ISBN 978-0-483-37421-8
PIBN 10681829

Forest Products Laboratory, U. S., celebrates its decennial ... 527
Formulas, steam ... 669
Fortieth anniversary meeting of A.S.M.E. ... (149), 712
Forty years ago, editorial ... 708
FOSTER, DEAN E. Effect of fittings on flow of fluids through pipe lines ... 616
FOSTER, FRANK A. Industrial training in China (C) ... 318
Foundations for machinery ... 671
Founders, four, of the A.S.M.E., portraits of ... (150)
FOWLER, GEORGE L. Scientific development of the steam locomotive (D) ... 40
Fractures, mechanism of, and reliability of materials ... 207
FRARY, HOBART D., necrology ... (164)
FREEMAN, JOHN R., addresses Civil Engineers on engineering conditions in Japan and China 593
FREEMAN, ROGER M. Armor-plate and gun-forging plant of U. S. Navy ... 657
FRENCH, H. J. Some applications of alloy steels in the automotive industry ... 501
Friction and spreading forces in rolling mills ... 11
FRITZ, JOHN (*See* John Fritz)
Fuel for motor transport ... 588
  liquid ... 348
  lubricant and motor, proper balancing of. 164
  uses, new, discussed by New York Section ... (25)
Fuels section under way ... (99)
Functions of engineering societies in industry ... (28)
Furnaces, metallurgical, at high altitudes, pulverized coal in ... 225
Future engineer—a man of affairs ... 254
Future of aviation ... 28

## G

Gages, plain limit, for general engineering work .. 420
Gaging in open channels, an improved weir for ... 83
GANTT, HENRY LAURENCE, necrology ... 70
GARDNER, H. C. The St. Lawrence river project 509
Gas, by-product producer (why not?), editorial .. 708
Gas, natural, efficiency of, used in domestic service ... 287
  natural, pipe lines for transporting ... 445
Gas, natural, saving the ... 643
Gases, dissolved, the separation of from water ... 273
Gasoline, motor, from heavier hydrocarbons ... 400
GAYLORD, W. W. Strength of thick hollow cylinders (C) ... 250
General Electric Co. presents $30,000 for research 588
German defenses on the coast of Belgium ... 319
German long-range gun ... 89
GIBBS, A. W. Steam vs. electric locomotives (D) 688
GIBBS, GEORGE. Scientific development of the steam locomotive (D) ... 40
  Steam vs. electric locomotives (D) ... 688
GILBRETH, FRANK B., now LL.D. ... (152)
GILES, GEORGE, necrology ... (103)
GODFREY, EDWARD. Fiber stress in a square plate (C) ... 475
GOMPERS, SAMUEL, remarks by at Fortieth Anniversary A.S.M.E. ... 714
GOSS, E. O. Strains in the rolling of metal (D) .. 42
Government activities in engineering ... 76, 134, 200, 257, 313, 365, 599
Gray-iron castings ... 437
Great Britain, research associations in ... 181
Great Lakes, extending ocean navigation to the .. 592
Gun, German long-range ... 89
Gun-forging and armor-plate plant of U. S. Navy 657
Gyro-compass, principles of ... 619

## H

HAAR, SELBY. Designing and machining of bearings (D) ... 42
HADFIELD, SIR ROBERT, prize ... 539
HALL, JAMES A. Federation dues and local societies ... 533
HALL, JOHN H., Steel castings ... 432
HAMPTON, F. G., LEH, C. F., and HELMICK, W. E. An experimental investigation of steel belting ... 369
HAND, FRANCIS L., necrology ... (14)
HARDIN, F. H. Steam vs. electric locomotives (D) ... 687
Hartford Branch (*See* Connecticut Section)
HARTMAN, JOSEPH W., necrology ... (140)
HARTNESS, JAMES, elected governor of Vermont by big majority ... 710
  to be next Governor of Vermont ... 594
Harvard Engineering School's new plan in technical education ... 306
Heat, dissipation of, by various surfaces ... 230
HEATH, CARL JEFFREY, necrology ... (103)
Heat-insulating value of cork and lith board ... 624
Heat treatments for alloy steels, some commercial 506
HECK, ROBERT, C. H. Steam formulas ... 669
HEILMAN, RALPH E. What may we expect of profit sharing in industry ... 26
HEINDLHOFER, KALMAN. Tests on rear-axle worm drives for trucks ... 613
HELMICK, W. E., HAMPTON, F. G., and LEH, C. F. An experimental investigation of steel belting ... 369
HEMSTREET, G. P. Industrial situation in relation to present conditions (D) ... 35
HENDERSON, ERNEST GEORGE, necrology ... 164
HERSCHEL, CLEMENS. An improved weir for gaging in open channels ... 83
  New type of hydraulic turbine runner (D) .. 37
HERSEY, MAYO D. Aeronautic instruments ... 263
HESS, HENRY. Lubrication of ball bearings (D) 42
  Technical library book and journal indexing (C) ... 419
  Unemployment, national taxation and profit sharing ... 185
HESSENBRUCH, GEORGE S. Research and industrial wastes ... 104
HIBBARD, H. WADE. Gray-iron castings (D) .. 439
High-speed machinery ... 644
HIRSHFELD, C. F. Research and social evolution ... 103
HOBBS, L. S., POTTER, A. A., CALDERWOOD, J. P., MACK, A. J. The heat insulating value of cork and lith board ... 624
Hoisting drums, rational design of ... 675
HOLLEY, ALEXANDER L., portrait of ... (150)
HOLLIS, IRA N. A great work possible in industrial education, editorial ... 592
Joseph A. Holmes Safety Association ... (24)
Holyoke, Mass., electric power for ... 70
Honors conferred on A.S.M.E. members ... (111)
HOOVER, HERBERT C., address by ... 722
  addresses Mining Engineers ... 194
  elected president of the American Engineering Council ... 720
  portrait of ... 720
  Washington award to ... 24
HOPKINSON, DR. EDWARD, discusses opportunity of organized mechanical engineers ... (28)
Horsepower of resistance, discussion of the (C) .. 126
Houses, making them truly livable ... 478
Housing, the financial problems of industrial ... 346
Houston Section ... (155)
HOYT, JOHN C. (portrait), secretary of organizing conference of F.A.E.S. ... 422
Human relations in industry, conference on ... 483
HUNT, CHARLES WARREN, appointed secretary emeritus of A.S.C.E. ... 201
HUNT, ROBERT W. medal fund ... 421
HUTTON, FREDERIC R., memorial to the late ... 71
  Memorial plaque of ... 71
Hydrocarbons, motor gasoline from heavier ... 400
Hydroelectric plant at Keokuk, the great ... 228

## I

ILLMER, LOUIS. Experiences with large center-crank shafts ... 610
Indexing, technical library book and journal (C) 419
India, engineering organization in ... 190
Indianapolis Section ... (48, 81, 164)
Industrial conference, report of available ... 538
  education, a great work possible in ... 592
  field, the need of research in the ... 487
  housing, the financial problems of ... 346
  management, the training of engineering students in ... 492
  research department opened at University of Michigan ... 252
  training in China (C) ... 305, 419
  unrest, causes of and remedy ... 17
  wastes research and ... 104
Industrial Relations Association of America, meeting ... 315, 425
"Industrial Unrest" not an A.S.M.E. pamphlet .. 190
Industry, employees' representation in management of ... 133
  functions of engineering societies in ... (28)
  profit sharing in, what may we expect of ... 26
  systems for mutual control of ... 24
Instrument specifications ... 181
Instruments, aeronautic ... 263
Insulating materials, thermal conductivity of ... 8
Insurance, war risk, special ruling issued by Government on reinstatement ... 259
Interdependence of countries and professions (Secretary's letter) ... (39)
Inter-professional conference ... (11)
Inventions, Government to develop ... 257
Iowa State college, engineering experiment station at ... 528

## J

JACKSON, A. C. Industrial situation (D) ... 35
JACKSON, E. R. The first transcontinental motor convoy ... 145
  Malleable castings (D) ... 440
Japan and China, engineering conditions in. John R. Freeman addresses Civil Engineers on ... 593
JEFFRIES, ZAY. Aluminum castings ... 427
John Fritz Medal, awarded Orville Wright ... 306
  presented to Orville Wright ... 364
JONES, EDWARD C., biographical sketch of ... (12)
Junior members, shall they vote? (C) ... 303
Juniors, should they have privileges of A.S.M.E. members? ... (45)

## K

Kansas State research council ... 181
KATTE, EDWIN B. Scientific development of the steam locomotive (D) ... 40
  Steam vs. electric locomotives (D) ... 688
KEMP, HENRY D., necrology ... (46)
KENNEDY, W. P. Motor-transport vehicles for the U. S. Army (D) ... 41
Keokuk, Iowa, the great hydroelectric plant at .. 228
  views of the Mississippi River Power Company's plant at ... 229
KIESEL, W. F. Steam vs. electric locomotives (D) ... 688
KIMBALL, DEXTER S., biographical sketch of ... (13)
  visits Atlanta ... (11)
KING, MATTHEW L., necrology ... (46)
KLOCKARS, CHARLES OSCAR, necrology and portrait ... 164
KOPP, CHARLES, necrology ... (165)

## L

Laboratory, new mechanical engineering, college of engineering, University of Illinois ... 110
Labor program, proposed national ... 191
  management and production, noteworthy papers on ... 594
  situation in Australia, development of ... 593
Land leasing bill, Congress passes general ... 200
  law, administering the new ... 257
LEE, CLIFFORD, necrology ... (14)
Legal papers, have engineers the right to prepare? 473
LEH, C. F., HAMPTON, F. G., and HELMICK, W. E. An experimental investigation of steel belting ... 369
LEIGHTON, M. O. Federated American Engineering Societies (C) ... 474
LEISERSON, WILLIAM L. Systems for mutual control of industry ... 24
LELAND, HENRY M., Doctor of Engineering degree conferred upon ... (112)
  portrait of ... (111)
Lewis Institute of Chicago and Portland Cement Association coöperating in research ... 528
Load-speed capacities of radial ball bearings (C) 534
(Lord) Leverhulme's six-hour day ... 72
Libraries, a census of special ... 426
Library notes ... 81, 142, 205, 261, 317, 368, 485, 548, 601, 655, 726
  technical, its relation to efficiency ... 479
Licensing of engineers ... 77
  bill for, passed by N. Y. legislature ... 535
  N. Y. law requires ... 574
Light and power stations, electric, census of central ... 599
Limits, engineering, the raising of ... 537
Liquid fuel ... 348
Lith-board, and cork, the heat-insulating value of ... 624
Local Sections (*See* Sections).
  active ... (136)
Locomotive feedwater heating ... 283
Locomotives, code for boilers of ... 187
  steam vs. electric—the electric side ... 684, 685
  steam vs. electric—the steam side ... 680
Long-range gun, the German ... 89
LORD, C. B. Tight-fitting threads for bolts and nuts ... 222
Los Angeles Section ... (10, 48, 101, 164)
LOW, F. R. How shall engineers organize? (C) 303
Lubricant, fuel and motor, the proper balancing of ... 164
LUCKE, CHARLES E. Professional sections (C). 303

## M

McBRIDE, THOMAS C. Locomotive feedwater heating ... 283
McCook Field and American aeronautics ... 441
  views from, shown by Col. T. H. Bane ... 30, 31
McDERMET, J. R. The separation of dissolved gases from water ... 273
McGEORGE, JOHN. Gray-iron castings (D) ... 440
McINTYRE, O. L. Pulverized coal in metallurgical furnaces at high altitudes ... 225
McMILLAN, L. B. Thermal conductivity of insulating materials (C) ... 304
  The value of sheet asbestos on hot pipes (C) 188

Machinery, foundations for ... 671
Machine tools—what are they? (C) ... 474
MACK, A. J., HOBBS, L. S., POTTER, A. A., CALDERWOOD, J. P. The heat-insulating value of cork and lith board ... 624
MACLAURIN, RICHARD C., president M.I.T., dies 132
Mail tunnel systems for large cities ... 313
Malleable castings ... 431
MALLET, ANATOLE ... 131
Management and production, can engineering students be given broad conception of (C) 640
industrial, the training of engineering students in ... 492
production and labor, noteworthy papers on 594
Management Section elects officers ... (136)
gets under way ... (122)
MANNING, JAMES H., necrology ... (125)
MANNING, DR. VAN H., resigns as director of Bureau of Mines ... 363
Marine propulsion, a new development in ... 307
MARSH, T. A., E.F.C. watet-tube boiler for wood ship ... 37
MASURY, ALFRED F. Motor-transport vehicles for the U. S. Army (D) ... 41
Materials Handling Section elects officers ... (153)
holds preliminary meeting ... (122)
Materials, reliability of, and mechanism of fractures ... 207
Mechanical Engineering Bulletin, slow delivery of ... 421
Mechanical engineering laboratory, new, of college of engineering, University of Illinois 110
Mechanism of fractures and reliability of materials ... 207
MEEK, ALDEN R., necrology ... (28)
MEEKER, DR. ROYAL. Employees' representation in management of industry ... 133
Membership, candidates for (*See* Candidates, etc.)
Members, new, of the Society ... (144)
Memorial, war service and members' ... (40)
Meriden Branch (*See* Connecticut Section).
MERON, FREDERIC L. Profit sharing in industry (D) ... 35
Metal castings ... 427
Metallic-electrode arc-welding process ... 572
Metric system, A.S.M.E. Council adopts resolution regarding ... (75)
Michigan, University of, industrial research department opened at ... 253
Mid-Continent Section ... 91
Military legislation, the engineer's attitude on ... 254
MILLER, A. H. Some commercial heat treatments for alloy steels ... 506
MILLER, B. F. Motor-transport vehicles for the U. S. Army (D) ... 41
MILLER, FRED J., A.S.M.E. president for 1920 .. 68
biographical sketch of ... (12)
Engineers in public service ... 643
portrait ... 1
MILLER, H. W. (LT.-COL.). The German defenses on the coast of Belgium ... 319
The German long-range gun ... 89
MILLER, SPENCER, honored for services on Naval Consulting Board ... (112)
portrait of ... (111)
Milwaukee Section ... 81, 101
Mining and agricultural appropriations, relative importance of ... 599
Mining Engineers, Herbert Hoover addresses ... 194
Mining engineers visit Lake Superior copper and iron industry ... 600
Minnesota Section ... (10, 25, 48, 101, 155)
Mississippi, design of ore fleet for upper ... 379
Mississippi Valley, river transportation in the ... 270
MITCHELL WILLIAM S. Design of ore fleet for upper Mississippi ... 379
MOLDENKE, RICHARD. Gray-iron castings ... 437
MONAHAN, LOUIS J., necrology ... (115)
MONTEFIORE, GEORGE, foundation contest ... (155)
MORGAN, D. W. R. Air pumps for condensing equipment (D) ... 39
MORRIS, JOHN FINDLEY, necrology ... (164)
MOSS, S. A. Superchargers for airplane engines 383
Motion, the measurement of minute ... 308
Motor convoy, the first transcontinental ... 145
fuel and lubricant, the proper balancing of .. 164
gasoline from heavier hydrocarbons ... 400
Motor Transport Corps, first transcontinental convoy of ... 478
MOTSINGER, HOMER N., necrology ... (140)
MUDGE, JAMES DOUGLAS, necrology ... (81)
MUHLFELD, JOHN E. Scientific development of the steam locomotive, closure to ... 40
MUHLFELD, JOHN E. Steam vs. electric locomotives the steam side ... 680
MUNRO, ROBERT E., necrology ... (156)
MURPHY, W. T. Lubrication of ball bearings (D) ... 12
MURRAY, THOMAS E., made Doctor of Science by Villa Nova College ... (111)
portrait of ... (111)
MURRAY, WILLIAM S. Notes on the superpower survey ... 642
portrait of ... 642
The "mysterious Mr. Smith" ... 129

## N

NASH, LEWIS H. Labor unrest (D) ... 35
National Advisory Committee for Aeronautics, meeting ... 75
National Electric Light Association, meeting of . 483
National Founders' Association, meeting ... 75
National Metal Trades Association, meeting ... 366
National organization of engineers, an inclusive (Secretary's letter) ... 3
National Physical Laboratory of Great Britain .. 638
National Public Works Department Association holds second convention ... 136
observations on work of ... 537
to hold conference ... 68
National Research Council elects new officers for 1920–1921 ... 538
Engineering Division of, issues report ... 650
National safety codes, conference on ... 69, 129
National Safety Council, annual congress of ... 652
Engineering Section of, holds meeting ... 366
National Screw Thread Commission report ... 337
National Service Committee, notes contributed by ... 76, 134, 200, 257, 313, 365, 599
Natural gas, saving the ... 643
Naval experimental station at Annapolis, research work at ... 50
Commercial testing at ... 51
Navigation, ocean, extending it to the Great Lakes ... 592
Navy, U. S., armor-plate and gun-forging plant of ... 657
Navy wage award on salaries of technical men, effect of ... 599
Necrology ... (14, 27, 46, 65, 80, 92, 103, 114, 125, 140, 156, 164)
NELIS, JOSEPH J. E.F.C. water-tube boiler for wood ships (D) ... 37
New members of the Society ... (144)
New Orleans Section ... (25, 48, 81, 101)
New York (Metropolitan) Section ... (10, 25, 48, 63, 81, 164)
New York Section discusses new fuel uses ... (25)
NOBLE, PATRICK, necrology ... (165)
Nolan Patent Office Bill H.R., 11,984, appeal to A.S.M.E. members on behalf of ... 707
Nominating Committee, changes in constitution and by-laws regarding ... (77)
Nominating committee for 1920 ... 11
invites suggestions ... (59)
note on ... (75)
report of, for 1920 ... (98)
NORRIS, JOHN H., necrology, portrait ... (115)
NORTON, C. H. Designing and machining of bearings (D) ... 42
Nozzles, calibration of for measurement of air flowing into a vacuum ... 607
Nuts, and bolts, tight-fitting threads for ... 222

## O

OATLEY, H. B. Scientific development of the steam locomotive (D) ... 40
Steam vs. electric locomotives, (D) ... 687
Officers for 1921 elected by A.S.M.E. membership ... (152)
Officers of the Society for 1920, the new ... 12
to be balloted on ... (110)
Ontario Section ... (10, 48, 81, 91, 164)
Ordnance department establishes new engineering staff ... 365
section to organize ... (162)
Ore fleet for upper Mississippi, design of ... 379
Oregon Section ... (48)
Organization and construction of dye houses ... 673
of engineers (C) ... 303
Organizing and financing a public-utility project, the cost of ... 557
Outdoor power plant ... 307

## P

Pacific coast, fuel and power conservation on the 710
PACK, CHARLES. Die castings ... 434
PADGETT, FRED W. Motor gasoline from heavier hydrocarbons ... 400
PAINE, F. B. H. The cost of organizing and financing a public-utility project ... 557
Pan-American Financial Conference, A.S.M.E. members at Second ... 190
PARISH, W. F. The proper balancing of fuel, lubricant and motor ... 164
Patent bill ... 257
legislation ... 200
PERKINS, GEORGE H. Steam use in textile processes ... 14
PERRY, THOMAS D. The engineer and the woodworking industry ... 448
Personals ... (13, 26, 47, 63, 80, 91, 101, 113, 124, 138, 155, 165)
Petroleum research ... 638
Philadelphia Section ... (10, 25, 48, 63, 81, 91, 101, 137, 155, 164)
Pipe lines, effect of fittings on flow of fluids through ... 616
for transporting natural gas ... 445
Pittsburgh Section ... (48)
PLANCHE, E. Economy of passenger automobiles at various speeds (C) ... 705
Plating, copper, effect of depth on carburization . 565
POLAKOV, WALTER N. Industrial situation in relation to present conditions (D) ... 35
POPE, F. H. Motor-transport vehicles for the U. S. Army (D) ... 41
PORTERFIELD, HENRY A., necrology ... (103)
Portland Cement Association and Lewis Institute of Chicago coöperating in research ... 528
POTTER, A. A. Specialized curricula in engineering colleges ... 127
CALDERWOOD, J. P., MACK, A. J., HOBBS, L. S. The heat-insulating value of cork and lith board ... 624
Power and light stations, electric, census of central ... 599
plant, the outdoor ... 307, 477
Power Section elects officers ... (122)
officers of ... (136)
Power supply, electric, of southern textile mills .. 161
Power Test Codes ... 718
General Instructions of ... 718
note on ... (99)
Presidents, three, at Worcester ... (23)
PRESSINGER, WHITFIELD P., necrology, portrait . (115)
Price levels in relation to value ... 553
Prices, are they coming down? ... 191
Printers, Mechanical Engineering changes ... 642
Prizes to be awarded by the Society ... 6
Production and management, can engineering students be given a broad conception of? (C) ... 640
labor and management, noteworthy papers on ... 594
Professional employment, conference on ... (6)
Professional sections (C) ... 303
Professional sections ... (6), (75)
activities ... (122)
a new activity ... (44)
standing committees on, now established ... (162)
Profit sharing in industry, what we may expect of ... 26
unemployment, national taxation and ... 185
Propeller, air, physical basis of design ... 213
Providence Engineering Society, meeting of . (154, 164)
Public questions, the engineer's relation to ... 336
Public service, engineers in ... 643
Public-utility project, the cost of organizing and financing a ... 557
Public Works, National Chamber of Commerce to vote on Department of ... 306
Public Works Association, National Department of, holds second convention ... 136
Public works, two attitudes on the department of 477
Pueblo Engineering Association, meeting ... (135)
Pulleys, compressed-spruce, an investigation of .. 451
Pulp and paper industry of New England ... 444
Pulverized coal in metallurgical furnaces at high altitudes ... 225
Pump, dredging, of novel construction ... 1

## Q

QUINN, C. H. Steam vs. electric locomotives (D) ... 687

## R

Radial ball bearings, load-speed capacities of (C) 534
Railroad engineers, classification of subordinate .. 314
Railroad Fire Protection Association, meeting ... 74
Railroad Section, completes organization ... (98)
meeting a huge success ... (162)
plans joint meeting ... (122)
presents first professional section papers ... (154)
RALSTON, A. L. Steam vs. electric locomotives (D) ... 687
RAND, CHARLES FREDERIC, retires as president of the U. E. S. ... 141
RANDOLPH, ISHAM, death of ... 540
RATEAU, A. The rectilinear flight of aeroplanes 268
Rateau, French Society presents A.S.M.E. honorary membership to Monsieur ... 481
RATEAU, M. AUGUSTE C. E., portrait of ... 482
Rational valuation—a comparative study ... 549

RAWDON, H. S. The electric-arc welding of steel: the properties of the arc-fused metal..... 567
RAYMOND, ALFRED, necrology.................. (103)
READ, T. T. Industrial training in China (C).. 419
Reclamation service........................ 200
Reclassification and compensation in the Government service.......................... 314
Rectilinear flight of aeroplanes................ 268
Registration of engineers, law for.............. 717
Reliability of material and mechanism of fractures.................................. 207
Research and industrial wastes............. ... 104
  and social evolution..................... 103
  associations in Great Britain............181, 638
  coöperative.............................245, 588
  department of scientific and industrial...... 700
  engineering..................50, 108, 181, 245, 300, 349, 414, 470, 528, 588, 638, 700
  Engineering Foundation reports progress in. 137
  gift, of $5,000,000 for promoting........... 191
  in the indusurial field, the need of......... 487
  laboratories in industrial establishments of the U. S............................. 588
REYNOLDS, PHILIP E. Air pumps for condensing equipment (D)......................... 38
RICE, CALVIN W. The Federated American Engineering Societies...................... 420
RICHARDS, CHARLES RUSS. The new mechanical engineering laboratory, college of engineering, University of Illinois............ 110
RICHARDSON, CHARLES G. Simplification of venturi-meter calculations (C)........... 304
RITTER, HENRY. Screwed pipe joints (C)...... 125
River transportation in the Mississippi Valley... 270
RIX, HAROLD PRESTON, necrology.............. (81)
Road bill, a new............................ 365
Road building, Government aid in............. 200
Road legislation............................ 314
ROBERTS, WILLIAM E., necrology............... (156)
Rochester Section........................(48, 81)
ROLLASON, G. M. Alloyed aluminum as an engineering material..................... 495
Rolling mills, friction and spreading forces in.... 11
Rust, economic loss caused by (C)............. 126

Safety Code program, progress of............. 716
Safety Codes, standardization of.............. 651
Saint Lawrence river project.................. 509
St. Louis, A.S.M.E. to meet in convention at, May 24–27............................ 252
  Preparing fine program for Spring Meeting 1920.................................. (39)
  Spring Meeting of A.S.M.E. at............ 309
St. Louis Section................(25, 48, 81, 155, 164)
San Francisco Section......(25, 48, 63, 81, 91, 101, 164)
SAUNDERS, W. L. Remarks by, at Fortieth Anniversary A.S.M.E. Meeting.......... 713
SCHADER, E. W. River transportation in the Mississippi Valley..................... 270
SCHMID, ALBERT, necrology.................... (46)
SCHRECK, H. Hvid engine and its relation to fuel problems (D)...................... 34
SCHROEDER, MAJOR R. W., record altitude flight of.................................... 255
SCHUHMANN, GEORGE, necrology............... (14)
Scientech Club of Indianapolis, note on......... (135)
Scientific American Supplement now issued as a monthly............................ 254
SCOTT, CHARLES F., remarks by, at Fortieth Anniversary A.S.M.E. Meeting.............. 713
SCOTT, EARL F., biographical sketch of......... (13)
Screwed pipe joints (C)..................... 125
Screw Thread Commission report, National..... 337
Screw threads, note on standardization of....... 362
Secretary's Letter....................(3, 23, 39, 59, 75, 90, 110, 121, 133, 152, 161, 420)
Section meetings......................(10, 25, 47, 62, 81, 91, 101, 137, 155, 163)
Sections (local):
  Conference of at 1919 A.S.M.E. Annual Meeting.............................. 9
  Conference of at 1920 A.S.M.E. Spring Meeting.............................. 100
  Activites planned for 1920–1921............ 136
See, James Waring, an appreciation............ 362
  necrology................................ 256
Self-supporting chimneys to withstand earthquake.................................. 157
SELLEW, R. W. Effect of radial play on life of ball bearings (C)...................... 706
  Load-speed capacities of radial ball bearings (C)................................... 534
Service vs. professional standing (Secretary's letter)................................. (23)
Shafting, standard sizes for.................. 251
Shafts, center-crank, experiences with large..... 611
Sheet asbestos on hot pipes, the value of......69, (C)................................... 188
SHELDON, Dr. SAMUEL, dies................... 595
SHEPARD, F. H. Steam vs. electric locomotives—the electric side....................... 685
SHIPLEY, ALBERT ROBOSSON, necrology......... (140)
SHORT, FRED A., necrology................... (125)
SIBERT, H. W. Calculation of stresses in a rectangular plate (C)..................... 249
SIFF, A. L. Industrial situation in relation to present conditions (D)................. 35
SIRRINE, J. E. Electric power supply of Southern textile mills....................... 161
SKINKLE, W. B. Friction and spreading forces in rolling mills........................ 11
SLADE, ARTHUR J. Motor-transport vehicles for the U. S. Army (D)................ 40
SMITH, H. R. Lubrication of ball bearings (D). 42
Social evolution, research and................ 103
Society affairs.........*See* Section Two of all months
Society of Automotive Engineers, meeting...... 547
  section policy........................... (40)
  technical papers at annual meeting of the... 130
Society of Industrial Engineers, meeting........ 315
Society of Naval Architects and Marine Engineers, meeting............................. 74
Society membership in other organizations...... 40
SPALDING, C. M. Comments on the proposed code of ethics (C)........................ 591
Specialized curricula in engineering colleges..... 127
SPILSBURY, EDMUND GYBBON, portrait and necrology............................... 424
Spreading forces and friction in rolling mills..... 11
Spring Meeting, of A.S.M.E., St. Louis, 1920, (39, 58, 60), 128, 190, 252, 309, 386
Standardization, engineering and industrial..544, 651, 716
Standardization of fire-hose couplings.......... 452
Standards Committee, American Engineering, actively engaged in organizing standardization work....................... 193
Standard sizes for shafting................... 251
Standard, tool-drafting (C)................... 535
Steam formulas............................ 669
Steam side—steam vs. electric locomotives...... 680
Steam use on textile processes................ 14
Steam vs. electric locomotives................ 680
Steel belting, an experimental investigation of... 369
Steel castings............................. 433
Steels, alloy, in the automotive industry, some applications of.......................... 501
  some commercial heat treatments for....... 506
Stoker for burning eastern coals............... 279
STOWE, L. R. A stoker for burning eastern coals 279
STREET, CLEMENT F., Scientific development of the steam locomotive (D).............. 40
Strength of thick hollow cylinders (C).......... 250
Stresses in a rectangular plate, calculation of (C). 249
Stress, fiber, determination of, caused by force fits (C).............................. 533
Student Branch meetings.................... 26
Students' Army Training Corps, history of the... 190
Students of engineering, can they be given broad conception of production and management? (C).......................... 640
  their training in industrial management.... 492
Sundry Civil Bill........................... 365
Superchargers for airplane engines............ 383
Superpower survey......................... 642
Surveys and maps, a new board of............ 134
  meets to perfect organization.............. 257
SWEET, JOHN E., portrait of.................. (150)
Swiss Economic Mission, note on.............. 189
  to study American economic conditions..... 421
Syracuse Section.......................... (164)

# T

Tariff legislation........................... 200
Taxation, national, unemployment and profit sharing................................ 185
Taylor Society, meeting.....................74, 366
Taylor system in Germany................... 135
TAYLOR, T. S. Dissipation of heat by various surfaces.............................. 230
  The flow of air through small brass tubes.... 334
  Thermal conductivity of insulating materials.............................. 8
  The value of sheet asbestos on hot pipes (C). 189
Technical inquiries......................... 24
  press comments on the F.A.E.S. Washington conference....................... 480
Testing of farm tractors..................... 101
Textile mills, electric power supply of southern.. 161
  processes, steam use in.................. 14
Thermal conductivity of insulating materials... 8
Thermal conductivity of insulating materials (C) 304
Threads, tight-fitting, for bolts and nuts........ 222
Three Presidents at Worcester................ (23)
THURSTON, ROBERT H., portrait of............ (150)
TRWING, L. L. Machine tools—what are they? (C)................................ 474
Toledo Section........................(81, 101)
Tool-drafting standard, why not a? (C)........ 535
TOUCEDA, ENRIQUE. Malleable castings....... 431
TOWNE, HENRY R., remarks by at Fortieth Anniversary Meeting of A.S.M.E............ 712
TOWNLEY, CALVERT (portrait), chairman of organizing conference of F.A.E.S........... 422
Tractors, the testing of farm.................. 101
Training of engineering students in industrial management........................ 492
Transactions, certain issues of for sale......... (136)
Transcontinental motor convoy, the first........ 145
Transportation, river, in the Mississippi Valley.. 270
Transporting freight in interchangeable containers.................................. 612
TREADWAY, LYMAN HAMBRIGHT, necrology....... (125)
Trucks, tests on rear-axle worm drives for....... 613
Tubes, boiler, the constitution and properties of.. 603
  the flow of air through small brass.......... 334
Tulsa, members enjoy visit to................ 398
Tunnel, vehicular, between New York and New Jersey ground broken for................ 644
Turbo-compressor calculations................ 151

Unemployment, national taxation and profit sharing.................................. 185
United Engineering Societies Library issues annual report............................. 142
United Engineering Society, report for 1919..... (41)
  Treasurer's report for 1919................ (62)
U. S. Chamber of Commerce, records its views on industrial relations.................. 539
Unrest, industrial, the causes of and the remedy.. 17

Valuation, and appraisal, of properties.......... 549
  rational—a comparative study............ 549
Value, price levels in relation to.............. 553
VAN DERHOEF, GEORGE N. Lubrication of ball bearings (D)........................... 42
Vehicular tunnel between New York and New Jersey, ground broken for................ 644
Venturi-meter calculations, simplification of.... 220
Venturi-meter calculations, simplification of (C).................................304, 419
Virginia Section........................... (48)

# W

Wage increases not necessarily on a straight-line curve................................... 477
WALDEN, A. E. The design of aspirators for sterilizing water........................ 562
WALKER, P. F. The need of research in the industrial field............................ 487
WALSH, EDWARD T., necrology................ (28)
War, dues of members in active service during the.................................... (39)
WARNER, WORCESTER R., and SWASEY, AMBROSE, celebrate fortieth anniversary............ (99)
Warner and Swasey Observatory dedicated at Case School of Applied Science.......... 717
WARREN, C. B. Simplification of venturi-meter calculations........................... 221
  Simplication of venturi-meter calculations (C).................................... 419
War Risk Insurance, Government issues special ruling on.............................. 259
War service and members' memorial............ (40)
Washington award to Herbert Hoover.......... (24)
Washington, D. C., Section......(10, 48, 81, 91, 164)
Washington State Section........(11, 48, 63, 81, 101, 155, 164)
Waterbury Branch (*See* Connecticut Section)
Water power act............................ 644
Water power bill, agreement expected on........ 257
  delay in................................. 365
  passes senate............................ 135
Water power problem, the important........... 536
Water powers, a plea for the development of our.. 536
Water, sterilizing, the design of aspirators for.... 562
  the separation of dissolved gases from...... 273
WATERS, EVERETT O. Rational design of hoisting drums............................. 675
WEBER, R. L. Proposed symbol for B.t.u. (C).. 249
WEBER, HENRY, necrology.................... (156)
WEICKEL, HENRY, necrology.................. (46)
Weir, improved, for gaging in open channels..... 83
Welding, autogenous, of boilers and pressure vessels................................ 70
  conference committee on.................. 532
  electric-arc, of steel: the properties of the arc-fused metal..................... 567
  electric, notes on....................... 567
WELLMAN, S. T., resolutions on death of........ (40)
Western Society of Engineers entertains boards of civil, mining, mechanical and electrical engineers.............................. 367
  holds annual meeting..................... 204

WEYMOUTH, C. R. Self-supporting chimneys to withstand earthquake............... 157

WHEELER, FRANK R. Air pumps for condensing equipment, closure on.................... 39

WHITE, ALBERT E. The constitution and properties of boiler tubes.................... 603

WHITE, WALTER J. A dredging pump of novel construction............................ 1

WHITE, WILLIAM MONROE. A plea for the development of our water powers ........ 536

WHITNEY, AMOS, dies ................ 540
portrait of .......................... 540

WILEY, WILLIAM H., biographical sketch of ..... 113

WILLIAMS, GARDNER S. Federation an opportunity for local societies (C)........... 332

WILSON, GEORGE S. An investigation of compressed-spruce pulleys.................... 451

Woodworking industry, the engineer and the.... 448

WOOLSON, OROSCO CHARLES, necrology........ (140)

Worcester Section.................. (11, 25, 48, 63)

Worm drives for trucks, tests on rear-axle....... 613

WORTHINGTON, HENRY R., portrait of......... (150)

WRIGHT, ORVILLE, awarded the John Fritz medal for 1920............................ 306
John Fritz medal presented to........... 364

WRIGHT, WILBUR, monument to, presented to Le Mans............................ 539
Monument of (photo).................. 539

Year Book for 1920........................ (25)

YOUNG, G. R. Motor-transport vehicles for the U. S. Army (D)...................... 41

YOUNGER, JOHN. Shall junior members vote? (C)............................ 303

Z

ZIMMERLI, F. P. The effect of depth of copper plating on carburization................. 565

# ENGINEERING SURVEY


- Abrasive, aluminous 169
- Abrasive wheels, testing 123
- Acids, fatty, in oil 357
- Adams, W. L. 298
- Adamson, Daniel 67
- Aerial transport of woolen goods 289
- Aerodynamic phenomena, wind-tunnel studies in 403
- Aerofoils, thick, properties of, aerodynamic 234
- Aeroplane engines, back pressure in exhaust line
  - engines and power loss of, muffler design 233
  - ranges and useful loads 351
  - woods, kiln drying of 291
- Air classification of pulverized material 581
  - classification, Raymond system of 581
  - compressors, overheating in 637
  - ducts, design of 463
  - filters, screenless 403
  - forces on circular cylinders 630
  - fresh-, problem, ozone as the solution of 244
  - mass of liter of 67
  - moist, graphical determination of weight of 170
  - specific heat of 124
- Air-cooled engine 583
- Aircraft carriers 693
  - engines, fuels for 173
  - engines, radiators for 463
- Aitchison, Leslie 59
- Alcohol as a by-product of distillation of coal 352
  - from coke-oven gas 633
- Alison, J. Melville 575
- Allen, Dr. H. S. 116
- Alloys, heat-resisting 236
  - impact tests on 519
  - of oxides 358
- Aloxite 169
- Alula wing 579
- American competition with British trades 290
- American Society for Testing Materials 408
- Ammonia compressor, feather-valve 66
  - vapor pressure of 234
- Andrew, J. H. 406
- Archbutt, L. 115
- Armstrong, A. H. 242
- Ash- and coal-handling equipment 241
- Ashes, volcanic 631
- Atkins, W. R. G. 467
- Automatic hydroelectric plants 518
- Automobile engines, slow running of, limits of 466
- Automobiles, frameless 525
  - Duplex car and engine 178


## B


- Babcock & Wilcox Company 629
- Back pressure in exhaust line and power loss of aeroplane engines 233
- Bacon, R. F. 405
- Balancing turbine rotors 460
- Bayley, F. 405
- Bean, W. R. 114
- Bearing, radio-thrust 468
- Benson, H. K. 352
- Berg, Eskil 62
- Bergstrom, Eric M. 513
- Bergsund original surface-ignition-type engine 295
- Beutell, A. 242
- Bicycle chain 467
- Blackstone engine 466
- Blake, A. E. 404
- Blast-furnace castings 352
  - gas, cleaning 692
- Block, Berthold 122
- Blower, turbo-, single-stage, 22,000-r.p.m. 630
- Boilers, flues of, heat transmission in 67
  - lap-seam, explosion of 65
  - locomotive, staybolts in, deflection of 123
  - longitudinal joint in, maximum length of 65
  - mercury, Emmet 325
  - Stevens-Pratt 453
  - Tests, Pennsylvania Railroad 629
  - waste-heat 526
- Bone, Prof. W. A. 633
- Boring, thrust, in earth 412
- Bowden petrometer 586
- Britain, women workers in 180
- British Association, fuel-economy report of 55, 633
- British trades, American competition with 290
- Bromley, Chas. H. 66, 175
- Brown, R. P. 297
- Buck, E. M. 682
- Buckingham 630
- Buckingham, Edgar 114
- Bureau of Standards 55
  - new engine-testing plant 408
- Burley, Geo. W. 117
- Burrows, Chas. W. 698
- Burt single-sleeve valves 583
- Bury, E. 352
- Caldwell, F. W. 403
- Camichel, C. 56
- Cams 467
- Canada, National Institute for 697
- Canfield, R. E. 352
- Car, frameless 525
- Carbonizing, pots and boxes used in 236
- Cardan shafts, strength of 524
- Carey oil-transmission system 696
- Castings, blast-furnace 352
  - steel, light 579
- Cement, portland, long-time tests of 631
- Centrifugal force in rope driving 576
- Chain, bicycle 467
- Chantraine metallurgical furnaces 405
- Chip-protection device 237
- Citröen gear for electric locomotives 413
- Clays, bond, American 55
- Coal and ash-handling equipment 241
  - Canadian, steaming tests with 241
  - consumption in terms of water evaporation 458
  - distillation of, alcohol as a by-product of 352
  - dust 405
  - powdered, in power stations 580
  - pulverized, in open-hearth practice 114
  - pulverizers 359
  - sub-bituminous, low-temperature distillation of 352
- Coffin, J. G. 351
- Coke breeze, burning on underfeed stokers 465
- Cold rolling, stresses caused by 404
- Cold-storage industries 360
  - plant 66
- Coleman, Frederick C. 360
- Combustion process 627
  - surface 404
- Competition, American, with British trades 290
- Compressors, air, overheating in 637
  - ammonia, feather-valve 66
  - reciprocating 410
- Concrete, reinforcing bars for 291
  - ship 239
- Condenser, mercury 525
  - tubes, corrosion of 290, 461
- Condron, T. L. 291
- Cone, Edwin F. 468
- Connecting rod, H-section 579
- Constantinesco, G. 633
- Constantinesco sonic waves 359
- Cooling, rapid, of steel cylinders 118
- Copper, intercrystalline brittleness of 171
- Corbet, Darrah 580
- *Corrientes*, S. S. 577
- Corrosion of condenser tubes 290, 461
  - of iron and steel 631
  - resistance, film or intergrain theory of 631
- Corrosive liquids, pumps for 634
- Cotton rope for power transmission 575
- Cottrell system 693
- Coxon automatic stop valves 64
- Cragoe, C. S. 234
- Crane, shipyard 353
- Crude-oil engine, Ruston and Hornsby solid-injection 521
- Cushing, W. C. 291
- Cycle, thermodynamic, for internal-combustion engines 361
- Cylinders, circular, air forces on 630
  - steel, rapid cooling of 118


## D


- Davey, Norman 57
- DeBuos, F. A. 465
- Deeley, R. Mountford 115
- deLoisy, E. 352
- Dezincification 290
- Diamond wire-drawing dies 118
- Dickinson, H. C. 463
- Die casting 173
- Dies, wire-drawing, diamond 118
- Diesel engine for motor vehicles 355
  - engine, Neptune 174
  - marine machinery 408
- Dijxhoorn, J. C. 692
- Distillation of sub-bituminous coal, low temperature 352
- Doherty, R. E. 410
- Dolder, Professor 123
- Dorman wave-power tools 633
- Drifting valves, Ripken, automatic, for locomotives 121
- Drilling machines, "Way," Foote-Burt 237
- Drills, twist, use of 585
- Drop forgings, defects in 579
- Dryden, Hugh L. 630
- Drying, kiln, of aeroplane woods 291
  - theory of 169
- Dunderdale, R. J. 579
- Duplex car and engine 178


## E


- Edwards, Junius David 114
- Electric locomotives, Citröen gear for 413
- Electrically welded ship *Fullagar* 524
- Emmet mercury boiler 525
- Engine, aeroplane, Zeitlin, variable-stroke 582
  - aeroplane, back pressure in exhaust line and power loss of, muffler design 233
  - air-cooled 583
  - aircraft, fuels for 173
  - Blackstone 466
  - Diesel, for motor vehicles 355
  - Duplex car and 178
  - internal-combustion, thermodynamic cycle 361
  - internal-combustion, valve steels in 59
  - Melton-Haury 235
  - motor-car, British twin-six 298
  - motor-car, Lanchester 586
  - motor-car, Turner-Moore 407
  - oil 627
  - power-recuperating 354
  - single-sleeve-valve 235
  - steam, reciprocating 242
  - submarine-boat type, Vickers 627
  - surface-ignition-type, Bergsund original 295
  - U-boat, built at Germania Works 627
  - testing plant, new, Bureau of Standards 408
- Ethylene 352
- European turbine manufacturers 513
- Explosion gaps 453


## F


- Fales, E. N. 403
- Fatigue and its effect on production 238
  - of metals under repeated stress 171
- Fatty oils 116
- Feiser, Myers J. 699
- Fennell, W. 407
- Ferroaluminum 173
- Ferroboron 173
- Ferromolybdenum 173
- Ferrouranium 173
- Ferrozirconium 173
- Fessenden 629
- Film or intergrain theory of corrosion resistance 631
- Films, theory of, Langmuir's 116
- Filters, air, screenless 403
- Firebricks, clay, porosity and volume changes 360
- Fitch, W. H. 114
- Flame propagation in complex gaseous mixtures 240
  - propagation in mixtures of methane and air 239
- Fletcher, J. E. 119
- Flow of liquids, velocity of 56
  - of oil in pipes 633
  - water-filament 516
- Flues, heat transfer in 628
- Flywheel effect 410
- Forest Products Laboratory 577
- Fowler, Geo. L. 123
- Francis turbines 513
- Frequencies, standardization of 242
- Fresh-air problem, solution of, ozone as 244
- Frey, Ernest 355
- Fry, Lawford H. 628
- Fuel injection 627
  - selection for electric railways 580
- Fuel-economy report of British Association 55, 633
- Fuel-oil situation in the East 56
  - installation 692
- Fuels, American 405
  - Canadian, steaming tests 241
  - for aircraft engines 173
- *Fullagar*, electrically welded ship 524
- Funck, Georges 354
- Furnaces, gas, heat-treating 353
  - metallurgical, Chantraine 405
- Fusain 405


## G


- Gage, V. R. 233
- Garvin 580
- Gas, blast-furnace, cleaning of 692
  - dynamical method for raising to high temperature 637
  - efflux through small orifices 114
  - furnaces, heat-treating 353
  - mixtures, air and, mass of liter of 67
  - producer 173

producer, for motor vehicles............ 177
supply, public...................... 88
turbine, Holzwarth................... 292
Geared turbines......................... 877
Gears, Citröen, for electric locomotives.... 413
cutting........................... 587
double-reduction.................. 577
erosion of........................ 460
reduction, mechanical, in warships.... 587
teeth of lubricating................ 587
turbine, high-speed................ 61
worm, David Brown & Son.......... 116
General Electric design of a superpower station 486
"Germ" process in lubrication........... 336
Germain Works, U-boat engine built at..... 627
Germany, metals in, war expedients of..... 464
Goddard, Robt. H.................... 179
Goldsmith, F. C..................... 468
Goodwin, H., Jr..................... 486
Gorsuch, W. S....................... 179
Graphite lubricants.................. 467
Green, B. H......................... 358
Groesbeck, E. C..................... 404
Ground vibrations, transmission of....... 693
Gun, 155-mm. Schneider.............. 411
Gyroscopic phenomena in synchronous-motor
drive.......................... 410

**H**

Hack saws.......................... 523
Halkett, R. R..................244, 406
Hassond, Edward K.................. 118
Hamon, W. A....................... 408
Hardening of carbon steel............. 580
Hardy, B. W....................... 116
Hardy, W. B....................... 360
Harris, R. H....................... 236
Harrison................... 113
Heat transfer through insulation, geometric
form and........................ 638
transfer in flues.................. 628
transmission in boiler flues......... 67
waste, utilization of............... 358
treating gas furnaces.............. 353
Helmholtz.......................... 630
Herck, J........................... 389
Herschel, Winslow H................. 463
Hess, H. L......................... 234
Heusi, Wilhelm..................... 124
Hirgl, J........................... 631
Hoehn, E........................... 169
Huk, W. H......................... 353
Holzwarth gas turbine............... 292
Holzwarth Hans..................... 292
Hood, British battle cruiser........... 297
Hopewell, tidal-power development at..... 115
Horwich superheater for locomotives..... 65
Hot stamping, American and British practice.. 583
Howd, H. M........................ 464
Howe suspended molding machine....... 235
Huygens, P. V...................... 237
Hutchins, Otis..................... 466
Hyde.............................. 115
Hydraulic and steam plants, combined operation
of.............................. 818
Section of Report of the National Electric
Light Association Committee on Prime
Movers......................... 517
turbine, Kaplan................... 816
valve, balanced.................. 294

**I**

Ignition, magneto and battery, system on
Liberty-12 aero engine............ 289
Impact tests on alloys............... 819
Indicator diagrams, low-pressure...... 295
Ingersoll Rand P-R oil engine......... 522
Interatom theory of corrosion resistance... 631
Iron, malleable, physical tests of...... 114
puddled, practice............. 119

Insane, A. S.................... 836
Isurl, A. S.................... 463
Jet apparatus................. 465
Johnson, N.................... 255
Johnson, J. P................. 63
Joint angular al............. 409
Jones............... 634
Jove.............. 627

**K**

Kanowitz, S. B................. 881
Kaplan hydraulic turbine........ 816
Kiving, C. F................. 894
Kiln drying of structural wood....... 891

Kleinschmidt, R. V............289, 463

**L**

Lanchester motor-car engine........... 386
Langdon, S. C...................... 171
Langmuir's theory of lubrication....... 116
Lathe, relieving attachment for........ 205
turning tools, cutting power of...... 117
Laxtonia 12-cylinder engine........... 208
Lead, intercrystalline brittleness of.... 352
LeChatelier, H..................... 352
Leduc, Prof. A..................... 67
Lee, Louis R....................... 299
Liberty-12 aero engine, magneto and battery
ignition system on............... 289
Lime, hydraulic, and volcanic ashes..... 631
Lippman, Gabriel................... 120
Locomobile, compound, with superheater.. 469
Locomotives, boilers for, staybolt deflection in.. 123
cylinders, parts of, cast iron for.... 631
drifting valve for, Ripken automatic.. 121
electric......................... 242
electric, Citröen gear for.......... 413
helping, ten-wheel................ 360
internal-combustion............... 356
superheater for, Horwich.......... 65
water-indicating devices for........ 693
Lugne............................ 692
Longobard, Michael................. 67
Loomis, George A................... 360
Lubricants, graphite................ 467
viscosity of, standard test for...... 408
Lubricating gear teeth............... 357
Lubrication........................ 115
"germ" process................... 336
Langmuir's theory of.............. 116
molecular layers in............... 116

**M**

McBride, R. S...................... 403
McCook Field wind tunnel............ 403
Machine tools, geometric progression of speeds in. 175
Machinery foundations............... 693
Magnetic testing apparatus........... 698
Mangnall, Capt. A. R............... 412
Marine machinery, Diesel............ 408
Mason, Walter..................... 239
Mass systems of aircraft propulsion..... 234
Melton-Haury engine............... 235
Mercury boiler, Emmet.............. 525
condenser....................... 525
Metals, fatigue under repeated stress.... 171
war expedients in Germany........ 464
Michell, A. G. A................... 115
Miller, C. P....................... 123
Mirrless, Bickerton and Day oil-injection method. 466
Molding machine, suspended, Howe...... 235
Mollier, Prof. R.................... 123
Molybdenum in alloy steel........... 172
Monel metal, working............... 522
Moore, Herbert F.................. 169
Motor vehicles, Diesel engine for....... 355
vehicles, producer gas for.......... 177
Motors, synchronous............... 410
Muffler design; back pressure in exhaust line and
power loss of aeroplane engines..... 233
Muller, Friedrich.................. 697
Myers, C. H....................... 234

**N**

National Electric Light Association, Committee
on Prime Movers of............... 459
Committee on Prime Movers, Hydraulic Sec-
tion, Report of.................. 517
National Institute for Canada........ 697
Neptune Diesel engine.............. 174
Nemcek, Otto...................... 632
Newbury, N. D..................... 62
Nickel, rolled...................... 468
Nilson, A. J....................... 465
Norton, F. H...................... 234
Nozzles, employment in turbines....... 123
steam........................ 697
Nutbelt.......................... 629
Nut lock, new type of.............. 237

**O**

Oberhoffer, P...................... 242
Obturators........................ 407
Oil engine, combustion process in....... 627
engine, Crossley................. 466
engine, Ingersoll-Rand P-R........ 522
engine, supplying liquid fuel to..... 465
transmission system, Carey......... 696
oil fuel installation.............. 693

"Oiliness".......................... 115
of lubricating oils............... 356
Oils, fatty........................ 116
Ollander, O....................... 352
Omnibuses, motor, stability of........ 413
Open-hearth practice, coal in, pulverized.. 114
Ozone air purification.............. 244
in ventilation.................. 406

**P**

Parsons, S. R...................... 289
Patel fuel-regulation valve for oil engines.. 465
Patent act, British................ 180
Payman, Wm....................... 240
Pelton wheel...................... 515
wheel, bucket design for.......... 520
Pennsylvania Railroad boiler tests..... 629
Periscopes......................... 543
Petot, Prof. A..................... 689
Petrometer, Bowden................ 586
Philo, Frank G.................... 692
Pickling of alloy steel............. 234
Pipe, flexible, Dorman............. 633
flow of oil in................... 515
lines.......................... 515
Piping, power.................... 63
power, standardization of......... 63
steam......................... 457
Pitman, Percy.................... 520
Pittsburgh district, water softening in... 587
Plug stoker....................... 632
Plywood webs, shear strength of...... 577
Poer, Joseph...................... 692
Porteven, M....................... 118, 380
Power generation, economics of....... 458
piping, standardization of......... 63
plants, standardization of......... 299
plants, steam, performance of...... 179
tidal, development at Hopewell..... 115
Pratt, Arthur D.................. 526
Pfatt-Stevens boilers.............. 483
Precipitation of salt by refrigeration... 122
Preger, Ernst..................... 403
Prentiss, F. L.................... 382
Press, adjustable-position.......... 696
Preston, Arthur C................. 633
Prime Movers, Committee of National Electric
Light Association on............. 439
Producer gas for motor vehicles...... 177
Projectile, laws governing punching of a plate
by a........................... 584
Pulverized fuel.................... 562
material....................... 581
Pulverizer, coal................... 590
oil-fuel sleeve type.............. 356
Pumps, centrifugal................ 692
for corrosive liquids............. 634
mercury, high-vacuum............ 242
roto-piston vacuum and pressure.... 633
Punching, high-speed, laws of....... 584
Pyrometer, signaling.............. 297

**Q**

Q-alloy of nickel-chromium family.... 236

**R**

Radiation, constants of............ 114
Radiators for aircraft engines....... 463
free-air, streamline casing for..... 289
head resistance due to............ 289
Rails, old........................ 291
Railways, electric, fuel selection for... 580
Rawdon, Henry S.............171, 352
Rayleigh, Lord.................... 630
Raymond system of air classification... 581
Reaction jet propulsion............ 234
Refrigeration, salt precipitation by.... 122
Regan automatic train-control system... 696
Relieving attachment for lathes...... 205
Research, scientific and industrial, in England.. 696
Reynolds......................... 630
Richardson, E. A. and L. T......... 631
Rings, piston..................... 407
Ripken automatic drifting valve for locomotives.. 121
Ripper, Prof. W.................. 117
Rippon, J. B...................... 115
Rocket action.................... 179
Rogers, W. O..................... 241
Rolling, coils, stresses caused by..... 404
Rope, cotton, for power transmission... 575
driving, centrifugal force in...... 576
Rosa, C. E....................... 635
Roto-piston vacuum and pressure pumps.. 633
Rowley, F........................ 579
Ruston and Hornsby solid-injection crude-oil en-
gine.......................... 521

## S

Salt precipitation by refrigeration............ 122
Sawmill refuse as fuel....................... 580
Saws, back................................. 522
Scavenging valve........................... 124
Scheel, Kar................................ 124
Schneider 155-mm. gun...................... 411
Scott, H................................... 351
Scragging springs, machine for.............. 635
Searchlights............................... 454
Shafts, Cardan, strength of................. 524
Shear strength of plywood webs.............. 577
Ships, concrete............................ 239
    merchant, war experience with.......... 526
    oil vs. coal for use on................ 124
    rivetless.............................. 524
Similarity, dynamical....................... 630
Sinnatt, F. S.............................. 405
Sled, motor................................ 299
Smith, A. R................................ 456
Smith, D. J................................ 177
Smith, J. W................................ 694
Smith, P. H................................ 466
Smith gas producers........................ 177
Snow, W. H................................. 585
Somerfeld.................................. 115
Sonic wave-power tools...................... 633
Southcombe, J. E........................... 116
Springs, helical, machine for testing....... 635
Stamping, hot, British and American practice in.. 585
Stamps, steam and friction.................. 585
Standardization of frequencies.............. 242
    of power piping........................ 63
    of power plants........................ 299
Stanton, Dr. F. H.......................... 115
Staybolts in locomotive boilers, deflection of.... 123
Steam engine, reciprocating................. 242
    low-pressure, control device........... 694
    nozzles................................ 697
    pressures, higher...................... 460
    plants, hydraulic and combined operation of. 518
Steel, alloy, molybdenum in................. 172
    alloy, pickling of..................... 234
    carbon, hardening of................... 580
    castings, light........................ 579
    copper................................. 692
    industry, truck transportation in...... 699
    nickel, critical ranges of............. 351
    physical properties of................. 406
    valve, in internal-combustion engines.. 697
    zirconium-containing................... 697
Stern, H................................... 405
Stevens, John A............................ 64
Stevens super-power central station......... 453
Stevens-Pratt boilers...................... 453
Stoker, mechanical, aboard ship............. 175
    Pluto.................................. 632
    underfeed, burning coke breeze on...... 465
Stoney, Gerald............................. 61
Storage-battery testing.................... 55
Strand, C. H............................... 631
Stromeyer, C. E............................ 65
Stuart, Geo. J............................. 63
Stutz, C. C................................ 175

Supercharging.............................. 354
Superheater, Horwich, for locomotives...... 65
Superpower central station, Stevens........ 453
    station, design of, General Electric... 456
Surface combustion......................... 404

## T

Talbot, Prof. A. N......................... 291
Tanner, J. Roy............................. 63
Taylor, C. S............................... 234
Taylor, Rear-Admiral D. W.................. 693
Taylor, Roger.............................. 63
Temperatures, high, dynamical method for raising gases to........................... 637
Testing, engine, new Bureau of Standards plant. 408
    magnetic, apparatus for................ 698
    plants for water turbines.............. 513
Tests, impact, on alloys................... 519
"Therm" (gas unit)......................... 633
Thom....................................... 524
Thoma, D................................... 409
Thornton, H. M............................. 353
Tidal-power development at Hopewell........ 115
Time, unit of, absolute.................... 120
Tiredness.................................. 238
Tools, lathe turning. Cutting power of..... 117
    sonic wave-power...................... 633
Tostevin, Engr-Commander H. B.............. 357
Train-control system, automatic, Regan.... 696
Transmission, oil, Carey system of......... 696
Transport, aerial, of woolen goods......... 289
Transportation, truck, in steel industry... 699
Turbines, employment of nozzles in........ 513
    Francis............................... 513
    gas, Holzwarth........................ 292
    geared............................... 577
    high-speed, gears for................. 61
    hydraulic, Kaplan..................... 516
    rotors for, balancing................. 460
    steam, development of................. 489
    steam, speed and power limits of..... 62
    water with two Pelton wheels of different diameters.............. 57
Turner-Moore motor-car engine.............. 407

## U

Uebelbronde................................ 356
Upton, G. B................................ 233

## V

Valve, drifting, automatic. Ripken, for locomotives............................... 121
    hydraulic, balanced................... 294

motor-car-engine, accentuated inclination of.. 407
    scavenging............................. 174
    single-sleeve, Burt................... 583
    steels in internal-combustion engines. 59
    stop, automatic, Coxon................ 64
Variable-stroke aero engine, Zeitlin...... 582
Vautrin, A................................ 238
Veil, S................................... 358
Ventilation, ozone in..................... 406
Vibrations, ground, due to reciprocating machinery 693
    ground, transmission of............... 693
Vickers submarine-boat type engine........ 627
Viscosity of lubricants, standard test for.. 408
    Saybolt, of blends.................... 463

## W

Walker, Wm. J.........................361, 467
War expedients as to metals in Germany.... 464
    Vessels, design of.................... 693
Watkinson, Prof. W. H..................... 637
Warships, mechanical reduction gears in... 337
Water consumption of power stations....... 458
    filament flow......................... 516
    indicating devices for locomotives.... 693
    power................................. 513
    purification.......................... 462
    softening in Pittsburgh district..... 587
    wheel for low-head plants............. 517
Waves, sonic, Constantinesco.............. 359
Webs, plywood, shear strength of.......... 577
Welding, electric, use on ship Faragut.... 524
Wheeler, Richard Vernon................... 239
Whistle valve, remote-control............. 454
Williams, Hugh R.......................... 522
Wills, H.................................. 385
Wilson, Austin R.......................... 519
Wilson, T. R. C........................... 291
Wind tunnel, McCook Field................. 403
Wind-tunnel studies in aerodynamic phenomena.. 403
Wing, Alula............................... 579
Wire drawing, diamond dies for............ 118
Wireless.................................. 454
Women workers in Britain.................. 180
Worm gear, David Brown & Son's type....... 116
    gears, in volute..................... 117
Wragg, A.................................. 406
Wuerth, Albert............................ 464

## Y

Yates, Raymond Francis.................... 123

## Z

Zeitlin variable-stroke aero engine....... 582
Zirconium-containing steels............... 697

FRED J. MILLER         PRESIDENT, 1920
THE AMERICAN SOCIETY OF MECHANICAL ENGINEERS

# MECHANICAL ENGINEERING

## THE JOURNAL OF THE AMERICAN SOCIETY OF MECHANICAL ENGINEERS

Volume 42        January, 1920        Number 1

# A Dredging Pump of Novel Construction

By WALTER J. WHITE,[1] NEW ORLEANS, LA.

THE port of New Orleans has long been denied its natural growth by restrictions imposed by the civil code of the state of Louisiana, which declares that the banks of all navigable rivers are public property and as such cannot be sold nor used for factory sites, shipyards, or other industrial enterprises. Attempts have been made to overcome this industrial handicap by proposals to build a navigable canal which would not be subject to the restrictions just enumerated. Some fifteen years ago Mr. Walter Parker proposed that such a waterway be built connecting the Mississippi River with Lake Pontchartrain, but the task was not undertaken until shortly before the war.

It is not the purpose of this paper, however, to deal with the history of this canal nor to present the many industrial advantages which its creation will bring to the port of New Orleans, but rather to present some of the outstanding engineering features incident to the completion of the work.

It may be of interest to note in passing, that in spite of the prevalent belief that the civil code has been a handicap to the general industrial development of New Orleans, even if there were no law standing in the way, manufacturing plants located upon the Mississippi would be exposed to many dangers occasioned by the encroachment of the river. Furthermore, any manufacturing plant due to the great range of level of the river would be burdened with the cost and inconveniences of loading and unloading of water craft and such fact would serve to make impracticable the operation of the plant. It is also generally conceded by experts in shipbuilding that operations of that kind could not be maintained on the banks of the river with any hope of success. It is of course obvious that an inner-harbor navigation canal (controlled by locks) would be entirely suitable for all such purposes.

During the construction of the canal the very unusual character of the dredged material necessitated a complete change in the construction of the dredge pumps, and because the en-

gineering features are particularly interesting, it is to this phase of the work that the writer wishes to direct attention; but before proceeding to examine the methods employed, a brief description of the canal itself will undoubtedly be of interest.

The surface elevation of that part of the city of New Orleans in which the canal is located is the same as mean gulf level. Occasional high tides, due to storms, occur when the water rises to an elevation of from three to five feet higher than the ground surface. It was therefore desirable that the dredge excavation be handled in such a way that the greatest benefit would be secured from the disposal of the material. This was accomplished by depositing the dredged material so as to form as nearly as practicable a continuous fill, which thus serves as a levee along both banks of the canal and which prevents any possibility of overflow during storm tides. The dredge excavation furthermore provides high, well-drained sites for shipyards and other industries. The length of the Inner Harbor-Navigation Canal from deep water in the Mississippi River to the 14-ft. contour in Lake Pontchartrain is 31,800 ft. The completed project is to have a clear channel width of 150 ft. on the bottom with a depth of 30 ft. below low water. In the development of shipyards and other industrial sites no encroachment on the channel width will be allowed, such locations being provided with launching basin, slips, piers, etc.

FIG. 4 THE DREDGE *Captain Huston*, USED IN THE EXCAVATION OF THE NEW ORLEANS INNER-HARBOR NAVIGATION CANAL

While it is true that the Mississippi River and Lake Pontchartrain will be connected, it is believed that the use of the word canal is misleading and very likely to be misunderstood. It should be remembered that Lake Pontchartrain is a shallow body of water when compared with the depth of 30 ft. in the inner harbor. The present plan contemplates the construction of an inner harbor, or basin having a constant level and connected with the deep-water channel of the Mississippi River, entrance from the river channel to the inner harbor being by means of one lock. After the harbor feature has been considerably developed, it is quite probable that extensive dredging operation will be undertaken to establish a deep-water channel from the north end through Lake Pontchartrain and Mississippi Sound to the Gulf,

---

[1]Supt. of Dredging, Board of Commissioners, Port of New Orleans. Assoc-Mem.Am.Soc-M.E.

Abstract of a paper presented at the Annual Meeting, New York, December 1919, of THE AMERICAN SOCIETY OF MECHANICAL ENGINEERS. All papers are subject to revision.

FIG. 2 VIEW OF CANAL SITE LOOKING SOUTH FROM THE RAILROAD

thereby justifying, in the fullest sense, the name Inner Harbor-Navigation Canal—an *Inner Harbor* having a constant level and a *Canal* connecting deep water in the Mississippi River with deep water through Lake Pontchartrain to the Gulf.

During the construction of the canal, traffic had to be maintained over railroads, electric-car lines, and highways which crossed the line of the Inner-Harbor. This feature and the necessity of utilizing the excavated material in levees and fills along the banks, and also the requirement that all excavated material be deposited well back from the edge of the bank, prohibited the use of dipper dredges or other types of dredges which dispose of material by means of scows or by depositing along edge of cut.

While planning methods by which this dredging could be done economically, and at the same time rapidly, several test pits were excavated which confirmed the belief that about 80 per cent of the area to be excavated was underlaid with large cypress stumps which at some time in the past had either subsided or had been covered by deposit from overflow. In addition to the submerged

stumps there was a heavy surface growth of cypress and other timber from Station No. 88 to Station No. 221. The topography of this country is best shown by Figs. 2 and 3. After clearing the standing timber from the area to be dredged tentative efforts were made to remove all stumps by shooting and grubbing, but it was found that this would greatly increase the cost of the work. During the progress of the work very little manual labor was expended on stumps and roots, although it was found to be advantageous to blast some of the larger surface stumps.

Notwithstanding the great number and character of stumps and roots, as evidenced by Fig. 4, it was decided for the reasons mentioned that the work could best be done by hydraulic pipe-line dredges of the cutter type and contractors were asked to submit bids on the work. Due to the difficulties mentioned, however, contract prices could not be secured and arrangements were therefore made to charter dredges, the Board of Port Commissioners accepting all responsibility for output. Four 20-in. dredges, and one 22-in. dredge, the *Texas*, were accordingly secured and work was commenced on May 15, 1918. Fig. 1

FIG. 3 VIEW OF CANAL AFTER EXCAVATION

shows one of these dredges, the *Captain Huston*. The dredges were powerful machines of good design in all respects, but it was fully realized that much trouble was to be expected on account of stumps and roots.

By utilizing Bayou Bienvenue four of the dredges were brought in to a point on the canal near Station No. 88, where one was placed in operation headed toward Lake Ponchartrain and three were started toward the Mississippi River; one dredge was also started south from Lake Ponchartrain. On account of large areas in suction and discharge pipes and pump, the 22-in. dredge *Texas* was placed in the most difficult section. After allowing a reasonable time for dredge and crew to become adjusted to conditions, it became evident that progress would be slow and the work would be expensive, due entirely to delays and decreased capacity occasioned by stumps and roots clogging pump throat and suction. The soil itself was easily handled by a hydraulic dredge and the pipe-line conditions were not difficult, but the amount and character of roots encountered and later successfully handled is beyond description. Fig. 5 shows the character of the material actually pumped through the dredge *Texas*. It was thought that the solution of this difficulty would be found in a pump impeller designed with large and easy passage areas and having no projecting webs on vanes or hub. The writer learned that Mr. A. B. Wood, mechanical engineer for the Sewerage and Water Board of New Orleans, had designed and patented a centrifugal-pump impeller for handling sewage containing trash. An impeller for a 12-in. pump was inspected, and while none had heretofore been used on dredges of the cutter type it was believed that the root problem was solved and Mr. Wood was engaged to design an impeller for the 22-in. dredge *Texas*. All costs in connection with the manufacture of the impeller were to be borne by the Board of Port Commissioners regardless of whether or not it proved a success.

From the first the results obtained with the new impeller were remarkable, the increase in output being between two and three hundred per cent. The increased yardage excavated by the dredge *Texas* is shown by the following comparative statement, covering thirty days' operation immediately before and immediately after

substituting Mr. Wood's impeller for the old type:

| | Old Impeller | New Impeller |
| --- | --- | --- |
| Excavating, hours | 482.43 | 394.00 |
| Clogged suction delay, hours | 130.75 | 71.50 |
| All other delay[1], hours | 106.82 | 254.50 |
| Advance, feet | 873 | 2,901 |
| Excavated material, cubic yards | 73,472 | 175,460 |
| Average yardage per excavating hour | 152 | 445 |

The operation and output of the 20-in. dredges *Captain Huston* and *Dixie* was very greatly facilitated and increased by certain modifications made in pump casings and impellers. Area in pump throat was increased by changing the shape of the throat ring and fitting a new piece of suction pipe. As previously mentioned, it was recognized that in order to secure maximum output in material of this character, it would be necessary to provide as large passages through pump and impeller as possible, and it was therefore determined that two-thirds of each alternate vane should be cut out of the impellers in these dredges, as shown in Fig. 6 and 7. The outer one-third of the vanes was left to preserve structural strength, but after observing the performance of the pumps it is the opinion of the writer that every alternate vane could have been entirely removed.

During the first three days that the dredge *Captain Huston* was in operation and prior to making alterations to the pump the yardage output of the dredge was practically nothing. This was due to the fact that the pump-suction throat was continuously clogged with stringy roots and stumps, which it was impossible for the pump to handle. Conditions could not have been worse and it was felt that any method whereby the pump throat and impeller could be opened up would result in an increased output. It is admitted that some uneasiness was felt as to how these alterations might effect the performance of the pumps, but results have more than justified what at first seemed to be a drastic and

[1] The large number of delay hours after installation of new impeller was due to time lost cleaning boilers and raising a sunken fuel barge. For the period covered in the tabulation it will be seen that although the installation of new impeller decreased the number of excavating hours, the average advance in yardage was much greater. From observation of the dredge in operation it was seen that the ability of the impeller to pass stumps and roots insured the maintenance of a satisfactory output.

hazardous measure. Pump and engine tests, typical curves of which are shown in Fig. 8, have shown that these alterations did not effect the capacity or efficiency of the pump, other than by greatly facilitating the passage of roots, etc., through the pump. While the writer understands that radical alterations should not be made without careful consideration, it is undoubtedly true that the performance of a dredge is to be judged largely, if not solely, by the unit cost of output, and where for any reason output is

The impeller designed for the 22-in. dredge *Texas* was of same diameter (80 in.) as that of the old impeller, but it had two instead of four vanes. An examination of Figs. 9 and 10 will show that, due to the peculiar shape of vane, the lodging or collection of roots or other material is made difficult, if not impossible, every encouragement being given by shape of vane and throat of impeller to the continuous flow of water containing irregular shape roots or other objects. The writer does not wish to convey

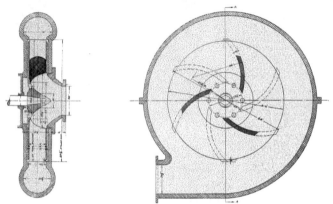

FIG. 6 DETAILS OF PUMP ON DREDGE *Dixie* SHOWING THE MODIFIED THROAT AND IMPELLER

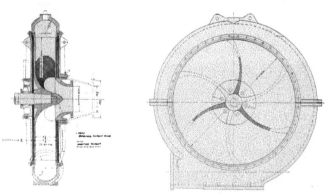

FIG. 7 DETAILS OF PUMP ON DREDGE *Captain Huston* SHOWING THE MODIFIED THROAT AND IMPELLER

not being secured any modifications of plant are justified, provided that the unit cost of output is controlled. In this particular instance there was involved the difference between failure and success. Assuming the same general character of excavation, a further modification along the lines indicated would, it is believed, result in a further increase in output.

the impression, however, that all of the large cypress stumps and roots were cut up by the cutters and handled through a 20- or 22-in. dredging pump. While suction ladders were built for heavy service and cutters were of good design and powerful, a great many of the stumps were undercut and allowed to sink to the bottom, where they were deposited below grade.

In general the operation and maintenance of the dredges followed the usual procedure for dredges of this type, and it is not believed that any novel feature other than in connection with modification of pump design can be presented. The 22-in. dredge *Texas* was equipped to use ball joints as connections between sections of pontoon pipe, and it is believed that the passage of roots was made easier than if rubber sleeves had been used.

Pump impellers having two, three, four, six, and seven vanes and varying in diameter from 68 in. to 96 in. were used in the different dredges engaged on this work. While it is the intention to present methods and actual results secured rather than to discuss the design and performance of centrifugal pumps and impellers, it is thought that some comparative observations made of the different pumps will be interesting. Tests to determine certain centrifugal dredge-pump characteristics while pumping water were carefully made. The results of these tests are given in Tables 1 and 2. Observers were selected and were rehearsed several times to make sure that they understood just what was required of them. All practicable precautions were taken so that accurate results would be secured. Readings were recorded as observed, the object being to eliminate any intention to interpret or analyze readings while test was in progress. A study of the readings does not reveal any wide range in any set of readings for a given condition. All the figures in Table 1 are a mean of a number of readings and the writer has confidence in the correctness of all observation and computations, and it is believed that the data presented will be of value in analyzing and studying the design and performance of centrifugal dredging pumps. At the time tests were made, the dredges were in good operating condition, but no special preparation of machinery was made. All of the pumps tested were of the single suction type and all impellers, with one exception, were shrouded on both sides. The pumps were directly connected to vertical triple-expansion engines.

Upon the completion of the tests made on the dredges *Captain Huston* and *Dixie*, both of which were equipped with pumps

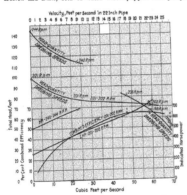

FIG. 8   CURVES OF DREDGE *Texas* EQUIPPED WITH IMPELLER
DESIGNED BY A. B. WOOD

having modified propellers, it was decided to test the pumps of the dredge *Pelican*, which was fitted with an open impeller. This pump, shown in Fig. 11, was tested with impeller in the pump in the normal manner with the vanes curved backward and also with the impeller reversed,[1] i.e., with the vanes

[1] In Trans.Am.Soc.C.E. vol 50, page 505, Oct. 1905, Colonel C. W. Sturtevant, Mem.Am.Soc.C.E., mentions a centrifugal pump having been operated with impeller vanes curved forward instead of backward.

curved in the direction of rotation. The results were surprising and are shown in detail in Table 1 and the curves of Figs. 12 and 13. So far as the operation of the dredge was concerned, both while pumping water and also while dredging, there was nothing

FIG. 9   REAR VIEW OF TWO-VANED IMPELLER DESIGNED BY
A. B. WOOD

to indicate to the closest observer that the pump was not assembled in the usual manner. After completion of tests it was found by comparison that there was a 10 per cent loss of

FIG. 10   VIEW SHOWING THE LARGE PASSAGEWAYS OF THE IMPELLER
DESIGNED BY A. B. WOOD

efficiency due to increased consumption of power with the reversed impeller. As a matter of interest it might be stated that the dredge *Pelican* was operated from June 1 to June 14, 1919, with the reversed impeller, nothing unusual being noticed in operation and no diminution in yardage being apparent.

In the tests of dredges *Texas* and *Captain Huston* the quantity of water discharged by the pump was measured by pitot tube and venturi meter. While testing the dredge *Captain Huston* the pitot tube was bent and was not used in the tests of the other dredges. During tests the venturi meter was located just astern of the dredge in a straight section of pipe, where it would not be influenced by elbows. Suitable arrangements were made so that the discharge pipe could be throttled, or closed off entirely, and in this way a variation in head and quantity was created. Piezometers were ¼-in. tee-handled pet cocks, which were screwed into

which reference has already been made; one 20-in. dredge chartered at a monthly rate for the bare dredge, all operating and maintenance costs being borne by the Board; and two 20-in. and one 22-in. dredges were chartered by the Board at a rate to include all operating and maintenance charges except fuel and shore pipe line, which were borne by the Board. In securing dredges for this work it was necessary to tow them to New Orleans from more or less distant points, two having been brought from Havana, Cuba, the Board assuming all towing charges and paying one-half the charter rate while dredges were in transit. A

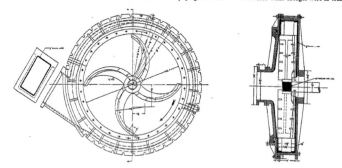

FIG. 11  DETAILS OF THE IMPELLER OF THE DREDGE *Pelican*

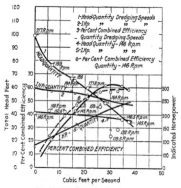

FIG. 12  TEST CURVES OF DREDGE *Pelican*

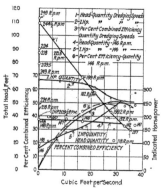

FIG. 13  TEST CURVES OF DREDGE *Pelican* WITH IMPELLER REVERSED

the pipe. Indicators were not calibrated, but are believed to have been in excellent condition. Cards were taken simultaneously on the three engine cylinders. It was unfortunate that during the drop in pressure test made on dredge *Dixie* it was impossible to obtain comparable peripheral speeds due to valve of intermediate cylinder being improperly set.

From May 15, 1918, to August 1, 1919, there were employed in the construction of the Inner Harbor-Navigation Canal four 20-in. dredges and one 22-in. dredge. One of the 20-in. dredges was owned and operated by the Board of Port Commissioners, to

careful record of costs was kept and the total yardages by months and the total and the unit costs of the same will be found in Table 3, page 80. These costs are divided under three headings:

ARBITRARY

  Cost of towing to and from New Orleans

  One-half rental paid while in transit

  Cost of 477,193 cu. yd. of preliminary and auxiliary dredging

  Cost of design and manufacture of two-vane impeller for 22-in. dredge

TABLE 1 DATA AND RESULTS OF TESTS OF CENTRIFUGAL DREDGING PUMPS

| Test Number | Steam Pressure, lb. sq. in. gage. | r. p. m. | Suction, ft. of water. | Discharge, ft. of water. | Total, Including Difference of Velocity Heads. | Flow in ft. per sec. by venturi meter. | Indicated hp. | Water hp. | Combined Efficiency, Engine and Pump | Peripheral Speed, ft. per sec. |
|---|---|---|---|---|---|---|---|---|---|---|
| **Texas — February 14, 1919** | | | | | | | | | | |
| 1 | 175 | 201 | − .94[a] | 76.02 | 77.03 | 20.16 | 423.4 | 175.8 | 41.5 | 70.16 |
| 2 | 175 | 201 | + .86[a] | 99.34 | 98.48 | 0.0 | 254.45 | 0.0 | 0.0 | 70.16 |
| 3 | 173 | 202 | −12.94 | 57.57 | 71.18 | 58.37 | 742.5 | 471.0 | 63.3 | 70.51 |
| 4 | 170 | 240 | − 1.34[a] | 112.04 | 113.47 | 22.86 | 681.9 | 293.0 | 43.0 | 83.78 |
| 5 | 175 | 244 | + .86[a] | 145.29 | 144.43 | 0.0 | 403.4 | 0.0 | 0.0 | 85.18 |
| **Captain Huston — February 23, 1919** | | | | | | | | | | |
| 1 | 200 | 141 | −30.20 | 25.83 | 54.26 | 52.38 | 657.9 | 322.1 | 49.0 | 59.06 |
| 2 | 198 | 161 | −15.23 | 85.11 | 99.56 | 35.17 | 706.7 | 397.8 | 56.3 | 67.44 |
| 3 | 200 | 165 | −12.74 | 93.65 | 105.69 | 33.05 | 729.5 | 396.0 | 54.1 | 69.12 |
| 4 | 200 | 166 | + .28[a] | 99.89 | 99.61 | 0.0 | 292.7 | 0.0 | 0.0 | 69.54 |
| 5 | 191 | 168 | −10.06 | 98.96 | 108.49 | 28.87 | 682.8 | 355.8 | 52.1 | 70.38 |
| 6 | 200 | 174 | −10.97 | 104.97 | 115.33 | 30.78 | 742.8 | 403.0 | 54.3 | 72.95 |
| 7 | 200 | 182.5 | − 6.56 | 123.67 | 129.88 | 23.47 | 756.8 | 346.0 | 45.7 | 76.48 |
| 8 | 200 | 184 | − 6.40 | 122.22 | 128.27 | 23.47 | 759.6 | 341.7 | 45.0 | 77.00 |
| 9 | 200 | 221.5 | + .28[a] | 186.94 | 186.66 | 0.0 | 426.0 | 0.0 | 0.0 | 92.78 |
| **Dixie — March 11, 1919** | | | | | | | | | | |
| 1 | 158 | 158 | −25.99 | 28.52 | 54.70 | 49.55 | 559.55 | 307.50 | 55.0 | 57.91 |
| 2 | 155 | 160 | −22.98[b] | 34.27 | 57.42 | 46.62 | 568.83 | 303.71 | 53.4 | 58.64 |
| 3 | 150 | 163 | − 7.68 | 58.36 | 66.10 | 27.29 | 446.78 | 204.66 | 45.8 | 59.74 |
| 4 | 156 | 163.5 | −18.48[b] | 44.22 | 62.85 | 42.0 | 551.42 | 299.56 | 54.3 | 59.92 |
| 5 | 145 | 166 | − 8.16 | 60.21 | 68.44 | 27.81 | 471.47 | 215.94 | 45.8 | 60.84 |
| 6 | 152 | 174.75 | − 9.32 | 66.68 | 76.08 | 30.35 | 545.4 | 261.8 | 48.0 | 64.05 |
| 7 | 155.7 | 175.5 | −10.74 | 64.71 | 75.54 | 32.33 | 577.42 | 277.25 | 48.0 | 64.32 |
| 8 | 154.5 | 176 | − 2.68[a] | 76.37 | 79.08 | 17.85 | 479.15 | 160.15 | 33.5 | 64.50 |
| 9 | 156 | 176 | − .03 | 78.91 | 78.95 | 8.95 | 407.72 | 80.11 | 19.65 | 64.50 |
| 10 | 158 | 176.5 | + .92[a] | 82.61 | 81.79 | 0.0 | 310.24 | 0.0 | 0.0 | 64.68 |
| 11 | 151 | 186 | − 3.60 | 83.76 | 87.40 | 19.47 | 547.96 | 193.06 | 35.3 | 68.17 |
| 12 | 150 | 197 | − .38[a] | 99.46 | 99.85 | 10.21 | 552.39 | 115.66 | 20.95 | 72.20 |
| 13 | 154 | 219 | + .82[a] | 127.17 | 126.35 | 0.0 | 578.31 | 0.0 | 0.0 | 80.26 |
| **Pelican — May 28, 1919** | | | | | | | | | | |
| 1 | 174.5 | 119.5 | + 3.99 | 8.93 | 23.16 | 30.06 | 139.95 | 78.9 | 56.3 | 38.06 |
| 2 | 175 | 132 | −17.40 | 10.58 | 28.26 | 32.90 | 184.33 | 105.2 | 57.2 | 42.04 |
| 3 | 175 | 145 | −21.96 | 11.78 | 34.10 | 37.04 | 246.10 | 143.0 | 58.1 | 46.18 |
| 4 | 182.5 | 145 | + .88[a] | 41.45 | 40.57 | 0.0 | 75.66 | 0.0 | 0.0 | 46.18 |
| 5 | 175 | 146.5 | −1[illegible].96 | 16.63 | 35.90 | 34.58 | 231.32 | 140.3 | 60.7 | 46.66 |
| 6 | 180 | 148 | − 3.87[a] | 36.38 | 40.32 | 16.61 | 162.82 | 76.0 | 46.6 | 47.14 |
| 7 | 175 | 149 | + .28[a] | 42.88 | 42.61 | 5.74 | 117.93 | 27.7 | 23.5 | 47.46 |
| 8 | 180 | 150.5 | − 1.57[a] | 40.06 | 41.67 | 11.99 | 147.15 | 56.6 | 38.5 | 47.93 |
| 9 | 182.5 | 159 | −11.34 | 34.03 | 45.55 | 25.96 | 236.30 | 133.8 | 56.7 | 50.64 |
| 10 | 182.5 | 164 | −12.91 | 35.03 | 48.14 | 28.01 | 249.94 | 152.8 | 61.2 | 52.23 |
| 11 | 182.5 | 177 | − 7.66 | 51.68 | 59.45 | 20.93 | 272.30 | 140.8 | 51.75 | 56.37 |
| 12 | 177.5 | 184 | − 4.0 | 60.61 | 64.69 | 15.97 | 276.22 | 117.0 | 42.4 | 58.60 |
| 13 | 175 | 186 | − 2.82[a] | 63.63 | 66.51 | 14.65 | 286.53 | 110.2 | 38.5 | 59.24 |
| 14 | 175 | 199 | + .08[a] | 76.78 | 76.71 | 6.79 | 283.2 | 59.0 | 20.75 | 63.38 |
| 15 | 177.5 | 217 | + .88[a] | 99.41 | 98.53 | 0.0 | 295.22 | 0.0 | 0.0 | 69.11 |
| **Pelican — May 31, 1919 — During this test the runner was reversed** | | | | | | | | | | |
| 1 | 175 | 118 | −13.49 | 7.98 | 21.69 | 28.85 | 151.78 | 70.9 | 46.75 | 37.58 |
| 2 | 176 | 137 | −19.67 | 10.53 | 30.52 | 35.1 | 255.12 | 121.3 | 47.6 | 43.63 |
| 3 | 189 | 141 | −21.75 | 11.28 | 33.39 | 36.9 | 288.57 | 139.3 | 48.35 | 44.91 |
| 4 | 172.5 | 142 | −17.53 | 15.08 | 32.89 | 32.85 | 256.88 | 122.3 | 47.7 | 45.23 |
| 5 | 177.5 | 146 | − 3.93[a] | 31.88 | 35.88 | 16.65 | 154.56 | 67.70 | 43.8 | 46.50 |
| 6 | 175 | 146 | + .88[a] | 37.98 | 37.10 | 0.0 | 75.85 | 0.0 | 0.0 | 46.50 |
| 7 | 175.5 | 147 | − 1.13[a] | 36.68 | 37.84 | 10.60 | 138.93 | 45.5 | 32.7 | 46.82 |
| 8 | 172.5 | 148 | + .88[a] | 38.18 | 37.30 | 0.0 | 79.48 | 0.0 | 0.0 | 47.14 |
| 9 | 172.5 | 150 | + .64[a] | 40.08 | 39.44 | 3.76 | 106.66 | 16.78 | 15 73 | 47.78 |
| 10 | 181 | 152 | −10.63 | 28.58 | 39.38 | 25.90 | 233.58 | 115.3 | 49.4 | 48.41 |
| 11 | 181 | 159 | −11.84 | 31.18 | 43.22 | 27.54 | 266.79 | 134.7 | 40.55 | 50.64 |
| 12 | 177.5 | 182 | − 7.14 | 50.18 | 57.44 | 21.38 | 307.91 | 139.0 | 45.2 | 57.97 |
| 13 | 187.5 | 199 | − 3.88[a] | 64.88 | 68.83 | 16.49 | 321.50 | 128.5 | 40.0 | 63.38 |
| 14 | 181 | 209.5 | + .88[a] | 79.68 | 78.80 | 0.0 | 194.58 | 0.0 | 0.0 | 66.73 |
| 15 | 180 | 218 | + .88[a] | 87.83 | 86.95 | .0 | 227.89 | 0.0 | 0.0 | 69.43 |
| 16 | 170 | 221.5 | − .43 | 88.78 | 89.23 | .17 | 327.96 | 82.50 | 25.18 | 70.55 |
| 17 | 185 | 224 | + .88[a] | 90.68 | 89.70 | .0 | 243.34 | 0.0 | 0.0 | 71.34 |
| 18 | 185 | 244.5 | + .88[a] | 109.78 | 108.90 | 0.0 | 334.57 | 0.0 | 0.0 | 77.87 |
| 19 | 185 | 249 | + .88[a] | 113.58 | 112.70 | 0.0 | 354.46 | 0.0 | 0.0 | 79.31 |

[a] Readings of suction gage incorrect because of whirl set up in pipe by pump impeller.

[b] Suction readings incorrect due either to pet cock being throttled down too much or getting temporarily choked up.

DISCHARGE-PIPE-LINE CONDITIONS FOR TESTS OF DREDGING PUMPS GIVEN IN TABLE 1

*Texas:* Tests 1 and 4, 8-in. valve on end of discharge pipe wide open. Tests 2 and 5, 8-in. valve on end of discharge pipe cut off. Test 3, pipe line 570 ft. pontoon and shore pipe.

*Captain Huston:* Test 1, Pipe full opening. Test 2, Iron plate bolted on discharge pipe. Tests 3, 7, and 8, 8-in. valve on end of discharge pipe wide open. Tests 4 and 9, 8-in. valve on end of discharge pipe cut off. Tests 5 and 6, Pipe line 3538 ft. pontoon and shore line.

*Dixie:* Test 1, Pipe full opening. Test 2, Wooden piece bolted on discharge pipe. Tests 3, 5, and 6, Pipe line 1583 ft. pontoon and shore pipe. Test 4, 2 pieces of wood bolted on end of discharge pipe. Test 7, Iron plate bolted on end of discharge pipe. Tests 8 and 11, 8-in. valve on end of discharge pipe wide open. Tests 9 and 12, 8-in. valve on end of discharge pipe partly closed. Tests 10 and 13, 8-in. valve on end of discharge pipe cut off.

*Pelican:* Tests 1, 2, and 3, Pipe full opening. Tests 4 and 15, 8-in. valve on end of discharge pipe cut off. Tests 5 and 10, Wooden piece bolted on end of discharge pipe. Tests 6 and 11, 3 wooden pieces bolted on end of discharge pipe. Tests 7 and 14, 8-in. valve on end of discharge pipe partly closed. Tests 8, 12 and 13, 8-in. valve on end of discharge pipe wide open. Test 9, 2 wooden pieces bolted on end of discharge pipe.

*Pelican* (with reversed runner): Tests 1, 2, and 3, Pipe full opening. Test 4, Wooden piece bolted on end of discharge pipe. Tests 5 and 12, 3 wooden pieces bolted on end of discharge pipe. Tests 6, 8, 14, 15, 17, 18 and 19, 8-in. valve on end of discharge pipe cut off. Tests 7 and 8, 8-in. valve on end of discharge pipe wide open. Tests 9 and 16, 8-in. valve on end of discharge pipe partly closed. Tests 10 and 11, 2 wooden pieces bolted on end of discharge pipe.

In all cases where length of discharge pipe is not given the length was less than 100 ft. from the stern of dredge. In general, it should be stated that the figures in Table 1 are a mean of a number of readings recorded during tests, which explains why the figures are worked out to two decimal places. The venturi meter had an inlet diameter of 22 5/16 in. and a throat diameter of 18 in. For readings at cut off, the elevation of water at rest in suction pipe was taken as suction head and for other readings of low quantity the suction head was taken from loss pressure curve plotted from high quantity readings.

TABLE 2 DATA ON CENTRIFUGAL DREDGING PUMPS

| | Texas | Capt. Huston | Dixie | Pelican | Pelican† |
|---|---|---|---|---|---|
| Diam. of impeller, in.... | 80 | 96 | 84 | 73 | 73 |
| Number of vanes....... | 2 | 3* | 3* | 4 | 4 |
| Diam. of suction pipe at suction gage, in....... | 23¾ | 19 11/16 | 20⅝ | 20¼ | 20¼ |
| Diam. of discharge pipe at discharge gage, in... | 22⅝ | 20 11/16 | 20⅝ | 25½x12⅝ | 25½x12⅝ |
| Triple expans. engine, in. | 14x21½x35 | 14½x22½x40 | 14½x22½x36 | 11⅝x18x29 | 11⅝x18x29 |
| Length of stroke, in..... | 18 | 20 | 18 | 18 | 18 |
| Water at rest in canal above center-line of pump, in............ | 10½ | 3¼ | 9¼ | ......... | ......... |
| Water in river above center-line of pump, in.... | ......... | ......... | ......... | 10½ | 10½ |
| Zero of suction gage below center-line of pump, in. | 19 | 20 | 16.68 | 1.68 | At center line of pump |
| Zero of discharge gage above center-line of pump, in............ | 20 | ......... | ......... | 7 | 7 |
| Zero of discharge gage below center-line of pump, in............ | ......... | 53⅝ | 47.66 | ......... | ......... |

*Originally 6, parts of 3 Vanes removed. †This test made with impeller reversed.

CHARTER

Charter cost of four dredges

ALL OTHER CHARGES

Fuel

Shore pipe lines

Labor, subsistence, repairs, material, insurance and supervision.

The writer was present and assisted in the very complete test of hydraulic dredges made for the Mississippi River Commission

(Continued on page 80)

# Thermal Conductivity of Insulating Materials

By T. S. TAYLOR,[1] EAST PITTSBURGH, PA.

*Despite the fact that numerous thermal-conductivity measurements have been made, but very few data are available for such materials as are used in the construction of electrical machinery. To further investigate this field the author accordingly conducted a series of tests based upon two methods, both of which are explained in detail in this paper. The results obtained are tabulated in the complete paper for such materials as fish paper, fuller board, cambric, mica tape, various kinds of woods, asbestos, plate glass, sheet steel, wool felt, etc., and a selection covering various miscellaneous materials is given in the present abstract.*

ALTHOUGH numerous experimenters have been interested in making thermal-conductivity measurements, but very few data are available for such materials as are used in the construction of electrical machinery. The most important work on such insulating materials thus far available was done by Symons and Walker,[2] but the materials tested by these authors were special and consequently the values obtained cannot be taken as applicable to similar materials used in the construction of electrical machinery in this country. The investigations herein described were therefore undertaken to obtain values of the thermal conductivity of insulating and other materials, the values of which are of direct interest to those concerned with the heat problem in electrical apparatus. Results have also been obtained for numerous other materials of general interest.

### THE NORTHRUP THERMAL BRIDGE

As a preliminary step it seemed worth while to try the "Thermal Bridge" method suggested by Prof. E. F. Northrup.[3] The bridge used is shown in Fig. 1. It consisted of two soapstone cylinders each 4⅝ in. in diameter, one being 8½ in. long and the other 3 in. long. Each cylinder consisted of an inner core 1⅝ in. in diameter surrounded by a concentric cylinder having a wall thickness of 1⅜ in. The two faces along $MN$ separating the two parts of the apparatus were ground so as to fit very closely. A spiral groove was cut in the top of the longer cylinder $E$ and a heater wire placed in this groove. Small holes, 1, 2, 3, etc., were drilled at right angles to the axis of the cylinders through the outer wall and into the central core as indicated. Copper-constantan thermocouples made of 0.005-in. wire were inserted in these holes for the purpose of determining the temperature gradient along the cylinders. The lower part of the apparatus $F$ was placed on a brass box which served as a cold-temperature reservoir when kept filled with water and ice or when the water was kept circulating through it. The heat generated at the top would flow down the core and outer wall through the junction $MN$ to the reservoir. The purpose of the core and surrounding wall was to insure a uniform flow of heat through the core. Felt was placed around the outer cylinder to further prevent undue loss of heat from the surface of the apparatus.

If the distance from thermocouple 1 to thermocouple 5 is made the same as that from 2 to 8, and the conditions are such that a uniform temperature drop exists along the core, the thermal conductivity of a material placed in $MN$ can be determined in terms of soapstone. First suppose the above conditions to exist when there is no specimen in $MN$. Then the temperature drop will be uniform along the apparatus, as can be tested by the thermocouples. When a sample is inserted in $MN$ and the temperatures of 1, 5, 2 and 8 are measured, the drop between 2 and

8 exceeds that from 1 to 5 by an amount equal to the drop through the sample. The soapstone equivalent of the sample is then readily calculated from the temperature drop through the sample and the drop per unit length along the soapstone.

[In the actual experiment, described by the author in his complete paper, the procedure was slightly different from that outlined above, from the fact that the distance from thermocouple 1 to 5 was not exactly the same as from 2 to 8. The use of the thermal bridge, however, proved to be unsuited to the work, for a series of tests gave results that varied considerably. Another method was therefore employed in which use was made of a "thermal meter." A description of this method follows.—EDITOR.]

### THE "THERMAL METER"

A sketch of the "thermal meter" used is shown in Fig. 2. It consisted essentially of an electric heater $H$ constituting two hot

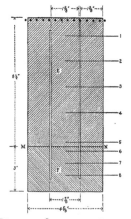

FIG. 1 DIAGRAMMATIC SKETCH OF THE NORTHRUP THERMAL BRIDGE

equitemperature surfaces or sources of heat and two cooling chambers $E$, $E$ (one on either side of the heater) or cold constant-temperature surfaces. The heat generated in $H$ passed laterally through the samples $I$, $I$, of a given material to the cold reservoirs $E$, $E$. The disks were made from two disks of soapstone 9 in. in diameter and ⅜ in. thick. Each disk had a spiral groove of 3/16-in. pitch cut in one face. A heater wire, No. 21 constantan, was wound and cemented securely in the groove of each disk and then the iron disks were cemented together with the sides containing the heater wire adjacent. The two heating elements in the two disks were joined together at the center by means of a peg in one disk being pushed into the spring contact in the other. Potential leads were brought out from each heating element at points 2 in. from the centers. It was later found that potential leads fastened to the heating elements at the points

[1] Westinghouse Research Laboratory, Westinghouse Electric and Manufacturing Company.
[2] See Journal of the Institute of Electrical Engineers (London), vol. 48, p. 674, 1912.
[3] See American Electrochemical Society Journal, no. 24
Abstract of a paper presented at the annual meeting, New York, December 1919, of THE AMERICAN SOCIETY OF MECHANICAL ENGINEERS. All papers are subject to revision.

where the wire started in the outer terminals of the spirals served equally well, since the temperature coefficient of resistance of the constantan wire is very small and furthermore the temperature of the heater was constant over its entire face. Extra turns of wire were wound around the outer edge of the heater in order to prevent the loss of the heat generated in the heating element proper through the edge of the heater. After several trials it was found that this procedure gave a heater which in use had a very constant temperature over its two faces, even up to the outer edge.

Two samples of the material to be tested were always used, each being 9 in. in diameter and from 0.1 to 0.75 in. thick, depending upon the nature of the material. One sample was placed on each side of the heater I, I, Fig. 2. Extra disks of lagging of the same material of the sample or something else suitable were placed on each side of the sample, as shown by the shaded portions in Fig. 2. This made the ultimate drop at high temperatures less than it otherwise would have been, and likewise gave a wider range of mean temperature. The faces of the cold reservoirs E, E, constituting the cold equitemperature surfaces, were made of heavy brass having a diameter of 10 in. The samples, heaters, and cold reservoirs were held securely together by means of bolts extending between these two plates. At first strong spiral springs were used around these connecting bolts to insure uniform pressure, but it was found later that equally satisfactory results could be obtained by merely turning down the nuts on

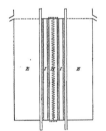

FIG. 2  DIAGRAMMATIC SKETCH OF THE THERMAL METER

the bolts till the samples, etc., were drawn tightly together. Thermocouples of 5-mm. copper-constantan wire were inserted on each side of the samples I, I, under test. Great care was taken in order to insure good contact between the sample and the thermocouple junction. Two couples were placed on each side of a sample, one at the mid-point and another about 1½ in. from the center. The electromotive force of the couples (the cold junction being always kept at 0 deg. cent.) was measured by means of a thermocouple potentiometer. The current in the heater was likewise measured by the same potentiometer by measuring the drop through a standard resistance placed in series with the heater. The potential drop per unit length of heater wire was also measured by means of the potentiometer. This necessitated placing a high resistance in parallel with the heater and then measuring the potential drop across a small fraction of this. The current was supplied by a storage battery and consequently remained quite steady. The cold equitemperature surfaces E, E, were maintained so by having water from the tap circulating through them continuously. The outer edges of the samples and heater were surrounded by felt in order to prevent undue loss of heat from the edges of the heater and samples.

In order to facilitate the work a second apparatus having the same parts as the one described above was constructed. The only difference being that the heater, etc., were of square cross-section. This heater was made of soapstone slabs 12 in. square. The heater wire (No. 21 constantan) was wound back and forth in

parallel slots 3/16 in. apart in each of the parts of the heater. This made its construction quite easy. The elements of the two parts of this heater were joined in parallel as its resistance would have been too high for the available voltage had they been in series. The potential leads of this heater were joined at the ends of a wire in a single groove, thus measuring the drop in a 12-in. length of the wire. Four sets of potential leads were inserted (two in each half) and the average of the four potential drops used in calculating the drop per unit length of heater wire. As for the round heater, extra turns of wire were put around its outer edges to insure a uniform temperature source. Great care was taken to have the same amount of resistance in each element of the heater, since they were joined in parallel. The faces of the cooling reservoirs were of cast brass 13½ in. by 13½ in. by 1 in., and being heavy they did not buckle when bolted together over the heater and the materials tested. Identical results were obtained with the two pieces of apparatus for a given material.

The samples of the materials tested were usually composed of one or more sheets. Extreme care was taken to eliminate the air between the surfaces of the component sheets as much as possible by the use of vaseline, shellac, carpenter's glue, etc. By the use of such materials the drop between component sheets was made quite negligible or at least of the same order of magnitude as the drop through the same distance in the material itself. This was due to the fact that the thermal conductivity of such materials as glue, etc., when dried differs but little in order of magnitude from that of the sheet materials dealt with. In making up a sample that material was used to make good contact between the components which lent itself most readily to the case in question.

## RESULTS OBTAINED BY THE THERMAL METER

Table 1 shows a typical set of results for samples made up of 7 sheets of 0.03-in. fuller board, 0.23 in. in thickness. By plotting the values of the mean temperatures as ordinates and the corresponding values of the thermal conductivity $k$ as abscissæ it is possible to get a measure of the temperature coefficients of thermal conductivity. The values given in Table 2 if plotted will give a straight line whose equation is

$$k_t = k_o (1 - at)$$

where $k_t$ and $k_o$ are the thermal conductiveness at temperatures $t$ and 0, respectively, and $a$ is the temperature coefficient. From the data given the value of $a$ for this sample is found to be about 0.0030.

In the manner thus outlined thermal-conductivity measurements have been made for a large number of materials. The values obtained are recorded in Table 3 of the complete paper and values pertaining to various miscellaneous materials will be found in Table 2 of this abstract. The results in both tables are expressed both in calories per cm. per deg. cent. per sec. and in watts per cm. per deg. cent. per sec. The values are also given for a definite temperature such as 20 deg. cent. as well as for an average range 20 deg. to 80 deg. The temperature coefficient is also recorded as determined in the previous paragraph. It is to be observed that all samples did not show a temperature coefficient. This is doubtless due to the changing characteristics

TABLE 1  THERMAL CONDUCTIVITY OF FULLER BOARD

| Hot-side temp., $t_1$, deg. cent. | Cold-side temp., deg. cent. | Mean temp. deg. cent. | Conductivity $k$, watts, per in. per deg. cent. per sec. |
|---|---|---|---|
| 29.15 | 18.10 | 23.62 | 0.00414 |
| 47.10 | 29.40 | 38.25 | 0.00432 |
| 59.45 | 37.15 | 48.30 | 0.00433 |
| 84.00 | 53.07 | 68.54 | 0.00458 |
| 102.60 | 65.05 | 83.82 | 0.00484 |
| 26.85 | 16.15 | 21.50 | 0.00407 |
| 40.87 | 24.20 | 32.54 | 0.00418 |
| 62.65 | 37.75 | 50.20 | 0.00436 |
| 115.20 | 70.30 | 92.75 | 0.00486 |

of the samples. Thus it is highly possible that the increase in the thermal conductivity due to an increase in temperature is counteracted by a corresponding increase in the thermal resistance due to the change in the surface contacts and likewise increase in air pockets. Besides measuring the thermal conductivity transversely for sheet material, measurements have been made as shown by Table 2 of the longitudinal conductivity along the laminations. The samples for this latter work were prepared for the one apparatus by winding up disks of the material 9 in.

TABLE 2   THERMAL CONDUCTIVITIES

| Material, thickness, inches | Thickness of sample, inches | Sp. gr. | Temp., deg. cent. | Cal. per cm. per deg. cent. per sec. ×10^-4 | Watts per in. per deg. Cent. per sec. ×10^-6 | Temp. Coef. ×10^-4 |
|---|---|---|---|---|---|---|
| HARD RUBBER[1] | 0.380 | 1.19 | 75–50 | 380 | 404 | |
| WHITE FIBER[1] | 0.383 | 1.22 | 20 | 663 | 705 | |
| | | | 20–80 | 695 | 728 | 12 |
| WOODS | | | | | | |
| White Pine[3] | 0.519 | 0.45 | 20–120 | 255 | 271 | |
| White Pine[4] | 0.732 | 0.45 | 30–80 | 613 | 652 | |
| White Oak[3] | 0.516 | 0.60 | 30–80 | 455 | 484 | 18 |
| White Oak[4] | 0.784 | 0.60 | 40–70 | 944 | 1003 | |
| Maple[3] | 0.733 | 0.72 | 20 | 1015 | 1078 | |
| | 0.733 | 0.72 | 20–80 | 1037 | 1100 | 8 |
| Maple[4] | 0.508 | 0.72 | 20–80 | 434 | 462 | |
| ASBESTOS | | | | | | |
| 4 in. Sheet[1] | 0.344 | 0.804 | 23–80 | 395 | 420 | |
| 0.075 in. Paper[1] | 0.306 | 0.98 | 20 | 345 | 367 | |
| 0.035[1] | 0.356 | | 20–100 | 375 | 399 | 24 |
| | | | 20 | 666 | 708 | |
| | | | 20–80 | 685 | 728 | 14 |
| ASBESTOS BOARD[1] | 0.507 | 1.93 | 20 | 1780 | 1890 | 14 |
| PLATE GLASS[1] | 0.252 | 2.49 | 20–90 | 1950 | 2080 | 14 |
| | | | 20 | 1785 | 1900 | |
| | 0.289 | 2.60 | 20–100 | 1945 | 2070 | 15 |
| | 0.289 | | 20 | 1905 | 2024 | |
| | 0.289 | | 20–120 | 2016 | 2142 | 12 |
| SOAPSTONE | 0.715 | 2.87 | 70–175 | 8000 | 8500 | |
| SIL-O-CEL | | | | | | |
| Brick[1] | 0.977 | 0.495 | 30 | 242 | 258 | 15 |
| | | | 30–150 | 262 | 279 | |
| Powdered | 0.955 | 0.15 | 30 | 208 | 222 | 31 |
| | | | 30–150 | 242 | 258 | |
| WOOL FELT | | | | | | |
| Dark Grey[1] | 0.98 | 0.15 | 40 | 149 | 159 | 76 |
| | | | 40–100 | 175 | 186 | |
| GRAPHITE | | | | | | |
| Solid | 1.04 | 1.58 | 50 | 105500 | 112200 | 12 |
| | | | 50–130 | 110200 | 117200 | |
| Powdered, through 20 mesh on 40 mesh | 0.476 | 0.70 | 40 | 2850 | 3030 | 48 |
| | | | 40–100 | 3200 | 3400 | |
| Powdered, through 40 mesh | 0.476 | 0.42 | 40 | 922 | 980 | 40 |
| | | | 40–100 | 1007 | 1080 | |
| Powdered, through 100 mesh | 0.476 | 0.48 | 40 | 438 | 467 | 34 |
| | | | 40–100 | 482 | 513 | |
| LAMP BLACK | | | | | | |
| Eagle Brand | 0.476 | 0.165 | 40 | 156 | 160 | |
| | | | 40–150 | 166 | 176 | 6 |
| Germantown | | | 30 | 265 | 282 | |
| | 0.476 | 0.73 | 30–150 | 298 | 317 | 23 |
| COAL DUST | | | | | | |
| IRON DUST AND SAND | 0.377 | 1.14 | 30 | 460 | 489 | 23 |
| | | | 30–150 | 517 | 550 | |
| SHEET STEEL | | | | | | |
| 0.0172 in M. A. Varnished[1] | 0.415 | ......... | 20 | 1370 | 1455 | 19 |
| | | | 20–80 | 1430 | 1520 | |
| | 1.48 | ......... | 40–100 | 103000 | 109500 | 6 |
| Unvarnished With Asphalt Paint on Sheets[2] | 0.420 | ......... | 40 | 101300 | 107700 | |
| | | | 40–100 | 4710 | 5020 | 10 |
| Unvarnished[3] 0.0172 in. | 0.416 | ......... | 20–80 | 4850 | 5160 | |
| | | | 40 | 1480 | 1570 | 25 |
| Same with Asphalt Paint[1] | 0.475 | ......... | 40–100 | 1580 | 1680 | |
| | | | 40 | 6360 | 6750 | 9 |
| W. A. SILICON STEEL 0.011 in. | | | 40–100 | 6520 | 6930 | |
| Varnished[1] | 0.419 | ......... | 40–80 | 1270 | 1350 | 16 |
| | 1.44 | ......... | 40–100 | 41800 | 44400 | 19 |
| Same Painted with Asphaltum[2] | | | 40 | 39500 | 42000 | |
| Unvarnished[3] | 0.422 | ......... | 40–80 | 4640 | 4920 | 10 |
| | 0.440 | ......... | 20 | 1340 | 1420 | |
| | ......... | ......... | 20–100 | 1470 | 1560 | 25 |
| Same painted as above[2] | 0.443 | ......... | 20 | 4290 | 4470 | |
| | | | 20–100 | 4520 | 4810 | 17 |

[1] Direction of head flow transverse.
[2] Direction of head flow longitudinal.
[3] Direction of head flow across grain.
[4] Direction of head flow along grain.

in diameter, and for the other apparatus by cutting the material into strips and then forcing them tightly together in a special press. By coating the edges of the strips while in the press with glue they could be held together in a square sample and later placed in the apparatus. It is interesting to note that the ratio of the longitudinal to the transverse conductivity is much greater for the mica combinations than for other insulating materials. This is due to the influence of the mica, whose longitudinal conductivity is so much hotter than its transverse. The same ratio has its least value for such materials as varnished cambric and black bias cloth. For these materials there is less difference

between the transverse and longitudinal construction than the mica compounds.

Results were also obtained for the transverse thermal conductivity of 0.0172-in. carbon sheet steel and for 0.014-in. silicon sheet steel. The values obtained are slightly greater than those obtained by other observers for similar materials. Attempts were made to detect the change of transverse thermal conductivity of iron stampings with pressure, but the apparatus did not lend itself readily to this since the exact pressure applied could not be determined. Since this work has become available and a study will be made in the near future of the influence of pressure on the transverse thermal conductivity of iron stampings, and this work will be reported later. It is interesting to note that the transverse conductivity of iron stampings can be increased some 3 to 4½ times by coating the sheets with asphalt paint before putting them together. Consequently if the punchings in electrical apparatus could be assembled in groups, having a gum or other suitable material between the constituents so as to have better contact, the heat generated could be much more readily conducted away. This will also be investigated when the work on iron stampings is resumed.

## SUMMARY OF RESULTS

Experiments in the thermal conductivity of insulating and other materials have led to the following conclusions:

a The "thermal bridge", recommended by Professor Northrup has not been found satisfactory for determining the thermal conductivity of sheet materials

b Two "thermal meters," one of circular cross-section and the other of square cross-section, have been found entirely reliable for the measurement of the thermal conductivity of sheet and other materials

c By putting vaseline, glycerine, glue, shellac or a similar material on the division between two surfaces the thermal drop due to such division can be largely eliminated. This is particularly true for poor conductors

d The thermal conductivity has been measured for a large number of materials both across and along the laminations. For the poor conductors the ratio of the longitudinal to the transverse conductivity varies from 2 for black bias cloth to 5½ for mica tape

e The temperature coefficient of thermal conductivity has been measured whenever the experimental results justified

f Of the electrical insulating materials tested, those containing mica have in general the better thermal conductivity

g As a thermal insulator soft pine is the best of the woods tested and is but little inferior to dark-grey felt

h The transverse conductivity of iron stampings can be increased some three or four times by the insertion of some suitable material between the stampings so as to make better thermal contact. This is for a pressure of about 50 lb. per sq. in. Nothing, however, destroying the electrical insulation could be used. By using something between sheets the ratio of the longitudinal to transverse conductivity could be reduced to 20 to 25 instead of 80 to 100

i In general the thermal conductivity of laminated products can be considerably increased by suitable impregnation so as to get rid of the air film

j Oil-soaking soft fuller board increases its thermal conductivity by about 50 per cent.

k The best thermal insulation for a given thickness should be made up from several thin sheets rather than from a single sheet. This effect is more pronounced the better the conductor.

l Results were obtained for longitudinal conductivity of iron stampings. 0.0172-in carbon sheet steel is about two and one-half times better than 0.014-in. silicon sheet steel. Carbon sheet steel has a longitudinal thermal conductivity about 80 times the transverse while silicon has but 32 times the transverse.

# Friction and Spreading Forces in Rolling Mills

### Description of Apparatus Designed for a Special Series of Investigations to be Made on the Experimental Rolling Mill at the Carnegie Institute of Technology

By W. B. SKINKLE,[1] PITTSBURGH, PA.

*In Mechanical Engineering for May, 1919, announcement was made of the decision of the officials of a number of leading steel works and firms building rolling-mill machinery to organize a Bureau of Rolling Mill Research and to install at the Carnegie Institute of Technology, Pittsburgh, for research purposes, an experimental rolling mill. This undertaking on so large and important a scale marks a distinct advance in coöperative research among manufacturers in this country. The present article by the director of the Bureau of Rolling Mill Research outlines the further development of these plans and in particular describes the equipment which has been designed for a special series of investigations on friction losses and spreading forces in rolls and roll housings as met in the actual operation of rolling mills. A feature of this apparatus is the use of what are practically frictionless support cylinders for measuring the forces acting, which provide hydraulic support in combination with the elements of a pressure gage.*

THE need of accurate knowledge of the forces acting on rolls and the friction losses on roll necks in different types of rolling mills has prompted the author to respond to the editor's request for an article, with a description of a device soon to be tried out on the experimental mills of the Bureau of Rolling Mill Research at the Carnegie Institute of Technology.

By means of this equipment it is expected that experimental data will be secured which will give positive, accurate knowledge, not only of the relative values of the various lubricants and bearing metals used on roll necks, but also of the spreading forces

acting on the rolls; from which the stresses in rolls and housings may be calculated to a far greater degree of accuracy than is at present possible.

The device consists of two hydraulic supporting cylinders marked Cylinder No. 1 and Cylinder No. 2 in Figs. 1 and 2, located in the bottom of the roll-housing window and so arranged that they can easily be removed and replaced by a filler.

The carrier for the bottom roll rests on the plungers of these cylinders in such a manner that the forces acting on the roll neck are recorded by the resulting hydrostatic pressure in these cylinders.

In Fig. 1 is a diagram showing the arrangement of these cylinders and their relation to the rolls. A bloom is shown between the upper and lower rolls with the lower roll carrier and cylinders. The full lines indicate the position of this carrier under static conditions (condition of rest). The conditions existing at the neck of the lower roll are greatly exaggerated for the purpose of illustration.

The total spreading force F, due to the bloom being pinched between the rolls, will cause a heavy force P to be applied[1] about as indicated in Figs. 1 and 2. Under static conditions this force would be transmitted from the body to the neck and would be applied to the bearing as indicated by the arrow A in Fig. 1. Such an application of forces would cause one half the load P

on each bearing to be carried by cylinder No. 1 and the other half by cylinder No. 2.

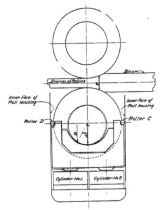

FIG. 1  ARRANGEMENT OF CYLINDERS AND THEIR RELATION TO THE ROLLS

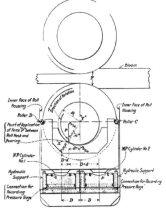

FIG. 2  DIAGRAM OF FORCES EXERTED DURING ROTATION

---

[1] Director, Bureau of Rolling Mill Research, Carnegie Institute of Technology.  Mem. Am. Soc. M. E.

This article was prepared for *Blast Furnace and Steel Plant* and is here published through the courtesy of that journal and of the author.

[1] It is not the purpose of this article to enter into a discussion of the complicated system of forces acting at the contact surfaces between rolls and bloom. Inasmuch as the rolls are able to move up and are restrained from horizontal movements in any direction by the bearings and roll housing, the resultant of all forces acting must be in a vertical direction.

The proportion of the total force $F$ which is distributed to each neck of the rolls will depend on the location of the pass in the roll body and can readily be figured by the well-known principles of beam reactions.

Where rotation starts, and the bloom is being rolled, the point of contact between roll neck and bearing changes to some other point, the force $P$ is then applied to the bearing as indicated by the dotted arrow $B$, and the lower roll carrier has a tendency to assume a position indicated by the dotted lines, using the roller $C$ as a fulcrum point; the roller $D$ lifting off its seat. The position indicated by these dotted lines in Fig. 1 of course is greatly exaggerated for the purpose of illustration. It is vitally important, however, that the bearing be perfectly free to rotate as indicated, and that no forces other than cylinders Nos. 1 and 2 be interposed to prevent this rotating tendency. (In actual practice it is expected that not more than 0.003 in. or 0.004 in. of movement will be obtained on these cylinders between no load and their maximum carrying capacity of 250,000 lb. each.)

When the contact point has moved to a place where it slides

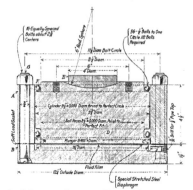

back as rapidly as it tends to climb, the angle of sliding friction has been reached and a condition of equilibrium is again established.

These conditions are shown and analyzed in Fig. 2. The pressures on the supporting cylinders will no longer be equally distributed, as in the case of static conditions, but will be much heavier on cylinder No. 1 than on cylinder No. 2.

In the analysis of these conditions the following symbols will be used:

$P$ = Total resultant force on the roll necks in pounds.

$R$ = Radius of the roll neck in inches.

$U$ = The angle whose tangent is the coefficient of friction.

$D$ = Distance from center line of roll housing to center line of supporting cylinder.

$d$ = Distance from center line of roll-neck bearing to the point of contact between the neck and bearing.

$N$ = Some fraction of $P$ greater than $\frac{1}{2}$.

$M$ = Some fraction of $P$ less than $\frac{1}{2}$.

If equilibrium in a "free body" is maintained the following conditions must be met:

1 The sum of all horizontal or vertical forces acting on the body must be equal to zero.

2 The sum of the moments of all forces acting on the body taken about any point must equal zero.

Fulfilling the first condition, as applied to the vertical forces

on the lower roll-neck bearing, we have $-P + NP + MP = 0$, whence $NP + MP = P$. Canceling the $P$'s we have

$$N + M = 1$$

This shows that for each increase of over half the total load carried by cylinder No. 1 there is a corresponding decrease in the load carried by cylinder No. 2 and that the sum of the pressures on cylinders Nos. 1 and 2 must be equal to the force $P$. This same fact can be shown by taking moments about the point $x$ where the roll neck applies its load to the bearing. In this case the force $P$ has no lever arm. Its moment would therefore be zero, and it may be neglected. Taking moments about $x$ we have

$$+ NP(D - d) - MP(D + d) = 0$$

which expands into

$$NPD - NPd - MPD - MPd = 0 \quad \ldots \ldots \ldots [1]$$

Taking moments about $Y$ the center of the roll neck we have

$$- Pd + NPD - MPD = 0 \quad \ldots \ldots \ldots \ldots \ldots [2]$$

The foregoing is true only if the hydraulic support cylinders are *frictionless*. If these hydraulic cylinders were of the ordinary hemp-packed plunger type the results obtained would be almost useless as the friction of the packing on the plunger would be very large and the exact amount of this friction would be very difficult if not impossible to determine.

It will therefore be necessary to produce a frictionless hydraulic support cylinder. This, the author believes, has been practically obtained in the cylinder shown in Fig. 3. While it is recognized that no machine can be absolutely frictionless, this cylinder so nearly approaches this ideal state that for all practical purposes it may be considered as frictionless.

The cylinder consists of an upper guide cylinder $A$ securely bolted to its base $B$ by sixteen $\frac{3}{4}$-in. bolts $G$. The base, which is really the cylinder proper, has a groove $9\frac{1}{4}$ in. inside diameter cut into it. The center portion of the base is then faced off to allow just a film of fluid not over $\frac{1}{16}$ in. in thickness to act on a special stretched-steel diaphragm $C$ which is securely held in place by the friction between the parts $A$ and $B$ of the cylinder and is made tight on its lower side by a soft lead gasket. By this means the fluid is completely confined in the cylinder, the only escape here being through the orifice leading to the recording pressure gage. The plunger $D$ does not come in contact with the fluid in the cylinder but rests on the upper side of the steel diaphragm $C$. These contacting surfaces must be very flat and true.

In order to avoid friction between the plunger $D$ and the side walls of the guide cylinder $A$ the plunger is turned $\frac{1}{8}$ in. smaller in diameter than the guide cylinder on its upper parts, and 0.005 in. smaller in diameter where it rests on the diaphragm.

In order to prevent lost motion, or the plunger from tilting and shearing the diaphragm, and further to prevent the plunger from getting "off center" and having a rubbing contact with the cylinder walls, two rings of fifty-six $\frac{1}{2}$-in. diameter steels balls each, marked $F$, are provided.

The inside diameter of the guide cylinder $A$ is ground to as nearly a perfect circle as possible and the grooves for the balls are ground and polished to an accurate fit.

When the plunger moves downward due to the application of a load these steel balls tend to lift off their seats and establish a rolling contact between the plunger and cylinder walls.

The only friction on this cylinder is the friction of these ball bearings, the internal friction of the steel in the diaphragm and the friction of the fluid in the pipes leading to their recording gages. These quantities are so small when compared with the maximum load of 250,000 lb. which each cylinder is designed to carry, that they may be neglected.

The displacement of fluid by the plunger when under load is equally small. Marks' Mechanical Engineers' Handbook, page 251, gives the following information: "A pressure of 1 lb. per sq. in. compresses liquids in volume as follows: Water 1 part in 300,000; mercury 1 part in about 4,700,000. For water in an iron pipe this corresponds to a compression of 2 in. per mile of length for a pressure of 10 lb. per sq. in."

It will easily be seen that such a compression of the fluid will not disturb the roll setting. The only other factors entering into the plunger displacement are the quantity of fluid necessary to move the needles of the recording gages and the expansion of the short pipes leading to these gages. Both of these factors are so small that they may be neglected.

The idea of this frictionless diaphragm cylinder is not new, but has had several stages in its development. It appears to have been originally invented by Mr. Emery, who first applied it to the well-known testing machine bearing his name. The general application of this invention was delayed by the fact that Mr. Emery measured the pressure in his hydraulic supports by means of a complicated balancing and weighing mechanism. The hydraulic support was made accessible to general use by the research work of Dr. A. Martens at the German Government materials-testing station, who measured the pressure in the hydraulic

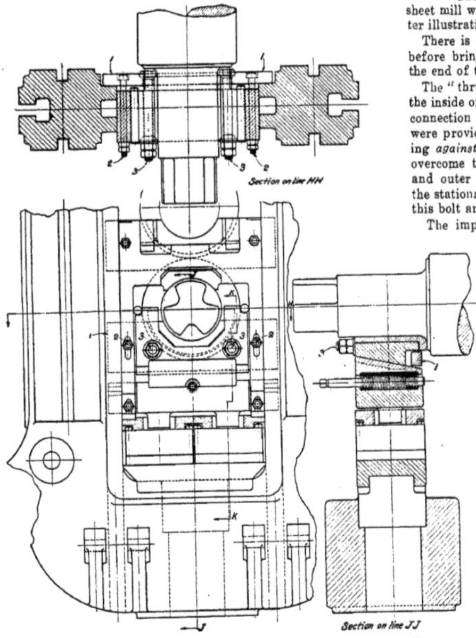

Fig. 4 Details of Mill Showing Bolt With "Ball Joint" to Provide Flexible Connection Between Thrust Plate and Lower Roll Carrier.

support by means of Bourdon pressure gages and who proved conclusively that the combination of hydraulic support and pressure gages forms the most accurate and desirable means for measuring without friction or lost motion, forces of any magnitude and of any suddenness of application.

Prof. W. Trinks, head of the department of mechanical engineering at the Carnegie Institute of Technology, first conceived the idea of applying cylinders of this type to roll necks sometime early in 1916, and made drawings of a special roll housing for this purpose.

When the author was placed in charge of the organization of the Bureau of Rolling Mill Research and of the designs of the experimental mill he carried the development one step further by designing the present cylinders in such a manner that they would fit in the window of an ordinary roll or pinion housing,

and could easily be replaced by a filler casting in case experiments not requiring their use were to be made.

The first cylinder of this type is now in the course of construction and will be carefully tested and calibrated under a 600,000-lb. compression testing machine before the final design is produced for the experimental mill.

Diaphragms varying in thickness from 1-64 in. to 3-64 in. are on hand and their relative merits together with the relative merits of mercury or water for the fluid will be carefully investigated before the final selections are made.

Fig. 4 shows these cylinders applied to the 18-in. three-high bar mill, which are somewhat different from those shown in Figs. 1 and 2. These last-mentioned figures were taken from the 24-in. sheet mill with the neck-brass "faked in" for the purpose of better illustration of the mechanical principles involved.

There is one more important item which should be considered before bringing this article to a close. That is the friction of the end of the roll-neck brass on the roll body.

The "thrust plate" marked "1" in Fig. 4 is held firmly against the inside of the roll housing by bolts 2 and 3 and, unless a flexible connection between the lower roll carrier and this thrust plate were provided, would offer a very substantial resisting force acting against the rotating tendency of the lower roll carrier. To overcome this bolts 3 are made with ball joints at their inner and outer ends. In this manner any retarding action due to the stationary thrust plate is avoided. Fig. 5 is a section through this bolt and clearly shows the construction.

The importance of careful experiments on roll-neck friction

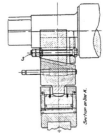

Fig 5 Details of Bolt Shown in Fig. 4

will be realized when it is stated that, according to the best information at present available, between 40 and 50 per cent of the total work of a blooming or bar-mill engine is absorbed in friction on the roll necks. These figures represent the best working conditions in rolling-mill practice. For mills such as plate, skelp or sheet mills rolling wide, thin sections, which are often cold, the work lost in friction on the roll necks is frequently between 70 per cent and 80 per cent of the total work of the engine or motor.

The designs of the experimental mills are being worked out with extreme care, which of necessity involves a slow rate of progress. They offer a great many problems similar to the one just described. It is expected that these designs will be completed some time during the next three or four months and will be given to the public very shortly thereafter.

The Carnegie Institute of Technology is throwing open to members of the Bureau nearly $400,000 worth of building and equipment, free of all charges for rentals, etc. More than half the money necessary to add this experimental rolling mill has already been definitely subscribed and the remainder is in sight. The completion of this experimental mill will make the Bureau of Rolling Mill Research what is probably the largest and most complete steel research laboratory in the world.

# Steam Use in Textile Processes

## By GEORGE H. PERKINS,[1] BOSTON, MASS.

*The following paper, presented at the Textile Session of the Annual Meeting of the A.S.M.E., and here given nearly in full, presents tables of comparative figures showing the importance of the use of steam for the manufacturing processes in the textile industry.*

*Analyses are given of the various uses to which steam is put and suggestions are made for research and improvement of economy.*

THE collective textile industries of the United States consume 9,662,600 tons of coal annually and require for their operation a total primary power of 2,495,000 hp. of which 66.8 per cent, or 1,665,900 hp., is developed by steam prime movers.

Tables 1 and 2 give figures compiled from the United States Census of Manufactures for 1914, showing the distribution of coal and primary power to the seven principal manufacturing groups included in the textile industry. These figures serve to emphasize the magnitude of the steam requirements of this industry, which are to be considered in this paper with particular reference to the heat demands of textile processes.

### RELATIVE ECONOMY OF TEXTILE INDUSTRIES

The relative fuel economy of the different textile groups is shown in Table 3, derived from data given in Tables 1 and 2. The items are listed in the order of comparative performance as figured on a common basis of power production and show clearly the effect on the overall economy of the extensive use of steam in the various processes of manufacture.

The cotton goods and cordage groups rank first, owing to their small demands for process steam. The relatively good economy shown by the woolen and worsted group may be attributed to a more general utilization of exhaust steam in processes and the better balance usually existing between the steam demands of prime movers and processes.

The dyeing and finishing group has a comparatively small power demand and a most extensive use for process steam. The relation of coal to power in the silk goods group appears out of proportion in comparison with the other industries and must be attributed, in part, to the wasteful use of steam and a more extensive use of the lower grades of fuel.

The validity of the results given in Table 3 may best be shown by citing a number of specific examples of individual plants of different types, giving the actual fuel consumption and power production. This is shown in Table 4. These figures check well with the averages for the respective groups which are given in Table 3.

### IMPORTANCE OF HEAT APPLICATIONS

The foregoing facts emphasize in general terms the importance of the problems of heat application to textile processes. These problems warrant far greater attention than they have yet received, both from the standpoint of production efficiency and from the consideration of fuel economy. But little progress has been made along either line in comparison with the marked advances that have taken place in the art of generating steam and power.

Many textile power plants are " saving at the spigot and wasting at the bung " and the average dyehouse, bleachery or print works is usually considered a necessary evil to be endured up to the limit of the boiler capacity of the plant. While production efficiency must be considered as of the first importance, the present fuel situation demands rigid conservation in the use of both coal and steam.

[1] Consulting Engineer, 34 Batterymarch St. Mem. Am. Soc. M. E.

### EFFECTS OF HEAT ON TEXTILE MATERIALS

The purposes for which heat is applied to textile materials in process are numerous and varied and only the principal ones will be summarized, as follows:

| Process | Material | Purpose or Effect of Heat |
|---|---|---|
| Scouring | Wool | To assist the action of the soap and alkali. |
| Carbonizing | Wools and shoddies | To free the acid from moisture to permit carbonization. |
| Combing | Worsted wools | To facilitate drawing through the pins. |
| Dyeing | All materials | To assist the fixation of dyestuffs. |
| Kier Boiling | Cotton goods | To assist action of liquor in softening impurities. |
| Drying | All materials | Evaporation and removal of moisture. |
| Tentering | Cotton and worsted goods | To dry goods under tension retaining width. |
| Steaming | Woolen, worsted and print goods | To shrink, remove glaze and to set fabrics and colors. |
| Soaping | Print and dyed goods | Warm water removes surplus color and improves appearance |
| Ageing | Print and dyed goods | To set or develop colors. |
| Washing | Woolen and worsted goods | Warm water removes oil and soap applied in previous processes. |
| Pressing | Woolen and worsted goods | With pressure, flattens surface for finish effect. |
| Calendering | Cotton goods | With pressure and friction, assists in producing luster. |

### CLASSIFICATION OF PRINCIPAL HEAT-USING PROCESSES

| Application | Machines of Similar Use |
|---|---|
| Direct contact of materials with steam or water vapor | Steamers for woolen goods, print goods, etc. Ageing machines for print and dyed goods Yarn conditioning machines |
| Direct contact of materials with heated surfaces for warming stock, pressing, polishing, drying, etc. | Worsted comb steam chests Worsted back washers Plate presses for woolen goods Rotary presses for woolen goods Slashers and dressers Cylinder and can dryers Calenders for cotton goods Polishers for twine, thread, etc. |
| Drying of materials with air heated by direct or indirect radiation | Hot air slashers and dressers Cotton and wool stock dryers Carbonizing dryers Cloth dryers Shrinking dryers Tenter frames Dry rooms for yarn Back washer dryers |
| Direct use of steam for boiling or heating liquids and use of warm water | Dye tubs and dyeing machines Scouring bowls for wool Bleaching kiers Cloth washers Crabbing machines Starch and size cooking Soapers |
| Indirect use of steam for heating liquids | Bleaching kiers with closed beaters and circulating pumps. Jacketed kettles for size, starch or dyestuffs. Tubs and dyeing machines with closed submerged coils. Prehenting of liquids for process or storage by closed heaters. |

The direct or indirect heating of rooms containing processes requiring careful temperature and humidity control may also be mentioned under the general heading.

### FACTORS AFFECTING PRODUCTION AND ECONOMY

Some of the various conditions commonly found, which affect either production or economy are as follows:

14

| Production | Economy |
|---|---|
| Wet steam. | Leakage from traps. vents, etc. |
| Poorly designed steam distribution system. | Wasted drips. |
| Incorrect arrangement of traps. | Overheating or overcooling through careless operation. |
| Deficient trap capacity. | Radiation losses through dryer housings, etc. |
| Delays in heating liquor, dyes, etc. | Waste of exhaust steam or condensing water. |
| Inefficient air circulation in dryers. | Lack of hot-water storage. |
| Deficient dryer radiation. | Failure to recirculate air in dryers. |

In addition to the data given above, there is usually little information available either of the steam requirements or

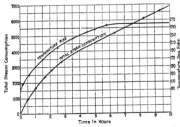

FIG. 1  TEMPERATURE RISE AND STEAM CONSUMPTION IN KIER TEST

demands. While considerable progress is being made in means and methods for correcting the conditions mentioned, there is still room for much improvement.

Many of the larger mills are making complete steam surveys of their plants periodically to check use against demand and also to enable steam costs to be properly apportioned to the various processes. Modern steam meters give reliable results and in-

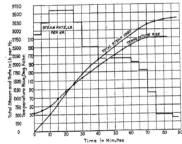

FIG. 2  TEMPERATURE RISE AND STEAM CONSUMPTION IN DYE-TUB TEST

variably reveal many surprising conditions which require attention and which cannot be disclosed by any other means.

## TEST RESULTS

The results of tests made at different plants are given in the following tabular form, and in Figs. 1 and 2, merely to illustrate common practice and not as standards of high performance. They will however be found useful for comparative purposes.

## SUGGESTIONS FOR INVESTIGATION AND IMPROVEMENT

It is hoped that the outline given of this subject will help to stimulate greater attention to this important matter.

The following suggestions are offered as concrete lines for useful and practical study and research:

*a* Further investigation of conditions of temperature, humidity and air circulation for drying various materials with maximum production and minimum heat consumption

*b* Development of more efficient arrangement of radiation and air circulation in dryers

*c* Automatic control of the production rate of dryers, through conditions resulting from drying results obtained

*d* Further development and application of temperature control devices to heat using processes

*e* More effective utilization of exhaust steam in processes, heating water storage, etc.

*f* Recovery of heat from spent and rejected liquors.

*g* Development of insulating materials to be used under conditions where equipment is subject to extreme moisture conditions or to mechanical injury

*h* Use of supplementary air circulation with can dryers.

The field is a broad one with many varied and complex problems, but the coöperative effort of manufacturers, machine builders and engineers should bring about much needed improvement in the present average practice.

TABLE 1  COAL CONSUMPTION AND STEAM-POWER REQUIREMENTS OF THE TEXTILE INDUSTRIES

| Industry | Anthracite, 1000 Tons of 2240 lb. | Per Cent of Total Anthracite | Bituminous, 1000 Tons of 2000 lb. | Per Cent of Total Bituminous | Steam Power, 1000 Hp. | Per Cent of Total Power |
|---|---|---|---|---|---|---|
| Carpets and Rugs | 74.7 | 3.1 | 193.7 | 2.7 | 29.5 | 1.8 |
| Cordage, Jute Twine and Linen | 91.9 | 3.8 | 195.2 | 2.7 | 65.8 | 4.0 |
| Cotton Goods.... | 313.5 | 13.0 | 3634.2 | 50.0 | 1011.3 | 60.7 |
| Dyeing and Finishing* | 490.6 | 20.5 | 896.6 | 12.4 | 111.5 | 6.6 |
| Hosiery and Knit Goods | 118.1 | 4.9 | 484.3 | 6.7 | 80.8 | 4.9 |
| Silk Goods | 1053.7 | 44.0 | 249.9 | 3.5 | 78.3 | 4.7 |
| Woolen, Worsted and Felt | 257.4 | 10.7 | 1608.8 | 22.0 | 288.7 | 17.3 |
| Total | 2399.9 | 100.0 | 7262.7 | 100.0 | 1665.9 | 100.0 |

*Exclusive of that done in textile mills.
Fuels other than coal have been omitted as the equivalent total is small.

TABLE 2  DISTRIBUTION OF PRIMARY POWER IN THE TEXTILE INDUSTRIES

| Industry | Total Primary, 1000 Hp. | Steam Power, 1000 Hp. | Water Power, 1000 Hp. | Internal-Combustion Engines, 1000 Hp. | Rented Power Electric, 1000 Hp. | Rented Power Other, 1000 Hp. |
|---|---|---|---|---|---|---|
| Carpets and Rugs | 44.0 | 29.5 | 4.1 | .... | 9.2 | 1.2 |
| Cordage, Jute Twine and Linen | 93.9 | 65.8 | 14.4 | 2.8 | 10.7 | 0.2 |
| Cotton Goods.... | 1585.9 | 1011.3 | 314.2 | 4.0 | 252.9 | 3.5 |
| Dyeing and Finishing | 130.1 | 111.5 | 9.9 | 0.7 | 7.2 | 0.8 |
| Hosiery and Knit Goods | 125.7 | 80.8 | 14.6 | 1.8 | 26.3 | 2.9 |
| Silk Goods | 117.0 | 78.3 | 7.6 | 1.8 | 23.8 | 5.5 |
| Woolen, Worsted and Felt | 396.4 | 288.7 | 76.3 | 2.7 | 25.0 | 5.7 |
| Total | 2493.0 | 1665.9 | 441.1 | 13.1 | 355.1 | 19.8 |
| Per Cent of Total | 100.0 | 66.8 | 17.7 | 0.5 | 14.2 | 0.8 |

### TABLE 3 COMPARATIVE FUEL ECONOMY OF THE TEXTILE INDUSTRIES

| Industry | Equivalent Bituminous 1000 Tons of 2000 lb.[1] | Per Cent of Total | Steam Power 1000 Hp. | Per Cent of Total Primary Power | Tons Per Hp. Per Year | Lb. Per Hp. Per Hr[2] |
|---|---|---|---|---|---|---|
| Cotton Goods....... | 3947.7 | 40.8 | 1011.3 | 63 7 | 3 99 | 3.12 |
| Cordage, Jute, Twine and Linen........ | 287.1 | 3.0 | 65.8 | 70.0 | 4.35 | 3 48 |
| Woolen, Worsted and Felt............ | 1866.2 | 19.3 | 288.7 | 72.3 | 6.50 | 5.20 |
| Hosiery and Knit Goods.......... | 602.4 | 6.2 | 80.8 | 64.5 | 7.45 | 5.96 |
| Carpets and Rugs... | 268.4 | 2.8 | 29.3 | 67.2 | 9 08 | 7 26 |
| Dyeing and Finishing | 1387.2 | 14.4 | 111.3 | 85.8 | 12.50 | 10.00 |
| Silk Goods ...... | 1303 6 | 13.5 | 78.3 | 67.0 | 16.70 | 13 36 |
| Total.......... | 9662 6 | 100.0 | 1665.9 | 66.8 | 5.80 | 4.64 |

[1]Anthracite ton of 2240 lb. included as equivalent of Bituminous ton of 2000 lb.
[2]Based on 2500 hr. operation per year.

### TABLE 4 ACTUAL ECONOMY OF INDIVIDUAL PLANTS

| Kind of Plant | Total Coal Tons | Total Steam Power Hp. | Coal Per Hp. Per Year |
|---|---|---|---|
| Cotton goods, no dyeing ....... | 10,500 | 2,800 | 3.75 |
| Cotton goods, with dyeing..... | 13,000 | 3,000 | 4.33 |
| Jute goods ................. | 9,400 | 2,200 | 4.28 |
| Worsted yarns ............. | 1,560 | 340 | 4.6 |
| Woolen goods.............. | 1,800 | 270 | 6.67 |
| Bleachery and finishing..... | 10,200 | 800 | 12.75 |
| Print works................ | 60,000 | 4,000 | 15.00 |

### TABLE 5 CLOTH DRYER TESTS

| | UNINSULATED HOUSING Woolen suiting | J-M No. 1 TYPE INSULATED HOUSING Woolen suiting |
|---|---|---|
| Weight of goods, oz. per yd. ...... | 13 | 12 |
| Weight of wet goods processed per hr., lb... | 990 | 1047 |
| Weight of goods dried per hr., lb....... | 523 | 609 |
| Yards dried per hr., lb........ | 644 | 747 |
| Weight of Water evaporated per hr., lb.... | 577 | 438 |
| Per cent moisture in wet cloth.......... | 41.9 | 41.9 |
| Heating surface, sq. ft............ | 2240 | 2240 |
| Steam pressure, gage............. | 42.5 | 42.5 |
| Temperature in dryer............. | 185 deg. fahr. | 158 deg. fahr. |
| Weight of steam condensed per hr., lb... | 875 | 7.8 |
| Steam condensed per sq. ft. of radiation, lb | 0.39 | 0.33 |
| Steam per lb. of dry cloth, lb.... | 1.67 | 1.21 |
| Steam per lb. of moisture evaporated... | 2.32 | 1.85 |
| Yards dried per lb. of steam............. | 0.735 | 1.01 |

### TABLE 6 CARBONIZING DRYER TEST

| | UNINSULATED HOUSING Woolen and cotton thread Waste | J-M No. 1 TYPE INSULATED HOUSING Woolen and cotton thread Waste |
|---|---|---|
| Weight wet stock processed per hr., lb.... | 402 | 585 |
| Weight stock dried per hr., lb.... | 266 | 391 |
| Weight of water removed per hr., lb.... | 136 | 194 |
| Per cent moisture in wet stock........ | 32.4 | 33.2 |
| Heating surface, sq. ft........ | 2440 | 1952 |
| Temperature in dryer.......... | 279 deg. fahr. | 261 deg. fahr. |
| Steam pressure, gage............ | 80 | 80 |
| Weight of steam condensed per hr., lb.... | 1044 | 432 |
| Steam condensed per sq. ft. of radiation, lb | 0.42 | 0.22 |
| Steam condensed per lb. of dry stock, lb.... | 3.93 | 1.11 |
| Steam condensed per lb. of moisture removed, lb. | 7.68 | 2 23 |

Note: The author wishes to acknowledge the contribution of the above tests on insulated and uninsulated cloth and stock dryers from the H. W. Johns-Manville Co.

### TABLE 7 TENTER FRAME TESTS

| | |
|---|---|
| Aver. Weight of cloth, yd. per lb.................. | 5.4 |
| Duration of test, hr.................. | 7.0 |
| Steam pressure, gage............... | 30 lb. |
| Outside temperature ............... | 70 deg. fahr. |
| Temp. of air supplied, dry bulb.... | 183 deg. fahr. |
| Temp. of air supplied, wet bulb.... | 85 deg. fahr. |
| Temp. of return air, dry bulb.... | 138 deg. fahr. |
| Temp. of return air, wet bulb.... | 84 deg. fahr. |
| Air supplied, cu. ft. per min.... | 8700 |
| Weight of cloth dried per hr., lb.... | 950 |
| Weight of Water evaporated per hr., lb.... | 106 |
| Per cent of water in wet cloth........ | 14 |

| | Without air recirculation | With air recirculation |
|---|---|---|
| Steam condensed per hr., lb.............. | 1122 | 433 |
| Steam per lb. of dry cloth, lb............ | 1.71 | 0.67 |
| Steam per lb. of Water evap., lb........ | 10.5 | 4.1 |
| Lb. of cloth dried per lb. of steam........ | 0.58 | 1.5 |
| Yd. of cloth dried per lb. of steam......... | 3.23 | 8.1 |

### TABLE 8 CAN DRYER TESTS

| Test No. | 1 | 2 | 3 | 4 | 5 |
|---|---|---|---|---|---|
| Wt. of goods, yd. per lb.......... | 1.45 | 2.5 | 5.15 | 8.10 | 10.85 |
| Kind of goods..... | Shoe Lining | Shoe Lining | Sheeting | Twill | Lawn |
| Steam pressure in cans, gage...... | 5.7 | 4.7 | 8.1 | 5.5 | 5.3 |
| Number of cans... | 47 | 47 | 47 | 23 | 13 |
| Size of cans....... | 23in.x100in. | 23in.x100in. | 23in.x100in. | 23in.x14½in. | 36in.x12¼in. |
| Material of cans... | Copper | Copper | Copper | Tinned Iron | Copper |
| Total can surface, sq. ft....... | 2350 | 2350 | 2350 | 1625 | 1265 |
| Cloth dried per hr., yd.......... | 2393 | 4420 | 10,135 | 17,750 | 12,730 |
| Cloth dried per hr., lb........ | 1650 | 1765 | 1968 | 2191 | 1173 |
| Per cent of water in wet cloth | 48.8 | 51.8 | 51.7 | 46.1 | 44.4 |
| Water evaporat. per hr., lb........ | 1570 | 1892 | 2112 | 1873 | 938 |
| Steam condens. per hr., lb........ | 2462 | 2625 | 3424 | 2530 | 1310 |
| Steam per lb. dry cloth, lb........ | 1 49 | 1.48 | 1.74 | 1.15 | 1.12 |
| Steam per lb. water evaporated, lb.. | 1 57 | 1.39 | 1.62 | 1 34 | 1 40 |
| Number of strings. | 2 | 2 | 2 | 4 | 4 |
| Average speed, yd. per min.,.... | 20 | 36.5 | 84 | 74 | 50 |
| Yards of cloth per lb. steam...... | 0.97 | 1.68 | 2.96 | 7.60 | 9.7 |
| Steam per hr. per sq ft. can surface ... | 1.05 | 1.12 | 1.46 | 1.56 | 1.04 |

### TABLE 9 KIER TEST (SODA BOIL)—SEE FIG. 1

| | |
|---|---|
| Size and kind of kier..........4 ton, pressure type, injector circulation | |
| Weight of dry cloth................... | 8000 lb. |
| Kind of cloth and wt. in lb. per lb. muslin 56 x 52...... | 7.25 yd. |
| Weight of Water entering goods.............. | 8000 lb. |
| Weight of liquor........... | 16800 lb. |
| Initial temperature ................... | 50 deg. |
| Final temperature ................... | 245 deg. |
| Steam pressure, gage................... | 25 lb. |
| Time of boil................... | 10 hr. |
| Total steam used................... | 8250 lb. |

#### APPROXIMATE HEAT BALANCE

| | Pounds of Steam | Per cent of Total |
|---|---|---|
| Heating wet goods .................. | 2264 | 27.5 |
| Heating liquor .................. | 2971 | 36.0 |
| Heating kief .................. | 238 | 2.9 |
| Radiation loss .................. | 1860 | 22.5 |
| Vent. evaporation and accounted for losses ... | 917 | 11.1 |
| Total .................. | 8250 | 100.0 |

### TABLE 10 DYE TUB TEST—SEE FIG. 2

| | |
|---|---|
| Gallons of liquor boiled................... | 1845 |
| Pounds of liquor boiled................... | 15390 |
| Initial temperature................... | 34 deg. fahr. |
| Final temperature................... | 210 deg. fahr. |
| Total temperature rise................... | 176 deg. fahr. |
| Time of total temp. rise, hin................... | 70 |
| Total steam used during total temperature rise, lb...... | 3230 |

# The Causes of Industrial Unrest and the Remedy

Labor Unions Have Promoted Inefficiency and Decreased Production—Employers Have Been Remiss
in Regard to Industrial Relations—Each Shop Must Work Out Its Own Salvation
Through Fair, Cordial and Sympathetic Co-öperation

By FREDERICK P. FISH[1], BOSTON, MASS.

EVEN under normal conditions, there is no phase of our modern social organization which is more complicated and exacts more from the intelligence, character and capacity of mankind than the machinery and methods of production and distribution by means of which our material needs are satisfied. But at the present time, conditions are not normal. Everything has been thrown out of adjustment by the Great War. There has been a destruction of property, the effect of which will embarrass us for years, and millions of men who would have worked and produced have been killed or incapacitated. But this is not all. The thought of the whole world has been disturbed. A condition almost approaching hysteria has been developed so that everywhere men are incapable of cool, sane judgment not only as to the facts of the existing situation but as to their needs and aspirations.

## Decreased Production and Its Causes

Leaving out of consideration all matters of governmental disturbance and confining our thought only to production, it seems certain that at the very time when there should be the greatest effort to produce, that we may make up for the losses of the last five years and establish our material resources on a sound basis, the attitude of workmen throughout the world is such that individually and as a whole their efficiency and productive power are much less than before the war. While in other parts of the world they are working under unfavorable conditions, that is not the case in this country. Here the sole reason for reduced efficiency is the atmosphere that has been developed, which results not only in strikes but in a definite attitude on the part of the worker that he will not produce to the extent of his reasonable ability. In spite of the fact that he is constantly demanding greater pay, he is insisting upon shorter hours and is disinclined to exert himself while at work. He almost cultivates a spirit of antagonism to the enterprise for which he works, forgetting that his interests and those of his employer are identical and that only by close coöperation can results be attained that will lead to the prosperity of either. The influence of labor unions, not only in the "closed shop," where with the acquiescence of the employer they largely dominate the situation, but in many of the so-called "open shops," in which are labor-union workmen influenced by the principles of the labor unions, is such as to promote inefficiency on the part of the workmen. Some deliberately refrain from effort, but I believe it to be the case with a large majority that, against their inclination, they fail to make proper effort because of fear. They do not dare to act contrary to the spirit instilled by the labor unions and labor-union men, which causes them to be treated as traitors to their class if they do all that they would gladly do by way of loyal devotion to their work.

Only this week the manager of a large plant told me casually of some of the conditions with which he has to contend. In one room where a large number of workmen are assembling a certain article, each and every one of them produces exactly fifty a day and twenty-five a half-day of the working period. It is, of course, absurd to suppose that the capacity of each and all of a

very large number of workmen should be so uniform as to lead to this result, unless there was a standard deliberately adopted and maintained. This same manager told of another circumstance which illustrates the same spirit. He has procured a machine which with one man will do the work heretofore done by eight men. Up to date he has been utterly unable to find a man in his establishment who will run that machine. The absurd fallacy that it is for the interest of wage earners that each should do little work, ineffectively, that more may be employed, prevails to a substantial extent under the influence of the labor unions.

In this plant to which I have just referred, wages have increased on the average 110 per cent since 1914, although the cost of living has not increased more than about 75 per cent, and the manager is satisfied that the efficiency of the individual workmen has been reduced during the same period about 40 per cent.

There is ample testimony that the situation is the same throughout the country.

Of course, it is the public, the consumer, that ultimately suffers from such failure on the part of the workers to recognize their duty as factors in production. For the moment at least, owing to the abnormal situation throughout the world, the employers are able to add to the price of their manufactured articles enough to make up for this absurdly unnecessary and vicious added cost of production, so that the public pays the bill. But such conditions cannot continue, and the time is surely coming when our industries will be unable to recoup themselves in this manner, with a resulting period of industrial depression which may well be the most serious in history.

## Reasons for the Present Unsatisfactory Employment Relations

What is the reason for these unsatisfactory employment relations and how can they be cured? To answer these questions, brief reference must be made to certain facts of recent industrial history, the importance of which should be clearly understood.

While the present unsatisfactory conditions have been accentuated by the war, they had existed, although to a lesser degree, for many years prior to 1914 in this country and even to a greater extent in England.

I believe the fundamental difficulty to be that industry has not had time to adjust itself to the extraordinary new conditions that have prevailed during the past forty or fifty years. While there is no more wonderful chapter in the history of the world than that of industrial progress during the period above mentioned, those who were in control of the industries and directed their progress and upon whom was the duty of assuring satisfactory employment relations under all conditions of effort, were so absorbed in the tremendous task of meeting the material advances in the arts, in responding to the vital changes in industrial methods and in securing the proper utilization of new and extraordinary resources that were for the first time available, that they overlooked and neglected the essential matter of properly adjusting the employment relation to new conditions.

The result of all this progress was great corporations and great manufacturing establishments such as were not even foreshadowed in former days, attended by the extraordinary development of power and of machinery on a large scale which was based upon marvelous inventions and striking achievements in engineering such as the world had never seen. Craft and hand-

[1] Chairman National Industrial Conference Board.

This contribution and the two succeeding ones, by Dr. Leiserson and Dean Heilman, were given at the Industrial Relations sessions of the Annual Meeting of THE AMERICAN SOCIETY OF MECHANICAL ENGINEERS, December, 1919. These three addresses, and the paper by Mr. A. L. De Leeuw on Wage Payment, published in the December, 1919, number, constitute a symposium on "The Industrial Situation in Relation to Present Condition," a general account of which is given elsewhere in the present issue. These addresses are here given in abstract and are subject to revision.

work were to a large extent eliminated. Vast numbers of workmen were brought together in individual factories, where their activities were specialized to such an extent that each man became trained to a high degree of efficiency in some line to which he devoted his energies exclusively. Only by organization of the most complicated character was it possible to deal with these new conditions; and organization took the place of that personal and direct control by management which had formerly prevailed. In such a radically new environment it was impossible for employment relations to remain as they were before. They inevitably required readjustment. Personal touch between the employer and employee, such as had existed from the beginning of history, was no longer possible. Where such personal relations exist there must be serious fault on one side or the other if they are not harmonious.

If organization is to take the place of such personal contact, it must be made to fit the conditions.

There can be no doubt that the achievements of those who have been responsible for our industrial development during the last fifty years are worthy of the greatest praise. They were secured by a wonderful exhibition of power, intelligence and capacity on the part of a large number of devoted servants of industry. But in the stress and strain of their efforts they could not cover the whole ground. Some aspects of the situation were neglected.

One illustration of this is familiar to us all. As business interests and business men became more and more efficient and powerful, evils appeared such as always attend a great and sudden social movement accompanied by strenuous effort and marked success. Those evils were due to the same weakness of human nature which has caused the bad to be mingled with the good in every great forward step in the progress of our race, whether in religion, in methods of government or in measures of social reform. They arose because management failed adequately to correlate industrial progress with the public interest.

The time came when the public dealt in its own crude way with a situation which it did not approve. Some twenty-five or thirty years ago it became persuaded that the "captains of industry," whom the public had idolized up to that time, and their associates, were developing the industries on lines that seemed contrary to the general interest. Big business became unpopular. The result was a public sentiment which, while it has undoubtedly brought about a needed reform in industrial methods and the relations of industries to the community, has surely been carried so far as to embarrass business to the disadvantage of society. Such a reaction was inevitable, for the people will and should resent the exercise of undue power on the part of

any class, no matter what may be its obligation to that class for real service.

Another direction in which management failed to do its full duty was in the employment relation. It proceeded in its strenuous efforts to advance the industries without recognizing that the conditions of employment were such that they required to be approached from a new point of view. Things were allowed to drift and finally came to a point, even before the war, which threatened trouble of a deep-seated and far-reaching character.

We now see clearly that enough attention was not given to the conditions under which men worked, to the principles upon which their compensation should be determined, to their health and comfort not only in the factory but at home. We recognize that the new and strenuous activities forced upon the workmen by modern machinery, which often determine the pace at which men shall work, required careful study that the length of the working day might be properly determined in view of the new environment in which the work was carried on. Enough attention was not paid to the effect upon the workmen of the intense specialization and concentration of their work, as compared with the greater variety and slower methods of former days. Due consideration was not given to the question whether or not the fundamental law of supply and demand, which will always be potent in determining the workman's compensation, might not properly be supplemented to some extent by other considerations. The serious matters of unemployment and the current practice, which seemed unavoidable, of hiring many men to secure a few, were not strongly dealt with.

If these subjects had been studied and the new problems properly worked out, there is no doubt in my mind that the feeling of dissatisfaction, almost of antagonism, on the part of the workmen that finally came into the industries would, to a large extent, have been avoided. It must not be forgotten that contented workmen, whose surroundings are satisfactory and whose health and comfort are maintained, are likely to work so efficiently as more than to compensate for the shortening of hours of work, if they are too long, and for the expense involved in giving them surroundings conducive to health and reasonable comfort; and that added efficiency is promoted by and should be recognized to some extent by increased remuneration, irrespective of what may be called the "market rate" for labor.

As discontent grew among the wage earners, and employers did not wake up to the necessity of trying to meet their aspirations, reasonable or unreasonable, as to better working conditions, and of educating them to sound views of the matter, it

---

*"The influence of labor unions, not only in the 'closed shop,' where with the acquiescence of the employer they largely dominate the situation, but in many of the so-called 'open shops,' in which are labor-union workmen influenced by the principles of the labor unions, is such as to promote inefficiency on the part of the workmen.*

*"They seem determined to carry on their war against civilized society, for that is what it really is, to the limit and to secure if possible the control of our industries by threats and force. If they should succeed by such methods, a law of brute force would be established which would mean the end of our Government and of society.*

*"There is no doubt, however, that the labor unions have also accomplished much good. Among other things they have focused the attention of industrial management and of the public upon those features of employment relations which needed study and revision. It may even be possible that the vicious methods of the strike and the boycott and the brutal and even criminal pressure brought by the trade unions upon non-union workmen and their families, as well as upon employers, have contributed to opening the whole question for proper consideration.*

*"There can be no specific general cure, no panacea for the evils that exist, either by way of legislation or of reorganization of our social or industrial methods; but the present unfortunate antagonism between employers and employees and the existing disturbance of the cordial feeling that should exist between them can be eliminated and right conditions restored only by definite personal effort to develop and maintain, in the individual industrial establishments of the country, suitable employment relations, based in each instance upon fair, cordial and sympathetic coöperation and a recognition of what is right and proper in the particular establishment."*

FREDERICK P. FISH.

was but natural that the trade and labor unions should succeed in bringing many workmen into labor organizations which seemed to them to promise relief from what they regarded as unsatisfactory working conditions. There is no doubt that the labor unions have accomplished much good. Among other things—and this is probably where they have been of chief value—they have focused the attention of industrial management and of the public upon those features of employment relations which needed study and revision. It may even be possible that the vicious methods of the strike and the boycott and the brutal and even criminal pressure brought by the trade unions upon non-union workmen and their families, as well as upon employers, have contributed to opening the whole question for proper consideration.

### ARBITRARY ATTITUDE OF LABOR UNIONS

But from the beginning the labor unions have adopted social and economic principles that were unsound and methods of propaganda and of action that should not be tolerated in any civilized society. They have worked for the suppression of production upon the false theory that the less men worked, the more men could be employed. On the same ground they have not been sympathetic to the introduction of improved machinery and have often opposed its use. They have undertaken to establish a closed organization which would prevent the employment of workmen who would not yield themselves body and soul to the dictation of labor-union leaders. In this way they have been the enemies of society and of unorganized labor. They have been opposed to apprenticeship systems and vocational education as likely to add to the number of competitors for the jobs of labor-union men. They have undertaken to enforce arbitrary and unreasonable rules in factories and to interfere with the proper execution by the management of its necessary prerogative of control. Not to speak of such crimes as those of the McNamaras, they have often violated the laws of the land as well as the principles of good morals. They have fostered and encouraged mob rule. They have tacitly and even directly encouraged assault and destruction of property as a means of winning the strikes upon which they relied for accomplishing their purposes. They have refused to be bound by the law and have too frequently, by the exercise of pernicious political influence based largely upon threats, obtained special exemptions from provisions of law than no one else could violate with impunity. They have practiced intimidation on the wives and families of workers and interfered with the inherent right which every man has, to choose his own way of getting a living, provided it is lawful.

They have used the fact that they were rich and powerful organizations, purporting to speak for a large number of men—who were, in theory, bound to work and vote as a unit—to further legislation for their own selfish purposes and not in the public interest. They have openly expressed disrespect for the law and courts and have shown that disrespect by their conduct. All these charges are capable of proof and are recognized as valid by those who are familiar with such matters.

There is no doubt that many of the labor-union leaders have been good men, working for what they regarded as the interests of the workmen. It is also a fact that some individual labor unions have been so reasonable in their activities that employers have found it possible to work with them with satisfactory results. But generally speaking, such has not been the situation. It is always the case in such organizations that a small element of active, virile and intelligent leaders, carried away by the desire for power and by their own ambitions, find themselves almost instinctively adopting such wrong and brutal methods as the exigencies of the case seem to require, on the theory that the end justifies the means. Moreover, their hands are forced by what they regard as the necessity of getting results. The majority of those who follow them are naturally good and law-abiding citizens, but in addition to the fact that pressure, even to the point of violence, is brought to bear upon them and their families, they become blinded in their devotion to the cause and fail to see things in true perspective; even if they do not approve of what is done in their name and ostensibly on their behalf, they have no power to control the situation. They are carried along by a stream too strong to permit the exercise of individual judgment.

Up to the time of the war, the American Federation of Labor, which stood for the recognized labor unions in this country, had succeeded in securing a membership of less than 10 per cent of the workers in industry. Considerably less than 10 per cent of the shops in the country were "closed shops," that is to say, shops in which the labor unions dominated. The "open shop" principle had been substantially maintained. This country therefore did not suffer as it would inevitably have suffered if the labor unions had been really a dominant force. In England, where they had been much more successful in securing the control of the employment relation, the consequences of their power and influence were clearly seen. Before the war, the individual American workman was turning out from two to three times the product turned out by the English workman. Because of his greater productive power, his wages were higher and his living conditions much better than in England. In the latter country, production was so reduced by labor-union rules and by the hostility to improved machinery, fostered by the labor unions, that England was rapidly losing her standing as a manufacturing nation, while at the same time wages and the standards of living of her workmen were extremely low. This was inevitable, for there was no large production from which good wages could be paid, large production being the only source from which increased returns to the workers can be derived.

When the war broke out, England found that under existing conditions she could not begin to produce what was required for fighting purposes and the support of her army and her people. There was only one thing to do, which was to persuade the labor unions to suspend their rules and abandon their characteristic methods during the period of the war. By great pressure from the Government they were induced to take this course, with the result that production was increased in England to an astonishing degree; but the labor unions made this sacrifice of what they regarded as their interests, only with the distinct understanding that, after the war, they should be allowed to reinstate themselves as a controlling force in industry. They are now trying to bring about this revival of the old condition of things. If they succeed, England as an industrial nation is doomed.

When we entered the war there seemed to be a fear, which was probably unfounded, that the organized workmen in this country had no patriotism and no thought of the public welfare. There seemed also to be a feeling on the part of those in power that the labor unions, represented by the American Federation of Labor, controlled the situation and must be placated in any event. There probably was no foundation for this feeling, but on this theory the entire influence of the Government during the war was on the side of yielding everything to the labor unions. The Government throughout the war seems largely to have followed the dictation of the labor unions. It certainly used its influence to force the employers of the country to yield to the organized workmen practically everything that was demanded, not only as to wages and working conditions but in the matter of unionizing the shops. This seems to have been the uniform policy from the date of the Adamson Bill, which was the first expression of it on a large scale.

The results have been that while every other section of the community, including unorganized labor, made substantial financial sacrifices to aid in winning the war, workmen in industry have been called upon to make no such sacrifice. Their wages have gone up far more than has the cost of living and today they are the conspicuous class in the community which has money to spend on luxuries to an extent never known before. And yet, during the war organized labor promoted an infinite number of strikes to the great embarrassment of the Government and loss to the public. Many and perhaps most of these strikes centered about the question of recognition of the labor unions by the employers.

### DETERMINATION OF LABOR UNIONS TO DOMINATE INDUSTRY

The great increase in power and influence which the labor unions have secured as the immediate result of the war, as well as their determination to dominate industry at all costs, is reflected in the steel strike, the coal strike and innumerable other strikes throughout the country, while one phase of their aspirations is illustrated by the strike of the Boston police. They threaten a railroad strike and even a general strike unless their demands, which are clearly unreasonable, are complied with. Either would cause intolerable hardship to the community. They seem determined to carry on their war against civilized society, for that is what it really is, to the limit and to secure if possible the control of our industries by threats and force. If they should succeed by such methods, a law of brute force would be established which would mean the end of our Government and of our society. There is no need to argue what a serious matter this would be for our industrial prosperity. But it also should be noted that if the labor unions succeed in this effort, they will establish an *imperium in imperio* in this country which will be utterly inconsistent with proper Governmental control and with the ideals of a free, democratic community.

There are at the outside something more than four million workmen who are members of the labor unions, not over 10 per cent of all the wage earners in the country. More than that, there is reason to believe that a substantial number of the men in the labor unions do not sympathize with the underlying aspirations of the leaders or the methods adopted by them. The recent election in Massachusetts is an indication that the workmen of this country are not in sympathy with such high-handed procedure. And there is much further evidence to the same effect.

In addition to the trade and labor unions associated with and in the American Federation of Labor, there are other labor unions of the same type, of which the railroad brotherhoods are the strongest and most important. There are also other associations of wage earners to which a brief reference must be made. Many of these, the so-called Independent Workers of the World being the most conspicuous, are distinctly revolutionary in character and criminal in act and intent. The normal American workman has no sympathy with their real aims and purposes, but some are deceived into believing that they will help to secure better conditions of living. The greatest strength of these lawless organizations is with the foreign-born worker who is not at all an American and whom it is easy to mislead. They are, in theory, opposed by the trade and labor unions partly because it is the policy of those last-named organizations to resent the intimation that they are aiming at the destruction of society and partly because their leaders are afraid that, by some wave of sentiment, the rank and file of the labor unions may be carried over into the other and more radical camp and new labor-union leaders selected in the place of those now in power.

It cannot be overlooked, however, that while the trade and labor unions have stood in opposition to the I. W. W. organization, and others like it, and have proclaimed their hostility to the destructive and revolutionary policy of such organizations, they themselves in their underlying principles, as well as in their practices and conduct, have laid the foundation for just such revolutionary and destructive ideas as those which their rivals have openly proclaimed and promoted.

It is incredible that at the present time there should be any danger of such public support of these revolutionary organizations as would make them a serious menace to our industries, to our Government and to our civilization. There is far greater danger that the trade and labor unions may play a conspicuous part in the future of our country, for at any rate until recently the people as a whole, carried away by sympathy, have been inclined to overlook their faults and even their crimes and to regard their activities with a certain degree of favor. The real question is this: Shall a series of organizations, welded into one great parent organization of enormous wealth and strength, be allowed to have and to exercise such power that it may at will paralyze our industries, and by methods such as are illustrated by the great strikes above referred to, and by threats—even threats to starve our people—attempt to secure its own selfish ends irrespective of the interests of the rest of the community? The answer to this question seems clear.

The same instinct on the part of the public which led to the more or less warranted attack upon big business some twenty years ago should bring about among our people a recognition of the tremendous danger involved in the improper and brutal exercise of power threatened by the labor unions. If public sentiment is once aroused, the end is certain. The American people will surely not submit to the domination of any class working solely in its own supposed interest for power such as no class should have.

The extraordinary efforts now made by labor unions have as their foundation no present grievance on the part of the working-man, for none of any real moment exists. The unions are simply taking advantage of the condition of dissatisfaction and mental disturbance on the part of the wage earners, the beginnings of which were before the war but which have been fostered and promoted by the circumstances of the war period. So long as that mental and moral condition continues, the field is open for labor-union propaganda and worse. If they succeed in their efforts the result will surely be inefficiency of production, inharmonious employment relations and the possibility of even worse developments along the lines of the aspirations of those who would destroy our civilization and our Government.

### HOW TO REMEDY THE EXISTING SITUATION

I have tried briefly to present certain phases of the existing situation and to show in part how present conditions arose. The great question, and it is the most serious with which we are now confronted, is how to deal with this existing situation. There are some, even among those who are intelligent and have given more or less attention to the subject, who seem inclined to believe that the solution of the present difficulties in the employment relation resides in the further development of the trade and labor unions and of their policies. They picture a condition of things in which all workers are organized and perhaps all employers in like manner organized. Then the labor unions on the one hand and the employer organizations on the other are to get together and by "collective bargaining" settle for all the industries all questions of wages, hours of labor, conditions of employment and other relations between employer and employee.

I am convinced that such a plan, if adopted, would be destructive of our industries and fatal to our prosperity and to our life as a free people. The whole history of the labor unions in this country and abroad shows that the result would be marked inefficiency, limited production and no real improvement in the relations of the wage earners and the employers. It is to me inconceivable that any one who has studied the subject should come to the conclusion that such a plan is even to be seriously considered.

Others talk loosely of the "democratization of industry," a vague phrase which seems to imply a more definite participation on the part of the workmen not only in the control of employment relations but even in management.

There is no reason to believe that the workmen themselves have any such aspirations and it is certain that the right and obligation of management to exercise its necessary function of direction and control must always be definitely recognized and maintained. Such suggestions are too vague for intelligent consideration.

### NO PANACEA AVAILABLE FOR EXISTING EVILS

It is my firm conviction that there can be no specific general cure, no panacea for the evils that exist, either by way of legislation or of reorganization of our social or industrial methods; but that the present unfortunate antagonism between employers and employees and the existing disturbance of the cordial feeling that should exist between them can be eliminated and right conditions restored only by definite personal effort to develop and maintain, in the individual industrial establishments of the country, suitable employment relations, based in each instance upon fair, cordial and sympathetic coöperation and a recognition of what is right and proper in the particular establishment.

Not only is the cost of labor far and away the largest item in practically all industry but men and women, unlike machinery, to a large extent control their own actions even when under discipline and must not only be trained but must be in the right frame of mind. If they work in a cordial spirit and with good will, the results are far greater than if their efforts are perfunctory. It would seem certain that an intensive study of this human relation in an industrial establishment would result to even greater advantage than the same amount of attention given to the material factors.

In this direction much has been done during the past few years and with good results. There has been a marked improvement in working conditions throughout the country. Sanitary and hygienic arrangements and devices protecting against danger have been introduced to a large extent. Shop hospitals and nurses are common in the industries. Definite and well-considered efforts have been made to promote the living and housing conditions of employees and to give them opportunity for healthy and stimulating physical and mental recreation.

### IMPROVED WORKING CONDITIONS BETTER EMPLOYMENT RELATIONS

There is no doubt that the employment relations have been bettered by these improvements in working and living conditions. They must have resulted in better health and greater efficiency and have developed a better feeling on the part of the workers.

Such methods should be adopted in every establishment to the extent to which they are justified by the conditions in that establishment, not only as a plain duty to the workmen but because they are a profitable investment leading to better service. Everything, of course, depends upon the spirit in which such efforts are made and accepted by the workmen. If they come as a natural, reasonable incident to the employment relation and not in such a way as to offend the just pride of the employee, their effect must be wholly good.

But such efforts are not enough. The workman is primarily interested in the more essential matters of wages, hours of labor, continuity of employment, and the fairness of the rules and regulations under which he works and of the spirit in which those rules and regulations are applied.

In all these matters he has inevitably personal views to which consideration should certainly be given, not to an extent which will interfere with the necessary control by management, but to an extent which will enable management to act intelligently and sympathetically. If in such matters the workmen can be made to feel that they are treated as men, with rights that are recognized and respected, they will surely be in a much better frame of mind and more likely to respond with a reasonable view as to that to which they are entitled.

It is not unnatural that while they must take orders and must work under rules that are essential to the carrying on of the business, they should feel dissatisfied if they have nothing to say as to their wages, hours of work and working conditions, but are in a position where they must take without protest what is offered to them or give up their job. No man likes to have another say to him, "Take it or leave it," without explanation or discussion.

It therefore seems certain that in each individual industrial establishment there should be the same scientific effort to study the men in their relations to the work as to analyze the materials, machinery and shop methods, in order that the best possible results may be produced. If instead of holding the workmen at arm's length, they could be taken into the confidence of the employer as to wages, hours and working conditions, it seems but reasonable to believe that a feeling of mutual confidence and willing coöperation would be developed which would be a most distinct asset to the business and likely to hold the men against any of the superficial inducements offered by trade and labor unions, which endeavor to excite antagonism on the part of the workmen toward the enterprise in which they are essential factors.

The burden of the effort to bring about such relations must be consciously and definitely assumed by the management in each individual establishment. From that management the initiative must come with the hope that the employees will respond.

### EACH EMPLOYER MUST SOLVE HIS OWN PROBLEM

No hard and fast rules on the subject can be laid down. Each employer should study his own problem and attempt to deal with it in view of the conditions in his own establishment, exactly as he works out his mechanical and technical problems in harmony with the requirements of his own work and with only general reliance on the experience of others.

Of the 450,000 manufacturing establishments in this country, the large majority are so small that it is still possible for the old-fashioned personal contact between employer and employee to continue. In such cases all that may need to be done is that the employer should treat his workmen as reasonable beings, whose inevitable interest in their own affairs entitles their views to frank sympathetic consideration. He should not hold them at arm's length. He should learn their ideas and their criticisms of their relations with him. He should endeavor to satisfy them of the reasonableness of his attitude in the matters that are of such vital importance to them. He should see that they understand the conditions of the business. He should let them know why the wages and hours of labor should be fixed at a certain standard and the reasons for the rules and regulations that are adopted. He should acquaint them with the difficulties under which he labors, in so far as those difficulties are material to a full understanding of the exact situation. He should invite their suggestions as to ways, large and small, by which shop methods may be improved or the product and quality of the goods increased. He should be willing to discuss with them, under proper conditions, all the questions in which employer and employee are jointly interested. He should listen to the complaints of the men, recognizing that if matters of complaint are dealt with promptly it is generally easy to dispose of them.

All this could be done by an intelligent, sympathetic employer without the expenditure of much time or effort, and there is no doubt that he would find a real satisfaction in the energy so expended.

Where the number of men employed is so large that personal contact is impossible, each employer should make every effort to work out a plan by which, in view of the conditions in his own establishment, he can get the equivalent of such personal contact. There is more than one way of bringing about this result. In some cases, it might be accomplished by the organization of a special department whose duty it would be to keep in close touch with the men and establish with them the confidential relations that are required.

### GOOD RESULTS FROM ESTABLISHING SHOP COUNCILS

Another plan that seems promising and which has already been adopted in several hundred establishments in this country, in some of which it is working with marked success, is the establishment of shop councils or some other form of shop or plant organization by which representatives of the employees, honestly selected by them from their own number, meet regularly with representatives of the employer to consider, in conference, the questions arising under the employment relation.

There is every reason why the employees should welcome such an opportunity to meet representatives of the employer. Such an arrangement would appeal to the sense of fairness of the employees and if the conferences were conducted as they should be on terms of perfect equality and with entire frankness, the results could not be other than beneficial. Experience seems to show, that it is only natural to expect, that representatives of the men selected for such purposes, under conditions which insure an honest election and a fair count, would be intelligent and reasonable. Through them the men would learn much that they never otherwise would know as to the troubles and aspirations of the management, the difficulties of carrying on the business and the reasons for the adoption of one policy or another in the employment relation.

If the workmen are not satisfied with some phase of the terms of employment or with some feature of the rules imposed upon them, or have a complaint against any particular foreman or other person in the establishment, or if an individual workman has some personal grievance, such conferences would bring the matters complained of to the attention of the management at an early stage, before the workmen have become demoralized by brooding on the subject, and, if dealt with promptly and fairly, there would be less danger that a small grievance would lead to serious trouble. Through their representatives, the workmen could be taught sound economic principles such, for example, as the fact that good wages and national prosperity as well depend upon efficient service and large production. They could be made to see that if it is reasonable that wages should respond to improved conditions of business, it is equally reasonable that if business declines, wages should be reduced. There would be an opportunity that would otherwise not exist to present to the workman the reasons why, from the point of view of his own selfish interest as well as that of the public good, a reduction of the hours of labor below the point of maximum production consistent with the health and reasonable comfort of the worker is a fatal mistake. At and by such conferences the workmen should be encouraged to make suggestions as to the improvement of shop methods, large or small. Many workmen are unquestionably capable of real help in this way. They could be made to realize that they were a vital element in the work whose coöperation in all ways was needed and expected. To such suggestions they would surely be inclined to respond.

It would seem that if in every one of the industrial establishments of the country a definite effort was made on the part of the management to bring about directly or indirectly such personal contact between the employer and employee as would result in a frank discussion of the employment relation and those matters concerned with the enterprise in which the employees are necessarily and vitally interested, and a definite recognition of the workmen as essential factors in the enterprise, antagonism might be eliminated and relations of loyalty and confidence established to an extent that would be most fruitful in good results.

It goes without saying that nothing can be accomplished by mere machinery. The right spirit must be developed on both sides; but it is unreasonable to suppose that it is impossible to develop that right spirit. In any event, it would, in my opinion, be a red-letter day not only for industry but for the public if all the employers of the land were to start in to bring about an improvement in employment relations by the very simple expedient of working, each in his own establishment, to see what could be done by personal contact between employer and employee to inspire confidence, harmony and coöperation.

### SELECTION OF EMPLOYEES' REPRESENTATIVES

In any such shop organizations as those above referred to, it is of vital importance that the representatives of the employees should be honestly chosen, and chosen from among the employees themselves and not from the labor unions. Otherwise, the labor unions would find it necessary to enlist as members only a small portion of the employees in any given establishment and that minority, acting as a unit, would be able to elect walking delegates or other members of the union as representatives of the whole body of employees. If that happened and the employer was obliged to meet these labor-union agents as the representatives of his men, his fate would be sealed. An "open shop" employer would surely find himself forced to the union shop, first, by the pressure of the negotiations, and, second, by the moral effect upon his own men, exaggerated by the skillful way in which the labor unions would utilize the fact that they and they alone through the prestige of their union were maintaining the rights and promoting the interests of the employees.

It is futile to contend that the employees in any given establishment cannot find among their own number men who are quite competent to represent them fairly in conference with representatives of the employer. In every shop there are many workmen quite capable of presenting their own views and those of their fellows in a most effective fashion. Such men are sure to be elected by the workmen as their representatives.

### ADVANTAGES OBTAINED FROM A WELL-ORGANIZED EMPLOYMENT DEPARTMENT

Another direction in which effort should be made by the management in each individual establishment to better the conditions of the employees and improve the employment relation is in the matter of unemployment and of hiring, placing and discharging. Unnecessary unemployment is one of the most serious incidents of modern industry. It undoubtedly has much to do with the discontent of the workmen. The subject has been given much less careful attention than its importance deserves.

Here again much can be accomplished by intelligent management in the individual industrial establishment. Nothing could be more absurd than the hiring of men without a careful examination into their fitness for the particular job for which they are hired. Their mental, physical and moral qualifications should be studied and they should be assigned to work in view of the qualifications which they are found to possess.

When employed there should be every certainty that they are not to be discharged without due cause. If a man proves unfit for the work to which he is assigned, an effort should be made to find other work in the establishment for which he is qualified. Every employee should feel assured that the management proposes to keep him at work during good behavior if it is possible to do so.

In this matter of hiring and discharge, it is not safe to trust the foremen unless they are carefully supervised. There is no doubt that one of the most serious needs in our industries is a better education of the foremen and like subordinate officials and such supervison that they are unable to make the mistakes and do the harm that the wrong men can easily do in such positions.

Experience seems to show that one of the greatest economies that can be practiced in an establishment of fairly large size is the organization of a department which shall be devoted to the hiring, placing and discharge of men on definite and scientific principles. Moreover, if workmen feel secure in their jobs, except for their own definite failure to do good work, they must inevitably have a greater feeling of content and a stronger inclination to serve well the employer who has given evidence that he proposes to continue them in his employ.

That stability of employment is a matter of the utmost consequence to the employer as well as to the employee, is illustrated by the fact that not infrequently the management of an establishment finds that it must hire a thousand or more men really to add two hundred to its force. The cost of hiring and training all these men is a severe burden on industry.

In this same connection it is also clear that the management of our industries should give far more attention than is given at the present time to reducing irregularity of employment. To this end much may be done in each individual establishment by way of spreading out the work and declining to go to extremes in pushing production in good times and reducing it when the market is bad. If employers as a whole were to take up the problem seriously and endeavor to work together for its solution, there is no doubt that cyclical as well as seasonal extremes of undue activity and undue depression could be to a large extent minimized.

Such schemes as profit sharing, bonus payments, the acquisition of stock by the employees and other devices for increasing the return to the wage earner are worthy of careful consideration; but in each case in view only of the conditions of the particular establishment. It may well be that in some instances the management may determine that one or the other of these plans is likely to give to the workmen a more permanent interest in the enterprise and therefore to remove antagonism and develop loyalty.

## WAGES AND WHAT SHOULD DETERMINE THEM

It seems probable, however, that the workmen for the most part are chiefly interested in the pay which they receive at stated intervals. While they should be encouraged to save and make investments, it may well be that in most cases they would not be greatly influenced by the chance of return at some period which seems to them remote.

While there is no doubt that the law of supply and demand will always, to a large extent, control in the matter of fixing wages, those engaged in industry are, I think, becoming more and more convinced that as a practical matter the operation of this law may well be modified in view of other considerations. In a normal industry each workman should surely be able to earn enough to live comfortably in accordance with the standards of a right-minded man of his position in life, and to save for a rainy day and for old age. If he cannot earn as much as this, he should not remain in that industry. If the industry cannot pay the normal workman this amount, it should not, in the long run, survive. It moreover seems clear that efficiency and loyalty should be recognized in the wages paid; also that the wages might well to some extent reflect the prosperity of the individual establishment.

In this matter also the burden is upon the management of each individual establishment to deal fairly and justly with its workmen in view of the conditions characterizing that establishment; but it is certain that more satisfactory results will be attained if consideration is given to the ideas of his workmen on the subject and an effort is made to make them understand why the wages proposed or paid are reasonable and proper.

So as to hours of work. There can be no possible reason for a fixed period for all industries. In each it should be determined by scientific study what number of hours a day will make possible the maximum production consistent with the health and reasonable comfort of the workmen. Even in the same industry the hours of work may well vary in different establishments. The locality, climate, general environment and living conditions are important factors in determining how long one may work without sacrifice of health or comfort. There is no comparison in this regard between, for example, the conditions of men in a shoe factory in the country and those similarly employed in a shop in the heart of a city like St. Louis.

Here again the solution of the problem is one for the individual establishment, to be determined by the management, preferably after conference with the employees who are very likely to assent to a sound conclusion when the conditions are fully explained to them.

There seems to be no doubt that overtime work under present conditions cannot be justified except in the case of a real emergency. As we all know, overtime work is insisted upon by many employees as a device for getting a larger return for their services. This is mere pretense, overtime being used as a plausible camouflage for an unstable and deceptive situation.

Much might be said as to the spirit of our institutions and how inconsistent with that spirit would be the domination of the industries by the labor unions; while, on the other hand, the "open shop" and the personal dignity of the relations between employer and employee, which is reflected in the "open shop," illustrate perfectly the principles of individual freedom and of fair play which are at the basis of our economic as well as of our political and social life.

## EMPLOYERS MUST DEAL OPENLY AND FRANKLY WITH THEIR WORKMEN

If I am right in my views, the opportunity for bringing about a satisfactory feeling between employers and employees, by definite and conscious effort on the part of the management in the individual shops, is so great that it seems to be the plain duty of every individual employer throughout the country to adopt the principle of dealing openly and frankly with his workmen in all matters which need to be understood on both sides if relations are to be cordial.

Management can never be asked to sacrifice its function of judgment and direction. If the so-called "democratization of industry" means the contrary of this, it is nonsense. The workmen would be the very first to recognize that ultimate control must be with the employer. Naturally, however, they cannot be satisfied if no opportunity is given them to state their point of view as to matters which are of the utmost importance to them in their life's work.

It is obvious that such relations as should exist between management and workmen are only possible in an establishment which is not tied up with the trade or labor unions. There alone are the workmen free from outside influence, which may be and often is used to stimulate antagonism and to breed dissatisfaction as well as to reduce efficiency.

## EFFICIENCY AND PRODUCTIVE POWER GREATER IN THE OPEN SHOP

In the "closed shop" it is in effect the labor union which negotiates with the employer as to wages, hours of labor and working conditions and its negotiations are carried on from the point of view of the labor union and not of the men; although even in a "closed shop," under some conditions, "shop councils" or meetings between the management and the employees would be possible. There are not, however, likely to be such relations in the "closed shop," particularly as the American Federation of Labor, at its annual convention in 1919, definitely resolved that "we disapprove and condemn all such company unions [works councils, etc.] and advise our membership to have nothing to do with them." The resolution further stated that "we demand the right to bargain collectively through the only kind of organization fitted for this purpose, the trade union." In the steel strike the "abolition of the company unions" was one of the fundamental demands of those promoting the strike.

It seems incredible that the people of this country, as a whole, if they really understood the situation, should not stand firmly for the "open shop," as distinguished from the "closed shop," not only because of the greater efficiency and productive power of men who are not bound by union rules but also because, in the "closed shop," personal and direct relations between employer and employee without outside interference are practically impossible.

Of course, an employer should be free to allow his shop to be organized and to deal with the trade unions, as representing his men, if he is willing so to do, just as he should have the right to work through a closed non-union shop; but certainly there should be no moral pressure brought upon him to force such action on his part.

There are many other questions which some regard as of great importance in relation to the present labor conditions. For example, how far if at all should the state interfere by the compulsory arbitration or by way of conciliation and mediation in labor difficulties? Something of the sort must come if labor-union domination is to characterize the future of industry.

On the other hand, if the development is on such lines that the employer and employee in each individual establishment take up in a sane fashion the problem of fixing the employment relations on a fair basis, it may well be that labor difficulties will be so reduced that no scheme of arbitration or conciliation would be worth while. Certainly, no one can look forward without perturbation to the interference of such boards. Experience has shown their incapacity and the unsatisfactory results of their efforts.

Another matter which would be of vital importance if labor-union domination were to prevail would be assurance of the responsibility of the unions and of their legal obligation to carry out the agreements that they make.

At the present time the labor unions are practically immune against attack even for their crimes. They are not clearly responsible to those whom they unlawfully injure. They may not be held on the contracts that they make. Their compliance with an agreement is almost altogether a purely voluntary act on their part. If they were to become the representatives of the men and were to act for the men in a substantial part of the

(Continued on page 72)

# Systems for Mutual Control of Industry

By WILLIAM L. LEISERSON[1], ROCHESTER, N. Y.

IT is difficult for us to change our conception of those conditions and relations as we have learned to know them from years of experience, said Dr. Leiserson, in introducing his discussion of shop committees, works councils and the coöperative or industrial democracy plans by which employers propose to give their employees some voice in the control of industry, as contrasted with trade-union collective bargaining by which organized wage earners attempt to force employers into a system of joint control of wages and working conditions.

"This is particularly true," he continued, "of conditions of employment and the relationship between employer and employee." He pointed out that the world war and the shortage of labor since the armistice was signed have materially changed the status of the employee, and that however much the employer may realize the changed industrial conditions he finds it immensely difficult to conceive of his employees as people who do not have to obey his orders, but instead have to be consulted in any action that may affect them.

"To talk about mutual or democratic control of industry or to establish plans for coöperative or industrial democracy without a clear realization of the revolution that has taken place in the status of the wage earner can lead only to confusion and to failure of the plans for mutual control of industry," Dr. Leiserson asserted. He then compared the present situation with that in the fourteenth century in Europe when an industrial revolution brought on by the Black Plague and characterized by much the same industrial ills that we suffer now, resulted in changing the status of the laborer from that of a serf to a wage earner. He then traced the history of the labor movement in this and other countries, covering much the same ground as in his former address, published in the November 1919 number of MECHANICAL ENGINEERING, p. 884, and continued: "In a democratic country with rising standards of living and large numbers in what

"In a democratic country with rising standards of living and large numbers in what we call the middle classes there is no great danger of labor completely dominating the situation and becoming the only masters of industry. But just because we are a democratic country with rising standards of living, and because our wage-earners are now approaching the status of members of the middle class, we cannot permit employers alone to remain the masters of industry as they have been in the past.

"Labor management means control, discipline, an industrial organization held together by rules, orders and authority, reaching down from an executive head at the top. Collective bargaining implies a questioning of that authority. It is based on the principle that an individual employee cannot effectively question the authority of the management. It says there must be no rules or orders affecting the lives and welfare of the wage-earners without the consent of those who must obey them.

"Here we have the age-long problem of reconciling authority with democracy. Law, order, government, we must have in industry as in all other human institutions, but the day of autocratic control is gone, at least in all western civilizations. To maintain law, order and government in the industries of these western countries there must also be democracy, the consent of the governed. Without such consent, without such democracy, our industrial governors find themselves unable to control their organizations, to have their laws, rules, regulations and orders obeyed. This is the essence of the industrial problem today."

WILLIAM L. LEISERSON.

have been in the past. There is more danger of Bolshevism from the employers' side in this country than from the workers' side. Dictatorship by the plutocrat is worst than dictatorship by the proletariat. We are warned against the latter and the people of the United States will never permit wage earners to take over our industries and control them absolutely. But we are not warned against the employer doing the very same thing and there is great danger that in our fear of Russian Bolshevism we will be led to side with those employers who attempt to fortify themselves in their old positions of sole and only masters and dictators of industry, Only the public through its regularly constituted government can be permitted to be masters of industry. Both employers and employees must be servants of the public. The mistake we are likely to make in dealing with this problem is to assume that because industrial democracy necessarily involves an organized labor force capable of acting as a unit through its representatives, therefore every organization of working people in industrial plants is an example of industrial democracy. As a matter of fact, however, a study of the employees' organizations now in existence will show that they classify themselves into three general groups and that the element of democracy is present in only two of these groups, but not to the same extent in each.

## EMPLOYEES' ORGANIZATIONS

"The first group of employees' organization plans may be called shop committees proper. These are merely advisory organizations of the working force selected either by the management or by the employees for the purpose of conferring with the foreman, with safety directors or personnel and service managers regarding various problems related to working conditions in the plants. The matters with which these committees are concerned are primarily safety and welfare work with a small number trying to extend their activities to include grievances. Although these committees are constantly hailed as examples of industrial democracy, they involve no element of collective bargaining or joint control over terms and conditions of employment. Complete authority is centered in the management, the committees merely giving advice and suggestions which may or may not be accepted by the management. The powers, functions and methods of operation of these committees identify them with the service work of the plants rather than with problems of bargaining, of wages, hours and shop discipline.

"The second group of organizations may be called 'employers'

[1] Chairman, Labor Adjustment Board.

Presented at the Annual Meeting, New York, December 2 to 5, 1919, of THE AMERICAN SOCIETY OF MECHANICAL ENGINEERS. The paper is here printed in slightly abridged form. All papers are subject to revision.

24

unions.' In this group are included all those plans which either explicitly say, as does the Midvale Steel and Ordnance Company plan, ' We recognize the right of wage earners to bargain collectively with their employers . . . . .' or which by implication recognize the principle of collective bargaining as do the plans of the Bethlehem Steel Company, the International Harvester Company and the Colorado Fuel and Iron Company. The employees' organizations in this group represent a long step in the direction of industrial democracy, for they involve getting the consent of the employees who are the governed in industry, in the making of industrial laws. As political democracy provides for the consent of the governed in the state, so industrial democracy will provide for the consent of the governed in industry. The difference between these 'employers' unions' and the ordinary unions known as 'organized labor' is that the former are initiated by the employers; they are confined to one company rather than connected with a national organization of employees and the management is not excluded from the meetings.

"The third group of organized workers are the ordinary labor unions, and they usually involve written or understood agreements between national unions of the employees and individual firms or associations of employers. The agreements invariably cover wages and hours and usually working conditions as well. The Hart, Schaffner & Marx plan and the contracts of a number of firms in Rochester with the Amalgamated Clothing Workers are examples of agreements with individual plants, while the conventions of the United Mine Workers and Coal Operators, and the Joint Boards of Control in the needle trades in New York City may be cited as examples of agreements with associations of employers.

"However much the employer whose workers are included in the third group may object to the principle of collective bargaining, he understands thoroughly that his agreement with the union involves that principle. In the second group, however, in the plans we have called 'employers' unions,' there is not always this clear understanding on the part of the employer of what his employees' organization involves. Where the principle of collective bargaining is present only by implication as is true in practically all the plans except that of the Midvale Steel Company, it quite often happens that the employer does not realize that his plan involves this principle. The employees, however, usually assume it does not mean anything unless it gives them the opportunity to bargain collectively with their employers. It is hardly necessary to point out that any misunderstanding like this of the purposes of a plan of employee representation is likely to cause trouble between employer and employee."

## COLLECTIVE BARGAINING WITH ORGANIZED LABOR

Dr. Leiserson then discussed the features of collective bargaining which he believed would contribute to its success, and those which would doom it to failure. He contended that the vast majority of such dealings between employers and employees must be carried on by organized labor for the reason that organized labor is growing very much faster and stronger than the competing movement of company organizations launched by the employers. "During the past four years," he said, "when the works council movement has made such great headway, very many more wage earners have joined the regular organized labor movement than are included in all the company plans put together." Also, that inasmuch as there can be no permanent raising of the status of the wage earners unless the whole industry is affected, "there can be no effective bargaining, particularly in a competitive industry, unless it be conducted on a national scale."

"Works councils, or a union confined to one plant or a few plants, can do little to give to the wage earner a larger share of the returns from industry as compared with what capital and management has been receiving. And this is what collective bargaining and democracy in industry aims at.

Competitive conditions will hold the single plants down to the level of the industry as a whole. Before the distribution of wealth produced by industry can be improved to give labor a larger share, collective bargaining on a national scale covering all markets, must be established. It is because the wage earner sees this that he is so easily led away from company unions by a national organizer.

"Labor management means control, discipline, an industrial organization held together by rules, orders and authority, reaching down from an executive head at the top. Collective bargaining implies a questioning of that authority. It is based on the principle that an individual employee cannot effectively question the authority of the management. It says there must be no rules or orders affecting the lives and welfare of the wage earners without the consent of those who must obey them. It joins the members of the industrial organization into a union and forms a democratic legislative body for the purpose of giving to those who have to obey the laws of industry a voice in determining what those laws shall be. Labor management implies power, authority or lawmaking in the hands of the employer. Collective bargaining means democratic representation in the process of that law making and democratic control of the exercise of that industrial power or authority.

"Here we have the age-long problem of reconciling authority with democracy. Law, order, government, we must have in industry as in all other human institutions, but the day of autocratic control is gone, at least in all western civilizations To maintain law, order and government in the industries of these western countries there must also be democracy, the consent of the governed. Without such consent, without such democracy, our industrial governors find themselves unable to control their organizations, to have their laws, rules, regulations and orders obeyed. This is the essence of the industrial problem today."

## AN ATTEMPT TO SECURE DEMOCRACY AND EFFICIENCY IN THE CLOTHING INDUSTRY

Dr. Leiserson then described an industry where an attempt has been made to put labor relations on a democratic and efficient basis, where collective bargaining with a Union prevails, and where the administrative machinery necessary to make collective bargaining work satisfactorily, has been created. This is the clothing industry, which employs over 200,000 people. It has four great market centers, New York, Chicago, Rochester and Baltimore. In all of these markets it is now almost completely organized and only a few firms are outside the agreement between the Clothing Workers Union and the employers' associations.

Dr. Leiserson said that both sides are organized, and that there can be no real collective bargaining if only the workers are organized and enforce their will on the employers, any more than when only the employers are organized. "In each of those cities," he said, "there is an agreement between the Employers Association and the local branches of the union. The agreements fix minimum wages, establish a working week of forty-four hours, provide for open or preferential shops, except with one association which granted a closed shop, prohibits strikes, and restricts the employer's rights to discharge by making him give a good reason for every dismissal.

"The local unions of the various trades in the industry elect delegates to a central body, known as a Joint Board. This board employs managers and deputies whose business it is to represent the employees in handling their grievances. In each shop the employees also elect representatives called Section Chairman and Shop Chairman.

"On the employers' side there is a similar organization. Each of the large plants employ a labor manager, who has charge of all bargaining and labor control, as well as employment and service functions. For the smaller plants, the employers' association employ labor managers who perform the same duties for groups of employers. In New York and Chicago, there is a chief labor manager employed by the association, and all the

(Continued on page 82)

# What May We Expect of Profit Sharing in Industry?

By RALPH E. HEILMAN,[1] CHICAGO, ILL.

THAT the subject of profit sharing is not one which is new to this country but dates back to the early 70's when there was a very widespread interest in the whole subject of profit sharing, was pointed out by Dean Heilman. Many experiments were introduced at that time and profit sharing was heralded as if a panacea for all industrial problems. Most of these experiments were soon abandoned, however, and profit sharing came to be regarded as a subject of interest mainly to the social reformer or theorist and of little interest to the business man. In the last few years, however, there has been a marked increase in the attention paid to the subject, which justifies a reëxamination of the profit-sharing plan in the light of modern business practice and methods.

### WHAT CONSTITUTES PROFIT SHARING

"At the outset," said Dean Heilman, "let us ask ourselves in what really consists profit sharing, because as a matter of fact there are many plans widely heralded throughout the country as profit-sharing plans, which do not, in any real sense, embody the fundamental principles of 'profit sharing.'

"Profit sharing is an agreement between an employer and his employees whereby the latter participate in some way in the profits of the business, either of the business as a whole or of some unit part of the business. This participation is on a predetermined basis definitely known and established. With that conception of profit sharing in mind, the question is asked, 'Why should any employer, firm, or corporation wish to introduce an arrangement of this kind and voluntarily share a portion of its profits with its employees?'" Continuing, the speaker discussed the various aspects of profit sharing as a business arrangement, and said in part:

"The typical wage earner seeks neither philanthropy nor charity, and likewise the employer does not wish to distribute gratuities to his employees. Therefore, if profit sharing is to justify itself, and to be introduced on any wide scale, it must be as a business arrangement which will prove mutually advantageous and profitable, both to the employer and employees.

### BUSINESS REASONS FOR INSTALLING A PROFIT-SHARING SYSTEM

"What, then, are the business reasons which would lead a firm to introduce this profit-sharing system and what are the advantages which its introduction would bring about? In the first place, it is claimed that profit sharing will increase efficiency and output.

"But only to a certain extent is this true, and only under special circumstances and subject to special limitations; in other words, generally speaking, more effective results are obtained from labor if compensation is on the basis of the measurable results of each individual so that he will be induced to put forth his best efforts. It is frequently difficult, if not impossible, however, to ascertain and apportion the results of the individual worker, since there are conditions under which labor presents a joint product, in which it is impossible to compute or disentangle one individual's work from the mass production of all the workers.

"Under these circumstances profit sharing may prove desirable and prove an additional incentive to increase production far in excess of what would be secured with the flat hourly rate. For example, consider the Procter & Gamble plant at Cincinnati,

[1] Dean, School of Commerce, Northwestern University.

Abstract of an address presented at the Annual Meeting, New York, December 1919, of THE AMERICAN SOCIETY OF MECHANICAL ENGINEERS.

which for 20 years has had a thoroughly successful profit-sharing plan in effect. Here the men are engaged in shifts, keeping the huge vats of soap in continuous operation, so that it would be impossible to separate the work of one shift from that of another and award any one shift over and above another.

"The same is true in most lines of work in the gas industry, which probably accounts for the fact that in England profit sharing is in vogue in connection with some of the gas plants in that country. Where work can be definitely measured, the best plan, probably, is to pay the individual on a basis of individual results. But where work cannot be definitely measured, then the profit-sharing plan may prove an effective supplement or addition to the regular flat-time wage.

"Secondly, profit sharing may be introduced for the primary purpose of promoting permanency and stability among the employed staff and as a consequence reduce the labor turnover. Ordinarily reliance is placed on what may be called two 'deferred features' to accomplish this result. The first of these is the preliminary or qualifying period which an employee is required to serve before he can participate in the profits. This may be six months, or two years as in the plant of the Dennison Manufacturing Company, or three years as in the case of the Burt plant, or seven years as in the case of the Sears, Roebuck & Company plant. Reliance is placed on this qualifying period to hold the employee until its expiration so that he may qualify to share in the profits.

"The second deferred feature is that the profits are only to be issued at stated periods, say, once in six months, or at the end of the year, with the proviso that if the employee leaves the service of the company before the profits are distributed, he foregoes his participation in the profits. Reliance is placed on this feature to hold the employee until the end of each period.

"The power of profit sharing to hold employees and promote stability of employment depends on the question as to whether the profits paid represent a net addition to the full current-market wage paid to similar classes of workers in other industries. If profit sharing is used to depress wages the employees would not desire to accept the profit-sharing plan, and it would defeat itself in that way; but if the profits to be paid represent a net addition to the going current-market wage, it has been demonstrated that in many cases the introduction of profit sharing will tend to reduce the labor turnover and promote permanency and stability among employees.

"Again, profit sharing is sometimes introduced with the hope that it will promote industrial peace, that it will do away with industrial strife, industrial disputes and strikes, and that thereby a larger degree of wholesome and cordial industrial relations will be promoted between employer and employee. Profit sharing has possibilities in this direction, but they have been exaggerated. It is significant that profit sharing is not now, and never has been, one of the demands for which organized labor has striven, and that in some instances strikes have immediately followed the introduction of profit sharing. Instead of giving an ironclad guarantee to industrial peace, profit sharing may give rise to new causes of controversy, as for example, controversy over the percentage or amount of profits which should be shared, or the method by which it should be computed.

"Nevertheless, under certain circumstances profit sharing may be expected to promote wholesome industrial relations, but if it is to be introduced for this purpose the most promising and fruitful field for its application is in those plants which have already introduced so-called shop committees, or internal unions of employees, or works councils. For here any degree of management sharing which is accorded to the employees may go hand in hand with profit sharing, the one strengthening and reinforcing the other, promoting a better understanding and a more whole-some, cordial and peaceful relations.

## VALUE OF PROFIT SHARING IN PROMOTING EFFECTIVE MANAGEMENT

The most important purpose for which profit sharing has been introduced, in the opinion of Dean Heilman, is to promote effective management upon the part of managerial executives, semi-executives, junior executives, and salaried employees all the way from foremen or sub-foremen up. Although the great bulk of the literature on profit sharing deals with its applicability to the rank and file of the wage-earning group, the most promising and fruitful field for the application of profit sharing rests with the salaried class, those holding the positions of larger importance and larger responsibility in business. People of this character ordinarily are paid a flat salary, because so far we have evolved no other satisfactory method of compensating this class of work-ers, but we are beginning to realize the inadequacy of a flat salary even for those who hold executive, semi-executive, or minor execu-tive positions. The flat salary usually fails to draw forth from a man the best exercise of his abilities, and it is not always the case that salaried employes find incentive through the hope of ad-vances do not come as fast as expected, or when the salary limit for the job is reached, there develops a lassitude, and often a lurking resentment in the back of the employee's mind which mars his efficiency.

One of the most important questions at the present time is how best to compensate those hired to manage an industry, because at the present time, in larger-scale industries, even those who occupy most important managerial places, are to a very consider-able extent hired employees. A partial answer to this question, according to the judgment of the speaker, lies in the application of profit-sharing schemes or restricted profit-sharing schemes. especially for those who occupy the positions of larger impor-tance and larger responsibility in these business organizations. Dean Heilman said:

"This principle has already been recognized by a considerable number of business organizations. Two organizations · which have successful profit-sharing plans, limited to the small upper group, are the Dennison Manufacturing Company, which includes about 200 employees out of a total of 3,000, and the Solvay Pro-cess Company, which divides the executive heads of its depart-ments into three groups according to their responsibilities. Then there is the plan of the Studebaker Company, grouping the em-ployees into four groups, in direct ratio to their responsibility and classification. The same thing is true of the Dold Packing Com-pany, which has its executives classified in two groups. A number of other companies might be mentioned which have their manage-rial employees classified into groups for profit-sharing purposes.

"This principle has been applied by the largest department and merchandizing establishments in the country to the depart-ment heads—in the latter case, the special or additional compen-sation generally takes the form of a percentage, not of the total operations of the business as a whole, but a percentage of the profits simply of the particular department of which the individ-ual happens to be in charge.

## EFFECTIVENESS OF PROFIT SHARING AS AN EFFICIENCY DEVICE

"The most important conclusion in this whole subject is that effectiveness of profit sharing viewed simply as an efficiency de-vice and peace arrangement for the purpose of increasing profits by sharing them, is in direct ratio to the rank of the participants and in inverse ratio to the size of the participating group. Stated in another way, the effectiveness of profit sharing purely as an efficiency arrangement is greatest among large groups, or if applied to the rank and file of wage earners, is greatest among small groups.

"What is the relation," asked the speaker, "of the effectiveness of profit sharing to the rank or responsibility of the participant? In the first place, the higher the rank, the greater is the oppor-

tunity which the individual has to exercise an important influence on profits. A purchasing agent, for example, by one shrewd purchase, can exercise a greater influence on the total profits of the business than a workman in the course of a year, no matter how faithful he may be in the performance of his duties.

"In the second place, the greater the responsibility, the easier it is for the employee to see the relation of his efforts to the profits of the business. It is easy for the sales manager to see that if he finds a new market for the product of the concern, he will thus increase profits. But it is a difficult thing for the elevator man to see that he can so run his elevator that he can produce any great increase in profit.

"In the third place the managerial or executive group are more familiar with business vicissitudes and uncertainties than the rank and file. This has an important effect when you come to a lean year. Many profit-sharing plans have gone on the rocks when they have struck a lean year. When the profits are small or non-existent, it often happens that the employee who has worked as hard as he did the previous year is suspicious that some advantage has been taken to deprive him of his portion of the profits. But in the case of the managerial employee he has foreseen the conditions, is familiar with the market conditions and market fluctuations, and instead of losing confidence in the profit-sharing plan, he may gain a larger sense of partnership in the enterprise.

"What is the relation between the effectiveness of the scheme and the size of the participating group? It would seem that a profit-sharing plan would operate successfully for the rank and file of wage earners only when applied in a small or medium-sized shop, because the larger the group of participants, the more difficult it is to educate them in the essentials of profit sharing, and the larger the group, the less effect each one can exercise on profits.

"These conclusions are verified by the history of profit sharing. In nearly every scheme of profit sharing for the wage earner which has been abandoned, it will be found that it was a scheme which included a large number of the rank and file indiscrimi-nately. Also many plans have been abandoned so far as the rank and file are concerned, but retained so far as the managerial and salaried group are concerned. Moreover, although many of the profit-sharing plans introduced for the benefit of the managerial group have been modified, none, so far as I know, have been abandoned as failures.

"Profit sharing is not a social panacea," concluded Dean Heil-man. "It will not solve the vexing problems which arise between employer and employee. It will not prove a substitute for good management on the part of the corporation, nor will it prove a substitute for the wage system. But within the wage system it has a place, a real place. In my judgment, we are destined to see a much wider and a much wiser use of profit sharing than in the past."

---

Although the Bureau of Standards dropped most of its ex-perimental work upon the use of metal to replace wood in air-plane construction, the future of metal construction appears promising, especially as the need for a fire-resisting airplane be-comes more generally understood. To avoid the difficulty which has been found in designing sheet metal parts so as to secure suf-ficient compressive strength one engineer has devoted much time and study to the use of steel tubes. A single metal tube has much less strength for its weight than a beam designed to concentrate most of the material in the flanges which are subjected to the higher stresses. The tubular design uses several small cold-drawn steel tubes for the flanges which are connected by diagonal lat-ticing of tubes of the smaller diameter. It is also possible to join tubes by inserting a snugly fitting ·tube of smaller diameter into another for a distance which will give satisfactory strength. Methods of fastening the two tubes together readily present them-selves. It is believed that encouragement and assistance given to metal construction by this Bureau is now about to result in material improvements which will come into general use.

# The Future of Aviation

Landing Fields and Wireless "Beacons" the Great Need—New Inventions Less Important—Future
Wars will See Armored Planes and Aerial Torpedoes

By COL. E. A. DEEDS,[1] DAYTON, OHIO

O N Thursday evening, December 4, at the Annual Meeting
of The American Society of Mechanical Engineers, the
night when the members of the Society and their friends
gather for their annual reunion and lecture, Col. E. A. Deeds,
the first speaker of the evening, was greeted by a large audience,
to which he presented in a most entertaining manner some of
the possibilities of the future of aviation both for military purposes and in its commercial applications. Following Colonel
Deeds, Col. Thurman H. Bane, Chief of Engineering, Division of
Air Service, and commanding officer of McCook Field, displayed
a series of lantern slides and moving-picture films of McCook
Field and the work accomplished there. These pictures were
released by the War Department for this lecture for the first
time and a number of the views are reproduced in the engravings
on the following pages. The remarks by Colonel Deeds are here
given in abstract and present some of the most interesting facts
which he brought out.

"When any one talks on the future of aviation we indulge in
flights of the imagination and we measure his ability as a speaker
by the number of thrills he can give us as he pictures the wonderful things which are likely to happen. This places one who
knows the industry at a disadvantage because the great thrills
of aviation have already taken place. I would rather, for
instance, take up tonight the work of the Wrights and the beautiful story of how these two brothers worked together; how their
savings were expended and how their sister readily gave hers, the
experiences around the fireside, and the sacrifices made—a story
which will never be duplicated. It is one of the finest examples
of courage and originality, interwoven with family devotion and
sacrifice, which will always stand as one of the great achievements of American invention.

"Then if we pass from the date when Orville Wright made
his first flight in 1903 to the time of the war, we will find a
struggling industry, many sacrifices, but here again, events that
never will be duplicated.

"The war itself gave a tremendous impetus to aviation; and
when we get enough perspective and are far enough removed
from it so that the public will be able to discern between the
patriotism and performance on one side, and propaganda and
politics on the other side, there will be found a real record of
achievement, more potent and more inspiring, probably, than
anything the future can give us, at least for some years to come.
The short period of the war gave aviation a greater impetus
than it would have received from fifty years of old peace-time
experiences along these lines. There was created during that time an
organization out of nothing, practically, that provided 16,852
airplanes, 40,420 engines, 642 balloons, several hundred flying
boats and many accessories that had to go with them—a tremendous undertaking, and accomplished with fairly good success.

"There were such accessories as the wireless telephone, the
oxygen apparatus, so that the aviator could breathe at high
altitudes, the bombs that went on the planes, the gun sights, the
electrically heated clothing, machine guns, fixed guns, synchronized
guns, 37,000 of which were provided, the pyrotechnic accessories
that went with the airplanes, the instruments of navigation and
others which created an industry in which there were approximately 300 concerns involved, and over 250,000 men working on
apparatus for the Air Service when the armistice was signed.

"Along with this development came the need for supplies—
the demand for raw material. There was particularly a large
demand for spruce timber. There were 27,000 or more boys in

uniform out on the coast who were engaged in supplying the
Allies with spruce for the building of airplanes, and because of
their activities we were never compelled to shut down a single
factory in this country or abroad, because these men supplied
all the lumber that was required. The chemical industries supplied
the dope that went on the wings, oil, and things of that sort, and
there is a long list of industries which furnished material and
supplies to advance the work.

"During these eighteen strenuous months there were developed
and produced 15,576 Liberty engines, representing a total of more
than 6,000,000 hp., an accomplishment which history unquestionably will put down some day as one of the great mechanical achievements of the war. The Liberty engine was the first across the
Atlantic, the first across the Continent, and by the sheer force of
its inherent merit, it has put itself among the foremost engines
of the world. It was a great accomplishment, done with great
credit to the manufacturers, who coöperated in a wonderful way
to make that work successful."

Colonel Deeds spoke about what the boys themselves did, stating that twenty-two thousand were trained to fly in this country,
and they flew approximately 66,000,000 miles in their training
alone, with the splendid record of only one fatality to every
200,000 miles, or the equivalent of eight times around the earth.
This is an indication of the comparative safety of this form of
transportation. That thousands of boys from all parts of the
country, brought into camp under military discipline, should go
through the most perilous part of flying, that of training, with
the loss of only one flier in a distance which is equivalent to flying
eight times around the world, is an achievement to be proud of.
With regard to the record of the boys in France, he said:
"The knights of old cannot picture anything more dramatic than
the young men who went out in these powerful machines,
going out alone, going up 10,000 or 15,000 ft. out over the enemy
lines, and meeting in hand-to-hand combat, as it were, the enemy.
Never will a page be written in history that is as brilliant and
thrilling, because in future wars we will look on these things more
as a matter of course, even though they will take a more important part.

"There has been during this war period a record of accomplishment which is more potent and more inspiring in many phases
than anything the future can give us for a long time. That does
not mean, however, that there is not a great future for our aviation. But it is going to have this disadvantage, that it will be
misunderstood and partially discredited by the public. Because
the airplane has not accomplished all the spectacular feats which
were prophesied for it, the public is now inclined to underrate
the possibilities of aviation. Too much credence should not be
given to those who picture the air full of airplanes on various
missions, moving on our hotels in flocks, surrounding our office
buildings, and carrying passengers to their occupations. The
future of aviation is bright enough viewed from a practicable
standpoint."

Looking to the future the speaker predicted that the line of
development in military aviation will take several forms. There
will be armored airplanes, protecting the pilot and observer, and
limited, of course, in size and thickness of armor, which will be
made of the more highly specialized and better-treated steel. Larger
planes should be developed, planes carrying cannon and more
guns, so that there will be no blind angle. These planes would
not maneuver, but go directly to their objective. The ground-
strafing machine, armored beneath, with machine guns pointed
down to cover the trenches, will undoubtedly create havoc.
The ability to reach high altitudes is another line of develop-

[1] Chief of Equipment Division of Air Service during the early development of the air program of the war. Mem. Am. Soc. M. E.

ment that is vital because, as Colonel Deeds put it, " About the only safe place in this game is to be up where the other fellow cannot get." That means that the air fighters of the future will have to be able to do their work at an elevation of 40,000 ft., in a temperature 50 deg. below zero.

Then, higher speeds will be necessary. Approximately 160 miles is the maximum speed today, not the advertised speed, but *the speed.*

There is also the aerial torpedo, that is, the pilotless aeroplane, that will go out 100 miles and then shed its wings and drop the whole fuselage, accurately hitting a target the size of a city.

" That," said the speaker, " is not a dream, and the problem was so well developed during this last war that there was a group of young men who were sorely disappointed when the armistice was signed, because they had expected to go over and discharge these pilotless machines against the enemy and create havoc in his ranks." That is a line of development which is sure to come.

There is the possibility of wireless control, too, but it has many difficulties, and may not be practicable.

The line of tactics that ought to be used by the military aviation division is a very important thing, and involves better communication, better relationship with other branches of the Army, the independent work to be done by the aircraft—a whole new line of work just touched upon by the war which must be interwoven now with our military tactics in every way. This can be done only by thorough, painstaking, experimental and development work. It is going to require the services of the engineer in the laboratory and of the flier in the field. Time, the greatest element in all of these undertakings, the thing we could not get in this last war, will be required to put this future aviation where it belongs in this country; that is, time enough for these engineers to do their work.

After outlining these various lines of development in military aviation, Colonel Deeds took up the subject of commercial aviation and said in part:

" At the present time the risks are too great, and the expenses too high, for private capital to go into commercial aviation. Government aid must be injected in some way into the commercial side of our aviation, if any progress is to be made. The worst thing that could happen to commercial aviation would be to place the entire responsibility upon the Government, but there ought to be some way worked out whereby moderate sums would be appropriated each year, to assist the commercial side of this industry until such a time as it can take care of itself.

" The most important thing for commercial aviation today is the simple thing of landing fields, because until we get them we can have no so-called commercial aviation. Municipalities should set aside certain grounds for this purpose and the Government should in some appropriate way assist the municipalities so that they can give these fields. The future should find every city of 10,000 inhabitants with these landing fields.

"Secondly, we need wireless beacons or lighthouses, as it were, scattered over the country, and placed with just as much care as we exercise in the case of our lighthouses for our water navigation.

"There is no reason at all, from an engineering standpoint, why we cannot have dotted over this country, at landing fields, and at frequent intervals, little wireless signal towers that will give out a proper signal, and enable the aviator to fly across this country and run no danger of being lost.

"We also need various instruments of navigation. We have not a compass upon which we can entirely depend. The gyro compass, or something of that kind, must be developed. We also need drift instruments, so that the aviator may fly straight in one direction without being carried off to one side. These are some of the practical, simple things that have to be done to make aviation safe. Commercial aviation will follow, if we have landing fields, wireless lighthouses and the proper navigation instruments.

"In addition to landing fields, wireless lighthouses, and proper navigation instruments there are other less vital needs. There is the matter of meteorology. There has been given to us a third dimension—we must study the condition of air currents at the highest altitudes and make maps that can be used for aerial navigation, a work which the Weather Bureau might well be handling now. Exploration of higher altitudes made during the War showed that 95 per cent of all the winds at an altitude above 10,000 ft. are from the west and east, both in this country and Europe. Explorations were made at 35,000 ft. and some as high as 60,000 ft. One interesting record was made at Chattanooga, November 5, 1918, where at 28,000 ft. there was found an air current of 154 miles an hour, which means this: that an airplane going 154 miles an hour would be standing still going one way, or it would be going at the rate of 308 miles an hour the other way.

"Among other things, licensing of pilots is vital, so that we get competent men in charge of the aircraft. The testing of machines, to be sure that we have proper machines going into the service, flying rules, and other regulations are important.

"The Post Office Department has greatly helped commercial aviation and it ought to be encouraged. The kind of service it requires, flying in all kinds of weather, under all conditions, is tending to develop more reliable engines for the airplanes. It is also securing data on the cost of operation which will be available to the commercial enterprises that are to follow.

"The designs of machines for commercial aviation will have to be changed. We have in the last two years devoted our energies in the direction of high speeds, quick maneuvering ability, and similar matters, but we should get down to lower landing speeds, more comfort, and enclosed airplanes. Our controls will be worked differently, and a number of other things will have to be changed, but none of these are very serious matters."

Colonel Deeds believes that the maximum size of the airplane, that is, the heavier-than-air craft for our commercial work, has been reached, unless we can get better landing facilities, and that the lighter-than-air type of aircraft will come into prominence for carrying the heavier loads.

The discovery and development of helium in this last War, which formerly was provided in the laboratory at the cost of $1700 per cu. ft., but now is available for less than 20 cents and is conserved by the Government, takes out of the lighter-than-air craft the one great hazard of fire, and with that hazard out of the way, an entirely new era is opening up. The lighter loads and greater speeds will be left to the smaller airplanes, perhaps, and heavier loads to the airships and to flying boats, which probably will be larger and more seaworthy.

In speaking of his own experiences and those of Mr. Kettering in flying, Colonel Deeds stated that the time required to go by plane from Dayton to Detroit is 2 hours; to Indianapolis, 1 hour; to Cleveland, less than 2 hours; to Washington, 3 hours; to New York, 4 hours; and to Wichita, Kansas, 7 hours. In addition to the great saving of time, the experience of flying is delightful. Colonel Deeds said that he believes in the privately owned airplane, but that he is restricted in his flying by the lack of landing fields and proper safeguards. Regarding casualties in flying, he said: " If you will take a list of the men who pioneered the automobile development, you will find that there is a long casualty list of men who have given their lives to that industry, and we have not any more of a casualty list in the airplane industry than in the automobile industry at this time."

Airplane development will come as that of the automobile came, first as a pleasure proposition and later as a utility proposition. Manufacturers are turning out a good many planes a week that are sold to individual fliers, but they have not the proper facilities for travel. Operating expenses are high but development of the industry will correct these. If engineers and manufacturers get together and are assisted by the Government, the airplane, while it probably will never attain the same degree of prominence that the automobile has attained, will nevertheless take a very important place in commerce. Aviation as an industry deserves the confidence and support of the conservative men of the country.

"Aviation will not reach the heights of the imagination which some have had about it," Colonel Deeds concluded, "but it will become a permanent part of the life of the people, and the thing that this country can do is to put the practical side into this industry for all civilization."

Views from McCook Field, Dayton, Ohio, Shown by Col. Thurman H. Bane, Chief of Engineering Division of Air Service, U. S. A.,
at Annual Meeting of the A.S.M.E.

Fig. 1—Airplane View of McCook Field, Showing Hangars, Headquarters, Supply Departments and Barracks; Fig. 2—High-Speed Wind Tunnel, Capable of a Velocity of 450 m.p.h.; Fig. 3—Dynamometer for Testing Air-Cooled Engines, With Wind-Tunnel Attachment in Order to Create
a Similar Condition of Air Flow; Fig. 4—Engine Dynamometer With Liberty-12 Motor on Test Stand Connected to Electric Dynamometer With

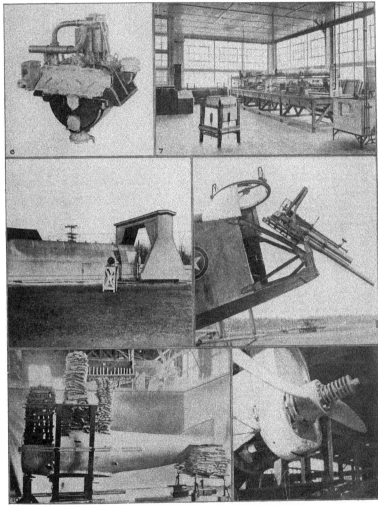

These Views Were Selected from Those Used by Colonel Bane to Illustrate the Present Development of the Military Airplane, and the Equipment at McCook Field

Fig. 6.—Dynamometer for Use in Connection With the Testing of Single Experimental Cylinders Shown Connected to Single-Cylinder Universal Engine; Fig. 7.—Chemical Laboratory, Materials Section; Fig. 8—Propeller Whirling-Testing Stand Capable of Delivering 800 hp. in Measuring Deflections, Flutter, Etc.; Fig. 9—Thirty-seven mm. Automatic Cannon on Flexible Mount as Placed in the Martin Bomber; Fig. 10—Method of Static Test of Fuselage. Factor of Safety Required Varies With the Size of Airplane; Fig. 11—Propeller Thrustmeter Placed on a DH-4 With Liberty Engine, Which Mechanically Records the Thrust of the Propeller.

# The Annual Meeting of the A. S. M. E.

Record Attendance of over 2100—Interest Centers on Committee Activities and Session on Industrial Unrest—Addresses on Aviation and Discussion of Valuation Problems Leading Features

ATTENDANCE at the Annual Meeting of The American Society of Mechanical Engineers found the Engineering Societies Building in holiday attire for the occasion, with the foyer on the first floor lavishly decorated with fir trees and palms and the other floors that were to be used appropriately arranged for the convenience of the Society's guests. Anticipating a large attendance, a Sub-Committee of the Committee on Meetings and Program, composed of local members having the social arrangements in charge, prepared to utilize all the available space in the building. Besides the Society's headquarters on the eleventh floor, the meeting rooms on the fifth and sixth floors and the auditorium were made use of, and on Thursday evening, the annual reunion night, members also gathered in the spacious foyer on the first floor. There are five public-meeting halls in the building, all of which were in use at times. The total attendance was 2116, of which 1346 were members and 770 guests, the largest at any meeting of the Society.

Although the convention nominally opened on Tuesday, December 2, many were present earlier for Council and committee meetings, Local Sections conferences, etc.; twenty-eight delegates were in attendance from the Local Sections of the Society.

There was widespread interest during the convention in the discussion upon the reports of the Aims and Organization Committee of the Society and the Joint Conference Committee of the Founder Societies, and in the conference of the Local Sections' delegates. A full account of these meetings will be found in Section Two of this number, with other matters of general interest pertaining to Society affairs.

Numerous committee meetings were also held, including those of the A. S. M. E. Power Test Codes and its Sub-Committee on General Instructions; Research Committee and its Sub-Committee on Fluid Meters; the Committee on Steel Roller Chains; the Committee on the Standardization of Shafting; and the Standardization Committee.

On the social side mention should be made in particular of the gathering at the ladies' reception and tea on Thursday afternoon, always one of the pleasantest occasions of the Annual Meeting.

The excursions to points of interest about the city were well attended, particularly that on Friday to the Curtiss aeroplane plant and flying field at Garden City where exhibition flights were given and special arrangements were made whereby members who so desired were able to undertake flights. As usual the meeting of representatives of student branches was held and following the convention on Friday evening there were numerous college reunions. An event of interest to many was an evening devoted to an account of the work of John Ericsson and Cornelius H. DeLamater, the founder of the famous DeLamater Iron Works, the occasion being the eightieth anniversary of the meeting of these two noted engineers and the thirtieth anniversary of their death. Several societies joined in these exercises.

## The President's Address and Reception

The formal opening was on Tuesday evening, when President Cooley gave his presidential address on the Society's activities and national scope of the work of the engineer, printed elsewhere in this number. Following his address, the announcement of the election of officers for the ensuing year was made as follows: *President,* Fred J. Miller; *Vice-Presidents,* E. C. Jones, R. H. Fernald and J. R. Allen; *Managers,* D. S. Kimball, L. E. Strothman and E. F. Scott; *Treasurer,* William H. Wiley.

In introducing the President-elect, President Cooley said that he had had the kind of experience which in his opinion qualified a man to be the President of The American Society of Mechanical Engineers. For 20 years he was a journalist, since which time he had been engaged in the business of factory management, an

expert on factory management. He served for five years on the Council of this Society, and for four years as Trustee of the United Engineering Society. His work as Trustee of the United Engineering Society had to do with the reorganization of its finances and other features, but principally the work of reorganization. During the past year he has had made a survey of the machinery—the office machinery—of the Society, and he has advised with regard to the development of MECHANICAL ENGINEERING, the Journal of the Society. He also rendered very distinguished service in the cause of his country during the war.

In replying, Mr. Miller said he regarded his election primarily an invitation to do more work, rather than a recognition of his service, and as such he gladly accepted it. His acquaintance with the history and methods of the Society made him fully appreciate that what should be done during the coming year would not depend so much upon the President as upon his fellow-members, and he would therefore bespeak the aid and coöperation of all those who desired to see the Society grow in numbers and in influence, keeping step with the best spirit and progress of the age.

The next event on the program was the conferring of Honorary Membership upon M. Charles de Freminville, consulting engineer, Creusot Works, France.

James Hartness, past-president Am. Soc. M. E., introduced M. de Freminville, who, he said, "is an engineer in the finest sense, a gentleman who understands the human element as well as the elements of machinery. He comes from a family, from an environment that combines the technical training with the practical life, that combines the science with the practice."

"M. de Freminville was one of the pioneers in introducing the corridor trains in France, and coming out of his long experience in engineering he has rendered distinguished service in the great war."

At the time when the long-range gun was shelling Paris, Mr. Hartness visited M. de Freminville in his home and went with him for a walk about the city to see some of the effects of the bombardment, and of the bombs which dropped from the aeroplanes. The spirit which M. de Freminville displayed at that time when his home and family were in such great danger made a deep impression on Mr. Hartness.

"While we are honoring one of the world's first engineers," he said, "we are also honoring a man who typifies that noble French spirit that took possession of the souls of the French people in that fight from the very first, and inspired their soldiers to 'carry on.'"

In his address of acceptance, M. Charles de Freminville expressed his deep appreciation of the tribute paid to him by the Society. Reviewing the history of the coöperation of America and France in their common ideal of civilization, he referred to the sewing machine as "the first object lesson which France had received from America," and the Westinghouse air brake as a second example. In return the Mallet compounding devices for the steam locomotive had been adopted in America and France had produced a steel car, the design of which inspired American engineers in the design of the present American steel car. The first low-level station was built in Paris, and it served as a model to the American engineers in the design of similar stations in New York. Again, coöperation between the two countries in automobile development had been very marked.

M. de Freminville also referred to the efforts made in both countries for the conquest of the air. In this connection he said:

While we were witnessing in France the successful experiments of Farman, about 1907, the news spread that two Americans, the Wright Brothers, had already obtained success in far greater achievements. This was very welcome, and this time France seemed to be the proper field for development of aviation. Count de Lambert flew over Paris some time later on a machine which he believed was designed by the

Wright Brothers, and he was promptly surpassed by Blériot crossing the Channel between Calais and Dover on his own machine.

M. de Freminville spoke of the work of Frederick W. Taylor and of the efforts of Professor Le Chatelier to introduce scientific management in France. The French public, he added, look upon the name of Taylor as a symbol of the efficient spirit that prevails in America.

He concluded by expressing the hope that America and France might come in closer contact by increasing interchange of students between the two countries. This interchange, he noted, has been highly beneficial in the past and would contribute greatly in the future to help the American and French engineers to develop a closer and more intimate friendship.

Following these exercises announcement was made that Honorary Membership was later to be conferred, in France, upon M. Auguste Camille Edmund Rateau, Chairman Board of Directors, Rateau, Battu and Smoot Company, France. Mr. F. R. Low was asked to outline the work of M. Rateau, which he did briefly, saying that he was a pioneer investigator in the field of the steam turbine and the turbo-compressor. He is author of basic tables and formulæ for the flow of steam through orifices, and author of methods of calculation and design of steam turbines now in practically universal use. He is the designer of numerous types of helicoidal fans. He is the inventor of the mixed-pressure steam turbine, and president of his own shops, manufacturing turbines and turbo-compressors as well as other types of high-speed rotating machinery, employing several thousand workmen. He has recently been active in the field of turbine propulsion of battleships, and in connection with the world war was author of notable contributions to the study of ballistics as well as inventor of devices to increase the efficiency of aeroplanes.

## Business Meeting

President Mortimer E. Cooley called the business meeting together on Wednesday morning and asked Ira H. Woolson to present the proposed amendments to the constitution. These were as follows:

C45 The Standing Committees of Administration of the Society shall be

    Committee on Finance
    Committee on Meetings and Program
    Committee on Publication and Papers
    Committee on Membership
    Committee on Local Sections
    Committee on Professional Sections
    Committee on Constitution and By-Laws

There shall be other Standing Committees of the Society as the By-Laws provide and the Council approves. The appointment, organization, duties and terms of service of all Standing Committees shall be as the By-Laws provide. The Chairman of all Standing Committees shall have a seat in the Council of the Society, but no vote. The Secretary of the Society shall be the Secretary of each of the Standing Committees.

C49 Professional Committees. The Council shall have power to appoint, upon a recommendation of the Society at a general meeting, or upon its own initiative, such Professional Committees as it may deem desirable, to investigate, consider, and report upon subjects of engineering interest. Any proposed expenses of such committees must be authorized by the Council before they are incurred.

C50 The Society may approve or adopt any report, standard formula or recommended practice and may print the same in the Transactions. It shall, however, not approve any engineering or commercial enterprise. It shall not consent to the use of its name or initials in any commercial work or business, except to indicate conformity with its standards or recommended practices.

C46 (To be eliminated. Superseded by C50 above.)

C51 Each committee shall perform the duties required of it in the By-Laws, or assigned to it by the Council. Membership on the Council or any committee of the Society shall terminate automatically on account of absence of any member, either wilful, or due to force of circumstances, as provided in the By-Laws.

Mr. Woolson then moved that the meeting approve the proposed amendments in the form in which they had been offered and order them submitted to letter ballot. The motion was discussed by G. K. Parsons who asked if the action would preclude the operation of the Nominating Committee newly created.

H. H. Vaughn wished to have the word "adopt" in C50 interpreted, and was asked to prepare such an interpretation with Robert I. Clegg, who offered his understanding of the meaning of the term. It was subsequently decided that the term would be interpreted in the by-laws.

L. P. Alford and Irving E. Moultrop also spoke on the motion, and R. H. Libbey sounded a note of warning against the ill-considered adoption of standards by the Society. R. H. Fernald called attention to a typographical error in C45.

The Secretary presented reports of Standing and Special Committees.

The preliminary Report of the Committee on Pipe Threads was presented by title by the Secretary and it was moved that the Society receive and order printed the report.

The Report of the Sub-Committee on Railway Locomotive Boiler Code was presented by H. H. Vaughn and members interested in the discussion of the Report adjourned to the Council room immediately.

Junior and Student prizes were awarded by the President to E. D. Whalen (junior) for his paper, "Properties of Airplane Fabrics," to C. F. Leh (junior) and F. G. Hampton (student) for their paper, "An Experimental Investigation of Steel Belting," to W. E. Helmick (student) for his paper on "An Experimental Investigation of Steel Belting." F. G. Hampton was present and received his prize in person. Upon a question by L. P. Breckenridge he informed the meeting that he and his collaborator were connected with Leland Stanford University, California.

Ira H. Woolson was called upon to introduce the chairman of the Sections Conference, A. C. Booth. In his remarks Mr. Woolson explained that the Society had taken action to create new machinery for the operation of the Nominating Committee in the nomination of officers of the Society. Legal advice had been obtained so that nominations had been made by the Sections delegates in conference, their action to be legalized by the adoption of a by-law to read as follows:

The nomination of the members of the Nominating Committee for the year 1920 made by the delegates of the Local Sections at their session held December 2, 1919, shall, if not revoked, be a sufficient nomination of this committee by the said delegates within the intendment of these by-laws.

The Committee on Constitution and By-Laws voted to accept this opinion, provided proper action was taken subsequently with regard to the details of the new machinery.

A. C. Booth, Chairman of the Sections Conference then reported the following list of members recommended for approval by the meeting to constitute a Nominating Committee:

Howard P. Fairfield ....................Worcester, Mass.
George K. Parsons ....................New York, N. Y.
G. A. Wechsler ....................Washington, D. C.
Elliot H. Whitlock ....................Cleveland, Ohio
George W. Galbraith ....................Cincinnati, O.
Chester B. Lord ....................St. Louis, Mo.
Robert Sibley ....................San Francisco, Cal.

Alternates were recommended as follows:

William W. Varney ....................Baltimore, Md.
H. G. Tyler ....................New York, N. Y.
Samuel W. Stratton ....................Washington, D. C.
C. M. Spaulding ....................Erie, Pa.
H. M. Norris ....................Cincinnati, O.
J. A. Hunter ....................Boulder, Col.
E. O. Eastwood ....................Seattle, Wash.

It was voted that the recommendation be approved. Ira H. Woolson then moved that Council be recommended to adopt the temporary by-law necessary to legalize the action.

Memorial exercises in honor of Professor F. R. Hutton, including the presentation to the Society of a plaque, were held, Past-Presidents W. R. Warner and Ambrose Swasey, H. H. Suplee, Professor Hutton's sons, Mr. M. S. Hutton and Dr. Lefferts Hutton, being asked to take seats on the platform. The Secretary read the presentation address by William H. Wiley, acting Chairman of the Memorial Committee.

Past-President W. R. Warner spoke reminiscently of Professor Hutton and unveiled the plaque, which was accepted in the name of the Society by the President.

F. R. Low presented a memorial of William Kent, which was prepared by a committee of which he was chairman and distributed in pamphlet form at the meeting.

Major Fred J. Miller read a memorial to the members who had given their lives in the service and the following resolution was passed with a rising vote:

Resolved: That this Society hereby expresses its greatest appreciation of and pride in the services of its members who gave their lives that freedom might be preserved among the nations of the earth and that a copy of these resolutions be forwarded to the family of each of the members.

L. C. Marburg, Chairman of the Committee on Aims and Organization, presented the Report of the Committee to the Society. The discussion of the report was so lengthy that it was continued at an adjourned meeting in the afternoon and will be found in Section Two of this issue.

Morris L. Cooke presented an amendment to the Constitution, C6, which was received by the Secretary for presentation at the Spring Meeting as required by the Constitution.

## Appraisal and Valuation Session

The Committee on Meetings and Program has received requests from time to time for papers and discussion upon the subject of Appraisal and Valuation of Property, a question in which many engineers are interested and about the principles of which there is a wide diversity of opinion. It was therefore decided to hold a session on this subject at the Annual Meeting, and a Committee, of which President Cooley was chairman, was appointed to arrange the details.

It subsequently developed that the American Society of Refrigerating Engineers, whose Annual Meeting occurs simultaneously with that of The American Society of Mechanical Engineers, were proposing to hold a similar session, and a joint session of the two Societies was accordingly planned for Wednesday, December 3. Papers contributed by the Refrigerating Engineers were upon Ice-Plant Depreciation, by George E. Wells, president of that society, and upon Depreciation of Insulation, by J. E. Starr. The A.S.M.E. contributed an Appraisal and Valuation Methods, by David H. Ray, an abstract of which appeared in MECHANICAL ENGINEERING for December. Supplementing this was a series of contributions, carefully developed by the Committee in charge, for a progressive discussion of certain fundamentals upon which a plan for valuation methods might be based. These papers were as follows: Fundamental Principles of Rational Valuation, by James R. Bibbins; Development of Project, by the late F. B. H. Paine; The Construction Period, by H. C. Anderson; Price Levels and Value, by Cecil Elmes; and Original Costs and Normal Values, by R. H. Nexsen.

There was very evident interest in the session, as it was largely attended and the discussion lasted until a late hour. In fact, the attention which the subject received and the evident approval of those present led to a motion recommending that a committee corresponding to the Boiler Code Committee of the Society be appointed to go into the subject of appraisal and valuation, making a study of it for several years if need be, trying, if possible, to bring order out of chaos in respect to the methods to be followed. This motion was carried.

Inasmuch, however, as no opportunity was afforded for the proper digesting and discussing of the several papers presented at this session, it has been further proposed that certain or all of the papers be distributed in advance of the St. Louis meeting to be held next spring, and that discussion upon them be solicited for that meeting. It is probable that this will be done and that a digest of the discussion given at the Annual Meeting will also be sent out. In view of the situation, therefore, no attempt has been made at this time to report the discussion on appraisal and valuation at the Annual Meeting. Should the matter be assigned to the St. Louis meeting, as proposed, a suitable review of the whole matter will undoubtedly then be published.

## Gas Power Session

The Gas Power Session was held on Wednesday afternoon, Dean Benjamin of Purdue University, presiding. The following papers were presented: The Hvid Engine and Its Relation to Fuel Problems, by E. B. Blakely; Combustion of Heavier Fuels in Constant-Volume Engines and in Engines of the Superinductive Type, by Leon Cammeu; Kerosene as a Fuel for High Speed Engines, by L. F. Seaton; and Oil Pipe Lines, by S. A. Sulentic. The first two of these papers appeared in abstract form in the December number of MECHANICAL ENGINEERING on pages 918 and 941, respectively; the last two in the November issue on pages 881 and 883.

Mr. Blakely's paper drew forth considerable discussion in which the following points regarding the Hvid engine were brought out: The standard types are 1½, 3, 6, and 8 hp., but larger units such as 15, 20, 30 and 60 hp. per cylinder are also being made. One marine concern is putting out a 25-hp. 4-cylinder block design with an electric starter; another is building 2, 4, and 6-cylinder units.

The cheapest grades of oil may be used for the Hvid engine as long as they are clean, free from grit, and will flow readily. The author told how crankcase oil obtained from a garage, after being strained through a piece of cheesecloth, was used by one concern to operate two of these engines. Undoubtedly this procedure saved several thousand dollars. Two fuel cups are supplied with each engine and about a half-hour is required to change over from a light-oil engine to a heavy-oil engine.

Carbonization in the cup may be attributed to the following causes: the cup being too hot when heavy oil is used; a cup designed for kerosene being used for heavy oil, or leakage occuring between the cup and the atmosphere. Other than these, the author knows of no other causes of carbonization.

Several of these engines have been running on test to see how long they will stand up and how long it will be before there will be trouble with the Hvid principle. Three 8-hp. engines have been running between five and six months. One of them has been in operation, both day and night, for a little over five months, with one shutdown, the cause of which is unknown, which occurred over two months ago. Since that time this engine has been running continually.

H. Schreck [1] took exception to the author's statement that the Diesel engine cannot be built for smaller units than 100 hp. He pointed out that a German firm has made a practice of building Diesel engines down to 8.5 hp. and has shipped them by hundreds to South America and Russia. In reply, the author stated that no small Diesel engines had been made in this country, and moreover that from the standpoint of initial cost, a small Diesel engine, with its complicated air compressors, could not compete with the simply constructed Hvid engine.

L. F. Seaton's paper on Kerosene as a Fuel for High-Speed Engines was read by title only.

In presenting his paper on the use of heavier fuels in constant-volume engines, Mr. Cammen pointed out that the gas turbine today is in the same state as aviation was before the advent of the Wright Brothers; moreover, the Wright Brothers of the gas turbine have not yet appeared. The United States, he said, is the greatest automobile-producing country in the world; nevertheless, during the past three years, more has been done abroad in the development of the new types of the internal-combustion engine, particularly as regards the superinduction types. Unless American engineers consider the effect of mass production and look to the future more carefully, we are apt to lose out in the automobile field just as, very unfortunately, we lost out in the development of the American invention, the aeroplane.

In the author's absence the paper by S. A. Sulentic was read by title only. In view of the fact that oil pipe lines have been so highly developed in this country, and further, that many available data exist in the hands of the builders and operators of these lines, Professor Lucke [2] expressed the opinion that such data should

[1] Consulting Engineer, New York City. Mem. Am. Soc. M. E.
[2] Prof. of Mechanical Engineering, Columbia University, N. Y. City. Mem. Am.Soc.M.E

be filed, either in the form of papers or discussions, and that they appear in the TRANSACTIONS of the Society. Professor Lucke also offered a resolution, which later was passed, that copies of Mr. Sulentic's paper be sent to engineers connected with pipeline and similar industries, with a request that they present written discussions verifying the conclusions arrived at, or substitute others for them, and that they add as many data as they can, in order that this valuable compilation become part of the TRANSACTIONS.

## Session on the Industrial Situation in Relation to Present Conditions

At the session devoted to the Industrial Situation in Relation to Present Conditions, held on the morning of Thursday, December 4, four papers were presented, which, by reason of the vital importance of the subjects discussed, drew an audience that filled every seat of the main floor of the auditorium. These papers were: The Causes of Industrial Unrest and the Remedy, by Frederick P. Fish, Chairman, National Industrial Conference Board; Wage Payment, by A. L. De Leeuw, Consulting Engineer; Systems for Mutual Control of Industry, by William L. Leiserson, formerly Chief, Division of Labor Administration, Working Conditions Service, U. S. Department of Labor; and What May We Expect of Profit Sharing in Industry? by Ralph E. Heilman, Dean, School of Commerce, Northwestern University, Chicago. Owing to the absence of the author, Dr. Leiserson's paper was read by Secretary Rice. The discussion of these papers and of a set of resolutions offered by the New York Section required a continuation of the session throughout the afternoon. President Cooley opened the morning meeting, later calling Vice-President Spencer Miller to the chair. Prof. D. S. Kimball presided at the afternoon session.

Discussion of the papers was opened by the Hon. George M. Barnes, M.P., representing the British Joint International Labor Conference and who also represented Great Britain at the Peace Conference at Versailles. Mr. Barnes, in the course of his remarks, said that the labor problem in its essence was not only that a man should get higher wages and more to eat and better clothes, but that he should take his place on terms of equality with all other classes of the community. Workmen were no longer illiterate and could hold their own in discussions on economic subjects. Employers on both sides of the Atlantic were now recognizing this fact and also that workmen had the right to organize and bargain for the disposal of their labor through agents. He did not think that profit sharing would appeal to the workmen unless it was applied to organizations of labor over wide areas on a national basis instead of to isolated establishments. Paternalism was well enough when shops were small and when every employer could know his men, but was no longer feasible in the present era of large-scale production where an establishment sometimes employed twenty to thirty thousand men. The best way to abolish the industrial unrest, in his opinion, would be to recognize that the workmen had equal rights with capital to organize on a large scale, and he believed that frankly recognizing this fact would contribute to increased production and to the benefit of all concerned. Mr. Barnes' remarks were greeted with hearty applause.

A. L. Siff[1] submitted a written discussion in which he stated that strikes, lockouts and the like were as unnecessary today and as unjustifiable as any of the more serious breaches of civil and criminal law. We had built up a system of courts and legal procedure to adjudicate all civil and criminal disputes and we should adopt the same system to adjust labor questions. He outlined an elaborate plan for such adjustments which included a conference council in each company, as well as courts of appeals and various industrial zones for the country at large.

Harry Alexander[2] wrote describing a profit-sharing plan in which the men are rated in six classes, varying amounts being

deducted from their normal share of profits. These deductions form an efficiency fund, 70 per cent. of which is divided between the two first-grade men, 20 per cent. among the three second-grade men and 10 per cent. among the five third-grade men. The men are rated 60 points for loyalty, 15 each for ability and judgment and 10 for speed. The plan, he said, had been in successful operation for many years.

Lewis H. Nash[1] believed that the present labor unrest was due to the fact that so many people had come here from foreign countries who consider work more as a degradation than an honor. He looked with great suspicion on any scheme that tended to limit the rights of an individual to run his business for the mutual benefit of himself and his employees without intermeddling by outside forces, either governmental, labor-union or other, because he believed individualism to be the basis of democracy.

A. C. Jackson[2] said that one question which should be considered was whether or not labor unions should be compelled to incorporate, so that an agreement made between a responsible corporation and a labor organization would be between two responsible parties and not as now between a responsible party and one irresponsible before the law; and further, that if such incorporation was made compulsory, the present exemption of labor organizations from the Sherman Anti-Trust Law should be removed.

William S. Bowen[3] said that the struggle between capital and labor now centered about the question of control in industry. At present it was controlled by property. Looking upon wages as the interest that labor received from its capital, he thought it was possible to devise forms of control certificates, issuable to both labor and investors, that would amalgamate the interests of both, and gave a brief outline of the provisions of these proposed certificates.

Frederic L. Meron[4], as his contribution to the discussion, presented particulars of a profit-sharing plan which he had recently published as part of a work on the Human Element in Organizations.

G. P. Hemstreet[5] said he had been urged many times to consider making partners of his workmen but felt that the union policy of limiting production hardly fitted a man to become a partner in a business. One thing that all could do was to see that their foremen became better acquainted with the men under them and treated them as human beings, for the foremen were the men who made or marred the reputation of a corporation with its employees.

Harrington Emerson[6], discussing Mr. De Leeuw's paper, said that what was most needed today was some authority or some arbiter who would determine for the employer and employee what was just; who could settle the unit of measurement for both sides so that it would not be guesswork. He knew of an experiment where 12,000 men were employed, in which such an arbiter had succeeded after a few years in increasing the amount paid out as wages over a million dollars annually. This man and his assistants drew up thousands of work schedules which had to be passed upon by officers of the company and accepted by the men before they became operative.

Walter N. Polakov[7] said that so long as goods continued to be produced primarily for profit instead of for use, we could not hope to attain a maximum production. What could and must be done in this connection was to settle on the definite amounts of goods necessary for consumption and then see to it that those amounts were produced with a maximum economy and a minimum loss. Today industry was absolutely uncoordinated and unregulated. If we had a competent council capable of determining how much

---

[1] Of the firm of Siff Bros. Co., Wholesale Clothiers, New York. N. Y.
[2] Pres. Harry Alexander, Inc., Contracting Engineer, New York, N. Y., Mem. Am. Soc. M. E.
[1] Pres. Nash Engineering Co., South Norwalk. Conn., Mem. Am. Soc. M. E.
[2] Gen. Supt. and Secy., Miller Lock Co., Frankford, Philadelphia, Pa., Mem. Am. Soc. M. E.
[3] 405 W. 118th St., New York. N. Y., Assoc-Mem. Am. Soc. M. E.
[4] Mechanical and Electrical Engineer, New York, N. Y.
[5] Gen. Works Mgr., Hastings Pavement Co., Hastings, N. Y., Mem. Am. Soc. M. E.
[6] Director. The Emerson Engineers, New York, N. Y., Mem. Am. Soc. M. E.
[7] Of Walter N. Polakov & Co., Consulting Engineers, New York, N. Y., Mem. Am. Soc. M. E.

and what goods were needed, as we had during the war, we might begin to see the light on other issues such as wage payment, organization of labor, etc., for these were capable of very simple solutions provided that those who were trying to settle them knew exactly what they were talking about and based their conclusions on facts, not on prejudices.

T. M. Ave-Lallement,[1] in an extended discussion of Mr. Fish's paper, challenged the statement that " some form of representation in industry is essential in order to make personal relationship possible under modern industrial conditions." He argued that representation in industry was a means of social control and not at all a means to the " resumption of personal relationship between employer and employed." What the workers meant definitely was a share in the control of industry through representatives of their own choosing. Piece-work, premium and bonus systems of remuneration were objected to by workers because they had no control of the determination of the rates employed. There remained as a further step in development the " collective labor contract," under which the workers contributed to the compound labor required in the production of an article agree collectively to do their work at a certain contract price for the whole job. This form of contract involved the complete determinateness of the piece-work contract, the principles of collective bargaining and the possibilities of acceptance by the workers of the principles of scientific management. Only in this way, in his opinion, could be much-to-be-desired maximum of production and minimum of cost be obtained.

RESOLUTIONS PROPOSED BY THE NEW YORK SECTION ON THE SUBJECT OF INDUSTRIAL UNREST

At the morning session, just prior to adjournment, L. P. Alford, representing the New York Section of the Society, took the floor and stated that the New York Section, during the preceding three months, had held a series of meetings to discuss aspects of industrial unrest. Acting in accordance with the wish expressed at one of these meetings, a brief declaration had been prepared on the subject, which was presented, discussed and approved at a largely attended meeting held on the evening of November 26, 1919, the motion of approval carrying a request that it be presented to the Council for their consideration and action. The Council had duly considered the declaration and referred it back to the Executive Committee of the New York Section, with a request that it be brought before the Society at the Annual Meeting. Mr. Alford then read the resolution together with two further resolutions which had been drawn up. At the afternoon session, following the discussion of the papers which had been presented in the morning, action was taken on these resolutions, and after a short debate they were adopted in the form given below.

RESOLUTION NO. 1—DECLARATION OF PRINCIPLES

Social and industrial unrest result from the fact that human relations have not kept step with economic evolution.

Competent directive management of essential enterprises is the logical solution. Such management must be free from autocratic control, whether by capital or by labor.

Sharp social or industrial disputes are no longer private. Society is affected, therefore such cases must be subject to the decision of authorities based upon intrinsic not arbitrary law.

Industry and public utilities must serve the people. There is no room for special privileges of capital or of labor. Strikes, irregular employment, or arbitrary acts of ownership or of management are harmful, not alone to the immediate parties but to society as a whole.

Productivity and public service are absolutely essential.

On account of the peculiarly intimate familiarity of engineers with industrial problems our responsibility is great.

Therefore, we, engineers and members of The American Society of Mechanical Engineers, declare that the following essentials are established by facts and experience, urge all of our members to uphold them, and invite other engineers to coöperate with us in having them unanimously recognized; viz:

    Every important enterprise must adopt competent productive management, unbiased by special privilege of capital or of labor, and disputes must be submitted to authorities based upon intrinsic law.

    Credit capital represents the productive ability of the com-

munity and should be administered with the sole view to the economy of productive power, that is, it should be granted only to those who are able to render valuable service.

RESOLUTION NO. 2

*Moved*, That a Committee on Agencies of Productivity be established with sub-groupings to study among others these major agencies:
1—Invested savings
2—Credit capital
3—Organized labor
4—Unorganized labor
5—Employers
6—Managers
7—Society as a whole
8—Economic and productive intelligence service.

RESOLUTION NO. 3

WHEREAS, The vital necessity of *immediate* constructive action in the field of industrial relations is beyond dispute; and

WHEREAS, Industrial " Society at Large " obviously stands in need of some organ or agency of continuous productive initiative; and

WHEREAS, A declaration of principles upon which to base constructive action has today been approved by the Society in Convention assembled; be it

*Resolved*, That we, members of The American Society of Mechanical Engineers, undertake to formulate a practical program for industrial relations betterment: and further,

That this Society undertake to formulate a practical plan of organization to assist in determining what immediate steps can be taken to put industrial relations on a better footing and coördinate and make effective the splendid work now being done by various agencies: and further,

That the Council of the Society be requested to appoint a committee of five members of The American Society of Mechanical Engineers to study and to make effective this program.

## Machine Design Session

With the exception of one paper, the Machine Design Session, held in the Auditorium on Thursday, was devoted to a discussion of various types of rotating machinery. Dean Cooley presided and the following papers were presented in the order named: Reliability of Materials and Mechanism of Fractures, by Charles de Freminville, Hon. Mem. Am. Soc. M. E.; Tests on Dredging Pumps Used in Inner-Harbor Navigation Canal, New Orleans, by W. J. White; A Perfected High-Pressure Rotary Compressor, by C. B. Lord; Turbo-Compressor Calculations, by A. H. Blaisdell, and A New Type of Hydraulic-Turbine Runner, by Forrest Nagler. The papers by Messrs. Lord and Nagler will be found in abstract form in the November and December issues of MECHANICAL ENGINEERING; those by M. Charles de Freminville and A. H. Blaisdell will appear at some future date.

Mr. White's paper, which was presented by Prof. W. B. Gregory, was illustrated by numerous lantern slides and many of the views showing striking features of the work of constructing the New Orleans Industrial Canal are reproduced in connection with the abstract, which appears elsewhere in this issue.

It is of interest to recall in this connection that a similar problem was encountered during the construction of the Panama Canal and this fact was alluded to by H. Cadwallader, Jr.[1], in the discussion. Mr. Cadwallader told of the difficulties of that work and how changes in dredging equipment increased the amount of material removed from 110,000 yd. per month to 225,000 yd.

During the discussion which followed Mr. Lord's paper, which consisted chiefly of a series of questions put to the author, it was brought out that the high-pressure rotary compressor he described will not be manufactured on a commercial basis because the principle involved is applicable alike to vacuum pumps, fluid pumps and air pumps, and it is thus obviously impracticable to manufacture the compressor in all its various forms. An organization has accordingly been formed for the purpose only of licensing other concerns to manufacture the particular type of compressor they require. For example, if a certain company requires a compressor for Diesel engines this new organization will supply them with drawings, act in a consulting capacity and give them a license to manufacture in their own shops the particular type of compressor they require.

In response to an inquiry Mr. Lord stated that the compressor

[1] 161 West 56th St., New York, N. Y. Late member of Research Staff, National Industrial Conference Board.

[1] Efficiency Engineer, Philadelphia, Pa. Mem.Am.Soc.M.E.

had shown a mechanical efficiency of 86 per cent and a volumetric efficiency of 92.7 per cent, but as to its adaptability to various types of work, such as the pumping of Mexican fuel oils or the compressing of casing-head oil, Mr. Lord was unable to give any definite data.

Turbo-Compressor Calculations, by A. H. Blaisdell, was presented, but by title only. The paper contains a discussion of the features which differentiate the turbo-compressor from the multi-stage centrifugal pump.

The final paper of the session was presented by its author, Forrest Nagler, of Milwaukee, Wis. The type of runner described in the paper is a decided step forward in the development of the hydraulic turbine and it is of interest in this connection to quote from the remarks of Clemens Herschel, a hydraulic engineer who was intimately connected with the early development of the turbine.

"The Francis wheel, so-called, is a misnomer," said Mr. Herschel. "He never invented anything of the sort that is now called the 'Francis wheel.' Of course, I knew him personally, and used to work with him. The so-called 'Francis wheel' was invented by a man by the name of George F. Swain. Of course improvements were made and in those days most of the development was done by what were called cut-and-dry methods. They would make a model and try it. If it did not work they would plug it or shave it down a little, and finally get it to suit."

## Discussion at Power Machinery Session

The discussion, of which a summary is here given, took place at the Power Machinery Session at the Annual Meeting and applies to the four papers which were read at the session: Emergency Fleet Corporation Water-Tube Boiler for Wood Ships, by F. W. Dean of the Emergency Fleet Corporation and Henry Kreisinger, engineer, U. S. Bureau of Mines; Flow of Water Through Condenser Tubes, by William L. De Baufre and Milton C. Stuart, of the U. S. Naval Experiment Station; Air Pumps for Condensing Equipment, by Frank R. Wheeler, of the C. H. Wheeler Mfg. Co., and The Thermal Conductivity of Insulating and Other Materials, by T. S. Taylor, non-member, of the Westinghouse Research Laboratory. Because of lack of room it has been impossible to give the discussion in complete form, but it is believed that what is presented here correctly records the most essential points that were brought out.

### EMERGENCY FLEET CORPORATION WATER-TUBE BOILER FOR WOOD SHIPS

In opening the discussion, Joseph J. Nelis[1] emphasized the radical departure for the merchant marine, accustomed to the Scotch marine boiler, in the introduction of the water-tube boiler on merchant ships. The chief differences between this boiler and the standard in use in the United States Navy existed in closer tube spacing, larger grate areas and the fact that there was no superheater. The boilers were designed with large grate areas to accommodate the use of Pacific Coast coals and natural draft, although all ships were equipped with fans. The tendency in the merchant marine, he said, was toward higher pressures, superheat and the use of turbines.

T. A. Marsh[2] wrote that the results of the experimental work by Mr. Kreisinger both as regards baffling and air supply over the fuel bed correspond with much that has been done in commercial practice by boiler and stoker companies during the past few years.

In the development of improved heat absorbers additional passes have been given to the gases in some of the standard boiler designs. Some of the first designs lacked proper areas, did not make due allowance for soot removal, and were difficult to maintain, but recent designs have made provision for overcoming these difficulties.

[1] Asst. Engr., Emergency Fleet Corporation, 140 North Broad St., Philadelphia, Pa.
[2] Chief Engineer, Green Engineering Co., East Chicago, Ind. Mem.Am. Soc.M.E.

In changing a boiler from a 3-pass to a 4-pass baffling, one must always consider the additional draft loss due to high resistance to the gases and the reduction in draft due to lower exit temperatures. It is necessary if some desired rating is required from the boiler, to insure that the draft after changing the baffles will be sufficient to assure adequate fuel consumption, otherwise in the effort to improve efficiency the capacity will be reduced.

Referring to air admission over the grates, there is no question but that the author is correct regarding the inability to force sufficient air through the fuel bed. Auxiliary air supplies have been introduced through bridgewalls, side walls, through arches, at the front of the furnace and in fact at almost every section of the furnace. This has proven of advantage in many instances, particularly with furnaces of limited capacity for combustion of the gases.

Mr. Marsh had seen several instances of increased efficiency due to the introduction of auxiliary air and was thoroughly in accord with the idea of installing systems for admitting air above the fuel bed provided they are installed and operated under the supervision of competent combustion engineers. There are many more installations suffering from a large excess of air than from a deficiency, and it is with great caution that we should advocate admitting air other than under the fuel bed.

Larger furnaces or stoker firing may afford a better solution to the problem. Each furnace requires its own specific investigation.

In a written discussion, Albert A. Cary[1] drew attention to the circulation within the boiler. Three hundred and eighty-eight 3-in. tubes generating steam, he said, had for their outlet to the steam drum 21 three-inch tubes. This disparity of area was not thus wholly expressed, but the relative volumes of water and steam must also be considered. By applying the usual formula for calculating the flow of steam through pipes, it could be shown theoretically that the twenty-one 3-in. tubes delivering steam from the rear header to the cross-drum of this boiler were not only ample in number and size, but that the steam was flowing at an absurdly low velocity. But it would be found that the difference in pressure between the rear header and the front cross-drum actuating the flow of steam was very small. Steam merely flows in these tubes as rapidly as steam is being taken from the boiler, and the pressure in the steam drum never falls very much below the pressure in the rear header.

During the course of his professional experience, he said, he had been called upon to perform investigations which led him to experiments to determine the true velocity of circulation within a water-tube boiler. A light-running propeller wheel mounted in a thin disk of metal was placed in the tube in such a manner as not to interfere with the circulation of the water. The number of revolutions of the propeller was recorded by an electric chronograph placed outside the boiler and varied with the velocity of circulation. While the experiments were unsuccessful so far as determining the actual velocities was concerned, he was able to discover that the direction of circulation was not as he had supposed. The upward circulation from rear to front headers was very rapid in the several lower layers of tubes. Above the vertical center of the bank of tubes there was little or no velocity of flow; while near the top there was a reversed circulation, from the front, back to the rear headers. This showed that the outlets from the top of the headers to the steam drum could not deliver the steam from the tubes as rapidly as it was generated, resulting in reversed circulation. With the tubes thus largely filled with steam their value as heating surface became diminished.

He discovered subsequently that these observations had also been made by M. Brull, the French engineer, with boilers of the Babcock and Wilcox type. In England he had seen the efficiency of boilers greatly increased by the use of steam-extraction tubes which delivered the steam directly to the steam drum as soon as it was generated.

L. P. Breckenridge[3] congratulated the Society upon having pre-

[1] 93 Liberty St., New York. N. Y.
[3] Professor of Mechanical Engineering, Sheffield Scientific School. Yale University. New Haven. Conn.

sented to it a paper by men so well qualified to do so as Messrs. Dean and Kreisinger, and pointed out that the complete combustion of fuel depended upon three things: the proper amount of air, mixed thoroughly with the fuel at the proper temperature.

R. Sanford Riley[1] thought that the water-tube boiler had made an effective entrance into merchant marine engineering practice, and that, due to the impossibility of obtaining skilled firemen at sea, the automatic stoker would soon come into use on ships.

John Van Brunt[1] said that underfeed stokers had been used on ships and that there was no mechanical reason why they should not come into universal use.

In the closure of the paper, Henry Kreisinger said that the Bureau of Mines experiments had showed that the circulation in boilers was as Mr. Cary had pointed out, very slight in the middle section of tubes and in opposite directions in the sets of tubes above and below. He did not imagine that the reversed circulation in the tubes would cause steam to remain in them for very long. He announced that the investigations published in the present paper would be completely reported in a publication of the Bureau of Mines.

### FLOW OF WATER THROUGH CONDENSER TUBES

The discussion on this paper, opened by Joseph J. Nelis, centered around the packing of condenser tubes in marine condensers. William de Baufre said that the present marine practice with the Navy was the rolled joint with bent tubes to take up expansion and contraction strains. Edwin B. Ricketts[2] thought that the corset-lace packing of standard stationary condenser practice should prove equally as serviceable at sea as on land and John F. Grace[2] said that he had inspected condensers on a recent destroyer where the corset-lace packing had been used.

Philip E. Reynolds[4] and G. L. Kothny[4] asked about the friction losses at the entrance of the tubes, and R. J. S. Pigott[4] wanted to know if any condenser manufacturer had attempted to eliminate much of this entrance and exit friction by chamfering the ferrules. August H. Kruesi[4] said that he had always specified chamfered ferrules, and John F. Grace drew on the blackboard a diagram of the chamfered ferrule made by his company.

In closing, Professor de Baufre said that the experiments showed a loss of about 3 ft. of tube due to the entrance and exit friction, and that further experiments should be performed to see if it would be possible to reduce this to, say, 1 ft. He had no data on the effect of bent tubes.

### AIR PUMPS FOR CONDENSING EQUIPMENT

By far the liveliest discussion of the Power Machinery Session was that on the paper by Frank R. Wheeler on Air Pumps for Condensing Equipment. After presenting the important features of his paper orally, Mr. Wheeler was followed by George J. Foran,[7] who read a lengthy discussion of the paper. Mr. Foran thought that the paper conveyed the impression that recent progress in the maintenance of high vacua was due to the introduction of the steam-ejector type of air pump, and gave evidence to show that such high vacua had been obtained by the rotative dry-vacuum pump in use for many years in commercial operation. He contended that every manufacturer of condensing equipment made ejector-type air pumps, and offered them for situations where he considered them advisable. His company had had no call for such pumps from the more important power plants.

[1] President, Sanford Riley Stoker Co., 25 Foster St., Worcester, Mass.
[2] Chief Engineer, Combustion Engineering Corporation, 11 Broadway, New York, N. Y.
[3] Asst. to Chief Operating Engineer, The New York Edison Company, 130 E. 15th St., New York, N. Y.
[4] Designing Engineer, Henry R. Worthington, Harrison, N. J.
[5] Vice-President, Croft-Reynolds Co., Inc., 95 Liberty St., New York
[6] Manager, Marine Department, C. H. Wheeler Mfg. Co., 18th St. and Lehigh Ave., Philadelphia, Pa.
[7] Supt., Raw Materials Department, Bridgeport Brass Co., Bridgeport, Conn.
[8] Engineer of Construction, General Electric Co., Schenectady, N. Y.
[9] Manager, Condenser Department, Worthington Pump and Machinery Corporation, 50 Church St., New York, N. Y.

He called attention of the members to a paper read before the National Edison Companies, in 1913, in which the hydraulic and rotative dry-vacuum pumps were compared. The power required by the hydraulic pump was several times that required by the rotative dry-vacuum pump, not 40 per cent as stated in the paper.

He submitted the performance curve of the Muller hydraulic pump (Fig. 1) with the statement that the pump had no small passages or sharp edges as stated by the author. He questioned accuracy of the figures given, and pointed out that the author had assumed an increased temperature corresponding to adiabatic compression which was true only for the steam ejector. The compression for dry air in the hydraulic vacuum pump was known to be isothermal. He then showed the effect of the actual conditions where the air was not dry, and that the hydraulic vacuum pump required less power than the ejector pump.

Mr. Foran criticized the author's statement regarding the mixture of air and water vapor, and said that the process should be

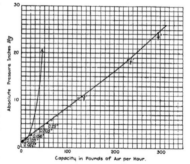

FIG. 1   PERFORMANCE CURVE OF MULLER HYDRAULIC PUMP

clearly indicated by which the method of determining the relative volumes was deduced, and he thought that with Fig. 5 (MECHANICAL ENGINEERING, December, p. 930) should have been shown also the curve showing volumes of 1 lb. of air at different pressures and temperatures of saturation. This curve, he contended, was applicable to the determination of volumes. He pointed out the discrepancy in figuring the volumes as had been done by the author.

With regard to Fig. 4, he pointed out that for the smaller capacities much less air allowance was shown than previously had been used, both in this country and abroad. On the other hand, the curve showed a much greater allowance than was required under actual conditions for the larger capacities.

Commenting on the statement that "the ratio of compression in the two stages (in the two-stage ejector pump) should be kept as nearly equal as possible," Mr. Foran pointed out that patents had been applied for such pumps under claims that:

a. Compression in first stage was greater than in second.
b. Compression in second stage was greater than in first.
c. Compressions in both stages were equal.

Philip E. Reynolds read a discussion in which he agreed with the conclusion of the author that the ejector type of air pump was the air pump par excellence. The fault of reciprocating pumps lay in the fact that the volumetric efficiency fell off very rapidly when high vacua were approached, that it was difficult to maintain high volumetric efficiency and to lubricate the pump.

He thought that the hydraulic air pump failed to meet the requirements in that it could handle only small quantities of air and was somewhat unstable at low vacua. The power required to obtain high vacua was very high. The openings were small and liable to become clogged; it required expensive equipment and the maintenance was high.

He said that the vacuum obtainable in a given condenser depended on the design and size of the condenser, the amount and temperature of the circulating water and the effectiveness of the air pump, and gave the basis for these statements. From an engineering standpoint, and with a given condensing equipment, the air pump which would remove the greatest volume of saturated air from the condenser would produce the best vacuum. However, there must be a compromise involving steam consumption, and the benefits derived from the increase in vacuum which might be obtained by increasing the volume of saturated air removed. It could readily be shown by simple mathematics, applying Dalton's law, that excessive volumes of saturated air might be removed from a condenser with only a negligible increase in vacuum. Just what volume it was best to specify to be removed depended on the design of the condenser, the temperature of the saturated air, steam consumption, benefits derived from increased vacuum, etc.

He said that the author's method of selecting the size of a pump was very simple, but was based upon the temperature of the saturated air leaving the condenser for determining which he gave no method excepting to state that it was assumed to be an average between the circulating inlet and outlet temperatures. This assumption could hardly be considered as accurate. It was not possible, he said, to obtain the temperature of saturated air by the usual mercurial thermometer and he had been led to regard the inconsistent readings obtained in attempting to determine this temperature as worthless as the basis of mathematical calculations.

He pointed out that the Croll-Reynolds evactor, shown in Fig. 10 of the complete paper, used two second-stage ejectors, each having different capacities. This gave the pump three capacities, a combination of the first with each of the second stages separately and the first with the two second stages combined for maximum capacity, thus obviating the use of two pumps.

The conclusion of the reading of the written discussions was a signal for advocates of the rotative dry-vacuum pump, the hydraulic vacuum pump and the ejector pump to vie with one another in debate, every speaker suffering the common fate of being misunderstood and of misinterpreting the meaning of his opponent's remarks.

Those taking part in the discussion were D. W. R. Morgan,[1] who contested that the openings of the hydraulic type of pump were not small, that the condenser manufacturer was interested in selling to the purchaser the type of pump best suited to his needs and that the capacity of the air ejector depended upon the vacuum; Paul A. Bancel,[2] who argued that the methods of computing the size of the air pump from Fig. 5 and from Dalton's law would give different results; John F. Grace, who suggested that the temperature in the dry air pipe could be found by means of a bare thermometer passed through a cork, and that air leakage with a view to its reduction was the great question to be investigated by the engineer; William G. Starkweather,[3] who thought that the question of air pumps came down to one of simplicity and reliability, and whether due to design or other factors, there was a demand for ejector pumps; G. L. Kothny, who devoted himself to a refutation of Mr. Foran's and Mr. Morgan's criticisms; and John H. Lawrence,[4] who also pointed out that Fig. 4 was misleading.

An amplification of the remarks at this session will appear in TRANSACTIONS.

In his oral closure, Frank R. Wheeler answered as many of the criticisms as the circumstances would permit. He called Mr. Foran's attention to the paragraph in the paper in which he stated that the Muller hydraulic pump was a distinct type to which the remarks in a later paragraph did not apply. With reference to air leaks, he was under the impression that he had made it clear that this was a great defect. In different installations, conditions would vary, so that the curves given in the paper might or might not apply. In any case, they were the curves used in determining sizes of pumps furnished by his company, two pumps being furnished for each installation. In answer to an objection that had been raised to the ejector pump in that it would not perform properly under reduced steam pressure, Mr. Wheeler pointed out that the ejector pumps were furnished with pressure-reducing valves which supplied them with steam at 105 lb. pressure, the pressure for which they were designed to operate.

An abstract of the paper on Air Pumps for Condensing Equipment will be found in the December, 1919, issue of MECHANICAL ENGINEERING, the figure numbers in the foregoing discussion referring to that abstract.

Owing to the fact that T. S. Taylor, author of the paper entitled The Thermal Conductivity of Insulating and Other Materials, was unable to be present, the paper was presented by title only. There being no one to discuss the paper, the meeting was adjourned.

## Textile Session

At the Textile Session, called to order by C. T. Plunkett, chairman of the sub-committee on Textiles, on Thursday afternoon, the following papers were presented: Teaching and Training Mill Employees, by Roy DeMitt; Steam Use in Textile Processes, by Geo. H. Perkins; Electric Power Supply to Southern Mills, by J. E. Serine; and Possibilities of Mechanical Handling of Materials in Textile Mills, by Kenneth Moeller. All of these papers with the exception of that by Mr. Perkins will later appear, in abstract form, in MECHANICAL ENGINEERING. Mr. Perkins' paper will be found on page 14 of this issue.

Mr. De Mitt[1] in presenting his paper on Teaching and Training Mill Employees, told of the methods employed in the southern states to educate mill operatives. Many of these mills run ten hours a day, and under the Federal Child Labor Act it is not permissible to work those under sixteen years of age over eight hours a day. After careful investigation it was decided to split the day into two five-hour shifts, and to compel the child operative to attend school for the remainder of the eight hours. This system, together with compulsory school attendance, has been extensively adopted throughout the South.

Four types of vocational school are employed: The vestibule school, in which would-be operatives are trained for their particular work when entering a mill; the all-day school, which the operative attends on alternate days; the part-time school, which the operative attends during part of each working day; and the evening school, which is attended outside of working hours. Short courses are offered because it is impossible to get the operative to sign up for a great length of time, and it has also been found advisable to teach subjects which will be of immediate use to the student rather than those of a more theoretical nature which might be of some practical use in the future.

The Smith-Hughes Act, providing for industrial education as well as other types, carries with it an appropriation of $7,000,000 per year. In the South the agricultural people get the bulk of this, although the state boards are usually in a position to render considerable financial assistance.

George H. Perkins presented his paper with the aid of lantern slides, and showed the useful results of a great deal of laborious research. Chas. T. Main[2] recalled the great dearth of information on the subject and emphasized the fact that it is very desirable to have such data for future reference.

Because of the absence of the author, J. E. Serine's paper was read by title only.

Kenneth Moeller's paper, illustrated by means of lantern slides, discussed different methods of transporting material from one part of a mill to another by means of conveyors. Among the

[1] Chief Engineer, Condenser Department, Westinghouse Elec. Mfg. Co., Machine Works, East Pittsburgh, Pa.
[2] Ingersoll-Rand Co., 11 Broadway, New York, N. Y.
[3] Starkweather and Broadhurst, 53 State St., Boston, Mass.
[4] Asst. M. E., The New York Edison Co., Irving Place and 15th St., New York, N. Y.

[1] Federal Board of Vocational Education, Atlanta, Ga.
[2] Consulting Engineer, Boston, Mass. Past-Pres. Am. Soc. M. E.

advantages of such systems are: that the trucking is done "on the ceiling" instead of on the floor, and that the mill requires about 20 per cent less operatives.

## Transportation Session

At this session, held on the morning of Friday, December 5, in the Auditorium, Prof. W. B. Gregory presiding, two papers were presented, namely: Scientific Development of the Steam Locomotive, by John T. Muhlfeld, and Motor-Transport Vehicles for the United States Army, by John Younger. Abstracts of these papers appeared in the November issue of MECHANICAL ENGINEERING.

### SCIENTIFIC DEVELOPMENT OF THE STEAM LOCOMOTIVE

Discussing this paper, F. J. Cole[1] wrote that much remained to be done in the way of designing boilers to safely and economically operate under pressure of 350 lb. and over. Hard riding, he said, did not always arise from improper counterbalancing, but might be due to worn driving-box bearings, etc. When more than four driving axles were used, flange wear should be taken into account and lateral-motion driving boxes or conjugated leading trucks employed. Attention should be given to details of domes and throttle valves so that water is not carried out into the steam pipes. The increasing cost of fuel seemed to warrant the use of many devices not considered economically desirable a few years ago.

Clement F. Street[2], discussing rule-of-thumb methods in design, said that scientific methods were all right in their place, but when it came to locomotive boilers, the thing to do was, as the late M. N. Forney had said, to "make the boiler just as big as you can." As to stoker-fired engines burning more coal than those hand-fired, that was the very purpose for installing the stoker—it was not intended as a fuel saver.

George L. Fowler[3] said that in regard to the disadvantage of using trucks, experience has shown that a symmetrical-wheelbase engine such as the Mikado type was much harder on the rails than a Consolidation engine, which has an unsymmetrical wheelbase and no trailing wheels on the back. He was unable to see what was to be gained in the way of evaporation by rapid circulation in a locomotive boiler, but it was an excellent thing as far as internal stresses due to expansion were concerned.

H. B. Oatley[4] wrote expressing his belief that a locomotive-boiler pressure of 500 lb. would be safely and economically used in the near future. Further, that the present method of transmitting power would be superseded by an improved type of construction that would result in the elimination of considerable unbalanced reciprocating and rotating weight and consequently in less rapid deterioration of rail, rail bed and driving mechanisms. Condensing operation would also be adopted.

Otto S. Beyer, Jr.,[5] said that in any discussion of the economy of electrified roads the possibilities of the super-power plant, located in the coal district itself and doing away with fuel haulage, should be considered. With rising labor costs it seemed to him that a premium would be placed more and more on speed in railway operation, which in turn would mean faster locomotives. The time element would also have to be taken into consideration in the matter of train loading. It was not a question of what a locomotive could haul up the maximum grade at the slowest possible speed.

Edwin B. Katte[6], referring to the author's remarks on the electric locomotive, said that the fact that nearly every large railroad in the country is now operating a part of its system

electrically would indicate that the capital cost is not prohibitive. As to interchangeability, steam locomotives did not habitually leave the divisions to which they were assigned, nor would electric locomotives. There was hardly an electrification nowadays but what had the advantage of two or three or possibly six or eight power stations feeding into its own transmission system by various routes. In regard to congestion of road, it was quite generally accepted that a yard or terminal or line would have its capacity increased 25 per cent or more by electrification. As to operating costs, it was not difficult to arrive at the cost of electric operation; the difficulty was to obtain the comparable cost of superseded steam operation. As to tie-ups, it had been the experience of most steam roads that during snow storms and blizzards the electrified divisions operated more nearly 100 per cent than the steam divisions.

George Gibbs[1] whote that electric traction was too big a subject to be disposed of in a few paragraphs. The author was mistaken in his belief that expected results had not been quite generally attained. More money had been made for the stockholders by reducing operating expenses, or by producing new business either by more attractive or more reliable service, or by an increase of capacity of the railways' facilities. He thought it entirely creditable to engineers that they had been able in the short period of, say, ten years' time, to develop electric locomotives which are reliable in service, have a moderate upkeep cost and which exceed in hauling capacity the most powerful steam locomotives.

Mr. Muhlfeld, in closing, said that in the boiler being designed for higher pressures the largest diameter of the shell, which was made of 1¼-in. plates, was 68 in. This was about as large a boiler as could be handled at present by locomotive builders. Mr. Oatley had referred to 500 lb. pressure. What had limited him in his argument in the paper to 350 lb. was the cylinder temperature, not boiler troubles; and tests made during the past three years had shown that a higher temperature than 750 deg. fahr. would be one that would destroy the cylinders, packing and bushings.

In the matter of burning coal, to which Mr. Street had referred, he would say that during the past five years he had made numerous tests of the cinders taken from the back of a stoker-fired locomotive and from accumulations in small boxes placed on top of a freight train, and had found that in nearly every case they were unconsumed particles of fuel. The coal was being put through the boiler but was not being burned.

In regard to questions propounded by Messrs. James H. S. Bates and Wm. H. Wood, he would say in reply to the former that the steam turbine had not as yet been adapted to the locomotive, and to the latter that fireboxes with flexible stays were being eliminated.

### MOTOR-TRANSPORT VEHICLES FOR THE UNITED STATES ARMY

The discussion of this paper was opened by Arthur J. Slade,[2] who said that the most intelligent opinion of the A. E. F. would undoubtedly disagree with the author's statement regarding operation of trucks off the roads. Any vehicle running on wheels required a supporting foundation, even caterpillar tractors, and in France before vehicles could keep pace with advancing troops it was always necessary to repair the worst spots in the roads. The matter of personnel was vital as with the most highly refined types of equipment, military motor transport in the mass could not function properly without systematic organization. He believed that the solution of the intricate questions involved could best be found by a commission of truly qualified experts who had been associated with the M. T. C. activities in the A. E. F. and who in coöperation with a board of army officers with similar experience could work out a sound constructive program that would justify the support of both the regular army and the general public.

[1] Chief Consulting Engineer, American Locomotive Co., Schenectady, N. Y. Mem.Am.Soc.M.E.<br>
[2] Consulting Engineer, New York. Mem.Am.Soc.M.E.<br>
[3] Consulting M. E., New York. Mem.Am.Soc.M.E<br>
[4] Chief Engineer, Locomotive Superheater Co., New York. Mem.Am.Soc.M.E.<br>
[5] Captain, U. S. A., Washington, D. C. Jun.Mem.Am.Soc.M.E.<br>
[6] Chief Engineer, Electric Traction, U. S. R. R. Administration. New York. Mem.Am.Soc.M.E.

[1] Consulting Engineer, Gibbs & Hill, New York. Mem.Am.Soc.M.E.<br>
[2] Lieutenant-Colonel, Aviation Section, Unattached. A. E. F. Mem.Am.Soc.M.E.

F. H. Pope[1] said that doubtless more rugged truck construction was necessary in military than in commercial vehicles, but army experience was apt to be misleading on this point as they had found the great cause for vehicle breakdown to be poor operation. As to the advisability of the 3-ton type, he would say that this type was urgently recommended for our army in 1916 by officers who had had actual motor-transport experience. The 1½-ton truck had its special value over a heavier truck in being more rapid, entailing less loss of capacity in bulky loads and being the truck par excellence for personnel carrying, but it was open to the objection made by Mr. Younger. Special vehicles, he believed, should not be allowed where standard types would answer the purpose. Motor cars and trucks should be districted by make or model, either in divisions or territorial districts. The 4-wheel-drive type had a very positive value, chiefly for artillery work, where their tractor properties could be utilized, and for cargo work over mountainous or difficult terrain. The caterpillar type, however, was the only one that was really serviceable in very difficult territory.

B. F. Miller[2] said that in the future the country would undoubtedly have a military policy, upon which would be based programs for any emergency that might arise, and he believed that motor-vehicle designers and engineers, manufacturers and military authorities should all have a part in perfecting such programs. The peace-time equipment of the army would probably be limited to that actually required for peace-time necessities but this equipment should be kept up to date by the yearly purchase and replacement of 20 per cent. of new equipment. He was of the opinion that we should not cling to the fetish of exclusive military design unless it could be shown to be capable of being produced more rapidly and economically than selected types of commercial design adapted to military requirements. Ford cars would be used in the future near the front lines and where the roads were bad, and light trucks would also be used to replace the animal-drawn vehicles of combat troops for cargo carrying. Four-wheel-drive trucks would always be required, not only for use with artillery and other hauling off the roads but also for cargo hauling. The use of trailers, in his judgment, should be limited to hauls that were obviously suitable and for mounting certain special equipment and bodies that were required to be moved infrequently.

William P. Kennedy[3] said he believed that standards must be effected by means of some compromise between military engineers and commercial engineers representing the most prominent manufacturers, who would discuss and lay down the principles that would be incorporated in the machines that were necessary to serve the army purposes. That is, the military organization could not expect to get all it would design or demand for its own interests and would have to accept a compromise between that and what the manufacturers considered it desirable to produce to remain in business and in such condition that they could produce quantities of trucks in a short time without great loss.

Alfred F. Masury[4] said that undoubtedly specially designed vehicles to meet the requirements of the army would be superior to any commercial vehicles, if for no other reason than on account of standard parts. The War Department, however, could not be expected to carry on hand in peace times a large stock of up-to-date war vehicles. It would therefore seem best to employ commercial vehicles so far as possible in the army and restrict the specially designed vehicles to those special uses for which truck and other special equipment was used. Commercial designers should be encouraged to incorporate in their designs features which might be of military use and which would also be of commercial use, such as provision for high gear ratios, etc.

G. R. Young,[5] in an extended written discussion, said among other things that one reason for standardization that should be strongly emphasized was the importance of reducing as much

as possible the varieties of makes and types of vehicles for the army. Motor vehicles for army use should be capable of taking great overloads, withstanding rough handling and operating under bad road conditions. To make any program a success it would be necessary in peace times to have the army supplied with, say, 40,000 to 50,000 motor vehicles. Immediately on the outbreak of a war, 20,000 or more must be immediately obtainable. In four to six months, under the gruelling conditions of field service, the motor vehicles first sent out would begin to wear out and break down in large numbers and would have to be replaced. This would call for, say, 50,000 vehicles, with further large increments until the close of the war. The peace-time estimate under a five-year replacement scheme would call for 8000 new vehicles a year. The question was, would manufacturers be sufficiently interested in a share of this business to construct these standardized vehicles with tools and fixtures for which the Government assumed the expense of keeping up to date?

Mr. Younger, in his closure, said that those in charge at Washington had to face facts as they were and their policies had been developed from day to day instead of during the period preceding the war. The paper had touched on important points that should be considered in the policy that should govern us during the next few years. There must be no compromise decisions in the future. The question of personnel raised by Colonel Slade was one of exceedingly great importance. The automobile industry had been successively called upon to furnish personnel to care for the tremendous expansion in aircraft, fighting-tank, agricultural-tractor and submarine-chaser activities. The truck-industry had to take what was left, which was one reason why many things appeared terrible overseas. While it was perfectly true that it had been found necessary to introduce new types of vehicles, if the war had lasted three years longer, which was expected at the last, he believed they would have been justified in their adoption.

As Mr. Kennedy had said, it was highly probable that a compromise would have to be effected, and one strong compromise suggested during the war was that vehicles of different makes be zoned. With a static battlefield as in France it would be easy to arrange these zones, but with a mobile battlefield as we had in Mexico other elements had to be considered. As to commercial-vehicle manufacturers supplying the vehicles in the future, if the army could dictate what kinds of guns and harness it was going to use, why should it not have the right to dictate, if it so desired, what type of motor vehicle it should use? One interesting thing in this connection was the question of roads. Millions of dollars were being spent on road construction and trucks would be designed more and more for use on highways, but it should be kept in mind that an army did not travel over good roads.

## Machine Shop Session

Four papers were read and discussed at the Machine Shop Session held Friday morning, December 5, under the auspices of the Sub-Committee on Machine Shop Practice. These papers were: Common Errors in Designing and Machining Bearings, by C. H. Bierbaum; Lubrication of Ball Bearings, by H. R. Trotter; Thread Forms for Worms and Hobs, by B. F. Waterman; and Electric Arc Welding, by H. L. Unland. The papers by Messrs. Bierbaum and Waterman will be found in the November issue of MECHANICAL ENGINEERING. Mr. Trotter's paper was published in the October issue.

Mr. Bierbaum's paper was a discussion of the disadvantages of tight-fitting bushings, the methods of finishing brasses to provide for expansion while running, and the methods of clamping bearings during tooling. The finishing of surfaces and the importance of using sharp tools with the proper rake was also considered.

In discussing Mr. Bierbaum's paper Frank L. Fairbanks[1] stated that while the principles set forth by Mr. Bierbaum were theoretically correct, it was nevertheless commercially impossible to obtain the conditions required by these principles; and even if it

[1] Colonel, Motor Transport Corps.
[2] Captain, U. S. A.; formerly Lieutenant-Colonel, Motor Transport Corps.
[3] Consulting Engineer, New York, N. Y. Mem.Am.Soc.M.E.
[4] Chief Engineer, International Motor Co., New York. Mem.Am.Soc. M.E.
[5] Captain, Engineer Corps; Chief, Engineering Branch, Motor Transport Corps.

[1] Chemical Engineer, Quincy Market Cold Storage & Warehouse Co., Boston, Mass., Mem. A.S.M.E.

were possible to maintain a full oil film between journal and bearing this would mean a projected area so large and a rubbing speed so low that it would be doubtful whether such machine could compete with one obtaining 750 to 1,000 lb. per sq. in. by letting the metal take some of the wear and tear.

Selby Haar[1] also discussed the paper and stated that because of the small clearance or radial air gap between stator and rotor, bearings for small alternating-current motors of the induction type must be manufactured with additional limitations to those mentioned by Mr. Bierbaum's paper. A looseness in the bearing of only a few mils, he explained, would cause a considerable percentage variation of air gap around the circumference of the rotor. He further remarked that inasmuch as motors are often shifted from one class of work to another it is necessary that motor bearings be more liberally designed than plain machine bearings.

C. H. Norton[2], discussing the questions of oil filament, oil grooves, and heat, stated that he had come to the conclusion that all oil grooves were of no possible value. He referred to a pamphlet published by the Texas Oil Co., in which he said it was concluded from a series of experiments on heat and speed that the fact that a bearing is heated to about 210 deg. is probably proof that oil is in it and the surfaces are separated by fixed oil and they become cooler as the oil film becomes thinner.

Of the lubricating of ball bearings, the subject of Mr. Trotter's paper, but little is known, according to the author, is chiefly due to the fact that there is no accepted method of determining the lubricating value of an oil or grease. In presenting his paper Mr. Trotter outlined a method which he has devised for this type of work. He also discussed the operating characteristics of a ball bearing and presented specifications for what he considered a satisfactory oil.

Henry Hess, in discussing the paper, suggested that the Research Committee of The American Society of Mechanical Engineers urge on Mr. Trotter the advisability of completing his investigations and that the Society aid by setting aside such funds as might be needed.

W. T. Murphy[3] said that 80 per cent of all trouble in bearings was caused by the presence in the lubricant of sand particles or other abrasives. Discussing the question of running anti-friction bearings without any lubricant, he observed that while such a procedure was theoretically correct, in practice the lubricant helped to overcome the slight friction that exists between the races of the bearings and the balls or rollers, and that there are cases where the use of grease or oil on such bearings is absolutely essential.

George N. Van Derhoef[4] challenged Mr. Trotter's conclusions as to the impossibility of there being an oil film between balls and races in a ball bearing. He observed that the research work so far conducted on oil films had been confined to bearings with much larger areas than there are in the case of balls traveling in the conventional raceway and also with surfaces with a different relative motion. He also pointed out that the difference in the coefficient of friction of lubricated and unlubricated ball bearings seemed rather to confirm the idea of an oil film existing in the case of the lubricated bearing.

H. R. Smith[5] claimed that the only practical method of gaging the lubrication qualities of an oil is by determining its viscosity value. In the case of a grease, the most important factors, he said, are the free acid—organic or mineral—or alkali, free fatty oil, solid matter, the character of the soap used and also the viscosity of the mineral oil employed in the manufacture of the grease.

Thread Forms for Worms and Hobs, by B. F. Waterman, was a discussion of both the theoretical and mechanical problems involved in the use of worm gearing. According to the author there has been a steady increase in the use of worm gearing and

accompanying it a corresponding increase in efficiency and durability.

In discussing Mr. Waterman's paper Henry J. Eberhardt[1] related an experience of the Newark Gear Cutting Machine Co. which led them to generate worm wheels having an angle of side of tooth of 14½ deg. In generating a worm wheel with a fly tool, according to the Noerstedt process, the fly tool having been made with straight sides and ground to 29 deg., included angle, the worm wheel thus generated would not run with a 29-deg. worm as it would not mesh. They, therefore, concluded that the only way to generate a worm wheel was to make a fly tool or thread on the hob of the same shape as the worm.

In the discussion of Mr. Unland's address on Electric Arc Welding, the question was brought up as to whether the Boiler Code Committee would accept and pass a boiler constructed and repaired by electric welding. In this connection Mr. Unland remarked that the question of repairing boilers had been previously discussed before The American Society of Mechanical Engineers, but as yet he did not know of any conclusions ever formulated in this respect by the Boiler Code Committee. He observed, however, that this type of repair is made and passed by all the shipping bureaus, Lloyds' and the American Bureau and others, on repairs to marine boilers, provided the company doing the work has the sanction of the underwriter. Mr. Unland stated that the majority of users are in favor of the direct current for electric welding, and that in this country the accepted practice is to use uncovered electrodes, while in England and on the continent of Europe the covered electrode is in favor.

## General Session

Four papers were presented at the General Session held on Friday morning, December 5. These were: Slow-Speed and Other Tests of Kingsbury Thrust Bearings, by H. A. S. Howarth; Octaval Notation and the Measurement of Binary Inch Fractions, by Alfred Watkins; Modern Electric-Furnace Practice as Related to Foundries in Particular, by W. E. Moore; and An Investigation of Strains in the Rolling of Metal, by Alfred Musso. Abstracts of all of these papers have appeared in MECHANICAL ENGINEERING, those by Messrs. Watkins and Moore in the November issue and those by Messrs. Howarth and Musso in the December.

Mr. Howarth's paper, illustrated by lantern slides, dealt with the application and tests of Kingsbury thrust bearings.

Mr. Watkins, whose paper was an argument for the adoption of a new method of measuring binary inch fractions, is a resident of England and his paper was presented by F. A. Halsey.[1] In the discussion following its presentation, Dean Benjamin brought out the fact that the question of measurement was not necessarily an academic one but rather a subject of broad interest to all engineers. Howard Richards, Jr., formerly associated with the University of China, advocated the metric system but Mr. Halsey maintained that the decimal system was hopelessly impossible as an instrument for reliable service; moreover, that the system advocated by Mr. Watkins was not a new one but merely a simplified method of using a system already existing.

The paper by W. E. Moore, on Modern Electric-Furnace Practice as Related to Foundries, was a discussion of the various features of the electric furnace which have made that type practically supreme in the field.

The paper by A. L. Musso, which was illustrated by lantern slides, dealt with one of the most important problems incidental to rolling-mill operation.

E. O. Goss,[3] in a written discussion of Mr. Musso's paper, gave the results of a series of experiments made to determine the percentage elongation and side spread of cold-rolled brass, and while the results obtained differed materially from those given in the paper, Mr. Goss was of the opinion that this wide difference was not entirely due to the difference in material.

[1] Electrical and Mechanical Engineer, New York, N. Y., Mem. A.S.M.E.
[2] Mechanical Engineer, Norton Grinding Co., Worcester, Mass., Mem. A.S.M.E.
[3] General Manager, Standard Machinery Co., Auburn, R. I., Mem. A.S.M.E.
[4] Asst. Con. Engr., Dodge Mfg. Co., 21 Murray St., New York. Mem. A.S.M.E.
[5] 47 South Twelfth Street, Newark, N. J., Junior A.S.M.E.

[1] Secretary, Newark Gear Cutting Machine Co., Newark, N. J., Mem. A.S.M.E.
[2] Commissioner, Am. Inst. of Weights and Measures.
[3] Vice-Pres., Leonille Manufacturing Co., Waterbury, Conn.

# President Cooley's Address

Annual Meeting, 1919, of The American Society of Mechanical Engineers

A Survey of the Society's Organization, a Review of the Last Year's Accomplishments, and an Appeal to the Engineering Profession for National Organized Effort

IT is my pleasant duty on this occasion to tell you something of the doings of the Society during the year while I have been its President. And I thought you would also be interested to hear something about the way in which the Society does its work. For most of you, I feel certain, my story will be like the opening chapters of a new book. It will, in fact, be from the opening chapters of an old book, familiar to you in appearance, but, I dare say, one you have scarcely ever opened. I mean the Year Book of the Society.

First, however, let me tell you how much I have enjoyed the year and how much I appreciate the honor you conferred on me when you entrusted me with the responsibilities of administering the affairs of the Society. It is truly a great responsibility and growing greater with each year's growth of the Society—growth both in its membership and in its activities.

I remember well those first years, back in the early eighties, when our members were only a few hundred, and our activities confined principally to two meetings, one in the spring, the other in the winter. The contributions were chiefly the papers read at the two meetings and the discussion they provoked. Today there are over 12,000 members, 72 committees engaged in the work of the Society; 17 committees representing the Society on other organizations; 31 local sections; and 47 student branches. These are directed by a Council of 21 members assisted by the chairmen of the six Standing Committees of Administration and a paid Secretary with 76 assistants. Thus has the Society grown in the 39 years since it was organized in 1880.

The machinery of this great national society is really somewhat complicated. I assume that you are as ignorant of it as I was when I took my place at the throttle valve. Ah! happy thought! How easy to explain when the right analogy comes to one. As you know a throttle valve has two functions, one to start, the other to stop. At first I thought its most important function had to do with starting. But then I was guileless. I did not know my colleagues on the Council. "Throttle valve" is just the right word. How many times we would in our discussions have settled the affairs of the Nation had something gone wrong with the stopping function of that throttle valve; but as it happened the valve never got out of order; only individuals got "out of order."

Probably only a few of you know what the Council is. I need not hesitate to speak openly and with freedom, as knowing me as well as they do, it would not have surprised me had there been not a single member of the Council present here this evening. The fact that a number are present is evidence of the discipline to which they have been subjected; or it may be evidence of the possession of a conscience still active in pointing the way of duty. But let that pass. The Council is composed of the President, six Vice-Presidents, nine Managers, the Treasurer, and five Past-Presidents. They are a fairly quiet and orderly, not to say, modest lot of men—starting with the President at the top. But those past-presidents at the bottom! What shall I say of those past-presidents? And to think that I, too, shall so soon graduate and become one of them. But so runs the machinery of the Society. The last five presidents have a place on the Council. Thus each president has five years in which to make amends for any bad judgment he may have exercised during his one year of being first in responsibility. While these 22 men do all the voting, they do not do all the talking. There are the six chairmen of Standing Committees to be reckoned with on the floor in debate. That makes 28. But that is not all—far from it. There is also the Secretary. He is a spouting fountain—of ideas for advancing the interests of the Society, and the watchdog of its activities. No one of you knows how great is his responsibility to keep the President functioning properly. The Treasurer of the Society is the only member of the Council who sits tight and says nothing.

Now that I have paid my respects to these gentlemen, my most immediate colleagues, let me say of them that it has been a rare privilege to be associated with such men in the great work of the Society. It is no mere honor as some have thought, to be a member of the Council. It means hard and conscientious work and vast quantities of it. If you could see these gentlemen sitting long hours around the Council Board, sometimes two full days at a time, you would realize how good servants of the Society they are. And let me tell you that were it more generally known how like horses these men have to work there would be fewer candidates, when the nominations for officers of the Society are made. The Council has held in all, during the past year, eleven meetings, one of them in Pittsburgh, two in Detroit and one in Indianapolis.

To my colleagues of the Executive Committee I would pay especial tribute. They have been ever faithful in responding to calls for some emergency action. And many times I fear they have thereby been greatly inconvenienced. I was told I must choose nearby men in order to get them together, and even then I would rarely have a quorum. But I wanted to go afield a bit in my selections for the Executive Committee and bethought me of a way to break down tradition. So I invited the Executive Committee to meet me at dinner. It worked like a charm. Not a man was absent. And at none of the many meetings we have held has there ever lacked a quorum. My heart goes out in gratitude to those loyal men who have so wholeheartedly helped the President over so many difficult places.

Next in our administration machinery come the six Standing Committees of Administration—Finance, Meetings and Program, Publication and Papers, Membership, Constitution and By-Laws, and Local Sections. I do not quite know how to tell you of my appreciation of the thirty men on these six committees. The amount of time and energy they give to the Society's work is enormous—appalling, I nearly said; but the word refers to the President, who must keep in touch with their several activities. The Society should know and appreciate the great care required and given so splendidly by the groups of five gentlemen composing each of our six standing committees of administration.

To aid in its work, the Meetings and Program Committee is assisted by six sub-committees, each charged with the responsibility of securing papers and preparing programs on particular subjects for sessions at the Annual and Spring Meetings of the Society. These sub-committees are really special committees in that they may be changed from time to time to meet the need of new subject-matter for presentation at the Society's meetings.

Next come the Standing Committees—House, Library, Public Relations, Research and Standardization. The Research Committee supervises all research activities, collaborates with kindred committees of other institutions and obtains results of researches conducted in other countries. It has at present seven sub-committees which are really special committees, each engaged in the investigation of some particular subject, such as the cutting action of machine tools, fuel oils, lubrication, fluid meters, worm gearing, bearing metals, and heat transmission. The latter acts jointly with a similar committee of the American Society of Refrigerating Engineers. The report of the Research Committee by its Chairman, Professor Arthur M. Greene, Jr., read at the Detroit meeting, merits wide attention. It deals with the subject comprehensively.

The Society has naturally a large number of special committees to deal with particular problems. These are formed as occasion

43

demands and discharged with the acceptance of their reports. Some of them are appointed annually and thus become standing committees for the year; such for example as the Committee on Junior and Student Prizes. Others are more in the nature of permanent standing committees, such as the Committee on Increase of Membership and on Student Branches. The latter is now called Committee on Relations with Colleges.

During the past year this Committee on Relations with Colleges has, under the direction of its Acting Chairman, Past-President Ira N. Hollis, made a comprehensive study of the Society's relations with educational institutions and of student and junior prizes. The report of this committee is very complete and well worth attention. It is the result of much hard work involving extended correspondence and study of other society organizations. The Council in accepting this report put itself on record as favoring prizes other than honorary membership, now accorded to persons of acknowledged professional eminence. Life membership, in addition to being acquired for a sum of money, could be awarded for the best contribution to mechanical engineering found in the papers for one year; Junior and Student Prizes, for papers of exceptional merit; a medal, for notable invention or some striking improvement in connection with the industries; honorable or special mention, for conspicuous contributions to engineering either of a practical nature or in literature; scholarships or fellowships, for exceptional attainments in college work; a medal or special mention, for exceptional achievement by Junior and Student members.

Rear-Admiral George W. Melville left a bequest of $1000 to the Society for the award annually of a medal for the best thesis on mechanical engineering; and another gentleman has this year made a similar contribution. The Society allowed him $1500 for expenses to France as the Society's official representative at the Franco-American Congress but he paid his own expenses and added of his own money $1000, making his contribution $2500. I refer to Past-President Charles T. Main.

The National Societies can well afford to further their relations with colleges. It is impossible to do too much toward helping to shape the ideals and ambitions of the youth of our country who are soon to take our places in the engineering world. The colleges would welcome the counsel of men in actual contact with the world's affairs. These engineers are giving little attention, or none at all, to the preliminary training of the young men they take on their staffs as assistants. They are, or should be, vitally interested; and they could if they would be of great help in shaping the curricula of engineering colleges. Whatever the practicing engineer may think of his brother the professor-engineer—and his thoughts are, I fear, sometimes none too flattering—he must not confuse his brothers with clairvoyants. While college professors may be conceded to have wonderful imaginations, they stop short of being occult. The advice of the practicing engineer would be eagerly welcomed. If he thinks of us of the near-cloth as being diviners, let him at least show his hand to be read.

It cannot be said in these days that the college man works for money. His real compensation comes from the success of his students after they leave college and enter the practical world. Their training must still, and for several years, be in progress. Their teachers only are different. If we are to achieve greatest success in our profession of engineering, that is, have it stand at the forefront in world affairs, you to whom are turned over the young men from our colleges must do your part. When you realize what your part is, you will be both willing and anxious to help the college man in his part of the work. The four years in college should not be separated, as by a wall, from the years which follow. There should in those after years be the same inspiration to high ideals of life—those quite apart from money. Our country—the whole world—never needed high ideals more than it does today.

But let me get on with my description of the Society's machinery. The Boiler Code Committee with its five sub-committees still occupies a front seat in the Society's activities. The fact that our Boiler Code has been incorporated in the laws of thirteen states and seven municipalities, is a source of great satisfaction and pride. It indicates a good work well done. The Boiler Code sub-committee on Railway Locomotive Boilers is very active. The railroads themselves through the Interstate Commerce Commission, have now sought the opportunity to coöperate in this work.

The Committee on Power Test Codes, reorganized in 1918, with 18 sub-committees, is a beehive of industry, and bids fair to rival the Boiler Code Committee for the Society's favor. Some of its sub-committees have already practically completed their work and submitted reports. Another year should get the important work of this committee well on toward completion.

Progress in various degrees is being made by other special committees. The Committees on Standardization of Feed Water Heaters, Shafting, Steel Roller Chains, have all reported progress. The Committee on Protection of Industrial Workers has recently completed jointly with the United States Bureau of Standards, an elevator code in which the Safety Committee of the National Association of Manufacturers participated. The special Screw Threads Committee and the Committee on Flanges and Pipe Fittings have been active. An international convention was recently called to meet in Paris to adopt a standard for pipe threads. This Society's official delegate is M. Laurence V. Benét, who resides in Paris; and to assist M. Benét technical advisors have been appointed from the Society's membership in various American industries engaged in the manufacture of pipe and fittings.

It will interest you to learn that standardization in general has recently taken a distinct step forward. The American Engineering Standards Committee is now, after being for several months the subject of active debate, an accomplished fact. While it is recognized that there are still several changes that could well be incorporated in its Constitution, it was considered better to get started and let experience determine the exact nature of those changes. The Committee starts off with representatives from the American Society of Civil Engineers, the American Institute of Mining and Metallurgical Engineers, The American Society of Mechanical Engineers, the American Institute of Electrical Engineers, the American Society for Testing Materials, and from the Government Departments of War, Navy and Commerce. The constitution provides for admitting to membership on the Committee, representatives from other societies and associations. This Committee does not create standards. That work is left to be done as now, by bodies especially interested in some particular standard. But the Committee gives its stamp of approval and serves as the medium through which standards are passed out to the world. One of the most useful functions of the American Engineering Standards Committee will be to coöperate with similar bodies in other countries in the formation of international standards; for instance, with the industries of Great Britain through the British Engineering Standards Association. When this Committee is functioning completely such matters as the coming meeting in Paris, of the Commission on an International Standard for Pipe Threads, would be handled by the American Engineering Standards Committee.

It will be remembered that last year at the Annual Meeting, President Main, after delivering his address, fled the country. I did not at first know the reason for his precipitate haste. I could scarcely imagine that it was because of the things he said in his address, although I now know that the President can say and do things amply justifying him in seeking safety. It was, however, an honorable mission that called him away. He went abroad as this Society's representative on a Committee of the four Founder Societies which was invited by the Société des Ingénieurs Civils and the Committee of the French Engineers' Congress, with the official approval of the French Minister of Armament, Public Works and Commerce, to study with French Engineers the rehabilitation of France after the war. Out of the service rendered by this Committee has come about the appointment of a Franco-American Committee in which this Society is represented by Charles T. Main and George W. Fuller.

Other exchanges of international courtesies that have afforded us satisfaction and pleasure during the year occurred in connection with the Ottawa meeting of the Engineering Institute of Canada in February and the Detroit meeting of this Society in June. Past-President Hollis went to Ottawa as our special representative and Fraser S. Keith, Secretary of the Engineering Institute of Canada, came to Detroit as special representative of the Institute.

The James Watt Centenary celebration was held in England on September 16, 1919. In response to an invitation to participate, this Society sent as honorary Vice-Presidents Messrs. J. Wilfrid Harris, Harry F. L. Orcutt, R. Sanford Riley, and Wilson E. Symons.

This Society has had the pleasure of participating in extending courtesies to members of several delegations from foreign countries who have come to this country on various missions. Official delegations have come from France, Belgium and Switzerland. In fact, our distinguished guest of the evening, M. de Freminville, is a member of the French Engineering and Economic Mission which was invited to this country by the Chamber of Commerce of the United States. M. Schneider, the illustrious head of this Commission, was recently entertained by our sister society, the Mining and Metallurgical Society of America.

An interesting and important part of the mechanism of the Society's machine is the groups of members, known as Society Representatives, who represent this Society on other bodies. There are 17 of these groups, and the men who compose them are those who have been conspicuous in the Society's work, or have qualifications especially fitting them for the particular duties involved. These other bodies and committees are the American Association for the Advancement of Science, American Engineering Standards Committee, Classification of Technical Literature Committee, Cost of Electric Power Committee, Engineering Council, Engineering Foundation Board, International Committee on Screw Threads, International Standard for Pipe Threads, John Fritz Medal Board, Joseph A. Holmes Memorial Board, National Research Council, Naval Consulting Board of the United States, Standards for Graphic Presentation, United Engineering Society —its Board of Trustees and its Library Board, United States Bureau of Mines Advisory Committee, Western Society of Engineers, Washington Award Committee. The names themselves sufficiently describe the character of work involved making unnecessary any detailed description. The reorganization of the National Research Council on a peace basis includes its affiliation with similar bodies of other nations. The Chairman of our own Committee on Research has a place on the National Research Council.

The Regular Nominating Committee and the Tellers of Election must also be mentioned. The Nominating Committee was in the old days chosen by the President alone. But President Jacobus in 1917 inaugurated the policy, followed since, of inviting the Local Sections to make nominations from which the President could choose. But a new plan is to be followed in the future.

The Nominating Committee is to be elected by the voting membership, and tomorrow morning the Society will decide on the machinery. The proposal is to utilize the Local Sections organization for conducting the primary for nominations, every voting member being assigned to a local section for that purpose. The nominations are to be confirmed at the Local Sections Conference to be held at the Annual Meeting, after which the election will take place at the Annual Business Meeting. Thus will be insured, it is believed, a thoroughly democratic election.

In 1918 there were twenty-one Local Sections. Now there are thirty-one, an increase of ten during the year. Three of them are on the Pacific Coast. One local section in New England has five branches. This growth of Local Sections and extension of section responsibilities is one of the marked features of the Society's development. Like a wide spreading tree, the Society must have its roots and branches widespread if it is to grow and thrive. The roots must reach out into new earth and the branches out into sunshine. It is my firm belief that by increasing the importance of its sections the Society will add to its strength and vigor and render the greatest service to its membership.

As I see future development of the profession as a whole, it is important that the members of local sections of our own and other national societies should be also members of sectional and local engineering societies which embrace in their membership all engineers whether they be members of national societies or not; and which include those who work with engineers as members of their staffs, and who may not possess the qualifications for membership in the national societies. This, happily, is the plan being followed. Cleveland is a good example. There is now an arrangement, approved for trial by our Council, whereby our members who are also members of the Cleveland Engineering Society may have part of their dues to this Society remitted; that is, there is a division of dues. No doubt, considerable latitude will be permitted Local Sections in order to encourage them to work out plans for improving their own organizations. With increase in importance of Local Sections work, there will come the necessity of providing more permanent local headquarters. It has been suggested, and I believe wisely, that financial assistance should be given to local building projects through the aid of societies owning the Engineering Building here in New York, and by loans from surplus reserves. This subject may well have our earnest consideration.

One of the Sections Committee's recommendations is a Traveling Secretary whose duty it shall be to visit Local Sections and assist them in developing their plans. It is a good suggestion; but I would add to it. The President's duties have become very heavy. Notwithstanding that I have given one-quarter or one-third of my time to the duties of the office, I am sure that I have not done more than one-half of what the President should do. Among other things, it has been impossible to respond to the many cordial invitations to visit Local Sections. It has also been impossible to respond to many invitations from other national bodies to be present and participate in their deliberations and ceremonial functions. You will agree with me that these are proper and important duties of the President, and the dignity of the Society requires that they should be performed.

What I would add, therefore, is that the proposed traveling secretary be the President's representative. He could be styled "Assistant to the President." He should be a man of parts and worth a salary of at least $5000 per year. With such assistance the President could perform many of his duties in whatever part of the country he might come from; that is, it would not be necessary for him to spend so much of his time at national headquarters.

This assistant could go to the President's home and stay a week or two at a time if necessary. He could represent the President at Local Section headquarters; or by lightening the burden of routine work, enable the President to go. Nor is this all. Many of the President's duties now necessarily fall on our Secretary, already overburdened with his own duties. The assistant to the President could relieve the Secretary of many of these, but not all of course. Certain very important duties could be performed by no one with more grace or with more complete regard for the proprieties and requirements, than by our Secretary.

In taking up the duties of President, it became evident to me after a time that pending the report of our Committee on Aims and Organization, it would be inexpedient to suggest or inaugurate any changes in the existing policies of the Society. It seemed rather a period during which the existing activities of the Society should be fostered and made as fruitful of results as possible. With this thought in mind, I undertook a study of committee work. As was to be expected with such a large number of committees, some of them were more active than others. Inactivity was not necessarily the fault of a committee. Some of the subcommittees were required to work only when called on. The work of other committees was temporarily suspended on account of the war. In a number of instances suspended committee work has been resumed.

One result of my investigation was to find frequently the same man serving on a number of committees, and only the older and more experienced members of the Society on many of the committees. While in some cases these conditions were warranted as tending to give the best results, it seemed to me that for the good of the Society it was desirable to increase the number of individuals engaged on committee work by limiting one's membership to a few committees and by adding some of the younger men. In this way more general interest would be stimulated in the Society's work and the younger men would be trained for more responsibility later. Accordingly, Local Section officers were requested to canvass their membership for men willing to accept committee appointments. There is now on file a considerable list of men not only willing to serve but who have promised if appointed to attend to the work.

Another result of my investigation was to find the membership of many of the committees confined to within reaching distance of national headquarters here in New York. It is absolutely necessary, in a number of cases, to have at least a majority membership living nearby. The Executive Committee, the Finance, the House, the Library and the Membership Committees are examples. It is not advisable, in general, to have members of a committee come from widely separated parts of·the country. It should be possible for them to get together without too great sacrifice of time consumed in travel, and without too great expense, which in most cases must be borne by the members personally.

It appeared that a solution of the difficulty might be found in choosing committee members with respect to localities. For instance, an automobile committee might be centered in Detroit and embrace Buffalo, Cleveland, Toledo, Jackson and Port Huron. A petroleum committee might be wholly from the Oklahoma or the California oil districts. A machine-tool committee might be wholly from the Ohio or the New England district, and similarly for committees on other subjects. A natural way with the increased importance of our local sections would be to choose the membership committee wherever practicable from the members of a particular local section. The local section itself could be held responsible for the work of the committee. The membership of such committees need not be confined to this Society. Indeed, it might be well, and in the interests of the Society as well as the work, to appoint some non-members. If they become greatly interested they might become members later. Committee work is of first importance in the Society, and should be well cared for. When it is of high order great credit comes to the members of the Committee and to the Society. It is the good work done by its members that enables the Society to occupy and hold its high place in the world. The honor of the Society is the honor of its members. The Society can give back only what its members put in both in kind and in quantity.

Of equal importance to that of Local Sections, if not greater, are the Society's publications. It is through them that the Society gives back to its members what they contribute in money and in work. Its journal has been rechristened and is now called MECHANICAL ENGINEERING. It is the medium through which the membership at large can keep in touch with all Society activities, and become promptly informed on important engineering topics discussed at the annual and spring meetings and at the Sections' meetings throughout the country. During the year the Society has acquired the ownership of "Engineering Index" which it has greatly enlarged, and in which now appear regularly reviews or titles of articles from over eleven hundred periodicals properly classified and indexed. The Council has recently authorized an increase in the scope of the Journal so that it can be of still greater service to the membership.

With the desire to take part in and help solve, so far as it could properly, the industrial troubles of the country, the Council early in the year directed the appointment of a committee of seven to investigate and report on the feasibility of the Society actively participating in industrial questions. This committee made its report in February, advising that inasmuch as the constitution of the Society set forth as its object "to promote the Arts and Sciences connected with Engineering and Mechanical Construction," it was inexpedient to take any active part as a society in industrial matters; but that it was highly important that the members as individuals, or in their business capacity, should take such part as they could.

Believing that it would be helpful to the membership to receive the results of investigations of industrial conditions the Council in April authorized affiliation with the National Industrial Conference Board and appointed a representative on the Board. The propriety of the Society's holding such relations being later questioned the Council in October directed the appointment of a committee to report on the question of retaining our membership. This Committee, while unanimously commending the work of the National Industrial Conference Board was of the opinion that the Society should not hold membership in any other society or organization. The Council, therefore, withdrew from the National Industrial Conference Board at its November meeting held on December 1.

By far the most important work of the Society during the year has been that of its Committee on Aims and Organization. Inasmuch as the report of this Committee has been sent out to the membership and is to be discussed and disposed of at this annual meeting, I have not deemed it necessary, or even proper, to do more than call your attention to it and to urge you to take part in its consideration when presented. Suffice to say that certain recommendations of the committee have already had the attention of the council since they were favorably considered by the Society at its spring meeting in Detroit. I refer particularly to the development of Local Sections, the encouragement of Professional sections and the enlargement of the scope of the Society's journal, MECHANICAL ENGINEERING. We are all most appreciative of the work of this Committee. Theirs was a hard task, and while they would have welcomed an opportunity to carry their work further and perfect it, the Council thought that the Society would do well if it agreed upon and adopted even a few of the large number of recommendations made.

A most important part of the work of our Committee on Aims and Organization was its coöperation with similar committees of the American Society of Civil Engineers, the American Institute of Mining and Metallurgical Engineers, and the American Institute of Electrical Engineers. A joint committee of these four Founder Societies has held several meetings to consider and agree upon certain fundamental things as the basis for closer coöperation in the future. Without in any way sacrificing their identity and purpose, and leaving the chief aim of each to be pursued independently, it has been agreed that a comprehensive national organization or council of engineers should be created to be composed of representatives of local, state and regional societies. Further, that this organization be so framed as to include allied technical societies; and that coöperation of these several bodies in technical activities, and affiliations of local organizations for general public-service and welfare work be encouraged. This joint committee also recommends that the bill to create a National Department of Public Works receive the active support of all engineers.

Every thoughtful engineer will recognize the importance of these recommendations. The engineer has in the past been too much of an individualist. He is at heart a idealist. He lives in the things he creates, and often his chief reward is the satisfaction he enjoys from work well done rather than pecuniary. It is a splendid quality of heart and mind that enshrines the engineer. It harks back to those earlier days when respect for one's achievements and personal character measured success rather than the dollars amassed. But to keep one's place in these modern days requires that the game be played in accordance with prevailing rules. Thus the engineer must perforce conform to these rules if he would exert his fullest influence. It is organized rather than individual effort that gets ahead today. Only the individual who can direct and control organized effort stands out as an individual. He is the general in command of forces.

The one criticism of the engineering profession. as I see it. is that there are too few generals. The positions of high command too often go to others who are rarely themselves engineers but who constantly make use of engineers in framing and executing their plans. That is to say, the engineer, while he has the knowledge and skill required to plan and build structures utilizing Nature's material forces for the benefit of mankind, has apparently given but little indication that he possesses the knowledge and skill required to handle human forces for the benefit of mankind. It would appear that that kind of leadership does not appeal to the engineer. He has made the instrument, even composed the music, which others have played to sway the multitude. In short, the engineer, if he has the qualities for leadership, has not been leading, except in his own field and amongst his own kind where of course he stands preeminent.

The world today needs a Moses to lead it out of the Wilderness. Moses must have been an engineer. That job of his in the Red Sea would indicate that he was an hydraulic engineer. Can the engineering profession aspire to such leadership as is today needed? No one knows. We think it can. But it would appear that the engineer has not been thought of in that connection. Why? Simply because he has in the past taken so little part in public affairs outside his profession. Here then lies the opportunity, the first step being to organize nationally and do the things recommended by our Joint Committee of the four national engineering societies.

One of the things recommended is the active support of the bill in Congress to create a National Department of Public Works. Such a department would at once bring the engineering profession before the eye of the world. It is such a normal and proper move that one would think it would carry itself. But it will not. You can be sure of that. It is forty years since the effort to secure such a department was first made. It failed as have all subsequent efforts for want of being properly pushed. And who is to push? You, gentlemen of the engineering profes-

sion. Not one, nor a few, but all of you and all together, and at this particular time.

I am now come to near the end of what I had in mind to say to you. There are other things that should be added to make my story of the year's doings complete. But something must be left for my successor in office to say next year. A president's address is not, I fear, the best medium of conveying a message to the multitude. It partakes more of the nature of an embalming process of ideas. It will find its mausoleum in the TRANSACTIONS. Some thousands of years hence it will be discovered, and the world of that day will comment on it much as we now do on the papyrus from the tombs of Egypt. But I have had great pleasure notwithstanding, in preparing my address.

Our worthy editor has for several weeks been keen to discover the title of my address. He was somewhat perturbed when I informed him that there was to be no title; that I proposed to be free, up to the last moment, to write what I pleased. The fact that I have succeeded in eluding him I consider some accomplishment, as you would agree did you know him as well as I do.

And now my adieus. It has been for me a happy year. I am grateful to you for your confidence, and your good will. It has brought forth the best that is in me. While I have many times failed to measure up to what I consider a proper specification for a president of The American Society of Mechanical Engineers, I have the satisfaction of feeling that I have stretched myself to the utmost in an effort to render all the service of which I was capable.

# WORK OF THE BOILER CODE COMMITTEE

*THE Boiler Code Committee meets monthly for the purpose of considering communications relative to the Boiler Code. Any one desiring information as to the application of the Code is requested to communicate with the Secretary of the Committee, Mr. C. W. Obert, 29 West 39th St., New York, N. Y.*

The procedure of the Committee in handling the cases is as follows: All inquiries must be in written form before they are accepted for consideration. Copies are sent by the Secretary of the Committee to all of the members of the Committee. The interpretation, in the form of a reply, is then prepared by the Committee and passed upon at a regular meeting of the Committee. This interpretation is later submitted to the Council of the Society for approval, after which it is issued to the inquirer and simultaneously published in MECHANICAL ENGINEERING, in order that any one interested may readily secure the latest information concerning the interpretation.

Below are given the interpretations of the Committee in Cases Nos. 228, 230, 252, 253, 256-261, inclusive, as formulated at the meeting of October 24, 1919, and approved by the Council. In accordance with the Committee's practice, the names of inquirers have been omitted.

### CASE NO. 228

*Inquiry:* Will wrought-iron material made by the puddling process be acceptable under the Boiler Code Rules for use in superheater construction where the requirements in the Code specify wrought steel or cast steel?

*Reply:* Where the Boiler Code specifies wrought steel or cast steel for superheater construction, the intent was to prohibit the use of cast iron. It is also the opinion of the Committee that puddled wrought iron that will meet the requirements in the specifications for wrought iron in the Code will be a suitable material.

### CASE NO. 230

*Inquiry:* In the case of a water-tube boiler of the water-leg type, using large tapered connections from the rear header to the drum, is it permissible to weld the flanged edges of the water-leg sheets instead of butt-strapping them, and also the longitudinal joint of the tapered connection piece if it is amply braced cross-

wise by staybolts? These welded joints would not be under tension, as the staybolts take all of the pressure load and the possibility of breakage of these staybolts is negligible.

*Reply:* It is the opinion of the Committee that this cannot be done and comply with the requirements of the Code. The safety of the structure may be dependent upon the weld to the extent that the interior stresses in the weld, which would be relieved if rupture should take place, may throw suddenly upon the nearest staybolts, excess stresses which may cause the staybolts to fail in series.

### CASE NO. 252

*Inquiry:* Is it necessary under the requirements of Par. 186 of the Boiler Code, where door-hole flanges are to be welded and the stress due to the steam pressure is to be carried by the adjacent stay-bolting, to consider any limitation on the staybolt pitch to secure proper distribution of the load from the door-hole flanges to the staybolting?

*Reply:* It is the opinion of the Committee that under the requirements of Par. 186 the door-hole flanges of the furnace and exterior sheets may be joined by autogenous welding, provided these sheets are stayed or otherwise supported around the door-hole opening, and provided that the distance from the flange to the surrounding row of stays or other supports does not exceed the permissible staybolt pitch, as per Par. 199.

### CASE NO. 253

(Annulled)

### CASE NO. 256

*Inquiry:* Can a vertical fire-tube boiler be constructed with an unstayed furnace from 36 to 38 in. in diameter with butt-strap longitudinal seams as required in Par. 239 of the Boiler Code?

*Reply:* It was not the intent of the Boiler Code that the requirements for longitudinal seams of furnaces between 36 and 38 in. inside diameter should apply to the unstayed furnaces of vertical firetube boilers. Such furnaces over 36 in. inside diameter should be fully staybolted, as it is necessary to use lap-riveted longitudinal seams in this type of furnace. Butt-strap seams may be used only in horizontal furnaces where the seams can be placed below the grate.

CASE No. 257

*Inquiry:* Will a specially constructed water gage with an automatic-valve-closing device, arranged to be operated by a balanced pressure attachment depending for action upon the breakage of the water glass, be considered under the rules of the A.S.M.E. Boiler Code as an automatic water gage, in which the automatic shut-off valves must conform to the requirement of Par. 427?

*Reply:* It is the opinion of the Committee that the requirements of Pars. 292 and 427 of the Boiler Code apply specifically to automatic shut-off valves on water gages and not to exterior automatic devices which may be provided to close the valves. If the automatic closing device is not incorporated into the construction of the shut-off valves, the requirements of these paragraphs do not apply.

CASE No. 258

*Inquiry:* a In large hot-water heating boilers which under the requirement of Par. 335 of the Boiler Code must be constructed under the requirements for power boilers, is it necessary that such requirements as covered in Pars. 291 and 315 shall apply?

b Was it the intent of Par. 349 of the Boiler Code to cover the possible contingency of the necessity of a water-relief valve discharging steam, provided the boiling point has been passed, by the bursting of the diaphragm, thereby increasing the capacity of the water-relief valve?

c Would it be assumed that the safety of the structure would be dependent upon the strength of a weld made in a header of a Hawley down-draft furnace attached to the shell by means of pipe connections from each header, and could such a header be welded and a patch strap then applied over the top of the weld, to take the strain of a joint which is in tension?

*Reply:* a It is the intent of the requirement in Par. 335 that only such portions of the rules for power boilers shall apply to hot-water boilers as refer specifically to constructional details. The requirements for some of the fittings and trimmings for steam power boilers are obviously not applicable, so that in such cases the rules for hot-water heating boilers shall apply.

b It was not the intent of Par. 349 that the diaphragm of a water-relief valve should break after the valve opens.

c It is the opinion of the Committee that this is a question of repairing boilers. The Code gives certain recommendations in regard to the use of autogenous welding for repairs, but does not deal with the general subject of repairing which must, in each case, be taken up with the local state or municipal inspectors.

CASE No. 259

*Inquiry:* Is it allowable under the requirements of the A.S.M.E. Boiler Code to fit a high-pressure steam boiler with a 4-in. blow-off connection, in case it is to be used initially for low pressure steam service, or will it be necessary under the requirements of Par. 308 to apply two or more 2½ in. blow-off connections for the return connections to the boiler?

*Reply:* It is the opinion of the Committee that the boiler could be fitted with the 4-in. blow-off connection when in use for hot-water heating and, when converted to a steam boiler, a reducing fitting could be used at the 4-in. opening to reduce to the pipe size necessary for the blow-off connection.

CASE No. 260

*Inquiry:* Is the requirement of Par. 185 of the Boiler Code, relative to planing down the girth joint, applicable to a special type of boiler formed of an h.r.t. shell with a water-tube element over which the gases pass before coming in contact with the h.r.t. shell?

*Reply:* Par. 185 applies only to horizontal return tubular boilers and does not apply to a combination design of fire and water-tube boiler as described. In such a boiler the girth joints are not as directly exposed to the heat of the fire as in the case of the h.r.t. type of boiler.

CASE No. 261

*Inquiry:* What is the stress allowable under the requirements of the Boiler Code for unwelded solid stays more than 20 diameters in length?

*Reply:* It is the opinion of the Committee that the stress should be that given in item c in Table 5 of the Boiler Code, which is 9500 lb. where the lengths between supports do not exceed 120 diameters.

## PROPOSED CODE FOR AIR TANKS AND PRESSURE VESSELS

For nearly a year a Sub-committee of the Boiler Code Committee has been engaged in formulating rules for the construction of air tanks and pressure vessels, to be known as Part I, Section IV, of the A.S.M.E. Boiler Code. It is the object of this section, which is devoted to pressure vessels other than fired steam boilers, to secure safe construction and at the same time to so establish the rules that the requirements are within practical working limits.

The work of the Sub-committee to which this task was assigned began in April, 1919, and the preliminary report which follows is the result of a careful and painstaking study. The Sub-committee consists of the following members:

E. R. Fish, *Chairman*    S. F. Jeter
Wm. H. Boehm    Wm. F. Kiesel, Jr.
C. E. Bronson    James Neil
E. C. Fisher

The preliminary report was ordered printed by the Council of the Society with a view of having it discussed by the membership and the public at large. Discussions should be addressed to the Secretary of the Boiler Code Committee, 29 West 39th St., New York, N. Y.

## A. S. M. E. BOILER CODE
### PART I—SECTION IV
### PRELIMINARY REPORT ON RULES FOR THE CONSTRUCTION OF UNFIRED PRESSURE VESSELS

NOTE: When this Section is incorporated in the Boiler Code, it will be desirable to add the words "AND OTHER PRESSURE VESSELS", at the top of the front cover and of the title page and on page 3 of the Code. Also the sentence in italics on page 3 will be changed to read, as follows:

*These rules do not apply to boilers and other pressure vessels which are subject to Federal inspection and control.*

Also to insert, after bracket in line beginning Part I, the following:

#### SECTION IV—UNFIRED PRESSURE VESSELS

##### DEFINITIONS

The vessels to which these rules apply are divided into two classes:

Class A—Vessels for containing liquids above the atmospheric boiling point, inflammable substances, or any gas, and limited in size to those over 6 in. in diameter, more than 1.5 cu. ft. in volume and carrying over 15 lb. pressure per sq. in.

Class B—Vessels for containing liquids, the temperatures of which are under control so as to be below atmospheric boiling point and limited in size to those over 9 in. in diameter, more than 4 cu. ft. in volume and carrying over 30 lb. pressure per sq. in. but not to exceed 100 lb. pressure per sq. in. For pressures over 100 lb. per sq. in. the rules for Class A vessels apply.

##### MATERIAL

These vessels may be constructed of any metal complying with the following rules:—

The maximum allowable working stress used in any of the

metals selected shall be determined by the following formula:—

For Class A Vessels:—
$$S = 0.0125 \ E \ (e + 8) \text{ but not more than } 0.4 \ E$$

For Class B Vessels:—
$$S = 0.0125 \ E \ (e + 16) \text{ but not more than } 0.65 \ E$$

where

$S$ = Maximum allowable working tension stress, lb. per sq. in.;

$E$ = Elastic limit of material used;

$e$ = Elongation of material used, in per cent, in 8 in.

Working stress in rivets in single shear shall be taken as 80 per cent of $S$;

Working stress in rivets in double shear shall be double that of single shear.

Staybolt values for allowable stress per sq. in. of net section shall be taken as follows:—

When $e$ is not over 10 per cent = 60 per cent of $S$;
When $e$ is from 10 to 20 per cent = 70 per cent of $S$;
When $e$ is over 20 per cent = 80 per cent of $S$.

Class A Vessels may be constructed of materials specified in the A.S.M.E. Boiler Code (Edition of 1918), in which case the formula above given need not be used.

Class A Vessels shall be built in accordance with the rules of construction, which apply, of the A.S.M.E. Boiler Code (Edition of 1918) except Par. 186 which shall read as follows:—

The ultimate strength of a joint which has been properly welded by the forging process shall be taken as 65 per cent of the tensile strength of the plate.

and Par. 269 to 328, inclusive.

Class B Vessels may be constructed of material specified in the A.S.M.E. Boiler Code (Edition of 1918) in which case the formula above need not be used.

For Class B Vessels, steel of untested tank quality may be used, assuming the tensile properties below:—

Tensile strength, lb. per sq. in.................48,000
Yield point, lb. per sq. in.......................25,000
Elongation in 8 in.............................20 per cent

All Class B Vessels shall be constructed in accordance with the rules of construction, which apply, of the A.S.M.E. Boiler Code (Edition of 1918) with the following exceptions:—

Par. 180—change FS = 4.

Par. 186—rewrite as follows:—

The ultimate strength of a joint which has been properly welded by the forging process shall be taken as 65 per cent of the tensile strength of the plate. The ultimate strength of a joint which has been properly welded by the autogenous process, may be used in Class B pressure vessels only and shall be taken as 50 per cent of the tensile strength of the plate.

Par. 187—change 36 in. to 48 in.

Par. 188—change 36 in. to 48 in.

Par. 190—omit entirely.

Par. 195—change formula as follows:—
$$t = \frac{5 \times P \times L}{2 \times TS}$$

Par. 210—holes may be punched full size.

Par. 253—rewrite as follows:—

All rivet holes may be punched full size.

Par. 254—omit entirely.

Par. 256—omit entirely.

Par. 257—rewrite as follows:—

Calking shall be done with a round-nosed tool.

Par. 262—rewrite as follows:—

Manhole plates may be of wrought steel, steel castings or of cast iron.

Par. 263—omit entirely.

Pars. 269-328—omit entirely.

No drain or blow-off shall be less than ¾ in. pipe size.

*Safety Appliances.* All pressure vessels shall be provided with such relieving, indicating and controlling devices as the industry in which they are employed or use to which they may be put shall require to insure their safe operation. All such devices shall be so located and installed that they cannot possibly be rendered inoperative. The relieving capacity of safety valves shall be such as to prevent a rise of pressure in the vessel of more than 6 per cent above the maximum allowable working pressure.

*Supports.* All vessels must be so supported as to equally distribute the stresses arising from the weight of the vessel and contents. Class A Vessels must be so arranged that the entire interior and exterior of the vessel may be thoroughly inspected. In the case of vertical vessels, the bottom head, if dished, of cylindrical vessels must be with the pressure on the concave side to insure complete drainage.

(Par. 325) Lugs or brackets, when used to support a vessel of any type, shall be properly fitted to the surfaces to which they are attached. The shearing and crushing stresses on steel or iron rivets used for attaching the lugs or brackets shall not exceed 8 per cent of the strength given in Pars. 15 and 16 of the Boiler Code.

*Corrosive Chemicals.* All pressure vessels which are to contain substances having a corrosive action upon the metal of which vessel is constructed should be designed for a pressure in excess of that which it is to carry to safeguard against early rejection.

Class A Vessels shall be stamped in accordance with the A.S.M.E. Boiler Code (Edition of 1918), Pars. 331-332.

Class B Vessels shall be stamped in accordance with the A.S.M.E. Boiler Code (Edition of 1918), Pars. 376 and 377.

$$\left\{ \begin{array}{c} \text{A.S.M.E.} \\ \text{B} \end{array} \right\}$$

*Location of Stamps.* Location of tank manufacturer's stamps to be as follows: plain cylindrical tanks, and all other tanks not of a steam boiler type shall have at least one stamp, all stamps to be plainly visible when the tank is completed. On a tank with a manhole, said stamps shall be close to the manhole opening; on a tank without a manhole, close to the handhole; and on a tank without either a manhole or handhole, in a conspicuous place.

The tank builder's stamp shall not be covered by insulating or other material.

According to a recent article in *Oil News* the British-Inter-Departmental Committee on the production and utilization of alcohol for power and traction has recommended that the British Government study the possibilities of this form of fuel.

In the British Empire, according to the committee, there are vast prospective sources of alcohol in the vegetable world, although in the United Kingdom itself production from these sources is now and is likely to remain small. Synthetic production, however, especially from coal and coke-oven gases, is promising. The Committee also states in its report that "to the enormous growth of road motoring during recent years, especially in the United States of America, there will now be added the requirements of high-grade petrol for aeroplanes and airships, to which no limits can be assigned. Whilst it is impossible for us to forecast the development of total petrol consumption for all countries and all purposes, facts are not wanting to indicate the likelihood in the not distant future of so great a pressure of demand as to cause at any rate a very high level of prices, and we are satisfied that close investigation should now proceed with the object of providing alternative supplies of motor fuels derived from new or supplementary raw materials.

"We are satisfied that the time has come for government action, which should pay due heed to both current and prospective prices for petrol, or other petroleum products, benzol, and alcohol motor-fuel or its admixtures.

"We have received exhaustive technical evidence from representatives of the Ministry of Munitions concerning the investigations made by them during the war in respect of the extraction of ethylene from coal and coke-oven gases, and concerning quantitative results so obtained. The testimony of witnesses and records of work done indicate that there is available in Great Britain a large potential source of power alcohol, but further investigations are necessary in this connection, particularly as regards the conversion of ethylene into alcohol, before definite figures as to quantities and price can be given."

# ENGINEERING RESEARCH
## A Department Conducted by the Research Committee of the A. S. M. E.

### Research on the Pacific Coast and at the Naval Experiment Station at Annapolis

DURING the past months the Committee on Research of the San Francisco Section has been active in sending out inquiries regarding research and the report below together with a letter of transmittal from Prof. W. F. Durand gives an excellent idea of research conditions in that portion of the United States.

When the Committee of the Society planned its present work it was not expected that we would receive information regarding research for private development purposes. This work in many cases represents a large expenditure of money and the Committee does not desire to publish any of this information until the proper time arrives. It has hoped that when such information is properly protected it will then be given to the profession as a contribution to our scientific knowledge.

The General Committee feels that the report of the San Francisco Committee is a most excellent one in bringing to our members problems on which work is being done and for which solutions are desired by engineers of the Pacific Coast.

The conditions for commercial testing at the Naval Experiment Station are described in a statement prepared from a note received from the Station. It is given here for the information of our members having apparatus or materials suitable for naval service.

LELAND STANFORD JUNIOR UNIVERSITY,
STANFORD UNIVERSITY, CAL.,
October 15, 1919.
PROF. ARTHUR M. GREENE, JR.,
Chairman, Committee on Research. A.S.M.E.,
29 West 39th St., New York, N. Y.

MY DEAR PROFESSOR GREENE:

After some delay caused by the special rush of work at the opening of our academic year, I am handing you herewith a brief résumé of the replies received to date from the circular letters which we sent out regarding problems under investigation or suggested for research in San Francisco and vicinity.

Out of nearly two hundred letters sent out, only about a dozen replies have been received of sufficient significance to justify listing with reference to the purposes of the survey.

From what I learned by private conversation, however, it is perfectly clear that this is only a small part of the investigation and research work which is actually going forward. Many people stated that they did not feel at liberty to state the character of the work on which they were occupied, simply for the reason that they considered it confidential in character and of possible commercial value, and were, therefore, not disposed to make any report of the matter which might give any indication to other people, possibly competitors, regarding the work on which they are occupied.

I anticipated precisely some such result as this, although it was impossible to tell before the event just how much we might be able to develop under conditions involving commercial competition and the safeguarding of private interests.

It is evident that the amount thus actually developed by no means represents adequately the amount of research and investigation work on which mechanical engineers in this vicinity are occupied, but under the circumstances it seems to represent the best showing that can be made so far as a public record or statement is concerned.

I must confess to some doubt as to the extent to which our Committee can be of very much significance in connection with such work of research and investigation, due primarily to the disinclination of people to talk about work of this character, and the very natural desire to safeguard in every possible way personal and private interests. However, something useful may perhaps develop, and we shall always be glad to coöperate with the work of the general committee in so far as we possibly can.

Sincerely yours,

(Signed) W. F. DURAND.

SURVEY OF ENGINEERING RESEARCH, SAN FRANCISCO AND VICINITY

PREPARED BY THE RESEARCH COMMITTEE OF THE SAN FRANCISCO SECTION OF THE AMERICAN SOCIETY OF MECHANICAL ENGINEERS

1 a Confidential metallurgical research work, the character of which cannot be made public.

b Suggestion of the following subjects as important for investigations:

  1 The development of a coke suitable for making pig iron on the Pacific Coast.

  2 A continuous process for producing spelter.

A large part of our spelter is produced in Kansas and Oklahoma where natural gas used to be cheap. The old-style distillation furnace is used. Gas is becoming more expensive now and these furnaces are costly and wasteful. Zinc ore is produced largely in the west and it is a long haul to Oklahoma and Kansas. During the war some ore was shipped from Australia to the Kansas smelters. Electrolytic zinc is not as satisfactory a solution as had been hoped for.

Fuel gas can be obtained by partially coking the coal in a by-product oven so that the by-products would pay for the gas. Or pulverized coal might be used. It should not be difficult to devise mechanical means so that a continuous distillation could be obtained. This would eliminate the wasteful charging methods and the loss of time and retorts. Considerable study could be devoted to the production of a more satisfactory retort.

2 a Confidential developments relating to rotary gas compressors and the application of same to the various industrial uses such as: vacuum pumps, air compressors and refrigerating apparatus.

  b Suggestion of development of heating buildings by reversal of thermodynamic cycle, using cheap electrical energy and low-temperature air as input with a delivery of high-temperature heat as output, measured in amount by the electrical input divided by the thermodynamic efficiency. Reference is made to Lord Kelvin's suggestion of possibilities along this line.

3 Investigation of certain forms of hydraulic dynamometers as a transmission device.

4 Development of a special or modified form of Diesel engine, cycle, intended to use kerosene and distillate. The special significance of this development lies in the possible extension of the use of kerosene or distillate for many purposes now requiring gasoline, and the resultant liberation of such amounts of gasoline for use in aeronautic and automobile engineering where the higher grades are especially desirable.

5 Development of proper type and mounting of ball bearings for deep-well and centrifugal pumps.

6 Development of a grade of steel suitable for pins for buckets on gold-dredging machinery. Pins vary in diameter from 3 in. to 7⅝ in. A few years ago a pin lasting nine months was considered as giving good service. With improved grades of steel, pins have now been in continuous service for three years.

7 Investigations on airplane propellers in connection with the National Advisory Committee for Aeronautics, the results of which are published in the annual reports of that Committee.

8 An investigation on the wear of gear teeth under loaded conditions, and including the influence of the following variable features:

  a Varying intensity of load
  b Varying materials
  c Varying tooth forms
  d Lubrication.

9 Investigation of the economic possibilities involved in the use of steel belting as a substitute for leather, and with special reference to the development of a satisfactory form of joint.

10 Research relating to the problem of the transfer of heat between fluids, vapors, and gases, and with special reference to the development of satisfactory commercial apparatus for such purposes, based on sound engineering principles. This research is intended to include a study of the transfer of heat between metal surfaces and fluids of widely varying viscosities and specific heats and including vapors and gases. In a detailed way the investigation will bear especially upon the transfer of heat between metallic surfaces and petroleum oils and water.

11 A general investigation relating to the problem of the development of a steam power plant, especially designed to meet the requirements of the self-propelled vehicle. This involves engineering research along the following lines:

  a A system of combustion, using for fuel kerosene and cheaper grades of fuel such as stove oil, etc. A successful system for this work requires complete combustion, as the formation of soot or carbon deposits would be objectionable.

  b In the construction of the combustion system, to make it entirely automatic and subject to control by boiler conditions, so that the temperature of superheat, the pressure of the steam and the supply of feedwater to the boiler, would all be automatically controlled, so as to maintain uniform conditions of operation. This involves the development of an automatic ignition device, so that by simply closing an electric switch, completing the circuit through the automatic control, the fire would be started under the boiler and brought under full control of the automatics to bring about the above results.

c A fuel and fire automatic control, so as to maintain within the boiler uniform temperature of superheat, of steam pressure and the necessary feedwater supply.

d The development of a new type of high-pressure boiler, especially developed to furnish steam at a high superheat, and at a maximum thermal efficiency.

e The development of a new type of steam engine which will have high economy, great flexibility of control and be able to utilize with maximum economy high-pressure steam with high superheat.

12 Investigations relating to improvements in electric furnaces, with special reference to the following applications:

a For the melting of steel scrap into steel castings.

b For the melting of steel scrap and ferroalloys into tool steels.

c The heat treating of steel castings, either annealing or tempering as the requirements may be.

d The melting of non-ferrous metals in commercial castings, particularly marine work.

## Commercial Testing at the Naval Experiment Station, Annapolis, Md.

The Bureau of Steam Engineering of the United States Navy conducts tests at its Naval Engineering Experiment Station of such engineering apparatus and materials as give promise of proving desirable for use in naval service. The requirements for making these tests are:

a That the material or device offered shall have definite application to naval needs.

b That the Government shall not be involved for any expense whatever either for the cost of the test, the material or apparatus tested, the shipment charges or damage to material or apparatus in consequence of the test.

c That the result of the test shall not be used for advertising purposes.

Application for test should be made in writing to the Bureau of Steam Engineering, U. S. Navy, Washington, D. C. If the nature of the article to be tested is not known or has not been explained in a personal interview the application must be accompanied by a complete description and blueprints, if necessary, for a thorough understanding.

If the test is authorized the applicant will be put into communication with the Superintendent of the U. S. Naval Academy, Annapolis, Md., who will furnish information as to any special requirements in connection with the test, such as number, size, and material of the article to be tested, steam pressure, etc., and will also designate the deposit which it will be necessary to make to cover the probable cost of test.

All necessary information having been received, the apparatus to be tested will be shipped, prepaid, addressed to the "Supply Officer," U. S. Naval Academy, Annapolis, Md., and plainly marked "For Engineering Experiment Station." A check will also be forwarded to the Superintendent of the Naval Academy, drawn to his order and covering the amount of deposit designated. Any unexpended portion of this deposit will be returned upon completion of the test.

The applicant will be allowed to have a representative present at all tests made on the apparatus submitted by him, and he or his representative may make suggestions or objections concerning the method of carrying out the test. Such suggestions will be followed so far as practicable and consistent with the purpose to obtain complete and conclusive data. The test authorities will be free to submit the apparatus or material to any test which in their judgment may be desirable to determine its suitability under any special requirements for use in the naval service, whether such tests have been contemplated by the applicant or not. When the applicant objects in writing to any proposed test on the ground that he believes the apparatus or material would be injured thereby, the particular test will not be made and the objection will be made part of the record of the test.

All applicants must agree in writing to accept the conditions

of the test set forth above and must further agree in writing that they will accept the results of the test as final and that they will not apply for a retest of articles or material until such changes have been made which will give reasonable promise of overcoming defects. This agreement in writing should be forwarded to the Superintendent of the Naval Academy together with the deposit mentioned in Par. 5.

Upon completion of test, a copy of the report thereon will be furnished the exhibitor by the Bureau of Steam Engineering.

For further particulars address Chief of the Bureau of Steam Engineering, Navy Department, Washington, D. C.

### A—RESEARCH RESULTS

The purpose of this section of Engineering Research is to give the origin of research information which has been completed, to give a résumé of research results with formulæ or curves where such may be readily given, and to report results of non-extensive researches which in the opinion of the investigator do not warrant a paper.

*Electric Power A6-19.* Electric Ranges. C. W. Pieper is the author of a Bulletin No. 2 of the Engineering Experiment Station at Purdue University on electric ranges. Seven electric ranges were tested. The time taken to preheat the ovens and to heat 3 lb. of water within the oven to boiling point varied from 12 min. to 45 min., the oven temperatures varying from 420 to 530 deg. fahr. On opening the oven doors for 45 sec. the temperature fell about 100 deg. fahr. This was regained in about two minutes. The ovens required varying amounts of power to care for radiation. The most efficient oven required 100 watts to hold it at 200 deg. F., 300 watts for 395 deg., 500 watts for 550 deg. and 700 watts for 680 deg., while the less efficient oven required 220 watts for 200 deg., 610 watts for 395 deg., 1220 watts for 550 deg. The surface burners, which are in the form of porcelain hot plates, gave from 51 to 69 per cent efficiency in the evaporation of water, while in heating water the efficiency varied from 8 to 51 per cent. To bake biscuits about 1 kw-hr. is required. Cake requires 14/10 kw-hr.; bread 13/10 kw-hr. The following table gives the results for power supply of families of different sizes when the cost of power is 11 cents for lighting and 6 cents for the first 10 kw-hr. and 3 cents per kw-hr. beyond this. The form of load curve is given in Fig. 1.

| Size of Family. | No. of Rooms. | Time Month & Day. | Range. kw-hr. | Light. kw-hr. | Total kw-hr. | Total Bill. | Net Range Bill. |
|---|---|---|---|---|---|---|---|
| 6 | 5 | 1-14 to 2-21 | 36 | 11 | 47 | $2.29 | $1.19 |
| 6 | 5 | 3-24 to 4-24 | 71 | 11 | 82 | 3.26 | 2.16 |
| 7 | 5 | 1-22 to 2-21 | 112 | 20 | 132 | 5.06 | 3.96 |
| 5 | 6 | 3-22 to 4-22 | 54 | 58 | 112 | 4.24 | 1.56 |
| 8 | 6 | 2-23 to 3-24 | 101 | 32 | 133 | 5.31 | 2.11 |
| 10 | 9 | 2-21 to 3-21 | 108 | 51 | 159 | 6.53 | 1.43 |
| 10 | 10 | 2-22 to 1-22 | 167 | 57 | 218 | 8.30 | 2.60 |

The Bulletin describes the various ranges tested and discusses the advantage of electric ranges. Engineering Experiment Station, Purdue University, Lafayette, Ind. Address C. H. Benjamin, Director.

*Lubricants A2-19.* Cutting Oils from Machine Tools. Diseases have been caused by infections in cuts and abrasions in the skin of machine operators when oil used on machine tools. This has been the subject of a report by Lieutenant-Commander A. G. Zimmermann of the Ordnance Department, U. S. Navy. His report quotes Bulletin No. 2 of the Department of Scientific and Industrial Research, No. 15 Great George St., Westminster S.W.-1, London, England. This Bulletin is a memorandum on Cutting Lubricants and Cooling Liquids, their Purposes, Properties, Composition, and the second part of the paper is on the Skin Diseases Caused by Lubricants.

Oil rashes are caused by:

(a) The plugging of glands of the hair follicles.

(b) Mechanical injury to the skin by metallic particles.

Any injury to the skin may allow germs to enter and cause septic infection. To prevent disease the workmen must keep perfectly clean, washing with hot water and soap and scrubbing brushes and not washing in the cutting compounds. The lubricant must be kept clean by constantly removing the metal particles by filtration and the machines must be kept clean by frequent removal of all old lubricant and dirt. The lubricants should be treated with disinfectants or antiseptics, although the use of these of these promote skin poisoning. It is well to sterilize the cutting oil by heat, and 300 deg. fahr. is suggested for the temperature of this heat. Any workers with septic infection of the hands should not be permitted to work. Workmen should be instructed to refrain from expectorating in the pans of the machines. The treatment for the blocking of the glands is frequent washing with soap and water followed by the use of zinc oxide and starch powder. Septic infections in cuts should be treated by suitable antiseptic dressings. Certain persons are more susceptible than others.

This report may be obtained from the Department of Scientific and Industrial Research, London, England.

Commander Zimmermann also quotes from the "Traveler's Standard" of July, 1918, prepared by the Engineering and Inspection Division of the Travelers' Insurance Company. This article speaks of the contamination of the oil from sputum. The report suggests repeated heatings to kill the germs and also repeated cleanings of the system. The dressings used for cuts should be such that the oil cannot further enter the broken skin. For that reason a collodion dressing may be quite useful. In many cases a rubber cot is advantageous.

One per cent of carbolic acid may be added as a germicide, but an excessive amount of acid will cause poisoning. It must be remembered that workmen have their hands wetted with the cutting fluid for hours at a time. Tanks used to hold the cutting oils should be covered so that animals may not drop in and become decomposed.

An investigation of oil samples from a machine tool showed the presence of bacteria found in sputum. It is important that notices against spitting in oil pans should be posted.

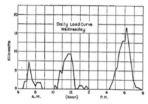

FIG. 1 FORM OF LOAD CURVE OF ELECTRIC RANGE

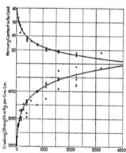

FIG. 2 BOTH CRUSHING STRENGTH AND MERCURY-CONTENT OF AN AMALGAM VARY AS LOGARITHM OF PACKING PRESSURE. LARGE DEVIATIONS FROM LOWER CURVE ARISE FROM CRUSHING SPECIMENS AT VARIOUS TEMPERATURES

*Marine Engineering A1-19.* The Propulsive Efficiency of Single-Screw Cargo Ships. A comparison of hulls of longitudinal coefficients of 0.74, 0.76, 0.78, 0.80 and 0.82 and of the same overall dimensions. The parallel middle body should be about one-third of the length as determined in a previous test, but in these tests the parallel portion of the ship was made longer. The model was tested by self-propulsion, using a floating form of dynamometer to get torque, while thrust was measured by a spring attached to the thrust bearing, and revolutions of the shaft were measured by suitable means. The model was also tested from the towing carriage. With speeds up to 7 knots there was little variation in power, but the resistance in the ships with the higher longitudinal coefficient the increase of power was much greater than the increase in displacement. This would not prevent the use of the ships with the longer parallel section where cargo capacity is

important. The general conclusion of the experiments is that for a well-designed hull with a propeller running at about 90 r.p.m., a hull efficiency of 1.09, a propulsive efficiency of 0.65 may be expected from single-screw cargo ships. This paper was read at the General Meeting of the Society for Naval Architects and Marine Engineers, Nov. 13, 1919, by Commander William McEntee, Construction Corps, U. S. Navy, Washington, D. C.

*Metallurgy and Metallography A20-19.* Copper Content in Steel. A report is made from the Bureau of Ordnance of the Navy Department that the Metallurgical and Testing Division have made experiments on steel in which the copper content was increased to as much as 2 per cent. The gun factory has been manufacturing nickel steel with copper contents between 0.4 and 0.7 per cent, and when properly handled these steels showed a marked superiority for most purposes over steel free from copper. The gun factory has not found that steel containing appreciable amounts of copper when properly made is more red-short than steel of similar quality free from copper. With poorly made steel red-shortness may be found.

In successfully manufactured steel containing an appreciable amount of copper, precaution must be taken to prevent the copper from liquating or separating during solidification as the solubility of copper is not as great in ternary alloys as in the case of other alloying elements. The usual practice in the gun factory is to keep the manganese content of the steel slightly higher when the alloy carries an appreciable amount of copper . The higher manganese content tends to hold the copper in perfect alloy and prevent the tendency to red-shortness due to the presence of slightly excessive amounts of sulphur. Steel with 2 per cent copper shows most excellent physical properties when liquation was prevented by rapid solidification of the ingot through a low pouring tem-

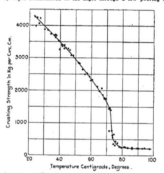

FIG. 3 EFFECT OF TEMPERATURE ON CRUSHING STRENGTH. TRANSITION REGION IS SHOWN BY THE RAPID FALL IN STRENGTH BETWEEN 70 DEG. AND 80 DEG.

perature and by use of a very heavy chill. Report by Lieutenant Commander A. G. Zimmermann. U. S. Navy. Address Rear Admiral Ralph Earle, Chief of Ordnance, U. S. Navy, Washington, D. C.

*Metallurgy and Metallography A21-19.* Bronze. Effect of Boiling Bronze in Oil. The Bureau of Construction and Repair of the U. S. Navy reported that test bars of manganese bronze were being boiled in oil to improve the physical characteristics. This matter was investigated by the Metallurgical and Testing Division of the Ordnance Department of the U. S. Navy under the direction of Lieutenant-Commander A. G. Zimmermann. Although there was no reason to expect any decided change under these conditions four block test bars were taken and two specimens were machined from each block. One of each was boiled for one-half hour in sperm oil at a temperature of 150 deg. fahr. The physical qualities of the treated and untreated bars are given below to show that this treatment has no effect on the physical qualities of manganese bronze:

| Not Treated. | | | Treated. | | |
|---|---|---|---|---|---|
| Yield Point. | Tensile. | Elong. | Yield Point. | Tensile. | Elong. |
| 46,100 | 76,400 | 21.75 | 50,400 | 76,400 | 22.00 |
| 42,900 | 75,200 | 25.50 | 47,900 | 74,900 | 23.50 |
| 44,400 | 77,400 | 23.25 | 43,900 | 75,600 | 23.50 |
| 49,400 | 79,500 | 18.75 | 47,900 | 79,800 | 16.50 |

Address Bureau of Ordnance, U. S. Navy. Rear-Admiral Ralph Earle, Washington, D. C.

*Metallurgy and Metallography A22-19.* Amalgams. The tests of dental amalgams formed by making an amalgam of mercury with an alloy of silver, tin and other metals has shown that the strength of the amalgam after hardening forms a logarithmic function of the pressure used in packing the amalgam and also the mercury content of this amalgam is a similar function. This is shown in Fig. 2.

The tests also show that there is a gain in strength due to the time of trituration, the maximum strength occurring at this time in six minutes. Temperature decreases the strength of an amalgam as shown in Fig. 3.

The amalgams change in volume during the time of hardening. This change in volume is called reaction expansion. They have been measured by a dilatometer capable of measuring to 1/500 micron. With heavy packing pressure the time to reach the maximum or minimum expansion is decreased. Increasing the trituration time accelerates the time of reaching the maximum pressure. Reducing the temperature retards the action. Time of packing does not have much effect. The amount of expansion is influenced by the time used in annealing the alloy for the amalgam. The greater the time of annealing the greater the reaction expansion. These results are of value to the mechanical engineer in that they are phenomena which occur in alloys not far removed from their melting points. These results are given in a paper by Arthur W. Gray, Ph.D., in the Transactions of the American Institute of Mining Engineers for December, 1918. Address Dr. Arthur W. Gray, L. D. Caulk Dental Company, Milford, Del.

*Properties of Engineering Materials A7-19.* Copper in Steel. See Metallurgy and Metallography A20-19.

### B—RESEARCH IN PROGRESS

The purpose of this section of Engineering Research is to bring together those who are working on the same problem for coöperation or conference, to prevent unnecessary duplication of work and to inform the profession of the investigators who are engaged upon research problems. The addresses of these investigators are given for the purpose of correspondence.

*Fuels, Gas, Tar and Coke B9-19.* Gasoline from Heavier Oils. A series of experiments have been made by E. W. Dean at the Bureau of Mines Station, Pittsburgh, Pa., in which a study was made of the cracking reactions for conditions favorable for transformation of heavier oils into gasoline. Bureau of Mines, Washington, D. C. Address Van H. Manning, Director.

*Fuels, Gas, Tar and Coke B10-19.* Oil Shales. The Bureau of Mines is investigating the retorting of oil shales and the crude oils obtained therefrom. This work is under the direction of W. D. Bonner and M. J. Gavin, Bureau of Mines, Washington, D. C. Address Van H. Manning, Director.

*Lubricants B9-19.* Analytical Distillation of Emulsified and High Boiling Oils by E. W. Dean, Pittsburgh, Pa. Bureau of Mines, Washington, D. C. Address Van H. Manning, Director.

*Metallurgy and Metallography B16-19.* Manganese Ores. The Bureau of Mines is investigating the mining and concentration of ores for the purpose of utilizing low-grade ores and the improvement of methods of mining. Bureau of Mines, Washington, D. C. Address Van H. Manning, Director.

*Petroleum, Asphalt and Wood Products B1-19.* Petroleum, Conservation and Production. The Bureau of Mines is conducting at its stations in Bartlesville, Okla, and San Francisco, investigations relating to petroleum conservation and production. The various headings are as follows:

Water Problems
Decline of Oil-Well Production
Oil Storage and Transportation Losses
Oil-Well Cements
Mode of Occurrence of Oil, Gas and Water in Underground Strata
Oil Casing, Perforation and Screen Pipes
Cement Oil Wells
Oil Camp Sanitation
Transporting Heavy Oils
Factors in Oil Production
Oil-Well Drilling on Government Lands
Transmission of Probable Depreciation
Methods of Increasing Production
Legislation
Construction and Operation of Oil Storage Facilities
Oil Shale
Circular Method of Oil-Well Drilling
Bureau of Mines, Washington, D. C. Address Van H. Manning, Director.

*Petroleum, Asphalt and Wood Products B2-19.* Examination of American Crude Petroleum. A study of physical and chemical properties of American crude petroleums by (1) routing comparison of relatively large number of samples. (2) A detailed study of a smaller number of representative composite samples by E. W. Dean, Bureau of Mines Station, Pittsburgh, Pa. Address Bureau of Mines, Washington, D. C. Van H. Manning, Director.

*Ventilation B1-19.* Ventilation of Metal Mines. The Bureau of Mines is reporting through Daniel Harrington the investigations on the ventilation of the Butte mines. The report will discuss the underground and surface conditions affecting the flow of air and its adequate distribution. This will be discussed in its application to general principles. Bureau of Mines, Washington, D. C. Address Van H. Manning, Director.

### C—RESEARCH PROBLEMS

The purpose of this section of Engineering Research is to bring together persons who desire coöperation in research work or to bring together those who have problems and no equipment with those who are equipped to carry on research. It is hoped that those desiring coöperation or aid will state problems for publication in this section.

### D—RESEARCH EQUIPMENT

The purpose of this section of Engineering Research is to give in concise form notes regarding the equipment of laboratories for mutual information and for the purpose of informing the profession of the equipment in various laboratories so that persons desiring special investigations may know where such work may be done.

*Kansas State Agricultural College D2-19.* Thermal Testing Plant. The refrigerating equipment: 2-ton experimental steam-driven refrigerating system of direct expansion, compression type and duplicate 1-ton electrically-driven system. Calorimeter room 9 ft. x 12 ft. x 10 ft., built of cork board and equipped with 600 ft. of 2-in. direct-expansion piping. The quantity of ammonia circulated together with temperature regulation is controlled through expansion valve and speed of compressor. Temperature regulated within ⅓ deg. fahr. Test boxes are cubes of 2 ft. in diameter. Three boxes are being used; one of 3-in. cork board, one of 3-in. waterproof lith board and one of 2-in. waterproof lith board. The boxes are heated electrically by special alloy resistance coils placed within each box. Three coils are used in each box. The current is supplied by 12-volt storage battery, being regulated by variable external resistance.

The temperature is measured by electric pyrometers accurate to 1/5 deg. fahr. These are thermocouples of No. 16 gage copperconstantan wire and are connected to the potentiometer indicator by a multiple switch. The cold end of the thermocouple is inserted in a vacuum bottle containing a liquid, the temperature of this liquid being measured by means of a mercury thermometer.

There is no fan within the test box. The heat through the top and bottom is figured separately from the heat through the sides by determining temperature difference of various surfaces.

It is hoped that investigations will throw light on heat transmission due to radiation as well as that due to convection. It is also planned to determine, if possible, the relation between temperatures at various points in the interior of the box and the heat transmitted per sq. ft. through the material.

*Massachusetts Institute of Technology D1-19.* Electrical Engineering Laboratory. The laboratories of the Department of Electrical Engineering of the Massachusetts Institute of Technology are four in number, their purposes and equipments being as follows:

*The Electrical Measurements Laboratory* is equipped for the general problems of electrical testing and electrical measurements including standardization. Insulating materials may be tested for dielectric strength of 100,000 volts and also special equipment for measuring losses of energy in dielectric when subject to alternating voltages.

*The Dynamo Laboratory* is equipped with instruments and apparatus necessary for problems relating to dynamos and special problems of losses in iron when subject to magnetization under various conditions of the form of test piece, the frequency of alternations of magnetism and variations of magnetic density.

*The High-Frequency Laboratory* is equipped with alternators giving alternating currents of 10,000 periods per second and 100,000 periods per second, also with kenetrons and other tuned devices for obtaining alternating currents of controllable frequencies and amounts. This laboratory offers excellent opportunity for studying special problems of radio telegraphy, and telephony as well as problems of skin effect in conductors when currents flow through them.

*The General Research Laboratories* are equipped with artificial lines representing certain conditions of very long-distance power-

MECHANICAL ENGINEERING THE JOURNAL
AM.SOC. M. E.

transmission lines and of telegraph and telephone lines and some of the characteristics of real lines have been traced out. The equipment includes instruments for measuring tractive effort, for studies of motor transportation and street-railway transportation, hot-wire ammeters, sensitiveness of oscillographs and the characteristics of diaphragms.

The Department has published twenty bulletins after their publication as papers presented to the American Institute of Electrical Engineers, numbered as follows:

1 The Economical Transportation of Merchandise in Metropolitan Districts.
2 Notes on the Cost of Motor Trucking.
3 Observations on Horse and Motor Trucking.
4 Relative Fields of Horse, Electric and Gasoline Trucks.
5 The Delivery and Handling of Miscellaneous Freight at the Boston Freight Terminals.
6 The Delivery System of R. H. Macy and Co. of New York.
7 Explorations over the Vibrating Surfaces of Telephonic Diaphragms as Derived from their Motional-Impedance Circles.
8 The Mechanics of Telephone-Receiver Diaphragms as Derived from their Motional-Impedance Circles.
9 Experimental Researches on Skin Effect in Conductors.
10 Tractive Resistances to a Motor Delivery Wagon.
11 Some Properties of Vibrating Telephone Diaphragms.
12 Experimental Researches on the Skin Effect in Steel Rails.
13 Skin-Effect Resistance Measurements of Conductors at Radio-Frequencies up to 100,000 Cycles per second.
14 Street Railway Fares; their relation to length of haul and cost of service.
15 Apparent Dielectric Strength of Varnished Cambric.
16 Magnetic Flux Distribution in Annular Steel Laminæ.
17 Electromagnetic Theory of the Telephone Receiver with special references to Motional Impedance.
18 A Rectangular-Component Two-Dimensional Alternating Current Potentiometer.
19 Current Distribution in Armature Conductors.
20 Alternating-Current Planevector Potentiometer. Measurements at Telephonic Frequencies.

*Mellon Institute D2-19.* The Mellon Institute of Industrial Research is equipped for engineering research as well as that in pure and applied science. The following equipment is available for engineering work:

Two 100-hp. Sprague electric dynamometers equipped with revolution counters operated by the same electric circuit.

A three-section pipe arranged so that there is no end transmission of heat from the electric heaters on the inside of the middle section.

One standard Dodge automobile 4-cylinder gas engine connected to 75-hp. Sprague electric dynamometer equipped with automatic test timing device actuated by fuel-weighing scale.

Gas producer of modern mechanical type

Complete grinding and pulverizing machinery

Well-equipped machine and woodworking shop

Sweetland filter press

Vacuum dryer

Furnaces for refractory and ceramic research

Lummus column still.

*Refrigeration D3-19.* Thermal Testing Plant. See Kansas State College, D2-19.

## E—PERSONAL NOTES

The purpose of this section of Engineering Research is to give notes of a personal nature regarding the personnel of various laboratories, methods of procedure for commercial work or notes regarding the conduct of various laboratories.

*United States Naval Experiment Station E1-19.* The work of the U. S. Naval Engineering Experiment Station at Annapolis. Md.. consists of the following:

*a* Determination of conformity to specifications issued by the Bureau of Steam Engineering for particular materials or machines.

*b* Establishment of an "improved list" for the purchase of types and makes of materials and appliances coming under the cognizance of the Bureau of Steam Engineering.

*c* Determination of the most satisfactory of a number of devices as regards efficiency, performance, or other operative peculiarities.

*d* Obtaining data for the basis of specifications for the purchase of material or the modification of existing specifications.

*e* Furnishing the Bureau of Steam Engineering with design data and methods of design for various kinds of apparatus for which the existing data and methods are either insufficient or unsatisfactory.

*f* Investigation of new machines or materials and, if necessary, the development of them into practical form.

*g* Discovering the cause of failure or of poor performance in order to recommend to manufacturers and to the Bureau of Steam Engineering, alterations which will fit the apparatus or material under test for use in naval service.

*h* Regular courses of instruction to classes of student officers from the Post Graduate School, U. S. Naval Academy, and to undergraduate midshipmen from the U. S. Naval Academy.

There are three laboratories: the Engineering Laboratory, the Chemical Laboratory and the Metallographic Laboratory. In the Engineering Laboratory the following items are included:

Internal-combustion engines,
Refrigerating apparatus,
Condensers, air pumps, blowers, fans and air compressors,
Steam turbines,
Heaters for feedwater, fuel oil and lubricating oil,
Evaporators, distillers, pumps, rotary engines,
Safety valves,
Steam traps,
Pump valves,
Sheet and rod packings.
Boiler fittings,
Blow valves,
Water columns,
Gage glasses,
Feedwater regulators,
Stop and non-return valves,
Pump governors,
Pressure-regulating valves.

The work of the Chemical Laboratory is divided into several parts:

*a* All oils submitted for use in the naval service are tested in this laboratory and from the results of the tests recommendations are made to the Bureau of Steam Engineering for their approval or rejection.

*b* Analyses are made of boiler compounds, fuels, boiler deposits, boiler waters and all other materials as requested by the Bureau of Steam Engineering.

*c* New processes for production of hydrogen for aviation are proposed and developed.

*d* All problems in lubrication for the United States Navy are taken up in this laboratory.

*e* The chemical features of many problems connected with the operation and maintenance of naval machinery are investigated.

The work done in the Metallographic Laboratory consists of the following:

*a* Investigation of the cause of failure of metals in naval service, including shafting, boiler tubes, boiler plates, condenser tubes, gland rings, studs, metallic packing.

*b* Research with the view of determining the suitability of new alloys or new metallurgical methods for adoption in the Navy. These include bearing metals, aluminum alloys, nickel, silver and aluminum solders, welding methods, etc.

*c* Inspection tests, including metallographic inspection of forgings intended for naval service.

*d* Miscellaneous tests involving investigations for the preparation of specifications, standardization of tests, etc.

*e* Instruction of post-graduate students and undergraduate students of the Naval Academy.

## F—BIBLIOGRAPHIES

The purpose of this section of Engineering Research is to inform the profession of bibliographies which have been prepared. In general this work is done at the expense of the Society. Extensive bibliographies require the approval of the Research Committee. All bibliographies are loaned for a period of one month only. Additional copies are available, however, for periods of two weeks to members of the A.S.M.E. or to others recommended by members of the A.S.M.E. These bibliographies are on file in the offices of the Society.

*Boilers and Accessories F3-19.* Feedwater Evaporating Equipment. A bibliography of 1¼ pages. Search 2731. Address A.S.M.E., 29 West 39th St., New York.

*Cement and Other Building Materials F1-19.* Sand, Its Occurrence, Properties and Uses. A 72-page bibliography prepared by the Pittsburgh Carnegie Library. Price 15 cents. Address Director of the Carnegie Library, Pittsburgh, Pa.

*Properties of Engineering Materials F2-19.* Determining, Producing and Maintaining Hardness of Metals. A bibliography of 28 pages. Search 2683. Address A.S.M.E., 29 West 39th St, New York.

*Properties of Engineering Materials F3-19.* Design and Strength of Riveted Joints. A bibliography of 8 pages. Search 2740. Address A.S.M.E., 29 West 39th St., New York.

# ENGINEERING SURVEY

A Review of Progress and Attainment in Mechanical Engineering and Related Fields, as Gathered from Current Technical Periodicals and Other Sources

SUBJECTS OF THIS MONTH'S ABSTRACTS

STORAGE-BATTERY TESTING
CADMIUM ELECTRODES FOR STORAGE-BAT-
    TERY TESTING
BUREAU OF STANDARDS ANNUAL REPORT
BOND CLAYS AND GRAPHITE CRUCIBLES
FUEL-ECONOMY REPORT, BRITISH ASSOCIA-
    TION
GAS-SUPPLY STANDARDS
FUEL-OIL SITUATION IN EASTERN STATES
FLOW OF LIQUIDS, DETERMINATION OF
    VELOCITY
RIVA WATER TURBINES
WATER TURBINE WITH TWO WHEELS OF
    DIFFERENT DIAMETERS

VALVE DESIGN IN INTERNAL-COMBUSTION
    ENGINES
VALVE STEELS
HIGH-SPEED TURBINE GEARS
WIRE ROLLING-MILL GEAR DRIVE
LIMITS OF SIZE OF SINGLE-SHAFT STEAM
    TURBINES
DESIGN OF VERY LARGE STEAM-TURBINE
    UNITS
POWER PIPING
STANDARDIZATION IN POWER PIPING
MAXIMUM LENGTH OF LONGITUDINAL
    BOILER JOINTS

COXON AUTOMATIC STOP VALVES
LAP-SEAM BOILER EXPLOSION
PLACING A NEW TURBINE IN SERVICE
HORWICH SUPERHEATER FOR LOCOMOTIVES
QUINCY MARKET COLD-STORAGE PLANT AT
    BOSTON
COMPRESSOR FOR QUINCY MARKET COLD-
    STORAGE PLANT
INTERCOOLER OF COMPRESSOR FOR QUINCY
    MARKET COLD-STORAGE PLANT
THE MASS OF THE LITER OF AIR AND GAS
    MIXTURES
HEAT TRANSMISSION IN BOILER FLUES

## BUREAU OF STANDARDS

STORAGE-BATTERY TESTING. In the operation and testing of storage cells, it is frequently important to know the individual potentials of the positive and negative plates. No standard method has been recognized. The results obtained with the cadmium electrode are often contradictory. This investigation has been made to determine the reliability of the cadmium electrode and the errors in measurement to which it is subject. The standard electrode used was the mercurous sulphate half-cell. It was found that the most serious error in using the cadmium electrode is due to polarization. By measuring the potential of the negative plate and computing the potential of the positive plate from this and the cell voltage, the most serious error of the cadmium electrode can be avoided, even when using an ordinary voltmeter. This investigation deals only with the accuracy of the cadmium electrode and does not discuss the cadmium readings with reference to the age or condition of the battery. (*Bureau of Standards Technologic Paper* No. 146, entitled The Cadmium Electrode for Storage Battery Testing, *g*).

### Annual Report of Bureau of Standards

ANNUAL REPORT OF THE DIRECTOR OF THE BUREAU OF STANDARDS. A review of the work of the National Bureau of Standards for the year ended June 30, 1919, is given in the annual report of the Director of the Bureau of Standards at Washington. The report describes the functions of the Bureau of connection with standards and standardization, and contains a chart and description of the several classes of standards dealt with. The Director also gives a clear idea of the relation of the Bureau's work to the general public, to the industries, and to the Government, and includes a special statement of the military work of the year. Brief statements are made upon practically all of the special researches and lines of testing completed or under way at the Bureau. The list of these topics occupies 12 pages in the table of contents. The Bureau is organized in 64 scientific and technical sections and 20 clerical, construction, and operative sections. During the year the Bureau has issued 51 publications, not including reprintings, 36 of which were new and 15 revisions of previous publications. In the several laboratories of the Bureau more than 131,000 tests were made during the year. The appropriations for the year, including special funds for war investigations, were approximately $3,000,000. A noteworthy event of the year included the completion of the industrial laboratory in which will be housed the divisions having to do with researches and tests of structural materials. The building also includes a commodious kiln house for use, among other purposes, of the ceramics division in the experimental production of new clay products and for general experimental purposes.

The report comprises 293 pages, and may be obtained as long as free copies are available by addressing the Bureau of Standards, Washington, D. C. (Abstract of *Miscellaneous Publication of the Bureau of Standards*, No. 40, *g*)

## ENGINEERING MATERIALS (See also Internal-Combustion Engines)

THE PROPERTIES OF AMERICAN BOND CLAYS AND THEIR USE IN GRAPHITE CRUCIBLES AND GLASS POTS, A. V. Bleininger. The properties of American bond clays are described in detail and expressed through characteristic numerical values, with special reference to their burning behavior. It is shown that materials equal in quality to those formerly imported from Germany are available, and that by suitable blending any desired combination of properties can be readily produced. The characteristics of natural and artificial graphite are described and means suggested for the control of crucible mixtures. The fact is brought out that the main advantage in the use of German glass-pot clay consists in its low shrinkage, and suggestions are made for obtaining similar conditions with the use of domestic materials and with increased resistance to corrosion. The compositions and the preparation of semi-porcelain and porcelain glass pots are given. The method of casting glass pots as practiced at the Pittsburgh laboratory of the Bureau of Standards is also described. (Abstract of *Technical Paper* 144, *U. S. Bureau of Standards*, *g*)

## FUELS AND FIRING

FUEL-ECONOMY REPORT OF THE BRITISH ASSOCIATION. The report consists of several parts. The first part outlines the work of the Committee since 1916. The second part deals with coal output and prices since 1913. The third part (on Coal Research) states that the recent experiments by Bone and Sarjant make it very doubtful whether the resinous and cellulose constituents of coal can really be separated by the aid of pyridine and chloroform. It would appear that pyridine affects the coal substance as a whole, especially at higher temperatures and pressures, while the presence of water and oxygen retards the action.

The last part of the report is devoted to the subject of future standards of public gas supply. The Committee believes that the consumer should be supplied with a suitable gas for lighting and heating, while the recovery of by-products should be secured in the general interest. It would not exclude the distribution of surplus of coke-oven gas through the pipe mains and the steaming of incandescent coke in continuous vertical-retort working; but it would not oblige a gas coke to convert all its coke into water gas. The Committee would not admit further that the relative value of a grade of gas was simply proportional to its calorific value. The composition of the gas was not an indifferent matter to the consumer—in particular, the percentage of $CH_4$ should not be reduced below 30; also, to strip the gas of benzenoid hydrocarbons seemed inadvisable, if the thermal basis were to be adopted. (*Engineering*, vol. 108, no. 2809, Oct. 31, 1919, pp. 574-576, *g*)

FUEL-OIL SITUATION IN THE EAST. An editorial article in *Power* touches upon the important change in the fuel used in large power plants in the Eastern States which has been gradually taking place in the last few years. One after another important plants have displaced coal in favor of fuel oil. This has been natural for several reasons. In the first place, the coal market was anything but attractive to the buyer, both as regards prices and conditions of supply. Also fuel oil is a more convenient fuel to handle, increases boiler and fuel efficiency and greatly cuts labor costs and troubles. On the other hand, the editorial calls attention to the very important fact that there appears to be no certainty that oil will be available for long periods as it is available today, even though three- to five-year contracts are possible. (*Power*, vol. 50, no. 17, Oct. 21, 1919, published Dec. 1, 1919, pp. 623-624, *g*)

## HYDRAULIC ENGINEERING

### Photographic Method of Determining Velocity of Flow of Liquids

DETERMINATION OF VELOCITY OF FLOW OF LIQUIDS, C. Camichel. An article describing the method of determining the velocity of flow of liquid by the observation of the travel of very fine particles carried along by the liquid.

This method was employed for the first time by Scott Russell. The author, who is the director of the Electrotechnic Institute of Toulouse, France, indicates briefly some of the results which have been obtained by applying this method to various hydrodynamic problems.

I *Flow of Water between Two Vertical Parallel Plates, such as Glass.* The $x$ axis is horizontal and parallel to the plates; the $y$ axis is horizontal and normal to the plates; the velocity of

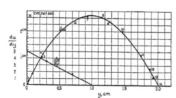

FIG. 1 CURVE DISTRIBUTION OF VELOCITIES IN FLOW OF WATER BETWEEN TWO PARALLEL PLATES

flow $u$ satisfies the relation resulting from the equation of Navier

$$\mu \frac{d^2 u}{d y^2} = \frac{dp}{dx} \quad \ldots \ldots \ldots \ldots \ldots \ldots [1]$$

where $p_0$ is the hydrostatic pressure; $p + p_0$ the pressure during the flow, $p$ being a function of $x$ only.

The following are some data obtained experimentally. The water flows between two parallel vertical glass plates located at a distance of 2 cm. from each other, the plates themselves being 30 cm. (11.8 in.) wide and 4 m. (13.1 ft.) long. The experiment consists in lighting up horizontally very fine particles suspended in the water by using a luminous batch of rays interrupted by a toothed wheel rotating at a constant speed, and in taking photographs by means of a camera placed at a point on the horizontal median plane.

A number of illustrations reproducing the photographs thus obtained are given in the original article, but they are not distinct enough for further reproduction. A very careful examination of these photographs shows that the reflection on the glass of the trajectories of the particles makes it possible to determine with considerable precision the values of $y$ measured from one of these walls. Table 1 gives data obtained from one such test. It is easy to see that the curve $(u, y)$ is a parabola and it is enough to plot a curve with ordinates $du/dy$ and abscissae $y$ to see that such a curve is a straight line. This has been done in Fig. 1.

If we take as a coefficient of viscosity of the water $\mu = 0.011$ in C.G.S. units for the temperature prevailing during the experiment (16 deg. cent.), we get

$$\mu \frac{d^2 u}{d y^2} = 0.0605 = \frac{dp}{dx}$$

The value of the loss of head $W$ for two points located on a line parallel to the axis of $x$ and at a distance of 80 cm. from each

FIG. 2 DISTRIBUTION OF VELOCITIES IN A GYRATORY MOVEMENT

other, we find to be $W = 0.05$ mm. Though it is difficult to find the correct value of such a small loss of head, it can be measured by means of the M. R. Threlfall manometer with a micrometric screw; this gives for $W$ the value of 0.07 mm., which is in sufficiently good accord with the formula above.

The above experiment has been carried out with the velocity

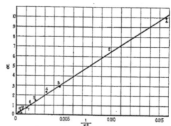

FIG. 3 DISTRIBUTION OF VELOCITIES IN A GYRATORY MOVEMENT

of flow of water reduced as far as possible, but similar experiments have been carried out for very much higher velocity.

II *Gyratory Movement about a Vertical Axis.* In studying rotary movements accompanying flow of liquids from orifices, Boussinesq has shown that if one limits himself to the consideration of the superficial particles of the liquid the velocity of flow is proportional to the distance $r$ from the axis, in the part where the trajectories of flow are approximately horizontal, so that

$$ur = C \ldots \ldots \ldots \ldots \ldots \ldots \ldots \ldots [2]$$

TABLE 1

| No. | y cm. | u cm. per sec. | Remarks |
|---|---|---|---|
| 1 | 0.026 | 0.15 | The period of oscillations is 0.17 sec. |
| 2 | 0.096 | 0.4 | |
| 3 | 0.23 | 1.03 | |
| 4 | 0.24 | 1.13 | |
| 5 | 0.53 | 2.02 | |
| 6 | 0.55 | 2.04 | |
| 7 | 0.94 | 2.54 | |
| 8 | 1.15 | 2.52 | |
| 9 | 1.27 | 2.34 | |
| 10 | 1.47 | 2.00 | |
| 11 | 1.73 | 1.11 | |
| 12 | 1.88 | 0.60 | |
| 13 | 1.94 | 0.3 | |

TABLE 2

| No. | a | r mm. | $\frac{1}{r^2}$ |
|---|---|---|---|
| 1 | 99° | 8 | 0.0166 |
| 2 | 63° | 10 | 0.0100 |
| 3 | 30° | 14.7 | 0.0045 |
| 4 | 23°3' | 17.8 | 0.0032 |
| 5 | 14°7' | 22.0 | 0.00207 |
| 6 | 10°5' | 25.5 | 0.00154 |
| 7 | 8°0' | 29.0 | 0.00119 |
| 8 | 4°9' | 39.0 | 0.00066 |
| 10 | 3°0' | 46.5 | 0.00046 |
| 14 | 2°2' | 57.8 | 0.00030 |

TABLE 3

| | $z_s - z$ | $2r$ | $\frac{1}{r^2}$ |
|---|---|---|---|
| 1 | | 60 mm. | 0.0011 |
| 2 | | 29.7 | 0.0045 |
| 3 | | 23.6 | 0.0072 |
| 4 | | 20.2 | 0.0098 |
| 5 | | 18.0 | 0.0123 |

FIG. 4  MERIDIAN OF GYRATORY MOVEMENT

This is roughly shown in Fig. 2 which demonstrates the distribution of velocities of flow in a gyratory movement about the vertical axis. Table 2 gives in numerical form the results of a similar experience; α designates the angle formed by the rays passing through two extreme positions of a particle at the beginning and end of a photographic exposure.

Fig. 3 represents the curve having for its abscissæ $1/r^2$ and for its ordinates α. It is a straight line. The angular velocity varies therefore inversely as the square of the distance of the particle from the axis, and hence the linear velocity varies inversely as the distance from the axis. in accordance with Formula [2]. If we apply the theory of Bernoulli and designate by h the depth $z_s - z$ of the free surface below the horizontal plane containing the limit of the vortex of radius $r_s$, which is very large, we have approximately

$$h = \frac{C^2}{2g}\frac{1}{r^2} - \frac{1}{r_s^2} = \frac{C^2}{2g}\frac{1}{r^2} \quad \ldots \ldots \ldots \ldots [3]$$

The experiment consisted in photographing the meridian of the gyratory movement, that is, the curve having for its abscissæ r and for its ordinates h, Fig. 4, and also the distribution of velocities shown in the original article in a figure not suitable for reproduction.

In this experiment 100 mm. of the caliber occupied 108 mm. on the photographic plate, the periods of interruption of the ray of light being 0.026 sec. The meridian of the gyratory movement is given in Table 3 for points which are not too close to the axis.

The curve with abscissæ $\frac{1}{r^2}$ and ordinates $z_s - z$ is a straight line for points which are not very close to the axis, as shown in Fig 5. The angular coefficient of this straight line is

$$\frac{C^2}{2g} = 4.1 \times 10^{-1}$$

and the distribution of the velocities gives

$$\frac{C^2}{2g} = 4.2 \times 10^{-1}$$

a value in good accord with what has been given above.

In the preceding experiment the gyratory movement was counter-clockwise. There is, however, no difference between the

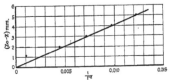

FIG. 5  MERIDIAN CURVE OF GYRATORY MOVEMENT

clockwise and the counter-clockwise movements. It is intended to carry out an investigation of the distribution of velocities in the neighborhood of the axis and in the interior of the liquids. (*Revue Générale de L'Électricité*, vol. 6, no. 21, Nov. 22, 1919, pp. 707-709 and 2 plates, 8 figs., etc.)

### Water Turbine with Two Pelton Wheels of Different Diameters

ITALIAN WATER TURBINES, Norman Davey. First of a series of articles entitled Notes on a Tour in Italy and France, devoted chiefly to the description of hydraulic machinery.

In the Riva (Milan) type of larger turbines particular attention is called to the oil-pressure thrust block. The end thrust of the rotor is taken up by oil pressure upon the surface of a disk attached to the turbine shaft; oil under pressure is circulated on both sides of the disk and an equilibrium of thrust is thus secured.

Mention is made of a rather unusual type of turbine which the author has seen in the process of construction. It is to be used to produce power for the Italian Electric State Railways. The machine consists of two Pelton wheels of different diameters, employing varying heads of water, the water in question being drawn from two different sources of supply. Of the two wheels which are upon the same shaft, one is operated by a head of 2000 ft. and the other of 670 ft. As a large variation of load occurs throughout the 24 hr., the plant is fitted with an automatic device by means of which the smaller wheel, which is only used when the load rises to a certain point, may be put to work. The diameters of the two wheels taken at the centers of the buckets are 6.4 ft. and 3.67 ft., respectively, and the speed 500 r.p.m. The larger diameter wheel develops 3500 hp. with one jet and the smaller wheel 2500 hp. with two jets.

The machine is shown in Fig. 6 where the oil thrust block can

be seen beyond the bearing to the left of the larger-diameter wheel.
The 6.4-ft. wheel has buckets 10 in. in diameter and the 3.67-ft.
has buckets 13 in. in diameter.

An interesting feature of the Riva turbine of the Pelton type

on the other end is fulcrumed the floating lever $B$. One end of
this floating lever operates the deflector $C$ and the other is connected with the governor. The fulcrum itself is connected with
jet $K$.

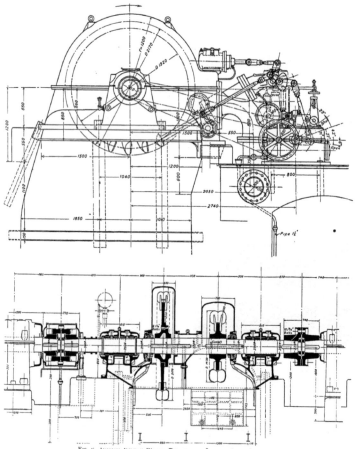

FIG. 6  DUPLEX PELTON WHEEL TURBINE FOR ITALIAN STATE RAILWAYS

is the method of governing speed, or rather the control of the
water jet. This is effected in two ways—first by deflection and
next by throttling (Fig. 7). In it $A$ is the lever working on a
fixed fulcrum. To one end of this lever is attached a dashpot $D$;

Suppose the load to be suddenly removed from the turbine and
racing to commence. The governor immediately reacts on the
floating lever $B$ and the deflector comes into action on the water
stream. Since resistance is offered in this operation to the turn-

ing of the lever $B$ about its fulcrum, the fulcrum itself becomes a moving point about the fulcrum of the lever $A$ and the jet $K$ begins to close. The arrangement is really an application of the floating lever principle as used in various differential gears to this particular purpose. (*The Engineer*, vol. 128, no. 3332, Nov. 7, 1919, pp. 456-457, 6 figs. in the text, $dA$)

## INTERNAL-COMBUSTION ENGINES

### Why Valves Fail in High-Speed Internal-Combustion Engines

VALVE FAILURES AND VALVE STEELS IN INTERNAL-COMBUSTION ENGINES, Leslie Aitchison. Discussion of the different troubles which arise with valves and the steps which should be taken in order to overcome them in the most likely manner, as well as data of tests carried out by the writer, who, during the war, was acting as metallurgist at the British Air Board.

As regards the removal of heat from the inlet valves, it is pointed out that this is done in three ways: First, by conduction down the stem of the valve; second, by direct radiation from the back surface of the head or the neck of the valve; and third, by direct conduction from the face to the valve seat, the third method being the most important.

For proper cooling of the inlet valve two things are essential:

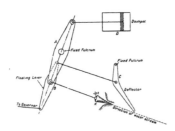

FIG. 7  RIVA SPEED-CONTROLLING ARRANGEMENT FOR PELTON WHEEL TURBINES

First, that the two surfaces shall come accurately together; and second, that the valve shall be closed at the moment of ignition, the second being the more important. The most prolific causes of troubles in valves are those produced by preignition in the cylinder. If the valve is not properly seated at the time of the explosion, there is a rush of hot gases around the valve, taking place at the time when the gases are near to their maximum temperature. The author has examined valves which failed in consequence of preignition and which must have been raised to a temperature of at least 1150 deg. cent. during the process.

Among the other factors which gave trouble is mentioned the work between the guides and the stem of the valves, which is of material importance in rotary engines.

The causes of failure of valves are classified and discussed. Only the most important will be mentioned here.

*Distortion of the head of the valve* may arise from various causes. In one engine it was traced to the fact that the stampings which were used had not been properly heat-treated, and the strains produced by the forging operation had not been removed. When the valve became heated in the running of the engine these strains were released and the valve warped appreciably, sufficiently so to prevent its seating properly. But distortion of the head of a valve may occur even with the proper heat treatment of the forging, and if the valve is so placed that the head is unequally heated this inequality of temperature may have a distorting effect.

Distortion in the head of the valve may further be due to excessive softening of the steel during running, as a result of which the valve is pulled out of shape by hammering against the seat. The causes of the softening are twofold—either the employment of a steel which is not sufficiently strong at high temperatures, or else the utilization of a design in which there is not sufficient metal in the valve head.

*Cracks in the valve face* are a fairly prolific source of trouble. They are usually due to faults of metal, but from whatever cause they may come they are dangerous.

*Excessive wear of the valve foot* is usually found in valves in which a blow of the tappet is delivered directly on to the foot of the valve, or in which the cam acts directly on the valve stem, no cap being fitted to take the blows or the wipe. It is not always easy to provide the proper steel in the foot of the valve to take care of this condition.

*Burning out of the head of the valve* may be due to several causes which may not be entirely attributable to the steel. It is much more nearly true to say that in all cases where burning out occurs, the engine has been running badly. One of the principal causes is the oxidation or scaling of the steel. Cracks in the valve face are also a very prolific source of burning out. If the valve does not sit properly, burning out is almost always a result.

*Breaking of the head or neck* is due to self-hardening. Various cases have been known in which the valve has worked satisfactorily for a certain length of time and then failed by breaking up of the head or the snapping of the neck of the valve. This is found to happen most frequently when starting up an engine which has previously run and then gone cold. Also, the valve which most usually fails in this manner is the one which is manufactured from air-hardening nickel-chromium steel. The valves during the first running have been heated to a temperature greater than, say, 750 deg. cent. and then have been allowed to cool down. As a result of this cooling, the valves have hardened to such a degree that the maximum strength of the steel has reached a figure of possibly 100 tons per sq. in.

This hardening is generally accomplished by the production of a certain amount of brittleness. When the engine is restarted the brittle valve breaks in the head, or else the head falls right off. To obviate this trouble, it is essential that a steel should be chosen which does not harden when cooled in the air from the temperature obtained by the valve during the running of the engine into which it is put.

### IDEAL VALVE STEEL

Steel which is to be used for the manufacture of gasoline-engine valves should possess certain properties and it is difficult to expect all these properties will be found in one and the same steel. These properties are:

1  The greatest possible strength at high temperatures
2  The highest possible notched-bar-value
3  The capacity of being forged easily
4  The capacity of being manufactured free from cracks, whether these arise in the manufacture of the steel bar or whether they are produced during the forging of the steel.
5  The capacity of being heat-treated easily, regularly and reliably
6  The least possible tendency to scale, and if scaling does occur the scale should be as adherent as possible.
7  The reliability to retain its original physical properties after frequent heatings to high temperatures followed by cooling to normal temperature, also after being heated to an elevated temperature for a considerable length of time
8  No liability to harden when cooled in air from the temperature which it will retain when used normally as a valve in an engine
9  The capacity of being heat-treated after forging so that it is free from strains liable to produce distortion
10  Sufficient hardness to withstand excessive wear in the stem
11  The capacity of being hardened at the foot of the stem with considerable ease if necessary

12 The capacity of being machined easily and satisfactorily by ordinary methods.

It does not appear that there is a steel which would possess all of these properties. Table 4 shows the compositions of the various steels which have been used, and there does not seem to be any reason for using such a large variety. In order to evaluate the adaptability of various steels for use in valves, a large number of tests have been carried out, mainly at high temperatures. These tests have been tensile tests chiefly, supplemented by notched-bar tests at high temperatures, and, in some cases, Brinell hardness tests and observation for scaling.

In view of the importance of tungsten steel, Fig. 8 is here reproduced; likewise Fig. 9, showing the resistance to scaling of the various types of steel.

The tests have shown that the principal factor in deciding the properties of a steel at high temperature is its type. By this is meant that all the tungsten steels, for example, have fairly similar properties which make them, as a type, distinct from chromium steels or nickel-chromium steels.

There is a considerable similarity as regards mechanical prop-

are liable to scale very considerably. Nickel-chromium steels scale to a greater extent than do the steels of any of the other types, while the high-chromium steels present the greatest resistance to scaling at high temperatures of any of the steels.

TABLE 4

| Steel | Carbon | Nickel | Chromium | Tungsten | Cobalt | Vanadium |
|---|---|---|---|---|---|---|
| a | 0.65 | ..... | 3.5 | 17.0 | .... | 0.8 |
| b | 0.60 | ..... | 3.5 | 14.0 | .... | 0.8 |
| c | 0.25 | ..... | 3.5 | 11.5 | .... | .... |
| d | 0.30 | 25.0 | ..... | .... | .... | .... |
| e | 0.25 | ..... | 13.0 | .... | .... | .... |
| f | 0.70 | ..... | 11.0 | .... | .... | .... |
| g | 0.80 | ..... | 7.0 | .... | .... | .... |
| h | 0.35 | 3.0 | ..... | .... | .... | .... |
| j | 0.60 | 3.0 | ..... | .... | .... | .... |
| k | 0.30 | 3.75 | 1.0 | .... | .... | .... |
| l | 0.30 | 4.25 | 1.4 | .... | .... | .... |
| m | 0.15 | 3.75 | 1.0 | .... | .... | .... |
| n | 0.10 | 5.8 | 0.25 | .... | .... | .... |
| o | 0.30 | ..... | ..... | .... | .... | .... |
| p | 1.0 | 0.5 | 11.5 | .... | 4.0 | .... |

The general properties of valve steels in use are next discussed and interesting data are given as to their heat treatment.

The writer recommends the selection of steels from one of the three types: tungsten steel, high-chromium steel and nickel steel,

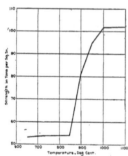

FIG. 8 STRENGTH OF TUNGSTEN VALVE STEELS AFTER COOLING FROM HIGH TEMPERATURES

erties at high temperatures between those steels which have a similar constituent.

Among the different steels in any type, variations in composition have a distinct effect upon the mechanical properties at high temperatures. This is particularly true of the high-tungsten and high-chromium steels in which variation in carbon content produces a very marked effect upon the tensile strength or the notched-bar value. This variation is also noticeable in the 3 per cent nickel and 3 per cent chromium steels that is not so marked as in the steels containing a larger proportion of alloying elements. In these steels a variation of 0.2 per cent of carbon will have a greater effect upon the strength than a variation of 5 per cent of tungsten and nearly 7 per cent of chromium.

Tungsten steels with high percentages of carbon (about 0.6 per cent) have the greatest tensile strength at high temperatures. This strength is nearly approached by that of the high-chromium steels containing a high percentage of carbon.

Steels containing a high percentage of chromium with a lower percentage of carbon (such as stainless steel) are distinctly weaker than either high-tungsten steels or the steels containing a similar proportion of chromium with high carbon.

Plain nickel or nickel-chromium steels are quite weak at high temperatures as compared with the other steels. Variations in composition among these steels have apparently very little effect. The influence of vanadium and cobalt appears to be negligible.

As regards scaling, tungsten steels scale comparatively little up to temperatures of about 850 deg. cent., but beyond that they

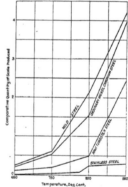

FIG. 9 COMPARATIVE CURVES OF RESISTANCE TO SCALING OF THE VARIOUS TYPES OF STEEL

eliminating entirely the nickel-chromium steels, and suggests the following allocation for the different steels:

All inlet valves.............................3 per cent nickel steel
Exhaust valves with a working temperature
    not higher than 600 deg. cent..............3 per cent nickel steel
Exhaust valves with a working temperature
    between 600 deg. cent. and 700 deg. cent..... high-chromium steel
Exhaust valves with a working temperature
    greater than 760 deg. cent................. tungsten steel

The article contains data of numerous tests on heat treatments which cannot be abstracted for lack of space. (*The Automobile Engineer*, vol. 9, no. 132, Nov. 1919, pp. 401-410, 13 figs., peA)

MACHINE PARTS AND DESIGN (See also Internal-Combustion Engines)

### Present Status of High-Speed Turbine Gearing

HIGH-SPEED TURBINE GEARS, Gerald Stoney. The paper starts with brief historical remarks on the use of high-speed gearing.

The present status of the matter is that gearing is used in all cases for large powers, except in the case of steam turbines driving alternators, where it is found that the satisfactory alternator can be made for speeds suitable for the steam turbine.

As regards types of gearing suitable with steam turbines, rope, chain and belt gearing are all impractical, and the only types used are hydraulic, electric and helical, only the latter being discussed here in detail. The helical gear, among other advantages,

FIG. 10  DIAGRAM OF A PINION OF A HIGH-SPEED TURBINE GEAR
WITH INDICATION OF STRESSES

possesses the highest efficiency, between 98 and 99 per cent, and is used especially in marine propulsion on an immense scale.

The design for helical gearing for use with steam turbines is discussed in brief. Tooth speeds up to 120 ft. per sec. are common and have, in many cases, been exceeded. What is exactly the limit of speed is uncertain, but so long as lubrication can be maintained there does not seem to be any reason why the figure

and is one that so far has been only partly worked out. In England, Messrs. Parsons, who were the pioneers in these gears, resolved not to exceed $a = 80$ and $b = 175$, but as experience was gained $b$ was increased first to 220 and now to 250.

The distortion of the pinion is next discussed. This is made up of two items—the twist of the pinion due to the torque it transmits, and the bending due to the load on the teeth. In estimating this it can be assumed that the load on the teeth is uniform along the pinion, and it may be further assumed that the effective diameter of the pinion is the pitch diameter.

At any point $x$ in Fig. 10, the torque $T$ is:

$$T_x = \frac{pxd}{2} = \frac{2fI}{d} = \frac{\pi fd^3}{16}$$

where $f$ is the shear stress at the circumference of the pinion and $I$ is the polar moment of inertia, or $\frac{\pi d^4}{32}$. Hence it follows that $f = \frac{8px}{\pi d^3}$. If $N$ is the modulus of rigidity, which for steel is 12,000,000 lb. per sq. in.,

$$\frac{dv}{dx} = \frac{f}{N} = \frac{dx}{\pi Na^3}$$

$$v = \frac{4p}{\pi N}\frac{x^2}{d}$$

which is a parabola. The twist in the first half of the pinion is:

$$y_1 = \frac{p}{\pi N}\frac{l^2}{d}$$

and the second half:

$$y_2 = \frac{3p}{\pi N}\frac{l^2}{d}$$

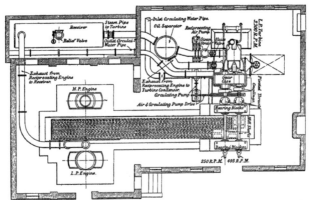

FIG. 11  COMBINATION OF GEAR TURBINE WITH A RECIPROCATING ENGINE FOR DRIVING A WIRE ROLLING MILL

given above should not be considerably exceeded. The load on the teeth is determined by the safe limit of bending stress on the tooth and of the pressures at the point of contact, as well as by danger of the oil-film failure. The bending stress can, in general, be ignored. As regards crushing stresses on the material of the tooth, the load per inch run will vary directly as the diameter, or $p = ad$, where $p$ is the load per inch run in pounds, $d$ the pitch diameter of the pinion in inches and $a$ a constant. Similarly, the stresses of the oil film will obey some such law as $p = bd^m$. The theory of the lubrication of gear teeth is a most difficult one

The alignment of the gears in the gear case is of the greatest importance, as also is the construction of the gear case. In this connection the author discusses the rigid gear-case construction originally introduced by Sir Charles Parsons and the floating-frame type introduced by John H. McAlpine in America.

Several applications of gearing are described and illustrated, such as, for example, a geared turbine driving a cotton mill, and the combination of a geared turbine with a reciprocating engine for driving a wire rolling mill having two shafts running at different speeds (Fig. 11). The compound reciprocating engine runs at

80 r.p.m. and is supplied with saturated steam at 160 lb. pressure, originally exhausting into a central condenser at about 14 in. vacuum giving 2100 hp. By allowing it to exhaust at 16 lb. abs. and leading the steam into a turbine, exhausting into a condenser with 28.8 in. vacuum (barometer 30 in.), the engine gave 1650 i.hp. and the turbine 1663 i.hp., or a total of 3313 i.hp., an increase of 58 per cent in power with no increase in steam used, giving an increased economy of 36.6 per cent. The exhaust turbine runs at 3700 r.p.m., and the geared shafts at 500 r.p.m. and 250 r.p.m. The difficulty of the uneven turning moment of the reciprocating engine and the even turning moment of the steam turbine straining the gear is met by the flexibility of the rope drive. For starting up, the reciprocating engine can be exhausted to atmosphere.

It may be noted that if the power supplied to each low-speed shaft is equal, there will be no bending of the pinion, and the only distortion will be that due to twist. (Paper before the Manchester Association of Engineers. Nov. 22, 1919, reprinted in *Engineering*, vol. 108. Nov. 28, 1919, pp. 729-733, 14 figs., *dg*)

## POWER PLANTS

### How Large Can Steam Turbines and Turbo-Generators Be Made Today?

PRESENT LIMITS OF SPEED AND POWER OF SINGLE-SHAFT STEAM TURBINES AND TURBO-GENERATORS. A symposium of three papers. The first, by Eskil Berg, starts by showing that the limit of a single-unit turbo-generator does not lie in the generator, but is confined to the steam turbine, and that the limiting feature of the latter is the last wheel. The paper is based mainly on the work done by the General Electric Company.

Questions of design of the large turbines and of the conditions limiting the size are discussed.

Efficient action of the blades in large-size turbines operating with high vacuum can be accomplished only by using the bucket speed that bears a proper relation to the steam velocity, which means that to get the largest capacity not only long buckets must be used, but they must be moved at a very high speed. On the other hand, to obtain good bucket action the buckets should not be more than about one-quarter as long as the pitch diameter of the wheel.

The use of a high steam speed in the last stage naturally implies that a relatively large proportion of the total steam energy must be utilized there, and such concentration of work into a single stage has its disadvantages, even if the best relations of velocities is maintained.

Since such a stage is doing a large amount of work, it is naturally less efficient than a stage of similar character doing less work.

In this way the last wheel of the turbine becomes one of the limiting features of the overall dimensions. Energy and efficiency curves (Fig. 12) of the last stage are given which show that at the most efficient point this stage absorbs 11.5 per cent of the total adiabatic available energy, and that the wheel efficiency is 66.25 per cent. But this is for an output of 21,000 kw., and in the same turbine when the load increases to 36,000 kw. the energy in the last stage is 20.9 per cent of the total energy and the wheel efficiency goes down to 54.2 per cent.

The conclusion which the author comes to is that for a given speed there is one particular size of turbine which can be designed to be most economical as to steam consumption, weight, space and price per kilowatt. Even if a size smaller than this is required, it would, in many instances, pay for the central station to install a larger unit, even though it would have to run at reduced load for some time before the station load increased sufficiently to utilize its full capacity. (*Proceedings of the American Institute of Electrical Engineers*, vol. 38, no. 11, Nov. 1919, pp. 1223-1231, 6 figs., *g*)

In the second paper, F. D. Newbury discusses the conditions determining the size of a turbine and conditions limiting it.

In the opinion of the writer, maximum output at any speed

when reduced to the simplest terms is attained when slot space is provided for the maximum possible ampere-turns (in either stator or rotor), and core cross-section is provided for the maximum possible flux. These conditions require the most effective rotor diameter and the maximum rotor and stator core length.

The most effective rotor diameter for a given speed is not necessarily the largest diameter, and to obtain maximum output at a given speed the rotor proportions must be chosen properly to balance mechanical stresses, rotor ampere-turns and flux.

The electrical factors in generator design are discussed by the writer, who believes, among other things, that a fundamental difficulty in setting down definite limiting outputs is the difficulty in arbitrarily setting limiting stresses.

On the other hand, limit of length of the rotor and stator cores is determined mainly by such considerations as cooling-air requirements, bearing proportions, by limits to weight imposed by transportation facilities and the ability to secure forgings of necessary diameter and weight. .

To give an idea of the character of these limitations, it may be stated that, for example, transportation facilities may impose a limit to size in the case of 6- and 8-pole 60-cycle generators. From the special nature of their design and the special skill and equipment required for winding and assembling, rotors should be completed at the builder's factory and shipped as a unit. The

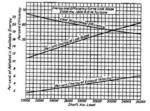

FIG. 12 ENERGY AND EFFICIENCY CURVES (LAST STAGE) OF A LARGE TURBINE

weight of the complete rotor of a 4-pole 1800-r.p.m. generator of 40,000-kva. capacity will be roughly 90,000 lb. This can be transported without difficulty, but the largest possible 12,000-r.p.m. rotor would weigh more than 200,000 lb. and would require rolling stock and trackage in some cases not now available.

Another general limitation to output of larger-diameter rotors is that imposed by the forging facilities of the country. At the present time it is not possible to obtain forgings of suitable physical characteristics weighing more than 50 or 60 tons nor much larger than 50 in. This limits the rotor made from a single forging to an output of, roughly, 50,000 kva. at 1500 r.p.m. By adopting the rotor construction involving 2-in. or 3-in. plates and upset flanged shaft ends, the limiting diameter may be increased sufficiently for the largest 1500- and 1200-r.p.m. outputs shown in Fig. 13. .

This figure shows in curve form limiting generator capacities at various speeds. At 1500 r.p.m. and higher the capacity is determined by the rotor and is inversely proportional to the r.p.m. squared. At lower speeds the capacity is limited by the stator and falls somewhat below the corresponding rotor limiting capacity, as indicated by the dotted extension of the rotor curve. The curve is an indication of the present boundaries based on existing commercial materials and current stresses and bearing proportions. In fact, the capacities shown by it are somewhat in advance of accomplished results.

The writer claims that mechanical forces due to short-circuit and damage caused by armature-winding failures are no greater in the very large generators, indicated by the above figure, than the present-day 20,000- and 30,000-kva. unit.

No opinion is expressed as to the wisdom of installing very large single-shaft units. If operating engineers desire units of 50,000 to 100,000 kva., there is no question but that such generators can be conservatively designed and constructed. *Proceedings of the American Institute of Electrical Engineers*, vol. 38, no. 11, Nov. 1919, pp. 1233-1242, *g*)

The paper by J. F. Johnson is restricted to a discussion of some of the factors which influence limits as applying particularly to turbines of the reactionary type. The author expresses the opinion that with the employment of high vacua the limit of power is determined largely by the area obtainable through the last stage, for the final expansion and passage of the steam prior to its entering the condenser.

The significance of this is shown by the fact that whereas a pound of steam when entering the first stage has a volume of less than 2½ cu. ft., when passing through the last stage it has a volume of approximately 395 cu. ft. if expanded to 28.5 in. vacuum, and 585 cu. ft. if expanded to 29 in., the ratio in the latter case being 1 to 234.

Because of this, in any discussion of limits of power it is neces-

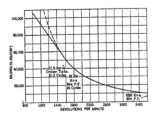

FIG. 13 LIMITING GENERATING CAPACITIES IN LARGE UNITS AT VARIOUS SPEEDS.

sary to assume conditions of pressure and superheat of the steam entering the turbine, the vacuum to which the steam is to be expanded in the blading and the efficiency of change of steam flow per unit of power.

The limiting factors are divided into three classes by the author: (1) Theoretical, including limiting steam velocities and effect on efficiency of velocity remaining in steam after leaving the last stage, and the area through the blades as affected by blade angle; (2) Physical, including methods of construction, materials, stresses, factor and safety against rupture, reliability factor, and limitations of transportation facilities; (3) Economical, including limits beyond which it may be physically possible but economically inadvisable to go, such as effect of size of structure or of character of materials employed on costs and time required to make inspection and repairs. The paper is largely based on the practice of the Westinghouse Company.

As regards theoretical limits with materials of infinite strength and rigidity, it would have been possible to build units of infinite capacity, but for a given diameter and blade height the capacity is limited by chosen maximum values of steam speed through the blades, in order to keep the energy available in the steam discharged to the condenser within permissible limits.

Throughout the entire turbine, with the exception of the last few stages, steam speeds only about 25 per cent in excess of the corresponding blade speeds are employed in order to secure maximum efficiency. In the latter stages, however, the volumes become so great that steam speeds are increased sometimes approximately 100 per cent in excess of the blade speed, in order to effect a compromise between maximum theoretical efficiency and physical dimensions.

The question of blade angle is discussed next.

As regards the physical limits, the chief ones limiting turbine

capacity are the physical characteristics of the material employed and the chosen limits to which these materials may be safely stressed.

Of course, the design of the rotor materially affects the stresses involved, but if the rotor design can be so modified as always to keep the stresses within necessary limits, then the stresses at the base of the limits or in blade fastenings determine the maximum capacity obtainable with a given speed.

There exist two interesting relations between the stress at the base of blades, steam-passage area through the blades, and rotative speed. For any given rotative speed and blade angle the steam capacity or steam area through the blades is directly proportional to the stress at the base of the blades, regardless of the diameter and blade height selected. This stress can only be modified by unevenly varying the cross-sectional area of the blades, such, for example, as thickening the blade near the base. Also for any given stress the area through the blades will vary inversely as the square of the speed, i.e., if at a speed of 1800 r.p.m. a given stress and area are obtained, then at 900 r.p.m. the area will be increased four times if the stress is kept constant.

In other words, as the writer shows in the original article, the area and stress are each equal to a constant times the product of mean diameter and blade height, and when the stress is constant this product varies inversely as the square of the revolutions per minute. The ratio of blade height to rotor diameter is therefore not a factor in determining physical limit of capacity, but only in determining efficiency cost and, to some extent, reliability of the turbine.

Increased capacity without decrease of rotative speed or increase in stresses may be obtained by employing multiple low-pressure stages, and the author gives a curve showing the limits of capacity of steam turbine with double-flow low-pressure stages.

In discussing the economic limits of turbines of large capacity, consideration is given to the fact that as yet such units are not required in sufficient quantity to warrant equipping and operating shops for their exclusive manufacture, and that they must be produced largely by the same processes and equipment as are used for smaller sizes which are built in greater quantities.

As the sizes become larger, a greater proportion of special equipment and processes becomes necessary, resulting in increased rates of cost unless accompanied by very material increase in production of quantity. Under present conditions this economic limit of capacity agrees closely with the physical limit of 1500-r.p.m. units.

Another factor tending to limit capacity of single units is the generating capacity loss resulting from suspension of service for inspection or repairs. For example, if a 30,000-kw. unit must be kept out of service ten days for a certain inspection or repair, a 60,000-kw. unit would have to be kept out probably fourteen days for a similar purpose because of the greater time required to handle the larger structure. Therefore, if two 30,000-kw. units were used and each held out of service ten days, the outage loss would be only five-sevenths as great as if a single 60,000-kw. unit were kept out fourteen days. (*Proceedings of the American Institute of Electrical Engineers*, vol. 38, no. 11, Nov. 1919, pp. 1243-1253, 6 figs., *g*)

## Standardization in Power Piping

POWER PIPING, J. Roy Tanner and Geo. J. Stuart. A paper discussing the subjects of piping systems, their design, selection of materials, and details of construction.

The steam pressures carried have gradually increased until the present extra heavy material used for high-pressure steam is designed for 250 lb. working pressure, and the tendency is still upward. Material designed for 350 lb. and superheat is actually in use but has not yet had wide enough application to be reduced to standard practice.

The steam turbine has changed practice by bringing superheated steam from occasional into extensive use. This made it necessary to find substitutes for cast iron and bronze in fittings.

The former was replaced by cast steel and the latter, to a great extent, by Monel metal.

The importance of standardization as a factor in the industrial position of this country today is nowhere better illustrated than in this industry under discussion. A few years ago each manufacturer had his own standard for the dimensions of flanges, fittings, and drilling templets. This compelled the large users of piping, such as steel mills, to adopt something which was very likely to be original. Also, about the first thing an engineer did after putting up his sign, was to calculate and design his own piping standard. There was a tremendous amount of brain power consumed which can now be diverted to other purposes. The effect of this confusion was that manufacturing in all that the word implies was impossible because changes of flange sizes or center to face had to be made at frequent intervals. This condition prevailed long after The American Society of Mechanical Engineers, in 1894, adopted, after conferring with the Master Steam and Hot-Water Fitters and some of the manufacturers, a standard flange templet which subsequently became the "A. S. M. E. Standard" used for lower pressures, along with the "Manufacturers' Standard" for extra heavy material, in 1901, and it was several years after this before the flange situation became clarified.

Some time after 1910 a group of manufacturers of valves and fittings formed an organization called the Committee of Manufacturers on Standardization of Valves and Fittings, and began the work of designing a completely standardized line of fittings, including center to face, and shell thickness. This work was completed and published and about a year later revised to meet the views of The American Society of Mechanical Engineers regarding the bolting of extra heavy flanges larger than 12 in. This revision was accepted at a conference in Washington by representatives of the United States Government. The American Society of Mechanical Engineers, the Master Steam and Hot-Water Fitters, and the Manufacturers' Committee. The American Society of Mechanical Engineers adopted the report of its Committee—compiled under the direction of the late Mr. H. G. Stott—and thus made effective the American Standard for Pipe Flanges, Fittings and Their Bolting, to become effective January 1, 1914. The report of this Committee was a masterpiece of complete analysis.

The American Standard consisted of specifications for "Standard" material up to 125 lb. pressure and "Extra Heavy" material up to 250 lb. pressure.

The American Society of Mechanical Engineers and the Manufacturers subsequently collaborated in standardizing hydraulic flanges up to 3000 lb. pressure, culminating in the adoption of a report in December, 1918.

Economy demanded the use of a lighter design than that specified for 125 lb., for use on low-pressure water, condenser service, etc., and The American Society of Mechanical Engineers has made a recommendation for lighter shells and smaller bolts than the American 125-lb. standard, and flanges of the same diameter, thickness, and bolt spacing. The Manufacturers' Committee, however, is considering the advisability of using a thinner flange for this service, owing to the foundry difficulties attending such a great change in section, and may request The American Society of Mechanical Engineers to reconsider this point.

Some work had been done by the Manufacturers' Committee to determine the feasibility of standardizing valve dimensions from center to face, but with no results. In the fall of 1918, when the Emergency Fleet Corporation was struggling with the problem of a design for the piping of fabricated ships, the standardizing engineer for the Emergency Fleet Corporation, Mr. John A. Stevens, who played a large part in formulating the Boiler Code of The American Society of Mechanical Engineers, attacked the problem of interchangeable dimensions for the valves to be used by the fleet, calling to Philadelphia engineering representatives of all the large valve manufacturers and, had it not been for the armistice, the fleet would probably by now have had a standard center-to-face dimension for every valve to be used by them on shipboard. It is therefore not impossible that at some future date we may even have interchangeable face-to-face dimensions of valves.

A large amount of standardization work has been done by the Power Piping Society, an organization of fabricators and erectors. The "Standard Specifications for Power Piping" adopted by the society in 1915 was the result of the experiences of a group of men who had installed a considerable part of the power piping in this country, and can therefore be accepted as safe modern practice.

Thus out of chaos has come some degree of order, for the benefit of the use of pipe, valves, and fittings. (*Proceedings of the Engineers' Society of Western Pennsylvania*, vol. 35, no. 7, Oct. 1919, pp. 311-332 [original paper] and 333-347 [discussion], *g*)

COXON AUTOMATIC STOP VALVES. Description of a series of devices of British manufacture designed to automatically shut off steam in the event of a burst in the steam main or a similar ac-

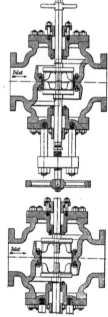

FIGS. 14 AND 15  COXON AUTOMATIC INSULATING VALVE FOR STEAM HEADERS

cident to the apparatus to which it is supplied. The valves described and illustrated comprise a boiler junction valve, a combination safety valve and stop valve and an insulating valve for steam headers. This latter is shown in Figs. 14 and 15. The casing contains two similar valves, closing in opposite directions. The lower one is forced to its seat by means of the hand wheel and screw, and acts as an ordinary stop valve when closed in this way. The valve spindle, however, is separate from the screw, and the valve can thus close independently of the hand wheel. The upper valve has no screw, and the spindle slides freely in its gland. When in use, both valves are normally held open by the steam pressure, but should the main burst on either side, the corresponding valve closes automatically and prevents steam flowing to the burst part. No by-pass is required to open the valve, because when the lower valve is shut it lifts the upper automatic valve slightly from its seat. When steam is admitted to the main, the upper valve also opens as soon as permitted by the operation of the hand wheel.

A feature of the valves described is that they have no springs and are so arranged that they can be readily tested at any time. (*Engineering*, vol. 108, no. 2809, Oct. 31, 1919, pp. 581-582, 9 figs., *d*)

WHAT SHOULD BE THE MAXIMUM LENGTH OF A LONGITUDINAL JOINT, C. E. Stromeyer. Attention is called to this article, the author of which is Chief Engineer of the Manchester Steam Users' Association for the Prevention of Boiler Explosions, and to an editorial on p. 624 of the same issue of *Power.* The article and the editorial are here abstracted together.

The writer of the article draws attention to the differences of engineering views on the two sides of the Atlantic. He claims that in England boilers have been made so safe that fatal boiler explosions rarely reach a dozen per year. He was shocked when on a recent occasion he heard an address by an American engineer who claimed to be qualified to speak about the practical working of boilers because he had been present at a dozen boiler explosions. Such an exciting experience is quite impossible in England.

The writer thinks, however, that there must be other causes than the use of long longitudinal seams to account for these numerous explosions and if these other causes be removed the question of length of joints will sink into insignificance.

The writer employed special strain indicators to measure the strains in the solid plates of boilers on either end of lap joints and also in the solid plates above and below the single-butt-strap joint of the hulls of iron ships, which construction was customary some thirty years ago. His observations showed that the strains in the solid plates close to the ends of lap joints were eight times greater than those in the solid plates away from the joints. Assuming that these joints had only 70 per cent of the strength of the solid plates, it is evident that the plates that adjoin lap joints are stressed about five times more severely than the joints themselves. These observations had been confirmed ·by unpublished experiments made in 1874 by the Manchester Steam Users' Association on a full-sized boiler whose solid plates ruptured under hydraulic pressure when the seams gave way.

The conclusion that is to be drawn from these two sets of experiments is that the breaking of joints is a source of weakness, that the longer we make the longitudinal joint for boilers the better. The author qualifies this statement, however, saying that it is intended to apply only to steels made by the acid open-hearth process. The Manchester Steam Users' Association refuses to sanction the use of basic steel in boilers.

As regards the experience of the Manchester Steam Users' Association, it is stated that since its foundation in 1854 it has issued over half a million certificates of safety and up to the time of the war it had had no accident entailing loss of life.

The editorial commenting on this article, asks:

" Why the Massachusetts Boiler Board considers both lap and strap joints dangerous in lengths exceeding 12 ft., while the Boiler Code Committee of the A. S. M. E. is apprehensive for lap joints only?

" Why the Board of Boiler Rules prohibits these joints on horizontal return-tubular. vertical-tubular, or locomotive-type boilers, and the A. S. M. E. only on horizontal return-tubular boiler?

" If they are dangerous in the shell of a locomotive-type boiler, why they are not equally dangerous in the drums of water-tube boilers? "

The editorial in *Power* also calls attention to the fact that both the Massachusetts Board and the A. S. M. E. Committee agree that the long joint is bad and should be avoided, while Mr. Stromeyer asserts that he is strongly in favor of long seams and prefers them to short ones, even if the plates are rolled parallel to the seams. (*Power*, vol. 50, no. 17, Oct. 21, 1919, published Dec. 1, 1919, pp. 626-627 (original article), and 624 (editorial comment) *gpA*)

EXPLOSION OF LAP-SEAM BOILER. Description of the explosion of a boiler as the result of a lap-seam crack which occurred at the boiler plant of the Sisters of St. Francis of Assisi at St. Francis, Wis. The boiler was 60 in. in diameter, 40 ft. long, and constructed of ⅜-in. plate of flange steel and was built in 1904. There were two sheets of the shell plates, each sheet forming a complete ring with the longitudinal seam in the usual place, above the water line.

The longitudinal seams were of the double-riveted lap-joint type, there being two rows of ⅞-in. rivets driven in 15/16-in. holes, with a rivet spacing of 3¼ in. The spring-loaded safety valve was set to blow at 95 lb. pressure. Information available shows that the boiler was clean and had received good attention in other ways. After the accident the safety valve was found to be perfectly free and unobstructed.

Examination of the exploded boiler developed the fact that the front course of the shell plate had failed from the head seam to the girth seam, a distance·of 6 ft. 9 in. along the rivet holes of the double-riveted lap seam. The crack probably started at a point about 5 ft. from the front head where the plate was not homogeneous and gradually extended to the front-head seam, working deeper into the sheet until separation occurred. (*Power*, vol. 50, no. 17, Oct. 21, 1919 [published Dec. 1], pp. 604-605, 2 figs., *d*)

PLACING A NEW TURBINE IN SERVICE, Roger Taylor, Mem.Am. Soc. M. E. An article of a strictly mathematical nature not suitable for abstracting. Attention is called to it because it presents

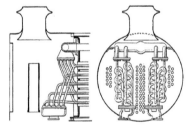

FIG. 16  HORWICH LOCOMOTIVE SUPERHEATER

data as to procedure followed in putting a steam turbine into service not easily found in print elsewhere. Such features as leveling, alignment, checking bucket clearance, adjustment of thrust and step bearings, valve setting and governor adjustments are considered. The article is the first of a series. (*Power*, vol. 50, no. 16, Nov. 22, 1919, pp. 578-582, 9 figs., *p*)

## RAILROAD ENGINEERING

*HORWICH SUPERHEATER FOR LOCOMOTIVES. A number of heavy 0-8-0-type locomotives at present being constructed at the Horwich Works of the Lancashire & Yorkshire Railway are fitted with the new type of superheaters, shown in Fig. 16.

In this superheater there are three separate headers, one at the top and two at the bottom of the smokebox, the latter being in direct communication with the cylinder steam chests. Saturated steam from the boilers enters the top header and branches right and left from a central chamber. Steam passes along small pipes enclosed in large flue pipes in the ordinary way, traversing the length of the boiler toward the firebox, returning toward the smokebox, then again toward the firebox and finally to the lower headers in the smokebox, thence to the smoke chest and cylinders. The large flue tubes are arranged in two vertical rows on each side, and in the particular boiler illustrated there are twenty elements. This arrangement gives a short direct connection between the lower headers and the cylinders for the passage of steam.

Owing to the fact that the saturated- and superheated-steam receptacles are separated from one another and the usual smokebox steam pipes dispensed with, any loss of superheat occurring during the passage of steam from the outlet ends of the superheater headers to the steam chests is reduced to a minimum.

It is also claimed that the parts are readily accessible and cleaning facilitated. (*The Railway Gazette*, vol. 31, no. 18, Oct. 31, 1919, pp. 556-557, 3 figs., d)

The low-pressure end has a header made of 16-in. pipe and carrying a 6-in. dry pipe (Fig. 17), the dry pipe being welded at each end to a plate, which, in its turn, is welded to the header. On the inside opposite the gas-admission side the dry pipe has a slit 2 in. wide for its entire length through which the gas passes into the dry pipe and thence to the suction pipe. To the bottom of the dry pipe is riveted a curved baffle plate which changes the direction of the gas flow from the coolers so as to throw out some of the entrained liquid. Both dry pipe and header are given a slight pitch toward the dead end, where liquid may drain into the header from the dry pipe and from the header into the coolers. The circular plate in the outlet end of the header has an opening of segment shape into the header, the opening being immediately above that into which the dry pipe is welded and so situated as to pass gas collecting at the top of the header to the suction pipe.

The conditions created by this and other means are such that

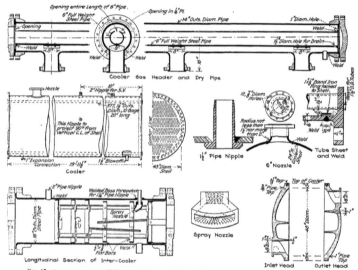

FIG. 17  COMPRESSOR OF THE T-WHARF PLANT OF THE QUINCY MARKET COLD STORAGE COMPANY, BOSTON, MASS.

## REFRIGERATING ENGINEERING

### Improved Compressor at Quincy Market, Boston

QUINCY MARKET COLD-STORAGE PLANT AT BOSTON, Chas. H. Bromley. Description of the so-called T-wharf plant of the Quincy Market Cold Storage and Warehouse Company. The station is of 500 tons rated capacity in one compound high-speed feather-valve ammonia compressor driven by a 600-hp. synchronous motor. The most remarkable part of it is this compressor.

Essentially it is of the standard Laidlaw type, but has a special arrangement of gas connections and gas intercooler.

The compressor has a 24-in. stroke, a low-pressure cylinder of 27-in. bore and a high-pressure cylinder of 15-in. bore. The low-pressure cylinder has 12 valves each with a free opening area of 9 sq. in., and the high-pressure cylinder six valves of the same area.

liability to slop over is remote under normal conditions, and, in any event, a slopover comes gradually and not in one smashing flow. There is a condition which tends to produce a slopover when the rate of feeding liquid to the brine coolers is suddenly and markedly reduced. The expansion-line nozzle inside the cooler projects the liquid upward at an angle on a line with the top far end of the cooler. Feeding may be so rapid that the liquid in the cooler will pile up at this end and thus leave more disengaging volume between liquid level and cooler top at the gas outlet end than existed before increasing the rate of liquid flow.

Another interesting feature of the compressor is the intercooler. The cooling medium is liquid ammonia taken direct from the liquid receiver. The principle of the intercooler is identical with that of the open feedwater heater, and the intercooler is provided with trays and baffles just as is the open heater. Liquid direct from the main liquid receiver is run through the intercooler, and what is not evaporated goes to the brine coolers. The gain is, of course, twofold: First, the gas is cooled, reducing the volume to be handled by the high-pressure cylinder; second, liquid in cooling by its own expansion uses up, say, 15 per cent of its cooling effect in cooling itself. Compared with its condition in the liquid receiver, this liquid passed through the intercooler has already lost this in a most useful way—the liquid as well as the gas has benefited by the heat exchange in the intercooler. As the intercooler pressure is 60 lb. and that in the brine coolers 5 lb., the liquid is trapped just before it reaches the coolers. It is convenient to use the last or No. 8 condenser as an intercooler, taking the gas before it goes to the liquid-gas intercooler.

The designer has cleverly obviated the use of a concrete crib at the harbor end of the suction pipe. This suction is 10 in. in diameter and at the harbor connects with a 10 x 8 x 8-in. Y. At each 8-in. outlet of the Y is a gate valve, a short nipple on the end of which is fitted with 12 ft. of 8-in. flexible rubber hose. On the end of the hose is a strainer made of 12-in. pipe, slotted to give capacity considerably greater than that of the 10-in. suction.

Several other interesting features at the new plant are discussed and illustrated in the original article. (*Power*, vol. 50, no. 18, Oct. 28-Nov. 4, 1919, pp. 638-644, 9 figs., *dA*)

## STANDARDS

THE MASS OF THE LITER OF AIR AND GAS MIXTURES. Critical abstract of a memoir on The Mass of the Liter of Air, published by Prof. A. Leduc in vol. 16 of the Travaux et Mémoires du Bureau International des Poids et Mesures.

Objections are made to the use of air as a standard of measurement. When this is done the composition and density of the air are tacitly assumed to be constant, the data having been reduced to normal temperature and atmospheric pressure. But the air density is not constant and the buoyancy is sufficiently affected by a barometric change of a few millimeters to upset balance established even between lumps of quartz and of platinum, each weighing one kilogram.

While the air density does not fluctuate widely, under ordinary conditions, such deviation as takes place is well within the limits of precise measurement.

The air measurements are also affected by the composition of the air. Thus, oxygen, being heavier than nitrogen, becomes concentrated in the lower atmosphere, this tendency, however, being counteracted by winds and convection currents. Further, air in laboratories may be seriously contaminated by water vapor and carbon dioxide, which are removed before analyzing the air but usually disregarded in density determinations.

There is another objection to the use of a mixed gas like air as a standard.

Dalton's law was formulated in ignorance of the fact that nearly all gases are more compressible than they should be according to the law of Boyle and Mariotte. Leduc's own experiments would lead to the following conclusions:

If we have equal volumes of two gases, $A$ and $B$, at equal pressure $P$ and allow the gases to mix, $A$ will not, in the doubled volume, be at the pressure $P/2$, but at the pressure $P/2 + \epsilon$, because the gas is more compressible than it should be; similarly $B$ will be at a pressure $P/2 + \epsilon_1$, and the pressure of the gas mixture will be $P + \epsilon + \epsilon_1$; but this ($\epsilon + \epsilon_1$) is not a constant and depends upon the nature of the gases. In one series of experiments to demonstrate this Leduc and Sacerdote connected two flasks of equal volumes, charged with various gases, by a three-way cock, and hastened the mixing of the two gases by cooling and heating the one or the other vessel and by shaking. Diffusion was very slow; sometimes it took days and weeks to obtain a mixture of constant composition (as shown by analysis) and pressure. In the case of two substances like $CO_2$ and $N_2O$ (which physically differ little) there was no distinct increase in pressure; in the case of two gases like $H_2$ and $SO_2$, physically dissimilar, there was a pressure rise of 3 mm. It is clear, therefore, that the nature of the gases has to be considered. What is less clear —from the brief account of the memoir, at any rate—is whether all possible chemical and other disturbing effects can be allowed

for in such experiments, and whether Leduc's version of Dalton's law meets the required conditions, namely, that the volume occupied by a mixture of gases is equal to the sum of the volumes occupied by each of the constituent gases at the same temperature and at the pressure of the mixture. As regards the atmospheric mixture of nitrogen and oxygen, the $\epsilon$ may practically be disregarded, Leduc finds; but the fact could not be presumed.

In the experiments made to determine the mass of the liter of air it was found that the rare gases, excepting argon, and hydrogen would only affect the 5th decimal of the air density, which is uncertain anyhow and may hence be neglected.

Samples of air taken in different parts of France and in Africa at various altitudes show that the oxygen proportion by weight varied between 0,2305 and 0,2325 and further the oxygen proportion was lower than the average in the north of France at sea level, which, it is suggested might account for the slightly lower oxygen values of Rayleigh.

The final result is that the liter of air at 0 deg. cent. and 760 mm. barometer weighs 1.2928 grams when $g = 980.665$. Under one megabar, taking the density of mercury as 13.5951 and $g$ as 980.97, the mass of the liter of air would be 1.2759 grams. (*Engineering*, vol. 108, no. 2809, Oct. 31, 1919, p. 569, *tA*).

## STEAM ENGINEERING (See also Power Plants and Railroad Engineering)

HEAT TRANSMISSION IN BOILER FLUES, Michael Longridge. In carrying out tests on a new type of boiler some years ago the author tried to determine whether heat transmission in boiler flues is proportional to the product of the density and speed of gas when the water circulation is well maintained. While the tests were not exhaustive enough to answer this question in a positive manner, enough data were secured to create a doubt as to its truth.

If the transmission per square foot of air per degree temperature difference were strictly proportional to the product of the density and speed of the gas the values of $H/\Delta \rho u$ (where $H$ is the average B.t.u. transmitted per sq. ft. of air; $\Delta$ mean temperature difference between gas and steam or water; $\rho$ mean density and $u$ mean velocity) would be constant in each case though the constant values might be different for the plug flue for the evaporator and for the economizer.

Actual test data show that in the case of the economizer the values are approximately constant and therefore approximately in accordance with the law $H/\Delta \rho u =$ constant.

In the case of the evaporator the variation is greater, and for this there are two reasons: First, greater uncertainty as to temperatures and radiation losses; second, the fact that owing to some of the tubes being bent the gas current was not evenly distributed over the cross-section of the flue.

In the case of the plug flue the variation is considerable, but here it must be remembered that the plug was quite hot at one end and more than red hot at the other end, and therefore the radiation from it was very great. As a result of this the heat absorbed by the flue plate was much greater than if the plate had received heat from contact with the hot gases only.

It appears, therefore, that the general law, namely, that heat transmission in boiler flues is proportional to the product of the density and speed of gas when the water circulation is well maintained, is well supported. (Abstract from an unpublished copy of a report made by the author who was Chief Engineer of the British Engine Boiler and Electrical Insurance Company, Ltd., of Manchester, England, on December 31, 1909. While not of recent date the report is of interest and this is its first publication, a copy having been secured through the courtesy of Daniel Adamson, Mem. Am. Soc. M. E.; *g*)

## CLASSIFICATION OF ARTICLES

Articles appearing in the Survey are classified as *c* comparative; *d* descriptive; *e* experimental; *g* general; *h* historical; *m* mathematical; *p* practical; *s* statistical; *t* theoretical. Articles of especial merit are rated *A* by the reviewer. Opinions expressed are those of the reviewer, not of the Society.

# MECHANICAL ENGINEERING

### THE JOURNAL OF THE AMERICAN SOCIETY OF MECHANICAL ENGINEERS

Published Monthly by the Society at
29 West Thirty-ninth Street, New York

FRED J. MILLER, *President*

WILLIAM H. WILEY, *Treasurer* CALVIN W. RICE, *Secretary*

PUBLICATION COMMITTEE:

GEORGE A. ORROK, *Chairman*         J. W. ROE
H. H. ESSELSTYN      GEORGE J. FORAN
RALPH E. FLANDERS

PUBLICATION STAFF:

LESTER G. FRENCH, *Editor and Manager*
LEON CAMMEN, ⎱ *Associate Editors*
C. M. SAMES, ⎰

W. E. BULLOCK, *Assistant Secretary*
*Editor Society Affairs*
FREDERICK LASK, *Advertising Manager*
WALTER B. SNOW, *Circulation Manager*
136 Federal St., Boston

Yearly subscription $4.00, single copies 40 cents. Postage
to Canada, 50 cents additional; to foreign countries $1.00 addi-
tional.
*Contributions of interest to the profession are solicited. Com-
munications should be addressed to the Editor.*

## Our New President and New Past-President

Major Fred J. Miller, the new President of The American
Society of Mechanical Engineers, brings to this important position
the fruits of many years of wide experience in large affairs. In
the introduction of the President-elect on the evening of the President's Reception at the Annual Meeting, it was stated that " for
20 years he was a journalist, since which time he has been engaged
in the business of factory management, an expert on factory
management. He served for five years on the Council of this
Society, and for four years as Trustee of the United Engineering
Society. His work as Trustee of the United Engineering Society
had to do with the reorganization of its finances and other features,
but principally with the work of reorganization. During the past
year he has had made a survey of the machinery—the office
machinery—of this Society, and he has advised with regard to
the development of MECHANICAL ENGINEERING, the Journal of
the Society. He had also rendered very distinguished service in
the cause of his country during the war."

The position of President of the A.S.M.E. has come to be in
reality a position demanding a large amount of time and personal
attention, not only to Society affairs almost overwhelming in
amount, but to public matters of broad scope in which the interests of engineers are involved. The Society is to be congratulated
on having for its leader during the coming year one whose experience and intimate acquaintance with the Society so well
qualify him to conduct affairs which need the sound judgment of
an able executive and officer. An account of Major Miller's life
will be found in Section Two of this number.

For the work of Dean M. E. Cooley, the retiring President,
the utmost gratitude is felt by all the members of the Society
and all who have been acquainted with his constructive genius.
Making the trip from Ann Arbor, Mich., to Society headquarters
each month, and sometimes oftener, and giving practically half
his time to his presidential duties, he has displayed a devotion to
the cause of the profession for which the Council and his fellow-
members in the Society have the fullest appreciation. His grasp
of Society activities is shown in his presidential address printed
elsewhere in this number. The manner in which he has handled
the many problems that have arisen, with combined forcefulness
and geniality, has been most happy and effective and has won the
admiration of his associates and his many friends.

## Engineering Index Issued in a New Form

The Engineering Index has now been conducted by The
American Society of Mechanical Engineers for a period of one
year, and as a result of this experience a change has been made
in the arrangement of the items comprising the Index, beginning
with this number of MECHANICAL ENGINEERING. Heretofore the
classified system has been employed whereby the items were
grouped under several broad headings, such as Mechanical,
Electrical, Civil and Mining Engineering, Marine and Railroad
Engineering, Organization and Management, General Science,
etc.

During the year suggestions have been solicited from the
membership for the improvement of the Index, with the result
that practically all of the criticisms or suggestions have pointed
in the direction of adopting a simple dictionary arrangement of
the items rather than the classified arrangement. This led the
Publications and Papers Committee of the Society to make an
investigation and secure the opinions of engineers, librarians,
and others who have given attention to the matter, which have
confirmed the opinions previously received to the effect that
the alphabetical arrangement, all things considered, is to be
preferred.

To turn from a classified system to an alphabetical one presents many problems, to solve which requires time and effort.
Undoubtedly many suggestions can be made for the further
improvement of the Index as it appears in the current issue of
this Journal, and it is hoped that those who consult the Index
will feel free to communicate with the editor with regard to any
deficiencies and give, if possible, constructive suggestions for
future improvement.

## Engineering Societies Employment Bureau Completes a Successful Year

During the past year the four Founder Societies have jointly
maintained in the Engineering Societies Building an Employment Bureau, open to all engineers, and especially extending its
aid to discharged engineer officers and men. During the year
the Bureau has interviewed over 20,000 men and has handled
nearly the same number of applications. At present there are
approximately 5000 men registered, many of whom, however,
are already employed and have entered their names in order to
advance themselves should opportunity offer. About 2500 positions are now available. Here is an opportunity for effective
coöperation. Business men and engineers who are seeking new
blood for their organizations can without question profit by communicating with the manager of the Bureau, Mr. W. V. Brown,
who should be addressed at 29 West 39th Street, New York, N. Y.
It should not be inferred that because of the smaller number
of positions available men are not in demand. On the contrary,
good men are constantly in demand and many of the positions
now available call for engineers of particular qualifications.
Those seeking engineering positions will therefore find it advantageous to enlist the aid of the Bureau.

## National Public Works Department Association Holds Conference in Washington

The National Public Works Department Association, which
grew out of a conference called by Engineering Council, and
held last April in Chicago, will meet at the Willard Hotel in
Washington, D. C., on January 13 and 14. The purpose of the
conference is to discuss the Jones-Reavis bill, which is now
being considered by Congress, and which proposes far-reaching
changes in the organization of the Federal Government. This
bill, to which reference has frequently been made in these

columns, proposes to create a Department of Public Works to replace the Department of the Interior. It is the primary object of the bill to assemble under one head all of the engineering activities of the Government and to place under the jurisdiction of existing appropriate departments such activities as are of a non-engineering character. The various details of the bill will be found in the August issue of MECHANICAL ENGINEERING (page 709).

The Tuesday session of the conference will be given over to a discussion of various committee reports and of the Jones-Reavis bill itself. The practical results achieved by the public works departments of several of the states will also be discussed. It is the hope of the National Public Works Department Association that every engineering organization in the country be represented in this conference, for the bill is one not alone of interest to engineers, but to the country at large.

## Annual Volume of Engineering Index to Be Alphabetically Arranged

The Engineering Index, which has been a monthly feature of MECHANICAL ENGINEERING during the past year, will later be issued in the form of an annual volume, according to the practice followed by the former publishers of the Index for many years past.

To meet the demand for an alphabetically arranged index, however, which a consensus of opinion clearly indicates is the most convenient form for reference purposes, the items have been entirely regrouped according to this plan, instead of following the classified arrangement of the Index as it has appeared in MECHANICAL ENGINEERING and in the previous annual volumes. By introducing many cross-references the engineer will be able to find under one heading all the references to any one subject.

For instance, under the new plan references to screw-thread gages will be found under the heading GAGES, Screw-Thread, instead of MECHANICAL ENGINEERING, GAGES, Screw-Thread. In other words, one will turn directly to the reference rather than attempt first to classify it as Electrical Engineering, Civil Engineering, Mining Engineering, Mechanical Engineering, etc.

The work of regrouping the 12,000 items which form the 1919 Index has been done entirely by engineers, men familiar with the terminology of the profession and thus eminently qualified to compile a work of this character. No expense has been spared to make the 1919 Index an accurate and reliable compilation of intrinsic merit. It will be a book for the desk of the individual engineer as well as for the reference library—one which every practicing engineer should own.

## Conference on National Safety Codes

Within sixty days a definite start will be made in the development of uniform national safety standards, beginning with those subjects on which rules are now being formulated or revised by state or other authorities. This seems certain as a result of the Conference called by the Bureau of Standards and held at Washington, D. C., December 8 and 9.

This meeting was attended by about 100 representatives of the trade associations, engineering societies, state industrial commissions and labor departments, government officials, insurance companies and bureaus, and large employers of labor.

The program included an exposition of the large number and variety of conflicting safety codes in existence which have been formulated by various state, insurance, and other organizations, and an outline of the efforts which have been made to secure uniformity. It was evidently the unanimous feeling of those present that such uniformity had become not only extremely desirable but almost imperative, in view of the embarrassment and expense to which employers and others are subjected when different standards are enforced by different authorities, and when these standards are changed at, in some cases, frequent intervals. The Conference voted unanimously to approve the plan of formulating safety standards under the general auspices of the American Engineering Standards Committee.

The vote of the Conference on December 8, already referred to, means that the formulation of uniform safety codes will be undertaken according to the regular procedure of the American Engineering Standards Committee as just described.

The Conference on December 8 further voted that to expedite early action, a general advisory committee should be formed to include representatives of all national associations, state commissions, and others legitimately interested, to survey the whole field of safety standards and recommend what standards should be undertaken first and what organizations should sponsor them. The conference recommended that such representative committee be organized by the National Safety Council, the Bureau of Standards, and the International Association of Industrial Accident Boards and Commissions, acting jointly. At an informal meeting of representatives of these three organizations, on December 9, plans were made for organizing the general advisory committee at once, and probably arranging for a meeting in January. Before this time, up-to-date information will be gathered as to what codes are now being written or revised in the various states and the general advisory committee will probably recommend that these subjects be given first attention. This committee will report to the American Engineering Standards Committee not later than February 1, and the definite assignment of sponsorships to the National Safety Council, the Bureau of Standards, and others will doubtless follow. The National Safety Council's participation in this work will be largely in the hands of the Engineering Section, in coöperation with the industrial sections especially interested.

## The Value of Sheet Asbestos on Hot Pipes

The following notes on recent research work carried on at the research laboratories of the Westinghouse Electric and Manufacturing Company, East Pittsburgh, Pa., have been contributed by T. S. Taylor:

While conducting some simple tests on the relative thermal resistance of layers of asbestos paper when applied loosely and tightly, respectively, some rather interesting results were obtained. In the preliminary tests, it was observed that a bare tin vessel containing hot water would cool less rapidly than the same vessel when wrapped tightly with one layer of 0.013 in. sheet asbestos. This observation, which is just the opposite to the usual ideas held by most people, was carefully checked by fixing up identical tin vessels, one being left bare and the other covered tightly with sheet asbestos, and then obtaining the cooling curves for each when placed under similar conditions. The uncovered one cooled much more slowly than the one covered with asbestos. However, to obtain results from conditions more nearly like those met in practice, heaters were made so as to slip into vessels 4 in. in diameter and 20 in. long and the temperature excess of the air at the center of the vessel above room temperature measured, for various inputs and surface conditions.

From these tests, it has been found that an asbestos-covered hot-air pipe loses 33 per cent more heat from its surface than a bare one. To reduce the heat loss of the covered pipe so as to be equal to that of the bare pipe would require a coating of asbestos, as usually applied, at least ¹⁄₁₆ in. thick. To make the heat loss of the asbestos covered pipe ¾ that of the bare one, would require a layer of asbestos about 1/9 in. to ⅛ in. thick. However, if the asbestos is put on loosely, three layers of 0.013 in. will reduce the heat losses to practically that of a bare pipe. Even with a dust-covered surface, such as usually exists on furnace pipes, the asbestos-covered pipe will lose at least 20 per cent more heat than the bare one.

The results show that it is not economical to cover hot-air pipes with the usual thin layer of asbestos, not only on account of the initial cost of labor and material but on account of heat losses later. Calculations based upon the data obtained indicate the possibility of reducing the heat losses so that there will be a saving of one ton of coal in sixteen. A more complete account of the experiments will appear shortly.

## Autogenous Welding of Boilers and Pressure Vessels

An important announcement has been received from the Steam Boiler and Fly-Wheel Service Bureau relative to the position it has taken in regard to autogenous welding in steam-boiler repair practice, at the suggestion of the Engineers' Committee of the Bureau. This Committee, which consists of representatives of all the insurance companies writing boiler insurance in the United States, has, in a report recently adopted unanimously, interpreted the application of Par. 186 of the A.S.M.E. Boiler Code on welding to boilers and other pressure vessels, for both new construction and repairs, in order to insure uniform practice in the acceptance of autogenous welding. Inasmuch as this report sets forth the position of the insurance companies with respect to the insurability of steam boilers repaired by autogenous welding, it is here reproduced with the permission of the Steam Boiler and Fly-Wheel Service Bureau.

By "autogenous welding" is meant any form of welding by fusion, that is, where the metal of the parts to be joined or added metal used for the purpose is melted and flowed together to form the weld. Such welding is accomplished by the oxy-acetylene, hydrogen or other flame processes, or by the electric arc; no distinction is made between any of these processes. The general rule to govern the acceptance of such welds in insured vessels is that prescribed by the Boiler Code of The American Society of Mechanical Engineers, Par. 186, as follows: "Autogenous welding may be used in boilers in cases where the strain is carried by other construction which conforms to the requirements of the Code, and where the safety of the structure is not dependent upon the strength of the welds."

The following illustrations will serve to point out where such work should be accepted or rejected.

1   Any autogenous weld of reasonable length will be permitted in a st: ybolted surface or one adequately stayed by other means so that, should the weld fail, the parts would be held together by the stays. It is necessary for the inspector to use judgment in interpreting the meaning of "reasonable length" as given above, since it may vary in different cases. In the average case it should be not more than 3 ft. Autogenous welding will not be accepted in unsupported surface.

2   The edges of the inner and outer sheets of vertical firebox boilers, or boilers of the locomotive type, may be joined by autogenous welding to form the door openings, if the surrounding surfaces are thoroughly stayed. This would also apply to other openings of a similar character in such surfaces.

3.   For low-pressure plate-steel boilers operated at a pressure not exceeding 15 lb. to the sq. in., or for higher pressures in unfired vessels subjected to water pressure only, rectangular headers may be autogenously welded at the edges if the sheets are properly held together by stays. Autogenous welding of cracks and fractures in cast-iron boilers will not be permitted.

4   Fire cracks in girth seams extending from the edge of the plate to the rivet hole may be autogenously welded provided the cracks are properly prepared by cutting out the metal at the crack in the form of a letter "V" to permit fusion through the entire thickness of the plates. Similar cracks in girth seams located between the rivet holes may also be autogenously welded, provided the cracks do not extend beyond the edge of the lap of the inner plate. In the latter class of cracks it is advisable to drill a hole not exceeding ¾ in. in diameter at the end of the crack before the weld is made. Cracks extending from rivet hole to rivet hole on girth seams cannot be welded. Calking edges of girth seams may be built up by autogenous welding between rivet holes and calking edge where the original section of the metal to be built up averages to be equivalent to one-fourth of the diameter of the rivet hole; and the portion of calking edge to be replaced does not exceed 30 in. in length in a girthwise direction. In all repairs to girth seams by autogenous welding the rivets must be removed over the portions to be welded and for a distance of at least 6 in. at each end beyond such portions. After repairs are made the rivet holes should be reamed before the rivets are redriven.

5   Stayed sheets which have corroded to a depth of not more than 40 per cent of their original thickness, may be reinforced or built up by autogenous welding. In such cases the stays shall come completely through the reinforcing metal so as to be plainly visible to the inspector.

6   Where tubes enter flat surfaces and the tube sheets have been corroded or where cracks exist between the tube ligaments, autogenous welding may be used to reinforce or repair such defects. The ends of such tubes may be autogenously welded to the tube sheets. The above-mentioned repairs for tube sheets and the welding in of tubes in the sheets, are not to be permitted where such sheets form the shell of a drum or boiler, as in the case of the Stirling type of boiler.

7   When external corrosion has reduced the thickness of plate around handholes to not more than 50 per cent of the original thick-

ness and for a distance not exceeding 2 in. from the edge of the hole, the plate may be built up by autogenous welding.

8   Pipe lines will be accepted where the flanges or other connections have been welded autogenously, provided the work has been performed by a reputable manufacturer and the parts properly annealed before being placed in position. Such welding when made with the part in place and unannealed will not be acceptable.

9   Autogenous welded-in patches in the shell of a boiler will not be acceptable regardless of the size of such patches. Autogenous welding of cracks in the shell of a boiler (except those specified in Par. 4) regardless of the direction in which they may lie will not be permitted, unless such welding is only for the purpose of securing tightness and the stresses on the parts are fully cared for by properly riveted-on patches or straps placed over the weld. The plates at the ends of joints may be welded together for tightness, provided the straps or other construction are ample to care for the stresses on the parts so welded.

10   Refinding or piecing of tubes for either fire-tube or water-tube boilers by the autogenous process will not be permitted.

## Electric Power for Holyoke, Mass.

Within recent months the Turners Falls, Mass., Company, which is the well-known hydroelectric development company on the Connecticut River and its tributaries, made a proposition to the city of Holyoke, Mass., for supplying the electrical energy to the present Holyoke municipal power-generating plant, the distribution of this power to be over the existing wires and equipment of the Holyoke plant and the latter to discontinue the generation of power. The consumption amounts to 60,000 kw. daily.

The limit of capacity of the Holyoke generating plant has nearly been reached, sufficient condensing water is not available and a very large new investment would be necessary in a few years if it should be decided to continue the manufacture of electrical energy. These considerations have aroused a great deal of interest among engineers in the Connecticut Valley. The newly formed Engineering Society of Western Massachusetts has taken an active part in the consideration of the matter and has participated in the public meetings which have been called by the city of Holyoke to discuss this subject. The result has been that S. M. Green, Mem. Am. Soc. M. E., consulting engineer of Springfield, Mass., was retained by the city of Holyoke to investigate the project and to render an impartial verdict.

John J. Kirkpatrick, manager of the Holyoke Gas & Electric Department and an active member of the Engineering Society of Western Massachusetts, has taken a leading part in the discussion of the problem and at the November meeting of the Engineering Society of Western Massachusetts in Holyoke, November 10, at the invitation of Charles L. Newcomb, the presiding officer, spoke upon the subject of municipal ownership, which had an interesting although not manifestly direct bearing upon this topic. Mr. Kirkpatrick said that while Holyoke was not imbued with socialistic tendencies, nevertheless it had been concerned in three very successful enterprises.

When but a town of 10,700 inhabitants, it spent $375,000 in the construction of a railroad between Holyoke and Westfield to connect with the New Haven road and the Boston & Albany road.

Again, in 1872 the town voted to establish its own water-supply system and voted $250,000 for this purpose.

In 1896 the Board of Aldermen passed the first vote favoring municipal ownership of the gas and electric plants. The law provides that a city shall not acquire a lighting plant until it has been so authorized by a two-thirds vote of its government passed in each of two consecutive years and thereafter ratified by a majority of the voters at an annual or special city election. In December, 1897, this ratification was secured. For the gas plant the city paid $432,295 and for the electric stations $377,252, both of which items included certain bonuses for mill powers. Subsequent development of the plants brought their total value up to $2,000,000.

Mr. Kirkpatrick said that the prices paid for street lights in Holyoke are the lowest in Massachusetts. If the city had paid for its lights during the past 16 years on the basis of the lowest price charged in any city or town in the state for metered electricity in buildings and for its street lights, it would have paid

$453,745 more than they actually cost the city. The property of the Gas and Electric Department is not taxable, but if taxes had been paid during these 16 years they would have amounted to only $358,134. While water power is used to a small extent, it is not an important element in the successful conduct of the plant. "The Gas and Electric Department, as it is known locally," he said, "is one of the most successful instances of municipal ownership in the country."

## Memorial to the Late Frederic R. Hutton

During the Annual Meeting of the A.S.M.E., as recorded in the account of the Business Session in this number, a tablet was unveiled of the late Frederic R. Hutton, for twenty-three years Secretary of the Society. This tablet, an engraving of which is reproduced herewith, was executed by Victor Brenner and now hangs in the reception hall of the Society headquarters. The tablet was presented with remarks by Major William H. Wiley, Treasurer of the Society, a close friend of Professor Hutton, and Acting Chairman of the Committee appointed for the preparation of this memorial. Major Wiley spoke as follows:

MEMORIAL PLAQUE TO THE LATE FREDERIC R. HUTTON

GENTLEMEN OF THE AMERICAN SOCIETY OF MECHANICAL ENGINEERS: In presenting this tablet, as Acting Chairman of the Memorial Committee appointed for that purpose, I was firmly of the belief that no words of mine would be fitting on such an occasion in speaking of a man so well known to the entire Society and whose work, as its Secretary and President, was a large part of the history of this organization, but the rest of the Committee unanimously differed from my judgment, hence I will yield to them and make a few brief remarks in tendering this memorial.

When Dr. Hutton became Secretary of the Society in 1883 the membership was 554. I was elected Treasurer in 1884 and the amount of money passing through my hands that year was, in round numbers, $6,500. With his usual energy shown in every enterprise with which he was connected, Dr. Hutton labored in season and out of season to increase the membership of the Society by every means in his power, and used wonderful tact in all such action. Naturally a wonderful success resulted from his efforts.

The Society showed its appreciation of him and of his work by electing him to the presidency in 1908, at which time the membership was 2,929 and the amount of money passing through my hands, as Treasurer, was nearly $54,000. I think I shall not say too much when I assert that a great part of the increase in membership was due to his untiring efforts. At the present time, the year ending September 30, 1919, the membership has increased to 11,082 and the

money passing through my hands, as Treasurer, $375,430. I mention these figures to show that the growth fostered by Dr. Hutton was maintained by his successor.

My associations with Dr. Hutton, as Treasurer, were naturally intimate, and most pleasant. His energy is known to all of us, and as his neighbor in the Catskill Mountains for over twenty years, I can say that he threw his efforts in that section to building up and increasing the activities of the Union Chapel with which he was connected. He was busy at all times, and devoted a great deal of valuable time to this work, and, like his actions for the interests of the Society of Mechanical Engineers, his efforts in the Mountains were marked by signal success. I could dwell at great length on the work of Dr. Hutton, both professionally and socially. He has been sadly missed in this Society as well as in Twilight Park, his summer home. Notwithstanding he went to the Mountains for a well-deserved rest, yet when an appeal for any charitable object reached him, he did not hesitate to volunteer to aid as a reader or a speaker or a contributor. I am just glancing at these matters in passing to pay a deserved tribute to one who did so much for others, but the speakers who are to follow me will go more into the details of his work.

I cannot close more fittingly than to apply to our late Secretary and President, the words inscribed on the tomb of a celebrated architect who was buried near one of the finest structures he had ever erected—"If you wish to see his work, look around you."

## Engineers Honor Memory of John Ericsson

Tribute to the memory of John Ericsson, the inventor of the *Monitor*, and to Cornelius H. DeLamater, the founder of the DeLamater Iron Works, where much of Captain Ericsson's work was done, was paid by fourteen professional organizations at a meeting held on the evening of December 3. The exercises, which took place in the auditorium of the Engineering Societies Building, New York, were in commemoration of the eightieth anniversary of the meeting of Mr. DeLamater and Captain Ericsson and of the thirtieth anniversary of their death, for both men after thirty years of devoted patriotic service died within the same year.

Addresses were made by Mr. H. F. J. Porter, the Hon. Charles Vezin, Jr., a grandson of Cornelius DeLamater; the Hon. W. A. F. Ekengren, the Swedish Minister to the United States, and Rear-Admiral Bradley A. Fiske, U. S. N. (retired). A feature of the meeting was the singing of Mme. Marie Sundelius of the Metropolitan Opera and of Samuel Ljungqvist, formerly of the Royal Opera, Sweden. The Swedish Singing Society Lyran of New York also rendered several selections.

H. F. J. Porter, the first speaker of the evening, presented an illustrated lecture on The Phœnix Foundry and the DeLamater Iron Works. He outlined the careers of the two men who for over half a century were associated in work of lasting value to the engineering profession and to the country in which they lived. Under the direction of Captain Ericsson, the first river, lake, and ocean vessels driven by propellers were launched near the Phœnix Foundry, at that time located at 260 West Street, New York City. Captain Ericsson was also instrumental, during his years of association with Mr. DeLamater, in bringing about many important developments in naval architecture and marine and commercial engineering. At the DeLamater Iron Works there was also initiated under his direction such naval equipment as self-propelled torpedoes, torpedo boats, torpedo-boat destroyers and submarines. Here, too, the engines of the *Monitor* were constructed, as well as the first rock drills, the first air compressors and the first ice machines.

The Phœnix Foundry has long since disappeared; the DeLamater Iron Works has given place to a Cunard line pier, and Captain Ericsson's house was recently torn down to make room for a garage. But these historic sites will at least be marked for future generations, for the organizations under whose direction the memorial meeting was held will erect in the near future suitable tablets at all three places.

Following Mr. Porter, the Hon. Charles Vezin, Jr., spoke briefly of the life and work of his illustrious grandfather, Cornelius DeLamater. The Hon W. A. F. Ekengren, Swedish minister to the United States, paid tribute to both men, and in reviewing the life and work of Captain Ericsson, said:

It is unnecessary for me to enter upon John Ericsson's career as

MECHANICAL ENGINEERING

an inventor and upon the services he rendered the United States and humanity. They are too well known and too well appreciated to be forgotten. But there is one feature in his noble character which I should like to mention. While John Ericsson was always faithful to the country which he had chosen as his and did not hesitate to offer her his services and his life, he loved and honored his mother country, Sweden, and never failed to render it a service when he was able to do so. His *Monitor* which he invented long before it came to practical test, was intended for the coast defense of Sweden and when the first ship of this type was built for the Swedish Navy he presented the Swedish Government armament for it. John Ericsson has spread fame and glory over the Swedish name. In that respect as in so many others he ought to be an example worthy of aspiration for subsequent generations of Swedes in this country.

The last speaker of the evening was Rear-Admiral Bradley A. Fiske, U. S. N., retired, who characterized John Ericsson as a genius and declared that his work the cause of civilization had been advanced. He said in part:

A more sudden and complete reversal of the conditions of a situation, cannot be found in history, or any one event more fraught with important consequences. The consequences that were to follow were perceived in their main features at the time, but their full value does not seem to have been realized. It was realized that the North could establish an effective blockade, but it does not seem to have been realized that this fact made it absolutely impossible for the South to win the war. Mahan was then an obscure young officer, and had not written his epoch-making book, The Influence of Sea Power on History: and the attention of the North was so wholly fixed on the movements of the armies, that no one seems to have apprehended the fact which is now so clear, that the weak point in the Confederate situation was their line of communication across the Atlantic Ocean, and that the victory of the *Monitor* made possible and even easy, the breaking of that line.

For this reason. sufficient advantage was not taken of the power the *Monitor* gave to the Union side, Ericsson's advice was largely disregarded. too few monitors were built, many faulty imitations were constructed that failed disgracefully: and almost the whole energy of the Northerners was devoted to widely scattered and relatively feeble military operations by soldiers whose only weapons were little muskets that they carried in their arms.

Even now. so imperfect a contrivance is the mind of man, the importance of the *Monitor's* victory is not realized, and it is deemed by many as of less importance than some of the land battles in which men were killed and wounded by the tens of thousands, with no special result beyond the killing and the wounding. Even great historians often fail to realize that the thing which makes a battle important is not the number of men engaged or killed or wounded, but the effect of the battle in winning or losing the war. Many of the most important battles have been comparatively bloodless, and many of the most bloody battles have been wholly indecisive.

But while the most important result of the *Monitor's* victory was that it assured ultimate victory in the war to the Union side, it assured another important result, the increased power of navies and therefore the increased security of civilized and wealthy countries as against countries less civilized and wealthy. In other words. it assisted the cause of civilization. and the upward progress of the race.

It did this to a degree so great that it is hard to designate any other achievement that did more. As the *Monitor's* victory was due almost wholly to the genius and character of her designer. we can truly declare that no one man can be designated as having done more for the well-being of mankind than that brave and forceful and honorable and patriotic inventive engineer. John Ericsson.

## Lord Leverhulme's Six-Hour Day

Lord Leverhulme, the wealthy English manufacturer who has recently visited America, is best known in this country as an advocate of the six-hour day. What his viewpoint is, the extent to which he believes the six-hour day applicable and why he advocates the short day Lord Leverhulme explained in an address before the Boston Chamber of Commerce, reported in its journal *Current Affairs* for December 8, 1919.

His plan for the six-hour day contemplates operating in two shifts and in consequence a proportionate reduction in overhead. He says: " You will find that you can pay the same wages for six hours' work as for eight hours provided you work in two shifts of six hours each and that your result in product will not be increased in cost, merely through the increased output and the reduced fixed charges; and that to the extent that the fixed charges are greater than the weekly wage bill there will be an automatic decrease in cost."

Lord Leverhulme admitted that he did not believe the six-hour day was applicable to all industries,—at least at the present time. He believes that it should be established in such industries as lend themselves for business reasons to its operations. If others follow it will be because the plan has succeeded by the sheer force of its own merit.

The strong appeal in the address was that portion devoted to a consideration of the place that domestic economy and home life should play in regulating the hours of labor in industry and it is probably this argument which would chiefly aid in selling the idea, at least in factory communities where parents and sons and daughters are all employed to a greater or less extent. He said one must remember that the housewife may have a husband and sons and daughters going to work and that their meals have to be provided for them when they return from their toil. They should go and come at times to fit in with the daily routine of the housewife, which answers the question so often asked, " Why not work two shifts of eight hours each instead of two of six hours each?" Starting early with the first six-hour shift, changing to the second at the middle of the day and completing the work early in the evening, enables a regular schedule of domestic economy to be maintained in the household, which cannot be done with the eight-hour shifts.

The guest of the Chamber of Commerce gave one warning. The plan must not be approached with the idea that it will make more money, although it may be based on the sound economic principles which make for success. The central idea of the scheme is the betterment of the conditions of industrial workmen, the making of better human beings and the realization that the road to success is along the line of our moral and spiritual resources.

## CAUSES OF INDUSTRIAL UNREST AND REMEDY

(Continued from page 23)

industrial establishments of the country, this situation would be intolerable. The employer would be bound by contracts he made, but the labor unions with whom he dealt would not be bound.

But in so far as the " open shop " is maintained, the employer, while he may suffer without redress from unlawful acts of the labor unions, is not called upon to make agreements with them at all and, therefore, insofar as contract relations are concerned, does not care whether or not they can be held responsible. It is his own men with whom he deals. While therefore it seems clear that from every point of view the law everywhere should be amended so that the labor unions should surely be responsible for their acts and for their contracts, the matter is of less importance if only the " open shop " can be maintained in full vigor.

The members of The American Society of Mechanical Engineers are a power in industry. Their influence pervades the industrial organizations throughout the land. Their interest in these great industrial questions is manifest. The membership of your Society in the National Industrial Conference Board is an evidence of this fact, for by becoming a member of that Conference Board your Society participates in a very definite, very honest and public-spirited effort to get at the real facts by investigation and research and to educate those engaged in industry as well as the public to a sound knowledge of the truth as to industrial questions, among the most important of which is that of the employment relation.

It is the duty of each member of this great body to study the question and to take a definite position as to what seems to be the single, important issue we are now facing. Shall our employment relations be developed on lines of trade-union domination. in which the employees are to be represented by the trade unions, working primarily for their own interests and the advancement of their own power on lines and by methods which to a large extent are economically, morally and legally unsound, or shall every effort be made to encourage the management of each industrial establishment to give itself wholeheartedly to the task of meeting its own employees in a friendly, sympathetic and conciliatory fashion, that by joint consideration of all questions of the employment relation, peace and contentment, coupled with the highest efficiency. may be attained?

# CHARLES DE FREMINVILLE ELECTED TO HONORARY MEMBERSHIP IN THE A. S. M. E.

TO the roster of illustrious men honored by The American Society of Mechanical Engineers is added the name of Charles de Freminville, engineer of the Schneider Works and member of the French Economic Mission.

Although so well known, the careers of famous men are always of interest and so a brief biographical sketch of this French engineer follows:

Charles de Freminville was born at Lorient, France, on August 16, 1856, from a family which has honored the engineering profession

CHARLES DE FREMINVILLE

for many generations. He obtained the degree of engineer at l'Ecole Centrale des Arts et Manufactures, an institution which all Americans who served in the artillery have learned to know and appreciate.

Soon afterwards he entered the service of the Paris-Orleans Railway, first as locomotive engineer and subsequently as designer and engineer. His knowledge of the English language determined his selection as the concern's representative at the Sharp Stewart Works of Manchester, England, in connection with an order placed at these works. This gave him occasion to study in all its details the practical operation of English railways.

Later on, by the request of the Orleans company, he visited America for the first time in 1883. He was accompanied by an intimate friend, Commandant Krebs, who together with Commandant Renard, played such a conspicuous part in the construction of the first dirigible La France, and who some years later also assisted in the construction of the first submarine Le Gymnote, under the direction of Gustave Zede. Mr. de Freminville was later to become the collaborator of Commandant Krebs during many years.

Having returned to France, the Orleans Railway Company commissioned Mr. de Freminville to design a new type of railway car following the lines of those used in the United States. He then designed an original type of car made almost entirely of steel, which was successfully adopted in Europe. American engineers noted this type of car with interest when they were compelled to renounce completely the utilization of wood in the construction of railway passenger cars.

In 1898 Mr. de Freminville again visited the United States with a view to studying the application of electric traction to the trains of the Orleans Railway Company.

Some time later he coöperated with his friend, Commandant Krebs,

who has assumed the direction of the Panhard and Levassor Works, with a view to manufacturing industrial types of automobiles, and particularly to organize production. His efforts were especially directed to the selection of metals for the construction of light machine parts which would nevertheless possess the required strength and rigidity. He had to make his selection among metals which were then new and the qualities and defects of which were little known and often misunderstood. The construction and operation of the automobile brought about a new conception of the resistance of materials.

In 1913, in response to an invitation of Dr. Frederick W. Taylor, Mr. de Freminville again visited the United States for the purpose of studying the scientific organization of workshops. This visit afforded him the opportunity of furthering his friendship with various engineers of the Pennsylvania Railroad and a number of automobile engineers of Detroit, and of visiting the laboratories at Harvard University where remarkable experiments on the action of high pressures on the resistance of materials had been performed by Professor Bridgman.

He returned to France and dedicated himself to the application and extension of the scientific principles of organization of workshops and, together with Prof. H. Le Chatelier, endeavored to attract the attention of French industrial manufacturers to the necessity of organizing their plants in order to obtain an intensive production. He emphasized always the great importance of the work done by Frederick W. Taylor, and undertook to show that it is necessary to take into account the human factor in the organization of the works.

When the war came he became engaged in the manufacture of projectiles. His attention was called to the solution of the many problems which arose from the fact that it was necessary to use steels coming from different sources; he also undertook to apply in the manufacturing operations the best methods of organization.

He was particularly successful in the reorganization of systems of work in the shipyards of Penhoet (St. Nazaire), which were converted into a school of application of scientific methods to the most varied kinds of work, such as the manufacture of machines and the construction of ships with all the specialties that such construction entails.

Mr. Eugene Schneider, desiring to make application of new methods in his establishments, which are by far the greatest industrial works in France, associated Mr. de Freminville permanently with the Schneider works after having made certain that under his influence the new ideas of organization were accepted by the personnel and effected important developments in production. Mr. de Freminville's task was singularly facilitated by the personal support of Mr. Schneider, who has known how to inspire his personnel with a faithfulness that is rarely found in industry.

Mr. de Freminville has been President of the Section of Transport of the Society of Civil Engineers of France. He is a member of the Council of the Society for the Encouragement of National Industry, of the International Association for the Standardization of Testing Methods, of the Taylor Society, etc.

## Resolution by Buffalo Section on Aims and Organization Report

At a meeting of the Local Section of the A. S. M. E. of Buffalo, November 18, 1919 the following resolution was presented and adopted:

*Resolved*, That the Buffalo Local Section of The American Society of Mechanical Engineers hereby approve the Report of the Committee on Aims and Organization as presented at the Spring Meeting; this Section considers that the work done by this Committee cannot but lead to broader opportunities and greater usefulness for our Society. This Section, however, cannot refrain from expressing its appreciation of the action taken in amending paragraph sixteen of the Report as presented at Detroit. We believe that The Journal can be made a means of far greater usefulness.

[The paragraph 16 referred to in the above resolution calls for the development and broadening of MECHANICAL ENGINEERING. The action at Detroit favored this, but advocated continuing the publication for the present as a monthly journal instead of publishing it weekly as recommended in the original draft of the report. The resolution above by the Buffalo Section, however, favors the full plan of development, which includes weekly publication.—EDITOR.]

# NEWS OF THE ENGINEERING SOCIETIES

Reports of Meetings of Electrical Engineers, Taylor Society, Engineering Society of Western
Massachusetts, Railroad Fire Protection Association, National Founders' Association, etc.

## New York Section of the American Institute of Electrical Engineers

A meeting of the American Institute of Electrical Engineers
was held at the Engineering Societies Building on the evening
of December 10 for the purpose of organizing a New York Section of the Institute. The necessary formalities were carried out;
by-laws were adopted, and the following officers were elected:
Chairman, H. W. Buck; Secretary-Treasurer, H. A. Bratt; Executive Committee, the Chairman, the Secretary-Treasurer, E. B.
Craft, W. S. Finlay and Farley Osgood.

There was no discussion regarding details of the work of the
Section. The by-laws adopted provide that the Executive Com-
mittee of five should be the governing body and should manage
the affairs of the Section. It is assumed that the scope of the
Section will be identical with that of the other sections of the
Institute throughout the country.

## Taylor Society

Endorsement of collective bargaining and of both trade unions
and shop unions, the one to supplement the work of the other,
was a prominent feature of the annual meeting of the Taylor
Society, held at New York on December 5 and 6.

H. K. Hathaway, consulting engineer in management, Philadelphia, spoke on Standards: Their Nature, Purpose, Necessity. He observed that standards of accomplishment are dependent upon standards of equipment, materials, methods and
product and that lower labor costs, which make higher wages
possible, are brought about by such methods as continuous
straight-line handling, elimination of waste motions, new machinery, keeping accurate records, reorganization of sales and
improvement in stock-keeping. He gave concrete examples of
ways of effecting economy by the installation of an automatic
tool-grinding machine in a machine shop, by standardizing tools,
by carefully regulating the tension of belts, etc.

Two committees were appointed, one to study "unemployment
within employment" and provide it to a minimum, and the
other to provide for the introduction of courses in scientific
management in university curricula.

## Naval Architects and Marine Engineers

The twenty-seventh general meeting of the Society of Naval
Architects and Marine Engineers was held in New York, November 13 and 14.

The president of the Society called attention in his address
to the recent surprising growth in American shipbuilding. He
surveyed the situation in the industry from 1900 when 80 steel
steam vessels of 168,000 tons were built as compared with Great
Britain's figures of 567 vessels of 1,341,000 tons to the fiscal
year of 1919 when the figures given, although they include only
steel steamers of over 1000 tons, reached the total value of nearly
2,500,000 tons.

One of the professional papers described in detail the propelling machinery of the U. S. S. Leviathan, which is the largest
vessel ever completed. The vessel is driven by the Parsons direct type of steam turbine arranged on four shafts and the turbine installation consists of eight separate turbines with their
respective rotors. The go-ahead units, of which there are four,
each driving a separate shaft, are arranged in triple
series; the go-astern units, of which there are also four, are
arranged to operate in parallel and consist of two independent

sets for the two port and the two starboard shafts, each unit
driving a separate shaft. The machinery installation is so designed that under any extraordinary conditions which may arise
at sea it is possible to operate the two turbines of any one shaft
entirely independent of the other units. Further, all intercon-
necting valves and piston are so designed as to permit carrying
out any change at sea without having to shut down for any appreciable length of time.

Other technical papers were: Methods Employed in the Con-
struction of Concrete Ships, by R. J. Wig; Submarines in Gen-
eral—German Submarines in Particular, by Commander E. S.
Land; The Application of Standardization and Graphical Meth-
ods to the Design of Cylindrical Boilers, by H. C. E. Meyer;
New Developments in High Vacuum Apparatus, by G. L.
Kothny; and Standard Oiling System for Geared Turbines, by
J. E. Schmeltzer and B. G. Fernald.

## Engineering Society of Western Massachusetts

Unusual interest both in the society itself and in the welfare
of the community was displayed at the November 10 meeting
of the Engineering Society of Western Massachusetts. This society was newly formed in the Spring of 1919 and an account
of its organization meeting appeared in the July issue of ME-
CHANICAL ENGINEERING, page 637. The officers of the Society
are Charles L. Newcomb, Mem.Am.Soc.M.E., president; C. C.
Chesney, vice-president; George E. Williamson, Mem.Am.Soc.
M.E., vice-president; Winfield E. Holmes, Mem.Am.Soc.M.E.,
secretary and treasurer. The headquarters of the Society are in
Springfield and its present membership, which includes engineers
of all classes, closely approaches three hundred.

At the fall meeting were gathered engineers from all over
Western Massachusetts. They spent the afternoon in visiting
some of Holyoke's plants of engineering and industrial interest,
where they were cordially received and shown all points of interest.

At the evening meeting, held at the Hotel Nonotuck, President
Newcomb delivered an interesting opening address of welcome
in which he touched briefly upon the engineering characteristics
of Holyoke and its origin, founded upon its early and maintained water-power development and upon its dam, over a
thousand feet in length, thirty feet high and with a potential
horsepower of 45,000.

The principal speaker of the evening was H. W. Rogers of the
Power and Mining Engineering Department, General Electric
Co., Schenectady, N. Y., who delivered an illustrated address
on Industrial Applications of Electric Power. Other speakers
were John J. Kirkpatrick, manager of the City of Holyoke Gas
and Electric Department, on Municipal Ownership of Public
Utilities, and W. E. Holmes, upon the need of the active interest
of members in the society.

A second meeting of the Engineering Society of Western
Massachusetts was held on December 16 in Springfield, Mass.,
at which Charles Spofford, of Boston, Mass., spoke on Some
Architectural Features of Boston Army Supply Base, and Fred-
eric H. Fay, also of Boston, on The Proposed Springfield-West
Springfield Bridge.

## Railroad Fire Protection Association

The best methods for handling and storing fuel oil and inflammable liquids were among the topics discussed at the annual meeting of the Railway Fire Protection Association held at Chicago

74

on November 18, 19 and 20. An extensive report specifying rules and regulations for the storage and use of fuel oil and for the construction and installation of oil-burning equipment was presented by the Committee on Oil-Burning Appliances.

A technical paper on Fuel Oil versus Coal for Locomotives, by A. M. Schoen, contained many recommendations concerning the various kinds of fuel oil, the methods of handling oil and the fire hazard involved. "The large tanks," according to Mr. Schoen, "should be at a distance of not less than 350 ft. from buildings or other combustible property, and this may well be made the minimum distance from the main line of the road as well, owing to the fact that should the tank be on fire and a strong wind blowing in the direction of the tracks, the burning gases and dense smoke could be driven a considerable distance, and if the tank is not sufficiently isolated the operation of the main-line trains might be affected for a considerable time. Also the tanks should be so located and protected that under no circumstances could the oil reach streams, either flowing or tidal, by means of which it might be carried into the proximity of destructible property. It is strongly recommended that the fueling of tenders by gravity pressure be avoided, in view of the danger of the entire contents of the tank escaping in event of accident. Most approved practices would require the oil to be pumped into the tender from reservoirs constructed below the grade level, which would permit of the flow being promptly shut off in event of accident to the piping or apparatus. These wells can be filled at regular intervals either by gravity feed from the large storage tanks or else direct from the tank cars. These underground reservoirs, if completely covered over, may be located at such points as may be most convenient for their intended purpose. Usually they should be constructed either of reinforced concrete or steel."

The report of the Committee on the Protection of Wooden Bridges and Trestles against fire recommended: That effective spark screens be provided in the front end of all locomotives; that ashpans and grates be made tight and kept in good working order; that special places be provided for the dumping of cinders and ashes, and that grates are not to be shaken down except at safe points; that all combustible refuse such as dry leaves, dead grass, weeds, brush and rubbish be cleared away from under and around all wooden bridges; that the decks of all wooden bridges, between the rails, be covered with No. 22 galvanized iron, this to prevent sparks from setting fire to the structure should they be dropped from the locomotive; and that all wooden bridges be coated with a fireproof or fire-retardent paint.

## National Founders' Association

The labor problems of the day were discussed at the annual meeting of the National Founders' Association, held in New York on November 19 and 20. President William H. Barr noted that 5800 strikes were called by the labor unions during the war, none of which occurred in non-union shops. He contended that more strikes are caused by the ill-advised conclusions or the hasty words of a foreman than through any other single cause, and advocated employee representation in order to reduce this possibility to a minimum by means of periodical meetings of foremen, discussion of policy, analysis of production plans and the emphasizing of the necessity of dealing justly with all employees.

Speaking on Industrial Unrest, George F. Monaghan, Detroit, counsel for the Association, suggested the employment of more efficient machinery and greater efficiency in management as remedies for the solution of present labor difficulties which he attributed to lack of production, the prevailing extravagance of all classes, socialistic agitation and profiteering on the part of both capital and labor. He also suggested that both employer and employee agree to adopt the present scale of wages and present profits until greater production can bring down the cost of living to the normal level.

The Industrial Association of Cleveland was described by H. P. Bope, vice-president of the Hydraulic Pressed Steel Co., Cleveland, as follows:

About five years ago ten of us got together to see how we could benefit our factory. First we realized the necessity to educate our foremen. Then we believed such a movement should not be confined to our factory, so we called together 50 leading business men of Cleveland and secured that number of members to form the association. There are 30,000 salaried employees in Cleveland and it was for their advancement as well as that of the employers that the movement was launched. A board of directors was formed, consisting of 25 employers and the same number of employees. We inaugurated monthly meetings with speakers and movies, a typical instance of the latter being a presentation of the manufacture of steel from the mining of the ore to the finishing of the steel product. Eight hundred men saw these pictures monthly. Last year in 20 meetings there was an average attendance of 1100 men—employers, superintendents and workmen. The object was to bring together the brains and the brawn and to distribute some of the brains into the brawn. The speeches were usually about 20 minutes long and were followed by an open forum. Cleveland is 500 per cent better off because of the industrial association. The foremen were broadened by mingling with the foremen of other shops and exchanging ideas and viewpoints. We have a magazine called Coöperation, which is very helpful in promoting friendly relations and affording education. We do not believe in unions—they are horizontal, whereas they should be vertical, or take in the executives and be local in their scope. We offset red propaganda by employing the vernacular of the laboring man in presenting economical truths.

Mr. Bope also referred to work being done to establish similar associations in Philadelphia, St. Louis and Detroit, and pointed out the advantages that could be derived by building a national association along the lines of the Cleveland Industrial Association.

Other papers were: Government Competition in Industry, by William H. Vandervoort; British and American Labor Problems, by Prof. J. Lawrence Laughlin; The Foreign Language Press, by Richard H. Waldo; and The Right of Individual Contract, by L. R. Clausen.

## National Advisory Committee for Aeronautics

At the annual meeting of the National Advisory Committee for Aeronautics, held in Washington on October 9, the general functions of the Committee were specified as follows:

1 Under the law the Committee holds itself at the service of any department or agency of the Government interested in aeronautics for the furnishing of information or assistance in regard to scientific or technical matters relating to aeronautics, and in particular for the investigation and study of problems in this field with a view to their practical solution.

2 The Committee may also exercise its functions for any individual, firm, association, or corporation within the United States, provided that such individual, firm, association, or corporation defray the actual cost involved.

3 The Committee institutes research, investigation, and study of problems which, in the judgment of its members or of the members of its various sub-committees, are needful and timely for the advance of the science and art of aeronautics in its various branches.

4 The Committee endeavors to keep itself advised of the progress made in research and experimental work in aeronautics in all parts of the world, particularly in England, France, and Italy, and will extend its efforts to the securing of information from Germany and Austria.

5 The information thus gathered is brought to the attention of the various subcommittees for consideration in connection with the preparation of programs for research and experimental work in this country. This information is also made available promptly to the military and naval air services and other branches of the Government, university laboratories, and aircraft manufacturers interested in the study of specific problems.

6 The Committee holds itself at the service of the President, the Congress, and the executive departments of the Government for the consideration of special problems which may be referred to it, such as rules for international air navigation, method of regulation and development of civil aerial transport, technical development policies of the military, naval, and postal air service, etc.

Dr. Charles D. Walcott was elected Chairman for the ensuing year, and Dr. S. W. Stratton, Secretary.

# GOVERNMENT ACTIVITIES IN ENGINEERING

*Notes Contributed by The National Service Committee of Engineering Council*[1]

SEVERAL bills of interest to engineers have been passed by the House and will be taken up by the Senate during the next session. These bills deal with such matters as oil, gas and coal land leasing; water power; national budget; tariff on ores, glassware, instruments, etc.; railroads; minimum wages, and distribution of motor vehicles to State highway departments. The National Soldiers' Settlement Bill has been reported to the House but has not as yet been acted upon. The details of many of these bills were published in the November issue of MECHANICAL ENGINEERING and there follow some additional data and comments of more recent date.

## HIGHWAY LEGISLATION

The original bill proposing to create a national highway system controlled by a Federal Highway Commission, which was proposed by Senator Townsend, has been practically abandoned because of the opposition which has developed. It is now believed that any plan for National Highways constructed directly by the Federal Government will be unpopular with the states because only a limited portion of the state will be benefited directly by such highways. Further, the Government with its road building activities would be in competition with the states and the counties. It is believed that it will be easier for the Federal Government to hold the states up to a standard rather than to attempt to maintain the standard through its own work. No plans have been made to secure consideration of the Townsend bill at the present session of Congress.

Another highway bill which proposes a National Highway Department and Highway Commission has been introduced into the House, which aims to do away with the defects which have become apparent in the Townsend bill. It will bring the control of highway work more closely under the control of the states, at the same time giving enough Federal supervision to insure coordination and standardization in the work. The states are to be divided into ten regional areas with corresponding regional congresses having direct control of the work in their region but under the general direction of the National Highway Commission.

For the purpose of carrying out the provisions of this new act and as an initial appropriation for the creation of a National Department of Highways, the new bill proposes to appropriate $10,000,000. The Secretary of Highways would receive $12,000 per year and each National Highway Commissioner would receive $6,000 per year.

## ESTIMATES FOR PUBLIC WORKS FOR 1921

The Secretary of the Treasury has approved and forwarded to Congress estimates for next year which total the unusually large sum of $283,921,810. The appropriations for public works during the fiscal year 1920 total $93,872,092. The estimates for work of particular interest to engineers and constructors are as follows:

Construction of Government buildings, including purchase of sites, $1,332,775; marine hospitals and quarantine stations, $5,204,621; fortifications, $117,793,330; military posts, $14,225,251; rivers and harbors, $53,659,265; navy yards and stations $20,606,000; Panama Canal construction and maintenance, $18,245,391; depots for coal and other fuels, $2,335,000; Reclamation Service, $20,134,000; $725,000 special appropriation for a fuel inspection system; $1,865,880 for the Bureau of Mines; $2,592,920, Geological Survey; $3,000,000 (approximate), Bureau of Standards.

The Technical News Bulletin of the Bureau of Standards describes progress in the following work: Radio Fog Signalling Experiments; Standardization of Electron Tube Symbols; Scientific Research on Electron Tube Amplifiers; Magnetic Analysis; Industrial Safety Standards; Photoelectric Spectrophotometry; Reinforced Concrete Investigation; Concrete Tanks for Storage of Oil; Oxyacetylene Welding and Cutting Apparatus; Locomotive Packing Ring Investigation; Weights and Measures Publications; Testing of Hydrometers; Thermostatic Return Line Valves for Heating Systems.

## NOTES ON THE BUDGET, TARIFF, COMMERCE. ETC.

The Senate Budget Committee hearings, started Dec. 15, indicate a desire for the McCormick Bill. This bill provides a budget bureau in the Treasury which will coöperate with a budget officer in each department. The final budget shall be approved and sent to Congress by the President. The House Budget bill simply proposes to transfer the present work of the Treasury Department to the President, who will then prepare and submit an alternate budget with recommendations.

The office of the Adjutant General announces that its present policy governing the selection and appointment of members of the Officers' Reserve Corps contemplates the establishment of a competent Reserve Commissioned Personnel, individual members of which will be qualified to perform fully duties of their respective grades when ordered into active service in case of emergency. Officers commissioned in Officers' Reserve Corps to Oct. 31, 1919, numbered 52,839.

Contracts for electrical machinery for the propulsion of three new battleships have been awarded by the Navy Department. The price for the three sets is $5,500,000. According to Navy officials, electrical propulsion has been successful. The *New Mexico*, which was the first electrically driven battleship, has given entire satisfaction, and the *Tennessee*, the second to be thus equipped, will shortly be put in commission.

The Bureau of Public Roads has announced that while expenditures during 1919 for hard-surface highways will set a new record with a total of $138,000,000, this figure is small in comparison with the computed available total for 1920 of $633,000,000, the spending of which promises to be dependent chiefly on the quantity of materials present limited railway facilities can transport.

The prohibition or control of importation of dyes and coaltar products as provided in the trading with the enemy act are continued until Jan. 15, 1920, by the passage by the Senate of the House joint resolution. The question of tariff bills was discussed on the Senate floor, with the decision that they must go over until after other legislation is considered.

The Commerce Committee of the Senate has been urged by the Chief Signal Officer of the U. S. Army to extend American cables as an aid to the development of the nation's foreign commerce. This is in connection with the testimony providing new cable lines across the Pacific.

A Corps of Civilian Engineers in the U. S. Army is proposed in a bill which is aimed specially to care for the resident engineers throughout the country and the Washington force, who have been underpaid, although in charge of some of the most important river and harbor work.

Senator Thomas of Colorado, during the closing hours of the recent session of Congress, introduced a bill for the establishment of a trained body of 200,000 railroad workers, to constitute a railroad army reserve force subject to call by the Secretary of War in the event of nation-wide railroad strikes.

[1] Engineering Council is an organization of national technical societies created to consider matters of common concern to engineers as well as those of public welfare in which the profession is interested. The headquarters of Engineering Council are located in the Engineering Societies Building, 29 West 39th Street, New York City. The Council also maintains a Washington office with M. O. Leighton, chairman of the National Service Committee in charge. This office is in the McLachlen Building, 10th and G Streets, Washington, D. C. The officers of Engineering Council are: J. Park Channing, Chairman; Alfred D. Flinn, Secretary.

# LICENSING OF ENGINEERS

## Committee Appointed by Engineering Council Recommends Uniform Registration Law

AT a meeting held on October 25, 1918, Engineering Council authorized the creation of a committee to make a thorough study and submit a report on the subject of licensing or registering engineers. Fifteen engineers of long experience in the various branches of the profession of engineering were selected from 13 states representing practically all sections of the United States, as well as mechanical, electrical, mining, hydraulic, municipal, sanitary, railway and structural engineering and also colleges of engineering.

The personnel of the committee is as follows:

Theodore L. Condron, Chicago, Ill., *Chairman*

| | |
|---|---|
| John W. Alvord, Chicago, Ill. | Caleb M. Saville, Hartford, Conn. |
| Bion J. Arnold, Chicago, Ill. | A. M. Schoen, Atlanta, Ga. |
| John H. Dunlap, Iowa City, Ia. | Francis C. Shenehon, Minneapo- |
| A. Lincoln Fellows, Denver, Colo. | lis, Minn. |
| Farley Gannett, Harrisburg, Pa. | Amos Slater, Seattle, Wash. |
| Arthur M. Greene, Jr., Troy, | Christopher H. Snyder, San |
| N. Y. | Francisco, Cal. |
| James H. Herron, Cleveland, O. | John W. Woermann, St. Louis, |
| John Klorer, New Orleans, La. | Mo. |

On December 18, 1919 this committee submitted its report to the Engineering Council, which included a "Recommended Uniform Registration Law," with the recommendation that the Council approve and endorse the bill for an act of legislation in each and every state for the regulation of the practice of engineering, architecture and land surveying. An abstract of the report is given below together with extracts from the recommended law, particularly those sections relating to the requirements for the registration of engineers.

On account of the overlapping of the work of engineers and architects in the design and erection of buildings and other structures and because of the close connection between this work and the surveying of the land and foundations upon which such structures are erected, it was deemed essential, as pointed out in the report of the committee, to provide in a single law for the registration of engineers, architects and land surveyors. By this means would not only life and health be protected but property rights as well, and to a fuller extent than if there were separate laws for each of these branches. The report was therefore carried through on this comprehensive scale, although the American Institute of Architects had previously prepared a "model law" to cover architecture and building.

## Abstract of Report on Licensing of Engineers

This preliminary investigation disclosed that very pronounced views were held by engineers throughout the country, both for and against state licensing or registration. The general sentiment one year ago was more opposed to such measures than it is today. The older members of the profession did not as a rule favor licensing nor did they feel there was need for state regulation of engineering practice, while among the younger men there was a feeling that licensing or registration by the states would add prestige to professional engineers and in many ways benefit the profession, as well as individual engineers.

The advantages claimed for state licensing or registration are the same as those presumably gained by the laws regulating the professions of law and medicine, namely, that those who are incompetent and unqualified professionally to practice are unable to obtain certificates or licenses and hence both the public and the profession are protected. On the other hand, those engineers who have already attained recognized professional standing feel they not only do not need the benefits claimed for such legislation, but they fear that state licenses or certificates of registration are apt to put the seal of state endorsement on men who do not deserve it and that the public would assume that a licensed or registered engineer was thereby certified by the state as fully quali-

fied, regardless of what might or might not be the requirements demanded before a license or certificate was granted.

However, the question has gone beyond the stage of debate, for already ten states[1] have enacted laws licensing or registering engineers, and other states are certain to enact similar laws during the present or coming sessions of their legislatures. In addition to these ten laws governing engineering practice, there are at least six states that require the licensing or registration of land surveyors, and in at least eighteen states[2] laws have been passed licensing or registering architects. Some of these ten laws are so drawn as to include both engineers and surveyors and some include engineers and architects and one or two include engineers, architects and surveyors. Moreover, these laws are not at all uniform and in several instances are likely to prove seriously embarrassing and annoying to engineers whose activities extend beyond the limits of a single state. Because of the nature of professional engineering work the practice of an engineer frequently extends over several states and therefore it is vitally important, if there are to be state regulations for engineering practice, that these regulations be made uniform so far as possible and that the engineering profession unites in wisely directing such legislation.

As stated, laws have been passed in at least 18 states for licensing or registering architects and the American Institute of Architects has endorsed and advocated such legislation, considering that both the architects and the general public are benefited thereby. Unfortunately some of the laws for licensing architects have been so drawn as seriously to interfere with legitimate engineering practice and the "model law" proposed and advocated by the American Institute of Architects contains definitions of "architecture" and "building" which, should such laws be passed and enforced, would prevent any one but a "registered architect" from planning or supervising the construction of any structure or any of the appurtenances thereto; consequently under this head would come a structure having simply foundations and girders, whether "with or without appurtenances." Under this model law proposed by the American Institute of Architects no one but a registered architect shall prepare plans for or supervise the construction of a building, and a "building" is thus defined in Section 19: "A building is any structure consisting of foundations, floors, walls, columns, girders, and roof, or a combination of any number of these parts, with or without other parts or appurtenances."

Since none but a registered architect shall have the right to design or supervise the construction of any structure or any appurtenance thereto, this matter becomes of vital importance to mechanical, electrical, sanitary and mining engineers, as well as to structural engineers.

In the State of Illinois a similar law for licensing architects was passed several years ago, the rigid enforcement of which made it necessary for engineers to unite in having a "structural engineers' license law" passed by the legislature, and now there are two laws in force in Illinois, one for architects, the other for structural engineers.

In some states the laws enacted and proposed are intended to regulate the practice of architecture, while in the other states laws have been enacted the purpose of which is simply to protect the term "architect," but not intended to regulate the practice of architecture. In Wisconsin and several other states no one may use the title "architect" without first obtaining from the state a certificate of registration as a "registered architect," but any one not an architect may prepare plans and supervise construction provided he does not style himself an "architect."

Therefore this committee at its meeting of October 13, 1919, adopted unanimously the following resolution:

[1] Colorado, Florida, Idaho, Illinois, Iowa, Louisiana, Michigan, Nevada, Oregon, Wyoming

[2] California, Colorado, Florida, Idaho, Illinois, Louisiana, Michigan, Montana, New Jersey, New York, North Carolina, North Dakota, Oregon, Pennsylvania, South Carolina, Utah, Washington, Wisconsin.

The following is the sense of this committee relative to the desirability of a law licensing or registering engineers:

*Resolved:* That the enactment of legislation to provide for the registration of professional engineers is desirable and necessary. Ten states have already enacted such legislation. Laws licensing architects have been enacted in several states and similar laws endorsed by the American Institute of Architects, are pending in several other states, which, if enacted, would prohibit engineers from continuing their customary and recognized practice.

The only basis on which the practice of any profession may be subject legally to state regulation is "in order to safeguard life, health and property." Land surveying does not involve matters that would ordinarily jeopardize life and health, but property rights are vitally affected by land surveying, and many states have deemed it essential to place restrictions and safeguards about the practice of land surveying. Land surveying is associated with both engineering and architectural practice.

This committee has therefore deemed it advisable and to the best interests of all concerned to include in one law, provisions for the registration of engineers, architects and land surveyors. It has recognized that the practice of engineering and architecture overlap in many instances, especially in connection with the larger projects of modern structures, where many branches of the arts and sciences are combined, involving architecture, structural, mechanical, electrical, sanitary, and other lines of engineering.

There are ample reasons why architects alone should judge as to the qualifications of those desiring to practice architecture and why engineers alone should pass upon the qualifications of those desiring to practice engineering. Hence a bill for legislation has been drafted by the committee along these lines and it is confidently hoped that the objections expressed by the American Institute of Architects against laws which might provide jointly for the registration of engineers and architects will be overcome by the terms of the bill herewith submitted.

### RECOMMENDED UNIFORM REGISTRATION LAW

This act is entitled "An act to regulate the practice of professional engineering, architecture and land surveying." Its sections deal with the enactment of the law, the appointment of a state board of registration for professional engineers, architects, and land surveyors, and the qualifications, expenses, certificates, powers, organization, meetings, receipts and disbursements, and records and reports of this board. Sections 9 and 13, extracts from which are here given, treat respectively of Application for and Issuance of Certificates, Exemptions, and Corporations or Partnerships, and the remaining sections take up questions of revocation and reissuance of certificates, the significance of the certificates, unlawful acts and penalties, public work, land surveying, and repeal of conflicting legislation.

#### APPLICATION FOR AND ISSUANCE OF CERTIFICATES

*Section 9:* The board shall, on application therefor, on prescribed form and the payment of a fee of ———— dollars ($————), issue a certificate of registration:

*Note:* The application required should include a complete statement of an applicant's education and a detailed summary of his technical work. The statements made should be under oath, and should be supported by the recommendations of not less than two professional engineers, architects or land surveyors as vouchers.

1 To any person who submits evidence satisfactory to the board that he or she is fully qualified to practice professional engineering, architecture or land surveying:

Provided, however, that no person shall be eligible for registration who is under twenty-five years of age, who is not a citizen of the United States or Canada, or who has not made declaration of his or her intention to become a citizen of the United States, who does not speak and write the English language, who is not of good character and repute, and who has not been actively engaged for six or more years in the practice of professional engineering, architecture or land surveying of character satisfactory to the board. However, each year of teaching, or of study satisfactorily completed, of engineering or architecture in a school of engineering or architecture of standing satisfactory to the board, shall be considered as equivalent to one year of such active practice.

Unless disqualifying evidence be before the board, the following facts established in the application shall be regarded as prima facie "evidence, satisfactory to the board," that the applicant is fully qualified to practice professional engineering, architecture or land surveying:

*Note:* When the law goes into affect a large percentage of practising engineers, architects and land surveyors will be registered to preserve the *status quo.* Long-continued practice, graduation from a technical school of approved standing with subsequent years of practice, or membership in high-grade technical societies, in the absence of disqualifying facts, is accepted as *prima facie* evidence of qualification, as stated below.

*a* Ten or more years of active engagement in professional engineering, architectural or land surveying work;

*b* Graduation, after a course of not less than four years, in engineering or architecture, from a school or college approved by the board as of satisfactory standing, and an additional four years of active engagements in professional engineering, architecture or land surveying work;

*c* Full membership in American Institute of Architects, American Society of Civil Engineers, American Institute of Chemical Engineers, American Institute of Electrical Engineers, American Society of Mechanical Engineers, American Institute of Mining and Metallurgical Engineers, American Society of Naval Architects and Marine Engineers, or such other national or state engineering or architectural societies as may be approved by the board, the requirements for full membership of which are not lower than the requirements for full membership in the professional societies or institutes named above.

Applicants for registration, in cases where the evidence originally presented in the application does not appear to the board conclusive or warranting the issuance of a certificate, may present further evidence which may include the results of a required examination.

*Note:* The standard of qualification is set high for two reasons: the public welfare will be better promoted by maturer competency; and the prestige attached to the term "Registered" will be more significant for the professional men themselves. In requiring the younger, less experienced men to serve somewhat longer as assistants or understudies to older men, no hardship is imposed which will not be compensated by the fuller return in recognition after registration.

In determining the qualifications of applicants for registration as architects, a majority vote of the architect members of the board only, shall be required; and in determining the qualifications of applicants for registration as professional engineers or land surveyors, a majority vote of the engineer members of the board only, shall be required.

#### EXEMPTIONS

*Section 13:* The following shall be exempted from the provisions of this Act:

Offering to practice in this state as a professional engineer, architect or land surveyor, by any person not a resident of and having no established place of business in this state.

*Note:* A professional card in a journal of national circulation is an "offer to practice" in any state in the Union. It would be manifestly unfair to compel a professional man to register in every state in which he may in this way, or by letter or otherwise, express his readiness to accept an engagement.

Practice as a professional engineer, architect or land surveyor in this state by any person not a resident in this state and having no established place of business in this state, when this practice does not aggregate more than fifteen days in any calendar year; provided, that said person is legally qualified for such professional service in his own state or country.

*Note:* It is a distinct advantage to the people of any state to be able to call in for consultation a specialist from any other state. Such practice may be brief and often of an emergency nature.

Practice as a professional engineer, architect or land surveyor in this state by any person not a resident of and having no established place of business in this state, or any person resident in this state, but whose arrival in the state is recent; provided, however, such a person shall have filed an application for registration as a professional engineer, an architect or a land surveyor and shall have paid the fee provided for in Section 9 of this Act. Such exemption shall continue for only such reasonable time as the board requires in which to consider and grant or deny the said application for registration.

Engaging in professional engineering, architecture or land surveying as an employee of a registered professional engineer, a registered architect or a registered land surveyor, or as an employe of a professional engineer, architect or land surveyor, authorized by paragraphs 2 and 3 of this section, provided that said practice may not include responsible charge of design or supervision.

Practice of professional engineering, architecture and land surveying solely as an officer or as an employee of the United States.

Practice of professional engineering, architecture or land surveying solely as an employee of this state or any political subdivision thereof, at the time this Act becomes effective and thereafter only until the expiration of the then existing term of office of such employee.

#### CORPORATIONS OR PARTNERSHIPS

A corporation or partnership may engage in the practice of professional engineering, architecture or land surveying in this state, provided the person or persons connected with such corporation or partnership in charge of the designing or supervision which constitutes such practice is or are registered as herein required of professional engineers, architects and land surveyors. The same exemptions shall apply to corporations and partnerships as to individuals.

# HENRY LAURENCE GANTT

ON November 23, 1919, H. L. Gantt died at his home in Montclair, N. J., after a brief illness. In his death the engineering profession has lost one of its foremost engineers, a pioneer in industrial management and a keen student of the human element in its relations to industry. He was associated with Frederick W. Taylor in his early work at the Midvale and Bethlehem Steel Companies and with this as a basis, and his personal ability as an organizer, he later established and successfully conducted his own consulting practice as an industrial engineer. His career was marked by original and thoughtful work, progressive in viewpoint, and effective in its accomplishment. In later years he developed a broad conception of industry as a national problem in which he regarded it essential that the man at the top should have the same close scrutiny and careful direction that has in the past been given his co-workers in the lower ranks. To these views he gave expression in a characteristically original manner in many addresses and written articles.

Henry Laurence Gantt was born May 20, 1861, in Calvert County,

HENRY LAURENCE GANTT

Maryland. He was graduated from Johns Hopkins as a Bachelor of Arts when only 19 years of age and taught for three years at the McDonough School in Baltimore County which he had attended as a boy. In 1884 he received the degree of Mechanical Engineer from Stevens Institute of Technology.

Following his connection with Frederick W Taylor, he conducted his consulting practice until his death and had as clients more than a score of prominent manufacturing plants. Among these were the Westinghouse Electric and Manufacturing Co., American Locomotive Co., factories of Remington Typewriter Co., Bethlehem Foundry & Machine Co., Saco-Lowell Shops, Ingersoll-Rand Co., Amoskeag Mfg. Co., Cheney Brothers (silk manufacturers), Brighton Mills, Sayles Bleacheries and Corticelli Silk Mills.

Mr. Gantt was active in The American Society of Mechanical Engineers and served on the Council as Manager, 1908-11, and as Vice-President, 1913-15; on the Meetings Committee, 1912-17; and on the Executive Committee of this Council, 1913-14.

His papers contributed to the Society and which enrich its TRANSACTIONS comprise the following list, of which his early paper on the bonus system brought prominently to the front new features of management developed during his early work with Mr. Taylor: Steel Castings; Recent Progress in the Manufacture of Steel Castings; Bonus System of Rewarding Labor; Graphical Daily Balance in Manufacture; Training of Workmen in Habits of Industry and Coöperation; Mechanical Engineer and the Textile Industry; Modifying Systems of Management; Measuring Efficiency; The Relation Between Production and Costs; Productive Capacity a Measure of Value of an Industrial Property; Expenses and Costs; Industrial Democracy.

During the German trip of the Society in 1913 Mr. Gantt was a frequent speaker at the various functions and at the Leipsic meet-

ing of the Verein deutscher Ingenieure, held jointly with our Society, aroused much interest by his discussion of Scientific Management.

Mr. Gantt was also the author of three books: Work, Wages and Profits: Their Influence on Cost of Living; Industrial Leadership (Yale Lectures); and Organizing for Work. In addition, he contributed many articles to the technical and daily press.

There grew up around Mr. Gantt several groups having as their central thought the principles which he enunciated. Among these was a circle of friends who regarded service to the community as the essential element in industry, and which it was believed could be attained by concentrating on the production of goods rather than on the production of profits. It was considered that industry required a "new machine" rather than a disrupting social movement, and the movement was thus given this unique designation.

During the war Mr. Gantt acted in a consulting capacity, first for the Ordnance Department by invitation of General Crozier, who recommended for the entire War Department the use of Gantt's production charts, employed by Mr. Gantt in his work and which had been used in the Ordnance Department. These charts are well known among engineers and constitute an important development in recording the progress of work. When General Crozier was retired as Chief of Ordnance a change in plans was made, but the charts were later used by the U. S. Shipping Board and Emergency Fleet Corporation in routing ships and in following up constructive work.

Mr. Gantt's vision of a better spirit in industry is well expressed by the following quotations from his latest book, Organizing for Work: "The community needs service first, regardless of who gets the profits, because its life depends upon the service it gets. The business man says profits are more important to him than the service he renders; that the wheels of business shall not turn, whether the community gets the service or not, unless he can have his full measure of profit. He has forgotten that his business system had its foundation in service, and as far as the community is concerned has no reason for existence except the service it can render."

At the last Annual Meeting of The American Society of Mechanical Engineers Mr. Harrington Emerson presented a tribute to Mr. Gantt, extracts from which here follow:

## A TRIBUTE BY HARRINGTON EMERSON

Only a few weeks ago Mr. Gantt and I bent our heads over the same declaration of beliefs and principles, agreeing even as to words in the form it should take, a declaration of principles intended for the President, for Congress and for the people of the United States.

The old Hindu philosophy states that a man's full life is divided into four periods:

He is first a learner;

Next a doer;

He conquers the position of executive, of leader;

He becomes in the fullness of his experience a counselor.

When we were both still mainly in the doing stage, there was keen and friendly rivalry between us. It lasted into the executive stage, but as we slowly qualified to counsel, the agreement between us became close.

I regarded him as the nearest man to me, the man whose opinions I most valued, the man to whom I was most willing to turn for counsel. As we grow older many friends drop away. The more closely do old men cling to the few tried-out companions who remain.

It is my custom to place in my files a size-up, an appreciation of great men with whom I come in contact. This I find about Mr. Gantt:

An energetic, dominant nature;

Courageous and fearless, in expressing his opinions, which he never gave without careful consideration;

Very decided when once his mind was made up;

Industrious and energetic in working out a problem;

Independent in thought and action, ready in speech and quick and impulsive in manner, a clear thinker and a man of positive and decided character;

Precise and exact in his work and systematic in his activities, with a thoroughness in all his investigations, worthy of imitation and praise;

One of the great leaders in Efficiency.

At the extraordinarily early age of eighteen he had already received the degree of Bachelor of Arts from one of our youngest and greatest universities, he had already been in touch with the great living minds and through them, with the great minds of the past, and without this double opportunity any man, however naturally great, is burdened in the race. But he went on. He taught. He went to Stevens Institute: he was the student for even more than the first score of years. For the next twenty years he was a doer, little known to the wide world, but accumulating personal experience, trying himself out on new and difficult problems.

Leadership, authority, rests on the possession of four entirely distinct major qualities and as to each I appraised Mr. Gantt. The first great quality is that of goodness, of character, of good will towards his fellow man. It was this that attracted me to Mr. Gantt before I personally met him.

At the December meeting, 1901, of this Society, he presented a paper on a Bonus System of Rewarding Labor; with that clearness of thought which always characterized him, he stated in his second paragraph the aim of the system, which turned out to be the aim of his own life: To harmonize the interests of the employer and of the employee: to give justice to the employee; to conform also to the best interests of the employer.

We have here Mr. Gantt's fundamental claim to leadership—*Harmony based on Justice.*

His last published utterance was in the New York *World*, Sunday, October 12, 1919. Speaking of industrial unrest, he said:

"No permanent peace is possible so long as one man is able to exploit the services of another for his own benefit."

In all his lectures and books in between these two dates he never deviated from the conviction of harmony based on justice. Therefore on this count of good will he was qualified as a leader.

The second qualification for authority, for leadership, is competence, is ability. The eminent position to which Mr. Gantt rose, his financial success, is the practical tribute paid to his competence and ability, but more than that his ability was recognized and deferred to by his own competitors in his younger days, his own friends in his last years.

The third quality essential to leadership is courage. Anybody who knew Mr. Gantt, who looked at his white head and heavy jaws and chin, knew that he was in the presence of a fighter.

He had physical courage;
He had mental courage;
He had spiritual courage.

There was never in anybody's mind any doubt as to where Mr. Gantt stood, whether he was leading you, at your side, or in behind you, you knew he was there and to be counted on.

The final quality essential to leadership, to authority, is personal charm.

We not only respected and depended on Mr. Gantt; we liked him. He could disagree without offending, and even in his wrath he was lovable. He furiously resented what he considered injustice, but he would come around into the wind as a full-sailed ship does turning a cape. Therefore we find in rugged form in Mr. Gantt that great quality of charm.

The noblest monument we can erect to him is to carry on his ideals. He had no faith in government machinery as government is at present organized.

His ideal held to and illustrated by his life should be ours: Harmony based on Justice.

## To Whom Credit Is Due

The social affairs of the Annual Meeting of the A.S.M.E., which were so great a success this year, as well as the professional features, were directly in charge of the Committee on Meetings and Program, the membership of which is Prof. Dexter S. Kimball, *Chairman*, Walter Rautenstrauch, A. L. DeLeeuw, W. G. Starkweather, and R. V. Wright. As a result of experience in conducting previous conventions, an organization was perfected this year whereby a member of the Committee on Meetings and Program, Mr. R. V. Wright, was made general Chairman of a Sub-Committee for the Annual Meeting, with various sub-divisions devoted to the different activities.

Of these sub-divisions the general Social Functions were in charge of N. F. Jacobus, *Chairman*; President's Reception,

Frederick A. Scheffler, *Chairman*; Dance, E. Van Winkle, *Chairman*; Reunion, F. T. Chapman, *Chairman*; Acquainteceship, Hazen G. Tyler, *Chairman*; Excursions, H. J. Marks, *Chairman*; Catering, Edric B. Smith, *Chairman*; Ladies, Mrs. H. C. Spaulding, *Chairman*, The President's Reception Committee was assisted by the Sections delegates who were present at the meeting, together with a number of the local members.

An effective department of headquarters was the Information Bureau to which Prof. Charles C. Sleffel of Columbia University gave his personal attention, proving to be a second "Mr. Foster."

All of the events of the meeting were conducted with smoothness and in a way that created general satisfaction and gave the greatest possible pleasure to the guests in attendance. For this, credit should not only be given to the Committee on Meetings and Program, and the special committees mentioned, but to the personal efforts of Mr. S. N. Castle, of New York, who gave the results of his experience in his work in past years for the benefit of the committee members, and materially assisted in organizing and coördinating the work.

## Index to Volume 41 of Mechanical Engineering

An index to Volume 41 of MECHANICAL ENGINEERING is now in the course of preparation, which, it is expected, will be issued early in February. A copy of this index will be sent to each member of the Society or subscriber who sends in a written request therefor. In order that no more copies than are necessary to supply the demand may be printed, requests for copies should be received at headquarters not later than February 1.

## DREDGING PUMP OF NOVEL CONSTRUCTION

*(Continued from page 7)*

by Colonel F. B. Maltby and a careful study of the latter's paper in Vol. LIV, Trans.Am.Soc.C.E., has been of material assistance in the preparation of this paper. Colonel Maltby has also very kindly offered suggestions and criticisms during the preparation of this paper.

In a hydraulic pipe-line dredge of the cutter type it is believed that all passages through pipes and pump should be designed so that clogging at any point by more or less irregularly shaped solids will be reduced to a minimum. As a result of tests and also based on observation of the dredges in operation, it is the conclusion of the writer that the modifications described did not decrease the efficiency or capacity of the main engines and pumps, but did greatly increase the output of the dredges as excavating machines.

TABLE 3 DREDGING YARDAGES AND COSTS

| Year 1918-1919 | Yardage Excavated | COSTS | | | | | | | | |
|---|---|---|---|---|---|---|---|---|---|---|
| | | Arbitrary | Unit Cost | Charter | Unit Cost | All Other Charges | Unit Cost | Total | Unit Cost |
| May, June and July | 383,643 | | | 74,887.19 | .196 | 12,143.72 | .031 | 87,030.89 | .227 |
| August | 291,256 | 9,524.07 | .015 | 90,547.20 | .245 | 11,053.66 | .034 | 111,124.93 | .294 |
| September | 578,055 | 18,902.72 | .022 | 63,993.60 | .183 | 44,602.56 | .054 | 127,498.88 | .239 |
| October | 734,056 | 24,003.03 | .027 | 62,486.40 | .147 | 53,717.97 | .061 | 140,208.00 | .235 |
| November | 427,522 | 13,979.97 | .028 | 54,566.40 | .143 | 77,431.04 | .082 | 145,977.41 | .233 |
| December | 597,488 | 16,534.82 | .028 | 66,628.80 | .138 | 61,053.99 | .086 | 147,217.61 | .232 |
| January | 498,340 | 13,293.72 | .029 | 64,145.76 | .136 | 80,726.85 | .097 | 160,168.33 | .262 |
| February | 455,159 | 14,883.70 | .029 | 57,874.80 | .136 | 45,607.02 | .095 | 118,365.52 | .261 |
| March | 222,508 | 7,266.01 | .029 | 24,807.55 | .133 | 35,857.67 | .101 | 67,931.23 | .263 |
| April | 167,481 | 5,478.63 | .029 | 19,027.80 | .133 | 22,203.91 | .102 | 46,708.34 | .264 |
| May | 94,777 | 3,099.20 | .030 | 12,724.80 | .133 | 6,028.60 | .101 | 21,852.60 | .264 |
| June | 130,386 | 4,263.62 | .030 | 13,915.20 | .132 | 8,376.64 | .100 | 26,555.46 | .262 |
| July | 141,233 | 4,618.32 | .030 | 14,275.20 | .131 | 3,528.24 | .068 | 22,421.76 | .259 |
| Preliminary and Auxiliary | 477,193 | | | | | | | | |
| Total | 5,199,107 | 140,846.41 | .027 | 619,680.68 | .119 | 462,331.87 | .089 | 1,223,060.96 | .235 |

All Yardages Computed from Sections Plotted from Soundings.  Above Total Yardage is 70% of Estimated Yardage for Entire Project.  Unit Cost Apply for Total Yardage Through Month Shown.

# LIBRARY NOTES AND BOOK REVIEWS

THE ANALYSIS OF MINERALS AND ORES OF THE RARER ELEMENTS. For Analytical Chemists, Metallurgists. and Advanced Students. By W. R. Schoeller and A. R. Powell. J. B. Lippincott Co., Philadelphia, 1919. Cloth, 6 x 9 in., 239 pp., tables.

This is a practical laboratory guide for those interested in the determination of the so-called rare elements, and is the first work, the authors state, in which the complete analysis of the minerals of these elements has received systematic treatment. The methods have been selected from various authorities, supplemented when necessary by methods devised by the authors. Full details are given in each case.

ASBESTOS—FROM MINE TO FINISHED PRODUCT. Asbestos and Mineral Corporation, New York, 1919. Cloth, 10 x 11 in., 194 pp., 60 pl.

A collection of sixty photographs of asbestos minerals, mines and manufacturing plants, forming a pictorial presentation of the mining of raw asbestos and its manufacture into finished goods. Brief explanatory notes accompany the illustrations.

CATALYSIS IN INDUSTRIAL CHEMISTRY. By G. G. Henderson. Longmans, Green & Co., New York. Cloth, 6 x 9 in., 202 pp.

This, the latest volume of the series of Monographs on Industrial Chemistry which is being published under the editorship of Sir Edward Thorpe, is intended to present broadly our present knowledge of the phenomenon of which it treats. The book reviews the existing literature on catalysis, paying particular attention to the many and varied applications now used in industry. The author hopes that it will also suggest the desirability of further investigations.

COMMERCIAL OILS—VEGETABLE AND ANIMAL. With special reference to Oriental Oils. By I. F. Laucks. First edition. John Wiley and Sons, Inc., New York, 1919. Cloth, 5 x 8 in., 188 pp.

This book is intended to give men without technical training, who are in the oil trade, the technical data and information required in everyday dealings, in concise form. The various commercial oils are listed. For each are given the sources, uses, and the physical and chemical characteristics. The values have been collected from the standard texts, supplemented, particularly in regard to oriental oils, by the results obtained in the author's laboratory.

THE CONDENSED CHEMICAL DICTIONARY. A reference volume for all requiring quick access to a large amount of essential data regarding chemicals, and other substances used in manufacturing and laboratory work. Compiled and edited by The Editorial Staff of the Chemical Engineering Catalog. The Chemical Catalog Company, Inc., New York, 1919. First edition. Cloth, 6 x 9 in., 525 pp.

This volume has been prepared to meet the need for a concise summary of the properties of chemicals. The arrangement of the book is alphabetical. The information given includes variant names, formulas, color and properties, constants (boiling and melting points, specific gravity, etc.), derivation, method of purification, grades obtainable, containers, uses, fire hazard and shipping regulations. Substances made in America are indicated. The volume is fully cross-indexed, and includes a number of useful tables of weights and measures, thermometric scales, etc. Synthetic dyes are omitted. While the compilers state that no attempt has been made to be exhaustive, but merely to give the outstanding facts concerning the chemicals ordinarily met in commerce, approximately six thousand substances are included.

ELEMENTS OF RADIOTELEGRAPHY. By Ellery W. Stone. D. Van Nostrand Co., New York, 1919. Flexible cloth, 5 x 8 in., 267 pp., 125 illus., 33 pl.

This book, by the instructor at the U. S. Naval Radio Station, San Diego, was written primarily for use in instruction at that school; it is believed, however, that it will be helpful in other schools and for self-instruction. The physical rather than the mathematical presentation of the subject is used. A knowledge of physics and mathematics is not necessary. Prominence is given to the development of the subject as illustrated by patents.

FIFTY YEARS OF IRON AND STEEL. By Joseph G. Butler, Jr. The Penton Press, Cleveland, 1919. Leather, 6 x 10 in., 145 pp., portraits.

This volume is one of a special edition, limited to one hundred copies, of an address delivered before the American Iron and Steel Institute in 1917, in which the author gives his personal reminiscences of the development in iron and steel making as observed during an active participation in the industry in eastern Ohio, which began in 1857. To this are added some brief data on the early history of iron and steel, and a chapter on the activities of the American iron manufactures during the great war. The volume is profusely illustrated with portraits of men of note in the industry.

FOREST PRODUCTS. Their Manufacture and Use. Embracing the Principal Commercial Features in the Production, Manufacture, and Utilization of the Most Important Forest Products Other than Lumber, in the United States. By Nelson Courtlandt Brown. First edition. John Wiley and Sons, Inc. New York, 1919. Cloth, 6 x 9 in., 471 pp., 120 illus., 1 pl., tables.

The object of this book is to provide a brief account of the chief commercial features involved in the manufacture and use of such products as wood pulp and paper, tanning materials, veneers, cooperage, naval stores, distilled products, charcoal, maple sugar and syrup, dye woods, rubber, cork, excelsior, shingles, mine timbers, etc., which will serve as a text for students and as a book of reference. The volume reviews a field in which, the author states, but little has been written on American practice. Brief bibliographies are appended to the chapters.

GEODESY. Including Astronomical Observations, Gravity Measurements, and Method of Least Squares. By George L. Hosmer. First edition. John Wiley & Sons, Inc., New York, 1919. Cloth, 6 x 9 in., 368 pp., illus., tables.

The author has endeavored to produce a textbook adapted to a course of moderate length which would make clear the underlying principles and emphasize the theory as well as the details of field work. The methods of observing and computing given are consistent with the practice of the Coast and Geodetic Survey.

HAND-BOOK OF FIRE PROTECTION. By Everett U. Crosby, Henry A. Fiske and H. Walter Forster. Sixth edition. D. Van Nostrand Company, New York, 1919. Flexible cloth, 5 x 7 in., 757 pp., illus., tables.

This work has long been a standard reference book for insurance engineers and inspectors, in which those methods of fire protection which have crystallized into good practice are collected and presented concisely. The present edition has been enlarged by new chapters and illustrations, and has been carefully revised.

HYDROLOGY. The Fundamental Basis of Hydraulic Engineering. By Daniel W. Mead. McGraw-Hill Book Co., New York, 1919. Cloth, 6 x 9 in., 647 pp., illus., maps, tables.

Lack of knowledge of the fundamentals of hydrology and of the importance of hydrological factors has been responsible, the author believes, for more failures in hydraulic-engineering projects than defects in structural design. To assist in removing this ignorance he has here set down some of the more important facts and principles, omitting everything that his long experience has not shown to be of practical importance. Carefully selected lists of references are appended to each chapter, to enable readers to complete their study of any phase of the subject which has not been sufficiently treated within the limits of the book itself.

A MANUAL OF PHYSICAL MEASUREMENTS. By Anthony Zeleny and Henry A. Erikson. Fourth edition. Cloth, 5 x 8 in., 261 pp., 129 illus., tables.

This manual is an outline of the laboratory course given to students of general physics at the University of Minnesota. One hundred and thirty-three experiments are included. Directions for assembling the apparatus and performing the experiments are given. Appendixes give the mathematical tables needed by users of the book.

THE METALS OF THE RARE EARTHS. By J. F. Spencer. Longmans, Green & Co., New York, 1919. Cloth, 6 x 9 in., 279 pp., diagrams, tables.

This volume supplies a review of our knowledge of these metals, in which their occurrence, separation, compounds and uses are described. An extensive list of references is included, which contains, the author believes, all the important papers published on the subject, as well as most of those of lesser importance.

OPPORTUNITIES IN CHEMISTRY. By Ellwood Hendrick. Harper and Brothers, New York, 1919. Boards, 5 x 8 in., 102 pp.

This little volume attempts to explain the relation of chemistry to our activities in general and the usefulness of chemical knowledge in various walks of life. It will serve to introduce the subject to those considering its possibilities as a profession.

OSMOTIC PRESSURE. By Alexander Findlay. Second edition. Longmans, Green & Co., New York, 1919. Cloth, 6 x 9 in., 116 pp., illus., tables.

The author of this monograph gives an account of the theories of osmotic pressure, the methods of determining it, and the general theory of solutions. An extensive bibliography is included. In preparing this new edition account has been taken of the results obtained by investigators during the past six years.

THE OUTLOOK FOR RESEARCH AND INVESTIGATION. With an Appendix of Problems Awaiting Solution. By Nevil Monroe Hopkins. D. Van Nostrand Company, New York, 1919. Cloth, 5 x 8 in., 241 pp., 1 pl., 6 portraits.

The object of this book is to stimulate a more general interest in American research, to indicate its necessity at this time, to explain the educational requirements for research workers and to point out the ways in which such work is done inefficiently at present. Appended to the volume is a list of suggested lines of research.

THE PHYSICAL CHEMISTRY OF THE METALS. By Rudolph Schenck. Translated and annotated by Reginald Scott Dean. First edition. John Wiley and Sons, Inc., New York, 1919. Cloth, 6 x 9 in., 239 pp., illus., tables.

This volume is the outcome of a course of lectures delivered in the Technischen Hochschule at Aachen before the engineers of the district. The purpose of the lectures was to show the use of chemical statics and to explain the applications of physical chemistry in the study of smelting operations and metallurgical processes. The translator has made additions to the original text when necessary and has revised the numerical data to agree with the accepted values.

PRACTICAL MATHEMATICS FOR HOME STUDY. Being the essentials of Arithmetic, Geometry, Algebra and Trigonometry. By Claude Irwin Palmer. First edition. McGraw-Hill Book Co., Inc., New York, 1919. Cloth, 5 x 8 in., 493 pp., illus., tables.

During the past fifteen years the author has taught mathematics in the evening school at the Armour Institute of Technology to classes of men engaged in practical pursuits of various kinds. For this purpose a course of instruction was published in four volumes, which has now been arranged in one volume, with particular reference to the needs of home students. In the present edition, a few new topics have been added, together with many solutions of problems. The book includes arithmetic, geometry, algebra, logarithms and trigonometry. It is intended for adult students and the three thousand exercises and problems are selected to illustrate actual practical problems.

PRACTICAL WINDINGS OF ALTERNATING CURRENT MACHINERY. By J. Herbert Wickman. J. H. Wickman, Milwaukee. Paper, 6 x 9 in., 36 pp.

This pamphlet is intended to give the necessary information to armature winders, without mathematical calculations or theoretical discussions. Directions for winding, reconnecting for other voltages or changing the windings from one phase to another are given.

PRINCIPLES OF DIRECT-CURRENT MACHINES. By Alexander S. Langsdorf. Second edition. McGraw-Hill Book Company, Inc., New York, 1919. Cloth, 6 x 8 in., 460 pp., 314 illus.

This book has been prepared to place before students of electrical engineering a reasonably complete treatment of the fundamental principles underlying the design and operation of all types of direct-current machinery. Instead of attempting to touch the "high spots" in the whole field of direct-current engineering, attention has been concentrated upon certain important features which are ordinarily dismissed with brief mention, but which, the author believes, are vital to a thorough grasp of the subject. The present edition has been thoroughly revised and enlarged in accordance with the results of experience in the use of the book.

## SYSTEMS FOR MUTUAL CONTROL OF INDUSTRY

(Continued from page 25)

labor managers together have what they call a Board of Labor Managers.

"When there is to be a general revision of wages or hours, or a new agreement is to be made, the Board of Labor Managers and the representatives of the Union Joint Board meet and work it out. When there are adjustments to be made, that apply to only one shop, the individual or group of employees affected takes it up with the labor manager of the plant and assistance may be given by the Section or shop chairman. If a satisfactory adjustment cannot be made, one of the managers or deputies of the Joint Board is called in. In the vast majority of cases, amicable adjustments are arrived at. Where such an agreement cannot be reached, the matter is left to impartial arbitration, for which permanent machinery is created by the industry itself—in effect an industrial court.

"The employers' association of each city, together with the joint board of the Union, select an impartial person to act as judge of all disputes. Technically he is the Chairman of a labor adjustment board, consisting of the labor managers and the union representatives who argue cases before him. His salary and the expenses of his office are paid half by the employers and half by the unions. He hears no cases of men who have stopped work. Strikes are illegal. The employer may suspend a worker, but if on appeal to the Chairman, this dismissal is found unjustified, the employee is reinstated and must be paid by the employer for the time he lost. This is the machinery of democracy. The institution of the impartial chairman or judge, is a recognition that law, order and government must prevail in industry, that peace must be maintained and the interests of the public conserved. But all this is not enough. With each market organized separately, employees complained that the same work was paying more in one place than in another, and there was a constant tendency to upset the wage agreements. The union, therefore, agreed with the employers, that a national plan of collective bargaining must be devised in order to stabilize labor conditions. A few weeks ago representatives were elected by the National Association of Clothing Manufacturers and the National Union, to form a joint industrial council for the entire industry, and they are now working on a national agreement to apply to all markets."

Dr. Leiserson stated that although this plan has not solved the whole labor problem, or eliminated all labor troubles, it has shown that war is not necessary to settle labor disputes. "The plan is beyond the experimental stage," said Dr. Leiserson, in conclusion, "It began five or six years ago in the Hart, Schaffner & Marx factories of Chicago, and the success of it in these factories, has spread it through the entire industry. It has taught the employers to respect union leaders whom they once thought were crooked advocators and it has made unions and their leaders have confidence in their employers and coöperate with them. It has enabled unions and employers to work together on such problems as the elimination of strikes, the introduction of machinery, time studies and standardization of occupations, increasing production and maintaining shop discipline. Disputes on all these questions have not been eliminated. They never will be; but they have been settled, in a peaceful, orderly manner according to laws of the industry, which labor and capital alike have had a share in framing. More than all this, however, the plan has enabled the unions to take on a peaceful character and to concern itself with the real problems of collective bargaining instead of emphasizing war policies as they have to do as long as employers fight them."

# MECHANICAL ENGINEERING
## THE JOURNAL OF THE AMERICAN SOCIETY OF MECHANICAL ENGINEERS

Volume 42        February, 1920        Number 2

# An Improved Weir for Gaging in Open Channels

## By CLEMENS HERSCHEL,[1] NEW YORK, N. Y.

*In the following paper are presented the results of a series of investigations on weirs which should mark an epoch in the measurement of water by that means. Asked by the Power Test Code Committee of The American Society of Mechanical Engineers to advise it with regard to weir measurements, and supplied by Engineering Foundation with the necessary funds to conduct the experiments, the author devised an entirely new weir which was tested in the Hydraulic Laboratory of the Massachusetts Institute of Technology. The tests revealed that for discharges of from 0 to 9.55 cu. ft. per sec. per ft. of weir length, the limits covered by the experiments, the quantity of water flowing, Q, was directly proportional to the difference, d, in two pressures, one measured just upstream of the weir—corrected automatically for the velocity of approach, and the other measured at the crest, the formula being: Q = 5.50 d. The weir, which bears no resemblance to the sharp-edged weir now in use, has a 2:1 slope of approach to the crest. The crest is made in the form of an arc of a circle and is hollow for observing the pressure at that point. The nappe of the stream is supported for a short distance beyond the crest by another 2:1 slope. The paper opens with a résumé of the history of weir measurements and contains descriptions and diagrams of the apparatus and tests of the new weir, with tabular and graphic reports of the results obtained.*

> " Daniel Bernoulli often said to me to eschew all those complicated formulæ; he believing that nature works according to laws that are too simple to lead to them; and if one finds such, that it is because one's computations have been based on false preliminary assumptions."—R. Wolf, Biographies Relating to the History of Civilization in Switzerland,[2] Vol. 3, (1860), p. 176.

T0 ANY ONE whose memory of hydraulic engineering in the United States, more especially in New England, runs back 60, or only 50, years, the standing at that time of the "Francis formula" for weir discharge will appear as little short of infallibility, or of perfection. Courts of justice recognized it as decisive; lawyers of eminence played with it as though it were one of the choicest morsels from the tree of knowledge; it was the last word in gaging the flow of water. And there was reason for such an estimate of its value at the time.

James Bicheno Francis was then easily the leading hydraulic engineer in the United States, with perhaps not over half a dozen of the species in the whole country. Grappling as a young man with the problem of insuring their lawful quantities, day by day, to each of the seven "Proprietors of the Locks and Canals on Merrimack River," at Lowell, Massachusetts, out of the common trough or canals which fed them collectively, he may be said to

have been the first one who thus brought order out of the disorder which that communistic distribution of water—naturally accompanying the American system of power distribution canals for manufacturing purposes, and now passing away in favor of the electrical distribution of power—always brings with it. For the purpose named he instituted, about 1850, the series of hydraulic experiments described in his book entitled Lowell Hydraulic Experiments; and he did this with all the accuracy obtainable at the time and place that has been named.

From another point of view just described were his experiments of a new order of importance. If any one will consult Weisbach's Experimental Hydraulik, or the later editions (1855-60) of Weisbach's Mechanics, he will be surprised at the Lilliputian dimensions of the apparatus then in vogue as the basis of engineering knowledge for use in the practice of hydraulics. Thus Weisbach's experimental reservoir had an area of 1.35 sq. ft.; and his ajutages were ordinarily about three hundredths of a foot in width, or diameter. Of course, we know that Atwood's machine, of extremely moderate dimensions, will measure the force of gravity exercised by the globe which we inhabit; but it requires more confidence in the "horse sense," no less than the technical ability of hydraulic engineers, than is possessed by the average mill owner or his agent to give him confidence in the work of the engineer he employs, should that engineer base it on experiments such as inspired Julius Weisbach or his immediate disciples. So that the experiments of Francis, conducted out of doors on a scale commensurate with their importance, at the lower locks of the Concord River, came to the interested "Proprietors" and others as a refreshing and new conquest of mind over matter. They were much more reliable because made on a natural scale; exact, as every one could appreciate, from the care taken in making the experiments and the computations following the experiments, and thoroughly reliable.

At that time arose the expression, "made with a standard weir," which has remained alive to the present day, although, as will presently be seen, there was no propriety whatever for calling any weir a "standard." The only standard water-gaging apparatus is a tank or reservoir. Weirs, orifices, venturi tubes and meters, and other water-gaging apparatus and methods, can only become truly rated by comparison with tank measurements, and thus thereafter competent to give reliable service. Such comparisons, moreover, establish the *degree* of reliability of the several methods that have been subjected to comparison, and from this point of view the weir has for many years been falling into deserved disrepute.

To treat the matter somewhat chronologically, there is the article in *Engineering News* for November 10, 1898, by the present writer, protesting against the multiplicity of formulæ that sharp-edged weirs and orifices, and present methods of observing heads upon them, bring in their train; and suggesting that observations be taken *at* the weir or orifice, and that weirs and orifices cut "*in* a thin plate" be abandoned.

In the Transactions of the American Society of Civil Engineers for 1914, the discussion on the paper of R. R. Lyman, Mem.Am.Soc.C.E., brought out clearly the number of methods there were in use to measure "head upon the weir," and their idiosyncrasies.

---

[1] Hydraulic Engineer, 2 Wall Street. Past-President, Am.Soc.C.E.

[2] " M. D. Bernoulli ma souvent dit de me défier de toutes ces formules compliquées; il croit que la nature est trop simple pour y mener; et que si on en trouve, c'est qu'on a fondé ses calculs sur des fausses hypothèses."—R. Wolf, Biographien zur Kulturgeschichte der Schweiz, Band 3 (1860), p. 176.

Abstract of a paper to be presented at the Spring Meeting, St. Louis, May 24 to 27, 1920, of THE AMERICAN SOCIETY OF MECHANICAL ENGINEERS. All papers are subject to revision.

These several methods go far beyond any forethought of the pioneer experimenters in that sort of work, and definitely confirm the rule that when any one formula is to be used to compute weir discharge, the weir and methods of taking observations of flow over it must imitate with slavish exactness the construction of the weir and methods of taking observations that were followed in constructing the selected weir formula, a matter of annoyance or difficulty as a rule.

Fteley and Stearns, for example, Trans. Am. Soc. C.E. for 1883, p. 5, say: "It is, however, almost impossible to cover all cases that may occur in practice."

Bazin, in his 1890 article, Part I, for that year, of *Annales des Ponts et Chaussées*, p. 4, and elsewhere, says that to attempt to experiment on all cases possibly to be met with in practice (instead of setting up a standard form to be used in such work) must be pronounced impracticable. And in his 1898 article, Part II for that year, p. 161, he says: "Each type of weir would therefore require a special study; and face to face with a complexity like this, one must abandon the idea of establishing formulæ for general application."

R. E. Horton, Trans. Am. Soc. C.E., 1914, p. 1300, tells us that "of the making of many formulæ there is no end." Also, p. 1326,

be automatically corrected for "velocity of approach," if one pleases, or corrected by computation before being used in the above formula, if preferred), and the other taken *at* the crest. In both cases *d* is obtained by measuring the elevation of a water surface in a pail by means of a hook gage, or otherwise as desired. This *difference* of pressures may also be obtained by means of one of the many devices already invented and on the market for measuring differences of pressure, in which case the register may indicate mechanically the quantity passing at any moment, sum it up for any given period of time, portray it graphically on a chart, or simultaneously do all three of these recording operations.

It will be observed that the process just described is not open to the criticism of R. E. Horton, mentioned above, who said: "No one equation, of the form he (Francis) uses, $Q = mh^n$, can fit accurately a series of experiments on weirs of the type (sharp-edged weirs) herein considered." It does this, moreover, with the added simplicity of making the exponent *n* equal to 1.

And the reason it does all this is because it is carried out with a weir that is not a "sharp-edged weir," the type that has rested like the incubus of an evil spirit upon the minds of all hydraulicians for full 200 years, ever since the Marchese G. Poleni first wrote about weirs in De Motu Aquae Mixto, Patavii, 1717. Bazin

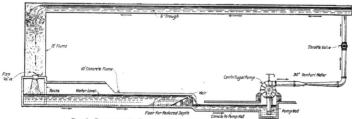

FIG. 1   DIAGRAMMATIC REPRESENTATION OF THE EXPERIMENTAL APPARATUS

"If Francis had used higher and lower heads, he would have discovered that no one equation of the form he uses, $Q = mh^n$, can fit accurately a series of experiments on weirs of the type herein considered."

Hamilton Smith[1] saw this, and very likely Francis did, for Francis was very particular to state that the formulæ he gives must only be used between stated limits of head, and only upon fulfillment of other stated conditions of weir construction and operation.

Finally, in the discussion on the Paper by Floyd A. Nagler, Trans. Am. Soc. C.E. for 1918, this subject now under consideration was again debated by interested hydraulic engineers, the present writer among the number; and the status of the whole question was summed up by F. P. Stearns, Mem. Am. Soc. C.E., himself an experimenter on weirs of more than 40 years' experience, when he said with startling, yet righteous, precision that "in the present state of the art the weir is not an accurate instrument for the measurement of water."

### AN ADVANCE IN THE STATE OF THE ART BELIEVED TO HAVE BEEN NOW ATTAINED

After this survey of the state of the art, and to give a résumé of what this article will endeavor to establish, it may not be improper to say at once that it treats of a weir whose discharge is represented by a straight-line formula from 0 up to 9.55 cu. ft. per sec. per ft. in length of weir, and thus continuing, presumably, indefinitely beyond that quantity. Its equation is $Q = 5.50d$; *d* being the difference of two observations of water height or pressure; the one taken just upstream from the weir (taken so as to

and others have thought that to attempt to set up a standard form of weir for measuring water was visionary, in view of the many variations in weir discharge so easily brought about. But this very thing was one of the objects of the experiments now under consideration, and time alone will tell whether or no it has not after all been accomplished.

### THE POWER TEST CODE COMMITTEE OF THE AMERICAN SOCIETY OF MECHANICAL ENGINEERS, AND ENGINEERING FOUNDATION

After this much said in anticipation, we will go back now to a chronological order of events. The American Society of Mechanical Engineers, having appointed a Committee to draft a revised form of Power Test Code in 1917, and invited the three other Founder Societies to appoint advisory committees to the same end, the writer was appointed as one such advisory member. It thereupon very soon appeared that the matter of weir measurements and of their accuracy was in a most deplorable state of confusion. On the proper representations having been made to Engineering Foundation that a new set of experiments on weir discharge were liable to lead to a considerable advance in knowledge in this branch of applied science, it made an appropriation for that purpose, to be expended by the writer.[1] The experiments about to be described have been the result of these preliminaries.

[1] Acknowledgment is also here gratefully made to the President and Faculty of the Massachusetts Institute of Technology, especially to Prof. Edward F. Miller; to Wm. F. Uhl, B. S. Rose and H. L. Woodruff of the office of Chas. T. Main and to Mr. Main; to Chas. W. Sherman, H. P. Eddy, Jr., and others of the office of Metcalf & Eddy and to their principals, all of Boston; and to Builders Iron Foundry, of Providence, R. I.; for most efficient aid rendered in various ways at various stages in the conduct of the experiments herein described; and done in a way that amounted to a contribution to enable all the described work to be done.

[1] Hydraulics, John Wiley & Sons, 1886.

### THE HYDRAULIC LABORATORY OF THE MASSACHUSETTS INSTITUTE OF TECHNOLOGY

The place selected for conducting the experiments was the Hydraulic Laboratory of the Massachusetts Institute of Technology. This laboratory had barely been completed when the United States entered the World War, and all the energies of the Institute thenceforth had been devoted to its notable war work to a degree that left improvement in and perfection of its school apparatus rightfully a secondary consideration. As a consequence, the hydraulic part of the laboratory on closer study was shown to have many remediable defects, which, no doubt, will be eliminated in course of time. These, again, affected both the difficulty of making the contemplated experiments and the precision at which results could be attained. But the choice had been made, many reasons existed for persevering in the place where the work had been planned, and thus the experiments were completed as they had been projected.

For a description of the Institute Hydraulic Laboratory reference is made to an account in the March, 1918, number of the *Journal* of the Boston Society of Civil Engineers, by Prof. Geo. E. Russell of the Institute. Briefly stated, the apparatus used in the present experiments and shown diagrammatically in Fig. 1 consisted of a 350-hp. Ball angle-compound engine, 240 r.p.m., direct-connected to a 30-in. Worthington centrifugal pump, having some 48 cu. ft. per sec. maximum capacity at 40 to 45 ft. lift. This pump takes its suction from a system of canals under the floor of the basement, which are or may be connected with the Charles River Basin.

Right here arose one great obstacle in the making of these hydraulic experiments. The canals that have been named at times took to "surging," causing oscillations of the pump-well surface, and these, in turn, were transmitted through the pump and could plainly be seen at the upstream end of the discharge pipe. They were also indicated by the 30-in. venturi meter set in line of the discharge pipe, next downstream from the pump. Two whole afternoons' work were rendered useless from this cause. But by experience and careful work any gross effect from this cause could be avoided.

The discharge pipe delivered into a riveted trough of steel, 5 ft. by 5 ft. and about 135 ft. long, at some 40 ft. higher elevation than the pump. In it was set the 16-in. by 5¼-in. venturi meter used to measure quantities smaller than about 8 cu. ft. per sec. (Sept. 25, 26 and 29), the larger quantities being measured by the 30-in. meter already spoken of.

Both these meters were read by means of mercury straight-tube manometers, as made by the builders of the meters. The scale of these manometers was based on a coefficient of discharge for the meters of 0.99. These were the facilities at hand, or readily procurable and installable; and though not so precise as a tank measurement would have been, were sufficiently so for the pioneer work to be done. It is not often, as yet, that one can get a reservoir to hold 46 cu. ft. per sec. for the period of a hydraulic experiment; and in the opinion of the writer the venturi meter is the next best thing. It certainly was at the hydraulic laboratory.

That the venturi meter is of a high order of accuracy has been shown by a compilation by the writer for the Power Test Code Committee of all attainable relevant experiments made with it during its existence of 30-odd years. A very significant series of tests, made by F. H. Shaw, Mem.Am.Soc.C.E. of a meter at Lancaster, Pa., show an average variation of 0.036 of 1 per cent for 44 tests covering a period of eight years.

To return to our present meter observations: The scale of the manometers read tenths of cubic feet per second; and hundredths could readily be estimated on both of the meters used.

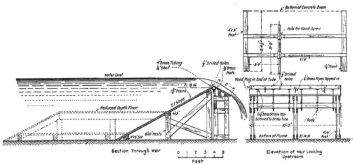

FIG. 2 THE NEW WEIR (THE SURFACE CURVE OF WATER DISCHARGING OVER THE WEIR IS DRAWN FROM MEASUREMENT.)

From the 5-ft. trough, the water fell into a vertical flume or wheel pit 12 ft. in diameter, made of riveted steel and 22.5 ft. high; and through a wooden adjustable flap valve in the floor of this flume, fell into a concrete tail-race pit, which, through racks and gridiron-valve gates, discharged into the concrete channel about 72 ft. long that led to the weir to be tested. The discharge of the weir fell back into the system of canals which fed the centrifugal pump, thus completing the cycle actuated by the Ball engine and pump.

The free discharge of the 5-ft. trough into the 12-ft. vertical flume could be rendered somewhat less turbulent by keeping the flume-water elevation as high as possible without waste, this again being crudely controllable by means of the wooden flap valve in the floor of the vertical flume. But this whole arrangement and the free discharge through the flap valve into the tail-race pit below were imperfect arrangements, acknowledgedly only temporary in character, and were the cause, together with the "surging" of the pump-well water above referred to, of many of the difficulties experienced in making these experiments. They also were the principal cause of most of the lack of precision that remains and is inherent in the final results.

It would have been gratifying to remove some or all of these crudities, and also to greatly extend the scope of the experiments, but the limits of the appropriation available forbade going beyond the work that was actually done.

To vary the discharge, the pump operating at a uniform speed as governed by a flywheel governor, and the boiler pressure also being fairly constant, a throttle valve in line of the pump discharge pipe was used. When the small 16-in. meter set in the 5-ft. trough was in action, the quantity and constancy of the flow leading to the meter and the efficient action of the pump, in spite of the small quantities lifted, could be regulated by opening a waste valve out of the pump discharge pipe. To a minute extent, certainly sufficiently to remove all air from the pump discharge pipe before it reached the 30-in. meter, the same office was performed by keeping in action a 2-in. steam siphon, set at the extreme upstream end of the pump discharge pipe and ordinarily used to exhaust air from the pump before starting it.

The concrete channel above referred to was nominally and almost exactly 10 ft. wide, and had at its downstream end a sharp-crested weir whose crest was about 5.35 ft. above the bottom of the channel. The new form of weir was so designed that it could be built over and around the old weir, and again removed without injury or displacement of what was found on the premises, and with its crest at nearly the same elevation above the bottom of the concrete flume.

### THE NEW WEIR

In its general outlines, the new weir, Fig. 2, was described in the *Engineering News* 1898 article above referred to. More particularly was it forecast in the Trans.Am.Soc. C.E. discussion of 1918 that has been named. So here was a case of an invention, or vision, if one pleases, in practical hydraulics that had to wait 20 years in the brain of the engineer for lack of public or private facilities to test it. An attempt to experiment as has now been done was made some 10 years ago, but was frustrated by the writer being informed by the manager of a richly endowed institution that its funds were only available for research in sciences that had no practical applications. The wonder is that any progress at all is made in this most useful of the arts. One endowment only exists as yet for the development specifically of applied science or engineering—the Engineering Foundation, begun by Ambrose Swasey, of Cleveland, Ohio; and it was most gratifying to the writer to have been allowed to endeavor to illustrate some of the work of which such an endowment is capable.

The fundamental idea followed in the design of the new weir was to have the water to be measured conducted over the weir in a gentle manner, and so to have it flow smoothly and regularly from the time it first encounters the weir construction until it leaves it. Instead of allowing the body of water to impinge with more or less violence, according to the velocity with which it approaches the weir, against a perpendicular wall in its path (the upstream face of the ordinary weir) it is gently led to the crest by a 2:1 slope. Instead of striking or being torn over a sharp edge at the crest, the crest is made in the form of the arc of a circle; and instead of bothering about air under the nappe, the nappe is supported on

FIG. 3   THE SHORT WEIR IN ACTION
Leakage on each side could be measured either by the new formula; or by carding it in a pail and weighing the amount caught in an observed period of time; or by both these methods simultaneously.

another 2:1 slope downstream from the crest. Moreover, the crest is made hollow, so that observations of the pressure, or lack of full pressure, whichever the water may elect to exercise, can be taken *at the crest*, not at a distance upstream from the crest varying according to the fancy of the experimenter.

If the quantity passing the weir turns out to be a function of this observed pressure, well and good. If not, we will see what virtue there is in the *difference* of water elevations or pressures, the one taken upstream from the weir and the other taken by means of the hollow weir crest.

The difference referred to proved to be the sought-for solution of the problem at hand, and hereafter it alone will be closely considered. A United States patent has been applied for, covering the weir construction herein described.

The orifice in the side of the concrete flume conducting water to the hook-gage pail for measuring water elevations upstream from the weir was made flush with the flume side, at a point 13.9 ft. upstream from the weir crest center, and 1.68 ft. below it in elevation.

Furthermore, to connect the water passing over the weir with the hollow of the weir crest, a series of ⅛-in. holes were carefully bored through into the latter. This was done by first tapping in a solid brass plug, and then boring a hole ⅛ in. in diameter through this plug. A natural inclination would be to bore these holes at right angles to the surface curve of the water, but as this varies with the quantity passing the weir, such an inclination is impracticable. There are only two other natural directions for such piezometer holes—either vertical, or at right angles to the upstream slope of the weir; and if the latter, naturally at the line of tangency of that upstream slope and the arc of the weir crest. The vertical holes are called "Orifices No. 1" and those perpendicular to the 2:1 slope, "Orifices No. 2". The original intention of the writer had been to make the crest tube so that it could be revolved around its horizontal axis, and thus find the best position of the piezometer orifices, but a cumbersome and unduly expensive experiment like this was at once abandoned, and in view of the success already attained, may now be disregarded by engineers.

Both kinds of piezometric orifices were tried, and the last-named proved much the superior for general use.

To vary the velocity of approach a false bottom was put into the channel for some of the experiments. These will be designated as having been made with "reduced depth" and others with "full depth."

The diagrammatic drawing, Fig. 1, will make clear the description of the weir test apparatus given in the foregoing paragraphs.

To increase the quantity per foot in length of weir, the available capacity of the pump was in many of the experiments made to pass over only about 3.3 ft. of the length of weir, as shown in Fig. 3, while other series of the experiments used a weir about 10 ft. long.

As these experiments had a practically useful end and aim —the establishing of a commendable method of weir measurements—end contractions were ruled out from the very beginning. They are of no use, ordinarily, and only complicate the situation. Even in the few cases of working with sharp-crested weirs, when side contractions are introduced so as to cause greater depths upon the weir with a diminishing quantity of water, or so as to cause a selected formula to apply, they could be avoided at the cost of a very little extra carpenter work. But with the straight-line formula about to be established and valid from zero up to the maximum found by the experiments herein described, there is no occasion whatever to change the length of the weir so as to accommodate the quantity of water about to pass it, and side contractions become either foolish or pseudo-scientific. None of the following experiments includes them.

This apparatus exists as yet only in theory and should be experimented with at the first favorable opportunity, because on drawing the water out of the flume at the close of our experiments a broken and leaky joint was discovered in the rubber pipe that carried the water pressure from the vertical tube to the hook-gage pail, so that the readings taken proved unreliable, or of no record value whatever.

### NEED AND SCOPE OF FURTHER EXPERIMENTS

A crew of six men, including three observers, with a captain, had been working with the apparatus from September 10 to October 3, 1919, and it was with great regret that the writer gave the order to dismantle the weir because so much remained to be discovered.

But October 3 was Friday, the last day of the week for us. The laboratory closed at 5 p.m., and on Monday the regular school term commenced. The appropriation for the work was nigh exhausted and it was high time we quitted the premises.

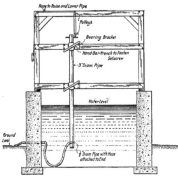

FIG. 4 APPARATUS TO OBSERVE THE WATER ELEVATION UPSTREAM FROM THE WEIR, AUTOMATICALLY CORRECTED FOR "VELOCITY OF APPROACH"

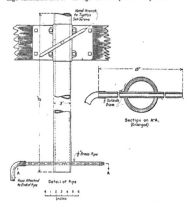

FIG. 5 DETAILS OF APPARATUS SHOWN IN FIG. 4 TO OBSERVE THE WATER ELEVATION UPSTREAM FROM THE WEIR

### AUTOMATIC CORRECTION FOR VELOCITY OF APPROACH

In the photographic view of the new weir in action, Fig. 3, a vertical brass pipe dipping into the water will be noticed in the background, which is also shown in detail in Figs. 4 and 5. This was intended as a means of observing the water elevation upstream from any part of the weir, automatically corrected for velocity of approach, by taking in the water pressure through a pitot orifice pointing upstream and transmitting it to a hook-gage pail in the usual manner from out of the downstream side of the vertical tube. A pointer above water, parallel to the pitot tubular orifice, which should point upstream, insures its correct position. Our experience leads us to advise that the vertical tube be not used as a reservoir, but only as a support for the pitot orifice.

The tube is held in its guides by two clamp screws, and may readily be set so as to bring the pitot orifice in the locus of mean velocity as it may be judged or found. In these experiments it was set at 6/10 depth of water, and 1/3 the width of the channel from either side. A scale of feet and tenths on the tube, reading up from zero at the pitot orifice, and aided by another scale or gage painted on the side of the concrete flume, reading up from zero at the flume bottom, readily permitted this. The apparatus is mounted on two beams crossing the water channel and at a convenient height above it.

There was, first of all requiring further investigation, the effect of the radius of the weir crest. Bazin had experimented on a weir composed of two 2:1 slopes meeting at a sharp angle and with a little pipe to indicate pressure "under the nappe," *let into the body of the weir slope upstream from the crest.* This can not enlighten us very much, but is the nearest approach to our experiments known to the writer.

Our weir crest had a radius of 0.198 ft.; and the outside surface was hard and smooth oil paint.

No doubt the radius of the weir crest is a factor in the application of the formula found for weirs of the construction shown in this article, but it remains for other sets of experiments to reveal the limits and detailed effects of this radius.

### THE EXPERIMENTS

With these preliminary statements, the record of the 40 experiments made is given in Table 1. Four of them have been marked doubtful on account of actual or supposedly defective observations and all those made September 9 and 30, also part of those on September 25, had to be rejected as worthless on account of the previously mentioned violent "surging" of the water discharged by the pump.

A separate set of gagings was made October 3 to compare the indications of the Institute's sharp-edged weir with the Institute's 30-in. venturi meter. Owing to the fact that this meter is set in line of the pump discharge pipe, only 12 ft. distant from the pump, it had been a question whether at that short range it would meter accurately, and the experiments of

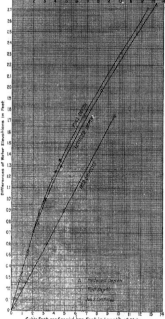

FIG. 6  DIAGRAM SHOWING THE RESULTS OF THE 40 EXPERIMENTS MADE

October 3 were made for the purpose of answering this question. The system for observing head on the weir, available, was that known as the Fteley and Stearns method. A metal plate was set in the side of the concrete flume, flush with it, and an orifice in this plate in the present instance was situated 6.8 ft. upstream from the crest and 1.68 ft. below it in elevation, connected with a hook-gage pail in the usual manner.

RECORD

Pump started about 12.30 p. m., discharging 30 cu.ft. per sec. at 1 p. m. All quantities corrected for velocity of approach.

| | | WEIR Cu. Ft. per Sec. | METER |
|---|---|---|---|
| Oct. 3, 1919, 2.05 to 2.10 p. m., both inclusive.. | | 38.13 | 38.30 |
| 2.50 " 2.55 " " " | " | 30.33 | 31.25 |
| 3.50 " 3.55 " " " | " | 17.91 | 18.80 |
| 4.40 " 4.45 " " " | " | 11.73 | 11.00 |

Having in mind the imperfections of the whole test apparatus from the canals in the basement, through the pump, riser, 5-ft.

TABLE 1  EXPERIMENTS WITH THE HERSCHEL WEIR

| Date 1919 | Length of Run P. M. | Velocity of approach, ft. per sec. | Depression = Piezometer height (corrected for velocity of approach) less weir indication. | | Quantity of weir per ft. of weir length. cu. ft. per sec. |
|---|---|---|---|---|---|
| | | | As measured | Corrected for v. of a. | |
| Sept. 10[1] | 2.45–2.50 | 0.70 | 1.332 | 1.340 | 4.53 |
| | 4.00–4.05 | 0.58 | 1.182 | 1.187 | 3.82 |
| | 4.30–4.35 | 0.48 | 0.991 | 0.995 | 3.06 |
| | 4.55–5.00 | 0.33 | 0.700 | 0.702 | 2.03 |
| Sept. 11[1] | 1.50–1.55 | 0.24 | 0.522 | 0.522 | 1.49 |
| | 2.20–2.25 | .... | 0.347 | 0.347 | 0.95 |
| | 2.50–2.55 | .... | 0.279 | 0.279 | 0.75 |
| | 3.20–3.25 | 0.24 | 0.508 | 0.508 | 1.47 |
| | 3.55–4.00 | 0.45 | 0.962 | 0.965 | 2.87 |
| | 4.45–4.50 | 0.62 | 1.240 | 1.246 | 4.07 |
| Sept. 15[2] | 2.15–2.20 | 1.24 | 1.319 | 1.349 | 4.48 |
| | 3.00–3.05 | 0.92 | 1.001 | 1.014 | 3.14 |
| | 3.50–3.55 | 0.72 | 0.790 | 0.798 | 2.34 |
| | 4.25–4.30 | 0.50 | 0.529 | 0.533 | 1.53 |
| | 4.55–5.00 | 0.30 | 0.305 | 0.307 | 0.85 |
| Sept. 17[1] | 2.30–2.35 | 1.62 | 2.683 | 2.723 | 12.53 |
| | 3.05–3.10 | 1.39 | 2.279 | 2.309 | 10.18 |
| | 3.40–3.45 | 1.00 | 1.725 | 1.741 | 6.87? |
| | 4.20–4.25 | 0.63 | 1.229 | 1.237 | 4.157 |
| | 4.50–4.55 | 0.40 | 0.864 | 0.868 | 2.57 |
| Sept. 18[1] | 1.55–2.00 | 2.87 | 2.603 | 2.733 | 13.38 |
| | 2.30–2.34 | 2.60 | 2.374 | 2.484 | 11.53 |
| | 3.05–3.10 | 2.01 | 1.915 | 1.977 | 8.31 |
| | 3.50–3.55 | 1.28 | 1.269 | 1.294 | 4.63 |
| | 4.35–4.40 | 0.75 | 0.773 | 0.782 | 2.45 |
| Sept. 25[4] | 3.10–3.15 | 0.40 | 0.478 | 0.482 | 1.33 |
| Sept. 26[5] | 1.43–1.45 | 0.47 | 0.491 | 0.495 | 1.42 |
| | 3.05–3.10 | 0.25 | 0.265 | 0.265 | 0.71 |
| | 4.10–4.15 | 0.13 | 0.120 | 0.120 | 0.36? |
| | 4.40–4.45 | .... | 0.101 | 0.101 | 0.24? |
| Sept. 29[5] | 3.15–3.20 | 0.42 | 0.249 | 0.253 | 0.13 |
| | 3.55–4.00 | 0.29 | 0.194 | 0.195 | 0.087 |
| | 4.25–4.30 | 0.18 | 0.149 | 0.149 | 0.051 |
| | 4.45–4.50 | 0.18 | 0.042 | 0.042 | 0.049 |
| | 5.05–5.10 | 0.08 | 0.022 | 0.022 | 0.021 |
| Oct. 1[4] | 1.30–1.35 | 2.22 | 1.658 | 1.738 | 9.55 |
| | 2.10–2.15 | 1.90 | 1.189 | 1.249 | 6.91 |
| | 3.10–3.15 | 1.34 | 0.865 | 0.895 | 4.93 |
| | 3.45–3.50 | 0.92 | 0.581 | 0.593 | 3.14 |
| | 4.20–4.25 | 0.73 | 0.438 | 0.447 | 2.42 |

[1] Long weir, full depth.    [4] Short weir, reduced depth.
[2] Long weir, reduced depth.    [5] Short weir, reduced depth. No. 2 orifices.
[3] Short weir, full depth.

trough, 12-ft. vertical flume, wooden flap valve, into the 10-ft. concrete flume, and over the weir, as already mentioned, it is quite possible that 38.13, 30.33, 17.91 and 11.73 cu. ft. per sec. were passing the weir on the average for 5 min., while 38.30, 31.25, 18.80 and 11.00 cu. ft. per sec. were being discharged by the pump on the average during the same 5 min. The difference could very well represent rise or fall in the mean elevations of the water in either the 5-ft. trough, the 12-ft. vertical flume, or in the 10-ft. concrete flume, or in two or more of these containers simultaneously.

The comparison which was made, alone, does not argue for superior exactness of gaging by either the weir or the meter —that must be determined by tank tests made with the two separately. It only showed that neither was materially inexact. And we know from data that have been given, that next to tank tests the venturi meter is an exact method for measuring water.

And with the result placed in evidence, that No. 2 orifices act according to the straight-line equation—

$$Q = 5.50d \text{ in English units}$$
$$Q = 1.675d \text{ in metric units}$$

this paper might fitly close.

Fig. 6 shows the results of Table 1 in graphic form.

(Continued on page 132)

*The Long-Range Gun in Action*

# The German Long-Range Gun

An Account of the German Super-Gun Bombardment of Paris from the Forest of Gobain, 70
Miles Distant, in the Spring of 1918, Together with Authentic Details Regarding
the Design of the Gun, Carriage, Emplacement and Projectile

By LIEUT.-COL. H. W. MILLER,[1] U. S. A.

AT 7.15 on the morning of March 23, 1918, just two days
after the Germans opened their offensive against the British 5th Army before Amiens, the people of Paris were
startled by an explosion of something that had fallen on the Quai
de Seine. The explosion was of such magnitude that it could be
heard over practically the whole of Paris. Fifteen minutes later
there was another explosion of the same magnitude, but this time
closer to the Seine, on the Rue Charles V; fifteen minutes later
another explosion occurred on the Boulevard de Strasbourg near
the Gare de l'Est. Until that time Paris had never been bombarded except from airplanes and zeppelins, and the first thought
of the people was that they were being bombarded from some
new type of aircraft that was being operated at such a height
that it was practically invisible.

The explosions continued to occur throughout the morning at
very uniform intervals of 15 min., and by evening 21 explosions
had occurred at the places shown in Fig. 1. After the first few
explosions, between seven and eight o'clock in the morning, business in Paris practically ceased. Stores were closed and part of
the Metro system ceased to operate; ticket offices in some of the
railway stations were closed, and great numbers of people could
be seen walking the streets looking skyward trying to locate the
planes that were dropping the supposed bombs.

## The Bombardment of Paris

Information of the extraordinary bombardment was telephoned
and telegraphed over practically the whole of France within a
few hours and was received everywhere with amazement. After
a few hours of the first day a sufficient number of the fragments

of the exploding agent were collected by officials in Paris to identify it as a projectile being fired from a gun, rather than a bomb
being dropped from an airplane. By noon representatives of the
Heavy Artillery Section of the Ordnance Department, located at
Tours, had made preliminary calculations in response to telephone
instructions from Paris giving the probable muzzle velocity at
which a projectile would have to start in order that it might travel
from a few kilometers within a point of the German lines nearest
Paris to the center of Paris. Inquiries were made likewise of the
Section as to the possibility that these projectiles could actually
be coming from a gun within the German lines. The reply was
that the probable muzzle velocity of the projectile, if it was
actually being fired within the German lines, was not less than
4500 ft. per sec. No such muzzle velocity had, to their knowledge,
ever been realized, but with a gun of sufficient length and with
a powder burning at a satisfactorily low speed it was considered
quite possible for the projectiles to be fired from within the
German lines. By the end of the first day officers of the French,
as well as the American, armies were quite certain that the projectiles were being fired from a newly designed long-range gun
located within the German lines, and operating at a probable
range of 110 km. (68.8 miles).

During the next few days some interesting theories were advanced in the various newspapers. In one case it was maintained
that the projectiles which arrived in Paris were being fired from
another much larger projectile, which actually served as a gun.
This larger projectile was said to have been fired from a gun
within the German lines, and upon attaining a certain height a
charge of powder within the larger projectile was automatically
ignited, firing the smaller one to a much greater distance. The
theory that the projectiles were being fired from guns concealed
in abandoned quarries or in heavily wooded regions near Paris
received considerable credence, and diligent search was made of

[1] Office of Chief of Ordnance. Artillery Division, Washington, D. C.
Released for publication by the War Department. Office of the Chief of
Ordnance, and contributed to Mechanical Engineering.

all such places to be certain that it was not true. A third theory
was that the projectiles were being fired from a pneumatic gun
located within Paris.

A plotting of all the bursts for the first day showed results that
were very puzzling. If the guns were being operated at comparatively short distances, the only way to account for the tremendous
dispersion was on a basis of actual laying of the gun for different
objectives. This theory did not seem plausible, however, because
if the gun was actually laid on given objectives, these objectives
were of relatively small importance. If, on the other hand, the
projectiles were being fired from a gun at a great distance, it was

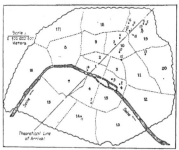

FIG. 1 MAP OF THE CITY OF PARIS, SHOWING POINTS OF BURSTS
ON FIRST DAY OF LONG-RANGE BOMBARDMENT (MAR. 23, 1918) IN
THE ORDER OF THEIR ARRIVAL

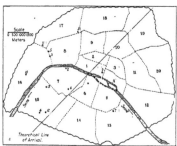

FIG. 2 MAP OF THE CITY OF PARIS, SHOWING POINTS OF BURSTS
ON LAST DAY OF THE LONG-RANGE BOMBARDMENT (AUG. 5, 1918)
IN THE ORDER OF THEIR ARRIVAL

seen at once that the dispersion was so great as to make it practically impossible to hit with any certainty any objective much
smaller than that portion of Paris within the walls. It was noticed that the dispersion in direction, that is, to the right or to the
left of the line on which the projectiles were arriving, was comparatively small, while the dispersion in range, that is, over or
under the probable point at which the projectiles were supposed
to arrive, was very great. Examination of the map, Fig. 1, shows
that on this first day, when the gun was new, the average dispersion in range was very small compared with the dispersion on
later days when the gun became worn.

Evidence received later indicates that at the end of this first
day's bombardment the first gun was about half worn out.

Eighteen projectiles had fallen within the walls of Paris, and
three outside; fifteen people had been killed and thirty-six
wounded. The destruction of property had been comparatively
small. By "comparatively" it is meant that the destruction was
small in comparison with that wrought by the 100- and 300-kg.
airplane bombs. Whenever the projectiles landed in the street
or in an open plot of ground, the hole made was seldom more
than from 12 to 15 ft. in diameter, and from 4 to 6 ft. in depth.
When the projectiles struck buildings it was not unusual to have
them explode in the interior without showing any serious signs
of damage on the outside.

On the 24th of March thirteen projectiles fell within the walls
of Paris and nine without, killing eleven people and wounding
thirty-four. On the 25th, four projectiles fell within the city
and two without. This was the end of the first gun. It was worn
out, and no more firing was done until the 29th, when the second
gun began its work.

Search of the files of the French War Office revealed the fact
that full drawings and plans had been on hand for quite a long
time for such a gun as the Germans were probably using. These
specifications had been submitted a number of years before and
had been discarded because of the excessive expense and the tremendous difficulties involved in manufacturing such a gun. It
was considered likewise by those who had turned down the specifi-

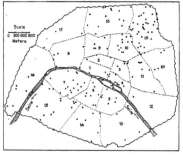

FIG. 3 MAP OF THE CITY OF PARIS, SHOWING POINTS OF BURSTS
OF THE 183 PROJECTILES THAT FELL WITHIN THE WALLS IN 44
DAYS OF BOMBARDMENT

cations that the value of such a gun was extremely questionable.
It was realized that its dispersion would be excessive, and that
with the powders with which the service was then familiar it
would be exceedingly difficult to secure a muzzle velocity sufficiently uniform to do effective work.

### DETAILS OF THE BOMBARDMENT

When the bombardment was over an examination of the periods
over which it had extended indicated that it had been divided into
three distinct series. The first extended from March 23 to May
1; the second from May 27 to June 11, and the third from July
15 to August 9. It seems certain that this division into series
was a part of the plan of the large offensives being waged about
that time. Table 1 on page 99 gives the bombardment by days
during this entire period and shows the number of projectiles
falling within and without the walls of Paris each day, and the
number of people killed and wounded. During the first few days
of the bombardment the first projectiles arrived between seven
and eight o'clock in the morning and continued to fall at intervals
of 15 min. through a portion of the day. On later days the
bombardment would begin at 12.40 noon.

Much has been said with reference to the effect of this bombardment on the people of the city of Paris. The writer's first direct acquaintance with the bombardment was during the third and fifth days. With considerable surprise it was observed that already the people were taking it quite philosophically; in fact, it could not be seen that they were paying much attention to it at all. At intervals of about fifteen minutes muffled explosions would be heard in different parts of the city, seldom two consecutive explosions in the same vicinity. Many people would stop for an instant and attempt to decide from what direction the sound had come, after which they would go on their way apparently unconcerned. It was, of course, still sufficiently novel to be of considerable interest, and people were talking about it everywhere. It is quite probable that the regularity with which the projectiles were arriving got on the nerves of some people, and that the bombardment was responsible for the departure of some of those who

the population left the capital and so increased the alarm caused by our successes." He is right, the bombardment did make a great impression; it made every one more angry, and alarmed very few. It is certain now that they could ill afford to use their manufacturing facilities for the making of such guns as these at a time when they were so desperately in need of heavy field guns to assist their armies in their big drives.

Long-range or super guns received consideration from the Allies for a very short period. There was a tendency at the time to favor construction of a great number of them, but a saner view soon prevailed and actual steps were taken for the construction of only a very few. Both the British and the French Governments began the construction of a limited number, some of which have now been finished. They built them, however, with a clear understanding that they could hope for but little more from them than the Germans were getting from their own. American ord-

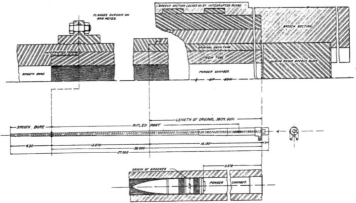

FIG. 4 DETAILS OF DESIGN OF THE GERMAN SUPER GUN (DIMENSIONS IN METERS)

were crowding the trains west and south from Paris. It seems more certain, however, that by the first of June the close proximity of the Germans to the city of Paris, together with the anxiety caused by another drive impending, added to the effect of the heavy bombardments from airplanes on every clear night, had far more effect than the gun. The visible destruction of property was so slight as to give little evidence to any one traveling about the city that the explosions they were hearing from time to time amounted to anything. The newspapers were very careful to avoid any discussion or even a lengthy reference to the bombardment from day to day, and neither gave the locations of the places where the projectiles had fallen nor the number of killed or wounded.

The Germans must certainly have known that their gun was not a profitable investment for the destruction of property, hence they must have continued the bombardment purely for its destructive effect on the morale of the Parisians and its beneficial effect on the morale of the Germans. This is likely the purpose which prompted the construction of the guns. In his book, My Thoughts and Actions, General Ludendorff says: " During the battle we had commenced bombarding Paris from near Laon with a gun having a range of 75 miles. This gun was a marvelous product of technical skill and science; a masterpiece of the firm of Krupp and its director, Rausenberger. The bombardment made a great impression on Paris, and on all France. Part of

nance officers feel that it would not profit us to construct more than two or three such guns at the very most, and probably none at all.

Within a very few days from the beginning of the bombardment it was possible, through careful examination of the direction of the passage of the projectiles through various buildings, to determine quite accurately the direction of their arrival and the probable place from which they were coming. This place was in the Forest of Gobain, southwest of Laon. Very soon thereafter the French Air Service was able to locate in this same forest three positions from any one of which, or possibly from all of which, the guns were being operated. Only one of these was sufficiently close to the German and the Allied lines to permit of any possibility of operating against it with the heavy guns then in service. The gun selected was a 34-cm., 45-caliber French gun on a railway mount. This mount was run up to a point very close to the lines and behind a hill of such size as to effectively conceal it from German line observers. It was well camouflaged by nets against the air observer and against German sound-ranging apparatus by placing two smaller guns at the right and left and several hundred yards behind it. These smaller guns were fired at intervals of one and two seconds before the larger gun, and evidently had the effect of so disturbing the German sound-ranging apparatus as to make it impossible to locate the heavy gun. After a half-day's firing from the 34-cm. gun, air photo-

graphs indicated that the emplacement had been destroyed. The
airmen undertook to demolish the other two emplacements, and
their photographs indicated that they had inflicted considerable
damage. Apparently they did not secure any direct hits, for the
gun or guns continued to fire without any long intervals between
shots.

After the advance of the Allies it was impossible to determine
from examination of the emplacement which had been nearest the
Allied lines whether a gun had actually been on this position or
not. A German artillery officer who worked with the Americans
for some time in the forward area after the armistice, and who

FIG. 5  CRADLE, COUNTERWEIGHTS AND ELEVATION ARC OF GERMAN
                              SUPER GUN

spoke with such certain knowledge of the long-range gun as to
indicate that he knew the details of design and had seen it, insisted that the position that had been destroyed by the Allies had
not had any gun on it.

Mention has already been made of the destruction of property
by these long-range projectiles. Except in a few places where
the damage was so exposed as to attract attention, no matter how
slight it might be, there was little evidence that the bombardment
had any effect. On March 29, the day on which the second gun
began to fire, and fired only four projectiles, the one projectile
which landed within the city of Paris struck and knocked out the
keystone in one of the arches in the roof of the church of St.
Gervais near the Hotel de Ville. The falling of the keystone
caused a large part of the arch and the roof to collapse, and most
unfortunately the church was quite full of people at the time.
Examination of Table 1 will show that this day holds the record
for casualties. On one other occasion a projectile burst in the
lobby of a hotel, killing a number of people but doing comparatively little in the way of material damage. Ordinarily an explosion would make only a comparatively small hole in the
ground.

Even during the active periods of the three series noted above
there were many days on which the gun did not fire. It is quite
certain that between the 25th and 29th of March the first gun
was being removed and a new one placed on the carriage. By
the end of the 25th of March 49 projectiles had arrived, and the
probable life of the guns is not more than 50 rounds. It is likely
that the guns were changed again between the 7th and 11th of
April and between the 21st and 24th of April.

After the armistice it was learned that the Germans had constructed a total of seven guns. These guns were first constructed
to a diameter of 21 cm., and after being worn out as 21-cm. guns
were rebored to 24 cm. All of the projectiles of the first two
series were 21 cm. in diameter, but during the last days of the
third series the projectiles were 24 cm. in diameter, indicating that
the entire seven guns had been worn out, and that probably the
gun that had commenced firing on Paris on March 23 as a 21-cm.
gun had been rebored and fired again as a 24-cm. gun. It was
learned also that the Germans were reboring the remainder of the
seven guns and were constructing additional guns at Essen. Repre-

sentatives of the Ordnance Department found in the Skoda Works
at Pilsen three more guns, which the engineer at these works said
were under construction as long-range guns at the time of the
armistice. In July, 1918, the Intelligence Service transmitted
information to the effect that one gun had been destroyed by a
premature explosion. No confirmation of this report has ever been
found, and it cannot be considered a certainty.

The third series began on July 15 with ten projectiles, followed
by four projectiles on the 16th. The bombardment then ceased
for three weeks, beginning again on August 5 and continuing daily
until the 9th. On this day twelve projectiles arrived, only two of
which fell within the walls of Paris. The famous bombardment
was finished. A comparison of the distribution of the bursts on
this last day (Fig. 2) with the locations on the first day (Fig. 1)
is very interesting. On the first day a considerable number of
the projectiles fell within a very small area in the northeast of
Paris. On this last day it will be observed that they are scattered
over the western section of Paris, and no two are very close together. The last projectiles of the entire bombardment fell between one and two o'clock on August 9. Already the Allies in
their successful drive north of the Marne and between Soissons
and Rheims were driving the Germans back so rapidly and had
made such progress as to put the long-range guns in serious
danger.

Fig. 3 shows the location of all of the bursts within the walls

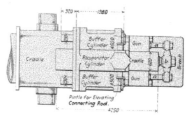

FIG. 6  RECOIL AND RECUPERATOR MECHANISM, BOTTOM PLAN

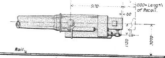

FIG. 7  CRADLE-RECOIL MECHANISM

of the city, a total of 183; 120 more fell outside of the city, making a total of 303 fired from seven 21-cm. guns, and probably one
gun rebored to 24 cm. The bombardment continuing actively over
forty-four days. Even before the last day of the bombardment
American forces operating north of the Marne had captured an
emplacement 9 km. (5 miles) north of Chateau-Thierry. This
emplacement will be described in detail later. It was 86 km.
(53.4 miles) from Paris, and it is understood that it was a new
emplacement from which the Germans had hoped to operate the
super guns more effectively.

### DESIGN OF THE GUN, CARRIAGE AND EMPLACEMENT

It has been mentioned before that the Germans used seven guns
of a caliber of 21 cm. or 8.15 in., that they had rebored at least
one of them to 24 cm. or 9.3 in., and that at the time of the
armistice three more guns were under construction at the Skoda

Ordnance Works at Pilsen in Austria. During the period of active warfare many reports were received with reference to the design of the gun, but never any information with reference to the carriage. The various reports on the gun agreed quite closely in the essential details.

In May 1919 a commission of ordnance officers was sent to the Skoda Works in Austria for the purpose of investigating their methods of constructing large and small ordnance. While there they secured additional data from the chief engineer of the plant with reference to the design of the gun, and saw the three guns which had been in process of construction on November 11, 1918. This engineer stated that when the first German gun began to fire on Paris, it was as much a surprise to the Austrians as it was to the Allies. Shortly thereafter he went to France to examine the gun and observed it in action. It was not until a few months before the armistice that the three guns were sent to Pilsen for conversion into long-range guns. Evidently the Germans had considered the bombardment a success, or sufficiently so to warrant the use of manufacturing facilities for the construction of a greater number of guns than they were able to handle at the Krupp Works at Essen.

The details of the design of the carriage for the gun were not learned until April 1919, although it had been quite certain since August 1918 that the emplacement found southwest of Fere-en-Tardenois was intended for the carriage of the long-range gun. Until July 1, 1919, no direct information had been received from the Germans with reference to the design of any part of the entire

## THE GUN

All of the long-range guns were constructed from worn-out 38-cm. (15-in.) 45-caliber (17.1 m. or 56 ft. in length) naval guns. The converted gun was in two parts, the main section 30 m. or 98.5 ft. in length, and the forward section 6 m. or 19.7 ft. in length. The 38-cm. gun was bored out and a very heavy tube with an inside diameter of 21 cm. was inserted, Fig. 4; 12.9 m. or 40.23 ft. projected beyond the end of the original gun, and over this projecting portion another hoop was shrunk and locked to the forward hoop of the old gun. This 21-cm. tube was rifled at a uni-

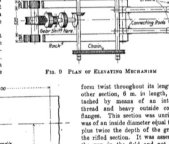

FIG. 9 PLAN OF ELEVATING MECHANISM

form twist throughout its length. The other section, 6 m. in length, was attached by means of an interrupted thread and heavy outside collars or flanges. This section was unrifled and was of an inside diameter equal to 21 cm. plus twice the depth of the grooves in the rifled section. It was assembled to the gun in the field and not removed until the gun had been worn out. The total weight of the original 38-cm. gun was 152,550 lb., and the weight of the reconstructed gun approximately 318,-000 lb. The 21-cm. liner was of such thickness that on being worn out at this caliber it could be rebored first to 24 cm. (9.3 in.) and later to 26 cm. (9.93 in.). Its probable life at any one caliber was not more than 50 rounds, and the maximum powder pressure did not exceed 3000 atmospheres or 44,000 lb. per sq. in. The design of the breech mechanism of the original 38-cm. gun did not require modification.

The long-range guns constructed by the British and French governments do not follow the German design to the extent of having a smooth-bore section on the front. The purpose of this feature was for some time in doubt and is worthy of some discussion.

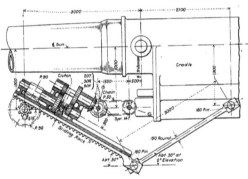

FIG. 8 SIDE ELEVATION OF ELEVATING MECHANISM

mechanism except the gun on which, as mentioned before, information had been received at various times before the armistice; the guns themselves were examined at the Skoda Works at Pilsen. Just why the Germans refused to talk about this gun is not known. In December 1918 and January and February 1919 a German engineer who was working with the American forces in the region northeast of Verdun helping identify long-delay fuses and assisting on other technical matters said that he was acquainted with the design of the gun and had seen it in operation. Very curiously, however, he refused to give any information with reference to the details of the design and was quite positive in his assertions that the Allies would never see any of the guns, and certainly not any of the carriages. This was difficult to understand in view of his perfect willingness to talk about the details of design of any other piece of ordnance that he was at any time asked about, and it was more curious in view of the fact that there is really nothing wonderful about the design of the long-range gun, its carriage or its emplacement.

The gun just described is approximately three meters longer than either the British or French long-range guns, both of which are rifled throughout their length. Two reasons might be given for the extra section.

Possibly additional linear velocity was imparted to the projectile as it traveled through the 6-m. smooth-bore section. It seems improbable, however, that this could be its primary purpose, inasmuch as the muzzle velocity was such a variable quantity. It is more probable that its purpose was to align the axis of the projectile more perfectly with the axis of the bore of the gun and reduce to a minimum the angular velocity of yaw as the projectile left the muzzle. When it is understood that this projectile was to mount to a height of about 24 miles and travel a horizontal distance of 76 miles, it can easily be appreciated that any tendency that the projectile might have to throw its axis out of alignment

FIG. 10  THE GUN CARRIAGE.  (THE GUN SHOWN IS THE 38-CM. 45-CALIBER GUN FROM WHICH THE
LONG-RANGE GUNS WERE MADE)

with its theoretical path would have disastrous results. Through
the impracticability of making projectiles fit perfectly in a gun
they have a tendency to hammer the walls as they travel down
the bore. This hammering action of the projectile is the result
of the operation of two forces, the one tending to increase its
linear velocity and the other to increase its rotational velocity.
The latter force is acting only while it is traveling through the
rifled section. It is invariably true that just as the projectile is
leaving the ordinary gun it is on one of its up, down or crosswise
hammer strokes, and that the axis has a tendency to yaw or devi-
ate from its theoretical path at a certain angular velocity. Cardboard screens placed in front of guns in proof firing invariably
show elongated holes, and not infrequently very decidedly so,
even at no greater distances than 100 ft. from the muzzle. This
tendency to yaw has a detrimental effect on the accuracy of even
our comparatively short-range guns, but no effective means have
yet been devised to neutralize it. With the extreme-range gun it
is so much more serious that it seems quite likely that the Germans adopted this method of neutralizing it, thereby reducing the
dispersion both in range and in direction. It is not improbable
that the German gun was first constructed without the smooth-
bore section, and that its shooting was found to be so erratic as to
require the addition of this feature.

This gun was operated at a muzzle velocity of from 1500 m.
(4760 ft.) to 1600 m. (5090 ft.) per sec. and at an elevation of
55 deg. This 55 deg. is worthy of comment, inasmuch as it has
ordinarily been supposed that nothing could be gained in range
by elevating a gun above 45 deg. It is known, of course, that if
a projectile is fired in a vacuum its maximum range is attained
when it starts at an elevation of 45 deg. When fired in the air,
however, the initial angle which permits it to attain its maximum
range depends both on the caliber and weight of the projectile
and the velocity at which it leaves the gun. With reasonably
large and well-designed projectiles, it may be said that for each
muzzle velocity (a) there is a density of air (b) into which the
projectile should enter at an approximate elevation of 45 deg.
For this projectile and the muzzle velocity of 1500 m. per sec. the
initial elevation of the projectile is 55 deg. It is very doubtful
whether any additional range can be secured with any projectile
at any muzzle velocity at an elevation above 55 deg.

Attention was called in the discussion of the first day's bombardment to the tremendous dispersion in range. Some of the

projectiles fell just within the
wall at the northeast side of
Paris, and some within the walls
at the southwest side. This is
accounted for by the fact just
mentioned that the muzzle velocity of the projectile varied as
much as 100 m. per sec. It was
impossible with the powder that
the Germans were using to secure
a muzzle velocity with a variation of less than 100 m. per sec.

### THE GUN CARRIAGE

*Cradle.* The cradle, Fig. 5, is
a cylinder of simple design having ribs at long intervals and of
a depth of only about 3 cm. The
walls of the cylinder have a
minimum thickness of 10 cm.
and a maximum thickness of 13
cm. over the ribs. The diameter
of the main trunnion is 46 cm.
and the length 33.5 cm. A col-
lapsible counterweight is attached to the top of the cradle.
In firing, the two sections of this
counterweight are raised and
locked together for the purpose
of raising the center of gravity
to such an extent that the gun may be elevated and depressed more
easily. This cradle is of a naval design, the front section being so
shaped as to close the opening in a turret. It is not improbable
that the cradle as well as most of the 38-cm. guns were removed
from German ships or coast fortifications.

*Recoil Mechanism.* The recoil mechanism is composed of two
hydraulic recoil cylinders and one spring-pneumatic recuperator
cylinder, all attached to the bottom of the cradle, Figs. 6 and 7.

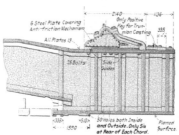

FIG. 11  CIRDER DETAILS

Each of the recoil cylinders is a separate cylinder carried in
brackets cast on the cradle, and each is provided with a buffer
approximately 32 cm. long. The plugs through which the cylinders are filled are on the ends of the buffers. The approximate
length of recoil is 1.3 m. (50.5 in.). This distance is, of course,
approximate and was arrived at after careful examination of the
cradle. A report examined in the office of the Chief of Artillery
of the Belgian Army at Brussels gives the length of recoil as
1.15 m. Inquiry disclosed the fact that this length of recoil was
also approximate.

Two rods *a, a,* screwed to the recoil lug of the gun, Fig. 6, and
extending to the rear, carry a heavy crosshead *b* to which the

Fig. 13  Sight, Recuperator Pump and Breech Mechanism

Fig. 15  Another View of Emplacement shown in Fig. 14

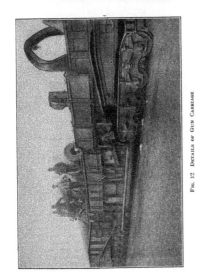

Fig. 12  Details of Gun Carriage

Fig. 14  Emplacement (Bois du Chatelet)

piston of the recuperator cylinder is attached. The recuperator
cylinder is likewise a separate cylinder carried in brackets cast
on the cradle. At the rear on both sides it is planed to serve as
a guide for the recoil lug to prevent rotation of the gun.

*Elevating Mechanism.* The elevating mechanism of this mount
is extremely heavy and unique in design. The cradle was evidently originally provided with the screw type of elevating mechanism, the screw being hinged to the cradle at the rear end.

The specification that the gun be provided with an elevation of
55 deg. necessarily led to the discarding of the screw mechanism.
The straight racks shown on Fig. 8 slide in ways which are parallel
to the inclined lower face of the forward end of the side girders.
It was necessary to put them in this position to secure sufficient
movement to attain to 55 deg. At the lower end they are connected with each other by a heavy shaft to which are attached
the two connecting rods running up to the bottom of the cradle
(Fig. 9). Provision is made for two handles on each side of

at the front and rear by heavy structural-steel transoms. They
are further reinforced in the front by the heavy cast-steel housing
for the elevating gear. The trunnion seats are of cast steel and
are simply bolted to the top chord with single key plates at their
rear. This single key plate has the positive backing of one top-
cover plate. The face of the horizontal section of the lower chord
of the side girders is planed and is provided with key plates at
each end, Fig. 11. Eight 2-in. holes are provided in each lower
chord at the front end and six at the rear for bolting the mount
to the emplacement.

## THE ANCHORAGE

When the German Army had retired from the salient between
Soissons and Rheims in August 1918, the emplacement shown in
Figs. 14 and 15 was found in the Bois du Chatelet, 9 km. directly
north of Chateau-Thierry. This emplacement was in process of

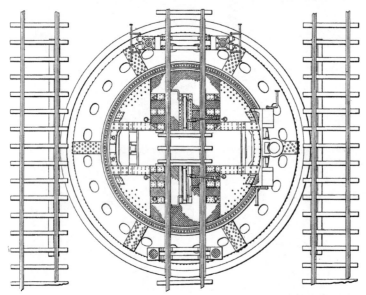

FIG. 16  PLAN OF EMPLACEMENT SHOWN IN FIGS. 14 AND 15

the car. each for two men, Fig. 9, and a two-speed transmission
is provided, permitting the mechanism to be operated at low or
high speed. The ratio of the low-speed gear is 4⅛ turns of the
handle for each degree of movement of the gun. It was not possible to determine the ratio of the high-speed gear, since the two
men making the examination could not operate it. It seems probable, however, that eight men could operate the mechanism at high
speed without undue difficulty.

*Traversing Mechanism.* See later paragraphs dealing with the
anchorage.

*Railway Car Body.* The railway car body shown in Figs. 10,
11, 12 and 13 is made up of two single-web side girders connected

installation, and had not been completely erected. The Germans
attempted to destroy it, but succeeded only in ripping loose a few
plates. At the time the emplacement was captured, no description
of any kind was available of the carriage for which it was intended. The writer's failure to find any emplacement in Belgium
from which the mount examined there could be operated led to
further examination of the emplacement in France, and it was
found that they fitted perfectly.

The emplacement is in two main sections, one a base and the
other a rotating section. The rotating section, which is shown
in Figs. 16 and 17, is about 28 ft. 6 in. in diameter and is supported
on 112 8-in. steel balls. The base of this section is about 35 ft. in

diameter and 6 ft. in depth. In Fig. 14 there can be seen at right angles to the direction of the track of the rotating section two girders, on the ends of which are key plates or pads. The plate which is labeled has six holes in it, and the plate on the opposite end of the girder has eight.

The mount is run on to the emplacement with the rotating section in the position shown in Fig. 14. It is then raised by the four jacks, which can be seen at the left of Fig. 14, and are shown in their exact positions in Fig. 16. When the mount is raised the trucks are removed and the rotating section of the emplacement turned through 90 deg. On lowering the mount on to the emplacement, the key plates on the bottom chords of the carriage are fitted to the corresponding key plates on the emplacement. The emplacement is shown in process of installation, with the gantry crane still in place, Fig. 19.

The carriage on which the gun is so mounted as to be capable of movement in a vertical plane only is traversed by rotating the top section of the emplacement. A complete circular traversing rack made up of angles and steel pins is bolted to the structural base (page 89). A pinion carried on a vertical shaft on the side of the rotating section which corresponds to the rear of the car-

mount from the trees on either side, and in Fig. 19 it can be seen that the gantry crane as well as the emplacement, while in the process of installation, were covered with tree branches.

*Trucks.* All of the trucks as well as the span bolsters are constructed entirely of structural steel as shown in Fig. 20. The front trucks contain five axles each, and the rear trucks four axles each. The journals are approximately 14 cm. in diameter by 22 cm. in length. The wheels are 95 cm. in diameter. On the front trucks three wheels are provided with two brakeshoes each; on the rear trucks the end wheels are provided with two shoes each and the inner wheels with one shoe each. The braking is done by hand only. The axles are equalized in pairs only. The center axles of the front trucks have no connection with the others. In Fig. 10 a portion of a circular rail can be seen on the top of the front truck on which the pads attached to the span bolster on each side bear.

### AMMUNITION SUPPLY SYSTEM

The provisions made for supplying ammunition are shown in Figs. 21, 23, and 25. These same provisions were used for the

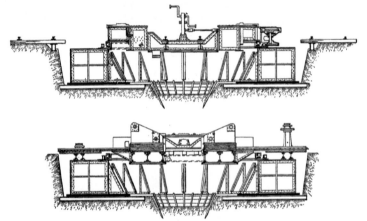

FIG. 17   ELEVATION OF EMPLACEMENT SHOWN IN FIGS. 14 AND 15

riage engages with this rack, and the mechanism is driven by two two-man handles on a horizontal shaft carrying a worm engaging with a worm wheel on the top of the vertical shaft, as shown at the right of the top view, Fig. 16. This shaft may likewise be driven by means of the four handles mounted on the two cases in the center of the rotating section and connected with the worm wheel and shaft through two shafts carrying two universal joints each. Apparently the mount was generally traversed by means of these latter handles, if one may judge of the scheme of operation shown in the illustration on page 89.

*Camouflage.* In the case of the emplacement found in the Bois du Chatelet, sockets were placed in the center of the approach track at intervals of about 30 ft., into which trees with trunks up to about 6 in. in diameter could be placed for the purpose of concealing these tracks. In the illustration of the gun in action, page 89, it will be observed that netting has been stretched over the

operation of the 38 cm. 45-caliber naval guns from which the long-range guns were made. The extension to the car body shown in Fig. 21 was used for the operation of the 38-cm. gun under certain conditions. This was not required and was removed when the long-range gun was operated on the carriage. In Fig. 25 a removable plate can be seen between the shot-truck rails. Above this removable plate is a light bridge, in the center of which is seen a ball on the end of the cable used in hoisting projectiles through the hole in the floor. This cable runs over a series of pulleys to a small drum in the box on the top of the left-side girder. The ammunition was supplied from the storehouse over the narrow-gage track shown in Fig. 23, the circular portion of the track being under the opening in the center of the floor of the mount. The gear in the box on the side girder is of the double-speed variety, the high speed being provided probably for rapid operating in the hoisting of powder.

FIG. 19  EMPLACEMENT IN PROCESS OF INSTALLATION

### AMMUNITION

The projectile used with this gun is shown in Fig. 24. Its weight is 120 kg., or 264 lb. The length of the main portion of the projectile is 490 mm., or approximately 19 in. To this was attached a false ogive or cap, the exact length of which is not known; it was probably as long, if not slightly longer, than the main portion of the projectile. The radius of the curve of this ogive was about 7 calibers or 58 in. The design of the interior as well as the exterior of this projectile is unique in many respects. It will be noted that the charge was separated into two parts by a diaphragm which has a number of holes in it. Two fuses are used, the one in the diaphragm and the other in the base. All of these projectiles were most carefully machined to elaborate specifications in order that the weight and dimensions might be practically exact and the center of gravity perfectly located; even after they were thus machined they were sorted out into lots according to their characteristics by weight, dimensions and location of the center of gravity, and the various lots of projectiles marked with center-punch marks. The necessary corrections in elevation and deflection were calculated for the projectiles of these various lots. The projectiles which fell in Paris were successively marked with three center-punch marks in a triangle or four center-punch marks in a square, etc.—projectiles of the same marking arrived consecutively.

The outside of the projectile was provided with two copper bands which served as gas checks, and in front of each of these copper bands the projectile was enlarged and rifled for a length of 70 mm. or 2.7 in. The shear from the extreme acceleration necessary to produce the required muzzle velocity was such that the ordinary copper bands could not be made to serve as rotating bands. The shearing strength of the copper is not sufficient to stand the strain that would have been put on the bands with so great an acceleration. The friction between the lands of the gun and the projectile was so great as to wear the gun appreciably on each round fired. This was evident from examination of the copper bands of successive projectiles. As a result of the necessity for providing rifled projectiles, it was necessary in placing

FIG. 20  TRUCK DESIGN

the projectile in the bore of the gun to fit the forward rifled section of the projectile into the beginning of the rifling of the gun, Fig. 4. The exact details of the design of the portion of the tube between the beginning of the rifling and the powder chamber are not known. It seems quite probable, however, that there is a smooth section of about 12 in. into which the portion of the projectile back of the first copper band fits, bringing the base of the projectile to the end of the powder chamber. The fact that the forward rifled section of the projectile was already fitted into the grooves of the gun made it certain that the rifled section at the

against the effects of inertia which might either provoke a premature explosion in consequence of the friction of the molecules of the charge against each other or might bring the charge beyond the critical density point, after which it probably would not detonate. At one time a theory was advanced to the effect that the diaphragm separated two liquids which mixed and detonated upon the fall of the projectile. There seems no good reason to believe that this could have been so.

### PREVIOUS DESIGNS AND LIMITS IN DESIGN

It has been mentioned already that investigation disclosed shortly after the beginning of the bombardment that an almost identical design had been proposed to the French government not

TABLE 1  BOMBARDMENT OF PARIS BY THE GERMAN
LONG-RANGE GUN

(March 23 to August 9, 1918.)

| Dates | | Number of Projectiles | | Killed | Wounded |
|---|---|---|---|---|---|
| | | Within the Walls | Outside the Walls | | |
| March | 23 | 18 | 3 | 15 | 36 |
| | 24 | 13 | 9 | 11 | 34 |
| | 25 | 4 | 2 | 1 | 3 |
| | 29 | 11 | 3 | 88 | 68 |
| | 30 | 18 | 3 | 10 | 60 |
| | 31 | 3 | .. | 1 | 1 |
| April | 1 | 3 | 1 | 8 | 8 |
| | 2 | 3 | 1 | .. | 3 |
| | 3 | 1 | .. | .. | .. |
| | 6 | 2 | 4 | .. | 3 |
| | 7 | .. | 1 | .. | .. |
| | 11 | 5 | 1 | 9 | 21 |
| | 12 | 5 | 2 | 2 | 14 |
| | 13 | 6 | 2 | .. | .. |
| | 14 | 3 | 1 | .. | .. |
| | 15 | 3 | 2 | .. | .. |
| | 16 | 2 | 1 | 17 | 114 |
| | 19 | 1 | 2 | .. | .. |
| | 21 | 2 | .. | .. | .. |
| | 24 | .. | 1 | .. | .. |
| | 25 | 6 | 4 | 1 | .. |
| | 26 | 1 | .. | .. | .. |
| | 30 | .. | 3 | .. | .. |
| May | 1 | .. | 1 | .. | 5 |
| | 27 | 7 | 7 | 4 | 20 |
| | 28 | 6 | 4 | 1 | 2 |
| | 29 | 5 | 5 | 1 | 7 |
| | 30 | 4 | 5 | 13 | 5 |
| | 31 | 1 | 3 | .. | .. |
| June | 1 | .. | 4 | .. | .. |
| | 3 | .. | 6 | 2 | 8 |
| | 4 | 4 | 1 | 4 | 16 |
| | 7 | 3 | 1 | 1 | 4 |
| | 8 | 1 | 2 | 3 | .. |
| | 9 | 1 | 3 | 1 | 9 |
| | 10 | 2 | 2 | 3 | 13 |
| | 11 | .. | 1 | .. | .. |
| July | 15 | 9 | 1 | 6 | 9 |
| | 16 | 4 | .. | 3 | 8 |
| August | 5 | 13 | 4 | 32 | 61 |
| | 6 | 12 | 7 | 8 | 39 |
| | 7 | 8 | 4 | 7 | 43 |
| | 8 | 1 | 4 | 1 | .. |
| | 9 | 2 | 10 | 3 | 6 |
| Totals........ | | 183 | 120 | 256 | 620 |

1 Shell struck an arch in the roof of the Church of Saint Gervais.  2 Night and day.  3 Night.

FIG. 21  VIEW SHOWING LOADING GEAR, TRUCKS AND BOLSTER

many years before. Further search of French and English records has brought to light the fact that the rifled projectile is quite old and seems to antedate the copper band considerably. In 1892 Sir Alfred Noble made a 100-caliber 6-in. gun which gave a muzzle velocity of 3700 ft. per sec.; in the same year the French government constructed a 10-cm. 80-caliber gun from which they attained almost the same muzzle velocity. It is not known what range was secured in either case. The forms of most standard projectiles at that time were not such, however, as to make it possible to secure the maximum range from the above-noted velocity.

FIG. 22  BREECH MECHANISM, RECUPERATOR GAGE AND TRAVELING LOCK

In taking up the design of such guns again the French and English services found that with guns of practicable length and muzzle velocities attainable with guns of such length, and with available powders, the projectile that would permit the greatest range to be realized would be approximately 8 in. in diameter and weigh from 225 to 250 lb. An appreciable change up or down in the diameter or the weight of the projectile would result in a loss of range. The gun designed by one of our allies uses a projectile weighing 240 lb. and a powder charge of 350 lb. One of the

rear would likewise fit as the projectile began to move up the gun.

It has been noted before that this projectile departed at a muzzle velocity of from 1500 to 1600 m. per sec. The velocity remaining on its arrival in Paris was approximately 700 m. per sec. and the time of flight about three minutes.

The analyses made of the various parts of the projectile are as given in Table 2.

Not a single one of the 303 projectiles that fell in or about Paris failed to detonate. Evidently the two fuses were responsible for this excellent record. The practice of separating the charges into two parts by a diaphragm is not new. The diaphragm had been used before in heavy French projectiles as a guarantee

TABLE 2 ANALYSES OF VARIOUS PARTS OF THE SHELL

| | Shell[1] | Cap Holder[2] | Cap | Diaphragm[3] | Copper Band | Dunkirk Shells |
|---|---|---|---|---|---|---|
| C | 0.43-0.85 | 0.58-0.73 | 0.10-0.18 | 0.60 | | 0.72 |
| Mn | 0.27-0.49 | 0.27-0.36 | 0.41-0.97 | 0.09 | | 0.26 |
| Si | 0.24-0.31 | 0.19-0.28 | 0.03-0.05 | 0.51 | | 0.23 |
| S | 0.04-0.10 | 0.03-0.07 | 0.05-0.11 | 0.04 | | 0.03 |
| P | 0.01-0.06 | 0.03-0.07 | 0.04-0.067 | 0.04 | | 0.04 |
| Ni | 1.82-2.32 | 2.10-2.39 | 0.00-0.11 | 0.0 | | 2.41 |
| Cr | 0.83-1.45 | 0.95-1.44 | 0.00-0.2 | 0.0 | | 1.32 |
| Pb | | | | | 0.12 | |
| Fe | | | | | 0.21 | |
| Cu | | | | | 99.67 | |

[1] Practically a casting. [2] Cast. [3] Forging.

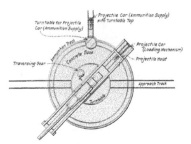

FIG. 23 PLAN OF AMMUNITION SUPPLY SYSTEM

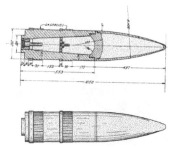

FIG. 24 DETAILS OF DESIGN OF PROJECTILE

FIG. 25 LOADING GEAR

to give a uniform pressure through a large portion of the period of the travel of the projectile in the bore of the gun. The powder made by one of our allies for their long-range gun is in flat strips resembling cardboard. These strips are three-ply, the various sections burning at different rates. The large sticks of powder mentioned above for the gun of the other of our allies is special, and is expected to live up to the characteristics just noted.

A somewhat recent report by the Bureau of Mines is authority for the statement that the metal-mining industry in this country is now passing through one of the most critical stages in its history, and is confronted by new problems upon the solution and adjustment of which depends, in a large measure, its future welfare and prosperity.

The chief factors, according to the Bureau, contributing to the cost of production are the cost of labor and supplies. It is conceded by our leading mine operators, managers, and engineers that any material reduction in the cost of labor or supplies, within possibly the next two or three years, cannot be anticipated; and it is not thought likely that pre-war prices will ever again prevail. This being true, the mine operator must look in some other direction to effect economies that will lower production costs. If this cannot be accomplished it will mean, in many instances, that mines and metallurgical works will have to be closed and abandoned, which will tend to disorganize the entire industry. How to lower production costs and still maintain the present high standard of wage is the problem that is engaging the serious attention of mine operators and managers over the entire country.

As a solution to this problem it is thought that the most promising field of investigation lies in the direction of systematizing and standardizing mining practice so as to effect economies in the use of both labor and materials. Already some of the large mining companies in the West have conducted extensive experiments and research in this direction, and the results obtained have furnished conclusive proof of the value of standardization of mining practice and mining equipment. For example, the Phelps-Dodge Corporation has made in the past year an exhaustive study of the various underground operations in their mines at Bisbee, Ariz., for the purpose of standardizing the mining practice. The results achieved have been most encouraging, and show that costs can be lowered even in the face of advancing prices in labor and material.

pieces of powder examined was about 2½ ft. in length and roughly elliptical in cross-section, the axes being about 1 in. and ½ in., respectively. It is interesting to note in this case that the powder charge has a weight about 50 per cent greater than that of the projectile. If the weight of the charge were increased, calculations indicate that most of the energy of the gas would be absorbed in the work of accelerating the charge rather than in accelerating the projectile, and no greater muzzle velocity would be obtained.

One of the vital requirements made of the powder is that it burn at a rate sufficiently slow to avoid excessive pressures and

# The Testing of Farm Tractors

## By CHARLES H. BENJAMIN [1], LAFAYETTE, IND.

*The future will unquestionably see a great development of the farm tractor, and the testing of these machines will of necessity be an important factor in that development. To assist in this work there has been installed in the Experiment Station at Purdue University a tractor-testing plant, the features of which are described in the present paper. Briefly, the plant consists of three elements, a treadmill, or moving platform, a traction dynamometer, and an absorption dynamometer, and is capable of testing tractors weighing as high as 12,000 lb. on the driving wheels and giving a drawbar pull of 5000 lb. at a speed of not over 5 m.p.h. A detailed description of the plant follows.*

THE latest machine to claim the attention of the investigator is the farm tractor. Its development has naturally followed that of the automobile and the motor truck, and the farmer, comparatively isolated as he is, finds the use of individual motive power almost a necessity. But, like all newly invented machines, the farm tractor is imperfect, and will not always do the work expected of it. Field tests throw some light on its good

experiment station, is capable of testing tractors weighing as high as 12,000 lb. on the driving wheels and giving a drawbar pull of about 5000 lb. at a speed not to exceed 5 m. per hr.; these tractors being either single- or double-wheel, front or rear drive, or of the caterpillar type, with maximum tread for the double-wheel tractor of 84 in. and a caterpillar length of 96 in., the face of the drivers not being greater than 26 in.

### DESCRIPTION OF THE PLANT

The plant is of the treadmill type having two caterpillar treads of 26 in. width, carried on a structural-steel frame so designed that with little difficulty it can be changed to accommodate the various styles of tractors.

The steelwork consists in part of a main frame made up of three 3-in. I-beams running across the plant resting on permanent supports at each end and on a movable support in the center, all supports being bolted to the foundation.

On the main frame are two rectangular frames which can be

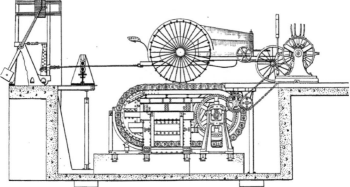

FIG. 1 SIDE ELEVATION OF THE COMPLETE PLANT FOR TESTING TRACTORS

and bad points, but unfortunately they do not always locate the source of poor performance. A laboratory test, on the other hand, if it fairly imitates what we may call standard working conditions, will enable us to compare one machine with another and thus secure information not given by the field test.

To carry on such laboratory tests, apparatus especially designed for the testing of farm tractors has been installed at Purdue University, and it is believed that a description of the installation will be of general interest. The plant itself consists of (1) a treadmill or moving platform, a sort of inverted "caterpillar" supported on driving wheels and idle rollers, and in its turn supporting the driving wheels or lags of the tractor; (2) a traction dynamometer to receive and measure the pull of the tractor; (3) an absorption dynamometer to control and measure the energy transmitted through the treadmill and its supporting wheels.

The plant, which is a permanent equipment for the engineering

moved along the 8-in. I-beams, thus taking care of the different treads of the tractors, or one can be placed in the center to accommodate the single-wheel type. The top of each rectangular frame is made up of four 3-in. I-beams for cross-members, which support the caterpillar tracks. These 3-in. beams can be removed and replaced by a weighing mechanism, so that the weight of the tractor on the driving wheels can be obtained while in operation.

At the rear end of each rectangular frame is a sliding semicircular track which transfers the caterpillar tread from the top to the bottom track, these tracks being forced endways by means of compression springs which maintain the desired tension on the caterpillar tread.

Under the rectangular frame is a separate track bolted to the foundation for the purpose of carrying the weight of the caterpillar tread on the return. Fig. 1 shows a side elevation of the complete plant.

The caterpillar tread consists of ninety-four 7-in. cast-steel links 26 in. wide, each link being carried on four 5-in. wheels equipped with bronze bearings. These wheels provide support

[1] Dean, Purdue University, Lafayette, Ind. Mem.Am.Soc.M.E.
Presented at the Mid-West Sections Meeting. Indianapolis, Ind., October 24 and 25, 1910, of THE AMERICAN SOCIETY OF MECHANICAL ENGINEERS.

for the links at all times except as they come around 48-in. pulleys at the front end, when the under sides of the links, which are curved to fit, transmit the pull to the 48-in. pulleys. To the upper sides of the links are bolted hard-maple blocks 26 in. long, 6 in. wide and 4½ in. thick, leaving 1 in. between the blocks. Lugs can be provided on the tractor driver wheels to mesh in the gaps and thus prevent slippage between caterpillar and wheel. Fig. 2 is a general view of the plant.

The 48-in. pulleys which take the pull from the caterpillar tread have glued and riveted to their faces strips of conveyor belting 26 in. wide and ½ in. thick to reduce slippage between the links and the pulleys. These pulleys are mounted on a 5-in. shaft which is carried on three pedestals equipped with roller bearings. Two of the pedestals can be moved along the shaft to take care of the different positions of the pulleys due to the change of tread

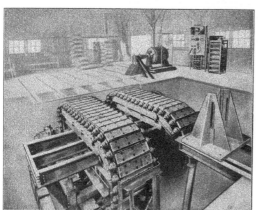

FIG. 2 GENERAL VIEW OF THE PLANT SHOWING THE ELECTRICAL EQUIPMENT

of the tractors. From the end of this shaft the power is transmitted by gears and silent chain to an electric dynamometer located on the main floor of the plant. All bearings in this transmission are either roller or ball type with final connection to the dynamometer by a flexible coupling.

### THE ELECTRIC DYNAMOMETER

The electric dynamometer which provides the brake load for the plant is of the Sprague Electric Company's standard design, rated at 100 hp. at 1200 r.p.m. and so wired as to operate either as a generator-dynamometer or as a motor-dynamometer. When operating as a generator-dynamometer the energy generated is absorbed by air-cooled rheostats, the resistance of which can be varied to accommodate the load. The rotating part of the dynamometer is supported on ball-bearing pedestals and by suitable linkwork connected to standard scales. These scales are so connected that the only readings necessary to determine the horsepower being transmitted are the weight on the scales and the revolutions per minute of the dynamometer.

### THE DRAWBAR DYNAMOMETER

The drawbar dynamometer is of the pendulum type, the pendulum and the recording arm being carried by a 5-in. drum hung on ball bearings. The pendulum weight and the recording arm are so proportioned that the record made by the recording pen shows 200 lb. drawbar pull for every inch of movement from the vertical

position. Tangent to the 5-in. drum, a flexible steel plate 0.025 in. thick is fastened connecting it with a reducing arm by a right-and-left-hand coupling nut. The reducing arm is pivoted on ball bearings and is accurately finished on the end to a radius of 36 in. This arm connects with the drawbar of the tractor at a radius of 12 in., thus giving a reduction for the dynamometer weight of three to one. To the 12-in. arm is attached a flat piece of steel 0.045 in. thick which is connected to the ¾-in. wire rope going to the drawbar of the tractor. This rope passes through a four-wheel guide which can be lowered or raised so that the pull on the tractor will be level at all times and so that there will be no chance for side pull on the dynamometer if the tractor should swing off the center line.

The recording apparatus for the drawbar dynamometer consists of a vertical brass table which can be moved along suitable guides to take care of the different loads. The 10-in. record paper is carried over the table from the lower roll to the upper, the winding mechanism being driven from the main 5-in. shaft of the plant so that the speed of the paper is in exact proportion to the speed of the tractor. Above the recording table are two recording arms which can be moved along a graduated scale, one arm with its pen marking the zero line for the test; i. e., if test is run at 2000 lb. drawbar pull, the zero line would be placed at 1500 or 1800 lb., thus giving a card which would be easy to measure. The second recording arm would be set at the desired load to be carried. This arm will have fastened to it a fine wire with a weight at the lower end which will keep the line plumb and straight at all times. As the operator looks at the recorder, he will see a line which will represent the load he is carrying on the tractor and by this he will be able to keep the electric-dynamometer adjustment correct. Along the side of the table is a sliding electric time recorder which can be set opposite the recording pen so that the time record will correspond to the load record.

A weighing mechanism will be placed under the guiding axle of the tractor so that the changing load on the guide wheel can be measured while the tractor is under load, thus giving a check on the force transmitted to the drawbar.

### TYPES OF EXPERIMENTS MADE

The experiments which can be made on a machine such as has been described are of a twofold character:

1 Those made with the tractor to be tested mounted on the moving platform and run in its usual manner, transmitting its energy through the platform, the wheels and the geared train to the generator which acts as a brake.

2 Those in which the engine and main driving shaft of the tractor are connected more or less directly with the generator either with or without the use of the transmission gears, the tractor wheels and the moving platform being "cut out."

The first may perhaps be termed the commercial and the second the scientific test. In the first we have field conditions and field operation imitated. The tractor is run at different speeds and under different loads and the economy or efficiency is determined in the gross, in terms of the drawbar pull and the gasoline consumption.

This would be followed by field or road tests to determine other practical characteristics of the tractor as they are affected by out-of-door conditions. The efficiency of the tractor considered merely

(Continued on page 132)

# Research and Social Evolution

By C. F. HIRSHFELD[1], DETROIT, MICH.

*Research, according to the author of the following paper, is the search for facts, and industrial research is the search for industrial facts. This, he points out, opens a broad and almost unlimited field for the research worker. The value of research lies in substituting methods of all kinds command large audiences and when past, have been the bases of judgments made by industrial leaders. Nor is the author pleading alone for what he terms material research but for economic research as well, and most particularly for sociological research, believing that our basic knowledge and control of human relations have not kept abreast of social evolution.*

THE study of history reveals a great similarity in the social and economic conditions following wars of great magnitude. There is always a period of topsyturvydom during which radicals of all kinds command large audiences and when individuals and nations act as though they had been relieved suddenly of all historical restrictions and bonds and were free to proceed unhampered in the pursuit of any sort of evolutionary or revolutionary scheme which met their individual or collective fancies.

This period always flows gradually into one in which the necessity of continuing an ordered social life is appreciated by the great majority, and the world, or that part of it affected by the war, settles down once more to the production of wealth and the pursuit of happiness upon a plane and according to methods not greatly different from those in use before the cataclysm.

The real lessons of the war and the lasting results thereof are seldom appreciated until they can be viewed in retrospect, and often through a vista of many years.

The present situation bids fair to duplicate the past. We are apparently just emerging from topsyturvydom and moving toward a more or less ordered existence. We are beginning to appreciate the fact that sociological surgery, like human surgery, has as yet certain limitations. We may perform operations of some sorts without seriously endangering a fairly normal patient, but when the severity of the operation exceeds a certain limit the patient is apt to succumb to shock, even after a most successful operation from the surgeon's viewpoint.

I feel that it is well to bear such facts in mind when discussing research and its application to industry in this country. The public at large seems to believe that industrial research did not exist in these United States before the war; that the phenomenal performance of such research in Germany and her allies was due to the development of such research in Germany; and that we are now about to take up work of this sort in characteristic American fashion, producing over night a dozen or more research laboratories for every one originally possessed by Germany. The facts are that industrial research was fairly common in this country before the war; that Germany's performance, in so far as it depended upon research, was largely based on the utilization of the results of research rather than upon the uncovering of new facts; and that to multiply research laboratories and facilities in this country at too rapid a rate would lead to chaos, waste and business failure instead of acting as a healthy stimulant to a slightly neurotic patient.

We must not expect the war to revolutionize over night our methods of producing and of doing business any more than we can expect it suddenly to "emancipate" labor. The fact that we are now trying to enunciate and codify principles governing the relations between employer and employee which have been in the process of development for decades should not lead us to attribute the principles themselves to the war. The disturbed conditions incident to the war merely serve to focus attention upon the problem. This is a convenient time for taking account of stock before we start out once more in ordered fashion to develop still further.

[1] Chief, Research Dept., The Detroit Edison Co., Jun. Am. Soc. M. E. Presented at the Mid-West Sections Meeting, Indianapolis, Ind., October 24 and 25, 1919, of THE AMERICAN SOCIETY OF MECHANICAL ENGINEERS.

Similarly, there has been growing in industry during the two decades just past an appreciation of the fact that refined methods of analysis produce results, and the war has simply served to focus attention upon the haphazard way in which industry has adopted and utilized such methods. This is unquestionably the time at which we can best study what we have achieved in the application of industrial research, and at which we can best plan the future development insofar as it is controllable. If you will look over the papers which have been presented before this organization and others of similar type during the past three years, you will discover that some of the best minds in the country are already engaged in just such a task.

## POPULAR IMPRESSIONS OF RESEARCH

There is another popular impression which I believe to be basically incorrect. Research is regarded as a name for the efforts of individuals who spend their working hours in physical and chemical laboratories delving into the so-called secrets of nature for the purpose of extracting hitherto hidden facts or formulæ which may be expected to have commercial value. To me this is entirely too narrow an interpretation. I regard *research* as the search for facts, and I regard *industrial research* as the quest for facts which have immediate *industrial* value. Research as now pictured by the multitude may be called *material* research. Industrial research must include many other varieties if it is to achieve its end, particularly economic research and sociologic research. Questions of economic and sociologic types are just as amenable to laboratory treatment as are problems in the proper method of heat-treating a given grade of steel or in the modifications required to take account of mass action in conducting chemical reactions on a commercial scale.

If research be regarded merely as the determination of facts, every department of industry is a proper field for research and possibly can be benefited thereby. In fact, if this definition be adopted, modern research is merely an organized or ordered development of the methods which have been in use for generations. Every executive and every individual in charge of industrial activities has had to make decisions, and, in general, has determined facts on which to base these decisions insofar as that has been possible. Unfortunately the determination of facts in many different lines by one individual and at the rate required to keep pace with the evolution of the industrial world is often impossible, and as a consequence most of our successful executives have fulfilled Elbert Hubbard's definition—they decided quickly and were sometimes right. The marvel is that they were right in such a large proportion of their decisions.

All that we can hope to do by research is to make available for such men the facts upon which to base a greater proportion of their decisions. They will thus have to depend upon judgment, instinct, intuition, or guess, as you like, for a smaller proportion of their decisions, and their work will be just that much nearer to an absolute basis.

## THE FUTURE OF RESEARCH

I do not like to pose as a prophet—I prefer to have some honor in my own country if it is obtainable. However, I believe it is patent that our industrial development on its material side is already outstripping the economic and sociologic development required to stabilize it. We are learning to produce in greater quantities and in greater varieties daily, and our attention is so focused on this work that we are giving insufficient thought to the problems of finance, distribution and utilization and to the desires, ambitions and hopes which we are creating.

The literature of the day is so full of articles giving the results of investigations in the field of heat treatment that no manufac-

103

turer of steel products can plead ignorance of the fact that exact information is available if he chooses to make use of it. On the other hand, where is that same manufacturer to learn the facts with respect to the advisability of making plant extensions during periods of rising as against falling prices, or during periods of decreasing business failures as against periods of increasing failures? Or where is he to determine facts with respect to the reasons why bonus systems or profit-sharing systems succeed in some places and fail in others? Or where is he to obtain information with respect to the relationship existing between the rate of increase of wages in his plant or in his industry and the rate of increase of living costs? Research in such fields is conducted to a limited extent and certain publications are issued for the purpose of distributing the results, but comparatively few industries seem to have recognized the necessity of applying this sort of research to the problems of industry.

Does it not appear that in the sudden public appreciation of the advantages to be gained from material research we are threatened with a one-sided development which can only lead to trouble unless corrected at the start? Does it not appear desirable for organizations such as this Society to bring to the attention of the public the fact that no decision of any sort need rest on guesswork if facilities for all the necessary types of research are available?

I am not now concerned with the way in which research is to be conducted or financed. For my present purposes it is immaterial whether the federal government pays for *this*, the state government pays for *that*, and the individual producers or a group of producers pay for something else.

I am concerned with developing appreciation of the fact that what we now call research is merely an extension and refinement of methods previously in use; with driving home the fact that material research is only one of the numerous varieties required to continue the present industrial development; and with spreading the doctrine that it is good business to obtain such information as is necessary to enable our industrial executives to base their decisions more largely upon determined facts and to a lesser extent upon so-called judgment. I feel particularly that the continued stable development of the industrial age depends almost entirely upon our ability to bring to our executives those facts which will enable them to stabilize the social structure within which industrialism is contained and has its growth. I see no way of doing this other than by painstaking research along economic and sociologic lines at such a rate as to keep pace with the results attained by the application of material research which industry itself may be depended upon to develop at an ever-increasing rate.

I admit that my program is large, but the problem with which we are faced is large. We of this country, in common with the rest of the civilized world, have developed to a point where the conditions we have created are leading us to question the adequacy and even the validity of what we have regarded as fundamental concepts in the relation of man to man and of nation to nation. Some of us attribute this condition to the rapid development of industry, others to the growth of social and class consciousness, and still others to the continued upward progress of man. Attribute it to what you will, it is a fact that parallel to our development of factories, modern methods of production and of ways and means of bending natural forces and laws to our ends, there has occurred a development in human relations about which we have much less basic knowledge and over which we can exercise correspondingly less control. Is not a rather large program justified by such a set of conditions?

# Research and Industrial Wastes

BY GEORGE S. HESSENBRUCH,[1] ST. LOUIS, MO.

*In the paper which follows the author introduces his subject by calling attention to the lack of interest in research which existed previous to the war. He divides research into two classes which he calls "pure scientific research" and "industrial research," and points out that the latter class has a direct financial return to the industry concerned. He makes a strong plea for research on the subject of industrial wastes and closes his paper by describing a research laboratory now contemplated for pursuing this type of investigation.*

UNTIL a few years ago, research in the United States was not looked upon with great favor, except by those few who realized and knew what it had accomplished and what it meant to science. The general opinion among the laity was that only a crank could be interested in research that no special benefit would be derived from it in a commercial way. Its unbounded limits were not recognized by the general manufacturer. No doubt many men in the industries often wished they knew why such and such was the case and what would happen if such and such was done and what might be developed. Research, however, was looked more upon as a kind of inventive genius with which only a few were gifted and not as a science.

Up to that time there were few laboratories devoted to this work and those few were not sufficiently equipped to accomplish very much. Furthermore, their endeavor was mostly along some specific and special line of investigation for a particular industry. But now largely as a result of the world war and the impulse created by it for development along all scientific lines and the consequent industrial activity, manufacturers have come to the realization that activity in research is one of the essentials in the progress of industry, that it has, in many instances proved highly profitable, opened up tremendous fields for industrial endeavor and given them a vast amount of accurate and definite information and data

on subjects that previously had only been guess work and chance.

The subject of research can be divided into two distinct classes, namely, "pure scientific research" and what might be called "industrial research." Pure scientific research in most cases does not show very great immediate financial gains and therefore must be undertaken for the pure love of science and knowledge. It requires an investigator of a highly trained mind and peculiar character for the devotion to its cause. He must be of a very analytical trend of mind, possess considerable knowledge, and have unlimited patience and courage. He must be unselfish and have the ability of great concentration. Pure scientific research does not, however, show very great financial gain. The unmeasurable value and importance of this branch is not realized, for with but a few exceptions has a life-long work in this branch brought a commensurate financial return to the one devoting himself to it. Nevertheless, this branch is of the greatest importance and any accomplishment or new discovery in it is of inestimable value. If it were not for pure scientific research we would have no foundation for science. Its discoveries bring forth and determine the fundamentals and principles upon which further development is based and make them available for industrial enterprise and exploitation. At this point the engineer and chemist steps in and applies these new fundamentals and principles, develops them and gradually produces for the world a new piece of apparatus, machinery or process such as in some instances have revolutionized an entire industry. The need of pure scientific research therefore is absolutely essential.

## INDUSTRIAL RESEARCH

Industrial research, although highly scientific in principle, might be classed as the branch with the greatest chance of financial returns and for this reason it attracts more attention, as it is more remunerative both to the worker and investigator as well as the investor and industry. It develops known facts still further, and

[1] President, Industrial Engineers Corporation. Mem.Am.Soc.M.E.
Presented at the Mid-West Sections Meeting, Indianapolis, Ind., October 24 and 25, 1919, of THE AMERICAN SOCIETY OF MECHANICAL ENGINEERS.

investigates established principles, fundamentals and materials with the sole object in view of making these principles, fundamentals or materials of greater commercial value and more remunerative and available for industrial purposes. The world war and its consequent activities have given this branch of research a great impetus during the last few years in the United States. A great number of the larger corporations are establishing or have established research laboratories. These laboratories are, however, mostly equipped for investigations along the special line of endeavor of the particular corporation. A few private laboratories have been established for the purpose of investigating any subject that may be presented to them by a manufacturer who does not have a laboratory of his own, or who cannot justify the expense of maintaining one, but who can afford to obtain more definite and accurate knowledge on the materials or methods of manufacture pertinent to his particular endeavor. These laboratories undertake this work for a fee and results of value are sometimes obtained; at other times, however, the investigation, although conscientiously carried forward, is bare of results except as to data.

The United States Government has realized the meaning of industrial research and is giving it its support, and many societies and organizations are collecting data as to laboratories, etc., and in this way industrial as well as pure scientific research is gradually receiving more and more attention.

### INVESTIGATION OF WASTES

In industrial research would be included the investigations of wastes. This subject, in the writer's opinion, is one of considerable importance, for in many instances the financial return from the wastes, when properly processed and obtained from the raw materials in a manufacturing plant, have been greater than the return from the actual manufactured article made in this same plant.

Today we are confronted from all over the United States by the cry of increased production. Every manufacturer is striving his utmost to help this movement. In fact, it is the solution or at least an amelioration of the evils that threaten us seriously. It is but natural for us to turn toward the utilization of wastes and waste materials, which is the most economical and beneficial way to help this movement. Reducing these enormous wastes or utilizing them in some way or other by suitable processing will not only save our national resources from being depleted more rapidly, but it will also teach the manufacturer to be circumspect. It will reduce costs, establish new industrial undertakings, give employment to more people, open up new channels of commerce, induce capital to invest in new enterprises, increase the production of our soil and finally bring about the one most important status, namely, independence.

One benefit, if we may call it that, arising out of the world war has been that the American manufacturer is today more willing to be convinced in regard to the loss he is incurring daily by the non-utilization of waste materials. A few years ago the writer had the sad experience of being laughed at when putting before several manufacturers a plan by which their waste material would net them a greater return than the article they were producing in their factories. It was only a question of the proper processing of the wastes to make a salable and valuable product, a product that had a big demand and which was being imported. The processing was demonstrated to them, but they frankly said they did not wish to go into the chemical business—they were manufacturers of a certain article and did not care to undertake a new line and would not be interested.

The writer does not wish to minimize the efforts that have been put forth or disparage the good work that has been accomplished by such institutions as the Mellon Institute in Pittsburgh, our own Government laboratories in Washington and Pittsburgh, the Rensselaer Polytechnic Institute and the many private laboratories throughout the country. They have all done good work and are accomplishing a great deal for the benefit of industry and the public in general, but he would like to see a concerted move in the direction of research of wastes and their utilization. In his opinion such research would be of great value, and a brief description of a laboratory for conducting such investigations will doubtless be of interest. Its scheme of operation is so novel and unique as regards the financial returns of the undertaking that the writer feels that every manufacturer, no matter how large or small will be glad to avail himself of the chance to have his waste materials investigated, and developed if possible. It is contemplated that the laboratory will be one of the best-equipped laboratories in the United States, and that the highest possible class of research workers only will be employed. No fees are to be charged and no investigations will be undertaken for a fee, nor will there be any testing of materials. The laboratory is to be devoted purely to wastes and the utilization of wastes. The data collected will be available to any accredited scientist, research worker, or institute free of charge and cannot under any circumstances be utilized commercially except by special sanction of the laboratory management. All processes will be properly patented and the patents will be the property of the laboratory for such disposition as the board of management may deem best advisable. The manufacturer who has sent in the waste will, after proper investigation, be advised in the form of a report of the possibilities. The report will show the investment required for the proposed undertaking, the estimated operating cost, the revenue, the present and future markets and all data such as are required for the sound and proper understanding of the proposed undertaking. If requested, plans and specifications will be furnished for the entire establishment and operation properly started and temporarily superintended. This latter part of course will call for the regular engineering fee as established in the engineering profession. The manufacturer will be assisted in any way it is possible for the laboratory to serve him. In short, every effort will be made to give the manufacturer the full benefit that expert knowledge and investigation in all of the engineering lines can give, the ultimate result in view being to see the establishment of an undertaking that will in some manner or form utilize this waste and bring about conservation.

It is fully realized that a large amount of capital will be required and that the laboratory in all probability will not become self-sustaining for a period of three or four years, but under the scheme as laid out by the originators of the project it is figured that after that time it will become remunerative financially as well as bring about a great benefit to the industries and the public in general. An additional benefit of inestimable value will be the large amount of accurate knowledge and data that will be obtained and placed at the disposal of all mankind, and it stands to reason that hand in hand with industrial research will go purely scientific research, the results of which will be available to all and will be given out for the benefit of science in general.

----

Following apparently satisfactory test-block studies of a synthetic airplane-engine fuel known commercially as Alcogas and composed of 38 parts alcohol, 19 parts benzol, 4 parts toluol, 30 parts gasoline and 7½ parts ether, the Post Office Department arranged for a test of the fuel under service conditions in the Air Mail.

Mail plane No. 35, a Curtiss Model R4 machine equipped with a high-compression Liberty 12 motor, was assigned for the work, the check plane, flying the opposite trips during the same period with high-test aviation gasoline, being mail plane No. 34, also a Curtiss Model R4 plane, equipped with a low-compression Liberty 12 motor.

Thirty-one trips were flown between New York and Washington, being 218-mile non-stop flights, on the regular air-mail schedule between August 4 and September 19, 1919. The flights made by the gasoline ship numbered nineteen.

The tests indicate a saving of 3.3 gal. of fuel per hr. in favor of the alcohol fuel. Noting the revolutions per minute, however, the saving is even greater, as alcohol fuel shows 1514.3 r.p.m. as against 1507.8 r.p.m with gasoline. This means that not only is there a saving of 3.3 gal. of fuel per hr., but that 6.5 r.p.m. are gained above that by the use of the alcohol fuel.

Alcohol fuel also shows a saving in lubricating oil. The average for this fuel was 4.4 qt. per hr. as against 4.98 qt. per hr. for gasoline, or a net saving of 0.58 qt. per hr. This saving is thought to be due to the greater thermal efficiency displayed by alcohol.

## FLANGES FOR CAST-IRON PIPE

A PAPER by Mr. John Knickerbacker, read before the American Water Works Association last June, invites closer co-operation from the Cast Iron Pipe Makers and the A.S.M.E. Committee on Standardization of Flanges and Pipe Fittings. Mr. Knickerbacker recalls the need for adoption of standard flanges for light cast-iron pipe, and presents diagrams visualizing the different standards and semi-standards now existing for each size.

It will be remembered that an A.S.M.E. Committee has had this subject, flange ends for all weights of pipe, under constant consideration since 1892, and it is to the 1918 Report of this Committee that Mr. Knickerbacker especially refers. Therefore the memorandum below by Arthur R. Baylis, Chairman of the A.S.M.E. Committee, is interesting, as explaining the difficulties under which the Committee was obliged to work.

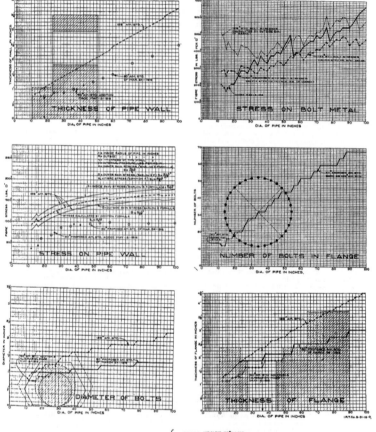

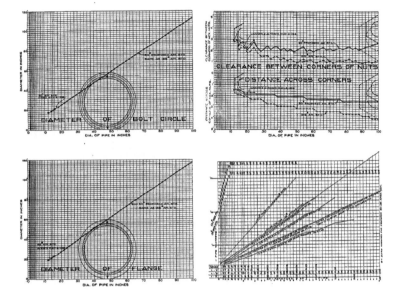

When the A.S.M.E. Committee on Standardization of Flanges and Pipe Fittings first commenced work on the new low-pressure standard, consideration was given to an entirely new set of dimensions propor-

TABLE 1   PROPOSED LOW-PRESSURE STANDARD FOR END FLANGES, BOLTINGS AND BODY THICKNESS—30 LB. WORKING PRESSURE

| Size | Diameter of Flange | Flange Thickness | Bolt-Circle Diameter | Number of Bolts | Size of Bolts | Body Thickness | Size | Diameter of Flange | Flange Thickness | Bolt-Circle Diameter | Number of Bolts | Size of Bolts | Body Thickness |
|---|---|---|---|---|---|---|---|---|---|---|---|---|---|
| 12 | 19 | 1¼ | 17 | 12 | ⅞ | 1 | 56 | 68¼ | 3 | 65 | 48 | 1½ | 1¾ |
| 14 | 21 | 1¼ | 18½ | 12 | ⅞ | 1 | 58 | 71 | 3¼ | 67½ | 48 | 1½ | 1¾ |
| 15 | 22¼ | 1¼ | 20 | 16 | ⅞ | 1 | 60 | 73 | 3¼ | 69½ | 52 | 1½ | 1¾ |
| 16 | 23½ | 1⁷⁄₁₆ | 21¼ | 16 | ⅞ | 1 | 62 | 75¼ | 3¼ | 71½ | 52 | 1½ | 1¾ |
| 18 | 25 | 1⁷⁄₁₆ | 22½ | 16 | 1 | 1¼ | 64 | 78 | 3¼ | 74 | 52 | 1½ | 1¾ |
| 20 | 27½ | 1⅜ | 25 | 20 | 1 | 1¼ | 66 | 80 | 3¼ | 76 | 52 | 1½ | 1¾ |
| 22 | 29½ | 1⅜ | 27¼ | 20 | 1 | 1¼ | 68 | 82½ | 3¼ | 78½ | 56 | 1½ | 1¾ |
| 24 | 32 | 1½ | 29¼ | 20 | 1 | 1¼ | 70 | 84½ | 3¼ | 80½ | 56 | 1½ | 1¾ |
| 28 | 34½ | 2 | 31¼ | 24 | 1 | 1¼ | 72 | 86½ | 3½ | 82½ | 60 | 1½ | 1¾ |
| 30 | 36½ | 2¼ | 34 | 28 | 1 | 1 | 74 | 88½ | 3½ | 84½ | 60 | 1½ | 2 |
| 32 | 38½ | 2¼ | 36 | 28 | 1 | 1 | 76 | 90½ | 3½ | 86½ | 60 | 1½ | 2¼ |
| 34 | 41½ | 2¼ | 38¾ | 28 | 1 | 1¼ | 78 | 93 | 3½ | 88½ | 60 | 1½ | 2¼ |
| 36 | 43½ | 2¼ | 40¼ | 32 | 1 | 1¼ | 80 | 95½ | 3½ | 91 | 60 | 1½ | 2¼ |
| 38 | 46 | 2¼ | 42½ | 32 | 1 | 1¼ | 82 | 97½ | 3½ | 93½ | 60 | 1½ | 2¼ |
| 40 | 48½ | 2¼ | 45½ | 32 | 1 | 1¼ | 84 | 99½ | 3¾ | 95½ | 64 | 1½ | 2¼ |
| 42 | 50½ | 2¼ | 47 | 36 | 1 | 1½ | 86 | 102 | 4 | 97½ | 64 | 1½ | 2¼ |
| 44 | 53 | 2¼ | 49½ | 36 | 1¼ | 1½ | 88 | 104½ | 4 | 100 | 68 | 1½ | 2¼ |
| 46 | 55½ | 2½ | 51¼ | 40 | 1¼ | 1½ | 90 | 106½ | 4¼ | 102½ | 68 | 1½ | 2¼ |
| 48 | 57½ | 2½ | 53¼ | 40 | 1¼ | 1½ | 92 | 108½ | 4¼ | 104½ | 68 | 1½ | 2¼ |
| 50 | 59½ | 2½ | 56 | 44 | 1¼ | 1⁷⁄₁₆ | 94 | 111 | 4¼ | 106½ | 68 | 1½ | 2¼ |
| 52 | 61½ | 2½ | 58½ | 44 | 1¼ | 1⁷⁄₁₆ | 96 | 113½ | 4½ | 108 | 68 | 1½ | 2¼ |
| 54 | 64½ | 2¾ | 60½ | 44 | 1¼ | 1⁷⁄₁₆ | 98 | 115½ | 4½ | 110 | 68 | 1½ | 2¼ |
| | 66½ | 3 | 62½ | 44 | 1¼ | 1⁷⁄₁₆ | 100 | 117¼ | 4½ | 113 | 68 | 1½ | 2¼ |

tioned to the specified working pressure and pipe diameters. Other considerations developed which influenced the Committee in arriving at the dimensions set forth in the present 50-lb. American standard.

The Committee of Manufacturers on Standardization of Fittings and Valves held an annual meeting on April 19, 1916, at which it discussed a proposed low-pressure standard. It finally authorized a subcommittee to conclude the matter in collaboration with the A.S.M.E. Committee. Their report, which maintains as far as possible interchangeability with the present 125-lb. American standard for flange- and bolt-circle diameters and number and diameter of bolts, is set forth in Table 1.

In determining flange thickness, pipe sizes were separated into groups, a common thickness of flanges being selected for each group, reducing to a minimum the number of stock bolts required, the diameters of bolts being standardized as well.

It does not appear that Mr. Knickerbacker's reference to foundry difficulties on the standardized flanges is well taken, as the permissible variation between thickness of flange and pipe body metal largely depends on the size and shape of the connecting fillet. Good safe castings can be made if the thin metal is gradually tapered into the thick.

In grouping the pipe sizes as mentioned above, it naturally follows that the smallest pipe in each group has a heavy flange while the largest pipe has a light one. When standards are formulated on the basis of interchangeability, with the object of reducing the number of patterns, templets and confusion in construction work, general proportion is of necessity somewhat sacrificed, and it is difficult to determine just where to draw the line. Sizes were selected for minimum thickness of pipe body metal closely following those of the American Water Works Association, class "B" pipe. The new standard is considered ample to meet all physical requirements.

The curves shown here are self-explanatory and a study of them largely explains the standards adopted. These curves were used during Committee discussions in preparing the new low-pressure standard.

Our Committee will be glad to coöperate further with any engineering organization interested in this subject.

(Signed)   ARTHUR R. BAYLIS, Chairman

A.S.M.E. Committee on Standardization
of Flanges and Pipe Fittings.

# ENGINEERING RESEARCH

## A Department Conducted by the Research Committee of the A. S. M. E.

### A—RESEARCH RESULTS

The purpose of this section of Engineering Research is to give the origin of research information which has been completed, to give a résumé of research results with formulæ or curves where such may be readily given, and to report results of non-extensive researches which in the opinion of the investigators do not warrant a paper.

*Apparatus and Instruments A1-20.* Electrical Measuring Instruments. An exhaustive paper on the accuracy of commercial electrical measurements requested by the Committee on Instruments and Measurements of the A. I. E. E. has been furnished and submitted to the Chairman of the Committee. It is scheduled for the Convention in February. It discusses the accuracy required, choice, installation, use and maintenance of instruments, conditions of use, features of design, sources of error and improvement. Address Bureau of Standards, S. W. Stratton, Director, Washington, D. C.

*Fuels, Gas, Tar and Coke A1-20.* Sulphur in Coals. A study of the forms in which sulphur occurs in coals. Bulletin No. 111 of the Engineering Experiment Station of the University of Illinois by A. R. Powell with S. W. Parr. The Bulletin gives the present status of the knowledge of sulphur in coal, the methods used in analyzing the changes in the form of sulphur in the coal, and also conclusions which have been abstracted.

1 The sulphur of coal occurs in four characteristic forms:
(*a*) The resinic organic type.
(*b*) The humus organic type.
(*c*) The pyritic or marcasitic inorganic form, and
(*d*) The sulphate sulphur.

2 Early methods of determining the forms of sulphur gave high results for pyritic sulphur and low results for organic. The methods developed in the Bulletin are more nearly accurate.

3 Some of the sulphur in coal is oxidized to the form of sulphate. Length of time and standing accomplished this.

4 The action mentioned in 3 is hastened by the presence of bacteria or some catalytic agent.

5 The amount of soluble sulphate formed is less than it would be if the amount of soluble iron were taken as a criterion. Some of the oxidized sulphur combined with the organic matter of the coal.

6 When coal is coked the sulphate sulphur is retained by the coke, the pyritic sulphur is partially volatilized, most of the resinic sulphur is left in the coke and the humus sulphur is partially volatilized.

7 The forms of sulphur in the coke were studied but not fully identified.

8 During the coking process secondary reactions between the constituents of the coal and the decomposed sulphur compounds affect the quantity of sulphur retained by the coke. Address Director of Engineering, Experiment Station, University of Illinois, Urbana, Ill.

*Heat A1-20.* Hot-Air Pipes with Thin Asbestos. In making simple tests on relative thermo-resistance of layers of asbestos paper when applied loosely and tightly, it was observed that tin vessels containing hot water would cool less rapidly than the same vessels when wrapped tightly with one layer of sheet asbestos 0.013 in. thick. It was found that it was necessary to have asbestos at least 1/16-in. in thickness to be equal in heat conductivity to the bare pipe. Address T. S. Taylor, Westinghouse Electric and Manufacturing Company, Pittsburgh, Pa.

*Metallurgy and Metallography A1-20.* Lead and Embrittlement. It has been found that the rate at which intercrystalline brittleness is brought about is proportional to the amount of impurities and to the concentration of acid in the solution in which the lead is placed. Practically all impurities which are found are lodged between the grains. The preferential attack by the corroding agents for these impurities and perhaps for the amorphous intercrystalline cement accounts for the brittleness. Investigations showed that specimens of exceptionally pure lead (99.99 per cent) when immersed in a neutral solution of an acetate for twenty-four days became appreciably embrittled by the formation of minute intercrystalline fissures. No evidence of the existence of allotropic form of lead similar to gray tin could be obtained. Address Bureau of Standards, S. W. Stratton, Director, Washington, D. C.

*Properties of Engineering Materials A1-20.* Rope. The Action of Acid on Rope with Protective Coatings. Tests of manila rope to determine the effect of asphaltum compound used as a protection to rope against action of muriatic acid show that rope so treated deteriorates as completely as untreated rope. Address Bureau of Standards, S. W. Stratton, Director, Washington, D. C.

### B—RESEARCH IN PROGRESS

The purpose of this section of Engineering Research is to bring together those who are working on the same problem for coöperation or conference, to prevent unnecessary duplication of work, and to inform the profession of the investigators who are engaged upon research problems. The addresses of these investigators are given for the purpose of correspondence.

*Aircraft B1-20.* Propellers. Investigation of airplane propellers in connection with National Advisory Committee for Aeronautics. Prof. W. F. Durand and E. P. Leslie, Stanford University, Cal.

*Apparatus and Instruments B1-20.* Recording Thermometers. A study of the permanence and accuracy of initial adjustment and sensitivity, of recording thermometers having an upper limit of reading of about 820 deg. fahr., together with a comparative examination of the mechanical excellence and convenience of manipulation and observation of the various types. The tests planned are such as to develop the factors of hysteresis, including pen friction, time lag, fatigue, constancy of zero adjustment and of the elastic properties of the Bourdon tube under frequently repeated temperature cycles. Work of the Sub-Committee on the Permanency and Accuracy of Engineering Instruments, F. J. Schlink, Chairman. Address F. J. Schlink, Physicist, Firestone Tire and Rubber Company, Akron, Ohio.

*Apparatus and Instruments B2-20.* Oxy-acetylene. An investigation of the apparatus for oxy-acetylene welding and cutting. Method of determining the specific-heat ratio of acetylene passing through acetylene meter has been completed. Bureau of Standards. Address S. W. Stratton, Director, Washington, D. C.

*Automotive Vehicles and Equipment B1-20.* Radiators. Heat tests of auto radiators. Sheffield Scientific School, Yale University. Address Prof. L. P. Breckenridge, New Haven, Conn.

*Friction B1-20.* Ball and Roller Bearings. The friction of ball and roller bearings. Sheffield Scientific School, Yale University. Address Prof. L. P. Breckenridge, New Haven, Conn.

*Heat B1-20.* Heat Transmission in Boiler Tubes. Sheffield Scientific School, Yale University. Address Prof. L. P. Breckenridge, New Haven, Conn.

*Heat B2-20.* Hot-Blast Heater Experiments. Sheffield Scientific School, Yale University. Address Prof. L. P. Breckenridge, New Haven, Conn.

*Heat B3-20.* Heat Transfer. Research relating to the development of satisfactory commercial apparatus for the transfer of heat between fluids, vapors and gases and to include a study of the transfer of heat between metal surfaces and fluids of widely varying viscosities. Prof. W. R. Eckert. Stanford University, Cal.

*Internal-Combustion Motors B1-20.* Modified Diesel Engine. Development of special or modified form of Diesel engine to use kerosene and distillate. Prof. C. N. Cross, Stanford University, Cal.

*Machine Design B1-20.* Ball Bearings. Development of proper type and mounting of ball bearings for deep-well pumps and centrifugal pumps. S. C. Kyle, Rialto Building, San Francisco, Cal.

*Machine Design B2-20.* Gear Teeth. Investigations of the wear of gear teeth under loaded conditions, including the influence of a varying intensity of load, materials, tooth forms, and lubrication. Prof. G. H. Marx, Stanford University, Cal.

*Machine Design B3-20.* Locomotive Packing-Ring Investigations. It has been found that low-sulphur air-furnace iron is more suitable for locomotive packing rings, bushings, valves and piston rings than cupola iron. It appears desirable that the present maximum allowable sulphur limit be lowered and that the minimum transverse strength of castings be slightly increased. It has not been possible to establish a relation between laboratory and service tests. A report on this investigation is now practically complete. Bureau of Standards, Washington, D. C. Address S. W. Stratton, Director.

*Metallurgy and Metallography B1-20.* Steel. Development of steel of suitable grade for pins of buckets on gold-dredging machinery. Pins have diameters from 3 to 7½ in. G. L. Hurst, Union Plant, Bethlehem Shipbuilding Corporation, San Francisco, Cal.

*Metallurgy and Metallography B2-20.* Electric Furnaces. Improvements in electric furnaces with special reference to the following applications:
   *a* The melting of steel scrap for steel castings.
   *b* The melting of steel scrap and ferroalloys into tool steels.
   *c* The heat treatment of steel castings for annealing or tempering.
   *d* The melting of non-ferrous metals for commercial castings in marine work.
   Address C. M. Gunn, 444 Market St., San Francisco, Cal.

*Properties of Engineering Materials B1-20.* Expansion of Marble. Two slabs of Tennessee marble and one slab of Vermont marble have been exposed to the weather at the Bureau of Standards. During the first three months a considerable change in dimensions was shown, after which the dimensions remained fairly constant One of the Tennessee slabs expanded 0.02 per cent and the other 0.13 per cent of its length, while Vermont marble contracted 0.04 per cent of its length. Bureau of Standards, Washington, D. C. Address S. W. Stratton, Director.

*Properties of Engineering Materials B2-20.* Tests of Materials before acceptance and to determine cause of failure and strength of electric welds. (Tests 1117-1119-1140-1171-1192-1203-1205-1223 and 1099.) Naval Experiment Station, Annapolis, Md. Address Bureau of Steam Engineering. Washington, D. C.

*Properties of Engineering Materials B3-20.* Bearing Metals. Tests of bearing metals. (Tests 1101-1114-1145-1092-869.) U. S. Naval Experiment Station, Annapolis, Md. Address Bureau of Steam Engineering, Washington, D. C.

*Properties of Engineering Materials B4-20.* Repeated Stress and Impact. Investigation of endurance and impact resistance of metals. Test 1220. U. S. Naval Experiment Station, Annapolis, Md. Address Bureau of Steam Engineering, Washington, D. C.

*Properties of Engineering Materials B5-20.* Magnetic Analysis. The Bureau of Standards is engaged at present on a study of the correlation between magnetic properties of steel and its structure and mechanical properties, and the development, methods and apparatus for the application of magnetic analysis in a practical way. A method has already been developed to determine the degree of homogeneity along the length of a relatively long piece. The quality of heat treatment is also being investigated by magnetic methods. Address Bureau of Standards, S. W. Stratton, Director, Washington, D. C.

*Pumps B1-20.* Air Chambers. Determination of proper size of air chambers upon reciprocating pumps. No. 1231. U. S. Naval Experiment Station, Annapolis, Md. Address Bureau of Steam Engineering, Washington, D. C.

*Railroad Tracks and Signals B1-20.* Rail Investigations. The Bureau of Standards is at present making a critical study of available literature for the purpose of classifying rail failures and test data preparatory to initiating an extensive series to discover cause and remedy for such accidents. Bureau of Standards, Address S. W. Stratton, Director, Washington, D. C.

*Steam Power B1-20.* Steam Power Plants for Automobiles. The Doble laboratories of San Francisco are investigating the development of steam power plants for automobiles under several divisions:
   *a* Systems of combustion using kerosene and cheaper grades of fuel without the formation of soot and carbon deposits.
   *b* The construction of the combustion system to make it automatic and subject to control by the boiler conditions.
   *c* A fuel and fire automatic control.
   *d* The development of a high-pressure boiler of maximum thermal efficiency.
   *e* The development of a new type of steam engine with high economy, flexibility of control and utilization of high-pressure steam with high superheat. Address Doble Laboratories, San Francisco, Cal.

*Transmission B1-20.* Hydraulic Dynamometers as Transmission Devices. H. B. Langille. University of California, Berkeley, Cal.

*Transmission B2-20.* Steel Belting. Investigations of the economic possibilities in the use of steel belting as a substitute for leather, with special references to development of satisfactory joint. Prof. G. H. Marx. Stanford University. Cal.

## C—RESEARCH PROBLEMS

The purpose of this section of Engineering Research is to bring together persons who desire coöperation in research work or to bring together those who have problems and no equipment with those who are equipped to carry on research. It is hoped that those desiring coöperation or aid will state problems for publication in this section.

*Boilers and Accessories C1-20.* High-Capacity Boiler. Design, construction and testing of boilers for high capacities. Address Sheffield Scientific School, Yale University, Prof. L. P. Breckenridge, New Haven, Conn.

*Fuels, Gas, Tar and Coke C1-20.* Pig-Iron Coke for Making Pig Iron on the Pacific Coast. Address Robert M. Hale, 220 Battery St., San Francisco, Cal.

*Heat C1-20.* Radiation. Heat radiation from gases to solids at high temperatures. Sheffield Scientific School, Yale University. Address Prof. L. P. Breckenridge, New Haven, Conn.

*Heating C1-20.* New Method of Heating by Reversal of Thermodynamic Cycle. Reverse a cycle to discharge high-temperature heat and also use discharge from engine driving cycle. Address J. G. De Remer, Rialto Building, San Francisco, Cal.

*Hydraulics C1-20.* Measurements. The measurement of the flow of fluids. Sheffield Scientific School. Address Prof. L. P. Breckenridge, New Haven, Conn.

*Metallurgy and Metallography C1-20.* Spelter. A continuous process for producing spelter. See letter from Prof. W. F. Durand in the Research Section of the January number of MECHANICAL ENGINEERING. Address Robert M. Hale, 220 Battery St., San Francisco, Cal.

## D—RESEARCH EQUIPMENT

The purpose of this section of Engineering Research is to give in concise form notes regarding the equipment of laboratories for mutual information and for the purpose of informing the profession of the equipment in various laboratories so that persons desiring special investigations may know where such work may be done.

*Sheffield Scientific School, Yale University D1-20.* Research Laboratory. The Research Laboratory of the Mechanical Engineering Department of Sheffield Scientific School is equipped as follows:

   Heating boilers for using anthracite and bituminous coal
   Exhaust fan and heater stacks
   Automobile-radiator testing laboratory
   Automobile testing drum
   Sprague electric dynamometer
   Material-testing machines up to 150,000 lb.
   Gas-explosion apparatus
   150-hp. water-tube boiler
   50-hp. two-stage air compressor
   Gas producer
   Gas and gasoline engines
   Steam engines and condensers
   Airplane engines of the following makes:
      Two 400-hp. Liberty motors
      Two 100-hp. Hall-Scott motors ·
      Two 200-hp. Curtiss V-2 motors
      Two 150-hp. Hispano-Suiza motors
      Two 100-hp. Curtiss OXX motors, and
      Two 100-hp. Gnome. motors.

Sheffield Scientific School, Yale University. Address Prof. L. P. Breckenridge, New Haven, Conn.

## E—RESEARCH PERSONNEL

The purpose of this section of Engineering Research is to give notes of a personal nature regarding the personnel of various laboratories, methods of procedure for commercial work or notes regarding the conduct of various laboratories.

*Connecticut Research Laboratories E1-20.* Through the Research Committee of the Local Sections in Connecticut the names of the following laboratories in the neighborhood of New Haven have been collected. Other lists are to follow:

   The Acme Wire Company. Address Dr. R. W. Langley, New Haven, Conn.
   American Brass Company. Address J. A. Coe, Waterbury, Conn.
   English and Hendricks Co. Address S. M. Bradley, New Haven, Conn.
   Farrell Foundry and Machine Co. Address Walter Perry, 2nd Vice-President, Ansonia, Conn.
   Groton Iron Works. Address H. O. Trowbridge, Groton. Conn.
   Greist Mfg. Co. Address H. M. Greist, New Haven, Conn.
   Malleable Iron Fittings Co. Address A. E. Hammer, New Haven, Conn.
   New Haven Gaslight Co. Address J. A. Norcross, New Haven, Conn.

Scoville Mfg. Co. R. S. Sperry. Waterbury, Conn.
Sargent & Co. Address Henry B. Sargent. New Haven, Conn.
Winchester Repeating Arms Co. Address Edwin Pugsley, New
    Haven. Conn.
*Sheffield Scientific School, Yale University E1-20.* The research work
    of the Sheffield Scientific School is given by the following titles
    of articles published:

Roll Pressure in Cold-Rolling Brass. W. K. Shepard and G. C.
    Gerner. Am. Mach., March 1915.
Hardness Tests of Cold-Rolled Steel, W. K. Shepard and C.
    T. Porter. Am. Mach., Feb. 18, 1915.
Hardness Tests on Brass, W. K. Shepard, Am. Mach., Feb. 8,
    1917.
Testing of House-Heating Boilers. L. P. Breckenridge and D.
    B. Prentice. Jl. A.S.M.E., Nov. 1916.
Experimental Determination of Moisture in Exhaust Steam, E.
    H. Lockwood. Journal H. & V. Eng., Apr. 1915.
Tests of a Hot-Air-Tube Heating System. E. H. Lockwood and
    A. C. Staley. H. & V. Magazine, Dec. 1914.
Tests of a Magazine Heater. E. H. Lockwood. H. & V. Maga-
    zine. May 1916.
Sampling and Analysis of Producer Gas. P. W. Swain.
    Jl. A.S.M.E., Dec. 1916.
Practical Testing of Motor Vehicles. A. B. Browne and E. H.
    Lockwood. Jl. S.A.E., Feb. 1917.
Power Losses in Pneumatic Tires, E. H. Lockwood. Jl. S.A.E.,
    Feb. 1917.

### F—BIBLIOGRAPHIES

The purpose of this section of Engineering Research is to
inform the profession of bibliographies which have been prepared.

In general this work is done at the expense of the Society. Extensive bibliographies require the approval of the Research Committee. All bibliographies are loaned for a period of one month
only. Additional copies are available, however, for periods of
two weeks to members of the A.S.M.E. or to others recommended
by members of the A.S.M.E. These bibliographies are on file in
the offices of the Society.

*Apparatus and Instruments F1-20.* Brightness Measurements.
    References relating to brightness measurements. A bibliography
    of one page. Address A.S.M.E., 29 West 39th St., New York.
*Leather and Glue F1-20.* Casein and Casein Glues. A bibliography
    of one page. Address A.S.M.E., 29 West 39th St., New York.
*Lubricants F1-20.* A bibliography of one page on oils and lubricants.
    Address A.S.M.E.. 29 West 39th St., New York.
*Transmission F1-20.* Gears. Worm and Helical Gears. A biblio-
    raphy of 18 pages. Address A.S.M.E., 29 West 39th St., New
    York.
*Wastes F1-20.* Incinerator Apparatus. The Manufacture of In-
    cinerators. A bibliography of one page. Search 2611. Address
    A.S.M.E.. 29 West 39th St., New York. (See also *Water,
    Sewage and Sanitation F1-20.*)
*Water. Sewage and Sanitation F1-20.* Incinerator Apparatus. The
    Manufacture of Incinerators. A bibliography of one page. Search
    2611. Address A.S.M.E., 29 West 39th St., New York. (See
    also *Wastes F1-20.*)
*Water, Sewage and Sanitation F2-20.* Underground Water in Arid
    Countries and Electrical Detectors. Search 2691. A bibliography
    of one page. Address A.S.M.E., 29 West 39th St., New York.

# THE NEW MECHANICAL ENGINEERING LABORATORY, COLLEGE OF ENGINEERING, UNIVERSITY OF ILLINOIS

BY CHARLES RUSS RICHARDS,[1] URBANA, ILL.

THE following description of the Power Laboratory of the
Department of Mechanical Engineering of the University of
Illinois presents the facilities for instruction and research which
are afforded by but one of the many laboratories of the College of
Engineering, and which are available for the research work of the
Engineering Experiment Station. Each department of the College
of Engineering is represented in the Engineering Experiment
Station and is conducting investigations with funds appropriated
for the purpose by the University, or with funds for special coöperative investigations provided by outside interests. The Station is prepared to undertake research work in architecture and
architectural engineering, ceramic engineering, industrial chemistry, civil engineering, electrical engineering, theoretical and
applied mechanics, mechanical engineering, mining engineering,
municipal and sanitary engineering, physics, and railway engineering. The results of completed investigations are published as bulletins of the Engineering Experiment Station, which are
distributed free upon request. At present 114 bulletins and 5
circulars have been published.

The following is a partial list of investigations which are being
conducted with funds provided by the University: Investigations
of the viscosity of glass; of certain ceramic materials and processes; of the drainage of land; of monolithic construction of highways; of riveted steel structures; of the electrical properties of
steel alloys; of acid-resisting alloys; of various properties of coal;
of the manufacture of gas; of reinforced concrete and other
materials of construction; of the explosion of gaseous mixtures;
of the gas-engine cycle; of twist drills and the drilling of metals;
of coal mining; of the subsidence of land from mining; of heat
transmission; of air-steam mixtures; and of the flow of air.

A number of very important coöperative investigations are now
in progress, the principal expenses of which are aid from funds
provided by the coöperating agencies. These include an investigation of the coking of coal in coöperation with Mr. A. T. Hert of
the American Creosoting Co.; an investigation of the stresses in
railroad track in coöperation with committees of the American
Society of Civil Engineers and the American Railway Engineering

Association; an investigation of the stresses in chilled car wheels
in coöperation with the Association of Chilled Car Wheel Manufacturers; investigations of problems concerned with coal mining
in coöperation with the Illinois Geological Survey and the U. S.
Bureau of Mines; an investigation of warm-air furnaces and
furnace heating in coöperation with the National Warm Air
Heating and Ventilating Association; and an investigation of the
fatigue phenomena of metals in coöperation with the Engineering
Foundation and the National Research Council. In these coöperative investigations the Engineering Experiment Station provides
such laboratory facilities as it may possess, administers the work,
and publishes the results obtained; while the cost of special apparatus and other expenses, and the salaries of special investigators, are paid from funds provided for the investigations.

## Description of Laboratory and Equipment

This is one of the largest laboratories of its kind in the country
with two main floor levels each approximately 120 by 125 sq. ft. in
area, exclusive of two small research laboratories 20 by 27 sq. ft.
Included in the same building are offices for a staff of from sixteen
to eighteen men, three computation rooms, a fairly large lecture room,
a stock room, a shop for mechanicians and a modern locker room.
The elevated main floor, Fig. 1, carries the principal equipment
and is of heavy reinforced concrete construction designed for a working load of 300 lb. per sq. ft. Numerous light and air wells afford
communication between this floor and the lower floor, which is of the
same area as the main floor, and sufficiently depressed below grade to
give a clear working height, in which there are no beams, of 8 ft. 6 in.
These features are all shown in Fig. 1. The roof is of the saw-tooth
type with three principal bays running east and west.

### STEAM-ENGINE AND TURBINE EQUIPMENT

The principal prime movers installed in the laboratory consist
of six engines and two turbines. There has been installed the following equipment, starting at the west end of this group:
A Chandler and Taylor cross-compound engine (9), made up by
connecting a 7-in. by 10-in. and a 10-in. by 12-in. simple throttling
slide-valve engine, served by a Dean surface condenser.
A 10-in. by 10-in. Ideal, automatic high-speed, non-condensing
engine (10), with balanced slide valve and inertia governor.
A 10-in. by 10-in. Ideal, automatic high-speed, non-condensing
engine (14), with a piston valve, and a throttling and centrifugal
shaft governor, either of which may be used.

[1] Dean, College of Engineering, and Director, Engineering Experiment
Station, University of Illinois. Mem.Am.Soc.M.E.

An 8-in. by 12-in. Meyer automatic engine (13), which operates non-condensing.

An 8-in. by 18 in. simple releasing gear, Murray Corliss engine (15), provided with a Dean surface condenser.

A 12-in. by 24-in. simple double eccentric, releasing gear, Allis-Chalmers Corliss engine (17) belted to a 100-kw. bipolar Edison d. c. generator. This engine has a disk speed-changing device on the governor, whereby the speed may be varied from 60 to 120 r. p. m. A Worthington surface condenser, together with reciprocating wet and dry-vacuum and circulating water pumps take care of the exhaust steam from this unit when the engine is operated condensing.

A Kerr single-velocity stage turbine (18) rated at 60 hp., served by the Worthington surface condenser in case the turbine is run as a condensing unit.

(24), capable of delivering 300 gal. of water per min. against a pressure of 150 lb. per sq. ft., and a De Laval two-stage turbo pump (23) having a capacity of 140 gal. per min. when operating against a head of 300 ft.

## MAIN STEAM PIPING

The high-pressure saturated-steam line is 8 in. in diameter, and enters the laboratory at the northwest corner. It runs underneath the elevated main floor along the west side for the entire width of the laboratory. Near the middle of the laboratory a connection is made by a long-sweep, vertical bend to the main high-pressure saturated-steam header running the length of the main floor. A separator and reducing valve are installed just after the bend at the point where the pipe again becomes horizontal. The steam pressure

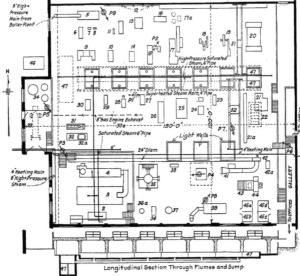

FIG. 1 MAIN FLOOR PLAN OF THE MECHANICAL ENGINEERING LABORATORY SHOWING THE LOCATION OF THE PRINCIPAL EQUIPMENT

A simple slide valve engine (19) designed by Professor Robinson, first professor of mechanical engineering at the University, and built in the mechanical engineering shops of the University.

A large Southwark triple-expansion engine with two direct-current generators (20), given to the University by the Commonwealth Edison Company of Chicago. This machine, formerly installed at the Fisk Street Station of that company, and the preceding unit (19), illustrating the development of the steam engine, will later be placed in a permanent museum.

In the University power house, adjacent to the laboratory, there is available for test work a 100-kw. Curtis turbo-alternator, equipped with the instruments necessary for the measurement of the power output. The turbo-alternator unit has a Wheeler surface condenser, a Wheeler horizontal crank and flywheel wet- and dry-vacuum pump, and a motor-driven centrifugal circulating pump. This condenser equipment is owned by the laboratory.

All engines and turbines, unless otherwise noted, are provided with absorption brakes, either of the Prony or of the rope type.

Equipment for the study of steam calorimeters, the calibration of pressure gages, and the calibration of indicators is located on the north wall (12) and on the adjacent bench.

### PUMPS

The equipment for test work on pumps consists of a compound Fairbanks-Morse direct double-acting, outside-packed plunger pump

in the steam header is controlled and kept constant by means of the reducing valve.

The piping from the experimental power boiler is so arranged that steam from the boiler may be discharged into the high-pressure saturated-steam main when operating without superheater, or into a second superheated-steam header running parallel to the saturated-steam header and directly underneath it, when operating with superheater. Both saturated- and superheated-steam headers are provided with sliding expansion joints, and are dripped through pot and dump traps.

### FLUMES AND SUMP

In order to care for the condensed steam discharged from the engines, surface condensers and pumps, two flumes (47) are provided. The piping from the experimental power boiler is so arranged that running nearly the entire length of the basement. The water may be dropped from the flumes through any one of six circular outlets in the bottom of each flume. These outlets discharge into a sump located at the west end of the flumes and underneath the basement floor. This construction is shown by the longitudinal section through flumes and sump. Provision is made for locating V-notch weir plates in the flume so as to measure the amount of water flowing.

At present the water from the sump is discharged to waste by means of an electrically driven centrifugal sump pump. The discharge will eventually be into a cooling tower (46), which is soon to be built.

### SMALL POWER PLANT GROUP

In order to demonstrate the operating characteristics and relation of the equipment of a steam plant, a small power plant has been erected. This plant consists of an Erie City hand-fired portable type of firebox boiler (5), rated at 25 boiler hp, capacity, operating under natural draft with a steel stack 20 in. in diameter, and 40 ft. high; a 7-in. by 10-in. Atlas steam engine (6), with a throttle governor, the output of which is absorbed by a Prony brake; a Cochrane open feedwater heater (7) of 50 boiler hp. capacity; an American single-acting simplex boiler-feed steam pump (8); and a Keystone water meter for measuring the amount of feedwater used.

### POWER BOILER FOR EXPERIMENTAL PURPOSES

The power boiler group (2), (3) and (4), consists of a Heine horizontal water-tube boiler (3) of 210 boiler hp. rated capacity,

FIG. 2 FLUMES, LOWER FLOOR. SEE FIG. 1, P-1

FIG. 3 POWER BOILER GROUP. SEE FIG. 1, P-3

equipped with a Green chain-grate stoker having a grate area of 38.2 sq. ft. and belt driven from a small vertical steam engine; a separately fired Foster superheater (2), having 3400 sq. ft. of super-heating surface capable of superheating 4500 lb. of steam per hr. through 200 deg. fahr., and a Sturtevant economizer and induced-draft fan (4), which draws the flue gases through the economizer when the economizer is being used, or through a bypass underneath the economizer when the dampers are thrown to cut the economizer out of service. The gases are discharged by the induced-draft fan into a steel stack 36 in. in diameter and 45 ft. high. In order to drive this fan, a small vertical Sturtevant steam engine is coupled directly to it. This engine also drives the mechanism for moving the scrapers which clean the economizer tubes.

A feedwater tank (1) is located near the boiler into which tank the feedwater is discharged, after having been measured in a cali-brated tank located directly above the feedwater tank. The feed-water can be delivered to the boiler from the feedwater tank either by a direct-acting boiler-feed pump located on the main floor directly

underneath the feedwater tank, or by means of three injectors located directly above the feedwater tank.

The arrangement of the boiler group is such that the operation of the boiler may be studied in any one of three ways; the boiler operating without either the economizer or the superheater, the boiler operating with either economizer or superheater, or the boiler operating with both economizer and superheater.

### HEATING AND VENTILATION

As already indicated, the south bay is largely devoted to work in the field of heating and ventilation. A four-unit steam-radiator testing plant (45-a,b,c,d) has been conveniently arranged for comparing the coefficient of heat transmission of various types of direct steam radiators. Steam can be supplied at any desired constant pressure so that simultaneous tests may be run on the radiators located in the four enclosures, and their comparative efficiencies and transmission coefficients easily determined.

An experimental hot-blast heating unit has been completely equipped for a variety of tests. The heater proper (43) consists of six sections of 1-in. pipe coils, each section containing 507 equivalent lin. ft. of pipe, making a heater equal to approximately 3000 sq. ft. of direct radiation. The condensation from each section can be weighed separately and the air flow and temperature rise measured, so that the effect of any air velocity on the heat transmission co-efficient is readily calculated. A Buffalo Forge Company's steel-plate fan (44) with a 100-in.-diameter fan wheel is operated by a direct-connected 6-in. by 6-in. vertical steam engine for drawing air through the unit. Pneumatically controlled bypass dampers con-trol the flow of air through or under the coils.

Near the above heater stands a Warren Webster & Company's

FIG. 4 WARM-AIR-FURNACE TESTING PLANT. SEE FIG. 1, P-4

air washer (41) of the humidifying type. This is equipped for tem-perature and humidity control with a Powers water thermostat operating on the steam and water mixer in the washer tank. Any desired initial air temperature can be secured by the Vento heating coil which is under the control of a Johnson Service Company's duct thermostat. The washer has a capacity of 3000 cu. ft. of air per min. at a free-area velocity of 500 ft. per min.

A No. 4 single-inlet Sirocco multiblade fan (42) is used to draw air through the above washer. This fan is mounted on a gallery and discharges the air into a metal duct 24 in. in diameter and 100 ft. long. Complete facilities are provided for making traverses of this duct at any point, so that accurate data on the air volume handled and coefficient of duct friction may be determined. By using the first 40 ft. of the duct and disconnecting the air washer, rating and effi-ciency tests of the Sirocco fan are easily made. The fan is operated by a Sprague electric dynamometer (42-a) of 10 to 25 hp. capacity with a speed range of from 1000 to 4000 r.p.m.

A heat-transmission testing plant (40) for finding the heat loss through various building materials has been set up in the form of a standard box with special walls 8 in. thick and 6-ft. by 6-ft. sides. An electrical heating element and numerous thermocouples at all surfaces supply and measure heat input and temperature gradients respectively. This plant has been used regularly, but has not yet been inclosed, as indicated on plans, to completely protect it from all disturbing influences.

A Kewanee down-draft brick-set smokeless steam boiler (30) with water-tube grate is the principal item in the heating-boiler equipment. This boiler is rated at 1900 sq. ft. of radiation with a water-grate area of 7.1 sq. ft. It is equipped for securing very complete data on efficiency and capacity, and can be fed with condensation from the hot-blast unit described above. The boiler was given to the University by the Kewanee Boiler Company.

A small Arco cast-iron round steam-heating boiler (37) rated at 800 sq. ft. of radiation is also included in the boiler equipment. This boiler has a grate area of 4.12 sq. ft. and was made by the American Radiator Company. In addition to the above, there is a Mercer cast-iron sectional steam-heating boiler (36) of 1275 sq. ft. rated capacity. This boiler is of the header type with a grate of 6 sq. ft. area, and was made by the H. B. Smith Company. The last two boilers were used in securing data for Bulletin No. 81 on House-Heating Boilers.

A single-sweeper vacuum cleaning machine (38), made by the American Rotary Valve Company, has been mounted in portable fashion to facilitate the testing of its efficiency, amount of air handled, and vacuum required for cleaning purposes. The exhauster is of the rotary, double-impeller type, belted to a 1-hp. motor.

Probably the most elaborate feature of this section of the laboratory is the warm-air-furnace testing plant (35) and its auxiliaries. This equipment has been erected for the purpose of measuring the efficiency and capacity of warm-air furnaces under conditions of service with leaders, stacks and registers at three floor levels. In this way it is not only possible to test the complete system, but also to determine the losses in various parts of a system when in operation.

In order to secure the necessary data, an elaborate system for temperature measurement, using thermocouples, has been worked out so that all readings can be taken at a central point. The air flow is measured at inlet and outlets by specially calibrated anemometers. These anemometers are checked before each test in two calibrating plants which reproduce the conditions of service in all details, such

box has two compartments, each similar to the one used by R. J. Durley in his experiments in the measurement of air. The machine is equipped with revolution counters, reducing motions and other apparatus requisite for making complete tests.

REFRIGERATION EQUIPMENT

The refrigeration plant consists of a York ammonia compression machine (21) and a Vogt absorption machine (21a). A freezing tank, with a capacity of 85 cans, each holding 100 lb., is built into the lower floor, and arranged so that the brine from either machine may be circulated through it for the purpose of making ice. This tank is of waterproof concrete construction, is built flush with the floor level, and is heavily insulated by means of cork blocks.

FIG. 6   HEATING AND VENTILATING EQUIPMENT, SOUTH TO NORTH.
SEE FIG. 1, P-8

FIG. 5   THE AUTOMOBILE ENGINE-TESTING PLANT.   SEE FIG. 1, P-5

FIG. 7   MAIN ELEVATED FLOOR, NORTH TO SOUTH.   SEE FIG. 1, P-9

as temperature of air, effect of register faces, movement of anemometer, and velocity of flow. This calibrating system is probably the most unique feature of the tests.

A small auxiliary plant equipped with a single leader and stack has also been set up for studying all types of leader and stack construction. This research work is being carried on in coöperation with the National Warm Air Heating and Ventilating Association.

AIR COMPRESSOR

The air-compressor equipment consists of an Ingersoll-Sargent compound steam and two-stage air compressor (25) having a rated capacity of 300 cu. ft. of free air per min. at a speed of 120 r.p.m. The rated delivery pressure is 80 to 100 lb. per sq. in. gage. The air cylinders are 12¼ in. and 18½ by 12 in. and the compressor is driven by a direct-connected steam engine having cylinders 12 in. and 22 in. by 12 in. The air from this compressor is delivered to a receiver on the lower floor of the laboratory, and from this receiver it is discharged into the atmosphere through a gage box and orifice. This

The York machine (21) is of the vertical single-acting duplex-compressor type, having cylinders 7½ in. in diameter and a stroke of 10 in. It is driven by a 11½-in by 10-in. Corliss engine and has a capacity of 10 tons of refrigeration per 24 hr. A set of cooling coils and a reheater is arranged on the lower floor of the laboratory, thus permitting the refrigeration capacity to be determined without the necessity of making ice, and eliminating the inconvenience of running long tests.

The Vogt machine (21a) has a capacity of 10 tons of refrigeration per 24 hr. It is of the standard commercial type modified slightly to adapt it to testing purposes. There are two anhydrous ammonia receivers equipped with gage glasses, thus facilitating the measurement of the anhydrous ammonia circulated. The piping to the absorbent, exchanger, weak-liquor cooler, and rectifier is arranged so that the cooling water may be circulated through them in any way deemed desirable for experimental purposes. The steam used for heating the strong liquor may be taken either from the low-pressure steam main (Continued on page 144)

# ENGINEERING SURVEY

A Review of Progress and Attainment in Mechanical Engineering and Related Fields, as Gathered from Current Technical Periodicals and Other Sources

### SUBJECTS OF THIS MONTH'S ABSTRACTS

| | | |
|---|---|---|
| EFFLUX OF GASES THROUGH SMALL ORIFICES | WORM GEARING, IMPROVED | SALT PRECIPITATION BY REFRIGERATION |
| CONSTANTS OF RADIATION OF UNIFORMLY HEATED ENCLOSURE | CUTTING POWER OF LATHE TURNING TOOLS | STAYBOLT DEFLECTION IN LOCOMOTIVE BOILERS |
| ATMOSPHERIC ABSORPTION OF HEAT | DIAMOND WIREDRAWING DIES | MULTI-STAGE NOZZLE STEAM TURBINES |
| MALLEABLE IRON, PHYSICAL TESTS | LONGITUDINAL INTERNAL STRESSES IN STEEL CYLINDERS | MOLLIER AIR REFRIGERATING MACHINE |
| PULVERIZED COAL IN OPEN-HEARTH PRACTICE | PUDDLED-IRON PRACTICE | WITH MULTI-STAGE NOZZLES |
| TIDAL-POWER DEVELOPMENT AT HOPEWELL, N. B. | ABSOLUTE UNIT OF TIME | SPECIFIC HEAT OF AIR AT VARIOUS TEMPERATURES |
| LUBRICATION | RIPKEN AUTOMATIC DRIFTING VALVE FOR LOCOMOTIVES | ABRASIVE WHEELS AND THEIR TESTING. |
| | FUEL OIL, USE AND HANDLING | |

## BUREAU OF STANDARDS

THE EFFLUX OF GASES THROUGH SMALL ORIFICES, Edgar Buckingham and Junius David Edwards. This paper contains a theoretical discussion of the results of experiments made during an investigation of the effusion method of determining gas density; and it is a sequel to the paper in which that investigation has already been described by one of the present authors. The effects on effusion of viscosity and thermal conductivity of the gas are studied, and formulæ for representing these effects are developed and compared with the observed facts. A physical interpretation is thus obtained for the most striking of the apparent anomalies observed in the behavior of gases flowing through very small orifices such as are used in apparatus for determining density of effusion. The mathematical work was, unavoidably, too cumbersome for satisfactory abstraction. The resulting agreement of the theory with the observations is exhibited graphically. It is fairly satisfactory but not perfect, and its imperfections suggest the directions in which further experiments should be made. (Abstract of *Scientific Paper No. 359, U. S. Bureau of Standards*, e)

THE CONSTANTS OF RADIATION OF A UNIFORMLY HEATED ENCLOSURE. Experimental data are given on atmospheric absorption. The paper gives also a recalculation of the coefficient of total radiation of a uniformly heated enclosure, or so-called black body, giving a value of

$$\sigma = 5.72 \times 10^{-12} \text{ watt cm}^2 \text{ deg}^{-4}$$

The effect of atmospheric absorption is discussed, and the conclusion arrived at is that, if corrections are made for atmospheric absorption, the value recently obtained at Naples is close to the average value, viz.,

$$\sigma = 5.7$$

In the second part of the paper the present status of the constant of spectral radiation is discussed. (Abstract of *Scientific Paper No. 357. U. S. Bureau of Standards*, by W. W. Coblentz, t)

## ENGINEERING MATERIALS

PHYSICAL TESTS OF MALLEABLE IRON. At the recent annual meeting of The American Society for Testing Materials, W. R. Bean, of the Eastern Malleable Iron Co., Naugatuck, Conn., presented data showing the possibility of making malleable iron in heavy sections.

Test pieces were cut from a malleable-iron bar 3 in. by 3 in. in section. These test pieces ran lengthwise with the bar, and were so selected that every portion of the cross-section was represented.

From data of these tests it would appear that the test pieces from the corners of the bar were the strongest as well as the most ductile, ranging slightly over 45,000 lb. per sq. in. in tensile strength.

The weakest samples were from the center of the bar, but even they gave a tensile strength of 39,859 lb. per sq. in., which is

quite good for iron in the center of such a heavy section of metal.

Data showing the effect of the skin on the strength of malleable iron were also given, and would indicate that there was little difference in minimum tensile strength or other test values before or after machining.

Interesting data are also presented on annealing of cast iron and the effects of carbon conditions. (*The Foundry*, vol. 47, no. 337, Dec. 15, 1919, pp. 906-908, 6 figs., e)

## FUELS AND FIRING

### What 18 American Steel Plants Say About Pulverized Coal in Open-Hearth Practice

PULVERIZED COAL IN OPEN-HEARTH PRACTICE, W. H. Fitch. A review of the experience of eighteen American steel plants. From this review it would appear that pulverized coal was applied in some cases deliberately and in other cases under the urgency of necessity, chiefly actual or threatened lack of the fuel regularly used.

Practically all of the furnaces under investigation were designed to use producer gas, though some of them passed to other fuels such as natural gas or fuel oil even before an attempt was made to burn powdered coal.

In some of the cases the results with powdered coal were decidedly unsatisfactory. Apparently this was due chiefly to the lack of proper design of the furnaces, although in the case of acid steel there appeared to be fundamental defects with pulverized coal.

In general, however, the investigation appears to have shown that pulverized coal is quite a proper fuel for open-hearth furnaces, provided only the conditions of its use are right.

In this connection the experience of plants G, H and I is illuminating. During the winter of 1916 there was a scarcity of natural gas, and the management became interested in finding a substitute, fearing the necessary supply of gas would not be obtainable. A careful investigation of the situation was made and an engineering company specializing in this line of work was given a contract to install pulverizing machinery, distributing systems and furnace fixtures.

As it was felt that furnace design was the principal factor in determining the success of operation with a new fuel, it was decided that it would be best to make as little change as possible to begin with, allowing for the result of experience of applying coal dust to one furnace to point the way and nature of changes necessary.

Considerable trouble in various ways was experienced at first, but due to a number of furnaces being in use it was possible to make minor changes at various times when the furnace was down for repairs, and in this way different arrangements of checker brick, ports and roof were experimented with. Changes that could be made easily to the furnace above the charging floor were lim-

114

ited, due to the necessity of making very radical changes in steel work, which, for the time being, were prohibitive.

Eventually, the slag chamber was found to be a very important factor, as its capacity decided the number of heats that could be made without stopping the furnace for cleaning.

TABLE 1

| | Producers | | Pulverizers | |
|---|---|---|---|---|
| Power................ | 75 kw. x 24 hr. x 1 c.$18.00 | | 122 kw. x 24 hr. x 1 c. | $29.28 |
| Gas and ash man...... | 18 x 8 x 60 c...... | 86.40 | | |
| Dryer man.......... | | | 3 x 8 x 60 c............ | 14.40 |
| Pulling man........ | | | 3 x 8 x 60 c............ | 14.40 |
| Unloading coal...... | 240 x 2 c........... | 4.80 | 180 x 2 c............. | 3.60 |
| Repairs............. | 240 x 10 c.......... | 24.00 | 180 x 10 c............ | 18.00 |
| Steam.............. | ................... | 44.40 | | |
| Circulating water.... | 200 gal. per min.... | 4.80 | | |
| | | 182.4 | | 79.68 |
| | $182.4 ÷ 240 = $0.76 | | $79.68 ÷ 180 = $0.45 | |

Several other companies have had the same experience with reference to the importance of the proper design of the slag chamber. It appears that the desirable design should embrace a slag chamber of large cubical content, having a removable bottom, and affording the gases an opportunity of expanding and precipitating the heavy oxides of molten ash. Also, there should be a regenerative chamber of large volume and provided with observation and cleaning doors at each end of the furnace, the checker brick being assembled to provide vertical flues of comparatively large area and the waste gases passed through the boiler.

The most economical results, however, are obtained by returning all of the waste heat available to the furnace, and a design of the furnace that will provide for the accumulation and removal of fine ash after passing through the slag chamber will have the effect of cleaning the gases before they enter the regenerative chamber. It will then be possible so to arrange the checker brick to form flue areas that will absorb the maximum quantity of heat from the outgoing gases and permit the incoming live air to be heated to the desired temperature.

There have been accidents and casualties in the use of pulverized coal, but the number is said to have been small and in many instances they were the result of negligence on the part of the operator.

As regards the comparative value of pulverized coal and gasified coal, Table 1 would appear to show that on the whole pulverized coal is cheaper. (*The Iron Age*, vol. 104, no. 26, Dec. 25, 1919, pp. 1323-1328, *gcA*)

## HYDRAULIC ENGINEERING

### Proposal to Develop 90,000 hp. from a Tidal-Power Plant in Canada

PROPOSED TIDAL-POWER DEVELOPMENT AT HOPEWELL, N. B. Data of a proposed development of unusually large size to be located at Hopewell, New Brunswick, where it is believed conditions are especially favorable to such an undertaking. The layout can be seen from Fig. 1, which is a roughly sketched map of the confluence of the Peticodiac and Memramcoke Rivers, which empty themselves into the tidal source, in this case the Shepody Bay.

There are two main dams proposed, the western one 49,000 ft. long and the eastern one 48,000 ft. long, and also a short wing dam of 900 ft. dividing the high-level from the low-level basins. It is proposed to have a highway and trolley line running over these dams to facilitate shore-to-shore communication.

The western dam would be provided with automatic flap gates opening upstream, which would allow the high-level basin to be filled at each high tide, and at the eastern dam there would be similar gates opening downstream to allow the low-level basin to empty on every ebb tide.

The cycle of operation beginning with low tide would be as follows: The water from the spillway would be discharged directly into Shepody Bay, which is the tidal supply, with the gradually

decreasing head as the tide rises. At the proper time the attendants close these gates and open other gates, allowing the spillway discharge to enter the low-level basin, into which the spillway will continue to discharge for about 6½ hr., that is, through the last 3½ hr. of flood tide and through the first 3 hr. of ebb tide, after which the cycle will be reversed and start all over again. During the flood tide the high-level basin is replenished and by this means continuous 24-hr. operation is secured.

In order to determine the commercial possibilities of a given location for tidal-power generation, it is the range of the tide and not the actual rise which has to be considered, and to determine this an analysis of the ranges which will occur in the course of a year is necessary. At Hopewell about 15 per cent of the time the tides exceed 42 ft. range, and for about the same percentage of time there are sub-normal neap tides ranging less than 32 ft.

The initial installation is proposed to be about 90,000 gross hp., for which it is believed there is ample water available. By steadi-

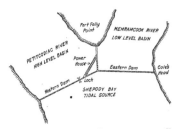

FIG. 1 LAYOUT OF THE PROPOSED TIDAL-POWER PLANT AT HOPEWELL, NEW BRUNSWICK

ly improving the ratio of the two basins, it is believed that the power output could be increased up to 200,000 hp. without making the level drop exceed the limit of 6 in. per hr. (*The Electrical Times*, vol. 56, no. 1469, Dec. 11, 1919, pp. 462-463, *d*)

## LUBRICATION

### The Physico-Chemical Phenomena Governing Lubrication

OILINESS OF LUBRICANTS—WHAT IT IS AND WHAT IT DOES

LUBRICATION. Abstract of a discussion on this subject before the Physical Society, London, in connection with a paper by R. Mountford Deeley on Oiliness and Lubrication.

The significant feature about this discussion is the large attention paid to the physico-chemical phenomena in connection with lubrication. In introducing the discussion, Dr. F. E. Stanton gave a brief account of the history of the physics of lubrication from the researches of Beauchamp Tower in 1833 to the work of Somerfeld and A. G. M. Michell and the later investigations of Harrison and Hyde.

An abstract of Deeley's paper will be given later. Among available data it appears that it presents a description of the apparatus used for testing oiliness at the National Physical Laboratory at Teddington, and data obtained therewith.

In discussing this paper L. Archbutt pointed out that the following two conditions should be clearly distinguished in considering lubrication problems: (1) Cases where the two solid surfaces were completely separated by a film of oil as with properly fitted journals revolving at high speed and abundantly fed with oil; and (2) cases where owing to the shape and condition of the surfaces, low speed, high pressure or inadequate oil supply, the oil film could not form completely or became broken so that the solid surfaces came into contact.

In the first instance the friction was entirely due to the viscosity
of the oil and the theory of Reynolds and the study of physical
and chemical characteristics of an oil could be used as guides.
In the second case the circumstances were entirely different and
trials had to be relied on. Oils alike in viscosity might have
different friction-reducing values and it is here that the property
termed " oiliness " becomes very important.

Mr. Deeley's machine seems likely to give a means of measuring that oiliness and his view that oiliness enables the oil or some
constituent of it to combine with the surface and form a lubricating film is in accord with Langmuir's view.

L. Archbutt recently found that under the same conditions of
speed and pressure and with the same oil, bearings of white metal
would carry double the load of bronze bearings without increase
of friction.

On Langmuir's theory the molecules of the oil and solid were
held together by residual or secondary valencies, and the more
chemically active the lubricant, the more firmly it would be
attached.

In this connection, the behavior of acids in lubricants is quite
significant. Free acids have so far been considered injurious in
lubricants, because they corroded and formed soaps thickening the
lubricant in the presence of moisture.

Southcombe found, however, that the very restricted amount of
fatty acids, while of no advantage in the first of the two cases
above referred to, had a very remarkable effect in the second of
the two cases, that is, the one in which the surfaces of journal
and bearing come into contact.

Experimenting with a Thurston machine at low speed, 7 ft.
per min., and pressure of 270 lb. per sq. in., Mr. Archbutt had
found that 1 per cent of the fatty acid of rape oil added to a
mineral lubricant lowered the coefficient of friction from 0.0047
down to 0.0033, and that 2 per cent of the acid did not effect any
further lowering. Of neutral rape oil 60 per cent was required
to produce the same effect; it was in that respect rather fortunate
that commercial rape oil always contained some fatty acid. The
important point was that a small percentage of fatty acid improved the lubricating power of a mineral oil as much as a large
percentage of fatty oil. But it was also noteworthy that he had
found a perfectly neutral rape oil to give lower friction than a
mineral oil of the same viscosity. In comparative tests made with
a commercial rape oil of 2.4 per cent of fatty acid $R$, with a perfectly neutral oil $R$, prepared from $R$, and with a mineral oil $M$
of the same viscosity, he had obtained the following friction coefficients: $M = 0.0078$, $R = 0.0050$, $R_1 = 0.0045$. These results, if
confirmed, would suggest that the oiliness or lubricating efficiency
of the unsaturated molecules of the neutral rape oil was nearly
as great as that of the free fatty-acid molecules, but that the
acid molecules were much more active in their influence on the
hydrocarbon molecules of the mineral oil. The size of the
molecules and the thickness of the molecular layer would be important factors, and Langmuir's researches, referred to in an
earlier paragraph, seemed to throw considerable light upon the
subject.

Dr. H. S. Allen, of Edinburgh University, in a note on
Molecular Layers in Lubrication, again referred to Langmuir's
theory of films. This theory might be expressed in its application to lubrication as follows: The metal covers itself with a
layer of oil one molecule thick resembling a piece of velvet firmly
glued to the metal with the pile outward, and two such velvet-
clad surfaces would glide over one another with but little friction.
If this be so, oiliness would depend neither on viscosity nor on
compressibility, but on chemical forces and on the nature of both
the lubricant and the metallic surface.

B. W. Hardy emphasized the highly complex nature of the
phenomena of lubrication, especially the chemical problems connected with the mixing of substances of different lubricating
powers.

J. E. Southcombe told of his experiments with the object of
finding out in what respect the fatty oils differed from mineral
oils. An abstract of his remarks will be given at a later date.
(*Engineering*, vol. 108, no. 2814, December 5, 1919, pp. 758-760.)

## MACHINE PARTS

### Improved Worm Gear Having Efficiency of Over 97 Per Cent

DAVID BROWN & SON IMPROVED WORM GEAR. Description of
a worm gear developed by a British concern. It is stated that in
a certified test made at the National Physical Laboratory, Teddington, England, an efficiency of over 97 per cent was secured.

The illustrations (Fig. 2) show the details of the character and
area of contact in the new F. J. gears. The view on the left shows
the worm in such a position that the drive is taken by the three
threads, $A$, $B$, and $C$, while that on the right shows the worm in
a different position, having been rotated through an angle of 45
deg., with the result that the two threads $A$ and $B$ are in simultaneous contact with the worm wheel. The area in dotted outline
across the teeth of the worm marks the zone of contact between
the gears when the worm is rotated in the direction indicated in
the end view. The intersections of the worm threads and the zone
give the lines of contact between the worm and wheel teeth, and
at no point outside this zone is there any contact between worm
and wheel. It is this contour of the zone of contact that was
arrived at by the mathematical calculation previously referred to,
and which was submitted to Faraday House.

In the left-hand view, the actual contact between the worm

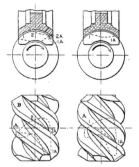

FIG. 2  ZONE OF CONTACT BETWEEN THE F. J. WORM AND WHEEL

thread $A$ and the wheel is marked by the dotted line 1-1$A$, that
between the worm thread $B$ and the wheel by the line 2-2$A$, and
on the worm thread $C$ by the line 3-3$A$. In the right-hand view
the worm has been rotated through an angle of 45 deg., the new
line of contact being again shown dotted, only in this case the
worm thread $C$ has passed beyond the zone of contact and has thus
ceased to make contact with the worm wheel.

The illustrations (Fig. 3) show the zone of contact for a four-
threaded worm of the ordinary shape in which the teeth of the
worm are straight-sided in section; that is to say, in the form of
a rack. As the linear section of this type of worm corresponds
to a straight-sided rack, and as the straight-sided rack forms
the basis of the involute spur gear, this shape of worm is generally
known as the involute worm. As before, the zone of contact is
shown in dotted line, but in this instance two teeth are in
contact in the left-hand view—$A$ and $B$—giving lines of contact
1-1$A$ and 2-2$A$, while in the right-hand view the tooth $A$ carries
the entire load along the line of contact 1-1$A$. The unsymmetrical
contour of the zone of contact will also be observed. More noticeable still is the difference between the irregular lines of contact
shown on the end views of the involute worm as compared with
the symmetrical lines of the F. J.

It must be understood that in both cases whenever the worm is revolved through an angle of 90 deg., the line of contact 1-1A will sweep across the zone of contact until it takes up the position marked 2-2A, because when a four-threaded worm is moved a quarter of a revolution (or 90 deg.), the thread A will naturally take up the position of the thread B. In the F. J. system it will be seen that the point 1 at the root of the worm (end view) falls into the position 2 after a quarter of a revolution; that is to say, the actual contact travels *in the same direction* as the worm through a distance of approximately 50 to 60 deg., hence about two-thirds of the motion is transmitted by a rolling action and only one-third by sliding. This means that a rolling action is introduced between the teeth of the gears, thereby reducing the ordinary rubbing velocity to a minimum. . Thus the gear teeth, instead of rubbing against each other at, say, 1000 ft. per min., really only rub at the reduced speed of about 350 ft. per min., this being due wholly to the rolling action of the F. J. tooth contact.

This is not so on the involute worm-gear system. In the left-hand view in Fig. 2 the point of contact 1A actually moves *in the opposite direction* to the rotation of the worm to the point 2A, thereby giving the rubbing velocity even more than its full the-

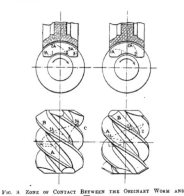

Fig. 3  ZONE OF CONTACT BETWEEN THE ORDINARY WORM AND WHEEL

oretical value, while rolling action, at the ends A of the lines of contact, is absolutely non-existent.

A peculiarly inefficient characteristic of the involute system is that the extremities of the lines of contact (see plan view of worm) travel from positions 1 and 1A *toward each other* until they meet, and vanish just beyond the points 2, 2A, as shown in the left-hand plan view (Fig. 3).

The plan view of the F. J. system (Fig. 2) shows the points A always moving from left to right, in the same direction as the rotation of the worm, while the left-hand plan view of the "involute" system shows the point A moving very slightly from right to left, actually against the rotation of the worm. The high rubbing velocity which occurs on the points AA, together with the converging lines of contact which induce a concentrated load, all tend to produce the worst possible conditions on the leaving side of the worm wheel. This explains the well-known fact that when a worm gear of the "involute" type is overloaded, "pitting" first takes place on this side of the wheel, and on this side only, while those portions of the teeth on the entering side of the wheel remain unmarked. (*The Automobile Engineer*, vol. 9, no. 133, December 1919, p. 449, 2 figs., d)

## MACHINE SHOP

### Tests on Best Speeds, Fits and Settings of Lathe Turning Tools

CUTTING POWER OF LATHE TURNING TOOLS, George W. Burley. In November 1913 Prof. W. Ripper and the present author read a paper under the same title as the present one before the Institution of Mechanical Engineers. The present article is a continuation of that article and deals with experiments made with lathe turning tools and in particular with such matters as relation between the cutting speed and the degree of finish obtained by the use of ordinary lathe finishing and light cutting tools and the influence of the various factors in tool design and position on the cutting power of the tools, power consumption, etc.

The paper, while of great interest, is too extensive for a detailed abstract. Because of this only the conclusions arrived at by the author are reproduced.

1 That there is no practical cutting speed below which it is impossible to obtain a satisfactory surface on plain-carbon steels by means of ordinary lathe finishing tools, whether these be made of plain carbon tool steel, ordinary (non-vanadium) high-speed steel, or superior (vanadium) high-speed steel. There is, however, a maximum limiting speed at or above which a satisfactory finish cannot be obtained on account of the tendency of the tool to pluck at and tear the surface, this tendency being related to the phenomenon of building up on the cutting edge of the tool. For the finishing of mild steel this limit is not very different from each of the above three varieties of tool steel and is within the range of 48 ft. to 58 ft. per min. For the finishing of hard steel this limit does depend somewhat on the variety of tool steel employed and is within the range of 23 ft. to 28 ft., 17 ft. to 21 ft., and 28 ft. to 34 ft. per min. for the three varieties, respectively.

2 The durability or life of a lathe finishing tool whether of plain carbon or high-speed steel, is for all cutting speeds below the limiting speed some function of the reciprocal of the cutting speed; in other words, an increase in the cutting speed below the limiting value is always accompanied by a decrease in the life or durability of the tool.

3 The most suitable angle of side rake (that is, the angle of side rake associated with maximum durability and cutting power) for a high-speed lathe roughing tool working on steel depends upon the physical properties of the steel. For mild-steel turning it lies between 20 and 25 deg., while for hard-steel turning it is of the order of 10 deg.; and if these angles are either increased or reduced there is always a depreciation of cutting power.

4 The color of the turnings formed by a high-speed lathe roughing tool when working on steel is not necessarily a true index of the condition of maximum cutting efficiency. Thus, in the case of hard-steel turning, the turning color which is associated with maximum cutting efficiency is a pale blue, while a mild-steel turning which is removed under the conditions of maximum efficiency is practically uncolored, apart, of course, from the natural gray color of the steel.

5 The net power consumption of a high-speed roughing tool is dependent, other conditions being constant, upon the amount of top rake on the tool, the relation between these two quantities being reciprocal in character, so that, within the limits of ordinary practice, a reduction in the top-rake angles of a tool is always accompanied by an increase in the net amount of power consumed. The law connecting the variations of the two quantities appears to be of the nature of a straight-line law for all qualities of steel machined. There are, therefore, no critical values of the rake angles in regard to power consumption as there are in regard to durability and cutting power.

6 The cutting power of a high-speed lathe tool is influenced by both the cross-sectional area of the shank of the tool and the nose radius, but the influence of the latter factor very largely predominates in all cases. Thus with a number of different sections of tool steel, an increase of the nose radius of 100 per cent produced an average increase in the cutting power of 45 per cent; whereas an increase of the cross-section of the shank of the tool of 500 per cent with a constant nose radius produced an average increase in the cutting power of only 8.5 per cent.

7   The effect of raising a roughing tool so that its cutting edge is slightly above the horizontal plane passing through the lathe centers is, generally, to increase the cutting power of the tool slightly and to reduce its net power consumption slightly, when compared with its normal position, that is, with its cutting edge at the same height as the lathe-center axis.

8   Forging the nose of a lathe cutting tool does not materially affect its cutting power and durability, there being practically nothing to choose between a completely ground tool and a forged and ground tool, otherwise identical.

9   The general direction and the active length of the cutting edge of a high-speed tool have a slight influence upon the net power consumption of the tool, the influence being such that, with any given depth of cut, if the active length of the cutting edge is increased by an alteration of the general direction of the edge, the net power consumed is increased under conditions of working otherwise identical.

10   There is no marked difference in the net amounts of energy required per cubic inch of material removed from mild-steel and hard-steel bars at high and low cutting speeds.

11   The increase in the cutting power of a high-speed roughing tool resulting from the use of a given stream of water as a cooling agent is greater with small cuts than with heavier ones, indicating that, with heavier cuts, heavier flows of coolant should be used. The velocity of flow of a stream of coolant does not very materially affect the improvement in the cutting power of a tool due to the use of the coolant, provided that the velocity is not such as to cause excessive splashing of the coolant.

12   The cutting efficiency of a high-speed roughing tool depends very largely upon the condition of the cutting edge of the tool,

opening, as indicated at $C$, Fig. 4. This opening is made by means of a diamond chip held by hand between the points of a pair of long-nosed pliers. After the conical hole has been cut, the drilling operation is started by means of a tool carried in the spindle of the machine, which is so designed that the table rotates and also has a vertical reciprocating motion imparted to it.

After the hole has been drilled about three-quarters of the way through the diamond, as indicated at $D$, the work is removed from the machine and turned over, after which it is reset and the conical opening is cut in the opposite side of the diamond, as indicated at $E$. Of course, this second conical opening has to be located exactly opposite the hole which has been drilled in the work. The next step is to drill from the bottom of this conical opening to make a connection with the hole entering the stone from the opposite side, $F$. It is very important to have the inside of the die absolutely smooth in order to prevent forming seams in the work and also to have the bell mouth of the die formed in such a way that the metal will pass into the die without undue frictional resistance. This is done by giving the die the shape shown at $G$. The inlet to the throat of the die is made slightly larger than the outlet at the back, which requires a great amount of dexterity to do. The dies are set in brass mounts.

The feature of diamond dies which makes them more economical in use than steel, cast iron or other comparatively inexpensive materials, is that they give a far greater amount of service before the tool is worn out, because of the extreme hardness of the diamond. This resistance against wear is also the means of maintaining greater uniformity in the diameter of the wire that is drawn through the die.

The time comes, however, when the die becomes too large to

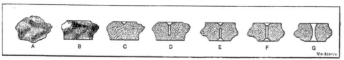

FIG. 4   ROUGH DIAMOND ($A$) AND THE CROSS-SECTIONAL VIEWS SHOWING SUCCESSIVE STEPS IN THE PROCESS OF MAKING WIRE-DRAWING DIES

and though a tool with its cutting edge blunted in the cutting process may continue to cut, it cannot be used in that condition for the purpose of starting a new cut. (Paper before the Institution of Mechanical Engineers, Dec. 19, 1919, *Journal of the Institution of Mechanical Engineers*. Compare also *Engineering*, vol. 108, no. 2817, Dec. 26, 1919, pp. 867-875, 91 figs., *eA*)

## MECHANICAL PROCESSES

### Diamond Dies and How They Are Made

MAKING DIAMOND WIRE-DRAWING DIES, Edward K. Hammond. Diamond dies are extensively used for drawing small wire made of all kinds of metals and for sizes from 0.08 in. down they are used practically exclusively. Because of this the following data as to the manufacture of such dies by the Vianney Wire Die Works at Trevoux, France, become of interest.

For the manufacture of dies diamonds are used which are of a grade unsuitable for use in jewelry.

The firm standardized the size of diamonds from which dies of different diameters are made, the diamonds themselves ranging from one carat to sizes as large as twenty carats.

As is generally known, the diamond is so hard that it can be cut only with another diamond or diamond dust. The first step is the flattening of a rough stone, which is done with laps charged with diamond dust made by pulverizing small-sized diamonds or chips from larger stones and then grading the dust by sifting it through sieves with various numbers of meshes per square inch. This is succeeded by several operations, comprising the cutting of the die opening. The first step is to chuck the diamond on the faceplate of a special drilling machine and cut a conical-shaped

produce wire that comes within the specified limit of tolerance. When this point has been reached the die is recut for a larger size, which is done by the use of a steel lapping tool charged with diamond dust. In connection with this operation the bell-mouthed opening must again be carefully finished to bring it tangent to the new throat which has been cut in the die and the inner surface must once more be carefully polished. The original article illustrates the special type of lathe used for refinishing diamond dies. (*Machinery*, vol. 26, no. 3, Nov. 1919, pp. 264-266, 5 figs., *d*)

## MECHANICS

### French Tests on Longitudinal Internal Stresses in Rapidly Cooled Steel Cylinders

INFLUENCE OF VARIOUS FACTORS ON THE RISE OF LONGITUDINAL INTERNAL STRESSES DUE TO RAPID COOLING OF STEEL CYLINDERS, M. Portevin. The longitudinal internal stresses were determined by measuring by means of a comparator the change of length which accompanied the removal in a lathe of concentric layers of metal on the cylinder.

The necessity of keeping the test pieces in a state of easy machining limited the initial temperatures of rapid cooling to the lower point of transformation of steel, which is around 700 deg. cent. (1292 deg. fahr.) in order to avoid the hardening which would follow other treatment.

The internal stresses were computed by using the formula of Heyn and Bauer. If $l$ and $l_n$ are the initial length of a test and its length up to the removal of the $n$th layer, $S''_n$ the area of a straight section remaining after the removal of the $n$th layer, and

$E$ the modulus of elasticity of the metal, then the stress $f_n$ which was initially present in the $n$th layer is

$$f_u = \frac{E}{l} \frac{S''_n (l_n - l) - S''_{n-1}(l_{n-1} - l)}{S''_{n-1} - S''_n}$$

For a given steel the internal stresses produced by rapid cooling are a consequence of the lack of isothermic condition of the mass, that is, of instantaneous inequality of distribution of temperatures during this operation. These variations of temperature at various points of a metal cylinder cooled by immersion into a liquid vary with the temperature of immersion, the external conditions of cooling and the diameter of the cylinder. In particular, the external conditions of cooling depend on the nature of the cooling liquid, its temperature and the duration of the immersion.

In the present instance the author investigates in a quantitative manner the results obtained by varying separately these different factors. The numerical results of the same investigation will be published separately.

*Temperature of Immersion.* The internal stresses rise very rapidly with the temperature of immersion when this latter increases from 200 deg. cent. (392 deg. fahr.) to 600 deg. cent. (1112 deg. fahr.). They are practically nil for temperatures below 200 deg. cent. (392 deg. fahr.) and the diagram of Fig. 5 gives for a steel cylinder the curves of internal stresses as a function of the square of the distance from the axis.

For reasons given above (necessity of keeping the metal machinable) the tests were not carried to higher temperatures, but in view of the viscosity of steel at high temperatures no material increase of internal stresses was to be expected.

*Nature of the Cooling Liquid.* Comparison of internal stresses arising through immersion in water and in oil at a temperature of 650 deg. cent. (1202 deg. fahr.) shows that in the latter case they are from two to four times smaller than in the former, according to the character of the steel and according to whether we consider the peripheral stresses of compression or the axial stresses of expansion.

*Temperature of the Water.* The internal stresses decrease with the temperature of the water and become practically nil for a temperature of 100 deg. cent. (212 deg. fahr.). The curves of Fig. 6 indicate this variation for a hard steel. It is found that when the temperature of the water is increased by 12 to 25 deg. cent. (21.6 to 45 deg. fahr.) a reduction of internal stresses is observed, small but unnoticeable, and in addition the internal stresses lose a good deal of their regularity, that is, they cease to maintain materially the character of revolution around the axis of the cylinder.

*Duration of Immersion.* In tests where the duration of immersion was, respectively, 1.5, 3.5, 8 and 13 sec., there was a net increase in the internal stresses for the case of cylinders immersed in the water at a temperature of 650 deg. cent. (1202 deg. fahr.). This is also apparent for the range of temperatures between 8 sec. and 13 sec., although by interpolating results of previous tests it may be estimated that beginning with 5 sec. the axial temperature falls below 300 deg. cent. (572 deg. fahr.). Fig. 7 gives the internal stresses as a function of the square of the distance from the axis for a semi-hard steel cylinder.

*Diameter of the Cylinder.* Tests carried out on semi-hard steel cylinders of 20 mm., 35 mm., 50 mm. and 60 mm. (0.7 in., 1.37 in., 1.96 in., 2.36 in.) in diameter, respectively, have given results in which the anomalies are apparently due to experimental errors, but which nevertheless indicate the decrease of internal stresses with the diameter, a decrease, it may be remarked. which is really not very rapid.

A knowledge of these results was necessary for clearing up the entire problem of tempering of steel. (Etude de l'influence de divers facteurs sur la création des efforts internes longitudinaux lors du refroidissement rapide de cylindres d'acier. *Comptes Rendus hebdomadaires des Séances de l'Académie des Sciences,* vol. 169, no. 21, Nov. 24, 1919, pp. 955-957, 3 figs., eA)

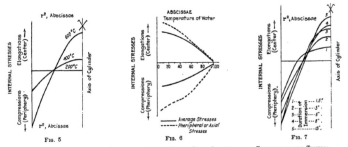

FIG. 5 VARIATION OF INTERNAL STRESSES IN RAPIDLY COOLED STEEL CYLINDERS AS A FUNCTION OF THE TEMPERA-TURE OF IMMERSION. FIG. 6 VARIATION OF INTERNAL STRESSES IN RAPIDLY COOLED STEEL CYLINDERS AS A FUNCTION OF THE TEMPERATURE OF THE WATER. FIG. 7 VARIATION OF INTERNAL STRESSES IN RAPIDLY COOLED STEEL CYLINDERS AS A FUNCTION OF THE DURATION OF IMMERSION

## METALLURGY

### Description of Modern Methods of Making Puddled Iron

PUDDLED-IRON PRACTICE, J. E. Fletcher. There has been a material revival of interest in puddling iron. Hall's process lies at the root of present and future puddling processes and is not likely to be superseded. It is conceivable that further attempts may be made to produce wrought iron directly from poorly silicious ores or by the decarburization of cheaply produced mild steel, but there are fundamental difficulties in the way of success affecting the character of the crystalline structure and the related welding and fibering properties of the resulting product.

There are three factors in puddling that affect the entire situation. These factors are:

1 The puddling furnace as we now know it is one of the most wasteful of metallurgical furnaces, using more fuel per ton of product than any other continuous process.

2 The labor cost per ton is abnormally high as is also the cost of furnace upkeep.

3 The rolling, shingling and material handling and transport plants are usually crude and excessively wasteful in power cost per ton of product.

The writer vigorously criticises the design of coal-fired puddling furnaces, claiming that the radiation losses in such a furnace and its waste-heat boiler amount to at least 20 per cent of the heat energy of the coal burned or equivalent to, say, 560 to 784 lb. per ton of puddled-iron heat, whereas in the steel furnace the radiation losses do not exceed 224 lb. per ton of steel cast.

The radiation losses in a puddling furnace per unit of metal charged average from 10 to 15 times as much as in the open-hearth steel furnace, and so on.

On the whole, the author comes to the conclusion that the small furnace used in puddling is a real obstacle to fuel economy as its relatively large hearth causes the excessive radiation areas.

The construction and proportions of the modern puddling furnace are not likely to be substantially altered or improved, but it is possible, to improve the grate and firebox and to provide

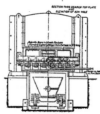

FIG. 8  END ELEVATION OF FURNACE ARRANGED FOR EASY REMOVAL
OF ASHES

efficient means for burning the poorer coals—in particular, to reduce the amount of unburned fuel that goes into the ashpit. Figs. 8 and 9 show sections of a furnace arrangement designed by the author.

The grate consists of a series of tubular bars perforated for air ejection, over which a number of specially shaped chilled cast.

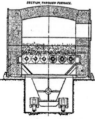

FIG. 9  SECTION THROUGH FURNACE SHOWING ARRANGEMENT OF
GRATES

iron disks are threaded. These form a kind of roller grooved longitudinally, so that when it is rotated it causes clinkers to crack and burned ashes accumulated in the troughs between the roller bars to roll into the ashpit.

The air required for combustion is forced through the tubes by steam jets and passes out through the perforations into the fuel bed and ashpit which is closed by the front plate of the ash-removal car. The air ducts between the grate disks are nozzle-shaped and the grates can be cleaned without opening the doors. Large clinkers can be removed through the large holes provided above the grate in the firebox back plate.

The application of gas firing to small puddling furnaces has

also been suggested and a furnace was built with a small producer at the firing end.

The process of puddling covers three stages, namely, the melting stage, occupying 30 to 40 min. depending on the class of pig iron used and the efficiency of the fuel; the boiling stage, occupying 30 min.; and the drying and balling, followed by fettling and charging periods, occupying about 40 min.

It appears further that the melting period demands, under the present conditions, from a little over 5000 lb. to close to 6000 lb. of coal per ton of metal melted and is the most wasteful stage of the puddling process.

Cupola melting which would be more economical as far as fuel consumption goes is not suitable as it yields a higher-carbon metal and very little reduction of the remaining metalloids, and because of this the boiling period in the puddling furnace is lengthened.

The author proposes a special method of procedure based on manipulation of the slags with a view to producing a low-sulphur metal from the cupola described in the original article.

The characteristic feature of the new process is the addition of manganese at a certain stage which causes a delay in the elimination of the carbon, while it accelerates the elimination of silicon and phosphorus. A liquid slag resulted and the metal poured flows easily and is well deoxidized.

The author also proposes the design of a plant consisting of an open-hearth tilting or fixed basin-lined furnace, the waste heat from which is used in attached efficient boilers fitted with auxiliary fuel-supply apparatus, this latter having the purpose of increasing the steam requirements during the charging and tapping stages. From the open-hearth furnace the metal is ladled to mechanical puddling furnaces fired with pulverized coal and having waste-heat boilers attached to them.

Each of the units composing the proposed plant are at work. The open hearth furnace is regularly melting iron on a pulverized-coal consumption of 448 to 560 lb. per ton of iron melted. Puddling furnaces similarly fired are operating on a consumption of 1300 to 1600 lb. per ton of puddled bar, when dealing with cold pig iron. When puddling molten charged material the coal burned should not exceed 800 to 1000 lb. per ton of product, six furnaces doing the work of the present ten.

If the roll trains and shingling machines or squeezers be operated by electrical power, and they can be, it is conceivable, that, given cheap electrical energy, such a combination of the duplex system with mechanical puddling furnaces and with electrical rolling and shingling equipment should furnish the most economical puddling plant possible. (Paper before the Staffordshire Iron and Steel Institute read Nov. 22, 1919, abstracted through *The Iron Trade Review*, vol. 66, no. 2, Jan. 8, 1920, pp. 152-156, 4 figs. *dg*)

## PHYSICS

ABSOLUTE UNIT OF TIME BASED ON NEWTONIAN LAWS, Gabriel Lippmann. The unit of time in general use, namely, the second, is an arbitrary and not an absolute unit. It is a submultiple of the solar day and can be obtained only by astronomic observation. If some fabulous being flew over from the solar system to Sirius, he would have lost the second on the way, at the instant when the earth's system disappeared from his sight. On the other hand, if the same being had some formula to express an absolute unit of time, he could reconstruct this unit no matter in what part of the world's system he might happen to be. In physics the use of the absolute unit is, however, of quite a practical value in that it permits the elimination of arbitrary numerical coefficients. Such coefficients are inconvenient in that they complicate the form of equations, introduce a possible source of error, and, what is still more important, in numerical computations tend to conceal the nature of relations existing between phenomena of different character. Without absolute units it would have been very difficult to create the electromagnetic theory of light, for example. It is therefore very desirable to secure a definition of a unit of time in absolute measurements.

The present author builds up an absolute unit of time on the basis of the Newtonian theory of gravitation. In accordance with this theory

$$f = k\,\frac{m\,m'}{r^2} \quad \ldots\ldots\ldots\ldots\ldots\ldots [1]$$

where $f$ is the static force of attraction between masses $m$ and $m'$ at a distance $r$ from each other, and $k$ is the Newtonian coefficient of attraction, the numerical value of which depends on the units of measurement selected, which may be such that $k = 1$. By this the unit of force is determined: It is the attraction which exists between two masses, each of which is equal to unity, separated by a unit of distance.

Hence, if we call $\gamma$ the intensity of the Newtonian field due to a mass $m$ at a distance $r$, we have $f = \gamma m'$, and hence

$$\gamma = \frac{m}{r^2} \quad \ldots\ldots\ldots\ldots\ldots\ldots\ldots [3]$$

Hitherto only static forces were considered and the element of time was absent.

But let us consider now the acceleration given to mass $m$ by the force $f$, assuming that the force is equal to the product of mass by acceleration. We have then

$$\frac{m}{r^2} = \frac{d^2 r}{d\theta^2} \quad \ldots\ldots\ldots\ldots\ldots\ldots [4]$$

where the value of the second member depends on the unit of time employed; conversely, Equation [4] defines the unit of force employed as the first member of the equation is completely defined. We may therefore write

$$\gamma = \frac{d^2 r}{d\theta^2} \quad \ldots\ldots\ldots\ldots\ldots\ldots [5]$$

The unit of force is therefore such that the acceleration is numerically equal to the intensity of weight, or the acceleration is equal to unity in a gravitational field equal to unity.

We may now calculate the dimensions of the absolute time We have to admit in the usual manner that the mass is measured by the product of a volume $L^3$ by density $\varrho$, this latter being taken for a substance of such type of water. Equation [4] becomes then

$$\frac{\varrho\,L^3}{L^2} = \frac{L}{\theta^2} \quad \ldots\ldots\ldots\ldots\ldots\ldots [6]$$

Hence

$$\theta = \frac{1}{\sqrt{\varrho}} \quad \ldots\ldots\ldots\ldots\ldots\ldots [7]$$

The absolute time is therefore a zero dimension in reference to lengths, which means that a unit of time is independent of the choice of units of length and is the same in the centimeter-gram system or the meter-ton system, or even the foot-cubic-foot system. As regards the coefficient $\varrho$, it has to be considered only if we wish to attempt to replace water as a typical substance by some other substance. For example, if we should desire to replace by mercury, the factor $\sqrt{\varrho}$ would be equal to $\sqrt{13.6}$ and the unit of time would be $\sqrt{13.6}$ times greater.

The author discusses in detail the methods of introducing this absolute unit of time into various physical computations. (*Annales de Physique*, 9th series, vol. 12, Sept.-Oct., 1919, pp. 215-237, *t*)

## RAILROAD ENGINEERING

RIPKEN AUTOMATIC DRIFTING VALVE FOR LOCOMOTIVES. Description of a drifting valve which is operated by the compression in the cylinder, known as the Ripken valve, and in use on the Minneapolis, St. Paul and Sault Sainte Marie Railroad.

The valve comes into operation before the cylinders are emptied of steam pressure after the throttle has been closed and is kept in operation and furnishes the needed amount of steam until the pistons cease to move. It is automatically closed just as the engine stops by the building up of pressure in the steam chest.

Only one valve per engine is used as the steam is passed directly to the branch pipe or the saturated end of the superheater header. This arrangement is advantageous as it reduces the amount of

steam required and also protects the superheater elements against overheating. It makes unnecessary the use of either air relief valves or bypass valves.

The original article illustrates and describes the construction of the valve, which is, however, well known generally. It also gives a series of indicator cards from an engine fitted with the Ripken valve, shown in Fig. 10.

Tests on a superheater locomotive with 25-in. by 26-in. cylinders supplied with steam from a 2-in. pipe showed that the valve when adjusted between 25 and 30 lb. pressure in the steam chest at 25 per cent cut-off at a speed of 60 miles an hour, prevented the formation of harmful vacuum even at such high speed. After a shut-off with the reverse lever in running position, the valve would open when the steam-chest pressure rose to 49 lb., which

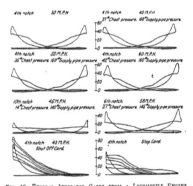

FIG. 10 TYPICAL INDICATOR CARDS FROM A LOCOMOTIVE ENGINE FITTED WITH THE RIPKEN AUTOMATIC DRIFTING VALVE

occurred just before the stop, allowing time for the superheater to empty itself before the engine came to rest. (*Railway Age*, vol. 67, no. 22, Nov. 28, 1919, pp. 1079-1080, 2 figs., *de*)

USE AND HANDLING OF FUEL OIL, A. M. Schoen. Paper presented before the Railway Fire Protection Association Convention at Chicago in November 1919, at the same time as the Report of the Committee on Oil-Burning Appliances, B. S. Mace, Chairman.

The Report of the Committee contains specific rules and requirements for the storage and use of fuel oil and for the construction and installation of oil-burning equipment.

A. M. Schoen's paper called attention to the fact that while oil has been used for locomotives in the West and on the Florida East Coast for a number of years, there would appear to be no well-defined standards accepted, and even the individual railroads appear to consider this fuel as not yet entirely permanent. It seems, however, to be time to formulate standard specifications in an advisory form, at least, that may be adopted by railroads preparing to adopt oil as fuel.

Not enough attention is being paid to safeguarding the storage of oil at points along the right of way, the loading and unloading of the oil, and the fueling of the tenders.

One of the obstacles to the determination of uniform standards lies in the wide variation in the flash test and viscosity test of the oils employed for fuel purposes in the different parts of the country.

A number of recommendations are made as to the design and location of large storage tanks and methods of fueling of tenders. (*Railway Age*, vol. 67, no. 22, Nov. 28, 1919, pp. 1069-1071, *gp*)

## REFRIGERATION

### Salt Precipitation from Mixed Solutions by Crystallization through Refrigeration

SALT PRECIPITATION BY REFRIGERATION, Berthold Block. In the chemical industry many salts have to be precipitated out of mixed solutions by means of crystallization. This is done by cooling the concentrated brines to the point where their ability to maintain the salts in solution is reduced.

Fig. 11 gives the solution curves of several salts in water, indicating how much of the salt at various temperatures may be contained in 100 grams of water (saturated solutions). As shown by these curves, the solubilities of various salts are unlike to an extraordinary degree and each salt will yield a greater or lesser precipitation out of a brine at a given temperature.

Thus, for example, in the case of Chilean saltpeter ($NaNO_3$) cooled from 90 to 20 deg. cent. (194 to 68 deg. fahr.), out of 165 grams held in solution at 90 deg. cent. so much is precipitated out that only 88 grams remain in solution.

The curves show that many brines can maintain quite considerable amounts of salt in solution even at 20 deg. cent. (68 deg. fahr.) and that substantial amounts of crystals are precipitated only when the brine is cooled to a considerably lower temperature.

Because of this, provision has to be made for certain salts to cool the brine to quite a low temperature and Fig. 12, for example, shows an installation built to precipitate out of its solution potassium chloride. In this case 3000 grams of brine at 20 deg. when cooled to — 10 deg. (14 deg. fahr.) will precipitate out in crystal form 200 grams of potassium chloride.

The cooling is effected in two stages: by means of a water cooler to + 20 deg. cent. (68 deg. fahr.), and then in an ammonia refrigerating machine to — 10 deg. cent. (14 deg. fahr.).

The ammonia compressor $A$ compresses the ammonia and forces it into a coil condenser $B$, in which ammonia under a pressure of 8 atmos. is liquefied by cooling water having a temperature of 15 deg. cent. (59 deg. fahr.). The liquid ammonia, the flow of which is controlled by a special valve, is then again vaporized in the coils $C$ of the crystallizing cooler $D$, the ammonia then going back to the compressor by means of pipe $E$.

The precipitation of the crystals is produced by cooling the brine in cooler $B$, the salt settling mainly on the cooling coils $C$. It is then swept off of them by special adjustable brushes, which also help to maintain the efficiency of the cooling action of the coil $C$. The crystals precipitated in this way sink through the brine and collect in the cone-shaped bottom of the cooler, whence they are from time to time discharged into the straining tank $O$ through a quick-acting valve $G$. It is necessary to have the valve of a quick-acting type in order to have the liquid pressure discharge the crystals containing brine without letting out too much liquid itself. The straining tank $O$ has on top a fine screen on which the brine falls. The thin liquid passes through and is strained off while the crystals remain on top of the screen and are collected therefrom from time to time.

In the installation described approximately 6000 calories (say, 24,000 B.t.u.) would have to be absorbed from the refrigerating plant per hour. In order to reduce the consumption of cold the installation is so arranged that the brine which leaves the crystallizing cooler $D$ at $H$ is conducted to the jacket of a second crystallizing cooler $J$ and on the way precools the fresh brine which enters at $K$ with the temperature of 20 deg. cent. (68 deg. fahr.).

Brine entering with the temperature of — 10 deg. cent. (14 deg. fahr.) takes up heat to the extent of leaving the cooler $J$ with the temperature of 5 deg. cent. (41 deg. fahr.), whereas the fresh brine which enters with the temperature of 20 deg. cent. (68 deg. fahr.) leaves with the temperature of 5 deg. cent. (41 deg. fahr.). The salt which is crystallized out in the meantime is swept away by brushes in the same manner as in the first cooler and precipitates out through the quick-acting valve $L$ into the same straining tank $O$. In this manner the consumption of cold is reduced to about one-half.

In computing the consumption of heat in refrigerating installations of this type it is necessary to bear in mind that in addition to the heat which is necessary to cool the brine itself, quite considerable amounts of heat have to be taken care of as heat is given up in the process of precipitation by crystallization at the instant

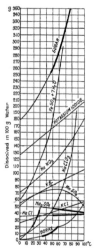

FIG. 11  SOLUBILITY CURVES OF VARIOUS SALTS

when the dissolved salt passes into the form of crystals, the amount of heat thus given up being the same as that which was previously consumed in order to dissolve the salt. In the case of cooking salt about 20 calories per kilogram (35 B.t.u. per lb.)

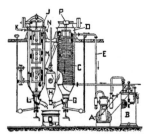

FIG. 12  EMIL PASSBURG INSTALLATION FOR PRECIPITATION OF SALTS OUT OF SOLUTIONS BY REFRIGERATION

were consumed, for potassium chloride ($KClO_3$) about 81.5 calories per kilogram (147 B.t.u. per lb.) and so on. (*Zeitschrift für die Gesamte Kälte-Industrie*, vol. 26, no. 8, August 1919, pp. 57-59, 3 figs., d)

## STEAM ENGINEERING (See also Railroad Engineering)

EMPLOYMENT OF NOZZLES INSTEAD OF CYLINDERS OR WHEELS IN TURBINES, Prof. R. Mollier. In 1904 Professor Dolder proposed in the *Schweizerische Bauzeitung* a new idea for the improvement of steam turbines, the main feature of which consisted in removing in the turbine only so much of the available energy of the steam jet that the steam might be able to regain its initial pressure in an expanded nozzle.

By these methods it was believed to be possible to shift the cycle at will into a range of high efficiency. The velocity in the turbine could be made as small as desirable and the compressor, which otherwise would be necessary with such a cycle, could be left out.

Professor Stodola denied the possibility of practically applying such a scheme. He pointed out the injurious effects of the resistances which would occur in all parts of the cycle, and brought forward as a deciding argument against the feasibility of such a cycle, the fact that the condensing steam always settles on the walls, which means loss of its kinetic energy.

The author states, however, that he himself was recently considering the same scheme in various forms; among others, its application in a reversed cycle to refrigerating machines, which he thought might be possible since with air or other non-condensable gases the fundamental objection brought forward by Stodola would not have to be considered. The scheme itself seemed to be highly attractive and if it could have been carried out in practice would have been of great benefit.

A closer investigation showed, however, that the presence of excessive resistance, which was only barely indicated by Stodola, offered an insuperable objection to the carrying out of this scheme.

It is the custom to carry out all kinds of changes of state of gases, in particular cyclic changes, in one or more cylinders with movable pistons and predetermined addition or abstraction of heat. There is, however, another way to do it and that is by substituting for the cylinder with movable piston a duct with variable cross-section through which the flow of gas takes place. If we neglect the question of frictional resistance the addition and absorption of heat will be the same as in the case of a cylinder with a movable piston.

In a cycle of the character occurring in the case of a duct of variable cross-section, there will be instead of work directed on to the piston a variation of kinetic energy of the gas. Should the cycle be continuous, we would then have to construct a circular duct, and the excess of energy created by the process would have to be abstracted from the gas at some point of the cycle, for example, by means of a turbine; or, if the cycle is of such a character that the kinetic energy of the gases is reduced, then at some place in the cycle this loss of energy would have to be made good by some means, for example, by a turbo-compressor. In both processes, however, the rule holds good that on the whole the amount of work either gained or expended by the gas is equal to the amount of heat that is either abstracted from or delivered to the same gas.

In cycles of the first class, that is, cylinders with movable pistons, the velocities of the gas are so small that the work necessary to overcome the resistance thus produced may be neglected entirely, and in a cycle of this character the integrals

$$\int p dv = -\int v dp \text{ and } \int T ds = -\int s dT$$

may be taken as fully representing the work gained during the cycle or the work expended to carry it out. In these equations $p$, $v$, $T$ and $s$ represent respectively the pressure, volume, absolute temperature and entropy of the gases.

In the cycle of the second class, that is, a duct with variable cross-section, the resistance cannot be neglected since in that case we are dealing with great, and often very great, velocities.

In a cycle of that character the work gained or lost cannot be expressed by the above integrals and the area of the cycle cannot be expressed in the p-v or T-s diagrams. As a matter of fact it is greater or less than that expressed by the above integrals, by the amount of the work of resistance $W$, so that the variation of energy is in that case expressed by

$$E = -\int v dp - W$$

Because of this, it appears that if in a heat engine the friction is equal to the integral of the area of work, the whole work produced by the engine disappears and the engine fails to work. The amount of heat introduced from outside is entirely consumed but at the same time a certain amount of heat passes from the hot body (working chamber of the engine) to the cold body (cooling jacket, however expressed).

The author considers in detail the amounts of resistances which are met with in the cycle proposed by Professor Dolder, whether it be for the case of a steam turbine or an air refrigerating machine, and shows that the resistances are so great as to make the ultimate efficiency of the cycle extremely low. (Ueber die Anwendung von Düsen an Stelle von Zylindern oder Kreiselrädern, *Zeitschrift des Vereines deutscher Ingenieure*, vol. 63, no. 35, August 30, 1919, pp. 830-833, 6 figs., *t*)

DEFLECTION OF STAYBOLTS IN LOCOMOTIVE BOILERS, Geo. L. Fowler, Mem.Am.Soc.M.E. As the length of fireboxes increased there was an increase in staybolt breakages and while there were many theories as to the cause of it, the amount of experimental data was quite limited.

In the present investigation an attempt was made to determine by actual measurement the amount of relative movement between the inner and the outer sheets of a locomotive firebox and also when the motion occurred and what was its general character.

The apparatus used was designed to resolve the motion of the inner sheet relatively to the outer one into its vertical and horizontal components and project them on a screen. These were afterward recombined to plot this relative movement in the form of a diagram.

The investigation is so extensive that only some of its conclusions can be reported here. The fundamental facts brought out were that the staybolt deflection is much greater in a flexibly stayed boiler than in a rigidly stayed boiler. It appears also that during the whole period of operation and probably until the boiler temperature has reached that of the atmosphere, the staybolt is in constant motion. The two types of boiler construction (Wootten and radially stayed) have quite different effects on staybolt deflection and firebox temperatures, and the action of the tubes has a marked influence on staybolt deflection at the front end of the firebox.

The tests also showed the great sensitiveness of the plates to changes of temperature. For example, a cold boiler may be filled with water of any temperature from cold to boiling and there will be no relative motion between the sheets. But let the fire be laid and a piece of lighted waste thrown in to ignite it, and it has, thus far, been impossible to get a reading before the sheets would show a movement, though this has been done within 10 sec. from the time of ignition of the waste.

The general impression left by the tests is that the Wootten firebox is much more rigid than the wide firebox when rigidly stayed with radial stays, and the difference might be still greater if it were given a complete installation of flexible bolts.

Because of the great sensitiveness of the sheets to changes of firebox temperature, it would appear that a boiler which is so built as to permit the sheets to expand under the influence of temperature changes will put less stress upon the staybolts and sheets than one where such motion does not exist.

The tests apparently left the observers with the conviction that the flexible bolt has superior possibilities as a means of reducing the probable stresses that the several parts of the boiler are called upon to sustain. (*Railway Age*, vol. 67, no. 23, December 5, 1919, pp. 1088-1094, 13 figs., *eA*)

## TESTING AND MEASUREMENTS

TESTING ABRASIVE WHEELS FOR EFFICIENCY, Raymond Francis Yates. Description of methods for testing abrasive wheels and some of the results obtained.

There are cases where it is necessary to determine the efficiency of various types and makes of abrasive wheels. In particular, the manufacturer of such wheels must have methods to determine which of his products gives the best efficiency.

Wheels are made in various grits, grades and bonds. By grit is meant the size of the abrasive particles that constitute the wheel, the particles being graded by the number of meshes to the square inch on a screen through which they pass. Thus, 10 would represent a very coarse wheel, while 200 would indicate a wheel composed of fine particles. Grade indicates the nature of the abrasive substance as regards its temper or hardness, and electrothermically it is now possible to produce abrasive substances of desired hardness. The bond of a wheel is the substance that holds the abrasive particles together in a solid form. It is extremely important and is a large factor in the determination of the efficiency of the wheel when cutting certain substances. For instance, if a wheel is cutting cast iron and its bond is not of the proper nature the particles may lose their sharpness and yet remain in place, or the bond may not be able to hold until they become sufficiently dull to break away from the rest of the wheel. In the first place very low cutting efficiency will result, as the dull abrasive particles hang on to the wheel and cause undue friction in rubbing over the surface of the iron. In the second case the wheel may cut freely and quickly, but will wear down very rapidly. (*American Machinist*, vol. 52, no. 3, Jan. 15, 1920, pp. 139-141, 6 figs., *ep*)

## THERMODYNAMICS

SPECIFIC HEAT OF AIR AT ROOM TEMPERATURE AND AT LOW TEMPERATURES, Karl Scheel and Wilhelm Heuse. An extensive paper of which only the conclusions can be given here on account of limited space available.

The specific heats of air were determined by means of two calorimeters and the average data of observations are presented in Table 2, where, in addition to the direct readings, are given the corresponding units in caloric masses obtained from electric mass by multiplying the latter with the heat equivalent of the watt-second, which is

$$0.23865 \frac{\text{g-cal}_0}{\text{watt-sec.}}$$

The tests confirm what has already been found by Swann, namely, that the specific heat of air at room temperature is considerably greater than that found by previous investigators, in particular by Regnault.

The value given by Swann in caloric mass units, namely,

$$0.24173 \frac{\text{g-cal}_{20}}{\text{g-deg.}}$$

was based on the value of 20-deg. calorie = 4.180 watt-seconds. From this may be computed the comparable value of the specific heat of air at 20 deg. in electric mass units

$$1.010 \frac{\text{watt-sec.}}{\text{g-deg.}}$$

which agrees perfectly with the results obtained by the present writers.

The value of the specific heat permits us also to determine the relation between the specific heats of air in accordance with the formula $c_p/c_v = x$. Usually

$$c_p - c_v = T \left(\frac{dp}{dT}\right)_v \times \left(\frac{dv}{dT}\right)_p$$

If one wishes to determine the values on the right-hand side of the equation, the van der Waals equation of state, then for atmospheric pressure and room temperature one obtains

$$c_p - c_v = \frac{R}{M} \times 1.0045 \frac{\text{watt-sec.}}{\text{g-deg.}}$$

where $R = 8.316$, which is the gas constant. and $M = 28.95$, which is the molecular weight of air. Hence

$$c_p = 1.009 \frac{\text{watt-sec.}}{\text{g-deg.}}$$

$$x = 1.400$$

When the temperature decreases, the specific heat of air increases. The increase is noticeable at −78 deg. cent. and reaches the value of about 5 per cent at −183 deg. cent. When the pressure is reduced to ⅔ atmosphere, the increase of specific heat is still noticeable at − 183 deg. cent., but has a smaller numerical value (3 per cent). It appears, therefore, that qualitatively in the neighborhood of its point of condensation, air shows the same behavior that Knoblauch and Jakob, and later Knoblauch and Mollier, have found for steam.

According to Linde the specific heat $c_p$ at pressure $p$ and absolute temperature $T$ may be represented by the specific heat $c$ of the ideal gas state independent of the temperature, that is, for the case of infinitely small pressure, when the following equation holds good:

$$\frac{c_p}{c} = \left(1 - \frac{ap}{T^3}\right)^{-\frac{1}{3}}$$

The constant may be calculated from experimental observations of the Joule-Thomson effect. If $p$ is measured in atmospheres, then for air $a = 21,500$. According to the tests of Linde this relation holds good for the case of small pressures down to the temperature of condensation of air.

With the aid of this formula it is possible to eliminate $c$ and to determine from the specific heat $c_p = 1.009$ at room temperature and atmospheric pressure the specific heat $c_p$ at equal or different pressure and at a lower temperature. One finds that under atmospheric pressure at − 78 deg. cent. $c_p = 1.013$ and from this value and the Linde equation the specific heat at − 183 deg. cent. is $c_p = 1.065$, while at a pressure of ⅔ atmosphere and − 183 deg. cent. $c_p = 1.045$, all values being in watt-seconds per g-deg. Within small fractions of one per cent these figures equal with the values found experimentally. (*Zeitschrift für Sauerstoff- und Stickstoff-Industrie*, vol. 11, nos. 9-10, May 1919, pp. *d*)

                            TABLE 2

| Temperature, deg. cent. | Specific Heat $c_p$ | |
|---|---|---|
| | in watt-seconds g-deg. | in g-calories g-deg. |
| +20 | 1.009 | 0.240, |
| −78 | 1.019 | 0.243; |
| −183 | 1.058 | 0.252, |

## VARIA

OIL VERSUS COAL FOR USE ON SHIPBOARD. The following statement in *Power* at the end of an article describing stokers for use on shipboard is highly significant.

The navies of the world will use oil regardless of price. So, too, will fast passenger ships. It is believed by many that there is not enough oil to warrant its wide use in the merchant marine for many years, although within recent weeks rich oil finds are reported from Colombia. Just now American coal is selling in Italy for $36 a ton. The freight rate is $22. So it is not bad business now to use oil-burning ships in the coal trade to Europe. Yes, the mechanical stoker for ships is here, and when oil again becomes high-priced or its supply alarmingly diminished, it will come into its own. (*Power*, vol. 51, no. 3, Jan. 20, 1920, pp. 86-89, 8 figs., *g*)

## CLASSIFICATION OF ARTICLES

Articles appearing in the Survey are classified as *c* comparative; *d* descriptive; *e* experimental; *g* general; *h* historical; *m* mathematical; *p* practical; *s* statistical; *t* theoretical. Articles of especial merit are rated *A* by the reviewer. Opinions expressed are those of the reviewer, not of the Society. The Editor will be pleased to receive inquiries for further information in connection with articles reported in the Survey.

# CORRESPONDENCE

CONTRIBUTIONS to the Correspondence Departments of MECHANICAL ENGINEERING by members of The American Society of Mechanical Engineers are solicited by the Publication Committee. Contributions particularly welcomed are suggestions on Society Affairs, discussions of papers published in this journal, or brief articles of current interest to mechanical engineers.

## Screwed Pipe Joints

TO THE EDITOR:

Very little attention is paid to the manner in which screwed pipe-thread joints are made. It is generally up to the pipe fitter to put red lead or some other kind of dope on the threads and then exert all his strength to screw the pipes together. When he comes to test the joints, however, he finds a goodly percentage of

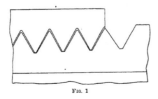

FIG. 1

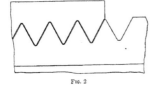

FIG. 2

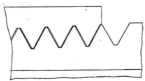

FIG. 3

them leaky. As a last resort he will apply a peening tool to close up the leaks, a thing that should not be done, as in most cases it only stops them for a short time. But the fault is not his, it is mostly the fault of the pipes and fittings—the way they are threaded.

In the first place, all pipe threads are supposed to be made to the Briggs Standard gage. It is almost impossible for the manufacturers to do this, as a tool made perfect to the Briggs Standard will not stay that way very long in practical work. It is most natural that the sharp point of the thread on the tap or die will

wear first, as shown by the male part of Fig. 1, and, as it would be utterly impractical to discard a tap or die before the shape varied from the Briggs Standard shape by a reasonable amount of wear, there ought to be some allowance made first in the making of the tool and then in the working gage. For instance, a perfect female pipe and male pipe that have been threaded with tools that have cut a good many threads before, as in Fig. 1, will never make a good joint except by applying main force, which should not be done. If, however, the bottom of the female thread were flattened somewhat, as shown in Fig. 3, the pipes would screw together with very little force; the sharp points on the male and female threads would be of such shape that they would float in place, the sides of the threads would come together and a good joint would result throughout, as shown in Fig. 2.

It is an easy matter to keep the tool so it will make a thread as shown in Fig. 4 for the male pipe, and in Fig. 5 for the female. The width of the flat part of the thread could vary from 1/10 to 1/8 of the depth of the sharp V-thread, and a joint would always

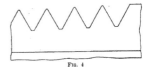

FIG. 4

FIG. 5

be made by floating the sharp point of the thread against it. As a matter of fact, however, the flat of the thread will be a round fillet, as the sharp point of a tool will soon wear round, and if made flat it will still wear round after it has performed a small amount of work.

Manufacturers have been reluctant to admit that the threads on fittings and pipes are not perfect to Briggs Standard, and it is high time to overcome that difficulty by making the working gage so it will gage the sides of the threads and not ride at the bottom when the thread is not perfect. I have seen pipes and fittings gaged with a Briggs Standard gage that were apparently all right, but when screwed together they would not make a joint for the simple reason that the female thread would be too large and the male thread too small. The crests of both the male and female work gages should be flattened to an extent not less than the U. S. Standard thread flat or one-eighth the depth of sharp V-threads; in fact, if one-sixth were taken off it would be still better.

HENRY RITTER.

Cincinnati, Ohio.

## Discussion of the Horsepower of Resistance

To THE EDITOR:

A measure of performance of tractive machines which is convenient for broad generalizations is

$$\frac{\text{weight} \times \text{speed}}{\text{horsepower}} \quad \text{or} \quad \frac{WV}{H}$$

For the battleship *Pennsylvania*, this coefficient has the value 47,000, based on shaft horsepower. A freight train with a resistance of 10 lb. per ton at 20 miles per hr. gives a coefficient of 75,000. In general, the value of the coefficient is $C = 750,000 \div R_T$, where $R_T =$ resistance in lb. per ton of weight.

A man on a bicycle may realize a value of $C$ around 40,000. For an average pleasure automobile, $C = 5000$. The lowest values of all are realized with airplanes, because these machines sustain as well as propel. The NC transatlantic hydroplanes give $C = 1900$ at 80 m.p.h. The usual range is from 1400 to 1800. High speeds always tend to reduce $C$, because power increases faster than speed.

Since for an airplane $H = dV^3(K_DA + K_P) \div 375$, the coefficient becomes

$$C = \frac{375 W}{dV^3(K_DA + K_P)} = \frac{375 W}{R} = \frac{375}{\tan \theta}$$

where $R =$ total resistance, lb., and $\theta$ is the gliding angle. The coefficient in question is therefore (in homologous units) simply the reciprocal of the gliding angle. It should have a high value, as the gliding angle should have a low value. Since $R_T = 2000 R \div W$, $C = 750,000 \div R_T$, as above.

Efficient airplane flight seems, therefore, to involve high altitude (low $d$), a proper wing form, and low parasitic resistance. The question of altitude is influenced, however, by the falling off of engine power at high altitudes.

W. D. ENNIS.

New York, N. Y.

## Economic Loss Caused by Rust

To THE EDITOR:

The actual economic loss and wastage of steel and iron tools, machines, vehicles, agricultural machinery and implements, parts and supplies of all kinds, due to rust from moisture condensation and to corrosion from fumes, etc., must be enormous.

Are any data or statistics available which express this loss?

INDUSTRIALIST.

New York City.

# WORK OF THE BOILER CODE COMMITTEE

*THE Boiler Code Committee meets monthly for the purpose of considering communications relative to the Boiler Code. Any one desiring information as to the application of the Code is requested to communicate with the Secretary of the Committee, Mr. C. W. Obert, 29 West 39th St., New York, N. Y.*

The procedure of the Committee in handling the cases is as follows: All inquiries must be in written form before they are accepted for consideration. Copies are sent by the Secretary of the Committee to all of the members of the Committee. The interpretation, in the form of a reply, is then prepared by the Committee and passed upon at a regular meeting of the Committee. This interpretation is later submitted to the Council of the Society for approval, after which it is issued to the inquirer and simultaneously published in MECHANICAL ENGINEERING, in order that any one interested may readily secure the latest information concerning the interpretation.

Below are given the interpretations of the Committee in Cases Nos. 262-266, inclusive, as formulated at the meeting of December 2, 1919, and approved by the Council. In accordance with the Committee's practice, the names of inquirers have been omitted.

### CASE No. 262

*Inquiry:* Does the term " equivalent pitch for cylindrical surface" as referred to in Par. 212c of the Boiler Code apply to the outside shell or to the inside shell?

*Reply:* It is the opinion of the Committee that this term in the formula of Par. 212c refers to the surface to be stayed, which in this case is the inside shell.

### CASE No. 263

*Inquiry:* a Is it the intention that Table 5 of the Boiler Code which specifies the maximum allowable stresses for stays and stay-bolts shall apply to iron as well as steel stays? Are sub-items a, c and e supposed to apply to iron as well as to steel?

b Is it assumed by the Boiler Code Committee that an ordinary expanded boiler tube has no holding power in the tube shell whatever? What holding power is a beaded tube or a welded tube supposed to have?

*Reply:* a Items a and c of Table 5 apply to iron as well as steel, but in accordance with the requirement of Par. 4, item e applies only to iron.

b It is the opinion of the Committee that tubes which are merely expanded in the tube sheets have a large initial holding power. However, under practical conditions of operation, where the seating of the tube in the sheet is disturbed, the holding power of tubes so attached is different. It was for this reason that the Code requires either beading, flaring or welding in of tubes. There have been many tests made on the holding power of tubes, the results of which may be found in technical literature and handbooks.

### CASE No. 264

*Inquiry:* What form of test will be required for the shell of a miniature boiler formed by the cupping and hot-drawing process from a sheet of firebox plate which meets the requirements of the Boiler Code? It is believed that the specifying of firebox boiler plate steel and subjecting each finished cupped and hot-drawn shell to an hydrostatic test, using the formula $P = 32,000t/D$, and the flattening test of a coupon cut from each cupped and drawn piece, this flattening test to go to 5 times the wall thickness, would be sufficient guarantee of the finished material and would be far more representative and better test than taking tests from the material before the cupping and hot-drawing operations.

*Reply:* It is the opinion of the Committee that if the firebox plate material from which the boiler shell is formed meets the requirements of the Boiler Plate Specifications of the Code, and if the shell is properly annealed after drawing, it may be considered as having met the requirements of the rules in the Boiler Code. The Boiler Plate Committee has been informed that for proper annealing, the shell should be subjected to a temperature of 1550 deg. fahr.

### CASE No. 265

*Inquiry:* Is it the intent of Par. 25 of the Boiler Code that chemical limits for carbon are applicable only to firebox plate and not to flange plate?

*Reply:* The Specification for Boiler Plate Steel in the Code, which is in conformity with the similar specification of the American Society for Testing Materials does not specify any chemical limits for carbon in steel plate of the flange grade.

### CASE No. 266

*Inquiry:* Does not the reply given in Case No. 206 conflict with the rules in the Boiler Code for calculation of structural members for the stiffening of unstayed portions on heads of boilers?

*Reply:* It was not the intent of the Boiler Code Committee in Case No. 206 to indicate that the stiffening effect of the members used to attach the braces to the head could be added to the value of the braces, in determining the maximum allowable working pressure for which the design was constructed.

# SPECIALIZED CURRICULA IN ENGINEERING COLLEGES [1]

## By A. A. POTTER, MANHATTAN, KAN.[2]

THOSE who are responsible for the conduct of engineering departments at universities or colleges are urged by some to develop specialists, and by others to make the curricula broad. Some alumni, particularly during their first few years after graduation, advocate courses to meet the needs of the special industries, others feel that engineering schools are now devoting too little time to basic principles and facts.

Mr. Benjamin G. Lamme, chief engineer of the Westinghouse Electric and Manufacturing Company, in discussing technical training for engineers before the Pittsburgh Section of the American Institute of Electrical Engineers in 1916, made the following remarks concerning specialization:

There has been quite a fad for specialization in engineering training. The writer's personal opinion is that specialization in college training is not advisable, except possibly in a very general way. There has been a false idea in many schools that if a man specialized along some individual line of work it would advance him more rapidly when he leaves school for active work. The writer almost never asks the student in what field he specialized. . . . In college training the time spent on commercially practical details is usually largely wasted, as it may give the student entirely wrong ideas. . . . The colleges should aim to develop analytical ability, imaginative faculty, ability to do independent thinking. . . . A broad general training is most desirable for the man who has ability to do something in the world.

The majority of engineering colleges are devoting their chief attention to curricula in civil engineering, electrical engineering and mechanical engineering. There seems also to be considerable demand for men trained in mining engineering, chemical engineering, architecture and architectural engineering. Several institutions are also offering curricula leading to degrees in agricultural engineering, ceramic engineering, flour-mill engineering, highway engineering, railway engineering and industrial engineering. Specialized curricula have also been proposed in business engineering, petroleum engineering, aeronautical engineering, etc.

An investigation has been made of the ground covered at twenty-four representative institutions in their civil, electrical and mechanical engineering curricula with results as follows:

About fifteen per cent of the total time is devoted to mathematics, twelve per cent to physics and chemistry, eight per cent to English and other modern languages, ten per cent to mechanics and hydraulics, three per cent to economics and business organization, five per cent to elementary engineering drawing, about thirty per cent to specialized engineering subjects, and about fifteen per cent to related engineering subjects.

The specialized subjects in a civil-engineering curriculum usually include: surveying, masonry and foundations, bridge design, railways, highways, drainage irrigation, water supply, sewage disposal, and concrete construction. The related engineering subjects include shop practice, elements of power engineering, and elements of electrical engineering.

In the electrical-engineering curriculum the specialized subjects usually include: theoretical electricity, direct- and alternating-current machinery, electrical measuring instruments, power engineering, telephony, electric railways, and electric design. The related engineering courses are usually thermodynamics, shop practice, machine design, surveying and factory organization.

In the mechanical-engineering curriculum the specialized subjects include: machine design, shop practice, thermodynamics, power engineering, power-plant design, heating and ventilation, refrigeration and factory organization. The related subjects are electrical machinery, structural design, and surveying.

An analysis of the various specialized curricula will reveal the fact that they differ mainly by name from those in civil, electrical or mechanical engineering. Such specialized curricula are usually chosen by the adventurous and poorer students who are attracted by good-sounding names.

Employers of graduates from engineering schools are usually not concerned with the name of degree the young engineer holds. The large electrical manufacturers do not differentiate between those trained as electrical and those trained as mechanical engineers. A representative of a telephone company stated that his company is willing to accept men trained in any engineering course, but that he does not care to select men from schools who claim to train telephone engineers.

If it were possible to predict the line of work the student will enter after graduation, certain specialization might be justified. Even in such cases a broad engineering training would be preferable, as the young engineer will find that his work will necessarily specialize him and will so absorb his energies as to leave him very little time to acquaint himself with related specialties. Records of graduates from engineering schools show that, while the majority follow some branch of engineering, comparatively few remain in the particular field which they had expected to enter when at college.

Dean M. E. Cooley, President of The American Society of Mechanical Engineers, in the September 25, 1919, issue of the *Engineering News-Record* makes the following statement:

Engineers are not wholly blameless for the position in which they find themselves—a responsibility which the engineering colleges must share. They have devoted altogether too much time to professional subjects. In their zeal to fit themselves for work in the field of their choice, non-technical subjects which make for breadth have been sacrificed. Thus they are narrow, are fenced in so that neither can they look out nor can others look in.

Close observation of the records of hundreds of graduates from engineering schools leads the author of this paper to the opinion that better results would be accomplished if the majority of engineering colleges would devote their energies to two curricula, one dealing mainly, in its applications, with static engineering and the other with dynamic engineering. In the first curriculum should be included the courses now in the civil-engineering curricula of our best schools and should familiarize the student with the broad principles underlying the design and construction of railways, highways, irrigation and drainage projects, masonry, concrete, sanitation, and business administration. In the dynamic-engineering curriculum should be included the fundamental courses now given in the electrical- and mechanical-engineering curricula, with considerable emphasis on factory management, power engineering and business administration.

Curricula in mining engineering, chemical engineering and architecture should be offered only by a small number of institutions. There may also be need for a few colleges to offer graduate courses or short courses which will prepare men for certain specialized work in the industries. Admission to such courses should be limited to men who have had considerable practical experience.

Experience indicates that broad engineering education will enable the college graduate to reach greater usefulness in his profession than will specialized technical training.

According to Albert Reeves, General Manager of the National Automobile Chamber of Commerce, 1,892,000 motor vehicles were built in the United States in 1919, of which 305,000 were motor trucks. Two hundred and sixty manufacturers were engaged in this production, employing 580,000 workmen. The wholesale value of these vehicles was, in round figures, $1,808,000,000. There are approximately 7,100,000 motor vehicles registered in the United States, of which 700,000 are motor trucks, these latter having displaced, it is estimated, over 3,500,000 horses.

[1] Address delivered before meeting of the Mid-Continent Section of THE AMERICAN SOCIETY OF MECHANICAL ENGINEERS at Bartlesville, Okla., October 30, 1919.

[2] Dean, Division of Engineering, Kansas State Agricultural College. Manhattan, Kan.

# MECHANICAL ENGINEERING

### THE JOURNAL OF THE AMERICAN SOCIETY
### OF MECHANICAL ENGINEERS

Published Monthly by the Society at
29 West Thirty-ninth Street, New York

FRED J. MILLER, *President*
WILLIAM H. WILEY, *Treasurer* CALVIN W. RICE, *Secretary*

PUBLICATION COMMITTEE:
GEORGE A. ORROK, *Chairman*                         J. W. ROE
    H. H. ESSELSTYN            GEORGE J. FORAN
            RALPH E. FLANDERS

PUBLICATION STAFF:
LESTER G. FRENCH, *Editor and Manager*
LEON CAMMEN, ⎫ *Associate Editors*
C. M. SAMES, ⎭
        W. E. BULLOCK, *Assistant Secretary*
            *Editor Society Affairs*
    FREDERICK LASK, *Advertising Manager*
WALTER B. SNOW, *Circulation Manager*
            136 Federal St., Boston
Yearly subscription $4.00, single copies 40 cents. Postage
to Canada, 50 cents additional; to foreign countries $1.00 addi-
tional.
    *Contributions of interest to the profession are solicited. Com-
munications should be addressed to the Editor.*

## Mechanical Engineers to Meet Next May at St. Louis

The Spring Meeting of The American Society of Mechanical Engineers is to be held at St. Louis, May 24 to 27, the first meeting being on Monday of the week of the convention in accordance with the plan followed so successfully at Detroit last year.

In view of the interest throughout the Mississippi Valley and in the territory contributory to it, one session will be devoted to a discussion of river-transportation problems, and the movement now under way for the development of transportation on the Mississippi River, with New Orleans as the port for shipment of goods to South American and other countries. Another session will be devoted to the subject of the Valuation and Appraisal of Industrial Properties, using as material the papers presented at the last Annual Meeting, upon which discussion is being solicited by the Committee on Meetings and Program. Papers are also in prospect on fuels, oil engines, measurement of the flow of water, transmission of heat, pumping engine tests, aeroplane design, etc. A paper is to be presented by the recently elected Honorary Member of the Society, A. C. E. Rateau, of France, on Turbo-Compressors for Increasing the Power of Aeroplanes at High Altitudes. It is also expected that papers will be presented by the Mid-Continent Section, representing the oil districts of Oklahoma, on subjects relating to the distribution and use of oil for industrial purposes.

St. Louis is a rapidly growing industrial center, with one of the most progressive Chambers of Commerce in the United States, engaged in educational work for the increase of production in agriculture and industry, both locally and throughout the state, and for the improvement of industrial conditions. A large housing undertaking is under way upon which there will be a contribution at the meeting. An immense plant being erected by the Chevrolet Motor Company is one of the influential factors in bringing about this movement. There are numerous manufacturing plants, some of large size and several of the most modern equipment, undertaking unusual lines of work of interest to members of the Society.

The parks of the city, with the outdoor municipal theater and the Art Museum in a location remarkable for its beauty and extent of view, are among the finest in the country. The hotel accommodations are good and modern, and altogether the city is an ideal place for a meeting of engineers, and a whole-hearted welcome is assured on the part of the local members and their many friends of the city, both within and without the profession. St. Louis is proud of her accomplishments industrially and as a civic community, and justly so. The Committee on Meetings and Program and the officers of The American Society of Mechanical Engineers are gratified at the opportunity which the city has extended to the membership for a convention which promises to be one of the greatest in the history of the Society.

## What Is Engineering?

The Committee on Development of the American Society of Civil Engineers, in its report to the Board of Directors presented on October 14, submitted the following definitions of engineering:

1 Engineering is the science and art of directing the great sources of power in Nature to the use and convenience of man.

2 Engineering is an art and a science. It is a science in so far as physical laws are its base, and an art in so far as in the application of these laws the things designed and constructed develop the spirit of progress, the creation of wealth, and the well-being of all peoples.

Engineering is generally divided into four major divisions, viz., civil, mechanical, electrical, and mining.

The practice of engineering requires knowledge of physical laws, forces, and the materials of Nature.

The professional engineer is one who by reason of special training, education, and experience is qualified to design and direct the construction of engineering work in one or more of the major divisions of engineering.

3 Engineering is the creative science and art of applying economically the materials and forces of Nature to the use and convenience of man.

4 Engineering is the science of industrial effort, and the science and art of applying this effort for the welfare of the public.

An engineer is one versed in the science and art of industrial effort made for the purpose of public welfare.

## Civilization Follows Transportation

Many years ago Professor Metchnicoff, of the Pasteur Institute, Paris, better known for his work in the field of biology and micropathology, published a book dealing with the history of ancient civilization.

He showed that civilization followed the course of the great rivers, which was only natural in the light of our modern knowledge. Great cities as well as even such crude manufacture as existed in the ancient world required the movement of goods, such as food and building and raw materials, and in a roadless world the rivers presented the only available cheap means of communication.

In this country the early settlements followed the same water paths. The Hudson and Ohio rivers, and later on the Mississippi, were the great liquid roads carrying the early American commerce and settlers.

As civilization moved, however, and the wealth of nations increased, the rivers were not sufficient to carry on the commerce of the world and canals were dug, at first as outlets for rivers into more convenient directions or as connecting links between important rivers, and later on as more or less independent means of communication. This represents the basic transportation tendency of the 18th century and the early days of the 19th century. The historical Manchester Canal in England is one of the best illustrations of this phase of the transportation problem in Europe, and the Erie Canal, which practically made an empire state of New York, serves as a good example of the application of this same principle to conditions throughout the entire United States.

But water transportation is at best only a makeshift. It is slow and, when created artificially, by no means cheap. In North America it is closed to users for several months in the year. Further, water carriers have to be large in order to be economical, which in turn means that the goods shipped have to be either in bulk character or have to be accumulated until there is enough to fill a steamboat. All this corresponds to the stage of economic development when industry was still moving slowly.

The next stage was the creation of the railroads, which have been the means of opening immense regions of the country devoid of rivers or canals. The great American Northwest, the mining regions of the central Rockies, and the wealthy states of the Pacific slope could not have been opened to settlement or commerce without the railroads.

In the 75 years of the railroads' existence, however, it has been found that they also have certain limitations. While they represent the most economical means of transportation over distances in excess of, say, 200 miles and for shipments not below a certain relatively moderate bulk, an insistent demand arose for an economical means of transportation of goods below the transportation range of the railroads, especially for distances less than 100 miles, or what might be called urban-circle deliveries. When a legitimate demand on a large enough scale arises, it is usually promptly satisfied, and the motor truck came to the relief of those who needed that type of transportation.

One more transportation question has been looming up for the last 25 years at least, and that is transportation between widely scattered localities where there are neither roads nor rivers, and where the comparatively small volume of traffic does not justify the construction of railroads. In the sparsely populated mining regions of South America, in the outlying settlements of Africa, in the islands of the Pacific, there was a demand for a way of rapidly conveying passengers and comparatively small amounts of goods and for exporting the products of local industries, such as rubber, gold, and other minerals. This problem cannot yet be considered to be definitely solved, but a promise of its solution lies in the recent rapid progress of the aeroplane and dirigible. The whole history of the improvement of the means of conveyance shows, however, that as means of transportation have improved civilization has followed in its way. As the rivers helped to build up ancient cities and the earlier towns of the Western Hemisphere, so the Erie Canal has helped to build the Empire State, the railroads have opened Minnesota, the Dakotas and Manitoba, the motor trucks are now opening up the farming communities throughout the country, and the aeroplanes will make possible the development of the rich mines of Central and South America and South Africa; and in the far distance loom still greater possibilities of which it would now perhaps be premature and but a flight of fancy to discuss.

## The "Mysterious Mr. Smith"

The story of how the new buildings and some of the equipment of the Massachusetts Institute of Technology in its new site on the Charles River Basin, Cambridge, Mass., were made possible by the munificent contributions of a certain "Mr. Smith" is generally known by all who have kept in touch with technical education in this country.

Who this Mr. Smith was, however, has been a zealously guarded secret, until at a recent jubilee dinner of the alumni of the Institute, to celebrate the successful completion of a drive for a $4,000,000 endowment, the real name of this benefactor of the M. I. T. was disclosed as Mr. George Eastman, the "Kodak King" of Rochester, N. Y.

In 1914 the Massachusetts Institute of Technology celebrated the fiftieth anniversary of its chartering and marked the occasion by deciding to move from its outgrown buildings which had become a landmark in the Back Bay district of Boston, to its present location in Cambridge. Later, President Maclaurin of the Institute, himself a physicist of renown, spent a day at the works of the Eastman Kodak Company, where he was greatly impressed

by the high order of scientific attainment in the laboratories and manufacturing departments of this great plant. Mr. Eastman was not at Rochester at the time and Dr. Maclaurin therefore wrote him asking for a conference with regard to the problems and opportunities of the Institute and its possibilities for the development of scientific work. They met at dinner in New York, and Mr. Eastman was so interested in the project for rebuilding Technology that he at once gave $2,500,000, expressing the desire that a comprehensive plan for the new buildings should be followed and the usual incoherent, piecemeal construction for such institutions avoided.

As the work progressed, $1,000,000 more was added, and just before the dedication of the new monolithic structure in which the Institute is now housed, he offered five dollars for every three that others contributed up to a limit of $2,500,000, an offer which was fulfilled. Still other donations were subsequently made and finally, when the drive for the $4,000,000 endowment above referred to was completed, Mr. Eastman matched it with a sum of equal amount. His total gifts to the Institute have amounted to $11,000,000, all of which, with the exception of a quarter of a million specifically for chemical engineering equipment was given anonymously.

Mr. Eastman preferred to keep the real name of the donor a secret in order to avoid notoriety and only disclosed it because of the legal requirements incident to the transfer of stock made necessary at the time of his latest gift of $4,000,000. Naturally there has been a great deal of guessing as to who was "Mr. Smith." Among those suspected were Frick, Rockefeller, Vail and DuPont, the latter a Technology alumnus who has contributed large sums to his Alma Mater under his own name. There also were many amusing incidents in this connection, of people who had announced definitely that they, or some one of their friends, were the man; and of wealthy suspects who met and accused each other of keeping under cover, and who separated believing that the other fellow was a colossal bluffer.

The authentic announcement of Mr. Eastman places him as one of the greatest supporters of technical education in America, and it is a pleasure to know that the gifts have come from one who, through his own experience, has full appreciation of the value of such education.

## Conference on National Safety Codes

A general conference on safety codes was called by the Bureau of Standards at Washington, for December 8. This conference was attended by representatives of a very large number of organizations interested in the preparation of standard safety codes, including engineering societies, trade organizations representing all phases of industry and City and State commissions. By a vote taken at this conference, the American Engineering Standards Committee was asked to request the National Safety Council, the International Association of Industrial Accident Boards and Commissions, and the Bureau of Standards acting together to appoint an Advisory Committee of fifteen for three purposes, viz.,

(1) to determine the order of priority in the development of safety codes

(2) to recommend to the A. E. S. C. suitable sponsor and joint sponsor bodies

(3) to make report, after preliminary study, not later than February 1.

This matter was taken up by the American Engineering Standards Committee at its meeting on January 3 and the Committee took favorable action on the recommendation of the Conference. A committee of fifteen was immediately organized and held a meeting in Washington on January 9, at which the following members were present:

F. D. Patterson, representing Pennsylvania Department of Labor and Industry.

W. C. L. Eglin, representing National Electric Light Association, Committee on Safety Rules and Accident Prevention.

M. G. Lloyd, representing Bureau of Standards, Safety Section.

Royal Meeker, representing U. S. Bureau of Labor Statistics.

L. W. Chaney, representing U. S. Bureau of Labor Statistics.

Dana Pierce, representing Underwriters' Laboratories.

W. P. Eales, representing The American Society of Mechanical Engineers.

E. B. Rosa, representing Bureau of Standards.

A. W. Whitney, representing National Workmen's Compensation Service Bureau.

S. J. Williams, representing National Safety Council.

P. G. Agnew, Secretary, American Engineering Standards Committee was also present.

The Committee first organized itself with Dr. E. B. Rosa as Chairman and Dr. Royal Meeker as Secretary. It then adopted the name of "National Safety Code Committee."

A total of 63 Safety Codes on as many subjects were scheduled by Dr. Rosa for consideration. Those for which The American Society of Mechanical Engineers was suggested as sponsor, are as follows:

Elevators and Escalators
Stationary Steam Boilers
Non-fired Pressure Vessels
Boiler Room Equipment and Operation
Steam Engines and Turbines
Internal Combustion Engines
Engine-Room Equipment and Operation.
Ladders
Locomotives

A number of other codes were discussed, some of which in all probability will be assigned to this Society, but no definite decision on these was reached at the time of this meeting of the Joint Advisory Committee.

## Technical Papers at Annual Meeting of Society of Automotive Engineers

A number of interesting professional papers were presented at the annual meeting of the Society of Automotive Engineers, held in New York, January 6 to 8. In a paper on Preignition of Spark Plugs, Stanwood W. Sparrow of the U. S. Bureau of Standards, described tests on specially prepared spark plugs, which indicate that one of the main causes of preignition is the lack of proper cooling of a spark-plug element, and in some cases improper design of the engine itself. The lack of proper cooling of the spark-plug elements may be due either to improper design of the plug or to subsequent causes preventing its proper preparation.

Major George E. A. Hallett, chief of Power-Plant Section, Engineering Division, Air Service, Dayton, Ohio, presented general data on the use of superchargers for aircraft engines. The conclusion to which the author comes is that supercharging in principle is of value to several classes of military aeroplanes, as well as to certain types of civilian planes. As a graphic illustration of the advantage of supercharging, it was pointed out that at 25,000 ft. altitude a supercharged 250-hp. engine will deliver as much power as a 1000-hp. engine without supercharging apparatus.

Interesting considerations on Dilution of Engine Lubricants by Fuel in Automotive Engines were presented by Gustave A. Kramer, of the U. S. Bureau of Standards. Among other things were described experiments to determine the effect of dilution by fuels on the viscosities of lubricants, the fuels being diluted either by an average grade of fuel gasoline or by commercial kerosene. The author suggested the use of an oil renovator as part of the engine itself.

The important problem of the use of low-grade fuels in the automotive engines was discussed briefly by J. G. Vincent, vice-president in charge of engineering, Packard Motor Car Company, Detroit, Mich., who suggested among other things, the use of a special manifold for preheating fuel with an automatic method of regulating the heat. The principle of the device is the employment of a difference in pressure existing on either side of the carburetor butterfly valve; a small amount of the combustible mixture is caused to pass through a passage which is in parallel with the main carburetor passage whereupon the mixture is burned in a suitable burner and the burned gases are allowed to mix with the incoming main supply to the engine above the throttle.

Notwithstanding the great progress made in the application of the constant-volume-type internal-combustion engine to use in road vehicles, an increased amount of attention has been paid recently to the problem of adopting steam to the same purpose. Louis L. Scott, chief engineer, Standard Engineering Company, St. Louis, Mo., presented a description of a Steam Automotive System which it is said has already been applied both to passenger cars and to trucks.

L. H. Pomeroy, consulting engineer, Cleveland, Ohio, discussed the trend of automotive-engine design in the light of the application of modern views of gasoline-engine thermodynamics, and noted recent improvements in mechanical efficiency, in engine balancing, in carburetion and the progress which has been made in the utilization of low-grade fuels.

Of special interest were also the following: Aluminum Piston Design, E. G. Dunn; The Measurement of Vehicle Vibrations, Benjamin Liebowitz; Needs in Engine Design, Frank H. Trego; Bettering the Efficiency of Existing Engines, Hugo C. Gibson; Mixture Requirements of Automotive Engines, O. C. Berry; Springs and Spring Suspensions, E. Favary; The Velocity of Flame Propagation in Engine Cylinders, R. K. Honoman and Donald McKenzie; Composite Fuels, Joseph E. Pogue.

The report of the Iron and Steel Division proposed specifications for low-tungsten steel for inlet and exhaust valves, nickel steel for case-hardened parts, malleable-iron castings for general automotive purposes, and high-chromium steel for exhaust valves in aeroplane engines. The composition recommended for low-tungsten steel expressed in percentage was: carbon, 0.50 to 0.70; manganese, not to exceed 0.30; phosphorus, not to exceed 0.035; sulphur, not to exceed 0.035; chromium, 0.50 to 2.00. Malleable castings were specified to be produced by air-furnace, open-hearth, or electric-furnace process; the minimum tensile strength permitted was 45,000 lb. per sq. in. and a minimum elongation of 7.5 per cent in 2 in. The suggested composition in percentage for high-chromium carbon steel was: carbon, 0.20 to 0.40; manganese, not to exceed 0.50; phosphorus, not to exceed 0.035; sulphur, not to exceed 0.035; chromium, 11.50 to 14.00: silicon, not to exceed 0.30.

## New York Automobile Show

The New York Automobile Show, held at the Grand Central Palace in the week of January 3 to January 10, left a mixed impression on an engineer who tried to gage the state of the industry from the exhibits.

The first impression was decidedly encouraging. Both the exhibits and the attendance at the show indicated a thoroughly healthy state of the industry. Practically all the old companies were represented, and there were a considerable number of newcomers who had either made good already, or who gave fair promise of doing so as soon as they had had time enough to arrive at this result.

The art of exhibiting automobiles has made a big advance since one or two years ago. The general disposition of the exhibits this year was good, and there was far less crowding than marked the exhibits of the previous years. More sectional models than ever were shown, operated by electric motors so as to give a chance to the public and the engineers to see the inside construction. This latter was a rather unexpected feature as in the last few years the study of engine development has been such that the ability of the engine to perform is now taken for granted. On the other hand, however, it may be that the technical education of the public has proceeded to such a point that a larger number than ever want to know the reason why and refuse to take anything for granted in mechanical lines.

More and more attention is being paid to body refinements, and the coach work of the cars shown at the Grand Central Palace exhibited a gratifying high form of development.

All this is highly encouraging, but some of the promises of the past two years do not seem to have been fulfilled, at least as far as one may judge from what was shown at the Grand Central Palace.

For nearly two years the automotive industry was in the service of the Government, engaged in assisting it in the strenuous work of building up in record time the great American war machine. This necessarily forced automobile engineers into entirely new lines of work. Some of them made shells, others aeroplanes and aeroplane motors; still others the fine instruments needed for the Ordnance and Navy Departments. It was confidently expected that this would bring a flood of new ideas into the automobile field just as it did in Europe, and from this point of view the show was to a certain extent disappointing.

There were exhibited no really new developments. The engines, the chassis construction (with some exceptions), the accessories still follow the conventional lines, and, notwithstanding even the impending shortage of gasoline, nothing at the show indicated that the industry is anything like prepared to meet this great condition.

It would be unfair, however, to exaggerate this fact, or to attach too great importance to it, for two reasons. In the first place, during the year and a half elapsed since the war the American automobile industry has gone through a period of tremendous strain. Following the great task of reconverting the industry back to the peace basis were the grave labor disturbances both in the automobile plants themselves, and still more so in the accessory plants supplying automobile plants with castings, bodies and parts, which created conditions anything but favorable to the undertaking of new developments on a large scale.

That in the face of all this the automobile industry has been able not only to meet the demands of the market, but to bring into the product a number of minor improvements, clearly shows the great vitality of the industry and the genius of the men directing it, and it would have been difficult to expect basic developments under such conditions in such a comparatively short time. That important research work is going on in practically all the larger companies is an open secret, but the results of this work were not shown at the Grand Central Palace, and from this point of view it cannot be said that the show represented the actual state of the industry.

While, however, no radical improvements were shown, a trained eye could easily discover the influence of war work in many directions. The finish of parts, the machining, and especially the attention paid to the balancing of moving parts, is far superior to anything that we have seen in the industry in the past.

What has been said about the Passenger Car Show at the Grand Central Palace, largely applies also to the Motor Truck Show which was going on simultaneously at the Jerome Avenue Armory.

Of the minor improvements, the following may be noted:

At the Passenger Car Show the spring suspension was employed in such a manner that a spring base of 130 in. is secured with a wheel base of only 100 in.

At the Commercial Car Show, the Shacht Motor Car Company exhibited a truck with a 10-speed transmission which is secured by using a regular five-speed transmission (four ahead and one reverse) and a sub-transmission doubling the range of the main transmission. So far this arrangement has been applied to 5-ton trucks only where apparently there may be the greatest demand for it.

The fundamental impression left by the thousands of exhibits at the show would indicate that the industry is a perfectly healthy one and that greater things may be confidently looked forward to in the very near future in the way of mechanical road transport.

This is of importance not only to the automobile industry generally but to the country at large, as transportation is a measure of the progress of civilization.

## Anatole Mallet

Anatole Mallet, Honorary Member of The American Society of Mechanical Engineers, died in Nice, France, in October, 1919. He was one of the few men whose name became the designation of a standard type of apparatus in a great industry. Born in 1837 at Carouge, and graduated in 1858 from the Central School of Arts and Manufactures in Paris, Mr. Mallet devoted all his attention for a number of years to problems in civil engineering. He was first connected with the Bureau of Direction of the General Company of Railroad Materials in France, and subsequently with the work on the Suez Canal. In 1864 he was engineer for the company which undertook the dredging of ports of Italy.

In 1867 he engaged for the first time in mechanical engineering, giving particular attention to steam engines with double

ANATOLE MALLET

expansion. The first application of this system to locomotives was introduced in 1876, when he introduced the first two-cylinder compound locomotive, which operated on the line from Bayonne to Biarritz. A fuel economy 20 per cent superior to that given by the standard types of the time was attained, which at once placed the inventor in the foremost ranks of locomotive designers.

The great success of the compound locomotive led to a material increase in the size of the tractive unit and to the development successively of three- and four-cylinder compound locomotives. In this way it became evident that the limit of size of a locomotive of rigid construction, especially on lines with sharp curves, would soon be reached, and the problem arose of building a locomotive of still larger size which would have the necessary flexibility. This problem was solved by Mallet in the design of the articulated locomotive, an invention which made his name familiar to every railroad engineer throughout the world. The most powerful locomotives now in existence are of the Mallet articulated type.

He was a member of the Society of Civil Engineers of France, The Society for the Encouragement of National Industry of France, and The Franklin Institute of Philadelphia. The French Society of Civil Engineers awarded him the Schneider prize in

1902, and the annual prizes in 1909 and 1911. He was made a Knight of the Legion of Honor in 1885 and promoted to Officer in 1905. The Institution of Mechanical Engineers of London awarded him a gold medal in 1915.

In addition to carrying on important engineering work, Mr. Mallet took an active part in the work of the French Society of Civil Engineers. From 1880 to within a few months of his death he was editor of the *Chronicle* of the Bulletin of the Society and furnished it with numerous technical notes and important memoirs, the last of which, treating of The Practical Evolution of the Steam Engine, earned for him the honors conferred by the Society.

## Death of the President of the Massachusetts Institute of Technology

Richard C. Maclaurin, President of the Massachusetts Institute of Technology, died of pneumonia at his home in Cambridge, Mass., on January 15. He was only 49 years old, but had gained an international reputation, not only as an educator and an administrator of large affairs, but as a physicist, a mathematician and a jurist.

Dr. Maclaurin was born in Edinburgh in 1870. His early boy-

RICHARD C. MACLAURIN

hood was spent in New Zealand, whence he returned to England to complete his preliminary education in English schools. He later was graduated from St. Johns College of the University of Cambridge, where he also received the advanced degree of Master of Arts, the thesis work for which was in higher mathematics.

He next spent several months in the United States and Canada studying educational institutions. Again returning to England he re-entered Cambridge, this time to study law, and was awarded the McMahon Law Studentship, the most highly valued in the university for law students.

In 1898 Dr. Maclaurin was appointed professor of mathematics in the University of New Zealand, became a trustee of the university, and took an active part in the organization of technical education in the colony. In 1903 he became Dean of the Faculty of Law in the university, which office he held for four years. In 1907 he was invited to take the chair of mathematical physics at Columbia University, New York, and a year later was made head of the department of physics.

In 1898 the degree of Doctor of Science was conferred upon Professor Maclaurin by Cambridge University for his researches in pure science, and again in 1904 he was honored with the degree Doctor of Laws by the same university for his achievements in the study of law.

In 1908 Dr. Maclaurin was appointed president of the Massachusetts Institute of Technology at a time when its finances were low, its building inadequate, and the call for the education at this institution of students from all over the world was very great. The new president succeeded in drawing together into a unified, enthusiastic body the corporation and faculty of the institute, and the alumni and students. By concerted action the movement was started which resulted in the new and greater Technology now located in its magnificent buildings on the Charles River in Cambridge, Mass. One of the fruits of this movement was a gift by T. Coleman du Pont of $500,000 toward the purchase of the site for the buildings. How Dr. Maclaurin, backed by the alumni, succeeded in raising the additional funds needed for the completion of the plans for expansion is told in another article in this number. The culmination of this was the announcement at a jubilee dinner by the alumni on January 10 of the success of the campaign for a $4,000,000 endowment, supplemented by a gift of an equal amount by George Eastman of the Eastman Kodak Company. Just previously, however, Dr. Maclaurin had been stricken by illness, and it was not permitted to him to tell the alumni personally of this greatest achievement of his life. A few days later he died—indeed a tragedy, his death coming at the height of his career, just as he had completed the foundation for the great educational work which the institute is destined to accomplish with its splendid equipment.

## AN IMPROVED WEIR FOR GAGING IN OPEN CHANNELS

(*Continued from page 88*)

One word more, however, the writer allows himself in relation to the formulæ for weir discharge. It had been his intention not to evolve any formula, but to present the result of his experiments in the form of a table, arranged somewhat as are tables of logarithms, so that quantities could always be read at a glance from the data observed. No one cares for the formulæ by means of which logarithms are computed; their use would not be furthered by a study of them. No more need a Power Test Code contain formulæ for weir discharge; a table would answer its purposes more fittingly.

But the exceeding simplicity of the weir formula appurtenant to the form of weir herein presented upset this plan. Such a weir does not need so much as a table to look up or to compute its discharge. All one has to do is to measure the length of the weir and then multiply it by 5.5 and by the difference in head.

But there remains the question, why 5.5? Why not 20, or any other quantity? And for the mitigation of the stated laudable curiosity, further experiments will no doubt be undertaken in the course of time; and much ink of disquisition may yet be set adrift.

## THE TESTING OF FARM TRACTORS

(*Continued from page 102*)

as a machine for converting gasoline into tractive force would remain as determined by the laboratory tests. These tests may be denominated consumer's tests.

The first and its corresponding field experiments will have determined the overall efficiency of the tractor as a power machine.

The second will differentiate these results and distribute the various factors of efficiency or inefficiency throughout the machine. It will serve to separate the performance of the engine from that of the transmission and other moving parts, and in so doing will perhaps be more satisfactory to the manufacturer. This we may denominate a producer's test.

# EMPLOYEES' REPRESENTATION IN MANAGEMENT OF INDUSTRY

The Economic Questions Involved, as Summarized by Dr. Royal Meeker, U. S. Commissioner
of Labor Statistics, at Conference on Industrial Relations Held at
Chicago, Together with Other Discussion

THE annual convention of the American Economic Association was held in Chicago on December 29 to 31 concurrently with those of the American Association for Labor Legislation and the American Sociological Society. The question of industrial unrest formed the topic of a great number of the addresses in the three programs. Particularly at a joint session of the Association for Labor Legislation and the Economic Association the discussion of this question was brought to a focus by Dr. Royal Meeker, United States Commissioner of Labor Statistics, who, in his address on Employees' Representation in the Management of Industry, asserted that neither the English Whitley scheme nor the various plans which have been introduced in American industries constitute anything more than an initial experiment in the direction of a solution of industrial unrest.

In Great Britain the Whitley system of organizing industries into national and district joint industrial councils and works committees was adopted at the instance of the Government. Up to date, national joint industrial councils have been set up in some fifty industries, besides the councils established in the Government departments for both manual and clerical employees. These national joint councils, according to Dr. Meeker, are analogous to the roof of a house suspended in mid-air with no supporting side walls or foundations upon which to rest. Only three industries have set up complete Whitley systems of joint industrial representation of employers and employees with a national council and district councils for the industry and works committees for individual shops, and these three industries are relatively insignificant, namely, match, rubber and pottery manufacturing.

Dr. Meeker divided the different types of shop committees and work councils in this country into three distinct groups: (1) closed-shop committees of union workers chosen exclusively by union men affiliated with national or international unions; (2) open-shop committees composed of workers chosen by the votes of all workers who have been employed in the shop a required length of time; (3) open-closed-shop committees chosen by all workers qualified to vote in a shop that is closed against trade unions and which may or may not have a local plant or corporation organization of the workers. There is, in fact, little fundamental variation, he said, in the published statements of the objects sought and the plans of organization as between group and group.

Comparing the different views of Americans and of the British, Dr. Meeker pointed out that respecting labor organization the position of organized employers and employees is exactly reversed in Great Britain. We today are where the British were about thirty years ago. The question of national unions versus plant unions was fought out in Great Britain and won by the workers. British employers were obliged to accept the result and bargained collectively with the representatives of the national unions. During the war the workers rebelled against this system, insisting that wages, hours and shop conditions should be negotiated for each shop by the local shop committees. The employers stood up valiantly for the established order, and insisted that they would have nothing to do with local shop committees, but would bargain collectively only with truly responsible and representative bodies, the executives of the national trade unions. American employers, equally valiant for the established order, will have nothing to do with irresponsible, unrepresentative officials of national trade unions and insist on bargaining collectively with representatives of the workers who know the local situation and who are chosen from the shop where a dispute is pending. One of the biggest questions to be settled, in Dr. Meeker's opinion, is whether employees' representation is to be local and under the direct control and domination of the employer, or whether it is to be nation- or world-wide and under the control of the workers themselves. or whether the general public

will insist on being a party to every collective agreement so as to prevent the employers and employees from agreeing too agreeably and charging the bill to the ultimate consumer.

If workers are to be admitted to actual participation, Dr. Meeker suggested they must participate to some extent in the losses as well as the gains of industry. A practical method of payment would be to guarantee for each position a minimum wage which must be paid regardless of any losses which the business may suffer. In addition to the minimum, a bonus should be provided varying according to the contribution of the workers in cutting down labor costs, in reducing costs of management, in decreasing spoilage of material, in decreasing wear and tear on machines and tools, in improving quality of product, in increasing business, or in any other way. This would obviate the objection to most bonus schemes that the worker is penalized or rewarded for the inefficiency or the good judgment of the managers.

Labor unrest, in the present-day meaning of the term, is, in Dr. Meeker's opinion, a natural and inevitable result of the industrial revolution, machine production, absentee ownership of industrial establishments, and the centering of industrial management in the hands of managers of finance whose offices are in the big banks and office buildings of the centers of finance and trade. Strikes in the modern sense have practically unknown until industrial units had grown to such a size as to erect formidable barriers between the workers on the one side and the responsible owners and managers on the other. Lack of interest in work, he said, grows out of absentee ownership, The absent industrial landlords, interested only or principally in dividends, have employed experts, scientific managers, to produce a substitute for the old-time workman's interest in his work. The theoretical ideal of maximum output with minimum expenditure of effort which can be figured out by mathematical formulæ and pictured on charts, is never attained in practice. A much more imperfect layout, which leaves much to the ingenuity and initiative of the individual workmen, in practice always achieves better results.

The present-day movement for industrial democracy, Dr. Meeker continued, is a partial recognition of the fundamental psychological phenomenon that industrial fatigue is not simply an engineering question to be stated mathematically in foot-pounds per hour or even a physiological question having to do with calories used up in the body. Work is hard primarily because it is uninteresting and monotonous, or easy because it demands ingenuity or skill. Paradoxical as it seems, the way to make work easier is to make it harder by requiring more of the workmen. In so far as scientific management has resulted in merely breaking processes up into component parts, segregating so far as possible the purely muscular and mechanical operations from the creative and planning functions, so-called "efficiency" has resulted in the most disastrous inefficiency.

Willard E. Hotchkiss, Chicago. took issue with various of Dr. Meeker's remarks and maintained that it has not been established as a fact that the workers object to repetitive jobs. Economists, he stated, have created a complete literature on the labor problem which is based on *a priori* conclusions. In his opinion little progress will be made in the direction of a solution of our labor difficulties until principles are worked out by inductive reasoning.

Prof. Ira B. Cross, University of California, raised the question whether the introduction of works councils did not constitute a dangerous economic policy. He cited the circumstance which attended the introduction of the works-council plan in a large American industry. The workmen, who were doubtful as to the advantages of the scheme, asked the advice of radical labor leaders and were told to support it on the theory that eventually it, in conjunction with similar councils in other industries. would form the basis for establishing a soviet form of government.

Prof. Frank W. Blackmar, University of Kansas, in an address on A Working Democracy, emphasized that neither an autocracy of capital nor an autocracy of labor can be permanently successful. A similar idea was discussed by A. W. Small, University of Chicago, who, in a paper on Democracy and Labor, read before the Sociological Society, described "laborocracy" as being as objectionable as any other kind of autocracy, and as inevitably doomed.

An optimistic view of the labor situation was presented by Matthew Woll, vice-president of the American Federation of Labor. He addressed the Sociological Society on Labor and Democracy, and the Economics Association on Labor and Immigration Questions. He asserted that the bulk of organized labor is loyal to America and its institutions and observed that while there have been a few spectacular labor disputes, like the steel strike and the coal strike, there have also been thousands of 'readjustments of wages and working conditions through direct negotiations between employers and organized labor without strikes.

John A. Fitch of The Survey, and S. K. Ratcliffe, a well-known English journalist, delivered interesting talks at a joint session of the Association for Labor Legislation and the Sociological Society, on the general subject of Democracy and Industrial Life.

Another general subject which was given much attention was the prospect for international labor legislation.

A more extensive account of these conventions is given in Iron Age for January 8, 1920, from which these notes have been prepared.

# GOVERNMENT ACTIVITIES IN ENGINEERING
*Notes Contributed by The National Service Committee of Engineering Council[1]*

## A New Board of Surveys and Maps

AN executive order recently issued by President Wilson creates a new Board of Surveys and Maps, the purpose of which shall be to coördinate the activities of the various map-making agencies of the Government, to standardize results and to avoid unnecessary duplication of work. This order is the outcome of a suggestion which Engineering Council made to President Wilson in a letter dated July 1, 1919, which pointed out that the official map work of the United States was being prosecuted by twelve separate and distinct Federal agencies, the majority of which are making maps for special purposes, and which are of little value for any other purposes. Furthermore, some states, counties, cities, and private organizations are also engaged in this work and since each of the Federal agencies and the others above mentioned follows methods developed in its own surveys, the result is precisely that which might be expected. There is no agency in the United States which has authority to coördinate the work.

The Council therefore advised the creation of a "Board of Surveys and Maps" consisting of a representative from each of the present Federal map-making agencies, together with representatives from well-qualified map-using agencies, which would be vested with authority to work out a plan of standardization and coördination of the work of the Government.

This letter was referred by the President to the Secretary of War with the request that a conference of representatives of the twelve map-making agencies named by Engineering Council be called to consider the suggestion in its several practical aspects. Representatives of these agencies and of the Forest Service and U. S. Hydrographic Office accordingly held a series of conferences at Washington, D. C., from September 15 to 29, inclusive, and the following engineering and scientific societies were invited to send representatives to the meetings to present their needs:

Engineering Council
The American Society of Civil Engineers
The American Institute of Mining and Metallurgical Engineers
The American Society of Mechanical Engineers
The American Institute of Electrical Engineers
The American Association of State Geologists
The American Association of State Highway Officials
The National Research Council
The Association of American Geographers
The American Geographical Society
The Geological Society of America
The National Geographic Society

Ten of these organizations were represented at the conference on September 22, but since the time available was not sufficient to formulate and present their views during the conference, it was

agreed that the above-named organizations would present their recommendations to the permanent body proposed by the conference.

The report of the conference, submitted to the President through the Secretary of War on September 30, 1919, includes a general survey of the subject; a statement of the map needs and of the map-making activities of each of the fourteen organizations represented at the conference; the practical program outlined by the Interdepartmental Committee on Aerial Surveying; suggestions for consideration by the proposed Board of Surveys and Maps; and estimates for the completion of primary control of the United States by the U. S. Coast and Geodetic Survey and for the completion of the topographic mapping by the U. S. Geological Survey.

The Board of Surveys and Maps, according to the executive order, will be composed of representatives of the fourteen organizations represented at the conference. The order further provides that the individual members of this Board shall be appointed by the chiefs of the various organizations named and shall serve without additional compensation.

The Board is also directed to make recommendations to the several Departments, or to the President, for the purpose of coördinating all map-making and surveying activities of the Government and to settle all questions at issue between executive departments relating to surveys and maps insofar as their decisions do not conflict with existing laws.

The Board is authorized to establish a central information office in the U. S. Geological Survey for the purpose of collecting, classifying and furnishing to the public information concerning all map and survey data available.

A permanent organization of the Board was effected on January 13 when representatives from each of the Departments met and elected O. C. Merrill, Chairman (Chief Engineer, Forest Service, Department of Agriculture); William Bowie, Vice-Chairman (Chief, Division of Geodesy, Coast and Geodetic Survey, Department of Commerce); and Colonel C. H. Birdseye, Secretary (Chief Geographer, Geological Survey of the Department of Interior).

It is hoped that this concentration of mapping work will permit larger proportionate appropriations for the country's mapping program under the direction of the Geological Survey. In bringing this matter to the attention of the President, Engineering Council realized the great value of making reliable maps of the whole country available to every branch of the engineering profession in the simplest and most direct way. This will be accomplished through the central information office provided in the Geological Survey to receive and distribute Government and public map data. Thus, if proper appropriations can be obtained and put to work under the Geological Survey, the engineers of the country as well as the Government departments will be greatly benefited.

In this connection it is pertinent to state that the National Service Committee of Engineering Council in coöperation with all state geologists are urging prominent engineers in every congressional district to bring pressure to bear upon their congress-

---

[1] Engineering Council is an organization of national technical societies created to consider matters of common concern to engineers as well as those of public welfare in which the profession is interested. The headquarters of Engineering Council are located in the Engineering Societies Building, 29 West 39th Street, New York City. The Council also maintains a Washington office with M. O. Leighton, chairman of the National Service Committee in charge. This office is in the McLachlen Building, 10th and G Streets, Washington, D. C. The officers of Engineering Council are: J. Park Channing, Chairman; Alfred D. Flinn, Secretary.

men to vote sufficient funds for topographic mapping. This has been accomplished by selecting a prominent engineer—members of the Founder Societies—in each congressional district and urging them by direct appeal to request their senators and representative to support a $600,000 appropriation for topographic mapping.

The state geologists have coöperated by sending these engineers complete information of the local conditions, so that they may be thoroughly conversant with their " back-home " requirements, and thus be more forceful with their men in Congress. Further, most of the state geologists are endeavoring to get increased state appropriations for this important work, which will have to be matched by an equal Federal appropriation in most cases before they become available for expenditure.

The estimates submitted by the Secretary of the Interior call for an increased mapping program to cost $600,000 annually, instead of $500,000 which was last year's estimate. It will be recalled that the last Sundry Civil Appropriation Bill provided $375,000 for topographic mapping, which was less by $50,000 than prewar annual appropriations for this work. This year's estimate contemplates a program that will complete topographic mapping in twelve years instead of the one hundred years which will be required if the old appropriation is not increased.

## Water Power Bill Passes Senate

Although it was stated in the last issue of MECHANICAL ENGINEERING that the Senate, during its next session, would take up, and possibly dispose of, several bills of interest to engineers, such action has not, up to the time of going to press, taken place.

The Senate, upon reconvening on January 5, after the usual holiday session, first devoted its attention to the Water-Power Bill, discussion of which lasted nearly a week for the debate both on the bill itself and on many of its amendments was very active. The bill finally passed the Senate with a majority of three to one, substantially as passed by the House, except as to certain conditions of renewal to take effect after 50 years' license. In reporting the bill to the Senate the Committee explained and clarified the definitions of " reservations," " navigable waters " and " costs." A proportionate license fee not exceeding twenty-five cents per horsepower is proposed, which is to defray costs of the administration of the act. Fair rentals are to be charged for Government dams and other Government property on water-power sites. The bill also provides that in case the United States does not exercise the right to take over the works, the license tendered shall be " on reasonable terms " and " which is accepted " by the original or the new licensee. This provision was thought necessary to make it certain that capital would invest under the law. A provision for severance damages also appears in the bill.

This bill now goes to a committee composed of members from both the House and Senate. This is for the purpose of settling on points of difference. The Senate is represented by Senators Jones, Nelson, Smoot and Fall (Republicans) and Bankhead and Myers (Democrats). The House conferees have not as yet been appointed. It is not expected, however, that further delay will be occasioned, and that not more than two weeks will be occupied in getting the bill through this Joint Conference Committee.

## Technical Work at Bureau of Standards

Bulletins dated January 9 on the progress of experimental work that has been started at the Bureau of Standards are as follows: (1) Accuracy of Electrical Measuring Instruments; (2) Electron Tubes in Radio Telephone Transmitting Sets; (3) Recent Radio Publications; (4) Industrial Safety Standards; (5) Measurement of Thermal Expansion of Various Materials; (6) Service Tests of Concrete Floor Treatments; (7) Consistency and Time of Set of Next Cement Determined by New Method; (8) Slag as an Aggregate for Concrete; (9) Flat-Slab Investigation; (10) Manufacture of Automobile Crankshafts; (11) Investigation of Electric Welding; and (12) of Chilled-Iron Car Wheels.

Because of the current nature of these investigations the results are not generally put in printed form, but complete data on the results of the work to date are always available on application to the Bureau of Standards or the National Service Committee in the Washington office of Engineering Council.

## Colonel Beach New Chief of Engineer Corps

Colonel Lansing H. Beach, U. S. A., has been appointed Chief of Engineers, replacing General William M. Black, retired. Colonel Beach's appointment came as a complete surprise to many, as he had not been publicly mentioned for the place. He is a native of Iowa and a graduate of West Point and has held the rank of Colonel since 1913. While Colonel Beach is best known for the work he has done in connection with Mississippi River projects, he has had the varied service which comes to all Engineer Corps officers who have spent most of their lives in its service. He has done extensive engineering work on Great Lakes projects and was in charge of the important improvements of the harbors of Baltimore and Jacksonville. More recently he has been stationed at Cincinnati in charge of the construction of the system of locks and movable dams being constructed in the Ohio River. Colonel Beach has been prominently identified with the development of inland waterways, as a member of the Mississippi River Commission and as a member of the Board of Engineers for Rivers and Harbors and as an advisor to the Waterways Division of Railroad Administration.

## TAYLOR SYSTEM IN GERMANY

The following advertisement of a firm of efficiency engineers taken from a German trade paper, gives a pithy expression of the benefits claimed for scientific management. The translation of the advertisement is also given in similar form.

---

**Für Arbeitgeber und Arbeitnehmer:**

Erhöhung des Einkommens und der Leistungsfähigkeit. Einschränkung jedes überflüssigen Kräfteverbrauchs.

Der einzige Weg zur Erhöhung der Löhne bei Verminderung der Produktionskosten.

**Für die gesamte Wirtschaft:**

Erhöhung der Produktion mit den jetzigen Einrichtungen.

Qualitäts-Erzeugung und Steigerung der Leistungsfähigkeit bei Verminderung der Erzeugungskosten. Erhöhung des tatsächlichen Arbeitswertes durch ergiebigere Ausnutzung aller aufgewendeten Mittel: Zeit, Kraft, Material und Geld.

**Für den Wiederaufbau Deutschlands die Rettung!**

**Das ist das Taylor-System!**

---

The Taylor System Means

**To the Employer and Employee:**

Higher Income and Efficiency

Reduction of Unnecessary Consumption of Effort

**The Only Way to Raise Wages and at the Same Time Reduce Cost of Production.**

**For the Entire Industrial System:**

Increase of Production With Now Available Facilities

Quality Production and Increase of Efficiency at Lower Cost of Production

Increase of the Actual Value of Work by a More Efficient Utilization of Employed Means, Namely, Time, Effort, Material and Money.

**For the Reconstruction of Germany—Salvation.**

# NATIONAL DEPARTMENT OF PUBLIC WORKS ASSOCIATION
## HOLDS SECOND CONVENTION
### Engineers Urge Congress To Pass Jones-Reavis Bill Consolidating Federal Public Works

NATIONAL interest in the bill now before Congress to replace the present Department of the Interior by a National Department of Public Works, and thus bring under one bureau all the engineering activities of the Government, was greatly intensified by a two-days' conference of the National Department of Public Works Association held at the Willard Hotel, Washington, D. C., January 13 and 14. Ninety-five delegates representing the leading engineering societies and organizations of the country, with a total combined membership of nearly 90,000, took part in the discussions and proceedings of the conference.

Four sessions were held. These were chiefly given over to reports of officers and committees and discussion of the Public Works Department bill now before Congress. A communication from Governor Lowden of Illinois dealing with the organization of the public works in that state was presented, and Representative Reavis, one of the authors of the bill, and Brigadier-General R. C. Marshall, chief of the Construction Division of the United States Army, also addressed the convention.

General Marshall pointed out that there are at present 27 separate and distinct Federal agencies engaged in the construction of public buildings, 18 in building roads, and 18 in work relating to hydraulics, river and harbor work. He was accordingly of the opinion that a Department of Public Works could accomplish a most constructive step in the history of Government work by placing all Federal construction under control of a single bureau with a single method of accounting and a single bureau of purchase.

M. O. Leighton, chairman of the Executive Committee, opened the convention by briefly reviewing the history of the Association since the inception of the movement by Engineering Council at a meeting held last April. As originally planned the Association was made up solely of technical men, but the movement has since become so general that its supports are now convinced that it should have active backing from those outside the profession. Chairman Leighton expressed these sentiments in his opening address as follows:

We here assembled are asking Congress to do the obvious thing. Give the Government a business-like organization. Coördinate the functions so that the processes of Government business shall dovetail. Cut out the wastes and the duplications. Abolish the rivalry between departments. We advocate a Department of Public Works not merely to secure technical symmetry in our Federal organization, desirable as that and may be, but our advocacy in its essential features is an attempt to stop some of the waste.

When this organization was set up at Chicago I think that none of us—certainly not the speaker—had an adequate idea of the scope of the movement. We saw a loose and inefficient public works organization, divided and subdivided into many different provinces. As technical men we knew how organically wrong that was. Of the wastes and inefficiencies we were well aware. With the necessity for a coördinated structure by virtue of which the technical and semi-technical fields of the Government could be rendered efficient and business-like we were profoundly impressed. But that our effort, our idea, our legislative bill would become the cornerstone of a structure embodying efficiency on all Departments of Government we could hardly foresee. As an organization our effort is still focused on a Department of Public Works and that alone. But we realize that with that principle established—that example set—reform in other provinces of Government business activity will occur by the mere logic of events. We are pioneers.

This is the reason why our project for a Department of Public Works strikes straight home to the business man, the manufacturer, the contractor, and the merchant, all of whom are represented here today. The technical men who met at Chicago last April to set up this organization built better than they knew. While the project retains all the virtues that appealed to us when it was launched—of technical excellence, of rational Government organization, of economy and efficiency, we now see that it reaches to national and to business prosperity, to the fiscal welfare of the nation, to the individual welfare of the productive business.

We may as well recognize and admit the fact that individual initiative in America will not continue to be that spontaneous thing that it has been in the past if the rewards are to be divided and a part of them are to be wasted in the support of a chaotic Government business

organization. The burdens of every business organization in the country are magnified by Government business inefficiency. We want no good and proper thing to be withheld from our Government nor in turn by our Government.

References to the efforts of engineering organizations to create a Department of Public Works have frequently been made in the columns of MECHANICAL ENGINEERING. The bill which creates this new Department is commonly known as the Jones-Reavis Bill, and the details of the changes which it proposes in the executive machinery of the Government were discussed in the August issue, page 709. Briefly the bill creates a Department of Public Works which is to replace the Department of the Interior and thus assemble under one head all the engineering activities of the Government. Such bureaus of the Department of the Interior as are of a non-engineering character will be transferred to existing appropriate departments.

The first convention in this movement was held in Chicago last April, and since that time various state committees have been at work within their own districts. A roll call of these various state committees held during the convention showed that considerable effective work has been done, chiefly, however, in the southern and western states. Two states reported that their entire congressional representation had pledged their support of the bill. The eastern states, and particularly those which represent the large industrial centers have, however, made but little progress toward obtaining congressional support. This condition of affairs was not altogether expected, and in the opinion of many, considerable effort will be necessary to overcome the opportunities thus lost.

Following the roll-call the Jones-Reavis Bill itself was discussed. The chief objections centered around Section 3 which provided for the transfer of the Bureaus of Education, Pensions, Indian Affairs and other similar non-engineering bureaus of the Department of the Interior to existing department of the Government. In order to place before the public its position on this and other controversial section of the bill the convention adopted the following resolution defining its position:

1 The National Public Works Department Association must, under its articles of organization, confine its efforts solely to the creation of a Department of Public Works.

2. Section 3 of H. R. 6649-S. 2232 (the Jones-Reavis Bill) has nothing to do with public works. The only reason for its presence in the bill is that some departmental disposition must be made of the non-engineering bureaus in the Interior Department.

3 It is manifest that the numerous organizations which become affiliated under the name of National Public Works Department Association for a common public-works purposes do not, by so doing, commit themselves with respect to any other question, national or otherwise. Therefore each affiliated organization may have and is entitled to express its views as to said Section 3 of the bill without in any way qualifying its approval of the remainder of the bill.

4 The National Public Works Department Association, as such, is therefore unable either to approve or to disapprove the specific assignments of bureaus in the said Section 3, and in advocating the bill before committees of Congress the agents of this Association are to be instructed to present this fact in an unmistakable way.

5 The National Public Works Department Association suggests to all affiliated organizations that they present to Congress their own individual views with respect to said Section 3, either through the legislative agents of this Association or otherwise.

6 The officers and directors of this Association are instructed to have the foregoing statements printed in suitable form and circulated as one of the campaign documents issued in support of this movement.

The second day of the conference was largely devoted to interviews with senators and congressmen, committees representing the several states having been appointed for this purpose the day previous. Reports of these interviews indicate that the proposed bill will doubtless be favorably reported, although it must not be inferred that success is altogether assured. On the contrary, the task of securing sufficient support to finally pass the bill has only just begun, and an undue optimism at this time may result in failure. The widest publicity possible should accordingly be given the measure and every engineer should make it a personal matter to lend his support to this exceedingly important bill.

The Committee on New Organization, of which Calvin W. Rice, Secretary of The American Society of Mechanical Engineers, is chairman, rendered a report in which it recommended that the scope of the appeal of the association be enlarged to include the business, professional and social bodies and that all such activities be accorded executive representation in the movement.

The Committee also suggested that there be a committee of the Association in every state; that each state committee use every existing organization which is in any way enlisted by this movement.

The Committee stated that while the present organization of the Association has been adequate and admirable up to the present time it should nevertheless be enlarged as indicated. What it needs is popular support in effort and in subscription.

The knowledge that the present officers have of the needs of the work should be accepted and such expansion should be granted as the officers and Executive Committee may recommend.

During the final session of the convention the financial needs of the Association were discussed at some length, and it was made very evident to those present that, if the Association was to continue, this issue must be met and settled. Heretofore the various member societies of the Association have contributed but little, the expenses of the work being largely met by individual subscriptions. The Committee recommended that each society be asked to assess its members at the rate of one dollar per man.

# ENGINEERING FOUNDATION REPORTS PROGRESS IN RESEARCH

THE second formal publication of the Engineering Foundation, scheduled to appear in October but delayed by the recent printers' strike in New York, has just been issued. Engineering Foundation is one of the three departments of the United Engineering Society, and was made possible by the presentation to that organization of a sum of money as a trust fund, the interest of which is to be used in the promotion of engineering research. This gift came from Ambrose P. Swasey, past-president and honorary member of The American Society of Mechanical Engineers. In 1914, in fulfilment of his desire to advance the work of the engineer and add to the accumulated facts which serve to guide him in his undertakings, he gave the United Engineering Society $200,000, the income of which should be used "for the furtherance of research in science and engineering, or for the advancement in any other manner of the profession of engineering and the good of mankind." In September 1918 Mr. Swasey added $100,000 to the endowment, making the present total $300,000.

The organization meeting of the Engineering Foundation Board was held on April 15, 1915. As at first constituted, the Board had eleven members, elected by United Engineering Society and each serving for a term of three years or less. By amendment of the By-Laws of United Engineering Society, April 25, 1918, the Board was enlarged to sixteen members, and by this amendment an executive committee of five members was also provided.

The Engineering Foundation Board is charged with the responsibility of administering funds placed at its disposal. It supports no project the merits of which have not been demonstrated by competent inquiry. In the development of such inquiry it is fortunate in possessing the coöperation of the National Research Council, an organization having world-wide scientific affiliations. While the Engineering Foundation Board reserves to itself the right to choose, initiate, and conduct under its own immediate direction such researches as may commend themselves to its membership, it is at the present time making extensive use of certain channels established by the National Research Council, notably the Council's Division of Engineering.

The National Research Council is an organization of American scientists, engineers and educators, established in April 1916 under the Congressional charter of the National Academy of Sciences, given in 1863. It comprises representatives of national scientific and technical societies, chiefs of technical bureaus of the Army and Navy, heads of governmental bureaus engaged in scientific research, representatives of other research organizations, and other persons whose aid may advance the objects of the Council. The Council is organized in thirteen divisions of two classes, six dealing with the more general relations and activities of the Council and seven with science and technology. The Division of Engineering, under this second class, includes twenty-eight members, at least five of whom are members of the Engineering Foundation, and one of whom is the Chairman of that body. A more detailed history of the National Research Council will be found in MECHANICAL ENGINEERING for July 1919, p. 631.

In June 1916 the Engineering Foundation Board decided to devote its resources to the National Research Council for one year, beginning September 1916. The close coöperation between the two organizations which resulted from this decision was found to be so mutually helpful that at a meeting of the Foundation Board in September 1917 it was resolved "to continue the coöperation between the two bodies in all practicable ways that may be now, or may become, mutually available, such ways consisting of the interchange of helpful suggestions, advice, information, office representation and similar facilities, and in addition a recognition of community of purpose that shall promote in the field of engineering research increasingly intimate relations between engineering and science."

In total, sixty members of the Founder Societies are members of National Research Council, and of these, eight are members of the Foundation Board. The total membership of the National Research Council is about 175. One member of the Engineering Foundation Board and its secretary, and 31 additional members of the Founder Societies, are members of the National Research Council in divisions other than the Division of Engineering and on its Executive Board. Engineers have therefore an influential position in the Council and a large measure of responsibility for its success.

The Engineering Division of the Council has 18 committees, each charged with a specific line of research and together covering a wide variety of subjects of interest to engineers and the industries. It is with this division, as stated before, that the Engineering Foundation is at present most closely associated.

In August 1919 the Executive Committee of the Foundation took action to support preliminary investigations in a number of subjects recommended by the Division.

Research in fatigue phenomena of metals is to be conducted in the laboratories of the Engineering Experiment Station of the University of Illinois under the direction of Prof. H. F. Moore. The Experiment Station shall have the right to publish the results in full as a bulletin of the station, but in addition shall prepare a brief, comprehensive statement suitable for publication in the journals of the engineering societies. A progress report of the Committee on Fatigue Phenomena of Metals, which summarized the facts and theories relating to fatigue failure appeared in the September 1919 issue of MECHANICAL ENGINEERING, p. 731.

From approximately fifty suggestions made to date, the Engineering Foundation Board has selected for investigation six subjects. These are: (1) wear of gears, (2) spray camouflage for ships, (3) directive control of wireless communication, (4) weirs for measurement of water, (5) establishment of a testing station for large water wheels and other large hydraulic equipment, (6) mental hygiene of industry. Investigations 1, 3, 4 and 6 are in progress; the other two have been completed.

An investigation concerning existing industrial research laboratories in the United States has resulted in assembling and classifying some information about each of approximately 250 laboratories connected with the industries, which give whole or part time to industrial research and development. This information is being prepared for publication under the joint auspices of the Engineering Foundation and the Division of Industrial Relations of National Research Council.

# CLASSIFICATION AND COMPENSATION OF ENGINEERS

## Engineering Council Approves Final Report of Committee

IN April 1919 Engineering Council organized a Committee on Classification and Compensation of Engineers, whose object it was to secure better conditions of employment for engineers by classifying engineering positions into several grades, and determining the relative salaries of each. The Committee decided to confine itself to a consideration of engineers engaged in railroad, state, county and municipal, and federal activities and a main committee with three sub-sections was accordingly appointed with chairmen and personnel as follows:

COMMITTEE ON CLASSIFICATION AND COMPENSATION OF ENGINEERS:
Arthur S. Tuttle, *Chairman*, Deputy Chief Engineer, Board of Estimate and Apportionment, New York, N. Y.
Francis Lee Stuart, Consulting Engineer, New York, N. Y.
John C. Hoyt, Chief, Division of Water Resources, U. S. Geological Survey, Washington, D. C.
Charles Whiting Baker, Consulting Editor. *Engineering News-Record*, New York, N. Y.
M. O. Leighton, Chairman, National Service Committee of Engineering Council, Washington, D. C.

*Railroad Section*:
Francis Lee Stuart, *Chairman*
Frank H. Clark, Consulting Engineer, New York, N. Y.
Bion J. Arnold, Consulting Engineer, Chicago, Ill.

*State, County and Municipal Section*:
Arthur S. Tuttle, *Chairman*
M. M. O'Shaughnessy, City Engineer, San Francisco, Cal.
F. W. Cappelen, City Engineer, Minneapolis, Minn.

*Federal Section*:
John C. Hoyt, *Chairman*
John S. Conway, Asst. Commissioner of Lighthouses, Washington, D. C.
Oscar C. Merrill, Chief Engineer, Forest Service, Washington, D. C.

From time to time during the year partial and preliminary reports dealing with the work of the sub-committees have been submitted and printed in MECHANICAL ENGINEERING. The main committee has now completed its investigation and submitted its final report to Engineering Council and the complete and definite classification for engineers in federal, state, county and municipal and railroad service was approved by Engineering Council at its meeting on December 18, 1919.

The extracts from this report which follow include a classification of engineers into eight grades, five professional and three sub-professional, and a statement of the necessary qualifications for each; a comparison of the results of this latest investigation with those reported in 1917 by a committee of the American Society of Civil Engineers; and a summary of the returns from the questionnaires, with average ages of incumbents, average salaries in federal, state and municipal service and average salaries recommended by the engineers in charge.

### FINDINGS AND CONCLUSIONS OF THE COMMITTEE

The investigations have shown the lack of any adequate or consistent employment policy with respect to professional services. This is evidenced by the following conditions which are believed to be largely responsible for the unsatisfactory status of men engaged in this class of work:

1 Absence of any uniform system of grading of positions
2 Lack of uniformity in titles of positions with respect to duties
3 Inequalities in compensation for positions of the same grade
4 Generally inadequate compensation for services rendered.

To the end that these conditions may be corrected and proper and equitable conditions of employment established, the following principles and practices are recommended by the Committee, though not yet acted upon by Engineering Council.

1 Positions should be classified in accordance with the type of work, and with the character of the duties to be performed and the qualifications necessary for their performance, as indicated by a system of grading.

2 Within the salary limits fixed for each grade, there should be a system of advancement through the grade based upon experience gained in the position and upon proof of increase in the proficiency of the employe in performing the duties of the grade.

3 Promotions from grade to grade should depend upon the existence of a vacancy in the higher grade and proof that the employe is qualified to fill the vacancy.

4 The determination of salary adequate to procure for and retain in engineering work a high class of employes, should take into account and properly weigh the following considerations:

a The capital invested, both in money and in time, in obtaining the requisite fundamental training
b The amount and character of experience and the degree of personal ability required
c The relative value of the classes of work to be performed
d The amount paid for similar service in other lines of work
e The amount necessary to enable the employee to maintain a standard of living commensurate with the general standards of the community for positions of similar dignity and responsibility.

5 In the interest of an adequate social policy, no position likely to be occupied by individuals of an age to assume family responsibilities should fail to pay an amount sufficient to permit the maintenance of the average family in reasonable decency and comfort.

6 In the interest of the employes as a whole and of the employer, a system should be established by which employes who fail to maintain satisfactory standards of service should be removed, transferred, demoted, or retired as may be equitable in the circumstances.

### CLASSIFICATION OF ENGINEERING POSITIONS IN FEDERAL, STATE, COUNTY, MUNICIPAL AND RAILROAD SERVICES

The grades proposed, which appear to be well adapted to use not only in all of the services represented but also for all other forms of engineering activities, have been divided into two classes, the Professional Service being deemed to include men who have received an engineering degree from an educational institution of recognized standing or who have obtained similar qualifications through practice of the profession and by mastering the fundamentals of engineering science, while the Sub-Professional Service includes assistants with at least a high school education, who enter upon the practice of the profession in the performance of responsible duties for which an engineering training is not essential, but who through experience and study may fit themselves for the higher grades. The classification is as follows:

#### PROFESSIONAL SERVICE

##### GRADE 1 CHIEF ENGINEER

*Duties*: To act in chief administrative charge of a technical organization, or of a main division thereof; to determine the general policies of the organization under the limitations imposed by law, regulation, or other fixed requirement; to have final responsibility for the preparation of reports, cost estimates, designs, and specifications and for the construction, maintenance, or operation of engineering works or projects; to have full charge of the collection and presentation of data for and the conduct of valuation proceedings; to conduct or direct the most comprehensive lines of engineering research.

*Qualifications*: Training and experience of a character to give substantial evidence of engineering knowledge and ability or of executive capacity of highest order along lines of work similar to those involved in the position to be occupied and of at least twelve years' duration, of which at least four years shall have been spent in duties of Engineer, or their equivalent, and at least five years in responsible charge of important work or projects. Fundamental training equivalent to that represented by professional degree granted upon the completion of a standard course of engineering instruction in an educational institution of recognized standing or, in absence of such degree, at least four years of additonal experience. The completion of each full year of such standard course shall be considered the equivalent of one year of such additonal experience.

## GRADE 2 ENGINEER

*Duties:* Under general administrative direction and within the limits of the general policies of the organization, to have responsible charge of and to initiate and determine policies for a major subdivision of an organization: to prepare for final executive action reports, cost estimates, designs, specifications, and valuation studies and data: to have immediate charge of the construction, maintenance, or operation of engineering works or projects of major importance: to conduct or direct major lines of engineering research: or to furnish for executive action expert or critical advice on engineering works, projects or policies.

*Qualifications:* Active professional practice or executive charge of work for at least eight years, of a character to demonstrate a high degree of initiative and of ability in the administration, design, or construction of engineering work or projects of major importance, of which at least three years shall have been spent in duties of Senior Assistant Engineer, or their equivalent, and at least three years in responsible charge of work. Fundamental training equivalent to that represented by professional degree granted upon the completion of a standard course of engineering instruction in an educational institution of recognized standing or, in absence of such degree, at least four years of additional experience. The completion of each full year of such standard course shall be considered the equivalent of one year of such additional experience.

## GRADE 3 SENIOR ASSISTANT ENGINEER

*Duties:* Under general administrative and technical direction, to be in responsible charge of an intermediate division of an organization: to exercise independent engineering judgment and assume responsibility in studies and computations necessary for the preparation of reports, cost estimates, designs, specifications, or valuations: to have immediate charge of the construction, maintenance, or operation of important engineering works or projects: or to conduct or direct important lines of engineering research.

*Qualifications:* Active professional practice or executive charge of work for at least five years, of which at least three years shall have been spent in duties of Assistant Engineer, or their equivalent, with at least one year in responsible charge of work. Fundamental training equivalent to that represented by professional degree granted upon the completion of a standard course of engineering instruction in an educational institution of recognized standing or, in absence of such degree, at least four years of additional experience. The completion of each full year of such standard course shall be considered the equivalent of one year of such additional experience.

## GRADE 4 ASSISTANT ENGINEER

*Duties:* Under specific administrative and technical direction, to be responsible for the conduct of the work of a minor subdivision of an organization; to collect and compile data for specific items of engineering studies; to take immediate charge of field survey projects and of the design and construction of minor engineering work; to lay out and develop work from specifications and to supervise the work of a drafting or computing force; or to conduct specific tests or investigations of apparatus, material, or processes.

*Qualifications:* Experience for at least two years in duties of Junior Assistant Engineer or their equivalent. Fundamental training equivalent to that represented by professional degree granted upon the completion of a standard course of engineering instruction in an educational institution of recognized standing or, in absence of such degree, at least four years of additional experience. The completion of each full year of such standard course shall be considered the equivalent of one year of such additional experience.

## GRADE 5 JUNIOR ASSISTANT ENGINEER

*Duties:* Under immediate supervision, to perform work involving the use of surveying, measuring, and drafting instruments: to take charge of parties on survey or construction work; to design details from sketches or specifications: to compute and compile data for reports or records: to inspect or investigate minor details of engineering work: or to perform routine tests of apparatus, material, or processes.

*Qualifications:* No experience required other than that involved in securing a professional degree upon the completion of a standard course of engineering instruction in an educational institution of recognized standing; but in absence of such degree, a high school education or its equivalent is required and at least four years' experience in the use of surveying, measuring or drafting instruments, or the computation and compilation of engineering data, together with evidence of a knowledge of the fundamentals of engineering science sufficient, with further experience, to qualify for the higher professional grades. The completion of each full year of such standard course of engineering instruction shall be considered as the equivalent of one year of experience.

## SUB-PROFESSIONAL SERVICE

### GRADE 6 SENIOR AID, OFFICE

*Duties:* To supervise the plotting of notes and maps, and to direct the work of a drafting or computing squad.

*Qualifications:* Experience for at least five years in tracing, lettering, drafting, and computing, of which at least three years shall have

been spent in the duties of draftsman. Education equivalent to graduation from high school. The completion of each full year of a standard course of engineering instruction in an educational institution of recognized standing shall be considered as the equivalent of the experience otherwise required, with the provision, however, that at least one year shall have been spent in the duties of draftsman.

### GRADE 6 SENIOR AID, FIELD

*Duties:* To direct work of field party on surveys or construction; to keep survey notes and engineering records; to supervise construction or repair work: to direct the work of computing surveys and estimates: to direct the work of making minor engineering computations.

*Qualifications:* Experience for at least five years in the use and care of surveying instruments, of which at least three years shall have been spent in the duties of instrumentman. Education equivalent to graduation from high school. The completion of each full year of a standard course of engineering instruction in an educational institution of recognized standing shall be considered as the equivalent of the experience otherwise required, however, that at least one year shall have been spent in the duties of instrumentman.

### GRADE 7 AID, OFFICE

*Duties:* To prepare general working drawings where design is furnished; to plot notes and prepare maps: to design simple structures; to make computations and compile data for reports and records: to check plans, surveys, and other engineering data.

*Qualifications:* Experience for at least two years in tracing, lettering, drafting and computing. Education equivalent to graduation from high school and familiarity with the use of the slide rule, and of logarithmic and other simple mathematical tables. The completion of each full year of a standard course of engineering instruction in an educational institution of recognized standing shall be considered as the equivalent of the experience otherwise required.

### GRADE 7 AID, FIELD

*Duties:* To run surveying instruments and to adjust and care for same; to compute surveys and estimates; to make minor engineering computations; to inspect incidentally construction or repair work.

*Qualifications:* Experience for at least two years in the duties of rodman. Education equivalent to graduation from high school and familiarity with the construction, operation, and care of surveying instruments. The completion of each full year of a standard course of engineering instruction in an educational institution of recognized standing shall be considered as the equivalent of the experience otherwise required.

### GRADE 8 JUNIOR AID, OFFICE

*Duties:* To trace and letter maps and plans; to make simple drawings from sketches and data; to make minor calculations.

*Qualifications:* Education equivalent to graduation from high school.

### GRADE 8 JUNIOR AID, FIELD

*Duties:* To run tape or levelling rod: to perform other miscellaneous subordinate duties in survey party in field or office, as directed.

*Qualifications:* Education equivalent to graduation from high school.

The completion of each full year of post-graduate work in the specific subject of study or investigation appropriate to a particular service or branch of service shall be considered the equivalent of one and one-half years of general experience, but such substitution shall not thus be made for more than four years of such experience or be considered as reducing the requirements in any grade of the number of years engaged in the conduct or direction of responsible work.

## RESULTS OF INVESTIGATIONS

An analysis of the federal, state, and municipal returns from questionnaires received up to the date of the committee report, December 15, 1919, shows them to be much better than those obtained in the investigation made by the American Society of Civil Engineers in the years 1913-1917. This is due to better coöperation on the part of responsible department heads. The financial status of 10,089 men has now been revealed as compared with a total of 6378 men who replied to the previous inquiry. It also appears that the respondents in the case of the previous investigation were largely confined to the classes receiving a maximum compensation, whereas the present returns cover a much broader range of salary.

Returns from men in federal service representing 16 engineering bureaus in civil establishments and 4 engineering bureaus in the Navy Department, and holding positions of decreasing responsibilities from chief administrative officer having full charge of organization, including determination of policy, to subordinate duty not requiring any special education, training or originality, show a corresponding gradual decrease in average salaries from approximately $8000 to $1000 in civil establishments and from $10.000 to $1400 in the Navy Department.

In state service 35 questionnaires from 27 states show a total of 1166 men in professional positions, ages ranging from 21 to 70. The average salary on July 1, 1919, was $2070 as compared with $1590 four years previous. This is an increase of 30 per cent, whereas a 72 per cent increase is recommended, making the average salary $2730. In sub-professional position there is represented 1056 men, ages 18 to 50. The average salary on July 1, 1919, was $1290 against $995 in 1915, an increase of 30 per cent. The engineers in charge advocate an average salary of $1470, a 48 per cent increase.

In municipal service, analysis of 66 questionnaires from 40 cities shows a total of 1478 men in professional positions, ages ranging from 25 to 70. The average salary on July 1, 1919 was $2370 as compared with $2230 in 1915, showing an increase of 6.5 per cent. The engineers in charge recommend an average salary of $3020, which is a 36 per cent increase. In sub-professional positions the returns show 1839 men, ages 17 to 65, average salary 1919, $1370 against $1160 in 1915, an increase of 18.5 per cent. The recommendation is for an average salary of $1700, a 47 per cent increase. It is thus disclosed that the average compensation for grades representing 68 per cent of the engineers in the federal service, other than the War Department, and 80 per cent of those in state and municipal service, is less than is required for the support of a family on a scale sufficient to provide for necessities (assumed at $2200 per annum), to say nothing of the expense of giving children anything like the education with which the breadwinner was equipped.

It would seem reasonable to assume that men in the "Junior Aid" grade have not reached an age which would require compensation sufficient to support a family, but that the salary limitation for men in higher grades should clearly be sufficient to meet this need. On this basis it is safe to state that with anything like suitable compensation, not more than about 25 per cent of the entire service, if properly organized, should receive less than $2200 per annum. The investigation made by the State and Municipal Section shows that the average age of the present incumbents of the lowest grades is actually about 27 years; if they are to be adequately provided for as permanent employees, the percentage of men to receive less than $2200 is clearly negligible.

A comparison with the pay of the industrial worker who serves under the engineer also bears testimony to the fact that, while admitted by none as to value, the actual compensation for brawn is to-day greater than for engineering brain. The need for setting up some scale of compensation for the engineer to correct this serious condition is, therefore, obvious, as is also that for the inquiry now being carried on by the Committee. Unless a radical improvement can be brought about, it seems evident that the profession cannot attract or retain men of the caliber required to command the respect in which it has heretofore been held by the public, and that so long as they are continuously struggling with the problem of making even a bare living, their efficiency will be minimized and their incentive to work with other than a purely selfish interest will be lacking.

That a serious condition of unrest exists in the municipal and state services is clearly evidenced by explicit statements in this respect in 44 per cent of the questionnaires returned, while in only 13 of these services were conditions as to morale reported to be satisfactory.

## RECOMMENDATIONS

From the investigation made by each of the sections of the Committee, it appears that the heads of sixteen engineering bureaus in the Federal Government (excluding the Navy Department) have recommended an increase in present compensation averaging about 59 per cent, while similar increases in the State and Municipal Services, as recommended by the engineering heads, average 26 per cent.

As a result of the inquiry now in progress under the direction of the Congressional Joint Commission on Reclassification of Salaries, the attention of all members of the federal service has been drawn to this question in such a way as to make it one of close study and analysis, which condition doubtless accounts for the comparatively modest recommendation made by the representatives of the state and municipal services, whose reply to our inquiry has been individual and who have doubtless been confronted with the necessity of recognizing relationship to other workers in the same services in whose interests no concerted move of this character has been attempted.

The Committee is impressed with the method which has been followed by the Federal Section in setting up a standard of compensation based on a readjustment to the new conditions as to cost of living, this being dependent on an award made on October 24, 1918 by the Shipping Wage Adjustment Board, which was designed to provide uniform national wage scale for all shipbuilding workers, including a scale of compensation for draftsmen and copyists. It is also understood that this wage scale, which is generally known as the Macy Scale, has since been adapted with modifications to the needs of certain bureaus in the Navy Department. In the accompanying diagram there is illustrated a comparison of the compensation fixed under the Macy Scale with the maximum rates suggested by the Federal Section. A further comparison has also been made with scales proposed by other organizations, which seem to justify the schedule of salaries now suggested by the Committee for discussion, which is as follows:

| Grade | Total Years Experience Required to Qualify | | Salary Range | |
|---|---|---|---|---|
| | With Professional Degree | Without Professional Degree | Minimum | Maximum |
| 8—Junior Aid | .... | 0 | $1080 | $1560 |
| 7—Aid | .... | 2 | 1680 | 2400 |
| 6—Senior Aid | .... | 5 | 2520 | 3240 |
| 5—Junior Assistant Engineer | 0 | 4 | 1620 | 2580 |
| 4—Assistant Engineer | 2 | 6 | 2700 | 4140 |
| 3—Senior Assistant Engineer | 5 | 9 | 4320 | 5760 |
| 2—Engineer | 8 | 12 | 5940 | No limit |
| 1—Chief Engineer | 12 | 16 | 8100 | No limit |

Applying the average proposed salary to the present incumbents of these positions as reported in the state and municipal services, it will be found that there would be a resulting increase in annual compensation totalling about $5,500,000 as compared with a total increase of $2,500,000 recommended by the service heads, against which, however, should be charged the economy growing out of the increased efficiency brought about by a restoration of morale.

The Committee is not prepared at this time to recommend the adoption of any definite schedule of compensation, and it is not at all clear as to the wisdom of fixing even a minimum limit on the highest grades of service or of keeping the maximum of one grade below the minimum of the grade above, all of which questions are now receiving its serious consideration, as is also the question concerning the provision to be made for advancing within the limits of a grade, for which a plan is suggested by the Federal Section. Under this plan it is proposed at yearly intervals, through ratings determined by a Personnel Board, to provide for awarding an increase in compensation to three-fourths of the men in a grade receiving less than the maximum compensation of the grade, who through fitness and industry have shown themselves to be of increased value to the service. Those in the highest third of this preferred list would each receive the maximum rate of increase, which under the proposed scale would be $480, while each of the next two-thirds would receive an increase of $240, the latter figure corresponding with what would then be the average increase for all of the men in the grade.

It is also the opinion of the Committee that, pending the completion of the investigation, the scale of compensation presented and the plan for promotion within a grade is adaptable to general use in all branches of engineering service. The Committee believes, however, that a general discussion of this question is desirable.

# CHARLES FREDERIC RAND RETIRES AS PRESIDENT OF THE UNITED ENGINEERING SOCIETY

FOR six years Charles Frederic Rand has been a member of the Board of Trustees of United Engineering Society, the first two as representative of the American Institute of Mining Engineers, and the last four in the presidential chair, and under his able and earnest guidance much has been accomplished. His term of office as president has just expired and upon the occasion of his retirement the Council of The American Society of Mechanical Engineers recognized his services by presenting to him framed resolutions appreciative of his accomplishments. Mr.

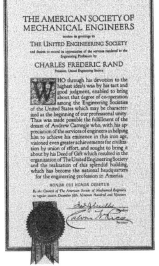

CHARLES FREDERIC RAND

W. M. McFarland, Mem.Am.Soc.M.E., and one of its representatives on the Board of Trustees of the United Engineering Society, in presenting the resolutions said in part:

It is always a pleasure to find that the faithful and efficient performance of duty has been appreciated by those for whose benefit it has been done, and it is a still greater pleasure when this appreciation is accompanied by its definite expression at a time when it will be a satisfaction and a slight reward to the one whose actions are commended.

With the realization of the fact that with this meeting the service of our esteemed associate, Mr. Charles F. Rand, as President of the United Engineering Society comes to a close by the termination of his service as a member, the representatives of The American Society of Mechanical Engineers brought this fact to the attention of the Council of that Society and took occasion to praise the devotion to the interests of united engineering which had been shown by Mr. Rand, first as Chairman of the Finance Committee and later as President.

The Council thereupon passed a suitable resolution expressive of their appreciation of Mr. Rand's work and directed that it be suitably engrossed and presented to him.

The pleasant duty of presenting these resolutions to Mr. Rand has been delegated to me, and it gives me very great pleasure to transmit to him this expression of the appreciation by one of the great American engineering societies for the splendid work which he has done in advancing the united interests of the engineers of America.

I feel that I am not only voicing my own sentiments but those of the entire Board when I express the deepest regret that the termination of his services will deprive us of the benefit officially of his ability, tact and wise judgment in the administration of the affairs of United Engineering Society, and of the great pleasure which it

has always been to be associated with him personally. I feel very sure, however, that he will always be ready to respond to any call for help which he can give us unofficially.

Mr. Rand is widely known as a mining engineer. He has done considerable work both in this country and abroad, and in 1913 he was decorated by the King of Spain for his notable efforts and accomplishments in advancing the mining industry of that country. Mr. Rand occupies an active and honored position in the affairs of the American Institute of Mining and Metallurgical Engineers and during the year 1913 served as president. He is now one of the directors of the Institute.

During his term of office the activities of the United Engineering Society have greatly expanded. The size of the Engineering Societies Building has been increased by the addition of three stories, two of which are now occupied by the American Society of Civil Engineers. There has thus been brought under one roof the headquarters of the great national engineering societies and this has made for more coöperation along every line.

The Engineering Societies Library has also greatly increased both as to size and the scope of its work. There have likewise been inaugurated two movements of considerable importance to the engineering profession at large. These are the Engineering Foundation for the advancement of research, created by

a gift of $200,000 from Mr. Ambrose Swasey, and Engineering Council, an organization devoted to a consideration of matters pertaining more particularly to the engineer's relation to public welfare. · Both of these organizations have contributed largely toward the advancement of the engineering profession, and Mr. Rand's participation in their activities has always been helpful and sympathetic.

# LIBRARY NOTES AND BOOK REVIEWS

## United Engineering Societies Library Issues Annual Report

The Library Board of the United Engineering Societies Library, which collection embraces the libraries of the American Society of Civil Engineers, The American Society of Mechanical Engineers, the American Institute of Electrical Engineers and the American Institute of Mining Engineers, has just issued its annual report for 1919.

The officers of the Board for the year 1919 were E. Gybbon Spilsbury, Chairman, Samuel Sheldon, Vice-Chairman, and Harrison W. Craver, Secretary. The four societies were represented on the Board as follows:

AMERICAN SOCIETY OF CIVIL ENGINEERS
Alfred D. Flinn                 Arthur P. Davis
John A. O'Connor                Willard Beahan
          Charles Warren Hunt

AMERICAN SOCIETY OF MECHANICAL ENGINEERS
Jesse M. Smith[1]               William N. Best
Walter M. McFarland             L. P. Breckenridge[2]
Andrew M. Hunt                  Calvin W. Rice

AMERICAN INSTITUTE OF MINING AND METALLURGICAL ENGINEERS
E. Gybbon Spilsbury             Alex. C. Humphreys
George C. Stone                 John H. Janeway
          Bradley Stoughton

AMERICAN INSTITUTE OF ELECTRICAL ENGINEERS
Samuel Sheldon                  W. I. Slichter
Edward D. Adams                 A. W. Kiddle
          F. L. Hutchinson

The year has been marked by an increased use of the Library in every direction. The number of readers during 1919 was 22,042, an increase of 6979, or 46 per cent over 1918. The attendance after six in the evening was 15 per cent of this, or 3470. The average daily attendance was 72.

Over 2000 telephone inquiries were received during the year and approximately 1500 letters which did not require elaborate investigation were received and answered. In other words, over 3500 persons not represented in the attendance figures nor in the records of the Service Bureau have been helped by the Library.

As in previous years, the Service Bureau has cared for the work requiring extended research, the preparation of translations, and copies of articles and bibliographies. During the year 552 searches, chiefly bibliographic, were prepared in answer to specific inquiries. Of these, the classification of searches by branches of engineering shows the largest number to have been made in the field of mechanical engineering; mining and metallurgical engineering ranking second. Some of these bibliographies were of great extent, requiring the time of one or two people for weeks and even months: and the work done exceeded that in 1918 by an amount greater than the increase of 44 in the number of searches indicates. The number of translations made was 71, containing 227,300 words. In 1918 there were 78 translations, amounting to 258,080 words. The orders for photographic copies of articles amounted to 2319 and required 23,951 prints. In 1918 there were 1150 orders for 6306 prints. This increase of 101 per cent in orders shows the wide appreciation of this form of assistance, which appeals particularly to those far from libraries. Copies have been made for engineers in every continent.

The resources of the Library, on January 1, 1919, were:

|                        | Volumes | Pamphlets | Total   |
|------------------------|---------|-----------|---------|
| Accessioned material   | 104,858 | 31,034    | 135,892 |
| Unaccessioned material | 13,962  | 7,941     | 21,903  |
| Total                  | 118,820 | 38,975    | 157,705 |

By December 31, 1919, all unaccessioned material had been examined and either accessioned or placed among duplicates. At that date the permanent collection contained a total of 152,091

[1] Resigned May. 1919.
[2] Succeeded Jesse M. Smith

volumes, pamphlets, maps and plans, and searches. There were also on hand 3958 duplicate volumes and 2076 duplicate pamphlets.

The receipts during the year were:

|                | Gift  | Purchase | Total |
|----------------|-------|----------|-------|
| Volumes        | 2,384 | 553      | 2,937 |
| Pamphlets      | 3,205 | 232      | 3,437 |
| Maps and plans | 118   | ...      | 118   |
| Searches       | ...   | ...      | 501   |
|                | 5,707 | 785      | 6,993 |

As in former years, the Library has received considerable important material by gift. The majority of the technical publications of American publishers have been sent for review in the publications of the Founder Societies, and there have also been many valuable gifts from individuals.

The technical work of the Library during the past year consisted principally in recataloging. A complete report of this work has been prepared by Harrison W. Craver, Director of the Library, and was published in the September 1919 number of MECHANICAL ENGINEERING, page 782. Arrangements for the work were made in May, and Miss Margaret Mann, formerly Chief Cataloger of the Carnegie Library of Pittsburgh, was engaged as cataloger. The work was begun on July 1, since which time the cataloging of new accessions and the recataloging of old material have proceeded simultaneously.

At the end of the year all unaccessioned material had been examined and either cataloged, listed as duplicates or discarded as valueless. The number of volumes completely cataloged was 6671, while 2210 duplicates had been listed, 374 searches had been cataloged and 1612 volumes added to the records as additions to existing sets. An index to the new catalog, containing 2920 entries, had been compiled and 6899 cards had been added to the catalog. The speed of the department has increased at a rapid rate as preliminary points have been settled and methods of procedure determined. The total work of the catalog department during the year, including that prior to the reorganization of the department, was 3009 volumes, pamphlets, maps and plans, and searches cataloged, and 5984 recataloged, totaling 8993. These figures, while they express the net results, are by no means an accurate index of the amount of work done. Much time is spent in collecting sets from their various locations, in correcting inaccurate records and in collating sets of periodicals and Government publications.

Changes in the staff have been less frequent than in previous years, and the effect of experience has been favorably shown in more helpful service.

The new system of cataloging the Library substitutes for all the old catalogs an author catalog and a classed subject catalog, the latter to be accompanied by a very full index. A classed subject catalog enters a book at that place in the catalog where all similar entries on the subject are collected. The author catalog will enable the reader to find a book when the name of the author is known, to find what books by any author are in the library, and also the various editions available; translations as well as their originals will be shown. The subject catalog will enable the reader to find a book when the subject is known, and will exhibit the resources of the Library in orderly form on every chief and most subordinate topics within its chosen field. The alphabetical index will comprise the names of the subjects of knowledge upon which the Library possesses important books or pamphlets, and it will at once direct the user to that section of the subject catalog where these books are listed and to the department of the Library where these works are shelved. When these catalogs are completed the catalog equipment of the Engineering Societies Library will be as complete, detailed and satisfactory as it is possible to make it, as well as a true and helpful guide for all reasonable methods of approach by readers.

142

# BOOK REVIEWS

L'ALUMINUM DANS L'INDUSTRIE. Métal pur Alliages D'Aluminium. By Jean Escard. H. Dunod et E. Pinat, Paris. 1918. Cloth. 6 x 10 in., 272 pp., illus., 18 francs.

The subjects discussed include the manufacture of aluminum, its physical and chemical properties, soldering, the influence of mechanical treatment on its properties, and its applications in metallurgy, electrical and mechanical industries, etc. The constitution, properties and uses of the various aluminum bronzes and alloys of aluminum with various metals, both common and rare, are described, and a final chapter treats of alloys with silicon, aluminum carbides and nitrides, and several alloys for special purposes. The volume gives a concise survey of present knowledge of aluminum.

APPLIED SCIENCE FOR METAL WORKERS. APPLIED SCIENCE FOR WOOD-WORKERS. By William H. Dooley. The Ronald Press Co., New York, 1919, $2 each.

These volumes present a course in the general principles of science, which gives particular attention to their applications in industry and is intended for use in vocational high schools, industrial schools, apprentice classes, etc. The general scientific matter in the two books is identical. Additional material relating specifically to the woodworking trades appears in one, while the other treats of metal-working in similar fashion.

BUILDING OF THE PACIFIC RAILWAY. The Construction-Story of America's First Iron Thoroughfare between the Missouri River and California, from the Inception of the Great Idea to the Day, May 10, 1869, when the Union Pacific and the Central Pacific joined tracks at Promontory Point, Utah, to form the Nation's Transcontinental. By Edwin L. Sabin. J. B. Lippincott Co., Philadelphia, 1919. Cloth, 5 x 8 in.. 317 pp., 10 pl., 4 portraits. 1 map, $2.

Mr. Sabin's object has been to tell how the Pacific railway came into being; to describe the actual building operations by which the Union Pacific, from the Missouri River, and the Central Pacific Railroad, from the Sacramento River, were constructed in six years instead of the allotted fourteen. The book is a popular, readable account of the men who built the roads and of the methods used.

CLAY-PLANT CONSTRUCTION AND OPERATION. By A. F. Greaves-Walker. Brick and Clay Record, Chicago, 1919. Cloth, 6 x 9 in., 212 pp., 79 illus., $4.

In writing this work the author has attempted to explain in understandable English some of the problems of the manufacturer of structural clay products. Technical terms, formulæ, and theories have been avoided, and practical facts alone have been presented.

ELECTRICAL ENGINEERING PAPERS. By Benjamin G. Lamme. This volume contains a collection of the author's most important engineering papers presented before various technical societies and published in engineering journals and elsewhere from time to time. 1919. Published by Westinghouse Electric & Mfg. Co., Pittsburgh. Pa. Cloth, 6 x 9 in., 773 pp., $2.50.

A collection of thirty-one papers and addresses which have appeared in various periodicals during the period from 1897 to 1918. These papers deal with problems of generator and motor design, the history of direct-current generators and of street-railway motors, and with technical training. The volume is published by the Westinghouse Electric and Manufacturing Company to commemorate Mr. Lamme's connection with it for thirty years.

ELEMENTARY PRINCIPLES OF AEROPLANE DESIGN AND CONSTRUCTION. A Textbook for Students. Draughtsmen, and Engineers. By Arthur W. Judge. James Selwyn and Co.. Ltd.. New York. 1919. Cloth, 6 x 9 in., 116 pp., 56 illus., 13 tables, $3.

Written to provide an inexpensive book, of an elementary nature, dealing with the fundamental principles of the design and, to a certain extent, the construction of airplanes. Follows the plan of the author's larger work, to which it may serve as an introduction.

FUNDAMENTALS OF CONTINUOUS CURRENTS OF ELECTRICITY. Book II of text-book series. School of Practical Electricity. By O. Werwath, assisted by Messrs. Vaughn, Hughes and Raeth. Electroforce Publishing Co., Milwaukee, Wis. 133 pp., illus., tables.

This book covers the fundamental principles of continuous current or direct electric current. It contains a nomenclature of electrical terms, discussion of Metric methods, basic electrical units, mechanical and electrical work and energy and detailed chapters on types of electrical circuits, electricity meters, resistance measurements, and electrical heating.

GUIDE TO THE STUDY OF THE IONIC VALVE. Showing its Development and Application to Wireless Telegraphy and Telephony. By William D. Owen. Sir Isaac Pitman and Sons, London. Cloth. 4 x 6 in., 59 pp., 12 illus., $1.

This little volume is intended to provide, for students of radio-telegraphy, an impartial, coherent record of the development of the ionic valve. References to the original sources of the information enable the reader to pursue his study of the subject.

INDUSTRIAL RECONSTRUCTION PROBLEMS. Complete Report of the Proceedings of the National Engineers, New York City, March 18-21, 1919. Paper, 6 x 9 in., 197 pp.

The papers presented at the conference dealt with a wide variety of problems connected with the reorganization of our industries on a peace basis. Consideration is given to questions of finance, commerce, production, labor, education, management, etc.

MACHINE TOOL OPERATION. Part I. The Lathe Bench Work and Work at the Forge. By Henry D. Burghardt. First edition. McGraw-Hill Book Co., Inc., New York, 1919. Cloth, 5 x 8 in.. 326 pp., illus., tables, $2.

The author of this work has used his experience as a teacher of machine work in a technical high school to prepare a textbook for those who desire a knowledge of the principles and elementary operations of machine-shop work, suitable for use in technical and trade schools and in apprenticeship courses. While designed primarily for use in connection with lectures by an instructor, it is also useful for self instruction.

THE MANUFACTURE AND TESTING OF MILITARY EXPLOSIVES. By John Albert Marshall. First edition. McGraw-Hill Book Co., Inc., New York, 1919. Cloth, 6 x 8 in., 261 pp., $3.

This volume contains a concise account of the explosives used for military purposes, in which attention is directed to those points which have a direct bearing on their manufacture, testing and storage. An extensive bibliography is included.

METAL WORKER'S HANDY-BOOK OF RECEIPTS AND PROCESSES. Being a collection of chemical formulas and practical manipulations for the working of all the metals and alloys; including the decoration and beautifying of articles manufactured therefrom, as well as their preservation. Edited by William T. Brannt. New enlarged edition. Henry Carey Baird and Co., Inc., New York, 1919. Cloth, 5 x 8 in., 582 pp., illus., tables, $3.

Five new chapters have been added to the present edition of this work, describing methods for welding with the oxy-acetylene flame, thermit and electricity, for galvanizing and for die casting.

MILLIONS FROM WASTE. By Frederick A. Talbot. J. B. Lippincott Co., Philadelphia, 1920. Cloth, 6 x 9 in., 308 pp., $5.

This work is written to indicate certain of the most obvious channels through which wealth incalculable is being permitted to escape, as well as to describe some of the highly ingenious efforts which are being made to prevent this wastage. While written essentially for the uninitiated, it will, the author hopes, prove of aid to those who are fully alive to the potential values of refuse. The volume is confined to those phases of the subject which are familiar to the average person.

MINERAL DEPOSITS. By Waldemar Lindgren. Second Edition. McGraw-Hill Book Co., Inc., New York, 1919. Cloth, 6 x 9 in., 957 pp., illus., tables, $5.

Mineral deposits are usually classified and described by the metals or substances which they contain; for instance, deposits of copper are described together, with little or no effort to separate them into genetic groups. This book is the outcome of a desire to place the knowledge of mineral deposits on the broader and more comprehensive basis of a consistent genetic classification and thus bring it into a more worthy position as an important branch of geology.

In this second edition, all of the chapters have been revised and several have been largely rewritten. A discussion of metallogenetic epochs has been added, as well as an index by elements. Some less essential descriptions have been excised in order that the book might not increase in bulk.

OIL ENGINES. Details and Operation. By Lacey H. Morrison. First edition. McGraw-Hill Book Co., Inc., New York, 1919. Cloth, 6 x 9 in., 472 pp., illus., tables, diagrams, $5.

This volume is intended for operators of oil engines. The details of construction of the more important oil engines manufactured in the United States are described and the proper methods of adjustment are explained at length.

THE PETROLEUM HANDBOOK. By Stephen O. Audros. The Shaw Publishing Co., Chicago, 1919. Flexible cloth, 4 x 7 in., 206 pp., illus., tables, $2.

In a book of pocket size, the author gives the fundamentals of each phase of the oil industry necessary to a clear understanding of the various operations entailed between the location of an oil well and the distribution of the refined products. The work is chiefly a compilation from the standard authorities, arranged for those who wish a brief, accurate account of the industry, devoid of unnecessary detail.

PRINCIPLES OF INDUSTRIAL ORGANIZATION. By Dexter S. Kimball. Second edition, revised and enlarged. McGraw-Hill Book Co., Inc., New York, 1919. Cloth, 6 x 9 in., 325 pp., 21 illus., 1 pl., 12 tables, $3.

This work, the first edition of which appeared in 1913, has been written to give young engineers a concise account of the salient facts relating to the most important economic and sociological problems with which they will be brought in contact, and to explain the origin and growth of the important features of industrial organization. The present edition has been revised, rearranged and enlarged.

PRINCIPLES OF REINFORCED CONCRETE CONSTRUCTION. By F. E. Turneaure and E. R. Maurer. Third edition, revised and enlarged. John Wiley and Sons, Inc., New York, 1919. Cloth, 6 x 9 in., 485 pp., illus., diag., pl., tables.

Like former editions, this work is intended "to present in a systematic manner those principles in mechanics underlying the design of reinforced concrete; to present the results of all available tests that will aid in establishing coefficients in working stresses; and to give such illustrative material from actual designs as will serve to make clear the principles involved." In the present edition the material has been considerably amplified and rearranged. Separate chapters have been devoted to the theory of flexure; bond and shear; and the design of beams. The experimental data have been reviewed. A new chapter on flat slabs has been added and the chapter on building construction extended.

RETAINING WALLS.—Based entirely on the Theory of Friction. By Pedro Dozal. Translated into English by R. T. Mulleady. First edition. M. A. Rosas, Buenos Aires, 1918. Cloth, 7 x 11 in., 161 pp., 62 illus.

The author presents a general theory in regard to the static pressures acting on a plane cutting an unbounded coherent, incoherent or liquid mass in equilibrium, and its application for the calculation of the pressures on retaining walls. The theory advanced differs from those presented by Rankine and other students of the subject and gives notably different results.

SCHOOL OF PRACTICAL ELECTRICITY. Book I of text-book series. By Oscar Werwath. Electroforce Publishing Co., Milwaukee, Wis., 1919. 60 pp., illus.

This practical text book, written by the president of the School of Engineering of Milwaukee in collaboration with several members of his faculty, sets forth the advanced methods of electrical instruction as taught at the School in a clear-cut, expository form, avoiding dry technicalities. It includes the history of electricity, detailed treatment of the telephone and telegraph systems, definitions of electrical terms, practical estimates and pointers regard-

ing electrical workmanship and installations of all kinds, and experimental outlines for the electrical laboratory.

SELENIUM CELLS. The Construction, Care and Use of Selenium Cells with special reference to the Fritts Cell. By Thomas W. Benson. Spon and Chamberlain, New York, 1919. Cloth, 5 x 8 in., 63 pp., 18 illus., $1.50.

The lack of definite information on the construction of selenium cells has led the author to publish the results of some of his experiments. After a brief review of various types of cells, the book describes in detail the manufacture, maturing and testing of the Fritts cell, and concludes with an account of some applications and suggestions on the care of cells.

TIMBER. ITS STRENGTH, SEASONING AND GRADING. By Harold S. Betts. First edition. McGraw-Hill Book Co., Inc., New York, 1919. Cloth, 6 x 9 in., 234 pp., illus., maps, charts, tables, $3.

This volume, for engineers, manufacturers and users of lumber and wood material, is intended to meet the lack of a work on wood as a structural material, similar to existing handbooks on steel, concrete, etc. The methods of testing the strength of wood and wood products, such as telephone poles, packing boxes and vehicle parts, are described and the results of numerous tests by the Forest Service are tabulated. A table based on 130,000 tests, showing the average mechanical properties of 124 American woods, is given. Proper methods of seasoning wood are fully discussed. The grading rules of the different manufacturers' associations are given and the book closes with extensive statistics on the lumber produced and used in the United States.

## MECHANICAL ENGINEERING LABORATORY, UNIVERSITY OF ILLINOIS

(Continued from page 113)

supplying the laboratory or from the exhausts from the ammonia and brine pumps. A 24-in. by 12-ft. generator is installed for this purpose. A 10-in. by 19-ft. brine cooler is placed on the lower floor, and is arranged so that the brine may be circulated either through the reheater or the brine tank.

Both machines are completely equipped with thermometer wells and weighing tanks, and are especially arranged for accessibility.

### INTERNAL-COMBUSTION ENGINES AND GAS PRODUCER

The section of the laboratory devoted to internal-combustion engines and the gas producer is located at the west end of the middle bay. At the extreme west of this bay is located a Smith gas producer rated at 75 hp. capacity. The producer (34) is complete with a gas-cleaning plant and is designed to operate continuously on bituminous coals.

In the opposite corner of the same bay with the gas producer there is to be installed an automobile-testing plant. This plant will be designed so as to be suitable for the testing of all sizes of pleasure cars and trucks up to five tons capacity. It is intended to measure the drawbar pull directly, and it will be possible to make various kinds of comparative tests on motor vehicle parts with this outfit.

While the automobile-testing plant is not yet installed, the laboratory has facilities to carry on tests of automobile, truck, tractor and even airplane motors. The principal unit of this equipment is a complete Sprague electric dynamometer (32a) of 150 hp. capacity. This outfit can be used either as an absorption or as a transmission dynamometer. At present a 4-cylinder Willys-Knight motor, a 4-cylinder Ford and a Peerless 6-cylinder motor (31) comprise the engine equipment for running tests with the dynamometer. A water brake designed by Dean C. R. Richards and built in the Mechanical Engineering Shops of the University of Illinois is shown in the plan view connected to the 6-cylinder 70-hp. Peerless motor (31).

The first large unit in the stationary gas-engine equipment is a single-acting, four-cycle horizontal 10-in. by 10-in. Otto engine rated at 25 hp. at 220 r.p.m., piped to the gas producer and supplied with the gas from the producer there.

There is also installed a smaller Otto single-acting four-cycle 5¾-in. by 12½-in. gas or gasoline engine (29) rated at 10 hp. at 300 r.p.m.; a 5-in. by 9-in. single-acting four-cycle Bogart engine (28), which is rated at 10 hp. when running at 315 r.p.m., and may be operated on either illuminating gas or gasoline; a Mietz and Weiss single-acting 9-in. by 10-in. engine (27), rated at 12 hp. at 340 r.p.m., and using kerosene as fuel with a hot-bulb type of ignition; and a Sargent 10-in. by 20-in. tandem double-acting four-cycle "complete expansion" engine (26), rated at 50 hp. at 250 r.p.m. and using illuminating gas for fuel. Mr. Sargent, the designer and inventor of this engine, is an alumnus of the College of Engineering, University of Illinois.

# MECHANICAL ENGINEERING

## THE JOURNAL OF THE AMERICAN SOCIETY OF
## MECHANICAL ENGINEERS

Volume 42        March 1920        Number 3

## The First Transcontinental Motor Convoy

An Account of the Successful Attempt Made by the War Department to Push Through a Large Train of Heavily Loaded Motor Trucks from Washington, D.C., to San Francisco, Together with Particulars of the Itinerary, Personnel, Operation and Maintenance Problems, and Conclusions Reached

### By E. R. JACKSON,[1] WASHINGTON, D. C.

THE Transcontinental Motor Convoy which started from Camp Meigs, Washington, D. C., at 8 a. m. on July 7, 1919, and finished at San Francisco, Cal., on September 6, 1919, was the first attempt to push through a large train of heavily loaded trucks from one end of the country to the other. Back of it were placed the resources of the War Department, and some measure of its success was due to the friendly co-operation of the local bodies of the communities through which the Convoy passed.

#### OBJECTS OF THE TRIP

The four purposes of the trip were as follows:

1 The War Department's contribution to the Good Roads movement for the purpose of encouraging the construction of through-route and transcontinental highways as a military and economic asset.

2 The procurement of recruits for the enlisted personnel of the Motor Transport Corps, or any other branch of the U. S. Army, young men to be enlisted with the train as candidates for the mechanical training schools opened under the direction of the Motor Transport Corps.

3 An exhibition to the general public, either through actual contact or resulting channels of publicity of the development of the motor vehicle for military purposes, which is conceded to be one of the principal factors contributing to the winning of the World War.

4 An extensive study and observation of terrain and standard army motor vehicles by certain branches of the army, particularly the Field Artillery, the Coast Artillery, the Air Service, the Corps of Engineers, and the Ordnance Department.

[1] Former 1st Lieut., Ord. Dept., U. S. A. Ordnance Observer.

For presentation at the Spring Meeting, St. Louis. May 24 to 27, 1920, of THE AMERICAN SOCIETY OF MECHANICAL ENGINEERS. All papers are subject to revision.

The writer was acting as Ordnance Observer for the Chief of Ordnance, attached to the Transcontinental Motor Convoy, and accompanied this truck train the entire distance from Washington, D. C., to San Francisco.

The Convoy departed from Camp Meigs at 8:30 Monday morning, July 7, 1919, proceeding to the Ellipse in Potomac Park, opposite the south front of the White House, where the Zero Milestone was dedicated by the National Highways Association in the presence of the Secretary of War, the Chief of Staff, general officers of the Army, and several Senators and Congressmen. Following his acceptance of the Zero Milestone on behalf of the President of the United States, Secretary Baker formally directed the Convoy to proceed overland to San Francisco, Cal., via the Lincoln Highway.

#### ITINERARY

Several days before it started the Convoy received from the Motor Transport Corps an itinerary giving scheduled control points, mileage and dates of arrival. Orders were given by the Secretary of War and Brig.-Gen. C. B. Drake of the Motor

FIG. 1 ROUTE OF THE FIRST TRANSCONTINENTAL MOTOR CONVOY—CAMP MEIGS, WASHINGTON, D.C., TO SAN FRANCISCO, CAL., JULY 7, TO SEPT. 6, 1919

Transport Corps, to hold to the schedule as closely as possible, and it is noteworthy that the Convoy arrived at Lexington, Neb., almost exactly half-way across the continent, on August 1 on time, thanks to the good roads of Pennsylvania, Ohio, Indiana and Illinois and to the fact that the gumbo roads of Iowa were dry and hard, there having been no rain in that state for several weeks previous to the arrival of the Convoy.

From this point west and until the Convoy reached the palm-lined boulevards of California the roads grew steadily worse except in the vicinity of Salt Lake City, Utah, and it became increasingly difficult to get the heaviest trucks through the soft, muddy roads and over the bad grades. The greater part of the Convoy trucks could not have reached San Francisco without the assistance rendered by the Militor and the 5-ton artillery tractor, produced by the Ordnance Department.

MECHANICAL ENGINEERING

THE JOURNAL
Am.Soc.M.E.

## PERSONNEL

The personnel of the Convoy consisted of Companies E and F of the 433d Motor Supply Train, comprised of 12 officers and 210 men; Service Park Unit No. 595, 1 officer and 38 men; Company E, 5th Engineers, 2 officers and 30 men; 1 Medical Detachment; 1 Field Artillery Detachment; and 17 commissioned officers detailed to duty as staff observers.

The majority of the enlisted men of the Motor Supply Train were raw recruits with little or no military training, many of whom had not driven a motor truck before. · Considering this fact, the success of the Convoy is quite remarkable, as the roads encountered through the mountains of the West required driving skill of the highest order. The men were poorly equipped at the start, due to the fact that many of them joined at the very last minute.

The performance of the various types of motor equipment is one of the chief points of interest to the engineering fraternity, inasmuch as it affords a most excellent opportunity for making valuable comparisons as to the ability and reliability of the several makes of motor vehicles and trailers. Practically all of the equipment was new when the Convoy left Washington, the exceptions being the Militor (the standardized four-wheel-drive motor vehicle of the U. S. Army) and the 5-ton artillery tractor, both of which were designed and built by the Ordnance Department, so that in so far as the condition of the vehicles was concerned, all had an equal chance of successfully completing the long journey across the continent.

While the cargoes carried varied somewhat from day to day, it may be said that in general the trucks hauled capacity loads. The most uncertain factor in the operation of these motor vehicles, especially during the first few weeks of the trip, was the

FIG. 2 ARTILLERY TRACTOR USED TO PULL BIG TRUCKS OUT OF BAD SPOTS. REPAIRING A BROKEN TRACK LINK

FIG. 4 TRUCK STUCK IN MUD SO BADLY THAT A 1¾-IN. MANILA ROPE AROUND WINCH OF MILITOR BROKE

FIG. 3 CLASS B TRUCKS, CHAINED TOGETHER, MAKING GRADE AFTER FIRST ONE HAD BEEN TOWED OVER TOP BY TRACTOR. EAST OF BIG SPRINGS, NEB., AUG. 6, 1919

FIG. 5 MACK TRUCK BREAKING THROUGH SMALL WOODEN BRIDGE, 31 MILES WEST OF GREEN RIVER, WYO.. AUG. 15, 1919

The other units were better trained and better equipped. The Engineer Corps and the Medical Detachment were especially efficient organizations. Much credit should be given to the Engineers for the extremely valuable work which they did in repairing roads and bridges, without which the expedition must have failed. The Medical Detachment handled all matters of camp sanitation and precautionary measures in a most satisfactory manner and as a result practically the entire personnel enjoyed excellent health during the whole trip.

## OPERATIONS

The Convoy was made up of 81 motor vehicles and trailers.

inexperience of the drivers, all of whom were new to convoy work. Even during the first half of the trip, where the roads were generally good, some drivers had considerable difficulty in keeping their trucks on the road. After the train officers had had ample opportunity to observe the comparative ability of the men for a few weeks, the best drivers were each assigned to one of the motor vehicles with an assistant driver and made fully responsible for the operation of their equipment. Immediately upon arrival at the control point for the night it was the duty of the driver to see that his truck was at once filled with gasoline, oil and water, cleaned, and any needed adjustments or repairs made. Three officers of the Motor Transport Corps inspected each vehicle every morning before the Convoy left camp, to see that this work had been properly done. This resulted in a very marked lessening of mechanical troubles on the road and a much more efficient operation of the entire equipment. As the Convoy proceeded across the country the men gained experience with their trucks and confidence in themselves, so that by the time the mountains and deserts were reached most of them were equal to the difficult tasks encountered there, and when the train reached San Francisco many of the men were really competent drivers, and their development as such is deserving of favorable comment. It is a truly noteworthy achievement that there were no trucks lost on the very dangerous mountain grades of the Rockies and Sierras.

Excellent work was done by the two pilots, who left camp on motorcycles each morning about one-half hour in advance of the

## MAINTENANCE

The maintenance and repair of the Convoy vehicles was the principal function of Service Park Unit No. 595, and the fact that only three trucks were retired from service en route bears tribute to the effectiveness with which they did their work. Included in this detachment were expert mechanics on motorcycles, motors, carburetors, magnetos, radiators, and tires, machinists, painters, welders, blacksmiths, and carpenters; each man a specialist in his own line of work. These men did very excellent work in general, frequently working all night in order to have some truck ready to run under its own power when the Convoy broke camp in the morning, and some very clever repairs were made with the facilities available. New connecting-rod bearings were fitted frequently, damaged cylinder blocks replaced, cracked

FIG. 6  NOON-STOP AT HIGHEST POINT ON LINCOLN HIGHWAY,
8020 FT. ELEVATION, BETWEEN LARAMIE AND MEDICINE BOW,
WYO., AUG. 11, 1919

FIG. 8  CONVOY PASSING HIGHEST POINT ON LINCOLN HIGHWAY
AT AMES MONUMENT IN WYOMING. 8020 FT. ABOVE SEA
LEVEL, AUG. 11, 1919

FIG. 7  MILITOR PULLING CLASS B TRUCK OUT OF DITCH NEAR
GRAND ISLAND, NEB., BY USING WINCH

FIG. 9  CLASS B TRUCK SO FIRMLY STUCK IN QUICKSAND THAT
EVEN THE TRACTOR TREAD SPUN

Convoy to see that the road was clear and to report any obstructions or deviations to the expeditionary commander. The simple road-marking system used by the pilots proved to be a great success. The markers consisted of salmon-colored paper isosceles right-angle triangles which were tacked on a conspicuous post or fence at each turn of the road or intersection. When the apex of the triangle pointed up it indicated that the route was straight ahead; apex pointing to the right indicated right turn; apex pointing left, left turn; and two triangles tacked up base to base, thus forming a square, indicated the location of the camp site. Without these markers considerable difficulty would have been experienced in following the route through cities, on detour, in the absence of official Lincoln Highway markers, or where those were interspersed with other highway markers.

crankcases repaired, new clutch assemblies installed, and even babbitt bearings poured on the roadside very satisfactorily.

The maintenance work was considerably hampered because of the necessity of carrying spare parts for so many different makes of trucks and the additional fact that the stock carried had not been intelligently selected. Large stocks of such parts as axles and drive shafts, among which there were no breakages, were carried, while no spares whatever were taken of many other parts, including radiators, on which there were frequent failures. In a few instances it was possible to replace a broken or worn part in one truck with a spare part for another make of truck, when careful calibration showed the dimensions to be practically the same. Another difficulty was due to the poor quality of the hand tools supplied, and better tools should be furnished the mechanics, as well as such special tools as are necessary to make certain adjustments properly.

### General Conclusions

In general, very few serious automotive troubles developed, but the experience gained on the trip and the information secured from an examination of each vehicle after arrival at San Francisco would seem to warrant the following statements:

*Cooling Systems.* The vertical tubular type of radiator, with bolted-on upper and lower tanks, is the most satisfactory on military vehicles, because of the facility with which it may be cleaned and repaired. The addition of a temperature regulator in the water line, which would restrict the flow of water until the motor had warmed up and such as is now standard equipment on many commercial trucks, would be most desirable. Such a device would be simpler and less expensive to install than radiator shutters, less liable to damage, and is automatic in its operation.

An endless flat fan belt, of leather or rubberized fabric, is most desirable for military motor vehicles, and an effort should be made to standardize the width and length of these belts. An extra endless belt should be carried in each vehicle at all times. The

The present acetylene headlights used on all trucks are entirely inadequate, and the carbide generators are unsatisfactory and poorly constructed. It is believed that an efficient system of electrical lighting should be adopted for all military trucks. Such a system was developed for the Ordnance Department during the war by Mr. R. M. Newbold and found satisfactory.

The use of mechanical or electrical self-starters on military vehicles is undesirable, as they merely add complications to the operation and maintenance of the trucks.

All service cars should be equipped with four electric spotlights for night repair work.

*Fuel Systems.* Except for a few broken gasoline lines the fuel systems now in use are generally satisfactory, although larger drain cocks in all gasoline tanks would be desirable.

*Transmission Systems.* A driving worm of 2½ pitch instead of 3½ pitch is recommended for all Class B trucks. Of the three Class B tankers, the one with the 3½-pitch worm was the one which was most frequently stalled.

It is recommended that steps be taken to have all future military motorcycles equipped with a complete protective guard around the drive-chain system. Some steps should also be taken

Fig. 10  Class B Truck, Towing Two-Wheel Liberty Kitchen Trailer. Stuck on Sandy Grade, and Towed over by Tractor

Fig. 11  Ordnance-Built Tractor Pulling Heavy Trucks out of Mud near Gothenburg, Neb., Aug. 3, 1919

"V" fan belts are not desirable, but if used, some form of a dust pan should be placed in front of the lower fan-belt pulley to protect the "V" belt.

*Motors.* Because of the high percentage of failures experienced with certain valve-tappet rollers and guides and the valve springs, prompt attention should be paid to the development of a more durable valve-lifter assembly, which will be interchangeable with that now in use. It is suggested that steel be used in the construction of the guide instead of cast iron, as at present, and that the valve springs be heat-treated more carefully.

The governors of all trucks in the Convoy were rendered useless during the first few days of the trip, resulting later in many burned-out connecting-rod bearings, due to speeding the vehicles. Some means should be provided to protect the governors against tampering on the part of drivers and mechanics.

A simplified carburetor would be advantageous on cars having sensitive automatic throttle and auxiliary air valve, as these are not thoroughly understood by the average mechanic.

Only the very best grades of oil should be used in motors, and a simple and positive means of determining the oil level should be provided on every truck.

A study should be made of the conditions which are responsible for the blowing out of cylinder-head gaskets in the motor of the Standardized Class B trucks, and such changes made as may be necessary to eliminate this trouble.

The mufflers on some of the trucks were found to clog up easily and this matter should be given careful attention.

*Ignition and Electrical Systems.* The Bosch magneto used on the Standardized Class B truck should be corrected to have a greater air gap between the pencil and the distributor block.

to provide for a unit power plant, eliminating the short drive chain. In future development steps should be taken to eliminate the chain drive entirely and to substitute shaft drive. There are no insurmountable difficulties to the inclusion of this type of drive in a motorcycle, but it cannot, however, be applied when the V-type two-cylinder motorcycle motor is used.

Chain-driven trucks cannot be operated satisfactorily where soft, sandy or muddy roads are encountered. They are essentially good-roads trucks.

As an automotive principle the four-wheel drive was surely vindicated by the Militor and the F. W. D. trucks. It is believed that the Militor is the most powerful wheeled motor vehicle constructed up to the present time.

*Vehicle Controls.* The steering gears gave very little trouble, but better lubricating facilities are desirable to secure the ease of steering which is so necessary on a motor truck, especially when operating on heavy roads, as some types are very hard to operate on rough roads.

On one class of trucks the brake rods are placed too close to the springs and there is frequently wear between these parts.

*Front Axles.* There were no front-axle troubles due to any inherent mechanical faults.

*Rear Axles.* No serious rear-axle trouble developed on the road other than stripping some teeth out of two ring gears in one make of car. All worm gears operated without any signs of overheating under the heaviest pulls. Increased ground clearance under most rear axles would be very desirable.

*Wheels and Tires.* All trucks for military use should have cast-steel wheels; and passenger automobiles for the army, wheels of the pressed-steel disk type.

Solid tires for trucks should be of the Giant type instead of the Dual type, as the Giant type gives better traction and apparently stands wear better than the Dual type. This was especially noticeable in the deserts, where a ridge of sand was pressed in between the two halves of the Dual tires, virtually eliminating the ground pressure of the tread, and thus reducing the tractive effort. Furthermore, the manufacturers have practically ceased production of Dual tires and are centering all their energies upon the construction of Giant tires.

*Chasses.* The frames of all vehicles completed the journey with no apparent damage. All frames were in alignment and only one case of deflection was observed. Rear bumpers should be placed on all trucks, and a stronger bumper, possibly with steel facing, should be used on military vehicles. Many towing hooks straightened under pull, and a much stronger hook should be developed.

*Bodies.* In two makes of 1½-ton trucks the body sills had to be cut out for wheel clearance and attention should be paid to this matter.

Tankers are overloaded and the center of gravity is too high. Slush was responsible for one overturning. Bumper should be in rear of faucets to protect them.

*Equipment.* Horns loosened up from vibration, and should be more securely attached to the body.

Tool boxes should be of heavier construction in order to withstand the very hard usage to which they are subjected in the field.

### RESULTS

It is the opinion of the Ordnance Observer that the First Transcontinental Motor Convoy, as the pioneer undertaking of its kind, was a successful operation, and that the experience gained during this trip will prove invaluable, not only to the Army, but to all users of motor trucks in general.

All along the route great interest in the Good Roads Movement was aroused by the passage of the Convoy, and it was reported that several states had voted favorably on large issues of road bonds. However, the officers of the Convoy were thoroughly convinced that all transcontinental highways should be constructed and maintained by the Federal Government, because the sparsely settled states of vast area in the Rocky Mountain region are absolutely unable to finance their own section of such a project. The Townsend Bill, now before Congress, provides for just such an

FIG. 12 DISTANT VIEW OF CONVOY GOING OVER THE GREAT SALT LAKE DESERT, AUG. 22, 1919

FIG. 13 CLASS B TANKER RUNS OFF ROAD ON OLD RAILROAD GRADE AND TURNS OVER 270 DEG. NEAR RAWKINS, WYO.

Whenever a train is started out for other than experimental purposes the trucks should be of the same type and tonnage, thereby eliminating the number of spare parts necessary to be carried.

For the mobile army light trucks not to exceed 2-ton capacity should be the practice. The heavier trucks should be used in the rear areas only.

With reference to shop trucks, no fixed tops should be used. The general scheme of using bows and paulins should be followed, because the fixed top is unnecessary, as it adds to the weight of the truck and often is an encumbrance in the passage of low bridges and low-hanging branches.

The equipment on the machine-shop trucks should be rearranged to provide greater accessibility, placing the motor-generator set where the lathe now is, across the forward end of the body, and locating the lathe transversely at the rear of the body. A double bench could then be placed across the body at the center, with the drill press on one end and the grinder on the other, and four vises of various sizes and types could be attached as convenient.

Some provision should be made on all trucks to permit both sides of the hood to be folded up and securely retained in that position. This would enable two men to work on opposite sides of the motor at the same time, and would make it possible to increase the circulation of air around the motor in very warm weather. A more sturdy form of hood clip should be developed.

The rope cleats and stake pockets all worked loose and a better means of fastening them should be used.

Floor boards are too weak and should be made of heavier gage stock. Laminated floor boards did not prove to be satisfactory.

undertaking under a Federal Highway Commission, and it is earnestly hoped that this measure may be favorably acted upon at an early date, in order that the nation may be ready for the commercial "Ship by Truck" movement.

Although the number of enlistments actually secured by the recruiting officer was not large, great interest in the various activities of the Army was stimulated throughout the regions covered by the Convoy, and reports from recruiting stations indicate that several hundred recruits were obtained as a direct result of the Convoy.

The interest of the general public in the Convoy was evidenced by a whole-hearted hospitality which never failed from beginning to end of the trip, and which was quite as spontaneous in the small towns as in the larger cities. Everybody was glad to see the trucks and the men, who were showered with every variety of refreshments all along the route. In the larger towns and cities luncheons and dinners were served, entertainments and dances given, and the clubs and many homes opened to the entire personnel. Great credit should be given to the Red Cross Canteen Service, the War Camp Community Service, the Knights of Columbus, the Y. M. C. A., the Chambers of Commerce, the Commercial Clubs, the Elks, and many similar organizations, who served the Convoy with distinction all the way "from Coast to Coast."

In the operation and maintenance of a motor-truck train many valuable lessons were learned, which warrant the following recommendations:

That extended use of triangular signs, bearing the designations of the unit using them, as road markers is invaluable.

That there should always be a very careful inspection of all vehicles for the proper amount of gasoline, oil, and water before departure on any run, no matter how short.

That a light touring car or light delivery car replace the present motorcycle with side car, which cannot negotiate soft, sandy or muddy roads successfully, and is a single-purpose vehicle.

That Class B trucks are too heavy for use in the field, except in the rear areas, on good roads and over strong bridges.

That only trucks of the four-wheel-drive type be used with artillery because of the ease with which they negotiate any sort of rough terrain, sand or mud, in any kind of weather.

That two Class A trucks be used instead of one Class B truck wherever there are unimproved roads or doubtful bridges.

That every Service Park Unit and Repair Base should have a Militor with winch for wrecking.

experienced in starting several motors on account of low-gravity gasoline.

To give an idea of the main troubles encountered under the strenuous conditions of the run the log on July 23 is reproduced in its entirety.

Departed Clinton, 6:30 a. m. White Staff Observation Car No. 111,506 delayed in starting on account of oil-pump plunger sticking. Class B No. 414795 stalled with oil-soaked magneto 6 mi. out. Class B spare parts truck No. 414668 taken in tow by Militor on account leaking cylinder-head gasket—would not pull on hills. This truck gave some trouble yesterday, but was not repaired in Clinton. Porcelain of one AC spark plug was broken. 3 mi. west of Lowden, Class B No. 80215 stopped to clean out gasoline line. Fender fell off 2-wheel kitchen. Starting crankpin breaks on one class B and another has trouble with back-firing in carburetor. Trailmobile kitchen breaks right rear spring through all leaves in center. Class B machine shop

FIG. 14 CLASS B SPARE-PARTS TRUCK RIGHTING OVERTURNED G.M.C. AMBULANCE No. 39309 EAST OF CHICAGO HEIGHTS. ILL., JULY 19, 1919

That the Militor should have a longer and stronger sprag, of the artillery type, similar to the spade of a gun-carriage trail; or perhaps two such sprags, one on each rear corner of the frame; that the cable reel should be placed on a squared shaft to facilitate rewinding and should be located ahead of the winch and guide sheaves which should have universal support.

That spare-parts trucks should have a stock-record card system, and that the arrangement for packing the parts be improved so that such articles as bearings may be protected with individual boxes or cartons.

That every effort be made to coöperate with carburetor manufacturers toward the production of a carburetor of the jet type suitable for use on motorcycles.

### DAILY LOG OF THE TRIP

A log was maintained in which the incidents occurring were written day by day. It makes interesting reading and gives a clear idea of the difficulties which the train had to overcome. A few of the incidents recorded may be of particular interest; for instance, those inserted in this abridged report.

As stated, the train departed from Camp Meigs, Washington, D. C., at 8:30 a. m., July 7, 1919.

Already on July 9 some of the trucks were held back by broken accelerator springs and sticky exhaust valves losing compression. Ignition troubles of various nature developed and various types of trucks also developed considerable magneto trouble.

Considering the personnel was untrained, the driving could be called excellent.

On July 12 on one truck the main-drive shaft had been twisted off in the transmission bearing by an inexperienced driver throwing in the reverse gear and dropping in clutch while the motor was turning over at high speed on down grade. The driver did it to use the motor as a brake.

On July 17, departed Delphos at 6:30 a.m. Difficulty was

FIG. 15 MILITOR WINCH PULLS SO HARD ON CLASS B TRUCK, STUCK IN SAND, THAT IT RAISES ITS OWN REAR WHEELS OFF THE GROUND AND STANDS ON THE SPRAG. TEN MILES WEST OF NORTH PLATTE, NEB., AUG. 5. 1919

FIG. 16 VIEW OF RIKER TRUCK BELONGING TO COMPANY E, 5TH ENGINEERS, TURNED OVER IN DITCH NEAR FULTON. ILL., JULY 22. 1919

truck No. 414,819 went through small culvert bridge and had to be jacked up and pulled out by Militor. Convoy escorted into Cedar Rapids by Lt. Gov. Moore, Mayor Rall. etc. Dinner served to entire command by Chamber of Commerce and Rotary Club. Made 87 miles in 10½ hr. Fair and warm. Good hard dirt roads. Arrived Cedar Rapids, Iowa, 5 p.m.

(Continued on page 205)

# Turbo-Compressor Calculations

### By ALLEN H. BLAISDELL, PITTSBURGH, PA.[1]

*After pointing out the features which differentiate the turbo-compressor from the multi-stage centrifugal pump, the author takes up the study of the pressure and volume changes occurring while a fluid is passing through one stage of the former and states the fundamental energy equations involved. He then discusses the number of stages to be used in a given case and follows this with an explanation of the method of laying out a compression diagram and of subdividing it according to the number of stages to be provided for.*

*The function of the impeller is next discussed and in this connection expressions are derived for the actual and effective impeller radii and the axial depth of channel of any radius. Blades and blade angles are also considered and a graphical method is described for calculating the theoretical head to be developed by an impeller.*

*The paper concludes with a brief study of the working of free-vortex and guide-vane diffusers, and of matters to be taken into account in designing in order to obtain a turbo-compressor of the highest possible overall efficiency.*

IT is well known that the turbo-compressor is quite similar in its nature to the multi-stage centrifugal pump. Whatever differences exist between the two types of machines is due to the fact that one handles a gaseous medium while the other handles a practically incompressible fluid, water. Both are high-speed machines.

The impellers of multi-stage centrifugal pumps are of the same size throughout, but those of turbo-compressors should be theoretically of continually diminishing dimensions because of the decrease in volume of the fluid as it is compressed from stage to stage. Actually the impellers are divided into two or more groups and each group calculated for its own average conditions. Since the capacities of these machines rarely exceed 50,000 cu. ft. per min., and in general range from 7000 to 10,000, the single-flow type of impeller is used in most turbo-compressors.

The only other important feature which differentiates the turbo-compressor from the centrifugal pump is the arrangement made for cooling the fluid as it passes through the compressor. In the reciprocating compressor practically all cooling is accomplished by means of intercoolers located between the different stages, the water jackets of the cylinders having but slight effect on the temperature rise of the fluid which is being compressed in the cylinders. In turbo-compressors the cooling is accomplished in two ways (a) by means of cooling coils between each stage, and (b) by means of carefully designed water jackets. The water-jacket method is much more effective in the case of the turbo-compressor than in that of the reciprocating machine because the fluid being compressed has a more intimate contact with the cooling surfaces.

The discussion will be divided into three main divisions:

A Thermodynamic Considerations
    a Study of the nature of the compression
    b Fundamental energy equations
    c Layout of compression diagram
B Pneumatic Considerations
    a Theory of fluid flow
    b Essentials of impeller design
    c Essentials of diffuser design
C Power Consumption and Efficiency.

## (A) THERMODYNAMIC CONSIDERATIONS
### NATURE OF THE COMPRESSION

Fig. 1 shows in section a portion of two stages of a multi-

---

[1] Instructor in Mechanical Engineering, Carnegie Institute of Technology. Assoc-Mem-Am.Soc.M.E.

Abstract of a paper presented at the Annual Meeting, December 1919, of THE AMERICAN SOCIETY OF MECHANICAL ENGINEERS. All papers are subject to revision.

stage centrifugal or turbo-compressor. The fluid being compressed enters the impeller at $a$ with a definite pressure, velocity and temperature. The impeller, rotating with a high speed, discharges the fluid with an increased pressure, temperature and velocity. Leaving the impeller the fluid enters the diffuser, where its velocity is greatly reduced and its pressure increased. The temperature will also be reduced somewhat, due to the cooling water which passes through the diffuser vanes. At $d$ the fluid leaves the diffuser and passes down between guide vanes, during which interval it is in contact with water-cooled surfaces $f$, and suffers a still further reduction of temperature and volume. At $c$ the fluid enters the succeeding impeller, where it goes through a similar cycle.

The change of state of the fluid being compressed will follow the law $pv^n = $ constant. The value of the exponent $n$ is governed by the kind of fluid being compressed, the moisture condition of the fluid and by the surrounding thermal conditions. When the fluid being compressed neither receives heat from, nor gives up heat to, surrounding bodies the compression is called adiabatic, and under these conditions $n = 1.41$ for air. When

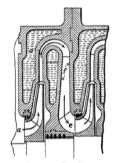

FIG. 1 SECTION OF PORTION OF TWO STAGES OF A MULTI-STAGE CENTRIFUGAL OR TURBO-COMPRESSOR

the nature of the compression is such that the temperature remains constant, then the compression is said to be isothermal and $n = 1$ for all gases.

It is very difficult to say exactly what pressure and volume changes occur while a fluid is passing through one stage of a turbo-compressor. The energy transfer to the fluid occurs in the impeller, which changes the angular momentum of each fluid particle, work being done upon the fluid as a whole. As a result of this energy transfer the fluid on exit from the impeller possesses an increased pressure and velocity. But this pressure is not the total pressure developed in the stage, and herein the action in a turbo-compressor differs somewhat from the reciprocating class of machines. In the reciprocating compressor work is done by the piston upon the fluid enclosed in the compressor cylinder, first compressing the gas to the discharge pressure and then ejecting the fluid from the cylinder. In the case of the turbo-machine the impeller does work upon the fluid which results in an increase not only of pressure but also of velocity, and if the amount of work is the same in both instances it is quite evident that the pressure rise in the impeller will not be so great as in the case of the reciprocating compressor, since a portion of the energy imparted to the fluid has the kinetic form. Consequently, in order to secure as large a pressure rise as possible, use is made of a diffuser wherein no further work is done on the fluid, but a portion of its velocity head is changed over into pressure head.

It is possible to study the pressure and volume changes in the impeller of a turbo-compressor by means of a $pv$-diagram just as in the case of the reciprocating machines. Thus, in Fig. 2, $ABCD$ may be considered as representing what occurs in an impeller. The line $BC$ will be the actual compression curve and must be located outside of the adiabatic $BF$. This is quite easy to understand, since there is likely to be but very little heat transfer through the walls of the impeller to or from the fluid because of the short interval of time which elapses while the fluid is passing through the impeller. Because of friction, impact, eddying and turbulent flow, extra work is done and this results in a compression which follows the polytropic curve $BC$ rather than the adiabatic $BF$. This means that the value of the exponent $n$ in

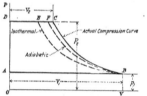

FIG. 2 DIAGRAM SHOWING PRESSURE AND VOLUME CHANGES IN THE IMPELLER OF A SINGLE-STAGE TURBO-COMPRESSOR

$pv^n = $ constant will be greater than 1.41. The temperature rise of the fluid must consequently be in excess of that due to adiabatic compression alone, and hence it is very essential that every effort be made to cool the fluid after it leaves the impeller, since by so doing the temperature and volume will be reduced before entrance into the succeeding stage and consequently the compression work of that stage will also be diminished.

Of the total energy imparted to the fluid in the impeller only a portion is in the form of pressure, the remainder being used up in increasing the velocity. Such being the case the expression for total work must contain a term to include the increase of kinetic energy as well as a term to cover the work of compression.

### NUMBER OF STAGES

The number of stages to be used depends upon the total pressure to be developed by the machine and the speed of the prime mover. Since the fluid being compressed is a gas and the terminal pressure for a single stage is proportional to the density of the gas, it is evident that a large number of stages will be required if very high delivery pressures are desired.

In selecting the proper number of stages a compromise has to be made between a few stages with large-diameter impellers and a large number with impellers of small diameter. If large impellers are used the frictional losses due to the fluid drag of the fluid on the impeller will be excessive, reducing the overall efficiency of the compressor. On the other hand, if small impellers are used the pressure rise per stage will not be so large and this means an increase in the number of stages, with a consequent increase in the cost of production, but the efficiency will be better than that of the first machine. In addition to reduced frictional losses there will also be a decrease in the compression work, and this factor helps to raise the efficiency above that of the machine with a small number of stages.

For equal amounts of compression work per stage the pressure developed will not be equal but will increase from stage to stage.

If the stages are divided into two or more sets, all impellers of a single group having the same diameter, then the same fact applies, but in this case the pressure rise will be the average for the group.

### LAYOUT OF COMPRESSION DIAGRAM FOR TURBO-COMPRESSOR

For illustrative purposes only, let us suppose that the diagram of Fig. 3 is that of a four-stage compressor. It is desired to study the peculiarities of the diagram and develop a graphical method of constructing such a diagram. The method to be explained was originated by M. de Stein.

The total energy transfer from all the impellers to the fluid passing through them is represented in the diagram by area $ABCDEFGHIK$. In other words, area $ABCD'$ is an equivalent $pv$-diagram for the first stage and represents not only the work of compression in the impeller but also the increase of kinetic energy. In the same way does $D'DEF'$ represent the energy transfer for the second stage. In other words, if $p_1$ is the inlet pressure to the first stage, then $p_2$ is the delivery pressure from the diffuser, and not the impeller. Consequently, the total pressure rise for the stage (impeller and diffuser) is $p_2 - p_1$. Such being the case, $p_2$ will be the delivery pressure of the machine if the kinetic energy in the fluid when discharged is neglected. The pressures $p_1$, $p_2$ and $p_3$ may be considered as the static pressures which exist between the diffuser of one stage and the entrance to the impeller of the following stage.

Good design necessitates that the areas $ABCD'$, $DEF'D'$, etc., shall be equal or that the energy delivered to the fluid shall be the same for each stage. The lines $BC$, $DE$, $FG$ are laid out as adiabatics. If the cooling is effective they will all lie in closer proximity to the isothermal $BM$ than the adiabatic $BN$. When there are a large number of stages, say eight or ten, it is allowable to assume the curve $BJ$ (passing through points $D$, $F$, $H$) as representing the equivalent compression curve.

To lay out the $pv$-diagram $ABJK$ for the machine requires that we shall know the initial pressure $p_1$, the initial volume $v_1$, in cu.

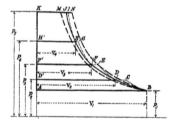

FIG. 3 COMPRESSION DIAGRAM FOR A FOUR-STAGE COMPRESSOR

ft. per lb., the delivery pressure $p_5$ and the value of $n$ in $pv^n = $ constant in order that the curve $BJ$ can be drawn.

Having laid out the isothermal and adiabatic curves it is now necessary to lay out the compression curve for the machine. In order to do this it is practically necessary to assume the position of point $J$ of the compression curve such that

$$v_J \eta_a (v_M - v_N) + v_N \quad \dots \dots \dots \dots [1]$$

where $\eta_a = $ adiabatic efficiency and $v = $ volume in cu. ft. per lb.

When the compression is adiabatic, if the final temperature is $T_N$ deg. abs., the adiabatic efficiency, or ratio between the theoretical internal work and the work actually furnished for the compression, can be expressed by

$$\eta_a = \frac{T_N - T_1}{T_J - T_1} \quad \dots \dots \dots \dots \dots [2]$$

where

    $T_J$ = final temperature in deg. abs. with adiabatic compression

    $T_N$ = final temperature in deg. abs. with actual compression

    $T_1$ = initial temperature in deg. abs.

If frictional resistances are considered, then the value of $\eta_a$ as calculated from [2] will be increased about 3 per cent.

## (B) PNEUMATIC CONSIDERATIONS

### DIVISION OF THE PRESSURE-VOLUME DIAGRAM

On completion of the actual compression diagram $ABJK$ (Fig. 5) it is next necessary to divide its area into the same number of parts as there will be stages or groups of stages in the compressor. In this instance it is four.

The total energy head developed by a single impeller or group of impellers can be stated in the form

$$H = \frac{p}{\delta_m} = K\frac{u^2}{g} \quad \ldots\ldots\ldots\ldots\ldots\ldots [3]$$

where

    $K = \eta_{ip} \times \theta$

and

    $\delta_m$ = mean density for a single stage, or group of stages

    $u$ = peripheral velocity of the impeller (or impellers) in ft. per sec.

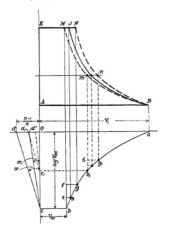

FIG. 4   METHOD OF LAYING OUT COMPRESSION DIAGRAM FOR TURBO-COMPRESSOR

    $\eta_{ip}$ = pneumatic efficiency for a single stage, or group of stages

    $\theta$ = form factor determined by the blade shape, angles, etc.

    $p$ = pressure rise for a single stage, or group of stages.

It is evident from Equation [3] that the compression diagram must be so divided into areas that the law $pv = $ constant shall hold true; that is, all the areas into which the $pv$-diagram is divided must be equal.

The following graphical method can be used with great facility in carrying out this subdivision, particularly when there are a good many stages to be considered.

Let $ABJK$ (Fig. 5) represent the compression diagram for the compressor. We will assume, as before, that the machine has four stages only and proceed to divide this diagram into parts such that the areas will all be equal.

Divide the distance $KA$ into any convenient number of equal lengths, say, six, marked by the figures 1, 2, 3, etc. Through these points draw horizontal lines 1-1, 2-2, 3-3, etc., cutting the diagram up into six trapezoidal areas which can be quite easily converted into equivalent rectangles as indicated in Fig. 6.

The areas of these rectangles are to each other as their bases. Reduce these bases by one-half, say, and add to each one of these the sum of the reduced lengths of the preceding ones. Through the ends of the lines thus drawn pass a smooth curve, $AabcdeC$. The length of any horizontal such as $KC$ represents to scale the area between it and the lines $JB$, $BA$, and $AK$. In other words, because of the construction the length $KC$ represents the area of the compression diagram. Such being the case, if we divide $KC$ into the same number of equal parts as there will be stages, or groups of stages (in this instance, four) and through the division points $H$, $F$, $D$, drop verticals intersecting the curve $AC$ at $I$, $G$, and $E$, respectively, then horizontals through $I$, $G$, and $E$ will

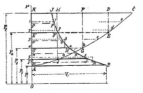

•   FIG. 5   PRESSURE-VOLUME DIAGRAM FOR COMPRESSOR

divide the compression diagram into four sections whose areas will be equal. The lengths $AI'$, $I'G'$, $G'E'$, etc., will represent the total pressure head to be developed by each stage, or group of stages.

If the line $AB$ represents the total volume of fluid taken into the compressor, then $mm$, $nn$, $oo$, $pp$, etc., will represent, to the same scale, the average volume for the various stages, or groups of stages. On the other hand, if $AB$ represents the specific volume of the fluid drawn into the first stage, then $mm$, $nn$, $oo$, $pp$, etc., will represent to the same scale the average specific volumes of the different stages.

If $Vm_1$, $Vm_2$, $Vm_3$, etc., are the mean specific volumes just defined, $Vm_1$ being the mean specific volume for the first stage, then the mean densities are

$$\delta_1 = \frac{1}{Vm_1}; \quad \delta_2 = \frac{1}{Vm_2}; \quad \delta_3 = \frac{1}{Vm_3}; \quad \text{etc.} \ldots\ldots [4]$$

It has already been stated that the pressures developed per stage are proportional to the fluid densities. Hence the ratio of the total pressure developed in one stage to that of the following or preceding stage must be the same as the ratio of the mean densities. If $p_1$, $p_2$, $p_3$, etc., are the discharge pressures per stage, we have

$$\frac{p_2}{p_1} = \frac{\delta_2}{\delta_1}; \quad \frac{p_3}{p_2} = \frac{\delta_3}{\delta_2}; \quad \frac{p_4}{p_3} = \frac{\delta_4}{\delta_3}; \quad \text{etc.} \ldots\ldots\ldots 5\|$$

where $p_1$ and $\delta_1$ are the pressure and density of the fluid on entrance into the first stage.

If a group of stages is made up of $n$ impellers and $p$ is the total pressure increase for the group, we can assume the average increase per impeller and diffuser (or stage) as given by

$$\Delta p = \frac{p}{n} \quad \ldots\ldots\ldots\ldots\ldots\ldots\ldots [6]$$

In reality the pressure increases will not be the same, as we already know, but the resulting error has no effect on the accuracy of the calculations as a whole. By using Equation [6] we virtually assume that the fluid is incompressible for a group of stages, or that the mean density of one stage is the same as those of the remaining stages of the group.

## IMPELLERS AND DIFFUSERS

In the design of impellers and diffusers for turbo-compressors we are forced to depend on the theory of centrifugal pumps, which at its best is far from being satisfactory or complete for centrifugal pumps even, and of course is far less so when applied to turbo-compressor calculations. However, by making certain assumptions it is possible to utilize the theory with quite satisfactory results.

In order to simplify the analysis of fluid flow it is necessary to make the following assumptions:

*a* That the fluid has stream-line motion, i.e., the fluid particles move in non-intersecting curves

*b* That the fluid will, since it is elastic, completely fill the channel of flow

FIG. 6 GUIDE-BLADE CONSTRUCTION. SHOWING EFFECTIVE IMPELLER RADII

*c* That the motion of the fluid is steady, i.e., each fluid particle has the same mean velocity as every other particle which is moving along with it through any section and the static pressure is the same at all points of a section which is at right angles to the fluid flow.

*d* That the fluid is non-compressible.

## THE IMPELLER

The function of the impeller is to impart a certain amount of energy to the fluid passing through it, with a maximum of efficiency. That is, friction, fluid impact, and turbulence should be reduced to a minimum.

On the assumption that the speed of rotation and the rate of fluid flow are known, there will be derived expressions for the actual and effective impeller radii, and the axial depth of the impeller channel at any radius. In addition to these there will also be given a brief discussion on blades and blade angles, together with the explanation of a graphical method for calculating the theoretical head to be developed by an impeller.

If we assume the impeller discharge to be zero, then from an equation given in the complete paper we obtain

$$W = \frac{144\, p_1 v_1}{\tau_{,i}} \cdot \frac{n}{n-1} \left[ \left( \frac{p_2}{p_1} \right)^{\frac{n-1}{n}} - 1 \right] \dots \dots [7]$$

where $\tau_{,i}$ = impeller efficiency, being practically identical with $\eta_i R_i$ but applying to the conditions of zero discharge only. This expression actually represents the work of compression performed in the impeller against the contripetal forces of the fluid particles. Since these forces vary from a minimum at the impeller entrance to a maximum at the discharge, the work done in forcing one pound of fluid from entrance to exit is equal to the average value of the centripetal force multiplied by the radial distance through which the pound of fluid is to move (with a negligible velocity). Such being the case, we have

$$F \times (R_2 - R_1) = W \dots \dots \dots [8]$$

where

$$F = \text{centripetal force} = \frac{N^2}{5908} (R_2 + R_1)$$

$R_2, R_1$ = actual outer and inner radii of the impeller in feet
$N$ = r.p.m. of impeller
$W$ = work of compressing 1 lb. of fluid in the impeller in ft-lb.
Substituting in [8] the value for $W$ given by [7] we finally get, if $R_1 = \sigma R_2$ and $\eta_{,i} = 0.75$,

$$R_2 = \frac{1095}{N} \sqrt{ \frac{p_1 v_1}{(1-\sigma^2)} \left[ \left( \frac{p_2}{p_1} \right)^{0.286} - 1 \right] } \dots \dots [9]$$

The value for $R_2$ given by [9] must be checked from the strength standpoint in order to fortify against the possible occurrence of dangerous stresses in the impeller at excessive speeds of rotation. In fixing upon final definite values for $R_2$ and $R_1$ care must be taken to see that the inlet area of the propeller is not too restricted by the actual fluid flow and causing an unnecessary loss of head at the impeller entrance.

The effective impeller radii $r_2, r_1$ (Fig. 6) are those at which the velocities $w_2, u_2, v_2, w_1, u_1, v_1$ are calculated and are not the same as the actual radii $R_2, R_1$. If an impeller is designed with velocity diagrams based on the radii $R_2, R_1$, it will be found that the impeller will not develop the desired head without an increase in the rated speed. It should be noted that if the impeller blades are radial at exit then the actual and effective radii (outer) will coincide. The determination of $r_2$ and $r_1$ is mostly a matter of judgment and will be taken up later on in the paper.

## IMPELLER BLADES

Just what shape of blade should be used is largely dependent upon the type of characteristic desired. In Fig. 7, *a* and *b*, are shown typical constant-speed characteristics $AB$ and $AC$. They represent in a general way how the pressure varies with the discharge from a single stage, curve $AB$ referring to curved-back

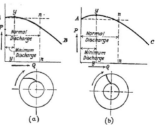

FIG. 7 TYPICAL CONSTANT-SPEED CHARACTERISTICS SHOWING HOW PRESSURE VARIES WITH THE DISCHARGE FROM A SINGLE STAGE

blades and $AC$ to radial blades. Both curves are supposed to show the combined effect of the impeller and diffuser. If the diffuser has been properly designed to suit the impeller and the casing, the characteristics $AB$ and $AC$ will clearly indicate the effect of the blade shape on the nature of the pressure characteristic. It will be noticed that the slope of $AB$ tends to grow steeper from the very first and the fact that it does rise at all is due to the effect of the diffuser. This so-called "drooping characteristic" is always typical when backward-curved blades are used. In the case of the radial blade the characteristic $AC$ has a decided rise from point $A$ and only starts downward some distance beyond $A$. Lines $yy$, in both figures, represent in a general way the relative locations of the maximum possible pressure and the corresponding magnitude of the discharge. With leaning-back blades the minimum possible discharge (accompanying the maximum possible pressure) from the compressor is about 30 per cent of the normal. With radial blades the minimum discharge ranges between 40 and 60 per cent of the normal, the normal discharge being indicated by lines $nn$ for both cases. The location of this lower discharge limit is of much importance, for if the volume is reduced any further, "surging" or "pumping" ensues, with consequent severe oscillations of the machine.

The radial type of blade is used to a large extent in turboblowers where the discharge pressure seldom exceeds 40 lb. per sq. in. gage. By using radial blades it is possible to use larger impeller diameters for a given speed of rotation. This means fewer stages with impellers of a simple but very strong construction.

The claim is sometimes made that the curving back of the blades at the outlet secures only a very small increase in the mechanical efficiency, but that this is accompanied with a relatively greater loss of manometric efficiency. This fact may be of importance in the case of fans and blowers, but it does not apply with equal emphasis in the case of turbo-compressors where the absolute velocity of discharge from the impeller is kept down to as small a value as possible by use of leaning-back blades in the impellers. Under any circumstances it is very difficult to convert much more than 50 per cent of the kinetic energy of the fluid, when discharged from the impeller, into pressure. When diffusers are used as a means of carrying out the conversion it is quite necessary that the absolute discharge velocity from the impeller shall be a minimum, otherwise there will be large pressurehead loss because of impact, eddying, etc., as will be more fully explained later on in the discussion.

If radial vanes are used, however, it is essential that they should be curved at the inlet to suit the relative velocity of the fluid to the vane at that point. Some manufacturers divide the blade into two parts, making the larger external portion straight and radial, a form that can be easily machined, and attaching to this a curved part at the inlet, which being small can be readily machined.

The total fluid head against which the impeller must work can be expressed in the following three ways:

$$H = \frac{v_z^2 - v_1^2}{2g} + \frac{144}{\delta_m}(p_z - p_1) + J(i_z - i_1) \quad \dots \dots \dots [10]$$

$$H = \frac{v_z^2 - v_1^2}{2g} + \frac{u_z^2 - u_1^2}{2g} - \frac{w_z^2 - w_1^2}{2g} + J(i_z - i_1) \dots \dots [11]$$

$$H = \frac{u_z(u_z - v_z \cos \alpha_z)}{g} + \frac{u_1(u_1 - v_1 \cos \alpha_1)}{g} + J(i_z - i_1) \cdot [12]$$

where

$H$ = theoretical head in feet of fluid (can be in terms of water head)

$v_z, v_1$ = absolute velocities of exit from and entrance to impeller, respectively

$u_z, u_1$ = peripheral velocities at exit and entrance to impeller, respectively

$w_z, w_1$ = relative velocities of discharge from and entrance to impeller, respectively

$i_z, i_1$ = sensible heat contents per lb. of fluid in B.t.u.

$J$ = 778

$\alpha_z, \alpha_1$ = angle made by vectors of $v_z$ and $v_1$ with those of $u_z$ and $u_1$

$p_z, p_1$ = absolute pressures in lb. per sq. in. abs. at exit and entrance, respectively.

Equations [10], [11] and [12] are, if the sensible-heat factor is neglected, well-known standard equations for centrifugal pumps and their derivation as given will be found in any textbook on turbine pumps. The increase in sensible heat is assumed to be that due to adiabatic expansion and friction in the impeller. If there are a large number of stages in the compressor the change in sensible heat per stage, due to adiabatic expansion, can be omitted when using the above equations. That due to friction is a very uncertain quantity and can be approximated only.

The head $H$ is the head which must be imparted to the impeller in order to obtain the actual head $H_A$; that is, $H$ must cover all pneumatic losses. If we let $H_L$ represent the loss of

head due to resistances to fluid flow, etc., and $\eta_p$ the pneumatic efficiency, then

$$\eta_p = \frac{H_A}{H} = \frac{H_A}{H + H_L} \quad \dots \dots \dots \dots \dots [13]$$

The pneumatic efficiency is analogous to the hydraulic efficiency of centrifugal pumps and is a measure of the pneumatic losses in the compressor. These losses cause a reduction in $H$ due to (a) the frictional resistances offered by the interior surfaces of the machine to the movement of the fluid, and (b) to too abrupt changes in the fluid path both in direction and cross-section.

In calculating the fluid velocities of entrance to and discharge from impeller, it is assumed, as mentioned before, that the fluid particles move through and discharge from the impeller with the same relative positions. In other words, the velocity at any cross-section of the impeller channel is the same for all the fluid particles. Of course, this can only be true for an impeller with an infinite number of blades of infinitesimal thickness. Actually, between any two blades the fluid stream has its maximum velocity at the center of the stream, while those particles adjacent to the blades are retarded in their motion due to the surface drag of the blade surfaces. From a practical standpoint, however, it is sufficient to assume that all the fluid particles have the same velocity as those of the middle filament. Such being the case, the effective radii, $r_z, r_1$ of the impeller will be different from actual exterior radii $R_z$ and $R_1$ (Fig. 6). The entrance- and exit-velocity diagrams should be laid out from the points $a$ and $b$, Fig. 6. If this is done there results an increase in $\eta_p$, bringing it closer to the value desired. Line 1–1 (Fig. 6) is drawn normal to the blade at point 2, while 3–4 is drawn normal to the blade at point 3. Points $x$ are located at the centers of these lines. It should be noted that the foregoing theory is to be applied with the expectation of good results, the impeller blades at entrance and exit must have relatively small divergence, thus insuring that the velocities across these sections will be fairly uniform.

### DIFFUSERS

On discharge from the impeller the energy content of the fluid, as previously shown, is largely in the potential and kinetic form. For a perfect impeller the division between the two would be nearly equal. Actually, however, this is not the case, the potential energy being smaller in value than the velocity energy. That is, the pressure head will be less than the velocity head. In order to convert part of this velocity head into pressure head, use is made of so-called diffusers. These can be made in two forms, the free vortex and the guide vane. As a matter of fact, the second or guide-vane diffuser is really at normal loads a freevortex diffuser, providing the guide vanes have been properly designed. Of course, this is not the case under ordinary conditions, particularly in a multiple-stage machine where it is desirable that the discharge from the diffuser shall be radial, or at least not far from radial.

On discharge from the impeller a fluid particle, if unhindered, will trace out a spiral path providing the space surrounding the discharge side of the impeller has sufficient radial depth. In other words, the fluid as a whole will have a free vortical motion and it can be shown that the path of any fluid particle is a logarithmic spiral, the equation of which is

$$r = e^{k\theta} \quad \dots \dots \dots \dots \dots \dots \dots [14]$$

where

$e = 2.7183$

$k$ = constant

$\theta$ = angle of radius vector with entrance position.

With free vortical motion of the discharge the maximum possible gain of pressure is as given by one of the equations in the complete paper. Due to instability of flow, friction of side walls, impact, etc., however, which cause the stream lines to be widely divergent from the theoretical, the gain of pressure given by this equation is not attainable in practice. If a free-vortex diffuser is used it is essential that its radial depth be sufficient for the vortical motion to develop and thus secure as large energy conversion as possible. The axial depth of the diffuser on the entrance side should be the same or slightly greater than that of the impeller discharge. This axial depth may remain constant or gradually diverge from inlet to outlet of diffuser. It is of course quite essential that the inner surfaces of the diffuser shall be as smooth as possible in order that friction shall be eliminated.

Because it is practically impossible to obtain non-turbulent motion in a free-vortex diffuser the use of guide vanes has been introduced. These are located in the diffuser channel and divide it up into several separate sections, through which the fluid can flow with a reduced tendency to become turbulent. As a consequence the conversion efficiency is higher than in the case of the diffuser without guide vanes. In practice it is difficult to transform much more than 50 per cent of the kinetic energy possessed by the fluid on entrance to the diffuser into pressure energy by the time the fluid is discharged. With a guide-vane diffuser the best conversion efficiency is obtained at normal load when the fluid leaves the impeller in such a direction as to enter the diffuser sections with a minimum of shock, eddying, etc.

The number of guide blades should be sufficient to give the desired uniformity of flow without undue friction, usually one-half to two-thirds as many guide blades as impeller blades. If the same number of blades are used in both there may result severe periodic impulses or vibrations due to the fact that the clear cross-section of the fluid channels at discharge of impeller and entrance to diffuser are increasing and decreasing in unison.

Theoretically the shape of the guide blades should be that of a logarithmic spiral to conform with the free path which the fluid tends to follow on leaving the impeller. Actually, the blade shape is governed by other requirements. A favorite form of guide-blade construction is that shown in Fig. 6, and is similar to those used on turbine pumps. The cavity inside the blades is connected with the cooling-water circuit and thus some reduction in the temperature of the fluid is obtained while it flows through the diffuser. The blade curves are customarily volutes or involutes on the entrance side of the diffuser and a combination of circular arcs and straight lines for the remainder of the blade length.

## (C) POWER CONSUMPTION AND EFFICIENCY

In order to compete successfully in the industrial field with the reciprocating compressor it is of great importance that the overall efficiency of the turbo-compressor be made as high as possible. That is, the difference between the compressor input (work delivered to the compressor) and output (work obtained from compressor) should be reduced to a minimum. The area of the $pv$-diagram represents the useful work done in the compressor, but it does not cover the pneumatic, mechanical and volumetric losses. If $\eta_p, \eta_m, \eta_v, \eta_t$ are the pneumatic, mechanical, volumetric and overall efficiencies, respectively, then the shaft horsepower must be given by the following expression neglecting the kinetic energy of discharge:

$$\text{Shaft Hp.} = \frac{0.261 \times L \times p_i v_i}{\eta} \times \frac{n}{n-1}\left[\left(\frac{p_i}{p_i}\right)^{\frac{n-1}{n}} - 1\right] \quad [15]$$

where

$\eta$ = varies between 0.55 and 0.65 and equals $\eta_p \times \eta_m \times \eta_v$
$L$ = lb. of fluid compressed per sec.
$p_i - p_i$ = total pressure rise in the machine in lb. per sq. in. abs.
$n$ = compression exponent and is calculated from an equation given in the complete paper.

The mechanical efficiency $\eta_m$ is the ratio between the total power actually delivered to the fluid and the shaft or brake horsepower. It is a measure of the power losses, made up of rotation loss (friction between outer surfaces of impellers and the surrounding fluid) and bearing loss or similar losses due to friction between metal surfaces. The rotation loss will vary between 10 per cent and 20 per cent of useful output of the machine, being determined by the density of the fluid, the speed of rotation, and the impeller diameters. For a single impeller it can be calculated very approximately from the following expression:

$$\text{Friction Hp.} = 0.44 \times 10^{-14} \times \mathfrak{F}_m \times N' \times D^5 \dots [16]'$$

where

$N$ = r.p.m.
$D$ = impeller diameter in ft.
$\mathfrak{F}_m$ = mean density of fluid for the stage.

Equation [16] is based upon experiments with constant-thickness disks in open air. It is readily seen from [16] why it is preferable to employ a large number of small-diameter impellers in place of a lesser number with large diameters. With a given discharge any increase in pressure must be accompanied by either an increase in speed or else an enlargement of the impeller diameter. Since the frictional resistance varies as the speed cubed and the impeller diameter to the fifth power, it is more economical to use high speeds with small impellers.

The volumetric efficiency is, of course, the ratio of the fluid actually delivered by the compressor to that taken into the compressor, and will vary between 96 and 98 per cent, a good average being 97 per cent. Most of this leakage will occur in the last stage between the inlet to that stage and the atmosphere, and is not materially affected by the number of stages.

Under ideal conditions the compression would follow the isothermal curve $BM$ on the $pv$-diagram, Fig. 3. The compression efficiency will be, when referred to the isothermal,

$$\eta_o = \frac{\text{isothermal work}}{\text{actual work}} = \left(\frac{n-1}{n}\right)\frac{T_i}{T_i - T_i} \log \frac{p_i}{p_i} \dots [17]$$

and will vary between 60 and 67 per cent.

The need of cooling the fluid as it passes through the compressor is emphasized by [17]. The heating of the fluid during compression renders it difficult to obtain efficiencies which are materially better than those attainable with reciprocating machines of modern construction. The amount of heat to be dissipated is more than that due to adiabatic compression since the mechanical energy loss due to short-circuiting, eddy currents, fluid impact, and fluid friction reappears as heat energy stored very largely in the fluid itself. By careful design of the cooling-water passages in the casing and diffusers, and by use of a large number of stages, it is possible to carry off most of this sensible heat. In spite of the use of a water-cooled housing, the temperature rise of the air in some turbo-air compressors has been over 200 deg. fahr., the inlet temperature being in the neighborhood of 70 deg. fahr. In the first few stages the cooling effect is small because of the small temperature difference between the air and the cooling water, but in the later stages the compression approaches very closely to being isothermal, as we already know, and the temperature of the air when discharged will range between 150 deg. fahr. and 175 deg. fahr. for discharge pressures of from 75 to 120 lb. per sq. in. gage.

What the Bureau of Mines has done for the great coal-mining industry, chiefly through investigations at the experiment station at Pittsburgh, Pa., has been published in numerous reports issued by the Bureau. Some of the more important accomplishments have been the development and introduction of permissible explosives for use in gaseous mines, the training of thousands of coal miners in mine-rescue and first-aid work, and the conducting of combustion investigations, aimed at increased efficiency in the burning of coal and the effective utilization of our vast deposits of lignite and low-grade coal. How vast are the deposits of low-grade ores being made available through the experiment stations is shown by the work assigned to the station at Minneapolis, Minn. The primary purpose of this station is to devise methods of utilizing low-grade iron ores. It has been estimated that the reserves of low-grade magnetic iron ores in Minnesota alone amount to some forty billion tons, but until recently these ores have been untouched because no process of treating them profitably has been devised. The Minneapolis station has already demonstrated that one process for utilizing the great deposits of manganiferous iron ore on the Cuyuna Range is metallurgically possible.

¹ See pp. 220 and 223, Bulletin of the Bureau of Standards, vol. 10, on Windage Resistance of Steam-Turbine Wheels.

# Self-Supporting Chimneys to Withstand Earthquake

### Nature of Stresses Developed by Seismic Shocks; Theory of Design of Chimneys to Withstand Such Stresses; Particulars Regarding Chimneys in Which Theory Has Been Applied

#### By C. R. WEYMOUTH,[1] SAN FRANCISCO, CAL.

*The studies of the Japanese Imperial Investigation Committee on the nature of the stresses developed in chimneys in consequence of earthquakes are summarized by the author, who also examines the conclusions of the committee in the light of data on failures of chimneys secured from inspections of wrecks caused by the San Francisco earthquake of 1906. An advancement to the theory developed by Professor Omori, Chairman of the Japanese Committee, on the design of chimneys to withstand earthquakes, is then presented. This assumes that the chimney, instead of being looked upon as rigid, should be considered as a body oscillating about the center of percussion. Reference is made in the paper to the design of various chimneys where the theory outlined has been applied.*

BECAUSE of the prevalence of earthquakes in Japan, and the immense damage resulting from them to buildings and other structures, the Japanese Government, a number of years ago, created the Imperial Earthquake Investigation Committee with a view to having it undertake a scientific study of earthquake phenomena, as well as of the effect of earthquakes on structures, etc. The Committee's reports would fill a good-sized volume. Many of these have been translated into English, and are obtainable at a few of the leading libraries in this country. Professor Omori, chairman of this committee, is the world's greatest authority on earthquake problems.

Although this subject is one of more or less academic interest, there are, however, certain localities in this country, mainly on the Pacific Coast, and many foreign localities, where the design of structures necessitates a careful consideration of the forces due to earthquakes. With the increasing tendency to invest American capital in foreign countries, American engineers will be required to design structures to withstand earthquakes.

A brief review of the work of Professor Omori, together with certain data pertaining to the San Francisco earthquake of 1906, a discussion of what the writer believes to be a limitation in a portion of Omori's work, and also an advancement in the theory of the design of chimneys to withstand earthquake are presented in this paper.

Earthquakes are due to various causes, the principal one being a slippage of the crust of the earth along fault lines. Earthquakes, of course, vary from mild trembles, barely perceptible and lasting but a few seconds, to violent shocks lasting several minutes and capable of demolishing very substantial structures. The principal movement is in a horizontal plane, but the vibration occurs in a number of directions in the same plane. There is generally a slight, but sometimes a severe, movement in a vertical direction, and also a severe twisting effect. On account of the combination of horizontal and vertical motion, earth waves visible to the eye frequently occur during earthquakes.

Earth movements are recorded at most astronomical observatories by means of seismographic instruments which record the displacement of the earth in three coördinates by measuring the amplitude, frequency, and duration of the earth wave.

By means of various earthquake records, values have been established for the maximum rate of acceleration of an earth particle, the maximum amplitude of an earth particle during a complete vibration, and the time for the complete period of vibration; and from these data it is possible to estimate approximately the stresses and strains in simple structures subject to earthquake shock. A sufficient number of earthquakes have been recorded and classified during recent years to make known the

maximum intensity of shock experienced in certain localities, and the best that an engineer can do is to design a structure based on recorded experience. There will, however, always be the doubt as to whether some future earthquake may not be more severe than those which have been recorded in the past.

The most severe earthquake recorded in Japan was the Mino-Owari earthquake of 1891, for which Omori gives the following estimates:

Displacement of an earth particle .................. 233 mm.
Complete period of the principal motion, being a wave
   cycle ................................................. 1.3 sec.
Duration of earthquake ................................ 4 min. 30 sec.
Maximum acceleration observed at different localities
   ranged from.............. 3,000 to 10,000 mm. per sec. per sec.

For the purpose of classifying various earthquakes Omori has devised an absolute scale of earthquake intensities as follows:

The following absolute scale of destructive earthquakes, or the relation between the maximum acceleration of the earthquake motion and the damage produced, has been deduced chiefly from analysis of the Mino-Owari earthquake, the intensity being arbitrarily divided into seven classes. It is to be noted that the scale applies principally to Japan.

1 Maximum acceleration, 300 mm. per sec. per sec. The motion is sufficiently strong to cause people generally to run out of doors. Brick walls of bad construction are slightly cracked; plasters of some old dozo (godowns) shaken down; furniture overthrown; wooden houses so much shaken that cracking noises are produced; trees visibly shaken; waters in ponds rendered slightly turbid in consequence of the disturbance of the mud; pendulum clocks stopped; a few factory chimneys of very bad construction damaged.

2 Maximum acceleration, 900 mm. per sec. per sec. Walls in Japanese houses are cracked; old wooden houses thrown slightly out of the vertical; tombstones and stone lanterns of bad construction overturned, etc. In a few cases changes are produced in hot springs and mineral waters. Ordinary factory chimneys are not damaged.

3 Maximum acceleration, 1200 mm. per sec. per sec. About one factory chimney in every four is damaged; brick houses of bad construction partially or totally destroyed; a few old wooden dwelling houses and warehouses totally destroyed; wooden bridges slightly damaged; some tombstones and stone lanterns overturned; shoji (Japanese paper-covered sliding doors) broken; roof tiles of wooden houses disturbed; some rock fragments thrown down from mountain sides.

4 Maximum acceleration, 2000 mm. per sec. per sec. All factory chimneys are broken; most of the ordinary brick buildings partially or totally destroyed; some wooden houses totally destroyed; wooden sliding doors and shoji mostly thrown out of the grooves; cracks 2 or 3 in. in width produced in low and soft grounds; embankments slightly damaged here and there; wooden bridges partially destroyed; ordinary stone lanterns overturned.

5 Maximum acceleration 2500 mm. per sec. per sec. All ordinary brick houses are very severely damaged; about 3 per cent of the wooden houses totally destroyed; a few tera, or Buddhist temples, thrown down; embankments severely damaged; railway lines slightly curved or contorted; ordinary tombstones overturned; ishigaki, or masonry walls, damaged here and there; cracks 1 or 2 ft. in width produced along river banks; waters in rivers and ditches thrown over the banks; wells mostly affected with changes in their waters; landslips produced.

6 Maximum acceleration, 4000 mm. per sec. per sec. Most of the tera, or Buddhist temples, are thrown down; 50 to 80 per cent of the wooden houses totally destroyed; embankments shattered almost to pieces; roads made through paddy fields so much cracked and depressed as to stop the passage of wagons and horses; railway lines very much contorted; large iron bridges destroyed; wooden bridges partially or totally damaged; tombstones of stable construction overturned; cracks a few feet in width formed in the ground, accompanied sometimes by the ejection of water and sand; earthenware buried in the ground mostly broken; low grounds, such as paddy fields, very greatly convulsed, both horizontally and vertically, sometimes causing trees and vegetables to die; numerous landslips produced.

[1] Chief Engineer, Chas. C. Moore & Co. Engineers. Mem.Am.Soc.M.E.

7 Maximum acceleration much above 4000 mm. per sec. per sec. All buildings, except a very few wooden-houses, are totally destroyed; some houses, gates, etc., projected 1 to 3 ft.; remarkable landslips produced-accompanied by faults and shears of the ground.

In the above scale of the seismic intensity, the earthquake motion has been assumed to be entirely horizontal. This supposition would not, except in places very near to the epicenter, cause sensible errors in the result.

A complete report was made of the San Francisco Earthquake by the California Earthquake Commission. This report, which has been published by the Smithsonian Institute, gives numerous charts, diagrams, photographs along fault lines and of wrecked structures, etc., and its study will be of value to any one desiring to pursue this subject further.

Seismographic records were almost valueless for the San Francisco earthquake because practically all of the instruments at the local observatories were put out of commission by the intensity of the shock, the full amplitude greatly exceeding their limits of movement. Omori visited San Francisco soon after the earthquake and, judging from the damage done, estimated the complete amplitude of the most severe shock to have been about 4 in., and the complete period of vibration of an earthquake wave about 1 sec. He also noted shear along the fault lines at different points from 16 to 20 ft.

Professor Lawson, a member of the California Earthquake Commission, estimated, from the Omori scale, that the acceleration for various formations in San Francisco had been as follows:

| Foundation | Acceleration mm. / sec.² |
|---|---|
| Serpentine | 250 |
| Made land | 1100 |
| Marsh | 3000 |
| Sandstone | 250 to 600 |
| Made land | 2900 |
| Sand | 600 |
| Sandstone | 400 |
| Sand (Mission Valley) | 1100 |
| Marsh | 3000 |
| Sundry solid rocks | 250 |

It will be noted from these data that earthquake shocks are least severe on structures built on rock foundations and most severe on structures built on loose or filled ground.

Omori discusses the force due to earthquake as impulsive or gradual, defining the former as a force applied to an elastic body so rapidly that it is finished in a time interval infinitely small in comparison to the period of natural vibration of the body, and the latter as a force applied so slowly that the body assumes a position of equilibrium without being thrown into vibration.

Within moderate limits of intensity of earthquake shock, Omori shows that structures are likely to fail which have a natural period of vibration materially less than the period of the earthquake, and structures are not likely to fail which have a natural period of vibration equal to, or greater than, the period of the earthquake. It is apparent that structures in the former class are short in comparison with their width, or diameter, and, until failure occurs, are accelerated throughout their mass by the earthquake vibration, inducing stresses resulting from the acceleration of the mass of the structure as a whole. Structures of the latter class are generally tall in comparison with their width, are more or less flexible, and, owing to their natural period of vibration with respect to the earthquake period of vibration, are able to yield in a measure to the earth movement, and are not accelerated as a whole at a rate equal to the earth's acceleration, and, consequently, are stressed to a less extent than in more rigid structures.

Omori brings out the well-known proposition that a force impulsively applied to an elastic body produces a stress double that caused by the same force when gradually applied, and further states that since brick fractures immediately the limit of elasticity is exceeded, the strength of a brick column against a force implied impulsively is half that against the same force applied gradually.

Having reduced the investigation of earthquakes to a basis of numerical acceleration and stated periods of vibration, Omori proceeds to show that for any given structure, such as a vertical prism resting on its base, the force necessary to accelerate the mass is to be computed from the well-known formula that the force is equal to the total mass of the structure, multiplied by the acceleration, divided by $g$; assuming, of course, that the mass is accelerated as a whole. This force may be considered as applied at the center of gravity, and it exerts a turning moment at the base of the prism equal to the computed force multiplied by the height of the center of gravity above the base.

In one of his reports Omori gives the results of investigation of forty-nine self-supporting brick chimneys. These failed at points from 24 per cent to 94 per cent of the height of the chimney, the average failure being at a point about two-thirds of the height of the chimney,. while practically none of the chimneys failed at the base. He points out that the point of failure corresponds approximately to the center of percussion of the chimney with reference to a horizontal axis at the base of the chimney. Unfortunately, no self-supporting steel chimneys nor reinforced-concrete chimneys were in existence in Japan during the severe earthquake shocks of 1891 and 1894.

Omori shows that the failure of chimneys at the center of percussion instead of at the base of the chimney indicates a tendency to oscillate, as it were, about the center of percussion. There is a whip-snapping effect of the upper portion of the chimney which may cause a condition of resonance. In chimneys which have failed portions of the top of the chimney have been thrown down, striking the ground never at a great distance from the base of the chimney.

The writer investigated the failures of a number of chimneys in the San Francisco earthquake of 1906, and found that brick chimneys failed in about the same manner as observed by Omori. The tallest chimney investigated was 240 ft. in height, and failed at a point about two-thirds its height. Many of the large brick chimneys in the San Francisco District failed, although a few on rock foundation and especially well built did not fail by having a portion of the chimney height demolished; but the chimneys that did not suffer in this way were at least cracked either vertically or horizontally.

There were no reinforced-concrete chimneys in San Francisco at the time of the earthquake, and but one large self-supporting steel chimney, this being 10 ft. in diameter by 120 ft. in height. This chimney, although brick-lined, was not injured by the earthquake. A 150-ft. self-supporting steel chimney at the Mare Island Navy Yard was uninjured by the earthquake. A 217-ft. brick-lined self-supporting steel chimney at Salinas, Cal., likewise was uninjured by the earthquake.

Omori made an interesting series of experiments with brick columns on a shaking table, duplicating in effect earthquake motions of varying intensities. The results of these experiments are too elaborate for presentation in this paper. They showed conclusively that estimates of the tensile strength of brickwork are exceedingly uncertain and undependable, owing to the variation in the character of workmanship in laying the brick. Omori has measured the natural period of vibration of a number of self-supporting brick chimneys. One chimney 50 ft. in height was found to have a natural period of vibration of about one second.

Recently a very tall concrete chimney has been erected in Japan for the Kuhari Mining Co., Saganoseki, in the Oita Prefecture, Kyushu. The main shaft has a height of 550 ft.; the inside diameter at the top is 23 ft. 6 in.; the base is circular and 95 ft. in diameter; the natural period of complete vibration of the chimney was observed by Omori to be 2.5 sec. The range of motion or double amplitude of the top of the chimney was 1 in. at a strong gale of 24 m. per sec., but it reached 7.7 in. at a hurricane wind blowing 35 m. per sec. The chimney is located on hard rock, on which it is estimated an earthquake movement would be comparatively weak, probably having an intensity of less than 500 mm. per sec. per sec. A 100-ft. reinforced-concrete chimney 3½ ft. inside diameter at the top was found to have a period of vibration of 0.8 sec.

Some interesting observations of a chimney at Providence, R. I., are given by Dr. D. S. Jacobus, in his paper on Counterweights for Large Engines.[1] The stack was built on a mass of concrete

[1] Trans. Am. Soc. M. E., vol. 26 (1906), p. 531.

which extended under the building structure. Referring to the vibration of the chimney due to the reciprocating engine, Dr. Jacobus states:

At a point near the extreme end of the foundation where the No. 1 engine was located the foundation was found to shake 0.008 in. Measurements made near the top of the chimney, which was erected to the height of about 175 ft. above the ground, showed that the maximum shake with the engine running at its ordinary speed of 90 r.p.m. was about 0.02 in. After measuring the vibration of the chimney with the engine running at its ordinary speed the engine was shut down and a marked result took place when its speed fell in harmony with the time of vibration of the chimney. When this occurred, the chimney shook to such an extent that the motion was beyond the range of the special instrument. The total movement of the pointer of the instrument was such that the chimney was shown to move more than one-eighth of an inch. In constructing the chimney the workmen had noticed that when they came to a height of about 130 ft. the vibration was much greater than it was after the chimney was built higher. This made it appear that at the height of 130 ft., at which there was the most shake, the time of the vibration of the chimney was in harmony with the number of revolutions made by the engine.

Professor Omori has also collected many data with respect to the effect of earthquakes on bridge piers.

Following the San Francisco earthquake and fire the city of San Francisco installed an elaborate high-pressure fire-fighting system, including two high-pressure salt-water pumping plants; the structures were designed to withstand earthquakes by utilizing the latest information obtainable from the San Francisco Earthquake Commission, an acceleration of 6.2 ft. per sec. per sec. being stipulated. Pumping plant No. 1 included four chimneys specified not to exceed 90 ft. in height, to minimize the earthquake risk, and measuring 64 in. inside diameter at the top. These chimneys were finally made of reinforced concrete and self-supporting.

As the writer's firm undertook to build a substantial part of this pumping plant he engaged a local engineer of prominence to make a special design of chimney to withstand earthquakes, and in the course of the next few years several other chimneys were designed in the same way. These designs followed Omori's theory, as outlined above; the bending moments due to earthquake were calculated for all sections, assuming the lateral displacement of the chimney to result from the earthquake acceleration imparted to the mass of the structure as a whole and the chimney shaft to remain vertical at all times. This theory gave a strength of chimney at the base materially greater than at the higher sections, and in view of Omori's observation of the frequency of breakage at two-thirds of the height, the reinforced amount was empirically increased at this section. No severe earthquakes have tested the accuracy of the design. Recently the writer had occasion to examine the subject further in connection with the erection of a chimney 220 ft. in height and 14 ft. in inside diameter at the top.

The chimney in question was constructed during the war time, and on account of the prohibitive price of self-supporting steel chimneys, decision was made in favor of reinforced concrete.

The writer concluded that while the chimney might have a tendency to sway somewhat about the base as an axis, the principal design should be made on the basis of accelerative rotation of the chimney about its two-thirds point, instead of a motion of accelerated translation parallel to itself, as heretofore assumed. The theory of rotation harmonizes with the observed failures of chimneys, whereas the theory of translation does not.

Considering the chimney as a rigid body and assuming the two-thirds point to be at rest and the acceleration of the earth to be known, then the angular acceleration about the two-thirds point becomes immediately known, and from the principles of moment of inertia the bending moment at different sections is easily computed. Such theory, however, would only apply if the chimney were under no strain at its base or foundation. If such rotation did exist, the chimney remaining rigid, then there must occur either a yielding of the soil under the chimney foundation, or a deflection or fracture of its base, or both. An investigation of the foundations of chimneys which had failed in the San Francisco earthquake did not lend support to this theory, and the writer finally concluded that the element of flexibility of the chimney shaft played a most important part in chimney design.

A further important consideration in chimney design is the deflection of the chimney from the base to its neutral point, assumed to be at two-thirds of its height. It is only on the assumption that a tall chimney will deflect within its elastic limit one-half the amount of the earth's vibration that its stability during a severe earthquake can be maintained without rupture.

The failure to find serious horizontal cracks near the base of the shaft or in the foundation of some large self-supporting brick chimneys can only be explained on the theory of deflection. This theory is not touched upon by Omori in any of his reports.

Fortunately it is comparatively easy to make an approximate estimate of the elastic deflection of a self-supporting chimney, but unfortunately the writer is not able to indicate the factor of safety to be applied to this deflection as compared with the earth's vibration.

The writer consulted E. P. Lewis, Professor of Physics, University of California, who submitted a statement as follows:

Since seeing you I have considered the questions raised in the memorandum which you gave me regarding the design of chimneys. As I told you, the problem seems so complex, and the factors so obscure, that I can do no more than make some suggestions regarding the physical principles involved.

In the first place, referring to Omori's view that the center of percussion of the chimney remains at rest under the impulse of an earthquake, Omori evidently makes this assumption by considering the case of an impulse at one end of the uniform rod on which no other forces act. In this case the center of percussion is a fixed axis of rotation, as is easily shown by considering that the resultant moment of momentum, due to the impulse, must vanish if taken about the point at which the impulse is applied. In the case of the uniform rod this is one-third the length of the rod from the opposite end. I think, however, that Omori's assumption is incorrect. The above applies to a body free from external forces after the application of the impulse, whereas the chimney remains fixed to its base, and the principle applies to a perfectly rigid body, which the chimney is not.

As a matter of fact, I am confident that something like this happens: The impulse comes along; an elastic wave runs up the chimney; during its first passage the center of percussion has a minimum motion; the earth movement ceases; the inertia of the chimney carries it forward until it is brought to rest by elastic reactions, which cause it to vibrate in the opposite direction, and any effect at the center of percussion vanishes. The amplitude of the motion of the top of the chimney (ventral segment of the vibration) probably is many times greater than that of the earth. If the earth vibrations are approximately synchronous with those of the chimney, the amplitude may become very great; if the frequency of the earth vibration is three times that of the chimney, "overtones" will appear with a nodal point one-third of its length from the top of the chimney (if of uniform section), thus adding to the possible effect due to the center of percussion.

A wooden or glass rod clamped to a massive base, which is then subjected to a shock or vibration, behaves as I have indicated above, and I am sure that a tall chimney behaves in qualitatively the same way.

I think you are right in considering that the vibratory motion of the chimney is of fundamental importance, and that account must likewise be taken of vertical displacements. Below [See Fig. 1] I sketch some of the phases of the motion of the chimney as I imagine it.

It occurs to me that especially violent effects might occur if succeeding non-periodic shocks come at such phases of the chimney vibration due to the first shock as to suddenly increase the acceleration.

I have given my ideas so far as they are clear to me. I do not speculate further because the problem is so difficult and obscure that I might easily suggest some error.

The sketches submitted by Professor Lewis are shown in Fig. 1. At a is shown a vibrating rod, fixed at the base, in its initial position; at b the effect of the center of percussion; at c the effect when the center of percussion vanishes; and at d the effect of a three-to-one overtone.

The ordinary chimney having much less flexibility than the experimental rod considered by Professor Lewis, it is the writer's opinion that the extreme effects which Professor Lewis mentions would not be experienced by a chimney with intensity of shock equal to that of the San Francisco earthquake.

In completing the design of the chimney in question, it was necessary for the writer to make certain assumptions, as follows:

1 That the minimum strength at any section must be equal, of course, to that required to withstand wind pressure; that an earthquake would not occur at a time of the hurricane wind velocity for which the chimney is designed; and that the addition of these forces would be unnecessarily conservative.

**2** That the minimum strength at any section to withstand earth-quake, should correspond to the equal acceleration of the chimney as a whole, considered as a rigid body, the axis of the chimney remaining vertical at all times; due to its flexibility, the chimney is not accelerated equally throughout, so that the writer assumed that this calculated moment should not be doubled, as customary for vibration.

**3** That the minimum strength at any section to withstand earthquake should correspond to that required for the angular acceleration of the chimney about the center of percussion, as explained above; allowance being made for the doubling of stresses due to the vibratory or whip-snapping action of the chimney; a factor of safety must still be applied on account of the uncertainty of such calculations, as further outlined below.

**4** That the vertical acceleration of the chimney should be considered as resulting from the vertical component of the earth wave. The stress computed in this manner will be generally found to be very small.

**5** That in the case of reinforced-concrete chimneys the temperature stresses must be computed for temperatures correspond-

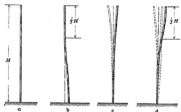

FIG. 1  VIBRATIONS OF AN ELASTIC ROD FIXED AT ITS BASE<br/>*a* Initial position. *b* Effect of center of percussion. *c* Effect when center of percussion vanishes. *d* Effect of a 3:1 overtone.

ing to the average overload on boilers, and this condition should be assumed as being coexistent with a possible earthquake, the temperature stresses being added to earthquake stresses.

**6** That the foundation or base of the chimney should be very carefully investigated with respect to the moments produced at the base of the chimney by the earthquake, and transmitted through the foundation.

**7** That the elastic deflection of the chimney through its two-thirds height, measured from the base, should be somewhat greater than the amplitude of an earth particle, the amplitude being taken as half the total movement of an earth particle.

As a check on the above theory of design, investigation was made along parallel lines of the brick-lined self-supporting steel chimney of the Spreckels Sugar Company, at Salinas, Cal., details of which are here reproduced, with the permission of the Spreckels Sugar Company, in Fig. 2. This chimney was erected in 1897, and was presumably designed to withstand wind pressure only. Having an inertia effect comparable with the chimney under consideration, and having successfully withstood the action of an earthquake of intensity approximately the same as that estimated for the reinforced-concrete chimney, it afforded a standard of comparison for testing the computed strength of the reinforced-concrete chimney. For any section, the ratio of the assumed elastic limit of the steel to the computed tensile stress, according to the above theory, gave a factor in all cases greater than unity, indicating either that, if the above theory is correct, the earthquake intensity could have been multiplied by this factor before the elastic limit of the steel would have been reached; or that the whip-snapping and inertia effects, due to earthquake, resulted in stresses greater than calculated, by a ratio numerically equal to this factor.

If the factor were not applied it does not necessarily follow that the reinforced-concrete chimney would be unsafe. It may, and probably does, indicate that the strength of the Spreckles chimney was somewhat greater than required to withstand the San Francisco earthquake.

The factors of safety, mentioned under [3], finally used, were as follows: Starting with a factor of 1 at the 20-ft. section (measured from the ground line) these factors were gradually increased up to 2.5 at the critical section, or two-thirds of the height, and this factor was continued to the top of the chimney. The amount of steel required for this increased safety in the upper or critical sections was relatively small and inexpensive.

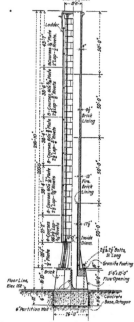

FIG. 2  SELF-SUPPORTING STEEL CHIMNEY, SPRECKELS SUGAR CO.,<br/>SPRECKELS, CAL.

On the other hand it seemed the part of wisdom to apply to the chimney such factors as would produce a stability against earthquake approximately equal to that of a chimney which had withstood earthquake shock.

Experience extending over a long period of time will alone indicate the proper factors to be applied in chimney design in order to insure the structure against earthquake acceleration. This, for the reasons stated above, was impossible in San Francisco. According to Omori, fifty years may elapse before another severe shock; the preceding one was in 1868.

The method used in computing the temperature stresses in the

(*Continued on page* 204)

# Electric Power Supply of Southern Textile Mills

By J. E. SIRRINE,[1] GREENVILLE, S. C.

IN that section of the South comprising the Carolinas, Georgia, Alabama, and Tennessee there are 840 cotton mills manufacturing cloth and yarn and 310 cotton knitting mills. Fifty-seven per cent of these plants are operating on purchased electric power and consume approximately 700,000,000 kw-hr. annually. In twenty years the number of mills has increased 311 per cent; and in the past ten years alone this increase has been 43 per cent. The present annual consumption of cotton is now 3,100,000 bales, or approximately 30 per cent of the entire amount grown in America.

The earlier textile mills were naturally located at water-power sites, and they accordingly suffered from lack of transportation facilities, access to labor markets and damage by flood. Transmission in these mills was entirely mechanical. The development of steam power made possible the more advantageous placement of the mills on transportation routes and in communities capable of supplying labor. These steam-driven mills were at first developed with mechanical transmission, and this equipment was developed to a high degree of perfection along well standardized lines.

The development of alternating current and the commercial production of the induction motor provided a new type of transmission which was quickly adopted, and complete mills were equipped throughout with these motors in the early 90's. It is interesting to note that some of the original motors are still in successful operation and giving first-class service.

The first textile mill to use transmitted electric power was probably the Columbia Duck Mill, in Columbia, S. C. At this plant the power was simply transmitted a few hundred feet from the generating plant on the power canal to the mill on the hill.

TYPICAL INSTALLATION OF SPINNING FRAMES EQUIPPED WITH INDIVIDUAL MOTORS

Following this came the Pelzer Manufacturing Company's installation where power was transmitted for approximately four miles at 3,500 volts and stepped down at the mill to 230 volts. These were both isolated plants. Probably the first commercial generating and transmitting project was the Portman Shoals plant of the Anderson Water, Light & Power Company, at Anderson, S. C. In this plant the power was generated at 10,000 volts and transmitted at this voltage to the Orr and other mills in the city of Anderson, S. C.

As the advantages of electric-motor drive became more generally recognized other companies were developed for the generation and transmission of electric power, and there are now approximately twelve large companies in the territory, with an interlocked network of transmission lines practically covering it.

[1] Mill Engineer and Architect.
Presented at the Annual Meeting, New York, December, 1919, of THE AMERICAN SOCIETY OF MECHANICAL ENGINEERS. All papers are subject to revision.

The hydraulic generating plants have an aggregate capacity of 500,000 kva., and in connection with these plants are operated steam auxiliary stations with a capacity of 150,000 kva., making a total developed capacity of 650,000 kva. The power companies own undeveloped water power sites with an aggregate probable capacity of 450,000 hp.

The hydroelectric developments utilize moderate heads, ranging from 12 to 580 ft. and averaging 40 to 60 ft. Storage reservoirs are being built on some of the streams to conserve the seasonal flow. In several cases a number of plants are built on the same stream and successive use of the water is obtained.

The steam auxiliary plants are now built with first-class construction throughout, and employ the best equipment and mechanical saving devices. In some of the earlier plants the mistake was made of depending on cheap and ample labor, factors which no longer exist.

A TYPICAL PICKER ROOM WITH INDIVIDUAL MOTOR DRIVE

The main transmission lines are usually designed for operation at 60,000 or 110,000 volts, and these lines are carried on steel towers. As a rule, twin circuits are used on the more important lines. For branch lines, voltages of 10,000, 22,000 and 40,000 are generally employed, and the circuits supported on wooden poles. All the standard systems are 60-cycle, 3-phase, so that interconnection and exchange of power between the companies is readily carried out.

The general map of the territory, page 162, shows the principal trunk transmission lines. It should be noted that there is a continuous trunk which can be interconnected from Goldsboro in the eastern part of North Carolina to Nashville in middle Tennessee, a distance by air line of 500 miles and by transmission line of 720 miles. This is equivalent to a transmission line between Bangor, Maine, and Philadelphia, which would serve the industrial territory lying between these cities.

There are two classes of electric power sold in the territory, one known as primary power, which is guaranteed for continuous supply, barring accidents; the other, known as secondary power, is furnished only when there is sufficient water in the streams to generate it, and when the power company cuts off this class of power, the mill plants are compelled to generate this power locally with prime movers.

The cotton mills give a steady all-day load which is highly desirable from the central-station standpoint. The maximum peak occurs at the starting hour in the morning when the machinery is cold. This peak is about 10 per cent above normal. The rest of the day the load is practically uniform.

At the mills the transmitted power is usually stepped down to 550 or 220 volts, and at these lower voltages is distributed throughout the plant.

(Continued on page 205)

161

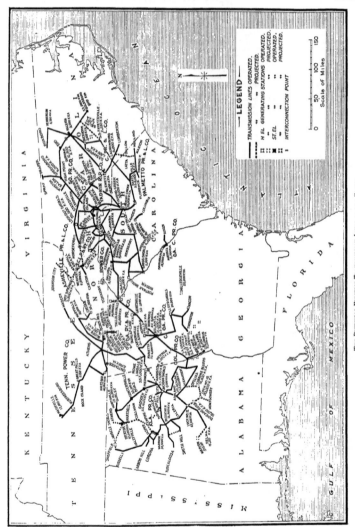

The Principal Electric Transmission Lines of the South

SOME TYPICAL EXAMPLES OF ELECTRIC POWER GENERATION AND DISTRIBUTION IN THE SOUTH

Fig. 1—Generator Room in the Georgia Plant of The Georgia Railway and Power Company; Fig. 2—Generator Room in the Coosa River Power House (Lock No. 12) of the Alabama Power Company; Fig. 3—40,500-kva. Distributing Station of the Alabama Power Company; Fig 4—110,000 Volt Transmission Lines of the Alabama Power Company; Fig. 5—Lock No. 12, Coosa River—Alabama Power Company; Fig. 6—Lookout Shoals, North Carolina—Southern Power Company.

# The Proper Balancing of Fuel, Lubricant and Motor

## A Study of the Dilution of Lubricating Oils in Aeronautic and Automobile Engines by the Gasoline Mixture and the Resulting Losses of Fuel, Power and Lubricant

By WILLIAM F. PARISH,[1] CHICAGO, ILL.

*Dilution of motor lubricating oil has always existed; this condition is caused by the gasoline mixture escaping past the piston rings, with a possible secondary cause—drainage of the diluted oil from the cylinder walls.*

*In the earlier years of highly volatile gasoline the gas escaping past the piston rings did not enter extensively into the body of the lubricating oil, and while mechanical losses of fuel took place in all four-cycle engines, this loss did not reflect itself in a problem of dilution. In the past few years, however, due to the heavy end components of the fuel, a larger amount after passing the rings, upon condensation, remained in the lubricating oil, thus creating a condition of dilution.*

*Not until the period of high-end-point gasoline did the matter of fuel loss become apparent, and then only in proportion to the amount held in solution in the lubricating oil. This condition, which has become especially accentuated in the past three or four years, has resulted in undue wear upon the engine parts as well as affecting the use of fuel, lubricant, and power.*

*The purpose of this paper is to call attention to this underlying trend in the situation and to suggest a method of application of motor oils to correct some of the present faults. It is felt that a true solution can be reached only in changes in mechanical equipment of the engine, and a plea is made for better engine design and greater economy in the use of gasoline and lubricating oils.*

**D**URING 1911, in conducting research work in connection with the use of various fuels and lubricant on a four-cylinder automotive engine and in examining the lubricating oils before and after use, it was noticed that the oil drawn from the crankcase after use had undergone a considerable physical change, the most noticeable thing being that the viscosity of the oil had been lowered during use until it was only one-half or one-quarter as much as the viscosity of the new oil. The flash or ignition point of the used oil was only a third of the original, in some cases being lowered from 470 deg. to 150 deg. fahr. Other just as remarkable changes had taken place in the gravity and cold test of the oil.

The automotive engineers who made the tests stated they had always noted a similar condition in the case of racing cars where, after a hard race, the oil drawn from the engine was as thin as water, even when it was cold.

Technical records were searched for similar information. The only reference regarding the lowering of the flash point and lowering of the body of the oil in automotive engines was found in an English paper, which stated that this condition had been noted but that no solution had been found. The matter was then taken up with the most prominent automotive manufacturers, the condition explained to them, and they were asked if they had noticed the changes that had taken place in the lubricant in their engines. Practically the only answer was that the oil in these engines had worn out and become thin.

An article prepared by the author and published in June, 1912, ascribed the changed condition of the motor lubricant to the leakage of gasoline vapor during the compression stroke, re-condensation of this vapor in the crankcase, absorption by the lubricant, and to a probable slight leakage down the walls of the cylinder of unconsumed heavy portions of the fuel which had mingled with the lubricant, producing a similar effect. This condition has always existed, but has become more pronounced and better understood during the last two years.

[1] Sinclair Ref. Co. Mem.Am.Soc.M.E.

Presented at a meeting of the Chicago Section of THE AMERICAN SOCIETY OF MECHANICAL ENGINEERS, January 23, 1920.

The tremendous growth of the automotive industry during the past few years, with the consequent burden placed upon the oil industry of producing the vast amount of fuel required, has brought about an unbalanced condition. Because of the enormous increase in consumption of gasoline—from 15,000,000 bbl. in 1910 to approximately 95,000,000 bbl. in 1919—the petroleum industry has been able to meet demands only by the manufacture of a less volatile fuel.

Navy Department specifications show that during the past few years automotive fuels have been gradually getting heavier, approximately from 360 deg. fahr. end point in 1916 to 428 deg. fahr. end point in 1919. (The end point is the temperature at which the final part of the fuel vaporizes.)

There is ample evidence that this increase in the use of heavy fuels will continue and it follows, therefore, that gas-engine designers must so design their engines as to consume successfully this less volatile product.

### LUBRICATION TESTS ON AN AERONAUTIC ENGINE

During the war, before writing specifications for lubricating oils and fuels for aeronautic and other internal-combustion engines, it was necessary to make a complete study of this subject in order that the fuel and lubricants could be so balanced that their combination in the engine would give the necessary lubricating results. These main tests resulted in the specifications for Liberty Aero Oil, and had much to do with specifications for aviation gasolines, including domestic, export and fighting grades. A Hall-Scott aeronautic engine, Type A-7-A, four-cylinder, 100 hp., was used for the special fuel tests of the series. This engine was operated under ideal test conditions, working with the same lubricant but with different fuels. Two of these fuels were made especially for the test and were distilled from the same crude.

The first fuel tested was made after the German specifications for aviation gasoline, and had a gravity of 75.2 deg. B., an initial boiling point of 110 deg. fahr. and a final boiling point of 230 deg. fahr.

The second fuel tested was made after the French specifications and had a gravity of 63.6 deg. B., an initial boiling point of 140 deg. fahr., and a final boiling point of 292 deg. fahr.

The third fuel complied with regular Navy specifications for motor-boat fuel under contract for the year 1917 and had a gravity of 61.3 deg. B., an initial boiling point of 135 deg. fahr. and a final boiling point of 385 deg. fahr. To this fuel a so-called gasoline energizer was added. This produced a fuel which, while it had an initial boiling point of 135 deg. fahr., had an end point of 422 deg. fahr., which in its last distillation range would be equivalent to the present-day commercial gasoline.

The method of conducting the tests was to run the engine continuously for five hours under test conditions, every observation in connection with the operation being recorded. At the end of every hour samples of lubricating oil were drained from the lubrication system of the engine and the same amount of new oil replaced. These samples were examined by a corps of very efficient oil chemists working at the Bureau of Standards, at the oil laboratory in Washington Navy Yard, at two commercial laboratories, and at the Naval Experiment Station at Annapolis.

The effect of the four different classes of fuel upon the same lubricant is clearly registered in the charts in Figs. 1 and 2. The lightest or most volatile fuel shown in test No. 3 caused a lowering of the viscosity of the lubricating oil during the first hour of 62 sec. (Saybolt universal viscosimeter). The oil, however, recovered its body, and at the end of the fifth hour was 87 sec. above its original viscosity or body.

In test No. 2, Fig. 1, with a slightly heavier fuel, there was a reduction in the body of the oil of 114 sec. during the first hour. The oil, however, recovered after the first hour, and at the end of the fifth hour had a viscosity of 55 sec. above its original body.

Test No. 1, with a semi-commercial fuel, showed a drop in the body of the oil in two hours of 304 sec., with a slight recovery during the balance of the test. There was a total reduction of 136 sec. at the end of five hours from the body of the original oil.

conditions. With the lightest volatile fuel, which would also be the case with the engine operating on producer, natural or furnace gas (some figures on this are given later), the tendency of the lubricating oil would be to build up in body, the same as in any other class of lubrication work where the lubricant does not come in contact with lighter hydrocarbons or lighter volatile products that can be combined with it. These tests also indicate quite clearly that the heavier the end point of the fuel, the greater is the effect of dilution and the more permanent is its character.

It must be borne in mind that in these experiments the aeronautic engine was operated with an exceedingly heavy lubri-

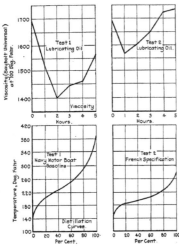

FIG. 1 CURVES OF GASOLINE DISTILLATION AND OIL VISCOSITY SHOWING CHANGE IN LUBRICATING OIL WHILE WORKING IN ENGINE

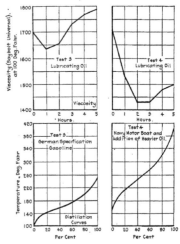

FIG. 2 CURVES OF GASOLINE DISTILLATION AND OIL VISCOSITY SHOWING CHANGE IN LUBRICATING OIL WHILE WORKING IN ENGINE

In test No. 4, Fig. 2, with the high-end-point gasoline, there was a drop in two hours of 280 sec. in the body of the lubricant, only a slight recovery and a total drop at the end of five hours of 213 sec. from the original body of the oil.

The sample taken at the end of the five-hour period of each one of these tests was tested in a distillation flask and the amount of heavy end of the gasoline recovered varied from 0.5 per cent to 1.2 per cent as indicated on the chart in Fig. 3. All of the physical characteristics of the oil, such as gravity, flash, fire, viscosity, pour test, acidity, carbon content, showed a considerable change as indicated in Fig. 5.

In Fig. 5 carbon conditions are indicated which are especially interesting as showing the laboratory variations on four samples of the same new oil when taken by the Conradson and the Waters carbon test. Particularly interesting is the sample by weight and residue, test No. A4, where practically commercial gasoline was being formed in the engine. This heavier carbon residue or deposit found in the used oil indicates the increased amount of carbon being formed in the engine while using the heavier fuel.

The results of the above tests indicate quite clearly that dilution exists with even the most volatile fuel, but only during certain periods of the operation of the engine and under certain fixed

cating oil, such as these engines require, and that carbureter and magneto adjustments were attended to by experts; that the engine was placed in perfect condition for each test and operated continuously without stopping, at high speed, for the entire five hours of the test.

### LUBRICATION TESTS ON AUTOMOBILE ENGINES

The following data were secured from the examination of lubrication conditions of a considerable number of cars and trucks during December, 1919, using commercial gasoline and lubricants as purchased in the Chicago market. In some cases the cars were new as received from the manufacturer, and in other cases old, or old and repaired.

In preparation for the tests, the motors were thoroughly cleaned and the crankcases filled with the oil to be tested. The oils so used covered a range of commercial medium and heavy grades as indicated by Government specifications for motor oils. The cars were allowed to operate in their regular business for periods of ten days, during which time no particular care was given them other than ordinarily given commercial vehicles, but at the end of each two-day period an eight-ounce sample was

drawn from each crankcase and eight ounces of the grade of oil
used on that particular motor was put in whenever additional oil
was necessary. The samples were examined for dilution by measuring all the product that could be distilled from the sample of
oil under 500 deg. fahr. The viscosity (Saybolt) of the sample
before removal of the diluent was taken at 100 deg. fahr. The
ash test was made in the Conradson carbon apparatus, and after
the carbon had been determined a gas flame was used for burning
the carbon out of the dish. The ash that remained was then
weighed and the percentage taken of the whole sample. This ash
was made up of iron, iron oxide and silica. The results of the
tests are given in Figs. 6, 7 and 8.

Fig. 6 shows the results on a 3½-ton Diamond " T " truck. This
was a new car which had not been run previous to this test. Viscosity at the end of ten days was reduced from 345 to 100 sec.
Dilution the tenth day amounted to 19 per cent and ash to 1.62
per cent.

Fig. 7 shows the results on a 5-ton Pierce-Arrow truck which
had run 5205 miles. On July 4, 1919, this motor was completely
overhauled, new piston rings being fitted. Viscosity dropped
from 358 to 90 sec. during the first six days, then recovered to
184 sec. at the end of ten days. Dilution in the case of this truck
seemed to be very dependent upon weather conditions; at the
end of the six-day period there was 20.6 per cent of dilution,
which was reduced to 8 per cent at the end of the ten-day period.

Fig. 8 shows the results on a 3½-ton Packard truck which had
run 12,551 miles. Overhauled May 5, 1919; new rings fitted.
(NOTE: This truck did not go out of the garage for the first two
days. Engine, however, was run during this time for a period that
would be equivalent to five miles' operation.) The viscosity held
practically constant for the first two days when the truck was not
operating in the open. After the second day there was a drop
in viscosity to 43 sec. Dilution at the end of the tenth day was
26.5 per cent.

### CAUSE OF DILUTION

With the present-day fuel and the present-day automotive
engine, all lubricants change considerably in body within a very

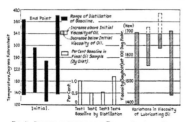

FIG. 3  RESULTS OF DISTILLATION TEST OF OIL AFTER FIVE-
HOUR RUN

short operating period, this new body being the result of dilution
and the dilution coming from the heavy end of the fuel.

The amount of dilution is influenced, to a great extent, by the
temperature of the surrounding air, the reason being that the
colder the engine, the greater the amount of fuel the oil will absorb
and hold. Conversely, the hotter the air or lubricating oil in the
engine, the less will be the amount of fuel held in dilution, as all
the fuel can be distilled off at high temperature.

In order to see if any of the dilution of a motor oil could be
caused by decomposition or cracking of the oil by being thrown
against hot surfaces, tests were made by the author with a com-

mercial car operated by pure benzol. From 4 to 8 per cent of
benzol was distilled from the motor oil after 100-mile runs. After
the benzol had been removed the oil possessed all of its original
characteristics. Benzol, wherever found, can be determined as
such by the chemist. These benzol tests determined satisfactorily
that dilution was being caused by the fuel, although an odor of
burned oil indicated some degree of decomposition.

The conclusions drawn from the above test with benzol are
confirmed by the results of a test on a Nash six-cylinder engine
operated with city gas for 10 hr. The original viscosity of the
oil at 100 deg. fahr. was 205 sec.; after ten hours' operation the
viscosity had increased to 255 sec., the flash and fire points remaining practically the same, showing that there had been no
change in the oil caused by cracking and reabsorption of the
lighter products, but that the absence of lighter products allowed
the oil to become heavier in body.

Dilution is caused mainly by leakage past the piston rings
during the compression stroke. There is a secondary leakage

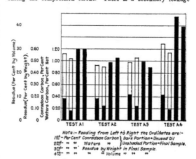

FIG. 4  RESULTS OF TESTS SHOWING VARIATION OF CARBON CON-
DITIONS IN FOUR SAMPLES OF NEW OIL

caused during the suction stroke, where there is a partial separation of heavy ends of the fuel from the gas entering the cylinder
and absorption by the film of oil on the cylinder walls, causing
some drainage on the lower exposed surfaces during the compression stroke. It is doubtful if any drainage or leakage takes place
during the exhaust stroke or during the power stroke. If we assume that the losses take place during the admission and compression strokes, we are dealing with one-half of the operating
period of the engine, and that will give all the opportunity necessary for the effects that have been noticed to take place.

Leakage of the gasoline or gasoline mixture is influenced by
the following mechanical or operation conditions:

  1 Piston clearance
  2 Richness of mixture
  3 Improper ignition
  4 Operating temperature of motor affected by weather
  5 Body of lubricant or thickness of film on walls of cylinders
  6 Condition of cylinder walls
  7 Condition of piston rings as affecting their all-around
    fit in the grooves and against the cylinder walls
  8 Intermittent operation
  9 Priming and using the choker when engine is cold in
    starting
 10 Choking when overloading
 11 Character of fuel
 12 Carburetor out of adjustment
 13 Vacuum system out of adjustment.

## LOSSES CAUSED BY DILUTION

These losses are of four kinds:

1 Loss of fuel
2 Loss of lubricant.
3 Loss of power
4 Wear of all parts.

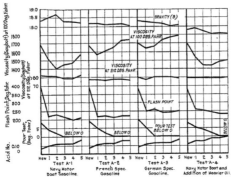

FIG. 5   CHART OF OIL ANALYSES AND CARBON CONTENT; ENGINE,
HALL-SCOTT, TYPE A7A, No. 3012, 100 HP.

## LOSS OF FUEL

This loss is not entirely due to the heavy end of the fuel or to the present heavy character of the fuel, as, irrespective of the nature of fuel, whether light and volatile or heavy, there will be a fuel loss during the compression stroke in a four-cycle engine. In the case of heavy-end fuel this loss is reflected to a magnified extent in the dilution effect upon the lubricating oil. In the case of a very much lighter fuel, say one of 72 deg. B., the loss by the rings during the compression stroke would be just as pronounced in percentage of the amount taken into the cylinder, supposing the sealing effect of the lubricant on the piston rings to be always the same, but the effect would not be as noticeable, due to the lubricant's inability to absorb and hold the lighter fuel as a diluent at operating temperatures.

The amount of fuel lost in this manner is in no way represented by the amount of dilution found in motor oils, as there is a limited point of saturation of a motor oil dependent upon the condition under which it is operating. The losses which constantly occur are disposed of by the engine in some other way than by being entirely absorbed by the lubricant.

The loss of fuel by leakage means that all the fuel placed in the cylinder is not converted into power.

## LOSS OF AND THROUGH THE LUBRICANT

The first effect of dilution is upon the viscosity or body of the lubricating oil, reducing this body to a point where it cannot be considered to be an efficient lubricant for a motor. This causes three conditions, which can be stated as:

1 Loss of lubricant
2 Loss of power
3 Wear of the machine

The first effect is the wastefulness of the lubricant, as the entire loss of lubricating oil in an internal-combustion engine is caused

by the lubricant creeping up the cylinder walls, mainly during the first cycle or admission stroke, and being later consumed in the combustion chamber. The lighter the oil is in body, the greater will be this loss. Further, there is another loss that is due to the present condition of dilution and the attempts made to offset it, and that is by the removal of lubricating oil from the engine at periods of from 300 to 800 miles, instead of the former periods of double this mileage, this mixture being thrown away.

## LOSS OF POWER

The basic rule in lubrication is illustrated best by considering the power necessary to operate a machine composed of a number of bearings. If the machine is lubricated perfectly with oil of the proper viscosity, the power will be at a minimum and speed at a maximum, and there should be practically no wear on the bearings. If, however, an oil of double or treble the viscosity is used as a lubricant, the power necessary to move the machine at the same speed will have to be materially increased, due entirely to the fluid friction of the lubricant itself. Under this condition there will be practically no wear, as the surfaces will be kept well apart. Attempting to lubricate the same machine with an oil that is entirely too light or thin for the work will also require the development of practically as much, or even more, power than in the case of the extra heavy oil, the resistance being due to metallic or solid friction, one of the results of which is, of course, abrasion with constant changing of the surfaces.

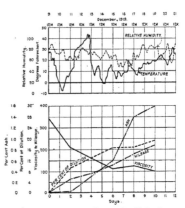

FIG. 6   VISCOSITY AND DILUTION TESTS OF OIL, 3½-TON DIAMOND
"T" TRUCK

The general rule as applicable to automotive engines when considering the dilution problem is well illustrated by a test on a tractor engine using kerosene for fuel where the output at the start of the test was 28 b.hp., gradually dropping to 24.5 b.hp. in eight hours, indicating a considerable increase in the frictional

horsepower. The viscosity (Saybolt) of the oil at 100 deg.
fahr. changed from 300 sec. at the start to 75 sec. at the finish of
the test, at which time there was 36 per cent of dilution in the oil.
Temperature of oil in crankcase increased from 98 deg. fahr. at
start to 135 deg. fahr. at end of run.

A lubricant actually working in a motor should have a body
at least equal to that possessed by a new heavy spindle oil (130
sec. viscosity at 100 deg. fahr.) and an oil of such light body must
be used in a forced-feed system where the pressure and constant
flow of the oil will make up for its lack of thickness of lubricating
film.

As a matter of fact, under present conditions of operation most
motor oils when in an engine, in the winter especially, are much
lighter than a lubrication engineer would ever think of applying
to the lightest piece of machinery he would be called upon to
lubricate, which is a light spindle in a textile mill. A description
of these spindles will permit comparisons to be made. They
rotate at about 10,000 r.p.m., and the side pull or pressure averages 2 lb. The pressure on the bearing side will not be above
10 lb. per sq. in. The high rotation speed of the spindle tends
to keep it away from the bearing surface and keep it in the center
of the bolster. About one ounce of oil is placed in the base of the
spindle, and this amount is added to every 15 to 45 days. It has
been proved by the most careful experiments that oil that has a
viscosity (Saybolt) of less than 60 sec. at 100 deg. fahr. will
allow these spindles to wear perceptibly in a month's time. The
oil remaining in the spindle base will become black and contain
a very considerable proportion of iron and iron oxide. Further,
the same spindles when lubricated with oil of from 70 to 80 sec.
viscosity do not wear, due to the fact that the oil is of sufficient
body to keep the surfaces apart.

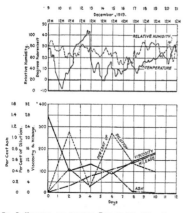

FIG. 7  VISCOSITY AND DILUTION TEST OF OIL, 5-TON PIERCE-
ARROW TRUCK

Therefore, the result of the lubrication of an automotive engine
with a mixture of lubricating oil and heavy ends of the gasoline
having a viscosity below that required for a textile spindle can
only result in excessive wear, and this is shown by the ash remaining from the complete distillation of the sample of mixture taken
from the motor.

### WEAR OF MACHINE

The effect of poor lubrication is indicated to some extent by
the readings of ash. These readings were taken on each sample
and represent the total amount of ash in that particular sample.
The ash is made up of oxide of iron, other metal, and of silica,

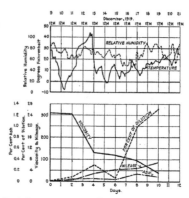

FIG. 8  VISCOSITY AND DILUTION TESTS OF OIL, 3½-TON PACKARD
TRUCK

the iron and other metal being from natural or unnatural wear
of the engine, the oxide of iron coming from the water that has
been in contact with the metal, and the silica coming from road
dust, though at the time of the year this particular test was made
the streets were well covered with snow and but little road dust
would be expected to enter the engine through the carbureter or
breather holes.

The wear of the machine is reflected in repair bills, in the necessity for reboring cylinders and refitting rings when an engine
becomes noisy and when compression is lost.

All of these matters are so well known as not to need explanation. They are largely the result of wear caused by the present
unbalanced condition of fuel, lubricant and engine.

### CONCLUSION AND RECOMMENDATIONS

As engineers we are now faced with a very definite problem.
It is necessary to reconsider the entire problem of the internal-
combustion engine on the basis of present-day fuel and its effect
upon the lubricant. There is but little more that the oil industry
can do in regard to the fuel situation. Considerable can be done
in regard to the application of motor oil; that is, the selection of
motor oil for these engines upon some other basis than has generally been prevalent in the past whereby a motor of any one
manufacturer was supposed to be properly suited by a certain
grade of lubricant for its entire life irrespective of its physical
condition. Modern lubrication engineering can, to a very considerable extent, rectify some of the damage done by this old
system of selecting lubricants, basing the selection upon more
scientific or practical grounds, taking into consideration all of the
factors which surround the problem.

The entire problem of lubricant, fuel and engine conservation
can only be solved by the automotive engineer or designing mechanical engineer providing means and methods whereby the fuel

(Continued on page 193)

# ENGINEERING SURVEY

A Review of Progress and Attainment in Mechanical Engineering and Related Fields, as Gathered
from Current Technical Periodicals and Other Sources

### SUBJECTS OF THIS MONTH'S ABSTRACTS

ALOXITE AA, NEW ABRASIVE
ABRASIVE MATERIALS AND SIZE OF GRAIN
DRYING, THEORY OF
DRYING, MOST ECONOMICAL TEMPERATURES
AIR, MOIST, GRAPHIC DETERMINATION OF WEIGHT
COPPER, INTERCRYSTALLINE BRITTLENESS
FATIGUE OF METALS UNDER REPEATED STRESSES

MOLYBDENUM STEEL
FERROALLOYS OF RARE METALS
DIE CASTING
AIRCRAFT FUEL TESTS
NEPTUNE DIESEL ENGINE
GEOMETRIC PROGRESSION OF SPEEDS IN MACHINE TOOLS
MECHANICAL STOKERS ABOARD SHIP
PRODUCER GAS FOR MOTOR VEHICLES

SMITH PORTABLE GAS PRODUCERS
DUPLEX AUTOMOBILE MOTOR
STEAM POWER PLANT STATISTICS
ROCKET ACTION IN THEORY
GODDARD ROCKET
ATMOSPHERE, INVESTIGATION AT EXTREMELY HIGH ALTITUDES
NEW BRITISH PATENT ACT

AERONAUTICS (See Internal-Combustion Engines)

## ABRASIVES

NEW ALUMINOUS ABRASIVE. Otis Hutchins. Description of an abrasive material known as Aloxite AA, developed by the Carborundum Company. The main application of the new abrasive is expected to be for the grinding of reamers, cutters and small tools and also for certain kinds of cylindrical grinding.

The chief structural characteristic of the new abrasive is the large size of the alumina crystals found therein.

The importance of this factor lies in the following: When the alumina crystals in the abrasive material are small a rupture of the grain may mean a tearing away of a few small crystals without fracture. When, however, the crystals are large and a rupture occurs in an abrasive grain it means that the crystal of alumina has been fractured and that new sharp cutting points have been produced. In this way, during the process of grinding, there is a breaking down of the abrasive grains in the wheel and the character of the new cutting edges is such as to facilitate an easy removal of the ground material. (*American Machinist*, vol. 52, no. 3, Jan. 15, 1920, pp. 113-114, 3 figs., d)

## DRYING

Drying Theory, Most Economical Temperatures, Graphic Determination of Weight of Moist Air

CONTRIBUTION TO THE THEORY OF DRYING, E. Hoehn. Discussion of the most economical temperature and most economical degree of saturation for the air at the outlet from the drying chambers; discussion of influence of air pressure; and graphical methods for the solution of problems in connection with processes of drying.

*Most Economical Temperatures.* The fact that the highest possible temperature of air is the most economical in drying is well known. This is due to the fact that the ability of air to hold moisture rises very much more rapidly than does its temperature.

A good deal less information is available, however, as to the temperature at which the air should leave the drying chamber. (Such air is referred to in what follows as "exhaust air.") As regards this latter, the impression seems to prevail that it is of advantage to have such air leave the drying chamber at the lowest possible temperature and the highest possible degree of saturation. The assumption as regards the necessity of employing a high degree of saturation to insure the economic operation of the drying process is correct, but the assumption as to the necessity of keeping the air at a low temperature rests on an erroneous conception of the character of the process. In order

to clear up this matter, the author investigates how many grams of water each heat unit (kg.-calorie) contained in the air (considered as a mixture of air and water vapor) can carry away.

The weight of 1 cu. m. of moist air is composed of $\phi\, G_d$ kg. of water and $0.465\, \dfrac{p - \phi p_d}{273 + t}$ kg. of dry air. In this formula $\phi$ is the degree of saturation in per cent; $G_d$ the weight in kg. of 1 cu. m. of saturated steam at the temperature $t$; $p$ the barometric pressure of the air-water vapor mixture in mm. of mercury; $p_d$ the steam pressure at the temperature $t$ (really the partial pressure of the steam content in the mixture) in mm. of mercury. Further, $p_i$ is the partial pressure of the air equal to $p - \phi p_d$ in mm. of mercury.

In this formula $G_d$ and $p_d$ for a given temperature can be found in steam tables; the degree of saturation $\phi$ can be measured by means of a hygrometer, or more precisely by a psychrometer. The heat content of 1 cu. m. of mixture at temperature $t$ (or 1 cu. m. of the mixture of the weight given above) is equal to

$$J = \phi G_d i + 0.465 \frac{p - \phi p_d}{273 + t} 0.24 t$$

where $i$ is the heat content of 1 kg. of steam at temperature $t$; the coefficient 0.24 is the specific heat $c_p$ of 1 kg. of air at constant pressure (all values in metric units).

It is clear from the above that the weight of water vapor corresponding to one heat-unit (kg.-calorie) content as part of the total heat content in 1 cu. m. of moist air is $\phi G_d / J$. It is desirable to plot this fraction as a function of temperature $t$ in order to see what would be its maximum and minimum values. To do this the author, instead of resorting to the complicated analytical treatment of the problem, simply computes the values of the numerator and denominator of the fraction and plots the quotients as coördinates with the temperatures as abscissæ. This is all the easier as both the parts of the fraction can be read off from existing tables, and as it does not matter whether the values are taken for the case of 1 cu. m. of moist air or for such a volume that the dry air contained therein would weigh 1 kg. (The tables are computed for the latter case.) If this computation be carried out for the barometric pressure of 720 mm. of mercury the curves of Fig. 1 are obtained.

This gives the desired picture showing the ability of the air to take up moisture as measured by heat content of the air. It appears that the curves come down at about 15 deg. cent. and then rise again, the greatest rise lying between 30 and 70 deg. Hence, the least favorable temperatures for drying purposes are located in the region of 15 deg. where the curves are at their lowermost points. In this region 1 kg.-calorie of heat contained in the air will enable the latter to carry off the least amount of water.

Attention is called also to the fact that Fig. 1 justifies the assumption that it is just at the average temperatures prevailing in the summer that the greatest consumption of heat is required by nature for the purpose of humidification of the air. From this point on, that is, from about 20 deg. cent., the absorption of moisture in drying becomes more economical the higher the temperature of the exhaust air is carried, assuming, of course, equal degrees of saturation. In the case of perfectly dried air ($\phi = 0$ per cent) the ordinate is zero and the curve reaches the axis of abscissæ. The highest location of the curves is at about 1.56 to 1.54 grams per large calorie. Each curve ends at its highest point, which latter corresponds to a definite state of mixture which is determined in the following manner. When in the weight

$$G = \phi G_d + 0.465 \frac{p - \phi p_d}{273 + t}$$

a partial steam pressure $p_d$ reaches the level of the total pressure $p$, that is, when $\phi p_d = p$, and hence the partial pressure of the air $p_i = p - \phi p_d = 0$, the second member of the right-hand side disappears; in other words, air disappears from the mixture and

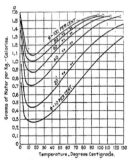

FIG. 1 CURVES OF MOISTURE CONTENT IN GRAMS AT VARIOUS DE-
GREES OF SATURATION AS FUNCTION OF TEMPERATURE $t$
WHICH CAN BE CARRIED OVER BY 1 CU.M. OF
MOIST AIR FOR EACH LARGE CALORIE OF
THE HEAT CONTENT OF THE AIR

only steam is present. If, at the same time, $\phi$ becomes equal to 100 per cent, we are dealing with a state in which the dewpoint is reached and this can occur only where the steam pressure is equal to the total pressure. This point may be called the limit dewpoint. When $\phi < 1$ we can find such a steam absorption $\phi$ on the basis of the temperature $t$, so that $\phi p_d = p$. Thus, with $\phi = 0.5$ the steam pressure $= p_d$ is 1520 mm. of mercury, at which at the barometric pressure of 760 mm. the steam drives the air entirely out of the mixture. This can happen at $t = 120.6$ deg. cent. It is obvious that the consumption of heat in the heat carrier is at its minimum when the heat carrier consists entirely of steam and no air is present. In Fig. 1 the curve $\phi = 100$ per cent at its interpolated continuation constitutes the upper limit of the region within which drying is at all possible. Fig. 1 gives also valuable information as to what degree of saturation is most economical. When very high degrees of saturation, such as 80 or 90 per cent, have been reached a further rise of saturation does not appear to improve matters very much. On the other hand, in the lower regions of saturation, such as 50 or 60 per cent, the amount of water carried away increases more rapidly with the heat consumption. On the whole, the curves of Fig. 1 would tend to indicate that the economy of the process is the same whether we operated with moist air of, say, 40 per cent saturation at 78 deg., or 60 per cent saturation at 63 deg., or 80 per cent saturation at 54 deg., or perfect saturation at 46 deg.: in all of these cases we have to expend 1 kg.-calorie in order to carry off

1.3 grams of water. The conditions most easily attainable mechanically and still satisfactory from an economic point of view would lie, as far as temperature is concerned, between 60 to 80 deg. cent. and 60 to 80 per cent of saturation, in which case we expend roughly 1 kg.-calorie of heat to take care of 1.4 grams of water, which means an expenditure of 0.7 kg.-calorie of heat for one gram of water, or 700 kg.-calories of heat for 1 kg. of water.

This is very nearly the same as the heat necessary to evaporate 1 kg. of water at atmospheric pressure, which is 640 kg.-calories. Under this condition the efficiency of drying is close to 90 per cent.

At 100 deg. cent. and $\phi = 100$ per cent, that is, when the air is entirely driven out of the mixture by the steam and the limit dew-

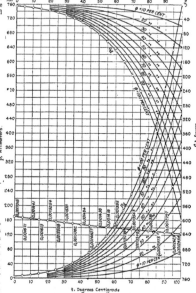

FIG. 2 CHART FOR COMPUTING THE WEIGHT OF MOIST AIR WHEN
ITS TEMPERATURE AND DEGREE OF SATURATION ARE KNOWN

point is reached, the end point of the curve shows a consumption of 1 kg.-calorie of heat to take care of 1.56 grams of water. The correctness of this figure can be proved in the following manner. Since 1000 kg.-calories of heat can take care of 1.56 kg. of water, to evaporate 1 kg. of water we have to consume 640 kg.-calories of heat, which is the value found in steam tables. A 100 per cent saturation at 100 deg. cent. is, however, unattainable in practice, since the drying process would have to operate at an efficiency of 100 per cent to make it possible, and such efficiency is not attainable.

*Graphical Determination of Weight of Moist Air.* As indicated above, the weight of 1 cu. m. of moist air in kg. may be calculated from

$$K_t = \frac{0.465}{273 + t}$$

$$G = \phi G_d + \frac{0.465}{273 + t}(p - \phi p_d)$$

We can represent these equations graphically by taking the first factor from the steam table and multiplying it by the value of the degree of saturation. This is done in Fig. 2 where the weight of steam $\phi G_d$ is plotted as ordinates from the upper axis of abscissæ downward, while the temperatures are plotted as abscissæ. As regards the second factor, both parts of it are reproduced graphically; $\frac{0.465}{273 + t} = K_t$ is at zero deg. fahr. = 0.0017015 and at 100 deg., 0.0012454. Any scale may be selected to plot this curve and it will be found in Fig. 2 as a practically straight line on the bottom.

At partial air pressure $p_z = p - \phi p_d$, the maximum barometric pressure $p = 760$ mm. is selected as the length of the highest ordinate and measured from the bottom upward; $p_d$, which is the steam pressure at complete saturation or a function of the dew-point curve, is transferred into the diagram from the steam tables and measured from the bottom upward; $\phi p_d$, which is the partial pressure of steam dependent on the degree of saturation, is to be found on one of the curves located below the dewpoint curve ($\phi = 100$ per cent).

Hence, the partial air pressure $p - \phi p_d$ is represented by the section of the ordinate located between the curve of total pressure $p$ (for example, barometric pressure of 720 mm.) and the corresponding curve $\phi p_d$, but the value of $p - \phi p_d$ is multiplied by the coefficient $K_t$ corresponding to the given temperature and the product, which is $K_t(p - \phi p_d)$, is to be added to the length of the ordinate $\phi G_d$ located on the same ordinate $t$ and measured from top downward.

The curves in Fig. 2 may be therefore used for a rapid computation of the weight of moist air for all cases and for barometric pressures up to 760 mm. pressure in C. G. S. units.

Other parts of the article may be abstracted at a later date. (Beitrag zuz Theorie des Trockens und Dörrens, E. Höhn, Zeitschrift des Vereines deutscher Ingenieure, vol. 63, no. 35, August 30, 1919, pp. 821-826, 5 figs., tp)

## ENGINEERING MATERIALS (See also Abrasives)

A PECULIAR TYPE OF INTERCRYSTALLINE BRITTLENESS OF COPPER, Henry S. Rawdon and S. C. Langdon. When copper is heated in a molten bath of sodium chloride for the purpose of cleaning and softening, it has been noticed that the copper becomes embrittled. Under these conditions the copper is more or less in contact with iron or steel, either from a stirring rod, forceps for handling the piece or the pot used to contain the salt, and a "galvanic couple" with the copper serving as cathode would be formed.

Experiments were made with small rods of copper and mild steel which were made either anode or cathode in a bath of molten salt and with an e.m.f. of approximately six volts. The brittleness of the copper rods was compared with untreated material by bending the specimens back and forth with one end firmly clamped.

The copper rod which was an anode was as tough and soft as the original and required practically the same number of bends to cause it to fracture. The cathode copper was found to be very brittle.

Microscopic examination showed the anode copper to be perceptibly smoother on the surface than the original material; the structure was that of annealed copper. In the cathode copper rod the crystalline boundaries on the surface are well defined, and the intercrystalline boundaries for an appreciable depth within the metal are made wider.

The probable explanation of embrittlement of the copper when it is made the cathode is that an appreciable amount of metallic

sodium is formed by the electrolysis which immediately alloys with the copper, the attack being selective toward the grain boundaries rather than forming an alloy layer of uniform thickness upon the outside. (Abstract of Technologic Paper of the Bureau of Standards, no. 158, e)

### Endurance Diagram as Criterion of Fatigue of Metals

FATIGUE OF METALS UNDER REPEATED STRESS, Herbert F. Moore. Discussion of fatigue failure of metals; criticism of some methods of determining it and discussion of a proposed standard method of the determination of fatigue resistance, namely, by means of an endurance diagram.

The author does not believe that elastic-limit figures help in any reliable way to judge as to the permissible fatigue stresses. He likewise does not attach much importance in this connection to a test consisting in subjecting samples of materials to a com-

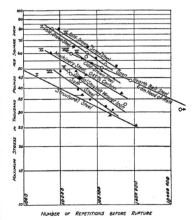

FIG. 3 ENDURANCE DIAGRAM FOR VARIOUS METALS TESTED UNDER
REPEATED STRESS

paratively small number of violent bends or twists and measuring the amount of the bending or twisting; though he does believe that such a test might be of value in testing material for crankshafts, in spite of the fact that it does not constitute a reliable index of the ability of a material to withstand long-time fatigue under working stress. In the author's opinion the best criterion for fatigue resistance is found in the endurance diagram, in which fiber stresses are plotted as ordinates and numbers of repetitions to produce failure are plotted as abscissæ. Fig. 3 shows several such diagrams. This kind of diagram can be advantageously plotted on logarithmic cross-section paper, in which ease the plotted points usually fall fairly closely along straight lines, at least up to one million repetitions of stress. The ordinates of such a stress endurance diagram show the fatigue strength for any given number of repetitions and the slope of the diagram gives an index of the relative fatigue-resisting strength for high stresses and for low stresses. Some uncertainty exists for the regions above four and five millions of repetitions, but this is true for all repeated-stress tests.

A fatigue failure of a machine part may start at some surface irregularity, especially at a sharp inward-pointing corner, a groove or even a deep scratch. Polishing the surface of test specimens has been found to greatly increase their resistance to fatigue, especially under low stress and long-time tests. Another method of increasing resistance to fatigue stresses is by using generous fillets at shoulders. (*Journal of the Western Society of Engineers*, vol. 24, no. 6, June 1919, pp. 331-336 (for the paper), and pp. 336-340 (discussion), 4 figs. in the paper, *ep*)

## Molybdenum in Chrome and Chrome-Nickel Steels

MOLYBDENUM IN COMMERCIAL ALLOY STEEL. Data of an investigation of these steels published by a commercial organization. Only the most important conclusions are reported here.

The addition of molybdenum to chrome steel increases the elastic limit to a more marked degree than would the further addition of chromium. But the brittleness is not only not increased, but actually decreased with the addition of molybdenum, the greater toughness being shown by the higher elongation and reduction of area.

In chrome-nickel steel molybdenum is claimed to increase the toughness of the material to a considerable extent.

Data are given on the properties of a chrome-vanadium steel, with additions of molybdenum up to the one present.

TABLE 1

| Carbon 0.25-0.32 | Manganese 0.71-0.76 | Chromium 0.65-1.04 | Silicon 0.11-0.22 | Molybdenum 0.32-0.46 |
|---|---|---|---|---|
| | | | | (On finished crankshaft) |
| Elastic Limit 131,700 | Tensile Strength 149,900 | Elongation Per Cent 17.7 | Red. of Area Per Cent 61.8 | Brinell Hardness 304 |

For spring steel, the type in Table 2 is said to have been found quite serviceable. This steel is in use on the Fifth Avenue buses of New York.

TABLE 2
(Molybdenum—0.25 to 0.40 per cent)

| Carbon 0.40-0.50[1] | Manganese 0.60-0.90 | Silicon 0.10-0.20 | Chromium 0.80-1.10 |
|---|---|---|---|
| Elastic Limit 180,000 to 210,000 | Tensile Strength 200,000 to 230,000 | Elongation Per Cent 12 to 15 | Red. of Area Per Cent 37 to 45 |

[1] Down to 0.35 for forgings. Up to 0.60 for rivet sets, etc.

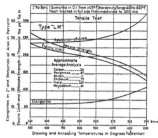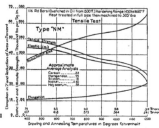

FIG. 4 THE RELATIVE PROPERTIES OF NICKEL-CHROME-MOLYBDENUM STEEL (LEFT) AND A NICKEL-MOLYBDENUM STEEL AT VARIOUS TEMPERATURES AND HEAT-TREATMENT CONDITIONS

As regards molybdenum in nickel steel, it is claimed that its addition increases the elastic limit, toughness and ductility, the effect being particularly pronounced when the steel is drawn at higher temperatures.

Molybdenum steels as a class permit a wide temperature range in which they can be heat-treated without detriment to their physical properties.

In the original article, figures selected from tests on commercial tonnages are given to illustrate this point, and it is claimed that, despite the extremely wide range in quenching temperatures, the variation in physical properties is slight. Tests made within a quenching range of as much as 300 deg. fahr. have shown the practical uniformity of results. In all steels so tested the content of molybdenum varied from 0.25 to 0.40 per cent.

It is also stated that molybdenum steels are less liable to warp in the course of manufacture.

Of the three grades of chrome-molybdenum steels, the one of medium-carbon content is probably the most important commercially, as it is used for parts in automotive equipment such as crankshafts, connecting rods, front axles, etc. Tests on approximately 4000 finished crankshafts forged from this type of steel showed the average properties of the metal to be as given in Table 1.

Chrome-nickel-molybdenum steel was used on Liberty-motor crankshafts, and Table 3 gives the results of tests made by a commercial company.

TABLE 3

| C. 0.305 0.236 | Mn. 0.69 0.50 | Si. 0.52 0.08 | Cr. 0.98 0.74 | Ni. 3.05 2.85 | Mo. 0.54 0.32 |
|---|---|---|---|---|---|
| Elastic Limit 130,000 | Ultimate Strength 142,000 | Elongation Per Cent 20.5 | Red. of Area Per Cent 65.0 | Izod 67 | Brinell 303 |

The crankshafts were treated as follows: Forged and then twisted through the required angle; shaft is then annealed at 1600-1650 deg. fahr. to relieve strains. Time at heat is about 1½ hr. Cooled in air. Heated in an electric furnace to 1475 deg. fahr., held ½ hr. and quenched in water. Drawn at 1150 for 2 hr. Cooled in air and straightened before the temperature has reached 800 deg. fahr.

Fig. 4 is of interest, as it gives the relative properties of a nickel-chrome-molybdenum steel (left) and a nickel-molybdenum steel (right) at various temperatures and heat-treatment conditions.

Molybdenum may be introduced into steel either as a ferroalloy or in the form of calcium molybdate. One of the large deposits of molybdenum in the form of molybdenum sulphide is located in this country at Climax, Colo., and is owned by the Climax Molybdenum Company. ("Molybdenum Commercial Steels," published by the Climax Molybdenum Co., abstracted through *Iron Age*, vol. 105, no. 6, Feb. 5, 1920, pp. 407-409, 3 figs., *ep*)

DEVELOPMENTS IN SOME OF THE RARER FERROALLOYS. The following extracts regarding some of the recent developments in the more uncommon ferroalloys are taken from British sources:

*Ferromolybdenum.* During the last couple of years the Canadian Imperial Munitions Board have produced 100,000 lb. of 70 per cent ferromolybdenum by the direct smelting of molybdenite concentrate at the works of the Tivani Electric Steel Co., Belleville, Ont. Vertical cylindrical furnaces of the single-phase type are employed, requiring 3000 to 4500 amperes at 50 volts per furnace. Each consists of an iron shell, lined with red brick, firebrick, silica brick, and carbon in succession, and resting on a concrete foundation. The lower electrode consists of a water-cooled bronze or copper block, from which iron rods project upward into the furnace bottom, and the upper electrode of a graphite or carbon rod. For the production of 70 per cent alloy from concentrate containing MoS, 75 and Fe 9 per cent, the charge consists of concentrate 100, lime 120, coke 10, and scrap steel 5 lb. Each furnace is tapped every 4 hr., the output being 1050 lb. of alloy per 24 hr. During the two years in which the plant was operated for the production of ferromolybdenum containing Mo 70, S 0.4, and C 4 per cent, the average composition of the product was Mo 70.43, S 0.38, and C 3.56 per cent.

*Ferrouranium.* Up to the present time the use of uranium in steel manufacture is not very fully developed or understood. The abundance of sodium uranate ($Na_2U_2O_7$), a by-product from the extraction of radium and vanadium from carnotite, provides the material from which the oxides are produced. The alloy is made by reducing the oxides with carbon according to the following equations, and then mixing with iron:

$$UO_2 + 2C = U + 2CO \qquad 3UO_3 + 5C = U + U_2C_3 + 2CO$$
$$U_3O_8 + 8C = 3U + 8CO \qquad U_3O_8 + 11C = U + U_2C_3 + 8CO.$$

The alloy usually contains from 25 to 35 per cent of uranium and under the microscope the uranium seems to be present, as the double carbide of iron and uranium, $Fe_3C.U_2C_3$, as a eutectic mixture of iron and carbon. It has been claimed that 0.5 per cent of uranium will replace several per cent of tungsten in tool steel, and the alloy is being used in making ternary uranium and quaternary uranium-tungsten steels.

*Ferrozirconium and Ferroboron.* Ferrozirconium is one of the latest of the ferroalloys to come into use. It is made by the reduction of zirconia, with aluminum, by the thermit process, to the element and then alloyed with iron. In Great Britain a 20 per cent ferrozirconium has been used to some extent in the place of ferrotitanium.

Ferroboron can be made by the reduction of a mixture of boron oxide and iron oxide with carbon in the electric furnace. Not much is known about this alloy, but experiments in France have shown that a very strong and tough steel can be made by using from 0.5 per cent to 2 per cent of boron.

*Ferroaluminum.* Ferroaluminum has been largely replaced in recent years by the metal aluminum, but it is suggested that a revival of its use would be desirable. It can be made by mixing scrap iron or iron ore with bauxite and reducing material. It formerly contained from 10 per cent to 20 per cent aluminum, but recent success in the production of ferroalloys indicates that a 50 per cent alloy could be produced. Aluminum is seven times as powerful as silicon and 28 times as strong as manganese in action upon the oxygen dissolved in steel, so that from one ounce to one pound of aluminum is sufficient to deoxidize a ton of steel. (*The Iron Age*, vol. 105, no. 7, February 12, 1920, p. 484, *g*)

## FOUNDRY

DIE-CASTING PARTS AND METALS. Brief statistical data of the American die-casting industry, the description of die-casting machines, metals used and methods.

The article brings out such recent developments in the die-casting industry as casting of parts in metals fusible at high temperatures, such as various bronzes and brasses, in particular, aluminum bronze.

It would appear that for low-fusible alloys, soft low-carbon steel makes a very good material for dies, both in respect to its ability to stand up to the work and to the comparatively low cost of production of dies. For use with alloys having high melting points, however, low-carbon steel is entirely unsuitable and materials of the order of tool steels have to be used, as they retain their hardness at temperatures of the metal when injected into the die. However, they make the production of dies both more difficult and more costly.

In addition to this threads cannot be cast on parts made from such materials as aluminum bronze, as can be done with white metal. This inability to cast threads is attributed to the failure of the small pieces of metal used for thread cores to withstand the high temperatures present for any length of time. (Serial article, *Raw Material*, vol. 2, no. 1, Jan. 1920, pp. 468-474, 13 figs., *d*)

## FUELS (See Internal-Combustion Engines, Motor-Car Engineering)

## GAS PRODUCERS (See Motor-Car Engineering)

## INTERNAL-COMBUSTION ENGINES (See also Motor-Car Engineering)

### Bureau of Standards Tests of Various Gasolines and Benzol Mixtures

FUELS FOR AIRCRAFT ENGINES. Symposium of three papers discussing the power characteristics of various fuels on the basis of tests made in the altitude laboratory of the Bureau of Standards and associated organizations.

The first paper (by H. C. Dickinson, W. S. James, E. W. Roberts, V. R. Gage and D. R. Harper, 3rd) presents data of tests of nine types of gasolines compared in turn with standard gasoline as a reference fuel.

Two of the fuels were representative American straight-run gasolines closely complying with the specifications for export gasoline. A third followed closely the specifications for fighting gasoline, while two others were slight variations of the fighting gasoline. These were imported from France. Two gasolines represented the average motor-car gasolines as sold by automobile filling stations, one of these being of eastern origin and the other a California gasoline. These fuels were tested quite thoroughly and among the properties the following were determined: Distillation curves, mean volatility, heat of combustion, critical solution temperature, chemical composition and magneto-optic rotatory power.

The reason of this latter test was as follows: In determining the properties of various organic mixtures it has been found that the presence of considerable proportions of compounds of the benzene group has a tendency to increase the rotation of the plane of polarization when a beam of monochromatic light is passed through the substance in a strong magnetic field. The presence of the benzene group has been thought by some to be associated with increased engine power. Thus, it was a conceivable hypothesis that the magnetic rotation might prove a convenient indicator of the power-producing ability of the fuel. This was worth while finding out, especially as this property of the gasolines could be measured very rapidly. Actually, no definite connection between the magneto-optic rotatory power and the engine performance of a gasoline was proved.

The tests on the whole have indicated that when the necessary carburetion changes have been made to suit each fuel, the difference in power-producing qualities of the various gasolines was quite slight and the extreme difference was of the order of the magnitude of 5 per cent, which, considering the permissible variation in this kind of work, is very little.

The paper by E. W. Roberts discusses the power characteristics of Sumatra and Borneo gasolines. It appears that the sample of Sumatra gasoline tested develops slightly more horsepower than American standard export aviation gasoline, while the sample of Borneo gasoline develops slightly less horsepower. Fractional distillation shows a lower volatility for the sample of Borneo gasoline than for the sample of Sumatra.

The fuel economy with Sumatra gasoline was higher than with American export aviation gasoline except at the highest altitudes. The fuel economy of the Borneo gasoline may be said to have averaged about the same as that of American export aviation gasoline.

As regards the method of working up data, one feature of the lines of horsepower plotted against barometric pressure was particularly helpful. It was found that the intercepts of these lines on the axis of zero pressure were quite evenly spaced. For example, in the equation of the line

$$y = ax - b$$

wherein $y$ = horsepower, and $x$ = barometric pressure in cm. of

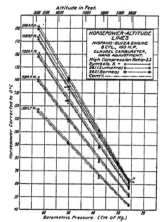

FIG. 5  DATA OF TESTS ON SUMATRA AND BORNEO GASOLINES AT
THE BUREAU OF STANDARDS

mercury, the values of $b$ increased by even increments for even increments of speed. The value of $b$ is a function of the friction horsepower at that particular speed. Since the friction horsepower is entirely independent of the fuel employed, the values of $b$ should be the same for each speed regardless of the fuel. Roughly estimating the equations of the various lines from the highest and the lowest horsepowers in each case showed that the values of $b$ were approximately the same for all fuels. Plotting these values of $b$ against speed and drawing an average line through these points gave the final values which were employed for the lines of Fig. 5.

The power characteristics of 20 per cent benzol mixture were also investigated by E. W. Roberts, the fuel consisting of benzol and gasoline. From the results of these tests the following conclusions have been drawn: This mixture shows very little gain in power as compared with aviation gasoline (export) at the

lower altitudes and the lower speeds. There is very little gain in horsepower below 10,000 ft. Above this altitude, and at high engine speeds, the engine shows considerable gain in power, which at 29,300 ft. and 2100 r.p.m. is 5.8 per cent.

While the methods of measuring fuel consumption were not entirely satisfactory, the fuel consumption of the 20 per cent benzol mixture appears to be about 4 per cent greater than $X$, at speeds and powers where it gives its best performance. (National Advisory Committee for Aeronautics Report no. 47, preprint from 4th annual report, 35 pp., numerous curves, eA)

### Neptune Diesel Engine; Novel Scavenging Valve Arrangement

BRITISH DIESEL ENGINE. Description of the Neptune Diesel engine built by Swan Hunter and Wigham Richardson, known as the constructors of the Mauretania.

The new engine is based essentially upon the Polar motor of the Atlas Diesel Engine Company of Stockholm, but embodies several new features. It operates on the two-stroke-cycle single-acting principle, but there is one scavenging pump for each working cylinder arranged directly below it; air at atmospheric pressure is drawn into it on the up stroke and compressed to about 3 lb. per sq. in. on the down stroke. The control of the air admission and its discharge into a scavenging trunk $C$, Fig. 6, is effected by means of the valve $A$, which is actuated from the camshaft by means of an eccentric and vertical link. From the scavenging trunks $C$ and $D$, which act as reservoirs, the scavenging air enters the working cylinders through ports at the bottom when these are uncovered by the piston on its downward stroke, the top

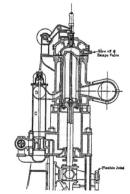

FIG. 6  CROSS-SECTION OF 1500-HP. NEPTUNE MARINE OIL ENGINE

of the piston being shaped in a special manner to help in the effective admission of the scavenging air and discharge of the exhaust gases.

The piston valve $A$ in combination with a change valve (not shown, but located back of $A$) serves another purpose. The scavenging cylinders are here used for starting purposes, compressed air at the low pressure of 100 to 150 lb. per sq. in. being admitted below the scavenging piston through the change valve and the piston valve $A$.

This indicates the object of the change valves. On starting up, these valves are placed in such a position that the admission pipe to the atmosphere is closed, but open to the compressed-air supply, so that compressed air passes through the piston valve $A$ into the scavenging cylinders.

The change valves are of the rotary type and are operated by means of a lever from the control station so that on starting up compressed air is admitted until the engine fires, after which the valves are immediately changed over to allow air to be drawn into the scavenging cylinders from the atmosphere. Only one valve, the fuel-admission valve, is required in each cylinder cover, as scavenging takes place through ports.

The scavenging cylinders, by the way, are supported on cast-iron columns at the front and by vertical steel columns at the back, with diagonal columns to provide stiffness. A pair of columns carries a pair of cylinders which permits ready standardization in 4-, 6- and 8-cylinder models.

A 6-cylinder engine, 18-in. diameter and 36-in. stroke, running at 115 r.p.m., develops 1500 i.hp. (*International Marine Engineering*, vol. 25, no. 1, Jan. 1920, pp. 29-32, 2 figs., *d*)

## MACHINE SHOP (See Abrasives)

## MACHINE TOOLS

GEOMETRIC PROGRESSION OF SPEEDS IN MACHINE TOOLS, C. C. Stutz, Mem.Am.Soc.M.E. Presentation of a method of proportioning the speeds of machine tools which may be applied to almost any problem of this type.

In designing machine tools of any type it is now an accepted practice to so proportion the speeds that they will form among themselves a geometric progression, so that each speed is increased from the preceding speed by the same multiplier.

By multiplying the slowest speed $a$ by the factor $f$, we obtain the second speed and so on through the whole series.

As a rule the number of speeds $n$ is such that it cannot be obtained by a cone pulley alone, and the machine is therefore either back-geared or triple-geared, or perhaps even quadruple-geared. In such a case it is a simple matter to obtain these gear ratios, as will be shown.

Assuming as an example a lathe with a four-step cone, ($d = 4$). This would give us four speeds with cone, four speeds with cone and back gears, four speeds with cone and triple gears; or 12 speeds in all, and the series (from the fastest to the slowest speed) would be: $af^{11}, af^{10}, af^{9}, af^{8}, af^{7}, af^{6}, af^{5}, af^{4}, af^{3}, af^{2}, af, a$.

It will be noticed that these three sections of the full series (each forming a series) bear the following relation to each other:

1 The second series can be obtained by dividing each member of the first series by $f^{4}$. In this case $f$ to the fourth power is the ratio of the back gears.

2 The third series can be obtained by dividing each member of the first series by $f^{4} \times f^{4} = f^{8}$. In this case $f$ to the eighth power is the ratio of the triple gears.

As the actual calculations are best carried out by logarithms we can summarize the above by writing

the log of ratio of back gears $= d \log f$ ............[2]

the log of ratio of triple gears $= 2d \log f$ ...........[3]

and the log of ratio of quadruple gears $= 3d \log f$ ...........[4]

Numerical examples of the application of these formulæ are given and also formulæ for the case of a machine with two crankshaft speeds and for dividing the speeds between the two crankshaft ratios. (*American Machinist*, vol. 52, no. 3, Jan. 15, 1920, pp. 117-118, *mp*)

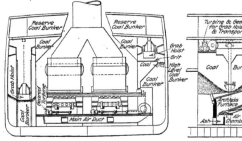

FIG. 7 RILEY MULTIPLE-RETORT STOKERS AND MECHANICAL COAL HANDLING AS INSTALLED BY EBITH, OF LONDON. INSTALLED ON NEW SHIPS ONLY

The writer assumes that the proper minimum speed $a$ and maximum speed $b$ for the given tool are given and that the total number of speeds desired $n$ is also known. From this he derives the various speeds in the following manner.

Let $n$, = total steps or intervals between total number of speeds
$f$ = ratio of one step or factor wherewith to multiply any speed to obtain the next higher speed, and
$d$ = number of steps in cone pulley.

Algebraically expressed, the various speeds therefore form the following series:

$a, af, af^{2}, af^{3}, af^{4}$ ............$af^{n-2}$ $af^{n-1}$

The last or fastest speed is expressed by both

$af^{n-1} = b$, or $f^{n-1} = \dfrac{b}{a}$

$$f = \sqrt[n-1]{\frac{b}{a}} \quad \ldots\ldots\ldots\ldots\ldots\ldots\ldots[1]$$

## MARINE ENGINEERING

### Mechanical Stoking Aboard Ship, Its Technical and Economic Aspects

MECHANICAL STOKERS ABOARD SHIP, Charles H. Bromley. Progress of the application of mechanical stoking aboard ship. It is claimed that the introduction of oil fuel holds back their wider adoption at present. The adoption of mechanical-stoker feed aboard ship is clearly related to the introduction thereon of the water-tube boiler, and this latter is opposed in some quarters because of its alleged inferiority to the Scotch boiler in respect to reliability.

The author points out, however, that the experience of American war vessels and the vessels of the Emergency Fleet Corporation has shown that when properly handled the water-tube boiler is highly reliable and instances are given to prove this contention.

Fig. 7 shows an installation of Riley multiple-retort underfeed stokers such as carried out by the Erith Engineering Company of London. In this case it has been adapted to the Babcock & Wilcox-type cross-drum water-tube boiler built with tubes 14 ft. long, and the boilers are installed athwart ship.

Owing to the moderate continuous load very high combustion charges are not required, the long-pattern stoker operating at comparatively low combustion rates. The stoker, of course, cleans itself of ash; the coal is mechanically handled, making on shipboard an installation closely similar to a modern stationary steam plant. Furthermore, the equipment does not occupy as much space as the usually installed Scotch boiler of equivalent steaming capacity, and the weight is much less also.

The Emergency Fleet Corporation follows a somewhat different line of design, as one can see from Fig. 8. The characteristic feature of this type of installation is the baffling found by experiment with the assistance of the Bureau of Mines and claimed to enable getting the most heat out of the flue gases and into the boilers.

Here also plenty of room is provided for ash deposit and removal at the refuse end of the stokers (the coal-handling machinery is not shown).

These boilers are very much lighter for a given steaming capacity than the Scotch. The Heine boiler of 3100 sq. ft. heating surface, 225 lb. pressure, hand-fired, without grates but inclusive

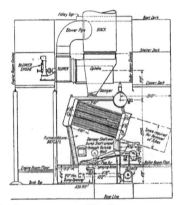

FIG. 8 GENERAL LAYOUT FOR WATER-TUBE BOILER AND RILEY MULTIPLE-RETORT STOKER FOR SHIPS OF EMERGENCY FLEET CORPORATION

of water, weighs 60 tons. The Scotch of 3032 sq. ft., 220 lb. pressure, weighs dry 68.48 tons and requires 29.62 tons of water, making a total of 98 tons, as against 60 tons for the water-tube. The 8000- to 10,000-ton ships of 11 knots speed need three such boilers. The saving in weight alone per ship is therefore appreciable.

While the mechanics of the application of mechanical stokers aboard ship has been fully developed by the engineers of the Emergency Fleet Corporation, the stokers will not be applied to the ships owing to the low price of fuel oil to the corporation. It is now somewhere in the neighborhood of 2½ cents a gallon, which puts coal out of the running. During the war, however, the Government had to pay from 7 to 9 cents per gallon of oil and it is now believed that the present low price will stay very long.

The navies of the world will use oil regardless of price. So, too, will fast passenger ships. It is believed by many that there is not enough oil to warrant its wide use in the merchant marine for many years, although within recent weeks rich oil finds are reported from Colombia. Just now American coal is selling in Italy for $36 a ton. The freight rate is $22. So it is not bad business now to use oil-burning ships in the coal trade to Europe. Yes, the mechanical stoker for ships is here, and when oil again becomes high-priced or its supply alarmingly diminished, it will come into its own. (*Power*, vol. 51, no. 3, Jan. 20, 1920, pp. 86-89, 8 figs., *dg*)

MILITARY ENGINEERING (See Testing and Measurements)

MOTOR-CAR ENGINEERING

Smith Gas Producer

PRODUCER GAS FOR MOTOR VEHICLES, D. J. Smith. General discussion of the advantages and disadvantages of the use of producer gas for motor vehicles, followed by a description of some attempts to do it on a practical scale—among others, those made by the author of the paper himself.

In the opinion of the author the ordinary producer is larger and heavier than it might be, and one of the reasons for this condition is that a very deep fuel bed is employed. To get away from this excessive weight the author resorted to the use of a thin bed of fuel which made it necessary to maintain two things: (1) a regular feed of fuel in small, measured quantities, and (2) continuous agitation of the whole fuel bed in order that no channels or holes in the fire might occur, and that all ash might be constantly sifted out and the fuel bed kept light and porous to give easy passage to the air and water vapor through it. In the producer used by the author the depth of the furnace chamber was approximately 12 in., the depth of fire being only 6 in., and it is stated that a producer with such a thin bed of fuel works very low, and that of anthracite, semi-anthracite and hard, non-caking steam coal only dust is found in the scrubber.

The feed and grate mechanism were driven by the engine of the vehicle, and a mechanical ash discharge was also added. This enabled a further reduction in the size of the producer by doing away with the necessity of providing an ashpan large enough to contain the ash formed during some hours' running.

As great flexibility in the operation of the producer was desired, it was found that water could not be fed directly to the fire which made it necessary to feed it as steam, and further, in order to keep the quality of the gas constant, the air passing to the producer had to be used as a regulating medium for the steam supply. A small vaporizer or boiler was therefore fitted to the producer heated by radiation through the fire and the passage of gas through a channel formed in it. (Figs. 9 and 10.)

The effect of varying level in the vaporizer due to the rolling of the vehicle on an uneven road was overcome by a device described in the original paper.

The producer as shown in Figs 9 and 10 was controlled only by the speed of the engine. A throttle was fitted to the air inlet to the vaporizer through which all air admitted to the producer had to pass. This throttle was coupled to the engine throttle, so that as the engine throttle was opened the throttle on the air inlet was partially closed. The result of this was to lower the boiling point of the water in the vaporizer, which then gave off steam freely owing to the drop in atmospheric pressure or slight vacuum to which it is subjected.

In addition to this, however, it was necessary to regulate the action of the producer also by the load on the engine, which was done by means of an arrangement in which the stroke of the fuel-feed device and that of the water pump were varied.

As regards the question of scrubbing, the device shown in Fig. 11 is claimed to have solved the problem of making it automatic. This device is in three sections: (1) the feed heater; (2) the cooling tubes; (3) the filter tubes.

The gas from the producer enters the top header at the left-hand end and passes down the internal tube of the feed heater, which at the lower end is coned, and this tends to increase the velocity of the gas. The gas then expands into a settling chamber or pot, and in doing so drops a very large percentage of dust in suspension. It then leaves the settling chamber by an annular passage and passes up and down two or more banks of cooling tubes. In these tubes and the headers more dust is deposited. Finally, the gas passes up two or more large-diameter tubes, which are fitted with fine-gauze filters of conical shape and filter fitting the bore of the tube at the lower end; to this a cross-handle is attached, which allows the filter to be easily withdrawn through the door in the top header. An arrangement is also provided whereby any filter can be closed off and withdrawn for cleaning

feed heater and filter is very small and can be easily accommodated on the dashboard on the side opposite to the producer. Many other forms of filters will, no doubt, be possible, but the author thinks it will be difficult to evolve a design which at once meets the requirements and is of less weight and size.

The material found in the filter is merely the fuel dust, which is quite dry and clean, and no trace of any tarry substance likely to cause trouble with the engine has ever been found.

The mixing valves and refractory lining of the producer are described in some detail. As regards the weight of the producer, it is stated that the plant which has a grate 12 in. in diameter, equivalent to 50 hp., weighs with its connections approximately 2.75 cwt. (308 lb.), corresponding to a weight of 6 lb. per hp., notwithstanding the fact that cast iron was used throughout. This

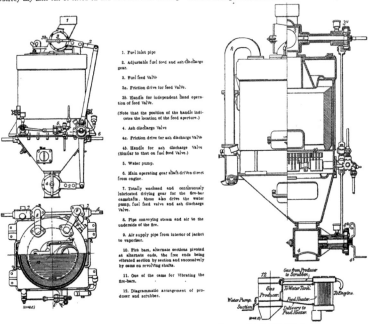

1. Fuel inlet pipe

2. Adjustable fuel feed and ash-discharge gear.

3. Fuel feed valve

3a. Friction drive for feed valve.

3b. Handle for independent hand operation of feed valve.

(Note that the position of the handle indicates the location of the feed aperture.)

4. Ash discharge valve

4a. Friction drive for ash discharge valve

4b. Handle for ash discharge valve (Similar to that on fuel feed valve.)

5. Water pump.

6. Main operating gear shaft driven direct from engine.

7. Totally unclosed and continuously lubricated driving gear for the fire-bar camshafts. these also drive the water pump, fuel feed valve and ash discharge valve.

8. Pipe conveying steam and air to the underside of the fire.

9. Air supply pipe from interior of jacket to vaporiser.

10. Fire bars, alternate sections pivoted at alternate ends, the free ends being vibrated section by section and successively by cams on revolving shafts.

11. One of the cams for vibrating the fire-bars.

12. Diagrammatic arrangement of producer and scrubber.

FIGS. 9 AND 10  GENERAL ARRANGEMENT OF THE SMITH PORTABLE GAS-PRODUCER PLANT

without stopping the action of the producer, but this is not likely to be required for vehicle work. The gas on leaving the filters is clean enough for all practical purposes and carries no more dust than is drawn into the air inlet of a carbureter when the roads are dusty. The cleaning of the filter is simple and only takes two or three minutes. The doors are opened, the gauze filters withdrawn and water is poured in at the top doors and all dust washed out; the gauze filters are then shaken and replaced. The water from the feed pump on the producer enters the feed heater at the bottom opening and leaves at the top of the vaporizer. This cools the gas considerably, and also recovers much valuable heat which would otherwise be lost. The combined

does not include, of course, the weight of the engine itself, but only that of the producer and its appurtenances.

As regards the engine design, the author states that the conventional gasoline engine is entirely unsuitable for use with producer gas. With such a cheap fuel a large, comparatively slow-running engine could be used with approximately the same characteristics as a stationary engine. A higher compression with a half-compression device is essential, as a pressure of 120 to 150 lb. gage would be necessary to obtain a fair economy with producer gas. The author recommends, for example, an engine such as the four-cylinder stationary type, 6½ in. diameter by 7½ in. stroke, running at 600 r.p.m.

As regards fuels, the author tested anthracite, semi-anthracite, coke, charcoal, peat, corncobs and straw, all of these fuels proving satisfactory under certain conditions.

In an editorial article on page 85 of *Engineering* for January 16, 1920, attention is called to the fact that in the paper no information is given as to the degree of reliability to be expected from the system described by Mr. Smith, a point which would be of interest to prospective users, especially if they had had experience with stationary producer plants. (Paper read before the Institution of Automobile Engineers, Jan. 8, 1920, abstracted through *The Automobile Engineer*, vol. 10, no. 134, January 1920, pp. 22-32, 9 figs., and *Engineering*, vol. 109, nos. 2819 and 2820, Jan. 9 and 16, 1920, pp. 59-64 and 92-95, 9 figs., *dA*)

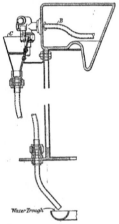

FIG. 11 SCRUBBER FOR THE SMITH GAS-PRODUCER PLANT

### Single-Sleeve Duplicated-Piston V-Type Motor

DUPLEX CAR AND ENGINE. Description of a light British car with an unusual type of engine.

The Duplex motor, Fig 12, is of the single-sleeve duplicated-piston type with the pistons running in conjointly twinned cylinders, and consists of a detachable headed monoblock of 8 cylinders of only 56 mm. (2.20 in.) bore by 75 mm. (2.95 in.) stroke.

The ports, inlet and exhaust are cut half around the outer side of each cylinder and are opened and closed by crankshafts through tiny connecting rods attaching to lugs at the bases of the sleeves (Fig. 13). These sleeves, which are of cast iron, carry three rings each, the top one wider than the other two, and as the ports are 5/16 in. deep on the exhaust side and ¼ in. deep on the inlet, the construction gives a large valve area with a comparatively slight sleeve motion (1⅛ in.).

The pistons carry their three rings in the middle of their trunks and the gudgeons are rather lower than usual, because of the connection in each pair to a common crankpin on the short four-throw crankshaft.

The crankshaft is mounted at the ends only in two roller bearings, the rollers being hollow to enable them to be link-caged and distanced somewhat like the rollers in a chain. The sleeve shafts,

on the other hand, are mounted in long plain bearings with bolt thrusts at the end and drive is by helical gearing from the front end of the crankshaft. The sleeve shaft also carries an eccentric actuating a horizontal piston pump internally spring-loaded, which draws oil from a large deep rearward sump by way of a non-return ball valve (right-hand side of Fig. 13) and delivers it downward again to a gallery lead in the casting for distribution to four transverse troughs from which dippers on the big ends effect splash lubrication to all interior surfaces.

It is claimed that this design permits obtaining rather interesting physical results; namely, a cylinder capacity of ap-

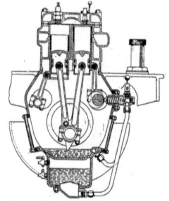

FIG. 12 DUPLEX SLEEVE-VALVE MOTOR, PLAN VIEW WITH HEAD BLOCK DETACHED, SHOWING THE SEMI-CIRCUMFERENTIAL PORT AREAS

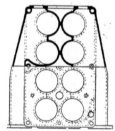

FIG. 13 DUPLEX SLEEVE-VALVE MOTOR, SECTIONAL END VIEW FROM THE FRONT

proximately 80 mm. (3.15 in.) by 75 mm. (2.95 in.) halved in area so as to produce a simultaneous effort under the medium long stroke or "internal combustion" conditions of 56 mm. (2.20 in.) by 75 mm. (2.95 in.). It is claimed that in this manner all the capacity of the short-stroke type is obtained for sustained high speed without its extravagance and jerky inflexibility at low and medium speeds. In fact, it is claimed that a motor is obtained in which the power curve is practically a straight line from the rated 10 hp. at 1000 r.p.m. up to 33 hp. at 3000 r.p.m.

As regards the valve mechanism, it is stated that the valve shafts having only frictional resistance to overcome are doing next to no work and their rotation absorbs little power, while the system affords easy timing. It is claimed the motor presents unusually good cooling facilities. (*The Auto Motor Journal*, vol. 25, no. 3/993, Jan. 15, 1920, pp. 66-68, 7 figs., *dA*)

## POWER-PLANT ENGINEERING

### Suggested Form of Reporting Steam Power-Plant Performance to Give Comparable Data

ESSENTIAL STATISTICS FOR GENERAL COMPARISON OF STEAM-POWER-PLANT PERFORMANCE, W. S. Gorsuch. Method for preparing statistical reports relating to power generation in steam power plants based on coal and load characteristics of the plant; suggested tabular forms for reports.

The principal difficulty with operating statistics of steam power plants as published by various state and government bodies is that they are inadequate for the purpose of determining the efficiency of a plant or making general comparisons between different plants. Bare figures of coal consumption per kw-hr. are given without information as to the quality of the coal and data as to the character of the load on the station. When load factors are given the figures are misleading, because they are computed upon different bases; thus, gross output is used in some cases and net output in others.

The author recommends that the following items be recorded in statistical reports relating to power generation in steam power plants and that they be expressed in a uniform manner. These items are said to reflect the influence of station design and arrangement of apparatus, methods of operation and management, load, quality and kind of fuel used and also indicate the character of the load imposed upon the station.

I  Coal Characteristics of the Plant:
    1 Average B.t.u. supplied to plant per kw-hr. net output
    2 Thermal efficiency of plant in per cent
    3 Average B.t.u. per dollar, coal as received (moist basis)
    4 Coal factor or pounds of coal per kw-hr. net output (moist basis)
    5 Average B.t.u. per pound of coal as received (moist basis)
    6 Cost of coal per ton (2240 pounds) delivered alongside of the plant
    7 Kind and character of coal
    8 Cost of coal in cents per kw-hr. net output.
II  Load Characteristics of the Plant:
    9 Average daily load factor of load
    10 Maximum load for the year
    11 Yearly load factor of load

Interval of maximum load 15 min. or 1 hour depending upon character of load.

    12 Kilowatt-hours net output for the year (kilowatt-hours sent out from the a.c. bus)
    13 Installed rated capacity, that is, the aggregate maximum continuous rating of the generators in kilowatts
    14 Average kilowatt-hours net output per kilowatt installed rated capacity.

Items 1, 2, 3, 8, 11 and 14 can readily be computed from items 4, 5, 6, 10, 12 and 13.

As regards the coal characteristics, generally speaking, the highest thermal efficiency is obtained when burning the highest quality of coal. On the other hand, a low grade of coal purchased at a low price may result in a better commercial efficiency. Therefore, as an index of the commercial efficiency of the plant, one has to have both the B.t.u. per kw-hr. net output and the B.t.u. per dollar.

The character of the load carried has a marked influence on plant efficiency since a fluctuating load and a load with sharp peaks and low load factor require more coal than an equally steady load or a load with a high load factor. The load conditions must therefore be known in order to compare the performance of the two plants, and these are best indicated by the ratio of the average load to the maximum load for the same

period; this is known as the load factor and is usually expressed in percentages.

There is, however, in the electrical industry no uniform method for determining the load factor. The writer recommends the use of the daily load factor which is the ratio of average net output during 24 hr. to the net maximum load multiplied by 100 in order to express the load factor in percentages. In comparing power-station statistics, which are usually on a yearly basis, the average daily load factor should be used because it is more representative of the operating conditions in the plant itself than the load factor based on the maximum of the month or year. The plant factor, which is the ratio of the average hourly load for the year to the rated capacity of the plant, is a measure of the utilization of the installed capacity. This is obtained by dividing 8760 into item 14.

For the purpose of illustrating the method and advantages for preparing statistical reports as described here, the following comparison is made between a few of the items of two plants, A and B, which have similar load characteristics but which use coal differing in quality and price:

| I Coal Characteristics of the Plant: | | Plant A | Plant B |
|---|---|---|---|
| 1 | Average B.t.u. supplied to plant per kw-hr. net output | 22,400 | 25,600 |
| 2 | Thermal efficiency of plant in per cent | 15.24 | 13.34 |
| 3 | Average B.t.u. per dollar, coal as received (moist basis) | 5,226,666 | 5,734,400 |
| 4 | Coal factor, or lb. of coal per kw-hr. net output (moist basis) | 1.60 | 2.00 |
| 5 | Average B.t.u. per lb. coal, as received (moist basis) | 14,000 | 12,800 |
| 6 | Cost of coal per ton (2240 lb.) delivered alongside plant | 6.00 | 5.00 |
| 7 | Kind of coal | Bituminous | Bituminous |
| 8 | Cost of coal in cents per kw-hr. net output | 0.428 | 0.446 |

| II Load Characteristics of the Plant: | | | | Plant A | Plant B |
|---|---|---|---|---|---|
| 9 | Average daily load factor of load | | | 50 | 54 |
| 10 | Maximum load for the year | Interval of maximum load, 1 hr. | | 90,000 | 80,000 |
| 11 | Yearly load factor of load | | | 40 | 38 |
| 12 | Kw-hr. net output for the year (kw-hr. sent out from the a.c. bus) | | | 315,460,000 | 217,248,000 |
| 13 | Installed rated capacity, i.e., aggregate max. continuous rating of the generators in kw | | | 125,000 | 100,000 |
| 14 | Average kw-hr. net output per kw. installed rated capacity | | | 2,525 | 2,173 |

(*Journal of The American Institute of Electrical Engineers*, vol. 39, no. 2, Feb. 1920, pp. 132-134, *p*)

## PRODUCER GAS (See Motor-Car Engineering)

## TESTING OF MATERIALS (See Engineering Materials)

## TESTING AND MEASUREMENTS

### Theory of Rocket Action; Goddard's Calculation of Rocket Capable of Passing Beyond the Range of Attraction of the Earth

A METHOD OF REACHING EXTREME ALTITUDES, Robert H. Goddard. Discussion of rocket action, the theory and design of rockets, and also the use of rockets for carrying recording instruments to the upper reaches of the atmosphere and beyond.

The problem considered in the paper was to determine the minimum initial mass of an ideal rocket necessary in order that on continuous loss of mass a final mass of 1 lb. would remain at any desired altitude.

An approximate mathematical method has shown that surprisingly small initial mass would be needed to carry a rocket to extremely high altitudes provided the gases were ejected from the rocket at a high velocity and also provided that most of the rocket consisted of propellant material.

The reason for this is, namely, that the velocity enters in the expression for the initial mass as an exponent, so that if the velocity of the ejected gases be increased fivefold, the initial mass necessary to reach the given height will be reduced to the fifth root of that required for the lesser velocity.

The solution of the problem of constructing a rocket of reasonable dimensions to reach extremely high altitudes necessitated the determination of the extent to which efficiency could be carried in a rocket of new design. The term efficiency as applied to rockets means the ratio of kinetic energy of the exploded gases to the heat energy of the powder, the kinetic energy itself being calculated from the average velocity of ejection—obtained indirectly by observations on the recoil of the rotors.

Tests have shown that the powder used in an ordinary rocket constitutes only a very small fraction of the total mass and the efficiency of the rocket is as low as 2 per cent. Experiments were performed with the object of increasing the average velocity of ejection of the gases. Charges of dense smokeless powder were fired in steel chambers provided with smooth tapered nozzles, the object of which was to obtain the work of expansion of the gases, much as is done in the DeLaval steam turbine. The efficiencies and velocities obtained in this way were very high, the highest efficiency being over 64 per cent and the highest average velocity of ejection being slightly under 8000 ft. per sec. Moreover, these velocities were proved to be real velocities and not merely effects due to reaction against the air resulting from firing the same steel chambers in vacuum and observing the recoil.

Since, however, a heavy steel chamber could not be used in a rocket, a reloading mechanism was designed in which successive charges were fired in the same chamber of light weight and small mass, so that most of the mass of the rocket could consist of propellant.

Regarding the heights that could be reached by the above method, it would appear that a mass of 1 lb. could be elevated to altitudes of 35, 72 and 232 miles by employing initial masses of from 3.6 to 12.6, from 5.1 to 24.3 and 9.8 to 89.6 lb., respectively, which is only a very small fraction indeed of what the initial masses would have to be in order to raise to the same heights a rocket of the Costen ship type.

In the course of the investigation several interesting methods were developed for deducing and measuring gaseous rebound: one to measure the force of the rebound, and another, called "impulse motors," for measuring the magnitude per unit area produced by the rebound in gases.

Among other things is discussed the question of calculating the minimum mass required to raise 1 lb. to an infinite altitude. The meaning of this term is as follows: Theoretically a mass projected from the surface of the earth with a velocity of 6.95 miles per sec. would, neglecting air resistance, reach an infinite distance after an infinite time, which in ordinary language means that it would never return to the earth. Actually, such a mass would not reach an infinite distance, but would come under the gravitational influence of some other heavenly body and follow the course prescribed by the resultant between the two forces acting upon it.

The construction of a rocket that would pass the sphere of gravitational attraction of the earth does not appear to present insuperable difficulties, the main difficulty lying in the finding of means of proving that such extreme altitudes as would be considered in this case had been reached. The only reliable procedure would be to send the smallest mass of flash powder possible to the dark surface of the moon when in conjunction in such a way that it would be ignited on impact. The light would then be visible in a powerful telescope.

Mathematical computations have shown that with the telescope of 1 ft. aperture a mass of flash powder of about a dozen pounds exploded on the surface of the moon would be very clearly visible, but to carry such a mass of flash powder to the moon a rocket of an initial mass varying between 8 and 16 tons would be needed.

The author of the paper plainly states that this plan of projecting a mass of flash powder to the surface of the moon, although a matter of much general interest, is not of obvious scientific importance. He adds that there are, however, developments of the general method under discussion which involve a number of important features not herein mentioned which could lead to results of much scientific interest. These developments involve many experimental difficulties but they depend upon nothing that is really impossible.

It appears, therefore, that contrary to the impression given by newspaper reports of the paper, the author does not consider that his rockets in their present stage can be employed for establishing a system of passenger transportation between our planet and its heavenly neighbors. (*Smithsonian Miscellaneous Collections*, vol. 71, no. 2, publication 2540, 1919, 69 pp. and 10 plates of figures, *teA*.)

## VARIA

New British Patent Act. This act was finally approved on December 23, 1919, and is now in force, except a few sections. It makes numerous alterations and amendments to the act of 1917, the most important of which are as follows:

The period of provisional protection has been increased from six to nine months. The duration of the life of a patent has also been extended and is now sixteen years instead of fourteen, as hitherto.

Provisions have been made to simplify the procedure for obtaining the prolongation of the life of the patent when it has been affected by war conditions.

An important alteration of the previous laws has been made in the way of extending the ground of opposition to the grant of a patent so as to include prior publications other than patent specifications, which makes it possible to now use references to newspaper or technical periodicals. But the examiner is not allowed to cite publications other than British specifications in opposition to the grant of a patent.

Under the new laws an inventor does not prejudice his right to a patent if he reads a paper containing a description therein before a learned society and has before such reading given notice of his intention to do so.

Extensive modifications not suitable for abstracting have been made in the closes referring to the working on a commercial scale of new inventions required under the patent laws. The system of granting compulsory licenses has also been modified in several material respects. On the whole, however, the old provisions as to the method of payment, etc., which made the British system so essential records. (Compare *Engineering*, vol. 109, no. 2819, Jan. 9, 1920, pp. 53-54, *g*)

Women Workers in Britain. Approximately 5000 firms in Great Britain have been asked by the Women's Industrial League for a statement of their experience of women as industrial workers. Over 1400 replies have been received, covering a wide field in engineering and other trades.

The replies show that 1422 firms were employing 79,700 women at the end of May, 1919, as compared with 43,200 before the war and 245,300 during the war. Of 382 firms who employed women before the war, 67 propose to increase the number, in consequence of the larger experience of women's work gained by them; while of 764 who employed women for the first time during the war, 228 propose to retain them. These firms do not comment on the present labor situation. A further 97 firms state that they would be willing to retain women but for trade union opposition. Taking all the figures, the estimate is that over 60 per cent of those who have tested women's work are ready to continue employing them. (*The Iron Trade Review*, vol. 66, no. 8, February 19, 1920, p. 558, *g*)

## CLASSIFICATION OF ARTICLES

Articles appearing in the Survey are classified as *c* comparative; *d* descriptive; *e* experimental; *g* general; *h* historical; *m* mathematical; *p* practical; *s* statistical; *t* theoretical. Articles of especial merit are rated *A* by the reviewer. Opinions expressed are those of the reviewer, not of the Society. The Editor will be pleased to receive inquiries for further information in connection with articles reported in the Survey.

# ENGINEERING RESEARCH

## A Department Conducted by the Research Committee of the A. S. M. E.

### Research Associations in Great Britain

THE proposed Research Associations in Great Britain which were described in the letter of Mr. Bancel in MECHANICAL ENGINEERING for August have been organized, and a recent circular issued from the Department of Scientific and Industrial Research of Great Britain gives the following list of Associations which have received licenses from the Board of Trade:

British Boot, Shoe and Allied Trades Research Association, Technical School, Abington Square, Northampton. Secretary, John Blakeman, M. A., M. Sc.
British Cotton Industry Research Association, 108, Deansgate, Manchester. Secretary, Miss B. Thomas.
British Empire Sugar Research Association, Evelyn House, 62, Oxford St., London, W. 1. Secretary, W. H. Giffard.
British Iron Manufacturers Research Association, Atlantic Chambers, Brazenose Street, Manchester. Secretary, H. S. Knowles.
British Motor and Allied Manufacturers Research Association, 39, St. James Street, London, S. W. 1. Secretary, Horace Wyatt.
British Photographic Research Association, Sicilian House, Southampton Row, London, W. C. 1. Secretary, Arthur C. Brookes.
British Portland Cement Research Association, 6, Lloyd's Avenue. London, E. C. 3. Secretary, S. G. S. Panisset, A.C.G.I., F.C.S.
British Research Association for the Woolen and Worsted Industries, Bond Place Chambers, Leeds. Secretary, Arnold Frobisher, B. Sc.
British Scientific Instrument Research Association, 26, Russell Square, W. C. 1. Secretary, J. W. Williamson, B. Sc.
British Rubber and Tyre Manufacturers Research Association, c/o Messrs. W. B. Peat & Co., 11, Ironmonger Lane, E. C. 2.
The Linen Industry Research Association, 3, Bedford St., Belfast. Secretary, Miss M. K. E. Allen.
Glass Research Association, 7, Senmore Place. W. 1. Secretary, E. Quine, B. Sc.
British Cocoa, Chocolate Sugar Confectionary and Jam Trades Research Association, 9, Queens Street Place, E. C. 4. Secretary, R. M. Leonard.

The Department of Scientific and Industrial Research of Great Britain has approved certain research associations which have not yet been licensed by the Board of Trade. These associations are given in the list below:

British Music Industries Research Association
British Refractory Materials Research Association
British Non-Ferrous Metals Research Association
Scottish Shale Oil Research Association.

The associations which are being considered by the Department of Scientific and Industrial Research are listed below:

British Launderers Research Association
British Electrical and Allied Industries Research Association
British Aircraft Research Association.

The silk manufacturers, the leather trade and the master bakers and confectioners are preparing articles of association for consideration by the Department.

Do the lists above suggest to our members the advisability of the formation of similar research associations in this country? The Heating and Ventilating Engineers, the National Canners' Association, the Association of Steel Manufacturers are all indicative of such work in America, but there is need for more coöperative work. The Research Committee feels that we will have accomplished much when those engaged in similar lines of manufacture can forget trade differences and coöperate in all active work for the development of knowledge in science and the arts.

### Kansas State Research Council

The State of Kansas has recently organized the Kansas State Research Council, which bears the same relation to the State of Kansas as the National Research Council does to the United States. Professor G. C. Shaad of the University of Kansas and Dean A. A. Potter of the Kansas State Agricultural College are on the Executive Committee of the Council.

### Research of the American Society of Heating and Ventilating Engineers

The first reports of the researches from the Research Laboratory of the American Society of Heating and Ventilating Engineers at the laboratory in Pittsburgh under the direction of Director John R. Allen were made at the annual meeting of the society in January. A paper on the total heat transfer through radiators and the amount of radiant heat from a radiator, as well as the tests on coal stoves made by Professor Allen and Mr. Rowley, were given. The excellent work of this society as shown in these reports is highly commendable, and the Research Committee of The American Society of Mechanical Engineers greatly appreciates their activity.

### Instrument Specifications

The Firestone Tire and Rubber Company are preparing a code of specifications for Bourdon tube pressure gages, recording thermometers, indicating mercurial thermometers and test gages. These are being prepared under the direction of Mr. F. J. Schlink, Physicist in Charge of the Instruments Inspection Department, and Chairman of the Special Committee on the Permanency and Accuracy of Indication of Engineering Instruments of The American Society of Mechanical Engineers.

#### A—RESEARCH RESULTS

The purpose of this section of Engineering Research is to give the origin of research information which has been completed, to give a résumé of research results with formulae or curves where such may be readily given, and to report results of non-extensive researches which in the opinion of the investigators do not warrant a paper.

*Apparatus and Instruments A2-20* The investigation on altitude effect of air-speed indicators is nearing completion and show conclusively that the compressibility of air may be legitimately omitted in determining the law of action of the usual venturi tubes; that viscosity must not be neglected in low-speed flights; that at low speeds the indication of the venturi is not directly proportional to the air density nor to the square of the speed but departs noticeably from this relation. Bureau of Standards, Washington, D. C. Address S. W. Stratton, Director.

*Cement and Other Building Materials A1-20* Crushed-Slag Aggregate. Crushed slag as a coarse aggregate produces concrete of as high or higher strength than gravel. The experiments already made are not extensive enough to determine the durability of slag, but so far as they have been carried on there have been no signs of disintegration due to sulphide sulphur. Slag sand does not produce an easily workable concrete when used as fine aggregate because of its lack of fine material. Bureau of Standards, Washington, D. C. Address S. W. Stratton, Director.

*Explosives and Explosions A1-20* Propagation of Flame in Pipes and Effectiveness of Arresters. The Underwriter's Laboratory of Chicago has issued a report on the above subject prepared by A. H. Nuckolls for the E. I. du Pont de Nemours and Company. The investigation was made by filling a pipe with a vapor-air mixture containing a known percentage of vapor and igniting by electric spark. The pressures developed were measured by recording gas-engine indicators and the effectiveness of arresters was determined by examining the products remaining in the pipe. The speed of propagation was not investigated. The results show that the pressure is dependent on the nature of the vapor and the relative proportion of air and is influenced by the size

181

and length of the pipe. Gasoline and ether produced maximum pressures from 75 lb. per sq. in. in ½-in. pipes to about 150 lb. per sq. in. in 6-in. pipes. Alcohol and acetone gave 50 to 100 lb. per sq. in. while amyl acetate and light oil gave about 70 lb. Rich and lean mixtures gave lower pressures than the mixture having the right composition for complete combustion. Pressures were found to increase with the diameter of the pipe but not with the length of the pipe. The propagation appeared to be progressive combustion. Arresters of the screen type were found to be more satisfactory than glass, wool or metal chips, especially when made in the cone shape, the length of the axis being twice the diameter of the base. Multiple coarse-mesh screens were ineffective. The effectiveness of arresters is influenced by the nature and proportion of the vapor, size and length of pipe and location of arrester. Flames from ether and gasoline are more difficult to arrest than those from other vapors. The larger the amount of air the more easily is propagation arrested. In large pipes the arresters are not as effective as in small pipes.

In connecting tanks the connecting pipe should not be over 4 in. in size; preferably it should be ½ in. or 1 in. The length should not be greater than 20 ft. Double cone screens should be employed. Flat screens may be used in 2-in. vent pipes.

This investigation should be continued over a wider range. Address Engineering Department, E. I. du Pont de Nemours Company, F. A. Wardenburg, Assistant Chief Engineer, Wilmington, Del.

*Heating A1-20* Coal Stoves. The efficiency of the standard design of base-burner coal stoves has been found to be about 70 per cent under ordinary operation. The two stoves tested showed an excessive amount of air leakage so that the percentage of $CO_2$ in the exhaust gases was about 4 per cent. After leaks were sealed the percentage rose to 10 per cent. The stove under ordinary conditions gave about 75 per cent efficiency, but at the higher rates of combustion it is impossible to get this efficiency due to the lack of radiating surface from which the heat can be given. The tests were made at the University of Minnesota by Professors John R. Allen and F. B. Rowley. Address John R. Allen, Bureau of Mines Building, Pittsburgh, Pa.

*Heating A2-20* Radiant Heat from Direct Radiators. Director John R. Allen and Prof. Frank B. Rowley have presented a paper to the American Society of Heating and Ventilating Engineers on tests made at the University of Minnesota to determine the amount of heat radiated from a direct cast-iron steam radiator. The radiator was placed within a tank from which the air was exhausted and the amount of heat radiated was determined from steam condensation, measuring the heat input in an electrically heated boiler. The experiments indicate that the Stefan-Boltzmann law holds. The amount per sq. ft. of radiation is equal to

$$Q = 0.137 R \left[ \left( \frac{T_s}{100} \right)^4 - \left( \frac{T_r}{100} \right)^4 \right]$$

In this formula $Q$ represents the heat in B.t.u. per sq. ft. per hour radiated from the radiator. $R$ is the proportion of the total radiator surface which is effective in radiating heat. This varies from 0.55 in one-column radiators, with a value of about 0.8 for wall radiators. The constant 0.137 was determined from physical experiments on clean cast iron, the value being less than this when aluminum paint is used. Table 1 gives the values of the constant $D$ which is 0.15 in the formula above.

TABLE 1.

| Condition of surface: | Temp. of tank surface. deg. fahr. | Temp. of steam. deg. fahr. | B.t.u. per sq.ft. radiating surface. | Value of constant $D$. |
|---|---|---|---|---|
| Cast iron, rusty........ | 70 | 215 | 180 | 0.142 |
| Cast iron, painted black. | 70 | 215 | 152 | 0.120 |
| Cast iron, clean....... | 70 | 215 | 189 | 0.148 |
| Aluminum paint........ | 70 | 215 | 131 | 0.104 |
| Physical experiment: | | | | |
| clean cast iron........ | 70 | 215 | 198 | 0.157 |

For practical work the authors suggest 0.137 to be used in problems. With steam temperature at 215 deg. fahr. and room temperature at 70 deg. fahr. the radiation amounts to 200 B.t.u. per sq. ft. of rated surface per hour. Address Director John R. Allen, Bureau of Mines Building, Pittsburgh, Pa.

*Heating A3-20* Heat Transfer from Direct Radiators. Director John R. Allen of the Research Laboratory of the American Society of Heating and Ventilating Engineers has presented a paper to the Society at its annual meeting January, 1920. His investigations show that the heat transfer from radiators is made up of two parts: that transmitted by radiation and that transmitted by conduction. The final formula is given as

$$Q = 0.157 R \left[ \left( \frac{T_s}{100} \right)^4 - \left( \frac{T_r}{100} \right)^4 \right] + (t_s - t_r)$$

In this equation the first terms give the heat delivered by radiation and the last term the heat by convection the constant for which is unity. In the equation $Q$ represents the B.t.u. per sq. ft. of total radiator surface per hour. $R$ represents the ratio of the radiating surface to the total surface. $T_r$ = the absolute

temperature of the room in fahrenheit degrees and $T_s$ the absolute temperature of the steam in fahrenheit degrees, while $t_r$ and $t_s$ are in degrees fahrenheit. The value of $R$ for ten-section radiators is about 0.55 for one-column radiators, 0.50 for two-column radiators, 0.38 for three-column radiators, 0.33 for four-column radiators and about 0.8 for wall radiators. These constants vary with the height of the radiator and with the

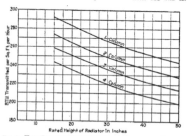

FIG. 1 HEAT TRANSMISSION FOR DIFFERENT HEIGHTS AND NUMBERS OF COLUMNS OF RADIATORS

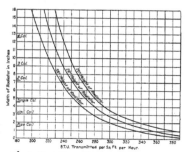

FIG. 2 HEAT TRANSMISSION FOR DIFFERENT WIDTHS OF RADIATORS

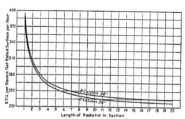

FIG. 3 HEAT TRANSMISSION FOR RADIATORS OF VARYING LENGTH

number of sections in the radiator, increasing with the decrease of height and decrease of number of sections. Curves and tables show the effect of heights of radiator, widths of radiator, lengths of radiator, temperature of steam and room, humidity, air circulation, final painting and position of radiator. The results are given in curves and tables in the original paper and the curves are reproduced below. The effects of grillwork or shields in front of the radiator are shown to reduce the heating capacity from 10 to 20 per cent, although when radiators are placed below seats reductions amount to 35 or 40 per cent. Address John R. Allen, Director, Bureau of Mines Building, Pittsburgh, Pa.

TABLE 2  EFFECT OF PAINTING ON TWO-COLUMN 38-IN RADIATOR

Steam Temperature 215 deg.  Room Temperature 70 deg. fahr.

| Condition of surface. | B.t.u. per sq. ft. |
|---|---|
| Cast Iron, bare.................. | 240 |
| Painted with aluminum bronze............. | 200 |
| Painted with gold bronze............... | 210 |
| Painted with white enamel.............. | 242 |
| Painted with maroon japan.............. | 240 |
| Painted with white zinc paint............. | 242 |
| Painted with no-luster green enamel....... | 230 |

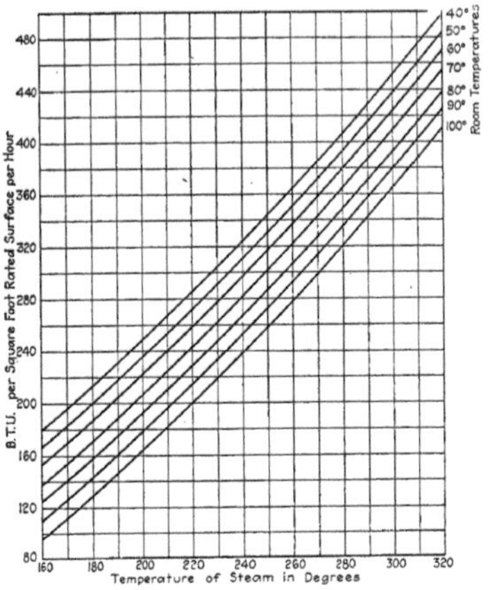

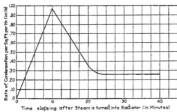

FIG. 4  HEAT TRANSMISSION FOR VARIOUS TEMPERATURES OF STEAM IN RADIATOR AND AIR IN THE ROOM

FIG. 5  RATE OF CONDENSATION WHEN STEAM IS TURNED INTO RADIATOR

**Heating** A4-20  Kiln Coils for Wood Drying. Return-bend heating coils gives a more uniform distribution of heat in kilns than coils built up of headers and should be used when possible. Forest Products Laboratory, Address Director, Madison, Wis.

**Machine Design** A1-20  Six-Throw Automobile Crankshafts. A report has been issued by the Bureau of Standards to certain manufacturers on the elimination of certain difficulties in making 6-throw automobile crankshafts. It was found that much of the trouble was of a mechanical rather than a metallurgical nature. Opportunities were found for improvements in the

metallurgical treatment of the process. Bureau of Standards. Washington, D. C.  Address S. W. Stratton. Director.

**Metallurgy and Metallography** A2-20  Microstructure of Steel Subject to Sudden Heating. Martensite structure has been observed in the superficial layers of metal in the bores of guns and pressure plugs. This has been probably due to rapid heating and high temperature followed by sudden cooling. This has been shown to be true by a sudden heating and cooling of the metal produced by the intermittent discharge of an electrode close to the metal surface. Bureau of Standards, Washington, D. C. Address S. W. Stratton, Director.

**Metallurgy and Metallography** A3-20  Spiegeleisen. Production of Spiegeleisen in Blast Furnaces, by P. H. Royster. War Minerals Investigations Series No. 6, Bureau of Mines. Address Van H. Manning, Director.

**Properties of Engineering Materials** A2-20  Fatigue Tests on Tubing. Tests in progress on iron and copper tubing under repeated stress indicate that pipes are seriously weakened by surface irregularities, especially by makers' names stamped on the pipe. Bureau of Standards, Washington, D. C. Address W. S. Stratton, Director.

**Properties of Engineering Materials** A3-20  Electric Welding of Metal. A comparison of tensile tests on welded metal plates shows that the mechanical properties of the material produced at the Bureau of Standards are uniformly lower than those of the Wirt-Jones test conducted by the Welding Research Sub-Committee. The personal element appears to be an appreciable factor in welding. Microscopic examination and hardness tests do not show any well-defined relation between physical properties and microstructure because of premature failure due to flaws. Bureau of Standards, Washington, D. C. Address S. W. Stratton, Director.

**Rubber and Allied Substances** A1-20  Gutta Percha. Weston A. Price, M.S., D.D.S., of Research Institute of the National Dental Association. in a report of laboratory investigations on tooth-filling materials discusses the expansion of gutta percha with temperature change and also the action of the evaporation of solvents on gutta percha. He gives a report of Dr. Dayton C. Miller of Case School of Applied Science. On heating and cooling a sheet of pink gutta percha the sheet contracted greatly in the direction which appeared to be that in which it was rolled, while expansion took place in the direction perpendicular to this. In heating from 22 deg. fahr. to 211 deg. fahr. the contraction of area was 14.7 per cent. A second piece 70 mm. by 70 mm. was heated from 22 deg. fahr. to 211 deg. fahr., changing in area to 63.0 mm. by 72.5 mm. The thickness increased from 0.843 mm. to 0.912 mm. This means a diminution of length of 10 per cent, an increase of breadth of 3 per cent and an increase of volume of about 1 per cent. The test under a rising and falling temperature showed a variation in the form of a hysteresis curve. On annealing in boiling water for 10 min. somewhat similar results were obtained.

Chloroform was found to absorb gutta percha. When the solvent was allowed to evaporate the volume of a thick solution just firm enough to hold its shape was 200 per cent of the volume when all of the solvent was driven off. Address Dr. Weston A. Price, Research Institute of National Dental Association, Cleveland, O.

**Wood Products** A1-20  Laminated-Wood Forms. Solid artificial limbs are difficult to obtain without checking. Built-up limb blanks have been proposed, made with water-resistant casein glue. Little or no waste is experienced in this way.

After 175 games, laminated maple bowling pins were returned for photographing. They were in excellent condition, but one gave way in a casein glue joint. This is not considered a defective pin, as solid pins often split. Hide glue and casein glue have stood up equally well. Birch pins are not so satisfactory. Two sets of maple pins are being made up to be combined with two sets of standard pins. Laminated gun stocks of black walnut are to be shipped to the Ordnance Department. Laminated shoe lasts are being tried by the Badger State Shoe Company. Forest Products Laboratory, Madison, Wis. Address Director.

**Wood Products** A2-20  Drying Schedule for Air-Seasoned Oak. Table 3 which follows for 4-in. by 4-in. plain-sawed oak with 15 per cent moisture content gives the same result as would be reached by 9 to 12 months of air seasoning.

After stock is dry, kiln is held at 145 deg. fahr. with 80 per cent humidity for 10 hours to balance moisture content and prevent warping.

**Wood Products** A3-20  Wood Preservatives. The Forest Products Laboratory has treated sap-pine ties and red-oak ties with sodium fluoride, zinc chloride and creosote. The sodium fluoride has high toxicity, is not injurious to metal and is convenient to handle. Sap-pine ties were placed in the mines at Birmingham, Ala., in 1914, while red-oak ties were placed on the tracks of the Baltimore and Ohio railroad. After five years of mine service the ties treated with sodium fluoride were found in as good condition as those treated with zinc chloride. Both showed little deterioration. The creosoted ties were in better condition while untreated ties were in advanced stages of decay. The railroad ties which were treated were practically all sound, while untreated ties had to be removed. Forest Products Laboratory, Madison. Wis. Address Director.

TABLE 8

| Stage of Kiln run in hours. | Temperature, deg. fahr. | Humidity, per cent. |
|---|---|---|
| 1 | 120 | 100 |
| 6 | 120 | 100 |
| 12 | 125 | 85 |
| 24 | 125 | 80 |
| 36 | 130 | 70 |
| 48 | 130 | 60 |
| 60 | 135 | 60 |
| 72 | 135 | 60 |
| 84 | 140 | 60 |
| 96 | 140 | 50 |
| 108 | 140 | 50 |
| 120 | 140 | 40 |
| 144 | 140 | 40 |
| 156 | 145 | 35 |
| 168 | 145 | 35 |
| 180 | 145 | 30 |
| 192 | 145 | 25 |
| Until dry | 145 | 25 |

Wood Products A3-20 Wood Pulp with Decay. Wood pulp infected with molds and fungi has been shown to be defective in producing dirty paper, in requiring extra sizing. in sticking to couch and press rolls, and gives trouble from excessive foaming. Infected pulp is free in its action and causes difficulty in carrying the necessary amount of water. Infected pulp yields 10 per cent less finished paper than clean pulp, in addition to reducing strength of the finished paper. Forest Products Laboratory, Madison. Wis. Address Director.

Wood Products A4-20 Shrinkage of Veneer. The veneers made from steam logs or boiled logs were tested and found to contain about the same amount of moisture before drying (65 per cent) and about the same amount after drying (10 per cent). The shrinkage of each kind was about the same. variation being no greater than that found in the pieces from the same treatment. Forest Products Laboratory, Madison, Wis. Address Director.

Wood Products A5-29 Moisture Content of Wood. Moisture content of wood has been found to be independent of the density of the wood. With a relative humidity of 38 per cent at 80 deg. fahr. in the oven the moisture content is about 11 per cent and with 88 per cent relative humidity the moisture content is 18.8 per cent. This is the same for specimens of varying densities. Forest Products Laboratory, Madison. Wis. Address Director.

## B—RESEARCH IN PROGRESS

The purpose of this section of engineering research is to bring together those who are working on the same problem for cooperation or conference, to prevent unnecessary duplication of work and to inform the profession of the investigators who are engaged upon research problems. The addresses of these investigators are given for the purpose of correspondence.

Aircraft B2-20 Wind Tunnels and Propellers. The National Advisory Committee for Aeronautics reports a research in progress by means of models to determine the best form of wind tunnel and the best type of propeller to use in drawing the wind stream through the tunnel. Various arrangements are being studied, especially with regard to their power consumption for a given wind speed and to the steadiness of velocity and direction of air currents. Address Joseph S. Ames, of the Executive Committee of the National Advisory Committee for Aeronautics, 4th St. and Missouri Ave., N. W., Washington, D. C.

Cement and Other Building Materials B1-20 Compressive Strength of Concrete. A paper is being prepared on the compressive strength of concrete based on the results of tests made by the Bureau of Standards at Lehigh University and Lafayette College in 1918 and at the Structural Materials Laboratory at St. Louis from 1904 to 1910. Bureau of Standards, Washington. D. C. Address S. W. Stratton, Director.

Internal-Combustion Motors B2-20 Ignition Apparatus. The Bureau of Standards has been developing a mathematical theory of the building-up of voltage in the secondary winding after the opening of the primary breaker. Data from nine magnetos have been received by the laboratory. The ratio of transformation and coupling coefficient has been determined and measurements of the mutual inductance are now in progress. Bureau of Standards, Washington, D. C. Address S. W. Stratton, Director.

Materials of Engineering B1-20 Thermal Expansion. A research in the determination of the thermal expansion for various substances. Several proposed outlines for the work have been drawn up. The work will keep the present laboratory force devoted to this subject busy for six months. Address Bureau of Standards, Washington, D. C. Address Director.

Metallurgy and Metallography B3-20 Hardness Relations. Relation of hardness, temperature of pouring and machining properties

of cast iron of given thickness and composition. Address Prof. W. W. Bird, Worcester Polytechnic Institute, Worcester, Mass.

Properties of Engineering Materials B6-20 Strain-Gage Test. A strain-gage test of the 350-ton fitting crane at the Philadelphia Navy Yard has just been made. The results of this test are to be made up in the form of a report. Bureau of Standards, Washington, D. C. Address S. W. Stratton. Director.

Properties of Engineering Materials B7-20 The Use of the Camera in Studying Physical Properties. The Bureau of Standards has been investigating photographic methods of studying elongations and lateral deformation of steel during tensile tests to determine the relation between elongation and lateral deformation. It is also developing a special camera for taking panoramic photographs of the interior of a gun barrel to study the effects of erosion. The lenses have been perfected and the camera is being constructed. Bureau of Standards, Washington, D. C. Address S. W. Stratton, Director.

Properties of Engineering Materials B8-20 Insulating Materials. The Bureau of Standards has carried on extensive investigations of the electrical and mechanical properties of insulating materials, giving particular attention to their behavior at radio frequencies. The research will include the effects of moisture and other tests to determine the practical usefulness of various materials. Special study has been made of bakelite and other materials of the phenol-methylene type. and materials such as glass, mica, celluloid, electrosote, paraffin, fiber, hard rubber, porcelain and wood will be included. Bureau of Standards, Washington, D. C. Address S. W. Stratton, Director.

Properties of Engineering Materials B9-20 Physical Constants of Materials Used in Refrigeration. The Bureau of Standards is investigating the physical properties of constants of materials used in refrigeration. The Bureau has recently determined the vapor pressure of saturated ammonia and densities of the liquid and saturated vapor. It has also determined the specific heats of sodium-chloride solutions. Bureau of Standards, Washington, D. C. Address S. W. Stratton, Director.

Protective Devices B1-20 Carbon Monoxide Detector. A. C. Fieldner, Supervising Chemist, G. St. J. Perrott, Physical Chemist. Bureau of Mines Station, Pittsburgh, Pa. So far the detector developed for gas-mask use by Lieutenant-Colonel Lamb and Captain Larson of the Chemical Warfare Service seems to be most promising. Bureau of Mines. Address Van H. Manning, Director, Washington, D. C.

Railroad Rolling Stock and Accessories B1-20 Chilled-Iron Car Wheels. Experiments are being made at the Bureau of Standards to determine temperatures at various portions of a wheel heated on the rim and to study the effects of this heating. Several wheels have already been tested at this plant. Work of a similar nature is being carried out at the University of Illinois, Bureau of Standards. Washington. D. C. Address S. W. Stratton, Director.

## C—RESEARCH PROBLEMS

The purpose of this section of Engineering Research is to bring together persons who desire cooperation in research work or to bring together those who have problems and no equipment with those who are equipped to carry on research. It is hoped that those desiring cooperation or aid will state problems for publication in this section.

Textile Manufacture and Clothing C1-20 Drying of Thin Gauze on Tentering Machines. The effect of humidity; the effect of velocity of air; the effect of heating cloth by radiation or by convection; the effect of drying selvages by installing steam pipes close to tenter fingers; best relative humidity of air at discharge for maximum drying rate. Address D. M. Ferris, Graduate Student, M.I.T., 37 Lee St.. Cambridge. Mass.

## D—RESEARCH EQUIPMENT

The purpose of this section of Engineering Research is to give in concise form notes regarding the equipment of laboratories for mutual information and for the purpose of informing the profession of the equipment in various laboratories so that persons desiring special investigations may know where such work may be done.

Harvard University D1-20 Cryogenic Laboratory. The equipment for the liquefaction of gases and for very low-temperature work includes two commercial air compressors for pressures to 3000 per sq. in., and two commercial types of liquefiers as well as special apparatus which is under construction. In the Refrigerating Laboratory ammonia and carbon-dioxide compression plants are about completed.

The Internal-Combustion Laboratory is being equipped, although much of the apparatus for this laboratory has not been purchased. Address Prof. Lionel S. Marks, Harvard University, Cambridge, Mass.

*Machine Tools D1-20* Cutting Action of Machine Tools. The University of Cincinnati is equipping its mechanical laboratory for the purpose of making investigations on the cutting action of machine tools. The university has recently received an Olsen efficiency-testing machine for files, milling cutters, lathe tools, hack saws and drills, and has also bought a number of Government machine tools to be used on research work. One of them is a large heavy-duty radial drill. A No. 3 planer milling machine will also be used for research work. The laboratory is to have a Charpy impact-testing machine and Olsen impact-testing machine and four Farmer fatigue-testing machines. Address Prof. A. L. Jenkins, University of Cincinnati, Cincinnati, Ohio.

*University of Cincinnati D2-20* Thermal Laboratory. The Mechanical Engineering Laboratory of the University of Cincinnati is equipped with an experimental gas engine made by the Foos Gas Engine Company, four small gas and oil engines, a 12 by 24-in. Hamilton Corliss engine, a Fairbanks gas producer and a number of automobile engines. Address Prof. A. L. Jenkins, University of Cincinnati, Cincinnati, Ohio.

### E—RESEARCH PERSONNEL

The purpose of this section of Engineering Research is to give notes of a personal nature regarding the personnel of various laboratories, methods of procedure for commercial work or notes regarding the conduct of various laboratories.

*Purdue University E1-20* The Experiment Station at Purdue University is about to issue a bulletin on the use of electric power in small pumping stations, by Professor Ewing, and a bulletin on the Coefficient of Discharge of Sewage Sprinkler Nozzles, by Prof. W. W. Greve. Prof. O. C. Berry has nearly completed a bulletin on the Investigations of Carburetors and Prof. R. V. Achatz has completed a circular on the Effect of Preservative Treatment of Various Kinds of Telephone and Telegraph Poles. Address Purdue University, Lafayette, Ind. Dean C. H. Benjamin, Director.

### F—BIBLIOGRAPHIES

The purpose of this section of Engineering Research is to inform the profession and especially the members of the A. S. M. E. of bibliographies which have been prepared. These bibliographies have been prepared at the request of members, and where the bibliography is not extensive, this is done at the expense of the Society. For bibliographies of a general nature the Society is prepared to make extensive bibliographies at the expense of the Society on the approval of the Research Committee. After these bibliographies are prepared they are loaned to the person requesting them for a period of one month. Additional copies are prepared which are available for periods of two weeks to members of the A. S. M. E. or to others recommended by members of the A. S. M. E. These bibliographies are on file in the offices of the Society and are to be loaned on request. The bibliographies are prepared by the staff of the Library of the United Engineering Society which is probably the largest Engineering Library in this country.

*Economics F1-20* Valuation of Street Railway Systems. A bibliography of 1½ pages. Search 2835. Address A.S.M.E., 29 West 39th St., New York.

*Hydraulics F1-20* Flow of Oil in Conduits. A bibliography of 10 pages. Search 2825. Address A.S.M.E., 29 West 39th St., New York.

*Metallurgy and Metallography F1-20* Microstructure of Electric and Acetylene Welds. A bibliography of 1½ pages. Search 2784. Address A.S.M.E., 29 West 39th St., New York.

*Petroleum, Asphalt and Wood Products F1-20* Flow of Oil in Conduits. A bibliography of 10 pages. Search 2825. Address A.S.M.E., 29 West 39th St., New York.

*Wood Products F1-20* Marine Borers. A bibliography prepared by the Forest Products Laboratory, U. S. Forest Service, Madison, Wis. This contains eight references to articles and treatises. Forest Products Laboratory. Address Director Forest Products Laboratory, Madison, Wis.

# Unemployment, National Taxation and Profit Sharing
## Reconciliation of the Needs and Demands of the Government, the Public, the Employer, and the Worker

### BY HENRY HESS,[1] PHILADELPHIA, PA.

THE present disturbance in our industrial position has been a long time in the brewing. This process may be likened to a period of slow boiling under a gentle heat that gave the cook no serious cause for alarm, even though it did occasionally tilt the cover of the pot.

But let us leave this simile to concern ourselves with basic causes and see whether their recognition may not lead us to a basic remedy. It will be necessary to go back to the beginning of the modern industrial era and note the change which that brought about.

Under the older feudal system the subsistence of the working element was more or less well taken care of by the overlords or by the masters in the various guilds; and as long as these overlords and masters fulfilled their obligations, matters went along smoothly. It was only when these obligations were forgotten and the duty of providing subsistence was omitted that matters became serious. In other words, when subsistence became endangered the social economy became endangered.

When such subsistence becomes endangered under our changed conditions, then our modern social economy is in danger of similarly breaking down.

There are many changes in the relative situation of the modern worker as compared with his brother of bygone ages, but the basic interest is the same; the worker must be assured of his subsistence and of of the improved character that the advance of general civilization and culture has brought about.

This is our problem today under our modern industrial system, as it has been the problem under the feudal system that has been displaced, and as it will be the problem under whatever future system may take the place of that of today, and so on to the end of time.

Now what one basic trouble stands out above all others? There are many minor ones that look large at the moment and large in the eyes of many; but which one applies to every man, woman and child worker? Unquestionably the fear that the means of subsistence may be withdrawn at any moment, which fear is the fear of unemployment. When we analyze the demand for increased wages we find that it resolves itself into the hope that the difference between increased and existing wages will permit the accumulation of a fund the interest of which will provide the basis for defraying the living cost during the recurring periods of unemployment.

The writer proposes to address himself to this problem, as he considers it the real problem, though frequently not definitely recognized as such; and to those others that are coexistent with it, such as the provision of the means for the maintenance of a proper government and all that that implies in activities that are best delegated to government. This then becomes the problem for our industrial system, since it is this which earns and creates wealth and then distributes it.

Men band themselves together in business organizations, which have gradually taken on the preferred corporate form; here capital and labor coöperate to increase the national wealth and to distribute it and to minimize the risks involved.

There are two groups in such corporations; one is generally

[1] 308 Bailey Building. Past Vice-President Am.Soc.M.E. Discussion prepared for the Industrial Relations Sessions at the Annual Meeting of THE AMERICAN SOCIETY OF MECHANICAL ENGINEERS, December 1919.

referred to as capital, the other as labor. The ideal grouping will equally divide the risks and responsibilities; and given such ideal grouping, the results or profits should be equally divided. Labor is today striving for such equal division, but is not yet willing to share equally the risk and responsibility. The more enlightened section of capital is willing to share with labor in proportion as labor fits itself to and does assume its share of the burden.

When that condition has been reached there will no longer be any problem that is basically serious; both sides will by then have forgotten such matters as unionism, the open shop, the closed shop, collective bargaining and whatever other standard people assemble under to wage industrial war and to inflict on one another the utmost injury of which they are capable; instead, they will mutually coöperate for mutually equal benefit.

Until that ultimate goal is reached man must live and live with

### HYPOTHETICAL NATIONAL BALANCE SHEET[1]

| | |
|---|---:|
| Common Capital | $50,000,000,000 |
| Preferred Capital | 10,000,000,000 |
| Borrowed Capital (corresponds to Bonds; to receive 5% interest and be retired in 20 yr.) | 30,000,000,000 |
| Wages and salaries of 12,800,000 employees (exclusive of firm members), $1060 per capita | 13,568,000,000 |
| Unemployment cost at 8% (approx.) | 1,085,000,000 |

| | | |
|---|---|---:|
| Gross income for 1917 (made up of Wages, Taxes, Interest, Dividends and Surplus for Manufacturing, Mining, Railroad and Public Utilities), $1,926 per capita | | $24,653,000,000 |
| Wages at $1060 per capita | $13,568,000,000 | |
| Bonded Indebtedness at 5% | 1,000,000,000 | |
| Retirement of borrowings per yr. | 1,000,000,000 | |
| Depreciation | 500,000,000 | |
| State and Local Taxes, 2/3 of 1917 | 1,000,000,000 | 17,068,000,000 |
| Balance Net Earnings before Federal Tax | | $7,585,000,000 |
| 1st Charge—Income Tax (1917) | | 405,360,000 |
| Balance | | $7,179,640,000 |
| 2d Charge—Cumulative Preferred Stock 6% | | 600,000,000[a] |
| Balance | | $6,579,640,000 |
| 3d Charge—Plant betterment (= twice depreciation) | | 1,000,000,000 |
| Balance | | $5,579,640,000 |
| 4th Charge—Profit Share 2% of Wage Bill[b] | | 271,360,000 |
| Balance | | $5,308,280,000 |
| 5th Charge—Corporations' share to unemployment | | 813,640,000[a] |
| Balance | | $4,494,640,000 |
| 6th Charge—Stockholders' Common share 4% dividend | $2,000,000,000 | |
| ½ as profit share | 1,000,000,000 | 3,000,000,000 |
| Balance | | $1,494,640,000 |
| 7th Charge—Excess Profits (2/3 of 1917 charge) | | 650,000,000 |
| Corporations' Surplus Balance is 1.69% | | $844,640,000 |

[1] Based on year 1917.
[b] To be applied to labor's share to unemployment.
[a] Annual fixed charge if industry is shut down.

reasonable decency and must adequately subsist during those periodic disturbances in the orderly process of increasing individual and national wealth.

So far there are probably very few but agree with these general statements; but how carry them into effect? They have been and are being carried out here and there by small bodies and with entire success and consequent satisfaction to both sides.

If these few concerns gather followers in sufficient number to comprise the many that make up the nation, then the sporadic few would have become the complete whole. The balance sheets that are made up by the many individual concerns could then be gathered into one national balance sheet.

The writer has prepared such a hypothetical balance sheet as well as the insufficient data available permit. Unfortunately the

Government has not included collection of data covering unemployment in its census. Reliance must therefore be placed on such fragmentary information as has been collected by other agencies over certain sections of the country.

Unemployment is probably larger the denser the industrial population. Therefore, applying the conditions of New York City to the entire country will probably give a result that is in the nature of an overstatement and is in consequence safe for the liberation of funds to cover this unemployment cost.

A very careful census for New York City is available and published by the U. S. Department of Labor, Bureau of Labor Statistics in its Bulletin No. 172 in April 1915.

Averaging a table on page 10 of that bulletin shows that of all male and female workers 60 per cent were unemployed for 65 days, which corresponds to an average constant unemployment of 10.5 per cent.

The Federal Government had found it necessary to gather reliable data in connection with the collection of taxes. A brochure published by Prof. David Friday and entitled Profits, Wages and Industrial Progress, has furnished much information useful in the set-up of the hypothetical balance sheet prepared by the writer.

An examination of that sheet will show that all of the usual features carried in a condensed balance sheet are provided for, such as taxes of various kinds, plant depreciation and plant betterment that keep the industry in productive health, interest and repayment of borrowed capital, dividends on preferred capital with a sufficient surplus to return this preferred capital, ordinary dividends and such new features as provision for unemployment and profit sharing of capital and labor.

It is believed that these figures are entirely conservative, since the 1917 actual surplus was $256.25 per capita or $3,280,000,000, which is nearly four times the amount the above hypothetical balance sheet leaves as surplus. Ten per cent unemployment is equivalent to 9 years' surplus accumulation or (9 x $844,600,000 =) $7,601,400,000. Deducting from this the cost of one year's shutdown, $1,413,640,000, leaves a balance of $6,187,760,000, which corresponds to a dividend on common capital for 10 years of 12.37 per cent or per year of 1.24 per cent, this equity increasing the annual normal common capital dividend to 5.24 per cent.

### Résumé

The proposal suggests the removal of a very potent cause of labor unrest by insuring labor against the economic loss of involuntary unemployment through the accumulation of a fund. This fund is contributed to by labor, which applies a portion of its participation in general profits.

The Government's need for funds, over and above the amount raised by income tax, is covered by lending to the Government this unemployment fund.

It is possible to analyze or to set up a balance sheet showing the tangible quantities of the nation's business, but that method cannot be applied to the more intangible factors that are nevertheless the more important ones. Avoidance of the many consequences of unemployment and the fear of that will relieve the community of a very heavy burden and tend to an increased output of work willingly rendered by the worker, who is not only relieved of this fear of unemployment, but is also assured of a fair participation of the resulting profits.

The employer or capitalist finds his reckoning in the general stabilization of his business, which stabilization is of the same character as the assurance against unemployment.

The general public finds its interest given in the stabilization of business conditions generally and in the relief from the burdens of various kinds, financial and social, which it must carry as the result of unemployment.

The Government finds its advantage in there being placed at its disposition, for a period of years, an amount which it will therefore not have to raise by taxation, and this again will ease its public relations, as tax gathering is probably the most-objected-to feature of government.

# Work of the Boiler Code Committee

*THE Boiler Code Committee meets monthly for the purpose of considering communications relative to the Boiler Code. Any one desiring information as to the application of the Code is requested to communicate with the Secretary of the Committee, Mr. C. W. Obert, 29 West 39th St., New York, N. Y.*

The procedure of the Committee in handling the cases is as follows: All inquiries must be in written form before they are accepted for consideration. Copies are sent by the Secretary of the Committee to all of the members of the Committee. The interpretation, in the form of a reply, is then prepared by the Committee and passed upon at a regular meeting of the Committee. This interpretation is later submitted to the Council of the Society for approval, after which it is issued to the inquirer and simultaneously published in MECHANICAL ENGINEERING, in order that any one interested may readily secure the latest information concerning the interpretation.

Below are given the interpretations of the Committee in Cases Nos. 267-272, inclusive, as formulated at the meeting of January 7, 1920, and approved by the Council. In accordance with the Committee's practice, the names of inquirers have been omitted.

### CASE NO. 267

*Inquiry:* a Is it not permissible to form door openings in the furnaces of locomotive type boilers with the overlapping flanges riveted so that the heads of rivets project into the door opening? Is it advisable that these flanges be welded instead of riveted?

b Is there any objection under the rules of the Boiler Code to the use of baffles over the feed pipe discharge into the side of the barrel of locomotive type boilers?

*Reply:* a The projection of rivet heads into the door opening is not prohibited by the Boiler Code, although it is the recommendation of the Committee that in case the flanges are riveted the heads in the door opening be countersunk. Welding of the flanges is permissible under the rules of the Boiler Code, but there is nothing in the Boiler Code that indicates preference for either riveting or welding.

b Inasmuch as the use of the internal feed pipe is required only for h.r.t. boilers, there is no objection to the use of a feedwater baffle over the discharge of the feed pipe It is, however, essential that the feed water shall not be discharged into the boiler, close to the riveted joints in the shell or to furnace sheets. (See Par. 316).

### CASE NO. 268

*Inquiry:* Inasmuch as the discharge of feed water into vertical fire tube boilers, in accordance with the requirement of Par. 316 of the Boiler Code, has caused serious difficulties from leaky tubes in the lower tube sheet, is it permissible to protect the feed water discharge through a feed water pocket riveted within the shell at a point above the water leg with handhole opening for cleaning.

*Reply:* It is the opinion of the Committee that such a method of delivery of feed water to the boiler is in full accord with the requirements of the Boiler Code.

### CASE NO. 269

*Inquiry:* If, under the provision made in Par. 212c of the Boiler Code, advantage is taken of the opportunity to increase the pitch of the staying for a cylindrical furnace, is it to be assumed that a portion of the increased load on the staybolt is to be supported by the resistance of the outer cylindrical shell to collapse, or must the staybolt be designed to carry the full load upon the stayed area of the furnace sheet?

*Reply:* It is the opinion of the Committee that the special provision made in Par. 212c for increased pitch, does not contemplate the division of the stress upon the stays, in view of the sup-

porting power of the outer cylindrical shell; it will not be permissible to increase the pitch under this rule, beyond the ability of the stays selected to carry the load.

### CASE NO. 270

(In the hands of the Committee)

### CASE NO. 271

*Inquiry:* Is it allowable under Par. 268 to utilize a flanged collar pressed into and flanged over the edges of the shell for a pipe connection, instead of riveting a flange to the shell, when the thickness of the plate is not sufficient to give the required number of threads?

*Reply:* It is the opinion of the Committee that such construction of pressed and flanged reinforcement for an opening in the shell, will be allowable under the requirement of Par. 268, for any pipe opening not exceeding 3 in. pipe size, provided the threaded portion complies with Par. 266 of the Code.

### CASE NO. 272

*Inquiry:* Is it permissible to form a dry pipe in a h.r.t. boiler by insertion of a 6 in. pipe threaded through a flange riveted to the rear head?

*Reply:* It is the opinion of the Committee that this connection of dry pipe is allowable provided: (1) That a flanged nozzle riveted to the head to receive a flanged fitting is employed; (2) That the dry pipe is detachably connected to the nozzle; (3) That the dry pipe is properly supported.

## CODE FOR BOILERS OF LOCOMOTIVES

In the August 1919 issue of MECHANICAL ENGINEERING (page 682) there was published the preliminary report of the Sub-committee of the Boiler Code Committee on Boilers of Locomotives, which is proposed as Part I, Section III of the Code. This report was the result of over two years careful investigation and study by this Sub-committee and the report was first submitted to the membership of the Society at the spring meeting at Detroit, Michigan, in June 1919. The report was accepted by the meeting and ordered published. Following the publication in August 1919, arrangements were made in accordance with the established practice of the Boiler Code Committee, for a public hearing on this question, which was held during the business session of the Annual Meeting on December 3rd, 1919, in order that all interested parties might be given an opportunity to discuss any features of the requirements in the proposed Code.

As a result of the hearing and the discussion, which lasted into the following day, December 4th, recommendations were accepted from the hearing, which resulted in the following revisions in the preliminary report. They are here published for the information of the membership. Anyone desiring to discuss the changes here proposed is requested to address the Secretary of the Boiler Code Committee, 29 West Thirty-ninth St., New York, N. Y.

## Revisions of Preliminary Report on Boilers of Locomotives

*Not Subject to Federal Inspection or Control*

Based on Report as published in MECHANICAL ENGINEERING (Journal of the American Society of Mechanical Engineers), August, 1919, page 682.

PAR. L-2

CHANGE WORDS IN SECOND LINE AS FOLLOWS:

excepting from tube sheets — to read — excepting front tube sheets.

PAR. L-18:

CHANGE HEADING OF TABLE 9 TO READ AS FOLLOWS:

Gage thickness of walls of fire tubes

PAR L-20:

CHANGE PARAGRAPH TO READ AS FOLLOWS:

The maximum allowable working pressure is determined by employing the factors of safety, stresses and dimensions designated in these rules.

The factor of safety used in design and construction of new boilers shall not be less than 4.5.

The factor of safety used in determining the maximum allowable working pressure calculated on the conditions actually obtaining in service shall not be less than 4.0.

The maximum allowable working pressure determined by conditions obtaining in service shall not exceed that for which the boiler was designated.

No boiler shall be operated at a higher pressure than the maximum allowable working pressure, except when the safety valve or valves are blowing, at which time the maximum allowable working pressure shall not be exceeded by more than 5 per cent.

PAR. L-21:

ELIMINATE LAST SENTENCE, WHICH READS AS FOLLOWS:

For new constructions covered in Part III, FS in the above formula = 4.

PAR. L-27.

ELIMINATE ENTIRE PARAGRAPH.

PAR. L-28.

RENUMBER THIS AS PAR. L-27.

PAR. L-29:

RENUMBER THIS AS PAR. L-28.

PAR. L-29:

INSERT NEW PARAGRAPH AS FOLLOWS:

Autogenous welding may be used in boilers in cases where the strain is carried by other construction which conforms to the requirements of the Code and when the safety of the structure is not dependent upon the strength of the weld.

PAR. L-30:

REVISE LAST SECTION OF PARAGRAPH TO READ AS FOLLOWS:

When boiler shells are cut to apply steam domes or manholes, the net area of metal, after rivet holes are deducted, in flange and liner, if used, must be not less than the area required by these rules for a length of boiler shell equal to the length removed. A height of vertical flange equal to three times the thickness of the flange shall be included in the area of the flange.

PAR. L-31

CORRECT FORMULA TO READ AS FOLLOWS:

$$P = C \times \frac{T^2}{p^1}$$

PAR. L-32:

CHANGE FIRST LINE OF SECOND SECTION TO READ AS FOLLOWS:

Staybolts behind permanent brickwork, frame braces, or grate bearers.

PAR. L-52:

ADD NEW SECTION TO THIS PARAGRAPH TO READ AS FOLLOWS:

The ends of arch tubes must be belled out to a diameter ¼ in. larger than the hole in the sheet to which they are connected.

PAR. L-58:

ADD NEW SECTION TO THIS PARAGRAPH TO READ AS FOLLOWS:

If, in any longitudinal section of an unstayed plate, more than two holes are so located that the strength of the ligament between them is reduced below the strength of ligament of the longitudinal seam, the plate must be reinforced to maintain the strength through that section.

PAR. L-60:

CHANGE LAST SENTENCE TO READ AS FOLLOWS:

than five per cent. above the specified boiler pressure.

PAR. L-73:

CHANGE FIRST LINE TO READ AS FOLLOWS:

Each boiler shall have at least one water glass provided with top and bottom shut-off cocks, and lamp

ALSO ADD NEW SECTION TO READ AS FOLLOWS:

Tubular water glasses must be equipped with a protecting shield.

PAR. L-75:

CHANGE LAST WORD TO READ " ENGINEMEN."

# CORRESPONDENCE

CONTRIBUTIONS to the Correspondence Departments of MECHANICAL ENGINEERING by members of The American Society of Mechanical Engineers are solicited by the Publication Committee. Contributions particularly welcomed are suggestions on Society Affairs, discussions of papers published in this journal, or brief articles of current interest to mechanical engineers.

## The Value of Sheet Asbestos on Hot Pipes

### Comments on Contribution in January Number

TO THE EDITOR:

In notes on research entitled Value of Sheet Asbestos on Hot Pipes, by Mr. T. S. Taylor, published in a prominent place in the January issue of MECHANICAL ENGINEERING, certain statements were made which, while entirely true in themselves, have been subject to considerable misconception.

There is nothing new or remarkable in the fact that a bright galvanized-iron surface will radiate heat at a much slower rate than a dull mat surface. Therefore, it is not surprising that placing over the bright surface a very thin layer of asbestos paper should increase the radiation. The paper was only a little over 1/100 in. thick, which is so thin that it could have but little insulating value, and the net effect was mainly the change in the character of the radiating surface and consequently the rate of radiation.

It should be borne in mind, however, that the high surface resistance to heat flow presented by a bright surface decreases rapidly as the surface becomes tarnished or if it is painted. Therefore, the radiation from such surfaces becomes much greater.

The principal cause for mistaking the true significance of the statements was the failure on the part of the average reader to distinguish between mere asbestos paper and asbestos insulation. This distinction was not mentioned in the article and it has been but natural that the casual reader should attribute to asbestos in general the poor results shown by the test for very thin asbestos paper.

The contribution is an arraignment and rightly so of the practice of covering furnace pipes with a thin layer of asbestos paper. The belief has been too prevalent that pipes covered with "asbestos" were properly protected, regardless of the thickness or quality of the asbestos. Therefore, instead of indicating any lack of usefulness of asbestos, the results of these tests properly interpreted bring out more forcibly the need for sufficient thickness of proper insulation.

New York, N. Y.          L. B. McMILLAN.

To the Editor:

I read with some concern an article in the January number of MECHANICAL ENGINEERING, page 69, on the value of sheet asbestos on heat pipes. This fact was reported in a paper read by the writer at the June 1919 meeting of the National Warm Air Heating and Ventilating Association, and has since been given wide publicity. I am therefore surprised to see this article appear.

It is interesting to note that the results obtained by this investigator agree very closely with the results reported by the writer in the proceedings of the meeting mentioned above. His work certainly serves to clinch our argument against the use of asbestos-paper covering on heat pipes.

I am sending under separate cover a copy of a paper by Prof. A. C. Willard, of the University of Illinois, Mem.Am.Soc.M.E., dealing with our testing program for warm-air furnaces. It contains some data on our tests of insulating materials for heat pipes, and the other information contained may be of interest to you.

Urbana, Ill.                                V. S. DAY, Research Assistant,
                                               University of Illinois.

In the report referred to by Professor Willard results are given of measurements made by Mr. Day of the temperatures across the air currents in a tin leader running from a hot-air furnace. Temperature traverses were made in a 10-in. leader of bright tin and in the same leader covered with one layer of asbestos paper weighing 10 lb. to the 100 sq. ft. These tests showed that the asbestos-paper-covered pipe lost heat more rapidly than the bright tin pipe, a fact which led to subsequent experiments on the heat lost from covered and uncovered pipe.

TABLE 1 RELATIVE HEAT LOSSES FROM THIN SHEET, METAL PIPES WHEN COVERED AND UNCOVERED

| Description of Drum | Wt. Steam Condensed in 10 hr. | Coef. of Emissivity |
|---|---|---|
| 1 Bright 1 C tin, 1 leader section.............. | 11.63 | 1.40 |
| 2 Same as No. 1, but covered with one sheet 10 lb. asbestos paper.................. | 17.80 | 2.30 |
| 3 Same as No. 2, but painted with gray enamel.. | 18.14 | 2.32 |
| 4 Same as No. 1, but painted with gray enamel, as in case No. 3........................ | 18.88 | 2.40 |
| 5 Black iron, No. 28 U. S. gage............... | 18.29 | 2.25 |
| 6 Galvanized iron No. 28 gage............... | 11.72 | 1.55 |
| 7 Same as No. 1, but covered with 3-ply air-cell asbestos paper ...................... | 5.86 | 0.70 |
| 8 Same as No. 1, with 5/16-in air space made by double wall of tin.................... | 6.32 | 0.78 |

For this purpose a special heat-transmission plant was built, supplied with low-pressure steam. Sheet-metal drums, all of the same size, were used in determining the heat loss. Five drums were tested at one time, with a sixth drum operated as a control in order to maintain uniform conditions. A few of the more interesting results are given in this table. In this table the " coefficient of emissivity " is based on the temperature difference, steam to air (steam pressure maintained at 1½ in. of mercury), and gives the loss in B.t.u. per sq. ft. of surface per hr. per deg. fahr., temperature difference.

As in the case of the tests by Mr. Taylor referred to in Mr. Day's letter, there was a large increase in heat loss when the metal was covered with a single thickness of asbestos paper without air space to retard the transmission of heat. Superior results were obtained, however, with air-cell asbestos paper and with a leader having a double wall and air space between, as shown in Tests Nos. 7 and 8 in Table 1.—[EDITOR.]

To the Editor:

It has been brought to the writer's attention that some readers of his notes on the above subject, contributed to the January issue of this magazine, have obtained very incorrect ideas concerning the proper interpretation of the results there given. Numerous letters relating to this contribution have been received,

but so far no one has given incorrect impressions of the results. The writer has learned indirectly, however, that some readers have interpreted the results to mean that since a little asbestos is a poor thing on hot-air pipes, more asbestos would be even worse. It is a little difficult to see how any one in reading the notes can come to such conclusions. While it is specified clearly that a thin layer of asbestos on hot-air pipes promotes the loss of heat to the extent of 33 per cent more than if the pipe was left bare, it is also made clear that three layers of thin asbestos 0.013 in. thick applied loosely will make the loss from the covered pipe the same as from the bare pipe; and furthermore it is pointed out that if the layer of asbestos is sufficiently thick, about ⅛ in., the loss through the covered pipe will be only 75 per cent of that through the bare pipe, thus resulting in a saving of 25 per cent. A thicker layer will cause even a less loss. This statement cannot be taken to hold indefinitely for various sizes of pipes, but it does hold within limits for pipes of such dimensions as 9 in. to 10 in. in diameter.

The reason why more heat will pass through a pipe when covered tightly with thin asbestos than when bare is that the effective area of the pipe as far as molecular dimensions and motions are concerned, is increased at least some two or three times by the layer of asbestos, while the thermal resistance due to the layer of asbestos is quite small. The result is that more molecules can come into close contact with the surface, and thus having their kinetic energy or temperature increased, more heat will be carried away. This increases the loss due to conduction and in particular that due to convection, which is the principal method by means of which heat is liberated at the temperatures here dealt with. The extra loss is thus due to the increase in molecular contacts and not to the difference in the radiating power of the two surfaces. Radiation at the temperature differences dealt with in furnace-pipe surfaces plays a very insignificant part in the loss of heat.

Now, since the effective molecular area of the surface of the pipe is so much increased by the asbestos surface over that of the smooth pipe, it is necessary to add enough asbestos or heat-insulating material to cause a temperature drop through this insulation so as to have its external surface at a considerably lower temperature and thus reduce the loss from the outer surface. Thus, if the layer is sufficiently thick the temperature of its outer surface will be low enough to overcome the advantage gained, as far as heat loss is concerned, by the rough surface. Consequently, in order to protect pipes, meet fire-insurance regulations, and also save loss of heat, hot-air pipes should have a much thicker coating of insulating material than is at present applied. This can be a thicker layer of asbestos, several layers of thin sheet asbestos, asbestos air-cell paper or any suitable material used for this purpose.

Pittsburgh, Pa.                                     T. S. TAYLOR.

## The Correct Statement of Averages

To the Editor:

While agreeing with Mr. Hess that an arithmetical average is very apt to be misleading (page 955, December MECHANICAL ENGINEERING), I do not feel sure that the mere statement of maximum and minimum figures is sufficient indication of the degree of reliability achieved.

Personally, I have found my own graphical method, which was described on page 226, vol. 104 of Engineering, the most reliable one I know, since it eliminates data which are obviously out of accord with those most often obtained in a series of experiments.

London, England.                                   C. H. WINGFIELD,
                                                      M. Inst. Mech. E.

The Swiss Economic Mission expects to sail on May 7, 1920, for a forty-two days tour of this country to study the general economic conditions. A delegation of about sixty engineers from the Mission will probably be in St. Louis at the time of the Spring Meeting.

# MECHANICAL ENGINEERING

THE JOURNAL OF THE AMERICAN SOCIETY
OF MECHANICAL ENGINEERS

Published Monthly by the Society at
29 West Thirty-ninth Street, New York

FRED J. MILLER, *President*

WILLIAM H. WILEY, *Treasurer* CALVIN W. RICE, *Secretary*

PUBLICATION COMMITTEE:

GEORGE A. ORROK, *Chairman*      J. W. ROE
H. H. ESSELSTYN      GEORGE J. FORAN
RALPH E. FLANDERS

PUBLICATION STAFF:

LESTER G. FRENCH, *Editor and Manager*

LEON CAMMEN, } *Associate Editors*
C. M. SAMES,

W. E. BULLOCK, *Assistant Secretary*
*Editor Society Affairs*

FREDERICK LASK, *Advertising Manager*
WALTER B. SNOW, *Circulation Manager*
136 Federal St., Boston

Yearly subscription $4.00, single copies 40 cents. Postage
to Canada, 50 cents additional; to foreign countries $1.00 addi-
tional.
*Contributions of interest to the profession are solicited. Com-
munications should be addressed to the Editor.*

## Spring Meeting of Mechanical Engineers at St. Louis

The attention of mechanical engineers is called to the Spring Meeting of The American Society of Mechanical Engineers to be held at St. Louis May 24 to 27. Two of the leading sessions will be devoted to subjects of local interest, although of national importance, those of the development of transportation on the Mississippi River, with a view to shipping products to foreign countries by water by the lower Mississippi and the Gulf of Mexico, and the other on recent developments in industrial housing in which there is a large project under way at St. Louis.

There will be papers on fuels, oil engines, measurement of the flow of water, transmission of heat, pumping engine tests, airplane design, superchargers for aeroplane engines and performance of automobile trucks, etc. All authors who expect to contribute papers are asked by the Committee on Meetings and Program to send their manuscripts to the Society by March 15. It is expected that the program of the meeting will be published in the April number of MECHANICAL ENGINEERING.

## A.S.M.E. Members in Second Pan American Financial Conference

That engineers are steadily coming to the front in the discussion of national and international problems was once again evidenced by the number of A. S. M. E. members who participated in the deliberations of the Second Pan American Financial Conference, which was in session in Washington for 10 days beginning on January 19. The Conference was attended by ministers of finance and other delegates from each of twenty Latin-American countries and representatives of the United States appointed by the Secretary of the Treasury. Among the leading Americans chosen were the following members of The American Society of Mechanical Engineers: Arthur L. Church, for Bolivia; Maurice Coster, for Columbia; Andrew Fletcher, for Chile; Hollis Godfrey, for Peru; John Hays Hammond, for Chile; George H. Harries, for Salvador; E. M. Herr, for Ecuador; Ira N. Hollis,

for Argentina; George L. Hoxie, for Panama; Chester B. Lord, for Dominican Republic; Charles T. Plunkett, for Dominican Republic; Calvin W. Rice, for Bolivia; W. L. Saunders, for Nicaragua; T. Stebbins, for Haiti; George F. Swain, for Panama; F. H. Taylor, for Uruguay; G. R. Tuska, for Mexico.

## "Industrial Unrest" Not an A.S.M.E. Pamphlet

With other members of the Society, the signers of this statement have received an anonymously written and distributed pamphlet entitled "Industrial Unrest," bearing the seal of The American Society of Mechanical Engineers and mailed with a New York City return address.

Inasmuch as we have been individually and officially active as members of the Executive Committee of the Metropolitan Section in bringing the subject of Industrial Relations before the Society, we feel compelled publicly to disclaim any connection with or knowledge of this publication.

We deplore the issue of this anonymous booklet bearing the imprint of the Society's emblem. Not only do we consider that an improper use has been made of the Society's emblem, but we resent the discourtesy to the Council of our Society perpetrated in sending out such propaganda on a subject that is now before the Council for action and on which a special committee appointed by the Council is soon to report.

    S. N. CASTLE      W. C. BRINTON
    G. K. PARSONS      W. S. FINLAY, JR.
    L. P. ALFORD      F. T. CHAPMAN.

## History of the Students' Army Training Corps

A review of the activities of the Committee on Education and Special Training has been published by the War Department. This records the history and general operations of the Students' Army Training Corps over which it had supervision from its inception on Feb. 10, 1918, until the completion of the settlement of the contracts made with colleges on June 30, 1919.

The report gives interesting information as to the personnel of the staff, the educational methods of intensive training in both vocational and collegiate divisions, the types of schools and equipment, costs, specialization curricula, rating and testing, course specialists, teachers, and the war-issues course.

Brig.-Gen. R. I. Rees was chairman of the Committee; Lt.-Col. Grenville Clark, secretary and executive officer. The activities were conducted by three major departments—the educational, the military and the business. The educational department was subdivided into three departments. Mr. C. R. Dooley was director of the vocational training division, the late Dr. R. C. Maclaurin, director of the collegiate training division, and Dr. Frank Aydelotte, director of the war aims division. The business administration was under the direction of Mr. E. K. Hall.

The Civilian Advisory Board included Dr. C. R. Mann, chairman; J. W. Dietz, secretary; Dr. J. R. Angell, Dr. S. P. Capen, Mr. Hugh Frayne, Dr. R. A. Pearson, Dr. Herman Schneider.

The demobilization of the S. A. T. C. was ordered on Nov. 23, 1918, and over three hundred thousand student soldiers in 680 educational institutions were returned to civilian status before Christmas, 1918.

## Engineering Organization in India

The Indian Industrial Commission, which was appointed to consider the development of engineering works and industries in India, included in its report a recommendation of the formation of a Society of Indian Engineers which should include all branches of engineering. Last December, at a representative meeting of engineers held in Calcutta, the formation of such a Society was realized. All the existing organizations in India have decided to corporate in the new Society.

Engineering Society developments in India have been interesting. In a vast country like India where the engineers are separated by great distances, exchange of experience and access to libraries is difficult and various efforts have been made to overcome these disadvantages.

It has been realized that the Government of India might on many occasions derive advantage if there were a recognized corporate body, representative of the engineering profession, which might be consulted upon such subjects as the education and training of engineers, the grants of concession for mining rights or water-power, the regulation of electrical supply undertakings, etc. The new Society of Indian Engineers hopes to be able to assume the responsibilities which would fall to such an organization.

## Gift of $5,000,000 for Promoting Research

The Carnegie Corporation of New York has announced its purpose to give $5,000,000 for the use of the National Academy of Sciences and the National Research Council. It is understood that a portion of the money will be used to erect in Washington a home of suitable architectural dignity for the two beneficiary organizations. The remainder will be placed in the hands of the Academy, which enjoys a federal charter, to be used as a permanent endowment for the National Research Council. This impressive gift is a fitting supplement to Mr. Carnegie's great contributions to science and industry.

The National Research Council is a democratic organization based upon some forty of the great scientific and engineering societies of the country, which elect delegates to its constituent divisions. It is not supported or controlled by the Government, differing in this respect from other similar organizations established since the beginning of the war in England, Italy, Japan, Canada and Australia.

All are familiar with the great achievements in research which have been accomplished in Germany; but in that country research has been backed and controlled almost exclusively by the government, banks and a few large firms and organizations closely allied to the government, and it was these same firms and government which have mainly benefited by the results of the research. The objectionable features of the autocratic régime of the country have been carried into the organization of the research work and have materially affected the handling of its results.

The National Research Council hopes to achieve scientific results at least comparable to those which the Germans have achieved, and to accomplish this by democratic methods in consonance with the spirit of this country and to create a movement toward research work that may ultimately become general among small as well as large organizations. It is intended that the work shall become so coördinated as to make the results beneficial to the industrial life of the country generally rather than to a few large concerns, and thus stimulate economic research throughout the nation.

The National Research Council was organized in 1916 as a measure of national preparedness, and its efforts during the war were mostly confined to assisting the Government in the solution of pressing war-time problems involving scientific investigation. Reorganized since the war on a peace-time footing, it is now attempting to stimulate and promote scientific research in agriculture, medicine and industry, and in every field of pure science. The war afforded a convincing demonstration of the dependence of modern nations upon scientific achievement, and nothing is more certain than that the United States will ultimately fall behind in its competition with the other great peoples of the world unless there be persistent and energetic effort expended to foster scientific discovery.

## Proposed National Labor Program

The three parties commonly regarded as closely related to the labor problem are the worker, the employer and the public. Mr. Henry Hess in this number (p. 185), recognizing that the problem is national in scope, proposes to add a fourth party—the Government—and advances a coöperative plan for reconciling the needs of all four. In this case, however, the role of the Government is not to be its usual one of adjuster of labor difficulties, but it is to play a more helpful and constructive part.

Mr. Hess believes that the trouble with labor is its inability to meet living expenses continuously; to save up against a day of unemployment and to accumulate funds against the period of old age. There are other issues, of course, but he holds that the fear of unemployment is basic.

On the other hand, he considers that an ideal condition would exist if labor and capital should divide equally their risks and responsibilities and share equally in the profits of industry. While this ideal is far from being reached, a few firms and their employees have advanced a long way toward this goal; and he feels that if such mutual coöperation should become general the groundwork would be prepared for a national plan whereby the demands, not only of capital and labor, but also of the public and the Government might be met.

His plan includes profit-sharing and provides for the accumulation of an unemployment fund to which labor would contribute a part of its share of the profits. The fund would be lent to the Government which would thus have placed at its disposal sums which it would not have to raise by taxation. Labor would benefit through ease of mind and insurance against time of need. Capital would benefit through stabilized business. The public would benefit for the same reason and because relieved of its burden of caring for the unemployed.

The engineering method in industry involves a study of facts and the formulation of a working plan before attempting to proceed with the work. In this proposal Mr. Hess has applied the engineer's method to a great economic problem, tempering it, however, by a recognition of the human characteristics involved. The proposal may or may not be practicable, but it is an attempt to solve a pressing national problem and so of timely interest.

## Are Prices Coming Down?

The industry of the country has been built up during the last few years on the idea of high wages, high prices of raw materials, and high prices of manufactured products. Many devices of production have been adopted in order to operate economically under such conditions as were created by war demands and war shortages. There are various indications, however, pointing to an impending change in economic conditions. Some of these are quite faint as yet, while others are fairly plain to all who would read.

A careful analysis of the situation, given by *The Annalist* (Monday, Feb. 9, 1920), would indicate that the long expected fall in prices may be at hand. This view is based on considerations of events in the commodity and security markets. Cotton, corn, pork, lard, and other commodities were marked down substantially under dealings at Chicago; and there was also a material fall in the prices of stocks. It is also influenced by the fall in European exchange and consequent decrease in the export of commodities. *The Iron Trade Review* (Feb. 12, 1920) gives important facts which bear very materially on the same situation.

The difficulties of exchange have affected the situation in two ways. In the first place, foreign purchasers find the prices of American goods at the present rates prohibitive and are making shifts to do with as little imported goods as possible. A number of projects in which American raw materials and machinery were to have been used have been discontinued, and in other projects orders were placed where the exchange situation was less difficult than in the American market. Thus, it has become known that a large order for locomotives which was to go to an American concern was placed in England.

The same exchange difficulties have brought about a situation where banks refuse to accept bills based on sales abroad, except for collection only, which means that the concern selling abroad has to furnish its own capital to handle the transaction. Considering the length of time which must elapse until a transaction is completed and paid for, this situation imposes an excessively severe hardship on the exporter. The *Iron Trade Review* cites a case where a $5,000,000 sale of iron and steel goods to an

Italian customer broke down in face of such a requirement on the part of American bankers.

The same situation affects the export of foodstuffs. Not only can American foodstuffs not be exported in the face of a 25 per cent handicap on the exchange, but European foodstuffs, such as Danish butter, are finding their way to United States ports.

At the same time it should be clearly realized that so far the fall of prices has mainly affected articles in which there has been no domestic shortage, and of which the prices have been maintained either by sales abroad or by storage holdings for foreign markets actually or prospective. Where the article is such that there is or is likely to be a serious domestic shortage prices are not only not coming down but have an upward trend. Such is largely the case with steel, and fuel oil and its various products.

On the whole it would appear that the trend of business affairs has clearly begun to favor the consumer, who, however, has no way of estimating just how far this will extend. The consumer sees the beginning of price recessions and an active demand which may be stimulated by lower prices, and the manufacturer has to view the same situation. The question is how these forces will act in relation to one another.

## Aeronautics in the United States

The impression appears to prevail that the state of aeronautics in America, as compared with that in Europe, is decidedly backward and in every way inferior. Developments on the two continents have been compared to the disadvantage of this country, both in regard to technical achievement and the government appropriations for aviation work.

The actual situation, however, from an American point of view, is far less disturbing than would appear. As regards appropriations, the situation is not good either here or abroad, although probably a good deal worse than in Europe. The original appropriations in England and France were enormous, the appropriation in England being over $300,000,000 for one year; but these appropriations have been very greatly cut down, partly because of the tendency to economize and also because of the growing feeling that important changes, mechanical and otherwise, in the aeronautical field are impending and that it might be well to await developments before incurring expenditures of such large sums.

In this country Congress has refused to appropriate any considerable sum for the present, and, in this respect, the situation is anything but encouraging.

The development of commercial aviation abroad has been of a more startling character and much better advertised than here. Thus, practically regular communication is maintained between London and Paris and a few other large cities. This is cited as showing the progress of aviation abroad. On the other hand, regular communication by aeroplane has been maintained in this country between New York and Washington for many months and is now being extended to Chicago. Our Forest Service is employing aeroplanes with great effectiveness for the detection and combating of forest fires. On the Pacific coast power transmission companies use aeroplanes to locate and repair faults and breaks in transmission and telephone lines, especially in the mountain regions. Ranchers find communication through the air a great boon in inspecting widely scattered herds of cattle and sheep over trackless ranges where an automobile is often of little service. In Texas, oil men have also found the aeroplane of great advantage, both in inspecting widely scattered properties and in quickly reaching new fields where indications of liquid wealth have been located.

It would be fair to state that from the point of view of organization and results achieved the American developments are, if anything, carried out on a greater scale than the European.

In the field of straight commercial aviation the United States probably leads. Thousands of planes are already in use by the public for every-day commercial work and such large companies as, for example, the Curtiss, are literally working day and night and still have considerable difficulty in keeping up with the orders accumulating on their books. An encouraging feature is

that a number of American aeroplanes and aeroplane parts have been sold or contracted for abroad, especially to South America, South Africa and China.

Comparing the number of aeroplanes sold to the public in their dollar value, America again leads. In making such a comparison it should be noted that only a very small number of war planes have been sold to the public and that the largest number of planes going for public use have been those built since the armistice and designed for peace purposes, a situation which is more favorable in every respect than that existing in Europe, where it is chiefly the discarded war planes which go for purposes of commercial aviation.

In this connection, rather significant statements have been made in British aeronautical papers in reference to the situation existing in Scandinavian countries. The conditions there favor the adoption of aeroplanes for communication between the islands in the North Sea and the inter-country and a considerable number of German planes were bought for this purpose. This was followed by a large number of accidents fatal and otherwise and when we learn that German discarded war planes have been bought with an extra engine for about $100 American dollars, it is very easy to conceive why their use has proved disastrous to aviators.

There have been several important technical developments in America to which the general public has not paid the attention deservedly due. One of the most significant of such developments was the tests carried out by the Navy Department, in connection with the Martin torpedo plane. This plane exceeded by far the requirements set by the Navy Department for its adoption and actually carried a torpedo weighing in excess of half a ton at a speed of 120 m.p.h. In fact, the results of these tests have been so encouraging as to point to the possibility of the torpedo plane entirely revolutionizing marine warfare. The torpedo plane is entirely an American invention and was proposed many years ago by a member of the American Navy.

Another important development has been refinement in the design and operation of the Liberty motor. In the months which have elapsed since the armistice a large amount of valuable work has been done on this motor whereby it has been brought to a high state of efficiency. The Liberty motor of today stands out as one of the great achievements of American engineering genius.

The wireless telephone for use in aviation is practically exclusively an American development and in the last year has been brought to an unusually high state of perfection.

In addition to these the most earnest consideration should be given to the important results achieved during the last two years in such research laboratories as McCook Field and the U. S. Bureau of Standards. Under the impetus of war and partly on funds provided by Congress during that period, a large number of important investigations on the stress of materials, fuels, electrical equipment of aeroplanes, etc., have been carried out, and now the American manufacturer of aeroplanes and accessories has a vast fund of information at his command; and what is, perhaps, more important, a number of men who have carried out investigations are cheerfully ready to give of their fund of knowledge to those wishing to use it in a legitimate way for the good of the country.

———

The first copy of *The Tech Engineering News*, published by the undergraduates of the Massachusetts Institute of Technology, has just been issued. It is a 16-page journal, 9 by 12 in. in size, published monthly, and its purpose is to provide a medium for the exchange of ideas between the Institute (the faculty, undergraduates and laboratory workers) and men of executive rank throughout the country. Its articles are intended to be of a character to interest practicing engineers as well as students, and to acquaint each with the needs and conditions of the other. The first issue contains a number of original articles of merit; some of them are strictly of a technical character, presenting new material, while others deal with the business side of the engineer's work, as in one contribution upon Possibilities for Civil Engineers in South America.

## American Engineering Standards Committee Actively Engaged in Organizing Standardization Work

The American Engineering Standards Committee, organized in 1918 by representatives of the national engineering societies to secure coöperation and prevent duplication in the work of standardization, now has a revised constitution, with by-laws and rules recently adopted by the coöperating societies. These are of wide scope and allow the direct or indirect participation of any organization interested in standardization. Provision is made for direct representation on the Committee of additional organizations or groups of organizations.

The Committee's activities are already taking on an international character. It has approved specifications for standard pipe threads, for which The American Society of Mechanical Engineers and the American Gas Association are sponsors, and are representing America on this subject at an international conference in Paris. Coöperation is in progress with the National Screw Thread Commission, by which direct coöperative work with the British, not possible by the official Commission, is being carried out. The Committee is also in active coöperation with the Canadians on bridge specifications, with the British on specifications for machine tools, and with the Swiss on specifications for ball bearings.

A large conference, in which practically all national organizations interested in industrial safety participated, has unanimously voted that all industrial safety codes should be prepared under the auspices of the American Engineering Standards Committee. This was reported in the February number of MECHANICAL ENGINEERING.

### HOW THE COMMITTEE OPERATES

Any organization may request the Committee to approve standards which it has formulated, or to approve committees that it has appointed, and by so doing becomes a sponsor society. Two or more organizations may act as joint sponsors. Approval of the standard is given when it is the substantially unanimous conclusion of a section committee made up of representatives of producers, consumers and general interests, and so selected that all interests concerned have adequate representation on the section committee.

It is understood that a standard must be referred to as that of the sponsor, using whatever title the sponsor has given it, followed by the statement—"Approved by the American Engineering Standards Committee." The approval may be given in one of three ways: "Recommended Practice," "Tentative Standard" or "American Standard." The "approval" of a standard does not mean that the Committee has itself worked over and approved each detail, but rather that the work has been carried out by a sectional committee adequately representing the industry concerned, and sponsored by one or more bodies of ability, experience and standing, so that the result may stand for what is best in American engineering practice.

Any standard or group of standards adopted by any organization or in process of preparation, prior to January 1, 1920, may be approved by the Committee, if in the opinion of the Committee it has either been developed by a properly constituted committee, or has, by actual practice, proved its right to become a standard.

In addition to the work of assistant in the selection of committees and certifying that their work has been done under proper conditions, the Committee will act as a bureau of information regarding standardization.

### PERSONNEL OF COMMITTEE

The Committee as it exists today consists of 24 members and includes representatives of the American Society of Testing Materials and of the Government Departments of Commerce, Navy and War, as well as of the founder Societies. Its former chairman, Prof. Comfort A. Adams, who has done painstaking work for the Committee since its inception, has been obliged to retire from this office because of the press of other affairs, and he is now succeeded by A. A. Stevenson, Past-President of the American Society for Testing Materials. Its secretary is Dr. P. G. Agnew, formerly of the U. S. Bureau of Standards, with headquarters in the Engineering Societies Building, 29 West 39th Street, New York. Until recently when it was possible to arrange for a permanent secretary, Prof. C. B. Le Page, secretary of the Standards and Technical Committees of the A. S. M. E., acted as volunteer secretary.

## PROPER BALANCING OF FUEL, LUBRICANT AND MOTOR

(Continued from page 168)

will be entirely consumed on top of the pistons. Effecting compression of the air and gasoline outside of the cylinder, and igniting the charge upon admission and during the power stroke, would overcome practically the entire amount of fuel-and-air-mixture leakage that takes place during the compression stroke of the present four-cycle motor. Other methods will undoubtedly be proposed and worked out successfully as soon as the seriousness of the present situation is fully established.

### CONSERVATION OF LUBRICATING OIL AND FUEL

In the Air Service during the war the necessity existed for the conservation of lubricating oil, and a simple system of reclamation which brought about a considerable saving in oil and money. The records show that 40 per cent of the oil issued and used in the aeronautic engines was sent back to the reclaimers, and of that amount 73 per cent, and as high as 95 per cent, was reclaimed and made available for reuse, and the test records show that this reclaimed oil was, if anything, a better lubricant than the original new oil. The engines were cleaner after its use, 50 per cent of the field flying engineers stated the reclaimed oil was preferred to new oil; the other 50 per cent said it was just as good as new oil.

Oil reclaimers are commercially successful. It is only necessary to have steam for their operation. They should be installed in every commercial garage. Reclamation of motor oil can be carried on without trouble and at a cost not exceeding ten cents a gallon. The oil so reclaimed will be just as good a lubricant as when new, but will be dark in color and have a slightly burnt odor, correction of which will require expensive and elaborate equipment and is not at all necessary.

Where reclaimers are installed oil should be removed every day. Motors should be made with larger drains and with easy methods of draining and filling. In fact, it should be as easy to drain and fill a modern motor with lubricating oil as it is to fill the radiator with water or the tubes with air.

Devices to remove the dilution in the oil in the motor or in the reclaimer do not solve the main problem of economy of use of gasoline. This is squarely up to the motor manufacturer or the designing engineer. The solution must come through some entirely different mechanical use of the fuel.

Then the importance of the care of gasoline and lubricating oil must not be overlooked, and the useless wasting of even the smallest amount of gasoline or oil per car must be avoided.

We are dealing with a disappearing asset. The Bureau of Mines has sounded repeated warnings as to the reducing reserves. The oil age which we are now living in will be shortened if we do not conserve our supplies by every possible means, and conservation is squarely up to the engineers of the country, as it is necessary first to secure engines that will utilize more of the heat units of the fuel, and are so designed that mechanical losses in the use of the fuel will not occur. And it is up to the millions of car users to prevent the small losses per car which in the aggregate mean million of gallons of a product which is of prime necessity to modern industry.

# Herbert C. Hoover Addresses Mining Engineers

### Discusses Railroads, Shipping, Industrial Relations, and Other National Problems in which Engineers Are Directly Concerned

A S we go to press the American Institute of Mining and Metallurgical Engineers is holding its annual meeting, the feature of which has been the inaugural address of Herbert C. Hoover, delivered upon assuming the presidency of the Institute for the ensuing year.

In beginning his address Mr. Hoover spoke of the work which the engineers have done for the nation during the war and of the greater task now facing them.

"Even more than ever before," he said, "is there necessity for the continued interest of the engineer in this vast complex of problems that must be met by our government. We are faced with a new orientation of our country to world problems. We face a Europe still at war; still amid social revolutions; some of its people still slacking on production; millions starving; and the safety of its civilization is still hanging by a slender thread."

Mr. Hoover referred specifically to some of the great problems in this country in which engineers are interested. Brief quotations from these references are given below, as reported in the daily press. Speaking of the railroads and shipping, he said:

"The war nationalization of railways and likewise the war nationalization of shipping are our two greatest problems in government control awaiting demobilization. There are many fundamental objections to continuation of these experiments in socialism necessitated by the war. They lie chiefly in their destruction of initiative in our people and the dangers of political domination that can grow from governmental operation.

"Already we can show that no government under pressure of ever present political or sectional interests can properly conduct the risks of extension and improvement, or can be free from local pressure to conduct unwarranted services in industrial enterprise.

"On the other hand, our people have long since recognized that we cannot turn monopoly over to unrestrained operation for profit nor that the human rights of employees can ever be dominated by dividends. The problem is easy to state. Its solution is almost overwhelming in complexity.

"As the result of war pressure we will spend over $2,800,-000,000 in the completion of a fleet of 1,900 ships of a total of 11,000,000 tons, nearly one-quarter of the world's cargo shipping. We are proud of this great expansion of our marine, and we wish to retain it under the American flag.

"Our shipping problem has one large point of departure from the railway problem, for there is no element of natural monopoly. Any one with a water-tight vehicle can enter upon the seas today, and our government is now engaged upon the conduct of a nationalized industry in competition with our own people and all the world besides. While in the railways government inefficiency could be passed on to the consumer, on the seas we will sooner or later find it translated to the national Treasury.

"Second, we may find it desirable to hold a considerable government fleet to build up trade routes in expansion of our trade, even at some loss in operation.

"Third, in order to create this fleet we have built up an enormous ship building industry. Fifty per cent of the capacity of our shipyards will more than provide any necessary construction for American account. Therefore, there is a need of obtaining foreign orders, or the reduction of capacity, or both.

"I believe, with most engineers, that with our skill in repetition manufacture, we can compete with any ship builders in the world and maintain our American wage standards; but this repetition manufacture implies a constant flow of orders.

## DEPARTMENT OF PUBLIC WORKS ESSENTIAL

"Our joint engineering committees have examined with a great deal of care into organization of our expenditure on public works and technical services. They report that the annual expenditure on such works and services now amounts to over $250,000,000 per annum, and that they are carried out today in nine different governmental departments. There is a great waste by lack of

national policy of coördination, in overlapping with different departments, in competition with each other in the purchase of supplies and materials and the support of many engineering staffs.

"They recommend the solution that almost every other civilized government has long since adopted—that is, the coördination of these measures into one department under which all such undertakings should be conducted and controlled.

## INDUSTRIAL RELATIONS A NATIONAL PROBLEM

"Another great national problem to which every engineer in the United States is giving earnest thought and with which he comes in daily contact, is that of the relationship of employer and employee in industry.

"We have until recently greatly neglected the human factor that is so large an element in our very productivity. The development of vast repetition in the process of industry has deadened the sense of craftsmanship and the great extension of industry has divorced the employer and his employees from the contact that carried responsibility for the human problem.

"I am daily impressed with the fact that there is but one way out, and that is to again reestablish through organized representation that personal coöperation between employer and employee in production that was a binding force when our industries were smaller of unit and of less specialization.

"Many of the questions of this industrial relationship involve large engineering problems, of which I know of no better example than that of the soft coal industry [a subject discussed by the A. I. M. & M. E. at their convention]. Broadly, here is an industry functioning badly from an engineering and, consequently, from an economic and human standpoint. Owing to the intermittency of production, seasonal and local, this industry has been equipped to a peak load of 25 or 30 per cent over the average load. It has been provided with a 25 or 30 per cent larger labor complement than it would require if continuous operation could be brought about.

"There lies in this intermittency not only a long train of human misery through intermittent employment, but the economic loss to the community of over a hundred thousand workers who could be applied to other production and the cost of coal could be decreased to the consumer. This intermittency lies at the root of the last strike in the attempt of the employees to secure an equal division among themselves of this partial employment at a wage that could meet their view of a living return on full employment.

## NATIONAL IDEALS SHOULD PREVAIL

"These are but few of the problems that confront us. But in the formulating of measures of solution we need a constant adherence to national ideals and our own social philosophy.

"In the discussion of these ideals and this social philosophy we hear much of radicalism and of reaction. In their present day practical aspects they represent, on one hand, roughly, various degrees of exponents of socialism, who would, directly or indirectly, undermine the principle of private property and personal initiative, and, on the other hand, those exponents who in various degrees desire to dominate the community for profit and privilege. They both represent attempts to introduce or preserve class privilege, either a moneyed or a bureaucratic aristocracy. We have, however, in American democracy an ideal and a social philosophy that sympathizes neither with radicalism nor reaction as they are manifested today.

"For generations the American people have been steadily developing a social philosophy as part of their own democracy—and in these ideals it differs from all other democracies. This philosophy has stood this period of test in the fire of common sense; it is, in substance, that there should be an equality of opportunity—an equal chance—to every citizen."

# The Engineer's Salary and Present Conditions

### Compensation Should Be Based on the Value of the Dollar—How This Value Has Changed and Is Likely to Change—Underpaid Service a Menace to the Public

I N MECHANICAL ENGINEERING for February there was published (page 138) an abstract of the report of the Committee of Engineering Council on Classification and Compensation of Engineers. Herewith is published the main portions of an appendix to that report dealing with the important question of compensation from two different points of view: First, what increase should be made in engineers' pay to compensate for the great reduction in value of the dollar which has taken place in the last five years? Second, how may the intrinsic value of engineering service be determined?

## I. THE DECREASE IN VALUE OF THE DOLLAR

A great deal of misunderstanding and injustice would be avoided in all our industrial relations were there a clear understanding of the fact that all prices today, whether of wages or salaries, or commodities, must be compared with the change which has taken place in the value of the dollar before it can be determined whether the price, measured by an absolute standard, has moved up or down. It should be obvious to every one that the value of the dollar is measured solely by its purchasing power. Whether it be a dollar received as wages by workmen, a dollar received as salary by an engineer, or a dollar received by a manufacturer in payment for goods sold, the actual amount of value which will be received in each case *will depend upon the average amount of other commodities which the dollar received will purchase.*

### CHANGES IN WHOLESALE PRICES

The "Index Numbers" which are regularly published by various statistical bureaus are an average of the *wholesale* market prices of a large number of standard commodities. In making up the index numbers the prices of the different commodities and classes of commodities are weighted in proportion to the per capita consumption of each. Thus a comparison of the index number at a given date with the index numbers at preceding and following dates shows the change in average wholesale market prices and such a comparison shows also the change in the purchasing power, or value, of a dollar in the wholesale market.

### TABLE 1 INDEX NUMBERS[1]

| Date | Dun's Index Number | Corresponding number indicating relative power of one dollar |
|---|---|---|
| Jan. 1, 1914 | 124,528 | 803 |
| Jan. 1, 1915 | 124,168 | 805 |
| Jan. 1, 1916 | 137,660 | 726 |
| Jan. 1, 1917 | 169,562 | 590 |
| Feb. 1, 1917 | 176,272 | 567 |
| Mar. 1, 1917 | 186,244 | 536 |
| Apr. 1, 1917 | 190,012 | 526 |
| May 1, 1917 | 208,435 | 480 |
| Aug. 1, 1917 | 218,779 | 457 |
| Jan. 1, 1918 | 222,175 | 450 |
| July 1, 1918 | 232,575 | 430 |
| Oct. 1, 1918 | 233,227 | 429 |
| Jan. 1, 1919 | 250,146 | 434 |
| Mar. 1, 1919 | 217,087 | 460 |
| May 1, 1919 | 222,193 | 450 |
| July 1, 1919 | 233,707 | 428 |
| Aug. 1, 1919 | 241,650 | 414 |
| Sept. 1, 1919 | 238,342 | 419 |
| Oct. 1, 1919 | 235,867 | 424 |
| Nov. 1, 1919 | 238,573 | 419 |
| Dec. 1, 1919 | 244,639 | 409 |

[1] From figures compiled by R. G. Dun & Co. from market quotations of wholesale prices of 300 standard commodities, showing variations in the cost of living and in the purchasing power of the dollar. In making up the figures the commodities are arranged in seven groups, as follows: (1) breadstuffs; (2) meat; (3) dairy and garden; (4) other food; (5) clothing; (6) metals; (7) miscellaneous. The price of each of these commodities is multiplied by a figure determined as the estimated per capita consumption of that commodity. The commodities listed under "miscellaneous" include such articles as coal, petroleum, building material, and drugs, so that the individual's expense for fuel and shelter is at least partially represented in the total. These index numbers published by R. G. Dun & Co. are believed to be the most reliable record obtainable of price changes in the wholesale markets.

This is illustrated in Table 1, in which the numbers in the last column, representing the buying power of a dollar, are the reciprocals of the index numbers. The index numbers in Column 2, it will be understood, are the amount of money required to buy a certain amount of certain standard goods at wholesale on a given date. Thus if the index number on a given date is, for example, $110, then on that date one dollar will buy 1/110 of this amount.

### CHANGES IN RETAIL PRICES

In order to determine the change which has taken place in the value of the dollar it is necessary also to investigate changes in retail prices as well as those in wholesale prices. Especially where wages and salaries are concerned, it is the changes in retail prices that determine the buying power of the dollar received.

In general, the changes in retail prices follow the general course of the changes in wholesale prices. There is, however, more or less variation in different localites. Where such variation occurs it will generally consist in a *greater* proportionate increase of retail than of wholesale prices. Under the stimulus of the abnormal demand of the past five years, which is chiefly responsible for the great increase in prices, it is common knowledge that the retailer has frequently advanced his prices faster than the wholesale dealer.

### CHANGES IN HOUSE RENTS

In the long run the prices paid for house rents and for personal service change to correspond with the change in commodity prices; but rents change more slowly than the market prices of commodities, so that allowance must be made for this "lag" in determining the change in value of the dollar. This is especially the case where such a sudden and rapid increase in market prices of commodities occurs as has taken place in the past five years.

Furthermore, the rise in rents varies in different cities. In some munition towns, rents rose with a bound in 1915, and so did prices for personal service. In most of the larger cities rents did not rise materially until 1919.

In order to determine accurately the change in value of the dollar in a given community, therefore, an estimate must be made of the change in retail prices compared with wholesale, and of the rate of change in rents and personal service charges.

The September Bulletin of the U. S. Department of Labor contains tables showing the results of the Department's investigation of the increase in the cost of living in 17 important industrial cities of the United States for December 1915, 1917, 1918, and June 1919 as compared with December 1914.

Averaging the figures for these 17 cities, it is found that the increase in the cost of living in June 1919 over December 1914 was 76.24 per cent. This so-called percentage of "increase in the cost of living" really represents the percentage of increase which has taken place in all prices, including rent and service, as well as commodities. Its reciprocal, therefore, represents the decrease in value of the dollar which has taken place.

### PRESENT PRICES COMPARED WITH 1914

In order to compare readily these Bureau of Labor figures with the Dun index numbers, Table 2 has been compiled, in which the prices of December 1914 are taken as a base and the percentage of increase is shown for each date to December 1, 1919.

As explained above, the chief reason why the "cost of living" figures have increased less rapidly than the increase in commodity prices is the delay in readjustment of house rentals. The

TABLE 2 PERCENTAGES OF INCREASE OVER DECEMBER
1, 1914 IN WHOLESALE PRICES AND IN COST OF
LIVING AT VARIOUS DATES

| Date | Dun's Index No for wholesale prices per cent of Increase over Dec. 1, 1914 | Bureau of Labor's per cent of Increase in Cost of Living Over Dec. 1, 1914; Average of 17 Industrial Cities. |
|---|---|---|
| Dec. 1, 1915 | 7.2 | 1.07 |
| Dec. 1, 1916 | 35.4 | 14.94 |
| Dec. 1, 1917 | 77.4 | 41.81 |
| Dec. 1, 1918 | 85.6 | 73.40 |
| June 1, 1919 | 83.7 | 76.24 |
| July 1, 1919 | 88.3 | ..... |
| Aug. 1, 1919 | 94.7 | ..... |
| Sept. 1, 1919 | 92.0 | ..... |
| Oct. 1, 1919 | 90.1 | ..... |
| Nov. 1, 1919 | 92.3 | ..... |
| Dec. 1, 1919 | 97.1 | ..... |

Bureau of Labor figures bring down the record only to June 1,
1919. Since that date there has been a great increase in house
rentals all over the country. There are no statistics to show
the amount of this increase and its effect on the cost of living,
but there is good reason to believe that this increase, coupled
with the general increase in charges for personal service that
has taken place during 1919, and the advance of prices by retail
merchants at a greater rate than the wholesale price increases,
is sufficient to make the percentage of increase in the cost of
living indicated by the Dun statistics a true record of the change
in value of the dollar at the present time.

There is still another item to be considered. The percentage
of increase in wholesale prices shown in Table 2 records the increase over the prices on December 1, 1914. A more accurate
comparison would be to take the average prices for the entire
year 1914 as the base with which to make the comparison. The
average of Dun's index numbers for the 12 months of 1919 was
$122.20. The average wholesale prices of commodities on
December 1, 1919, were therefore double the average prices for
the year 1914. The value of the dollar today, therefore, in the
wholesale market is just half what it was during 1914, and it
is less than it was at any time during the war.

CHANGES IN "COST OF LIVING" AND IN "VALUE OF THE
DOLLAR" ARE NOT THE SAME

The customary reference to the general increase of prices
above reviewed as "the high cost of living" has tended to confuse the minds of many people. Some employers have argued
that the high cost was not their affair. They have
declared the real trouble to be "the cost of high living." On
the other hand engineers or other professional workers, and
many classes of salaried men, have hesitated to press claims for
increased pay on the ground that changes in the cost of living
make it difficult to live on their incomes. Such men rightly feel
that their living expenses are their own private affair.

When, however, it is clearly understood that what has taken
place is a change in the value of the dollar, the claim for an increase in the rate of pay measured in dollars rests on entirely
different ground. The proper and dignified position for the
engineer is to assume that his work should receive at least the
same compensation in absolute value that it received five years
ago and that, therefore, the compensation measured in dollars
should be increased by whatever amount is necessary to offset
the decreased value of the dollar.

There can be no denial of the justice of this claim, even though
the difficulty of satisfying it to the full extent in many departments of engineering work is recognized. The compensation of
many engineers is dependent upon laws and ordinances, custom
and precedent. Great inertia must often be overcome to effect
a change. In many cases the compensation of the engineer, like
that of many other public servants, is dependent upon revenue
raised by taxation; and the difficulties in increasing tax rates
to correspond to the great decrease in the value of the dollar
are known to everyone.

Engineers engaged in business on their own account have to
meet the difficulty of raising their fees to offset the changed
value of the dollar, a task especially difficult in fields of engineering where work is inactive and the competition for it is keen;
yet without such increase they cannot adequately raise the pay
of their own employees. This illustrates anew the need for
emphasizing the change in value of the dollar, rather than the
change in living costs. The former is at once recognized to bear
directly on fair prices for goods sold and for the fees charged
by professional men as well as on wages and salaries.

How LONG WILL HIGH PRICES STAY?

There has been reluctance to raise salaries to correspond to
the changed value of the dollar because of the idea that prices
were to drop back with the conclusion of the war. So far from
this being the case, the above quoted records show that following
the lull in business after the armistice prices have risen above
even the war time scale and are now at the highest point ever
reached. Business has largely readjusted itself to the changed
conditions and the activity in some lines exceeds that registered
during the war.

At the bottom of the changed price conditions is the surplus
of demand over supply. The world urgently needs more food
and coal and steel and cotton, more of the goods made from
them and from other raw materials than are now being produced.
The competition among buyers that sent prices soaring in the
early years of the war is still an active force to maintain prices.

The only two things which can restore prices to their former
level are increased production or decreased consumption. World
wide disorganization of industry and of government, deficient
capital and deficient transport facilities all tend to reduce production. The world of consumers, long held down to a war
diet and war clothing, now eagerly seeks to replenish its larder
and wardrobe and to repair and renew its stock of buildings
and machinery.

The outlook is that it will take years to again organize the
world's equipment for production and distribution, including
finance, transportation by land and sea, and merchandising, so
that the demands of consumers may be met as before the war
and prices be brought back to former level.

The present price level is not considered merely temporary
by such of our Government agencies as the Department of
Labor and the Federal Reserve Board or by such economists as
Irving Fisher and J. S. Holden. Substantial relief from the
high cost of living, therefore, cannot reasonably be expected
through a decrease in prices; it must be met by increases in
salaries.

From such considerations as these, the Committee feels justified
in urging that a readjustment of compensation should be based
on the assumption that the present scale of prices is to continue
for an indefinite time.

II. THE INTRINSIC VALUE OF ENGINEERING
SERVICE

It will be generally agreed that the salary of an engineer
ought to be at least sufficient to enable him to live in the manner
which his position and responsibility call for, and in addition
to repay within a reasonable time the investment in time and
money he has made in gaining the education and experience
which is necessary for his work.

Unfortunately there has been for fully a decade a tendency
to lower the pay of engineers. The law of supply and demand
has operated to reduce the pay of engineers in many branches
of the profession far below the standards above defined.

By paying too low a rate for engineering service, the inevitable tendency has been to lower its quality. This has been
especially marked in the case of engineers in Federal, State, and
Municipal service. Here the inertia which prevails in all public
affairs has prevented the engineers from receiving more than a
trifling part of the increase in pay, measured in dollars, that is
required to offset the shrinkage in the dollar value.

The obvious result has been to drive out of the public service
the best and ablest men, who can obtain better positions elsewhere, and to leave only the men who by reason of age or
inferior ability cannot make such a change.

## CHEAPENED ENGINEERING SERVICE MEANS WASTE AND DANGER

It cannot be too strongly emphasized that the public loses through cheap engineering service many times the amount it may seem to save through lower salaries. The professional engineer in a responsible position in designing, constructing, or executive direction of important work should have initiative, sound judgment, broad knowledge, and executive ability. Lack of these qualities often results in great loss of money, often by needlessly increasing the cost of work of which the public never knows.

This matter deserves emphasis here because where readjustment of salaries has taken place to compensate for the changed value of the dollar it has been common to confine the increase to the lower paid men and to do little or nothing for the men receiving salaries above $2500 to $3000. There is no longer an excuse for this, as the above review amply proves. The measure of seeming economy, also, is very small; because the engineers in the higher positions are few in number compared with the rank and file of professional workers.

### ROUTINE WORKERS IN THE PROFESSION

It is frankly recognized that there is another class of technical work of a routine order which calls for little in the way of initiative, originality, or judgment. Much of this routine technical work in the field, the office, the shop, or the laboratory can be and is being done by boys and young men with limited education and no more training than that afforded by a correspondence school or a few months study in a trade school. Most of this work does require, however, a degree of reliability and fidelity which deserves fair compensation. The best guide to fair rates of pay for this class of technical workers is found by comparison with the standard rates of wages paid to skilled workers in the trades. These workmen are now generally receiving rates of pay much higher than the routine technical worker and in many cases higher than even the engineer who carries large responsibility for design or administration.

The Federal Government is now paying thousands of its highly trained clerical and technical force less than a living wage. Except for the temporary bonus of $240 a year for positions paying salaries of $2,500 or less, no attention has been paid to the constantly diminishing purchasing power of the salaries paid to this class of employees. On the other hand the Government has given full recognition to increased living costs in fixing the wages of organized labor.

A "shipfitter" in the Navy Yard, for example, receives $1,750 a year while he is learning how to do his work. After three months of apprenticeship he gets $2,000. If he is made a "straw boss" in charge of 12 or more men, he gets $2,450, and if a "sub-foreman" in charge of 30 or more men, he gets $2,900. A blacksmith (heavy fire) gets $2,400. A "hammer and machine forger" (heavy) gets $3,700.

In many instances the amount paid for skilled labor is greater than the amount paid to the trained Government engineer. Over 40 of the labor crafts were awarded a rate of wage of $2,000 and more by the Labor Adjustment Board.

The skilled laborer is not required to know how to read or write, and he may receive full pay after an experience varying from two weeks to six months. The Government engineering employee, on the other hand, to get an equivalent amount of pay, must have had from two to eight years' experience, if he is not a technical graduate, and in many instances will not be admitted at all without a technical degree and then only with from two to four years' practical experience.

Even though a temporary oversupply of men trained in engineering work may make it possible to keep salaries for these brain workers below the wage of laborers, the inevitable result will be a dissatisfied working force, which carries out the daily routine without energy or good will, and the public's work will not be done with efficiency or economy.

# Civil Engineers Discuss Development Problems

### Features of Annual Meeting Are Discussions of Development Committee Report and Recommendations of Special Committee Appointed to Review It

THE Sixty-Seventh annual meeting of the American Society of Civil Engineers was held in New York on January 21 and 22. Chief interest centered upon the comprehensive report of the Committee on Development presented to the Board of Direction in October. The report of the Joint Conference Committee of the four Founder Societies was appended to the Development Committee's report and also came under consideration.

With regard to the technical activities of the Society the Development Committee recommended in part:

That local "sections" hold not less than four stated meetings per year and encourage joint meetings with kindred societies; that periodic joint meetings of the four National societies be encouraged; that the fortnightly meetings of the parent society be discontinued; that the semi-annual meetings of the parent society be supplemented by the addition of a spring and a fall meeting, held successively in different sections of the country in coöperation with local sections, and that at these meetings as well as at the Annual Meeting and the Annual Convention attention be given to technical and broad economic problems, and that at the Annual Convention at least half the time should be scheduled for the serious consideration of technical matters; that the present system of securing papers which depends wholly on voluntary offers and results in a limited variety of subjects be supplemented by papers that will fit into a definite logical program, directed by the Publication Committee and announced in advance for a whole season; and that papers describing works be brief on the narrative side, with more attention given to stating the technical lessons of successes and failures; that

so far as practicable the subjects of the papers should cover current progress, and development of the various important branches of civil engineering and simultaneous discussion of important papers by the local section should be encouraged; that the monthly *Proceedings* should be extended to include editorial work, abstracts and reviews of important engineering articles and subjects of popular interest to the great majority of engineers; that coöperation with the American Engineering Standards Committee be approved and that it be continued and extended but that the idea of developing this Committee into another association be deprecated; that the Board of Direction be requested to consider the advisability of compiling a book of recommended principles of practice and standards to follow the development of engineering; and that to carry out the intent of these recommendations effectively the technical editorial staff be under the direction of the Publication Committee.

A special committee of the Board of Direction appointed to study the report of the Development Committee and review its suggestions, offered the following criticisms of the Joint Conference Committee Report, which, as it has been stated, was included in the report of the Development Committee:

1  The Joint Conference Committee's report does not sufficiently confine the voting power of the proposed new National Organization to the corporate members of the present National Engineering Societies, and of such other allied technical societies as are clearly predominately professional in their aim, and who are equally careful with the present national societies in the admission to their membership.

2  The proposed new organization does not have close enough connection and coördination with the officers and boards of direction of

the present national societies and such other national and regional societies as it may see fit to admit to its membership.

3  The Committee's plan creates a large amount of new, somewhat complicated, and expensive machinery that appears to us to be unnecessary, and which it may be reasonable to doubt can be made to work smoothly for some time to come, at least insofar as the present experience of engineering societies has gone.

4  The plan for the new organization is so drawn as largely to deprive the present national societies of any considerable influence in large public or national technical problems in which they are naturally interested and in which they are preeminently capable of having voice.

5  The plan commits the welfare of the engineering profession to a new and untried organization without history, precedent, or experience, from which semi-professional, and even semi-commercial local societies are apparently not only not barred, but possibly may become from time to time a predominating influence.

6  The representation in the new organization accorded to the four national societies is numerically insufficient in view of their numbers, experience and standing.

In view of these and other minor alleged objections, the special committee suggested that the officers and Boards of Direction of the present national and their affiliated engineering societies could form, ex-officio, a National Engineering Council for the purposes of professional unity, which would, in effect, continue the work of the present Engineering Council, and that to the council thus formed might be added, if desired, delegates from state and local councils to be formed throughout the country. [In this connection see also the account in this number, of the Conference of the governing boards of the Founder Societies on January 23.]

The Board of Direction then voted that the main question in the Development Committee's report be submitted to the society for action by letter ballot as soon as possible. The resolutions of the Board included a suggested form of the questionnaire to be sent to the membership. The wording of the questions was modified in subsequent discussion and was finally adopted as follows:

1  Shall the American Society of Civil Engineers adopt the principle of becoming an active national force in economic, industrial and civic affairs?

2  Shall the Society actively coöperate with other engineering and allied technical societies in promoting the welfare of the engineering profession?

3  Shall the Society coöperate in the formation of the comprehensive organization provided for in the Joint Conference Committee report?

4  Shall the recommendations of the committee regarding technical activities be approved?

5  Shall the Society be divided into sections to embrace all members?

6  Shall the directors in each geographical district be nominated and elected by vote of the corporate members resident therein?

7  Shall a portion of the Society dues be allotted to local sections?

8  Shall the present nominating committee be abolished and candidates for the offices of president, two vice-presidents, and treasurer be nominated by representative of local sections in annual conference?

9  Shall the dues of all non-resident members above the grade of junior be increased to provide for greater activities of the Society, said increase not to exceed $5 per annum?

A resolution was also passed authorizing the Board to add any questions that in its judgment were needful to develop the opinion of the members.

# First Step Toward Federation of Engineering Societies

## A Conference to Be Called Immediately of National and Local Societies to Formulate Plans for a Comprehensive Organization

A N important all-day conference of the officers of the four national engineering societies—the first such meeting ever held—was called in New York on January 23 by Engineering Council to discuss matters relative to the Council and its public work and to consider the broad question of engineering organization as reported in the recommendations of the Joint Conference Committee of the Founder Societies. The Trustees of the United Engineering Society, the members of the Joint Conference Committee and of Engineering Council were also present, the whole gathering numbering ninety-five leading engineers from all sections of the country.

The feature of the meeting was the passage of a resolution as a recommendation to the several societies approving the recent report of the Joint Conference Committee and calling for a conference in the near future of representatives of engineering organizations throughout the country to formulate plans for a federation of engineering societies. This is probably the most important and significant step which has been taken toward securing concerted action by the engineers of the country in respect to their common interests and their participations in public affairs. The wording of the resolution was as follows:

Whereas, This conference of national engineering societies has considered the recommendations of the Joint Conference Committee of the four Founder Societies, therefore, be it

Resolved, That the conference adopts in principle that report and requests the Joint Conference Committee to call, without delay, a conference of representatives of national, local, state and regional engineering organizations to bring into existence the comprehensive organization proposed.

Voted unanimously that if this conference be called together again, it be called for a meeting in Chicago on April 20.

The conference on January 23 was called to order by Vice-Chairman D. S. Jacobus in the absence of J. Parke Channing, Chairman of Engineering Council, who was ill. Dr. A. R. Ladoux, Vice-President, Am. Inst. Min. & Met. Engrs., was elected chairman of the Conference.

### FUTURE OF ENGINEERING COUNCIL DISCUSSED

A paper prepared by Chairman Channing, read at the conference, reviewed the main facts regarding the present organization and work of Engineering Council, covering important activities which have been reported from month to month in the columns of MECHANICAL ENGINEERING.

The financial needs of Engineering Council were outlined by Charles T. Main, Past-President Am.Soc.M.E. He stated that $45,000 was needed to continue the activities modestly for 1920, and $25,000 to reimburse Chairman Channing for money advanced during 1919 to inaugurate and maintain the Washington office: total $70,000. Total expended in 1919, $49,600. Resources for 1920 now visible are: balance from 1919, $900; appropriations by Member Societies, $12,600; contributions in response to 50,000 appeals to engineers throughout the country, $5,000; total $19,-000. Needed, approximately $50,000. Whatever reorganization of, or substitution for Engineering Council may be made, the latter must continue as now constituted through 1920, and should be at the end of the year a well organized activity.

Resolutions were passed at the conference calling for the support of Engineering Council so that its present activities need not be curtailed. These were voted as recommendations to the several societies and asked for a contribution of $5,000 for the year 1920 in place of $3,000 as now provided; and further for $1,000 from the Am. Soc. for Testing Materials instead of $600. It was asked that appeals for contributions be made at $2 per person in each of the societies for the support of welfare work during the interval before the recommendation of the Joint Conference Committee shall have been made effective.

### REPORT OF JOINT CONFERENCE COMMITTEE

Following these matters there was a general discussion throughout the afternoon of the report of the Joint Conference Committee. This report, an abstract of which was published in Section 2 of MECHANICAL ENGINEERING for January, 1920, is in two parts, dealing respectively with public activities of engineers and with activities of common interest to the four Founder Societies. The conference devoted its attention to the former, which presented tentative plans for "a single comprehensive organization to secure united action of the engineering and allied technical professions in matters of common interest to them." Certain specific objects which such an organization might have were pointed out in the report to be as follows:

1 To render the maximum of service to the nation through unity of action.

2 To give the engineers of the country a more potent voice in public affairs.

3 To secure greater recognition of the services of the engineer, and to provide for his advancement.

4 To promote *esprit de corps* among the members of the profession.

5 To provide the machinery for prompt and united action on matters affecting the profession, among which are:

Registration of engineers;

Washington office for engineering societies;

Department of Public Works of the Government;

Conservation of national resources;

Publicity;

Classification and compensation of engineers;

General employment bureau;

Engineering education;

International affiliation of engineers;

Industrial relations.

Commenting on these the Committee says:

"In the greater vision resulting from the world war it is apparent that no one alone can successfully solve the problems with which he is confronted. He needs the coöperation of his fellows. National problems demand united effort. The Joint Conference Committee believes that the greatest value of the proposed organization is in the united effort for the service of the nation, from which effort will result the greatest service to the individual."

### NEW ORGANIZATION TO BE FROM BOTTOM UP

The tentative plan for the proposed organization called for the following component parts:

1 Local Affiliations, preferably under the auspices of local engineering societies or clubs, as follows:

a "Local Associations" or "Sections" of the national engineering or technical societies;

b Local engineering societies; and

c Other local engineers and members of allied technical professions and associates.

These local affiliations when secured are to be headed up in:

2 A National Council, consisting of representatives of national engineering and technical societies and of representatives of local, state or regional affiliations or organizations.

State councils were also recommended wherever and whenever the local conditions warrant.

The proposed basis of representation on the National Council is for each local, state or regional affiliation or organization whose membership is not otherwise represented than through the national engineering or technical societies to have one representative to the National Council for a membership of from 100 to 1000 inclusive, and one additional representative for every additional 1000 members or major fraction thereof.

Each national engineering or technical society is to be entitled to one representative for a membership of from 200 to 2000 inclusive, and an additional representative for every additional 2000 members or major fraction thereof.

### ENGINEERING COUNCIL MIGHT BE MERGED

In reference to the present Engineering Council, whose work has been so effective in the many matters of general concern among engineers, the Committee states:

"In preparing this plan the Committee has recognized that there exists in Engineering Council a tool which is engraving an honorable record on the pages of professional history, but its limitations are well known and its poverty is chronic. If desired, Engineering Council can be moulded into this organization by making it more democratic and founding it on direct representation of all engineers, rather than appointment as at present.

The great object is to provide an effective body, widely and truly representative, modestly yet adequately financed, which will be neither autocratic nor aristocratic, which will at all times stand as the representative and defender of the profession in matters affecting its honor, welfare and common interest."

J. W. Alvord, Chairman of a committee appointed by the Board of Directors of the Am.Soc.C.E. to study the report of its Committee on Development and make recommendations with regard to it, presented the conclusions of this Committee, saying that the problem is first to consolidate the engineering profession within itself by knitting together the Founder Societies and other national societies equally broad in purpose and careful in membership; second, to affiliate semi-technical societies in such a way that they will not control. He said that the new organization proposed by the Joint Conference report is quite separate and distinct from the Founder Societies and that he would propose to depart from that Committee's plan by consolidating into the new National Council the directors and officers elected by the Founder Societies. To this and other similar proposals by various speakers objection was raised on the ground that an organization such as proposed by Mr. Alvord would not be democratic, builded from the bottom up, in a way to satisfy the urgent demands of the various organizations of engineers both large and small.

### DISCUSSION ON JOINT CONFERENCE REPORT

At the conference the plans of the Joint Conference Committee, certain of which have been referred to above, were presented by J. V. W. Reynders, who said that:

"First, the movement should be democratic, and so far as possible membership of the Engineering Council should not be appointed on the new Council. Second, the plan should not be fathered by the Founder Societies more than necessary in order to avoid hurting the susceptibilities of those not affiliated with these societies. Any nation-wide movement will fail unless it is the outcome of a justifiable, as well as an enduring principle. There must first be an unmistakable and widely-felt sentiment. There must be plainly recognizable signs of powerful initiative. In the national societies are sufficient members to insure success of this movement. This is a constituency upon which we can build without fear of consequences."

Further discussion, in which a large number participated, hinged upon the form which the proposed organization should take; that is, whether it should be in some way related to the present Engineering Council through the reorganization of the latter or otherwise organized through the present governing boards of the national societies; or whether it should be based on initial action taken by membership of the leading societies both national and local as proposed by the Joint Conference Committee. The drift of the discussion can be summarized by brief reference to remarks by Dr. Jacobus, who said the great handicap on Engineering Council has been the feeling that it is organized from within. The moment we say we are going to reorganize Engineering Council it will spread a feeling abroad that Council is going to fix things to suit itself. It should come from without, not within.

Another question raised was the matter of representation on the National Council of the proposed organization, whereby the local representation is much larger than that of the national societies. To this R. L. Humphrey of the Joint Conference Committee replied that the members of the local societies are so often members also of the national societies that the delegates from the local organization would usually be members of the national societies as well and that therefore the latter would be amply represented. He believed that it would add to the feeling of democracy if the men from the local affiliations had a larger proportion of representation on the National Council. It was specified in the report, however, that the local representation on the Executive Board would be more restricted.

# Government Activities in Engineering

*Notes Contributed by The National Service Committee of Engineering Council*[1]

## Congress Passes General Land Leasing Bill

After ten years of waiting, the General Land Leasing Bill has passed both houses of Congress, run the conference gauntlet and was finally accepted by the Senate on February 11th.

It is estimated that by the terms of the bill, the oil, gas, coal, phosphate, sodium, and oil-shale lands made available for leasing and development by private enterprises are approximately 75,000,000 acres of the public domain. This land is largely in the western states.

The conference report reserved the right of the Government to extract helium from all gas produced from such lands. A Senate provision was put back in the bill permitting individuals, or associations of individuals, to obtain limited licenses or permits to secure a supply of coal for strictly domestic needs.

The relief section of the bill relates to very valuable oil-producing lands, which are now involved in litigation. The purpose of the House language, which was finally retained, was to prevent the claimant or holder of excess acreage from disposing of such excess. This excess under the terms of the House bill would revert to the Government to be leased by competitive bidding. At the same time it does not prevent one, holding not more than the maximum amount allowed, from disposing of any part thereof. It also recognizes an exchange of interest in lands made prior to January 21, 1920, provided the exchange does not increase or reduce the acreage held in excess of the allowed maximum. Thus the status quo of the excess holder is not changed. It was thought best not to interfere with sales of oil lands which were made by claimants holding less than the maximum amount allowed. The Senate provision requiring substantial improvements to have been made prior to an oil-land withdrawal in Alaska was reinserted in the bill.

The Senate provision provided maximum charges by the Government of 20 cents per ton for coal, and 25 per cent of the value of oil produced. The House bill contained no restriction on these charges and the conference agreed to the House provisions. It was finally agreed that 70 per cent of the proceeds from past sales, bonuses, royalties, and rentals was to go to the reclamation fund and 20 per cent to the State. Of the proceeds from future receipts, $52\frac{1}{2}$ per cent is to go to the reclamation fund and $37\frac{1}{2}$ per cent is to go to the State. Ten per cent of both past and future funds is to be turned into the United States Treasury.

It will be recalled that this bill provides for leasing lands on the public domain in multiples of 40 acres, and the total allowed to any one claimant is not to exceed 2560 acres.

## Reclamation Service

The Reclamation Service has submitted supplemental estimates for $5,000,000 to be expended during the fiscal year 1921. The Secretary of Interior's letter of transmittal pointed out the limitation on the Reclamation Service due to the money available in the Reclamation Fund. At the time the regular estimate was submitted it was anticipated that not more than $7,873,000 would be available. When it became certain that the General Land Leasing Bill would become a law, it was possible to add at least $5,000,000 more to the estimate, because the reclamation fund is to receive 70 per cent for past production and $52\frac{1}{2}$ per cent for future production derived from the bonuses, royalties, and rentals provided in the General Land Leasing Bill. It is estimated that this will make between $5,000,000 and $7,000,000 additional available in the reclamation fund.

[1] Engineering Council is an organization of national technical societies created to consider matters of common concern to engineers as well as those of public welfare in which the profession is interested. The head-quarters of Engineering Council are located in the Engineering Societies Building, 29 West 39th Street, New York City. The Council also maintains a Washington office with M. O. Leighton, chairman of the National Service Committee in charge. This office is in the McLachlen Building, 10th and G Streets, Washington, D. C. The officers of Engineering Council are: J. Parke Channing, *Chairman*; Alfred D. Flinn, *Secretary*.

## Patent Legislation

Legislation covering the administration of patent law, the status of the Patent Office, and our international relations in patent work have been very active in Congress. The House has passed a bill ratifying a Patent Convention which will enable holders of patents in this country and South American countries to have their patents registered and cleared through a common office at Havana. The registration fee is very small—only a fraction of what it formerly cost to have these patents registered.

The Senate and House bills authorizing the Federal Trade Commission to accept and administer inventions, patents, and patent-rights have been reported to both Houses. This is for the purpose of benefiting and encouraging the public in an effort to stimulate inventions and to encourage their industrial use. This bill provides for coöperation as between individuals, Government, and other agencies, and with scientific agencies of the Government. The House patent bill providing for increase of force and salaries in the Patent Office has been rewritten and reported to the House calendar. It provides that the Commissioner of Patents shall receive $6000 annually, the First Assistant to the Commissioner $5500, and the Assistant Commissioner of Patents $4500. Five examiners-in-chief are also provided with a salary of $5000 each. Salaries of other officials are raised in proportion.

The bill proposing to establish a Court of Patent Appeals has been re-written by a sub-committee of the House Patents Committee and will be introduced to the House calendar shortly. A bill proposing to make a separate establishment of the Patent Office is now under consideration by another sub-committee.

## Tariff Legislation

Despite the fact that the Senate had threatened not to take up piece-meal tariff legislation, much of which had been passed by the House, some important tariff bills relating to magnesite, cobalt, chrome, zinc, tungsten, porcelain and scientific instruments have been ordered favorably reported by the Senate Finance sub-committee to the full committee. Some of these emergency tariff bills have been reported without change.

The magnesite bill provided one-half of a cent per pound on commercial ore; three-fourths of a cent per pound on calcined, deadburned and grain magnesite, and 10 per cent ad valorem and three-fourths of a cent per pound on magnesite brick. These are the same as the duties provided in the House bill. Laboratory, optical and scientific glassware also carry the same ad valorem duties as provided by the House, namely, 60 per cent on laboratory apparatus; 45 per cent on optical glass; and 45 per cent on scientific instruments.

The zinc bill was amended so that zinc-bearing ores of all kinds are to pay a duty of two cents per pound on the zinc content instead of one cent, as provided by the House bill. The bill was also amended to carry the cobalt duty of twenty-five cents. The tungsten bill was amended so as to provide for a duty of $1.50 on the tungsten content of imports of alloy steel.

## Government Aid in Road Building

A new bill has been introduced by the Chairman of the House Military Affairs Committee which is intended to liberalize the conditions under which surplus motor vehicles may be transferred to the State Highway Departments. The bill provides for transfer of equipment other than motor vehicles, and is in accordance with the recommendations of the Bureau of Public Roads, which recently expressed its desire that the following materials be itemized in the bill, which in other respects is similar to the bill that has passed the Senate. The list of materials includes industrial railway equipment, conveyors, gravity, power, donkey engines, corrugated metal roofing, steel and iron pipes, etc.

# Charles Warren Hunt Appointed Secretary Emeritus of the American Society of Civil Engineers

AFTER serving the American Society of Civil Engineers for twenty-eight years, three as Assistant Secretary and twenty-five as Secretary, Dr. Charles Warren Hunt, at this own request, has been relieved from active duty. Dr. Hunt will continue to serve in an advisory capacity. His relation will be that of Secretary Emeritus, this office having been created for him by the Board of Direction of the American Society of Civil Engineers by a resolution passed January 21, 1920. The text of the resolution follows:

WHEREAS, Dr. Charles Warren Hunt, after 25 years of devoted and efficient service as Secretary of the American Society of Civil Engineers, during which he has contributed to its upbuilding and prosperity to a degree which places the Society under lasting obligation to him, has indicated his desire to be relieved from active duty, and

WHEREAS, He possesses an intimate knowledge of the business of the Society and an extensive acquaintance with its membership which render his advice and assistance of the greatest value to the Society and its Board of Direction, the benefit of which they can ill afford to lose, therefore be it

Resolved, That Dr. Charles Warren Hunt be and hereby is appointed Secretary Emeritus of the American Society of Civil Engineers at a salary of $9000 for the coming year, and a salary of $6000 per annum thereafter with such duties as the Board of Direction may assign to him.

Throughout his administration, Dr. Hunt has kept closely in touch with the business details of the Society's office and its organization, and has conducted its affairs on sound business lines. The high professional standards of the Society have been maintained, and under his guidance and strong personality the Society has grown, not alone in numbers, but in professional prestige as well.

Dr. Hunt was born in New York City on May 19, 1858 and when only eighteen years old was graduated from New York University with the degrees of B. S. and C. E. This was in 1876, and for the next sixteen years he was engaged in the practice of his profession. His experiences were varied and included work on municipal parks, water works, railroads, and docks.

It was while associated with the Brooklyn elevated railway companies in 1882 that Dr. Hunt was elected Assistant Secretary of the society. This was a newly created office and while its acceptance meant financial loss he saw an opportunity to serve his fellow men and he did not hesitate to accept. From that time on the events in his career are intimately connected with the history of the American Society of Civil Engineers.

Bringing to the society the enthusiasm of youth, Dr. Hunt had well conceived ideas of the functions of a national engineering society. He believed it necessary for a technical society not only " to advance engineering knowledge and practice" but " to preserve the high character and professional qualifications of its membership; to maintain the dignity and standing of the organization; to keep in touch with and take proper action on all matters in which the relation of the Profession to the public is involved; and to do whatever is possible for its members individually and in general." This was Dr. Hunt's creed and to the realization of such ideals he devoted himself unstintingly.

Among the many developments which took place during Dr. Hunt's administration was the creation of Local Associations of the society. The formation of such associations had several times been considered, but it was not until 1905 that definite action was taken. In that year two associations were formed, one at Kansas City and the other at San Francisco. Their great success at once led to the formation of similar bodies in other cities, and at the present time the society has a total of twenty-five.

The house at 57th Street, which for over twenty years served as headquarters of the society, was also erected during Dr. Hunt's term of office. It was a building suited in every way to the needs of the society and only because of a desire to more thoroughly coöperate with the other national engineering societies was it vacated in 1916, when the American Society of Civil Engineers became one of the Founder Societies with headquarters in

the Engineering Societies Building, 29 W. 39th St., New York.

To Dr. Hunt the society is indebted for the cataloging of its Library. He early realized the importance of a library index and after two years of careful and concentrated effort completed the work which made the library of great use to the profession. Dr. Hunt was also a pioneer in the making of engineering " searches," a work which has now become so important in all engineering reference libraries.

CHARLES WARREN HUNT

During Dr. Hunt's term of office the publications of the Society have undergone several changes. In 1892, in addition to the regular Transactions, the Bulletin dealing with current events and containing short abstracts of the technical papers was begun. In January, 1896, the publishing of the monthly Proceedings was started, the technical matter contained in these being subsequently collated and published in volumes of Transactions. This method of handling society notes and technical papers was new in society publications, but it soon won universal approval and has been retained ever since. In 1899 the society began the publication in the Proceedings of a " list of current engineering articles of interest." This service, started in a modest way, was greatly appreciated by the membership and has been retained ever since as a monthly feature.

Dr. Hunt has always taken the greatest interest in matters pertaining to the profession at large. In 1903, when The American Society of Civil Engineers was invited by the Directors of the Louisiana Purchase Exposition to undertake the arrangements for an International Engineering Congress, he personally supervised and directed the work of organizing the congress, the securing, translating and publishing of papers, the reception and entertainment of foreign representatives, and the details of the many meetings.

Dr. Hunt has long served as a representative member on such bodies as the John Fritz Medal Board of Award, the Engineering Foundation, and the United Engineering Society. He has given freely of his time and energy for the advancement of the entire engineering profession, and his active participation in the affairs of the profession will be greatly missed.

# NEWS OF THE ENGINEERING SOCIETIES

American Society of Heating and Ventilating Engineers—The Engineering Institute of Canada—
A. S. M. E. Akron Meeting—Engineering Society of Western Massachusetts

## Engineering Society of Western Massachusetts

A meeting of the Engineering Society of Western Massachusetts was held at Springfield on January 20, at the Springfield Gas Light Company's plant, State and Water Streets. A number of the members visited the plant in the afternoon. Following a dinner in the office building of the plant, an evening session was held at which Arthur S. Hall gave a description of the plant and Harold C. Andrew, an efficiency expert, spoke on Chemical Control of the Gas Industry.

## American Society of Heating and Ventilating Engineers

The twenty-sixth annual meeting of the American Society of Heating and Ventilating Engineers was held in New York on January 27, 28, and 29. The reports of the officers and council reflected the increased growth of the society during the past year. The society's Journal, beginning with the month of January, will be published monthly instead of quarterly. The Research Bureau of the society is now a functioning institution, as was evidenced by the data on heat losses from direct radiation presented by Director John R. Allen and given in part in the Engineering Research section in this issue.

The results of four years' experience in prevention of corrosion of pipe in the deactivating plant of a 12-story apartment building in New York City, were presented by F. N. Speller and W. H. Walker. The principle employed is to fill a storage tank with suitably prepared steel lathing designed so as to pack closely with an exposed surface of approximately 100 sq. ft. per cu. ft. of space. In passing the water heated to 160 deg. fahr. through this metal, all the oxygen is fixed in the form of hydroxides which are readily removed by filtration through a sand filter. The result is water inactive toward iron or other metals, free from oxygen but otherwise unaltered in composition except for the presence of a small amount of free hydrogen resulting from solution of iron in the deactivating tank. The results have been entirely satisfactory. It is said the pipes carrying deactivated water have shown no measurable corrosion.

E. R. Knowles described various types of installations using pulverized fuel, and quoted figures indicating to what extent this form of fuel is being utilized. It appears that in the United States there have been used to date over 50,000,000 tons of coal in powdered form; there are about 6,000,000 tons of powdered coal now used annually in cement making; about 2,000,000 tons in copper roasting and smelting; about 2,000,000 tons in steel manufacture and about 200,000 tons for general power.

An Advance in Air Conditioning in School Buildings, by E. S. Hallett, carried the general note of progress into the field of school ventilation and presented a system of using ozone in a way apparently so reasonable and convenient that it undoubtedly points out an economical solution to the problem of improving health conditions in schools.

National consciousness was appealed to in two papers. The first, The Coöperative Movement, by James P. Warbasse, president of the Coöperative League of America, discussed this great world movement of coöperation, tracing its history through humble beginnings to the present prosperity which it now has in England, Belgium and other countries, and pointing out the steady progress it is making in the United States. The other paper, Work of the Construction Division of the War Department, by Major R. W. Alger, gave an account of the difficulties under which this division worked and the almost incredible things accomplished under such handicaps. It concluded with a plea for constructive planning rather than destructive criticism in

order that the country might be prepared should another great emergency overtake it.

Other papers were: Relations of Boiler Heating-Surface Area to Boiler Capacity, by P. J. Dougherty; Oil as a Fuel for Boilers and Furnaces, by H. H. Fleming; Oil Fuel, by F. W. Staley; Fuel-Oil Equipment, by John P. Leask; Oil Fuel versus Coal, by David Moffat Myers; Oil Fuel, by W. C. McTarnahan; and Color Schemes for Distinguishing Plant Piping, by H. L. Wilkinson.

## A Record Meeting of Canadian Engineers

The Annual and Professional Meeting of The Engineering Institute of Canada held in Montreal January 27, 28 and 29 was easily the most successful in the history of The Institute. In view of the fact that the percentage of members lost in the war was very appreciable, the attendance of over 450 was especially good. The unusual attractiveness of the social features is explained by the statement that this is the first year that ladies have been invited to attend and participate in the events of the week. The professional papers were both varied and important.

The program for the Annual Meeting, held on the first day of the convention, included, in addition to the business sessions, an address on Modern Highway Problems, the address of the retiring president, Colonel R. W. Leonard, and the inauguration of the incoming president, R. A. Ross. At a luncheon held for members and guests, greetings from The American Society of Mechanical Engineers were conveyed to the gathering by Calvin W. Rice, Secretary of the Society, who stressed the importance of all technical engineers becoming members of The Engineering Institute.

During the past year, The Institute has been coöperating with other engineering societies in an effort to secure for Canadian engineers a uniform registration law similar to that recently recommended by Engineering Council in this country. The monthly Journal of The Institute is now well established, provincial divisions including 18 branches are now actively working, and the membership has increased greatly in excess of previous years. During his recent visit to Canada the Prince of Wales accepted honorary membership in The Institute and Government official recognition of The Institute was accorded by the Minister of Labor in granting it representation at the Industrial Conference at Ottawa in September.

The Professional Meetings of the convention opened Wednesday, January 28, in one of the most important sessions, a discussion as to whether engineering training should be specialized or conducted in a general manner. Papers were presented by Profs. R. F. Ruttan and A. S. Eve, of McGill University. Dr. Ruttan spoke favorably of the plan adopted by the Massachusetts Institute of Technology whereby the student while still under the direction of the instructors leaves during the winter of his Junior year and for the period of from two to three months in each of four or five leading companies' works under actual conditions in the field, returning the following winter and completing his college course. The general feeling evidenced by the discussion was that the engineering graduate should be a man competent to "take up a thing and see it through."

Engineering activities in the province of Quebec were dealt with at the Thursday morning session of the convention. The water power resources, the public health administration and the highways of the province were the subjects discussed. In the afternoon three interesting papers were presented, The Policy of the Air Board of Canada, Pulp and Paper Industry and The Forests of Quebec.

Several important points touching upon the situation in the engineering world today were brought out in the address of the retiring president, Colonel Leonard. He emphasized the need for the engineer's clear thinking, always working from cause to effect, to help settle the labor unrest of the present time and quoted from an address which W. R. Ingalls gave before the Canadian Mining Institute to the effect that the increase in production which has come about during the last century is not because men have grown any stronger but because mind has taught labor how to become more effective and has provided it with machines and organization. Further progress depends on the ability of the leaders of labor to instill the sound principles of an intelligent democracy into their followers so that they may have a sense of their duties and responsibilities to the public as well as their rights. He said, "If solutions are possible whereby the lot of the workingman can be improved, the engineer—whose training and life work are devoted to the solution of practical problems—is the most capable of finding them."

Colonel Leonard urged coöperation among all classes and nations. He not only outlined various opportunities for Canadian engineers in Canada but importuned them to turn to engineers of other nations for advice and thus increase their usefulness. This suggestion, as applied between the engineers of the United States and Canada, would be for the greatest good of the development of the engineering profession in both countries.

We congratulate The Engineering Institute of Canada on its wonderful prospects and are grateful for the words of an appreciation in its annual report for the address last year of our past-president Dr. Hollis, and for the benefit which it obtained from the coöperative arrangements with our Society with respect to the Engineering Index.

## A. S. M. E. Meeting at Akron

Cleveland members of the A.S.M.E. and members of the Cleveland and Akron Engineering Societies held an all-day convention in Akron on Tuesday, January 27, headquarters being established at the Firestone Club House, where the professional session was held.

A paper, illustrated with slides, was presented by John Schaub, superintendent of the Akron plant of the Wellman-Seaver-Morgan Company, on the subject Rubber-Working Machinery. The paper, while utilizing as examples the heavy type of machinery common to rubber-working plants, was more correctly a discussion of the application of quantity-production methods to the machining of parts usually considered large enough to be run singly through the shop as separate jobs.

In the Wellman-Seaver-Morgan plant mixing mills are carried through the machine shop in units of five for the larger castings, such as housings, mill boxes and driving gears, while the smaller parts are worked in lots of two or three that number. In the design of these parts such clearances are specified that practically all the patterns may be used for machine molding.

Before planing the mill housings for their final fit, they are run through a milling machine and the larger portion of the surplus cap stock taken off. Then, when the housings are set on the planer, in lots of ten, there remains but a finishing cut. Machine time is cut appreciably by using two set-ups and two different machines for roughing and finishing. Mill bearings are gang-planed in lots of twenty, after which the pad for the bearing cap is milled and the individual bearing set for the final boring.

Mill guides are first planed, then dropped into a drill jig for drilling, after which they are placed on a special angle fixture on the horizontal boring-mill table, the drilled holes being used as a register to hold the guides in position. A complete set of guides with the guide tees adjusted is bored at one setting. The boring bar is fitted with two tools, one for each mill-roll radius, so that when the guides which fit on the large roll have been bored the operator moves the table sufficiently to place the other pair of guides in line with the boring bar.

A bonus system for machine-time saving is in successful operation, and all saving on the estimated time is split with the workman on a fifty-fifty basis, who receives as additional compensation one half of the time saved, at his hourly rate.

Calenders are also built in lots of five and erected on a cast-

iron floor to insure correct alignment. A dummy roll, considerably lighter than the service roll, is used for scraping in the bearings to a perfect fit, eliminating much of the laborious work usually connected with roll fitting.

The calender is divided into units for assembling smaller parts, such as wind-ups, let-offs and roll-raising mechanisms, being put together by special assembly groups, after which the completed sectional units are brought to the erection floor for assembly.

A. O. Austin, of the Ohio Insulator Company, Barberton, Ohio, presented a series of slides illustrating the Ohio Insulator plant and gave a brief description of the purpose and operation of each machine shown.

Following a luncheon at the Firestone club house the convention divided into three main inspection groups. One section of about thirty members visited the plant of the Ohio Insulator Company, under the guidance of Mr. Austin, and had an opportunity to view the manufacture of large porcelain insulators, ranging in height from two inches to almost five feet. A demonstration of high-tension insulator testing was given, a voltage of 330,000 volts and a frequency of 35,000 being used. This voltage produced a standing spark 36 in. or more in length. During the test a novel feature was introduced by subjecting the insulators under test to all the conditions met in a severe rainstorm.

The group electing to inspect the Wellman-Seaver-Morgan plant was taken first to what is known as the "Old Plant" building and there shown two 52,000-hp. hydraulic turbines under construction, both for use at Niagara Falls. The magnitude of these turbines is shown by the fact that the cast-steel metal casings have a water intake 7 ft. 6 in. in diameter. Other machinery under construction included rubber-mixing mills, calenders, washers, steam tire-vulcanizing presses, condensers for marine engines still being built for the U. S. Merchant Marine, and also the first lot of "Akron" tractors to be sent through the plant.

Passing through the galvanizing room, containing a tank capable of holding 100 tons of metal, and one of the largest in the country, the party completed its tour by inspecting assembly methods used in constructing mechanically stirred gas producers.

For the rubber-working industry the Goodyear Tire and Rubber Company acted as host, groups of eight men and a guide being sent through the plant, following, as closely as possible, the flow of the rubber through the plant from the raw material warehouse to the finished product. The manufacture of two types of inner tubes was shown, of lapped tubes—formed from sheet rubber much as lap-welded pipe is formed from skelp—and of laminated tubes, these latter being made up from a thin sheet of prepared rubber which is rolled around a mandrel of the proper size until it is covered by two or more complete layers. Both types when steam-vulcanized.

Guides for the inspection groups were taken from the organization in the Goodyear plant known as the "Flying Squadron," made up of high-school and college graduates, working out a kind of engineering apprenticeship with a view to being located eventually in the department most attractive to them individually.

The evening dinner was served at the Goodyear plant, with Dean Ayer, president of the Akron Engineering Society, as toastmaster. Dr. D. C. Tremaine, from the National Industrial Speakers' Bureau, discussed the topic, Human Nature in Industry. Introducing the factor of the human element in industry, the speaker brought out its present status, as compared to that of business, and emphasized the need for easy access to the manager's office by the worker. The responsibility for present-day business is not one-sided, but rests equally with worker and management. The growing importance of each man to the organization and to the industrial life of America was shown in its bearing on the relation of increased production to the high cost of living.

While the day's program was styled a "convention," there was nothing in the papers presented, the talks, inspection trips or the evening dinner to render it in the least bit formal. Primarily, the day was planned to afford an opportunity for engineers of Cleveland to meet engineers of Akron, and for them to enjoy together a glimpse into some of the industries which are typical of the manufacturing progress of the Rubber City.

## Western Society of Engineers Holds Annual Meeting and Discusses Projects for an Engineering Building and Affiliation of Sections with the National Societies

The fiftieth annual meeting and dinner of the Western Society of Engineers was held in the Hotel Morrison, Chicago, January 28. Four hundred members of the Society were in attendance.

Major General Leonard Wood, U. S. A., the principal speaker of the evening, declared that the establishment of a National Department of Public Works would be of great benefit to this country. He said the Engineer's Reserve Corps should be maintained at maximum strength, for it had proved its ability in the great war, both at home and abroad.

Engineers should take an active part in public affairs, he asserted. All colonial development of the United States has been dependent on the work of the engineer, but after their labors in establishing lines of communication and transportation improving sanitary conditions and establishing a basis for commercial activity, they have been content to let the remainder of the work be taken over entirely by the administrative forces of the country.

A. Stuart Baldwin, retiring president, gave a masterful address bringing out the need for more coöperative action among the members of the Society.

There must be a more stern and conscientious application of the "Golden Rule," if civilization is to stave off a period of decadence in the future, he said. The engineer must analyze the forces that today are a menace to the peace and prosperity of the world and are directly antagonistic to "Do unto others as ye would have them do unto you."

In beginning his address, Mr. Baldwin said that a development committee had been appointed to study the status of engineers in Chicago and vicinity, and thereafter enter upon a campaign for bringing together the engineers from all branches of the profession, organizing them into one great society and so coördinating their activities as not only to gain professional prestige but to function as a strong and useful member of the body politic. The Society could look forward at no distant date to doubling its present membership.

The next important move should be the construction of an adequate building in Chicago as an engineering center. "With so large a membership," he said, "the work of the directors and executives must to a much greater extent be de-centralized and thrown upon the heads of the Sections, each of which must become to a certain extent a self-contained engineering society, in its own specialty, all coöperating for the common good. If each Section were the working local branch and representative of the national society with which the most of its members were affiliated and if the national societies would father these sections and become their working associates, the Sections would become tremendously strengthened and not only serve to strengthen the Western Society of Engineers, but act as a reservoir from which the national societies would gain strength and inspiration. The national societies all have need for local work on the part of their members. The opportunities for getting together nationally occur but a few times during the year, whereas local sections should be able to meet monthly or even more frequently, if necessary. It will undoubtedly be found that those members of a national society who are able to get together locally most frequently for papers, discussions and consideration of society matters, will be the most influential in the affairs of the national society."

Frederick K. Copeland, Mem.Am.Soc.M.E., the new president, laid down the ambitions of the society for the coming year. Briefly, they are: (1) Increase membership; (2) Organize more sections to aid in professional development of engineer; (3) Pay an important and decisive part in civic affairs; (4) Develop the social activities of the engineer beyond the confines of engineering alone.

"Each and every one of us must show our loyalty and willingness, by giving of our time and our energies, to take part in this great development of the Western Society of Engineers," he concluded.

Mr. Copeland is president of the Sullivan Machinery Co. and for a number of years has been active in the affairs of the Western Society. He was admitted to membership in 1892. In 1919 he was appointed first vice-president to fill the unexpired term of Kempster B. Miller. Besides being active in the affairs of Engineering Council, Mr. Copeland is a director of the National Department of Public Works Association. He is president of the General Committee of Technical Societies of Chicago.

The rapid growth of the Western Society of Engineers has made it necessary that additional facilities as well as space for the administrative force of the organization be obtained. Two additional rooms adjacent to the present quarters of the society on the 17th floor of the Monadnock Building have been obtained and will be available May 1st. The tentative plan is to devote the present quarters to the library, reading rooms, auditorium and committee rooms. The records of the society will be moved to the new quarters as far as will be practicable. Such a move as this will add materially to the advantages to be obtained from the library in that more shelving space will be available, it will be possible to add to the comforts of the reading room and committees may work in rooms in direct contact with the reference files of the society.

H. T. ELLISTON.

## SELF-SUPPORTING CHIMNEYS TO WITH-STAND EARTHQUAKES

(Continued from page 160)

reinforced-concrete chimney is that given in Turneaure and Maurer's Principles of Reinforced-Concrete Construction, except that the temperature reinforcement was increased materially beyond that indicated in this textbook. The allowed tensile stresses of the reinforcing steel, as well as the compressive stresses in the concrete, due to earthquake, were normal. The temperature stresses alone were slightly above normal for concrete, and less than normal for steel. When these were added to the earthquake stresses, the final result gave a stress safely within the elastic limit for steel, but considerably above what would be considered a safe allowance for the compressive stress of concrete. Investigation of the temperature stresses shows that the above condition is unavoidable, and the conclusion is that while a chimney so designed would not fail by falling during an earthquake, it is easily believable that the combination of earthquake and temperature stresses would result in serious fractures at critical portions of the chimney.

A very important point in the design of such chimneys is the assumption of the difference between the temperature of the gases within the chimney and the temperature of the outer shell of the chimney. Some tests made by the writer indicated that this temperature difference is greater than was assumed in the example given by Turneaure and Maurer, and further data along these lines would be of much interest to engineers.

The writer was unable to find any new material of light weight and of low conductivity to be used in lieu of the ordinary concrete or brick lining of a reinforced-concrete chimney; for manifestly it is highly important to minimize the mass of the chimney subject to vibration. It was finally necessary to assume a brick lining. The question then arose as to the height to which this should be carried to minimize temperature stresses in the unlined upper portion of the chimney, which, from experience with earthquakes, is the weakest section of the chimney. It was considered impracticable to carry the chimney lining to this critical section; which indicates a manifest defect of the reinforced-concrete chimney for withstanding earthquakes.

It has recently been suggested to the writer by Mr. C. A. G. Weymouth that in the building of reinforced-concrete chimneys to withstand earthquakes it would be advantageous to use terra-cotta concrete, as developed by the Concrete Ship Section of the Emergency Fleet Corporation. As is well known, terra-cotta concrete materially reduces the weight per cubic foot, and in addition has a high compressive strength. It should be pointed out, however, that any saving in the weight of the chimney shaft, as compared with the ordinary concrete construction, would, of necessity, be made up by an increase in the weight of the base of the chimney to give the necessary stability against overturning from wind pressure.

The requirement under [7] for the deflection of the chimney shaft should be made with some margin of safety. No data are at present available, but it should be borne in mind that any increased elasticity of the chimney should not be secured at the sacrifice of strength actually necessary to resist moments in the lower portion of the chimney shaft.

It is hardly necessary to state that the chimney designer must give due consideration to the shear produced by earthquake, as well as the shear due to wind pressure; other stresses, however, are usually the determining factors.

It is of course obvious that a chimney to withstand earthquakes should be constructed of materials of maximum strength, of minimum weight, and with maximum elasticity. So far, this combination seems to be best supplied in the use of self-supporting steel chimneys, regardless of whether the conditions of operation require a chimney lining. In California, at least, the earthquake evidence testifies eloquently in favor of the use of steel chimneys.

In closing, the writer will state that while the ultimate solution of the earthquake chimney problem appears impossible in an accurate sense, it is nevertheless to be hoped that some of our engineering physicists will give further study to the rather special questions which it involves.

## FIRST TRANSCONTINENTAL MOTOR CONVOY
### (Continued from page 150)

This was in running through Iowa where the roads were very good. A very different class of troubles developed when running through the western section of the trip as indicated, for example, by what happened during such an average day as August 13.

Departed Rawlins 6:15 a. m. 22 mi. west detour was made around dangerous bridge over dry creek bed. At Creston Station a Class B truck slipped off the road and was helped back by another Class B. Also Class B spare parts truck ran off road on abandoned Union Pacific R. R. grade and was rescued by Militor. Old right-of-way single truck most of way and many high, dangerous fills and a number of deep sand holes were encountered. Near Latham Station Class B water tanker No. 80216 ran off road on abandoned railroad grade and rolled over 270°, resting on left side. It was righted by 2 Class B's and proceeded under its own power in 20 min. No serious damage. Class B tank trucks appear to be somewhat top-heavy, and driver claimed that this truck was very hard to steer and almost impossible to keep on road. Trailmobile kitchen broke spring and bent axle. At Wamsutter, F. W. D. No. 415767 stopped to tighten up front universal joint and it was also found to have a broken radiator support stud. Two G. M. C.'s, Garford, Class B and Dodge had carburetor trouble. 3 mi. east of Tipton Station, Class B Machine Shop No. 414,319 ran off road into soft sand hole and had to be towed out by Militor. Right wheels of Mack Blacksmith Shop No. 4 sheared through 16-3″ x 10″ bridge floor planks, and narrowly averted dropping into 12′ ravine. Bridge reinforced and truck rescued in 55 minutes. Six other bridges reinforced by Engineers and it seems that timber in this dry atmosphere loses its elasticity and becomes very brittle. Camped on Red Desert, on barren sandy plain. No inhabitants or buildings other than railroad personnel and property. Nearest natural water supply 16 miles. Fair and warm. Fair dirt roads. Made 58 miles in 10¾ hr. Arrived Tipton Station, Wyo. 5 p. m.

Even that, however, cannot be called the worst and to judge of the troubles that the Convoy had to handle it may be well to read, for example, the log for August 30 to September 1.

Departed Fallon, 6:30 a. m. Followed direct road to Carson City, via the Lahontan Dam. Heavy trucks had some difficulty in getting through soft sand holes about 2 miles out of camp. Five or more trucks were chained together and men not driving helped by pushing. Delay ½ hr. White Reconnaissance Car No. 111505 had to replace spark plug on account of broken porcelain. Required 11 hrs. to move Convoy from Fallon to a point 12 mi. west, on account several stretches of unstable, dry sand up to 1½ ft. deep, and wet quicksand. All heavy vehicles, including Cadillacs had to be pulled and pushed through by combined efforts of tractor and men, over wheel paths made of sage brush. Two Class B tankers sank to a depth of 5 ft. in sand, requiring extensive excavation and construction of false work of railroad ties to aid Tractor in rescue. Dodge Light Delivery No. 26991 broke generator drive chain, damaging oil pump drive. One small bridge broke down. Convoy met by Gov. Boyle, etc.

Chicken dinner. Co. E arrived 11 p. m. Fair and warm. Poor gravel roads. Made 66 miles in 20 hrs. Arrived Carson City, Nevada, 2:30 a. m. (31st).

Aug. 31. Carson City. Sunday rest period. Tractor towing Mack with bad clutch, arrived 11:30 a.m after resting on road until daylight. Inspected and adjusted equipment. Changed several tires, pressing done at Reno. Transportation provided and personnel taken to Hot Springs for bathing. Union religious services held at camp on Capitol grounds, at which Governor Boyle made address of welcome. At officers' meeting instructions were given relative to operation precautions to be taken on King's Grade climb tomorrow. Fair and warm.

Sept. 1. Departed Carson City 6:30 a.m. Convoy arranged with Temporary control established at base of King's Grade for inspection of steering gears, brakes, tow chains and wheel blocks, also for spacing vehicles 100 yds. apart. Full supply of gasoline, oil and water verified. Experienced drivers only allowed to drive and one man on each vehicle stood ready to block the wheels at each halt. All passengers required to be on the alert for any emergency. Nevada State Highway Dept. suspended all eastbound traffic from 6:30 a.m. until after Convoy had crossed the Sierras. Motorcycle riders required to maintain vehicle spacing and inspect bearings every 4 miles. Reached altitude of 7030 ft. at summit over narrow, winding road of sand and broken stone, cut out of, and in places built up on, mountain side. Total climb 14 mi. made in 6 hr., slow progress being necessary to prevent accident. Grades 8 to 14 per cent. Crossing Sierras without accident may be considered noteworthy achievement for heavy vehicles. Class B broke fan belt. One Trailmobile kitchen overturned and was wrecked, and was shipped to San Francisco. Mack broke connecting rod, pushing piston through top of the cylinder and cracking crank case, near Glenbrook, where upon "chow" was served, and was left there to make emergency repair by cutting out damaged cylinder, piston being welded and cemented in place. At California-Nevada line the War Camp Community Service and Mayor's Comm. of San Francisco gave fine barbecue and enthusiastic welcome. Mr. Celio, owner of Meyer's Ranch, served refreshments in evening around huge camp fire, while the Firestone representatives furnished movies and smokes. Scenery throughout day of greatest beauty, especially at Lake Tahoe. Fair and cool. Fair gravel roads. Made 34 miles in 13½ hrs. Arrived Meyers, Cal., 8 p.m.

On the whole it may be well to repeat what has been said before as to the results, namely, that the pioneer undertaking of its kind, the First Transcontinental Convoy, was a successful operation and the experience gained during this trip will prove invaluable not only to the Army but to all users of motor trucks in general.

## ELECTRIC POWER SUPPLY OF SOUTHERN TEXTILE MILLS
### (Continued from page 161)

Early practice employed large group motors, but comparatively little benefit was derived from their use. The practice now is to use small motors ranging in size from 0.5 hp. on looms to 5 and 7.5 hp. on spinning frames directly applied to the machines. The resulting benefits, principally in production, are a great deal more than sufficient to offset the extra cost for this type of installation. A great many of the old mills, either with mechanical or with large group motor drives, are changing to these individual motor drives. In some departments where it is still customary to use group motors, a number of small units are used, instead of a few large ones. The illustration on page 161 being typical.

Some of the mills so equipped have been in operation for four or five years, and the results justify the expectancy of an increased production, averaging probably 8 per cent, together with a very low maintenance cost. Incidental to the main advantage of increased production, are such minor but well worth while advantages as greater cleanliness, better lighting conditions, and flexibility of operation.

A very strong argument for purchased power is the flexibility it provides for the growth of industrial plants. A plant can be built requiring only a very small amount of power, with the feeling that unlimited extensions can be carried out without fear of power limitation.

The treatment accorded by the power companies to their customers has in general been quite liberal, and the rates have not been increased in line with the increased cost of other commodities.

The illustrations show typical examples of generating plants, distribution lines, substations, and motor installations.

# LIBRARY NOTES AND BOOK REVIEWS

APPLIED CALCULUS. Principles and applications. Essentials for Students and Engineers. By Robert Gibbes Thomas. D. Van Nostrand Company, New York. 1919. Cloth, 5 x 7 in., 490 pp., 45 exercises, 166 illus., $3.

In this college text-book, the author has aimed to make clear the basic principles of calculus and to show that fundamental ideas are involved in familiar problems. Effort has been directed toward the attainment of a working and fruitful knowledge of the elements of the subject and an ability to use it efficiently.

ATLAS AMÉRICA LATINA. General Drafting Company, Inc., New York, 1919. Cloth, 11 x 16 in., 196 pp., $20.

The detailed maps which form the most important part of this atlas have been prepared from the best authorities. They cover Mexico, Central and South America and the West Indies and are drawn to a uniform scale of fifty miles to the inch. Towns, railways, administrative divisions and physical features are clearly shown and a geographic index is provided.

The atlas also contains maps showing steamship routes, ocean currents, natural vegetation, prevailing winds, summer and winter rainfall and temperatures, principal products, principal minerals and the language divisions of Latin America, as well as charts showing the foreign trade of each country and descriptive articles concerning their possibilities and conditions.

AVIATION. Theorico-Practical Text-book for Students. By Benjamin M. Carmina. The Macmillan Co., New York, 1919. Cloth, 8 x 5 in., 172 pp., 92 diagrams, table, $2.

The author discusses the theory of flight and the construction, rigging and maintenance of airplanes, and gives suggestions on flying. The treatment is non-mathematical throughout, but an appendix is included which gives the mathemataical analysis of the laws governing flight.

DER BAU DES DIESELMOTORS. By Kamillo Korner, Julius Springer, Berlin, 1918. Cloth, 11 x 7 in., 350 pp., 500 diagrams, 30 marks.

The present work is devoted to the practical details of Diesel engine construction and omits all discussion of theoretical considerations. The various elements of engines are discussed in detail and examples of the usual designs are given.

CHILTON TRACTOR INDEX, Vol. 3, No. 1. Chilton Company, Philadelphia, 1920. Paper, 10 x 7 in., 478 pp., $1.

The Tractor Index is a directory of the farm tractor industry of the country, which gives in tabulated form the specifications of the tractors on the market, lists of tractor and power farm machinery manufacturers, makers of parts and accessories and farm trucks. In addition to these directories, the volume includes a collection of data and articles on power farming, farm lighting plants, tractor testing, tractor standards and other topics of interest to users and dealers.

CONTROLLERS FOR ELECTRIC MOTORS. A Treatise on the Modern Industrial Controller, together with Typical Applications to the Industries. By Henry Duvall James. D. Van Nostrand Company, New York, 1919. Cloth, 8½ x 5 in., 307 pp., 259 illus., $3.

The aim of this volume, which is based on a series of articles published in the Electric Journal during 1917 and 1918, is to provide students, operating engineers and users of electrical apparatus with a good general account of their principles of operation.

THE ELEMENTS OF ASTRONOMY FOR SURVEYORS. By R. W. Chapman. J. B. Lippincott Company. Philadelphia, 1919. Cloth, 5 x 7 in., 247 pp., 36 diagrams, $1.75.

Although there are several excellent books on surveying that deal more or less thoroughly with astronomical observations, the author of this work believes that there is a need for an elementary work suitable for the student and for the surveyor who is taking up astronomical observation for the first time. The present work is an attempt to provide an elementary exposition, not only of the practical methods of observation and computation, but also of the main principles that must be thoroughly understood if the surveyor is to be master of his profession.

GOVERNORS AND THE GOVERNING OF PRIME MOVERS. By W. Trinks. D. Van Nostrand Co., New York, 1919. Cloth, 6 x 9 in., 264 pp., 140 illus., $3.50.

This volume aims to present the principles and essentials of the subject in such a manner that the reader may be able to judge existing and future types of governors as well as the properties of prime movers with regard to regulation. It has been written to provide a text-book on the general subject of governors and governing, upon which, the author states, there is no book of any consequence in English. A selected bibliography is included.

THE GREAT LAKES RED BOOK. A list of over 1,000 Vessels of the Great Lakes, with the Name of Owner, Captain and Engineer of each Vessel. The Marine Review, Cleveland, 1919. Paper, 5 x 3 in., 159 pp., $1.

Gives a list of the fleets upon the Great Lakes, as well as owners, captains and engineers. There is also a table showing capacity of ore-carriers and a port directory.

MINING MATHEMATICS SIMPLIFIED. By James Wardlaw, Sr., Scottdale, Pa. Cloth, 7 x 5 in., 189 pp., $2.50.

This volume gives simple arithmetical solutions for the problems which confront the practical miner. The author is engaged in preparing miners for examinations for certificates as fire bosses, foremen, etc., in the Pennsylvania bituminous coal-field and has written this text for their use.

THE NEW AMERICAN THRIFT. The Annals of the American Academy of Political and Social Science. Vol. 87, No. 176, Philadelphia, 1920. Paper, 248 pp., $1.

Contents: Introduction; Thrift for the Individual and the Family; Thrift for the Nation; American Needs for Capital, Typical Examples; Thrift in Resources and Industry, Typical Examples; The Investment of Savings; The Promotion and Practice of Thrift in Different Countries; Suggestions for Promoting Thrift.

A symposium by a number of bankers, economists and business men on the desirability of thrift in national and individual life. The problem is broadly presented and discussed in its various phases by experts.

A TEXTBOOK OF RAND METALLURGICAL PRACTICE, Vol. 2. By Ralph Stokes, Jas. E. Thomas, G. O. Smart, W. R. Dowling, H. A. White, E. H. Johnson, W. A. Caldecott, A. McA. Johnston and C. O. Schmitt. Second edition, revised. Charles Griffin & Company, Ltd., London, 1919. Cloth, 9 x 6 in., 462 pp., 467 illus., 25 shillings.

This work was prepared in 1911 by a body of technical men actively engaged in current metallurgical practice upon the Witwatersrand, to provide a textbook giving a detailed account of the methods in actual use in that district, both mechanical and chemical.

The present volume, which discusses the design and construction of reduction plants and the transport of materials is by C. O. Schmitt. No radical change in practice, has occurred since the publication of the previous edition, but considerable progress in the development of the use of heavy stamps, and tube mills, and a corresponding advance in the methods of classification and amalgamation have been made. This revised edition has been prepared in the light of these improvements.

# MECHANICAL ENGINEERING
### THE JOURNAL OF THE AMERICAN SOCIETY OF MECHANICAL ENGINEERS

| Volume 42 | April 1920 | Number 4 |
|---|---|---|

## Reliability of Materials and Mechanism of Fractures
### Behavior of Materials Under Test and in Actual Use Discussed by Distinguished French Engineer
#### By CHARLES DE FREMINVILLE[1], PARIS, FRANCE

*It is a common experience that the behavior of materials in practice does not always agree with their behavior under test; designs worked out from data secured from test pieces have not always proved satisfactory, and actual machines have had to be modified from the experience of practice. The differences between test predictions and actual results are set forth in the following paper, and it is shown that the reasons for the discrepancy between test and actual conditions can be determined by an analysis of the mechanism of fractures of the test pieces and of actual specimens. Such an analysis reveals the fact that the mechanism of the fracture follows certain laws which can be formulated. The paper also deals with methods of designing members so that stresses in them do not conflict with the laws of mechanism of fracture. Pieces can be so designed and so assembled that the stresses in them largely avoid the conditions of fracture, and it is significant that when such design and assembling is carried on, the behavior of the actual machines is much more similar to the behavior of the test pieces themselves.*

W HEN the author was invited to read a technical paper before The American Society of Mechanical Engineers the first subject which came to his mind was, The Reliability of Materials, chiefly because of his previous experience in the use of metals and with the testing of metals. If that word "reliable" impressed itself upon his mind, it was certainly because it is a word of which he had realized the full meaning in America, where the people are always looking for something reliable: reliable information, reliable products, reliable men; a country in which he found so many reliable friends, and which has proved herself so prominently and so thoroughly reliable in her immense effort to save the mutual ideal of civilization of France and America, which cannot conceive Science associated with an ideal other than Liberty.

Reliability is the quality we desire in the metals we use, but it is very difficult to ascertain it through a short test. The reliability of materials is like the reliability of friends. Whatever brilliant qualities they may exhibit, their reliability is known only in the long run, so that in the case of an apparent failure, the qualities of a metal, like those of a friend, should not be questioned too quickly. It is not enough to know the intrinsic qualities of the metal; a very careful study of the circumstances under which the failure occurred must also be made, and it is this fact which has made the study of fractures and of the mechanism of fractures one of great interest to the author.

During the early days of the automobile industry the problem of resistance of materials began to be considered in a new light. The first motor cars that were built were designed with factors of safety then generally accepted for the resistance of medium-

carbon steel. But the engineer soon came into contact with the sportsman and was tempted to put into the gearing a stress much greater than the one for which it had been made. One of the problems to be solved, therefore, was to secure a material which could safely be used for gears.

At that time steel makers had produced, among certain other new products, ternary steels, which showed brilliant promises of new qualities if the results of the laboratory tests to which they had been submitted could be taken as a proof of their reliability.

Viewing the question of material for motor cars in a general manner, the constituent parts appeared to belong to two distinct classes. The first included such parts as gears, crankshafts, connecting rods, etc., which had to withstand severe normal shocks. Also only a small amount of wear was allowable and the class of accidents to which they were liable was limited, being chiefly those of seizing or an encounter with stray nuts or bolts.

The parts of the second class did not have much stress imposed upon them in ordinary service, but they did have to meet with what can be termed "normal accidents," and were therefore expected to bend to a certain extent without ceasing to fulfill their duty. To this class belonged front axles and, more generally, all the parts connected with the steering mechanism. For these, the "normal accident" can happen very frequently. It is almost impossible to say that the wheel of a car will not come rather abruptly into contact with a curbstone, or even with the wheel of another car. In such case something must yield or break. It is of course but natural to hope that something will yield and that this yielding has been foreseen; also that it will take place on a piece easily seen and easily replaced, and that the material of which the piece is made will not cease to be reliable.

The impact test, to which attention was directed at that time in France by M. Considère and later by M. Frémont, seemed to be a very convenient method for ascertaining the characteristics of the parts of each class. The impact test, however, was devised primarily for soft steel and had to be adapted for very hard steel. But by making use of the impact test it was possible to select for the parts of the first class a grade of steel which only fractured after a very small amount of deformation, and this could be considered as a very excellent quality for the parts of this class.

The parts of the second class were expected to bend under the impact test to a very great extent, and it was easy to find wrought iron or mild steel possessing this characteristic, although the reliability of these materials for the use considered is a matter open to discussion. But here again the problem became complicated, because there were those who seemed reluctant to admit that some of the parts had to bend easily under certain circumstances and that a limited amount of toughness was not desirable, inasmuch as the steel maker pretended to have grades of steel uniting in a rare degree ductility with toughness.

The tensile tests of these steels showed a high ultimate strength and a great elongation. The impact test showed that a consider-

[1] Architectural and Industrial Engineer. Consulting Engineer, Schneider Works. Hon.Mem.Am.Soc.M.E.

Presented at the Annual Meeting, New York, December 1919, of THE AMERICAN SOCIETY OF MECHANICAL ENGINEERS. All papers are subject to revision.

able amount of energy had been absorbed by the test piece, which gave a great amount of distortion. It is evident that if it had been possible to have a diagram of the impact test, showing at every moment what was the amount of stress corresponding to a given deflection, the matter would have been made very much clearer. But since there was no machine that would give such a

by drop tests and quick-bend tests on notched test pieces. In these diagrams the ordinates of curves $A$ are proportional to loads corresponding to deflections shown on the abscissa scale. In curves $B$ (which are the integrals of curves $A$,—ordinates $y_1y'_1$, $y_2y'_2$, etc., being proportional to surface $Y$ and so on) the amount

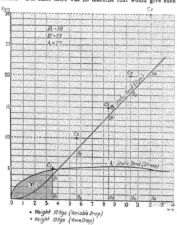

FIG. 1 DIAGRAM FOR COMPARING THE RESULTS OF STATIC BEND
AND DROP BEND TESTS

diagram, a substitute was found in a static bending machine registering the amount of energy taken in the deformation process, for it was possible to ascertain that in all the cases which had to

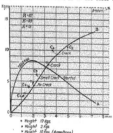

FIG. 2 DIAGRAM FOR COMPARING THE RESULTS OF STATIC BEND
AND DROP BEND TESTS

be investigated, the number of kilogram-meters absorbed was practically the same as in a quick bend or in a drop test. So far as the appearance of the fracture was concerned there was no perceptible difference in either case, which showed that the metal which seemed to possess ductility and toughness did not possess these characteristics to the extent believed.[1]

Figs. 1 and 2 are diagrams for comparing the results obtained

FIG. 3 RESULT OF SHARP SHOCK ON A GLASS VESSEL

FIG. 4 DEVELOPMENT OF THE FRACTURE RESULTING FROM A SHOCK
ON A THIN SHEET OF GLASS

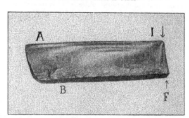

FIG. 5 FRACTURE SURFACE IN A SLAB OF GLASS

of energy consumed in the bending test is given for each value of the deflection. The ordinates $y'_1C_1$, $y'_2C_2$, etc., are proportional to the amount of energy absorbed by the drop test for the same

[1] The author has recently learned that the same idea of interpreting a drop test by means of a diagram given by a static bend test has been successfully used in this country for testing rails, and he believes that it can be used in many instances.

FIG. 6 SMALL SPLINTERING FOCUS—GLASS

FIG. 7 SPLINTERING FOCUS—GLASS

FIG. 8 FRACTURE IN JUDEA BITUMEN

deflection, as measured in the bend test, and can be easily compared with $y_q y'_q$ $y_q y'_r$. The energy absorbed in the drop test appears to be only slightly greater than that absorbed in the quick-bend test, but the process has followed the same course in both cases, and the difference may be attributed to errors in the experimental work.

Fig. 1 is a diagram of a test of wrought iron or mild steel in which the amount of stress necessary to obtain a given deflection remains constant over a long range. This is the type of material which can be expected to give good results for parts of the second class.

Fig. 2 is a diagram of a test of a steel with a certain amount of ductility, in which the energy consumed for a given deflection appears to be greater than in the case of Fig. 1, and which was offered for parts of the second class. It should be noted that in

FIG. 9 FRACTURE IN GLASS

Fig. 2 curve $A$ is absolutely different from curve $A$ of Fig. 1, as it shows that a maximum amount of the energy is absorbed after a small deflection has been obtained. This maximum is the point where the piece begins to show a crack, which of course did not exist in the test piece whose diagram is shown in Fig. 1.

Parts of the second class made of this metal proved a decided failure. Even without experiencing the "normal accidents" which they were expected to withstand, they showed a great tendency to crack without the slightest deformation, which method of fracture has been called *fissillité* by the late M. Brustlein, of the Jacob Holtzer firm, Unieux, France.

By looking closely into the fracture of the broken test piece, which was supposed to possess ductility and toughness, it could be seen that most of the distortion, or flowing of the metal, had taken place not before the fracture was started, but during the process of extension of the fracture, and this gave to the amount of energy recorded an altogether different meaning. In this case the failure of the material could be traced to a bad interpretation of the test, but other failures were encountered which could not be traced to the quality of the metal used. Typical among these

FIG. 10  FRACTURE IN JUDEA BITUMEN

was the sudden snapping of levers, under conditions which could be specified but which were difficult to explain. The impact test, like others, could not give a clue as to reliability in certain circumstances.

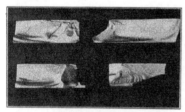

FIG. 12  FRACTURES IN TOOL STEEL.

### THE STUDY OF FRACTURES

The uncertainties in the behavior of material, the propagation of a crack without any distortion of the piece, the sudden snapping of a rod without any perceptible alteration in the neighbor-

FIG. 13  RENTS FORMED UNDER THE MAIN SURFACE IN GLASS

hood, even the changes which were produced in brittleness under heat treatment, appeared to be difficult to account for, and it

seemed to the author to be worth while to add to the theoretical views of the question by making, as far as possible, a thorough study of the fracture itself, which would tell, in some cases, the story of the failure.

It is frequently believed that fracture is a form of failure of such peculiar character as to escape all rules. However, the attention of the author was first attracted by the very regular fractures of sandstone, which is used extensively in France for pavements, and which so resemble one another that they seem to be real diagrams of what has happened during the rupture. Moreover, a certain number of peculiarities of the fractures were found to be common to both sandstone and steel. It was also noticed that these same features were found in glass fractures, in which they could be observed with great accuracy; also that they were still easier to observe in Judea bitumen, where the breakage was produced very easily. A further study has shown that the fractures developed in an identical manner in these various materials.

It is not possible in the limited time at the disposal of the author to go thoroughly into the subject of the study of fractures, and as a result attention can only be called to some of the main features. Fig. 3 shows a fracture developed in a glass vessel by a sharp shock. In this should be noted the symmetry of the surfaces originated and the peculiar disposition of the starting points of these surfaces from both ends of a small line to which the surfaces are tangent. Fig. 4 shows the development of the fracture resulting from a similar shock in a thin sheet of glass. Here great symmetry is seen, the network also originates from a small line, and some of the surfaces extend entirely through the glass, some only partially.

FIG. 11  SCHEME OF FRACTURE IN JUDEA BITUMEN

FIG. 14  SCHEME OF TEARING, WITH CRACKS THAT ACCOMPANY IT

Fig. 5 is taken from a slab of glass broken in the same manner. It shows one of the surfaces dividing the slab into pieces, which was also partially cut by other surfaces. In this particular case the blow was given at $I$ on the top of the slab, and it will be seen that the small line, noted in the previous figures as the origin of all the cracks, is situated at $F$ at the opposite side of the surface. It presents a very remarkable appearance and the general disposition of the surface induces us to believe that the fracture originated in that region.

Figs. 6 and 7 show in great detail other specimens of very peculiar surfaces, which illustrate very clearly that there is a starting point of the fracture. These examples enable us to see that one of the main surfaces cutting the test sample has not expanded at one stretch, but is formed by surfaces overlapping each other like the blades of a fan. This is made still more evident by the fracture shown in Figs. 8 and 9. These surfaces overlap each other at distances often smaller than a hundredth of an inch, and by making use of the method called recoupement (recutting) by the geologists, it is possible to follow the successive order in which the various elements of the fractures have developed.

Fig. 11 shows in greater detail one of these surfaces, and enables us to point out some of its special features. Part of the surface is bright and part dull. A study of the bright part can be made from Fig. 10, which shows a standard fracture in Judea bitumen. It is very easy to follow the surfaces overlapping each other, as is shown by the scheme in Fig. 11. The general arrangement of the surfaces bears a very close resemblance to the surface of water as it runs in a culvert, passes over a dam, or expands in a shallow pond.

The same arrangements of fractures, overlapping surfaces, foci, etc., which are easily observed in glass and Judea bitumen are also met with in tool steel, as shown in Fig. 12.

We find that the dull part gives the impression that it has been rent or torn away, while the bright surface seems to have been carefully cut. Fig. 13 shows how the dull surface is originated by small rents coming from under the surface. If now we come back to the focus of Figs. 6, 7 and 8, and look for the point where it has been originated we find that it also lies under the surface.

Fig. 14 shows what takes place in the tearing of a tough substance such as gelatine. The cracking *bb* travels ahead of the parting at *a* and prepares the way for its advance.

Fig. 15 is another and striking example of the tearing of a tough material. The parting at the edge of the tool has been made possible by a "cracking" similar to that shown in Fig. 14. The sketch in Fig. 16 shows the scheme of the formation of the steel chip.

But how can a bright surface be produced? How can the cutting process be carried on without the sub-surface cracking? How can surfaces be cut which run parallel within such small distance as the ones we have noted? The only explanation which seems to be satisfactory is that the cutting power is a vibratory motion of the surfaces of the fracture, so long as they do not join each other.

 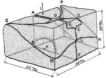

FIG. 16  SCHEME OF STEEL CHIP

FIG. 18  OUTLINE OF THE FRACTURE OF A BITUMEN BLOCK

Fig. 17 shows a very typical arrangement of the fracture surfaces in a bitumen block and Fig. 18 gives an outline of the different parts. The same arrangement is constantly found in steel fractures, Fig. 19 showing a similar fracture in a piece of a steel rail broken under a drop test.

### STRESS AND STRAIN

What are the conclusions to be drawn from this brief study of the mechanism of fractures? How can it pretend to be a contribution to the science of resistance of materials? In a general way, by showing the succession of the phenomena which have taken place during the spreading of the fracture, it affords an experimental proof that the straining effects are more detrimental to the material than mere stress. This is in accordance with what has been shown by a purely mathematical study of the question.

In case of an impact fracture, be it in a piece of glass or in a piece of steel, we find the starting point of the fracture under the surface, although the maximum stress must certainly be located on the surface itself.

In the case of a piece subjected to tension only if a comparison is made with the fractures already considered it will show that fracture has originated in the interior of the test piece (see Fig. 20). The surface of the test piece itself sustains a very important stress, but without being subjected to the strains resulting from the contact of layers undergoing uneven stresses, flowing, so to speak, to give the reduction of area.

The predominance of the internal or straining action is no less

striking in the case of a rent where the straining effect travels in advance of the stress which could be originated at the separating

FIG. 15  STEEL CHIP

FIG. 17  FRACTURE OF A BITUMEN BLOCK

FIG. 19  PIECE OF STEEL TIRE BROKEN IN A DROP TEST

point. (See Fig. 14.) This condition should induce the metallurgist to pay more attention to the causes which produce straining in other words, to the tensions which may exist in the pieces after being cast or forged. This is illustrated by the changes in brittleness which occur in glass undergoing heat treatment. The heat treatment does not bring about any change in its molecular arrangement, but the glass coming out of the furnace is very brittle, and after being properly annealed is so much less so that it is in the so-called unbreakable state; that is to say, it can sustain shocks and bending to a much greater extent than ordinary glass.

The first condition is one of very uneven tension; the second, one without tension; and the third, one which has a beneficial tension.

FIG. 20 TEST PIECE RUPTURED UNDER TENSION; FRACTURE ORIGINATED ALONG THE AXIS

This unevenness of tension also exists in steel, irrespective of the chemical or other changes in structure. An example of the effect of these internal tensions is given by Fig. 21, which shows a fracture in a rail. It is evident that the fracture was started in the neighborhood of the neutral axis, a fact which could not be accounted for if the piece had been under normal conditions;

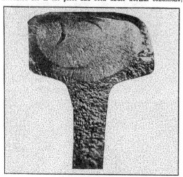

FIG. 21 TENSION FRACTURE IN A STEEL RAIL

therefore the causes of a fracture of that kind must be traced to a tension most likely produced in the rolling mill. This case illustrates the mode of cutting which produces bright surfaces in glass, and which can only be explained by the presence of a vibratory motion of the surfaces when they are apart from one another.

In conclusion the author trusts he has created the impression that a methodical study of fractures is of the greatest importance, especially as it leads to a knowledge of the causes of failures which cannot always be attributed to the intrinsic qualities of the metal itself.

## Large Gas-Well Fire Extinguished with Explosives

The successful application of explosives in extinguishing a fire at one of the Standard Oil Co.'s wells near Taft, Cal., reported in a recent issue of the *American Gas Engineering Journal*, suggests the possibility of greatly simplifying the present expensive methods of extinguishing gas-well fires, oil-well fires and oil-reservoir fires.

Drilling operations with a rotary drill were in progress in well No. 7, in the Elk Hills, in the Midway Oil Fields. The well was at a depth of 2135 ft. in 10-in. casing when a supposed water sand was encountered. To test this sand for water the mud in the well was bailed down to relieve pressure and to allow any water present to accumulate in the well. During the course of bailing the well blew out under tremendous pressure and gas began to flow at the rate of 125,000 cu. ft. per min. An attempt was made to get the well under control by closing a 12½-in. gate valve with which the casing was equipped, but the shale and sand carried by the stream cut out the seat of the valve while the gate was being closed and the well continued to blow. Soon afterward the friction caused by shale and sand had produced a red-hot heat on the casing, and the gas, igniting, burst into a column of fire, gushing 300 ft. skyward. Its maximum width of about 50 ft. was reached after the flame had traveled about 100 ft. upward. The gas emerged from the casing with such force that for a distance of 12 ft. about the casing there was no flame. The derrick, 100 ft. in height, was immediately consumed and all that remained was a twisted mass of iron and steel. Metal at the casing melted, while within a radius of 50 ft. it was raised to a white heat. Everything within hundreds of feet was burned and baked. It is reported that the fire could be seen from a distance of 80 miles and that the roar from the escaping gas could be heard at a distance of ten miles.

Attempts were made to extinguish the fire, first with ten boilers, and subsequently with sixteen and twenty boilers. With the twenty boilers there were twenty-one 3-in. lines and nine 4-in. lines and eleven rotary pumps throwing steam, water, mud and carbon tetrachloride. Nearly half a million dollars were expended in vain attempts to extinguish the fire. In the meantime the casing above ground had blown off, the well had formed a crater and the flame was down to the ground level.

As a last resource it was decided to suspend a package of Giant blasting gelatin upon cables in such a manner that the package could be safely brought to and exploded at the edge of the fire at a distance of about 40 ft. above the ground. Two derricks approximately 40 ft. high were built on either side of the burning well and so far away that the heat would not destroy them. A 9/16-in. steel cable was attached to the top of one of them, then laid around the well outside of the heat zone to the other derrick where it was run through a pulley at the top and the end attached to the drum of a small winding engine underneath. The explosives, bound in one package and covered with heavy sheetings of asbestos, were securely attached to this cable at the same distance away from the first derrick as the burning well was. The engine under the second derrick began drawing the cable taut, which pulled the explosives toward the well and at the same time hoisted them. When they had reached the proper spot they were exploded electrically. All of these operations had to be done very quickly so as to bring the explosive charges to the necessary point before they were ignited or exploded by the heat from the burning well.

When the package exploded the fire was divided into three sections. The section at the point of explosion was moved bodily away from the column of gas in a horizontal direction, the upper section was blown upward and beyond danger of reignition while the lower section was snuffed out, the flow of gas being temporarily checked.

An idea of the enormous amount of gas flowing from this well may be formed by noting that it could conveniently supply eight cities of the size of San Francisco, which alone consumes 22,000,000 cu. ft. of gas each day.

# Physical Basis of Air-Propeller Design

By F. W. CALDWELL[1] AND E. N. FALES,[1] DAYTON, OHIO

*Description of wind tunnel at McCook Field, Dayton, Ohio; phenomena discovered by its use; analysis of propeller theory and application thereof to fan design in the light of facts established experimentally; methods for visualizing air flow.*

*The wind tunnel at McCook Field is notable first for the high velocity of air used (500 m.p.h.) and second, for means adopted to visualize the air flow. This latter is done by maintaining the humidity of the air in the tunnel at a certain predetermined level and by providing opportunities for its condensation where it will show the character of the air flow.*

*The study of air flow by these means gave new data on vortex formation created by the presence of aerofoils and made it possible to form a clearer conception of the conditions surrounding the operation of propellers.*

*The value of such investigations as can be made with the McCook Field equipment lies in the fact that hitherto the theory underlying flight phenomena has been purely rational and not directly applicable to engineering design because of the absence of empirical measurement of flight vortices. Because of this, it was impossible to predetermine the performance of aerofoils of new shapes, speeds and sizes without first building a model and determining the particular coefficients applicable to the new design experimentally. Data obtainable in a wind tunnel having provision for visualizing the air flow are, however, capable of supplying this deficiency.*

*The authors show the application of some of the data obtained to the design of blowers and fans.*

MEMBERS of The American Society of Mechanical Engineers as well as most of the older engineering professions have shown a keen interest in aeronautical problems and in the progress of the scientific side of aviation. This interest is no doubt partly due to the romance of seeing realized mankind's age-long desire to imitate the birds in their flight; and partly to the patriotic hope that America will become predominant in a field which owes its existence to two American inventors.

This interest is very valuable to aeronautical engineers, as we feel that association with the older professions ought to enable us to keep at least one foot on the ground. At the same time it is felt that the information gained in aeronautical research may prove helpful in some ways to the mechanical engineers.

It is the purpose of this paper to summarize our conceptions of flight phenomena as developed by the use of the high-speed wind tunnel at McCook Field. The understanding of the fundamental aerodynamical phenomena upon which flight depends has been rather vague throughout the past, simply because of the invisibility of the air. Our knowledge of these phenomena has been based on deductions made from observing the forces produced. (The variation of the air resistance has been well understood for various types of wings, and also for solid bodies of geometrical shape.) To evaluate these forces, however, the results only have been analyzed—not the phenomena themselves. Measurement of the dynamic and static pressures in a given air flow has contributed somewhat to our knowledge of what goes on in the air. The use of threads, solid particles, such as smoke, etc., has also been universal for determining the direction of given portions of an air flow. Such methods, however, have barely scratched the surface of the problem and have been of small value to the aerodynamical analysis of the mathematicians.

A means of actually visualizing the air is therefore of greatest interest; and while its immediate application has been to the specific case of studying air flow past aerofoils, many other uses suggest themselves wherever air in motion constitutes an en-

[1] Air Service, Engineering Department, McCook Field.
Abstract of paper to be presented at the Spring Meeting. St. Louis. May 24 to 27, 1920, of THE AMERICAN SOCIETY OF MECHANICAL ENGINEERS. Advance copies of the complete paper may be obtained gratis on application. All papers are subject to revision.

gineering or scientific factor. Thus, for instance, the study of meteorology involves a knowledge of the circulation of air currents on a large scale over the earth's surface; this knowledge, too, is imperfect because of the invisibility of the medium concerned, and can be greatly improved by application of means for visualization. Moreover, the flow of air in ventilation and other engineering processes is a question which may frequently require analysis by a method of visualization.

It is thought that the general engineering profession will quickly deduce from the experiments recorded in this paper principles which can be of service in their more commercial fields. Therefore it has been our aim to summarize the main features of these experiments without going too deeply into the technical details necessarily of interest only to the aeronautical engineer.

The experiments have resulted in discoveries which may be briefly summarized as follows:

The flow of air, whether in a duct or past an aerodynamic body, is subject to parasite energy losses connected with accidental whirls and eddies, and also to internal motion on account either of

FIG. 1 PHOTO OF TIP VORTEX, CAMERA AXIS BEING COINCIDENT WITH VORTEX AXIS

the shape of the containing duct or the aeronautical body. Analysis of these motions in the air can be facilitated by visualizing them. The writers' method for producing visualization involves condensation of the moisture contained in the air. This moisture, when turned into visible vapor by suitable means, takes up shapes dependent upon the motion and pressure gradient of the air and justifies us in considering that the air itself is thus visible. (See Fig. 1).

When the air is made to flow past an aerodynamic body, special shapes of flow result and can be visualized by our method. It has been shown in our experiments that when the speed changes, the characteristics of the internal motion in the air alter. This alteration has been a source of much uncertainty to aeronautical engineers in the past; but the combination of visual study of these changes of air flow, together with quantitative analysis of the forces involved, has made it possible to throw much light on the underlying structure of the air as it applies to dynamic flight and propulsion.

In our experiments, then, we introduced small wooden wings (aerofoils) into the air current of the wind tunnel and studied their performance; intending to interpret the results for the purpose of propeller design.

## WIND-TUNNEL APPARATUS USED FOR OBSERVING AEROFOIL COEFFICIENTS AND AIR FLOW

The McCook Field wind tunnel (Fig. 2) was designed by the writers for advanced study of aerofoil coefficients used in propel-

213

FIG. 2  McCOOK FIELD WIND TUNNEL, GENERAL EXTERNAL VIEW

lers. It is a departure from the accepted type of wind tunnel in two ways: first, it has a speed higher than elsewhere attained for similar purposes—500 miles per hour at 200 hp.; second, it makes possible the visualization of aerodynamic phenomena.

It seems wise to summarize briefly the general function of a wind tunnel, which is the laboratory apparatus of the aeronautical engineer. An airplane, when once in the air, is very difficult to analyze; neither the pilot nor an observer on the ground is able to gage accurately the speed, angular attitude, nor the air forces acting. So far, no thoroughly accurate instruments for accomplishing full-flight analysis have been perfected. Projects have been proposed for attaching a full-sized airplane to a moving electric locomotive properly equipped to record accurately all the forces produced; but such a plan has drawbacks. It is found that the best, most accurate and cheapest way to analyze airplane characteristics is by means of the wind tunnel. This is a confined artificial blast of air which is made to blow against a model wing or airplane. The model is supported in the air blast by a thin rod reaching into the wind tunnel from outside, where it connects to a delicate scale. Thus all forces created on the model are transmitted to the scale where they can be measured.

It is found that the forces due to the air current blowing against fixed objects are the same as though the object were moved through still air, provided the air current is smooth and without eddies. Elimination of eddies is therefore an important requirement in a wind tunnel.

The forces found in a wind tunnel apply to full-sized machines, and wherever aeronautical design is carried on the wind tunnel is essential. Besides the McCook Field wind tunnel there are in America about a dozen others. In Europe there are many more, the British National Physical Laboratory employing seven. The largest in existence is now being built at Paris, 13 ft. in diameter and of 1000 hp. capacity. The typical wind tunnel has a stream of air 15 to 50 sq. ft. in cross-sectional area, flowing at a velocity of from 30 to 90 miles per hour.

## THE QUESTION OF WIND-TUNNEL SPEED

The speed and size of a wind tunnel have a recognized importance in the interpretation of the results obtained. The usual coefficients of aeronautical engineering are determined by tests on a model aerofoil of from 3 to 6 in. chord at velocities of 45 to 130 ft. per sec. The data obtained must be applied in the full scale of areas a hundredfold greater in the case of wings, and to velocities ten times greater in the case of propellers. Thereby may be a discrepancy between model and full-scale coefficients, and this is spoken of as "scaling effect."

*Scaling Effect.* While scaling rules and empirical factors worked out in practice have enabled us to produce very fair results, there is decided room for improvement. Indeed, so meager is the available information on the scaling effect, as treated under the "law of dynamic similarity," that a growing tendency has appeared among aeronautical engineers to regard

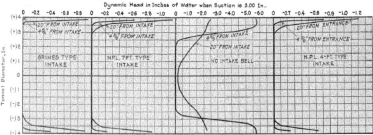

FIG. 3  CHART OF VELOCITY TRAVERSES FOR VARIOUS INTAKES

the classical coefficients $K_x$ (coefficient of drag) and $K_y$ (coefficient of lift) as inadequate.

It has been almost universally the practice to write

$$L = \rho/g K_y A V^2$$
$$D = \rho/g K_x A V^2$$

where $L$ is the lift (vertical lifting force), $D$ the drag (resistance), $A$ the area of the supporting surface, $V$ the velocity of advance, $\rho$ the density of the air in weight units, $g$ the acceleration due to gravity, and hence $(\rho/g)$ the density of the air.

It is well known as the result of experience that the values of $K_x$ and $K_y$ vary somewhat with velocity and also with the size of the surface under consideration. If $l$ represents one of the linear dimensions of the surface and $v$ the coefficient of kinematic viscosity, it is assumed, according to the law of dynamic similarity, that values of $K_x$ and $K_y$ are functions of $Vl/v$. Since $v$ is usually considered constant, it is customary to compare the product $Vl$ for experiments in a given medium.

The wind tunnel designed by the writers permits the attainment of a $Vl$ product of 60 ($V$ in ft. per sec., $l$ in ft.). While this figure is twenty times less than the corresponding full-sized wing values, it is only four to eight times less than full-sized propeller values.

As a matter of interest, tip speeds of some of the propellers in actual use are given below:

| Airplane | Engine | M.p.h. |
|---|---|---|
| USD-9 | Liberty-12 | 650 |
| VE-7 | Hispano-Suiza, 150 hp | 545 |
| Thomas-Morse | Le Rhone, 80 hp | 380 |
| Verville Chasse | Hispano-Suiza, 300 hp | 600 |
| Roché XB-1-A | Hispano-Suiza, 300 hp | 625 |
| Curtiss JN-4 | Curtiss OX-5 | 420 |
| DH-9 | Rolls-Royce, 375 hp | 430 |

It would be ideal if a wind tunnel could be built large enough for testing propellers and airplanes of full size and speed. Such apparatus would, of course, be enormous and of prohibitive cost. We must therefore be content with wind tunnels whose $Vl$ product is smaller than the full-flight value. In designing a tunnel

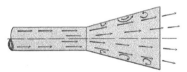

FIG. 4  DIAGRAM SHOWING EDDIES IN CONE OF TOO LARGE ANGLE

we may have either large size and low speed, or vice versa. In general a large wind tunnel requires a large, expensive hall, while a high-speed tunnel requires high horsepower.

Thus in the McCook wind tunnel we have attained a velocity equal to full-flight values, and the size of model is correspondingly small. The results obtained, however, demonstrate greater significance than were the same $Vl$ product obtained with a larger model and smaller velocity. By means of the high velocity, discoveries of outstanding importance have been made, namely, the means of visualizing flight vortices, and the identification of changes in these vortices with simultaneous changes in the aerofoil coefficients.

### DESCRIPTION OF THE McCOOK FIELD WIND TUNNEL

The tunnel has the form of a venturi tube $18\frac{2}{3}$ ft. long, built of turned laminated wood. Its proportions and novel features have been the subject of exhaustive investigation and experiments on scale models.

*Intake Bell.* Thus the intake bell has a radius of curvature of

$1\frac{5}{8}$ times the throat diameter, such as best to take care of the *vena contracta* and insure best velocity distribution and parallelism in the air flow (see chart, Fig. 3).

*Throat.* The throat is unusually short—1.3 diameters—with resulting economy of power. The short throat and the location of the model close to the intake are justified by the study of the

FIG. 5  McCOOK FIELD WIND TUNNEL, SHOWING OPERATING PANELS

chart in Fig. 3, and is made possible by the elimination of the usual "honeycomb" and by proper intake-bell design.

*Expanding Cone.* The cone leading from throat to fan has an angle of 7 deg., the maximum value established by Eiffel. We have found that in cones of larger divergence the air flow does not fill the cone, a space of whirling and eddying air being left between the blast and the containing walls (see Fig. 4), with resultant loss of energy. The principle involved is that flowing air surrounded by solid walls loses less energy by surface friction than if surrounded by inert air.

*Straighteners.* To eliminate velocity fluctuations a novel straightener is used—the conventional "honeycomb" being eliminated. The straightener is placed 4 ft. down the cone and has four radial blades 4 ft. long, so located as to prevent inflow whirl by obstructing the formation of the pressure apex. An auxiliary straightener (Fig. 5) with flat radial vanes is located outside the intake; and together the two reduce velocity fluctuations from 15 per cent to 3 per cent, without appreciable power loss.

### POWER PLANT AND FAN MEASUREMENTS

Velocity is measured in terms of dynamic head minus static head. For measuring dynamic head there is an "impact" tube at the throat somewhat off the center of the tunnel axis. For measuring static head a perforated plate is used, set in flush with the walls. A differential manometer records the velocity head, the whole apparatus being analogous to the conventional pitot tube and Krell manometer.

Temperature of the air passing through the throat is calculated on the assumption that expansion is adiabatic from atmospheric pressure to pressure corresponding to condensation, and is polytropic below the latter pressure. Correct knowledge of throat temperature is, of course, essential, and it is necessary to develop a special method of thermometry for reading it. Present methods are inapplicable to its direct measurement, for a thermometer introduced into the air stream occasions more or less adiabatic compression of the air striking it, with consequent rise of temperature at the point of impact. (See chart, Fig. 6.) The most advantageous position for the thermometer is with the bulb down stream, where it is subject chiefly to skin friction rather than impact. Further investigations are being made of the matter.

The balances are of two types. The first one measures lift and drag on two separate instantaneous reading Toledo scales

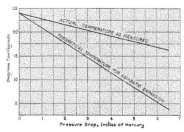

Fig. 6   Chart of Theoretical and Measured Temperatures

and is mounted upon a portable carriage (see Fig. 7). The spindle for the model projects horizontally and axially from this carriage into the mouth of the wind tunnel, carrying the model at its free end. The spindle terminates in a thin flat bar, the latter clamping a graduated disk which is rigid with the model at the center of the span. Three advantages of this type of balance are: (1) That instantaneous readings make it possible to synchronize balance and velocity observations and to practically eliminate the effect of velocity fluctuations; (2) The air forces can be qualitatively studied; as, for instance, in the case where a given setup has two values of $K_y$ or $K_x$. The balance can then be seen to change from one reading to another; (3) The method of support affords a highly accurate means of skin-friction observation.

The conventional supporting spindle, as adopted at the National Physical Laboratory, cannot be utilized under the conditions of these tests. In order to definitely delimit the effect of the supporting member, further developments are proposed where this effect will virtually be eliminated from the tests. The effect of the center support on the lift coefficient is not considered serious. This conclusion is based on experiments run at other laboratories where the effect of the support has been determined; also on a comparison of the present series of experiments with tests made elsewhere on a larger model supported at the end, for the same $l'l$ values. The effect of the center support on drag, however, is known to be large; the air-flow disturbance is visualized in Fig. 8, where it appears as a white tuft above the support.

*Power Plant.* The power plant consists of a Sprague dynamometer, capable of delivering 200 hp. for one-half hour at 250 volts and 1770 r.p.m. without overheating. The 5-ft. fan (Fig. 9) is made with a solid-center disk 40 in. in diameter, and has 24 blades 10 in. long. At the upstream side of the 40-in. disk a bell of equal diameter is fixed in the tunnel so that the air is let up to the annular discharge opening with a minimum of eddies. The operating efficiency of the whole tunnel is 75 per cent higher than has been usual for the determination of aerofoil coefficients in other wind tunnels. By efficiency is understood the ratio of kinetic energy of air stream at throat of tunnel,

minus the energy absorbed by the fan, all divided by the energy of the air stream at the throat.

Careful study of fan and cone design results not only in reduced losses but also in reduced noise. The question of noise has in the past been a serious objection to speeds greater than 70 miles per hour in wind tunnels. The roar of the fan is analogous to that of an airplane propeller, which usually makes more noise than does the unmuffled motor exhaust. For wind-tunnel use the combination of fan and cone adopted has brought about a considerable improvement, as indicated in the following tabulation:

From the operator's position:
| | | |
|---|---|---|
| Fan is noiseless at...... | 50 | m.p.h. |
| Starts to roar at......... | 60 | m.p.h. |
| Conversation easy at.... | 125 | m.p.h. |
| Conversation slightly forced at .......... | 155 | m.p.h. |
| Possible 12 in. apart at. | 240 | m.p.h. |
| Possible 4 in. apart at.. | 300 | m.p.h. |

*Fan.* The fan represents a very interesting union of airplane-propeller principles and blower-design principles, combining the high pressure usually associated with centrifugal blowers with the high efficiency of the airplane propeller. While essentially a propeller fan of aeronautical design, its 16-in. pressure head is such as would ordinarily be expected only from a centrifugal blower.

This application of airplane-propeller principles to blower design is significant to the ventilation engineer, who may by such means produce in wood, at comparatively small cost, blowing apparatus of high efficiency, high horsepower, high head, and reduced noise.

### Application of Propeller Design to Commercial Use

As a further illustration of the benefits to the ventilation engineer of airplane-propeller principles, the following may be cited as of interest:

Fig. 7   McCook Field Wind Tunnel, Balances and Intake

It has been found possible, by impressing special properties, to extend the usefulness of one of our most common and universal devices—the electric cooling fan. In the ordinary electric cooling fan referred to, the air flow is analogous to that of an airplane propeller rotating at a fixed point on the ground, and is shown in the sketch, Fig. 10-A.

The blast converges as it leaves the blades, reaching a maximum velocity a short distance from the fan, and gradually dissipating its energy as the flow lines expand along a small-angle cone. Thus, in such a blast the energy is concentrated in a narrow stream. Numerous experimenters have conceived the idea of placing vanes in front of these fans to diffuse the air and make it spread outward, but no practical success has been secured.

The problem is one of many minor ones which have found their solution as a side issue to the war activities of the Air Service. By judicious application of aerodynamic principles, a cooling fan has been developed which delivers a fan-shaped expanding blast, while moving as much air as the older conventional type of fan. See sketch, Fig. 10-B.

This fan directs its stream, not along a single axial line, but along a sector embracing 90 or even 180 deg. The advantages are obvious:

*First,* the energy of the blast is rapidly dissipated over a large area, causing a large, gentle circulation in place of a small, violent one.

*Second,* the blast spreads over an angle equal to that attained by the conventional oscillating type of fan, and accomplishes the same beneficial results without the unpleasant features associated with intermittent action.

*Third,* such a fan can produce a truly conical blast of which the fan-shaped expanding blast is only a special case; thus, the advantage of large cross-sectional area is secured from a small high-speed fan; and for purposes of overhead or ceiling use such a device is considerably cheaper than the conventional low-speed, large-diameter ceiling fan.

*Fourth,* the device involves only a simple, inexpensive addition to the conventional fan, may be readily applied to it, and causes no additional noise.

### VISUALIZATION OF FLIGHT VORTICES BY THE WRITERS' METHOD

The method of visualizing air flow discovered by the writers together with C. P. Grimes offers a solution of the fundamental problems of aerodynamics. This problem is the quantitative empirical measurement of the phenomena of fluid dynamics as applied to flight and air flow.

The accepted theory upon which flight has its physical basis is purely rational. It has not yet been directly applicable to engineering design, because empirical measurement of flight vortices has never been made. Therefore, the aeronautical engineer's

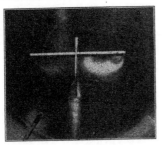

FIG. 8   TIP VORTEX, HIGH-LIFT RÉGIME

use of aerodynamics is largely according to the cut-and-try method. He cannot, on the drafting board, depart from known shapes, speeds or sizes without first building a model and determining the coefficients applicable to his new design.

To illustrate this point it is only necessary to refer to the simplest case, that of an airplane wing. We can measure the

coefficient of force on a small model of this wing to an accuracy of 1 per cent. But we do not know definitely how the accuracy is changed by scaling up to full size, or to full speed. We cannot, without tests, predict the change of coefficient to be expected when the wing shape is altered, or the angle of attack, or the position with reference to other surfaces.

Aerodynamical theory will serve practical use when supported by empirical data. In the past, flight vortices have never been

FIG. 9   McCOOK FIELD WIND TUNNEL, FAN END

measured, nor even visualized to a usable extent. Analysis of air flow has been confined to the use of smoke or powder set loose in the air to indicate lines of flow; or of threads used as wind vanes. Or we have been driven to analogies derived from study of fluids of different viscosities and densities, such as water. Or, further, we have sought by measurement of static pressures in the air surrounding a body to deduce the lines of flow. But these methods have given small encouragement to the practical application of the vortex theory to engineering use.

The writers' method bids fair to supply the missing link between aerodynamical theory and design. It depends upon the fact that the moisture in the air condenses out as fog when the temperature is reduced to the dewpoint, provided there is a solid or liquid nucleus to start the condensation. In the McCook Field wind tunnel the temperature drop is brought about through expansion of the air during acceleration due to 100 in. of water suction. Relative humidity of the atmosphere can be artificially raised if too low. The necessary nucleus for condensation is provided by the model itself.

Flight vortices become readily visible by the writers' method and can be photographed with the aid of searchlights. Several efforts were made to take the pictures with a plate camera but these were not very successful. Finally a good moving picture was taken and some of the films enlarged. While these films showed up very well on the screen the detail was not very clear in enlargement, so that, in addition to the searchlights which were provided with nitrogen-filled incandescent lamps, two carbon arc lights were set up in order to give a greater amount of blue light. The results obtained from the motion-picture camera with the carbon-arc lighting were fairly satisfactory, and a number of enlargements from the motion-picture films are reproduced in this paper. Fig. 1 is an enlargement of a moving-picture film looking downstream, and shows the left half-span of a small model aerofoil, with tip vortex trailing downstream from the tip and edge vortex-sheet below the model. The edge vortex-sheet is made up of vortices with axes parallel to the rear edge of the model; they build up and run off at such frequency as to appear continuous. It is somewhat inferior to visual observation. To the naked eye the tip vortices are in line with the wing tips

and are clean-cut, perfect circles. They extend downstream a
distance of several dozen chord lengths from the rear corner
of each wing tip, enlarging in diameter as the distance increases,
and converging slightly in the horizontal plane (see Fig. 11).
In the vertical plane the tip-vortex axis takes a decided downward
angle, intermediate between the horizontal and the line of the
flat sheet of edge vortices.[1]

At slow speeds condensation is not obtainable, but the vortex
phenomena can be corroborated by injecting a jet of visible
steam into the air current. Such a jet, when of proper saturation, affords a better indication of flow direction than does smoke
or thread.

Adequate analysis of the flight vortices is now being made with
stroboscopic apparatus. The shape, size and direction of the

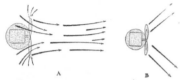

tip vortices can be easily noted and seem fairly susceptible of
measurement. The periodic run of the edge vortices is too quick
for recognition by the naked eye, or even for identification by
the moving-picture camera; it requires stroboscopic analysis.

The observed vortices differ for different aerofoil setups and
different speeds. For instance, the observed tip vortex at 18 deg.

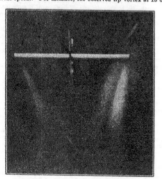

FIG. 11  PLAN VIEW OF FLIGHT VORTICES

has less than one-half the diameter manifest at 8 deg., while the
line of edge vortices is less noticeable at 18 deg.

Again, the character of the general vortex phenomenon undergoes remarkable change at the critical speed. In the high-lift
régime the general shape is like a trough whose floor (edge

[1] For an excellent mathematical discussion of the shape and arrangement of the tip vortices and the trailing vortices, see report No. 28 of the
National Advisory Committee for Aeronautics (U. S.). On page 44 the
author, Dr. George de Bothezat, has shown in an interesting way that the
axis of the tip vortices is intermediate between the direction of movement and the sheet of trailing edge vortices.

vortex-sheet) slopes downward from the trailing edge, and whose
walls (tip vortices) are increasingly high as the distance downstream increases. The cross-section is roughly like a shallow U.

At higher speeds, however, in the low-lift régime, the observed
phenomenon is suddenly altered. Following out the above homely
analogy, we may imagine that the "walls" of the trough remain
substantially as before. The "floor," however, splits longitudinally, curls upward, and extends the two limbs, now free, to a
point well above the level of the tip vortices. This is shown

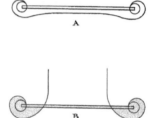

FIG. 12  DIAGRAM OF HIGH- AND LOW-LIFT VORTICES

in enlargements of two motion-picture exposures which were
intended to record the sequence of the change. These photographs, however, were not altogether satisfactory, and are therefore replaced by the two sketches of Fig. 12.

Fig. 12-A is a diagram of the end view of the high-lift phenomenon as it appears distinctly to the naked eye. The right-hand
side of the aerofoil in Fig. 8 approximates this condition, the
left-hand side having already gone over to low-lift flow. Fig.
12-B is a diagram of the low-lift phenomenon; the left-hand side
of the aerofoil shows this fairly well, the right-hand side being
in transition stage. Fig. 11 is a three-fourths view of the low-
lift flow, and also represents other features mentioned above.

An interesting variation of the flight vortices is furnished by
replacing the aerofoil by a flat plate normal to the wind. Here
the phenomenon can be seen as a "stream-line" fog surface,
converging towards a point half a dozen diameters downstream.

Fig. 8 illustrates the distance above the aerofoil to which the
flight-vortex phenomena may extend, and shows their tendency
to merge with other whirls attributable to the wind-tunnel walls.
The extent of the phenomenon may be four or more chord
lengths above the aerofoil; this further develops a discovery
made by the writers in 1911, when it was shown experimentally
that the air flow above a wing is disturbed to a distance of at
least four chord lengths.[1]

### VISUALIZATION OF UNOBSTRUCTED AIR FLOW

When the model is removed the vortices and eddies of flow
through the unobstructed throat may be observed by looking into
the intake or through the transparent shield of the observation
section. The condensation is more pronounced behind the impact
tube and thermometer bulb than elsewhere, since these are obstacles to the flow and therefore constitute nuclei for condensation.
A projecting cotter pin 1/16 in. high at the wall causes a perfect
vortex, which shows up against the white, foggy background as
a black circle.

The general appearance of the air flow, which may be considered typical of all air flow, is as follows: A cross-section at
the throat shows a seething mass of fog specters, denser at the

[1] See The Center of Pressure Travel on Airplane Surfaces and Birds'
Wings, Mass. Inst. of Technology, 1911. Reference may also be made in
this connection to the work of J. R. Pannell, dealing with experimental
evidence as to the extent of circulation about an aerofoil.

wall than at the center, though occasionally the entire disk fills up with fog to the point of opaqueness. The specters have in the cross-sectional plane a gentle movement like the flame of an alcohol stove, showing the constant readjustment of equilibrium. Vortices and S-shaped whirls continually form and, after moving about, lose themselves in the general confusion. In a diagonal view they take the appearance of long, foggy fibers, stretching down the tunnel like wooden moldings. The axes of whirl, of course, are longitudinal. Under proper humidity and lighting conditions the whole becomes a beautiful iridescent sight, violet and purple hues predominating.

*Lift Coefficients.* Coefficients of lifting force and resistance have been determined for a series of model aerofoils 6 in. long, 1 in. in chord length, camber varying from 0.08 to 0.20. The cross-sectional shape is that upon which the Engineering Division

dynamic phenomena resulting from air flow past a solid body. When the speed of air flowing past an aerofoil increases, there is first a régime of relatively low-lift effect, then at higher speeds an efficient lift effect such as applies in flight, then at still higher speeds a drop back to a second low-lift effect. As the angle of camber increases, the high-lift régime becomes discontinuous and is succeeded by the low-lift régime; the transition point is spoken of in conventional graphs as the critical, or stalling angle, or the "burble point."

All of the sections show, at certain angles, two speeds at which the flow is unstable and discontinuous. At the point of discontinuity occurring at the lower speed, increase of speed shows an increased lift coefficient and a decreased drag coefficient so that the lift-drag ratio ($L/D$) is enormously increased. At the point of discontinuity occurring at the higher speed, increase of

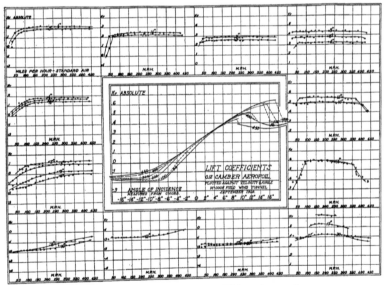

FIG. 13  CHART OF LIFT COEFFICIENTS OF 0.12 CAMBER AEROFOIL

has standardized for propeller use. A chart of lift coefficients for one of these aerofoils (0.12 camber) is presented, and shows the variation of lift coefficient over the entire range of speeds and angles tested. See Fig. 13.

The outstanding conclusion to be drawn from the tests is that we have more than one régime of air flow to deal with in aerofoil study, and that these régimes are separated by conditions of discontinuity. The characteristics usually associated in aeronautical engineering with a practical aerofoil do not apply outside the small range of cambers, speeds and angles utilized in flight. Beyond this range the flow about the aerofoil no longer produces the familiar results in terms of lift and drag but becomes analogous to the flow about a body of irregular shape. Efficient lift of an aerofoil is only a single case of several distinct aero-

speed shows a decreased lift coefficient and an increased drag coefficient, so that the lift-drag ratio is enormously decreased. Thus these sections have a definite speed range, at each angle, within which the flow is efficient and produces a high lift-drag ratio, and a fairly constant lift coefficient. It may be called the régime of high $L/D$, and includes the phenomena appertaining to practical flight. This speed range has been definitely measured for the higher angles, but apparently it goes beyond the speed obtainable in the tunnel for the lower angles.

It will be noted that the $K_y$ curves are drawn discontinuous to correspond with discontinuity in the type of air flow. In some cases the graphs show a third curve intermediate between the high-régime curve and the low-régime curve. This third inter-

(*Continued on page 260*)

# Simplification of Venturi-Meter Calculations

By GLENN B. WARREN,[1] SCHENECTADY, N. Y.

*In this paper the author describes a method which he has devised for the simplification of the calculations involved in the venturi-meter measurement of the flow of compressed air, and which reduces the work of one determination to a simple slide-rule computation requiring but a few settings. The method, which is described in considerable detail, is based on the venturi-meter formula as given by Herbert B. Reynolds in Trans. Am. Soc. M. E., Vol. 38, p. 799, and can be easily applied to any venturi meter used for the measurement of a gas.*

**D**URING the winters of 1918 and 1919 the writer had occasion to measure with considerable accuracy the amount of compressed air used in connection with some research work which was being carried on at the University of Wisconsin. The venturi meter presented itself as the logical method, but aside from the mechanical difficulties of building manometers to satisfactorily stand pressures of more than 100 lb. per sq. in. the difficulties presented by the calculations where several hundred volume determinations would have to be made

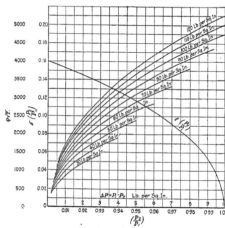

5000    0.20
4500    0.18
4000    0.16
3500    0.14
3000    0.12
2500    0.10
2000    0.08
1500    0.06
1000    0.04
500     0.02
0       0

$\Delta P = P_1 - P_2$  Lb. per Sq. In.

$\left(\dfrac{P_2}{P_1}\right)$

Fig. 1  Flow Curves for 1.6-in. × 0.498-in. Venturi Meter

were of considerable importance. Accordingly, a method of shortening the calculations was devised which reduced the entire work of one determination to a simple slide-rule computation, requiring but a few settings, and which at the same time did not affect the accuracy to any appreciable extent. Although the method of shortening these computations may seem somewhat laborious in itself, when it is remembered that with but a few changes it can be applied to any venturi meter used for the measurement of a gas, the result can be seen to have abundantly justified the means.

For Presentation at the Spring Meeting, St. Louis, May 24 to 27, 1920, of The American Society of Mechanical Engineers. All papers are subject to revision.
[1] 25 North Ferry Street.

The venturi-meter formula is given by Dr. Lucke as follows:

$$W = CA_2 \left(\frac{P_2}{P_1}\right)^{\frac{1}{n}} \sqrt{2g \frac{n}{n-1}\frac{P_1}{V_1}\left[\frac{1-\left(\frac{P_2}{P_1}\right)^{\frac{n-1}{n}}}{1-\left(\frac{A_2}{A_1}\right)^2\left(\frac{P_2}{P_1}\right)^{\frac{2}{n}}}\right]} \dots [1]$$

where $W$ = weight of vapor or gas passing per second
$C$ = coefficient
$A_1$ = area of entrance
$A_2$ = area of throat.

The units in which $W$ is expressed are dependent, of course, upon the units used in all the other quantities of the formula.

Herbert B. Reynolds in Trans. Am. Soc. M. E. for 1916, p. 799, gives the following formula, which can be derived from Equation [1] by the substitution of the proper units:

$$Q = C\frac{3514\,A_2T_s}{P_s}\sqrt{\frac{n}{(n-1)G}\left(\frac{P_1}{\sqrt{T_1}}\right)\left(\frac{P_2}{P_1}\right)^{\frac{1}{n}}}$$

$$\times \sqrt{\frac{1-\left(\frac{P_2}{P_1}\right)^{\frac{n-1}{n}}}{1-\left(\frac{A_2}{A_1}\right)^2\left(\frac{P_2}{P_1}\right)^{\frac{2}{n}}}} \dots\dots\dots\dots\dots [2]$$

where   $Q$ = cu. ft. per min. at 32 deg. fahr. and 14.7 lb. per sq. in.
$C$ = coefficient (experimentally determined)
$A_1$ = area of entrance, sq. ft.
$A_2$ = area of throat, sq. ft.
$T_s$ = absolute temperature of standard air = 492 deg. fahr.
$P_s$ = absolute pressure of standard air = 14.7 lb. per sq. in.
$n$ = constant = $C_p/C_v$ (= 1.402 for air)
$G$ = specific gravity of gas (= 1.0 for air)
$P_1$ = pressure of air at entrance, lb. per sq. in. absolute
$P_2$ = pressure of air at throat, lb. per sq. in. absolute
$T_1$ = temperature of air at entrance, deg. fahr. absolute.
3514 = constant determined by above choice of units

The *Sibley Journal of Engineering* (vol. 29, pp. 90-95) presents this formula in still another form, which has been derived by means of the Fourier series and is in better shape for calculation, where the ratio $(P_2/P_1)$ is very nearly unity, or in other words, for low rates of flow. This formula is:

$$W = CA_2\left(\frac{T_s2gn}{V_sP_s(n-1)}\right)^{\frac{1}{2}}\sqrt{\frac{P_1}{T_1}}\left(x\left[\frac{n-1}{n(1-a)}\right]\right)^{\frac{1}{2}}$$

$$\times \left(1-\frac{[3+a]x}{2n[1-a]}\right) \dots\dots\dots\dots\dots [3]$$

where $V_s$ = volume under standard conditions = 12.38 cu. ft. per lb. for air; $a = (A_2/A_1)^4$; and $x = \Delta P/P_1$, where $\Delta P = P_1 - P_2$; the other symbols being the same as before.

Any one of these three formulæ presents a rather difficult computation as far as solution is concerned, this being especially the case in formulæ [1] and [2] because of the last radical. The log-log slide rule enables the solution to be made quite readily, excepting for values of $P_2/P_1$ greater than 0.98, after which it is necessary to resort to very complete logarithmic tables, or else to the third formula.

### METHOD USED TO SIMPLIFY COMPUTATIONS

The first step in the process of reducing these formulæ to forms which could easily be handled consisted in determining the values of $A_1$ and $A_2$ which would give a venturi meter having the desired range of capacity without making the difference between $P_1$ and $P_2$ either too great or too small for easy and accurate reading. In this series of tests it was necessary to have a range of capacity from 25 to 250 cu. ft. of free air per min. at a pressure in the venturi tube of about 100 lb. per sq. in. gage. Inasmuch as the flow is nearly proportional to the square root of the pressure

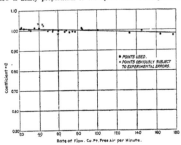

FIG. 2  CALIBRATION CURVE FOR 1.6-IN. × 0.498-IN. VENTURI METER

difference, as can be seen from formula [3], a range of possible rates of flow of 10 to 1 means a range of readable pressure differences of nearly 100 to 1 in magnitude. This necessitated the combination of a mercury and a water manometer, which will be described later.

It was found by a trial and error method that a venturi meter having an entrance diameter of 1.6 in. and a throat diameter of 0.5 in. would answer the requirements. As actually constructed, the throat diameter was found to be 0.498 in. This gave a ratio of $A_2/A_1$ nearly equal to 0.10.

Upon substituting these numerical values for $A_1$, $A_2$, $P_2$, $n$, and $G$ in formula [2], there is obtained the following expression:

$$Q = 297.6C \left(\frac{P_1}{\sqrt{T_1}}\right) \left[ \left(\frac{P_2}{P_1}\right)^{0.714} \sqrt{\frac{1-\left(\frac{P_2}{P_1}\right)^{0.286}}{1-0.01\left(\frac{P_2}{P_1}\right)^{1.43}}} \right] \quad \ldots [4]$$

It can be seen that the part in brackets is independent of everything but the ratio $P_2/P_1$; in other words, it is a function of this ratio. Then if we let

$$f\left(\frac{P_2}{P_1}\right) = \left[ \left(\frac{P_2}{P_1}\right)^{0.714} \sqrt{\frac{1-\left(\frac{P_2}{P_1}\right)^{0.286}}{1-0.01\left(\frac{P_2}{P_1}\right)^{1.43}}} \right] \quad \ldots [5]$$

and plot a curve of $f(P_2/P_1)$ as ordinates against values of $P_2/P_1$, it follows that the finding of $Q$ will merely amount to multiplying several known quantities on the slide rule, since $f(P_2/P_1)$ can be found from the curve, thus:

$$Q = C \frac{297.6P_1}{\sqrt{T_1}} f\left(\frac{P_2}{P_1}\right) \quad \ldots [6]$$

In practice it is the quantity $P_1 - P_2 = P$ which is measured by the manometers, and the most accurate method of finding $P_2/P_1$ from this is as follows:

$$\frac{P_2}{P_1} = \frac{P_1 - \Delta P}{P_1} = 1 - \frac{\Delta P}{P_1} \quad \ldots [7]$$

It can easily be seen that this last expression can be evaluated much more accurately on the slide rule than can the second.

The values of $f(P_2/P_1)$ for values of $P_2/P_1$ from 1.00 to 0.90 are given in Table 1 and are shown plotted out in the form of a curve in Fig. 1. The values of $f(P_2/P_1)$ for values of $P_2/P_1$ between 0.98 and 1.00 were found by evaluating the last two parentheses in formula [3]. It is rarely advisable to use a gas

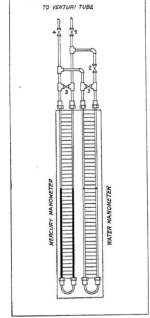

FIG. 3  ARRANGEMENT OF MANOMETERS, PIPING AND VALVES

venturi meter at values of $P_2/P_1$ less than 0.90, inasmuch as the total drop in pressure across the meter becomes too great, and also because the reserve or overload capacity of the meter is so greatly restricted. This set of values is worked out for a ratio of $A_2/A_1 = 0.10$, but the effect of a slight change in this ratio upon the values of $f(P_2/P_1)$ is so slight that these values may be used for meters having a ratio differing considerably from 0.10. However, a set of values can be worked out for a tube having a ratio other than this.

The family of curves shown in Fig. 1 was worked out to facilitate rapid slide-rule calculation of the flow during a test. In order to plot these, the temperature in formula [6] was transposed to the left-hand member, thus:

$$Q\sqrt{T_1} = 297.6CP_1f\left(\frac{P_2}{P_1}\right) \quad \ldots [8]$$

and curves for numerous values of $P_1$ plotted against values of

(Continued on page 260)

# Tight-Fitting Threads for Bolts and Nuts

By CHESTER B. LORD,[1] ST. LOUIS, MO.

*Although in general the thread forms now in use are quite satisfactory, perfection has by no means been reached. There is still considerable search for what the author of the following paper terms "a thread that will not loosen," and in the preliminary portion of his paper he discusses the fundamental principles involved in the manufacture of threads. The problem to be solved, he states, is as follows: "Without sacrifice of strength, without increase of rejection, without additional manufacturing costs, find a method whereby a male and female thread of the same lead and, pitch diameter may be made after repeated loosenings to fit tight without the aid of a locking device." The reasons for departing from accepted practice are presented and discussed, and as a result of experimental work, the author draws the following conclusions: (1) The cause of stripped threads is lack of room into which the metal can flow; (2) the pitch diameter should be the same in both threads; (3) The lead should be the same; (4) The thread angle should differ by not more than 10 deg.; (5) The limits for the inside diameter of nut need not be adhered to closely, as the inner part of the nut thread holds very little, if any; (6) The outside diameter of plug and pitch diameter of both plug and nut are important and should be adhered to fairly closely.*

WHAT is the cause of our periodic dissatisfaction with threads when in general they are so satisfactory? What other machine element is so easily made or is so satisfactory as regards strength? Why are there so many different kinds of threads when all are equally satisfactory, or rather unsatisfactory? Furthermore, is the dissatisfaction founded on performance or merely upon theory? Also, is the form or angle of thread a matter of importance, or merely an excuse for mathematical gymnastics? The only answer the writer has been able to elicit in reply to these various questions is that we are looking for a better thread; which statement, however, is rather indefinite and usually simply means a thread that will pass the gage. Of course, the real object of the search has been to find a thread that will not loosen.

In the past we have attributed our troubles to the fact that our fits were not close enough—the engineer's alibi for a poor design; but fundamentals cannot be violated in mechanics any more than elsewhere in nature, and we are attempting to violate two by insisting upon our present methods of inspection: (a) that interchangeable manufacture is a matter of percentage, which depends upon tolerance and cost; and (b) that a force fit is not possible between two parts the surfaces of which are complements one of the other.

Having in mind the first fundamental, it is obvious that the chances of securing a perfect fit are limited by the cost, and the second fundamental would seem to render this entirely hopeless. It is therefore proper to conclude that a good fit is usually due to error, and that if changing both male and female threads produces no relative change, changing one thread must of necessity do so. It is the object of this paper to demonstrate that by making this latter change threads can be produced that are interchangeable practically regardless of tolerance, that will not loosen, and are cheap to manufacture.

A physician always diagnoses a case before prescribing, so let us do likewise. The loosening of a thread fit is caused by vibration or repeated shock, the chief result of which is to flatten and burnish the parts of the thread that are in contact. This produces a slight looseness and the nut tends to follow the thread incline until it again fits. The same performance is repeated until finally the nut reaches an obstacle too large to flatten, or else the bolt and nut vibrate in unison and there is no further

relative movement. This same phenomenon occurs in the case of a bolt screwed in a tapped hole.

The writer can remember in his shop days picking out and numbering nuts and bolts, and every mechanic knows that he cannot take the bolts out of a cylinder head on an engine of good make and put them back again indiscriminately. They must go back into the holes from whence they came. All have demonstrated to their satisfaction the fact that a tight thread will not loosen by vibration, and that one with tolerance will, unless it is prevented from loosening by a lock washer or its action is limited by some type of nut lock. We cannot lessen our tolerance because the tap wears small and the die wears large, and the lesser cannot contain the greater.

## THE PROBLEM TO BE SOLVED

To gain a better conception of the problem to be solved, consider a board cut as shown in Fig. 1. We cannot obtain a forced

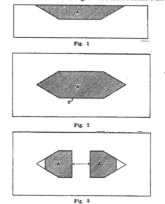

Fig. 1

Fig. 2

Fig. 3

FIGS. 1-3 APPLICATION OF PRESSURE IN THREAD FITS

fit between the two pieces because one surface is the complement of the other. If we apply a force to A, we are no better off because the pieces cease to fit the instant the pressure is released. We may therefore state as a rule that where two surfaces are complementary to one another a tight fit cannot result without some means of maintaining pressure. But if we cut the board as shown in Fig. 2, theoretically removing no material, we can replace A without force, and any pressure exerted will not make a tight fit unless we distort A or drive a wedge at C. This is analogous to a perfect thread, and driving the wedge at C is equivalent to introducing a slight difference in lead. If, now, we cut off the ends of A as shown in Fig. 3, and apply a force in the direction shown, we can obtain a forced fit because only the angles are complements and because we have a method of maintaining pressure. We can also even distort A because we have room for it to expand. We have only then to provide for three things: a method of making the parts in contact absolutely complementary; the introduction of sufficient metal; and a method of maintaining pressure. With our present type of thread we can only meet one of these—namely, that of partly maintaining

[1] Consulting Mechanical Engineer, Research Engineering Co., Mem. Am.Soc.M.E.

Abstract of a paper to be presented at the Spring Meeting, St. Louis, May 24 to 27, 1920, of THE AMERICAN SOCIETY OF MECHANICAL ENGINEERS. Copies of the complete paper may be obtained gratis upon application. All papers are subject to revision.

pressure. This is demonstrated in Fig. 4, which shows in an exaggerated manner the effect of off lead.

While two slightly varying leads make a better fit, both as regards gaging and in actual use, this practice is not to be commended. Using a different lead to secure a fit is doing imperfectly on one side of the thread what the different-angle method does perfectly on both sides, because by having the leads identical and the thread supported on both sides, we secure a uniform finish instead of a distortion. Where the leads are different, the amount of distortion necessary to secure a fit increases with each thread. Thus, if the lead of a 20-thread stud is 0.05 in. and we make it 0.052 in., it will be 0.002 in. off center on the second thread and 0.018 in. on the tenth thread. This is entirely possible, and superior to a so-called perfect thread as regards fit, but a distortion, unmechanical, and unnecessary.

U. S. thread, because we have filled out the female thread tolerance as well as that of the male. We will also have uniform pressure on all flanks; the maximum possible material at the pitch line; a hard, smooth surface analogous to a case-hardened one; a fit that will remain snug despite repeated removals and that may be screwed together by ordinary means; and yet we still retain an interchangeable bolt and nut according to U. S. standards. The nut has not been changed or distorted in any way, but has simply served as a finishing roll.

The writer has stated in a previous article that, aside from threads, nowhere else in machine work do we expect micrometer limits on a roughing cut; and the question naturally arises whether an operation similar to that described would not be an effective finishing and sizing operation for commercial work. This would be the equivalent of *making* them fit the gage, and

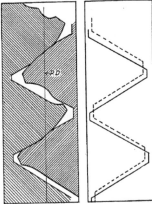

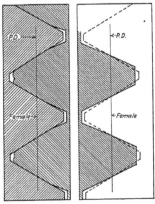

FIG. 4  THE EFFECT OF LEAD (GREATLY EXAGGERATED)

FIG. 5  DIFFERENT-ANGLE THREADS (GREATLY EXAGGERATED)

This is what is done with railroad fish-plate bolts where the specifications state the minimum foot-pounds at which the nut and bolt may be assembled. The impossibility of meeting these requirements in quantity production is recognized by purchasing agents and most engineers, and so the lead is slightly changed. This, however, is merely a subterfuge and really defeats the purpose of the specifications in that it permits of a poorer and weaker thread than would otherwise be possible.

We are thus confronted with this problem: Without sacrifice of strength, without increase of rejection, without additional manufacturing costs, find a method whereby a male and female thread of the same lead and pitch diameter may be made after repeated loosenings to fit tight without the aid of a locking device. This, according to specifications, calls for a full thread at contact points, pressure applied continuously on all flanks, and maximum strength at the pitch line. This means the addition of surplus metal to the male (which is the only one affected) sufficient to fill out the female threads, which would be an impossibility were it not for the ductility and elasticity of steel. If we add this surplus metal we will find that we can more than fulfill the required conditions by changing the angle of the male thread to a lesser one than that of the female, having the two intersect on the pitch line.

When a male thread of this type and a standard mate are screwed together, we will have transformed a male of lesser angle to one of larger angle or of wider base than the standard

would greatly reduce the cost due to rejections. We would, of course, still have variable nuts and the necessity of sizing them.

### REASONS FOR DEPARTING FROM ACCEPTED PRACTICE

Let us now see what authority and precedent we have for making so radical a departure from accepted practice. To do so, let us consider how threads are made, especially by rolling. Generally speaking, thread rolling is that process whereby the diameter of a part is increased at certain points by a corresponding reduction of diameter at other points, due to pressure alone. It is applicable to both flat and round surfaces, but for the purpose of this article we will consider only round surfaces.

In turning screws and bolts to size the diameter is held, generally speaking, to the pitch diameter. The displacement of metal from the root and lower flank forms the addendum under the process. Cutting a thread on a bolt with a die is a somewhat analogous operation, the similarity increasing as the die becomes duller. For a die-cut 1-in. bolt it will generally be found that with a diameter of 0.990 in. a fuller thread may be secured than with one of 1.000 in., the explanation being that with the die we secure a combined cutting and rolling operation. Due to lack of clearance, if the diameter is too large, part of the metal is pressed into the bottom of the die and with such force that it drags and is torn off, thus leaving a thread of smaller outside diameter than would have been otherwise secured. This same phenomenon occurs when a nut is too tight.

If we require a holding fit on a shaft, do we use tolerances that allow of the shaft being several thousandths smaller than the hole it is to fit? Quite the contrary. We not only specify the fit but also the minimum pressure allowable to assemble the two parts, and we do this both for single units and for quantity production. We might term a shaft and rotor spider a nut and bolt with microscopic threads and assembled with a right-angled instead of a helical motion. Why not fit our bolts the same way, making our tolerances plus instead of minus and using a force fit we can depend upon when it means no change in the tools for assembly?

If we look at a finished commercial thread through a microscope, it will be seen that the edge is serrated and that slivers stand up all over its surface. By running a nut over it once we but slightly change its appearance, but by repeatedly doing so—always using a tight nut—we may finally burnish the thread so that it will not thereafter change its size and will have a surface somewhat comparable to a case-hardened one.

Fig. 5 shows diagrammatically the different-angle thread with the angles of the male greatly exaggerated to demonstrate the principle. We are complying in this case with all the conditions we have just been discussing: We are making the lesser contain the greater, angularly speaking; we are applying force from both directions, at right angles to the axis of the bolt; we have made the angles complementary—not merely two equal angles; we are securing the maximum strength at and near the pitch line, and transforming a thread with a lesser angle than that of the standard male to one with a larger angle, thus filling out the space perfectly and so doing away largely with vibration; and we are securing, whether under pressure or not, contact on all flanks, whereas the standard thread when under pressure secures contact on possibly one-half the flanks, both because it is compressible and because it does not fill the female thread. The only problem to be solved, therefore, was to find an angle of such slope that it could be formed without distortion of the nut or requiring too much force to screw home. To demonstrate this, threads as small as No. 10-32 were used, and as being of possible interest, the writer presents a brief outline of the engineers' report of the experimental work.

### RESULTS OF EXPERIMENTAL WORK

Diagrams of the different threads were first laid out on a 100 to 1 scale so as to determine approximately the most suitable angles to be tested; the Löwenherz thread with an angle of 53 deg. 8 min. being used as a basis. The nut was to have the regular Löwenherz thread with same diameter and pitch as in the 155-mm. shell adapter used by the U. S. Government.

Threads of 44 deg. and 45 deg. for the plugs seemed most favorable and accordingly the following cold-rolled-steel plugs were made up, with nuts having the same pitch diameter and lead as the plugs but a thread angle of 53 deg. 8 min.:

1 Angle of thread 44 deg. 54 min., pitch diam. 0.8748, lead 0.0787
2 Angle of thread 45 deg. 2 min., pitch diam. 0.8748, lead 0.0787
3 Angle of thread 44 deg. 0 min., pitch diam. 0.8748, lead 0.0787

*Test No. 1.* Nut No. 1 and plug with thread angle of 44 deg. 54 min. were screwed together *without* a lubricant. They were started about a half a thread by hand, and then an 8-in. wrench was used for about four threads. The plug was then so tight in that a 10-in. wrench was required to turn it to full depth. After a couple of backward turns the plug stuck so tight that a 20-in. wrench would not move it. The nut was then sawed open and removed from the plug and about one-third of a thread of the plug was taken off in a piece of the nut. A magnifying glass showed that the threads in both plug and nut were drawn and cut out of shape where there was a tendency for them to overlap, due apparently to too much metal and no lubricant.

*Test No. 2.* Nut No. 2 and plug with thread angle of 45 deg. 2 min. were screwed together *with* a lubricant. They were started by hand for about one-half turn, then an 8-in. wrench was used for five or six threads, and a 10-in. wrench for the remainder. The plug came out slightly easier than going in. After this had

been repeated three times the plug could be screwed in by hand. The maximum and minimum plug gage for the nut showed no change in the thread of the nut. Under a magnifying glass it was seen that the metal had flowed to the top of the plug thread from about the pitch diameter outward. The plugs were screwed into the nuts fifty times and there was still what could be termed a "snug fit."

*Test No. 7.* Plugs were tried out with commercial 1-in. nuts. The thread angle of the plugs was 58 deg., and the pitch diameter 0.9228 in., with one plug this diameter plus 0.001 in., and one minus 0.001 in. This diameter allowed the plug thread in the layout to overlap the entire thread of the nut instead of only half, as in the previous cases. The nuts used were picked out of stock for size with a standard 1-in. thread gage. The smallest plug went into the nut easily by hand. The largest two went together easily with an 8-in. wrench. After being twice screwed together with the wrench they went together with a snug fit by hand. The magnifying glass showed that the thread from near the pitch diameter outward had been drawn and compressed slightly.

*Test No. 9.* Plugs of ½-in. diameter with pitch diameters of 0.4684 in. and 0.4699 in. (same as ½ S. A. E. nuts) and a thread angle of 50 deg. were tried with commercial ½-in. S. A. E. nuts. The difference in the two pitch diameters made practically no difference in the fits, as they both readily went in with an 8-in. wrench. After they had been screwed together four times, they would go together by hand, but without shake. After they had been screwed together 75 times there was still what could be termed a "snug fit."

*Test No. 10.* An attempt was made to compare the strength of an S. A. E. standard ½-in. thread with a 50-deg. thread of the same size. The plug with the standard thread on one end and the special thread on the other end was used with standard nuts. A pull of 14,000 lb. was gradually applied and the metal began to give way, which prevented an additional load. During this pull, observations were made to determine if there was any "give" in either of the threads, but both remained the same throughout. The nuts were removed and there was no apparent distortion of the threads.

*Test No. 12.* In this test ¼-20 plugs with pitch diameters of 0.2165 in. and 0.2181 in. with 50-deg. angle were tried out with ¼-in. U. S. standard nuts, one being a commercial nut and the other of a standard size but made in our toolroom—the tap being 0.250 in. in diameter. Both nuts were tried with a ¼-in. standard plug gage, both being apparently the same size. The nut made in the toolroom was screwed on the maximum plug and went on about one and a half times its length and then stuck and would not go either way. It was finally removed by hammering it on the sides. The threads were rolled and torn from about the pitch diameter outward, but there was not that tendency for the metal to roll upon the outside of the thread as in the previous tests, the outside diameter being only 0.251 in. as against its original 0.250 in.

The commercial nut went on the minimum plug with an extremely tight fit, but it came off very readily with the wrench and left quite a different thread from the previous one. The thread was not torn at all but rolled out to almost a perfect V-thread with outside diameters of 0.2555 in. as against the original 0.250 in. The outside diameter of the tap for the commercial nut must have been over 0.250 in. to allow this metal to flow out to 0.2555 in. and not jam the nut. If the toolroom tap had been sharper on the flats or its outside diameter greater, there would have been room for the metal in the test plug to flow out to a larger diameter and avoid tearing the thread, and consequent jamming of the nut. Two facts are clearly demonstrated: first, necessity of room for metal to flow; second, one of the limitations of the thread gage.

*Test No. 13.* Plugs corresponding to standard 10-32 with 50-deg. angle were tried out with standard 10-32 nuts; pitch diameters 0.170 in. and 0.1716 in. They were a little too tight a fit to go together by hand, but after being screwed together once with a wrench they went together by hand snugly.

(Continued on page 289)

# Pulverized Coal in Metallurgical Furnaces at High Altitudes

By OTIS L. McINTYRE,[1] NEW YORK, N. Y.

*The following paper describes experiments with pulverized coal which led to the installation of apparatus for its use in the blast furnaces, reverberatories, and sintering machines of the Cerro de Pasco Copper Corporation at La Fundicion, Peru. Information is given of the method of determining the firing performance of pulverized coal; the behavior of the various mixtures; experiences with different air pressures; and the effects of ash. The experiments were conducted at an elevation of 14,000 ft. and this fact greatly adds to the value of the data presented, for heretofore but little attention has been paid to the effects of altitude upon the burning of pulverized coal.*

THE Cerro de Pasco Copper Corporation at La Fundicion, Peru, uses about 65,000 tons of coke per year, of which about 85 per cent is local coke made at the smelter, and 15 per cent is imported. This latter is very expensive, and of course both classes of coke enter largely into the smelting costs; consequently, about two years ago it was decided to determine what could be done in the way of using pulverized coal in the

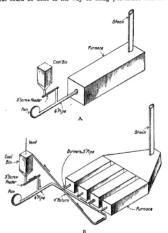

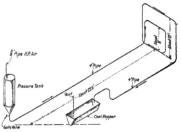

Fig. 1 Layouts of Equipment Used in Furnace Tests of Pulverized Coal

Fig. 2 Layout Used to Test Feasibility of Conveying Pulverized Coal Under Direct Air Pressure

various departments of the smelter. The preliminary work consisted in determining the general combustibility of the local coals in pulverized form. These coals are obtained from two mines operated by the company and have the following general analysis:

| Ash | Volatile Material | Fixed Carbon |
|---|---|---|
| 26.8 | 40.05 | 33.15 |

This coal was dried by hand on steam hot pans to less than 1 per cent moisture, and then ground in a 4-in. by 4-in. Marcy

For presentation at the Spring Meeting, St. Louis, May 24 to 27, 1920, of THE AMERICAN SOCIETY OF MECHANICAL ENGINEERS. All papers are subject to revision.

[1] Mech. Supt., Cerro de Pasco Copper Corpn. Mem.Am.Soc.M.E.

mill, the product being stored in barrels until a sufficient quantity had been pulverized to run a test. An average screen test of this pulverized coal was about as follows:

| + 60 mesh | 8.00 per cent |
|---|---|
| +100 mesh | 8.00 per cent |
| +200 mesh | 14.00 per cent |
| −200 mesh | 70.00 per cent |

The equipment used in the test is shown by Fig. 1-A and consists of coal hopper, a 3-in. feed screw driven by variable-speed motor, and a No. 2 Sturtevant blower supplying the air. The burner was a standard 6-in. pipe projecting about 12 in. into the furnace, which was approximately 4 ft. by 4 ft. by 16 ft. and constructed of firebrick. A number of tests were run with this equipment and though no pyrometric measurements were taken, observation of the furnace showed the results to be satisfactory.

The tests were first made with pure pulverized coal, and then with mixtures of coal and coke breeze, varying from 10 to 35 per cent of coke breeze, which gave practically the same results as the pure coal. The layout was then changed, Fig. 1-B, to test the practicability of using more than one burner with a single feeder. This test was run with the 4-in. return pipe both open and closed, the results indicating that satisfactory operation could be obtained by either method with a properly proportioned pipe system.

The next test made was in the sintering of fine ores on a standard Dwight-Lloyd sintering machine. These machines are oil-fired, and if coal could be substituted it would show a considerable saving. The equipment used in this test was the same as was shown in Fig. 1, except that a 1-in. screw feeder, a smaller fan, and a 2-in. pipe burner were used. This test produced a satisfactory sinter, though some trouble was encountered in the primary ignition of the coal, and the standard oil muffle proved to be too small.

The next experiment was to test the feasibility of conveying pulverized coal under direct air pressure. The layout used is shown in Fig. 2. Pulverized coal was placed in the pressure tank and air at 20 to 25 lb. was then admitted through the ¾-in. pipe at the top of the tank. The 4-in. valve at the bottom was then opened and the coal passed through the 4-in. piping system to the coal hopper. In this way 4000 lb. of coal was transported in 1½ to 2 min. The loss through the vent pipe varied from 100 to 200 lb. This can be taken care of by using dust collectors on the hopper, or an exhaust system which would return this waste coal to the main hopper.

The foregoing tests showing up so well, it was decided to erect a larger experimental pulverizing plant. There were available for this purpose one set of 18-in. by 36-in. rolls, one 4-ft. by 4-ft. Marcy mill and two 6-ft. by 4-ft. Allis-Chalmers ball granulators. The drier consisted of five passes of 16-in. by 12-ft. screw conveyor, mounted in a brick housing on top of the reverberatory flue, and through which part of the flue gases were by-passed. The layout in plan is shown in Fig. 3.

After completing this plant it was decided to make the first experiment on the blast furnaces, so No. 5 blast furnace was selected for the purpose and was equipped on one side only, as shown by Fig. 4. The coal was ground at the experimental plant and transferred to the No. 5 furnace in a hopper car, being weighed in transit. A number of tests varying from 8 to 12 hr. were run with this equipment. The furnace air pressure averaged 34 oz., and auxiliary air for injecting coal about 22 lb. The coke charge was reduced first 25 per cent, and then 50 per

shifters. By using dust-proof bearings and a better-designed injector, we expect to eliminate this trouble.

Keeping the tuyeres open is absolutely essential to the safe and efficient operation of this process, and as it is a manual operation it must be handled by the operators. During these tests, tuyeres were "punched" every fifteen to twenty minutes on signal. On one occasion a tuyere became badly blocked, the feed was cut off and the tuyere cap opened. The blast from the furnace blew out a dense cloud of coal dust and molten material. The dust was ignited and burned on the outside of the furnace for 20 to 30 sec. with an intense flame about six feet long, the tuyere acting as an ordinary coal burner.

In view of the difficulty of keeping the tuyeres open and the connections airtight, it is probable that the most satisfactory place to inject the coal into the furnace would be through a separate opening in the jackets, between and preferably somewhat above the tuyeres.

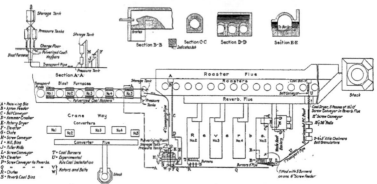

FIG. 3   LAYOUT OF PRESENT CERRO DE PASCO COPPER CORPORATION'S SMELTER, SHOWING LOCATION OF COAL-PULVERIZING PLANT

cent. These tests showed up so well that it was decided to equip the other side of the furnace with coal feeders and run a test of of several days' duration. This was done and the results were entirely satisfactory. During these tests the auxiliary air was taken from the converter air line, which varied from 12 to 16 lb. pressure.

The following quantities will give an idea of the proportion of coke and pulverized coal used:

| | 14-Hr. Run | 50-Hr. Run |
| --- | --- | --- |
| Normal charge of coke. lb. | 31,000 | 114,000 |
| Actual charge of coke, lb. | 17,000 | 61,800 |
| Pulverized coal fed to furnace, lb. | 8,900 | 41,000 |

The analysis and screen tests of the coal used were practically the same as noted above. The furnace performance through all tests was carefully observed and was found to be fully equal to that when operating on the normal coke charge. Two difficulties were encountered on the blast-furnace test, namely, keeping some to the feeders in operation and keeping the tuyeres open. It was observed that in some of the feeders there was a slight back pressure, due probably to partially blocked tuyeres. This did not affect materially the feeding, but forced some coal dust into the feeder bearings which mixed with the oil and finally bound the bearings so that it became necessary to shut down that particular feeder and clean the bearings. This was easily done without shutting down the other feeders, as the gears on the main shaft were mounted on feathers and provided with

The No. 5 reverberatory was selected for the final test. All four reverberatory furnaces in use are identical: they are old style, designed for hand-firing and about 18 ft. by 58 ft. inside the bridge wall. No. 5 was fitted up as shown at U, Fig. 3. The coal was discharged from the last mill into a hopper and dropped into a 7-in. pipe where it was picked up by an air jet and conveyed to the coal hopper, a distance of about 150 ft. and rise of about 30 ft.; the top of the hopper was constructed similar to a cyclone dust collector. From the hopper a 6-in. variable-speed screw feeder fed the coal into the suction side of a No. 9 Sturtevant Monogram blower; this in turn discharged the mixture of coal and air into the feed piping from which branched five 6-in. pipe burners into the furnace, the excess air and coal returning to the hopper.

This test was disappointing from the actual results, but when the following difficulties which were encountered are corrected, the furnace will, beyond question, show a much higher efficiency than the hand-fired furnace. First, the coal could not be dried sufficiently, the average moisture being in excess of 1.5 per cent. This introduced handling troubles. The plant would not grind sufficient coal to the required fineness, the average screen analysis being:

$$+ \; 65 \; \text{mesh} \quad 22.8 \; \text{per cent}$$
$$- \; 65 \; \text{mesh} \quad 8.5 \; \text{per cent}$$
$$-100 \; \text{mesh} \quad 25.6 \; \text{per cent}$$
$$-200 \; \text{mesh} \quad 42.4 \; \text{per cent}$$

The discharge from hopper to feeder was too small, and the coal continually caked and bridged. The screw feeder was too short so that the coal flushed badly at times; also the discharge from the feeder was too far from the fan so that the coal accumulated in the suction pipe and had to be removed with an air jet. Under these conditions it was obvious that uniform feeding, which is essential to efficient operation, was impossible.

This test covered about nine days, and was run for about two days with the return pipe open. Sometime during the second day the return pipe was blocked, due to overfeeding, and it was decided to continue the test without opening the run pipe, the only difference being an apparently heavier feed at the burner farthest from the fan. With a properly designed piping system, there seems to be no reason why a series of burners cannot be operated from a single feeder with or without a return. The last day's run of this test was made with a mixture of 75 per cent coal and 25 per cent coke breeze, which gave results equal to straight coal.

The following table shows a comparison between the average performance of reverberatories Nos. 2, 3, 4 and 5 over the same period:

| No. | Charge Smelted Per Hr. | Coal Used per Hr. | Smelting Ratio | Hr. Run | Time Lost |
|---|---|---|---|---|---|
| Aver. of 2-3-4 | 5.35 | 2.00 | 2.67 | 262 | .... |
| 5 | 4.63 | 1.99 | 2.33 | 225 | 37 |

These results are really not so bad when the troubles encountered are considered and it is remembered that this furnace was not designed for pulverized coal and that it cools very rapidly during any shutdown, and considerable time is required to bring it up to the smelting temperature again.

As ash accumulations are an important factor in reverberatory smelting with pulverized coal, close observation was made of these accumulations, and the following samples taken:

1 Ash and slag float on the bath: comes out when skimming in small and large pieces, sometimes has to be broken to pass the skimming door, is easily handled when furnace is hot, but is tough and sticky when furnace is cool.

2 Ash in boiler cross-flue: Spongy mass of ash and some slag accumulates in fairly large quantities in cross-flue between furnace and waste-heat boilers; is soft and easy to remove when first deposited, but if allowed to remain, is very difficult to remove. (See sections E-E and D-D, Fig. 3.)

3 Ash on sides of roof of furnace: Almost pure ash, lightweight and brittle when cold, appears to accumulate on sides and roof of furnace until too heavy to stick, when it drops and floats on the bath.

4 Ash in reverberatory flue, very similar to No. 2 (See Section C-C, Fig. 3).

Section B-B, Fig. 3, is taken at right angles to the bridge wall of the furnace and shows the large mass of ash, slag, unburned coal and coke which accumulated here and materially affected the operation of the furnace. This was deposited on the bridge wall and gradually built up in the shape of large horns in front of each burner, sometimes reaching within 8 or 12 in. of the burner.

Quite a large quantity of ash was deposited each shift on the boiler tubes, but was easily blown off by compressed air once or twice a shift. It was estimated that at least 50 per cent of the total ash was disposed of in the manner described, while the remainder was deposited in the main flue and went up the stack. The average ash analysis was as follows:

| SiO₂ | Al₂O₄ | FeO | CaO | MgO | Cu |
|---|---|---|---|---|---|
| 49.9 | 28.4 | 11.0 | 2.5 | 0.26 | 1.24 |

The high copper content is probably due to calcines lodging on the ash and slag float. Two samples showed Cu 2.92 and 1.91, respectively.

As a result of these experiments a modern 250-ton coal-pulverizing plant was designed and is now in course of erection. Blast furnaces, reverberatories and sinter plant will be equipped for pulverized coal, and the experiments will be continued to ascertain the equipment most suitable for local conditions, which will then be used at the new smelter now being constructed.

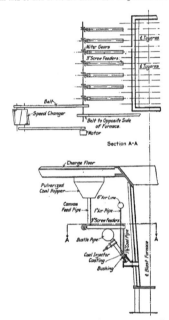

FIG. 4 EXPERIMENTAL PULVERIZED-COAL EQUIPMENT OF No. 5 BLAST FURNACE

Fig. 3 shows the layout of the present smelter in plan, with various elevations, flow sheet of coal-pulverizing plant, and the location of this plant relative to the general smelter layout. In conclusion, it may be of interest to note that these experiments and tests were carried out at an elevation of 14,200 ft.

---

The Historical Report of the Chief Engineer, Including All Operations of the Engineer Department, American Expeditionary Forces, 1917-1919, published by the War Department, consists of a main body and appendices. The main body constitutes the history of the engineers in France and is divided into three parts as follows: Part One deals with the organization and development of the department from the time of the appointment of the first chief engineer, A. E. F., May 18, 1917; Part Two is devoted to military engineering, including staff and army operations; Part Three relates to engineer activities in the services of supply, construction and forestry. The appendices consist of technical discussion, departmental and special service reports, regimental histories, and material of a similar detailed character.

# The Great Hydroelectric Plant at Keokuk

## Excursion to This Plant Is Planned for Members and Guests of The American Society of Mechanical Engineers in Connection with Their Spring Meeting

THE Mississippi River Power Company's power plant at Keokuk, Iowa, is by far the greatest low-head hydroelectric development ever undertaken. The entire structure is built of monolithic concrete and includes a dam, power house, navigation lock, and dry dock. Measured from end to end it is 10,560 ft., or two miles long, and is believed to be the largest monolith of its kind in the world.

It is located at the foot of the Des Moines Rapids, which extend from Montrose to Keokuk, a distance of twelve miles. In this distance the river bed has a fall of 23 ft., and the surface fall varies from 23 ft. at low water to 16 ft. at high water, which provides a normal working head for hydraulic machinery of

will provide room for 30 units each comprising a 10,000-hp. single-runner vertical Francis turbine connected to a vertical 9000-kva., 11,000-volt., 25-cycle, three-phase generator operated at 577 r.p.m. The present structure is only half of that called for by the plans, and provides for only 15 main generating units. For the substructure foundations, excavation was carried 25 ft. below the surface of the blue limestone bed of the river. From the forebay the water passes through the gate openings in the gatehouse section of the building, thence entering four branch intake tubes for each 10,000-hp. turbine. These four entry openings, 22 ft. by 7.5 ft., deliver the water to the scroll chamber at the sides and rear of the turbine setting. By the design of

GENERAL VIEW OF THE GREAT HYDROELECTRIC PLANT AT KEOKUK

32 ft., varying from 21 ft. to 39 ft. according to the stage of the river. The discharge of the river, at the site of the power station, where it is approximately a mile wide, is 200,000 cu. ft. per sec. at low water and 372,500 cu. ft. per sec. at the flood stage, which discharge made possible the project.[1]

A 4650-ft. spillway dam of the gravity section type, which is surmounted by a bridge, extends from Hamilton, Ill., on the east side of the river, to the power house which is near the west side of the river. It consists of 119 arched spans, each having 30-ft. openings and 6-ft. piers. Each of these 30-ft. spillways may be closed by a gate or wier; in other words there are 3570 ft. of spillway. The upstream side of the spillway section is vertical: the downstream side is rounded off into an ogee curve which discharges the flow quickly into the river below.

In order that drift and ice may be excluded from the forebay, a 2340-ft. fender pier was constructed which extends upstream from the power house and curves toward the west shore. There is a 300-ft. opening between the shore abutment and the end of the pier to allow the passage of river traffic. When navigation is closed, or when there is a large amount of floating matter in the river, this opening is closed by means of a floating boom.

The completed power house will be 1700 ft. by 123 ft. and

[1] These figures are the results obtained by 20 years of observation.

the scroll chamber, 39 ft. in diameter and molded in concrete to follow the mathematical curvature required, the water is impinged upon the turbine blades from all sides with equal force and velocity. The draft tubes, 18 ft. in diameter at the rotor, enlarge rapidly as they assume a horizontal direction to empty into the tail race. Because of this design the water, as it enters, moves at a speed of 14 ft. per sec. and is discharged into the tail pool at about 4 ft. per sec.

A short dam, extending downstream from the power house, a navigation lock, and a dry dock at the west shore, form the forebay of the power house. The lock is a single 40-ft. lift, the chamber of which is 400 ft. by 110 ft., and may be filled in 10 minutes and emptied in 12 minutes. One of the features of this lock is the upstream gate, which may be described as a single leaf with a void chamber. This gate is operated as follows: To open it, the chamber is filled with compressed air, and because of the buoyant effect it floats into place; to close it, the compressed air is allowed to escape, and it sinks into the chamber provided for it because of its weight.

At the present time this station provides 110,000 hp. to St. Louis, East St. Louis, Alton, Hannibal, Quincy, Burlington, Ft. Madison, and adjacent territory; hence, it serves a population of about 1,120,000. With the 15 generators at the station it is possible to develop 13,500 kva.

VIEWS OF THE MISSISSIPPI RIVER POWER COMPANY'S PLANT, KEOKUK, IOWA

Fig. 1   Interior of Lock Operator's House
Fig. 2   Switchboard Room, Chief Operator's Board in the Foreground
Fig. 3   High-Tension Room
Fig. 4   Main Generating Room
Fig. 5   View Showing Power House and Dam
Fig. 6   Steamboats Passing through Navigation Lock, Upper Level

Photographs obtained through the courtesy of Hugh L. Cooper & Co., Consulting Engineers, Mississippi River Power Company, 101 Park Ave., New York City

# The Dissipation of Heat by Various Surfaces

By T. S. TAYLOR,[1] PITTSBURGH, PA.

*Although it has long been the popular opinion that by covering hot-air pipes with sheet asbestos the heat loss was thereby reduced, some recent experiments by the author of the following paper indicate that quite the opposite is the case. Using bare tin as a standard of reference, he found that tin covered with 0.33 mm. (0.013 in.) of sheet asbestos will dissipate about 37 per cent more heat; asbestos-covered tin having a layer of dust, 32 per cent more; and tin with a layer of dust only, 7 per cent more. Calculations based on a series of tests indicate that it would require a covering of about 0.2 in. of sheet asbestos to make the loss from the covered pipe equal to that from the bright uncovered pipe. A thickness of 0.4 in. of asbestos, however, would result in a saving of 25 per cent over that of bare pipe, and about seven layers of 0.013 in. sheet asbestos loosely applied on a bare pipe would effect a saving of about 75 per cent. The author also presents figures, derived from tests, showing the effect of air velocity upon the dissipation of heat. Curves are given showing the watts dissipated per unit area and air velocity for definite surface temperature excesses of from 10 to 70 deg. The effect of the angle of incidence of the air is also discussed in the light of experimental work and it is shown that the maximum dissipation of heat takes place when air is blown over the object at an angle of from 40 to 45 deg.*

N OT long ago, while conducting a series of tests to determine the relative thermal insulating properties of asbestos, it was observed that warm water placed in a plain tin vessel cooled more slowly than when placed in a similar vessel covered with thin asbestos. Since this observation was in direct contradiction to popular opinion, it seemed worth while to make some more definite tests on the relative dissipation of heat by such surfaces in still air. To conduct such tests tin vessels were accordingly constructed with a lid at one end and having a diameter of 10 cm. and length of 50 cm. A cylindrical heater was made by winding No. 21 constantan wire longitudinally on an asbestos-board framework so as to slip readily into the vessels. The heater was so constructed that the same amount of heat would be developed per unit area of surface of the vessels, both sides and ends. This made it possible to maintain the temperature within the vessel at various values above the surrounding air temperature.

The outer surfaces of the vessels were as follows: plain bright tin; tin covered tightly with 0.33-mm. (0.013 in.) sheet asbestos; tin covered loosely with three layers of 0.33-mm. asbestos; tin aluminum-painted; galvanized; and various dust-covered surfaces. A thermometer inserted through the side of each vessel provided means for measuring the temperature at the center of the vessel, and thermocouples of 0.005-in. copper-constantan wire were attached to the outer surface of each vessel so as to measure their surface temperatures. The vessels were always placed horizontally in such positions in the room as to be free from unnecessary convection currents.

Observations were taken of the amounts of electrical energy, measured by ammeter and voltmeter, necessary to maintain various differences between the temperature within the vessel as determined by the thermometer and the surrounding temperature. Surface and room temperatures were also taken at the same time.

The room temperature was taken at points sufficiently distant from the vessel to be uninfluenced by it. In this manner results were obtained showing the watts dissipated per unit area for various temperature excesses, internal above room, for different surfaces. Curves 1 and 2 in Fig. 1 show the results obtained for tin covered tightly with 0.33-mm. sheet asbestos and bare tin, respectively. Curves 1' and 2' give the relations in watts per sq. cm. per deg. cent. plotted against temperature for the corresponding surfaces. It is seen that the dissipation of heat per unit area per degree of temperature excess increases almost uniformly with the temperature difference over the range of temperatures used. Thus the curves 1 and 2 can be represented approximately by the relation, $W = AT + BT^2$; where $W$ is the watts dissipated per unit area, $T$ the temperature excess (internal above surrounding air) and $A$ and $B$ constants for each surface.

Table 1 gives the values of the heat dissipated by the various

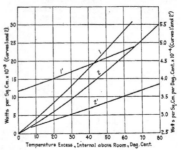

FIG. 1 HEAT DISSIPATED FROM TIN TIGHTLY COVERED WITH 0.33-MM. SHEET ASBESTOS

surfaces at corresponding temperature excesses. In addition to the surfaces listed in Table 1, results were also obtained for galvanized surface and tin *loosely* covered with three layers of sheet asbestos when dust-covered. Table 2 gives the relative

TABLE 1   HEAT LOSSES FROM VARIOUS SURFACES
All values are expressed in watts per sq. cm. ×10⁻⁴.

| Temperature excess internal above room deg. cent. | No. 2—Bare Tin | No. 1—Tin covered tightly with 0.33-mm. asbestos | No. 1 Covered with dust | No. 3—Tin covered with dust | No. 5—Galvanized sheet iron | No. 6—Tin covered with 3 layers asbestos paper | No. 7—Tin, aluminum-painted |
|---|---|---|---|---|---|---|---|
| 5 | 1.33 | 1.87 | 1.75 | 1.40 | 1.50 | 1.50 | 1.60 |
| 10 | 2.70 | 3.80 | 3.62 | 2.90 | 3.00 | 3.00 | 3.20 |
| 15 | 4.09 | 5.90 | 5.60 | 4.55 | 4.70 | 4.65 | 4.85 |
| 20 | 5.70 | 8.00 | 7.70 | 6.20 | 6.36 | 6.20 | 6.66 |
| 25 | 7.18 | 10.20 | 9.85 | 7.95 | 8.20 | 7.85 | 8.60 |
| 30 | 9.09 | 12.70 | 12.06 | 9.85 | 10.11 | 9.80 | 10.45 |
| 35 | 10.99 | 15.15 | 14.35 | 11.75 | 12.15 | 11.35 | 12.35 |
| 40 | 12.88 | 17.51 | 16.90 | 13.85 | 14.20 | 13.15 | 14.35 |
| 45 | 14.90 | 20.15 | 19.50 | 15.90 | 16.35 | 14.90 | 16.30 |
| 50 | 16.90 | 22.20 | 22.10 | 18.10 | 18.50 | 16.85 | 18.70 |
| 55 | 19.03 | 25.45 | 24.70 | 20.35 | 20.85 | 18.80 | 20.90 |
| 60 | 21.24 | 28.35 | 27.60 | 22.50 | 23.00 | 20.75 | 23.15 |
| 65 | 23.40 | 31.40 | 30.25 | 24.87 | 25.30 | 22.70 | 25.45 |
| 70 | 25.55 | | | 27.15 | 27.65 | 24.70 | 27.75 |
| 75 | 27.90 | | | 29.35 | | 26.70 | 30.20 |

[1] Mellon Institute, University of Pittsburgh; formerly of the Westinghouse Electric & Manufacturing Company.

[2] Since compiling these results the writer has learned that somewhat similar results have been observed independently by V. S. Day at the University of Illinois and were noted at the meeting of the National Warm-Air Heating and Ventilating Association, Columbus, Ohio, June 11, 1919. Observations showing the results outlined in the earlier portions of the present paper were first taken about February 1, 1919. Nothing further was done at that time owing to the pressure of other work. Recently, however, occasion permitted further work and the results confirmed the original ones. A preliminary notice of the present results appeared in MECHANICAL ENGINEERING for January, 1920, p. 69.

Abstract of a paper to be presented at the Spring Meeting, St. Louis, May 24 to 27, 1920, of THE AMERICAN SOCIETY OF MECHANICAL ENGINEERS. Copies of the complete paper may be obtained gratis upon application. All papers are subject to revision.

230

amounts of heat dissipated as compared with bare tin for corresponding differences in temperature. Tin covered tightly with one layer of 0.33-mm. sheet asbestos will dissipate 37 per cent more heat in still air than the bare tin. Even when both are covered with dust, such as that usually found on furnace pipes, the asbestos-covered surface will lose 23 per cent more heat than the bare one. The effect of dust on the pipes is to increase the loss of the bare pipe and decrease the loss of the asbestos-covered one. It requires at least three layers of 0.33-mm. sheet asbestos applied loosely on a bare pipe in order to dissipate no more heat than the uncovered bare pipe. This is readily shown by Table 2.

It will be observed from a study of Table 2 that the ratio for

TABLE 2  HEAT DISSIPATED BY VARIOUS SURFACES AS COMPARED WITH THAT DISSIPATED BY BARE TIN No. 3.

| Temperature excess internal above room deg. cent. | No. 1—Tin covered with 0.33-mm. sheet asbestos | No. 2—Dust covered | No. 3—Dust covered | No. 5—Galvanized sheet iron | No. 6—Tin covered with asbestos loosely layers three times | No. 5—Dust covered | No. 6—Dust covered | No. 7—Tin, aluminum-painted |
|---|---|---|---|---|---|---|---|---|
| 5 | 1.41 | 1.32 | 1.05 | 1.13 | 1.13 | 1.18 | 1.04 | 1.20 |
| 10 | 1.41 | 1.34 | 1.07 | 1.11 | 1.11 | 1.19 | 1.02 | 1.18 |
| 15 | 1.43 | 1.37 | 1.11 | 1.14 | 1.13 | 1.18 | 1.05 | 1.18 |
| 20 | 1.40 | 1.34 | 1.09 | 1.11 | 1.09 | 1.17 | 1.03 | 1.17 |
| 25 | 1.40 | 1.33 | 1.08 | 1.11 | 1.06 | 1.18 | 1.02 | 1.17 |
| 30 | 1.40 | 1.33 | 1.08 | 1.11 | 1.05 | 1.18 | 1.02 | 1.15 |
| 35 | 1.38 | 1.30 | 1.07 | 1.10 | 1.03 | 1.17 | 1.02 | 1.12 |
| 40 | 1.37 | 1.31 | 1.04 | 1.10 | 1.02 | 1.19 | 1.01 | 1.12 |
| 45 | 1.35 | 1.31 | 1.07 | 1.10 | 1.01 | 1.20 | 1.01 | 1.11 |
| 50 | 1.34 | 1.31 | 1.08 | 1.10 | 1.00 | 1.20 | 1.00 | 1.11 |
| 55 | 1.33 | 1.30 | 1.07 | 1.10 | 0.99 | 1.20 | 1.00 | 1.10 |
| 60 | 1.34 | 1.30 | 1.06 | 1.08 | 0.98 | 1.20 | 0.99 | 1.09 |
| 65 | 1.34 | 1.28 | 1.06 | 1.08 | 0.97 | 1.19 | 0.99 | 1.09 |
| 70 |  |  | 1.06 | 1.08 | 0.97 |  | 0.98 | 1.09 |
| 75 |  |  | 1.05 |  | 0.96 |  | 0.97 | 1.08 |
| M'e | 1.37 | 1.32 | 1.07 | 1.10 | 1.05 | 1.24 | 1.01 | 1.13 |

all surfaces with respect to the bare pipe becomes somewhat smaller with increasing temperature excess. This indicates that if the temperature difference are high enough, very little difference would exist between the amounts of heat dissipated by each. This condition would not exist, however, until the surface temperatures were such that the heat would be lost chiefly through radiation, which for the present temperature range plays but little part.

### HEAT LOSSES FROM HOT-AIR PIPES

One very interesting feature about these results is their application to the loss of heat by hot-air furnace pipes. From the results in Tables 1 and 2 it is quite evident that hot-air-furnace pipes lose more heat when coated with the usual sheet asbestos than when left bare. Furthermore, this difference is too great to be merely given a casual consideration, and the following brief discussion will emphasize the point. Let us consider in the first place what thickness of covering would be necessary in order to insure no more loss of heat by the covered pipe than by the uncovered one. For the same temperature excess, say, 40 deg. cent. internal above surrounding air, the covered pipe loses $17.52\times10^{-3}$ watts per sq. cm. while the bare pipe loses $12.70\times10^{-3}$. This is seen by reference to curves 1 and 2 in Fig. 1. Experiments show that a covered pipe (No. 1) to lose only $12.70\times10^{-3}$ watts per sq. cm. would require an outer-surface temperature excess of 12.8 deg. cent., and that when losing $17.52\times10^{-3}$ watts per sq. cm. its surface temperature excess is 17.2 deg. cent. Hence sufficient insulation must be added to reduce the surface temperature to the lower value, or there must be enough asbestos added to produce a drop of 4.4 deg. (17.2 − 12.8). The thickness of asbestos necessary is given by the following equation:

$$12.7\times10^{-3}\times0.239=\frac{4.4\times0.00035}{d}$$

where 0.239 is the factor to reduce watts to calories and 0.00035 the thermal conductivity of asbestos paper in calories per cm. per sec. Solving the equation for $d$ gives a value of 0.51 cm., which is practically 0.2 in. While this is an approximate solution it shows that considerably more thickness of insulation should be applied to hot-air pipes in order to make them as efficient as if they were left bare.

It is interesting to speculate as to the possible saving that would result by leaving the pipes bright and uncovered. Suppose there is a temperature excess, internal above surrounding air, of, say, 40 deg. cent. (72 deg. fahr.). As is shown above this corresponds to a loss of $17.52\times10^{-3}$ watts per sq. cm. or 0.113 watt per sq. in. from the covered pipe. If we have 10 pipes 10 ft. long and 10 in. in diameter, that is, approximately 36,000 sq. in. of surface, the total loss would be $0.113\times36,000=4068$ watts. The total loss per day would be $4068\times24\times3600$, or $3.52\times10^{9}$ joules. One pound of coal has a heating value of approximately 12,500 B.t.u. $=1.32\times10^{7}$ joules. Consequently the loss in pounds of coal per day would be $3.52\times10^{9}\div1.32\times10^{7}$, or 26.6. This would be equivalent to about 75 bu. during the heating season. The loss through a bare pipe would be equivalent to $100\div137$ (see Table 2) of this value, or about 54 bu. These considerations indicate, therefore, that the pipe system in question covered with 0.33-mm. sheet asbestos will lose during a winter season a quantity of heat equivalent to that obtained from 20 bu. of coal more than the same system would lose if left uncovered.

The explanation of the larger loss through a pipe when covered

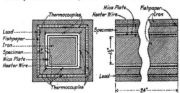

FIG. 2  APPARATUS USED IN MEASURING HEAT LOSSES

with a thin layer of asbestos is due to the fact that the asbestos surface is some three or four times as great as the plain tin surface so far as molecular dimensions are concerned. The loss being due chiefly to air contact, it is readily seen that the greater the surface for the molecules to come into contact with, the more heat will thus be liberated. The radiating power of the asbestos also being larger than that of tin, will contribute an additional amount to the advantage of the asbestos as far as heat loss is concerned. The loss due to radiation, however, at these temperature differences is quite small compared with that lost by convection currents. Since the asbestos surface facilitates the loss of heat due to its increasing the effective molecular contacts, the surface of the asbestos and also the outer surface of the tin will thus have their temperatures considerably decreased and the result will be that more heat must pass through the tin and asbestos as a consequence of this condition. Therefore, when the surface of the pipe is thus changed and the heat losses increased for a given temperature gradient, it is necessary to overcome this by increasing the thickness of the asbestos layer to such an extent that the thermal resistance of the pipe and asbestos or insulation will cut down the heat flow to the desired amount. That is, the radiation resulting from increasing the effective area must be counteracted by increasing the thermal resistance through the addition of a greater thickness of insulation.

### INFLUENCE OF AIR VELOCITY ON DISSIPATION OF HEAT

The work described under this heading was primarily undertaken for the purpose of securing data useful to engineers in designing electrical apparatus. The results obtained are for the surface of a typical end coil of a turbo-generator, but they nevertheless are of value to anyone who is interested in the problem of air cooling. The apparatus upon which the wrapper, composed of treated cloth and tape, was placed was the same as had been previously used in measuring the thermal conductivities of coil wrappers (see *Electrical World*, February 14, 1920, p. 369). An iron bar 1½ in. by 1½ in. by 24 in. was used as a core in order to secure rigidity. A layer of ₁₆-in. heater mica was pressed over the core and a heater wire of No. 21 constantan wound over the mica so as to have eight turns to the inch. The

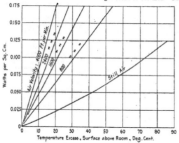

FIG. 3   EFFECT OF AIR VELOCITY ON DISSIPATION OF HEAT

space between the turns was filled with asbestos cement and a second layer of ₁₆-in. heater mica plate was then pressed over the entire apparatus. After having been thoroughly dried out by sending a current through the heater wire while the entire heater was held between clamps, the apparatus was wrapped with the insulation according to definite specifications. Two thermocouples were placed on the wrapper on each side of the heater, one at the center of each side and another at the edge at corresponding positions along heater. The thermocouple wires (0.005 in. copper and constantan) were run out along the heater its entire length, the copper ones to one end and the constantan to the other. A sketch of the heater is shown in Fig. 2.

The heater after having been thus wrapped and arranged was placed about eight inches in front of the outlet of a blower and parallel to the opening so that the air stream fell at right angles upon it. The ends of the heater were covered with wool felt to prevent loss of heat therefrom. The outlet from the blower was of such dimensions (2¼ in. by 2¼ in. by 24 in.) that the coil was completely within the air stream. Baffle plates placed in the air channel made it possible to secure a symmetrical distribution of the air stream. A small pitot tube made from a hypodermic needle was used to measure the velocity. The differences of pressure were read on a differential draft gage. Observations were also made of the air temperature and barometric pressure. The velocity of the air ($V$) was calculated for standard conditions, 760 mm. pressure and 25 deg. cent. temperature, by use of the formula

$$V = \sqrt{\frac{2\,ghd}{d'}}$$

where $g$ is the acceleration due to gravity, $h$ the height of the liquid in the differential gage, $d$ the density of this liquid, and $d'$ the density of the air.

Measurements of the air velocity at the point in the air stream where the coil was situated showed but little variation from that for corresponding points at the opening. At least, whatever variation did exist was of the same order of magnitude as the experimental error. It was therefore assumed that the average of the velocity would be a fair value to take as the velocity of the air blowing over the coil.

The current in the heater was maintained constant and a constant number of watts were thus dissipated per unit area. Observations were taken of the excess of the surface temperature of the coil (determined from the average of the eight thermocouples on its surface) above the temperature of the impinging air for various air velocities and it was found that the heat liberated per degree excess increases approximately uniformly with the velocity over the range of velocities investigated.

In the above manner relations were determined for various amounts of heat liberated up to 0.186 watt per sq. cm (1.2 watts per sq. in.). From these values relations were obtained between the watts dissipated per sq. cm. and the corresponding temperature excess of the surface above air temperature for various air velocities. The curves in Fig. 3 show the watts dissipated per sq. cm. and corresponding temperature excesses for air velocities of 0, 800, 1600, 2400, and 4000 ft. per min. For still air it is seen that the amount of heat liberated per degree of temperature excess increases with increasing temperature excess. On the contrary, it is seen that the watts liberated per unit area per degree excess is practically constant for all air velocities other than natural convection currents. The watts dissipated per unit area varies uniformly with the temperature excess for constant air velocities. However, it is not safe to assume from these

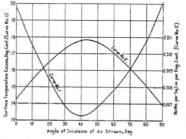

FIG. 4   EFFECT OF ANGLE OF INCIDENCE ON DISSIPATION OF HEAT

experiments that such linear relationships continue to hold indefinitely as the watts dissipated increases.

### EFFECT OF ANGLE OF INCIDENCE OF AIR STREAM

All the foregoing results were obtained for perpendicular incidence or when the angle of incidence of the air stream was zero. From results that had been obtained previously on the cooling of a very small coil of wire (⅛ in. in diameter) when placed in an air current, it was seen that the amount of heat dissipated for a given temperature excess was different for different angles of incidence of the air stream. It therefore seemed worth while to make some tests to determine the way in which the temperature excess of the surface of the coil wrapper varied with the angle of incidence for a definite amount of heat dissipated per unit area and constant air velocity. This was done for a dissipation of 0.145 watt per sq. cm. (0.938 watt per sq. in.) and an air velocity of 3267 ft. per min. Curve No. 1, Fig. 4, shows how the temperature excess changes under the above conditions as the angle of incidence of the air stream changes from 0, that is, perpendicularly, to 90 deg. or parallel to the coil. Curve No. 2 shows how the watts per sq. cm. deg. cent. changes with the angle of incidence. The curves are quite interesting in that they show the relative cooling effects of air at various angles of incidence. It is seen that the temperature excess for the particular air velocity of 3267 ft. per min. and a dissipation of 0.145 watt per sq. cm. at an angle of incidence of air stream of about 40 to 45 deg. is only 67 per cent of what it is for an angle of incidence

(Continued on page 259)

# ENGINEERING SURVEY

A Review of Progress and Attainment in Mechanical Engineering and Related Fields, as Gathered
from Current Technical Periodicals and Other Sources

SUBJECTS OF THIS MONTH'S ABSTRACTS

MUFFLER DESIGN FOR AEROPLANE ENGINES
BACK PRESSURE IN EXHAUST LINE AND
POWER LOSS OF ENGINES
MÉLOT SYSTEM OF REACTION JET PRO-
PULSION
AERODYNAMIC PROPERTIES OF THICK AERO-
FOILS
VAPOR PRESSURE OF AMMONIA
PICKLING AND PHYSICAL PROPERTIES OF
ALLOY STEELS
HOWE SUSPENDED MOLDING MACHINE
MELTON-HAURY SINGLE-SLEEVE-VALVE EN-
GINE

POTS AND BOXES FOR CARBONIZING
HEAT-RESISTING ALLOYS
NICKEL-CHROMIUM ALLOYS
NUT LOCK
MULTIPLE-WAY DRILLING MACHINES,
FOOTE-BURT
CHIP-PROTECTION DEVICE ON MULTIPLE-
WAY DRILLING MACHINES
FATIGUE AND ITS EFFECT ON PRODUCTION
FATIGUE AND ACCIDENTS
CONCRETE SHIP, PRESENT STATUS
PROPAGATION OF FLAME IN GAS MIXTURES

ASH- AND COAL-HANDLING EQUIPMENT
CANADIAN FUEL AND STEAMING TESTS
HIGH-VACUUM MERCURY PUMP
STEAM AND ELECTRIC PROPULSION FOR
RAILROADS
STANDARDIZATION OF ELECTRIC FREQUEN-
CIES
FUEL UTILIZATION ON RAILROADS
STEAM AND ELECTRIC LOCOMOTIVES COM-
PARED
AIR PURIFICATION BY OZONE
OZONE-GENERATING APPARATUS

## AERONAUTICS

### Muffler Design; Back Pressure in Exhaust Line and Power Loss of Aeroplane Engines

INVESTIGATION OF MUFFLING PROBLEM FOR AEROPLANE EN-
GINES, G. B. Upton and V. R. Gage, Members Am.Soc.M.E. Data
of tests carried out under the auspices of the National Advisory
Committee for Aeronautics on a Curtiss aeroplane engine and
several stationary and other engines, using a fan dynamometer.
The paper describes in considerable detail the methods of measurements and the formula used. As regards the relation between
back pressure and power output, it appears to have been found
that for moderate back pressures the power loss is substantially
proportional to the back pressure, while for higher back pressures
the power loss mounts rapidly, apparently at such a rate that a

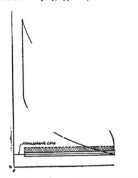

FIG. 1 INDICATOR DIAGRAM SHOWING LOSS OF POWER THROUGH
BACK PRESSURE AT EXHAUST VALVE

back pressure of even less than 10 lb. per sq. in. would stop the
engine.

A possible explanation of this changing effect of back pressure
as the back pressure increases may be found by considering the
indicator card. This is schematically shown in Fig. 1. For small
back pressures we may expect the main effect to be a lifting of
the exhaust line of the card by an amount substantially equal to

the increase of back pressure. The result would be a loss of indicated mean effective pressure equal to the back pressure, because
the elevation of the exhaust line would extend through the whole
stroke. The loss of brake mean effective pressure will be smaller
than the loss of indicated mean effective pressure in the ratio of the
mechanical efficiency of the engine to unity.

At higher back pressures the exhaust gases are held back in
greater amounts in the cylinder, leaving the clearance space, at
the end of the exhaust period, filled with an abnormal weight of

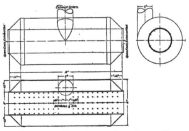

FIG. 2 TANGENTIAL-WHIRL-CHAMBER TANK MUFFLER RECOM-
MENDED BY THE NATIONAL ADVISORY COMMITTEE FOR AERO-
NAUTICS FOR THE LIBERTY AEROPLANE ENGINE

hot, dead gases. These reëxpanding, interfere with the incoming
charge in various ways, lessening the amount of the fuel mixture
drawn in. The decrease of charge quantity will result in a decrease
of the mean effective pressure which is added to the decrease of
mean effective pressure due to lifting of the pressure of the exhaust line.

Probably it is the decrease of charge which is the principal reason for the possibility of stalling the engine by fairly completely
choking the exhaust pipe and before complete closure is reached.

It was also found, as regards the relation between the brake
horsepower and the actual back pressure, that "the back pressure
increases as some exponential function of the horsepower, when the
conditions of the exhaust passages remain unchanged" [quoted
literally.—EDITOR.].

Data are presented to demonstrate that the choking of the exhaust by sharp turns, pipe fittings, etc., gives the same results as
choking by a muffler.

In the course of the experimental work carried out some peculiar
phenomena were noted. One such was the abnormal power drop,

233

considering the back pressure, at certain critical speeds. It was found that this abnormal power loss could be avoided by a very small change of speed either way from the critical.

The critical speed changed or disappeared with change of exhaust manifold. The supposed cause of this abnormal power loss at the critical speed is a reflected wave of exhaust gas killing the clearance of some cylinder just before its exhaust valve closed.

A number of mufflers were tested.

The authors recommended a tentative design for the Liberty 12-cylinder engine, that shown in Fig. 2, where air is supposed to pass through the inner tube to some extent aiding in cooling. (*National Advisory Committee for Aeronautics*, Report No. 55, preprint from Fifth Annual Report 1919, 38 pp., 27 figs. and numerous tables, *e*)

### Blast Engine with Reaction Jet Propulsion

MÉLOT SYSTEM OF AIRCRAFT PROPULSION. At the recent Paris aeronautical show there was exhibited a trial machine embodying the Mélot principle of propulsion. It is stated that the inventor experimented during the war at the Laboratoire du Conservatoire des Arts et Métiers in coöperation with the French Ministry in Munitions.

The machine consists of a burner with fireproof lining. Into this an explosive mixture is injected and ignited in the first instance by a spark plug, after which combustion continues uninterruptedly. The burned gases are exhausted through a blast

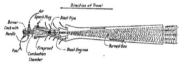

FIG. 3 MÉLOT SYSTEM OF PROPULSION FOR AIRCRAFT

pipe and four blast "engines," which are really graduated nozzles, one large and three small.

At the inlet mouth of each of these nozzles air is drawn in by suction, and the whole mixture of air and burned gases is exhausted by the last and largest blast engine and makes a direct thrust against the air at the rear of the engine (Fig. 3).

The machine is said to give 30 hp. for a relative speed of 50 m. (164 ft.) per sec. The weight per horsepower is said to be 0.5 kg. (1.1 lb.), and it is obvious that if the machine is capable of functioning for any length of time, its cost ought to be quite low because of the absence of valves, pistons and other parts of the conventional engine. (*Aeronautics*, vol. 18, no. 331 (New Series), Feb. 19, 1920, p. 157, 2 figs., *d*)

THE AERODYNAMIC PROPERTIES OF THICK AEROFOILS, F. H. Norton. A report dealing with the results of a series of tests conducted at the Massachusetts Institute of Technology wind tunnel, with a view to developing aerofoil sections thick enough to permit of internal bracing and the use of cantilever wings without any external bracing of the airplane wing truss. The sections tested were based on the Durand 13, and were varied in section form, in thickness along the span, and in chord along the span. Tapering both in thickness and in chord was found to be highly beneficial to efficiency, and some of the wings developed have $L/D$ ratios practically as high at angles corresponding to very high speeds of flight as the best of the wing sections for normal type, together with very much higher maximum lift coefficients. In particular, it was found that the substitution of a thick tapered wing for R. A. F. 6 on a 3600-lb. fighting biplane with Liberty engine would increase the maximum speed by 18 m.p.h., due to the saving in parasite resistance by entire elimination of the interplane bracing. Report No. 75 of the *National Advisory Committee for Aeronautics*, preprint from Fifth Annual Report 1920, pp. 5-26, 29 figs., *te*)

## BUREAU OF STANDARDS

VAPOR PRESSURE OF AMMONIA, C. S. Cragoe, C. H. Myers and C. S. Taylor. The previous measurements of the vapor pressure of ammonia are briefly reviewed and tabulated. A detailed description is given of the apparatus and method employed in the present measurements throughout the temperature interval — 78 deg. to + 70 deg. cent. Seven samples of thoroughly purified ammonia were used. Special tests showed less than one part in 100,000 by volume of non-condensing gases and less than 0.01 per cent by weight of other impurities. The methods of purification and filling manometers are briefly described. The phenomenon of hysteresis was observed near the normal boiling point of ammonia with a commercial sample containing a small amount of air, which indicated the necessity of very complete removal of dissolved gases for any accurate measurements of vapor pressure by the static method. Lags in coming to equilibrium were encountered and studied in order to determine the most advantageous procedure in establishing equilibrium. The normal boiling point of ammonia was determined by the static and also the dynamic method, the mean of the results by the two methods being —33.35 deg. cent. Two empirical equations were found to represent closely the results in the temperature range covered experimentally and also the latest determination of the critical data for ammonia. The results of 122 measurements in the interval — 78 deg. to + 25 deg. cent. made with direct observations of mercury columns agree with the empirical equations within 1 mm. of mercury. The results of 28 measurements in the interval + 15 deg. to + 70 deg. cent. made with an accurately calibrated piston gage agree with the empirical equations within about 3 mm. of mercury. As a final result the vapor pressure of ammonia is expressed in the range — 80 deg. to + 70 deg. cent. by either of the following equations:

$$\log p = 30.256818 - (1914.9569/\theta) - 8.4598324 \log \theta$$
$$+ 2.39309 \times 10^{-3}\theta + 2.955214 \times 10^{-6}\theta^2$$
$$\log p = 12.465400 - (1648.6068/\theta) - 0.016386466 + 2.403267$$
$$\times 10^{-5}\theta^2 - 1.168708 \times 10^{-7}\theta^3$$

where $p$ is expressed in mm. of mercury and $\theta$ is degrees absolute (deg. abs. = deg. cent. + 273.1). (*Abstract of Scientific Papers of the Bureau of Standards* No. 369, *e*)

## ENGINEERING MATERIALS

### Does Pickling Affect the Quality and Machinability of Steel?

EFFECT OF PICKLING ON ALLOY STEEL, H. L. Hess, Mem.Am. Soc.M.E. Pickling is extensively used, as it facilitates the discovery of seams and surface defects generally, and removes all furnace and rolling scale. Various solutions are used, such as highly diluted baths of sulphuric or hydrochloric acids.

A series of tests were carried out by the metallurgical department of the Hess Steel Corporation, Baltimore, Md. In these tests 1 lb. of a special pickling compound (not otherwise specified) was dissolved in 3 gal. of water, and the mixture held at a constant temperature of about 200 deg. fahr. The tests were made on round pieces (1½ in. in diameter and 3 in. long) of chrome steel with approximately one per cent carbon and 1.50 per cent chromium.

As the object was to determine the lasting effect of the pickling on the metal, various subsequent treatments were used in order to neutralize fully or partly the effect of the pickling, the pieces being washed in cold, hot or lime water and tested either immediately or after periods varying from 7 to 28 days. The tests were made for hardness without cleaning, polishing or filing the surfaces in any way previous to the test, this being done in order to avoid the possible removal of a skin which might have appeared as a result of the pickling.

On the whole, it was found that the hardness results do not seem to be affected by the pickling treatment, although there was a slight indication that the unpickled steel has a somewhat softer surface than the pickled specimens.

Careful machining tests showed absolutely no difference in machinability, and it would appear that pickling affects only the skin, and even that to an effect not noticeable in machining.

Further tests were made to find out, if possible, whether pickling would have any effect upon the surface of the material and whether this effect were traceable to any appreciable depth. In these tests each piece of steel was subjected to a careful microscopic examination of the surface before and after aging and at various magnifications. The photomicrographs do not show any distinct difference between the pickled and unpickled bars.

The general conclusion drawn from these tests is that pickling in itself, as well as pickling followed by various treatments, does not interfere in any noticeable way with the quality or machinability of the steel. (*Iron Age*, vol. 105, no. 9, Feb. 26, 1920, pp. 593-594, 6 figs., *e*)

## FOUNDRY

### Howe Suspended Molding Machine

SUSPENDED MOLDING MACHINE. Description of the installation and use of suspended molding machines at the works of the Standard Malleable Iron Company, Muskegon Heights, Mich., where it is claimed that their use effected an increase of production per molder of from 15 to 30 per cent, depending on the class of work done.

The machine hangs on a rail which extends lengthwise over

the aisle to the wall parallel with the course of the molding machine. At the close of molding operations for the day, the sand machine cuts the sand in each pile in the foundry. It moves from aisle to wall and back in the brief space of 1½ min. as compared with 30 min. of hand labor employing several men.

Each of the molding benches is equipped with adjustable shelves and a long side bench, the latter being attached to the machine by a bracket and supported on a wheel. (*Iron Age*, vol. 105, no. 10, Mar. 4, 1920, pp. 665-666, 3 figs., *d*)

## FUELS (See Railroad Engineering)

## HANDLING OF MATERIALS (See Power Plants)

## HEAT TREATING (See Machine Shop)

## INTERNAL-COMBUSTION ENGINES (See also Aeronautics and Physics)

### Single-Sleeve-Valve Engine with Outside Sleeves

MELTON-HAURY ENGINE. Description of a new engine, known as the Melton-Haury, whose characteristic feature is the single sleeve located outside of the cylinder, which means that the usual amount of water-cooling area is available.

The single-sleeve valve (Fig. 5 on the following page) is a cylindrical structure with port openings of large area. This cylindrical structure is split longitudinally and must be sealed

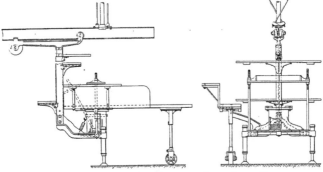

FIG. 4  HOWE SUSPENDED MOLDING MACHINE WITH SQUEEZER ATTACHMENT

the milling floor and is easily movable from place to place. It is equipped with pneumatically controlled legs (Fig. 4) by which it can be held firmly in a stationary position regardless of the irregularities of the floor. These legs are in reality plungers fitted in pneumatic cylinders and are controlled by a stopcock.

In its simplest form the machine is really a movable molding bench. With benches fixed at the walls, the molder was forced to carry his work over the floor to the aisle, running the risk of shifting the molds. The suspended bench enables him to commence work close to the aisle near his sand pile and gradually move toward the wall as his day's work progresses. The machine also enables the molder to handle all his own molds, even such where ordinarily a helper would be needed to carry them out.

Another advantage of this device is that the molder's afternoon work is not interfered with by the heat radiating from the hot sand dumped from the morning molds.

It is claimed that the machine facilitates the use of a sand cutter. The sand is dumped from the flasks in a pile extending from

in that direction. This is accomplished by a dovetailed edge that effectively seals the joint.

The cylinders are removable and are machined inside in the same manner as poppet-valve engines, giving also a minimum of friction surface on the outside, the valves traveling against this outside surface with the bearing surface integral with the block.

The valves are positively operated by individual valve shafts driven by helical gears, these shafts being located directly below the driving lug on the sleeves in order to reduce any side thrust or undue strains.

The construction is of interest because it permits the cylinder to be well cooled while the sleeve valve is amply lubricated. On the exhaust side, the exhaust manifold is provided with hot points protruding downward into the intake manifold against which the carburetor is attached. The gases, being preheated, pass through the block between the cylinders into the distributing manifold inside of the engine block. (*Commercial Car Journal*, vol. 18, no. 6, Feb. 15, 1920, pp. 23-24, 3 figs., *d*)

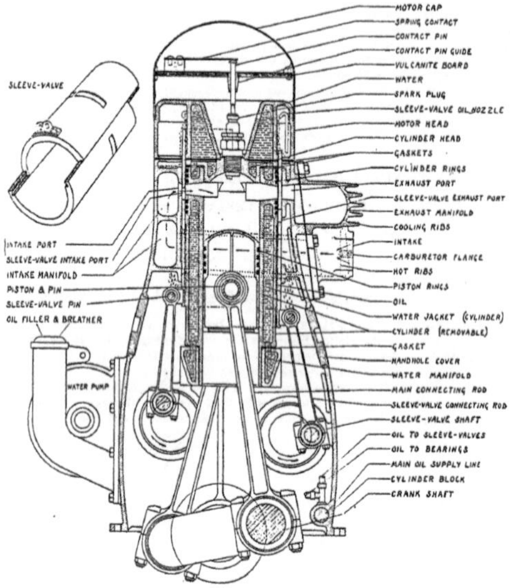
SLEEVE-VALVE
INTAKE PORT
SLEEVE-VALVE INTAKE PORT
INTAKE MANIFOLD
PISTON & PIN
SLEEVE-VALVE PIN
OIL FILLER & BREATHER
WATER PUMP
MOTOR CAP
SPRING CONTACT
CONTACT PIN
CONTACT PIN GUIDE
VULCANITE BOARD
WATER
SPARK PLUG
SLEEVE-VALVE OIL NOZZLE
MOTOR HEAD
CYLINDER HEAD
GASKETS
CYLINDER RINGS
EXHAUST PORT
SLEEVE-VALVE EXHAUST PORT
EXHAUST MANIFOLD
COOLING RIBS
INTAKE
CARBURETOR FLANGE
HOT RIBS
PISTON RINGS
OIL
WATER JACKET (CYLINDER)
CYLINDER (REMOVABLE)
GASKET
HANDHOLE COVER
WATER MANIFOLD
MAIN CONNECTING ROD
SLEEVE-VALVE CONNECTING ROD
SLEEVE-VALVE SHAFT
OIL TO SLEEVE-VALVES
OIL TO BEARINGS
MAIN OIL SUPPLY LINE
CYLINDER BLOCK
CRANK SHAFT

FIG. 5 MELTON-HAUBY ENGINE

## MACHINE SHOP

### Design and Materials for Pots and Boxes Used in Carbonizing; Heat-Resisting Alloys; Q-Alloy of Nickel-Chromium Family

POTS AND BOXES USED IN CARBONIZING, H. H. Harris. While heat-treating furnaces and processes have been consistently developed in the last score of years the pots and boxes used in the cyanide and lead-hardening processes have changed very little.

Three factors govern the service received from carbonizing pots and boxes, namely, design, method of manufacture, and material. As regards design, the author states that while in some plants there have been considerable improvements, through the industry at large a great majority of carbonizing boxes are of a design as well adapted for packing soap in as they are for carbonizing, which affects the results obtained in a very undesirable manner.

To be good a box must be designed to permit tight sealing with some refractory or other material, in such a manner, however, that the clay should not enter the box and mix with the compound.

Proper dimensions of the box are also important. The thickness of the box should be as thin as possible in order to prevent warpage and still sufficient to permit proper flow of metal in the casting.

Materials for pot and box manufacture may be grouped into six classifications: cast iron, cast steel, pressed and rolled steel, "trick" materials, alloys and materials calorized.

Cast iron is both the cheapest first-cost material and the poorest. It oxidizes rapidly, gives a non-uniform service, becomes distorted easily, forms scale which mixes with the carbonizing compound and is likely to spoil the work, and finally is affected by the cyaniding mixtures.

Cast steel is generally much superior to cast iron. It costs about twice as much per pound, but may give much longer service. The principal objection to cast steel is that the grade of steel used for pots and boxes is usually of inferior quality, sometimes even semi-steel being offered for steel. The method of casting the pots and boxes is also often unsatisfactory.

Judging, however, from the average of various types and conditions studied in many of the largest plants in the country, the author does not believe that any material has shown on the average a uniformly longer life per dollar invested than east steel, with the exception of a nickel-chromium alloy which he refers to as "Q-alloy."

Pressed- and wrought-steel pots were quite generally used some time ago before the advance in price of this material. Its advantage is its light weight and consequent small heat consumption. Its disadvantages are its high price and the comparatively few shapes in which it can be secured. Some companies manufacture their own annealing boxes from wrought steel, riveting or welding them together. Some of these are said to give satisfaction in extreme temperatures, but not under general conditions. Wrought-steel riveted or welded pots for cyanide, chloride or lead conditions have never been successful owing to leakage.

By "trick" materials, the author means products misrepresented by their makers or sold under a trade name which does not correctly indicate their nature. Alloys for heat-resisting purposes are a new development and may be said to be still in their infancy. At this time there are more than 35 patents covering heat-resisting alloys.

The latest development in the alloy field is a special nickel-chromium alloy, the analysis of which the author is not yet authorized to divulge. This material differs from nickel-chromium alloys on the market in that it retains many of the physical characteristics of the cold metal at a temperature of 2800 deg. fahr. and rings like a bell when struck with a hammer at this temperature. This new alloy is known as Q-alloy, grade X.

The only true scale of value by which a user can judge most competitive production materials consumed in service is by comparing their life under service conditions with their initial cost, and determining how many units of service each renders per dollar investment. In computing pot and box cost a unit is a heat-hour, which is an hour in the furnace under heat. Cost per heat-hour is arrived at by dividing the number of heat-hours received into the initial cost of the pot or box; for instance, if a cast-iron box weighing 100 lb. and cost 5 cents per lb. runs 100 heat-hours, the cost is 5 cents per hour. A steel box at 12 cents per lb. should run at least 300 hours under the same conditions, making a cost to the user of 4 cents per heat-hour. Under certain circumstances parallel to this Q-alloy has been known to run 7000 hours, making a cost to the user of 2 cents per heat-hour. Where an alloy affords a saving in cost per heat-hour it minimizes warpage and allows a thinner section to be used, thereby saving fuel in heating the box and its contents.

Thousands of tons of metal per annum are being consumed in the fires of heat-treating furnaces. This metal is paid for by companies every one of which could use the money so expended to the betterment of their product rather than writing it off in the profit-and-loss column and passing the tariff on to the ultimate consumer. American industry quite frequently puts up with undue waste to obtain production and perhaps the greatest waste in the metal-working industries, which can be directly attributed to ignorance and neglect, is the waste of metal consumed in the heat-treating processes. This waste can never be entirely eliminated, but it may be greatly reduced. (*The Iron Age*, vol. 105, no. 11, Mar. 11, 1920, pp. 729-731, *de*)

## MACHINE PARTS

A NEW NUT LOCK. A new form of locking washer for bolts and nuts has just been brought out by the Palnut Company, Limited, of 6 Great St. Helens, London, E. C. 3, and, although we have not yet had an opportunity of testing its efficacy thoroughly, the principles on which it is designed appear to us to be sound. The washer is a simple stamping which takes the form of an inverted dish, with the edge turned up in such a way as to make a hexagonal rim. The center of the dish is pierced for the passage of the bolt, and there are half a dozen radial slots which divide the dish up into sector-like teeth connected together round the rim. The teeth are so stamped that when the washer lies on the top of a nut their apices follow the contour and project well into a single thread on the bolt. All that is necessary to put the washer in action is to screw it down tight above the nut to be locked. There is a natural tendency for the dish to flatten out when it comes into contact with the nut when forced down by a spanner. As the sector-shaped teeth cannot expand outward during the flattening process, they dig inward into the thread of the bolt, and thus obtain, so it is claimed, a very secure hold. On the other hand, if the nut tends to slack back, it will produce the same flattening effect, and the grip of the washer will be increased so long as it does not turn with the nut. In order to prevent that happening, that part of the washer which beds on the nut is formed with a smooth, rounded surface, with the object of insuring that the friction between the nut and washer shall be reduced to a minimum. The washer can, of course, easily be removed by means of a spanner. There is just one little matter in which we think that the washer might be improved. It is that it should be stamped plainly with the word top, or some other indication as to which way up it is to be used. The device would be valueless if put on the bolt upside down, and the perverseness of mechanics in the use of lock nuts is proverbial. (*The Engineer*, vol. 129, no. 3343, Jan. 23, 1920, p. 102, *d*)

## MACHINE TOOLS

### Multiple-Way Drilling Machines, Heavy-Duty Type, with Chip-Protection Device on Inverted Head

THE FOOTE-BURT "WAY" DRILLING MACHINES, J. V. Hunter. Multiple-way drilling machines are a comparatively recent development. The early types had a single vertical column supporting the work jig on which ways were planed for the travel of the saddles of both the upper and inverted drilling heads. The side heads were carried by comparatively light wing brackets bolted to each side of the column, which gave the machine a somewhat spidery appearance. About 1912-1913 these machines were improved and built more substantially. Among other things, the numerous individual oil cups of the older type were supplanted by gang sight-feed oilers with feed pipes to the various bearings; and a year or two later a provision was introduced to avoid the troubles caused by chips and dirt which work into the bearings of the inverted spindle head.

This head (Fig. 6) is now protected at the top by a sloping shield *A*, which tightly surrounds the spindle and prevents the accumulation of chips.

Below the shield are two plates, *B* and *C*, spaced a fraction of an inch apart where they surround the spindles and bolted together with an airtight flange around the outer edges. At each end of these plates is an air-hose connection; from one of these a blast of air can be blown, the exhaust taking place through the other, and all fine dust which enters this place is blown out, thus preventing its reaching the inner bearings. In addition to this each spindle is surrounded by a tapered cone *D*, which fits over

a stationary sleeve *E* that extends through the top shield to the plate *B*. The sleeve *E* is stationary, so no wear occurs between it and the shield.

The appearance of the latest model of a four-way drilling machine is shown in Fig. 7. Here the base is wide and heavy, supporting a full-width column of such proportions that no side arms are required. The single oil pump in the base handles practically the entire oiling.

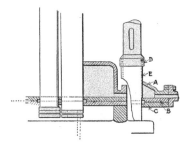

FIG. 6  CHIP-PROTECTION DEVICE USED ON INVERTED HEAD OF THE FOOTE-BURT MULTIPLE-WAY DRILLING MACHINE

In addition to this, drill levers *A* (Fig. 7) automatically reverse the feed to draw out the drills rapidly on the completion of their cut.

One of the latest machines carries 61 active spindles. In addition to the four-way there is a three-way drilling machine, such

FIG. 7  LATEST MODEL OF FOOTE-BURT FOUR-WAY DRILLING MACHINE

as would be necessary for drilling the automobile-engine oil pan. It is simpler than the four-way machine because of the absence of the inverted spindle head, but does not materially differ from it otherwise. (*American Machinist*, vol. 52, no. 10, March 4, 1920, pp. 485-487, 8 figs., *d*)

## MANAGEMENT

### Fatigue, Its Character and Forms; Tiredness; Fatigue and Accidents

FATIGUE AND ITS EFFECT ON PRODUCTION, A. Vautrin. Discussion of the nature of fatigue, its forms and degrees, and its influence on productivity of labor, accidents, etc. The distinction is established between fatigue as an objective phenomenon and the subjective feeling of fatigue, or tiredness.

It has been known for some time and experimentally established by Kraeplin that all work done by man, whether physical

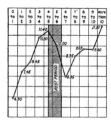

FIG. 8   DISTRIBUTION OF ACCIDENTS IN THE CHEMICAL INDUSTRY DURING THE WORKING DAY IN THE YEARS 1897-1907

Hours Worked and Percentage of Employees Injured During Each Hour

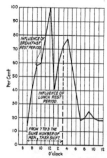

FIG. 9   DISTRIBUTION OF ACCIDENTS IN A GERMAN CABLE FACTORY THROUGHOUT THE WORKING DAY

or mental, is accompanied by a series of reactions which ·either favor its continuation or oppose it. The most important of the reactions opposing the tendency to continue working is known as fatigue, and may be of a physical or mental character or a combination of the two.

As regards its physical nature, it appears that all work produces in the body substances (chemical or organic) which exert gradually a narcotic effect on the central organs of the nervous system. In fact, Weichhardt claims to have found present in animals a poison produced by fatigue which he calls kenotoxin.

Fatigue may be either physical or mental. In physical work certain groups of muscles are set in action, thereby producing a corresponding excitation in the central nervous system. Muscular fatigue takes place, therefore, partly on the muscular periphery and partly at the brain center, and the more demand is made on

the brain center, the greater the final fatigue. Purely mental work creates a demand on the large brain, which means a gradual exhaustion of the gray matter and a certain demand on the nervous ganglia connected therewith. In either case, however, sooner or later exhaustion of the organs brought into action follows.

Moreover, there is a close connection between physical and mental fatigue, and it is a well-known fact that when the body is overtired for any considerable length of time, this also reduces the ability of the brain to do mental work. The reason for this is clear. All work may be physiologically considered as a process of consumption which calls into play not only the organs directly producing the work, but, more or less, all the energy reserves of the body. Hence bodily exhaustion will gradually lead also to mental exhaustion, and an overburdening of one part of the body by products of fatigue will necessarily sooner or later affect the operating ability of all the other parts of the body.

Fatigue is only a moderate degree of the lowering of the producing ability of the body due to exhaustive work. A greater degree thereof is known as exhaustion, and the difference between the two is stated as follows: Fatigue leads to the lowering of the ability to perform work; exhaustion represents a state at which the performance of work becomes entirely impossible. Physiologically, exhaustion represents such an accumulation of the products of fatigue that the body is unable to reconstruct itself for the time being.

A clear difference must be made between fatigue, which is an objective phenomenon and represents the actual failing in the ability to perform work, and the subjective feeling of fatigue, namely, tiredness. Whereas fatigue is produced by the actual physiological processes of exchange of materials in the body, tiredness may be the result of various conditions and circumstances often lying entirely outside of the effort to produce work. Fatigue is a physiological phenomenon; tiredness, psychological. It may be due to lack of interest in the work performed, outside happenings in the life of the workman, etc.

FIG. 10   FREQUENCY OF ACCIDENTS IN THE VARIOUS DAYS OF THE WEEK COMPUTED PER 100 INJURED OR KILLED PERSONS

The degree of tiredness depends mainly on the condition of the workman, and the same amount of effort will produce a greater tiredness in a weak man than in a strong man, in a man poorly nourished than in one well nourished, and in boys and women than in grown-up men.

Fatigue and Accidents. The paper presents a number of facts in confirmation of the claim often brought out before that there is a distinct connection between fatigue and accidents, although the author makes it clear that he does not consider accidents as being due to fatigue exclusively.

From his data it would appear that in the hours when the

fatigue of the workman is still slight there are scarcely any accidents, while in the sections of the working day or week, when the fatigue has grown to an appreciable extent, the number of accidents is more than double the average. Figs. 8 to 10, together with tabular data in the article, would indicate that interruption of the working day by rest periods reduces accidents very materially, and that of all the days of the week, Saturday, the day when fatigue is at its climax, is the worst day from the point of view of accident occurrence. This is particularly apparent in plants which do not employ the Saturday half-holiday.

The following statistics were collected in Lower Frankonia in 1895:

Length of working day, hours ...... 9¼     10½     13     over 13
Accidents per 100 workmen .........1.1     2.0     13.2     17.0

In printing establishments the accidents for the years 1910 to 1913 were on an average distributed in the following manner throughout the days of the week:

| | | | |
|---|---|---|---|
| Sunday | 136 | Thursday | 543 |
| Monday | 535 | Friday | 580 |
| Tuesday | 583 | Saturday | 585 |
| Wednesday | 530 | | |

(*Technik und Wirtschaft*, published by the Verein deutscher Ingenieure, vol. 12, no. 11, Nov. 1919, pp. 748-758, 3 figs., *g*)

## MARINE ENGINEERING

### Concrete Ships Still an Experiment, Reports American Concrete Institute

PRESENT STATUS OF THE CONCRETE SHIP. The original program of concrete-ship construction of the Emergency Fleet Corporation was reduced after the armistice to a total of fourteen vessels, and in October 1919 contracts for two 7500-ton cargo vessels were cancelled.

At the present date there are in service three 3500-ton cargo vessels and one 3000-ton cargo vessel, three 7500-ton tankers and twenty-one 500-ton canal barges built by the Railroad Administration under the supervision of the Emergency Fleet Corporation. The remaining vessels are in various stages of completion and are expected to be in commission by next summer.

In general, it is stated that in carrying out this program no construction problems were encountered which were not successfully met. The experience of the vessels in service thus far indicates that for cargo vessels there is ample structural strength, and that the barge is a much simpler problem in concrete than in usual materials.

On the other hand, it was not found that reinforced-concrete hulls could be built with greater speed than steel hulls, the average time of constructing the concrete hull being seven months.

In service concrete ships stood up quite well. In fact, there is generally less vibration in concrete ships than in corresponding steel ships, and also a considerable increase in the period of roll, which is desirable and is apparently due to the fact that these vessels have a larger moment of inertia around longitudinal axis than steel ships, due again to the mass of the concrete shell being considerably greater than the mass of the shell in a steel ship.

On the other hand, experience seems to indicate that these vessels are unable to withstand severe concentrated blows on the shell without the shattering of the concrete.

Impact, which in the case of a steel ship would probably only cause indentation to the plates, in the case of the concrete ship is apt to cause a shattering of the concrete over the area adjacent to the point of impact.

Repairs, especially in the case of barges, are, however, relatively simple and can be effected with little loss of time and at small cost.

The dead-weight capacity of the concrete ship was found to be lower than expected, the ratio of dead-weight capacity to displacement being on the average little more than 0.50. Furthermore, in the case of a steel ship and a concrete ship having the same dead-weight capacities, the concrete ship because of the greater weight of the ship itself must have greater dimensions than the steel ship, and in consequence must have greater hold spaces. For heavy-weight cargoes such as steel, coal or oil, in which the dead-weight capacity is reached before the hold spaces are filled, steel has obviously an advantage over concrete as a material of construction, assuming that construction and operating costs are equal. For bulky cargoes, such as ordinary goods, cotton, fruit or other materials for which the space required exceeds about 70 cu. ft. to the ton, the concrete ship will actually carry more dead weight than the steel ship for the reason that the hold spaces of the steel ship will be filled before dead-weight capacity is reached.

The report expresses the opinion that no definite conclusions should be drawn as yet from the experience of these vessels; and that the only general conclusion therefrom seems to be that it is possible to construct ships of concrete in about the same time, and for approximately the same cost as corresponding steel ships, which would indicate that with more experience in the art it will be possible to reduce both the cost and the time for construction. As regards the length of life of concrete ships, no sufficient data are available, and the brief experience of vessels afloat has disclosed no serious inherent weakness to shorten the life of a concrete ship. (Report read at the Convention of the American Concrete Institute, at Chicago, Feb. 16-18, 1920, by Committee on Reinforced Concrete Barges and Ships, H. C. Turner, Chairman. Abstracted through *Engineering News-Record*, vol. 84, no. 10, Mar. 4, 1920, pp. 463-464, *g*)

## PHYSICS

### Rate of Propagation of Flame in Mixtures of Methane and Air; Photographic Analysis of the Flame; Detonation Wave

THE PROPAGATION OF FLAME IN MIXTURES OF METHANE AND AIR, PT. I—HORIZONTAL PROPAGATION, Walter Mason and Richard Vernon Wheeler. In a previous paper the authors discussed the initial "uniform movement" of flame in gaseous mixtures. The uniform movement, however, is only one phase in the propagation of flame and is of comparatively short duration. The speeds attained by the flame during its régime are slow compared with the speeds during other phases in the propagation of flames in mixtures wherein no detonation wave has developed.

A knowledge of the speeds of flame in such mixtures as methane and air is of considerable importance, for example, in connection with the safe working of coal mines and also indirectly for internal-combustion-engine design.

The present paper describes experiments relating to phases other than the uniform movement during the horizontal propagation of flame in mixtures of methane and air. The experiments were carried out in tubes of different dimensions and materials, measurements of speeds having been made by the "screen wire" method. Supplementary information was obtained by photographic analysis of the flames.

The experiments were carried out first on ignition at the open end of a tube closed at the other end; next, ignition at the closed end of a tube open at the other end; and finally, ignition at one end of a tube open at both ends.

Taking the case of ignition at the open end of a tube closed at the other end, the initial phase of propagation of flame constitutes the "uniform movement."

The linear duration of this phase is controlled by such factors as influence the establishment of resonance in the column of gases in the tube, among these being the speed of the flame and hence the composition of the inflammable mixture, length, diameter and uniformity of bore of tube. Eventually, as a direct outcome of the establishment of resonance, the flame front acquires a periodic undulatory motion, leading sooner or later to violent vibrations which vary considerably in amplitude but remain periodic.

During the vibratory movement the oscillations of the flame are of wide amplitude and the mean speed of translation of volume is considerably enhanced.

The vibratory movement is also an excellent example of the effect of agitation or turbulence in accelerating the translation of flame through a gaseous mixture. The effect is a mechanical one. During each forward impulse the flame is drawn rapidly through previously unburned mixture by reason of the motion acquired by the resonating column of gases. In a certain degree also the forward motion of the flame is assisted by the expansion in volume of the burning gases, especially when the flame is at some distance from the open end of the tube, so that the escape of the expanded gases there is retarded.

The two other cases are discussed in the same manner. The most important one probably is that of ignition at one end of a tube open at both ends.

In that case it was found that the speed of the flame in all mixtures except the limit mixtures increases continuously over the whole distance traveled, and it seemed possible that the detonation wave might be developed if the flame could travel far enough.

In tests with a steel tube 15.25-m. long, it was found that the flame had acquired a vibratory character after traveling over the length of the tube. The length of the tube was then increased to 90 m. in the expectation that a greatly increased distance of travel would produce regular recognizable vibrations of large amplitude. The results confirmed this expectation. The propagation ultimately become strongly vibratory, but the early stages of the propagation were deeply modified by the increased length given to the tube. Instead of increasing rapidly in speed from the beginning as with the shorter tube, the flames now travel from the point of ignition at a constant and comparatively slow speed over a distance of between 12 and 15 m. and then began to vibrate. The vibrations acquired their greatest amplitude about half way along the tube (as was the case with the shorter tube) and continued throughout the remaining distance.

Of these three conditions under which the ignition of mixtures of methane and air has been effected, the one which would appear to lead to the most disastrous results in industry is the third, that is, ignition at one end of a tube or gallery open at both ends.

The fastest speed of flame acquired in any experiment was about 60 m. per sec. and was of short duration. The experiments have not, however, shown that a detonation wave cannot be developed in mixtures of methane and air at normal temperatures and pressure. On the contrary, the experiments with the long steel tube indicated that such an eventuality is possible. (*Journal of the Chemical Society*, vols. 117 & 118, no. 687, Jan. 1920, pp. 36-47, 2 figs. in text and 3 plates, *eA*)

THE PROPAGATION OF FLAME IN COMPLEX GASEOUS MIXTURES, Wm. Payman. It is customary to describe the inflammation of a gas mixture containing, for example, hydrogen and oxygen as the "burning of hydrogen in oxygen." This phrase is partly a relative one, and it is equally correct to regard the combustion as the burning of oxygen in hydrogen. Thus, the upper limit of inflammability of hydrogen in oxygen is the lower limit of inflammability of oxygen in hydrogen.

Mixtures of a combustible gas with air can be considered in a similar way and burning of hydrogen in air may be considered as burning of oxygen in a mixture of nitrogen and hydrogen.

For purposes of thoroughly investigating the mode of combustion of complex inflammable gas mixtures, it is desirable to examine their behavior with "atmospheres" other than air, the simplest problem being the combustion of a pure inflammable gas such as methane in pure oxygen.

The present research deals, therefore, with mixtures containing more oxygen than air and with mixtures with pure oxygen, the subject under investigation being mainly the initial uniform movement of flame which is supposed to be mainly effected by the conduction of heat from the burning to the adjacent unburned layer of gas mixture as different from what is known as "detonation wave."

If we neglect losses of heat to the walls of the containing vessel, the speed of propagation of flame during the uniform movement

can be regarded as depending mainly on two factors, namely, (1) the rate of conduction of heat from layer to layer of the mixture, which in turn depends on the difference in temperature of the burning and the unburned gases and on their thermal conductivities, and (2) the rate of reaction of the combining gases, which for a given combustible gas will vary with the composition of the mixtures (presumably according to the usual laws of mass action) and with the temperature produced by the reaction. A third factor might be added, namely, the ignition temperature of the mixtures, but this is perhaps dependent on the other factors.

The mixture of hydrogen and air for complete combustion, that is to say, the mixture having the greatest heat of combustion, contains 29.6 per cent of hydrogen, but the mixture in which the speed of the uniform movement of flame is greatest contains about 38 per cent, or nearly 10 per cent in excess.

Two series of determination of the speed of the uniform movement of flame in mixtures of oxygen with an atmosphere of nitrogen and methane and one of nitrogen and hydrogen have indicated that a displacement of the maximum speed mixture is toward mixtures containing an excess of oxygen. Further tests with combustible gases burning in air have shown that the addi-

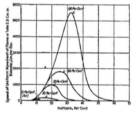

FIG. 11   CURVES OF SPEEDS OF UNIFORM MOVEMENT OF FLAME IN
VARIOUS MIXTURES OF METHANE AND "ATMOSPHERES"

tion of either combustible gas or oxygen to the basic mixture results in an increase in speed of the flame; and furthermore, in accordance with the laws of mass action, the displacement is greater on the addition of oxygen than on the addition of methane, due to the fact that the combination of one molecule of methane with two molecules of oxygen results in complete combustion. On the other hand, combustion, according to the same laws, should be less with the oxygen than with the hydrogen since two molecules of hydrogen combine with one molecule of oxygen, and experiment shows this deduction to be correct.

The speeds of uniform movement of flame in mixtures of methane with atmospheres containing 13.7, 22, 33, 50, 66 and 100 per cent of oxygen have been determined, some by means of an automatic commutator and single recording stylus, some by delicate Deprez indicators and the fastest speeds photographically. The results of the determinations are given in tables and also in Fig. 11.

The most striking results are those for mixtures of methane with pure oxygen. Here the maximum speed of the uniform movement of flame is attained with the mixture containing methane and oxygen in combining proportions $(CH_4 + 2O_2)$. The result is a sharp distinction from what obtains when the detonation wave is developed in mixtures of methane and oxygen, for the mixture in which the speed of the detonation wave is greatest contains equal proportions of methane and oxygen.

The retarding effect of addition of methane to the mixture for complete combustion $(CH_4 + 2O_2)$ is well illustrated by photographs given in the original article. In such a case, either methane or oxygen acts exactly like an inert gas, such as nitrogen, and in fact, methane, having the highest specific heat of the three, has the greatest retarding effect.

In the latter part of the article the writer gives the method of calculating mixtures so as to obtain given uniform speeds of flame movement. (*Journal of the Chemical Society*, vols. 117 & 118, no. 687, Jan. 1920, pp. 48-58, 6 figs., *eA*)

## POWER PLANTS

### Ash- and Coal-Handling Equipment

ASH- AND COAL-HANDLING EQUIPMENT, W. O. Rogers. In an article entitled Power Drives for Rolling Mills, the writer describes an interesting system of ash- and coal-handling equipment installed at the new power plant of the Republic Iron and Steel Company at Youngstown, Ohio. Ashes from the stokers of the boilers are deposited in an individual three-outlet iron ash hopper lined with red brick. The outlet of the ash hopper is provided with gates (Fig. 12) operated on rollers and actuated pneumatically by a plunger controlled by a four-way valve. The ashes

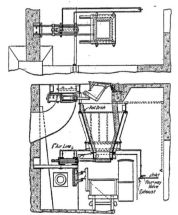

FIG. 12 ·ASH HOPPER AND ASH GATES AT THE ROLLING-MILL POWER PLANT OF THE REPUBLIC IRON AND STEEL COMPANY, YOUNGSTOWN, OHIO

are dumped into motor-operated cars having a capacity of 50 cu. ft. The loaded car is run to a hopper from which the ashes have elevated to the top of the hoist and automatically dumped into a chute leading to a bin from which they are loaded into railroad cars for removal.

The operation of the ash hoists is of interest. When the ash handler desires to elevate a load of ashes that has been dumped into the hopper, he presses a starting switch to start the 50-hp. d.c. motor. When the 70-cu. ft. capacity skip nears the top, a resistance coil is cut in on the circuit, which causes the motor to slow down. When the ash bucket reaches the dumping position, the circuit is automatically cut and the bucket stops long enough to discharge its contents. By means of a relay switch on the control board, after the bucket has remained at the dumping position a sufficient length of time to dump, the current is automatically applied to the motor, which has been reversed by means of a limit switch, and the bucket is sent to its original loading position ready to receive another load. The cycle comprises a period of 2½ minutes.

Coal is delivered from railroad cars into a rack hopper, from which it is conveyed to the crusher by means of an apron-type

conveyor 65 ft. long, center to center, and 36 in. wide. This conveyor has a capacity of 180 tons per hour at a speed of 40 ft. per min. Power to operate it is taken from the crusher shaft transmission by means of a chain drive to the hoist countershaft of the conveyor drive.

The crusher is of the four-roll type, each roll being 36 in. in diameter and 36 in. face. One pair of rolls is held rigidly in position, while the other pair is provided with heavy coil springs strong enough to press the coal and yet permitting of giving in case a foreign substance is encountered. The rolls and apron conveyor are operated by a 100-hp. compound-wound motor at 850 r.p.m. The capacity of the crusher, with run-of-mine bituminous coal crushing to 90 per cent ¾-in. size, is 180 tons per hour.

The article contains also a description of the boiler house and furnaces with their stokers. (*Power*, vol. 51, no. 8, Feb. 24, 1920, pp. 291-296, 5 figs., *d*)

### Tests of Canadian Fuels; Distribution of Heat Losses; Factors Affecting Efficiency of Boiler Furnaces

RESULTS OF 41 STEAMING TESTS CONDUCTED AT THE FUEL-TESTING STATION, OTTAWA. The paper gives results made on tests with Canadian coals and as such is mainly of local interest. Some remarks are, however, of general interest and are reported here.

The tests were conducted in a Babcock & Wilcox water-tube boiler having 677 sq. ft. of heating surface. Some of the tests were conducted on a grate area of 23 sq. ft. with an air space of ¼ in. between the bars; others on a grate area of 21 sq. ft. with an air space of ½ in. between the bars.

As regards losses, the following remarks are found in the report. The principal heat loss was that due to the total heat of the flue gases, which, in this case, does not include that due to the uncondensed steam. The variation of the loss due to the escaping hot gases for approximately the same boiler output is due almost entirely to the change in the amount of excess air. A figure is introduced showing the relation between the ratio of the flue-gas total heat loss to the heat usefully employed for steam raising and the ratio of the total air supplied to the air whose oxygen content is combined with the fuel (this relation is expressed as a straight line). This air ratio is calculated from

the flue-gas analysis, and is equal to $\dfrac{21}{21 - O_2 \dfrac{79}{N_2}}$ where $O_2$ and $N_2$

represent the volumes of oxygen and nitrogen in the flue gas,

and $\dfrac{21}{79}$ the ratio of oxygen to nitrogen in the atmosphere. The

expression $\dfrac{21}{100}\left(\dfrac{100 + XL}{X(1+L)}\right)$ represents the ratio of air supplied

to air required in terms of the carbon dioxide content of the flue gas ($X$) and the factor $L$, which depends upon the chemical con-

stitution of the coal, and is equal to $\dfrac{3}{C}\left(H - \dfrac{O}{8}\right)$, where

H, O, and C represent the relative weights of the hydrogen, oxygen, and carbon contents of the fuel. This expression holds good only for complete combustion of a fuel whose sole constituents—which pass off with the flue gas—are carbon, hydrogen and oxygen.

As regards the variation in excess air, it is found that the air ratio was less when using a grate of 21 sq. ft. area with ½-in. air spaces than when using a grate of 23 sq. ft. area and ¼-in. air space for the same fuel, and approximately the same rate of steaming. On the other hand, it was found that the excess air increased with an increase of rate of combustion for the majority of coals.

While a change from the grate with small to one with larger air openings reduced the excess air loss, it was found that the loss due to unburned solid fuel increased for the trials with the larger air space and that this loss tended to increase considerably with an increasing ash content.

Briefly then, burning coal on a grate with ¼-in. air spaces, tends to cause the loss due to solid unburned carbonaceous material to be less and the loss due to the escape of flue gases to be greater than when using a grate with ½-in. air spaces.

The principal reasons for the variation in the radiation, and unaccounted-for loss, where the rate of steaming remains the same, are variations in the heat transmitted from the hot gases and incandescent fuel through the boiler setting, and in the variation of the undetermined combustible content of the products of combustion.

A number of figures are plotted in the original report showing the total loss due to radiation, unburned gases and the unaccounted-for loss. These figures are plotted on a base representing the ratio of the air supplied to that used for combustion. While no attempt has been made to plot a curve showing any general law, it would appear that the loss tends to decrease with an increased air supply.

An increase in the air-supply ratio may be expected, therefore, to be accompanied by a decrease in the loss due to incomplete combustion and by an increase in that due to the increased mass of gas escaping at a high temperature; and the efficiency of the boiler, based on the solid combustible consumed, will show the net result of the change.

In general, it would appear that the predominant effect of the increase in the air ratio for any particular fuel is to lower the efficiency, but this must not be considered in the light of an absolute law. (*Bulletin No. 27, Mines Branch, Department of Mines, Canada*, Ottawa, 1920, pp. 3-83, 41 charts, *eA*)

BOILER AND FURNACE TESTING, Rufus T. Strohm. Recommendations for simple tests of a particular character, the purpose of which is to find out how efficiently the boiler is working.

The test is of the nature of an evaporation test and comprises the determination of the total weight of coal used during the test, the duration of which is recommended as 8 hr.; the total weight of water fed to and evaporated by the boiler; the average temperature of the feedwater; the average steam pressure in the boiler.

The methods of determining these factors and simple formulæ for calculating results are given, as well as recommendations for using the results. (*Bureau of Mines Technical Paper* 240, reprint of Engineering Bulletin No. 1, U. S. Fuel Administration, Washington, 1920, *p*)

## PUMPS

### High-Vacuum Mercury Pump with Provision for Storing Gases Exhausted

HIGH-VACUUM MERCURY PUMP WITH PROVISION FOR THE PRESERVATION OF THE EXHAUSTED GASES, A. Beutell and P. Oberhoffer. Both mercury pumps of the Sprengel type and such mechanical pumps as the Gaede have the disadvantage of possessing no provision for conveniently collecting and preserving the gases exhausted from a high vacuum. This becomes of importance now that a considerable amount of interest has been aroused in the subject of gases occluded in minerals and metals. In the present article a pump is described with which it is possible to collect the gases exhausted from a high vacuum. It is stated that the new arrangement does not materially affect the rate of operation of the pump; thus, the 500-cc. container may be exhausted to a pressure of 0.0007 mm. in 10 min. and to a pressure of 0.00011 mm. in 20 min. Furthermore, if it is not desirable to collect the exhausted gases, the operation of the pump is not affected in any way.

As seen from Fig. 13, the pump is of the water-jet type, but the exhausted gases are not directed as usual into the water jet at *N*. Instead, there is inserted the small tank *E* between the downward tube *Fa* and the upward tube *St*, this container being equipped with a three-way cock. The upward tube *St* is provided at its lower end with an expanded section, the purpose of which is to

prevent the flow of air from *L* into the container *E*. This made it necessary to increase the dimensions of the container *N*, the purpose of which is to take up the mercury displaced by the gas collected in *E*.

Because of this the amount of mercury used in the present pump is about 100 cc. as compared with about 20 cc. for similar pumps previously used and having no provision for the collection of exhausted gases.

Instead of the asbestos stopper formerly used for regulating the air admission at *L*, there is now welded in there a thick-walled capillary tube protected against penetration of dirt into it by an-

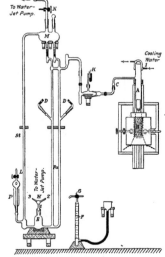

FIG. 13   BEUTELL AND OBERHOFFER HIGH-VACUUM PUMP WITH PROVISION FOR THE PRESERVATION OF EXHAUSTED GASES

other glass tube closed at one end. The three-way cock *N* at the top of the tube likewise employs a capillary tube instead of the normal asbestos stopper for the purpose of safely admitting air to the pump against the vacuum therein.

The operation of the pump does not materially differ from that of similar pumps previously described. In addition, the article describes an apparatus for gas analysis. (*Chemiker-Zeitung*, vol. 43, no. 125, Oct. 16, 1919, pp. 705-706, 4 figs., *d*)

## RAILROAD ENGINEERING

### Steam and Electric Propulsion for Railroads Compared; How the Railroads Utilize Their Fuel; Standardization of Frequencies; Electric Locomotives

THE LAST STAND OF THE RECIPROCATING STEAM ENGINE, A. H. Armstrong. By "the last stand of the reciprocating steam engine" the author, who is Chairman of the Electrification Committee, General Electric Company, refers to the steam operation of railroads.

In brief, he claims that where applicable the electric locomotive is more reliable and more economical from the point of view of fuel consumption. He makes a reservation, however, to the effect that many roads of lean tonnage should not be immediately electrified, as they would render no adequate return upon the large capital investment required.

To illustrate the present situation he gives Table 1, showing the subdivision of the tonnage passing over the tracks of our railways (The source of the table is not indicated).

The first four items, representing nearly 85 per cent of the total ton-miles made during the year 1918, may be regarded as fundamentally common to both steam and electric operation. By introducing electric propulsion the last four items are considerably cut down, and it is claimed that approximately 12 per cent or about 150 billion ton-miles at present hauled by steam engines over the roads will be eliminated. This means that if *all* American railways were completely electrified they could carry one-fifth more revenue-producing freight tonnage with no change in present operating expenses or track congestion.

Another table would indicate that a quarter of all the coal mined in the United States is consumed by the railways, and the author claims that this is done with extreme wastefulness. This claim is

TABLE 1 TOTAL TON-MILE MOVEMENT
(All Railways in United States—Year, 1918)

| | Per Cent | Ton-Miles |
|---|---|---|
| 1 Miscellaneous freight cars and contents | 42.3 | 515,000,000,000 |
| 2 Revenue coal cars and contents | 16.23 | 197,000,000,000 |
| 3 Locomotive revenue, driver weight only | 10.90 | 132,300,000,000 |
| 4 Passenger cars, all classes | 16.13 | 196,000,000,000 |
| Total revenue, freight and passenger | 85.56 | 1,040,300,000,000 |
| 5 Railway coal | 5.00 | 60,600,000,000 |
| 6 Tenders, all classes | 6.50 | 78,800,000,000 |
| 7 Locomotive railway coal | 0.39 | 4,700,000,000 |
| 8 Locomotive, non-driving weight | 2.55 | 31,000,000,000 |
| Total non-revenue | 14.44 | 175,100,000,000 |
| Grand total (all classes) | 100 | 1,215,400,000,000 |

TABLE 2 COAL REQUIREMENTS FOR U. S. RAILROADS IN 1918 REFERRED TO ELECTRICAL UNITS

| | |
|---|---|
| Total ton-miles, 1918 | 1,215,400,000,000 |
| Watt hours per ton mile | 40 |
| Kw-hr. total movement | 48,700,000,000 |
| Coal required at 7 lb. per kw-hr., tons | 170,000,000 |

TABLE 3 COAL SAVING BY ELECTRIFICATION

| | |
|---|---|
| Total ton-miles, steam | 1,215,400,000,000 |
| Reduction by electrification | 146,000,000,000 |
| Total ton-miles electric | 1,069,400,000,000 |
| Kw-hr. electric at 40 watts | 42,776,000,000 |
| Coal on basis 2½ lb. per kw-hr. tons | 53,500,000 |
| Equivalent railway coal, 1918, tons | 176,000,000 |
| Saving by electrification, tons | 122,500,000 |

based on tests made upon the Rocky Mountain Division of the C. M. & St. P. Ry. to determine the relation existing between the horsepower-hours of work done in moving trains and the coal and water consumed on the steam engines in service. From these tests it would appear that the actual coal consumed under the engine boiler in 24 hr. divided by the actual work performed by the engine is equivalent to 10.18 lb. per hp-hr. at the driver rims. These tests have also shown that the coal consumed while doing useful work was raised 30 per cent by stand-by losses. In electrical constants this was found to be equivalent to 7.56 lb. of coal per kw-hr. at power supply on a basis of 55 per cent efficiency.

From this the author proceeds to an estimate of the cost in pounds of coal to do the same work electrically, taking the condi-

tions existing on certain western railroads. To do this, the ton-mile values of Table 1 are reconstructed into those of Table 2. From this he proceeds to the construction of Table 3 showing the saving of coal by electrification again on the basis of Table 1.

The startling conclusion arrived at is that approximately 122,-500,000 tons of coal, or more than two-thirds the coal now burned in our 63,000 steam engines, would have been saved during the year 1918 had the railways of the United States been completely electrified along lines fully tried out and proved successful today.

The author claims this estimate is probably too conservative as no allowance has been made for the extensive utilization of water power which can be developed to produce power more cheaply than by coal in many favored localities. On the other hand, one should bear in mind his previous statement that many roads of lean tonnage would render no adequate returns upon the large capital investment required by electrification. It would appear, therefore, that essentially the question is more economic than technical.

Of interest in this connection is a passage at the end of the paper referring to the study of the congested tracks of the Baltimore and Ohio Railroad between Grafton and Cumberland. Company coal movement in coal cars and engine tenders constituted over 11 per cent of the total ton-mileage passing over the tracks.

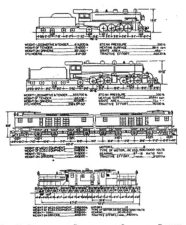

FIG. 14 DIAGRAMMATIC COMPARISON OF STEAM AND ELECTRIC LOCOMOTIVES

In other words, due to the very broken profile of this division, the equivalent of one train in every nine is required to haul the coal burned on the engines.

It is assumed that with the electric-locomotive operation the three tracks now badly congested with present steam-engine tonnage could carry 80 per cent more freight.

Attention is called to Fig. 14 which gives an interesting comparison between steam and electric locomotives. (*Journal of the American Institute of Electrical Engineers*, vol. 39, no. 3, Mar. 1920, pp. 209-218, 11 figs., d)

The United States Bureau of Mines announces that it has developed a novel method for giving a danger warning in mines in which compressed air is used throughout the workings. An ill-smelling substance is injected into the compressed-air line and within a few minutes the odor is spread throughout the mine.

STEAM ENGINEERING (See Railroad Engineering)

VENTILATION

### Ozone Air Purification

Ozone as the Solution of the Fresh-Air Problem, E. S. Hallett. Description of installations tested out in the public schools of St. Louis, Mo. The Head of the Hygiene Department of the Board of Education of St. Louis came to the Building Department with the complaint from one of the downtown schools that the air was so bad in some rooms that teachers threatened to resign on the advice of their physicians. The writer proposed to the Hygiene Department to test the application of ozone.

The ozone apparatus was set up in the air passage between the air washer and fan, and regulated to produce just sufficient ozone to be barely detected by the odor on entering the building, but not enough to make one conscious of an odor. The result was the actual disappearance of all the stuffy conditions and bad smells complained of. The teachers stated that the conduct of the children as to lessons and behavior was noticeably better. Colds and coughs nearly disappeared. No contagious diseases developed during the six weeks' trial, although the influenza was epidemic at this time. In fact, during the period of an influenza epidemic, the attendance was 30 per cent higher than the general average attendance for this school.

The experiment was then transferred to a colored school having the plenum system with the Zellweger air-washing fan and with complete recirculation of the air. The ozone machine was set up just back of the fan, the ozone acting upon the water of the air washer as well. In this test the pupils and teachers were weighed weekly and a close inspection made by the staff physician of the Hygiene Department. About 75 per cent of the children gained in weight on an average of about one pound, about 5 per cent lost weight, while the rest suffered no change. Several very stout girls weighing about 170 lb. each lost from 5 to 8 lb. weight (total duration of test not stated). No indication of illness or discomfort was noted. No contagious diseases occurred in this school and colds were noticeably less.

The coal consumption was measured, and in comparison with days of equal outside temperature the coal used was almost exactly one half.

Agar plates were exposed in a room filled with pupils and showed an average count of bacteria of 225, which was extremely low, indicating that the ozone had destroyed the active germs of the air.

These and other tests during a year's period indicated that ozone destroys all odors resulting from the respiration, bodies and clothing of the children, and produces a mild exhilaration resembling that of a sea breeze or the air on a morning after a thunderstorm. It appears from limited data to be a preventive of influenza, and it is believed that its introduction in ventilation would probably remove the necessity for open-air schools now common in most cities.

The maximum concentration should be too low to give an ozone odor, and if used up to this concentration is safe for ventilation. Persons not used to ozone in air must be employed for detecting the air, as the sense of smell for ozone quickly declines when exposed to it.

The writer developed a standard which may be used in determining in advance the proper concentration for any given volume of air movement or for a given number of occupants in a room.

This standard was developed after ascertaining that with a given voltage and with a given thickness of dielectric the amount of ozone generated was proportional to the number of brush discharge points of the generator.

The most satisfactory apparatus uses 4000 volts alternating current from a static transformer, all included with the ozone generator unit, which uses a micanite plate dielectric 0.040 in. in thickness and aluminum points spaced approximately $\frac{1}{2}$ in. apart.

It was observed that 600 brush points made just enough ozone for 1000 cu. ft. of air from the blast fan.

This test was with rooms filled with 45 to 50 children much below the average in cleanliness. For rooms occupied by fewer people, the brush points or voltage should be reduced. If conditions are to remain constant, some points should be disconnected, but with varying conditions a controller should be installed to regulate the voltage by taking taps out of the primary of the transformers.

Where the air is recirculated in whole or in part, the ozone must be cut down to the point where no ozone odor is noticeable. In fact, the revitalizing of the air of the average school room when recirculating 90 per cent of the air will be effectively done with half the maximum stated above. The writer believes that the delay in the use of ozone in ventilation has been due to trials made with too high concentration and to the absence of any information on a means of control.

Ventilation for its final effect depends on the purity of the air surrounding the building, and if the building, such as a school or hospital, is located in an unhealthy neighborhood, ventilation is apt to do more harm than good. Ozone, by killing the germs in the air, practically creates clean air and is independent in its effect on the location of the building to be ventilated.

The writer proceeds to show a heating system designed in the light of these tests. It has no air washer. (Paper presented at the Annual Meeting of the American Society of Heating and Ventilating Engineers, Jan. 27-29, 1920, under the title, An Advance in Air Conditioning in School Buildings. Compare *Heating and Ventilating Magazine*, vol. 17, no. 2, Feb. 1920, pp. 25-29, 1 fig., *ed*)

The Detection of Invisible Objects by Heat Radiation. In 1918 the United States Army started to devise methods for detecting men and inanimate objects which were at a higher temperature than adjacent things by means of the emitted radiation. A thermopile was placed at the focus of a parabolic mirror. The radiation from the warmer object fell on the mirror and was reflected to the thermopile where it produced an electric current which was observed by means of a galvanometer. It was possible to detect a man 600 ft. away. "A man lying in a depression in the ground at a distance of 400 ft. was detected unfailingly as soon as he showed the upper part of his face above ground." Secret signaling could be carried on by covering and uncovering the face. This apparatus was sent to the A. E. F. in August, 1918.

An aeroplane at an altitude of 3500 ft. could be picked up by such an instrument and its course followed. A wisp of cloud produced confusion by causing as large a deflection as the aeroplane did, but the two possible causes were different in the way in which the galvanometer deflection began.

In an address before the Section on Physics of the American Association for the Advancement of Science, St. Louis, December 1919, Prof. Gordon F. Hull called attention to the rather peculiar fact that in the official manual for the U. S. rifle the value of the ballistic coefficient of the ordinary service rifle bullet of 0.30 in. caliber is given as 0.3894075, "as determined experimentally at the Frankford Arsenal." Commenting on this he said: " The experimental skill which can determine to an accuracy indicated by several places of decimals a quantity as highly capricious as the so-called ballistic coefficient, is of rather questionable value."

## CLASSIFICATION OF ARTICLES

Articles appearing in the Survey are classified as c comparative; d descriptive; e experimental; g general; h historical; m mathematical; p practical; s statistical; t theoretical. Articles of especial merit are rated A by the reviewer. Opinions expressed are those of the reviewer, not of the Society. The Editor will be pleased to receive inquiries for further information in connection with articles reported in the Survey.

# ENGINEERING RESEARCH

## A Department Conducted by the Research Committee of the A. S. M. E.

### Coöperative Research

AT PRESENT seven associations of manufacturers are supporting research at the Mellon Institute at Pittsburgh. The name of the association, the number of companies and the amount subscribed for the fellowship foundation in each case is given in the following table:

| | Number of Companies | Amount |
|---|---|---|
| Magnesia Association of America..... | 4 | $ 6000 |
| Asbestos Paper Manufacturers Assn.. | 10 | 3500 |
| Rex Spray Companies............. | 9 | 3500 |
| Laundry Owners National Association | 2500 | 5000 |
| Leather Belting Exchange Corporation | 40 | 5600 |
| Refractories Manufacturers Association | 84 | 10000 |
| Container Club .................... | 24 | 3500 |

The investigations on magnesia insulation are for the purpose of determining its heat-insulating value. The investigations for the Asbestos Paper Manufacturers are for the purpose of standardizing the specifications for asbestos textiles and asbestos millboard. The Rex Spray fellowship is for the purpose of improving the manufacture of insecticides. The Laundry Owners' fellowship is devoted to the investigation of water, soap, permanency of colors, stains and dyes. The Leather Belting fellowship is for the purpose of determining the comparative power-carrying capacities of leather belts with substitute belts. The fellowship of the Refractories Association is for the purpose of determining standard tests to be used in manufacturing. The Container Club fellowship is in connection with the development of a paper tester and the testing of fiber board with this instrument.

### A—Research Results

The purpose of this section of Engineering Research is to give the origin of research information which has been completed, to give a résumé of research results with formulæ or curves where such may be readily given, and to report results of non-extensive researches which in the opinion of the investigators do not warrant a paper.

*Apparatus and Instruments A3-20* Gage for Small Pressures. The Wahlen gage was developed by W. G. Wahlen of the Engineering Experiment Station of the University of Illinois. It is sensitive to 0.0001 in. of water. It is composed of a rigid base on leveling screws with two large glass bulbs communicating through a special inverted U-tube. One of the bulbs is controlled by a micrometer caliper. Alcohol of density 0.8195 is used in the bulbs and part of the U-tube. It is colored with aniline dye to give a red color. The upper part of the U-tube is filled with a mixture of turpentine and ligroin of density 0.810. The gages are first balanced at zero with both ends open. The pressure connection is then made and the movable carriage is adjusted to bring the meniscus in the U-tube to its reference point. By using a U-tube with legs of different sizes accurate observations may be made. This gage may be used with pitot tubes for the calibration of instruments or the measuring of low air currents. Address Engineering Experiment Station, University of Illinois, Urbana, Ill. C. R. Richards, Director.

*Apparatus and Instruments A4-20* Radiation Effects on Thermocouples. In determining the temperature of the air in hot-air leaders of a warm-air furnace, the Experiment Station of the University of Illinois has found that the radiation from the thermocouples affects the temperature reading shown by them. The temperature of a leader was determined by a thermocouple when the leader was uncovered and when the leader was covered with hair felt. If the temperature of the air going to the leader was found to be the same in each case, the difference shown in the leader would represent the loss due to radiation. This was found to be 26 deg. fahr. when the temperature observed was 195 deg. fahr. Engineering Experiment Station, University of Illinois, Urbana, Ill. Address C. R. Richards, Director.

*Buildings A1-20* Vibration. A preliminary report by the Aberthaw

Construction Company on the Effects of Vibration in Structures has been issued. This gives a statement of the research from 1133 replies in connection with this investigation. The report mentions the natural frequencies of vibrations, the method of recording vibrations, method and type of construction, the reduction of efficiency and production caused by vibration, experiments in various forms of mills, methods of absorbing vibration, methods of tracing vibration, types of building used and methods of design. Address Aberthaw Construction Company, 27 School St., Boston, Mass.

*Fuels, Gas, Tar and Coke A2-20* The Comparative Value of Fuels, by M. K. Thornton, Jr., issued by the Texas Engineering Station, College Station, Tex. Address J. C. Nagle, Director.

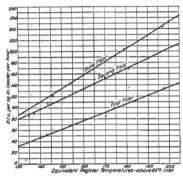

Fig. 1  Curves Showing Effect of Register Temperature on Leader Capacity

*Fuels, Gas, Tar and Coke A3-20* Wood. The fuel value of wood is given by the Forest Products Laboratory. Two pounds of dry wood of any non-resinous species has the heating value of 1 lb. of good coal. Resin gives about twice as much heat as the wood. As the average amount of resin is 15 per cent, this increases the heating value 30 per cent. One cord of hickory, oak, beech, birch, hard maple, ash, elm, locust, long-leaf pine or cherry is equal to one ton of coal; and 1½ cords of short-leaf pine, Western hemlock, red gum, Douglas fir, sycamore and soft maple are equal to one ton of coal. Two cords of cedar, redwood, poplar, catalpa, cypress, basswood, spruce or white pine are equal to one ton of coal. Forest Products Laboratory, Madison, Wis. Address Director.

*Heating A5-20* Warm-Air Furnaces. Bulletin No. 112 of the Engineering Experiment Station of the University of Illinois, on the Progress of the Warm-Air Furnace Research, by A. C. Willard, has been issued. The report describes the method of measuring temperature and allowing for the radiation effect, the method of measuring air at low velocities with anemometers and pitot tubes using the Wahlen gage accurate to 0.0001 in. of alcohol. Researches indicate that a square inch of leader pipe on the first floor has a far less heating and carrying value than that for second- or third-floor leader pipe. The tests indicate that the capacity of a warm-air furnace system is greatly reduced by transporting leaders to the upper floors. The temperature at register faces should be 175 deg. fahr. The amount of heat delivered per square inch of leader area is given by the three curves in Fig. 1. The efficiency of the furnace as installed is shown by the curve in Fig. 2, below which is shown the arrangement of the installation. Engineering Experiment Station. University of Illinois, Urbana, Ill. Address Director C. R. Richards.

*Hydraulics A1-20* Sewage Sprinkler Nozzles and Coefficient for Discharge. The discharge from sewage sprinkler nozzles can be represented by the equation $Q = CA \sqrt{(2gh)}$ after the head on the sprinkler is greater than 1½ ft. At lower heads the coefficient decreases. The exponent of the head is practically 0.5. Seven different nozzles were tested with the following coefficients:

| | |
|---|---|
| Chase square nozzles | 0.648 |
| Taylor round nozzles | 0.756 |
| Taylor square nozzles | 0.696 |
| Worcester round nozzles, 13/16 in. | 0.660 |
| Worcester round nozzles, 11/16 in. | 0.675 |
| Merritt square nozzle | 0.598 |
| Priestman-Beddoes round nozzle | 0.569 |

*Internal-Combustion Motors A1-20* Spark Plugs. The Fifth Annual Report of the National Advisory Committee for Aeronautics will contain a report on the properties and preparation of ceramic insulators for spark plugs. Report No. 53, in four parts, gives the results of research and development work conducted by the Bureau of Standards at the instance of the National Advisory Committee for Aeronautics. A discussion is given of the methods for measuring the resistance of insulators at high temperatures and the method finally adopted for this work is described. Re-

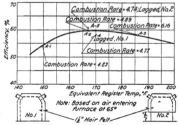

FIG. 2 EFFICIENCY CURVE FOR A FURNACE OPERATING AT VARIOUS REGISTER TEMPERATURES AND RATES OF COMBUSTION

sults of tests of a large number of insulating materials at high temperatures are tabulated and their significance discussed. The preparation and composition of ceramic insulators is outlined and compositions are given of new porcelains having superior properties. Results of an experimental investigation into the corrosion effects and permanency of various cements for spark-plug electrodes are discussed. The four parts of this Report are entitled as follows:

Part I   Methods of Measuring Resistance of Insulators at High Temperatures, by F. B. Silsbee and R. K. Honaman.
Part II   Electrical Resistance of Various Insulating Materials at High Temperatures, by R. K. Honaman and E. L. Fonseca.
Part III   Preparation and Composition of Ceramic Bodies for Spark Plug Insulators, by A. V. Bleininger
Part IV   Cements for Spark-Plug Electrodes, by H. F. Staley. Address National Advisory Committee for Aeronautics, Washington, D. C.

*Light A1-20* Reflectometer. The Bureau of Standards has developed a new reflectometer after extensive experiments. The instrument is to be used to determine the amount of reflection from walls to rooms. It consists of a small metal sphere from which a segment has been removed leaving about nine-tenths of its original surface. It is painted inside with a diffusing white paint and a beam of light is projected through a small hole in the wall to the surface which is to be tested. The test surface may be compared with the standard surface or with a flat surface painted with the same paint as the sphere. Bureau of Standards, Washington, D. C. Address S. W. Stratton, Director.

*Lubricants A1-20* Solid Lubricants. The Department of Scientific and Industrial Research of Great Britain has recently prepared a memorandum on solid lubricants. This can be obtained for sixpence from H. M. Stationery Office at No. 20 Abington St., Westminster S. W. 1, London, England. The subject is dealt with under the following headings:

1   Characteristics of Solid Lubricants
2   Action of Solid Lubricants
3   Analyses of Lubricating Graphites
4   The Grading of Graphite
5   Hot Bearings
6   Methods of Applying Solid Lubricants

7   Drawbacks to the Use of Colloidal Solid Lubricants
8   Observations on Results Obtained by the Use of Solid Lubricants

*Mining A1-20* Concentrating Minnesota Wash Ores. A new machine for concentrating Minnesota wash ores. Bulletin No. 6 of the Engineering Experiment Station of the Minnesota School of Mines. Address W. R. Appleby, Director, Minneapolis, Minn.

*Mining General A2-20* Panel Systems of Coal Mining, by C. M. Young. A graphical study of the percentage of extraction by the panel system of coal mining. The paper shows that the greatest extraction with room 300 ft. long and 30 ft. wide on 50-ft. centers is only 57.05 per cent and with 40-ft. rooms 68.48 per cent. The Bulletin shows the reasons for low extraction, the advantages of high extraction and the method of investigating the percentage extraction. Engineering Experiment Station, University of Illinois, Urbana, Ill. Address Director.

*Properties of Engineering Materials A4-20* Boiler Plate at Elevated Temperatures. A series of tests on cold-rolled boiler plate at temperatures has just been made. The curves from the tensile tests show the same variations as those obtained with temperature increases and the results are available for those interested in the subject. The summary gives the tensile strength, transverse strength, absorption, porosity and weight per cu. ft. Bureau of Standards, Washington, D. C. Address S. W. Stratton, Director.

*Properties of Engineering Materials A5-20* Granite and Indiana Limestone. Investigations on the properties of commercial granites of the United States has been commenced and tests have been completed on twelve varieties used in the eastern part of the country. Seventy-two samples of Indiana limestone have been studied and the results are available for those interested in the subject. The summary gives the tensile strength, transverse strength, absorption, porosity and weight per cu. ft. Bureau of Standards, Washington, D. C. Address S. W. Stratton, Director.

*Properties of Engineering Materials A6-20* Over-Sanded Concrete. In order to make certain concrete mixtures more workable it has been proposed to introduce additional sand. The Bureau of Standards has investigated a 1:2:4 limestone concrete and a 1:3:3 mixture of the same substances. After 28 days the latter mixture gave 988 lb. per sq. in. while the former gave 768 lb. per sq. in. The latter sample took 487 lb. of cement per cu. yd. while the former took 508 lb. per cu. yd. This shows the advantage of over-sanding. Further tests are under way. Bureau of Standards, Washington, D. C. Address S. W. Stratton, Director.

*Wood Products A6-20* Gluing. When large glue joints are to be made, high strength of the whole joint is obtained by heating the wood for 10 or 15 min. to a temperature of 120 deg. fahr. before gluing. Forest Products Laboratory. U. S. Forest Service, Madison, Wis. Address Director.

## B—RESEARCH IN PROGRESS

The purpose of this section of Engineering Research is to bring together those who are working on the same problems for coöperation or conference, to prevent unnecessary duplication of work and to inform the profession of the investigators who are engaged upon research problems. The addresses of these investigators are given for the purpose of correspondence.

*Aircraft B3-20* Model Tests. The theory of model tests of flying boats. Address Curtiss Research Laboratory, Curtiss Engineering Corporation, Garden City, L. I.

*Aircraft B4-20* Static Pressure Gradients. The theory of static pressure gradients in wind tunnels; method of eliminating pressure gradients and short method of correcting for pressure gradients. Address Curtiss Research Laboratory, Curtiss Engineering Corporation, Garden City, L. I.

*Aircraft B5-20* Pressure Gages. Absolute pressure gage for wind-tunnel work. Address Curtiss Research Laboratory, Curtiss Engineering Corporation, Garden City, L. I.

*Aircraft B6-02* Cable Air Resistance. Tests on resistance of multiple airplane cables behind each other at varying amounts of separation. Address Curtiss Engineering Corporation, Garden City, L. I.

*Apparatus and Instruments B3-20* Absolute Pressure Gage for Wind-Tunnel Work has been designed and constructed at Curtiss Research Laboratory. Address Curtiss Engineering Corporation. Garden City, L. I.

*Chemistry, Inorganic B1-20* Fixation of Atmospheric Nitrogen. An investigation of a number of factors for increasing the efficiency of the arc process in the fixation of atmospheric nitrogen, by Prof. E. A. Loew and F. K. Kirsten, Engineering Experiment Station. University of Washington, Seattle, Wash. Address C. E. Magnusson, Acting Director.

*Electrical Communication B1-20* Multiplex Radiotelephony. An investigation for sending five radio telephone messages from same antenna and reception of these through single antenna. F. M. Ryan, J. R. Tolmie, Roy Bach, A. Kalin, graduate students in

Electrical Engineering. Engineering Experiment Station, University of Washington, Seattle, Wash. Address C. E. Magnusson, Acting Director.

*Electrical Instruments B1-20* Order and Amplitude of Harmonics in Voltage Waves. A method for determining the order and amplitude of harmonics in voltage waves by indicating meters. A wave analyzer is being constructed to apply this method. Prof. L. F. Curtiss, Engineering Experiment Station, University of Washington, Seattle, Wash.

*Electrical Instruments B2-20* Oscillograph. A convenient timing device for use with the oscillograph. A convenient method for producing time waves of any desired frequency. J. R. Tolmie, Engineering Experiment Station, University of Washington, Seattle, Wash. Address C. E. Magnusson, Acting Director.

*Electric Power B1-20* Trunk Transmission Lines. An investigation on the need for trunk transmission lines on the Pacific Coast by C. E. Magnusson, Engineering Experiment Station, University of Washington, Seattle, Wash. Address C. E. Magnusson, Acting Director.

*Glass and Ceramics B1-20* Enamels. The Bureau of Standards is investigating the formation of fish-scaling in enamels on sheet steel and also the subject of the relation of the composition of the enamel to solubility in acids. They have made over 4000 sample enamel plates comprising 21 different enamels in the production of fish-scaling and 33 enamel compositions have been tested for the solubility in acids. One of the most important factors in producing fish-scaling is too severe heat treatment in firing the enamel. The composition of the enamel and the physical and chemical characteristics of underlying metal together with the method of melting the mixture and the shape and weight of the mixture are important factors. The work on the acid-resisting properties indicates that oxides which are chemically similar in nature do not have similar effects when incorporated in enamel compositions. Thus calcium chloride is not so good as barium oxide. Bureau of Standards, Washington, D. C. Address S. W. Stratton, Director.

*Heat B1-20* Evaporation from Liquid Surfaces. Address Curtiss Engineering Corporation, Garden City, L. I.

*Internal-Combustion Motors B3-20* Available Fuels. An investigation of available fuels for internal-combustion engines in the future using suggested manifold and heater for the application of high-boiling-point fuels. J. G. Vincent, Packard Motor Car Company, Detroit, Mich.

*Properties of Engineering Materials B10-20* Cables. Tramway Cable Loaded at One Point, by S. Herbert Anderson. An investigation to obtain equations which completely describe the conformation taken by a flexible chain or cable anchored with one end at a higher level than the other and loaded at any one point such as a tramway cable or logging skyline. Equations have been obtained for easy computation of tension at any point, maximum tension for given loading and position of load for maximum tension. Equation contains three constants or parameters, one of which is expressed in terms of the other two. Constants can be determined from weight of cable per unit of length, amount of load, and heights of end points above load. University of Washington, Seattle, Wash. Address C. F. Magnusson, Director.

*Textile Manufacture and Clothing B1-20* Methods of Determining the properties of cotton fibers with respect to their manufacturing properties. These methods are to be used in connection with determining the effect of the present manufacturing properties in the attainment of higher-speed textile machinery and more economical use of cotton fibers. Textile Research Company, 34 Batterymarch St., Boston, Mass. Address E. D. Whalen, Manager.

*Transmission B3-20* Wood Pulleys. An investigation of pulleys prepared from spruce. Pulleys built up in the same manner as hardwood pulleys. Density of the spruce fibers increased by compression at right angles to grain to one-half of original volume. Tests being made are those for frictional driving power and investigations of mechanical strength, and are in charge of Prof. G. S. Wilson, Engineering Experiment Station, University of Washington, Seattle, Wash. Address C. E. Magnusson, Acting Director.

## C—Research Problems

The purpose of this section of Engineering Research is to bring together persons who desire coöperation in research work or to bring together those who have problems and no equipment with those who are equipped to carry on research. It is hoped that those desiring coöperation or aid will state problems for publication in this section.

*Automotive Vehicles and Equipment C1-20* Resistance of Automobile Bodies. Tests on resistance of automobile bodies at various speeds. Address Curtiss Engineering Corporation, Garden City, L. I.

## E—Research Personal Notes

The purpose of this section of Engineering Research is to give notes of a personal nature regarding the personnel of various laboratories, methods of procedure for commercial work or notes regarding the conduct of various laboratories.

*Heating E1-20* The work of the Engineering Experiment Station at Pennsylvania State College in connection with heating and ventilation is being conducted along the lines of obtaining the coefficient of heat transmission by means of the "Hot Box." This method gives the effects of conduction as well as the surface effects. The work is under the direction of Prof. A. J. Wood. Address Dean R. L. Sackett, Pennsylvania State College, State College, Pa.

*Heating E2-20* Warm-Air Furnace Research, by A. C. Willard. Among the various problems to be carried on in the researches connected with warm-air furnace heating may be mentioned the following:

Transmission, efficiency and capacity with various register temperatures with draft not to exceed 0.2 in. of water. The length of firing period is also to be investigated. The heat- and air-carrying capacities and heat losses in various kinds and sizes of leaders and stacks are to be investigated. Losses in boots and registers are to be determined as well as the length and pitch of leaders. Comparison of relative efficiency and capacity of various furnaces of same grate area but of different amounts and arrangements of heating surface. Comparison between hard- and soft-coal furnaces. A study of cold air and recirculation. A standard method of rating furnaces. The effects of modifying the design and installation. Address Director C. R. Richards, Engineering Experiment Station, Urbana, Ill.

*Iowa State College, Engineering Experiment Station E1-20* The Engineering Experiment Station was organized in 1904 and receives an appropriation of $35,000 per annum. The complete equipment of the Engineering School is available for the Experiment Station. The application of mechanical engineering to agricultural machinery, including tractors, receives considerable attention. Iowa State College, Agriculture and Mechanical Arts, Ames, Iowa. Address Anson Marston, Director.

*Texas Engineering Experiment Station E1-20* The Texas Engineering Experiment Station, College Station, Texas, has issued twenty-two Bulletins, fourteen of them dealing with highway engineering, one with chemistry, one with cotton cultivation, one with purchasing, one with geology, three with bones and one with fuels. Address J. C. Nagle, Director, Texas Engineering Experiment Station, Agricultural and Mechanical College of Texas, College Station, Texas.

## F—Bibliographies

The purpose of this section of Engineering Research is to inform the profession of bibliographies which have been prepared. In general this work is done at the expense of the Society. Extensive bibliographies require the approval of the Research Committee. All bibliographies are loaned for a period of one month only. Additional copies are available, however, for periods of two weeks to members of the A.S.M.E. or to others recommended by members of the A.S.M.E. These bibliographies are on file in the offices of the Society.

*Chemistry, General, F1-20* Chemical Warfare. A bibliography of 12 pages prepared by Dr. Clarence J. West of the Arthur D. Little Co., Inc. Address Arthur D. Little, Inc., Cambridge, Mass.

*Chemistry, Organic F2-20* Alcohol. The Production of Alcohol from Sulphite Waste Liquors. A bibliography of 6 pages prepared by Dr. Clarence J. West of the firm of Arthur D. Little, Inc. Address Arthur D. Little, Inc., Cambridge, Mass.

*Electrochemistry F1-20* Electric Furnaces, Design and Construction. Search 2880. A bibliography of 5½ pages. Address A. S. M. E., 29 West 39th St., New York.

*Fuels, Gas, Coal and Coke F1-20* Colloidal Fuels and Low-Temperature Distillation of Coal. A bibliography of 1½ pages. Search 2878. Address A. S. M. E., 29 West 39th St., New York.

*Heat F1-20* Utilization of Waste Heat for Generating Steam. A bibliography of 3 pages. Search 2741. Address A. S. M. E., 29 West 39th St., New York.

*Hydraulics F2-20* Sprinkler Nozzles, by F. W. Greve and W. E. Stanley. A bibliography of 6 references on sewage sprinkler nozzles. Address A. S. M. E., 29 West 39th St., New York.

*Metallurgy and Metallography F2-20* Electric Furnaces. See Electrochemistry F1-20.

*Research F1-20* Industrial Research. A bibliography of 8 pages on a reading list on Industrial Research, prepared by Dr. Clarence J. West and Edward D. Greenman, of Arthur D. Little, Inc. Address Arthur D. Little, Inc., Cambridge, Mass.

# Work of the Boiler Code Committee

*THE Boiler Code Committee meets monthly for the purpose of considering communications relative to the Boiler Code. Any one desiring information as to the application of the Code is requested to communicate with the Secretary of the Committee, Mr. C. W. Obert, 29 West 39th St., New York, N. Y.*

The procedure of the Committee in handling the cases is as follows: All inquiries must be in written form before they are accepted for consideration. Copies are sent by the Secretary of the Committee to all of the members of the Committee. The interpretation, in the form of a reply, is then prepared by the Committee and passed upon at a regular meeting of the Committee. This interpretation is later submitted to the Council of the Society for approval, after which it is issued to the inquirer and simultaneously published in MECHANICAL ENGINEERING, in order that any one interested may readily secure the latest information concerning the interpretation.

Below are given the interpretations of the Committee in Cases Nos. 273-282, inclusive, as formulated at the meeting of February 26, 1920, and approved by the Council. In accordance with the Committee's previous practice, the names of inquiries have been omitted.

FIG. 3  VERTICAL SINGLE-FLUE BOILER WITH CORRUGATED FIRE BOX

## CASE No. 273

*Inquiry:* Does Par. 296 of the Boiler Code require that the tee or lever-handled cock be placed immediately under the steam gage where a long connecting pipe is used and the locked-open valve is permitted close to the boiler?

*Reply:* It is the intent of the requirement in Par. 296 that the tee or lever-handled valve shall be located near to the steam gage so that it will be readily evident to any one observing the gage, even though the locked-open valve is used at or near the boiler.

## CASE No. 274

*Inquiry:* Is it permissible to use upon an h.r.t. boiler with the third return type of setting, an extended nozzle formed of a short length of pipe screwed into flanged fittings at the boiler connection and the outer end, in order that it may reach well above the boiler brickwork?

*Reply:* The construction proposed will not meet the Code requirements, where the pipe is over 3 in. pipe size and the working pressure exceeds 100 lb. per sq. in. A double-flanged nozzle riveted to the boiler shell should be provided, one form of

which is illustrated in Case No. 232, or a standard pressed-steel nozzle may be used in lieu of the one illustrated by Case No. 232. If such standard nozzle is used, it should be protected by insulation.

## CASE No. 275

*Inquiry:* Is autogenous welding permissible for the longitudinal joint of the fire box of a form of vertical boiler as shown in Fig. 3, where the furnace section is, after welding, heated and corrugated by rolls? The corrugations are rolled to a depth of 1½ in. on 8 in. centers, and the weld shows no fracture or distress after either corrugating or flanging.

(In the hands of the Committee)

## CASE No. 276

*Inquiry:* In the design of a cast-steel water box to be set in the side walls of furnaces and to be subjected to full boiler pressure, is it necessary to apply to the sections containing flat surfaces, the formula in Par. 199 of the Boiler Code, using C = 120, or is it necessary to use this formula with the value of C = 156?

*Reply:* It is the opinion of the Committee that the construction of pressure parts of the type referred to is provided for by Pars. 9 and 247. If the Secretary of the Boiler Code Committee can be notified when the specimen is ready for test, steps will be taken to have a representative of the Committee present.

## CASE No. 277

*Inquiry: a* An interpretation is requested of the application of Par. 212a of the Boiler Code with reference to any curved stayed surface subject to internal pressure. Does this refer to both the outer and furnace sheets of vertical tubular boilers?

*b* If under the requirements of Par. 239 of the Boiler Code relative to furnaces under 36 in. in diameter, the design of the furnace does not permit of operation without staying, is there any rule in the Code for the staying in this case, or must the furnace sheets be stayed as flat surfaces, using Table 4 for the pitch?

*c* Is it the intent of Par. 212c of the Boiler Code that the increased pitch allowed, may be used for the same working pressure and thickness of plate as indicated in Table 4? It is the understanding that Table 4 is based on the formula given in Par. 199.

*Reply: a* The term " curved stayed surface subjected to internal pressure " in Par. 212a of the Code is intended to refer to any surface in a boiler structure that is subjected to pressure on the concave side. It therefore refers only to that part of the outer shell of a vertical tubular boiler which is stayed.

*b* Where a furnace under 36 in. in diameter requires staying, it should be stayed as a flat surface as provided for in Table 4, except that the pitch may be increased as indicated in Par. 212c.

*c* It is the intent of Par. 212c that the increased pitch there permitted, may be used for the same steam pressure and thickness of plate as specified in Table 4.

## CASE No. 278

(In the hands of the Committee)

## CASE No. 279

(In the hands of the Committee)

## CASE No. 280

*Inquiry: a* Is it permissible under the requirements of the

Boiler Code to use a blow-off valve of the type used on locomotive boilers, operated by a lever lift, for stationary boilers?

*b* Is it considered safe to use superheated steam of 125 lb. pressure and 125 deg. of superheat, or total nominal temperature of 478 deg., with a piping system having extra-heavy cast-iron fittings and medium-weight cast-iron valves?

*Reply: a* It is the opinion of the Committee that the blow-off valves required by Par. 311 for stationary boilers, may be of the lever-lift type, provided they are of extra-heavy construction, and so designed that they may be operated without shock to the boiler.

*b* Attention is called to Par. 12 of the Code which states that cast iron shall not be used for nozzles or flanges attached directly to the boiler for any pressure or temperature, nor for boiler and superheater mountings such as connecting pipes, fittings, valves and their bonnets, for steam temperatures of over 450 deg. fahr. While the Code only covers the parts therein specifically mentioned, this paragraph clearly indicates the judgment of the Committee as to the safety of the construction in question.

### CASE No. 281

*Inquiry:* Is it the intent of Par. 306 of the Boiler Code that

every superheater shall be so fitted with a drain that it can actually be completely drained? Many superheaters are fitted with drains which are, however, unable on account of their positions, to completely drain the apparatus.

*Reply:* It is the opinion of the Committee that every superheater should be so fitted with a drain as to substantially free the superheater from water when the drain is opened.

### CASE No. 282

*Inquiry:* Is it the intent of the Boiler Code Committee that the diameter at the base of the threads on the threaded ends of through rods or braces for h.r.t. boilers, shall be equal to or greater than the nominal diameter of the rod? Fig. 14 and Pars. 208 and 211 would seem to infer that it should be at least equal, but isn't it preferable to make it greater, so that the point of greatest weakness in the rod may not be in the threaded portion where permanent set due to strain would tend toward fracture?

*Reply:* It is the opinion of the Committee that it will be desirable that the threaded ends of through rods or braces for h.r.t. boilers, shall be sufficiently upset so that the minimum diameter at the base of the threads is in excess of the nominal diameter of the rod.

# CORRESPONDENCE

CONTRIBUTIONS to the Correspondence Departments of MECHANICAL ENGINEERING by members of The American Society of Mechanical Engineers are solicited by the Publication Committee. Contributions particularly welcomed are suggestions on Society Affairs, discussions of papers published in this journal, or brief articles of current interest to mechanical engineers.

## Proposed Symbol for B. T. U.

TO THE EDITOR:

For some fifteen years I have used the following symbol as a substitute for the three letters of B.T.U.:

You will note that the three letters are combined and are speedily produced in two strokes without the use of periods. Having found it very convenient, I am submitting it to you for possible publication in the hope that others interested might find it equally helpful.

R. L. WEBER.

Kansas City, Mo.

## Calculation of Stresses in a Rectangular Plate

TO THE EDITOR:

There are several formulæ in the handbooks for the solving of stresses in rectangular plates, but so far as the writer knows, each one of them is either empirical in its nature or has been mathematically derived from an assumption which is admittedly not true, but which, it is hoped, approximates the truth more or less closely. The writer, therefore, starting with such an assumption, presents a method for calculating the stresses in a rectangular plate supported at its four edges and loaded perpendicularly to its surface, and hopes that some light may be thrown upon the subject which will definitely determine whether or not the assumption and the method based upon it are correct.

It is a well-established fact of mechanics that in an ordinary beam the stress at any point, due to the combination of several loads upon the beam, is equal to the algebraic sum of the stresses caused at that point by each of the several loads taken independently. The writer assumes that this same method can be used to solve rectangular plates by first finding for a given point the stress, considering the plate as a simple beam supported on only two of its opposite edges, and then finding the stress for the same point, considering the plate as a simple beam supported on its other two opposite edges, and finally by combining these two stresses to obtain the magnitude and direction of the stress in the rectangular plate when supported at all four edges. As these two stresses are at right angles to each other they must be combined vectorially, which makes the resultant stress equal to the square root of the sum of the squares of the two simple beam stresses.

In solving these two simple beams we cannot take the same loading for each simple beam as we have on the rectangular plate itself, because then upon combining the two simple beams, the combined loading would be greater than the loading on the rectangular plate itself. What we must do is to take such loading on each of the two simple beams that the summation of their loading equals the loading on the rectangular plate.

In order to take the simplest possible case, we will consider a square rectangular plate of length $L$, uniform thickness $d$, with a uniformly distributed load upon its surface of $w$ lb. per square unit of surface, and with all four of its edges supported freely.

Obviously, as the plate is perfectly symmetrical, we should make the loading on each simple beam half the value of the loading on the rectangular plate, i. e., ½ $w$ lb. per square unit of surface. Following this through, we find that the stress at the center of the square plate is $\dfrac{3\sqrt{2}wL^2}{8d^2}$, and its direction is along the diagonal of the square. The stress at the middle point of each edge is $\dfrac{3wL^2}{8d^2}$, with a direction along the edge itself. The stress at each of the four corners is zero. The stress for other points in the plate could be found similarly.

An additional assumption is needed for the solving of rectangular plates that are not square, but the writer does not think it worth while to take up in detail this more complicated case until the assumption for the solving of square plates has been proved. He therefore gives only the result, as follows: Taking the same conditions for the rectangular plate as for the square plate, except that the rectangle has sides of length $a$ and $b$, the stress at the center becomes $\dfrac{3\pi a^2 b^2 \sqrt{a^2 + b^2}}{4 \delta^2 (a^2 + b^2)}$.

The writer concludes with the statement that if this method is correct it can be used as the basis of the calculation of such problems as determining the thickness of cylinder heads, steamchest covers, and other problems involving flat plates.                                    H. W. SIBERT.
Galesburg, Ill.

## Strength of Thick Hollow Cylinders

TO THE EDITOR:

In the discussion on my paper on Valves and Fittings for High Hydraulic Pressures, presented at the Annual Meeting of the A. S. M. E. in 1918[1], the point was made that the formula used for the strength of hydraulic pipes and cylinders was difficult to apply, especially when contrasted with the greater simplicity of Barlow's formula. The method which I used was based on the results of experiments by Cook and Roberts, published in London *Engineering*, December 15, 1911, which, after considerable study, I was convinced offered the best basis for design.

[1] Trans. Am.Soc.M.E., vol. 40, page 977.

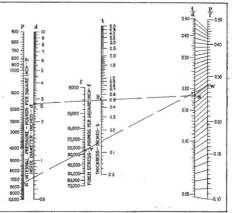

FIG. 1 STRENGTH OF THICK HOLLOW CYLINDERS

FORMULA: $t = -r \pm \sqrt{P r^2 /(0.6 f - p) + r^2}$, in which $r = 0.5\,d$.
EXAMPLE: Given $P = 7500$ lb. per sq. in.; $f = 27,000$ lb. per sq. in.; $d = 2\frac{3}{4}$ in. *Solution:* Draw line *wx* from 7500 on $P$-scale through 27,000 on $f$-scale, cutting $P/f$-scale at 0.278; next draw *xyz* to $t/d$-scale, cutting it at 0.183; then draw *zyx* to 2¾ on $d$-scale, cutting $t$-scale at 0.50 +, which gives required thickness.

Recently, having occasion to calculate some piping, I found that I could make a chart, using the principles of the formula given in my paper, by which the strength of almost any cylinder could be obtained without calculation. I am enclosing a print of this chart, and also one for the application of Barlow's formula which I find many designers use. It will be noted that these two charts are the same except for the relation lines between the $P/f$ and $t/d$ scales at the right side of the chart. The charts are self-explanatory and will be found time savers and sufficiently accurate for any ordinary case.
                                    WM. W. GAYLORD.
Torrington, Conn.

A national association of present and former officers of the Corps of Engineers and civil engineers who have served in any branch of the army—Engineers, Ordnance, or Signal Corps; Infantry, Cavalry or Artillery—is being organized in Washington by a committee appointed by the Chief of Engineers. The objects of this society are to promote the science of military engineering, and to foster the co-operation of all arms and branches of the service, and of civilian engineers in that science. Provisions have been made for an annual meeting, the date of which is to be fixed with reference to that of the American Society of Civil Engineers, in order that members may attend both meetings. The annual dues will not exceed $5 per year. Further information regarding the society may be secured from Col. G. A. Youngberg, Office of the Chief of Engineers, U. S. Army, Washington, D. C.

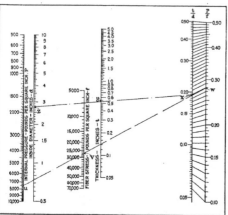

FIG. 2 STRENGTH OF THICK HOLLOW CYLINDERS

BARLOW'S FORMULA: $P/f = t/R$, or $P/f = 2t/(d + 2t)$, in which $P =$ internal pressure; $f =$ fiber stress; $t =$ thickness; $r =$ outside radius; $d =$ inside diameter.
EXAMPLE: Given $P = 7500$ lb. per sq. in.; $f = 27,000$ lb. per sq. in.; $d = 2\frac{3}{4}$ in. *Solution:* Draw line *wx* from 7500 on $P$-scale through 27,000 on $f$-scale, cutting $P/f$-scale at 0.270; next draw *wx* to $t/d$-scale, cutting it at 0.193; then draw *xyz* to 2¾ on $d$-scale, cutting $t$-scale at 0.53, which gives required thickness.

# Standard Sizes for Shafting

### Committee of the A.S.M.E. Recommends 14 Sizes of Transmission Shafting and Large Reduction in Number of Sizes of Machinery Shafting

THE desirability of reducing the number of sizes of shafting and in consequence the number of parts of power-transmission equipment that must be carried in stock has long been recognized. It remained for the conditions of the war, however, to bring about definite action in this regard, from the standpoint of the conservation of materials. The activities of the Committee of The American Society of Mechanical Engineers on War Industries Readjustment brought to light the fact that an immense amount of steel is continuously tied up in manufacturers' and dealers' stocks of shafting and that a corresponding amount of cast iron is also held in stock in the form of hangers, bearings, couplings, collars, bushings, pulleys, etc. At the suggestion of the chairman of the Committee on War Industries Readjustment, therefore, a committee was formed to investigate the subject of the standardization of shafting sizes. The personnel of this committee is as follows:

Cloyd M. Chapman. *Chairman*     Geo. N. Van Derhoef
Hunter Morrison                  Louis W. Williams
Russell E. Nelles

This Committee was confronted with two distinct but closely related problems, viz., the standardization of the diameters of shafting used for the transmission of power, such as lineshafts, countershafts, etc., and the standardization of the diameters of shafting used by machinery manufacturers in making up their product. The first of these problems seemed to be the simpler of the two. While a large number of sizes of transmission shafting are now listed and stocked, it was believed that a comparatively few of these are in extensive general use. Accordingly, a letter was sent to 36 of the largest manufacturers and dealers in transmission shafting asking for statistics on the consumption of each size of shafting handled by them. Some 20 of the largest concerns in the industry furnished complete statements of their sales over periods of time chosen by themselves. These data were reduced by the committee to a uniform basis of percentages. The amount of each size sold was expressed as a percentage of the total sales, both on a weight basis and on a lineal-foot basis. From these data, plotted in the form of a diagram, it was very evident which of the sizes were popular and generally used and which were more rarely called for. A tentative list of 12 sizes was prepared from this diagram and sent to forty-six other dealers in transmission shafting and shafting supplies from whom twenty replies were received.

In the letters to these firms, the committee expressed the opinion that the custom of using shafting $\frac{1}{16}$ in. under the unit sizes is so firmly and so nearly universally established in this country, that it would be unwise to attempt to adopt sizes in even inches and fractions as standard. It was pointed out, also, that certain sizes stand out preëminently as "popular sizes" and that others are sold in relatively small quantities. It seemed very feasible to select a series of standard sizes which would meet the popular demand and give a sufficient selection of sizes for general purposes and at the same time reduce the number of sizes now listed by the trade from some 50 or 60 down to 12 or 15.

The response to these letters was hearty and practically unanimous in opinion. The transmission-shafting users and dealers, almost to the last one, approved the plan of standardization and the sizes suggested were very generally approved except that the diameters $1\frac{11}{16}$ in. and $2\frac{7}{16}$ in. were in many cases requested to be included. After due consideration the committee decided to include these two sizes in the original list, making the 14 sizes now adopted as standard.

The second problem was a more intricate one. The number of sizes now produced by the rolling mills for use in machinery is very large. Almost every sixty-fourth of every inch up to three inches is drawn. This means excessive equipment at the mills and large stocks in the warehouses. If a reasonable number of these sizes could be eliminated or classed as "Specials" and a comparatively few sizes selected as standard or stock sizes, a great saving would thus be effected and a valuable service performed.

In order to get the opinions of leading consumers of shafting for machinery purposes, the committee decided to lay the plan before some 225 large consumers of this material and invite their comment upon its desirability or feasibility and their advice as to the size interval between standard diameters which should be considered. It was explained that it was not intended that the adoption of certain sizes as standard should make it impossible to secure any other size required on special order; but that the general elimination of a great number of the sizes now in use and the consequent greatly increased production of the standard sizes could only tend to a reduction of mill costs and capital invested in manufacturers' equipment and in stocks in warehouses. Both of these savings should have a lowering effect upon the price to the consumer and the problem was, therefore, truly one of conservation.

In the case of machinery shafting the users were equally unanimous in their approval of the plan to standardize sizes, but recommendations as to size interval varied greatly. However, these recommendations, in so far as they were definite and specific enough, were tabulated and a diagram constructed showing the relative popularity of the various size increments for each inch of diameter.

With these data accumulated and sifted down to usable form, the committee felt that it was in a position to present its information and preliminary deductions to representatives of other interested organizations. Accordingly, invitations were issued to twelve societies and associations requesting them to consider the proposed lists of standard sizes and to appoint representatives to confer with the committee before its report was finally formulated. The seven organizations listed below responded and the standard sizes which follow have the unanimous approval of these representatives and, as far as can be learned, of their associations.

American Hardware Manufacturers' Association
American Railway Engineering Association
American Supply & Machinery Manufacturers' Association
National Association of Manufacturers of the U. S. A.
National Association of Purchasing Agents
National Machine Tool Builders Association
Southern Supply & Machinery Dealers Association

The committee then considered that it had completed the first part of the work to which it had been assigned, so on January 14 submitted to the Council a progress report in which it recommended the approval and adoption of the following lists of sizes as standard for the Society:

*Transmission Shafting:*

15/16 in; 1-3/16 in; 1-7/16 in; 1-11/16 in; 1-15/16; 2-3/16 in; 2-7/16 in; 2-15/16 in; 3-7/16 in; 3-15/16 in; 4-7/16 in; 4-15/16 in; 5-7/16 in; and 5-15/16 in.

*Machinery Shafting:*

Size intervals extending to 2½ in., by sixteenth inches; from 2½ in., to 4 in., inclusive, by eighth inches; from 4 in., to 6 in., by quarter inches.

The Council approved the report and accepted the recommendations.

In the opinion of the committee the adoption of standard sizes of shafting will mean that in the future there will be a gradual elimination of odd sizes from makers' lists and from dealers' stocks, and for new construction only standard sizes would be selected.

Before undertaking the standardization of the shafting formulæ and the dimensions of shafting keys and keyways the committee plans to reorganize itself and add to its membership.

251

# MECHANICAL ENGINEERING

### THE JOURNAL OF THE AMERICAN SOCIETY OF MECHANICAL ENGINEERS

Published Monthly by the Society at
29 West Thirty-ninth Street, New York

FRED J. MILLER, *President*

WILLIAM H. WILEY, *Treasurer*    CALVIN W. RICE, *Secretary*

PUBLICATION COMMITTEE:

GEORGE A. ORROK, *Chairman*        J. W. ROE
H. H. ESSELSTYN       GEORGE J. FORAN
RALPH E. FLANDERS

PUBLICATION STAFF:

LESTER G. FRENCH, *Editor and Manager*
FREDERICK LASK, *Advertising Manager*
WALTER B. SNOW, *Circulation Manager*
136 Federal St., Boston

Yearly subscription $4.00, single copies 40 cents. Postage to Canada, 50 cents additional; to foreign countries $1.00 additional.

*Contributions of interest to the profession are solicited. Communications should be addressed to the Editor.*

## Editorial Associates on Mechanical Engineering

C. E. Davies, a graduate of Rensselaer Polytechnic Institute, and for several years in production work with the Remington Typewriter Company, Syracuse, N. Y., has joined the staff of MECHANICAL ENGINEERING as one of the Associate Editors. Mr. Davies was in the Ordnance Department at Frankford Arsenal for 15 months during the war, and left the service with the rank of Captain.

E. W. Tree, who has been with the Society during the past year and was formerly an instructor in the Electrical Engineering Department of Brooklyn Polytechnic Institute, has also been appointed an Associate Editor. Of the other members of the Staff, Leon Cammen, for many years Associate Editor, will continue in this capacity and as in the past prepare the Engineering Survey; and C. M. Sames, Associate Editor for four years, will have the direction of the Engineering Index.

## A.S.M.E. to Meet in Convention at St. Louis, May 24 to 27

A tentative program for the Spring Meeting is printed in Section 2 of this number of MECHANICAL ENGINEERING. In connection with it, announcement is made of an opportunity to visit the immense hydroelectric plant at Keokuk, on the Mississippi River, illustrations and description of which appear in this number. Particulars regarding the excursion are contained on page 60 of Part 2, or Society Affairs Section, of this issue.

Papers to be presented include several groups upon certain special subjects, such as aeronautics, the appraisal and valuation of property, and foundry practice. One session will be devoted to scientific papers, and another to miscellaneous papers, mainly on power-plant subjects. The St. Louis Local Committee will have a session of their own, with papers of local interest, principally upon new developments in river transportation, and on industrial housing, a large project for which is under way at St. Louis.

Two papers of unusual interest come from Government sources, one of which will be by Lieut.-Colonel H. W. Miller, Artillery Division, U. S. A., who recently contributed the German Long-Range Gun article to MECHANICAL ENGINEERING. Colonel Miller has investigated heavy artillery abroad and will speak on the formidable German defenses on the coast of Belgium. Lieut. E. R. Jackson, Ordnance Department, U. S. A., Tank, Tractor and Trailer Division, will show moving pictures of the transcontinental trip made by a large number of army trucks, an extended account of which was published in the last number of MECHANICAL ENGINEERING.

The Local Committee has arranged several excursions and social affairs, listed in the tentative program, which are most attractive. Members of the A. S. M. E., and their friends, who can do so, are cordially invited to attend.

## "Conservation Through Engineering"

Past-Secretary of the Interior, Franklin K. Lane, who upon retiring from that office characterized official Washington as "rich in brains and in character, but poorly organized for the task that belongs to it" has given considerable attention to the engineering features of the work of that department. This is evidenced by the frequent and comprehensive references made to engineering work in the last annual report of the Secretary of the Interior, and by the specially-printed extracts of that report which have been published under the title quoted above.

This report deserves the careful attention of all engineers. The recent coal strike and the attention given it by the Interior Department is covered in considerable detail with pertinent questions asked and answered as to the means of conserving coal and meeting future emergencies of this kind. Under the heading, White Coal and Black, the question of water power is briefly treated. Petroleum from the standpoint of its production from oil shale and its use as fuel and generating power is also discussed with a comprehensive summary. It will be recalled that Mr. Lane was especially active with broad plans for soldier settlement legislation,—it was he who first introduced soldier settlement plans. Other phases of the reclamation and development program are covered in an interesting manner. Conservation Through Engineering is a public document known as Bulletin No. 705, and may be obtained from the Superintendent of Documents, Government Printing Office, Washington, D. C., at ten cents per copy.

## A University for the Industries as Well as for the People

The University of Michigan, long a source of help and inspiration to the people of its own state, as well as to its many students and friends in other sections, is now arranging for coöperation with the industrial and technical interests of the state along the lines of scientific research.

A new department of Industrial Research is to be organized in connection with the Colleges of Engineering and Architecture. The department will be under the administration of a director, who will be assisted by an administrative committee of the University and by an advisory board composed of men representative of the industries of the state. Investigators and assistants will be assigned to the Research Department as required, and the work in this department will be open to graduate and undergraduate students under suitable conditions. The industry or group of industries having researches undertaken is to pay for these researches, and the University reserves the right to make public at such times and in such ways as it deems best the results of any work done on any problem by the Research Department. The University is to set aside for the miscellaneous expenses of the department a sum of $10,000 a year, and it is expected that the industries of the state will provide initial funds, in part, at least, for meeting the preliminary expenses incident to the laboratory equipment required for the new undertaking.

This action by the University has followed a report recently made by the Michigan Manufacturers' Association recommending such coöperation. A great deal of credit is also due to the Chicago Engineering Alumni of the University who prepared a report on this subject in 1916 and whose long-sustained efforts to bring the industries into closer relations with the University are now to bear fruit. These Chicago alumni have done a great deal in various directions to advance the cause of research as evidenced by resolutions presented at the last Spring Meeting of The American Society of Mechanical Engineers, largely through their efforts, in which a broad program was outlined for research generally throughout the country, and in which the colleges of the country were an important element.

## Annual Volume of Engineering Index to be Issued May 15

The 1919 annual volume of The Engineering Index will be available about May 15th, the earliest date at which it is possible to complete printing. This publication has been compiled from the monthly issues of The Engineering Index which have been a feature of MECHANICAL ENGINEERING during the year just passed. The items as they appeared from month to month were classified under general headings designating the broad divisions of engineering, but an extended investigation of the merits of different systems of indexing showed that an alphabetical or dictionary arrangement was greatly preferred by the users. Accordingly, in the preparation of the annual this system has been followed. The 12,000 items comprising the index have been entirely rearranged according to this system by a staff of engineers, and while this has entailed many weeks of labor it is believed that the result will fully justify the effort.

## Diesel-Electric Drive for Ships

The Diesel engine has regularly been used in submarines as the propulsive power for driving the propeller when on the surface, and also for charging the storage batteries through a generator, thus providing an electric drive when submerged. In the meantime a method for driving the propeller from a prime mover, through the intermediary of an electric generator and a motor, was developed in this country and first applied in the form of a steam turbine-electric drive such as now used on the U. S. battleships.

The many good features of the electric drive for ships, in combination with steam turbines, have been so fully demonstrated that it was to be expected that the same system would be tried in combination with Diesel engines. The Diesel engine, like the turbine, is capable of extraordinarily high economy, and, like the turbine, it should be operated under the conditions which its constructive features demand, rather than under the conditions imposed by the speed limitations of the propeller.

What is said to be the first trial of a Diesel-electric drive for ships is reported in the current issues of the marine papers as being on the yacht *Elfay*, a schooner of 152 ft. length overall, 30 ft. beam and 313 gross tons capacity. The yacht is driven by a 6-cylinder Winton Diesel oil engine rated at 115 hp., at 425 r.p.m., direct-connected to a 75-kw. Westinghouse generator, which in turn is direct-connected to a 9-kw. exciter. The motor for the propeller is rated at 90 hp., at 360 r.p.m. The entire boat is operated and controlled by electricity, no steam being used or required. All the auxiliaries are electric-driven and all movements are controlled from the bridge without having to signal the engine-room.

The yacht is said to make 8½ knots on 7½ gal. of fuel per hour; and since she can carry 2400 gal. of fuel, she is able to run 2000 miles on her propelling equipment alone. Although the *Elfay* is by no means a typical merchant vessel, it is believed that if the new drive proves successful on her it will be extended to purely commercial ships. In fact, consideration is now being given to the application of Diesel-electric power to some of the wooden freight vessels of the Ferris type now offered by the U. S.

Shipping Board, where there is not sufficient headroom in the hold for boilers and engines of the usual type. In an article in *International Marine Engineering* for March, the project is outlined for placing Diesel engines on the main deck of the ship, using as many units as are required for the power to be generated, these engines being of the stationary type, but light enough for marine use.

## Recent Development in American Aeronautics

The aeronautical show held in New York during the second week of March brought out no very startling and novel developments, but it clearly demonstrated that the aeronautical industry in America is developing along sane and conservative lines.

Contrary to statements repeatedly made in the daily press, the aeroplane and aeronautic-engine plants are neither closing their doors nor dismissing their engineering personnel. On the contrary, they are doing fine work and are gradually placing the industry in the position to which it is entitled in view of the fact that the aeroplane was conceived in this country and made to fly for the first time as a result of the genius and perseverance of American engineers.

In this connection attention may be called to an important publication issued by the director of Air Service, Washington, D. C., namely, a Report on First Transcontinental Reliability and Endurance Test, conducted by the Air Service, U. S. A., in October, 1919. The official conclusions arrived at on the basis of a very careful analysis of all that happened are of unusual interest.

As regards motors, it is stated that on the whole the performance of the Liberty motor can be considered quite satisfactory although one is constrained, after very careful consideration of all the circumstances, to arrive at the conclusion that the Liberty motor cannot run full out for more than 40 hours, though with careful nursing to avoid excessive vibration it will run over 100 hours without overhaul.

A Hispano-Suiza motor should act well and the test showed that with some care the high-compression Hispano-Suiza motor will stand up against fearful strain.

It is not entirely pleasant to find that on the whole the German Mercedes 160-hp. low-compression motor gave a performance which the official report describes as "wonderful." No spare parts were needed and only a few minor adjustments to the magnetos, jets and plug points were made.

As regards planes, the test has shown conclusively that the DH-4 is a very good machine. On the other hand, the SE-5A made a remarkable showing in the test. Although as a single-seater plane it was not designed for the long, continuous flights of a trans-continental trip, it gave less trouble than the DH-4. The Martin bomber, which was the only type of multi-engined machine entered, showed an extremely fine performance and despite its great load kept up with the lighter craft.

In this flight there were several fatalities, the causes of which are carefully analyzed in the report. In general it would appear that while extensive precautions are taken in a test of this nature, it is nevertheless practically impossible to avoid accidents. One of the rules of the test stated that aeroplanes entered must be capable of a speed of over 100 m.p.h. These aeroplanes designed for war purposes are not as safe as the less speedy craft which would be flown for commercial purposes. It is another question whether such a rule should have been made for such a test, as obviously even in war times high-speed aeroplanes would hardly be used for transcontinental flying, except under very unusual circumstances.

The test strongly accentuated the importance of the pilot and demonstrated clearly that he is not merely an aerial chauffeur.

Considering the show and the report of the transcontinental flight, together with the high-altitude flight of Captain Schroeder (of which more is said elsewhere in this issue) as the three outstanding events of March, 1919, one can hardly help coming to the conclusion that aeronautics is quite alive in this country and very gratifying results are being obtained.

## The Engineer's Attitude on Military Legislation

Despite the fact that Congress has failed to enact legislation dealing with the subject of military training, and that public opinion seems at least to be divided on the question, the engineers through Engineering Council have gone on record as favoring a sound and effective development of both army and navy. This recommendation of Engineering Council comes through its Military Affairs Committee which was appointed in December, 1919, to consider the relation of engineers to the future military activities of the United States. Col. William B. Parsons is chairman of the committee, which is composed of both civilians and ex-officers of all branches of the service. The committee's recommendations to Engineering Council which were unanimously adopted by the executive committee of that body are in part as stated in the following paragraphs:

We believe it of vital national importance that sound military legislation should be enacted during the present session of Congress. Effective provision for national defense with an adequate trained army and navy sufficient to discharge our national and international obligations is essential to security and stability, which must be present if our country is to go forward in constructive activity and achieve its possibilities as a nation.

The careful selection and training of personnel and their assignment, when reservists, to organized reserve units are vitally important to the efficiency of the technical services of the army and navy, and therefore essential to the proper preparedness of the individuals of the engineering and allied professions and trades for the discharge of their constitutional obligations in national military service.

The full utilization of specially and technically qualified men and specialized industries has proved impossible where an organization was built up only after war had become imminent or had been declared.

We believe national security and stability can be assured only by application of the foregoing principles, in particular the principle of universal military training and the creation of organized reserves, above all in the technical branches. We, therefore, recommend to Engineering Council that the President, and Vice-President of the United States, the Secretary of War, the Secretary of the Navy, the Speaker of the House and the Chairmen and Members of the Military and Naval Committees of the Senate and the House be urged to incorporate the foregoing principles into any bills passed by either body for army or navy organization, and particularly the principle of universal training.

## The Future Engineer—A Man of Affairs

Engineering started as an art; then it developed into a science. Later it was realized that the engineering of men was as important as the engineering of things and the science of management came into being. Finally, the management of affairs, of men and things taken together collectively, is receiving a great deal of consideration and there is a general movement in the technical schools and colleges toward the teaching of business administration to engineering students.

How general this movement is has been shown in a recent report of the U. S. Commissioner of Education based on the replies to letters sent out to a number of the larger colleges and universities requesting "a statement of present practice and any change contemplated in courses of study in engineering and commerce." Fifty-five replies were received, a study of which brings out some interesting facts.

Approximately a third of the replies received came from colleges or universities which offer subjects in economics and business administration in connection with their engineering courses. In general the curricula provide business training in greater or lesser amounts for engineers who expect to enter positions concerned with the management or administration of manufacturing, construction, and transportation enterprises, which demand a knowledge of business as well as of scientific and engineering principles.

There are several institutions in which the combined business and engineering course is already well developed. For over 20 years Stevens Institute of Technology has given its students instruction in the commercial or business side of engineering. Dartmouth College has offered for four years a course in engineering management. A unique feature of this course is the large requirement in outside reading covering thirty or more books on the subject. The University of North Dakota has offered for three years a course in general engineering, providing about 60 per cent of the work along fundamental subjects in sciences, mathematics and technical engineering, and about 40 per cent in subjects relating to both business and engineering careers.

Massachusetts Institute of Technology gives a course in engineering administration which combines instruction in general engineering studies and in the methods, economics and law of business. The course includes (1) the instruction common to all courses, in literature, language, and history, and in chemistry, physics and mathematics; (2) a choice of engineering studies classified under three options: civil engineering, mechanical and electrical engineering, chemical engineering; and (3) a selected group of subjects in business and economics.

Other colleges which do not cover as much ground in their prescribed course, offer these subjects as electives or, in a few cases, in a special fifth-year course devoted to business administration.

There is only one institution reported in which no course is offered in either commercial or industrial engineering. A few commercial schools include engineering subjects in their courses.

The balance of the reports come from institutions which although they do not now offer a combined course, nevertheless are in favor of doing so as soon as they feel that the demand is great enough. Some of them are now at work on new engineering curricula covering a four, five, or six-year combined course. Columbia University, for instance, plans to have in the near future a course in manufacturing and industrial engineering, and Tulane University, New Orleans, now has under consideration a revision of its engineering curriculum to provide proper business training for engineers.

## Scientific American Supplement Now Issued as a Monthly

The *Scientific American Supplement*, edited by A. Russell Bond, is now issued as the *Scientific American Monthly* and has been greatly broadened in scope. Besides containing important announcements of distinguished technologists appearing in foreign as well as domestic publications, it furnishes translations of the complete texts of significant articles in European books, etc., and has several departments to which various organizations contribute. The National Research Council and the Bureau of Standards each supply material for a department, and a similar section is to be edited by the U. S. Bureau of Mines. The American Society of Mechanical Engineers furnishes regularly a review of important articles in the field of mechanical engineering and a similar service covering the field of mining and metallurgy is being rendered by the American Institute of Mining and Metallurgical Engineers. Reviews are also being given in the fields of electrical engineering and industrial chemistry. The journal is unique in that it carries no advertising, all the pages being text pages. While it is realized that the rapidly rising cost of paper and printing may make it impossible to issue a publication with 96 pages of text, such as this journal contains, from the income received solely from subscriptions, the attempt is nevertheless being made to conduct the journal on this plan.

The *Scientific American Supplement* was founded in 1876 at a time when there was aroused by the Centennial Exposition a great public interest in science. It was established to meet the growing demand for information of a higher technical character than could properly be published in the more popular *Scientific American*.

# Major Schroeder's Record Altitude Flight

### Notes on Major Schroeder's Accomplishment—Technical Difficulties Encountered—Military Value and Commercial Uses of Supercharger

On February 27, Major R. W. Schroeder, chief test pilot of McCook Field, Dayton, Ohio, drove an aeroplane to a higher altitude than has ever before been reached. Official figures show that an altitude of 36,020 ft. was attained and that the official record of 30,300 ft. reached by Rolland Rholfs on July 30, 1919, and the unofficial record of 33,137 ft. credited to Adjutant Casale, the French pilot, were broken.

The public is familiar with the story of how Major Schroeder set out to reach the altitude of 40,000 ft.; of his battle against a temperature of 67 deg. below zero and a wind of 100 m.p.h.; of how his oxygen gave out; of his fall of five miles; and of his miraculous landing at McCook Field. Further information regarding the flight, prepared for MECHANICAL ENGINEERING by Col. Thurman H. Bane, Commanding Officer of McCook Field and released by the Air Service, is given below.—EDITOR.

M AJOR SCHROEDER'S successful altitude flight was the result of a long series of important tests in engineering development by the Engineering Division of Air Service, U. S. A. This series of tests has disclosed problem after problem which had to be overcome in some manner before further advances could be made. Some of these troubles were:

*a* Obtaining a suitable propeller
*b* Keeping the mixture ratio of the carburetor constant throughout a wide range of altitude and therefore varying the differences in proportion between the throat of the carburetor and the atmosphere surrounding it
*c* Delivering fuel to the carburetor against its varying pressure
*d* Cooling at high altitude and raising the boiling point of the water
*e* Providing a drain valve to let the water out of the radiator at high altitude in case the engine should stop, so that the engine and cooling system would not be ruined by the water freezing; and further, of so arranging the valve that it would not itself freeze and thus become inoperative
*f* Securing goggles which would not freeze over in the intense cold at great heights; (such goggles were invented and designed by Major Schroeder)
*g* Developing a special instrument to show the pilot how to handle his exhaust bypass gates, or, in other words, how to control the supercharger pressure in his carburetors without the need of making any calculations.

The Engineering Division has been most fortunate in having Major Schroeder to pilot the airplane throughout the preliminary supercharger development tests which were made, because, in addition to being a surpassingly good pilot, he is an excellent engine mechanic and is one of the few pilots who can really sit in an airplane and know what an engine is doing and what it needs. In this way he has helped the development immensely.

Considerable trouble has always been encountered from preignition when running with a supercharger, due to the fact that the air delivered to the carburetor is at very high temperature. Future designs of superchargers will provide additional intercooling between the compressor and the carburetors. The fuelfeed system, prior to Major Schroeder's flight, had operated quite satisfactorily, but in spite of this fact it was necessary for Major Schroeder to close the vents in his gasoline tanks and pump some pressure in them with a hand air pump in order to help the fuel pumps deliver the fuel at the extreme altitudes.

To reduce the preignition which it was expected would be encountered on this flight, a specially prepared fuel was provided by Mr. Thomas Midgeley, Jr., who has been developing "antiknock" fuels for Mr. Charles F. Kettering, Mem. Am. Soc. M. E. This fuel proved to be of great assistance in this flight as it caused the motor to run much more smoothly than it would otherwise have done.

The supercharger used by Major Schroeder was the old original General Electric supercharger designed by Dr. Sanford A. Moss, Mem. Am. Soc. M. E., and originally tested on Pike's Peak in 1918; and it is not surprising, therefore, that no good means had been provided for blowing off the exhaust gases which issue from the turbine. The exhaust gas has bothered the pilot to a certain extent, due to the fact that it sweeps past his face. No way has been found to date to entirely carry the exhaust clear of the occupants of the machine. On this record flight it seems that the gases expanded more rapidly as they issued from the turbine in the thin air at great altitude and that they swept past the pilot's face in even larger volume. Judging from the doctor's report, it seems that the carbon-monoxide poisoning gave Schroeder more trouble than the lack of oxygen.

In military work an automatic oxygen-feed apparatus is provided for the pilot which regulates the amount of oxygen in proportion to altitude so that it is not necessary for him to think of making any adjustments. Major Schroeder had been in the habit of using a simple rubber tube extending from the neck of the oxygen flask to his mask in such a manner that he could adjust the flow by hand, as be has often had trouble with the oxygen pipe freezing up and flow stopping at high altitudes. On his record flight he knew he would be up for a long time, and he desired to use the automatic apparatus as long as he could. He believed it would work to about 29,000 ft., and therefore took one bottle of oxygen, connected through the automatic oxygen feed and one connected direct. However he found that the automatic apparatus did not work at all and he therefore had to start using his emergency bottle at about 18,000 ft. and realized that it might run short, although he thought it would last long enough for him to accomplish his record. Nevertheless, due probably to the large amount of exhaust gas he was breathing, Major Schroeder had to use an excess amount of oxygen, which, of course, resulted in his reaching the end of his supply sooner than he expected.

It is noteworthy that the instrument which shows the pilot what pressure his supercharger is delivering to the carburetor showed a pressure something close to that of sea level, even when he was at the highest point of the flight. The operation of the supercharger was excellent throughout the flight and it was found to be in good condition afterward.

Major Schroeder states that he actually reached warmer temperature at the top of his climb. The coldest temperature record was about —67 deg. fahr. and two or three thousand feet higher, at the top of his climb, the temperature was 4 deg. higher. He encountered the usual strong west wind which he has in every case encountered at altitudes above 25,000 ft. He believes the velocity of this wind to be close to 175 m.p.h., judging by the rate at which it drifted his machine eastward, even though he was headed west and climbing at an extraordinarily high air speed due to the use of the supercharger. It is to be regretted that figures on these speeds cannot be given for publication.

It is an interesting fact that after a certain low temperature is reached the exhaust gases issuing from the engine become snow white from the freezing of the vapor in them, and from that time on long, white clouds formed by this exhaust are visible from the ground on a clear day such as the one on which Major Schroeder's flight was made. This results in ice forming on all the wires and struts coming in contact with the stream of the exhaust.

When Major Schroeder's oxygen supply finally failed, he raised his goggles in order to see more clearly in his endeavor to "coax" some more oxygen from one or the other of his tanks, and at this moment unconsciousness suddenly overtook him—but not before he reached for his switch and put the machine into a spiral. He intended to make one steep spiral which would bring him down to about 20,000 ft. above the ground, where he expected to recover; but, although he believed after his fall that he had succeeded in doing this, as a matter of fact the plane fell into a "shot pigeon" down to about 3000 ft. above the ground, where Schroeder regained consciousness, "righted" the plane and, although he was still semi-conscious and could scarcely see at all due to the chilling of his eyes, had the presence of mind to open the vents in his gasoline tanks so that the engine would continue to get fuel and run. It happened that he succeeded in getting only one switch instead of two which are present in all Liberty ignition systems, therefore his engine had been running practically wide open throughout the fall. This kept the water from freezing in the cooling system and gave him the use of the engine after his recovery from the fall.

The daily papers have given the details of his marvelous landing in spite of his almost total blindness, so nothing will be said here about that; however, it is of interest to note that the gasoline tanks which probably had a plus pressure of several pounds in relation to the surrounding atmosphere at the top of the climb had collapsed,—that is, three out of four of them had collapsed, one of them almost totally, due to the fact that at the bottom of the fall conditions had changed so that there was a minus pressure inside of several pounds. This is why it was necessary for Major Schroeder to open the vents in the tanks in order to be able to deliver fuel to the engine to get to a suitable place to land.

The military value of the supercharger will be very great. It will greatly increase the speed of airplanes at high altitudes and enable them to go much faster than they can near the ground. It will be useful for photography at extremely high altitudes, because the photographer will not be hampered by attack if the plane goes high enough. An airplane with supercharged engine will be valuable to carry dispatches or a high-ranking officer over great distances in a very short time. Superchargers, when applied to heavy bombers, will enable this type of machine to reach a ceiling well above enemy anti-aircraft gun fire, and in fighting planes superchargers will greatly increase speed and climb.

Commercial use of superchargers will be to enable heavy passenger- or express-carrying airplanes to climb over the highest mountains or over thunderstorms with the use of comparatively low-powered and low-priced engines; without the supercharger, very large engines would have to be installed in order to have sufficient power to sustain the airplane at high altitudes. This is largely unnecessary if superchargers are used. It is felt that passenger-carrying airplanes can be provided with a supercharger and an airtight cabin for the passengers so that the supercharger can keep the air in this cabin at a density and temperature which will make it practically comfortable for all passengers, and at the same time the airplane can fly at extreme altitudes at very much greater speeds; and speed, after all, is one of the chief advantages of air travel over other kinds.

## James Waring See

James Waring See, a charter member of the A. S. M. E. and a widely known engineer, died at his home in Hamilton, Ohio, on January 31, 1920. Mr. See was better known to the readers of mechanical journals as "Chordal," a pen name used by him for much of his writing. About 1880 he began a series of articles in the *American Machinist* under the title of "Extracts from Chordal's Letters." These articles were of the shop by a shop man and it would be difficult for a person of the present generation to understand the deep impression which they made on the mechanical men of the country and the widespread interest which they created.

Mr. See was born in New York City on May 19, 1850. He received his earliest education in a country school at Rutland, N. Y., though he later attended school at St. Louis, Arcadia and Springfield, all in Missouri.

At the outbreak of the Civil War Mr. See was employed in the Springfield military hospital as an assistant in the dispensary and in the operating ward. After the battle of Carthage he was made telegraph messenger to the Federal forces in and

around Springfield. Upon the ending of hostilities Mr. See served an apprenticeship as machinist at the Springfield Iron Works and it was there he laid the foundation for his future career. At the completion of his apprenticeship he found employment in various shops located from St. Louis to Yankton, S. D., eventually settling at Omaha, Neb., where he started a shop of his own.

The story is told of him that being disappointed in a lathe he bought from the Niles Tools Works he wrote to that firm to the effect that if he could not design a better lathe he would eat it. Alexander Gordon, head of the Niles works, testily replied that if he was able to design a better lathe than the one in question, he wished he would come to Hamilton and do it.

Long after Mr. Gordon had forgotten the incident there appeared in his office a tall, lanky individual who said with a drawl, "My name is See and I came up here to design that

JAMES WARING SEE

lathe for you." He was put off with one excuse after another until in disgust he appropriated an empty drawing board and did design a lathe that made Mr. Gordon sit up and take notice.

He was at once employed by the Niles Works, filling positions as foreman, chief draftsman and chief engineer, respectively. In 1876 Mr. See opened an office in Hamilton, Ohio, as a consulting mechanical engineer and his keen insight into mechanical matters soon brought him a wide practice in connection with some of the largest machine establishments in the United States and Europe.

With the invention of the telephone, Mr. See became greatly interested in its practical workings and was the inventor of several valuable devices connected with central-station apparatus. In recognition of his work he was made an honorary member of the Telephone Exchange Association. For a time he was editor of the *Telephone Exchange Reporter*. He built the first telephone line in Hamilton. Also in connection with Alexander Gordon and James K. Cullen, he built an electric-light plant at the Niles Tool Works, the first in Hamilton.

As Mr. See's business increased, he became a patent attorney and as such developed an enviable reputation as an expert in patent litigation, and was called on to give expert testimony in more than three hundred cases.

By appointment of Governor Campbell, Mr. See acted as one of the commissioners for Ohio to the World's Fair held at Chicago in 1893.

# Government Activities in Engineering

*Notes Contributed by The National Service Department of Engineering Council* [1]

## Agreement Expected on Water Power Bill

As stated in MECHANICAL ENGINEERING for February the Water Power Bill has been in the hands of a committee composed of members from both the House and Senate. This was for the purpose of settling on points of difference, and while it was expected that not more than two weeks would be occupied in getting the bill through this Joint Conference Committee, nevertheless on March 9th. when the committee adjourned for a period of two weeks, there still existed two important points of difference. This adjournment was taken to await the return to Washington of Representative Taylor of Colorado and to permit Senators Jones and Smoot to devote some time to the Merchant Marine and Executive Bills, both of which are now pending.

These differences pertain to license charge and recovery and the proposal to have an army engineer as an executive secretary. The strong opposition to this latter plan will probably result in the placing of a civilian in that office with the authority to have an engineer officer detailed to assist in the work. Those in immediate touch with the status of the Water Power Bill regard the situation as very satisfactory and it is anticipated that a committee agreement may be shortly expected. The power project for Great Falls on the Potomac River, which carries an appropriation of $25,000,000 and which was placed in the bill as a rider by the Senate, was eliminated in the conference. No further details are at present available concerning the other items of the bill.

## New Board of Surveys and Maps Meets to Perfect Organization

The various developments leading to the creation of a new Board of Surveys and Maps which will coördinate the various map making agencies of the Government were presented in considerable detail in MECHANICAL ENGINEERING for February (p. 134). Since then the members of the Board have met with representatives of non-federal organizations for the purpose of determining the policy of the new Board and the manner in which it shall function. This meeting, called on March 9th, was called by the chairman of the Board, O. C. Merrill, Chief Engineer of the Forest Service, Department of Agriculture.

The report of the committee of representatives of the non-federal organizations was presented by M. O. Leighton of Engineering Council, who stated that while efficiency and economy might be certain occasions when it would be more advantageous for a particular department to do its own mapping. The consideration of the Board in such cases would have to be in a large degree judicial and representatives of the interested department should be disqualified from participating, in decisions affecting that department. It was further suggested that a plan should be adopted by which coöperating non-federal agencies could have a voice in the determination of the manner and the place of disbursement of their contributed funds.

The report of this non-federal committee also commented upon the following special features of the conference report: the preparation of skeleton maps of various scales; the methods of distribution as affected by the size of editions, costs, etc.; and the preparation of maps suitable for educational institutions. The need of a general topographic map of the United States and the means by which its preparation may be expedited was covered from the highway, railroad, and military standpoints by governmental and military officers.

## The Patent Bill

Due much to the efforts of Edwin J. Prindle and his colleagues, the Nolan Patent Bill (H. R. 11984) was put on the House calendar for March 5th, and was passed on the same date. Mr. Prindle is chairman of the Patent Committees of both Engineering Council and National Research Council. It will be recalled that extensive hearings were held before the House Patents Committee so that the case presented to the Rules Committee consisted in urging immediate consideration of the Bill.

About sixty engineers, members of the Founder Societies or representatives of organizations using the Patent Office, supported Mr. Prindle. When the hearing was over, the chairman of the Rules Committee told these men that it had been the most satisfactory hearing held by his committee in over ten years and that the thorough disinterested manner in which the engineers had presented their case assured early action in the House. This assertion was borne out when the House passed the bill a few days later. The bill is now before the Senate Patents Committee who intend to hold short hearings.

## Administering the New Land Leasing Law

The provisions of the Land Leasing Bill, to which reference was made in the last issue (p. 200), went into effect when the President signed it on February 26th. Its administration was handed over to the Interior Department, and to correct the misunderstandings that have arisen in connection with this new law, that department immediately prepared a set of regulations which have now been approved by the Secretary of Interior and are available for distribution.

Inquiries received indicate that many had gained the impression that leases on proven oil lands of great value were to be had merely by the filing out of an application prior to anyone else. The regulations point out that this is not the case, because proven oil lands of the Government are largely covered by claims of various kinds that have been in litigation for some time and which must be submitted for adjustment within six months. Until such claims are acted upon none of the lands can be leased and then only to the extent to which such claims are rejected. Further, if the Government grants a lease for their oil or coal land, except under the relief sections or as the result of a permit, it will only be by competitive bidding of which ample notice will be given to the public.

The only part of the act which is self-operative and not subject to the regulations of the Department is the section which provides for securing a preference right for oil prospecting permit by posting a notice on the ground. This applies only to lands not in the geologic structure of a producing oil field.

## Government to Develop Inventions

In order that the public, the industries, and the various departments of the government may obtain the full benefit of scientific inventions and patents of departmental employees, a bill has been introduced by the Senate Committee on Patents, proposing to give the Federal Trade Commission authority to arrange for the development of useful inventions, patents, and patent rights in the broadest way. This procedure is to be under such regulations as the President shall prescribe.

The Commission is authorized to collect fees and royalties for licensing the inventions, patents, and patent rights in such amounts as the President may direct and to deposit these in the United States Treasury; the necessary percentages however may be reserved for the remuneration of the inventors of meritorious ideas. The appropriations of Governmental departments are made available for the payment of fees charged in connection with the granting of patents under this Act.

[1] Engineering Council is an organization of national technical societies created to consider matters of common concern to engineers as well as those of public welfare in which the profession is interested. The headquarters of Engineering Council are located in the Engineering Societies Building, 29 West 39th Street, New York City. The Council also maintains a Washington office with M. O. Leighton, chairman of the National Service Department, in charge. This office is in the McLachlen Building, 10th and G Streets, Washington, D. C. The officers of Engineering Council are: J. Parke Channing, Chairman; Alfred D. Flinn, Secretary.

257

# NEWS OF THE ENGINEERING SOCIETIES
Meetings of The American Institute of Mining and Metallurgical Engineers, the American Institute of
Electrical Engineers and the Engineering Society of Western Massachusetts

## American Institute of Mining and Metallurgical Engineers

The annual meeting of the American Institute of Mining and Metallurgical Engineers was held in New York, February 16 to 20. The retiring president, Horace V. Winchell, outlined the civic activities of the Institute for the year. The new president, Herbert C. Hoover, an abstract of whose address was published on p. 194 of the March issue of MECHANICAL ENGINEERING, spoke on the national problems in which engineers are directly concerned.

The leading technical feature was a conference of coal operators from all over the country, who assembled to discuss the stabilization of the coal industry. At the opening of the conference, President Hoover, at whose suggestion it was called, outlined a program of study and work for the Institute and a constructive plan for the better working of the bituminous coal industry. Dr. Van. H. Manning, Director of the Bureau of Mines, talked on Problems of the Coal Industry. Dr. George Otis Smith, Director of the Geological Survey, in Fluctuation of Production of Coal—Its Causes and Effects, presented a statistical analysis of the rate of output over a period of years, indicating the relative effect of shortage of transportation, shortage of labor, lack of market and other factors in producing intermittency in the operation of coal mines. H. H. Stoek, professor of mining at the University of Illinois, discussed the Storage of Bituminous Coal and Its Possibilities of Stabilization. Transportation as a Factor in Irregularity of Coal Mine Operations, by S. L. Yerkes, vice-president, Grider Coal Fields Co., Birmingham, Ala., contained exact data as to the effect transportation facilities have on coal production, use of cars for storage, effect of more equipment, effect of lower rates in spring and summer, reduction of cross hauling, and long hauls by railroads of their own coal.

Numerous papers of scientific interest were presented at the technical sessions on Iron and Steel, Non-Ferrous Metallurgy, Milling and Smelting, Oil and Coal, and Industrial Organization. P. W. Bridgman described an experiment in one-piece gun construction. During the war, the Navy undertook the construction of an experimental gun embodying features designed to lessen the cost and time of production. It was demonstrated by actual construction and firing tests that a gun could be made from a single forging, producing the required reduction of internal stresses by a preliminary application of hydrostatic pressure so high as to strain the material considerably beyond its yield point. The procedure was to start with a single forging of approximately the dimensions of the finished gun, subjected to internal pressure in one or more stages, depending on the external shape, high enough to stress the metal permanently and thus to raise the elastic limit by producing compression in the inside layers and tension in the outside layers. The technique of controlling the pressures required, which were of the order of 100,000 lb. per sq. in., rested chiefly in employing a packing which automatically became tighter at higher pressures.

Tensile Properties of Boiler Plate at Elevated Temperatures was presented by H. J. French. At the request of a committee of the Engineering Division, National Research Council, a study was undertaken of the properties of boiler plate at various temperatures up to about 900 deg. fahr. Tests were made of $\frac{1}{2}$-in. plates of firebox and marine grades. The test specimen was heated by means of an electric tube furnace, two spiral resistors in series being used. In both grades of plate, increase in temperature from 70 to 870 deg. fahr. was accompanied by distinct changes in strength and ductility. The tensile strength at first decreased a few thousand pounds per sq. in., reaching a minimum at about 200 deg. fahr. This was followed by an increase up to about 550 deg. fahr., where the tensile strength reached a maximum about 10 per cent greater than the normal room temperature, after which an-

other and final decrease occurred. The percentage elongation in 2 in. decreased rather slowly up to about 200 deg. fahr., after which it dropped more rapidly, until a minimum was reached at about 470 deg. fahr. The reduction in area had a minimum at a slightly higher temperature than the elongation. The proportional limit at first increased slightly and showed a maximum in the neighborhood of 400 deg. fahr. for the firebox plate and the highest values between 200 and 300 deg. for the marine plate.

Of striking importance were also the following: Intercrystalline Brittleness of Lead, by Henry S. Rawdon; Blast-Furnace Flue Dust, by R. W. H. Atcherson; Critical Ranges of Some Commercial Nickel Steels, by Howard Scott; the Coefficient of Expansion of Alloy Steels, by John A. Mathews; Physical Changes in Iron and Steel Below the Therman Critical Range, by Zay Jeffries.

## The American Institute of Electrical Engineers

The eighth midwinter convention of the American Institute of Electrical Engineers was held in New York from February 18 to 20. A feature of uncommon interest was a symposium on superpower plants and transmission for the Northeast seaboard zone. W. S. Murray, consulting engineer, New York City, explained that the super-power plan provided a means by which a present estimated machine capacity of 17,000,000 hp. in a region between Boston and Washington and inland from the coast 100 to 150 miles, now operating with a load factor not exceeding 15 per cent, could be lifted to a load factor of more than 50 per cent and possibly 60 per cent, and a means by which, conservatively speaking, one ton of coal would do the work of two or three. The railroads within the above zone and those carrying coal into that zone would thereby be relieved of transporting one-half the amount of coal required for power and lighting purposes. That is, the value of machine capacity from a utilization standpoint would be increased from threefold to fourfold and coal resources for the purposes named be conserved twofold. It has been estimated that a saving of $300,000,000 a year would result from the establishment of this super-power system. In one of the other contributions it was pointed out that generating stations of from 200,000-kw. to 300,-000-kw. capacity, employing generating units of from 50,000-kw. to 75,000-kw. capacity, would involve no difficulties of design, construction or operation, and that from such stations a steam consumption rate of 10 lb. per kw-hr. or less and a total station rate of less than 1½ lb. of good quality coal per kw-hr. output should be obtained. Hope was expressed that a bill appropriating money for this national need would be presented before Congress by July 1, otherwise the plan would have to be financed by backing it with Government bonds or interstate or private capital.

The Measurement of Projectile Velocities, by Paul E. Klopsteg, physicist, Leeds and Northrup Co., and Major Alfred L. Loomis, Ordnance Department, was the title of a paper discussing the requirements imposed by proving-ground practice upon a chronograph intended for general ammunition testing. The number of Boulangé instruments in possession of the Ordnance Department was entirely insufficient for testing the immense quantities of ammunition contracted for by the Government during the war, and a new type of instrument was accordingly developed and adopted as standard ordnance chronograph, which was designated "Aberdeen chronograph." It is an assembly of standard parts with a few necessary special ones and rapid production was thereby made possible. Many comparative tests as to accuracy were made during the process of development, and invariably the average dispersion with the Boulangé chronograph was found to be from two to four times as great as with the Aberdeen. In routine firing, the three Aberdeen chronographs tested agreed within one or two feet per second on a velocity of 1700 ft. per sec., and only rarely did the maximum dispersion at this value become as great as 5 ft. per sec.

The economic and sociological aspects of daylight saving were discussed by Preston S. Millar, who stated that this wartime measure had reduced the total output of certain central stations and of one gas company by about 3 per cent during the seven summer months, reduction in output for lighting alone having averaged 8 per cent. Applying these data to the country as a whole, Mr. Millar estimated an annual saving by the public of $19,250,000 in expenditure for artificial light and a reduction of about 495,000 tons per annum in consumption of coal. On the other hand, he estimated the economic losses resulting in consequence of interfering with agriculture, dairying and truck gardening at over $1,000,-000,000. He therefore concluded that the very obvious solution of the problem appeared to lie in diversification of hours of industry. Since advancement of clocks, while serving the interests of one part of the population, has proved so disadvantageous to another part as to compel return to correct time, it seemed obvious, Mr. Millar pointed out, that those who benefit by advanced time in summer should adjust their habits as desired without disturbing the practice of the remainder of the population.

Participation of engineers in public affairs was urged by President Townley in his address. A sub-committee of the Standards Committee of the Institute submitted a proposal to standardize symbols for use in electrical diagrams, and offered for criticism and approval a tentative list of the more fundamental symbols. A dinner-dance at Hotel Astor constituted the social function of the convention.

## Engineering Society of Western Massachusetts

The February meeting of the Engineering Society of Western Massachusetts was held on February 24 at the High School of Commerce Auditorium, State Street, Springfield, Mass. Mr. W. E. Hodge, who is deputy in charge of street lighting in the city of Springfield, spoke on Street Lighting as an Engineering Problem, and Mr. John L. Harper, vice-president and chief engineer of the Niagara Falls Power Co., spoke on Niagara Falls Power Developments. Both lectures were illustrated by slides.

## Government Issues Special Ruling on Reinstatement of War Risk Insurance

Under a new and very liberal ruling, of far-reaching importance to millions of former service men, War Risk (term) Insurance, regardless of how long it may have been lapsed or canceled, and regardless of how long the former service man have been discharged, may be reinstated any time before July 1, 1920.

The only conditions to be met are: (1) The payment of two monthly premiums on the amount of insurance to be reinstated; and (2) The applicant must be in as good health as at the date of discharge, or at the expiration of the grace period, whichever is the later date.

The new ruling is the most important liberalization of War Risk Insurance since the passage of the Sweet bill, and is designed for the special benefit of service men who failed to reinstate their insurance prior to the new law, and who have been dischargd for more than 18 months.

Ex-service men may still reinstate their lapsed term insurance at any time within 18 months following the month of discharge by complying with the same conditions. Within three months following the month of discharge reinstatement may be made by simply remitting two months' premiums without a formal application or statement as to health.

Reinstatement may also be made after 18 months following discharge, as follows: If the insurance has not been lapsed longer than three months by complying with the conditions outlined above. From the fourth to the eleventh month, inclusive, after lapse, by complying with the same conditions, and in addition submitting a formal report of examination made by a reputable physician substantiating the statement of health to the satisfaction of the Director of the Bureau.

War Risk (term) Insurance may be converted into United States Government Life Insurance, now or at any time within five years after the formal termination of the war by proclamation of the President, and such Life Insurance, including Ordinary Life, Twenty Payment Life, Thirty Payment Life, Twenty Year Endowment, Thirty Year Endowment, and Endowment at Age of 62, may now be paid in a lump sum at death, if such method of payment is designated by the insured.

## Annual Report of the A.I.M.E.

In the report of the president of the American Institute of Mining and Metallurgical Engineers for 1919 it is gratifying to note the establishment of the Robert W. Hunt Prize for the best paper on the subject of iron and steel presented to the Institute for publication. Rules for this prize have been adopted and a medal is being designed by the distinguished artist, Mr. Emil Fuchs. Two bronze tablets have been erected at the headquarters of the Institute, one in memory of Dr. James Douglas and the other commemorative of the American Institute of Metals which was affiliated with the A. I. M. E. in the year 1918.

The Institute has also taken an advanced stand with regard to professionalism and established a precedent worthy the attention of other professional organizations in the expulsion of members believed to be unworthy the honor or privilege of membership. The report states that two members have had the privileges of membership withdrawn—as many as during the past six years—and comments as follows: "If an engineer is unworthy because of his lack of integrity, it is better that he be expelled forthwith, and members are urged for the good of the profession to bring to the attention of the Directors any cases of unprofessional conduct on the part of A. I. M. E. members. We believe that the increase in number of cases reported this year is due to the establishment by the members of a higher qualification for membership, and therefore a greater appreciation of the honor and responsibilities of membership."

## TIGHT-FITTING BOLT THREADS

(Continued from page 224)

In conclusion it may be stated that the tests would seem to indicate the following:

1 The cause of stripped threads is lack of room into which the metal can flow

2 The pitch diameter should be the same in both threads

3 The lead should be the same

4 The thread angles should differ by not more than 10 deg.

5 The limits for the inside diameter of nut need not be adhered to closely, as the inner part of the nut thread exerts very little, if any, holding power.

6 The outside diameter of plug and pitch diameter of both plug and nut are important and should be adhered to fairly closely.

## DISSIPATION OF HEAT BY SURFACES

(Continued from page 232)

of 0 deg. Then, since the watts dissipated varies directly with the temperature excess for a constant air velocity, it is seen that about 45 per cent more heat will be dissipated per degree under the above conditions of air velocity and heat supplied for an angle of incidence of about 42 deg. than for either perpendicular or parallel incidence. It is further seen that the least heat will be dissipated for 90 deg. incidence or when the coil is parallel to the air stream.

Tests were also made for perpendicular incidence of the air stream for various positions of the coil about its axis. No noticeable difference was observed other than what might well be ascribed to experimental error. This is quite significant since it shows the factor of importance to be the relative position of the axis of the coil in the air stream and not the relative position of the coil about its axis. Such a condition is quite likely to hold for all objects completely within the air stream.

## VENTURI-METER CALCULATIONS

(Continued from page 220)

$P_1 - P_2$, that is, actual manometer readings. Interpolation could then be resorted to, and by dividing the value of $Q\sqrt{T}$, found in the ordinate by the square root of the inlet temperature the flow in cubic feet of free air per minute could be obtained. The accuracy would depend upon the care taken in the interpolation. It was generally within 2 per cent, however, which was sufficient to answer the purpose for which this set of curves was devised.

### CALIBRATION

After consideration and trial of several methods of calibration, it was decided to check the meter against thin-plate orifices, using the empirical formula

$$Q = \frac{405A\,(P_1{}^2 - P_2{}^2)^{0.48}}{\sqrt{T}} \quad \dots\dots\dots\dots [9]$$

where $Q$ and $T$ are the same as before; $A$ is the area of the orifice in sq. in.; $P_1$ is the pressure in lb. per sq. in. abs. in front of the orifice; and $P_2$ is the pressure in the same units beyond the orifice. This formula, like formula [2], is due to Mr. Reynolds, and has been checked by him by means of a large number of very accurate experiments. Three orifices were used, having diameters of $\frac{3}{8}$ in., $\frac{1}{4}$ in. and 0.191 in., respectively. All the plates were $\frac{1}{16}$ in. in thickness and made of brass.

The results of this calibration are shown in Fig. 2. The coefficient seems to decrease with an increase in flow. This is not in accordance with the results obtained by several other investigators, but the orifices were later checked up and further runs made with substantially the same results. Another calibration was made after more than a year of operation of the meter, and, despite some slight corrosion and roughening of the walls, the coefficient was found to be very nearly the same. It is the writer's opinion that for most experimental work one would be very safe in assuming the coefficient of a carefully constructed venturi meter when metering compressed air to be practically unity, providing the ratio $P_2/P_1$ was not less than 0.90.

TABLE 1   VALUES OF $f(P_2/P_1)$ FOR VALUES OF $P_2/P_1$ FROM
1.00 TO 0.90

| $\frac{P_2}{P_1}$ | $f\left(\frac{P_2}{P_1}\right)$ | $\frac{P_2}{P_1}$ | $f\left(\frac{P_2}{P_1}\right)$ |
|---|---|---|---|
| 0.90 | 0.1600 | 0.97 | 0.0917 |
| 0.91 | 0.1534 | 0.98 | 0.0753 |
| 0.92 | 0.1456 | 0.985 | 0.06535 |
| 0.93 | 0.1369 | 0.99 | 0.0535 |
| 0.94 | 0.1276 | 0.995 | 0.0370 |
| 0.95 | 0.1170 | 0.9975 | 0.02688 |
| 0.96 | 0.1057 | 1.0000 | 0.00000 |

### MANOMETERS

As stated before, it was necessary to use two manometers, one filled with water and the other with mercury. This was necessary in order to read accurately values of $\Delta P$ from about 11 lb. per sq. in. to 0.10 lb. per sq. in. A valve and piping arrangement with which it was possible to quickly change from one manometer to the other while under pressure, and at the same time keep the liquid in the manometers from being blown out, is shown in Fig. 3. Its operation is apparent. It is necessary, however, to be sure that the valves connecting the two legs of each manometer are open when making any change of valves Nos. 1, 4 or 5, or whenever valves Nos. 4 and 5 are closed, and also that valve No. 2 remains open at all times when the water manometer is not in use.

It was found that rubber washers cut from soft sheet rubber made the most satisfactory packing for the stuffing boxes on the air, water and mercury ends of the manometers. Brass stuffing boxes and U's were used, excepting on the lower end of the mercury manometer, where wrought iron was used. Glass tube 3 mm.

outside and 1 mm. inside diameter was found to be satisfactory. At first an attempt was made to make the manometers from a single piece of glass bent into shape, but it was discovered that the internal strains set up by the heating were so great as to render the tubes unreliable when used at this pressure.

In the final set-up it was necessary to replace the $\frac{1}{8}$-in. piping which connected the venturi meter with the manometers with $\frac{1}{2}$-in. piping, and to provide small water traps or pockets at the point where these pipes were connected to the venturi meter. This was done in order to prevent water, which was frequently in the compressed air, from working its way over into the manometers and thus rendering the readings inaccurate. A large number of experimental runs, otherwise perfect, were spoiled before this fact was discovered.

The total drop in pressure across this venturi tube for rates of flow under 100 cu. ft. per min. was only a fraction of a pound per square inch, while at 200 cu. ft. per min. it was only about 2.5 lb. per sq. in. when $P_1$ was 100 lb. per sq. in. gage. At rates of flow greater than this, however, the drop increased rapidly.

## PHYSICAL BASIS OF AIR-PROPELLER DESIGN

(Continued from page 219)

mediate line undoubtedly represents different types of flow on the two parts of the aerofoil. This is possible because of the center support which divides the aerofoil into two parts. At the point of discontinuity corresponding to the second critical speed the lift reading becomes unsteady and the flow phenomena become unstable and jump from one type to another until the new form is established. Fig. 12-$A$ shows the unstable flow of the transition, and Figs. 12-$B$ and 8 show the flow of the final low-drift régime.

The discovery of this second critical speed is one of the novel and significant features of the experiments. Simultaneous observation of the balance and the flight vortices made the discovery possible; affording proof that the two types of flight vortices can be identified with the two values of the lift coefficient, one belonging to a high $L/D$ régime, the other to a low $L/D$ régime.

Various experimenters have in the past given evidence of discontinuity of the flow past aerodynamic objects. As applied to aerofoils, Mr. Orville Wright conducted in 1918 a particularly interesting series of experiments, the results of which unfortunately have not been published, wherein new discoveries were brought out regarding the discontinuity of flow about propeller sections. He found that at certain angles thick sections manifested a dual value of the coefficients, the angle at which discontinuity occurred depending on whether the angle was increasing or decreasing at the time when the readings were taken. He interpreted the phenomenon as due to a change of air flow. He found two distinct values of the lift coefficient and of the drift coefficient throughout the range of instability; the high lift value always corresponding to a low drift value and vice versa. It was possible to make the values change back and forth between the two points at will by disturbing the air current.

The writers have been able to verify conclusively Mr. Wright's assumption that his results involved a change of air flow; wherever discontinuity has been encountered in our experiments it has always been found that the drag and lift became discontinuous simultaneously, and that the high value of lift always coincided with the low value of drag.

### PRACTICAL SIGNIFICANCE OF THE RESULTS

Reference has already been made to the prospect, opened by these experiments, of developing a usable physical thory of flight. The immediate practical results have also been important, especially as regards propeller design. For example, we have demonstrated the existence of a limited tip speed.

When the critical speed at which the change in flow takes place is reached, there is in some cases a violent chattering of the model and support; and the nature of the vibrations readily suggests a connection with the fluttering sometimes observed in propellers. The speed encountered is equal to the tip speed of slow-speed propellers, but is considerably inferior to the tip speed of the very large fast-turning propellers used on the Liberty engine.

Many static tests carried out by the writers on propellers have shown an effect which appears to be related to the discoveries made during the series of aerofoil experiments. It is well known that the ratio of propeller thrust to propeller torque must be independent of speed if both thrust and torque are proportional to the square of the revolutions. It has been found in practice, however, that the ratio of thrust to torque decreases greatly with revolution speed when high tip speeds are reached, and that it sometimes shows a rapid decrease accompanied by violent fluttering. The analogy to the results of the aerofoil tests is obvious and it is felt that considerable study is desirable in order to connect these facts in a rational manner.

We have found by practical experience that if we do not go below a value of $V/ND$ (velocity/r.p.m. of engine $\times$ diameter of propeller) of 0.65 we get a very fair propeller efficiency. As we have gradually increased the speed of our planes we have gone on increasing the r.p.m. of the engine and the diameter of our propellers so that the value of $V/ND$ has remained about the same for the great majority of propellers in actual service.

We have always assumed that there was no limit to this development aside from the characteristics of the plane and engine. That is, we have made the assumption that we could double our propeller speed just as soon as we were able to double our plane speed and strengthen our engine enough to stand the stresses involved.

It now appears, however, as though there is a limit to propeller speed aside from the value of $V/ND$ or, to use more familiar terms, aside from pitch ratio.

Unfortunately, even the speed obtainable in the McCook Field wind tunnel is not great enough to measure the limiting velocity for relatively thin sections when set at low angles, consequently we are only able to infer that it exists.

### CONCEPTION OF LIMITING STRESS IN FLUID DYNAMICS

Mathematical studies of first importance, which are now classical, on the nature of the flow about an aerofoil have been developed by Helmholtz, Kirchoff, Lord Rayleigh, Lanchester, Prandtl, Kutta, Karman, Greenhill, Lewis and others. Dr. Georges de Bothézat has put forward some very interesting ideas about the effect of stresses in the fluid on the nature of the air flow, and by his theory it is proper to consider in fluid dynamics the same sort of stresses with which we are familiar in solid statics.

It is Dr. de Bothézat's conception that the type of flow which establishes itself is governed by the stresses set up in the air passing the aerofoil. The unit stresses increase as the velocity rises. It is easy to conceive that a given type of flow is possible only so long as the shearing stress, developed in the fluid, does not exceed a certain magnitude which depends on the value of the viscosity coefficient. When the stress reaches a certain critical value, rupture occurs in the air-flow structure; adjacent layers of air begin to slide past each other. Since there is no longer flow similarity, the aerofoil characteristics must change in the manner described earlier in this paper.

# BOOK NOTES

AMERICAN GAS WORKS PRACTICE. Standard Practical Methods in Gas Fitting. Distribution and Works Management. By George Wherle. Progressive Age Publishing Co., New York, 1919. Cloth, 6 x 8 in., 741 pp., illus., tables, $4.00.

This work is intended as a general reference book on gas-works practice in this country, with special emphasis on gas-fitting practice. Approximately, one-half of the book is given to the latter topic, and the methods used in street and house distribution, standards adopted, etc., are fully described.

APPLIED MOTION STUDY. A Collection of Papers on the Efficient Method to Industrial Preparedness. By Frank B. Gilbreth and L. M. Gilbreth. The Macmillan Co., New York, 1919. Cloth, 5 x 8 in., 220 pp., plates, 1 chart, $1.50.

This collection describes the application of motion study in various fields of activity and the methods by which it is applied. It also gives the results obtained in various cases and suggests the fields in which it may be used with benefit.

CHILTON AUTOMOBILE DIRECTORY, January 1920. Published quarterly by the Chilton Co., Philadelphia. Cloth, 6 x 10 in., 1000 pp., $5.00.

This quarterly directory of the automobile industry contains a classified list of the American manufacturers of passenger and commercial motor-cars, automobile equipment, parts and machinery for their manufacture. The arrangement is alphabetic and the classification is quite detailed.

In addition, the book contains a directory of automobile trade associations, a table of the serial numbers of American motor cars, the standards of the Society of Automotive Engineers and various engineering tables.

ENGINEERING MACHINE TOOLS AND PROCESSES. A Text-Book for Engineers, Apprentices, and Students in Technical Institutes, Trade Schools and Continuation Classes. By Arthur G. Robson. Longmans, Green & Co., New York, 1919. Cloth, 6 x 9 in., 307 pp., illus., table, $4.80.

This work presents a course in the systematic study of machine work and machine tools which is practical rather than theoretical in character. The methods and machines described are those used in British shops and the volume is intended for use in that country.

FATIGUE STUDY. The Elimination of Humanity's Greatest Unnecessary Waste, a First Step in Motion Study. By Frank B. Gilbreth and Lillian M. Gilbreth. The Macmillan Co., New York, 1919. Cloth, 5 x 8 in., 175 pp., plates, $1.50.

This book is a study of fatigue in workmen and its prevention. It aims to determine what fatigue results from various types of work and how unnecessary fatigue may be eliminated and necessary fatigue reduced to a minimum. Numerous appliances and methods are described.

HIGHWAY INSPECTORS' HANDBOOK. By Prevost Hubbard. First edition. John Wiley & Sons, Inc., New York, 1919. Flexible cloth, 4 x 7 in., 372 pp., 55 illus., diagrams, tables, $2.50.

The author has endeavored to present most of the important details of highway construction and maintenance, as briefly as possible, in such form as to be quickly available to the inspector, who wishes to be told what to do rather than what others have done under similar circumstances. Considerable explanatory matter has been included, and diagrams have been used freely to present data in convenient form for field use.

MANUFACTURE AND USES OF ALLOY STEEL. By Henry D. Hibbard. John Wiley & Sons, Inc., New York, 1919. Cloth, 6 x 9 in., 110 pp., $1.25. (Wiley Engineering Series.)

In this monograph the author has tried to give briefly information of present value relating to the manufacture and uses of the various commercial alloy steels, with the hope of stimulating the demand for them and extending their practical use. The steels considered are tungsten, chromium, manganese, nickel, silicon, nickel-chromium, chromium-vanadium and high-speed tool steels. Bibliographies are given for each.

THE PRINCIPLES OF ELECTRICAL ENGINEERING AND THEIR APPLICATION. Vol. 2. By Gisbert Kapp. Longmans, Green & Co., New York, 1919. Cloth, 6 x 9 in., 388 pp., 173 diag., $6.

The present textbook is intended for all general engineering students and also as a handbook for general engineers. For the latter, the author attempts to provide a work which will give him the fundamental principles of the subject and describe their application in practical engineering, without burdening him with minute details of design. It will, the author hopes, enable the general engineer to determine whether and how any particular piece of electrical plant can be used or adapted for a particular purpose.

Volume 1, dealing with Principles, appeared in 1916. The present volume treats of the applications of these principles in electrical machines.

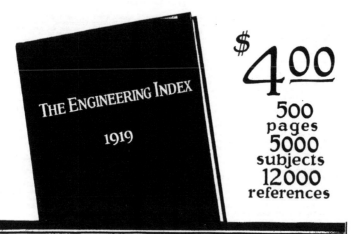

$$\$4^{\underline{00}}$$

500
pages
5000
subjects
12000
references

# Have You Ordered *Your* Copy?

The 1919 edition of THE ENGINEERING INDEX — a master key to the important engineering developments of the year — will soon be off the press.

It should find a preferred position on every engineer's book shelf. The engineering press of the world, comprising more than 1100 publications, representing twelve different languages, has been systematically searched during the past year for the more than 12,000 references which have been compiled, classified, and alphabetically arranged.

### Edition Will Be Limited

The number of copies to be printed will be determined strictly by the number of advance orders and requests for examination, which are received.

To insure against your being disappointed, your order or request should be sent in without delay.

In case you desire to examine the volume before purchase, it will be sent subject to your approval. Sign and mail the coupon.

# MECHANICAL ENGINEERING

## THE JOURNAL OF THE AMERICAN SOCIETY OF
## MECHANICAL ENGINEERS

Volume 42          May 1920          Number 5

## Aeronautic Instruments

A Paper on the General Principles of Their Construction, Testing and Use, Based on the
Work of the Aeronautic Instruments Section of the Bureau of Standards

By MAYO D. HERSEY,[1] WASHINGTON, D. C.

*This paper, which is a contribution from the Aeronautic Instruments Section of the National Bureau of Standards, affords a brief sketch of the general principles of such instruments. The discussion takes up successively matters relating to the construction, the testing, and the practical use of instruments for aircraft. In each case principles rather than details are emphasized. In a few instances particular instruments are described or the results of particular investigations stated, in order to illustrate the principle involved. The paper, however, does not pretend to serve as a final or comprehensive report on the subject.*

*That the general field of aeronautic instruments is a less familiar one to the engineer at large than it logically ought to be, is probably due to the confidential character of the work which has gone forward during the war. Now that attention is again directed to civilian aeronautics the subject of instruments takes on great importance for two chief reasons: first, because the efficiency and safety of air navigation, especially in clouds, depend on the reliability of the instruments which are used; and second, because the whole future development of aeronautic engineering rests upon the measurement of aircraft performance, which requires suitable instruments of precision.*

T HE life cycle of any aeronautic instrument starts with the recognition of some distinct physical measurement that has to be made in flight; then, broadly speaking, it passes through the three main stages of construction, testing, and use; and ends by revealing the need for some particular improvement or new development, at which point the cycle begins over again.

As a convenient way of arranging the available information, we will consider in succession these different stages or processes, emphasizing only the general principles involved, but illustrating the latter by their application to particular instruments such as the aneroid barometer or tachometer.

### 1 CONSTRUCTION OF INSTRUMENTS

In classifying instruments with regard to the purpose for which they are needed a fundamental distinction may be drawn between service instruments and experimental instruments. Service instruments have to be light and rugged, fool-proof, and reliable under all conditions of flight, and must not occupy too much space in the cockpit. Experimental instruments must be extremely accurate and the self-recording feature is desirable.

A further distinction may be drawn between instruments for balloon and airship use and those for airplanes. Historically, instruments for lighter-than-air craft were the first to be developed; but owing to the more extensive use of airplanes during the war, the scientific elaboration of airplane instruments stands now far in advance of airship instruments. The chief characteristics of airship instruments are the relatively lower range of

Abstract of a paper to be presented at the Spring Meeting, St. Louis, May 24 to 27, 1920, of THE AMERICAN SOCIETY OF MECHANICAL ENGINEERS. Advance copies of the complete paper may be obtained gratis upon application. All papers are subject to revision.
[1] Chief, Aeronautics Instrument Section, U. S. Bureau of Standards.

speed and altitude over which the mechanism has to function accurately, and the correspondingly weaker forces available to operate the movement. In addition, there are some extra instruments needed on airships, for measuring conditions relating to the air and hydrogen supply.

Finally, all aircraft instruments may be classified with respect to the nature of the measurement made. From this viewpoint the most important groups are: (1) altitude instruments; (2) speed indicators; (3) direction indicators; (4) tachometers; (5) gasoline indicators; (6) thermometers and pressure gages for the power plant; (7) timepieces; (8) oxygen apparatus; and (9) additional instruments for lighter-than-air craft.

### ALTITUDE INSTRUMENTS

Taking up these nine groups in order, altitude instruments, broadly considered, may be regarded as covering the following species:

   *a* Altimeters
   *b* Barographs
   *c* Thermographs and aerographs
   *d* Statoscopes
   *e* Rate-of-climb indicators
   *f* Night altitude indicators.

The altimeter is in principle a common aneroid barometer with a movable altitude scale instead of a fixed pressure scale. For airplane use, however, the aneroid has to be made uncommonly well; otherwise the instrument will have serious errors caused by the peculiar conditions of flight, and the pointer may vibrate so much that it can hardly be read.

A typical altimeter is seen in the upper left-hand part of the instrument board picture, Fig. 1. The elements of its construction are probably familiar, comprising a stiff steel spring coupled to a thin corrugated-metal vacuum box, together with a delicate transmission mechanism which multiplies the movement of the box. A dilation or contraction of the box only 0.002 or 0.003 in. causes a motion of 1 in. at the tip of the pointer. Various minor adjustments are provided, and usually a bimetallic bar forms part of the lever system, serving in some measure to compensate for the influence of temperature.

Barographs usually contain a tier of vacuum boxes, one on top of another from two to five in number, with a pair of steel springs inside each box. The multiplying factor of the transmission mechanism is decidedly less than in altimeters, so that the pen arm has correspondingly greater power for tracing a firm ink line on the revolving chart.

Thermographs are intended for recording the temperature of the atmosphere at different altitudes. They have been built both with bimetallic elements and with delicate Bourdon tubes containing mercury or ether. Aerographs are combination recording instruments comprising hair hygrometers for humidity determination along with the barometric and thermometric elements.

Statoscopes serve to tell whether a balloon or airship is slightly ascending or descending. They are remarkably sensitive, too much so to be of much use on airplanes. The statoscope in its simplest form is a thermally insulated vessel of air, connected to the external atmosphere by a glass tube containing a drop of red liquid which moves freely one way or the other in response to any minute change of external pressure.

Rate-of-climb indicators give a direct reading of the vertical speed in feet per minute. The most usual types are based upon the pressure difference set up between the two ends of a capillary tube connecting a thermally insulated air container with the external atmosphere. What is actually observed is the manometer indication of that differential pressure.

The night altitude indicator is an optical projection instrument of the range-finder type, particularly valuable for seaplanes flying low over the water.

## SPEED INDICATORS

The pressure type of air-speed indicator shows the impact pressure of the air due to the motion of the aircraft and so enables

type) are usually based on some rotating-vane or Robinson cup principle and connected, either directly or through electric transmission, with a tachometer movement.

Absolute or ground-speed indicators have been successfully made in the visual type which gives its result by combining an observation of the angular velocity of the line of sight with a knowledge of the altitude.

## DIRECTION INDICATORS

The third general group, direction indicators, may be subdivided as follows:

    *a* Compasses and turn indicators
    *b* Inclinometers and banking indicators
    *c* Angle-of-attack indicators
    *d* Side-slip or yaw indicators
    *e* Drift indicators
    *f* Course-finding instruments.

Compasses may conceivably be either magnetic or gyroscopic, but the latter have not yet been successfully developed for airplanes. Magnetic compasses in common use are of the short-

FIG. 1 TYPICAL AIRPLANE INSTRUMENT BOARD, SHOWING ALTIMETER, TIMEPIECE, AIR-SPEED INDICATOR, COMPASS,
TACHOMETER, PRESSURE GAGES, AND RADIATOR THERMOMETER

the pilot to keep above his stalling speed and below the danger point due to excessive speed. Such instruments consist of two parts: the indicator proper on the instrument board in the cockpit (see Fig. 1), and a pressure head mounted out on some strut where it will catch the full force of the undisturbed atmosphere going by. The indicator part measures the differential pressure generated by the pressure head and transmitted to it through some length of tubing. Liquid manometers were used for the indicator early in the war, but later were superseded by various diaphragm constructions. Not only corrugated metal diaphragms, but even rubber and doped silk have been found satisfactory for this purpose.

The pitot tube and venturi tube have been tried out in a great variety of forms as pressure heads for air-speed indicators. A combination of the two gives the greatest differential pressure and was adopted in this country in preference to the pitot tube, in order to accelerate production of indicator mechanisms during the war.

Air-speed indicators of the anemometer type (or true air-speed

period type. Turn indicators have been made in both aerodynamic and gyroscopic forms. The former type measures the differential pressure between two static heads placed one at each wing tip. It is therefore a purely atmospheric instrument which might read zero if the aircraft were resting in a large portion of the atmosphere itself undergoing rotation. The gyroscopic forms show absolute rotation: one type is a very sensitive indicator showing whether the aircraft is or is not turning, and whether to the right or left, while the other type is somewhat less sensitive but actually measuring the rate of turning.

Inclinometers are to be distinguished from banking indicators in that they measure the inclination of the aircraft to the ground, while banking indicators show only the departure of the aircraft from the proper banking position. Gyroscopic inclinometers are usually of the long-period pendulum type. They function as a simple pendulum several thousand feet long, and so are free from the effects of transient accelerations, but not from persistent acceleration in the same direction. The most common inclinometers are liquid-filled tubes in the form of a large, triangular closed circuit, and are used for fore-and-aft observations. The commonest form of banking indicator is a modification of the familiar spirit level, bent into an arc of about 20 deg. each side of the zero. Mechanical pendulum types also are in use.

Angle-of-attack indicators show the angle between the wing chord, say, and the direction of motion. Both weather-vane and differential pressure types have been used, the latter like the side-slip indicator below.

Side-slip or yaw indicators show whether the aircraft is heading exactly in the direction of motion (relative to the atmosphere) or not. It operates by the differential pressure set up between symmetrically placed openings on the two sides of a sphere, attached to some strut so as to head directly forward. The two openings are connected to the two sides of a diaphragm movement with dial and pointer.

Drift indicators show the speed of drift compared to the speed of advance and involve the rotation of some reference line by the observer until it is parallel to the apparent trajectory of objects going by on the ground.

Course-finding instruments are merely mechanical devices for determining the course and distance made good, by reference to data already available.

### Tachometers

Tachometers constitute the fourth general group. All the various types serve for the sole purpose of indicating the revolutions per unit time of the propeller shaft, with the exception of one instrument which indicates also the angular acceleration of the propeller shaft, that is, its rate of speeding up or of slowing down. The diverse types of construction may be subdivided respectively as: Chronometric; centrifugal; magnetic; electric; air-viscosity and air-pump; and liquid.

The chronometric or clockwork type, the most accurate but most complicated, actually counts the number of revolutions in a fixed time.

The centrifugal type, most commonly used on aircraft, works on the principle of the change which takes place in the position of a revolving mass due to centrifugal force acting against the restoring force of a spring.

The magnetic type operates by the rotation of a permanent magnet near an electrically conducting disk which is mounted on the pointer spindle and controlled by a spring.

The electric type consists of a magneto, geared to the engine shaft, which transmits the resulting voltage to a millivoltmeter graduated to read revolutions per minute.

The air-viscosity type is similar to a torsion viscosimeter, measuring the product of viscosity by speed of rotation. Here the viscosity of the fluid, air, is so nearly a constant that the deflection of the spring and pointer is almost directly proportional to the speed of rotation. The air-pump type is based on the force of impact of the air against a movable vane resisted by a spring; the pump is geared to the engine shaft so that the deflection depends on the speed of rotation.

The liquid type involves the observation of the maximum height of a liquid column or deflection of Bourdon gage which can be maintained by a centrifugal pump geared to the engine shaft. By comparing the height of this column with that of a similar one heavily damped an indication of angular acceleration is also obtained.

### Gasoline Indicators

Gasoline indicators, the fifth group, comprise, broadly speaking, depth gages and flow meters. Depth gages show the available supply and are of two types, those based on a simple float principle and those which measure the hydrostatic pressure near the bottom of the tank. Flow meters show the rate of consumption of gasoline at any instant. They have been based on the venturi principle and also on the principle of utilizing the force of impact of the stream of liquid against a movable element resisted by a spring or by gravity.

### Thermometers and Pressure Gages

The sixth group comprises thermometers and pressure gages for the radiator, the lubrication system, and the gasoline tank, respectively. These thermometers are of the long-distance transmission type with liquid-filled or vapor-filled bulbs. The indicator movement, installed in the cockpit, consists essentially of a Bourdon tube element, as do also the air and oil pressure gages.

### Timepieces

The seventh group, timepieces, presents no great novelty aside from the reversing stop watches used in connection with certain bomb sights; but they have to be rugged enough to withstand airplane service and provided with suitable dials.

### Oxygen Apparatus

Oxygen apparatus for the aviator (group 8) may be either of the compressed-oxygen, liquid, or chemical type. Only the first has been put into regular use in this country. These outfits are provided with automatic regulators which control the flow of oxygen from the storage tank to the aviator's breathing attachment. The regulators are designed to deliver an amount of oxy-

FIG. 2 AVIATOR'S OXYGEN REGULATOR

gen which increases at the higher altitudes in accordance with a prescribed formula. The mechanism is operated by diaphragms. An example of such a regulator is shown in Fig. 2.

### Balloon and Airship Equipment

The ninth and last group comprises balloon and airship equipment not already included among the foregoing. Special pressure gages are made for indicating the amount of water ballast carried. Delicate manometers are made for measuring the air pressure in the balloonetts and the hydrogen pressure in the envelope; liquid types have been used, but sensitive diaphragms are better. Electric-resistance thermometers are used to indicate the temperature of the gas in the envelope. Hydrogen detectors based on osmotic pressure action are valuable for locating leaks.

### Design, Development and Production

The design of aeronautic instruments is at present almost wholly on a cut-and-try basis. The Bureau of Standards has made some progress, however, in placing instrument design on a rational basis. Consideration has been given to such general problems as: the stiffness of elastic systems where two bodies such as a spring and diaphragm are coupled together; the effects of temperature and elastic lag on coupled systems; bimetallic bars for temperature compensation; the balance of

moving parts to resist angular acceleration and vibration; the securing of a uniform scale by suitable design of transmission
mechanism; and the treatment of aerodynamic problems and the
general action of damping fluids by the method of dimensions
described before this Society by Dr. Buckingham at the Buffalo
meeting in 1915 in his paper on Model Experiments. There is
hardly space here even to outline those results, which it is hoped
may form the substance of later communications.

Experimental development has to follow the preliminary design
of an instrument before it can be put into quantity production.
Among the recent development projects of the Aeronautic Instruments Section at the Bureau of Standards may be mentioned the
following, several of which were undertaken at the request of the
Air Service or the National Advisory Committee for Aeronautics:
*a* The conversion of barographs from a low-altitude to a high-
altitude range, together with the elimination of ink.
*b* The conversion of altimeters from an altitude scale to a pressure scale, to adapt them for aircraft performance testing.
Improvements were also made in the method of mounting the
main spring of commercial altimeters to diminish the ob-

*g* A working model of a type of dynamical ground-speed indicator which would become available if there existed a suitable
gyro-stabilizer to hold it horizontal. This model consists of
a large steel ball free to roll back and forth in a glass tube
filled with a viscous liquid. It can be shown mathematically
that if the fluid resistance is directly proportional to the
speed of the ball through the tube, while the tube is held in a
horizontal fore-and-aft position, then the displacement of the
ball at any instant from its initial position in the tube is directly proportional to the absolute ground speed of the aircraft.

The development work on any given model of an instrument
must be dropped and the design rigidly fixed before quantity
production can be efficiently carried on. In this country the
quality of products made in quantity has to be artificially held up
to standard by means of specifications and inspection. This situation is not so commonly met with abroad, especially in France,
where every individual involved, from the mechanic in the shop
up to the head of the concern, takes a professional artistic pride
in the quality of his output. This difference is not the fault of

FIG. 3  TESTING OUTFIT FOR AIRPLANE TACHOMETERS

served differences between readings with increasing and decreasing altitudes.
*c* The development of a precision altimeter with a large dial
giving a ¼-in. movement of the pointer for each 100 ft. of
altitude, a special design being made for the elastic system
to diminish elastic lag.
*d* The development of a direct-reading rate-of-climb indicator
based on the capillary-tube-leak principle, with the use of a
liquid entirely dispensed with through the use of sensitive
diaphragms.
*e* Development of a reduced scale working model of a proposed
gyro-stabilizer on the long-period pendulum principle,
possessing mechanical features of extreme simplicity.
*f* The application of the moving-picture camera for securing a
record of instrument readings during the performance test
of an aircraft in flight. The whole outfit—instruments,
illumination, camera and timing system—forms a unit which
takes the place of the human observer in the airplane.

the American manufacturer, and there are some advantages in
our system, yet the fact remains; and this spectacle of successful
instrument production without specifications formed one of the
most astonishing impressions which the writer gathered from his
trip to Europe during the war.

## 2  TESTING OF INSTRUMENTS

The testing of any aeronautic instrument consists in a direct
comparison with some standard, the instrument being operated
over its full scale; the whole procedure should then be repeated
under different conditions varying in flight. The most important of these conditions are:

    *a* Extreme low temperatures
    *b* Change in pressure or density of air
    *c* Acceleration and inclination
    *d* Vibration
    *e* Time elapsed during the flight.

A typical installation of testing apparatus, in this instance for the testing of tachometers, is shown in Fig. 3. The results of such a test are ordinarily given out in the form of tables or curves showing the correction of the instrument from point to point of its scale. By the correction is meant the amount that has to be added algebraically to the reading of the instrument in order to give the true value of the quantity measured. Typical correction plots or curves of this sort are seen in the first two charts shown on Fig. 4.

Factory inspection tests can be reduced to a comparatively simple routine after the first few instruments of each pattern have been inspected. The purpose of such tests is to control the quality of the output.

The short test for service instruments is an attempt to secure, in a minimum time, such data regarding the maximum calibration correction or temperature effect as may be needed for actual refer-

course on the nature of the work for which the instrument is to be used, and on the flight conditions which will prevail. For instruments such as aneroid barometers, which involve serious errors due to elastic lag or temperature lag, there can be no more reliable method of determining the corrections necessary for the interpretation of the readings than to make a flight-history test by actually reproducing in the laboratory the identical variation of pressure, temperature, etc., which was experienced from time to time during the flight. This is particularly important in connection with the use of barographs for competitive altitude flights. It is now well known that barographs subject to any appreciable amount of elastic lag will read a higher altitude the greater the time elapsed during the flight.

It is a short step from some of these more complete testing methods to the general subject of the experimental investigation of sources of error in instruments. Such investigations have

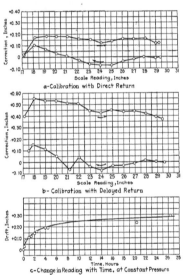

a-Calibration with Direct Return

b- Calibration with Delayed Return

c-Change in Reading with Time, at Constant Pressure

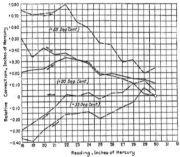

d-Calibration of Same Aneroid at Three Temperatures.

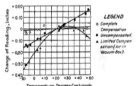

*LEGEND*
o *Complete Compensation*
• *Uncompensated.*
▲ *Limited Compensation (Air in Vacuum Box).*

e-Calibration for Temperature at Constant Pressure.

FIG. 4 TYPICAL INVESTIGATION RESULTS

ence in connection with the installation or use of instruments on aircraft.

The Bureau of Standards general test is the one which is furnished on instruments submitted to the Bureau without special instructions. It is a comprehensive standard form of test affording a complete calibration curve together with, at least, numerical values for the more common errors due to temperature and other well-known conditions.

The Bureau makes more thorough tests on sample instruments of new type whose mechanism is not familiar. Such additional observations usually include some form of accelerated life test, such as a five-day run to determine the endurance of the instrument under vibration. The steadiness of the pointer and the accuracy of the instrument with and without vibration are observed in such a test.

The special tests on experimental instruments depend of

formed one of the most interesting features of the work of the Bureau of Standards and are illustrated by the examples given in Fig 4.

In this figure five charts are shown based on experiments with high-grade commercial aneroids. The first chart (a) shows the ordinary calibration of an aneroid changing the pressure at the rate of an inch of mercury in five minutes. (At sea level one inch of mercury corresponds to about one thousand feet of altitude. At twenty thousand feet one inch corresponds to about two thousand feet of altitude). The calibration curve would be distinctly different at some different speed, due to elastic lag in the corrugated-metal vacuum boxes. For the same reason, as shown in chart b, the discrepancy between the readings going up and coming down is much greater when the aneroid is held at the maximum altitude for a considerable time interval. This drift, or change of reading at constant pressure, is shown in chart c as a function of the time elapsed. These curves may be taken as a criterion of the elastic quality of the material in the instrument; and several hundred such have been determined in the course of the Bureau's investigations. Chart d shows the increase in the width of the hysteresis loops with rise of temperature, and it also shows the decrease in the sensitivity of the movement when an instrument is chilled. This last effect is largely due to the change of elasticity with temperature, and cannot be fully compensated for by means of the ordinary bimetallic lever. The remaining chart (e) shows what can be done in the way of temperature compensation. Of three superimposed curves one represents a typical uncompensated aneroid. The graphs show the change of reading with temperature at a constant pressure corresponding to sea-level conditions. A second curve shows a partial degree of compensation obtained by admitting air to the vacuum box. The remaining curve shows satisfactory compensation by means of a bimetallic bar for sea-level observations; but this same aneroid may have a serious temperature error when the movement has shifted into the new position which it will occupy at twenty thousand feet.

Still further characteristics of aneroid barometers have been studied at the Bureau in addition to those shown in Fig. 4, and similar investigations have been extended to all of the other aeronautic instruments.

### 3   USE OF INSTRUMENTS

Many practical airplane pilots disclaim the use of instruments, and certainly it is desirable that the aviator should be trained to become just as independent of any mechanical aid as possible. Some of the times when airplane instruments are really of the most importance are the following:

a For various military and naval operations such as bomb dropping
b Long-distance navigation
c Flying at night and in clouds
d Preparing to land on a perfectly smooth body of water where the height is deceptive
e Flying in formation with instructions to hold to a prescribed air speed and altitude
f Photographic and survey work.

The need for instruments is still more evident in airship work. Here it is necessary to control the position of the ship, especially near the ground, with comparatively greater precision; nor is it possible for the pilot to act so quickly in response to his unaided senses as it might be in an airplane.

The following precautions may be suggested for getting the best results in the practical use of service instruments:

a Select for installation in the first place only those instruments which can be certified by the Bureau of Standards as suitable for the intended use.
b In any event see that the instrument is tested immediately prior to installation to avoid the occurrence of large errors due to secular changes or mechanical injury.
c When accuracy is required in flight observations the aviator must remember where appreciable lag errors are stated to exist, whether due to elastic lag or temperature lag or other causes, that some time will be required for the instrument to completely respond to any given change, and that the results will be different with increasing and decreasing scale readings.
d At least once in six months it is desirable that instruments should be overhauled, during which time they may perhaps be replaced by a reserve supply of fresh instruments properly tested and adjusted.

The problem of salvaging has been studied by the Bureau of Standards and the statistical results of such work are of the greatest value as a basis for future improvements in instrument construction. The actual salvage proposition consists of dismantling a certain number of damaged instruments and putting them together again so as to form a smaller number of good instruments. The study of expedients for readjusting the mechanism forms an integral part of such work. This can only be done in the testing laboratory.

In concluding this paper, a recapitulation may well be made of some of the outstanding problems of instrument development, the solution of which is much to be desired:

a A satisfactory gasoline depth gage is needed.
b Barographs and other recording instruments of better quality than any now existing are needed. Better clockwork, and a much more open scale, and more complete freedom from errors due to friction, temperature, elastic lag, and lack of balance are desirable.
c A wholly satisfactory form of compass for airplanes is not as yet available. Besides the northerly turning error, so-called, to which all single-pivot compasses are necessarily subject, there are troubles due to the action of the damping fluid.
d Air-speed indicators suitable for the low speeds of airship flight are required.
e A gyro-stabilizer is needed, different from existing types in that it must be able to stabilize a body which is not rigid, but contains a freely moving mass. Such a stabilizer is necessary before the full possibilities of any dynamical type of ground-speed indicator can be realized.

The material in this paper is of course not due to any one individual, but is based upon the work of the entire Aeronautic Instruments Section of the Bureau of Standards. Later, the detailed official reports prepared by the various members of the staff concerned, should be available through the publications of the Bureau and the National Advisory Committee for Aeronautics.

# The Rectilinear Flight of Aeroplanes

## Basic Principles of Analytical Theory Advanced by Eminent French Engineer and Scientist and Their Application—Calculation of Maximum Cruising Radius

BY A. RATEAU,[1] PARIS, FRANCE

IN his paper, of which the following is an abstract, M. Rateau presents the basic principles underlying his theory of the rectilinear flight of aeroplanes, and further indicates how they may be applied in the solution of certain important aeronautical problems. In doing this he establishes at the outset characteristic relations connecting the main factors governing the operation of the propeller.

The paper is based on the assumption that a constant-compression, not supercharged, engine is used, the carbureter being

equipped with an altitude regulator, in order to maintain the mixture at the optimum ratio all the time, and that further the propeller is of constant pitch, is not subject to deformation, and is held on the motor shaft.

The same theory may apply to a propeller with gear reduction, but the author does not wish to introduce into his equations the complications created by the use of gear reduction between the motor and propeller.

One of the most essential elements affecting flight is the specific weight of air at the different altitudes, and the author establishes a formula connecting the specific weights of air at the ground and at various elevations.

As regards the engine, the most important factor is not the

[1] Mem. Institut de France; Hon.Mem.Am.Soc.M.E.

Abstract of paper to be presented at the Spring Meeting, St. Louis, May 24 to 27, 1920, of THE AMERICAN SOCIETY OF MECHANICAL ENGINEERS. Advance copies of the complete paper may be obtained gratis on application. All papers are subject to revision.

power output, which varies with the speed, but the couple which it sets up, and if the carbureter is properly adjusted this couple should vary but little with the speed of rotation in revolutions. It is, however, tremendously affected by the specific weight of the air sucked in, though not proportional to it.

On the other hand, the consumption of fuel and oil in an engine varies with the speed of rotation and with the specific weight of air. As regards fuel consumption, if the carbureter is properly adjusted it varies in proportion to the total weight of the air sucked in, and hence in proportion to the speed of the engine and to the specific weight of air. The oil consumption depends on the character of the oil and the temperature of the air with which the engine is supplied, and increases with the speed of the engine more rapidly than the speed itself.

The consumption of both fuel and of oil is affected by a number of factors in such a manner that it is impossible to keep track of them in an exact manner. It may be said to be proportional to $\omega + o$ where $\omega$ is the specific weight of air and $o$ is a constant, the numerical value of which is about 0.08 in the case of non-rotary motors lubricated by castor oil and a little larger, say, about 0.1, in the case of mineral oil, and still larger for rotary engines.

Considerable attention is devoted to the characteristic functions of the propeller. In this connection only the effective pitch of the propeller and not the so-called "nominal pitch" is taken into consideration. While the effective pitch is of a physical character and therefore cannot be measured directly on the propeller, it is nevertheless of the nature of a constant characteristic of the propeller, no matter what the speed of the latter may be.

The following basic formulæ are established by the writer:

Theory shows that the tractive force $F$ of the propeller in kilograms and its counteracting couple $\Gamma$ in kilogram-meters are expressed by the relations

$$F = b\omega n^2 f(\sigma)$$
$$\Gamma = \frac{bH}{2\pi}\omega n^2\varphi(\sigma) \quad \Big\} \quad \dots\dots\dots\dots\dots[1]$$

where $f(\sigma)$ and $\varphi(\sigma)$ are the functions of slip and of certain coefficients of the propeller mentioned later, and where $b$, which the author calls the factor of thrust, has for its expression in the case of aeroplane propellers with narrow blades—

$$b = \frac{2k\pi}{g}HM\dots\dots\dots\dots\dots\dots[2]$$

where $\pi$ is the ratio of circumference to diameter; $g$ the gravitational constant in metric units ( $= 9.81$); $k$ a numerical coefficient of the order of 2, and $M$ the moment, referred to center of propeller, of the projection of the active surface of the wings on a plane perpendicular to the axis. The units in the case of $H$ and $M$ are the meter and the square meter.

The function $f(\sigma)$ is of the form $\sigma(1-e'\sigma)$ where $e'$ is the coefficient depending on the ratio $p$ of the pitch $H$ to the peripheral diameter $D$ of the propeller and decreases in a certain proportion to the decrease of the ratio above referred to. When $p$ is less than unity (and this is the case with practically all aircraft propellers), $e'$ is less than 0.01. Since, on the other hand, the slips $\sigma$ as a rule vary only from 0.2 to 0.4, it can be seen why the corrective term $e'\sigma$ varies only about 0.016 in particular applications and hence may be entirely neglected when subtracted from unity. This is what the author does. At the same time, whenever necessary, the complete formula makes it possible to carry out the calculations with the necessary degree of precision.

The function $\varphi(\sigma)$, which is of basic importance, has the form

$$\varphi(\sigma) = a + \sigma - e\sigma^2\dots\dots\dots\dots\dots[3]$$

where $e$ always has a value in the neighborhood of 0.5 and $a$ of 0.03, and are variable from one propeller to another in accordance with its shape and roughness of the surface. It is only this $a$ that is affected by losses in the propeller due to shocks, friction and turbulence of the fluid; it is connected with the coefficient $\varepsilon$ (loss of relative velocities across the propeller) by the formula—

$$a = \varepsilon\left(\frac{p}{\pi v} + \frac{\pi v}{p}\right)^2\dots\dots\dots\dots\dots[4]$$

where $v$ is the ratio of a certain average radius to the radius of the periphery.

If it be assumed that $e = 0.5$, then

$$\varphi(\sigma) = a + \sigma - \frac{\sigma^2}{2}\dots\dots\dots\dots\dots[5]$$

and hence

$$\psi = \frac{\varphi(\sigma)}{f(\sigma)} = 1 - \frac{\sigma}{2} + \frac{a}{\sigma}\dots\dots\dots\dots\dots[6]$$

This function $\psi$ of the slip is one that constantly occurs in the author's aeroplane formulæ.

The efficiency of the propeller $\rho$, if by $v$ is denoted the velocity of advance, is

$$\rho = \frac{Fv}{2\pi n\cdot\Gamma}\dots\dots\dots\dots\dots[7]$$

If $v$ is replaced by $nH(1-\sigma)$, and $F$ and $\Gamma$ by their expressions taken from Equation [1], then

$$\rho = \frac{f(\sigma)(1-\sigma)}{\varphi(\sigma)} = \frac{1-\sigma}{\psi}\dots\dots\dots\dots\dots[8]$$

Attention is called to the fact that in this expression of the efficiency of the propeller, in which $\psi$ may be used in the form given by Equation [6], only the slip $\sigma$ and the coefficient of losses $a$ are used.

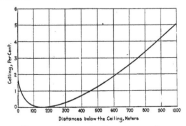

FIG. 1  VARIATION OF FUEL AND OIL CONSUMPTION PER KILOMETER OF FLIGHT WITH VARIATION IN DISTANCE OF PLANE FROM ITS CEILING DURING FLIGHT.  (COMPUTED FOR AEROPLANE-TYPE)

All these novel and fundamental formulæ are excellently confirmed by experiments as precise as one could wish for in a subject of such complexity as the one under consideration.

### CHARACTERISTIC FUNCTIONS AND COEFFICIENTS OF AN AEROPLANE

For a given aeroplane the essential variables are its weight $P$ in kilograms and its incidence $\alpha$, that is, the angle in degrees formed by the wings with the trajectory.

Formulæ are established connecting in horizontal rectilinear flight the weight of the aeroplane $P$ and the resistance $R$ to forward motion, where both $P$ and $R$ are expressed in terms of specific weight of air $\omega$ and velocity of movement of the plane $v$ in the form of

$$P = Y\omega v^2\dots\dots\dots\dots\dots[9]$$

and

$$R = X\omega v^2\dots\dots\dots\dots\dots[10]$$

where the coefficients $Y$ and $X$ for each given plane are functions of the incidence $\alpha$. As such they are expressed in the form of

$$Y = Y_o(1 + \eta\alpha) \quad \Big\} \quad \dots\dots\dots\dots[11]$$
$$X = X_o(1 + \xi\alpha^2)$$

where $Y_o$, $X_o$, $\eta$ and $\xi$ are the four specific coefficients of an aeroplane which have to be determined by measurement or computation.

As an example, the author considers a supposititious airplane

(Continued on page 316)

# River Transportation in the Mississippi Valley

### Recent Progress in Construction and Operation of Barges and Towboats—Need for Waterways Transportation—Developments Under the Federal Control Act

BY E. W. SCHADEK,[1] ST. LOUIS, MO.

*A review of recent progress in the construction of barges and towboats for river transportation, particularly in the Mississippi River Section, and of provision made for their operation. The towboats are of the screw-tunnel type with new features. Traffic by rail in the Mississippi Valley has long since reached the capacity of the railroads. The need for water transportation facilities was recognized by the Government during the war, and under the Federal Control Act appropriations were made for the construction and operation of river fleets. When the vessels are completed and placed in service, a good start will have been made in utilizing our national waterways for the development of industries and commerce.*

THAT great part of our country lying between the Allegheny and Rocky Mountains, known as the "Mississippi Basin," contains more than one-half of the country's population and produces three-fourths of its food products and one-half of its manufactured products. The Mississippi Valley district includes 21 states and has an approximate area of 1,725,000 square miles, with 120 cities having a population of more than 25,000 each. It furthermore contains a number of great ports and in addition the country's greatest waterway.

The capacity of the railroads of the Mississippi Valley has long since been reached. Every one knows of the annual shortage of cars during the crop-moving season, but few realize how greatly this interferes with the flow of commodities from farmer to consumer. To provide ample rolling stock, trackage and terminals to take care of this peak load, however, would impose on the railroads too great an investment in equipment for a comparatively short period of the year, so that the only remedy for the situation is the improvement of waterways as recommended many times by the engineers studying this situation. It has been computed that the capital saving in rolling stock of the extra equipment alone would pay for the entire cost of the waterway improvement within a comparatively few years.

Our government has spent many millions of dollars upon the river improvements in the last 40 years, but hardly one-half of the necessary work has been accomplished up to the present time. The Newton Bill, pending in Congress, calls for completion of the improvements, and covers:

9-ft. channel in Ohio River from Pittsburgh to Cairo
9-ft. channel in Mississippi from Cairo to New Orleans

[1] The Ruemmell-Dawley Mfg. Co. Mem. Am. Soc. M. E.
Abstract of a paper to be presented at the Spring Meeting. St. Louis, Mo., May 24 to 27, 1920, of THE AMERICAN SOCIETY OF MECHANICAL ENGINEERS. Advance copies of the complete paper may be obtained gratis upon application. All papers are subject to revision.

8-ft. channel in Mississippi from Cairo to St. Louis
6-ft. channel in Mississippi from St. Louis to St. Paul
6-ft. channel in Missouri River from Kansas City to Mississippi River

## PROVISION MADE DURING THE WAR FOR NEW FLEETS AND THEIR OPERATION

During the World War the Government recognized the importance of adequate waterways in time of need. The Federal Control Act appropriated funds for new fleets and for immediate operation upon inland waterways in which were comprised the following developments:

An appropriation for the Mississippi-Warrior section provided for the construction of a new fleet on the lower Mississippi River, to consist of six tunnel towboats, 200 ft. by 40 ft. by 10 ft. in dimensions, and 40 new steel barges each 230 ft. by 45 ft. by 11 ft. The larger number of these barges is completed and the balance, according to reports, will be ready before June 1. Deliveries of the six new steel towboats were scheduled for March, June, and July, so that the complete fleet is expected to be in operation by the end of July.

The barges under construction for the lower Mississippi River have a capacity 2000 tons each, on 8 ft. draft. which is twice the capacity of the largest barges operated in the present fleet. One barge has a capacity equal to 100 cars. The towboats for this new fleet are of 1800 hp. each, with a capacity to tow downstream five of these barges, making a total of 10,000 tons, or 500 carloads.

An appropriation of about $5,000,000 provides for the construction of a new fleet on the Warrior, Alabama and Tombigbee Rivers, to consist of four steel, self-propelled barges of the double-tunnel type and three tunnel towboats, each 140 ft. by 24 ft. by 6 ft. As the bulk of the tonnage on these rivers consists of coal from Alabama fields to New Orleans and Mobile, 20 steel cargo retainers of 10 tons capacity each were constructed to stimulate other cargo for northbound trips, and derricks are being installed at terminals to facilitate handling.

In the very near future there will not be a city or town on the Warrior and Tombigbee River without port facilities. Plans are being drawn and estimates made of the cost of terminal facilities at Demopolis and Jackson, and also at Spocari, six miles from Demopolis. Two turning basins are to be established on the Warrior River, one directly opposite the port of Birmingham and the other at Locust Fork. These basins, dredging work for which

FIG. 1 U. S. SNAGBOAT *J. N. Macomb.* STEEL HULL BUILT IN 1874
Showing snag drawn up between catamaran bows, to be sawed up and destroyed.

Fig. 2  Assembly of Floors, Center-Line Bulkheads and Side Frames on One of the Boats at the Early Stage of Construction

Fig. 3  Two 3000-Ton Upper Mississippi River Steel Barges in Tow of U. S. Dredge H. S. Taber, Below Dam, Keokuk

Fig. 4  Scotch Marine Boilers Built in St. Louis, Here Shown Barely Clearing the Grand Avenue Viaduct

Fig. 5  Showing the Boat Fairly Advanced with the Forward-Cargo, Center-Cargo and Back-Cargo Space Practically Completed and With the Engine-Room and Boiler-Room Sides Up

will start shortly, will enable the boats to turn without backing.
Craft with high stacks which cannot pass under Ensley Bridge
will use the Birmingham basin.

The New York Barge Canal is to receive 20 modern power self-
propelled barges each 149 ft. by 20 ft. by 12 ft. and 51 canal
barges each 150 ft. by 20 ft. by 12 ft. Some concrete barges are
also being built, but these are too heavy for shallow draft.

The Inland Waterways shipbuilding program above outlined
would not be complete without reference to the $3,860,000 appro-
priation of the Emergency Fleet Corporation for the Upper
Mississippi River fleet, which will consist of 19 of the most up-
to-date barges, each 300 ft. by 49 ft. by 10 ft., and four stern-
wheel towboats 230 ft. by 58 ft. by 8 ft. This fleet is being built
under the direction of the U. S. War Department, Engineers
Office, St. Louis District, headed by Major Wilbur Willing, and
Mr. Wm. S. Mitchell, Assistant Engineer, who are directing the
designing and construction.

### PARTICIPATION IN BOATBUILDING BY ST. LOUIS

When the Government program of building river fleets became
known to some of the leading citizens of St. Louis they decided
that this city should also be represented in the shipbuilding program. A shipbuilding company was accordingly organized and
incorporated under the name of St. Louis Boat and Engineering
Company, of which Mr. Edward A. Faust and Mr. Albert
Ruemmeli are the leaders.

On account of the war and later the post-war conditions, the
necessary equipment and arrangement of the shipyard was delayed
to such an extent that the company could not start operation
early enough to secure any of the orders for the construction of
the Lower Mississippi River fleet; but when the proposition for
building the Upper Mississippi River fleet came up, the company
was ready and received an order for two of the powerful stern-
wheel towboats.

The U. S. Railroad Administration awarded the contract for
the construction of four self-propelled barges for the Tombigbee,
Alabama and Warrior Rivers to the St. Louis Boat and Engineering Company, and these boats are now being built in the first St.
Louis construction shipyard. They are of the twin-screw steel
tunnel type, 280 ft. long, 49 ft. wide and 10 ft. deep, and are
built for a maximum draft of 7 ft. They are of single-deck,
shallow-draft, flat-bottom type, built entirely of steel. The machinery enclosure is located aft, and the cargo space is all on deck,
consisting of two steel houses for package freight, one forward
and one aft, and two steel guards amidship forming coal-cargo
space. The forecastle head is raised to accommodate part of the
crew, and other crew quarters are located above the machinery
enclosure. The pilot house is raised and the captain's room is
located above. The center-line bulkhead extends from keel to
deck and from forward collision bulkhead to engine-room bulkhead, all watertight. The cross bulkheads in these barges are
six in number.

The propelling machinery consists of two 400-hp. vertical,
triple-expansion condensing engines, and of two marine-type
water-tube boilers for 250 lb. working pressure, arranged for
forced draft, for which a fan of 10,000 cu. ft. of air is provided.
All auxiliaries are arranged for running condensing to collect the
condensate for feedwater, and the make-up water is carried along
in tanks, two of which are forward underneath the crew quarters
and the other four are next the engine room. The total capacity
of these six tanks is 135 tons of water. The surface condenser
has ¾-in. outside diameter tubes with 1100 sq. ft. of condensing
surface. The propellers are four-bladed, two in number, each 6
ft. 4 in. in diameter, 6 ft. 9 in. pitch.

The most interesting feature of the construction of these self-
propelled barges is the selection of propeller drive instead of the
paddle-wheel drive which, up to the present time, has been in
use on our rivers. It is known that the propeller drive is 25 per
cent more efficient than paddle-wheel drive, permitting more compact arrangement of machinery and reducing not only the size and
initial cost of the boat, but also its operating expense.

### SCREW-TUNNEL VS. STERN-WHEEL TOWBOATS

The origin of the screw-tunnel type of boat is found in the
English patent of Yarrow and Company for a hinged flap forming the adjustable tunnel for a screw propeller. In this country
the few attempts which have been made to replace the stern-
wheel towboats with screw-tunnel boats have met with little success because the latter type of boat has not been found effective
for towing purposes. While making curves with tow, towboats
must flank or reverse the rotation of the wheel or propeller, in
order to throw water against the rudders so as to turn the tow,
which precedes the boat, and prevent its grounding on the bank
of the river. The rudders of a stern-wheel towboat are connected
in parallel and at flanking the boat will throw against the rudders
all the water the wheel is capable of handling, making a powerful
flanking.

The present tunnel towboats, besides having the ability to steer
and flank, are also able to twist (where one engine runs forward
and one astern). Their flanking ability with both engines running
astern, however, is less than one-half that of stern-wheel towboats,
and twisting, combined with flanking, reduces the total ability to
handle the tow to about one-third that of stern-wheel boats. All
this is well known to river "rats," hence their skepticism about
tunnel towboats.

The new-type tunnel boats now being built will overcome the
defects of the present-type tunnel boats by introducing independent control of starboard- and port-side rudders. A tunnel towboat
has one rudder forward of each propeller and one aft of each
propeller, and these rudders up to the present time have been
connected in parallel. The new tunnel boats will have rudders,
one forward and the other aft of propeller on port side, controlled by one steering mechanism; and in the same way rudders
on the starboard side will be controlled by another steering
mechanism. All is ingeniously constructed so that the rudders
can be operated in parallel or independently, either by power or
by hand, from pilot house or forecastle.

By independent control of the two sets of rudders, the total
ability to handle a tow combines that which comes from steering,
twisting, flanking and side thrust of propellers, so that, when
steering is not sufficient for handling the tow, the ability of flanking is increased by the ability to twist and side thrust, making the
total ability to handle the tow as great as for the stern-wheel towboat, when this ability is based on the proportions of available
horsepower propelling the boats.

The ingenious design of this rudder control was introduced and
designed by the well-known firm of naval architects, Cox and
Stevens, of New York. They have been engaged by the U. S.
Railroad Administration to design and direct the construction of
the new fleets as previously outlined.

### IMPORTANCE OF WATERWAYS TRANSPORTATION

When all the vessels comprising the U. S. Railroad Administration program are placed in service, a fairly good start will have
been made in utilizing our national waterways for the development of the industries and commerce of the Mississippi Valley.
The importance of the waterway transportation to the development and welfare of any state or city is clearly shown in the
project for which the state of Illinois appropriated $20,000,000.
The project is to connect the terminus of Chicago drainage canal
in the Desplaines River at Lockport with the Illinois River at
Utica, a distance of about 60 miles. When this is completed the
route from Chicago to the Gulf will be realized. Time for construction of the new canal is estimated as 5 years and the carrying
capacity of this new waterway 60,000,000 tons a year. The capacity of the Government's new fleet for the Lower Missispi River
between St. Louis and New Orleans is 1,000,000 tons. This gives
us an idea what is still to be accomplished in building new fleets
to fully utilize the waterways transportation.

There are now 600 firms in St. Louis selling in foreign markets
and they did about $75,000,000 worth of foreign trade business
in 1918, increasing this about 25 per cent during the year 1919.

# The Separation of Dissolved Gases from Water

## Description of a Method Therefor, Together with Particulars Regarding Its Use in Inhibiting Corrosion in Boilers and Economizers and in Increasing the Conductivity of Surface Condensers

By J. R. McDERMET,[1] PITTSBURGH, PA.

*This paper covers the results of a research on the separation of dissolved gases from water, both on a laboratory and a commercial-plant scale. Fundamentally the process consists of the rapid injection of heated water into a region of vacuo, and an explosive boiling of it at the expense of the heat of the liquid available to the vacuum, with a simultaneous recovery of the heat liberated to the vacuum by a heat exchanger or condenser cooled by the incoming water, preliminary to its heating. It is possible to deduce from a résumé of existing knowledge the results to be expected in the elimination of corrosion from boilers and economizers fed with oxygen-free water. Test data are submitted to show the economies to be gained from the surface condensation of steam free from non-condensable gases both in increased conductivity and higher vacua, and chemical analyses suggest the elimination of carbonates dissolved in water as bicarbonates in excess of their true solubility as carbonates.*

THE idea upon which the method of separating dissolved gases from water described in the present paper is based was originated by Mr. W. S. Elliott of Pittsburgh and sponsored by him for development under the industrial fellowship system of the Mellon Institute of Industrial Research, Pittsburgh, Pa. It appeared at the conclusion of the laboratory investigation that the process evolved from the idea successfully fulfilled three requirements which were regarded as necessary for commercial utility: It was able to handle water in quantities sufficient for the feeding of power-plant boilers with a minimal operating expense and was particularly applicable to this service; it was capable of removing all of the dissolved gases in a commercial sense; and it involved no methods or apparatus with which the average operating engineer was not familiar. The scale of experimentation was therefore enlarged, but the basic idea underwent no change.

The solution of gases in water obeys Henry's law, which states that a solvent in contact with a gaseous mixture dissolves the component gases in proportion to the partial pressures which they exert, provided they do not unite chemically with the solvent. Carbon dioxide which unites chemically with water deviates from this law, but is itself of variable content in atmospheric air. With this reservation in mind, it is possible to say that if extraction is complete, the gas mixture extracted from water is precisely of the same composition as air from which solution was attained. At incomplete extractions proportionality factors between the components appear due to different degrees of solubility. The process with which this paper deals might therefore with perfect propriety be called "air separation."

The initial laboratory installation consisted of an instantaneous water heater drawing its water under pressure from a supply main, and delivering heated water under thermostatic control to the spray head of a separator. The separator consisted of an airtight steel tank with a series of copper return-cascade pans supported by a central steel column, over which the spray head delivered water uniformly in a thin sheet. The separator tank was connected with a barometric tail-pipe column at the bottom to remove the treated water and with an air-pump connection at the top. Vapors and gases were withdrawn from the air-pump connection through a small condenser cooled by the supply water to the heater, and equipped with a trap for retaining condensate. Non-condensable gases were withdrawn from the condenser by a variable-speed motor-driven air pump and passed to a displace-

[1] Research Engineer, the Elliott Company. Assoc-Mem.Am.Soc.M.E.
Abstract of paper to be presented at the Spring Meeting, St. Louis, May 24 to 27, 1920, of THE AMERICAN SOCIETY OF MECHANICAL ENGINEERS. Advance copies of the complete paper may be obtained gratis on application. All papers are subject to revision.

ment air bell, adapted from a standard gas-meter prover, for collection and measurement. Thermometer wells were installed to indicate all significant temperatures and weighing tanks to collect and measure treated water. The condensate trap was so valved that it could be cut in and out of the system for the purpose of collecting the condensate over stated periods, and the air bell collected all non-condensable gases. While the provisions for determining all data but volume of non-condensable gases proved adequate and satisfactory, it was immediately discovered that the air-bell data were not of consistent accuracy in spite of the laboratory tightness of the complete system. It was accordingly necessary to turn to direct testing of the water which had passed through the process. The laboratory arrangement is indicated in the diagram of Fig. 1.

Three methods, developed previously by experimenters in other fields, were utilized in testing for the dissolved gas content, after careful checking had established their accuracy; (a) The iodimetric titration method of Winkler for dissolved oxygen; (b) Freeman and Preusse's modification of Reichardt's method of boiling; and (c) a displacement with either carbon dioxide or hydrogen. As experience soon demonstrated, the degree of separation of dissolved gases was so complete that the proportions of constituent gases were the same as for atmospheric air. When this fact was definitely confirmed by a long series of pipette analyses, reliance was placed entirely upon the chemical titration method of Winkler.

At the conclusion of about four months' intermittent testing the apparatus was taken apart and examined. Steel tank, copper pans and condenser gave evidence of very accelerated corrosion. Cognizance of this virulent action was taken in the design of the first commercial apparatus, and the separator was made entirely of cast iron and non-corroding bronzes, which materials have withstood the corrosive tendencies for over four years of experimental operation.

The laboratory experiments also developed a further fact which was the cause of much confusion at the outset. Water which has been passed through this process and completely degasified is never in equilibrium with the temperature as far as solubility is concerned. Such water has an enormous avidity for redissolving small amounts of air and it is only by the most painstaking manipulation that accurate results are possible in the determination of the extent of separation. In the practical application of the process, absolutely rigid exclusion of air must be maintained if the full benefits of the process are to be secured.

Empirical design factors governing separator volumes, pan areas, and the proportioning of the injection nozzle were adopted from the laboratory installation and embodied in the first commercial installation which was put in service in the Jeannette, Pa., works of the Elliott Company of Pittsburgh in the autumn of 1915. This installation has been in routine operation since that time. The separator proper was designed for 1000 boiler hp. at the outset, and consisted of a 100-hp. Elliott cylindrical heater shell with modified water injection, different proportions of water-storage volume and pan area, and a change of materials to resist corrosion. This design of separator has been retained intact, but the auxiliary arrangement has been modified three times. At the outset a 300-hp. open feedwater heater was installed to supply hot water, and vacuum was maintained by a rotary motor-driven dry-air pump in series with a condenser. Condensate from the condenser was allowed to drain back into the separator, and the air pump discharged non-condensable gases directly into the atmosphere; otherwise the scheme of connections was identical with that in Fig. 1.

An increase in boiler capacity later made necessary the installation of a 1000-hp. heater of the same shell type as the separator, and the entire outfit was repiped and made into an integral vertical unit as shown in Fig. 2. Simultaneously the mechanical air pump was replaced by a steam ejector and the condenser was mounted on the atmospheric side to recover the ejector steam, condensate drainage being now into the heater instead of the separator.

It was felt at the beginning that condenser conductivities would be very low on such vapor mixtures as are withdrawn from the separator of an air-extraction apparatus, and a condenser of a relatively large capacity was installed. This condenser was removed from the apparatus when opportunity permitted and tested under the same circulation rates and with steam-air mixtures of

in thin sheets exposed to a vapor mixture corresponding in pressure to the vacuum maintained. Under steady conditions of operation at a pressure reduction in the separator of 10 in. of mercury and admission temperature of 212 deg. fahr., this vapor mixture corresponds to an air content roughly 0.11 per cent by weight but 0.07 per cent by volume. The water vapor is undoubtedly in equilibrium with its liquid, and as such is relatively dense and moist. With a separator water-storage volume corresponding to two minutes' operation at rating, it is impossible to detect at any rate of operation a temperature depression in the vapor mixture or in the water leaving the separator below that corresponding to the vacuum. Water, vapor, and total pressure are therefore in equilibrium. The most

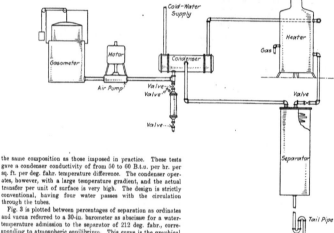

FIG. 1  LABORATORY SEPARATION EQUIPMENT

the same composition as those imposed in practice. These tests gave a condenser conductivity of from 50 to 60 B.t.u. per hr. per sq. ft. per deg. fahr. temperature difference. The condenser operates, however, with a large temperature gradient, and the actual transfer per unit of surface is very high. The design is strictly conventional, having four water passes with the circulation through the tubes.

Fig. 3 is plotted between percentages of separation as ordinates and vacua referred to a 30-in. barometer as abscissæ for a water-temperature admission to the separator of 212 deg. fahr., corresponding to atmospheric equilibrium. This curve is the graphical average of a large number of determinations made both with the laboratory and the commercial-size apparatus. Percentage of separation is defined for the purposes of these curves as the ratio of the quantity of gases removed to that actually present in the raw water, and the calculations were made in terms of volume. Calculation in terms of weight would, however, have had identical values.

The mechanism of the separation appears to depend upon three factors; namely, the natural decrease of solubility with temperature, the mechanical agitation of boiling, and the reduction of pressure. The reduction of solubility with heating amounts roughly to 70 per cent of the volume at normal temperatures, and separation to this extent was arbitrarily credited to the heater.

Water at 212 deg. fahr. injected suddenly into the region of vacuum of the separator has considerable heat available to the vacuum and boils with explosive violence at the expense of the heat of the liquid, behaving exactly as though it were superheated. Vapor bubbles form upon minute entrained air bubbles as nuclei, and in passing off into the region of reduced pressure mechanically take some portion of the dissolved air with them.

By far the major portion of the separation, particularly in the elimination of the final traces of dissolved gases, is accomplished by the reduction of partial pressure and represents a simple application of Henry's law. The water flows through the separator

logical conclusion to be deduced from this is that, with the continuous sweep of the wet, dense water vapor from the separator, the air component of the vapor pressure has been reduced by entrainment substantially to zero, and that the process has furnished a very simple and efficient means of securing such a reduction of pressure. Experiments by the writer have confirmed the results of other investigators that it is impossible to secure complete separation within a reasonable time by pressure reduction alone in excess of 80 per cent of saturation, provided the water is refrigerated to prevent vapor evolution. This is primarily a question of apparatus, since vacuum pumps of types which permit of this experiment are invariably of low volumetric capacity at extreme vacua.

It was found early in the experimental work that water in pipe lines and in small streams unpolluted by sewerage is likely to be supersaturated with air. It was accordingly necessary to analyze water both before and after entering the process to get the measure of the results accomplished. Entering water temperatures fell continuously within the range of 38 deg. fahr. to 70 deg fahr., but did not in the least conform to solubility-table values of dissolved gases. It is significant of the physical phenomena underlying the method that at low degrees of separation (70 to 95 per cent) the test results indicate a definite degree of separation rather than the reduction of dissolved gas content to a definite small value. At separation values much in excess of 95 per cent, the quantity of gases remaining is so exceedingly small that accuracy of testing is a limiting feature in extending such a conclusion into this range. It is the writer's frank opinion that the flat asymptotic portion of the curve in Fig. 3 represents complete separation of gases, but it is impossible to prove it.

Fig. 4 presents data on oxygen separation plotted as a function of separator vacuum with a constant separator temperature gradient of 22 deg. fahr., and illustrates the effect of variation of pressure. The separator temperature gradient of 22 deg. fahr. was selected as the minimum value that could give maximum separa-

FIG. 2 INTEGRAL VERTICAL-UNIT COMMERCIAL INSTALLATION

tion. This curve is plotted in terms of dissolved oxygen because dissolved oxygen only was determined, the methods of testing for dissolved air being too inaccurate to apply over the major part of the range. It is, however, probably of equal applicability to dissolved air.

The actual operation of a separator equipment is altogether as simple and straightforward as it can conceivably be. At a constant vacuum and admission temperature the water delivered by the separator is in exact equilibrium with the vapor-pressure temperature of the vacuum, and the increase of pumping head is exactly measured by the pressure head corresponding to the vacuum. The amount of heat in the vapor mixture and the amount of condensed vapor to be handled by the condenser are calculated from the heat drop in the separator, and is constant at any specified rating. The circulation ratio in the condenser, which handles all the supply water in its cooling circuit, is constant for all rates of flow. Neglecting radiation, heat losses are substantially zero. The arrangement of exhauster and condensing

equipment can be made sufficiently flexible to meet all conditions of application. Where a steam ejector is used, however, it is desirable to subdivide condensing capacity, using one section to condense the vapors withdrawn from the separator to relieve the ejector of the duty of evacuating them, and the other section to recover the heat of the ejector steam. The theory of the steam ejector seems best satisfied by the assumption of an entrainment and velocity compression of an inelastic fluid, making no concessions for the intrinsic energy of the fluid compressed, and a very slight one for differences of entrainment with variations in density. An ejector handling the vapor mixture from the separator under a 10-in. mercury vacuum has approximately five times the steam consumption of an ejector evacuating only the air. There are no moving parts in the apparatus except the float level control valves in heater and separator.

Careful tests were made upon the two heaters which at different times supplied the separator. The heaters were of a standard induction open-heater type, with spray pans for agitating the water in its passage through the steam. The vent openings, which were standard with the manufacturer, were freely connected to the atmosphere. Both of these heaters were able to deliver their rated capacity (300 and 1000 boiler hp.) of water with a 2-deg. fahr. temperature drop on atmospheric steam. Water was taken from below the water line of both of these heaters by a slow, large volume displacement into special apparatus and was tested for air without releasing the pressure or lowering the temperature. These tests gave reductions of dissolved air of only from 10 to 20 per cent, but when a similar sample was collected by rapid displacement in an open bottle, results indicated from 60 to 70 per cent at 212 deg. fahr. and a maximum of 78 per cent at 224 deg. fahr. Within this range of temperature, water within the 300-hp. heater viewed through a glass window in a blank-flanged connection against a submerged incandescent lamp bulb, appeared to be milky with entrained air bubbles. While any generalization from such meager data as these would be presumptuous, the writer cannot help but feel that from the standpoint of dissolved-air separation the heating of water from mixture with steam is an essentially different proposition from heating from a submerged and wetted source in the matter of temperature gradient within the body of the liquid. These conclusions apply specifically to a heater of the induction type, and probably would be very different for a "through" heater with its greater venting.

This work was undertaken as an industrial-research problem with the expectation that it would have two distinct fields of usefulness: the elimination of boiler and economizer corrosion in boiler plants, and of that of the entire distribution system in central-station heating, and the improvement of conductivity in surface condensation of steam. It developed in the course of experiments that it had additional merit in the removal of dissolved and half-combined carbon dioxide. The matter of corrosion has always received wide attention from the engineering profession and a vast amount of investigational work has been done on it. It was accordingly assumed at the beginning that a recapitulation of existing knowledge was all that was needed to establish the pernicious effect of dissolved oxygen. In the field of surface condensation and the removal of scale-forming materials, however, it was realized that direct experimental proof must be submitted to carry any conviction.

### CORROSION

Of the gases dissolved in water, nitrogen is altogether inert, hydrogen is inert except in the ionic form, while oxygen and carbon dioxide are the agents producing boiler and economizer corrosion. Similarly, among the dissolved solids magnesium chloride and soluble nitrates are to be included. Some waters, dependent upon locality, contain hydrochloric acid, and are included in the same category. Irrespective of the direct agent, however, all cases of boiler or economizer corrosion are ultimately either those of oxygen oxidation or of electrolytic currents. The ionic notation is the best vehicle for explaining the mechanism of the oxidation process, and in the complete paper recourse is had thereto to demonstrate that dissolved oxygen and dissolved carbon dioxide, both of which are removed by the process, are responsible for the dangerous features of boiler and economizer corrosion: the former by destroying the inhibitive polarization and displacing the equilibria of the inevitable slight solution of iron in water; the latter by increasing the concentration of hydrogen iron and with it the initial tendency of the iron to dissolve.

Carbonic acid contributes to the acidity of boiler waters and, as a corollary, to the electrolytic action, and, being volatile, mingles with the steam. In this sense whatever portion of it that returns with the condensate is cyclic in its effects. From its ability to dissolve ferrous carbonate as ferrous bicarbonate it tends to concentrate all the products of pipe-line corrosion in heating systems in the boiler itself, and makes the apparent boiler corrosion much greater than it really is. · Certainly it is a very undesirable constituent of boiler waters; but not nearly so dangerous as dissolved oxygen.

A high-pressure steam boiler is so complicated a chemical factory that there can be no general panacea for its ailments, but it is thought that the removal of dissolved oxygen and carbon dioxide is a decidedly effective step in the elimination of corrosion. Since it is a basic attack at the root of the evil, it will contribute to the success of all methods of eliminating corrosion which engineering experience has found good, but as long as boilers are made of iron and fed with water, some corrosion is inevitable. Boiler compounds, alkalinity, electric potentials, etc., are involved with such an infinitely variable condition of equilibria within the boiler that their success in practice can only be partial, while the effect of the removal of dissolved gases is aimed at the equilibria conditions themselves and is self-regulating.

The internal corrosion of economizer tubes is a phenomenon closely parallel with that of the corrosion of boilers. The corrosion here is more severe and localized and the corrosion products

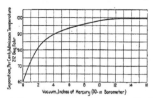

FIG. 3   RELATION BETWEEN AIR SEPARATION AND VACUUM

more alarming to the eye, although this is due primarily to a superficial change in the water of crystallization of the resulting hydrated iron oxides. The increasing cost of coal is bound to increase the use of economizers, but unfortunately both the cast-iron and the steel-tube types are unfavorably regarded, the former because of its uncertain factor of safety and the latter from its liability to corrode. Necessity, however, has already developed the steel-tube economizer closely along the type of header and tube construction used in boilers, and economizer boiler units are now available which show a decided increase in evaporative efficiency and reduction of cost over straight boiler units. The removal of carbon dioxide and oxygen will undoubtedly prolong the life of cast-iron economizers greatly, and will make the installation of steel-tube economizers economically practical and conservative. On this basis the gas-removal process will have technical significance in the future greater than it has at present.

### SCALE PRECIPITATION

Carbon dioxide exists in water as carbonic acid ($H_2CO_3$), which is capable of combining with carbonates to form bicarbonates; thus,

$$Ca(CO_3) + H_2(CO_3) = CaH_2(CO_3)_2,$$

The carbonates of calcium and magnesium are soluble in the proportions of 23 and 106 parts per million by weight as carbonates, or, in other words, in water free from carbon dioxide, but soluble as bicarbonates in any proportion in which they are apt to exist in natural waters. Bicarbonates of either of these elements are metastable and readily part with their carbonic acid under the conditions existing in the separation equipment, being reduced to carbonates and precipitated in excess of their solubility values. Sodium carbonate and bicarbonate are soluble in all proportions, although half-combined carbon dioxide is also removed.

The air-separation equipment has been tried out on two waters; one from the water works of the city of Pittsburgh, which had been drawn through a filtration plant from the Allegheny River; the other from a small local stream that supplies the Jeannette,

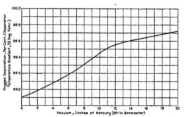

FIG. 4   RELATION BETWEEN OXYGEN SEPARATION AND VACUUM

Pa., works of the Elliott Company. Both waters were from the same water shed and had the same general analysis.

|                    | Allegheny City of Pittsburgh Parts per million | Surface Stream Jeannette, Pa. Parts per million |
|--------------------|----------------------------------------------|-----------------------------------------------|
| $CaCO_3$ ............... | 24.92   | 15.7 |
| $CaSO_4$ ............... | 130.53  | 18.2 |
| $MgSO_4$ ............... | 4.59    | 15.7 |
| $NaCl$ ................. | 3.63    | 3.5  |
| $Na_2SO_4$ ............. | 1.87    | 8.6  |
| $FeSO_4$ ............... | 9.14    | 3.0  |
| $Al_2SO_4$ ............. | 19.07   | 18.3 |
| $NaOH$ ................ | 4.94    | 0.0  |
| Organic volatile ...... | 27.00   | 12.  |
| Total ................ | 225.09  | 93.  |
| Total as determined ... | 214.0   | 87.  |

The effect of the process in eliminating calcium carbonate was initially discovered by the soap titration test of Clark on a cooled and decanted sample of water which had been put through the process. It was difficult, however, to determine the dilution of the water with exhaust steam in the heater, and results from the method were only indicative. Careful gravimetric tests over the period June 1 to November 1, 1919, supplemented by careful volumetric titrations with phenolphthalein and methyl orange indicators, gave a uniform series of results running from 18 to 25 parts per million of $CaCO_3$, which agreed closely with solubility values. While exhaust-steam dilution appeared also in these results, the reduction was greater than could possibly be accounted for by dilution, and agrees closely with solubility values. The water after passing through the separation process invariably gave a pink color with phenolphthalein, although it gave no coloration before, indicating thereby the elimination of free carbon dioxide. Technical boiler water analysis is, however, so much a matter of hypothesis that such results as the foregoing may be regarded as confirmatory but not conclusive.

Calcium carbonate precipitates out of the hot solution of the process as a colloid, which coagulates very rapidly into a flocculent precipitate and becomes granular by the time it arrives in the boiler, where it is deposited as sludge. Scale taken from the boilers during the operation of the separation equipment gave only traces of calcium carbonate, although the sludge from the bottom of the boilers contained a sufficient amount of it to be effervescent when treated with acid. No trouble was ever experienced from scaling of the boiler feed lines.

### SURFACE CONDENSATION

The surface condensation of steam involves three distinct physical steps: the conduction of heat through a definite thickness of metal—the tube wall; and two boundary conditions, the transfer of heat from a vapor to a metal, and the transfer of heat from a metal to a liquid. In the surface condenser this latter condition is dependent upon the velocity, viscosity and temperature of the cooling water and the laws of the heat transfer are not definitely understandable. The other boundary condition is complicated in practice because the area exposed to steam flow is surrounded by a rarefied air film so that in reality two heat transfers are involved: from the steam to the air film and from the air film to the metal surface. The elimination of this film of non-con-

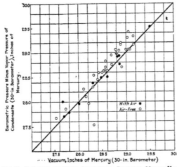

FIG. 5 RELATION BETWEEN BAROMETRIC PRESSURE MINUS VAPOR PRESSURE AND VACUUM

densable gas and the attainment of an ideal boundary condition is all that the air separation can possibly attain. It is to be expected that it will increase the conductivity by eliminating the resistance of the air film and increase the vacuum by reducing the partial vapor pressure; but the intricacy of the experimental data is very apt to obscure the simple physical fact involved.

In the experiments on surface condensation the boilers were fed with water which had been put through the air separation process and which was practically air free; and high-pressure steam was conducted to a cast-iron baffled cone bolted to the steam inlet of the condenser, expanded to condenser pressure, and allowed to enter the condenser with considerable velocity but no impingement. Two surface condensers of very dissimilar types were tested under these conditions. The first was of an antiquated type designed for engine service. The second was one furnished by the C. H. Wheeler Manufacturing Company after a design which had been built for the Bureau of Mines for experimental purposes, and was the type of unit that would ordinarily be furnished for a 100-kw. turbine. It consisted of 228 tubes, ⅝ in. O. D., No. 18 B. W. G., with an active length of 8 ft., arranged in three water passes; calculated area, 301.7 sq. ft. It was furnished with a steam baffle and air-pump suction baffle, and was without rain plates, steam lanes or differential spacing. It was equipped with a motor-driven centrifugal condensate pump with condensate-water-sealed glands, and a 16-in. by 18-in. Rotrex dry-vacuum pump. This pump was separately motor-driven at constant speed, was arranged for separate sealing water injection, and at its operating speed had a volumetric capacity of 63.3 cu. ft. per min.

Vacuum was measured at the air-pump suction and at the nozzle, a special gage being made by welding a connection to the top of a U. S. Weather Bureau mercurial barometer. The air entering the top of the mercury column was chemically dried, the vacuum was never released, and the piping was all of copper tubing. This gage had a reading accuracy of 0.01 in. of mercury, and it is felt that the accuracy is consistent.

A thermometer well was inserted into the air-pump suction in the hope that it would furnish a check on the gage. The curious situation developed that, while a thermometer well drilled in a flange of the pipe wall which was insulated with hair felt maintained within 0.1 deg. fahr. of the temperature of the vacuum, the thermometer well projecting into the pipe showed a temperature several degrees higher at extremely high vacuum (i. e., 0.01 to 0.05 in. of mercury of the theoretical absolute) and equaled the temperature of the pipe at lower vacuum. The only plausible explanation was that the air particles were carried along in the rarefied vapor stream with a translatory motion but with a pressure so close to zero that all sensible molecular vibratory motion had ceased.

The temperature of the condensate was measured by a thermometer well immediately below the exit from the condenser, mercury-filled and equipped with a certified thermometer reading to 0.1 deg. fahr. With the centrifugal pump the condensate removal was continuous. Heat transfer was determined from the temperature rise and weight of cooling water. Cooling water was delivered under constant gravity head through a regulating valve, and after going through the condenser passed through a baffled weir box and over a 90-deg. weir. This weir was sharp-edged, cut from ⅛-in. brass plate and mounted against a rigid iron frame with a commercially plane surface.

A method of checking this heat transfer was adopted which is curious in that it was successful, although the apparent odds against it were rather high. Steam was admitted to the expansion valve on the condenser after having been throttled from boiler pressure to 85 lb. gage by an automatic reducing valve, and in its passage from the boiler to the reducing valve the quality was measured. The condensate for an entire test was weighed in a large weighing tank. The sealing water from the air pump was weighed, its temperature rise measured, the amount of condensed steam calculated, and the amount of heat withdrawn with the air-pump vapors determined. The expansion of the steam into the condenser was assumed as one of constant heat content. This obviously did not represent the actual nature of the expansion, yet it was difficult to conceive of an appreciable quantity of heat being lost in the process. This heat value of the steam was corrected for initial quality, heat withdrawn by the condensate and by the air-pump vapors, but no correction was applied for radiation from or to the condenser shell. From this calculation heat transfer was computed and compared with that obtained from cooling-water measurement. The agreement between these values is the part that is curious. In a total of 52 tests the maximum difference is 35 per cent, the agreement perfect in two instances, and the average difference 7 per cent. The conductivities varied from 129 to 949 B.t.u. per sq. ft. per hr. per deg. fahr. difference in temperature.

In order to provide for as precise a comparison as possible between tests with air and those without air, a series of tests was run in which air was not extracted from the water before it entered the boiler, although it was heated up to atmospheric temperature in the open heater of the separation equipment. Air was not admitted directly into the condenser because no means of measuring the quantity admitted was known which would be of consistent accuracy with the remainder of the experiments. Such tests as either of these series are evidently not representative of operating conditions, in that they were performed upon a condenser of air-tightness far in excess of anything attained in commercial operation. They are an attempt toward an ideal condition of condensation, however, and are offered as being of specific interest because they apply to an actual condenser.

On account of the volume of the data of these tests, some 3400 tabular values, it is not practicable to publish the complete log in the present paper. However, a specimen log of two tests condensed to about one-third of its original volume is given in Table 1 for the use of those who wish to make a numerical comparison between the two groups, and the remainder of the data is presented in the form of curves, between such groupings of coördinates as are conventional in the representation of condenser tests.

It appears to be impossible to raise conductivity without recourse to methods that are freakish. Most of the heat transfer in the air-free tests took place in the upper tube banks; the transfer of heat was enormous, but without the additional refrigerating effect of the lower tubes it would have been impossible to secure

TABLE 1   SPECIMEN LOG OF TESTS (CONDENSED)

| Date ................................................. | Separated 8/28/17 | With Air 8/23/17 |
|---|---|---|
| Duration, hours ..................................... | 1 | 1 |
| Barometric pressure, inches Hg................. | 28.96 | 28.98 |
| Condensate, lb...................................... | 2,988 | 3,201 |
| Condensate, lb. per sq. ft. per hr.............. | 9.91 | 10.61 |
| Condensate temperature, deg. fahr............. | 89.1 | 87.1 |
| Temperature cooling in, deg. fahr............. | 67.5 | 64.2 |
| Temperature cooling out, deg. fahr............ | 82.8 | 80.4 |
| Heat rise, cooling, B.t.u.......................... | 14.96 | 16.16 |
| Weir head, ft........................................ | 0.682 | 0.682 |
| Weir volume, cu. ft. per min.................... | 5.852 | 5.852 |
| Specific Weight, lb. per cu. ft.................. | 62.20 | 62.22 |
| Weight cooling, lb. per min..................... | 3638 | 3640 |
| Weight cooling, lb. per hr....................... | 218,322 | 218,424 |
| Total heat to cooling, B.t.u..................... | 3,266,097 | 3,535,460 |
| Mean temperature difference log. deg. fahr... | 76.7 | 73.6 |
| Vacuum in condenser, inches Hg.............. | 28.60 | 28.57 |
| Temperature of Vacuum, deg. fahr............ | 89.5 | 90.2 |
| Absolute pressure in condenser, lb. per sq. in..... | 1.40 | 1.43 |
| Heat transfer, B.t.u. per sq. ft. per hr. per deg. fahr, .............................................. | 846 | 706 |
| Vapor pressure of condensate, inches Hg...... | 28.62 | 28.70 |
| Cooling-water velocity, ft. per sec............ | 8.47 | 8.47 |
| Cooling-water ratio ............................... | 73 | 68 |

a satisfactory vacuum. With the necessary refrigerating area the apparent conductivity was decreased.

The following data are significant of the range covered by the tests as a whole:

| | Maximum | Minimum |
|---|---|---|
| Condensation rate, lb. per sq. ft. per hr........ | 18.87 | 2.82 |
| Cooling-water ratio ........................... | 174.5 | 8.7 |
| Cooling-water velocity, ft. per sec........... | 11.94 | 2.77 |
| Cooling-water temperature, deg. fahr. ......... | 90.9 | 33.0 |
| Cooling-water rise in temperature, deg. fahr.... | 122.5 | 6.8 |

Fig. 5 is plotted between barometer minus the vapor pressure of the condensate as ordinates and vacuum as abscissæ, a solid line representing equilibrium being drawn thereon. The refrigerating effect is very decisively indicated. The condenser circulation was contraflow, three-pass, and in the air-free tests no transfer could be detected in the first pass. Simultaneously the vacuum drop in the condenser fell from 0.2 in. of mercury for 5000 lb. condensation per hour and 29-in. vacuum air to zero at all rates up to 5700 without air. Unfortunately these data were not systematically taken, since it was not known at the outset just where the advantages of air separation would appear. Ordinarily with steam-air mixtures and very efficient air removal from the condenser the phenomenon is exactly reversed, the maximum rise and the maximum transfer taking place in the lower tubes. It was possible by closing a valve in the air-pump suction to operate the condenser easily for an hour with a drop of 0.05 in. of vacuum at vacua as high as 29.06 in. When the valve was closed the condensate temperature immediately came into equilibrium with the vacuum. When, however, the loss of vacuum amounted to 0.1 in., the further drop was very rapid, amounting to a complete loss in about 10 minutes. The law of loss of vacuum was not definitely determined, but it was never a straight-line function of the time. Under the conditions of condensate removal there was never any opportunity for "drowned" tubes.

The tightness of the condenser was maintained by three methods: filling the steam space with water under pressure and watching for leaks; filling with compressed air under a pressure of about 10 lb. per sq. in. and painting with soap suds; and pump-down tests with a laboratory high-vacuum pump. If pump-down tests showed a leakage greater than 0.1 in. mercury per hour, tests were never performed. Loss of vacuum under pump-down

conditions was always a straight-line function of the time over a period of several hours. Curiously, however, tests on the condenser operating with air-free steam showed higher vacuum than it was possible to secure on pump-down tests, and the loss of vacuum in such operation followed an entirely different law.

Fig. 6, giving relation between condensate depression (temperature of the vacuum minus temperature of the condensate) and conductivity, shows again and more pronouncedly than Fig. 5 the effect of refrigeration in the lower tube banks.

## CONCLUSIONS

A careful consideration of the data of these tests in all their phases has led the writer to the following conclusions, the basis for forming which he has unfortunately not been able to make altogether clear:

*a* A comparison between individual tests on air-free steam and steam with air which are sufficiently alike in the variable quantity values or which it has been possible to reduce to a satisfactory basis of comparison, indicates an increase in conductivity of about 20 per cent, figured upon the entire area of the condenser.

*b* It is possible to reduce the air component of the vapor pressure in a tight condenser to within an immeasurable amount of zero, and simultaneously eliminate all dead areas in the condenser due to air blanketing.

*c* The fundamental problem in condensation of air-free steam is not to conduct the heat but to get rid of it after it is conducted.

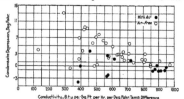

FIG. 6   RELATION BETWEEN CONDENSATE DEPRESSION AND
CONDUCTIVITY

The economical limit of permissible hydraulic losses will limit conductivities to substantially the values that are used in commercial practice today.

*d* In the condensation of air-free steam two processes are involved: the actual condensation and the refrigeration of the liquid, which take place in different zones and are distinctively different processes. One process disposes of the heat and the other produces the vacuum, and each is therefore as indispensable as the other. In the zone of condensation heat transfers reach enormous values. In the zone of refrigeration both heat transfer and heat available are very low, and the conduction is from a liquid to a liquid. In this zone a very close spacing of tubes is permissible, and any change in design which would substitute water films for water droplets would be a constructive improvement. It is possible to isolate from the data of the tests the fact that the zone of condensation is limited at least to 30 per cent of the tube area, and the reduction of vacuum loss indicates it to be much less. The ratio of temperature rise between the different circulating-water passes is radically altered from operation with very slight amounts of air.

*e* From the radically different rates at which heat is transmitted in the different zones, it is impossible in tests on condensers of different types with air-free steam to arrive at any data from which to formulate or verify laws on the conduction of heat. Neither is it possible to apply such laws without an empirical assumption on the distribution of heat between the various tube banks. These tests, while they truly represent the condensation of air-free steam, contribute no information on numerical values of actual conductivities.

# A Stoker for Burning Eastern Coals

Particulars Regarding a Newly Developed Stoker of the Conveyor-Feed Type in Which Forced
Draft is Employed and High Rates of Burning Are Obtained

By L. R. STOWE,[1] ST. LOUIS, MO.

*It is well known that the chief difficulty in burning eastern
bituminous fuels on chain-grate stokers is due to the fact that they
cake or coalesce under the arch, and thus prevent the entrance of air
through the fuel bed and check combustion. As a result of two years
of experimental work a stoker has been developed by the Laclede-
Christy Clay Products Company which, it is claimed, may be used
satisfactorily with eastern bituminous fuels. This stoker, which the
author describes, is of the conveyor-feed type and operates with
mechanical draft. High ignition temperatures are employed, the air
supply is graduated from feeding end to discharge end, and the fire
is thickened at the point of ash discharge, all of which is said to result in the prevention of caking, the assurance of a $CO_2$ content of
from 10 to 12 per cent and an intimate mixture of air with the combustible gas. Further, rates of burning up to 60 lb. per sq. ft. of
grate area per hour are obtained, as compared with the 25 to 35 lb.
on chain grates, and the percentage of combustible in the ash is
reduced 50 per cent.*

F OR the past two years the Laclede-Christy Clay Products
Company has been experimenting on a new forced-draft
stoker for use with the various kinds of fuel found in this
country. The stoker used employs the conveyor method of feeding, i. e., progressively moving chain-grate conveying elements
feed the coal through the furnace and discharge the refuse to the
ashpit. Fig. 1 shows the test boiler and stoker.

The feature of the conveyor feed that commends its use is its
handling of the fuel bed in a positive, definite fashion. By
striking off the fuel evenly under a feed gate and then literally
carrying it on through the furnace and over the point of ash
discharge, this mechanism prevents in a large measure the evils
of less positive feeding devices, namely, specifically uneven fires,
balled-up spots, and immovable clinker formations. These advantages of conveyor feeding are very attractive to the stoker
designer, and as a result of this and of the fact that the under-
feeds have quite thoroughly demonstrated the great advantages
of mechanical draft, many builders are now experimenting on
and developing various forms of forced-draft conveyor-feed
stokers. In fact, work in this direction characterizes the efforts
of the stoker industry at the present time, and the following
particulars regarding the experimental work of the Laclede-
Christy Company and the results it has attained should therefore
prove of timely interest.

For many years it has been quite clearly established that nearly
every requirement of a successful stoker is lacking when burning
caking fuels on chain grates with natural draft. The coal cakes
almost immediately upon entering the furnace, high rates of burning cannot be secured, the combustible content of the ash is high,
the admission of the air is not properly graduated to the requirements of the various stages of combustion, and maintenance of
links is excessive. This is indeed a complete indictment, and with
the sole exception that the automatic features are always in
evidence, it may be said that the performance of chain grates on
eastern bituminous coals violates every requirement of a successful
stoker.

It is well known that the caking property of eastern coal is
the most serious obstacle in the way of the successful use of this
fuel on chain grates. In endeavoring to burn Pocahontas and
similar coals it is observed that tarry oils flow through the fuel
bed as a result of the application of heat at moderate tempera-

[1] Stoker Dept., Laclede-Christy Clay Products Co.
Abstract of a paper to be presented at the Spring Meeting, St. Louis,
Mo., May 24 to 27, of THE AMERICAN SOCIETY OF MECHANICAL ENGINEERS.
Advance copies of the complete paper may be obtained gratis upon appli-
cation. All papers are subject to revision.

tures. It is further observed that the coal under the arch is
caked or coalesced by these tarry oils, and forms a sticky blanket
that prevents the entrance of air through the coal and hence
checks combustion and heat generation under the arch.

In attempting to remedy this situation the most natural thing
to do is to employ mechanical means to break up the caked fuel
bed, which is the course almost invariably taken by the inventor.
In fact, the idea is prevalent that " A mechanical agitation of the
fuel bed, for the purpose of breaking up the coke, is necessary
or advisable when burning eastern coals." So many statements
have appeared similar to the one just quoted that mechanical
engineers have come to accept it as a basic fact. The results
obtained at the experimental plant, however, have disproved this

FIG. 1 THE TEST BOILER AND STOKER

contention and in doing so have made possible the application of
conveyor feeding to the burning of eastern coals.

In the experiments with this new stoker the ignition temperatures are extremely high and at these temperatures considerable
amounts of air are forced through the fuel bed, thus providing
for the absorption of sufficient oxygen in a short time to destroy
the initial caking tendency. Moreover, it is reasonable to suppose
that these unusually high temperatures at the same time cause a
quite different procedure in the destructive distillation than that
which obtains with lower temperatures.

The absence of coke formation has almost as marked an influence on the last half of the fire as on that portion under the
arch, because since coke is not formed in the distillation zone
the cemented coke lumps cannot be present in the final stages of
combustion. The individual pieces of coal retain their identity
exactly the same as when Illinois coal is used, with the result that
the fuel bed has the soft, incandescent and fluffy appearance

which every chain-grate fireman realizes means economy and high rates of burning. The absence of coke formation in the final stages of combustion will be mentioned again.

### THE NEW STOKER

A description of the final apparatus decided on is advisable at this point for the purpose of showing how the high initial temperatures are attained. Fig. 2 is a view of the grate surface, its general incline being 20 deg. The arrows indicated by $A$ point to narrow chain-grate or conveying elements, about 4 in. wide and placed on 7-in. centers. In the intervening 3-in. spaces are placed stationary tuyeres indicated by arrows at $B$. At the lower end of the grate surface the tuyeres $B$ terminate in short pivoted bars $C$ which are made automatically to rise above, then drop below the surface of the conveyor chains. This action takes place slowly. $D$ is a feed gate, adjustable to give various thicknesses of fire exactly as on a chain grate. Partly overhanging the vibrating bars $C$ is the bridge wall $E$, shown cut away in the center. An ignition arch is shown at $F$. This latter provision conforms to chain-grate practice except that with the greater incline of the grate the arch is farther away from the fire. Fig. 3 shows a cross-sectional side elevation of the stoker and boiler and Fig. 4 details of the conveying elements marked $A$ in Fig. 2.

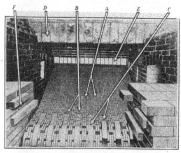

FIG. 2 VIEW OF THE GRATE SURFACE AND FURNACE TESTED

Close study of chain-grate practice has revealed much in the art of building up high ignition temperatures. A great expanse of bridge wall extending from a point horizontal with the arch to or nearly to the fuel bed is known to be of utmost value in directing heat on to the incoming fuel. This is the surface indicated by $A$ in Fig. 3. The steep incline of the stoker in question permits this dimension $A$ to be greatly extended over that possible with the ordinary chain grate. Moreover, it exposes the fuel bed more directly to the intense heat reflected from the bridge wall and the flame in front of it.

The form of arch was also taken from chain-grate practice and its design and placement has, of course, contributed its part toward the building up of the high fuel-bed and ignition temperatures that play so important a part in preventing the caking and ultimate coking of the fuel.

In view of the troubles known to exist with eastern fuels when using a stoker that conveys the fuel undisturbed into the furnace, the author deems it advisable to give a further description of the stoker in question in the light of what is required today in the economical burning of coal for the generation of steam, touching particularly on those features that have been prominently wrong with chain grates. The following outline presents the entire situation at a glance, and later in the paper each of the items therein is briefly touched upon.

PRESENT-DAY REQUIREMENTS OF AN AUTOMATIC STOKER COMPARED WITH THE PERFORMANCE OF CHAIN-GRATE AND CONVEYOR-FEED STOKERS—EACH USING EASTERN COALS.

| Requirements of the Present-Day Stoker | Results Obtained With Eastern Coals | |
| --- | --- | --- |
| | With Chain Grates | With Conveyor-Feed Stoker Described |
| 1 { It should completely burn all combustible that arises from the grate and perform this task under the great handicap of using no more air than is required. | Thin fires spotted with upstanding coke result in excess air and low $CO_2$ due to the fact that the air supply is not graduated down to a minimum at the rear of the stoker and to influences attending the coking of the fuel bed. | Graduation of air supply and thickening of fire near point of ash discharge and prevention of coke assure a consistent 10 to 12 per cent. $CO_2$ content with positive assurance of an intimate mixture of air with combustible gas. |
| 2 { It should bring about low combustible in the ash. | Combustible in the ash from 30 to 50 per cent. | Combustible in the ash from 16 to 24 per cent. |
| 3 { It should burn coal at high rates of burning—not only, to obtain high capacities but to widen the range of economical rates of burning as well, that great flexibility may be assured. | Rates of burning from 25 to 35 lb. of coal per sq. ft. of grate surface per hour. | Rates of burning up to 60 lb. of coal per sq. ft. of grate surface per hour. |
| 4 { It should be as nearly wholly automatic as possible in operation. | Successful in the matter of automatic operation. | Successful in the matter of automatic operation. |
| 5 { It should not entail unreasonable expense in the upkeep of the grate surface or of the furnace brickwork. | No unusual maintenance cost for furnace brickwork, but a very high one for grate links. | Use of mechanical draft imposes a more severe condition for furnace brickwork, requiring a higher grade of refractories, but maintenance of ironwork is not excessive—in fact, may be considered quite satisfactory |

It is of interest to note that the difficulties with chain-grates on eastern coal have been violations of the essential requirements of the modern stoker, so it is not surprising that the opinion prevails that success can not be obtained with eastern coals using the conveyor feed.

### AIR DISTRIBUTION

It will be generally admitted that one of the prime considerations, if not indeed the most important, in requirement No. 1 above, is that each particle of combustible (whether solid, vapor, liquid or gas) as it leaves the grate shall be accompanied initially and primarily by its own quota of oxygen-giving air, and it will also be admitted that these combustible particles leave faster from some parts of the fuel bed than from others. The necessity of considering the most vital subject, air distribution, is therefore indicated.

Samples of the furnace gases from all parts of the fuel bed have been analyzed from time to time as succeeding designs of stokers have been put into use. A survey of these analyses pointed out that the greatest amount of air is required at the forward end of the stoker and that only a limited supply is advisable just in front of the point of ash discharge. The individual tuyeres are only 14 in. long and are made in several interchangeable patterns, differing only as to the percentage of their air space, which ranges from 2 per cent to 22 per cent. It is to be noted, therefore, that great latitude in the matter of air supply is possible for any stage of combustion in these experiments.

Table 1 shows the air-space distribution finally decided upon. This arrangement gives the most uniform condition of furnace gas thus far attained.

There is a rather gradual decrease in the amount of air space from the initial to the final stages of combustion between the two extremes indicated in the table. It is to be noted therefrom, more-

TABLE 1  AIR-SPACE DISTRIBUTION ADOPTED
FOR CONVEYOR-FEED STOKER

| Air Space in Each Element at Feeding End | | Air space in Each Element at Discharge End | |
|---|---|---|---|
| Conveying elements...... | 8 per cent | Conveying elements..... | 8 per cent |
| Tuyere elements......... | 22 per cent | Tuyere elements........ | 2 per cent |
| 8 per cent of 60 per cent. = | 4.8 per cent | 8 per cent of 60 per cent. = | 4.8 per cent |
| 22 per cent of 40 per cent. | 8.8 per cent | 2 per cent of 40 per cent.. | 0.8 per cent |
| Total air space........ | 13.6 per cent | Total air space........ | 5.6 per cent |

Conveyor elements—about 60 per cent of total grate surface.
Tuyere elements—about 40 per cent of total grate surface.

over, that it has been found that more than twice as much air is required for the zone of destructive distillation as for the bed of well-spent fuel, and hence the advantage of graduating the air supply is readily apparent.

### LOW COMBUSTIBLE IN THE ASH

One of the best ways of reducing combustible in the ash with the conveyor-feed type of stoker is to build up high ignition temperatures. In the initial stages, the coal burns from the top down, while during the latter stages the burning is reversed, the bottom of the fuel bed, where the air is richest in oxygen, being most active. The stronger or hotter this reversal is effected, the quicker this reversal is effected, and if an early burning from the bottom up is established the combustible content of the ash will be materially decreased. The high ignition temperatures of the improved stoker have previously been referred to and their influence has unquestionably aided in reducing the combustible of the ash to a satisfactory amount.

With the caking and ultimate coking of eastern coals that occur with chain grates, the air breaks through the continuous mass of coke at the fissures or crazed places. The burning starts at these fissures and continues into the bed for its entire depth. This burning is at first quite rapid when the fissures are small, but as they increase in size the burning of the remaining coke slows down, due not only to the cooling influence of the bare spots but as well to the lessening intensity of the draft occasioned by the continual increase of the effective air openings. By reason of the original coked condition of the fuel bed, then, the fire in its final stages is characterized by bare spots and upstanding coke. To avoid exceedingly low CO, it is necessary to run these bare spots out of the furnace even though the spots of coke must be carried along with them. In other words, an effort to slow down the feed and burn the coke would result in a stack loss greater than the attending saving from burning the combustible from the ash.

With the coking of the fuel bed prevented and with each particle of coal preserving its identity in the last stages of combustion, the air filters through the innumerable interstices of the coal and causes the fuel bed to burn down evenly, giving that soft, intensive and highly luminous fuel bed that makes for high CO₂ in the gas and low combustible in the ash.

As to the retarding mechanism shown in Fig. 2, the short pivoted bars C in their lowest position allow the adjacent conveying elements to convey the ash from the furnace, but in their raised position they not only form an obstruction to the movement of the ash but raise the ash above the influence of the conveyor chains. The position of these bars with respect to the chains can be altered at the will of the operator to bring about any desired retarding influence.

The feeding at the upper end of the stoker is not affected by the retarding bars, and consequently while the ash is being positively held back of the point of discharge, additional ash-laden fuel is just as positively being conveyed toward the point of discharge by the positive feed from above. These actions result in a thickening of the fuel bed and in the squeezing out of any bare

spots that may have formed. Thus it is seen that not only is the combustible given a longer period of time in which to be completely burned, but the fuel bed is thickened so that the air entering through it may be effectively used.

It appears to the author that a particular virtue of mechanical draft lies in the burning of the combustible from the ash, for at the start of this process we are dealing with an extremely poor fuel of which the ash content may be considered to be as high as 60 per cent. It is with this in mind that the stoker in question has been so arranged that the full pressure of the wind box is made effective throughout the last stages of combustion. The experiments have been so conclusive on this point that if in further experiments it is decided to graduate the air pressure under the grate, the greatest pressure will be provided where the ash content is the highest. Of course, because of the relatively small amount of combustible in the fuel bed less air is required, so that, as explained above, the air spaces under this part of the fuel bed are materially smaller than elsewhere.

Four matters, then, are of importance in burning the combustible from the ash of eastern coals with the stoker in question.

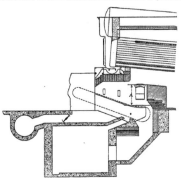

FIG. 3  SECTIONAL ELEVATION OF BOILER AND STOKER
(The boiler has 2500 sq. ft. of heating surface and the stoker 40 sq. ft. of grate surface, giving a ratio of 62½ to 1.)

These are: (1) The influence of high ignition temperatures on the combustible content of the ash; (2) A fuel bed that may be said to be free-burning as compared to one characterized by a coking tendency; (3) The holding back of the ash and attending thickening of the fuel bed in the final stages of combustion; (4) Greater intensity of draft through the bed of nearly spent fuel.

### HIGH RATES OF BURNING

Enough has already been said regarding the new stoker and its performance to make it evident that rates of burning with this type will be materially increased over those obtainable on chain grates using eastern coals with natural draft.

For instance, it is readily apparent that when the plastic, caking condition of the fuel bed is replaced by one that is free-burning and intensely active, the change will be very favorable to higher rates of burning. It is quite evident, too, that the application of forced draft will be an aid in this direction. It may be mentioned here that the chief means employed to obtain high CO₂ under forced draft have been (a) the air-distribution arrangement previously mentioned; (b) the retarding influence of the bars C together with the somewhat thicker depth of the general fuel bed over chain-grate practice; (c) the increased chemical activity of the fuel bed due to the higher fuel-bed temperatures;

and (d) the impinging of the air on the fuel bed through the employment of miniature air spaces.

### AUTOMATIC OPERATION

Having satisfied the first three requirements, this stoker must justify the use of the word "automatic"; that is, it must operate with the minimum of effort from the attendant. In this connection it is simply necessary to recall the basic principle of the conveyor-feed stoker—a positive feed from coal hopper to ashpit.

The only thing that interferes in the least with this entirely automatic and continuous operation is the clinker formation of the side walls. While this formation is no more serious than

FIG. 4  DETAILS OF THE CONVEYING ELEMENTS MARKED "A" IN FIG. 2

with the chain grates, experiments are nevertheless under way in the experimental plant to minimize this evil.

As regards eastern coals, however, it may be of interest to note that they require less slicing of the side walls than do Illinois coals, probably because of the lower ash content and the higher fusing temperatures of the ash. Further, the eastern coals have a higher heat value and a greater firing stability than do Illinois coals, and consequently require less attention. And because of these advantages the firemen of the experimental plant much prefer on this new equipment eastern fuels to those of the Central West.

### LOW MAINTENANCE

All of the factors relating to the actual burning of the coal have now been discussed and there remains only the consideration of the matter of low maintenance.

Excessive maintenance of chain grate links when using eastern coals may in part be ascribed to the uneven burning of the fuel bed at the rear of the grate and to the low ash of these fuels. These two conditions combine to lay bare and expose the grates to the heat of the furnace.

Further, in an effort to build up a high $CO_2$ content in the escaping gases and thus exclude the thief, too much air, the boiler dampers are apt to be partially closed. This procedure may be

effective in increasing the CO and in raising the furnace temperatures, but it makes for a sluggish movement of the air over the faces of the links, thus reducing the cooling influence. It further causes the heat to penetrate down into the grate links instead of being carried briskly out of the furnace. An action of this kind, especially where stokers too large for the load are provided, produces in any stoker a condition under which no grate can permanently endure.

With a stoker of the type described the thickening up of the fuel bed, especially near the bridge-wall end, together with the high rates of burning, permits one with a relatively small area being employed. This same thicker fuel bed, too, may probably be considered the most important factor in protecting the links from the heat of the furnace.

The use of mechanical draft and small air spaces with the attending high velocity of air over the sides of the grates exerts a great cooling influence on the links even at moderate loads. Moreover, the links have been especially designed for this service. They are short, and much deeper than chain-grate links. They are free from any appreciable corrugations or flutings near the top face. Cooling veins extend deep down the sides of the links.

In the successive designs of stokers which led up to the one finally adopted, no one was operated for any appreciable time, but the links of the latest designs run so cool under both forced and natural draft that their low maintenance can easily be predicted. The maintenance of the tuyeres can only be determined by their extensive and prolonged use. But no deterioration of these or of the retarding bars has thus far been noticed and it is thought that the same influences that save the links will be more or less operative with the tuyeres, though due allowance must be made for the one important difference that the tuyeres remain continuously in use while the links as in the chain grate are used intermittently.

Regarding the maintenance of the brickwork, the higher rates of burning and the higher furnace temperatures of course impose more severe conditions. The maintenance due to mechanical draft, however, is not such as might at first thought be expected, for only a moderate air pressure (less than 2 in. water column) is used, and with the very small air openings occurring uninterruptedly across the entire width of the stoker, any impinging influence of the flame is lost quite near the surface of the fuel bed.

The problem of more severe furnace conditions has thus far been met by employing refractories of a higher grade. The side walls and bridge wall have been made with standard 9-in. shapes and the writer is advised that the higher refractoriness of the brick used is in part due to its higher percentage of alumina and its low content of fluxes.

Regarding the ignition arch, both the process and the mixture are new in stoker work. The material is highly refractory and porous and its performance under the harder usage bids fair to rival the ordinary arch maintenance in chain grates.

The essential points of this paper have been sufficiently marked to require no summary, but it may be said in closing that those connected with the experimental plant have derived considerable satisfaction in attaining the degree of success noted through the employment of such simple means.

In a lecture on Some Obscure Points in the Theory of the Internal-Combustion Engine, given recently at a meeting of the Junior Institution of Engineers, Prof. F. W. Burstall said that as regards the combustion of gases in a metallic envelope, it would appear that most of the phenomena connected with afterburning can be explained when the variation of the specific heat with temperature is taken into consideration, but there still remains doubt as to whether or not combustion is complete at the point of maximum temperature. He feels that nearly all the experimental work which has been done on the internal-combustion engine has a most serious drawback from the point of view of the student of thermodynamics, namely, the number of variable quantities involved is so great that it becomes almost impossible to deduce any exact laws from a series of experiments. (*The Engineer*, March 12, 1920.)

# Locomotive Feedwater Heating

## Description of a New Type of Heater and a Discussion of the Advantages of Locomotive Feedwater Heating

By THOMAS C. McBRIDE,[1] PHILADELPHIA, PA.

*This paper describes a system of locomotive feedwater heating which makes use of the so-called open type of heater in which the exhaust steam comes into direct contact with the cold feedwater. The author describes the construction and operation of the heater and discusses the coal saving, water saving, and possibilities of increased capacity due to locomotive feedwater heating. These are dealt with first from a theoretical standpoint, and second from the practical standpoint, based upon the reports of actual tests. Tables and curves based on tests are presented and analyzed, and the paper is concluded with a discussion of the question as to whether the exhaust steam taken by the feedwater heater would cause trouble with the draft. From actual operating conditions the author states that "there was a surplus of exhaust steam left available for the draft."*

THE use of feedwater heaters on locomotives dates back to the beginning of the locomotive itself. There are records of the application of an open or injection type heater to a locomotive in England in 1827, and an English patent of 1828 describes a tubular heater for the same purpose. Ross Winans of Baltimore applied tubular heaters to two types of Baltimore & Ohio railroad locomotives in 1836. It is worthy of note that the desirability of the feedwater heater for the locomotive was recognized at these early dates, although it must be appreciated that the advantages possible at that time were proportionately greater than at the present time because of the low-steam pressure then in use.

There is much of interest in the history of locomotive feedwater heating in a paper, Feedwater Heaters and Their Development, by J. Snowden Bell, which was presented at the fiftieth annual meeting of the American Railway Master Mechanics Association, in June, 1918. Every conceivable type of heater and all possible sources of heat, even the hot ashes, seem to have been proposed or tried, and in view of the amount of study that has been given to the subject it seems rather remarkable that locomotive feedwater heating is not now in more general use. Possibly the reason lies partly in the fact that it is only in comparatively recent years that the essentials of heater performance and pump operation have been understood.

There are two possible sources in a locomotive from which otherwise waste heat may be obtained to heat the feedwater. These are: (*a*) the waste gases in the smokebox or stack and (*b*) the exhaust steam.

The exhaust-steam heater may be either of the open or injection type in which the exhaust steam comes into direct contact with the cold feedwater so that the water condensed from this exhaust steam is added to and mixed with this feedwater; or, of the closed or surface type in which the heat from the condensation of the exhaust steam passes to the feedwater through thin sheets of metal, generally in the form of thin brass tubes. Practically all of the development work to date that has been at all successful has considered only the surface type. There are a great many heaters of this type in service in Europe, but there are objections to it for locomotive use. It is a complex and delicate structure. It wastes the water condensed from the exhaust steam with its heat or raises complications as to saving it, and requires enough exhaust steam to heat all of the feedwater. On the other hand, the open or injection type of heater necessarily recovers both the water condensed from the exhaust steam and its heat so that both are returned direct to the boiler with the water taken from the tender. Less water is therefore taken from the tender and less exhaust steam is required for the heating.

[1] Sales Manager, Worthington Pump and Machinery Corporation. Mem. Am.Soc.M.E.

Abstract of a paper to be presented at the Spring Meeting, St. Louis, Mo., May 24 to 27, 1920, of THE AMERICAN SOCIETY OF MECHANICAL ENGINEERS. Advance copies of the complete paper may be obtained gratis upon application. All papers are subject to revision.

### Description of a New Type of Heater

With these thoughts in mind, the author has devised a system of locomotive feedwater heating which makes use of a heater of the open type which is manufactured and marketed by the Worthington Pump and Machinery Corporation. The system follows marine practice in heaters of this type, and the pump and heater have been combined into a single unit which can be conveniently attached to the side of the locomotive boiler as is the practice with air-brake compressors. Fig. 1 shows a unit with a capacity of 60,000 lb. of feedwater per hour in service on a Mikado locomotive and Fig. 2 a sectional view of the heater and pump. In these illustrations the heater is shown at the left, the feed pump at the right.

Fig. 1 Pump and Heater Applied to Mikado Locomotive

The heater is a simple cast iron box, the upper part of which is supplied with exhaust steam taken from the locomotive through holes cut in the back of the cylinder saddle and a pipe which has in it a stop check valve and an oil separator, the latter having a small hole in it so that there is a continuous drip for the oil.

The pump is of the vertical type attached to the side of the heater by the various pipe connections. The steam cylinder of the pump is at the top, and it has two water cylinders clearly shown in the illustrations, both operated from the same piston rod. The water cylinder in the middle of the pump takes cold water from the tender and delivers it through a port in the side of the heater to a spray valve in the extreme top of the heater where it is sprayed into the exhaust steam that fills the top of the heater. This condenses as much of the exhaust steam as the water needs to heat it, and the mixture, dropping to the bottom of the heater, is taken by the other pump cylinder and delivered to the boiler through the usual check valve.

The pump is driven by saturated steam taken from the locomotive boiler through a 1-in. steam pipe and controlled by a throttle valve in the cab within convenient reach of the engineer. The exhaust steam from the pump is led into the back of the stop check valve where it mixes with the exhaust steam from the locomotive and passes with it through the oil separator to the heater.

The adoption of the open heater for the locomotive may seem a radical step, but from this description it will be evident that no new elements are used. The heater itself· is of a general type that has been in use in marine service for at least twenty years. The open heater with the oil separator to remove the oil from the exhaust steam before it enters the heater because of its merits has been almost universally adopted for stationary plants. One of these heater and pump units was thoroughly tested out on the locomotive test plant of the Pennsylvania Railroad at Altoona, Pa., in 1917 and has been in operation on road service since October, 1918. Three more units have been in operation for

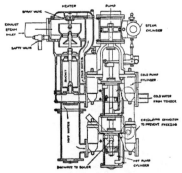

FIG. 2 SECTION OF THE HEATER

approximately a year on pooled locomotives and have given no trouble other than the usual pump maintenance. Nothing has developed indicating that the open heater as a type is not well adapted to the locomotive and the coal and water savings have been so marked as to be evident to the crew.

### ADVANTAGES OF LOCOMOTIVE FEEDWATER HEATING

It is not possible to make an exact general statement of the advantages to be obtained from locomotive feedwater heating because so much depends on the particular locomotive to which the heater is to be attached, the capacity at which it is worked, the temperature of the water in the tender, and the basis on which the advantages obtained are stated.

We are liable to base our notions as to the advantages of feedwater heating on the results obtained in stationary practice where the coal saving is very properly considered as equivalent to the reduction in the amount of heat necessary to evaporate the water because of the higher temperature at which the water is delivered to the boiler due to the heater. In the case of the locomotive, however, with its very complicated relations of the different operations that are going on and the very wide range of capacity through which the boiler is operated, there is generally a much greater coal saving than in stationary practice. Three factors enter into the final result. These are as follows:—

1. The reduction in the amount of heat required to evaporate the water in the boiler, which might be called the *heat saving due to the feedwater heater*.

2. The reduction in the amount of heat required results in a greater reduction in the amount of coal required because of the better efficiency of the boiler at the lower rate of combustion. These two factors give the *coal saving based on water evaporated*.

3. The exhaust steam taken by the heater from the exhaust ports of the locomotive reduces slightly the back pressure in the locomotive and should show some advantage. All three factors combined will give the *coal saving based on indicated horsepower*.

An analysis of these three factors will show the coal and water saving referred to the steam that is used in the locomotive cylinders and, therefore, the minimum saving since it does not include the saving in generating the steam used by the auxiliaries. It is necessary to make this analysis partly theoretical, assuming certain constant operating conditions in order that a fair comparison may be made of the effect of the heater on different locomotives and at different locomotive capacities. It can only apply to the coal burned while running since the heater can have no influence while standing or at terminals. The percentage of coal saving due to the heater, and to a lesser extent the percentage of water saving when based on ton miles, the month, or the year will be less than the percentage while running by an amount determined by the use made of the locomotives.

### HEAT SAVING DUE TO FEEDWATER HEATER

The first factor, the heat saving, is determined by the amount the feedwater is heated by exhaust steam taken from the locomotive. This factor will evidently be greatest when the tender water temperature is low and when the exhaust steam pressure in the heater is highest, as this pressure determines the temperature to which the feedwater can be heated. This pressure will be highest when the locomotive is working hard, and when the exhaust steam for the heater is taken from that part of the heater valve chests or cylinder saddles where its pressure is greatest.

Test-plant results and actual daily operation on the road with superheated steam locomotives show that the feedwater heater described can be depended on to effect a reduction of at least 10 per cent in the summer time and 12 per cent in the winter time in the amount of heat required to evaporate the water and superheat the steam. This heat saving will not be constant at these figures but will vary with the capacity at which the locomotive ·is

TABLE 1 HEAT SAVING DUE TO FEEDWATER HEATER

| | | | | |
|---|---|---|---|---|
| Assumed steam pressure in branch pipe, lb. per sq. in. | 200 | 200 | 200 | 200 |
| Assumed superheat, deg. fahr. | None | None | 150 | 150 |
| Assumed temperature water in tender, deg. fahr. | 40 | 70 | 40 | 70 |
| Heat content per pound of steam, B. t. u. | 1199.1 | 1199.1 | 1284.6 | 1284.6 |
| Heat content per pound of water, B. t. u. at 40 degrees | 8.0 | | 8.0 | |
| at 70 degrees | | 38.0 | | 38.0 |
| Heat required to generate one pound of steam B. t. u. | 1191.1 | 1161.1 | 1276.6 | 1246.6 |
| Heat saving in B.t.u. when water is heated from 40 to 215 degrees | 175.0 | | 175.0 | |
| 70 to 215 degrees | | 145.0 | | 145.0 |
| Heat saving, per cent | 14.7 | 12.5 | 13.7 | 11.6 |
| Total heat required by locomotive with feed pump, as compared to injector operation, per cent | 102 | 102 | 101.75 | 101.75 |
| Heat saving in locomotive with heater as compared to injector operation, per cent | 15.0 | 12.75 | 13.94 | 11.80 |
| Heat required, per cent, as compared to injector operation | 87.0 | 89.25 | 87.81 | 89.95 |
| Heat saving as compared to injector operation, per cent | 13.0 | 10.75 | 12.2 | 10.05 |

worked because the steam pressure in the heater determines the temperature to which the feedwater can be heated. These figures can, however, be conservatively accepted as representing average working conditions at which the greater part of the coal is burned.

The tabulation given in Table 1 is based on heating the feedwater to 215 deg. This is quite conservative as it has been found easily possible to obtain this temperature with this heater even with tender-water temperatures of 40 deg. and on locomotives worked at less than two-thirds of their maximum steaming capacity. The tabulation shows that greater heat saving is possible with saturated than with superheated steam locomotives; therefore the percentage of coal saving will be greater with saturated- than with superheated-steam locomotives.

From the tabulation it will also be noted that in the case of the feedwater heater for the locomotive, because of its comparison to injector operation, it is not possible to apply the usual short rule customary in stationary practice stating that a certain number of degrees of heating of the feedwater results in 1 per cent heat saving, but rather that it is more exact to state, assuming the conditions of the table, that in the saturated steam locomotive, after the first 23.8 or 23.2 deg. heating, each 11.9 or 11.6 deg. of heating represents 1 per cent heat saving, and that in the superheated steam locomotive after the first 22.3 or 21.8 deg. of heating each 12.8 or 12.5 deg. of heating represents 1 per cent heat saving, the first figures in each case applying with 40 deg. and the second figures with 70 deg. water temperature.

European practice, as far as the author is informed, involves the use of tubular heaters differing only in details of their construction and discharging to the track the water condensed from the exhaust steam in the heater, thus losing both this water and its heat, necessarily with some additional steam that must be blown through the heater to prevent it becoming air-bound. These locomotives require 2 per cent more water than when operated with the injector because of the steam used by the feed pump and discharge to the track from 15 to 18 per cent of this with from 2.6 to 1.8 per cent of the heat of the saturated and 2.4 to 1.6 per cent of the heat of the superheated steam locomotive in the water alone for 40 and 70 deg. water respectively. There are a few surface heaters in operation on locomotives in this country in which this water is saved by pumping it back to the tender through filters to remove the oil, but in doing so approximately 80 per cent of its heat is lost by radiation from the tender. This loss of heat from the locomotive with the surface heater does not count directly against the coal saving since it is supplied by the exhaust steam, but it does count indirectly since it decreases the amount of exhaust steam left available for the draft. The open heater is more economical in the use of exhaust steam and leaves more available for the draft with the possibility of lower back pressure and saving because of this.

The locomotive with the open heater or the surface heater saving the water condensed from the exhaust steam requires the extra 2 per cent steam to operate the feed pump, but from 15 to 18 per cent of this 102 per cent is recovered, so that the net water saving is from 13.3 to 16.4 per cent respectively for the assumed summer and winter conditions as compared to injector operation.

## COAL SAVING BASED ON WATER EVAPORATED

To combine these results with the second factor in order to obtain the coal saving based on water evaporated it is necessary to consider particular locomotives since the second factor involves the change in the efficiency of the boiler with varying boiler capacity. For this purpose the curves of Fig. 3 have been prepared. The curves on the left marked A are on the basis of 10 per cent heat saving; those on the right marked B on the basis of 12 per cent heat saving. The upper solid curves marked "Injector" are test plant results of a Consolidation, Atlantic, and Mikado type locomotive respectively and show the total amount of water evaporated per hour against dry coal fired. The dotted curves marked "Heater" for each of these three locomotives show the amount of water that would have been evaporated by the same amount of coal with a feedwater heater reducing by 10 per cent and 12 per cent respectively the amount of heat required to evaporate the water and superheat the steam. The vertical distances between the solid and dotted curves represent the amount of coal saved and these quantities are shown in the lower set of curves. This lower set of curves, therefore, shows the coal saving based on water evaporated due to the feedwater heater that follows a reduction of 10 per cent and 12 per cent respectively in the amount of heat necessary to evaporate the water and superheat the steam.

It would be difficult to credit this result were it not warranted theoretically by the second factor mentioned above, and had it not been confirmed in test plant results with a wide margin to spare. A study of the upper curves in connection with the lower curves of both figures will show that the very great coal saving possible is accompanied by an increase in steaming capacity due to the feedwater heater and that it should be possible to work the locomotive with the heater at greater steaming capacities than are possible with the injector, but how much greater would depend largely on the particular locomotive.

Referring to the lower set of curves of Fig. 3 and assuming that the consolidation locomotive would be operated normally at a steaming rate of say 25,000 lb. per hr., it is noted that the 10 and 12 per cent heat savings assumed as summer and winter conditions would result in 14 and 16.6 per cent coal savings, respectively. Assuming that the Atlantic type of locomotive would also normally be operated at a steaming rate of 25,000 lb. per hr., the coal saving would be 12.8 and 14.5 per cent respectively. Assuming that the Mikado locomotive is hand-fired and operated at a steaming rate of 37,500 lb. per hr., the coal saving would be 12 and 14.2 per cent, respectively.

On the other hand, it is altogether possible that advantage should be taken of the feedwater heater either in increased speed or increased hauling capacity, and the same amount of coal burned with it as would have been burned if the locomotive had been operated with the injector. In this case, the summer-time 10 per cent heat saving would result in a $\frac{100}{100-10}$ per cent steaming capacity, or 11.1 per cent increase; and the 12 per cent heat saving would result in a $\frac{100}{100-12}$ per cent steaming capacity, or an increase of 13.6 per cent. But as compared to injector operation of the first two factors as shown in Fig. 3, leading up to the Consolidation locomotive would be 15.5 and 19 per cent, in the Atlantic locomotive 13.3 and 15.5 per cent, and in the Mikado locomotive 13 and 15.3 per cent respectively.

The feedwater heater does not appear to as great advantage in the Mikado locomotive as in the other type locomotives because

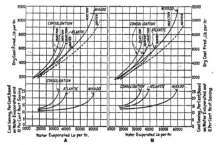

FIG. 3   CURVES SHOWING THEORETICAL COAL SAVING BASED ON
WATER EVAPORATION

it is not as fully loaded. If the Mikado locomotive was operated at a steaming rate of say 50,000 lb. per hr., the assumed 10 and 12 per cent coal saving would have resulted in a 15 and 17.3 per cent coal saving respectively, and if advantage had been taken of the feedwater heater for the increased capacity burning the same amount of coal as with the injector, the coal saving would have been 18 and 21.2 per cent respectively.

### COAL SAVING BASED ON INDICATED HORSEPOWER

Any modification which the third factor makes in the combina-tion of the first two factors as shown in Fig. 3, leading up to the coal saving per indicated horsepower, must be but slight and depend on the particular locomotive. Saturated steam locomo-tives should show a further gain of a few per cent from reduced back pressure. Test plant records show some reduction in this back pressure in superheated steam locomotives accompanied by a reduction in superheat and indicate that, at least until further tests are made, this factor should not be considered for super-heated steam locomotives.

Table 2 and the curves of Fig. 4 are test-plant results of a Mikado locomotive, comparing its operation with the injector and

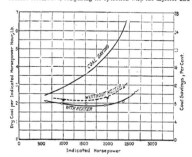

FIG. 4 CURVES SHOWING COAL SAVING BASED ON INDICATED HORSEPOWER

with the feedwater heater described in this paper. They show a coal-saving curve of the same general character, but slightly higher than the curve for the Mikado locomotive of Fig. 3-B, which is based on water evaporated, although the tests were made under practically identical operating conditions as to tempera-tures and pressures.

The coal-saving curve of Fig. 4 has been reproduced in Fig. 5, and a curve added showing the reduction in the amount of heat required to evaporate the water and superheat the steam because of the heater. The cross-hatched area below this curve shows the saving in heat due to the heater. The cross-hatched area above this curve shows the additional saving, presumably all due to the better efficiency of the boiler with the reduced amount of heat required from the fire because of the heat recovered by the heater.

### EXHAUST STEAM AVAILABLE FOR THE DRAFT

Question has been raised as to whether the exhaust steam taken from the locomotive by the feedwater heater would cause trouble with the draft. This question could be analyzed from the partly theoretical curves of Fig. 3, but as actual operating condi-tions are of more interest, curves have been added to Fig. 5, show-ing the total exhaust steam taken by the heater and the exhaust steam taken from the exhaust ports of the locomotive. The distance between these two curves represents the exhaust steam furnished by the feed pump to the heater. It is noted that the

curve of exhaust steam taken from the exhaust port is below the curve of coal saving. The feedwater heater, therefore, reduced the amount of coal burned at a greater rate than it reduced the amount of exhaust steam, so that there was a surplus of exhaust steam left available for the draft, no necessity for reducing

TABLE 2 COAL SAVING AND INCREASE IN EFFI-CIENCY WITH FEEDWATER HEATER

| Test Nos. | 1 and 5 | 2 and 6 | 3 and 7 | 4 and 8 |
|---|---|---|---|---|
| Speed in m. p. h........................... | 14.6 | 14.6 | 22.0 | 22.0 |
| Indicated Horsepower. | | | | |
|   Without heater......................... | 990 | 1549 | 2001 | 2388 |
|   With heater............................ | 965 | 1534 | 1949 | 2373 |
| Coal per indicated horse-power hour, lb.. | | | | |
|   Without heater......................... | 2.2 | 2.3 | 2.3 | 2.9 |
|   With heater............................ | 2.0 | 2.0 | 1.9 | 2.2 |
| Coal saving by feed heating, per cent........ | 9.1 | 13.1 | 17.4 | 24.7 |
| Drawbar horsepower. | | | | |
|   Without heater......................... | 824 | 1289 | 1600 | 2019 |
|   With heater............................ | 783 | 1293 | 1679 | 2055 |
| Thermal efficiency of locomotive, per cent.. | | | | |
|   Without heater......................... | 7.0 | 6.7 | 6.7 | 5.4 |
|   With heater............................ | 7.6 | 8.0 | 8.6 | 7.8 |
| Increase in efficiency with heater, per cent.... | 8.5 | 19.4 | 28.3 | 44.4 |

the size of the blast nozzle and consequent possibility of reduced back pressure in the locomotive cylinders because of this heater. For instance, at 2000 i.hp., the reduction in the amount of coal burned was 18 per cent, and in the amount of exhaust steam 12½ per cent. As compared to injector operation, 82 parts of coal had to be burned and 87½ parts of exhaust steam were left avail-able for the draft to burn it.

### CONCLUSIONS

The feedwater heater on the locomotive may be used to reduce the coal and water consumption working the locomotive at the same capacity as with the injector or to increase the capacity of the locomotive, or partly for both, as occasion demands. It will

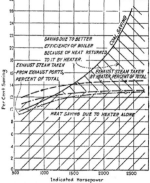

FIG. 5 ANALYSIS OF COAL SAVING BASED ON INDICATED HORSE-POWER WITH EXHAUST STEAM LEFT AVAILABLE FOR DRAFT

save coal and water all the time, and permit of increased capacity when needed.

The feedwater heater might be considered as alternate to large size of locomotive boiler and grate as a means to coal economy,

(Continued on page 316)

# Efficiency of Natural Gas Used in Domestic Service

### Results of Tests on Commercial Burners Under Kitchen Conditions, Showing Effect of Gas Pressure and Distance Between Burner and Vessel on the Efficiency

By ROBERT F. EARHART,[2] COLUMBUS, OHIO

*The following paper describes a series of efficiency tests on natural gas which the author made in the physics laboratory of Ohio State University. The operating conditions under which the tests were made were similar to those which commonly obtain in domestic service. Eight pounds of water were heated, in a granite-ware kettle, from the temperature of the tap to the boiling point. The definition of efficiency adopted is: heat units employed in raising temperature of water divided by total available heat units in gas consumed. The tests show that, under low-pressure conditions, an efficiency greater than 45 per cent can be secured when the burner is close to the vessel and that an equal efficiency can be obtained with the flame up to a distance of two inches from the vessel, under high-pressure conditions. It is further shown that with reasonable care in adjusting these distances the efficiency is independent of the rate of gas consumption and that under the varying conditions which now exist, it is advisable that the distance between burner and vessel should be between 1 and 1½ in. This will render operation possible under low pressures with fair efficiency and will lie within the range of maximum efficiencies which may be secured at moderate pressures and without undue sacrifice of time. A final experiment shows that a slight increase in efficiency is obtained by using an aluminum instead of a granite-ware container.*

O F the three divisions of the natural gas industry: production, distribution and utilization, utilization now offers the most inviting field for improvement and conservation. The general topic to be studied is the relation between the thermal content and efficiency of a gas. The term "efficiency" as applied to the utilization of gas in the kitchen of the average home is a very indefinite term. Leaving out of account the operation of a furnace, the ordinary household processes are boiling, baking, frying, and perhaps ironing. In several of these it is impractical to formulate a rigorous definition of efficiency. In baking, the B.t.u. actually put into the loaves of bread should form the numerator of the efficiency fraction. The number of B.t.u. which must be expended to bring the oven to the proper temperature and the number radiated during the actual baking process cannot properly be regarded as units directed into performing useful work. The process of frying involves a similar difficulty in developing a satisfactory definition of efficiency.

The boiling process involves less difficulty. The intent, usually, is to direct a sufficient number of heat units against a vessel so that the liquid contained therein may be raised from room temperature to the boiling point. In formulating our particular definition for efficiency, only those heat units used in raising the temperature of water are regarded as performing useful work. The heat used in raising the temperature of the containing vessel is not reckoned as useful work; i. e., the water equivalent is neglected.

Among the factors which determine the efficiency of a particular gas are (1) the type of burner, (2) the pressure of the gas, (3) the size of the mixer, (4) the rate of flow of the gas, and (5) the distance from the burner to the bottom of the containing vessel. The one factor which seems to combine these is

[1] This article is one of a series on the physical properties of natural gas. Former articles have appeared as follows:
Deviation of Natural Gas from Boyle's Law, R. F. Earhart and S. S. Wyer, Trans. Am.Soc.M.E., vol. 38, p. 283.
Ratio of Specific Heats and Coefficient of Viscosity, R. F. Earhart, Trans. Am.Soc.M.E., vol. 38, p. 979.
[2] Ohio State University.
Abstract of paper to be presented at the Spring Meeting, St. Louis, May 24 to 27, 1920, of THE AMERICAN SOCIETY OF MECHANICAL ENGINEERS. Advance copies of the complete paper may be obtained gratis upon application. All papers are subject to revision.

the location of the tip of the flame with reference to the containing vessel. When the tip of the flame impinges on the bottom of the vessel, the various separate factors are so determined as to yield a maximum efficiency. The deviation from this best position produces considerable variation.

### METHODS AND APPARATUS USED IN TEST

In these tests, 8 lb. of water were placed in a granite-ware preserving kettle. The water was raised from the temperature at which it was drawn from the tap, to the boiling point. The vessel measured 8 in. across the bottom, weighed 2.03 lb. and was provided with a tin cover. A thermometer was suspended in the vessel, passing through a small hole located 1½ in. from the center of symmetry of the cover. The vessel was supported by an iron ring-stand touching it only at the periphery, so that there was no obstruction between the burner and the bottom of

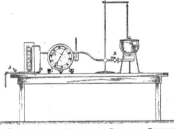

FIG. 1 ARRANGEMENT OF APPARATUS FOR DETERMINING EFFICIENCY OF GAS

the vessel. The arrangement is shown in Fig. 1. The gas was metered through a Wright wet meter, the indicator of which made one complete revolution of the dial for 0.1 cu. ft. and could be read accurately to 0.001 cu. ft. During a test, the time of each passage of the pointer over the zero of the scale was accurately noted and recorded. The rise in temperature of the water in the vessel was noted at two-minute intervals up to within a few degrees of the boiling point and then, at one-minute intervals. These data could then be plotted and the curves extrapolated when changes in boiling point due to variation in atmospheric pressure sometimes made it desirable to compare exactly equal ranges in temperature. Plotting the gas-consumption curve and the temperature curve of the water made it possible to ascertain the exact amount of gas consumed during any given rise in temperature of the water. The meter having been accurately calibrated and the temperature and pressure of the gas observed, its volume was reduced to standard conditions, viz., 30 in. pressure at 60 deg. fahr. The rise in temperature of the 8 lb. of water being observed, the number of B.t.u. per cu. ft. of gas developed was computed, this figure serving as the numerator of the efficiency fraction. The total B.t.u. content was determined by means of a constant-flow gas calorimeter manufactured by the American Meter Company and loaned by the Ohio State Utilities Commission for these tests. The gross B.t.u. content was determined; also the net B.t.u. content, which is the gross value less the amount of heat given up by the product of combustion in condensing. The usual procedure was to run (1) a series of two tests with the granite kettle; (2) a series of three tests with the gas calorimeter; and (3) another test with the granite vessel.

Four types of burners, shown in Fig. 2, were used for these tests. All burners were selected because of their common use in widely distributed types of stoves and no appliances nor means of adjustment not ordinarily supplied as part of the equipment were used. Burners of types A and B were of the simplest pattern. The gas issued from them as small independent bunsen flames of equal height. These simple burners where the envelop of the flame surface is flat, gave efficiencies a little higher than the other types. As has been stated, the essential feature for maximum efficiency is that the tips of the flames impinge on the

provided the gas is properly aerated and proper flame contact is attained. Data upon which this conclusion is based were obtained on gas whose calorific value was about one-half that of the gas used in the present tests. The present experiments, however, are in agreement with this conclusion.

Some few measurements were made with the burner at a distance of ½ in. from the vessel. Operating difficulty at this distance was experienced, as the flame produced was unsteady and likely to flash out, being smothered by its burned gases, and unpleasant fumes resulting. The thermal condition seemed to be good notwithstanding the bad hygienic condition. An oversupply of gas with the burner at ¾ in. gave rise to the same condition and considerable care in adjustment was required, especially when the vessel was cold.

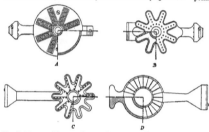 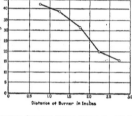

FIG. 2   TYPES OF BURNERS USED IN DETERMINING EFFICIENCY OF GAS

FIG. 3   EFFECT OF CHANGING BURNER DISTANCE WITH CONSTANT LOW GAS PRESSURES

bottom of the vessel and this is best accomplished when the separate jets of burning gas are of the same height.

Twelve simple mixers which could be used on any of the burners were available. The largest of these corresponded with an opening made by a No. 28 B & S gage drill; the smallest, to a No. 60 drill. The latter is impractical for low pressures and the larger size may be used in a very wasteful manner at high pressures. In some cases a special mixing device was tried. This consisted of a mixer opening, back of which was a secondary mixer having six holes. Its only advantage is the possibility of securing a greater flow of gas under low-pressure conditions than could be obtained with the simple mixer. This advantage means flexibility of adjustment rather than efficiency.

In these tests no refined appliances were used to secure adjustments, but effort was made to utilize to the best advantage the air and gas adjustments provided as part of the regular equipment.

### EFFECT ON EFFICIENCY AT VARYING DISTANCES, PRESSURES AND GAS CONSUMPTIONS

The results secured under a variety of conditions indicate that under low pressures an efficiency greater than 45 per cent can be secured. Equal efficiencies may be obtained up to a distance of 2 in. from burner to vessel if higher pressures are used and the rate of consumption is increased. This has a practical advantage because the operation can be performed more quickly with the same efficiency. Efficiency of this order, however, could not be secured when the distance between burner and vessel exceeded 2 in.

A bulletin published in May, 1918, by Mr. S. S. Wyer and Miss Edna White, called particular attention to the fact that a maximum efficiency could be obtained with short gas flames under low-pressure conditions. Results obtained by the writer under similar conditions confirm this point. In a report of the Research Committee of the Institute of Gas Engineers of England, published in the *Gas Journal*, October 29, 1918, it is stated that efficiency is independent of gas consumption within a considerable range,

### EFFECT OF LOW GAS PRESSURE

Tests were run, using the same common-size mixer throughout, to learn how low gas pressure affected efficiencies. When the pressure head falls below 0.3 in. of water, household operations are carried on with difficulty. At the start of the test, the pressure of the gas was reduced to 0.3 in. the burner set ¾ in. from the bottom of the vessel and the mixer valve adjusted for maximum efficiency. The burner was then lowered in steps of ½ in., readings being taken at each step without additional adjustments. When the distance was increased to 2¼ in. the water could not be

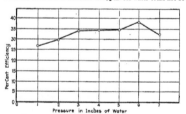

FIG. 4   EFFECT OF PRESSURE VARIATION WITH MIXER VALVE WIDE OPEN, BURNER DISTANCE CONSTANT

raised to the boiling point. Beyond a certain stage, radiation equaled input and no further rise in temperature could be obtained. At 2¼ in. the maximum temperature was 207 deg. fahr. If we regard the heat used to raise the temperature of the water to 207 deg. fahr. as usefully employed, the efficiency was 19.8 per cent. At a distance of 2¾ in. the maximum temperature attained was 197.5 deg. fahr. When we distort our definition of efficiency

(Continued on page 316)

# ENGINEERING SURVEY

A Review of Progress and Attainment in Mechanical Engineering and Related Fields, as Gathered from Current Technical Periodicals and Other Sources

SUBJECTS OF THIS MONTH'S ABSTRACTS

HEAD RESISTANCE OF AEROPLANE RADIA-
TORS
AEROPLANE RADIATORS WITH STREAMLINE
CASINGS
AERIAL TRANSPORT OF WOOLEN GOODS
AMERICAN COMPETITION WITH BRITISH
TRADES
CONDENSER TUBE CORROSION
CORROSION OF BRASS, FIVE TYPES
OLD RAILS AND TRANSVERSE FISSURES

AEROPLANE WOOD, KILN DRYING
REINFORCING BARS FOR CONCRETE
HOLZWARTH GAS TURBINE DEVELOPMENTS
ELECTROLYTIC IRON FOR GAS-TURBINE
BLADES
BALANCED HYDRAULIC VALVE
HYDRAULIC RECOIL AND RECUPERATOR
GEAR VALVES FOR GUN CARRIAGES
BERGSUND ORIGINAL SURFACE-IGNITION
TYPE ENGINE

LOW-PRESSURE INDICATOR DIAGRAMS FOR
CONSTANT-VOLUME ENGINES
RELIEVING ATTACHMENT FOR LATHE
BRITISH BATTLE CRUISER "HOOD"
SIGNALING PYROMETER
MOTOR SLED
LAXTONIA TWIN-SIX MOTOR-CAR ENGINE
TWIN-SIX ENGINES, ANGLE AND "V" AND
VIBRATION
STANDARDIZATION OF POWER PLANTS

## AERONAUTICS (See also Engineering Materials)

### Head Resistance of Radiator Cores—Resistance Due to Nose Radiators—Streamline Casing for Free Air Radiators

HEAD RESISTANCE DUE TO RADIATORS. The present report consists of three papers, of which the first, by R. V. Kleinschmidt and S. R. Parsons, treats of the Head Resistance of Radiator Cores. The main conclusions arrived at are as follows:

The shape and size factors are insignificant for the range covered (8 by 8 in. to 16 by 16 in. and 12 by 24 in.), there being no perceptible variation in the resistance per square foot.

The effect of inclining the surface of the radiator to the air stream (yawing) is, in general, to increase the resistance considerably for angles up to at least 30 deg. to 45 deg. Cores of special construction vary greatly in this characteristic.

The horsepower absorbed by the radiator in being lifted and pushed through the air is high, being, at 100 m.p.h., roughly 5 per cent to 20 per cent or more of the total power developed by the engine. It thus appears that a small gain in radiator performance may be very appreciable at high speeds.

The investigation would indicate that a high head resistance is not in itself a detriment to a radiator if properly placed in the aeroplane. A radiator placed in the open air as between the planes or beside the fuselage must be of low head resistance. In the case of nose radiators the real limit is the frontal area which may be occupied. On the other hand, if the radiator is placed in the wings, head resistance is of little detriment.

A table giving the characteristics of various radiator cores and a table of comparisons of head resistance of radiator cores are given in the original article.

The second paper, comprising a preliminary report on Resistance Due to Nose Radiators, by R. V. Kleinschmidt, is based on wind-tunnel tests over model fuselages which show qualitatively that at any given plane speed the total resistance of a fuselage with the front nose radiator is increased by increasing the air flow through the radiator.

It would further appear that the resistance of a fuselage with streamline nose is increased more by removing the streamline nose and substituting a radiator than it is by adding an equivalent unobstructed radiator and retaining the streamline nose.

A brief report on the Effect of Streamline Casing for Free Air Radiators, by S. R. Parsons, would indicate that head resistance can be decreased very materially in this way, but the accompanying decrease in mass flow of air through the radiator and consequently in heat transfer nullifies the advantage gained, with the possible exception in the case of very high speeds. (*National Advisory Committee for Aeronautics*, Report No. 61, 1920, Preprint from the Fifth Annual Report, 22 pp., illustrations, *ep*)

COMPARATIVE MERITS OF MAGNETO AND BATTERY IGNITION SYSTEM ON LIBERTY-12 AERO ENGINE. Data of tests in which the Dixie magneto represented one system and the Delco the other system.

The engine was connected to an electric cradle dynamometer in the customary manner. Tests were first conducted to determine the difference in fouling of spark plugs when used with either of the two systems. To determine this the engine was run with each ignition system for one-half hour at idling speed, and then at one hour at normal speed under full throttle operation, allowing one gallon of lubricating oil to flow into the intake system during each run. To do this the center pet cock of each of the four intake manifolds was connected with the corresponding outlet from a calibrating container filled with lubricating oil, the amount of flow being regulated by means of a valve.

The results of tests appear to be as follows: When the engine was run at idling speeds, the plugs fouled equally, no matter what system of ignition was used. The difference in power output on these runs is accounted for by the fact that the run with the Delco system was made with the spark fully retarded, while the run with the Dixie system was made with the spark fully advanced. During the one-hour runs at full power none of the plugs ceased firing with either system, the oil merely burning on the plugs to form carbon. The resultant carbon deposits were about the same with either system. The power runs show that the magneto ignition gave a little more power both on the one-hour run and the run with set spark. To offset this, however, it should be noted that on the last power run with the battery ignition and spark set at speed the power obtained was much higher than that which was obtained on any previous run, either with magneto or battery ignition. It is therefore concluded that spark-plug fouling and power output are the same for both systems, within the limits of experimental error. (*Air Service Information Circular, Engineering Division, Air Service*, McCook Field, Feb. 13, 1920, vol. 1, no. 17, 2 pp., *ec*)

AERIAL TRANSPORT OF WOOLEN GOODS. Owing to the congestion of traffic at Hull Docks and to the dockers' strike at Amsterdam, the transport of goods from the Midlands to Holland, which has been taking the preposterous time of from six to eight weeks, has, for the meantime, become quite impossible. For this reason Messrs. Heatons (Leeds), Limited, who do considerable business with Holland in ladies' clothing, decided to despatch by aeroplane a quantity of goods which they were anxious to deliver in Holland in time for the spring season. A Blackburn "Kangaroo" commercial aeroplane was therefore chartered, from the North Sea Aerial and General Transport Company, of Leeds, to convey 22 packages, with an aggregate weight of over 1000 lb. to Amsterdam. The machine was loaded at Brough aerodrome, near Hull, on Friday, March 12, and proceeded to Lympne, near

Folkestone, for examination by the air-port authorities and thence to Soesterberg aerodrome at Amsterdam, where it arrived safely on the following day. The Dutch customs authorities arranged to examine the cargo immediately on its arrival, so that the packages could be despatched to their final destinations with little or no delay. The machine was expected to return with a valuable cargo of dyes of a class urgently required in Yorkshire and Lancashire. It is hardly necessary to point out that the commercial value of the service would be greatly augmented if the visit to Lympne for customs examinations could be avoided. Lympne is, at present, the only recognized point of departure for Continental aerial traffic, and this fact is obviously a severe handicap on aircraft traveling from the northern counties to Holland and the Baltic. Evidently the traffic would be greatly facilitated by Government recognition of an air station on the North Sea coast, since a direct flight across the North Sea could be made in half the flying time, and in much less than half the total time, owing to the fact that the long southward diversion renders a night's stay at Lympne unavoidable. (*Engineering*, vol. 109, no. 2828, March 12, 1920, p. 344, *g*).

## CORROSION (See Engineering Materials)

## ECONOMICS

AMERICAN COMPETITION WITH BRITISH TRADES. This subject is discussed in *The Engineer* (London) in an editorial entitled The Question of American Competition.

The writer calls attention to the sales efforts of American concerns in Europe; such as the fact that a leading American locomotive firm is alleged to have sent a representative to The Hague where he offered to quote prices not only on locomotives and other railroad equipment but also on all kinds of machinery, including machinery for treating sugar, rubber, coffee and cocoanuts in tropical countries.

On the other hand, it is stated that in the calendar year of 1919 American exports decreased while imports increased, the reduction being due to the fall in the quantity of semi-finished steel sent out of the country for the account of the Allied nations. The imports increased through purchases of scrap and pig iron. (The statistics refer, of course, exclusively to the metal and machinery trades.)

If the American statistics are contrasted with those for Great Britain, it is found that the British exports showed a recovery for the first year after the end of active hostilities. The exports rose from 1,618,000 tons in 1918 to 2,254,000 tons in 1919, being an advance of 636,000 tons, giving a net balance of 360,000 tons to the advantage of Great Britain. The figures in the case of machinery are also hopeful for Britain, although a strict comparison between the latter and the United States cannot be made. The British exports advanced in tonnage by 65 per cent as compared with the preceding year.

As regards the future, it is stated that American steel and machinery makers have secured a great hold on international markets under the abnormal circumstances of a world-wide war, and thereby have become enormously enriched. Furthermore, though peace prevails, the conditions throughout the world have not become normal and probably will not do so before several years have elapsed. There is a great international demand for iron, steel and different classes of machinery in various countries and it is believed that Great Britain will obtain its fair share of this world trade as in former years. In fact, figures appear to show that it has already been able to make a fresh and favorable start in the resumption of its former position in the markets of the world. Though the present high prices tend to facilitate American competition in external markets, except where the rate of exchange exercises an unfavorable influence upon prospective customers, nevertheless the amount of work available in the world is so enormous that international competition for the purpose of securing it should be inconsiderable for a long time to come. (*The Engineer*, vol. 129, no. 3349, Mar. 5, 1920, pp. 249-250, *g*)

## ENGINEERING MATERIALS

### Classification of Types of Corrosion of Brass

THE CORROSION OF CONDENSER TUBES. Abstract of the Fifth Report of the Corrosion Committee of the Institute of Metals, confined to the corrosion of condenser tubes, mainly 70:30 brass.

The first section deals with what has been called the diagnosis of condenser-tube corrosion and describes the procedure to be followed in withdrawing and preparing a tube for examination, and also symptoms indicating corrosion. Conditions existing within the plant, in particular, water supply, are stated to be of great importance.

The second section is devoted to a consideration of certain features in the structure of condenser tubes and attention is directed to the presence on the tubes of a surface layer consisting of structureless, highly distorted metal. This layer has a greater resistance to corrosion than the underlying crystalline metal so that whenever this layer is penetrated corrosion will proceed at an increased rate. The thickness of this layer is usually of the order of 0.01 mm. (0.00039 in.) and indications have been obtained that its composition may be different from that of the underlying metal.

In the third section the five main types of brass-condenser-tube corrosion are considered in detail. These types are:

*Type I, General Thinning.* This may be considered as an accelerated form of the complete corrosion which normally occurs in saline solutions, the tube being gradually and uniformly reduced in thickness. Rapid general thinning, however, is essentially a fresh-water phenomenon and is usually associated with the presence of free acid in the water supply. Tests have shown that even with dilute solutions of hydrochloric acid such as 3 parts in 100,000, in six weeks tubes had lost from 3 to 4 per cent in thickness. Proper neutralization of the acid is an effective remedy for this trouble. It is difficult, however, to detect the acid, especially if it appears in the water supply only intermittently.

*Type II, Deposit Attack.* This results in pitting and is based on the following chemical reaction. In the presence of sodium-chloride solution the cuprous oxide formed on a brass surface gradually changes to cuprous chloride. This latter is usually swept out of a condenser tube by the circulating water, but under various conditions may adhere at different parts of the tube surface. When such adherence has occurred, conditions allow of the further gradual change of the insoluble cuprous chloride under the influence of oxygen to soluble cuprous chloride and cuprous oxide. The action of cupric chloride solution on brass is very rapid and, for example, a piece of brass tube 2 in. long, placed in a strong cupric chloride solution at ordinary temperature, was completely disintegrated and partially replaced by a pseudomorph in copper in two days.

*Type III, Dezincification.* In dezincification, in the case of a 70:30 brass, true parting of zinc and copper does not occur but the so-called residual copper is always redeposited copper. There is therefore no real but only apparent dezincification. The layer type which is characterized by the disintegration of the brass tube and redeposition of copper over large areas has been found to occur with both salt and fresh water, and several ways in which it may occur are indicated.

*Type IV, Plug or Local Dezincification.* This type of corrosion proceeds beneath a deposit and is stimulated by the presence of foreign bodies. It differs, however, in many ways from Type II. So far as is known, it occurs only in sea water and is always associated with appearing quite salt consisting of colloidal zinc oxychloride, also containing some carbonate. The dezincifying action is thought to be due to a small concentration of hydrochloric acid contained within the colloidal white salt. The behavior of different tubes is not always the same and the reason for it is not clear.

*Type V, Waterline Attack.* In brass tubes only partially immersed in sea water increased corrosion takes place not at the waterline itself but above it—sometimes as high as 2 cm. (say,

0.8 in.) above the air-sea-water surface. Further, the attack is not uniform but is concentrated at areas where salt deposits have formed. This type of attack may occur at the inlet end of the tube and is prevented by eddying effects from being swept away by the water flow.

The fourth section of the report contains an account of preliminary work on the electrolytic protection of condenser tubes. The particular question investigated was that of the efficiency of electrolytic protection in preventing deposit attack. It is considered that a small electric current slows down, but does not necessarily inhibit corrosion. By raising the current density sufficiently it is possible to prevent all corrosion, even in a cupric chloride solution, but the current must be quite heavy. (Official Abstract of the Fifth Report on Corrosion to the Corrosion Committee of the Institute of Metals, cp. *The Engineer*, vol. 129, no. 3350, Mar. 12, 1920, pp. 267-268, *eA*)

## No Fissures in Old Rails

OLD RAILS. A report by W. C. Cushing, Chief Engineer of Maintenance, Pennsylvania Lines West of Pittsburgh, gives an interesting sidelight on the cause of the appearance of transverse fissures.

This report covers an investigation of rails which had been in service from 10 to 45 years; some of the oldest were of wrought iron, others rolled from iron and steel pile, while others 10 to 25 years old were of bessemer steel.

In the old wrought-iron rails, head strains corresponding to 24,000 to 27,000 lb. per sq. in. compression were found, which appeared the more serious because the material had a low elastic limit, 26,670 to 48,760 lb. per sq. in. The piled rails showed maximum head strains of 24,000 lb. per sq. in., and the elastic limit of the material ranged from 33,700 to 58,830 lb. per sq. in. Still higher internal strains, however, were found in steel rails that had been in service 20 to 25 years, several measurements of 36,000 to 33,000 lb. per sq. in. being made, as against elastic limits of 46,500 to 58,450 lb. per sq. in. These high values of permanent strain were measured at the surface of the head and on the side opposite the gage side in nearly every case. One abnormal result was a reading of 57,000 lb. per sq. in., tensile stress, in the flange of an 85-lb. rail that had been in service nearly 20 years (from 1898 to 1917); Mr. Cushing calls this "extraordinary and inexplicable."

From a knowledge of the service to which these rails were submitted, one would expect to find them filled with transverse fissures. In fact, rails adjacent to them and undergoing precisely the same condition of service had been found to contain such fissures, which would indicate that the difference in the behavior of the two classes of rails must be sought in the process of manufacture. (Report No. 84 to the Rail Committee of the American Railway Engineering Association, abstracted through *Engineering News-Record*, vol. 84, no. 11, March 11, 1920, p. 527, *g*)

## THE EFFECT OF KILN DRYING ON THE STRENGTH OF AEROPLANE WOODS, T. R. C. Wilson. This paper is one of a series of monographs contributed by the Forest Products Laboratory, maintained at Madison, Wis., by the Forest Service, U. S. Department of Agriculture, in coöperation with the University of Wisconsin.

It embodies the data of a series of tests which included 26 species of wood, approximately 100 kiln runs and over 100,000 mechanical tests.

These tests were made the basis of the specification (Signal Corps No. 20500) for the general kiln-drying process for aeroplane stock (now with some slight modification specification 20500-A of the Bureau of Aircraft Production).

The general conclusions of the tests reached are:

1 That wood may have its strength properties, particularly toughness or resistance to shock, quite seriously damaged without any *visible* evidence of such damage. Hence, *appearance of the* material cannot, where maximum strength is essential, be accepted as the sole basis of judgment of the effect of a drying process on wood.

2 That the effect of a given process is not the same on all species of wood.

3 That apparently a given process may be entirely safe for some but quite detrimental to other material of a species.

4 That proper kiln drying produces material equal in all strength properties to that resulting from air drying under the most favorable conditions.

5 That specification 20500-A of the Bureau of Aircraft Production (Table 1 or 2 as specified) can in most cases be expected to produce material fully equal to air-dried.

6 That best results (with respect to strength properties) on Douglas fir will result from the use of somewhat milder drying conditions than those laid down in Specification 20500-A (Table 1). Table 2 of this specification (temperatures 105 deg. fahr. initial to 135 deg. fahr. final and relative humidities 85 per cent initial to 40 per cent final) is recommended for drying Douglas fir.

7 That in some species there is apparently no relation between drying temperatures up to 180 deg. fahr. and the strength properties of the dry material. Such a conclusion, however, needs further confirmation, and temperatures higher than those of Specification 20500-A have not been recommended.

This work has been done under the necessity of getting results as quickly as possible and with the primary object of checking the safety of the general kiln-drying specifications when applied to the drying of airplane lumber on a commercial scale, together with the more or less incidental object of ascertaining if conditions adapted to more rapid drying could be used. Under these circumstances it has not been possible to investigate the subject in the fundamental way which it merits. Completion of the work already begun is contemplated and it is hoped to be able later to carry out experiments to determine the effect of the various factors involved, both separately and in combination, and to ascertain the maximum temperature and minimum humidity that can be safely applied at any stage of the drying process. It is hoped also to carry on comprehensive investigations of the closely related subject of the bending of wood, for the purpose of determining the steaming or other processes best adapted to successful bending and to secure more accurate knowledge of the effect of such processes on the strength of the wood.

An interesting part of the table is the discussion of kiln drying as it affects the various species of woods. The methods used in the tests and the kilns are illustrated in the original article. (*National Advisory Committee for Aeronautics*, Report No. 68, 1920, preprint from the Fifth Annual Report, 69 pp., 22 figs. and 9 plates, *ep*)

REINFORCING BARS FOR CONCRETE. Discussion at the annual convention of the American Railway Engineering Association at Chicago in March, 1920.

Specific prohibition of twisted bars for the reinforcement of concrete, as provided in the masonry committee's specifications, met with some objection. Hot-twisted bars are dangerous, according to T. L. Condron, while Mr. Condron and Prof. A. N. Talbot (Mem.Am.Soc.M.E.) agreed that cold-twisted bars give no advantage over plain or deformed bars commensurate with the extra cost of the twisted bar. A suggestion that the twisting is in itself a test of the good quality of the steel was characterized by Mr. Condron as "salesman's talk." A proposal to limit the prohibition of hot-twisted bars was voted down.

Copper-plated bars, copper-alloy bars and a rough enamel coating for bars were suggested by the committee on electricity for use in reinforcing concrete subject to electrolytic action. This was accepted as information without discussion. It was pointed out that in reinforced-concrete ships protection has been based upon density and richness of concrete, and high quality of materials and workmanship, while experiments have been made with non-metallic waterproofing paints. (Abstracted through *Engineering News-Record*, vol. 84, no. 13, March 25, 1920, p. 615, *g*)

## GAS TURBINES

Holzwarth Gas Turbine, Thyssen Type; Large Com-
pression and Power Output, Nozzle Design, Blade
Design, Electrolytic Iron for Blades; Steam-
Turbine Practice Applied to Gas-Turbine
Design

DEVELOPMENT OF THE HOLZWARTH GAS TURBINE SINCE 1914,
Hans Holzwarth. In 1914, a few months before the beginning
of the war, tests were started on a 1000-hp. vertical gas turbine
built by the firm of Thyssen & Co., in Mülheim (Ruhr). These

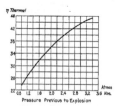

η Thermal

FIG. 1 THERMAL EFFICIENCY OF THE HOLZWARTH GAS TURBINE
(ASSUMED TO BE FREE FROM LOSSES) AS A FUNCTION OF THE
CHARGE PRESSURE PREVIOUS TO THE EXPLOSION (PRODUCER-GAS
FUEL OF HEAT VALUE 500 KG-CAL. PER CU. M.; TEMPERATURE OF
MIXTURE, 77 DEG. CENT.

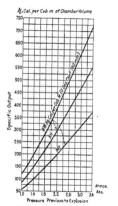

Kg. Cal. per Cub. m. of Chamber Volume

FIG. 2 SPECIFIC OUTPUT OF THE HOLZWARTH GAS TURBINE (AS-
SUMED TO BE FREE FROM LOSSES) AS A FUNCTION OF THE CHARGE
PRESSURE PREVIOUS TO THE EXPLOSION FOR VARIOUS HEAT
VALUES OF THE MIXTURE

tests were discontinued during the war, but were started over
again in 1918. The Thyssen turbine does not differ in external
appearance from the gas turbine previously built by the author
at Mannheim, but differs from it in several particulars of design.

*Higher Charge and Explosion Pressures.* These were adopted
to increase the output per cubic unit of chamber capacity and the

thermal efficiency. The average explosion pressure for continuous
operation was raised to from 12 to 14 atmos. abs. as compared
with 5 to 6 atmos. abs. of the Mannheim turbine.

There is fundamentally nothing in the way of raising this
pressure still higher by increasing the charge pressure above 2.3
atmos. abs. The curves of Figs. 1 and 2 show the influence of
the charge pressure and explosion pressure on the efficiency and
specific output of the turbine.

*Reduction of the Period of Expansion.* The shorter the time
taken by expansion the less the loss of heat to the chamber and
nozzle walls, and the greater the amount of heat transformed into
kinetic energy, and hence the greater the efficiency. At present
the duration of the period of expansion in the gas turbine is
around 0.1 sec. In large gas engines running at 90 r.p.m. the
expansion lasts about 0.33 sec., while in aircraft engines, in which
the heat consumption is not materially inferior to that in Diesel
engines, it is 0.02 sec. at 1500 r.p.m. Fundamentally there is nothing to prevent reducing the expansion period in the gas turbine
still more. To do this, however, the nozzle cross-section must be
increased, and this in turn leads to an increase in the length of
the turbine blades and in flow resistances.

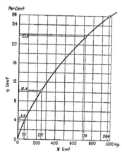

Per Cent

FIG. 3 TOTAL EFFICIENCY (AS MEASURED ON THE WHEEL PE-
RIPHERY) AS A FUNCTION OF THE POWER OUTPUT AT THE WHEEL
IN THE HOLZWARTH GAS TURBINE

*Changes in the Shape of Blades and Blade Fastenings.* Decrease in the period of expansion provides an increase in pressure on the blades. Furthermore, unlike what takes place in steam
turbines, the jet in explosion turbines acts intermittently for short
periods, and is of the character of a blow. Because of this the
blade shapes and fastenings used, for example, in Curtis turbines
would not be satisfactory. New blades, therefore, had to be developed which were planned to withstand the added stresses much
better.

As regards the material of the blades, many kinds of alloyed
and unalloyed hard and soft steels were tested, with rather unsatisfactory results, and finally a soft electrolytic iron was adopted.
With proper heat treatment blades of this iron proved to be very
satisfactory for use in gas turbines. Table 1 shows the physical
properties of this material from tests made in the testing laboratory of the Technical High School at Stuttgart.

While blades made out of harder or alloyed steels rapidly show
changes in the structure of the metal, cracks and fissures, electrolytic iron in gas-turbine blades appears to be quite capable of
withstanding wear and also resistant against surface corrosion and
erosion, provided, however, no substantial amounts of wet steam
or water are present in the gases of combustion.

*Nozzle Shapes.* DeLaval nozzles and nozzles with parallel exit
walls were tested as well as several other shapes. The advantage
of nozzles with parallel walls, which lies in their offering a greater

cross-section of exit, is counterbalanced by an increase in the deflection of the jet, and on the whole it was found that steam-turbine experience applies within reason to the design of gas turbines. The most satisfactory results were obtained with DeLaval nozzles with minimum possible angle of exit, just as in steam-turbine practice.

Practical experience has confirmed theoretical expectations to the effect that the reduction of heat gradient during the expansion does not materially affect the efficiency of the gas-turbine nozzles and blading. For example, if the instantaneous maximum jet velocity at the beginning of expansion is in the neighborhood of 1500 m. per sec., more than 95 per cent of the mechanical energy of the gases of combustion is available at the wheel up to the time when the jet velocity decreases to about 1000 m. per sec. This

TABLE 1 PHYSICAL PROPERTIES OF ELECTROLYTIC IRON USED FOR GAS-TURBINE BLADES

| | Tested at | |
|---|---|---|
| | Room temperature | 450 deg. cent. |
| Yield Point..................... kg. per sq. cm. | 4204 3185 | 1975 |
| Breaking Load............... kg. per sq. cm. | 4510 | 2675 |
| Elongation........................ Per cent | 27.2 | 50.2 |
| Contraction..................... Per cent | 73 | 88.4 |

means that more than 95 per cent of the energy is available at jet velocities varying 19 per cent either way from the average jet velocity for which the nozzle and blading have been designed. Practical experience confirms these theoretical considerations. Time after time useful efficiencies of about 55 per cent (i. e.,

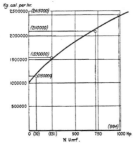

FIG. 4 TOTAL HEAT SUPPLIED AS A FUNCTION OF THE POWER OUTPUT AT THE WHEEL OF THE HOLZWARTH GAS TURBINE

about 55 per cent of the output attainable in a gas turbine free from all losses) have been indicated on the periphery of the wheel.

In steam-turbine practice the upper and lower limits in the heat-drop process are determined on one hand by the initial pressure and temperature of the steam, and on the other hand by the vacuum in the condenser. Furthermore, the peripheral velocity of the rotor is also predetermined, and all that the designer can do is to rearrange the stages in one way or another. In gas turbines the lower limit of the heat drop is predetermined and for practical purposes so are the peripheral velocity and number of stages. On the other hand, however, the designer is free to vary the upper limit of the heat drop, and through this the important ratio $\frac{\text{jet velocity}}{\text{peripheral velocity}}$ without materially affecting thereby the thermal efficiency of the turbine.

*Change in Nozzle Valves.* The employment of higher charge

pressures and the increase in the cross-section of the nozzles required some changes in the design of the nozzle valves, which are described in a very general manner.

*Tests.* The tests were carried out in the presence of various railroad officials in December 1919. The fuel used was coke-oven gas with a heat content of 3860 kg-cal. per cu. m. (432 B.t.u. per cu. ft.). The turbine was driving a dynamo delivering current to a water rheostat. The main test lasted four hours and gave the results shown in Table 2.

TABLE 2 DATA OF TESTS OF HOLZWARTH-THYSSEN GAS TURBINE

| | 1 | 2 | 3 | 4 |
|---|---|---|---|---|
| Gas consumption, reduced to 0 deg. cent., and 760 mm. mercury pressure, cbm. per hour........ | 300 | 400 | 550 | 630 |
| Heat supplied, cal. per hour........ | 1,150,000 | 1,530,000 | 2,110,000 | 2,415,000 |
| Power output at wheel N-umf, hp... | 70 | 251 | 724 | 984 |
| Heat consumption, kg-cal. per hp.-hour............ | 16,430 | 6,090 | 2,915 | 2,450 |
| Efficiency η umf, per cent........ | 3.9 | 10.4 | 21.8 | 26 |

Figs. 3 and 4 give the amounts of heat delivered to the turbine per hour, and the efficiency $\eta_{umf}$ refers to the power output $N_{umf}$ delivered by the wheel. The work consumed in compressing the blast air and gas amounted to about 5.7 per cent of the exhaust heat. If the efficiency of the compressor be assumed to be 70 per cent, of the condenser steam turbine 18 per cent and utilization of exhaust heat 60 per cent, then the total efficiency for the exhaust-heat plant and the blower system is $0.70 \times 0.18 \times 0.60 =$ 7.6 per cent, in which case the power consumption for the blowers can be well taken care of by exhaust heat.

*Construction.* Gas turbines of the horizontal type appear to be preferable in several important ways to vertical turbines, and it

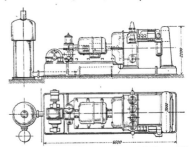

FIG. 5 HOLZWARTH OIL-FUEL GAS TURBINE, 500 HP. AT 3000 R. P. M.

is intended in the future to build only horizontal gas turbines. Fig. 5 shows the general outlines of a 500-hp. oil-fuel gas turbine now under construction to run at 3000 r.p.m. and drive a direct-current generator. The blower and exciter are built in on the same base as the main engine, and directly connected thereto is the exhaust-fuel-heated boiler delivering steam for the blowers. This installation is to go to the Prussian Railway Administration. The same article gives an illustration of a 12,000-kw. gas turbine, which seems, however, to be as yet only in the design stage.

It may be added that during the war there were several rumors of a gas turbine of extremely high efficiency having been developed by the Augsburg-Nuremberg Machine Company. In fact, it was stated that such a turbine was to be applied to drive an aeroplane. Thus far no information has become available to confirm the truth of these rumors. (*Zeitschrift des Vereines deutscher Ingenieure,* vol. 64, no. 9, Feb. 28, 1920, pp. 197-201, 14 figs., dA)

## HYDRAULIC MACHINERY

### Balanced Hydraulic Valve for Hydraulic Recoil and Recuperator Gears for Gun Carriages—Hydraulic-Valve Construction for Minimum Leakage

BALANCED HYDRAULIC VALVE. As a result of experience during the recent war in connection with hydraulic recoil and the recuperator gear for gun carriages, in which construction and design finally reached a stage when hydraulic pressure could be maintained in a group of cylinders for several weeks at a stretch with a loss of less than 5 lb. pressure. Mr. Pearson of the Erith Works, Vickers, Ltd., has designed the hydraulic valve illustrated in Fig. 6.

The chief points about this valve are that it has been designed to eliminate leakage at the moving parts, and that it is balanced both at the inlet and exhaust ends. Ease of effecting inspections or renewals has been also a main consideration.

showing the machine in connection with the exhaust, the valve head is near the open end of the bush, and the ports of the latter are uncovered, allowing connection via the interior of the sleeve, to the exhaust pipe. In Fig. 6-B, both connections from the inlet and to the exhaust are blocked. The pressure valve works in a chamber fitted with hydraulic rubber cup packings held apart by a bulb-headed spacing tube of frame, the sides of which are cut away. The exhaust sleeve is held in place and packed also with hydraulic rubber. The exhaust-valve heads are fitted with rubber packings pressed down on to the bulb-headed rings. The rubbers used in this valve are of a special composition, which thoroughly proved its great durability and efficiency in connection with gun carriages throughout the war. They are far superior to the ordinary leathers used in connection with hydraulic installations. The pressure packings and frame, and the exhaust sleeve and packings are held in place by their respective covers, which are fastened to the body by four studs and nuts. It will thus be seen that all parts may be quickly removed for cleaning and inspection; in fact, this can be accomplished and the parts re-

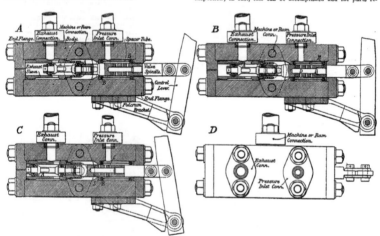

FIG. 6  BALANCED HYDRAULIC VALVE NOTABLE FOR ITS VERY SMALL LEAKAGE

The body of the valve is cut from a rectangular block of steel and bored out to take valve bushes; inside these bushes works the valve itself. In Fig. 6-A the valve is shown with the connection from the machine in communication with the exhaust. In Fig. 6-B the position of the valve cuts off the machine both from the pressure inlet and exhaust, maintaining therefore pressure on the machine and keeping it stationary. In Fig. 6-C the valve is shown putting the machine connection in communication with the pressure inlet. It will be noticed that the inlet valve is tubular, and constructed with two series of ports $C$ and $D$ (Fig. 6-B and C). The pressure inlet connects with the space surrounding this valve, and when in the position shown in Fig. 6-C the pressure supply is put through to the machine by way of these ports, the set $C$ acting as inlets and $D$ as outlets to the machine connection. At the exhaust end the ports are in the bushing, and not the valve. In the position, Fig. 6-C, with the supply and machine coupled up, the exhaust-valve head is beyond the exhaust ports in the bush, and the exhaust is cut off. In Fig. 6-A,

placed in less than 5 min., or in about 10 min. if new rubbers are required. The pipe connections can be bored and fitted to different sides as convenient and two or more valves can be coupled for combination working. It will be noticed that the ring packings are arranged so as to be more effective in preventing leakage. A small hole is drilled through the exhaust-end cover so that the double-headed piston exhaust valve shall be in equilibrium. The control lever can be easily moved by hand, and as the valve is balanced in all positions, will remain in whatever position it is placed.

In addition to the illustrations reproduced in this abstract the article gives two halftones, one showing a general view of the valve and the other a view with the control lever dismantled and also showing the interior components and links for the lever. The whole is an interesting illustration of a minor mechanical detail developed to a very high stage of perfection under the stress of war demands. (*Engineering*, vol. 109, no. 2826, Feb. 27, 1920, pp. 281-282, *d*)

IGNITION ((See Aeronautics)

INTERNAL-COMBUSTION ENGINES (See also Aeronautics, Gas Turbines, Motor-Car Engineering)

BERGSUND ORIGINAL SURFACE-IGNITION-TYPE ENGINE. Description of an engine now built in Sweden but which, it is stated,
is likely to be introduced soon into the American market. Its design does not essentially differ from that of similar engines now
known.

Fig. 7 shows a section. The cycle is as follows: Fuel oil is
injected by the pump into the hot-bulb " T " via the injection
nozzle H immediately before the piston reaches top center, and
is sprayed on to the surface of the projecting surface V, which is
heated when starting by transmission of heat from the double-
lipped disk I, which is made red-hot by the blow lamp R. After
the engine has been running a short time, the blow lamp is ex-

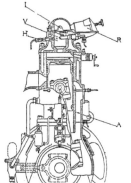

FIG. 7  BERGSUND ORIGINAL SURFACE-IGNITION-TYPE ENGINE

tinguished and the projection R and the ignition plate are kept
hot by the temperature of combustion. It will be noticed that
both the hot bulb and cylinder head are watercooled.

The projecting piece V has a double duty as the fuel is sprayed
on it at an acute angle in an almost solid stream, and is split by
it into fine particles. And, in order to avoid chance of misfire at
light loads, the lipped ignition disk $V_1$, termed the light-load ignition surface, is fitted just above it, and this enables the engine
to run for long periods without load, or without the use of the
blow-torch.

Fuel is fed to the injector by the pump A which is actuated by
a cam on the governor. The governor regulates the position of
the cam in such a manner that the stroke of the pump is suited to
the load of the engine, hence the quantity of the fuel being varied.
Before entering the cylinder, the fuel oil is passed through to
separate filters, one of which may be taken out and cleansed while
the engine is working. All troubles from the choking-up of the
fuel-oil valves and the nozzle are obviated by this arrangement.
(Motorship, vol. 5, no. 4, April, 1920, pp. 324-326, 13 figs., d)

CALCULATION OF LOW-PRESSURE INDICATOR DIAGRAMS, E. C.
Kemble. The investigation deals with the theoretical study of
the pressure variations in an internal-combustion engine cylinder
during the low-pressure part of the cycle.

The method of calculation is based on certain assumptions

which are all to be regarded as rough approximations to the
truth, such that it is believed that their adoption does not introduce any material errors.

In order to apply the method, it is necessary that pressures in
the intake and exhaust manifolds at points adjacent to the valves
and the effective clear valve openings shall be known functions
of the time. By the effective clear opening of a valve is meant
the product of the minimum sectional area of the passage through
the valve and the coefficient of efflux.

The use of the method is of course restricted by the above
requirements, since the manifold pressures are usually not
known, but to a certain extent the actual varying pressure may
be replaced by its average value which will give results that are
interesting though not strictly correct quantitatively.

The subject of the value of the coefficients of efflux is discussed
in considerable detail, the author being inclined to the opinion
that the coefficients for steady flow should not differ very greatly
from those applicable to intermittent flow obtained in practice,
provided that the pressure used in making the calculations are
those actually existing at points close to the valves.

Differential equations for the rate of change of the pressure in
the cylinder during the exhaust and suction strokes are set up
and a method for the graphic integration of these equations is
indicated. Efflux coefficients taken from experimental data are
used in the application of the theory to a typical motor and a
theoretical low-pressure indicator diagram for this motor is
obtained.

It is also stated that a method of computing theoretical low-
spring diagrams has been developed which is capable of further
development and refinements not justified by the experimental
data now available. At present the absolute values deduced from
the theory cannot be relied on, but future experiments may show
how to modify the theory so that it will give quantitatively correct values. (National Committee for Aeronautics, Report No.
50, Preprint from the Fifth Annual Report, 13 pp., 7 figs., tA.)

MACHINE TOOLS

Attachment Converting Ordinary Lathe into Relieving
Lathe

RELIEVING ATTACHMENT FOR LATHES. It is usually considered
to be uneconomical to make milling cutters and other small tools
in a general engineering shop, but occasions arise when it is desirable to do so.

All the equipment necessary to make such tools is usually
available in shops, but difficulty arises in connection with the
work of backing off or relieving the teeth of the cutters. The
teeth can, of course, be formed by milling with special cutters, but
the ordinary shop cannot have all the cutters required for different shapes and sizes of tools. A relieving lathe which might be
employed is likewise an expensive tool of very little general
utility and it would not pay to have one to take care of an
occasional odd job.

From this point of view the attachment described here and
built by Milton, Ltd., London, is of interest. It is intended to
take care of just the kind of job that can usually be done only
on a relieving lathe or by means of special tools.

Drawings, Fig. 8, A to E, illustrate the construction of the attachment. It is mounted on a gibbed plate which is screwed on
to a lathe saddle. Into this plate is fitted a steel casting having
a hole through which the work passes, or the arbor on which the
work is mounted between the centers of the lathe. There are
also two oscillating tool posts carrying tools which can operate
simultaneously on the front and back of the work, though for
relieving the front tool only is employed. Provision is made for
adjusting the tools in a vertical direction to allow for any alteration in height which may have been introduced by grinding,
likewise for the transverse adjustment of the tools. The tools
themselves fit in slots in the tool posts and are each clamped in
position by a bolt and nut.

The tool posts are rocked by means of two eccentrics mounted on a central shaft passing through the upper part of the main casting as shown at *B*. These eccentrics, which are connected to the tool posts by means of short links shown at *A*, are set at an angle of 180 deg., so that angular motion of the shaft will cause both the tool posts to advance into or recede from the work together. Mounted loosely on the opposite end of the central shaft bearing is a gear wheel, and on a sleeve extending from the boss of this wheel is screwed a ratchet-toothed cam seen in the figure at *A* and *D*. The cam is driven from the nose of the lathe by means of a shaft having two universal joints. The cam rotates at the same speed and in the same direction as the work and must have the same number of teeth as the tool being relieved. The actual diameter of the cam and the shape of the teeth are immaterial, but the teeth may be formed with feather edges.

bell-crank lever is communicated to the central shaft, and also, through the eccentrics, to the tool posts. Two short springs fitted between projections on the forked parts of the tool posts and the main casting are provided in order to take up any lost motion due to wear of the links, etc.

It will be obvious that the position of the block in the segmental groove determines the magnitude of the angular oscillations of the central shaft, and consequently those of the tool posts, so that, by adjusting the block, the amount of relief can be varied at will quite independently of the shape of the cam. The actual amount of relief is indicated by a scale engraved on the arm of the bell-crank lever, as shown at *D*, the figures giving the amount of relief in millimeters. The only other feature of the mechanism now remaining to be explained is the

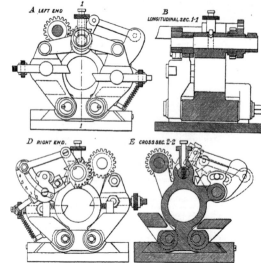

FIG. 8 MILTON RELIEVING ATTACHMENT FOR LATHES

means provided for limiting the backward movement of the tools in normal working, while allowing them to be brought back clear of the work before reversing the motion of the saddle to commence a new cut. This arrangement is best shown in the cross-section *E*, from which it will be seen that a bolt fitted with a sliding sleeve is screwed into the middle of the central shaft. A tongue formed on the sliding sleeve fits into a slot in the main casting, and normally limits the movement of the shaft by coming into contact with the end of the slot, but when it is desired to withdraw the tools from the work the tongue is pulled out of the slot by raising the sleeve. This allows the spring attached to the bell-crank lever to pull the latter back into the position indicated by the dotted lines in *E*. In this

Reamers and cutters in which the teeth are unequally spaced can be relieved by means of the attachment, but in such case the teeth of the cam must, of course, be spaced to correspond with those of the work.

Referring now to *D*, it will be noticed that the teeth of the cam engage with an adjustable pawl mounted on one arm of a bell-crank lever, which is pivoted on the main casting. As the cam rotates in anti-clockwise direction the arm of the bell-crank lever will be pulled downward until the pawl escapes from the tooth of the cam, when it will be pulled up into engagement with the next tooth by the action of a light helical spring seen in both illustrations. The combined effect of the cam and spring is thus to impart an oscillating motion to the bell-crank lever. On the other arm of the latter is formed a segmental groove, and in this groove a block, to which a short link is pivoted, can be clamped in any position. The other end of the link is connected to an arm keyed on to a central shaft, above referred to, and in this way the oscillating motion of the

position the tools will be quite clear of the work and the pawl clear of the cam; the lathe can then be reversed without altering the adjustment of the tools in the tool holders. This is a great advantage when relieving screwed hobs, and is also convenient for ordinary screwing without relief.

An important advantage of the pivoting tool holder in comparison with the horizontal slide of the ordinary relieving lathe, is that, in the former, the tools can be adjusted so that the cutting angle remains constant at all parts of the cut, and consequently greater cutting efficiency is attained and a tool with much less clearance can be used. That it is necessary to tilt the tool for this purpose will be obvious when it is remembered that the normal relief to the curve at the cutting edge of the tooth being formed coincides with the radius of the plain cylindrical blank, since no metal is removed from this part of the tooth, whereas, at the inner end of the curve, the normal is inclined by an amount depending upon the amount of relief. In an ordinary relieving lathe, in which the tool moves horizontally, no allowance is made for this feature, and the cutting angle may therefore vary by 15 deg. or 20 deg. from one end of the relief curve to the other. Another advantage of the pivoted tool holder is that the friction is considerably less than that of a sliding holder, and consequently a greater proportion of the power supplied is available for the cut in the former system. It is also pointed out that the fact that the tool holder is operated from a point well above the tool gives great rigidity in cutting.

By illustrations and descriptions the original article shows the large variety of work which can be done by means of this attachment. (*Engineering*, vol. 109, no. 2828, Mar. 12, 1920, pp. 345-346 and 348, 11 figs., *d*)

## MARINE ENGINEERING

### Largest and Fastest Battle Cruiser Afloat in Any Navy

BRITISH BATTLE CRUISER "HOOD." Some data of the new battle cruiser *Hood*, built by John Brown & Co., Clydebank, which is said to mark the close of a period in naval history. The ship was intended to be one of a group of four, but is the only one to reach completion and is today the greatest unit afloat in any navy in existence.

The *Hood* was built under conditions of unusual difficulty. The builders were furnished with sketches of preliminary design in April, 1916, and the ship was launched on August 22, 1918, a highly creditable performance considering that the first keel plate was not laid until September, 1916.

In many features the ship heralds a new departure. The difference in structure that strikes the eye first (as compared with the existing battle cruisers) lies in the "bulge" or "blister" which secures the ship against effective attack by a torpedo. This blister surrounds the main hull of the ship, and has been carefully proportioned in order to reduce to a minimum the additional resistance it causes.

The use of the blister involved many interesting problems of design. Thus, the subdivisional bulkheads have been so arranged that the energy of an external explosion is absorbed at the spot, thus preventing the piercing of the main hull should the outer hull be blown in by the explosion.

The armor has been disposed so as to provide a protection against shell fire to the vitals of the ship. The deck over the magazines is 3 in. thick. The side armor is fitted in three tiers, the lowest of which is 12 in. thick amidships, tapering to 6 in. and 5 in. forward and aft. Above the main 12-in. plate the middle and upper tiers are respectively 7 in. and 5 in. thick, while the conning-tower armor is 11 in. in thickness.

The main armament of the ship consists of eight 15-in. guns, all on the middle line. Each pair of guns is mounted in an armored barbette. The barbette armor is 12 in. thick. The gun shields are 15 in. thick in front and 11 in. thick at the sides.

The secondary armament consists of twelve 5.5-in. guns, five of which are mounted on each side of the vessel on the forecastle deck, the remaining two being mounted on the top of the superstructure. All these 5.5-in. guns are protected by gun shields forming an integral part of the mounting. In addition there are four 4-in. anti-aircraft guns also mounted on the superstructure.

The propelling machinery comprises four distinct sets, each driving one shaft, arranged in three engine rooms. The turbines are of the Brown-Curtis type and each set of machinery consists of compound ahead turbines driving the propeller shafting through single reduction gearing. For use at cruising speeds a cruising turbine of the Brown-Curtis type is provided. This has a separate casing and can, at will, be connected to, or disconnected from, the main high-pressure turbine shaft by suitable clutches. At cruising speed this turbine is operated in series with the high-pressure and low-pressure turbines of the set.

The reverse turbine is fitted with the low-pressure casing and rotates in a vacuum when the ship is running ahead. Each of the four sets of machinery is complete in itself with its own condenser and auxiliaries and can be used independently.

Each high-pressure ahead turbine rotor consists of ten wheels,

of which the first and second are velocity-compounded having two rows of blades each, and the remainder are simple impulse wheels. Each low-pressure ahead-turbine rotor consists of eight wheels, of the simple impulse type. Each cruising-turbine rotor has four velocity-compounded wheels with three rows of blades on the first wheel and two rows on each of the others. The astern turbine for each shaft is incorporated in the low-pressure ahead-turbine casing and consists of two velocity-compounded wheels, each carrying three rows of blades.

The turbine casings are of cast iron with the nozzle-control-valve seatings cast in the body. The rotor spindles are of forged steel, having a tensile strength of 34 tons to 38 tons per sq. in. The wheels are of forged steel, of similar quality, and phosphor bronze has been used for the blades, which are of the usual impulse type. The blades have dovetailed roots, and the tips of the blades are supported by phosphor-bronze shrouding. The diaphragms are of cast iron, with the nozzle blades of nickel steel (about 3½ per cent nickel) cast in. The diaphragms in the cruising turbines and the first diaphragm in the high-pressure turbines are solid, the remainder of the diaphragms in the high-pressure turbine, and those in the low-pressure turbines, being split. The spindle is made steamtight where it passes through the diaphragms by segmental serrated brass rings, kept up to their work by garter copper-coated steel springs in the cruising and the first diaphragm of the high-pressure turbines, and by sherardized coach springs in the other diaphragms of the high-pressure and in the low-pressure turbines. The glands are packed, as is usual in Brown-Curtis turbines, with carbon rings made in segments and held in position by springs. The high-pressure ahead forward-end gland has eight carbon rings, the after end five, the low-pressure ahead four rings and the astern turbine four rings. The cruising turbine is fitted with five rings in each of the glands.

The turbines are designed to develop a total horsepower of 144,000 with about 210 lb. pressure in the control chest and 28 in. vacuum in the condensers. Steam is supplied by 24 Yarrow-type small-tube boilers burning oil fuel and located in four compartments. The working pressure is 235 lb. per sq. in. and the heating surface per boiler 7290 sq. ft.

The original article is illustrated by a number of plates giving an idea of the stupendous size of the unit itself and its various details. (*Engineering*, vol. 109, no. 2830, Mar. 26, 1920, pp. 397-399 and 4 pp. of plates, 12 figs., *dA*)

## MILITARY ENGINEERING (See Hydraulic Machinery)

## MEASURING INSTRUMENTS

THE SIGNALING PYROMETER, R. P. Brown. The ordinary pyrometer which either indicates on a temperature scale or records the temperature on a chart is satisfactory where the employees are capable of reading the temperature scale or understand the movement of the pointer. During the last few years, however, an inferior grade of labor has had to be employed, often unable to understand the instrument, and it has become necessary, therefore, to use some other means to indicate to the furnace operator when the temperature is correct or too high or too low.

In some plants this is done by signaling the temperature by lights from one central station. Three lights are grouped at each furnace: green, to indicate that the temperature is too low; white, correct; and red, too high. This system requires, however, the maintenance of a central operator to flash the signals and usually an elaborate switchboard and wiring system from the furnaces to the central station.

Recently, however, an instrument has been developed to operate the signal lights from a pyrometer automatically. In this instrument, whose construction is very similar to that of the ordinary indicating pyrometer, a contact table adjustable throughout the whole scale range carries three contacts corresponding to the three lights, the central contact operating the white light, usually covering a space of 20 deg. on the scale, amounting to 10 deg. above or below the desired temperature. The pointer on the pyrometer is automatically depressed once a minute on the tungsten contacts.

An automatic three-position relay is operated by the closing of the contacts in the instrument. Let us assume that the furnace is started up and the low contact is closed, corresponding to the green light, and the relay complete the circuit through the green light. The contact in the instrument is only closed a few seconds to operate the relay, but the green light continues to burn until the temperature rises to the desired point, when the relay shifts, causing the white light to burn and extinguishing the green light. The proper light continues to burn at all times, although the contact in the instrument only closes for a few seconds once a minute. (*Journal of American Steel Treaters' Society,* vol. 1, no. 4, Jan. 1919, pp. 131-132, *d*)

## MOTOR-CAR ENGINEERING

MOTOR SLED. Description of a motor sled now being perfected by a company in Seattle, Wash.

The propelling force is obtained by applying the principle of the ship propeller to two revolving drums turning in opposite directions and having secured to them a series of spirals' that force the sled ahead just as a bolt is forced through a nut when turning. The speed of the sled is changed by a special transmission through a conventional gear shift.

The fact that the propelling drums rotate in directions opposite to one another makes it necessary to employ a special gear drive. In this case it has three gears, the drive pinion from the engine directly meshing with only one of these gears. The lower two gears again mesh with one another and drive through separate shafts to centrally located sprockets and then through chains to the main-drive sprockets located in between the two drums.

Finally, the drums must be made perfectly cylindrical in shape in every transverse section and must be perfectly smooth.

It is claimed that there is absolutely no slip between the snow surface and sled.

In other words, for a given number of revolutions of the drum with a certain pitch of the skates, the actual travel is equal to the theoretical length of the travel.

The drums are made of pressed steel fastened by means of flanges and spiders to the drive shaft. The steel skates spirally located on the drum are of a very shallow depth—probably not over ⅝ in. high.

In addition to this, there are two vertical guides fastened on to the superstructure of the car, the object of which is to keep the drum shafts parallel in a vertical plane and also to allow the drums and drum shafts which are supported on a central pivot to oscillate around this point in order to conform to the road bed. (*Motor Age,* vol. 37, no. 12, Mar. 18, 1920, p. 30, 1 fig., *d*)

### Laxtonia Twin-Six Motor-Car Engine; Angle of "V" and Vibration of Engines

BRITISH TWIN-SIX MOTOR-CAR ENGINE. Description of the 12-cylinder touring-car engine designed by W. L. Adams of the Laxtonia Engineering Company, Peterborough, England.

The engine has several interesting features. The two blocks of six cylinders (Fig. 9) have an angle of 30 deg. between them. Such an arrangement is apt to increase vibration but makes the engine more compact, which, in this case, is further enhanced by the fact that the cylinder blocks are cast with the main portion of the crankcase as a single unit in aluminum.

The 30-deg. setting also simplifies the design as it enables a single overhead shaft midway between the two blocks to operate all the overhead valves, both inlet and exhaust, through the medium of rockers.

The crankshaft (Fig. 10) has only three throws and two journal bearings of the roller type. Two bearings are considered to be sufficient because the crankpins and journals are of 2¾ in. diameter and the individual impulses are claimed to be of comparatively small magnitude.

A detachable cylinder head is fixed over each block of cylinders and carries the overhead valves. Each block of cylinders is staggered in relation to the other, to the extent of 21/32 in.,

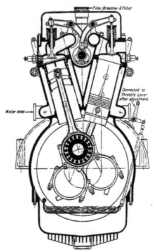

FIG. 9  SECTION OF THE LAXTONIA 12-CYLINDER ENGINE

which allows the connecting rods of opposing pairs to lie side by side without their being offset in relation to the piston centers.

The overhead camshaft formed in one piece with its 24 cams

FIG. 10  CRANKSHAFT OF THE LAXTONIA 12-CYLINDER ENGINE

is driven by a train of three spur pinions enclosed at the front of the engines and lies in a trough formed above the main casting. The oil pump delivers oil through the branch pipe to this trough and the rotating cams serve to distribute lubricant to all the overhead gear, which is enclosed by a detachable aluminum cover.

While the engine has a capacity of 6900 cu. cm. (421 cu. in.) the two cylinder blocks are but 12 in. overall.

While it is admitted that the uniformity of turning effort in this engine falls below that of a 60-deg. twelve, its torque curve approaches that of the 12-cylinder Liberty engine, which has its cylinder blocks set at an included angle of 45 deg. There are two peaks in the torque curve of every 120 deg. of crankshaft revolution, the minor peak occurring 30 deg. in advance of the main peak. (*The Engineer*, vol. 129, no. 3349, March 5, 1920, pp. 252-254, 4 figs., *d*).

## POWER PLANTS

### Principles of Standardization of Power Plants; Suggested Types of Standard Units

STANDARDIZATION OF POWER PLANTS, Louis R. Lee. Standardization is recommended in the interests of plant efficiency, which the author defines as the ratio of the gross input to the gross output. The gross output may be expressed in pounds of steam or in kilowatt-hours. In the first case the efficiency ratio should

be effected only when they will not reduce the possible saving in operation and maintenance.

Fig. 11 shows several types of assembly of the complete unit and the different possible building arrangements. In *B* the plant shows the boilers, economizers and auxiliaries are housed in the main building, while the induced-draft equipment, coal bunkers and ash discharge are outdoors.

In *C* and *D* a reduction in size of the main building is secured by providing each economizer unit with a self-containing waterproof covering, this covering also extending over the aisles between economizers.

It is claimed that a comparison of any of these building arrangements with the best present-day practices will show much saving in cubical contents of the building when compared on a rate of horsepower basis. There is also a great saving in material for supporting equipment particularly the economizers and draft equipment.

The standard unit as designed by the author consists of a cross-drum boiler set singly with suitable stoker furnace and provided with a direct-connected economizer. The draft equipment consists of a forced-draft fan and an induced-draft fan, both motor-

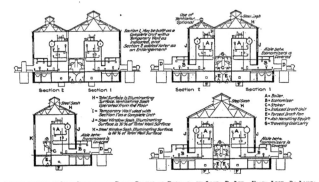

FIG. 11   SECTIONAL VIEWS OF SOME STANDARDIZED BOILER PLANTS AS DESIGNED BY LOUIS R. LEE.   UPPER LEFT, *B*; LOWER LEFT, *C*;
UPPER RIGHT, *D*; LOWER RIGHT. *E*

show the cost per thousand pounds of steam and in the latter case per kilowatt-hour, the cost of input covering all that is actually put into the plant, including cost of fuel, labor for operating, labor for maintenance and fixed charges on the total investment.

One of the most essential parts of all power plants is the boiler plant. The boiler complete with furnace draft equipment and feedwater supply is essentially a complete plant and it is only necessary to multiply the output of such a unit to secure any desired output. By making the unit a complete fractional part of a plant extensions are taken care of with minimum trouble and cost.

To make estimating of the cost of the installation easier and also to effect savings the author recommends devising the unit design with the limited selection of characteristic equipment bearing in mind the effect that the price of the fuel has on the installation.

Unless necessitated by local conditions such as occur in congested city districts, it is recommended in the design to use more ground area rather than high elevation. To secure low operating cost all equipment must be so located that all parts are quickly accessible and the disassembly is easily effected for making repairs and maintenance. Savings in first cost of construction must

driven. The induced-draft equipment together with stack is located outdoors where space permits, and inspection of the fan is obtained through a doorway opposite each unit aisleway for which the width of 10 ft. is recommended as standard.

A central observing and controlling station is provided at each side of the boiler unit, these stations being arranged where desired, right and left hand, so that one operator may observe and control two units. Data as to the equipment and details of the apparatus used are also given. (*Power*, vol. 51, no. 10, Mar. 9, 1920, pp. 375-378, 2 figs., *gd*)

## RAILROAD ENGINEERING (See Engineering Materials)

## WOOD (See Engineering Materials)

## CLASSIFICATION OF ARTICLES

Articles appearing in the Survey are classified as *c* comparative; *d* descriptive; *e* experimental; *g* general; *h* historical; *m* mathematical; *p* practical; *s* statistical; *t* theoretical. Articles of especial merit are rated *A* by the reviewer. Opinions expressed are those of the reviewer, not of the Society. The Editor will be pleased to receive inquiries for further information in connection with articles reported in the Survey.

# ENGINEERING RESEARCH

A Department Conducted by the Research Committee of the A. S. M. E.

A—RESEARCH RESULTS

The purpose of this section of Engineering Research is to give the origin of research information which has been completed, to give a résumé of research results with formulæ or curves where such may be readily given, and to report results of non-extensive researches which in the opinion of the investigators do not warrant a paper.

*Apparatus and Instruments A5-20* Accuracy of Chronographs. The use of photographic records to determine the uniformity of motion of clockwork used to drive recording mechanism and to determine the uniformity of chronographs has been developed by the Bureau of Standards. Bureau of Standards, Washington, D. C. Address S. W. Stratton, Director.

*Apparatus and Instruments A6-20* Anemometer. An air-speed indicator of the Robinson cup type with centrifugal indicating mechanism has been calibrated. This instrument is of German manufacture and is very compact and well adapted for use on dirigibles. It is independent of changes in air density and gives actual air speeds. Bureau of Standards, Washington, D. C. Address S. W. Stratton, Director.

*Apparatus and Instruments A7-20* The Leucoscope Used in Pyrometry. The Bureau of Standards has found that the leucoscope of Helmholtz is applicable to pyrometry and spectral energy distribution in a light source or furnace. These data have been communicated to the Optical Society of America and will appear in its Journal. Bureau of Standards, Washington, D. C. Address S. W. Stratton, Director.

*Apparatus and Instruments A8-20* New Instrument for Measuring Small Air Velocities. Thesis by A. E. Reinhard and J. R. Cain, Rose Polytechnic Institute. Use of the standard pitot tube with instrument for measuring the difference of pressures.

Two shallow, airtight tin boxes about 6½ in. square have square openings about 6¼ in. on a side cut in the tops. Within each opening a square aluminum plate, 5⅝ in. on a side, is located so as to be central and in the plane of the top of the box. Thin gold leaf is then cemented to the aluminum plate and to the top of the box so as to make an airtight receptacle with a flexible top.

A suitable wooden base is provided with a column for supporting an equal-arm balance, which is sensitive to one grain. The air boxes are fastened to the base so that the centers of the boxes are vertically below the respective ends of the balance lever, and the aluminum plates are connected by suitable cords with the arms of the balance. Suitable stops are also provided for limiting the movement of the diaphragm so as to prevent tearing the gold leaf.

When the air boxes are connected respectively to the dynamic and the static sides of the pitot tube, the net effect upon the balance is due to the pressure equivalent of the air velocity which is being measured. The scales are first balanced with both air boxes open to atmosphere. Then connections are made to the pitot tube and weights added until a balance is again obtained. The added weight divided by the effective area of the diaphragm gives an absolute measure of the difference of pressure.

The instrument was found to be sensitive enough to record a velocity of 1½ ft. per sec. At 3 ft. per sec. the probable error was about 3 per cent. This instrument is not suited to high velocities. It was used successfully to measure velocities between 1½ and 15 ft. per sec.

Address Prof. F. C. Wagner, Rose Polytechnic Institute, Terre Haute, Ind.

*Internal-Combustion Motors A2-20* Monel Metal for Engine Valves. The Bureau of Standards has recently used exhaust valves made of monel metal in a 180-hp. Hispano-Suiza engine during 130 hr. of running time. The tests were of 6 hr. duration. During this period 90 hr. of the test were devoted to the study of lubricating oils while the engine ran at 1800 r.p.m. and developed 130 to 140 hp. At the end of 45 hr. the valves were found to be badly pitted and the valve seats were found to be in the same condition. During the next 45 hr. the power was the same as before and after this the valves were opened for 40 hr. service under severe conditions while the engine was used to study preignition. High temperatures were experienced and the electrodes of the spark plugs were fused, the melting point being in the neighborhood of 1500 deg. cent. In spite of this severe test the

valve seemed to hold up as there was no drop in power. The valves were found to remain red for nearly a minute after the valves in the other cylinders had become black. The pitted appearance of the valves does not seem to affect the gas tightness. Bureau of Standards, Washington, D. C. Address S. W. Stratton, Director.

*Chemistry, Organic A1-20* Camphor. A new step in the process of manufacture of synthetic camphor from pinene (turpentine) has been studied by the Institute of Industrial Research. Camphor is a necessary constituent in the manufacture of celluloid and allied compositions, and is used in the manufacture of moving-picture films. Up to the present time camphor is an almost exclusive Japanese monopoly. Institute of Industrial Research, 1845 B. St., N. W., Washington, D. C. Address Allerton S. Cushman, Director.

*Metallurgy and Metallography A4-20* Zirconium. A new process for treating zirconia and zirconite ores so as to get them into a condition in which they are readily soluble in acids for repreciptation and purification, has been developed by the Institute of Industrial Research. Since zirconia is of great present interest on account of its extraordinary refractoriness at high temperatures, this process may prove of value. The Institute of Industrial Research, 1845 B. St., N. W., Washington, D. C. Address Allerton S. Cushman, Director.

*Metallurgy and Metallography A5-20* Mechanical Hardness and Grain Size of Carbon Steels. The research just completed at the Bureau of Standards shows no simple relation between grain size and hardness in carbon steels such as exists in alpha-brass and other constituent alloys. Grain size is a factor of minor importance. The amount of distribution of structural constituents determines the hardness to very great extent. The work already completed throws some light upon the magnitude of the stressing of low-carbon steel by cooling and the work in order to affect subsequent grain growth upon annealing below the critical temperature. Bureau of Standards, Washington, D. C. Address S. W. Stratton, Director.

*Petroleum, Asphalt and Wood Products A1-20* Iroline Process. A new process known as the Iroline Process for treating heavy petroleum oils and distillates in order to produce light, low-boiling hydrocarbons of a gasoline-like nature, has been developed by the Institute of Industrial Research. This process has been given a trial on the plant scale of operation in a specially designed plant capable of treating 1000 bbl. of distillate per day. The results obtained were satisfactory. The Institute of Industrial Research, 1845 B. St., N. W., Washington, D. C. Address Allerton S. Cushman, Director.

*Refrigeration A1-20* Ammonia Vapor Pressure. A paper has been published in the Journal of the American Chemical Society, vol. 42, p. 206, giving a summary of the research made at the Bureau of Standards on the vapor pressure of ammonia from —78 deg. cent. to +70 deg. cent. Seven samples of thoroughly purified ammonia were used. Special tests showed less than one part in 100,000 by volume of non-condensing gases and less than 1/100 per cent by weight of other impurities. The commercial samples of ammonia were found to give hysteresis due to the presence of impurities. The normal boiling point was determined by static and dynamic methods. Two empirical equations were found to represent the results for the temperature range covered with close agreement to actual results. Bureau of Standards, Washington, D. C. Address S. W. Stratton, Director.

B—RESEARCH IN PROGRESS

The purpose of this section of Engineering Research is to bring together those who are working on the same problem for coöperation or conference, to prevent unnecessary duplication of work and to inform the profession of the investigators who are engaged upon research problems. The addresses of these investigators are given for the purpose of correspondence.

*Internal-Combustion Motors B4-20* Kerosene Vaporizer. The development of a kerosene vaporizer for use on gasoline engines to be inserted between an ordinary carburetor and intake manifold in which wet mixture is added by passing part of exhaust gas through tubular heater. From investigations so far conducted the device appears to show more power and better efficiency than that obtained from gasoline with the ordinary carburetor. There is freedom from knocking, no contamination of the lubricating oil and no greater sooting or carbonizing. Freedom from smoke requires careful handling. Address Prof. E. F. Church, Polytechnic Institute, Brooklyn, N. Y.

*Heating B5-20* Test on Vapor-Pressure Regulating Valves for Heating Systems. A study of the fluids used in various types of vapor-pressure control valves with tests of their action in use on heating systems. Address Prof. E. F. Church, Polytechnic Institute, Brooklyn, N. Y.

*Metallurgy and Metallography B4-20* Expansion of Low-Carbon Chrome-Nickel Steel. A study of the expansion of low-carbon chrome-nickel steel during heat treatment is being carried out at the Michigan Agricultural College at East Lansing. Address Prof. H. B. Dirks.

*Metallurgy and Metallography B5-20* Expansion and Contraction of Steel by Heat Treatment. The expansion and contraction of steel by heat treatment as in the problem of the shrinkage of oversized holes and the expansion of undersized pins and disks is being carried out in the Michigan Agricultural College, East Lansing, Mich. Address Prof. H. B. Dirks.

*Properties of Engineering Materials B11-20* Effect of Temperature on Tensile Strength of Muntz Metal, Bronzes and Steel. Tests on the ultimate strength and ductility of various materials at temperatures from 20 deg. to 550 deg. cent., including effects of annealing, cold-rolling, extrusion and such actions in relation to strength at high temperatures. Address Prof. E. F. Church, Polytechnic Institute, Brooklyn, N. Y.

## D—RESEARCH EQUIPMENT

The purpose of this section of Engineering Research is to give in concise form notes regarding the equipment of laboratories for mutual information and for the purpose of informing the profession of the equipment in various laboratories so that persons desiring special investigations may know where such work may be done.

*Michigan Agricultural College D1-20* Mechanical Engineering Department.

Exhaust fan and heater stacks
Material-testing machines for tension, compression and torsion
12-in. by 30-in. Nordberg Corliss engine
8-in. by 12-in. Skinner automatic cut-off engine
75-kw. Terry turbine direct-connected to d. c. generator
12-in. by 7½-in by 14-in. Sullivan tandem compound steam-driven air compressor
6-in. by 6-in. Ingersoll-Rand Imperial-type air compressor
Condensers, pumps and injectors
8½-in. by 12-in. tandem compound Elyria gas engine
Horizontal and vertical gasoline engines
50-hp. cradled motor
Single and three-stage centrifugal pumps
16-in. Pelton water wheel
Venturi meter and weirs
Hardening and tempering furnaces, scleroscopes and Brinell hardness-test machines
Leeds and Northrup pyrometer equipment
Leeds and Northrup transformation point apparatus
Bausch and Lomb microscopes and cameras for microphotographs
Electric furnaces with pyrometers
Gas-fired crucible furnaces
One-ton cupola
Motor-driven machine tools

## E—RESEARCH PERSONAL NOTES

The purpose of this section of Engineering Research is to give notes of a personal nature regarding the personnel of various laboratories, methods of procedure for commercial work or notes regarding the conduct of various laboratories.

*Petroleum Asphalt and Wood Products E1-20* Petroleum Stills. Mr. Henry Kreisinger has just completed an investigation on the furnace conditions and heat transmission in petroleum stills for the Bureau of Mines. This work will soon be ready for publication. Bureau of Mines, Washington, D. C. Address Van R. Manning, Director.

The Bureau of Mines has received the following monthly reports of investigations for March, 1920:

Manufacture of Carbon Black from Natural Gas, by Roy O. Neal
Records of Individual Wells, by A. W. Ambrose
Recent Articles on Petroleum and Allied Substances, by E. H. Burroughs
Comparison of British and American Coal-Mining Conditions, by G. S. Rice

Decrease in Coal-Mine Accidents, by W. W. Adams
Sulphur in Coal and Coke, by Alfred R. Powell
Diatomaceous Earth, by W. C. Phalen
Marble in Guatemala, by Oliver Bowles
Asbestos, by Oliver Bowles
General Aspects of the Leasing Law, by James R. Jones
Employee Representation in Mining Enterprises, by T. T. Read
Observations with the Geophone, by Alan Leighton
Automobile Exhaust Gases in Vehicular Tunnels, by A. C. Fieldner.

Bureau of Mines, Washington, D. C. Address Van R. Manning, Director.

## F—BIBLIOGRAPHIES

The purpose of this section of Engineering Research is to inform the profession of bibliographies which have been prepared. In general this work is done at the expense of the Society. Extensive bibliographies require the approval of the Research Committee. All bibliographies are loaned for a period of one month only. Additional copies are available, however, for periods of two weeks to members of the A.S.M.E. or to others recommended by members of the A.S.M.E. These bibliographies are on file in the offices of the Society.

*Calorimetry, Pyrometry, Thermometry F1-20* Treatise on Pyrometric Practice. The Bureau of Standards has completed a comprehensive treatise on modern pyrometry. This volume will consist of 300 pages, with a large number of illustrations. It will be ready in nine months and may be obtained at cost from the Superintendent of Documents, Washington, D. C. Bureau of Standards, Washington, D. C. Address S. W. Stratton, Director.

*Mining, General F1-20* A Glossary of the Mining and Mineral Industry, by Albert H. Fay of the Bureau of Mines, 754 pp. Bureau of Mines, Washington, D. C. Address Van R. Manning, Director.

*Light F1-20* Nomenclature of Chromatics and Colorimetry. The Bureau of Standards has prepared a draft of a report on nomenclature and standards of colorimetry. It was submitted to the Optical Society of America. The draft has 50 pages and the following table of contents:

Introduction
Nomenclature, including General Terminology
Fundamental Psychologic Terms; Outline of the Methods of Practical Colorimetry; Classification of the Methods of Measurement Contributory to Colorimetry; Terms Relating to Transmission; the Physical Terms of Homo-heteroanalysis and their Correlation with the Attributes of Color Standards, including Standards of Spectral Energy Distribution; Transmission Standards
Bibliography
Usage of Terms Relating to the Attributes of Color and their Correlations in Stimulus
Usage of the word " Light."

The report may be borrowed from the Bureau of Standards, Division 4, Section 3, Washington, D. C.

*Petroleum, Asphalt and Wood Products F2-20* The Bureau of Mines has prepared a bibliography of the recent articles on petroleum and other allied substances, by E. H. Burroughs. Bureau of Mines, Washington, D. C. Address Van H. Manning, Director.

*Textile, Manufacture and Clothing F1-20* Drying of Textiles on Tentering or Stentering Machines. A bibliography of one-half page. Search 2904. Address A.S.M.E., 29 West 39th St., New York.

*Transmission F2-20* Roller Bearings. A bibliography of 10 pages. Search 2654. Address A.S.M.E., 29 West 39th St., New York.

In experiments on gravitation reported by L. Majorana (*Comptes Rendus, Acad. des Sciences*, vol. clxix, Nos. 15 and 17, 1919), a lead sphere of mass 1274 g. suspended from the right arm of a balance was counterpoised by a similar sphere on the left arm. Both spheres were surrounded by walls of the same shape. The first sphere had its center coincided with the center of a cylinder with vertical axis which could be filled with mercury of mass 104 kg. The liquid was kept from touching even the protecting wall about the lead sphere. A vacuum was maintained within the balance case and all thermal, electrical, magnetic, and electromagnetic effects were eliminated. As the result of a lengthy series of experiments, it was found that the lead suffered a loss of weight of 0.0009 mg. when it was surrounded by the mercury, that is, a loss of $7.10^{-10}$ of its own weight. This is of interest, since hitherto it has been believed that the force of gravity acts through matter just as if it were a vacuum.

# Work of the Boiler Code Committee

*THE Boiler Code Committee meets monthly for the purpose of considering communications relative to the Boiler Code. Any one desiring information as to the application of the Code is requested to communicate with the Secretary of the Committee, Mr. C. W. Obert, 29 West 39th St., New York, N. Y.*

The procedure of the Committee in handling the cases is as follows: All inquiries must be in written form before they are accepted for consideration. Copies are sent by the Secretary of the Committee to all of the members of the Committee. The interpretation, in the form of a reply, is then prepared by the Committee and passed upon at a regular meeting of the Committee. This interpretation is later submitted to the Council of the Society for approval, after which it is issued to the inquirer and simultaneously published in MECHANICAL ENGINEERING, in order that any one interested may readily secure the latest information concerning the interpretation.

Below are given the interpretations of the Committee in Cases Nos. 253, 268, 270, 279, 283-290, inclusive, as formulated at the meeting of March 18, 1920, and approved by the Council. In accordance with the Committee's previous practice, the names of inquirers have been omitted.

### CASE NO. 279

*Inquiry:* Is it permissible under the requirements of the Boiler Code to construct steel heating boilers in diameters up to 72 and 78 in. with shell plates ¼ in. in thickness? Is it the opinion of the Committee that boilers of the h.r.t. type so constructed have sufficient shell plate strength to permit of proper lug attachments for support of the shell?

*Reply:* The section of the Code on Heating Boilers is incomplete in not specifying the minimum plate thickness to be used in the shells and tube sheets, and below which the thickness shall not be made in any case. The Boiler Code Committee recommends that until a revision be made to cover this class of boiler construction, the minimum thickness of shells and heads for various shell diameters of steel plate heating boilers shall be as follows:

| Diameter of Shell, Tube Sheet or Head | Minimum Thickness Allowable Under Rules | |
|---|---|---|
| | Shell, in. | Tube Sheet or Head, in. |
| 42 in. or under | ¼ | 5/16 |
| Over 42 in. to 60 in. | 5/16 | ⅜ |
| Over 60 in. to 78 in. | ⅜ | 7/16 |
| Over 78 in. | 7/16 | ½ |

### CASE NO. 283

(In the hands of the Committee)

### CASE NO. 284

*Inquiry:* Is it the intent of Pars. 269 and 270 of the Boiler Code that a boiler requiring safety valve relieving capacity greater than 2000 lb. per hour shall have two safety valves, each one of which alone will properly discharge the steam, or does it require two safety valves, the combined capacity of which will properly discharge the steam?

*Reply:* For capacities greater than 2000 lb. per hour, the Code requires two or more safety valves on each boiler, the combined capacity of which shall meet the requirements of Par. 270.

### CASE NO. 285

*Inquiry:* Is it permissible under the rules of the Boiler Code to use standard extra-heavy cast-iron flanges and fittings on pipe connections between boilers and attached-type superheaters, and on the ends of superheater-inlet headers for pressures up to 250 lb. per sq. in.? It is pointed out that neither the inlet pipe connections nor the superheater inlet flanges would be subjected to other than saturated steam temperatures.

*Reply:* It is the opinion of the Committee that the flanges referred to may be made of cast iron, provided the temperature of the steam does not exceed 450 deg. fahr., as specified in Par. 12.

### CASE NO. 286

*Inquiry:* Does Par. 321 of the Boiler Code require that all water column connections to boilers must be fitted with crosses for cleanout purposes, or is it permissible in a connection of a water column to a steam drum as shown in Fig. 5 to omit the crosses, provided provision for cleanout is afforded by the gage glass connection on the opposite side of the water column?

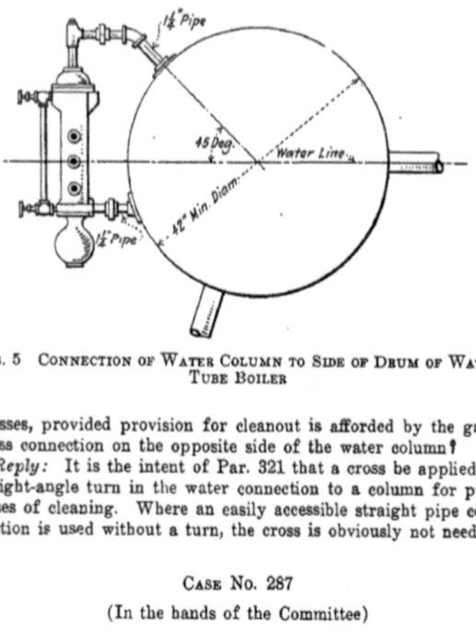

FIG. 5 CONNECTION OF WATER COLUMN TO SIDE OF DRUM OF WATER TUBE BOILER

*Reply:* It is the intent of Par. 321 that a cross be applied at a right-angle turn in the water connection to a column for purposes of cleaning. Where an easily accessible straight pipe connection is used without a turn, the cross is obviously not needed.

### CASE NO. 287

(In the hands of the Committee)

### CASE NO. 288

(In the hands of the Committee)

### CASE NO. 289 (Revision of Case No. 254)

*Inquiry:* Does the reply to Case No. 254 apply only to flat plates or also to the holes in the shells of drums?

*Reply:* This Case applies to flat sheets. It is not permissible to burn tube or other holes into the shells of drums unless such holes are strengthened with flange or other reinforcements.

### CASE NO. 290

*Inquiry:* In determining the ratios of lengths between supports to diameter in Table 5, is the diameter at the root of the thread to be used, or the diameter of the body of the stay.

*Reply:* It is the opinion of the Committee that the diameter of the body of the stay is to be used in determining this ratio.

# CORRESPONDENCE

CONTRIBUTIONS to the Correspondence Departments of MECHANICAL ENGINEERING by members of The American Society of Mechanical Engineers are solicited by the Publication Committee. Contributions particularly welcomed are suggestions on Society Affairs, discussions of papers published in this journal, or brief articles of current interest to mechanical engineers.

## Shall Junior Members Vote?

To THE EDITOR:

I have read with a great deal of interest the results of the Philadelphia Section and Mr. M. L. Cooke's plea that junior members of the Society should have the right to vote. There is just now a comparatively large class of young engineers growing up—men upon whom we are going to have to place responsibilities because of the shortage of executives—and it is interesting to note how these young men are bearing their burdens.

During the war, when it was necessary to examine the ages of men very closely, I found to my surprise that in the automotive industry if every man had been physically and otherwise able to be drafted between the ages of twenty-one and thirty-one, the automotive industry would have lost almost all of its engineers.

Now the automotive industry has certainly had a very progressive engineering personnel. The work is complicated, the mechanisms are delicate and, on the average, the engineering work involved is of a very high order and, as stated above, this has been done almost entirely by comparatively young men. There are many instances that come to my mind wherein men under thirty-one are occupied in positions of high responsibility in large firms, doing very capable and successful engineering work. During the war, we must not forget, the airplane was largely developed by young men in the draft age, as were the tank and other automotive apparatus.

I feel it desirable to recognize the fact that age should not be one of the primary requisites in deciding whether a man should be a member or not, but the man's own ability, the nature of the work he is doing, and the actual experience he has put in should be the decisive factors. There are many men of the age of twenty-five who have much greater engineering instincts and greater capacity for engineering work than men who are admitted as members simply because they are thirty-two years of age and have as well, of course, other qualifications. I, therefore, look with much interest toward any method which will allow comparatively young men who are capable of assuming responsibility, to do so.

JOHN YOUNGER.

Cleveland, Ohio.

## How Shall Engineers Organize?

To THE EDITOR:

I wonder to what extent the current activity for the organization of a united engineers' organization is spontaneous with the membership, and whether the rank and file are really interested in it and are giving it the thought and attention that such a fundamental movement deserves.

Does the average member realize and believe in the possibilities of a general organization for increasing the opportunities for service and the recognition and emolument of the engineer?

In the general opinion, should such an organization consist:

(a) Of one big association of engineers of all kinds, the local branches of which would impress engineering thought and experience upon local administrations and activities, hold their members together by frequent meetings of general engineering interest, bring the engineer into prominence as a factor in the commu-nity and improve his social and material condition; the collective thought of such locals finding its expression in a national convention and its execution through national officers and commit-tees? Or

(b) A federation of existing engineers' societies, local, regional and national, through local committees or councils composed of representatives of such societies as are invited to join, and a federal council composed of representatives from such local councils and from regional and national societies? Or

(c) Some compromise between or combination of such plans?

If a new organization, is it to be confined to professional engineers or open to all who have a legitimate interest in and connection with engineering "from," as it has been expressed, "the blueprint boy up"?

If a federation, is it to include only the professional societies or may it include those whose activities are concerned with the operative and fellowship sides of engineering and, through technical methods, with the direct betterment of the material conditions of their members?

I should like to see some expression from the membership in advance of the meeting which is to determine the form and scope which the proposed organization is to have.

F. R. LOW.

New York, N. Y.

## Professional Sections

To THE EDITOR:

The new plan of The American Society of Mechanical Engineers for the establishment of Professional Sections meets with my hearty approval, and I am glad to learn that the idea the nucleus of which I was responsible for some years ago as a member, and later as the chairman of the Committee on Meetings, is working out in this way. It seems to be the logical sequence to our old mechanism of the sub-committees of the Committee on Meetings that sectional meetings on special topics should be arranged for.

The Society can never attain a degree of usefulness commensurate with the interests represented by its members by conducting single general meetings to be attended by everybody or nobody. This is a natural and inevitable result of the diversity of machinery with which we are concerned, and which is responsible for a corresponding difference in the nature of the problems which engineers have to solve. New machinery of each class is produced in proportion as those engineers concerned with it establish a community of interest in the problem, and the Society owes a debt to the machinery industry to provide a medium of exchange whereby these groups of engineers may have an opportunity to exchange ideas. For example, a group of engineers concerning themselves with the development of the motorship industry can derive but little help by attending a meeting where the subject under discussion is stationary steam plants, nor is there any help to be derived by engineers concerned with chemical factory machinery in attending a meeting where the subject under discussion is motor cars.

For these reasons professional sections must be organized for the discussion of the problems typical of the classes of machinery represented by the various groups of engineers who will attend the meetings of such sections. Any one group of engineers held together by a bond of common interest really consists of two groups: those concerned with the production of the machinery, and those concerned with its proper and economic use. Together they may, by discussion, come to a real and proper appreciation of the problem; and by exchanging data materially assist in the solution of that problem. This does not mean that there will be any interference with legitimate commercial competition, but it does mean that the business represented by all will be logically and properly advanced not only for the common welfare of the machinery interests, but of the public, which is entitled to regular improvements as a mark of progress.

Unless professional sections of this kind are organized and properly conducted, the inevitable result must be the organization of separate societies; perhaps such separate societies will be organized anyway. In some cases this has happened, notably in the case of the Refrigerating Engineers and the Automotive Engineers, to cite only two examples. Other subdivisions will form unless provision is made for the development of their interests within our Society. While in special cases the interests may be so very large as to make it quite impossible properly to care for them by professional sections in our Society, I am firmly of the belief that professional sections should be maintained anyway. My reasons for this are that there are certain fundamentals underlying the design, production, and in some cases, also, the economic use of all machinery which are apt to be lost 'ight of in a special society; and also that there are many thing reduced in the development of one line of machinery or m chanical appliances which could be adopted with minor or no modifications by quite another group. If, therefore, these groups are kept quite separate, as would be the case if they met in separate societies, as the Automotive and Chemical Societies are separate from those of the Lawyers' Society, professional progress would not be as rapid as if they were to come in contact at the same time and place, where there would always be a harking back to the questions of a community of interest and basic scientific fundamentals. This means that the professional sections in the A.S.M.E. could accomplish something which separate societies could not accomplish. These professional sections will not prevent the organization of separate societies when the interests involved become large enough to justify such action, but it will prevent their premature or unwise organization. Furthermore, the maintenance of professional sections dealing with the same topics as separate societies may become, through proper coöperation, the centralizing means of establishing communities of interest among all machinery interests and the professional interests representing them.

Some of the existing separate societies which have been organized may well take this matter under serious consideration, because it is believed that some of them will do well to disband and become professional sections of the A.S.M.E. The typical cases in which this is believed to be true are those in which the separate societies are finding difficulty in securing either proper financial support or have a small and not uniformly high grade of membership, professionally speaking, and have difficulty in securing proper papers of high professional character. The interests represented by these groups can be far better served by professional sections than by struggling independently.

On the other hand, those separate organizations, such, for example, as the Society of Automotive Engineers and the Marine Engineers and Naval Architects, which have reached a high degree of society prosperity in finances, membership and number and quality of papers, enthusiasm of meetings and such matters, should welcome, and in my opinion would welcome, a parallel professional section in the A.S.M.E., and would be willing to support it as a means of coördinating related interests and fundamental professional matters, the professional section acting as a centralizing link.

As you know, I am a member of several organizations and have been a student of society affairs for quite a long time, and I am quite sure in my own mind that the professional-section idea cannot be overrated in importance, although it can of course be misused. With the Council supplying broad-minded, helpful advice and actively concerned with the problem of promoting the development of American machinery along the most effective professional lines, the professional-section idea may well be expected to produce results never yet attained, and react to the very great credit of our Society as a national institution.

New York, N. Y.

CHARLES E. LUCKE.

## Thermal Conductivity of Insulating Materials

To THE EDITOR:

In his paper presented at the Annual Meeting and published in MECHANICAL ENGINEERING for January on page 8, Mr. T. S. Taylor has given considerable valuable information on the thermal conductivities of electrical insulating materials. Only a few heat-insulating materials are considered, however, and none of the more efficient ones.

In the case of electrical insulating materials high thermal conductivity is a desirable quality as it facilitates the dissipation of heat from electrical conductors, while obviously in the case of heat-insulating materials it is low conductivity that is desirable. Furthermore, the results are expressed in units convenient to the electrical but not to the mechanical engineer. Therefore it would seem that the paper would have been more likely to receive attention of the engineers most interested had the title been more explicit.

The units in which the results are expressed are not those ordinarily used in this country in connection with heat transmission. In order to convert gram-calories per square centimeter per centimeter thick per deg. cent. per second into B.t.u. per sq. ft. per inch thick per deg. fahr. per hour, the values must be multiplied by 2903. The values in column 5 of Table 2 are evidently gram-calories per square centimeter per centimeter thick per deg. cent. per second $\times 10^6$ and not $\times 10^{-5}$ as stated. Therefore the equivalent conductivities in B.t.u. per sq. ft. per inch thick per deg. fahr. per hour are obtained by multiplying values in column 5 by 0.002903.

Substitution of temperature coefficients in the equation given by Mr. Taylor would give decreasing values of conductivity as temperature is increased. Conductivities of poor conductors of heat generally increase as temperatures are increased, therefore apparently the equation should read $k_t = k_o (1 + at)$.

L. B. McMILLAN.

New York, N. Y.

## Simplification of Venturi-Meter Calculations

To THE EDITOR:

A paper on the Simplification of Venturi Meter Calculations, published in the April 1920 issue of MECHANICAL ENGINEERING, describes a method which its author has devised for the simplification of the calculations involved in the venturi-meter measurements of the flow of compressed air, and which reduces the work of one determination to a simple slide-rule computation requiring but a few settings. His method is based on the venturi-meter formula as given by Herbert B. Reynolds, Mem. Am. Soc. M. E. in Trans. Am. Soc. M. E., vol. 38, p. 799. In using this or any other modification of previously devised formulæ it is wise to check at least one point in the problem by a true formula, as some of the less complicated formulæ are limited in their application.

In 1898 F. G. Gasche, of the Illinois Steel Company, developed the thermodynamic formula for measuring gas and steam through the venturi meter. Nine years later E. P. Coleman, Mem. Am. Soc. M. E., in a paper entitled The Flow of Fluids in a Venturi Tube,[1] gave a formula equivalent to the author's Equations 1 and 2. The method which the author has independently worked out of handling what might be termed the most disagreeable factor of the equation is the same method which has been employed by several others. For instance, Mr. Gasche developed an elaborate series of curves and diagrams for calculating gas and steam flows

[1] Trans. Am. Soc. M. E., vol. 28, p. 483.

through venturi meters and among these were curves involving the ratios of the absolute pressures between throat and inlet, also specific heats, corresponding to the argument employed by the author in the solution of Equation 5. In 1910 Beebe and Herrick in experiments on the measurement of air through a venturi meter at the University of Wisconsin also reduced this factor to a curve.

It is probably the author's intention to confine his various conclusions to the results obtained with tests on a single small venturi meter; for a more universal application, the following exceptions should be noted:

a There is no difficulty commercially in making manometers to withstand working pressures exceeding 100 lb., those on the market being adapted for pressures up to 250 lb. or even higher.

b A friction loss of 2.5 lb. through a 1.6-in. by 0.5-in. venturimeter tube under 100 lb. gage pressure and discharging air at the rate of 200 cu. ft. per min. is larger than would be obtained by a venturi tube of standard proportions, which would show about 1.5 lb. at this rate. It would therefore seem that the venturi tube employed in the test was of some special design, that the downstream piezometer was placed too near the outlet end, or that the loss in pressure was incorrectly indicated. Tests conducted by the Builders Iron Foundry of Providence, R. I., in 1911 on a 2-in. by 1-in. venturi-meter tube measuring air showed the friction loss to be about 1/5 of the difference in pressure between the inlet and throat. Beebe and Herrick in their tests on a 2-in. by ½-in. venturi tube also show the friction loss to be a straight-line function of the differential between inlet and throat and give the value as $F = H/5.467$, where $F$ is the friction loss and $H$ is the differential pressure. In a series of comparative tests made by W. F. Schell in July, 1914[1] between various types of meters measuring natural gas including 4-in., 10-in. and 12-in. venturi meters, it was found that "the ratio of the differential pressure to the pressure loss through the venturi meter is approximately eight to one." Messrs. Bacon and Moulson, in a comprehensive series of calibrations at the Illinois Steel Co. on venturi meters 24 in., 36 in. and 48 in. in diameter, found the drop in line pressure caused by the insertion of the meter tube to be $F = H/9$, the outlet pressure being observed about one and one-half diameters beyond the outlet end of the meter tube. Tabulating these results we have

| Size of Meter Tube Inlet and outlet diameter, in. | Friction Loss Equals differential between inlet and throat divided by |
|---|---|
| 2 | 5 |
| 4 to 12 | 8 |
| 24 to 48 | 9 |

This shows conclusively that the author's determination of friction loss was not representative.

| $\frac{A_2}{A_1}$ | $n$ | $P_1$ lb. per sq. in. | $P_2$ lb. per sq. in. | $\frac{P_2}{P_1}$ | $f\left(\frac{P_2}{P_1}\right)$ |
|---|---|---|---|---|---|
| 0.2500 | 1.402 | 114.7 | 104.7 | 0.9128 | 0.1548 |
| 0.1111 | 1.402 | 114.7 | 104.7 | 0.9128 | 0.1513 |
| 0.0625 | 1.402 | 114.7 | 104.7 | 0.9128 | 0.1507 |
| 0.2500 | 1.270 | 114.7 | 104.7 | 0.9128 | 0.1396 |
| 0.2500 | 1.402 | 89.7 | 79.7 | 0.8885 | 0.1721 |
| 0.2500 | 1.402 | 64.7 | 54.7 | 0.8484 | 0.1970 |

c For experimental work the reduction of the author's tabular arrangement of Equation 5 to curve form, using a coarse scale, would be valuable. A wider application to venturi meters in general would involve a large family of curves due to the variations which occur in the values $A_2/A_1$, $n$, and $P_2/P_1$. For instance, for the measurement of gas each size of standard venturi meter tube has four throats, the diameters of which are approximately 3/12, 4/12, 5/12 and 6/12 of the inlet diameter. For $n$ there would be a variation from 1.402 for air to 1.135 for dry saturated steam. $P_2/P_1$ might go much below the limit of 0.90

[1] Proc. Natural Gas Assn of America, vol. 8

mentioned by the author; for instance this value becomes 0.8454, assuming 50 lb. gage pressure at the inlet of a venturi meter tube connected to a mercury manometer indicating a differential of 10 lb. The wide range in the value of $f(P_2/P_1)$ for such changes in value of the terms under the radical sign is illustrated by the following table. Since this function enters as a factor in the final equation any variation in its value involves an equal variation in the calculated weight or quantity discharged through the venturi meter.

Providence, R. I.          CHAS. G. RICHARDSON.

## Industrial Training in China

TO THE EDITOR:

The writer, a member of the A.S.M.E., has been a resident of China for the past ten years and as a result of his observations is impressed with the opportunity which exists for the enlightenment of the Chinese in industrial and educational matters. He believes that a great service could be rendered in this direction by the Society or some of its members, which would benefit not only China but America as well.

At present, the Chinese have very little knowledge of what can be accomplished by the use of the best engineering processes and with modern equipment of tools and machinery. In the case of machine tools, which are at the foundation of manufacturing, a large number of those in common use are poor oriental copies of ancient European types. The Chinese believe these to be as good as any, or the best to be had, and that the work accomplished on such machines is the best that can be done. They are broadminded and willing to learn, and can be taught the value of up-to-date methods and machinery only by actual demonstration that new methods will pay them better than those in present use. They have had no opportunity to learn that such a benefit will result from the use of modern equipment.

China also needs to be shown what good engineering schools really are and how they should be managed. In the present condition of the country it is almost inevitable that such schools should be undervalued by those in governmental power and made more a matter of political jobbery than a factor in the upbuilding of the country. Concrete examples are needed of good American technical and trade schools, controlled by American capital and equipment; for technical education has not yet come to be so highly appreciated by those in political control in China as to gain the financial support necessary for the large outlays for equipment and maintenance which such schools require.

A suggestion for help in this direction might be taken from the activities in China of such institutions as Yale, Princeton, Harvard and Oberlin. These have selected certain lines of helpful work out here and are carrying them on successfully in certain large and definitely determined sections of the country. Why should not some large, representative society or societies, like the A.S.M.E., undertake to help in this technical education?

A step in this direction has been taken by British engineering societies and firms through the establishment of Hongkong University, which is well equipped with machinery and tools contributed by numerous far-seeing manufacturers. The Germans equipped a technical school out here before the war. The French are now making a very successful effort to reach the engineering student class of young men here. There is a very energetic and well-organized society in France promoting such efforts. Special low rates of passage are offered and great friendliness shown to graduates of Chinese trade schools of what would correspond to the American high-school grade. Living wages are paid these men who, in most cases, would not otherwise be in a position to take advantage of such an opportunity.

An effort to establish an American trade and technical school is being made by Peking University, an American missionary institution controlled by broad-minded men with practical and far-sighted vision. They realize that one of the greatest needs, perhaps the greatest, for the upbuilding of China today, morally

(Continued on page 318)

# MECHANICAL ENGINEERING
### THE JOURNAL OF THE AMERICAN SOCIETY
### OF MECHANICAL ENGINEERS

Published Monthly by the Society at
29 West Thirty-ninth Street, New York

FRED J. MILLER, *President*

WILLIAM H. WILEY, *Treasurer*   CALVIN W. RICE, *Secretary*

PUBLICATION COMMITTEE:

GEORGE A. ORROK, *Chairman*       J. W. ROE
H. H. ESSELSTYN     GEORGE J. FORAN
RALPH E. FLANDERS

PUBLICATION STAFF:

LESTER G. FRENCH, *Editor and Manager*
FREDERICK LASK, *Advertising Manager*
WALTER B. SNOW, *Circulation Manager*
136 Federal St., Boston

Yearly subscription $4.00, single copies 40 cents. Postage
to Canada, 50 cents additional; to foreign countries $1.00 addi-
tional.

*Contributions of interest to the profession are solicited. Com-
munications should be addressed to the Editor.*

## Meeting in Memory of Andrew Carnegie

A meeting in memory of the life and work of Andrew Car-
negie was held on April 25 in the United Engineering Societies
Building under the joint auspices of the Authors' Club, the
Trustees of the New York Public Library, the Oratorio Society,
St. Andrew's Society and the United Engineering Society. The
program included music by the Oratorio Society and addresses
by Dr. John H. Finley and the Hon. Elihu Root. Dr.
William P. Merrill of the Brick Presbyterian Church gave the
invocation and tributes were read from Hon. William H. Taft,
Sir Oliver J. Lodge, Viscount Bryce and Lord Morley.

## John Fritz Medal

At a meeting to be held in the Engineering Societies Building at 8.30 p. m., May 7, the John Fritz Medal for 1920 will be
presented to Mr. Orville Wright, Honorary Member of The
American Society of Mechanical Engineers. The speakers of
the evening are to be General O. Squier, Chief Signal Officer,
U. S. A., and Colonel E. A. Deeds, of Dayton, Ohio, who was
prominent in the Aircraft Production Board during the war.
Mr. B. B. Thayer, present chairman of the Board of Awards,
will preside and Prof. Comfort A. Adams, chairman of the Engineering Division of the National Research Council, will present
the award. All members of the Society are cordially invited to
attend.

## American Railway Engineering Association
## Joins Engineering Council

The American Railway Engineering Association, whose excellent technical work in many branches of railway construction
and maintenance is well known, has accepted an invitation from
United Engineering Society to become a member society of Engineering Council. The Association has about 1,650 members
and its headquarters are at 431 South Dearborn Street, Chicago,
Illinois. Its President is Mr. Harry R. Safford, and its Secretary Mr. E. H. Fritch.

The Association has named as its representative upon Engineering Council its President, Mr. Safford, who is a member of
the American Society of Civil Engineers and Engineering Institute of Canada. Mr. Safford was recently appointed Assistant
to President Hale Holden, of the Chicago, Burlington and
Quincy Railroad Company, the Colorado and Southern Railway
Company, the Fort Worth and Denver City Railway Company,
and the Wichita Valley Railway Company. He was formerly
Chief Engineer of the Grand Trunk Railway, and is well known
in Canada, as well as in the United States.

Six engineering organizations with a combined membership
of 45,000 are now represented in Engineering Council. These
are: American Society of Civil Engineers, The American Society of Mechanical Engineers, American Institute of Mining and
Metallurgical Engineers, American Institute of Electrical Engineers, American Society for Testing Materials, and the American Railway Engineering Association.

## National Chamber of Commerce to Vote on
## Department of Public Works

The question of whether or not the Federal Government should
establish a Department of Public Works has recently been submitted for a referendum vote to the 1,305 member organizations
of the United States Chamber of Commerce. The result of this
referendum will determine the National Chamber's attitude on
this question when it comes up for Congressional action.

The questions submitted by the committee appointed by the
National Chamber to make a study of this matter are as follows:

1. Shall there be established by the National Government a
Department of Public Works?

2. Shall such department be established by suitable modification of the existing Department of the Interior, excluding therefrom the non-related bureaus or offices, and by the change of
name of the Department of the Interior to the Department of
Public Works?

3. Shall such department be created by the creation of an
entirely new department?

Engineering Council, which is one of the member organizations
of the National Chamber, has gone on record as favoring the
adoption of the Jones-Reavis Bill (S-2232-H.R.649) now pending in Congress. This bill provides that the name of the executive department, at present designated as the Department of the
Interior, be changed to the name of the Department of Public
Works, and that the head thereof shall continue to be a member
of the cabinet under the official designation of Secretary of Public Works. The details of this bill, which would transfer much
of the work now transacted under the Department of the Interior to other departments and would also place additional bureaus under the new department, have been discussed at length
in past issues of MECHANICAL ENGINEERING, notably in August,
1919, issue, p. 709, and the February, 1920, issue, p. 136.

## Putting the Technic in Technical Education

The Harvard Engineering School has adopted a new plan of
instruction for the junior year of the engineering course whereby
students will be given an opportunity to combine classroom work
with six months of active engineering practice and industrial
training.

The courses for the freshman, sophomore and senior years will
remain approximately as they are now, but during the junior
year the students who desire the industrial training will spend
three separate periods of two months each in the outside work.
The schedule will be so arranged that these students will also receive the regular amount of classroom instruction.

In general, this plan lies along the lines of the highly-developed
and successful plan of industrial coöperation which was initiated
by Dean Schneider at the University of Cincinnati, where it has
been carried on so successfully for many years, and which also

has been applied in a modified form at the University of Pittsburgh. This plan has been modified still further to meet the different conditions and needs at Harvard, where the technical work of putting it into operation will be in charge of Mr. H. O. Drufner of the University of Cincinnati.

The plan has received the support of the Associated Industries of Massachusetts, comprising some 1400 industrial and engineering concerns, and these industries have indicated their willingness to take on as many students as the Engineering School desires to place. It is expected that the men who go into this work will be paid current wages for the periods in which they work in the plants, and that they will be able to earn sufficient money to pay their expenses during these periods, so that the experience, if not actually profitable, at least will not be a financial burden.

Professor Hector J. Hughes, chairman of the Administrative Board of the Engineering School, explains the purpose of the new plan as follows: "One of the first problems which the staff of the new Engineering School set itself to solve was to find an effective way of getting the new School and its students into closer relations with industrial and engineering work before they graduate. The need for such relations has been increasingly evident in the past few years. The object of such coördination is manifold: to stimulate interest in the classroom work; to keep the teaching staff well-informed of the needs of industry and how to train engineers to meet them; to give the students some intimate knowledge of the great problems of labor and industry which they have to meet after they graduate, and thus to anticipate to some extent the period of initiation which all students must go through and better to fit them to begin their careers; to give them an opportunity to discover how intricate and interesting the basic industries are and to what extent scientific knowledge may be used in work which is too frequently looked upon as non-technical; in other words, to find out how many kinds of careers are open to technically trained men and how wide is the opportunity for such men. Another object of the new plan is to stimulate the interest of the industries themselves in the adaptation to their special needs of education in engineering. To sum up, the object of such coördination is to give our students the chance to find themselves."

## The Outdoor Power Plant

It has been the accepted practice in engineering design that a hydroelectric power plant must be sheltered by a house. The question, however, whether there is really any need for the expensive buildings that have always been erected for housing generating equipment is of increasing importance in view of the present economic situation. This question of the practicability of the outdoor generating station is pertinently discussed by H. W. Buck in the March issue of the *General Electric Review*.

The modern, vertical-shaft, internal revolving-field generator, he argues, is essentially a waterproof structure. The vital parts of the machine are all on the inside protected by a massive casing of cast iron, and the openings in the upper spider are usually plated with steel for purposes of ventilation so that the top is protected. With slight modifications in its design, the standard vertical-shaft alternator can therefore be made proof against all stresses of weather. The waterwheels themselves naturally need no housing.

Thrust bearings are housed in heavy steel or iron casings and need little protection from weather. The various auxiliaries, including governors, connected with a hydroelectric plant can easily and advantageously be installed under the main generator floor in the various compartments naturally existing in the substructure of such a plant and there protected from the weather.

It thus appears pertinent to inquire why millions of dollars are expended to house machinery which does not need housing at all.

As an example of a scheme eliminating the expensive superstructure, Mr. Buck described the general design of the Muscle Shoals Development prepared for the War Department by Viele, Blackwell & Buck, Engineers, in August, 1918. This design comprised the outdoor installation of fifteen 20,000-kw. generating units. It was proposed to install within the generating structure itself only the water turbines, generators, excitors, governors and the various pumps. The usual iron frame of the stationary armature was to be omitted and the armature laminations and coils were to be attached to the surrounding concrete. The switchboard panels were to be located in a relatively small pilot house which would also contain the low-voltage oil switches and busbars. All of the transformers and high-tension equipment would be installed in the open, adjacent to the pilot house. In this arrangement the switchboard operators would be housed and the generator attendants also protected in the substructures. The plant auxiliaries would also be housed. All of the vital parts of the alternators could be inspected from below the machines. Only the occasional inspection of thrust bearings would have to be made in the open. In order to perform under protection from the weather the work of installation and repair on the generating units, it was planned to install a housed-in travelling gantry crane of sufficient length to cover a single unit, which would travel on rails the total length of the generator structure.

In the case of this Muscle Shoals plan the estimated saving in construction cost by omitting the superstructure was very large, amounting to over $700,000 after paying for the pilot house and gantry crane.

It is true that this Muscle Shoals scheme was designed to meet the weather conditions of Alabama, where the climate is mild, but there is no reason why it should not be possible to make an outdoor generator snow-proof as well as rain-proof. The operator would be normally housed in any case in the pilot house or between decks in the substructure, and would be required to go on deck at occasional intervals only.

## A New Development in Marine Propulsion

Five years ago the writer of this had the pleasure of listening to a paper on the Application of Electricity to Marine Transportation by William P. Donnelly, Mem. Am. Soc. M. E., presented before the Brooklyn Engineers' Club, dealing with a plan which the author had developed for the generation of electricity upon a vessel designated as the "power boat" and transmitting it by means of a flexible, waterproof cable to one or more other vessels and thereby utilizing the energy for propulsion through propelling apparatus operated by electric motor.

Mr. Donnelly at that time had designed and constructed an electrically-propelled yacht, *Dawn*, 50 ft. long and 12 ft. beam, with a 60-hp. gasoline engine, a 40-kw. electric generator and a 20-hp. motor for driving the propeller, which allowed a surplus of from 20 to 30 hp. beyond that required for driving the boat itself, to be used for the propulsion of a second boat. The yacht was designed primarily for experimental purposes, one of the features being the elaborate instrument board with fifteen dials for ammeters, voltmeters, tachometers, gages, etc.

The author pointed out that through the use of a power boat in connection with cargo vessels at sea, canal barges or pleasure boats, there would be a material saving in cargo or passenger space and expense of equipment. In the case of passenger boats it would be possible to reduce materially the fire risks, the discomfort from excessive heat in warm weather through the elimination of the steam boilers, and annoyance from excessive vibration due to the elimination of reciprocating engines. Further, it would extend the general use of electricity for refrigeration, heating, cooking, etc., adding to the convenience and cleanliness of the boats.

In an article in *Marine Engineering* for April, 1920, Mr. Donnelly publishes the sequel to his earlier paper in which he describes a second yacht, the *New Era*, completed last year and designed for pleasure purposes. The lines laid down were the same as for the *Dawn*, but with the power plant eliminated. A 20-hp. electric motor for propulsion was installed in the same manner as on the *Dawn*, below the floor of the forward cabin, thus leaving the entire cabin space free of machinery. Incidentally the yacht was made non-sinkable, following out the scheme developed by Mr. Donnelly and tried out during the war on one of the transports. This was accomplished by placing the floor of the cabin 6 in. above the water line and filling in the vacant space below with air-tight copper tanks, and with balsa wood in the irregular parts of the hull. The electric current was received from the generating plant of the *Dawn* through a flexible cable, as described above, which ran from a reel on the bow of the *Dawn*, which was to serve as the power boat, to the stern of the *New Era*, which was to serve as the pleasure craft—in effect a floating home.

Last November a trial trip was made from New York to Schenectady, by way of the Hudson and the New York Barge Canal, the *New Era* leading and the *Dawn*, or power boat, following. The trip was successful, and demonstrated the possibilities of this method of propulsion either for pleasure or for commercial uses. It showed the great degree of comfort which could be had in touring with electric power in practically unlimited quantity so far as immediate necessities were concerned, and with the relegation to the power boat of the power plant, crew, fuel, stores and the many things which must go with one in cruising.

The broader application of electricity to marine transportation yet to be put into actual use is obvious. In place of the small yacht described we can conceive of a large power boat convoying up the Hudson one or more palatial passenger craft devoid of all the objectionable features of such boats; or similarly supplying the power for taking up freight vessels or barges. These freight carriers could be left at the dock at the terminal point and the same power boat used if necessary to bring back other vessels while the first were being unloaded. In his recent article the author points out the possibility, also, of using electrically propelled car floats conducted by a power boat floating between them and shows a schematic illustration of three electrically propelled barges moving one behind the other through a canal, and followed by a fourth barge containing the power-generating plant.

## The Measurement of Minute Motion

A little instrument recently described by an Indian scientist well-known in scientific circles in this country, Sir Jagadis Chunder Bose, and called by him the "crescograph" promises to render a service to science in some ways similar to that which has been rendered by the microscope, and particularly the ultra-microscope developed by Siedentopf and Ziegsmondy.

The instrument, full details of which have not yet reached this country, appears to comprise means sensitive enough to be influenced by movements as minute as, for example, the growth of a leaf of a plant. It magnifies these movements millions of times and projects them on a screen so that the growth of an ordinary plant can be made actually visible.

The inventor of this instrument developed it in the course of many years of investigations on the responsiveness of plants to external influences, and the crescograph has already given very important and but little suspected information on plant life.

It is obvious, however, that its usefulness is by no means limited to the particular object for which it has been developed, and one can easily see many useful fields of its application in engineering. The following few topics are mentioned simply by way of illustrating some of its possibilities and calling the attention of engineers to what may become a new and important tool of modern research.

Physics tells us that metals expand with increase of temperature and within certain limits it would appear that such expansion is proportional to increase of temperature. Whether, however, this is really so for very small ranges of temperature is by no means established and yet would be well worth knowing. The crescograph would easily show variations in the length of even a short metal bar with changes in temperature as low as 1 deg. fahr.

Another problem the crescograph could solve is what law is followed in the expansion of bodies of irregular section; for example, elliptical or streamline. The next subject on which we

have practically no information at all is whether wood,—in particular, laminated wood,—is affected by changes in temperature and humidity of the ambient medium and to what extent.

We all know that a casting changes in shape within some period after it has been taken out of the mold,—a fact which often leads to important distortions and always has to be taken into consideration. We are entirely ignorant, however, as to the chronology of this change in shape and dimensions and, in fact, do not know when it begins and how it proceeds and how long it takes for it to run out with castings of various shapes. It may be hoped that a crescograph or an instrument properly developed for this purpose may one day enable us to take, let us say, a casting for a large gas engine, put on it the sensitive feelers of the instrument and within half an hour or so determine by looking on a screen in a dark room whether the casting has definitely set or is still "alive."

Numerous tests have shown that a bar of iron or steel in a magnetic field undergoes certain modifications of its dimensions, depending on the character of the metal, its location with respect to the field, strength of the field and certain other factors. In fact, attempts have been made to use this method for practical purposes in "setting" the metal.

The most interesting phenomenon, however, is the behavior of a bar of magnetic material, such as iron or steel, in an alternating magnetic field and it would have been extremely interesting to determine just how and to what extent the variations in the size and shape of the bar respond to the variations in the density of the field. There are numerous highly interesting factors here of which we are absolutely ignorant, such as the relation between the hysteresis in the magnetization of the field surrounding the bar and the hysteresis in the movements of the bar itself, factors which very materially affect practical applications of such methods.

With today's methods we are incapable of knowing what happens to the bar on every cycle of the alternations in the magnetic field and can guess at it only from the final results,—a method crude and unfruitful. With a crescograph we might be able to see the metal contract and expand on every cycle and as we can see by means of other instruments already in existence the variations in the density of the magnetic field, we should have no trouble in establishing the laws governing this highly interesting phenomenon.

It would appear, therefore, that here, as in other cases, the instrument developed for one purpose might serve to help us out in many ways for which it was never intended, and with the spirit of research abroad in the land as it is today, there is scarcely any doubt that the possibilities of the principle of the crescograph will be rapidly realized and a new world of infinitesimally small motion, which at present lies entirely beyond our ken, will be open to us.

## New Engineering Periodical in Three Languages

*Engineering Progress*, the first issue of which has just been received, is a review of engineering published in Germany in three parallel editions—German, English and Spanish. It treats of German machinery and products in the various engineering fields in a semi-popular, semi-commercial manner and as far as possible in non-technical language. It is one of a very few papers devoted more particularly to the needs of the purchasers of machinery than to the students of design. In view of the present controversy as to the use of the metric and English units of measurement, it is of interest to note that although metric units are in common use in Germany, *Engineering Progress* uses the English units and does so without giving their metric equivalents.

This new periodical is published jointly by the three most prominent German engineering organizations, the Society of German Engineers, the Association of German Electrical Engineers, and the Society of German Metallurgists. The editor is Dr. C. Matschoss, well-known to many mechanical engineers in this country in connection with the trip of the members of The American Society of Mechanical Engineers to Germany not long before the war.

# A. S. M. E. Spring Meeting at St. Louis

## Headquarters, Hotel Statler—Dates, May 24-27—Excursions to Keokuk and Tulsa

ST. LOUIS RIVER FRONT

EN route to the meeting, members will have an opportunity to visit the Keokuk Hydraulic Power Development. Special cars or a train will be arranged to bring the members together on Sunday, May 23, at Keokuk. The opportunity to see this remarkable plant and development is one that every engineer should grasp.

### THE TECHNICAL SESSIONS

The papers to be presented at the Meeting Sessions have been grouped under the headings of Appraisal and Valuation, Aero-

fore give the members an opportunity to complete the discussion and consider the new papers presented. In this day of inflation and profiteering, the stand of the engineering profession on questions of appraisal and valuation is of extreme importance. This is rapidly being realized and the subjects to be discussed will make this session one of great value to every engineer.

Tuesday morning is given over to a session of special interest to St. Louis engineers and to a consideration of the Elevator Code. The St. Louis Session will discuss the housing problem, vital to every community today, and the river-transportation situation, upon which the development of St. Louis so largely

ST. LOUIS BRIDGES

FREE BRIDGE

EADS BRIDGE

nautics, Foundry, Scientific, and Power and Combustion. A session with papers presented by St. Louis members covering miscellaneous subjects has also been arranged.

The Appraisal and Valuation Session with which the meeting will be opened on Monday afternoon is a continuation of a session on the same subject held at the 1919 Annual Meeting, where such great interest was shown that time did not permit full discussion. This session at the Spring Meeting will there-

depends. Papers will also be given by St. Louis engineers on a new form of stoker and on the subject of screw threads.

The Aeronautic Session, to be held simultaneously with the Foundry Session, comes Wednesday morning. The authorities at McCook Field have revealed a method by which the effect of unknowns may be reduced in the design of propellers. This method is set forth most interestingly in the paper by Messrs. Caldwell and Fales, the presentation of which will be aided by

309

motion pictures of a phenomenon seldom exhibited—air in motion. The eminent French engineer, A. Rateau, who so ably assisted his country in airplane development during the war, is contributing at this session a paper of basic theoretical value to the airplane designer. Further, the work of the Bureau of Standards in the Design, Construction and Inspection of Aeronautic Instruments is well set forth in the paper by M. D. Hersey. The fourth paper for the meeting is of a research nature, covering the Flow of Air through Small Tubes.

The Foundry Session will contribute valuable data to the machinery designer. The various characteristics and limitations of the different types of metal castings will be given by men thoroughly familiar with their subjects.

Thursday morning is given over to the Scientific Session and the Power and Combustion Session. The Scientific Session is held jointly with the American Society of Refrigerating Engineers and the American Society of Heating and Ventilating Engineers, and a paper on the Thermal Conductivity of Heat Insulators is to be contributed by M. S. Van Dusen for the first-named society. The record of a research made for the Engineering Foundation will be presented in a paper by Clemens Herschel which gives the results obtained in his development of a weir for open-channel measurement of the flow of water. The other papers to be considered at this session relate to venturi-meter calculations and to the dissipation of heat by various surfaces.

The Power and Combustion Session will consist of papers on the burning of pulverized coal at high altitudes, on the description of a new development in locomotive feedwater heaters, on a method of separating dissolved gases from boiler feedwater, and on the efficient use of natural gas in domestic service.

A complete list of the papers is given here so that the membership may order the copies they desire for consideration before the meeting. The coupon facing page 263 of this issue should be used in ordering advance papers.

### PAPERS TO BE PRESENTED AT MEETING

APPRAISAL AND VALUATION METHODS, David H. Ray
FUNDAMENTAL PRINCIPLES OF RATIONAL VALUATION, James Rowland Bibbins
DATA ON THE COST OF ORGANIZING AND FINANCING A PUBLIC-UTILITY PROJECT, by the late F. B. H. Paine (contributed by Dean M. E. Cooley)
THE CONSTRUCTION PERIOD, H. C. Anderson
PRICE LEVELS AND VALUE, Cecil Elmes
THE HOUSING PROBLEM, Nelson Cunliff
INDUSTRIAL HOUSING, Leslie H. Allen (Springfield, Mass)
RIVER BARGES AND TERMINALS, Wm. S. Mitchell
MISSISSIPPI VALLEY RIVER TRANSPORTATION ACTIVITIES, E. W. Schadek
BURNING EASTERN COALS SUCCESSFULLY ON A CONVEYOR FEED TYPE STOKER, L. R. Stowe
TIGHT-FITTING THREADS FOR BOLTS AND NUTS, C. B. Lord
PHYSICAL BASIS OF PROPELLER DESIGN, F. W. Caldwell and E. N. Fales (illustrated by moving pictures)
ANALYTICAL THEORY OF AIRPLANES IN RECTILINEAR FLIGHT AND CALCULATION OF MAXIMUM CRUISING RADIUS, A. Rateau, Hon. Mem. Am. Soc. M. E.
AERONAUTIC INSTRUMENTS—GENERAL PRINCIPLES OF CONSTRUCTION, Mayo D. Hersey
FLOW OF AIR THROUGH SMALL TUBES, T. S. Taylor
WEIR FOR GAGING THE FLOW OF WATER IN OPEN CHANNELS, Clemens Herschel
SIMPLIFICATION OF VENTURI-METER CALCULATIONS, Glenn B. Warren
DISSIPATION OF HEAT BY VARIOUS SURFACES, T. S. Taylor
PULVERIZED COAL IN METALLURGICAL FURNACES AT HIGH ALTITUDES, Otis L. McIntyre
EFFICIENCY OF NATURAL GAS USED IN DOMESTIC SERVICE, Robert F. Earhart
A METHOD FOR SEPARATION OF DISSOLVED GASES FROM WATER AND SOME OF ITS USES, J. R. McDermet
LOCOMOTIVE FEEDWATER HEATING, Thos. C. McBride
GRAY-IRON CASTINGS, Richard Moldenke
MALLEABLE CASTINGS, Enrique Touceda
STEEL CASTINGS, John H. Hall
ALUMINUM CASTINGS, Zay Jeffries
DIE CASTINGS, Charles Pack
BRASS AND BRONZE CASTINGS, C. H. Bierbaum
ELEVATOR CODE, (Committee Report)

### ELEVATOR CODE AND RULES FOR AIR TANKS AND PRESSURE VESSELS

The Elevator Code will be presented for public discussion at a session to be held Tuesday morning. This code is the result of several years' work on the part of the Committee on the Protection of Industrial Workers of The American Society of Mechanical Engineers, assisted by the U. S. Bureau of Standards, elevator manufacturers' associations, casualty and fire insurance companies and related engineering societies and elevator specialty manufacturers. The code covers the construction, operation and maintenance of elevators, dumbwaiters and escalators.

The Boiler Code Committee will present for preliminary open-meeting discussion the rules for the construction of air tanks and pressure vessels, published in the January, 1920, issue of MECHANICAL ENGINEERING. This presentation will be held Wednesday morning simultaneously with the Aeronautics and Foundry Session.

### ENTERTAINMENT

A series of entertainments and excursions have been laid out to insure a very agreeable passing of time and an opportunity to see the points of greatest interest in St. Louis. The excursions include trips to the plants of the Commonwealth Steel Company, Busch-Sulzer Diesel Engine Company, Heine Safety Boiler Company and the large power and gas plants, as well as to the parks and botanical gardens. One afternoon will be given over to a boat ride up the river to the Chain of Rocks. The evening entertainments will be enhanced by addresses to be given by Lt.-Col. H. W. Miller and E. R. Jackson. Lt.-Col. Miller will tell of the remarkable Belgian coast defences erected by the Germans after their occupation. These fortifications rivaled Heligoland in impregnability and remained intact under frequent British bombardment during the war. Mr. Jackson's talk will cover the Motor Convoy from Washington, D. C., to San Francisco in the summer of 1919, which he accompanied as Ordnance observer. A joint entertainment with the other societies holding meetings in St. Louis and with all of the local engineers will be held in the mammoth Municipal Open Air Theater in Forest Park on Wednesday evening.

### TULSA TRIP AND MEETING

Those going to Tulsa will leave St. Louis on Thursday evening, May 27, arriving in Tulsa in the morning. The Mid-Continent Section of the Society has planned a fine program of trips and meeting sessions to be held jointly with the Oklahoma Section of the American Chemical Society. The trips will include an inspection of the large oil refineries and a visit to the oil fields where wells in various stages of development will be seen. The following papers will be presented at the Tulsa meeting:

THE PREPARATION OF MOTOR GASOLINE FROM HEAVIER HYDRO-CARBONS, Prof. Fred W. Padgett, the University of Oklahoma
THE INFLUENCE OF TIDAL FORCES ON A FLOWING OIL WELL, J. E. Stillwell, Bartlesville, Okla.
THE MANUFACTURE OF SYNTHETIC GASOLINE FROM CRUDE MINERAL OILS BY THE AID OF HYDROSILICATES, F. C. Thiele, Oklahoma City, Okla.
MUNICIPAL WATER WORKS, T. R. Hollway, Tulsa, Okla.
META-NITROPHENYL ETHERS, Dr. Hilton I. Jones, of Oklahoma A. and M. College
WHAT IS GASOLINE; LIGHT PETROLEUM PRODUCTS DEFINED, D. E. Foster, Tulsa, Okla.
MID-CONTINENT GASOLINE, Dr. Charles K. Francis, Tulsa, Okla.

Dean M. E. Cooley of the University of Michigan will also address the meeting.

### RESERVATIONS AND EXCURSION ARRANGEMENTS

The St. Louis headquarters of the meeting will be the Hotel Statler, where the meeting sessions will be held. Reservations at the Statler should be made at once by communicating directly with the hotel. It is requested that all those who have any intention of attending the meeting will fill out the coupon facing page 263 of this issue and send it in at once.

# John Alfred Brashear

ENDING a long and active career of notable achievement, Dr. John A. Brashear, famed throughout the world as a scientist and maker of precise astronomical instruments, and known to a multitude of devoted friends as "Uncle John," died at his home in Pittsburgh on April 8, in his eightieth year.

During his later years he was the recipient of several honorary degrees, but when asked at one time what degree he would prefer, he replied, "Doctor of Humanity," which fitly expresses the crowning characteristic of his life, evident in all his relations with his fellow-men. It would be difficult to recall the name of any person in recent years who has received the outward expression of real devotion from so many men, women and children, of all stations of life and of all nationalities, as did John A. Brashear. On several anniversaries he received testimonials of appreciation and respect, literally by the thousand, and it is a source of gratification that he could have lived to receive such abundant recognition of the value attached to a life of service.

He was born at Brownsville, Pa., November 24, 1840, attended the Brownsville common school, learned pattern-making, and at the age of 22 was established, apparently for life, as a millwright in a Pittsburgh rolling mill. He remained in millwork for 20 years, after which he engaged in his life work of astronomical and physical instrument making.

From his earliest boyhood the study of astronomy had been his hobby. A traveling astronomer exhibiting the wonders of the heavens to passersby at five cents a look gave to John Brashear, then a lad of seven, his first glimpse of the wonders of the heavens. This made a deep impression on his young mind and he began to dream and study about the stars. He longed for a telescope of his own, and when he became married early in his industrial life he and his wife set out to make one. They fitted a workshop with a small engine and lathe at their humble home on the south side of the city where they made the tubes and ground the lenses, often working until midnight in absorbed joy in the work after a long, hard day at the mill and household tasks. Their first glass took three years to grind. Dissatisfied with its limitations, they set to work on a larger one only to have it break in the silvering process after two years of patient grinding and correcting.

The next night Brashear returned from the mill completely discouraged. His wife was not in the kitchen, so he wandered out into the workshop, where he found a fire under the boiler, steam up and a fresh piece of glass in the lathe. "We'll make a new one," his wife said, and they did. This proved to be the turning point in Brashear's life and was the beginning of his really great work in the field of science. Through the interest which he attracted from Langley, of flying-machine fame, then head of the Allegheny Observatory, he was enabled to move to Allegheny and set up a shop for lens grinding,—the beginning of the firm of John A. Brashear Company, Ltd., known today wherever astronomical instruments are used. Here have been made many

of the most important optical parts in the world. It has been the policy of the firm to have no patents and no secrets. Whatever it accomplished it has given freely to the world. His son-in-law, Mr. James B. McDowell, joined him in the work and to him Dr. Brashear gave a large share of credit for the success of the undertaking.

Of the products of the Brashear Company reference should be made to the following:

A 72-in. parabolic mirror, Dominion Observatory, Victoria, Canada. The illustration on the next page shows Dr. Brashear at work on this glass in his workshop.

JOHN ALFRED BRASHEAR

Two 37-in. parabolic mirrors for Lick Observatory, Mt. Hamilton, Calif., and University of Michigan, Ann Arbor, Mich.; and one 30-in. parabolic mirror for Allegheny Observatory, Pittsburgh, Pa.

Several plain and parabolic mirrors and object glasses ranging from 18 in. to 30 in. in diameter for Yale University, New Haven, Conn.; Allegheny Observatory, Pittsburgh, Pa.; Swarthmore College, Swarthmore, Pa.; Chabot Observatory, Oakland, Calif.; Dominion Observatory, Victoria, Canada; and University of Pennsylvania, Philadelphia, Pa.

Some twenty-five object glasses, ranging from 9 in. to 18 in. in diameter, for various other observatories throughout the world.

A 16-in. photographic doublet for the Königstahl Observatory, Heidelberg, Germany, and a 10-in. photographic doublet for Bruce Telescope, Yerkes Observatory, Williams Bay, Wis.

Perhaps his most important achievement was in connection with the design and development of the spectroscope for astronomical uses. Many of the spectroscopes in the principal observatories in the world have been made in the Brashear workshop, including that for the 38-in. telescope of the Lick Observatory. The excellence of the work done by Professor Keeler at this observatory is freely attributed to Dr. Brashear's skill and genius.

In 1884 Dr. Brashear was commissioned by Professor Rowland of Johns Hopkins University to make the speculum metal plates for his diffraction gratings. Mr. Ambrose Swasey, in his presidential address before the A. S. M. E. in 1904, referred to the extraordinary accuracy of these plates and of others made by Dr. Brashear for Professor Michelson in connection with the determining of the standards of length. Mr. Swasey said:

The Rowland engine was made especially for ruling diffraction gratings of speculum metal, and with it a metal surface has been ruled with 160,000 lines, there being about 29,000 to the inch, and as many as 43,000 lines to the inch have been ruled.

The gratings mostly used have from 14,000 to 20,000 lines to the inch, and with such exactness is the cutting tool moved by the screw that the greatest error in the ruling does not exceed one millionth of an inch.

The production of these gratings, which has enabled the physicist in his study of the spectrum to enter fields of research before unknown, has not only called for the highest degree of perfection ever

attained in the spacing of linear distances, but it has also called for
a refinement most difficult in the optical surfaces upon which the
lines are ruled. To Mr. Brashear was given the problem of producing
such surfaces, and notwithstanding the many difficulties encountered
in working and refining the speculum metal plates, he has made many
hundred plates with surfaces either flat or curved with an error not
to exceed one-tenth of a wave length of light, or one four-hundred
thousandth of an inch.

As the established standards of length, which are the yard of Great
Britain and the meter of France, are made of metal, and liable to
destruction or damage, Professor Michelson conceived the idea of
determining the lengths of these standards in wave lengths of light,
which would be a basis of value unalterable and indestructible.

For the purpose of carrying out these experiments, the interfero-
meter was constructed, an instrument which required the highest
order of workmanship and the greatest skill of the optician. Again
Mr. Brashear proved to be equal to the occasion, and made for the
instrument a series of refracting plates, the surfaces of which were
flat within one twentieth of a wave length of light, with sides
parallel within one second. This was the most difficult work ever
attempted in the refinement of optical surfaces.

No event in Dr. Brashear's life gave him greater satisfaction,
nor elicited from him
greater enthusiasm
than the completion of
the 72-in. glass for the
Dominion Observatory,
and its safe arrival at
Victoria. The disk of
glass was cast in Bel-
gium and weighed
when completed for
shipment from Pitts-
burgh 2¼ tons. The
final grinding and pol-
ishing was done by
hand in the basement
of the Brashear work-
shop where tempera-
ture conditions could
be kept constant and
where the great tube
with its double walls
was located for making
the final tests. The
correcting of the sur-
face had to be done by
hand by local polish-
ing. After working on

DR. BRASHEAR AT WORK ON THE 72-IN. DISK FOR THE
CANADIAN REFLECTING TELESCOPE

the glass for a few hours it had to be lifted to the testing
tube where it remained for 10 hours to attain uniform temperature. When the glass finally arrived at Victoria in its
special car Dr. Brashear wrote: "The glass reached its destination in less than six days after I bid it good-bye on the car in
Pittsburgh, a trip of 2200 miles. You bet Uncle John and his
son-in-law and some others are happy over this news, after three
years and eight months." The Dominion Observatory was dedicated on June 11, 1918.

Dr. Brashear's scientific work brought recognition and about
the time of his removal to Allegheny he was given an appointment in the University of Western Pennsylvania of which the
Allegheny Observatory was a department. From 1898 to 1900
he was acting director of the Allegheny Observatory and raised
$300,000 for the building and equipment of a new observatory in
Riverview Park. He always kept in touch with the development
of this observatory, and through his efforts one department has
been put in every possible way at the disposal of the public, as
well as for astronomical and astrophysical research.

For twenty years past Dr. Brashear has been a trustee of the
Carnegie Institute, for fifteen years of the Carnegie Institute of
Technology and for twenty years of the University of Pittsburgh,
of which latter he has also served as chancellor. Several years
ago a friend placed in his hands an endowment fund of $250,000
to be used for the advancement of teachers and teaching in the
public schools, as a result of which to date over seven hundred
teachers have been sent to different parts of the country for rest

and study, bringing back with them new ideas and greater enthusiasm.

Dr. Brashear was elected to membership in The American Society of Mechanical Engineers in 1891 and was made an honorary
member in 1908. He was a manager of the Society from 1899
to 1902. In 1911 he was elected one of the Society's representatives on the John Fritz Medal Board to serve for four years. He
was president of the Society for the year 1915, during which time
he twice toured the country, visiting the sections and speaking
before many engineering and other organizations. His magnetic
personality made him always in demand as a speaker. Later he
addressed large audiences of the Society on Photography of the
Stars and once on The Science of the Beautiful in Commonplace
Things.

He was a fellow of the American Association for the Advancement of Science and the Royal Astronomical Society of Great
Britain; he was a past-president of the Engineering Society of
Western Pennsylvania and the Pittsburgh Academy of Arts and
Sciences; he was a
member of the British
Astronomical Associa-
tion, the Société As-
tronomique de France,
Société de Belgique,
the American Philo-
sophical Society, the
Astrophysical Society
of America, the Wash-
ington Academy of
Sciences, the National
Geographic So-
ciety, and an honorary
member of the Royal
Astronomical Society
of Canada.

He was honored with
the degree of LL.D. by
Washington and Jef-
ferson College, by
Wooster University
and by the University
of Pittsburgh, and with
the degree of Sc.D. by
Princeton University
and the Western Uni-
versity of Pennsylvania; also with the degree of Doctor of Engineering by Stevens Institute of Technology.

A few years ago he was voted Pennsylvania's foremost citizen
by a committee of prominent men appointed by the governor of
the state. Governor Brumbaugh had been asked by the officials
of the Panama-Pacific Exposition to name the most distinguished
citizen of Pennsylvania. Having consulted men of prominence
throughout the state, the Governor unhesitatingly named John
Brashear. When word of this honor was brought to "Uncle
John," he was at work in his laboratory and his only comment
was: "I guess you're joking." Then he went on with his work.

Dr. Brashear's seventy-fifth birthday was celebrated with a
dinner that will long be remembered in Pittsburgh. Alexander
Graham Bell, Charles M. Schwab, Henry Clay Frick, Rear-
Admiral Robert E. Peary and many other men of high achievement joined with Uncle John's neighbors to pay him homage.
One thousand people sat down at dinner and the table was piled
high with telegrams of congratulation—from the president of the
United States, from men of science all over the world and from
thousands of his less-famous but as well-loved neighbors. That
night Dr. Brashear's friends made up a purse of fifty thousand
dollars, the interest of the fund to go to him as long as he lived
and after that "to carry on, but not to take the place of," the
work he had begun.

Scarcely less notable was a public reception tended him at
Pittsburgh on the eve of his 76th birthday and of his departure
for an extended trip to the Orient, with Messrs. Ambrose Swasey

and John R. Freeman, Past-Presidents A. S. M. E. Concerning this, the following extract from a letter written at the time by his secretary, Miss Martha C. Hoyt, gives a just estimate of the affection in which he was held by the people of his own city.

It was a truly wonderful sight, such as Pittsburgh will probably never see again. In the receiving line were the Jew and the Christian, Protestant and Catholic, well-known public men and private citizens, the rich and the poor. There were even people whom the doctor has rescued from privation. In the line which came to greet him were the most prominent educators, social and financial leaders, down to the dirty little newsboy with his papers under his arm, who conducted himself during his progress down the receiving line with as much dignity as the more experienced members of society. There was a deaf and dumb man who knew him when a young man and brought his congratulations and best wishes written on a paper so he would not take too much of Dr. Brashear's time presenting them, and an Italian laborer who was attending night school in the neighborhood and came running in to say what his teacher had taught him for the occasion: "I wish you many happy returns of the day." He was radiant with his success at being understood. There were delegations from the Carnegie Technical Schools, the University of Pittsburgh, the School for the Blind, the Newsboys Home, and the Loendi Club, a group of colored men in whom he has been interested and who have had their share of his loving service.

Among the many benevolent acts which were typical of the life of Dr. Brashear was the founding three years ago of the Brashear Settlement Association which has its headquarters in the old Brashear home, where the doctor spent his younger years. He suggested that the house should be called "Inspiration," for the humble old-fashioned structure with the rude shop in the rear would point out to these lads that from humble beginnings one may rise to greatness through hard work.

Upon the last celebration of the anniversary of his birth in November, at his request, money which would have been used by his numerous friends to purchase gifts for him was given to the Brashear Settlement to enable it to further the work which it had instituted. It was Dr. Brashear's wish that the work should live after his death, and acting upon this the officers of the Association will endeavor to make it a permanent tribute to his name. On this same anniversary a "birthday shower" of postal cards was arranged by the officers of the Brashear Settlement House and it was planned and carried out for thirty thousand Pittsburghers to send cards of greeting to Uncle John. At this time Dr. Brashear was confined to his home by illness.

Memorial Hall, Pittsburgh, was the scene of Dr. Brashear's funeral services, which were simple and carried out in accordance with his wishes expressed shortly before his death. Thousands from all walks of life were present to show the love they had for "Uncle John." Not only was the great hall filled, with hundreds standing, but many hundreds more for whom there was no room lingered outside. The ashes of Dr. Brashear, together with those of his wife, Phoebe Stewart, who died a few years ago, will rest in the base of the great telescope erected in Allegheny Observatory. The inscription, written by Dr. Brashear, reads, "We have loved the stars too fondly to be fearful of the night."

# Government Activities in Engineering

*Notes Contributed by The National Service Department of Engineering Council*[1]

## Congress To Study Reorganization of Federal Executive Departments

When the Public Works plan was set in motion by the introduction of the Jones-Reavis Bill in the Senate and House, comparatively little attention was paid to it by the legislators on Capitol Hill. As the importance of this measure was impressed upon them through their constituency back home, through personal calls from prominent engineers, and through the definite message that was the result of the hearings before Senator Smoot's Committee on Public Lands, there was a general awakening as to the existing conditions throughout the executive departments. With this awakening came a desire for more knowledge as to the exact conditions to which the enormous wastes in Government operations could be traced.

And under existing demand for a material cut in Government expenditures, there were produced two definite plans, one in the House and one in the Senate, proposing to investigate all the activities of the executive branches of the Federal Government with a view to their reorganization on a business basis. This was substantially the job which the engineers had taken upon themselves with regard to the engineering functions of the Government.

The most recent proposed legislation on this subject is the Senate Joint Resolution No. 191, introduced by Senator Smoot. It proposes a "joint committee on reorganization" to consist of three members from the Senate and three members from the House. This joint committee is to make a survey of the administrative services of the Government to secure pertinent information as to their distribution among the several Government departments, including their overlapping and duplication of authority, with a view to redistribution of activities under proper correlation. The regrouping under each executive department is to be such that only services having close working relation with each other administering directly to the primary purpose of that department are to be included therein.

The committee is to report its findings, prepare and submit bills or resolutions for the coördination of Government functions and their most efficient and economic conduct, and will employ experts in this kind of work, expenses to be equally divided between the Senate and House.

The wording of this resolution is obviously entirely in line with the plan for a National Department of Public Works, and needless to say it has the sanction and active support of the National Public Works Department Association.

## Mail Tunnel Systems for Large Cities

When the Senate and House conferees agreed on the Post Office appropriation bill, it contained a provision for a commission, which is to study among other things, the advisability of establishing a system of tunnels to handle mail in congested centers—especially in New York City.

The commission is to consist of a chairman and four members of the Committee on the Post Office and Post Roads from both the Senate and House and a postal expert appointed by the Postmaster-General. This commission is to have authority to appoint an advisory council of seven to be composed of those having special knowledge of the work in hand—preferably representatives of commercial organizations, and engineers who are acquainted with existing conditions and requirements. Further, the commission is authorized to employ such engineers and special experts as it is deemed necessary to assist in its investigation.

The appointment of the advisory council to this commission is in line with the recommendations that Engineering Council has made to the Post Office Committee of Congress. It is especially fitting that this legislation should carry special provision for the use of engineering opinions in the accomplishment of a joint congressional committee's work.

[1] Engineering Council is an organization of national technical societies created to consider matters of common concern to engineers as well as those of public welfare in which the profession is interested. The headquarters of Engineering Council are located in the Engineering Societies Building, 29 West 39th Street, New York City. The Council also maintains a Washington office with M. O. Leighton, chairman of the National Service Department, in charge. This office is in the McLachlen Building, 10th and G Streets, Washington, D. C. The officers of Engineering Council are: J. Parke Channing, *Chairman;* Alfred D. Flinn, *Secretary.*

## Road Legislation

It will be recalled that last year's Post Office appropriation bill carried a rider appropriating $200,000,000 of the Federal funds for road construction throughout the several states. This year the bill became law without any provision for highway legislation, solely because it had not been seriously urged that increased appropriations or further legislation extending Federal aid should be included.

When the conference report was finally accepted on the floor of the Senate, Senator Townsend, the sponsor of the bill to establish a national highway system under a highway commission, gave notice that he intended to proceed with hearings on this bill and would soon be prepared to present to Congress more concrete information for an improved highway policy. The need of investigation on this subject was emphasized because it was contended that the continuity of good roads is necessary for general welfare in time of railroad strikes and when for other reasons the rail transportation facilities do not meet the needs of the public.

Some discontent exists among legislators because they are not pleased with the way that Federal money is being expended in highway construction. This is largely due to conditions in the states where the Federal aid money is divided among counties in the same way the Government apportions it to the state, that is, according to population, territory, and road mileage. Short stretches of good roads are the result instead of complete through highways built for permanency. Senator Townsend said, "I maintain that the present methods of expending Federal aid funds are inexcusable."

Senator Townsend urges connecting road systems between the states and a provision for permanent maintenance. He seeks a national system, built and maintained by the Federal Government for the good of the whole country and built on such a basis that they will inspire the states to build good road systems connecting the Federal roads. The Townsend bill provides that the national system of highways be composed of two roads in each state, one running east and west and one running north and south.

The Post Office appropriation bill carried a provision whereby tractors may be loaned by the War Department to the states for use in highway construction. When the conferees finally agreed to this provision the Secretary of War was vested with authority to use his discretion in loaning such tractors as could be spared for road construction work. All expenses for repairs and upkeep as well as loading and transportation charges are to be paid by the state.

The Military Affairs Committee has also favorably reported the Reavis Bill, H. R. 13329. This provides for transfer of additional machinery and equipment to the Bureau of Public Roads for use in highway construction, and will permit of the use of mobile machine shops and other similar army equipment. According to reports recently submitted by the Bureau of Roads, this equipment will be very acceptable and will be immediately put to good use.

## Important Flood Legislation

The Minnesota River and the Big Stone Lake, which are on the boundary between Minnesota and South Dakota, are proposed for flood protection in a bill which has just passed both House and Senate. This legislation will protect property worth millions of dollars when it is enacted into law and put into effect. Minnesota flood-control work has been delayed up to the present time because that state is forbidden by its constitution to spend funds for such internal improvement. It is proposed to use the Big Stone Lake as a reservoir to hold back the flood waters, this being accomplished by the construction of a dam at the foot of the lake. The entire cost of improvement is to be borne by those benefited according to provisions of the bill. The project has the sanction of prominent engineers and the approval of the Engineers Corps of the War Department.

## Reclassification and Compensation in the Government Service

What has been known as the Keating Commission on Reclassification and Compensation of Federal Employees has been at work for fifteen months investigating every branch of the personnel in the Government Service, with special reference to compensation. On several occasions the national engineering societies have been called on for information which they were able to compile and for general advice on technical service. One of the largest committees formed for this purpose was able to advise as to existing salaries in the technical-commercial field.

The Commission found in its investigations the same kind of troubles that appear to be chronic in our Government. For instance, pay for the same positions requiring the same qualifications show wide variations from department to department, and in many cases within the same department. This was especially in the war expanded departments. The report shows that the Government has no standard guide in fixing salaries appropriate to the character and importance of the work, so that there is little or no means of rewarding special training or marked ability. In many cases there is no uniform system of promotions nor reward for long service. These factors were shown to lead to many ambiguities in appropriations for Government service and to a serious growing discontent of Government employees, especially technical employees. Wide variation was everywhere apparent from accepted commercial practice.

Detailed recommendations for first classifying all Government employees and then obtaining a standard compensation in these various classes covered over 200 pages of the report, while nearly a thousand additional pages were used in what appears to be a vast dictionary of Government positions. This report covers every kind of technical position in the Government.

It is contemplated that this report will form the basis for a bill proposing at least some helpful changes in Government positions but it is hard to tell when such proposed legislation will receive attention before either the House or Senate.

## Classification of Subordinate Railroad Engineers

The Interstate Commerce Commission has held lengthy hearings on the classification of engineer assistants in the railroad employ, and as a result has issued regulations classifying subordinate officials as provided in the Transportation Act. These engineers are grouped into a class as "Engineers of Mechanics." This class includes Civil Engineers, inferior in rank to Engineers of Maintenance-of-Way, Chief Engineers or Division Engineers. It includes draftsmen, engineers on maintenance-of-way work, and other engineers of mechanics, who are not vested with authority to employ, discipline or dismiss subordinates.

Representatives of the American Association of Engineers, whose membership includes many engineers in this class, presented arguments in favor of a separate grouping for subordinate technical engineers. It was shown that twenty thousand technical engineers in the employ of the railroads would be included in such classification, whereas only five per cent of that number are engineers that come within the category of executives. No other class of railroad employees have a more trusted relationship with their companies than do the subordinate engineers, assistant engineers, architects, rodmen, chairmen, and draftsmen. Because of the fact that this class of engineers have previously had no classification they have received practically no consideration from the railroads in the point of increases in salary, promotion, schedules, etc. For the same reason, also, it has been hard for engineering organizations representing them to obtain improved conditions.

It will be recalled that the Engineering Council went before the old Board of Wages and Working Conditions under the Railroad Administration to assist in getting a classification and increased compensation for the men who have now been classed "Engineers of Mechanics." Some good was accomplished but conclusive results could not be obtained because this group of engineers were not then known as a unit on the railroads.

# NEWS OF THE ENGINEERING SOCIETIES

Meetings of the Industrial Relations Association of America, the Society of Industrial Engineers
and the American Railway Engineering Association

## Industrial Relations Meeting in Chicago

The Industrial Relations Association of America, formerly the National Association of Employment Managers, will hold its Second Annual Convention in the Auditorium Theater, Chicago, from May 19 to 21. The program will cover the various phases of employment and industrial relations work. Two sessions will be held daily on broad subjects such as Man and Industry, Community Conditions Affecting Labor Stability, The Foreman of the Present and Future, Organized Labor in Industry, and Incentives and Production. Subject Luncheons and Section Dinners have been arranged for the discussion of the details of the work. Some of the subjects to be discussed at these affairs are Americanization, Shop Committees, Insurance Plans, Developing Plant Spirit, Radicalism, and Financial and Non-financial Incentives. The speakers will be manufacturing executives and labor leaders of prominence, while the chairmen of the sessions will be employment managers of experience and ability.

## Society of Industrial Engineers

The practical application of the principles of industrial engineering was the keynote of the spring meeting of the Society of Industrial Engineers held at the Bellevue-Stratford Hotel, Philadelphia, on March 24-26. The feature of the meeting was a paper by C. E. Knoeppel, president of the New York industrial-engineering firm bearing his name. He chose as his subject, The Most Effective Type of Industrial Organization, and used as an illustration of the ideal, the human body. Everywhere in the human body it is found that while each organ performs some definite function, the coöperation of all is necessary to the life of the person. While two or more organs perform totally distinct functions, they coöperate for some major or compound function in which their individual functions are completely merged.

Mr. Knoeppel outlined the managing organization of the human body as follows: The chief executive—the cerebrum—exercises judgment, reason, cognition, and volition; the first assistant—the cerebellum—coördinates and harmonizes the voluntary organs and thus guides the performance of the organization; the second assistant—the medulla oblongata—coördinates and harmonizes the work of the involuntary organs in order that the proper service may be secured; the intelligence, formulative and standardization bureau—the cranial nerves—control and link up the action and routine division with the coördinating division.

A better example of organized control cannot be found anywhere in the whole wide world, declared Mr. Knoeppel; its component parts are of a finer kind than we shall ever approximate, and the functions and their relations are coördinated more smoothly than we shall ever be able to arrange human relations. We can, however, pattern our industrial organizations after this perfect model with the full expectation of securing both economy in the expenditure of energy and efficiency in the attainment of results.

Dexter S. Kimball, dean of the college of engineering, Cornell University, spoke on How Industry Can Assist the Educator, and pleaded for greater coöperation from the industries with the student and asked that students be given more chances for summer work.

L. P. Alford, editor of *Industrial Management*, speaking on The Value of Team Work Among Engineers, pointed out that engineers were aloof, self-centered specialists often regarded as caring for nothing but their narrow professional aims. He urged a nation-wide, all-embracing society expressive of the ideals, purposes and work of all engineers.

John Calder, manager of industrial relations for Swift & Co., Chicago, insisted that the key to many intangible uncertainties in industry is found in the education and training of the foreman, as already carried out in a large number of up-to-date plants. He related the practice of Swift & Co. Their forty classes range from Boston to Portland, Ore., with 50 to 400 foremen in each class, and about 5000 executives in all. Each class is split up into inner groups of from 15 to 20 men, and for each team an executive is selected as team leader. Every class meets as a whole once in two weeks at a lecture, and each class meeting concludes with an open forum in which questions are asked and answered publicly. In most plants, the men have found the training so interesting and their group meetings so helpful that they have effected permanent organizations or clubs, to continue the benefits of the discussion.

Other papers were presented on industrial management in army construction work, industrial safety, fatigue study, cost systems, effect of the industrial engineer on the earning power of industry, and the need of industrial preparedness. About forty companies displayed exhibits which consisted of pictures, charts, and labor-saving devices.

## American Railway Engineering Association

The twenty-first annual convention of the American Railway Engineering Association was held in Chicago from March 16 to 18. Fundamental additions to existing knowledge of the stresses developed in railway track were contained in the second progress report of the committee on stresses in railway track, a joint committee with the American Society of Civil Engineers. It develops from the results obtained in numerous tests with different types of engines that the action of the driving-wheel counterbalance multiplies the static load stresses in the rail. The unbalanced rotating masses at the several drivers subject the rail to excess vertical forces, often larger than the computed effects. In some cases the excess stress chargeable to the unbalance is much more than the static rail stress. With respect to the tie, the nature of the service given by it is defined by measurements of tie bending and distribution of the rail load to the ballast.

The committee on electricity included in its report a study made by a sub-committee of the electrified section of the Chicago, Milwaukee & St. Paul Railway. This experience has demonstrated that the transmission of railway electric power over long distances is efficient, economical and reliable at 110,000 volts, with 3,000 volts on the contact wire. Furthermore, the introduction of electric locomotives equipped with electric brakes has made possible the haulage of heavier trains at higher speeds on heavier grades, such as mountain divisions, and with greater safety and reliability under all climatic conditions. These facts, it was observed, pointed out the way in which hydroelectric power could be still further utilized for the electrification of other railways.

New steel bridge specifications presented by the committee on iron and steel structures were adopted by the convention. Rounded ends for through-plate girders are specified. This is said to be important in the case of derailment of cars. Specific permission for the welding of minor defects in castings is not required.

There was considerable discussion concerning specifications, proposed by the committee on masonry, for billet-steel concrete reinforcement bars specifically prohibiting the use of twisted bars. It was stated that hot-twisted bars are dangerous and that cold-twisted bars give no advantage over plain or deformed bars. The specifications were adopted.

315

## EFFICIENCY OF NATURAL GAS USED IN DOMESTIC SERVICE

(Continued from page 288.)

as in the previous case, the efficiency falls to 15.4 per cent. These efficiencies are represented by a curve in Fig. 3. In many stoves adapted to the use of natural gas, 2¾ in. is the distance from burner to the place upon which a vessel is supported. It is apparent not only that there is a provoking loss of time in carrying on heating operations but the gas consumed under such conditions is used very inefficiently.

### EFFECT OF PRESSURE VARIATION IN PRESSURE WITH MIXER VALVE WIDE OPEN

Fig. 4 is an efficiency curve obtained when the mixer valve was opened wide and the burner set 2¾ in. from the bottom of the vessel. The pressure was varied from 1 in. water head to 7 in., in steps of 1 in. At 7 in. pressure the gas impinged so strongly on the bottom of the vessel that the flame floated around the edges of the vessel. The results are shown in Fig. 4. After the final reading in this set was taken, viz., at 1 in. pressure, another reading was taken under the same flow conditions but with the burner 1¼ in. from the kettle. This increased the efficiency from 26.5 to 41.8 per cent. In justice to stove manufacturers it should be stated that their devices are arranged for pressures which vary from 2 to 4 oz., i. e., from 3½ to 7 in. of water, approximately.

A final experiment was made to determine the comparative merits of a granite-ware vessel and an aluminum preserving kettle of the same general size and shape, the general conditions of all the tests being the same. The water equivalent of the containers was not included. The results show a small margin in favor of the aluminum vessel.

As a result of these experiments it may be concluded that efficiencies of 45 per cent or more can be obtained when the burner is set within 2 in. of the vessel, a condition which can be secured easily by adjusting the flame height. Under the varying pressure conditions which now exist, a distance between the limits of 1 in. and 1½ in. is preferable. At this distance the ordinary operations are fairly efficient even at low pressures, while at moderate and high pressures maximum efficiencies may be secured without excessive loss of time.

The author is under obligation to the Public Utilities Commission for the loan of equipment, to the Ohio Experiment Station and to several gentlemen connected with the national gas industry, especially to Mr. S. S. Wyer.

## RECTILINEAR FLIGHT OF AEROPLANES

(Continued from page 269.)

which he calls Aeroplane-Type, and for which he gives certain numerical values of these four coefficients closely approximating those of the Bréguet army airplane having 50 sq. m. of wing surface.

The main equations of an airplane in horizontal flight are established and an equation is derived indirectly or directly comprising the ten specific coefficients of the airplane, propeller and engine, clearly showing the influence of each of them on the final performance of the airplane. Among other things, it determines through the medium of the specific weight of air the altitude of flight possible with any given incidence; also it determines the ceiling of the airplane in terms of a function involving its specific coefficients.

It also makes possible determining the altitude of flight at which the consumption of the fuel or oil is a minimum, which the author proceeds to do in detail. There are two problems in this connection which can be and are solved practically simultaneously, namely, the altitude at which the plane must fly in order to reduce the consumption of fuel and oil to a minimum, and the cruising radius which may be reasonably expected with airplanes of present construction.

The formulæ derived make it possible to construct a table show-

ing variations of consumption with various angles of incidence and various positions of altitude. Fig. 1 plotted from data of this table shows the variations of fuel and oil consumption at various positions below the ceiling of the plane, which is in this case $Z = 5106$ m. In this figure as abscissæ are plotted the distances in meters of flight at the ceiling, and as ordinates the increases of consumption in percentages with respect to the minimum equal to 0.2997 kg. per hp-hr. for an incidence of $\alpha = 4.6$ deg.

From this curve it appears that the minimum consumption is secured at an altitude only 151 m. below the ceiling. At the ceiling the excess consumption amounts to 2 per cent; at 450 m. below the ceiling to 1 per cent; at 615 m. to 2 per cent and at 1000 m. to 5.1 per cent. These are the limits of altitude within which the pilot should remain in order not to exceed the minimum possible consumption of fuel or oil by 1, 2 or 5 per cent.

A very important problem to the discussion of which the author next proceeds is that of the cruising radii of planes of the present time, because he theoretically establishes the limits attainable in long-distance flights. The conclusion to which he comes is that tests such as the transatlantic flight from Newfoundland to Ireland have shown that a non-stop flight of 3000 km. (1865 miles) is already possible.

He further shows that with the present constructions 5000 km. (3100 miles) are certainly possible and 6000 km. (3725 miles) probable, but that the attainment of a cruising radius of 7000 km. (4350 miles) is quite doubtful.

A somewhat involved formula is given expressing the maximum cruising radius of a plane in terms of a number of variables involving the altitude of flight, fuel and oil consumption, efficiency of propeller, etc.

In the final count it is the ratio $P_o/P_1$, which is the ratio of total weights of the plane at the beginning and end of the trip, that has most of the influence on the cruising radius. It is easy now to build planes that will lift a load of fuel and oil equal to their own weight, including the weight of the reservoirs, pilot and all of its equipment and accouterments. A lift of 1.5 times the weight of the plane in fuel and oil load is possible, one of twice that weight less certain, and above this quite difficult to accomplish. From these data it would appear that roughly 7000 km. is about the limit that one may hope to attain in the way of a cruising radius under present conditions of aeronautical engineering.

In order to do better than this it will be necessary to build airplanes with a higher fineness ratio or airplanes capable of rising from the ground with a load twice as great as their own weight. In addition to this it will be necessary to provide means making it possible for the pilots to exist comfortably for dozens of hours at altitudes varying from 5000 m. (16,400 ft.) to perhaps 8000 m. (26,250 ft.), which is quite a problem in itself.

## LOCOMOTIVE FEEDWATER HEATING

(Continued from page 286.)

if for any reason both are not considered advisable or possible for still further economy.

The feedwater heater offers a solution of the pressing problem of these times, when so many fairly old and small locomotives have outgrown their usefulness because they are not large enough for present-day demands. The feedwater heater will not only increase the capacity of these locomotives, or what might be called their economical operating capacity, at a comparatively low cost and with little change in them, but it will also show a larger percentage of coal saving because of their full loading than would be obtained with present-day larger locomotives.

The feedwater heater offers the peculiar advantage of being of most assistance to the locomotive just at the time when the locomotive needs assistance. It offers its greatest saving in fuel or increase in capacity in the winter time; furthermore these advantages are not affected by reduced steam pressure or superheat, but continue in spite of them and at a proportionately greater rate.

# LIBRARY NOTES AND BOOK REVIEWS

AMERICAN CIVIL ENGINEERS' HANDBOOK. Editor-in-chief, Mansfield Merriman. Fourth edition. John Wiley & Sons, Inc., New York. 1920. Flexible cloth, 4 x 7 in., 1953 pp., tables, $6.

The fourth edition of this well-known reference work follows the plan of the previous editions and is the achievement of a board of eighteen associate editors, under the direction of Professor Merriman. The volume has been thoroughly revised, a collection of mathematical tables included and new sections added on electric railways, irrigation and drainage. Nearly four hundred pages have been added to the work. Because of the comprehensiveness of the book it has been entitled a "Handbook" instead of, as formerly, a "Pocket Book."

AMERICAN MACHINISTS' HANDBOOK. A Reference Book of Machine-Shop and Drawing-Room Data, Methods and Definitions. By Fred H. Colvin and Frank A. Stanley. Third edition, thoroughly revised and enlarged. McGraw-Hill Book Co., Inc., New York. 1920. Flexible cloth, 4 x 7 in., 758 pp., illus., diagrams, tables, $4.

This handbook has been prepared to present in convenient form such data as will be of value to practical men in the various branches of machine work. The present edition has been thoroughly revised. Seventy-seven pages have been added and much other new material included by the elimination of the less important matter from the previous edition.

DESIGN AND CONSTRUCTION OF HEAT ENGINES. By William E. Ninde. First edition. McGraw-Hill Book Co., Inc., New York, 1920. Cloth, 6 x 9 in., 704 pp., illus., diagrams, tables, $6.

The object of this book is to supply in one volume the material most essential to the well-equipped, independent designer of heat engines, and to give this material in the form most convenient for use in class-room and practical work by a separate treatment of the different phases of the subject. The book is the outgrowth of twenty years' experience, and a study of the literature, drawings and practical data. The volume is confined to the steam engine, steam turbine and internal-combustion engine.

THE DESIGN OF SCREW PROPELLERS. With Special Reference to their Adaption for Aircraft. By Henry C. Watts. Longmans, Green & Co., New York, 1920. Cloth, 6 x 9 in., 340 pp., illus., charts, diagrams, plate, tables, $8.

During the war the author was in charge of technical work in connection with propellers for aircraft at the Admiralty and Air Ministry of Great Britain, and many of the propellers used were designed by him or under his supervision. This volume records the results of his experience. It is intended as a guide to practical design and gives slight attention to mathematical theory of the behavior of screw propellers. A chapter on the design of windmills is included.

FLOW AND MEASUREMENT OF AIR AND GASES. By Alec B. Eason. J. B. Lippincott Co., Philadelphia; Charles Griffin & Co., Ltd., London, 1919. Cloth, 6 x 9 in., 252 pp., illus., diagrams, tables, 25 shillings.

The engineering problems investigated in this book arise in connection with pneumatic tubes-and compressed-air, gas-lighting and ventilating systems. The author has investigated, by study of the literature and by experiment, the friction of gases and the coefficient of friction in pipes, the question of suitable meters for gas and air, and the working of pneumatic tubes; and has attempted to coördinate the results of various tests and formulas, so that the reason for variations may be appreciated. Full references to the sources of information are given.

GRAPHIC PRODUCTION CONTROL. By C. E. Knoeppel, assisted by various members of the author's firm and staff. The Engineering Magazine Co., New York, 1920. Cloth, 6 x 9 in., 477 pp., illus., diagrams, charts, $6.

This treatise on the use of graphic charts in shop control is intended to provide a complete description of the proper methods of making graphic charts and of applying them to industrial problems. The general principles of graphic control, the preliminaries of its installation, the practical operation of the system, etc., are discussed in detail.

HENDRICKS' COMMERCIAL REGISTER OF THE UNITED STATES FOR BUYERS AND SELLERS. Twenty-eighth annual edition. S. E. Hendricks Co., Inc., New York, 1919-20. Cloth, 8 x 10 in., 2541 pp., $12.50.

This new edition, like its predecessors, is devoted to the electrical, engineering, hardware, iron, mechanical, mill, mining, architectural, quarrying, chemical, railroad, steel, contracting and kindred trades. The firms included are listed alphabetically and by products. A subject index and an index of trade names are also included.

INORGANIC CHEMICAL SYNONYMS and Other Useful Chemical Data. By Elton R. Darling. D. Van Nostrand Co., New York, 1919. Cloth, 4 x 7 in., 100 pp., $1.

The author has collected the various synonyms that have been used in scientific and trade literature to designate the chemicals in common use, and has arranged them so that the substance in question can be accurately identified.

THE LABOR MARKET. By Don D. Lescohier. The Macmillan Co., New York, 1919. Cloth, 5 x 8 in., 338 pp., 9 charts, $2.25.

This work is intended for general readers, employers, legislators, employment officials, college teachers and students of the employment and labor problem. The factors of supply and demand, labor market, machinery and special employment problems are discussed. The author advocates national control of the problem of employment. An extensive bibliography is included.

METALLOGRAPHY. Part I. Principles of Metallography. By Samuel L. Hoyt. First edition. McGraw-Hill Book Co., Inc., New York, 1920. Cloth, 6 x 9 in., 256 pp., illus., diagrams, tables, $3.

This volume, the first of a series of three based on the author's lectures at the University of Minnesota, deals with general principles and with some of the more important methods used in general investigations in the metallographic laboratory. Future volumes will discuss the metallography of the metals and alloys and the applications of metallography to the metallurgical and engineering industries.

MOTOR VEHICLE ENGINEERING. Engines (for Automobiles, Trucks and Tractors). By Ethelbert Favary. Second edition. McGraw-Hill Book Co., Inc., New York, 1920. Cloth, 6 x 9 in., 333 pp., illus., plates, tables, $3.50.

The author states that many chapters of this edition have been entirely rewritten, that the practical data have been brought up to date and that the work has been generally revised. The book treats of automobile engines and is intended to give a concise, simple statement of the every-day information needed by automotive engineers.

THE ORGANIZATION OF INDUSTRIAL SCIENTIFIC RESEARCH. By C. E. Kenneth Mees. First edition. McGraw-Hill Book Co., Inc., New York, 1920. Cloth, 6 x 8 in., 175 pp., $2.

Contents: Types of research laboratories; coöperative laboratories; position of the research laboratory in an industrial organization; internal organization of industrial research laboratories; staff of a research laboratory; building and equipment of the laboratory; direction of the work; design of a research laboratory for a specific industry.

This volume is a contribution to the current discussion of the relation of scientific research to industry. The author is concerned with a study of the best methods of organizing research work for industrial purposes and of the conditions under which such work should be conducted, and has made an effort to give definite suggestions upon these points. The book is intended for those who plan to undertake research work rather than as an exposition of its theoretical advantages.

317

ORGANIZING FOR WORK. By H. L. Gantt. Brace & Howe, New York, 1919. Cloth, 7 x 5 in., 113 pp., $1.25.

Our civilization depends, according to Mr. Gantt, upon the effectiveness with which our combined industrial and business system works, and recent revolutionary attempts to overthrow it are due to the failure of the present system to recognize fully its responsibility to the community. The author believes that there must be a return to the principle that the first aim of business is to render service to the community and that this result can be peacefully obtained by the use of familiar methods, whose use for the purpose he discusses in the present book.

PRINCIPLES OF METALLOGRAPHY. By Robert S. Williams. McGraw-Hill Book Co., Inc., New York, 1920. Cloth, 8 x 5 in., 158 pp., 75 illus., tables, $2.

This is a brief introduction to the subject written for students of general science or engineering, who do not specialize in metallography, but who will use it to a limited extent in connection with their professional work, and for general readers who wish an introduction to more specialized books. Particular attention is given to the applications of metallography.

THE ROMANCE OF A GREAT FACTORY. By Charles M. Ripley; introduction by Dr. Chas. P. Steinmetz. The Gazette Press, Schenectady, N. Y., 1920. Cloth, 6 x 9 in., 200 pp., illus., $2.

This book, which deals with electrical subjects, especially the contributions made by the Schenectady Works of the General Electric Company, opens up a new line of thought. It portrays the romantic or human-interest side of the present-day electrical industry. The thousand-word introduction by Dr. Steinmetz and the large number of illustrations, photographs and artists' drawings of electrical scenes in and out of the shop are prominent features of the book.

STATISTICS IN BUSINESS. Their Analysis, Charting and Use. By Horace Secrist. McGraw-Hill Book Co., Inc., New York. 1920. Cloth, 6 x 8 in., 137 pp., illus., diagrams, maps, tables, $1.75.

This volume has been prepared to serve as a handbook for executives in the application of business statistics to problems which arise. It aims to present briefly and concretely the reasons why statistics should be used in business analysis and to illustrate how and with what effect they may be applied to the solution of business problems. The discussion is of a practical nature. Especial attention is given to the use of charts and graphs. Examples of good and bad usage are given.

A SYSTEM OF PHYSICAL CHEMISTRY. By William C. McC. Lewis. Second edition, three volumes. Longmans, Green & Co., New York, 1918. Cloth, 6 x 9 in., diagrams, $4.50.

Professor Lewis's work is intended as a general textbook for use by those who already possess some knowledge of both physics and chemistry and as a fairly comprehensive account of our present knowledge. It is well equipped with references to the original literature and so is useful for reference use.

A third volume, on the quantum theory, has been added in this edition. Other changes, necessitated by the advances made since the previous edition have caused considerable expansion of the book.

TECHNISCHER LITERATURKALENDER. 1918. München-Berlin. R. Oldenbourg. No date. Cloth, 6 x 8 in., 640 columns, 1 portrait, 12 marks plus 20 per cent increase.

This volume is a concise biographical dictionary of living German authors of technical and scientific books, compiled by the Librarian of the German Patent Office. Over five thousand writers are included. The information given includes their addresses, education, age, occupations, technical specialties, publications, etc.

WHITE LEAD—ITS USE IN PAINT. By Alvah Horton Sabin. First edition. John Wiley & Sons, Inc., New York. 1920. Cloth. 5 x 8 in., 133 pp., tables, $1.25.

In his small work Dr. Sabin has provided a brief, non-technical account of the subject for the instruction and guidance of those whose need and use of white-lead prompts them to seek knowledge about it of a simple but reliable sort. The various methods of making white-lead are given, together with the properties of the usual substitutes; methods of mixing paint for different purposes, etc.

## Industrial Training in China

(Continued from page 305.)

as well as materially, is such industrial development as will put the great mass of the people as many steps as possible from starvation.

This University would like to coöperate with some such institutions as the A.S.M.E. and other engineering bodies connected with and interested in the promotion of engineering in its various branches. They are at present planning to include in the University a really great American technical school which shall have as great an influence in shaping the engineering progress of China as possible. The proposed enlargement of scope contemplates a far-reaching plan and requires a broad and substantial foundation in the way of equipment, staff and endowment. They wish to make the instruction the most thoroughly practical kind possible.

Another phase of the educational situation is that of the Chinese students who study abroad. From one school in Paotingfu alone, fifty or more students have gone to France in the past two years to get further technical training and experience. Some six thousand Chinese students are expected to go to France under these conditions this year, and this is only the beginning. The students from Paotingfu receive 15 francs a day for their work and have to work continuously for two years before they can attend school. They have training in the French language before leaving China. Students who are willing and anxious to earn their way in such a manner are the kind who will do constructive and important work on their return home.

The students going to France, in a great majority of cases, would prefer to go to America, if they were permitted to earn their way. Considerable effort is being made to make this possible. There should be some well-organized society for encouraging such a condition in America, as there is now in France. It would have most beneficial effects on both China and America. The writer holds invitations from thirteen of the leading machine-tool manufacturers of America for some fifty Chinese students to come to their shops for practical experience, and at the same time, continue their studies. The students who wish to fill these positions are anxiously waiting for some decision of the State Department which will allow them to enter the country. They have had some shop experience, have studied English and are graduates of what would correspond to the American high school. Can the A.S.M.E. use any influence in bringing about a change or modification of laws permitting the entrance of such men to America? The obstruction at this point seems to be the fact that such students work for wages. The American commercial organizations over here are all anxious to see such a modification brought about as will permit the admission of such students. The sooner that can be accomplished, the sooner the tide will be directed in our favor.

During the past summer, the National Machine Tool Builders' Association has contributed to Chinese industrial schools over twenty machine tools, either as gifts or sales at half price. Such things as this contribute vastly to good feeling between the two countries. The students who have the use of these superior machines will naturally have a great appreciation of the generosity of the givers which in years to come will react favorably for the country and firms contributing. There is room for much more good work of the same kind, not only in machine tools, but in small tools and other equipment in the various lines of industrial training such as textile work, tanning, agriculture, mining, etc.

The writer plans returning to America this spring as a delegate from the American Chamber of Commerce of China to the Convention of the National Foreign Trade Council to be held in San Francisco in May. He plans also to visit the East and hopes to have the opportunity of discussing these matters further with A.S.M.E. members.

Paotingfu, China.                FRANK A. FOSTER.

# MECHANICAL ENGINEERING
## THE JOURNAL OF THE AMERICAN SOCIETY OF MECHANICAL ENGINEERS

| Volume 42 | June 1920 | Number 6 |

# The German Defenses on the Coast of Belgium

### A Description of the System of Fortification Employed, Together with Particulars Regarding the Guns, Fire-Control Stations, Shelters, etc., and Details of the Mechanisms of the Various Batteries

#### By LT.-COL. H. W. MILLER,[1] U. S. A.

ON October 15, 1914, the Germans occupied Ostend and all of the Belgian coast north of that point. During the next few days their offensive gained for them the coast for a distance of about 10 miles farther south, including the city of Nieuport. It is understood from inhabitants of the city of Ostend that on the morning of October 15 the last boat loads of refugees left the harbor for England. The line of the farthest advance of the Germans in 1914 ran in a snake line south from the city of Nieuport, placing the cities of Dixmude, Poleappelle, and Ypres within the German lines. They were unable to hold all of their gains against the offensives of the Belgians and British during the next few months, and their line was pushed back to a point between Nieuport and Ostend, at Westende, about 8 miles south of Ostend. This line became a part of what has been known familiarly as the Hindenburg Line. From Westende

line. Parts of the city are very old, dating back at least a half-dozen centuries. In this part of the town, which is mainly along the old harbor, the streets are narrow and characteristically crooked, the houses of quaint construction, and it does not require much imagination to make one believe that he has gotten back into the time of Columbus. It is understood that the old city of Ostend was heavily fortified and withstood several lengthy sieges. The last of these fortifications were removed in 1865, and only the slightest traces of them are still remaining. Other parts of the city are of quite recent construction and likely date from 1898, since which time considerable development has taken place. The streets in this section of the town are much wider, and straight, and the buildings of modern construction.

For many centuries Ostend figured prominently in the ocean trade between Belgium and the Indies and for a long period the

Fig. 1  German Railway Artillery Used for Coast Defense

it passed somewhat to the east of Nieuport and directly through Dixmude, about a mile to the east of Poleappelle, 5 miles to the east of Ypres, resting on what is known as Zonnebeck Ridge. This line was heavily fortified by the Germans, and their men were provided with adequate shelters. Even in the great offensives of 1918 this end of the line remained practically stationary. Four years after the original occupation, almost to a day, the Germans were driven out of Ostend, on October 17, and from Zeebrugge farther north, on the 18th.

Ostend, which figures as practically the center of these coast fortifications, numbered about forty thousand inhabitants at the beginning of the war. It is the second seaport in Belgium and stands about at the middle of Belgium's forty-two miles of coast

company known as the Ostend Company was very prominent in foreign shipping. The city is now equally prominent as a port and as practically the most attractive summer resort on the coast of France or Belgium. It is the home port of a large fleet of fishing vessels and the port of entry for a great part of Belgium's medium-tonnage ocean commerce. In 1900 work was begun on the harbor with a view to extending it back a distance of about two miles and providing a large basin and docking facilities for a considerable number of boats of medium tonnage. It is not known whether this program has been completed or not; apparently not, however, since the distance inland to the present locks is certainly not two miles. Extensive and heavy quays are provided all along the harbor for any boats that can negotiate the entrance. The quays nearest the entrance are the oldest and are now used almost entirely by the fishing boats, while the larger boats proceed farther up the harbor. The entrance to the harbor is quite narrow, being apparently not over two or three hundred

[1] Office of Chief of Ordnance, Artillery Division, Washington, D. C. Released for publication by the War Department, Office of the Chief of Ordnance. Delivered as a lecture at the Spring Meeting, St. Louis, Mo., May, 1920, of THE AMERICAN SOCIETY OF MECHANICAL ENGINEERS. Subject to revision.

feet in width. At the time of this inspection the hull of the British boat *Vindictive* was lying along and parallel to the northern side of the entrance, taking up about one-third of the space.

There is a promenade or "digue," constructed entirely of granite, on top of the dune in front of the city, which extends from the entrance to the harbor south for a distance of about two miles. The beach from a point near the entrance to the harbor for a considerable distance south is almost unsurpassed. There is a long line of fine summer hotels along the beach among which is

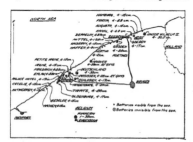

the famous Royal Palace Hotel, which has been very seriously damaged by the Germans, and in front of which one of the 17-cm. batteries and one of their fire-control stations was installed.

The Kursaal or Casino at the northern end of the principal bathing section of the beach was badly damaged by the Germans during their stay. The floors were removed, the windows knocked out, and the building generally damaged. Other portions of the city were badly damaged by bombs dropped from planes. The writer was told by the proprietor of the hotel at which he was staying that most of the damage seen about the town had been done by young German aviators, who, returning from a fruitless bombing trip, usually released a few of their bombs on some prominent object in the city. The writer witnessed an interesting conference between a number of prominent citizens, at which plans were discussed for the awarding of some simple medal to those citizens who at considerable risk of their lives had rendered service in putting out fires started by these bombs and in rescuing wounded people from the damaged buildings. Those buildings which to all intents and purposes were intact had been almost completely robbed of their contents.

The locks leading from the head of the harbor into the canal to Bruges had been blown up, but whether by Allied air bombs or by the Germans on evacuating the area it is not known. They were being repaired in March 1919. An interesting bit of the Flemish language was seen on a sign at the end of a temporary foot bridge thrown across the locks. This sign read: "Est ist verboden te brugg over te gann."

## THE COAST AND COUNTRY

The coast of Belgium is low and sandy, and the slope of the shore is very gentle. Apparently there are only three points at which it is possible for ships to make an entrance, and it would be exceedingly difficult for even large rowboats to make a landing elsewhere, because of the fact that they would be grounded at a point several hundred yards beyond the shore line. All along the coast there is a sand dune which is about 50 ft. above the mean level of the ocean, and from 20 to 30 ft. above the average level of the land. This dune is not so pronounced below the city of Ostend as above. The Germans very systematically planted grass in these dunes in an attempt to prevent their shifting. It is quite probable that the sand gave them a great deal of difficulty in the maintenance of their artillery.

The land behind the dunes is comparatively level, so level, in fact, that it is inclined to be swampy in many places unless special care is taken in its drainage. In consequence, one finds drainage canals and ditches in all directions.

As previously mentioned, the Germans were very careful in building up fortifications along their line to provide concrete shelters for their men which made them comparatively comfortable and dry. It will be remembered that both armies paid considerable attention to the damming up of canals in the attempt to drown each other out, and there is a newspaper record of one case in which about 40,000 Germans were drowned. The accuracy of this statement, however, is doubted.

In March 1919 the canals and ditches about Nieuport were just being cleaned and repaired. Water was then flowing off some of the land that had been inundated for several years. In all directions platforms on stakes were visible in the grass, and in some cases board walks built on piles could be seen running for long distances. All sorts of platforms, shelters, etc., had been improvised. Usually where the line crossed roads, heavy concrete machine- and field-gun shelters had been constructed on both sides, these leaving spaces about 12 ft. broad in the middle, which were heavily barricaded with barbed wire. In numerous places the bones of both horses and men could be seen sticking out of the mud where they had fallen, and where it had been impossible to reach them and give them decent burial.

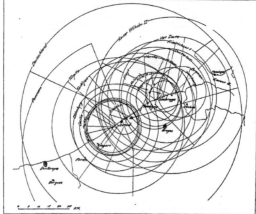

## INSTALLATION OF THE FORTIFICATIONS

It is quite probable that shortly after their capture of Ostend and the occupation of the coast of Belgium, the Germans began the installation of small-caliber coast guns. It does not seem probable that they could have paid any particular attention to the installation of the extremely heavy guns at once, although they may have done so. As mentioned before, it is impossible for boats of any size to make a landing except at Zeebrugge and Ostend, and inasmuch as these two places had at first no particular value to the Germans, it is likely that they concluded that the Allies would not make any serious attempt to take them, since they would have to maintain considerable forces there and would have to use a considerable portion of their transport facilities in supplying them. When the German General Staff gave serious consideration to the plan for their major submarine campaign, it is probable that the harbors of Ostend and Zeebrugge loomed up prominently in their plans and that at once they set to work to fortify these two points with heavy guns.

The closest point on their own coast that could well be used as a base for submarines is Wilhelmshaven. They did not expect to operate their submarines to any great-extent in the North Sea, hence there was nothing to be gained and a great deal to be lost by having their base so far from their scene of operations. The distance from Ostend to Wilhelmshaven is about 350 miles. With a base at Ostend for supplies and a well-protected point farther inland for minor repairs, they were saved a great deal of time and could more effectively annoy the Allies in the destruction of their shipping. The harbor of Ostend is ample for sheltering a fleet of submarines of whatever size, and it is connected by a canal with the city of Bruges, fourteen miles inland, where the submarines might be

FIG. 4 BATTERY POMMERN (380-MM. GUN) AT LEUGENBOOM, BELGIUM

repaired in the already existing basins. In recent years another canal has been constructed connecting the city of Bruges with the coast at the little village of Zeebrugge, fourteen miles north of Ostend. The village of Zeebrugge has no railroad facilities and there is no real harbor, boats entering there being lifted by the locks into the canal and proceeding to the docks at Bruges. It is probable that the submarines which required maintenance entered by Zeebrugge, since they could proceed more rapidly and easily inland than from Ostend. The extent of the repair facilities provided at Bruges is not known.

As soon as they decided to make use of the port of Ostend and the entrance at Zeebrugge, as well as the basins at Bruges for the supply and maintenance of their submarines, it became imperative that the Germans protect these two points against the raids the

Allies might undertake in the attempt to bottle up the harbors or seriously damage the locks. To this end they started constructing the emplacements for their heavy guns at both points in 1915.

To the best of the writer's knowledge the installation of all the primary and secondary defenses, with the exception of those known as railway batteries, was complete in 1916. While the writer was with the British Fourth Army at Ypres in March 1918, he was told by the British Ordnance officers that they had received very authentic reports to the effect that the Germans were in-

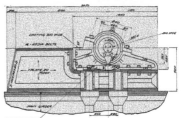

FIG. 5 TRUNNION-BEARING DETAIL OF GERMAN 380-MM. MOUNT, BATTERY POMMERN.

stalling railway batteries at various places along the coast. They had been given to understand that the guns for these railway batteries had been removed from the German Second Navy. The 28-cm. guns on railway mounts that were operated in the Batteries Preussen, Hannover and two other batteries not included on the map were probably some of these guns. The emplacements for three of these railway-mount batteries did not make their appearance in the air photographs until the spring of 1918.

### LOCATIONS OF THE ARMAMENT

As a consequence of the practical impossibility of making any landing except at Zeebrugge and Ostend, no attempt was made to fortify the coast at other points. The heaviest of the fortifications were centered about Ostend, with others not much less effective about Zeebrugge. The primary armament, including the heaviest guns, was generally located at a distance of from two to five kilometers from the shore. The 38-cm. gun of the Battery Pommern at Leugenboom, the farthest point inland, was likely placed primarily for land service. This gun could operate against both Dunkirk and Ypres, and it did operate very extensively against the city of Dunkirk. It is probable, though not certain, that this gun took some part in the shelling of the city of Ypres when it was being reduced to ruins in 1915. The existence of this gun was well known to the men of the army about Ypres, with which the writer was associated for a time in March 1918.

The distribution and caliber of all of the 30 batteries on the coast is shown in Fig. 2. All of the batteries shown by dots were located on top of the dunes and were visible from the sea. Those batteries indicated by dots enclosed in rings were behind the dunes and had to depend upon the stations located in the dunes for their observation. The Palace Hotel Battery of 17-cm. guns was installed on the broad promenade just in front of the Royal Palace Hotel. This battery was particularly conspicuous. Battery Tirpitz just south of the city of Ostend was located in very swampy land, and it is understood that a great number of piles had to be driven to render the concrete emplacement stable. Considerable

FIG. 6  BATTERY DEUTSCHLAND
Note camouflage, wickerwork, and barbed-wire entanglements.

attention was paid to the draining of the region around this battery and Battery Oldenburg. All of the other inland batteries were installed on comparatively dry and solid ground.

### FIRE CONTROL

Fire-control stations, particularly of those batteries located behind the dunes, were installed in the dunes on either side of Ostend and Zeebrugge. The stations for the batteries about Ostend were located, one on the promenade in front and to the north of the Royal Palace Hotel near the Palace Hotel Battery, and the other in the dune near the Battery Petite Irene. These stations are located in pairs and determine the position and range of a target by the method of triangulation. Specially devised rapid-operating plotting boards are provided to convert the readings of the observers into ranges and azimuths. Observations are made at regular intervals and the resulting data phoned from the plotting room to the various batteries which it controls. A third auxiliary station was found on the southern edge of the city. This probably was to be used in the event that the station in front of the Royal Palace Hotel, which was very conspicuous, might be damaged. This station was camouflaged as a house, having windows and doors painted on it. The men stationed in it observed the boats at sea through an 8-in. slit at the top. The stations located in the dunes were entirely covered with sand and it was practically impossible to see them from the sea. From some British officers in Ostend it was learned that the German batteries had been able to land heavy projectiles on the decks of some of their monitors at 40,000 yd. Somewhat similar feats are reported for the heavy guns of some of the German ships. This is exceptionally fine shooting and attempts were made to determine the methods of fire control employed in operating such ranges. No information to the writer's knowledge has been secured to date. The German ships as well as the fire-control rooms of the Belgian coast fortifications had been stripped of their fire-control apparatus before being surrendered or abandoned.

### SHELTER

In all of these fortifications the Germans fully lived up to their reputation for being strong on concrete shelters. Numerous such shelters for the men were found in the dunes and about the inland batteries. A partially constructed shelter was found just beside the locks at the upper end of the harbor of Ostend. If all the reinforcement which was exposed were covered with concrete, the roof of the shelter would be about 1.5 m. or 60 in. thick. Accord-

ing to their usual custom, several meters of earth would be placed on the top of this, making a very effective shelter.

### SCOPE OF SERVICE

The range and scope of service of the various batteries is shown in Fig. 3. All of the guns, except the 38-cm. gun Pommern at Leugenboom, were capable of being traversed 360 deg. The range of 55 km. reported for the Battery Deutschland seems extreme. Various reports conflict in this respect, some crediting the batteries with a range of 42 km., some with a range of 47 km., and the last report, given in the *Bulletin Renseignements de l'Artillerie* of January and February 1919, with 55 km. The ranges given in Fig. 3 are in accordance with the report in the publication just mentioned.

### ALLIED ATTACKS ON OSTEND AND ZEEBRUGGE

During the spring of 1918, April 23 to be exact, an attempt was made to block the harbors of Zeebrugge and Ostend. The mole at Zeebrugge was pierced and an obsolete cruiser, loaded with

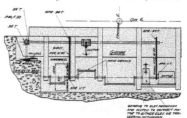

FIG. 7  TRAVERSING MECHANISM (ELEVATION) OF GERMAN 380-MM.
MOUNT, BATTERY DEUTSCHLAND

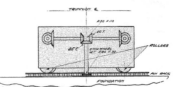

FIG. 8  TRAVERSING MECHANISM (REAR ELEVATION) OF GERMAN
380-MM. MOUNT, BATTERY DEUTSCHLAND

concrete, was placed across the entrance of Ostend harbor. The boats engaged in these undertakings received terrific punishment, but apparently accomplished their mission. It seems probable that had the Allies cared to make the sacrifice, it would have been possible to force either harbor, although it is not very likely that they could have retained possession of them for any great length of time. At the time of the inspection of these defenses, the writer observed the *Vindictive* still at the entrance to Ostend harbor. It had been raised and moved to a position as close as possible to and parallel with the north side of the entrance.

In the attack upon Zeebrugge the *Vindictive* was under the fire of a 150-mm. German gun at ranges of from 200 to 500 yd. for approximately one hour. Portions of the superstructure of the boat were injured, but the *Vindictive* was not prevented from fulfilling its mission of landing a force of marines from two ferry boats which it had in tow. Certain facilities on the mole and in the harbor of Zeebrugge were destroyed and the mole was broken. This mole is simply a wall or breakwater extending into the ocean for a distance of about 100 yd. on the southern side of the entrance to the harbor.

### DESTRUCTION FROM SHELL FIRE AND BOMBS

In spite of the fact that a majority of the batteries were located on top of the dunes and in plain sight of the sea, there is no evidence that any of them were damaged by shell fire. It is probable that no firing was done against the Palace Hotel Battery because of the damage that would be done the large buildings round about. It is understood on good authority that the Allied ships paid constant attention to both those batteries installed on the dunes, as well as those located behind the dunes. As early as 1916 the exact locations of practically every battery behind the dunes were determined from airplanes. Some of the pictures are shown in Figs. 12, 20, and 30. It is understood that the monitors came within a few kilometers of the coast at night, camouflaging their gun fire by blinding flares. It is reported that on practically all occasions when the ships shelled the coast fortifications during the day, heavy smoke screens were at once set up by the Germans which evidently afforded effective protection. Just back of the Battery Irene, between the dune and road, a large number of steel pots or cylinders 18 in. in diameter and 24 in. in height were found, which had been used by the Germans in setting up their smoke screens. The writer saw many of the holes made by airplane bombs and shown in Fig. 30, but could not find any single case in which either the inland guns or the guns in the dunes had been struck by shell firing from sea or by bombs dropped from airplanes.

The British monitors controlled some of their fire by the scheme of triangulation. Knowing that the Germans would at once set up a smoke screen between the batteries fired upon and the monitors firing, they were in the habit of placing one boat of inconspicuous construction a great distance off to serve as observer. This observing boat was so located as to be able to see behind the smoke screen. For firing on the heavier batteries behind the dunes, it was of course necessary to operate by indirect fire.

It is significant to note that when it became necessary for the Germans to evacuate this area in August 1918, the only artillery that they were able to get out was the railway artillery. From the inhabitants of Ostend it was learned that all of the batteries in that vicinity had been operated quite continuously against the Allied land forces for some days before the evacuation.

### FINAL DESTRUCTION OF THE BATTERIES

Just before the area was evacuated, all the guns were destroyed with the exception of the 38-cm. gun at Leugenboom. One method seems to have been employed on the guns. In each case the rotating band was removed from one projectile, which was rammed into the bore of the gun; a second projectile was then rammed in and the gun fired. When the rear projectile struck the forward projectile it detonated, and in every case, except in the Battery Pommern as noted above, the breech of the gun was completely blown off and in most cases the carriage practically wrecked. In some cases the forward projectile likewise detonated, swelling or tearing off the muzzle of the gun; in other cases it was simply projected a short distance out of the gun. The guns and carriages after being wrecked were practically valueless, and no attempt was made to keep them in condition.

### PURPOSE OF THE INVESTIGATION

Two investigations were made of these defenses. One of these was for the purpose of studying the tactical distribution of the entire armament. Until this war it has apparently been an unde-

cided question as to whether any point on a coast or any section of a coast line can be so fortified as to be impregnable to attack from sea. During this war there were three sections of coast line that were so fortified as to be considered practically impregnable, except at a prohibitive cost, to attack from the sea. These were the section of German coast at Kiel, defended by mine fields and the fortifications at Heligoland; the Turkish center of Constantinople protected by the fortifications of the Strait of Gallipoli; and the Belgian coast protected by the fortifications of the only two landing points, Ostend and Zeebrugge. The disastrous attempt and failure to force the Straits of Gallipoli instilled in the Allies a wholesome respect for the difficulties involved there, and to the best of the writer's knowledge no real attempt was made to force the defenses of Heligoland and Kiel. It is probable that the defenses at Ostend could have been forced if the Allies had been willing to pay the price, but apparently they were not willing to do so. The investigation on the tactical distribution of this

FIG. 9   PEDESTAL AND PINTLE BEARING OF THE BATTERY DEUTSCHLAND

FIG. 10   380-MM. PROJECTILE AND SHOT TRUCK

latter armament was made by Majors Armstrong and Norton, Coast Artillery Corps, U. S. A., and their report was published in the *Journal of the U. S. Artillery* during the months of June, July, and August, 1919.

The second investigation was made by the writer for the purpose of determining whether the Germans had followed fixed policies in the designs of the various parts and mechanisms of their heavy guns and carriages, for example, cradles, carriages, armor, elevating and traversing mechanisms, etc., and for the purpose of making a detailed study of these parts and mechanisms. With ordnance designers of our own country and of the countries with which we have been associated in this war, there have been a number of questions on which there has been and still is some difference of opinion. Some believe that the cradles of heavy guns should be heavily braced by ribs, while others feel that there is no reason why they should not be simple smooth cylinders. Some designers are in favor of a front-pintle type of carriage

FIG. 11 BATTERY DEUTSCHLAND
Air photo taken August 9, 1916.

with simple friction bearings, or at best, roller bearings with racers of small diameter and more or less crude wheels or large rollers at the rear of the carriage on which it may be traversed. Others believe in a central-pintle type of carriage with large roller paths. It is seen at once that there is a vast difference in the difficulties involved in the manufacture of carriages designed under these different principles. Cradles with heavy ribs on the outside are difficult to cast and do not lend themselves to rapid manufacture. If smooth cylinders will answer just as well, it does not seem wise to hamper the manufacturers with the other design. Front-pintle carriages are in general easier to manufacture than the center-pintle type, hence unless there is something very vital to be gained in the central-pintle type of carriage with its large and difficult-to-manufacture roller paths, it would seem that the first design should have preference.

## DETAILS OF THE MECHANISMS OF THE VARIOUS BATTERIES

### BATTERY POMMERN

*Gun.* This battery, Fig. 4, consists of but one gun, Model 1914, Krupp No. 15 L. Its weight is 77,530 kgs., overall length 17.13 m. and length from breech block to the muzzle 16.13 m., giving it an effective length of 42 calibers. It seems to be rated in some reports as a 45-caliber gun. This rating is evidently based on its overall length. There are 100 grooves, and the twist of the rifling is to the right 1 cm. in 10. The diameter of the powder chamber is 42.5 cm., and the outside diameter of the breech 1 m. The breech block is of the usual Krupp sliding-wedge type. The design of the gun is likewise identical with that of the 38-cm. guns of the Battery Deutschland, Fig. 6. In this figure it will be observed that the breech section has been blown away. The scheme of at-

taching the breech section by the interrupted-ring method is well shown. The recoil lug has a bearing at the bottom on the two sides of the recuperator cylinder for the purpose of preventing rotation of the gun.

*Cradle.* The cradle is a cylinder of simple design, having ribs at long intervals and a depth of only about 3 cm. The walls of the cylinder have a minimum thickness of 10 cm. and a maximum thickness of 13 cm. over the ribs. The diameter of the main trunnion, Fig. 5, is 46 cm. and the length 33.5 cm.; the diameter of the small trunnion 20 cm. and the length 19 cm. The anti-friction mechanism is of the rolling-wedge type. This cradle was provided with a counterweight identical with that shown on Fig. 6, and its front is so designed as to close the opening in the armor turret. Its total length is 3 m. and it is provided with brackets for the recoil and recuperator cylinders on the bottom. The cradle is lined with a bronze liner, approximately 6 mm. thick and 1 m. in length both at the front and rear. A portion of this liner may be seen in Fig. 6.

*Recoil Mechanism.* The two recoil cylinders are carried in brackets on either side of the center on the bottom, and the filling plugs are on the ends of the buffers. The recuperator cylinder is carried in the center in similar brackets and likewise has a filling plug on its forward end. Its rear end, which is planed on both sides, serves as a guide for the breech lug, to prevent rotation of the gun.

*Elevating Mechanism.* See the description of the elevating mechanism for the 38-cm. guns of the Battery Deutschland. The only difference between the two is that in this battery there is no provision for hand operation. The report found in the office of the Belgian Chief of Artillery stated that this gun had been originally provided for hand operation only, but that within the last year of the war it has been equipped with motors for electrical operation. The maximum elevation obtainable is 45 deg.

*Traversing Mechanism.* See the description of the traversing mechanism of the Battery Deutschland. There is provision for both hand and power operation of this mechanism. The wheel for hand operation, however, is located on the left of the carriage, very close to and on a level with the traversing rack. It is 1 m. in diameter. The maximum traverse provided for is about 157 deg. The center line of this field of fire passes approximately through Dunkirk.

*Carriage.* See the detailed description of the carriage for the 38-cm. guns of the Battery Deutschland. The only difference between these carriages is in the armor provided on the guns of the Battery Pommern. This will be described later under the heading of Protection.

*Emplacement.* The emplacement for this carriage is of exceed-

FIG. 12 BATTERY KAISER WILHELM II, NEAR KNOCKE

ingly massive concrete construction. The diameter of the central pit or well, in which the carriage is placed, is approximately 22.439 m. The form of the pit is not a complete circle, but is so shaped as to allow the carriage to be traversed 157 deg. The depth to the level of the traversing rack is 3 m. and the additional depth to the floor on which the center pintle rests 1.5 m., making a total depth of 4.5 m. On either side there are practically identical concrete structures, one of which is shown to the right in Fig. 4, for the housing of ammunition and personnel. One structure is for projectiles, the other for tools. Between these two structures at the front there is a concrete parapet 2 m. high and 3 m. thick. The earth slopes gradually away from the top of the parapet at the front, dropping about to the level of the main floor of the emplacement. The thickness of the roof of the structures is 3 m., the total height above the floor being 5.5 m.

*Ammunition-Supply System.* As just noted under the previous heading, the projectiles were stored in the concrete storehouse on the right and the powder in the storehouse on the left. Three weights of projectiles were used. The ammunition was supplied entirely by hand, shot trucks being used. The arrangement of the storehouse is in general, similar to that for the 305-mm. gun of the Battery Kaiser Wilhelm II. (See Fig. 19), the projectiles being piled two high. The report of the Belgian Chief of Artillery states that the projectiles were rammed by twelve men and that the rate of fire was 1 shot in 5 min. with electrical operation of the elevating mechanism, and 1 shot in 10 min. with hand

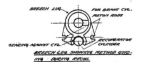

FIG. 13  BREECH LUG. SHOWING METHOD OF GUIDING DURING RECOIL, BATTERY KAISER WILHELM II

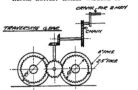

FIG. 14  TRAVERSING GEAR. BATTERY KAISER WILHELM II

operation of the elevating mechanism. The report further states that according to reports of people living at Leugenboom, the personnel originally provided for operation of the mount when it was operated by hand was one captain, two lieutenants, ten non-commissioned officers, and one hundred and sixty men. After provision was made for electrical operation of the various mechanisms, the personnel was reduced to one captain, two lieutenants, five or six non-commissioned officers, and seventy men. The shells fired by the 38-cm. gun are as follows:

| Name | Weight kg. | Fuse | Range km. |
|---|---|---|---|
| H. E. 38 cm. Sp. Gr. L/4.1 Bds. with base fuse.................................... | 730 | Sp. Grn. R. | 42 |
| H. E. 38 cm. Sp. Gr. L/3.6 m. Bds. (m Haube) with false ogive and base fuse... | 695 | Sp. Grn. R. | |
| Do.: without false ogive.................. | 600 | | 44 |
| H. E. 38 cm. Sp. Gr. L/4 Bds. A. Kz. (mHaube) with double fuse at head and base false ogive............................ | 342 | Sp. Grm. K. | |
| Do.: Without false ogive................. | | | 48 |

The charges are three in number:

| | | | |
|---|---|---|---|
| 1 Hutzenkartasch | Cartridge containing | 87 kg. |
| 2 Vorkartasche | Charge containing | 96 kg. |
| 3 Vorkartasche | Charge containing | 118 kg. |

It seems then that the gun can fire with two charges:

*a*  87 kg. + 118 kg. = 205 kg.
*b*  87 kg. + 118 kg. + 96 kg. = 301 kg.

*Protection.* To protect the personnel operating this gun against aircraft bombs and aircraft machine-gun fire, the carriage

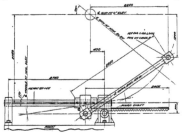

FIG. 15  ELEVATING MECHANISM (ELEVATION) OF GERMAN 305-MM. MOUNT, BATTERY KAISER WILHELM II

was covered with 6-cm. flat armor. This plating extends to within a few centimeters of the floor of the pit. The hole in the front through which the gun projects, is sealed by the small shield on the front of the cradle.

*Discussion.* The parapet has been blown away in the center (See Fig. 4). There is evidence to indicate that the Germans used their characteristic method in attempting to destroy this

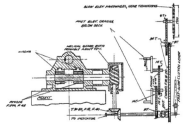

FIG. 16  ELEVATING MECHANISM (SECTION) OF GERMAN 305-MM. MOUNT, BATTERY KAISER WILHELM II

gun, but that in this case neither projectile detonated until they struck the parapet in front. Neither projectile could be found, hence it is assumed that both detonated. The gun was damaged only to a slight extent at the muzzle where some fragments of the projectiles were blown back into the bore, scoring it deeply.

### BATTERY DEUTSCHLAND: 38-CM. GUNS

*Gun.* See the description of the 38-cm. gun of the Battery Pommern. The design of these four guns is identical with the design of the Model 1914 Krupp Gun, No. 15-L. The gun, Fig. 6, is Model 1916, No. 36-L. Its weight is given as 77,562 kg. This is Gun No. 4 of the battery; No. 3 is likewise a Model 1916, and numbered 35-L. Gun No. 2 is Model 1916, No. 41-L, and Gun No. 1 is Model 1914, No. 9-L.

Each of these guns had been destroyed by the characteristic German method of placing one projectile in the bore of the gun and firing another projectile into it. This procedure had resulted, in this case, in the blowing off of the entire breech and in the wrecking of the carriage. There was little evidence that any of these guns had been used to any great extent. The lands and grooves were in well-nigh perfect shape at least. It was evident that such damage as was visible had not been caused through ordinary fire. The width of the lands is 6 mm., and the depth of the grooves 3 mm.

*Cradle.* The cradle on each of these four guns is identical in design with that of the 38-cm. gun of the Battery Pommern. The counterweights on the Battery Pommern and Battery Deutschland cradles are identical. The one point of difference between the cradles of three of these guns and the cradle of the gun of the Battery Pommern is in the lack of any provision for the reduction of the friction of the trunnions. In Fig. 6 it will be observed that there are no auxiliary trunnions and no friction-reducing mechanism. There is no friction-reducing mechanism on Guns Nos. 2, 3,

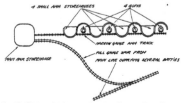

and 4, but Gun No. 1 has one identical with that of Battery Pommern, Fig. 5. This omission of anti-friction devices on Guns Nos. 2, 3, and 4, and which by the way are Model 1916 guns, can hardly be attributed to the lack of time for the installation, since the small amount of machine work necessary in the shop might have been handled without difficulty, and the work finished in the field if desired. Apparently they preferred to rely on a surplus of man power in elevating their guns.

*Recoil Mechanism.* The recoil mechanism is composed of two recoil cylinders and one spring pneumatic recuperator cylinder, all located at the bottom of the cradle. It is identical in design with the recoil mechanism of the Battery Pommern gun.

*Elevating Mechanism.* The elevating mechanism on each of these guns, as well as on the 38-cm. gun of the Battery Pommern, is composed of two large telescoping screws, the larger screw passing through a nut carried in an oscillating bearing at the bottom of the carriage. The larger of the elevating screws can be seen about in line with one of the elevating and traversing handles at the bottom of the carriage in Fig. 6. The larger of these two screws is about 37 cm. in diameter and the smaller about 20 cm. The large nut in the oscillating bearing is driven through cross-shafts either by the motor or by the hand mechanism which is used likewise for the traversing mechanism.

The maximum elevation at which the gun is operated is not known. There were no elevating arcs remaining on any of the cradles that would indicate to what extent they could be elevated. It is assumed, however, that the maximum elevation obtainable is the same as that in the Battery Pommern gun, which is 45 deg. The box just to the rear of the camouflage near the bottom of the carriage, Fig. 6, is the housing for the clutch connecting the elevating mechanism with either the motor or the hand drive. All of this mechanism is duplicated on the left side of the carriage. The motor driving the elevating and traversing mechanisms is on the same platform as the transmission box, but, as will be seen,

it is practically hidden by the wire camouflage which has fallen.

*Traversing Mechanism.* The various details of the traversing mechanism are shown in Figs. 7 and 8. Provision is made for its operation by hand as well as by motor. This hand mechanism is operated by eight men.

The rear of the carriage is supported on two heavy rollers 96 cm. in diameter by 23 cm. on the face which are carried on 21-cm. spindles. A complete circular steel bearing plate is bolted to the first shelf of the pit. This plate is 12 cm. thick by 1 m. in width, and a traversing arc made of three angles and a series of 5-cm. steel pins is bolted to its outer edge. The star traversing pinion, 21 cm. outside diameter, meshes with this rack and is driven by either the motor or the hand mechanism. The radius to the center of the traversing rack pins is 10.439 m.

*Carriage.* The carriage is entirely of structural steel (see Fig. 6). It is of the front-pintle type, the front being carried on steel balls. The ball path is 2 m. in diameter and the balls, which are separated by a bronze distance ring, are 15 cm. in diameter. The pintle, which is not much less than 2 m. in diameter, is a part of the base ring which is bolted to the top of the pedestal, Fig. 9. The pintle projects into the racer a distance of about 12 cm. with which it comes into direct contact on firing, thereby transmitting the horizontal component of the shock of recoil into the pedestal and concrete base. As noted under Traversing Mechanism, the rear of the carriage is supported on two rollers 96 cm. in diameter. These rollers have a face of 23 cm. and are carried on steel spindles 21 cm. in diameter. The distance between the centers of these spindles is approximately 3.3 m. The rollers appeared to be perfectly cylindrical, with rounded edges. The construction of this carriage for one of their heaviest guns is exceedingly simple. The only heavy castings are those for the pedestal, the base ring, the elevating oscillating bearing, and the trunnion bearing. The only parts of the entire mechanism which obviously require fine machine work are the trunnion bearings, the heavy steel balls, and the ball paths.

*Emplacement.* With the omission of the heavy concrete struc-

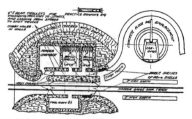

tures described under Battery Pommern for ammunition and personnel, the emplacements for the Batteries Pommern and Deutschland are nearly identical. Battery Pommern was finally fitted up for electrical operation only, and most of the mechanism for hand operation is removed. In the lower part of the Battery Deutschland pit a wood floor made up in sectors raised the floor to such an extent that the men operating the traversing and elevating mechanisms would be on the proper level. The pit was filled with water at the time of the inspection to such an extent that these sectors had come loose and were floating. The emplacement provided for 360 deg. traverse of the carriage as against 157 deg. for the Battery Pommern.

*Ammunition-Supply System.* See the description of the ammunition-supply system for the Battery Pommern. There was no evidence of any provision for any mechanical handling of the projectiles except in the lifting of them from the floor to the storehouse and placing them on the shot trucks. Between Guns Nos. 2 and 3 of this battery there was found a practice shell-ramming tray. This tray comprises a steel trough about 10 ft. long with an ordinary railway-car buffer at the end. A shot truck carrying a projectile with a false ogive is shown in Fig. 10.

*Protection.* It is significant to note with reference to the three of the newest of these guns, that is, Nos. 2, 3, and 4, that there was absolutely no protection provided either for the gun carriage or personnel. Evidence of their scheme of camouflage can still be seen in Fig. 11. This camouflage was unable to hide the guns from the air photographers, as this photograph (Fig. 11), taken on August 9, 1916, shows the emplacements quite plainly. In spite of this fact there was no evidence to show that any of the guns or emplacements had ever been damaged either by shell fire from sea or by bombs from the air. Several holes that had been made either by bombs or by shells were visible in the fields in front of the guns. Gun No. 1 was protected with 6-cm. armor in the same fashion as the gun of Battery Pommern. Gun No. 1 of this battery and the 38-cm. gun of the Battery Pommern are identical in design throughout. Both are Model 1914 guns, and it is probable that the hand elevating and traversing mechanisms of the Pommern gun were identical with those of the Deutschland guns before the electrical equipment was provided.

*Discussion.* Significant points with reference to this battery are that guns Nos. 2, 3, and 4 are not provided with either trunnion anti-friction devices or positive protection in the form of steel armor.

### BATTERY KAISER WILHELM II: 305-MM. GUN

*Gun.* With each of these four guns, Fig. 12, the scheme of destruction was so effective that it was impossible to find a breech after a half-day's search. It is assumed that they were of a model at least as late as 1916. The *Bulletin de Renseignements de l'Artillerie* states that they were 50 calibers in length. One breech block found later was of the usual Krupp sliding-wedge type. The guns are rifled with 88 grooves and the twist of the rifling is to the right 1 cm. in 10. Projectiles found in the storehouse had two rotating bands, from which it may be assumed that the pitch of the rifling is uniform.

*Cradle.* The cradles for all of these guns are smooth cylinders, the walls of which are 10 cm. thick. The brackets at the bottom of the cradle provided for the recoil and recuperator cylinders are quite similar to those of the 38-cm. gun. The front of the cradle is provided with a shield (see Fig. 12), which closes the opening in the armor. The cradle is likewise provided with a heavy counterweight, which can also be seen in Fig. 12. The anti-friction device is of the rolling-wedge type. It is significant that in this case the auxiliary trunnion is but slightly less in diameter than the main trunnion. In practically all other cases observed, the diameter of the small trunnion was approximately one-half of the diameter of the large trunnion.

*Recoil Mechanism.* The recoil mechanism for this gun is in general a duplicate of that for the 38-cm. gun. The breech lug, with the method of attaching the recoil pistons, and the bearing of the lug on the recuperator cylinder to prevent rotation of the gun, are shown in Fig. 13. In contrast to the 38-cm. gun, in which the recoil pistons pass through holes in the recoil lug and the recoil lug bears on two planed sides of the recuperator cylinder, it will be observed in this case that the lug is slotted on the sides to receive the recoil pistons, and the bearing on the recuperator cylinder is circular instead of flat.

The outside diameter of the recoil cylinder is 38 cm. and the length is 1.68 m. The diameter of the piston is 15 cm. and the length of recoil 1.37 m. These cylinders are both smooth forgings, with flanges at the front bearing against the cast bracket on the cradle. There was no evidence of a counter-recoil buffer. On the front end of the cylinders there are filling plugs similar to those found on the 38-cm. cradles. The single recuperator cylinder is likewise a smooth forging, with one large flange on its forward end. It is of the combined air-spring type and is ap-

proximately 45.0 cm. in diameter by 4.25 m. in length. This cylinder likewise has a filling plug at the front end.

*Elevating Mechanism.* The elevating mechanism for these guns is in principle the same as that for the 38-cm. railway mount. The designs of its details, shown in Figs. 15 and 16, are, however, quite different. Apparently it was operated entirely by hand. There are two two-man handles below the deck of the carriage for rapid elevation, and above the deck a single handwheel of large diameter for final and careful setting. The clutch for shifting from the low gear to the high is located below deck. Motion is transmitted from the handwheels to the horizontal shaft at the bottom and through the bevel and helical gears to the two pinions on the main horizontal elevating shaft. The two straight racks, which in this case are horizontal, are connected to the crosshead, which slides on a single round shaft as a guide (see Fig. 13). This crosshead in turn transmits motion to the gun through the two connecting rods. The two racks have guides on the top, which slide in ways in the forward support of the main horizontal guide shaft (see Fig. 16). It was not possible to elevate the

FIG. 10  BATTERY · KAISER WILHELM II
Air photo taken August 9, 1916.

gun, and the ratio of the gearing is not known. This mechanism, together with its companion mechanism on the 38-cm. railway mount, is quite unique among the mechanisms observed on all German artillery. The reasons for the design seen on the railway mount seem obvious, but the same reasons do not hold on this carriage. Certainly a simpler mechanism of equal efficiency could have been provided. The maximum elevation obtainable is 45 deg.

*Traversing Mechanism.* The traversing mechanism, Fig. 16, is operated both by hand and motor. The roller track is set in the concrete emplacement. In this case, contrary to the design of the same mechanism for the Batteries Pommern and Deutschland, there is no traversing rack provided in connection with the roller track. Motion is transmitted through chains and gears from the motor or handwheel to the traversing pinion, which meshes with the two gears bolted to the faces of the traversing rollers. The simplicity of this mechanism is very striking, especially for guns of this size. Apparently it worked satisfactorily, or it would not have been retained in a gun of such value. It is certainly worthy of serious consideration for similar carriages for our own service. An azimuth circle is embedded in the vertical walls of the concrete emplacement about two feet from the top. An indicator is provided on the rear of the carriage with a vertical wire quite close to this azimuth circle. The switchboard and seat for the traversing operator are located beside this indicator.

*Carriage.* The carriages are constructed entirely of structural steel. The side girders are in three main sections comprising a central section of uniform depth, a top brace for the trunnion support, and a bottom section for the pintle. The pintle bearing is of the ball type, the balls being about 15 cm. in diameter. In this case the racer serves as a pintle, having a positive bearing against the base ring for the transmission of the horizontal component of the force of recoil into the foundation.

*Ammunition - Supply System.* The general plan of the Battery Kaiser Wilhelm II is shown in Fig. 17. Storehouses are located on the left of each gun and the main storehouse is 100 m. to the left of the battery. The narrow-gage line, shown extending from t h e main storehouse, passes through each auxiliary storehouse. A typical plan of one gun and its storehouse is given in Fig. 18. The double narrow-gage l i n e s shown passing through the storehouse are the same as those shown in Fig. 17. These storehouses are of excellent construction and are typical of the design of the storehouses for most of the heavy batteries inspected. The plan for the Battery Deutschland is almost identical. The external plan for the Battery Pommern is slightly different, but the interior arrangement is well nigh the same.

Projectiles are transported from the main storehouse to the individual storehouses on narrow-gage trucks. From these trucks they are carried by means of an overhead trolley through openings in the walls into the projectile rooms, where they are stacked two high. Later they are picked up and carried by the same trolley into the corridor just outside the shell room and placed on the shot trucks. The table of this truck is quite broad and the shell is placed on one side, where it is held by two arms until it is to be rammed. The truck is provided with a shelf, presumably for powder. There was no evidence of any scheme of handling the ammunition except by hand. The rammer found beside one of the guns is of such a length as to indicate that the projectile was probably rammed by eight men. The guns are loaded at zero elevation. Another practice shell-ramming tray was found beside one of the emplacements.

*Emplacement.* The design of the emplacement for these guns is shown in Fig. 12. It will be observed that all the space about

the gun leading to the ammunition storehouse is floored with concrete. In this respect the emplacement differs from that of the Battery Deutschland, in which the concrete work did not extend much beyond the vertical walls of the pit. It is not known just why the walls of the pit are higher in front than in the rear, as there seems to be no particular reason for it. The depth of the pit to the roller path is 2.4 m., and the total depth 3.6 m.

*Protection.* Each of the four guns of this battery is protected with 6-cm. flat armor. The guns and carriages were likewise elaborately camouflaged to hide them from the sight of the airmen, but air photographs taken on August 9, 1916, one of which is reproduced in Fig. 19, indicate that it was perfectly possible to see the emplacements. In spite of this there was no evidence that any of these guns, carriages, or emplacements had ever been damaged by shell fire from sea or by bombs from airplanes. Some holes were visible some distance in front of the emplacements, w h i c h may have been made by either bombs or shells. A few similar shell holes were visible some distance to the rear.

FIG. 20 DETAILS OF THE 280-MM. GERMAN RAILWAY MOUNT
*A* Elevating Racks                    *C* Base, Ball Bearing and Swiveled Racer
*B* Trunnion Anti-Friction Mechanism   *D* Emplacing Jack

BATTERY PREUSSEN: 28-CM. GUN RAILWAY MOUNT

*Gun.* The gun on t h e mount examined (see Fig. 1) is a 40-caliber naval piece of Model 1914 of 42 calibers total length. The tube is rifled with 80 grooves having a uniform twist to the right of 1 cm. in 10. The breech block is of the ordinary Krupp sliding-wedge type and is fitted with a mechanical firing mechanism. As with all other large guns having this type of breech, this gun uses semi-fired ammunition.

*Recoil Mechanism.* The recoil mechanism is of the hydro-pneumatic type, and comprises one pneumatic recuperator cylinder mounted at the bottom of the cradle in the center and two hydraulic cylinders likewise mounted on the bottom of the cradle and on either side of the recuperator cylinder.

*Elevating Mechanism.* The elevating mechanism, shown in Fig. 20-A, comprises two straight racks engaging with two pinions enclosed in floating housings and attached to one shaft. This shaft is connected by means of a worm, wheel and shaft, and on the right side of the gun with elevating handwheel shown near the elevation quadrant. This is the only railway mount so far observed in any of the armies in which the attempt has been made to use this straight rack, which is much more easily machined than any of the usual curved elevating racks. On the under side of the floating housings two rollers are carried on which the back of the rack rides and which hold the rack in perfect mesh with the pinions. These two racks are attached by means of heavy pins to the rear of a cradle. The cradle trunnions are provided with an anti-friction device of the type shown in Fig. 20-B.

*Traversing Mechanism.* Two traversing mechanisms are provided on the mounts used in these batteries. The one is provided to permit the mount to be fired from its wheels on a standard track. This mechanism permits the car body to be rotated about the pintle or kingpin of the front truck, and affords a total traverse of 2 deg., that is, 1 deg. on each side of the center line. The other mechanism permits the mount to be rotated about a central pivot attached to the car body. This mechanism affords 360 deg. traverse and is the only one with which we are concerned in the use of the mount for coast defense. It comprises a center pivot and two rear support rollers. The center pivot, Fig. 20-C, comprises a base 1.78 m. in diameter which bolts to the ground platform by 20.6-cm. bolts. This base contains a pintle about which the mount rotates and supports a ball bearing with sixteen 15-cm. balls on which the swiveled racer carrying the mount rests. The racer is carried by means of its trunnions in steel bearings bolted to the bottoms of the side trunnions. Thus the car body can swivel in a vertical plane as well as rotate in a horizontal plane about the base. The rear end of the mount is supported on a track in the foundation by two rollers 60 cm. in diameter by 10 cm. on the face, the housing of which is bolted to the side girders. These rollers are set on a radius of 4.5 m. from the center pivot. To the face of each roller a spur gear is bolted. A single pinion placed between these spur gears and meshing with both of them is driven from the hand mechanism on the side of the mount. Twenty-four turns of the handle give one revolution of the roller. An azimuth circle was found painted on the steel roller track. It seems unlikely, however, that this was ever used except for approximate settings of the mount, for without much doubt the mount was laid exactly in azimuth by means of a panoramic sight and an aiming point.

*Gun Carriage.* The gun is carried in a cradle of simple cylindrical design, which is supported by means of its trunnions in bearings attached to the side girders of the car body. The cradle is provided with a heavy counterweight just above the trunnions to raise the center of gravity of the tipping parts sufficiently to permit easy elevating and depressing. Each of the trunnions is provided with an anti-friction device of the design shown in Fig. 20-B. The car body is built up of two single-web structural-steel side girders carried by a series of structural-steel transoms and deck plates, and the car platform is covered with light armor. Both the traversing roller housing and center pinion bearings are bolted to the bottom of the side girders and serve to stiffen it. Jacking beams, which in the case of the mount examined had the jacks attached (see Fig. 1), are carried under the forward and rear ends of the car body. The car-traversing mechanism forms a part of the upper center plate of the rear truck.

*Emplacement.* The construction of the emplacement, which is typical of the 16 emplacements found along the coast, is shown in Fig. 21. The two short sections of rail are carried by a steel plate which rests on top of the base for support as the mount passes over. This steel plate was blown off when the Germans were demolishing the emplacement. The standard-gage track extends a short distance beyond the emplacement to receive the forward truck when removed from the mount. The mount is run into place with the emplacement arranged in general, as shown in Fig. 21. The small sections of track of the center are then removed in order that the pivot may be lowered on to the structural base by means of the lowering screw and bolted fast by 21 bolts. Four special jacks of the design shown on Fig. 20-D and attached to the jacking beams located under the front and

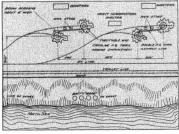

rear of the mount are then laid down and the mount is raised by them sufficiently to permit the trucks to be removed, as well as the two sections of track connecting the center with the rim of the emplacement. The entire mount is then lowered until the rear rests upon the central track by means of the rollers, and the forward end rests on the 16 steel balls in the central pivot.

The general location of the emplacement of this battery with reference to the main railway line on which the mounts were brought in, and likewise with reference to the coast line, is shown in Fig. 22. Two spur lines are run in from the main line. It is not quite clear why this was done. It would obviously have been possible to run short lines to the two guns on the left from the spur line in the center, which line, as will be seen from the figure, has been used only for the two guns on the right.

*Ammunition-Supply System.* The location of the ammunition storehouses and the narrow-gage connections with the emplacements is shown in Fig. 22. In each case the standard-gage line runs at the rear of the simple concrete storage houses which are provided with two doors both at the front and rear. The powder is kept in the one storehouse and the shells in the other. The double line of 30-cm. track connects the storehouses with the two emplacements. These narrow-gage lines connect with a turntable just inside the wicker protection. A complete circle of narrow-gage track is provided about each emplacement, making it possible to supply shells to the mount in any position. There was no evidence of any scheme of storing the projectiles or powder closer to the mount than the storehouses. Evidently a sufficient number of shot trucks are provided for these narrow-gage lines so that the shells can be provided directly from the storehouses at the maximum

FIG. 21  EMPLACEMENT OF BATTERY PREUSSEN

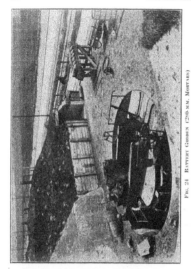

FIG. 24  BATTERY GROBEN (280-MM. MORTARS)

FIG. 26  BATTERY IRENE (150-MM. GUNS)
Note stranded German submarine.

FIG. 23  BATTERY TIRPITZ, NEAR OSTEND

FIG. 25  BATTERY GOEBEN (170-MM. GUNS) AT ZEEBRUGGE

firing rate of the gun. The ammunition storehouses are well camouflaged, but in spite of this, air photographs taken in 1918 clearly indicate the positions of the guns as well as the storehouses.

*Protection.* The gun crew was protected to a certain extent by the light metal cab on the mount. The mounts were likewise screened with wire and brush, the remains of which can be seen in Fig. 21. In spite of this camouflage, air photographs showed quite clearly the position of the emplacements. There was no evidence of damage from shell fire or air bombs on the emplacements or storehouses. A heavy concrete earth-covered headquarters shelter with telephone connections from the observation stations in the dunes and telephone connections to the various guns was provided between the pairs of guns. In the case of each emplacement there is a cable through the central pintle, base which evidently carries the telephone connections between headquarters and the mount.

### BATTERY TIRPITZ: 28-CM. GUN

*Gun.* Battery Tirpitz is a four-gun battery, and all of the guns are of Model 1911. The gun of the four which was examined in detail, shown in Fig. 23, is Model 1911, No. 5. Its weight is 33,875 kg. The length of the tube from the face of the breech block to the muzzle is 11.220 m., and the overall length of the gun is 11.950 m. There are 80 grooves, and the twist of the rifling is to the right 1 cm. in 10. The diameter of the powder chamber is 30 cm. The breech is of the usual Krupp sliding-wedge type. As with the 38-cm. gun found on the railway mount, the tube of this gun has lengthened from firing, thereby separating the forward hoop a noticeable distance from the hoop to the rear. Apparently these hoops are not locked to each other. The lower extension of the breech lug is machined to bear against the planed inner surfaces of the lower recuperator cylinders, thereby preventing rotation of the gun. The gun does not show signs of very great wear, both the lands and the grooves being quite sharp.

*Cradle.* The cradle is a smooth cylinder with walls 10 cm. thick and is lined with bronze liners about 6 mm. thick by 1 m. in length at both the forward and rear ends. The trunnions are located noticeably close to the forward end. The anti-friction device is of the rolling-wedge type. There were no unusual features in this design, and no sketch has been made.

*Recoil Mechanism.* The recoil mechanism comprises two recoil cylinders located on the top and bottom of the cradle and four recuperator cylinders. The two recoil cylinders are smooth forgings carried in the cast brackets on the top and bottom of the cradle. Their length, not including the buffer, is 1.22 m., and the length of the buffer is 24 cm. The piston rods, 11.5 cm. in diameter, pass through holes in the extensions of the breech lug. The length of recoil is approximately 76 cm. The four recuperator cylinders are placed symmetrically above and below; the upper cylinders are combined spring and air recuperators, while the lower are spring only. A common air line is connected to the valves on the forward ends of the upper two cylinders. It is believed that the air is supplied from bottles since there was no evidence of an air pump about the carriage. Both above and below, two rods connect the extensions of the breech lug with the crossheads attached to the piston rods of the recuperator cylinders. These rods taper from 4 cm. at the crossheads to 5 cm. at the breech lug, and are turned down to 4 cm. through the breech lug. This design strikes one as being close to the limit of safety as it places these rods, which are not of very great diameter, always under compression.

*Elevating Mechanism.* The elevating mechanism is operated by hand power only and the ratio of the gearing is five turns of the handwheel for 4 deg. of elevation. The elevating rack is double, although cast in one piece. Identical pinions on the same shaft mesh with these racks. The range of elevation is from zero to 45 deg.

*Traversing Mechanism.* The traversing mechanism is quite similar to that found on the guns of the Battery Kaiser Wilhelm II. Details of this traversing mechanism are shown in Fig. 27. As with the 305-mm. battery, there is no traversing rack attached

to the roller path. The operation is by hand only from a large handwheel located on the left side of the carriage and about on a level with the roller path. In the case of the 305-mm. carriages, the single pinion meshed with spur gears attached to the face of the two large rollers. In this case, the motion from the large handwheel is transmitted directly to only one of the four rollers on which the rear of the carriage is supported. Although the mechanism was seriously damaged, it was possible to traverse the carriage just far enough to indicate that one man could operate the mechanism without difficulty.

*Carriage.* It was not possible to secure such a photograph of these mounts as would show satisfactorily the construction of the carriage. In general, the design is not unlike that of the carriages for the 305-mm. guns. There are two main girders, each of which is in three sections: a central section of uniform depth, a top section for the trunnion support, and a bottom section carrying the pivot. It is made of standard structural plates and angles throughout. The racer is attached to a heavy yoke which is supported by its trunnions in heavy trunnion bearings attached to

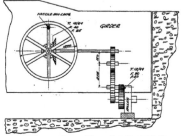

FIG. 27 BATTERY TIRPITZ, 280-MM. GUN TRAVERSING GEAR

the carriage. The design of the racer connection with the carriage is not at all unlike that found on the 28-cm. railway mount.

*Emplacement.* The emplacement is quite similar to that already described for the 305-mm. guns. The traversing roller path is practically identical and the general dimensions of the pit are nearly the same. Again, there is a raised section of concrete in front of the gun, the purpose of which is not apparent. It is understood that unusual difficulties were encountered in constructing these emplacements, inasmuch as the ground is quite swampy, and it was necessary to drive numerous piles in order that the emplacements might be sufficiently stable.

*Ammunition-Supply System.* The ammunition-supply system for this battery is quite similar to that shown and described for the 305-mm. gun. The shot trucks are of the same design.

*Protection.* Each of these four guns is armored with 6-cm. flat armor. In spite of the camouflage, the remains of which can be seen in Fig. 23, legible air photographs were secured on August 9, 1916 (see Fig. 28). The camouflage is carried on a framework attached to the carriage and rotates with it. Most of the concrete about the guns is completely demolished, but there is no evidence that would indicate to the writer that any of the destruction had been effected by shell fire from the sea. It is certain at least that none of the carriages had been so struck or damaged. There are numerous holes to the front and rear of the batteries that were probably made by air bombs.

### BATTERY GRODEN: 28-CM. MORTAR

*Top Carriage.* These mortars are supported directly by means of trunnions on a structural-steel top carriage of the design shown in Fig. 24. The carriage was supported on the inclined sides of the main carriage by means of four rollers, two on each side.

*Recoil Mechanism.* The recoil mechanism comprises two recoil cylinders, the pistons of which were attached to the forward ends of the main carriage body; the cylinders are carried in the sides of the top carriage. The length of recoil is estimated as 1 m., the gun returning to battery by force of gravity. On the front of the main carriage, there are four buffers, each made up of seven pairs of Belleville washers 15 cm. in diameter by 5 mm. in thickness.

*Elevating Mechanism.* The elevating mechanism comprises a single circular rack bolted to the gun and meshing with a single pinion on a horizontal shaft in the top carriage. The handwheel operating the elevating mechanism is located on the right side of the carriage on a platform rotating with the carriage. On the end of the horizontal shaft in the top carriage, there is a worm wheel meshing with the worm carried on a long shaft parallel to the inclined rails on which the top carriage rolls. The worm is

simply keyed to the shaft, moving with the top carriage as it recoils. This is the only elevating mechanism observed in which a slip friction device is provided. This slip friction device provides for the slipping of the worm on the end of the horizontal shaft through the top carriage. It includes two sets of Belleville washers.

*Traversing Mechanism.* The traversing mechanism is made up of one rack attached to the base ring, a traversing pinion, vertical shaft, and simple spur-gear mechanism leading to the handwheel. There is nothing unusual in the design and from the condition of the carriage, it was impossible to traverse it to learn the ratio of the gearing. The extent of traverse is 360 deg.

*Carriage.* It is believed that the carriage is shown in such detail in Fig. 24 as to make any lengthy description unnecessary. The diameter of the ball path is 3.5 m. and the balls are about 10 cm. in diameter. The carriage is constructed throughout of standard structural steel.

*Emplacement.* The pit is 6 m. in diameter and 1.5 m. in depth. In front of each mortar there is a concrete parapet about 5 m. high having a slope of about 45 deg. A part of this parapet is shown in Fig. 24. It is evidently a continuation of the concrete storehouse which is covered with earth.

*Ammunition-Supply System.* The ammunition-supply system is also shown in Fig. 24. Between the shell table and the road a narrow-gage line can be seen which leads to the main storehouse. At the end of the shell table on the concrete floor there are guides to place the shot truck in loading projectiles from the table. It is assumed that the mortars were loaded at zero elevation, although this is not certain. The shot truck is similar in design to those used with the 38-cm. guns and shown in Fig. 10. Since the powder charge is not of great weight, it is probable that it is carried in a two-man tray.

*Protection.* The only protection afforded these mortars has been mentioned under the heading Emplacement. There was no evidence of any camouflaging, although it is assumed that some sort of camouflage was provided. It is quite certain that no amount of camouflage could effectively conceal them. There is no evidence, however, that any of them had ever been either struck or damaged by shell fire or bombs.

### BATTERY GOEBEN: 17-CM. GUN

*Gun.* All of the guns of this battery (see Fig. 25) are Model 1914, and the gun examined in detail is Model 1914, No. 71-L. Its weight is 10,701 kg., the length of the tube 6.44 m., and the total length of the gun 6.980 m. There are 52 grooves and the twist of the rifling is to the right, 1 cm. in 10. The diameter of the powder chamber is 19 cm. and the breech is of the standard Krupp sliding-wedge type.

*Cradle.* The cradle is identical in design with the cradle found with the 21-cm. railway mount. There are two hydraulic buffer cylinders, one on the top and the other directly below the cradle, and each is attached to the cradle by a heavy pin about which it can rotate in a vertical plane. Four spring cylinders are arranged about the cradle symmetrically, two above and two below. The diameter of the trunnions is 22 cm. and the length 15 cm.

*Recoil Mechanism.* The outside diameter of the recoil cylinder is 35 cm. and the estimated length of recoil 40 cm. As noted before, the recuperator cylinders are four in number and they are 1.6 m. in length. There are two columns of springs in each cylinder, the mean diameter of the outside springs being 17 cm., diameter of wire 2.4 cm., and pitch 5 cm. The mean diameter of the inside spring is 11 cm., the diameter of the wire 1.5 cm. and the pitch 3 cm. Tension rods connect the crosshead at the front of each spring column with the recoil lug.

*Elevating Mechanism.* The elevating mechanism includes a circular rack bolted to the side of the gun near the breech. The pinion meshing with this rack is driven by an electric motor carried on the part of the carriage below the floor.

*Traversing Mechanism.* A complete circular rack is attached to the base ring. The traversing motor is carried on the section of the top carriage which extends below the deck and connects with the traversing rack through two sets of bevel gears and one spur gear. The entire original gun and turret mechanism, without any modifications, were installed on the concrete emplacement as shown in Fig. 25.

*Carriage.* The one detail of the carriage which seems of interest to describe is the traversing ball bearing. A portion of the light shield protecting the bearing has been torn away approximately at the center of the turret, and almost above the long exposed anchor bolt. It was found, on removing this shield, that the carriage is provided with a double ball path, the diameter of the outer path being 3.2 m., and the inner path 2.8 m. The balls in the outside bearing are about 5 cm. in diameter and in the inside bearing, 7.5 cm.

*Emplacement.* These guns, all placed in the shore dunes at Zeebrugge, are easily visible from the sea. The concrete emplacement shown in Fig. 25 is about level with the top of the dune. A railway line, evidently constructed by the Germans, runs along the top of the dune just in front of the gun. The thickness of the concrete wall at the rear is about 1.5 m. This wall increased in thickness on the sides to several meters, flaring off to the right and left at the front. There was a door leading into the operating room under the turret just under the two short anchor bolts seen in the center of the picture.

*Ammunition-Supply System.* Ammunition was supplied to the gun through the floor just mentioned in the emplacement and was hoisted by the usual type of electrical turret hoist. Apparently there was no provision for supplying ammunition in any other way.

*Protection.* All of the guns were protected as shown in Fig. 25. This turret was made up of 10-cm. armor.

*Destruction.* Apparently the usual scheme of destroying the gun was not employed in these cases. Instead, a number of projectiles were detonated inside the turret and in the operating room below. All but one of the emplacements was destroyed as badly as the one shown in Fig. 25. In each case the roof of the turret was blown off. The explosions which destroyed the emplacements do not seem to have destroyed the turret mechanism as one would have expected. In the case of the gun shown there were about half a dozen unexploded projectiles in the turret and twice that number in the operating room below.

### BATTERY IRENE: 15-CM. GUN

*Gun.* All of the guns of the Battery Irene, Fig. 26, are of the Model 1900, and the gun inspected is Model 1900, No. 478-L. The weight is 4861 kg., the total length 6 m., and the length of the tube to face of the breech block 5.57 m. The number of grooves is 44 and the twist of the rifling is to the right 1 cm. in 10. The diameter of the powder chamber is 18 cm. The breech block is of the standard Krupp sliding-wedge type.

*Cradle.* The cradle is not unusual in design. The trunnions are provided with an anti-friction device. Two lugs are cast on the bottom of the cradle near the rear to which the recoil piston is attached by means of a pin. Two other lugs are provided to which the spring recuperator cylinders are bolted.

*Recoil Mechanism.* Attention is called to the rather unique design of the recoil mechanism. This is the only case observed in which the cylinder is carried in the breech lug, the piston being attached to the cradle by means of a pin passing through the two lugs just mentioned. The recuperator cylinders are two in number. Two 3-cm. tension rods connect the breech lug with the crosshead at the forward end of the spring cylinders. These recuperator cylinders are faced off on the inside at the rear to serve as a guide for the breech lug to prevent rotation of the gun.

*Other Details.* There are no features regarding elevating mechanism, traversing mechanism, carriage, emplacement, ammunition-supply system, or protection that are worthy of description.

### CONCLUSIONS

*Guns.* Most of the guns inspected are about 42 calibers in length, measuring from the face of the breech block, or about 45 calibers in total length. All guns are provided with the standard Krupp type of sliding-wedge breech block. In all cases, except with the 28-cm. mortars, which are Model 1892, the twist of the rifling is uniform to the right, 1 cm. in 10. There are no guns older than 1904. Quite a number of the guns are as late as 1916. If the data given in the *Bulletin de Renseignement de l'Artillerie* of January-February, 1919, can be accepted, it is certain that the Germans were securing ranges from these guns at ordinary muzzle velocity far in excess of the ranges that we are securing from our guns. This is probably through the improvement in their projectile design.

*Cradles.* All cradles constructed may be termed smooth cylinders, and the maximum thickness of the walls is 10 cm. The cradles of the larger guns are provided with bronze liners about 6 mm. thick and 1 m. long, at both forward and rear ends. In all cases, the cradles are provided with the simplest types of brackets for the attaching of the recuperator and recoil cylinders.

*Provisions to Prevent Rotation of the Gun.* All breech lugs are fastened to the gun by the interrupted-ring method. All lugs are so shaped as to bear either between two recuperator cylinders or on two sides of a single recuperator cylinder to prevent rotation of the gun. There is no evidence in any case of the use of the spline, typical in American design, for the prevention of rotation of the gun.

*Hydraulic Buffers.* In most cases two hydraulic buffers are provided. Apparently there is no fixed policy of balancing these, since in many cases the two cylinders are located on the bottom of the cradle. In some cases the hydraulic cylinders are provided with extensions on the forward end which are evidently counter-recoil buffers. In a number of other cases no such extensions are visible.

*Recuperators.* The designers seem to have favored the combined air-spring recuperators for the heavy guns. There is no attempt at balancing them, and the number of cylinders varies from one to four. All of the 380- and 305-mm. guns are provided with only one recuperator cylinder each. As noted under the heading Provisions to Prevent Rotation of Gun, at the bottom of the preceding column, in every case the recuperator cylinders are used as guides for the extension of the breech lug to prevent rotation of the gun.

*Gun Carriages—Traversing Mechanisms.* In all cases the heavy gun carriages are constructed entirely of structural steel, and with the exception of the 28-cm. mortar carriages, are of the front-pintle type. The pintle bearings in all cases are of the ball type. The rear of the carriages is carried on heavy rollers (two to four in number, 10 to 20 cm. across the face and 0.60 to 1.00 m. in diameter) running on circular tracks set in the concrete emplacement. In some cases (38-cm. guns) the traversing pinion meshes with a rack attached to the roller path. In other cases (380- and 305-mm. guns), the traversing pinion meshes with gears bolted to the face of the rollers.

*Elevating Mechanisms.* The variations in the design of the elevating mechanisms found in German coast carriages as well as railway carriages is very striking. On the 38-cm. railway mount found at Brussels, and the 305-mm. carriages of the Battery Kaiser Wilhelm II, there are double straight racks, but the dimensions and designs differ quite radically. With the 38-cm. guns of the Batteries Pommern and Deutschland, there are double telescoping screws. On the Battery Tirpitz there is a double curved rack attached to the bottom of the cradle. On the Battery Groden 28-cm. mortars and the guns of several other batteries there are single curved racks attached to the bottom of the cradle. On the Battery Goeben 17-cm. guns circular racks are attached to the sides of the cradles.

*Ammunition-Supply System.* In all cases the ammunition is conveyed from the storehouses into the gun by hand. The shot trucks are all of extremely simple design, and the projectiles in all cases are rammed by hand.

*Ammunition Storage.* With all of the heavier batteries, the layout of the ammunition storehouses is as shown in Figs. 17 and 18. In each case the storehouses are designed to house projectiles and powder for the main guns as well as ammunition for the anti-aircraft guns provided for the protection of each of the big batteries.

*Protection.* It is very significant that there is no evidence of a policy of providing heavy protection for these large-caliber and valuable guns. Apparently all of the guns which are provided with the 6-cm. flat plate armor have been removed from other coast fortifications where they had been previously provided with the same armor. All of the guns were elaborately camouflaged, but air photographs taken in 1916, 1917 and 1918 show quite clearly the positions of all of the guns, ammunition storehouses, approach tracks, etc.

In spite of the lack of protection and the clear evidence of the position of the batteries from air photographs, there is no evidence that any of the guns were ever damaged, or even hit, by shell fire or by bombs from airplanes. It is understood that the coast fortifications were shelled constantly by the heavy guns of the Allies' monitors. The position of the guns were known, but either the smoke screens that were at once put up by the Germans were unusually effective, or the systems of fire control that were employed were defective. The reasons for the failure of the aviators to obtain any satisfactory results are not certain. They dropped many bombs in the vicinity of the various batteries. It is probable that the accuracy of the anti-aircraft guns provided with all of the large batteries was such as to compel the aviators to operate at a very great height.

# The Flow of Air Through Small Brass Tubes

By T. S. TAYLOR,[1] PITTSBURGH, PA.

*A study of the flow of air through brass tubes ⅝ in., ⅞ in. and 1½ in. in diameter, respectively, has been made by means of small pitot tubes. Under the conditions of the experiment, it has been observed that the velocity distribution does not become constant in tubes of these dimensions until the air has passed through a length of about 200 cm. The ratio of the average velocity to the maximum velocity at the center has been found to have a value of from 0.82 to 0.85 for all velocities for each of the tubes tested. Tests were made of the influence of dust and oil on the walls of the tubes and very interesting results obtained. A small quantity of dirt irregularly distributed greatly diminishes the total air flow for a given static pressure and also produces a marked change in the velocity distribution, the average velocity being made considerably less with respect to the maximum velocity.*

T HE investigations discussed in the present paper were undertaken as a preliminary study of the general ventilation problem, dealing in particular with the flow of air through tubes of various sizes and shapes. At present, literature on ventilation contains very little information along this particular line that is useful to the designing engineer. Quite a little attention has been given to the mathematical consideration of the flow of liquids through tubes, but so far no special experimental study has been made of the factors influencing the flow of air through small tubes.

In the first place, no very satisfactory device has as yet been made for measuring gas flow through such tubes as are dealt with in the present experiments. The tubes thus far tested in

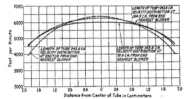

FIG. 1 VELOCITY DISTRIBUTION AT THREE POINTS IN 1½-IN. TUBE

this investigation were of seamless brass having internal diameters of ⅝ in., ⅞ in., and 1½ in. respectively. In order to determine the velocity distribution of the air across these tubes a hot-wire anemometer with thermocouple was developed. The method of use consisted in measuring the electrical energy required in the heating wire to maintain the temperature of the thermocouple a definite amount above the temperature of the air in which it was placed. The device proved to be a very satisfactory one for laboratory experiments. For shop use, however, it is not satisfactory as it requires accurate manipulation and is not "fool-proof." Attention was therefore turned to the usual pitot-tube method. In the preliminary study it was only necessary to have a tube for measuring the total pressure, as the static pressure could be measured by means of a small tube in the side of the main tube through which the air was passing. The pitot tubes used were made from hypodermic needles

[1] Mellon Institute, University of Pittsburgh, formerly of the Westinghouse Electric and Manufacturing Co., East Pittsburgh, in whose Research Laboratory the work herein described was done.
Abstract of a paper presented at the Spring Meeting, St. Louis, May 24 to 27, 1920, of THE AMERICAN SOCIETY OF MECHANICAL ENGINEERS. Copies of the complete paper may be obtained upon application. All papers are subject to revision.

about 5/64 in. in external diameter. The velocity pressure was measured in the usual way by means of inclined oil gages. The use of a special low-pressure differential gage was found unnecessary.

The air current was furnished by means of a No. 3 V blower made by the American Blower Co., and was driven by a d.c. variable-speed motor. To prevent vibration of blower and motor being transmitted to the tube, a short rubber hose was used for connection between blower outlet and the brass tube used in the experiment. From the readings of the inclined gage the velocity of the air was calculated by the usual formula

$$V = \sqrt{2ghd/d'}$$

where $g$ is gravity, $h$ the difference in level in the gage, $d$ the density of the medium in the gage, and $d'$ the density of the air. All velocities were calculated and reduced to standard conditions of 760 mm. pressure and air temperature of 25 deg. cent.

## VELOCITY DISTRIBUTION

A study was made of the velocity distribution at various positions along a 1½-in. brass tube 243.6 cm. (8 ft.) long. This was done with a pitot tube about 1/40 in. (0.068 cm.) in external diameter inserted into the brass tube. The curves in Fig. 1 show the distribution of velocity for three different positions. The velocity at the center of the tube is seen to increase as the distance from the end nearest the blower increases. The average velocity for each of the cases was 5300 ft. per minute. This average velocity was obtained in the following manner. The cross-section of the tube was divided into several small concentric zones and the area of each zone calculated. The area of each zone was then multiplied by its average velocity. The sum of the products for all zones was then divided by the area of the cross-section of the tube, which gave the average or mean velocity of air flow. In other words, this is equivalent to finding the volume described by revolving the curve for the velocity distribution about its axis and dividing this volume by the area of the tube. These results show clearly that the velocity distribution of air flowing through tubes does not reach a steady state in the distance in which it is usually assumed that it does. It is customary to assume a steady distribution to exist in a length of tube equivalent to 10 times its diameter.

In order to test this last point a little more, a second tube 407 cm. (13.3 ft.) long and 1½ in. in diameter was used. It was found that the velocity distribution had attained a definite condition and was the same for all points distant 200 cm. from the blower end of the tube. Another rather interesting point was observed in regard to the ratio of the average velocity to the maximum velocity for various velocities. The results in Table 1

TABLE 1 SHOWING RATIO OF AVERAGE VELOCITIES TO MAXIMUM VELOCITY AT VARYING AIR VELOCITIES IN 1½ IN. BRASS TUBE

| Blower Speed r.p.m. | Static Pressure in Inches | Maximum Velocity ft. per min. | Average Velocity ft. per min. | Average Ratio. =Avg. Max. |
|---|---|---|---|---|
| 435 | 0.0465 | 872 | 751 | 0.862 |
| 739 | 0.0913 | 1410 | 1186 | 0.853 |
| 983 | 0.146 | 1912 | 1610 | 0.842 |
| 1098 | 0.176 | 2144 | 1804 | 0.842 |
| 1490 | 0.322 | 2973 | 2480 | 0.834 |
| 1857 | 0.437 | 3784 | 3216 | 0.830 |
| 2710 | 0.980 | 5373 | 4460 | 0.830 |
| 3150 | 1.385 | 6330 | 5460 | 0.863 |
| General Mean.................................. | | | | 0.847 |

show that the average velocity is the same percentage of the maximum velocity for all velocities from 800 ft. per min. to 6300 ft. per min. These results were all taken in the long tube, 407 cm., at a point 240 cm. from the end adjacent the blower. The values of the static pressure were taken at the point along the tube when the velocity distribution was determined, and thus represent the pressure which is necessary to maintain the air flow through the remainder of the tube, or 167 cm. The fact that this ratio of the average velocity to the maximum is approximately the same for all velocity distribution, seems quite interesting. Similar results were also obtained for tubes 7/8 in. and 5/8 in. in diameter, respectively. The general mean for the ratio of the average to the maximum velocity was found to be 0.830 for the 7/8-in. tube and 0.870 for the 5/8-in. tube. These results would indicate that there is little difference between these ratios for different-sized tubes provided the surface conditions are identical. These results indicate also that the velocity distribution, for steady state, across smooth brass tubes having similar surface conditions is practically the same for all tubes and all average velocities between 0 and 6000 ft. per min. It is interesting to note that the velocity near the wall of the tube is so large as compared with the

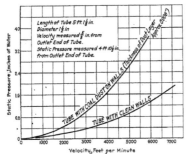

FIG. 2 EFFECT OF DUST IN TUBE

maximum velocity at the center. The velocity must therefore drop off very rapidly close to the wall of the tube. This drop occurs in such a short distance that it cannot be observed with a pitot tube 1/40 in. in diameter.

### EFFECT OF DUST IN TUBES

Having determined the above data on the air flow through smooth tubes, the next step was to see what effect would be produced by the addition of dust to the walls of the tubes. Oil was poured through brass tubes 1½ in., 7/8 in. and 5/8 in. in diameter respectively, each being a little over 5 ft. long. Coal dust was then sifted through the tube and only that which was allowed to remain which was not readily jarred out when the end of the tube was struck against some object. The velocity across each tube was then measured near the outlet end for various air velocities. The thickness of the dirt layer is only approximate. The dust and oil were wiped out of the tubes after measurements were taken and the thickness of the layer determined from its mass and density. The results obtained are summarized in Table 2.

The curves in Fig. 2 show the difference in the static pressure necessary to force air at various average velocities through clean and dirty tubes as specified in the figure. They show that a very small but irregular distribution of dirt upon the surface of smooth tubes will have a very decided effect upon the amount of air that will pass through the tube for a given static pressure. The effect is not merely a diminution in the effective area of cross-

TABLE 2  INFLUENCE OF LAYER OF DIRT ON VELOCITY DISTRIBUTION IN TUBES

| Diameter in. | Clean Tube, Avg. V/Max. V. | Dirty Tube Avg. V/Max. V. | Thickness of Layer, in. |
|---|---|---|---|
| 1½ | 0.832 | 0.75 | 0.006 |
| 7/8 | 0.833 | 0.65 | 0.016 |
| 5/8 | 0.823 | 0.80 | 0.008 |

section but must give rise to turbulency since the ratio of the average velocity to the maximum is so much influenced by the presence of the dirt. The diminution in this ratio due to the dirt must depend chiefly upon the manner in which the dirt is distributed. At any rate the distribution must be of considerably more importance than the amount per unit area, as can be seen by reference to Table 2 and Fig. 2. The values of the static pressure to maintain an average velocity of 3000 ft. per min. through the tubes when clean and dirty as outlined above are given in Table 3.

TABLE 3  STATIC PRESSURE IN INCHES OF WATER NECESSARY TO MAINTAIN AVERAGE VELOCITY OF 3000 FT. PER MIN. IN TUBES 5 FT. LONG

| Diameter of Tube in. | Static Pressure, in. of Water | | Thickness of Layer, in. |
|---|---|---|---|
| | Clean Tube | Dirty Tube | |
| 1½ | 0.37 | 0.94 | 0.006 |
| 7/8 | 0.80 | 3.20 | 0.0016 |
| 5/8 | 1.00 | 5.00 | 0.008 |

While the results of the present experiments should be considered as only preliminary, they nevertheless bring out some rather interesting points. The striking features about them are to be found: (a) in the way in which the velocity distribution changes along a tube; (b) in the relative relation of the ratios of average velocity to maximum velocity for different tubes as well as different velocities; and (c) in the influence that dust collected upon the wall of a tube exercises upon the total flow of air under a definite static pressure.

In conclusion, I take pleasure in acknowledging my indebtedness to the Westinghouse Electric and Manufacturing Company, in whose research laboratory the work was done, for furnishing the facilities for carrying out the experiments, and in particular to Mr. C. E. Nolan for his able assistance throughout the work.

---

An instrument for determining the direction of flow of fluids in pipes is described by Mr. J. S. G. Thomas in a paper entitled A Directional Hot-Wire Anemometer of High Sensitivity, especially Applicable to the Investigation of Low Rates of Flow of Gases, recently presented to the Physical Society. Essentially the instrument consists of two fine platinum wires arranged parallel and one behind the other in close juxtaposition, transversely to the direction of flow of gas in the pipe. The wires constitute two arms of a Wheatstone bridge, the remaining arms being formed of a resistance of 1000 ohms and an arm capable of adjustment. The operation is dependent upon the fact that that platinum wire experiences the greater cooling effect upon which the stream of gas is first incident, since it exercises by its presence a shielding effect upon the second wire. Such a hot-wire anemometer affords a ready means of ascertaining the direction of flow of fluids in various units of a complicated network of gas or other mains. The author states that subsequent experience with anemometers of this type has shown that they possess special characteristics which make them particularly useful in the investigation of very low rates of flow.—*The Engineer* (London), April 16, 1920.

# THE ENGINEER'S RELATION TO PUBLIC QUESTIONS

Address of President Miller at the Banquet Given by the Western Society of Engineers to the Governing Boards of the Four National Societies

ON another page will be found an account of the two-day conference held April 19 and 20 in Chicago, when the members of the governing boards of the A.S.C.E., A.I.M.E., A.I.E.E. and A.S.M.E. were the guests of the Western Society of Engineers. President Miller's remarks at the banquet given in the Hotel La Salle on the evening of the 19th follow:

It gives me the greatest pleasure to meet along with my fellow-engineers from the outlying districts of Chicago, extending from the Pacific to the Atlantic, with your Society of Western Engineers, which, I am told, is, in point of age, the third society among engineering associations in this country. In fact, it is about as old as I am.

Now, an individual can attain age simply by continuing to exist, but no society can do that. A society or an organization of any kind, not subsidized, cannot attain the age that I have attained unless it does something that is creditable and worth while, and we all know that the Western Society of Engineers has much more to its credit than mere existence since its foundation.

The American Society of Mechanical Engineers, which I have the honor to represent, has always shown a considerable spirit of progress and coöperation. I think its members generally recognize that the engineer is by tradition and training an innovator. His work renders valuable that which before perhaps had no value, and, on the other hand, sometimes renders of no value that which previously had value. These are necessary consequences of the engineer's work. The engineer, this being the case, should, I think, be willing to take his own medicine and societies, no matter how old they may become, should retain that attribute of youth which enables us to consider and to study and perchance to adopt new ideas.

Our society, immediately following the appointment by the Civil Engineers of their committee to consider the problems now before the engineering societies concerning coöperation, appointed a committee which we called the "Aims and Organization Committee." That committee, after considerable conscientious and hard work, submitted a report advocating considerable changes in our Society, particularly along the lines of democratization of its administrative features, and strongly recommended coöperation with other engineering societies. None of us can be certain of what will happen in the future, but I think it is easy to predict that the time will come when there will be in Chicago, by the growth of engineering in this neighborhood, a headquarters building of engineering somewhat similar to the United Engineering Societies Building in New York. In that building probably will be housed all the engineering interests of this section, and I believe it is easy to perceive and to predict that it will eventually house branch offices of the principal engineering societies of the country; and, if we allow our imagination to go a little further, we might imagine that by the development that may take place in the future those branch offices may, as sometimes happens in other organizations, become really the principal offices of those associations.

### THE ENGINEER AND PUBLIC QUESTIONS

It is generally expected that in our joint activities we shall attempt to deal more than ever before, and especially more effectively than ever before, with matters of public concern. Our relation to those problems will necessarily be of a twofold nature: for first, we are citizens of a great country, with the duties and responsibilities that pertain to that citizenship; and after that we are engineers, with duties and responsibilities not only to ourselves and those dependent on us, but to the high and honorable profession to which we belong.

With respect to purely engineering matters engineers can and will speak with the authority of experts, but all public questions connected with engineering will have, or nearly all will have, an aspect of public policy, and I think we will do best to remember, in discussing these matters, that we will do best to confine ourselves to the purely technical side of those questions, for we must remember that where a question has a public aspect, other citizens of the United States, not engineers, will have as much right and will be as much expected to hold and to express views upon those matters as we; and our testimony as engineers will have the more effect if we carefully refrain from giving the impression that we assume any extraordinary or special ability regarding the public side of those questions that may come before us. We can be, in a sense, expert witnesses, testifying for what we believe to be right, but with all the well-understood limitations of the expert witness (Laughter); and our influence on the expert or on the technical side of these questions will be the greater in proportion as we confine ourselves to that side of the matter.

It is obvious that our own members will have various views regarding the public questions which arise, and for that reason I think we should also be very cautious about affiliations of our engineering societies with other organizations not composed of engineers nor having engineering objects, because if we attempt to commit our members to questions not dealing strictly with engineering matters, then our own members will have widely varying views on those matters, and by attempting to commit our members to them we shall, I fear, promote dissension and possibly disintegration. I think it is well we should bear that in mind and be cautious about such affiliation and joint action.

### THE ENGINEER AND PUBLICITY

Finally, if we are to serve the public, the public must know much more about the engineers' work than it has been able to discover heretofore, and that means publicity, a thing which the engineers have not cultivated as they should, and which I think needs a great deal more attention than has been given it. Recently, the annual banquet of a society of engineers in a New England city was held, and at that banquet the governor of the state, the mayor of the city, the representatives of the four engineering societies, and others, spoke. The governor of the state and the mayor of the city spoke remarkably well regarding the coöperation of engineers with public matters in the service of the state and of the municipality. The mayor of the city particularly said that certain activities carried on under the advice and with the help of engineering associations of that locality had been carried on much better than ever had been possible before. Further, he stated flat-footedly that he believed the time had come when it must be recognized that the activities of a municipality involving engineering matters could no longer be satisfactorily conducted under the control of politicians, but must be conducted under the guidance and control of engineers—a very significant thing to be said by the mayor of a great city of this country.

The day after that meeting took place the papers of the city were examined to see how they had treated this occasion. One had in its headlines the name of a prominent lawyer who had spoken well at the meeting on an important matter of international politics, but no reference to engineers nor to an engineering meeting. The article went on to say that the lawyer had said certain things on this subject at a meeting of engineers held the night before. That was all the publicity that the engineers of the city secured from the occasion, although the speeches were filled with matter of the utmost importance to the public if it could have been gotten to them.

Now, it has seemed to me that the engineers need to cultivate that particular thing, and that we shall do better if the local societies of engineers can select some one of their number who can, in coöperation with a newspaperman knowing the newspaper game, translate into human and public terms what it is that the engineers are trying to do, why they are trying to do it, and what it means to the public. The newspaper game is a good deal like any other game. One who takes part in it and succeeds must know the game; he must know what will interest the city editor and in what form it will interest him, and engineers are not trained for that work. They see only the engineering side of it. The newspaper man can see the human side of it, and with the assistance of the engineer I believe a combination can usually be made in any locality that will secure for engineers at least a part of the publicity to which they are entitled, and which they ought to have.

This will not be selfish advertising. The public needs our coöperation, but in order to secure that coöperation it needs to understand what we are doing. It is a matter of human experience that when groups of people wish to coöperate or to work together, they cannot do so effectively until each group understands what the other purposes to do, how it proposes to do it, and why it proposes to do it, and that means publicity, which I recommend be given much more attention than engineers have been in the habit of giving it heretofore.

---

A new chemical reagent which has proved to be one of the most striking solvents discovered in recent years and contains possibilities of extensive industrial use has recently been worked out in the chemical laboratories of the University of Wisconsin. The discovery was announced at the meeting of the American Chemical Society in St. Louis recently by Prof. Victor Lenher, the chemist who conducted the investigation.

The new solvent is technically known as selenium oxychloride, and the selenium from which it is made is a by-product in the electrolytic refining of copper, a substance for which there is now no market, although hundreds of tons of it are annually going to waste.

The reagent is an excellent solvent for unsaturated organic substances. The unsaturated hydrocarbons, such as acetylene, benzine toluene, etc., dissolve readily in it, while the paraffin hydrocarbons, such as gasoline, kerosene and the so-called mineral waxes, vaseline and paraffin, are unaffected. The vegetable oils are acted upon, many with violence, and linseed oil forms a thick, exceedingly mucilagenous and rubberlike mass.

Bakelite, redmanol, the waterproof insoluble casein glue used in airplane construction, pure rubber, vulcanized rubber, asphalt and bitumen all dissolve with ease. The bituminous material in soft coal can be dissolved out, leaving a carbonaceous residue, and in all probability some information as to the character of the naturally occurring asphalts and bitumens can be obtained by the use of this new solvent.

# National Screw Thread Commission Report

### This Commission, Composed of Engineers and Government Representatives, Is the First to Be Appointed by Congress for Standardization Work. The Report Covers Standards for Threads and Thread Gages, Including Tolerances

AS a result of the efforts of the American Society of Mechanical Engineers, the Society of Automotive Engineers, the Bureau of Standards and prominent manufacturers of specialized thread products, the National Screw Threads Commission was appointed by Act of Congress, July 16, 1918, to investigate and collate standards for screw threads to be adopted by manufacturing plants under the control of the army and navy, and also for adoption and use by the public. The work of organizing the Commission was expedited by the Committee on Coinage, Weights, and Measures of the Department of the Interior, and it was under their auspices that the act was modified on March 3, 1919, to take care of changes in the requirements of the personnel and to extend the life of the Commission until March 21 of this year; and again on March 1, when the life of the commission was extended for two years.

The Commission is composed of two representatives of the army, two representatives of the navy, and four civilians nominated by the engineering societies of America, with Dr. S. W. Stratton, Director of the Bureau of Standards, as chairman. The members serve without compensation with the exception of the regular salaries received by the army and navy members and that received by the Director of the Bureau of Standards.

The work for which the Commission was created, which, in the opinion of Dr. Stratton, is one of far-reaching effect, has been largely completed by the issuance of a progress report, a copious abstract of which is given herewith. The report is very complete in its text and includes, in addition, a large number of diagrams and tables. A few of the latter are included in the abstract.

It was believed best to keep the Commission in existence for two years more, so that it may pass upon any questions of adjustment or modification of the proposed standards. The appointees of The American Society of Mechanical Engineers serving on the Commission are James Hartness and F. O. Wells.—EDITOR.

## ABSTRACT OF REPORT

IT is the desire of the Commission to make available to American manufacturers the information contained in its progress report for immediate use, rather than delay making a report in order to consider more fully the possibilities of international standardization of screw threads. It is the opinion of the Commission, however, that international standardization of screw threads is very desirable and that the present time is most opportune for accomplishments in this direction.

In 1864 a committee appointed by The Franklin Institute, Philadelphia, prepared a report upon its investigations of a proper system of screw threads, bolt heads and nuts, which recommended the thread system developed by William Sellers, known at present as the Franklin Institute, Sellers Thread, or more generally as the United States Thread. The accomplishment realized in the general adoption of the United States Standard Thread was brought about largely by the great need of standard threads by American railroads. Certain other industries, however, such as the automobile, machine-tool, and light machine industries, were in need of additional standard threads, and to fulfill these needs a thread system having finer pitches was recommended by the Society of Automotive Engineers; and a machine-screw thread series which provided smaller-sized screws than the United States Standard Thread was recommended by The American Society of Mechanical Engineers.

### THREAD SERIES RECOMMENDED

It was the aim of the Commission, in establishing thread systems for adoption and general use, to eliminate all unnecessary sizes, in addition to utilizing, as far as possible, present pre-dominating sizes. While from certain standpoints it would have been desirable to make simplifications in the thread systems and to establish more thoroughly consistent standards, it is believed that any radical change at the present time would interfere with manufacturing conditions and involve great economic loss.

The testimony given at the various hearings held by the Commission is very consistent in favoring the maintenance of the present coarse- and fine-thread series. The coarse-thread series is the present United States Standard thread series supplemented in the sizes below ¼ in. by the standard established by The American Society of Mechanical Engineers for machine screws. This series includes only those sizes which are essential and is recommended for general use, in engineering work, in machine construction where conditions are favorable to the use of bolts, screws, and other threaded components where quick and easy assembly of the parts is desired, and for all work where conditions do not require the use of fine pitch threads. (Table 1.)

TABLE 1   NATIONAL COARSE-THREAD SERIES

| Identification | | Basic Diameters | | | | Thread Data | |
|---|---|---|---|---|---|---|---|
| Numbered and Fractional Sizes | Number of Threads per Inch | D Major Diam., Inches | E Pitch Diam., Inches | K Minor Diam., Inches | Metric Equivalent to Major Diam., mm. | p Pitch, Inches | h Depth of Threads Inches |
| 1 | 64 | 0.073 | 0.0629 | 0.0527 | 1.854 | 0.0156250 | 0.0101 |
| 2 | 56 | 0.086 | 0.0744 | 0.0628 | 2.184 | 0.0178571 | 0.0116 |
| 3 | 48 | 0.099 | 0.0855 | 0.0719 | 2.515 | 0.0208333 | 0.0135 |
| 4 | 40 | 0.112 | 0.0958 | 0.0795 | 2.845 | 0.0250000 | 0.0162 |
| 5 | 40 | 0.125 | 0.1088 | 0.0925 | 3.175 | 0.0250000 | 0.0162 |
| 6 | 32 | 0.138 | 0.1177 | 0.0974 | 3.505 | 0.0312500 | 0.0203 |
| 8 | 32 | 0.164 | 0.1437 | 0.1234 | 4.166 | 0.0312500 | 0.0203 |
| 10 | 24 | 0.190 | 0.1629 | 0.1359 | 4.826 | 0.0416667 | 0.0271 |
| 12 | 24 | 0.216 | 0.1889 | 0.1619 | 5.486 | 0.0416667 | 0.0271 |
| 1/4 | 20 | 0.2500 | 0.2175 | 0.1850 | 6.350 | 0.0500000 | 0.0325 |
| 5/16 | 18 | 0.3125 | 0.2764 | 0.2403 | 7.938 | 0.0555555 | 0.0361 |
| 3/8 | 16 | 0.3750 | 0.3344 | 0.2938 | 9.525 | 0.0625000 | 0.0406 |
| 7/16 | 14 | 0.4375 | 0.3911 | 0.3447 | 11.113 | 0.0714286 | 0.0464 |
| 1/2 | 12 | 0.5000 | 0.4500 | 0.4001 | 12.700 | 0.0763931 | 0.0500 |
| 9/16 | 12 | 0.5625 | 0.5084 | 0.4542 | 14.288 | 0.0833333 | 0.0541 |
| 5/8 | 11 | 0.6250 | 0.5660 | 0.5069 | 15.875 | 0.0909091 | 0.0590 |
| 3/4 | 10 | 0.7500 | 0.6850 | 0.6201 | 19.050 | 0.1000000 | 0.0650 |
| 7/8 | 9 | 0.8750 | 0.8028 | 0.7307 | 22.225 | 0.1111111 | 0.0722 |
| 1 | 8 | 1.0000 | 0.9188 | 0.8376 | 25.400 | 0.1250000 | 0.0812 |
| 1 1/8 | 7 | 1.1250 | 1.0322 | 0.9394 | 28.575 | 0.1428571 | 0.0928 |
| 1 1/4 | 7 | 1.2500 | 1.1572 | 1.0644 | 31.750 | 0.1428571 | 0.0928 |
| 1 3/8 | 6 | 1.5000 | 1.3917 | 1.2835 | 38.100 | 0.1666667 | 0.1083 |
| 1 3/4 | 5 | 1.7500 | 1.6201 | 1.4902 | 44.450 | 0.2000000 | 0.1299 |
| 2 | 4 1/2 | 2.0000 | 1.8557 | 1.7113 | 50.800 | 0.2222222 | 0.1443 |
| 2 1/4 | 4 1/2 | 2.2500 | 2.1057 | 1.9613 | 57.150 | 0.2222222 | 0.1443 |
| 2 1/2 | 4 | 2.5000 | 2.3376 | 2.1752 | 63.500 | 0.2500000 | 0.1624 |
| 2 3/4 | 4 | 2.7500 | 2.5876 | 2.4252 | 69.850 | 0.2500000 | 0.1624 |
| 3 | 4 | 3.0000 | 2.8376 | 2.6752 | 76.200 | 0.2500000 | 0.1624 |

The fine-thread series is made up of sizes taken from the standards of the Society of Automotive Engineers. This series is recommended for general use in automotive and aircraft work, for use where the design requires both strength and reduction in weight, and where special conditions require a fine thread, such as, for instance, on large sizes where sufficient force cannot be secured to set properly a screw or bolt of coarse pitch by the use of an ordinary wrench. (Table 2.)

The adoption, in practically its present form, of the American Briggs Standard Pipe Thread sizes as recommended by The American Society of Mechanical Engineers and all other organizations in any way interested in pipe threads was heartily favored by the great majority. This also applies to the fire-hose-coupling pitches as established by the National Fire Protection Association.

The National Fire Hose Coupling Threads Series is for fire-hose couplings from 2½ in. to 4½ in. in diameter, the basic sizes and dimensions of which correspond in all details to those recommended by the National Fire Protection Association, by the

337

**TABLE 2 NATIONAL FINE-THREAD SERIES**

| Identification | | Basic Diameters | | | Thread Data | | |
|---|---|---|---|---|---|---|---|
| Numbered and Fractional Sizes | Number of Threads per Inch | Major Diam., Inches | Pitch Diam., Inches | Minor Diam., Inches | Metric Equivalent to Major Diam., mm. | Pitch, Inches | Depth of Threads Inches |
| 0 | 80 | 0.060 | 0.0519 | 0.0438 | 1.524 | 0.0125000 | 0.00812 |
| 1 | 72 | 0.073 | 0.0640 | 0.0550 | 1.854 | 0.0138889 | 0.00902 |
| 2 | 64 | 0.086 | 0.0759 | 0.0657 | 2.184 | 0.0156250 | 0.01014 |
| 3 | 56 | 0.099 | 0.0874 | 0.0758 | 2.515 | 0.0178571 | 0.01160 |
| 4 | 48 | 0.112 | 0.0985 | 0.0849 | 2.845 | 0.0208333 | 0.01353 |
| 5 | 44 | 0.125 | 0.1102 | 0.0955 | 3.175 | 0.0227273 | 0.01476 |
| 6 | 40 | 0.138 | 0.1218 | 0.1055 | 3.506 | 0.0250000 | 0.01624 |
| 8 | 36 | 0.164 | 0.1460 | 0.1279 | 4.166 | 0.0277778 | 0.01804 |
| 10 | 32 | 0.190 | 0.1697 | 0.1494 | 4.826 | 0.0312500 | 0.02030 |
| 12 | 28 | 0.216 | 0.1928 | 0.1696 | 5.486 | 0.0357143 | 0.02319 |
| 1/4 | 28 | 0.2500 | 0.2268 | 0.2036 | 6.350 | 0.0357143 | 0.02319 |
| 5/16 | 24 | 0.3125 | 0.2854 | 0.2584 | 7.938 | 0.0416667 | 0.02706 |
| 3/8 | 24 | 0.3750 | 0.3479 | 0.3209 | 9.525 | 0.0416667 | 0.02706 |
| 7/16 | 20 | 0.4375 | 0.4050 | 0.3725 | 11.113 | 0.0500000 | 0.03246 |
| 1/2 | 20 | 0.5000 | 0.4675 | 0.4350 | 12.700 | 0.0500000 | 0.03246 |
| 9/16 | 18 | 0.5625 | 0.5264 | 0.4903 | 14.288 | 0.0555556 | 0.03608 |
| 5/8 | 18 | 0.6250 | 0.5869 | 0.5528 | 15.875 | 0.0555556 | 0.03608 |
| 3/4 | 16 | 0.7500 | 0.7094 | 0.6688 | 19.050 | 0.0625000 | 0.04060 |
| 7/8 | 14 | 0.8750 | 0.8286 | 0.7822 | 22.225 | 0.0714286 | 0.04640 |
| 1 | 14 | 1.0000 | 0.9536 | 0.9072 | 25.400 | 0.0714286 | 0.04640 |
| 1 1/8 | 12 | 1.1250 | 1.0709 | 1.0167 | 28.575 | 0.0833333 | 0.05413 |
| 1 1/4 | 12 | 1.2500 | 1.1959 | 1.1417 | 31.750 | 0.0833333 | 0.05413 |
| 1 1/2 | 12 | 1.5000 | 1.4459 | 1.3917 | 38.100 | 0.0833333 | 0.05413 |
| 1 3/4 | 12 | 1.7500 | 1.6959 | 1.6417 | 44.450 | 0.0833333 | 0.05413 |
| 2 | 12 | 2.0000 | 1.9459 | 1.8917 | 50.800 | 0.0833333 | 0.05413 |
| 2 1/4 | 12 | 2.2500 | 2.1959 | 2.1417 | 57.150 | 0.083333 | 0.05413 |
| 2 1/2 | 12 | 2.5000 | 2.4459 | 2.3917 | 63.500 | 0.083333 | 0.05413 |
| 2 3/4 | 12 | 2.7500 | 2.6959 | 2.6417 | 69.850 | 0.083333 | 0.05413 |
| 3 | 10 | 3.0000 | 2.9350 | 2.8701 | 76.200 | 0.100000 | 0.06498 |

Bureau of Standards, and by The American Society of Mechanical Engineers. This series is to be used on all couplings and hydrant connections for fire-protection systems and for all other purposes where hose couplings and connections are required in sizes between 2½ in. and 4½ in. in diameter. (Table 3.)

**TABLE 3 NATIONAL FIRE-HOSE COUPLINGS**
BASIC MINIMUM COUPLING DIMENSIONS

| Nominal Size | Number of Threads per In. | Pitch Inches | Depth of Thread Inches | Major Diameter mm. | Major Diameter Inches | Pitch Diam., Inches | Minor Diam., Inches | Allow. Inches |
|---|---|---|---|---|---|---|---|---|
| 2.5000 | 7.5 | 0.13333 | 0.0955 | 78.550 | 3.0925 | 2.9670 | 2.8715 | 0.03 |
| 2.0000 | 6.0 | 0.16667 | 0.1243 | 82.837 | 3.4550 | 3.3307 | 3.4063 | 0.03 |
| 2.5000 | 6.0 | 0.16667 | 0.1243 | 108.712 | 4.2800 | 4.1556 | 4.0313 | 0.03 |
| 4.5000 | 4.0 | 0.25000 | 0.1765 | 147.320 | 5.8000 | 5.6235 | 5.4470 | 0.03 |

BASIC MAXIMUM NIPPLE DIMENSIONS

| Nominal Size | Number of Threads per In. | Pitch Inches | Depth of Thread Inches | Major Diameter mm. | Major Diameter Inches | Pitch Diam., Inches | Minor Diam., Inches | Allow. Inches |
|---|---|---|---|---|---|---|---|---|
| 2.5000 | 7.6 | 0.13333 | 0.0955 | 77.788 | 3.0625 | 2.9670 | 2.8715 | 0.03 |
| 2.0000 | 6.0 | 0.16667 | 0.1243 | 82.075 | 3.6250 | 3.5008 | 3.3763 | 0.03 |
| 2.5000 | 6.0 | 0.16667 | 0.1243 | 107.950 | 4.2500 | 4.1256 | 4.0013 | 0.03 |
| 4.5000 | 4.0 | 0.25000 | 0.1765 | 146.050 | 5.7500 | 5.5735 | 5.3970 | 0.03 |

The National Hose Coupling Threads Series is for hose-coupling threads from ¾ in. to 2 in. in diameter. These are to be used on all couplings and connections where sizes between ¾ in. and 2 in. in diameter are required. (Table 4.)

The National Pipe Thread Series is described in detail in the section on National (American) Pipe Threads.

## FORM OF THREAD

The form of thread profile, known previously as the United States Standard or Sellers Profile, which was recommended by the Commission, is known throughout the report as the National Form of Thread. This profile is to be used for all screw-thread work except when otherwise specified for special purposes. The basic angle of thread between the sides of the thread measured in an axial plane is 60 deg. The line bisecting this 60-deg. angle is perpendicular to the axis of the screw thread. The basic flat at the root and crest of the thread form is 1/8 p or 0.125 p. The basic depth of the thread form is $0.649519p = 0.649519/n$, where $p$ is the pitch in inches and $n$ the number of threads per inch. A clearance is provided at the minor diameter of the nut by removing the thread form at the crest by an amount equal to 1/8 to 1/4 of the basic thread depth. A clearance at the major diameter of

the nut is provided by decreasing the depth of the truncation triangle by an amount equal to 1/3 to 2/3 of its theoretical value.

For threads cut on fire-hose couplings, the form of the thread profile recommended by the National Fire Protection Association, The American Society of Mechanical Engineers and the Bureau of Standards, and previously known and specified as the National Standard Hose Coupling Thread, is specified in this report as follows:

The basic angle between the sides of the thread measured in an axial plane is 60 deg. The line bisecting this 60-deg. angle is perpendicular to the axis of the screw thread. The crest and root of the basic thread are flattened or truncated from a sharp V form as follows:

| | | | |
|---|---|---|---|
| Threads per inch | 4 | 6 | 7 ½ |
| Depth of truncation, inches | 0.02 | 0.01 | 0.01 |

**TABLE 4 NATIONAL HOSE-COUPLING THREADS**
BASIC MINIMUM COUPLING DIMENSIONS

| Nominal Size | Number of Threads per In. | Pitch Inches | Depth of Thread Inches | Major Diameter mm. | Major Diameter Inches | Pitch Diam., Inches | Minor Diam., Inches | Allow. Inches |
|---|---|---|---|---|---|---|---|---|
| 3/4 | 11 1/2 | 0.08696 | 0.0565 | 27.242 | 1.0785 | 1.0160 | 0.9565 | 0.01 |
| 1 | 11 1/2 | 0.08696 | 0.0565 | 33.150 | 1.3051 | 1.2486 | 1.1922 | 0.01 |
| 1 1/4 | 11 1/2 | 0.08696 | 0.0565 | 41.908 | 1.4499 | 1.5934 | 1.5369 | 0.01 |
| 1 1/2 | 11 1/2 | 0.08696 | 0.0565 | 47.976 | 1.8885 | 1.8323 | 1.7759 | 0.01 |
| 2 | 11 1/2 | 0.08696 | 0.0565 | 60.013 | 2.3628 | 2.3063 | 2.2498 | 0.01 |

BASIC MAXIMUM NIPPLE DIMENSIONS

| Nominal Size | Number of Threads per In. | Pitch Inches | Depth of Thread Inches | Major Diameter mm. | Major Diameter Inches | Pitch Diam., Inches | Minor Diam., Inches | Allow. Inches |
|---|---|---|---|---|---|---|---|---|
| 3/4 | 11 1/2 | 0.08696 | 0.0565 | 26.988 | 1.0635 | 1.0060 | 0.9495 | 0.01 |
| 1 | 11 1/2 | 0.08696 | 0.0565 | 32.696 | 1.2951 | 1.2386 | 1.1822 | 0.01 |
| 1 1/4 | 11 1/2 | 0.08696 | 0.0565 | 41.654 | 1.6399 | 1.5634 | 1.5269 | 0.01 |
| 1 1/2 | 11 1/2 | 0.08696 | 0.0565 | 47.722 | 1.8785 | 1.8223 | 1.7659 | 0.01 |
| 2 | 11 1/2 | 0.08696 | 0.0565 | 59.761 | 2.3528 | 2.2963 | 2.2398 | 0.01 |

## CLASSIFICATION OF FITS

For general use, four distinct classes of screw-thread fits were established. These four classes, together with the accompanying specifications, given below and in Tables 5, 6, 7 and 8, are for the purpose of insuring interchangeable manufacture of screw-thread parts throughout the country. The examples given under each class of fit are for the purpose of illustration only. It is not the intention of the Commission to arbitrarily place a general class or grade of work in a specific class of fit. Each manufacturer and user of screw threads is free to select the class of fit best adapted to his particular needs.

The following general specifications apply to all classes of fits: In order to conform to the general ideas of standardization, the pitch diameter of the minimum threaded hole or nut should correspond to the basic size, the errors due to workmanship being permitted above the basic size. The maximum length of engagement for screw threads manufactured in accordance with any of the classes of fit specified herein shall not exceed the quantity as determined by the formula $L = 1.5D$, where $L$ is the length of engagement and $D$ the basic major diameter of the thread.

The specifications established for the various classes of fit are applicable to the National Coarse Threads, the National Fine Threads, the National Hose Threads, Straight Pipe Threads, and to any special thread required in manufacture which is not intentionally tapered.

### CLASS I, LOOSE FIT

The loose-fit class of screw threads is defined and specified as follows: This class is intended to cover the manufacture of strictly interchangeable threaded parts where the work is produced in two or more manufacturing plants. In this class are included threads for artillery ammunition and rough commercial work, such as stove bolts, carriage bolts, and other threaded work of a similar nature, where quick and easy assembly is necessary and a certain amount of shake or play is not objectionable. National Straight Pipe Threads and National Hose Coupling Threads are to be produced in this class of fit only. National Fire Hose Threads are to be produced in this class in accordance with special allowances and tolerances for fire-hose-coupling threads. The pitch diameter of the minimum nut of a given diameter and pitch will correspond to the basic pitch diameter as specified in the tables of thread systems given herein, which is computed from the basic major diameter of the thread to be manufactured. The pitch diameter of the minimum nut is the theoretical pitch diameter for that size.

The tolerance on the nut will be plus, applied from the basic size to above basic size; the tolerance on the screw will be minus, applied from the maximum screw dimensions to below the maximum screw dimensions. The allowance provided between the size of the minimum screw, which is basic, and the size of the maximum screw for a screw thread of a given pitch, and the tolerance allowed on a screw or nut of a given pitch will be as specified in Table 5.

TABLE 5   CLASS I—LOOSE FIT
ALLOWANCES AND TOLERANCES
SCREWS, NUTS AND GAGES

| 1 | 2 | 3 | MASTER GAGE TOLERANCES | | | 7 |
|---|---|---|---|---|---|---|
| No. Thds. per Inch | Allowances, Inches | Extreme or Drawing Pitch Diam. Tolerances, Inches | Diameter, Inches | Lead, (+ or -) Inches | 1/2 Angle, (+ or -) Degrees | Net Pitch Diameter Tolerances, Inches |
| 80 | 0.0007 | 0.0024 | 0.0002 | 0.0002 | 30' 00" | 0.0020 |
| 72 | 0.0007 | 0.0025 | 0.0002 | 0.0002 | 30' 00" | 0.0021 |
| 64 | 0.0007 | 0.0025 | 0.0002 | 0.0002 | 30' 00" | 0.0022 |
| 56 | 0.0008 | 0.0028 | 0.0002 | 0.0002 | 30' 00" | 0.0024 |
| 48 | 0.0009 | 0.0031 | 0.0002 | 0.0002 | 30' 00" | 0.0027 |
| 44 | 0.0009 | 0.0032 | 0.0002 | 0.00002 | 30' 00" | 0.0028 |
| 40 | 0.0010 | 0.0034 | 0.0002 | 0.0002 | 30' 00" | 0.0030 |
| 36 | 0.0011 | 0.0036 | 0.0002 | 0.0002 | 30' 00" | 0.0032 |
| 32 | 0.0011 | 0.0038 | 0.0002 | 0.0002 | 20' 00" | 0.0034 |
| 28 | 0.0012 | 0.0042 | 0.0003 | 0.0002 | 15' 00" | 0.0034 |
| 24 | 0.0013 | 0.0044 | 0.0003 | 0.0002 | 15' 00" | 0.0040 |
| 20 | 0.0015 | 0.0051 | 0.0003 | 0.0002 | 15' 00" | 0.0045 |
| 18 | 0.0016 | 0.0057 | 0.0004 | 0.0002 | 10' 00" | 0.0049 |
| 16 | 0.0018 | 0.0063 | 0.0004 | 0.0003 | 10' 00" | 0.0055 |
| 14 | 0.0021 | 0.0070 | 0.0004 | 0.0003 | 10' 00" | 0.0062 |
| 13 | 0.0022 | 0.0074 | 0.0004 | 0.0003 | 10' 00" | 0.0066 |
| 12 | 0.0024 | 0.0079 | 0.0004 | 0.0003 | 10' 00" | 0.0071 |
| 11 | 0.0026 | 0.0085 | 0.0004 | 0.0003 | 10' 00" | 0.0077 |
| 10 | 0.0028 | 0.0092 | 0.0004 | 0.0004 | 5' 00" | 0.0084 |
| 9 | 0.0031 | 0.0100 | 0.0004 | 0.0004 | 5' 00" | 0.0092 |
| 8 | 0.0034 | 0.0111 | 0.0004 | 0.0004 | 5' 00" | 0.0103 |
| 7 | 0.0039 | 0.0124 | 0.0004 | 0.0004 | 5' 00" | 0.0116 |
| 6 | 0.0044 | 0.0145 | 0.0006 | 0.0005 | 5' 00" | 0.0133 |
| 5 | 0.0053 | 0.0169 | 0.0006 | 0.0005 | 5' 00" | 0.0157 |
| 4 1/2 | 0.0057 | 0.0184 | 0.0006 | 0.0005 | 5' 00" | 0.0172 |
| 4 | 0.0064 | 0.0204 | 0.0006 | 0.0005 | 5' 00" | 0.0192 |

## CLASS II-A, MEDIUM FIT (REGULAR)

This class is intended to apply to interchangeable manufacture where the threaded members are to be assembled nearly, or entirely, with the fingers, and where a moderate amount of shake or play between the assembled threaded members is not objectionable. This class includes the great bulk of fastening screws for instruments, small arms and other ordnance material such as gun carriages, aerial-bomb-dropping devices, and interchangeable accessories mounted on guns; also machine screws, cap screws, and screws for sewing machines, typewriters and other work of a similar nature.

The pitch diameter of the minimum nut of a given diameter and pitch corresponds to the basic pitch diameter as specified in tables of thread systems herein given and is computed from the basic major diameter of the thread to be manufactured. The major diameter and pitch diameter of the maximum screw of a given pitch and diameter correspond to the basic dimensions as specified in tables of thread systems herein given, which are computed from the basic major diameter of the thread to be manufactured.

The tolerance on the nut will be plus, applied from the basic size to above basic size; the tolerance on the screw will be minus, applied from the basic size to below basic size. The allowance between the size of the maximum screw and the minimum nut will be zero for all pitches and all diameters. The tolerance for a screw or nut of a given pitch will be as specified in Table 6.

## CLASS II-B, MEDIUM FIT (SPECIAL)

This class is intended to apply especially to the higher grade

TABLE 6   CLASS II-A—MEDIUM FIT (REGULAR)
ALLOWANCES AND TOLERANCES
SCREWS, NUTS AND GAGES

| 1 | 2 | 3 | MASTER GAGE TOLERANCES | | 6 |
|---|---|---|---|---|---|
| No. Thds. per Inch | Extreme or Drawing Pitch Diam. Tolerances, Inches | Diameter, Inches | Lead, (+ or -) Inches | 1/2 Angle, (+ or -) Minutes | Net Pitch Diameter Tolerances, Inches |
| 80 | 0.0017 | 0.0002 | 0.0002 | 30' 00" | 0.0013 |
| 72 | 0.0018 | 0.0002 | 0.0002 | 30' 00" | 0.0014 |
| 64 | 0.0019 | 0.0002 | 0.0002 | 30' 00" | 0.0015 |
| 56 | 0.0020 | 0.0002 | 0.0002 | 30' 00" | 0.0016 |
| 48 | 0.0022 | 0.0002 | 0.0002 | 30' 00" | 0.0018 |
| 44 | 0.0023 | 0.0002 | 0.0002 | 30' 00" | 0.0019 |
| 40 | 0.0024 | 0.0002 | 0.0002 | 30' 00" | 0.0020 |
| 36 | 0.0025 | 0.0002 | 0.0002 | 20' 00" | 0.0023 |
| 32 | 0.0027 | 0.0002 | 0.0002 | 20' 00" | 0.0025 |
| 28 | 0.0031 | 0.0003 | 0.0002 | 15' 00" | 0.0025 |
| 24 | 0.0035 | 0.0003 | 0.0002 | 15' 00" | 0.0027 |
| 20 | 0.0036 | 0.0003 | 0.0002 | 15' 00" | 0.0030 |
| 18 | 0.0041 | 0.0004 | 0.0003 | 10' 00" | 0.0035 |
| 16 | 0.0045 | 0.0004 | 0.0003 | 10' 00" | 0.0037 |
| 14 | 0.0049 | 0.0004 | 0.0003 | 10' 00" | 0.0041 |
| 13 | 0.0052 | 0.0004 | 0.0003 | 10' 00" | 0.0044 |
| 12 | 0.0056 | 0.0004 | 0.0003 | 10' 00" | 0.0048 |
| 11 | 0.0059 | 0.0003 | 0.0003 | 10' 00" | 0.0051 |
| 10 | 0.0064 | 0.0004 | 0.0004 | 5' 00" | 0.0056 |
| 9 | 0.0070 | 0.0004 | 0.0004 | 5' 00" | 0.0062 |
| 8 | 0.0076 | 0.0004 | 0.0004 | 5' 00" | 0.0068 |
| 7 | 0.0085 | 0.0004 | 0.0004 | 5' 00" | 0.0077 |
| 6 | 0.0101 | 0.0006 | 0.0005 | 5' 00" | 0.0089 |
| 5 | 0.0116 | 0.0006 | 0.0005 | 5' 00" | 0.0104 |
| 4 1/2 | 0.0127 | 0.0006 | 0.0005 | 5' 00" | 0.0115 |
| 4 | 0.0140 | 0.0006 | 0.0005 | 5' 00" | 0.0128 |
| Allowances, in all cases. * zero | | | | | |

of automobile screw-thread work: It is the same in every particular as Class II-A, Medium Fit (Regular), except that the tolerances are smaller. (Table 7.)

TABLE 7   CLASS II-B—MEDIUM FIT (SPECIAL)
ALLOWANCES AND TOLERANCES
SCREWS, NUTS AND GAGES

| 1 | 2 | 3 | MASTER GAGE TOLERANCES | | 6 |
|---|---|---|---|---|---|
| No. Thds. per Inch | Extreme or Drawing Pitch Diam. Tolerances, Inches | Diameter, Inches | Lead, (+ or -) Inches | 1/2 Angle, (+ or -) Minutes | Net Pitch Diameter Tolerances, Inches |
| 80 | 0.0013 | 0.0002 | 0.0002 | 30' 00" | 0.0009 |
| 72 | 0.0013 | 0.0002 | 0.0002 | 30' 00" | 0.0009 |
| 64 | 0.0014 | 0.0002 | 0.0002 | 30' 00" | 0.0010 |
| 56 | 0.0015 | 0.0002 | 0.0002 | 30' 00" | 0.0011 |
| 48 | 0.0016 | 0.0002 | 0.0002 | 30' 00" | 0.0012 |
| 44 | 0.0016 | 0.0002 | 0.0002 | 30' 00" | 0.0012 |
| 40 | 0.0017 | 0.0002 | 0.0002 | 30' 00" | 0.0013 |
| 36 | 0.0018 | 0.0002 | 0.0002 | 20' 00" | 0.0014 |
| 32 | 0.0019 | 0.0002 | 0.0002 | 20' 00" | 0.0015 |
| 28 | 0.0022 | 0.0003 | 0.0002 | 15' 00" | 0.0016 |
| 24 | 0.0024 | 0.0003 | 0.0002 | 15' 00" | 0.0018 |
| 23 | 0.0026 | 0.0003 | 0.0002 | 15' 00" | 0.0020 |
| 18 | 0.0030 | 0.0004 | 0.0003 | 10' 00" | 0.0022 |
| 16 | 0.0032 | 0.0004 | 0.0003 | 10' 00" | 0.0024 |
| 14 | 0.0036 | 0.0004 | 0.0003 | 10' 00" | 0.0028 |
| 13 | 0.0037 | 0.0004 | 0.0003 | 10' 00" | 0.0029 |
| 12 | 0.0040 | 0.0004 | 0.0003 | 10' 00" | 0.0032 |
| 11 | 0.0042 | 0.0004 | 0.0003 | 10' 00" | 0.0034 |
| 10 | 0.0045 | 0.0004 | 0.0004 | 5' 00" | 0.0037 |
| 9 | 0.0049 | 0.0004 | 0.0004 | 5' 00" | 0.0041 |
| 8 | 0.0054 | 0.0004 | 0.0004 | 5' 00" | 0.0046 |
| 7 | 0.0059 | 0.0004 | 0.0004 | 5' 00" | 0.0051 |
| 6 | 0.0071 | 0.0006 | 0.0005 | 5' 00" | 0.0059 |
| 5 | 0.0082 | 0.0006 | 0.0005 | 5' 00" | 0.0070 |
| 4 1/2 | 0.0089 | 0.0006 | 0.0005 | 5' 00" | 0.0077 |
| 4 | 0.0097 | 0.0006 | 0.0005 | 5' 00" | 0.0085 |
| Allowances, in all cases. * zero | | | | | |

## CLASS III, CLOSE FIT

This class is intended for threaded work of the finest commercial quality, where the thread has practically no backlash, and for light screwdriver fits. In the manufacture of screw-thread products belonging in this class, it will be necessary to use precision tools, selected master gages, and many other refinements. This quality of work should therefore be used only in cases where requirements of the mechanism being produced are exacting, or where special conditions require screws having a precision fit. In order to secure the fit desired, it may be necessary, in some cases, to select the parts when the product is being assembled.

The pitch diameter of the minimum nut of a given diameter and pitch will correspond to the basic pitch diameter as specified in tables of thread systems given herein, which is computed from the basic major diameter of the thread to be manufactured. The major diameter and pitch diameter of the maximum screw of a given diameter and pitch will be above the basic dimensions as specified in tables computed from the basic major diameter of the thread to be manufactured, by the amount of the allowance (interference) specified in Table 8.

The tolerance on the nut will be plus, applied from the basic size to above basic size; the tolerance on the screw will be minus, applied from the maximum screw dimensions to below the maximum screw dimensions. The allowance (interference) provided between the size of the minimum nut, which is basic, and the size of the maximum screw, which is above basic, and the tolerance for a screw or nut of a given pitch will be as specified in Table 8.

TABLE 8 CLASS III—CLOSE FIT
ALLOWANCES AND TOLERANCES
SCREWS, NUTS AND GAUGES

| 1 | 2 | 3 | MASTER GAGE TOLERANCE | | | 7 |
| No. Thds per Inch | Interference or Negative Allowances, Inches | Extreme or Drawing Pitch Diam. Tolerances, Inches | Diameter, Inches | Lead, (+ or -) Inches | 1/2 Angle, (+ or -) Degrees | Net Pitch Diameter Tolerances, Inches |
|---|---|---|---|---|---|---|
| 80 | 0.0001 | 0.0006 | 0.0001 | 0.0001 | 15' 00" | 0.0004 |
| 72 | 0.0001 | 0.0007 | 0.0001 | 0.0001 | 15' 00" | 0.0005 |
| 64 | 0.0001 | 0.0007 | 0.0001 | 0.0001 | 15' 00" | 0.0005 |
| 56 | 0.0002 | 0.0007 | 0.0001 | 0.0001 | 15' 00" | 0.0005 |
| 48 | 0.0002 | 0.0008 | 0.0001 | -0.0001 | 15' 00" | 0.0006 |
| 44 | 0.0002 | 0.0008 | 0.0001 | 0.0001 | 15' 00" | 0.0006 |
| 40 | 0.0002 | 0.0009 | 0.0001 | 0.0001 | 10' 00" | 0.0007 |
| 36 | 0.0002 | 0.0009 | 0.0001 | 0.0001 | 10' 00" | 0.0007 |
| 32 | 0.0002 | 0.0010 | 0.0001 | 0.0001 | 10' 00" | 0.0008 |
| 28 | 0.0002 | 0.0011 | 0.00015 | 0.0001 | 7' 30" | 0.0009 |
| 24 | 0.0003 | 0.0012 | 0.00015 | 0.0001 | 7' 30" | 0.0009 |
| 20 | 0.0003 | 0.0013 | 0.00015 | 0.0001 | 7' 30" | 0.0010 |
| 18 | 0.0003 | 0.0015 | 0.0002 | 0.00015 | 5' 00" | 0.0011 |
| 16 | 0.0004 | 0.0016 | 0.0002 | 0.00015 | 5' 00" | 0.0012 |
| 14 | 0.0004 | 0.0018 | 0.0002 | 0.00015 | 5' 00" | 0.0014 |
| 13 | 0.0004 | 0.0019 | 0.0002 | 0.00015 | 5' 00" | 0.0015 |
| 12 | 0.0005 | 0.0020 | 0.0002 | 0.0002 | 5' 00" | 0.0016 |
| 11 | 0.0005 | 0.0021 | 0.0002 | 0.00015 | 5' 00" | 0.0017 |
| 10 | 0.0006 | 0.0023 | 0.0002 | 0.0002 | 2' 30" | 0.0019 |
| 9 | 0.0006 | 0.0024 | 0.0002 | 0.0002 | 2' 30" | 0.0020 |
| 8 | 0.0007 | 0.0027 | 0.0002 | 0.0002 | 2' 30" | 0.0024 |
| 7 | 0.0008 | 0.0030 | 0.0002 | 0.0002 | 2' 30" | 0.0026 |
| 6 | 0.0009 | 0.0036 | 0.0003 | 0.00025 | 2' 30" | 0.0030 |
| 5 | 0.0010 | 0.0041 | 0.0003 | 0.00025 | 2' 30" | 0.0028 |
| 4 1/2 | 0.0011 | 0.0044 | 0.0003 | 0.00025 | 2' 30" | 0.0036 |
| 4 | 0.0012 | 0.0048 | 0.0003 | 0.00035 | 2' 30" | 0.0042 |

### CLASS IV, WRENCH FIT

This class is intended to cover the manufacture of threaded parts ¼ in. in diameter or larger which are to be set or assembled permanently with a wrench. Inasmuch as for wrench fits the material is an important factor in determining the fit between the threaded members, there are provided herein two subdivisions for this class of work, namely, A and B. These two subdivisions differ mainly in the amount of the allowance (interference) values provided for different pitches.

Subdivision A of Class IV, Wrench Fit, provides for the production of interchangeable wrench-fit screws or studs used in light sections with moderate stresses, such as for aircraft and automobile-engine work.

Subdivision B of Class IV, Wrench Fit, provides for the production of interchangeable wrench-fit screws or studs used in heavy sections with heavy stresses, such as for steam-engine and heavy hydraulic work.

The pitch diameter of the minimum nut of a given diameter and pitch for threads belonging in either subdivision A or subdivision B will correspond to the basic pitch diameter as specified in tables of thread systems given herein, which is computed from the basic major diameter of the thread to be manufactured. The major diameter and pitch diameter of the maximum screw of a given diameter and pitch for threads belonging in either subdivision A or subdivision B will be above the basic dimensions as specified in tables of thread systems given herein, which are computed from the basic major diameter of the thread to be manufactured, by the amount of the allowance (interference) provided.

The tolerance on the nut will be plus, to be applied from the basic size to above basic size; the tolerance on the screw will be minus, to be applied from the maximum screw dimensions to below maximum screw dimensions. At the present time [when this report was released] the Commission does not have sufficient information or data to include the values for tolerances and allowances for wrench fits. It is hoped, however, that sufficient information resulting from investigation and research will enable the Commission to decide, at an early date, the allowance and tolerance values for these two classes of wrench fits included herein, which will be applicable to the various materials and which will meet the requirements found in the manufacture of machines or products requiring wrench fits.

## TOLERANCES

There are specified herein, for use in connection with the various fits established, three different sets of tolerances (as given in Tables 5, 6, 7, and 8), which represent the extreme variations allowed on the work.

The tolerance limits established represent in reality the sizes of the " Go " and " Not Go " master gages. Errors in lead and angle which occur on the threaded work can be offset by a suitable alteration of the pitch diameter of the work. If the " Go " gage passes the threaded work, interchangeability is secured and the thread profile may differ from that of the " Go " gage in either pitch diameter, lead or angle. The " Not Go " gage checks pitch diameter only and thus insures that the pitch diameter is such that the fit will not be too loose.

The tolerances established for Class I, Loose Fit, and Class II. Medium Fit, permit the use of commercial taps now obtainable from various manufacturers. For Class III, Close Fit, in which it is desired to produce a hole close to the basic size, it is recommended that a selected tap be used.

The pitch-diameter tolerances provided for a screw of a given class of fit will be the same as the pitch-diameter tolerances provided for a nut corresponding to the same class of fit.

The allowable tolerances on the major diameter of screws of a given classification will be twice the tolerance values allowed on the pitch diameters of screws of the same class.

The minimum minor diameter of a screw of a given pitch will be such as to result in a basic flat ( = ⅛p) at the root when the pitch diameter of the screw is at its minimum value. When the maximum screw is basic, the minimum minor diameter of the screw will be below the basic minor diameter by the amount of the specified pitch-diameter tolerance.

The maximum minor diameter may be such as results from the use of a worn or rounded threading tool, when the pitch diameter is at its maximum value. In no case, however, should the form of the screw, as results from tool wear, be such as to cause the screw to be rejected on the maximum minor diameter by a " Go " ring gage, the minor diameter of which is equal to the minimum minor diameter of the nut.

Attention is called to the fact that the minimum threaded hole or nut corresponds to the basic size; that is, the minimum nut is basic for all classes of fit. This condition permits the use of taps which when new are oversize and which are discarded when the hole cut is at the basic size.

In order to secure the desired fit the screw size is varied; the maximum screw corresponds to the basic size for the medium-fit class, is slightly above basic size for the loose-fit class. The tolerances specified in column 7 of Tables 5 and 8, and column 6 of Tables 6 and 7, are the net tolerances which are in no way reduced by permissible manufacturing tolerances provided for by master gages. These master-gage tolerances are provided for by being added to the net tolerances. Thus the extreme or drawing tolerances are the net working tolerances increased by the master-gage increment of equivalent diametrical space required to provide for the master-gage tolerances. The limits established for the extreme tolerances should in no case be exceeded. The application of gage tolerances in relation to tolerances allowed on the work can be best understood by considering that the extreme tolerances represent the absolute limits over which variations of the work must not pass. The manufacturing tolerances required for master gages are then deducted from the extreme working tolerances, producing the figures specified as net tolerances. Further reduction of the extreme tolerances is caused by the manufacturing tolerances required for the inspection gages and working gages.

It is essential that the proportion of the tolerances used by the workmen producing the work at the machine be well within the net tolerance limits. The net tolerance limits as established by the master gages may be considered as the largest circle of a target, the space occupied by the master-gage tolerances representing the width of the line establishing the largest circle. The marksman always aims to hit the bull's-eye. Any mark inside of the largest circle or cutting the circle scores. Any mark outside of the largest circle does not score. The same is true in producing work—the careful manufacturer will aim to produce work which is in the center of tolerance limits. The bull's-eye in this case, which is the working tolerance used at the machine, will be considerably less than the net tolerance and the result will be that a very large percentage of the work will be accepted, and spoiled or rejected work will be reduced to practically nothing. If the net tolerance limits are used as working limits at the machine, there will be a larger percentage of rejections due to differences in gages and wear of both tools and gages.

## GAGES

The following general specifications which refer in particular to gaging systems which have been found satisfactory by the army and navy for the production of interchangeable parts are built upon the following assumptions:

*a* Approved limit master gages do not reduce the net working tolerance.

*b* Permissible errors in angle of thread specified for "Go" gages tend to reduce the net working tolerance, while similar permissible errors on the "Not Go" gage tend to increase the net working tolerance. These two factors, therefore, balance each other.

*c* Permissible lead errors specified for the "Go" gage reduce the net working tolerance, while permissible lead errors on the "Not Go" gage tend to increase the net working tolerance.

*d* In order to realize the full net working tolerance, the permissible diametrical variation specified for both "Go" and "Not Go" gages (gage increment) is placed outside of the net tolerance limits. The extreme tolerance equals the net tolerance plus gage increment.

*e* The "Go" gage should check simultaneously all elements of the thread (all diameters, lead, angle, etc.).

*f* The "Not Go" gage should check separately the elements of the thread.

### GENERAL SPECIFICATIONS

The following specifications are included for the use of manufacturers where definite information is lacking but are not to be considered mandatory:

*Classification:* Thread gages may be included in one of four classes, namely, Standard Master Gages, Limit Master Gages, Inspection Gages, and Working Gages.

*Standard Master Gage:* The standard master gage is a threaded plug representing as exactly as possible all physical dimensions of the nominal or basic size of the threaded component. In order that the standard master gage be authentic, the deviations of this gage from the exact standard should be ascertained by the National Bureau of Standards, and the gage should be used with knowledge of these deviations or corrections.

*Limit Master Gages:* Limit master gages are for reference only. They represent the extreme upper and lower tolerance limits allowed on the dimensions of the part being produced. They are often of the same design as inspection gages. In many cases,

however, the design of the master gage is that of a check which can be used to verify the inspection or working gage.

*Inspection Gages:* Inspection gages are for the use of the purchaser in accepting the product. They are generally of the same design as the working gages and the dimensions are such that they represent nearly the net tolerance limits on the parts being produced. Inasmuch as a certain amount of wear must be provided on an inspection gage, it cannot represent the net tolerance limit until it is worn to master-gage size.

*Working Gages:* Working gages are those used by the manufacturer to check the parts produced as they are machined. It is recommended that the working gages be made to represent limits considerably inside of the net limits in order that sufficient wear will be provided for the working gages, and in order that the product accepted by the working gages will be accepted by the inspection gages.

### INSPECTION AND WORKING-GAGE SETS

*For Screws:* The following list enumerates the inspection and working gages required for producing strictly interchangeable screws as specified for National Coarse Threads, National Fine Threads, or other straight threads:

*a* A maximum or "Go" ring thread gage, preferably adjustable, having the required pitch diameter and minor diameter. The major diameter may be cleared to facilitate grinding and lapping.

*b* A minimum or "Not Go" ring thread gage, preferably adjustable, to check only the pitch diameter of the threaded work.

*c* A maximum or "Go" plain ring gage to check the major diameter of the threaded work.

*d* A minimum or "Not go" snap gage to check the major diameter of the threaded work.

*For Nuts:* The following list enumerates the inspection and working gages required for producing strictly interchangeable nuts, as specified for National Coarse Threads, National Fine Threads, or other straight threads.

*a* A minimum or "Go" thread plug gage of the required pitch diameter and major diameter. The minor diameter of the thread plug gage may be cleared to facilitate grinding and lapping.

*b* A maximum or "Not Go" thread plug gage to check only the pitch diameter of the threaded work.

*c* A "Go" plain plug gage to check the minor diameter of the threaded work.

*d* A "Not Go" plain plug gage to check the minor diameter of the threaded work.

### LIMIT MASTER GAGES REQUIRED FOR CHECKING WORKING OR INSPECTION GAGES

*Used on Screw:* The following list enumerates the limit master gages required for the verification of the working or inspection gages as previously listed for verifying the screw:

*a* A set plug or check for the maximum "Go" thread ring gage, having the same dimensions as the largest permissible screw.

*b* A set plug or check for the minimum or "Not Go" thread ring gage having same dimensions as smallest permissible screw.

*c* A maximum plain plug for checking the minor diameter of both the "Go" and "Not Go" inspection thread ring gage.

*Used on Nut:* The following list enumerates the limit master gages required for the verification of the working or inspection gages as previously listed for verifying the nut:

*a* A minimum or "Go" threaded plug to be used as a reference for comparative measurements and corresponds to the basic dimension, or standard master gage.

*b* A maximum or "Not Go" threaded plug to be used as a reference for comparative measurements and corresponds to the largest permissible threaded hole.

*c* A minimum plain ring gage to check the major diameter of the "Go" and "Not Go" master threaded plug unless suitable measuring facilities are available for this purpose.

### DESIGN AND CONSTRUCTION

The following specification will be helpful in the design and construction of gages used for producing threaded work:

*Material:* Gages may be made of a good grade of machinery steel pack-hardened, or of straight carbon steel of not less than 1 per cent carbon; or preferably of an oil hardening steel of approximately 1.10 per cent carbon and 1.40 per cent chromium.

*Handles and Marking:* Handles should be made of a good grade of machinery steel plainly marked to identify the gage.

*Plain Plug Gages:* All plain plug gages should be single-ended. Plain plug gages of 2 in. and less in diameter should be made with a plug inserted in the handle and fastened thereto by means of a pin. Plain plug gages of more than 2 in. in diameter should have the gaging blank so made as to be reversible. This can be accomplished by having a finished hole in the gage blank fitting a shouldered projection on the end of the handle, the gage blank being held on with a nut.

The "Go" plain plug gage should be noticeably longer than the "Not Go" plain plug gage, or some distinguishing feature in the design of the handle should be used to serve as a ready means of identification, such as a chamfer on the handle of the "Go" gage.

*Plain Ring Gages:* Both the "Go" and "Not Go" gages should have their outside diameters knurled if made circular.

The "Go" gage should have a decided chamfer in order to provide a ready means of identification for distinguishing the "Go" from the "Not Go" gage.

*Snap Gages:* Snap gages may be either adjustable or non-adjustable. It is recommended that all snap gages up to and including ⅛ in. be of the built-up type. For larger snap gages, forge blanks, flat plate stock or other suitable construction may be used.

Sufficient clearance beyond the mouth of the gage should be provided to permit the gaging of cylindrical work.

Snap gages for measuring lengths and diameters may have one gaging dimension only, or may have a maximum and minimum gaging dimension, both on one end, or maximum and minimum gaging dimensions on opposite ends of the gage. When the maximum and minimum gaging dimensions are placed on opposite ends of the gage, the maximum or "Go" end of the snap gage will be distinguished from the minimum or "Not Go" end by having the corners of the gage on the "Go" end decidedly chamfered.

*Plug Thread Gages:* All plug thread gages should be single-ended. Thread plug gages 2 in. and less in diameter should be made with a plug inserted in a handle and fastened thereto by means of a pin.

Plug gages of more than 2 in. in diameter, unless otherwise specified, should have the gaging blank so made as to be reversible. This can be accomplished by having the finished hole in the gage blank fitting a shouldered projection on the end of the handle, the gage blank being keyed on and held with a nut.

"Not Go" thread plug gages should be noticeably shorter than the "Go" thread plug gages, in order to provide a ready means of identification, or the handle of the "Go" gage should be chamfered.

End threads on plug thread gages should not be chamfered, but the first half-turn of the end thread should be flattened to avoid a feather edge.

Inspection and working thread plug gages should be provided with dirt grooves which extend into the gage for a depth of from one to four threads.

The length of thread ($L$) parallel to the axis of the gage should, for all standard "Go" thread plug gages, be at least as much as the quantity expressed in the formula: $L = 1.5D$, where $D$ is the basic major diameter of the thread. For threaded work of shorter length of engagement than 1.5D, the length of thread on the "Go" gage may be correspondingly shorter.

"*Not Go*" *Thread Gage for Pitch Diameter Only:* All "Not Go" thread plug gages should be made to check the pitch diameter only. This necessitates removal of the crest of the

thread so that the dimension of the major diameter is never greater than that specified for the "Go" gage, and also removing the portion of the thread at the root of the standard thread form.

*Ring Thread Gages:* All ring thread gages should be made adjustable.

The "Go" gage should be distinguished from the "Not Go" gage by having a decided chamfer and both gages are to have their outside diameter knurled if made circular.

The end threads on ring thread gages should not be chamfered but the first half turn of the end thread should be flattened to avoid a feather edge.

*Length of Thread:* The length of thread parallel to the axis of the gage should, for all standard "Go" ring thread gages, be at least as great as the quantity determined in the formula, $L = 1.5D$. For threaded work of shorter length of engagement than 1.5D, the length of thread on the "Go" gage may be correspondingly shorter.

"*Not Go*" *Ring Gage for Pitch Diameter Only:* "Not Go" ring thread gages should be made to check the pitch diameter only. This necessitates removal of the crest of the thread so that the dimension of the minor diameter is never less than that specified for the maximum or "Go" gage, and also removing the portion of the thread at the root of the standard form.

### TOLERANCES

There are specified herein for use in the production of National Coarse Threads, National Fine Threads, National Hose Coupling Threads, and for other straight threads, and for National Pipe Threads, several tables of gage manufacturing tolerances.

Table 9 will be found practicable for all plain plug, ring and

TABLE 9   MANUFACTURING TOLERANCES ON PLAIN GAGES

| Manufacturing Tolerance allowed on Work | Allowable Tolerance for Master Gages | | Allowable Tolerance for Inspection Gages | | Suggested Tolerance for Working Gages | |
|---|---|---|---|---|---|---|
| | Minimum Gage (+) | Maximum Gage (−) | Minimum Gage (+) | Maximum Gage (−) | Minimum Gage (+) | Maximum Gage (−) |
| 0.002 | 0.0000 | 0.0000 | 0.0001 | 0.0001 | 0.0003 | 0.0003 |
| | 0.0001 | 0.0001 | 0.0003 | 0.0003 | 0.0005 | 0.0005 |
| 0.002 to 0.004 | 0.0000 | 0.0000 | 0.0002 | 0.0002 | 0.0004 | 0.0004 |
| | 0.0002 | 0.0002 | 0.0004 | 0.0004 | 0.0007 | 0.0007 |
| 0.004 to 0.006 | 0.0000 | 0.0000 | 0.0004 | 0.0004 | 0.0007 | 0.0007 |
| | 0.0003 | 0.0003 | 0.0007 | 0.0007 | 0.0011 | 0.0011 |
| 0.006 to 0.010 | 0.0000 | 0.0000 | 0.0006 | 0.0006 | 0.0010 | 0.0010 |
| | 0.0004 | 0.0004 | 0.0010 | 0.0010 | 0.0015 | 0.0015 |
| 0.010 to 0.020 | 0.0000 | 0.0000 | 0.0010 | 0.0010 | 0.0015 | 0.0015 |
| | 0.0005 | 0.0005 | 0.0015 | 0.0015 | 0.0021 | 0.0021 |
| 0.020 to 0.030 | 0.0000 | 0.0000 | 0.0020 | 0.0020 | 0.0026 | 0.0026 |
| | 0.0006 | 0.0006 | 0.0026 | 0.0026 | 0.0033 | 0.0033 |

snap gages used in connection with a measurement of screw-thread diameters. In addition to the master-gage tolerances, suggested tolerances for inspection and working gages are also given.

Table 10 will be found practicable for both standard and limit master thread gages for thread work designed in accordance with the manufacturing tolerances for Class I, Loose Fit and Class II, Medium Fit, made to Tables 5, 6 and 7.

Tables 11 and 12 contain suggested manufacturing tolerances for inspection thread gages with a small allowance for wear for use in quantity production of Class I, Loose Fit and Class II, Medium Fit thread work, made to 5, 6 and 7.

Table 13 contains the tolerances suggested for both standard and limit master thread gages for work designed in accordance with manufacturing tolerances for Class III, Close Fit thread work, made to Table 8. As the component tolerances for this class are relatively small, it is believed that the working gages will be required to be held within the gage tolerances shown in this table.

### APPLICATION OF GAGE TOLERANCES

*Tolerances for Plain Gages:* For plain plug gages, plain ring gages and plain snap gages required for measuring diameters of screw-thread work, the gage tolerances specified in Tables 10, 11, 12 and 13 should be used.

Attention is called to the fact that the tolerances on thread diameters vary in accordance with the number of threads per inch on the screw or nut being manufactured. In manufacturing a plain plug, ring or snap gage, in absence of information as to the number of threads per inch of the screw to be made, or for gage dimension other than thread diameters, the tolerances for plain gages given in Table 10 may be used.

*Tolerances on Lead:* The tolerances on lead are specified as an allowable variation between any two threads not farther apart than the length of thread engagement as determined by the formula, $L = 1.5D$.

*Tolerances on Angle of Thread:* The tolerances on angle of thread as specified herein for the various pitches are tolerances on one-half of the included angle. This insures that the bisector of the included angle will be perpendicular to the axis of the thread within proper limits. The equivalent deviation from the

was read before the Institution of Civil Engineers of Great Britain.[1] In 1886 representatives of various manufacturing concerns and a committee of The American Society of Mechanical Engineers jointly adopted this system in detail and master gages were made. This standard has since been used in the United States and Canada.

At various conferences in the recent past, representatives of American manufacturers and The American Society of Mechanical Engineers established additional sizes, certain details of gaging, tolerances, and special applications of the standard. These were tabulated in a much more complete manner than was originally done by Mr. Briggs, by a special committee known as the Committee of Manufacturers on the Standardization of Valves and Fittings, and published at the suggestion of The American

TABLE 10 DIMENSIONS ON MASTER THREAD GAGES FOR LOOSE-FIT AND MEDIUM-FIT WORK
(This applies to both Standard and Limit Master Gages)

| Number of Threads Per Inch | Allowable Variation in Lead between any two Threads not farther apart than Length of Engagement (+ or −) | Allowable Variation in one-half of Angle of Thread Gages (+ or −) | Allowable Tolerances on Diameters of Minimum Thread Gages (+) | Allowable Tolerances on Diameters of Maximum Thread Gages (−) |
|---|---|---|---|---|
| 4 to 6 | 0.0000 5 | 5' 00" | 0.0000 0.0006 | 0.0000 0.0006 |
| 7 to 10 | 0.0004 | 5' 00" | 0.0000 0.0004 | 0.0000 0.0004 |
| 11 to 18 | 0.0003 | 10' 00" | 0.0000 0.0004 | 0.0000 0.0004 |
| 20 to 28 | 0.0003 | 15' 00" | 0.0000 0.0003 | 0.0000 0.0003 |
| 30 to 40 | 0.0003 | 20' 00" | 0.0000 0.0002 | 0.0000 0.0002 |
| 44 to 80 | 0.0002 | 30' 00" | 0.0000 0.0002 | 0.0000 0.0002 |

TABLE 12 SUGGESTED MANUFACTURING TOLERANCES FOR WORKING GAGES FOR LOOSE-FIT AND MEDIUM-FIT WORK

| Number of Threads per Inch | Allowable Variation in Lead between any two Threads not farther apart than Length of Engagement (+ or −) | Allowable Variation in one-half of Angle of Thread (+ or −) | Allowable Tolerances on Diameters of Minimum Thread Gages (+) | Allowable Tolerances on Diameters of Maximum Thread Gages (−) |
|---|---|---|---|---|
| 4 to 6 | 0.0006 | 5' 00" | 0.0015 0.0025 | 0 .0015 0.0025 |
| 7 to 10 | 0.0005 | 10' 00" | 0.0010 0.0020 | 0.0010 0.0020 |
| 11 to 18 | 0.0004 | 15' 00" | 0.0008 0.0015 | 0.0008 0.0015 |
| 20 to 28 | 0.0003 | 20' 00" | 0.0006 0.0012 | 0.0006 0.0012 |
| 30 to 40 | 0.0002 | 30' 00" | 0.0005 0.0010 | 0.0005 0.0010 |
| 40 to 80 | 0.0002 | 45' 00" | 0.0004 0.0006 | 0.0004 0.0006 |

TABLE 11 SUGGESTED MANUFACTURING TOLERANCES FOR INSPECTION GAGES FOR LOOSE-FIT AND MEDIUM-FIT WORK

| Number of Threads per Inch | Allowable Variation in Lead between any two Threads not farther apart than Length of Engagement (+ or −) | Allowable Variation in one-half of Angle of Thread Gages (+ or −) | Allowable Tolerances on Diameters of Minimum Thread Gages (+) | Allowable Tolerances on Diameters of Maximum Thread Gages (−) |
|---|---|---|---|---|
| 4 to 6 | 0.0006 | 5' 00" | 0.0006 0.0015 | 0.0006 0.0015 |
| 7 to 10 | 0.0005 | 10' 00" | 0.0004 0.0010 | 0.0004 0.0010 |
| 11 to 18 | 0.0004 | 15' 00" | 0.0004 0.0008 | 0.0004 0.0008 |
| 20 to 28 | 0.0003 | 20' 00" | 0.0003 0.0006 | 0.0003 0.0006 |
| 30 to 40 | 0.0002 | 30' 00" | 0.0003 0.0005 | 0.0003 0.0005 |
| 44 to 80 | 0.0002 | 45' 00" | 0.0002 0.0004 | 0.0002 0.0004 |

TABLE 13 MASTER GAGE TOLERANCES FOR CLASS III, CLOSE-FIT WORK
(This applies to both Standard and Limit Master Gages)

| Number of Threads per Inch | Allowable Variation in Lead between any two Threads not farther apart than Length of Engagement (+ or −) | Allowable Variation in one-half of Angle of Thread Gages (+ or −) | Allowable Tolerances on Diameters of Minimum Thread Gages (+) | Allowable Tolerances on Diameters of Maximum Thread Gages (−) |
|---|---|---|---|---|
| 4 to 6 | 0.00025 | 2° 30' | 0.0000 0.0003 | 0.0000 0.0003 |
| 7 to 10 | 0.0002 | 2° 30' | 0.0000 0.0003 | 0.0000 0.0003 |
| 11 to 18 | 0.00015 | 5° 00' | 0.0000 0.0002 | 0.0000 0.0002 |
| 20 to 28 | 0.0001 | 7° 30' | 0.0000 0.00015 | 0.0000 0.00015 |
| 30 to 40 | 0.0001 | 10° 00' | 0.0000 0.0001 | 0.0000 0.0001 |
| 44 to 80 | 0.0001 | 15° 00' | 0.0000 0.0001 | 0.0000 0.0001 |

true form caused by such irregularities as convex or concave sides of thread, rounded crests, or slight projections on the thread form, should not exceed the tolerances allowable on angle of thread.

*Tolerances on Diameters:* The tolerances given for thread diameters, in Tables 10, 11, 12 and 13, are applied in such a manner that the tolerances permitted on the inspection and working gages occupy part of the extreme tolerance. This insures that all work passed by the gages will be within the tolerance limits specified on the part drawing as represented by the limit master gages. The tolerances given also permit the classification and selection of gages so that if a gage is not suitable for a master gage it may be classified and used as an inspection or working gage provided that the errors do not pass outside of the net tolerance limits. The application of the tolerances on diameters of thread gages is exactly the same as explained herein for plain gages.

## NATIONAL (AMERICAN) PIPE THREADS

The American Pipe Thread Standard, also known as the American (Briggs) Standard, was formulated by Mr. Robert Briggs, an American engineer, prior to 1882. After his death a paper containing detailed information of the system he devised

Society of Mechanical Engineers, in October, 1919, under the title, A Manual on American Pipe Threads. This report contained the first mention of tolerances in connection with pipe threads. It was endorsed by the American Gas Association and later was adopted by the National Screw Thread Commission with only such changes as were necessary to bring it into conformity with the remainder of its report.—EDITOR.

## OUTLINE OF STANDARD

The National (American) Pipe Thread Standard establishes the following: Outside Diameter of Pipe; Diameter of External (Male) Thread; Diameter of Internal (Female) Thread; Profile of Thread; Pitch or Lead of Thread; Length of Thread; Taper of Thread; Engagement (by hand) of External and Internal Threads; Construction and Use of Gages; Tolerances; Use of Taper Threads; Use of Straight Threads.

The dimensions of National (American) Pipe Threads are expressed in inches to one-one hundred thousandth (0.00001) of an inch, and in millimeters to one-thousandth (0.001) of a millimeter.

While this is a greater degree of accuracy than is ordinarily

---

[1] Recorded in the Excerpt Minutes, vol. LXXI, sessions of 1882–1883, part 3.

used, the dimensions are so expressed in order to eliminate errors which might result from less accurate dimensions.. The relation between the inch and the meter used in calculating the dimensions in the tables is that established by law in the United States and on record in the Bureau of Standards, Department of Commerce, Washington, D. C., viz.,

1 meter = 39.37 inches, exactly
25.40005 millimeters = 1 inch.

### SPECIFICATIONS OF THREAD

The standard outside diameter of pipe is given in Column G of Table 14 of dimensions. These diameters should be very closely adhered to by pipe manufacturers.

The pitch diameters of the taper thread are determined by formulæ based on the outside diameter of pipe and the pitch of thread. These are as follows:

$$A = G - (0.05G + 1.1)P$$
$$B = A + 0.0625F$$

where

$E$ = Length of effective thread
$G$ = Outside diameter of pipe
$P$ = Pitch of thread.

The taper of the thread is 1 in 16, measured on the diameter. The normal length of engagement between taper external and internal threads when screwed together by hand is shown in column F, Table 14. This length is controlled by the construction and use of the gages.

In these specifications the lead of the screw is expressed in terms of the number of threads in one inch and the number of threads in 254 millimeters (10 in.).

### TYPICAL SPECIFICATIONS FOR SCREW-THREAD PRODUCTS

*Material:* The material used shall be cold-drawn Bessemer Steel Automatic Screw Stock.
*Composition:*

Carbon, 0.08 to 0.16 per cent

TABLE 14—DIMENSIONS OF NATIONAL (AMERICAN) PIPE THREADS

| Nominal Size | | A | | B | | E | | F | | G | | Depth of Thread | | Number of Threads | |
|---|---|---|---|---|---|---|---|---|---|---|---|---|---|---|---|
| Inches | M.M. | Inches | M.M. | Inches | M.M. | Inches | M.M | Inches | M.M. | Inches | M.M. | Inches | M.M. | Per Inch | Per 254 M.M. |
| ⅛ | 3 | .36351 | 9.233 | .37476 | 9.519 | .2038 | 6.700 | .180 | 4.572 | .405 | 10.287 | .02963 | .753 | 27 | 270 |
| ¼ | 6 | .47739 | 12.126 | .48989 | 12.443 | .4018 | 10.206 | .200 | 5.080 | .540 | 13.716 | .04444 | 1 129 | 18 | 180 |
| ⅜ | 10 | .61201 | 15.545 | .62701 | 15.926 | .4078 | 10.358 | .240 | 6.006 | .675 | 17.145 | .04444 | 1 129 | 18 | 180 |
| ½ | 13 | .75843 | 19.264 | .77843 | 19.772 | .5337 | 13.556 | .320 | 8.128 | .840 | 21.336 | .05714 | 1.451 | 14 | 140 |
| ¾ | 19 | .96788 | 24.579 | .98886 | 25 117 | .5457 | 13.861 | .339 | 8.611 | 1.050 | 26.670 | .05714 | 1.451 | 14 | 140 |
| 1 | 25 | 1.21363 | 30.826 | 1.23863 | 31.461 | .6828 | 17.343 | .400 | 10.160 | 1.315 | 33.401 | .06956 | 1.767 | 11½ | 115 |
| 1¼ | 32 | 1.55713 | 39.551 | 1 58338 | 40.218 | .7068 | 17.953 | .420 | 10.668 | 1.660 | 42.164 | .06956 | 1.767 | 11½ | 115 |
| 1½ | 38 | 1.79609 | 45.621 | 1.82234 | 46.287 | .7235 | 18.377 | .420 | 10.668 | 1.900 | 48.260 | .06956 | 1.767 | 11½ | 115 |
| 2 | 50 | 2.26902 | 57.633 | 2.20627 | 58.325 | .7565 | 19.215 | .436 | 11.074 | 2.375 | 60.325 | .06956 | 1.767 | 11½ | 115 |
| 2½ | 64 | 2.71953 | 69.076 | 2.76216 | 70.159 | 1.1375 | 28.892 | .682 | 17.323 | 2.875 | 73.025 | .10000 | 2.540 | 8 | 80 |
| 3 | 76 | 3.34063 | 84.852 | 3.38850 | 86.068 | 1 2000 | 30.480 | .766 | 19.456 | 3.500 | 88.900 | .10000 | 2.540 | 8 | 80 |
| 3½ | 90 | 3.83750 | 97 473 | 3.88881 | 98.776 | 1 2500 | 31.750 | .821 | 20.853 | 4.000 | 101.600 | .10000 | 2.540 | 8 | 80 |
| 4 | 100 | 4.33438 | 110.093 | 4.38713 | 111.433 | 1.3000 | 33.020 | .844 | 21.438 | 4.500 | 114.300 | .10000 | 2.540 | 8 | 80 |
| 4½ | 113 | 4.83125 | 122.714 | 4.58594 | 124.103 | 1.3500 | 34.290 | .875 | 22.225 | 5.000 | 127.000 | .10000 | 2.540 | 8 | 80 |
| 5 | 125 | 5.39073 | 136.925 | 5.44929 | 138.412 | 1.4063 | 35.720 | .937 | 23.800 | 5 563 | 141 300 | .10000 | 2.540 | 8 | 80 |
| 6 | 150 | 6.44809 | 163.731 | 6.50597 | 165.252 | 1.5125 | 38.417 | .958 | 24.333 | 6.625 | 168.275 | .10000 | 2.540 | 8 | 80 |
| 7 | 175 | 7.43984 | 188.972 | 7.50234 | 190.560 | 1 6125 | 40.987 | 1.000 | 25.400 | 7.625 | 193.675 | .10000 | 2.540 | 8 | 80 |
| 8 | 200 | 8.43359 | 214.214 | 8.50003 | 215.901 | 1 7125 | 43.497 | 1.063 | 27.000 | 8.625 | 219.075 | .10000 | 2.540 | 8 | 80 |
| 9 | 225 | 9.42734 | 239.455 | 9 49797 | 241.249 | 1 8125 | 46.037 | 1.130 | 28.702 | 9.625 | 244.475 | .10000 | 2.540 | 8 | 80 |
| 10 | 250 | 10.54531 | 267.851 | 10 62094 | 269.772 | 1.9280 | 48.895 | 1 210 | 30.734 | 10 750 | 273.050 | .10000 | 2.540 | 8 | 80 |
| 11 | 275 | 11.53906 | 293.002 | 11.61938 | 295.133 | 2.0250 | 51.435 | 1 285 | 32.639 | 11.750 | 298.450 | .10000 | 2.540 | 8 | 80 |
| 12 | 300 | 12.53281 | 318.334 | 12.61781 | 320.402 | 2 1250 | 53.975 | 1.350 | 34.544 | 12.750 | 323.851 | .10000 | 2.540 | 8 | 80 |
| 14 O.D. | 350 | 13.77500 | 349.886 | 13 87262 | 352.365 | 2 250 | 57.150 | 1.562 | 39.675 | 14.000 | 355.601 | .10000 | 2.540 | 8 | 80 |
| 15 O.D. | 375 | 14.76875 | 375.127 | 14.87419 | 377.805 | 2.350 | 59.690 | 1.687 | 42.850 | 15.000 | 381.001 | .10000 | 2.540 | 8 | 80 |
| 16 O.D. | 400 | 15.76250 | 400.368 | 15.87575 | 403.245 | 2.450 | 62.230 | 1.812 | 46.025 | 16.000 | 406.401 | .10000 | 2.540 | 8 | 80 |
| 17 O.D. | 425 | 16.75625 | 425.609 | 16.87800 | 428.626 | 2.550 | 64.770 | 1.900 | 48.200 | 17.000 | 431.801 | .10000 | 2.540 | 8 | 80 |
| 18 O.D. | 450 | 17.75000 | 450.851 | 17.87500 | 454.026 | 2.650 | 67.310 | 2.000 | 50.800 | 18.000 | 457.201 | .10000 | 2.540 | 8 | 80 |
| 20 O.D. | 500 | 19.73750 | 501.333 | 19.87031 | 504.707 | 2.850 | 72.390 | 2.125 | 53.975 | 20.000 | 508.001 | .10000 | 2.540 | 8 | 80 |
| 22 O.D. | 550 | 21.72500 | 551.816 | 21.86562 | 555.388 | 3.050 | 77.470 | 2.250 | 57.150 | 22.000 | 558.801 | .10000 | 2.540 | 8 | 80 |
| 24 O.D. | 600 | 23.71250 | 602.290 | 23.86094 | 606.069 | 3 250 | 82.550 | 2.375 | 60.325 | 24.000 | 609.601 | .10000 | 2.540 | 8 | 80 |
| 26 O.D. | 650 | 25.70000 | 652.781 | 25.85625 | 656.750 | 3.450 | 87.630 | 2.500 | 63.500 | 26.000 | 660.401 | .10000 | 2.540 | 8 | 80 |
| 28 O.D. | 700 | 27.68750 | 703.264 | 27.83150 | 707.431 | 3.050 | 92 710 | 2.625 | 66.675 | 28.000 | 711 201 | .10000 | 2.540 | 8 | 80 |
| 30 O.D. | 750 | 29.67500 | 753.746 | 29.84687 | 758.112 | 3.850 | 97.790 | 2.750 | 69.850 | 30.000 | 762.001 | .10000 | 2.540 | 8 | 80 |

in which

$A$ = Pitch diameter of thread at end of pipe
$B$ = Pitch diameter of thread at gaging notch
$G$ = Outside diameter of pipe
$F$ = Normal engagement by hand between external and internal threads
$P$ = Pitch of thread.

The angle between the sides of the thread is 60 deg. when measured in the axial plane, and the line bisecting this angle is perpendicular to the pipe axis for taper or straight threads.

The crest and root are truncated an amount equal to 0.033 P. The depth of the thread, therefore, is 0.8 P.

The length of the taper external thread is determined by a formula based on the outside diameter of pipe and the pitch of the thread. This is as follows[1]:

$$E = (0.8G + 6.8)P$$

[1] The above formulæ are not expressed in the same terms as the formulæ originally established by Mr. Briggs; however, both forms give identical results.

Manganese, 0.50 to 0.80 per cent
Phosphorus, 0.09 to 0.13 per cent
Sulphur, 0.075 to 0.13 per cent.

*Method of Manufacture:* Bolts and nuts may be either rolled, milled, or machine-cut, so long as they meet the provided specifications, and are to be left soft.

*Workmanship:* All bolts and nuts must be of good workmanship and free from all defects which may affect their serviceability.

*Finish:* All bolts and nuts shall be semi-finished; that is, the bodies are to be machined, under side of head and nut faced, and upper face of head and nut chamfered at an angle of 30 deg., leaving a circle equal in diameter to the width of the nut.

*Form of Thread:* The form of thread shall be the "National Form," as specified herein, and formerly known as the United States Standard or Sellers Thread.

*Thread Series:* The pitches and diameters shall be as specified in Table 1, and known as the National Coarse Thread Series.

*Class of Fit:* Class II-A, Medium Fit (Regular).

*Dimensions:*

  a  Nominal Size: ½ in.
  b  Number of Threads Per Inch: 13.
  c  Length Under Head: 3 in. ± 0.05 in.
  d  Minimum Length of Usable Thread: 1 in.
  e  Diameters: See Table XI, in complete report.

*Tolerances and Allowances:* See Table 6.

*Nuts:*

  a  Form: Hexagonal.
  b  Thickness: ½ in. ± 0.01 in.
  c  Short Diameter (Across Flats): ⅞ in. ± 0.01.

*Heads:*

  a  Form; Hexagonal
  b  Thickness: 7/16 in. ± 0.01 in.
  c  Short Diameter (Across Flats): ⅞ in. ± 0.01 in.

*Gages:* The gages used shall be such as to insure that the product falls within the tolerances as specified herein for Class II, Medium Fit (Regular).

The following gages are suggested and will be used by the purchaser:

For the Screw:

  a  A maximum or "Go" ring thread gage
  b  A minimum or "Not Go" ring thread gage to check only the pitch diameter.
  c  A maximum or "Go" plain ring to check the major diameter.
  d  A minimum or "Not Go" snap gage to check the major diameter

For the Nut:

  a  A minimum or "Go" thread plug gage
  b  A maximum or "Not Go" thread plug gage to check only the pitch diameter
  c  A "Go" plain plug gage to check the minor diameter
  d  A "Not Go" plain plug gage to check the minor diameter of the thread.

*Inspection and Test:* Screws and nuts shall be inspected and tested as follows:

At least three bolts and nuts shall be taken at random from each lot of 100, or fraction thereof, and carefully tested. If the errors in dimensions of the screws or nuts tested exceed the tolerance specified for this class, the lot represented by these samples shall be rejected.

*Delivery:* Unless otherwise specified the assembled bolts and nuts are to be delivered in substantial wooden containers, properly marked, and each containing 100 lb.

### FUTURE WORK OF COMMISSION

The problems of standardization so far considered by the Commission have been those of most pressing importance to manufacturers and users of screw-thread products.

It is the intention of the Commission to continue the work of gathering information in regard to special problems still to be considered. The following list includes the more important screw threads which require standardization:

  a  Threads cut on brass tubing
  b  Instrument threads
  c  Acme, square, buttress and other special threads.

In addition to the standardization of various thread systems, the Commission believes that it would be of great advantage to American manufacturers to have established standards for stock. tools and other appliances used in the production of screw threads, such as the following:

  a  Taps
  b  Dies
  c  Sizes of bar stock for producing cut threads
  d  Sizes of bar stock for producing rolled threads
  e  Dimensions of bolt heads and nuts
  f  Standardization of sheet-metal and wire-gage sizes
  g  Standardization of tap-drill sizes.

The recent war has demonstrated the need of interchangeability of articles manufactured in this country with those manufactured abroad and it is known that manufacturers and authorities of Great Britain, France, and other foreign countries are awake to the situation and, in fact, have already taken steps toward the international standardization of screw threads and other manufactured articles. Furthermore, international standardization is of great importance in connection with the development of foreign trade.

In July, 1919, the Commission sent to Europe a delegation of its members to confer with British and French engineering standards organizations, and while no definite agreements were reached in regard to international standardization of screw threads, it was apparent in both France and England that the engineers and manufacturers in these countries are anxious to coöperate with the United States in this work. The time is very opportune for accomplishments along this line, and it is the opinion of the Commission that, as a result of the war, it should in time be possible to reach an agreement on an international standard thread. Such an international standard should be established by giving consideration to the predominating sizes and standards used in manufactured products, as well as to the possibilities of providing a means for producing this international screw thread by the use of either the English or the Metric system of measurement.

## Opposed-Piston Diesel Engine

A type of marine Diesel oil engine operating on the opposed-piston principle, recently developed by Messrs. Cammell Laird and Co., of Birkenhead, England, has been described in various periodicals, notably London *Engineering* of Jan. 30, 1920. The chief advantage claimed for it is that its weight, and therefore its cost, for a given power is lower than with any existing type, while the space occupied is smaller. Each unit comprises two parallel cylinders in each of which are two opposed pistons. The usual three cranks between the main bearings in the opposed-piston engine have been reduced to two by cross-connecting the top pistons to the crossheads attached to the adjacent bottom pistons. This arrangement, which is a characteristic feature of the engine, relieves the framing of most of its stresses except those imposed by the oblique rods extending from the upper pistons and enables a lighter framing to be employed than with ordinary motors. Each unit is equal in effect to a single-cylinder double-acting engine and the same effect is obtained with two connecting rods, two cranks, two cylinders and four pistons, as would require two cylinders, four pistons, six connecting rods, and six cranks with the ordinary opposed-piston type of engine.

In other respects the lines of the ordinary opposed-piston design have been followed. The fuel is injected by means of a centrally placed fuel valve, just before the pistons reach their nearest central point, the combustion chamber being thus formed between the faces of the two pistons. At the bottom of the cylinder are arranged ports around the circumference through which scavenging air is admitted when these ports are uncovered by a bottom piston in the course of its stroke. Exhaust takes place through a similar series of ports at the top of the cylinder, these being uncovered in their turn by the upper piston just before the end of its stroke. The scavenging air thus passes through the whole cylinder and a very good scavenging effect is obtained, giving the engine a high efficiency for a two-cycle type, the fuel consumption being 0.42 pound per brake horsepower hour. The scavenging pumps are of simple design. The air supply, both for the injection of the fuel and for starting purposes, is obtained from a three-stage vertical air compressor driven by the engine. The pumps for the circulating water and lubricating oil are driven from eccentrics. The engine is to be built in standard sizes of 1000 hp. for installation in single or twin screw cargo ships.

In view of the small space occupied by this engine, it is stated that the steam engine and boilers can be taken out of existing steamships and replaced by an engine of this type without altering the shafting or propellers. The low speed of revolution consequent upon the adoption of opposed pistons—the piston speed at 110 r.p.m. is only some 450 ft. per min., or equivalent in effect to 900 ft. per min.—makes this possible.

# The Financial Problems of Industrial Housing

By LESLIE H. ALLEN,[1] SPRINGFIELD, MASS.

*This paper treats industrial housing as a financial problem. Our housing shortage is said to be due partly to the fear of a financial panic and partly to the fact that, high as rents are, they are not high enough to show an adequate return on present-day construction costs. The relation of rents to capital invested, the calculation of proper rents, and methods of financing house construction are discussed in some detail. The financial difficulties which face those desiring to purchase new homes are also taken up and selling plans that will meet the purchasers' needs and their ability to pay are outlined. The paper closes with a discussion of the scheme of coöperative housing, which the author suggests may be the solution of America's housing problems.*

T HE close of the year 1919 finds our industrial housing problem still unsolved. The expected fall in prices has not come and does not seem likely to come. The speculative builder seems to be definitely out of the field. Rents are still rising but have not risen high enough to show a proper return on present-day construction costs. The shortage of materials and freight cars has caused a notable increase in building costs during the past few months, so that today a house will cost 35 per cent more than it did last November. Some employers who built houses last year are using these as an inducement to attract help from other localities. Those who did not build are coming to the conclusion that they cannot hold their help without building this year, and in the meantime their production and sales are falling off because they are short of help.

Manufacturers are still hesitating and doubting the wisdom of starting their housing programs because of the fear of a financial panic or a period of depression and fall in prices. They persist in ignoring the presence of our Federal Reserve Bank which was created for the very purpose of stabilizing our financial situation. They also ignore the fact that organized labor is demanding a larger share of the earnings from industry than it has received before and is going to see that it gets it.

## Present Renting Difficulties

The second difficulty is that the employer thinks he cannot rent his new house at a figure that will show an adequate financial return. It is unfortunate that the unattractiveness of the house as an investment is due to the fact that receipts from rent are usually not enough to pay a fair interest and allow for depreciation.

The rents charged were fixed many years ago; the manufacturer has not ventured to raise them to correspond with rising costs or increased wages, because of the odium that attaches to such action when the employer is also the landlord. In consequence, we find that most mills are not receiving enough rents to take care of the cost of maintenance and taxes. These low rents aggravate the housing shortage because they influence rentals and real-estate values all through the town. Other parties may get slightly higher rents, but not enough more to make property owning an attractive investment, and this discourages them even in normal times and is now an absolute barrier to real-estate development by others.

## Calculation of Proper Rent

At this point it may be well to discuss what is a proper rent for a house. Assuming a lot and a house built today cost $6000, of which $500 is the cost of the lot and $5500 the cost of the house, the annual charges to be covered by rent would be as follows:

The prevailing interest on first mortgages is 6 per cent. Taxes vary according to locality, but a total charge of $100 will probably cover taxes, insurance and water.

[1] With Fred T. Ley & Co., Inc.
Abstract of a paper presented at the Spring Meeting, St. Louis, Mo., May 24 to 27, 1920, of The American Society of Mechanical Engineers. Copies of the complete paper may be obtained upon application. All papers are subject to revision.

At the end of twenty years the house, even after renewing the roof, heater and plumbing, will not be worth as much as its original cost. It will have depreciated through wear and tear at least 30 per cent on the first cost, the figure varying according to the quality of material and workmanship in the building and the amount of care taken of the premises by the tenant. If we assume in this case an average of 30 per cent on $4000 (the cost of the structure above the cellar) we obtain $1200 or a charge of $60 per annum. A tabulation of these charges worked out on an annual basis follows:

| | | | |
|---|---|---|---|
| Outside painting | .... $100÷ 3 | ....................... | $ 33.33 |
| Inside painting | ..... 140÷ 7 | ....................... | 20.00 |
| Roofing | .......... 100÷20 | ....................... | 5.00 |
| Heater | ........... 150÷20 | ....................... | 7.50 |
| Plumbing | ......... 350÷20 | ....................... | 17.50 |
| General Repairs | .... Average | ....................... | 15.00 |
| Depreciation, at 2 per cent, on remainder of superstructure | | ....................... | 60.00 |
| | | | |
| Total maintenance and depreciation | | ............. | 158.33 |
| Taxes and insurance | | ............................ | 100.00 |
| Interest on investment | | ......................... | 360.00 |
| | | | |
| Total | | ................................. | $618.33 |

This figure, amounting to slightly over 10 per cent on the investment, does not allow anything for management. It is equal to a rent of $51.50 per month and is as low as it is safe to figure with favorable conditions. The calculations take no account of possible rise in land values or in cost of building materials which might offset depreciation.

## Methods of Financing House Construction

The third big problem that confronts the employer in his housing problem is how to finance the large expenditure needed. The ultimate purchaser usually cannot make a large initial payment and his earnings do not permit very large monthly payments on the amount he borrows. It is very seldom that more than 10 per cent of the purchase price is offered as a first payment, and repayment of principal borrowed on mortgages is stretched out by employers over periods of from 10 to (in one case) 34 years.

In most of the plans now in use the employer has to take back a second mortgage for at least 40 per cent of the total cost.

The building and loan associations of America find themselves with insufficient funds available for home buying and home building because their only source of working capital is derived from weekly cash deposits paid in by the association members. To relieve these conditions the Federal Building Loan Act has been introduced in Congress to Encourage Home Ownership and to Stimulate the Buying and Building of Homes. These bills would create a system of Federal Building Loan Banks operating under the general supervision of the Treasury Department, and make available, for the purpose of dwelling construction, a considerable portion of the two billion dollars now tied up in the mortgages held by the building and loan associations throughout the country.

The combination of manufacturers into a local housing corporation which shows a willingness to take the risk of building houses has proved of considerable encouragement to the prospective home purchasers. The housing corporations also gain the advantage of economical purchasing and quantity production in the erection of houses.

There are many localities where united action by manufacturers is not possible and here the employer is advised to organize his own subsidiary realty company. It is unquestionably of benefit to him that he should not be both employer and landlord.

The suggestion has been made in several quarters that new house construction should be exempt from local taxation for a period of five or ten years. The benefits of this in communities where the housing shortage is very acute would at once be felt, as local taxation usually amounts to one-quarter of the rental cost of a house. It is a method often used to attract new manufacturing industries to a town and precedent for such procedure is therefore established. It may be objected, however, that the building of a number of houses adds a far greater burden to a town's expenses than an expenditure of equal amount in factory construction, as the houses usually call for an expenditure on sewers, water, etc., out of a town's funds and an additional tax on the town's school facilities, fire-protection service, police, etc. The exemption from local taxation, therefore, can only be considered an emergency measure. It is not economically sound and should be used only as a means of making all classes of a community contribute toward the relief of a town's housing shortage where other methods of enlisting their support have failed.

There is a wide difference of opinion among employers as to the wisdom of renting or selling homes to employees. The rental of houses acts as a check on the tenant leaving the factory of the company who owns the house. Rental is further necessary for young married couples not in a position to buy. The selling of houses to the workman is of advantage because the married man in his own home makes a better workman, becomes a better citizen and acquires a permanent interest in the community. For the selling scheme to be satisfactory the purchaser must be guaranteed that his investment is a liquid one. The employer must be protected if values rise and if the employee leaves his employ before the house is entirely paid for. Various plans for determining the paid-in value of the house are used which insure fairness to the workman and the employer.

A selling plan that will meet the purchaser's needs and his ability to pay is of great importance. The ordinary workman does not understand the business of home ownership, so if the selling plan is to be a success most of the thinking and planning and financing must be arranged for him. The first payment may be as low as 2 per cent, but a man worthy to purchase a house should be able to put down at least 5 per cent. Interest charges are the large item in his annual carrying charges and these will be decreased by a large initial payment. The best plan is for the man to arrange a monthly payment to take care of all outgoings and amortization of the principal.

The following examples show various methods of financing the sale of a home. It is assumed that the purchaser is earning $2500 a year, and should be able to pay about $600 a year (or $50 a month for carrying charges and amortization). In each case the price of the house and lot is assumed to be $6000 and the amount of the first payment $600, leaving $5400 to be financed. Smaller or larger transactions can be figured pro rata.

The method of dealing with the mortgage finance corporations is to place a first mortgage for 50 or 60 per cent of the value and a second mortgage for the difference between that and the first payment. This would work out as follows:

| | |
|---|---:|
| First mortgage, $3000 at 6 per cent | $189 |
| Second mortgage, $2400 at 6 per cent, amortized in 12 years | 288 |
| Taxes, insurance and water | 100 |
| Minor repairs and painting (average) | 30 |
| Total ($50 per month) | $598 |

When the second mortgage is amortized the workman can clear off his first mortgage in about eight years more if he continues to pay in $50 per month.

Where the purchaser's income is large enough a building and loan association mortgage for 75 or 80 per cent of the price can be taken out, and the difference between this and the first payment would have to be taken by a second mortgage. This would work out as follows:

| | |
|---|---:|
| Coöperative bank mortgage, $4000 at 6 per cent | $240 |
| Coöperative bank dues to amortize mortgage in 12½ years | 240 |
| Second mortgage of $1400 at 6 per cent amortized in 12 years | 168 |
| Taxes, insurance and water | 100 |
| Minor repairs and painting | 30 |
| Total ($65 per month) | $778 |

Under this method the property would be entirely paid for in 12½ years, but the monthly payments are so high that very few workmen are able to purchase in this way.

Some firms prefer to sell on a purchase agreement by which title does not pass until the house is paid for. The tenant gives a note or else land contract for the amount owing on this. He pays interest annually, and a certain minimum payment on principal, which he is encouraged to increase if possible. The advantage in this method is that a tenant is more likely to make extra efforts to pay off the principal because he sees his interest payments being reduced as often as he makes payments. In times of prosperity or high wages he can do this easily, whereas if hard times come he can reduce his annual payments to the specified minimum. The annual payments and maturity length would work out as follows:

| | |
|---|---:|
| First mortgage or purchase agreement at 6 per cent (amortized in 24 years) | $432 |
| Taxes, insurance and water | 100 |
| Minor repairs and painting | 30 |
| Total ($47 per month) | $562 |

## AMORTIZATION OF MORTGAGES

The rate of interest is such a big item in the annual cost of a house that some firms are charging less than market rates in order to help their men. One national concern is taking back a first mortgage of 95 per cent of the net cost at 4½ per cent, the principal being amortized in 30 years. At first sight it may seem that this is unbusinesslike, but the plan is undoubtedly a wise one that will benefit both parties to the deal. The difference between 4½ per cent and the prevailing interest rate represents the annual cost to the company of keeping a steady workman on the job and keeping down labor turnover and may save it several times that sum.

As labor turnover is supposed to cost from $40 for unskilled men up to as much as $250 for skilled workmen, it would seem, from this point of view, a real economy, as the loss of interest would amount to only $60 per annum.

The importance of making proper provision in any sales plan for the amortization of both first and second mortgages should be emphasized. We recognize the fact that a house will not and does not last forever, yet this fact seems to be lost sight of in the usual negotiation for mortgages. The financial difficulties of many mortgagees may be traced to the fact that mortgages placed on a conservative basis soon become very poorly secured if the mortgage debt is not reduced. Whenever a man buys an automobile he has to reduce his debt very quickly; in buying pianos or furniture on the installment basis, his repayments on principal are usually as large or larger than the amount of interest he is paying on his debt. The same idea ought to be adopted in the purchase of a house, as even the best-built house has plumbing, roofing, hardware and many other items that are wearing out. Most men will readily see the advantages of amortization. The continually reducing charges for interest and the progressive increase in repayments of principal are very attractive. With money at 6 or 6½ per cent a mortgage can be reduced very quickly on a moderate amortization payment.

## COÖPERATIVE HOUSING

Much interest is being shown in copartnership or coöperative housing, which is being tried in several New York City apartment houses and is the plan adopted by the English garden city companies. Under this system a company is organized to purchase and develop real estate and the stockholders are admitted as tenants to the property. The usual scheme is for each man to purchase stock to the value of half the cost of his house and land, the other half being carried by a mortgage on the whole property. His rent is sufficient to amortize the mortgage in 7 to 12 years, and then the housing company can either reduce rents or pay large dividends.

To apply this system to the housing of the working classes

would necessitate the placing of blocks of stock in the hands of employers and charging enough rent to pay for the purchase of the stock by tenants from the block-holders in monthly or annual installments instead of selling the whole of the stock to the tenants at the start. The rent paid would be large enough to take care of installment payments on the stock, as well as the amortization of the bonds. A rental of 12 per cent of the cost of the development would amortize the bonds and place the whole of the stock in the hands of the tenant in about 27 years.

The tenant in signing a rent agreement will acquire the right to receive annually second preferred stock ·to the amount of one-sixth of the rental paid, this stock being turned over to him in quarterly installments of two shares each. In addition to this he will be given one share of common stock (which carries with it voting privileges) as soon as his holdings in preferred stock reach $100, and one additional share for each additional $100 he acquires.

At the end of twenty-five years the whole of the second preferred stock will be in the hands of the tenants and half of the common stock will be held by them, the treasury stock being exhausted for this purpose.

In the twenty-seventh year the bonds will be entirely redeemed, and if the rent remains at the same figure there would thereafter remain $69,000 per annum available for dividends on the common stock. Probably $5 per share would be used for this purpose and the remainder carried to surplus, so that each tenant would receive $370 in dividends, viz., $245 on his preferred stock and $125 on his common stock.

The original guarantors of the preferred stock who have been holding common stock all this time, without receiving a dividend, would receive a compensation for the risk taken in lending their credit, in the shape of $25,000 per annum dividend.

This plan has a saving incentive that would appeal to a great many workmen. Some arrangement could be made by which they could transfer from smaller to larger houses as their families increased in size.

It will be desirable to guarantee to the tenant a market for his preferred stock in case he wishes to vacate the house and leave the town, so as to remove from his mind the fear of loss which is otherwise likely to deter him from entering the scheme. It is not probable that it will be necessary to maintain this guarantee very long, as in a few years the value of the assets behind it, as the bonds are retired, will make it worth more than the purchase price.

The advantage of such a plan to the employer is that all he has to do is to guarantee an annual dividend of small amount and his funds are not otherwise tied up.

The advantage to the tenant is that he has the freedom of a tenant and yet shares the profits of the landlord and he acquires by installments a liquid investment in a housing property. He has all the rights of a householder except the speculative selling privilege and he is relieved of the fear of loss through depreciation or forced sale.

In higher-class developments where purchasers are better provided with funds, the preferred stock can be sold outright to them at the commencement, and the rental can be reduced by 2 per cent.

The plan has so many obvious merits that it is certainly worth trying out on a large scale, although it is feared that the constitutional dislike of the American workmen to any form of co-operative merchandising may prove a bar to its success. It may be that one or two successful demonstrations, however, will show that the successful solution of America's housing problems lies in this direction.

Our housing problems can be solved only by action. We should strive to put the business of housing on a self-supporting basis, for we have not solved the problem if our housing does not pay its way. We may make some mistakes in the steps we take, but if all those who are trying to remedy the present state of affairs will work together, we can overcome all the difficulties and solve our problems to the lasting benefit of our country and our homes.

# LIQUID FUEL

In the last few months evidence has been accumulating to show the serious situation with respect to liquid fuels, in this country and abroad.

In England, the government has already realized the urgency of the situation and its importance for the future and has decided to render assistance to the industry through the purchase of bonds of one of the largest oil companies, the so-called Shell Group, by which means it will be supplied with working capital.

In America the question has been left largely to private initiative, with the only exception that a new Oil Land Leasing Bill may open new fields for increasing the production of oils.

The consumption of mineral-oil fuels in motor cars and in marine and stationary engines has grown so tremendously in the last five years that unless new sources of fabulous wealth are discovered the end of our liquid-fuel supplies is clearly in sight. In a recent statement the Director of the U. S. Geological Survey, George Otis Smith, estimated that the present known supplies will last only in the neighborhood of twenty years.

It is of interest, therefore, to review briefly what has been done recently to supplement mineral oils or to increase their usefulness. The work has been carried on in several directions, of which the following may be mentioned without, however, indicating the relative importance of the various remedies by the order in which they are enumerated.

During the war the Submarine Defense Association developed the so-called colloidal fuel, a mixture of oil and pulverized coal where the coal is held in suspension in the oil through the presence of a material called by its inventors the "fixateur." It is claimed that this material may be used for oil-fired furnaces in about the same manner as oil alone, and that for a given weight of oil it has a very much higher heat capacity, than oil alone, due to the presence of suspended coal.

Next, in particular reference to motor cars, may be mentioned the possible use of engines of the Diesel type, reference to which is made in an article in the Engineering Survey of this number. Numerous statements have been made in this direction without conspicuous success, which the author of the article referred to explains by claiming that Diesel-engine manufacturers are too busy to attempt the problem, while motor-car manufacturers do not know enough about Diesel engines. The use of Diesel engines on motor cars would relieve the situation to the extent of permitting the use of unrefined oils, instead of only such upper fractions of distillation as gasoline and kerosene.

Mention should be made, also, of various products of coal distillation such as benzol, products of its hydrogenation, and motor spirits. The production of these depends on the availability in this country of coals with high nitrogen content and such coals do not appear to be very numerous.

The last, and possibly the most promising solution of the liquid-fuel problem, in particular for motor vehicles, appears to be the use of alcohol, and it is interesting to note in this connection (1) that fuels with alcohol base have recently been used with great success in motor vehicles in this country, in Europe, and especially in South Africa; and (2) that new sources for the production of alcohol are constantly being discovered. In addition to the generally known vegetable sources, of recent years methods have been developed for the production of alcohol from sawdust, from refuse of cane-sugar manufacture, from calcium carbides, and recently, at the Skinningrove Works, from coal gas. As the largest and best-known field of consumption of alcohol in this country has been destroyed for the time being, the use of alcohol for power purposes would not meet with the competition which it would formerly have had and the price of alcohol should not be high, especially as compared with the prices of gasoline now prevailing and expected in the near future. There are certain difficulties in the way of wide use of alcohol as a fuel, which, however, are of a non-technical nature and will not be discussed here. It may be stated, however, that the discovery of a cheap, positive and reliable denaturant would go a long way toward making possible the wide adoption of alcohol for power purposes.

# ENGINEERING RESEARCH

A Department Conducted by the Research Committee of the A. S. M. E.

A—RESEARCH RESULTS

The purpose of this section of Engineering Research is to give the origin of research information which has been completed, to give a résumé of research results with formulæ or curves where such may be readily given, and to report results of non-extensive researches which in the opinion of the investigators do not warrant a paper.

*Apparatus and Instruments A9-20* Air Analyzer for Cements. Technologic Paper No. 48 of the Bureau of Standards describes an air analyzer for determining the fineness of cements and describes measurements to determine the size of the particles passing through a 100- or 200-mesh sieve as compared with maximum-size particles blown off by Nos. 1, 2 and 3 nozzles of air analyzers. In reporting the results of tests it has been at all times desirable to know the maximum-size particle passing through a 300-mesh sieve. The following table gives the mean results of two dimensional measurements:

| 100 mesh = 0.00793 in. | No. 1 nozzle = 0.00216 in. |
| 200 mesh = 0.00428 in. | No. 2 nozzle = 0.00154 in. |
| 300 mesh = 0.00301 in. | No. 3 nozzle = 0.00076 in. |

Bureau of Standards, Washington, D. C. Address S. W. Stratton, Director.

*Cement and Other Building Materials A2-20* Finely Ground Cement. The Bureau of Standards has completed a series of comparative tests on cement of normal fineness in which 86 per cent passes through a 200-mesh sieve and on finely ground cement in which 98 per cent passes through a 200-mesh sieve. The two cements were tested neat, in 1 : 3 sand mortar and in concretes of 1 : 1½ : 2, 1 : 2 : 4 and 1 : 3 : 6. The following results were obtained:

1 The strengths of the concrete from the fine cement were regularly and consistently greater than those made with the normal cement so far as the tests have been completed.

2 The percentage increase in the strength of the fine cement concrete varied as follows:

| | 2 days | 7 days | 28 days | 3 mo. | 6 mo. |
|---|---|---|---|---|---|
| Minimum gain | 90 | 50 | 30 | 21 | 14 |
| Average gain | 121 | 66 | 56 | 41 | 42 |

3 The strength increases in lb. per sq. in. were greater in the mixes containing the greater proportion of cement.

4 The approximate savings for the same strength at 28 days if consistencies and aggregates were the same were found to amount to 1.2 to 2.1 bags of cement per cu. yd.

5 When used in 1 : 3 mortar the fine cements produced a more marked percentage increase than when used in concrete, but the percentage increases in neat mixtures were of the same order as those in the concretes.

6 The fine cement requires no more water than normal cement.

7 The lumps of very fine material were not broken up under the sieving action, but when tested by the air analyzer these were broken up and the results gave a true indication of the fineness. Hence it is recommended that air analyzers be used in determining the fineness of very fine cements.

Bureau of Standards, Washington, D. C. Address S. W. Stratton, Director.

*Cement and Other Building Materials A3-20* Adherence of Gypsum to Concrete. The Bureau of Standards has made the following determinations from a research on the adherence of gypsum to concrete:

1 There seems to be no detrimental reaction between the two materials; the tensile strength of the mixture of gypsum and cement is very nearly equal to their combined strengths, dependent upon the proportions of the two ingredients used.

2 The suction of the surface to which the plaster is applied is an extremely important factor; for example, if gypsum plaster is applied to a dry concrete surface, the suction of the concrete will take so much water out of the gypsum that it will prevent its proper hardening.

3 The expansion of neat cement, when wet, is of an entirely different order from the expansion of neat gypsum, and a bond between the two materials can be permanently maintained only when enough sand is added to both materials to reduce the expansion for both.

The above-mentioned research work has been supported by the Gypsum Industries Association and a report covering it has recently been made to them. Address Bureau of Standards, Washington, D. C. S. W. Stratton, Director.

*Electric Power A1-20* Corona Discharge. Bulletin No. 114 of the Engineering Experiment Station of the University of Illinois on Corona Discharge by E. H. Warner and J. H. Kunz has been issued. The Bulletin is divided into eleven chapters and contains a bibliography of 2½ pages. The titles of the chapters are as follows:

  I Introduction (with statement of phenomenon and description of apparatus

  II General Appearance of Corona about a Wire in a Cylinder (The appearance of the corona with direct current and wire positive in air; negative in air; positive in hydrogen and negative in hydrogen with direct current is shown, as well as the appearance of the corona with alternating current)

  III The Starting Point of the Corona

  IV Characteristic Curves

  V Additional Factors Affecting the Starting Point and the Corona Current

  VI Alternating-Current Rectification

  VII Distribution of the Potential in the Corona Tube

  VIII The Pressure Increase in the Corona

  IX Ozone Formation in the Corona

  X The Influence of a Series Spark on the Corona

  XI Other Types of Corona Discharge.

Engineering Experiment Station, University of Illinois, Urbana, Ill. Address Dean C. R. Richards, Director.

*Electric Power A3-20* Power Factor. A mathematical study of the theoretical side of the effect of power factor on economy for the purpose of defining power factor in a quantitative way has been made by the Bureau of Standards and will be presented at the June Convention of the American Institute of Electrical Engineers. The purpose of this study was to aid the Joint Committee of the Institute and the National Electric Light Association in the study of this question. The paper points out the type of definition which most logically fits a number of different cases and shows the essentially conflicting character of some of the requirements which a single definition should satisfy. It is suggested that an additional quantity called the balance factor will enable these requirements to be made in a more satisfactory manner. Bureau of Standards, Washington, D. C. Address S. W. Stratton, Director.

*Fuels, Gas, Tar and Coke A4-20* Natural Gas. The University of Kansas Experiment Station has issued Engineering Bulletin No. 11, by H. C. Allen and E. E. Lyder, on a chemical survey of the natural gases of Kansas and Oklahoma. The Bulletin contains 101 pages and discusses the gas and oil development in the two states, the theory of accumulation, the properties of natural gas, the methods of determining constituents of natural gas and its heating values, and gives a survey of the fields and analysis of typical gases. The variations in composition and pressure, earlier tests, analyses of other gases of North America and foreign countries and comparative costs between natural gas and coal are also discussed.

The Bulletin describes the modification of apparatus for analysis of gases and shows a great variation in the percentage of hydrocarbons and nitrogen present. The maximum percentage of $CO_2$ was found to be slightly over 4 per cent. The report shows that the shallower gases are poorer in quality than the deeper ones from the same locality. Analytical data show the effect of mixing poorer gas with good gas. The principal causes of complaint from the domestic consumer are low pressure, poor quality and variation in quality. No air is intentionally mixed with the gas by the companies. The report shows possibilities for the recovery of gasoline by absorption processes and of utilizing geological means in oil and development work.

Address W. A. Whitaker, Director of the Division of State Chemical Research, University of Kansas, Manhattan, Kan.

*Fuel Utilization A1-20* Fuel Research. The report of the Fuel Research Board of the Department of Scientific and Industrial Research of Great Britain for the years 1918-1919 is now on sale. It deals with the following subjects: The immediate importance of fuel economy; oil fuel for the navy and the mercantile marine; the fuel-research station and functions; survey of the national coal resources from the physical and chemical points of view; work at the fuel-research station; domestic heating; air pollution; pulverized coal; peat inquiries; use of alcohol as fuel; gas standards.

The report contains appendices on Fuel Economy and Low Temperature Carbonization, and Summary of Reports on the Efficiency of Cooking Ranges. The price is 1s. 8½d. by post and copies may be obtained from H. M. Stationery Office, Imperial House, Kingsway, London, W.C.

*Fuels, Gas, Tar and Coke A5-20*  The United Gas Improvement Contracting Company has developed a device by which a water-gas set may be operated with promptness and celerity by a proper adjustment and arrangement of valves. This automatic control is placed in a small dustproof case of cast iron mounted on cast-iron legs and occupying 6 sq. ft. of floor space. It is operated by 1/10-hp. electric motor. It is interlocking and foolproof and permits one man to control several sets of apparatus and the gas maker may handle his own fuel and do other work. It permits shorter cycles. United Gas Improvement Contracting Company, Philadelphia, Pa. Address J. M. Rusby, Engineer of Tests.

*Metallurgy and Metallography A6-20*  Melting Point of Slags. The Bureau of Standards is preparing for publication a paper on a recent investigation of the melting point of various slags typical of those which occur in different lines of metallurgy. Bureau of Standards, Washington, D. C. Address S. W. Stratton, Director.

*Metallurgy and Metallography A7-20*  Mechanical Working of Metals. The Bureau of Standards has shown that it is possible to melt nickel in an atmosphere of hydrogen so as to prevent the formation of slight traces of oxide which causes crumbling in the working of the metal. The hydrogen also acts as a dioxidiser. Wires as small as 0.05 mm. diameter have been drawn from some of the pure nickel thus melted. Bureau of Standards, Washington, D. C. Address S. W. Stratton, Director.

*Metallurgy and Metallography A8-20*  Behavior of Hardened Steel upon Heating. The Bureau of Standards has recently investigated the effect of time in tempering hardened steels at relatively low temperatures. It has been found that the transformation which is suppressed by rapid cooling of the metal in hardening occurs upon reheating to 200 or 250 deg. cent. By heating for a long time at a lower temperature results similar to that obtained in a shorter time by higher temperature is produced. Bureau of Standards, Washington, D. C. Address S. W. Stratton, Director.

*Metallurgy and Metallography A9-20*  Aluminum-Alloy Castings. Blowholes, Porosity and Unsoundness in Aluminum-Alloy Castings. Technical Paper 241 of the Bureau of Mines, by Robert J. Anderson, 31 pages.

The Bulletin has the following sectional headings: Unsoundness in general and factors affecting it; definitions of blowholes, porosity and unsoundness; general factors affecting soundness of castings; gases in aluminum; solidification of metals; analogy with steel; effect of casting temperature; effect of method of melting; effect of rate of melting; effect of method of molding; effect of design of castings; effect of quality of ingot; so-called deoxidation of aluminum; description of experiments; metallography of unsoundness; radiography of castings; miscellaneous consideration; conclusions and publications on metallurgy.

The general conclusions show that broad generalizations cannot be drawn from the results of the experimental work, the experience of foundries and the contradictory literature. However, it is possible to recognize the existence of a large number of variables that may conduce to unsoundness and blowholes. The number of blowholes present is a function of the pouring temperature: the higher the pouring temperature, the greater the amount of blowholes and the more unsound the casting. Unsoundness varies with the temperature to which the charge is heated. The higher the temperature the more unsound the casting, irrespective of the pouring temperature.

Unsoundness is a function of the length of time of melting. The longer the melting is held the more unsound the castings. Melting and pouring should be carefully supervised. Close pyrometric control is necessary. It is better to have the molding floor wait for the metal rather than for the metal to wait for the floor.

Address Bureau of Mines, Washington, D. C. Van H. Manning, Director.

*Metallurgy and Metallography A10-20*  Magnetic Uniformity of Steel. See *Properties of Engineering Materials, A9-20*

*Properties of Engineering Materials A7-20*  Bakelite. Modulus of Rupture. Computed by formula $S = 8PL/2BD^2$. Determined from ½ by ½ by 5-in. specimens. With span of 2 in., $S = 12,500$ lb. per sq. in., with span of 3 in., $S = 12,460$ lb., and with span of 4 in., $S = 13,170$ lb.

Moduli of Elasticity. Determined in above specimen by formula $E = WL^3/4ABD^3$; for 2-in. span $E = 760,000$ lb. per sq. in., for 3-in. span, 887,000 lb.; for 4-in. span, 944,000 lb.

These tests were repeated, giving about the same stress as before, but slightly higher moduli of elasticity for 3-in. and 4-in. span.

Address General Bakelite Company, Gilbert L. Peaks, Electrical Engineer, Perth Amboy, N. J.

## B—RESEARCH IN PROGRESS

The purpose of this section of Engineering Research is to bring together those who are working on the same problem for coöperation, to prevent unnecessary duplication of work and to inform the profession of the investigators who are engaged upon research problems. The addresses of these investigators are given for the purpose of correspondence.

*Apparatus and Instruments B4-20*  Wave Meter  The development of a design of heterodyne wave meter for undamped wave work by Prof. R. W. Goddard, New Mexico College of Agriculture and Mechanic Arts, State College, New Mexico. Address A. F. Barnes, Dean of Engineering.

*Electrical Communication B2-20*  Wave Meter. See *Apparatus and Instruments, B4-20.*

*Electrical Communication B3-20*  Spark Frequency and Power. An investigation of the relation of spark frequency on power input to damped wave systems of wireless with given wave lengths and capacity condensers. Prof. R. W. Goddard, New Mexico College of Agriculture and Mechanic Arts, State College, New Mexico. Address Dean A. F. Barnes.

*Electric Power B2-20*  Heat by Induction and Eddy Currents. An investigation on heat by conduction and eddy currents in high frequency currents by Prof. R. W. Goddard, New Mexico College of Agriculture and Mechanic Arts, State College, New Mexico. Address Dean A. F. Barnes.

*Fuels, Gas, Tar and Coke F11-20*  Volatility of Motor Fuels. An investigation to determine by vapor-tension apparatus the minimum temperature to establish a stable mixture of air and gasoline vapor in combining proportions. Address Frank A. Howard, Standard Oil Company, 26 Broadway, N. Y.

*Fuels, Gas, Tar and Coke F12-20*  Crushed Fuel. A new method of burning crushed fuel is being developed at Purdue University. Fuel is introduced after crushing to small-size grains through tuyeres at one side of a circular hearth. The coal is introduced by means of an air current through a fan blower, additional air for combustion being introduced by side openings. High capacity has been obtained with ease of control and regulation. The cost of operation is much less than that of the pulverized coal methods. Address Prof. G. A. Young, Purdue University, Lafayette, Ind.

*Internal-Combustion Motors B5-20*  Mixture Requirements. The mixture requirements of internal-combustion engines. Investigation by O. C. Berry, Engineering Experiment Station, Lafayette, Ind. Address C. H. Benjamin, Director.

*Metallurgy and Metallography B6-20*  Refractory Crucibles. The Bureau of Standards is working on the method of making crucibles from highly refractory oxides and minerals. Titanium oxide, zirconium dioxide and carborundum firesand have been used. Bureau of Standards, Washington, D. C. S. W. Stratton, Director.

*Petroleum, Asphalt and Wood Products B3-20*  Volatility of Motor Fuels. See *Fuels, Gas, Tar and Coke, F11-20.*

*Pumps B2-20*  Electrically Driven Pumps in Small Water Works. Investigation by G. C. Blalock and D. D. Ewing at the Engineering Experiment Station, Purdue University, Lafayette, Ind. Address C. H. Benjamin, Director.

*Wood Products B1-20*  Preservation. Preservative Treatment of Wood Poles. An investigation by R. V. Achatz, Purdue University, Lafayette, Ind. Address C. H. Benjamin, Director.

## E—PERSONAL NOTES

The purpose of this section of Engineering Research is to give notes of a personal nature regarding the personnel of various laboratories, methods of procedure for commercial work or notes regarding the conduct of various laboratories.

*Purdue University E2-20*  The Engineering Experiment Station of Purdue University has just issued a Bulletin describing their work, the equipment and the experiments in progress. A copy of this may be obtained by addressing Dean C. H. Benjamin, Director, Purdue University, Lafayette, Ind.

The Engineering Experiment Station is planning to include the following subjects for study during the coming year:

A   Friction and Endurance of Lubricating Oils
B   The Slippage and Windage of High-Speed Belts and Pulleys
C   Composition of so-called Standard Liquid Fuels
D   Combustion of Crushed and Pulverized Coal
E   Effect of Load and Power Factor on Central-Station Rates
F   Farm Lighting Plants
G   Adaptation of Circular Energy for Service in the House and on the Farm
H   Electrification of Grain Elevators.

The Station is prepared to do commercial testing at the established rates.

The plans for industrial coöperation are similar to those employed by other laboratories. Industrial fellowships may be founded by individuals or corporations, the university supplying the necessary facilities for the work. The Station is established primarily to afford opportunities for scientific research which would be of direct benefit to the people of the state and the nation.

# ENGINEERING SURVEY

A Review of Progress and Attainment in Mechanical Engineering and Related Fields, as Gathered from Current Technical Periodicals and Other Sources

SUBJECTS OF THIS MONTH'S ABSTRACTS

AEROPLANE CRUISING RANGES AND USEFUL LOADS
TRANSFORMATIONS IN COMMERCIAL NICKEL STEELS
INTERCRYSTALLINE BRITTLENESS OF LEAD
ALCOHOL as BY-PRODUCT OF DISTILLATION OF COAL
CASTING FROM BLAST FURNACE INTO MOLD
LOW-TEMPERATURE DISTILLATION OF SUB-BITUMINOUS COAL
HEAT-TREATING GAS FURNACES
ROTOPLAN TOOL-HARDENING FURNACE
SHIPYARD CRANES (HOE)

POWER-RECUPERATING ENGINES
DIESEL ENGINES FOR MOTOR VEHICLES
INTERNAL-COMBUSTION LOCOMOTIVES
LUBRICATING OILS AND FREE FATTY ACID CONTENT
GERM PROCESS OF LUBRICATION
MECHANICAL REDUCTION GEARS IN WAR-SHIPS
OIL SPRAYER FOR LUBRICATING LARGE REDUCTION GEARS
ALLOYS OF OXIDES
WASTE HEAT FROM STEEL FURNACES
COAL PULVERIZERS

POWER TRANSMISSION BY SONIC WAVES
SONIC WAVES AND LAWS OF ALTERNATING CURRENTS
CLAY FIREBRICK AT FURNACE TEMPERATURES
COLD-STORAGE INDUSTRIES AND THEIR SCIENTIFIC PROBLEMS
BIOLOGICAL PHENOMENA IN COLD STORAGE OF FRUITS AND MEATS
THERMODYNAMIC CYCLE OF INTERNAL-COMBUSTION ENGINES

## AERONAUTICS

### Graphical and Mathematical Investigation of Aeroplane Ranges and Useful Loads

A STUDY OF AEROPLANE RANGES AND USEFUL LOADS, J. G. Coffin. The report is in three parts, the first dealing with numerical and graphical analyses of the conditions for maximum range, fuel consumption, determination of the maximum load for a given objective and consequences of flying at maximum speed, together with conditions for flying at minimum power. Graphical methods are chiefly used.

In the second part practically the same factors as were considered in the first part are investigated again, but this time by mathematical methods. Equations are deduced such as time-weight, time-distance, weight-distance and useful-load-objective distance. Proof is presented to the effect that the load-objective curve is a straight line. The third part considers the effect of wind on range.

The report is of particular interest when taken in connection with the paper on Rectilinear Flight of Aeroplanes, by A. Rateau, Hon. Mem. Am. Soc. M. E., published in the May, 1920, issue of MECHANICAL ENGINEERING, with this distinction, however, that while Professor Rateau makes the attainment of the maximum cruising radius of an aeroplane dependent on the altitude of flight, this latter element is introduced into calculations by the author only incidentally. Doctor Coffin, however, also attaches importance to flight at high levels and cites its advantages in the following manner:

*a* The motor running full open will probably use less fuel per horsepower than has been assumed for throttle, say, in the ratio of 0.6 to 0.7.

*b* The motor running at a higher speed can develop slightly more power with proper adjustment, which will increase the height, and therefore the speed.

*c* A very good third reason is that the duration of the flight will be considerably lessened and this together with

*d* The increased safety due to high altitude and greater flying speed lead to the conclusion that: For bombing purposes the aviator should fly at a certain predetermined constant angle of attack; he should allow the plane to rise as the load diminishes.

Since the work consumed in rising to the higher level is at least partially returned when the machine glides down at the end of the trip without power, these works have not been considered.

The following results have been attained mathematically as regards flight in calm air:

1 The machine should fly at a constant angle of attack, the angle corresponding to the minimum value of $W/R$, where $W$ is the weight and $R$ the total resistance.

2 It is practically immaterial whether the machine flies high or low as far as range is concerned.

3 There is an advantage in flying high in that the time is much reduced.

4 The resistance is proportional to the weight at a given altitude.

5 The result of flying at maximum speed is a very much diminished range, or for a given range a very much diminished useful load.

6 The result of flying at minimum power is to slightly reduce the range.

7 The times of flight at the same level for flying at best range speed and at minimum power speed are practically the same.

8 The condition for best range is shown.

9 The weight-time curve is deduced.

10 The range-time curve is deduced.

11 The weight-range curve is deduced.

12 The effect of altitude has been taken into account.

13 The time is greatly diminished for flying at corresponding levels.

14 The theory checks closely with the ordinary methods of Part I.

Part 3 gives a theoretical solution of the effect of wind on range. First, a proof of a method for determining the $L/D$ and air speed for the machine under any wind conditions is given. A new method is shown wherein but one $P$-$V$ curve is required for any load and any wind speed.

Variations in $L/D$ for changes in load and wind speed are derived and checked against the usual methods, and the weight-distance formula is derived as modified by winds.

The results of interest for flight in winds are:

1 The angle of attack changes but slightly when flying against winds of reasonable strength, and but very slightly when flying with winds of any strength.

2 The altitude of flight affects the range. The reason being that higher speeds are attained at higher altitudes and the ratio of air speed to wind speed changes. However, as wind speeds change with altitude it does not seem worth while to go into the matter more fully.

3 Other things being equal, it is slightly advantageous to fly high, especially as to time of flight. (*National Advisory Committee for Aeronautics*, Report No. 69. Preprint from 5th Annual Report, Washington, D. C., 1920, pp. 6-29, 11 figs., *t*)

## BUREAU OF STANDARDS (See also Refractories)

NOTES ON THE CRITICAL RANGES OF SOME COMMERCIAL NICKEL STEELS, H. Scott. The transformations in some commercial nickel steels have been studied by means of thermal analysis. The $Ac_1$ transformation, the temperature of which is of direct practical

351

value in heat-treatment specifications, is also the most difficult to definitely locate. It is found that slow rates of heating give better definition of this point' and that the end of the $Ac_1$ range, as interpreted from the thermal curves, corresponds to the minimum temperature at which all the ferrite is in solution. The effect of rate of temperature change and of nickel content on the critical ranges of the steels, which approximate 0.40 per cent carbon content, is shown in curves, and the effect of nickel and carbon on the end of $Ac_1$ is deduced from the data obtained and those of other observers whose work is discussed. The effect of nickel on the $A_1$ transformation is to lower $Ac_1$ by 10.5 deg. cent. and $Ar_1$ by 21.5 deg. cent. for each 1 per cent nickel over the range 0 to 4 per cent nickel. The eutectoid ratio is also decreased by 0.042 per cent carbon for each 1 per cent nickel added over the same range. (Abstract of *Scientific Paper of the Bureau of Standards*, No. 376, *e*)

THE INTERCRYSTALLINE BRITTLENESS OF LEAD, Henry S. Rawdon. Sheet lead sometimes assumes a very brittle granular form during service, due to corrosion. An explanation which has been offered by previous investigators for this change in properties is that it is due to an allotropic transformation, the product resulting from the change being analogous to the well-known " gray tin." Contact with an electrolyte, particularly a weak acid solution of a lead salt, has been claimed to be the agency by which the transformation is brought about.·

Metallographic examination of the granular " allotropic " lead shows that each grain has the characteristic properties of the ordinary form of lead. The intercrystalline cohesion of the grains for one another, however, has been so weakened that the material has a granular appearance.

The rate at which the intercrystalline brittleness is brought about is proportional to the amount of impurities and to the concentration of acid in the solution in which the lead is placed. Practically all the impurities which are found in lead are lodged in between the grains. The preferential attack by the corroding agent for these impurities, and perhaps also for the " amorphous intercrystalline cement," accounts for the brittleness produced. Investigation showed that specimens of exceptionally pure lead (99.993 per cent), when immersed for 24 days in a neutral solution of lead acetate, became appreciably embrittled by the formation of minute intercrystalline fissures. No evidence of the existence of an allotropic form of lead similar to gray tin could be obtained. (*Scientific Paper of the Bureau of Standards*, No. 377, *t*)

## CHEMICAL ENGINEERING

ALCOHOL AS A BY-PRODUCT OF THE DISTILLATION OF COAL, E. de Loisy. On December 15, 1919, H. Le Chatelier presented to the Academy of Sciences, in Paris, a short paper written by the present author describing the process of synthetic production of alcohol as a by-product of coal distillation. By a queer coincidence on the same day E. Bury and O. Ollander presented a paper on the same subject to the Cleveland Institution of Engineers, describing the same process as developed at the Skinningrove Works.

Briefly, the process is based on the following facts: It has been known for a long time that illuminating gas contains about 2 per cent of ethylene, and it has been known that ethylene can be converted into alcohol. The question was, however, (1) how to separate the ethylene from the gas, and (2) the best method for its conversion into alcohol.

As regards the former problem, it was solved during the war when vast quantities of ethylene became necessary for the manufacture of so-called mustard gas in connection with the Chemical Warfare Service. A process was developed for segregating ethylene in coke-oven gas by absorbing it in charcoal, the gas having been previously washed in lime water to free it from hydrogen sulphide and carbon dioxide. In fact, during the war a continuous process was developed such that at one stage cold charcoal absorbed the gas and liberated it at another stage at a temperature

of about 350 deg. cent., whereupon it was ready to be cooled and used over again.

The next stages in the conversion into alcohol comprise the absorption of ethylene by concentrated sulphuric acid, which led to its transformation into acid ethyl sulphate or sulphovinic acid, and the treatment of the latter which could be carried out in two ways—either by oxidation of the acid by ozonized air or electrolysis with the view of obtaining acetic acid, or by hydrolysis, in which case alcohol is obtained.

The process is of interest in view of the apparently growing demand for power alcohol. (*Revue de Métallurgie*, vol. 17, no. 2, Feb. 1920, pp. 56-62, *ep*)

## FOUNDRY

DIRECT BLAST-FURNACE CASTINGS, F. L. Prentiss. An article (Making Ore Pile Part of Automobile Plant) describing the blast-furnace installation at the Ford Motor Company intended to produce the iron needed in the manufacture of Ford cars.

One of the main interesting features of this plant is the attempt, only partly realized, to pour hot metal directly from the furnace to the molds in the foundry eliminating remelting the pig iron in the foundry cupolas. Mr. Ford, believing that this could be accomplished, some time ago set his engineering and foundry departments to work on the problems involved. In the process as it has been worked out the cupolas will not be eliminated entirely, but the metal from the blast furnace and the metal from the cupolas will be mixed in definite proportions. It is stated that tests have proved that high-grade castings possessing all the qualities required for machining will be produced by this process. (*The Iron Age*, vol. 105, no. 19, May 6, 1920, pp. 1295-1302 and one sheet of illustrations, 11 figs., *dA*)

## FUELS AND FIRING (See also Chemical Engineering; Power Plants)

Decomposition Point in Sub-Bituminous Coal—University of Washington Investigation

LOW-TEMPERATURE DISTILLATION OF SUB-BITUMINOUS COAL, H. K. Benson and R. E. Canfield. Data of an investigation carried out at the Laboratory of Industrial Chemistry, University of Washington, Seattle, Wash., the investigation being on a semi-commercial scale.

The sample of coal was black in color with a brown-black streak. It had a rather dull luster, was massive in texture, without joints, and had a conchoidal fracture. The approximate analysis showed in air-dry sample a sulphur content of 0.36, volatiles 39.4, and nitrogen 1.47 per cent.

The most striking result of the investigation is the well-defined decomposition point between 350 and 400 deg. cent. This marks a maximum in the yield of tar oils, and an abrupt rise in the quantities of nitrogen and methane. A decrease in paraffins also occurs at this point, suggesting the possibility of the cracking of the oils.

Rather interesting data have been secured on the yield of ammonium sulphate in pounds per ton of coal (Fig. 1).

Attention is called to the following general conclusions:

About 3.5 per cent of the coal may be obtained as raw oils. These raw oils are a mixture of coal tar and petroleum-like oil, with the former predominating.

The yield of light oils decreases rapidly as the temperature increases, that of the paraffin oils less rapidly, while the yield of the medium oils remains fairly constant.

About 5.3 lb. of paraffin wax per ton of coal may be obtained at 350 deg. cent.

The gas given off up to 600 deg. cent. is small in volume and low in heat value, but relatively high in illuminants.

The residue at 350 deg. has a calorific value of 12,700 B.t.u., which is an increase of 22.8 per cent over the coal as mined, and of 14.7 per cent over the dry coal. (*The Journal of Industrial and Engineering Chemistry*, vol. 12, no. 5, May 1920, pp. 443-446, 4 figs., *e*)

## FURNACES

HEAT-TREATING GAS FURNACES, H. M. Thornton. In the course of a paper on Gas in Relation to Increased Output and National Economy, the author presents some interesting data on the use of gas for heat treatment and· describes some of the furnaces used in England.

Among these are the so-called L. P. G. A. (low-pressure gas and air) used for annealing high-speed steel in Sheffield. This furnace is of the "over-fired" type, which means that the gas supply at normal city pressure is led through ports at one or both sides, where it comes into contact with the air (supplied at approximately 2 in. water gage) obtained from a small fan. This air is well preheated by being taken through the opposite side of the furnace in fireclay tubes and then passed along the bottom of the furnace in close proximity of the hot waste products.

Combustion takes place inside the working chamber round the furnace walls, the usual combustion chamber being absent. The flames produced keep up well and sweep round the arch. The products of combustion then pass to the opposite side of the furnace, are carried under the floor and up the other side away to the flue. All the waste heat possible is utilized in preheating the air, thus promoting economy and efficiency.

With pressure inside the furnace, air cannot enter through the door or any other orifices even when they are opened for any purpose.

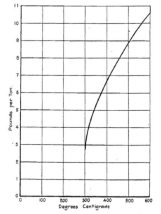

Another of the furnaces mentioned is the Rotoflam tool-hardening furnace, Fig. 2. In this furnace the chamber is circular and is heated by gas and air-blast burners, two of which are used— one at the top and one at the bottom of the chamber, and equidistant on the circumference. The flame from the burners encircles the inside walls of the furnace and maintains a temperature suitable for any class of high-speed steel.

There is no direct flame contact, all the heating being done by radiation. Therefore there is no excessive heat at any one point, and a gradual soaking heat is insured. There is no flue or chimney in this furnace, the exhaust gases being expelled automatically through the mouth. This arrangement makes it impossible for any

free air from outside to enter the chamber and oxidize the steel

The next furnace described is the Brayshaw, also used for heat-treating. In this the upper chamber is heated by waste heat from the lower and serves for preheating tools, etc., before they are put into the lower furnace, raising them gradually to a medium temperature preparatory to the final quick heating required to bring them up to the high temperature necessary for hardening.

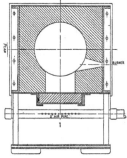

FIG. 2 ROTOFLAM TOOL HARDENING FURNACE (SECTIONAL VIEW)

The main advantage claimed in the paper for gas-heated heat-treating furnaces is that they permit such a close control of conditions as to reduce the waste due to improper heat treatment to very low figures.

Several other furnaces are described and comparative data given on the cost of coal and gas firing for certain purposes. (*Journal of the Royal Society of Arts*, vol. 68, no. 3517, Apr. 16, 1920, pp. 346-366,. 17 figs., *dp*)

## GAS ENGINEERING (See Furnaces)

## HEAT TREATMENT (See Furnaces)

## HOISTING MACHINERY

### Principles of Design of Shipyard Cranes and Description of Hok Crane

SHIPYARD CRANE, W. H. Hok. Description of a crane used in a Swedish yard (Lindholmen Shipbuilding and Engineering Company of Gothenburg).

A large number of European yards adhere to the ordinary mast-and-derrick shipyard crane. Invariably the hoisting winch, whether steam or electric, was placed on the ground level with the attendant trouble of having the hoisting rope leading from the winch to the derrick mast always entangled in bars, plates and all sorts of rubbish. This arrangement also necessitated having signalmen placed here and there, as the winch operator could not as a rule see what he was doing.

Other yards are equipped with expensive overhead traveling cranes, such as gantry cranes, or with cantilever cranes common to two contiguous berths and running either on rails laid on the ground or on a high gantry erected between the berths. Revolving cranes are also used, either of the high-power type traveling on rails laid on the ground or a small revolving crane running on rails laid on a gantry erected between berths.

The author of the paper investigated the crane arrangements on some two score of plants and found that, on the whole, nobody seemed satisfied with the crane arrangements he possessed, and the cranes not only were not standardized, but it was almost always the case that a different system of cranes was tried at almost

every building berth in the same yard and for every new berth
laid down.

The conclusions to which the author came as regards the general principles of crane construction are, as follows:

1  The mast-and-derrick arrangement is quite satisfactory, provided it can be so arranged that all side staying of masts can be done away with.

2  The operator's platform should be placed high above the ground on, for instance, the level of the principal weather deck of the ship, so that the operator can see what he is doing, thus obviating the necessity of using signalmen.

3  The lead from the hoisting winch to the derrick mast should not be taken along the ground among staging uprights, shoring, plates and bars, and various rubbish strewn about the ship. The lead from the winch should be free from all obstructions.

4  A space or passageway is desirable between contiguous ships to enable building material to be brought down between ships and hoisted on board from the nearest point at the ship's side, and not from the ship's end only. Such an arrangement

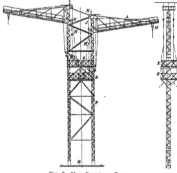

FIG. 3  HOK SHIPYARD CRANE

covers many more chances of rapid building than an arrangement based, for instance, on the material being taken in hand by the hoisting gear at the end of the ship only, because the former arrangement offers so many more points of attack on the ship than the latter.

Guided by these conclusions the author designed the stationary crane shown in Fig. 3. This type is chiefly composed of a stationary main structure and two swinging arms, the former consisting of two latticework masts placed about 15 ft. 3 in. apart from center to center, and rigidly connected to one another by cross-stays and trusses making the main structure stable in the thwartship direction, thus obviating the necessity of fitting side stays. The author calls such a structure consisting of two derrick masts rigidly connected to one another by cross-stays and trusses a *derrick frame*, in contradistinction to the solitary derrick mast. Each derrick frame carries two derricks or arms $A$ which can swing about 120 deg. to each side of a vertical thwartship plane through the derrick frame, i. e., well past the center line of railway $B$ laid down in the passageway between contiguous ships. The derrick frame is held in place by fore-and-aft wire stays only. For this purpose two sets of stays are fitted, the lower ones $C$ partly for giving rigidity to the cranes in a fore-and-aft direction and partly for preventing collapse of a whole group of interconnected cranes in case the top and end stays $R$, or some of the top stays $D$, fitted from crane top to crane top of the group should give way.

At a suitable place above the ground two winch platforms $E$ are built into the derrick frame, one above the other. Access to them is given by ladders inside the frame legs $F$.

Each crane arm and the load are controlled from the platforms $E$ by a single ordinary alternating-current electric shipyard winch $G$, which means that they are controlled by one electric motor only. Hoisting and lowering are done with the winch center barrel in the ordinary way, and slewing with the extended winch ends by taking a couple of turns round the appropriate extended winch end with a loose end of a tackle $H$, actuating the wire $J$, which is carried round and fastened to the rim of the horizontal wheel $K$ on top of the crane.

A spiral spring $N$ is introduced above the slewing tackle for the purpose of taking up the inertia of the crane arm when it arrives at the extreme end of the swing, in this way preventing damage and unnecessary straining of the connections in case of rough usage or ignorance on the part of the man handling the crane.

The method of racking motion is described in some detail, as well as the method of erecting these cranes. (Paper read before the Institute of Engineers and Shipbuilders in Scotland, abstracted through *Marine Engineering and Canadian Merchant Service Guild Review*, vol. 10, no. 4, April 1920, pp. 84-87, 6 figs., d)

## INTERNAL-COMBUSTION ENGINEERING (See also Thermodynamics)

### Power-Recuperating Engines: Principles of Operation and Design of Valve Gear

A POWER-RECUPERATING ENGINE, Georges Funck. Discussion of the design of engines that would maintain power at altitude with supercharging. The writer discusses the operation of standard engines under various conditions on the basis of their entropy.

The power-recuperating engines are classified into two main groups, the first consisting of those having a cycle of operations such as to maintain a maximum pressure and the second a cycle to maintain a constant compression temperature, taking into account for both the atmospheric temperature variation.

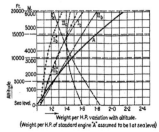

FIG. 4  CURVES OF WEIGHT PER HORSEPOWER OF ENGINES OPERATING ON VARIOUS CYCLES AS COMPARED WITH A STANDARD ENGINE AT SEA LEVEL

They may further be divided into certain subdivisions. Thus, case (a) represents an engine designed with a combustion space of such a size that its ratio to the total cylinder volume gives a maximum compression ratio required for the altitude at which the engine has been designed to work.

In order to vary the effective actual compression ratio as required by the altitude, it is proposed to close the inlet valve before the piston reaches the end of the stroke which amounts to having an engine with a variable compression but a constant-expansion stroke.

This cycle is really the Atkinson cycle, with the modification that the effective compression ratio is variable with the altitude while the expansion ratio is constant, until the standard cycle is attained at the predetermined altitude. In this connection Fig. 4 is of interest as showing that the weight of such an engine per unit output.would be quite large at sea level.

The simplest way to perform this cycle mechanically appears to be to provide a variable inlet charge cut-off by altering the timing of the inlet valve, which could be done in several ways; for example, the valves could be operated by means of rockers mounted on eccentric pins as illustrated in Fig. 5.

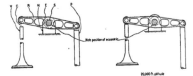

FIG. 5  VALVE OPERATION OF ATKINSON-CYCLE POWER-RECUPER-
ATING ENGINE

By rotating on its housing $H$ the eccentric $E$ which carries the swivel pin $S$, the clearance $C$ between the rocker $R$ and the valve $V$ is altered and the valve lift and duration of opening is modified, thus effecting the inlet charge cut-off.

Case (b) represents an engine so designed that the volume of combustion space is adjustable. Attempts to do this were made by introducing a second piston in the upper end of the cylinder, so located that its position can be modified at will, thereby re-

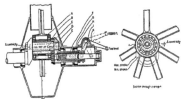

FIG. 6  RADIAL POWER-RECUPERATING ENGINE

ducing or enlarging the combustion space. This involves a complicated and inefficient valve gear. Another proposition has been made to make the whole cylinder movable up and down, which is again too complicated for immediate consideration.

The variable-stroke method has also been proposed. The design of a variable-stroke engine of the multi-throw type of crankshaft presents mechanical difficulties. It is, however, comparatively simple in connection with radial engines. Thus, Fig. 6 shows the design of a radial power-recuperating engine.

Over the crankpin $C$ is fitted an eccentric sleeve $D$, which carries the connecting rod $A$ mounted on ball bearings in the usual manner. The eccentric sleeve can be rotated partially round the crankpin and held in any desired position. It is obvious that by this arrangement the stroke of the piston can be altered at will. Only a small eccentricity is required to give the desired result. Even in the extreme case $B$ of group 1, to which the curves relate, an eccentricity of 5 per cent of the shortest stroke is all that is necessary to effect power recuperation up to 20,000 ft.

In order to operate this eccentric sleeve from outside, a gear wheel $B$ is attached to the sleeve. Into this gear meshes another wheel $E$ carried on the spindle $F$, the latter being provided with a multi start quick thread as shown. A sleeve $H$ with a cor-

responding internal thread engages the thread of the spindle $F$. This sleeve is prevented from rotating by one or more keys $G$ located in the crankshaft in such a manner that an axial sliding motion can be imparted to the sleeve $H$ by means of the lever $I$. It will be seen that by operating the outside lever $K$, which is coupled direct with lever $I$, the eccentric sleeve $D$ can be rotated to any desired position, and maintained there while the crank is revolving, modifying the effective throw of the crank and thereby the piston stroke.

As the eccentricity is very small, the reaction of the connecting rods while under load on the eccentric sleeve is small, and no undue force would be required to operate the mechanism while the engine is running. Should there be any difficulty, however, the throttle may be closed for the short period of the change, which would be effected in steps according to the altitude. However, in proportioning the lead of the multi thread and the levers correctly, this process of operation would hardly be necessary. For the same reason the different positions of the eccentric affect the timing of the engine only to a small degree, and it should be possible to design a cam so that a satisfactory timing for all altitudes is arrived at; or such a timing may be evolved which gives the best result at the altitude at which the engine is called upon to run normally.

The article is illustrated by numerous pressure-volume diagrams worked out for the various cycles at three different levels; namely, sea level, 10,000 feet altitude, and 20,000 altitude. (*The Automobile Engineer*, vol. 10, no. 137, April 1920, pp. 145-153, pA)

## Details of Diesel-Engine Design for Use on Motor Vehicles

THE SMALL DIESEL ENGINE FOR MOTOR VEHICLES, Ernest Frey. The author, who is chief engineer of the Oberursel Motor Company, claims that the chief reason why Diesel engines have not yet been adopted for use on motor vehicles lies in the fact that Diesel-engine manufacturers did not pay sufficient attention to this particular application, while motor-car manufacturers did not have the experience and knowledge of design necessary to solve the problem of designing a small Diesel engine. Inherently, however, he believes that this can be done without much difficulty.

As regards injection, either air or solid injection may be applied. Because of the greater weight of the Diesel engine as compared with the conventional automobile engine, the crankcase has to be built lighter than usual if possible. To do this, the author recommends casting the upper part of the crankcase integral with the cylinder block. The cylinders themselves should be equipped with thin steel liners.

The crankshaft should be equipped with ball or roller bearings so as to compensate for the somewhat lower mechanical efficiency of the Diesel engine. As such bearings do not have to be wide, plenty of room is available for wide connecting-rod bearings, which is important in view of the great piston pressures employed. Furthermore, for the sake of safety it is recommended that each crank have a bearing on both sides, so that a four-cylinder engine would have five bearings. This should be combined with pressure lubrication.

As regards the pistons, it is recommended that a deep depression be made toward the center of the piston head, in order to compress the air of combustion under the fuel valve which should be located as closely as possible to the center. The piston then at its upper dead center will have only a very small amount of clearance (say, 2 mm.) at its periphery against the cylinder head.

In reference to the compression pressure, it is stated that the exponent of the compression pressure curve rises with the speed of rotation; on the other hand, however, because of the small dimensions of the motor-car-type Diesel engine, the area available for heat transmission is quite large as compared with the cylinder volume. The following values of this exponent $\epsilon$ are recommended:

for $n =$ 800 to 1000, $\epsilon =$ 18
for $n =$ 1000 to 1400, $\epsilon =$ 20

The following valve timing is recommended: The fuel valve opens at 5 deg. ahead of the dead center and stays open until 40 deg. past the dead center; the air inlet valve opens 5 deg. ahead the dead center and stays open until 40 deg. past the dead center; the exhaust valve opens 45 deg. ahead of the dead center and stays open until 10 deg. past the dead center; and the starting valve opens 2 deg. ahead of the dead center and stays open until 120 deg, past the dead center, the cycle of operations being distributed over the four strokes in the usual manner.

Where air injection is used the design of the high-pressure compressor (whether two or three-stage) requires particularly careful attention. For the sake of simplicity the two-stage reciprocating compressor should be selected. Because of the great compression (as high as ninefold in a single stage) it is absolutely necessary to employ an efficient intercooler, which may conveniently be arranged concentrically with the high-pressure cylinder. It is also recommended that the cooler be provided at its lowest point with a drain cock or valve in order to drain off, from time to time, the oil or water that may accumulate therein. A calculation is given, indicating by means of an example how the dimensions of the compressor should be computed.

Compressed air is recommended for starting the engine. As regards the fuel pump, it is stated that this pump and its injection nozzle constitute the most important parts of the Diesel engine. In the face of numberless attempts to solve the problem, the author states that for multi-cylinder high-speed engines the only possible construction is that in which each cylinder is equipped with its own pump supplying to it the necessary amounts of fuel, no matter how small these may be.

Since the fuel pump is not capable of automatically lifting the fuel to it, it is necessary to supply the fuel from the fuel container under a small pressure, say, 0.3 atmos. gage, which pressure may be created by a separate small air pump. Care should be taken that no air gets into the suction chamber of the fuel pump, and as it is impossible to prevent small air bubbles from penetrating into the pump together with the fuel, provision should be made for maintaining a vacuum in the pump suction chamber. Above all, the pump itself must be so designed that not the slightest air bubble can persist in the compression chamber.

Fuel pumps for multi-cylinder engines consisting of one plunger and so-called distributors, which latter are supposed to take care that each cylinder gets its proper quota of fuel, have been given up as unsatisfactory. Likewise, the pumps in which the injection pressure is regulated instead of metering the fuel amounts themselves have been found unsatisfactory. Such a system of regulation even though possibly suitable for single-cylinder motors always gives an unequal fuel distribution in multi-cylinder motors, since the flow resistance in individual nozzles is never equal even when the greatest precision has been used in the production of these nozzles.

As regards the injection valves, it was found that the best results were obtained from needle valves equipped with strong springs, the methods of controlling these valves being different for solid injection from those used in air injection.

The article is to be continued. (Der Motorwagen, vol. 23, no. 2, pp. 30-33, 1 fig., tp)

THE INTERNAL-COMBUSTION LOCOMOTIVE IN STANDARD RAILWAY PRACTICE. An editorial raising the question as to why so little has been done toward the production of a locomotive operated by an internal-combustion motor for general service.

The internal-combustion engine is now fully established as an efficient, reliable and satisfactory source of motive power for stationary purposes. It has also been successful in marine work and has been applied for certain special purposes in railway traction but not in locomotives for general purposes.

There are several directions in which the large-power internal-combustion locomotive may possess distinct value in heavy railway practice. It eliminates the expensive steam boiler with its heavy maintenance costs and all the expensive appurtenances in the way of water supply and coaling-plant installations. Further-more, no fuel expenditure is involved when the machine is not at work.

In addition to this are advantages of a directly engineering character, as for instance, the considerable power capacity which can be concentrated within relatively moderate dimensions, and the great heat and power efficiency of the well-designed modern internal-combustion engine with its reliability and freedom from breakdowns.

The editorial ascribes the lack of development in this direction to the fact that railway companies cannot undertake the necessary experimental work, while private firms have not the opportunity for experiment and trials under service conditions. The chief explanation is said to be found, however, in the actual conditions of railway traffic operation. The steam locomotive is peculiarly adaptable for variable working conditions. The internal-combustion engine requires being operated under conditions favorable to itself. Adaptations to suit speeds, gradients, light and heavy loads, etc., are easily made with steam. But the internal-combustion locomotive must either depend largely upon gearing or special transmission or be working disadvantageously whenever requirements vary much below the standard for which it is designed, and furthermore, it does not easily carry overload.

It is this elasticity in service which is so characteristic of steam-locomotive practice and which is so essential under modern traffic conditions that constitutes the greatest obstacle to the introduction of the internal-combustion engine as a practicable factor in heavy railway working. (The Railway Engineer, vol. 41, no. 483, Apr. 1920, pp. 138, gc)

## LUBRICATION

### Physico-Chemical Bases of Lubrication—Measurement of Oiliness—Free Fatty Acids as Affecting Lubricating Properties of Oils

THEORY OF PRACTICE OF LUBRICATION: THE "GERM" PROCESS. Discussion of the theory of lubrication, in particular the so-called "oiliness" of lubricating oils; methods of its measurement and influence of the presence of free fatty acids on the lubricating properties of oils.

The "oiliness" of lubricating oils has been observed for some time, but hitherto no methods have been found to measure it and no consistent theory has been offered to account for it.

Uebbelohde pointed out years ago (Journ.Am.Soc.M.E., June 1912, p. 963) that only a liquid which wets or spreads over the solid can constitute a true lubricant, but this did not give a basis of differentiation between various lubricants as all of them wet solids.

In order to determine the oiliness of a lubricant it would have been necessary to measure the surface tension between the oil and the solid metal bearing, which, unfortunately, we do not know how to do. Because of this, the present authors decided to measure the surface tension of the oil against the immiscible liquid in the hope that this procedure might furnish some criterion of oiliness.

The liquid selected was water and the measurements have shown several interesting facts.

In the first place, it was found that the interfacial tension of vegetable and animal oils against water is much lower than in the case of a mineral oil. Furthermore, there was a distinct difference between the tensions in the case of mineral and saponifiable oils independent of their viscosity, density, etc., and this difference appeared to be in conformity with the lubricating properties of the oils.

Further, experiments proved that the lowering of the interfacial tension against water in the case of fatty oils was due to the presence in them of small amounts of free fatty acids. In fact, it was found that when free fatty acids are removed from the saponifiable oils the tension rises, and that when they are added to mineral oils the tension can be lowered. Analyses have shown that fatty acids are present to a certain extent in practically all oils.

These experiments led to the following conclusions:

1  Capillary effects hitherto ignored in lubrication play a fundamental part
2  A neutral glyceride possesses a tension similar to that of a
   neutral mineral oil, and
3  The addition of a relatively minute amount of a fatty acid to
   a neutral mineral oil reduces the tension to that of a commercial animal or vegetable oil or compounded lubricating
   oil.

Interfacial tension affects lubrication in the following way:
The permanency of films depends on a diminished interfacial
tension between the oil and the metal in contact therewith. If
such a film is broken, it will unite the faster the lower the interfacial tension.

Extensive tests were carried out to determine the behavior of
various oils and it was found, for example, that 1 per cent of the
free fatty acids of rape oil added to a mineral oil are as effective
in reducing the value of the frictional coefficient as is the addition
to the mineral oil of 60 per cent of neutral rape.

The next subject taken up in the paper is that of the colloidal
characters of the fatty acids. In recent years it has been shown
that while the lower members of the fatty-acid group possess
relatively little colloidal character, the higher members are highly
colloidal and there appears to be a gradation in these properties as
one ascends the scale, lauric acid occupying a sort of intermediate
position.

The fatty acids which occur in commercial oils are not pure
chemical individuals, but are mixtures in various proportions of
a number of fatty acids, higher or lower members predominating
in accordance with the character of the oil. Thus, cocoanut oil is
characterized by containing appreciable percentages of the lower
members of the series, while rape oil rarely contains anything
but the higher members. The behavior of the oils is determined
by the fatty-acid groups which predominate in them and it is possible to reproduce the capillary properties in any particular animal or vegetable oil by adding suitably chosen fatty acids to
mineral oil.

The above considerations have a practical importance. Thus, in
a steam engine using saturated steam there is a tendency for condensation to occur in the cylinder and valves. In such a case the
presence of a substance in the oil which lowers the surface tension
against water will assist in the formation of oil films by enabling
the oil to spread more readily or by reducing the tendency of the
water to wash the oil film off.

In certain classes of lubrication where the oil is brought into
contact with water, it may be desirable that the oil shall either
separate itself rapidly from the water (demulsification), or, conversely, that it shall mix or emulsify with the water, but the emul-
sification is dependent upon the colloidal properties of the oil
while demulsification is brought about by a greater concentration
of hydrogen ions. Consequently, by varying the types of fatty
acids present in the oil, it is possible to control this particular
property, oils containing higher members of the fatty-acid group
possessing an emulsifying tendency while the ones containing the
lower members possess a demulsifying tendency.

The authors use the expression "germ process" to describe the
production of oils made by using one or more fatty or other acids
with one or more mineral oils. As a matter of fact, no germs
whatsoever are known to have anything to do with this process.

Specifications are given for oils used in the lubrication of
various classes of machinery. (Paper presented before the Society of Chemical Industry, *Journal of the Society of Chemical
Industry*, 1920, pp. 51-60, T, 2 figs., et)

## MACHINE PARTS AND DESIGN

### Mechanical Reduction Gears in Warships Have "Made Good" in the British Navy

MECHANICAL REDUCTION GEARS IN WARSHIPS, Engr.-Com-
mander H. B. Tostevin, R. N. Data on the experience and practice on this subject in British warships. Up to about 1912 only
comparatively moderate amounts of power on shipboard were

transmitted by gearing in the British Navy, but in the torpedo-
boat destroyers *Leonidas* and *Lucifer* laid down in 1912 it was
arranged to transmit the whole power, equal to 22,500 hp.,
through two sets of gearing. The two boats were put into commission in August 1914 and received a very severe service test,
which they passed in a highly satisfactory manner.

Even before the completion of these destroyers, however, two
light cruisers of 40,000 hp. were arranged to have all-geared
units, one vessel with four shafts and the other with two. Both
installations again proved quite satisfactory.

The paper includes a table showing the total horsepower and
number of all-geared sets fitted and being fitted in warships. This
total reaches the very respectable figure of 7,828,000 shaft hp.

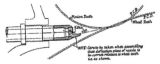

FIG. 7  SPRAYER FOR LUBRICATING GEARING TEETH

transmitted through 652 gears, the power per gearing set ranging
from 1750 hp. up to as high as 36,000 hp.

As regards the gearing design, helical gearing appears to be
used with the angle of obliquity of 14½ deg. Fine-pitch gears
give more silent running at high speeds and a normal pitch of
7/12 in. has been adopted for all but the very largest installations.
While, however, the pitch and obliquity have remained the same,
the proportion between addendum, dedendum and pitch, respectively, and the shape of the root and tip have been changed from
time to time with growing experience.

The question of the lubrication of the teeth is very important

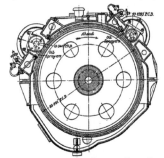

FIG. 8  SECTION THROUGH A GEAR CASE FOR A TORPEDO-BOAT DE-
STROYER SHOWING RELATIVE POSITIONS OF PINION AND WHEEL
AND LUBRICATING ARRANGEMENT FOR THE TEETH

and oil is usually supplied through nozzles (¼ to ⅜ in., of about
5-in. pitch) discharging the oil under a pressure of from 5 to 10
lb., the jet being fan-shaped so that the whole length of the teeth
is lubricated.

Figs. 7 and 8 show a type of oil sprayer fitted, together with
its arrangement on the gear case.

A difficulty in the early history of gear cutting was caused by
the fact that the master worm wheel which rotated the gear wheel
during the cutting operation was not quite accurate; this difficulty
led to an addition to the machines of a "creep" mechanism, by
means of which the job is rotated at a higher speed than the table.
Any recurring error in the worm wheel which might be copied

on the job being cut is, with such a machine, no longer in the direction of the axis, but is distributed in helices on the gear wheel or pinion. With improvements in the accuracy of this master worm wheel and the maintenance of the machine in good condition with all parts well lubricated during a cutting operation, the advantage derived from the use of the " creep " has been lessened, and from a list of ten firms that have cut the majority of naval gears it is a matter of interest that on the wheel machines only four, and on the pinion machines six firms, use " creep." As regards the tendency of the worm wheel to wear out of truth, inquiry has shown that in a number of machines that have been working constantly for at least three years the accuracy has not been affected.

In all naval work the turbine spindles, pinions and gear wheels are supported on rigid bearings and the alignment is determined by accurate machine work in boring the gear housings and fitting the bearings. No gears of the floating-frame type have been fitted and the system is not favored by the writer.

Several important questions which cannot be abstracted because of lack of space, are, in particular: the action between the teeth when transmitting power, the pinion and the wheel as tending to increase their distance between centers, and the speed of the teeth.

Some of the troubles experienced with gearing are described, such as, for example, the pitting of the faces of the teeth and corrosion of gears due to the use of improper lubricating oil. In an appendix an abstract from Admiralty specifications for gearing for turbines is given. (Paper read before the Institution of Naval Architects, Mar. 26, 1920. Abstracted through *Engineering*, vol. 109, no. 2832, Apr. 9, 1920, pp. 474-480, 14 figs., *dA*)

## MARINE ENGINEERING (See Hoisting Machinery; Machine Parts and Design)

## METALLURGY

ALLOYS OF OXIDES, Miss S. Veil. It is of interest to determine the combinations which may be formed between oxides compressed and heated according to methods analogous to those applied to metallic alloys.

The problem is both difficult and delicate in view of the slowness with which phenomena of diffusion take place between solid bodies, and the properties which can be investigated and measured with any degree of precision are few in number.

Among others, interesting results have been obtained with a mixture of oxide of chromium and oxide of cerium.

By varying the proportions of the constituent parts, the author has been able to investigate concurrently, on one hand, the electric conductivity at high temperatures, and, on the other hand, the coefficient of magnetization at ordinary temperatures.

The electrical conductivity has been measured on material pressed into the form of small rods heated in a platinum-resistance-electric furnace. The coefficient of magnetization was determined by means of the Curie and Cheneveau balance.

The results of both series of tests are presented in the form of diagrams which cannot be consistently interpreted except on the basis of admitting the existence of definite combinations between the oxides under consideration. Furthermore, while the two methods of investigation are different from each other, they give results consistent between the two.

The author indicates the different combinations of oxides which would appear to have either a certain or probable existence. (*Comptes Rendus des Séances de l'Académie des Sciences*, vol. 170, no. 16, Apr. 19, 1920, pp. 939-941, 2 figs., *et*)

## MOTOR-CAR ENGINEERING (See Internal-Combustion Engineering)

## POWER PLANTS

### Waste-Heat Utilization in an American Steel Plant

USING WASTE HEAT FROM FURNACES, B. H. Green. Description of an installation in a steel plant utilizing waste heat from

furnaces, the arrangement being such that steam from waste heat is supplemented by that from coal-fired boilers and used for electrical power generation.

The waste heat is obtained from ten 75-ton open-hearth furnaces fired with producer gas or tar. The furnaces each deliver at the boiler an average of 65,000 lb. of waste gas per hour at a temperature ranging from 900 to 1300 deg. fahr., but averaging

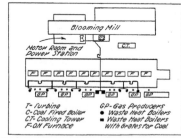

FIG. 9  PLAN SHOWING LOCATION OF WASTE-HEAT BOILERS IN AN OPEN-HEARTH STEEL PLANT

fairly well around 1100 deg. during weekly operation. The open hearths are shut down over Sundays and because of this little steam is generated on Mondays and Tuesdays, the output increasing until the maximum is reached on Thursdays and Fridays.

The boiler plant had to be fitted into the existing steel-plant operation, the available space for the boilers being very limited indeed. Fig. 9 shows how this problem was solved. Vertical two-pass boilers (Wickes type) were installed, as in this particular location the arrangement in this type of boiler permitted a

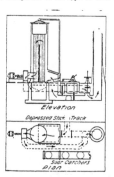

FIG. 10  PLAN AND ELEVATION SHOWING ARRANGEMENT OF UNITS IN LARGE BOILER PLANT AND VERTICAL-SLIDING WASTE-GAS VALVES USED AT THE BOILERS

shorter gas travel between the furnace and boiler. A total space width of about 16 ft. was all that was available for the boilers between the gas-producer soot catchers and a depressed stock track, and Fig. 10 shows how this location was utilized.

There were other problems that had to be considered in determining the location of the boiler. Thus it was desirable to keep the stack warm and to allow its draft to assist the induced-draft fans used in connection with the boilers.

Fig. 10 also illustrates the type of vertical-sliding water-gas valves used at the boilers. These valves are two in number and counterbalance each other as shown, thus rendering the raising or lowering a comparatively easy matter for one man on the winch. A 4-ft.-diameter saucer valve with vertical screw arrangement is also used to shut off the boiler gas passage at any time when the boiler is shut down and the furnace is discharging directly into the stack.

The boilers averaged 250 boiler hp. each during actual operating hours after the initial troubles had been overcome.

The fans have a capacity of from 75,000 to 80,000 lb. of gas per hour when running at a speed of about 600 r.p.m. and produce a draft of 4 in. of water with a gas temperature of 450 deg. at the fan breeching, which gives a draft of about 1¼ in. at the furnace valve damper. The boiler fans should be of sturdy construction with substantial bearings and shafts, as in this location, among the gas producers, great quantities of fine ash sift in through every conceivable crack. Heavy rigid bearings were adopted with substantial dust collars applied externally and clamped tightly so as to hold the felt rings snugly against the shaft. The shafts were made of large size, in order to avoid damage due to possible warping or bending of the fan wheel by reason of heat or by the accumulation of ash on the blading of the fan, which would tend to produce unbalancing. All fan bearings are water-cooled.

Gas explosions are experienced occasionally in the boilers upon reversal of the furnaces, but it has been found that these can be avoided to a great extent by the careful handling of the reversing valve by the furnace operator. It is possible that a careful timing of throw-over of gas and air valve with relation to each other might eliminate explosions entirely by allowing either the free air or the unburned gases to escape into the boiler alone and not in an explosive mixture of the two together.

The original article describes also the auxiliary coal-fired boilers and the piping used in the plant. (Paper before the Cleveland Section of the Association of Iron and Steel Electrical Engineers. Compare the *Iron Trade Review*, vol. 66, no. 15, Apr. 8, 1920, pp. 1065-1068, 3 figs., *d*)

COAL PULVERIZERS. Description of the Aero-Pulverizer, a self-contained unit which first reduces coal to the requisite degree of fineness for burning in furnaces and then feeds it into the furnace.

The advantages of such a unit are that the coal is used as soon as it is powdered and no provision has to be made for storage, and that unless the raw coal is excessively wet it does not need to be dried prior to grinding. It is claimed that it can be ground with a moisture content as high as 6 per cent.

As shown in Fig. 11, the casing is divided into compartments, the number of which varies with the size of the plant. In each compartment there revolves at high speed a disk paddle equipped with a series of hard steel paddle blades. The paddles are all keyed to one horizontal shaft passing from end to end of the machine. The final compartment contains a series of fine blades which revolve with the shaft carrying the paddles. At the end of the machine, remote from the fan, there is an adjustable air inlet, and the amount of air sucked in through the fan and hence the fineness of the dust delivered from the pulverizer are determined by altering the area of the orifice at that inlet. The coal is fed into the pulverizer from a hopper in the form of a sleeve and the amount can be regulated by simply raising or lowering the sleeve.

The diameter of the paddles and their blades is somewhat less than that of the compartments in which they revolve, there being about ⅝ in. clearance all around. There is therefore no grinding action between the blades and the casing, the pulverizing being done entirely by a beating or impact action. As more coal is fed into it the first compartment is gradually filled until the partly

powdered coal overflows the division piece into the second compartment, and so on to the last. During the whole process air is being sucked through the machine by means of the fan and the most finely ground coal is carried away by the current. This is the reason why the strength of the current determines the fineness to which the coal is ground.

A small compartment can be seen on the hinged back casing between the three pulverizing compartments and the fan compartment. It is for the purpose of admitting more air to the

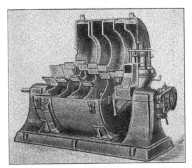

FIG. 11  AERO-PULVERIZER, A COAL-PULVERIZING AND FEEDING MACHINE

apparatus, this additional air serving the purpose of carrying the powdered coal into the distribution system to support combustion of the coal. It is claimed that this machine has been successfully employed for the firing of the furnaces of various types as well as water-tube boilers. (*The Engineer*, vol. 129, no. 3351, Mar. 19, 1920, pp. 306-307, 3 figs., *d*)

## POWER TRANSMISSION

### Constantinesco Sonic Waves, Their Nature and Application—Power Transmission by Rapid Impulses in Liquids

POWER TRANSMISSION BY SONIC WAVES, J. Herck. Description of the Constantinesco system of power transmission.

All methods of power transmission may be divided into three basic classes: First, rigid transmission from solid body to solid body, as in gearing, belt and other friction-drive methods; next, transmission by fluid under pressure, no matter what the nature of the fluid may be—water, air or even electricity (direct current); third, the so-called "sonic" waves of M. Constantinesco, a Roumanian engineer, who is said to have spent close to twenty years in experimental work connected with the development of his idea.

The employment of sonic waves is based on experimental work which is said to have demonstrated that, contrary to popular impression, liquids are essentially compressible.

Let us consider a conduit having a piston at each end, $A$ and $B$, and entirely filled with a liquid, the pistons being leak-proof. If, now, the piston $A$ is given a sudden impulse the liquid will be compressed, the volumetric compression storing up in the liquid a quantity of energy proportional to the coefficient of elasticity of the liquid. Furthermore, since the deformation produced is assumed to have been sudden, a wave is created in the liquid and moves through it with a velocity equal to $\sqrt{(E/m)}$, where $E$ is the coefficient of elasticity of the liquid and $m$ the mass per unit of volume. This velocity is the same as the velocity of propagation of sound in the given liquid.

If the length of the conduit be properly selected, the wave will cause the piston $B$ at the other end to move, and if a vibratory motion be imparted to the piston $A$, the piston $B$ will receive the same motion, the conditions being somewhat analogous to what takes place with single-phase alternating current in electricity.

The intercalation of an auxiliary reservoir creates a capacity having the same effect as a condenser in an electric circuit; a spring in the circuit, by its inertia, performs the same functions as a self-induction coil, and resistance in the conduit to the passage of the wave reminds one of an electric reactance, while the average pressure produced may be compared to the voltage in an alternating-current circuit.

The analogy between sonic waves and transmission of power by alternating current can be carried still further. If we take three conduits that are interconnected and have impulses spaced 120 deg. apart, we obtain three waves spaced likewise 120 deg. apart and have something similar to three cables carrying the three-phase alternating current—in this case only sonic instead of electric. All that is necessary to create it is three pistons placed star-wise, while at the receiving end two arrangements may be employed, depending on whether it is desired to have a synchronous or asynchronous motor.

It is interesting to note further that the transmission of energy in sonic waves is governed by laws remarkably analogous to such laws of electric circuits as Ohm's law, Joule's law, etc.

The Constantinesco principle has been applied for numerous purposes during the war when the inventor was working for the British Admiralty. Among these may be mentioned the hydraulic hammer with the "single-phase" sonic motor; another type of hammer for chipping stone; drills with two-phase asynchronous sonic motors; servo-motors for use on aeroplanes, which are very powerful for their small size; and what is of particular interest, a device for oil injection on Diesel engines. This device has been applied by the British Admiralty to a 1000-hp. Diesel engine and is said to have given entire satisfaction.

Another interesting application of the same principle has been made in connection with bomb throwers capable of hurling a bomb weighing 220 lb. to a distance of close to 5000 ft., and that without either fire or noise. In this device the energy of a small cordite cartridge is absorbed in the compression of a liquid which restores it by pushing the bomb under constant pressure over the entire length of the gun barrel, this affecting the efficiency of the explosion very favorably.

It is stated that experimental work is being carried on to apply the same principle to the power transmission between the engine and propeller of an aeroplane.

The best-known application of the Constantinesco principle is that of synchronizing the propeller and the machine gun on aeroplanes, extensively used by the Allied armies during the recent war. (*Bulletin Technique du Bureau Veritas*, vol. 2, no. 4, Apr. 1920, pp. 69-73, 5 figs., *d*)

## RAILROAD ENGINEERING

New Ten-Wheel Helping Engine on the Midland Railway, England, Frederick C. Coleman. Description of a locomotive of a new type recently built at the Derby Works in England. It has four cylinders, each 16¾ in. diameter by 28 in. stroke, cast in pairs, one steam chest to each pair. In these cylinders cross-ports have been introduced which makes it possible for one piston valve to supply both cylinders on one side of the engine. The front piston-valve head serves the front port of the outside cylinder and the back port of the inside cylinder, and vice versa.

Owing to severe gradients on the road all the wheels are braked, the front three pairs being operated by a steam cylinder and shaft placed just behind the driving axle, and the two hind pairs of wheels by a cylinder and shaft placed under the draft casting. In addition a hand brake has been fitted on the engine to act on all wheels. (*Railway Review*, vol. 66, no. 16, Apr. 17, 1920, pp. 657-658, 1 fig., *d*)

## REFRACTORIES

The Porosity and Volume Changes of Clay Firebricks at Furnace Temperatures, Geo. A. Loomis. This paper deals with the permanent changes in porosity and volume of clay firebricks when reburned to temperatures at or above those to which they were originally fired. These were measured for a series of temperatures to determine what relation, if any, might exist between these changes and the deformation of the same bricks under load at furnace temperatures. The possibility of such a relation is suggested by the fact that contraction of clay on heating and decrease in porosity are, to a certain extent, indications of the amount of softening of the mass due to the action of fluxes present, and hence indicative of decreased resistance to deformation under pressure or decreased viscosity. Softening-point determinations were also made to determine what relation these might bear to the results of the load test.

The results of tests on a large number of clay firebricks from various parts of the country show that bricks which withstand a load test of 40 lb. per sq. in. at 1350 deg. cent. without marked deformation, show no marked changes in porosity or volume up to 1425 deg. cent. Bricks which do not withstand the test generally show appreciable contraction or expansion, accompanied by considerable decrease in porosity. Bricks which showed overburning and the development of vesicular structure below 1425 deg. cent. by marked expansion or increase in porosity, invariably failed under load. In general, bricks which show a decrease in porosity exceeding 5 per cent or a volume change exceeding 3 per cent (amounting to approximately 1 per cent in linear dimensions) when refired to 1400 deg. cent. fail to pass the load test.

No definite relation could be determined between the softening point of a brick and its ability to withstand pressure at high temperatures. All bricks softening below cone 28 failed completely in the load test. Some showing quite high softening points also failed, probably due to the use of an inferior bond clay in the mixture or too small an amount of bonding material. (Abstract of *Technologic Paper of the Bureau Standards*, No. 159, *e*)

## REFRIGERATION

### Biological Phenomena in Fruits and Meats in Cold Storage

Scientific Problems of Cold-Storage Industries, W. B. Hardy. The author, Secretary of the Royal Society and Director of Food Investigation, discusses the wider biological aspect of food preservation. Essentially, food preservation by means of cold storage means an attempt to stop certain organic processes which would otherwise lead to the decomposition of the food.

It appears, however, that while this is being achieved other processes may be initiated and ultimately lead to undesirable changes in the properties of the foods. The author expresses this by saying, in particular in reference to fruits and vegetables, that if these latter are to be preserved in any semblance of their normal selves they must be kept alive, and their preservation as living organisms must in principle depend on the selection of some agent which will lengthen the normal duration of their life.

An apple, like every portion of living matter, is an internal-combustion motor constructed to work over a rather remarkably large range of temperature. It is true that the moving parts are small, being, in fact, chemical molecules. Once plucked, an apple is like a clock wound up and will go only for a certain period. It will die when the chemical changes it is wound up to perform are completed and will also die if the normal progress of those chemical changes is interfered with sufficiently.

To lengthen the duration of the life of a fruit, we take advantage of two features of living matter considered as a machine, namely, the temperature coefficient of the vital processes, and the fact that its rate of working is determined within limits, first by the supply of oxygen and next by what an engineer might call the back pressure of the products of its own chemical changes. The temperature coefficient is of the same nature as that affecting a large number of chemical changes where the processes are being retarded within limits by cold and accelerated by heat. Hence the use of cold storage to keep fruit.

The final products of chemical changes in living organisms are carbonic acid and water. If the analogy between living matter and an internal-combustion engine is sound, prevention of escape of the carbonic acid should, broadly speaking, produce in fruit an effect comparable to that produced when the free escape of the exhaust from a gasoline engine is hindered. The engine is slowed down, and this is what actually happens to fruit. Apples or other fruit placed in a chamber from which the carbonic acid cannot escape do actually ripen much more slowly than similar apples placed where there is free escape. The combustion motor may also be slowed down (and the ripening of fruit delayed) by diminishing the supply of oxygen, and this is the scientific principle which underlies the familiar practice of pitting potatoes. However, interference with normal processes brings about a train of new phenomena.

If the exhaust of an internal-combustion engine is choked or if it is starved of air, the whole series of chemical changes which occur in the cylinder are modified, as our sense of smell informs us. Fruit behaves in precisely the same way. If its normal respiration is interfered with, the intrinsic chemical changes are not merely delayed but modified and may be modified to such an extent as to render the fruit totally unfit for an article of food. Combustion in the fruit, as in the gasoline engine, is incomplete. In the former, the products of incomplete combustion take, among other things, the form of alcohol. Thus, strawberries preserved in this way are apt to develop amyl alcohol and in such quantities as to give them the flavor of a pear drop.

The cold storage of meat is discussed by the author from the same biological standpoint. (*The Cold Storage and Ice Association*, vol. 16, no. 1, 1919-1920, pp. 23-35 and discussion 36-45, *tA*)

## THERMODYNAMICS

### New Thermodynamic Cycle Applicable to Internal-Combustion Engines

NEW THERMODYNAMIC CYCLE, Wm. J. Walker. Description of a thermodynamic cycle especially applicable to internal-combustion engines.

The author claims that the usual cycles based on the formula

$$\text{Thermal efficiency} = \eta = 1 - \frac{1}{r^{\gamma-1}} \dots \dots [1]$$

where $r$ is the adiabatic compression ratio, give an impression that engine efficiencies are limited wholly by the compression ratios with which they operate, which is not necessarily always the case.

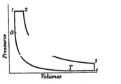

FIG. 12 DUAL COMBUSTION CYCLE FOR INTERNAL-COMBUSTION ENGINES

In the present investigation the thermodynamic cycle shown in Fig. 12 has been chosen to represent the most general type. Cycles of this type are termed by the writer "dual combustion cycles," on account of the fact that the heat is imparted to the working fluid by internal combustion both at constant volume and constant pressure. The problem which the writer considers is the determination of the maximum efficiency condition for a cycle com-

prised between given pressure and volume limits, assuming heat to be supplied either at constant volume, constant pressure, or both.

The operations concerned in the cycle of Fig. 12 are:

From $T$ to 0, adiabatic compression
From 0 to 1, heat given at constant volume
From 1 to 2, heat given at constant pressure
From 2 to 3, adiabatic expansion
From 3 to 4, rejection of heat at constant volume
From 4 to $T$, rejection of heat at constant pressure.

The efficiency of the cycle is given by

$$\eta = 1 - \frac{\text{Heat rejected}}{\text{Heat received}}$$

$$= 1 - \frac{(T_3 - T_4) + \gamma(T_4 - T)}{(T_1 - T_0) + \gamma(T_2 - T_1)}$$

After several operations the author obtains the following equations:

$$\eta = 1 - \frac{A - \gamma'}{B - r^{\gamma}} \dots \dots \dots [2]$$

and

$$(\gamma - 1)r^{\gamma} - Ar^{\gamma-1} + B = 0 \dots \dots \dots [3]$$

where $A$ and $B$ are defined by

$$A = c^{\frac{1}{\gamma}}\left\{\varrho^{\gamma} + (\gamma - 1)\right\}$$

and

$$B = c\left\{1 + \gamma(\varrho - 1)\right\}$$

and $c$ is a constant for given pressure limits and is equal to $ar^{\gamma}$.

The solution may be obtained either graphically from Equation [3] or by curves from Equation [2] showing the relationship between $\eta$ and $r$ for different values of $\varrho$. These curves are shown in Fig. 13, the value of $\gamma$ being taken as 1.3.

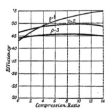

FIG. 13 CURVES SHOWING THE RELATIONSHIP BETWEEN THERMAL EFFICIENCY AND ADIABATIC COMPRESSION RATIOS FOR DIFFERENT ENGINE TYPES

The curve where $\varrho = 1$ is the efficiency curve for the Atkinson cycle. The author points out, however, that this cycle is the limiting case of a series of cycles, all of which have a definite maximum efficiency at a compression ratio less than the compression ratio required to compress the gases to the maximum pressure of the cycle. (*Engineering*, vol. 109, no. 2832, Apr. 9, 1920, p. 467, 2 figs., *tA*)

## CLASSIFICATION OF ARTICLES

Articles appearing in the Survey are classified as *c* comparative; *d* descriptive; *e* experimental; *g* general; *h* historical; *m* mathematical; *p* practical; *s* statistical; *t* theoretical. Articles of especial merit are rated *A* by the reviewer. Opinions expressed are those of the reviewer, not of the Society. The Editor will be pleased to receive inquiries for further information in connection with articles reported in the Survey.

# MECHANICAL ENGINEERING

### THE JOURNAL OF THE AMERICAN SOCIETY OF MECHANICAL ENGINEERS

Published Monthly by the Society at
29 West Thirty-ninth Street, New York

Fred J. Miller, *President*

William H. Wiley, *Treasurer*  Calvin W. Rice, *Secretary*

PUBLICATION COMMITTEE:

George A. Orrok, *Chairman*     J. W. Roe

H. H. Esselstyn     George J. Foran

Ralph E. Flanders

PUBLICATION STAFF:

Lester G. French, *Editor and Manager*

Frederick Lask, *Advertising Manager*

Walter B. Snow, *Circulation Manager*
136 Federal St., Boston

Yearly subscription $4.00, single copies 40 cents. Postage to Canada, 50 cents additional; to foreign countries $1.00 additional.

*Contributions of interest to the profession are solicited. Communications should be addressed to the Editor.*

On the date when this number of MECHANICAL ENGINEERING is issued, the Spring Meeting of the A.S.M.E., for which its members and friends at St. Louis have made such careful preparation, will have been concluded and a goodly number, it is hoped, will be en route to Tulsa for the meeting in that city and the inspection of the great oil fields. An account of both meetings will appear in MECHANICAL ENGINEERING for July. Long may the memories of these meetings linger in the minds of those who are so fortunate as to attend!

### Standardization of Screw Threads

The Congressional Screw Thread Commission, an abstract of whose report appears elsewhere in this number, has accomplished a fine piece of standardization work worthy of the careful consideration of all manufacturers of machinery. The Commission has adhered to the recognized thread standards already in use in this country, namely, the United States standard and A.S.M.E. machine-screw standard for coarse threads; the S.A.E. standard for fine threads; the Briggs standard for pipe threads; and the standards recommended by the National Fire Protection Association and the Bureau of Standards for fire-hose threads.

It is well and good to specify standardized sizes and forms, but such are of very little avail unless means are provided for the carrying out of these standards in practical industrial work. It is here, therefore, that the chief value of the Commission's efforts are to be found. A great deal of the report is devoted to the question of tolerances and the subject of gaging. Every effort has been made to establish tolerances which are reasonable and workable, and the Commission has gone to the extent of recommending different limits for screw threads for different classes of machinery, ranging from the heavy and more crude machines to the light and delicate types requiring the finest workmanship.

It is due to the Commission that a fair trial of these tolerances be made by manufacturers, as only by such earnest coöperation can the industries of the country derive the benefit which is sure

to come to them if workable screw-thread standards can be brought into general use.

Members of the A.S.M.E. are acquainted with the painstaking work previously accomplished by their own committee on screw thread tolerances; and it is a pleasure to know that this committee's investigations were extremely useful in the Screw Thread Commission, particularly in the facts which they brought out with respect to the influence of variations of lead on screw-thread fits, which are equally as important in determining limits as the question of variations in pitch diameter.

### Modern Coast Defenses

The naval and coast-defense problem of this country receives considerable light from the description of one of the most powerful coast defenses in the world given by Lt-Col. H. W. Miller in this issue's leading article. The fact that the Allied fleet with all its strength was not able to reduce the German batteries after three years' bombardment proves again the superiority of land fire over gun fire from floating vessels, while the extensive use of anti-aircraft guns prevented the accurate dropping of bombs. The location of the large batteries at considerable distance from the shore increased the difficulty of the already difficult problem for the naval gunner of hitting a target of low visibility from an unsteady platform. The adoption of railway mounts and use of smoke screens for coast defense by the Germans for their Belgian fortifications are points that can well be used by the engineers who will lay out the future coast defenses of this country. The simplicity of construction and scheme of location of these defenses are also important to the United States.

### Special Bulletin on Society Affairs

A part of the matter which ordinarily would have been contained in Section Two of this number of MECHANICAL ENGINEERING, including the Positions Available of the Employment Bulletin, has been issued to the membership of the A.S.M.E. ten days in advance of the publication date of this Journal, in the form of a special bulletin. The reason for this is explained in a brief article in the bulletin referred to under the heading " How Do You Like It? " which is quoted herewith.

A " speed up the service " policy has been advocated by the Committee on Local Sections of the A. S. M. E. and backed by the Council. There are many things about the work and accomplishments of the various committees of this and other societies which you as a member ought to know about and *know about promptly*. Part Two of MECHANICAL ENGINEERING. issued each month, tells the story but does not tell it soon enough. It requires nearly a month to publish and distribute a journal of the size of MECHANICAL ENGINEERING. In the case of the Employment Bulletin particularly this is a serious matter. Those seeking positions are entitled to prompt advice about new openings. By means of a small Bulletin like the present one, issued occasionally to supplement Part Two of the regular Journal, this much-to-be-desired quick news service can be given. This first copy of the proposed Bulletin has been brought out at the request of the Committee on Local Sections as a " Spring Meeting Special," and for the purpose of trying out the idea among the membership.

Would you like to have it continued—occasionally?

Write the Secretary.

### James W. See—An Appreciation

Few of the younger generation know much of one of the outstanding figures in the world of mechanical engineering and one of the most appreciated members of our American Society of Mechanical Engineers. The main facts of Mr. See's career are correctly enough given in the condensed biography in the April number of MECHANICAL ENGINEERING; but were that biography expanded a thousand fold, as it might well be, it would still remain but a cold recital did it not take account of Mr. See's, or, to give his old nom de plume, Chordal's essential humanity, clearheadedness and logic so rarely found coupled, as they were with him, with an underlying kindness that was always at the disposal of any seeker for help. The writer for one will never forget when, in his early days, many a problem that appeared to him almost unsolvable before he took it to Chordal, became so apparent after a few minutes' talk, that he was invariably surprised at himself for thinking it at all abstruse or difficult; yet never did Chordal convey the impression of imparting information. Recalling this in later years, after a riper experience, the writer is convinced that See had that quality of ingrained common sense that penetrates at once and with no apparent effort to the root of any problem. All those who were privileged to come into personal contact with See, and there were many, felt that, even though their ways might have diverged widely over the years, with his passing the world, and not only the engineering world, had lost one of its kindliest and most helpful figures.—HENRY HESS.

## Engineering Foundation Seeks Large Endowment

The Engineering Foundation, created by an endowment of $300,000 from Ambrose Swasey, past-president and honorary member of The American Society of Mechanical Engineers, is now actively seeking additions to this endowment which will raise the total to at least a million dollars.

Through Mr. Swasey's generous gifts the Foundation has maintained since 1915 a liaison between engineers, as represented by the founder and other societies, and scientific workers, as represented by the National Research Council. Practical means for coöperation in research thus exist for the engineers in the numerous branches of the profession to join with the workers in other fields of science, in the attack upon problems of common interest and in the exchange of knowledge.

Engineering Foundation seeks to build up its endowment to dimensions worthy of the profession, for progress will be made approximately in proportion to the funds available. Besides problems relating to the materials and forces of engineering, many acute social and economic questions of our day need the patient study of scientists and technologists.

It is the purpose of the Engineering Foundation to stimulate, coördinate and support research work in existing scientific and industrial laboratories, coöperating, in so far as may prove advantageous, with the National Research Council. It does not plan to build laboratories and conduct research work directly.

Mr. Charles F. Rand, past-president of United Engineering Society, and of the American Institute of Mining and Metallurgical Engineers, is chairman of Engineering Foundation.

The office of the Foundation is in Engineering Societies Building, 29 West 39th Street, New York, and further information may be had by addressing that office. A booklet giving an account of the Engineering Foundation and its work will be mailed upon request.

## The Engineering Index for 1919 Just Issued

The 1919 annual volume of The Engineering Index, the first to be compiled by the staff of The American Society of Mechanical Engineers, is now ready for distribution. This year's volume is a book of 528 pages and contains over 12,000 references to the engineering and allied technical publications of the world. These have been selected from over 700 periodicals representing some twelve languages, thus making the index a comprehensive guide to the engineering literature of 1919.

In addition to containing more references than any previous annual volume, the items of this year's Index are arranged in alphabetical or dictionary form, instead of being grouped under the divisions of engineering as has heretofore been the custom. This feature makes it possible for one to turn directly to the subject claiming his particular attention, instead of being put to the necessity of first classifying it as civil engineering, electrical engineering, mechanical engineering, etc. The price of the work is $4.

The Engineering Index is the continuation of a work started in 1884, when Prof. J. B. Johnson, of Washington University, St. Louis, began to regularly index for the Journal of the Association of Engineering Societies the articles appearing in the leading current periodicals of that time. The undertaking proved

highly successful, and at the end of five years Professor Johnson compiled and published the first volume.

As the work developed it seemed advisable to combine it with a somewhat similar service then being rendered by the Engineering Magazine Company (New York), and accordingly in 1895 the second volume, while edited under the direction of Professor Johnson, was published by that concern. From then on until the close of 1918 The Engineering Index was regularly published by the Engineering Magazine Company.

During these years the Index was greatly enlarged in its scope and the number of periodicals indexed was increased to about 240. At the close of 1918 The Engineering Index was acquired by The American Society of Mechanical Engineers. It was combined with the Selected Titles of Engineering Articles then appearing in the Society's monthly Journal, and the number of periodicals indexed increased threefold. The Index is a regular feature of MECHANICAL ENGINEERING.

## Resignation of Dr. Manning as Director of Bureau of Mines

Dr. Van H. Manning, Director of the Bureau of Mines, Department of the Interior, resigned on June 1 to become Director of Research of the recently organized American Petroleum Institute, the most important body of petroleum men of the country. In his letter of resignation to President Wilson, Dr. Manning said:

In leaving the Government service there comes to me, as it has over and over again, the thought that although this Government spends each year many millions of dollars in useful scientific work for the benefit of the whole people, the monetary recognition of its scientific and technical servants is not sufficient to enable them to continue in the service for the people. This has been especially true within the last few years when it has been impossible for many men to remain in the Government service.

With the marvelous expansion of the industry in this country and the growing necessity of science to industry, the scientific bureaus have been utterly unable to hold their assistants against the competition of industry which is taking their highly trained men at salaries the Government does not pay or even approach.

These words of warning bring out prominently the serious handicap under which the work of the Governmental departments is now being carried on. There is no more pressing need under the present conditions of unrest than the stabilization at least of the departments of the Government where accomplishment is dependent upon the maintenance of a strong organization.

In regard to Dr. Manning personally and his connection with the Bureau of Mines, Prof. O. P. Hood, chief mechanical engineer of the Bureau of Mines, writes as follows and it is a pleasure to publish this testimonial from one of Dr. Manning's most prominent associates:

"Mr. Manning's handling of the Bureau's work has been characterized by administrative energy and decision. His habit of quick decision has kept things moving. There has been no radical change of policy in the Bureau's objectives but a steady healthy growth and enlargement of fields of usefulness. The Petroleum and Natural Gas Division has been largely developed during his administration, and the new work which he enters is but a continuation of similar effort.

"Mr. Manning has greatly developed schemes of coöperation with individuals and industries whereby governmental and private organizations having the same objectives in research, work together and make available to the public information which would otherwise be had by few.

"His foresight in marshalling the chemical investigators of the country around a small section of the Bureau's work when chemical warfare was forced upon us is characteristic. What finally developed into the Chemical Warfare Service was initiated and administered for fifteen months as a part of the Bureau of Mines. Similar foresight and energy was shown in utilizing the Bureau for war-time service in several fields.

"His methods of management have been such as to win the loyal support of the technical and administrative men who regret to see him leave, but who rejoice that he finds more remunerative work."

Dr. Frederick G. Cottrell, the new director succeeding Dr. Manning, became connected with the Bureau of Mines in 1911 and during the past year has been assistant director. Several years ago he evolved what is known as the Cottrell process for the electrical precipitation of fume and fine particles suspended in the gases from metallurgical furnaces and cement works. This process has been applied successfully in many large plants in different parts of the world. In a desire to encourage scientific research, Dr. Cottrell turned over his patent rights to a non-dividend paying corporation, known as the Research Corporation, a body formed for that purpose. A fundamental requirement in the incorporation is that all net profits shall be devoted to the interests of scientific research.

Aside from his work on smelter smoke Dr. Cottrell has been deeply interested in and intimately connected with work on the separation and purification of gases by liquefaction and fractional distillation. During the world war and subsequent thereto the development of the Norton or Bureau of Mines process for the recovery of helium from natural gas has been his special care, and it was chiefly through his efforts that a plant for recovering helium on a large scale for military aeronautics has been erected near Petrolia, Texas.

## Cruising Radius of Aeroplanes

In view of the somewhat loose talk of crossing the Pacific by aeroplane in a non-stop flight, it may be of interest to call special attention to the recent paper by Dr. A. Rateau, Hon. Mem. Am.Soc.M.E., in the May issue of MECHANICAL ENGINEERING and the report made by Dr. J. Coffin to the National Advisory Committee for Aeronautics, of which a brief abstract (preliminary to publication in the Annual Report of the Committee) appears in the present issue.

The two investigators came to values which do not materially differ from each other. Doctor Rateau investigates the question of maximum cruising radius more directly than does Doctor Coffin, and the conclusion to which he comes is that an aeroplane of the present type with a non-supercharged engine can cover 5000 km. or 3100 miles certainly, 6000 km. or 3700 miles possibly, but that it is very doubtful whether it can fly 7000 km. or, say, 4350 miles.

Even these cruising radii, however, are predicated on the condition that the flight should occur very near the ceiling of the given aeroplane, which means at altitudes ranging in the proximity of 20,000 ft. We are all familiar from the experience of Major Schroeder and of Rohlfs with the discomforts of flight at very high altitudes and it is rather difficult to expect that a flier would have the endurance to remain for a period of many hours at a stretch at these extremely high altitudes without the artificial stimulants or artificial protection from the elements which hitherto have alone made possible high-altitude flying.

It should borne in mind in this connection that the figures of Rateau are based on the assumption that the aeroplanes would not carry a single pound of weight not absolutely necessary for the purpose of covering the greatest possible distance. The presence of oxygen tanks or similar devices would mean materially added weight and would cut the cruising radius to a corresponding extent. Enclosed compressed-air chambers for aviators would also cut down the cruising radius, though in ways different from the addition of compressed-air tanks. It would therefore appear that with the Brown and Alcock flight from Newfoundland to Ireland across the Northern Atlantic, we have already come fairly close to the maximum practical range of the present aeroplane and can look forward to increasing it only through essential changes in the construction of either the aeroplane or the engine, or possibly through the development of a fuel of greater heat capacity than any we have at our disposal today.

An important limitation to the formulae developed by Doctors Rateau and Coffin for the maximum cruising radius of aeroplanes, is that they apply only to flight in calm air.

The remarkable high-altitude flights of Major Schroeder and of Rohlfs have shown that the air at high altitudes is anything but calm. The flights of Major Schroeder have indicated the prevalence of steady "trade" winds of unsuspected velocities, rising to possibly as high as 200 miles per hour, or even more, and it has been stated that the War Department was planning to send a dirigible from San Francisco to the Eastern seaboard, utilizing these high-velocity winds to accomplish the transcontinental trip in what has been estimated to be from eight to ten hours.

It is obvious that if such should be the case and the winds relied on for carrying the airships,—whether heavier or lighter than air is immaterial at such tremendous speeds,—the cruising radius would be correspondingly extended. It would consequently appear not impossible to cross from San Francisco to London in the air in a period no longer than taken by special trains to carry a passenger from Chicago to New York by rail.

## John Fritz Medal Presented to Orville Wright

On May 7, the sixteenth presentation of the John Fritz Medal was made to Orville Wright, Hon. Mem. A. S. M. E., for noteworthy work in the development of the airplane. The ceremony took place in the auditorium of the Engineering Societies Building, Charles F. Rand, former president of the United Engineering Society and past-president of the American Institute of Mining and Metallurgical Engineers presiding in the absence of Benjamin B. Thayer, past-president of the American Institute of Mining and Metallurgical Engineers and chairman of the board of award.

The first speaker was Major General George O. Squier, Chief Signal Officer, U. S. A., who as an officer of the Signal Corps in 1908 presided over the board of officers that prepared the specifications and supervised the acceptance tests of the Wright planes, and in this connection became very closely connected with the Wright brothers. He related the incidents connected with the first flights at Fort Myer and gave the history of the first Wright flights abroad. He paid tribute to the concentration and thoroughness, reticence of speech and capacity for work of the two Wright brothers and called attention to the fact that these characteristics made them great as engineers. He made it clear that the painstaking pioneer work of the Wright brothers furnished the foundation for the rapid and sure development of the airplane and in closing saluted Mr. Wright as the most distinguished engineer in the world.

Edward A. Deeds, former Colonel in the Signal Corps, member of the Aircraft Production Board and a lifelong friend of the Wright brothers, told of the early work of the two boys in Dayton, in a delightfully informal and intimate way. He spoke of their inspirations, their early successes and especially their failures, which made it necessary for them to develop by laborious research the first basic theory for their future work. He dwelt on the thoroughness with which the design of their first planes was consummated, with the result that present designs differ little from the originals of the Wright brothers. In closing he spoke particularly of the remarkable team work with which Wright brothers worked and risked their lives, and of the wonderful sacrifices made by the entire Wright family, to the end that man might fly.

In presenting the medal, Comfort A. Adams, past-president of the American Institute of Electrical Engineers, said in part:

"It is with particular pleasure that we honor one who, through years of patient, painstaking and discouraging research, in the face of almost insuperable obstacles visited with danger to life and limb, finally succeeded in developing a machine that would actually fly. We wish you to know that we are not unmindful of that other whose life was sacrificed in this cause, and that, in spirit at least, this medal is awarded to the Wright brothers."

Mr. Wright responded very briefly, expressing his especial appreciation at the receipt of this mark of great honor from the engineers of the country.

# Government Activities in Engineering

*Notes Contributed by The National Service Department of Engineering Council* [1]

## Senate Considering Water Power Bill

The Water Power Bill was reported back to the Senate and House after agreement by the conferees and the House promptly accepted the conference report after a short debate by a decisive vote of 259 to 30. In the natural course of events this practically assured the early passage of the bill into law but quite a formidable filibuster has developed in the Senate. It now appears that the bill is in serious danger unless the conferees make further changes in it.

The filibuster, it is understood, will be led principally by the Senators from New England who object to the implied federal control of business. It will be recalled that the bill has had a treacherous path through many sessions of Congress and that it was lost by a narrow margin in the filibuster and legislative jam that came at the end of the last session. Those in favor of the measure, however, are in hopes of forcing a vote and bringing the measure to favorable final action before Congress adjourns.

It will also be recalled that after the passage of this bill by both Houses considerable delay resulted in conference, due to the inability of the conferees to agree on points of difference concerning principally the appointment of an executive secretary, definition of navigable waters, and limitations for awards on severance damages incurred on recapture of property at the end of the fifty-year license period. The original provision for severance damages as provided in the first House bill was restored by the conferees. Some are of the opinion that the definition of "navigable waters" as given in the conference report will make it impossible for any one to build a dam for any purpose in any creek or small stream which empties into a navigable river without first obtaining a license from the Federal Power Commission.

## The Sundry Civil Bill For 1921

At present the Sundry Civil Bill, introduced on the floor of the House on January 29, is undergoing discussion in that body. This bill carries many items of interest to engineers, those dealing with the investigation of the Boston-Washington super-power project and an increased appropriation providing for early completion of the topographic map of the United States, being of special importance.

It will be recalled that Engineering Council took a particularly active part in urging an appropriation for the power investigation so that that work could go forward at once. The Council's committee composed of William S. Murray, Prof. D. C. Jackson, Prof. L. P. Breckenridge and M. O. Leighton, appeared before the House Committee on Appropriations in behalf of this measure. Mr. Leighton also appeared for Engineering Council in behalf of the increased topographic mapping appropriation. Both of these matters have come up for consideration on the House floor and have remained in the bill, although there has been some opposition to the power investigation in the House discussion. The former appropriates $125,000 to the Geological Survey to carry on the investigation. The mapping appropriation carries $330,000 as compared with $325,000 for last year. In addition to the latter amount, $100,000 will be available from the Army for topographic mapping work.

Other items of interest to engineers that have remained in the Sundry Civil Bill after House discussion are: California Debris Commission, $15,000; enforcement of anti-trust laws, $150,000; new mining experiment stations, $175,000; petroleum and natural gas investigations, $135,000 (in all for the Bureau of Mines, $1,-277,542); Bureau of Standards, $87,272; Reclamation Service, $7,898,000; mineral resources work in Geological Survey $125,000 (in all for the Geological Survey $1,655,700); Emergency Shipping Board, $70,000,000; maintenance of Panama Canal, $9,281,-851; National Advisory Committee for Aeronautics, $200,000; flood control on Mississippi, $6,670,000; and prevention of obstructions and deposits in New York harbor $109,260.

## Ordnance Department Establishes New Engineering Staff

The Ordnance Department has established a staff of highly trained engineers under the direction of General C. L'H. Ruggles as chief of Technical Staff, whose function it shall be to study all of the technical problems of the Ordnance Department which seem worthy of development.

Through the National Service Department of Engineering Council, General Ruggles has requested that engineering societies throughout the country consider the following problems the solution of which will be beneficial to both the Ordnance Department and to many manufacturing industries:

*Machinability of Metals.* This question directly affects the production and output of a shop. Much time can be lost due to indiscriminately mixing articles of different machinability which are to go through the same machine operations. Maximum production can only be secured through the control of the character of the metal sent through the machine shop.

A test to determine machinability and thereby regulate the distribution of parts to be machined should be developed into a simple practical shop test. This test should result in separating materials of different machinability into different groups and each group could then be sent to the machines best adjusted to handle it.

*Nomenclature of Metals.* The methods of nomenclature of metals vary from shop to shop, from industry to industry, from society to society, and from country to country. The British Institute of Metals has been at work on the problem for a number of years. Several engineering societies in America have made individual attempts to solve the problem. The Bureau of Standards has a system of its own for designating metals. To date nothing good enough has been developed to support either as an international or a national standard.

It is believed that by proper correlated and concerted effort a simple comprehensive system can be evolved which will be broad and expansive enough to allow all non-ferrous alloys and metals to come within its scope. It is further believed that by similar methods a system can be evolved for ferrous metals.

Suggestions for the solution of these problems will be welcomed by the Secretary.

## A New Road Bill

The Senate Committee on Post Office and Post Roads began hearings on the Townsend Road bill on May 4. This bill proposes to create a Federal Highway Commission having charge of all Federal road work. The hearings will attempt to develop whether or not Congress should discontinue its present system of Federal aid to roads, or whether to appropriate larger sums for use after 1921, and whether it is to make a change in all its road policies.

So far the hearings have not developed sentiment in favor of stimulation of road construction at this time, due to the great need for labor on farms and in the industries. Some of the witnesses have argued that this is a good time to formulate definite and more coherent road-building policies. Practically every witness has favored the construction of special highways to be built and maintained by the Federal Government.

From the engineers' standpoint, the purpose of the Townsend Bill is most commendable, but it is pointed out that the plan for a National Department of Public Works contemplates that the Bureau of Roads will be included in the Public Works Department, which means that the formation of road policies and the administration of those policies will come under suitably qualified direction that will not only accomplish all that the Townsend Bill proposes, but at the same time will coördinate this work with the other similar functions of the Government without creating another separate commission.

[1] Engineering Council is an organization of national technical societies created to consider matters of common concern to engineers as well as those of public welfare in which the profession is interested. The headquarters of Engineering Council are located in the Engineering Societies Building, 29 West 39th Street, New York City. The Council also maintains a Washington office with M. O. Leighton, chairman of the National Service Department, in charge. This office is in the McLachlen Building, 10th and G Streets, Washington, D. C. The officers of Engineering Council are: J. Parke Channing, *Chairman*; Alfred D. Flinn, *Secretary*.

365

# NEWS OF THE ENGINEERING SOCIETIES

Meetings of the Engineering Section of National Safety Council, Taylor Society, American Gear
Manufacturers' Association, National Metal Trades Association and the Western Society of Engineers

## Standardization of Industrial Safeguards

The first conference of the Engineering Section of the National Safety Council was held in New York on April 27. David S. Beyer, vice-president and chief engineer of the Liberty Mutual Insurance Co., and one of the founders of the Engineering Section, explained the activities which this new section of the National Safety Council will undertake. . About ten years ago the states began to pass laws on safety in industry which naturally were very general and which were left in most cases to the interpretation of inspectors. Later on efforts were made to formulate adequate and reasonable standards, but even these were not always perfect and were more or less in conflict with one another, and with the requirements of insurance companies. It is with a view to harmonizing the conflicting points of view of the engineers, the insurance companies and officials in industry that the Engineering Section of the National Safety Council has been formed. The work will be undertaken under the general control of the American Engineering Standards Committee, and the Engineering Section will engage in the active codification of industrial safeguarding.

## Taylor Society

A meeting of the Taylor Society was held at Rochester on May 6, 7, and 8, under the auspices of the industrial management council and the manufacturers' council of the Rochester Chamber of Commerce.

One of the professional papers, Promulgation of Standards by the Taylor Society, by William O. Lichtner, contained an interesting suggestion in regard to enlarging the field of activities of the society. It proposed to standardize the terminology involved in the operation of the Taylor system, to define clearly the functions and executive titles of a standard oragnization, to prescribe definite policies on bonus payments, base rates and total earnings, and to compile a list of reference books on industrial management. As a preliminary step in this direction and with a view to stimulate discussion, Mr. Lichtner submitted a list of standard definitions of such terms as apportioning stores, available future, board ticket, base rate, day work, employee's time sheet, idle machine ticket, job, etc., and a bibliography on industrial management containing over 100 references. He also outlined some general principles as a basis upon which a definite policy of remunerating employees could be constructed. These principles refer to the adoption of a standard time for all major operations and setting a base rate for each operator.

Honorary membership was conferred on Carl G. Barth, of Buffalo, N. Y., who is the third person thus honored by the society. The other two honorary memberships were conferred on the late Frederick W. Taylor and Henri Le Chatelier.

## American Gear Manufacturers' Association

Important committee reports on standardization work were presented at the fourth annual meeting of the American Gear Manufacturers' Association, held in Detroit from April 29 to May 1. The report of the general standardization committee called the attention of the Association to the American Engineering Standards Committee organized as a clearing house for the standardization work of this character. It referred to the acceptance by the association of joint sponsorship with The American Society of Mechanical Engineers for the standardization of gears of all kinds under the American Engineering Standards Committee. A sectional committee is to be formed by these organiza-

tions to formulate gear standards which, if found to meet requirements, will in time be known as the American standard.

The bevel and spiral gear committee submitted for the consideration of the members a table giving the maximum addendum for bevel gears based on a back cone radius of 1 in. and covering 14½-deg. and 20-deg. obliquity, with ratios of from 1 to 1 up to 1 to 8.

The report of the sprocket committee contained tables of dimensions for roller-chain sprockets, relative position of wheels, speeds, and approximate speed ratios and sprocket diameters for single-width roller chain wheel with chain of different pitches. The standardization of herringbone gears and of worm gears was discussed in connection with the preliminary work in those directions contained in the reports of the respective committees.

The hardening and heat-treating committee submitted a list of 15 kinds of forged and rolled steel suitable to the gear industry. The first steel was a basic open-hearth or bessemer steel which was suggested as a good steel for a cheap class of work, being somewhat superior to screw-machine stock and possessing good machine, heat-treating and case-hardening qualities. The percentage analysis of this steel is: Carbon, 0.15 to 0.25; silicon, maximum, 0.25; manganese, 0.60 to 0.90; sulphur, 0.06 to 0.09; and phosphorus, maximum, 0.06. A number of the other steels listed conformed to the specifications of the Society of Automotive Engineers, and some had slightly different specifications. The suggestion was made in the discussion of this report that the Society of Automotive Engineers' specifications be followed as closely as possible to avoid confusion.

The Hump Method of Steel Treating was the subject of an address by G. W. Tall, Leeds and Northrup Co., Philadelphia. Mr. Tall stated that the cost of using the Hump process was 1½ to 2 cents per pound for hardening and 2 cents for drawing, or about the same as with the use of oil or lead pots. During the discussion, Mr. Peterson, Packard Motor Car Co., asserted that steel treated by the Hump process shows an elastic limit of 222,000 lb. per sq. in. as compared with 201,000 lb. per sq. in. when pieces are treated in a lead pot.

Other professional papers were Gears from the Purchasers' Standpoint, by D. G. Stanbrough, Packard Motor Car Company, and Routing of Gears and Machine Parts Through the Factory, by J. A. Urquhart, Brown & Sharpe Manufacturing Company.

## National Metal Trades Association

The need for increased production in industry was emphasized at the twenty-second annual convention of the National Metal Trades Association, held in New York on April 21 and 22. President J. W. O'Leary insisted that systems of payments based on production are "the only corrective of the prevailing fallacious idea that wages ought to be based on the cost of living." He said that careful studies of the consumption requirements of the market open to industry both in the United States and the world were necessary before it could be properly determined whether the work day could be shortened or must be lengthened.

With the aid of numerous charts, M. W. Alexander, general manager of the National Industrial Conference Board, Boston, compared the advance in wages to the high cost of living. The National Industrial Conference Board, he asserted, had found from authentic sources that to March 1920 the average cost of living had increased 95 per cent from the July 1914 level. In that time clothing showed the greatest advance and was still tending upward. He demonstrated, however, that the high cost of living has lagged behind the cost of wages in leading industries. A chart giving a comparison of changes in hourly and weekly earnings with changes in the cost of living, taking the condition in 1914 at 100 points as a basis, showed that hourly earnings in the metal-manufacturing industry increased from 100 to 216 points, weekly earnings from 100 to 233 points, and the cost of living from 100 to 195 points. Another chart giving a comparison of index numbers of changes in average weekly earnings of male workers in different industries with those applying to the cost of living, showed that earnings in the metal trade rose from 100 to 223 points, in the cotton industry from 100 to 244 points, in the wool industry from 100 to 254 points, in the silk industry from 100 to 216 points, in the boot and shoe industry from 100 to 212 points, while in the cost of living it rose from 100 to 195 points.

Dr. Richard H. Waldo, of the Inter-Racial Council, New York, asserted that there is today in the United States a shortage of labor amounting to more than 5,000,000 workers. He further said that the policy of the American Federation of Labor to keep the supply of labor down to the smallest possible amount would ultimately result in workers being detained in foreign countries to manufacture goods for the United States. He urged that the National Metal Trades Association exert its efforts in connection with securing the adoption by the United States of a well-considered policy of immigration.

There was a general discussion of the question of shop representation with arguments both in favor of and against such representation. In each instance the arguments set forward were warmly approved, a circumstance which clearly indicated that the question is still an open one among the members. That such is the case was admitted by President O'Leary, who observed that the question was too broad to be settled easily and expected that it would be discussed again at the next convention.—Abstracted from report in *Iron Age*, April 29.

# Western Society of Engineers Entertains Boards of Civil, Mining, Mechanical and Electrical Engineers

IT is evident to those who have kept in touch with the work of the engineering organizations that the last two years have witnessed profound changes in the relationship of the societies to each other. Coöperation between all the societies there has always been, but this recent period has seen the birth of a real movement for society unity, the embryo of which has already taken shape and is sufficiently fashioned to enable its advocates to " show it in public " in Washington at the Organizing Conference to be held there June 3 and 4.

Sympathetic with this movement for society unity, the Western Society of Engineers invited the governing boards of the four national engineering societies to be its guests in the western metropolis on April 19 and 20, and incidentally, as the account of the meeting will show, provided the psychological opportunity for securing the unity of at least these four societies in the plans for the Organizing Conference.

Three of the boards—the A. S. C. E., A. I. M. E. and the A. S. M. E.—were able to hold regular meetings and transact their usual business while in Chicago. The Electricals had just had their regular meeting in Boston, but members of their board, headed by their President, Mr. Calvert Townley, attended informally.

The meeting of the A. S. M. E. Council was held on Monday morning, and was the regular April meeting. An account of it is given elsewhere under the usual heading of Council Notes.

On the evening of the 19th, the members of the boards were the guests of the Western Society at a banquet at the Hotel La Salle. A. Stuart Baldwin, Past-President of the W. S. E., presided, and introduced the guests of the evening. The officers of the national societies and of the Western Society occupied seats on the platform.

Arthur P. Davis, president of the A. S. C. E., Arthur Fletcher, director of the A. I. M. E., representing Herbert Hoover, president, Fred J. Miller, president of the A. S. M. E., and Calvert Townley, president of the A. I. E. E., were each introduced in turn and made appropriate addresses on engineering organization, emphasizing the main purpose of such organization as development of the engineer for service to the public.

President Fred J. Miller spoke particularly of the public work of the proposed engineering federation, expressing his opinion that engineers should speak as expert witnesses only, and should not forget that the laymen have as much right to their opinion as the engineers in matters of public policy. He spoke also of the need for engineering publicity, developed by the societies appointing staff members to collaborate with editors of the daily press in translating what the engineer is doing into terms the public could understand. His address appears on page 336.

Mr. Townley voiced as the essential problem the development of team work among engineers, to the end of unselfish public service, the ideal of professional organization.

Mr. Davis stated that the great changes which have come into men's minds in the past five years must be reflected in an expansion and recasting of the engineering organizations to keep abreast of them.

Mr. Fletcher laid stress on the engineer's fitness for solving present-day problems, as exemplified in the service rendered to the public by Mr. Hoover.

The meeting was concluded by a masterly address by Dr. Theodore G. Soares, head of the Department of Theology, University of Chicago, who declaimed the traditional modesty of engineers. He held that the engineer must be more than an expert witness. He must prescribe, and order, and see that his orders are carried out. Engineering has a first place in modern life, and grave responsibilities now rest on the engineer.

On the Tuesday the program continued with a luncheon at the University Club, at which were present members of the board of W. S. E., members of the boards of the four national societies, Chicago representatives on Engineering Council and also the Chicago Committees of the local sections of the national societies. Mr. Townley presided at this meeting until he had to leave to catch a train, when his place was taken by Mr. Miller.

E. S. Nethercut, Secretary of the W. S. E., opened with a lucid description of the origin and growth of his society. He enumerated the present extensive committee activities of the organization, in which the field of public service was in no way neglected. He discussed the relation between the local society and the section of the national societies, by means of which effective coöperation had been secured and mutual development consummated.

E. S. Carman, member of the Joint Conference Committee of the National Societies, and Chairman of the Local Sections Committee of the A. S. M. E., was called upon by the presiding officer to describe the plan of society federation proposed in the report of the Joint Conference Committee. Mr. Carman traced the history of coöperative movements, and indicated how many of those in the societies felt that the time for the final step in organization had now come.

During the two days at Chicago, the Civil Engineers were concerned with reconciling the vote of their membership upon so-called " Question 3 " with the desire of the Board to participate in the Washington Conference, and they invited into session the members of the boards of the other societies. The Board eventually evolved a plan enabling the Society to participate in the Organizing Conference in Washington, June 3 and 4.

The President of the Western Society, Frederick D. Copeland, was unavoidably absent, as was also Arthur L. Rice, Chairman of the A.S.M.E. Chicago Section, who was performing railroad strike duty as a member of the Home Guard.

# LIBRARY NOTES AND BOOK REVIEWS

AIRCRAFT YEAR BOOK. Issued by Manufacturers' Aircraft Association, Inc. Published by Doubleday, Page & Co., New York, 1920. Cloth, 6 x 9 in., 333 pp., illus., $2.

The first issue of this annual review of the industry appeared in 1919. The present issue reviews the progress up to date in various fields of aeronautical activity. Aircraft in commerce and in warfare, technical developments between 1914 and 1919, and cross-country flying are discussed and a detailed story of the recent achievements of the firms composing the Association is given. The book also contains the text of the convention relating to international air navigation, the report of the American Aviation Commission, a chronology of the events of 1919 and appendices giving information on governmental activities.

COURS DE MÉCANIQUE RATIONNELLE avec de Nombreuses Applications a l'Usage des Ingénieurs-Cinématique-Statique-Dynamique. By L. Legrand. Ch. Béranger, Paris and Liége, 1920. Cloth, 6 x 10 in., 618 pp., illus., 48 francs.

The author of this textbook believes that there is need for a work which will present the subject in strictly scientific manner, but which will draw its illustrations from the realm of industrial mechanics, rather than from celestial mechanics, as is usually done in theoretical treatises, and offers the present book for this purpose. He has attempted to supply a complete course in which an engineer will find the theory illustrated by problems which arise in the practice of applied mechanics in various industries.

EFFICIENT BOILER MANAGEMENT. With notes on the operation of reheating furnaces. By Charles F. Wade. Longmans, Green & Co., New York, 1919. Cloth, 6 x 9 in., 280 pp., illus., tables, $4.50.

The author of this work endeavors to explain, in their proper sequence, the elementary scientific principles underlying the various subjects combined in boiler management and the systematic practical application of these principles to obtain the greatest efficiency. The book is intended to fill the gap between the treatises upon the chemistry of combustion, etc., in which practical applications are omitted, and the practical textbooks on boiler plants, which give little attention to the fundamental principles governing their operation.

ELEMENTS OF STEAM AND GAS POWER ENGINEERING. By Andrey A. Potter and James P. Calderwood. First edition. McGraw-Hill Book Co., Inc., New York, 1920. Cloth, 5 x 8 in., 297 pp., illus., $2.50.

The object of this treatise is to provide a clear, concrete statement of the principles underlying the construction and operation of steam and gas-power equipment. It is intended for the use of students of engineering with power-plant equipment before they take up the study of thermodynamics and design, and for those responsible for the operation of power plants.

HOW TO MAKE AND USE GRAPHIC CHARTS. By Allan C. Haskell with an introduction by Richard T. Dana. First edition. Codex Book Co., Inc., New York, 1919. Cloth, 6 x 9 in., 540 pp., diagrams, $5.

The object of this book is to call attention to the many functions which graphic methods can accomplish and to indicate the suitability of the different methods of charting for various purposes. After describing the theory and construction of the types of charts, the author gives many examples of those used to aid in organization and management, in analyzing costs and operating characteristics, in recording tests, in predicting trends and tendencies, in computing, designing and estimating. Bibliographies are given with most of the chapters.

MENSURATION FOR MARINE AND MECHANICAL ENGINEERS. (Second and First Class Board of Trade Examinations). By John W. Angles. Longmans, Green & Co., New York, 1919. Cloth, 5 x 7 in., 162 pp., illus., diagrams, $1.75.

This textbook is intended to enable students to pass the examinations of the Board of Trade (Great Britain) for licenses as marine engineers, but will be useful, the author hopes, to engineer-

ing students in other lines. A feature is made of fully solved examples, illustrating the practical applications of the theoretical principles involved in the text.

OPPORTUNITIES IN ENGINEERING. By Charles M. Horton. Harper & Brothers, New York. Paper, 5 x 8 in., 90 pp., $1.

The tremendous power which engineers wield in world affairs has inspired the author to set forth in this book the opportunities for constructive work which lie before the man who selects engineering as his profession. He also describes the type which, being best fitted for the work, is most likely to succeed and gives some hints for the guidance of the student who is choosing his vocation, as well as some examples of what has been done by those already in the work.

ORGANIZING FOR WORK. By H. L. Gantt. Harcourt, Brace & Howe, Inc., New York, 1919. Cloth, 7 x 5 in., 113 pp., $1.25.

Our civilization depends, according to Mr. Gantt, upon the effectiveness with which our combined industrial and business system works, and recent revolutionary attempts to overthrow it are due to the failure of the present system to recognize fully its responsibility to the community. The author believes that there must be a return to the principle that the first aim of business is to render service to the community and that this result can be peacefully obtained by the use of familiar methods, whose use for the purpose he discusses in the present book.

SAFETY FUNDAMENTALS. Lectures given by Safety Institute of America (Maintaining the American Museum of Safety). Safety Institute of America, New York, 1920. Cloth, 5 x 8 in., 228 pp., illus., plates, $2.

Contents: The body which gets hurt.—The injured body and its treatment; (a) Protective clothing for men; (b) Suitable work garments for women in industry.—Safe heads and good eyes.—Guarding machinery.—Arrangement of machinery and working places.—Heating and ventilation.—Illumination.—Nature's forces for and against workmen.—Safety education and shop organization.

These lectures were delivered during 1919 before an audience of factory inspectors employed by the City of New York, the States of New York and New Jersey, and insurance companies in and near New York. They are intended to enlarge the knowledge and increase the experience of inspectors with respect to the various fundamentals that affect the mind and body of the workmen.

THE STORY OF ELECTRICITY. Vol. 1. Edited by T. Commerford Martin and Stephen Leidy Coles. The Story of Electricity Company, M. M. Marcy, New York, 1919. Cloth, 11 x 8 in., 661 pp., $25 for vol. 1 and 2.

The authors of this volume have prepared an account in popular language of the development of the electrical industry, with particular reference to American achievement. After an introductory chapter on the beginnings of electrical science, the invention and growth of the telegraph, telephone, central station, electric railway, etc., are described. Chapters are devoted to the great electrical companies. The various chapters are accompanied by biographical sketches and portraits of engineers of prominence, past and present. Numerous well-selected illustrations add to the value of the work.

TECHNO-CHEMICAL RECEIPT BOOK. Containing Several Thousand Receipts and Processes. Covering the Latest. Most Important and Most Useful Discoveries in Chemical Technology and Their Practical Application in the Arts and the Industries. Compiled and edited by William T. Brannt and William H. Wahl. New enlarged edition. Henry Carey Baird and Co., Inc., New York. 1919. Cloth. 5 x 8 in., 516 pp., illus., tables, $2.50.

The principal aim in preparing this work has been to give a compendious collection of approved receipts and processes of practical applications in the industries. The receipts have been principally derived from German sources and most of them have been tested practically. The present edition has been revised and various receipts added.

# MECHANICAL ENGINEERING
## THE JOURNAL OF THE AMERICAN SOCIETY OF
## MECHANICAL ENGINEERS

| Volume 42 | July 1920 | Number 7 |

# An Experimental Investigation of Steel Belting

Particulars Regarding Belt Material and the Joints Used—Description of Apparatus Employed in
Tests—Determination of Relationships Existing Between Horsepower Transmitted,
Velocity of Slip on Pulleys, and Coefficient of Friction

BY F. G. HAMPTON, C. F. LEH, AND W. E. HELMICK, STANFORD UNIVERSITY, CAL.

*At the Annual Meeting of The American Society of Mechanical
Engineers, held December 2 to 5, 1919, in New York, Student and
Junior prizes were awarded to the authors of the following paper. It
treats of an investigation undertaken by them at Leland Stanford
University as a partial requirement for the degree of engineer. Part
I was written by Messrs. Hampton and Leh in 1918, and Part II by
W. E. Helmick the year following. The first section deals with a
description of the apparatus employed, the character of the belting,
and a discussion of the results obtained in investigating the coefficients of friction and velocity of slip. Part II deals more particularly
with the slip of the belting, which the original investigators recognized should be more carefully studied.*

### PART I GENERAL INVESTIGATIONS

ATTENTION has been called from time to time to the successful use of thin ribbons of steel for belting, and the
purpose of the experimental work herein described was
to determine as far as time would permit the characteristics of
operation and the general laws controlling the performance of
this means of power transmission. The literature on the subject
of steel belts is very limited, and so far as is known there has
been little or no experimental work done.

The controlling features of the design of an apparatus for testing steel belting may be enumerated as follows:

1 The machine must be so constructed that very high speeds may
be obtained, and also means provided to vary the speed over
as large a range as possible

2 Since very high efficiencies are expected, it is necessary to
provide means for measuring, as accurately as possible, the
losses of power which occur in the belt

3 The apparatus must be so constructed that the tensions in the
tight and loose side may be accurately determined

4 On account of the relatively small slip it is necessary to provide
specially constructed apparatus to accurately determine this
variable

5 For the sake of convenience it is necessary to have the machine
built so that belts can be changed easily and different lengths
used.

#### DESCRIPTION OF APPARATUS

Two special high-speed pulleys were constructed upon which
the belt to be tested were run. Since the speed at which cast
iron can be safely run is far below that which was desired, it was
necessary to select a stronger material and so construct the pulleys
that they could be faced with some other material than steel.

The pulleys used were built in the department shops by the
authors and consisted of two boiler-plate disks, a cast-iron hub,
and a wooden rim. The pulleys were made relatively large in
order that high peripheral speeds could be obtained with relatively low shaft speeds. The facing consisted of sheet cork, which

was chosen as the best material on account of the friction properties and its durability.

One of these pulleys, which was to be the driven pulley of the
machine, was keyed directly to the shaft of a 100-hp. Sprague
dynamometer (see Fig. 1), which was used in testing only as
a means of absorbing the power delivered to the driven pulley,
and, although record was kept of the readings of the dynamometer, they were only used as a check and did not enter into the
calculations of the test.

The other pulley, the driver, was mounted on a short countershaft held in two bronze bearings which were cast and finished
specially for the purpose, the bearings being held in a pair of
standard shaft hangers which were bolted to the frame. This
shaft carried the driver pulley on one end, and on the other end

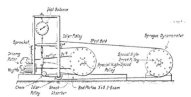

FIG. 1 DIAGRAMMATIC SKETCH OF STEEL-BELT TESTING MACHINE

the pulley with which the shaft was driven by a motor through
a leather belt.

In order to measure the tensions in both sides of the steel belt,
two special idlers were hung behind the driver in such a way that
each one completely reversed the direction of the belt on both
the tight and loose sides between the driver and the driven
pulleys.

In the original design of the machine the straight sections of
the belt were all to be kept parallel by using a small driver, a
large driven pulley and idlers of such size that the sum of the
diameters of the driver and two idlers would be equal to the
diameter of the driven pulley. This arrangement was discarded
because of the work which would have been necessary on the
available apparatus and the belts were run with their straight
sections at an angle. The manner in which this angularity was
corrected for will be discussed later.

The idler pulleys were hung on swinging frames so that they
were free to swing in the plane of middle of the driver and driven
pulleys' faces, but were constrained from moving in any other
direction. They were carried on short shafts with high-grade ball
bearings in order that friction would be reduced to a minimum.

The design of these pulleys was also controlled by the high speed at which they were expected to run, and it was necessary to use a stronger material than cast iron.

Castings were made which formed the hub and a thin, solid web, and rims were made by cutting sections from a piece of heavy lap-welded 10-in. pipe. The castings and steel rings were finished all over, the proper allowance being made for a shrink fit, and the rims heated and shrunk on to the webs, after which they were again machined all over and accurately balanced.

Since it was not considered desirable that the steel belt come in contact with the steel surface of the idlers, even though slippage at this point would be very improbable, the pulley faces were covered with the same variety of sheet cork that was put on the large pulleys.

Throughout the test no trouble was encountered with the cork faces of either the idler pulleys or the large pulleys, and, although speeds as high as 12,000 ft. per min. were attained, there seemed to be no tendency for the cork to fly off.

In order that the pull exerted by the belt on the idlers could be accurately measured, yokes were put on the idler shafts around the idlers, and to these were attached link chains which passed horizontally over ball-bearing sprockets so that, when the swinging frames were in their normal position, the chains were tangent to the arc of their swing and in the same plane. The chain from the idler, over which the tight side of the belt passed, hung vertically downward after passing over its sprocket, and on it was suspended a stem with a plate on its lower end and upon which weights could be placed. The chain from the other idler extended vertically upward after passing over its sprocket, and was attached to the lower part of a dial balance which measured the pull on the loose-side idler.

After starting the test it was found necessary to introduce some kind of a shock absorber to reduce the vibration of the needle on the dial balance, and for this purpose a spring balance of 200 lb. capacity was used. It was also necessary to adjust the length of this chain on account of the variations arising from difference in tension, and to do this a turnbuckle was put in between the dial and spring balances.

The swinging frames, guides, sprockets, and the dial balance were all carried on a rigid superstructure composed of iron bars and angles, and this was firmly bolted to the same I-beams which carried the shaft hangers holding the countershaft, and also the driving motor, so that there could be no relative movement between the various parts of the assembly.

The bedplates which supported this unit consisted of two 8-in. I-beams 16 ft. long, which were set at right angles to the shaft of the Sprague dynamometer and firmly bolted to the floor and shimmed with neat cement. These bedplates formed a guide upon which the driver unit was supported so that this part of the apparatus could be easily moved along the bedplate to accommodate belts of various lengths and without having to disturb the relations between the several elements of the unit.

The motor which was used to supply the driving power was a variable-speed, three-phase induction motor with speeds of 600, 900, 1200, and 1800 r.p.m., and rated at 4, 6, 8, and 12 hp. for these respective speeds. The motor was mounted on an adjustable base so that the leather belt driving the countershaft could be tightened, and the controller was mounted on the sliding frame near the motor.

The motor, which is of the most desirable type for the purpose, was not nearly large enough, and as a consequence the range of experimentation was limited by the available power supply and not by the transmission properties of the steel belt as it should have been.

Since it was necessary to make as accurate a determination as possible of the slip which occurred in the belt at all times, two similar devices were made which would electrically control two speed counters, one mounted on the center of the driver shaft and the other on the driven shaft. Two Veeder revolution counters were used for this purpose, being fixed to the shafts so that their spindles rotated on the same centers as that of the shaft; the remainder of the counter floating on the spindle and remaining

stationary. On the stationary part was suspended a bar which carried an electromagnet so arranged that when current passed through the coils the movement of the armature would actuate a small lever, which slid the floating part of the counter axially so as to engage the dog clutch and count the revolutions of the spindle. Springs returned the counter to the original position and disengaged the clutch when the circuit was opened.

This arrangement gave a fairly accurate measure of speed because both instruments were made alike; and since the coils were in the same circuit and the springs of the same strength, it should give very dependable results. The difficulty, however, was not in the determination of the absolute speed, but of the difference of speed, and, since the counters only registered to the nearest revolutions and the clutches were only two-jaw clutches, there is a possibility and also a probability of an error of one revolution either way, or two revolutions, and when the slip is small, say, only 2 r.p.m., may involve an error in slip of 100 per cent, although the error involved in determining absolute speed is only a very small fraction of one per cent. In order to make accurate measurements of slip, it is evident that a differential counter should be used of such design that it would record to at least 0.10 r.p.m.

The entire testing apparatus as herein described is in itself a transmission dynamometer which measures its own losses. Having measured the tension in both the tight and the loose sides of the steel belt, and the peripheral velocity of both driving and driven pulley, the power delivered to the driving end of the belt may be computed from the value of the net pull, which is the difference between the two belt tensions and the peripheral speed of the pulley face. In a similar manner the power delivered to the driven pulley may be computed from the net pull and the peripheral velocity, the tension in the belt being uniform between pulleys, and the friction or slippage losses may be obtained from the net pull and the velocity of slip.

It is assumed that no work is done or power consumed in bending the belt, and this assumption is substantiated by the fact that the material is perfectly elastic within the limits worked, and whatever power is required to bend the belt is given back to the system when it returns to its normal condition.

In order to further verify this assumption, pieces of the belting were caused to vibrate at a very high rate of speed by holding them on the teeth of a rapidly turning sprocket. Had there been any appreciable internal friction losses present there would have been a noticeable rise in temperature in the part of the steel which was subjected to such rapid bending. Although the pieces became warm at one end from the friction of contact with the sprocket, and on the other end from impact on the material with which they were held, the part which was subjected to the severest bending remained cool.

This leaves the only losses sustained in transmitting power by steel belting to be those due to slippage of the belt on the cork surface, and since this can be reduced to practically nothing by increasing the belt tension, it appears that practically 100 per cent transmission efficiency may be obtained. The experimental results consistently verify this statement.

### BELT MATERIAL AND CORRECTION FOR ANGULARITY

The material which was used for the steel belts in this work was what is known in the commercial world as clock spring. It is very high-carbon steel drawn and rolled, apparently ground to size, hardened, and drawn to a dark-blue color. A rough test showed a tensile strength of slightly over 300,000 lb. per sq. in. and an elastic limit nearly as high. The material in pieces 0.01 in. thick receives no permanent set when bent around a radius of $\frac{1}{2}$ in. and snaps with a clean break when bent around a radius of $\frac{1}{16}$ in.

The chief difficulty in using such material for belting is to get a joint which will develop a sufficient proportion of the strength of the material. Considerable work was done in this connection, and, although no prediction of the durability of the joints used can be made, on account of the short duration of the test, they proved to be entirely satisfactory within the limits used, and under conditions which were probably much more severe than would be found in an actual commercial installation.

Because of the arrangement of the pulleys, and the consequent reversal of the curvature which any part of the belt receives when passing over them, it was impossible to use the standard fastening, which consisted of a strap of a curvature equal to the smallest pulley and to which the two ends of the belt were made fast by means of screws or rivets. A later discussion of the subject of joints will describe the various ones used and show why the plain butted joint held with silver solder was the best one for the purpose.

In order to correct for the angularity of belt pull on the idler pulleys and its effect upon the readings of the dial balance and the dead weights, a method was used which calibrated the exact belt tension directly against the readings of both the dial balance and the dead weights.

A belt was put on which, instead of being spliced, had its two ends joined by a link composed of a turnbuckle and an accurate spring balance. Weights were put on the hanging platform, the swinging arms brought to their normal positions by adjusting the turnbuckle below the dial balance, and readings were taken of the dial balance and the spring balance while slowly rotating the driven wheel first in one direction and then in the other. This method divided the friction, and the average of the two readings gave the true readings which could be plotted against one another and a curve drawn from which the true belt tension could be read for any reading of the dial balance or the dead weights.

### DESCRIPTION OF TESTS

On account of the limited time it was only possible to cover a small portion of the work necessary for a complete test of steel belting, and accordingly the field in which speed, horsepower, and tension are the chief variables was selected for investigation.

A series of runs was made at constant speed and, keeping one value of tension at a constant value, the load was varied from no load to an upper value limited either by the available supply of driving power or by the slippage in the steel belt.

Since the tension in the tight side, $T_1$, was kept constant for any run, it may readily be seen that for an increase in transmitted horsepower the factor $T_1 - T_2$ could be made larger only by decreasing $T_2$, and that when this quantity was reduced to a very small value the loose side of the belt would become very unstable. After covering the range of horsepower possible, the tension was changed and another run made with varying horsepower.

Having covered the desirable range of tensions, the speed was then changed and another series of runs at the new speed was made. In making a run, the apparatus was adjusted to operate under conditions of practically no load, the swinging frames on the idlers were brought into their normal position by adjusting the turnbuckles and readings were taken of r.p.m. of driver and driven pulleys, the dial balance indicating the tension in the loose side and the dead weights the tension in the tight side. In timing the r.p.m., 2-min. intervals were used in order to reduce the error due to starting and stopping the counters.

The instruments indicating the other readings were found to be exceptionally constant, and since each reading was checked by both the parties making the run, only one reading for the most part was entered on the log sheets. By taking readings of the Sprague dynamometer beam, data were made available for a calibration of this machine at low capacity values and which also served as a rough check on the other readings.

An accurate calibration was made of all the instruments used in the test, including the two spring balances, stop watch, and lead weights. The readings taken can be depended upon to have a high degree of accuracy, with the exception of the difference of r.p.m. in driver and driven pulleys. Since this is a very small value and the error involved is in starting and stopping the counters, the percentage of error in the indication of absolute speed is practically zero.

Since the readings of the tight-side tension were made with weights, they are as accurate as the weights. No difficulty was encountered in reading the dial balance to the nearest pound, and

since the idlers were adjusted to their normal position for every change, the readings relative to the belt tensions should be accurate to within one per cent.

The arc of contact was measured for each belt length used. As can be seen, when the distance between centers of driver and driven pulleys is varied there will be a slight change in the arc of contact. Since this change was so small and was used only in connection with the value of slip which as explained before had a large percentage of error, the change in the arc of contact was not considered.

The following will show how the various values used were obtained from the results of the tests:

*Revolutions per Minute.* The difference in the readings of the positive speed counters at the beginning and at the end of the timed interval will give the r.p.m. of the driver and driven pulleys, while the difference between the values thus obtained will give the r.p.m. slip.

$V_s$. On account of a slight difference in the circumference of the driver and driven pulleys, it was necessary to make a correction. This was made by constructing a diagram from which the velocity of slip could be read directly for any r.p.m. slip and any speed.

$V_d$. The driver velocity is obtained by multiplying the driver r.p.m. by its circumference.

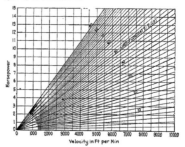

FIG. 2 DIAGRAM FOR COMPUTING HORSEPOWER FROM VELOCITY AND DIFFERENCE IN TENSION

$T_1$. The tension in the tight side of the belt is read directly from a calibration curve.

$T_2$. The tension in the loose side is also read directly from a calibration curve.

$t_1$. The unit tension in the tight side is found by dividing $T_1$ by the cross-sectional area of the belt, or

$$T_1/0.01 \times 0.75 = T_1/0.0075$$

*Driver Hp.* was obtained from the relation:

$$\text{Hp.} = \frac{\text{force} \times \text{distance moved per min.}}{33,000} = \frac{(T_1 - T_2)V_d}{33,000}$$

A chart was made (Fig. 2) from which the horsepower at any speed or any value of $T_1 - T_2$ may be read directly. This was done by plotting the lines $\text{Hp.} = KV_d$ for values of $K$ corresponding to the required values of $T_1 - T_2$.

*Hp. Loss.* This can be obtained from Fig. 2 by using $V_s$ in place of $V_d$ and dividing both horizontal and vertical scales by the same power of ten.

*Efficiency* is computed by taking the ratio of the power input to the belt to its power output:

$$\frac{\text{Hp. driver} - \text{Hp. loss}}{\text{Hp. driver}}$$

$t_c$. The centrifugal tension is determined from the relation:

$$t_c = \frac{W V^2}{g}$$

Fig. 3 shows a plot of this using the proper units and from which $t_c$ for any belt speed may be read.

u The value of the coefficient of friction was calculated from the relation:

$$\log \frac{t_1 - t_c}{t_2 - t_c} = 0.4343\ u\theta$$

where $u$ = coefficient of friction
$\theta$ = angle of contact in radians.

### DISCUSSION OF RESULTS

On account of the uncertainty of the test values of $V_s$ it was first necessary to find, if possible, how this value varied with the others and to discover the law which connected its variation with

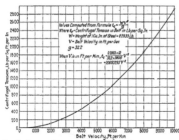

FIG. 3   CURVE OF CENTRIFUGAL TENSION FOR BELT SPEEDS UP TO
10,000 FT. PER MIN.

that of other quantities involved. The first step in this process was to plot the test values of $V_s$ against the corresponding values of hp. This was done over the whole range of the test, one sheet for each constant, $T_1$. Table 1 and Fig. 4 show results and curves for $T_1 = 100$ lb.

After careful consideration of the location of these points, it seemed that the curves for all speeds for any $T_1$ seemed to follow a straight line beginning at the origin and then to break sharply and leave this line, the point of departure depending upon the speed, and being different for different values of $T_1$.

It was also noticed that for high values of $T_1$ the straight line from the origin was much steeper than for lower values. Working on this assumption straight lines were drawn which represented the mean of all points which seemed to locate the line. Although there is a wide variation in the location of these points, it must be remembered that the percentage of possible error in the value of $V_s$ is very large, in fact large enough to justify the moving of practically any of the points to the mean line.

Since it was evident that some simple law existed between the various curves, as shown by the change in slope of the main straight line from the origin for different values of tension, values of $V_s$ were plotted against their corresponding values of tension, at constant horsepower. This was done by taking the values of $V_s$ from each curve at 10 hp. and plotting them against the value of tension corresponding to the curve.

The relation as shown from the plot proved to be a very close approximation to a straight line and accordingly the line was assumed and the lines previously located to represent the mean of the plotted points were so changed as to make their inclination to the hp. axis conform to the law. This correction was considered desirable and even necessary on account of the uncertainty in the values of $V_s$, which has been discussed before.

Data were now available for the construction of a characteristic diagram in which lines of constant hp. were laid down on $V_s$ and $T_1$ axes. (See Fig. 5.) It was noticed on the hp.-$V_s$ curves that

TABLE 1   DATA AND RESULTS OF TESTS ON 0.75 x 0.01-IN. STEEL BELT
($T_1 = 100$ lb.)

| No. Run | R.p.m. Driver | R.p.m. Driven | slip ft. per. min. | $V_d$ ft. per. min. | $T_2$ lb. | Hp. Driver | Hp. Loss | Eff. per cent | u | $V_{ss}$ |
|---|---|---|---|---|---|---|---|---|---|---|
| 21 | 243 | 242 | 4 | 2075 | 46 | 0.16 | 0.0002 | 99.9 | 0.023 | 0.1 |
|  | 242 | 240 | 4 | 2055 | 34 | 0.90 | 0.0008 | 99.9 | 0.122 | 0.8 |
|  | 240 | 239 | 4 | 2048 | 19 | 1.82 | 0.0031 | 99.8 | 0.317 | 1.7 |
|  | 238 | 217 | 176 | 2032 | 3.5 | 2 64 | 0.1200 | 81.4 | .... | 44.0 |
| 16 | 364 | 363 | 4 | 3110 | 46 | 0.235 | 0.0001 | 99.9 | 0.024 | 0.3 |
|  | 363 | 361 | 4 | 3090 | 33.5 | 1.41 | 0.0013 | 99.8 | 0.127 | 1.4 |
|  | 360 | 359 | 4 | 3080 | 20.5 | 2 61 | 0.0043 | 99.7 | 0.300 | 2.5 |
|  | 358 | 357 | 4 | 3060 | 12.5 | 3.32 | 0.0065 | 89.6 | .... | 3.0 |
|  | 357 | 295 | 493 | 3030 | 0.5 | 4.41 | 0.7200 | 83.5 | .... | 246.5 |
| 13 | 485 | 484 | 0 | 4140 | 45.5 | 0.38 | 0.0001 | 99.9 | 0.097 | 0.4 |
|  | 480 | 479 | 0 | 4100 | 32.5 | 1.97 | 0.0022 | 99.9 | 0.136 | 2.3 |
|  | 476 | 472 | 0 | 4060 | 20.0 | 3.50 | 0.0037 | 99.8 | 0.322 | 3.1 |
|  | 466 | 410 | 473 | 3990 | 0.0 | 5.88 | 0.7000 | 88.8 | .... | 227.0 |
| 1 | 727 | 725 | 5 | 6210 | 36.2 | 2.31 | 0.0010 | 99.9 | 0.118 | 2.1 |
|  | 723 | 721 | 5 | 6180 | 28.5 | 3.75 | 0.0035 | 99.9 | 0.223 | 3.5 |
|  | 715 | 711 | 22 | 6110 | 14.5 | 6.30 | 0.0089 | 99.5 | 0.601 | 5.8 |
|  | 709 | 701 | 87 | 6060 | 7.2 | 7.58 | 0.0706 | 99.1 | .... | 10.0 |
|  | 702 | 636 | 338 | 6060 | 2.5 | 8.37 | 0.4990 | 94.1 | .... | 179.0 |
|  | 698 | 607 | 729 | 6060 | 1.7 | 8.45 | 1.0300 | 87.9 | .... | 303.0 |
| 33 | 1102 | 1099 | 5 | 9400 | 33.0 | 4.31 | 0.0033 | 99.9 | 0.231 | 3.7 |
|  | 1065 | 1062 | 8 | 9085 | 21.5 | 7.41 | 0.0105 | 99.8 | 0.690 | 6.5 |
|  | 1030 | 1047 | 9 | 8905 | 13.5 | 9.47 | 0.0170 | 99.8 | .... | 8.5 |
|  | 972 | 939 | 265 | 8205 | 2.2 | 11.44 | 0.6570 | 96.7 | .... | 128.0 |

there seemed to be some symmetrical relation between the points at which the curves for different speeds broke away from the characteristic straight line. Accordingly all of these points were transferred to the $V_s$ and $T_1$ diagrams, the result being a series

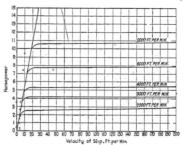

FIG. 4   CURVES SHOWING RELATION BETWEEN VELOCITY OF SLIP
AND HORSEPOWER FOR $T_1 = 100$ LB.

of surprisingly well-defined lines radiating from very near the origin and cutting the lines of constant hp.

Since the quantity $T_1 - T_2$ is more valuable for use in connection with this work than the quantity $T_1$, it was found desirable to provide some means of connecting the relation of $T_1 - T_2$ to the other variables.

To do this the values of $T_1 - T_2$ were computed for various points on the diagram, and after marking the values in their proper places contours were run so as to show the lines of constant $T_1 - T_2$. The intersections of constant-hp. lines and constant-speed lines were used for this purpose and the corresponding values of $T_1 - T_2$ were computed from the following relation:

$$T_1 - T_2 = \frac{33,000 \times \text{Hp.}}{V_d}$$

This diagram may be used in connection with the design of a steel belt within the range of the values used, and of course applies only to steel belting of the same size as used in the test and run over cork-faced pulleys.

For any belt speed the maximum horsepower which can be transmitted with a given tension and the corresponding velocity of slip and $T_1 - T_2$ can be determined. For instance, if we wish to transmit 10 hp. we have a large range of values of both $T_1$ and belt speeds from which to make a choice, but when either of these values is fixed, the other assumes a definite value.

It must be remembered in using this diagram that owing to the method of its construction some of the quantities represent not the absolute values, but the limiting values corresponding to the others which have been fixed. Thus if we assume a belt speed of 6000 ft. per min. with the original assumption of 10 hp. transmitted, it will be seen that $T_1$ will have a value of about 135 lb. This means that with the conditions which were assumed (10 hp. and 6000 ft. per min.) any value of $T_1$ less than 135 lb. would be below the limiting value and that excess slip would be expected. It would be advisable, however, to use a somewhat higher tension, since in this way the velocity of slip would be materially decreased.

On the other hand, if we wished to find what speed would be necessary if we assumed $T_1$ to be, say, 200 lb., from the diagram it appears that a speed of about 3600 ft. per min. would be the least speed at which 10 hp. could be transmitted without excess slippage, and at this condition the slip would be about 12½ ft. per min. Here again it would be advisable to either increase the speed or the tension, in order to work safely below the critical point, because the relations on the diagram represent conditions which correspond to the points on the straight lines of the $V_s$ and hp. curves at which the holding power of the belt on the pulley breaks down and a very great increase in slippage begins.

By extending this diagram to cover a large range of hp., belt velocity and tension, and applying certain coefficients to compensate for changes of belt size and other variables which may occur, a complete working diagram for steel-belt design would be obtained.

The next relation which was wanted was that of the coefficient of friction to the velocity of slip. The method used in calculating

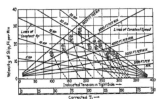

FIG. 5  CHARACTERISTIC DIAGRAM FOR STEEL BELTING

This diagram was constructed from the data and computed values obtained from the tests. It shows the relations which exist between the several variables involved, particularly as these relations are controlled by the points which determine the limiting conditions; that is, the point where excess slippage begins. The dotted lines represent lines of constant $T_1 - T_2$.

these values has been previously explained, the values of $V_s$ used being the corrected values obtained by the method before described.

Fig. 6 shows the curves obtained, and it was found that not only does the coefficient of friction vary with the velocity of slip but also with the belt speed, increasing with increased velocity of slip and with a decrease of speed.

A rather interesting feature was discovered in connection with the computation of the values of the frictional coefficient. In using the quantity

$$\log \frac{t_i - t_e}{t_s - t_e}$$

it was found that for some readings the values of $t_i - t_e$ became negative, and hence the expression

$$\frac{t_i - t_e}{t_s - t_e}$$

also became negative.

Investigation of the curves showed that such values appeared to develop at or near the points where the hp. and $V_s$ curves broke away from their common characteristic line.

Very little can be determined for these values from the experimental data taken, and it is evident that such points are outside of the practical range for the use of steel belting. The phenomenon can, however, be explained as follows:

Whenever the value of $t_e$ became greater than the existing value

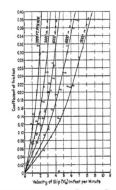

FIG. 6  CURVE SHOWING RELATION BETWEEN VELOCITY OF SLIP AND COEFFICIENT OF FRICTION

of $t_i$, which is the condition under which $t_i - t_e$ is negative, it is evident that there was no force acting to keep the belt in contact with the pulley on the loose side. Accordingly, the arc of contact was diminished and conditions brought about which gave rise to the unstable and uncertain conditions which showed up in the experimental work.

In obtaining the values of the coefficient of friction an attempt was made to compute them from the relation:

$$\text{Ft-lb. lost due to friction} = u\, p\, r\, \theta\, V_s$$

where  $u$ = coefficient of friction
$p$ = pressure per unit length of belt on pulley face
$r$ = radius of pulley
$\theta$ = angle of contact in radians
$V_s$ = velocity of slip.

Assuming that the total power loss due to friction is also equal to the product of the velocity of slip by the difference in tension,

$$(T_1 - T_2)V_s = u\, p\, r\, \theta\, V_s$$

and since

$$p = \frac{T_1 + T_2}{2r\theta}$$

$$V_s(T_1 - T_2) = u\frac{T_1 + T_2}{2r\theta} r\theta V_s$$

or

$$u = \frac{2(T_1 - T_2)}{(T_1 + T_2)}$$

Fig. 7 shows the curves plotted from this relation, but on account of their being independent of $V_s$, values of $u$ were obtained in another manner.

### DISCUSSION OF BELT JOINTS

One of the serious difficulties which had to be contended with was the development of some method of joining the ends of the belt in such a way as to develop a fair proportion of the tensile strength of the steel belt and at the same time not to interfere with the smooth operation of the belt and to be of such a nature that the joint will be sufficiently durable.

The following requirements are those which determined the most desirable joint to use:

*a* In order that there shall be the least possible amount of concentrated strain it is necessary either to have the joint extend along the belt as small an amount as possible, or to have it so constructed that its flexibility is approximately equal to that of the original belt. If the joint is of such construction that there is a distinct change in the flexibility in adjacent sections, the stresses would be so concentrated at the junction point that destructive stresses would cause an early failure of the material.

*b* On account of the design of the testing machine it was necessary that the joint would be capable of withstanding reversed stresses and that nothing should interfere with using both sides of the belt in contact with the pulley wheels. This condition, while

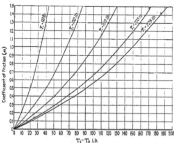

FIG. 7  CURVES SHOWING RELATION BETWEEN COEFFICIENT OF FRICTION AND DIFFERENCE OF TENSION

being essential for the test, is not necessarily applicable to all installations, because for ordinary straight drives only one side of the belt is run in contact with the pulleys.

*c* The material of the belt is a very high-carbon steel carefully tempered and any joint which depends upon the use of solder must be made in such a way that the original temper is either not drawn or that it is drawn to only such a degree that the material is not damaged.

It can be readily seen that a joint could not be retempered without tempering the entire belt, because there would be a section each side of the heated part which would have received enough heat to draw its temper and still would not be hot enough to harden. Efforts were made to develop a joint which satisfied the foregoing requirements, and after much experimentation it was found that the most successful manner of joining the belt ends, so far as the test was concerned, was to make a butted joint with the edges beveled about 60 deg. and then secured by means of silver solder. Care was taken to confine the annealing effect of the soldering flame to as small a length along the belt as possible and only enough solder was used to make the joint complete. With the 3/4-in. belt it was possible to confine the annealing effect to a very small length (as little as 1/2 in.), and to so distribute the solder that it was not necessary to do any filing in order to make the joint ready for service.

Joints of this kind were used throughout the test with apparently no signs of weakening or failure and showing no undesirable effects on the running stability of the belt.

An attempt was made to weld the belting by means of an oxy-acetylene torch, but on account of the extreme thinness of the material and the high temperature required it was unsuccessful.

In an attempt to make a joint which would develop a large percentage of the original strength of the material, one of the ends to be joined was cut to a sharp V and the edges serrated, and the other end so cut as to interlock with it. It was expected that when this joint was silver-soldered the strain would be taken up by the material instead of by the solder as it would be with the simple butt-soldered joint. This method was not a success, because when heated to a sufficiently high temperature to flow the solder it buckled and twisted and could not be made flat.

Riveted joints were also tried and proved fairly satisfactory. The joints were made in the V-shape in order to distribute the rivets and both ends were cut out so that the tendency to stiffen the joint would be reduced to a minimum. They were made by the use of a punching die into which one end of the belt was clamped and 15 No. 50 holes were punched. The other end was punched in the same die, being put through from the opposite end so that the holes in the two pieces exactly matched. The rivets used were No. 51 punchings from phosphor bronze about 0.03 in. thick. These were placed in the holes and headed down lightly on both sides, care being taken not to stretch the material by upsetting them too much. A joint of this type was run for a considerable length of time and then a very small crack was discovered in the edge. It is the opinion of the authors that with sufficient care and experimentation a very satisfactory riveted joint could be developed.

### PART II  INVESTIGATION OF SLIP

IN carrying out the work described in Part I both Hampton and Leh realized that certain relationships existed between the horsepower transmitted by the belt, velocity of slip of the belt on the pulleys, and the coefficient of friction, but owing to the very large error in their observed data on velocity of slip, they were unable to find such expressions as have been developed below.

The difference of r.p.m. of the driver and driven pulleys as observed by the above authors was subject to a large error in that the revolution counters used did not read less than unity. One revolution would correspond to a slip of 8.625 ft. per min. were the observation period one minute long, as the circumference of both pulleys was 8.625 ft.

Hampton, realizing that more accurate apparatus was necessary to measure the slip, designed and partly constructed a differential revolution counter that was completed by the author. This revolution counter would indicate one one-hundredth of one revolution difference, with a probable error of 5 per cent.

### THE DIFFERENTIAL REVOLUTION COUNTER

The positions of the idler pulleys reverse the direction of the belt, and hence the direction of rotation of the driver and driven pulleys is opposite. The mechanism of the differential, as is commonly known, is such that the two wheel shafts rotate in opposite directions and with equal r.p.m. when there is no rotation of the ring gear. The driver and driven shafts are connected 1 to 1 to the wheel shafts by light chains and sprockets. Hence, if the driver and driven pulleys are rotating with the same speed, the ring gear is stationary. The pulleys having equal circumferences, no slip would be in evidence when this were the case. However, as soon as the belt began to slip, the ring gear would rotate in proportion to the difference of peripheral velocity of the two pulleys.

The ratio of the ring gear to the pinion is 1 to 20, and that of the differential gears is 1 to 1, making the ratio of the wheel shaft to the ring pinion 1 to 10. A disengaging coupling is connected to the ring pinion, by means of 1-to-1 sprockets and light chain, and is operated by the solenoid which in turn is controlled by the knife switch. This coupling has twelve teeth on one side and a knife edge on the other, making the probable error 1/24 or 4.16 per cent. In order that the chain be kept tight, a friction brake was put in between the sprocket mounted on this coupling shaft and its supporting standard.

The revolution counter used was one specially constructed for this particular use. The unit wheel being divided into tenths gave a ratio of 1 to 100 from the wheel shaft to the counter. Hence the difference of r.p.m. of the driver and driven shafts was determined quite accurately, that is, to one one-hundredth of a revolution, which in terms of the pulley circumference is 0.0863 ft.

It was found after running some time that the vibration of the chains connecting the pulley shafts to the revolution counter caused them to become stretched in places, effecting a slight rotation of the ring gear first in one direction and then in the opposite. This appeared to vary with the speed, but, as nearly as could be determined, a probable error of from 1 to 2 per cent might be introduced.

Through all tests this machine served its purpose very accurately and was a great help in adjusting the load on the Sprague dynamometer, for the point at which the belt began to slip excessively could be observed at a glance. Without this machine the data on the slippage of the belt would be so inaccurate that the results of the tests would have been useless.

### BELT MATERIAL AND JOINTS

The material used for belting in this test was clock-spring steel ¾ in. by 0.01 in. It is manufactured in this country from Swedish high-carbon charcoal steel, drawn, rolled, ground to size and tempered to a dark blue. It can be obtained in widths of from ½ in. to 3 in. and 0.01 in. thick and costs from 6 to 18 cents per ft.

Tests of this material show the ultimate strength to be over 300,000 lb. per sq. in. and the elastic limit slightly less. The belt used in this test had no permanent set when bent around a radius of ¾ in. but it was found it would rupture when bent around a radius of $\frac{1}{16}$ in.

When the belt bends around a pulley, as much work is put into bending it as appears when the belt straightens out on leaving the pulley. This is not true of any other kind of belt, as power is required to bend it and again to straighten it out.

The joint used on the belt tested was a silver-soldered lap joint. It proved very satisfactory throughout all tests and showed no

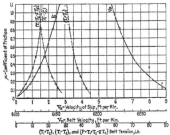

FIG. 8 CHARACTERISTICS OF A 0.75-IN. BY 0.01-IN. STEEL BELT

signs of necking-in where the temper had been drawn. These joints when tested have an efficiency of about 60 to 65 per cent in pure tension, the rupture taking place on either side of the lap where the metal has been softened.

Owing to the low efficiency of the silver-soldered and brazed joints, a dozen or more riveted joints were constructed. Phosphor-bronze rivets were used and one triple-riveted lap joint by test had an efficiency of 84 per cent, rupture occurring by shearing all the rivets. This joint could not be used on pulleys smaller than 30 in. in diameter.

Other similar lap joints averaged 76 per cent by test when cold-drawn iron wire was used for rivet material.

Owing to the idler pulleys reversing the motion, one side of the belt runs on the driven pulley and the other side on the driver pulley, consequently it was necessary to have both sides of the belt joint as smooth as possible.

The type of joint used in an installation of steel belting in one of the shops was of the bent cover-plate type and would have been quite successful had the pulley on the motor been of larger diameter. This cover plate was bent to conform with the curvature of the smaller pulley, and the ends were bent up on a radius of ¾ in. It appeared that the tangential acceleration of the joint as it came on to the motor pulley kept the joint from following the pulley curvature. This induced flexure in the belt, just ahead of the first row of rivets, and the combined stress of the tension and flexure being in the neighborhood of the elastic limit caused rupture after 15 to 20 hr. operation.

It is the opinion of the author that this type of cover-plate joint would be quite satisfactory for use on pulleys of not smaller radius than 15 in.

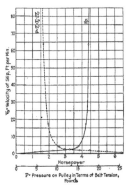

FIG. 9 CHARACTERISTICS OF A 0.75-IN. BY 0.01-IN. STEEL BELT

### DESCRIPTION OF TESTS

The calibration of the weights and spring balance was made just after the differential revolution counter was completed. The work checked quite well with the calibration made by Hampton and Leh.

Several preliminary runs were then made to determine the capacity of the motor and the variations of speed with load. A preliminary curve run was then made. A tension of 23.6 lb. in the tight side of the belt was selected in order that the tension in the loose side could be made practically zero, allowing an excessive slip. The speed selected would also bring in the characteristics of the motor and allow a certain change in the centrifugal tension of the belt. The curves resulting from this set of values and shown in Figs. 8 and 9 will be discussed below.

The four speeds of the motor with the arrangement of pulleys then on the machine would give belt speeds of about 3150, 4770, 6250 and 9450 ft. per min. In order that a wider range of belt speeds could be obtained, a large pulley and a small pulley could be placed on the driver-pulley shaft, and with these pulleys belt speeds of 2080 and 14,500 ft. per min. could be obtained.

Seven runs were made with each velocity for various tensions in the tight side of the belt. The smallest is 23.6 lb. and corresponds to 50 lb. on the weight pan, while the largest is 111.6 lb.

and corresponds to 225 lb. on the weight pan. The interpolated results for $T_1 = 99.0$ are given in Table 2.

The run was always begun at the highest horsepower possible, and from five to six observations made as the hp. varied to zero. The load could be held quite constant throughout each run when the Sprague dynamometer was separately excited, but near the end of the test period, when the generator was self-excited, some difficulty was experienced in making all the observations. Before a run was begun, the surfaces of the cork pulleys were always cleaned in order that the conditions of the test might be the same.

The following values were observed and will be discussed below:

$(T)_1$ = tension in lb. on the idler over which the tight side of the belt passes; observed by the weights on the weight pan

$T_1$ = corrected tension in the belt in lb. as taken from the calibration curves

$(T)_2$ = tension in lb. on the idler over which the loose side of the belt passes; observed by the reading of the spring balance

$T_2$ = corrected tension in the belt in lb. as taken from the calibration curves

$N$ = r.p.m. of the driver pulley; observed from the reading of the electrically operated Veeder revolution counter

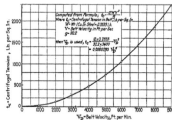

FIG. 10　CURVE OF CENTRIFUGAL TENSION FOR BELT SPEEDS UP TO 9000 FT. PER MIN.

$u$ = difference of r.p.m. of the driver and driven pulleys as observed from the differential revolution counter.

A reading of the brake arm on the Sprague dynamometer was also taken and used as a check on the power transmitted.

### NOTATION

$T_1$ = actual tension in the tight side of the belt in lb.

$t_1$ = actual unit tension in the tight side of the belt in lb. per sq. in.

$T_2$ = actual tension in the loose side of the belt in lb.

$t_2$ = actual unit tension in the loose side of the belt in lb. per sq. in.

$T_c$ = centrifugal tension corresponding to the velocity at which the belt is passing around the pulleys in lb.

$t_c$ = actual unit centrifugal tension in lb. per sq. in.

$T_1 - T_2$ = difference in belt tensions in lb. or the useful force in the transmission of power.

$V_d$ = belt velocity in ft. per min.

$V_s$ = computed velocity of slip of the belt relative to the pulley in ft. per min.

$V_{sc}$ = velocity of slip of the belt in ft. per min. corrected from the curves where Hp. $= 0.17 T_1^{1.45} \times V_s$

$P$ = actual tension in belt producing pressure on pulley, lb.

$p$ = pressure in lb. per sq. in. per unit length of belt on the pulley face

$W$ = weight of 1 cu. in. of steel in lb.

$g$ = gravitational force

$r$ = radius of driver and driven pulleys in ft.

$u$ = coefficient of friction

$\theta$ = angle of contact in radians the belt makes on the pulley face.

TABLE 2　DATA AND RESULTS OF TEST ON 0.75 x 0.01-IN. STEEL BELT

| Run No. | $T_1$ | $T_1*$ | $P$ | $V_d*$ | $H_p.$ | $u$ | $V_s$ |
|---|---|---|---|---|---|---|---|
| R-4 | 99.0 | 22.6 | 120 | 2018 | 4.67 | 0.482 | 1.37 |
| .... | | 23.4 | ... | 3000 | 6.98 | 0.481 | 2.05 |
| .... | | 29.0 | ... | 4800 | 9.76 | 0.440 | 2.88 |
| .... | | 35.6 | ... | 5730 | 11.03 | 0.376 | 3.24 |
| .... | | 56.4 | ... | 8920 | 12.76 | 0.236 | 3.76 |
| | 99.0 | 43.0 | 140 | 2033 | 3.45 | 0.269 | 1.01 |
| .... | | 45.0 | ... | 3070 | 5.25 | 0.260 | 1.48 |
| .... | | 50.0 | ... | 4610 | 6.84 | 0.234 | 2.00 |
| .... | | 57.0 | ... | 6040 | 7.69 | 0.162 | 2.26 |
| .... | | 79.0 | ... | 9300 | 5.64 | 0.092 | 1.66 |
| | 99.0 | 63.0 | 160 | 2068 | 2.255 | 0.145 | 0.66 |
| .... | | 65.0 | ... | 3100 | 3.195 | 0.137 | 0.93 |
| .... | | 70.0 | ... | 4880 | 2.980 | 0.117 | 0.87 |
| .... | | 78.0 | ... | 6200 | 3.950 | 0.084 | 1.16 |
| | 99.0 | 83.0 | 180 | 2075 | 1.005 | 0.087 | 0.30 |
| .... | | 85.6 | ... | 3163 | 1.286 | 0.046 | 0.84 |
| .... | | 90.8 | ... | 4740 | 1.178 | 0.026 | 0.35 |
| .... | | 98.0 | ... | 6250 | 0.189 | 0.004 | 0.05 |

* Interpolated values.

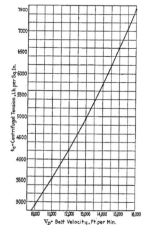

FIG. 11　CURVE OF CENTRIFUGAL TENSION FOR BELT SPEEDS UP TO 16,000 FT. PER MIN.

### DISCUSSION OF CURVES

The computations necessary in working out a set of values involved a great number of operations. An attempt was made to plot curves of various functions that would constantly enter into certain equations.

The values of $t_c$ and $T_c$ were first computed and plotted as shown in Figs. 10, 11, and 12. The equation for unit centrifugal tension is

$$t_c = \frac{12 W V_d^2}{g}$$

where $W = 0.2833$ lb.
$g = 32.2$

and in this equation $V_d$ is in ft. per sec. If $V_d$ is to be used in ft. per min., the equation becomes

$$t_c = \frac{12 \times 0.2833 \times V^2_d}{32.2 \times 3600}$$
$$= 0.0000293 \, V^2_d \text{ lb. per sq. in.}$$

As $T_c = t_c \times A$, the multiplying factor then becomes
$$T_c = 0.0000293 \times 0.01 \times 0.75 \times V^2_d$$
$$= 0.00000022 V^2_d$$

Computations of all observed data were then made and the results tabulated, the equations employed being given below.

Circumference of driver and driven pulleys = 8.625 ft.

$$V_d = \left( N - \frac{n}{2} \right) \times 8.625 \text{ ft. per min.}$$

$$t_1 = \frac{T_1}{0.0075} \text{ lb. per sq. in.}$$

$$\text{Hp.} = \frac{(T_1 - T_2) V_d}{33,000}$$

where $T_1 - T_2$ = force in lb.

$V_d$ = distance

$$\text{Efficiency} = \frac{\text{Hp.} - (T_1 - T_2) V_{ec}}{\text{Hp.}} \times 100 \text{ per cent}$$

$$u = \frac{\log_{10} \dfrac{t_1 - t_c}{t_2 - t_c}}{0.43436}$$

$$V_s = \frac{n}{2} \times 8.625 \text{ ft. per min.}$$

The above equation for $u$ is derived in Smith and Marx' Machine Design.

Curves of all relations that might possibly exist were then plotted as are shown in Figs. 8 and 9.

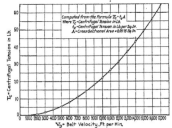

FIG. 12 CURVE OF CENTRIFUGAL TENSION FOR BELT SPEEDS UP TO 17,000 FT. PER MIN.

The most noticeable relation is that of velocity of slip and horsepower. This curve shows practically straight-line variation up to a certain point where the velocity of slip increases very suddenly. Inspection of the data shows that at this point $t_2 = t_c$ and hence there is no pressure on the pulley where the belt leaves the driven pulley. As this is the case, it would be expected that the velocity of slip would become excessive.

The curve of $P$ as a function of slip shows that the pressure decreases with an increase of slip up to a point where the pressure decreases but slightly with a large increase of slip. As can be seen from these two curves, the pressure on the pulley and the horsepower are related, but for the present the most striking relation appears to be in the curve of horsepower and slip.

The relation between the coefficient of friction and velocity of slip apparently exists in Fig. 8. This is a misconception, for the curve should not pass through zero, since the coefficient of friction of rest between cork and steel is not zero. All that this

curve indicates is that the coefficient of friction at zero velocity and zero slip is indeterminable, due to other variables. It may also be noted that the belt velocity is not constant and that the pressure varies.

A relation between $u$ and $T_1 + T_2$ was suspected, but after some investigation it appeared that $T_c$ entered into this factor. It was then that the pressure curve was plotted, and although it has no significance as it appears here. it was necessary in the work that was to follow.

## METHOD OF OBTAINING RESULTS

The first relation that was attempted was that between hp. and

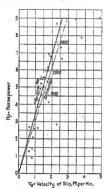

FIG. 13 CURVES SHOWING RELATION BETWEEN VELOCITY OF SLIP AND HORSEPOWER

$V_s$. It was noticed that for a constant value of $T_1$ all the points up to certain limits would fall on a straight line. These curves were plotted, the curve for $T_1 = 99$ lb. being shown in Fig. 13. A dashed line was then drawn such that the estimated slope would as nearly as possible coincide with the mean average of the points. In these curves this line is a light dashed one and noted by the word "Estimated." It must be remembered that in these curves all points in the vicinity or above that where $t_2 = t_c$ are not plotted. The figures on the right show the belt velocity and hp. at which $t_2 = t_c$.

When this was done for all values of $T_1$, the slope was found and tabulated with the corresponding value of $T_1$. The equation of a straight line passing through zero is $Y = mX$, where $m =$ slope. With this in mind the slope $m$ was plotted as a function of $T_1$, and resulted in a parabolic curve. To determine the equation of this new curve the data were plotted on logarithmic cross-section paper. The result was a straight line which had 0.17 for its intercept and a slope of 0.65, indicating that the slope $m$ was equal to $0.17 \, T_1{}^{.65}$.

From this it was a simple matter to deduce the equation—

$$\text{Hp.} = 0.17 \, T_1{}^{.65} V_s$$

As all of the points of the estimated slope $m$ did not fall on the logarithmic curve, another line, the slope of which was computed from the above equation, was drawn on the curves and noted by the equation.

The values of the velocity of slip were then corrected by the use of these curves and the corrected value was tabulated as $V_{sc}$.

Fig. 13 was then plotted to give an idea of the relation of hp. to $V_{sc}$ at various values of $T_1$. The lines of constant $T_1$ are shown as solid straight lines. The light dashed lines are lines of constant

$V_d$ and the points of intersection with the lines of constant $V_d$ were computed from the equations—

$$\text{Hp.} = (T_1 - T_2) \times V_d$$
$$T_2 = T_e = 0.00000022 \, V^2_d$$

The relation of $u$ to $V_{ec}$, $V_d$ and $P$ was more difficult to determine owing to the large number of variables. An attempt was made to select groups of values from the observed data where two of the values remained constant. This could not be done for enough values to find a relationship between the other two variables, and for a time the solution looked much more difficult. It seemed that new data must be taken or that interpolations must be made from the data already at hand. As the taking of new data, such that two of the variables remain constant, was impractical for but one observer, interpolated data had to be computed. The values of $T_2$ and $V_d$ were interpolated from curves and the observed data.

A large number of groups were selected, and in each group two and sometimes three sets of values of the two variables could be obtained when the other two remained constant. These values were plotted on logarithmic cross-section paper and were found to be straight lines having approximately the same slope. For example, the relationship between $u$ and $V_{ec}$ is shown in Table 3 for one value of $P$ throughout the set of groups.

TABLE 3 RELATIONSHIP BETWEEN $u$ AND $V_{ec}$

| Run No. | Constants | | Variables | | Relation |
|---|---|---|---|---|---|
| | $P$ | $V_d$ | $u$ | $V_{ec}$ | |
| R-1 | 70 | 5000 | 0.143 | 1.20 | $u = 0.1 \, V_{ec}^{1.33}$ |
| R-2 | .. | .... | 0.360 | 2.75 | |
| R-1 | .. | 4000 | 0.190 | 1.15 | $u = 0.155 \, V_{ec}^{1.33}$ |
| R-2 | .. | .... | 0.465 | 2.25 | |
| R-1 | .. | 3000 | 0.271 | 1.00 | $u = 0.223 \, V_{ec}^{1.33}$ |
| R-2 | .. | .... | 0.515 | 1.80 | |
| R-1 | .. | 2000 | 0.240 | 0.75 | $u = 0.386 \, V_{ec}^{1.33}$ |
| R-2 | .. | .... | 0.550 | 1.25 | |

Data similar to those in Table 3 were obtained for pressures of other values and the relation found. The exponents in each were averaged and found to be 1.37, it then being stated that

$$u = V_{ec}^{1.37} \times f(P) \times K \times f(V_d)$$

where $K$ is some constant.

The characteristics of the curves plotted of $u$ as a function of $V_d$ and $P$ when $V_{ec}$ was constant, showed that $V_d$ and $P$ varied inversely according to some power. Interpolations and computations similar to the above were made, and it was found that $u$ varied inversely as some constant multiplied by the square root of $P$. Again the same methods were followed, and it was found that $u$ varied inversely as some constant multiplied by $V_d^{1.4}$.

This gave the equation

$$u = \frac{V_{ec}^{1.37} \times K}{P^{1.4} \times V_d^{1.4}}$$

and 75 values of $K$ were solved from the equation

$$K = \frac{u \times P^{1.4} \times V_d^{1.4}}{V_{ec}^{1.37}}$$

The values varied from $2.71 \times 10^4$ as a minimum to $4.04 \times 10^4$ as a maximum. The majority, however, were in the neighborhood of the average, which was $3.1992 \times 10^4$.

The final deduction of the relation of $u$ to the other variables is given by the equation

$$u = \frac{V_{ec}^{1.37} \times 3.2 \times 10^4}{P^{1.4} \times V_d^{1.4}}$$

## CONCLUSIONS

The first equation found in this investigation is useful in determining the tension in the tight side of the belt, when a given amount of power is to be transmitted and a reasonable velocity

of slip (not greater than 3 ft. per min.) is assumed. The wear on the pulley facings is in proportion to the velocity of slip, and hence it is advisable to keep this value as low as possible to prevent wear. Knowing this tension, the belt may be cut to length by computing the total elongation due to the tension, and the distance that the cork will be depressed due to this tension. The belt may then be joined and placed on the pulleys.

The second equation, while not as practical as the first, is interesting from the technical nature with which the different factors vary the value of the coefficient of friction as the range of power transmission is passed over. The coefficient of friction of rest between cork and steel at a given pressure would not appear to be zero as this equation would indicate were the velocity of slip zero.

The efficiency of transmission was always so near 100 per cent that it was not tabulated in the data. Only in extreme cases did it fall below 98 per cent, and the author spent but little time on this computation.

BIBLIOGRAPHY

Tests by Eloesser, in Germany, forming the basis of the subject.
*Engineering News*, Jan. 9, 1908.
*Engineering News*, Oct. 14, 1909. Article by Shroeter.
*Power*, May 3, 1910. Article by Krall.
Tests by General Electric Co. Not published.
*American Machinist*, Nov. 21, 1912. Article by Conkhete.
*Machinery*, Feb., 1913. U. S. Consul F. D. Hale.
*Machinery*, May, 1914. Editorial.
*American Machinist*, May 14. 1914. Article by Chubb.
Steel Belting. A Thesis by Hampton and Leh, Stanford Univ., Cal.

A paper on Recent Advances in Utilization of Water Power, by Eric M. Bergstrom, of London, read before the Institution of Mechanical Engineers, January 23, 1920, directs attention to the main lines along which the development of this highly-specialized branch of engineering has taken place, and indicates briefly its general effect on the utilization of water power. In considering these developments as described in the paper, Mr. Bergstrom summarizes the outstanding features as follows:

*a* The exclusive use of two types of turbines only, namely, Francis reaction turbines for low and medium heads and Pelton impulse wheels for high heads

*b* The extension of the use of Francis turbines under heads approaching 800 ft. and Pelton wheels in single stage up to 5500 ft.

*c* The exclusive adoption of balanced wicket gates for the regulation of Francis turbines and the circular nozzle with combined deflector and needle regulation for Pelton wheels

*d* The standardization of turbine runners and increased specific speed permitting the use of single vertical units of large output under low heads

*e* The general increase of output per unit, the maximum output at present being 31,200 hp.

*f* The general increase of the overall efficiency of about 6 per cent to 10 per cent

*g* The exclusive use of oil-pressure governors

*h* The efficient regulation (by means of differential surge tanks) of turbines using long pipe lines

*i* The employment of large-diameter pipe lines under high heads, resulting in an appreciable reduction of the initial cost of development.

These developments, coupled with the improved construction of impounding dams in addition to multifarious improvements in details to insure effective safeguards and reliability in operation, constitute as such an enormous advance from a purely technical point of view, but of even greater significance has been the unlimited field thereby opened up for the application of electricity and the extended scope given to the utilization of water power in the service of civilization; issues of the greatest magnitude have been raised affecting the economic life and future prosperity of nations, and from being a question to be left solely to private enterprise, the control and development of water power now rank among the vital problems of national interest.

# Design of Ore Fleet for Upper Mississippi

Economy of River Transportation of Ore as Shown by Experimental Trips on Upper Mississippi
River—Particulars Regarding Design and Construction of Barges and
Towboats—Terminal Facilities Projected, etc.

By WILLIAM S. MITCHELL,[1] ST. LOUIS, MO.

*This paper deals principally with the design of an ore fleet for the upper Mississippi River. The author describes two experimental trips which were conducted during the war to try out the possibilities of such transportation on the Mississippi. The results of these experimental trips warranted the construction of a fleet of 24 barges and 4 towboats for traffic of ore and fuel on the upper Mississippi River. The writer, as engineer of the U. S. Engineer Office at St. Louis, was in charge of the design of these vessels and describes the difficulties encountered and the manner in which they were handled. He also gives information concerning the fleets for the lower Mississippi River and the Warrior River, and the terminal facilities being developed in cities and towns along these rivers.*

I N the Spring of 1917, after this country had entered into the world war, it was quickly realized that the transportation of materials for the manufacture of munitions and equipment presented one of the most important problems to be solved, especially as the movement of large bodies of troops and their impedimenta soon began to crowd and embarrass the railroads to the point of breakdown. For relief, in part, the use of the waterways was early suggested, and in May of that year a decision was reached by the War Department, with this idea in view, to aid in transportation on the Upper Mississippi River of iron ore from Minnesota ore fields to the coal fields of the Middle West, to increase, in blast furnaces there, the production of pig iron for the manufacture of steel near St. Louis. This effort started with a small improvised fleet composed of six steel barges of the U. S. Engineer District at St. Louis, each of 600 tons capacity, and a steel-hull stern-paddle-wheel towboat of ordinary western river type, from the Mississippi River Commission. This fleet was assigned to a lessee to try out the possibilities of the upper river for such traffic, and, loaded with about 3000 tons of Illinois coal to be sold at St. Paul to meet in part the expense of the undertaking, left St. Louis July 31, at the beginning of the annual low-water season, and proceeded without incident upstream, arriving at destination in the middle of August.

At St. Paul considerable delay ensued in unloading at the water front the unusual quantity of coal, and in obtaining and loading ore for the return trip. An extra quantity of ore having been delivered by error of the railroad and ore companies, four more small wooden barges were borrowed at St. Paul from the Engineer Department, for its transport, and were added for the downstream trip to the tow, already rather too large and heavily laden for the prevailing river stage, which had been falling steadily until the conditions almost reached the recorded low limits for a navigable season.

The whole fleet, with about 3400 tons of ore, left St. Paul August 28, and although the draft of the barges was at the extreme limit permitted by the channel depths, they were floated over the shoals and bars with fair success; but the towboat, built for the deeper water of the lower river, was of slightly greater draft, passing that limit, and grounded at nearly every river bar encountered. Then, too, as the boat was nearly touching bottom at many other places, it lost control of the tow, and the barges also were frequently grounded in the soft river sand. Other smaller and lighter-draft boats were brought to the assistance of the struggling fleet, and after two months of incessant effort the vessels arrived at St. Louis.

Meanwhile a similar second fleet had been collected at St. Louis and, loaded somewhat more lightly with coal, had started north,

but because of the fallen river it was unable to proceed much beyond the mouth of the Illinois River, to which stream it was diverted, and after discharging cargo at Hennepin it returned to St. Louis.

The seemingly disastrous southbound trip from St. Paul unfortunately created a strong impression of unsuitability of river transportation to modern needs; but as a matter of fact, that trip, made at such an extraordinarily low stage, developed practically and commercially the fact of an excellent, well-improved river channel for nearly 700 miles between the cities named, available to at least 4 ft. depth for wide tows of barges during the low-water season, increasing at high stages to 8 and 9 ft., which guarantees to vessels built specially for such shoal depths a capacity for transportation far beyond that of any railroad. As the route by rail from the Minnesota ore fields to St. Paul and thence by river 675 miles to St. Louis is about 200 miles shorter than the prevailing route by rail to the lakes and thence by water to southern lake ports and again by rail to St. Louis; and as the ore is always at hand in full-capacity quantities for the down trip, and the coal, eagerly sought in the north, is equally plentiful for the upstream haul, there is offered an ideally balanced freight-traffic, commercially profitable even with these low-grade (tariff) commodities, which, if continually carried in large quantities, cannot fail to effect great economy in the transportation costs of both.

## UPPER MISSISSIPPI FLEET

As a war measure, to increase the production of pig metal and in contribution to the relief of the acutely felt shortage in ore and coal cars, the United States Shipping Board, Emergency Fleet Corporation, made an allotment of $3,360,000 for the construction of 24 barges and 4 towboats, to be entirely of steel, and specially designed for this through traffic of ore and fuel on the shallow Upper Mississippi River. The number of barges was afterward reduced to 19 and the funds increased to $3,860,000 to aid in the construction of terminal loading and unloading apparatus.

### DESIGN AND CONSTRUCTION OF BARGES

The design of the vessels fell to the writer's charge under the U. S. Engineer Office at St. Louis, under the War Department, and in this work most valuable counsel and assistance was freely given by Mr. M. H. Von Pagenhardt, Naval Architect, formerly with the Kansas City-Missouri River Navigation Co., and by Mr. J. W. Gerell, Naval Architect, Mississippi River Commission, both at St. Louis.

As is well known, large barges offer (a) greater carrying capacity per unit of fleet area and draft, (b) lower cost for construction and upkeep, and (c) less resistance in propulsion with consequent reduction in transportation cost per unit of capacity than a fleet of smaller vessels aggregating the same capacity. With these considerations in view, the dimensions of the barges were fixed just within the physical limits set by bridges and locks in the river reach to be navigated.

The dimensions chosen for the barges were 300 ft. length, 48 ft. width or beam, and 10 ft. depth. The length, 300 ft., of the locks on the rapids at Moline and Le Claire fixed the length of the component barges. The minimum channel clearance for the bridges, with one exception, is about 150 ft., thus the width of a unit fleet or tow was limited to about 144 ft. A width of 48 ft. allowed three barges to be grouped about each towboat, the most advantageous arrangement in towing, without necessity of breaking tow when running through the many bridges. The depth provided the necessary longitudinal strength and stiffness of hull. With these dimensions it was found that such a barge would

[1] Principal Assistant Engineer, U. S. Engineer Office.
Abstract of a paper presented at the Spring Meeting, St. Louis, Mo., May 24 to 27, 1920, of THE AMERICAN SOCIETY OF MECHANICAL ENGINEERS, NEW YORK. All papers are subject to revision.

carry about 850 tons of cargo on a draft of 4 ft. during low water, increasing at high stages to 2500 tons on 8-ft. draft when going upstream, and to 3000 tons on 9-ft. draft when floating over the bars coming downstream. It was estimated that the annual average load would be 1500 tons per barge, making 4500 tons per tow each way, and also, that about 16 round trips for each boat could be made during the boating season of 7 to 8 months, giving a unit annual capacity to each towboat of 144,000 tons, or, 576,000 tons to be moved by the fleet annually between the cities. The barges, therefore, were laid down on this basis.

It was found that Minnesota ore at about 115 lb. per cu. ft. could be readily carried as a deck load to the capacity limits assigned by the construction of necessary bulwarks 5 ft. high entirely around the barge. Also this would permit the use of the full unbroken deck as an upper chord to the barge considered as a girder, giving the desired longitudinal strength of structure with minimum weight of metal. But the full deck had to be abandoned as the upstream capacity cargo of coal could not be adjusted to these conditions, because the much greater bulk of coal per ton (about 55 lb. per cu. ft.) would require piling so high above deck as to be dangerous, and this compelled sinking the latter in the form of a great hopper, or open hold, 5 ft. deep, the barge thus losing longitudinally the strength of the sunken part of the deck which becomes the hopper bottom near the neutral axis of the vessel.

Longitudinal strength of structure was therefore given by wide (6 ft.) deck stringers on each side of the hopper, increased by a steel combing 1 ft. high above deck, running entirely around the hopper and adding to the depth of the latter.

The final dimensions of the cargo hopper are: an unbroken length of 256 ft., 36 ft. width, and 6 ft. depth. The hopper is drained at the corners and middle of each side by 4-in. drain pipes passing through the barge at about the 4-ft. draft line, and controlled from the main deck by proper valves. The hopper bottom is of ¼-in. plate, supported on longitudinal 6-in. channels spaced 12-in. centers at deck reinforces, to withstand inevitable careless handling of heavy clamshell unloaders.

The open hopper, of course, is suitable for carrying any other materials or commodities not requiring protection from the weather, either in bulk or manufactured, as clay products, brick and tile, structural steel and machinery, lumber, ore, stone, etc.

The form of the barges, in section, thus given by the hopper, is that of two side pontoons, each 6 ft. wide by 10 ft. deep, connected by a double bottom 36 ft. wide by 5 ft. deep, and in the bottom and pontoons, which latter act also as expansion tanks, there is stowage space for 1500 tons of oils or nearly 2000 tons of bulk molasses "blackjack" now beginning to move to this country from the West Indies via New Orleans. Liquid cargoes are to be loaded and discharged by a 6-in. pipe system, with suction and discharge valves in each barge compartment, operated independently by a large steam pump, and also by hand pumps, installed in one fore peak of the barge, steam supply being had from the towboat or from steam plant on the loading or discharging dock. The pipe system can be used also as a bilge system.

In addition there is a smaller pipe system, with fusible plugs and nozzles, passing under deck for steam supply to each compartment for fire smothering, and the compartments are entered through water-tight deck hatches, fitted with ladders, and are vented for oil evaporation by goose-neck pipe vents through the decks of the side pontoons close to the hopper combing, the mouths of the vents being carefully screened by fine-mesh copper netting, and closed against water ingress by balsa wood ball float valves soaked in linseed oil and varnished to prevent weather checking.

The barges are fitted at each end with hand capstans, hawsers, steel anchor cable, cable reels and stoppers, heavy stockless anchors handled from davits with blocks and falls, towing bitts and numerous kevels, chocks, side fenders, etc.

The straight body of the barge is about 168 ft. in length, the ends are alike and each, with long, easy rake of 42 ft., is molded for 66 ft. in length, with about 1 ft. sheer, and forms a spoon-shaped bow, giving remarkably easy lines and resultant least re-

sistance to towing, yet the block coefficient being kept above 80 per cent, giving large displacement and capacity. The loaded economical speed is calculated at 5¼ m.p.h., and at this speed and 3½ ft. draft the resistance was found on trials to be only 1.7 lb. per ton of displacement (total weight of barges and cargo, 1200 tons), or fully one-third less than the resistance of standard short-rake, scow-bowed barges.

Owing to the great combined width of bottom (144 ft. ) of a tow of three barges, with attendant heavy "drag" over shoal water, they are fitted with short wooden swinging side fenders in order to spread them slightly apart, thus offering spillways between the vessels, and for further relief, escape of drift, etc., the bilges are rounded with 5 ft. radius below the average draft line.

The barges are divided by 12 cross-bulkheads, and 1 central longitudinal bulkhead, or keelson, between bottom and hopper, into 2 fore peaks and 22 water- and oil-tight compartments, each controlled, as stated, by pipe lines and valves, vents, hatches, etc. Each barge when empty weighs about 550 tons, on 19½ in. draft, and the average cost at prevailing high prices is about $110,000 per barge.

The barges have been constructed under separate contracts at three yards. They have been largely "fabricated" in the contractors' and associated shops, from templets taken from full-size mold lines, the fabricated parts being then erected on the launching ways. At this writing seventeen barges have been launched and completed, and the remaining two are expected to be finished in June.

### DESIGN AND CONSTRUCTION OF TOWBOATS

The design of the towboats presented a problem even more complex than that of the barges. The power required to move a unit tow of three fully loaded barges, carrying 7500 tons of freight, upstream against the strong current of high water, is about 3000 hp., a greater power than has been heretofore provided (or developed) on a river towboat, yet the great weight of such a power installation, had to be mounted upon a hull of very light draft necessitated by the shallow depths at low water stages of the river. Nor can boats, as can barges, utilize fully in draft even the meager depths thus offered, but require, especially at the stern, a marginal or surplus depth below the hull to insure vitally essential backing power and maneuvering ability obtained by throwing the race water from the wheel under the hull. This requirement limited the low-water draft of the boat to slightly less than 4 ft. at the bow and about 3 ft. at the stern, but as these extreme conditions obtain only during the low-water season, with its greatly reduced currents and necessarily lighter loading of the barges, the full power of the wheel is not then required to be developed, and also the fuel supply to be carried may be greatly reduced, these conditions aiding to secure the lighter draft desired. At high stages the increased quantity of fuel to be carried, and possibly some water ballast in hull compartments provided therefor, can be relied upon to increase the draft 18 in. and 12 in., respectively, at bow and stern, and thus secure at will wheel immersion sufficient to develop the full power of the propelling machinery.

Three types of towboats and propelling wheels and appropriate machinery were considered; namely, hull with twin screw propellers in tunnels; side-wheel towboats; and the stern-paddle-wheel towboat of western river type, which was found to best fulfill the exacting conditions of the service planned. Objections to this type were the great overhanging weight of the paddle wheel, making difficult the fore-and-aft trim of the hull, and its rather inferior steering-ahead qualities as compared with side paddle wheels or twin screws; but its great backing and "flanking" (backing with strong sidewise movement) power give it unquestioned superiority over the other types.

The selection of the proper type of paddle wheel was also important. The radial fixed paddle wheel was chosen because of its simplicity of construction, ease of repair and, as compared to other types, light weight. The wheel designed probably will be one of the largest that has ever been installed in a stern-wheel towboat. While it weighs less than 60 tons, it will nevertheless

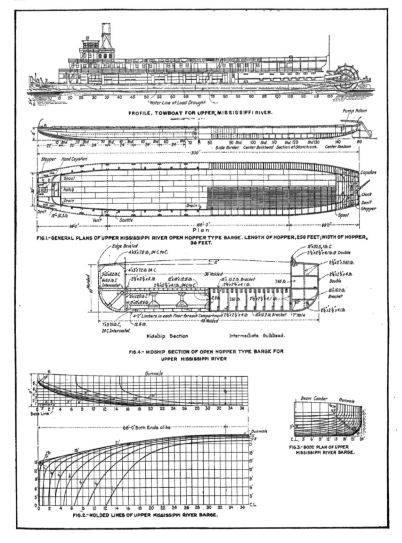

PROFILE, TOWBOAT FOR UPPER MISSISSIPPI RIVER.

Plan

FIG.1.—GENERAL PLANS OF UPPER MISSISSIPPI RIVER OPEN HOPPER TYPE BARGE. LENGTH OF HOPPER, 256 FEET; WIDTH OF HOPPER, 36 FEET.

Midship Section            Intermediate Bulkhead.

FIG.4.— MIDSHIP SECTION OF OPEN HOPPER TYPE BARGE FOR
UPPER MISSISSIPPI RIVER

FIG.3.— BODY PLAN OF UPPER
MISSISSIPPI RIVER BARGE.

FIG.2.— MOLDED LINES OF UPPER MISSISSIPPI RIVER BARGE.

be 22 ft. in diameter with 12 wooden paddles, each 40 ft. long by
3½ ft. wide, borne on 7 wheel flanges. The r.p.m. will be 22 to 25,
to develop the full power of thrust of 50,000 lb. required. The
shaft will be of forged nickel steel, carefully annealed and machined to dimensions. It will be 46 ft. 2 in. in length, 25 in. outside diameter, but with 7 octagonal seats on which the wheel
flanges will be keyed, hollow-bored to 19 in. diameter, with the
ends reinforced interiorly and integrally to 6 in. thickness under
the crank grips and journals, tapering into 3-in. walls under and
between the outside flanges.

The engines are twin tandem compound condensing, 24-in.
high-pressure and 48-in. low-pressure cylinders, and 8 ft. stroke,
and are designed for an initial pressure of 250 lb., with steam
superheat of about 75 deg. In order to carry the great weight of
the engines well forward of the stern of the boat, they are fitted
with 2 sets of pitmen, crossheads and slides, separated by about
17 ft., an arrangement which has been tried recently with success
on other government boats in the St. Louis District. The surface
condenser is of usual type and has 3000 sq. ft. of cooling surface.

The boilers on each boat are two in number, of Heine cross-
drum marine type, with 3½-in. tubes 14 ft. long; each boiler
having 4250 sq. ft. of heating surface, 100 sq. ft. grate surface,'
and with stack superheater (Foster type) and feedwater economizer, fan blower leading directly above to an individual stack
and with turbine forced draft in the furnace which is fitted with
grates for coal fuel, but arranged also for ready installation of
burners, air preheaters, etc., for oil fuel, as it is highly probable
that tar fuel, which can be obtained from local coking plants,
will be carried in the oil compartments in the hull.

The auxiliary machinery will include heavy steam capstans,
band capstans, settling tanks and neutralizing and clarifying
plant for boiling water, domestic water supply, electric-lighting
plant, refrigerating plant, steam steering engines and gear, oil
and bilge pipe lines and the usual feed, fire and deck pumps, injectors for emergency boiler feed, coal-handling apparatus,
machine tools for repairs, motor-boat tender for sounding, etc.

The hull will be fitted with a heavy steel spud and hoisting
engine, as for a dredge, for anchoring the boat and tow and to
afford a firm anchorage against which to pull to release barges
grounded on the shoals, which so frequently happens in all shallow
rivers.

Also, each boat is equipped with two sets of balanced rudders,
one set of 4 rudders fitted, as usual, at the stern transom, forward
of the wheel, and the other set of 2 rudders fitted aft of the wheel,
each set acting alternately in the strong race water from the
wheel when backing or going ahead, respectively, the double installation greatly increasing the steering and maneuvering power
of the craft.

It was originally planned to install towing engines and tow
barges behind the boat on cable hawsers as done in Europe, especially when running through narrow bridges, but on account of
the expense of installation these engines were omitted, as were
also the large rudders necessary for such towing. However, the
foundations for the engines on the towboats and the rudder
attachments on the barges were retained, and in future, if desired,
the remainder of the required astern towing plant may be installed.

Each hull for all this machinery is 230 ft. long, 58 ft. beam
and 8 ft. deep, and will displace about 1100 tons when under
steam. As in the barges, so also in the boat hulls the sunken deck
is used, forming a hopper 3 ft. deep from the boiler room nearly
to the after transom, thus lowering the height of the deckhouses
and reducing the wind surface of the boat, which will aid greatly
in maneuvering heavy tows through the bridge spans, narrow
channels and shoal water to be encountered. The deckhouses and
accommodations also are of steel, and fireproof as far as is possible without undue weight.

A novel feature of the hull construction will be stiffening and
strengthening trusses on each side running from the boilers to
the wheel beams aft of the shaft, supporting the wheel and also
supporting on the upper chords the entire steel middle deck (and
superstructure), which thus incorporated into the top members

of the trusses becomes an important component in the longitudinal strength of the boat.

Each boat will be completely outfitted for a full crew of 40
men, and will cost about $375,000 when complete, outfitted and
turned over to the lessee.

Fabrication for all the boats is now well advanced and the
erection of two hulls at Stillwater is progressing rapidly. It is
expected that both will be in operation during the coming fall
season, and that the remaining two will be ready by next spring.

### OTHER RIVER FLEETS

As a reflex of this attempt to stimulate and revive commercial
navigation on the upper Mississippi River, a similar movement
was at once started to utilize the lower Mississippi River between
St. Louis and New Orleans for general commercial transportation,
and the United States Railroad Administration furnished funds
to construct a fleet of 6 towboats and 40 large—2000-ton—barges,
fitted with steel protective deckhouses, for that service, but without awaiting the construction of these vessels the Railroad Administration began operations in the summer of 1918 with a fleet
of vessels obtained by purchase and lease from private and governmental owners. The new fleet is under construction, a number
of the barges already have been delivered by the builders, and one
of the towboats is soon to be in operation. The barges are of
the scow-bow type and the boats are of the tunnel twin-screw type
and about 1500-hp., that type being chosen for the boats as the
low-water channel depths in the lower river are thought to be
sufficient for such craft. It is expected that interesting comparison of the relative efficiencies of the two types of towboats and
barges in the two river sections may be made after one or two
years of operation.

The Great Warrior River also is to have a fleet of power barges
now building as another of the sequelæ to the upper-river revival;
and numerous inquiries from individuals and corporations indicate
that many private barge and steamboat lines may be expected
as the success of the Government lines becomes apparent.

### TERMINAL FACILITIES

As a result of widespread and searching public discussion and
of investigations by the War Department, covering the entire
field of transportation, it was early seen that modern, up-to-date
docks and terminal facilities, warehouses and yard trackage, all
equipped and arranged for convenient, quick and economical
interchange of freight between water and rail carriers, are essentially, almost vitally, necessary adjuncts to successful and effective
use of the rivers.

In promotion of such facilities a circular letter was issued in
October 1917 by the Board of Engineers for Rivers and Harbors
(advisory to the Rivers and Harbors Committee of the House of
Representatives in Congress), which was distributed by the War
Department to all municipalities concerned, calling upon cities
and towns located on navigable streams to construct and maintain
proper river terminals, open to all on equal terms, as a pre-
requisite to further appropriations for river improvement, and
in encouragement of a revival of river-borne traffic. In response
thereto the cities on the Mississippi River and its tributaries, generally speaking, have made earnest efforts to comply by planning
and arranging for terminals and wharves to such extent that every
important city on the river between St. Louis and Minneapolis,
and on the Illinois River, has either built terminals or has prepared financially to do so, as necessity therefor arises. Particulars of these developments are given in the complete paper.
This indicates an intense public interest in the question of river
traffic, as, practically, commercial navigation on these rivers is
now at its lowest ebb.

In the foregoing sketch of that part of the present active effort
to revive water transportation, possibly centering at St. Louis
and on the Mississippi River, may be seen the very general public
interest in the use of the river, and it is hoped that the seed sown
by the Government in its aid to the movement may have fallen
on fertile ground and will return the economic relief and harvest
for which the Mississippi Valley has hoped so long.

# Superchargers for Airplane Engines

### The Necessity for Supplying Air to the Carbureter at Sea-Level Pressure at High Altitudes— The Moss Supercharger for This Purpose, Its Design, Development and Performance

By SANFORD A. MOSS,[2] WEST LYNN, MASS.

A N airplane flying at high altitude is in an atmosphere of comparatively low density. For instance, at 20,000 ft. altitude the density is practically half that at sea level. This means that a given volume contains half as much actual air by weight. The cylinders of an airplane engine are therefore charged with an explosive mixture which has about half the value of a charge at sea level. The engine actually delivers about half of its sea-level power at 20,000 ft.

Both the low temperature and the decreased pressure at high altitude have effect in fixing the high-altitude density. Both the decrease of temperature and the decrease of weight of the charge affect the carburation at high altitude. The fixed clearance volume and the decreased initial pressure give a decrease of compression pressure resulting in a loss of efficiency. There is, therefore, a combination of causes which gives as a net result a decrease in engine power very nearly proportional to the decrease in density.

At high altitude the resistance of the air to the motion of the airplane is decreased directly in proportion to the decrease of density. The power required for a given airplane speed is therefore greatly reduced. However, the engine power has been so reduced that the usual net result is a considerable decrease in airplane speed. When the engine power is maintained at the sea-level value, there is, however, a considerable increase of speed at high altitude.

Filling the cylinders of an internal-combustion engine with a charge greater than that which would normally occur, is called "supercharging." Methods of doing this have engaged the attention of a great many experimenters.

The centrifugal compressor is an apparatus similar to the fan blower except that the shape of the impeller blades and the passages leading air to and from the impeller are so arranged as to give an efficiency very much greater than that of the usual type of fan blower, so that the apparatus forms a satisfactory means for compressing air to appreciable pressures. A line of single-stage centrifugal compressors has been developed for compressing air from 2 to 5 lb. per sq. in. above atmosphere, to be used for many industrial purposes; as well as a line of multi-stage machines for compressing air and gas up to pressures of 30 lb. per sq. in. above atmosphere.

The turbo-supercharger is a combination of a gas turbine and a centrifugal compressor, arranged as part of an airplane gasoline engine. The hot products from the engine exhaust are received upon the turbine runner and furnish power whereby is driven a centrifugal compressor mounted on the same shaft, which compresses air for supply to the carbureters. A more detailed description is given later, as well as particulars regarding its development.

In the latter part of 1917 the National Advisory Committee for Aeronautics requested the coöperation of the General Electric Company in the development of the turbo-supercharger in the United States. Our work was originally started at the suggestion of Dr. W. F. Durand, then Chairman of the Committee, who knew of our long experience with gas turbines and centrifugal compressors. It has since been carried on under the supervision at various times of Col. J. G. Vincent, Col. T. H. Bane, Major H. C. Marmon, Major G. E. A. Hallett and Major R. W. Schroeder. Major Hallett has had charge of the development since the armistice, and he has given considerable study to the matter of superchargers in general.

[1] Condensed from an article on The General Electric Turbo-Supercharger for Airplanes, published in the June 1920 issue of the General Electric Review, to which journal MECHANICAL ENGINEERING is indebted for advance proofs of the article and for the illustrations here used.
[2] Engineer, Turbine Research Department, General Electric Co. Mem. Am.Soc.M.E.

## THE TURBO-SUPERCHARGER CYCLE

Fig. 1 shows an airplane engine equipped with a turbo-supercharger. The exhaust of the engine is received by an exhaust manifold which leads it to a nozzle chamber carrying nozzles which discharge it on to the buckets of a turbine wheel. On the same shaft as this turbine wheel is the impeller of a centrifugal compressor. This compresses air from the low-pressure atmosphere to approximately normal sea-level pressure and delivers it to an air-discharge conduit which supplies the carbureters.

The turbine nozzles are of such area as to maintain within the exhaust manifold and nozzle box a pressure approximately equal to that at sea level. The difference between this pressure and the altitude low pressure gives a pressure drop for the exhaust gases which furnishes the power that operates the system.

Due to the respective temperatures this power input suffices to

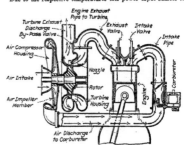

FIG. 1 DIAGRAMMATIC SKETCH OF AN AIRPLANE ENGINE EQUIPPED WITH A SUPERCHARGER

give the desired compression and also to supply the inevitable losses. However, in order to avoid back pressure on the engine, above the normal sea-level value, both turbine and compressor must be designed with utmost attention to efficiency.

With an efficient arrangement the engine when at high altitude exhausts at normal sea-level pressure and receives its air at the carbureter at normal sea-level pressure. Hence, normal sea-level power is delivered at all altitudes up to the maximum for which the supercharger is designed, so that the plane speed will increase uniformly as the altitude density decreases.

## MECHANICAL PROBLEMS OF SUPERCHARGING

The General Electric superchargers thus far constructed have been designed to give sea-level absolute pressure at an altitude of 18,000 ft., which requires a compressor that doubles the absolute pressure of the air. This pressure ratio, with the quantity of air involved, means about 50 shaft hp. input for the compressor. The design of a complete power plant of this size, to suit an existing airplane engine, with such weight and location as will not impair the flying characteristics of the plane, has of course offered many problems. The possibility of driving the compressor of the supercharger by engine power instead of by the exhaust gases suggested itself. Much experience with the operation of the gas turbine, however, led the writer to prefer its problems to those of the driving mechanism of a supercharger operated from the engine. The turbine involves merely the addition to the compressor of a single extra wheel, designed for the conditions, with no extra bearings. The engine-driven scheme involves a 50-hp. transmission with a multiplicity of gears, bearings, clutches, belts, and the like. These offer more or less drag on the engine when the supercharger is not in use at low altitudes, and very serious problems of acceleration when the supercharger is to be thrown into action, since the engine will be then running at its full speed of about 1800 r.p.m.

The exhaust manifold and nozzle box have proved to be a very efficient exhaust muffler and conductor. Such a muffler and conductor is needed in any event, and the design of means for withstanding the increased pressure difference of the turbo-supercharger has been successfully accomplished.

### POWER FOR TURBO- AND ENGINE-DRIVEN SUPERCHARGERS

An efficient turbo-supercharger theoretically deducts from the indicated horsepower of the airplane engine an amount corresponding to the difference between sea-level absolute pressure and altitude pressure. There is this additional back pressure during the exhaust stroke. The theoretical power available for driving the turbo-supercharger is greater than this, however, owing to the fact that there is available not only the energy due to the direct pressure difference mentioned, but also the energy of perfect expansion from the higher to the lower pressure. If there were no turbo-supercharger the engine would waste this energy in sudden pressure drop as the exhaust valve opens. The turbine can utilize this energy. The sum of these two amounts of avail-

(with muffling advantages) and delivery to the turbine wheel. As already mentioned, practical success to date is in favor of the turbo-supercharger and the writer feels that this is really due to its innate superiority.

Engine-driven superchargers with positive-pressure blowers have been proposed. These have the additional disadvantage that with the desirable pressure ratios of about two to one there is an appreciable compression loss due to the fact that the machine only displaces air and has no direct means for compression.

Supercharging engines of various kinds, in which the engine crankcase or the engine cylinders themselves are arranged for additional compression, have been shown to give excessive weight and complication as compared with a turbo-supercharger.

### DEVELOPMENT OF THE TURBO-SUPERCHARGER

The machines used thus far have been designed to give sea-level pressure at 18,000 ft. altitude, which corresponds to a pressure ratio of about two. The rated speed for these conditions is 20,000 r.p.m. Sea-level pressure has readily been obtained up to 22,000 ft. altitude. The control is entirely by hand operation of waste gates, which permits of free escape of some of the exhaust gases.

The entire apparatus, exclusive of exhaust manifold and air-discharge conduit, weighs about 100 lb. The exhaust manifold and air conduits have nearly the same weight as equivalent parts with no supercharger.

The turbine and compressor wheel have diameters somewhat

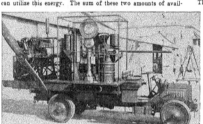

FIG. 2 MOTOR TRUCK PREPARED FOR EXPEDITION TO SUMMIT OF PIKE'S PEAK FOR TESTING LIBERTY MOTORS

FIG. 3 SHOWING WAY IN WHICH MOTOR TRUCK OF FIG. 2 WAS LEFT AFTER END OF DAY'S WORK

able energy, multiplied by the efficiency of the turbine wheel, gives the shaft power delivered to the compressor.

For an engine-driven supercharger compressor there is greater engine indicated power due to a lower exhaust pressure. However, the shaft power for the supercharger compressor must be transmitted through the engine connecting rod and crankshaft, with losses, and then through the supercharger driving mechanism with additional losses. The total shaft power thus subtracted from the engine, multiplied by the efficiencies of these two transmissions, gives the shaft power delivered to the compressor. This is the same as for the turbo-supercharger. For a Liberty motor of about 400 hp. and sea-level power at 18,000 ft. altitude, this power is 50 hp.

The comparison then is as follows: The turbo-supercharger subtracts from the engine indicated power, adds power of expansion which would not otherwise be used, and has turbine wheel losses. The engine-driven supercharger puts this indicated power through the engine (with some additional loads on the pins and bearings) and has engine and transmission losses.

With usual efficiency there is probably not a great difference between the gross subtraction from engine power in the two cases. There is then the disadvantage of transmitting the supercharger power through the engine pins and bearings, as well as through some mechanism between engine and supercharger, to be compared with the collection of the hot gases under pressure

less than a foot. The present design has been hampered by necessity for accommodation to existing engines and planes. It is proposed, however, to construct apparatus in which engine and supercharger are integral, with all parts arranged for the full possibilities of the combination.

In the combination under consideration the airplane, propeller, engine, radiator, cooling system, and supercharger are so intimately associated that no adequate tests can be made without the complete system in operation at full speed at altitude. During the initial development of the Liberty motor a testing expedition had been sent to the summit of Pike's Peak, and it was decided to repeat this performance with the supercharger. Fig. 2 shows the motor truck that was prepared for the expedition and Fig. 3 the way it was left after each day's work. The Liberty motor carrying the supercharger was mounted on a cradle dynamometer, with scales and all arrangements for accurate measurement of power, gasoline consumption and the like. In fact, a complete testing laboratory was provided. The motor truck was shipped by rail to Colorado Springs, and then proceeded by its own power to Pike's Peak summit on the "Pike's Peak Auto Highway," a well-constructed but very tortuous mountain road 28 miles long. The summit has an altitude of 14,109 ft. and it is the highest point in the United States easily reached by road.

The testing work at the summit lasted through September and half of October, 1918. The usual difficulties with experimental work were, of course, encountered with the addition of many delays, due to the cold and snow, and distance from repair shops. The apparatus was finally arranged to give good mechanical operation and it was found possible at the existing altitude of 14,000 ft. not only to supercharge so as to give full sea-level power, but also to overcharge so as to cause the engine to preignite.

It was agreed that results of the tests warranted the immediate installation of the supercharger on an airplane, and arrangements for doing this were in progress when the armistice caused a cessation of the work. After the armistice, careful reëxamination of the situation resulted in resumption of the work in the early part of 1919. Various rearrangements were made in view of the experience gained at Pike's Peak and the apparatus was finally installed on an airplane.

It soon developed that a very appreciable increase of power was easily obtained when the supercharger was opened up. The whole airplane installation was not properly arranged to take advantage of this power, however, and changes were necessary in the radiator, cooling system, propeller system, gasoline tank,

FIG. 4  RIGHT TO LEFT: MAJOR R. W. SCHROEDER, LIEUT. GEO. W. ELSEY AND DR. MOSS

pump system, etc. Changes in these parts have been made from time to time, and this work is still in progress. As the work proceeds more and more power is developed by the engine. Changes have also been made in the supercharger itself.

Fig. 4 shows Major R. W. Schroeder, who has made all of the flight tests to date, together with Lieut. George W. Elsey, who has made all of the flight observations to date. The aviators are of course clothed for the intense cold of high altitudes and carry the parachutes that are now regularly used by the U. S. Air Service in experimental work.

### SUPERCHARGER PERFORMANCES

The supercharger which has been used to date was primarily desired for high speeds at altitudes of 18,000 to 22,000 ft. The Le Pere plane on which the installation was made had a ceiling of about 20,000 ft. with two men, and a speed at this altitude of 70 miles per hour. With the supercharger in use, a speed of about 140 miles an hour has been attained at 22,000 ft. As already pointed out, this has been attained with various parts of the plane installation in a partially developed state. Theoretical computations have been made showing that much higher speeds at high altitudes are to be expected, and the flight tests to date indicate that the theoretical expectations will be fully realized.

The making of high altitude records has been very attractive and the supercharger has, of course, been used for this purpose as well as for the speed courses mentioned. Successively higher altitudes have been reached as experience has been gained regarding the manipulation of oxygen, gasoline, and other details.

On February 27 Major Schroeder made a flight alone, attaining an actual height above the ground finally computed at 36,130 ft. (6.85 miles). The lowest temperature reached was minus 67 deg. fahr. At the maximum altitude his oxygen apparatus failed and he became unconscious and lost control of the plane, which fell almost vertically. As he neared the earth he partly recovered consciousness and, at an altitude of about 3000 ft., succeeded, in a half-dazed semi-automatic way, in righting the plane and making a good landing in his own field, again becoming unconscious. He was taken to a hospital in a serious condition, but has since almost completely recovered. The supercharger, engine, and plane were in perfect working order after the flight.

At the maximum altitude attained, recording instruments showed that the plane was still climbing at the rate of about 125 ft. per minute and it was estimated that an altitude of 40,000 ft. would have been attained if the oxygen apparatus had not failed.

[In the May issue of the *General Electric Review* there also appears a reprint of a paper on Superchargers and Supercharging Engines, by Major George E. A. Hallett, U. S. A., presented at the annual meeting of the Society of Automotive Engineers, January 7 and 8, 1920. Major Hallett, who is Chief of Power-Plant Division, U. S. Air Service, deals at some length with the various methods employed in supercharging and refers to the work of the U. S. Air Service on the Rateau type of turbo-compressor, under the supervision of Mr. E. H. Sherbondy, prior to that undertaken by Dr. Moss. Commenting on the working of the Moss supercharger, he says:

It would naturally seem at first thought that the extremely low temperatures always found at great altitudes would make possible the easy solution of cooling problems, but in reality the low density of the air reduces its heat conductivity and capacity for heat absorption to such a point that a supercharged engine developing sea-level power at 20,000 ft. requires a little *more cooling surface* than it does when developing normal power at sea level.

The Liberty engine and many others run best with a water temperature of about 170 deg. fahr. To maintain the cooling water at this temperature in the reduced atmospheric pressure at 25,000 ft. it is necessary to use several pounds of air pressure in the radiator to prevent the water from boiling away. Very effective radiator shutters are needed when the engine is throttled to make a descent from altitudes of over 20,000 ft. to prevent the water in the radiator from freezing before warmer air is reached.

Contrary to expectations, the Moss turbo-compressor now being tested at McCook Field does not complicate the pilot's controls. On a normal engine the pilot handles the throttle and the altitude carburetor control which thins down the mixture as he ascends. With the turbo-compressor the altitude control becomes unnecessary up to the altitude at which the engine can no longer deliver sea-level power but is used, as with a normal engine, if the plane is driven higher.

As to the future of the supercharger, Major Hallett says:

The uses of the supercharger for military service can be divided into: first, for airplanes in which it is desired to reach extreme altitude; second, for airplanes in which it is desired to increase the rate of climb and horizontal speed and therefore maneuverability at altitudes where it is intended to fight; and, third, for airplanes which carry large loads such as bombers, which normally are handicapped by having a very low ceiling and whose entire usefulness would, if larger engines were installed to pull them to a higher ceiling, be lost on account of the large amount of fuel and other material that would have to be carried, thus decreasing their radii of action.

In the first case it is believed that a special supercharger can be built that will make feasible much greater altitudes than any that have been attained with the present General Electric turbo-compressor; and it is considered essential that we have airplanes capable of reaching very great heights. In the second case, it is pointed out that military machines not fitted with superchargers engines, when fighting at an altitude of 20,000 ft. or more, are so near their ceiling that their rate of climb, speed, and maneuverability are comparatively poor, but the use of a supercharger seems to overcome this difficulty easily.

The use of superchargers in commercial airplanes of the future is assured because superchargers will make possible far *more miles per hour* and *more miles per gallon* with a given engine and airplane, and speed is the main advantage of air over other kinds of transportation. It is thought by many qualified judges that by flying at a sufficient height with a supercharged engine and a suitably designed airplane, a speed of 200 m.p.h. can be maintained.]

# Spring Meeting of The American

### Whole-Hearted Hospitality Extended by St. Louis — Engineers Present From Many Sections
### Professional Sections Started — Discussion of Local Engineering Problems
### Joint Session With Society of Refrigerating Engineers

THE 1920 Spring Meeting was the first convention of the Society to be held in St. Louis in twenty-four years and so open-hearted was the hospitality, so well carried out were the excursions and entertainments, and so thoughtfully was every detail worked out for the comfort of the guests that every member in attendance enthusiastically demanded an early repetition of a St. Louis Spring Meeting. It was successful in every phase, as in addition to the social features the professional sessions were well attended and pertinent discussions contributed.

The meeting gained momentum quickly, the business session starting promptly at 2 p.m. on Monday, being followed by the professional sessions as scheduled. The total registration was 475 members and 330 guests, most of whom arrived for the first session and remained to the end. All of the professional sessions were held at the Hotel Statler, the headquarters of the meeting.

#### Opening Reception

On Monday evening a reception for visiting members was held in the Hotel Statler. Clarence H. Howard, president of the Commonwealth Steel Company of St. Louis, extended a warm welcome and spoke about the development of the Junior Chamber of Commerce and the importance of business training for young men and boys. President Miller responded with the thanks of the Society for the hospitality shown up to that time and anticipated during the remainder of the convention. He also spoke of the problems in industry and earnestly pleaded for closer cooperation between workmen and managers. Mr. E. R. Jackson displayed several reels of the official moving-picture record of the First Transcontinental Motor Convoy. The pictures were accompanied by comments from Mr. Jackson on the various events taking place on the trip, and were received with interest and close attention. Mr. L. C. Nordmeyer, chairman of the St. Louis Section, presided at this meeting. The members of the St. Louis Section proved delightful hosts and everyone thoroughly enjoyed the dancing which followed.

#### Entertainment

The Spring Meeting banquet, held Tuesday evening, May 25, in the large dining room of the Missouri Athletic Association, was attended by over 700 members and guests. A fine musical program was presented during the dinner, after which Lt.-Col. H. W. Miller, of the Ordnance Department, presented a paper on the German Defenses on the Coast of Belgium. Colonel Miller's paper, which appeared in the June MECHANICAL ENGI-

NEERING, is a valuable contribution to the information necessary for the solution of the coast-defense problem of this country. The very formidable fortifications thrown up by the Germans on the coast of Belgium kept the Allied fleets at bay for three years and the lessons that may be learned from this paper are of great interest to every one who understands the great problem before the War College in the selection of proper methods of coast defense.

A special entertainment for all engineering societies holding meetings in St. Louis and all members of St. Louis engineering societies was arranged at the beautiful new municipal open-air theater in Forest Park. The interesting program of music and interpretative dancing was presented from a picturesque natural stage, which, viewed from the outer circle of the amphitheater, gave an inspiring perspective of lights and color. St. Louis may well be proud of this concrete structure, holding 7000 people and having perfect appointments.

On Thursday afternoon the excursion steamer St. Paul was chartered for a boat ride up the Mississippi. This trip was a decided success in that it afforded a fine opportunity for making acquaintances and for that reason was voted the most successful entertainment of the meeting. Music and dancing were enjoyed on the boat.

The ladies attending the convention were particularly well taken care of and were afforded very interesting trips through the residential section, through Forest Park and visits to the Art Museum, Shaw's Garden, Washington University and to Brentmoor, Westmoreland and Portland Places.

#### Trips of Inspection

The trips of inspection planned by the St. Louis Committee covered a wide variety of industries and engineering works for the visiting members. On Tuesday afternoon the plant of the Commonwealth Steel Company at Granite City, Ill., was visited by a large number who were very much interested in the new methods shown for making steel castings. The 122,000-kw. plant of the Union Electric Light and Power Company at Ashley Street was also visited Tuesday afternoon, followed by a call at the docks of the St. Louis Boat and Engineering Company where steel barges are under construction for the United States Government.

Wednesday morning the trips included the plants of the General Motors Corporation, Wagner Electric Company, Everhght Piston Ring Company, and Heine Safety Boiler Company.

# Society of Mechanical Engineers

of the Country—Members Enthusiastic and Meeting Pronounced a Great Success—Two Important Feature—Appraisal and Valuation Session Excites Interest— and Society of Heating and Ventilating Engineers.

Lunch was served at the Bevo Plant of the Anheuser-Busch Association, following which the party moved on to the works of the Busch-Sulzer Brothers Diesel Engine Company, the Mississippi Valley Iron Company, and the coke plant of the Laclede Gas Light Company.

### PUBLIC HEARINGS

On Tuesday morning, May 25, at 10 a.m., the Committee on Protection of Industrial Workers opened to public discussion the Elevator Code covering the construction, operation and maintenance of elevators, dumbwaiters and escalators. Many useful criticisms and suggestions were advanced which will be considered in the work of the Committee. At 2 p.m. on Tuesday the Boiler Code Committee held a public hearing on the Rules for Air Tanks and Pressure Vessels. At this session the discussion centered largely on the value of welding the seams of the vessels under consideration.

### KEOKUK EXCURSION

Eighty members of the Society availed themselves of the opportunity to inspect the hydroelectric developments of the Mississippi River Power Company at Keokuk. The special cars arrived at Keokuk Sunday morning and were met by representatives of the Power Company and after an Iowa breakfast the party was split up into small sections which were effectively guided across the locks and docks to the power house and dam. Every detail was explained carefully and with consideration by the power-house attendants. Mr. L. E. Dickinson was in charge of the plant for the day and to his courtesy and thoughtfulness the success of the excursion was undoubtedly due. After the power-house inspection the party motored about the adjoining country and after dinner went aboard the sleeping cars and arrived at St. Louis on Monday morning.

### TULSA TRIPS

Following the meeting in St. Louis, about fifty of the attending members left for Tulsa, where, Friday and Saturday, a joint meeting of the Mid-Continent section of The American Society of Mechanical Engineers was held with the Oklahoma section of the American Chemical Society. A full account of this meeting and the trips in connection with it is given elsewhere in this issue.

### LOCAL COMMITTEES

The entire arrangement for the entertainments, excursions and publicity for this meeting was under the jurisdiction of the Gen-

eral Committee, the personnel of which was as follows: M. L. Holman, Hon. Chairman; Louis C. Nordmeyer, Chairman; Lewis Gustafson, Vice-Chairman; Fred. E. Bausch, Secretary; William P. Samuel, Treasurer; Ernest L. Ohle, H. Wade Hibbard, Chester B. Lord, Edward W. Schadek, Victor J. Azbe, and Edward R. Fish.

Emphasis must again be placed on the hospitality extended by the St. Louis Section and the thoughtfulness with which every detail was worked out. The work of the Women's Committee, of which Mrs. Ernest L. Ohle was chairman, in anticipating every desire of the wives of the visiting members and guests was particularly noteworthy.

In the publicity work the St. Louis Convention and Publicity Bureau was very helpful and its representative, Mr. Francis E. Turin, worked in very close coöperation with the Committee.

## Business Meeting

THE matters coming before the business meetings of professional societies are often regarded by the members as uninteresting. However, at the semi-annual business meeting in St. Louis, held on the Monday afternoon, the attendance was quite large and the discussion general, indicating a good interest in the various details presented for consideration.

Vice-President Fred R. Low opened the meeting, and after a few minutes President Miller took the chair and introduced the subjects of the meeting.

The first business presented by the Secretary was the report of the Tellers of Election, John H. Lawrence, Mancius S. Hutton and R. K. MacMaster, on amendments to the Constitution. This report was adopted on the motion of Mr. Low and the amendments declared carried. They provided for

*a* Addition of Standing Committee on Professional Sections.

*b* Approval or adoption of reports, standard formulæ or recommended practices.

*c* Termination of membership on the Council or any committee on account of absence.

The amendments were published in MECHANICAL ENGINEERING, January, 1920, p. 33.

Mr. H. J. C. Hinchey then presented the report of the Special Committee on Code of Ethics. This report with the Code incorporated in it will be published in MECHANICAL ENGINEERING. The report was referred to Engineering Council or to such organization as may be formed for joint action by the Engineering Societies on matters of common interest to form the basis of a possible general Code of Ethics. The Committee was continued to take such action as might seem advisable looking toward the adoption of a Code by the Society in case no other suitable action was taken.

A report of the Boiler Code Committee on Boilers for Locomotives was presented by title and those interested were referred to a special session held for its discussion.

Under new business President Miller introduced an amendment to the Constitution shortening the time required to complete amendments to the Constitution, which was discussed but no modification made. This amendment will follow the usual procedure.

The question of coöperation by the Society with the National Public Works Department Association was introduced and discussed, and the next day the following resolution was passed indorsing this movement:

WHEREAS: The American Society of Mechanical Engineers be-

At the adjournment of the business meeting the session on Appraisal and Valuation convened immediately.

## Two Professional Sections Started

WELL-ATTENDED conferences looking to the organization of Machine Shop Practice and of Power Professional Sections under the new authority of the Council were held at the Spring Meeting, and at both of them the sentiment toward the organization of the sections was unanimous.

The Machine Shop Practice conference was held on the Monday morning and was called to order by Prof. H. P. Fairfield who was the Secretary of the former Sub-committee of the Meetings Committee on this subject.

The meeting instructed Professor Fairfield to appoint a committee of three to proceed with the organization of the section, and he selected R. E. Flanders, C. B. Lord and himself. The Committee met immediately and decided on a form of petition to

ANALYSIS OF JUNIOR MEMBERSHIP PRESENTED BY MR. M. L. COOKE AT THE BUSINESS MEETING

lieves thoroughly in the principles of functional organizations and its members, as engineers, are particularly concerned with reorganizations leading to the substitution of efficiency and economy for looseness and waste, therefore, be it

*Resolved:* That The American Society of Mechanical Engineers in convention assembled fully indorses the effort which the National Public Works Department Association is making to achieve these results and specifically urges upon its local sections that in conjunction with the Local Engineering Societies these sections support this movement fully and effectively.

Mr. Morris L. Cooke introduced an amendment to the Constitution to extend the privilege of voting to Junior Members. He cited a resolution passed by the Philadelphia Section making such a recommendation. He quoted figures to show that the Junior Membership was approximately 23 per cent of the total and that that part of it below the age of twenty-seven was only 8 per cent.

Mr. Cooke also spoke of the necessity of affording proper outlet for the ideas and activities of the younger members and stated that the required combination consists of the sagacity and experience of the older men and all that young men give. Mr. Cooke's analysis of Junior Membership is shown in the above table.

The proposal was discussed by S. N. Castle, M. M. Case, H. B. Dirks, Max Toltz, C. B. Lord, J. H. Bernhard and J. M. Spitzglass, and Mr. Cooke's proposal was accepted by the meeting and his amendment will therefore go through the usual procedure.

the Council and reconvened the general session on the Wednesday when signatures to the petition were taken.

On the Monday the Council voted to refer this petition when received by the Secretary to the Executive Committee of the Council for action.

The conference on the Power Section was held on the Wednesday morning when Prof. R. H. Fernald, member of the Special Committee on Professional Sections representing this subject, called the meeting to order.

Practically the same procedure was followed as in the meeting of the Machine Shop Practice Section and Professor Fernald appointed F. R. Low, J. R. Bibbins, and A. L. Rice as the Organization Committee. This Committee drew up the petition and it has already been received by the Secretary and is being balloted on by the Executive Committee of the Council.

To date 350 members of the Society have registered in the new Machine Shop Practice Section and 850 members in the Power Section. Both of these Sections expect to conduct sessions at the forthcoming Annual Meeting under the auspices of the Committee on Meetings and Program.

During the discussion at these conferences it transpired that there was a little doubt as to the difference between a professional section and a professional session of a general meeting. The Professional Sections are intended to be organizations of a permanent character of groups of members of the Society who will devote continuous attention to their development of their subjects and will report either at Annual Meetings or at Local Section meetings or at meetings of their own or through the publications of the Society. Professional sessions at general meetings are simply divisions of the program of the professional papers presented to the Society determined upon by the Committee on Meetings and Program for convenience in conducting the Annual and Spring Meetings; the names of these sessions depending on the character of the papers on hand and the topics selected by the Meetings and Program Committee for the conventions.

### ADVANTAGES OF PROFESSIONAL SECTIONS BROUGHT OUT

The organization of Professional Sections is a new one and it might be well to quote some of the views of the members expressed at the Spring Meeting on the development of the Society's work into this channel. Here are some of the opinions:

There is great value in what we sometimes call "verandah talks" in little groups with no reporter present, and so long as we keep within the limits of the Society's activities prescribed by the Council, I see no harm in these.

The Professional Sections will give the particular phases of engineering to which they are devoted more standing and dignity than was possible under the old arrangement of Sub-committees of the Meetings Committee.

There are two tendencies in the Society: One is to narrow everything down to the purely technical, and the other is to open everything out to include general things of interest. The more we force ourselves into the narrow technical side of engineering the less good we will be to the community and the less influence we will have.

This Professional Section movement is a very important one in the life of the Society, but we must guard against duplication of effort which will have a tendency to weaken what is destined to become a very important development.

In the Professional Sections so far we are simply limited by the matter of expense. When that is overcome we shall have a perfectly free hand as to presenting papers and holding sessions at any time and anywhere in the country, with the restriction that if we desire sessions in the regular Annual and Spring Meetings, we must come on the regular program prepared by the Meetings Committee and do things under their instructions.

The general idea prevails throughout the country that in order to get into The American Society of Mechanical Engineers one must possess a M. E. diploma or be a college graduate. This is not the actual requirement, but nevertheless the qualifications for membership are high. One object of the Professional Sections is to interest a number of men in the work who are not immediately eligible for membership in the Society, but who can join these Sections as Affiliates.

There are a great many men in the country who know all about every phase of the work of the various Professional Sections being organized, but there are no organizations in their particular little fields which interest them. These men should be brought into our organization.

## Appraisal and Valuation Session

THE five papers submitted for this session had been previously contributed as discussion at the 1919 Annual Meeting and because of lack of time for complete consideration, it was decided to present the papers again at the Spring Meeting in 1920. The wisdom of this action was apparent from the interest shown and the value of discussion advanced. No agreement was reached as to methods of valuation and because of the grave importance of the subject to the engineering profession, resolutions were passed at the close of the Session which will keep the subject a vital issue before the Society.

This session, presided over by Dean Charles Russ Richards, Manager of the Society, was well attended and after three hours, was adjourned to the next day when, under the gavel of Vice-President F. R. Low, the discussion extended over two hours.

The session was opened by reading the following brief of Dr. David H. Ray's[1] paper on Appraisal and Valuation Methods:

The outstanding feature of the paper is the method of integration of value by weighted average. This method cannot easily be applied by accountants because they do not know how to determine the coefficients.

[1] Professor of Industrial Engineering, California Institute of Technology, Pasadena, Cal.

---

Members of St. Louis Local Committees, Whose Efforts Contributed So Fully to the Splendid Success of the Meeting

**General Committee**
M. L. Holman
   *Hon. Chairman*
Louis C. Nordmeyer
   *Chairman*
Lewis Gustafson
   *Vice-Chairman*
Fred. E. Bausch
   *Secretary*
William P. Samuel
   *Treasurer*
Ernest L. Oble
H. Wade Hibbard
Chester B. Lord
Edward W. Schadek
Victor J. Azbe
Edward R. Fish

**Reception Committee**
Roderick H. Tait
   *Chairman*
A.S.M.E. St. Louis Membership

**Women's Committee**
Mrs. Ernest L. Oble
   *Chairman*
Mrs. Edw. R. Fish
   *Vice-Chairman*
Wives of the Associated Engineering Societies Membership

**Hotel Committee**
William H. Reeves
   *Chairman*
Joseph W. Peters
   *Vice-Chairman*
E. Julius Boehmer
William E. Doll
Robert F. Wiselogel

**Publicity Committee**
Victor J. Azbe
   *Chairman*
Edmond Siroky
Walter Siegrist

**Finance Committee**
Philip DeC. Ball
   *Chairman*
John F. O'Neil
William M. Duncan
Hugo Wurdack
Louis C. Nordmeyer

**Program Committee**
Ernest L. Oble
   *Chairman*
H. Wade Hibbard
   *Vice-Chairman*
Robert G. Clyne
Dwight T. Farnham
George S. Hessenbruch

**Entertainment Committee**
Fred. E. Bausch
   *Chairman*
Frank N. Jewett
Arthur S. Hawks
Charles F. Frede
George B. Evans
Leonard A. Day
Stevenson A. Dobyne
Edward H. Tenney

**Information Committee**
Edward R. Fish
   *Chairman*
Nathan Kohn
Fred Key
Lyman C. Huff
William A. Hoffman
Edward L. Dillon
Ray L. Howes

**Transportation Committee**
Walter E. Bryan
   *Chairman*
James E. Allison
G. Horton Blackman
Franz A. Berger
Henry F. Gauss
Allen C. Staley
B. Frauenthal

**Meetings and Program Committee in General Charge of the Society's Conventions**
Dexter S. Kimball      W. G. Starkweather
   *Chairman*      R. V. Wright
A. L. De Leeuw      J. W. Roe

---

The second feature is the development of an "aleatory factor," which applied to modified original cost or reproduction cost, gives going value, must be based on the bed rock of theoretical economics and developed by experience and judgment in industrial engineering.

The basic economics is contained in John Stuart Mill's Principles of Political Economy, volumes I and II, a striking paragraph being:

"The result of this long investigation may be summed up as follows. The economical progress of a society constituted of landlords, capitalists and laborers, tends to the progress enrichment of the landlord class; while the laborer's subsistence tends on the whole to increase and profits to fall."

Normally by general economic law profits tend to become less, and laborers normally find themselves in a tighter position with the years, for subsistence; in other words the mechanism of fundamental world economics normally reduces profits and increases cost of living (and demands for wages).

The way out of the dilemma is mechanical research and development. Here the skilled judgment of the industrial engineer comes into play and he must allocate the values and profits according to the various factors concerned in production and develop an aleatory factor which will differ with the class of business and its location.

It is to be noted that accounting systems often carry land valuation at cost until sale or appraisal as our modern economic industrial line up tends broadly "to the progressive enrichment of the landlord class." this may or may not be a factor in fair valuation. It however requires consideration in the formulation of a just aleatory factor.

In general therefore the determination of the aleatory factor is to be determined by an allocation of the elements of risk in the project considered along broad economic lines and modified by the special peculiarities of the business and its situation.

In presenting his paper on Price Levels in Relation to Value, Cecil F. Elmes[3] quoted John Stuart Mill's definition of value as "the value of a thing is its general power of purchasing, the command which its possession gives over purchasable commodities in general." Price he defined according to the New Standard Dictionary as "the amount demanded by an owner in return for an article as the condition of sale. . . . the quantity of labor which its possessor will take in exchange for it."

After showing the results of historical studies of commodity values and wages Mr. Elmes stated his conclusions as follows:

a   It is of the utmost value to all of us and certainly to those engaged in the public utility business that precise use of the terminology of economics be insisted upon.

b   Low commodity prices while superficially desirable may be in a broad sense the reverse of a blessing. Inasmuch as they signify a shrinkage in demand in commodity markets, they may be an indication of depression or stagnation in business.

c   High prices—so long as they do not run into panic or famine prices—may equally be by no means an economic evil. Panic and famine prices are and should be within the control of the community and their evil effects can be largely eliminated by a wise use of that control.

d   A great and progressive movement in the conditions of the laboring man is without question a large factor in the high prices of the present day.

e   Every public-spirited citizen, having the welfare of the community at heart, and conversant with the evil of low wages, desires to see labor get just recognition and every incentive to progress and development.

f   Outside of catastrophic disturbances which we cannot undertake to predict, a belief in lower commodity prices in the immediate future is hard to justify unless lower wages are taken for granted.

g   The general laws of economics apply to public utilities just as much as to other lines of commerce, and every procedure and decision based upon the theory that economic laws apply to everything else except to public utilities is essentially unsound and vicious.

h   In determining the rates of a public utility (and for that matter in any proceeding as between any two litigants before any tribunal) the word "dollar" should not be used to mean two different things, a high figure to one party and a low figure to the other party in one and the same controversy.

Rational Valuation—A Comparative Study, was the title of the paper presented by J. R. Bibbins[4]. The author called particular attention to the complex stage that the valuation problem had reached, and to the fundamental similarity between industries and utilities from the various aspects of valuating physical properties. He gave the results of an interesting historical study of an electric-railway property covering thirty years of the horse-drawn life, and its period of transition to electric propulsion up to the present day. This showed the rates earned over the period, the increase in net investment and the effect on deficit and surplus of varying rates of return. After considering the financial plans for rate-making purposes, and making a plea for a careful consideration of the present vague terminology of valuations, Mr. Bibbins closed as follows:

Is it not clear that the vagaries and inconsistencies of appraisal procedure could have been largely avoided thus:

1   By closer study of past history

2   Economic research into all fundamental facts

3   Reasonable forecast of future income and liabilities

4   More complete appraisal reports dealing with every aspect of reasonable value

5   An open mind—not trying to find the answer before we start

6   Acceptance of accrued depreciation as a physical fact to be handled like any other actuarial problem

7   Cost to reproduce theory used judiciously to reflect the original cost as nearly as possible

8   Institution of perpetual inventories and values based upon well-maintained and preserved records

9   Development of flexible or sliding-scale methods as bases for rates, service and return on investment

10   Adopting for utilities the practice of progressive industry in

preserving its physical and economic integrity. Pay down "while the paying is good."

11   An attempt to educate rather than to obstruct or defeat Public Policy?

The paper entitled Data on the Cost of Organizing and Financing a Public-Utility Project, prepared by F. B. H. Paine,[1] was reviewed by Mr. Bibbins. Mr. Paine's paper presents the data accumulated in a report on a particular problem. He outlines the preliminary steps in the development of a public utility involving an expenditure of money. He analyzes these expenditures and upon the basis of his investigation concludes that the proper percentage for each of the following three heads is as follows:

The development of the project............2.5 per cent
The cost of money.....................5.0 per cent
The promoter's remuneration.............5.0 per cent

Prof. H. C. Anderson[2] was unable to be present and his paper, The Construction Period, was also reviewed by Mr. Bibbins. This paper was presented to show in specific terms the relation of the construction period to the assignment of portions of construction cost known as overhead. These items of cost are items which must go into the property but which cannot be assigned as rails and ties are assigned. Professor Anderson's paper discussed proper allowance to be made for interest and taxes during the period of construction, items frequently ignored in early days of appraisal and too often exaggerated lately. His conclusion was that the proper interest charges can only be determined by a thorough study of the particular property under consideration with a view of determining: (1) The magnitude of the property, (2) the method used in finding the reproduction cost, (3) the proper construction period, and (4) knowing the construction period, what is the proper method to use. In determining the percentage to be allowed for taxes the author is of the opinion that the proper amount depends: (1) on the length of the construction period, (2) what percentage of the construction cost the assessor will find, and (3) the tax rate.

Mr. Elmes' paper was opened for discussion first and J. N. Chester,[3] commenting on the inflation of the dollar shown by Mr. Elmes, suggested that instead of attempting to change the valuations with corresponding changes in money value, the rate of return on money be changed. Mr. Chester stated that it was his belief that when the purchasing power of money decreased the compensation should increase and that this was the simplest form in which experts' opinions and lawyers' pleadings might be placed before courts and commissions as means of compensating capital.

Harold Almert[3] discussed the importance of considering that the dollar in which the property was valued was also the dollar with which the service must be paid for. He also expressed the desire that the engineer should familiarize himself with the matter of appraisal and valuation so that he should be in a position to guide the courts rather than be guided by tradition and precedent.

James E. Allison,[4] commenting on Mr. Chester's idea of adjusting the rate of return, mentioned some of the difficulties involved. He spoke of the values that a utility has, based on its desirability and based on an identical property reproduced new. He accentuated the point that getting a valuation without considering the elements of return is a very difficult procedure.

Mr. Bibbins called attention to the important part of Mr. Elmes' paper, that the dollar should not mean two things to two different people. He called attention to the difficulties thus far brought out in the discussion of both methods suggested: (1) holding value steady and varying the rate to absorb fluctuation in purchasing power, and (2) boosting property value and holding to return cost of money and rate of return on money at present in vogue. Mr. Bibbins was of the opinion that changing property would cause some embarrassment later, even though the present-day prices might continue.

---

[1] Sanderson and Porter, Chicago, Ill.
[3] Consulting Engineer, Chicago, Ill. Mem.Am.Soc.M.E.

[4] Deceased. Formerly Consulting Engineer, Paine, McClellan & Campion, New York.
[5] Professor of Mechanical Engineering, University of Michigan, Ann Arbor, Mich. Mem.Am.Soc.M.E.
[6] Chester and Fleming, Pittsburgh, Pa. Mem.Am.Soc.M.E.
[7] Consulting Engineer, Chicago, Ill. Mem.Am.Soc.M.E.
[8] Consulting Engineer, St. Louis, Mo. Mem.Am.Soc.M.E.

E. H. Whitlock,[5] offered the solution that commodities, wages, and prices of all other articles should more or less be based on the increase in money circulation per capita.

A discussion by Bion J. Arnold[10] was read. Mr. Arnold made a strong plea for complete valuation surveys, using the various elements necessary to satisfy the purpose of the valuation, and for a consideration of the same principles in valuation for utilities that are used in industries in determining cost of service, which has been neglected in the past. He stated that in his belief much of the antagonism between corporations and the public may be ascribed to application of various intangible values without regard to the character of the project, such as contractual relations under which it is organized, efficiency of administration, etc. He decried the use of expediency in present controversies over current pricing in appraisals. In this connection he stated that current pricing is directly applicable in some instances for sales, renewals, etc., but warned against the confounding of fluctuating purchase price of money with basic capital induction and the consequent embarrassing situation of being obliged to apply this same theory, current pricing, to properties at a time when prices have slumped far below normal or cost. He pointed out that if high prices are to remain, capital investment must rise gradually, due to high prices of renewals, capital investment reaching ultimately a new higher level which, however, cannot be reached suddenly as some desire for all purposes except taxation. He emphasized the importance of amortization of that part of the capital investment which is liable to be dissipated through future developments as a factor of cost of production as great as the cost of current maintenance of the property. In closing, Mr. Arnold made a particular plea for a classification of the terms used in valuation and appraisal.

C. E. Grunsky[11] stated that in his opinion, in the matter of valuation procedure the engineering profession should establish the guideposts of progressive thought courageously, fairly and without undue dependence upon leaders' precedents, giving due weight to the decisions of the courts and commissions but only in so far as these are sound in the light of true engineering economics. As a solution of the present difficulties caused by changing values of money, Mr. Grunsky suggested the establishment of a definite commodity unit to supplement money, which will be based not on metals but on things which in civilized communities are necesary to meet the wants of man, based on definite quantities of staples such as food and fuel, iron and such of the many other things which a family requires to live in ordinary comfort. The commodity unit would be incomparably more stable than the dollar. Mr. Grunsky further suggested that the change of value of the money unit as measured by the commodity unit might be used to secure a correlated increase or decrease of public utility earnings, thus making frequent revaluation for current prices unnecessary.

In a discussion of the contributions of Messrs. Arnold and Grunsky, Mr. Allison questioned the value of principles adopted by engineers in an attempt to upset the dicta of the courts and expressed himself as being opposed to a codification of laws of appraisal and valuation by the Society. Mr. Chester spoke again on the impossibility of solving the problem of proper rate return by simultaneously raising the base and the rate.

In response to the questions raised on discussion, Mr. Elmes replied that the unquestioned right of the state to regulate commerce within its borders applied as much to a fish store as it did to a street railway and that one law or one rule of fair dealing should apply to both. Touching on the point of the relation of expansion of currency to present rise in prices, Mr. Elmes said that a study of the 1896 dollar showed that in 1919 its purchasing power had shrunk to about thirty cents, inflation of currency alone reducing the purchasing power to thirty-nine cents. He stated that in his opinion the inflation of the dollar would not go down to any serious degree in the near future and quoted

Prof. Irving Fisher as saying that it looked like fifty years before we could hope for a return of prices prevalent a year ago.

A written discussion presented by B. J. Denman[12] was read. Mr. Denman confined himself entirely to the question of valuation of public utilities and called attention to the fact that the reason for the decrease in the value of securities bearing a fixed rate of return, has been the lessened value of the return on account of the decreased purchasing power of the dollar. He also showed that the common stocks of utilities decreased in value, due to the inability of utility companies to promptly pass on to their consumers increased costs of operation. Mr. Denman made a plea for fair treatment of public utilities, and in this connection emphasized the importance of "Going Concern" value, which he believed should include the following elements: Value of established business; early losses; cost of establishing business; unamortized superseded property; cost of consolidation; good will. Mr. Denman was also of the opinion that the terminology of valuation should receive careful consideration, and when formulated be given wide publicity.

In a written discussion R. H. Nexen[13] called attention to the difficulty of obtaining true cost of commodities, because of wide fluctuations in market value, and gave the results of an investigation to determine the true weighted average price paid for copper from 1904 to 1913, and also to determine whether an average base price of wire and cable could be determined for this period, using as a base the price paid for fabricated wire and market quotations of Lake ingot copper during this period. He showed a very close agreement between the average method and the actual method.

Mr. Almert spoke concerning the variation in copper prices, due to the condition of the buying market rather than to economic conditions affecting the cost of production. He questioned the inclusion in the valuation of items which had been spent in original costs of installation but which are not at present included— as, for example, a bridge built by a new railroad which is to be dedicated to a municipality.

Max Toltz[14] quoted from his experience in the building of steam roads the fact that supplementary expenses were added to the value of the road. Mr. Toltz expressed himself as being in favor of a careful study by all engineering societies of the problem of appraisal and valuation.

In answering Mr. Almert's question, Mr. Bibbins stated that he would follow the procedure of examining every possible phase of proper investment in the past, excluding nothing that pertains to a functioning property in public use, and then determine from the past records the ability of the company to amortize some of the old charges. Mr. Bibbins also stated that amortization should take place after a fair return has been paid. In closing Mr. Bibbins presented the following resolution which was adopted by the meeting:

IT IS MOVED that a Committee on Present Valuation and Appraisal be appointed by the President of the Society to review, digest and crystallize the essential features of the various papers and discussions already submitted to the Society for presentation in concise form to the next Annual Meeting for further discussion and definition; supplemented by such additional papers on fundamentals as may be submitted or requested by the Committee.

## Local Session

AT the Local Session, Tuesday morning, May 25, over which E. S. Carman of Cleveland, presided, the following papers were presented: The Housing Problem, Nelson Cunliff; Industrial Housing, Leslie H. Allen; River Barges and Terminals, William S. Mitchell; Mississippi Valley River Transportation Activities, E. W. Schadek; Burning Eastern Coals Successfully on a Conveyor Feed Type Stoker, L. R. Stowe, and Tight-Fitting Threads for Bolts and Nuts, C. B. Lord. Mr. Allen's paper was published in abstract form in the June number of MECHANICAL ENGINEER-

[5] Gen. Mgr., Am. Fire Clay & Products Co., Cleveland, Ohio. Mem.Am. Soc.M.E.
[10] Consulting Engineer, Chicago, Ill.
[11] Consulting Engineer (C. E. Grunsky Co.), San Francisco, Cal.

[12] Vice-President United Light and Rwys. Co., Davenport, Ia. Mem.Am. Soc.M.E.
[13] New York City.
[14] Toltz, King & Day, Inc., St. Paul, Minn. Mem.Am.Soc.M.E.

ING; those of Messrs. Schadek and Stowe in the May number, and Mr. Lord's in the April number.

Mr. Cunliff's paper presented a problem of nation-wide importance. Because of the fact that the Priority Board declared that water works took precedence over housing the demand for dwellings throughout the country is very acute at the present time. Government statistics show that during the war there were 1,040,000 marriages, people eligible for new homes, and that there were but 27,000 houses built during that period.

The people of St. Louis have formed a home and housing organization, the purpose of which is the erection and sale of homes to the people of that city. The lives of the purchasers are insured so that in case of death the property goes to their estate which relieves the housing association and the Chamber of Commerce of being put in the position of foreclosing on widows and orphans.

The homes built are 3-, 4-, and 5-room two-story houses with hardwood floors in the lower story, and cost from $4,400 to $8,800. In all cases these houses are sold for the actual cost, which can only be figured out after the building has been completed.

Max Toltz,[1] discussing the paper, told of the housing problem in St. Paul. He pointed out that there is a shortage of 3,500 dwellings and two-thirds as many 30 to 40-flat apartment houses at the present time. Four-room frame houses cost $4,200 (the same house was built in 1917-1918 for $1,900) and brick houses $5,500 to $6,200. The home-building association that had been organized has not had much success because the people could not afford to buy houses in spite of the fact that the association gave 80 per cent of the value of the house as a first mortgage. In South St. Paul where the new Armour plant has been completed, there is a shortage of 800 houses. Inasmuch as nobody else will undertake to build these houses, the Armour Company will be obliged to build and rent them to its employees.

Regarding business houses, such as manufacturing plants and jobbing houses, of which there is a very urgent need, it is expected that nothing will be done for the next twelve months, not only because of the high prices of labor and material, but because of the fluctuation in prices. Building estimates are given with a statement that the cost sheet is "good only for today and tomorrow," and for this reason $17,000,000 worth of work is held up at the present time. As soon as material and labor can be secured at a fixed price, it is expected that a great deal of building will be done.

In closing, Mr. Cunliff pointed out that Government figures show that building costs for residences and flats have increased 215 per cent since 1914. People have been educated to pay a very considerable advance in the cost of clothes, shoes and food, but not to pay 215 per cent increase in the cost of the homes in which they live; therefore something will have to be done to bring them face to face with the fact that building costs are up and are going to remain up. The work of the St. Louis Home-Building Association is to establish people in permanent homes which they will be proud of, proud to live in, and proud to possess.

Mr. Allen's paper on Industrial Housing because of his absence was read by title only. In a written discussion Henry Wright[2] strongly advocated the building of intelligently planned flats. It was impossible to produce a well-constructed detached house of even four or five rooms so as to sell it on a basis of $50 per month. It was, however, possible to build a given amount of living space in the form of a flat at from 60 per cent to 65 per cent of what it would cost for the identical space in an individual house. A double house could be built for 10 per cent less than a single house; a cross-party-wall house, providing for four individual two-story houses, would save at least 15 per cent per unit; and houses built in rows would save 10 per cent on the end houses and 20 per cent on the middle houses. In all of these cases fuel consumption would be reduced in about the same proportion as the reduction in cost.

John R. Fordice,[3] formerly a terminal engineer of the Mississippi-Warrior Barge Line, who has made a special study of terminal conditions along the Mississippi River, in discussing Mr. Mitchell's paper, called attention to the fact that with the exception of St. Louis and New Orleans, there are no river ports providing the wharves and machinery for properly handling the cargoes which the river service is expected to carry. Among several reasons which he enumerated for the lack of terminals are: the great cost, lack of faith in the enterprise, and a lack of proper engineering and advice. The river conditions vary greatly each season; a bar may form one year where deep water existed the year before, and the course of the river may change and deep water may exist today where a town was situated last year. For these reasons the towns do not feel that they should attempt to build permanent structures.

To provide a terminal which will meet these conditions, a "terminal barge" was designed by the speaker. This barge is 250 ft. long with a 75-ft. beam and has two railway tracks, each of which holds five freight cars. It is provided with two cranes to handle heavy, unwieldy pieces of freight, eight conveyors to handle small packages, and pumps to handle liquids. This barge will anchor at the foot of a railway incline, the cars will be run on and the cargo will be transferred when necessary to river boats and barges with little difficulty. For local freight, a barge might be designed for horse-drawn vehicles and auto trucks which could be provided with removable bodies. The cargo will be unloaded either through hatches in the roof or through doors at the side, according to the stage of the river and the character of the cargo; therefore it should be easy to transfer cargo, of whatever nature, quickly and cheaply. At the present time none of these barges are in operation, but it is expected that one, now nearly completed, will be tried out in St. Louis in the near future.

Charles Whiting Baker,[4] in a written discussion, took up the subject from an economic standpoint. He pointed out that an investment of $157 per ton, which would be required for three barges and one tow boat, is a much larger investment cost than would be required for cars and a locomotive to move the same amount of freight. The crew—40 men to move 4500 tons of freight—is much larger than the train crew necessary to move such a tonnage by rail. The risk of transportation by river or canal is greater than that by ocean or lake and many times greater than that of rail transportation, and risk is an important element of cost. If private investors were asked to put their money into such an enterprise, such an estimate would have to be presented; therefore it follows that the United States Shipping Board, which is investing the taxpayers' money. should investigate this matter carefully and determine whether or not the investment will prove profitable.

In answer to a question raised by Melvin Overstrud[5] as to whether or not any investigation has been made of the feasibility of running from Stillwater, Minn., south instead of from Minneapolis and St. Paul, John N. Chester[6] asked whether or not such investigations were profitable. He pointed out the tremendous waste of money in maintaining the inland waterways of the country from which so little return is realized. He recalled the drydock at Keokuk, practically idle, and the slip locks which, with the exception of a few excursion boats, served only a small packet, which uses it twice a week at an expense of $5000 to the Government. In addition to this he mentioned the tremendous expense of the construction and upkeep of the 53 dams of the Ohio River and of the New York State Barge Canal, for all of which there was but a very small return.

John H. Bernard[7] reviewed European practice (Germany and Holland) in regard to waterways, and pointed out that because of the development in these countries the railroads have to be

[1] M. E. Toltz, King & Day, Inc., St. Paul, Minn. Mem.Am.Soc.M.E.
[2] Adviser in Allotment and Community Planning, New York City.
[3] St. Louis, Mo.
[4] President, Baker Economic Transport Corporation, New York City. Mem.Am.Soc.M.E.
[5] Asst. to President, Twin City Forge & Foundry Co., Stillwater, Minn. Jun.Am.Soc.M.E.
[6] Chester & Fleming, Pittsburgh, Pa. Mem.Am.Soc.M.E.
[7] Transportation Engineer, New York City. Assoc-Mem.Am.Soc.M.E.

protected against them. One of the main reasons why there is not more water transportation in the United States is because of the lack of suitable terminals. Railroads spend a large percentage of their income on their terminals, whereas the percentage spent by barge lines is practically negligible.

J. R. Bibbins[*] discussed the results of a comprehensive investigation made in New Orleans to determine the probable distribution of the freight movements in this country. Taking Gibraltar, Pernambuco and the Panama Canal as the basing points for export rates, it was found that for non-European ports, traffic originating south and west of a curve running from Savannah east of St. Louis and up through the northwest should move via the Gulf: for the east coast of South America all freight originating west of the line passing near Chicago should move via the Gulf; and for the west coast of South America and Pacific ports all freight west of the curve passing through Buffalo should move via the Gulf. But the fact was that the freight did not move that way; and the reason therefor was one of the biggest fundamental problems of industrial economics that existed in this country.

Theodore Brent[*] told of the activities of the Mississippi-Warrior River division during the war. He pointed out that within the last twenty years there has been no boat service on the lower Mississippi until the Government proposition was tried. During the régime of the U. S. Railroad Administration railroad river rates were secured by means of which the man living inland received the benefit of saving of what was considered, up to the present time, the proper relation between rail and rail and water rates. The fact had been demonstrated that river rates could be made that would be at least 20 per cent below rail rates, and not only save money for shipper and consumers, but pay the cost of operation as well.

Mr. Schadek accompanied the presentation of his paper by lantern slides, which showed the types of boats being built under the governmental program. He called attention to the Newton Bill now pending in Congress, which provides for the dredging of channels in the Ohio, Missouri and Mississippi Rivers, and referred to the fact that because of the importance of river transportation, the state of Illinois had appropriated $20,000,000 for the purpose of making improvements on the Chicago drainage canal.

John H. Bernhard expressed his opinion that the river terminals are not built properly. He emphasized the fact that freight must move continually and rapidly, and that removing freight by means of a vertical lift is not the best method. He pointed out the advantages of self-propelled barges in crooked rivers, and as the introducer of this type of barge expressed his regret at the present time, as he felt they could not be successful in such rivers as the Warrior, which had a number of bends of 24 ft.

Charles Whiting Baker submitted a written discussion to the effect that water transportation could in no way compete with rail transportation. Ninety-five per cent of the freight moved by rail, he wrote, is carload freight which is loaded by the shipper and unloaded by the consignee. It costs more to move a ton of freight by wagon or motor truck than to haul it 200 miles by rail. This handling is nearly always necessary with boat transportation, whereas by rail most of the carload freight is taken from and delivered to the private sidetrack of the mill, mine or warehouse. Hopper-bottom cars, which drop their load directly into the boiler-house bins, have an advantage that the floating barge cannot possibly overcome. He maintained that the burden of proof was on the waterway advocates to demonstrate, by estimates that would bear critical analysis, that freight can be moved at enough lower cost by water than by rail to induce capital to undertake the building and operating of boats. Until this was demonstrated, business would not be restored to the rivers, no matter how much more the Government spends on their channels.

[*] Consulting Engineer, Chicago, Ill. Mem.Am.Soc.M.E.
[*] U. S. R.R. Administration. Mgr. of Mississippi-Warrior River Division.

R. A. Hiscano,[10] in a written discussion of the Mitchell and Schadek papers, pointed out the value of river transportation in the East. The last two months had shown conclusively that the statement made that the rail transportation systems were entirely able to take care of the country's business was far from the truth. If it had not been for the boat lines operating out of New York City to Hudson River, Long Island Sound, and New York Bay points, conditions during the past year would have been considerably worse than they were. The lack of room in New York and vicinity in which to expand rail terminal facilities made it imperative that the waterways be utilized.

Mr. Stowe illustrated his paper by means of lantern slides. R. H. Kuss[11] and R. Sanford Riley[12] paid tribute to the originality and intelligence with which the paper was worked up. Each, however, commented on the lack of data and expressed the opinion that more should have been given. Mr. Riley pointed out that hitherto experiments in mechanical stoking had perhaps neglected to give sufficient attention to the chemical side of the problem. Such a procedure as Mr. Stowe followed, he said, opens up an entirely new field and is a real contribution to the art.

Written discussions by Donald H. Hampson,[13] Wilfred Lewis,[14] and H. L. Van Keuren[15] were submitted for C. B. Lord's paper. All were impressed favorably with the idea of putting the Sellers nut on a Whitworth bolt, and welcomed this scheme as a very good solution of a perplexing problem. Mr. Lewis, however, pointed out that this scheme would develop a rivet-like joint which it would be necessary to either twist or cut apart. For the most part, nuts are put on with the idea of taking them off if necessary, therefore considerable experimental work must be done to determine the safe limits in the materials used.

## Aeronautic Session

FOUR noteworthy papers were presented and discussed at the Aeronautic Session held Wednesday morning, May 26, Dean Charles Russ Richards presiding. These papers were: Physical Basis of Air-Propeller Design, F. W. Caldwell and E. N. Fales; Analytical Theory of Airplanes in Rectilinear Flight and Calculation of Maximum Cruising Radius, A. Rateau; Aeronautic Instruments—General Principles of Construction, Testing and Use, Mayo D. Hersey; and Flow of Air Through Small Brass Tubes, T. S. Taylor.

Messrs. Caldwell and Fales' contribution was exceptional, their paper, The Physical Basis of Air-Propeller Design, being given by Mr. Fales with the aid of moving pictures which showed very clearly eddies and vortices in air moving against an aerofoil in the wind tunnel. This paper appeared in abstract form in the April issue of MECHANICAL ENGINEERING.

Walter C. Baker[1] submitted a written discussion of the paper in which he stated that if it was possible for an end of a wing to fail while the other continued to lift, as stated by the authors, numerous accidents might be looked for as one would expect an airplane to turn over sideways under such circumstances. He thought that further information of interest could be furnished by the authors concerning the need for the straighteners and just why they were located with reference to the wind tunnel as they were. It seemed probable to him that the nuclei for fog condensation in the model wing experiments were not of a chemical or electrical nature, but were certain microscopic regions of low pressure existing in the centers of axes of various small eddies.

M. C. Stuart[7] wrote calling attention to the advantages of the specific-speed method of dealing with geometrically similar air propellers in connection with their design, and referred to a

[10] General Manager, Catskill Evening Line, New York City.
[11] M. E., Malcolmson Briquet Engng. Co., Chicago, Ill. Mem.Am.Soc.M.E.
[12] Pres. Sanford Riley Stoker Co., Worcester, Mass. Mem.Am.Soc.M.E.
[13] M. E., Morgan E. Wilcox Mfg. Co., Middletown, N. Y. Assoc.Mem.Am.Soc.M.E.
[14] Pres. The Tabor Mfg. Co., Philadelphia, Pa. LifeMem.Am.Soc.M.E.
[15] Gage Engineer, Wilton Tool & Mfg. Co., Boston, Mass. Assoc.Mem.Am.Soc.M.E.
[1] Consulting and Manufacturing Chemist, Winthrop C. Durfee, Boston, Mass. Mem.Am.Soc.M.E.
[7] M. E., U. S. Naval Engineering Experiment Station, Annapolis, Md. Mem.Am.Soc.M.E.

paper on the subject which he had contributed to the Journal of The American Society of Naval Engineers in May, 1918.

Morgan Brooks' said that he was impressed with the definiteness at which the discontinuity was fixed by the experiments described, and also by the superiority of the motion-picture method used by the author over the old method of point-by-point tests. There was some suggestion toward the end of the paper in regard to discontinuity having an influence on very high-speed flight. We were now reaching speeds of 600 to 700 ft. per sec. at the tip of the propeller, perhaps more, and as it had always seemed to him that discontinuity must come at the velocity of sound in air, that is, 1100 ft. per sec., in designing we might have to take account of discontinuity.

Mr. Fales, in closing, said in reply to Mr. Durfee, that the energy at the tips of an airplane was a very small percentage of that in the air flow and he doubted whether the amount would be enough to upset the airplane. The use of a straightener to decrease the pulsations in the wind tunnel was the result of experiments made by Mr. C. P. Grimes and himself and was suggested originally by the flow out of a bathtub. It was found that it was possible to stop the whirl by means of such straightener as had been used.

Replying to Professor Brooks, he said that the moving pictures and photographs were vastly inferior to the actual view of the phenomenon in the wind tunnel itself. The effect of the discontinuity had already changed over the design to some extent, but he doubted whether this phenomenon would affect an airplane very greatly.

Dr. Rateau's paper on The Analytical Theory of Airplanes in Rectilinear Flight and Calculation of Maximum Cruising Radius, was presented by Prof. George O. James of Washington University, who outlined the mathematical reasoning of the author, from which the conclusion was drawn that the maximum cruising radius of an airplane today is about 4350 miles. An abstract of Dr. Rateau's paper appeared in the May issue of MECHANICAL ENGINEERING.

Walter C. Durfee, discussing the paper, wrote that since the sole object of attaining the high altitudes mentioned by the author, so far as cruising radius was concerned, was to cut down the tractive force while maintaining a perfect carburetion of fuel, it would seem that the same thing could be otherwise arranged for, as by cutting out some of the engines in case there were several; or the plane could be flown constantly at its most economical speed, allowing the surplus tractive force to be stored as potential energy of elevation, which could be used up from time to time by shutting off the engine and allowing the machine to glide. The cruising radius would then be somewhat greater as compared with the high-altitude radius, because the flight might be accomplished at a low level where the internal losses of the motor would be a small proportion of the total expenditure of energy.

J. G. Coffin[1] wrote that the author went much more into the detail of the subject than practical considerations necessitated. He had treated the question of aeroplane ranges in a paper prepared in 1918 and which had just been issued as Report No. 69 of the National Advisory Committee for Aeronautics. In this paper his aim had been to introduce the least number of complications possible and derive a practical method for giving an estimate of possible ranges. Conclusions similar to his had since been published in French, English and Italian reports. The main factors to be taken care of were evidently the propeller efficiency, the fuel consumption per horsepower-hour, and the $L/D$ ratio for the complete plane loaded to plane empty.

E. N. Fales[2] submitted an extended written discussion in which he gave calculations, based on data taken from actual flight tests made in February at McCook Field, showing that speeds of 300 m.p.h. were possible with the Thomas-Morse M. B. 3

aeroplane tested, providing its propulsive power were maintained constant by supercharging the engine (a 300-hp. eight-cylinder Hispano-Suiza) and applying an adjustable-pitch propeller. This would mean New York to St. Louis in four hours.

Morgan Brooks said that near the close of the paper the author added one per cent to the cruising radius by planing down after all the gasoline had given out, but he had not allowed for the extra fuel it required to originally reach the altitude. Having allowed nothing for rising he thought the author was not entitled to claim the extra distance for coming down without gasoline.

In presenting his paper on Aeronautic Instruments, Mr. Hersey outlined the important work done by the Bureau of Standards in this department and supplemented it with slides showing additional types of instruments tested by the Bureau. This paper, slightly abridged, was published in the May issue of MECHANICAL ENGINEERING.

Alexander McAdie,[3] discussing the paper, wrote that the typical altimeter as described by the author would show an error of 8 per cent at 25,000 ft. elevation and the outstanding need was for a temperature correction. It was a fact that Doctor Jeffries, who crossed from England to France through the air in 1785, could tell his maximum elevation (6600 ft.) with his mercurial barometer with much greater accuracy than a present-day flier with an altimeter.

The atmosphere had been quite well explored and pressure, temperature and humidity conditions determined for heights far beyond airplane or airship limits. The lowest temperature in the air thus far obtained was — 130 deg. fahr. at 17,000 m. elevation over the equator. A temperature scale successfully used for several years at the Blue Hill Observatory had its zero at absolute zero and 1000 for the freezing point. This eliminated minus signs and degree symbols and was known as the 4 K system.

It seems to be impracticable for an aviator to determine the true velocity while both machine and air were in motion, but by sending him from a certain point a combined wireless and sound signal, a close approximation of his distance from the source could be made. The interval between the arrival of the two signals made simultaneous would give the time of travel of the sound or aerial wave, and the proper correction for temperature and density, the distance.

Elmer A. Sperry,[4] in a written discussion, called attention to the unreliability of the magnetic compass in fog, cloud and night flights. It was highly reliable in straight flight, however, but this was extremely difficult when no landmark could be seen. The gyroscopic turn indicator he had developed had been made to indicate clearly a turn about a radius as great as five miles, distinguishing clearly between this and straight flight; and by thus safeguarding the magnetic compass very excellent navigation had been secured. The great advantage of dealing with true geometric meridians and the high degree of accuracy attained by the gyroscopic compass made it very desirable for large aircraft.

Albert F. Hegenberger[5] wrote concerning the work of the Engineering Division, Air Service, in developing aeronautical instruments, especially those for military work. A new sensitive barograph which had been functioning successfully for several months would record a change in altitude of about 30 ft., and a mechanical rate-of-climb meter now being tested out used a sylphon for the sensitive diaphragm element and had worked very well. Aeronautical instruments, however, were still very far from satisfactory. Gyroscopic compasses thus far tested had been affected so greatly by bumps and similar accelerations as to be worthless. Turn indicators of the gyrostatic type should be improved to prevent oscillations of the hand, which were very confusing. The static-head type was free from this vibration, but was affected by side slips and skids. A great many instrument problems would be solved if some means could be developed for indicating the true vertical and horizontal. Increasing reliance was being placed on instruments as they continued to im-

[1] Professor of Electrical Engineering, University of Illinois, Urbana, Ill. Life Mem.Am.Soc.M.E.
[2] Director of Research, Curtiss Aeroplane & Motor Corpn., Garden City, L. I. N. Y.
[3] Air Service, Engineering Department, McCook Field, Dayton, Ohio.
[4] Director, Blue Hill Observatory (Harvard University), Readville, Mass.
[5] President and Engineer, The Sperry Gyroscope Co., Brooklyn, N. Y. Life Mem.Am.Soc.M.E.
[6] Second Lieut., A. S. A., Air Service Representative, Dayton, Ohio.

prove and it was not at all improbable in the near future that large aeroplanes and planes with enclosed cockpits would be flown entirely by instruments.

J. A. Hoogewerff[1] wrote that Mr. Hersey did not sufficiently stress the point that it was most necessary for the air pilot to have a thorough knowledge of the instruments available for his use and respect for them based on that knowledge. He should be sufficiently informed to make due allowances for instrumental errors and inaccuracies, but he should also realize that good instruments are far more accurate than human judgment and that to depend on judgment alone might be fatal.

Thomas D. Cope,[a] in a written communication, stated that an experimental gasoline depth gage tried out at Langley Field had given excellent service and reliable indications. Its new features consisted of a ball-and-socket endless chain running over pulleys and actuated by a float. The motion of the upper pulley was communicated through a ground joint to a rack and pinion and thus by a wire to a dial and pointer on the instrument board.

In closing Mr. Hersey said, apropos of Admiral Hoogewerff's remarks on the need for training pilots, that it might be of interest to those present to know that the Aeronautic Instrument Section at the Bureau of Standards had conducted two schools during the war in which scientific courses were given to naval officers to fit them to take charge of instrument work in the Navy. This was done at the request of the Chief of the Bureau of Operation. He felt that the discussions that had been presented were all useful contributions to the subject of the paper.

Mr. T. S. Taylor's paper on the Flow of Air through Small Brass Tubes appeared in June MECHANICAL ENGINEERING. Mr. Taylor was unable to be present and his paper was presented by title.

H. Wade Hibbard[b] said that while Mr. Taylor's tests showed that for a tube 1½ in. in diameter it required a length of flow of 50 diameters to reach a stable condition, he would state that in large conduits the case was different. He had found that in transporting sawdust and shavings by air blast through a 3-ft. metal conduit 800 ft. long, if a piece of metal accidentally fell into the material being transported, it would dent the conduit its entire length, and that the dents would be apparently as numerous toward the end as at the beginning.

Mayo D. Hersey asked if the author had compared his experimental results with the well-known formula of Dr. Lees. This was recognized in England as covering the law of the flow of fluids through any form of pipe, and it would be of value to have the comparison made.

## Castings Session

THE Castings Session convened on Wednesday morning, May 26, with F. O. Wells, Manager of the Society, presiding. The session was organized by the former Sub-Committee on Foundry Practice as a symposium on the various types of castings and the papers, including information as to what may be obtained from commercial foundries, are of great value to designing engineers. The symposium consisted of the following papers: Malleable Castings, Enrique Touceda; Die Castings, Charles Pack; Aluminum Castings, Zay Jeffries; Steel Castings, John H. Hall; Gray-Iron Castings, Richard Moldenke, and Brass and Bronze Castings, C. H. Bierbaum. Dr. Moldenke was present as the representative of the Foundry Committee and guided the discussion which culminated in an expression of desire to hold further meetings to consider the foundry problems of the engineer. Mr. Pack was also requested to present a treatise on die castings for the Annual Meeting.

The castings papers, together with an account of the discussion which they elicited, will appear in abstract form in a future number of MECHANICAL ENGINEERING.

## Scientific Session

THE Scientific Session on Thursday morning was a joint session with the American Society of Refrigerating Engineers and the American Society of Heating and Ventilating Engineers, both of which organizations were meeting in St. Louis for their spring conventions. Mr. Matthews, president of the American Society of Heating and Ventilating Engineers, presided at the meeting and five papers were presented, three being contributed by The American Society of Mechanical Engineers, and one each by the Refrigerating, and Heating and Ventilating Engineers. These papers and authors were as follows: An Improved Form of Weir for Gaging in Open Channels, by Clemens Herschel; Simplification of Venturi-Meter Calculations, by Glenn B. Warren; The Dissipation of Heat by Various Surfaces, by T. S. Taylor; The Thermal Conductivity of Heat Insulators, by M. S. Van Dusen (contributed by Am. Soc. R. E.); and Ship Ventilation, by F. R. Still (contributed by A. S. H. & V. E.).

Of the papers presented by The American Society of Mechanical Engineers, that by Mr. Herschel appeared as the leading article in the February issue of MECHANICAL ENGINEERING (p. 83), while the papers by Messrs. Warren and Taylor will be found in abstract form in the April issue, on pages 220 and 230, respectively.

Mr. Herschel's paper, presented by F. R. Low,[1] outlined the results of a series of investigations on weirs which mark a distinct step forward in the measurement of water by that means. The investigations were undertaken at the instigation of the Power Test Code Committee of The American Society of Mechanical Engineers and the necessary funds were supplied by Engineering Foundation. The new weir which is described in the paper bears no resemblance to the sharp-edged weir now in general use, for its crest is made in the form of an arc of a circle and is hollow for observing the pressure at that point. Mr. Herschel's tests showed that for discharge of from 0 to 9.55 cu. ft. per sec. per ft. of weir length, the limits covered by the experiments, the quantity of water flowing, $Q$, was directly proportional to the difference, $d$, in two pressures, one measured just upstream of the weir, and the other measured at the crest, the formula being $Q = 5.50 \, d$.

In the discussion which followed A. M. Greene, Jr.,[2] stated that he was unable, by means of the formula given, to compute the results which were tabulated by the author. He therefore asked that Mr. Herschel furnish at least one set of the complete data so that he might be able to follow through the computations and accordingly check the values given. Professor Greene also asked why Mr. Herschel selected the position for the hook gage as stated in his paper when he claimed that hydraulic engineers do not know where to put the hook gage in the standard form of sharp-crested weir.

Dean R. L. Sackett,[3] in discussing the use of weirs now commonly employed, stated that the weir itself ought not to be condemned, but rather the methods employed in its use, for in the past there has been no agreement either upon a standard type or upon the position at which to measure the head. The new weir is not the only one in which sharpness varies as the head, as recent laboratory tests indicate that a sharp-crested weir of a standard curved type in vertical position does have a constant coefficient within the range of experiments made. It might therefore appear, Dean Sackett said, that the new weir is better for large volumes of water than the old type, but in the measurement of water used for cooling purposes or return water in power plants, where the discharge would vary in its head, would be of great advantage.

C. G. Richardson[4] said that to his personal knowledge Mr. Herschel would be greatly pleased if the technical schools and colleges would study this problem and carry on investigations

[a] Rear-Admiral. U. S. N.; Supt. U. S. Naval Observatory, Washington, D. C.
[10] Assistant Professor of Physics, University of Pennsylvania, Phila.
[11] Professor of Mechanical Engineering, University of Missouri, Columbia, Mo. Mem.Am.Soc.M.E.

[1] Editor. Power, New York. Vice-President Am.Soc.M.E.
[2] Professor of Mechanical Engineering, Rensselaer Poly Inst., Troy, N. Y. Mem.Am.Soc.M.E.
[3] Dean of Engineering. Penn. State College, State College, Pa. Mem.Am.Soc.M.E.
[4] Chief Mechanical Engineer, Taft-Pierce Co., Woonsocket, R. I. Mem. Am.Soc.M.E.

THE JOURNAL
AM. SOC. M. E.

along the line he had suggested. He further stated that F. M. Connet,[1] in discussing with him the features of the new weir, had expressed the opinion that it seemed to be a special form of venturi tube and it should, therefore, be expected that the quantity flowing would vary as the one-half power of $H$, where $H$ represents the difference in head between the inlet and throat of that special venturi tube. It seems possible, however, that the departure from the formula so that $Q$ varies directly as the head may be due to the departure of the stream line from the true venturi principle as it passes over the contraction or, one might say, the throat of the tube.

The second paper of the session, Simplification of Venturi-Meter Calculations, by G. B. Warren, was presented by title only. It describes a method devised by the author which reduces the calculation of venturi-meter measurements to a simple slide-rule computation requiring but a few settings. Written discussions of Mr. Warren's paper were presented by Thomas G. Estep,[2] M. C. Stuart,[3] and C. G. Richardson.[4] Additional discussion will also be found in the Correspondence Section of this and the May issues of MECHANICAL ENGINEERING.

Mr. Estep in his discussion called attention to the fact that the equations developed by Mr. Warren are simple derivations of fundamental equations for the frictionless adiabatic flow of elastic fluids and credit is therefore due to older investigators than those mentioned by Mr. Warren. He also presented a formula which he had developed, with the suggestion that if it is desired to simplify the calculation that an impact tube be used instead of the upstream static opening, retaining, however, the static opening at the throat. The differential pressure then represents the total velocity at the throat instead of simply the change in velocity as when the two static openings are used. He also stated that in his opinion Mr. Warren's calibration shows that H. B. Reynolds'[5] empirical equation for a thin-plate orifice is a good one, but it does not prove the accuracy of the venturi meter.

M. C. Stuart stated that several years ago he had developed a method of simplified calculations almost identical with that presented in the paper, and in his discussion gave details of his method.

M. S. Van Dusen, of the Bureau of Standards, next presented his paper on the Thermal Conductivity of Heat Insulators, which dealt with the development of the methods for measuring the insulating properties of materials. Mr. Nichols, in discussing Mr. Van Dusen's paper, called attention to the necessity in heat-measurement work of settling upon a definite and very closely regulated method for absolute conductivity measure. Messrs. Cramer, McGeorge, F. R. Still, Voorhees, and L. R. Whitney also discussed the paper, commenting chiefly upon the excellent scientific work being done at the Bureau of Standards whose research work is of value to the entire country and not limited in application to any particular class of industry. This sentiment was the result of a suggestion by one of the discussers that the large industrial corporations might very profitably finance scientific investigations being carried out under the direction of the Bureau of Standards.

The paper by T. S. Taylor on the Dissipation of Heat by Various Surfaces, dealt for the most part with the value of sheet asbestos as a covering for hot-air furnace pipes. In addition to the oral discussion which follows, there were two written discussions, one by A. C. Willard and V. S. Day, of the University of Illinois, and one by L. R. Ingersoll, of the University of Wisconsin.

The discussion by Messrs. Willard and Day dealt chiefly with the first part of Mr. Taylor's paper. They pointed out that in general Mr. Taylor's results confirmed the work of a similar nature already completed at the University of Illinois and re-

ported in Bulletin No. 117 of the Engineering Experiment Station as the Emissivity of Heat from Various Surfaces, by V. S. Day. In fact, when the results of the two investigators are compared on the basis of the excess of surface temperatures of the cylinders above the surrounding room temperatures, a satisfactory agreement results. Messrs. Willard and Day also pointed out that the only definite basis for estimating the amount of heat dissipated by thin metal surfaces which separate air or gas at high temperatures from air or gas at low temperatures, is the difference in temperature between the surface itself and the temperature of the cooler surrounding air. The values given by Mr. Taylor are used as a basis for a calculation, as the result of which they conclude that Mr. Taylor's values of the coefficient of emissivity when surface temperatures are used as a basis are only 5 to 6 per cent lower than Mr. Day's values for bright tin cylinders.

L. R. Ingersoll, in commenting upon the paper, stated that the results are by no means striking to one familiar with the main laws of heat transference. The effect would be exaggerated if the pipe were of very small diameter, for in this case the addition of insulation would add appreciably to the radiating surface. From a practical standpoint, however, furnace pipes are so generally dirty and dusty that to leave off the customary asbestos-paper covering in the expectation of saving heat would be a practice of doubtful wisdom.

L. B. McMillan[10] called attention to the fact that the author's conclusion that a thin layer of asbestos paper applied loosely to a bright surface increased the loss of heat from the surface, is undoubtedly correct and in accord with former tests of the same general nature. The increase in heat loss, he said, is entirely due to the change in character of the exposed surface. The addition of a thin layer of asbestos adds some resistance, but not enough to offset the decrease in surface resistance. In the case of furnace pipes the loss from the bright surface is so great that it would be wasteful to leave them bare. The important factor, therefore, is what one thickness of insulating material will save as compared with another. He also called attention to the superior insulating qualities of cellular construction and stated that actual tests at the University of Illinois on surfaces covered with three layers of air-cell asbestos showed a saving of twice that obtained by the same material by the author's calculations.

Mr. McMillan also stated that experiments have established the fact that radiation from a galvanized surface increases very materially as the surface becomes tarnished. Tests reported in Engineering, October 19, 1917, showed that galvanized iron exposed to the weather for one year offers practically no more resistance to radiation than does black iron.

E. R. Hedrick[11] called attention to the fact that the laws that were presented were for entirely different circumstances than the laws for the dissipation of heat when gas or other material is passed through the pipe, for in that case a very important consideration, especially as there is a considerable drop in temperature, is the passage of material down the pipe. However, since the experiments were conducted in a box of uniform size, agreement between the laws that are represented and the laws for the transmission of heat which apply for a material being passed through a pipe, is not expected. Mr. Hedrick also commented on the formula and curves given by the author, and stated that he had not succeeded in doing more than check the formula presented and in his opinion it did not hold. He also stated that the observations were limited because they had been carried on for a difference of only 10 deg., which in his opinion constituted an insufficient basis for argument for ordinary heat transmission.

J. D. Hoffman,[12] in commenting upon the value of Mr. Taylor's paper, stated that he considered it of especial value because of the fact that it confirmed the work that had been done at the University of Illinois and the results obtained were also in close

[1] Chief Engineer, Builders Iron Foundry, Providence, R. I. Mem.Am. Soc.M.E.
[2] Assistant Professor Mechanical Engineer, Carnegie Inst. Tech. Mem. Am.Soc.M.E.
[3] Mechanical Engineer, U. S. Naval Engineering Experiment Station, Annapolis, Md. Mem.Am.Soc.M.E.
[4] Sales Engineer, Builders Iron Foundry, Providence, R. I. Mem.Am.Soc. M.E.
[5] Trans. Am. Soc. M. E., vol. 38 (1916), p. 790.

[10] Construction Engineer, H. W. Johns-Manville Co., New York, N. Y. Assoc-Mem.Am.Soc.M.E.
[11] Professor of Mathematics, University of Missouri, Columbia, Mo. Mem.Am.Soc.M.E.
[12] Professor, Director of Practical Mechanics, Purdue Univ., Lafayette, Ind. Mem.Am.Soc.M.E.

accord with the work done by Mr. Soll, of the American Radiator Company in his investigation on blast heating.

The final paper of the session, on Ship Ventilation, was presented by F. R. Still [a] of the American Society of Heating and Ventilating Engineers. This paper, which appeared in the April issue of the *Journal of the American Society of Heating and Ventilating Engineers*, page 363, takes up the subject of ventilating passenger ships, refrigerating ships, cattle ships, general freighters, and lake and river steamers. There was no discussion.

## Power Session

THE interest among mechanical engineers in the subject of power was evidenced by the announcement at the Power Session of Friday morning, by F. R. Low, that over 700 members had expressed a desire for a Power Section and that (as announced elsewhere) at a meeting previously held steps had been taken for the formation of such a Section. Prof. Robert H. Fernald presided at the Power Session and four papers were discussed as follows:

The first paper, by Otis L. McIntyre, New York, dealt with experiments on pulverized coal which led to the installation of apparatus for its use in the metallurgical furnaces of the Cerro de Pasco Copper Corporation at La Fundicion, Peru. These were unique in that they were conducted at an altitude of 14,000 ft., a condition under which there have been practically no data available. It should be said, also, that the paper itself is unique in that the difficulties encountered in the development of the apparatus are plainly stated. The pros and cons of pulverized fuel are all of interest to the engineer, but it has usually been easier to get at the pros than the cons in papers prepared on this subject.

W. N. Best,[1] the only one to discuss this paper, styled it as "excellent." His only criticisms were on the number of burners used and lack of combustion space provided. With pulverized coal as few burners as possible should be used—one burner will give better results than two or four because the flame and temperature can be more easily controlled by the operator. A combustion chamber is not only required for good combustion of the fuel, but will prevent the piling up of the unconsumed fuel before reaching the bath of molten metal.

The second paper, on Efficiency of Natural Gas Used in Domestic Service, by Robert F. Earhart, Columbus, Ohio, touched a vital point in the fuel situation in this country. The supply of natural gas is no longer sufficient for both industrial and domestic purposes; and under the present wasteful methods it will not long be sufficient even for the latter use. Anything contributing to greater economy in the consumption of gas in the thousands of homes in the natural-gas regions, therefore, is of the greatest importance. The paper reports the results of a series of efficiency tests on natural gas made in the physics laboratory of Ohio State University. The conclusions relate mainly to the location of the burner with respect to the utensil on the stove and show that with proper adjustment an efficiency in excess of 45 per cent can be secured.

G. I. Vincent,[2] in a letter on the subject, wrote that whereas it was known in a general way that the position of the burner under the vessel in domestic appliances materially affected the efficiency, investigators had not sufficiently emphasized the fact that the efficiency of gas-burning appliances and cooking vessels of the present design is very low. Only a small part of the heat goes to cook the food immersed in the water. The saving from properly designed and located burners would hardly be a drop in the bucket compared with what might be accomplished by revolutionary changes in the design of household utensils and gas-burning appliances. While the latter might be possible, however, the former presents an almost impossible task. Another saving

would be accomplished if the gas supply could be cut off as soon as a cooking process was completed or a vessel removed from the stove.

A general discussion followed, chiefly upon the relation between the gas pressure and the mixture of air and gas, and the possible use of automatic devices for varying the air supply with the pressure, or with the height of flame, such as used in gas annealing furnaces. Suggestions to this end were made by H. B. Dirks, H. R. Godeke, and H. S. Dickerson.

John Borge[3] said there are a number of burners on the market in which the air shutter is geared to the gas part so that opening part way would admit a certain amount of air and opening further would admit correspondingly more air. Investigations show, however, that there is not much justification in such an elaborate combined valve.

There was also discussion by Victor J. Azhe and others on efficiency, or inefficiency, in larger installations than required in domestic kitchens, such as in bakeries, etc. One instance cited was of a large plant using gas to the value of $200 a day, but at an efficiency of 20 per cent when it should have been 50 per cent with suitable equipment and adjustments.

The next paper, on Locomotive Feedwater Heating, by Thomas C. McBride, described a recently-developed system of feedwater heating for locomotives which makes use of the open type of heater in which the exhaust steam comes in direct contact with the cold feedwater. Mr. McBride said that the practice of feedwater heating for locomotives is followed abroad far more extensively than in this country. In 1918 a single road in France had 61 heaters in service and he had been informed that all the German locomotives turned over to the French under the terms of the armistice had feedwater heaters.

F. N. Speller[4] said that information upon the amount of gas separation in the type of heater described by the author would be of interest as an indication of whether there would be less corrosion in the boiler through the use of an open heater. An open heater of adequate size has a marked effect in retarding corrosion in stationary boilers. A serious objection to closed heaters is the rapid deterioration of the heater itself through the action of corrosive gases in solution. He further referred to heater tests on a western road which showed rapid deterioration.

George M. Basford[5] sent a comment to the effect that while the paper covers certain phases of the subject which have not been well understood and gives new data of value, it is unfortunate that the author confined his treatment to the open type of heater. He had presented his subject in a somewhat less favorable light, in some particulars, than he might have done if he had included results obtained with the closed type of heaters on a number of railroads in this country and abroad.

Grafton Greenough[6] wrote that there is every evidence that the necessity is upon us for the reappearance of the locomotive feedwater heater which, as stated by the author, has lain dormant for over three-quarters of a century. This necessity is twofold: First, the sizes of our most powerful locomotives are seriously restricted by physical conditions and any device which will enable us to obtain greater horsepower without increasing the size of the locomotive boiler should receive attention; and second, there is urgent need for conserving labor, money and the transportation facilities required in producing and delivering coal. The locomotive feedwater heater is a practical agency whereby the size of the locomotive boiler may be reduced and a saving in fuel effected. He predicted that before long it would be regarded as an established device in common with means for superheating and compounding, needful to bring about greater efficiency and economy in the locomotive engine.

The final paper of the session, on the Separation of Dissolved Gases from Water, by J. R. McDermet, awakened a great deal of interest and elicited considerable discussion. Valuable data were given, based on extended experiments leading to the prac-

---

[a] Vice-President, American Blower Co., Detroit, Mich. Mem. Am. Soc. M. E.

[1] Pres., W. N. Best, Inc., New York. Mem. Am. Soc. M. E.

[2] Engineer, Syracuse Lighting Co., Syracuse, N. Y. Mem. Am. Soc. M. E.

[3] Pres. Borge Incinerator Corp., New York City. Assoc-Mem. Am. Soc. M. E.

[4] Metallurgical Engineer, National Tube Co., Pittsburgh, Pa.

[5] Pres. Loco. Feed Water Heater Co., New York. Mem. Am. Soc. M. E.

[6] Vice-Pres. in Charge, Baldwin Loco. Wks., Philadelphia.

tical development of the process originated by W. S. Elliott of
Pittsburgh. Fundamentally it consists of the rapid injection of
heated water into a region of vacuo, and an explosive boiling of
it at the expense of the heat of the liquid, with a simultaneous
recovery of the heat liberated by a heat interchanger or condenser
cooled by the incoming water, preliminary to its heating.

F. N. Speller, in a written discussion, said that the oxygen
content of natural water varies from 5 to 10 cc. per liter, according to seasonal temperature. It is desirable to remove dissolved
gases as far as possible in an open heater. Quoting the performance of three open heaters in different plants, he said that
one heater operating at a temperature of 170 deg. fahr. left in
the water an average of 2.55 cc. per liter; another at 204 deg. an
average of 0.75 cc. per liter; and the third at 209 deg., 0.70 cc.
per liter.

Experience indicates, however, that to prevent corrosion the
feedwater used with steel economizers should carry less than 0.20
cc. per liter. Experimental work done by his company and by Dr.
W. H. Walker of the Mass. Inst. of Technology indicates that
open heaters as usually operated will not reduce the oxygen low
enough to stop corrosion. The residual oxygen, after passing
through a properly vented open heater, may be reduced below
the danger point by further mechanical deaeration, such as passing the water over baffles at about the boiling point in a separate
chamber at lower pressure, or by injecting the superheated water
into a low-pressure chamber with a condenser in series, as
described by Mr. McDermet; or the residual oxygen passing
through the heater may be quickly and completely removed by
chemical treatment, or a combination of both systems as most
convenient. His company, he said, had used the latter principle
for five years. It is based on passing the heated water under
pressure through a large mass of expanded steel scrap, which
he termed " deactivating" the water in distinction to mechanical
deaeration.

C. M. Garland[7] wrote that with coal at $6 a ton there are
few plants that cannot afford to install the most efficient boiler-
room equipment obtainable. This means the installation of economizers. The economizer, however, has not yet reached the
stage of development which might be expected, due mainly to the
corrosive action of dissolved gases in water. This action has
made it necessary to use cast-iron tubes, which must be of large
diameter and for high pressures of heavy weight, making the
economizer bulky, clumsy and to a large extent impractical.

If the author's investigations and apparatus lead ultimately
to the elimination of corrosion in economizer tubes, he will have
eliminated one of the most difficult problems in economizer

design. Steel-tube economizers can be designed to occupy about
one-eighth the space allotted to economizer equipment and will
greatly reduce the cost of such installations.

Geo. H. Gibson,[8] in a written communication, said that while
the author had credited the method described to W. S. Elliott,
it had been pointed out by Wm. H. Walker, in a paper before
the Iron and Steel Institute in 1909, that pitting might be
entirely avoided by removing air from the feedwater before its
introduction into the boiler and that this might best be done by
the employment of an open feedwater heater connected to the dry-
vacuum pump of a condenser. Also he referred to a similar
proposal for the elimination of corrosion in the famous 350-mile
Coolgardie pipe line for the Goldfields Water Supply, of Perth,
Western Australia, in 1909.

Mr. Gibson contended that effective separation of air can
be effected without the complication of vacuum appliances and
quoted results obtained by heating water in an open feedwater
heater. Under atmospheric pressure the oxygen content has been
reduced slightly below 0.2 cc. per liter with the water at the
outlet at 210 deg. fahr. It was found essential that the heater
outlet temperature be near 212 deg., the oxygen content increasing
rapidly at lower temperatures. The maintenance of high temperature in a heater operating under atmospheric pressure requires careful attention to water distribution and to venting.

R. N. Ehrhart,[9] in reply to Mr. Gibson, with regard to the
utility of the open heater in preventing corrosion, said that both
the author and Mr. Speller had realized the necessity of taking
out the oxygen, which the open heater will not do. Some companies have attempted to check corrosion by raising the feed
temperatures, say, from 130 to 140 deg. to 210 deg., but the consensus of opinion is that this will not produce the results and
that air-extraction apparatus is absolutely essential.

E. H. Tenney[10] pleaded guilty to being one of those who had
raised feedwater temperatures and said that there had since been
no indication of corrosive action in the economizer. In the
Mississippi Valley, where his company's plants are located, the
water is highly impregnated with dissolved organic matter, which
gives another important problem to be dealt with in addition to
that of dissolved gases. This is a troublesome subject which he
believed could well be taken up for consideration by the Society.

---

[7]Manager Power Dept., Allen & Garcia Co., Chicago, Ill.  Mem. Am.
Soc. M. E.
[8]Consulting Engineer, Geo. H. Gibson Co., New York.  Mem. Am.
Soc. M. E.
[9]Elliott Co., Jeannette, Pa.  Mem. Am. Soc. M. E.
[10]Chief Engineer of Power Plants, Union Electric Light & Power Co.,
St. Louis, Mo.  Mem. Am. Soc. M. E.

## Members Enjoy Oil Field Visit

THE Mid-Continent Section of the A. S. M. E. and the Oklahoma Section of the American Chemical Society held a joint
meeting at Tulsa on May 28 and 29, the occasion being noteworthy as the first joint meeting of a chemical and an engineering
society.

Following the St. Louis meeting, about 50 members of the
A. S. M. E. left for Tulsa, arriving there Friday morning. They
were conducted to the Hotel Tulsa by the Local Section to join
them at breakfast at the hotel. Mayor Evans delivered the
"keys of the city with all the ceremonies," and promised the
visitors that they would either leave the city millionaires or
"broke." President Miller responded in his usual gracious
manner.

The first part of the morning was spent in riding around the
beautiful, well-laid-out city of attractive homes and imposing
office buildings. At 11 o'clock the local members with their guests
gathered at the Cosden Refinery, where, under the guidance of
F. E. Holsten, manufacturing superintendent, and W. H. Pape,
mechanical superintendent, the intricacies of the $20,000,000 re-
finery were explained. The ingenious devices installed to promote
the safety of operation for the employees and the various pro-

cesses of distillation and filtration were gone over thoroughly
The plant has a capacity of about 25,000 barrels per day.

Luncheon was served at the refinery of the Texas Company.
Mr. W. K. Holmes, superintendent of the refinery, conducted
the party through the spotlessly clean plant which attracted
very many favorable comments.

A short business session held in the City Hall auditorium was
followed by a professional session, at which the following papers
were presented:

THE PREPARATION OF MOTOR GASOLINE FROM HEAVIER HYDROCAR-
    BONS, Prof. Fred W. Padgett
MODERN WATER PURIFICATION, W. R. Holway
REFINING OF CRUDE MINERAL OILS AND THE ACTION OF ABSORP-
    TIVE CLAYS ON SAME, F. C. Thiele
FLOW OF FLUIDS THROUGH PIPE LINES AND THE EFFECT OF PIPE-
    LINE FITTINGS, D. E. Foster
SOME INVESTIGATIONS IN BRIQUETTING OF OKLAHOMA BITUMINOUS
    COAL, Prof. James C. Davis
MID-CONTINENT GASOLINE, Dr. Chas. K. Francis
COAL VS. OIL COST PERFORMANCE CHART, L. C. Lichty

Professor Padgett's paper is presented on another page of this
issue, and the other papers will receive consideration in later
issues of MECHANICAL ENGINEERING.

SECTION OF OIL FIELDS SOUTHWEST OF TULSA

A banquet was given at the Hotel Tulsa in the evening, Louis Bendit, the dean of engineers in Tulsa, serving as toastmaster. Frank Greer gave a rapid-fire sketch of the remarkable development and marvelous resources of Tulsa, the Magic City, and Oklahoma, the Billion-dollar State. President Miller spoke of the development of coöperation among engineers and of the problems of relation between employer and employee, emphasizing the fact that there must be correct relations for proper production. Dean P. F. Walker, of the University of Kansas, presented the report of the Mid-Continent Section Research Committee which will be abstracted in the August issue of MECHANICAL ENGINEERING. Secretary Rice spoke on the subject of the engineer and his citizenship responsibility and laid out the broad lines on which the activities of the engineer must be organized.

On the following day the party was organized early for a 60-mile automobile trip to the casing-head gasoline plant of the Gypsy Oil Company, crude-oil pumping stations, and the Glenpool oil fields. Mr. H. P. Porter, chairman of the Mid-Continent Section and superintendent of the Gypsy casing-head plant, revealed the mystery of making gasoline in his plant. A most interesting instance of the complete utilization of heat energy was afforded at this plant. After the high-pressure gas has passed the usual condenser and had the normal amount of liquid removed, it is led to a cross-compound engine where it serves as an expansive vapor in the same manner as steam is commonly employed. In this expansion the gas is reduced to a very low temperature, this action being made cumulative in a counterflow system of pipes whereby it is made to refrigerate itself on the way to the engine and in this way extracts from itself an amount of high-gravity gasoline which could not have been recovered in the ordinary high-pressure condenser using water as the cooling system.

A specially arranged feature of the day was the shooting of an oil well. The party watching at a respectful distance saw the 80 quarts of nitroglycerine poured carefully into the tube and lowered 2200 ft. to the bottom of the well. They waited while the casing was being withdrawn but only a few venturesome spirits stayed on, in spite of a foreboding storm and the approaching train time, to see the final shot. Those who did stay witnessed the remarkable spectacle of a fountain of oil mounting 60 ft. above the top of the derrick—but missed their train.

The entire meeting and program of excursions was a brilliant success and every detail of arrangement was worked out very carefully by Messrs. H. P. Porter, J. H. McEwen, D. E. Foster and Earle A. Clark, the officers of the Mid-Continent Section, to all of whom the thanks of the Society are due.

OIL REFINERIES AND STORAGE TANKS OF WEST TULSA. ARKANSAS RIVER IN DISTANCE

# Motor Gasoline from Heavier Hydrocarbons

### Particulars Regarding Its Preparation by Various Cracking Processes, Together with a Discussion of the Theory of Cracking and Notes on the Present Status of the Industry

By FRED W. PADGETT,[1] NORMAN, OKLAHOMA

*The following paper is one of the group prepared by the Mid-Continent Section of the A.S.M.E. for the Tulsa, Okla., meeting held immediately after the Spring Meeting of the Society at St. Louis, and which was attended by a good-sized delegation of those who were present at the latter meeting. It is mainly concerned with those methods known as "cracking" processes which, when applied to the heavier hydrocarbons present in petroleum, produce mixtures within the range of motor fuel. It has as its objects the discussion of the theory of cracking, the description of standard processes, and the consideration of the present status of the industry. It will be found a convenient and valuable summary of the various processes, particularly useful for reference purposes.*

THE growing importance of the internal-combustion engine during the past decade has brought with it the attendant problem of fuel supply. While the production of motor fuel in the past has been ample, the data indicate that in the near future we shall be faced with the problem of augmenting the supply, or see a depressing effect upon the automobile industry on account of shortage. The Bureau of Mines in a recent report on refining statistics shows that the gasoline production in 1919 increased 10 per cent over 1918. On the other hand, the number of automobiles registered increased 22 per cent during the same period and the increase of 1920 will be of like order. The question therefore presents itself as to how it will be possible to meet the future demand for motor fuel even though gasoline stocks were increased during the past year. The problem is not now a serious one, but may become so in a very few years.

There are several sources from which additional supplies of motor fuel may be expected. Benzol and toluol from by-product coke-oven plants may be blended with gasoline to give a very satisfactory product, while industrial alcohol is a product to fall back on, if necessary. Crude oil may also be derived from such kerogenous substances as oil shale and there are enormous deposits of the raw starting material in various parts of the country, especially in Colorado, Wyoming, and Utah, from which the crude oil may be extracted by low-temperature carbonization or a combination of this with ultimate high-temperature effect in the presence of steam to recover the by-product ammonia.

Petroleum consists essentially of a mixture of hydrocarbons which are in the main totally miscible with one another. Organic nitrogen and sulphur compounds are also present in varying amounts and occasionally oxygen compounds, as well as water and inorganic material in disperse form. When this complex of naturally occurring hydrocarbons is treated by dry distillation at atmospheric pressure, the hydrocarbons of lowest boiling point are vaporized first and come into the condenser followed by those of higher boiling point as the temperature rises. Varying degrees of fractionation will obtain depending upon the design of still and method of operation. From the point where gases first appear at the condenser outlet until the temperature in the still reaches approximately 625 deg. fahr., the products secured are in the main natural—that is, they exist in the crude oil as such. Above this temperature, however, especially if the distillation is carried out slowly, what is known as "cracking" takes place and a distillate is secured, which upon redistillation will yield more gasoline and kerosene, their quality, however, in most cases, not being comparable with that of the natural products.

If instead of carrying out the distillation above the 625 point at atmospheric pressure in an ordinary still, one of special design is substituted, and if the operation is conducted at a pressure varying from 50 lb. to 150 lb., profound decomposition or crack-

ing occurs and a yield of gasoline of 15 per cent or more may be obtained, based on the original charge of residue, upon fractionating the pressure distillate secured. The residue remaining, and the pressure distillate which has thus been freed from gasoline, may again be subjected to the process and a further yield of gasoline secured, this being less than in the first case and decreasing upon subsequent treatments.

This process of cracking is a special application of the well-known phenomenon which occurs when organic compounds are heated to a high temperature, decomposition taking place and the molecule being broken down into compounds of lower molecular weight, even with the formation of the elements carbon and hydrogen.

Again, if a sample of the same residual oil be placed in a vertical still having a fractionating column or "tower" interposed between the still and the condenser, and the distillation is carried out at atmospheric pressure, with stirring, in the presence of anhydrous aluminum chloride, the latter acts as a catalyst to cause the decomposition of the oil into gasoline, giving a yield of 15 per cent or more. This process is an illustration of the fact that chemical compounds, working at atmospheric pressure, may effect a result similar to pressure distillation.

## COMMERCIAL PROCESSES FOR THE PRODUCTION OF GASOLINE FROM HEAVIER HYDROCARBONS

The various processes for producing motor fuel from heavier hydrocarbons may be roughly classified as follows:

1 *The Pressure Still.* A "two-phase" cracking system. Examples are the Burton and Coast processes.

2 *The Pipe Still.* Generally a "single-phase" cracking system. Examples are the Greenstreet, Hall, and Rittman processes.

3 *The Use of Catalytic Agents at Atmospheric Pressure.* This may include reaction both in the single-phase and the two-phase systems. An example is the aluminum chloride process, which is likely two-phase to some extent.

4 *Combinations and Modifications of Two or More of the Above Methods.* The cracking may be in either single-phase or two-phase systems. Examples are the Dubbs, Jenkins, and Bacon processes.

5 *Processes Involving Principles not Included Under the Above Four Headings, but Which May Involve Similar Apparatus or the Use of Pressure.* These methods may have either single-phase or two-phase cracking systems. Examples are the proposed processes of Cherry,[2] Coast,[3] Ellis,[4] and many other processes equally unique. In the first a bipolar high-voltage, oscillating, silent electric discharge is thrown across the vapors; in the second, hot gases are passed through the oil or into contact with oil spray; and in the third, air is supplied to the cracking chamber where, by combustion with a portion of the oil, heat necessary for the cracking of the remainder is generated.

During the past eight years numerous patents involving almost every conceivable form of apparatus have been granted for the purpose of securing hydrocarbons within the gasoline range from those of higher boiling point.[5] Many of these patents seek to obviate the trouble incident to coke deposition, while others have the object in view of producing a gasoline which is comparable

[1] Assistant Professor of Chemistry, University of Oklahoma.
Presented at a meeting of the Mid-Continent Section, Tulsa, Okla., May 28 and 29, 1920.

[2] U. S. Pat. 1,229,886, Jan. 12, 1917, and 1,327,023, Jan. 6, 1920.
[3] U. S. Pat. 1,252,401, Jan. 8, 1918, and Cosden-Coast, 1,261,215, Apr. 2, 1918.
[4] U. S. Pat. 1,295,825, Feb. 25, 1919.
[5] Some of the principles involved are as follows: The use of molten metal as a cracking agent; oil may be sprayed against an electrically heated plate within an autoclave; a perforated basket charged with metallic turnings may be disposed in a furnace; oil vapors may be subjected to progressively increasing and then progressively decreasing temperatures; superheated steam; various methods utilizing electric discharge across vapors; emulsified oil as starting material; etc., etc.

in quality with the natural product. Still others are along entirely new lines.

As early as 1865 James Young, the Scotch industrial chemist, secured a patent for the production of illuminating oil by means of pressure at about 20 lb. per sq. in.

The Pintsch gas process for the production of illuminating gas from gas oil dates back to the seventies in its commercial applications in the United States. As usually carried out at atmospheric pressure, the oil is passed by gravity into cast-iron retorts heated to bright redness. In the upper chamber of the retort oil is vaporized, while in the lower one cracking takes place. The vapors then traverse a tar seal, condenser, purifier, and meter, and are stored in a gas holder at a pressure near that of the atmosphere. The gas is finally compressed to 12 or 14 atmospheres in heavy welded tanks ready for delivery. One gallon of gas oil will produce from 60 to 75 cu. ft. of gas and when the product is compressed at the last stage a liquid known as "Pintsch gas drips" collects to the amount of about 2 or 2½ gal. per 1000 ft. of gas compressed. This liquid, while quite volatile, is very unsaturated and entirely unsuitable as motor fuel.

*Benton's Apparatus* for the production of illuminating oils from residual oils was patented in 1886 but was never used commercially. A pipe coil within a furnace is heated to a suitable temperature while the oil is forced through at a pressure of 285 lb. per sq. in. or more. The products of the reaction are released into a vapor chamber, which in turn is in communication with a condenser. At the inlet end of the coil is placed a safety valve and at the outlet a stop-cock for regulating the pressure on the coil.

In 1890 a patent was granted *Dewar and Redwood* for an apparatus which differed from that of Benton, to be used in cracking Russian residual oil to illuminating oil, but was never applied commercially because of the introduction of the starting material as fuel and its subsequent rise in price. A retort disposed in a furnace has one end in communication with a dome, the latter also being connected to a condenser. Safety valve, pressure gage, charging line and receiver are also provided, and the condenser and receiver are held under pressure. It is proposed to fill the retort partially, to distill under pressure and to collect the distillate in the receiver. Dewar and Redwood do not state what pressure gives the best results.

*The Burton Process.* Burton's first patent was granted in 1913 and the process is extensively used at the present time by the various Standard companies. In 1918 it was estimated that 10 per cent of the gasoline produced was by cracking, a large proportion of this being through the Burton still. The development of this process marks one of the milestones in the history of petroleum refining and its originator merits the distinction of being the first to demonstrate that pressure distillation could be done safely and practically on a large scale. In addition, Burton realized the limitations of the type of apparatus and did not attempt, in the original design, to make the operation continuous.

The original claims of Burton consisted in the distillation of petroleum under a pressure of 4 to 5 atmospheres and a temperature of 650 to 850 deg. fahr., maintaining the pressure upon the volatile products until they had traversed the condenser.

The standard Burton apparatus includes a still of 200 bbl. capacity or more, of conventional shape, so constructed that high pressures may be used. Beyond the condenser is located a safety valve which prevents the pressure from rising above a desired maximum and permits the escape of the permanent gases. Next in line is a valve where the pressure upon the distillate is released before it enters the run tank.

Modifications of the original Burton process call for introduction of the oil in the vapor line,[*] amounting to semi-continuous operation, the use of false bottoms[*] upon which the main portion of the carbon collects, and lastly a still containing internal flues.[*]

*The Design and Operation of Pressure Stills.* At the temperature at which the pressure still operates steel begins to decrease in tensile strength. It is therefore evident that a fundamentally important consideration lies in the material and method of construction of the stills. As an outgrowth of this need there are now firms which specialize in the construction of stills for severe usage. The walls of the stills are ½ in. to ⅞ in. thick and hammer-welded, this method of construction being claimed as superior to both riveting and to acetylene or electric welding for the purpose in view. In the operation of pressure stills the operator has before him thermometers and pressure gages, which inform him of the conditions prevailing within the various units under his control.[*] In the operation of some pressure stills using gas oil as the starting material, the distillation is continued at the average rate of 1 to 2 per cent an hour and the run is completed in 48 hours, giving a yield of 35 per cent of 450 deg. fahr. end-point gasoline. During distillation the bottom of the still is watched and if "hot spots" develop due to superheating from deposited carbon, close surveillance is maintained in order to determine if it will be necessary to close down the still before the run is completed. After each run the deposited carbon is completely removed from the still before recharging.

*The Greenstreet, Hall and Rittman Processes.* These processes involve cracking mainly in the vapor phase. By the Greenstreet method[*] a mixture of oil vapors and steam is passed through a coil of pipe which is heated, the products next being permitted to expand in drums of considerable size. The pressure is released at a point between the expanding drums and the condenser. The Hall process[*] in a broad way resembles that of Greenstreet but has the fundamental difference in that steam is not used and the pressure is released as the products leave the coils and enter the expanding drums. In the Rittman process[*] the cracking takes place in an upright tube containing a rod and chain, the latter being thrown against the sides of the tube when the former is revolved. The oil is fed into the tube at the top while the vapors are taken off separately from the tar pot at the bottom. All of these processes include the use of pressure.

*The Aluminum Chloride Process.*[*] The most conspicuous developments under this method have been carried out by McAfee of the Gulf Refining Co.[*] The oil is heated and stirred in a still in the presence of anhydrous aluminum chloride or other anhydrous salt of aluminum. Before treatment the oil must be freed from water and a quantity of the catalyst equal to a maximum of 8 per cent by weight of the charge is added before the distillation is begun. Fractionating towers are interposed between the still and the condenser so that the higher-boiling vapors may be returned to the still along with aluminum chloride which has been vaporized and carried out. The distillation is continued slowly at a temperature of 500 to 550 deg. fahr. over a period of 24 to 48 hours. By this method a yield of gasoline of 15 per cent or more may be obtained from residual oils.

Very interesting in connection with this process is the effect upon the residue remaining in the still and the character of the distillate secured. At the end of the distillation it is found that the aluminum chloride is in mechanical combination with granular coke which is easily removed, and that a heavy oil free from asphalt, and with its viscosity in no way impaired, may be separated from the coke and then utilized for the production of high-grade lubricating oils, the recovery of paraffin wax, or the manufacture of petroleum. The distillate secured by aluminum chloride treatment is water white and possesses a pleasant odor, and in order to secure the finished gasoline it is only necessary to wash the "rerun" distillate with dilute alkali and water.

By the action of anhydrous aluminum chloride on residual petroleum, sulphur compounds are destroyed and possibly nitrogen and oxygen compounds such as naphthenic acids as well.

The one difficulty suggesting itself as being inherent in this

*U. S. Pat. 1,199,464, Sept. 26, 1916.
[*]Humphreys Patents 1,122,002 and 1,122,003, Dec. 22, 1914, and 1,119,700, Dec. 1, 1914.
[*]Clark Patents 1,119,496, Dec. 1, 1914, 1,129,024, Feb. 16, 1915, and 1,132,163, March 16, 1915.
* Liquid meters are sometimes placed on the run lines.
[*] See Petroleum Age, 6 (1919), 66.
[*] See Lomax, Jour. Inst. Pet. Tech., 3 (1916), 36.
[*] See Bureau of Mines Bulletin 114 and Technical Paper 161.
[*] See Jour. Ind. Eng. Chem. 7 (1915), 737.
[*] Ibid.

process is that of recovering the aluminum chloride, or of manufacturing it cheaply. McAfee has patented the two following
methods for the recovery of the catalyst:

By the first method the aluminum chloride present in the coke is
dissolved in water. The solution is then evaporated and heated
moderately, when the hydrated salt is broken up into aluminum
and hydrochloric acid. The alumina, when mixed with carbon
and calcined in the presence of dry hydrochloric-acid gas, reacts
to form aluminum chloride, hydrogen, and oxides of carbon.

By the second method the residual coke is heated in an
atmosphere of chlorine, when the anhydrous salt is volatilized and
carried away by the chlorine vapors.

Finally, it is of interest to introduce one of McAfee's tables on
the treatment of 34 deg. Baumé Oklahoma crude petroleum, showing the comparison of the aluminum chloride process with that of
the customary method.

| | Per Cent of Crude Oil: | |
| | $AlCl_3$ Process | Usual Process |
|---|---|---|
| Gasoline | 34.82 | 12.50 |
| Kerosene | 29.47 | 41.00 |
| Gas oil | 5.36 | 35.00 |
| Residual oil | 14.07 | 9.00 |
| Sum of products | 83.72 | 97.50 |
| Loss | 16.28 | 2.50 |
| | 100.00 | 100.00 |

While the above table is not representative of the actual amount
of gasoline procured at the present time from the average Oklahoma crude petroleum, still the data may serve for making comparisons. At this time the average Oklahoma crude is being refined to about 20 per cent of straight-run gasoline and 17 per
cent of kerosene.

*Other Processes.* Three methods which have recently come into
prominence because of favorable reports concerning their operations, are those of *Dubbs, Jenkins,* and *Bacon.*

The design and operation of the *Dubbs plant* is given in the
report of a committee of the Western Refiners' Association to its
members. Extracts from this report are quoted as follows:
"The plant proper consists of a cracking coil made up of ten
lengths of 4-in. extra heavy lap-welded pipe, each 20 ft. in
length, jointed on the ends by return bends. This coil of 4-in.
pipe is placed horizontally in a furnace in two rows, six of the
pipes being in the lower row and four in the upper row. The
furnace temperature is maintained at approximately 1540 deg.
fahr. The outlet of the 4-in. coil is connected to an expansion
chamber which consists of four 20-ft. lengths of 10-in. common
extra heavy pipe. These pipes are connected in series by means
of return bends so as to form an expansion chamber approximately 80 ft. long. These 10-in. pipes are positioned horizontally
in a chamber and are not heated, but insulated on the outside to
prevent loss of heat by radiation.

"The raw oil is fed into one end of the 4-in. coil by means
of a force pump, and as the oil passes through it is heated to
about 820 deg. fahr. and is then discharged from the other end
into the 10-in. expansion pipes, which are maintained approximately half full of oil. The vapors are liberated from the oil
and pass up through connecting goose necks to a manifold; then
enter vapor lines leading to a spiral vapor condenser, and finally
into a water-cooled condenser. A pressure of about 135 lb. per
sq. in. is maintained on the entire apparatus. The unvaporized
portion of the oil in the 10-in. coil is continuously drawn off
from the end of the last unit."

In regard to the Dubbs process, it should be noted that the
cracking takes place in a two-phase system.

The summary at the top of the next column gives the results of
a test run of 168 hours' duration.

In the *Jenkins process* an apparatus is used resembling the
water-tube boiler. The tubes in the latest installation are $2\frac{1}{2}$ in.
outside diameter, and cold-drawn. The furnace is of the vertical
baffle type with Dutch oven, no heat coming in direct contact
with either the longitudinal or traverse drums. The operation of
the process is continuous and the oil is circulated mechanically.
The safe limit of operation (with the type of starting material

| | | |
|---|---|---|
| Total gas oil treated | 20,952 gal. | |
| Total uncondensable gas | 25,851 cu. ft. | |
| Fuel used [a]—equivalent of 3677 gal. of 14 Bé. fuel oil. | | |
| Products (per cent yield of original oil charged): | | |
| Gasoline (440 end point) (58.59 Bé.) | 26.29 per cent | |
| Kerosene (40-41 Bé.) | 14.11 per cent | |
| Pressure distillate bottoms (31-32 Bé.) | 27.19 per cent | |
| Residuum (13-14 Bé.) | 24.71 per cent | |
| Loss | 7.70 per cent | |
| | 100.00 | |

being used) has been found to be when a volume of oil equal to
14 times the changing capacity has been passed into the apparatus. Working under these conditions it is claimed that the
still has long life and that the time of cleaning is only five minutes
per tube. In closing down after a run the reduction of temperature is accomplished gradually by forcing more charging oil into
the apparatus under complete mechanical circulation, thus eliminating undue stresses or strains which might be caused by a
sudden temperature drop. Mr. Jenkins contributes the following data in connection with a run of 46 hours, using pressure
distillate gas oil[b] from Homer crude as the starting material:

| | | |
|---|---|---|
| Total oil circulated | 17,130 gal. | |
| Overhead pressure distillate | 11,600 gal. | |
| Fuel used—48,000 cu. ft. of city gas (natural). | | |
| Products (per cent yield of original starting material): | | |
| Gasoline (118 I.B.P., 400 end point) | 24 per cent | |
| Kerosene | 29 per cent | |
| Steam still bottoms | 14 per cent | |
| Pressure still bottoms (52 vis. Saybolt at 100 deg. fahr. and when treated, $4\frac{1}{2}$ color) | 27 per cent | |
| Total loss | 6 per cent | |

The pressure used in the above run varied from 105 to 110 lb.
gage; temperature, 700 to 710 deg. fahr.

The *Bacon process* was developed under the direction of Raymond F. Bacon and Benjamin T. Brooks at the Mellon Institute
of Industrial Research. As originally designed some difficulties
were encountered during operation because of coke deposition on
the tubes; but according to the latest information certain improvements have made the process appear very promising. The
"cracking" is carried out in a vertical tube 20 ft. in length and
6 in. to 19 in. in diameter, the oil level being maintained at the
top of the heated zone and the process operated continuously.
The oil inlet and vapor outlet are located at the top of the
tube. The lower end of the tube is connected to a
drum into which the tar and coke settle and are drawn off
from a pipe at the bottom. When operating an apparatus of
this type, coke accumulates very slowly on the metal surface in
the cracking zone and in addition a large heating surface is exposed to the oil. The pressure specified is from 60 to 300 lb. per
sq. in., but preferably 100 lb. The yields of gasoline (56 Bé.)
claimed for various oils are as follows:

| | |
|---|---|
| Oklahoma gas oil (32 Bé.) | 45 per cent |
| Mexican fuel oil (12 Bé.) | 50 per cent |
| California fuel oil (14 Bé.) | 47 per cent |
| Caddo heavy crude (12-14 Bé.) | 48 per cent |

*Deposition of Coke During Cracking.* It has already been
mentioned that one of the most serious difficulties encountered
during "cracking" is the deposition of coke. Unless removed
from the walls of the apparatus very soon after its formation, the
coke produced becomes very hard and seems to unite with the
metal. Superheating (hot spots), consequent softening of the
metal and finally "blowing out" may be the result of this condition. A majority of the processes include, among other things,
methods of eliminating the coke from the heating surface. Thus
the Burton process involves a modification in which false bottoms
are used, the Rittman process a revolving rod and chain, the
Bacon process vertical heating surface in the presence of liquid,
and so on." One patentee specifies the use of steel balls which
are passed intermittently through the "cracking" tube.

[a] Consumed during the run.
[b] It should be noted that this starting material has already been once through the process.
[c] In some refineries it is the practice simultaneously to agitate the oil slightly and to scrape the bottom of the still by means of a sweep with chains attached.

(Continued on page 426)

# ENGINEERING SURVEY

A Review of Progress and Attainment in Mechanical Engineering and Related Fields, as Gathered from Current Technical Periodicals and Other Sources

### SUBJECTS OF THIS MONTH'S ABSTRACTS

WIND-TUNNEL STUDIES OF AERODYNAMIC PHENOMENA
MECHANICAL SCREENLESS AIR FILTERS
RASCHIG RINGS IN VALVES
STRESSES CAUSED BY COLD ROLLING
SURFACE COMBUSTION, MIXTURE REGULA-TION IN
AUTOMATIC MIXTURE PROPORTIONERS FOR SURFACE-COMBUSTION FURNACES
FUBAIN AND COAL DUST
CHANTRAINE METALLURGICAL FURNACES
POLYMULTIPLE FLAME COMBUSTION IN METALLURGICAL FURNACES

OZONE CONCENTRATION IN VENTILATION
INITIAL TEMPERATURE AND PHYSICAL PROPERTIES OF STEEL
OBTURATORS IN INTERNAL-COMBUSTION ENGINES
TURNER-MOORE MOTOR-CAR ENGINE
ACCENTUATED VALVE INCLINATION IN MO-TOR-CAR ENGINES
BUREAU OF ST NDARDS ALTITUDE TESTING LABORATORY
RATE OF FLAME PROPAGATION, MEASURE-MENT OF

VISCOSITY OF LUBRICANTS, A. S. T. M. PROPOSED STANDARD TEST FOR
DIESEL MARINE MACHINERY, TRENDS OF DESIGN
UNIVERSAL JOINT, MECHANICS OF
FLYWHEEL EFFECT, SYNCHRONOUS MOTORS CONNECTED TO RECIPROCATING COM-PRESSORS
SCHNEIDER 155-MM. GUN
THRUST BORING IN EARTH
STABILITY OF MOTOR OMNIBUSES

## AERONAUTICS

### WIND-TUNNEL STUDIES IN AERODYNAMIC PHENOMENA AT HIGH SPEED, F. W. Caldwell and E. N. Fales.

A great amount of research and experimental work has been done and fair success obtained in an effort to place airplane and propeller design upon an empirical basis. However, one cannot fail to be impressed with the apparent lack of adequate knowledge toward establishing flow phenomena upon a rational basis, such that they may be interpreted in terms of the laws of physics.

With this end in view it was the object of the authors to design a wind tunnel differing from the usual type especially in regard to large power and speed of flow, which involves features whose suitability cannot be predicted. After all available information has been secured on full-size and model wind tunnels in various parts of the world, there remains much obscurity about the airflow phenomena. It is the assumption of Dr. George de Bothezat that the type of air flow which establishes itself is governed by the stresses set up in the air passing the aerofoil. The unit stresses increase as the velocity rises, and it is easy to conceive that a given type of flow is possible only so long as the shearing stresses developed in the fluid do not exceed a certain magnitude, which depends on the value of the viscosity coefficient.

Experimental investigation of the flow has heretofore been rather unsuccessful because of lack of adequate methods. The writers laid out the design of the McCook Field wind tunnel to investigate the scaling effect due to high velocities of propeller aerofoils. During the course of the experiments, however, it was found possible to visualize the air flow by the following method: The velocities of the air flow discovered by the writers offer a solution to one of the fundamental problems of aerodynamics. This problem is the quantitative empirical measurement of the phenomena of fluid dynamics pertaining to flight and air flow. The method described in the report for visualizing air flow depends upon the fact that the moisture in the air condenses as a fog when the temperature is reduced to the dewpoint, provided that there is a solid or liquid nucleus to start the condensation. In the McCook Field wind tunnel the temperature drop is brought about through expansion of the air during acceleration, due to a drop of pressure of 100 in. of water. The relative humidity of the atmosphere can be artificially raised if too low, and the necessary nucleus for condensation is provided by the model tested. Flow vortices become readily visible, and the report contains many photographs showing the air flow past an aerofoil under different conditions.

In this connection, attention is called to the paper by the same authors presented at the Spring Meeting of the A. S. M. E., and given in abstract form in MECHANICAL ENGINEERING for April 1920, p. 213 (National Advisory Committee for Aeronautics Report No. 83, 52 pp., 42 figs., e.A)

## AIR MACHINERY

### German Screenless Air Filters

MECHANICAL SCREENLESS AIR FILTERS, Ernst Preger. There are four ways of filtering air. One is to pass it through a screen so fine that the dust will be kept out. The second way is to pass it through a liquid, which in this case acts as a screen of infinitely fine mesh. The third method is to pass the air through an electrical screen, the mesh of which is constituted by lines of electric force. This is the Cottrell system of dust precipitation, which is very effective under certain conditions. The fourth way is to interpose in the path of the flowing air mechanical obstacles which will not interfere to any very material effect with the flow of air itself, but will cause precipitation or separation of the heavier particles such as dust. A number of devices belonging to this fourth class are described in the present article.

In the air filter of the Balcke Machine Co. of Bochum, Germany, Fig. 1, the air flows between several rows of vertical

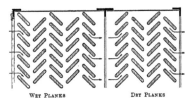

WET PLANKS          DRY PLANKS

FIG. 1 BALCKE AIR FILTER (SECTION THROUGH THE PLANKS)

wooden planks staggered and located at angle of 45 deg. with respect to each other, as shown in the drawing. Because of the frequent changes of direction the dust particles are projected against the wooden planks and washed off them by the water flowing over their faces. Because of this arrangement, it is necessary to clean the planks. Fig. 2 shows a Balcke cleaner capable of handling 24,000 cu.m. (847,300 cu. ft.) of air per hour.

Water is raised to the top of the cleaner by pump A and is distributed over the planks N by the trough B and hopper C. The water accumulates at D and may be used for purposes where the presence of dirt in the water is not injurious.

Back of the wet planks N are located several rows of dry planks P, the purpose of which is to catch the water particles carried off from the wet half of the filter. It has been found in

THE JOURNAL
AM. SOC. M. E.

practice that these planks remain very nearly dry, a good indication that the use of the wet filter causes only an insignificant increase in the humidity of the air. Measurements have shown that with fairly dry entering air there is an increase of humidity of from 3 to 5 per cent, and when the humidity of the entering air is considerable, there is practically no increase of humidity.

Air filters of this type are mainly used for cleaning air employed for purposes of cooling, since there the variation of humidity is of lesser importance than, for example, with air

Fig. 2 Balcke Air Filter in Section

compressors. As advantages for this type of filter are claimed the low resistance in the filter, the elimination of filter cleaning and the absence of fire danger from the accumulation of dry dust.

The next mechanical air filter described is the Viscin filter of the German Air Filter Construction Co. of Berlin. The active part of the Viscin filter consists of a frame filled with so-called Raclig rings of about 14 mm. (0.56 in.) height and the same diameter. These rings are wetted with a liquid called viscinol, which is

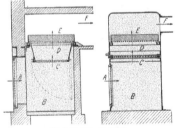

Fig. 3 Moeller Air Filter

claimed to be very sticky and little subject to deterioration. This type is not described in detail as its operation is evidently dependent on the use of a particular liquid used in wetting the rings.

The Moeller filter, Fig. 3 (K. & Th. Moeller, Brackwede, Germany), consists also of a frame filled with Raclig rings, but in this instance the rings are not wetted. The air enters through a perforated sheet-iron wall A, its velocity in the chamber B is considerably reduced and there occurs further a sharp change in direction of flow which causes the precipitation of the larger particles of dust. The medium-size particles of dust are held back

by the screen C. E is a horizontally located container filled with Raclig rings which hold back the finest particles of dust. The air comes out at F. Because of the three-stage method of separating the dust, namely, at D, C and E, a high degree of filtering is obtained notwithstanding the fact that no wetting of rings is resorted to. On the other hand, the dimensions of the air filter have to be made larger than in wet filters. Also the filters have to be cleaned every two or three months.

Strictly speaking, the Moeller filter does not belong to the class of screenless filters as a screen is used. It performs, however, secondary functions. (Stofflose Luftfilter, *Werkstatts Technik*, vol. 14, no. 3, Feb. 1, 1920, pp. 71-73, 16 figs., d)

## BUREAU OF STANDARDS

A Note on Stresses Caused by Cold Rolling, H. M. Howe and E. C. Groesbeck. The experimental data here presented show that a given amount of reduction by rolling causes less residual stress in the metal rolled if it is brought about by a large number of light drafts than if by a smaller number of heavy ones. This is shown by means of the curvature induced in each of a pair of superposed strips when they are simultaneously reduced in thickness by rolling, with varying reductions.

The result is attributed to the greater skin friction between the metal rolled and the face of the rolls with heavy than with light reductions. Because of this greater skin friction more of the reduction occurs through the backward forcing of the deeper-seated layers, and less through the elongation of the surface of the metal in contact with the rolls. (Abstract of *Technologic Paper of the Bureau of Standards* No. 163, e)

## FUELS AND FIRING

### Mixture Regulation in Surface-Combustion Furnaces

Surface Combustion, A. E. Blake. An extensive paper discussing the principles and describing the application of surface combustion. Only some passages of particular interest can be abstracted here owing to lack of space.

The factor of the highest importance in the perfected surface-

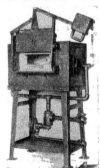

Fig. 4 Surface-Combustion Furnace with Low-Pressure Automatic Proportioner

combustion process is the production of perfect mixture or of any desired mixture to produce essential conditions within furnaces. There are several types of such automatic proportioning or mixing devices. Fig. 4 shows one type attached to the manifold upon the furnace. It is called the low-pressure system. In this case air is supplied at a maximum pressure of about 1 lb.

per sq. in. Two cocks are provided and the air supply is governed by the upper of these cocks. The air from the chamber passes through a venturi tube to a manifold supplying the burners. Within the throat of the tube is located an adjustable nozzle and the location of this nozzle determines the proportion of air and gas in the mixture. In operation air creates suction upon the nozzle. The gas comes from a governor, the function of which is to deliver the gas at constant atmospheric pressure at all times regardless of the rate of flow. There is a diaphragm which is moved by atmospheric pressure and causes the opening of the gas entry port. Changing the air supply automatically changes the gas supply and during operation the gas cock is never touched.

In the original article another type of automatic proportioner, the so-called high-pressure type, is shown and described. No decisive steps have been taken as yet in the matter of preheating air and gas either separately or combined. This is partly due to the fact that the highest temperatures which furnaces will withstand can be secured easily without preheating, and also to the fear that preheating may cause premature ignition.

The presentation of the paper was followed by an interesting discussion which cannot be abstracted for lack of space. (Advance copy of a chapter in a book, "American Fuels," by R. F. Bacon and W. A. Hamor (McGraw-Hill) presented before the Engineers' Society of Western Pennsylvania. Abstracted from the *Proceedings of the Engineers' Society of Western Pennsylvania*, vol. 36, no. 3, April, 1920, pp. 145-174, 27 figs. in the original paper, and discussion pp. 175-204, *g*)

COAL DUST AND FUSAIN, F. S. Sinnatt, H. Stern and F. Bayley. Investigation of the composition of fusain and the amount present in the dust produced during the working of the coal at the colliery, with particular reference to Lancashire coal fields.

The investigation is being continued with a special object of determining, among other things, the properties of the fusain with special reference to the temperature at which it will ignite and the influence of fusain upon the general properties of the coal or coal dust in which it occurs.

Fusain is a jet-black powder which causes a black and dirty smudge on the skin. It has a needle-like structure different from particles of bituminous coals of the same degree of fineness. It ignites with considerable ease at a low temperature and continues to smolder at a dull red heat. It may smolder for a long time without the production of flame or the evolution of compounds having an odor, which adds to the dangers associated with the composition of fusain, as practically no indication is afforded of the substance being on fire.

The following experiments indicate the manner in which fusain burns: A briquet measuring 4 in. by 4 in. by 9 in. was made by mixing fusain with water; it was dried at a temperature of about 105 deg. cent. and allowed to stand for one week. The briquet was then placed on a stone slab and the tip of one corner heated to redness by means of an ordinary match. The briquet was allowed to remain undisturbed, and in about an hour the whole of it was consumed without the production of smoke or flame; there was no evidence of combustion except a faint, pleasant aromatic odor resembling that of pitch pine. The mass smolders internally, for when the surface is removed a glowing dull-red zone is exposed. The residual substance, in certain experiments, was a finely powdered coke—in fact, it appeared as if the volatile matter only had been consumed; in others the residual substance was the ash. A briquet of similar dimensions containing 50 per cent of fusain and 50 per cent coal acted in the same manner, and again practically no odor or smoke was produced.

When fusain is carbonized a coke is produced having an appearance practically identical with that of the original substance. This coke is jet black and is quite deficient in caking power.

It is of interest to note that dust from a dust collector of which 95 per cent by weight would pass through a sieve of 1/90 mesh was found to contain about 49 per cent fusain. (*Lancashire and Cheshire Coal Research Assn. Bulletin* No. 5, 1920, 19 figs., *ep*)

## FURNACES

### Polymultiple Flame Combustion in Metallurgical Furnaces

CHANTRAINE METALLURGICAL FURNACES. Description of a new type of furnace in which combustion is carried on by means of a large number of separate jets of flame.

In conventional metallurgical furnaces the firing is effected by means of flame obtained through a simultaneous admission of the air and gas. With these systems the high temperatures needed for the metallurgical process can be maintained only in certain definite sections of the furnace, which, because of this, has a very variable régime throughout the combustion chamber and this leads to trouble, both as a result of the irregularity of distribution of heat and because of oxidation. It is claimed that the system described in this article, which the author calls "polymultiple flame" combustion, avoids these defects.

The characteristic features of the Chantraine method are:

1 There is a reservoir of hot air, usually produced by making the roof of the combustion chamber double and providing for an interspace through which pass the conduits leading the air to the combustion chamber from the hot-air regenerator.

2 A number of orifices are provided in the lower roof of the hot-air reservoir, these orifices acting as tuyeres to admit air to the combustion chamber.

3 A sheet of combustible gas floats under the perforated roof of the combustion chamber; this sheet of gas is fed in slowly and continuously through gas conduits from the gas producers.

4 A continuous apparatus is provided for recuperating the combustion-chamber waste heat and to employ this latter for a preliminary preheating of the secondary air.

The operation of the system is based essentially upon the following: The air under pressure, whether natural or artificial, passes through the multiple tuyeres and then through the sheet of gas slowly and continuously fed to the roof of the combustion chamber; in doing this, the air creates a bundle of flaming darts, a compact bundle directed toward the material to be heated. The products of combustion are e hausted solely by the pressure in the combustion chamber.

The theory of this system of combustion is as follows: If a jet of gas passes into the air, as happens in conventional systems, combustion occurs on the periphery of this jet and this periphery produces heat radiations, the majority of which pass into the air which is diathermanous while the commercial combustible gas is athermanous.

On the other hand, if a jet of air is made to penetrate into the atmosphere of ordinary combustible gas so as to produce the same number of calories per second, the periphery of the jet also emits heat and light radiations and a great majority of the heat radiations go into the air of the jet, preheating it all the more the less air there remains for the oxidation of the gas, which latter is athermanous. As a result of this, the tip of the flame of air in a gas atmosphere is many times hotter than the tip of the flame of gas in an air atmosphere and the combustion even with the low initial temperature is practically perfect with the minimum access of air, while in the opposite process (gas going into an air atmosphere) the excess of air must be very large.

Furthermore, in conventional furnaces the radiation spreads equally over the roof and the side walls as well as over the floor of the furnace and the material located thereon. In polymultiple flame furnaces the sheet of gas floating against the roof protects it as well as the walls against the radiations from the bottom of the furnace and the melting material. Also since the roof of the furnace is equipped with a number of orifices through which a constant flow of air takes place, the refractory material of the roof is never permitted to attain a temperature high enough to soften it or melt it.

Uniformity of temperature distribution in the furnace is secured because each jet projects on to the part of the flow whereon it projects a certain number of calories.

The analyses of exhaust gases obtained when this system was

employed have given contents of carbon dioxide very closely
approaching the theoretical maximum. This fact is explained in
the following manner. As the tips of the jets of air are very
highly superheated, they consume all the oxygen which they contain during the passage through the fairly stagnant and thick
sheet of combustible gas, the temperature of which increases as the
flame nears the floor of furnace.

Fig. 5 shows the general construction of the Chantraine furnaces. As regards the temperatures obtainable with this system, it
is stated that the temperatures at the tip of the individual jets
are close to the theoretical temperature of combustion. Since the
jets cannot radiate their heat through the sheet of gas through
which the air flows, all the heat is retained and is used to raise
the final temperature of combustion.

In ordinary furnaces a certain amount of metal is lost
through oxidation. In the Chantraine system the metal is
separated from the oxidizing agents by a thick and uninterrupted
sheet of neutral gases (products of combustion) the oxygen of
which has been completely neutralized during the passage of the
flame through a very hot sheet of reducing gases. This oxygen

epidemic of 1918 the writer accidentally discovered that very few
motormen of the street-car service contracted the disease, which
was ascribed to the action of ozone emitted by the car controllers.

It was also found that employees in offices having ozone air
purification proved to be remarkably immune to influenza. This
is explained by the fact that ozone increases the oxyhaemoglobin
content of the blood and thereby increases the resistance of the
body to all disease. (*Journal of the American Society of Heating
and Ventilating Engineers*, vol. 26, no. 4, May 1919, pp.
423-425, *gp*)

## HEAT TREATMENT

THE EFFECT OF INITIAL TEMPERATURE UPON THE PHYSICAL
PROPERTIES OF STEEL, J. H. Andrew, J. E. Rippon, C. P. Miller,
and A. Wragg. Discussion of the effects which variation in initial temperature has upon the position of the resulting transformation point in certain nickel, chromium, and nickel-chromium
steels.

The volume change at the normal transformation points as

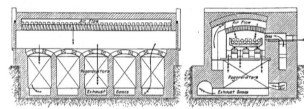

FIG. 5  CHANTRAINE "POLYMULTIPLE FLAME" METALLURGICAL FURNACE

itself having been heated to an exceeding high temperature.
Furthermore, the pressure existing in the combustion chamber
prevents the possibility of the entrance of outside air. It is
claimed that a fuel economy as compared with conventional types
of furnaces may be secured as high as 40 per cent, with lower
metal losses and lower maintenance costs. This construction is
particularly suitable for reheating furnaces, for steel-melting
furnaces, etc. ( *Le Génie Civil*, vol. 76, no. 14, April 3, 1920, pp.
336-337, 2 figs., *dA*)

## HEATING AND VENTILATION

THE SIGNIFICANCE OF ODORLESS CONCENTRATION OF OZONE IN
VENTILATION, E. S. Hallett. This paper is essentially a companion to the paper by the author abstracted in the April issue
of MECHANICAL ENGINEERING (p. 244).

The author states that chemical methods of determination of
ozone cannot be conveniently utilized for ventilation purposes—
(1) because the apparatus must be used often under conditions
where chemical tests are not available, and (2) because the concentration of ozone for ventilation is too low for such tests.
Therefore the odor remains the only indicator of excessive concentration, notwithstanding the fact that only a rough approximation of measurement is thereby made available.

Ozone in ventilation is not a stimulant and its refreshing action and the freedom from languidness are only normal conditions due to freedom from depression resulting from odors and
excessive heat. Furthermore, the effects of ozone do not wear
off rapidly, and are not psychological merely, as excellent results
have been observed in schools where teachers and pupils were
unaware of the presence of ozone.

Interesting data are presented concerning the influence of
ozone on the degree of susceptibility to influenza. During the

shown by the dilatometer has been found to be composed of two
changes working in opposition, namely, allotropic change and
change due to the carbide. The latter change in steels of high
carbon content is not completed at the ferrite or carbide (SE)
line, but continues to a temperature depending on its composition, particularly carbon content, increasing with this. This
quantity has been shown to be a measure of the molecular concentration of the carbide in solution, and controls the temperature
at which the allotropic transformation takes place.

This variation in the molecular concentration is explained by
the assumption that carbide dissociates at and above the temperature of the normal transformation point.

Nickel-chromium steels behave in much the same way as carbon
steels, excepting that in the latter the time factor necessary to
produce the same changes in the carbide is of a much lower order.

Martensite has been shown to be the product formed when
the allotropic change has been depressed by dissociated carbide
to a temperature at which the latter is able to exist as such in
the alpha iron formed. At the temperature of the normal transformation point martensite is not formed as an intermediate
phase between austenite and pearlite.

By tempering austenite in such a way that the carbide becomes
partially associated, the allotropic change is able to take place on
recooling, forming martensite, thus explaining the phenomenon
of secondary hardening.

It has been shown that in the carbon steels experimented with,
quenched at 1000 deg. cent., the iron in the product is in the alpha
state. In alloy steels gamma iron is present in various proportions, particularly with high carbon content. In extreme cases
alpha is entirely absent. (Paper before Iron and Steel Institute,
Annual Meeting, May 6-7, 1920, abstracted from advance proof,
82, pp., 71 figs. and 1 plate of photomicrographs, *eA*)

## INTERNAL-COMBUSTION ENGINEERING (See also Marine Engineering)

### Obturators in Internal-Combustion Engines

OBTURATORS VERSUS PISTON RINGS. Experiences in air-cooled engines on aircraft led in some cases to the replacement of the ordinary piston rings by an obturator, which is essentially the equivalent in metal of the cup leather used for packing hydraulic cylinders. The air-cooled cylinders of rotary engines, it was found, suffered distortion when at work owing to the leading side of the cylinder being more effectually cooled than the trailing side. As a consequence, piston rings often failed to hold the pressure satisfactorily. Any increase in their number would have involved a longer and heavier piston, and the obturator was introduced in consequence. This, as explained in an interesting and valuable paper by Mr. W. Fennell, M. I. E. E., which was read at a recent meeting of the Diesel Engine Users' Association, consisted of a flexible L-shaped ring forced against the cylinder wall by fluid pressure. At the outset the life was short, but improvements made at the works of "Engineering and Air Lamps" increased the life from 10 hours to 60 hours, and in exceptional cases to 250 hours. Brass was the metal first employed, but a special phosphor bronze is now used. The obturator was placed near the top of the piston, but being flexible and forced by the pressure into close contact with the relatively cool wall of the liner, it did not burn. Nevertheless, it is now considered preferable to fix it some little distance below the piston head. As originally fitted the rings were split similarly to a piston ring, but a lap joint is now used and with this the obturator works well, even if the liner be worn out of truth. In consequence of the satisfactory behavior of these obturators, Mr. Fennell determined to fit one to a three-cylinder Sulzer-Diesel engine rated at 140 bp. The liners of this engine had worn badly, being far from circular and tapering to the extent of 0.1 in. It was impossible under the conditions to obtain new

liners, and piston rings failed so rapidly the engine could not be run save at a prohibitive cost. It was decided accordingly to fix the three top rings of the piston and to place an obturator in the fourth groove. After a run of 100 hours without loss or blowing by, the obturator was examined. The wear was less than 1 mil, and the engine was set to work again, without replacing it. This obturator lasted 380 hours and then failed, due to wear at the lap joint. Here the motion was considerable, owing to the taper of the liner. Further experiments showed that even with these badly-worn liners the obturator might be counted on to last 300 hours, which was longer than the piston rings would stand. Further tests, Mr. Fennell stated, are in progress, and they indicate that with the liners in proper condition the life of the obturator should be well over 1000 hours. (*Engineering*, vol. 109, no. 2833, April 16, 1920, pp. 520, editorial, *g*)

### Motor-Car Engine with Hollow Crankshaft and Strongly Inclined Valves

TURNER-MOORE MOTOR-CAR ENGINE. Description of the new Turner-Moore 3½ by 5-in. four-cylinder engine having some interesting features. One of these is the hollow crankshaft. It is 2 in. in diameter at all bearings. There are no oil tubes, all of the leads being drilled. The engine also has a tunnel in which are housed a camshaft and bearing points of the tappets. This construction permits the operation of the camshaft and cams in an oil bath.

Fig. 6 shows the valve arrangement with its accentuated inclination of the valves, the purpose of which is to secure better

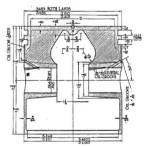

FIG. 7   TURNER-MOORE MOTOR-CAR ENGINE (SECTION SHOWING WEB CONSTRUCTION AND OIL RELIEF IN PISTONS)

scavenging inasmuch as it permits the exhaust gases to be swept off the top of the piston directly into the exhaust passage. The inclination also permits of increased water space around the valve-stem guides and in addition gives a shorter combustion chamber.

On the pressure-feed lubricating system there is a ball check and relief valve at the end of the oil line, the overflow from which lubricates the timing gears. The cylinder walls are lubricated by oil mist thrown from the connecting-rod bearings, and the remaining bearings are taken care of by direct-pressure feed.

The pistons are semi-steel and as shown in Fig. 7 are heavily webbed, the web being split at the center as to avoid distortion due to expansion. The piston has two rings at the top end. Below the lower ring there is an oil groove 1/32 in. deep around the entire piston, and at the point directly above the piston pin there is a slot cut in the side of the piston, the lower end of which is in connection with the drilled lead to the top of the piston-pin bearing. The effect of this is said to be to force oil to

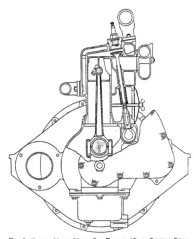

FIG. 6   TURNER-MOORE MOTOR-CAR ENGINE (CROSS-SECTION SHOWING INCLINED VALVES)

the top of the piston bearing on every up-stroke. (*Automotive Industries*, vol. 62, no. 17, Apr. 22, 1920, pp. 946-947, 5 figs., *d*)

### Bureau of Standards Altitude Testing Laboratory and Apparatus to Measure Rate of Flame Propagation

BUREAU OF STANDARDS NEW ENGINE-TESTING PLANT, R. S. McBride, Description of the altitude testing laboratory at the Bureau of Standards. The purpose of the laboratory is to test aeronautical engines, reproducing artificially the surrounding conditions under which they have to operate at various altitudes.

During the war period the laboratory was housed in temporary buildings, but a new laboratory has just been completed. This is a substantial brick and concrete building about 50 ft. by 100 ft. The altitude chambers are so designed that conditions can be reproduced in them such as are likely to be encountered at altitudes up to 40,000 ft. These chambers are large enough to take the largest aeroplane engine.

The operation of the altitude laboratory requires a rather complicated system of controls for the engine, the air system and the dynamometers. In the new laboratory all these have been brought nearer together so that a single operator has complete control of the system without moving from his regular position of observation. It is thus possible to quickly adjust conditions for

FIG. 8 TESTING A SINGLE-CYLINDER LIBERTY ENGINE FOR RATE OF FLAME TRAVEL

a test, carry through the observations and be ready for a new adjustment of conditions.

Other highly interesting equipment provided includes, for example, the arrangement shown in Fig. 8 to permit the study of the rate of flame propagation within the cylinder of an engine. This has been made possible by a triple spark-plug system which operates roughly as follows: At the time the flame of burning gas-and-air mixture passes over the spark plug the gases are ionized and a reduction in electrical resistance results. This permits the passage of a spark across the gap. The spark plugs are connected to a spark chronograph (which is being adjusted by the operator at the right in Fig. 8) or an oscillograph and the relative time of the passage of the flame past these three points can thus be measured (*American Machinist*, vol. 52, no. 20, May 13, 1920, pp. 1061-1063, 4 figs., *d*)

### LABORATORIES (See Internal-Combustion Engineering)

### LUBRICATION

PROPOSED STANDARD TEST FOR VISCOSITY OF LUBRICANTS. The test was proposed as tentative in 1919 to the American Society for Testing Materials, but it has not yet been finally adopted.

According to this proposal viscosity shall be determined by

means of the Saybolt standard universal viscosimeter described in a general manner and illustrated in the original text.

Viscosity shall be determined at 100 deg. fahr., 130 deg. fahr. or 210 deg. fahr., and for viscosity determinations at the first two temperatures either oil or water may be used as the bath liquid.

The discharge is measured by a standard receiving flask shown in cross-section in the original paper and the time in seconds for the delivery of 60 cc. of oil is the Saybolt viscosity of the oil at the temperature at which the test was made. (*Report of Committee D-2 on Lubricants of the American Society for Testing Materials*, preprint, 5 pp., 2 figs., *gp*)

### MACHINE PARTS AND DESIGN (See Internal-Combustion Engineering)

### MARINE ENGINEERING

THE GENERAL TREND OF DIESEL MARINE MACHINERY. According to an editorial in *Engineering*, there is at the present time no problem concerning the minds of marine engineers to a greater extent than that of the application of the Diesel oil engine to marine propulsion.

Statistics, as far as they are available, and without any data as to German construction, show that in the field of ships of greater deadweight capacity than 2000 tons considerably more than 150 are on order and in course of construction at the present time and are destined to be fitted with marine Diesel engines.

In the list of countries where this kind of work is done Denmark occupies the first place, the United States of America the second, and Great Britain the third, but it is claimed that the position of Great Britain is rapidly being improved.

A review of the type of engine being constructed goes to show that the simple single-acting 4-stroke-cycle engine is largely preferred today, and of the vessels built only some 16 per cent are to be fitted with engines of the 2-stroke-cycle type. Moreover, almost half of the 2-cycle engined ships are single-screw, whereas 90 per cent of the 4-stroke cycle machinery is of the twin-screw variety. Of all the motor vessels over 80 per cent are fitted with twin screws.

The 2-stroke-cycle engines comprise the opposed-piston types at present exclusively built in Britain, the supercharging port scavenging engine originated in Switzerland and being built now chiefly in Switzerland and Italy, and the valve scavenging engine which is now retained by only one French constructor. In America very few 2-cycle Diesel engines are in the course of construction, and Sweden, Denmark and Holland are solidly in favor of the 4-stroke cycle engine.

In Britain not only is the greatest variety of type evidenced, but also the largest engines are being constructed, the most powerful, the 4-cylinder opposed-piston two-stroke engine of 750 b.hp. per cylinder. The cylinder diameters are 580 mm. and the stroke 1160 mm., and the engine develops at 70 revolutions a total of 3000 b.hp. in the four cylinders. The largest 4-cycle engine being constructed in Britain, and also elsewhere, has cylinder diameters of 740 mm. by a stroke of 1150 mm., and runs at 115 r.p.m., developing over 3500 b.hp. in eight cylinders. The 2-cycle engines are of four and six cylinders—generally of four cylinders —and the 4-cycle engines have six cylinders, with a few exceptions where eight cylinders are employed to give the desired high power.

The British policy of trying out all types of engine is certain to lead in the end to a very favorable position, and is a very bold one and to be commended in view of the leeway which had to be made up and the lead unquestionably held by the earlier constructors of 4-cycle engines. The unfortunate and unlikely event of failure with any of the newer types cannot now, in view of the varied and long experience already piled up to the good, be taken as evidence sufficient to condemn this prime mover in respect of its suitability for marine work. (*Engineering*, vol. 109, no. 2836, May 7, 1920, pp. 617-618, editorial, *g*)

## MECHANICS (See also Transportation)

### The Mechanics of the Universal Joint

FORCES ACTING IN A UNIVERSAL JOINT, D. Thoma, D. E. Universal joints are used to connect shafts, the axes of which intersect at an angle of 90 deg. or less. If the shafts transmit no moment of torsion, the universal joint does not oppose the free variation of the angle of intersection of the two shafts and permits a free oscillation of one shaft axis relatively to the other about the center of the joint. This being so, it is easy to neglect mistakenly the fact that this resistance-free ability of oscillation persists only until the moment when the shafts begin to transmit a torsional moment, and that if the angle of intersection of the two shafts is other than zero, the universal joint as a rule must transmit the bending moment in addition to the torsion moment.

Even a superficial consideration will show that the universal joint actually exerts a bending moment on the shafts. Whatever may be the axis of the moment transmitted between the shafts through the means of the joint, it forms with the axial direction of at least one of the two shafts an angle equal to or more than half the angle between the shafts, so that the moment, at least with respect to one of the shafts, has a component normal to the axis of that shaft, which means a bending moment.

Since the conditions with respect to the two shafts are substantially symmetrical as a rule, both shafts will have to undergo the effect of a bending moment.

The universal joint coupling is a machine element having an extensive application, and it may be worth while, therefore, to investigate more closely the forces acting thereon. This is practically of interest in connection with certain tests on the critical speed of rapidly rotating shafts where the universal joint plays an important part.

The investigation of the forces presupposes the movements of the joint itself and of the shafts. Let us assume, as in Fig. 9, a system of rectangular coördinates such that both shafts lie in the x-y plane and that the positive z axis is located normally to the plane of the drawing and is directed upward. The angles of rotation of the shafts are measured from the position shown in Fig. 9, which may be called position A. Fig. 10 shows the general location of the coupling: The shaft I is rotated from the position A through an angle φ and the shaft II through an angle ψ in the direction of the arrow. We shall consider φ as an independent variable and shall determine ψ as a function of φ.

Let P be some point on the axis of the part of the universal joint belonging to the shaft I, and l its distance from O which is the point of intersection of the axes of the shafts and of the arms of the universal joints. The projection of OP shown in Fig. 10 is then equal to $l \cos \varphi$ and hence the coördinates of P are

$$x = l \cos \varphi \cos \alpha$$
$$y = - l \cos \varphi \sin \alpha$$
$$z = l \sin \varphi$$

From this it is found that the cosines of the angles $\alpha_1$, $\beta_1$, and $\gamma_1$ which are enclosed between the axis of the arm of the universal joint belonging to shaft I and the axes of the coördinates are

$$\left.\begin{array}{l} \cos \alpha_1 = \cos \varphi \cos \alpha \\ \cos \beta_1 = - \cos \varphi \sin \alpha \\ \cos \gamma_1 = \sin \varphi \end{array}\right\} \quad \dots \dots \dots [1]$$

In a similar manner are found the cosines of the angles $\alpha_2$, $\beta_2$ and $\gamma_2$ which are enclosed between the axis of the arm of the universal joint belonging to shaft II and the axis of coördinates, these cosines being

$$\left.\begin{array}{l} \cos \alpha_2 = - \sin \psi \\ \cos \beta_2 = 0 \\ \cos \gamma_2 = \cos \psi \end{array}\right\} \quad \dots \dots \dots \dots [2]$$

The directions of the two arms of the universal joint are not, however, independent of each other, the arms being connected with each other in such a manner that they are always at right angles to each other. The usual formula of analytical geometry of space gives therefore the relation—

$$\cos \alpha_1 \cos \alpha_2 + \cos \beta_1 \cos \beta_2 + \cos \gamma_1 \cos \gamma_2 = 0$$

If we insert into this formula the values from Equations [1] and [2], we obtain after transforming—

$$\sin^2 \psi = \frac{\sin^2 \varphi}{\cos^2 \alpha + (1 - \cos^2 \alpha) \sin^2 \varphi} \dots \dots [3]$$

This situation may also be expressed in the form

$$\tan \psi = \frac{1}{\cos \alpha} \tan \varphi \dots \dots \dots \dots \dots [4]$$

which will be used later on in this article and which is very simple.

*The Action of the Forces.* For the sake of simplicity it may be assumed that the joint transmits only moments but not individual forces. This assumption is usually satisfied at least if the joints are properly manufactured and installed, or if the journals of the arms have sufficient axial play in the bearings to take up the inequalities of manufacture, and furthermore we may assume that at least one of the two shafts is displaceable in an axial direction and will not take up any forces directed along its axis. The universal joint itself is considered as lacking either in mass or weight.

It is further assumed that shaft I is the driver. The joint transmits then the moment taken from shaft I unchanged in either magnitude or direction to shaft II. The arms of the joint are rotatable in the forks on the shafts and furthermore are

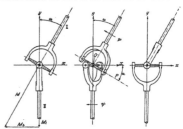

FIG. 9 UNIVERSAL JOINT IN POSITION A
FIG. 10 UNIVERSAL JOINT IN INTERMEDIATE POSITION
FIG. 11 UNIVERSAL JOINT IN POSITION B

assumed to be rotatable without friction; the axis of the moment which the joint takes up from shaft I and the fork located thereon is, therefore, normal to the arm held in that fork; the axis of the moment equal thereto in magnitude and direction which the part of the joint transmits to the fork on shaft II, and thereby to shaft II itself, is normal to but inverse in direction with respect to the arm of the joint located in that fork. The axis of the moment transmitted by the joint is thereby determined as the normal to the two arms of the joint, or, as one may express it otherwise, is the normal to a plane passing through the two arms.

In the position B (Fig. 11) this normal coincides with the axis of shaft II and the universal-joint coupling transmits to shaft II a pure torsion moment. In the position A (Fig. 9) the normal deviates most from the axis of shaft II, viz., by the angle α. Hence there the joint transmits to shaft II a bending moment $M_b = M_t \tan \alpha$ in addition to the torsion moment $M_t$. The forces acting on shaft I in both positions are just the reverse of this.

In order to estimate numerically the moments occurring in the intermediate positions, we shall designate the angles between the normals to the arms of the joint with the axes of coördinates by letters λ, μ, and ν. We shall further designate the total moment transmitted to shaft II by M, the component of M in the x direction by $M_x$ (= bending moment about x-axis) and the component (occurring in intermediate positions) in the z direction by $M_z$ ($=$ bending moment about the $z$-axis). It is then obvious that

$$M_t = M \cos \mu, \quad M_x = M \cos \lambda, \quad M_z = M \cos \nu$$

In addition to the torsion moment $M_t$, the transmission of which is the purpose of employing the joint, there are transmitted to shaft II also the bending moments—

$$M_x = \frac{\cos \lambda}{\cos \mu} M_t$$

and

$$M_z = \frac{\cos \nu}{\cos \mu} M_t \qquad \Bigg\} \quad \dots\dots\dots\dots [5]$$

It is necessary then to determine the values of $\cos \lambda$, $\cos \mu$ and $\cos \nu$. But according to the definition given above $\lambda$, $\mu$ and $\nu$ are the angles which the normals to the arms of the joint ($\alpha_1$, $\beta$, and $\gamma_1$ and $\alpha_2$, $\beta_2$, and $\gamma_1$) make with the axes of the coördinates; these angles are then determined by known equations of analytical geometry, viz.:

$$\cos \lambda = \cos \beta_1 \cos \gamma_1 - \cos \beta_1 \cos \gamma_1$$

and similar equations for the other two angles. In these equations account has been taken of the fact that the directions given by $\alpha_1$, $\beta_1$, $\gamma_1$ and $\alpha_2$, $\beta_2$, $\gamma_1$ are themselves at right angles to each other. If we insert into these equations the values from Equations [1] and [2] we obtain

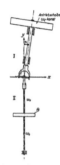

FIG. 12   UNIVERSAL JOINT DRIVE SUCH AS HAS BEEN INVESTIGATED
BY O. FOEPPL

(*Antriebschiebe* $\omega_1 = Konst.$, Driving Pulley $\omega_1 =$ Constant.)

$$\cos \lambda = -\cos \varphi \cos \psi \sin \alpha$$
$$\cos \mu = -\sin \varphi \sin \psi - \cos \varphi \cos \psi \cos \alpha$$
$$\cos \nu = -\cos \varphi \sin \psi \sin \alpha$$

where Equation [4] holds good for the relation between $\varphi$ and $\psi$. By inserting these values into Equation [5] we obtain the comparatively simple formulae—

$$M_x = \frac{\sin \alpha \cos \alpha}{\cos^2 \alpha + \tan^2 \varphi} M_t \quad \dots\dots\dots\dots [6]$$

$$M_z = \frac{\sin \alpha \tan \varphi}{\cos^2 \alpha + \tan^2 \varphi} M_t \quad \dots\dots\dots\dots [7]$$

As an example of the application of the above, may be investigated the system shown in Fig. 12, which is of particular interest because of the observations therein published by O. Foeppl in the *Zeitschrift des Vereines deutscher Ingenieure*, 1919, p. 867.

The moment of inertia of the driving-belt pulley is so large as compared with the moment of inertia $\theta$ of the small flywheel pulley that the angular velocity of shaft I, which is $\omega_1$, may be

considered as being constant. The angular velocity of shaft II is $\omega_2 = \omega_1 \dfrac{d\psi}{d\varphi}$ and its angular acceleration is

$$\frac{d\omega_2}{dt} = \omega_1 \frac{d}{dt}\left(\frac{d\psi}{d\varphi}\right) = \omega_1 \frac{d}{d\varphi}\left(\frac{d\psi}{d\varphi}\right)\frac{d\varphi}{dt}$$

and since $\dfrac{d\varphi}{dt} = \omega_1$

$$\frac{d\omega_2}{dt} = \omega_1^2 \frac{d^2\psi}{d\varphi^2}$$

The moment of torsion transmitted to shaft II by the joint in order to cause it to move is therefore

$$M_t = \theta \frac{d\omega_2}{dt} = \theta\omega_1^2 \frac{d^2\psi}{d\varphi^2} \dots\dots\dots\dots [8]$$

The bending moments $M_x$ and $M_o$ can then be derived from Equations [6] and [7]. In order to carry out the computation we must first determine $\dfrac{d^2\psi}{d\varphi^2}$ by double differentiation of Equations [3] and [4]. By substituting this value in Equation [8] and by substituting the value of $M_t$ thus obtained in Equations [6] and [7], we obtain for $M_t$, $M_x$ and $M_z$ rather complicated expressions which may, however, be materially simplified when the angle $\alpha$ is not very large, as then one may neglect the expression $(1 - \cos \alpha)^2$. This is permissible even for fairly large angles as, for example, $(1 - \cos 8.5 \deg.)^2 = 0.00012$. If this is done we obtain

$$M_t = -2 \theta \omega_1^2 (1 - \cos \alpha) \sin 2 \varphi$$
$$M_x = -\theta \omega_1^2 \sin \alpha (1 - \cos \alpha) (\sin 2 \varphi + \tfrac{1}{2} \sin 4 \varphi)$$
$$M_z = -\tfrac{1}{2} \theta \omega_1^2 \sin \alpha (1 - \cos \alpha) (1 - \cos 4 \varphi)$$

From the last two equations it appears that the shaft II receives through the universal joint powerful impulses tending to produce bending with a frequency double and quadruple the number of revolutions, and these impulses create new critical speeds and may entirely overshadow other effects. (*Schweizerische Bauzeitung*, vol. 75, no. 17, Apr. 24, 1920, pp. 187-188, 4 figs., t.A)

## Gyroscopic Phenomena in Synchronous-Motor Drive of Reciprocating Compressors

FLYWHEEL EFFECT FOR SYNCHRONOUS MOTORS CONNECTED TO RECIPROCATING COMPRESSORS, R. E. Doherty. From an analysis of mechanical and electrical conditions of operation of synchronous motors connected to reciprocating apparatus it appears that there is a very close relation between the flywheel effect and the proper design of the motor.

Two motors of similar design are, for example, installed on different types of compressors; one operates and the other does not. It appears, for example, that the motor gives trouble from surging and hunting. Then the flywheel is removed and the operation becomes satisfactory. In another instance, however, in which surging or hunting occurs, the flywheel is added and the trouble is cured. It would appear, therefore, that there can be either too much or too little flywheel and the author offers a method of solving this problem.

To do this, an analogy is offered first, such as Fig. 13, where two trains are assumed to be running side by side at constant speed and abreast of each other.

Under these conditions the driving force of the trains is exactly equal to the opposing forces of friction, windage, etc., and the position may be called "stable position."

Assume, however, that at an instant designated by $O$ the driving force on train $A$ is suddenly increased by an amount indicated in the figure and that after a time period $t$, this excess is suddenly removed. The presence of this added force will affect the velocity and relative displacement of trains $A$ and $B$ in the following manner. The excess force $a$ being consumed entirely in the acceleration of the train may be used to represent graphically the product of acceleration $\times$ time, that is, excess velocity. But the result of its application will be that while train $B$ reaches point $m$, Train $A$ will have gone further, viz., to point *n*. The excess displacement *mn* is represented by the maximum ordinate at the end of curve *c*, and the idea is that it is possible from the curve of unbalanced force impressed upon a mass to find the change in both velocity and displacement simply by integrating the area of the loops in the unbalanced force diagram. The first integration gives velocity and the second gives displacement.

From this the author proceeds to the consideration of a case where, instead of the momentary excess force described above, a periodically varying force is applied. The resultant curve of velocity will then be as shown at *e* and the displacement curve at *f* shows the position of train *A* first ahead and then behind the " stable " position.

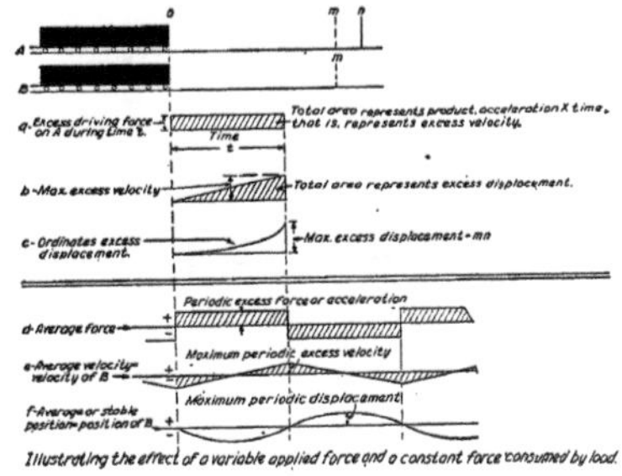

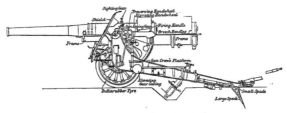

Illustrating the effect of a variable applied force and a constant force consumed by load.

FIG. 13 DIAGRAMS SHOWING HOW PERIODIC RELATIVE DISPLACE-MENT OF TWO TRAINS IS DERIVED FROM CURVE OF UNBAL-ANCED DRIVING FORCE

One part of the problem of determining the proper flywheel effect for synchronous motors direct-connected to reciprocating compressors is illustrated in this way,—that it is required to find what mass train *A* must have in order to limit to a definite amount the plus and minus displacement, as shown in *f*, when a given periodic variation in the driving force is impressed upon train *A*.

A synchronous machine driving an air compressor will have a pulsating speed on account of the pulsating torque of the compressor, and the variable torque on the motor will cause it to displace first ahead and then behind "reference positions" which might be marked on a similar motor running at constant speed. For electrical reasons, however, it is necessary to place limits to this displacement or deviation, and the real problem is to determine how much flywheel is required to keep the angular dis-

placement within given required limits which can be worked out from the turning-effort diagram of the compressor.

The writer explains in considerable detail the processes of obtaining the turning effort curve for a reciprocating compressor which is done by integrating the areas under the plus and minus loops of a tangential force curve along lines indicated in connection with Fig. 13. This integration gives the ordinates of the displacement curve which makes it possible with the given flywheel effect in the motor and a given tangential curve, to determine accurately what angular displacement would occur under the assumed conditions; or, conversely, what flywheel effect would be required to limit the angular displacement to a definite value.

Integration of the loops of the crank-effort and velocity curves by planimeter is, however, a very laborious process, and the writer recommends an analytical method which consists in expanding the unbalanced component in the crank-effort diagram, which to proper scale is the acceleration diagram, into a Fourier's series of sines and cosines, and integrating this equation twice for displacement. By proper substitutions the maximum angular displacement may be expressed in electrical degrees.

The author analyzes in considerable degree the electrical phenomena in a synchronous motor as affected by, and affecting, the flywheel effect. This part is not suitable for abstracting. (*A. S. R. E. Journal*, Nov. 1919, vol. 6, no. 3, pp. 159-177, 9 figs., *t*)

## METALLURGY (See Heat Treatment; Furnaces)

## MILITARY ENGINEERING

### Schneider 155-mm. Gun

SCHNEIDER 155-MM. GUN. Description of the Schneider gun of the heavy field-artillery type allowing of firing at a flat trajectory and high-angle fire up to a positive angle of 40 deg.

The original article describes both the 1917 and 1918 patterns. The maximum range exceeds 10 miles with the projectiles weighing close to 95 lb. and having a muzzle velocity of close to 2200 ft. The gun fires from the carriage without any preparation or platform. Three rounds can easily be fired per minute and this rate can be increased to five rounds. The gun is shown in firing position in Fig. 14. It is of steel and consists of a tube, a jacket and a rear block forming an equalizing weight. The breech is closed by an interrupted breech mechanism with plastic obturator op-erated by hand. Firing is by percussion such that the device does not act until the breech is completely closed.

A table is given showing particulars of this gun. (*Engineering*, vol. 109, no. 2833, April 16, 1920, pp. 512-513, 7 figs., *d*)

FIG. 14 SCHNEIDER 155-MM. GUN IN FIRING POSITION

## MOTOR-CAR ENGINEERING (See Internal-Combustion Engineering)

## MUNITIONS (See Military Engineering)

## RESEARCH (See Internal-Combustion Engineering)

## SPECIAL MACHINERY

### Machine for Boring Horizontal Holes in Clay

THRUST BORING IN EARTH. Description of a machine invented by Capt. A. R. Mangnall and made by a British company. It is claimed that this machine will pierce a hole 4 in. in diameter and 150 ft. long horizontally through heavy clay in half an hour. It is intended for cable laying, pipe laying, drainage or similar work. The machine is in a sense an outcome of the war. Several attempts were made to devise an apparatus which could be started from British trenches and bore under those of the enemy holes for mining and countermining. The action of the machine can best be compared with that of thrusting a stick into clay. The machine consists of a hydraulic cylinder of steel tube, carried

is again placed horizontally, and the extension piece is attached to the end of the pilot by a simple pin joint. The pressure is put on again, and the pilot driven forward another 4 ft. Then the same operation is repeated, extension piece after extension piece being added until the pilot breaks through into a pit at the far end—150 ft. away. The pilot is then uncoupled and lifted out of the pit, and the extension pieces are withdrawn and uncoupled one by one. The whole operation of raising the gun, inserting an extension piece, lowering the gun, fixing the pin joint, and thrusting the piece home occupies something less than one minute.

The joint is extremely simple and ingenious. The extension pieces are tubes about 2½ in. diameter for a 4-in. hole, with muff couplings at the leading end. Through each coupling is

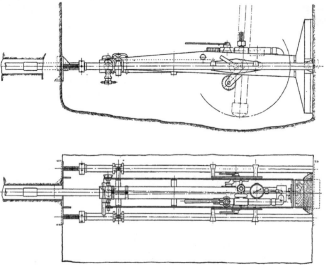

FIG. 15  MANGNALL EARTH THRUST BORER—GENERAL ARRANGEMENT

on trunnions in a light steel frame. Inside , the cylinder is a short piston with hinged guiding fingers.

This "gun" complete on its framework is dropped into a pit dug in the ground to the depth required, a plate at the back of the framework or carriage pressing against a few timbers to distribute the load. The cylinder is connected by flexible pressure pipes to a little three-throw pump driven by a petrol engine. The exhaust water returns to the pump from which the pump draws. The operation of thrust boring is begun by turning the gun up to a vertical position and dropping the "pilot" into it. The gun is then turned down and clamped in position, and hydraulic pressure is admitted behind the piston, pressing the pilot horizontally 4 ft. into the soil. The stroke being completed, the admission valve is closed and the exhaust valve opened—by one and the same lever—the gun is raised to the vertical and the first extension piece pressed into it. The exhaust water is then returned to the sump. The three guide fingers referred to keep the base of the extension piece in a central position. The gun

drilled a pair of opposite holes; through the other end of each piece is drilled a corresponding pair of holes, but of smaller diameter. The pins are turned to fit the smaller holes freely, but at each end are reduced in diameter so as to leave a shoulder. The operator brings the holes opposite each other, slips the pin in position, and dabs a little bit of clay round the end. It is found that the clay locks the pins effectively and that they never fall out. The withdrawal of the extension pieces is easy, since their diameter is a good deal less than that of the hole. The method generally employed is to attach a rope to the last piece and carry it round a pulley. Two or three men on the surface can then pull the whole chain of extension pieces back until the next joint is reached. The operation is repeated until the operator in the pit can pull the rods back without assistance. An alternative method is to turn the gun up to a vertical position, and by means of suitably arranged ropes and pulleys employ it like the cylinder of an hydraulic crane. (*The Engineer*, vol. 129, no. 3354, April 9, 1920, pp. 369-371 and 376, 7 figs., *dA*)

STEEL MILLS (See Bureau of Standards)

TESTING (See Internal-Combustion Engineering)

## TRANSPORTATION

THE STABILITY OF MOTOR OMNIBUSES. In view of the growth of the use of motor omnibuses for city transportation, an editorial in *The Engineer* in reference to the safety of this type of vehicle becomes of considerable interest. It would appear that within the past year there have been a number of cases in England where motor omnibuses overturned and there is a growing suspicion that the motor omnibus is not quite as safe a vehicle of travel as has been assumed hitherto.

Among the causes of this alleged condition are mentioned the carelessness of drivers who learned to take undue risks in France, the bad condition of the roads and the degree of overcrowding now permitted in England in the interior of the omnibuses.

These causes do not appear, however, to be capable of explaining some of the accidents and it would seem as if the real cause lay in the design of the omnibuses. As regards this latter, the motor omnibus used on the London streets is a vehicle weighing in service condition some 9850 lb., of which weight a little over 64 per cent is in the chassis. It is capable of carrying without overcrowding 36 people, representing on an average a load of 5400 lb. so that when fully loaded the vehicle and load weigh something in the neighborhood of 6½ tons.

Its center of gravity in service condition, but without load, lies low down, being only about 44.5 in. above the road surface and just a little above the level of the tops of the rear wheels. The gage of the wheels being 70 in., it follows that the vehicle will recover itself if it is tipped up until the rear axle is lying at 38 deg. to the horizontal. When the omnibus is fully loaded the center of gravity rises to 62 in. above the road surface so that in this condition the maximum heel from which the vehicle will recover itself is 29.5 deg. These figures do not take account of the fact that the body is mounted flexibly on the chassis. In an actual test the yielding of the springs permitted the vehicle to be tipped when empty until the center line of the body was inclined at 46 deg. to the vertical.

It would appear, therefore, that the motor omnibus is quite stable when stationary. When it is in motion the angles of heel mentioned above are a measure of its stability so long as its course is a straight line. The critical test for the stability of the vehicle arises when centrifugal force comes into play during the rounding of a corner. Taking 30 ft. as the radius of the curve and 7 miles per hour as the speed when rounding the corner, it would appear that the stability of the omnibus is reduced by a little over 18 per cent, and if the vehicle carries passengers on the top only the loss of stability is a trifle over 19 per cent. Under these circumstances the critical condition of equality between the upsetting moment and the righting moment will be reached if the heel amounts to 23¼ deg. with the vehicle fully loaded or 22½ deg. if the passengers are entirely on the top. These angles would be reached only if the wheels on the inner side of the curve were lifted off the ground to a height of about 27 in., a condition hardly likely to be encountered, and even this aggregation of unfavorable conditions is counteracted to a certain degree by the fact that the body of the vehicle is mounted on springs.

The stability of the modern motor omnibus may therefore be considered as satisfactory and while the vehicle could be made more stable this is hardly necessary and perhaps not even desirable, as increased stability could be obtained only at the cost of reducing maneuverability and this might lead to an increase of accidents on crowded streets.

It would appear, therefore, that such accidents of overcrowding as have occurred have been due to exceptional conditions present in each individual case.

It might be added in this connection that such accidents as overturning of an omnibus are practically unknown in New York City. (*The Engineer*, vol. 129, no. 3355, April 16, 1920, pp. 401-402, editorial, *g*)

CITRÖEN GEAR FOR ELECTRIC LOCOMOTIVES. Description of the application of Citröen gears for electric locomotives by the Oerlikon Works Company.

This is a result of tests over a period of many years. The locomotive put on the line in 1911 is still in service. It was, however, a comparatively slow-speed type. In 1913 another type was placed in service to run with commercial speeds as high as 70 km. (43.5 m.p.h.) and operate on rather heavy grades at speeds of about 30 m.p.h.

A somewhat strange phenomenon was noticed on that locomotive. Rather heavy vibrations occurred when the speed was between 24.5 and 24.8 m.p.h. and reached their maximum at speeds between 24.8 and 25.4 miles, but disappeared when the speed was between 25.4 and 25.7 m.p.h. The same phenomenon occurred during the slowing down and was maintained so long as the speed remained between 24.5 and 25.7 miles.

At first the trouble was thought to be due to an inaccurate division of the teeth, but had that been the case the oscillation would have increased between the speed of 25.4 and 43.5 miles in proportion to the square of the speed. These vibrations were so serious that they led in some instances to fracture of the coupling rods. It was clear that the vibrations arose through resonance. One of the causes of the resonance could be looked for in the play of the crankpin and coupling-rod bearings. Such a play, when it exists, has in itself an influence which is constant and is independent of the speed; this influence becomes serious only when it is amplified by resonance—that is, when another periodical cause is superimposed upon it. Here, apparently, this second cause has a fixed period which corresponds precisely to the speed of 24.8 to 25.4 miles, but just what its nature is does not appear to be clear yet, notwithstanding much experimental work.

The load is distributed on the axles by means of equalizers, as in the case of steam locomotives. The locomotive complete weighs in running order 108.6 metrical tons; the five central pairs of wheels and axles weighing each 2040 kg., and the two end ones 1530 kg. each, while the suspended load is 95,340 kg. On the other hand, the maximum designed load on a driving axle is 17,000 kg.; on this basis, and reckoning further on, a maximum load of 11,800 kg. for each end axle, the suspended load does not exceed 14,960 kg. for each of the five driving axles and 10,272 kg. for each of the two end axles. But the weight of a complete motor is made up as follows: Rotor, 4,715 kg.; pinion, 449 kg.; shaft, 459 kg.; stator, 6715 kg.; intermediate shaft and toothed wheel, 2410 kg.; crankplates, 400 kg.; making up a total of 15,150 kg. in round numbers. Since half the weight of each of the two motors is taken by the middle axle, as also a part of the weight of the frame plates and body, it is clear that the load over this middle axle must be in excess of that authorized by the regulations, so long as the rigidity of the frame is insufficient to transfer part of this load to the other axles, since the central load is distributed solely by special springs on the corresponding axle, without an equalizer. Consequently, if the conditions under which the frame plates undergo flection be gone into by using the Clapeyron formula, it is necessary to take into account the flection of the frame plate between the axles preceding and following the middle one, otherwise the middle axle would be loaded above the regulation figure.

The frame plate can thus only be considered as a girder resting on six bearings instead of seven, bearings which are on a level, while the central axle bearing exerts an upward force of 14,960 kg. (*Engineering*, vol. 109, no. 2839, May 26, 1920, pp. 712-714, 7 figs., *dt*)

## CLASSIFICATION OF ARTICLES

Articles appearing in the Survey are classified as *c* comparative; *d* descriptive; *e* experimental; *g* general; *h* historical; *m* mathematical; *p* practical; *s* statistical; *t* theoretical. Articles of especial merit are rated *A* by the reviewer. Opinions expressed are those of the reviewer, not of the Society. The Editor will be pleased to receive inquiries for further information in connection with articles reported in the Survey.

# ENGINEERING RESEARCH

### A Department Conducted by the Research Committee of the A. S. M. E.

### Endowment for Engineering Foundation

ENGINEERING FOUNDATION was organized to care for the generous gifts of Mr. Ambrose Swasey, the income from these gifts being devoted to engineering research. Since its organization the funds have been used to aid the National Research Council and others in performing research directly connected to engineering.

Potential benefits for the whole nation are very great, but these benefits cannot be gained without expenditure of effort and materials. Research workers must be supported. Equipment, materials, working places and traveling facilities must be provided. Since the benefits accrue to the profession, the industries and the public in general, support in large measure should come from general funds, such as those provided by endowments.

Although engineers, like other professional men as a class, are not wealthy, some individual engineers have large means. Engineering Foundation seeks to build up its endowment to dimensions worthy of the profession. Engineers connected with industrial and financial organizations having great resources can aid by convincing proper officials of corporations that the continued prosperity of our industries depends upon continued progress of research. Since the commercial and industrial establishments of the country reap the larger proportions of the financial profits arising from scientific and technological work, these establishments should contribute liberally to the support of research.

There are many problems relating to the materials and forces of engineering on which further knowledge is needed. Progress will be made approximately in proportion to the funds made available. But there are other kinds of problems which concern the engineer. No longer may one declare, as did Prof. J. H. Johnson a generation ago, that "Engineering differs from all other learned professions in this, that its learning has to do only with the inanimate world, the world of dead matter and force." Many acute social and economic questions of our day need the dispassionate, impartial, patient study of scientists and technologists. To these questions must now be applied the scientific method of collecting facts by thorough study, and the engineer's capacity for planning and performing, instead of ill-considered "reforms."

Occasionally experimental work is undertaken in accordance with a well-conceived plan as a necessary or desirable adjunct to the main operation. In such cases the exigencies of the main operation sooner or later interrupt the experimental work; or the men who have it in hand leave the force; or the information is gained but never written up; or the statement is buried in some report of limited circulation; or greater familiarity with research methods and a broader conception of the problem could, with small additional expense, have secured much more valuable results and have made them more generally useful.

The services described in the foregoing paragraphs, and many others, could be performed by Engineering Foundation, if adequate funds could be placed at its disposal. The Foundation does not plan to build laboratories and conduct research work directly, but rather to stimulate, coördinate and support research work in existing scientific and industrial laboratories, coöperating, in so far as possible, with the National Research Council.

Mr. Charles F. Rand, of 71 Broadway, New York, Past-President of United Engineering Society, and of the American Institute of Mining and Metallurgical Engineers, was elected Chairman of Engineering Foundation March 19, to succeed Dr. W. F. M. Goss, resigned. With the collaboration of Mr. Swasey, Mr. Rand is actively seeking additions to the endowment which will swell the total to at least a million dollars in the near future. Mr. Swasey's gifts amount to $300,000.

The office of Engineering Foundation is in the Engineering Societies Building, 29 West 39th St., New York. Further information may be had by addressing this office or the chairman. A booklet giving an account of the Engineering Foundation and its work will be mailed upon request.

### A.—RESEARCH RESULTS

The purpose of this section of Engineering Research is to give the origin of research information which has been completed, to give a résumé of research results with formulæ or curves where such may be readily given, and to report results of non-extensive researches which in the opinion of the investigators do not warrant a paper.

*Cement and Other Building Materials A4-20* The Use of Wet and Dry Sand for Concrete. Concretes made from Potomac sand and gravel in various proportions from 1:1½:3 to 1:3:6 with various extremes of flowability have shown that when the aggregates are proportioned by volume measure it is found that wet aggregate requires from ½ to 1 bag of cement per cu. yd. more than concrete with dry aggregate. Sand is generally wet or at least moist when used so that the full difference may never be apparent in field practice, yet the use of sand from a pile which has just been wet from the rain will result in the employment of more cement for a given volume of concrete than would have been the case had the work been done on a dry day. Marked improvement in the working qualities of concrete will be noted under the usual conditions when the relative volume of sand is increased and the gravel proportionately reduced. Bureau of Standards, Washington, D. C. Address S. W. Stratton, Director.

*Heat A2-20* The Efficiency of Insulation Used in Cold-Storage Work. The Bureau of Standards has conducted a number of experiments on the coefficient of conduction through insulating materials. This has been done by the flat-plate method, in which the coefficient of conduction rather than the coefficient of transmission has been determined. Preliminary results have been transmitted to the June Meeting of the American Society of Refrigerating Engineers and issued by them in their Journal and also issued by the Bureau of Standards as a scientific paper. Address S. W. Stratton, Director. Bureau of Standards, Washington, D. C.

*Leather and Glue A1-20* Service Tests on Chrome-Tanned and Oak-Tanned Sole Leather. Twenty-five service tests have shown that chrome leather outwore oak leather by about 23 per cent per unit thickness. Bureau of Standards, Washington, D. C. Address S. W. Stratton, Director.

*Metallurgy and Metallography A11-20* Heat Treatment of High-Chromium Steel. Experimental work on samples of high-chromium steel containing C 0.20 per cent, Mn 0.38 per cent, Si 0.70 per cent, and Cr 13.2 per cent has been completed and a report submitted to the Journal of the American Society of Automotive Engineers. The samples were heated to various temperatures and quenched in oil. The following results are shown:

a Hardness by Brinell and Shore instruments increases with the temperature until 1950 deg. fahr. is reached. Maximum range of hardness is generally obtained by quenching from this temperature up to the highest heat used, but in some cases this hardness actually decreases due to the retention of the solid solution.

b Quenching from about 1750 deg. fahr. develops the best combination of strength and ductility which is not coincident with maximum hardness. Quenching from this or a lower temperature does not retain all the carbide in the solution, as is the case in samples quenched from considerably higher temperatures, notably 2100 deg. fahr. and 2250 deg. fahr.

c Ductility as measured by elongation and reduction is very low when quenching occurs from 1850 deg. fahr. or above.

d Short-time tempering at temperatures above 800 deg. fahr. decreases brittleness. Ductility is increased to a greater extent in sample quenched from 1750 deg. than those at higher temperatures. Tempering above 800 deg. fahr. decreases the strength and hardness in a marked manner, although ductility is increased. Address S. W. Stratton, Bureau of Standards, Washington, D. C.

414

*Properties of Engineering Materials A8-20* Strength of Steel. The relation between the elastic strengths of steel in tension, compression and shear is the subject of Bulletin No. 115, Engineering Experiment Station of the University of Illinois, by Prof. F. B. Seely and W. J. Putnam. The Bulletin presents the results of experiments on soft, mild and medium carbon steels and vanadium, nickel and chrome-nickel steels. The elastic strengths in tension, compression and shear are given for each of the six grades. The shearing strength in torsion was found from solid cylindrical specimens and thin-walled hollow specimens. The main object of the investigation was to determine carefully the elastic shearing strength of ductile and semi-ductile and find the ratio of the elastic shearing strength to the elastic tensile strength.

Ductile rods or specimens from rolled shapes when tested in tension showed a sudden or abrupt elastic breakdown. The elastic shearing failure of solid cylinders is also abrupt with well-defined yield point. This seems to indicate that tensile failure is really a shear failure. The hollow thin-walled specimens do not show such an abrupt elastic failure in shear, and therefore the shearing breakdown is more gradual than the corresponding tensile breakdown. The following abridged conclusions are made at the end of the Bulletin:

1 The correct value of the elastic shearing strength and yield point in shear of steel as measured by the proportional limit or the useful limit point may be determined from a torsion test of a hollow thin-walled cylindrical specimen.

2 The correct value of the elastic shearing strength of steel as measured by the proportional limit or the useful limit point may be taken, with reasonable accuracy, as eighty-five hundredths (0.85) of the elastic strength obtained from a torsion test of a solid cylindrical specimen. The correct value of the yield point for the more ductile materials is slightly more than eighty-five hundredths (0.85) of the yield point obtained from a test of a solid torsion specimen.

3 A solid cylindrical torsion specimen may be used to obtain test results. The character or progress of the breakdown of the elastic action may be obscured in the test of the solid specimen. Test results obtained from a solid specimen may also be used to judge of the general properties, quality and reliability, of the materials.

4 The correct value of the elastic shearing strength of steel as measured by the proportional limit or the useful limit point is from fifty-five to sixty-five hundredths (0.55 to 0.65) of the elastic tensile strength and may be taken with reasonable accuracy as six-tenths (0.6) of the elastic tensile strength. The maximum shear theory of the failure of elastic action of ductile steel (sometimes called Guest's Law) is, therefore, not an accurate statement of the law of elastic breakdown, since Guest's Law assumes that the elastic shearing strength is one-half (0.5) of the elastic tensile strength. The maximum-shear theory as expressed by Guest's Law, however, is of much use in obtaining approximate results.

5 Contrary to the general belief, the breakdown of the shearing elastic action of ductile and semi-ductile steel as found in these tests is gradual; more gradual, as a rule, than the failure of elastic action in a tension specimen.

6 The amount of hot-rolling received by steel may have a marked effect upon the relation of the elastic tensile and compressive strengths. For the material from $\frac{3}{4}$-in. and $\frac{7}{8}$-in. bars, the elastic compressive strength is somewhat greater (about 5 per cent) than the elastic tensile strength, whether the elastic strength is measured by the proportional limit, useful limit point or the yield point. For the material from the $3\frac{1}{2}$-in. bars, the compressive proportional limit is very much greater (27 per cent) than the tensile proportional limit, while the compressive yield point is somewhat less (4 per cent) than the tensile yield point.

7 It appears, therefore, that the choice of the criterion of elastic strength may be of considerable importance in some cases, and that the usual tensile test of material may require supplementary compressive tests or a better correlation between test data and the amount of treatment received during rolling before the elastic properties of the material as obtained from tensile test data can be relied upon.

8 The elastic strength and ultimate strength of steel in tension, in compression, and in shear is affected little, if any, by the direction of rolling in the case of a slab 2 in. thick, although the ductility, as measured in tension tests by the percentage of elongation and percentage of reduction of area, is materially less when the stress is perpendicular to the direction of rolling than it is when the stress is parallel to the direction of rolling.

University of Illinois, Engineering Experiment Station, Urbana, Ill. Address C. R. Richards, Director.

*Properties of Engineering Materials A9-20* Magnetic Uniformity of Steel. The Bureau of Standards has recently tested 54 bars 1 in. in diameter, 12 ft. long for magnetic uniformity. The bars were intended for experiments on heat treatment of carbon steel. The determination was made by passing the test specimen at a uniform rate through a magnetizing solenoid energized by direct current and noting the deflection on a sensitive instrument connected to a system of test coils located within the magnetizing solenoid. With uniform material no deflection occurs while variations in uniformity gave deflections of which a photographic record was made. Five bars were found to give relatively large deflections at certain points. Regions thus located were cut out and not used in the investigation. Bureau of Standards, Washington, D. C. Address S. W. Stratton, Director.

*Welding A1-20* Electric Arc Welding. In an investigation about to be published by the Bureau of Standards it has been discovered that fusion welds differ from all other types in that the metal of the weld is essentially a casting. No refinement of this metal is possible in the method as ordinarily carried out. The characteristic properties of the arc-fused metal have been determined in this study. The work shows that skill, care and patience on the part of the operator is a matter of vital importance. Bureau of Standards, Washington, D. C. Address S. W. Stratton, Director.

## B—RESEARCH IN PROGRESS

The purpose of this section of Engineering Research is to bring together those who are working on the same problem for coöperation or conference, to prevent unnecessary duplication of work and to inform the profession of the investigators who are engaged upon research problems. The addresses of these investigators are given for the purpose of correspondence.

*Automotive Vehicles and Equipment B2-20* Dynamics of Load Suspension on a Motor Vehicle. A study of the forces involved in the action of tires, axles, springs and body of truck when passing over obstructions in the path of the wheels. An analysis of curves obtained by driving a truck at constant speed across the field of a camera in the dark, lights on the body and axles tracing the motion of those parts on the camera plate. This includes the determination of the force of impact on the obstruction and the pavement beyond the obstruction. Address Asst. Prof. W. E. Lay, Charge of Automobile Engineering, University of Michigan, Ann Arbor, Mich.

*Automotive Vehicles and Equipment B3-20* Fuel Economy of a Motor Vehicle. Much has been said about the apparent improvement in operation of the automobile engine on a summer evening. Preliminary tests have indicated the possibility of a daily cycle in the variation of fuel-mileage economy. The attempt is being made by means of a series of 24-hour runs to determine upon which variable or combination of variable conditions the change in fuel mileage depends. Address Asst. Prof. W. E. Lay, Charge of Automobile Engineering, University of Michigan, Ann Arbor, Mich.

*Cement and Other Building Materials B2-20* Building Columns Under Fire. The Underwriters' Laboratory of Chicago is working on a report of a test of 105 columns under fire. These columns have been loaded to capacity and in this condition were subjected to the standard fire temperatures. The work is being done in coöperation with the Bureau of Standards, National Board of Fire Underwriters and the Associated Factory Mutual Fire Insurance Companies. It has been in progress for three years and the report will soon be made. Address Underwriters Laboratories, 207 E. Ohio St., Chicago, Ill.

*Fuels, Gas, Tar and Coke B1-20* Carburetion of Fuels. See *Internal-Combustion Motors B7-20*.

*Heating B2-20* Sample House in Saskatchewan. Seven different types of house built as 6-ft. cubes of different types of building construction have been erected at the college at Saskatoon, Sask. The temperature in this locality reaches 50 deg. below zero with no wind and 15 to 20 deg. below zero with a strong wind. Recording thermometers of the Leeds and Northrup type are being used. The amount of heat to warm these houses is being determined. Address Prof. A. R. Greig, Saskatoon, Sask.

*Heating B3-20* Direct-Indirect Systems. The Mechanical Engineering Department of the Iowa State College has been making tests on direct-indirect systems of heating. Address Prof. M. P. Cleghorn, Iowa State College, Ames, Ia.

*Heating B4-20* Loss of Heat from Underground Mains. The Mechanical Engineering Laboratory of the University of Michigan is coöperating with the Ric-wil Company of Cleveland, Ohio, and the Research Bureau of the American Society of Heating and Ventilating Engineers in attempting to secure data from experiments upon a bare pipe buried in the ground whereby the paths of heat flow can be traced. Address Prof. J. E. Emswiler, University of Michigan, Ann Arbor, Mich.

*Heating B5-20* Velocities in Hot-Water Heating Systems. The Mechanical Engineering Department of the University of Michigan is conducting some experiments to determine the velocity of water when the radiator is located upon the first, second and third floor above the heater when natural circulation of the water is employed. Address Prof. J. E. Emswiler, University of Michigan, Ann Arbor, Mich.

*Metallurgy and Metallography B7-20* Tensile Properties of Boiler Plates. Boiler plate rolled at a blue heat (300 deg. cent.) has been used to determine the properties at elevated temperatures and the effect of blue work. The effect of low-temperature annealing is to be determined. Special apparatus has been constructed to determine the effect of the rate of loading on the elastic limit at elevated temperatures, using a motion-picture camera. Bureau of Standards, Washington, D. C. Address S. W. Stratton, Director.

### C—RESEARCH PROBLEMS

The purpose of this section of Engineering Research is to bring together persons who desire coöperation in research work or to bring together those who have problems and no equipment with those who are equipped to carry on research. It is hoped that those desiring coöperation or aid will state problems for publication in this section.

### D—RESEARCH EQUIPMENT

The purpose of this section of Engineering Research is to give in concise form notes regarding the equipment of laboratories for mutual information and for the purpose of informing the profession of the equipment in various laboratories so that persons desiring special investigations may know where such work may be done.

*Delaware College D1-20* The small laboratory at Delaware College is equipped with two single-expansion engines, a condenser, a small air compressor, a 30,000-lb. testing machine, a hydraulic ram, two gas engines, and a centrifugal pump. Address Merrill V. G. Smith, Delaware College, Newark, Del.

*General Bakelite Company D1-20* The Physical Laboratory of the General Bakelite Company at Perth Amboy, N. J., contains the following apparatus for use in their research work:
  One 100,000-volt, 60-cycle testing transformer with facilities for making tests on standard A. S. T. M. molded samples in air and oil at full voltage.
  One 30,000-lb. Olsen Universal testing machine.
  Insulating resistance outfit with direct current at 15,000 volts with galvanometer of sensitivity of $5 \times 10^{-10}$ for resistance of five million megohms.
  Olsen heat-distortion testing machine for molded insulation materials according to A. S. T. M. standards.
  Special apparatus for determining plastic quality.
  Auxiliary apparatus for standard laboratory work.
  Address Gilbert L. Peak, Electrical Engineer, General Bakelite Company, Perth Amboy, N. J.

### E—PERSONAL NOTES

The purpose of this section of Engineering Research is to give notes of a personal nature regarding the personnel of various laboratories, methods of procedure for commercial work or notes regarding the conduct of various laboratories.

### F—BIBLIOGRAPHIES

The purpose of this section of Engineering Research is to inform the profession of bibliographies which have been prepared. In general this work is done at the expense of the Society. Extensive bibliographies require the approval of the Research Committee. All bibliographies are loaned for a period of one month only. Additional copies are available, however, for periods of two weeks to members of the A.S.M.E. or to others recommended by members of the A.S.M.E. These bibliographies are on file in the offices of the Society.

*Electrochemistry F2-20* Electric Furnaces. The electric furnace as applied to metallurgy. A reading list from 1900 to 1919 by Clarence J. West, a paper presented to the American Electrochemical Society at Boston, April 8 to 10, 1920. See Vol. 37 of the Transactions of the American Electrochemical Society. A bibliography of 92 pages of general articles on electrodes, iron, steel alloys, ferrosilicon, ferromanganese, ferrochromium, ferrotungsten, ferromolybdenum, non-ferrous metals, brass, copper, nickel, tin and miscellaneous metals. Address Arthur D. Little, Inc., Cambridge, Mass.

*Fuels, Gas, Tar and Coke F4-20* Fuel-Oil Use in Locomotives. A bibliography of one page. Address A.S.M.E., 29 West 39th St., New York.

## Cleveland Engineers Hold Joint Meeting

### Cleveland Engineering Society Celebrates Its Fortieth Anniversary with Engineers' Club of Dayton, A.S.M.E. and Other Local Sections Coöperating

The fortieth annual meeting of the Cleveland Engineering Society was held on June 8 with an attendance of nearly four hundred. A delegation of seventy-eight members of the Engineers' Club of Dayton was present and contributed to the success of the meeting. The Cleveland A. S. M. E. Section had charge of the arrangements for the several trips of inspection which were made during the day, which included visits to a number of the leading industrial plants of the city.

During the morning while registration was progressing, motion pictures were shown at the rooms of the Cleveland Engineering Society on the fourteenth floor of the Hotel Statler. These illustrated operations in the plant of the Cleveland Automatic Machine Co., with talks by E. E. Blundell on the Cleveland Automatic Turret Machines and by H. T. Simmons, of the Wellman-Seaver-Morgan Co., on An Improved Car Dumper.

The Cleveland University Club was the scene of a banquet in the evening, at which inspiring addresses were given by Col. Edward A. Deeds, of Dayton, Ohio, and Dr. E. J. Cattell, of Philadelphia, Pa. Colonel Deeds spoke on The Benefits of Engineering Associations and he pointed out the many opportunities awaiting the engineering profession through coördinated action on the part of the local and national engineering societies. He stated that he was thoroughly convinced that an analysis of the reasons for the remarkable development of many of our large industrial cities, such as Cleveland, would undoubtedly show that the local engineering organizations had played a very important part in the growth of such places. He emphasized the fact that such a city as Dayton, with no séaport facilities, was largely dependent upon the industrial activity of the community for its progress, and through the establishment of a strong local engineering organization like the Engineers' Club of Dayton, provision was made for fostering the research and experimental work so essential for the continued growth and prosperity of the city. In closing he extended a hearty and most cordial invitation to the engineers of Cleveland to visit Dayton next year, because he believed that the exchange of ideas and the mingling of engineers of one locality with those of another advance the profession to a much greater extent than is appreciated.

Honorary membership in the Cleveland Engineering Society was conferred upon Dr. Dayton C. Miller, of the Case School of Applied Science, for his achievements in developing instruments for measuring the distance and direction of sound, and whose research work in sound waves and their photography has been notable. The other honorary members include such notable engineers and scientists as Charles H. Benjamin, Charles F. Brush, Gustav Lindenthal, Albert A. Michaelson, E. W. Morley, W. H. Searles, Ambrose Swasey, and Worcester R. Warner.

Mr. Edwin S. Carman, retiring president of the Cleveland Engineering Society, gave a brief outline of the work accomplished by the organization during his term of office and emphasized especially the degree to which it has succeeded in bringing about coöperation with the local sections of the various national engineering organizations and other local engineering groups. All such activities in Cleveland are now concentrated in the rooms of the Cleveland Engineering Society. The new president, Mr. William J. Carter, a consulting civil engineer, assured the organization of his best service and promised to do everything to develop the activities of the society along public service lines and in the leadership in such matters as suitable legislation in the state of Ohio as to the licensing of engineers. The Cleveland Engineering Society has long taken an active interest in municipal engineering matters, and the community has come to look to it for guidance in this direction. Col. Elliott H. Whitlock made an excellent toastmaster and the general atmosphere of the meeting was such as to justify the statement that it "was the best ever."—E. H.

# WORK OF THE BOILER CODE COMMITTEE

*THE Boiler Code Committee meets monthly for the purpose of considering communications relative to the Boiler Code. Any one desiring information as to the application of the Code is requested to communicate with the Secretary of the Committee, Mr. C. W. Obert, 29 West 39th St., New York, N. Y.*

The procedure of the Committee in handling the cases is as follows: All inquiries must be in written form before they are accepted for consideration. Copies are sent by the Secretary of the Committee to all of the members of the Committee. The interpretation, in the form of a reply, is then prepared by the Committee and passed upon at a regular meeting of the Committee. This interpretation is later submitted to the Council of the Society for approval, after which it is issued to the inquirer and simultaneously published in MECHANICAL ENGINEERING, in order that any one interested may readily secure the latest information concerning the interpretation.

Below are given the interpretations of the Committee in Cases Nos. 253, 268, 275, 278, 283, 287, 288, 291-298 inclusive, as formulated at the meeting of April 22, 1920, and approved by the Council. In accordance with the Committee's practice, the names of inquirers have been omitted.

## CASE No. 253 (Previously annulled)

*Inquiry:* Is it permissible under the requirements of the Boiler Code to use for boiler construction, tubes or flues that have been formed into the desired lengths by welding together at their ends, short lengths of such tubing by the electric pressure welding process, provided the metal upset at the weld is immediately upset down to the proper thickness while under the heat of the weld?

*Reply:* It is the opinion of the Committee that while there is no objection to the safe-ending of tubes for fire-tube boilers by a suitable forge or pressure welding process, the suggested method of making the tube from a number of short pieces would be an objectionable practice. The Committee does not consider the practice of safe-ending of tubes to be a suitable one for water tube boilers.

## CASE No. 268 (Reopened)

*Inquiry: a* Inasmuch as the discharge of feedwater into vertical fire-tube boilers in accordance with the requirement of Par. 316 of the Boiler Code has caused serious difficulties from leaky tubes in the lower tube sheet, a reconsideration is requested of Case No. 145, in which the opinion of the Boiler Code Committee is given that the discharge of feedwater in the water leg of a vertical fire-tube boiler is not permissible.

*b* In case it is ruled that such discharge of feedwater into the water leg of a vertical fire-tube boiler is not allowed, is it permissible to protect the feedwater discharge through a feedwater pocket riveted within the shell at a point above the water leg, with handhole opening for cleaning?

*Reply: a* It is the opinion of the Committee that Par. 316 does not permit of discharging the feedwater in the water leg of a vertical fire-tube boiler, and that difficulties with leaky tubes may be avoided by the use of an internal feed pipe which does not discharge into the water leg.

*b* It is the opinion of the Committee that such a method of delivery of feedwater to a vertical tubular boiler through an internal pocket or baffle above the water leg, with handhole opening for cleaning, is in full accordance with the requirements of the Boiler Code.

## CASE No. 275

*Inquiry:* Is autogenous welding permissible for the longitudinal joint of the fire box of a form of vertical boiler as shown in Fig. 3, where the furnace section is, after welding, heated and corrugated by rolls? The corrugations are rolled to a depth of 1½ in. on 8 in. centers, and the weld shows no fracture or distress after either corrugating or flanging.

*Reply:* It is the opinion of the Committee that the construction referred to fully complies with the requirement of Par. 186

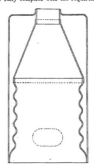

FIG. 3  VERTICAL SINGLE-FLUE BOILER WITH CORRUGATED FIRE BOX

of the Boiler Code, and the thickness of the furnace for a given maximum allowable working pressure should be determined in accordance with Par. 243.

## CASE No. 278

*Inquiry:* What would be considered the lowest permissible water level for a vertical water-tube boiler of the type shown in Fig. 4? Par. 430v of the Boiler Code to which reference is made

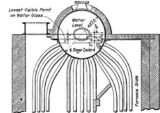

FIG. 4  LOCATION OF THE LOWEST PERMISSIBLE WATER LEVEL

in Par. 291 relative to the lowest permissible water level, leaves the location rather problematical as there may be a difference in opinion as to what the lowest permissible water level may be.

*Reply:* It is the opinion of the Committee that for a water-tube boiler of the design illustrated, a location for the lowest permissible water level such as referred to in Par. 430 i would be a suitable and safe construction.

## CASE No. 283.

*Inquiry:* In the construction of a water-tube boiler in which the tubes enter the tube sheet at an angle as great as 20 deg. from the normal, may the requirements of Par. 252 of the Boiler Code be fulfilled if the least projection of the tube be not less than ¼ in., or more than ½ in.?

*Reply:* It is the opinion of the Committee that Par. 252 applies to tubes which enter normally to the sheet and that where

417

the tubes are at an angle, the least projection of the tube should be not less than ¼ in. or more than ½ in., which may result in the tube projecting more than ½ in. at other points around the tube. The rule given in Par. 251 for flaring should apply to the true diameter of the tube hole measured in a plane at right angles to the tube end and not to the maximum diameter of the elliptical opening in the tube sheet.

### CASE No. 287

*Inquiry:* Is it permissible to burn off the edges of boiler plates and use them without any finish where sheared edges would be permitted by the rules in the Boiler Code?

*Reply:* The Code does not prohibit the practice suggested.

### CASE No. 288

*Inquiry:* Where a manhole is applied to the head of a dome and an opening for access to the boiler is placed in the shell under the dome, is it necessary to reinforce this opening in accordance with the requirements of Par. 260?

*Reply:* It is the opinion of the Committee that such an opening should be reinforced in accordance with the requirements of Par. 260. The reinforcing effect of the base of the dome may be included in calculating the reinforcement.

### CASE No. 291

*Inquiry:* In the case of a special arrangement of spacing of tube holes in the shells of the drums of a water-tube boiler with the pitch unequal in every second row, but shifting the adjoining rows to an exact stagger as shown in Fig. 6, is it

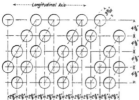

Fɪɢ. 6   Sᴘᴇᴄɪᴀʟ Sᴘᴀᴄɪɴɢ ᴏғ Tᴜʙᴇ Hᴏʟᴇs ɪɴ Dʀᴜᴍ ᴏғ Wᴀᴛᴇʀ-Tᴜʙᴇ Bᴏɪʟᴇʀ

proper to calculate the efficiencies of ligaments under the requirements of Pars. 192 or 193 of the Boiler Code?

*Reply:* It is the opinion of the Committee that the efficiency of the tube sheet should be obtained by employing the method outlined in Pars. 192 and 193 of the Code for diagonal ligaments. All possible methods of failure should be assumed and the lowest value obtained through any one of the methods, should be used.

### CASE No. 292

*Inquiry:* In the case of a triple-riveted butt and double strap joint with the pitch of rivets in the first and third rows one-half of that of the second row, as shown in Fig. 7, what is the proper method of determining the back pitch between the first or inner and second or middle rows of rivet holes?

*Reply:* It is the opinion of the Committee that the back pitch required between the first or inner and the second or middle rows of rivet holes in such a design of joint, shall be governed by the rules of Par. 182 for determining the back pitch between the second and third or outer rows of rivets.

### CASE No. 293   (In the hands of the Committee)

### CASE No. 294

*Inquiry:* Is the intent of Par. 331 of the Boiler Code that at least one complete set of the four stamps specified for boiler plate in Par. 36 remain visible after the completion of the boiler,

or if the slab or melt number is invisible, are the remaining markings sufficient to meet the Code requirements?

*Reply:* It is the opinion of the Committee that all of the information which the Code requires to be stamped on the plates at the mills must be obtainable from the finished drum structure for each plate. All four stamps need not be visible at a given point if the fragments from various sets can be pieced together so as to give the desired information.

### CASE No. 295

*Inquiry:* How may an inspector in the field know the range of tensile strength of boiler plate?

*Reply:* Attention is called to the fact that the range for all classes of steel specified in the Code is 10,000 lb. per sq. in. and the tensile strength stamped on the plate indicates the minimum of this range, which minimum under the rules must not exceed 55,000 lb. per sq. in.

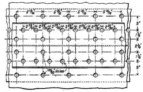

Fɪɢ. 7   Tʀɪᴘʟᴇ-Rɪᴠᴇᴛᴇᴅ Jᴏɪɴᴛ ᴡɪᴛʜ Wɪᴅᴇ Rɪᴠᴇᴛ Sᴘᴀᴄɪɴɢ ɪɴ Iɴ-ɴᴇʀ Rᴏᴡs

### CASE No. 296

*Inquiry:* Will a water-tube boiler of a type largely used in marine service, the heating surface of which is constructed mainly of iron pipe screwed into malleable iron and steel fittings, meet the requirements of the Boiler Code, if the pipe used is a special quality of redrawn lap-welded iron pipe of puddled stock and tested to 1000 lb. hydraulic pressure?

*Reply:* There is nothing in the Boiler Code to cover the use of iron pipes for the tubes of such a boiler with screwed joints. Until such a time that the specifications for piping have been formulated, the Boiler Code Committee would consider that if the other requirements of the Boiler Code are met, special redrawn pipe not to exceed 1½ in. standard pipe size made from lap-welded iron of puddled stock and tested to 1000 lb. hydraulic pressure, may be used for a working pressure not to exceed 200 lb. per sq. in., provided the wall thickness is at least 50 per cent greater than the wall thickness required by the Code for tubes of water-tube boilers.

### CASE No. 297

*Inquiry:* Will the requirements in the Boiler Code for automatic water gages be met under the terms of Par. 427e, if the shut-off valve in the upper fitting has a projection which pushes the ball away from the seat for a distance of ⅛ in., causing the ball to drop to a position considerably greater than ¼ in. away from its seat?

*Reply:* It is the opinion of the Committee that the intent of the requirements of the Code will be met in the construction described.

### CASE No. 298

*Inquiry:* Is it permissible in the case of a forged steel steam outlet riveted to the boiler shell to use a wrought steel flange screwed to the outer end of the endneck, the neck being properly threaded and peened over into a beveled part cut away from the flange?

*Reply:* It is the opinion of the Committee that this method of forming a steam outlet connection meets the requirements of the Rules in the Boiler Code.

# CORRESPONDENCE

ONTRIBUTIONS to the Correspondence Departments of MECHANICAL ENGINEERING by members of The American Society of Mechanical Engineers are solicited by the Publication Committee. Contributions particularly welcomed are suggestions on Society Affairs, discussions of papers published in this journal, or brief articles of current interest to mechanical engineers.

## Simplification of Venturi-Meter Calculations

To THE EDITOR:

In regard to my article on the above subject in the April issue of MECHANICAL ENGINEERING and Mr. C. G. Richardson's comments in the May issue, I wish to state the following:

Mr. Richardson is probably entirely right in his statement regarding the drop which should take place across a well-designed and constructed venturi meter. As I was not interested particularly in this drop at the time of making the experiments referred to, I took no pains to measure it accurately and the 2½ in. drop mentioned was not intended to be more than an approximation.

Mr. Richardson's statement as to the permissible ratio between the pressure at the throat and the initial pressure being as low as 0.84 is of course correct, but I would not advise designing a venturi meter having such a ratio of pressure at its rated capacity, inasmuch as any flow in excess of such rated capacity would increase this ratio to such an extent that it would probably go beyond the range of the pressure-measuring apparatus used in conjunction with the meter, whereas if a meter is designed for its rated capacity with the ratio nearer 0.90, it will be possible at times to exceed this flow without increasing the pressure differences very greatly out of proportion to the increase of the flow.

The tabular values of the function of $P_1$ and $P_t$ were plotted out to a very large scale for use during the experimental work and are shown in my article only to a reduced scale.

GLENN B. WARREN.

Schenectady, N. Y.

## Industrial Training in China

To THE EDITOR:

The letter appearing on page 305 of MECHANICAL ENGINEERING for May 1920, entitled Industrial Training in China, is of great interest to me, not only on account of my three years as a university professor in China, but because I am generally interested in relations between China and the United States. Mr. Foster has a distinct end in mind, namely, the encouraging of American manufacturers to supply mechanical equipment for schools in China, but in the course of making his argument, quite inadvertently no doubt, he presents a somewhat inaccurate picture of the present status of technical education in China. I am very sure the general reader of his letter will get the impression that there are no technical schools in China worth mentioning, except Peking University. The name of the latter is incorrectly given, by the way, Peking University being a Methodist school which, so far as I know, has never given any engineering courses. The writer doubtless has in mind the Peking Union University, which is just being organized and therefore has its record to make as far as technical education is concerned. On the other hand, there have existed in China for a number of years technical schools that of course have their drawbacks but which on the whole compare very favorably with the state colleges of the less populous states of the West. As to the character of work done, to cite a specific instance, the graduates of Pei Yang University in engineering have in the past found no difficulty in entering American universities of the first class as graduate students and taking their master's degree in one year and their doctor's degree in two years.

I am glad to be able to say that in the majority of cases the instructors in these schools have been Americans and that in nearly all instances they are of a grade that will compare favorably with those in American colleges.

The present situation in regard to the tendency of Chinese students to go to France rather than the United States is even more serious than Mr. Foster indicates in his letter, and it is highly desirable that an effort should be made to promote both the coming of Chinese students to the United States for study and the study of engineering in China.                T. T. READ.

Washington, D. C.

## Technical Library Book and Journal Indexing

To THE EDITOR:

I have just read an abstract of a report made by Mr. Harrison W. Craver, Director of the U. E. S. Library Board.

As one who is much interested in the entire subject and more distinctly so in certain phases, I am pleased to be at one with Mr. Craver in his advocacy of the Dewey Classification and its later extension, the Brussels Classification, and for the same reasons that Mr. Craver has so well expressed.

Mr. Craver states that "the catalog cannot be expected to act as an index to all the books in the library." Undoubtedly this is correct as applied to the present books with their present indexes, but I hope that the time will arrive when all books and all technical journals will be indexed by the same method that is suggested for the books in the library. When that time arrives the book or journal indexes will be logical extensions in detail of the general library index. Nor is this by any sense a utopian dream. The Dewey or Brussels Index is, in actual practice, so simple that it would require probably less time than the preparation of the usual indexing method employed.

Before I advocated this to the publishers of certain technical journals I tried it out myself and made up an extensive index covering the automotive industry in detail. This and the existing Dewey Index applying to mechanical subjects I applied to a year's issue of certain automobile publications and to a year's issue of the American Machinist. I found that the entire work occupied less than one-half hour's time daily for a week. Surely an index of this character will be far more useful than the usual haphazard one now found in technical journals under names or titles that have frequently no relationship whatever to the subject itself. The situation is not quite as bad in most technical books, but it is bad enough.

Those engineers who realize that much useful information is to be found in the technical journals which rarely or very late finds its way into books, and who therefore collect such data, would find their task enormously facilitated by the adoption of this indexing method. This work would be further facilitated if they would secure duplicate copies of the journals that they find most useful, in that they could then gather all of the information contained under any given index number in a group and so file it without the necessity of destroying part of any article that might happen to be closely printed with another article under another index number.

The adoption of this plan would also make technical journals much more useful in a library such as the United Engineering Societies Library, in that the index of each journal would be a logical continuation of the Library's index.

It is to be hoped that Mr. Craver will use his and the United Engineering Library's undoubtedly very great influence with publishers and editors to adopt this plan in their own interest as well as the interest of the Library and that of the users of books and journals.

Philadelphia, Pa.                HENRY HESS.

# MECHANICAL ENGINEERING
### THE JOURNAL OF THE AMERICAN SOCIETY OF MECHANICAL ENGINEERS

Published Monthly by the Society at
29 West Thirty-ninth Street, New York

FRED J. MILLER, *President*

WILLIAM H. WILEY, *Treasurer*   CALVIN W. RICE, *Secretary*

PUBLICATION COMMITTEE:

GEORGE A. ORROK, *Chairman*   J. W. ROE

H. H. ESSELSTYN   GEORGE J. FORAN

RALPH E. FLANDERS

PUBLICATION STAFF:

LESTER G. FRENCH, *Editor and Manager*

FREDERICK LASK, *Advertising Manager*

WALTER B. SNOW, *Circulation Manager*
136 Federal St., Boston

Yearly subscription $4.00, single copies 40 cents. Postage to Canada, 50 cents additional; to foreign countries $1.00 additional.

*Contributions of interest to the profession are solicited. Communications should be addressed to the Editor.*

Those of us who dwell in the past must have memories, enjoyable and fine. With generous, engineering comradery and unbounded St. Louis hospitality we may build a memory of the 1920 Spring Meeting of the A.S.M.E. that will long stand.

The short elapsed time does not allow us to record the parts of the meeting that years from now will remain in mind; but so successful was the well-planned and carefully-executed program of the St. Louis meeting in its appeal to both old-timer and novice that material for memory building abounds.

As suggestions, think of any of the interesting plants visited, recall the beautiful Shaw's garden, the picturesque open-air theater, the jolly ride up the river, or the quiet talk of experience and mutual confidence with an old friend and the making of several new friends. Whatever the individual reason, St. Louis will always be remembered as the place of a wonderfully successful Spring Meeting.

## The Federated American Engineering Societies

The outstanding event of the past month has been the conference in Washington of the representatives of about seventy-five (75) of the leading national, regional and local engineering and allied technical organizations.

This conference was the culmination of the work of several committees in local and national societies during the past few years, resulting in a joint conference committee of the four national engineering societies of Civil, Mining and Metallurgical, Mechanical, and Electrical Engineers.

To say that the conference was a complete success does not begin to describe it; the get-together movement of the professional men in engineering, architecture, chemistry and allied professions is an accomplished fact.

Interest was keen from the minute the conference opened until it closed with inspiring addresses from such men as Homer L. Ferguson, Past-President of the United States Chamber of Commerce; Robert S. Woodward, President of the Carnegie Institution; and James R. Angell, Chairman of the National Research Council.

The reason for the success was the sincere desire of all the delegates to subordinate personal interest, even of the organizations which they represented, to the common good. And the cause of the common good was expressed more fully and perfectly than I have ever yet heard it expressed: *"Service to others is the expression of the highest motive to which men can respond and duty to contribute to the public welfare demands the best efforts that men can put forth. Therefore, it shall be the object of this organization to further the interests of the public through the use of technical knowledge and engineering experience, and to consider and act upon matters common to the engineering and allied technical professions."*

In the long run, success not only of individuals but of organizations is dependent upon the fundamental motive's being unselfish. This is not inconsistent with the greatest possible personal rewards, both to individuals and to groups, which are the natural result of the indispensable services rendered. In other words, the man who can produce the greatest results is the man who can command a high salary. When engaging a man absolutely indispensable to the success of any given enterprise, one of the commonest inducements is that "salary is no object."

With the splendid spirit which actuated the entire conference, the Joint Conference Committee was invited by the delegates to arrange for a fall meeting to complete the organization, the constitution and by-laws having been adopted. In the meantime, until the perfection of the new organization, the Engineering Council was invited to continue to conduct the work which it has started and to take up any other matters which should be acted upon by the Federated American Engineering Societies when that body shall be organized.

With good generalship, the members of Engineering Council (of which there were a goodly number present, in their capacity as members or representatives of some of the societies) secured from their body a formal invitation to all of the organizations represented at the Washington Conference to attend by delegate the next meeting of Engineering Council, on which occasion such delegates would be accorded the privileges of the floor and participation in all debate.

One of the officers of one of the leading foundations is obligated among his duties to investigate and report on all organizations conducted under national, state or municipal auspices. And I firmly believe that if this officer were to investigate in the same manner the Federated American Engineering Societies and reflect on its objects he could report that it is destined to be one of the most useful organizations—that everyone working with it, whether by serving on its committees or otherwise, will be inspired to give his best to the nation and to the community in which he lives and thus enhance the standing of his profession.

CALVIN W. RICE,
*Secretary.*

## Plain Limit Gages for General Engineering Work

The new Sectional Committee of The American Society of Mechanical Engineers recently organized to undertake the standardization of plain limit gages held its organization meeting in the rooms of the Society on Friday, June 11. The Committee met as a whole in the morning and divided into sub-committees for the afternoon session. As a result it has already mapped out its work for the next three months.

This Committee, as will be recalled, was organized by the Society at the request of the American Engineering Standards Committee, and it is planning to coöperate with a similar committee recently organized by the British Engineering Standards Association. The personnel of the Committee is as follows:

E. C. PECK, *Chairman*, Gen. Supt., Cleveland Twist Drill Co., Cleveland. Ohio.

L. D. BURLINGAME, *Vice-Chairman*, Industrial Supt., Brown & Sharpe Mfg. Co., Providence, R. I.

H. W. BEARCE, *Secretary*, Gage Dept., Bureau of Standards, Secy. National Screw Thread Commission, Washington, D. C.

P. W. ABBOTT, Lincoln Motor Company, Detroit, Mich.

JOHN BATH, President, John Bath & Co., Inc., 8 Grafton St., Worcester, Mass.

EARLE BUCKINGHAM, Engineer of Standards, Pratt & Whitney Company, Hartford, Conn.

FRED H. COLVIN, Assoc. Editor, *American Machinist*, 10th Ave. & 36th St., New York City.

W. A. GABRIEL, Ch. Draftsman and Designer, Elgin National Watch Co., Elgin, Ill.

F. O. HOAGLAND, Vice-Pres. & Works Mgr., The Bilton Machine Tool Co., Bridgeport, Conn.

EDWARD H. INGRAM, Works Manager, The Cleveland Milling Machine Company, Box 350, Cleveland, Ohio.

J. O. JOHNSON, Major attached to office of Chief of Ordnance. War Department, Washington, D. C.

A. W. SCHOOF, Gage Engineer, Greenfield Tap & Die Corporation, Greenfield, Mass.

G. T. TRUNDLE, Consulting Engineer, Engineers' Building, Cleveland. Ohio.

H. L. VAN KEUREN, The Van Keuren Company, 362 Cambridge Street, Boston, Mass.

## Swiss Mission to Study American Economic Conditions

The group of Swiss engineers, architects, manufacturers, merchants, and agriculturists known as the Swiss Economic Mission, arrived in New York on May 27. The fact that this organization intended to visit this country to study economic conditions here was made known to the Society last autumn by the Hon. Hans Sulzer, of the Swiss Legation of Washington, D. C. The members, although not official representatives of the Swiss government, are progressive professional men of eminence.

The fact that Switzerland feels its indebtedness to this country for its supply of foodstuffs, especially cereals, during the war, was expressed by Professor Hilgard, of Zurich, at a dinner given to the entertainment committees of the American Society of Civil Engineers and The American Society of Mechanical Engineers at the Hotel Pennsylvania, June 3.

The party has been conducted on a general sight-seeing tour throughout the eastern and central sections of the country, with a brief stay in Canada. The entertainment for the mission was largely arranged for by the Local Sections of the A. S. M. E. in conjunction with the local chambers of commerce of the various cities visited. The party will sail for Liverpool on or about July 5.

## Austrian Engineers Appeal for Help

The Austrian Society of Civil Engineers and Architects, through its president, secretary and editor, has appealed to The American Society of Mechanical Engineers for relief for about 500 of the 4000 members who are without occupations and who are in distress.

Under the plan through which food drafts are sent to Austria through the American Relief Administration, 115 Broadway, New York City, the Austrian society is ready to deposit at a bank a sum in Austrian crowns equivalent to six times the present exchange for every dollar sent by Americans. This organization is further ready to distribute food not only among its own members but among all Austrian technical men who may call for help.

The Council of the A.S.M.E. received this appeal at its last meeting and refers it through the columns of MECHANICAL ENGINEERING to those of the members who may wish to extend succor by sending food drafts.

Drafts may be obtained at any bank in this country and should be made available to the Austrian Society of Civil Engineers and Architects, Wien, 1, Eschenbachgasse 9, mailing them to the American Relief Administration. This system of food drafts is apparently very sure and any member sending help in this manner may be certain of its receipt.

## The Robert W. Hunt Medal Fund

The eightieth birthday of Captain Robert W. Hunt was commemorated by the American Institute of Mining and Metallurgical Engineers by the presentation to him of the Hunt Medal for meritorious contributions to the art of making steel. The medal was designed and executed by Emil Fuchs, a medalist of international fame, and is considered, both from the point of execution and of symbolic significance, one of the most beautiful of any connected with the engineering profession.

The presentation ceremony took place on May 27 in the auditorium of the Engineering Societies Building. A. A. Stevenson opened the meeting and introduced John J. Cone, one of Captain Hunt's partners, who gave an account of the captain's career and his notable contributions to steel metallurgy. He then presented to the Institute on behalf of the associates and

employees of Captain Hunt the Robert W. Hunt Fund, which consists of $7500, as an endowment fund to provide an annual award, and $1000 for the expenses of providing the medal. Herbert Hoover accepted the Fund on behalf of the Institute and, following a short speech by Charles M. Schwab, announced that

the Iron and Steel Committee had awarded the first medal to Captain Hunt himself for his contributions to the steel industry, particularly the development of the steel rail. The ceremonies closed with a short speech of acceptance by Captain Hunt.

## Slow Delivery of Mechanical Engineering Bulletin

The U. S. Post Office Department threw in low gear in the distribution of the "Mechanical Engineering Bulletin" issued to the A.S.M.E. membership in May. This bulletin was planned by the Sections Committee of the Society with the expectation that it might be issued at intervals and give prompt service in respect to Employment Bulletin items and important news notes. That the "prompt service" idea was missing in this first issue, at least in some cases, is evident from the fact that copies were not received in Cleveland until three weeks after the mailing date (May 19). This is an example of the many difficulties now experienced by those who are sending out large quantities of second and third-class mail matter. If it should be decided to continue the bulletin, the most that can be said is that we hope the Post Office Department may gradually get back into high-gear operation.

# Organizing Conference Plans Federation of Engineering Societies

### Delegates from 71 Engineering Organizations and Allied Technical Societies Adopt Constitution Creating "The Federated American Engineering Societies"—New Organization to Represent Profession in Matters of Common Concern and in National and State Affairs

IN a two-days' conference at Washington, D. C., June 3 and 4, which was marked by the intense enthusiasm of the attending delegates, the representatives of 71 engineering and allied technical societies took the first step toward the creation of an organization which should speak with a single voice for the engineering profession of the United States. This Organizing Conference, as it was known, was called by the Joint Conference Committee of the Four National Engineering Societies, who for some time past have been concerned with the problem of democratization of the profession and of effecting a closer coöperation between existing engineering and allied technical organizations. The conference thus called was a success in the highest degree, for with a fine disregard of unimportant details it adopted a constitution and by-laws of an organization to be known as "The Federated American Engineering Societies," a body composed of national, local, state, and regional engineering organizations which is to function through an "American Engineering Council."

CALVERT TOWNLEY
Chairman of the Conference

The opening session of the conference was called to order by Richard L. Humphrey, of the American Society of Civil Engineers, the chairman of the Joint Conference Committee. He reviewed the history of that committee and explained the basis of representation of delegates at the conference. He also spoke briefly of the need of a federation of engineering societies and of the value to the profession of such a body. Following Mr. Humphrey, Lt-Col. C. W. Kutz, U. S. A., engineer commissioner of the District of Columbia, welcomed the conference on behalf of the engineering societies of Washington.

The delegates then proceeded with the election of officers and Calvert Townley, past-president of the American Institute of Electrical Engineers, was chosen chairman, and John C. Hoyt, of the American Society of Civil Engineers, secretary. Dr. F. H. Newall, of the American Association of Engineers, then presented a resolution to the effect that the delegates of the conference have the full power to make rules governing its procedure. He also suggested that the conference proceed at once with the business of the day rather than adhere to the program as arranged by the committee. To these suggestions the conference readily assented.

Major G. S. Williams, of the Grand Rapids Engineering Society, thereupon at once offered the following resolution:

*Resolved,* That it is the sense of the convention that an organization be created to further the public welfare wherever technical knowledge and engineering experience are involved and to consider and act upon matters of common concern to the engineering and allied technical profession; and

*Resolved,* That it is the sense of this convention that the proposed organization should be an organization of societies or affiliations and not of individuals.

This resolution was at once seconded and a spirited debate immediately ensued. The representatives of the American Associa-

tions of Engineers were the first to discuss the resolution. They argued for the creation of an organization composed not of societies or affiliations but of individuals, and they pointed out the advantages such a society would have for the individual engineer. They urged that their association be made the medium through which the entire engineering profession should speak in matters of federal and state concern and they appealed to all engineers to join their ranks.

Following the delegates of the American Association of Engineers, the representatives of the large national and the smaller local and regional organizations expressed their opinion as to the resolution, and it soon became evident that the conference was overwhelmingly in favor of a federation of existing societies and not of accepting any existing society to represent them nor of the formation of a new society composed of individuals.

The delegates who spoke in favor of the resolution urged its passage upon the grounds that an organization of organizations was the great need; a body which could represent the profession in such matters as concerned it as a group and not as individuals, as the interests of the individual, they claimed, were being well cared for by the existing national and local societies. The small local societies, particularly, were uniformly in favor of the resolution, and their representatives pointed out the value of a central organization through which they might be represented in federal and state matters, in which at the present time they are unable to participate.

When the resolution was put to vote, the societies voted in the order of their arrangement on the program, which was an alphabetical one. This placed upon the American Association of Engineers the necessity of voting first, and although they had argued against the resolution, nevertheless, sensing the wishes of the majority, they announced that their seventeen delegates voted aye. This announcement came as a great surprise to the conference, who at once voiced their approval by hearty applause. As the balloting continued it became evident that the resolution would easily pass, and when the secretary announced the final vote it showed that 119 delegates had voted aye, none had voted in the negative, and of the 71 societies present only five had not voted. This important decision of the conference taken so soon after it had convened showed the intense interest of the delegates in the movement and made clear their ability to act along broad, general lines without allowing themselves to be confused with details. The passage of the Williams resolution established beyond doubt the ultimate success of the movement and marked the first step toward the creation of the new organization.

JOHN C. HOYT
Secretary of the Conference

Following the vote on the resolution, Chairman Townley appointed three committees to deal with the details arising in connection with the conference. These were committees on constitution and by-laws, resolutions, and program. R. L. Humphrey was appointed chairman of the constitution and by-laws committee; Philip N. Moore, chairman of the resolutions committee; and

Robert H. Fernald, chairman of the program committee. These committees at once retired and the conference proceeded with the program as arranged by the Joint Conference Committee. This included addresses on the Functions of the Engineers in Public Affairs, by Arthur P. Davis, president of the American Society of Civil Engineers; Philip N. Moore, past-president of the American Institute of Mining and Metallurgical Engineers; and Leroy K. Sherman, president of the American Association of Engineers.

On Friday morning the conference began a consideration of the constitution and by-laws as prepared by the committee appointed the day previous. These were presented by Richard L. Humphrey, chairman of the committee, and after considerable discussion of their many details, both constitution and by-laws were passed in very nearly the same form as received from the committee. The vote on the constitution was particularly interesting in that it was indicative of the probable participation of the various societies in the new movement. Eight societies went on record as "present but not voting," among them being the American Association of Engineers. The final vote showed 88 delegates casting affirmative votes, ten absent and none against.

Following the announcement by Chairman Townley of the passage of the constitution and by-laws, L. P. Alford, chairman of the delegation representing The American Society of Mechanical Engineers, announced that the delegates who represented the Society were empowered to commit the organization, and that as soon as The Federated American Engineering Societies were ready to receive applications The American Society of Mechanical Engineers would apply for membership. The Society is thus the first organization to join the new movement.

The constitution, in accordance with the instructions of the conference, is now in the hands of an editing committee, and later MECHANICAL ENGINEERING will present it in its complete and finished form. The main features of the Constitution are:

*Membership.* Membership is to consist of national, local, state and regional engineering and allied technical organizations and affiliations, classified as follows:

(1) National engineering and allied technical organizations
(2) Local, state, or regional engineering or allied technical organizations other than local associations, sections, branches or chapters of national organizations

(3) Affiliations consisting of local sections of national organizations, local engineering societies or clubs and local engineers.

*Management.* The management of the new organization is to be vested in a body known as the American Engineering Council and its Executive Board. This Council is to consist of representatives of member societies each of which shall be entitled to one representative for a membership of from 1 to 1000, and one additional representative for each additional thousand members or a major fraction thereof. This Council is to hold an annual meeting. The elected officers of the Council will be a president, to hold office for two years, four vice-presidents, to hold office for two years, and a treasurer, to hold office for one year. An executive officer who shall also be secretary is also provided. The Executive Board is to consist of thirty members of the Council, and this Board is to conduct the business of the new organization under the direction of the Council. The six officers selected by the Council are to be members of this Executive Board, and the balance of twenty-four shall be selected by the National Societies and local, state and regional organizations according to districts. The president and secretary of the Council shall be, respectively, the chairman and secretary of the Executive Board.

*Funds.* The funds for the use of the new organization are to be obtained in the following manner: Each national member society represented on the Council is to contribute annually $1.50 per member. Each local, state or regional organization represented is to contribute annually $1 per member.

*Local Affiliations and State Councils.* The Council is empowered to encourage the formation of local affiliations to consider matters of purely local public welfare with which the engineering profession is concerned. It is also authorized to create state councils consisting of representatives of the local affiliations within the state. These councils are also for the purpose of considering matters of public welfare as associated more particularly with state matters.

Evening sessions were held at the New Willard Hotel on both days of the conference. On Thursday evening the delegates were addressed by Homer L. Ferguson, past-president of the United States Chamber of Commerce; James H. McGraw, president of the McGraw-Hill Publishing Co. of New York; and George Otis Smith, director of the United States Geological Service, Washington, D. C. On Friday evening addresses were given by Samuel M. Vauclain, president of the Baldwin Locomotive Works, Philadelphia, Pa.; Robert S. Woodward, president of the Carnegie Institution, Washington, D. C.; and James R. Angell, chairman of the National Research Council, Washington, D. C. All of these addresses were highly inspirational to the delegates assembled, and it is to be regretted that space will no' afford even a brief comment on them.          E. W. T.

## ORGANIZATIONS PARTICIPATING IN CONFERENCE

*American Association of Engineers*
*American Association of Petroleum Geologists*
*American Ceramic Society*
*American Concrete Institute*
*American Electric Railway Engineering Association*
*American Electrochemical Society*
*American Institute of Electrical Engineers*
*American Institute of Mining and Metallurgical Engineers*
*American Ordnance Association*
*American Railway Engineering Association*
*American Society of Agricultural Engineers*
*American Society of Civil Engineers*
*American Society of Heating and Ventilating Engineers*
*American Society of Mechanical Engineers*
*American Society of Naval Architects and Marine Engineers*
*American Society of Refrigerating Engineers*
*American Society of Safety Engineers*
*American Society for Testing Materials*
*American Water Works Association*
*Associated Engineering Societies of St. Louis*

*Associated Engineers of Spokane*
*Association Railway Electrical Engineers*
*Boston Society of Civil Engineers*
*Brooklyn Engineers' Club*
*Cleveland Engineering Society*
*Colorado Society of Engineers*
*Connecticut Society of Civil Engineers*
*Detroit Engineering Society*
*Duluth Engineers' Club*
*Engineers' and Architects' Club of Louisville*
*Engineers' Club of Baltimore*
*Engineers' Club of Columbus*
*Engineers' Club of Minneapolis*
*Engineers' Club of Philadelphia*
*Engineers' Club of St. Louis*
*Engineers' Club of Trenton*
*Engineering Council*
*Engineering Society of Akron*
*Engineering Society of Buffalo*
*Engineers' Society of Pennsylvania*
*Engineers' Society of Western Pennsylvania*
*Florida Engineering Society*
*Grand Rapids Engineering Society*
*Illinois Society of Engineers*
*Illuminating Engineering Society*
*Indiana Engineering Society*
*Institute of Radio Engineers*
*Iowa Engineering Society*

*Kansas Engineering Society*
*Los Angeles Joint Technical Society*
*Mining and Metallurgical Society of America*
*Minnesota Surveyors' and Engineers Society*
*Mohawk Valley Engineers' Club*
*Nashville Engineering Association*
*National Fire Protection Association*
*National Safety Council, Engineering Division*
*New England Water Works Association*
*Oregon Technical Council*
*Providence Engineering Society*
*Scientoch Club*
*San Francisco Joint Council of Engineering Societies*
*Society of American Military Engineers*
*Society of Automotive Engineers*
*Society of Engineers of Eastern New York*
*Society of Industrial Engineers*
*Society for the Promotion of Engineering Education*
*Taylor Society*
*Technical Club of Dallas*
*Topeka Engineers' Club*
*Vermont Society of Engineers*
*Washington Society of Engineers*

## EDMUND GYBBON SPILSBURY

ON May 20, 1920, Edmund G. Spilsbury, mining and metal-lurgical engineer of international reputation, died suddenly after many years of useful service to the profession and to the engineering societies.

Mr. Spilsbury was born in London, England, in 1845. He was educated in Belgium and was graduated from the University of Louvain in 1862. He then took up practical work in Germany, entering in 1864 the service of the Eschweiler Zinc Co., Stolberg, one of the largest miners and smelters of lead and zinc in the world. The following year he was given charge of that company's mines and works in Sardinia. In 1867 he became connected with McClean and Stillman, of London, in charge of the construction of the lock gates for the Surrey Commercial Docks; in 1868 he accepted the position of designing engineer with J. Casper Harkort and was assigned to the detail work of the Keulenberg Bridge in Holland, the Danube Bridge at Vienna and the

EDMUND GYBBON SPILSBURY

Rhine Bridge at Düsseldorf. Later he returned to the Eschweiler Company as chief engineer, and in 1870 came to the United States for the Austro-Belgian Metallurgical Co. to investigate the resources of this country in lead and zinc.

After two years in this country Mr. Spilsbury decided to take up practice here, in the course of which he introduced the Harz system of ore dressing for the zinc ores of Pennsylvania and New Jersey. During this period he was also engaged in explorations on the northern shore of Lake Superior and in Colorado, Montana, Utah and California. From 1873 to 1875 he was engaged as general manager of the Bamford (Pa.) Smelting Works; in 1879 he designed and built the Lynchburg Blast Furnace and Iron Works; he was also consulting engineer for the Coleraine Coal & Iron Co., Philadelphia. In 1883 he became general manager of the Hailer Gold Mine in South Carolina and in 1887 became connected with Cooper, Hewitt & Co., New York. From 1888 to 1897 he was managing director of the Trenton Iron Co., Trenton, N. J. While manager of these works he introduced as

specialties of their business the Elliot locked wire rope and the Bleichert system of aerial tramways.

His practice as a consulting mining engineer and metallurgist took him into many parts of Europe, Africa, the United States, Mexico and South America. During the winter and early spring of 1920 he spent a number of weeks in Brazil on a mining project for clients in the United States and had returned to the United States only a few weeks prior to his death.

Mr. Spilsbury was the author of a number of important technical papers, mostly on mining processes and on mine appliances. He also wrote entertainingly in a lighter vein, and the following paragraphs from "Technical Reminiscences," published in *Mining and Scientific Press* in 1915, recall early interesting experiences.

Early pioneer work in Sardinia was most interesting and instructive. In the first place, the island was almost as much a terra incognita as our own West a century ago. And yet the civilization of Sardinia was one of the oldest in the world. The Phoenicians, the Egyptians, and later the Carthaginians and the Romans, had all left their marks, in old excavations, dumps, slag piles and more especially in the tombs cut out of solid rock. In the early 60's the northern and eastern parts of the island were dangerous, as well as difficult of access. The tribes occupying this mountainous district did not recognize the Italian Government and were generally classed as bandits. During the second year of my stay one evening I heard a disturbance outside my camp and the clatter of a number of horses. Thinking that I was about to be raided by the bandits. I called the two foremen who were in the house and, getting our guns, we started to see what the trouble was. Opening the door, I saw a small cavalcade. The leader was a very tall and slim young gentleman, dressed in full hunting costume. He addressed me by name, stating that he had a letter of introduction and asking if he and a friend could stay the night. I thought I had never seen a taller or lankier fellow, but withal, a most pleasant face. He handed me a letter from one of the directors of our company, asking me to make the stay of the bearer as pleasant as I could, as he was a friend of theirs, being the Duke of Brabant, afterward Leopold the Second, King of the Belgians. The Duke had come to Sardinia on a mountain sheep hunt. Years afterward, when he had succeeded his father, I always received a hearty welcome whenever I met him either in Brussels or Paris.

And again we find these paragraphs telling of the changes which have taken place in mining practice:

I wonder how many mining engineers of the present day can call to mind changes and advances in the practice of mining which have taken place in the last half century? Practically every appliance beyond the pick and shovel, hand hammer and drill, has been introduced during that short period. When I first started my practical course in underground work, we knew none of the aids now considered requisite to work economically. We had no machine drills to save the physical labor of hand work. Dynamite or its precursor, nitroglycerine, had not yet been invented; fuse was still unknown; even steel was more or less of a luxury, and our hand drills were still made of iron with steel bits welded on to them. The mushrooming of the head of the drill under the hammer blows was a constant source of trouble and injury to the hands of the miners. The only explosive known was the large-grained black powder.

It seems hardly credible today, and yet it is a fact that the contract price for driving or stoping ground was influenced by the condition of the rye crop. In seasons of drought, the height of the rye was much curtailed and the length of the straw between joints would be so short that the time occupied in jointing and filling the stalks which were used as fuses, greatly lessened the driving capacity of the miners. The straw primers naturally limited the possible depth of the hole and three feet was considered a good average. These straw fuses were very rapid and it was necessary to ignite them by a slow-burning sulphur wick.

Mr. Spilsbury was intensely interested in the work of the national engineering societies. He was a member of the American Institute of Mining and Metallurgical Engineers, of which he was president in 1896, the American Society of Civil Engineers, the Institution of Mining and Metallurgy of Great Britain, the Mining and Metallurgical Society of America, and the American Electrochemical Society; he became a member of The American Society of Mechanical Engineers in 1890. In 1916-17 he was president of the Engineers' Club of New York. He was a trustee of the United Engineering Society from 1916 until his death; a member of the Engineering Societies Library Board from its organization in 1913 and a member of the Engineering Foundation Board from 1916. He was also a member of the National Research Council. In all these associations he was an active and useful member and served on numerous committees and contributed freely of his time and ability.

# Industrial Relations Association of America

### Earnest Convention at Chicago Considers Vital Problems — "Educating with Truth" a Solution of Industrial Relations — Union Officials Boast of Increasing Production

THE Second Annual Convention of the Industrial Relations Association of America was held in Chicago at the Auditorium on May 19, 20 and 21. Organized in 1918 as the National Association of Employment Managers, its rapid growth and the appreciation of the breadth of the field led to the change in name.

To understand the spirit that permeated this gathering, one must picture one of the old-time preaching services lasting several days, people coming from all the countryside, listening to hour sermons with divine expectation, and sitting through long experience meetings, all in the hope of guidance in solving their daily problems. The Chicago convention of the I. R. A. A. was just such a meeting. Managers came to Chicago from east and west, from small plant and from large plant, 2500 in all, with a determination to absorb every portion of the great feast of information provided. Glimpses of receptive notebooks and poised pencils gave an impression accentuated by the earnest faces and quick shouts of "louder," if the speaker dropped his voice even momentarily, that these people were hard against the big problem in industry today, that of the relation of workmen and employer, and that they were seeking every atom of help from the knowledge and experience contributed at the convention.

The plan of the convention was based on the principle that every one of the twenty-five hundred in attendance should be given opportunity to check up his individual stock of experience with that of those attacking similar problems or engaged in similar industries. Thirty-three round-table subject luncheons were arranged at which topics such as Employment Office Methods, Radicalism, Americanization, Developing Plant Spirit, Housing and Introducing the New Worker, were considered under the guidance of capable leaders. At dinner those working in the same industry met for a consideration of the problems peculiar to that industry. The main sessions of the convention were given over to a treatment of the important phases of the industrial relations problem by well-known and properly equipped speakers.

In the opening address Cyrus H. McCormick, works manager of the International Harvester Co., laid down the principle that truth and its wide dissemination was the only firm basis on which lasting coöperation between employers and employees can be built. He stated the duty of the employment manager to be the education of the employees and emphasized the importance of the proper fulfilling of this duty.

Incentives and Their Relation to Production were discussed by L. C. Marshall, dean of the School of Commerce and Administration, University of Chicago, and Frank J. Raymond, of the Inter-Racial Council of New York.

The importance of having each workman understand the part he is playing in the complete scheme of manufacture and the effect of the individual's effort on the finished product, was handled by H. B. Bole, vice-president, Hydraulic Pressed Steel Co., Cleveland, and Harry N. Clarke, president, The Cort-e-Scope Co., Cleveland. They outlined methods successfully used in giving their men a perspective of the entire operation of their plants.

C. A. Lippincott of South Bend, Ind., told of the housing scheme developed by the Studebaker Co. as a solution of the problem of community conditions affecting labor stability. The Studebaker plan is to build the houses and sell them to their workmen, loaning the money without interest. Mr. Lippincott paid his respects to the professional uplifters with the words, "Man does not need a gratuity. The activities of some welfare associations seem to intimate to the person to whom so-called succor is to be given, that he is in the mire and needs uplifting, such as they only can render."

In her talk on the Status of Women in Industry, Mary Van Kleck of the Russell Sage Foundation scouted the idea that women should receive special treatment as a peculiar problem in industry and spoke forcefully in favor of considering women's place in every phase in industry. Treating the subject of employee representation, Miss Van Kleck pointedly questioned the exploitation of industrial representation as a means of forcing false ideas of loyalty down the throats of employees, and stated as her real opinion that representation honestly offered by the employer would have a wonderful effect in bringing out leadership and developing true democracy.

The vital problem of the foreman and his relation to the organization and to the workman received special consideration. Dudley R. Kennedy, of Philadelphia, formerly in charge of industrial relations at Hog Island, touched the spot in his talk: "It's high time we stopped blaming the Reds and the agitators for the prevailing conditions and hang the blame where it belongs. If the radicals can do more with our men in two hours at night than we can, having them at our command for eight hours a day, then something is radically wrong. The fault lies at the top. "Industry must open its books to the workers. Also we must equip our foreman to answer the absurdities of the soapbox orators. Production has fallen off, not primarily because of derelictions among the workers, but because leadership isn't present, and it is up to management to provide the leadership." · Mr. Kennedy's talk was emphasized by Leroy Kramer, vice-president of the Willys-Overland Co., of Toledo, who said that the foreman must be closer to the management of the plant than he is, so that he may be in a position to give the workers the truth, rather than permit them to get half-truths or untruths from irresponsible agitators.

One of the peaks of enthusiasm of the convention was reached at the close of E. J. McCone's speech stating his position on the open-shop question. Mr. McCone, general manager of the Buffalo Commercial, recently testified before the Senate committee that every newspaper in the country, except three, are subject to censorship of the typographical union. His subject was Organized Labor in Industry, and he made the plea for democracy in unionism and the right of an individual to join a union or not. Mr. McCone spoke with an eloquence and enthusiasm that, at his closing words, brought the convention to its feet with applause and cheers. Sidney Hillman, President of the Amalgamated Clothing Workers of America, speaking on the same subject, uttered a new note in the call of organized labor. He stated the principal question of the day to be, "Shall humanity be starved to death through lack of production?" and pleaded for an industrial relationship that will establish a coöperation leading to greater production. Mr. Hillman outlined the work his union had done in bettering the conditions of the workmen in the clothing industry and quoted one of the employers who admitted that the organization had increased production in his shop. In closing, he made an earnest plea for a careful study of the industrial situation by the best minds of the country so that decisions in the future would be based on facts rather than the opinions advanced by glib talkers.

Sherman Rogers, of New York City, a former lumber jack and a man who has spent his life with the workmen, testified that what the workman wanted most today was the truth. He expounded forcefully his theory that placing the cards flat on the table was the only way to settle present conditions and he illustrated, from his experience, how Bolshevism had been checked and strikes forestalled by plain statements of the facts in the case.

What the workman thinks about modern employment methods and welfare work was illuminatingly explained by Whiting Williams, of Cleveland, who told how he was treated when he doffed his white collar and was hired and fired in shops and factories throughout the country. Severe though his comments were on the way men are treated, they were justly and graciously received by the convention. He showed the importance of a broad point of view in handling industrial relations and the necessity for dignifying the employment manager's job.

Broadly stated, according to the principal speakers the solution of the problem of satisfactory industrial relations lies along the open recognition by the employer that his leadership depends on the confidence he places in his men and in the way he takes his men into his confidence.

The convention relaxed from its inspiring work for one evening when it presented "An Industrial Jamboree," a satirical burlesque on the Second Industrial Conference. All the characters were represented and their testimony before the conference of disagreements brought peals of laughter from the convention.

Plant visitation trips included the Stock Yards, the International Harvester Co., General Electric Co., Illinois Steel Co. and other smaller plants that have achieved success with their industrial relations work.   C. E. D.]

## MOTOR GASOLINE FROM HEAVIER HYDROCARBONS

(Continued from page 402)

It is becoming the practice to utilize heavy distillates for the purpose of cracking rather than the residual oils. In this way the amount of coke produced during the operation is materially lessened. For example a heavy asphaltic oil may be distilled to coke at atmospheric pressure and certain of the distillates used for the purpose of cracking.

*Composition and Refining of "Cracked" Distillates.* Unrefined "straight run" or "natural" gasoline distillates are composed of varying proportion of paraffins, naphthenes, aromatics, small percentages of olefines and a quantity of organic nitrogen and sulphur compounds generally small, but depending upon the character of the original crude petroleum. Oxygen compounds such as naphthenic acids may also be present in traces. Some of the straight-run gasolines which have been secured where direct steam was used in the distillation process necessitate little or no treatment with sulphuric acid.

The gasoline distillates produced by "cracking" contain diolefines in addition to the compounds mentioned above while the proportions of aromatics are larger. Cracking in the vapor phase only, because of the higher temperatures generally employed, produces larger proportions of unsaturated hydrocarbons and aromatics than in the case of the pressure still.[19] Cracked distillates upon standing may separate a resinous product which is derived, no doubt, from the diolefines and possesses explosive properties when heated.[19] The disagreeable odor of these distillates is attributed by some to the unsaturated hydrocarbons, particularly the diolefines. Brooks and Humphrey,[20] however, record that this disagreeable odor is to be attributed more to the presence of organic nitrogen, oxygen and sulphur compounds.

Sulphuric acid acts both chemically and physically upon unrefined distillates, colloidal coloring matter being precipitated and dissolved by the acid, unsaturated hydrocarbons, nitrogen, sulphur, and oxygen compounds entering into reaction, the three last mentioned being by no means completely removed. The rise in temperature which is especially marked in the case of unrefined "cracked" distillates when treated with sulphuric acid is no doubt largely influenced by the diolefines. According to Brooks and Humphrey, the lower olefines react to form alkyl esters, tertiary and secondary alcohols, some polymerization also taking place. Above the hydrocarbon $C_{12}H_{24}$, polymerization takes place to form di- and tripolymers which contain but one double bond. Brooks and Humphrey made the interesting discovery that the higher olefines such as those found in the residual oils and in lubricating oils are but slightly reactive with concentrated sulphuric acid. In the case of gasoline distillates these polymers dissolve mainly in the hydrocarbon layer, raising the final boiling point. Likewise, neutral esters of sulphuric acid remain in the gasoline and are not removed when the product is

washed with alkali solution; upon standing the product is likely to precipate a resinous, viscous layer due to these esters.

It is well known that the "cracked" distillates are more difficult to refine successfully than the "natural" products. The loss is likely to be large and the resultant product may be colored. Ellis and Wells[19] report that by hydrogenation of "cracked" gasoline was secured a water-white product, but they do not give the specific method used nor the constants for the gasoline before and after treatment.

Brooks and Humphrey[20] recommend that the unrefined product be treated with not more than 6 per cent of 85 to 90 per cent sulphuric acid, followed by *redistillation.*

*Experiments Undertaken by the Author.* In order to determine practical refining conditions for "cracked" distillates, as far as *color, odor,* and *stability* are concerned, the writer and his students have carried out some experiments upon "cracked" distillates from commercial pressure stills. Constants on the unrefined pressure gasoline distillate and the natural gasoline distillate, both from the same crude petroleum mixture, were determined as follows:

|  | Crude Straight-Run Gasoline | Crude Pressure Still Gasoline |
| --- | --- | --- |
| Gravity, deg. Bé. | 57.7 | 54.3 |
| Refractive index | 1.4142 | 1.4252 |
| Heat of reaction with sulphuric acid (1.84 sp. gr.), deg. fahr. | 1 | 11 |
| Loss of sulphuric acid (room temp.), per cent | 9 | 16 |
| Loss of sulphuric acid (32 deg. fahr.), per cent | 5 | 8 |

The experimental work undertaken upon the refining of cracked distillates has developed in such an interesting way that the work may be considered as only begun. The best results from the standpoint of both color and odor were secured by direct treatment of the pressure still gasoline with concentrated (1.84) acid at 32 deg. fahr. followed by redistillation. The color of the product did not deepen after the sample had been exposed to the light (sunlight part of the time) for approximately 30 days.

Other most interesting results have been secured but it is not advisable to discuss them at this time until further experiment and verification.

## A Census of Special Libraries

Because of the fact that there does not exist an adequate directory of special libraries, the Special Libraries Association is now actively engaged in determining the number of such institutions throughout the country. During the war, when army camps and military centers were being furnished with libraries for research and educational work, the lack of a satisfactory list of information centers was keenly felt. There were a large number of occasions when army men could have used a directory of institutions or corporations having special libraries if such a publication had been in existence. It is to fill such a want that the present census is being taken.

A special library has been defined as a good working collection of information upon a specific subject or field of activity; it may consist of general or even limited material serving the interests of a special clientele, and preferably be in charge of a specialist trained in the use and application of the particular material. Libraries coming within the above qualifications are earnestly requested by the Special Libraries Association to supply the following information:

1  Name of institution or company
2  Name by which library is known
3  Name of librarian or custodian
4  Can it be classified as any of the following: financial; business; legal; engineering or technical; institutional; municipal; reference; agricultural?
5  Does it serve a special clientele?
6  Would your librarian be willing to assist other special libraries to a reasonable extent?

These data should be sent to the Chairman, Library Census Committee, care of General Electric Company, Schenectady, N. Y.

---

"Where steam is used with the oil vapors, it is claimed that this condition does not obtain and that a sweeter product is obtained.
"See Ellis and Wells, Jour. Ind. Eng. Chem. 7 (1915), 1029.
"Brooks and Humphrey, Jour. Am. Chem. Soc. 40 (1918), 822.

# MECHANICAL ENGINEERING
## THE JOURNAL OF THE AMERICAN SOCIETY OF MECHANICAL ENGINEERS

| Volume 42 | August 1920 | Number 8 |

## Metal Castings[1]

### A Group of Papers on the Usefulness and Limitations of Various Types of Castings Contributed by the A. S. M. E. Sub-Committee on Foundry Practice

MECHANICAL engineers are called upon to decide or advise on the selection of the proper kind of casting for a given piece of construction and should therefore have available reliable information as to the present state of the art in foundry practice. The following papers are intended to outline the special properties of the particular kinds of casting of which they treat and to explain where they are to be preferred to other forms of cast metal. They also review what is actually being done so that engineers and designers may know what they should be able to secure from first-class commercial foundries at the present time. In the preparation of the papers the endeavor has been to avoid giving information on the art of making castings, the specifications or methods of testing castings or the production of the metals from which the castings are made.

construction. In May, 1917, at the request of Messrs. Manley and Cardullo of the Curtiss Aeroplane Company, an investigation was made of the properties of aluminum castings then being used for aeroplane crankcases and other parts. It was found that the quality was very poor, the physical properties being neither uniform nor sufficiently high.

Immediate steps were taken to introduce technical control in the foundries to insure against wrong metal compositions and abuse of metal during the melting operations, and also to insure the maintenance of the proper pouring temperature. The results were very satisfactory, as is evidenced by the production record of Liberty engine castings in which a minimum tensile strength of 18,000 lb. per sq. in. was specified and the rejection for failure to pass specifications was practically nil. Even in such

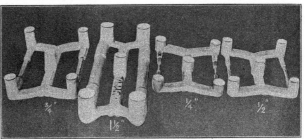

Fig. 1 METHOD OF CASTING ROUND TEST BARS OF VARIOUS DIAMETERS

## ALUMINUM CASTINGS
### By ZAY JEFFRIES,[2] CLEVELAND, OHIO

THE aluminum-castings industry can hardly be said to be more than fifteen years old. It is true that castings were made prior to 1905, but in such small quantities as to be insignificant in comparison with the present magnitude of the industry. The annual production of aluminum at the present time is about three hundred million pounds. The percentage of this used in castings is not definitely known, but it is estimated at about one-half the world's production.

The aluminum-casting art has developed largely by rule-of-thumb methods and even as recently as 1917 the quality of castings was so uncertain that aeroplane manufacturers and aviators had to be reassured of the reliability of the metal for aeroplane

[1] Abstracts of papers presented at the Castings Session of the Spring Meeting, St. Louis, May, 1920, of THE AMERICAN SOCIETY OF MECHANICAL ENGINEERS. All papers are subject to revision.
[2] Director of Research Laboratory, Aluminum Castings Company, Cleveland, Ohio.

highly stressed parts as those in the Liberty engine, no failures in service have come to the author's attention.

### ALUMINUM ALLOYS FOR CASTINGS

The principal aluminum alloy used for castings in the United States consists of about 92 per cent aluminum and 8 per cent copper and is generally known as Aluminum Company of America's No. 12 alloy or S. A. E. specification No. 30. This alloy when cast in a ½-in. test bar in green sand and tested without machining off the skin should give an average tensile strength of about 20,000 lb. per sq. in. and an average elongation of about 1.5 per cent in 2 in. A modification of this alloy having somewhat better physical properties is now finding considerable favor in castings for the automotive industry. This alloy has an analysis of 7.5 per cent copper; 1.5 per cent zinc; 1.2 per cent iron, and the remainder aluminum. The tensile strength of this alloy will average about 21,000 lb. per sq. in. and the elongation will be somewhat greater than that of No. 12 alloy.

S. A. E. Specification No. 32 is another alloy of aluminum and copper which finds considerable use, especially in parts which need to be water-tight. This alloy contains from 11 to 13.5 per cent copper, the remainder being commercial aluminum.[1] Its tensile strength when the test bar is cast in green sand about ½ in. in diameter and broken without machining off the skin should show an average of 20,500 lb. per sq. in. and the elongation will be practically nil. This alloy is used for carburetors, pumps, manifolds, radiators, engine cylinders, etc.

Another aluminum-copper alloy containing about 5 per cent copper is used for cast automobile bodies. This alloy has an average tensile strength of about 18,000 lb. per sq. in. and an average elongation of 3.0 per cent.

specific gravity of this alloy is 3.0 and its tensile strength when cast in sand should be over 25,000 lb. per sq. in. and the elongation should exceed 1 per cent. In England the test bars of this alloy are cast in chill molds and the minimum physical-property requirements are a tensile strength of 11 long tons (24,640 lb.) per sq. in. and an elongation of 4 per cent. This alloy is used in England as a substitute for our No. 12 alloy.

An alloy containing 10 per cent zinc and 2.5 per cent copper is put out in ingot form in England by the British Aluminum Company. This alloy should have a tensile strength of more than 22,000 lb. per sq. in. and an elongation of more than 2 per cent in 2 in.

Aluminum Manufactures, Inc., use an alloy containing 7 per

FIG. 2 GRAIN SIZE IN A ROUND TEST BAR OF 8 PER CENT COPPER ALLOY, 1¼ IN. IN DIAMETER, CAST IN GREEN SAND. MAGNIFICATION 50 DIAMETERS

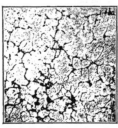

FIG. 3 GRAIN SIZE IN A ROUND TEST BAR OF 8 PER CENT COPPER ALLOY, 1½ IN. IN DIAMETER, CAST IN GREEN SAND. MAGNIFICATION 50 DIAMETERS

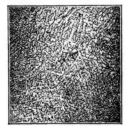

FIG. 4 MICROGRAPH OF ALUMINUM ALLOY CONTAINING 8 PER CENT COPPER, SHOWING GRAIN SIZE PRODUCED BY CASTING CYLINDER 1¼ IN. DIAM. IN GREEN IRON MOLD. MAGNIFICATION 50 DIAMETERS

FIG. 5 MICROGRAPH OF ALUMINUM ALLOY CONTAINING 8 PER CENT COPPER, SHOWING GRAIN SIZE PRODUCED BY CASTING CYLINDER 1¼ IN. DIAMETER IN GREEN SAND. MAGNIFICATION 50 DIAMETERS

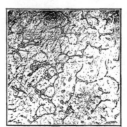

FIG. 6 SPECIMEN SHOWN IN FIG. 4 BUT AT HIGHER MAGNIFICATION TO SHOW STRUCTURE OF NETWORK SURROUNDING THE GRAINS. MAGNIFICATION 250 DIAMETERS

FIG. 7 SPECIMEN SHOWN IN FIG. 5. HIGHER MAGNIFICATION BRINGS OUT THE STRUCTURE OF THE NETWORK AROUND THE GRAINS. MAGNIFICATION 250 DIAMETERS

For pistons cast in permanent molds an alloy containing 10 per cent copper, 1.25 per cent iron, 0.25 per cent magnesium is extensively used. This alloy gives a good wearing surface both for the piston against the cylinder wall and for the bearing of the wristpin in the bosses. The pistons and camshaft bearings for the Liberty engine were made from this alloy, the former being cast in permanent molds.

While the aluminum-copper alloys have been used mostly in the United States, light alloys containing zinc have been used more extensively abroad. The most common alloy in use in England for sand castings contains 13.5 per cent zinc and 2.75 per cent copper, the remainder being commercial aluminum. The

cent zinc, 2.75 per cent copper and 1.5 per cent iron. This alloy has an average tensile strength of about 27,500 lb. per sq. in. when cast in green sand and an average elongation of about 4.5 per cent.

Zinc is added to aluminum up to 33 per cent. As the zinc increases up to this amount the tensile strength increases and the elongation decreases. The 33 per cent zinc alloy has been used considerably in the past but it is too brittle to find extensive employment in the industries in competition with more ductile alloys which also have lower specific gravities.

At the Westinghouse Electric and Manufacturing Company an alloy composed of 95 per cent aluminum and 5 per cent magnesium is used for starting-motor frames and other automobile work where extreme lightness and good machining qualities are

[1] Commercial aluminum always contains iron and silicon in amounts of 0.25 per cent and upward according to the grade of ingot.

essential. The specific gravity of this alloy is 2.47, the average tensile strength about 27,000 lb. per sq. in., and the elongation about 3 per cent.

At the United States Navy Yard, Washington, D. C., an alloy containing 2 per cent copper and 1 per cent manganese is used for all aluminum-casting work. It is used for parts for gun sights, optical instruments, loading-tray sights for naval mounts, mine-laboratory experimental work, parts for nitrating machinery for use in the manufacture of powder, etc. This alloy has a tensile strength of about 18,000 lb. per sq. in. and an elongation of about 8 per cent. Its proportional limit is very low, often not exceeding 2000 lb. per sq. in.

An alloy containing 98 per cent commercial aluminum and 2 per cent manganese is also used for certain aluminum castings which should resist the corrosive action of chemicals used in rubber molding.

### PHYSICAL PROPERTIES OF ALUMINUM AND ITS ALLOYS

*Hardness.* The Brinell hardness of pure aluminum is about 25 and that of the hardest aluminum alloy is more than 120. Where hardness is the controlling factor, any value between 25 and 120 can be produced.

*Coefficient of Expansion.* The coefficient of expansion of pure aluminum is about 0.000025 per deg. cent. The coefficient of expansion of the alloys varies between 0.000022 and 0.000027. Alloys containing about 5 per cent magnesium have the higher coefficient of expansion and the high-copper alloys show the lower values.

*Heat Conductivity.* The heat conductivity of pure aluminum is about 0.50 $K$ or about 48 per cent of the conductivity of copper at ordinary temperatures, and varies but slightly with temperature. The heat conductivity of aluminum alloys is less than that of the pure metal, varying between 25 per cent and 48 per cent of the conductivity of copper, the general rule being that the greater the percentage of alloying elements, the lower the heat conductivity.

*Electrical Conductivity.* The electric resistivity of pure aluminum is about 2.94 microhms per cubic centimeter. The electric resistivity of the alloys is higher than that of pure aluminum, ranging up to 5.70 microhms per cu. cm.

*Melting Points.* The melting point of pure aluminum is 658 deg. cent. The lowest melting points of the useful alloys range around 580 deg. cent.

*Crystallization Shrinkage and Allowance for Pattern Shrinkage.* The crystallization shrinkage of pure aluminum is about 5 per cent by volume, and a further 5 per cent decrease in volume takes place on cooling from the melting point to room temperature. The crystallization shrinkage is somewhat less in the alloys. The pattern shrinkage allowed for all aluminum alloys is 5/32 in. per ft.

*Specific Gravity.* The specific gravity of pure aluminum is about 2.7 and that of the alloys is not far from that calculated from the specific gravities of the various constituents. An alloy containing 95 per cent aluminum and 5 per cent magnesium has a specific gravity of about 2.47 and one containing 67 per cent aluminum and 33 per cent zine a specific gravity of about 3.3. These represent minimum and maximum values for the usable aluminum-base alloys.

*Modulus of Elasticity.* The modulus of elasticity of aluminum is about 10,000,000 lb. per sq. in. Values as low as 8,500,000 and as high as 11,000,000 have been obtained.

### GENERAL OBSERVATIONS ON ALUMINUM CASTINGS

*Classes.* Aluminum castings may be divided into three classes according to the method of manufacture, namely,

1 Sand Castings
2 Permanent Mold Castings
3 Pressure Die Castings.

Sand castings are so named because they are cast in either green or dry sand. The method of molding is very similar to that used for other metals except that the gates, risers and chills must be used to suit the particular properties of aluminum and its alloys. Such parts as crankcases for gas engines, oilpans, manifolds, steering-wheel spiders, transmission housings, differential carriers, etc., are usually cast in sand.

Permanent-mold castings are made in iron or other metal molds and consequently have a greater degree of chill than the sand castings. The physical properties of the various alloys vary in proportion to the degree of chill. Molds for the permanent-mold castings are provided with ample gates and risers for feeding, consequently the permanent-mold casting is non-porous as distinguished from the pressure die casting, which is always porous in the interior. As a general rule the permanent-mold process is limited to small parts with simple shapes.

Pressure die castings are considered in a paper on Die Castings presented at this session by Mr. Charles Pack.

*Maximum and Minimum Weight.* Aluminum sand castings have been regularly made weighing as high as 480 lb. each, but one of 150 to 200 lb. is usually considered a large casting. The

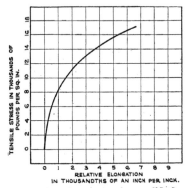

FIG. 8   STRESS-STRAIN DIAGRAM OF ALLOY CONTAINING 92 PER CENT ALUMINUM AND 8 PER CENT COPPER

minimum weight may be taken as 1 oz., but neither this weight nor 480 lb. should be considered as an extreme limit.

In the permanent-mold process the maximum weight is about 20 lb. and the minimum weight about 1 oz. Permanent-mold castings have been made weighing as high as 150 lb. but these large castings should be considered as special.

*Section Thickness.* On gas-engine oilpans the walls should be 3/16 in. or more in thickness and on crankcases a wall thickness of 7/32 in. or greater is common practice. Fillets should have large radii, especially where heavy sections join thin sections. The best thickness on pipes, manifolds, etc., is ⅛ to 5/32 in. The minimum thickness on permanent-mold castings is from 3/32 in. to ⅛ in.

*Finish.* On large sand castings ⅛ in. should be allowed for finish and on bench work ¹⁄₁₆ in. A finish of 1/32 in. should be allowed for disk grind.

The finish on permanent-mold castings up to 3 in. in diameter, for example, pistons, should be about 0.075 in. on the diameter, and the finish on the piston head should be 0.045 in. The finish on permanent-mold pistons over 3 in. in diameter should be about 0.093 in.

*Tolerances on Dimensions and Weights.* In sand castings a tolerance of 1/32 in. in thickness should be allowed and in permanent-mold castings the tolerances are plus or minus 0.010 in. The weight tolerances on sand castings using metal-pattern equipment are plus or minus 3 per cent and with wood patterns plus or minus 5 per cent. The weight tolerances on permanent-mold castings are plus or minus 2 per cent.

*Machining.* For roughing cuts speeds of 500 to 700 ft. per min. and a feed of 1/16 in. to 5/32 in. are recommended, according to the nature of the work. No lubrication is recommended for roughing. For extra fine finish a speed of about 125 ft. per min. is recommended, but speeds up to 600 ft. can be used and satisfactory finish obtained. For turning, the finishing feeds should be from 0.01 in. to 0.125 in., and for planing, up to 0.25 in. A good lubrication mixture for finishing is 70 parts kerosene and 30 parts lard oil.

Aluminum alloys take a fine finish by grinding. Good results have been obtained using a No. 40 grain crystolon wheel and a mixture of 10 per cent lard oil and 90 per cent water for lubrication. The grains of the crystolon wheel even when dislodged do not stick in the metal.

### Heat Treatment of Aluminum-Alloy Castings

Aluminum castings, especially those containing copper, may have their volumes permanently changed by heat treatment. Also the hardness may be changed. These facts are taken into consideration in connection with the heat treatment of aluminum-alloy parts such as pistons which will have to operate at an elevated temperature. The pistons are given a heat treatment which will cause permanent growth to such an extent that no additional change of volume will take place during normal running.

### Effect of Thickness of Section on Physical Properties of Aluminum Alloys

In general, the tensile strength and elongation of the aluminum alloys decrease as the thickness of the section increases. In the case of round test bars cast as shown in Fig. 1, the decrease in tensile strength in an alloy containing 8 per cent copper and 92 per cent aluminum is from approximately 25,000 lb. per sq. in. and 3 per cent elongation in the ¼-in. test bar, to 14,000 lb. per sq. in. and less than 1 per cent elongation in the 1½-in. test bar, with proportional intermediate values for the ½-in. and ¾-in. bars.

It appears that this alloy on account of its low elongation does not show up so well when tested in a large piece as would be indicated from the results obtained from small test bars machined from the larger pieces. For example, small test bars cut from the 1½-in. bars averaged about 19,000 lb. per sq. in. and the elongation was more than 1 per cent, whereas the large bars tested 14,000 lb. per sq. in. and less than 1 per cent elongation. The tensile strength of one of the zinc alloys described earlier decreases at an approximately linear rate from about 30,500 lb. per sq. in. in a ¼-in. test bar to 19,500 lb. in a 1½-in. test bar. Small bars cut from the 1½-in. bar tested about the same as the 1½-in. bar itself. The elongation of the zinc alloy decreases also with increase in thickness of section. It is thought that the greater ductility of this alloy is responsible for the similarity of results between the small test bars cut from the large bars and the large bars themselves.

The lower strength and ductility of the large sections are due to a combination of variation in grain size and soundness of the casting. The smaller sections are as a rule better fed during solidification and since they solidify more rapidly than the thicker sections, they are also finer-grained. Figs. 2 and 3 show the variation in grain size in the 8 per cent copper alloy when cast in ¼-in. and 1½-in. bars.

### Effect of Rate of Chill on the Physical Properties of Aluminum Alloys

As a general rule, the more quickly aluminum alloys solidify the smaller the grain size and the higher the tensile strength and elongation. No. 12 alloy, for example, when cast in large sections in sand may have a tensile strength as low as 14,000 lb. per sq. in. and an elongation of less than 1 per cent. When cast in small sections in a chill mold its tensile strength may be 31,000 lb. per sq. in. and its elongation as high as 6 per cent. Figs. 4, 5, 6 and 7 show the effect of degree of chill on the structure of an alloy containing 92 per cent aluminum and 8 per cent copper.

### Fatigue Resistance of Aluminum Alloys

In sand-cast bars of No. 12 alloy the fatigue resistance as measured on the White-Souther machine is about 14,000 lb. per sq. in. maximum stress for 500,000 reversals and about 8500 lb. per sq. in. maximum stress for 16,000,000 reversals. The proportional limit of this alloy is about 5000 lb. per sq. in. The stress-strain curve departs very gradually from a straight line after the proportional limit is exceeded, as is shown in Fig. 8. It is believed that the proportional limit of the aluminum alloys does not represent a breakdown of the main metallographic constituent as in a single-component metal like wrought iron, but that it simply breaks some of the unfavorably-situated, brittle aluminum compounds and thus causes slight permanent set. The fatigue life of the aluminum alloys is quite unusual, inasmuch as the safe limit seems to be equal to or higher than the proportional limit as measured by the extensometer. Chill-cast alloys are even better than the sand-cast alloys for fatigue resistance.

In a very comprehensive paper entitled Aluminum Alloys for Aeroplane Engines, by Prof. F. C. Lea, published by the Royal Aeronautical Society in 1919, fatigue tests on various aluminum alloys are reported, together with stress-strain diagrams. These tests show a minimum fatigue range in the Wohler test of 12,000 lb. per sq. in. (— 6000 to + 6000) for 12,000,000 reversals on an alloy containing 12 per cent copper. The proportional limit of this alloy was 5000 lb. per sq. in., although the amount of permanent set up to 6000 lb. was very slight.

### General Considerations Regarding Use of Aluminum Castings

As an engineering metal aluminum is in a class by itself because of its low specific gravity. Its use is determined by many factors, among which are the following:

*a* Where lightness is of prime importance. Aluminum in sheet form has about seven times the stiffness of steel for equal weights. For a given rigidity in any metal structure aluminum will weigh less than any other metal used in engineering. For such parts, therefore, as aeroplane engine crankcases, camshaft housings, oil-pans, etc., where lightness and rigidity are of prime importance, aluminum alloys find extensive use.

*b* In the manufacture of gas-engine crankcases it is found that the aluminum castings can be machined about three times as rapidly as cast-iron parts, and consequently the first cost of plant installation and labor cost of machining is greatly decreased by the substitution of aluminum for cast iron. The lightness of the crankcase is the chief advantage because the weight of the engine per horsepower is reduced.

*c* Where the lightness of reciprocating parts, like pistons, is of great moment, such as in high-speed gas engines, the use of aluminum is very beneficial. The aluminum piston also has the advantage of high heat conductivity, which is sometimes of greater importance in a gas engine than reduction in weight.

*d* For reducing unsprung weight in motor vehicles, aluminum is particularly valuable. This includes such parts as differential carriers, rear-axle housings, brake shoes, hub caps, wheels, etc.

*e* Aluminum alloys, especially with moderate pressures, function nicely as bearings against hardened steel. As an example, the Liberty Engine camshaft was run in aluminum bearings, as were also the rocker arms. Aluminum alloys, however, do not bear well against soft steel.

*f* For equal volumes aluminum is cheaper than brass or bronze. Aluminum alloys find extensive use as a substitute for brass and bronze castings where freedom from atmospheric corrosion is of prime importance and where the strength requirements are not too severe.

### Selection of Alloy

For aluminum castings not highly stressed, No. 12 alloy should be used. Where freedom from leaks is the main requirement, the 12 per cent copper alloy is good and the more-ductile zinc alloy should be used for highly stressed parts. Present-day engineering materials are required to stand abuse rather than normal use. Ductility is essential if an alloy is to stand abuse, and this is the main reason for the use of the aluminum alloy containing 2.75 per cent copper, 1.5 per cent iron and 7 per cent zinc.

[The complete paper includes a brief discussion of the metalography of aluminum alloys, and of the aging of aluminum-alloy castings.—EDITOR.]

## MALLEABLE CASTINGS

### By ENRIQUE TOUCEDA[1], ALBANY, N. Y.

IN general, the selection of the proper kind of a ferrous casting for use in a given piece of construction is dependent upon a number of considerations, chief among which may be mentioned the following:

a Ability to successfully withstand the abuse to which it will be subjected in service

b High static and dynamic strength, which implies minimum sections for the strains involved, this in turn implying low cost for raw castings, as such are not only less costly in direct proportion to their strength, but their appearance is enhanced thereby

c Ease of machining, accompanied by a good surface on faced and turned parts, and freedom from unsoundness; which implies beauty of finished castings and low cost due to less loss from defective castings

d Smoothness of surface and trueness to pattern.

In discussing the foregoing a·fact should be remembered that is frequently overlooked, viz., that the gray-iron, the malleable-iron and the steel casting have each a legitimate field of their own quite sharply defined by virtue of certain peculiarities possessed by each particular product. If a bedplate is desired for a certain large and ponderous machine, so dimensioned as to possess sufficient mass to enable it to absorb vibration, it would be unwise to specify that it be made of malleable cast iron. The malleable-iron founders are not at the present time equipped to cast and anneal castings of such size, and as cheapness and weight—obviously not high strength per square inch—are the dominant requirements in this case, the malleable-iron casting should not be specified. The same remarks hold good in the case of the steel casting, except of course that it is practical to make such castings as large as any made of gray iron.

### GRAY-IRON CASTINGS vs. MALLEABLE-IRON CASTINGS

The truly legitimate field of gray iron, in the writer's opinion, should and will in time be restricted to the production of that size of casting that cannot successfully be produced in malleable iron and is too costly to produce in steel, while it is peculiarly well fitted for such castings as are designed to sustain static compression, for which purpose the gray-iron casting is without question superior to either of the other two. The average ultimate tensile strength of the gray-iron casting as measured by the test bar does not exceed 20,000 lb. per sq. in., while it possesses practically no ductility. The writer can state with positiveness that in the case of at least sixty of the malleable-iron foundries the average ultimate strength and elongation of their product, as measured by the standard test bar, are 51,000 lb. and 12.50 per cent, respectively.

It can safely be stated that if a gray-iron and a malleable-iron casting are designed to perform the same function, the design of each being based strictly upon its physical properties, in view of the great difference in their relative strengths, coupled with the fact that a larger safety factor must be used in the case of the former due to its inherent brittleness, the latter can safely be made more than one-third lighter. Inasmuch as this difference in weight will in large measure counterbalance the difference in cost per pound with the substitution of a ductile casting for one that is brittle and is actually more easy to machine, the writer believes that when this fact is fully under-

stood and its significance appreciated, many castings now made of gray iron will be replaced by malleable, with the result that the gray-iron-casting field will of necessity be narrowed to this extent. It is therefore logical to conclude, as between these two products, that if the part can be successfully cast of malleable iron, the following statements hold true:

a For equal physical properties it is as cheap per raw casting

b It is as cheap per machined casting

c It will have a more pleasing appearance when assembled

d It will be lighter in weight

e Transportation charges will be less

f The weight per square foot on the floor will be less

g Breakage during transportation will be eliminated.

### STEEL CASTINGS vs. MALLEABLE-IRON CASTINGS

If the same line of reasoning be used as between the malleable iron and the steel casting, it can be stated at the start, that neither can successfully trespass in certain directions in the field of the other, for each, as will be explained, has certain limitations that control the situation. There is hardly a limit to the size or weight in which the steel castings can be made, while as previously pointed out there is a limit in the case of malleable iron. On the other hand, there are castings that can be very nicely made of malleable iron that cannot well be made of steel, using the same pattern.

The American Society for Testing Materials standard specifications for steel castings deal with two classes: Class A, for which chemical but no physical requirements are specified and which need not be annealed unless so specified; and Class B, which is divided into three grades, hard, medium, and soft, it being specified that all of these grades shall be properly annealed, and also conform to certain chemical and physical requirements. While the ultimate strength of both Class A castings and the soft grade of Class B castings may average higher than that of malleable iron, their yield point will average lower by some 4000 lb. Consequently for equal stiffness the malleable-iron casting has an advantage in the case of equal sections.

Inasmuch as the A. S. T. M. specification for Class A castings states that they need not be annealed unless so specified, it is to be expected that the majority of these castings do not receive that treatment. This means the presence of internal strains in such castings, minimized to a certain extent, however, if baked molds are used. The writer knows of quite a number of large steel-casting companies that turn out a large daily tonnage of castings that are never annealed and assumes that in such cases they are made in accordance with specifications other than those of the A. S. T. M., and after considerable inquiry he has come to the conclusion that 70 per cent of the steel castings produced in this country are not annealed.

It happens that the heat treatment given the malleable-iron casting during the period in which the white iron is being converted into the malleable product, is carried on at a temperature just slightly in excess of the critical range, a temperature which yields practically the finest crystallization that the metal can attain, while the cooling from this temperature is effected at the rate of about ten degrees per hour. In consequence of this the malleable-iron casting is not only fine-grained but is invariably free from internal strains. On the other hand, not only are unannealed steel castings in a state of internal strain, but they possess a coarsely crystalline structure corresponding to the temperature of solidification. In view of all this, the writer believes that if a given casting be made of malleable iron it will stand more abuse in service than will a similar unannealed steel casting.

If the cost of the raw casting is the dominant factor, then unquestionably preference must be given to the malleable-iron casting, as the difference in cost is considerable. If based upon the cost of the finished casting, the difference between the cost of the two will be much greater, due to the greater ease with which the malleable casting can be machined, which difference will vary directly in proportion to the amount and character of the machining operations.

[1] Consulting Engineer.

The malleable-iron casting is characteristically free from blow-holes, while this is acknowledged to be one of the shortcomings of the steel casting; consequently, there will be less scrap in the case of the former. If rust-resisting properties are a consideration, as in the ease of castings for refrigerator cars, the malleable-iron casting should be given preference over steel.

### PROPERTIES AND LIMITATIONS OF MALLEABLE-IRON CASTINGS

In commercial practice the engineer who has placed his contract with a reliable foundry can depend upon a uniformity of product that will rarely have a yield point lower than 31,000 lb. per sq. in. and frequently as high as 33,000, an ultimate strength less than 50,000 lb., or an elongation less than 10 per cent. He can quite safely depend upon the integrity of the casting, for the reason that the founder has finally learned through costly experience that freedom from shrinkage depends not only upon correct gate emplacement, but more particularly upon the use of large shrink heads, so located as to eliminate such defects. Castings have been produced commercially as long as 5 ft., with sections at some parts as thick as 3 in. In regular practice castings varying in weight from 300 to 500 lb. are made daily, while it would be difficult to place a limit on how small they can be run.

The wearing properties of this metal correspond to what can be expected of wrought iron or dead soft steel, while as is well known its machining properties can hardly be excelled. As far as the ordinary uses of malleable cast iron are concerned, the writer feels quite certain that no beneficial results can follow from its heat treatment. In a limited number of cases, however, such as in the production of cheap hatchets, cleavers, and axes, malleable iron can be made file-hard by heating to just above the critical range and quenching.

### COMPARATIVE COSTS OF GRAY-IRON, STEEL AND MALLEABLE-IRON CASTINGS

A general statement relative to the comparative cost of manufacture of the three products mentioned must obviously involve the consideration of many different cases and conditions. If it were not for the fact that in the case of the manufacture of malleable iron the annealing (which lasts from six to seven days and is a rather costly part of the process) must be taken into account, an approximate estimate could be made by figuring out the cost of the metal in the ladle for each of the three products; but the varying cost of molding in each case and the fact that the sprue can be easily knocked off from both gray and malleable-iron castings while it must be machined or cut off from the steel casting, render it rather difficult to make a comparison of the costs of the raw castings. In general, however, gray-iron castings run heavier than do those made of malleable iron. Consequently there is less sprue and therefore less remelt. Also there is less variation between the composition of the same and that of the charge and as a consequence a much larger percentage of scrap can be used in the mixture when making gray-iron than is possible in the case of the malleable castings. Assuming the highest fuel ratio in each case, 10 of iron to 1 of coke for the cupola, and 3 of iron to 1 of soft coal in the air furnace, it will be seen that the cost for fuel is less in the former than in the latter case. The same holds true in connection with furnace maintenance. Aside from cleaning and clipping, the gray-iron casting is finished when the metal fills the mold, while the white-iron castings must be cleaned, taken to the annealing room, packed in saggars, and then charged into the annealing oven and heat-treated for a period in most instances of seven days. The saggars must then be removed from the annealing oven, the castings again cleaned and sorted, to the cost of which must be added that of the fuel used for heat treatment, oven maintenance, supervision, and overhead. It can therefore be very easily seen that there must be considerable difference between the cost of production of these two products. The writer believes that on an average it will easily cost 1 cent a pound more to produce malleable-iron than it will the gray-iron castings, and at least 30 per cent more to produce steel castings than malleable.

# STEEL CASTINGS

BY JOHN H. HALL,[1] HIGH BRIDGE, N. J.

THE more observing users of castings, having learned during the war that it is possible to produce a far better steel casting than had formerly been supplied, have in some cases profited by their recently acquired knowledge, and are demanding and securing the better product. The testimony of some of the very best makers of small steel castings, however, is to the effect that the buyers of their wares still adhere to their old practice of requiring first of all a freely machining casting and will not tolerate improvement in the quality of the steel if it interferes in the slightest with ease of machining. This attitude is unfortunate and really short-sighted, as in the long run it can have but one result. Those who maintain it are sure, sooner or later, to suffer from the competition of concerns making and machining their own castings, and willing to sacrifice a little to better quality. It is true at the present time that the very best steel castings made are used for the most part by those who make them, and for certain classes of product that are subjected to very severe service. Castings are now being made that show all-around physical properties nearly twice as good as those of the ordinary commercial steel casting, but the average machine-shop proprietor absolutely cannot be persuaded to use them, simply because they do not machine quite so freely as those he has been accustomed to in the past.

### RANGE OF WEIGHT OF STEEL CASTINGS PRACTICALLY UNLIMITED

There are practically no limitations to the maximum and minimum weights of steel castings that can be produced commercially. The foundryman using a small bessemer converter, a crucible furnace or a small electric furnace can and does make castings with sections as light as $\frac{3}{8}$ in., and in some cases even $\frac{1}{4}$ in., provided their size and intricacy is not such as to make this impossible. On the other hand, the open-hearth furnace will pour castings as heavy as are ever called for, some poured in the last few years weighing 150 tons or even more, and with maximum sections of several feet. In fact, the steel foundryman is not so much concerned with the actual thickness of any particular section as with the combination of thick and thin sections in one and the same casting. When absolutely necessary, he can execute a casting of great intricacy in which thick and thin sections occur, often in exasperating combinations; but it is an undoubted fact that if the designer of castings would take counsel more often with the foundryman, he would obtain advice that would help him to secure a far better casting, and one that would more successfully meet the service it is called upon to stand.

### PHYSICAL PROPERTIES OF STEEL CASTINGS

The physical properties of the average commercial steel casting depend largely upon whether or not the casting has been annealed. Perhaps 50 per cent of the steel castings made in this country are shipped without annealing. This statement is not made on actual knowledge and is consequently subject to criticism as too high or too low, depending upon whether the critic is or is not a steel foundryman. The American Society for Testing Materials specifications for steel castings call for two classes, A and B. Class A castings "need not be annealed unless so specified." The carbon of these castings is specified not over 0.30 per cent, and the phosphorus not over 0.06 per cent (at present 0.07 per cent is allowed). Such castings will test about as follows:

| | |
|---|---|
| Tensile strength, lb. per sq. in. | 60,000 |
| Yield point, lb. per sq. in. | 25,000 |
| Elongation in 2 in., per cent | 10 to 15 |
| Reduction of area, per cent | 15 to 25 |

The A. S. T. M. specifications do not call for tensile or bending tests on Class A castings. These specifications are intended to cover only such castings as are not subject to very severe service, and in which the chief characteristics desired are soundness and comparatively easy machining.

[1] Taylor & Wharton Steel Co.

The writer is aware that some engineers are willing to accept unannealed steel castings for quite severe service, even for parts of locomotives and cars. The properties of the metal, as shown by the ordinary static tensile and bending tests, are considered good enough to justify the use of the unannealed steel. Of course the only steel so accepted is quite low in carbon, and accordingly fairly tough, at least in static tests.

From a fairly long acquaintance with cast steel, however, the writer is forced to state that in his opinion this policy cannot be justified. Our knowledge of endurance or "fatigue" tests teaches us that a coarsely crystallized metal such as unannealed cast steel cannot have a really high resistance to severe alternating stresses, and nothing like as high as the same metal will have after the coarse structure has been broken up by annealing and replaced by a finely crystalline condition.

It is when we come to consider resistance to sudden heavy shocks, however, that the danger of using unannealed cast steel for severe service strikes us most forcibly. Tests made by the writer have shown that under heavy shock cast steel even as low in carbon as 0.10 per cent breaks almost without bending, and with very little absorption of energy. In other words, the metal is truly brittle under impact test. After ordinary annealing the resistance to shock is greater, and the test piece bends through a greater angle before rupture. If the steel under test has been heat-treated by rapidly cooling and reheating, the shock resistance and angle of bend are increased to an even greater extent.

These tests were made on a Frémont testing machine; but the writer has made many tests on full-sized castings that amply justify him in his opinion that the use of unannealed cast steel for service calling for ability to resist sudden heavy shocks cannot be justified unless the sections of the piece can be made very heavy for the strength desired—in other words, unless weight per unit of strength can be disregarded.

The A. S. T. M. Class B castings must not exceed 0.05 per cent phosphorus or sulphur. (At present this has been temporarily increased to 0.06 per cent.) No other chemical specifications are given, the analysis other than phosphorus and sulphur being left to the judgment of the foundryman. Annealing, however, is specified. These castings are divided into three grades: Hard, Medium, and Soft, which must give physical properties as follows:

| | Hard | Medium | Soft |
|---|---|---|---|
| Tensile strength, lb. per sq. in. | 80,000 | 70,000 | 60,000 |
| Yield point, lb. per sq. in. (=0.45 T. S.) | 36,000 | 31,500 | 27,000 |
| Elongation in 2 in., per cent | 15 | 18 | 22 |
| Reduction of area, per cent | 20 | 25 | 30 |
| Cold bend around 1-in. pin, deg | .. | 90 | 120 |

Reliable foundries can meet these specifications with practically no trouble beyond the expense necessarily due to meeting the requirement that castings be kept together by heats as poured, and the cost of making the test specimens and carrying out the tests. Most specifications as now written call for so many test pieces that the cost of producing the castings is considerably increased. This is particularly true when it is specified that a test shall be attached to each and every casting over a certain weight. The writer has known cases in which this specification was literally insisted upon where as many as fifty almost identical tests were pulled from a heat of open-hearth steel. The A. S. T. M. specifies a test attached to every casting weighing over 500 lb. unless the sections are such that a satisfactory test piece cannot be attached to it. In that case, and for castings under 500 lb., two test bars may be cast to represent a heat, or "in the case of small or unimportant castings, a test to destruction on three castings from a lot," may be used. This test shall show the material to be "ductile, free from injurious defects, and suitable for the purpose intended." This provision for testing small castings to destruction in place of using test bars is an excellent one and should be used more than it is.

## ANNEALING AND HEAT TREATMENT

In order to meet specifications similar to those of the A. S. T. M., annealing is essential. The annealing practice followed is almost invariably such as to give the castings the greatest possible softness and ease of machining, and no heat treatment as

the term is usually interpreted is needed to meet these specifications if the steel is well made in the first place. Castings annealed in this manner should machine very freely or fairly freely, depending upon whether they are of "soft" or "hard" grade, as prescribed. If the first cut is deep enough to get well under the scale left by annealing and if the casting is free from adhering sand, it should give no trouble in machining.

Heat treatment as usually understood, that is, annealing the castings at a proper temperature, cooling more or less rapidly, and then heating a second time, produces castings with very much better physical properties than those of plain annealed cast steel, and at the same time produces a somewhat harder and much-finer-grained steel which has far better wearing qualities and much better resistance to fatigue than is possible in the case of castings which are simply annealed. In extreme cases, by using heat treatment and special analysis, it is possible to produce castings that will show as good tensile strength as "Class B Hard" of the A. S. T. M. specifications, and equal or exceed the elongation, reduction of area and bend of "Class B Soft" castings. The yield point especially is raised by heat treatment, and the wearing properties are improved. The resistance to fatigue is also greatly increased, and, most important of all, the power to resist sudden shock is enormously improved.

The cost of such heat treatment, if substituted for plain annealing when the castings are made, need not be higher than that of plain annealing, provided the foundry is properly equipped to carry out heat treatment. It must be confessed, however, that the average foundry today is not so equipped and in many cases does not have on its staff a man who thoroughly understands heat treatment. In the writer's opinion this state of affairs is due as much to the buyer of castings as to the maker, because the buyer as a rule is not at all interested in the superiority of heat-treated or special-analysis steel castings, as soon as he finds that such castings machine a little less freely than the common variety. The writer has had violent protests from customers because some of his castings gave a long, curling chip in the lathe, as a forging does, instead of the usual crumbly chip of the common steel casting, and the fact that the physical properties of the castings were about as follows:

| | |
|---|---|
| Tensile strength, lb. per sq. in. | 85,000 |
| Yield point, lb. per sq. in. | 55,000 |
| Elongation in 2 in., per cent | 25 |
| Reduction of area, per cent | 50 |
| Cold bend around 1-in. pin, deg | 180 |

did not interest the customer in the least; and yet he had come to the foundry in which the writer was employed, and asked particularly for a casting to resist extremely severe fatigue and shock!

### COST OF STEEL CASTINGS

The cost of steel castings, of course, varies with the size, weight and intricacy of the pattern to be executed. At the present time heavy castings such as are suitable for production in open-hearth foundries can be obtained for from 8 cents per lb. up, depending upon the difficulty of the job, while small and difficult castings command a price of about 12 cents per lb. up. No upper limit for price can very well be given, as very small or difficult castings, on which a piece price is often quoted, may go as high as $1.00 per lb. or even higher; but the foregoing figures are close enough for comparative purposes.

### WHERE STEEL CASTINGS SHOULD BE SPECIFIED

Steel castings should be used for all cases where a forging is not suitable or is prohibitive in cost, and where the service to be rendered calls for strength and toughness combined. To enumerate the uses of steel castings would be virtually impossible. It is enough to say that cast iron is suitable for service in which the stresses are chiefly compressive, or where the stresses to be endured can be taken care of by increasing the size of the casting, regardless of the resulting weight. In fact, in many cases where cast iron is used, as in bed plates of engines and heavy machines, great weight for a given strength is even desirable. Where moderate strength and toughness, combined with fairly good resistance to wear and fatigue, are desired, a malleable-iron casting is plenty good enough. When, however, the service demands all the strength, toughness, and resistance to wear and fatigue that are obtainable in a cast metal, a steel casting should be specified; and if the service is unusually severe, the engineer should use the best cast steel he can buy, and should not allow himself to be influenced in his choice by the fact that the metal does not machine as freely as inferior cast steel or malleable iron.

## DIE CASTINGS

### By CHARLES PACK,[1] BROOKLYN, N. Y.

DIE CASTINGS may be defined as castings made by forcing molten metal, under pressure into a metallic mold or die. It is erroneous to assume that all die castings have similar properties, since it is apparent that the properties of the die casting will depend upon the nature of the alloy used. The die-casting process is best adapted to alloys of comparative low fusing points which may, for convenience, be divided into the following groups:

*Group A* Zinc Alloys, consisting essentially of zinc alloyed with tin, copper or aluminum

*Group B* Tin Alloys, consisting essentially of tin alloyed with copper, lead or antimony

*Group C* Lead Alloys, consisting essentially of lead alloyed with tin or antimony

*Group D* Aluminum Alloys, consisting essentially of aluminum alloyed with copper.

No general rules can be laid down governing the design and application of die castings since the art depends largely upon the skill of the designers, and quite frequently a part may be considered as impractical from a die-casting standpoint which, if measured by given standards, may be redesigned and die-cast very successfully. Nevertheless the writer will endeavor to outline briefly the general properties of the alloys used, their fields of application and their limitations.

### GROUP A—ZINC ALLOYS

*Typical Alloy:*

Zinc............87.5 per cent     Copper...........4.0 per cent
Tin.............8.0 per cent      Aluminum.........0.5 per cent

*Properties:*

Color ...............................................Silver white
Weight per cu. in. ...................................0.253 lb.
Melting point ......................................780 deg. fahr.
Initial fusing point .................................275 deg. fahr.
Tensile strength ......................16.100 lb. per sq. in.
Elongation ....................................2 per cent
Compressive strength .........................27.670 lbs.
Pressure required to shorten bar 1 in. diameter ..10 per cent
Hardness number (Brinell) .........................04.6

*Casting Limits:*

Maximum weight for casting ........................8 lb.
Minimum limit of wall thickness .................1/10 in.
Small castings ....................................1/16 in.
Variations from drawing dimensions per inch of diameter
   or length ...................................0.001 in.
Cast Threads, Minimum Number, External....24 per inch
   Internal ............depends on conditions, often cast
Cast Holes: Minimum diameter ................0.031 in.
   Depending largely upon the depth and thickness of casting.

Draft: Cores, 0.001 in. per inch of length or diameter.
   Side walls, 0.001 in. per inch of length.

*General Design.* Sections of castings should be as uniform as possible. Sharp corners should be avoided and fillets added wherever permissible. Undercuts in castings should be avoided wherever possible.

*General Remarks.* Alloys of this type are corroded by any alkaline or aqueous solutions of any salts. Castings may be polished to a high luster, but soon tarnish when exposed to ordinary atmospheric conditions. Castings made from this alloy may be readily plated with nickel, copper, brass, silver or gold. When properly plated such castings will retain their luster as well as those made from brass or bronze.

[1] Doehler Die Casting Co.

*Applications.* Castings made from this alloy should not be used for parts that are subjected to severe stress or sudden shock in service. They are used extensively for parts of phonographs, calculating machines, drinking-cup, cigar, candy, stamp and gum vending machines, magneto housing, automobile-body trimmings, pencil-sharpening machines, time-recording devices, stamp-affixing machines, and for many other devices of a kindred nature.

### GROUP B—TIN ALLOYS

*Typical Alloys:*

|       | Tin Per Cent | Copper Per Cent | Lead Per Cent | Antimony Per Cent |
|-------|------|------|------|------|
| No. 1.................... | 90 | 4.5 | 0 | 5.5 |
| No. 2.................... | 86 | 6 | 0 | 8 |
| No. 3.................... | 84 | 7 | 0 | 9 |
| No. 4.................... | 80 | 0 | 10 | 10 |
| No. 5.................... | 61.5 | 3 | 25 | 10.5 |

Alloy No. 1 is a so-called "genuine babbitt" metal and was used very extensively during the war for main-shaft and connecting-rod bearings on all American-made aeroplanes and motor trucks. No. 2 is somewhat harder and is used extensively for bearings of internal-combustion engines. No. 3 is somewhat harder than alloy No. 2 and is the S. A. E. standard for high-grade internal-combustion-engine bearings. No. 4 is in general use for light bearings on stationary motors. No. 5 is a bearing metal for light duty and is used on a large number of moderate-priced automobiles for mainshaft and connecting-rod bearings.

In addition to the five compositions mentioned, hundreds of similar alloys may be made having various specific properties. A study of these alloys, however, would prolong this paper unduly, and in the opinion of the writer, is beyond its scope. The die-casting process, it may be said, is applicable to any of the alloys of this group and it may be left with the engineer to specify the alloy best suited to his requirements.

*General Properties:*

Maximum fusing point ....................450 deg. fahr.
Weight per cu in. ................depends on lead content

*Casting Limits:*

Maximum weight for casting......................10 lb.
Limit in wall thickness..........................1/32 in.
Variations from drawing dimension per inch of
   diameter or length........................0.0005 in.
Cast Threads, minimum number:
   External ..................................27 per inch
   Internal ...........depends on conditions, often cast
Cast Holes: Minimum.................0.031 in. diameter,
   depending on depth and thickness of casting
Draft: Cores, 0.0005 in. per inch of length and diameter.
   Side walls, 0.001 in. per inch of length.

*Applications.* Tin alloys find their largest field of application in their use as bearings for internal-combustion engines. They are also used for parts of soda fountains, cream separators, milking machines, surgical apparatus, galvanometers parts, player pianos, etc., where a tensile strength of over 8000 lb. per sq. in. is not essential and where resistance to corrosion is of importance. They are not affected by water, weak acid or alkaline solutions, and when free from lead, are extensively used for food-container parts.

### GROUP C—LEAD ALLOYS

*Typical Alloys:*

|       | Lead Per Cent | Tin Per Cent | Antimony Per Cent |
|-------|------|------|------|
| No. 1........................ | 83 | 0 | 17 |
| No. 2........................ | 90 | 0 | 10 |
| No. 3........................ | 80 | 10 | 10 |
| No. 4........................ | 80 | 5 | 15 |

Alloy No. 1 is generally known as C. T. (Coffin Trimming) metal, due to its extensive use in the manufacture of coffin trimmings. This alloy is also a good bearing metal for light duty and is used for thrust washers and camshaft bearings on light internal-combustion engines. No. 2 is somewhat softer and more ductile than No. 1. No. 3 is used extensively for light bearing duty, being somewhat tougher and stronger than Nos. 1 and 2. No. 4 is somewhat harder than No. 3 but less ductile. Many similar alloys may be compounded, all of which may be die-cast readily.

*General Properties:*
    Weight per cu. in..................depends on lead content
    Maximum fusing point......................600 deg. fahr.
*Casting Limits:*
    Maximum weight for casting......................  15 lb.
    Minimum wall thickness........................1/32 in.
    Variation from drawing dimensions per inch of
        diameter or length............................0.001 in.
    Cast Threads, Minimum Number:
        External ..............................24 per inch
        Internal.............depends on conditions, often cast
    Cast Holes: Minimum diameter..................0.031 in.
        depending on depth and thickness of casting
        Draft: Cores, 0.0005 in. per inch of length and diameter
        Side walls, 0.001 in per inch.

*Applications.* Lead alloys may be used where a metal of non-corrosive properties is desired and where a tensile strength of not over 8000 lb. per sq. in. will suffice. They are used extensively for fire-extinguisher parts, low-pressure bearings, ornamental metalware, and many parts that come in contact with corrosive chemicals. They should not be used for parts that may come in contact with foods or that may be handled often in service, since the poisonous properties of lead and lead alloys are well known.

The main advantage of these alloys lies in their comparatively low cost, but their high specific gravity must be considered, some lead alloys having a specific gravity double that of the zinc alloys.

During the war lead alloys were used for all hand-grenade fuse parts and many millions of these parts were made. Lead-alloy die castings were also used for thermite grenades, offensive grenades, trench-mortar fuse plugs and many other parts where a non-corrosiveness was an essential requirement.

### Group D—Aluminum Alloys

*Typical Alloy:*
    Aluminum, 92 per cent; Copper, 8 per cent.
*Properties:*
    Color ............................................silver white
    Weight per cu. in.............................0.115 lb.
    Melting point ...........................1150 deg. fahr.
    Tensile strength .....................21,000 lb. per sq. in.
    Elongation ................................1.5 per cent
    Hardness number (Brinell)..........................60.5
*Casting Limits:*
    Maximum weight for castings.....................  5 lb.
    Minimum wall thickness .......................1/16 in.
    Variation from drawing dimensions per inch of
        diameter or length ........................0.0025 in.
    Cast Threads, External: Minimum number......20 per inch
        Threads are cast oversize 0.01 in. to be chased to size.
        Internal threads rarely cast.
    Cast Holes: Minimum diameter 0.093 in. and not deeper than
        1 in. Larger cores may be cast much deeper; smaller
        holes may be spotted to facilitate drilling
    Draft: Cores, 0.015 in. per inch of diameter or length.
        Side walls, 0.005 in.
    Cores of less than ¼ in. diameter to have 0.005 in. draft per
        in of length and diameter.

The composition described above is well known in the arts as No. 12 alloy and is used very extensively for automobile and aeroplane parts. By varying the copper content harder or softer alloys may be obtained, all of which may be die-cast successfully.

*Applications.* Aluminum die castings find wide employment in the manufacture of parts of automobiles, such as spark and throttle control sets, magneto parts, battery ignition and lighting systems, speedometers, etc. They are also used for parts of vacuum sweepers, phonographs, milking machines, vending machines, etc.

### Brass and Bronze Die Castings

Die castings made from various types of brasses and bronzes were put on the market as early as 1910, but have never been successful commercially. At the present time there is only one die-casting manufacturer producing brass die castings in any appreciable quantity.

It is a comparatively simple matter to produce a small quantity of sample brass die castings, but no material has yet been found for die-making purposes, which will withstand the continuous action of molten brass and at the same time retain its shape,

surface and size. The die casting of brass and bronze must be considered as in the experimental stage at the present time, with little or no immediate prospect of the solution of the problem.

### Developments Due to the War

The most important development in the art of die casting during the war was the perfection of the process of die casting aluminum and its alloys. A suitable steel was developed for making the dies for this process that would withstand the action of molten aluminum without cracking, a problem the solution of which was essential to the development of the industry.

The part that this development played in the winning of the war will be readily appreciated when it is stated that at the cessation of hostilities there were being produced about one million aluminum die castings daily in this country for parts of gas masks, machine guns, aeroplanes, motor trucks, motor ambulances, surgical instruments, canteens, field binoculars, and many other appliances of war.

### Comparative Cost of Die Castings

The cost of die castings cannot be computed on the pound basis since it depends on the design of the piece, the number and position of the cores, the quantity to be produced and certain other factors. For comparative purposes it may be stated that at the present time tin-alloy castings are the highest in cost, being followed by those of aluminum alloy, zinc alloy and lead alloy in the order named.

In considering the use of die castings it is well to bear in mind that on a pound basis die castings are far more expensive than iron sand castings where the machining cost is not considered. As the zinc alloys, whose properties are similar to cast iron, cost from $200 to $275 per ton in ingot form, it is apparent that the substitution of a die casting for an iron casting can only be considered when the machining cost is sufficient to compensate for the difference in cost of the raw materials.

# BRASS AND BRONZE CASTINGS

By CHRISTOPHER H. BIERBAUM,[1] BUFFALO, N. Y.

THE terms "brass" and "bronze" are not as definite in their meaning as might be desired. The term bronze was formerly applied to an alloy of copper and zinc. Usage, however, has changed and it is now generally applied to an alloy of copper and tin, while the term brass is used to denote an alloy of copper and zinc. This change of nomenclature has led to a great deal of confusion. Even today we speak of crankpin, crosshead, or engine brasses when we have bronzes in mind; that is, a copper-tin alloy with comparatively little, if any, zinc in its composition. In general, it may now be said that without a modifying adjective brass is an alloy of copper containing a large amount of zinc; and bronze is an alloy of copper containing a relatively large amount of tin. In this country no general standard of terminology for these alloys has as yet been adopted; the British Institute of Metals has adopted one, but theirs does not seem entirely satisfactory for our purpose. For the present, in this discussion, these alloys will be designated according to the prevailing commercial usage in this country; and other alloys will, as far as possible, be designated in the order of their preponderant constituents. Thus, the copper-tin-lead alloy is one containing, by weight, more copper than tin, and more tin than lead.

It is very generally understood that in designing machine parts which are to be made of cast materials a larger factor of safety is necessary than that required for rolled or forged material, although the compositions may be very similar or even identical. This in itself is not sufficient. The fact that the strength in any given casting varies in its different parts must also be taken into consideration, since it is impossible to make a casting with varying sections of uniform strength.

[1] Vice-President, Lumen Bearing Co.   Mem.Am.Soc.M.E.

## PRECAUTIONS TO BE TAKEN IN SPECIFYING BRASS AND BRONZE CASTINGS

In general, the engineer should be cautioned against specifying alloys upon which he has only such knowledge as has been derived from laboratory tests alone, since it is impossible to determine, by ordinary experiment, all of the economic conditions which enter into the founding and tooling of a casting and the requirements of a subsequent satisfactory use of an alloy for any given service. To illustrate: Years ago the favorite Government composition known as gun metal (88 per cent copper, 10 tin and 2 zinc) was unreservedly specified for valves, steam connections and fittings. At the present time, however, no successful valve manufacturer would think of using this alloy for this purpose, although, from purely laboratory results this and similar high-tin alloys may appear to meet the requirements. The safe procedure is either to investigate and follow standard practice, or determine definitely why it should not be followed; or else to leave the composition and the guarantee of results to a responsible manufacturer and hold him accountable.

Aluminum should never be used in an alloy which is to be tight against leakage of either a gaseous or fluid substance. Aluminum has the peculiar property of forming a characteristic skin upon the surface of its alloys, due to its great affinity for oxygen. Appreciably less than one two-hundredth of one per cent of aluminum produces an appearance upon the surface which is very distinctly that of a brass or bronze casting. In fact, this surface film is often found in castings where the amount of aluminum present is so small that the chemist may be in doubt as to whether even a trace of it exists. In pouring a casting containing aluminum this surface film becomes entrained in the convecting currents as the molten metal fills the mold and often extends through the entire thickness of the wall of the metal and thus gives rise to porosity which causes leakage. For the foregoing reasons it is always desirable, when pouring, that the molds should fill with as little bubbling and splashing as possible. This applies to all alloys in general, but to those containing aluminum in particular. The "horn gate" or some form by means of which the molten metal enters at the lowest point in the mold is very desirable.

In the alloying of copper and tin a low-fusing crystal forms known as the tin-copper eutectoid. This crystal is hard and brittle and in general should not be present in bronzes used for machine parts; and more particularly so if these parts are subjected to temperatures approaching that of high-pressure steam. This crystal always forms in ordinary foundry practice when tin is present in amount equal to 9 per cent of the copper content. In slow cooling the amount of this crystal is increased and slow cooling also causes it to appear with a lower percentage of tin. High-tin bronzes should neither be used under high temperature nor where they are subjected to severe shock. Without doubt this eutectoid crystal constitutes the most valuable bearing crystal for bearing metals to be found in any of the bronzes.

Considerable experimental work has been done at different times and in different laboratories on the subject of heat treatment of foundry bronzes and brasses which are to be used without subsequent forging or tooling. These experiments have been applied particularly to the copper-aluminum alloys and to the copper-tin alloy known as gun metal. Nothing, however, has as yet developed to bring this process of heat treatment into general commercial importance.

Brass-foundry products do not compete directly with cast iron, cast steel or steel forgings in machine construction. They are specified by the designer only when certain properties are necessary, such as appearance, bearing value, chemical composition, and the like. These non-ferrous products never can compete directly in cost with the ferrous products for the reasons that copper is always higher-priced than iron and the price of tin is always subject to foreign speculation; while few of the minor constituents compare favorably in price with iron.

### BRONZES

*Manganese Bronze* is the term generally applied to a composition which, properly speaking, is a brass instead of a bronze. It

is an alloy of copper containing a high percentage of zinc, aluminum and iron, in which a small amount of manganese has been used. The beneficial effects in this alloy due to the deoxidizing and iron-carrying properties of the manganese are very striking. Beneficial results are obtained even though no more than a mere trace of manganese finally remains in the resulting product. In none of the brass-foundry products are the high merits of an alloy so much due to the results of skilful and intelligent foundry practice as in the case of this particular alloy. The composition is relatively cheap when the cost of its constituents is considered.

The cost of the final product, however, is much higher than its composition would indicate, due to the following requirements necessary for the production of satisfactory castings: The introduction of a small percentage of iron is necessary for proper results, and care and skill with proper equipment are indispensable for this operation; secondly, the amount of metal required for gates, heads and risers necessary for this alloy is unusually large, so much so that in the production of ordinary automobile castings only from 17 to 39 per cent of the entire metal melted finally appears in the form of finished castings; and lastly, the loss in melting is unusually large. For machine construction this alloy, of all brass-foundry products, is of first importance. In strength it successfully competes with steel castings and steel drop forgings although the price is from three to five times that of malleable castings. This, and its non-corroding properties make it a most desirable material in automobile and marine construction. As a foundry product it is a very desirable material as it has a very smooth surface. The labor for finishing it is in many cases so light compared with that required for finishing steel castings that it can often compete with steel castings. This is especially true in automobile construction. It has the necessary strength for brackets, standards and fixtures, so that it can replace both steel castings and forgings for these purposes. Manganese bronze has a field of usefulness in experimental work in that it can be produced with less delay than steel forgings, stampings or malleable castings, requiring as it does so little finishing. By varying the two principal constituents of manganese bronze the desired physical characteristics can be obtained. An increase of the percentage of copper increases its elongation, whereas an increase in zinc increases its ultimate tensile strength but decreases its elongation. When this material is made up with the necessary precautions as to purity and foundry methods, very excellent commercial results can be obtained. For example, a tensile strength of 90,000 lb. per sq. in. with 30 per cent elongation can be secured in a standard test coupon, whereas many producers consider 65,000 lb. tensile strength with 15 per cent elongation as fair.

*Aluminum Bronze* is an alloy of copper and aluminum containing from 4 to 11 per cent of aluminum and the remainder copper. In recent years, however, this alloy has been made up with an addition of from 1 to 6 per cent of iron. The introduction of iron produces twofold results: First, it causes the alloy to solidify with smaller crystals, that is, it produces a finer grain; and secondly, it adds a third hard crystal formed by the union of aluminum and iron which is much harder than the other two crystals. This is the only brass-foundry product which can compete commercially with manganese bronze. With care it can be so made up as to have a somewhat higher ultimate strength than manganese bronze, although it does not possess a higher yield point or elastic limit. The drawback to this alloy is the inherent difficulty of making solid castings. The tendency to have blow-holes is very characteristic, and even extreme skill will overcome this difficulty only in a degree. Aluminum bronze may be used for bearing purposes and, recently, when containing iron, has had a limited use for motor-truck worm drives. The only advantage of this latter alloy over a copper-tin-bronze composition is its cheapness; for bearing purposes, however, it cannot compare with a copper-tin alloy.

*Acid-Resisting Bronzes.* These are some of the brass-foundry products which formerly were of more importance than they are today, owing to the fact that stoneware and specially modified cast iron are now largely replacing these uses. There are, however, some uses for bronzes to resist sulphuric and sulphurous acids in mine service, pulp mills, and other industries. For these purposes two alloys are most serviceable; a copper-tin-lead alloy containing substantially 86 per cent copper, 12 per cent tin and 2 per cent lead; and aluminum bronzes containing a small iron percentage.

*Copper-Tin Alloys.* First among these alloys highest in its tin content may be mentioned bell metal, a composition containing from 16 to 25 per cent of tin. As the name implies, it is used for the manufacture of bells, gongs, steam whistles and the like. It is a resonant metal. For industrial purposes the compositions used are those having the lower percentages of tin, say from 16 to 18 per cent. The higher percentages of tin, especially between 20 and 25 per cent, are used for bells in which high tone quality is required. The production of these bells requires considerable skill and constitutes a special industry. In general, however, it may be said that for all purposes the constituents for these alloys should be extremely pure and every precaution against oxidation should be taken during melting and pouring, since the quality of the product is altogether dependent upon the purity of the alloy.

The next alloy of this type to be considered is a bronze used in turntables and movable bridges. This composition is given in the tentative A.S.T.M. specification as " Class A, Bronze Bearing Metals for Turntables and Movable Bridges." It is also given as " Grade A " in the American Railway Engineering Association specification bulletin of July, 1918. It contains 20 per cent of tin, and phosphorus not to exceed 1 per cent. It is very hard, due both to its tin and its phosphorus contents. For best service conditions it should not be used except when placed between hardened steel plates. It is intended for pressures over 1500 lb. per sq. in. and slow-moving bearings.

A composition similar to the foregoing is one given as " Class B Bronze " under both of the immediately foregoing specifications and containing 17 per cent of tin with a permissible phosphorus content of 1 per cent. It does not seem wise to permit this amount of phosphorus in a bearing metal of this kind, however, if to be used as specified between " soft steel plates." Better service conditions can be obtained by the production of the necessary hardness through increase of tin content and reduction of phosphorus. An alloy containing 19 to 20 per cent tin and only a trace of phosphorus would give better service when used on soft steel. In a case of this kind where a copper-tin bronze is used between unhardened steel surfaces the amount of phosphorus present should be very limited owing to the exceeding hardness of the copper phosphide crystals.

The next alloy of importance as to its tin content is a bronze containing 11 per cent of tin and a maximum of 0.2 per cent of phosphorus. This alloy is the standard composition now in very general use for worm wheels in motor-truck drives and in the motor-car reduction worm gears. A feature of considerable importance in the making of this alloy seems to be that of producing the proper amount of chilling effect in the cooling of the castings. In some cases zinc has been added to an amount of 2¼ per cent; results show, however, that the addition of zinc is undesirable although it produces an additional hardness of the alloy as indicated by the Brinell and scleroscope tests. This additional hardness rather detracts from its bearing value, and, at the same time, makes it more difficult for tooling. It seems self-evident that the above alloy, as it now stands, could be improved by the addition of a small amount of lead,—not sufficient to produce undue weakening,—and that this would result in a threefold benefit: easier tooling, increased accuracy of tooled surfaces, and a higher bearing value.

*Copper-Tin-Zinc Alloys.* The most popular alloy of this class is the one known as ordnance bronze or gun metal. Its composition is 88 copper, 10 tin and 2 zinc. This alloy has come to be a very generally known composition in that it is a general-average bronze. However, with a more complete metallurgical knowledge of the relative values of different compositions, it becomes of less importance, in that it is specifically neither a bearing bronze of high merit nor a bronze especially adapted for machine parts when maximum economy of alloy is considered. The high tin

content produces in this alloy a high percentage of the tin-copper eutectoid, with the resulting disadvantages when the alloy is used for other than bearing purposes. The use of ordnance bronze for the construction of machine parts is, for two reasons, a wasteful one: It contains an unnecessary amount of tin, a high-priced constituent, and more serviceable alloys can be produced by the use of a lower percentage of tin. For machine parts, steam connections and valves, a composition having, say, 90 per cent copper, 6½ tin, 2 zinc and 1½ lead is a superior alloy for the reasons that it can be produced as a more economic foundry product, its price is less, its physical properties are superior, and lastly, its machining qualities are superior. On the other hand, the zinc in ordnance bronze distinctly detracts from its bearing value.

*Copper-Tin-Lead Alloys.* These alloys are strictly bearing compositions containing not in excess of 82 per cent of copper and a tin maximum of 11 per cent with a lead maximum also of 11 per cent. This range of alloy corresponds to " Class C " of the A.S.T.M. tentative specifications for Bronze Bearing Metals for Turntable and Movable Railway Bridges. These alloys are usually deoxidized with phosphorus which has a tendency to suspend or diffuse the lead, since in all these alloys the lead is held in mechanical suspension without any chemical union. The most generally used of this series contains 80 per cent copper, 10 tin and 10 lead.

To enumerate all of the copper-tin-lead alloys given by various authorities as bearing bronzes would constitute an almost infinite series. The whole matter, however, can be summed up in a few words: Lead is a desirable constituent in a bearing alloy but has the drawback of weakening the bronzes, therefore, for light service, or under conditions where a supporting oil film cannot always be maintained, as in car or truck bearings, the compositions containing the higher proportions of lead should be used. Where the punishment is severe the alloys should have a lower percentage of lead. As a rule it is economy to use as much lead as possible, since lead is a cheaper metal and it also adds materially to the bearing value of the alloy.

*Copper-Lead-Tin-Zinc Alloys* are those containing from 3 to 12 per cent of lead, from 4 to 6 per cent of tin and from 1 to 10 per cent of zinc. This series of compositions includes several used for bearing purposes, especially those having the high lead limits. The addition of zinc in this alloy is for the purpose of reducing the cost; it does not add to the value of the alloy except in its physical strength and it detracts from its bearing value. The series includes the popular red-brass composition; 85 per cent copper, 5 tin and 5 zinc; and also a cheaper composition containing 77 per cent copper, 10 lead, 3 tin and 10 zinc used for low-pressure valves and plumbers' supplies.

# GRAY-IRON CASTINGS

By RICHARD MOLDENKE,[1] WATCHUNG, N. J.

IN working out problems in machine design the mechanical engineer is constantly confronted with requirements for those parts he wishes to make of cast iron, which makes it necessary for him to be fully informed in regard to the latest developments in the foundry industry. Where to use a casting and what metal or alloy to make it from should be pretty well known by now, but what class of cast iron to use and what kind of a foundry to get it from may be matters open to doubt in these days of rapid changes. Therefore, even though he depends upon the foundryman to deliver him the kind of casting he needs, the engineer should have some knowledge of the underlying causes for the peculiar properties exhibited by cast iron when put into the form of castings.

## THE CARBON SITUATION IN CAST IRON

To understand the nature of the complex material known as cast iron it is necessary to go behind the returns of an "ultimate analysis," as it were, in order to get the "rational" analysis, or the grouping up of the several elemental constituents. Here it

[1] Consulting Metallurgist. Mem.Am.Soc.M.E.

will be seen that the total carbon is present in two forms, part of it being a mechanically mixed crystalline graphite and the remainder being chemically combined. The amount, distribution and physical condition of the graphite thrown out of solution at the moment of set are the more particular items of importance to the engineer in connection with the strength and machinability of the casting. While the total carbon of the metal is the same throughout the mass, the division into mechanically mixed graphite and combined carbon will be such that the highest proportion of the former will be found in the center of the section, together with the least of the latter form, and vice versa. Since the presence of combined carbon in increasing proportion means corresponding increase in hardness, the metal at the surface of a casting will be harder than in the center. Further, the oxide of iron formed through contact of molten iron with a damp mold surface and the fluxes in the molding sand, if not properly covered with adhering graphite, on interacting form an enamel which is very hard on the cutting tool, and hence this should be made to dig under the skin of the casting and lift the surface metal away as it cuts along under lubrication of the graphite present and not be allowed to glide over the skin.

With the foregoing explanation of the carbon situation in cast iron, and remembering that graphite is in mechanical admixture only, and consequently is an element of weakness by cutting up the continuity of the iron mass, it will be understood that fundamentally cast iron is a steel of given combined carbon content, with the other elements present in much higher percentages than one would expect in any steel. For instance, if the percentage analysis of the chips taken from the first eighth inch below the surface of a flat plate gave silicon 2.00, manganese 0.70, phosphorus 0.45 and sulphur 0.12; with graphite 3.20 and combined carbon 0.30; then that part of the plate was for practical purposes a 30 carbon steel containing much graphite, double the usual sulphur and ten times as much phosphorus as a steel ought ordinarily to contain. The fact that the proportions of the two forms of carbon vary from surface to center conclusively shows that while the total carbon of a cast iron analysis may be perfectly reliable, the items "graphite" and "combined carbon" in such an analysis are worthless unless the actual point from which the sample has been taken is indicated.

Silicon as the chief determining factor of the condition of the carbon present may run from as low as 0.35 per cent in deeply chilled castings up to 3.25 per cent in the very lightest of hardware and novelty work. The absence of silicon allows the carbon to remain in chemical combination with the iron. As the percentage of silicon increases the iron loses its power to hold the carbon in combination and it is thrown out as graphite at the moment of set. The higher the silicon the greater the graphite percentage. A factor affecting the silicon content to be employed is the rate of cooling of the metal when a mold is filled with molten iron. The thinner the section the greater the power of the iron to hold the carbon in chemical combination in spite of the silicon present. Hence high silicon is necessary to prevent thin sections from turning out hard, and since the foundryman must deliver his casting in condition to machine properly, he naturally regulates the silicon of his mixture to care for the thinnest section. If he were to aim for strong iron with a reasonable machinability for the heavier sections of the casting to be made, making his mixtures accordingly with high-steel-scrap percentages and comparatively low silicon, the thin ribs and brackets of the casting would come out hard and brittle, if not actually white in fracture, and the greater contraction of these in cooling from the molten state would cause bad warping and internal strains. The mechanical engineer should therefore study the metallurgical side of the foundry requirements so that he may design his work with the least abrupt changes in section, avoid sharp corners and distribute his metal to the best advantage—in other words, give the casting a chance.

The total-carbon percentage has come to be a most important factor in the iron-castings industry. Of recent years the comparatively low-total-carbon iron made in the air furnace, and called "gun iron" from its use previous to the days of the steel cannon, has been paralleled by cupola metal in which great quantities of steel scrap are added to the mixture. Even in the air furnace, the former long periods of refining to reduce the total carbon are shortened by adding all the way up to 30 per cent of steel scrap to the mixture of pig irons and home scrap. The product gets the erroneous name of "semi-steel" from those who are not informed on the properties of this grade of cast iron—for it is only a high-grade cast iron, with none of the properties of steel. Its importance to the engineer lies in the fact that the reduction in total carbon through steel additions can be carried to a point where the ultimate strength is nearly doubled and where the chips taken from such a casting curl up almost as well as those from hard steel, for which reasons it serves very well where it is not essential that steel be used. The foundryman, however, has a problem before him in running these high steel-scrap heats. Only the best of cupola and mold-gating practice will yield castings free from blowholes, cracks and serious internal shrinkages. The silicon percentage must be watched particularly, so that separation of graphite and combined carbon may leave the latter low enough for proper machinability. The effect of low total carbon (it may be brought down to 2.25 per cent, or just above the line of division into steel), if the combined carbon is kept normal, is to give an otherwise lower graphite percentage. Necessarily this graphite is finer in crystallization, and there being less of it than normal, the planes of separation in the solid metal are smaller and fewer, hence the greater strength. Moreover, there is much less phosphorus present as the result of the low phosphorus of the casting.

Manganese in the castings should not be allowed to run below 0.50 per cent and not over 1.00 per cent for normal machinable work. Phosphorus should be as low as possible, considering the pig irons that must be used, and for important and thick work had better run below 0.40 per cent. Very thin castings, such as art and architectural details in structures, which really correspond to the "stove plate" of the foundryman, require very nearly 1.00 per cent of phosphorus to come out in perfect detail. Table 1 gives recommended compositions of castings for many purposes. It is not wise for the engineer to specify the chemical composition of the castings he requires, but if strength is important it should be specified according to the standards of the American Society for Testing Materials.

### SPECIALIZATION IN IRON FOUNDING

The foundry industry is pretty well divided up into specialized lines of work. If the engineer needs chilled castings he turns to the makers of chilled rolls. Many foundries, however, other than roll makers produce crusher castings, balls for grinding, shoes and dies for stamp mills. Another important branch of foundry work is the making of cylinders. For the highest grade, foundries having air furnaces should be selected, particularly if they make a specialty of cylinders, piston rings and the like; and they will be found among the great foundry and machine works of the country building heavy machinery and engines, particularly of the marine type. Castings to withstand high pressures, such as those of air, steam and ammonia, are made by foundries specializing in air compressors, pumps, refrigeration machinery, and the general valve and fitting trade. For light parts to serve for ornamental purposes there are many stove shops to which recourse may be had. In general, however, where reciprocating parts or sections of intricate character and for severe service strains are required, it is safer to go to the large concerns making well and favorably known lines of product. The foundries connected with such establishments have the proper facilities and the trained men to undertake the work, and will be in position to advise improvements in design to give the metal a better chance for the purpose intended. Foundries of the class just mentioned usually have need of several mixtures of iron to supply the parts required for the line of machinery turned out. Hence, they usually run a few charges of fairly low silicon, high steel scrap-metal for such castings as air-compressor cylinders, rolling mill parts, and other work of high strength and density. The next charges may then care for ordinary machinery castings, and finally a few extra soft or high-silicon charges may care for pulleys and the very light-section product. Sometimes the process is reversed, all light, soft work being run first and the harder classes at the end of the heat.

### RECENT TENDENCIES IN GRAY-IRON FOUNDRY PRACTICE

The passing of the war has given us time to realize more fully the train of evils that have been its legacy to the foundry. While

a quarter of a cent is thus added to the cupola-melting cost per pound of metal "duplexed," where current is cheap.

Molten cupola metal, with possibly 0.12 per cent sulphur, may thus be brought down to about 0.05 per cent, and even lower. Moreover, the metal is highly superheated—cupola metal itself is intensely hot as compared with furnace iron—and thoroughly de-oxidized. The castings made are therefore much better and sounder than the ordinary run, equaling charcoal-iron castings

TABLE 1   RECOMMENDED ANALYSES FOR VARIOUS CLASSES OF CASTINGS

| Castings | Light | | | | | Medium | | | | | Heavy | | | | |
|---|---|---|---|---|---|---|---|---|---|---|---|---|---|---|---|
| | Si | Mn | S | P | T.C. | Si | Mn | S | P | T.C. | Si | Mn | S | P | T.C. |
| Acid-resisting | 2.00 | 0.75 | 0.05* | 0.20* | 3.25* | 1.80 | 1.25 | 0.05* | 0.20* | 3.25* | 1.00 | 1.25 | 0.05* | 0.20* | 3.23* |
| Agricultural | 2.50 | 0.60 | 0.06 | 0.75 | 3.75 | 2.25 | 0.70 | 0.08 | 0.70 | 3.50 | 2.00 | 0.80 | 0.10 | 0.60 | 3.25 |
| Air Cylinders | 1.90 | 0.70 | 0.08 | 0.50 | 3.40 | 1.50 | 0.80 | 0.09 | 0.40 | 3.25 | 1.00 | 0.90 | 0.10 | 0.30 | 3.00 |
| Annealing Boxes | | | | | | 0.65 | 0.20 | 0.08 | 0.20 | 3.50 | | | | | |
| Automobile Cylinders | 2.25 | 0.65 | 0.06 | 0.40* | 3.25* | 2.00 | 0.75 | 0.08* | 0.40* | 3.25* | | | | | |
| Balls for Grinding | | | | | | 0.75 | 0.50† | 0.15* | 0.40* | 3.75† | 0.50 | 0.50† | 0.15* | 0.40* | 3.75† |
| Bedplates | 2.00 | 0.70 | 0.08 | 0.60 | 3.75 | 1.75 | 0.75 | 0.10 | 0.50 | 3.50 | 1.50 | 0.80 | 0.12 | 0.40 | 3.25 |
| Boiler Castings | | | | | | 2.00 | 0.80 | 0.06* | 0.30* | 3.50* | | | | | |
| Car Wheels | | | | | | | | | | | 0.65 | 0.50 | 0.08 | 0.33 | 3.50 |
| Chilled Castings | | | | | | | | | | | | 1.00 | 0.08* | 0.20* | 3.50* |
| Chills | | | | | | 1.00 | 0.50 | 0.05 | 0.20 | 3.00 | | 1.25 | 0.06 | 0.20 | 3.25 |
| Crusher Jaws | | | | | | 1.00 | 1.00 | 0.04 | 0.20 | 3.50 | 0.80 | 1.25 | 0.06 | 0.20 | 2.75 |
| Dies for Hammers | | | | | | 1.50 | 0.80 | 0.04 | 0.20 | 3.00 | 1.25 | 0.80 | 0.06 | 0.20 | 2.75 |
| Dynamo Castings | 2.80 | 0.50 | 0.05 | 0.75 | 3.75 | 2.75 | 0.50 | 0.05 | 0.50 | 3.50 | 2.15 | 0.50 | 0.06 | 0.50 | 3.25 |
| Electrical Work | 3.00 | 0.50 | 0.03 | 0.60 | 3.75 | 2.00 | 0.60 | 0.06 | 0.50 | 3.50 | | | | | |
| Engine Frames | | | | | | 2.00 | 0.60 | 0.08 | 0.30 | 3.25 | | | | | |
| Fire Pots, Grates | 2.25 | 0.60 | 0.05 | 0.20 | 3.50 | 1.50 | 0.60 | 0.08 | 0.20 | 3.25 | 1.25 | 0.70 | 0.08 | 0.30 | 3.25 |
| Fly Wheels | 2.00 | 0.60 | 0.05 | 0.50 | 3.50 | 1.60 | 0.60 | 0.06 | 0.40 | 3.25 | | | | | |
| Friction Clutches | 2.40 | 0.60 | 0.10 | 0.70 | 3.75 | 2.00 | 0.70 | 0.12 | 0.50 | 3.50 | | | | | |
| Furnace Castings | 2.40 | 0.60 | 0.05 | 0.60 | 3.75 | 2.15 | 0.80 | 0.06 | 0.50 | 3.50 | | | | | |
| Gas-Engine Cylinders | 2.00 | 0.70 | 0.08 | 0.40 | 3.25 | 1.50 | 0.80 | 0.09 | 0.30 | 3.00 | 1.25 | 0.90 | 0.10 | 0.20 | 2.85 |
| Gears | 2.25 | 0.60 | 0.08 | 0.70 | 3.75 | 2.00 | 0.80 | 0.09 | 0.60 | 3.50 | 1.50 | 1.00 | 0.10 | 0.50 | 3.25 |
| Glass Molds, Pipe Balls | | | | | | 2.50 | 0.60 | 0.04 | 0.20 | 3.25 | | | | | |
| Grate Bars | | | | | | 2.00 | 0.60 | 0.05 | 0.20 | 3.50 | | | | | |
| Gun Iron | 2.50 | 0.70 | 0.08 | 0.80 | 3.75 | 1.50 | 0.50 | 0.05 | 0.30 | 3.25 | 1.00 | 0.60 | 0.05 | 0.30 | 3.00 |
| Hardware | | | | | | 2.00 | 0.80 | 0.06 | 0.20 | 3.23 | 1.30 | 1.00 | 0.06 | 0.20 | 3.00 |
| Heat-resistant Iron | | | | | | 1.50 | 0.80 | 0.06 | 0.40 | 3.25 | 1.00 | 1.00 | 0.06 | 0.20 | 2.85 |
| Hydraulic Cylinders | | | | | | | | | | | 1.25 | 0.80 | 0.06 | 0.20 | 3.75 |
| Ingot Molds | | | | | | | | | | | 1.50 | 1.00 | 0.10 | 0.50 | 3.25 |
| Machinery Castings | 2.50 | 0.60 | 0.08 | 0.70 | 3.75 | 2.00 | 0.80 | 0.09 | 0.60 | 3.50 | | | | | |
| Mine Wheels | | | | | | 0.90 | 1.00 | 0.10 | 0.20 | 3.00 | | | | | |
| Ornamental Castings | 2.75 | 0.60 | 0.06 | 0.90 | 3.75 | 2.25 | 0.70 | 0.08 | 0.80 | 3.50 | | | | | |
| Pipe (Water) | 2.25 | 0.60 | 0.06 | 0.80 | 3.75 | 2.00 | 0.90 | 0.08 | 0.70 | 3.50 | 1.50 | 1.00 | 0.10 | 0.60 | 3.50 |
| Piston Rings | 2.00 | 0.70 | 0.05 | 0.60 | 3.50 | 1.75 | 0.80 | 0.06 | 0.50 | 3.25 | | | | | |
| Pulleys | 2.40 | 0.50 | 0.05 | 0.70 | 3.75 | 2.15 | 0.60 | 0.07 | 0.60 | 3.50 | 1.90 | 0.70 | 0.09 | 0.50 | 3.25 |
| Radiators | | | | | | | | | | | | 1.00 | 0.08 | 0.30 | 3.00 |
| Rolls (Chilled) | | | | | | | 0.80 | 0.06 | 0.40 | 3.25 | | | | | |
| Soft Castings | 2.00 | 0.50 | 0.05 | 0.60 | 3.75 | 2.40 | 0.60 | 0.08 | 0.50 | 3.50 | | | | | |
| Soil Pipe | 2.25 | 0.60 | 0.08 | 0.80 | 3.75 | 2.00 | 0.80 | 0.10 | 0.60 | 3.50 | | | | | |
| Steam Cylinders | 2.00 | 0.60 | 0.08 | 0.50 | 3.50 | 1.60 | 0.60 | 0.09 | 0.40 | 3.50 | 1.25 | 1.00 | 0.10 | 0.30 | 3.50 |
| Stove Plate | 2.50 | 0.50 | 0.06 | 1.00 | 3.75 | 2.25 | 0.60 | 0.08 | 0.80 | 3.50 | | | | | |
| Valves | 2.25 | 0.60 | 0.07 | 0.50 | 3.25 | 1.75 | 0.80 | 0.08 | 0.40 | 3.00 | 1.25 | 1.00 | 0.09 | 0.30 | 2.85 |
| White Iron Castings | | | | | | 0.75* | 0.20† | 0.23* | 0.75* | 2.50† | | | | | |

*Below.    †Above.

there was a tremendous expansion of other industries due to war demands, the foundries simply pushed production to the utmost consistent with ability to get raw materials and men. Impossible prices for pig iron resulted in the use of so high a proportion of scrap in the mixtures that when the castings made during this period of stress eventually laid into the scrap piles of the country—probably within the next twenty years—a serious problem of very high sulphur and oxidized iron will confront the foundryman. Instances of 90 per cent scrap in the mixture were common and castings with over 0.20 per cent sulphur equally so.

The problem will have to be faced, and there are two possible solutions. The method heretofore employed in such a case is the increased use of pig iron in the mixtures. Before the war the common practice was to use not over 40 per cent scrap with 60 per cent pig. Now it is reversed. Consequently, instead of having the sulphur run about 0.08 per cent maximum in important work, it is more nearly 0.12 per cent. Ordinary work today touches 0.18 per cent without much comment. It is true that even a higher sulphur content will not militate against machinability provided the melting practice is of the very best, but the castings are more subject to danger from shock and casting strains.

The other method for correcting the high-sulphur tendency in daily work at present and for that to come is a desulphurization process in which the regular cupola method is employed, the molten metal being transferred into an electric furnace having a basic lining. In this way the heavy current variations are avoided, and by taking advantage of the comparatively cheap cupola-melting cost, that of the electric furnace for refining only is very materially reduced. It is a question whether more than

made in the air furnace. The first cost of the installation is high, but it soon pays for itself in the quality of work turned out. While developed during the war period, it is none the less a logical sequence in the world advance of the art of making iron castings and should be welcomed as a satisfactory way out of a very bad situation ahead.

In conclusion, the mechanical engineer is urged to turn his attention more specifically to foundry operation. The technical staffs of all the foundries of continental Europe, from manager down to assistants in the several production and testing departments, are all graduate engineers. The consequence is that castings are made strictly for the purpose intended, and not merely to get by the machine shop safely, as is unfortunately so often the case here. Close coöperation between men trained in the science with those trained in the art of making castings can only result in good.

## DISCUSSION AT CASTINGS SESSION

### GRAY-IRON CASTINGS

H. Wade Hibbard,[1] who was the first to discuss the paper on Gray-Iron Castings, called attention to the author's statement that "cast iron is ideal for frames for machinery of all kinds, and as it can be locally chilled it gives splendid wearing surfaces," and stated that it had always been his impression that a special grade of cast iron was needed if a chilled surface was desired, and that the grade of cast iron used for frames for machinery was not the grade commonly employed. To this Doctor Moldenke

---

[1] Prof. Mech. Eng., Univ. of Mo., Mem. Am.Soc.M.E.

replied that "chilled" may refer to one of two things. It may refer to an iron that has been turned perfectly white or an iron which is merely cooled sufficiently rapid to become coarse-grained.

John McGeorge[2] referred to the author's statement regarding the transfer of the melted iron to the electric furnace having a basic lining and asked if the use of such a lining did not eliminate the phosphorus. Doctor Moldenke stated that the phosphorus was not eliminated, and that the carbon must first be removed, also that the sulphur must be eliminated.

Mr. T. A. Bovie inquired as to the strength of the iron and Doctor Moldenke replied that it depends upon the proportion of graphite and that of course, in general, as the total carbon content is decreased the strength is increased.

### MALLEABLE CASTINGS

Captain Hughes, who opened the discussion on the paper on Malleable Castings, called attention to the author's statement that the Ordnance Department was unfamiliar with the physical characteristics of malleable iron. He stated that the Ordnance Department was planning at the present time to make drop bombs out of malleable iron but has been unable to obtain satisfactory castings, especially in the larger sizes, and that furthermore the Department has had great difficulty in inducing the malleable-iron castings companies to undertake the necessary research work. He requested information as to where the Department might possibly secure satisfactory castings.

Doctor Moldenke gave as his reason for unsatisfactory castings, the fact that a white-iron casting has a contraction from the molten to the solid of over 12 per cent, and unless it is possible to take care of that contraction in the feeding of the metal, the outside of the casting will be sound but the inside be porous. He also spoke of Professor Touceda's work in assisting the smaller foundries to manufacture a uniform product.

Professor Hibbard stated that thick sections in malleable castings have been feared by the designer and objected to by the manufacturer and he was therefore interested in the author's statement that "castings have been produced commercially as long as 5 ft. with sections at some parts as thick as 3 in." "With certain of our malleable-iron foundries," he said, "the ability to produce malleable-iron sections as thick as 3 in. or more has been known for several decades, but since malleable iron before it is annealed is essentially white and very brittle, by putting a chilling piece into the sand mold where the section is thick, that chilling piece thus renders it more capable of receiving the annealing process." Professor Hibbard also wondered if our foundries are acquainted with this method of causing a thick piece to lend itself readily to the annealing process. Doctor Moldenke replied to Professor Hibbard's inquiry and stated that to his knowledge this particular method was commonly known. Doctor Moldenke also stated that from his experience he had found that malleable castings could withstand shocks better than steel castings. "There is nothing better," he said, "than malleable under shock."

Replying to an inquiry as to whether or not any work had been done toward the casting of malleable iron in permanent molds. Doctor Moldenke said that in making railroad tie plates in the permanent or iron molds he invariably found that the material so made was weaker than that made in a sand mold. He related his experiences in the production of the tie plates and called attention to the difficulties encountered, particularly those arising from the rapidity with which the metal chilled in the iron mold. He stated that in his opinion permanent molds would never amount to very much. "We ought to aim in a different direction," he said, "and make molds of a substance that will not be injured by one, two, three, four or even ten molds."

Lieut. E. R. Jackson[3] requested information dealing with the use of malleable iron in connection with parts for tractors and motor trucks. "Out of twenty of the standardized trucks that we took across the country," he said, "there wasn't a single valve tap that completed the trip. They all had to be replaced." He

referred to Professor Touceda's statement that malleable iron can replace such parts, but said that to date they have not been satisfactory in the Army. Doctor Moldenke replied to Lieut. Jackson's inquiry and stated that the difficulty was probably due to the fact that when a malleable casting is skinned a large portion of the carbon content is removed. "There is a difference," he said, "between the malleable casting of America and the malleable casting of Europe." In Europe it is the practice to almost completely remove the carbon and as a result they obtain practically a steel, which, of course, wears very well. Over here, however, our castings do not have the carbon removed and consequently do not wear so well.

### STEEL CASTINGS

Professor Hibbard was the first to discuss the paper on Steel Castings. He called attention to the author's statement that "when the service demands all the strength, toughness and resistance to wear and fatigue that are obtainable in a cast metal, a steel casting should be specified." He stated that perhaps he had an undue prejudice against wearing or suitability of cast steel on a rubbing surface and that in his own practice he did not employ steel castings with finished surfaces for rubbing surfaces.

Lieut. Jackson endorsed Mr. Hibbard's remarks, stating, however, that the Ordnance Department has found forged steel parts quite satisfactory for such work.

Mr. McGeorge related an experience in the production of large castings which did not shrink although a very considerable shrinkage was expected. The cores were very hard and they did not crush. "The following castings were annealed," he said, "in a large furnace which we built for that purpose. We then obtained the proper shrinkage."

Doctor Moldenke said that while he was not very well posted on steel castings, he had had similar experience with cast iron. "If the core is too hard," he said, "either the metal must give or it cracks. That is one of the reasons why in our foundry practice, when we make a pattern, we first make a casting and break it and measure it, for we may have to change that pattern."

Professor Hibbard also commented on this phase of the work and called attention to the methods employed by the Commonwealth Steel Company.

Captain Hughes stated that at the present time a very interesting case of steel castings failing to shrink the anticipated amount is to be found in the making of recoil bands for 16-in. guns. "We used to make them solid," said Captain Hughes, "and then in order to get greater weight, we cored them out and filled them with lead. We made some castings which weighed about 25 tons, and were about 129 in. long and 77 in. wide. On the 129-in. dimension they were about 1¼ in. long, throwing off all holes, ribs and other parts of the casting, so that it is necessary to make new castings, which is due to this same phenomenon of being unable to shrink due to hard cores. We annealed these castings afterward and realized the physical properties desired which are those of the U. S. Cast Steel, No. 3, which, as you know, are very high."

[EDITOR'S NOTE: The time devoted to the discussion of the papers by Messrs. Moldenke, Touceda, and Hall consumed practically the entire time set aside for the Foundry Session, and those present at that session accordingly decided that the discussion of papers by Messrs. Pack, Jeffries, and Bierbaum should be in written form and included in the general account of the meeting. To date, however, but one discussion has been received. This is given below.]

### BRASS AND BRONZE CASTINGS

ROBERT J. ANDERSON[3] (written). The paper by Mr. Bierbaum mentions the fact that in the case of manganese bronze sprue returns are exceptionally large—so much so that in the manufacture of ordinary automotive castings only about 17 to 39 per cent of the entire metal melted finally appears in the form of finished castings. In the other papers comprising the (Continued on page 480.)

[2] Special Engineer, Ohio Body and Blower Co., Cleveland, Ohio, Mem. Am. Soc. M. E.
[3] Tank, Tractor and Trailer Div., Ordnance Dept., Washington, D. C.
[3] Bureau of Mines, Experiment Station, Pittsburgh, Pa.

# McCook Field and American Aeronautics

Inception of McCook Field and Nature of the Work Carried on—The Problem of the Variable-Pitch Propeller—Development of Measuring Instruments—Research Work in Altitude Flying— Present and Future Potentialities of the Field

By LEON CAMMEN,[1] NEW YORK, N. Y.

A MODERN aeroplane is far more than a few sticks of wood and canvas thrown together with an engine and propeller stuck in. In many respects the design of an aeroplane is a real course in post-graduate engineering, even for those who have done extensive work in other lines. The aeroplane is the lightest craft for the amount of power that it carries, moves at greater speed than any other passenger-carrying vehicle, and is subject to a greater variety of stresses and influences than any other structure know to man. Moreover, these stresses and influences, the character of which is as yet little known, are most difficult to determine.

Classical economics used to say that all production involves three basic elements: land, labor and capital. In aeroplane construction more perhaps than anywhere else the fourth element of production plays a tremendously important part, and that is—knowledge.

Many delays in the execution of the American aeroplane program at the beginning of the war were due to the fact that we, in this country, did not have the accumulation of technical data, either theoretical or experimental, that was necessary to carry out the vast program undertaken, and that many of these data could not be secured even from abroad, where, notwithstanding all war regulations, they were a jealously guarded secret of commercial companies.

It takes a vast amount of painstaking work in experimenting, measuring, calculating and devising to secure data useful in aeroplane construction. Moreover most investigations bring in results in the form of tables, equations of stresses, or little matters of design which cannot be patented and made the exclusive property of some particular individual or company. Because of this, commercial aeronautic companies, such as there are, either do not undertake work of this character at all, or, if they do, guard the results as their trade secrets.

[1]Associate Editor, MECHANICAL ENGINEERING.

It is obvious, however, that if aeronautics is to be developed in this country there must be some place where investigations into matters pertaining to this new art can be carried on without any regard to immediate commercial returns. In other words, we have to have what practically amounts to a national laboratory for aeronautical research. In this direction a certain amount of work is being carried out in various institutions, but the place where such work is carried out probably on the largest scale is at McCook Field, Dayton, Ohio.

### THE INCEPTION OF McCOOK FIELD

The field happened to be located at that particular spot more or less accidentally. Since the days of Wilbur and Orville

FIG. 1  AERIAL VIEW OF McCOOK FIELD

Wright, Dayton has, to a certain extent, been the aeronautical Detroit of the United States, and about the time of our entry into the war several companies located in that particular city were engaged in aeronautical work or in work having an important bearing on the art of flying. The War Department did its best to coöperate with them, and the original McCook Field was the result.

It was quite a modest endeavor at first; but with our entry into the world war and the great appropriations granted by Congress for aeronautics, McCook Field rapidly assumed its present great proportions and, in fact, became the center of the main engineering activities of the Air Service of the Army.

The aerial view of the Field, Fig. 1, gives an idea of the scale on which this enterprise was carried out under the stress of war conditions when so much depended on the Air Service arm of our fighting forces.

The chief work on the development of the Liberty engine from its beginning to the present stage was carried out at McCook Field and many other investigations of great importance have passed through various stages of development in the same great organization.

### NATURE OF THE WORK CARRIED OUT AT MCCOOK FIELD

The work carried out at McCook Field is of an extremely varied nature and covers a wide variety of subjects; from trial flights in the course of which new equipment is being tested to the development of curves of performance of various apparatus, instruments of measurement, new devices and new laws in branches of engineering bearing on man flight.

During the war all that was done at the Field was naturally covered with the deepest veil of military secrecy, and even now research is carried on along lines and on subjects which are not yet ready for publication. A tremendous amount of work has been done, however, which can now be revealed. It may be stated, in this connection, that as far as possible the authorities at McCook Field never tried to make more of a mystery of their work than was absolutely called for either by the exigencies of war or the necessity for reticence about work yet in progress.

The Bulletin of the Airplane Engineering Department, U. S. A., and the Technical Orders of the Engineering. Division, Air Service, Dayton, Ohio, available now to every American engineer who can show a bona-fide desire to use them for the good of the country, represent a very valuable source of information on research in the aeronautical field.

It was at the McCook Field, in coöperation with the Forest Research Laboratory, that the whole subject of veneers, plywood, and glue in aeroplane construction received the development which, in this connection, gave the United States the foremost position that it now occupies in the world. There were several other problems that were carried to completion at the Field. One of the first that may be mentioned is that of supercharging. Since the early days of the war it was noticed that engines, as they then were, lost a large proportion of their power at high altitudes. A partial solution was offered by the supercharger as built by Dr. A. Rateau, Hon. Mem. Am. Soc. M. E., in France, but further work was necessary before the apparatus could be made practical. The theoretical part of this work was carried out in the Altitude Laboratory of the Bureau of Standards, but it remained for the engineers of McCook Field to do the practical work—in coöperation with Mr. E. H. Sherbondy, of the De Laval Company, and especially Dr. Sanford A. Moss, Mem. Am. Soc. M. E., of the General Electric Company. The Moss supercharger was the one which was finally brought to the practical stage and important flights were made with it from McCook Field. The exact value of the supercharger on aeroplanes may not yet be clear, but there is scarcely any doubt as to its representing an important advance in the engineering of high-altitude flight. Another development of importance, the magnitude of which it is scarcely possible to realize today, is represented by the so-called variable-pitch propeller.

### THE PROBLEM OF THE VARIABLE-PITCH PROPELLER

The efficiency of an aeroplane propeller depends on so many variables that it is practically impossible to design a propeller that will be equally efficient under all conditions. The propellers used today represent nothing but compromises designed to give the best efficiency under the average conditions under which the aeroplane is supposed to fly. This, however, deprives the aeroplane of a good deal of its flexibility and, for example, makes one designed to fly at moderate altitudes highly efficient at high altitudes, and vice versa. If we could make a propeller with a variable pitch and vary the pitch from the seat, it is obvious that we would have a means for adapting the propeller to the varying conditions of flight, and would give the aeroplane somewhat the same flexibility that the change gear gives to the automobile of today. The problem, though comparatively simple theoretically, proved to be a very puzzling one when it came to practical developments. It must be remembered in this connection that the variable-pitch propeller gear must be of light weight and yet have a high factor of reliability, and that the latter is made particularly difficult because of the fact that the amount of power transmitted through the gear is quite large—for example, with the Liberty motor, 400 hp. or more as compared with some 60 to 80

hp. that the change gear of an ordinary large automobile has to carry. A still further problem lay in the fact that a wooden propeller had to be made in two sections engaged with the variable-pitch gear and that means had to be found to hold the wooden parts in such a manner as to withstand the tremendous stresses induced by the centrifugal forces acting on the propeller and otherwise.

It would not be an exaggeration in any way to say that the development of a variable-pitch propeller installation represents as difficult a job in mechanical design as any that had to be tackled in the development of modern aeroplanes and its practical solution may be viewed with pride by the American engineering fraternity. This all the more so, as it may be expected that ultimately the solution of the variable-pitch propeller problem may lead not only to the improvement in the overall efficiency and flexibility of operation of the aeroplane but also to the solution of one of the most vital problems in aeronautics of today—the problem of insuring safety in landing.

To maintain itself in the air an aeroplane has to fly at a certain minimum speed. If that speed goes below the necessary minimum limit, the aeroplane will fall to the ground like a plummet, and this is what makes landing so difficult and dangerous. A high speed, varying from 30 miles for large and slow-going planes to some 60 miles or even more for the small, extremely fast machines, must be maintained up to the very instant the wheels touch the ground, and it is obvious that unless some way is found for braking an aeroplane, it will continue to move for quite a considerable time at the same speed after reaching the ground. An aeroplane, however, is not designed to run at very high speeds on the ground unless it is very smooth and even, and running on rough ground is apt to cause trouble either by breaking the landing gear or upsetting the plane. As a matter of fact, the extremely high-speed scout planes can land safely only on specially prepared landing fields.

The situation is about the same as if automobiles without brakes were permitted to travel at, say, 50 m.p.h., and it is easy to see how dangerous a pursuit motoring would have been under these conditions. There has been no lack of effort to provide braking for aeroplanes, but, unfortunately, practically all attempts in this direction have failed for one reason or another. However, there is what appears to be an entirely practical way of handling this problem and that is by using the aeroplane propeller itself as a brake, which means reversing the propeller at or immediately after the instant of landing on the ground. This requires that means for reversing the propeller and controlling its operation in the reverse condition should be available, and that is what the McCook Field work on variable-pitch propellers promises to give us.

It may be said that for work of this and similar character on propellers there are available at the Field unusually good facilities, and actual-size propellers can be tested in various ways with instruments necessary to determine the main features of propeller performance.

### DEVELOPMENT OF MEASURING INSTRUMENTS AT MCCOOK FIELD

Another field in which a large amount of highly interesting work is carried on is that of instruments. In the early days scarcely any instruments were used in power-plant engineering work. Coal was weighed after a fashion and in some plants pressure gages were employed. The thermometer was also used, though grudgingly, and in a few laboratories rather crude dynamometer facilities were available on a limited scale.

It was the introduction of electric power generation and transmission which really laid the foundation for the modern view as to the importance of measuring instruments. The transmission and application of electric power was not feasible without instruments, such as ammeters, voltmeters and wattmeters, and when it came to alternating currents, frequency meters, etc. In this way the amount of power delivered to the generators was measured and known and this induced a feeling of curiosity as to what was happening at the steam end, which led to the more extensive use of the indicator first and to the development of

FIG. 2 VIEW IN HANGAR SHOWING MACHINES UNDER TEST      FIG. 3 VIEW IN HANGAR SHOWING MACHINES UNDER TEST

FIG. 4 ENGINE OVERHAUL STANDS AND BENCHES      FIG. 5 VIEW SHOWING INTERIOR OF ENGINE MUSEUM

FIG. G FOKKER WITH LIBERTY-G ENGINE

the present varied and highly efficient instruments for measuring and recording water and coal consumption, exhaust-gas analysis, temperatures at the various stages and other elements. But even these relatively highly organized methods of measurement appear crude when compared with the vast metrical work that is being carried on in connection with aeronautics. There man has projected himself into a world of which very little is known, and what little information is available is either incomplete or misleading. Everything about the air composing the upper strata of the atmosphere as well as about the vehicle in which man goes up into the air has to be learned anew, which means that new and precise instruments have to be developed.

The work of developing these instruments is being carried on vigorously at McCook Field and recently, for example, a new instrument was developed permitting the recording of altitude with a precision many times greater than has hitherto been possible. The time has not arrived, however, for telling more than what has just been stated regarding this branch of work, which, by the way, has a particularly great importance for the American aeronautical industry, as comparatively little is known about foreign developments in instrument manufacture since the armistice.

A good deal of the design work in connection with development of planes and their testing is also carried on at the Field, which has good facilities therefor. Likewise, important work is carried on in connection with the design of military accessories, but of this branch the less said the better.

### RESEARCH WORK IN ALTITUDE FLYING

There is, however, one line of research which deserves particular attention. At the present writing a member of the McCook Field Flying Force holds the world's record for altitude flight, and, what is most important, altitude flying as carried out at the Field does not have the sole purpose of establishing records; as a matter of fact, the sporting element is kept entirely in the background.

For the Field, altitude flying is a grim business. They are developing the machinery to make altitude flying possible if the exigencies of war or aerial commerce should make it necessary. The flying is done with supercharged engines, which necessitate a number of inventions to make this new type of internal-combustion motor fully reliable to operate under the unusual and strenuous conditions prevailing at a height of several miles above sea level.

Then again, military authorities want to know just what is going on in the air at such altitudes, and it is already apparent that some of the phenomena occurring there are of vital and unexpected importance. They are not telling all they have learned and they should not. But one of the things which may be told is that winds of unsuspected velocities, ranging perhaps as high as 250 or more miles an hour, have been discovered blowing from west to east. Whether winds blowing in the other direction with similar velocities can be found remains yet to be seen, but even with the knowledge which we now have it appears perfectly feasible to go up in the air in California and land around New York within, say, 10 to 12 hours from the start.

### PRESENT AND FUTURE POTENTIALITIES OF McCOOK FIELD

McCook Field, however, is more than a research organization of the War Department. If it is not so already, it nevertheless has a nucleus of the possibility of becoming what might be called a national post-graduate research laboratory in aeronautics.

There is one room in the Field (Fig. 5) which makes an engineer connected with aeronautical research think. This spacious, splendidly lighted room has on its floor practically every aeronautical engine of importance—Liberties of several types, the big Sunbeams, Gnome engines as dainty as a Swiss watch, a sturdy German Mercedes. All of these can be studied there and a visit may save months of research and trouble to any one connected with the design of aeronautical engines. It is certainly the most complete engine museum in this country, perhaps in the world, and one cannot help thinking what a splendid thing it

would have been for American aeronautical engineering if side by side with the engine museum were other halls containing similar collections of propellers, wing sections and aeronautical accessories, and these backed by the courteous and expert staff on the Field. It may be mentioned, in this connection, that while there are sections of the Field closed to visitors as they should be, any person who can show a legitimate reason for inspecting those parts of the Field that do not fall under this restriction can easily obtain access to them. There is but little red tape about admission and the engineers and employees appear to be ready to help out any one expressing a fair reason for coming there. This is all the more worthy of note as the present staff is none too numerous for the work being done.

There is a good deal of talk now about Government assistance to aeronautics. The experience of France and England, for example, has shown that the art is yet too new and the commercial possibilities of aeronautics too limited and uncertain to develop without some form of Government assistance. Whether this will prove to be the case also in this country is a matter worthy of the most serious consideration. There is one form of Government assistance, however, the value of which will hardly be questioned, and that is assistance along the lines of research which no commercial concern can undertake.

McCook Field as it stands, together with the Aeronautical Laboratory of the Bureau of Standards, the Forest Laboratory at Madison, Wis., the testing fields of the Navy, already represents a powerful stimulus to the development of the art and science of aeronautics in this country.

A visitor to the Field cannot help being impressed with the fact that the actual coefficient of utilization of the Field is at present very low. The organization was conceived on a big scale in accordance with the vast American aeronautic problem of war time, and only a skeleton of it is maintained, and yet there is room and need for an organization even larger than that of the original McCook Field plan. In the field of aeronautical research undoubtedly good work has been done in this country, but the art is advancing by leaps and bounds and unless we keep up with it we are in danger of being left behind just as we were in the years 1914-1917.

It might be well, perhaps, to remember that the world's conditions are still extremely unsettled, and that it would really be a bold man who would dare to prophesy that this country will be confronted within the life of the present generation with an emergency similar to that of February, 1917. The development of aeronautical engineering as carried on by institutions like McCook Field is not a cheap proposition. On the other hand, however, it does not require any sums that might in the least prove to be burdensome to a country as wealthy as ours, and the value of such an investment would be repaid a thousandfold in many ways, especially if we should have the misfortune, together with the rest of the world, to be faced with a grave war emergency in the future.

## The New England Pulp and Paper Industry

The pulp and paper industry of New England probably ranks third in money value of product; the textile and shoe and leather trades leading, but, from the traffic viewpoint, pulp and paper with the immense raw material tonnage involved in their manufacture easily lead all our industries. A "balanced" book paper mill, for example, involves about seven tons inbound for each ton of final product—a "balanced" mill meaning a plant producing its own sulphite and soda pulp and its own bleach and alkali by electrolysis. Thus a "balanced" mill with 250 tons daily product involves a total shipping tonnage of 600,000 tons per year.

With certain undoubted advantages for the manufacture of paper, New England also has some serious disadvantages. The mills have long hauls inbound on coal and many important raw materials of heavy tonnage, and outbound on much of their manufactured product. New England produces many times its own requirements of paper, and while the home market is important, and New York City is much the largest paper market on the continent, the center of consumption is very far west of there.

# Pipe Lines for Transporting Natural Gas

### The Construction, Repair and Operation of Jointed and Welded Lines, Including Particulars Regarding the Couplings Used, Pipe-Laying Machines, Strength of Welded Pipe, Pressure Regulation, etc.

By CLIFFORD E. BROCK[1], BARTLESVILLE, OKLA.

D URING the early development of the natural-gas industry many different styles of pipe and joints were used. The first line, a small wooden one, was laid in Pennsylvania in 1872. Small screw lines were next used, and these were followed by leaded-joint lines, which later developed into the rubber-gasket joint of the present day.

A coupling for plain-end pipe consists of a center ring *A*, two followers *B*, and two rubber gaskets *C*, as shown in Fig. 1. When laying coupling line, skids are placed across the ditch and the pipe, according to size and weight, is either carried or rolled on these skids. One follower, one gasket, and the center ring are placed on the first joint, and the other follower and gasket are put on the next joint, the latter being fitted into the center ring on the first joint. Bolts are then placed through the followers and are drawn up against the center ring, forcing the gaskets against the center ring and pipe and thus making a tight joint. After a section of line has been laid, it is lifted off the skids by means of short scaffolds or horses and lowered into the ditch as the skids are removed (see Fig. 2). Great care must be exercised in lowering coupling lines, or the joints will be strained and leaks develop.

Many different styles of couplings have been used, starting with cast-iron couplings, which make excellent joints but are too heavy and easily cracked if strained when the pipe is lowered

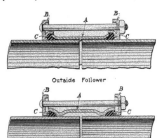

Outside Follower

Inside Follower

FIG. 1 TYPES OF COUPLINGS FOR PLAIN-END PIPE

into the ditch. Practically all couplings used today, however, are made of steel and with either outside followers, which fit over the center ring, or with inside followers, which fit under the center ring. See Fig. 1.

It is the writer's belief that the outside-follower coupling is the most practical, for the reason that it is impossible for the rubber to blow out between the follower and the center ring; while with an inside follower a part of the rubber is pushed over the follower and there is nothing to prevent the rubber from blowing out between the follower and the center ring, in which case the gas flow cannot be shut off temporarily with wooden wedges while repairs are being made.

[1] Supt. Gas Pipe Lines, Empire Gas & Pipe Line Co., Jun. Am.Soc.M.E.
Presented at a meeting of the Mid-Continent Section of THE AMERICAN SOCIETY OF MECHANICAL ENGINEERS, Bartlesville, Okla., October 30, 1919.

## REPAIRING COUPLING LINES

Leaks on coupling lines occur from three different causes, namely, rubber blowing out, cracked center rings, or broken followers; and in case the leak is of any consequence it can be repaired without interrupting the service by installing an emergency sleeve over the coupling.

This emergency sleeve (Fig. 3) is made in six pieces, two large halves and four half-followers, which, when joined together, make a joint similar to a coupling. The large halves also have inside rubbers which make them gastight, as well as a relief valve located in the top half through which the gas flows while the repair is being made. When the repair work is completed the valve is

FIG. 2 LOWERING COUPLING LINE INTO DITCH

closed, making a tight joint. The sleeves are left on until it is convenient to shut the line out to remove them and repair or replace the coupling.

With the development of natural gas in the Mid-Continent field it was found that the rubbers heretofore used in coupling lines were affected by the gasoline in the gas, and after a short time were eaten out. About six years ago gasoline-proof rubbers were put on the market, and practically all lines laid with these rubbers have given very little trouble. These rubbers have a tendency to enlarge when saturated with gasoline, and in this way make a very tight joint. On a pipe-line system in California it was found that after a gasoline plant was installed, leaks occurred along the line due to shrinkage caused by the rubbers drying out, and it became necessary to bypass the gasoline plant occasionally and allow the rubbers to become saturated with gasoline until they had swelled sufficiently to shut off the leakage.

The great advantage in laying coupling line is the speed with which it can be laid. There is practically no limit to the amount of any size of plain-end pipe that can be laid in one day, except the number of men that can be employed on the work. While screw pipe will carry higher pressures than plain-end pipe, the following results obtained in tests show that plain-end lines will carry the high pressures employed with a good margin of safety:

| Size, in. | 6 | 8 | 10 | 12 | 16 | 18 |
|---|---|---|---|---|---|---|
| Weight, lb. | 17 | 24 | 32 | 42 | 48 | 53 |
| Pressure, lb. per sq. in. | 450 | 400 | 400 | 350 | 325 | 300 |

## WELDED LINES

In the natural-gas business a large amount of screw pipe and casing becomes unserviceable after being used once or twice, unless it is rethreaded and new collars put on. To eliminate this expense by utilizing casing unfit for further employment in drilling, experiments in laying welded pipe lines were started in 1916.

Several small lines were laid in the Mid-Continent field, and gas men were so well pleased with the results that attempts at welding large lines were made. A 10-in. line was laid by the Wichita Pipeline Company from Drumright to Cushing, on which very few leaks have occurred since its completion. The Oklahoma Natural Gas Company laid a 10-in. line from Dilworth to Enid with good results, allowance being made for expansion and contraction by digging a wide ditch and blocking the line first on one side of the ditch, then on the other. Later the blocks were burned, leaving sufficient slack in the line to care for all the expansion or contraction.

In 1917 the Wichita Natural Gas Company laid a 12-in. line from the El Dorado field to Valley Center, Kan., but this line did not give as good results as the 10-in. line mentioned. A great many breaks occurred, and in places this line would pull apart three or four inches.

Another 12-in. line laid by the Empire Gas and Pipeline Company from Bigheart to Hominy, Okla., also gave considerable trouble, breaking as often as sixteen times in twenty-four hours.

All of the above-mentioned lines were laid without expansion

FIG. 3 EMERGENCY SLEEVE FOR REPAIRING LEAKS ON COUPLING LINES

sleeves or couplings, and after watching closely the operation of these lines, it is the opinion of the writer that it is impracticable to lay welded lines larger than 10 in. without allowing for expansion and contraction.

The process of laying welded lines must necessarily be slow on account of the time required to make a weld and the scarcity of good welders. The field is full of welders who are graduates of welding schools where they are taught to weld castings and flat steel, but these men must be taught how to weld pipe after they reach the job.

Large lines are usually welded in sections of six or eight joints before lowering into the ditch, and snub holes dug to weld the line solid. This process saves much time and digging, and gives the welder a chance to roll the pipe while welding rather than lay on his back in the ditch to weld the bottom of the joint.

The two principal welding irons used are vanadium steel and Norway iron, both of which are good, but better results are obtained with the latter on account of its being a softer metal.

The greatest drawback to welded lines is the great inconvenience and danger in repairing a break. It is almost impossible to repair a cracked weld with high pressure in the line, for two reasons: First, if the crack is very large, the gas pressure will be too great to permit the use of an emergency sleeve; second, if the top half of the sleeve is accidentally dropped on the pipe, the weld is likely to give away entirely, allowing a large volume of gas to escape and blow the top half of the sleeve and the men out of the ditch. In order to avoid such danger the writer has found it advisable to instruct all field men to always shut out and drain a welded line before attempting repairs, which of course would be disastrous to the service if the line were the only source of supply to the market.

The writer does not recommend welded lines in sizes larger than 8 in. without using a coupling or expansion sleeve every six or eight joints, but lines smaller than 8 in. can be laid successfully, and in connection work, such as Y's, tees, ells, etc., welded

work can be depended upon and, if properly done, a great saving effected.

Excellent results have been obtained with the welding process in building meter stations, regulator stations, gasoline plants and refineries, in the construction of which it is necessary to do a large amount of pipe fitting, which, when threaded nipples and flanges are employed, is slow and expensive.

In pipe-line work it is more practical to use acetylene and oxygen in tanks on account of the necessity of moving often, but in stationary work an acetylene generator can be used and the acetylene generated at about half the cost when supplied in tanks.

Table 1 gives results of tests made by the Kansas City Testing Laboratory, and shows the relative strength of the weld to the pipe.

TABLE 1 STRENGTH OF PIPE WELDS

WELDING ROD TESTS

| Description | Elastic limit, lb, per sq. in. | Ultimate strength, lb, per sq. in. | Elongation, per cent in 8 in. | Reduction of area, per cent. | Per cent of elastic limit of pipe sample. | Per cent of ultimate strength of pipe |
|---|---|---|---|---|---|---|
| Vanadium No. 1.......... | 46230 | 56870 | 11 | 73 | .... | .... |
| Vanadium No. 2.......... | 73300 | 79000 | 10 | 75 | .... | .... |
| Norway No. 1.......... | 42600 | 50300 | 35 | 60 | .... | .... |
| Norway No. 2.......... | 63620 | 59200 | 20 | 75 | .... | .... |

PIPE AND WELD TESTS

| Description | Elastic limit, lb, per sq. in. | Ultimate strength, lb, per sq. in. | Elongation, per cent in 8 in. | Reduction of area, per cent. | Per cent of elastic limit of pipe sample. | Per cent of ultimate strength of pipe |
|---|---|---|---|---|---|---|
| Pipe Sample No. 1........ | 48000 | 56360 | 22 | 40 | .... | .... |
| Welded Sample No. 1... | 46350 | 54660 | 8 | 21 | 96 | 97 |
| Welded Sample No. 2... | 38800 | 38800 | 0 | 0 | 81 | 69 |
| Welded Sample No. 3... | 37600 | 41000 | 3 | 12 | 88.3 | 72.6 |
| Welded Sample No. 4... | 31940 | 31940 | 0 | 0 | 67 | 54.5 |
| Pipe Sample No. 2........ | 42280 | 60080 | 22 | 40 | .... | .... |
| Welded Sample No. 1... | 39400 | 44620 | 3 | 16 | 93.5 | 74.3 |
| Welded Sample No. 2... | 38680 | 42800 | 3 | 12 | 91.5 | 71.4 |
| Welded Sample No. 3... | 38080 | 51190 | 5 | 18 | 90.1 | 85.3 |
| Welded Sample No. 4... | 33950 | 47730 | 6 | 11 | 80.2 | 74.1 |

### SCREW LINES

Screw lines are laid by two different methods, by hand and by machine. In laying pipe by hand, one of the gang, known as the "stabber," starts the joint pipe into the collar of the preceding joint, and rolls it by hand for several threads to see that it is not cross-threaded. The joint is then screwed up as far as possible with a snubbing rope, and tongs are used to tighten it. These tongs are provided with a square key let into the shorter jaw crosswise, the exposed corner serving to grip the pipe, and this key can be very easily turned to expose a fresh corner or replaced. Before laying screw pipe the threads should be oiled with a heavy oil or pipe dope, which will aid in screwing up the joint and will also preserve the threads and make the joints easy to break when reclaiming a line.

Screw-connection work where only flange unions are used is a very particular job, requiring great patience and accuracy, but when completed makes the best connections. However, a large amount of screw-connection work is finally connected with a solid steel sleeve similar to a plain-end coupling but of greater length, to allow for expansion and contraction.

Several machines are manufactured by means of which it is possible to lay screw lines by power. This method is much faster than laying with tongs, but requires careful supervision to avoid screwing up a joint cross-threaded or burning the threads by running up a joint too fast.

### PIPE-LAYING MACHINES

The two pipe machines most used in the Mid-Continent field are the "Buckeye" and "Mahoney," or California pipe machine. The Buckeye machine moves on caterpillars, and the pipe is laid from its rear end. The Mahoney machine is not equipped with caterpillars but moves on the pipe by means of a traveling gear operated by a clutch. When a joint has been laid a jack is placed under its head end and the machine moves forward to the end of the joint by means of the traveling gear. The machine is also equipped with a jack on which it is raised by power to take the weight off the pipe while another joint is being screwed up. Two sets of tongs are located on the front end, one for back-ups to hold the joint previously laid, and another to tighten the next joint. When it is desired to make an over-bend in the pipe, it is merely necessary to place a jack under the pipe at the point where the bend is to be made and run the machine ahead far enough to give the desired weight to make the bend. An attachment may also be purchased for these machines with which it is possible to make a sag bend in the pipe. This is done by the use of a 10-ton hydraulic jack operated by power from the motor. A force of fifteen or twenty men can operate the Mahoney machine, and where the country is fairly level a great deal of time can be saved, as it is possible to lay from thirty to forty joints an hour.

Screw pipe, while used in laying small gas lines, is used exclusively in laying oil lines, as a much higher pressure can be maintained on a screw line than on a coupling or plain-end line, and after a time rubbers in a coupling line will become affected by the oil. There is also considerably less friction in a screw line than in a coupling line, due to the different design of the joints.

### General

In laying gas pipe lines there are many details which aid efficient and continuous operation of a-line.

When connecting wells in a field it is always good policy to install a check valve at the point where the well line connects to the main line, to avoid a loss of gas in case of a break in the well line, or gate or tubing in the well.

In fields where salt water or gasoline is encountered with the gas it is necessary to install traps or drips to catch the fluid before it reaches the main line, as fluid affects measurement of the gas, regulation of pressure and operation of compressing stations, and at times even reaches the consumer's meter. A great many different types of drips have been designed, but all perform in practically the same manner—that is, they allow the gas to expand and decrease in velocity, which causes precipitation of the entrained fluid.

When constructing a main pipe line it is advisable to install a main-line gate every four or five miles, with a relief valve and pressure-gage tap on each side of the main-line gate. When a line is so laid it is possible, in case of a break, to shut out a short section of the line and exhaust the gas through the relief valves without draining the entire line and discontinuing service to consumers on sections of the line not affected by the break. With the gage taps it is possible for the man supervising the repair to observe the pressure being maintained on either side of the break.

In constructing river crossings for pipe lines which must give continuous service, it is always advisable to lay more than one crossing. The lines in the river can be of smaller pipe than the main line, but the total capacity should equal that of the main line. River crossings should be laid in deep water when possible, as the greater depth will lessen the danger of having the line torn out by drift. A line should never be laid straight across a stream, but swung in an arc upstream, the radius of the arc varying with the length of the crossing. In this manner the resistance of the line is greatly increased and many breaks avoided.

River lines are usually laid of screw pipe and each collar covered with a river clamp (Fig. 4), which removes all strain from the joint and adds weight to the line.

Occasionally welded river crossings are constructed, and in this case a split emergency sleeve, or a river clamp which has a rubber packing similar to that of an emergency sleeve, should be used, as it would be impossible to repair a welded crossing without first removing it from the water or without constructing cofferdams.

### Operating a Pipe-Line System

In operating a gas pipe-line system it is good practice to carry the lowest pressure with which it is possible to handle the market, and in cases where small towns are located some distance from the main line and are supplied by smaller lines, regulators are installed at the main line and the pressure reduced rather than carry main-line pressure to the town limits.

In the operation of a natural-gas system it is impossible to turn in all gas available and allow it to go where it will, as the consumers on one part of the system would have an oversupply of fuel, while those on another part would suffer from shortage. To avoid such unequal service, gas companies maintain what is usually known as a dispatcher's office. Here pressure reports are received by telephone from all important points on the system at certain intervals during the day, and by carefully tabulating and observing these pressures, it is possible by regulation of the gas to maintain sufficient pressures throughout the entire system. In case of line breaks or line shut-outs for any reason, the field men are required to notify the dispatcher's office before a line is taken out of service in order that the service may be maintained by supplying from another source. In fact, the duties of a pressure dispatcher parallel those of a train dispatcher, and all wells, lines and regulators are operated upon orders from him.

In the operation of a pipe-line system the entire effort of the organization is devoted to giving continuous service, and in order to do this men are stationed at various places along the system.

FIG. 4   CLAMP USED TO COVER COLLARS OF SCREW-PIPE RIVER CROSSINGS

The lines are patrolled continuously by line walkers, who repair small leaks as they find them and report to their immediate superior leaks which they cannot repair. In such cases a crew is taken to the leak at the earliest possible moment to make the repairs.

It is seldom realized that while many homes are comfortably warmed by natural gas, it is often necessary for pipe-line repair gangs to be out night after night working to keep the lines together and prevent gas shortage. In fact, the gas consumer seldom hears of a break or leak on the gas line from which he receives his supply, due to the efficiency which must necessarily be maintained in a pipe-line organization.

---

The methods employed during the war of applying the electro-deposition of iron to the repair of motor vehicles were described recently before the Institution of Automobile Engineers by Mr. B. H. Thomas, who instituted the process. It was found possible to deposit a layer of iron up to about 2mm. in actual thickness on any simple cylindrical surface of wrought iron or steel, mild or cast. If the work is properly done this layer is firmly adherent, and it is practically impossible to chip it away with hammer and chisel from the basis metal. It is deposited direct on the surface without preliminary coppering, and can be subjected to red heat without apparent deterioration. It can be carbonized and hardened in the ordinary way. The layer presents an extremely smooth surface, and the thickness put on in a given time can be predicted with considerable accuracy. It can be filed or ground, and takes a high finish. It is perfectly satisfactory on such parts as brake and clutch shaft ends and starting handle spindles, and there is no obvious reason why it should not be satisfactory for building up worn journals.

# The Engineer and the Woodworking Industry

### A Great Industry in Which There Exists an Urgent Need of Engineering Skill

By THOMAS D. PERRY,[1] GRAND RAPIDS, MICH.

The lack of the application of engineering principles and practice to the woodworking industry has been felt very strongly by a group of Grand Rapids members of The American Society of Mechanical Engineers. One of these, Mr. Thomas D. Perry, has prepared the following article, outlining the scope of woodworking field and presenting some of the engineering problems involved, with a view of opening the subject for discussion and as a preliminary to a possible session on woodworking at a later meeting.—EDITOR.

WOODWORKING, one of the oldest civilized trades, is now one of the largest industries in the United States. It is doubtful whether any other major group of modern manufacturers gives evidence of less scientific knowledge of its products. The reason for this is quite apparent.

Agriculture has been very slow in adopting scientific means to improve the efficiency of its workmen because son could easily

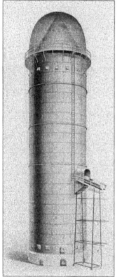

FIG. 1 THE MUSKEGON REFUSE BURNER

follow father with no special training. Carpenters' sons could do likewise; hence, a few branches of rudimentary woodworking were not at all difficult to acquire. In each case one or two or a half-dozen men could complete the finished product and no extensive organization, capital or costly machinery was necessary.

The problem of the metal worker was far different. The

[1] Vice-President and Manager, Grand Rapids Veneer Works.

necessary processes were complex and expensive. They demanded skill for the many necessary steps between the iron ore and the rolled boiler plate or the seamless boiler tube. It was therefore quite natural that the engineer was attracted to the intricate and involved problems of the metal worker, and that the metal worker, in his turn, felt the need of the trained scientist.

A survey, no matter how superficial, will demonstrate that, while the woodworker may not have needed the engineer in the past, he certainly needs him now. Raw material is increasingly more valuable; and wasteful methods are increasingly more culpable; labor requires higher wages; mechanical methods are in demand; manufacturing units are larger; need for coördination is more pressing; and output is more standardized. Proper adjustments to meet these conditions will permit larger and more intensive production.

The woodworker is by no means wholly to blame for not desiring or utilizing engineering knowledge. The engineer seems to have considered that such simple problems as those of the woodworker were beneath his dignity and not worth while. It follows, therefore, that if the woodworking industry and the engineering profession are to be of mutual benefit, a broader aspect and a complete readjustment of attitude are necessary.

It may be of interest to enumerate a few of the branches of the woodworking industry and suggest a few of the problems that need better solutions. The fundamental division is between primary woodworkers, converting timber and logs (natural growths) into lumber or some other form of raw material intermediate between the tree and the final product; and secondary woodworkers, who take the lumber and make it into pianos, furniture, caskets, house trim, etc. The problems are quite unlike, although there may be a few exceptions where the primary manufacturer still produces the finished product, as in the case of cooperage and flooring and many kitchen implements made directly from the log.

The sawmill operator has the problem of bringing logs to the mill, which includes hill and valley logging, land and swamp skidding, rail and water hauling, all more or less complicated by the rotation of the seasons and the variations of the weather. Considerable engineering skill has been shown in some of these problems, mostly, however, in the direction of machine design. The utilization of waste in such plants stands out as a gigantic travesty on the principles of economy. It is frequently said that steam power does not cost the sawmill man anything because he has plenty of waste, therefore, why watch the economies of steam consumption!

Nearly every sawmill has a large waste burner (Fig. 1) to consume the refuse so that the burning of the waste will not be an undue hazard to mill buildings and lumber yards. Temperatures of 800 deg. are not unusual in these burners. The better class of waste is put under the boilers and the remainder in the waste burner, by means of which untold potential power is dissipated. The waste from trimming trees in the woods is left where it offers fuel for forest fires. In a few instances the author has known of the utilization of mill and woods waste for power production such as the manufacture of artificial ice in southern states and for making wood fiber for wrapping papers. These few instances, however, are exceptions to the rule of universal and unjustifiable waste in lumber production.

The rotary cutting, slicing and sawing of veneers is subject to all of the above problems, and in addition has several refinements such as preparing, cutting and handling thin and delicate sheets, selecting for figure, preventing discoloration, keeping flat, etc.

Another question, which needs the attention of the engineer is that of artificial drying. The problem of establishing the conditions of heat and circulation are not so difficult, but the tendency of the wood to dry too rapidly on the surface and establish

FIG. 2  TYPICAL VIEWS IN WOOD-TREATING AND WOODWORKING ESTABLISHMENTS

(a) Typical view of wood-treating plant, showing at the left a horizontal cylinder for treating ties and timber, at the right a vertical cylinder for treating paving blocks, with a gantry crane between them for handling cross ties and lumber, and at the extreme right a steam storage locomotive. (b) Rotary veneer cutting from basswood log at the plant of the Grand Rapids Veneer Works, Grand Rapids, Mich., using Coe lathe. (c) Typical interior of modern Grand Rapids furniture factory, showing center trucking aisle with machines located near windows. (d) Typical sawmill installation on the Pacific coast. (Copyright Cress Dale Photo Co., Seattle). (e) Interior of a Grand Rapids vapor kiln of the box type; one of a battery of 14 at the plant of the Packard Motor Car Co., Detroit.

internal strain is usually overlooked. Fig. 3 shows a typical specimen of lumber drying: the cross-section of this board shows a wet center and a dry surface. The establishment and maintenance of a proper degree of enveloping humidity is the most difficult part of the drying operation, usually not appreciated by engineers. Of the various systems on the market each has its advantages: the blower types for veneer, finishes, glued work, etc., where there is no great thickness to be dried; the condensing kiln for the very thick and green lumber where plenty of time is available and careful operation possible; the ventilating type for speed and efficiency in the average plant of the majority of lumber producers and consumers. Artificial drying for piano actions, phonograph horns, and high-grade furniture requires a more refined process than that of general lumber drying. The opportunity for research and practical investigation along these lines is very great.

Cooperage, or the making of barrels, kegs, tubs, pails and casks, is in a very primitive condition and would completely discourage any engineer who might assume the problem of developing it. In a tight barrel the staves are first baked dry in a kiln, then machined to size and then assembled in barrel form and hooped

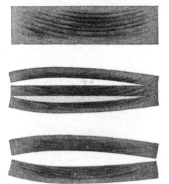

Fig. 3 Cross-Sections of Unevenly Dried Board with More Moisture in Center than on Outside Surfaces
The two lower sections show the effect of resawing this board and releasing the internal strain. The big problem in lumber is to dry without internal strain.

for one-third of their length. The next operation is steaming to soften and render the assembled staves pliable so that the bulge of the barrel can be produced. When the hooping operation is completed the barrels, without heads, are set over little fires to again bake them dry. This process is only typical of others as crude.

The impregnation of woods with various preservatives has received some engineering encouragement, perhaps because the creosoting of ties has been so intimately connected with railroad engineering. Very light woods used for life preservers need an "encysting" process to prevent rapid mold and rot. Woods used for insulation require similar retardents against decay. Fireproofing treatments to make wood shingles less inflammable are in the same class. Sterilizing treatments are necessary to destroy various fungus growths in warm climates. such as the blue stain in pine timbers. Woodwork to be sent to oriental countries must be made unattractive to the multitude of insect borers which attack nearly all untreated woods. The solution may be partly chemical, but it is certainly well within the broader engineering scope.

The making of wood pulp is an example of wood utilization developed through engineering skill, but it happens to be a process that is complicated and demands large production units.

The strength of timber of various species has been the subject of many tests under varying conditions, but a recognized standard compilation does not exist. The airplane has brought this subject prominently to the front, because the life of an airman or the success of his military mission would frequently depend on the reliability of wood parts. This phase of strength, recognized in only two of many uses, offers a vast field for consistent investigation and comparative study. What relation has density, or frequency of growth rings, or heart and sap wood to strength and elasticity? The authorities are few and their conclusions far from final. The tip of an airplane propeller moves at the rate of five miles, or 26,000 ft., per min. How can wood be made to endure this strain?

Wood is used in automobiles for spokes, felloes, bodies, top bows, dash boards, foot boards, running boards, battery parts, and sometimes for the all important work of distributing the weight of body and engine on the axles. Authorities vary widely as to the merit of wood and steel wheels, and as to the relative comfort and durability of wood and steel trusses from body to axles. No conclusive authority exists for intelligent guidance.

Production methods or industrial engineering are hardly developed as yet in the woodworking industry. Factory equipment and systems are largely lacking in productive efficiency. The application of production processes which have proved successful in other lines than woodwork is hardly to be regarded as more than experimental in such a new field. The average woodworker has very little regard for the production engineer at the present time. However, there are developing here and there unrelated and uncoördinated examples of scientific factory production. What better service could be rendered by engineers and engineering publications than to assist in standardizing and disseminating the best results? A few firms are specializing in this work, but there is available no supply of trained men from which to recruit their staffs of workers.

The public schools of this country have been struggling with the sloyd or so-called manual training in woodworking branches for many years, but where is an institution that produces even the rudiments of a woodworking training? Even the Massachusetts Institute of Technology has next to nothing to offer except patternwork and tests on strength of lumber. The forestry schools, notably the New York State College of Forestry at Syracuse, are making a beginning, but employers do not realize the need of sending their men for even short courses. Recent attempts to draw from the whole state of New York for a six weeks' course in kiln engineering resulted in less than a dozen pupils for the second year's trial, and the first year brought even less response, which did not warrant carrying the course through.

The foregoing is far from exhausting the list of the many opportunities in woodworking for suitably trained engineers. The demands for recreation and amusement require golf clubs, ball bats, billiard equipment, bowling-alley equipment, dominoes, checkers, chessmen, and gun stocks, all of which call for manufacturing efficiency in diverse lines.

The domestic equipment of the average home embraces matches, toothpicks, clothespins, cutlery handles, churns, caloric cookers, hollow woodenware, rolling pins, potato mashers, broom handles skewers, and many other articles. Each is made in sufficient quantity to occupy many factories and employ hundreds and often thousands of workmen.

Airplanes and automobiles are not the only vehicles of transportation that use wood in large measure. There are wagons, buggies, freight cars, passenger cars, boats and ships in which strength and durability are absolutely essential in material, design and fabrication.

The field for the engineer in woodworking is almost unlimited, but the development of such a new and untried line will take education, patience and adaptability on the part of all who are vitally interested in the trades that employ so large a proportion of our citizens.

# An Investigation of Compressed-Spruce Pulleys

By GEORGE S. WILSON[1], SEATTLE, WASH.

WITH the view of determining the practicability of manufacturing pulleys from native spruce, a series of tests was conducted in the Mechanical Engineering Laboratories of the University of Washington. These tests extended over a period of nine months from July 1, 1919, to April 1, 1920, and the investigation included the following:

*a* Determination of some of the factors influencing the carrying capacity of compressed-spruce pulleys

*b* A comparison of compressed-spruce pulleys with other commercial types.

The work was undertaken at the request of Watts & Walsh, of Tacoma, Wash., manufacturers of compressed-spruce power-transmission and conveying equipment, to whom much credit is due for assistance given in conducting the tests.

The tests were made with single-ply leather belting 4 in. wide by 0.225 in. thick or 0.9 sq. in. in cross-section. The pulleys were nominally 18 in. in diameter and 6 in. face. A constant speed of 2000 ft. per min. was maintained.

In general, the tests consisted of sets of observations on belt slips (slip includes creep) at different initial belt tensions and varying brake loads. Five different initial tensions were used in each set (37½, 75, 112½, 150 and 187½ lb per sq. in. of belt

Fig. 1 Machine Used in Pulley Tests at University of Washington

section), and the load varied from a minimum to the maximum that the belt would carry or the capacity of the driving motor would allow.

The machine used in the tests was built for the Mechanical Engineering Department of the University of Washington. Fig. 1 shows the general arrangement, *A* being the driving pulley, *B* a universal coupling, *C* the tension-weighing device, *D* the arc-of-contact meter, *E* the driving test pulley, *F* the driven test pulley, and *G* the Prony brake and tension-adjusting device.

The slip meter used in the tests consists of two cams, on the driving and driven shafts, respectively, each of which makes contact with a spring brush once each revolution. When the cams are in phase

Abstract of a paper presented at a meeting of the Washington State Section of The American Society of Mechanical Engineers, Seattle, Wash., April 19, 1920.
[1] Associate Professor of Mechanical Engineering, University of Washington. Mem. Am. Soc. M. E.

an electric circuit is closed through the contacts and a lamp flashes once each revolution. As the driven shaft lags behind the driving shaft, however, the cams gradually get out of phase and the light goes out until the driven shaft has dropped one full turn behind the driving shaft, when it starts flashing again. To determine the slip a continuous counter, operated by the driving shaft, was started at the first light flash and allowed to run until the light had died out; when the light again began to flash, the counter was stopped. The reading obtained was the number of revolutions made by the driving shaft for the loss of one revolution of the driven shaft.

After considerable preliminary experimenting it was decided that a 50 per cent end-grain spruce pulley built with about ¼-in. laminations and compressed under 250 tons total pressure, with

Fig. 2 Compressed-Spruce Pulley Employed in Tests
(¼-in. laminations compressed under 250 tons total pressure)

the face properly crowned, smoothly finished and coated with high-grade shellac, should be used in a comparative series of tests with steel, split wood, paper and cast-iron pulleys. Fig. 2 shows the pulley selected.

In the comparative tests the driving motor used was not sufficiently powerful to load the pulleys to their maximum capacity, but the curves in Figs. 3 and 4 show the results within the load

### TABLE 1 HORSEPOWER CAPACITIES OF PULLEYS PER INCH WIDTH OF BELT

| Pulley | Slip, per cent | | | Comparative transmitting capacity at 2 per cent slip |
|---|---|---|---|---|
| | 1 | 1.5 | 2 | |
| Cast iron | 0.58 | 0.69 | 0.74 | 100.0 |
| Steel | 0.71 | 0.78 | 0.81 | 109.4 |
| Wood, split | 0.63 | 0.77 | 0.82 | 110.8 |
| Paper | 0.62 | 0.76 | 0.83 | 112.1 |
| Compressed spruce | 0.76 | 0.81 | 0.84 | 113.5 |

limits. It will be noticed that for belt slip below 2 per cent the variation in the capacity of the different pulleys is small, with the exception of the cast-iron pulleys, which have the lowest capacity of any of those tested.

Beyond the 2 per cent slip range the variation in capacity is probably greater. In all cases the capacity will be affected by the amount of crown on the pulley, the finish of the pulley face, and the condition of the belt and belt lacing.

While Figs. 3 and 4 furnish the engineer a definite measure

451

of pulley capacities, Table 1, which gives the horsepower per inch
width of belt, will probably better serve the needs of the average
pulley user. This table is calculated from Nagle's formula, assuming a belt contact of 180 deg., a belt thickness of 0.225 in.,
a belt speed of 2000 ft. per min. and a stress of 200 lb. per sq.
in. of belt section; values of the coefficient of friction were taken
from Fig. 4.

Since the weights of pulleys may be a factor in some installations, the pulleys tested (all 18 in. diam. by 6 in. face) were
weighed, with the following results:

| Pulley Weight, lb. | Cast Iron 65 | Steel 40 | Wood, Split 21.5 | Paper 41.5 | Compressed Spruce 22 |
|---|---|---|---|---|---|

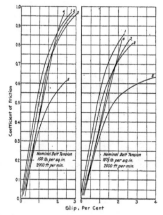

FIGS. 3 AND 4  FRICTION-SLIP RELATIONS FOR VARIOUS TYPES OF
PULLEYS

Pulleys: 1, Compressed Spruce; 2 Steel; 3, Wood, split; 4, Paper; 5,
Cast Iron. Belt speed, 2000 ft. per min.

Table 1 and the foregoing weights show that the compressed-
spruce pulleys as now developed are at least as serviceable as any
of the ordinary commercial types.

## STANDARDIZATION OF FIRE-HOSE COUPLINGS

THE state-wide movements in several states of the Middle
West toward the adoption of National Fire Hose Couplings,
reveal the fact that this problem of standardization is fast becoming one of nation-wide importance. Failure to recognize it as
such in the past has resulted in enormous fire losses which have
been due in no small measure to the lack of uniformity in fire-
hose couplings in adjacent cities and towns.

The first real movement for standardization may be said to
date back to the great Boston fire of 1872. Following this con-
flagration several of the near-by cities, profiting by the experience, adopted the Roxbury thread coupling which was the
Boston Standard at that time. Later New York City adopted a
thread which became the local standard through general adoption
by most of the New Jersey and New York State municipalities
within a radius of from 50 to 75 miles. Local standards have
been adopted in a number of places which have given a feeling of
security and complacency. Experience has shown, however, that
such an attitude is not entirely justified because of the fact that

outlying towns are little better off than if there were no standards
at all.

The first organized effort toward the adoption of a universal
standard thread was made by the International Association of
Fire Engineers during the period 1873-1883. It was not, however,
until a Committee on Hose Couplings and Hydrant Fittings of
the National Fire Protection Association was appointed in 1905
that the work was effectively started. Through the efforts of this
committee the couplings known as the National Standard promptly received unqualified approval and adoption by all of the leading organizations concerned with fire protection and by many
municipal fire departments. Coincidentally The American Society of Mechanical Engineers appointed a sub-committee on fire
protection whose report (Transactions No. 1398, Volume XXV,
1913) describes the methods by means of which the standards
were adopted, the specifications for hose couplings, and the
methods for converting non-standard couplings.

In December 1919 an illustrated pamphlet, Standardization for
Fire Hose Couplings and Fittings, containing an extended treatment of the subject of this standardization, and a complete description of the resizing or standardization tools, was issued by
the National Board of Fire Underwriters of New York City.
Upward of 12,000 copies of this pamphlet have been distributed
among water works and fire-department officials, state fire
marshals, insurance commissioners and other interested organizations. This pamphlet, copies of which are still available, contains a description of a set of tools devised for the purpose of
reforming to standard size, threads which are slightly different
from standard.

There are about 8000 cities or towns in the United States and
Canada that have fire-hose fittings in service. Statistics show
that 15 per cent of them have National Standard thread
couplings; 70 per cent have couplings that may be resized and
made standard, and the remaining 15 per cent are either using
couplings which have six threads to the inch or the various
types of snap or clutch couplings.

Because of the progress in adopting the National Standard
Coupling made by the state of Indiana, this state has been recognized as a pioneer in the field of state-wide adoption. As a
result of the satisfactory demonstration of a set of resizing or
standardization tools on the undersized fittings of the town of
Thornton, the oversized fittings of Lebanon and a variety of
threads in the headquarters of the Indiana Fire Department, the
organized movement for state-wide adoption was initiated. According to the latest statistics, 13.5 per cent of Indiana towns are
already using standard threads, and 70 per cent have threads
which may readily be made standard, but in the case of the remaining 16.5 per cent it is believed that the best results will be
obtained through complete replacement of the fittings now
in use.

It has been estimated that the above-mentioned 70 per cent of
Indiana towns and cities can be completely standardized in eight
months by keeping in continuous use six complete sets of resizing
or standardizing tools. It can readily be seen that the cost of
the entire work is small as compared to the advantages to be
derived from the elimination of confusion and delays due to the
existing misfits.

Similar state-wide standardization movements have been instituted in Kentucky, West Virginia, Minnesota, Michigan and
Illinois, and the added interest made manifest at recent demonstrations of the resizing tools and the number of inquiries received at the office of the National Board of Fire Underwriters
in New York, clearly indicate that this long-desired movement
toward standardization and conservation will, in the near future,
attain nation-wide magnitude and importance.

The American Society of Mechanical Engineers has supported
this movement from its inception and, as a body, is much interested in the recent strides toward its general adoption. Members who reside in localities where standardization movements
have not already been initiated, are urged to take upon themselves the matter of arousing their respective communities to the
importance of the subject.

# ENGINEERING SURVEY

A Review of Progress and Attainment in Mechanical Engineering and Related Fields, as Gathered from Current Technical Periodicals and Other Sources

SUBJECTS OF THIS MONTH'S ABSTRACTS

STEVENS SUPER-POWER STATION
GENERAL ELECTRIC DESIGN OF A SUPER-POWER STATION
ECONOMICS OF POWER GENERATION
REPORT OF COMMITTEE ON PRIME MOVERS OF THE NATIONAL ELECTRIC LIGHT ASSOCIATION
EXPLOSION GAPS IN LARGE POWER STATIONS
STEAM TURBINE, DOUBLE-CYLINDER EXTRACTION AND LOW-PRESSURE
STEAM TURBINE, POWER-PRODUCING, PRESSURE-REDUCING
SUPER-POWER ZONES IN NEW ENGLAND

HIGH-PRESSURE, HIGH SUPERHEAT STEAM PLANT
POWER PLANTS, DISTRIBUTION OF FUEL AND WATER CONSUMPTION
BALANCING TURBINE ROTORS
GEAR EROSION
CONDENSER TUBE CORROSION
HIGH STEAM PRESSURES
PULVERIZED FUEL
WATER PURIFICATION IN POWER PLANTS
RADIATORS FOR AIRCRAFT
AIR-DUCT DESIGN (CHART)
VISCOSITY OF OIL BLENDS, SAYBOLT

METAL SUBSTITUTION IN GERMANY DURING WAR
COKE BREEZE ON UNDERFEED STOKERS
HEAVY-OIL ENGINES, FUEL INJECTION
AUTOMOBILE ENGINES, LIMITS OF SLOW RUNNING IN
GRAPHITE LUBRICANTS
CAMS FOR RADIAL AND ROTARY ENGINES
BICYCLE CHAINS
ROLLED PRODUCTS OF 99 PER CENT NICKEL
BEARINGS, RADIO-THRUST, LOAD CHARACTERISTICS
LOCOMOTIVE, COMPOUND, WITH SUPERHEATER

## The Stevens Super-Power Central Station

OF the many remarkable changes in the life of the people and in the industries of the world during the past 30 or 40 years, none has been more profound than the rapid increase in the consumption of mechanical power. A man of two-score years ago consumed very little mechanical power. His food and supplies to a large extent came from nearby neighborhoods, brought by horse-drawn vehicles. He did comparatively little traveling, his home was lighted by kerosene lamps, and in most towns his water was pumped by hand from wells or drawn from springs. Many hand-made articles were in daily use and those which were produced in factories involved hand operations which have since been superseded by machine operations requiring the use of power.

The man of today, however, at every step is directly or indirectly a large consumer of power, from the moment he gets up and turns on the water for his bath to the $time$ when on retiring he turns off the light and thus relieves the load on the lighting plant of his own town. Not only has there been a great increase in the multiplicity and variety of manufactured articles, but there has been a corresponding growth in transportation and illumination. Raw materials, food supplies and produce are brought from all parts of the world, and manufactured products are as widely distributed. The great systems of electric transportation and electric illumination have come in, taking power from immense central stations, and the automobile is made by the million in plants with thousands of power-driven machine tools.

It is obvious that with this enormous increase in power requirements some means must be provided for generating the power on a scale commensurate with the demand and adequate to the requirements of the future. This means larger stations, and the tendency toward bigger units is strengthened by the growing cost of fuel and labor, as it is a well-established fact that the larger units are more economical in both regards than the smaller ones.

Nevertheless the concentration of very large volumes of power generation in one unit involves certain not inconsiderable difficulties and dangers. Items that would be comparatively small in the central station of moderate size become very big indeed in the giant installations, such as the proposed Stevens super-power central station, and furthermore, like the proverbial carrying of all the eggs in one basket, it involves great dangers in the way that a breakdown of a single big super-power central station would affect the life of the community to a far greater extent than a similar breakdown of only one of several scattered and possibly interconnected units.

It is obvious that on one hand there is a real field for the very large power-generating plants. On the other hand, the design of such plants involves a superlative degree of skill and care both to secure the highest economies in operation possible for this type of construction and to insure the degree of reliability consistent with the vital place which the super-power central station is to play in the life of the community which it serves.

From this point of view the two articles by John A. Stevens, Mem. Am. Soc. M. E., in *Power* (May 25 and June 1, 1920) are worth the most careful consideration.

The basic features of the Stevens plant are the use of a definite large size of unit, namely, 30,000 kw.; the separation of all boiler houses from their turbine rooms by "explosion gaps," the separation of the turbine rooms from each other by explosion walls; and further, the use of a particular type of turbine as the driving unit.

Thirty thousand kilowatts has been selected as the standard size of the unit as being the size within the zone of maximum economy in steam-turbine engineering practice and also as being about the limit of size that can be supplied with four sections of boilers served with one central stack. This size of unit, of which about eight would be used in the station, determines the general arrangement of the station.

The description of the station may be well started with that of the boiler plant. It is proposed to use Stevens-Pratt boilers, of 1716 hp. normal rating, arranged in units of four with one central stack which is from 300 to 350 ft. high in order to discharge the gases and dust at high altitudes. As shown in Fig. 1, the boilers will be installed in practically an independent building with an outdoor passageway between it and the turbine room. The reason for this is safety. Notwithstanding the high degree of skill in the design of large modern boilers, explosions and minor accidents, such as tube failures, are to be expected. To provide for such cases and also to guard against the results of short-circuits in generators an "explosion gap" (so-called) is left between the boiler room and the turbine room, localizing the damage to the plant where the accident occurred.

The same regard for safety has led to the division of each group of turbine rooms themselves by protecting walls which reach to the crane height. With such a wall 25 or 30 ft. high there is some chance for the operators to escape in the case of bursting of steam mains or fittings unless the explosion is very disastrous. Still further, each 60,000-kw. capacity power house should be separated from the next unit of the same size by a space of 20 to 40 ft. in the clear. See Fig. 2.

As regards heating the air which supports the combustion, a very high temperature is not advocated, but only a moderate degree of heat, steam being extracted from high-pressure stages to supply the air heaters.

On the firing floors stands are provided for the main motor-actuated stop valves, which may be controlled electrically from the switch house or from any part of the plant. Also, a master valve at or near the boiler plants, with remote motor control, is desirable, to be used in the event of serious disaster in the turbine room.

For economizers a type is recommended representing a combination of cast-iron sections and steel sections through which the water is pumped in series.

Among features not usual in ordinary plants but suggested for the Stevens super-power station may be mentioned periscopes for observation of the furnaces and for reading the water-gage-glass levels, a remote-control whistle valve, and even a wireless apparatus with aerials connected to the stacks, also searchlights

determined by the appropriateness of the selection of the power-generating units. From this point of view it is of interest to note that in the present instance a rather novel turbine arrangement is proposed for use, which the author terms a double-cylinder extraction and low-pressure turbine.

In addition to the power generated by super-stations, considerable amounts of steam have to be used either for district heating wherever there is a field for it or for manufacturing and heating processes. Until recently, however, no turbines were built for operating non-condensing, or against a back pressure of over 60 lb., though in some instances, as, for example, in the rubber industry, considerably higher pressures (80, 90 or even 100 lb.) are desired; and these are what the double-cylinder extraction and low-pressure turbine is intended to supply.

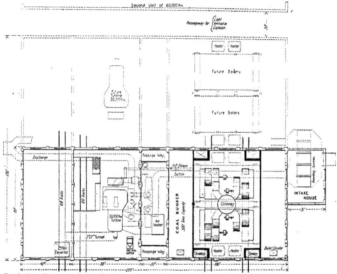

FIG. 1 PLAN VIEW OF A THIRTY THOUSAND-KILOWATT INSTALLATION SHOWING FOUR 1710-HP. STEVENS-PRATT BOILERS AND ONE 30,000-KW. TURBINE. DOTTED LINES INDICATE DUPLICATE INSTALLATION, MAKING A TOTAL OF 60,000 KW. FOR A COMPLETE UNIT

above the elevators for protection of the station in case of disturbances. Telephones are provided at all important points, for example, at or near the boiler-drum manholes, near the fans, etc., so that any accident or defect can be reported instantly. Above the umbrella on the stack and at the base of the umbrella, pipes are provided to wash off the dust that may collect on the plate-glass roof, while underneath this roof a shade is installed on rollers so that on a hot summer day the operating room may be shaded and made comfortable.

As a precaution against damage to the generating units a check valve made of heavy plate is placed at the warm-air discharge to the stoker fan, so that in the event of a minor explosion in the tube area of the boilers the dust and soot will not be blown into the main turbine generator.

No matter how carefully and skillfully a plant of the size of the Stevens super-power central station may be designed or built, from the commercial point of view its ultimate success will be

As shown in Fig. 3, such a turbine consists of high-pressure and low-pressure units mounted on one shaft with, say, a 10,000-kw. generator. This machine is said to operate automatically with steam extracted from zero gage to, say, 80 lb. pressure, or from 0 to 244,000 lb. extracted at a predetermined pressure. It will operate entirely condensing with only a slightly reduced economy from that of a standard machine, and it will operate also for any intermediate extraction of steam or generation of power. At the same time large amounts of steam may be admitted to the low-pressure casings in the way of returns from vulcanizing machinery, which should be put through regenerators or separators at 1 to 2 lb. pressure for the precipitation of the acids. The control of this machine is from the sustained or maintained pressure in the steam line to the manufacturing systems. In other words, the machine is controlled from its by-product rather than from its electrical end; but at the same time, if more service is demanded from the electrical end than the proportionate

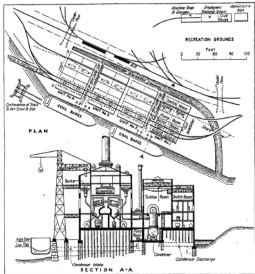

FIG. 2 PLAN AND ELEVATION OF A TWO HUNDRED AND FORTY THOUSAND-KILOWATT
SUPER STATION

Note the explosion gaps between each 60,000-kw. unit.

For smaller units the author recommends a type of turbine which he terms a power-producing, pressure-reducing turbine (Fig. 4). This turbine is so valved and arranged that it will automatically float on the line and deliver quantities of steam in proportion to the demand, giving a proportionate amount of by-product power. Furthermore, as these turbines are manufactured in small units they may be located at a distance from the boiler plant, which will do away with the necessity of running large and expensive low-pressure mains throughout the plant from the central station.

Summing up, Mr. Stevens states that in his opinion the power plant of the future will be along the basic lines presented in his articles, and he suggests that power contracts be made with the permission to manufacture what power is desired by by-product turbines—that is, by-product to power production—and by such station designs as he has shown in the articles. Continuing, he says:

Locating these super-power stations in large industrial communities would enormously reduce the amount of coal necessary to handle the power of such communities, for the coal consumed per unit in the super-stations, broadly speaking, is about one-half that of the isolated plant, and by making use of the by-product turbines for manufacturing and heating steam, enormous further reductions in the use of coal are made, to say nothing of a relay power at the customer's plant or at least a part relay to the purchased power.

These large stations require such vast amounts of condensing water that they can be located only on such rivers as the Connecticut or Merrimack, or at shore ports

amount of steam passing through to the manufacturing process would admit, the automatic valve mechanism will open to the low-pressure or condensing turbine the necessary amount of steam to follow the electrical demands.

As regards economies, it is stated that a machine of this type

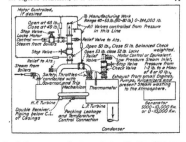

FIG. 3 DIAGRAMMATIC SKETCH OF DOUBLE CYLINDER EXTRACTION
AND LOW-PRESSURE TURBINE

may be installed to keep the entire output when running non-condensing or part non-condensing and part condensing, at a rate of 7 lb. of steam or less per kilowatt-hour at the generator leads chargeable to power in kilowatts.

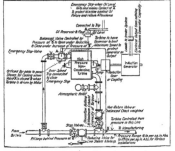

FIG. 4 DIAGRAMMATIC SKETCH OF TURBO-GENERATOR ARRANGED TO
OPERATE AS A POWER-PRODUCING, PRESSURE-REDUCING TURBINE

on rivers, lakes or oceans, where the intake canal and discharge canal, or conduit, from the condensers should be 1000 ft. or more apart. Of course, the question of currents, tidal effects and many other conditions enter into this problem, but this statement is made for guidance of designers in planning a station location. It would not be advisable to start any super-station where the ultimate development possible would be less than 100,000 kw. It would be better not to locate in any position where an ultimate development of 240,000 kw. or 300,000 kw. was not possible. This, however, would be controlled largely by the cost of transmission lines, franchises, money values, and innumerable other conditions that enter into a project of this size.

When otherwise practicable, it will be well so to choose locations for central stations that later full advantage may be taken of economies to be brought about by the recovery and utilization of distillates and other chemical products from coal. This is a matter at present little thought of but likely to be of increasing importance for the future.

To illustrate further what prominent engineers of the country think of the possibilities of the great super-power stations of the

FIG. 5  MAP SHOWING APPROXIMATE SUPER-POWER ZONE

future, Mr. Stevens refers to the super-power plan presented at the Midwinter Convention of the A. I. E. E., New York, February 19, 1920. The zone embraced by this plan (see Fig 5) comprises the area between Boston and Washington, and extends inland from the coast 100 to 150 miles. The transmission lines from the hard and soft coal fields, Niagara Falls, Cedar Rapids on the St. Lawrence, and from the Maine hydroelectric plants which Mr. Stevens suggests are shown in Fig. 6. He says:

From investigations and research I have found that there is a 300,000-kw. load, not including the steam-railroad electrification, within a radius of fifty miles from Newburyport or Beverly, Mass., which could be tied in with the great water-power developments at Manchester, N. H., Lowell and Lawrence, Mass. and with the great development and proposed developments of the Edison Company, of Boston, Mass.. the conservation in this power zone alone amounts to about 500,000 long tons of coal per year, conservatively speaking, with all the attendant advantages in the conservation of coal, money and men. The project considers the use of the river banks and railroad rights-of-way for transmission lines, and placing a station of that kind in the northeastern range of the proposed super-power

zone illustrated by Fig. 5 would reinforce the loop transmission line suggested.

Further, there is a vast amount of coal in Nova Scotia which, if used as pulverized fuel, would make two methods of shipment to this point by water—Nova Scotia coal and coal from the Virginia fields. This would greatly relieve the congested freight situation.

A cross-tie in this loop in New England and somewhat paralleling the Connecticut River and its hydroelectric and large steam developments is suggested. This would further substantiate and augment the eastern part of the super-power zone loop in New England and would also further relay the great power demands of New York City and vicinity. There are other untold resources in New England for such a project as has been outlined herein.

The time will come when no company or individual will be allowed to operate any plant at a thermal efficiency of less than a predetermined percentage; neither will any company be allowed to operate any turbines or equivalent machinery condensing when they could be operated non-condensing and the exhaust steam used in heating and

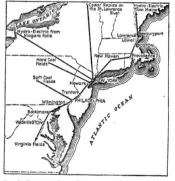

FIG. 6  SUGGESTED POWER-TRANSMISSION LINES FOR SUPER-POWER ZONE

manufacturing, or when purchased power from a large super-station can be advantageously made available. In other words, as coal and oil, in fact, all known fuels, are non-renewable, natural resources, the general public will be forced to use the utmost care in their consumption. (Power, vol. 51. no. 21, May 25, 1920, pp. 832-836, 3 figs; and no. 22, June 1. pp. 879-8844. 11 figs.)

## General Electric Design of a Super-Power Station

THE high cost of fuel today and the prospect that it may go still higher in the future have given a powerful impetus toward the development of central-station design in which coal will be used with the greatest economy.

This tendency has been materially assisted by the rapidly growing demand for power. Both of these factors have created a situation where, in many instances, it becomes possible to build power plants of practically any size that will permit their economical operation, the question of getting rid of superfluous power being practically absent through the fact that there is no superfluous power available today and very little prospect of there being any in the near future.

The design described by H. Goodwin, Jr., and A. R. Smith was developed for a particular condition where some of the fundamental considerations were high fuel cost, moderately good load factor, extreme river floods and high-voltage underground distribution. It contains, however, features of great interest and capable of general application. Some of the apparatus shown,

particularly boilers, economizers and preheaters, are only proposed designs.

The idea underlying the entire design of the proposed super-power station is that great economy can be obtained from large-capacity generating units operating at high steam pressure, especially when a number of these are located in one station.

In the present instance, a steam pressure of 350 lb. and super-heat of 350 deg. fahr. are proposed to be used, which means that both the boilers and the turbines have to be adapted to comparatively new conditions.

The boilers (Figs. 1 and 2) will have a rating of some 1600 hp., or 16,000 sq. ft. of heating surface, and will have the two banks of tubes separated with the superheater in between. The reasons for this arrangement are, first, to permit using a two-pass boiler and still keep the economizers on the main floor, which means a greater number of tubes in height; and second, to get a high amount of superheat without an excessive amount of superheater surface.

The high pressures and superheats employed make it necessary to simplify the steam piping as much as possible so as to make the entire piping system more flexible and avoid accidents. Hence there are no steam headers in the general sense of the word, but there is a transfer header for equalizing pressures and transferring steam between boiler rooms. The steam piping is also materially simplified by the fact that there are no steam-driven auxiliaries except the few house or auxiliary turbines, of which more will be said later.

The economizers will be of wrought-iron tube construction with headers inclined with relation to the tubes. They will be cleaned with steam soot blowers, and it is anticipated that there will be no moist soot deposit, because the water entering the economizers will first be heated to 150 deg. or 160 deg. with exhaust steam, this bringing the temperature well above the dewpoint of the gases.

The flow of water in both the vertical tubes and the headers

The very high steam temperature involved in the present design makes it practically impossible to use any small steam-driven auxiliaries, such as boiler-feed pumps, etc. On the other hand, the auxiliaries, especially feedwater pumps, should be of the great possible reliability, which means that it would not be advisable to drive them electrically from the main buses. The installation of low-pressure boilers to operate the boiler-feed pumps would not be advisable either.

In order to meet these conditions it has been decided to use one 2500-kw. auxiliary or house turbine for each pair of main turbines. Such a unit can be very well adapted to the steam conditions contemplated, and will have sufficient capacity to supply all auxiliaries through a separate busbar.

As regards condensers, the main condensers will be of standard design. The house turbines will each be provided with a low jet condenser which will normally produce about 15 in. of vacuum. The circulating water for this condenser is the condensate from

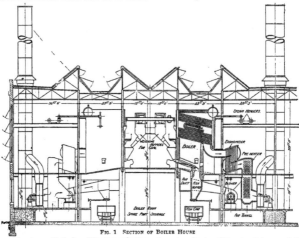

FIG. 1 SECTION OF BOILER HOUSE

is of the thermo-siphon type, with the flow of the gases and water in the economizers in countercurrent.

The air going to the boilers comes through a tunnel (Figs. 1 and 2) from the turbo-generators, which means that it is preheated. In addition to this, air may pass through heating tubes in preheaters divided into two units per boiler, the arrangement being such that the boilers nearest to the auxiliary room will burn the heated air discharged from the generators, whereas the boilers at the far end will burn the air from outside. In this connection a remark is made to the effect that it is believed that preheaters are perfectly practicable for use in stationary plants and can be made to show a good return on the investment if the cost of coal is at all high.

The use of economizers and preheaters makes induced draft unavoidable in view of the increased draft losses and the low temperature of gas entering the stacks. The draft fans might be of the ordinary belt type, or possibly of the multi-vane type, as the temperature is low and the pressure comparatively high. The induction type of stack employing high-pressure blowers may be substituted and, in general, provision should be made for the case of failure of fans.

the main units, in addition to which some of this water may be recirculated.

In a plant of the proposed size, the question of coal- and ash-handling equipment becomes of great importance. As regards the latter, it is proposed to dispense with all kinds of ash-conveyors and instead provide under each boiler ash hoppers of such capacity as to contain 12 or 24 hours' storage, so that ashes need to be removed only once or twice during the day, the ash hoppers emptying directly into standard railroad cars.

The coal-handling equipment has been designed so as to have the bulk of the coal unloaded from barges by means of traveling crane towers at the dock and transported by two belt conveyors directly to the four receiving hoppers and crushers. From the receiving hopper the coal is delivered directly to the outside storage or through a crusher and skip hoist to the overhead outside beams.

The duplicate belt conveyors provide a large coal-handling capacity when needed, and also act as a reserve in case one conveyor should break down. All belts are dead-ended and there are no traveling trippers.

The intention is that one operator located in a control cab

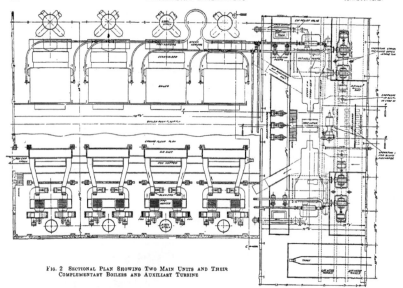

FIG. 2  SECTIONAL PLAN SHOWING TWO MAIN UNITS AND THEIR
COMPLEMENTARY BOILERS AND AUXILIARY TURBINE

above each receiving hopper will operate the revolving gantry crane, the crusher, the skip hoist, etc. Another operator will be located on each electrically operated larry to transport and weigh the coal from the overhead bins to each boiler. The emergency coal storage handled by the locomotive traveling crane will be operated only when the excess coal is being stored or reclaimed.

It is proposed that there be one spare larry which can be readily run into any one of the four firing aisles to replace any defective larry. It will be observed that the revolving gantry crane reclaims the coal from the circular storage without moving the bridge; thus this method is very rapid when reclaiming coal, although in distributing the coal the bridge will have to be moved

slightly from time to time, but to minimize this movement outside shoots are shown on the four sides of each receiving hopper tower. The revolving gantry cranes overlap so that coal can be transferred from one pile to another, and, furthermore, the design shown can be partially built and extended from time to time without interfering with operation or without changing existing structures.

As yet no data are presented as to the character of the main driving units. It is stated, however, that the total generating power will be 245,000 kw., 300,000 kva. capacity, on 66,000-volt distribution. (*General Electric Review*, vol. 23, no. 5, May 1920, pp. 355-418, 15 figs)

## Economics of Power Generation

THE growing cost of fuel and labor forces power-generating stations to keep much closer track than ever before of every item of their expenditures, with the result that some rather unsuspected relations have been discovered.

Thus, a recent analysis of the log-sheet records of a typical electric generating station in England showed that the coal consumption might be regarded as being made up of two separate items, namely, a constant quantity per shift, which in this instance was 14,560 lb., and an additional amount for every kilowatt-hour produced—in this case 2.2068 lb.

This held good not only for heavy enough loads but also for the Sunday and early-morning loads, when the station was running only at a small percentage of capacity.

The corresponding investigation with respect to the water consumption of the station showed that this also might be regarded in a similar way, namely, as a constant quantity of

60,000 lb. of water per shift plus an additional amount of 14.37 lb. for every kilowatt-hour generated, this covering, however, only the amount of water delivered to the boilers.

These results may be expressed in the form of equations—

$$C = 14,560 + 2.2068\,K \quad \ldots\ldots\ldots\ldots [1]$$
$$W = 60,000 + 14.37\,K \quad \ldots\ldots\ldots\ldots [2]$$

where $C$ represents the total number of pounds of coal burned per shift, $W$ the number of pounds of water evaporated per shift, and $K$ the output per shift expressed in kilowatt-hours. This permits expressing either the coal consumption in terms of water evaporation or vice versa as shown by the following equations:

$$C = 5346 + 0.1536\,W \quad \ldots\ldots\ldots\ldots [3]$$
$$W = 6.512\,C - 34,810 \quad \ldots\ldots\ldots\ldots [4]$$

From Equation [1] it is evident that the greater the output the less important becomes the constant quantity of 14,560 lb.

of coal, the station tending more and more nearly to a coal consumption of 2.2068 lb. per kilowatt-hour as the load increases. This is the limiting value of coal consumption for the particular plant. As is the case with all limits, it can never be actually attained, but will be approached as the output becomes greater. The same applies to the question of steam consumption.

These equations would further indicate that within the capacity of the plant any additional load given to the station may be considered to be generated for a net expenditure of 2.2068 lb. of coal and 14.37 lb. of water per kilowatt-hour, although the average coal and water consumption, if calculated by usual methods, would be considerably greater. This determines the cost of service on extra loads, which do not necessitate any increase in wages, capital charges, or items of expenditure, except coal and water.

Equation [4] is of interest as it shows that the water evaporated per shift is 34,810 lb. less than it would be if every pound of coal evaporated 6.512 lb. of water. Obviously, the greater the output the less important becomes the constant, which has to be deducted from the weight of water evaporated and in the limit 6.512 lb. would be turned into steam for every pound of coal burned. The low value of evaporation is in this instance ascribed "entirely to the nature of the material which is sold as coal nowadays," in the words of the author, which show that the British have to contend with pretty much the same conditions as we have in this country.

An interesting application of the above equation is made to the measurement of the so-called stand-by losses, which is done in the following manner:

Referring to Equation [1], it will be seen that in the particular station under consideration, if no power at all were generated for a shift, there would nevertheless be required 14,560 lb. of coal to be burnt during the time to keep the station in a condition to take the load when required. The purpose of this amount of coal is to make good all the heat which is lost in the plant, whether by radiation, chimney losses, or condensation,

and irrespective of whether the losses are in the engine room or the boiler room. Turning now to Equation [3], we see that if no water were to be evaporated during the shift, i. e., if the boiler-feed valves and junction valves were shut and only sufficient fuel were burnt to maintain the boiler house in a state to take up the load when called upon, the coal consumption would be 5346 lb. per shift. Hence this amount may be considered as equivalent to the boiler room losses, while the difference between 14,560 and 5346, which is 9214 lb., may equally be regarded as representing the engine-room losses. The writer is aware that this method of reasoning is open to criticism, but, from a considerable experience of power-plant economics, he is inclined to believe that the deductions are of practical value and possess a very large measure of truth.

The constant water loss of 60,000 lb. per shift, given in Equation [2], may seem high, but it is well in line with corresponding figures from other power stations. It must be remembered that the no-load condition in the engine room implies that the output has fallen to zero, but not that the plant is shut down. Now when it is realized that even a small turbine may have a no-load consumption with ordinary conditions of steam and vacuum, of something like 5000 lb. of steam per hour, or 40,000 lb. per shift, the total stand-by loss of 60,000 lb. of steam per shift does not appear excessive. The balance of 20,000 lb. per shift has to provide for running the turbine auxiliaries, the boiler-feed pumps, and the stoker engines as well as for leakage and condensation in the whole system. The stand-by steam has obviously to be generated by the stand-by coal, so that the evaporation at no load is found to be equal to 4.12 lb. of steam per pound of coal. If, however, we subtract the 5346 lb. of coal required to make good the boiler-room losses as deduced above, the evaporation works out to 6.512 lb. of water per pound of coal, which is what would be expected from Equation [4]. (Robt. H. Parsons, in *The Times Engineering Supplement*, vol. 16, no. 548, June 1920, p. 190).

# Report of Committee on Prime Movers of the National Electric Light Association

THE following excerpts have been made from the report of the Committee on Prime Movers of the National Electric Light Association, Technical and Hydro-Electric Section, which was presented at the forty-third convention of that society, May 18-22, 1920. The report is very much in detail, comprising 130 large-size pages, and covers the following subjects, considered in the main with respect to recent developments and current practice: Steam Turbines, Condensers, Boilers and Superheaters, Economizers, Grates and Stokers, Burning Lignite, Pulverized Fuel, Coal By-Products, Fuel Oil, Coal and Ash Handling, Power-Station Auxiliaries, Boiler- and Turbine-Room Instruments, Ventilation of Station Buildings, Higher Steam Pressures, Oil Engines, Hydraulic Prime Movers, and Hydraulic Operation and Combined Operation of Hydraulic and Steam Plants. Considerable space is devoted to reports from manufacturers on improvements or changes in design in steam boilers, condensers, grates, stokers, etc. In the quotations which follow, however, attention is mainly given to new developments of a broader nature and to operating features. These quotations, also, relate only to steam plants using coal as fuel.

## STEAM-TURBINE DEVELOPMENT

A record of the units which have been manufactured since the last report indicate that 30,000 kw., 1,800 r.p.m., 60 cycles, and 35,000 kw., 1,500 r.p.m., 25 cycles, can be considered for the present as the maximum standard sizes of single-shaft, single-barrel units for large installations.

During the past year a considerable amount of information has been obtained from the operation of these large-sized units as well as from some of the smaller machines, and a period of adjustment and improvement in both design and construction,

following the experience which has been gained, would seem to be imperative if these large-sized machines are to be brought to a requisite degree of reliability and operating efficiency.

The records of operation of single-shaft units have not by any means been satisfactory. Since the last report there have been several additional wheel failures which have occurred to units of this type and on sizes ranging from 5000 to 30,000 kw. capacity.

The General Electric Company has made a number of marked changes in the design of its large-sized turbines, the principal changes being the stiffening of the center bearing and modifications in the design of turbine disks, buckets, diaphragms and packing.

The changes in the turbine disks consists of a thickening of the disk in an attempt to reduce the possibility of injurious vibration; elimination of the weight-balancing holes, and the change in the number and location of steam-equalizing holes so as to move them well in from the rim; and changing from an even number to an odd number of holes with rounded and polished edges at the wheel surface.

The changes made in the buckets, particularly in the last stages, consist of the addition of a reinforced section, carrying it nearly to the center of the bucket so as to stiffen the construction materially. In addition, the very long buckets are reinforced by the use of a bracing strip, in order to minimize any vibration which might occur.

The changes to the diaphragms consist of replacing some of the cast-iron sections by cast-steel construction where stresses indicate this to be necessary. Additional clearance space around the diaphragm is also provided with the idea of allowing for the removal of a large part of the condensed water, which has , contributed to serious erosion of the blading in certain stages. A radical departure has been made in the type of diaphragm packing. The new type differs from the earlier construction in that the stationary, inwardly projecting short teeth of the labyrinth have been changed to much longer teeth which are mounted on the revolving element. It is claimed that this improvement makes it practically impossible, when properly installed, for any rubbing to distort the shaft due to the generation of heat at the point of contact.

An analysis of the cause of the serious wheel failures during the past year has resulted in the accumulation of sufficient data to indicate that the trouble has been due in the past to a vibration of the wheel structure. This vibration, which has been described as a fluttering of the web of the wheel in combination with the high working stresses of the material, gives rise to the formation of fatigue cracks. These cracks, which occur not only in the wheel but also in the bucket, when allowed to continue, result in a complete rupture. This action was accentuated in the disks where sharp tool marks or rough surfaces are present, or where holes for weight- or steam-balancing purposes are drilled in the wheel. Considerable investigation is still necessary, however, in connection with the subject of vibration of wheels before the full facts can be known and a proper design be forthcoming that will eliminate any danger from this source of trouble.

No marked changes have been introduced during the past year by the Westinghouse Electric & Manufacturing Company in the design of its single-cylinder units up to 30,000 kw. rating, although a number of minor improvements have been made. To meet the demand for larger capacities than 30,000 kw., the company has standardized on combinations of single-cylinder units of 20,000 to 22,000 kw. maximum rating in tandem, cross-compound and triple-cylinder arrangements.

While the combination of compound turbo-generators seems to be increasing in favor, and while test results indicate a higher thermal efficiency for such a combination as compared with single units of the same aggregate capacity, sufficient data are not at hand to show conclusively whether the practical operation of units bears out the theoretical advantages which are claimed.

### BALANCING TURBINE ROTORS

While it has been standard practice for years in large manufacturing companies to balance statically all high-speed rotating parts, it is only recently that equipment has been developed that will readily permit of correct dynamic balancing of rotating elements.

One type of machine, the Carwen static-dynamic-balancing machine, has proved very effective in the balancing of moderate-size rotating elements. The principle of operation of this machine depends upon the creation of an artificial unbalance in some of its own moving parts, which exactly counteracts the unbalance of the body to be tested. When this artificial couple or unbalance becomes the exact counterpart of the unbalance in the body being tested, the whole unit ceases to vibrate and the quantities of the unbalance are indicated on suitable dials located on the machine.

The Akimoff combined static and dynamic balancing machine, although built along somewhat different lines, has also a bedplate hinged in such a manner as to vibrate in two planes. No compensating couple is used in this machine. Instead, a balancing clamp is attached to one end of the object to be balanced and rotates with it. The position of the clamp and the location of a sliding weight attached to the clamp are changed until the resultant of the unbalanced force is neutralized and the vibration reduced to zero.

Another device which has recently been designed and built for measuring the amplitude of the vibration of machines while in operation—called the "Vibroscope"—has given very satisfactory results. In this device the frame or vibrating part of the instrument is so arranged that it may be conveniently attached to the machine whose amplitude of vibration is to be determined. The upper leg of the frame is slightly inclined, and has attached to it by a thin, flexible spring hinge a heavy, flat-plate pendulum supporting a micrometer microscope as its outer end. This pendulum plate has a very slow period and becomes the "steady point" of the combination. Directly beneath the microscope at the end of the horizontal part of the frame is a narrow slit illuminated by an electric light. By looking through the microscope with the instrument at rest, the width of the slot is adjusted to, say, five divisions on the illuminated scale; then with the machine running, any apparent increase in the width of the light slot is due to vibration. The amplitude of the vibration of the machine under test can be directly determined by proper calibration of this amplitude with relation to the fundamental amplitude of vibration of the vibroscope.

In connection with vibration and the fatigue of metals the following statement has been obtained from Prof. C. A. Adams, Chairman of the Engineering Division, National Research Council:

A very comprehensive research into the fatigue of metals is being conducted at the University of Illinois under the auspices of the Engineering Division, National Research Council, and under the direct supervision of Prof. H. F. Moore, with the assistance of an advisory committee of the Engineering Division.

The amount appropriated by the Engineering Foundation Board for this research is $15,000 per year for two years. In addition to this, the University of Illinois contributes, in the services of Professor Moore and other overhead expenses, the equivalent of at least $6,000 per year.

The chief object of the research is to establish as far as possible such fundamental laws of the fatigue of metals as will make possible the intelligent design of apparatus and structures which are likely to be subjected to alternating stresses.

The investigation will determine whether the stressing of the metal during vibration is above or below the elastic limit, and if below, how much time may be expected to elapse before rupture occurs.

It is expected that this experimental work by the National Research Council will furnish a great deal of valuable data in connection with the design of turbine blading and disks.

### GEAR-DRIVEN TURBINE UNITS—EROSION OF GEARS

The record of performances of small gear-driven units as experienced by operating companies has, in a number of instances, been far from satisfactory, the troubles being due in many cases to the abnormally excessive wear on the reduction gears.

An analysis of other gear troubles experienced indicates a number of reasons for their occurrence which summarize as follows: In some cases the oiling systems are incorrectly designed, or an improper grade of lubricating oil is used. Other cases of gear troubles are traceable to excessive or uneven tooth pressure, end play and poor meshing. The remedy for this class of trouble is purely of a mechanical character and readily corrected if detected in time. Mention has also been made of cases where the wear and tear on the gears has been aggravated by unbalanced rotating parts, or by foundations which have not been sufficiently rigid to maintain properly the correct alignment of the gear-teeth faces.

The manner and point of application of the oil to the gear are of prime importance. It is evident, therefore, that the oil must be applied at a point on the gear where it will most effectively reduce the friction to a minimum. The oil should be of a viscosity best adapted for the class of work intended, used in sufficient quantity, and applied to the entering side of the gear in a continuous film. Excessive oil does not improve gear lubrication and is likely in some cases to reduce the efficiency of certain types and design of gears, with a consequent increase of temperature. Sufficient space, however, should be provided in the gear casing so that when the oil has reached the maximum safe operating temperature it may be allowed to cool, rest and settle, to permit liberating as much as possible of the air or water globules it has collected during the period of work under actual churning conditions. Maximum oil temperatures allowable in best operating practice vary from 110 to 160 deg., according to the grade of oil and class of gear.

Examination of a number of gears from units which have been in operation over various periods of time, which were operated under proper conditions of lubrication, and where the alignment was satisfactorily maintained, shows a decided pitting below the pitch line in the gear and slightly above in the pinion, this line being the point at which the maximum tooth pressure occurs. A number of interesting experiments have been made during the last year, which seem to indicate that a considerable amount of gear erosion is due to causes other than those directly resulting from mechanical action.

Through the courtesy of Mr. S. K. French, of the American International Shipbuilding Corporation, a considerable amount of interesting information has been secured, based on the results which he has obtained in the investigation of gear operation in connection with gear drive for ship propulsion. It would appear that the pitting is due to the action of the oxygen content of the air in the oil film, which is entrapped under high pressure between the gear-teeth faces.

When the gear teeth come in contact with each other, the oil film on the gear is submitted to an action similar to that of a hammer stroke, more sliding than direct. The globules of air, which have been churned with the oil, are entrapped in this oil film, and are likewise submitted to a pressure sufficient to generate enough heat to cause an oxidization or burning of the metal in the tooth face. A number of experiments have already been carried out following this assumption. These experiments consisted in, first, adding a large quantity of oxygen to the gear casing; second, in supplying the gear casing with a mixture of nitrogen and $CO_2$. In the first series of tests, excessive pitting was noticed within a few hours' running, while in the operation of the gears in a mixture of nitrogen and $CO_2$, or where only a small amount of oxygen—approximately six per cent or less—was left, continuous operation over a long period of time indicated absolutely no pitting. Although the results cannot be regarded as conclusive, there is reason to believe that valuable information will soon be forthcoming, aiding in the solution of the problem of the more successful operation of gears. It is well known that certain classes of oils have a greater affinity for air and water than others of a different base or grade. These oils, after being churned with air or water and then allowed to settle, free themselves very slowly. It is evident that if this is true, the settling action, the size of the gear casings, the capacity of the oiling system, and the grade of lubricating oil must be given most careful consideration.

### HIGHER STEAM PRESSURES

Boiler manufacturers seem to feel certain that steam generators can be produced which will work satisfactorily at pressures as high as 1000 lb., but they all recognize the fact that such steam generators must be designed along lines different from those now in use. Present designs of water-tube boilers are variously estimated as adaptable to pressures of from 300 to 450 or even 500 lb.

Turbine builders express themselves as capable of producing turbines for any pressures for which boilers can be developed, but they place an upper limit on temperature. This limit is variously set at from 650 to about 800 deg. fahr.

Manufacturers of valves and fittings believe that they can produce such materials for steam pressures up to 1000 lb., provided the maximum temperature is limited to between 700 and 800 deg. fahr.

The majority of the manufacturers of power-plant equipment and the better known power-plant designers all seem to feel the desirability of using higher pressures as a means of continuing the development toward higher thermal efficiency which has been so rapid during the past two decades.

Power-plant executives will welcome anything which will assist in counteracting the effect of the ever-rising prices of fuel and labor.

This may all be summarized by saying that such information as is available indicates the possibility of building plants to operate at pressures not in excess of 400 to 450 lb. at temperatures not in excess of 700 to 800 deg. fahr, if it can be shown that the thermal gain actually realizable is sufficient to more than offset the increased first cost and probably increased maintenance. The boilers used would probably be heavy versions of present-day designs and not necessarily the most economical for such pressures.

However, it is not to be assumed from this summary that such an advance can be made in any easy fashion. Almost every piece of apparatus will require redesign and the highest type of engineering skill will be required to prevent costly failures.

In this connection attention is called to the fact that the A.S.M.E. standardization of flanges extends to a maximum of 250 lb. only.

At the present time 300 lb. with a total temperature of from 650 to 700 deg. fahr. may probably be taken as a conservative limit, and even under such conditions a high degree of operating skill is required even when the best obtainable apparatus is installed. Smaller plants which cannot ordinarily afford the most skilful operators will probably find the economic balance at a pressure in the neighborhood of 200 to 250 lb. and a total temperature not in excess of 600 deg. fahr. for some time to come.

As a result of these visits it can be stated that real operating difficulties have been experienced with valves and fittings operating at pressures in the neighborhood of 300 lb.

Welded joints in the steam line, such as were described in a previous report, are satisfactory so long as the line remains in continuous service. If cut out of service frequently, leaks develop at a rapid rate.

On large valve bodies by-pass valves and bonnet joints have given more trouble than usual.

Leaks developing under this pressure must be repaired at once because cutting of the metal occurs very rapidly. Attention is called to the fact that this corresponds to the experience of the builders of high-pressure steam automobiles, who found it extremely difficult to construct a tight throttle valve because of the rapidity with which the metal is cut away by the steam.

Blow-off valves have given more than the usual amount of trouble because of leakage, and similar statements can be made with respect to water column blow-off valves, valves on drips, and other small valves.

### INFLUENCE OF MICROSTRUCTURE OF METAL ON THE RESISTANCE OF CONDENSER TUBES TO CORROSION

The experience of the Southern California Edison Company with condenser tubes lasting over a period of eight and a half years at the Long Beach steam plant, has indicated that the microstructure of the metal has a great influence upon the resistance of the tube to corrosion.

As an illustration of the difference in tubes operating under similar conditions, there are three condenser units at Long Beach: No. 1 on a 12,000-kw., No. 2 on a 15,000-kw., and No. 3 on a 20,000-kw. turbine; all vertical machines, all taking water from the same tunnel and operating under practically the same temperatures.

In No. 1 condenser 21 tubes failed during 725 hours of operation; in No. 2 condenser 13 tubes during 2156 hours; and in No. 3 condenser 1892 tubes during 6232 hours. Thus there failed respectively 29, 6, and 336 tubes for 1000 hours of operation. The No. 3 tubes were found to be very coarsely crystalline, whereas both No. 1 and No. 2 tubes were of fine grain. Replacing the coarse-grained tubes with others of a fine structure cured the trouble.

Such microphotographic study as it was possible to make indicated that dezincification occurred round the crystal boundaries or possibly in the matrix surrounding the crystals. When the crystalline structure is coarse there is more tendency to local pitting; with fine-grained tubes the dezincification proceeds uniformly until the tube finally breaks from brittleness.

As a precaution against electrolysis the tube sheet was bonded to the condenser head and shell.

At times there was experienced trouble with erosion of the ends of the tubes which the water enters, this erosion extending in about three inches, probably until a steady condition of flow takes place. It has always been associated with dredging operations near the water intake and is attributable to fine sand in suspension. All the tubes are of Admiralty mixture, not tinned.

It would be of great advantage to condenser-tube users if some simple physical test could be applied to all tubes to in-

dicate their fitness as regards microstructure, as actual microscopic study takes too long and is too expensive.

## PULVERIZED FUEL.

The past year has witnessed an increased number of powdered-
coal-burning plants installed and a quite rapidly growing interest on the part of engineers in this development. The handling
of powdered coal in large quantities has been successfully carried
on in steel plants and cement mills, the coal being transmitted
for a distance as great as 2000 ft. from the powdering plant.
On the other hand, several companies are now placing on the
market unit pulverizing equipments, one or two units being
installed for each boiler. So far as the mechanical details are
concerned, either the centralized powdering plant or the unit
system seems to be highly satisfactory.

One of the earliest and largest plants with pulverized-coal
equipment, both in combined capacity and capacity of individual
boilers, is that built by Stone & Webster for the Western Avenue
steam-heating station of the Puget Sound Traction, Light &
Power Company at Seattle, Washington.

The Seattle installation was the outcome of extensive experiments, and the equipment is especially adapted for handling
lignite washery sludge from a dump of about 200,000 tons at the
Renton Mine, just outside of that city.

The boiler plant comprises ten boilers of an aggregate rated
capacity of 4100 hp. The preparation plant includes: Raw-
coal bunkers, roll-type crusher, crushed-coal bunker, rotary driers,
drier-coal bunker, pulverizing mills and pulverized-coal
bunkers, with the necessary conveying and transferring equipment. From the pulverized-coal bunkers the fuel is delivered
by screw feeders to the burners. As the driers are fired with
pulverized coal, and the entire preparation equipment is operated under a slight vacuum, the installation is practically dustless throughout.

The sludge burned reaches the plant carrying about 25 per
cent moisture, with a heating value of 7500 B.t.u.; ash on the
dry basis, 20 to 25 per cent. With this fuel boilers are operated
on a steam-heating load at 150 per cent to 175 per cent rated
capacity.

There are other installations in various parts of the country
operating on somewhat less severe fuel conditions serving boiler
plants of capacities around 2500 hp. which are also showing
very satisfactory performance. Among these may be mentioned
the 2340-hp. installation of The Milwaukee Electric Railway
& Light Company at Milwaukee, Wisconsin, burning a mixture
of Eastern Kentucky and Youghiogheny slack; the 1875 hp. of
the British Columbia Sugar Refinery Company at Vancouver,
B. C., burning the Nanaimo slack from Vancouver Island; the
2840 hp. of Morris & Company at Oklahoma City, Oklahoma,
burning bituminous slack from the Arkansas-Oklahoma fields;
and the 1800 hp. of the M. A. Hanna Coal Company at Lytle,
Pa., burning anthracite slack.

Speaking broadly, however, the boiler installations, as now
in operation, were not designed primarily for pulverized coal,
but are adaptations of settings originally constructed for use
with some other form of fuel and, on that account, are necessarily subject to many limitations inherent in previously existing
local conditions. In spite of this handicap, the commercial
practicability of pulverized coal as a boiler fuel has been
thoroughly demonstrated. A number of installations are now
under construction, designed throughout for the use of pulverized
coal and thus freed from the limitations referred to, on which
superior results may be anticipated. Proper volume and arrangement of furnace and simplicity in preparation and conveying
equipment are the most important requisites of the successful
installation.

Weighing the results so far obtained, we believe that the scope
of application of pulverization to the steam power plant may
be defined about as follows:

1　For a wide range of low-grade coals which only with great
difficulty, if at all, can be handled on existing forms of me-

chanical stoker, pulverization offers a satisfactory means of
efficient combustion.

2　For relatively higher-grade coals and very high in price, the
slightly improved efficiency obtainable by combustion in
pulverized form may frequently be sufficient to warrant
adoption where size of the installation and annual load factor are such as to make small percentages in combustion efficiency of large moment in operating cost.

3　For reasonably good grades of coal available at moderate
price, it is doubtful if the expense and complications of the
pulverizing equipment can be justified—except in very large
installations or in case of a very high annual load factor.

With regard to the comparatively recent development of unit
pulverizing equipments, one or more for each boiler, the following report from the Erie City Iron Works is quoted:

We have been conducting experiments in the burning of powdered
coal under a power boiler. The pulverizing equipment is that manufactured by the Aero Pulverizing Company. The pulverizer is direct-
connected to a variable speed motor. The coal is fed directly from
our coal bunkers into a funnel-shaped hopper at the feed end of the
pulverizer. The fuel is burned in a water-jacketed furnace applied
to one of our old-style horizontal water-tube boilers. In connection
with the test data which we give we wish to call attention to the
fact that the boiler has been in service a great many years and the
setting is far from being air-tight, thereby reducing the overall
efficiency, which would otherwise be greater.

We are at present experimenting with the pulverizer itself and
expect in the near future to make improvements therein which will
result in a considerably higher degree of fineness, which in turn
should result in a higher overall efficiency being obtained. We also
expect to produce this greater degree of fineness of the pulverization
at a lower power consumption.

A synopsis of the test referred to which was made in accordance
with the Short Form Boiler Test advised by the Boiler Test Committee of the A.S.M.E. is as follows:

| | |
|---|---|
| Percentage of ash in dry coal | 11.47 |
| Calorific value of 1 lb. of dry coal | 13.136 B.t.u. |
| Equivalent average per hr. from and at 212 deg. per | |
| sq. ft. of water heating surface | 5.6 lb. |
| Percentage of rated capacity developed | 162.6 % |
| Equivalent evaporation from and at 212 deg. per lb. of | |
| dry coal | 10.08 |
| Equivalent evaporation from and at 212 deg. per lb. of | |
| combustible | 11.38 |
| Efficiency of boiler and furnace based on combustible | 74.4 % |
| Furnace temperature just within furnace | 2.487 deg. |
| Stack temperature | 508 deg. |
| Speed of pulverizer | 1,480 r.p.m. |
| Lb. of coal as fired per hr. | 2.391 |
| Kw-hr. per ton of wet coal pulverized | 31. |
| Per cent moisture in coal before pulverizing | 6.44 |
| Per cent moisture in coal after | 1.78 |
| Per cent through 200 mesh screen | 41.6 |
| Per cent through 100 mesh screen | 66.1 |
| Per cent through 65 mesh screen | 79.8 |
| Per cent on 65 mesh screen | 20.3 |
| Volatile in coal | 30.3 % |
| Fixed carbon in coal | 52.52 % |
| Ash | 10.73 % |
| Sulphur | 2.45 % |
| Draft in furnace | 0.03 in. |
| Draft in damper | 0.44 in. |
| $CO_2$ | 16.41 % |
| $O_2$ | 2.10 % |
| CO | 0.11 % |
| Excess air | 10.14 % |

The coal was fed into the furnace at the front of the boiler
through a Y connection, each branch of the Y being 12¼ in. in
diameter, this Y being directly connected to the pulverizer. No
trouble was experienced due to Y connection clogging up.

## WATER PURIFICATION—THE USE OF EVAPORATORS

In marine service it has long been common practice to supply
pure make-up water for the boilers and other purposes from
sea water by means of evaporators. The use of evaporators for
supplying make-up water for generating stations has come into
use only during the past four or five years. One of the first
widely discussed installations of evaporators was that of the
Buffalo General Electric Company. The following is from a
letter from H. M. Cushing, Mechanical Engineer, giving information upon the operation of the evaporators:

We have installed two 30,000-lb.-per-hour, Griscom-Russell three-
unit, double-effect evaporators, taking steam at 200 lb. saturated. This

is obtained from a 275-lb. 275-deg. high-pressure line, by means of a reducing valve and desuperheater. The first equipment was installed in November, 1916, and the second one in April, 1919.

Our make-up runs between 5 per cent and 7 per cent. We have some reserve capacity in the evaporator equipment. For the year 1919 we evaporated 14,000,000 gal. of water.

In regard to economy, when one considers the one factor of cost of treated water per 1000 gal. of water used, the evaporator system will figure most expensive, but, although we have operated our boilers continuously from eight to ten hours a day at 300 per cent of rating, often operating at 400 per cent of rating for an hour at a time, we have had practically no boiler trouble, and none whatever traceable to the heavy loads which the boilers have been called upon to carry. We attribute this to the use of the evaporator system for obtaining our boiler make-up. We have had no pitting in our boilers or elsewhere on the feedwater system, although both the manufacturers and ourselves have searched diligently for signs of it.

The advantages and disadvantages of evaporators are given by the Committee in their report and are here summarized in brief form:

*Advantages:*

1 Practical elimination of boiler blowdown, which ranges from one to three per cent of the total feedwater.

2 Improvement in heat transfer due to elimination of scale.

3 Operation of the boilers over longer periods of time without shutdown.

4 Low operating cost of evaporation, which is automatic.

5 Almost entirely and permanently eliminates bad feedwater troubles.

*Disadvantages:*

1 Distortion of plant heat balance unless the entire electrical and steam-driven auxiliary system is laid out for the use of evaporators. The low thermal loss of an evaporator system depends on the fact that the latent heat of the vapor from one "effect" is used either as live steam in the next or for heating the feedwater above 210 deg. in a closed feedwater heater, or to 210 deg. in an open feedwater heater. Each "effect" used reduces the amount of vapor to be disposed of and consequently the amount of heat-balance distortion, but increases proportionately the first cost of the equipment. The "two-effect" system with vapor from the second effect going to the feedwater heater seems to be the more common layout for generating stations.

2 High first cost. If the proportion of make-up to total feedwater runs much over 8 or 9 per cent, either the heat-balance distortion is very great or the installation cost due to additional effects and perhaps a closed heater makes the cost of installation correspondingly high. Consequently there is an economic point somewhere around 10 per cent of make-up beyond which it does not pay to put in evaporators except in very special cases.

3 Leakage in condensers, sealing glands, etc., reduce the advantages of an evaporator system.

## Short Abstracts of the Month

### AERONAUTICS

RESULTS OF TESTS ON RADIATORS FOR AIRCRAFT ENGINES, H. C. Dickinson, Mem. Am. Soc. M. E., W. S. James and R. V. Kleinschmidt. The present report is the fourth of a series of reports on airplane radiators, incorporating the results of experimental work at the Bureau of Standards under the joint auspices of that Bureau, the Aviation Departments of the Army, and the National Advisory Committee for Aeronautics.

PART I shows in tables and curves the results of measurements of geometrical characteristics of 59 types of radiators, together with such physical properties as heat transfer, head resistance, air flow through the core, power absorbed, and figure of merit. In most cases the properties are shown for speeds running up to 120 miles per hour.

The terms used in describing the radiators and their performance are defined, the more evident relations between the properties and characteristics are stated, and applications of the results to the design of a radiator are pointed out.

It is shown that the most efficient type of radiator tested, for use at high speeds and mounted in "unobstructed" positions on the aircraft, is one whose water tubes are flat hollow plates, placed edgewise to the air stream, and continuous from front to rear of the radiator.

PART II shows in curves the pressure heads required to produce given rates of flow of water in twelve sections of radiator, each of a different type, briefly described and shown in photographs, but each with a core 8 in. square. The methods used in the tests and in computation are stated in detail, and a method is developed for estimating, by the use of twelve sets of auxiliary curves, the pressure head required for a given rate of water flow in each type of radiator, when of any size, on the assumption that losses of head at inlet and outlet of the water tubes are negligible in comparison with the resistance in the tubes of the 8-in. sections. (Synopsis of Report No. 63, *National Advisory Committee for Aeronautics, 1920, e*).

### AIR MACHINERY

DESIGN OF AIR DUCTS, J. C. Recommendations and chart for the determination of velocity in air ducts. The writer claims that in many cases the designer improperly figures the main duct as equal in area to the aggregate of the several branches, which gives to the main duct excessive proportions. Also an arbitrary velocity for the main duct is assumed which is considerably higher than that in the branches, with the result that the velocity for the main is excessive, the branches nearest the fan do all the work, and unnecessary use of metal is involved in making the duct larger than is really required.

In dust-extraction work it is important to maintain a velocity in the main duct equal to if not slightly greater than the velocity in the branches, so that no particle of dust is released from suspension until the dust separator is reached.

To solve this sort of problem the author plotted a chart (Fig. 1), the method of using it being as follows:

Suppose we require the size of main duct necessary to deal with fifty 5-in.-diameter branches. Let the eye follow up the line indicating 50 branches until the curve for 5-in. branches is encountered. If we now look along the horizontal line through the intersection of the said line and curve we obtain 24 in. diameter as the size required on the scale on the left-hand side of the chart. Now, fifty 5-in. branches total 981.75 sq. in., while the 24-in.-diameter main gives only 452.39 sq. in. This difference is brought about by the fact that there is a relatively greater frictional resistance offered by the smaller-diameter pipes owing to the factor (perimeter divided by sectional area) increasing as the diameter becomes less and less.

While the chart was prepared for 100 branches, the size of the main necessary to take the larger number of branches may be found (provided the main does not exceed 60 in. in diameter) by dividing the number of branches into equal groups not exceeding 100 each, then finding the size of pipe required for each group, and finally looking up the size of main capable of dealing with these larger pipes. (*Mechanical World*, vol. 67, no. 1743, May 28, 1920, p. 346, 1 fig., *pA*).

### BUREAU OF STANDARDS

THE SAYBOLT VISCOSITY OF BLENDS, Winslow H. Herschel. It has long been recognized that the viscosity of a mixture or blend of two oils is less than the arithmetical mean between the viscosities of the component oils. This is true whether viscosities are expressed in Saybolt seconds, poises, or any other known unit of viscosity. For definite chemical compounds which are inert with respect to each other, a formula has been found which gives the viscosities of mixtures with great accuracy, but this formula cannot be used with oils whose molecular weights are not known.

For the purpose at hand, the best available formula was found to be

$$\log \mu = v_1 \log \mu_1 + v_2 \log \mu_2 \dots \dots \dots [1]$$

where $\mu$, $\mu_1$, and $\mu_2$ are the viscosities, in poises, of the blend and of the component oils, and $v_1$ and $v_2$ are the percentage volume concentrations. Expressed in words, Equation [1] shows that the viscosity of a blend is the weighted geometrical mean between the viscosities of the component oils, that is, the logarithms of the viscosities are additive.

If it is assumed that a Saybolt viscosimeter is used, the viscosity in poises may be obtained by the formula—

$$\mu = [0.00220\ t - (1.80/t)]\gamma\ \dots\dots\dots [2]$$

where $t$ is the time of flow, or Saybolt viscosity, in seconds, and $\gamma$ the density of the oil at the temperature of test, in grams per cc.

It was found that the viscosities of blends, as determined by test, agreed in some cases fairly closely with the viscosities estimated from Equations [1] and [2], but the estimated value was always too low when there was a considerable difference between the viscosities of the component oils. For given component oils, the error increased as the amount of the lighter oil increased, reaching a maximum in a blend containing about 60 per cent of the lighter oil. . It was also found that the percentage of error

## ENGINEERING MATERIALS

### How Shortages of Metals Were Met During the War in Germany

WAR EXPEDIENTS IN THE FIELD OF METALS IN GERMANY, Albert Wuerth. The shortage of many metals in Germany during the period of the war forced manufacturers to resort to some interesting expedients which the present article describes. It further discusses the particularly important point as to the extent to which war practice in this field is likely to be maintained with the coming of peace conditions and restoration of the availability of the usual materials.

*Engine and Pump Manufacture.* Crankshaft housings for submarine engines have to be built of a special bronze having very high mechanical qualities, such as high tensile strength and high percentage of elongation. Since the beginning of 1915, due to the shortage of metals, these crankshaft housings were made of

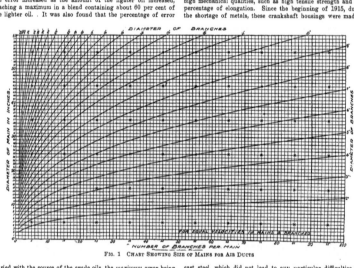

FIG. 1   CHART SHOWING SIZE OF MAINS FOR AIR DUCTS

varied with the source of the crude oils, the maximum error being obtained if a heavy naphthene-base oil was blended with a light paraffin-base oil, and the minimum error being found with a blend of heavy paraffin and light naphthene oils.

In order to carry the difference in viscosity to an extreme, a heavy naphthene-base oil was mixed with aviation gasoline, and the error in estimating the viscosity of the blend was found to be 221 per cent. But even in this case the viscosity determined by experiment was not as low as would have been estimated on the assumption that the reciprocal of viscosities are additive, a rule sometimes advocated as the true law of mixtures. Within the range of viscosities actually employed in blending operations, it was found that the maximum error in the viscosity calculated from the logarithmic rule would be 33 per cent for naphthene-base oils or 25 per cent for paraffin oils. A diagram was obtained for estimating the correction factor to be applied to viscosities of blends as calculated from the logarithmic rule. It is believed that with the help of this diagram the viscosity of a blend may be estimated with an error not greatly in excess in determining the viscosity of the component oils. (Abstract of *Technologic Paper of the Bureau of Standards* No. 164, ep)

cast steel, which did not lead to any particular difficulties and, indeed, with the construction of larger sizes of submarine engines, proved to be an advantage, as it gave to the housing a greater rigidity. On the other hand, the machinability of the material proved to be poorer than that of bronze.

All the screws, bushings, etc., which before the war were made of bronze or copper castings, had to be of iron and sherardized in order to protect them against rust. (This was done by the Schoop process.) On the other hand, parts which used to be made of aluminum castings were during the war made of thin sheet iron and then galvanized.

Considerable trouble was experienced with piping. The former copper piping was replaced by steel piping galvanized inside and out. The zinc coating did not prove satisfactory, however, as it was rapidly eaten away by sea water and salty sea air. The pipes therefore had to be protected by lead, which, however, did not adhere well to the steel and chipped off easily under mechanical influences. It proved necessary to protect the piping by two coats of metal—zinc first and lead afterward. This required great skill in manufacture and a close control of the temperatures at which the metals were applied.

Vigorous efforts were made to conserve tin. Bronze bushings or parts under heavy stress (such as steering gears) used to be made of a bronze with a 15.5 per cent tin content. During the war these bronze bushings were made of an alloy containing about 30 per cent less tin and were not cast in sand, but in a metal mold so as to give them a greater hardness. This was based on investigations made in 1910 by Heyn and Bauer, which would indicate that the hardness of a copper-tin alloy is really a matter which may be controlled either by variation in tin content or by the proper control of temperature conditions during cooling, and, in particular, velocity of cooling. It should clearly be borne in mind that this hardening by rapid cooling is possible only with bronzes and not with brasses.

Bronze bushings with a tin content of 10 per cent were machined and after having been set in their seats were hardened on their surfaces by having a conical mandrel driven in under high hydraulic pressure.

Parts not subject to heavy stresses were made of zinc castings, and it appears that the strength of such parts was materially affected by the heat treatment which they used to undergo.

An effort was made to produce a bearing metal containing 52 per cent tin, 42 per cent lead, 2 per cent copper and 14 per cent antimony, but it has not as yet been carried to a successful conclusion.

*Jet Apparatus.* Considerable trouble was experienced in attempts to find substitute metals for use in jet apparatus, by which are meant nozzles and their equivalents through which water, steam, gases or acids are discharged under pressure. Here, in addition to other conditions, the tendency to corrosion and erosion had to be met. This latter particularly when the material discharged through the nozzle contained sand or suspended solids.

In particular, in injectors, attempts were made to substitute cast-iron nozzles for bronze, but it was found that the former did not stand up well when used either with steam or hot water. The inner wall of the nozzle was strongly corroded, which interfered with the efficiency of the apparatus. Tests were begun with various aluminum alloys for use in nozzles, but have not yet been concluded, and promising results were obtained with aluminum bronze containing 90 per cent copper and 10 per cent aluminum. This, however, would bring a saving only in tin and not in copper.

The water-atomizing nozzles in humidifying apparatus used to be made of bronze. During the war an attempt was made to make the housings out of so-called zinc bronze, leaving the general shape of the device the same as formerly and using bronze nozzles. It was found that housings so made would not stand up in service and broke easily and developed leaks. The shape was then altered and the main housing made of cast iron with screwed-in zinc bushings for water inlets.

In air nozzles made of zinc bronze unevenness of the surfaces rapidly developed, which made the atomizing incomplete and irregular. The reason for this was found to be the formation of zinc hydroxide, and the result of it was that much greater skill in attendance was necessary in order that good atomizing might be obtained.

*Oil Preheaters.* In Germany, during the war, the development of oil preheaters was of considerable interest. In ante-bellum days the jackets and covers of oil preheaters were made of bronze. Attempts were made to use steel castings as substitutes for bronze in covers, which led to certain mechanical troubles. The covers were then made with heavier walls and out of cast iron, but broke easily.

The design was then entirely changed, the brass tubes replaced by steel tubes and the cover made of thick wrought-iron disks, while the jackets were made of steel blocks. This latter did not prove satisfactory because in the machining of the blocks close to 90 per cent of the material was wasted which made the process expensive and not economical, in addition to which the riveting of the parts proved to be difficult. The jackets were made out of sheets and held in place by a combination of riveting and welding. (*Stahl und Eisen,* no. 13, April 1, 1920, pp. 421-426, 14 figs., serial article, not concluded, *d*)

## FUELS AND FIRING

BURNING COKE BREEZE ON UNDERFEED STOKERS, F. A. DeBoos. Coke breeze, by which name the fine abrasions which are a by-product of coke manufacture are known, amounts to approximately 6 per cent of the total coke produced. In general, 40 per cent will pass through ½-in. mesh screen and 60 per cent through a 1½-in. mesh. These sizes, however, are too small for blast-furnace work, and, in particular, for the finer particles ranging from dust to sizes around ⅜ in., the demand is very slight.

The material has a good heating value, averaging on a dry basis 10,000 to 11,000 B. t. u., but is difficult to burn efficiently. Various attempts have been made to burn it either mixed with coal or on specially designed stokers of the chain-grate type, but there are objections to either of these methods. Furthermore, in many plants the amounts of coke breeze available are not sufficient to give the full power demanded, and coal-fired boilers must be used. In such installations there is a demand for a stoker which will burn varying mixtures of coke breeze and coal and which will have the ability not only to give good ratings and good efficiencies with such mixtures but will also be capable of securing extremely high ratings with proper efficiencies when burning all coal.

The author calls attention to the requirements for burning coke breeze and coal mixtures and also results of tests made at the Michigan Limestone and Chemical Co., Rogers City, Mich., with Jones A-C multiple-retort-type stokers.

In these tests a mixture of 225 lb. of coke breeze to 100 lb. of coal was burned with an efficiency of 65.6 per cent at 155.5 per cent rating; also with a mixture of 325 lb. coke breeze to 100 lb. coal, 141.1 per cent rating was developed with 59.1 per cent efficiency.

During the summer of 1919 tests were also made by the Ford Motor Co., in the course of which a mixture of two parts coke breeze to one part coal was first used, and then changed to three parts coke breeze to one part coal. As a result of these tests they are now burning at the Ford Hospital, Detroit, Mich., a mixture of two parts coke breeze and one part coal quite successfully and at a considerable saving in cost. The coke breeze is stored in a bin separate from the coal and the charge is mixed by dropping alternate charges of coke breeze and coal into the traveling bin. The mixture then descends through a short chute into the hopper of the stoker.

When coke breeze and coal mixtures are burned light air pressure is necessary, as otherwise the fine particles of breeze are blown up into the baffling of the boiler or down on to the dump plate with the objectionable result that holes will develop in the fire.

This makes it necessary to keep the fuel bed fairly thin, as otherwise heavy clinkers will develop and make it impossible to use the light air pressure. Also coke breeze-coal mixtures must be held together so that the particles of coke may "freeze", thus permitting the air pressure to facilitate combustion. (*The Blast Furnace and Steel Plant,* vol. 8, no. 6, June, 1920, pp. 362-364, 2 figs., *pe.*)

## INTERNAL-COMBUSTION ENGINES

### Fuel-Injection and Regulation Methods in Heavy-Oil Engines

THE APPLICATION OF LIQUID FUEL TO OIL ENGINES, A. J. Nilson. The author discusses the subject of injection and devotes considerable space to the description of various systems for its regulation. Of interest is the Pasel design (Fig. 2) for cutting out three-quarters of the pulverizer rings and giving the fuel a less tortuous path.

A groove is cut in the valve spindle down to the level of the bottom pulverizer ring and the entry to this groove or channel at the upper end is blanked by a sleeve having two ports opposite one another and coinciding with the groove in the valve spindle. This sleeve is operated from the outside and can be turned so as to close the aperture as may be required.

Another method of controlling the injection shown operates electrically and was designed by Mirrless, Bickerton and Day. Several solid-injection systems are also described. Of particular interest is the system used in the Blackstone engine, where duplex injection is employed, which is a compromise between the true Diesel air-injection engine and the solid-injection vaporizer type of the semi-Diesel. In this system (Fig. 3) blast air at a pressure of 350 lb. per sq. in. is taken to a pair of fuel inlet valves or atomizers, one of which opens with a slight

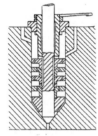

FIG. 2  PASEL BLAST-AIR INJECTION REGULATION VALVE (THE VARIABLE-RESISTANCE TYPE)

lead on the other. The leading valve delivers a small charge of oil to a hot bulb and maintains this automatically and continuously at the required temperature for igniting the main charge at all conditions of load. The jet of flame from this hot bulb is arranged to escape into the main cylinder through an orifice which deflects the flame obliquely into the spray from the main fuel valve, causing the charge to burn rather than explode. The initial compression in the main cylinder is only 150 lb. per sq. in. and the engine might be described as a low-compression Diesel with pilot ignition were it not that the ignition

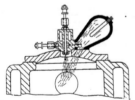

FIG. 3  BLACKSTONE DUPLEX-INJECTION TYPE OF CYLINDER HEAD

originates from the heat of a hot bulb and not as a result of air compression.

The Crossley oil engine is also described in considerable detail (compare MECHANICAL ENGINEERING, December, 1919, p. 896).

The subject of fuel ignition is discussed at some length and a good deal of attention is paid to the employment of tar oils as fuel for Diesel engines.

In this connection the choking of the pulverizer with black deposit is considered and the sleeve pulverizer introduced by P. H. Smith is described. It is stated that with this type of fuel valve the accumulation of black deposit is not as objectionable as in the standard type. The same trouble may be dealt with by running the engine for about fifteen minutes on gasoline every

five or six hours, which clears out most of the deposit and renders it possible to run for long periods on tar oil without dismantling the fuel valve.

It is also stated that lignite oils are perfectly satisfactory in any Diesel engine which will run on a petroleum gas oil and no special apparatus is required as in the case of coal-tar oils. (*Journal of The Institution of Petroleum Technologists*, vol. 6, no. 22, April, 1920, pp. 141-165, 16 figs., and discussion pp. 165-185, 3 figs., *dg*).

## Analysis of Conditions Affecting Slow Running of Automobile-Type Engines

THE LIMITS OF SLOW RUNNING, A. Johnson. It is well-known that as regards the ability to work efficiently at a low rate of revolution, the ordinary gasoline engine compares unfavorably with the steam engine. This is the reason for the use of change-speed gear drives in automobiles and numerous attempts to produce a slipping clutch (electric transmission).

As regards the mechanics of slow running, it may be noted that the low rate of speed is favorable to the induction of a large quantity of explosive mixture and thus to the attainment of a high pressure at the beginning of the compression stroke. On the other hand, the compression stroke itself is performed slowly and therefore the mean effective pressure is comparatively low. The balance of these two opposing causes, other things being

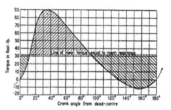

FIG. 4  TORQUE CURVE FOR A 4-CYLINDER ENGINE RUNNING LIGHT

equal, determines the speed at which the mean effective pressure is a maximum.

Fig. 4 shows an approximate torque curve for a 4-cylinder engine, 4-in. bore by 5-in. stroke, running light. The internal friction in this engine is estimated to be equal to a constant torque of 32 ft-lb. When the crank is at either dead center the torque is zero and does not become equal to the frictional load until the crank has turned through 12 deg. From 12 deg. to 93 deg. past the dead center the engine torque is in excess of the load, and from 93 deg. to 192 deg. it is again insufficient to overcome the resistance.

Because of this the revolving mass of the engine is accelerated between 12 deg. and 93 deg. and decelerated between 93 deg. and 192 deg.

The shaded area above the line of mean torque in the diagram represents work done by the engine in accelerating the flywheel or equivalent revolving mass, and the shaded area below that line represents work given up by the flywheel to compensate for the defect of engine torque. These two areas must be equal if the rate of revolution is to return to the same value at each 12 deg. of crank advance from either dead center, and considering the total lower area, it will be found by measuring the diagram that the mean height represents a torque of 28 ft-lb. exerted during the crank's passage through 99 deg.

The work done or given up by the flywheel is therefore $28 \times (99/360) \times 2\pi = 48.38$ ft-lb., which is also the amount of work previously done by the engine in accelerating the flywheel under the assumption made above.

Assigning certain arbitrary values to the dimensions and weight of the flywheel or its equivalents, the writer obtains a mean rate of revolution for a given engine of 33 r.p.m., which is the absolute minimum rate of revolution for the engine and flywheel, assuming, however, that at such a slow speed the pressure in the cylinder would not vanish by cooling.

This rate of revolution is, however, the mean of zero and 66.43 r.p.m., which is the rate of revolution attained by the crank at 93 deg. past the dead center, and it is obvious that such extreme variations of rate within half a revolution cannot be permitted. If an arbitrary limit be set to the effect that the variation of speed shall not exceed 10 per cent of the maximum, it will be found that the mean rate of revolution is roughly 145 r.p.m., which is the lowest rate at which the engine will run light with a specified degree of regularity.

To determine whether the same flywheel running on the same limit of regularity will permit the engine under full load to run at a reasonably slow speed, it is assumed that in the full-load torque all the ordinates of the light-load diagram are multiplied by five. This, according to the author's calculation, brings the mean r.p.m. at slow speed under full load to about 324 r.p.m. This is probably slower than the engine will ever be required to run and will, in many cases, be lower than the speed of maximum mean effective pressure and therefore lower than the lowest speed at which the engine can run at full load. (*The Automobile Engineer*, vol. 10, no. 139, June 1920, pp. 231-232, 1 fig., *t*)

## LUBRICATION (See also Bureau of Standards)

GRAPHITE LUBRICANTS, W. R. G. Atkins. In a brief paper the author describes the various forms of graphite lubricants and comes to the following conclusion:

It is obvious that to try to use graphite as a lubricant instead of oil would be useless, as solid friction is always greater than liquid friction. The oil used must be able to keep the metallic surfaces apart, either by its oiliness, its power of creeping with unbroken film over the metal, which necessitates a low surface tension, or by its viscosity, whereby it is carried along by the rotating portions, and wedged in between them and the fixed portions. To use an oil of viscosity more than sufficient to do this is wasteful, as energy is lost in shearing oil. Liquid lubricants, however, are greatly influenced by rise of temperature, and a decrease in viscosity of about 3 per cent per degree centigrade is quite a usual value for thick oils between 60 and 90 deg. cent. At cylinder temperatures all oils have very much the same viscosity. Under such conditions liquid oil films are apt to break, so the presence of a solid lubricant like graphite, which is almost entirely unaffected by such conditions, is highly desirable. Graphite, then, may be considered as a lubricant of special value for cylinders and valves of internal-combustion engines. It is also of value for use on bearings, especially when subjected to heavy pressure, in which case the oil film may fail. A graphite film in a bearing, besides reducing the solid friction of metal to metal, also provides a skin which is renewed from the oil, this skin being worn away instead of the metal. After graphite has been in use in an engine for some time the amount in the oil (always a very small quantity) may be further reduced, for once the graphite surface is formed a very small quantity suffices to maintain it. It is difficult to explain why the use of graphite reduces the consumption of lubricating oil, but it undoubtedly does so. The film deposited is too thin sensibly to alter the clearances. The explanation which appears the most probable is that by reducing the friction and the eddy currents in the oil due to surface irregularities, the temperature of the oil film is, on an average, reduced, and its viscosity being on that account greater, its rate of flow is diminished. It is also possible that, on the cylinder walls, where oil probably vaporizes with the fuel, a certain residue of graphite remains, and in this case the clearances between piston rings and cylinder walls may be slightly reduced. Apart, however, from any explanation of its action, the saving is an established fact, and in certain cases it appears that power is also saved. (*The Automobile Engineer*, vol. 10, no. 139, June 1920, pp. 239-240, *g*)

## MACHINE ELEMENTS AND DESIGN

### Cam Design for Radial and Rotary Engines

CAMS FOR RADIAL AND ROTARY ENGINES, Wm. John Walker. The writer attempts to determine by a general method whether or not any multi-lobed cam will, by being geared to run at an appropriate speed, perform its functions correctly. The method given by the author is applicable to the problem of designing a cam disk geared to revolve in the same direction as the crankshaft.

In Fig. 5 is illustrated a tappet distribution for a 9-cylinder engine, but the tappets have been numbered to correspond to an engine having any number of cylinders, $n$. The cam is shown with only one lobe, so that the figure is equally applicable for a cam revolving in either the same or in the opposite direction to that of the crankshaft.

The author shows, among other things, that the correct angular distance between two consecutive lobes is $4\pi/(n-1)$ radians, and also that the total number of lobes of the cam disk must be $(n-1)/2$.

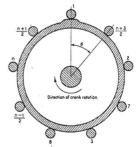

FIG. 5 TAPPET DISTRIBUTION FOR A 9-CYLINDER RADIAL AND ROTARY ENGINE

If the cam is geared to revolve in the same direction as the crankshaft, it must be geared to revolve at $1/(n+1)\times$ the speed of rotation of the crankshaft. Also the angular distance between two consecutive lobes must be $(n+1)/2$.

Thus, in the case of a 9-cylinder radial stationary engine, the two possible multi-lobed cams are either (1) the 4-lobed cam geared to run at ⅛ crankshaft speed in the direction opposite to that of the crankshaft, or (2) a 5-lobed cam geared to run at 1/10 the speed, and in the same direction as, the crankshaft. The same applies to the case of the rotary engine, the cylinder speed being considered instead of the crankshaft speed. (*The Automobile Engineer*, vol. 10, no. 139, June 1920, p. 241, 1 fig., *t*)

### Emergency Design of Bicycle Chain to Meet Present Shortage

NEW BICYCLE CHAIN. Last May a conference of jobbers in bicycle supplies was called together at the plant of the Diamond Chain Manufacturing Co., Indianapolis, to discuss the serious bicycle-chain shortage brought about by all the difficulties with which the majority of manufacturers have been confronted for the past several years. This shortage was the controlling factor in bringing about the situation where it is practically impossible to secure chains for replacements.

From reports made it would appear that the mills from which the rollers and bushings of the usual roller chains are obtained had been in constant labor trouble since July, 1919, and were closed completely from the end of September to December during the steel strike, while the production of some types of chains was not resumed until the middle of January. In addition, the working forces were seriously handicapped by the influenza epidemic during January and February, when for a period of about six weeks the absences among operators ranged from 25 to 45 per cent.

Under these conditions the Diamond Chain Manufacturing Co. recommended the employment of a new type of chain, or rather

FIG. 6 THE ONE-INCH PITCH 3/16-INCH LEAF BLOCK CHAIN OF THE DIAMOND CHAIN MANUFACTURING CO.

the revival of a type formerly used but produced by different methods.

With roller chains it is difficult to create increased production. The limiting factor is in the manufacture of bushings and rollers as the equipment for these parts is even now being driven to the limit. Material for these parts in increased quantity is not obtainable and the special machinery required for manufacture could not be produced under many months.

To meet this serious situation the leaf block chain (Fig. 6) has been designed. It is made up of a series of stampings and therefore does not require special machinery to produce it as is necessary in the production of roller and solid-block chains. It can also be assembled much quicker than either roller or solid-block chains. In general design it is similar to chains produced many years ago before the solid-block chain came on the market and is recommended especially for replacement business until the industry has picked up with the demand upon it. (*Motorcycle and Bicycle Illustrated*, May 13, 1920, pp. 37-39, 2 figs., d).

## MECHANICAL PROCESSES

ROLLED PRODUCTS OF 99 PER CENT NICKEL, Edwin F. Cone. Ninety-nine per cent nickel can now be rolled into the various shapes into which mild steel is rolled; and not only that, but forgings have been attempted with success. It appears that nickel products can be fabricated with no more difficulty than those of

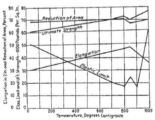

FIG. 7 THE PHYSICAL PROPERTIES OF 99 PER CENT ROLLED NICKEL.

mild steel, and as the same apparatus is used for working the nickel as is used for steel, it is possible to alternate the rolling of steel and nickel products without any changes or adjustments in the machinery. This is done at the plant of the Nickel Alloys Company at Hyde, Clearfield County, Pa.

Pure nickel products are of value wherever non-corrodibility and ability to maintain an article in an antiseptic condition are of importance; for example, in dairy machinery, dyehouse equip-

ment, etc. The new product is said to be malleable, can be welded to iron, steel or to itself, has a melting point of 1485 deg. cent. and the physical properties shown in Fig. 7.

The raw material is refined and specially treated in small 2- to 3-ton open-hearth furnaces. The hot metal is poured into ordinary ingot molds at a temperature of approximately 3200 deg. fahr. These ingots are later broken down under hammers or in rolls after the usual preheating. Sheets are rolled down in packs of 8 to 32 sheets to a thinness of 0.001 in.

The following data in addition to the curve are available as to the physical properties of the material:

*Transverse test on a 1-in. diameter hot-rolled bar:*
     Elastic limit.................... 27,500 lb. fiber stress
     Ultimate strength............... 84,300 lb. fiber stress
     Modulus of elasticity............16,500,000 lb. fiber stress
*Torsion test of a 1-in. rod, hot-rolled:*
     Elastic limit.................... 12,250 lb. fiber stress
     Ultimate strength............... 80,800 lb. fiber stress
     Degrees of twist per running inch.. 320 deg.
     Complete twist in 24 in.......... 21 revolutions

The metal is of great flexibility and does not appear susceptible to the formation of scale, even under the most trying conditions. (*The Iron Age*, vol. 105, no. 25, June 17, 1920, pp. 1713-1716, 9 figs., dA)

## MECHANICS

### Load-Carrying Capacity of an Angular-Contact-Type Ball Bearing

LOAD CHARACTERISTICS OF RADIO-THRUST BEARINGS, F. C. Goldsmith. The article is a reply to one by F. W. Gurney which appeared in a previous issue of the *American Machinist*. The author is chief engineer of the New Departure Manufacturing Company.

The part which is of particular interest is that in which the author establishes an equation for the load-carrying possibility

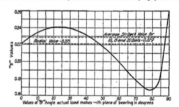

FIG. 8 CURVE SHOWING VARIABLE FACTOR AGAINST VALUES OF LOAD ANGLE IN A CERTAIN TYPE OF BALL BEARING

of an angular-contact type of ball bearing in terms of certain known quantities. The notation which the author uses is:

$P$ = total resultant load on one ball row
$P_o$ = load on bottom ball in direction of contact
$\theta$ = angle between line of load application and plane of balls
$\alpha$ = angle between balls in vertical plane
$\beta$ = angle of contact, and
$\eta = \omega_\eta = 0$

where $\omega$ is angle between line of load application and contact lines.

With this notation the author derives the formula—

$$P = P_o \left\{ \cos(\theta - \beta) + \frac{2}{\cos^{1.5}(\theta - \beta)} \right.$$
$$[(\sin\theta \sin\beta + \cos\theta \cos\beta \cos\alpha)^{2.5} +$$
$$(\sin\theta \sin\beta + \cos\theta \cos\beta \cos 2\alpha)^{2.5} + \ldots$$
$$\left. (\sin\theta \sin\beta + \cos\theta \cos\beta \cos\eta\alpha)^{2.5}] \right\}$$

from which one can see that for any one value of the load angle $\theta$ all other quantities are known. By solving this equation for bearings containing 10, 15 and 20 balls and letting θ vary through 90 deg. one can arrive at a variable factor which has a fixed value for each corresponding value of θ. Plotting these values of the variable factor against degrees of the angle of load application, the author obtains a curve from which, the angle of load application being given, one can pick the proper value of one of the principal bearing load factors.

In Fig. 8 is given such a curve showing the variable factor $K$ in ordinates against values of the load factor θ as abscissæ for a popular line of thrust angular-contact type of ball bearings, the values of $K$ being the average of bearings having 10, 15 and 20 balls.

In this figure, as a dotted line, is shown also the reciprocal of the number 0.229, which is the average value of the constant from the Stribeck equation for radial or annular ball bearings having

## POWER PLANTS

COMPOUND LOCOMOBILE WITH SUPERHEATER. Description of the locomobile built by Marshall, Sons & Co., Ltd., Gainsborough, England.

The engine is a 2-cylinder compound mounted on a cylindrical tubular boiler. This latter is 6 ft. 1 in. in diameter and has a length of .16 ft. It is fitted with a single flue 4 ft. 3½ in. in external diameter and 7 ft. 7 in. long. The tubes are 129 in number and 2½ in. in external diameter, giving a total heating surface of 769 sq. ft., of which 79 sq. ft. represent the flue surface.

A smokebox superheater, Figs. 9 and 10, provides 600 sq. ft. while a horizontal feed heater through which the exhaust from the low-pressure cylinder passes on its way to the condenser furnishes 65 sq. ft. of heating surface. The boiler pressure is

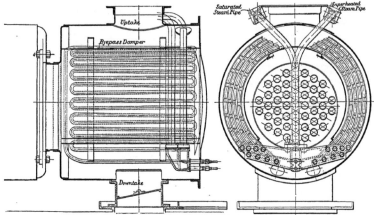

FIGS. 9 AND 10 SUPERHEATER OF MARSHALL, SONS & CO. COMBINED LOCOMOBILE

10, 15 and 20 balls. This is done to bring out the fact that for some load angles this bearing will carry a greater load than an annular bearing of the same size can carry, even as a pure radial load.

The following may be said as to the use of the curve in question: All speed-load catalog ratings are given as pure radial loads, and the corresponding value of $K$ would be 0.221. Now, supposing the angle of load application happens to be 25 deg., the corresponding value of $K$ would be 0.24, and in selecting the proper size of bearing from the catalog rating we would multiply the resultant load by 0.221 over 0.240 or

$$\text{Catalog load} = \frac{0.221}{0.240} \times \text{resultant load} = 0.92 \times \text{resultant load}$$

In case the load angle proved to be 85 deg.,

$$\text{Catalog load} = \frac{0.2210}{0.1675} \times \text{resultant load} = 1.32 \times \text{resultant load}.$$

After a careful study of this curve we can see that when the load angle is less than 48 deg. we really have in this angular-contact-type ball bearing a very excellent load-carrying device.

The author also gives an interesting curve sheet (Fig. 10 of the original article) such that after the radial and thrust loads have been determined one can readily arrive at the proper bearing size. (*American Machinist*, vol. 52, no. 23, June 3, 1920, pp. 1190-1196, 10 figs., *t*)

180 lb. per sq. in. and in tests reported in the original article it was found that the coal consumption per brake horsepower per hour was 1.7 lb. running non-condensing and 1.39 lb. condensing with the heat value of coal of 11,500 B.t.u. The engine is furnished with a jet condenser, together with an Edwards-type vertical air pump. The feed pump is fixed to the air-pump casting and works off the same eccentric rod. The feed is passed through the tube at the feed heater. An injector is fitted to the boiler as a stand-by.

In connection with Figs. 9 and 10 it may be added that they represent the general arrangement of the superheater, but not the actual arrangement adopted for this size of engine. (*Engineering*, vol. 109, no. 2840, June 4, 1920, pp. 767-768, *d*)

## STEAM ENGINEERING (See Power Plants)

## CLASSIFICATION OF ARTICLES

Articles appearing in the Survey are classified as *c* comparative; *d* descriptive; *e* experimental; *g* general; *h* historical; *m* mathematical; *p* practical; *s* statistical; *t* theoretical. Articles of especial merit are rated *A* by the reviewer. Opinions expressed are those of the reviewer, not of the Society. The Editor will be pleased to receive inquiries for further information in connection with articles reported in the Survey.

# ENGINEERING RESEARCH

## A Department Conducted by the Research Committee of the A. S. M. E.

### Research Association in Great Britain

IN addition to those listed in the March issue of MECHANICAL ENGINEERING (p. 181), the following Research Associations have received licenses from the Board of Trade under Section 20 of the Companies' Act of 1908:

The Glass Research Association, 50, Bedford Square, W. C. 2, Secretary, E. Quine, B. Sc.

The British Non-Ferrous Metals Research Association, 29, Paradise St., Birmingham, Secretary, E. A. Smith, A. R. S. M., M. Inst. M. M.

The following Rssociations have been approved, but not licensed:

British Music Industries Research Association
The British Refractory Materials Research Association
The Scottish Shale Oil Research Association
The British Leather Manufacturers' Research Association
The British Launderers' Research Association.

The following Associations are under consideration:

The British El<ctrical and Allied Industries Research Association
The British Aircraft Research Association.

The following Associations are preparing memorandum and Articles of Association:

Silk Manufacturers
Master Bakers and Confectioners
Cycle and Motor Cycle Manufacturers
Users of Liquid Fuels

### Research Laboratories In and About Washington

The following list of laboratories has been prepared by Mr. C. A. Briggs, Associate Physicist of the Bureau of Standards:

*Industrial Research Laboratories:*
Institute of Industrial Research, 19th & B Sts., N. W., Washington, D. C.
National Canners' Association, 1739 H St., N. W., Washington, D. C.

*Endowed Laboratories:*
Geophysical Laboratory of the Carnegie Institution
Department of Terrestrial Magnetism of the Carnegie Institution.

*Laboratories of Educational Institutions:*
American University
Bliss Electrical School
Catholic University
Gallaudet College
Georgetown University
George Washington University
St. John's College
Trinity College
Washington Missionary College.

*Departmental Laboratories:*
Bureau of Chemistry—10 laboratories
Other Bureaus of the Department of Agriculture—22 laboratories
Coast and Geodetic Survey—3 laboratories
Bureau of Standards
Bureau of Mines—5 laboratories
Navy Department—17 laboratories
Smithsonian Institution—4 laboratories.

### A—RESEARCH RESULTS

The purpose of this section of Engineering Research is to give the origin of research information which has been completed, to give a résumé of research results with formulæ or curves where such may be readily given, and to report results of non-extensive researches which in the opinion of the investigators do not warrant a paper.

*Apparatus and Instruments A10-20* Thermal Expansion. A stretched-wire apparatus for measuring thermal expansion has been developed by Dr. Arthur W. Gray. This instrument is simpler in form than the one designed by him for the Bureau

of Standards. The Report in *Chemical and Metallurgical Engineering* for Nov. 26, 1919, gives the form of apparatus together with results from its use. The instrument is simple and gives accurate readings. Address Dr. Arthur W. Gray, L. D. Caulk Company, Milford, Del.

*Cellulose and Paper A1-20* Gas Masks in Pulp and Paper Mills. Gas masks of a nose-breathing type with canisters containing special chemicals are used in the digester house of the Forest Products Laboratory at Madison, Wis. This mask enables the operator to make ready repairs of $SO_2$ leaks under conditions otherwise unbearable. They are of great assistance in bleach rooms. Address Forest Products Laboratory, Madison, Wis.

*Electric Power A3-20* Safe Use of Alternating Type of Coal-Cutting Equipment. L. C. Ilsley and E. J. Gleim have recently made a report to the Bureau of Mines on machines, circuits, protection and switches for alternating currents applied to coal-cutting equipment. Bureau of Mines, Washington, D. C. Address F. G. Cottrell, Director.

*Fuels, Gas, Tar and Coke A6-20* Evaporation from Crude Oil. J. H. Wiggins, Assistant Petroleum Engineer of the Bureau of Mines, has recently made a report showing a yearly loss of nearly $27,000,000 from evaporation of crude oil in the Mid-Continent oil field. This amounts to 3 per cent of the total gasoline produced in the United States. The major portion of this evaporation occurs on the lease when the oil is still fresh. It is possible to eliminate two-thirds to four-fifths of this evaporation by protecting oil from free contact with air. Bureau of Mines, Washington, D. C. Addr ss Director.

*Fuels, Gas, Tar and Coke A7-20* The Peat Resources of Ireland. Special Report No. 2 of the Fuel Research Board of the Department of Scientific and Industrial Research of Great Britain by Professor Pierce F. Purcell is to be issued. The subjects treated are: Origin and formation of peat deposits, occurrence of trees in bogs, physical properties of peat, elimination of water, area and fuel content of Irish peat deposits, peat deposits of other countries, hand winning, mechanical winning, uses, utilization, allowable moisture for producer work, reclamation and cultivation of bogs. Price by post, 11d, obtained at Imperial House, Kingsway, London. W. C. 2.

*Leather and Glue A2-20* Effect of Oils on Strength of Glues in Plywood. Plywood near machinery and tanks containing oils may be used with little danger of the joints becoming weak through the action of the oil or gasoline. Plywood panels glued with animal, vegetable, and casein glues were immersed for nearly a year in engine oil and gasoline. All the glues weakened somewhat during the early part of the test, the animal and vegetable glues showing the greatest reduction in strength. The total loss was small, however. In only two or three instances did the strength of the casein or the blood albumen glues fall below 150 lb. per sq. in. The shear strength of the glue of 100 to 125 lb. per sq. in. was found in most cases. During 45 weeks the wood absorbed 60 per cent of its original weight of gasoline. Forest Products Laboratory, Madison, Wis. Address Director.

*Machine Tools A1-20* Shearing Steel at High Temperature. Series No. 20 of the Engineering Experiment Station of the University of Missouri presents a paper by Guy D. Newton on Energy Necessary to Shear Steel at High Temperatures. Experiments were tried with a rotary shear and with a straight shear. The metal was heated in a gas furnace and temperatures were taken by a Hoskins thermoelectric pyrometer inserted in the specimen. The shearing energy was determined by the change in speed of the flywheel. The results are shown in Figs. 1 and 2.

*Ventilation A1-20* Tests of Siphoning Ventilators. A series of comparative tests have been conducted by A. A. Potter, J. P. Calderwood and A. J. Mack, of the Engineering Experiment Station, Kansas State Agricultural College, to study the effectiveness of ventilators. The results of these tests are shown in Table 1. All ventilators used in the tests were 10-in. Ventilators No. 1 and 2 were of the siphoning type (see Fig. 3), the wind velocity being utilized to create an additional current of air through the ejector tube. No. 3 ventilator was not intended to produce any siphoning effect.

A wind tunnel 26 in. square and 15 ft. long was utilized. A variable-speed fan was inserted in one end of the tunnel capable of reproducing wind velocities up to 5 miles per hour. The ventilator to be tested was placed in the other end of the tunnel. A 10-in. pipe extending through the bottom of the tunnel was inserted to receive the various ventilators.

In performing the test, a ventilator was inserted in the tunnel and the fan regulated to produce the proper conditions. The velocity of the air in the tunnel and that induced through the ventilator was measured by means of an anemometer.

During the test each ventilator was subjected to as nearly as possible the same condition. The speed of the fan being regulated to secure the desired result. The velocity of the air passing over the ventilator and that siphoned through the tunnel when no ventilator was inserted would induce a current of air through the uncovered pipe leading into the tunnel, this data was recorded and is referred to in Table 1 under the head " No Ventilator."

In deducing any conclusions from these results it should be remembered that under actual conditions the ventilator action would be more pronounced. The tests, however, are of value in that each ventilator was subjected to similar conditions and the results show their comparative value.

The increase in velocity of the air in the ventilator above that produced when no ventilator was used gives a direct measure of the siphoning action. No. 3 ventilator shows actually a retarding of the air in passing through the ventilator, while Nos. 1 and 2 ventilators show a decided siphon action.

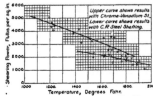

FIG. 1  ENERGY REQUIRED TO SHEAR STEEL AT HIGH TEMPERATURES
WITH A ROTARY SHEAR

Ventilator No. 1 was at a slight disadvantage in these tests, as it was larger in size than the other ventilators tested, and consequently obstructed the tunnel to a greater extent. This created a larger air pressure near the ventilator, increasing the tendency for the air to pass downward reducing the siphoning effect.

Address Director. Engineering Experiment Station, Kansas State Agricultural College, Manhattan, Kan.

*Water, Sewage and Sanitation A1-20*  The Bureau of Mines has recently received a report on the Collection and Examination of Rock Dust in Mine Air. This method consists in drawing air through a glass tube containing a layer of sized granulated sugar which filters the dust particles, using a calibrated air pump to measure the air. The sugar is then dissolved in distilled water, leaving the undissolved matter for examination. The sugar is submitted to an examination for insoluble matter before using it in the sampling tubes.

After dissolving the sugar a dust count is made from 1 cc. of the solution which has been made up to about 500 cc. by distilled water. The work was done by W. A. Selvig, F. D. Osgood and A. C. Fieldner. Bureau of Mines. Washington, D. C. Address F. G. Cottrell. Director.

*Water, Sewage and Sanitation A2-20*  Sulphur Dioxide as a Factor in the Smoke Problem of Salt Lake City. G. St. John Perrott has recently made a report to the Bureau of Mines on the sulphur dioxide contained in the atmosphere at Salt Lake City. The results show that the concentration is at no time to be an important factor in polluting the atmosphere and that most of the sulphur dioxide comes from the combustion of fuel in the city. At 12 noon the concentration is 0.1 part per million, while the greatest concentration occurs before 10 a. m. The concentration is higher on smoky days than on clear days and it decreases with the wind. ' The average concentration in December and January was about 0.15 part per million, while in March it was 0.01 part per million. Six parts per million have been observed in London, 0.6 part in Berlin and 1.8 parts in San Francisco. Samples taken by an airplane at 3000 ft. elevation, 2000 ft. elevation and 1000 ft. elevation showed a trace of sulphur dioxide. Bureau of Mines, Washington, D. C. Address F  G. Cottrell, Director.

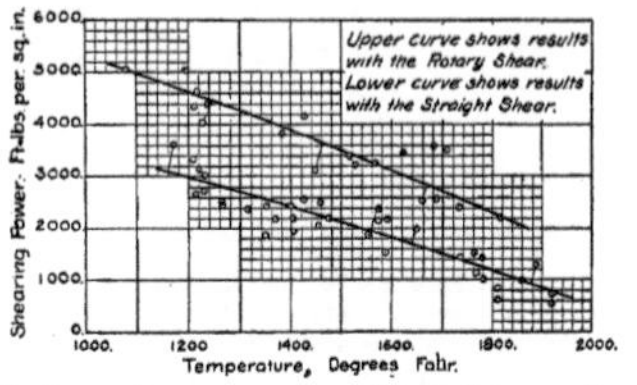

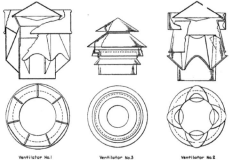

FIG. 2  ENERGY REQUIRED TO SHEAR CHROME-VANADIUM STEEL AT
HIGH TEMPERATURE WITH ROTARY AND STRAIGHT SHEARS

FIG. 3  TYPES OF VENTILATORS TESTED

TABLE 1  RESULTS OF TESTS ON VENTILATORS AT THE ENGINEERING EXPERIMENT STATION, KANSAS STATE AGRICULTURAL COLLEGE

| Velocity of air, miles per hour. | Velocity of Air in Ventilators in Feet Per Minute. | | | |
|---|---|---|---|---|
| | No ventilator. | Ventilator No. 1. | Ventilator No. 2. | Ventilator No. 3. |
| 1 | 26 | 35 | 30 | 17 |
| 2 | 55 | 70 | 61 | 36 |
| 3 | 85 | 106 | 93 | 55 |
| 4 | 112 | 143 | 123 | 73 |
| 5 | 140 | 178 | 153 | 90 |
| 6 | 168 | 213 | 183 | 107 |
| 7 | 197 | 250 | 213 | 126 |
| 8 | 224 | 286 | 244 | 144 |
| 9 | 252 | 319 | 274 | 162 |
| 10 | 280 | 235 | 305 | 179 |

*Wood Products A7-20*  Preservation. Light creosote oils properly injected into wood will prevent decay until the wood wears out or until it checks so badly that untreated portions are exposed. The heavier creosote oils used 25 or 50 years ago gave similar results. Forest Products Laboratory, Madison, Wis. Address Director.

*Wood Products A8-20*  Drying Periods.  Table 2 gives the approximate time for kiln drying under mild conditions of 1-in. boards in green condition and partially air-dried to reduce the moisture content to 6 per cent of the weight of the dry wood. For stock greater than 1 in. and not more than 3 in. in thickness the drying time is proportional to the thickness. Quarter-sawed stock requires from 25 to 35 per cent more time. If the drying is carried on to 10 or 14 per cent instead of 6 per cent, the time is reduced to 1-4 or 1-3 less than before.

Address Director, Forest Products Laboratory, Madison, Wis.

## B—RESEARCH IN PROGRESS

The purpose of this section of Engineering Research is to bring together those who are working on the same problem for coöperation or conference, to prevent unnecessary duplication of work

TABLE 2   APPROXIMATE TIME REQUIRED TO KILN-DRY, UNDER MILD CONDITIONS, 1-IN. PLAIN SAWED STOCK TO MOISTURE CONTENT OF 6 PER CENT, BASED ON WEIGHT OF DRY WOOD

| Species | Days drying time | |
|---|---|---|
| | Green from the saw | Partially air-dry (25 per cent moisture content) |
| HARDWOODS: | | |
| Swamp oak......................... | 45 to 50 | 20 to 25 |
| Northern oak....................... | 30 to 40 | 17 to 20 |
| Walnut, cherry..................... | 22 to 30 | 13 to 15 |
| Mahogany, beech.................... | 16 to 22 | 9 to 12 |
| Tupelo, gum........................ | 20 to 26 | 10 to 14 |
| Birch, ash, sycamore............... | 15 to 21 | 9 to 12 |
| Poplar, basswood, chestnut, butternut, elm, cherry....................... | 8 to 10 | 4 to 6 |
| Maple, hickory..................... | 17 to 23 | 9 to 13 |
| CONIFERS: | | |
| Western larch...................... | 9 to 12 | 4 to 6 |
| Cypress, redwood................... | 10 to 18 | 6 to 8 |
| Douglas fir, yellow pine, incense cedar, spruce............................ | 4 to 6 | 3 to 4 |

and to inform the profession of the investigators who are engaged upon research problems. The addresses of these investigators are given for the purpose of correspondence.

*Fuels, Gas, Tar and Coke B2-20* Volatility of Motor Fuels. An investigation to determine by vapor-tension apparatus the minimum temperature to establish a stable mixture of air and gasoline vapor in combining proportions. Address Frank A. Howard, Standard Oil Company, 26 Broadway, N. Y.

*Fuels, Gas, Tar and Coke B3-20* Crushed Fuel. A new method of burning crushed fuel is being developed at Purdue University. Fuel is introduced after crushing to small-size grains through tuyeres at one side of a circular hearth. The coal is introduced by means of an air current through a fan blower, additional air for combustion being introduced by side openings. High capacity has been obtained with ease of control and regulation. The cost of operation is much less than that of the pulverized-coal methods. Address Professor G. A. Young, Purdue University, Lafayette, Ind.

*Heat B5-20* The Heat Transmission of Cork and Lith Board. A series of experiments in preparation for the winter meeting of the Society. Address Prof. J. P. Calderwood, Kansas State Agricultural College, Manhattan, Kan.

*Internal-Combustion Motors B6-20* Condition of Air Flow Through Radiators. Investigations on the condition of air flow through airplane cooling radiators. Twenty-four specimens have been tested so far. The problem of direct cooling of the engine is being studied and progress is also being made in determining the temperature distribution in the air cells of certain cooling radiators. A comprehensive technical paper is being prepared. Address Dr. S. W. Stratton, Bureau of Standards, Washington, D. C.

*Internal-Combustion Motors B7-20* Carburetion of Fuels. Determination of the effect on the power and thermal efficiency of a tractor engine of varying the temperature of the fuel and air as it enters the carburetor. Quite elaborate apparatus is used to determine the temperature and pressure in the whole fuel intake system and to maintain exactly the conditions desired. Address C. W. Good, Department of Mechanical Engineering, University of Michigan, Ann Arbor, Mich., or Asst. Prof. W. E. Lay, Charge of Automobile Engineering, University of Michigan, Ann Arbor, Mich.

*Steam Power B2-20* Corrosion of Steam Pipes. An investigation of the corrosion of steam pipes at Iowa State College. Results so far seem to indicate the presence of $CO_2$ as the cause of corrosion. Address Prof. M. P. Cleghorn, Iowa State College, Ames. Ia.

#### C—RESEARCH PROBLEMS

The purpose of this section of Engineering Research is to bring together persons who desire coöperation in research work or to bring together those who have problems and no equipment with those who are equipped to carry on research. It is hoped that those desiring coöperation or aid will state problems for publication in this section.

#### D—RESEARCH EQUIPMENT

The purpose of this section of Engineering Research is to give in concise form notes regarding the equipment of laboratories for mutual information and for the purpose of informing the profession of the equipment in various laboratories so that persons desiring special investigations may know where such work may be done.

*Pennsylvania College D1-20* The Department of Mechanical and Electrical Engineering of Pennsylvania College, Gettysburg, Pa., has the following equipment:
One Riehlé 100,000-lb. testing machine.
One Riehlé 1000-lb. cement-testing machine.
One gage tester
Miscellaneous electrical and mechanical equipment.

#### E—RESEARCH PERSONAL

The purpose of this section of Engineering Research is to give notes of a personal nature regarding the personnel of various laboratories, methods of procedure for commercial work or notes regarding the conduct of various laboratories.

#### F—BIBLIOGRAPHIES

The purpose of this section of Engineering Research is to inform the profession of bibliographies which have been prepared. In general this work is done at the expense of the Society. Extensive bibliographies require the approval of the Research Committee. All bibliographies are loaned for a period of one month only. Additional copies are available, however, for periods of two weeks to members of the A.S.M.E. or to others recommended by members of the A.S.M.E. These bibliographies are on file in the offices of the Society.

*Fuels, Gas, Tar and Coke F5-20* Fuel Oil. A bibliography of one page. Address A.S.M.E., 29 West 39th St., New York.

*Internal-Combustion Motors F1-20* Spark Plugs. A bibliography of one page. Address A.S.M.E., 29 West 39th St., New York.

*Petroleum, Asphalt and Wood Products F3-20* Petroleum and Allied Substances. The Bureau of Mines has compiled by E. H. Burroughs a list of recent articles on petroleum and allied substances. This bibliography is prepared each month. Address Bureau of Mines, Washington, D. C., F. G. Cottrell, Director.

*Petroleum, Asphalt and Wood Products F4-20* The Natural Hydrocarbons such as gilsonite, elaterite, wurtzilite, grahamite, ozokerite and others by Raymond B. Ladoo, Mineral Technologist, Bureau of Mines. One-page bibliography of 22 items. Address Bureau of Mines, Washington, D. C., or A.S.M.E., 29 West 39th St., N. Y.

*Water, Sewage and Sanitation F3-20* Process used in U. S. for Treatment and Filtration of Sewage Waters. A bibliography of three pages. Search 2955. Address A.S.M.E., 29 West 39th St., New York.

## WORK OF THE A. S. M. E. BOILER CODE COMMITTEE

*THE Boiler Code Committee meets monthly for the purpose of considering communications relative to the Boiler Code. Any one desiring information as to the application of the Code is requested to communicate with the Secretary of the Committee, Mr. C. W. Obert, 29 West 39th St., New York, N. Y.*

The procedure of the Committee in handling the cases is as follows: All inquiries must be in written form before they are accepted for consideration. Copies are sent by the Secretary of the Committee to all of the members of the Committee. The interpretation, in the form of a reply, is then prepared by the

Committee and passed upon at a regular meeting of the Committee. This interpretation is later submitted to the Council of the Society for approval, after which it is issued to the inquirer and simultaneously published in MECHANICAL ENGINEERING, in order that any one interested may readily secure the latest information concerning the interpretation.

Below are given the interpretations of the Committee in Cases Nos. 299 to 306 inclusive, as formulated at the meeting of May 24, 1920, and approved by the Council. In accordance with the Committee's practice, the names of inquirers have been omitted.

### CASE No. 299

*Inquiry:* In using Par. 212 to determine the maximum allowable internal working pressure that may be allowed on a staybolted surface under the Code Rules, it is stated that the weakening effect produced by drilling the staybolt holes must be considered. Cylindrical surfaces of the character referred to in this paragraph usually contain a riveted joint that weakens the structure more than does the drilling of staybolt holes. Should not the weakening effect of such joints be considered in the applying Par. 212?

*Reply:* Where the curved staybolted surface referred to in Par. 212 contains a riveted longitudinal joint or other construction except hand-holes, and the strength of the surface through such joint or other construction is less than through any line of staybolt holes in the same direction, the weakening effect of the joint or other construction instead of that produced by the drilling of the staybolt holes, is to be considered in making the calculations for pressure by this paragraph.

### CASE No. 300 (Annulled)

### CASE No. 301

*Inquiry:* Is it the intent of the Boiler Code under Par. 182 that the back pitch of tubes and rivets on drums which have a relatively high ratio of shell thickness to diameter of drum, shall be measured on the inside surface, the outside surface, or the median line of the shell?

*Reply:* It is the opinion of the Committee that in measuring the back pitch of tube holes or rivets, the measurements should be made on the flat plate before rolling, so that in checking up these measurements after construction, the result would correspond to the dimensions on the median line, or the mean of the measurements on the outside and inside surfaces.

### CASE No. 302

*Inquiry:* Is it the intent of Par. 278 of the Boiler Code which requires a full-sized direct connection to the boiler for each safety valve, that the exact meaning of this term shall be applied to small brass pop safety valves with male inlet connections: It is impossible to maintain the nominal threads on the inlet of these valves if the inlet opening is maintained full-sized in a male connection.

*Reply:* It is the opinion of the Committee that the rule given in Par. 278 of the Code applies to the connection leading to the safety valve and not to the safety valve itself. There need not be a full-sized opening through the valve itself, as the stamping of the capacity thereon by the manufacturer indicates in any case what is actually guaranteed for the valve by the maker.

### CASE No. 303

*Inquiry:* Is it the intent of the Boiler Code Committee in its references to commercial lap-welded pipe in Case No. 218 that the weld therein may be considered as conforming to the requirements for welded joints in Par. 186 of the Boiler Code?

*Reply:* For calculations of the maximum allowable working pressures to be permitted in open-hearth lap-welded pipe used in connection with boilers, it is the opinion of the Committee that the weld may be assumed as meeting the requirements of Par. 186.

### CASE No. 304

*Inquiry:* What amount of cooling and shrinking under pressure is necessary in the driving of rivets under the requirements of Par. 256 of the Boiler Code? Is it consistent with the intent of this paragraph to release the pressure while the rivet shows any degree of heat redness in color?

*Reply:* It is the opinion of the Committee that the pressure should be maintained upon the rivet after it is driven until no part of the head shows red in daylight.

### CASE No. 305
(In the hands of the Committee)

### CASE No. 306

*Inquiry:* Is it the intent of Par. 20 that the minimum thicknesses of tube sheets there specified for h.r.t. boilers is applicable also to vertical fire tube boilers and to locomotive type boilers?

*Reply:* It is the opinion of the Committee that while this paragraph refers specifically to h.r.t. boilers, it is equally applicable to boilers of the vertical fire tube and of the locomotive types, and its application to these two latter types is recommended by the Committee.

## Have Engineers the Right to Prepare Legal Papers?

The New York Court of Appeals recently handed down an opinion which seemed to restrict the drawing of legal documents to persons possessing the right to practice law. This raised question in the minds of some engineers as to whether engineers and architects could continue in New York State to prepare contracts and certain other legal or semi-legal documents as has hitherto been their practice. It is the opinion of one legal authority, however, that engineers having special knowledge of the engineering work required and the professional skill enabling them to express it are not precluded as an incident in the practice of their profession from preparing "contracts for engineering work as has become the common practice for years." By so doing they are not holding themselves out as practicing law and are not practicing law.

The decision rendered by the Court of Appeals, Judge Crane writing the opinion, is that of People vs. Alfani (227 N. Y., p. 334). In that case there was evidence, as the chief judge said, "consisting of defendant's sign and repeated acts which permitted the trial court to find that the defendant held himself out to the public as being entitled to and did practice law." Defendant had a sign up advertising himself as a notary public and as a person drawing "all legal papers." The evidence showed that he did draw a bill of sale and chattel mortgage and advised as to the necessity of filing the mortgage in the county clerk's office and that he charged and received $4 therefor. When the client was leaving he inquired of the defendant, "In case I have any trouble of any kind and I need any legal advice, can I come back to you?" to which the defendant replied, "Yes."

In People vs. Title Guarantee & Trust Co. (227 N. Y., p. 366) a similar question arose in the prosecution against the trust company for drawing deeds and similar instruments in connection with their business of insuring titles to real estate, and the court held that such action did not violate the statute. The court says:

We know that in cities constantly men engaged in the real-estate business and banks have prepared for their customers such instruments without doubt or criticism.

The legislature when it enacted not only Section 280 of the Penal Law, which we have been considering, but also Section 270 relating to the practice of law by an individual without being admitted and registered, was charged with the same knowledge of prevailing customs and practices with which we are chargeable. Its members knew, oftentimes doubtless by practical and personal observation and experience, that laymen throughout the state were rendering such services as are here involved. Not only by practice and custom, but by inherent privilege they had the right to do this unless forbidden by statute, and if the legislature intended to prohibit a widespread practice and establish a new rule, it was its duty to say so clearly and unmistakably in the statute relating to the practice of law and rendition of legal services by individuals. It did not say so, and in my opinion there is not to be found in that section of the Penal Law any provision against the rendition of such services by an individual. We think the same idea is emphasized in Section 280, that an individual who is not admitted to practice must not assume the character of an attorney at law. He is forbidden to practice or appear "as an attorney at law or as an attorney and counselor at law" or to make it a business to practice "as an attorney at law or an attorney and counselor at law" or to hold himself out to the public as being entitled "to practice law as aforesaid or in any other manner" or "to assume to be an attorney or counselor at law." But there is nothing which can fairly be regarded as indicating an intention to abolish an existing and widespread practice and to prevent a layman as such and without any simulation of or pretense to the character of an attorney from drawing a simple instrument as instructed by his customer and not involving or predicated upon any legal advice then given.

# CORRESPONDENCE

C ONTRIBUTIONS to the Correspondence Department of MECHANICAL ENGINEERING are solicited. Contributions particularly welcomed are discussions of papers published in this Journal, brief articles of current interest to mechanical engineers, or suggestions from members of The American Society of Mechanical Engineers as to a better conduct of A. S. M. E. affairs.

## Announcement

The Publication and Papers Committee of The American Society of Mechanical Engineers desires to afford members of the Society every opportunity to express themselves on the conduct of the Society and on other matters related to its affairs. Owing to the infrequency of meetings, communications to the editor of MECHANICAL ENGINEERING seem to offer the most prompt and direct means of placing the views of individuals before the entire membership. The Committee desires to encourage the writing of such communications as one evidence of the interest of the members.

Having in mind the reasonable limitation of space it is the intention to publish all such communications submitted. Where the communication cannot be published in full the author's name and address with a quoted sentence or a synopsis giving the purport of the letter will be published. It will be assumed that members will write in a terse and considerate style appropriate to our publications.—EDITOR.

## Federated American Engineering Societies

To THE EDITOR:

To be successful the Federated American Engineering Societies must be an instrument of service. Any engineering society that approaches the question from any other point of view is bound to be both disappointed and destructive. Manifold benefits will surely be derived by the member societies, but they are merely the rewards of service.

There is so much to be done for the public weal that only engineers can do that there can be no choice for you and me as to our participation or non-participation in the Federation. If we include ourselves among the "recreated engineers" of the present it all comes down to a question of choosing the most effective instrument of service. The world's experience amply shows that there is only one such instrument—union.

The reputation for indifference and to ineffectiveness in public affairs long since achieved by engineers is not of their deliberate choosing. Their aggregate effort is probably as great as that of many other professional classes. The principal difference lies in the fact that the engineers' efforts have usually been "out of phase." The Federated American Engineering Societies is essentially a synchronizer.

Many engineering societies will find that the constitution now drawn up will not conform to their ideas in all respects. No set of men is wise enough to draw up an instrument that will find universal approval. Let those who raise objections remember (a) how easy it is to object; (b) how easy it is to let petty complaints divert the mind from the master principle. This is a time for getting together. If we wait for a perfect constitution we shall always remain apart. Besides, a bad constitution will reveal itself in operation, and the cure is plain.

Some local societies think that in the Federation idea lies loss of identity. Let us not forget that there are two kinds of identity, one of which is "gladly lost." As for the other, it is strengthened by broad association, guidance and sustenance. No one has observed that the suckling child loses its identity, nor the adult who in his temporary weakness is revived by a blood transfusion.

To the Joint Conference Committee it may be said that adverse action by many of the societies should not be considered as a defeat of the project. There will be many societies which genuinely cannot adopt the plan without reservations in certain

relatively small respects. If such reservations do not lessen the responsibility of such societies to the Federation then they should be counted among the members and the necessary adjustments can be made by the new organization at its leisure.

M. O. LEIGHTON,
*National Service Representative,*
Washington, D. C. *Engineering Council.*

## Select Strong Men for Directors of the New Federation!

To THE EDITOR:

If an engineer believes that his profession should take an active part in public affairs and that it should organize effectively to protect its own interests, then he should lose no opportunity to urge general approval of the plans for the new Federation of Engineering Societies to be known as the American Engineering Council, which were presented at the Washington conference of engineers held on June 4 and 5.

Criticism of the machinery of the new organization is no longer pertinent or helpful. Nor is the detail of the machinery anyway quite so important as has been sometimes claimed. The question of greatest importance is: How and by whom is this new machinery to be operated? What the new American Engineering Council is able to accomplish for the benefit of the profession and the public will depend very much on the quality of the thirty-odd engineers who are to be members of what may be called the Board of Directors of the profession.

Will the members of the local engineering societies and of the national societies take a live interest in the selection of the delegates who are to make up this new body? Will they use their influence to have men chosen who are able and willing to make the personal sacrifices of time and money necessary to attend the meetings of the Council, and who have in addition the ability and breadth of mind to enable them to judge wisely in all matters submitted to them?

It is upon the answers to such questions as this that the success or failure of the new plan for coöperative effort by various branches of the engineering profession will turn.

The experience gained in the operation of the present Engineering Council during the three years since its establishment has proved that there is real and important work to be done. It is work that demands time, energy and ability. It requires no small outlay of money, which the profession must furnish; but the profession must see to it also that men are selected for this position not because of a desire to honor some personal friend, but solely because the voters know the candidate chosen to be competent for the job and willing to faithfully discharge its obligations if elected. CHARLES WHITING BAKER.
New York, N. Y.

## Machine Tools—What Are They?

To THE EDITOR:

At a meeting of The American Society of Mechanical Engineers in 1885 one of the subjects for discussion was the power requirements for machine tools. In the course of this discussion one of the speakers, complaining about the indefiniteness of the question, said, "What a machine tool is, is slightly in doubt," and this question has been with us ever since. The dictionaries define it as a machine that uses tools, and the Custom House officials have accepted this and include woodworking tools under this classification. Users, dealers and manufacturers, however, do not accept the definition and the whole subject is as far from being settled as it was in 1885.

The question of the proper use of any word proceeds along lines of advocacy of certain use by various men, based on derivation, literal interpretation, or ancient authority; while the general acceptance of these words by those who have frequent occasion to use them is the actual test of their proper use. "Whatever is, is right," is sound etymological doctrine.

Edward H. Knight in his mechanical dictionary, published in 1872, gives the following definitions:

*Machine:* An instrument of lower grade than an engine, its motor being separate. It is distinct from a tool, as it contains within itself its own guide for operation.

*Machine Tool:* A machine in which the tool is directed by guides and automatic appliances. Among tools (sic) of this class for metals are lathes, punches and shears. Machine tools for wood are sawing machines, planing machines, etc.

It is the habit of dictionary makers to copy each other, and it is not improbable that our present-day dictionary interpretation of the word is derived from the above. It must also be said that to define a machine tool as a tool-using machine is a very natural and obvious definition to give. There is some doubt, however, if the original and common use of the words is as indicated above.

Mr. Cameron Knight, author of The Mechanician and Constructor, published in 1869, in the course of a very complete definition of shop terms, includes the following:

Tools include every implement small and large, simple or complex, which is used to produce or operate upon the work in the course of progress. A center punch in a turner's pocket is a tool, and the lathe before him is a tool.

It is an easily verified fact that in the text and advertising columns of the early mechanical papers such as the *Scientific American* (1846), the *Mechanics' Magazine* (1836), and the *American Artisan* (1862), the term "machine tools" is used but seldom, and then exclusively by editorial writers. The first observed use was in 1862. "Tools" is the word that was in universal use during the Civil War period to indicate what we now call machine tools. Advertisements for the sale of the after-war surplus were either under this caption or under the heading "machinist's tools." This use of the word persists into the eighties. In fact, in the discussion referred to in the opening paragraph, the word "tools" was not infrequently used. There can be no reasonable doubt, therefore, that although the expression "machine tools" was at least recognized as far back as 1860, "tools" was the word in common and everyday use until the eighties.

In 1879, Frederic B. Miles formed a company with the name "Machine Tool Works." This is the first noted instance of the use of this term by a shop man. The above company advertised the manufacture of steam hammers, planing machines, and lathes.

On first thought we may be inclined to define a machinist as a machine-using workman, and a machine shop as machine-using shop, but this definition will admit weavers and textile mills to these classifications. It is evident that a machinist is a machine-*building* workman, and a machine shop is a shop that builds machinery. By an analogous reasoning a machine tool is not a tool-using machine, but a machine-building tool; that is, a tool or machine for building machinery. This definition conforms to the best modern definition of the term, viz.: Machine tools are machines which, when taken as a group, will reproduce themselves. This excellent definition has the disadvantage of being general rather than specific, and as a secondary definition I would suggest the following: A machine tool is any metal-working tool whose waste is in the form of chips. This is specific and can be easily applied to any particular machine. It is not necessary to define a metal "chip," but it may be advisable to recall that the sparks from a grinding machine are chips, very small chips to be sure, but chips. This definition excludes sheet-metal-working machinery and metal-forming and forging machines, but these are probably not classed as machine tools by a

majority of shop men, and in any case it is necessary to draw the line somewhere, and this seems to be the best place.

This classification will then include all metal-cutting machinery whose action is a progressive cutting away of surplus stock, a gradual reduction in size until the finished dimensions are reached. A considerable part of such machines use a single-point cutting tool, and in these cases as well as the case of milling cutters the angles of these cutting tools are of common interest to machinists and machine-tool builders. The design of these machines has progressed along similar lines and the introduction of single-pulley drives with all-geared feeds and speeds has found ready acceptance among manufacturers. Lathes, milling machines, grinders, drills, boring machines, etc., are now so equipped.

A press, when used for piercing sheet metal, has very little in common with any of the above-mentioned machines. They are all metal-cutting machines, and here the similarity ceases. The metal is not in the same form, and the cutting tools have nothing in common with other metal-cutting tools as to construction or cutting angles. When a press is used for forming it has nothing in common with lathes, etc., other than the fact that they are all metal-*working* machines, such as power hammers, bulldozers, swaging machines, etc.

If we admit that the terms "machine tools" and "metal-working machinery" are coextensive, or go further and define a machine tool as a tool-using machine, we are then confronted with the lack of any definite and restricted term to apply to 90 per cent of our machine-shop equipment. It seems desirable to have such a term, inasmuch as a steam hammer has no more in common with a lathe than a blacksmith has with a machinist. It is probably a fact that 90 per cent of those who are variously engaged in the machine-tool industry will say that a universal grinding machine is a machine tool, but would not be quite as sure about a punch press, a steam hammer or a mowing machine, and this great majority of opinion should be the deciding element in this question of terminology.

L. L. THWING.

Boston, Mass.

## Fiber Stress in a Square Plate

TO THE EDITOR:

In your April issue, p. 249, H. W. Sibert gives a method for finding the fiber stress in a square plate supported on the edges and loaded with a uniform load normal to its surface. I believe his method of finding this stress is incorrect, and that therefore his results are incorrect. He finds a bending moment of constant intensity in a direction at right angles to a rectangular axis through the center of the plate and equal fiber stress in that direction from center of plate to the middle of a supporting edge. It is not possible for any stress to exist parallel with the support at any point along that support, for stress, of necessity, means a curving of the plate; and no curving could take place while the plate is in contact with its straight and rigid support.

It is further incorrect to combine these two rectangular stress intensities, as Mr. Sibert has done, by taking the square root of the sum of the squares, for the diagonal of the square on which the combined force acts is increased (over the side of the square) in exactly the same ratio, so that the intensity would remain exactly the same in a diagonal direction.

In my book "Concrete," p. 287, I have given a method for solving the problem which is practically correct and about as simple as it can be made. In that solution two rectangular strips are taken across the middle of the plate, each one unit wide. The load on the middle square unit is equally divided between these two strips, but other intersecting strips take less and less of the load $w$, because, as the edge is approached, these strips cannot bend and take their equal share of the load. The load on the critical strip is the full $w$ per unit at the support, and the curve of increase from center to support is assumed to be a parabola. This gives $7wL^2/96$ as the bending moment in the middle of the plate against Mr. Sibert's value of $wL^2/16$.

EDWARD GODFREY.

Pittsburgh, Pa.

# MECHANICAL ENGINEERING

### THE JOURNAL OF THE AMERICAN SOCIETY OF MECHANICAL ENGINEERS

Published Monthly by the Society at
29 West Thirty-ninth Street, New York

FRED J. MILLER, *President*

WILLIAM H. WILEY, *Treasurer*  CALVIN W. RICE, *Secretary*

PUBLICATION COMMITTEE:
GEORGE A. ORROK, *Chairman*      J. W. ROE
H. H. ESSELSTYN      GEORGE J. FORAN
RALPH E. FLANDERS

PUBLICATION STAFF:
LESTER G. FRENCH, *Editor and Manager*
FREDERICK LASK, *Advertising Manager*
WALTER B. SNOW, *Circulation Manager*
136 Federal St., Boston

Yearly subscription $4.00, single copies 40 cents. Postage
to Canada, 50 cents additional; to foreign countries $1.00 addi-
tional.

*Contributions of interest to the profession are solicited. Com-
munications should be addressed to the Editor.*

## The Federated American Engineering Societies

### Steps to Be Taken to Complete the Organization—Its Effects on the Work of Existing Local and National Societies

E. S. CARMAN, CHAIRMAN OF
LOCAL SECTIONS COMMITTEE

THE Federated American Engineering Societies, organized in Washington June 3 and 4, marks one of the milestones in this, the engineering societies' epoch of organization.

It was apparent several years ago that in order for the engineering profession to continue to occupy a high rank in the catalog of professions, a means must be provided whereby all engineers, regardless of branch, could act together in considering matters of common interest to the profession at large and also to the public.

In the organization of the Federated American Engineering Societies the means has been provided. Therefore, without further statement as to the desirability of having such an organization, let us accept the facts as they are, viz., that the organization has been started, and that it is now more important to discuss what will be necessary to complete its organization and properly begin its activities in order that the Federation may be most effective in representing the interests of the engineers of the country.

The organizing conference at Washington placed in the hands of the Joint Conference Committee of the four Founder Societies and Engineering Council the *ad interim* authority until the organization should be fully completed. The next steps then to be taken are:

1 Consideration by local, state, regional and national societies and applications made for charter membership (all organizations invited to attend the Washington Conference are eligible) in the F. A. E. S.

2 Appointment of societies' representatives on American Engineering Council.

3 Conference of representatives to elect officers and complete organization.

4 Activities to be immediately begun by the officers and organization.

As an obligation which they owe to the profession, every engineering society in the United States should be fully acquainted with the constitution and by-laws of the F. A. E. S. and be prepared to make an early decision and become a charter member.

The effect that the F. A. E. S. will have on other societies, and how it will affect their present activities can best be summarized as follows:

*Local Societies*

1 Participation and expression in all national, state or regional affairs of interest to the engineering profession.

2. Assistance in local affairs (when desired) by the national organization.

3. Greater local activities by complete local affiliation of all local and national societies' local sections.

*State Societies*

1 Participation and expression in all national, state or regional affairs of interest to the engineering profession.

2 Assistance in state affairs by local, regional and national societies.

3 Uniform recommendation in the various states regarding license and registration laws, road building, public utilities, etc.

*Regional Societies*

1 An opportunity to assist local, state and national societies in matters of common interest.

2 Receive assistance from local, state and national societies in furthering matters of common interest.

*National Societies*

1 Provide a place through local affiliations where the individual members may meet and have an opportunity of expression in local, state and regional affairs.

2 Provide a means of further coöperation with all engineering societies and coördinating such activities as
     Standardization
     Research
     Education
     Ethics
     Legislation
     National public engineering projects.

3 A more complete coöperation between the several national societies.

The question as to whether or not the engineer desires to have a part in local, state and national public affairs is answered today in the affirmative, as we now find him so engaged. He has accepted his local citizenship responsibilities and is now desirous of having a means of expression in exercising his national citizenship obligations.

The engineer today is alert and aggressive and eager to accept the responsibility that comes with the greater activities.

The F. A. E. S., as stated before, is only a beginning of society coöperation and of necessity must proceed cautiously and yet effectively.

Is it too much to expect that as the years come and go and complete confidence has been established by successful performance, the final result will be: One large, all-inclusive American Engineering Society with national professional branches and organizations, local branches and clubs in all of the many cities, and only one payment of dues by the members to entitle them to all privileges of local, state, regional and national organizations?      E. S. CARMAN,

*A. S. M. E. Representative,*
Cleveland, Ohio.         *Joint Conference Committee.*

## Wage Increases Not Necessarily on a Straight-Line Curve

Early this summer Mr. Charles Piez, Mem. Am. Soc. M. E., submitted to the United States Railroad Labor Board, for the Board of Directors of the Illinois Manufacturers' Association, a statement of principles which he believed should prevail in deciding upon wage increases. "The practice," he says, "of basing wages on the cost of living, without taking into account the work performed for the wages is, in our opinion at the very bottom of the present disturbed and unsatisfactory labor condition. . . . If we accept the figures submitted that $1700 represents the lowest subsistence level and $2500 the lowest comfort level for an American family, and base the scale of wages in the transportation service on the assumption that every man engaged in this service, no matter where he lives, whether married or single, is entitled to a wage based on a scale with those figures as minimums, would it not be fair to assume that every wage earner in every character of employment everywhere in this country is entitled to a scale based on that same minimum?"

Mr. Piez contends that if such a universal demand were made and acceded to, based on a straight time basis rather than on a production basis, the farms, the mines and the industries would be wholly unable to meet the cost. "Wages can't be paid unless they are earned and we can't get out of the common pot more than we put into it."

He further says there is undoubtedly justice in the demand for an increase among railroad workers and that there is no intention of arguing against a fair increase; and that particularly there should be an increase among the lower-paid classes as the high cost of living has affected them the most adversely. Nevertheless, "The divorce of wages from production has been one of the calamities of the war, for it has created in the mind of the wage earner the delusion that irrespective of output, performance, or character of service rendered, he is entitled to live on a certain scale. Statisticians spend their time in developing elaborated lists of family requirements, instead of determining how wage increase based on these requirements can be paid.

"And so it happens that during the most critical time in the industrial history of the country, when the consumption of commodities has expanded, and world production of commodities has been greatly impaired, we have established a shorter working day, have abolished piece and premium forms of payment, and have imbued the wage earner with the idea that he is entitled to a good living if he but spends eight hours at a job."

## Two Attitudes on The Department of Public Works

As a member of the National Chamber of Commerce, Engineering Council was instrumental in that having that body present to its member organizations a ballot designed to record the attitude of the Chamber on the question as to whether or not a Department of Public Works should be brought into being.

The propositions, which were submitted to 478 organizations, and the results of the balloting on each were as follows:

1 Shall a Department of Public Works be established by the national government?—827½ votes in favor—639½ votes opposed.

2 Shall a Department of Public Works be established by a suitable modification of the existing Department of the Interior, excluding therefrom the non-related bureaus and offices and by change of name from Department of the Interior to Department of Public Works?—676½ votes in favor—549½ votes opposed.

3 Shall a Department of Public Works be established by creation of an entirely new department?—283 votes in favor—993 votes opposed.

In accordance with these votes, the Chamber is not committed to either the first or the second proposition, for the reason that two-thirds of the votes cast were not in favor of either. As to the third proposition, the Chamber is placed by the voting in a position of opposition; for more than one-third of the total voting strength of the Chamber was recorded and more than two-thirds of the votes thus cast, representing more than twenty states, were recorded in opposition.

In this connection it is gratifying to note the interest of men of affairs in this movement as reflected in one of the planks of the Republican platform. As this plank was initiated and brought about by engineers it will be of interest to quote its phraseology:

We advocate a thorough investigation of the present organization of the Federal departments and bureaus with a view to securing consolidation, a more businesslike distribution of functions, the elimination of duplication, delays and overlapping of work and the establishment of an up-to-date and efficient administrative organization.

It is further understood in Washington that the Smoot-Reavis bill proposing a general reorganization of all departments along the lines suggested by the engineers for the Department of Public Works carries the provision that the first department to be so reorganized will be the Department of the Interior and its conversion to a Department of Public Works.

## The Outdoor Power Plant

In the May issue of MECHANICAL ENGINEERING, p. 307, reference was made to an article on the practicability of the outdoor generating station, written by H. W. Buck, and published in the *General Electric Review* of March, in which the advantages of such construction were pointed out. There has now appeared in the May number of that review an article by Henry G. Reist, Mem. Am. Soc. M. E., calling attention to the difficulties to be encountered in the outdoor plant, one of the most serious of which is the necessity for the construction of the generators so that the windings will be kept dry and at the same time will be properly ventilated. He says:

It has always been assumed that the windings of electric machines should be kept dry. It is difficult to maintain this condition and get proper ventilation in a generator exposed to the weather, without special construction and additional expense. If the generator is allowed to become cold when not in use there is likely to be condensation of moisture on coils under certain atmospheric conditions. To prevent this, many pieces of electric apparatus are provided with special heaters of some sort to keep the machine warm when it is idle.

In case machines are exposed in very cold weather there is danger of difficulty from the oil becoming too thick. This danger would apply especially to self-oiling thrust bearings placed at the top of the machines, since these might heat at starting before the oil became sufficiently thin to circulate freely.

Automatic stations, which operate without an attendant, should be inspected from time to time by a patrol. In case of very cold or disagreeable weather it is probable that such inspection would be superficial.

On synchronous machines placed out of doors the installation of exciter units presents difficulties. In small machines exciters are frequently belted and on larger machines they are either direct-connected or driven by motors. Whatever arrangement is used, considerable inconvenience will be caused by lack of a protecting roof. In some installations these parts, together with the governing mechanism, can be placed in chambers in the masonry below the generator floor; but in other cases, due to danger from high water, this could not be done, and with small low-head machines such masonry construction would add considerably to the expense.

I would suggest that instead of building special weatherproof outdoor generators, standard generators be used, and that simple, inexpensive semi-portable shelters be erected over them. In the case of several machines in one installation, I think it would be desirable to have a low house extending over the line of machines, with a gantry crane bridging the house. the house, or at least its roof, to be made in sections which can be moved on a track parallel with the line of generators to telescope with the roof over the adjacent generator. or each generator to be covered with a unit roof which may be lifted bodily from its place by means of a crane. I believe there would be sufficient heat generated to keep such a powerhouse comfortable, and that by suitably arranging the ventilators warm air could be passed over any generator that was not in use, thus obviating the difficulties which might be experienced with the generators out of doors.

The simple house suggested would provide a shelter for the exciters, waterwheel governors, and the patrol on his visits, and would seem to offer all the advantages of a more expensive power house. It is probable that a protecting house, as described, could be constructed for little more than the extra cost of a generator exposed to the weather.

## Across the Continent by Land and Air

The first transcontinental convoy of the Motor Transport Corps which crossed from Washington, D. C., to San Francisco, Cal., last summer has given, as shown in the paper presented by Lieut. E. R. Jackson, Mem.Am.Soc.M.E., at the Spring Meeting of The American Society of Mechanical Engineers, a large amount of valuable information both as to the state of the roads and as to the various makes of automotive equipment and methods of handling them.

It is gratifying to find that an earnest effort is being made to add to this store of knowledge. On June 14 the second transcontinental convoy left the East on its long, and at times weary and dangerous, journey to the Pacific coast, and we may expect to hear in a few months how the experience of last year helped on the present trip.

An interesting enterprise is now being carried out in another direction also. There were two long flights made last summer: one from Long Island to the Pacific and back, carried out in the form of a race, and the other, the "flight around the rim," which was intended as a test of endurance of machines and accessories. Both of these flights gave a considerable amount of valuable information and showed what can be done by the Army pilots.

This summer conditions have on the whole been less favorable to Army aeronautics, but an attempt is being made to carry out an all-air flight from Long Island to Nome, Alaska, which is of interest not only because of the great distance to be covered, but because in its final laps this flight will probably become a precursor of a regular system of communication by air which will bind more closely to this country a wealthy but somewhat neglected possession.

In this connection, attention may also be called to the vigorous extension of the net of aerial mail transportation by the Post Office Department, which has recently asked for bids for several new routes which will afford air delivery of mails to towns in the South and West.

## Making Houses Truly Livable

Taking our living conditions all around, it would appear that while we may have progressed in some ways during the last 50 or 100 years, and have learned to keep our houses clean (improved sewerage), supplied with good water, and efficiently lighted, in other directions we have not made very much progress as compared with the houses of the times of the first white settlers on this continent.

It is generally recognized in anthropology that one of the greatest milestones in the development of the human race was passed when men learned to heat their houses. This is what made the habitation of the so-called moderate-climate zone possible, and it is there that most of the progress of the human race has been achieved.

The human race has not, however, passed the next stage of development in this direction, which is to make our houses and offices truly livable the year round. We have learned how to combat cold, but not how to handle heat,—at least we have not yet learned the wisdom of actually doing it.

One of the reasons why tropical countries are so hard on the white man as to make his development in those sections a very uncertain problem, is because a white man cannot live and preserve his efficiency year in and year out at a temperature above blood heat. Even in the so-called temperate zone we have conditions for several months in each year which very decidedly reduce the efficiency of our working forces and sometimes quite harmfully affect the health of our population.

If we are heating our houses in the winter, why, with our present knowledge of refrigeration, can we not cool them in the summer? As a matter of fact, this has already been done in numerous instances. Thus, a refrigerating plant was installed in the New York Stock Exchange a few years ago, with the result that a temperature of 65 deg. fahr. is maintained on the floor of the exchange no matter what may be the outside atmospheric temperature.

The development of the small automatic refrigerator for preserving perishable goods which works practically without requiring any attendance, bears an important promise for the future. And if automatic refrigeration can be installed to keep meat and butter fresh and clean at a cost of a few cents a day, there is no reason whatsoever why a similar but larger machine could not be installed to do the same service for human beings, who need a constant temperature to be in good condition just as much as perishable foods. This applies especially to places like hospitals, offices, etc.

At least two obstacles have in the past held back the general introduction of cooling systems. The first is the higher relative cost and expense of operation of a refrigerating plant compared with a heating system, which is offset in part, however, by the fact that the temperature range required in cooling is only about half that required in heating. The second is the difficulty of dealing with the humidity often present in the warm air and which, at reduced temperatures, becomes exceedingly uncomfortable. If the air is cooled its vapor is condensed, and unless proper care is taken it will collect on the cooler surfaces of a building. However, ways have already been found to deal with this side of the problem and no doubt other ways will be found as soon as an earnest effort is made to apply artificial house cooling on a wide scale.

Probably the chief and real obstacle to the general introduction of cooling systems in buildings has been the fact that people can get along without them by the use of such local expedients as electric fans or arrangements for the proper ventilation and circulation of air. People cannot get along without heating systems, however, and they accordingly have become universal and have reached their high state of efficiency.

The American public, at least, is accustomed to having the things it wants regardless of cost, as evidenced by the high percentage of the population who have secured automobiles within the period of a few years, and as the development of our buildings progresses in the matter of comfort and convenience, there will undoubtedly be a greater demand for the same degree of comfort during the warm weather as during the cold weather, and we may expect the mechanical engineer eventually to have a large and fertile field for his abilities in the design, construction and operation of small cooling systems, comparable in cost and size to our small heating systems.

## Ten Ways to Kill a Branch of an Engineering Society

(1) Don't come to the meetings. (2) If you do come, come late. (3) If the weather doesn't suit you, don't think of coming. (4) If you do attend a meeting, find fault with the work of the officers and other members. (5) Never accept office, as it is easier to criticize than to do things. (6) Nevertheless, get sore if you are not appointed on the Committee, but if you are, do not attend Committee meetings. (7) If asked by the chairman to give your opinion on some matter, tell him you have nothing to say. After the meeting tell everyone how things ought to be done. (8) Do nothing more than is absolutely necessary, but when members roll up their sleeves and willingly, unselfishly use their ability to help matters along, howl that the branch is run by a clique. (9) Hold back your dues as long as possible, or don't pay at all. (10) Don't bother about getting new members. "Let George do it."—*The Engineering Institute of Canada.*

## Listing Forms for Directory of Consulting Engineers in 1920 Volume of Condensed Catalogues

It is planned to further extend and improve the Directory of Consulting Engineers in the forthcoming issue of the Condensed Catalogues, and members actively engaged along consulting lines are invited to write to the Society for special forms covering listings in this section of the volume. In developing the Directory it is particularly desirable that all the specialized lines of practice should be indicated so far as possible, and the coöperation of the members to this end will be welcomed.

## CHARLES ETHAN BILLINGS

Charles Ethan Billings, chairman of the Board of Directors and former president of the Billings & Spencer Manufacturing Co., Hartford, Conn., died on June 4, 1920, after a long illness.

Mr. Billings was born in Wethersfield, Vt., on December 5, 1835. After receiving a common-school education, he served a three-year apprenticeship in the gun department of the machine works of the Robbins & Lawrence Co., Windsor, Vt. In 1856 he entered the employ of Colt's Patent Firearms Manufacturing Co. as a toolmaker and die sinker. During the early stages of the Civil War he was employed in the gun factories of E. Remington & Sons, Ilion, N. Y.

From 1863 to the close of the war, Mr. Billings was making drop forgings for the U. S. Government. He was one of the

CHARLES ETHAN BILLINGS

first men in the country to use drop hammers in the manufacture of arms, and it was he who perfected the forgings that were universally used in the manufacture of pistols for many years afterward.

When the war was at its height and many large contracts for pistols were coming in from the Government, it was found necessary often to run the shops at night. The continual pounding of the drop hammers was a source of great disturbance to the surrounding neighbors and they voiced their indignation in no uncertain terms. It was not long before a delegation composed of the mayor and some prominent citizens waited upon Mr. Billings to complain of the noises that issued nightly from the shop and to demand that they cease. Mr. Billings stated briefly that he had orders to run his shop night and day until the contract was completed. "Orders," exclaimed the mayor, "Be pleased to understand, sir, that all orders in this town come from me. I should like to ask you, sir, who gave you these orders?" "My orders," replied Mr. Billings, "come from Abraham Lincoln." The noisy drop hammers continued their work until the peace of Appomattox.

At the close of the Civil War, Mr. Billings returned to Hartford where he became superintendent of the Weed Sewing Machine Co. for three years. In 1869 the Billings & Spencer Co. was organized, C. M. Spencer being the other member of the firm. The company was incorporated to manufacture drop forgings and develop various improvements and inventions in the numerous small parts of machinery which, through the inventive genius of Mr. Billings, became of great value to the world. He was the inventor of a score of articles now in general use such as drills, chucks, pocket knives, wrenches, etc.

Mr. Billings was a life member of the Society, and one of the first to join the organization, which he did in 1880. He served as vice-president and president from 1893 to 1895.

He was also very active in the civic affairs of Hartford, where for twelve years he was president of the Board of Fire Commissioners. For several terms he served as councilman and as alderman. He was president of the State Savings Bank of Hartford and trustee of the Hartford Trust Co. He leaves a wife, two sons and one daughter.

## The Technical Library As Related to Plant Efficiency

Never in the history of industry in the United States has so forceful an effort been directed toward the mental and physical elevation of the individual, with the ultimate purpose of greater and improved production through increased efficiency, as at the present time. Never has so much intensive attention been devoted to the fundamental and advanced training of executives and labor in general in industrial establishments as in these days when economic conditions demand a most comprehensive understanding of the technicalities that complex production. From president to apprentice, thorough knowledge of the task in hand is an absolute essential to the success of the industry. Even the unskilled laborer must be more than a machine performing routine work merely as routine, without intelligent comprehension of, and interest in, his labor.

Upon the uncontested premise that greater efficiency means increased and better production and that efficiency is the result of intelligent interest, study and knowledge, rests the claim that the industrial establishment which invests in the installation of a technical library within its plant, will soon outstrip its competitors who have not had the foresight to adopt the plan. The American Library Association, coöperating with the Special Libraries Association, two national organizations of high standing, plans under its enlarged program to provide estimates, outline the selection of books best suited to the industry, and explain the working system that will bring the best results,—in fact, to coöperate in every possible way with the concern which is to have the library installed.

"It is true," declared a famous university professor, "that knowledge is power, but the best kind of knowledge does not consist of knowing facts, but in knowing where to go and find facts when you need them."

The progressive executive knows the value of forwarding the mental and physical welfare of his employees and realizes that their condition is reflected in the condition of his industry. He knows that the higher the plane of the workman's intelligence, the worthier his output, and he adopts every feasible method of promoting that efficiency which is derived from mental advancement. He recognizes that by study the worker becomes more valuable to himself and to the concern, and he opens every possible avenue that leads to that end. Once the interest in serious, purposeful reading is awakened, it continues, the reader attains a higher position, his work is bettered and the general welfare of the industry enhanced.

It is therefore to the great advantage of the employer that the habit of reading and study be encouraged to the fullest extent, and to this end the technical library is the best possible coöperative agent. But the technical library is not designed for the benefit of the working forces alone. While it is a wonderful medium for the promotion of increased efficiency through study and comprehensive reading, it is the source, if properly conducted, of information for the executives as well. It is the fountain head to which they can turn for a solution of practically every problem that presents itself; it is the clearing house from which are available the minutest details of shop and office direction; it is the source of time- and money-saving knowledge, obviating the necessity of experimenting, with its waste of material, effort and financial outlay. It contains treatises on the most modern methods, the newest appliances and devices, system and management—in fact, the yield of the best brains of many persons for the instant use of the individual.

# News and Notes ot The Federated American Engineering Societies

## Technical Club of Dallas Joins the New Organization

. One of the outstanding features of the Organizing Conference was the enthusiastic interest of the delegates representing the local, state, and regional organizations. It is not surprising, therefore, that these local organizations should be among the first to apply for membership in The Federated American Engineering Societies. Already the Technical Club of Dallas has made application and in doing so filed its claim as being the first local organization to apply. It thus evidenced its enthusiastic desire to take an active part in the formation of the Federation. Congratulations to the Technical Club of Dallas.

## Engineering Council Gives Its Endorsement to the Federation

At its last regular meeting held on July 17 Engineering Council passed resolutions endorsing the new Federated American Engineering Societies and offering its assistance in the conduct of the work necessary before the new organization can function. Inasmuch as the constitution of the Federation calls for the creation of an American Engineering Council with representatives from the various member societies, the action of the present Council is one of great importance, as to all practical purposes it means that the American Engineering Council is now functioning with six societies backing it. The resolutions of the Engineering Council follow:

*Voted*, That Engineering Council heartily endorse the plan of organization of The Federated American Engineering Societies and the American Engineering Council, adopted by the Organizing Conference of technical societies in Washington June 3 and 4, and authorize its Executive Committee to proffer and perform on the part of Council such assistance as may be practicable in completing the work of the Organizing Conference of the Joint Conference Committee of the Founder Societies in establishing the American Engineering Council.

*Voted*, That Engineering Council authorize its Executive Committee to deal with any question of coöperation with the Joint Conference Committee of the Founder Societies, relating to the permanent organization of The Federated American Engineering Societies, which may come up during the summer.

*Voted*, That the Secretary be instructed to invite to future meetings of Engineering Council delegates of the societies participating in the Organizing Conference in Washington, June 3 and 4, and editors of technical journals who may be interested.

## Three More National Societies Favor the Federation

The last issue of MECHANICAL ENGINEERING carried the announcement that The American Society of Mechanical Engineers were the first to voice their intention to join the newly organized Federated American Engineering Societies. Since then three other national engineering societies have met in convention and favorably considered the question of also joining the new organization.

At the annual meeting of the American Institute of Chemical Engineers held in Montreal June 28th-July 3rd, the question of the Institute becoming a member of The Federated American Engineering Societies was discussed and referred to its Council for definite action at its meeting on July 25th.

The report of delegates to the Organizing Conference was read at the meeting of the Board of Direction of the American Institute of Electrical Engineers at the annual convention at White Sulphur Springs, W. Va., June 30 and the following resolution adopted:

RESOLVED: that it is the sense of this Board that the American Institute of Electrical Engineers should join The Federated American Engineering Societies, but that as there is a small attendance at this meeting and a new Board be constituted, commencing with the administrative year on August 1, action be deferred until the August meeting of the Board and that a letter be sent to the members of the incoming Board, with a request that they give careful consideration to the matter and be prepared to act at the next meeting.

The report of the delegates to the Washington Organizing Conference was read at the meeting of the Board of Direction of the American Institute of Mining and Metallurgical Engineers on June 25, was favorably discussed and referred to the Finance Committee to devise and report on means for meeting the financial requirements.

The report of the delegates to the Washington Conference representing the American Society of Civil Engineers will be presented at the Annual Convention of that Society, at Portland, Ore., August 10-12, 1920.

## Some Technical Press Comments on the Washington Conference

The youngest profession has voluntarily accepted its responsibility in human affairs and is now reaching forward to a career of essential service. This is unmistakably the meaning of the action of the organizing conference of engineers held in Washington, early last month. . . . . the object of the Federated American Engineering Societies is to give expression to the exalted motive of service for the public good. It stands to the credit of the profession that this was done without a dissenting vote, although the forces of materialism were active and vigorous in expressing their opposition in debate.

Truly one of the great milestones in engineering history has been passed. From now on the youngest of the professions will begin to wield a power in public affairs which is destined to be a controlling factor in the working out of economic and industrial conditions.—*Industrial Management*, July, 1920.

Seldom does the habitué of technical and general gatherings absorb so many inspirational thoughts tumbling fast through a medium electrified with a sharp clash of intense purpose as at Washington last week. This alone would mark the recent organization of The Federated American Engineering Societies. Not only were the brief and pithy addresses from eminent engineers in various walks of life full of allusion to the widespread value of engineering thought and achievements, but even the sharpest discussion was based entirely upon the best and most effective method of giving service. Service was the keynote—not struck by the first speaker of the meeting and then forgotten, but animating the proceedings to the end.—*Chemical and Metallurgical Engineering*, June 9, 1920

This organization by engineers is one that will result in a large measure of good to the profession of engineering. It is a practical demonstration of views long and stoutly maintained by this journal

and exactly in line with definite progress toward a closer and more valuable working association of engineers throughout the country.—*American Architect*, June 1920.

The greatest step, or at least the greatest potential step, in the history of engineering in the United States, was taken on June 3 and 4 at the Joint Engineering Conference in Washington when the Federated American Engineering Societies was organized. This body was created for the avowed purpose of public service. The American Electrical Railway Engineering Association should assume its full share of coöperation in the federated societies.—*Electrical Railway Journal*, July, 1920.

No gathering of engineers in America has ever quite equaled in potential importance that which took place at Washington, D. C., on June 3 and 4. With the entry into The Federated American Engineering Societies of the component organizations that are expected to support it, there would be 100,000 engineers banded together for public service and the advancement of the profession in the United States. It was intimated that the American Association of Engineers would continue to carry on in its own way. The outcome of this experiment in federation will be watched with the greatest interest by engineers in Canada. It is not unlikely that similar action will before long become a matter for serious consideration and discussion amongst technical men in this country.—*Canadian Engineer*, July 1, 1920.

The Federated Engineering Societies are to be congratulated on their recognition of the principle of publicity. When the engineering world shall speak out habitually and authoritatively through the professional and lay press alike, to recount its achievements, to outline its projects, and to tender its advice on all matters where it has

480

knowledge and a collective concern, it will both perform an imperative duty and help individual engineers to gain the great degree of recognition, the high scale of remuneration, and the broad sphere of influence that should be theirs.—*Electrical World*, June 19, 1920.

One of the most important meetings of engineers that has been held in the United States, assembled in Washington, June 3-4, to take up the question of forming a federation of the engineering societies of the country. The importance of this conference can hardly be overestimated. It has been said of engineers that they cannot coöperate, or at any rate they do not, and that this has been the reason why engineers have not had the influence and standing in the community to which the importance of their professional work entitles them.—*Power Plant Engineering*, July 1, 1920.

Organization of The Federated American Engineering Societies was effected at the organizing conference of national, local, state and regional engineering and allied technical organizations at the Cosmos Club, Washington, D. C., June 3 and 4. Thus is brought into being the greatest national engineering society in the world.—*Mining and Metallurgy*, July, 1920.

In laying the foundation last week for a national federation of engineering societies, the Organizing Conference at Washington, D. C., took action which will add a new chapter to American engineering history. The two-day meeting of delegates representing more than sixty national, state and local societies . . . . . . . . . . . . . . . . . . . . . . . . . . . . was unique in its purpose in the widely representative character of its participants and in the spirit of coöperation and service which animated every session and which, at times, seemed to transform the delegates from mere technical men concerned chiefly with the details of engineering construction and design into apostles of a new order of professional solidarity.

Never before have engineering organizations throughout the country enjoyed the opportunity that is now presented to them of affiliating to secure united action.

The conference was truly remarkable in its representative character, as indicated by the list of societies and delegates . . . . . . . . . . . . Engineers have been criticised, and just so, for their professional aloofness. During the past decade or so the profession has become highly specialized and where, in former times, we had only "engineers" we now have civil, mechanical, electrical, mining, industrial, automotive, and scores of other classifications. This specialization has tended to segregate engineering into groups working more or less independently. At the conference, however, engineers of all kinds met and worked together.

The federation of engineering societies, as did the conference which created it, offers to technical men, for the first time in their history, an opportunity of presenting a *united* front on matters of civic and economic import.—*Engineering News-Record*, June 10, 1920.

June third and fourth were epoch-making days in the history of American Engineering. On those dates one hundred and twenty-three representatives of sixty-one engineering societies met at Washington and instituted an organization through which the united engineering talent of the country can function for the service of the public and the progress of the profession. The stage is now set, the program prepared in which the engineer is to take a larger part. Not a more active part, for his has always been the working role,

but more prominence given to his part. Less work behind the scenes and in the supporting cast and more of the headlines. The engineer has been trained and used to think and do rather than talk. He has been content to be the servant of the administrator at a price—*Power*, June 15, 1920.

An epoch-making step in engineering organization was taken in the Washington conference of technical society representatives which decided upon the formation of a federation of these technical associations. This is the first all-embracing movement of the engineers to place their activities upon a full professional basis comparable with that of the other learned professions of law, medicine and the clergy; and as such is of the greatest importance. For a number of years the several larger national engineering societies have been affiliated through the Engineering Council and other channels, but the work thus done is limited in many ways. It is a happy development, therefore, that gives us the larger agency which is now assured.—*Coal Age*, June 10, 1920.

Increased prestige and opportunity for service came to A.A.E. out of the conference in Washington June 3 and 4 called by the Joint Conference Committee of the national technical societies. Little if any occupation of the field occupied by A.A.E. is contemplated by the Federation. On the one hand is the Federation composed of societies and created for the public good; on the other hand is the democratic organization of engineers devoted primarily to helping the individual engineer to get on in life. A.A.E. was apparently recognized as the servant of the profession in the welfare field. There can therefore be nothing on the part of A.A.E. but cordial good wishes to the Federation.—*Professional Engineer*, July, 1920.

An event of immense importance to the engineering world took place in Washington, June 3 and 4. The large number of great engineering organizations represented would alone stamp this meeting as of tremendous import, but the object sought places it above that of any engineering gathering ever held. If the plan projected becomes workable, the engineer will at last be given the recognition due him for his public-spirited, though hitherto too modest, service. The American people *need* successful engineers in public office, along with successful business men and manufacturers, as we called to your attention in our recent editorial: "Not Politics—But Common Sense." The American people need engineers to take active interest in the great national problems. The Engineers need the American people to take an interest in THEM—a human, urging interest. It was to bring about the satisfying of this mutual need that the Federation was first proposed.—*American Machinist*, June 17, 1920.

The Organizing Conference has met and fulfilled its mission.— But if the new organization undertakes to trespass on the jurisdiction of A. A. E., in a spirit of competition rather than coöperation, by duplicating the work that A. A. E. began and has continued with such marvelous success, then justification for the continuation is difficult to find. There is abundant room for two comprehensive organizations of engineers, provided that one busies itself with joint technical matters and the other with social. A. A. E. has avowed its purpose to adhere to the latter. For any other organization to become a competitor in this work would be an act to invite dissolution rather than consolidation, an act to inspire the belief that after all engineers learned nothing from their war experience.—*Engineering World*, July, 1920.

# French Society Presents A.S.M.E. Honorary Membership to M. Rateau

THE Société d'Encouragement pour l'Industrie Nationale conferred Honorary Membership on M. Auguste C. E. Rateau at its meeting on May 8. In a letter of greeting to the Société, Past-President M. E. Cooley, who was in office at the time the Society conferred this award both on M. Rateau and on M. de Freminville, another of his countrymen, thanked the French organization for extending this courtesy to our Society. Dean Cooley wrote:

GENTLEMEN:

I have the honor to convey to the Société d'Encouragement pour l'Industrie Nationale the greetings of The American Society of Mechanical Engineers, and to express the great pleasure that our President and Council feel in the gracious coöperation of the Société d'Encouragement pour l'Industrie Nationale in conferring honorary membership in The American Society of Mechanical Engineers on M. Rateau, who, much to their regret, was unable to be present at our annual meeting in December, 1919, to receive the certificate of membership in person.

Permit me also to say that this willingness of our confrères in France to assume the responsibilities and perform the functions of a sister society in America roused in us anew the spirit of friendship

which had its birth in the splendid service rendered America by your gallant and intrepid Lafayette.

It was our desire in conferring honorary membership on MM. Freminville and Rateau, in addition to recognizing men of great distinction in the profession of engineering to give France additional evidence of America's appreciation of your country and your countrymen in stemming the recent tide which so nearly overwhelmed civilization and which unchecked might have precipitated the world into another dark-age period.

In the belief that France and America will always be found standing together in the cause of right against wrong, in peace as well as in war, I have the honor to remain.

Most respectfully
(*Signed*) MORTIMER E. COOLEY.

PRESENTATION ADDRESS OF M. DE FREMINVILLE

M. de Freminville, member of the Council of the Société d'Encouragement, and honorary member of The American Society of Mechanical Engineers, had been authorized by the Society to present at a public meeting of a great French technical association, the certificate of honorary membership which The American Society had conferred upon M. Auguste Rateau. In presenting the certificate to M. Rateau, and thereby conferring upon him Honorary Membership in the Society, M. de Freminville expressed the sentiments of American engineers in desiring to bestow this honor and reviewed the features of M. Rateau's professional accomplishments which led to his selection for honorary membership. He said in part:

MY DEAR M. RATEAU:

At this time when economic relations are so difficult to regulate among the countries of the world, and when there is so much clashing of individual interests, it is most pleasing to meet with men who desire above all to show their sincere attachment to France and their desire to collaborate with her. This is what the members of the Council of The American Society of Mechanical Engineers have desired to emphasize in conferring upon you honorary membership in their Society, a distinction extremely rare, and appreciated in America as the greatest that can be conferred upon an engineer at home or abroad.

This honor was bestowed upon you at the Annual Meeting of the Society, held on December 2 last in New York. During my recent

M. AUGUSTE C. E. RATEAU

visit to the United States I had the great pleasure of attending that meeting, and I remember well the reception which was accorded me, in the course of which I was requested to present this certificate to you.

It was very natural that the choice of the Mechanical Engineers should fall on a man whose scientific accomplishments, as well as his record of practical achievements, are universally recognized. But permit me to assure you that our friends in America had the intention of honoring in you the singular genius of France for which they have the greatest admiration and the greatest respect.

The Americans, emboldened by the immense resources of their country, are remarkable for their great conceptions. Not being influenced by customary traditions, they have realized that the surest means of enhancing the value of their possessions is to appeal to the teachings of science. However, in spite of their love of doing things in a rush, they have not accepted conclusions without going back to their sources and without studying the works of the pioneers who have opened new fields of endeavor, and it is this policy which has made known to them the French men of science whose names are written on the walls of this society which so many of them have visited. Whatever the originality of their works, the Americans like to connect them with those of French scientists. They have never ceased to follow the development of science in our country and know that you have taken a great part in its development; that you are one of the pioneers in the manufacture of steam turbines and turbo-compressors and high-pressure centrifugal pumps; that you are the author of rules and formulæ for the steam flow through orifices of various forms and the originator of methods for the calculation and design of steam turbines, now in common use; that you have made important contributions to the theory of the hydraulic turbine and of water hammer in conduits under pressure and have studied numerous types of centrifugal and helicoidal ventilators; that you are the inventor of the steam turbine operating under mixed pressure and of the rational utilization of exhaust steam, and that you manage with genius the works in which your turbo-compressors and other high-speed turbine machinery are being built. Moreover, that you have taken an active part in the application of turbines to warships, that important researches in ballistics are due to you and that you are the inventor of the supercharger which greatly increases the efficiency of aeroplanes.

Although the Americans admire scientific attainment, they are first

of all a people who attain results, and the enumeration of your accomplishments shows that under this head you are also entitled to their admiration.

## RESPONSE BY M. RATEAU

M. Rateau, in accepting the certificate of honorary membership, responded to M. de Freminville's speech of presentation, saying in part:

It is a source of great satisfaction to me to be so highly honored by The American Society of Mechanical Engineers, but in a great measure it is also a source of pride to all the engineers in France who are being honored in my person by the American society.

At this hour when we are confronted with the tremendous work of reconstruction of the devastated regions, it is certainly a great consolation to know that our friends across the Atlantic think of us, and are to be associated with us in that task. And not alone in the work of reconstruction can we work together, but there are very many questions, the solution of which it would be highly desirable to discuss in common, such as for example, as patents, industrial research, bibliographic reports, standardization of mechanical quantities, technical education, etc., etc.

The scientific organizations of the Allied countries and of the United States have recently formed an international council of scientific research, the principal object of which is the coördination of the activities in the different branches of science and its applications. Now, applied science needs as much collective effort and organization as pure science, for there is in this field great need of coördinating the work that is now being done by the various laboratories without any joint purpose, and I take this occasion to point out that a central organization instituted by the principal societies of engineers in America and in the Allied countries, would be well qualified to occupy itself with this important matter.

In the matter of education we Frenchmen could profit much by visiting the United States and learning what is being done there in this regard. Indeed, we have had accounts of recent developments in America, such as that which M. de Freminville has given us on the progress of the Taylor System, and the work of Omer Buyse, which contains an extensive exposition of the methods of teaching in the schools of the United States. But, in spite of these efforts, the American methods remain as yet unknown by the greater part of our educators and people.

There is in France now a well-directed endeavor to reform our methods of teaching, but nothing much has yet been accomplished. Our ideas, our methods, except in special cases, are the same as they were fifty years ago. However, the time of pure speculative learning is past, and in order that we may not be overcome and ruined by our rivals in industry, in commerce, and in agriculture, it is necessary for us to have energetic men of action, and not dilettantes.

All our teaching, remarkable as it is from many points of view, seems to result in destroying initiative and energy. It cultivates, it is true, the various essential faculties of man; memory, intelligence, rectitude, logic, exact reasoning, elegance of thought and expression, etc.; but are there no abuses? Is the memory not overcharged with a number of useless details which dim the fundamental truths; and then is our teaching sufficiently comprehensive? Is an attempt made to develop also, general ideas, spirit of observation, independence of spirit, intuition, and above all courage for responsibilities, and a number of other qualities useful to the men of action? Very little, indeed.

Forgive me for making a personal reference. I have been a professor of pure and applied science and, having started young in the profession, I have followed the usual errors. Now that I have seen and studied teaching methods in other countries, have visited many technical schools and other laboratories in America, England and Germany, and have given considerable thought to their methods, I can truly say that if I were to undertake teaching again, I would adopt radically different methods with the conviction that my students would derive a much greater benefit.

Why should we not send our best men, those capable of becoming professors, to foreign countries, particularly America, where they can see for themselves the results obtained by foreign methods of teaching? A few months would be sufficient and they would be admirably received there.

Instruction trips have been organized for young engineers, but it would be more profitable still to organize trips for men capable of comparing our methods with those of other peoples, and of deriving a great profit from their study. I for one am sure that it would be time and money well spent.

Reciprocally, others can learn from us what is best in our methods. These, however, are fairly well known. Our schools of the last century, particularly l'Ecole Polytechnique, that admirable creation of the great men of the Convention, have served as models to numerous institutions outside of the boundaries of France, with such modifications as have been necessitated by special local conditions, and the progress of technical science. It is now our turn to look about. To examine and to study the achievements of our contemporaries with a view to benefiting by them, can never be a sign of inferiority, but exactly the reverse.

M. Baclé, vice-president of the Société d'Encouragement, in the name of that society presented his compliments to M. Rateau as well as to M. de Freminville for their election to honorary membership in The American Society of Mechanical Engineers, a distinction as rarely accorded to strangers as that of honorary membership in the Institution of Naval Architects of England.

M. Baclé read the letter sent to the Société d'Encouragement by President Cooley and the Council of the Society. He said that the members of the former body were very much impressed with the marks of sympathy and friendship of their American confrères, to whom he presented the thanks of the Société and his hope that their bonds of friendship might be still further strengthened.

### BIOGRAPHY OF AUGUSTE RATEAU

M. Auguste Rateau was born on October 13, 1863, at Royan (Charente-Inférieure), France. His successful career in college at L'Ecole Polytechnique—where he graduated as honor man—and later at the Technical School of Mines, completed in 1888, presaged the distinguished life of service he has lived as an engineer.

M. Rateau typifies the eminently practical technician. He has excelled as much in the application of theoretical mechanics to the study of engineering phenomena as in the practical utilization of the machines he has perfected theoretically. For twenty years professor of engineering, he became celebrated by his numerous theoretical and experimental researches, particularly concerning the mechanics of fluids. It was during this time, in 1899, that he won the Fourneyron Prize, which had been offered by the French Academy of Sciences to any one suggesting any improvement in the theory of nozzles and confirming his results by experiments.

Some time later M. Rateau's renown had reached America. In 1904 the University of Wisconsin conferred upon him the degree of Doctor of Laws, "in recognition of his achievements as a mechanical engineer, as a contributor to the science of the flow of fluids, as a distinguished inventor of steam-turbine engines, and as an author of standard books in engineering."

In 1906, on the occasion of the celebration of the fiftieth anniversary of the Society of German Engineers, the Technical University of Higher Studies at Charlottenburg conferred upon M. Rateau the degree of Doctor of Engineering. A similar tribute was paid him by the Société de l'Industrie Minérale in 1908, and also recently by the University of Birmingham, England, and by the Institution of Mining Engineers of England.

M. Rateau's technical works comprise: Theoretical and experimental studies of the mechanics of fluids—air, steam, water—and of the machines operated by these fluids—fans and centrifugal pumps, hydraulic and steam turbines, nozzles, air propellers, etc.; aerodynamic studies (1909); turbo-compressors for aeroplane engines (1910); and the theory of aeroplanes (1919).

M. Rateau's practical contributions to industry are attested by the numerous patents bearing his name, which have been used to great advantage in industrial countries. These are spoken of in detail in M. de Freminville's address, which appears above.

---

# NEWS OF THE ENGINEERING SOCIETIES

Conference on Human Relations in Industry, Annual Convention of National Electric Light Association, General Meeting of the American Iron and Steel Institute

---

## Conference on Human Relations in Industry

There will be held this year at Silver Bay, Lake George, New York, August 27 to 29, a conference on Human Relations in Industry. Over six hundred representatives of industry are expected to attend, including some of the most prominent industrial leaders. Among the speakers who have already accepted are Allen T. Burns, Americanization Study of the Carnegie Foundation for the Advancement of Teaching; F. J. Kingsbury, president, Bridgeport Brass Company; Clarence H. Howard, President, Commonwealth Steel Company; C. J. Hicks, Assistant to the President, Standard Oil Company; R. B. Wolf, Consulting Engineer; S. J. Carpenter, Lumber Manufacturer; L. P. Alford, Editor, Industrial Management; and Timothy Healy, International Brotherhood of Stationary Firemen.

A similar conference held at Silver Bay last year, attended by five hundred prominent industrial leaders and others, created a demand for a larger conference this year. The greatest problems of the day are industrial. The most important factor in industry is the human factor. Men dealing with the human factor want to get together, exchange experiences, and discuss these matters with experts in the field; a conference at Silver Bay affords the ideal opportunity. Members of the Society who would like to attend should reserve accommodations through Mr. Fred H. Rindge, Jr., 347 Madison Ave., New York City.

## National Electric Light Association

THE forty-third annual convention of the National Electric Light Association was held at Pasadena, Cal., on May 18-22. The deliberations forcefully indicated that the public utilities of the country are no longer facing the question of obtaining business, but rather that of obtaining funds with which to finance the extensions necessary to meet the rapidly increasing demands for electric service. Coupled with the appreciation of the seriousness of the financial situation was the realization that utilities are essential public servants and that their salvation lies in close relationship with public sentiment.

Numerous committee reports and technical papers were presented. One of the most important, complete and comprehensive reports was that of the committee on water-power development. It contained statistics of present and potential water power in the United States. The latest authentic estimate shows there is approximately 50,360,000 hp. of potential undeveloped water power in the United States, and only 9,823,540 hp. of developed water power. Of the potential maximum power, 68.6 per cent lies in California, Oregon, Washington, Arizona, Colorado, Idaho, Montana, Nevada, New Mexico, Utah and Wyoming. Of the total maximum potential water power of the United States, 74.3 per cent represents undeveloped power coming under federal jurisdiction. In the western states 94 per cent of the maximum potential water power is in this class. In other words, practically all of the future development of water power in the western states, covering approximately 70 per cent of the total potential water power of the country, will be dependent upon federal action in the matter of issuance of workable permits for the development of these projects. Only 16.6 per cent of the maximum potential water power has been developed up to the present time. For the western states this proportion is only 6.5 per cent. In the case of central and eastern states, particularly the New England states, a comparatively large portion of the potential water power has been already developed.

Figures tabulated by districts showed that the per capita use of electricity in the far-western states is 2.21 times that of the remainder of the nation. Electric power rates average 1.65 cents per kw-hr. in the western section, as compared with 2.19 cents per kw-hr. elsewhere in the country.

Reference was made to projected developments in eastern states. More than 1,000,000 hp. is contemplated in Alabama, North Carolina, Georgia and Maryland. Electrification of the western trunk-line railroads will require development of a portion of the water power in that territory. Data obtained from 52 power companies of the far-western states, showing their loads during the past ten years, and also their estimated loads up to and including 1928, were obtained and compiled for presentation in the report in connection with graphic charts. They show that during the next ten years the expected development of power in the West will rise to the vast total of 1,776,-260 kw. At the present prices for material and labor this improvement will involve an investment of over $700,000,000.

That increasing costs and the scarcity of labor are resulting in the more general and more rapid commercialization of electricity and electrical appliances, was stated by John G. Learned, chairman of the commercial section. Electric furnaces for steel making, he said, are now almost a necessity and many have been installed this year for melting brass. Further advances have been made in the application of electrical energy in the reduction of copper, zinc, and aluminum, and in the manufacture of nitrates and chlorine compounds. Electricity for ice making has been developed to such a degree in some places that entire communities are dependent upon central-station service for ice. Industries are employing small electrical appliances to facilitate production.

An address on The Use of Central-Station Power in Electric Furnaces was delivered by Robert M. Keeney, director of the department of metallurgical research, Colorado School of Mines. He pointed out the great possibilities for central-station service in the electric-furnace field, stating that over 200 companies in this country have furnace loads that are considered very desirable. Electric melting of steel and scrap to produce low-grade pig iron has a big future in the western states. There are 875 electric furnaces in the world, of which 363 are in service in the United States, of a total capacity of 1600 tons, with an input of 600,000 kva. The production of electric steel in this country totaled 500,000 tons last year. Electric furnaces have a poor load factor, but this is due mainly to poor operation, the tendency in design being to increase the demand input per ton. Heavy reactances are being eliminated. Speaking further of the relation of the electric furnace to central stations, Mr. Keeney said that small furnaces operating 24 hours per day should be encouraged. He also referred to the development of brass melting by electricity.

## American Iron and Steel Institute

THE seventeenth general meeting of the American Iron and Steel Institute was held in New York on May 28. The opening address by Judge Gary dealt with business conditions in the industry. He referred to the panic of 1907 when the industry had successfully confronted the difficulties which arose in that critical period, and expressed his confidence that now in another period of stress the iron and steel men were ready to do their full duty again.

Joseph G. Butler, Jr., Youngstown, Ohio, spoke on the contribution of the iron and steel industries to the war, stating that a conservative estimate made from the latest information at hand placed the amount of steel contributed to the combined armies and navies of the Allied countries at not less than 100,-000,000 tons.

The technical papers were of the customary high order. George Otis Smith, director, United States Geological Survey, spoke on the rapidly diminishing oil resources of the United States and on the relation between the steel and oil industries. It appears that in terms of oil the decade 1910-1920 has witnessed a transition from over-supply to over-demand. Ten years ago the wells of the United States were adding to our reserve stocks 15,000,000 bbl. in a year; in the last nine months our stored petroleum has been drawn upon to the extent of 15,000,000 bbl. The first three months of 1920 have established a rate of domestic production that if unchecked will mean a total for the year of 415,000,000 bbl., or nearly twice the output of 1910, and a rate of consumption that would make the year's requirements more than 400,000,000 bbl., or one-seventh more than last year's

consumption. An estimate of the petroleum resources of the world, to which Mr. Smith referred, gives 60,000,000,000 bbl. as the total figure, with 7,000,000,000 bbl. as the quantity remaining available in the ground in the United States. Thus, if our present rate of consumption of nearly half a billion barrels per year continues unchecked, our estimated supply cannot last fifteen years longer.

Mr. Smith also discussed the possible remedies. The conversion into motor benzol of the light oils obtained as coke by-products in steel manufacture cannot be regarded as promising enough motor fuel even to meet the present increase in demand. Steel plants now in operation and under construction have a capacity of less than 2,000,000 bbl. a year. A better hope lies in exploiting the oil-shale resources of the United States, as their oil content is fairly comparable with the petroleum reserves of the world. But as the oil thus obtained could not be as cheap that that flowing from the wells, it is not at all unlikely that the next generation of American business men will see their trade rivals across the Atlantic turning the wheels of industry and commerce with cheaper oil than will be available in the American markets. Concluding his paper, Mr. Smith urged discouragement of the use of fuel oil as a substitute for coal in stationary power plants, and pioneering by American oil companies in other countries, particularly in South America, which continent is credited with a third more oil reserves than the United States.

The report of progress in an investigation on the fatigue of metals under repeated stress, being conducted under the auspices of the National Research Council, Engineering Foundation and the Illinois Engineering Experiment Station, was presented by H. F. Moore, research professor, and J. B. Kommers, associate professor of engineering materials, University of Illinois, Urbana, Ill. It has been ascertained that the cause of failure under repeated stress is not due to crystallization of metal, as it has been thought, but to the breaking up of crystals within the metal. Metals while being subjected to fatigue were observed under the microscope, and it was found that the crystals of which the metal was composed would allow deformation to occur by movement along certain gliding planes within them. Fatigue failure would follow a path through the crystal grains themselves rather than along their boundaries, and this was true even though in going from one crystal to another the plane of failure would change its direction, because of the different orientation of the crystals. The primary cause, therefore, of fatigue failure is localized deformation. This deformation in any particular crystal is very small in amount, and apparently even very accurate and sensitive extensometers cannot detect the deterioration which is going on.

Tests conducted with apparatus of extreme sensitiveness showed that the action of materials within the ordinary elastic limit is not perfectly elastic. For instance, when a specimen is first loaded in tension, and then the load is reduced to zero, the return path of the stress-deformation curve does not coincide with the original path, but lies slightly lower, the deformation seeming to lag behind the stress, so that when the stress is zero the deformation of the specimen is not yet back to zero. If now the specimen is loaded in compression, and the load reduced to zero, it is found that a loop has been formed by the stress-deformation curve. This loop is called a mechanical hysteresis loop, from analogy with magnetic hysteresis. The area of this loop represents energy which has been absorbed by the specimen, but which has not been given back again. In each cycle of stress, therefore, there is a small amount of work done because of the inelastic action of the material. It is planned to complete a series of fatigue tests of several typical steels accompanied by very careful static tests, in order to secure data as to the relation of fatigue-resisting power to the ordinary static elastic properties.

Other technical papers were: The Microscope and the Heat Treatment of Steel, by Albert Sauveur; The Acid Open-Hearth Process, by B. E. L. De Maré; Pipeless Rolled Products from Annular Blooms, by C. A. Witter; and Welfare Work in the Steel Industry, by C. L. Close.

# LIBRARY NOTES AND BOOK REVIEWS

AIRMAN'S INTERNATIONAL DICTIONARY. Including the most important technical terms of aircraft construction; English, French, Italian and German. Edited by Mario Mele Dander. Charles Griffin & Co., Ltd., London, 1919 (J. B. Lippincott Co., Philadelphia). Cloth, 4 x 6 in., 227 pp., 6s.

The terms in this dictionary are classified under systematic headings, so that those relating to a particular part of the airplane are brought together and an alphabetical index is provided.

ASBESTOS AND THE ASBESTOS INDUSTRY AND OTHER FIREPROOF MATERIALS. By A. Leonard Summers. Sir Isaac Pitman & Sons, Ltd., London, New York, n. d. Cloth, 5 x 7 in., 107 pp., illus., plates, 2s. 6d.

This little book gives a non-technical account of asbestos, its sources, varieties and uses, that is intended for users of this material and for general readers, rather than for those in search of detailed information.

AUTOMOBILE CONSTRUCTION AND REPAIR. With Questions and Answers. By Morris A. Hall. American Technical Society, Chicago, 1920. Flexible cloth, 6 x 8 in., 711 pp., illus., $3.50.

In this book the various parts of the automobile are described in detail, and explicit instructions for their adjustment and repair are given. The volume is intended for owners, chauffeurs and repairmen.

CEMENT. By Bertram Blount, assisted by William H. Woodcock and Henry J. Gillett. Longmans, Green & Co., New York, 1920. Cloth, 6 x 9 in., 284 pp., illus., plates, tables, $6.

This monograph, like its companion volumes, is intended to give a connected, balanced account of the cement industry in the light of our scientific knowledge of the principles underlying it. Processes are explained so far as they are necessary to elucidate principles, but no attempt is made to present minute technical details.

DIE HELMOLTZSHE WIRBELTHEORIE FÜR INGENIEURE. By G. Bauer, R. Oldenbourg, München & Berlin, 1920. Paper, 7 x 10 in., 146 pp., illus., diagrams, 14 marks.

The hydrodynamic problems met in such fields as airplane and ship building call for an acquaintance with the theory of vortex motion, but this theory, although it has for some time taken a considerable and useful place in electrical theory, is less well known by those engaged in other fields, it is the author's opinion that the chief cause of this ignorance is the lack of any work from which the engineer can obtain an understanding of the theory of vortices with a minimum of effort, a lack that the present monograph is intended to supply. The method adopted is that of a commentary on Helmholtz's famous paper of 1858 on vortex motion. The paper is presented in full, with extensive illustrations and explanations. Helmholtz's two answers to Bertrand's papers on the motion of fluids are also included.

DYKE'S AUTOMOBILE AND GASOLINE ENGINE ENCYCLOPEDIA. Treating on the Construction, Operation and Repairing of Automobiles and Gasoline Engines; also Trucks, Tractors, Airplanes and Motorcycles. By A. L. Dyke. Twelfth edition. A. L. Dyke, St. Louis, 1920. Cloth, 7 x 10 in., 940 pp., illus., charts, $6.

This well-known book has undergone its annual revision, and has also been enlarged by the addition of material on motor trucks, tractors, motorcycles and airplanes. With its wealth of illustrations and its clear practical information on all subjects connected with automobile operation and repair, it is well named an encyclopedia of the subject.

FORGE PRACTICE AND HEAT TREATMENT OF STEEL. By John Lord Bacon. Third edition, revised and enlarged by Edward R. Markham. John Wiley & Sons, Inc., New York, 1919. Cloth, 5 x 8 in., 418 pp., illus., tables. $1.75.

This is a textbook to accompany a course in forge practice, which explains the principles and provides a suitable set of exercises, but does not attempt to give minute instructions for making tools. The present edition has been enlarged by the inclusion of instruction in the heat treatment of steel.

GRAPHIC PRODUCTION CONTROL. By C. E. Knoeppel, assisted by various members of the author's firm and staff. The Engineering Magazine Co., New York, 1920. Cloth, 6 x 9 in., 477 pp., illus., diagrams, charts, $6.

An inaccurate review of this book was published in the May issue of MECHANICAL ENGINEERING. The book is not intended, as was stated, "to provide a complete description of the proper methods of making graphic charts and of applying them to industrial problems." It reveals a system for the control of production using graphics, charts, diagrams, dispatch boards, control boards and other forms of mechanism to visualize the facts by which control is secured.

HANDBOOK OF NATURAL GAS. By Henry P. Westcott. Third edition. The Metric Metal Works, Erie, Pa., 1920. Cloth, 5 x 8 in., 725 pp., illus., plates, charts, tables, $3.75.

The intention of this work is to supply a one-volume manual covering the entire industry in a practical way. It begins with a brief statement of the geology of the gas fields, followed by a history of the industry, and then proceeds successively to the properties of gases, field work, measurement of gas wells, pipeline construction and capacity-gas compression, measurement, regulation, distribution and consumption. Casinghead gasoline, the carbon black industry, helium and acetylene welding are also discussed. The present edition has been revised and enlarged.

HARDENING, TEMPERING, ANNEALING AND FORGING OF STEEL, including Heat Treatment of Modern Alloy Steels. A Complete Treatise on the Practical Treatment and Working of High and Low-Grade Steel, Comprising the Selection and Identification of High and Low-Carbon Steel, and Modern Chrome-Nickel and Chrome-Vanadium Alloys; the most Modern Hardening, Tempering, Annealing and Forging Processes, the Use of Gas Blast and Electric Furnaces, and Heating Machines. By Joseph V. Woodworth. Fifth edition. The Norman W. Henley Publishing Co., New York. 1919. Cloth, 6 x 9 in., 321 pp., illus., tables, $3.

The author has prepared a compendium of practical information on the subjects mentioned, based on trade literature and personal experience, for the use of mechanics and other users of steel. The present edition has been revised and enlarged, and a section treating of the newer alloy steels has been added.

INTERNAL-COMBUSTION ENGINES. Their Principles and Application to Automobile, Aircraft and Marine Purposes. By Wallace L. Lind. Ginn & Co., Boston, (Copyright 1920). Cloth, 6 x 8 in., 225 pp., illus., chart, $2.20.

This is a brief, practical, up-to-date text for students, covering the theory and operation of these engines, but omitting the question of design.

INTERNATIONAL COMMERCE AND RECONSTRUCTION. By Elisha M. Friedman. With a foreword by Joseph French Johnson. E. P. Dutton & Co., New York, copyright, 1920. 432 pp., 5 x 8 in., tables. $5.

Beneath the military campaigns of the war, a silent economic process has been changing the character of the commerce of the world. Profound changes have taken place, which this book attempts to trace in detail, and the principles of which it aims to discuss. It recites some of the facts essential to the formation of a new commercial and financial view and advocates a definite trade policy for the retention of our commercial gains from the war years.

INVENTIONS. Their Development, Purchase and Sale. By William E. Baff. D. Van Nostrand Co., New York, 1920. Cloth, 8 x 5 in., 230 pp., $2.00.

The author of this work discusses the marketing of inventions. The subject is considered from the various points of view of the inventor, the capitalist or investor, and the manufacturer, and advice given to each as to the methods for making a correct judgment of the value of any invention.

A KINETIC THEORY OF GASES AND LIQUIDS. By Richard D. Kleeman. First edition. John Wiley & Sons, Inc., New York, 1920. Cloth. 5 x 8 in., 272 pp., illus., $3.

The object of this book is to formulate a kinetic theory of certain properties of matter, which shall apply equally well to matter in any state. By considering atoms or molecules as mere centers of forces of attraction and repulsion, instead of as elastic spheres and by modifying the definition of the free path of a molecule so that the exact nature of molecular interaction is not involved, it is possible to lay the foundation of a general kinetic theory which applies to matter in any state, and which furnishes a number of important formulas.

PIEUX ET SONNETTES. By Edward Noë and Louis Troch. Gauthier-Villars et Cie, Paris, 1920. Paper, 7 x 10 in., 348 pp., illus., 20 francs plus 50% temporary increase.

This work is a theoretical and practical discussion of piles and pile-driving, intended for engineers and architects. The first chapter treats of wooden piles, the second of metallic piles and the third of the different varieties of concrete piles, their advantages and disadvantages. Chapter four discusses metal-sheet piling. The theory of pile-driving is then studied, followed by a description of the various types of pile-drivers and of the method of driving by means of a water jet. The volume ends with a study of the equilibrium of piles and notes on the proper choice of piles.

RÉSISTANCE DES MATÉRIAUX ET ÉLASTICITÉ. Cours Professé a l'Ecole des Ponts et Chaussées. By Gaston Pigeaud. Gauthier-Villars et Cie, Paris, 1920. Paper, 8 x 6 in., 772 pp., illus., 64 francs.

This textbook represents the instruction given at L'Ecole des Ponts et Chaussées, and its arrangement, the author states, has been largely determined by the necessity of quickly preparing the students to apply the subject in the field of girders. At the beginning there is, therefore, no attempt to introduce the study by a preliminary exposition of the theory of elasticity, and throughout there is an avoidance of abstract principles. After a brief review of the principles of mechanics and graphic statics, the theory of girders is explained. This is followed by a discussion of the various uses of the girder. The theory of elasticity is then introduced and occupies the concluding third of the book.

THE RUNNING AND MAINTENANCE OF THE MARINE DIESEL ENGINE. By John Lamb. Charles Griffin & Co., Ltd., London; J. B. Lippincott Co., Philadelphia, 1920. Cloth, 4 x 7 in., 231 pp., illus., plates, 8s, 6d.

This is a practical manual, intended to prepare marine engineers with no previous experience with oil engines, to qualify for positions on motorships. For this purpose the author describes the usual types of marine Diesel engines, the defects which have developed during operation and the methods that he has used to remedy them.

TECHNICAL WRITING. By T. A. Rickard. First edition. John Wiley & Sons, Inc., New York. Cloth, 5 x 8 in., 178 pp.. $1.50.

Mr. Rickard's primary object is to awaken the interest of engineers in the importance of the proper use of language. His volume is a discussion of the subject, in which particular attention is given to the usual faults of technical writing as he has observed them during his long experience as an editor. His rules are simple and definite, and their use is illustrated by numerous examples, so that the book will prove a useful guide to authors.

A TEXTBOOK OF INORGANIC CHEMISTRY. Edited by J. Newton Friend. Vol. IX, Part I. Cobalt, Nickel, and the Elements of the Platinum Group. By J. Newton Friend. Charles Griffin & Co., Ltd.. London; J. B. Lippincott Co.. Philadelphia, 1920. Cloth, 6 x 9 in., 367 pp., tables, 18s.

This installment of this important reference work covers the eighth group of the Periodic Table, with the exception of iron. Like the previous volumes, it gives a concise yet full account of our present knowledge of these elements and their inorganic compounds, accompanied by numerous references to the leading works and monographs that treat of them.

THOMAS' REGISTER OF AMERICAN MANUFACTURERS AND FIRST HANDS IN ALL LINES. Eleventh edition. Thomas Publishing Co.. New York, 1920. Cloth, 9 x 12 in.. $15.

Contents: Finding list and index; lists of manufacturers classified according to business; manufacturers of the U. S. arranged alphabetically; leading trade names, brands, etc.; makers, boards of trade and other commercial organizations; leading trade papers; manufacturers' representatives; export and import

houses; steamship lines and forwarding agents; overseas importers, merchants, etc.

The eleventh edition of this widely known directory has been revised and enlarged to the size of 4500 pages. It covers manufacturers and dealers in all classes of materials, and contains a classified list of manufacturers, with indications of their approximate ratings; an alphabetical list of the more important firm names, and a list of brand or trade names of manufactured articles.

TRAITÉ DE MÉCANIQUE RATIONNELLE. By Paul Appell. Fourth edition, vol. 1. Statique—Dynamique du Point. Gauthier-Villars et Cie, Paris, 1919. Paper, 7 x 10 in., 619 pp.

The treatise of which this book forms the first volume is a comprehensive one, covering the course given for many years in the Faculté des Sciences of Paris. The first volume includes the theory of vectors, statics, and the dynamics of particles. The present edition has been thoroughly revised and references to recent writings on the subject have been added. No previous knowledge of mechanics is assumed. The work is designed to give a thorough understanding of present knowledge of mechanics to those studying the subject in preparation for an engineering career.

TIME STUDIES FOR RATE SETTING. By Dwight V. Merrick, with a foreword by Carl G. Barth. The Engineering Magazine Co., New York, 1919. Leather, 6 x 9 in., 366 pp., 137 figs, $6.

The author of this work presents in amplified form the principles covering time study for rate setting, describes various mechanisms that have been found helpful in making and using such studies and presents some details of practice. An example of the application of time studies to a line of machine tools is included, together with detailed times for a number of other kinds of work.

TRANSIENT ELECTRIC PHENOMENA AND OSCILLATIONS, THEORY AND CALCULATION OF. By Charles Proteus Steinmetz. Third edition, revised and enlarged. McGraw-Hill Book Co., Inc., 1920. Cloth, 6 x 9 in., 696 pp., $6.

This volume, which is to some extent a continuation of the author's Theory and Calculation of Alternating Current Phenomena, deals with the transient phenomena of the readjustment of stored electrical energy which is necessitated by a change in circuit conditions. The present edition has undergone extensive revision and expansion and has practically been rewritten. A new section, entitled Variation of Circuit Constants, has been added. The method of symbolic representation has been changed from the time diagram to the crank diagram.

ZINC AND ITS ALLOYS. By T. E. Lones. Sir Isaac Pitman & Sons, Ltd., London and New York. Cloth, 5 x 7 in., 127 pp., illus.. plates, 2s, 6d.

This is a concise, readable account of the occurrence, recovery and uses of zinc, intended for persons engaged in the zinc trade or in search of general information about the metal.

## DISCUSSION OF CASTING SESSIONS

(Continued from page 440)

symposium on castings, I do not find any references to sprue returns, casting losses, or melting losses. These considerations are of especial importance in the selection of an alloy for the production of castings, and in fact they might properly be regarded as governing factors in a sense.

Mr. Bierbaum does not give any data on casting losses in brass and bronze, and I wish to inquire whether he has any figures. In the founding of light aluminum-alloy castings, about 10 to 12 per cent may be taken as an average figure, on the basis of an investigation recently made by the Bureau of Mines. Average figures probably do not mean much, since casting losses necessarily vary a great deal, depending upon the type of castings made, upon local foundry conditions, and upon other determining factors. However, it would be useful for comparative purposes to know what the average casting losses in brass and bronze are. I do not here refer to melting losses, but have specific reference to actual scrap (casting) losses incurred because of rejections of castings in which foundry defects exist.

## The Need of Research in the Industrial Field

### The Study of the Economics of Location—The Scope of the Work and the Engineer's Part Therein—Industrial Research in the Middle West—The Power Problem in the West

By P. F. WALKER,[1] LAWRENCE, KAN.

*The following paper is a discussion of the problems which await the research engineer in the field of industry. The author first outlines the various studies which he considers of fundamental importance. He lays particular emphasis upon the study of the economics of the location of an industry, pointing out in this connection the unfortunate conditions which now exist in Kansas and the neighboring states solely because this problem is as yet unsolved. The solution of the problem, the author states, is distinctly an engineering one, and he accordingly next presents a discussion of the engineer's part in the field of industrial research. He also outlines the scope of the work which in his opinion should be undertaken. The concluding portion of the paper is an abstract of a report prepared by the Research Committee of the Mid-Continent Section of The American Society of Mechanical Engineers, of which the author is chairman. It deals with the status of industrial research in the Middle West and covers all phases of industry. It is divided into three groups: (1) the food group, (2) the fuel group, and (3) the miscellaneous group, the latter comprising such industries as mining, cement manufacture, metal manufacture, etc. The paper is a comprehensive summary of the status of industrial research in the Middle West and a concise statement of the great need for such research throughout the entire United States.*

INDUSTRIAL research is a term now coming into frequent use. There appear to be, however, about as many variations in the meaning assigned as there are persons using it, so it may not be out of place to briefly discuss its several phases before giving attention to the particular kind of work and study to be discussed in detail. This is not for the purpose of imposing the writer's personal definition on any one, but merely for the sake of clearness. Strictly speaking, all engineering research is industrial. So also is all research in either pure or applied science laboratories, if the results are of a nature such that they can be given direct application in industrial operations or designing. One of the most apparent tendencies is to make a distinction between the laboratory of a public institution and the laboratory of a manufacturing or other business firm, calling the work done in the latter industrial research. As a matter of fact the same kind of work may be going on in both. The private firm, it is true, is looking for results applicable to its own business, but the results of the other may also be applicable there. It is a distinction based on motive, and form of organization, rather than on kind of research.

Another distinction is sometimes made by placing the study and development work carried out with commercial equipment already in operation in the business, in a different category than similar work being done for the purpose of experimentation with equipment set up to simulate operating conditions. Both, presumably, are technical studies, and the only possible difference lies in the facilities which may be provided in the latter case for securing observations. Such a distinction would have the effect of branding industrial research as inferior in scientific character and perhaps superior in commercial value. This is obviously an improper sort of classification.

A distinction based on the character of the work is made by including in the term "industrial" the study and analysis of organization and management problems, as distinguished from technical work based on physical science. This view is well represented in the interesting article by Mr. Hirshfeld in the February 1920 number of MECHANICAL ENGINEERING. Industry is now in that period of development where the most marked increases in the rate of production come from development of the works personnel, rather than from refinements in mechanical equipment. Many of these developments are far past the experimental stage in some branches of industry, but work still remains to be done to show results in many others.

### THE STUDY OF THE ECONOMICS OF LOCATION

But there is still another line of study that can be called industrial research, different in kind from work based on physical science. It has to do with the economics of location of industrial enterprises and with the possible development of natural resources in any given locality. In order to make a clear statement of the methods that may be followed in such work certain general principles will first be given a brief discussion. In passing, it may be observed that the war called attention to the importance of developing certain essential industries in this country, and this fact while somewhat aside from the main line of thought, is of prime significance.

The distribution of goods is the greatest industry in the world. In the United States it has become especially large in comparison to other activities, because of the great extent of country and the high per capita wealth which stimulates buying. The absence of a peasant class means that in every nook and corner of the land people are in the market for-the highest classes of every line of commodities. This means that shipments of standard goods follow population groupings in approximately uniform quantity.

What this means is indicated by the following illustrations. In a simple everyday commodity like starch the state of Kansas calls for the shipment of 2000 tons annually, the haul being mainly from Illinois and eastern Iowa, an average distance of perhaps 500 miles. This means one million ton-miles of traffic imposed on the railroads, although Kansas ships out many thousands of tons of corn.

A single small city in Kansas, Garden City, on the main line of the Santa Fe railroad, ships thousands of tons of alfalfa meal and dried sugar pulp for dairy feed to the southern states. The town does not have even a small dairy-products plant, hence it must in turn ship in its butter, notwithstanding the fact that it is in the midst of a cattle country. In varying degree this statement applies equally well to scores of towns in several states. The dairy-feed traffic probably amounts to several hundred millions of ton-miles annually, near half of which could be eliminated by a scientific distribution of manufacturing enterprises.

A condition more vitally wrong exists in the handling of lead and zinc ore. Forty per cent or more of the output of the United States comes from a small area near the corner junction of Oklahoma, Missouri, and Kansas, a few miles east of the oil and gas belt. The concentrated ore is first shipped to smelters in the gas field, a proper procedure because it saves transportation of fuel. But from here the greater part of the spelter in semi-

[1] Dean, School of Engineering, University of Kansas. Mem. Am.Soc. M.E.

487

crude form is shipped to New Jersey, and finally the finished metal, in sheet or other stock forms, is again shipped, some to the states of its origin, and much of it to states much nearer that origin than to the rolling mills of New Jersey.

Many other instances of unnecessary traffic might be cited. An unavoidable freight producer is the group of standard food products—meat, fruit, and cereals. The origin of this traffic is in the grain belt west of the Mississippi River, on the stock ranges of the West and Southwest, and in the fruit districts of the Pacific and Gulf states. Granted that, in the main, the movement is always toward the markets, a greater saving will nevertheless result when the center of population moves further westward from its present location in Indiana and this will be accomplished as the trans-Mississippi states experience the industrial development which the full utilization of their resources makes reasonable.

As a matter of fact, the agricultural states are facing a disturbing situation which has a reflex effect on the country as a whole, threatening the food supply of the nation. These states are near a standstill in population, if not actually losing ground. Out of the 105 counties of Kansas, sixteen prevented the state from declining in population during the decade ending in 1915. These sixteen are the ones having the important manufacturing industries and hence the largest cities. The agricultural counties lost in the aggregate. The same thing holds in Iowa, and doubtless in other states for the same period. This means an undeveloping home market. It means further the absence of those large masses of labor usually found in industrial centers and so necessary during wheat harvest. The seriousness of this lack of labor at the peak of the agricultural load is not realized by people far removed from the grain states. In 1919 the farmers in western Kansas counties were paying eight and even ten dollars per day, and usually for incompetent labor, and even at that were short-handed. This condition so increased the cost of production that in many cases in spite of the high price of wheat net incomes were almost wiped out. The result has been that the farmers are reducing the acreage to an amount that they can handle with their own labor plus the neighborhood exchanges. This is menacing the food supply. Many a farmer is also without funds to pay his 1919 bills, because the railroads have not been able to move the wheat. The writer has talked with farmers in several of the largest wheat counties of Kansas, and knows that these statements are in accord with the facts.

Some one has said transportation is the neck of the bottle that limits the capacity of the country to produce. But while waiting for the railroads to recoup themselves, why should not industrialists bestir themselves in an effort to check the ever-growing load of freight by a more systematic plan of production? When viewed from the standpoint of a national industrial system industry has grown in a haphazard fashion as to location. Business instinct has been logical and true within the recognized sphere of determining factors, and master minds have built up our successful enterprises. But the controlling factor has been cost of production, based on the assumption that transportation facilities would be adequate to bring the materials and carry the goods to market, irrespective of distance. Sometime a limiting condition will be set on this assumption, and it may be that the time is nearer than has been supposed. It is not too soon to begin to consider a national industrial policy, based on economy of combined production and distribution, in which the associated problems of food and fuel supplies will be given their due consideration.

## The Engineer's Part in Industrial Research

The basic principles on which such a coördinate system must be established are scientific in character. It is an engineering problem that is involved, in which the economics of transportation and production are in the foreground. It is the business of the engineering profession to recognize the need and to proceed with the study and analysis of the vital facts. Too often engineers hesitate to enter upon projects wherein financial and economic situations are the first to be met, even when the forces most directly influencing those situations are technical and such

that the man with engineering experience can most readily pass judgment upon them. Production and transportation, taken separately, are matters subject to scientific analysis, with which engineers have long dealt. In combination they are no less susceptible to such study and the need for a scientific system is becoming more and more evident.

The investigations in which the writer is now engaged form but a slight part of the complete survey and study indicated by the preceding statements. An attempt is being made to analyze the industrial possibilities of the state of Kansas. It is being done with the coöperative support of the governor's office, and of the state department of industry and labor. Whenever it should become necessary the authority of those administrative agencies may be invoked to open the doors to information respecting individual enterprises, but diplomacy has been uniformly successful in securing all that is needed. Another coöperating agency is the Kansas Engineering Society, of whose standing committee on manufacturing industries the writer is chairman. Also the writer's position as head of the research committee of the Mid-Continent Section of The American Society of Mechanical Engineers is of contributory value, particularly as regards the petroleum industry. Likewise the fact of his being head of the engineering branch of the state university fixes his status as practically that of a state official in such matters. These points of a somewhat personal nature are mentioned merely to show the means at command for making a consistent statistical study of existing conditions.

## The Scope of the Work

The work under way naturally divides itself into three branches:
1 A survey of operating conditions in existing enterprises
2 A study of industrial needs of the state
3 A study of the vital economic factors in such kinds of manufacturing as Kansas resources and conditions make it feasible to consider.

Along with these branches of individual effort a certain form of educational campaign has been conducted. The means for this has been an association with the chambers of commerce of the cities, by speaking before gatherings, conference with committees, and providing secretaries with statements of facts as to individual industries. This is akin to promotion work, and as a matter of fact some real promoting is being fostered at the present time. This is special, however, and essentially a private enterprise.

In the survey of existing enterprises the usual routine of the census enumerator is amplified. Volume of business, capital, quantity or value of raw material, employees and payroll, and power, are interesting and valuable items, but in this study the following facts are more to the point:
1 Markets and marketing methods. The section of country served, the nature of the demand for the product, relations with other industries, and trend of development are points of information sought
2 Sources of raw material. Actual original source and the freight rates or other cost of delivery to the plant
3 Transportation facilities, for both raw and finished stock
4 Details of power and fuel supply, cost, etc.
5 Labor conditions. Questions of supply and quality of labor, aptitude of residents of the community for specific kinds of work: also the local situation as to balance in employment opportunities for men and women
6 Reasons why the enterprise was located as it was. Sometimes this reveals only some personal preference, but often there is a well-considered reason
7 Business and social conditions of the town and community.

The second branch of the work—the industrial needs of the state—presents several distinct features. For different regions the problems are different. Many think of the state of Kansas as one great wheat field with a few cattle ranges interspersed. As a matter of fact there is a wide diversity of natural products. In the southeastern counties are the mining interests in coal, lead, and zinc. Just west of that and extending half way across the state is the petroleum district. North of the western end of the oil belt in central Kansas are large deposits of a good grade of salt. In several counties, some northeast and some south of the salt deposits, there are large quantities of workable gypsum. In many sections the clays and shales are furnishing materials for cement, brick, tile, and other ceramic products. Corn, as well as wheat, is grown in all excepting the far-western counties, but almost nothing is done now in the manufacture of corn products. Sorghum cane and its products hold important possibilities, but lightly touched upon as yet. The dairy-products business holds most promising possibilities and is destined to become an important factor soon. Large portions of the seemingly dry western counties have abundant ground water within pumping distance of the surface. Irrigation has been practiced successfully for the growing of sugar beets and alfalfa, and much is to be expected from further developments. Canning and preserving of vegetables and fruit is an industry that has not yet begun to touch its possibilities. These and several other industries are of great promise, many of them being needed to change agricultural methods from extensive wheat growing to an intensive system with diversified crops, the better to realize the food-producing capacity of the soil. Industries producing lines of goods in metal, leather, paper, and cheaper fabrics, already represented in several cities, are capable of extensive development. Any line of manufacturing which produces goods expensive to ship in finished form, from raw material that can be shipped in bulk on a low freight tariff, can be operated successfully in this region. These comments illustrate the line of investigation without covering at all adequately the possibilities of this or any other region.

The third phase is the logical one of analyzing carefully many different lines of manufacturing to determine which ones are best adapted to the conditions in any given locality. Those requiring large amounts of power or process fuel, like the smelters and cement mills, must cling to the coal and oil districts. For others the power demand is small and they may follow markets or raw material. Those for which the value added by the manufacturing process is large can go anywhere, other factors being equal and provided the labor supply is in sight. It is here that we have the final application of all that has preceded, namely, the fitting of the enterprise to the locality. No task is more interesting than the one of providing a locality with an industry or an industry with a locality.

It is submitted that this is a form of service to the state or other district that warrants the attention of engineers. It is a step toward the great enterprise of establishing a comprehensive industrial system that shall take into intelligent consideration all of the factors of production, markets, economy in transportation, economic distribution of population, and maximum utilization of the food-producing power of the country. We have been prodigal in the exploitation of natural resources, but the end of this is coming. To become a true industrial nation means the extension of scientific methods to the broader problems of industry.

### INDUSTRIAL RESEARCH IN THE MIDDLE WEST

As a definite example of this form of service to the state the writer desires to present, as the concluding portion of his paper, an abstract of a report prepared by the Research Committee of the Mid-Continent Section of The American Society of Mechanical Engineers. This report was based upon that society's action in calling for information regarding industrial scientific investigations in the Mid-Continent Section territory; (a) On such topics as may be so well worked out as to warrant a final statement of results; (b) On problems now in process of solution or development. A further point was the formulating of information as to facilities available for carrying on research work in the several industries and educational institutions in the same territory. Still another was to take such steps as might be possible toward the securing of funds to be applied in the support of investigations having as their object the improvement and development of the varied industrial interests of that section of the country.

In compliance with this request the committee has applied itself to the carrying out of the program, particularly as to the gathering of data pertaining to investigational and development work now in progress. Circumstances have required, however, that the writer who is chairman of this committee should formulate the report without conferring with other members on matters other than the general intent and scope of the work to be undertaken. Opinions expressed herein are therefore largely his personal ones, for which he assumes responsibility. In general it is to be understood that the ideal that guides in this work is to promote the best interests of the people of this large Mid-Continent area, which in turn touch the interests of the entire country in a most vital manner. Engineering is a profession of service. It is given to engineers to bring the energies existing in nature to the use and convenience of man, and the first concern of those who study the industrial needs of an inhabited region is the economic betterment of the population along lines that are permanent and in accord with a sound national policy. Individual industries are to receive the direct study, of course, and the benefits of such development as might be stimulated thereby, but the motivating impulse lies in the general aim just expressed.

### THE SCOPE OF THE INDUSTRIES

The industries of the territory which have received consideration may be grouped as follows:

A—The food-products group, including
  1 The beet-sugar industry
  2 The dairy-products industry
  3 The canning and preserving industry
    NOTE: The two largest industries of this group, meat packing and flour milling, are not included because they have reached such a stage of development as to make generalized study unnecessary, excepting as to choice of motive-power equipment in flour mills.

B—The fuel group, including
  1 The petroleum and petroleum-products industry
  2 Coal mining.

C—Miscellaneous, including
  1 The zinc and lead mining and smelting industries
  2 Cement manufacturing
  3 Metal manufacturing
  4 Other manufacturing, considered as to conditions influencing possible development.

### THE FOOD-PRODUCTS GROUP

Research work among the industries producing foodstuffs is of that kind having to do with economic conditions influencing their establishment. This is a territory that produces much and is capable of producing much more. The soil is made to yield only a fraction of what intensive methods would make possible, this being strikingly true of the western portions of Kansas and Oklahoma. To bring about the change there is demanded a development in the field of engineering.

The great need of the section of country referred to is for power. Irrigation is necessary to make sure the crops, and this would be provided from the unfailing sheet of ground water if cheap power were obtainable for pumping. In some sections the start has been made, but the mounting cost of fuel oil is making continuance difficult and uncertain.

### THE FUEL GROUP

On the coal-mining branch comment has already been made with respect to possible development of a coking and by-products industry. In mining itself the problems are largely with respect to labor and social conditions. Investigations are contemplated by the mining staff of the University of Kansas on such questions as the percentage of extraction being secured under the mining systems now in vogue, and on economic advantages arising from electric underground haulage. It is understood that similar studies are under way in Oklahoma. Engineering problems are much more serious in the Arkansas coal fields than in either

Kansas or Oklahoma, but no information has been secured as to specialized work in that field.

The petroleum industry occupies the center of the stage in this territory, and necessarily comes in for a major share of attention. It has often been said that there is very little scientific engineering work carried on in the oil field. At one time in the history of the industry, undoubtedly this was true. Conditions now are very different, however, and while there are still found those regions where in the flush of new development the methods employed are wasteful in the extreme and based largely upon rules of thumb and traditions handed down through the old rough-and-ready type of oil men, there is at the same time being applied in the representative districts where well-organized companies are in control, a degree of engineering talent and skill not to be surpassed in any industry of the country. Margins are not what they once were, and, in spite of the apparently high rates of return, the safety of the excessively great investments in what may in many instances prove to be short-lived enterprises demand a close and accurate study of those conditions which lead to economy. To the same end there has been necessary a large amount of study and research in the development of processes and of specialized equipment that have made possible the effective handling of an elusive substance upon a basis profitable for the investor and adequate for the market. The uninitiated would be surprised on discovering the amount of attention being given to these scientific problems which have a direct bearing upon the commercial issues of business.

It is necessary to appreciate the demand which is being placed at this time upon the producers of high-grade liquid fuels. This demand is increasing rapidly and the public press is employed frequently by those who would express their convictions that a crisis is being approached when supply will no longer be sufficient to meet demand. It is said by many that the peak of oil production has been reached. It is not in the province of this committee in its report to express opinions on this point, but it is not out of place to remark that the apparent limitations to production in the Texas fields need not occasion as much alarm as it seemingly has in some quarters. There is evident a renewed interest in pushing forward new developments in older fields of Oklahoma and Kansas, particularly in the way of reaching sands considerably deeper than those which have been producing for many years. Results obtained are highly gratifying and there is ample evidence of the existence of a great storehouse of petroleum that as yet has been drawn upon but lightly. This is of particular significance in the present connection because it gives assurance of the continuance of the industry over a period of years during which there may be put into effect the many advanced methods now being perfected in both the producing and treating of oil. It is this which gives special significance to the scientific work that is being carried on.

Our topic divides into two parts, representing the major fields of activity. These are, first, the production of oil, and second, the handling and refining of the oil when once it has been brought to the surface.

*Production Problems.* It is of first importance that there should be drawn from the producing sand the largest possible percentage of the oil contained therein. How it may be increased is one of the greatest objects of study, and many suggestions have been made covering a wide range as to feasibility of methods. Perhaps the most extreme suggestion is that some day we shall be mining the oil sand, bringing it to the surface for the application of extractive processes more frequently associated with the handling of oil shale. To the practical oil man, however, the first step is the perfecting of producing methods. With lessened cost of production, wells can be operated down to a correspondingly lower output and the sand more completely drained. With lessened cost of drilling, wells may be more numerous and contribute to the same end. With this in view reasearch activities have been directed toward the following:

1  The development of more efficient types of deep-well pumps
2  The development of more efficient types of power-generating equipment as applied at the well

3  The application of new forms of power equipment, notably the introduction of electric power for deep-well and surface pumping and also drilling
4  The recovery of oil from the troublesome water-oil emulsion so commonly met with in most producing fields
5  The perfection of methods for extracting the gasoline otherwise carried to waste in vapor with natural gas, commonly designated as plants for manufacture of casing-head gasoline
6  The problem of water supply, frequently a serious one where quantity and quality of the water are vital factors
7  The development of accurate metering devices for both oil and gas.

A great amount of time, talent, and capital has been expended upon each of these problems, upon any one of which a complete paper might well be written. To the writer the most interesting and seemingly the most striking of these are the application of electric power in the field, the perfection of dehydrating plants for the recovery of oil from emulsion, and the casing-head gasoline plants.

The accomplishment of the first of these three is dependent upon a supply of electrical power. This condition is met in most admirable manner in the El Dorado field in Butler County, Kansas, where a high-voltage power line leads directly from the plant of the Kansas Gas and Electric Company of Wichita to a substation maintained by the Empire Gas and Fuel Company. From this substation current is distributed to the entire field. A great amount of specialized design was necessary in order to produce machinery adapted to the service, but this has been accomplished in a most creditable manner. It should be observed in passing that continuity in operation is a most important factor in preserving the productivity of an oil well.

No one problem has called for more careful study than has the design of dehydrating plants for the recovery of oil from emulsion. Probably the most effective system has not yet been evolved. The one in most common use involves the centrifuge principle, wherein the last step in extraction is brought about through centrifugal force. Efficiency of this system is such that less than one per cent of water remains. One process which has been demonstrated to be successful involves the principle of passing the water-and-oil mixture through a strong magnetic field, breaking the oil film and permitting the water to separate at once by gravity. Developments in this line are rapid, so that one sees much discarded equipment which has become obsolete in the presence of improvements.

In the field of production of casing-head gasoline there is active discussion as to the relative merits of the two systems, designated as the compression and the absorption methods. Experience seems to indicate that each one has its field of application. Apparently the gas less rich in condensable vapor is handled more effectively by the absorption method.

To further illustrate the investigational work which the successful operation of an oil property makes necessary, the following researches accomplished by one company are mentioned. These tasks have been accomplished through detailing to the work regular members of the operating staff of a single department.

1  The conducting of an extensive series of tests for volumetric efficiency of compressors drawing gas from wells under widely varying pressures. This was especially well done when considered as an isolated piece of work, and made possible the proportioning of equipment to demand on a scientific basis
2  The carrying out of extended tests on total gasoline content of gases before and after the application of the casing-head gasoline process
3  The determination of shrinkage of volume of gas while being thus treated by the compression method
4  The determination of evaporation losses from blends of casing-head gasoline with heavier oils when handled at different temperatures.

*Handling and Refining Problems.* In the transportation of oil by pipe line much analytical work has been done which cannot be reported upon in detail. Records of this work are in the files of the larger companies, and all that can be said is that engineering talent of a high order is being employed in maintaining service.

Problems that are obviously calling for solution in the development of equipment are those relating to the determination of constants for the flow of heavy oils in pipes, and the development of more efficient pumping machinery. Here again the question of power supply enters as a factor. With this mere mention the question of pipe-line transmission will be dismissed.

The problems of the refinery are many and complicated. Many of them come within the range of the chemist rather than of the mechanical engineer. Development work to perfect the process of cracking of oils to increase the yield of gasoline still continues, many companies being engaged in this work.

With the larger refining companies the following are problems continually discussed and for which the positive final solutions have not been found:

1　Determination of the exact amounts of heat required to vaporize oils of different densities and characteristics
2　The rate of conduction of heat through metal surfaces where oil is one of the media involved in the transfer
3　The proper design of furnaces for stills when handling (a) crude oil; (b) residues under pressure
4　The coefficient of expansion of oils
5　Fuel-economy problems, including the design of the most efficient oil burner
6　Devising of a method for the mechanical separation of wax from lubricating-oil stock
7　Successful methods for the treating of sulphur crudes
8　The development of synthetic processes for by-products, such as ammonia, dyes, perfumery, etc.

Progress in refining methods in the petroleum industry is of peculiar importance. Developments which have been made in recent years have greatly increased the production of the lighter and more valuable fuels, without which the market today would be in a much more critical condition. The chemist and the engineer are working together to improve existing methods and to devise new ones, and the end is not yet. All of the eight unsolved problems mentioned are subjects of study from which a mass of valuable data will be forthcoming in the near future.

### THE MISCELLANEOUS GROUP

The lead- and zinc-mining industry is one of the most important activities in the Mid-Continent territory. What is needed most is a stabilization of the demand for zinc plate. This metal is adaptable to many uses outside of the galvanizing and alloy business, and some pioneer work along this line would be productive of results. Some of the same conditions hold for the cement-manufacturing industry. This is a line of business, however, for which an active organization directs the necessary advertising and maintenance of market conditions.

Only one line of research is mentioned, namely, a close study and analysis of power distribution and costs, which has been in progress for some time at the plant of the Lehigh Portland Cement Company at Iola, Kan. That work was interrupted by the war, but will be renewed during the next year.

Metal manufacturing in the territory has been added to by the installation of several new plants which are engaged mainly in the oil-well supply business. One thing greatly needed is a source of supply of a good grade of metallurgical coke. Important investigations on possibilities of coking the mid-western coals are under way in Illinois and Kansas, and it is hoped that some help will be afforded the foundrymen in the not far distant future.

### THE POWER PROBLEM IN THE WEST

In preceding sections reference has been made to the need of cheap power for several important industries. This is a matter that is holding back industrial development in a large area. Coal in the region west of a north-and-south line not far from Tulsa is scarce and high in price. Under ordinary conditions oil is too valuable a fuel for use under steam boilers, and its availability for internal-combustion engines is becoming limited while the high price is having a marked effect on cost of generation of power in small units. Present assured markets make the promoting of an extended system of interlocking generating plants markedly hazardous.

Conditions indicate a possible revival of the producer-gas enterprises that were checked in development by the rapid extension of the petroleum fields about fifteen years ago. The lignites of the northwest hold the balance as regards original sources in this matter. The successful outcome of such a proposition rests with the engineering profession, and a reward is in store for those who provide the solution for this on a large scale. Before it can be accomplished, however, it will be necessary to construct a new line of north-and-south railroad connecting the northwest with this great semi-arid region in the western portions of Kansas, Oklahoma, and Texas—a bit of railroad construction that is well worth considering. The region in question has untold possibilities for the maintenance of the food supply of the nation.

### RESEARCH AGENCIES IN THE SECTION TERRITORY

The exclusively research institution in the territory is the Oil and Gas Experimental Station of the United States Bureau of Mines located in Bartlesville. The station has been active on several phases of the oil industry, its principal lines of study at the present time being:

(a) Dephlegmating towers for refinery stills
(b) An investigation of the underground conditions of the Hewitt Field
(c) Increased recovery from oil properties
(d) Evaporation losses from storage tanks
(e) The recovery of gasoline from uncondensed still vapors in refineries.

It has also taken up the question of water supply and the effect of water impurities on oil products.

There are three state universities and two state colleges of agriculture and mechanic arts within the limits of the territory. Research facilities exist at all of these institutions, and in at least three of them active steps are being taken to engage in investigations of direct interest to the petroleum industry. Mention has been made of work in progress at the University of Kansas. In addition a productive study has been made during the year just ending of the treatment of oil shale. Plans are now definitely made for taking up in the fall two of the problems mentioned in the refining section: namely, the conduction of heat through metal plate with oil as one of the media of exchange, and the latent heats of evaporation of oils of varying gravity. At the Oklahoma institutions preparations are being made for an extended study of chemical problems.

But the significant element in research in this territory is the part taken by the organized staffs of the larger oil companies. All have engineers and chemists whose time is given in considerable part to new problems, and if all that is being accomplished in this way could be brought to light the extent of the work would be a matter of much surprise. In this organized work the Empire Gas and Fuel Company holds a conspicuous place with its full-fledged department of engineering research. An independent agency like this committee can gain access to the results of such activities, and by adopting a conservative policy as to publication such as will retain it in good standing with the companies which cannot be expected to throw everything open to the public which it has learned through the expenditure of much effort and money, it can accomplish much for the good of the profession of engineering and of the industry.

The time is ripe for bringing to bear upon this important industry the best and united efforts of all technical and scientific men. Quoting from a recent publication of the Bureau of Mines: "The time has arrived when, with crude oil higher than it has been for 50 years, inefficiency in the production and manufacture of petroleum should not be countenanced. Petroleum and its products should be reserved for those uses for which it is peculiarly adapted and for which there are no substitutes."

# The Training of Engineering Students in Industrial Management

BY BRUCE W. BENEDICT,[1] URBANA, ILL.

*The author of the following paper presents the details of a pioneer experiment in the method of teaching industrial management to engineering students. The field of engineering is broad in its scope and purpose and there is a constant call for men trained to perform a multitude of diversified tasks. The technical schools, however, cannot attempt to train men for special tasks, but the presentation of the fundamentals of engineering must nevertheless be governed by the needs of present engineering practice rather than by traditional academic considerations. With this thought in mind the University of Illinois has organized its school shops upon a commercial basis, the details of the system employed and the manner of conducting the work being described by the author. The plan is an exceedingly simple one and is based upon a recognition of the fact that the engineer is primarily a manager of human enterprises rather than a technician making plans for others to execute. The paper is not intended to present the specifications for a complete course in industrial management designed to develop executives and managers from untrained men within the course of a few weeks, but it does outline an experiment in the development of engineering leadership which it is hoped will be fully realized.*

I F we accept the proposition that "the ultimate aim of engineering is reduction in cost of the elements of living through the development of improved facilities for changing raw materials into usable products," we have a definite guide for the training of engineers, and the prospective engineer has an equally definite ideal to direct him in preparing for and carrying on his life's work. It is not certain that technical schools generally have been guided by ideals which embrace the broad function of engineering as just defined, nor is it clear that the engineering profession as a whole is, or has been, inspired by adequate conceptions of its task. Engineers to a great extent have narrowed their field of activity to design and construction. Their genius has been directed toward the building of engineering works, but with the completion of the physical structure they have taken themselves off to similar tasks elsewhere, leaving operation to other hands. Operation is a prime function of engineering. It is the final link in the cycle which makes engineering work of value to man, and engineers must, if they see their task in its true perspective, assume the responsibility not only of creating improved production facilities but also of operating them.

The limited conception of the field of engineering held by engineers in general is due in no small part to the influence of the technical school. No one questions the high sense of duty and the devotion of engineering educators, but because of the limitations imposed by inadequate financial resources, the absence of the human factor in laboratory and classroom exercises and the detachment of college environment from affairs of the world, the technical school has been unable to construct a curriculum that reflects the larger ideals of engineering as expressed in the opening paragraph. Lacking a real engineering setting and having no direct need for consideration of those economic and human factors which enter so largely into the effective management of engineering, the technical school has to a large degree ignored them in its treatment of engineering subjects. Gradually the curriculum has been developed around the idea that proficiency in the solution of mathematical and scientific problems constitutes the proper initial training for engineering work. Unquestionably the preparation for theoretical and material problems of the engineering profession has taken precedence over the larger task of developing the fundamental qualities of leadership not only from the reasons mentioned above but also from the belief that knowl-

edge of practical affairs is to be acquired in practice after graduation. Allowing for the fact that experienced teachers can and do mold the most rigid curriculum into practical and inspiring forms, we cannot avoid the conclusion that the conventional scheme of technical training, as organized and applied, tends to produce technical advisors rather than engineering leaders.

Industry absorbs the bulk of engineering talent and young men entering this field of endeavor must be prepared for exceedingly complicated tasks, involving, in addition to mathematics and science, a thorough knowledge of economics, of production and of men. Technical courses taught on the basis of pure theory without reference to the latter factors certainly do not meet the need of a large majority of young engineers for conditions confronting them at this time.

Without lowering in any degree the pressure on fundamental theory, instruction must emphasize the importance of the factors of time, cost, production, safety, and of the human element. Even the fundamentals, mathematics and science, must be given' new life by teachers who look at these subjects as tools of the engineer, instead of as academic exercises for training the mind.

## THE SCOPE OF THE ENGINEERING FIELD

The field of engineering is broad in scope and purpose. It calls for an increasing force of trained men to perform a multitude of diversified services. Notwithstanding appeals for men with particular training, the technical school cannot attempt the preparation of engineers for special tasks. It must seek to promote knowledge of engineering fundamentals to the exclusion of technical skill required in specialized branches of engineering and of industry. But the presentation of fundamental matter should be governed, as previously suggested, by the needs of present engineering practice rather than by traditional academic considerations. It goes without saying that all will not agree with this proposal to liberalize present curriculums. The reasons formerly considered as fundamental are in opposition to the views expressed, but it is questionable whether the arguments employed to sustain them are entirely effective under present conditions.

We may consider the engineer as a technician or as an operator; we may carefully define his ideals and his tasks; we may prepare him for a certain type of service, but the composite result will be naught if our subject cannot produce and produce cheaply. Now the engineer is preëminently a producer, a producer of something useful to man, and it is obvious that his training must be organized to fit him for this task. The technical school is responsible for what is roughly the second training period, and if it fails to contribute its full share to the effective preparation of the engineer for the services demanded of him, we may expect the pressure of need eventually will force an adjustment in methods of training. To this, the ultraconservative will not agree, since to him it is a mistake to change the existing order. But educators are beginning to see more clearly that methods of training to be effective must respond to the developments in engineering practice, and also that a change in method necessarily does not effect' a correct application of basic principles. In recent years technical education has undergone a number of striking modifications, which indicates a growing effort on the part of the technical school to keep pace with developments in engineering practice. Among these we note: Direct coöperation of technical schools and industry in administering joint courses of instruction; establishment of specialized courses in technical schools for employees of engineering and industrial concerns; organization of post-graduate courses with highly technical aims by industrial concerns with the assistance of the technical school; establishment of new colleges of industrial

[1] Manager of Shop Laboratories, University of Illinois.
Abstract of a paper presented at the Indianapolis Mid-West Sections Meeting of THE AMERICAN SOCIETY OF MECHANICAL ENGINEERS, Oct. 24 and 25, 1919.

organization, commercial engineering, chemical engineering, ceramic engineering, operation and management, etc.; introduction of courses covering a wide variety of technical subjects; and lastly, the promotion of a closer working relationship between practicing engineers and the technical school. These developments have not lowered the status of engineering education; on the contrary, they have greatly enlarged the usefulness of the technical school and brought to it a larger measure of support from those interests directly concerned with engineering affairs.

While progressive educators recognize the fact that technical education is entering upon an era of more effective service, they are not agreed as to definite lines of development. With a problem so complex as this, agreement as to policies and details should not be looked for; in fact, concurrence of views in respect to this matter is an impossibility from every standpoint. The cause of technical education will be advanced most rapidly by the development of distinctive policies and a frank exchange of ideas about them. It is this thought that prompts the effort to describe the plan of training engineers in the principles of industrial management at the University of Illinois. Although it is recognized that a large majority of engineering graduates enter industrial work, a special course of study for training engineers in industrial management has not been established at this institution. This policy may be modified in the future, but the belief still prevails that the best preparation for the practice of engineering, in all branches, is a thorough grounding in the fundamental subjects of mathematics and of science, coupled with a certain amount of specialization in particular engineering subjects. The main engineering courses seem to satisfy, for the time at least, these general requirements. Consequently no attempt has been made to provide a special curriculum for men expecting to enter industrial work. On the other hand, it was felt that engineers should be given definite training in the principles underlying the management of men engaged in productive enterprises, which led to the adoption of a new policy in administering the traditional shop-work courses.

### THE COMMERCIAL ORGANIZATION OF TECHNICAL-SCHOOL SHOPS

It is not the intent of this paper to discuss the changes in methods and ideals which have been made in the shop courses at the University of Illinois, as these possess little interest in comparison with the more vital topic: What is being done now and with what results? The plan is a practical application of the theory of engineering as expressed in the opening paragraph. It recognizes the engineer primarily as a manager of human enterprises rather than as a technician making plans for others to execute. And again the plan represents the new spirit in engineering education of which mention previously was made, as it was developed on the theory that training for engineering tasks must join with the study of fundamental subjects, actual experience with processes of production. To provide a medium for carrying out the latter part of this program, in the year 1912 the shop laboratories were organized on the lines of a commercial plant for manufacturing gas engines. The plant consists of five departments: pattern, foundry, forge, machine, and assembly. Each department, except the latter, is in charge of a superintendent who is responsible for both operation and instruction. An assistant superintendent, mechanics and toolroom attendants complete the departmental organizations. An assistant manager looks after plant operation. Final authority on all matters is vested in the office of manager, who is responsible to the head of the department of mechanical engineering. The instructional staff serves as a shop committee and acts in an advisory capacity on important matters.

In the plant itself the most effective production methods within the existing limitations are in use or being installed. All shop operations are standardized and covered by detailed instruction cards. Jigs and fixtures and special tools are generally employed. Machinery is effectively grouped and maintained in operating condition, and equipped with safety guards and devices. Modern tool stockrooms with stocks of standard tools and supplies are maintained in each department. Approved methods of routing, dispatching, inspection, and transportation of parts are effective throughout the plant. Control and storage of materials is effectively carried on by accepted methods. In respect to methods and facilities for the production of work, the shop laboratories may be considered as representative of the better type of small commercial plants manufacturing similar products. It has been the aim to duplicate the form of the commercial manufacturing plant along with its problems of organization, production, and management. This result is not secured in all respects, since the actual industrial environment is lacking, but the essential requirements of a production laboratory are established in quite definite form. In effect, the University of Illinois has a manufacturing plant, equipped with facilities, methods and materials for the production of a useful product, but without an operating organization. The latter is supplied from the ranks of students specializing in mechanical, electrical and railway engineering during their sophomore and junior years.

This group at present numbers approximately 500 men. It is untrained, but in the aggregate it represents a wealth of human talent. The instructional plan is a simple one of training this group of embryo engineers to operate the plant and produce gas engines efficiently and economically. Production is sought not for the usual commercial reasons, but as a laboratory exercise for training men in the principles of management and in the mechanical processes of production. The aim is to stimulate the spirit of leadership by placing the responsibility of operation squarely before the student. In this feature, the plan departs from the usual method of academic instruction, and in a considerable degree it fulfills the requirements previously mentioned for training that will develop leaders rather than technicians.

### THE DEVELOPMENT OF THE INSTRUCTIONAL PLAN

Essentially the industrial process is a simple one of changing materials from one form to another by the application of power through the agency of tools, under human control and direction. Specifically, the instructional plan is developed around the theory that preparation for industrial management is accomplished through a study of the three important elements of the industrial process just mentioned: (a) the worker or the producing unit; (b) the equipment or the unit worked with; and (c) the materials or the unit worked upon. The meaning of these basic elements and their relation to each other is brought out by treatment of the following topics under three headings:

*The Worker.* Types of organization, functions of the executive, relations to industry, physical facts, effects of environment, methods of reward, costs of service.

*The Equipment.* Selection and specifications, maintenance methods, operation methods, planning methods, production methods, standardization methods, testing and experimental methods, and costs of production.

*The Materials.* Studies in construction and suitability, purchasing methods, storage methods, dispatching methods, ordering and inventory methods, transportation and delivery methods, costs of materials.

As arranged, these topics form a syllabus of the course of instruction, which has determined the details of the practical working plan, a brief description of which follows.

### THE CLASSIFICATION OF THE STUDENT BODY

The entire body of students previously referred to is automatically divided by registration into four groups, one to each of the main departments of the shops. Each of the department groups is subdivided into staff assistants and shop workmen. Assignments to these duties are made arbitrarily, but each student is moved periodically from one task to another so that during the entire course he performs all of the staff functions and the required mechanical, or shop, operations. The schedule is arranged so that the time is divided equally between the staff and shopwork. Duty assignments are for periods of two weeks or four weeks, depending upon the work to be covered, and they are alternated between the office and shop, according to expediency. From the performance of the duties of production assistant, a student, in the foundry, for instance, may be found during the following work period making molds of pistons on one of the molding machines, or vice versa. At the beginning of the term demonstrations of shop operations and of staff duties are given to all students, which are expected to prepare them for any subsequent work they may be called upon to do, with the aid of carefully prepared standard practice instructions.

The staff group performs the executive and supervisory work of the plant, and the shop group serve as workmen at the bench, on machines, in the core room, at the hardening furnaces, or wherever their services are required. Job orders are issued by the staff group to the shop group, according to the requirements of the production schedule adopted at the beginning of the school term. There are four sections of students in each department operating independently on different periods of the week, which introduces considerable rivalry in the matter of output. Naturally those sections with the most effective staff work secure the greatest production and the highest shop efficiency.

There are seven distinct divisions in the staff work: production, safety, standard practice, materials, mechanical, experimental, and accounting. Every student is expected to serve the allotted time in each division, but only once throughout his entire shop course. He may obtain his training in production, for instance,

Fig. 1 General Layout of Shop-Management Courses for Engineering Students
Showing organization of student operating staff and the main features of the training in executive and administrative work; the main shop departments and the most important items of training in mechanical shop practice.

in the foundry, and in standard practice after entering the machine department. He will perform, however, all of the staff functions before completing his shop course.

## THE WORK OF THE STUDENTS

The required work in each staff division is shown by the diagram of Fig. 1, a study of which, it is believed, will substantiate the statement that the various items are fundamental in the training of engineers for the practice of industrial management. In general, the student obtains from this part of the course a rather definite impression of routing and dispatching parts through the shop, standardizing work operations, rate setting, the custody of materials and supplies, safeguarding dangerous equipment, upkeep of machinery and tools, inspection of parts, cost keeping, as well as the supervision of men engaged in the production of work.

For the conduct of staff functions, effective but simple systems embodying the use of shop records, forms and production facilities have been adopted in somewhat different detail by the various departments. Production boards are employed in this work, and the effect of graphical records such as this upon the more intelligent students is very marked. Those with inborn executive ability immediately display their special qualifications for organization and leadership by increasing output over less gifted predecessors. The records of this group of students are repeated in all departments. There is no doubt about them; they are born to lead, and they do lead, although they may not measure up to the highest academic standards in other courses.

## THE DEPARTMENTAL ORGANIZATION

Departments are organized on similar lines and they administer instruction after a common plan, but the difference in character of their mechanical processes, and of the materials used, makes it possible and also advisable to vary departmental emphasis on certain features of the instruction. In the pattern department the subject of planning and work analysis is given special consideration; in the foundry and machine departments production of parts is emphasized; and in the forge department experimental work is given prominence. By this arrangement the more important factors of shop management are brought forcibly to the student's attention. As the time allotted to the shop courses is entirely inadequate for a thorough presentation of the subject of industrial management, instruction is planned to throw the important elements into bold relief with the object of creating lasting impressions of them upon the mind of the student. It follows that certain features of the work, which were formerly considered of some importance, are given no more than passing attention, but this is immaterial if the larger principles of management are effectively presented to, and thoroughly understood by, the group of future leaders.

Charts, diagrams, graphical records, and illustrations of all kinds are freely employed throughout the various stages of the instruction. It has been clearly demonstrated that facts and impressions are more easily and accurately acquired through the eye from illustrations than by any other method. In addition an effective graphical representation appeals to the imagination and stimulates the desire to work and to produce results.

As the object of the course is the training of engineers in the principles of management, the staff portion of the instruction is strongly emphasized, but not to the exclusion of the other portion relating to manual shop-work. The time devoted to each of these phases of plant operation is divided equally. A general idea of the manual operations performed during the entire course will be gained by referring to the diagram of Fig. 1. Briefly, this work includes patternmaking and preparing stock, molding, core making, pouring metals, and finishing castings, forging, heat treating and case carbonizing, machine-tool operation, assembling, finishing and testing. All of these operations are performed on standard gas-engine parts. Operations are standardized as to method and time. Jigs and special tools are employed wherever they apply. Jobs are scheduled by a staff assistant according to the production program, so the student workman must produce a given volume of work that will pass inspection if he obtains a passing grade in the course. A record of actual time on each operation is recorded by time clocks on job orders. The ratio of this time to the standard time allowed by the instruction card gives the efficiency of production of individuals on separate jobs. No attempt is made to teach students the so-called fundamental shop operations. Machine and tool operation is explained, and accurately prepared instructions for doing single jobs are issued, from which the student is expected to go ahead and produce. Generally, it may be said, that the average student does this in satisfactory fashion.

The above-mentioned plan is not perfect. Its faults, however, are superficial rather than fundamental. It is not intended to fulfill the specifications of a complete course in industrial management, nor is it expected to possess the power of developing competent executives and managers from untrained men within the period of a few weeks; but as a pioneer experiment in technical education it presents possibilities of usefulness in the development of engineering leadership which, it is hoped, will be fully realized.

# Alloyed Aluminum as an Engineering Material

By G. M. ROLLASON,[1] CLEVELAND, OHIO

*Upon its discovery aluminum was hailed as a panacea for all the ills of the metallurgist and its use was proposed for every imaginable purpose from armor plate to chemical chambers. But because it failed to do all that was expected of it a general distrust was created which has been very hard to live down. Recent developments, however, especially those which took place during the war, are rapidly dispelling this idea. In his paper the author first treats of unalloyed aluminum and its light alloys. He traces the improvement in the development of commercial alloys, presenting in connection therewith photomicrographs of the alloys commonly used. He also discusses the subject of the casting of aluminum as well as the methods employed in both the cold and hot rolling of the metal. The possible future uses of aluminum and its various alloys are also briefly considered by the author, who ventures the opinion that perhaps it would not be too optimistic to say that while we have passed through the Stone Age and the Bronze Age, and are now living in the Iron Age, that the future holds in store for us an Aluminum Age.*

THE physical properties of aluminum which render it useful for certain general applications are well known, but its special properties and more particularly those of its alloys which render the metal useful for strictly engineering purposes are not, however, so thoroughly understood as the present diversified uses in engineering would seem to imply. At the same time, as is natural during the development of a comparatively new material, there is a tendency to expect sometimes too much and sometimes not enough.

Attempts to use aluminum under unsuitable conditions, however, are not now as prevalent as they were ten years ago. It should nevertheless, be realized that the possible field of application is limited by the specific properties of the metal. Aluminum and its alloys have their true field just as clearly indicated by their properties and price as are the fields of copper and its alloys or iron and its alloys. A consideration of the fields of application where manufactured aluminum has found its way by natural growth and scientific development will indicate those fields where its application is most legitimate.

Fig. 1 shows both the annual production of aluminum metal in the United States from 1895 to the beginning of the war period, and the total number of motor vehicles produced yearly through a period over which the increase of aluminum production has been very marked. The world's total annual aluminum production at the present time may be estimated about 300,000,000 lb.

During the war period development of aircraft was of course responsible for considerable increase in the application of light alloys. The parallel growth in the two production curves confirms the actual facts in that a very large percentage of the aluminum production finds its way into automotive vehicles, which is by far the greatest field for engineering application of the metal. It is to be expected, however, that development of aerial navigation in the future will take a very large share of the increased light-alloy production.

## UNALLOYED ALUMINUM

The characteristic and dominant properties of aluminum are of course very well known, the most important being its low specific gravity. Next in importance come the relatively high conductivities, both thermal and electrical. This is true not only of the commercially pure metal, but of its light alloys, that is, those containing aluminum as their principal ingredient. In these the properties of aluminum itself, such as specific gravity and conductivity, are dominant, but the mechanical properties can be very greatly improved by alloying.

[1] Aluminum Castings Co.
Abstract of a paper presented at a meeting of the Metropolitan Section of THE AMERICAN SOCIETY OF MECHANICAL ENGINEERS, February 10, 1920.

The engineering applications of the commercially pure metal are fairly limited and somewhat beyond the scope of this paper, but the physical properties of the pure metal are interesting as a basis of comparison for those of the alloys. In the cast or annealed conditions these properties are:

Tensile strength.....................13,000 lb. per sq. in.
Elongation..............................25-35 per cent
Modulus of elasticity....8,500,000-10,000,000 lb. per sq. in.
(Cast iron, 20,000,000; machined steel, 30,000,000)
Brinell hardness.....................................25
Sclerescope hardness...............................5-6
Specific gravity....................................2.7
Specific electrical conductivity...........61 (copper, 100)
Specific thermal conductivity at 64 deg. fahr., 116(copper, 222)
Extreme ductility and softness make figures for compressive strength meaningless.

The commercially pure metal in the cast condition has practically no engineering application. In the wrought or cold-worked state, however, there is of course a very large commercial use. Domestic use of spun, stamped, or rolled aluminum ware is very familiar and the characteristic lightness and heat conductivity are again the controlling properties.

Rolled aluminum sheets and drawn wire are also in extensive use. The rolled sheet is used largely in automobile bodies, gaso-

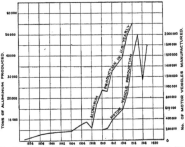

FIG. 1  PARALLEL GROWTH OF ALUMINUM AND AUTOMOTIVE INDUSTRIES

line tanks, etc., where the improvement in properties due to working is utilized for mechanical strength, rigidity, etc. The metal is rolled hot at about 400 deg. cent. from the ingots or slabs to a sheet ¼ in. to ⅜ in. thick and further reduction down to gage thickness is done by cold-rolling. Annealing of cold-worked aluminum, that is, recrystallization, will take place slowly at around 200 deg. cent., but for practical purposes the range is around 350 deg. cent. and higher. In addition to the rolled sheet, stamped or pressed aluminum is used in the automobile industry for fenders, beading, hoods, etc., and has found some application in airplanes and dirigibles for parts subjected to light stresses. There are, however, alloys of aluminum which are of much greater importance and possibilities than the pure metal, in particular duralumin, which will be referred to later.

Hard-drawn aluminum wire is used to some extent for transmission of electrical power, particularly in Europe, and by virtue of the combination of high conductivity and specific gravity. The following is a comparison of copper and aluminum wire:

| | Copper | Aluminum |
|---|---|---|
| Specific conductivity.................... | 100 | 61 |
| Cross sectional area for equal conductivity.. | 100 | 164 |
| Weight for equal conductivity............. | 100 | 50 |
| Tensile strength of hard-drawn wire........ | 60.000 | 28,000 |

Specific gravity.............................  8.9      2.7
Specific tenacity.............................  6,740    10,400

The last figure, specific tenacity, is an index of comparison which has been used to some extent by English metallurgists. It simply amounts to expressing strength in terms of weight, and the figures quoted are the unit tensile strength divided by the specific gravity. The term was developed in order to provide a reasonable index of comparison for useful light alloys with more common materials. The commoner industrial metals, iron, copper, zinc and nickel, for instance, have so nearly the same specific gravity that their tensile strength can be compared section for section, but where the section can be enlarged for equal or better strength with saving in weight it is necessary to get down into more strictly comparative terms. The substitution of aluminum wire for copper has of course been influenced by the relative market prices of the two materials and has only been extensive during periods when the copper market was abnormally high.

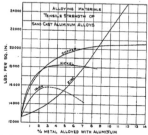

FIG. 2 TENSILE STRENGTH OF SAND-CAST BINARY ALUMINUM ALLOYS

With normal market conditions of the two metals, a price of 15 cents per pound for copper corresponds to 25 cents per pound for aluminum. In cases where aluminum can be substituted bulk for bulk the possibilities for the application of aluminum are therefore very obvious. There are some other applications of the unalloyed metal; for instance, aluminum is worked hot by extrusion under hydraulic pressure, the temperature used for this being about 400 deg. cent. It is fabricated in this way into structural shapes, rods and tubes, and sections up to 6 in. in diameter with wall thicknesses as small as ⅛ in. have been made. Continuous tubing may also be made by this method. This tubing finds some application in pneumatic conveyors for department-store service, also in the chemical industry where, for resistance to certain types of corrosion, the commercially pure metal gives a very satisfactory performance.

### LIGHT ALLOYS OF ALUMINUM

The possibilities for engineering application of aluminum alloys are due to the fact that the lightness of aluminum can be combined in the alloys along with marked improvement of mechanical properties such as tensile strength. The term "light alloys" is one which has been finding favor during the past few years, particularly in connection with war work, and may be said to embrace those alloys having aluminum as their base and whose specific gravity ranges from 2.65 to 3.0. This division is purely arbitrary, but it is interesting to note that when there is enough heavy metal in the alloy to give a specific gravity higher than 3, the mechanical properties at the same time change to such an extent as to render the material unsuitable for many engineering purposes. The alloying of a metal lighter than aluminum, such as magnesium, will bring the gravity down below 2.65, but there are not many alloys in use where magnesium is alloyed in such an amount as to bring the gravity down without simultaneous additions of other heavier metals to produce a counteracting effect.

The outside limits for aluminum alloys in commercial use are set on the light end by an alloy containing 5 per cent magnesium, with specific gravity of 2.47, and on the heavy end by an alloy containing 33 per cent zinc, with a specific gravity of 3.3.

Commercial aluminum always contains iron and silicon in amounts ranging from 0.25 per cent and upward according to the grade of ingot, and while there is a tendency to regard iron and silicon as impurities, nevertheless by their judicious control they can be made to act as useful alloying ingredients.

The commoner metals which are used to alloy with aluminum are copper, zinc, magnesium, and nickel. There are many others which have been used in aluminum-base alloys, most of which, however, have hardly passed the experimental stage. Among these may be mentioned chromium, molybdenum, tungsten and vanadium. In general the effect of progressive additions of the alloy metals to aluminum is to render the base metal correspondingly harder, stronger, and less ductile. It should be explained that "ductile" does not refer here to the property of being easily drawn into wire, but merely to the measured percentage elongation, that is, the amount of possible deformation without fracture.

Three types of aluminum alloys are manufactured for engineering purposes, namely,

1 Casting
2 Cold-working
3 Forging or hot-working.

It is estimated that a half of the world's production of aluminum goes into castings, that is, it is used in the alloyed form, and since a large part of the unalloyed aluminum is put to miscellaneous and non-engineering uses, the majority of engineering

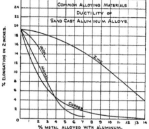

FIG. 3 DUCTILITY OF SAND-CAST BINARY ALUMINUM ALLOYS

uses and mechanical applications must consist of castings. The casting alloys will therefore be considered first.

### DEVELOPMENT OF PROPERTIES BY ALLOYING

Figs. 2 and 3 show the variation in tensile strength and ductility of aluminum, cast under carefully controlled conditions, produced by progressive additions of copper, zinc, nickel, and iron as alloying ingredients. Copper gives increased strength and decreased ductility up to about 11 per cent, at which point the alloys become quite brittle. The effect of zinc is less marked than that of copper in amounts up to 8 per cent especially as regards ductility; above 8 per cent the zinc alloys continue to become stronger and less ductile after the effects of copper have practically reached their maximum. This difference in behavior of copper and zinc alloys is due to difference in structure. (See later paragraphs under the heading Metallography.) Iron gives useful strengthening properties in amounts up to 2 per cent, but above that the effects are doubtful owing to rapid rise of the melting point and other disturbing conditions. This effect of iron, unlike that of the other metals, is additive, that is, it can be superimposed on that of the other metals. Nickel in behavior is intermediate between copper and iron.

Magnesium is a valuable addition to aluminum or its alloys when used judiciously. Of all metals it has perhaps the most pronounced effects on the properties of aluminum when added in small amounts, probably owing to presence of silicon. It is a very pronounced hardener and especially if the silicon content of the aluminum ingot is high the effect of magnesium even in amounts as low as 0.1 per cent will be to render the alloy stronger, harder and more brittle.

When about 12 per cent of copper or 15 per cent of zinc is

it cuts down the ductility of the alloy, which in this case shows only about 1½ per cent elongation. Though the solid solution is more ductile and would show more deformation without break, the brittle eutectic network prevents it.

Zinc, as shown in Fig. 5, is more soluble than copper and the solid solution will contain about 40 per cent of zinc without the appearance of a separate constituent. In general appearance this alloy is quite similar to pure aluminum. The effect of the zinc is to make the crystals of solid solution stronger, but there is no

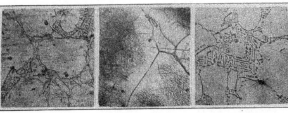

FIG. 4 TYPICAL 8 PER CENT COPPER-ALUMINUM ALLOY      FIG. 5 TYPICAL 15 PER CENT ZINC-ALUMINUM ALLOY      FIG. 6 ALUMINUM ALLOY CONTAINING Mg₂Si COMPOUND

added, the specific gravity reaches about 3 and it will be noted that when this figure is reached the alloy has reached a state of comparative brittleness and low ductility which renders it not only of limited usefulness in the cast state, but incapable of being cold-worked. It will be seen that it is comparatively easy to choose mixtures for alloying composition which will give high tensile strength, but to obtain a combination of high strength and ductility it is necessary to use careful selection and to give attention to constitution. The most important alloying materials are copper and zinc; that is, for casting and working purposes aluminum-copper alloys, aluminum-zinc alloys and aluminum-copper-zinc alloys cover most of the field. At the present time the aluminum-copper alloys find a good deal of favor in this country, while aluminum-zinc alloys seem to be preferred in Europe. This condition, however, has changed from time to time, since there have been periods when aluminum-zinc alloys have been in extensive use in this country. It is probable, as investigation and development progress, that successive improvements will involve more complex mixtures.

### METALLOGRAPHY OF ALUMINUM AND ITS ALLOYS

While the development of alloys, particularly casting alloys, in the past has been more or less a matter of cut and try, the metallography of these alloys is becoming quite well understood and study of constitution is leading to the development of properties in light alloys which have not been brought out through many years of experiment under the older methods.

Fig. 4 shows a micrograph of a typical 8 per cent copper-aluminum alloy. It contains two constituents, one being the background or matrix of solid solution, that is, so-called mixed crystals of copper dissolved in aluminum; the other constituent which appears as a network around the grain boundaries is a eutectic of the solid solution and a compound CuAl₂. This network is very hard and brittle and serves to reinforce the alloy, making it harder and stronger. At high temperatures, just below the melting point, the solid solution will contain about 4 per cent of copper. In an alloy like this, with 8 per cent copper, the other 4 per cent goes to form the eutectic network around the grain boundaries. It is to be noticed that the network is quite complete, that is, it forms a closed structure, and since it is brittle

separate brittle constituent with cellular structure as was the case with copper so that for a given tensile strength the zinc alloy will have a greater elongation. Fig. 6 shows the appearance of magnesium. The magnesium will form a solid solution up to a certain point and then separate out, a compound Mg₂Al₃ being formed, but since all aluminum ingot contains silicon, the result is generally somewhat different. The magnesium combines with silicon to form a compound Mg₂Si, which is practically insoluble and very brittle. Its effect is the same as that of the copper compound, but very much more pronounced; that is, an addition as

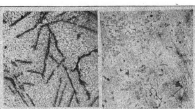

FIG. 7 ALUMINUM ALLOY WITH NEEDLES OF FeAl₃ COMPOUND      FIG. 8 ALUMINUM-COPPER-ZINC-IRON ALLOY

low as 0.25 per cent magnesium has a marked hardening effect.

Iron, manganese, and nickel form compounds with the aluminum such as FeAl₃, which show up in the form of long needle-shaped crystals as in Fig. 7. These compounds are practically insoluble, and even a small amount of iron will cause the appearance of the needles. The effect of these needles is to break up the continuity of the copper or other network, and in the case of a break the iron needles make the path of rupture longer. In this way small additions of iron give both greater strength and greater ductility.

Fig. 8 shows a ductile casting alloy containing copper, zinc, and iron. The zinc does not show up, since it stays in solution. The copper network is just beginning to appear, since there is about 3 per cent copper present, or just in excess of the amount which will stay in solid solution under the normal cooling conditions. The iron needles are distributed around and perform their function of giving additional strength and ductility.

### COMMERCIAL CASTING ALLOYS

The development of useful physical properties by alloying selected metals with aluminum has made possible the commercial use of a number of widely varying alloys. A good many of these are either straight binary alloys or binary alloys with very slight impurities or additions of third or fourth metal. The light aluminum alloys which are used for sand casting in America center for the most part around different copper-aluminum combinations.

The principal aluminum alloy used for castings in the United States consists of about 92 per cent aluminum and 8 per cent copper, and is generally known as Aluminum Company of America's No. 12 alloy or S. A. E. specification No. 30. This alloy when cast in a ½-in. test bar in green sand and tested without machining off the skin should give an average tensile strength of about 20,000 lb. per sq. in. and an average elongation of about 1.5 per cent in 2 in. A modification of this alloy having somewhat better physical properties is now finding considerable favor in castings for the automotive industry. This alloy has an analysis of 7.5 per cent copper, 1.5 per cent zinc, 1.2 per cent iron, and the remainder aluminum. The tensile strength of this alloy will average about 21,000 lb. per sq. in. and the elongation will be somewhat greater than that of the No. 12 alloy.

Where greater ductility is required the copper content is cut down to about 5 per cent, resulting in a tensile strength of about 18,000 lb. per sq. in. and an average elongation of 3 per cent. This alloy is used for castings such as those for automobile bodies which require to be pressed or bent into their final shape. This alloy is somewhat more difficult to cast than No. 12, due to the lower copper content and consequently higher shrinkage.

Another alloy which is well known under the designation S. A. E. Specification No. 32, contains approximately 12 per cent copper and remainder commercial aluminum. The high content of copper renders it somewhat easier to cast, as it cuts down the solidification shrinkage of the alloy and makes in general for denser and sounder castings. As a result of this the alloy is selected for use in castings where tightness against leaking is the principal requirement. It has the disadvantage of being quite brittle, due to the large amount of copper constituent present. The casting fails to show a marked improvement in tensile strength over the common 8 per cent copper alloy owing to the fact that in the structure of the latter the reinforcing network of copper compound is completely closed (See Fig. 4), and no further strengthening is produced by the excess amount given when the copper is increased up to 12 per cent.

As previously mentioned, there is a tendency in Europe, and especially in England, to favor the zinc-aluminum combinations, and the alloy which corresponds to our No. 12 is known as L-5 and contains 13.5 per cent zinc, 2.75 per cent copper and the remainder commercial aluminum. This has a specific gravity of about 2.95, or higher, which is relatively high as compared with No. 12 alloy at a specific gravity of 2.83. Its tensile strength when cast in sand is over 25,000 lb. per sq. in., but the elongation is not much over 1 per cent.

Another alloy which finds favor in England contains 10 per cent zinc and 2.5 per cent copper. Here some ductility has been gained over L-5, that is, it gives over 2 per cent elongation, but at a sacrifice of tensile strength, since the specification only calls for 22,000 lb. per sq. in.

The foregoing are some of the aluminum alloys most widely used in sand castings. There are, however, a few special uses which demand special alloys; for instance, commercially pure aluminum, in general, is much less subject to corrosion than any of its alloys, but in cast form is too soft for much practical use. In this connection an alloy of 98 per cent commercial aluminum and 2 per cent manganese finds some application, that is, it re-

sists corrosion practically as well as the unalloyed metal, while the small addition of manganese gives the necessary strength and hardness in the castings.

### PERMANENT-MOLD CASTINGS

By far the greater part of the cast aluminum used is in the form of ordinary sand castings, although a considerable tonnage of aluminum is manufactured in permanent molds. In this country the manufacturers of aluminum castings in permanent molds have not so far undertaken the casting of very large pieces, but castings up to 150 lb. in weight have been made in chill molds on a semi-commercial scale.

The chill-cast aluminum finds its principal outlet in the form of gas-engine pistons and this has led to the development of specialized alloys. Permanent molding, though a much newer art than sand casting, is susceptible of more scientific control and actually approaches a more exact science. The feasibility with which any alloy can be cast into a chill mold depends upon the simplicity of the desired casting design, and for casting complicated shapes the alloy selected should be such that its solidification shrinkage is relatively small.

For the permanent-mold casting of fairly intricate shapes such as aluminum pistons, bearing caps, etc., an alloy containing 10 per cent copper, 1.5 per cent iron and 0.25 per cent magnesium, remainder commercial aluminum, is used fairly extensively. This has the property of running well in permanent molds, the shrinkage being kept low by the relatively high percentage of copper. Fig. 9 shows the difference in structure obtained by casting this

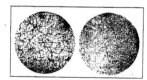

Cast in Green Sand　　　Cast in Permanent Mold

FIG. 9 SAND- AND CHILL-CAST PISTON ALLOY

alloy in a permanent mold and in sand. It gives a fine-grained, dense structure with high strength and hardness, making it suitable for wearing or bearing surfaces and easy to machine. In a properly made chill casting this alloy will have a tensile strength of 28,000 lb. per sq. in. and 2 or 3 per cent elongation, but these properties can only be attained by careful attention to all the factors going to make up a good casting.

Table 1 enumerates the properties of some of the alloys whose composition has been given above.

TABLE 1  PROPERTIES OF CERTAIN ALUMINUM ALLOYS

|  | No. 12 | S.A.E. 32 | British L-5 | Lynite 145 |
|---|---|---|---|---|
| Tensile strength, lb. per sq. in....... | 20,000 | 24,500 | 25,000 | 27,500 |
| Elongation, per cent................ | 1.5 | 0.5 | 1.5 | 4.5 |
| Brinell hardness.................... | 65 | 70 | 80 | 65 |
| Scleroscope hardness ............... | 14 | 16 | 18-20 | 14 |
| Specific gravity.................... | 2.84 | 2.90 | 2.93 | 2.89 |

### STRESS-STRAIN CHARACTERISTICS—VALUE OF DUCTILITY

In the selection of a casting alloy for certain applications, ductility must rightly receive consideration. For parts which are subjected to normal stress, rigidity and a reasonable amount of strength are the only essentials. In many cases the factor of safety is also very high, since the dimension of the cast piece is controlled by ability to cast or manufacture—for example, the familiar aluminum crankcase casting. However, where a cast-aluminum member is designed for a heavily stressed part the resistance to normal stress is not the only matter to be considered. The factor of safety might be made such that the ordinary stress range could be taken care of, but if this were exceeded and there were no ductility in the material, complete failure would be the result. It is in connection with abnormal stresses in engineering parts and with abuse as opposed to normal use that ductility must be regarded as a very valuable and essential property for cast aluminum or similar engineering materials. There are also cases where, apart from the performance of the finished part, ductility is absolutely essential to the process of manufacture. Sand castings must of necessity have liberal tolerances for the dimensions. With these variations assembling of parts will involve some rough handling and knocking together and there are instances on record where assembling shops could not use the ordinary brittle aluminum alloys when it was attempted to substitute aluminum for bronze castings, but where a ductile aluminum alloy of high strength met with good success.

In general it may be said that the cast aluminum alloys are not very elastic materials, that is, if the proportionality of stress to deformation is considered to be the test of elasticity. However, though the proportional limit of cast aluminum alloys is low, the yield point is relatively high and in comparing the stress-strain curve of a cast aluminum alloy with that of machine steel, for instance (see Fig. 10), there is a range of relatively high stress, about 13,000 lb., for aluminum alloy and 34,000 lb. for the steel when both have taken on small amounts of permanent deformation, but at this point for further slight deformation of the aluminum alloy it is necessary to put on a relatively large additional stress, while the steel will deform greatly under quite a small additional stress.

The fatigue life of ordinary aluminum alloys under repeated reversal of stress (White-Souther test, for example) is not very well recorded. The performance, however, in such few tests as are on record seems very promising. There appears to be a safe working range well above the proportional limit. For instance, in ordinary No. 12 alloy the proportional limit is around 5000 lb. per sq. in. and tensile strength about 20,000 lb. per sq. in., but this alloy has shown a fatigue life of 16,000,000 reversals under a stress of 8500 lb. per sq. in. Machine steel, on the other hand, for a life of 16,000,000 cycles will take a stress of only 11,000 lb. per sq. in. when its tensile strength is over 60,000 lb. and its proportional limit correspondingly higher than that of the aluminum alloy. The fatigue properties of the aluminum alloys, however, are not thoroughly accounted for as yet, and while such results as the above have been obtained, there are only a limited number of them on record. In accounting for such performance it might perhaps be explained in terms of structure by the fact that the aluminum alloy is a combination of brittle and ductile materials, and whereas in the straight tensile pull the brittle materials are pulled out of place with a relatively small amount of total deformation and their strengthening effect is lost, under the reversal of stress in the fatigue test these constituents remain in their original position and the alloy retains the benefits of their presence.

### POSSIBLE EXTENSIONS OF THE ENGINEERING APPLICATION OF ALUMINUM

The automotive field at the present time forms by far the largest engineering outlet for aluminum alloys. There are some possible applications which are based very specifically on the properties developed in aluminum alloys which may be mentioned, but these possibilities at the present time are not to any extent realized. For instance, in any field of light high-speed machinery the inertia of reciprocating parts and consequent vibration and shocks can be cut down by judicious substitution of aluminum alloys for heavier materials provided their physical properties are properly understood and can be relied upon. Also in high-speed pulleys where centrifugal stresses run high, cast aluminum can be substituted for cast iron to good advantage.

Where rigidity is the prime consideration, the figures controlling the substitution of aluminum for ferrous materials are interesting. The modulus of elasticity of aluminum and its alloys may be taken as 10 million lb. per sq. in. with gray iron at 20 million and machine steel at 30 million. The rigidity is proportional to the modulus of elasticity and to the section modulus. To substitute aluminum for gray iron with equal rigidity, therefore, the section modulus must be doubled, and this can be done with a 50 per cent decrease in weight for a cross-section of equal rigidity.

### THE WORKING OF ALUMINUM AND ITS ALLOYS

Aluminum alloys are not used to any great extent for cold-working. Cold-worked aluminum is nearly always unalloyed aluminum. The more ductile alloys are fashioned and shaped to some extent in the cold, for instance, in automobile body work, and during the war a very large number of fuse bodies for shrapnel were made by cold-stamping an aluminum alloy containing 3 per cent copper.

The rolling, drop forging and general hot-working of aluminum

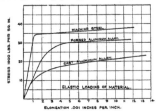

FIG. 10 STRESS-DEFORMATION CURVES OF ALUMINUM ALLOYS AND MACHINE STEEL

alloys have recently undergone considerable development and this has been largely due to development of particular alloys. At the present time the most remarkable alloy of aluminum is undoubtedly duralumin, which was developed in Germany by A. Wilm in 1903. A series of patents originating in Germany were brought out from the years 1903 to 1914. This alloy is notable for the fact that while it is fabricated in the hot condition, that is, well above the lowest annealing temperature, its final metallurgical condition and properties are controlled and improved by heat treatment and are very different from what the alloy composition gives in the cast or simple annealed form. Duralumin came into a good deal of prominence due to its extensive use in aircraft during the war. It was used to a large extent for framework and joints on dirigibles, being rolled and formed into structural shapes and extruded into tubes and sheets and a large number of other uses. The composition of the alloy varies somewhat, but the elements essential to the development of the characteristic properties are copper in amounts of 3.5 to 5.5 per cent and magnesium from 0.50 to 1 per cent. Additions are also made of such elements as iron or manganese in small amounts and even in some cases chromium or molybdenum.

In the manufacture of duralumin parts the general cycle of operations consists of:

1　Pouring of the ingots
2　Hot-rolling to a slab or bar
3　Hot-working to the final shape
4　Heat-treating and quenching
5　Aging

The ingots are poured with the metal at as low a temperature as possible, that is to say, just enough above the melting point to prevent cold shuts, a special type of tilting ingot mold being used. The ingot is then hot-rolled to a slab or bar for final working, the temperature being kept with advantage pretty close to 500 deg. cent. The final fabrication may be made by hot-rolling, hot-forging, hot-stamping, etc., according to the nature of the shape it is desired to produce. These operations are carried out within a carefully controlled temperature range, between 450 and 500 deg. cent. and the material is carried to its final shape.

The hot-worked material is possessed of properties greatly improved over what would show in the cast ingot, but the full development of its usefulness is only obtained by a specific heat treatment. The alloy is heated to a temperature of 500 to 540 deg. cent. for a period of ½ to 1½ hr., depending upon the size of the piece, and immediately quenched in cold water.

After this heat treatment and quenching the properties are still further improved, but are not fully developed until a process of aging is gone through. During the aging the alloy takes on a

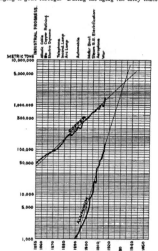

FIG. 11  WORLD'S PRODUCTION OF COPPER AND ALUMINUM PLOTTED LOGARITHMICALLY TO SHOW RELATIVE RATES OF GROWTH

further increase of tensile strength and elongation. This aging is analogous to what takes place with the cast aluminum-copper alloys, but its effect is much more pronounced. The very best properties can perhaps be produced after heat treatment by an artificial accelerated aging at elevated temperatures around 150 deg. cent. As an example of the manner in which the properties are developed by the various stages of manufacture, the following are fairly representative figures of an alloy of general duralumin composition: The chill-cast ingot shows a tensile strength in the neighborhood of 30,000 lb. per sq. in. and elongation below 8 per cent. After rolling, hot-forging and annealing, the tensile strength runs from 45,000 to 50,000 lb. and the elongation from 10 to 20 per cent. The exact balance between strength and ductility at this point depends upon the amount of cold work put on to the piece. After heat-treating at 500 deg. cent. or above and quenching, the properties show a marked increase: 55,000 to 60,-000 lb. per sq. in. tensile strength, and 25 to 30 per cent elongation. On a freshly quenched piece these figures would not be so high, but they develop after aging for two or three weeks. If the heat-treated material is given an artificial aging by exposure to a temperature of 150 deg. cent. for 48 hr. the tensile strength will

be from 2000 to 5000 lb. higher than what is obtainable by natural aging. Extensive investigations have been carried out at the United States Bureau of Standards and elsewhere on the structural and other reasons controlling development of properties by working, heat treatment and aging and a fairly satisfactory explanation has been arrived at. Metallographic considerations would show that the changes brought about are due in general to the change in solubility of the copper-aluminum compound as the temperature is lowered from 500 deg. cent. to ordinary temperatures.

## THE APPLICATIONS OF DURALUMIN

The possibilities for applications for an alloy of these properties and low specific gravity are of course very numerous. The rolling of structural shapes and of frameworks for permanent or portable structures has an immediate future. Drop-forged duralumin is finding its way into use in the automotive industry in the form of connecting rods, rocker arms, etc., as well as gears. Forged duralumin connecting rods have been submitted to prolonged test in special testing machines and in the motor on the block and on the road.

The fatigue life of the metal is very favorable indeed. A sample connecting rod made for use in a standard passenger car was run in a special machine consisting of a motor-driven crank and weighted crosshead. The speed was about 1500 r.p.m. and the load on the crosshead would correspond with about 50 per cent overload on the motor. The duralumin rod gave a life of 353 hr. and was still functioning when a casting on the crosshead failed in fatigue. A standard drop-forged steel rod for the same car showed a life of only 25 hr. under the same conditions, for example, with 50 per cent overload and another between 40 and 50 hr. at normal load both the steel rods failing in fatigue at the big end. The steel rod was of smaller section, but about double the weight of the forged aluminum rod. Some of these rods also have stood up well in block and road tests in the motor to the extent of over 10,000 miles of running. A notable feature of a number of these tests has been that at the big end bearing, the rod has been run direct on the crankshaft without any bushing whatsoever. Forged and heat-treated duralumin, in fact, is showing up as a very satisfactory bearing material when the wear is against a sufficiently hard shaft. Rods have been run against both case-hardened and heat-treated shafts. Owing to the relatively high expansion of the aluminum alloys, however, it has been found necessary to allow a little more end play at the big end bearings than is standard practice with steel rods.

The immediate future is going to tell considerably more about the actual performance of these and other applications of duralumin. Credit must be given the inventor of this alloy for having worked it out from fundamental scientific facts, and from the spectacular point of view this alloy represents a great achievement in developing properties in aluminum by alloying. In this case aluminum with a strength of 13,000 lb. has produced an alloy with a strength of over 60,000 lb. without loss of ductility, but the specific gravity has increased less than 10 per cent. What other similar possibilities may lie in the future it is not safe to say, but let us remember the short space of time that has been spent in reaching even the present state of alloy development. A picture which is fairly familiar to the aluminum industry is that when we speak of the beginnings of iron or bronze or most other commercial alloys we go back thousands of years to prehistoric times, but when we speak of aluminum we go back as far as our own generation only. The abundance with which aluminum occurs in nature and the reception which the engineering world has given it insure a rapidly increased production for years to come.

Fig. 11 shows the history of the annual rates of production of copper and aluminum. By the use of logarithmic coördinates and by extrapolation the prediction can be made that somewhere about 1935 or 1940 the yearly production of aluminum will equal that of copper and afterward exceed it. Perhaps it would not be too optimistic to say that while we have passed through the Stone Age and the Bronze Age and are living in the Iron Age, the future holds in store for us the Aluminum Age.

# Some Applications of Alloy Steels in the Automotive Industry

By H. J. FRENCH,[1] WASHINGTON, D. C.

*The following paper deals with the application of various types of nickel-chromium steels in the automotive industry. Particular reference is made to the uses of steel containing 1 per cent nickel and 1 per cent chromium and also 3 per cent nickel and 3 per cent chromium. Data are also given for these two steels showing the tensile properties and hardness developed in small-size rounds subjected to varying heat treatments. A somewhat detailed mention is made of nickel-chromium steels and treatments used for gears and airplane-engine crankshafts. General specifications governing this latter part for both rotary and stationary engines are also given, and important difficulties frequently met with in the production of crankshafts are noted. "Streaks" and "temper brittleness," both encountered in the use of nickel-chromium steels, are briefly discussed and curves showing the tensile properties and hardness of "stainless steel" under varying oil-quenching and tempering treatments are presented. There are also shown the results of cutting tests made with cast high-speed-steel milling cutters and a comparison is given with high-speed cutters made by present-day ordinary methods. Micrographs are included to show the refractory nature of the carbide in cast high-speed steel.*

ALLOY steels are very widely used in the automotive industry both in the manufacture of airplanes and automobiles, and without doubt have been one of the leading factors in the development of the industry. Adequately to present this subject, however, would require more space than is here available, and so this paper has accordingly been limited to a discussion of structural nickel-chromium and carbon-chromium steels. The results of a test of a cast high-speed steel cutter are also given because of the interest of all engineers in efficient cutting media.

Some common types of nickel-chromium steels having varied applications in automotive work are given in Table 1. The nominal percentages of the alloying elements are shown and the seven steels listed constitute a highly interesting series. They find use in production of such parts as axles, connecting rods, crankshafts, gears, steering knuckles, high-tensile bolts, and a variety of small parts either forged or machined from hot- or cold-rolled or drawn bars.

In general, as the alloy contents of this series increases more care in working and treatments is required, but under suitable conditions of manufacture a variety of properties may be obtained.

Steels containing nickel and chromium in the ratio of 2 to 1 seem to be in greatest favor, but any of the steels listed are capable of developing excellent combinations of strength and ductility greatly superior to plain carbon steels of similar carbon content. Steel No. 1, Table 1, will serve as a good example for discussion.

## 1 Per Cent Nickel and 1 Per Cent Chromium Steel

This steel, containing approximately 0.40 per cent carbon, 1 per cent nickel and 1 per cent chromium, was largely used in the production of various parts of a well-known rotary engine under widely different physical specifications. Three such parts with the tensile requirements and heat treatments used are listed in Table 2.

While both cams and valve lifters are of small size, the use of this steel for the former is not recommended, for it is exceedingly difficult to meet the severe requirements of the specifica-

tions. However, several thousand cams made of this steel were successfully treated and, according to all available information, proved entirely satisfactory in service, but the desired properties and uniformity as shown in Table 3 were obtained only under most closely controlled conditions. The steel used was produced

### TABLE 1  NICKEL-CHROMIUM STEELS

| No. | Carbon, per cent | Nickel, per cent | Chromium, per cent | Ratio, Nickel to Chromium |
|---|---|---|---|---|
| 1 | | 1.00 | 1.00 | 1 to 1 |
| 2 | | 1.25 | 0.60 | 2 to 1 |
| 3 | | 2.00 | 1.00 | 2 to 1 |
| 4 | 0.10-0.50 | 2.50 | 1.25 | 2 to 1 |
| 5 | | 3.00 | 1.00 | 3 to 1 |
| 6 | | 4.00 | 1.00 | 4 to 1 |
| 7 | | 3.00 | 0.50 | 6 to 1 |

### TABLE 2  NICKEL-CHROMIUM STEEL

Carbon, 0.40 per cent, nickel, 1.0 per cent, chromium, 1.0 per cent

| Part | Requirements | | | Heat Treatment (All parts normalised before machining) |
|---|---|---|---|---|
| | Ultimate Strength, lb. per sq. in. | Yield ratio, minimum | Elongation in 2 in., per ct | |
| Connecting Rods | 145,600 to 156,800 | 0.75 | 18 to 15 | [1]1550 deg. fahr.—oil [2]950-1000 deg. fahr.—oil |
| Valve Lifters | 190,000 to 202,000 | 0.75 | 14 to 12 | 1525 deg. fahr.—oil 780 deg. fahr.—oil |
| Cams | 247,000 to 268,000 | Not specified | 12 to 10 | 1525 deg. fahr.—oil—Tempered 30 min. in oil at 380 deg. fahr. |

[1] Quenched in oil from 1550 deg. fahr.  [2] Tempered at 950 to 1000 deg. fahr. and quenched in oil.

### TABLE 3  HEAT-TREATED NICKEL-CHROMIUM STEEL

Composition (per cent). C, 0.420; Mn, 0.400; Si, 0.170; P, 0.024; S, 0.031; N, 1.120; Cr, 1.00.  Size: Approx. 0.4 in. diam.

| Ultimate Strength, lb. per sq. in. | Elongation in 2 in., per cent | Reduction of Area, per cent | Brinell No. |
|---|---|---|---|
| 261,500 | 11.5 | 44.2 | 555 |
| 256,100 | 11.5 | 46.7 | 495 |
| 259,000 | 11.5 | 47.8 | 555 |
| 260,800 | 11.5 | 47.1 | 512 |
| 263,700 | 12.5 | 47.1 | 512 |
| 261,700 | 11.0 | 41.2 | 512 |
| 263,200 | 12.0 | 47.1 | 555 |
| 259,000 | 10.0 | 46.7 | 512 |
| 267,700 | 11.5 | 43.5 | 512 |
| For 9 tests | | | |
| 267,700 (max.) | 12.5 | 47.8 | 555 |
| 256,100 (min.) | 10.0 | 41.2 | 495 |
| 261,430 (avg.) | 11.4 | 45.7 | 524 |

in the electric furnace, and the parts were small, enabling treatments to be carried out in small electric heating units. Attention is called to this because there is without doubt a great deal of high-alloy-content steel used where cheaper and lower alloys would equally well serve the purpose, though under conditions

[1] Met.E.; Metallurgical Division, Bureau of Standards, Washington, D. C. Presented at a meeting of the Washington Section of THE AMERICAN SOCIETY OF MECHANICAL ENGINEERS, March, 31, 1920. Published by permission of the Director, Bureau of Standards.

not as severe as those cited, thereby saving considerable expense
to the industry.

The tensile properties and hardness of this steel in small sizes
under varying but single heat treatments are shown in Fig. 1.
Variation in tempering temperatures between about 650 and
1350 deg. fahr. gives a variation in properties approximately as
follows:

Yield point, lb. per sq. in............. 200,000 to  85,000
Tensile strength, lb. per sq. in........ 220,000 to 117,000
Elongation in 2 in., per cent..........     11.5 to   25.5
Reduction of area, per cent............     45 to      72
Brinell hardness ......................    440 to     220
Shore hardness ........................     60 to      32

### 3 Per Cent Nickel and 1 Per Cent Chromium Steel

Steel containing about 3 per cent nickel and 1 per cent
chromium, or often somewhat lower percentages of this element,
is used for crankshafts, connecting rods, etc., when higher
combinations of strength and ductility are required than may be
developed in the steel previously discussed. The tensile properties and hardness of such steel containing about 0.25 per cent

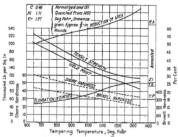

Fig. 1  Tensile Properties and Hardness Numbers of a Heat-
Treated 1:1 Nickel-Chromium Steel

carbon when subjected to various single heat treatments, are
shown in Fig. 2. The approximate variation in properties obtained by varying the tempering temperature between about 450
and 1300 deg. fahr. follows:

Yield point, lb. per sq. in............. 190,000 to  87,000
Tensile strength, lb. per sq. in........ 220,000 to 117,000
Elongation in 2 in., per cent..........     14 to      26
Reduction of area, per cent............     55 to      70
Brinell hardness ......................    440 to     240

If comparison of these values with those given for the 1-to-1
nickel-chromium steel is made it must be remembered that the
higher-alloy steel contains much less carbon.

### Crankshafts

The question of crankshafts for which this steel as well as
others are used is one of such importance, especially for airplane engines, that discussion from the standpoint of the part
itself now appears justified.

Despite the variety of design in both engines and parts, crankshaft specifications for most types of airplane engines may be
grouped into two classes, as shown in detail in Table 4. There
are exceptions to the values as given, but the table generally
applies. Ordinarily the crankshaft requires skilled metallurgical
control in all stages of manufacture and the best steels and treatments are none too good. The treatments used to develop the
required physical characteristics vary, but with the steels
ordinarily used a relatively high temperature—tempering above
1000 deg. fahr. is necessary, which is of course desirable. Both
water and oil quenching are used, as shown in Table 5, which

also shows analyses, treatments and properties obtained on
several types of crankshafts. Where water quenching is used
the shafts are usually removed from the bath after a given length
of time and before they are cold, and immediately tempered to
avoid cracking.

It is essential that the crankshafts be thoroughly annealed after
forging and before machining in order to minimize distortion
which sometimes occurs during the machining operations. Cold
setting after heat treatment also gives trouble and in such cases
the crankshafts should be rough-machined before heat treatment.
Hair-line seams or streaks and "blue brittleness" are two additional factors which, while they are often encountered in production of crankshafts, are general and will be briefly discussed
later.

In order to avoid such difficulties, care in melting, working, and
thermal treatments is required. Acid open-hearth or cold-melt
electric is undoubtedly the best steel to use. Excessive aluminum
additions should be avoided and casting temperatures should
preferably be under pyrometric control. Sufficient discard should
be made to remove segregation and pipe, the amount depending
upon casting practice, and all ingots and blooms should be carefully surfaced. The steel should be worked sufficiently so as to
remove all traces of cast structure and heating whether for
forging or heat treatment should be slow and uniform.

### Heat-Treated, Cold-Drawn Nickel-Chromium Steel

Nickel-chromium steels have also found application in their
heat-treated, cold-drawn condition where close adherence to size
as well as high physical characteristics are required. Such steel
is particularly useful in production of parts carrying threads

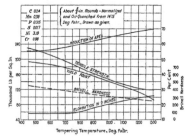

Fig. 2  Tensile Properties and Hardness Numbers of a Heat-
Treated 3:1 Nickel-Chromium Steel

where oxidation from heating after machining is objectionable.
Ordinarily the bars are hot-rolled to a size close to that desired
and then either normalized or quenched and tempered after which
they are cold-drawn. Table 6 illustrates such drawing practice
and shows the effect of cold work on the tensile properties. Increase in elastic limit and tensile strength with decrease in ductility as measured by elongation and reduction in area results.
Heavy final cold reductions induce brittleness and should be
avoided, and in addition impair machining qualities.

### Gears

Nickel-chromium steels of various analyses are also widely used
for gears. The type of steel recommended depends largely upon
the service desired and this, of course, together with the analysis
chosen influences the treatment. The steels used, however, may
be divided into two groups, viz: (a) tempering steels and (b)
case-hardening steels. Typical analyses of these two classes are
shown in Table 7.

TABLE 4  CRANKSHAFT FORGINGS FOR AIRCRAFT ENGINES

| Percentage Composition | | | | | Physical Properties | | | | |
|---|---|---|---|---|---|---|---|---|---|
| C | Ni | Cr | P | S | Ultimate Strength, lb. per sq. in. | Yield Ratio, Minimum | Per Cent Elong. in 2 in., Min. | Per Cent Red. of Area, Min. | Min. Impact Values-Izod 120 ft-lb. machine |
| STATIONARY ENGINES | | | | | | | | | |
| Max. 0.36 | Max. 4.00 | 0.50 to 1.30 | Max. 0.045 | Max. 0.05 | ¹132,000 to 137,000 | 0.75 | 17 | 44 | 35 |
| ROTARY ENGINES | | | | | | | | | |
| Max. 0.35 | Suitable proportions | | Max. 0.045 | Max. 0.05 | 106,000 to 128,000 | 0.75 | 17 | 50 | 40 |

¹ For each 4500 lb. increase above the minimum ultimate the required Izod minimum is reduced 1 ft-lb.

TABLE 5  CRANKSHAFT STEELS AND TREATMENTS

| Engine | Specifications | Yield Point, lb. per sq. in. | Tensile Str., lb. per sq. in. | Per Cent Elong. in 2 in. | Per Cent Red. Area | Brinell Number | Impact, ft-lb. |
|---|---|---|---|---|---|---|---|
| A | Carbon..0.35 per cent Manganese...0.61 per cent Nickel..3.09 per cent Chromium.0.50 percent Normalised.1600 deg. fahr. 1550 deg. fahr. oil 1030 deg. fahr. oil | 131,250 139,400 | 137,350 147,400 | 20.0 17.5 | 59.0 56.3 | 286 302 | |
| B | Carbon..0.48 per cent Manganese...0.69 per cent Nickel..1.74 per cent Chromium.0.85 percent Normalised.1450 deg. fahr. Drawn..925-1125 deg. fahr. | 127,700 124,500 130,350 | 143,150 141,000 145,700 | 20.0 20.5 18.5 | 58.8 56.4 57.2 | 302 302 302 | Izod 120 ft-lb. machine 47 46 46 |
| C | Carbon.0.35-0.45 per cent Manganese.0.50 - 0.80 per cent Nickel..1.00 - 1.50 per cent Chromium..0.45 – 0.75 per cent. | 107,500 117,600 | 125,500 132,200 | 22.0 22.0 | 61.6 58.7 | 286 286 | Olsen machine 99 99 |

TABLE 6  HEAT-TREATED, COLD-DRAWN NICKEL-CHROMIUM STEEL

Composition (per cent) ; C, 0.37 ; Mn, 0.68 ; Ni, 1.60 ; Cr, 0.59 ; P, 0.010 ; S, 0.034.

| Sizes, in. | Treatment | Elastic Limit, lb. per sq. in. | Ultimate Strength, lb. per sq. in. | Elongation in 2 in., per cent | Reduction of Area, per cent |
|---|---|---|---|---|---|
| 1- 11/32 | 1540 deg. fahr. oil, Tempered 930 deg. fahr. | 87,650 | 109,400 | 23.2 | 55.7 |
| 1- 5/16 | Cold - drawn  1/32 in. | 108,300 | 140,800 | 19.0 | 57.1 |
| 1- 11/32 | 1540 deg. fahr. oil, Tempered  930 deg. fahr. | 86,200 | 104,800 | 24.5 | 66.1 |
| 1-5/16 | Cold - drawn 1/32 in. | 108,200 | 125,900 | 20.0 | 61.3 |

TABLE 7  PERCENTAGE ANALYSES OF NICKEL-CHROIMUM GEAR STEELS

| | C | Mn | Ni | Cr |
|---|---|---|---|---|
| Tempering Steels | 0.50 | 0.45 | 1.75 | 1.00 |
| | 0.50 | 0.60 | 3.00 | 0.75 |
| | 0.40 | 0.45 | 3.50 | 1.25 |
| | 0.30 | 0.45 | 4.50 | 1.50 |
| Case-Hardening Steels | 0.15 | 0.45 | 1.25 | 0.60 |
| | 0.15 | 0.45 | 1.75 | 1.00 |

The treatment of the first class, comprised of tempering steels, is similar to treatments previously outlined and consists in normalizing or annealing the blanks before machining, after which the gears are quenched and then tempered at comparatively low temperatures (about 500 deg. fahr.). High hardness and strength together with fair ductility can be obtained.

The second class mentioned, comprised of low-carbon case-hardening steels, requires different treatment. The machined gears are carburized in suitable carburizing material and then usually double-quenched for refinement of core and of case. The result is a material of dual nature. The core, low in carbon, has high ductility and relatively low strength, while the case, high in carbon, is extremely hard and has little ductility.

Where the highest combination of strength and ductility is not essential the gears may be carburized at somewhat lower temperatures but above the critical ranges of the steel under treatment, and then quenched but once for hardening the case. With such treatment the coarse structure of the core produced during carburization still remains but a hard wearing surface is obtained.

In general tempering steels are applicable to clash gears while the higher hardness obtained in the case of the case-hardening steels makes them admirably suited for constant-mesh gears. Both types are used interchangeably, however.

### HAIR-LINE SEAMS

Two phenomena encountered in the use of nickel-chromium steels which warrant discussion are "hair-line seams," also called manganese sulphide streaks, slag inclusions, etc.; and "blue brittleness," referred to as "Krupp Krankheit" and more properly termed "temper brittleness."

The first is in many instances undoubtedly due to inclusions in the ingot and consists of manganese sulphides or silicates. Probably iron silicate and iron and possibly nickel sulphides are also present. These streaks or cracks, which are not to be confused with quenching cracks, vary in depth, length and in numbers, but are most frequently present at the surface of the finished forgings. Often removing some surface metal will cause them to disappear, only to be replaced by others, whereas sometimes the removal of a small surface layer will cause them to disappear entirely. This suggests that in some cases they may originate by being driven in from the outer surface during forging. It is therefore good practice to leave more than the ordinary amount of metal for finish-machining. However, their absence from the finish-machined forging is no proof that they do not exist in the interior. Whether such streaks are deleterious is open to discussion, but in the writer's opinion they should certainly be regarded with suspicion when present in considerable numbers, particu-

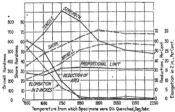

Fig. 3  Effect of Varying Quenching Temperature on the Physical Properties of a High-Chromium Steel.

larly if at or near sharp corners in machined forgings subjected to high stresses.

### Temper Brittleness

The second phenomenon mentioned is evidenced in the low impact values obtained in slow cooling from the tempering heat, between temperatures of 400 to 1100 deg. fahr., and more particularly 700 to 1000 deg. fahr.  By quenching from these tempering

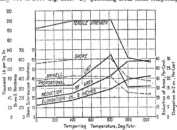

Fig. 4  Effect of Varying Tempering Temperature on the Physical Properties of a High-Chromium Steel, Oil-Quenched from 1750 Deg. Fahr.

temperatures, normal impact values are obtained. Table 8 illustrates this effect.

So far, no relation between low impact values and resistance to fatigue or other ordinary tests has been found, but until proved otherwise heats showing temper brittleness should be specially treated or discarded. At least they can be questioned seriously when intended for important parts. This whole subject is now under investigation by a committee[1] appointed by the National Research Council.

[1] Committee on Physical Changes in Iron and Steel below the Thermal Critical Range, Dr. Z. Jeffries, Chairman.

### TABLE 8  TEMPER BRITTLENESS OF CRANKSHAFT STEEL
Composition (per cent): C, 0.41; Mn, 0.50; Ni, 3.13; Cr, 0.82

Treatments:
1 Normalizing.......1670 deg. fahr.—air-cooled
2 Hardening.......1510 deg. fahr. for 1 hour—oil quenched
3 Tempering ......Heated 30 min. at 1100-1125 deg. fahr. and cooled as follows:
    (a) oil...........50 ft-lb. energy absorbed
    (b) water.........48 ft-lb. energy absorbed
    (c) still air.......43 ft-lb. energy absorbed
    (d) furnace (rate from 1050-775 deg. fahr. averages 2 deg. per min.) 8.7 ft-lb. energy absorbed

### Carbon-Chromium Steels— Stainless Steel

Steels without nickel containing various percentages of chromium also find interesting applications in the automotive industry.  For many years the use of chromium was restricted to high-carbon steels on account of difficulties in producing carbon-free ferrochrome.  With improvements in production of ferroalloys some interesting steels have come into use.

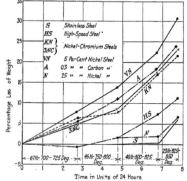

Fig. 5  Resistance to Oxidation of Various Steels

One of these is steel containing about 0.20 to 0.40 per cent carbon and 11 to 15 per cent chromium, produced under the name "stainless steel." Originally used for cutlery it has found successful application in valves for airplane and automobile engines where resistance to the action of hot gases and good strength and ductility are required.  Figs. 3 and 4 show the effect of varying quenching and tempering temperatures on the tensile properties and hardness, while in Fig. 5 is shown the resistance to oxidation compared with common types of alloy and carbon steels.[1]  The steel presents its maximum resistance to corrosion and oxidation only when properly hardened and finished (polished or ground). It is air-hardening in small sizes and may be quenched in oil, air, or water.  Quenching is best carried out from about 1650 to 1750 deg. fahr. and for valves the steel is usually tempered at relatively high temperatures—about 1250 deg. fahr.  Properties obtained after such treatment are shown in Fig. 4.

### Cast High-Speed Steel Cutters

Before closing this discussion it will probably be of interest to touch upon another phase of the application of alloy steels in the automotive industry, namely, as cutting media.  The follow-

[1] From tentative report of Iron and Steel Committee of the Society of Automotive Engineers, appearing in Jl. S. A. E., vol. 5, no. 3, Sept. 1919, pp. 262-263.

FIG. 6  CAST HIGH-SPEED STEEL MILLING CUTTER USED IN TEST;
DIAMETER, 3.75 IN.

known as dendrites, imbedded in a matrix of softer material.

One cutter having a double set of teeth joined by a curved cutting edge was tried on rail steel at two different cutter speeds and depths of cut. One set of teeth burned immediately at a speed of 100 ft. per min. and 0.200 in. depth of cut, while the second set of teeth were worn so as to be useless after cutting 2 in. at a cutter speed of 45 ft. per min. and 0.100 in. depth of cut. Examination of the structure in Fig. 8 reveals the cause of this poor service. The treatment given has not brought the coarse carbide dendrites into solution, the cast structure persisting. Comparison with the structure of a properly hardened forged tool in Fig. 9 will make this clear.

Another cast cutter was annealed at 1650 deg. fahr. for one hour and cooled in the furnace, but did not machine well and was reheated at this temperature for five hours more and cooled in the furnace. After machining to standard milling-cutter shape it was heated to 1250 deg. fahr. in 3¼ hr., then raised to 2150 deg. fahr. in 10 min., and oil-quenched. After grinding it was subjected to test in comparison with a high-speed steel cutter of standard make, the results being given in Table 9.[1]

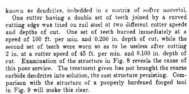

Cutting test made by M. A. Grossman, formerly of the Bureau staff. Micrographs prepared in metallographic section of the Bureau.

(Continued on page 547)

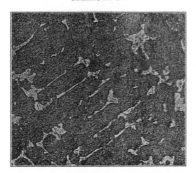

FIG. 7  STRUCTURE OF STEEL USED IN CUTTER OF FIG. 6 AS CAST,
MAGNIFICATION, 500 DIAMETERS

FIG. 8  CAST HIGH-SPEED STEEL—HEAT-TREATED. MAGNIFICATION,
500 DIAMETERS

ing will be confined, however, to the results of tests made on some cast high-speed steel milling cutters, which in the past have generally been made from bars or forgings. This latter method when properly carried out has the advantage of decreasing machine work as well as of saving costly material, and produces cutters of equally high if not better efficiency than those machined from bars.

Fig. 6 shows one of the cast high-speed steel cutters used in the test about to be described and having the following analysis:

Carbon ................................... 0.85 per cent
Tungsten ................................. 19.40 per cent
Chromium ................................. 4.20 per cent
Vanadium ................................. 0.80 per cent
Manganese ................................ 0.05 per cent
Iron ..................................... 74.60 per cent

Total ............................ 99.90 per cent

It will be noted that both carbon and tungsten are somewhat higher than proportions usually present in high-speed steel tools produced in the usual manner, and examination of the photograph shows a remarkably smooth surface, generally free from imperfections. The structure of this steel as cast is illustrated in Fig. 7, which shows the characteristic tree-like crystals of carbide,

FIG. 9  FORGED HIGH-SPEED STEEL—HEAT-TREATED. MAGNIFICATION, 500 DIAMETERS

# Some Commercial Heat Treatments for Alloy Steels

By A. H. MILLER,[1] PHILADELPHIA, PA.

*The object of alloy-steel heat treatment, and indeed with a very few exceptions all heat treatment, is to produce a grain size as small as possible, with a degree of hardness suitable for the purpose intended. The three variables which must be controlled for a successful heat treatment are temperature, time, and rate of cooling, and together with these the influence of mass must not be neglected. The author of this paper deals with the heat treatment of alloy-steels used for structural purposes, especially of nickel and nickel-chrome steel. He describes a series of tests which were conducted to determine the effect of the various heat treatments on samples of the same chemical composition, and the varied results are illustrated by a series of photomicrographs which show the effect of different adjustments of the above-mentioned variables.*

ALLOY steels, both for tool and structural purposes, have had an increasing application for a number of years. This paper is confined to a discussion of steels for structural purposes, and will further limit itself to their heat treatment. It will apply directly to the two alloy steels which are probably used to a greater extent than all others combined; namely, nickel and nickel-chrome steels. It is to be borne in mind, however, that the statements to be made in regard to these two alloys are almost equally applicable to all of the structural alloy steels, provided temperature changes are made which correspond to the changes in the critical temperature of other alloys.

In speaking of heat treatments, a fundamental thought must always be kept in mind: All fabricated steels are submitted to a heat treatment. The differences between steels known as heat-treated and others commonly spoken of as untreated is merely that the treated steels have supposedly received a preconceived, carefully-carried-out treatment, whereas the so-called untreated steels have received a variable and generally unknown treatment which is the result of casting, forging, and cooling at an unknown and variable rate from the casting or forging temperature.

### The Time Element in Heat Treatment

In the heat treatment of alloy steels the three variables which must be controlled for a successful heat treatment are temperature, time, and rate of cooling. The influence of mass on these three variables must never be neglected; moreover, it must be borne in mind that an increase in mass may increase the treatment temperature, should increase the length of time held at temperature, and will inevitably alter the rate of cooling.

Too little attention is generally paid to the time element of the heat treatment, whereas it actually is of very great importance. The illustrations, Figs. 1-7, are a series of photomicrographs of a nickel-chrome steel of the following composition: carbon, 0.35-0.40 per cent; nickel, 3 per cent; and chromium, 0.75 per cent; and show the microstructure in a typical forged condition, and after annealing at a proper annealing heat for varying periods of time. It will be noted that this series gives the time held at the annealing temperature from zero (meaning that the piece was brought to temperature and the furnace was immediately shut down) to 10 hr. A study of the photomicrographs shows that the ferrite as contained in the cell outlines of the forged specimen was not dissolved and uniformly diffused until the piece had been held at the annealed temperature for ½ hr. Fig. 7 shows that there had been a slight growth of the austenite crystals between the time of completed uniform solution at ½ hr. and the end of the run, 10 hr.

This series also shows that a new cell system may grow in steel simultaneously with the breaking up of the previously existent

[1] Research Department, Midvale Steel and Ordnance Co.
Presented at a meeting of the Washington Section of THE AMERICAN SOCIETY OF MECHANICAL ENGINEERS, March 31, 1920.

system. The pieces, representative micrographs of which are shown, were all cut from the same bar, and were treated by placing them together in a furnace controlled by a thermocouple, withdrawing them one by one at the end of the specified time and plunging each immediately into a box of well-aerated lime.

The reason that a considerable length of time is required to produce a uniform structure is probably as follows: After the steel is raised to a temperature above the critical temperature, the iron is in the gamma form, in which iron carbide is soluble. There is, however, a certain length of time required for this solution, and, more than that, a certain added length of time is necessary to allow the solution to become homogeneous, just as, in dissolving a lump of sugar in water, a certain length of time is required to complete the solution, and a certain further length of time for the water to become uniformly sweet. Analogously, if the iron carbide be dissolved in the gamma iron and this solution does not have time to become homogeneous before it be recooled, the ferrite will naturally separate out on cooling at the point where the greatest concentration existed in the solution.

### Procedure in Alloy-Steel Heat Treatment

The object of alloy-steel heat treatment, and indeed, with a very few exceptions, all heat treatment, is to produce a grain size as small as possible, with a degree of hardness suitable for the purposes intended, by the simplest possible means. Thus the ill-controlled and generally very poorly forged structure must first be broken up and a fine uniform structure established. In steels which are sensitive to heat treatment, of which the nickel and nickel-chrome steels are excellent examples, this object is best achieved in several steps, each of which is designed to break up the structure resulting from the previous step and bring the material into a more nearly ideal condition.

If the forging conditions are bad, as is the case in most forging processes, especially that of drop forging, a treatment of numerous steps may be necessary. As an example of the most drastic the following is given:

1. Anneal from approximately 1450 deg. fahr.
2. Quench from 1600 deg. fahr.
3. Quench from 1400 deg. fahr.
4. Draw at 1250 deg. fahr.
5. Quench from 1400 deg. fahr.
6. Draw at such a temperature as will give the desired hardness.

This heat treatment is not of unheard length, as it is quite conceivably necessary in many cases. As a matter of fact, in manufacturing pieces which will not subsequently be forged by the purchaser, steel companies very frequently give all of the preliminary steps of this treatment to their regular product. It must be well understood, however, that this number of steps is necessary only to guard against lack of uniformity, due to one piece out of a great number having possibly been subject to a poor forging heat. If the forging temperature can be accurately regulated, however, many of the steps in this treatment can be eliminated.

In much commercial work, with good forging practice, a simple anneal at 1450 deg. fahr., followed by a quench just above the critical temperature and a draw, will put the steel in excellent preliminary condition, at which point the steel can be machined to its final shape. If conditions are such that the steel must be extraordinarily hard (as, for instance, in automobile gears), a final quench with a draw at about 400 to 600 deg. fahr. is then given.

It must be borne in mind when laying out treatments that the time at which the steel is held at temperature during any treatment, whether it be an anneal or a quench, is of quite as great

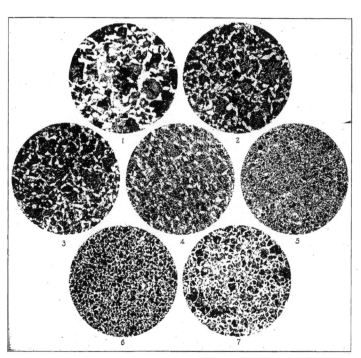

FIGS. 1 TO 7  EFFECT OF TIME AT NORMALIZING HEAT  (1450 DEG. FAHR.)  FOLLOWED BY SLOW COOLING
Fig. 1  As forged.  Fig. 2  Not held, cooled in lime.  Fig. 3  Held 5 min.  Fig. 4  Held 10 min.  Fig. 5  Held 15 min.  Fig. 6  Held
30 min.  Fig. 7  Held 10 hr.  x 80.

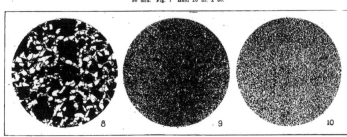

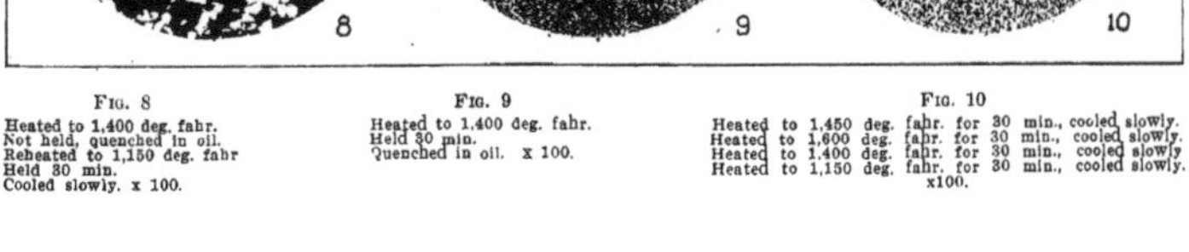

FIG. 8
Heated to 1,400 deg. fahr.
Not held, quenched in oil.
Reheated to 1,150 deg. fahr
Held 30 min.
Cooled slowly. x 100.

FIG. 9
Heated to 1,400 deg. fahr.
Held 30 min.
Quenched in oil.  x 100.

FIG. 10
Heated to 1,450 deg. fahr. for 30 min., cooled slowly.
Heated to 1,600 deg. fahr. for 30 min., cooled slowly.
Heated to 1,400 deg. fahr. for 30 min., cooled slowly
Heated to 1,150 deg. fahr. for 30 min., cooled slowly.
x100.

importance as the temperature. This is illustrated by photo-micrographs, Figs. 8-10, of two pieces cut from the same bar as those previously shown, both of which were placed in the furnace together. One of these pieces was drawn from the furnace and quenched immediately it had reached the quenching tempera-ture (in this case 1400 deg. fahr.). The other was allowed to remain in the furnace for ½ hr. and was then quenched. It will be seen that the ferrite areas in the first case had been slightly or incompletely broken up, whereas in the second case they were very completely dissolved.

These photomicrographs differ from a corresponding one in the first series of annealed samples in that there is shown no new grain growth within the old partially broken-up system. This, of course, is due to the fact that in the second case the time ele-ment necessary for the separation of the ferrite during cooling was not sufficient.

### RESULTS OBTAINABLE IN HEAT-TREATING ALLOY STEELS

From results obtained in the careful heat treatment of nickel-chrome steels, a series of curves, Fig. 11, has been prepared which show the physical properties of a nickel-chrome steel resulting from proper preliminary treatment and varying drawing tem-peratures.

The type composition only is given in this figure, because it is

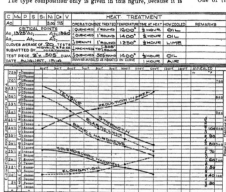

FIG. 11   PHYSICAL PROPERTIES OF A NICKEL-CHROMIUM STEEL RESULTING FROM PROPER PRELIMINARY TREATMENT AND VARYING DRAWING TEMPERATURES

a mean of the results of about twenty bars from several heats of slightly varying compositions. The nickel steels of the same approximate carbon content give results which are somewhat inferior to this nickel-chrome curve, whereas the results of an-other type of nickel-chrome steel, of 3½ per cent nickel and 1½ per cent chromium, would be slightly superior.

Starting from the extended heat treatment just described, the development of the cheapest and simplest treatment which will give good results is a matter of intelligently eliminating or alter-ing steps of the ideal heat treatment as conditions permit. For instance, in a certain case where important drop forgings were manufactured from the grade of nickel-chrome steel shown on the curve, the actual treatment to which pieces were subjected is as follows: The pieces were forged under a drop hammer, and were dipped immediately afterward into a tank of oil which was maintained close to the forge. The pieces were kept in this oil for about four minutes, removed at a temperature between 700 to 900 deg. fahr., and were buried in ashes as a precaution against

cracking. Then the pieces were subjected to a single quench at 1400 deg. fahr. and were drawn at 1200 deg. fahr., in which condi-tion they were machined and received no further treatment. The uniformly excellent results obtained (each of the pieces was sep-arately tested) showed that this very simple treatment had been entirely effective. A little thought will show that the reason for this was that the drop forging was not excessively high, and that the growth of large-cell outline was prevented by the quench after forging was completed. The single quench and draw were sufficient to completely refine the steel from the fair condition which was thus produced.

There is this to be observed in all cases of quenching of alloy or indeed any other steels: Following the quench, the piece quenched is in a condition of great strain and is liable to crack. This liability to crack persists until the piece has been drawn, and it is therefore wise to draw the piece as soon as possible after the quenching.

In cases where a drastic quench is advisable it is better to remove the piece from the quenching medium before it becomes entirely cold. By this procedure the great proportion of the con-demnations due to cracking are avoided.

### FIBER FRACTURE A CRITERION OF PROPER TREATMENT

One of the significant effects of a correct heat treatment on alloy steels, which is indeed a criterion as to the efficiency of the treatment, is the production of a peculiar type of fracture in a broken piece, known as "fiber." This fiber fracture is abso-lutely distinctive, and cannot be mistaken by one who is even slightly skilled in inspection. It is produced in all of the well-melted, shock-resisting alloy steels by proper heat treatments, and is so closely related to impact test values that failing impact tests can almost invariably be selected from broken impact test bars by the absence of this type of fracture. The ease of producing fiber by heat treatment is a criterion of the value of an alloy for shock-resisting properties. So im-portant is the presence of this feature that armor plate, which must withstand shock test of the more severe character, is never knowingly shipped without it.

In its report covering its activities for the year 1919, The National Physical Laboratory (Eng-land) reviews its accomplishments in many fields of science, among them being its work in metal-lurgy. The most important work in the Metal-lurgy Department of the laboratory was in con-nection with the production of light alloys, espe-cially for aircraft and aeroplane engines. Alloys were developed to meet special requirements for castings in general, and for parts such as pistons working at high temperatures; wrought alloys were produced for use in the construction of rigid airships, aeroplane spars, etc.; and the rolling of light alloys into thin sheets to serve as a substitute for fabric in covering aero-plane wings was successfully accomplished. Special investi-gations were made with the object of providing substitutes for alloys for munitions when difficulties of supply arose. Among these was a substitute for the use of antimony in hardening lead for shrapnel bullets. Another case in which a substitute was re-quired was the aluminum tip forming part of the standard small-arms bullet at the beginning of the war. The possibility of re-placing the aluminum tips by bodies of precisely similar size, shape, and weight was thoroughly demonstrated, with the col-laboration of a pottery in the Manchester district, where the diffi-culty of preparing small pottery bodies with the necessary ac-curacy of shape and dimensions was successfully overcome. Investigations were made on steel for torpedo air vessels and tur-bine gearing, and methods of hardening and case-hardening war material were improved.

# The St. Lawrence River Project

An International Project Which Promises Unlimited Benefits in Solving the Transportation Problem
of Central North America as Well as the Fuel and Power Crisis of the Eastern Seaboard

By HORACE C. GARDNER,[1] CHICAGO, ILL.

*For centuries the rapids of the St. Lawrence have been regarded as a drawback to the development of the surrounding country, and in the sense that the people of a century ago comprehended the matter they are still a disadvantage. But with the existing demands for power and inland transportation systems already breaking down under the stress of traffic, the river and its rapids should be looked upon as a resource and one of nature's best gifts, for they furnish the solution to both power and transportation problems. It is to this proposition that the author of the following paper has addressed himself. If developed, the improved St. Lawrence, together with the Great Lakes, would constitute, the author states, a waterway from the heart of North America direct to the continent of Europe. And as a result an ocean steamer could be loaded at a Great Lakes port and not discharge its cargo until it reached a European port. This, of course, would relieve the congestion which now takes place on the railroads every year during the season when the crops of the West and Middle West are shipped to the seaboard. It would also materially lessen the car shortage which occurs during that period, and moreover it would lessen the expense of shipping. The problem, the author states, is not a new one. It is over a hundred years old, and during the past seventy-five years the Canadian Government has been making improvements, most of which, however, are not large enough for the present demands. At the present time the Great Lakes-St. Lawrence Tidewater Association is endeavoring to encourage coöperation between the United States and Canada in the development of the St. Lawrence both for navigation and power. The International Joint Commission, which consists of three representatives of the United States Government and three representatives of the Canadian Government, are also holding hearings at various places in both countries to learn the sentiment of the people. The paper is an interesting discussion of an important problem which is bound to receive considerable attention in the immediate future.*

I N 1534 Francis I, King of France, commissioned that most famous French navigator of the time, Jacques Cartier, to attempt the discovery of the much-sought northwest passage to India. Accordingly in the spring of that year Cartier sailed from St. Malo and spent the summer in and about the St. Lawrence gulf, during which time he circumnavigated Anticosti Island. He actually entered the mouth of the St. Lawrence River, but he was not sure as to whether it was a river or an arm of the sea. In 1536 Cartier made another expedition and explored the St. Lawrence for 550 miles, or as far as the present site of the city of Montreal. The encouraging results of these expeditions brought about others, and eventually the five fresh-water inland seas that have come to be known as the Great Lakes were disclosed to the world, which, together with the St. Lawrence, constitute one of the finest, and in many respects one of the most remarkable, fresh-water basins and drainage systems on the globe.

The entire distance from the headwaters of the St. Louis River, in Minnesota, which flows into Lake Superior, to the mouth of the St. Lawrence River, is 2100 miles. The lakes themselves, constituting enormous reservoirs, cause the great river to have the most even flow of any of the earth's large streams. The variations between mean flow and maximum and minimum is but 25 per cent of the mean in contrast with thirty times the minimum at the mouth of the Ohio, and practically the same extremes in the Mississippi at Keokuk.

But the great disadvantage of the river is its rapids, the greatest of which is the Lachine, which ends at Montreal. The total fall between the outlet of Lake Ontario and mean tide at St.

Lawrence is 221 ft. For centuries these rapids have been regarded as an enormous disadvantage, but with the present need for power and modern methods of development the people of both the United States and Canada are sure to reach within a decade or two a condition of thankfulness that the rapids exist.

It will perhaps be interesting to speculate as to what would have been the order of development of population and industry on the North American continent had the St. Lawrence been an ordinary stream without the great rapids and had the Great Lakes been connected to this outlet by intermediate channels with only the normal fall of the usual river. The St. Lawrence proper was explored for its full length in 1536, and it will be recalled that settlements were made on the Atlantic Coast as follows: Jamestown, Virginia, in 1607; New Amsterdam (Manhattan Island) in 1615; and Plymouth, Mass., in 1620. It would therefore seem evident that had it been possible to conduct water transportation by the vessel of that day from the Great Lakes basin directly to Europe, actual colonization would have been under way on both shores of Lake Erie and Lake Ontario while the English were establishing themselves on the Atlantic Coast and the Dutch in eastern New York. The impetus then would have been in the Great Lakes basin; and, with the superior fertility of these lands in mind, it seems probable that populous centers would have been established on the shores of the Great Lakes, with settlements and developments extending westward, and with the less fertile marginal lands along the Atlantic Coast in a much more backward state. But nature provided otherwise, and the rapids of the St. Lawrence proved so much of a barrier to commerce that it took a century for the advance of civilization to reach Detroit and another century to reach Ft. Dearborn; and it was not until recently that the people awakened to the supreme necessity of providing adequate facilities so that ocean-going ships may pass up the St. Lawrence and navigate these fresh-water seas in the heart of the continent.

The territory included within the heavy lines on the map of the United States, shown in Fig. 1, is economically nearer to the ports of the St. Lawrence and Great Lakes than to any seaport east, south or west, either in actual miles or because of mountain ranges to be crossed. This territory is correctly called the heart of our country. It embraces something more than a third of our continental area, with more than a third of the population and more than a third of the wealth. All who know the United States are aware of the fact that in this area lies potentially much more than one-third, and probably much more than one-half, our productive capacity. According to the latest figures available its present productivity may be stated as follows, in terms of percentage of the total:

| | | |
|---|---|---|
| Corn ..................... | 65 | Apples ..................... 21 |
| Wheat .................... | 74 | Beet sugar................. 53 |
| Flax, practically............ | 100 | Wool ..................... 47 |
| Cattle, over................. | 50 | Coal mined............... 36 |
| Hogs ..................... | 57 | Coal reserves............... 72 |
| Horses ..................... | 60 | Copper ..................... 80 |
| Butter ..................... | 54 | Lead ..................... 46 |
| Eggs ..................... | 54 | Iron ore................... 85 |
| Cheese ..................... | 57 | Zinc ..................... 74 |

On the map of Fig. 1 it will be noted that various centers are indicated by numbered circles. These are the centers of production of various products, comprising practically all of the fundamental necessities of life.

With one exception—the center of population—the locations are the crossings of median lines. There are two methods of determining such centers. One is to calculate and determine a point at which, if all of the given product were assembled and piled in one pile, the least possible transportation would be necessary. The

[1] Gardner and Lindberg. Mem. Am. Soc. M. E.
Abstract of a paper presented before the Chicago Section of THE AMERICAN SOCIETY OF MECHANICAL ENGINEERS, June 8, 1920.

center of population as shown on the map was so determined by the Census Bureau. The other method, and the one used with respect to all of the centers except population, is to determine a north and south line, east and west of which half of a given product is produced; then determine an east and west line, north and south of which half is produced. It is at the crossing of these median lines that the center is located. In determining the centers used in this paper, other than the center of population, the latter method was employed.

To theorize further on the two methods, the determination of the crossing of median lines gives quite as useful and satisfactory results as the first method, but it would not necessarily do so with respect to everything, as, for example, citrus fruits. These are grown in Southern California, to a small extent in Arizona, and in Florida. The actual center would probably lie in the eastern part of Southern California. If Florida produced approximately the same quantity as California, a north and south line bisecting the total production might lie almost anywhere between California and Florida, and would certainly come in a territory of no production. Fortunately, most of our fundamental staples are produced over large areas, and the method of determining the crossing of median lines is as accurate and satisfactory as any that can be devised. Moreover, in considering the significance of the location of such centers, the normal direction of freight movement, which is always from the region of greatest production toward the region of less or no production, must be kept in mind. Grain, foodstuffs and feed move always toward the east, while manufactured products, for the most part, move toward the west.

The centers as shown on the map have been determined by the use of the latest figures available. The center of population was determined according to the census of 1910. Many of the most important figures are founded on figures of the Census Bureau bearing the date of 1914, but these centers do not shift rapidly.

### THE PRESENT FREIGHT SITUATION

And yet with all this potential and actual wealth both the United States and Canada have suffered, for business has been hampered by the lack of transportation. There are many reasons for this condition, but the fundamental one is due to the fact that such an enormous part of the total production must travel long

distances by rail and the railways are congested progressively as one moves eastward. The greatest transportation demand comes every year at the crop-moving season. Taking the experience of past years, it is found that the distribution of car supply is such that some sections of the country have too many and others too few cars to handle the business. As a rule, cars are well distributed in the spring, but when the crop-moving season begins there is always an abnormal movement of the car supply toward the seaboard. At the height of this season the distribution is so abnormal that the supply of cars east of the Alleghany Mountains is about 120 per cent and in the central part of the country about 80 per cent of the normal supply. This condition prevails for several months. It occurs, of course, because of the large amount of tonnage which the railroads must carry during the fall and early winter. There is a very natural disposition upon the part of the railroad lines to hold empty cars for westbound loading, hoping that they may be relieved of the necessity

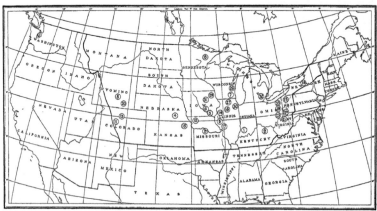

FIG. 1 CENTERS OF PRODUCTIVITY IN UNITED STATES

of moving cars without load in great numbers and over long distances. To some extent this is justified, because the demand for merchandise always increases after the crops are moved and the farmers have the increased buying power that comes therefrom. To some extent, no doubt, the railways delay empty-car movements longer than they should to postpone this dead expense. They are particularly prone to do so when there is a shortage of motive power and of man power, also a shortage of fuel, all three of which conditions have prevailed within the past six months.

During the latter part of 1919 and up to the present time, instead of working toward a normal car distribution as might be expected during that season of the year, conditions have become increasingly worse, not only because of the three conditions just named but also on account of the shortage of vessel room for export cargoes and of local strikes among the railway switchmen and dock laborers. In fact, vessels have lain at anchor in New York Harbor for weeks waiting for a place to berth. The whole situation has become almost intolerable and in many cases shippers have been put to great and unexpected expense. Three or four cases of excessive charges on export consignments shipped through the port of New York within the past few months are as follows:

Case 1—6 barrels, weight 521 lb.:

| | |
|---|---|
| Freight to New York | $4.94 |
| Cartage to warehouse | 4.50 |
| Cartage to steamer | 4.50 |
| Overtime for delay of truck | 7.20 |

Case 2—8 crates, weight 96 lb.:

| | |
|---|---|
| Freight to New York | $0.96 |
| Cartage to warehouse | 6.00 |
| Cartage to steamer | 6.00 |

Case 3—156 packages toys:

| | |
|---|---|
| Freight to New York | $22.56 |
| Freight to warehouse | 78.00 |
| Cartage to steamer | 14.25 |

Such conditions are necessarily ruinous to export trade because shippers cannot possibly satisfy their customers, nor can they afford to take the risk of making prices that would under ordinary conditions afford them a reasonable profit. No doubt a considerable part of our high cost of living comes from exactly such conditions. At our various ports and at many other interior points our domestic shipments have been delayed as well as subjected to many unusual expenses.

What is the remedy for this intolerable state of affairs? The answer is short—better transportation. There are, however, many ramifications of the problem. First, our railways must be rehabilitated, provided with an ample number of cars and adequate motive power, and must build up their personnel to take much better care of business. Fundamentally, however, it is wrong to impose upon the railways the enormous peak load of tonnage that comes every year during the crop-moving season, requiring them either to build up their freight-carrying capacity equal to the peak and consequently to have much of their rolling stock and power idle for several months in the year, or to repeat just what is being experienced at the present time; that is, to allow empty cars to pile up in the eastern zone and suffer the consequent shortage in the Central West.

A study of figures showing our exportations of grain and heavy freight shows that if we could load vessels at the ports of the Great Lakes and sail them direct to European ports, practically all of the excessive peak load could be absorbed and incidentally a tremendous car-mileage could be avoided. For example, wheat, which constitutes the greatest tonnage item of all our exports, comes principally from the Dakotas, Minnesota and Kansas. Think of the car-mileage involved in transporting this wheat to the Atlantic seaboard that could be saved if the vessels were loaded at the convenient ports on Lake Superior, Lake Michigan, and at Detroit and Toledo. Furthermore any railway system upon which tonnage of this class originates that could deliver directly to docks without leaving its own rails, could have much better control of disposition of empty cars; they could be rushed west for reloading not only without delay but without any of the red tape and trouble incident to business that must move over two or more railway systems.

The real remedy, then, for our annual freight blockade would seem to be to open the Great Lakes so that ocean-going vessels may freely enter and make use of the various ports, at many of which our railway systems already center. This being an acknowledged fact, it is, indeed, rather puzzling to know why we have suffered for so many years so greatly without taking real steps to remedy the difficulty.

### PAST IMPROVEMENTS AND PROGRESS

In applying our remedy such improvements of the St. Lawrence River and of the canal across the Niagara peninsula must be made that vessels of ocean-going type and large size can be admitted. The subject is almost a century old; in fact, in an academic way it was thought out and discussed much more than a century ago, and three-quarters of a century ago a series of canals were built around the various rapids of the St. Lawrence River and across the Niagara peninsula. These, while adequate for their time, were soon outgrown. They were improved somewhat so that vessels drawing up to 14 ft. of water might come and go. However, with the rapid increase in the size of ships these canals soon proved too small and have constantly hampered

commerce. Something over twenty years ago the subject became quite active and the Canadian Government undertook surveys and estimates as to the practicability and cost of improving the Ottawa River and the building of a canal across the divide to the French River and down that stream to Georgian Bay. Even then their ideas as to size were such as would not now receive serious consideration. These surveys and estimates showed, however, that canals with the depth and lock capacity for vessels of 20 ft. draft would cost practically $100,000,000. There was at that time in Canada a considerable sentiment in favor of building a canal entirely through Canadian territory. About the same time consideration was given in this country to the idea of building a canal from Albany to Lake Ontario. Fundamentally the plan was, of course, to meet the proposed all-Canadian canal with an all-American canal. Studies and estimates were made in detail. One of the propositions considered was for a canal with a depth of 30 ft. The estimate at that time gave the cost as about $200,-000,000. Nothing was done, however, and we were soon in the midst of the Spanish-American War, following which the Panama Canal attracted the attention and energy of the country.

About two years before the European War began, Canada undertook on her own account the construction of a new Welland Canal across the Niagara peninsula; the depth through the earth sections to be 25 ft. and the miter sills of the locks 30 ft. The expected cost was $50,000,000 and the work was well under way when, in the summer of 1914, Canada found herself at war. The work was courageously continued for about one year, when for financial reasons and lack of man power it was stopped. Within a year after the armistice Canada resumed work, and it is expected this canal will be finished within three years. It will be adequate for the typical ocean freighter of today, in fact, for all of the commercial fleets of the world with the exception of a very few large ships, which are principally used on the North Atlantic.

What about the St. Lawrence Rapids? How can they be avoided or what may be done to enable vessels to come up and use the new Welland Canal? About a dozen years ago, Gen. W. H. Bixby, then Chief of Engineers of the United States Army, pointed out in an address that a full investigation would probably show that the St. Lawrence might best be improved by building a series of great dams, drowning out the rapids and converting them into lakes, putting in locks so as to pass vessels of any size desired. This suggestion, which by the way General Bixby does not claim to have originated, was of course not based upon definite surveys and full engineering consideration, but it seems to be the general opinion that, upon full study, this method will be found the best. One of its great advantages is that it would incidentally develop a large amount of power.

### RECENT PROGRESS

In the very early days of 1914 a committee representing commercial and other bodies in the Central West visited Washington and took up with the Department of State the question of uniting with Canada in the making of adequate improvements in the St. Lawrence so that both countries might have the use of channels and facilities suitable for ocean-going ships; also the development of the incidental power. The Department of State officially brought up the question with the Canadian Government and it was pending when the war came on and then of course nothing was done. A little over a year ago the subject was again taken up and an organization known as Great Lakes-St. Lawrence Tidewater Association was effected, the purpose being to create in this country a sentiment in favor of uniting with Canada for the improvement of the St. Lawrence both for navigation and power. This association has working headquarters in Duluth and it became active at once. Generally, a most sympathetic feeling was experienced and Congress was appealed to and very promptly passed a joint resolution referring to the International Joint Commission the problem of (a) an investigation as to the need for the improvements; (b) the best method; (c) the probable cost, and (d) the proper and equitable division of the cost between our country and Canada.

The International Joint Commission is composed of three members from the United States and three from Canada. It is a standing Commission that has been in existence for eleven years. It was created by virtue of a treaty between the two countries, and has jurisdiction of all boundary waters and problems in connection therewith.

Immediately after our Congress passed the joint resolution the Canadian Government, by order in Council, took the same action. Following this each government appointed a conferee to consider the exact instructions to be issued and to lay down for the International Joint Commission the exact scope of the work to be done. These conferees agreed promptly on the series of problems and questions. The two governments also each appointed an engineer charged with the duty of deciding upon the engineering questions; that is, the best method for making the improvement and the probable cost. The engineer for the United

of the entire Atlantic coast of the United States and the Atlantic coast of Europe. The following comparisons of distance should be of interest: Rochester is nearer to Liverpool than New York; Buffalo is but a few miles farther; Toledo is nearer Glasgow, Belfast and all the Scandinavian and Baltic ports than New York; Detroit is but a few miles farther. From all our ports on Lakes Ontario and Erie it is 250 miles farther to Liverpool, London and the northern French ports and all North Sea ports via New York than via the St. Lawrence; to Glasgow, Belfast, Scandinavian and Baltic ports this differential is 500 miles. To the Mediterranean ports, however, the handicap via the St. Lawrence is but slight and not worth mentioning. It is readily seen, therefore, that distance is in favor of the St. Lawrence route.

Another reason is that there are but 45 miles of rapids, and if these be drowned out by the building of the dams, there would be a minimum of restricted channel.

FIG. 2  RELATION OF ATLANTIC COASTS OF UNITED STATES AND EUROPE

### HYDROELECTRIC RESOURCES

Another tremendous benefit to be derived is that of hydroelectric development. The normal mean flow of the St. Lawrence River at its outlet from Lake Ontario is 240,000 sec-ft. There are a number of affluents but they are all of minor consequence, except the Ottawa, which enters the St. Lawrence just above Montreal, and hence are not to be considered as amounting to much so far as the power development is concerned. There is between the head of the rapids near Ogdensburg, N. Y., and the foot of the lowest rapid just at the city of Montreal, a total fall, as before stated, of 221 ft. This is divided into several rapids, and in the river proper above Montreal there are two lakes: Lake St. Francis, about 30 miles long, and Lake St. Louis, about 16 miles. Of course these lakes would not be interfered with. Some channeling has already been done in them, but more is necessary. The levels will probably be raised somewhat by the series of dams already spoken of, and the pools that would be caused by them, together with these two lakes, as well

States is Colonel Wootten, an army engineer stationed at Detroit. The engineer representing the Canadian Government is Mr. W. A. Bowden. These engineers are actively at work. The International Joint Commission is now engaged in a series of hearings as to the need for the improvement; they have had meetings and presentations have been made by commercial bodies and others at various points in Canada and in the states of North and South Dakota, Montana, Idaho, Wyoming, Colorado, Nebraska, Iowa, Minnesota, Wisconsin, Michigan, Ohio and New York—additional hearings to follow in October. It is hoped that the Commission will reach an agreement and file its report with the two governments next December.

### ADVANTAGES OF ST. LAWRENCE RIVER ROUTE

To the American mind, of course, very naturally comes the question, Why the St. Lawrence? Why not an all-American route up the Hudson and into Lake Ontario? There are, however, several very good reasons for favoring the St. Lawrence. In the first place, nature shaped the course of the St. Lawrence so that it runs from the heart of the North American continent directly toward the principal ports of Europe. This fact, however, is somewhat difficult to realize from the ordinary map.

Fig. 2 shows, as well as can be shown on a flat surface, the relation of not only of the St. Lawrence River and its outlet but

as the whole body of Lake Ontario, would give a reservoir capacity that would be ample, so that the river flow might be almost absolutely controlled and the power production timed to meet the demand. In fact, there is now under consideration by the two governments the question of controlling not only the flow of the St. Lawrence but of the Niagara River as well, the purpose being to compensate for the diversion of water through the Chicago Drainage Channel. The Sanitary District Trustees have offered to meet the cost, which is comparatively small. This question of control of the flow of the St. Lawrence will no doubt be merged with the improvement of the river for navigation and power. Deducting from the 240,000 sec-ft., say, 10,000 sec-ft. for diversion through the Chicago Drainage Channel, and reckoning on the remaining 230,000 sec-ft. at full head, the power development on the 100 per cent basis would amount to 5,750,000 hp. If 70 per cent can be realized, the development will amount to over 4,000,000 hp. Offhand, this would seem to be practicable.

The first 113 miles of the river from its outlet from Lake Ontario are international, that is, they constitute the boundary between the two countries. The fall in this section, which is confined to about 45 miles, is 92 ft. and the power possibilities on the same basis, would be practically 1,660,000 hp. Normally half of this would go to Canada and half to the United States.

(Continued on page 527)

# SURVEY OF ENGINEERING PROGRESS

## A Review of Attainment in Mechanical Engineering and Related Fields

SUBJECTS OF THIS MONTH'S ABSTRACTS

| | | |
|---|---|---|
| RECENT UTILIZATION OF WATER POWER | COMBINED OPERATION OF HYDRAULIC AND | HACK SAWS |
| EFFICIENCY OF FRANCIS WHEELS | STEAM PLANTS | ELECTRICALLY WELDED SHIP "FULLAGAR" |
| TENDENCIES IN DESIGN OF FRANCIS | IMPACT TESTS ON ALLOYS | CARDAN SHAFTS, STRENGTH OF |
| WHEELS | PELTON-WHEEL RECONSTRUCTION | FRAMELESS AUTOMOBILE |
| KAPLAN HYDRAULIC TURBINE | RUSTON AND HORNSBY CRUDE-OIL ENGINE | EMMET MERCURY BOILER |
| HYDRAULIC SECTION OF REPORT OF NA- | INGERSOLL-RAND P-R OIL ENGINE | WASTE-HEAT BOILERS |
| TIONAL ELECTRIC LIGHT ASSOCIATION | MONEL METAL WORKING | WAR EXPERIENCE WITH MERCHANT SHIPS |

## Recent Utilization of Water Power

In the years 1890-1898 a power station was erected at Chevres, near Geneva. It consisted of 15 Jonval turbines of the multiple-runner type, each having an output of 1200 hp. under a maximum head of 26.5 ft. and at a speed of first 80 r.p.m. and then 120 r.p.m. In its day, which was not so long ago, this plant was representative of the highest stage of hydraulic engineering as it existed. If this plant were built today and equipped with high-capacity Francis turbines, a *single*-runner turbine at 120 r.p.m. would give an output per unit of 2500 hp., or, conversely, with an output of 1200 b.hp. the maximum speed would be approximately 175 r.p.m., which would mean a very material saving in the cost of the electrical-generator equipment. With each unit consisting of a double Francis turbine, a maximum output of 5000 b.hp. would be secured at a speed of 120 r.p.m., or with the original output of 1000 b.hp. a speed of 290 r.p.m. would be obtained.

To present the recent advances and present state of water-turbine development was the purpose of the paper recently presented before the Institute of Mechanical Engineers (Great Britain), by Eric M. Bergstrom, from which the above facts are taken.

Under the heading of Francis turbines Mr. Bergstrom calls attention to the early development of standard turbines which could be sold at an exceedingly low cost. This was followed by an increase in size of units specially designed for the conditions of installation which gave considerably greater economies than could be realized from standard machines. Following this development came the recognition of the value of "specific speed," which many leading manufacturers now use as the basis for the manufacture of standard runners, a series of runners corresponding to a certain specific speed being selected as standards to meet the conditions most frequently met with.

The important factor in connection with modern turbine practice is that exhaustive tests have been made giving a complete knowledge of the efficiency and other characteristics of each standard wheel made, so that knowledge of its behavior under varying conditions will enable a runner to be selected which will meet most efficiently a given set of requirements. The standardization of the modern turbine has been developed from these improved methods of systematic tests which provide definite knowledge of the chief characteristics of the turbine. The early American standardization practice was empirical and did not encourage further developments.

It is a general practice for European turbine manufacturers to install testing plants within their own works. This enables them to accurately analyze not only standard runners, but any new special designs. The importance of tests of research work has best been shown by the evolution of the Francis runner, the development of which has been possible by exhaustive trials and intelligent applications of the results obtained.

The development of the high-capacity runner has enabled designers to use a higher value of specific speed during the last ten years. Where the limit of specific speed was approximately 75 (330 metric system) in 1909, today turbines have been installed with a specific speed of approximately 95 with good efficiencies resulting. It is now possible to obtain runners of specific speed of 100, together with even a higher maximum efficiency than secured previously with a specific speed of lower value and with only small sacrifice in efficiency at part gate.

A design of runner recently constructed by the Escher Wyss and Co. gave good results at speeds corresponding to a specific

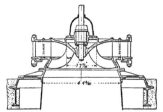

FIG. 1 NEW TYPE OF TURBINE, BUILT BY ESCHER WYSS & CO.

speed of from 85 to 112. The main feature of this new turbine, shown in Fig. 1, is the large space between the guide vanes and the entrance edge of the runner.

Two examples of notably high efficiencies are given: the American installation at the Appalachian Power Co. with a 6000-hp. single open vertical unit operating under a 49-ft. head at 116 r.p.m., specific speed 69.5, with an overall efficiency of 93.7 per cent; the Forsse Hydroelectric Plant, Sweden, with a 3750-hp., double enclosed horizontal unit, operating under a 57-ft. head at 250 r.p.m., specific speed 80, giving a maximum overall efficiency of 94 per cent. Other efficiencies ranging from 87 per cent up are given of plants installed in Norway, Switzerland, Canada, and the United States.

It follows logically that these most notable achievements in maximum efficiencies are the results of the more careful and correct runner design referred to before. Improvements in the design of casing, guide apparatus, suction casings and suction tube are based on a better understanding of the conditions of flow in various parts of the turbine, thus eliminating as far as possible impact losses and eddy losses in the water during its passage through the turbine.

Another significant fact is the long range of gate openings for which an efficiency of over 80 per cent is obtained. In 1911 results of experiments were published showing the average efficiency of tests then on record to be 85 per cent. This value has been greatly exceeded, and a general increase of 10 per cent above efficiencies obtained twelve years ago can now be recorded.

The low-pressure Francis turbines used for heads up to 75 ft. are in Europe generally located in open flumes to which the water is conducted through an intake channel, the plant being arranged with vertical or horizontal shaft. The American arrangement for this class of turbine is generally a single

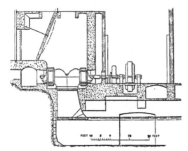

Fig. 2 Hydroelectric Power Station at Forshulten, Sweden

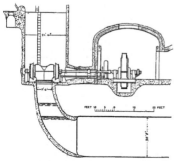

Fig. 3 Mockfjaerden Hydroelectric Power Station, Sweden

vertical turbine, and the most notable developments in hydro-electric plants have been with this type of runner.

The power plant at Forshulten, Sweden (Fig. 2) is an example of a typical horizontal low-pressure turbine plant. The units, of which there are six in operation, each with an output of 3000 hp., are arranged with double runners on a horizontal shaft. They operate at 187 r.p.m. under a net fall of 42.6 ft., the wheels being placed in open concrete pits protected by sluice and strainer racks, the two runners discharging into a common cast-iron suction casing. The shaft is supported at each runner on two outside ring-lubricated bearings, in addition to a babbitt-lined automatically grease-lubricated bearing inside the suction casing. The practice of providing the turbines with lig-

num-vitæ underwater bearings has now been discontinued in favor of outside ring oil-lubricated bearings, the bearing on the inlet side being made accessible through an inspection tunnel as in the present case or through a vertical steel funnel protruding above high-water mark.

More than usual interest is presented by the arrangeemnt adopted at Mockfjaerden Hydroelectric Power Station, Sweden, on account of the power house being situated underground and using open turbines under the relatively high fall of 72 ft. The arrangement of the plant is seen from Fig. 3, and comprises four units each having an output of 5000 b.hp. at 225 r.p.m. with double runners on horizontal shafts. Each turbine is placed at the bottom of a concrete-lined shaft and discharges into the common tailrace tunnel. The powerhouse is blasted out of solid rock, the floor level being approximately 70 ft. below the surface, whereas the switchboard and transformers are housed in a separate building on ground level and communicate with the power house below through an inclined shaft. The alternators are directly connected to the turbine shaft and, due to the underground situation, special arrangements had to be made to insure sufficient cooling. For this purpose the alternators are totally enclosed and cold air is forced down a separate shaft and distributed through underground ducts to each alternator, the hot air escaping from the top of the housing and being expelled through the vertical shaft communicating with the surface.

Medium-pressure Francis turbines are classified by the author as those operating between the heads of 75 to 150 ft. A unique installation of this type of turbine is the Porjus Power Station, Sweden. As in the case of the Mockjaerden Power Station, this plant is also situated underground, but the turbines are enclosed in steel casings and placed at the bottom of the intake shafts about 160 ft. below ground level. The vertical shafts are cut through solid rock and provided with liners of steel pipes with an internal diameter of 11 ft. 6 in.; and with flanged connection to the turbine casing. There are five units with an average capacity of 12,500 hp. each under a net fall of 163 ft. running at 225 r.p.m. The turbines are of the double type with two runners, discharging into the common suction casing. The power house is also blasted out of solid rock, and is 36 ft. wide and 310 ft. long, communicating with the turbine chambers through the short tunnels which accommodate the shaft extension connecting turbines and alternators.

The roof is supported on a strong concrete arch, and by the provision of false walls and roofs leaving a space between the rock and the walls through which warm exhaust air from the generator is allowed to pass, all damp is prevented from penetrating into the power house. The generators have a normal output of 11,000 kva. and 10,000 to 11,000 volts 3-phase current. The necessary switchgear and transformers are also in this case placed in a separate building on ground level, a shaft providing communication between this building and the power house below, through which the heavy parts of the machinery can be lowered; in addition to which there is lift accommodation both for passengers and goods. The line voltage is 80,000, the power being utilized for railway traction and for mining purposes.

As an instructive example of the arrangement of the medium-pressure turbine with horizontal shaft and spiral casing, Fig. 4 shows a section through a unit of the Massaboden Hydroelectric plant used in connection with the Simplon Tunnel in Switzerland. Each unit is capable of developing 3500 b.hp. under a net head of 142 ft. at 500 r.p.m. The turbine is equipped with two runners cast back to back in one piece, the outside bearing being arranged with thrust collars to take up any unbalanced thrust in an axial direction.

The noteworthy fact with medium-pressure plants as in the case of low-pressure turbines, is that recent developments seem to favor single-runner units on account of higher mechanical overall efficiency and less foundation work, coupled with lower initial cost. The decision in each case must be made on its own merits and with a careful consideration of local conditions.

The outstanding feature in the development of the high-pressure Francis turbine is its adoption for use with increasing heads. Today Francis turbines utilizing a head of from 500 to 600 ft. are not uncommon, the highest fall for which a Francis turbine has been designed being approximately 745 ft.

The most important improvement in the Pelton wheel is in the system of regulation which was necessary for new conditions of electrical transmission, and the demand for accurate and reliable automatic governing. The inherent defect of the method of by-pass regulation being liability to stick and excessive wear, together with difficulty of insuring synchronizing action, resulted in the introduction of the deflecting nozzle. In recent years the combined spear and deflecting nozzle has come to the front and has now been adopted in most modern plants. A number of different designs of this method of operation can be traced to one of the three systems shown diagrammatically in Fig. 5. In each case the operating cylinder of the oil-pressure governor operates the deflector and spear simultaneously when opening, but by sudden closing of the governor the deflector will in the first instance cut into the jet and divert the water from the wheel until the spear by slowly overcoming the dashpot resistance, regulates the water supply corresponding to the load, when the deflector will be brought back into a position just tangential to the reduced jet. The free movement of the deflector, independent of the spear in the closing direction, is in each case permitted by the "lost motion" existing in the mechanical connection between the deflector and spear.

The principal development in the design of pipe lines has been the use of concrete for low-pressure lines where its low initial cost and durability give it a decided field of usefulness. In some places concrete pipes have been used as a pressure line when a special form of reinforcement and treatment to exclude leakage has been adopted. Wood-stave pipe line is of value because of its durability, low cost, and low coefficient of friction, and is therefore of practical advantage for the construction of the upper portion of pipe lines where suitable wood can be procured.

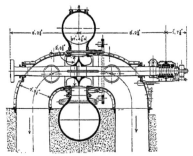

FIG. 4 ONE OF TWO TURBINES OF 3500 HP., MASSABODEN

The welded-steel pipe lines signify the most important developments in recent years because of the progressive utilization of high heads requiring close study of pipe-line construction. Welded pipe lines are exclusively used for high-pressure installations on account of their superior strength and absence of rivets which obstruct the flow of water. Welded pipes are heated by means of water gas and welded under high-speed, mechanically driven hammers which produce a weld of approximately 95 to 97 per cent of the strength of the full plate. After welding the pipes are annealed to remove all internal stresses. The foregoing process is only suitable for material up to about 1¼ in. thick, as above this thickness the heat would not penetrate sufficiently

to produce a uniform welding heat. For larger plate thickness, the "wedge-welding" method is resorted to, the edges being brought together and a separate bar inserted, forming the weld. With this method pipes up to a thickness of 1¾ in. can be satisfactorily welded. The material used in welded pipe lines is best open-hearth steel, with a tensile strength of 56,000 lb. per sq. in. and a elongation of 20 to 25 per cent in 8 in. On account of pressure surges plate thicknesses are calculated with a factor of safety of from 4 to 5, based on the strength of the weld equal to 100 per cent. Pipes are made in lengths of an average of 18 to 20 ft., using riveted joints for medium pressures and flange or expansion joints for high pressures. In the riveted, or so-called

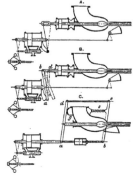

FIG. 5 SYSTEMS OF REGULATION FOR PELTON WHEELS

"bump" joint, the pipe ends are swelled so that the rivet heads do not obstruct the free area of the pipe.

A further improvement in the art of welding is reinforced welded pipe, consisting of a number of solid forged rings shrunk on to the outside of the pipe and adding further strength to the pipe with a reduction of plate thickness. This innovation in pipe design permits, for high heads, the use of a larger-diameter pipe without exceeding the maximum limit of plate thickness to obtain a reliable weld, thus reducing the number of pipe lines for large installations and the initial cost of installation. Although the present method of manufacture permits only short lengths of reinforced pipe to be made, these may be welded together with a circumferential weld into lengths of about 18 ft. In high heads this method of pipe-line construction makes an appreciable reduction in the initial cost of the development.

In his conclusion, Mr. Bergstrom shows the results of an estimate of the available horsepower in the principal countries of Europe and America. In the United States 28,100,000 hp. are shown as available, 24.9 per cent of which has been developed; 2,000,000 in Switzerland, of which 25.5 per cent has been developed; 4,000,000 in Italy, of which 24.4 per cent has been developed. Lowest percentages of development have been in Great Britain, with 8.3 per cent of 963,300 hp.; Spain, with 8.8 per cent of 5,000,000 hp.; and Austria-Hungary with 8.8 per cent of 6,460,000 hp. These figures apply, as stated above, only to Europe and the United States of America, and do not include the vast resources in white coal of Australia, Asia, Africa and the South American continent. (*Journal of the Institution of Mechanical Engineers*, Feb. and Mar. 1920, pp. 55-151, 51 figs. Also in *Engineering*, Jan. 30 (pp. 140-143), Feb. 6 (pp. 191-195) and Feb. 13 (pp. 227-232), 1920, from which journal the illustrations here used have been reproduced.)

# The Kaplan Hydraulic Turbine

THE Kaplan turbine, designed by Professor Kaplan of Brünn, Austria, is a development of the Francis type. In a Francis turbine the water acquires a certain predetermined velocity and direction, and then flows under the turbine wheel where it is deflected by the runner blades in such a manner as to pass into the suction pipe as far as possible without loss. The Francis turbine belongs to the class of pressure turbines; that is, the water possesses a certain excess pressure before it enters into the turbine wheel, and the pressure head is only partly converted into velocity in the guide apparatus.

The efficiency of the turbine depends, therefore, upon the blade angle and this latter in its turn is determined by the necessity of providing for the entrance of water as free of shock as possible. This again is accomplished when the runner blades are inclined with respect to the circumference of the runner in the direction of the flow of the entering water. In other words, if the water (Fig. 1) enters from the guide apparatus into the runner with the velocity of $w_i$, then it has only a velocity $v_i$, relatively to the runner moving with the

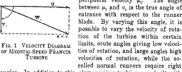

FIG. 1 VELOCITY DIAGRAM OF MEDIUM-SPEED FRANCIS TURBINE

peripheral velocity $\mu_i$. The angle between $\mu_i$ and $v_i$ is the true angle of entrance with respect to the runner blade. By varying this angle, it is possible to vary the velocity of rotation of the turbine within certain limits, acute angles giving low velocities of rotation, and large angles high velocities of rotation, while the so-called normal runners require right angles. In addition to this, slow-running turbines are built with runners of as large a diameter as possible, and fast-running turbines with runners of small diameters. In this way various types of turbines with their characteristic design are secured, and the original article gives illustrations of two such turbines, one for slow speed and one for high speed.

It has been usual heretofore in designing the runner blades to divide the turbines into so-called elemental turbines and to determine the proper dimensions of cross-sections for each line of flow, assuming the flow to be as free as possible from losses. It is claimed that Professor Kaplan has entirely deserted all these rules based on the so-called theory of "water-filament flow." He, in his theory of turbine design, attaches as much importance to the question of friction, which has been hitherto practically neglected, as has been done by others to the character of flow. According to his theory, other conditions being similar, it is the number of blades that has the determining importance. If it is too small, the flow of water suffers, and if it is too large, then the efficiency is reduced.

In an effort to create a turbine of maximum speed Professor Kaplan in the first place shortened as much as possible the length of the blades. Furthermore, he found that a large runner clearance does not harm in any way. This led him to locate the blades further and further back of each other until he obtained a runner having only actual flow. Finally, it is claimed that he succeeded by proper selection of the suction pipe in reconverting a large amount of the energy of the water at discharge into pressure.

All these changes led to the design of the new form of turbine such as is shown diagrammatically in Fig. 2. Contrary to what happens in a Francis turbine, the water flows throughout through the runner in an axial manner and the deflection of the water in the runner is eliminated. The guide apparatus is designed in the same manner as in the conventional Francis turbines, but the vanes are so arranged that the water discharges not only along the longitudinal edges, but also at the front edges. The guide-wheel cover has been made flat. According to the prevailing opinion, the water left to itself after its exit from the guide apparatus should lose most of its velocity through turbulence in the annular space formed

by the flat cover, and at first glance it would appear that the dead space at $E$ would unfavorably affect the flow of water. The two lines of flow $S_i$ and $S_i$ show, however, in what a powerful manner the water is deflected between the guide wheel and the runner.

It is claimed that the Kaplan turbine has several advantages of design, of which the following may be mentioned:

The turbine shaft can have its bearings located nearer the center of gravity of the runner than is the case in the Francis turbine. This results in a stable, comparatively light construction, avoiding the difficulties due to unbalancing which are encountered in Francis high-speed turbines.

The runner is somewhat smaller in diameter than the draft tube at its narrowest place. This makes it possible to insert the runner into the casing both from the draft-tube side and from the guide-wheel cover side, contrary to what is the case with high-speed turbines, which can be installed only from the draft-tube side, which is not always easy to do.

The diameter of the runner boss can be kept comparatively small; likewise the free blade area $F$ is considerably smaller than in the Francis turbine. The weight of the Kaplan runner is therefore very much smaller, in fact, it is claimed to be

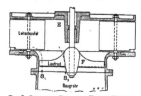

FIG. 2  SECTIONAL VIEW OF A KAPLAN TURBINE
(Leitschaufel, guide vane; Laufrad, Runner; Saugrohr, Draft tube.)

on the average only about one-fifth of the weight of a Francis wheel of the same diameter. Furthermore, the difficult method of holding the blades is eliminated, as a result of which it is claimed that a Kaplan runner can be made in about one-fourth the time required to make a Francis runner.

The principal advantage of the new construction is claimed to lie, however, not in these advantages affecting construction, but in the fact that with a Kaplan turbine speeds become attainable which are beyond the capacity of the Francis turbine, in addition to which the efficiency of the turbine at various loads varies in a more advantageous manner than with the conventional turbines. These points are further elucidated in the following manner:

It is important to have hydraulic turbines run at as high a speed as possible, because this permits a more economical utilization of the electrical generators usually connected to them; and it appears that the most economical speed of rotation for the generator is from 200 to 300 r.p.m. for power outputs of from 500 to 5000 hp., and from 400 to 600 r.p.m. for power outputs of from 100 to 500 hp.

These speeds are beyond the range of Francis turbines, unless the head of water is very large and the volume of water small. This results in uneconomical plants for low water heads, and in illustration a calculation is given showing that in a low-head plant with a turbine having an output of 5000 hp. and running at 83 r.p.m., the generators at peace-time prices would cost 150,000 marks (say \$37,500), while if the turbine could be run at 250 r.p.m. the cost of the generators would be only 100,000 marks (say \$25,000). Furthermore, the high-speed turbine, all else being equal, would be built much cheaper than the low-speed machine.

The low-speed machine is uneconomical from another point of view: namely, that where large volumes of water have to be handled a large number of units have to be used, since otherwise the velocity would have to be excessively low and the units excessively large. Numerous small units, however, are always more expensive than a few large ones.

A good insight into the question of the speeds available with various turbine types is obtained by employing the conception of "specific speed," which may be called "unit speed." In this the volume of water handled and the head of water for a given turbine are eliminated, and a unit speed determined for a given type of runner. By such a unit speed is understood the speed at which a machine similar in construction to the given turbine would have to run when delivering 1 hp. under a head of 1 m. If a turbine delivering $N$ hp. with a head of $H$ runs at $n$ revolutions per minute, its unit speed of rotation is

$$n_s = n \frac{\sqrt{N}}{H\sqrt{H}}$$

The following unit speeds are given for the conventional types:

    12 to 50 for Pelton wheels
    50 to 100 for Francis slow-speed wheels
    100 to 200 for Francis normal-speed wheels, and
    200 to 300 for Francis high-speed wheels.

As compared with these, a very much higher figure is possible with the Kaplan turbine: The first of these turbines built in Sweden had unit speeds of from 500 to 600, the turbine tested at the Technical High School in Brünn, 900, and in recent tests speeds of 1200 to 1600 have been obtained with good efficiency.

### EFFICIENCY OF TURBINES

Francis turbines have a very high efficiency, exceeding 80 per cent. This efficiency depends, however, on the type of the turbine, high-speed turbines having usually poorer efficiency than low-speed. Fig. 3 shows how the efficiency varies at various loads for a Francis turbine running with a unit speed of about 200. As is usual, this turbine has been designed to work with an average volume of water equal to three-fourths of the maximum volume, and at 0.75 load it attains its highest efficiency, which falls off materially with smaller load.

If this is compared with the brake diagram of a Kaplan turbine built in Sweden (Fig. 4), it can be seen that notwithstanding the high unit speed, namely, 750, the efficiency is not only very good in itself, but remains practically the same over a range of from about one-half to full load.

In this connection the curve shown in Fig. 5 is of considerable interest. This was secured from a brake test of a turbine installed in lower Austria and designed to operate at a unit speed of rotation of about 700. The wheel has a diameter of only 600 mm. and was designed to deliver about 35 hp. with 1.1 cu. m. of water per second and a head of 3 m. As shown by the curve, the efficiency rises from full load to one-half load, and even at 40 per cent. of the specified volume of flow it is

still close to 83 per cent. Brake tests were made on the turbine under the same load but with various heads of water and various speeds of rotation and these tests have also shown a remarkable uniformity in efficiency.

This lack of sensitiveness of the Kaplan turbine to variations in the volume of flow and in the head is particularly important in low-head installations, as it makes it possible to build the turbine for the maximum available volume of flow and then operate on smaller volumes and yet attain a high efficiency. In

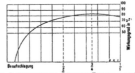

FIG. 3  EFFICIENCY OF A FRANCIS TURBINE $n_s = 200$

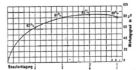

FIG. 4  EFFICIENCY OF A KAPLAN TURBINE $n_s = 750$

FIG. 5  EFFICIENCY OF A KAPLAN TURBINE $n_s = 700$
(*Beaufschlagung*, load; *Wirkungsgrad*, Efficiency; 1*lr./sk.*, liters per sec.)

addition to this, it is not necessary to go to the trouble of providing high-water reserves as the Kaplan turbine can very well stand overloads.

An interesting feature of this type is that in many cases it is very easy to convert a conventional Francis turbine into a Kaplan turbine, for only the runner has to be changed, the guide apparatus remaining the same. (*Zeitschrift des Bayerischen Revisions-Vereins*, vol. 24, no. 9, May 15, 1920, pp. 71-73, 7 figs.)

# Hydraulic Section of Report of the National Electric Light Association Committee on Prime Movers

THE important feature of the past year's development is the increased size of water wheels installed. The growth of operating systems makes feasible this increased size, individual units of approximately 40,000 hp. capacity having been installed and larger ones being contemplated. One large manufacturing company reports the starting of a total of nearly one-quarter of a million kilowatts in large hydroelectric generators in units of over 12,000 kw. each. In addition there are a large number of smaller vertical direct-connected wheels being installed as additional or replacement units.

### PROGRESS AND DEVELOPMENT OF HYDRAULIC PRIME MOVERS

The tests of the large water wheels in place show remarkably

high efficiency, which makes it profitable in a great number of cases to replace old installations with new. The new large units are uniformly distributed over the United States. Many of the older hydraulic developments which used dams and canals, supplying widely scattered individual water wheels, are now planning a single hydroelectric station to utilize the entire fall and transmit the power electrically.

Several large units have been started and continuously operated for over a year without a single shutdown on account of wheel troubles.

A new type of water wheel for low-head plants using an exceptionally high-speed runner has been developed by Forrest Nagler of the Allis-Chalmers Company.

Four 15,000-hp. tangential wheels for operating under an effective head of 1008 ft. have been ordered by the Great Western Power Company. These wheels contain 21 buckets, each 36 in. wide and 24 in. deep, and water is supplied from 13 nozzles. They are to operate one each overhung from the ends of a 22,200-kva. generator shaft running at a speed of 171.4 r.p.m. It may be of interest to state that several water-wheel manufacturers designed and recommended Francis-type wheels for this head at a speed of either 600 r.p.m. or 750 r.p.m. The operating engineers, however, were afraid that the erosion on the clearance passages and gate mechanism might be rather severe when operating under a pressure of over 430 lb. per sq. in. This, coupled with the fact that it would require a turbine-type generator with smooth-core field which was not suited for charging conditions on their long transmission line, led them to choose the impulse wheel.

There are at present under construction some 800-ft. Francis wheels for the Southern California Edison Company, to be run 600 r.p.m. and deliver a maximum horsepower of 22,500.

Automatic hydroelectric plants are receiving considerable attention. These plants operate without attendants, and the gate openings and output are regulated according to the water supply available, by means of a float control from pond level. These plants may be tied into a local power system or be run in connection with a steam station which takes care of the fluctuation of load and varying flow of stream.

There is a radical tendency, where climatic conditions are favorable, to confine the power-station building to the housing of control apparatus only, locating generating apparatus, step-up transformers and high-tension equipment outdoors.

General practice is divided between the centralization of auxiliaries and a tendency to make each large unit self-contained with its own exciter, governor, pumps, etc. A satisfactory solution of this problem is accomplished by having individual speed-control sets for each of the main units and by having an emergency pipe connection from each individual set to each adjacent set. Each main oil-pressure set is of sufficient size to operate two main units under emergency conditions. Thus all of the advantages of the central oil-pressure system are combined with the advantage of the individual system.

## HYDRAULIC OPERATION AND COMBINED OPERATION OF HYDRAULIC AND STEAM PLANTS

The combined operation of hydroelectric and steam-electric generating stations has become a subject of vital interest to the central-station industry. It will be increasingly important in the future, since the marked trend toward interconnection will bring more and more steam plants into parallel operation with hydraulic plants, and will make possible the development of water powers which cannot now be utilized economically. Yet intensive study of the subject is comparatively new. Individual companies and systems have made more or less careful analyses of their own conditions and problems, but there has been little information exchanged or published. During the past year the committee has considered or has had brought to its attention a large number of problems. These are listed below with the idea of having before the committee an outline to be followed in future work.

The problems of combined operation and hydraulic operation may be classified under the three heads outlined below: namely, (1) problems of the combined system; (2) problems of the steam plant; (3) problems of the hydraulic plant. Some problems of purely hydraulic operation are included in the third group, although they do not depend on combined operation.

### I—PROBLEMS OF THE COMBINED SYSTEM:

A—Parallel operation of coöperative systems:

1 Contract relations between hydraulic and steam plants
2 Advantages and disadvantages of interconnection
3 Division of load between hydraulic and steam plants
   (a) High-flow operation
   (b) Low-flow operation

4 Division of wattless current between hydraulic and steam plants
5 Frequency regulation and governing
6 Relative reliability of hydroelectric and steam-electric service.

B—Parallel operation of non-coöperative systems:

In this group there are other problems which are chiefly electrical and do not fall within the scope of the Prime Movers Committee, but it should be mentioned that a complete study of combined operation would include the following problems: Increased damage to property and interruption to service which may result from greater capacity feeding in so short-circuits; protective methods; the use of reactance to limit short-circuit currents; the effects of such reactance on voltage and synchronizing power; voltage regulation; synchronizing power of tie lines.

### II—PROBLEMS OF THE STEAM PLANT:

1 Standby service
2 Minimum load on steam turbines
3 Time required to start up steam turbines
4 Shutting down steam plant during "off-peak" hours
5 Banking practice as related to combined operation
6 Novelties and difficulties in operating
7 Station design as affected by combined operation.

### III—PROBLEMS OF THE HYDRAULIC PLANT:

1 Efficiency measures to secure maximum output or water economy.
2 Measurement of water
3 Dispatching of water and hydrographic methods
   (a) When the water is used by only one plant
   (b) When the same water is used by two or more plants
4 Ice troubles and methods of fighting same
5 Power requirements and operation of auxiliaries
6 Planning of maintenance, inspection, and repairs
7 Operating rules and procedure
8 Novelties and troubles in operating
9 Station design as affected by combined operation.

From this analysis of the problems of the combined operation the report considers the points in detail, outlining the information required for further reports and giving some data that have been discovered in the solution of the subdivisions of the problems. The report states that every system must constitute a separate problem, but it is believed that many problems and conclusions will have a common application to a large number of systems. In abstracting the remainder of the report, material will be included only where new information has been submitted:

### I—PROBLEMS OF THE COMBINED SYSTEM: A—PARALLEL OPERATION OF COÖPERATIVE SYSTEMS

1 Contract Relations Between Hydraulic and Steam Plants. The ideal in every combined system should be minimum overall cost of delivery to the ultimate consumers. The solution of the problem, where hydraulic and steam stations are separately owned, lies in some sort of flexible arrangement on a coöperative or cost- and profit-sharing basis in which rigid limits of power and energy are disregarded and the maximum output of the hydraulic plant is utilized as far as possible. Such a solution furthers the conservation of our natural resources in which the central-station industry is vitally interested.

3 Division of Load Between Hydraulic and Steam Plants. Under this head the report calls attention to some instances where steam generation is necessary in spite of the fact that the hydraulic plant is carrying the load. Such steam generation may be necessary for maintenance, economy or safety of steam-plant equipment, or reliability of service. Unnecessary steam generation may be caused by faulty governing or load dispatching. Under a discussion on the unnecessary steam generation due to faulty governing or load dispatching, the report points out the necessity for a graphic meter giving total station out-

put for each generator station, and the necessity of means of communicating output figures from all generating stations to one central point.

The importance of not having the hydraulic capacity of the turbine limited by the generator capacity is also emphasized.

4 *Division of Wattless Current Between Hydraulic and Steam Plants.* In the case of a hydraulic plant operating in parallel with the steam plant, the problem consists of comparing aggregate steam costs for high load factor but increased steam generation due to hydro losses, with steam costs for a poor load factor but a smaller amount of energy output. If the hydro plant has large storage or high head and little tailrace backwater, so that the reduction in head due to low load factor is not great, it will usually be found that it is most economical for the hydro plant to take the poor load factor. The result may be reversed if the percentage loss of delivered output from the hydro plant is considerable and if the percentage increase of steam costs due to lower load factor and variations is negligible.

5 *Relative Reliability of Hydroelectric and Steam-Electric Service.* The report is quite positive in its statement that in recent years particularly, the individual hydraulic prime mover is more reliable than the individual steam turbine. The present-day single-runner vertical-type water-wheel-driven generator equipped with a Kingsbury bearing seems to give almost perfect continuity of service. It has been nearly always found that hydro service can be restored before the steam plant can pick up a large proportion of the load even with a number of boilers held ready under "hot bank." As a result, with modern conservative construction of transmission lines, there is very little pure stand-by service in the East and Middle West. On the Pacific Coast, where use of fuel oil makes quick firing possible and transmissions of great length with considerable inherent voltage drop, extensive stand-by service is more common.

## II—PROBLEMS OF THE STEAM PLANT

1 *Stand-by Service.* It is stated that in coal-burning plants underfeed stokers with forced draft and fan and ram speed control, or chain grates with forced draft may be found most satisfactory for quick starting. Future developments will be along the line of using oil and coal under the same boiler, and of adapting pulverized fuel for large central-station use.

The example of the municipal plant of Zurich, Switzerland, is cited as demonstrating that during off-peak hours stand-by boiler pressure may be supplied electrically from the hydroelectric plant. In this case, the use of electricity for which there was no market was more economical than paying for banking coal.

2 *Minimum Load on Steam Turbines.* Motoring, even with a certain amount of steam entering the turbine, dropping below a certain minimum turbine load, and sharp fluctuations of load are regarded as injurious to large high-speed steam turbines by many operating engineers in Eastern plants.

On the Pacific Coast, however, the steam turbines have been floated on the line with practically no load, and in some cases, as synchronous condensers, with less than no load steam entering the turbine without apparent damage.

4 *Shutting Down Steam Plant During "Off-Peak" Hours.* To avoid the difficulties caused by accumulation of water in pipe lines and leaky joints, occurring in the steam plant during shutdowns, some plants operate at least one boiler at near normal rating during the off-peak hours and use the steam thus generated to operate a small unit, circulating the steam through all the steam mains before it reaches the unit. In other cases the generating plant is shut down entirely, but steam is blown off for some time near the turbine inlet before the unit is started up.

5 *Banking Practice as Related to Combined Operation.* A steam plant carrying the peak loads of a combined system operates at a poor load factor and therefore has an increased amount of banking. The report gives the result of a 42-hour banking test on a 822-hp. Stirling boiler with underfeed stok-

ers as 0.112 lb. of 13,500-B.t.u. coal per rated boiler hp. banking hour. The pressure being allowed to drop about 15 lb. below normal, coal was added after 16¾, 23¾ and 40 hours. The steam generated after the bank had nominally begun was approximately measured and an equivalent amount of coal deducted. In a similar test on oil-burning boilers a value of 0.046 lb. of oil per hp-hour was found. This is equivalent in heating value to 0.063 lb. of 13,500-B.t.u. coal.

6 *Novelties and Difficulties in Operating.* The percentage changes in load on steam plant are very greatly increased when the hydro plant takes flat load during high flow. This condition makes a foreknowledge of load variations essential to avoid excessive blowing off through safety valves or reduction of hydraulic output. To assist in estimating load in the boiler room, the Ashley Street Station of the Union Electric Light and Power Company of St. Louis operating in parallel with the Mississippi River Power Company and equipped with a number of varying sizes and types of generating units of which a kilowatt load does not give an accurate idea of the boiler room, has adopted a novel unit called the "kilber." The kilber is 1000 lb. of steam per hour. The boiler room is regulated by the number of kilbers required, an estimate which is furnished from the switchboard at half-hour intervals.

## III—PROBLEMS OF THE HYDRAULIC PLANT

1 *Efficiency Measures to Secure Maximum Output or Water Economy.* The report contains a detailed consideration of the operation of the Holtwood Station for information under the above heading. This station operates most of the time in parallel with one, frequently two, and rarely with three, steam-generating stations. Energy is transmitted at 70,000 volts to Baltimore, Md., 40 miles distant, and to Lancaster, Pa., 15 miles distant.

The Holtwood Station is on the Susquehanna River, the flow of which varies from about 3000 to about 700,000 cu. ft. per sec. The average height of the dam is about 55 ft. and flashboards of a height of 2½ ft. or 4½ ft., according to the season, are usually installed on the crest. Except during high floods and extreme low flow, the working head varies from about 47 ft. to 62 ft. The area of the pond at the elevation of the dam crest is less than four square miles. There are eight main turbines of the vertical-shaft, inward and downward flow Francis type. The total of the turbine ratings is 118,000 hp. (ratings are not, however, at the same head). The actual maximum sustained capacity of the plant is about 83,000 kw. Seven of the turbines are set in wheel pits with two runners per shaft. The eighth and newest unit is a single-runner turbine set in a reinforced-concrete scroll case.

Tailrace conditions are such that, with no flow over the dam, the tailrace elevation at the power house is about 14 ft. higher at full load than at no load. (*Report of Committee on Prime Movers*, presented at 43d Convention of the National Electric Light Association, May 18-22, 1920, pp. 105-128.)

## Short Abstracts of the Month

### ENGINEERING MATERIALS

STUDY OF IMPACT TESTS ON ALLOYS, Austin B. Wilson. Impact tests are of interest as comparatively little is known of the properties of metals under this type of stress. In the present case tests were made of three distinct classes, namely, McAdam impact-shear tests with unnotched bars, Fremont direct impact tests with notched bars, and Landgraf-Turner alternating-impact tests.

Several series of tests of each kind were made. At first machined bars were used, but the tests indicated that it was not necessary to machine the bars. Second, the bars were heat-treated by quenching from 920 deg. cent., reheating to 600 deg. cent., and furnace-cooling. This, it was thought, would increase the resistance to impact-shear, but such was not the case. In

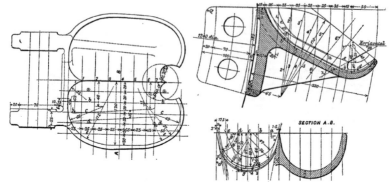

FIG. 1  LAYOUT FOR PELTON TURBINE BUCKETS

general, it was found that the aluminum bronzes tested excelled any of the other bronzes of the series in resistance to impact by shearing.

Aluminum bronze also showed high resistance to direct impact where *notched* effect was present; likewise it surpassed other bronzes in resistance to failure through fatigue.

Data of the tests are presented in the original article in the form of tables but no analyses of the materials tested are given. (*Foundry*, vol. 68, no. 352, August 1, 1920, pp. 616-617 and 622, 3 figs., *e*)

## HYDRAULIC ENGINEERING

### Pelton-Wheel Bucket Design

PELTON WHEEL RECONSTRUCTION, Percy Pitman. Description of a reconstruction carried out during the war period. The work was difficult as it had to be carried out on the turbine as it stood

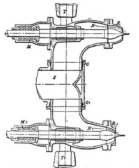

FIG. 2  PERCY PITMAN GOVERNOR FOR PELTON WATER TURBINE

in the power house and only the top cover could be removed, which made the position in which to work very cramped.

The turbines were of the Pelton type made by Ganz and Com-

pany of Budapest, and were installed in 1907, to work with a head of water of 1010 ft. Each Pelton wheel consisted of two 4-ft. diameter steel disk runners and each wheel was fitted with 22 staggered buckets. Both the buckets and the needle and nozzle

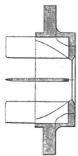

FIG. 3  EXPERIMENTAL NOZZLE FOR USE IN PELTON WATER TURBINE

were of an extremely bad shape and were continuously breaking, which was particularly dangerous in view of the high peripheral velocity employed.

The portion of the paper describing the new installation is of particular interest because it presents a complete layout for the new buckets, a thing not often found in engineering literature.

The bucket, Fig 1, is of the usual Pelton shape with the exception that it is slightly curved on the dividing wedge, and also on the bottom face of the bucket in a radial direction. The purpose of this is to give a larger radius of effective action after the bucket has passed the theoretically best position; that is, when the bucket face is at right angles to the entering stream.

The drawings show the points from which the various radii are struck. The methods of setting out shown are applicable to almost the whole range of impulse waterwheels, providing load conditions, head of water, size of nozzle, and speed required are taken into consideration.

The author designed a separate governor for the automatic regulation of the needles to enable the plant to take care of load changes, and at the same time economize water. This is shown in Fig. 2. The bad stream lines made by the existing steel nozzles $R$ are shown by dotted lines $N$. The new bronze nozzles $R_1$ give a much straighter path $N_1$ through the nozzles. $T$ and $T_1$ are the trunnions and $G$ and $G_1$ the glands, while $S$ is the T-shaped supply pipe. The springs $M$ and $M_1$ shown in Fig. 2 were designed to improve matters, but unfortunately up to the present the plant could not be spared the necessary length of time to carry out such a desirable improvement. The old Ganz buckets, needle and nozzle with 420 lb. pressure on the gage gave 970 kw. corrected, and the new buckets with the existing steel nozzle and spear (which was not made by the writer) gave 1082 kw., and after allowing a substantial correction factor of 4 per cent, it showed an increase of 70 kw., or a net increase in efficiency of 7.3 per cent.

Experimental nozzles were also made of bronze, into which were dovetailed four rustless-steel blades which were ground and highly polished up to a thin knife-edge on the inside, as shown in Fig. 3. These nozzles proved to be a big improvement, producing a jet of extraordinary solidity and transparency, the water for about 2 ft. issuing almost like a glass rod. The electrician in charge of the power station claimed that there was an increase of efficiency of 5 per cent due to this improvement.

Ordinarily the swinging or rotative character of the stream produced great vibration which seems to have produced fatigue of some sort in the steel blades, causing them to break off short at different intervals after they had been in use for some months.

An intensive analysis of the defects of the turbine as it existed previous to reconstruction is given in the original article. (*Engineering*, vol. 109, no. 2843, June 25, 1920, pp. 851-853, 13 figs, *dp*)

## INTERNAL-COMBUSTION ENGINEERING

### Cold-Starting Oil Engine

SOLID-INJECTION RUSTON AND HORNSBY CRUDE-OIL ENGINE. The Ruston and Hornsby engine is of the so-called cold-starting type, and depends upon the heat of compression for the ignition of the charge. It is claimed that a fuel consumption of below 0.4 lb. of fuel oil per b.hp. at full load has been obtained without difficulty.

The engines are built of the single-cylinder type in 11 sizes, ranging from 15 b.hp. to 170 b.hp., and of the double-cylinder type in 5 sizes, from 100 b.hp. to 340 b.hp.

The fuel pumps and valve gear for both cylinders are driven from a single lay shaft running along the right-hand side of the engine, the eccentric operating the inlet, and exhaust valves being keyed on the extreme end of this shaft. The motion of the eccentric is transmitted to the valves by means of rolling levers (Figs. 4 and 5) which give a more rapid opening and closing motion than would be obtained by the eccentric motion alone. The small handwheel seen near the eccentric is for the purpose of relieving the compression by raising the exhaust valves slightly from their seats.

When working with the lowest grades of tar oil, it is necessary, in order to insure proper ignition of the charge, to inject just ahead of the main charge a very small quantity of crude or other more inflammable oil. This only amounts to

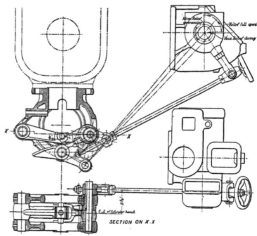

SECTION ON X.X

FIGS. 4 AND 5 FUEL PUMPS AND VALVE GEAR

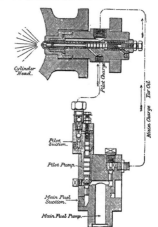

FIG. 6 ATOMIZER AND DOUBLE FUEL PUMP, RUSTON AND HORNSBY HEAVY OIL ENGINE

about 5 per cent of the total fuel consumption at full load. It is called the pilot oil and is forced to the atomizer independently of the main supply by means of a separate cam-driven pump.

The construction of the atomizer and double fuel pump is

shown in Fig. 6. The single main pump primarily deals with the main supply of tar oil. It takes in fuel through the main suction valve and delivers it to the atomizer through the discharge valve in the usual way. The by-pass valve controlled by the governor is not shown.

Above the main suction valve is the pilot valve, which consists of two plungers of different diameters, the larger one being acted upon by the pressure of the main fuel. At every stroke of the main pump the upper and smaller plunger is raised against the action of the spring by the pressure of the fuel oil on the lower one. The upper plunger therefore acts as another pump on its own account, and by means of its own suction and discharge valves it forces the pilot oil through the atomizer separately from the main supply.

With such an arrangement the amount of pilot oil delivered at every stroke can be regulated by varying the stroke of the pilot pump. Also, the pilot oil is necessarily injected immediately in advance of the main fuel oil, which is the required condition. (*Engineering*, vol. 110, no. 2844, July 2, 1920, pp. 7-10, serial article, 22 figs., *d*)

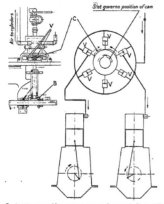

FIG. 7  AIR-STARTING MECHANISM OF THE INGERSOLL-RAND P-R OIL ENGINE

### American Heavy-Oil Engine, Intermediary Between Diesel and Semi-Diesel

INGERSOLL-RAND P-R OIL ENGINE. These engines are of the heavy-oil type and yet neither straight Diesel nor hot-bulb. The P-R is of the low-compression self-igniting type, but without hot plate or hot spot.

The engine is started by compressed air through a distributor (Fig. 7) which consists of six air valves *V*, one for each of the cylinders. Below these air valves is a cam *C* which is operated by a gear *B* from the side shaft. When the air is turned on, the six valves are driven down by the air against the action of springs and are closed, with the exception of one valve which strikes the top of the cam. This valve being held open by the cam, permits the air to pass to the cylinder which it controls. The engine turns in a given direction and the cam likewise turns, opening each valve in succession. To reverse the engine, it is stopped, the cam is moved to its opposite position, and the air valve opened.

When the engine is absolutely cold, the ignition at the start is assisted by means of a small electrical starting element located in the combustion chamber and heated to a dull red. This heating element is used only for a few revolutions and then discontinued. It does not act as a hot point or hot bulb, and after the engine is warmed up, the heating element may be removed and the engine started again without it.

Ignition of the fuel when the engine is in operation is produced by combination of a combustion chamber of peculiar shape, arrangement of spray nozzles, and timing of fuel injection. As regards this latter, it is stated that the intake valve is closed by the valve spring, the piston in its turn compressing the air from the cylinder into the combustion chamber to a pressure of approximately 200 lb. per sq. in. Injection of the fuel starts near the end of the stroke and is completed before the piston has reached the end of its travel. The solid system of injection is used. (*Automotive Manufacturer*, vol. 62, no. 3, June 9, 1920, pp. 7-12, 6 figs., *d*)

## MACHINE SHOP

### Hints for Working Monel Metal

WORKING MONEL METAL, Hugh R. Williams. Monel metal cuts like no other metal. In fact, so far as cleaving action of the cutting tool is concerned, it is more like brass than like steel. It is removed in ribbons and because of its toughness, high-speed-steel cutting tools with keen edge and decided rake have to be used.

For monel-metal cutting only the better grades of high-speed steel should be employed. As a general rule, monel metal can be machined dry, though cutting lubricants and cooling solutions may be necessary for fine work. Table 1 gives a list of such lubricants for monel metal.

Monel metal can be machined effectively at a wide range of cutting speeds from a slow speed of 8 ft. or 10 ft. per min. with a heavy cut and feed to as high as 250 ft. per min., with a light cut and feed, provided ample power is available. As a rule, on general work a speed of 50 ft. or 60 ft. per min.,

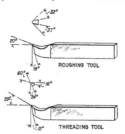

FIG. 8  TOOLS FOR WORKING MONEL METAL

with a ⅛-in. cut and a 1-32-in. feed, will be found to be satisfactory, but if a high finish is desired the depth of the cut should be decreased and a higher cutting speed employed, care being taken to keep the tool sharp.

The following instructions are given for polishing and grinding monel metal:

*Castings.* (1) Use a solid stone, of which there are several grades and makes. The Norton Company's grade "Q," grain No. 20; and Carborundum Company's grade "G," grain No. 16, are very satisfactory. (2) A rag, wood, or canvas wheel coated with No. 40 emery. (3) A rag, wood, or canvas wheel coated with No. 120 emery. (4) A rag, wood, or canvas wheel coated with No. 120 emery and finished with an ordinary buff, using buffing compound.

*Hot-rolled Rods.* (1) A rag, wood, or canvas wheel coated with No. 90 emery. (2) A rag, wood, or canvas wheel coated

TABLE 1. CUTTING LUBRICANTS FOR MONEL METAL

| Operation | Lubricant |
|---|---|
| Boring (fine) | Mineral lard oil, lareul, lard oil and 10 per cent turpentine |
| Cutting-off | Soluble oils, solul, borax, and aquadag. |
| General | Mineral lard oil (light), lareul, solul, soluble oil s exsnol. |
| Milling | Soluble oils, solul, oakite. |

with No. 120 emery. (3) A rag, wood, or canvas wheel coated with No. 120 emery and finished with an ordinary buff, using buffing compound.

*Sheets.* (1) A rag, wood, or canvas wheel coated with No. 90 emery. (2) A rag, wood, or canvas wheel coated with No. 120 emery. (3) A rag, wood, or canvas wheel coated with No. 120 emery and finished with an ordinary buff, using buffing compound.

The original article also discusses in detail the welding, annealing and pickling of monel metal. (*Mechanical World*, vol. 67, no. 1746, June 18, 1920, pp. 400-401, 3 figs., *p*).

## MACHINE SHOP

### Tests on the Behavior, Efficiency and Life of Hack Saws

HACK SAWS, THEIR SELECTION AND USE. The ability to cut in the shortest time is one of the important factors determining the value of a hack saw, but the rate at which the saw cuts may

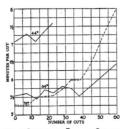

FIG. 9 EFFECT OF PRESSURE ON TIME PER CUT AND EFFICIENCY OF SAW

be made such as to become destructive. The endurance of the saw is also a factor.

The gradual lowering of efficiency as the saw is used and the effect of successive increases in the pressure are displayed in Fig. 9. The curve for 44 lb. shows how rapidly the saw was running beyond the limit of efficiency, while that for 64 lb. makes clear how the tendency was immediately restrained by the application of more weight. This figure also illustrates the effect of weight or pressure on the rate of cutting and the efficient endurance of the saw, identical blades being used on the same material in this test and different weights employed, the starting weight remaining constant throughout the test in each case.

In the case of the 44-lb. weight, the weight used was below that recommended for the blade employed. The second solid line shows the improvement in performance of a similar saw using the proper weight of 64 lb., while the dotted line illustrates what happens when a slightly excessive pressure is employed. Insufficient weight resulted in an excessive time per cut and on the whole the chart would indicate that, first, it is better to exceed the proper weight a little at the outset than to use too little pressure, and second, no matter how nearly correct the weight is at the outset, after a certain number of

cuts have been made the pressure must be increased, not only for the sake of reducing the time per cut to a point within the limits of efficiency, but also to prolong the life of the saw.

The effect of a slight excess in weight on the average time of cuts made before the blade is dulled is clearly shown by the two lower curves in Fig. 9. For the first fifteen cuts the advantage lies clearly with the heavier weight, but from that point on, it is apparent that the first saving in time was made at the expense of the general average for the entire series of cuts.

Another series of tests illustrated by curves in the original article indicates the destructive action of an excessive pressure. While the hack saw must be made to withstand a great amount of abuse, there are limits beyond which it will not go. Where a hack saw is forced to cut at a greatly excessive speed, the gain in time per cut may be offset by loss in saws, spoiled stock, etc. On the other hand, however, insufficient pressure affects the life of the saw almost as rapidly and in the same manner as when too much weight is used. In this case, moreover, the teeth of the blades are destroyed by slipping and sliding over the work rather than by cutting. It would appear,

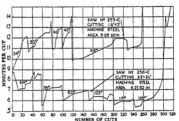

FIG. 10 EFFECT OF PROPER AND IMPROPER REGULATION OF WEIGHT ON CUTTING SPEED AND LIFE OF SAW

therefore, that using too little pressure is almost as inefficient and costly as using too much, and has not even the doubtful advantage of saving time at the expense of the blade and stock as is the case where too much pressure is employed.

Data are presented on the comparative advantage of flexible and all-hard blades. From this it would appear that the all-hard blades designed for use in machines require less time per cut than flexible blades, while the total number of cuts made by each blade is practically identical and the saws failed in almost the same manner. It would therefore appear that the use of flexible saws in power machines is attended by a loss of time in cutting without any corresponding gain in the life of the blade.

Data are also presented comparing the results of starting with an excessively light weight combined with proper subsequent regulation on one hand, and on the other hand with results obtained by using a slightly excessive weight to start with and then regulating it properly as the number of cuts progresses.

These results are presented graphically in Fig. 10 and from them it would appear that on the first blade (No. 255-C) an excessively light weight was used. The time per cut was excessive until about the thirtieth cut, when the weight was increased with a result in drop in time per cut from 30 min. 35 sec. to 10 min. 35 sec. The blade, however, was dulled so much by sliding over the work that it was not until the weight had been still more increased that the proper time per cut was reached.

In the second test (No. 256-C) a weight of 64 lb. was employed, whereas only 43 lb. is recommended. While overweighted at the start, the blade cut well within the limits of efficiency for nearly 40 cuts.

In presenting the data for these tests, the paper calls attention to the fact that the relative efficiencies of blades can be better determined by the number of square inches cut, rather than merely the number of cuts made. Thus in the above two tests, the first saw cut, before failing, 1365 sq. in. and the second saw 1742 sq. in.

Tests presented graphically in Fig. 11 are of considerable interest as they show the effect of the lubricant. Starrett No. 250 flexible blades were used and all were started under the same pressure of 24 lb. The line *AB* shows the performance of a saw run without lubricant at 65 strokes per min., an ex-

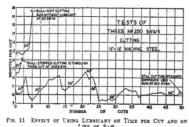

FIG. 11  EFFECT OF USING LUBRICANT ON TIME PER CUT AND ON LIFE OF SAW

cess of 15 strokes over the recommended speed. This saw failed when partly through the fifth cut.

The saw of curve *CD* was run dry at 100 strokes per minute, or double the proper speed, and failed very rapidly. The saw *EF* was run with the lubricant compound at 100 strokes per minute, and completed 50 cuts. The test was discontinued before the saw failed.

The following formula is given for making the lubricant compound: To a quart of sal soda thoroughly dissolved in 10 gallons of cold water add 4 quarts of equal parts of mineral and lard oil, and mix thoroughly. The compound is ready for use 10 or 12 hours after it is made.

The original article discusses the question of cost per cut and gives several tables, among others tables for combined cost per hour for saw, labor and overhead at different costs of labor and saw blades. The question of cornering work is also discussed in detail and various methods of proper and improper cornering are illustrated. (*Canadian Machinery*, vol. 24, no. 4, July 22, 1920, pp. 91-97, 19 figs., *peA*)

## MARINE ENGINEERING

### British Rivetless Ship

ELECTRICALLY WELDED SHIP "FULLAGAR." Description of the electrically welded, motor-driven ship *Fullagar*.

This ship is a coaster 150 ft. long by 23 ft. 9 in. beam, and 11 ft. 6 in. deep to the main deck. It is an entirely welded ship and there is not a rivet in her from one end to the other. Of course, this may not prove to be absolutely the necessary way of construction, but it was a good way of demonstrating the suitability of electric welding for shipbuilding purposes, although it did not demonstrate that it is necessarily the cheapest way of construction.

The hull is constructed of steel of the usual ship qualities, tested in accordance with Lloyd's rules in the ordinary way.

As a working rule, it was arranged that as far as possible no overhead welding should be called for on the job. Complete elimination of such welding, however, was impossible, but in almost every case it was found that joints could be planned so that such overhead welding as was done should be of a light type and should be reinforced by heavy welds put on in a downward direction.

Illustrations are given of the various types of welds used. Of

these, the system of "broken" welding is of particular interest. For joints not required to be watertight, the welding is put on in short lengths with unwelded spaces between instead of being continuous. The only reason for this is economy of labor and material. A sample of such continuous welding is shown at the connections between the main frames and the shell plates in Fig. 12A. Also, in Fig. 12B an arrangement is shown where the welds are placed alternately on either side of the frames.

In the welds shown in Fig. 12C (lower end of the main frames) the frame is welded all around to the plate. Service holes are also shown. These are, of course, employed in connection with erection, but in order to conserve the all-welded construction, they are not riveted after use, but are plugged and welded over.

The whole of the welding has been carried out by the quasi-arc

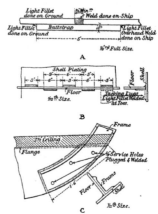

FIG. 12  WELDS USED ON THE ELECTRICALLY DRIVEN SHIP *Fullagar*

system, previously approved by the Lloyd's Register of Shipping, which latter has also embodied in the ordinary rules of the Society a set of rules for its application.

The body is fitted with a Cammellaird-Fullagar opposed-piston oil engine of about 540 b.h.p., working on a single screw. (*Engineering*, vol. 110, no. 2844, July 2, 1920, pp. 25-26, 14 figs., *dA*)

## MECHANICS

THE STRENGTH OF CARDAN SHAFTS, A Johnson. A mathematical discussion of the forces acting on a Cardan shaft. In this connection attention is called to a paper by Thom in a Swiss engineering magazine abstracted in MECHANICAL ENGINEERING, July 1920, pp. 409-410.

Besides the ordinary twisting moment, a Cardan shaft is subject to a bending moment due primarily to its own weight. It is in the position of a beam supported, but not fixed, at the ends, and uniformly loaded with *w* pounds per unit length.

The equation which defines the critical length of a shaft as given in reference books is:

$$2l = \pi \left( \frac{gEI}{wa^3} \right)^{\frac{1}{4}} \dots \dots \dots \dots [1]$$

where $2l$=whole length of shaft
$g$=acceleration due to gravity
$E$=Young's modulus

$I$ = moment of inertia of section
$w$ = weight of shaft per unit of length, pounds
$a$ = angular velocity of shaft in radians per second.

This formula unfortunately lends itself to ambiguity owing to the selection of incompatible units or careless interpretation of symbols. It is true only when all the dimensions are in *feet*, and consequently when $E$ is Young's modulus in pounds *per square foot*, or 144 times the ordinary tabular value of $E$ in pounds per square inch. Similarly when all four dimensions are taken in inches, the most convenient unit, it is necessary to remember that $g$ is not the abstract number 32.2, but represents a concrete quantity of *feet*, and therefore when 1 in. is the unit the formula for the limiting length of a shaft becomes:

$$2l = \pi \left( \frac{12gEI}{wa^2} \right)^{\frac{1}{4}} \dots \dots \dots [2]$$

where $2l$ = whole length of shaft, inches
$g$ = 32.2
$E$ = Young's modulus in pounds per square inch
$I$ = moments of inertia of section
　 = $\pi D^4/64$.for a solid round shaft when $D$ = diameter of
　　shaft in inches
$w$ = weight in pounds per inch length
$a$ = angular velocity of shaft in radians per second.

The author derives the following equation for the greatest bending moment $M$:

$$M = \beta^2 (Q - 2N) EI$$
$$= \beta^2 \frac{12g}{2a^2} \left( \frac{1}{\cos\beta l} - \frac{2}{(e^{\beta l} + e^{-\beta l})} \right) EI \dots \dots [11]$$

where $\beta = \left( \frac{12gEI}{wa^2} \right)^{\frac{1}{4}}$ and $Q$ and $N$ are constants.

Now the bending moment $M = fz$, where $f$ is the greatest stress per square inch of the material, and $z$ is the strength modulus of the section, which for a solid round shaft of diameter $D$ inches is $\pi D^3/32$.

Therefore we have from [11]:

$$f \frac{\pi D^3}{32} = \beta^2 \frac{12g}{2a^2} \left( \frac{1}{\cos\beta l} - \frac{2}{(e^{\beta l} + e^{-\beta l})} \right) E \frac{\pi D^4}{64}.$$

or

$$f = DE \frac{\beta^2}{4} \frac{12g}{a^2} \left( \frac{1}{\cos\beta l} - \frac{2}{(e^{\beta l} + e^{-\beta l})} \right) \dots \dots [12]$$

where $\beta^2 = \sqrt{\frac{wa^2}{12gEI}}$ as before.

From these equations the author derives Equation [2] by noting that the bending moment and the stress per square inch become infinite when $\cos\beta l = 0$, that is, when $\beta l = \pi/2$.

Thus the author has found that in a shaft of which the limiting length calculated according to Equation [2] was 32 in., a length of 30 in. showed a stress due to bending of only 1400 lb. per sq. in. as calculated from Equation [12], which would indicate that the limiting length of a shaft may be approached closely without danger of stress.

The author believes that if the statements embodied in Equations [11] and [12] have any real physical basis, a shaft need not be in danger at a speed higher than the critical speed if it could be restrained from whirling until the critical speed were well past. (*The Automobile Engineer*, vol. 10, no. 140, July, 1920, pp. 266-267, *tm.*)

## MOTOR-CAR ENGINEERING

FRAMELESS CAR. Description of a non-conventional design of a car recently patented by the Lancia Company. The body (Fig. 13) consists of a stamped sheet-steel shell and combines the function of a body with that of a frame. At the front of this shell is attached horizontally a form of stirrup, to which in turn is attached a semi-elliptic transverse spring, shackled at each end to the forks of a straight front axle.

Fore-and-aft location of the front axle relative to the body is obtained by the use of a radius rod on each side, running from

the bottom of the axle back to the body. Suspension for the rear axle consists of long quarter-elliptic springs, the butt ends of which are attached to the body at a point where a channel rib forming part of the front-seat support crosses and stiffens the body.

An essential feature of this construction is the use of a channel-section rib or tunnel running longitudinally from the front

FIG. 13　LANCIA FRAMELESS CAR

portion to the back and forming also a compartment open at the bottom in which the propeller shaft can move. Clearance for the rear axle is given by a similar tunnel running transversely across the back of the body. These channels may be made by stamping the sheet metal forming the lower portion of the shell to form a longitudinal member, thus improving the stiffness of the body without, to any extent, reducing the footroom capacity of the seating accommodation.

No information is given as to how accessibility to the engine is secured in this design. (*Autocar*, vol. 64, no. 1287, June 19, 1920, p. 1160, 2 figs. *d*).

## POWER GENERATION

EMMET MERCURY BOILER. The basic idea of the Emmet mercury boiler has been known for several years. It is to so use mercury as to increase the temperature range through which the engine operates. Mercury permits it to do so because of its higher boiling point as compared with water. On the other hand, however, if a boiler or turbine were to operate with mercury in the same manner as they are operated today with steam, the amount of mercury needed would be extremely large and there would be considerable danger of leaks, undesirable both because of the high cost of mercury and still more so because of the fact that mercury vapor is poisonous.

To avoid this, Mr. Emmet has designed a binary-vapor plant using mercury vapor and steam. The products of combustion from the furnace pass upward through part of the tubes which form the heating surface of the mercury boiler and pass on through tubes containing water. The mercury tubes are connected to the lower mercury header and to the mercury header which takes the place of the steam drum in an ordinary boiler. Mercury vapor at about 10 lb. gage pressure is collected in the header and passes to the mercury turbine. Owing to the high density and low velocity of the mercury vapor this turbine may be a single-stage machine of reasonably low speed and may have short buckets. The wheel may be placed inside the mercury condenser for simplicity.

The mercury condenser is a vital part of the unit. It consists of a tank supporting another tank which may be called a steam drum. A number of straight tubes extend from the bottom of the steam drum into the condenser. The exhaust from the mercury turbine is condensed on the surface of these tubes and as the boiling point of the mercury at 28 in. vacuum is 455 deg. fahr., steam is generated inside the tubes. This steam is led to a superheater and finally to the place where it is to be used, while the steam condensate is returned to a condensed-steam receiver and from there through a feedwater heater to the steam boiler.

The mercury condensate is drained from the bottom of the mercury condenser to the lower mercury header and thence to the mercury boiler.

In order to utilize mercury economically and to use a minimum amount of it, a special type of boiler had to be designed. In this boiler flattened tubes are used. Leaks are said to be prevented by great care in the design and construction of the fittings and pipes leading to and from the mercury boiler.

The original article states that Mr. Emmet has claimed that by the addition of this device to an assumed good modern power station with an increase of 15 per cent in the amount of fuel used, the same amount of steam can be supplied to the steam turbine as under present conditions, and the mercury turbine will generate power equal to about 66 per cent of the power generated by the steam turbine.

The experimental equipment was in operation for a short time last summer with a load of over 1000 kw. on the mercury turbine, and its operation showed that the economies predicted were fully realized.

In an editorial article in the August 3 issue of *Power* it is stated that there is a persistent rumor that an Emmet mercury-vapor plant is being built for the Hartford (Conn.) Electric Light Company. This would be of particular interest, as it was in the plant of this company that the first steam turbine of commercial size was installed for electric power generation. (*Power*, vol. 52, no. 5, Aug. 3, 1920, pp. 167-168 and 1 full-page colored diagram, *dA*)

WASTE-HEAT BOILERS, D. S. Jacobus and Arthur D. Pratt, Members Am. Soc. M. E. In modern practice with waste-heat boilers no special effort is made to minimize the frictional resistance of the gases flowing through the boiler by providing a large flow space.

Nowadays, waste-heat boilers are provided having a high draft resistance and in which the baffling is so arranged as to give a relatively small area of flow space through the passes so as greatly to increase the velocity of the gases. On the other hand, a suction fan is used to overcome the higher draft losses due to the increase in velocity of the gases, and the heating surface of the boiler is arranged in series in such a manner that in connection with the added draft loss under the increased velocity, the desired draft will be available at the outlet of the furnace supplying the waste heat as well as the outlet of the boiler.

In waste-heat boilers the heat-transfer rate should be materially higher than in direct-fired boilers. In the latter a temperature is developed greatly in excess of the average waste-gas temperature and a considerable proportion of the total heat absorbed by a boiler is absorbed through direct radiation to the tubes which are exposed to the radiant heat of the furnace.

The absence of this important factor may be compensated in the waste-heat boiler by raising the gas velocity. In this connection it should be borne in mind that the heat-transfer rate is dependent upon the gas velocity and not upon the number of passes, except in so far as the number of passes affects the velocity.

The best arrangement to use in most cases is the three-pass, but the number of passes depends generally on several considerations discussed in detail in the original paper.

Economizers are frequently installed in connection with waste-heat boilers.

The authors recommend a combination in which a three-pass boiler is used. The gases pass from the boiler through the vertical rear circulating tubes directly into an economizer which is made up of horizontally inclined tubes extending transversely to the boiler setting. The gases pass downward over the economizer tubes while the water has an upward flow, the countercurrent being obtained between the water and the gases, and the water in the economizer flowing continuously upward so as to avoid any pockets should steam be formed in the economizer.

Single-pass waste-heat boilers without induced draft were among the earliest to be used and there still appears to be a considerable field for such boilers, even under modern conditions. They can, however, be successfully used only in connection with a furnace requiring a low draft. A single-pass boiler may be used to advantage where a low weight of gas is available and may be best to install in the case of an isolated plant where there is outlet for the power developed expect for use in the plant. (*Proceedings of the Engineers' Society of Western Pennsylvania*, vol. 36, no. 4, pp. 221-234, and discussion pp. 235-242, *g*)

## VARIA

WAR EXPERIENCE WITH MERCHANT SHIPS, Prof. J. J. Welch. A paper discussing damage to vessels of various types caused by torpedoes or mines, and suggesting certain changes in design which might make the ship more fit to withstand external explosions.

As regards the nature and extent of damage caused by a torpedo or mine, attention is called to the fact that when a vessel is struck in way of a cargo hold, the hatches are usually blown off, which gives a certain relief of the pressure set up by the explosion. It would appear as probable that large hatchways afford an improved measure of protection to a ship.

The sizes of the holes made by torpedoes varied in the vessels which subsequently reached port from about 40 ft. by 23 ft. to 16 ft. by 14 ft. It would appear that a vessel should be reasonably safe against attack from a single torpedo if it were made capable of remaining afloat with any two holds open to the sea, while the minimum length of hold should not be less than, say, 40 to 45 ft., so as to avoid the possibility of two bulkheads being injured at the same time.

The first effect on a damaged ship was to produce a heel temporarily to the damaged side, due to the influx of water on that side, which heel was eliminated or reduced as the water became uniformly distributed across the ship. In some cases the loss of metacentric height, due to the loose water admitted to holds, was sufficient to render the ship initially unstable, and heeling resulted from this cause; such heeling being diminished, or the ship even regaining the upright, as the water in the hold reached the level of the water outside. This follows from the well-known fact that the loss of initial stability is generally greatest when the quantity of water admitted is relatively small. Treating the question as one of loss of buoyancy, with constant displacement, the position of C. G. of ship remains fixed throughout, while the variation in location of metacenter depends upon the rise of the center of buoyancy and the length of B. M. At an early stage of the inflow, the position of C. B. remains practically unaffected, while the diminution in the length of B. M. is given by $\mu i/V$ with the usual notation, $\mu$ being the fractional permeability of the damaged compartment. The value of $i$, the total amount of inertia of water plane area in damaged compartments, is more or less constant at all heights, and hence the loss of G. M. may be considerable in the early stages, and less important as the water rises in the compartment and so raises the position of the center of buoyancy of ship. For this reason it is more important to have watertight doors low down in the containing bulkheads of damaged compartment shut at the time of the casualty than similar doors at a higher level. The final result may be the same, but by confirming the longitudinal extent of the free water low down, the ship may continue upright, and if the ship has ultimately to be abandoned, the upright position will facilitate the getting away of the boats and so conduce to the saving of life.

An examination of the steamers torpedoed showed that very few capsized, but a fairly large number retained a list after damage. Some of these ships listed in a lightly laden condition, due to the fact that many vessels of relatively fine form have only very moderate metacentric heights when lightly laden, so that the loss of stability due to the inflow of water might have been sufficient to give a negative G. M. with consequent heeling; in one case, indeed, the loss of stability due to this cause was so great as to result in the capsizing of the vessel. On the other hand, vessels in full form carrying their breadth well down often have very large metacentric heights when lightly laden and many such vessels in this condition have continued afloat upright with two or more large compartments flooded. (Paper read before the Institution of Naval Architects, July 6, 1920, abstracted through *Engineering*, vol. 110, no. 2845, July 9, 1920, pp. 37-40, 2 figs., *dg*)

## CLASSIFICATION OF ARTICLES

Articles appearing in the Survey are classified as c comparative; d descriptive; e experimental; g general; h historical; m mathematical; p practical; s statistical; t theoretical. Articles of especial merit are rated A by the reviewer. Opinions expressed are those of the reviewer, not of the Society. The Editor will be pleased to receive inquiries for further information in connection with articles reported in the Survey.

## United States Forest Products Laboratory Celebrates Its Decennial

E XTENSIVE conservation of our national wealth of wood through more efficient utilization was the keynote sounded throughout the Decennial Celebration at the Forest Products Laboratory at Madison, Wis., July 22 and 23. Over 200 visitors from all parts of the country were present, representing every line of wood-using industry, including 59 wood-using associations and companies, 18 lumber-manufacturing associations and companies, the deans of 12 forestry schools, the United States Forest Service, and other visitors and friends. They came to pay tribute to the laboratory's ten-year record of service to American industry.

The Forest Products Laboratory is a government institution of industrial research in wood and all wood products maintained by the United States Forest Service in coöperation with the University of Wisconsin. In this work 220 engineers, wood technologists, manufacturing specialists, and assistants are employed in developing new uses for wood and improving present manufacturing methods. Investigations and experiments are undertaken both independently and for individuals and companies on a coöperative basis.

The committee in charge of the celebration consisted of Governor Philipp of Wisconsin, honorary chairman; H. F. Weiss, Burgess Laboratories, Madison, and former director of the laboratory, chairman; C. P. Winslow, director, Forests Products Laboratory; H. J. Thorkelson, business manager, University of Wisconsin, and Don E. Mowry, general secretary, Madison Association of Commerce.

The opening session of all the celebration convened at ten o'clock on the morning of July 22. The program of addresses included Legislative Measures for Forest Conservation, by Governor Philipp; Translating Knowledge into Power, by President Birge of the University of Wisconsin; and The Forest Products Laboratory, by C. P. Winslow, present director of that institution.

After luncheon the program continued with inspection of the work and exhibits of the laboratory. Guides in charge of small parties showed the visitors the various lines of endeavor in which the laboratory is saving millions of dollars a year to the people of the country.

Nearly 500 prominent lumbermen, manufacturers, and users of forest products, and members of the Forest Products Laboratory, attended the banquet in the evening, at which Lieut-Col. W. B. Greeley, Chief Forester of the United States, spoke on Forests and National Prosperity; and Mr. Max Mason, research specialist of the National Council of Defense, gave an illustrated talk explaining in detail the submarine detector which he perfected during the war and which was successfully used in European waters.

At the Friday morning session D. C. Everest, secretary and general manager, Marathon Paper Mills Company, spoke on Some Problems of the Pulp and Paper Industry; H. E. Howe, chairman, Research Extension Division, National Research Council, on America's Place in Industrial Research; and W. A. Gilchrist, representing the National Lumber Manufacturers' Association, on Some Problems of the Lumber Industry.

In the Forest Products Laboratory the Government has established an institution which is doing much direct good for all of the wood-using industries. Director Winslow in speaking of its purposes and work illustrated the manifold uses of wood in connection with the every-day life of the people. He brought out the great problems of conservation and utilization of forests, of cut lumber and of finished product. He stated that it was the broad purpose of the Forest Products Laboratory to aid the nation in solving these problems.

Mr. Winslow gave statistics showing by conservative estimate that the work of the Forest Products Laboratory effected an annual increase in production and decrease in waste aggregating $30,000,000. These figures, he said, should prove the value and importance of industrial research.

## THE ST. LAWRENCE RIVER PROJECT

(Continued from page 512)

It would seem, therefore, that the minimum power that may be developed from the St. Lawrence and distributed through northern New York and northern New England would be about 830,000 hp. It is not unlikely, however, that in the final bargaining between the two countries regarding the whole matter the United States may assume some of the expense of the improvement below the international section of the river and in return receive more power, inasmuch as the demand is relatively greater in the United States.

In a preliminary way the Canadian Government recently estimated the cost of the St. Lawrence improvement for the international section (including preparation for power installation but not the machinery and equipment for power development and distribution) at $60,000,000 and for the all-Canadian section of the river $50,000,000, not to include power development. The total, $110,000,000, may upon more detailed examination prove to be too small and $200,000,000 may be necessary. Suppose that the United States pays one-half of the total, $100,000,000, and gets one-half of the 4,000,000 hp. No matter how low the rental price on this may be, great profits should be derived and in a few years the entire cost amortized. Then there is the more fundamental question of conservation. A horsepower-year has been reckoned as equivalent to a coal saving as high as 30 tons; however, the author considers this too much and assumes 20 tons instead. The total coal to be saved is a very important item, and if 40 million tons of fuel can be saved, the additional tonnage necessary to transport it from the coal fields to the territory where the electric power would be used will be saved with it. This element of saving is variously reckoned at from 10 to 25 per cent of the primary coal tonnage. Considered internationally the saving would be predicated on the full 4,000,000 hp. and might amount to 100 million tons a year.

Again, there is the question of man power. How many miners are required to mine 100 million tons of coal, and how many mine laborers and others are employed in and about the mines? How many railway employees would be necessary to transport the coal and to haul the empty cars back to the mines? How many firemen and engineers in steam-driven power plants would be saved for other productive effort? The problem has many ramifications, but from whatever angle it is considered, a possibility of saving is at once apparent. The improvement of the St. Lawrence river is amply justified, even if considered solely as a conservation measure. Indeed, a careful study of the problem will show that there is no excuse for the people to neglect the taking of adequate steps to develop this power.

Returning to the question of transportation, if the United States is to continue shipping abroad yearly 300 million bushels of grain produced in the Middle West, how much of the shipping cost can be saved? Mr. Julius Barnes, who has just been relieved of his duties as Chief of the U. S. Grain Corporation, states that it may safely be estimated at from 5 to 6 cents per bushel, which would be $15,000,000 to $18,000,000 per annum. There are also other goods to be shipped in and out. That the total saving would be enormous and sufficient to justify the expense is undoubted.

Consider also the effect on immigration. At the present time the United States and Canada are suffering from a lack of man power. At the present time immigrants, particularly from the northern countries—Holland, the Scandinavian countries, and the British Isles—are very desirable. The cost of coming to our shores is now several times what it used to be, and when they land at some Atlantic port they must either stay in the already congested areas there or pay a big charge for railroad transportation. Some of the northwestern states have a large percentage of Scandinavian-born people. Practically all of the Scandinavians who come to this country settle in Wisconsin, Minnesota, and the Dakotas. Others will no doubt do the same. If they could come directly to some one of our Great Lakes ports, consider how much easier it would be for them and how much less expensive, not to mention the benefits of such immigration to the Central West.

# ENGINEERING RESEARCH

A Department Conducted by the Research Committee of the A. S. M. E.

## Engineering Experiment Station, Iowa State College

THE Engineering Experiment Station of the Iowa State College was established in 1904 and since that time it has issued 57 bulletins. Twenty-one of these have been devoted to sewage, drainage, drainage tiles and sewer pipes, nine to bricks and roads, nine to the use of Iowa coal for power and house heating and to mechanical-engineering subjects. In addition to these there have been bulletins on the holding power of nails, cement, concrete, electric power on the farm, topographical surveys, lighting, structural work and paints. For further information address Dean Anson Marston, Director, Engineering Experiment Station, Ames, Iowa.

### Coöperation

The Portland Cement Association and the Lewis Institute of Chicago are coöperating in their structural materials research laboratory. A number of bulletins have been issued as a result of this work. The Advisory Committee from Lewis Institute includes Profs. Duff A. Abrams and Philip B. Woodworth, while the Portland Cement Association is represented by Chairman F. W. Kelley of Albany, N. Y., and Ernest Ashton, of Allentown. Pa.

### Research Résumé of the Month

#### A—RESEARCH RESULTS

The purpose of this section of Engineering Research is to give the origin of research information which has been completed, to give a résumé of research results with formulæ or curves where such may be readily given and to report results of non-extensive researches which in the opinion of the investigators do not warrant a paper.

*Cement and Other Building Materials A5-20* Modulus of Elasticity of Concrete. Bulletin No. 5 of the Structural Materials Research Laboratory, Lewis Institute, Chicago, Ill., by Stanton Walker, is on modulus of elasticity of concrete, with an appendix applying the results obtained to flexure of reinforced-concrete beams. Results show that the stress-deformation curve is of the form $S = Kd^a$ and the modulus of elasticity varies with the stress in a manner represented by the equation $E = CS^n$, where $S =$ unit stress, $d =$ unit deformation and $E =$ modulus of elasticity.

There is no true elastic limit since the stress-deformation diagram is a curved line. There is a point where the equation above for stress and deformation no longer holds and this corresponds to the yield point used by some writers. This occurs at from 50 to 90 per cent. of the ultimate strength.

Four moduli of elasticity are noted: The initial tangent called $E_i$; the tangent modulus at from 5 to 50 per cent of the compressive strength called $E_1$; the secant modulus from 5 to 50 per cent of the compressive strength $E_2$; and the cord modulus $E_0$, which is determined by taking points on each side of the point at which the tangent modulus is found. The tangent modulus and the cord modulus are approximately the same. The initial modulus is given by the equation $E_i = 33,000 S^m$, where $m = 0.625$. The modulus at 25 per cent of the compressive strength is given by $E_{1/2} = 60,000 S^m$, where $m = 0.5$. The modulus of elasticity and strength increases as the aggregate becomes coarser, as the quantity of cement in the batch increases, as the age of the concrete increases and as the time of the mixing increases. The quantity of water has a great effect on the modulus and the strength. These decrease as the water is increased. There is no marked difference between concretes made of high-grade pebbles, crushed limestone, crushed granite or blast-furnace slag. Address Prof. Duff A. Abrams, Lewis Institute, Chicago, Ill.

*Cement and Other Building Materials A6-20* Effect of Fineness of Cement. The effect of fineness of cement is discussed by Professor Duff A. Abrams in Bulletin No. 4 of the Structural Ma-

terials Research Laboratory, Lewis Institute, Chicago, Ill. The bulletin contains 81 pages and is the result of tests on 6125 concrete cylinders, 4935 cement-mortar cylinders and 4065 briquets, as well as over 5000 other tests. Some of the conclusions reached are as follows:

1. There is no necessary relation between the strength of concrete and fineness of cements if different cements are considered.
2. In general the strength of concrete increases with the fineness.
4. Fine grinding of cement is more effective in increasing the strength of lean mixtures than rich ones.
5. Fine grinding is more effective on short-time tests.
12. The fineness of cement has no appreciable effect on the yield or density of concrete.
14. The unit weight of cement decreases with fineness.
23. Tension tests of briquets do not give a correct measure of the relative merits of different cements as determined by compression tests of mortar and concrete.

Address Prof. Duff A. Abrams, Lewis Institute, Chicago, Ill.

*Cement and Other Building Materials A7-20* Storage of Cement. Bulletin No. 6 of the Structural Materials Laboratory of Lewis Institute, Chicago. Ill., on the Effect of Storage of Cement, by Prof. Duff A. Abrams, was issued in June. The report includes compression tests on about 1000 6-in. by 12-in. concrete cylinders and 1000 2-in. by 4-in. cylinders of 1:3 standard sand mortars and about 500 miscellaneous tests. Tests have been under way for 3½ years. The following conclusions are noted:

1. Indications of the 1:5 concrete cylinders and 1:3 Ottawa sand-mortar cylinders are comparable.
2. Compression tests showed a deterioration in strength with storage of cement for all samples. The deterioration was greatest in the samples stored in open shed in yard, sample stored in basement of building showed less deterioration and those in laboratory showed still less.
3. After three months' storage in shed in yard cement had 80 per cent. of its original strength; after six months, 71 per cent.; after one year 61 per cent, and after two years 40 per cent. The deterioration was probably greater in these tests on from 8 to 12 sacks than it would be found in a larger amount of cement stored under similar conditions.
5. For periods up to 1½ years there is no marked difference in the quality of cement stored in cloth and paper bags.
6. Only a slight advantage was found from protection of cement in cloth sacks which were covered by a thin layer of portland cement or hydrated lime.
8. Storage of cement prolongs the time of initial and final setting.
10. The normal consistency was slightly affected by storage.
12. The deterioration of cement in storage appears to be due to absorption of atmospheric moisture causing partial hydration which exhibits itself in reducing the strength of the concrete and prolonging the time of setting.

Address Prof. Duff A. Abrams, Lewis Institute, Chicago, Ill.

*Cement and Other Building Materials A8-20* Strength of Drain Tiles. Bulletin No. 57 of the Engineering Experiment Station of Iowa State College, on supporting strength of drain title and sewer pipe under different pipe-laying conditions, by W. H. Schlick, was issued in April 1920. The bulletin contains 68 pages divided into six sections as follows:

1. The problem of design of drains and pipe sewers with safe supporting strengths.
2. General description of investigation.
3. Earth and pipe-laying methods.
4. Concrete-cradle pipe-laying methods for firm soils.
5. Concrete-cradle pipe-laying methods for yielding soils.
6. Pipe-laying methods with concrete cradles whose values are independent of the nature of the soil.

Tests were made on 24-in. sewer pipes, 24-in. clay drain pipes, 18-in. sewer pipes and drain pipe and 12-in. sewer pipe. The tests for the ordinary earth beddings showed the ordinary supporting strength of pipes to be slightly exceeded, while special laboratory tests showed that the effect of eliminating hub holes in trenches decreased the strength of the pipe from 15 to 27 per cent. With concrete beddings the strength of the pipe was increased from 70 to 100 per cent. in firm soils and from 32 to 68 per cent. in yielding soils. In general the concrete bedding will increase the strength of the tile about 100 per cent.

Address Engineering Experiment Station Ames, Iowa, Anson Marston, Director.

*Cement and Other Building Materials A9-20* Concrete and Cement. The bulletins issued by the Structural Materials Research Laboratory of the Lewis Institute carried out through the coöperation of the Lewis Institute and the Portland Cement Association are as follows:

- Circular 1 Colorimetric Test for Organic Impurities in Sands, by Duff A. Abrams and Oscar E. Harder (1917). Out of print.
- Bulletin 1 Design of Concrete Mixtures, by Duff A. Abrams (1919).
- Bulletin 2 Effect of Curing Condition on the Wear and Strength of Concrete, by Duff A. Abrams (1919).
- Bulletin 3 Effect of Vibration, Jigging and Pressure on Fresh Concrete, by Duff A. Abrams (1919).
- Bulletin 4 Effect of Fineness of Cement, by Duff A. Abrams (1919).
- Bulletin 5 Modulus of Elasticity of Concrete, by Stanton Walker (1920).
- Bulletin 6 Effect of Storage of Cement, by Duff A. Abrams (1920).

Address Prof. Duff A. Abrams, Lewis Institute, Chicago, Ill.

*Fuels, Gas, Tar and Coke A8-20* Heating Values. The heating value of a gas containing sufficient hydrocarbons to burn yellow may be obtained by burning the gas in a burner so constructed that the gas and air thoroughly mix and determining the relative amount of air to gas at the instant at which the yellow tip disappears. From a great number of experiments at different pressures and with gas of different heating values it was found that the air-gas ratio was always the same at the instant at which the yellow tip disappeared. The fact that the amount required for complete combustion should be a measure of the heating value has been known for some time. These experiments, however, were conducted to prove the fact and to develop a simple form of burner and apparatus by which the mixture of air and gas could be readily determined at the disappearance of the yellow tip. The mixture is expelled by rising water from an apparatus which gives a constantly changing air-gas ratio and then water. The water is stopped when the flame passes through the yellow tip and the stationary level is a measure of the heating value of the gas. The laboratory has devised the formula B.t.u. $= 158R + 276.5$, where $R$ is the value of the air-gas ratio and the heating value is the heating value of 1 cu. ft. The slope of the line (158) is approximately equal to the gram molecular heat of combustion of the radical $CH_2$. The constant of the equation 276.5 is the heating value per cu. ft. of $C_2$ burned to $CO_2$ on the assumption that $C_2$ is a gas.

Address Edward J. Brady, Physical Laboratory United Gas Improvement Company, 3101 Passayunk Ave., Philadelphia, Pa.

*Metallurgy and Metallography A12-20* Arsenic in Steel. The effect of arsenic in steel has been determined at the naval gun factory by P. E. McKinney, chemist and metallurgist.

"In view of the fact that the question has frequently arisen as to the effect of varying percentage of arsenic in steel," says Mr. McKinney, "it was deemed expedient to make a few experiments in connection with the regular manufacturing operation on the effect of this element.

"Two series of experiments were made, the first consisting of a comparison between a plain converter steel and steel from the same heat to which had been added 0.1 per cent arsenic. The second series was identical except that an addition of 0.5 per cent arsenic was made.

"After adding the final addition to a regular converter heat, a 3¾-in. diam. by 32-in. long split ingot mold was top-poured from a bull ladle for this plain test. Then about 3 in. of steel was poured into a hot bull ladle, of about 100 lb. capacity, to cover bottom, the metallic arsenic mixed with several times its weight of thermit and wrapped in paper, was thrown in ladle, the ladle was then filled with steel, mixed and top-poured into a similar mold. This constituted the arsenic test ingot. Both series were handled in the same way.

"The ingots were stripped the following morning, sent to forge shop, heated and forged longitudinally from bottom end to ¾-in. square, and cut into 6-in. lengths—a convenient size for test bars. They were not soaked or annealed before forging but both series worked excellently while being forged. All of the 6-in. lengths were annealed at 1400 deg. fahr., and then heat-treated as shown in Table 1. The ingots whose heat numbers are followed by the letter A are those to which arsenic was added:

TABLE 1. EFFECT OF VARYING THE PERCENTAGE OF ARSENIC IN STEEL

FIRST SERIES

| | | | Heat No. | C | Si | P | S | Mn | As | Remarks |
|---|---|---|---|---|---|---|---|---|---|---|
| Chemical analysis | | | C-1082 | 0.18 | 0.26 | 0.025 | 0.021 | 0.60 | 0.031 | No arsenic added |
| | | | C-1082A | 0.10 | 0.24 | 0.023 | 0.030 | 0.62 | 0.089 | 1% arsenic added |
| Annealed (deg. fahr.) | Oil-quenched (deg. fahr.) | Drawn (deg. fahr.) | Heat No. | P. E. L. | T. S. | Elong. % | Red. % | Rupture | Bend. | Fracture |
| 1400 | .... | .... | C-1082 | 48300 | 75300 | 32.50 | 62.30 | 138900 | .... | Perfect cup |
| 1400 | .... | .... | C-1082A | 43200 | 72200 | 33.50 | 62.79 | 125520 | .... | ½ cup |
| 1400 | 1450 | 500 | C-1082 | 48400 | 92900 | 26.00 | 57.22 | 162100 | 180° | 8/10 cup |
| 1400 | 1450 | 500 | C-1082A | 53500 | 88100 | 28.50 | 60.56 | 164300 | 180° | Perfect cup |
| 1400 | 1450 | 1000 | C-1082 | 58600 | 88400 | 28.25 | 60.56 | 157800 | .... | ¾ cup |
| 1400 | 1450 | 1000 | C-1082A | 53400 | 85300 | 31.25 | 64.42 | 179200 | .... | 4/5 cup |
| 1400 | 1450 | 1200 | C-1082 | 53500 | 84500 | 32.25 | 65.43 | 161100 | .... | 95/100 cup |
| 1400 | 1450 | 1200 | C-1082A | 53500 | 80000 | 32.00 | 67.74 | 162500 | .... | ¾ cup |

SECOND SERIES

| | | | Heat No. | C | Si | P | S | Mn | As. | Remarks |
|---|---|---|---|---|---|---|---|---|---|---|
| Chemical analysis | | | C-1123 | 0.13 | 0.25 | 0.019 | 0.048 | 0.57 | 0.065 | No arsenic added |
| | | | C-1123A | 0.12 | 0.33 | 0.018 | 0.045 | 0.57 | 0.310 | 0.5% arsenic added cold |
| Annealed (deg. fahr.) | Oil-quenched (deg. fahr.) | Drawn (deg. fahr.) | Heat No. | P. E. L. | T. S. | Elong. % | Red. % | Rupture | Bend | Fracture |
| 1400 | .... | .... | C-1123 | 35500 | 67700 | 32.00 | 68.00 | 147300 | .... | ¾ cup |
| 1400 | .... | .... | C-1123A | 39600 | 67800 | 31.50 | 67.50 | 139000 | .... | ½ cup |
| 1400 | 1450 | 500 | C-1123 | 45800 | 75100 | 31.00 | 62.20 | 143800 | 180° | 3/5 cup |
| 1400 | 1450 | 500 | C-1123A | 45800 | 77600 | 25.00 | 58.50 | 141300 | 180° | ½ cup |
| 1400 | 1450 | 1000 | C-1123 | 45800 | 72100 | 30.15 | 65.40 | 141500 | .... | 3/5 cup |
| 1400 | 1450 | 1000 | C-1123A | 49600 | 74000 | 33.50 | 64.60 | 143700 | .... | 4/5 cup |
| 1400 | 1450 | 1200 | C-1123 | 42700 | 69300 | 32.00 | 70.40 | 157300 | .... | 3/5 cup |
| 1400 | 1450 | 1200 | C-1123A | 48300 | 73400 | 33.50 | 69.50 | 153600 | .... | ¾ cup |

"The results of these tests show practically no difference between the steel containing no arsenic and that to which arsenic has been added, and if anything the result of test shows slight superiority in favor of the steel containing arsenic. While these tests were made on small ingots and the test piece received considerable longitudinal forging work, the results of these preliminary tests would not indicate that arsenic has the detrimental effect to steel attributed to it by some authorities.

"There is no noticeable difference in the properties of steel containing arsenic as compared with that to which no arsenic additions were made. In the pouring or forging the steel acted normal in every respect."

While it is evident from these experiments that 0.3 per cent of arsenic is not injurious, as far as static testing can disclose, the fact must not be lost sight of that it is extremely hard to get rid of arsenic after it is once present in steel, and if the steel is used for scrap purposes after its usefulness has ceased, there is a constant automatic augmentation of the arsenic content which will in time get beyond the limits desired. It would also be interesting to note the effect of arsenic upon shock-resisting qualities of the steel which is of major importance where ordnance work is concerned.

The Gun Factory hopes to have an impact machine installed in the very near future, and further experiments along this line will be carried out.

Address Bureau of Ordnance. U. S. Navy, Washington, D. C.

*Metallurgy and Metallography A13-20 Tests for Defects in Spring Steel.* Numerous breakages were obtained in ¾-in. round silico-manganese spring steel during a 24-hr. solid clamping test of spring steel. Deep seams were found in torsion tests. After various heat treatments seams developed. To guard against the receipt of such bars a comparison test has been devised in which a piece, the length of which is 1½ times the thickness of the bar, is compressed to a length equal to the thickness of the bar. Internal seams open widely under such a test. Metallurgical and Testing Division. Naval Gun Factory, Washington, D. C., Address Chief of Bureau.

*Metallurgy and Metallography A14-20 Alloy Steels.* An investigation including an extensive micrographic examination of specimens indicates that zirconium, titanium and aluminum are not truly alloying elements but act merely as scavengers. When not eliminated in slag they are present in the steel as inclusions. In small amounts they may help produce soundness but otherwise they can not do much good and may even be detrimental. Bureau of Standards, Washington, D. C. Address S. W. Stratton, Director.

### B—Research in Progress.

The purpose of this section of Engineering Research is to bring together those who are working on the same problems for coöperation or conference, to prevent unnecessary duplication of work and to inform the profession of the investigators who are engaged upon research problems. The addresses of these investigators are given for the purpose of correspondence.

*Metallurgy and Metallography B8-20 Bearing Metal.* Compressive tests at 25, 50, 75 and 100 deg. cent. have been made on five different compositions of babbitt metal, four of which correspond to proposed specifications of the S. A. E. The best pouring temperature and the effect of aging as well as the micromechanism of failure will be investigated. Bureau of Standards, Washington, D. C. Address S. W. Stratton, Director.

*Textile Manufacture and Clothing B3-20 Cotton Research Company.* The work of the Cotton Research Company is divided between researches in the mill and in the laboratory. The mill researches are as follows:

Relative value of speeds of beaters and blows per inch in picking processes;

The best speed for doffers and drafts to use in carding different cottons;

Comparisons of yarns made from various varieties of cotton with and without combing process;

Comparison of spinning with self-weighted and lever-weighted rolls;

Size mixtures most suitable for various types of warp yarns;

Connection between atmospheric conditions and relative weights throughout manufacture of yarns.

The laboratory researches are devoted to the following:

Analysis of cotton fibers and raw cotton throughout manufacture;

Value of fabrics from yarns containing various amounts of twists per inch;

Effects of humidity on sizing and breaking of fabrics.

Address Cotton Research Company, 1020 Washington St., Boston, Mass. E. D. Walen, Manager.

### C—Research Problems

The purpose of this section of Engineering Research is to bring together persons who desire cooperation in research work or to bring together those who have problems and no equipment with those who are equipped to carry on research. It is hoped that those desiring cooperation or aid will state problems for publication in this section.

*Internal-Combustion Motors C1-20 International Harvester Co.* The problems covered by Laboratory No. 5 of the International Harvester Company devoted to gas-power engineering is concerned with two sets of problems:

A Development of present manufactured products.

B Development for future market.

Under A there are two main divisions:

I Materials of Construction.

II Working Processes.

Under Materials of Construction the following sub-division occurs:

1 Characteristics of present products

2 Changes to eliminate failure in service

3 Changes for improvement in operation

4 Changes to facilitate production or manufacture

5 Development to meet future conditions.

Under Working Processes the following subdivisions are found:

1 Determining the performance, characteristics and laws of the present working process.

2 Changes to meet present difficulties or failures in service.

3 Development to secure improved operation, production or economy.

4 Development to meet future conditions.

*Machine Design C1-20 Ball Bearings.* Recent investigations on the arrangement, strength and serviceable features of ball bearings for flat footstep bearings is desired by W. S. Aldrich, American Bridge Company, Gary, Ind.

### D—Research Equipment

The purpose of this section of Engineering Research is to give in concise form notes regarding the equipment of laboratories for mutual information and for the purpose of informing the profession of the equipment in various laboratories so that persons desiring special investigations may know where such work can be done.

*International Harvester Company D1-20 Gas-Power Engineering Laboratory.* Laboratory for the development of research work on trucks, tractors, stationary engines of high and low compression, lighting outfits, various accessories and other products. Road tests are also operated by the Company.

The laboratory is situated in a building with a sawtooth roof, equipped with traveling crane, piping for water, gasoline, kerosene and compressed air and wired for electric power. Fuel is supplied to engines from individual tanks on scales. Exhaust is cared for by overhead exhaust line connected to fan blower. Ventilator stacks with openings at floor level are equipped with an exhaust fan. A motor-generator set supplies direct current for battery charging, field and armature current for d.c. motors and generators and field current for the Sprague dynamometer.

*Equipment*

100-hp., 1050 to 3500 r.p.m. Sprague dynamometer fitted with electric contactors for fuel weighing and recording revolutions.

Electrically indicating tachometer.

Electrically controlled revolution counter.

Hand tachometers.

Revolution counters and stop watches.

Toledo scales, electrically controlled for time element and hydrometers are used in fuel measurement.

Venturi meters recording water meters, sharp-edged orifices and pressure gages are used for water measurement.

Bristol recording thermometers, nitrogen-filled mercury thermometers.

Leeds & Northrup recording and indicating potientiometer.

Hoskins pyrometer and thermocouples are used for heat measurement.

Crosby indicators, diaphram indicators of the Bureau of Standards and manograph are used for power measurement.

An air-equalizing tank with sharp-edged orifice and Ellison gage.

Venturi tubes, pitot tubes and anamometers, displacement gas meters, recording and indicating barometers.

Thermometers and psychometers are used for air measurement.

Five test stands with direct-current generators are used for absorbing power.

Five stands with prony brakes are used for testing engines. Three of these are equipped with balances and supported brakes. Several other test stands are used for small and large engine work.

A brake band testing outfit is arranged to measure belt tension, slippage and durability as well as the coefficient of friction and endurance of brake and clutch linings. An apparatus for determining the endurance of fan belts under various tensions and temperatures is installed in this section of the laboratory.

The friction losses in gear-transmission lubricants are determined by a motor driving a generator through gear transmission.

The laboratory is planning apparatus to determine the tractor drawbar pull, tractive effort, torque reaction about driving axle, power delivered by belt pulley and general performance data from tractors.

The laboratory is equipped to determine the heat dissipation of radiators at various rates of air and water flow.

The laboratory is equipped for the development work on high-compression engines. A 1-hp. Hvid thermoil engine and a 10-hp. high-compression engine are installed.

The capacity, endurance and performance characteristics of generators and storage batteries or other electrical apparatus may be determined by the equipment in the laboratory. Apparatus for determining the characteristics of steam and steam engines are found in the laboratory.

Calorimeters for determining the heating values of solid and liquid fuels are found in the Chemical-Physical laboratory of the Company.

The laboratory is equipped for complete and thorough investigations of automotive apparatus, the characteristics of engine auxiliaries and development work on specific problems.

*Research Personnel:* .

Chief Engineer, Steam Research Engineer, several development engineers and office force of computers, a test-force foreman, chief tester, testers and mechanics.

Address International Harvester Company, 606 South Michigan Ave., Chicago, Ill. E. A. Johnston, Manager, Experimental Department.

#### E—RESEARCH PERSONNEL

The purpose of this section of Engineering Research is to give notes of a personal nature regarding the personnel of various laboratories, methods of procedure for commercial work or notes regarding the conduct of various laboratories.

#### F—BIBLIOGRAPHIES

The purpose of this section of Engineering Research is to inform the profession and especially the members of the A. S. M. E., of bibliographies which have been prepared. These bibliographies have been prepared at the request of members, and

where the bibliography is not extensive, this is done at the expense of the Society. For bibliographies of a general nature the Society is prepared to make extensive bibliographies at the expense of the Society on the approval of the Research Committee. After these bibliographies are prepared they are loaned to the person requesting them for a period of one month. Additional copies are prepared which are available for periods of two weeks to members of the A. S. M. E. or to others recommended by members of the A. S. M. E. These bibliographies are on file in the offices of the Society and are to be loaned on request. The bibliographies are prepared by the staff of the Library of the United Engineering Society which is probably the largest Engineering Library in this country. .

*Mechanics F1-20* Impact and Alternating Stress Tests. Impact. A bibliography of 10 pages. Address A. S. M. E., 29 West 39th St., New York.

*Molecular Physics F1-20* Capillarity and Surface Tension. A bibliography of 1 page. Address A. S. M. E., 20 West 30th St., New York.

*Properties of Engineering Materials F1-20* Impact and Alternating Stress Tests. Impact. A bibliography of 10 pages. Address A. S. M. E., 20 West 39th St., New York.

## WORK OF THE A. S. M. E. BOILER CODE COMMITTEE

*T*HE *Boiler Code Committee meets monthly for the purpose of considering communications relative to the Boiler Code. Any one desiring information as to the application of the Code is requested to communicate with the Secretary of the Committee, Mr. C. W. Ubert, 29 West 39th St., New York, N. Y.*

The procedure of the Committee in handling the cases is as follows: All inquiries must be in written form before they are accepted for consideration. Copies are sent by the Secretary of the Committee to all of the members of the Committee. The interpretation, in the form of a reply, is then prepared by the Committee and passed upon at a regular meeting of the Committee. This interpretation is later submitted to the Council of the Society for approval, after which it is issued to the inquirer and simultaneously published in MECHANICAL ENGINEERING, in order that any one interested may readily secure the latest information concerning the interpretation.

Below are given the interpretations of the Committee in Cases Nos. 269 and 307 to 312 inclusive, as formulated at the meeting of June 24, 1920, and approved by the Council. In accordance with the Committee's practice, the names of inquirers have been omitted.

#### CASE No. 269 (Reopened)

*Inquiry:* If, under the provision made in Par. 212 of the Boiler Code, advantage is taken of the opportunity to increase the pitch of the staying for a cylindrical furnace, is it to be assumed that a portion of the increased load on the staybolt is to be supported by the resistance of the outer cylindrical shell to collapse, or must the staybolt be designed to carry the full load upon the stayed area of the furnace sheet?

*Reply:* The special provision made in Par. 212c for increased pitch is made possible by the additional strength afforded by the convexing of the plate. It is permissible under this rule to increase the spacing of the staybolts to $p$, in the formula, whereas the required cross-sectional area of the staybolts should be based on $p$.

#### CASE No. 307

*Inquiry:* Is it permissible to so locate the supporting lugs on h. r. t. boilers where more than four lugs are required and under Par. 323 of the Code, must be set in pairs, that those in each pair come close together, or must the horizontal distance between the center lines of rivets attaching the adjacent lugs to the shell be at least equal to the vertical spacing of rivets that is required for lug attachments in Par. 323, as shown in Fig. 8?

*Reply:* There is no requirement in the Code specifying the distance apart of the lugs forming pairs as required by the last sentence of Par. 323. It is the opinion of the Committee, however, that in locating lugs in pairs on the shells of h. r. t.

boilers, it will be in conformity with the spirit of the second sentence of Par. 323 if the lugs of the pair are so spaced that the horizontal distance between the centers of the rivets which come nearest the adjacent edges of the lugs is at least 6 in., and not more than 12 in.

FIG. 8  SPACING OF SUPPORTING LUGS IN PAIRS ON H. R. T. BOILERS

#### CASE No. 308

*Inquiry:* Under what rules in the Boiler Code should the top head of a vertical submerged-tube type of fire-tube boiler,

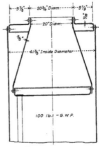

FIG. 9  DESIGN OF TOP HEAD OF VERTICAL SUBMERGED-TUBE TYPE OF FIRE-TUBE BOILER

such as shown in Fig. 9, be calculated to determine whether or not it requires bracing?

*Reply:* There is no rule in the Code specifically applying to such construction. However, in Par. 216, an allowance is made for surfaces located between tubes and shells. It is the opinion of the Committee that it would be entirely safe to permit similar allowances in this case, that is, the distance between supported points could be made 3 in. greater than the permissible spacing of staybolts for the corresponding plate thickness and pressure given in Table 4.

### CASE NO. 309

*Inquiry:* Is it necessary to furnish test reports of the steel used in the tubes or flues of special type boilers which are formed of 18 in. lap welded steel tubing, 15 ft. long, where the wall thickness is ½ in. and the weld meets the requirements of Par. 186?

*Reply:* It is the opinion of the Committee that the material used in the manufacture of lap-welded low-carbon steel tubing should meet the stipulations prescribed in Pars. 23 to 39 inclusive of the Code, which will make it necessary to furnish mill test reports of the material.

### CASE NO. 310
#### (In the hands of the Committee)

### CASE NO. 311

*Inquiry:* Is it necessary under Par. 188 of the Boiler Code to use butt-stray joints in the construction of very small drums, say, 10 or 12 in. in diameter, for pressures exceeding 100 lb. per sq. in? Such construction does not appear to be practical for such small drums and neck pieces sometimes used to connect drums or shells.

*Reply:* Lap-riveted construction is prohibited by the Code Rules, but it is the opinion of the Committee that this should cause no hardship as lap-welded or seamless pipe or tubes could be used, provided such tubes or pipes are constructed from material which in its initial form of plate or skelp conforms to one or the other of the specifications for open hearth steel given in the Boiler Code. No test would be required on the completed tube. (See Case No. 255).

### CASE NO. 312

*Inquiry:* Is it necessary in the construction of small Star-type water-tube boilers for steam heating which are to carry more than 15 lb. pressure at times, to drill the inside and outside ends of staybolts? It is believed that it was the intent of the Committee to cover in this requirement the water legs at front and back ends which are considered as headers.

*Reply:* If the grate area is more than 15 sq. ft., the staybolts are less than 8 in. in length, and the pressure exceeds 15 lb., it will be necessary to drill the staybolts in order to comply with the Code requirements.

## Conference Committee on Welding

The Boiler Code Committee takes pleasure in announcing that the American Welding Society has appointed a Committee to confer with the Sub-Committee of the Boiler Code Committee on Welding. The above is the result of an invitation extended by the Council of the A.S.M.E. at the request of the Boiler Code Committee. It is the desire of the Boiler Code Committee that the Committee of the American Welding Society shall coöperate with the Sub-Committee of the Boiler Code Committee in discussing the rules now in the Code and in proposing any revisions or new rules that may be embodied in the Code at the next revision period. The personnel of the Committee appointed by the American Welding Society is as follows:

A. S. KINSEY, *Chairman*, Stevens Institute of Technology, Hoboken, N. J.

C. A. ADAMS, National Research Council and Director of American Bureau of Welding, 29 West 39th St., New York, N. Y.

A. M. CANDY, Westinghouse Electric & Mfg. Co., East Pittsburgh, Pa.

ALEXANDER CHURCHWARD, Wilson Welder & Metals Co., 253 36th St., Brooklyn, N. Y.

J. H. DEPPELER, Metal & Thermit Corporation, 92 Bishop St., Jersey City, N. J.

R. E. KINKEAD, Lincoln Electric Co., Cleveland, Ohio.

VICTOR MAUCK, John Wood Mfg. Co., Conshohocken, Pa.

STUART PLUMLEY, Davis-Bournonville Co., Jersey City, N. J.

H. S. SMITH, Prest-O-Lite Company, 30 East 42nd St., New York, N. Y.

R. E. WAGNER, General Electric Co., Pittsburgh, Pa.

# CORRESPONDENCE

CONTRIBUTIONS to the Correspondence Department of MECHANICAL ENGINEERING are solicited. Contributions particularly welcomed are discussions of papers published in this Journal, brief articles of current interest to mechanical engineers, or suggestions from members of The American Society of Mechanical Engineers as to a better conduct of A. S. M. E. affairs.

## Federation an Opportunity for Local Societies

TO THE EDITOR:

The engineering profession of the country has been groping for many years after a suitable means of impressing its personality upon the public quite aside from the effort to secure betterment of individual conditions which has led to the remarkable development of the American Association of Engineers. The means sought for is at last at hand in the organization created on June 3 and 4 at Washington.

Every task accomplished is a source of satisfaction to those who participate in the performance, and so every society represented at the Washington conference has felt a thrill of pride at having been a part of the first movement offering a chance and a definite program for united service. Those societies who were so unfortunate as not to be represented, as the facts become known to them, will begin to wonder why they were so indifferent. This is particularly true of the local society which finds in its local affairs only intermittent opportunities to expend its energies. Though local matters are the first that should receive engineers' attention, they are the soonest settled, while as the broader fields of service are approached the necessity for longer and more persistent effort rapidly develops. The opportunity afforded by the new organization to every local society to make its voice heard and take a part in the great affairs of the nation will certainly stimulate activity and interest in their members.

The effect of the organization on the national societies will be probably less apparent than on the smaller associations. They have long imagined that they were performing a great work and that they will view the new organization as more or less of a side issue. Through it they will be able to exert a more effective influence on national and state questions, but the reaction will be of less importance than in the case of local and state organizations. Those technical matters in which the national societies are now coöperating seem unlikely to be affected, as the energies of the new organization will be absorbed in the solution of broader problems.

When it comes to such questions as a National Department of Public Works, the attitude of government toward the transportation industries of the country, the safe-guarding of national resources, and the protection and development of the laboring class, the new organization affords the long-looked-for opportunity for that group of our citizens best qualified by education and experience, to place before the public and the legislators its recommendations and its criticisms.

GARDNER S. WILLIAMS.

Ann Arbor, Mich.

### Federation Dues and Local Societies

To the Editor:

We all recognize that The Federated American Engineering Societies should do a great deal to stimulate the professional spirit of the engineer, as it will emphasize his relations to the public and to national affairs. However, it is even more important for the engineer that he take an interest in state and local affairs of an engineering nature. This can only be done if the idea of the Federation is carried down into the smaller units so that in every locality sections of the national societies and other engineering organizations are affiliated together and these affiliations are also carried out for the various states. This aspect is of course recognized in the constitution of The Federated American Engineering Societies, but I feel that active steps should be taken to further it at the very beginning of the organization.

The only difficulty for the average local society in coming into the Federation is the question of expense. Where a society has been very active in increasing its membership and has kept its dues at a low figure, say, $5.00, so as to reach the largest number, membership in the Federation will take 20 per cent of its annual budget. Furthermore, where the society is largely made up of members of national societies, some objection may be made to an increase in dues on the ground that payment has already been made through the national organizations. On the other hand, it is extremely important that as far as possible these local societies should become members of the Federation. It seems to me, therefore, that before these societies are asked to definitely decide the question of joining the Federation, a careful estimate should be made of the activities and expected expenditiures and also the anticipated revenue of that organization. The possibilities of increased activities or of smaller payments should also be carefully pointed out. The possession of this information will help materially in the discussion of the finance committees and councils of local societies.

I hope that it will be possible to get practically every engineering organization of the country into The Federated American Engineering Societies.

JAMES A. HALL.

Providence, R. I.

### A Criticism of the New Federation

To the Editor:

The constitution of The Federated American Engineering Societies makes two distinct and notable contributions to the progressive upbuilding of engineering as a profession, first through the avowal that the prime object of the organization is "to further the public welfare," and again in its stand "for principle of publicity and open meetings." This latter provision sounds a new note in quasi-public organizations.

But elsewhere in this constitution and by-laws one fails to discern any desire to get away from the traditional type of organization which has heretofore trammelled American engineering bodies especially in their work for the public.

Of vastly greater importance, however, than any possible defects in the type of organization adopted is the complete failure to recognize the necessity for enlisting the interest—aye, the enthusiasm—of the rank and file of the profession if we are to accomplish even a tithe of the task which the times demand of us.

In bringing about a union of all the engineering societies in the United States we are seeking fundamentally to effect three things:

1 The solidarity of the profession
2 To have our group thus unified, accept its public responsibilities, and
3 Gradually to develop and then to use the collective initiative thus created for the accomplishment of great public purposes.

The engineers of the United States today number at least 100,000—possibly 200,000. Within a decade or two it is not impossible that this may be increased to 500,000. Anything short of an effort to bring every one of these engineers individually into the fray by giving them a direct vote and otherwise, would be in the first place to go counter to every modern tendency in democratic theory, but of even greater immediate importance, it would deprive the new Federation of the most obvious means of educating and gradually enthusing the rank and file. This is just the kind of opportunity which was missed in the failure to provide that the members of the American Engineering Council should be elected by the votes of the members of constituent organizations rather than by the boards of these organizations.

In fact, the whole conception of the new Federation as expressed in its scheme of organization appears to be short-sighted almost to the point of insuring failure. This is not so much because it creates a typical super-organization but rather in its failure to recognize that power of the kind that makes for righteousness and the public welfare is grounded in the activities and devotion of the whole body—whatever that body may be, a ball team, a technical organization or the whole people. The effort in creating the Federation seems to have been to provide a board which could speak for the engineering profession rather than so to integrate the whole profession that when it speaks it will

1 Have something to say, and
2 In saying it carry reasonable conviction.

Still we must hope that this last effort at unity in the profession has something real to contribute, for deep in our hearts we engineers have a mighty purpose to fulfill—a purpose which cannot and will not long be thwarted and which only needs organization to make it effective.

MORRIS LLEWELLYN COOKE.

Philadelphia, Pa.

### Determination of Fiber Stress Caused by Force Fits

To the Editor:

Having read with interest Mr. Wm. W. Gaylord's communication on the Strength of Thick Hollow Cylinders in MECHANICAL ENGINEERING for April, I am led to submit the two accompanying alignment charts which I have recently constructed for the determination of stress in thick hollow cylinders as applied to force fits. These charts are based on Birnie's formula for thick hollow cylinders, adapted to the case where the external pressure is zero—given in Fig. 2—in which

$f_t$ = maximum fiber stress

$e$ = allowance, or difference between inside diameter of hub and diameter of shaft

$E_s$ = modulus of elasticity of shaft

$E_h$ = modulus of elasticity of hub

$R_s$ = radius of shaft

$R_h$ = outside radius of hub.

Fig. 1 has been constructed to obtain the value of $A$ in the formula and Fig. 2 to obtain the value of $f_t$, $A$ having been previously determined. Their use can best be explained by the solution of a definite problem.

Assume a steel shaft $2\frac{1}{2}$ in. in diameter ($R_s = 1\frac{1}{4}$ in.) to be forced into a cast-iron hub $5\frac{1}{2}$ in. outside diameter ($R_h = 2\frac{3}{4}$ in.) with an allowance ($e$) of 0.0025 in. Assume also that $E_s = 30,000,000$ for steel and $E_h = 28,000,000$ for cast iron.

*Solution:*

1 Lay a straight edge across the two points $R_s = 1\frac{1}{4}$ in. and $R_h = 2\frac{3}{4}$ in. of Fig. 1 and read $A = 1.854$ on the $A$-scale.

2　Obtain point $u$ by laying a straight edge across the two points $A = 1.854$ and $E_h = 28,000,000$ on Fig. 2 and noting where it intersects the vertical ungraduated scale.

3　Connect points $e = 0.0025$ in. and $R_s = 1\frac{1}{4}$ in. by the line $st$.

4　Draw through point $u$ a line $uv$ parallel to $st$ and read on the horizontal scale $f_t = 21,000$ lb. per sq. in.　An algebraic solu-

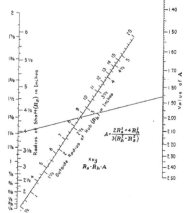

Fig. 1　Fiber Stress Caused by Force Fits.　Chart for Determining Value of $A$ in Birnie's Formula (see Fig. 2)

tion of the two equations gives $A = 1.8542$ and $f_t = 20,970$ lb. per sq. in.

It will be noticed in Fig. 1 that for small values of $R_s$ and $R_h$ (below $R_s = 2$ in. and $R_h = 5$ in.) the points determining the straight line are very close together, and therefore the position of the line cannot be accurately determined.　To obtain more accurate values of $A$ in this region, a second scale has been added to the $R_s$- and $R_h$-axes.　The outside graduations of these axes

are for values of $R_s$ from $\frac{1}{4}$ in. to 2 in. and $R_h$ from $\frac{1}{2}$ in. to 5 in., while the inside range is for $R_s$ from 2 in. to 6 in. and $R_h$ from 4 in. to 20 in.　In either case the straight line cuts the value of $A$ on the right-hand scale.

It will also be noticed that in Fig. 2 the $f_t$- and $R_s$- scales have been so constructed that they are the same length and their grad-uations of the same length.　If, then, these two scales be inter-changed, which simply means interchanging the numerical values, the scales being identical, a chart will be obtained for inter-secting index lines instead of parallel index lines.　The secondary values on these two scales are for that purpose.　The solution of the problem for intersecting lines is shown by the dotted lines and is obtained as follows:

1　In Fig. 2 draw line $qur$ as before.

2　Connect points $e = 0.0025$ in. and $R_s = 1\frac{1}{4}$ in., $R_s$ now being on the horizontal scale (line $sw$).

3　Draw a straight line through the intersection ($y$) of the line $sw$ and fixed diagonal $DB$, and point $u$ and at $x$ read $f_t = 21,000$ lb. per sq. in. on the secondary values of vertical scale $BC$.

The use of intersecting index lines has certain advantages and disadvantages, as has also the use of parallel index lines.　The chart of Fig. 2 has the advantage of using either or both together, taking the mean value of the two results.　The original pencil-line charts give very accurate results, but their accuracy, of course, decreases through reproduction on a smaller scale.

The $A$-scale of Fig. 2 need not have been graduated below $A = 1.333$, for $A$ can never be less than that value for positive values of $R_h$ and $R_s$.　The scales may also be extended to any desired limits.

It will be noticed that a chart of this type has one very great advantage over that of the ordinary logarithmic chart, in that the scales are all natural scales, i.e., not logarithmic, and therefore the graduations do not close up and become very small at one end.　For example, a fiber stress of 29,300 lb. per sq. in. is just as easy to read from the chart as one of 11,300 lb., and if the scale were continued to 100,000 lb. it would still be just as easy to read, which would not be true of a logarithmic scale.

S. R. Cummings.

Cambridge, Mass.

## Load-Speed Capacities of Radial Ball Bearings

To the Editor:

The purpose of this communication is to arouse interest in a subject, often incorrectly treated, with a view of pointing out a correct solution.　Ball bearings are in such general use and under such varied conditions that the engineer must have an accurate formula for calculating their strength.

There are at present two widely used formulæ for figuring the load-speed characteristics of radial or annular ball bearings.　One is the Hess-Bright or D.W. F. formula—

$$W = knd^x$$

In which　$W$ = load in lb.

　$n$ = number of balls

　$d$ = diameter of balls in eighths of an inch

　$k$ = a constant varying with the condition and type of bear-ing, as also with the material and speed.

The second formula is that worked out by Professor Goodman—

$$W = knd^x / (RD + Cd)$$

In which　$W$ = load in lb.

　$n$ = number of balls

　$D$ = diameter of outer ring race in inches

　$d$ = ball diameter in inches

　$R$ = r.p.m.

　$K$ and $C$ = constants having the values given below:

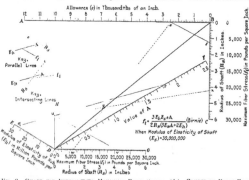

Fig. 2　Chart for Determining Maximum Fiber Stress ($f_t$) Caused by Force Fits

| | $C$ | $k$ |
|---|---|---|
| Thrust bearings, flat races | 200 | 500,000 |
| Thrust bearings, grooved races | 200 | 1,000,000 to 1,250,000 |
| Radial bearings, flat races | 2000 | 1,000,000 |
| Radial bearings, grooved races | 2000 | 2,000,000 to 2,500,000 |

As correctly stated by the authors of the Hess-Bright formula, the capacity of a bearing increases in direct proportion to the number of balls and the square of the ball diameter. The capacity decreases with the speed and varies with the type and material of the bearing. All of the variables last mentioned have been combined by some method in the constant $k$. From the writer's experience it would seem that a value of $k$ must be worked out for every speed. This makes it a somewhat overworked constant and harder to assign a value to it than to figure the load-speed characteristics on a more rational basis.

Let us look at the Goodman formula with these points in view. We find that the load increases with the number of balls.

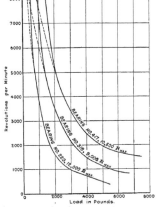

Fig. 3   Variation of Load Composition of Representative Ball Bearings with Speed in R. P. M.

and, since $d^2$ is expressed in the numerator and $d$ in the denominator, with the square of the ball diameter. The correctness of these facts is agreed to by all ball-bearing formulæ. We also find that the capacity decreases with the speed, which is quite obviously true. It also decreases with an increase in the diameter of the outer ring race, or in other words the diameter of the enclosing circle around the balls. We will see the logic of this assumption if we consider two bearings identical in number and size of balls but one having an outer ring race of greater diameter than the other. If the balls are evenly spaced there will be a greater length of arc between balls in the larger bearing than in the small one, and this naturally gives rise to a greater chance for deflection in the outer ring. We have left two constants which vary as shown in the table only for certain specific conditions, a value therefore being easily given them.

It appears from the foregoing that the Goodman formula is by far the more logical. There is, however, one consideration that this formula, and in fact all formulæ, neglects, namely, the maximum speed at which a bearing may be run. As shown in Fig. 3, which graph was obtained by plotting the Goodman formula for a few representative bearings, the curves reach zero load or

their maximum speed at infinity. The curves shown by solid lines are the values given by Goodman and the additional dotted lines are a modification of this formula which will be described.

A formula has been derived by the writer for the maximum safe speed of a bearing under no load. This has been worked out from the standpoint of centrifugal force and a large factor of safety is used. It reduces to—

$$R_{max.} = K\sqrt{1/(D-d)}$$

In which $R_{max.}$ = maximum safe speed under no load
　　　　$D$　= diameter of outer ring race in inches
　　　　$d$　= ball diameter in inches
　　　　$K$　= a constant to which a value of 25,000 has been assigned.

This value is correct for radial bearings having raceways of a radius = 0.54 of the ball diameter, made of alloy steel hardened clear through, and of the best accuracy known to the art.

To use this formula the values given by the Goodman formula are plotted. The maximum speed is figured from the above formula and plotted on the coördinate of zero load. Through this point a line is drawn tangent to the curve previously plotted. The result obtained is shown in the graph. This results, for large bearings, in a considerable reduction in the load-carrying capacity at the higher speeds.

The writer is indebted to Mr. Harry R. Reynolds for assistance in the preparation of this communication.

R. W. Sellew.

New Britain, Conn.

## Why Not a Tool-Drafting Standard?

To the Editor:

Many of our members who have been connected with the metal-manufacturing industry for many years can remember when tool drafting consisted mostly of verbal information, with perhaps a rough sketch of what was desired. This has been supplanted by more modern methods and the writer believes that tool designing and tool drafting have reached a stage today worthy of a consideration of some sort of codified standardization.

During 18 years of experience as tool maker and tool designer the writer has had the opportunity to observe many different uses of lines, letters, figures and symbols to express an idea, the majority of which failed to give a complete conception of the tools wanted. Mechanical engineers and chief draftsmen who in times of stress have had recourse to outside engineering firms to design and draw up the tooling of a new product, will vouch for the time lost in attempting to adapt drawings made by these outside firms to the system in vogue in their own drafting department. Such difficulties could be overcome if there was available some standard according to which the work might be done.

The writer suggests, therefore, that the newly created machine-shop section of the Society investigate the possibilities of an A. S. M. E. tool-drafting code whereby the best methods of expressing an idea by means of lines, letters, figures, and symbols would form some sort of universal language, the economical value of which would soon be felt.

Fred C. Friedli.

Elmira, N. Y.

## New York State Legislature Passes Bill for Licensing Engineers

The bill licensing engineers in the State of New York, adopted by the legislature in April 1920, became a law on May 14, 1920, when it was signed by Governor Smith. The bill is based upon Engineering Council's recommended uniform registration law, with certain modifications and changes conforming to the provisions of e isting law, a bill requiring the licensing of architects already having been enacted. A discussion of the bill as recommended by Engineering Council may be found on pages 77 and 78 of the January, 1920 issue of Mechanical Engineering.

# MECHANICAL ENGINEERING

Published Monthly by The American Society of Mechanical
Engineers at 29 West Thirty-ninth Street, New York

FRED J. MILLER, *President*

WILLIAM H. WILEY, *Treasurer* CALVIN W. RICE, *Secretary*

PUBLICATION COMMITTEE:

GEORGE A. ORROK, *Chairman*          J. W. ROE
    H. H. ESSELSTYN          GEORGE J. FORAN
          RALPH E. FLANDERS

PUBLICATION STAFF:

LESTER G. FRENCH, *Editor and Manager*
    FREDERICK LASK, *Advertising Manager*
          WALTER B. SNOW, *Circulation Manager*
                186 Federal St., Boston

Yearly subscription $4.00; single copies 40 cents. Postage
to Canada, 50 cents additional; to foreign countries $1.00 addi-
tional.

*Contributions of interest to the profession are solicited. Com-
munications should be addressed to the Editor.*

## A Plea for the Development of Our Water Powers

THE people, of a right, should look to the engineer to point the way in the utilization of nature's forces which will result in the greatest relief from human toil, as they have looked in the past to the medical profession for relief from human ills. The people's faith in the good to come from the engineer's work is well expressed in the generous support given the Engineering Educational Institutions.

There has been installed recently at Niagara Falls a combined hydroelectric unit of 40,000 horsepower capacity. The operation of this unit requires two men per shift. It has been calculated that to produce the same amount of power by small isolated steam plants, as would be the case were the water power not developed, there would be required over 800 men to mine, hoist, break, screen, load, transport by railroad, unload, store, rehandle, and fire under boilers the coal necessary to develop by steam an equal amount of power. The operation of this one unit conserves not only one train load of coal per day, but also conserves the man power which would be required to produce it. This striking example shows the great amount of man power which can be conserved for other fields of endeavor by developing our water powers.

W. M. WHITE

May it not be that the shortage of labor to accomplish the tasks before us today has been brought about by the failure to follow that path in the development of our natural resources which leads to the minimum of human toil?

Some time ago a member of the British Parliament pointed out in open meeting that the American workman has twice the horsepower per man behind him that the British workman has, and gave this as the reason for the ability of the manufacturers in the United States to pay to labor wages twice that paid in England and yet compete successfully in the World's markets in

manufactured products. We have the power behind the workman, but the major portion of it is steam power, and we are paying a dear price for that power in men absorbed.

The economic advantage which the United States has heretofore enjoyed by reason of adequate power is being sought by other nations. For example, Japan only recently began the development of her water powers, but has already developed one and a half million of her total eight million water horsepower.

Failure to develop our water horsepowers has been seriously retarded due to a wrong conception of conservation, as manifestly there can be no true conservation as long as the power in the rivers is running to waste.

The shortage of labor which has increased the price of coal and the shortage of fuel oil, forces to our attention the necessity for the development of a large amount of our water powers during the next decade.

Laws have just been enacted governing the development of water power on Government Lands which will greatly influence hydro-electric development, particularly in the Western States. It seems a fitting task of the engineer to urge and aid the development of our water powers under the new law.

There has been developed in the United States today approximately nine million horsepower from water power, and there is yet available for development by this same means over fifty million horsepower; the development of which would save the labor of one million men.

When the war came on we were lamentably short in certain chemicals, chemical compounds and alloys. That small portion of our water powers which had been developed proved a great asset in our national crisis. That great portion of our resources in water power which was yet undeveloped was of no value in meeting the great emergency. The Government took control of the developed powers and directed their energy to war work solely. Water power centers became great producing communities in all things chemical. The energies of the great developed water powers were let loose in electric furnaces, and in electrolytic cells for making good the shortage in chemicals, chemical compounds, and alloys, but even this was not sufficient to meet the emergency. Huge steam plants were hurriedly constructed to supply additional power for this purpose, but this great construction work required labor not only for the construction, but for the operation of such plants which required enormous quantities of coal, thus absorbing man power at a time when every potential fighting unit should have been available for the front.

The development of our fifty million water horsepower now running to waste would mean not only a tremendous economic advantage to the people but a potential asset which might prove of incalculable value in case of war requiring the concentration of the man force of the nation for other than domestic uses.

Early development of the water powers would be the greatest move which could be made in the direction of true conservation.

Milwaukee, Wis.                    WM. MONROE WHITE.

## The Important Water Power Problem

A study of the severe transportation problem has resulted in offering as a solution, vitally interesting to the engineer, the elimination of coal movements by developing the power at the mines, and uniting in one big system the power lines of the large industrial centers. Of equal interest in this problem, and of great value in the conservation of our fuel resources, is the problem of hydraulic power development.

In our desire to bring this problem home to the mechanical engineers of the country, several pages in the Survey of Engineering Progress in this issue are given to a digest of notable engineering articles covering the development of hydraulic machinery. Our signed editorial this month also bears on this problem, pointing out a conservation in man power that may be gained by greater utilization of water power.

The use of our hydraulic resources is a nation-wide question which commends itself to mechanical engineers, whether closely associated with its development or not.

## The Raising of Engineering Limits

Throughout the entire field of engineering there has recently been apparent a tendency to raise the limits of working far beyond what was considered either safe or commercially practicable only a few years ago.

In the General Electric design of a super-power station, described in the August issue of MECHANICAL ENGINEERING, it is proposed to use a steam pressure of 350 lb. and superheat of 350 deg. fahr., the implication being that pipe, valves and fittings can now be commercially produced that will not be excessive in price and yet will safely stand these high mechanical and thermal stresses. In fact, 350 lb. pressure and 350 deg. fahr. superheat are by no means the limiting values under consideration, for 600 lb. pressure has been seriously mentioned as a possibility of the near future, if not a commercial value for use today.

In the field of refrigeration we have very nearly reached the limit of possibility, as Prof.. Kammerlingh-Onnes in his work has come within 3 or 4 deg. of the absolute zero of temperature. It is true that such an excessively low temperature has not as yet been applied in commercial work, but only for the reason that thus far no process has been devised where it could be used commercially.

In the field of pressures other than steam, the limits have also been raised materially. Only a few years ago engineers spoke with somewhat bated breath of pressures of 200 atmospheres used in the various Linde processes for the separation of gases. Today commercial applications are being made of pressures at least five times as high, and pressures of 1200 and 1500 atmospheres are seriously under consideration.

The ability to raise the commercially reliable engineering limits of pressure, temperature, compression, etc., is undoubtedly due primarily to two causes. In the first place, our knowledge of the behavior of materials and our ability to produce materials of uniform quality have increased tremendously. Only fifteen years ago a tensile strength of 150,000 lb. per sq. in. was considered to be about the limit of strength of commercial metals, while 250,000 lb. does not represent the possibility of some of the special alloys of today. Moreover, with our present methods of testing materials and of producing alloys, and especially with our present methods of heat-treating them, it is possible to insure that a metal part will stand up under given enormous stress where there could be no such certainty only a few years ago.

In the second place, the improvement in our methods of machining has had a powerful influence in the same direction. In this connection an interesting illustration may be cited from the field of internal-combustion engineering. The Doxford oil engine uses a system of solid injection which is effected by means of a plunger pump exerting a pressure of 6000 lb. per sq. in. The remarkable part about this pump, however, is that it uses no packing. It appears, therefore, that it has become possible to produce a plunger and cylinder with such an accurate fit that a pressure of several tons to the square inch may be generated. It is easy to see what a substantial advance this represents in the precision of machine-shop processes.

This advance, in turn, has become possible through the improvement in the machine tools themselves, and especially through the fact that they are not only built better and of better materials, but are far more massive and rigid than they used to be a few years ago.

In the 50's of the last century, when the scientific world was still being taught the full realization of the principle of the conservation of energy, a favorite illustration used to be the lighting of a match. It was said that the world represented such a closely balanced unit that the release of the potential heat-energy content in the match was immediately reflected throughout the universe. The burning of the match raised the temperature of the surrounding atmosphere that produced a flow of air upwards which resulted in a tiny wind, etc.

Practically the same close co-working of phenomena is to be observed in engineering. The improvement of materials, tools and methods of machining makes possible the use of temperatures,

pressures and compressions greater than any that have been used before. This, in its turn, calls attention to the economies and possibilities of the higher limits and calls for new efforts to improve still further the materials, processes, and machinery of production.

As a matter of fact, some very remarkable results in machining to close limits have been attained in the production of gages, as, for example, the Johanssen and Hoke precision gages. A remarkable part of this work is the ability of the gage makers to produce surfaces of such flatness that when two blocks are put together they adhere to each other with great force. The production of such surfaces today is of value mainly in such a limited field as gage manufacture, but it would not be at all surprising, as stated by Mr. Johanssen, if a step further in the same direction would bring about a rather revolutionary process of joining metals. Gage blocks are already machined with such precision that when the surfaces of two such blocks are put together it takes a force of hundreds of pounds to tear them apart. One step further and it might become possible to produce blocks with two such perfect surfaces that when brought together they would exert upon each other their full molecular attraction, and, to all practical purposes, become one. In other words, welding by heat, which is nothing but bringing two metals to a state in which molecular attraction becomes possible, might be replaced by machining the two parts to a condition where the same result would be brought about.

Such an accomplishment admittedly would be difficult and probably impossible under our present knowledge. It is mentioned, however, as an illustration of the great new fields that loom ahead of the modern engineer and that have been brought within the range of possibility by the extension of our knowledge of producing and handling engineering materials.

## Observations on the Work of the National Public Works Department Association[1]

The technical men who started the public works movement ought to know the accomplishments to date so that they may plan wisely for the future. This is a period of recess in legislative matters and we are given an opportunity to derive lessons from the events of the past. The National Public Works Department Association therefore submits this review with confidence that the supporters of the movement will have no reason to regret their participation therein.

Like all other movements of this kind, there are stages in which there is much visible progress and others in which development can be discerned only by those in intimate touch. We are now in one of the latter stages in which the principal effort is the preparation for achievement at the fall session of Congress. There could be no final success without these interim developments for they are as necessary to the final enactment of the measure by Congress as is the President's approval of the finished document following its passage.

We have won decisively in the first phase of this campaign, which is the establishment of the underlying principle. If every indication can be believed a Department of Public Works will be created. The leaders of Congress are in agreement on the principle and some of them are working on its details during the present legislative recess. The candidates for president and vice-president have signified their approval and in this they are in step with the great leaders of thought and action the country over.

The second phase of the campaign, which consists of securing the right kind of a department of public works, will be more laborious than the first. It involves the minute consideration of

[1] The National Public Works Department Association is a league composed of individuals, associations and of national, state and local societies, having an aggregate membership of over 100,000 business men, engineers, architects, constructors, manufacturers, chemists, geologists and economists. Its purpose is to organize under one department the many and varied public works functions of the Federal Government. M. O. Leighton is chairman of the Association, and C. T. Cheney, secretary. The Association's headquarters are in the McLachlen Bldg, 10th and G St., Washington, D. C.

details, the sifting of evidence and the presentation of the results in convincing form. The supporters of this movement should not make the mistake of believing that the campaign is over merely because the principle has been accepted. It is obvious that the wrong kind of a Department of Public Works will be as bad if not worse than the present organization of governmental functions.

The most troublesome as well as the most inexcusable setback has been the failure of the movement in the referendum of the United States Chamber of Commerce. This was plainly due, first to the lack of adequate publicity, and second, to the inaction of the local engineering bodies which were depended upon to take care of the issue in their own local commercial organizations. Our finances would not permit us to handle the first, but we had every good reason to expect diligence on the part of " our own crowd " in the second. Evidence is constantly being accumulated that the important commercial bodies which voted adversely on that referendum did so in ignorance of the character, intent, and purport of the movement.

The principles underlying the public works movement—economy and efficiency and businesslike conduct of the Government's affairs—are not confined to the public works functions. The need is apparent in every line of Government functions. The Republican Party, for example, has adopted the principle as one of the planks in its platform. Reorganization of all departmental functions is as certain and steadfast a part of the Congressional legislative program as any other feature now before the public. It has been remarked in places that the public works department advocates may have done their job too well and have put forward so logical and persuasive a set of facts and principles that they must perforce be applied to the entire government structure with the possibility that the creation of the public works department may be delayed pending reform of all government functions. We are assured, however, that there will be no danger of such a result if the technical men of the country faithfully continue their activities on behalf of a public works department.

One of the most persuasive arguments against a proposal to delay the public works feature pending the announcement of a general reorganization bill is that previous efforts toward general reorganization have failed, first, because of the size and complexity of the job and, second, because it encountered so many points of opposition in the Federal bureaus. It is the opinion of such leaders in Congress as Senators Smoot of Utah and McCormick of Illinois. of Speaker Gillett of the House and of Representatives Madden of Illinois and Reavis of Nebraska that success will depend on the selection of one set of functions and the logical adjustment of them so that the underlying principle may be established in statutory form. It is agreed that the first and most logical set of functions is that which comprises the nation's public works.

From the present until the convening of Congress on the first Tuesday of December next, the engineers, architects, and technical men of other branches throughout the country have a rare opportunity to further this movement, because Congress is in recess and the representatives are at home. If each man who is convinced of the necessity for. this reform would embrace this opportunity to instruct his representative in Congress and to enlist his active coöperation in the furtherance of this legislation, progress at the next session of Congress would be rapid and certain. For the present, members of Congress are almost out of reach of the Association's officers. The field of intensive effort has shifted " back home " where the technical men of the country may render "first aid."

Washington, D. C.                                        M. O. LEIGHTON.

## Report on Industrial Conference Available

The American Society of Mechanical Engineers has on file at headquarters a limited number of copies of the Report of the Industrial Conference called by the President on December 1, 1919. The Society will be glad to furnish copies to those interested.

## Charles Whiting Baker Retires from Journalism to Enter New Field

Charles Whiting Baker, Mem.Am.Soc.M.E., formerly editor-in-chief of *Engineering News* and later consulting editor of *Engineering News-Record*, announces his retirement from engineering journalism, in which he has so long been identified. to become managing director of a new service for the engineering and technical industries. Realizing the difficulty often experienced in bringing together those who have engineering properties to sell, and possible purchasers who have had experience in the fields of these industries and are qualified to manage them, Mr. Baker is to establish an engineering business exchange where owners and qualified customers may conveniently meet.

Mr. Baker has been connected with engineering journalism for thirty-four years, during twenty-two of which he was editor-in-chief of *Engineering News*. Readers of this publication and of *Engineering News-Record* are quite. familiar with the illuminating quality of his writings and the breadth of view so often expressed in his editorial comment. As editor of a journal of civil engineering, he had to deal with much construction work of a public character, and his editorials have consistently discussed the broad aspects of such enterprises in which the public is vitally concerned. He has never been one of those who regard the field of engineering as restricted to the materials, the laws, and the forces of nature. On the contrary, he has been a pioneer in advocating the close association of the engineer with matters concerning the public interest and the human relations of society—matters which are now so much discussed in connection with the formation of the Federation of the American Engineering Societies.

In wishing Mr. Baker unqualified success in his new undertaking, we must also tell him that in the years to come we shall greatly miss his frequent and valued contributions to the technical press—we hope they still may be occasional.

## Minnesota Proposes State Federation of Architects and Engineers

Representative architects and engineers from all parts of Minnesota recently met at Duluth and took the first step toward the organization of a State Federation of Architects and Engineers. The unanimous sentiment of those in attendance, reports *The American Architect*, was in favor of such a federation. Only by unification into one state-wide organization will the engineers and architects of Minnesota have the power and weight of numbers behind them to force attention to matters of public concern, having to do with problems of engineering and architecture, or with the regulation of affairs affecting their joint interests.

It was pointed out by Max Toltz, chairman of the meeting. that there are about 4800 engineers, architects, and draftsmen in Minnesota. Many of these men are not identified with any existing organization. It is not the purpose of the men who are back of the proposed federation to supplant any existing organization.

## National Research Council Elects New Officers

The National Research Council, which was organized in 1916 under the direction of the National Academy of Sciences, has just elected new officers for the year beginning July 1, 1920. These are as follows: Chairman, H. A. Bumstead, professor of physics and director of the Sloane Physical Laboratory, Yale University; First Vice-Chairman, C. D. Walcott, president of the National Academy of Sciences and secretary of the Smithsonian Institution; Second Vice-Chairman, Gano Dunn, president of the J. G. White Corporation, New York; Third Vice-Chairman, R. A. Millikan, professor of physics, University of Chicago; Secretary, Vernon Kellogg, professor of biology, Stanford University; Treasurer, F. L. Ransome, treasurer of the National Academy of Sciences.

The National Research Council was one of the potent factors in development of scientific and research problems during the war. Originally created to deal with such problems, it was reorganized in 1918 on a peace-time basis and is now functional through five divisions, namely those of Physical Sciences, Engineering, Chemistry and Chemical Technology, Geology and Geography, and Biology and Agriculture.

## U. S. Chamber of Commerce Records Its Views on Industrial Relations

Overwhelming approval of a platform setting up twelve principles of industrial relations has been given by the membership of the Chamber of Commerce of the United States in a referendum vote, the result of which were recently announced. The vote taken was on the report of a special committee of the Chamber's board of directors. This report went deeply into the subject of the employment relation and recommended among other things recognition of the right of open-shop operation and the right of employers and employees to deal directly with each other without participation by outside interests.

At the same time the Chamber's membership in another referendum vote has approved a report of its Committee on Public Utilities recommending that strikes by public-utility employees should be explicitly prohibited and that tribunals should be created by law to adjudicate in decisions binding on both parties differences between public utilities corporations and their employees. The vote on the two referenda was the largest ever recorded by the Chamber on any subject.

## Sir Robert Hadfield Prize

Sir Robert A. Hadfield, D.Sc., D.Met., F.R.S., has placed in the hands of the Institution of Mechanical Engineers (England) the sum of £200, the income therefrom to be awarded at the discretion of the Council of the Institution as a prize, or as prizes, for the description of a new and accurate method of determining the hardness of metals, especially of metals of a high degree of hardness.

Competitors should familiarize themselves with the ordinary tests of hardness, such as are described in the Report of the Hardness Tests Research Committee (*Proceedings* of the Institution of Mechanical Engineers, 1916, pages 677 to 778). What is desired is the description of a research for or an investigation of some method of accurately determining hardness, suitable for application in metallurgical work in cases in which present methods partially fail.

The Council of the Institution will consider annually all communications received, and may then award a prize or prizes. But in January, 1922, the offer of prizes will be withdrawn, and any unexpended balance of the Prize Fund will be diverted to any other purposes to be determined at the discretion of the Council. Communications should be accompanied by scale drawings of any new apparatus described, or by a model or an example of the apparatus itself. If the communication describes a new invention likely to be of commercial value, it is desirable that provisional protection should have been obtained before it is submitted for consideration. Communications should be addressed to The Institution of Mechanical Engineers, 11 Great George Street, Westminster, London, S.W. 1.

## Monument to Wilbur Wright Is Presented to Le Mans

Tribute to the life and work of Wilbur Wright, American pioneer in aviation, was recently paid by the unveiling in Le Mans, France, of a superb monument. The occasion of the unveiling, on July 17, was a notable one, being attended by a large number of prominent Frenchmen and Americans.

Speeches were made by Baron d'Estournelles de Constant, president of the Wilbur Wright Committee; M. Castille, Mayor of Le Mans; Rear-Admiral T. P. Magruder, representing Hugh C. Wallace, the American Ambassador to France; Comte Henri de La Vauls, M. Stephen Lausanne and Mr. Myron T. Herrick, former United States Ambassador. The dedication exercises concluded with the presentation of the Grand Cordon of the Legion of Honor to Louis D. Beaumont, donor of the monument, a governor of The Aero Club of America, and a prominent patron of aviation during the war.

The Wilbur Wright monument, facing the east in the Place des Jacobins, is the work of the French sculptor Paul Landowski and the architect Bigot. Standing 40 ft. high, its base is carved with figures in bas-relief of Wilbur and Orville Wright and of Léon Bollée. It also carries the names of the precursors of aviation from the time of Dedalus, listing the heroes who gave their lives for the development of the science. It up-

THE WILBUR WRIGHT MONUMENT

holds at its summit the splendid figure of a man reaching to the skies.

Admiral Magruder, who represented the American Ambassador, spoke in French. He said in part:

The names of Wilbur and Orville Wright have added themselves to those of the great American inventors, Fulton, Howe, Morse, Edison, and I am proud of the thought that my fellow-citizens have contributed to give the world a finer life than before. Americans are a practical race. What I admire among you, my friends of France, is that you know how to join the ideal with the practical. One will never forget the names of Latham, Blériot, Pégoud, aviators before the great war. During the war, Wright's invention permitted other Frenchmen to make themselves ever famous—Garros, Gilbert, and the incomparable master of the air, Guynemer.

Under the Admiral's supervision a stalwart American marine placed a wreath at the foot of the statue, the gift of the townspeople of Wilbur Wright's home, Dayton, Ohio.

Former Ambassador Herrick recalled the prominence of his home state, Ohio, which has given Edison and the Wrights to civilization, and rehearsed the steps which brought Wilbur Wright to France in 1908. He also lauded the names of American inventors from the time of Fulton and Franklin, revealing their close relation, in their final success, to France and French assistance, and he concluded with a fervent appeal for the fostering of a common Franco-American spirit.

## Amos Whitney Dies

Amos Whitney, one of the most conspicuous figures in the early growth of the New England machine-tool industry, a member of the A.S.M.E. since 1913, died on August 5, at the Poland Springs House, Poland Springs, Me. Mr. Whitney was born 88 years ago on October 8, 1832, in Biddeford, Me. He came from distinguished Colonial and English ancestry. In this country the family has continuously held a prominent place, many of its members showing decided mechanical tastes as Eli Whitney, inventor of the cotton gin, Baxter D. Whitney, the Winehendon machine builder, and Amos Whitney.

He was apprenticed at the age of 14 to the Essex Machine Co. of Lawrence, Mass. Six years later he moved to Hartford where he was employed at Colt's Armory, at that time the headquarters for many New England mechanics.

There he met Francis E. Pratt, who left shortly afterward to take charge of the Phoenix Iron Works, and Asa A. Cook. In 1853 Mr. Cook also went to the Phoenix Works as a contractor,

AMOS WHITNEY

taking young Whitney with him as a full partner. While there Mr. Whitney became closely associated with Mr. Pratt, who designed the "Lincoln" milling machines. Whether the idea of the milling machine came originally from Windsor, Vt., as some claim, or originated in Hartford, is difficult to prove at this time; but at any rate the machine is known as the Lincoln milling machine all over the world. In 1860 they determined upon entering business together. They rented a small room where, in addition to their regular employment, they started the manufacture of a little machine for winding thread known as a spooler. Within two years the business had increased to such an extent that they gave up their positions at the Phoenix Works and in 1865 erected their first building on the present site. In 1869 the Pratt & Whitney Co. was formed with a capital of $350,000, later increased to $500,000, and in 1893 reorganized with a capitalization of $3,000,000. In 1893 Mr. Whitney was made vice-president; later he was president, in which office he continued until January 1901, when the control of the company was acquired by the Niles-Bement-Pond Co. Mr. Whitney remained as one of the directors.

Mr. Whitney's unbounded optimism was well displayed when the Pratt & Whitney Co. went through its first panic. It kept right on making standard machine tools, but selling almost nothing, until all the available storage room was filled. Then a large space was hired from the Weed Sewing Machine Co. and when

this was filled another large space was hired in Colt's "West Armory" and this in turn was filled with finished machinery. It is well to note as a matter of history that when this immense stock finally did begin to move, it was practically sold out in 30 days.

Mr. Whitney took an important part in the development, in this country, of standard measuring instruments, one of the first moves being a determined effort to secure a standard inch block. His company purchased at considerable expense a standard rectangular bar, 1 in. square and 12 in. long, which had been used as a standard of measurement. Twelve 1-in. cubes were then made as accurately as possible and tested by the 12-in. piece. It was found that the twelve 1-in. cubes were not as long as the single bar, supposed to be exactly 12 in. long. Careful measuring and comparison with such standard instruments as were available led the company to believe that the individual inch-blocks were more nearly accurate than the long piece, and this was afterwards proved by the Rogers-Bond comparator, which was developed in the Pratt & Whitney Works.

At the time of his death Mr. Whitney was president of the Gray Telephone Pay Station Co., and treasurer of the Whitney Manufacturing Co., Hartford, which was organized by his son, Clarence.

## Death of Isham Randolph

Isham Randolph, consulting engineer and for fourteen years chief engineer of the Sanitary District of Chicago, died on August 2 in that city. Mr. Randolph was born in Clarke County, Virginia, in 1848. At the age of 20 he started work as an axeman for the Winchester & Strasburg Railroad, thus beginning a long career in which railroads played so large a part.

From 1880 to 1885 he was chief engineer of the Chicago & Western Indiana Railway and the Belt Railway of Chicago, at the end of which period he began private practice in Chicago. The following year he became chief engineer on the construction of the Chicago, Madison & Northern Railway and Freeport & Dodgeville Railway for the Illinois Central Railroad. In 1888 he resumed general practice in Chicago, and was consulting engineer for the Union Stockyard & Transit Co., the Calumet Terminal Railway, and the Baltimore & Ohio Railroad. From 1893 until 1907, when he resigned he was chief engineer of the Sanitary District of Chicago. He was retained, however, as consulting engineer until the end of 1912.

In 1905 Mr. Randolph was appointed by President Roosevelt on the Board of Consulting Engineers for the Panama Canal, and he was one of the five members of the board whose minority report was accepted by the President and Secretary of War, approved by the Panama Commission and adopted by Congress. Again, in 1908, he was one of the six engineers who accompanied President-elect Taft to Panama to consider the advisability of changing the plans for the Canal.

He was chairman of the Internal Improvement Commission of Illinois which made plans for a canal from Lockport to Utica, Ill., of the State Conservation Commission, of the State Rivers and Harbors Commission and of the Chicago Harbor Commission. He made an engineering study, report and plans for a commercial harbor for Milwaukee and was consulting engineer. From 1893 until 1907, when he resigned, he was chief engineer for Toronto and Baltimore on track-elevation work. He served as a member of the Toronto Water Supply Commission. He was also consulting engineer for the Little River Drainage District of Cape Girardeau, Mo., and reviewed the plans and estimates for the Lake Erie and Ohio River Barge Canal. He was chairman of the Florida Everglades Engineering Commission and made a complete report on the drainage of the swamp lands of Florida. The report was accepted by the Governor. He designed the harbor system for Miami, Fla. Upon appointment by the Queen Victoria Niagara Falls Park Commission he designed the Obelisk dam above the Horse Shoe Falls.

Mr. Randolph was past-president of the Western Society of Engineers, and a member of the American Society of Civil Engineers and the American Institute of Consulting Engineers.

# The Federated American Engineering Societies

THE Joint Conference Committee, which is acting for The Federated American Engineering Societies in accordance with the instruction of the Organizing Conference held in Washington, D. C., June 3-4, 1920, has sent out to approximately 150 engineering and technical organizations an invitation to become a charter member of the new Federation. An invitation has also been extended to the same societies to appoint delegates to the first meeting of the American Engineering Council which is being planned for this fall and of which more extended notice will be given later.

Copies of the constitution and by-laws of the Federation have also been sent to the above societies. This constitution, which is printed below in full has just been released for publication in accordance with the instruction of the Washington Conference who referred it to an editing committee for revision. MECHANICAL ENGINEERING presents this constitution and by-laws with the hope that its readers will study the form of the new organization which is being set up, for as time goes on changes will doubtless be made, and if the Federation is to prosper it is necessary that these modifications be made in the light of widespread criticism. To this end MECHANICAL ENGINEERING will welcome correspondence dealing with all matters pertaining to the Federation.

## Constitution

### PREAMBLE

ENGINEERING is the science of controlling the forces and of utilizing the materials of nature for the benefit of man, and the art of organizing and of directing human activities in connection therewith.

As service to others is the expression of the highest motive to which men respond and as duty to contribute to the public welfare demands the best efforts men can put forth, Now, THEREFORE, the engineering and allied technical societies of the United States of America, through the formation of The Federated American Engineering Societies, realize a long-cherished ideal—a comprehensive organization dedicated to the service of the community, state, and nation.

### ARTICLE I. NAME

The name of this organization shall be The Federated American Engineering Societies.

### ARTICLE II. OBJECT

The object of this organization shall be to further the public welfare wherever technical knowledge and engineering experience are involved, and to consider and act upon matters of common concern to the engineering and allied technical professions.

### ARTICLE III. MEMBERSHIP

SECTION 1. *Scope.* The membership may consist of national, local, state and regional engineering and allied technical organizations and affiliations, classified as follows:

1 National engineering and allied technical organizations
2 Local, state, and regional engineering and allied technical organizations other than local associations, sections, branches, or chapters of national organizations
3 Affiliations consisting of any one or a combination of the following constituents:
(a) Local sections, branches, chapters, and associations of members of national organizations included under (1)
(b) Local engineering and allied technical societies and clubs, not of national scope
(c) Local engineers and members of allied technical professions and their associates.

SECTION 2. *Qualifications.* The qualifications for membership shall be as provided in the By-Laws.

SECTION 3. *Application for Membership.* Application for membership shall be made in the form and manner prescribed in the By-Laws.

SECTION 4. *Termination of Membership.* The membership of any constituent organization may be terminated by it or by the American Engineering Council in the manner provided in the By-Laws.

### ARTICLE IV. MANAGEMENT

SECTION 1. *Managing Body.* The management of this organization shall be vested in a body to be known as "American Engineering Council," and its Executive Board.

### AMERICAN ENGINEERING COUNCIL

SECTION 2. *Functions.* The American Engineering Council shall consist of representatives of Member-Societies selected as hereinafter provided. This council shall coördinate the activities of state councils and of local and regional affiliations whenever these activities are of national or general importance or may affect the general interests of engineers.

SECTION 3. *Representation.* Each national, local, state, and regional organization and affiliation shall be entitled to one representative on the Council for a membership of from one hundred to one thousand inclusive and one additional representative for each additional one thousand members or major fraction thereof; provided, that in the determination of the representation of local, state, and regional organizations and affiliations no count shall be taken of any organization which is represented individually or through another local, state or regional organization or affiliation; and, provided further, that no organization shall have more than twenty representatives on the Council.

SECTION 4. *Selection of Representatives.* Representatives on the Council shall be selected as stipulated in the By-Laws.

SECTION 5. *Meetings.* The Council shall hold an Annual Meeting. Other meetings may be called by the Executive Board and shall be called by it upon the written request of twenty-five representatives on the Council.

SECTION 6. *Officers.* The elected officers of the Council shall consist of a President, to hold office for two years, who shall be ineligible to reëlection, four Vice-Presidents, to hold office for two years, two to be elected each year, and a Treasurer, to hold office for one year. These officers shall be elected by a letter ballot of the representatives on the Council as provided in the By-Laws. There shall be an Executive Officer who shall also be Secretary appointed by and holding office during the pleasure of the Executive Board. He shall not be a member of the Executive Board but may be a representative on the Council.

### EXECUTIVE BOARD

SECTION 7. *Functions.* There shall be an Executive Board of the Council constituted as hereinafter provided and charged with conducting the business of the organization under the direction of the Council.

SECTION 8. *Membership.* The Executive Board shall consist of thirty members, of whom six shall be the officers elected by the Council and twenty-four shall be selected, a part by the national societies, and the remainder by the local, state and regional organizations and affiliations according to districts, as provided in the By-Laws; provided, that the number of members representing the national societies on the Executive Board shall bear as nearly as may be the same ratio to the number of members representing the local, state and regional organizations and affiliations thereon, as the number of representatives of the national societies on the Council bears to the number of representatives of the local, state and regional organizations and affiliations thereon.

SECTION 9. *Electoral Districts.* For the purpose of facilitating the selection of the district members on the Executive Board the Council shall, as provided in the By-Laws, divide the United States into districts based upon an equitable representation, having regard to both its membership and area.

SECTION 10. *Officers.* The President and Secretary of the Council shall be, respectively, the Chairman and the Secretary of the Executive Board. There shall be two Vice-Chairmen, elected by the Board from its members.

### ARTICLE V. UNEXPIRED TERMS

Vacancies in the offices of the President, the Vice-Presidents, and the Treasurer and in the Executive Board and among the representatives on the American Engineering Council shall be filled as soon as feasible by the agencies originally selecting the incumbents. The officers and representatives thus chosen shall serve for the unexpired terms.

### ARTICLE VI. FUNDS

SECTION 1. Funds for the use of the organization shall be contributed as follows:
(a) Each national Member-Society shall contribute annually one dollar and fifty cents ($1.50) per member
(b) Each local, state and regional organization and affiliation Member-Society shall contribute annually one dollar ($1.00) per member. No portion of such funds shall be applied to the use of local affiliations or state councils.

SECTION 2. The American Engineering Council may receive and administer gifts, bequests or other contributions for carrying out the purposes of the organization.

### ARTICLE VII.  LOCAL AFFILIATIONS

SECTION 1. *Object.*  The American Engineering Council shall encourage the formation of local affiliations to consider matters of local public welfare with which the engineering and allied technical professions are concerned, as well as other matters of common interest to these professions, in order that there may be united action and that suggestions and advice may be offered to the Council.

SECTION 2. *Constitution.*  Each local affiliation desiring membership in this organization shall submit its constitution and by-laws, and all subsequent amendments thereof, to the Executive Board of the Council for the approval of such portion or portions thereof as may affect its eligibility or its relation to the work of the Council.

### ARTICLE VIII.  STATE COUNCILS

SECTION 1. *Object.*  State councils, consisting of representatives of local affiliations within the state or otherwise representative of the majority of engineers and members of allied technical professions in the state, if members of this organization, shall consider state matters of public welfare with which the engineering and allied technical professions are concerned, as well as other matters of common interest to these professions, in order that there may be united action in state affairs.

SECTION 2. *Constitution.*  Each state council desiring membership in this organization shall submit its constitution and by-laws, and all subsequent modifications thereof, to the Executive Board of the American Engineering Council for the approval of such portion or portions thereof as may affect its eligibility or its relation to the work of the Council.

### ARTICLE IX.  DELIMITATION OF AUTHORITY

Local organizations and affiliations, state organizations and councils, if members of this organization, and the American Engineering Council shall deal with local, state, and national matters, respectively, and shall be autonomous with respect thereto. It shall be the duty of the American Engineering Council to interest itself in the activities of local and regional organizations and affiliations and state organizations and councils if such activities are of national scope, or affect the general interest of the engineering and allied technical professions; provided, that nothing herein stated shall be construed as preventing the discussion by any local and regional organization and affiliation or state organization and council or by the American Engineering Council of any matters of interest to engineers and members of allied technical professions, or action by that Council on local or state matters where there is no local or regional organization or affiliation or state organization or council.

### ARTICLE X.  PUBLICITY

This organization shall stand for the principle of publicity and open meetings, under such provisions as may be provided in the By-Laws.

### ARTICLE XI.  AMENDMENTS

SECTION 1.  An amendment to this constitution may be proposed by the Executive Board.

SECTION 2.  An amendment may be proposed, in writing, by at least twenty-five representatives on the American Engineering Council. Such amendment shall be considered first by the Executive Board, which shall report on it to the Council, approving or disapproving or modifying it or submitting an alternative amendment. This report shall be accompanied by the original proposed amendment.

SECTION 3.  An amendment proposed as provided in Sections 1 and 2 shall be considered at a meeting of the Council after it has been submitted to its members at least ninety days in advance of the meeting. At such meeting, provided a majority of the representatives be present, the amendment may be rejected, or otherwise submitted to the members of the Council for letter ballot within thirty days thereafter, with such modifications as may be adopted by a majority of those present. The amendment shall require a two-thirds affirmative vote for its adoption.

## By-Laws

### CHAPTER I.  MEMBERSHIP

SECTION 1. *Qualifications.*  Any society or organization of the engineering or allied technical professions whose chief object is the advancement of the knowledge and practice of engineering or the application of allied sciences and which is not organized for commercial purposes is eligible for membership.

SECTION 2. *Admission.*  The Executive Board shall submit to a letter ballot of the American Engineering Council, each application for membership made to it on a prescribed form, accompanied by a

statement of its findings as to eligibility and the number of representatives to which the applicant would be entitled. The applicant shall be admitted by a majority vote of the Council, provided, that not more than twenty-five per cent of the members of the Council shall vote in the negative.

SECTION 3. *Termination of Membership.*  1.  Any Member-Society may terminate its membership on June 30 or December 31 of any year after giving at least three months previous written notice to the Secretary of its intention thereof, provided, however, that the financial obligations of such organization are discharged to the said June 30 or December 31, respectively.

2.  On complaint brought by any three members of the Council, and transmitted in writing to the Secretary giving reasons why the membership of any Member-Society should be terminated, the Committee on Membership and Representation of the Executive Board shall investigate said complaint, inform itself of all matters pertaining thereto, and report its findings to the Executive Board, which may either dismiss the proceedings or make recommendations to the Council. The Council, by a two-thirds vote of those present at a meeting, may dismiss the proceedings or order a letter-ballot.

3.  A two-thirds affirmative vote shall terminate membership.

### CHAPTER II.  MANAGEMENT

SECTION 1. *Terms of Representatives.*  Representatives on the American Engineering Council shall serve for two years; provided, that after the first election, if there is more than one representative from one organization, approximately one-half the number of the representatives shall be elected each year.

SECTION 2. *Announcement of Representatives.*  Each organization represented shall send to the Secretary of the Council on or before August 15 of each year the names of its representatives who are to serve for the term beginning January 1 of the following year.

SECTION 3. *Votes of Representatives.*  Representatives on the Council and members of the Executive Board shall each have one vote in meetings of these bodies.

SECTION 4.  1. *Meetings.*  At all regular meetings of the Council the order of business shall be as follows:

   (a) Roll call of representatives
   (b) Approval of minutes of previous meeting
   (c) Report of Secretary
   (d) Report of Treasurer
   (e) Report of President
   (f) Report of Executive Board
   (g) Report of committees
   (h) Unfinished business
   (i) Special business
   (j) New business.

2. *Rules of Order.*  Unless otherwise provided, Robert's Rules of Order shall govern the proceedings of all meetings of the Council.

3. *Quorum.*  A majority of all the representatives on the Council shall constitute a quorum at any of its meetings.

SECTION 5. *Nomination and Election of Officers.*  1. The Secretary shall send to each member of the Council at least ninety days in advance of the Annual Meeting, nomination blanks for offices to be filled at that meeting. Nominations received within thirty days shall be canvassed by the tellers appointed by the Executive Board and reported by them to the Board. The Board shall place upon the ballot the names of the three candidates receiving the highest number of votes.

2. The Secretary shall mail to each member of the Council, at least thirty days before the Annual Meeting, the ballot containing the names of the above selected nominees for each office.

3. Ballots received before 7 a. m. of the first day of the Annual Meeting shall be canvassed by tellers appointed by the Executive Board, and the result shall be certified to the President of the Council, who shall announce the result of the election at the Annual Meeting. A plurality of votes shall elect; in case of a tie vote the Annual Meeting shall immediately select by ballot one of the two candidates.

SECTION 6. *Duties of Officers.*  1. The terms of all officers elected at an Annual Meeting of the Council shall commence on the adjournment of that meeting.

2. The officers shall have the usual duties pertaining to their respective offices except as may otherwise be provided in the Constitution and By-Laws.

3. It shall be the duty of the President to represent the Council on any formal occasion.

4. The Vice-Presidents in the order of their seniority of election and age shall, in the absence or disability of the President, discharge his duties.

5. The Treasurer shall receive all moneys and shall deposit them in the name of the Council with a bank or trust company approved by the Executive Board. He shall invest all funds not needed for current disbursements as ordered by the Executive Board. He shall pay all bills covering expenditures authorized by the budget or the Executive Board, by checks countersigned by the Chairman of the Finance Committee or some other member thereof. He shall make an annual report and such other reports as may be prescribed by the Executive

Board. He shall give a bond at the expense of the Council, in amount and with surety satisfactory to the Executive Board.

6. The Executive Officer shall be appointed and his compensation fixed annually by the Executive Board and he shall hold office during its pleasure. He shall be the Secretary of the Council and of its Executive Board. He shall manage the business of the Council under the direction of the Executive Board and shall perform such duties as may be assigned to him by the Council or the Executive Board. He shall be the custodian of the property of the Council. He shall collect all moneys due the Council and transfer them to the custody of the Treasurer. He shall scrutinize all expenditures and use his best endeavors to secure economy in the administration of the business of the Council. He shall certify to the accuracy of all bills or vouchers on which money is to be paid. He shall give a bond at the expense of the Council in amount and with surety satisfactory to the Board. He shall pay the current expenses of the office, and for this purpose he shall have at his disposal a suitable sum of money to be fixed by the Board, which amount shall be periodically replenished under the authority of the Finance Committee upon the presentation of an account of disbursements in the form required by it. He shall mail to the Member-Societies bills for their annual contribution thirty days prior to the beginning of the fiscal year. He shall perform such other duties as may from time to time be assigned to him by the Council or the Executive Board.

SECTION 7. *Selection of Executive Board.* 1. The Secretary shall submit to the Executive Board, at its September meeting, a list of Member-Societies in good standing with their respective memberships.

2. The Executive Board shall thereupon determine the number of its members for the next ensuing administration year to be selected by national societies, and by the representatives of the local, state, and regional organizations and affiliations, as provided in the Constitution, and it shall prescribe the boundaries of the districts from which the representatives of the local, state, and regional organizations and affiliations are to be selected.

3. The Secretary, within two weeks after the September meeting of the Executive Board, shall mail to the proper officer of each Member-Society a copy of his report on membership and a statement of the Board's action with respect to the membership of the Board, and its delimitation of districts.

4. Members of the Executive Board representing national societies shall be selected by the said societies in such manner as they may determine; the secretary shall be advised of this selection.

5. Members of the Executive Board representing local, state, and regional organizations and affiliations shall be nominated and elected from each of the several districts by the representatives therefrom on the Council, at a meeting duly called for the purpose, provided, three-fourths of all representatives of said district are present or represented; the selection shall be reported to the Secretary.

SECTION 8. *Attendance at Meetings.* If any member of the Executive Board or any of its committees shall fail to attend the meetings of said Board or committee for a period of five consecutive months, not including July and August, he shall cease to be a member of said Executive Board or committee. The vacancy thus created shall be filled as provided for in the Constitution in Article V. on Unexpired Terms.

SECTION 9. *Duties of Executive Board.* 1. The Executive Board shall organize within thirty days after the adjournment of the Annual Meeting of the Council.

2. It shall hold regular monthly meetings except during July and August. Regular meetings shall be held on the second Monday of each month except that a regular monthly meeting shall be held in connection with the meeting of the Council.

3. Special meetings may be called at the discretion of the President and shall be called at the written request of five members of the Executive Board.

4. The Secretary shall mail the notices of each regular meeting at least fifteen days in advance thereof and shall mail notices of each special meeting, stating its purpose, at least ten days or shall telegraph them at least six days in advance, to each member of the Board. No business other than that for which it has been called shall be transacted at a special meeting.

5. A quorum of the members of the Executive Board shall be fifteen members.

6. The Executive Board, unless otherwise provided, shall appoint all special committees of the Council and of the Executive Board. The membership of such committees may be drawn from the membership of the Council or of the Member-Societies.

7. The Executive Board shall, whenever practicable, provide for the whole or a part of the expenses of members or of representatives attending its own meetings and those of the Council.

SECTION 10. *Appointment of Committees.* The following committees shall be appointed from the membership of the Board annually by the incoming President with the approval of the Executive Board, each member to serve one year or until his successor is appointed:

(a) On Procedure
(b) On Constitution and By-Laws
(c) On Publicity and Publications
(d) On Membership and Representation
(e) On Finance
(f) On Public Affairs.

SECTION 11. *Duties of Committees.* 1. The Committee on Procedure shall act for the Executive Board in the interim between its meetings, and shall perform such other duties as may be assigned to it.

2. The Committee on Constitution and By-Laws shall report on all proposed amendments, with any modifications thereof it may deem desirable, referred to it by the Executive Board.

3. The Committee on Publicity and Publications shall, when so requested, prepare all public statements and shall have the direction of publications of the Council.

4. The Committee on Membership and Representation shall report to the Executive Board on the eligibility of each applicant for membership and the number of representatives to which it would be entitled. It shall review and report at least ninety days before each regular meeting of the Council the number of representatives to which each Member-Society is entitled. It shall review the existing electoral districts and report thereon to the Executive Board, at least once every two years. It shall report on all questions regarding registration and credentials of the representatives on the Council.

5. The Finance Committee shall have supervision of the finances of the organization. It shall report an annual budget to the Executive Board at least sixty days prior to the end of each fiscal year. The Chairman or some other member of the Committee shall countersign all checks for the payment of money.

6. The Committee on Public Affairs shall report to the Executive Board on all public affairs with which the engineer is concerned or which affect the relation of the engineer to the public.

7. Special committees shall report to the Executive Board.

## CHAPTER III. FUNDS

SECTION 1. *Fiscal Year.* 1. The fiscal year shall begin on January 1 of each year.

2. The contributions of each Member-Society shall be payable in advance in semi-annual payments, on January 1 and July 1 of each year, and shall be based upon a certified statement of its membership as of January 1.

3. A Member-Society failing to pay its semi-annual contribution within six months after it is due shall be suspended or dropped from membership at the discretion of the Executive Board.

4. Funds shall be disbursed, as authorized by the budget or the Executive Board, by checks signed by the Treasurer and countersigned by the Chairman of the Finance Committee or some other member thereof.

## CHAPTER IV. PUBLICITY

SECTION 1. The privilege of attendance at all meetings of the American Engineering Council, of the Executive Board, and of Committees, when not in executive session, shall be extended to any proper person, but this privilege does not include the right to speak or vote. Any proper person shall have the right to inspect and make true copies of the official records of all meetings of the Council, the Executive Board, and Committees.

SECTION 2. The Committee on Publicity and Publications may employ a Publicity Secretary whose duty under the direction of the Executive Board, shall be to prepare and supply to the engineering, technical, and general press information concerning The Federated American Engineering Societies, or the engineering world, and to coöperate with the editors of engineering and technical publications in disseminating information in regard to the organization and its activities. This Committee may appoint a coöperating board of engineering editors to counsel and assist in any or all of its activities.

## CHAPTER V. AMENDMENTS

SECTION 1. An amendment to these By-Laws may be proposed by a representative on the American Engineering Council or by a member of the Executive Board. The Executive Board shall consider all proposed amendments and mail a copy of such amendments together with its report thereon to each representative on the Council at least sixty days prior to the date of a regular meeting.

SECTION 2. These By-Laws may be amended by an affirmative two-thirds vote of all representatives on the Council present at a regular meeting thereof, provided, that the proposed amendment shall have been mailed to each representative at least thirty days in advance of such meeting.

SECTION 3. Sections 2, 4 (paragraph 1), 5, 6 (paragraphs 3, 4, 5 and 6), 7, and 10, of Chapter II, of these By-Laws, and any By-Law adopted subsequent to ———— may be amended by a vote of three-fourths of the members of the Executive Board at any regular meeting; provided, that the proposed amendment shall have been presented in writing at a regular meeting of the Board, and transmitted to each member of the Council at least thirty days previous to the date of its adoption.

# Engineering and Industrial Standardization

## A Review of Standardization Work Now in Progress, Both in This Country and Abroad, as Reported by the American Engineering Standards Committee and Foreign Standardizing Bodies

STANDARDIZATION in all its various phases is now receiving, both in this country and abroad, the closest study, and through the coöperative efforts of engineering societies and associations is rapidly being brought to the point where duplication of effort and the promulgation of conflicting standards will be largely avoided.

The American Engineering Standards Committee which is directing the standardization work in the United States is now engaged with the development of a great variety of standards for engineering and industry. The history of the committee will be found in MECHANICAL ENGINEERING for July 1919 (p. 638), and it is thus sufficient to state at this time that while the original committee was composed only of representatives of the American Society of Civil Engineers, the American Institute of Mining Engineers, the American Institute of Electrical Engineers, The American Society of Mechanical Engineers, the American Society for Testing Materials, and the Government departments of War, Navy, and Commerce. Later the constitution was changed so as to permit of the election to membership of other bodies interested and taking part in standardization work. Under this revised constitution five additional bodies have been admitted, thus making a total of 13 organizations now participating in the work of the Committee. These are:

1. Fire-Protection Group, composed of
   National Fire Protection Association
   National Board of Fire Underwriters
   Associated Factory Mutual Fire Insurance Companies.
2. National Electric Light Association
3. National Safety Council
4. Society of Automotive Engineers
5. Electrical Manufacturers' Council, composed of
   Electric Power Club
   Associated Manufacturers of Electrical Supplies
   Electrical Manufacturers' Club.

### STANDARDIZATION WORK IN THE UNITED STATES

By the creation of the American Engineering Standards Committee the joint action of the above-mentioned organizations along the lines of standardization has been greatly facilitated and the formulation of standards in many fields has been made possible. Under its rules the Committee selects the sponsor or joint sponsors for a given standard, who in turn organize the Sectional Committee which develops the standard. For example, The American Society of Mechanical Engineers and the American Gas Association were selected as sponsors for the standardization of pipe threads. Another society, The American Society for Testing Materials, was made sponsor for standards dealing with tests for portland cement and specifications for fire tests for materials and construction.

This spring the Sectional Committee on Structural Shapes for Ships was organized by the American Steel Manufacturers, The American Society of Mechanical Engineers and the American Society of Naval Architects and Marine Engineers acting as sponsors. The American Bureau of Shipping and the United States Navy are also represented on this Sectional Committee and real progress has been made.

Another subject receiving the attention of the Committee is that of the standardization of shafting. The American Society of Mechanical Engineers has been asked to become sponsor for this work since the Society has already done a considerable amount of work on a set of standard diameters for transmission and machinery shafting. It is proposed that the work when carried on by the Sectional Committee shall be broadened to include the standardization of the method of determining what diameters of transmission shafting should be used

for given loads, the dimensions of shafting keys and keyways, and the setting of dimensional tolerances.

### COÖPERATION WITH FOREIGN COUNTRIES

The American Engineering Standards Committee is also seeking to coöperate with similar organizations in Great Britain, France, Switzerland, Belgium and Holland. The Swiss, for example, have requested coöperation on Standards for Ball Bearings, and the formation of a Sectional Committee is now under way. Sectional Committees for Screw Threads and Cylindrical Limit Gages have also been formed. These Committees will act in coöperation with British Committees dealing with the same subject although so far as screw threads are concerned the American Engineering Standards Committee has a satisfactory understanding with National Screw Thread Commission, an abstract of whose latest report was published in 'MECHANICAL ENGINEERING for June (p. 337). The British have also asked for coöperation on Standards for Gears, and The American Society of Mechanical Engineers and the American Gear Manufacturers Association have accepted sponsorship for this work.

Another request has come from The Swiss Standards Association, who have addressed a communication to the American Engineering Standards Committee, proposing the international standardization of the widths across flats on nuts and bolt heads. The proposal covers the range of ¼ in. (6mm.) to 3 in. (80mm.) diameter of bolts. The numerical values proposed are a compromise between the United States Standard, the British and Whitworth, and the metric "Systeme Internationale."

### STANDARDIZATION ORGANIZATIONS ABROAD

In the accomplishment of such international standardization as is mentioned above foreign standardizing bodies will play an important part, and in order that the American Engineering Standards Committee might intelligently work with these foreign associations, its Secretary, Dr. Paul G. Agnew, during a recent trip to England and the continent secured information regarding the attitude of European countries toward international standardization and the means they possessed for bringing it to pass. In his report of his trip Dr. Agnew described the forms of standardizing organizations existing in Belgium, France, the Netherlands and England. Brief abstracts of these descriptions follow.

#### ASSOCIATION BELGE DE STANDARDISATION

This association was organized and began work in 1919. It was organized by the Comité Central Industriel de Belgium with the support of the national engineering societies and industrial associations. The Comité Central, upon which 80 associations are represented, is a strong organization which is playing an important role in the reconstruction of Belgian industry.

There are 37 members of the main committee, 4 representing the Comité Central, 13 representing the national engineering societies, and 20 representing various industrial associations. The latter cover a wide field, among them being the chemical, paper and cotton industries. The government is not represented on the main committee, but is represented on the working committees.

The Belgian association has no arrangement corresponding to sponsorships, the actual standardization being carried out under the general direction of the main committee, and the detailed administration is handled by the central office. Members of the working committees represent the organizations interested in the particular subject in hand, and are named by the organizations. The usual size of these committees is from 10 to 15.

Owing to the industrial position of Belgium, and its commercial relations to other countries, they are planning to make, and in their work so far have made, extensive use of the standardization work of other countries, making such modifications as will adapt it to their needs. Comprised in the work already carried out are a standard series of bolts and rivets. Work now in hand includes reinforced concrete, mechanical power transmission (pulleys, bearings, etc.), water-pipe fittings, specifications for steel bridges, economics in coal consumption, and steel sections. The secretary of the Association is M. Gustave Gerard.

### COMMISSION PERMANENT DE STANDARDISATION (FRANCE)

The French Commission differs from other national standardizing bodies in that it is strictly an official organization, and supported wholly by government funds. It was established by decree of the President of the Republic in June 1918. It is attached to the Ministry of Commerce, Industry, Posts and Telegraphs, Maritime Transport and the Merchant Marine. The Commission itself, which performs much the same functions as our Main Committee, has 24 members, 9 representing various government departments, one representing the Academy of Sciences, and the remainder, national engineering and industrial bodies. Officially, they are appointed by the Minister of Commerce, but they are nominated by the organizations which they represent.

The actual technical work is in charge of 14 committees covering such subjects as: Specifications for metal products; specifications for materials of construction; railways and tramways; machine elements; metallurgy; textiles; heat engines and hydraulic turbines, and materials for the chemical industry.

Special provision was made in the decree establishing the Commission for the coöperation of the government departments in the preparation of specifications, and the subsequent use of the specifications by the departments, but not in such a way as to hamper the departments by restrictions which are too ironclad. The matter is receiving further study in order to enlist the coöperation and support of the many officials in the departments.

The regular procedure is for a committee to submit a tentative specification to the Commission, and the central office submits it to all interested bodies for criticism, (societies, industrial syndicates, government departments, etc.), and the technical press is advised of the matter. The criticisms are digested by the central office, and decisions reached by the committee for final approval of the Commission.

### HOOFDCOMMISSIE VOOR DE NORMALISATIE IN NEDERLAND

The "Main Committee for Standardization in the Netherlands" was organized in 1916, by the Society for the Encouragement of Industry and the Royal Institute of Engineers. Each of these societies names two members of the Main Committee, the total membership of which is 15. The other 11 members are designated by the Committee itself, the term being 3 years. The president, vice-president and treasurer form a small executive committee which meets frequently. There are four regular meetings of the Main Committee a year. As in the case of Belgium, the relatively small size of the country and its industrial and commercial position necessarily affect their standardization work.

Industrial associations (syndicates) do not seem to be very largely developed in Holland. Often there are not more than three or four concerns in an industry. It would appear that these facts have had an influence on the methods adopted in handling the appointment of working committees, etc. When a member of the Main Committee has the necessary qualifications for the chairmanship of a working committee, he is chosen for it. The other members of the standing committees are chosen by the Main Committee.

The central office does a much larger part of the technical work of standardization than is the case with the other national standardizing bodies. The office opened in July, 1918. There are a total of 16 on the staff. There are two technically trained engineers besides the secretary, one electrical and one mechanical, and it is planned to add a civil engineer and a marine architect in the immediate future. There is a drafting room with an experienced man at the head. A very complete information service is maintained on the various phases of engineering standardization. An elaborate index system covers not only formally published standards, but also technical articles bearing on subjects in which standardization work is going on, or in which it is considered likely that such work will be undertaken. Mr. E. Hymans is the secretary.

The office prepares the material for all committee meetings. In taking up a new subject, for example, all available standards and proposals, both in Holland and in other countries, are shown in comparative tabular or graphic form. But the office goes farther and draws up what may seem to it as a reasonable proposal for the consideration of the committee. Of course, the office has no voice in decisions, and is not to attempt to influence decisions.

The form of publication is unique. Standards are issued as single sheets, perforated for binding in loose-leaf folders. The essential information is given in compact form, any necessary explanatory notes being given on the back of the sheet. The general style is that of a working drawing, and the idea is that they shall be issued directly to the draughtsman in the plant. They are lithographed from drawings, and printed on special paper that does not soil. The price is 15 Dutch cents ($0.06) per sheet. New sheets are being issued at the rate of about one per week.

### BRITISH ENGINEERING STANDARDS ASSOCIATION

The British Engineering Standards Committee was organized in 1901 by five leading technical societies. It was incorporated in 1918, and the name changed to Association. The general direction of affairs is in the hands of the Main Committee, which constitutes the sole executive authority. There are 24 members, 19 of whom are appointed by the technical societies and the remaining 5 are chosen by the appointed members, two being from the Federation of British Industries.

Sectional committees are responsible for the technical work in different subjects. Working under the sectional committees are subcommittees, which handle smaller subdivisions of the work. Still smaller subdivisions are handled by small committees termed panel committees, or briefly, panels, to avoid the use of the word subcommittee. Most of the detailed work is done in the panels. There are some 275 of these committees, with 1370 members in all. The members of all the different committees constitute the membership of the Association. A general meeting of the Association is held once a year. The government is represented on most of the working committees, but not upon the Main Committee. There are now 40 on the staff of the central office.

The steps in setting up a committee for a new project are as follows: A request must first come either from an association or a government department. A conference is then called at which all interests to be affected by the proposed work are represented. The purpose of the conference, which is presided over by the chairman of the Association, is to determine whether standardization is to be undertaken. In case the decision is that it shall be undertaken, the chairman of the sectional committee is named by the Main Committee. The members of the Sectional Committee are designated by organizations directly interested in the work of the committee. Generally the members so designated add a very few men of recognized standing to complete the committee.

Great stress is laid by the British upon this procedure which has been evolved as the result of 19 years' experience. They feel that wherever they are responsible industrial associations, it is emphatically best to deal with the industry through the association, and to have the members of the sectional committees appointed by the associations, rather than to have them "hand-picked" by a central body. The addition of a few men by the members who are representatives, makes it possible to obtain the services of outstanding men whose services would not otherwise be secured. The financial support of the Association comes from a variety of sources. In the last year contributions were received from 7 government departments, 47 municipalities, 34 railways, 18 industrial associations, and 98 individual firms.

During the war the government requested the Association to take over the formulation of specifications for aircraft material, and this is being continued. Official governmental representation on the International Aircraft Standards Commission is essentially the same as that in the organization created by the Association for standardization in the domestic industry. A similar arrangement exists in the case of the national committee of the International Electrotechnical Commission. A recent development of the Association has been in the field of shipbuilding. There are two sub-committees—one on ships' fittings and one on machinery. An unusual feature of this work is that there is a geographical division between the Tyne and the Clyde, the work being carried on largely independently in the two regions. Perhaps the most important work in this line is the agreement which has been reached upon standard tail shafts.

### INTERNATIONAL STANDARDIZATION

In concluding his report Dr. Agnew briefly touched upon the question of international standardization. "One heard a great deal of the necessity of international agreement in various lines of engineering result," he writes. "A suggestion has been made by the Swedish organization that there should be a regular publication in which the work of the projects under way in the different organizations could be abstracted. On the continent there has been some discussion of the desirability of a general international organization, modeled somewhat after the national organizations. While some steps have been taken in this direction, there seems to be a general opinion that such an organization would be unwieldy and liable to fall of its own weight. The reason for this view is that in most, if not in all countries, national standardizing organizations are not yet sufficiently developed, and industry itself is not sufficiently well organized, to permit of a general international engineering standards organization. A much less formal approach to the problem, consisting merely of an informal conference of the secretaries of the various national organizations, seems to have received considerable discussion. This last was favorably considered in a meeting of the Main Committee of the British Association, and it is probable that informal inquiries will be addressed to the various national bodies to determine what their attitude might be toward such an undertaking."

# NEWS OF THE ENGINEERING SOCIETIES
American Drop Forge Association and American Society for Testing Materials Hold Annual
Meetings — Society of Automotive Engineers Meet in Semi-Annual Convention

## American Drop Forge Association

TECHNICAL papers on accident prevention, standardization, company stores, pulverized coal and the lubrication of the steam drop hammer, were presented at the annual meeting of the American Drop Forge Association, held at Atlantic City, N. J., June 17, 18 and 19.

Accident prevention with special reference to forge shops was discussed by G. A. Kuechenmeister, Dominion Forge and Stamping Co., Walkerville, Ont., who described scale and treadle guards in use by the Dominion company, explaining how the use of a sweat pad in connection with goggles had been a factor in getting the goggles used, and outlined means for stimulating the interest and enlisting the coöperation of all employees in safety work.

A paper on the standardization of die blocks was read by C. B. Porter, president of the Sizer Forge Co., Buffalo, and chairman of the standardization committee of Forgemen's Exchange. Standardization of sizes of blocks into the fewest number of sections and lengths, he explained, would facilitate prompt deliveries more than any other factor, because it would allow orders from several customers to be combined and made in lots. He suggested that the users make up a list of standard sections with a few standard lengths for each section. The maker could then easily determine the percentage of replacements that usually occur in each grade and size of block and could add the proper number of stock order blocks to each lot order to compensate for any rejections, and any spares that would thus be made in excess of the order could be immediately applied to future loss. The sizes of blocks, steel in blocks, method of forging, annealing and hardening treatment, were also taken up in detail by Mr. Porter.

The operation of the stores department of Cleveland Hardware Co. was described by Edgar E. Adams. Every month a statement is rendered to each foreman showing how much in dollars the supplies used in his department have cost. When this policy was first established it had little effect, but soon the different foremen began to compare amounts they were paying for supplies and a saving began to be noticed. In time the expense of shop supplies was cut down fully 50 per cent.

C. F. Herington, Bonnot Co., Canton, Ohio, quoted comparative cost figures of burning oil fuel and pulverized coal. One company operating ten stills has been spending $18,000 per month for fuel oil. They are about to install a pulverized-coal plant and expect to make with it a saving of $130,000 per year. The cost of the plant completely erected is to be $85,000, so that they will save that cost in about nine months.

Answering an inquiry as to what the minimum daily consumption of fuel oil would be that it would pay to convert to pulverized coal, Mr. Herington said that about 3000 gal. In such case, he added, with present prices of oil, the pulverized-coal plant would pay for itself in less than a year.

The problem of supplying the proper amount of lubricant to the steam drop hammer was discussed by Harry Johnson, Ingalls-Shepard Forging Co., Harvey, Ill. The success attained with any lubricating system in connection with steam cylinders, he explained, depends on the layout of the steam line and the method employed in draining. In a test conducted on a 12,000-lb. steam drop hammer with the object of obtaining an oil that would lubricate it economically and would leave exhaust steam sufficiently clean for heating purposes, best results were obtained with a product composed of a full filtered cylinder stock compounded with an acidless tallow oil.

An arrangement of a 2-qt. four-feed motor-driven mechanical lubricator used for supplying this oil was described by Mr. Johnson. The oil was discharged into the steam through a siphon atomizer. One drop of oil per minute was supplied for each 2 sq. ft. of wearing surface involved in the circular area of the cylinder and in that portion of the piston rod that passed through the piston-rod packing.

## American Society for Testing Materials

The annual meeting of the American Society for Testing Materials was held at Asbury Park, N. J., on June 22-25. The committee on steel, A-1, proposed new tentative specifications for commercial bar steels, covering ordinary commercial carbon bar steels, rounds, squares and hexagons of all sizes, and flats not over 6 in. wide and hot-rolled or cold-finished. The classification is as follows:

GRADES AND CHEMICAL COMPOSITION OF BAR STEELS

| Grades | Carbon, per cent | Manganese, per cent | Phosphorus, per cent | Sulphur, per cent |
|---|---|---|---|---|
| Open-hearth: | | | | |
| Dead soft........... | 0.05-0.12 | 0.55 max. | 0.05 max. | 0.06 max. |
| Soft............... | 0.08-0.18 | 0.55 max. | 0.05 max. | 0.06 max. |
| 15-25 carbon....... | 0.15-0.25 | 0.60 max. | 0.05 max. | 0.06 max. |
| 20-30 carbon....... | 0.20-0.30 | 0.70 max. | 0.05 max. | 0.06 max. |
| 25-40 carbon....... | 0.25-0.40 | 0.70 max. | 0.05 max. | 0.06 max. |
| 35-50 carbon....... | 0.35-0.50 | 0.70 max. | 0.05 max. | 0.06 max. |
| Bessemer: | | | | |
| Welding............ | 0.12 and under | 0.60 max. | 0.115 max. | 0.08 max. |
| Soft............... | 0.15 and under | 0.70 max. | 0.115 max. | ........ |
| 15-25 carbon....... | 0.15-0.25 | 0.90 max. | 0.115 max. | ........ |
| 25-40 carbon....... | 0.25-0.40 | 0.90 max. | 0.115 max. | ........ |
| 35-50 carbon...,... | 0.35-0.50 | 0.90 max. | 0.115 max. | ........ |
| 40-60 carbon....... | 0.40-0.60 | 1.00 max. | 0.115 max. | ........ |
| Screw Steel: | | | | |
| Bessemer screw...... | 0.08-0.16 | 0.60-0.80 | 0.09-0.13 | 0.075-0.15 |
| Open-hearth screw.... | 0.15-0.25 | 0.60-0.90 | 0.06 max. | 0.075-0.15 |

At the 1910 meeting, by recommendation of committee A-1, the note which had been added to 43 specifications by action of the Society in 1918, raising the rejection limits for sulphur in all steels and for phosphorus in acid steels 0.01 per cent above the values given in the specifications, was removed from 29 of those specifications, and with respect to the remaining 14 specifications consideration of the removal was deferred until the meeting in 1920. Acting on an agreement that open conditions still obtain, the committee recommended that the removal of the note be again deferred until the meeting of 1921.

The other recommendations of the committee were largely revisions to established and tentative specifications.

Among the technical papers were those dealing with the following subjects: A new method for testing galvanized coatings, offered by Dr. Allerton S. Cushman, president Institute of Industrial Research, Washington; a description of the processes of manufacture, heat treatment and physical properties of large power-forged chain as made at the Boston Navy Yard, presented by Carlton G. Lutts, physical metallurgist of the hull division at the Boston yard; suggestions for more intensive investigations in delimiting the scope and kind of application for the various materials of engineering, contained in President J. A. Clapp's address; a strong brief for molybdenum as an alloying element in structural steel, presented by G. W. Sargent, and a discussion on so-called shattered zones in steel rails by Dr. J. E. Howard, engineer-physicist Interstate Commerce Commission, Washington.

## Society of Automotive Engineers

The Society of Automotive Engineers held its semi-annual convention on June 21-25 at Ottawa Beach, Mich.

At the fuel session Dr. W. S. James presented the findings of the Bureau of Standards on intake-manifold experiments. The program of the work, which was undertaken by the Bureau at the request of the committee of the society on utilization of present fuels in present engines, included measurements of engine performance under conditions of both steady running and rapid acceleration with different temperatures of the intake charge secured by three different methods: (a) the hot-air stove supplying heated air to the carburetor; (b) the uniformly heated intake manifold; and (c) the "hot-spot" manifold. Fuel economy was determined for both part and full-throttle operation. The engine which was used was of modern six-cylinder automobile design.

A preliminary series of runs was made with an intake manifold having the tee section of Pyrex glass and motion pictures were taken of the phenomena observed. In but one of the runs the inside walls were completely dry, and this condition was attained only with great difficulty by the direct application of the flames from two blazing torches to the glass tee. The greatest amount of liquid was present in the manifold at 650 r.p.m. and full throttle, the layer being at least 1-16 in. deep in some places. The liquid collected mainly on the lower side of the manifold between the tee and the intake parts.

The main series of runs was made with an exhaust-jacketed intake manifold. Fuel consumption tests were conducted at both full and half load at speeds of 650 and 1200 r.p.m. These conditions were selected as covering the range of average driving from 15 to 25 m.p.h. Acceleration runs were made for which a dynamometer load was selected to correspond with that of a car at about 45 m.p.h. full throttle. The results thus far obtained were summarized by Mr. James as follows:

1 At constant speed, mixture ratio, and power outputs, the fuel consumption in pounds per brake-horsepower-hour is independent of the temperatures and methods of heating the intake charge within the range tested.

2 The rate at which an engine will accelerate with a given mixture ratio, or carburetor setting, is markedly affected by the amount of heat supplied and its method of application. Within the limits of this work the greater the amount of heat supplied to the charge and the higher its temperature at the intake port the more rapidly the engine will accelerate.

Another speaker at the fuel session was O. H. Ensign, president Ensign Carburetor Co., Los Angeles, Cal., who described a carburetor that has been in use for the last five years on the Pacific Coast, and was designed to handle an intermediate grade of petroleum distillate, known as engine distillate, which has there been supplied to users of motor vehicles. The principle involved for metering the fuel and air is the drop in pressure that occurs at the center of a whirling mass. The suction air whirls in a special chamber at the center of which terminates a pipe connecting with a cylinder provided with small orifices and surrounded by the fuel. The decrease in pressure or suction at the center of the whirl draws fuel through the various passages into the whirl chamber where it thoroughly mixes with rotating air. The device, Mr. Ensign explained, is permanent in its performance and high in fuel economy.

A symposium on engine design was also presented at the fuel session. It consisted of the following papers: Notes on the Use of Heavy Fuel in Automotive Engines, by H. M. Crane; Engine Design for Maximum Power and Economy of Fuel, by C. A. Norman; Saving Fuel with the Carburetor, by W. E. Lay; and Some of the Factors Involved in Fuel Utilization, by P. S. Tice.

That power farming offers a wide field for engineering research was emphasized by the speakers at the farm-power session. O. B. Zimmerman read a paper entitled Analysis of Fundamental Factors Affecting Tractor Design. The paper dealt chiefly with the effect of grade upon speed and drawbar pull and the effect of rolling resistance upon tractor output. Numerous graphs were included to show the variation of some of the factors with others. Referring to the amount of interesting reliable research work that can be done with advantage in the tractor and plow industry, Mr. Zimmerman said:

When it is realized what the actual expenditure of energy each year on the United States farms really is and that the majority of it can be done by mechanical power; that a very large part of it can be performed by a properly designed tractor and that estimates of this energy expenditure show it in actual horsepower to be greater in quantity than all other energy expenditures of industry in the United States, including mines, manufacturing, lighting, etc., we must admit that every available resource of engineering skill and scientific research must be expended and quickly, to raise the tractor and implement industry to one where the most exact and careful design will be evident from every possible economic standpoint. Basic data must be established along many lines; special test equipment and apparatus must be designed and built; whole series of experiments must be carried out at various seasons and in different soils; these latter must be studied in various conditions of moisture content; plow and engine design features must be shifted and standards adopted toward which the whole industry will then be enabled to work for early sensible economic solutions in the general interest of the nation as a whole.

There were two other professional sessions, on production and on transportation. Some of the papers read at these sessions were: Production Control and Systems of Accounting, by A. C. Drefs; The Workman, an Element in Production, by A. F. Knobloch; Interdepartmental Production Contests, by R. R. Potter; Motor-Bus Transportation, by G. A. Green; and The Relation of the Motor Transport Corps to Commercial Transportation, by Col. B. F. Miller, M. T. C., U. S. A.

## ALLOY STEELS IN THE AUTOMOTIVE INDUSTRY

*(Continued from page 505)*

It will be noted that in this particular test the cast cutter, while run at a slightly lower speed than the standard cutter (93 ft. per min. vs. 109 ft. per min.) took a more severe cut. It cut

TABLE 9 CUTTING TEST OF CAST HIGH-SPEED STEEL MILLING CUTTER

| Material | Speed, ft. per min. | Feed, in. per min. | Depth, in. | Length cut, in. | Condition of cutter |
|---|---|---|---|---|---|
| Cast iron | 60 | 0.30 | 0.125 | 8.2 | Good, slight wear |
| Soft steel | 60 | 0.30 | 0.219 | 1.5 | Good, slight wear |
| | 71 | 0.30 | 0.219 | 2.5 | Slight wear, but cutting smoothly |
| Rail steel | 60 | 0.30 | 0.200 | 1.0 | Good |
| | 93 | 0.30 | 0.200 | 0.8 | Good |
| | 93 | 0.69 | 0.200 | 4.0 | Good |
| | 93 | 1.31 | 0.200 | 2.0 | Cutter began to "go," the wear increasing. Millings "blue heat" |

A high-speed steel cutter of standard make was tested on the same rail. It had previously been worn slightly, corresponding approximately to the wear resulting from the cast-iron and soft-steel cuts on the cast cutter.

| Rail steel | 70 | 0.30 | 0.200 | 1.0 | Good |
|---|---|---|---|---|---|
| | 109 | 0.30 | 0.200 | 0.8 | Good |
| | 109 | 0.69 | 0.200 | 1.5 | Cutter began to "go," one corner burning and wearing down. Millings hot but still "white" |

4 in. at 0.69 in. per min. before failure, whereas the standard cutter failed after cutting 1.5 in. at a lower feed of 0.69 in. per min.

Attention is invited to the fact that this test is not presented as being conclusive evidence that cast high-speed steel cutters are either superior or inferior to tools produced from bars or forgings, but is of interest chiefly in showing the refractory nature of the carbides in cast high-speed steel. It indicates that ordinary heat treatment is not sufficient to bring these carbides into solution, and further that good service may be obtained. Probably such tools may be applied to special lines of work.

# LIBRARY NOTES AND BOOK REVIEWS

APPLIED AERODYNAMICS. By G. P. Thomson. The Norman W. Henley Publishing Co. Cloth, 7 x 10 in., 292 pp., illus., charts, diagrams, plates, $12.50.

This book is the work of a former officer in the Royal Air Force, who, during the War, was in touch with both the demands from the Front and the attempts made to fulfil those demands from home sources. The results presented are drawn from data accumulated during the War, some of which has been or will be published in the Reports of the Advisory Committee for Aeronautics, and the author has also had free access to the staff and records of the Aircraft Manufacturing Company. The work is concerned only indirectly with the mechanical side of aircraft design, but it specializes upon aerodynamics as a branch of engineering, though an account is also given of such progress as has been made on this side of physics from an engineer's point of view.

A number of suggestions for further inquiry and research are made, such as the problems relating to the interaction of aeroplane fuselages, nacelles and the airscrews attached to them. The importance of full-scale work is emphasized, and the problem of stability is dealt with at length. The question of how heavily the wings of an aeroplane should be loaded is discussed and the subject of abnormally large aeroplanes is also taken up. The information given was up-to-date at the time of writing and pains have been taken to point out the cases in which there was any considerable probability of error.

THE COAL CATALOG, combined with Coal Field Directory for the year 1920. Containing Explanatory Articles on Rank, Usage, Analysis, Geology, Storage and Preparation of Coals. Compiled and published annually by Keystone Consolidated Publishing Company, Pittsburgh. Cloth, 9 x 12 in., 1133 pp., illus., maps, tables, $10.

The second edition of the Coal Catalog, in addition to describing and listing the coal mines of the United States, with the output and operating companies, has added much new data such as the fusion points of ash from the various coals, the specific gravities and weights of coals and the so-called "fuel ratios." A list of national and local associations is included as well as the wholesale dealers in the larger towns and cities. A Directory section follows each state.

DIRECT-CURRENT MOTOR AND GENERATOR TROUBLES, Operation and Repair. By Theodore S. Gandy and Elmer C. Schacht. First edition. McGraw-Hill Book Co., Inc. New York, 1920. Cloth, 6 x 8 in., 274 pp., illus., tables, $2.50.

The chief object of this book is to give simple, effective methods for finding and remedying troubles of direct-current motors and generators. In addition, the selection, operation, care and repair of direct-current machinery are analyzed from the operator's point of view. The theory underlying the design of these machines is omitted, as the book is intended for operators, not for designers.

DRAINAGE ENGINEERING. By Daniel William Murphy. First edition. McGraw-Hill Book Co., Inc., New York, 1920. Cloth, 6 x 9 in., 178 pp., illus., plates, tables, $2.50.

This is a general treatise on the drainage of agricultural lands, which indicates the various questions to be considered in connection with a drainage project, and presents the principles involved in the design and construction of drainage works. The author was formerly a drainage engineer in the U. S. Reclamation Service.

ENGINEERING FOR LAND DRAINAGE. A Manual for the Reclamation of Lands Injured by Water. By Charles Gleason Elliott. Third edition, revised. John Wiley & Sons, Inc., New York, 1919. Cloth, 5 x 8 in., 363 pp., illus., maps, tables, $2.50.

This book is prepared to present the essential features of drainage engineering as practiced in this country today and is adapted to the use of professional engineers and students. This edition has been revised and enlarged. The discussion of the hydraulics of flow has been rewritten and new tables for the discharge of tile drains are given. A diagram to facilitate the

use of Kütter's formula in the design of ditches and canals has been added, as well as new material.

DIE MAGNETISCHE INDUKTION IN GESCHLOSSENEN SPULEN. By Arthur Scherbius. R. Oldenbourg, München & Berlin, 1919. Paper, 7 x 10 in., 91 pp., illus., 7 marks.

This monograph treats of the possibilities, both theoretical and technical, of phase transformation by transformers and machines without commutators. The author discusses these possibilities with special reference to the limits within which technical applications may be possible and gives particular attention to the saturated transformer.

MANUAL FOR THE OIL AND GAS INDUSTRY under the Revenue Act of 1918. By Ralph Arnold, J. L. Darnell and others. John Wiley & Sons, Inc., New York, 1920. Cloth, 6 x 9 in., 190 pp., charts, tables, $2.50.

This manual was first issued in 1919 as a bulletin by the Oil and Gas Section of the Bureau of Internal Revenue. The demand having exceeded the supply, this private reprint, in which certain features have been brought up to date, is published.

The book is intended to assist the taxpayer of the oil and gas industry in preparing his Federal tax returns correctly and expeditiously. It is divided into three parts. Part one deals directly with the law and regulations, part two with depreciation, while part three describes methods of estimating underground reserves, especially by means of production curves, a collection of which are included.

A METALLOGRAPHIC STUDY ON TUNGSTEN STEELS. By Axel Hultgren. John Wiley & Sons, Inc., New York, 1900. Cloth, 6 x 9 in., 95 pp., illus., 43 plates, tables, $3.00.

As the question of tungsten steels has been studied by several investigators whose views have not always agreed, it is the purpose of this work to bring about a clearer understanding by presenting certain new facts and problems and by critically examining previous results and opinions. The subject is divided into two sections: (1) The transformations of tungsten steel during different heat treatments and the structures thereby formed; (2) Carbides in tungsten steels. Each section contains a review of the results and opinions of previous investigators and the author's own results and conclusions, based on his metallographic investigations at the Institute of Technology of Charlottenburg, the Royal Institute for Testing Materials, Stockholm, and in the Laboratory of the SKF Ball Bearing Co., Gothenburg, Sweden. The paper was written in Swedish in 1918. In its present form it includes a translation of the Swedish paper as well as an appendix containing a critical review of the investigations on tungsten steels of Honda and Murakami.

POPULAR OIL GEOLOGY. By Victor Ziegler. Second edition. John Wiley & Sons, Inc., New York, 1920. Flexible cloth, 5 x 7 in., 171 pp., illus., chart, maps, tables, $3.

The second edition of this work on oil geology has been partially rewritten and new material along the more theoretical lines of geology has been added. The principles important in the examination of prospective oil land have been emphasized and the work of the oil geologist described in some detail. The work is intended to make intelligible to the layman the fundamental principles of oil geology and is written in as clear and simple language as possible.

PRACTICAL CHEMISTRY. Fundamental Facts and Applications to Modern Life. By N. Henry Black and James Bryant Conant. Macmillan Co., New York, 1920. Cloth, 5 x 8 in., 474 pp., illus., plates, $2.

It is the purpose of this book to present the fundamental facts of chemistry and to show by the application of those facts to experiences of everyday life that the subject is real and practical. The economic significance of chemistry of growing allied industries has been stressed and the chemistry of growing things has also been considered. Only such topics have been included as young people can grasp and find useful. The book is adapted for class use as a textbook.

## Appraisal and Valuation of Properties

Rational Valuation; Price Levels in Relation to Valuation; Cost of Organizing and Financing
Public Utilities; Interest and Taxes During Construction; as Discussed
at St. Louis Meeting of A.S.M.E.

I N response to frequent requests from engineers for a discussion of the appraisal and valuation of property, sessions for this purpose have been held at the last two general meetings of The American Society of Mechanical Engineers. In arranging for these, a special effort has been made to contribute results of definite value in the broad field of appraisal and valuation. By such discussions as these it is hoped to bring about practical coöperation among members of the engineering profession, so that new light and more effective guidance may be given to the courts and commissions who in the end are obliged to rely upon technical findings in reaching equitable judgments.

At the two sessions mentioned, five papers were presented, of which one, Methods of Appraisal and Valuation, by David H. Ray, has already appeared in MECHANICAL ENGINEERING.[1] Abstracts of the other papers appear below. These papers are as follows: Rational Valuation, J. R. Bibbins; Price Levels in Relation to Valuation, Cecil Elmes; Cost of Organizing and Financing a Public Utility Property, F. B. H. Paine; Methods of Computing Interest and Taxes During Construction, H. C. Anderson.

The purpose of the first paper is to emphasize the breadth of view which a complete valuation should hold in order to be of the greatest usefulness, especially with reference to the facts of history. The author makes a comparative study of valuation practice illustrated by a complete economic analysis of a large utility from the period of the Civil War to the present time,

yielding in substance the history of the five-cent trolley fare.

Similarly, the second paper is a study of commodity prices. The author traces the changes that have taken place in prices and wages in this country and England since very early times. He believes that greatly reduced commodity prices are not to be expected unless lower wages are to be taken for granted. He further points out that the laws of economics as outlined in the paper must apply to public utilities as well as to other things, and that in any proceeding the word "dollar" should not be used to mean two different amounts, a high figure to one party and a low figure to the other in one and the same controversy.

The third paper discusses the preliminary steps which must be taken before a project is finally launched. These embrace the development of the project and the securing of sufficient capital funds for preliminary outlays and for the actual, construction of the utility.

Finally, the fourth paper takes up the proper allowances to be made for interest and taxes during the period of construction—items frequently ignored in the early days of appraisals and for which latterly excessive figures have often been used.

At the first session at which the papers were presented, the time was given almost entirely to the reading of the papers. In MECHANICAL ENGINEERING for July 1920, there is a summary of the discussion of the papers which took place at the second session held at St. Louis last May.

## Rational Valuation—A Comparative Study

By JAMES ROWLAND BIBBINS,[2] CHICAGO, ILL.

V ALUATION should be characterized as cold fact and engineering judgment combined. We should at least agree on facts. To clear the decks, let us eliminate what may be regarded today as the established mechanics of the valuation method. All of the factors noted below are probably admissible as facts, although the specific rates of construction overhead (items 7-10) may be subject to discussion and variance according to the kind of property, the locality and the historical period during which it was built. Then follow the principal contestable factors.

ADMISSIBLE FACTS

1 Extent of property owned and establishment of unencumbered title
2 Physical inventory in detail or by type of percentage-method according to accuracy desired. (Engineering approximations favored)
3 Cost extension based on original voucher costs or representative normal average costs. (Current costs as of date used in special cases)
4 Age of original structures still in use and property superseded
5 Normal life and rate of renewal encountered in shorter-lived property under peculiarly local conditions, such as ties, investment in an established property?
6 Initial expenses, legal and otherwise, incident to organization and construction
7 Engineering and superintendence prior to and during construction—i.e., cost of construction organization
8 Specific carrying charges, such as interest, taxes, insurance during construction
9 Discounts on construction finds not covered by normal interest rates

10 Accrued depreciation existing as a physical fact, not considering, for the moment, its application.

CONTESTABLE FACTORS

1 Purpose of the valuation as affecting "value":
  (a) Taxation (b) Accounting (c) Condemnation (d) Rate-making (e) Purchase or Transfer (f) Capitalization, and recently, during high prices (g) Insurance
2 Original Cost vs. Cost-to-Reproduce New theory under
  (a) Original conditions of construction, or
  (b) Conditions today with modern substitute plant
3 "Original Conditions" logically associated with actual Cash Costs (actual investment). Likewise current prices today with substitute plant where obsolete elements are considered
4 Appreciation in values since original construction, principally in real estate and earning value, recently in all material and labor.
5 Accrued depreciation on physical elements due to age, wear, inefficiency, and being out of date (obsolescence)
6 Franchise value, early losses, going concern, cost of developing the business, good-will, etc., asset or liability, or neither?
7 When does a business drop from a "going concern" to a "losing venture" and necessarily subject to lower return?
8 The range of applicability of current or "spot" prices: (a) Boom prices; (b) Panic prices. Has a spot value (as of today) any weight in an absolute determination of capital investment in an established property?
9 Should the true cost of service include at all times as an element of operating expenses, the wearing down or depletion due to accrued depreciation?

SCOPE OF A COMPLETE VALUATION

From the above it will be apparent that to attempt intelligent discussion of "Value" involves sidetracks and pitfalls in hopeless

[1] December, 1919, p. 931.
[2] Supervising Engineer, The Arnold Co., Mem. Am. Soc. M. E.

number, unless the issue is most clearly defined and limited. For it is a fact that, by reason of contractual requirements, public necessity, and the rights of society of past, present, and future generations, not to mention the investors, business with its risks and its responsibilities has reached an exceedingly complex stage. So that it no longer suffices to value a great property as a boy's jack-knife, "unsight—unseen." The writer submits instead that a proper valuation must view the subject from all possible angles, such as:

a The actual true plant investment cost as built, from historical studies

b The cost to reproduce plant as it exists today under reasonably normal prices—i.e., neither boom nor panic prices, which is a check on (a)

c The cost to develop the physical property to its present state of efficiency; called Plant Development Expenses

d The money loaned for commercial development, called Deferred Earnings during development

e The present shrinkage in physical plant from aging or other form of depreciation

f The contractual and moral responsibilities of the owners to the public and vice versa

g The history of development, net income for fixed charges and true adjusted return on net plant investment.

In other words, valuation especially for rate-making purposes has risen from simple arithmetic to the high level of an economic science requiring the utmost power of analysis, accounting judgment and vision. It demands the facts from the past, the conditions of the present and the outlook for the future.

Upon this substantial foundation may then be built up the case in question. In fact, a rate case rests upon a so-called three-legged stool, the economic stability of which rests in turn upon (1) Basic Value Assigned, (2) Rate of Return allowed on that value, (3) Disbursement of Income. To shorten any leg overturns the stool.

Time was, with the phrase "Public Utility" yet unborn, when properties were conceived, built, and operated distinctly for personal or corporate profit, without restriction, just like the industries furnishing the equipment. The major difference between utility and industry rested upon that intangible asset (or liability) called a franchise. Thus came to be established the personal point of contact with the public; in railways the most intimate, in telephones less so, and in light, power, and gas the least. And right here, unconsciously, industry became the public's business, as time and bitter experience have proved.

But here was the rub; when the utility industry awoke to the true cost of service based upon proper cost accounting, the balance sheet footed red. With income practically fixed, costs rising, depreciation mounting and right to do business expiring in the face of new conditions imposed by a more intelligent public, a readjustment became inevitable.

Meanwhile, private industry boomed, under no restraint as to right to do business and enjoying flexible rates to absorb variations in cost of service, with no limit except that imposed by competition. Hence earning power became the principal basis of capitalization as it has been to an unfortunate extent in the early acquisition of utilities. The difference was that, while perhaps justified in industry, it was not justified in the case of utilities, owing to the specific incumbrance of the franchise.

The economist, John H. Gray, has shown[1] that Federal regulation was devised to limit risk by preventing speculation and thereby to stabilize values accordingly—i.e., a police power which has already been exercised in the cases of utilities, banks and insurance companies, but only now taking form in the case of the private industries. Also he has demonstrated that the real difference between public and private property is that one is subject to the "servitude" or incumbrance of specific public service, the other is not, and that this incumbrance entirely vitiates franchise value, which itself is dependent upon rates fixed by public authority. Here we run around in a complete economic circle.

Is any one wise enough to deny the possibility of ultimate

[1] Trans. Am. Soc. M. E., 1916, p. 1247.

Federal regulation over industry such as that of the Interstate Commerce Commission over common carriers, and that such control will be based on the following principles:

a Actual investment cost contributed by the owners

b Rates calculated upon this investment

c Variable risk, requiring varying charge rather than variable value.

In other words, to stabilize values by eliminating the speculative element more and more, in essential industries.

It appears to the writer that fundamentally utilities and industries are not dissimilar as to the various aspects of valuation relating to the physical properties; and that only the speculative element is shrouded in doubt, which at least is minimized in the utilities subject to public regulation and protection.

### THE VALUE OF ECONOMIC RESEARCH

At the very foundation of rate making lies the determination of the true Cost of Service, or of product. It is no exaggeration to say that much of the controversy over valuations could have been avoided entirely had it been possible for the contending parties to view the complete facts concerning which there could be little or no argument.

Concerning the valuation basis, it is a matter of common knowledge that the partisan viewpoint has performed wondrous acrobatics in shifting from actual investment to replacement cost, then back to original cost, and again to cost to reproduce new on current prices. In earlier usage, "cost to reproduce new" was generally held to mean cost to reproduce in brand-new condition the property inventoried, under construction conditions and prices prevailing. Later, the definition was broadened somewhat in general usage so as to mean "cost to reproduce new," under substantially "original" conditions, the property inventoried, using normal prices or prices averaged over a term of years where wide fluctuations occurred. Quite recently, the scope of valuation has undergone still further modifications in the effort to satisfy the condition where the owners are held entitled to receive "a fair and reasonable return on the actual investment," which requires the determination of the historical value, i.e., the total actual cash investment resulting from efficient administration in (1) producing the property, (2) operating it efficiently, (3) extending it as required and (4) maintaining it at the physical standards necessary for reasonable requirements of the service. This later interpretation of "cost to reproduce" obviously attempts to reproduce the original cost and in fact, were no historical records available, such a value would probably represent closely the original cost of a property built in times of normal conditions and prices.[1]

Looking, then, into the future of valuation methods, can we escape the necessity of visualizing original cost at least as the starting point of valuation, in any direction of enterprise, and with the facts available as revealed by economic research, will there not be far less to squabble about?

### OPENING THE DOOR OF THE PAST

In order to crystallize the principal thoughts presented in this paper, a typical case may be used for purposes of illustration.[2] The exhibits represent a comprehensive study of all phases of development of a typical railway property from its inception, keeping an open mind as to what constitutes fair value for rate-making purposes today and taking into consideration essential economic elements of past, present and future. This is a deliberate attempt to illuminate the complex problem confronting society as a whole in fairly distributing the burden of development and operation of its public utilities. The study presents what may be termed the problem of three generations and illustrates, it is hoped, in unmistakable terms, the practical difference between renewals (or completed depreciation), accrued deprecia-

[1] The single exception is true appreciation of realty values by time and development. Some deny the right to appreciation over actual cost. However, it seems clear that land is a somewhat different classification from other physical property. And there is also involved an exceptional factor —viz., that the public, by taxation, recognizes Value at least to the extent of the taxables reported; it could not justly deny Value and tax it too.

[2] Presented by courtesy of Bion J. Arnold, Consulting Engineer, Chicago.

tion (or unmatured renewals), and amortization (a purely actuarial process designed to liquidate a debt). Further, the study shows the essential need of a definite amortization policy under all franchises of finite term and especially under short-term franchises. And here is the cause, usually neglected, of most of the contention over the admissibility of depreciation; i.e., whether the principal of the funded debt can be paid back, at expiration of franchise, either directly or by previous amortization.

### HISTORY OF THE FIVE-CENT FARE—FIG. 1

*Net Investment*, the upper curve, shows the progress and growth of a typical property and should be held in mind as an approximate measure, of the relative size of the property and volume of operations for the early and late periods of development. This represents the actual property in operation as would be determined by inventory at any year, at the original cost, i.e., all superseded and abandoned property is deducted. It will be noted that the horse-car operations are relatively unimportant, as the horse-property was only about one-fourth the size of the electric property today, and would probably average not more than one-fifth.

*Annual Rate Earned* curve indicates the annual net income balance (applicable to fixed charges, amortization, and surplus) in per cent of the average net investment for the year. In computing this return, the superseded and abandoned property is

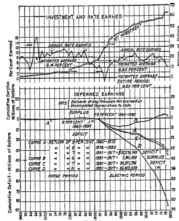

FIG. 1 HISTORICAL INVESTMENT AND DEFERRED EARNINGS

charged off in the year superseded. As a matter of fact, practically no amortization reserves have been built up in this property to absorb accrued depreciation by reason of the evident insufficiency of income of past years. The annual rate earned is then more apparent than real.

*Deferred Earnings*, curve D, shows the cumulative surplus and deficit in this net income balance, or, in other words, the relative possibility of the electric lines as a whole earning an assumed fixed rate of 8 per cent on the net investment for fixed charges, amortization, and surplus. Consequently, the downward tendency of curve D, after electrification was completed, really indicates the true trend of electric-railway operations in comparison with the income realized from the rates of fare actually established and in force. These deficits were compounded at this rate

as they had to be funded; likewise the surpluses. Curve C corresponds to a rate of 7 per cent and Curve B to 6 per cent, for the electric property. Curve A covers the entire history horse and electric operation corresponding to a rate of 7 per cent electric period and 8 to 9 per cent during the horse period, the rate of the early years recognizing the higher cost of money commensurate with the risk involved.

The zero line of Fig. 1 obviously represents par; i.e., for a continuing, stable business there should be no cumulative surplus or deficit at a rate of return, taken over a long period, adequate for the support of this investment. Now the hard facts which

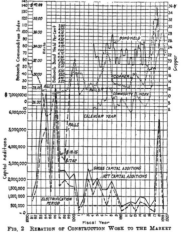

FIG. 2 RELATION OF CONSTRUCTION WORK TO THE MARKET

the owners of these properties face today as drawn from this typical economic study may be summarized as follows:

Apparent Rate Earned, 1860-1890 ......... 6.85 per cent
Apparent Rate Earned, Electric Period, 1891
    to 1919 ........................... 6.60 per cent
Actual Rate Earned, Electric Period, corrected
    for actual deficits and crediting surpluses 6.10 per cent
Actual Rate Earned, Electric Period, further
    corrected for assumed amortization of Ac-
    crued Depreciation to date ............ 5.27 per cent
Actual Rate Earned, Electric Period, finally
    corrected to cover Renewal Fund pay-
    ments necessary to meet future Renewal
    Liability (basis, pre-war normal prices).. 5.03 per cent
All rates expressed as weighted averages.

Thus the actual economic history of electric operation in this property is seen to be on a true basis of about 5 per cent, which would necessitate in financing 4 per cent bonds and 6 per cent stock, assuming the same amount of each issued on the property. Obviously, enormous discounts could not very well be avoided on such a basis.

### RELATION OF CONSTRUCTION WORK TO THE MARKET—FIG. 2

The lower curves trace the rate of capital investment per year in the electric property (and the extensive supersession) as compared with the upper curves of market price levels. Evidently the major construction and reconstruction work was not carried out during periods of e.cessively high prices, which has an important bearing upon the study in substantiating the Original Cost appraisal basis. (The apparent exception for 1918 capital additions is explained by the fact that the abrupt increase for the year is due to taking over property formerly rented.)

### ADJUSTMENT CURVE OF DEFERRED EARNINGS—FIG. 3

This curve is designed to interpolate rates and earnings between the curves $B$, $C$, and $D$ of Fig. 1, so that one may find an expression for the particular rate which he considers "fair" for such a property. The curve measures the deferred earnings of the electric period resulting from the assumption of any given apparent rate earned. Thus, this property averaged 6.60 per cent, but without any provision for funding deficits or meeting Accrued Depreciation. The curve shows that to carry these deficits an annuity of 0.50 per cent was required, reducing the actual rate earned at part to 6.10 per cent (amortization would further lower it 0.83 per cent and renewal liability 0.24 per cent, or to 5.03 per cent final return). It will now be apparent that the financial policy of a utility as compared with an industry

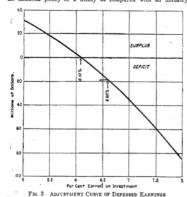

FIG. 3   ADJUSTMENT CURVE OF DEFERRED EARNINGS

may be extremely handicapped by the deficiency in income resulting from a too rigid bargain for franchise and fares.

### ESTIMATED FUTURE RENEWAL LIABILITY—FIG. 4

How many properties make a serious attempt to ascertain with any degree of accuracy their future renewal maturities, such as illustrated in Fig. 4? At the left, the actual costs of past maintenance and renewals are shown, rising to a maximum, equivalent to 20 per cent of the gross earnings. Then came the retrenchment of war years, and the actual physical renewals were even less than the money renewals because of the higher cost. Here is a renewal program based on original cost and reflecting exactly the depreciation schedules of the appraisal except that maturities were spread so as to lessen the severity of the peaks.

The principal result shown is that the future requires about $1,300,000 per year to maintain the integrity of the property. Obviously, the only sensible way to meet this condition is by a renewal fund just sufficient to carry the property over the maximum peak of the future. To amass such a fund by the year indicated, 1923-24, would have required an annuity of 0.24 per cent during the entire electric period. This explains the term used, renewal liability.

### THE THIRD GENERATION. PAY-AS-YOU-GO, OR PAY-DOWN—FIG. 5

We now come to the crux of the entire problem under dis-

cussion, a problem as broad as that of society and social justice. In view of the rapid advancement of the art and the impossibility of foreseeing the ultimate result of rigid franchises granted long years ago, what financial plans with respect to rate making can be devised which will hold the maximum of justice and fairness to investor and public and distribute most equitably the cost of service and cost of development without economic shock? Three typical plans will illustrate.

Plan $A$ assumes a "fair deal" as to franchise expiration—i.e., either a continuing or indeterminate franchise requiring necessarily no amortization of either funded debt nor depreciation; in other words, the "pay-as-you-go plan," renewing the property only as fast as the maturities indicated by Fig. 4. Here each generation takes care of itself; pays for renewing plant actually worn out by its predecessors, and passes on to the future its own burdens. In a normally expanding and renewed property, however, this plan is the simplest and cheapest.

Plan $B$ assumes a definite expiration of franchise and actual liquidation of the business. Obviously there is no other alternative than an amortization fund to validate the principal of the investment when the property is taken over by the city at "actual present value" (and there is little reason to believe that this will be any other than Original Cost New less Accrued Depreciation, despite present-day contentions for boom prices). Here, then, the rate must include an amortization charge depending upon the contractual term of the franchise. This plan holds practically the only fair economic solution if Public Policy has set definitely towards Depreciated Value as a basis of rates.

Plan $C$ interprets the result of a determination to adopt depreciated value at once as a rate basis. This means that 20 to 30 per cent of the investment must be liquidated, provided suffi-

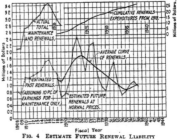

FIG. 4   ESTIMATED FUTURE RENEWAL LIABILITY

cient returns have not been earned in the past over and above fair return to absorb this shrinkage. But Fig. 1 has shown this not to be the case. There is then no other reasonable alternative but to transfer the amortization charge to the future as in Plan $B$, basing rates on a gradually reducing capital value resulting from the progressive amortization. As a result the public would be supporting, in 1942, an investment no greater than in 1898, although the actual property would be 50 per cent larger.

Assume that depreciated value is today in fact imposed by courts and commissions. Then the following alternatives hold in this particular property:

    *a* The past generation has accepted service at less than cost and the franchise bargain was a bad one and the shrinkage a dead loss, or

    *b* The present and future might accept a double amortization charge sufficient to refund at some future date (which is unthinkable) both the shrinkage and deferred return thereon.

But the concluding fallacy is exposed also in Plan $C$, for in this property an inventory each year between 1923 and 1938 would show an actual depreciated value in excess of the assumed agreed value today (neglecting for simplicity capital additions). The conclusion is plain—that depreciated value as a rate basis must be reappraised annually, or determined currently by a perpetual inventory and inspection (depreciated by the usual methods); otherwise, confiscation plainly enters into the controversy, as shown graphically by Fig. 5. Obviously such a method of current valuation would result in the maximum expense and complexity. If we had made sensible bargains in the past, with sufficient rates for a "pay-down plan," depreciated value might be acceptable in principle. That is, if past riders had actually paid for the property *they* wore out, in full; but this was rarely the case.

The economic conclusions from the study are either (1) that the electric railway has been a losing venture, and therefore should be scaled down to a new par securities basis, or else (2) that the public's responsibility has been unintelligent in permitting malnutrition while accepting service below cost. The responsibility is certainly mutual although rarely considered so, except by the aggrieved party.

### FACING THE FACTS

Is it not clear from the foregoing that the vagaries and inconsistencies of appraisal procedure could have been largely avoided thus:

1 By closer study of past history
2 Economic research into all fundamental facts
3 Reasonable forecast of future income and liabilities
4 More complete appraisal reports dealing with every aspect of reasonable value
5 An open mind—not trying to find the answer before we start
6 Acceptance of accrued depreciation as a physical fact to be handled like any other actuarial problem
7 Cost to reproduce theory used judiciously to reflect the original cost as nearly as possible
8 Institution of perpetual inventories and values based upon well-maintained and preserved records
9 Development of flexible or sliding scale methods as basis for rates, service and return on investment
10 Adopting for utilities the practice of progressive industry

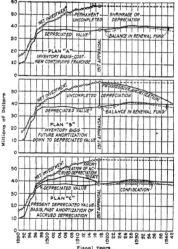

FIG. 5 FINANCIAL PLANS FOR RATE-MAKING PURPOSES

in preserving its physical and economic integrity. Pay down "while the paying is good"
11 An attempt to educate rather than to obstruct or defeat Public Policy.

# Price Levels in Relation to Value

### BY CECIL F. ELMES,[1] CHICAGO, ILL.

I N most rate controversies and appraisals, the opposing engineers are commonly able to agree within very reasonable limits on all purely engineering features. It is when these are set aside and economic questions are broached that sweeping differences appear. Engineers labor faithfully to reach an agreement within twenty thousand dollars on the "bare bones" value of a street railway, or of an electric utility, and then proceed to differ by perhaps two million dollars as to the overhead costs which should be applied to this figure. Their disagreement is not upon engineering but upon economic features, and the results to the public utility business as a whole are so serious that I have no apology to make for devoting myself in what follows to the subject of economics.

Let us start with a precise definition of two of the words in the title of this paper—"Value" and "Price." John Stuart Mill defines the value of a thing as "its general power of purchasing, the command which its possession gives over purchasable commodities in general."

Particular attention is called to the word "general" in this definition. All discussion of value by the great economists is predicated upon the assumption of a free exchange of commodities, with freedom of communication and transport and willing buyers and sellers. The inquiry is never as to the peculiar value of a particular article, such as an heirloom or the last morsel of food among the survivors of a shipwreck. The broad

[1] Sanderson & Porter.

conception of value as set forth by Mill is no more intended for such individual application than an actuary's tables of mortality are intended for use in forecasting the date of a particular individual's death.

The New Standard Dictionary defines "Price" as "The amount demanded by an owner in return for an article as the condition of sale . . . the quantity of labor which its possessor will take in exchange for it."

What relation, in practice, the price demanded by the owner will bear to the value as defined by John Stuart Mill depends upon the circumstances of the case. The likelihood of the two being identical increases, generally speaking, with the available supply of the article, and the readiness with which it may be obtained from any one of a number of sources. The commonness of an article makes it difficult or useless to set an exorbitant price upon it.

In determining the value of any piece of property which is the subject of litigation, courts have always been swayed, and quite properly so, by consideration of the price which the article would fetch if placed upon sale, then and there, under conditions fair to both a willing seller and a willing buyer.

### WHY PRICES FLUCTUATE

Price, as defined above, can be arrived at in terms of money or of labor. In the economic civilization of today we always quote prices in money. If we possessed an ideal currency not liable to fluctuation in its intrinsic worth, prices of commodities would vary in accordance with the laws of supply and demand, the demand for any commodity tending to cause a rise in its price, and conversely a lack of demand leading to the commodity being offered at a lower price. However, the currency of this and every other country is manufactured out of metals which are themselves commodities subject to the laws of supply and demand and to consequent variations in market price.

Another variable element which must be considered in dealing with fluctuations in market price is foreign exchange, or the ratio at which the currency of one country is traded for the currency of another.

### Low Prices

The idea is quite prevalent that we will all be more prosperous when the present high prices give way to lower levels. The fallacy of this hope lies in the fact that lower prices usually result from a lack of demand for commodities, and a lack of demand means a slowing up in commerce, a depression in business, and its result—a lack of employment for labor. Let any one who is clamoring for low prices refer back to such a time as the early 90's of the last century, when commodity prices were at their lowest. The financial panic of 1893 was an effect of prolonged hard times rather than a cause and the charts shown hereafter make it clear that a profound industrial depression overspread the country.

### High Prices—Panic or Famine Prices

If low prices, then, are no blessing, high prices in the same way are by no means a curse. In general, they indicate prosperity, of which the times we live in afford abundant proof. In spite of all clamor against high prices, not only are commodities of every kind being eagerly bought, but it is the judgment of retail merchants that the most expensive wares command the readiest market. At no time within the recollection of the present generation have luxuries been more sought, or has their distribution been more widespread.

So long as the free exchange of commodities goes on, high prices are, generally speaking, an indication of prosperity. It is only when the free exchange of commodities comes to a breakdown point that this ceases to be true. The definitions and conceptions of economics are based upon a free exchange of commodities, and they must not be projected into an analysis of the conditions which result when the exchange ceases to be free, or when the ordinary processes of commerce collapse. In such a case we have panic or famine prices. The most common occasions of panic or famine prices are crop failures, the monopolistic act of an individual merchant or group of merchants, or the warlike act of a belligerent in cutting off supplies.

### Wages

One of the vital commodities whose price must be considered in any general discussion of price fluctuation is labor. In what follows, labor will frequently be dealt with in terms of the commodities which may be purchased with it. No serious student, however, can afford to treat labor solely as a commodity and wages solely from the standpoint of a price for that commodity. It is useless to talk about selling a particular article at a price which will not allow a living wage to the workers producing it. As intelligent and progressive citizens, we are forced to postulate a decent living wage in every industry and to regard any commodity price incompatible with this requirement as basically unsound.

Statistics compiled upon the wages of bricklayers in the state of New York, machinists in the state of Pennsylvania, and laborers in the state of Illinois during the last sixty or seventy years, show the first have multiplied by more than four; the second by five; and the third by nearly seven. This relates only to the monetary increase without reference, for the moment, to how much the increased wages will buy.

In view of the fact that labor enters so largely into the cost of every commodity, we have in the foregoing a reason to anticipate

a higher level of commodity prices today as compared with the commodity prices of sixty or seventy years ago. The increased efficiency and productiveness of the wage earner will act, of course, as an offset to this. Nevertheless, anyone urging lower commodity prices, is evolved by any one of several methods of committing himself to a demand for lower wages. And anyone who demands that property be valued today on the basis of what it would have cost to produce it in some bygone year should be ready to squarely meet the issue of whether he is not thereby committed to championing the wages current at that time.

### Commodity Price Indexes

In studying the upward or downward movement of commodity prices as a whole, economists have evolved various systems of commodity price index numbers. An index number, applied to commodity prices, is evolved by any one of several methods of averaging actual market prices of the important articles of commerce. A list of basic commodities is selected which the economist regards as being of sufficiently wide range and commercial importance to fairly reflect the commerce of the day. At regular intervals, say every month, an index number is arrived at which is to represent the combined fluctuations of the commodity market. Sometimes the index number is obtained by simply averaging the unit market prices of all the commodities on the selected

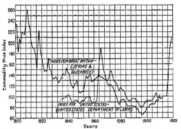

FIG. 1 INDEX NUMBERS OF WHOLESALE COMMODITY PRICES IN THE UNITED STATES AND GREAT BRITAIN IN THE NINETEENTH AND TWENTIETH CENTURIES

list. Sometimes the prices of the individual commodities are weighted to correspond with their relative importance as to: (a) quantity consumed; (b) vital significance to the life of the community.

There is vastly more iron sold, for instance, than indigo. It would be a criticism against an index number of fluctuations in the indigo market modified it as profoundly as changes in the cost of so basic a commodity as iron. Weight for weight there might be as much paving brick sold as wheat, yet there is no comparison in its importance to human life. An index number which is as largely affected by a change of the market price of paving brick as of wheat is open to criticism. Each commodity price index is therefore largely dependent upon the judgment of the statistician who evolves it and in this way quite a number of commodity price indexes have become current.

Perhaps the most authoritative index in the United States is that published by the Bureau of Labor Statistics of the United States Department of Labor, which will be frequently referred to here. The Bureau has published an index number going back as far as 1860, and in the data compiled for the Aldrich investigation the same index number was carried back as far as 1840. Over 320 commodities are now considered in compiling this index number. Other important index numbers for the United States are those compiled by *Dun's Review* and by Bradstreet; Gibson's index and that of the *Annalist* are also widely used. In England, the London *Economist's* index and that of Sauerbeck are, perhaps, two of the most important. French commodity prices **may** be studied from the index number of the *Statistique Générale*.

In Fig. 1 the degree of resemblance will be observed between the index number of the Department of Labor from 1840 to 1919 and the Sauerbeck index for England, the latter being carried back to 1800 by means of tables compiled by W. S. Jevons, the distinguished economist. In this diagram the present practice of the Department of Labor has been followed of using the average commodity price index number for the year 1913 as "100."

The Department's practice in this respect is convenient, inasmuch as it enables us to use the current index numbers as published month by month by the Department to show directly the change in current purchasing power of the dollar as compared with prices prevailing shortly before the war. For instance, the last published number is 226 for the month of August, 1919. That is to say, commodities which could have been purchased for $1.00 during the year 1913 would have cost in August last $2.26.

### ANALYZING PAST PRICES AND WAGES

If we desire to arrive at intelligent conclusions regarding the probable future movements of prices, a careful analysis of past price movements is well worth our while, but at the outset it is necessary to utter a word of caution. Amateur statisticians often attempt to predict the future by literally projecting forward such curves as those in the charts accompanying this paper. Nothing is easier than to take a pencil and ruler and project some of these curves forward into the future. Nothing is more meaningless, or, generally speaking, more misleading. Facts of great value can be obtained from a study of past prices but not by such juvenile proceedings.

The real value of such charts is for the purpose of determining the *underlying causes* of the price changes which occurred in the past. Only when we do this and note their effect can we speculate intelligently on the possible future of commodity markets. There are three groups into which we may divide the factors most profoundly affecting commodity prices:

1   First, the events of contemporary history, wars, revolutions, pestilences and famines. Any study of commodity prices for the last seventy years in the United States which failed to take an account of the Civil War would be a waste of time.

2   Second, the commercial and industrial development of the individual commodities. Any study of brick buildings in England which disregarded the fact that there was a prohibitive tax on brick down to quite a modern period could scarcely have any value.

3   Third, those social and economic movements which have profoundly influenced the welfare of man, without necessarily involving war or revolution. A study of the increase in cost of erecting a brick building which failed to disclose that the wages of bricklayers have in some places multiplied by eight in the last seventy years could not possibly be complete.

[Following this the author takes up these three factors in turn, showing how each in its way has influenced commodity prices. In studying commodity prices in the light of contemporary history, he shows a chart of price indexes in England from the year 1600 to the year 1918 in which are recorded fluctuations due to the great economic and social upheavals occurring during this long period. (Fig. 2.) In relation to the second factor of commercial and industrial development, a study was made of the cost of iron during the 500 years since its discovery, an epoch which the author divides into 12 periods, in each of which there have been economic or industrial conditions which have had a marked influence on the prices of iron.

Finally, in dealing with the third factor, that of economic movements which have influenced the welfare of men, he prepared a chart, reproduced in Fig. 2, in which are plotted the wages of English workmen in certain trades during the 300-year period considered in connection with Fig. 1, in his study of the fluctuation of commodity prices. The chart in Fig. 2 expresses the wages both in dollars per day and in terms of the commodities which might be purchased therewith. "It appears to indicate,"

the author states, "(a) a prolonged period from 1600 on for almost a century, during which the purchasing power of wages was lower than it has ever been since; (b) a noteworthy improvement setting in toward the end of the seventeenth century and culminating in the period 1730-1740; (c) another long period of depression in the second half of the eighteenth century and extending beyond 1810, practically to the time of the fall of Napoleon; (d) a great and continuous upward trend in the buying power of wages culminating about 1900."

Turning to history for an explanation of the causes of the fluctuation of the purchasing power of wages in these several periods, the author attributes the low purchasing power in the first to the state of abject poverty existing in England at that

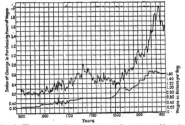

FIG. 2   WAGES OF AN ENGLISH ARTISAN (CARPENTER OR MASON)
FROM THE YEAR 1600 TO THE YEAR 1918

time—a period when it is estimated there were paupers and beggars to the number of 1,333,000, or more than one in five of the population. Following this came the adoption of sound currency and an improvement in conditions until the third period,

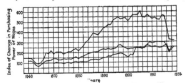

FIG. 3   RELATIVE CHANGE IN THE PURCHASING POWER OF WAGES OF CERTAIN CLASSES OF WAGE EARNERS IN THE UNITED STATES FROM THE YEAR 1860 TO THE YEAR 1920

that of the Seven Years' War, and the development under the German kings of England of a tyrannous government which led to the Revolution. During that period the factory system was in full swing with an oppressive system and unhealthy conditions prevailing, a day's work consisting of 15 or 16 hours and young children regularly working in the mills 14 hours a day. The fourth and last period covers the improvement of the status of the workingman which is one of the great features of the nineteenth century.]

### AMERICAN WAGES IN THE LIGHT OF COMMODITY PRICES

Turning to modern American wages, compared with the corresponding commodity prices, as plotted in Fig. 3, conditions are in striking contrast to the grim picture presented in former times in the history of English labor. The unit or index employed to measure the relative change in the purchasing power of a workingman's day's labor is not important. It is the rate of change in this index which is significant, the fluctuation in the purchasing power of his day's work from year to year. If that is ascending, it is fair to assume that his conditions are improved. If it shows a steady decline, it is reasonable to suspect that his conditions are becoming worse. The reader must be cautioned in regard to peak or high points. While apparently indicating prosperity to the individual artisan in continuous employment, they take no account of the fact that the bulk of unemployment may be great and actual distress is the result.

No claim is made that the curves in Fig. 3 are in any way complete for the purpose of such a study of the labor situation. That is not the primary purpose of this paper. Nevertheless, even with the small amount of data here presented, it will be noted how the workingman's conditions have improved, even allowing for the fact that the curves in Fig. 3 do not show to the full many recent wage increases. Taking Fig. 3 as it stands, however, labor today appears to have increased its purchasing power over conditions in, say, 1870 in a ratio of roughly fifteen to thirty. That is, the purchasing power of a man's working day yields approximately twice as much in commodities as it did in 1870. The standard of living of the workingman has greatly improved. This is a highly desirable condition. Without a doubt this country as a whole has greatly benefited from raising the living standard of the wage earning portion of the population. Further, no one will claim today that a lack of demand for labor is to be reckoned as nullifying the improvement in buying power of the wages an artisan receives.

### PREDICTING THE FUTURE COURSE OF PRICES

I have emphasized the fact that any efforts to predict future prices by projecting forward from a curve or graph of former prices by means of a ruler are foolish. Nevertheless, the inquiry as to what may be anticipated in the future will always be raised, and we are thoroughly entitled to use our best judgment in hazarding a prediction.

A guess as to the future price of an individual commodity must necessarily allow for progress and invention in that particular industry. A guess as to the future movements of commerce as a whole must be based upon some reasonable assumptions as to the general direction of human and economic progress. Great historical events, such as wars and revolutions, may be expected to produce big effects in the future, as in the past, but we are practically unable to predict them at all, and therefore cannot make allowance for them in future prices. Setting such greater disturbances aside, we may still do something towards forecasting future prices in the light of events of today.

The predominant element in making up the price of any commodity today is labor. Therefore, if any man claims that prices are to come down, without at the same time predicting great advances in the way of invention to increase the efficiency of labor, he is really arguing that the price of labor also is coming down. Any prophets who forecast lower commodity prices in the near future (particularly some of those whose prophecies are used as a basis for making rates on public utilities) should be forced to admit that they are in effect taking future lower wages for granted and when they do this they must be made to recognize that the burden of proof of any such claim is upon themselves.

### RELATION OF COMMODITY PRICES TO PUBLIC UTILITIES

At first sight it might seem unnecessary to call attention to the relation of public utilities to the prevailing commodity prices of today. It ought to be self-evident. The public-utility man is just as much a merchant as any other merchant. How can he fail to be affected by conditions that affect every living human being? And yet, the events of the past two or three years, in fact events of the last decade, show that this apparently self-evident fact is not evident at all to a large section of the public. An immense number of people seem to be educated to the idea that the butcher, baker, and candlestick maker are all living with them in the year 1919, but public utilities, in some mysterious way are expected to live half in the present and half in the pre-war past. A dollar bill handed to them for the payment of a 1919 electric-light bill is to be able to transact as much business for them, to purchase as much raw material and labor as it could have done in, say, the year 1913—although it is fully recognized that it has no such occult power in the hands of the customer who tenders it to them.

I will hazard a guess that future generations may look back with curiosity to some of the practices now in use in valuing railroad property throughout the country. As fast as a railroad completes construction today, with labor and materials purchased and erected today, paid for at the prices of today, the rate-making body proceeds to place a "valuation" upon the newly completed construction work based not on what it actually cost, not on what any living human being could do it for today, but on the assumption that a magician built the work with labor and materials called up out of the past at the prices current on June 30, 1914. Whatever legitimate use a "valuation" of the railroads of the country "as of June 30, 1914" may be thought to have, it would surely seem that this order of "theorizing" a value constitutes nothing less than a perversion of economic thought.

### SETTING A PRICE ON THE SERVICE OF A PUBLIC UTILITY

Assuming then, the prices we find actually prevailing in the market today, how shall we set a price upon the service of a public utility? Exactly as we determine the price of any other merchandise. What shall we pay for it? We should pay for it what the service is worth. How shall the worth be measured? In terms of the currency used to pay for it. If the service was

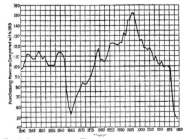

FIG. 4 FLUCTUATION IN WHOLESALE COMMODITY PURCHASING POWER OF THE DOLLAR FROM THE YEAR 1840 TO THE YEAR 1919

worth $100 in 1913, and if $226 in August, 1919, will buy as much of commodities in general as $100 bought in 1913, then public utility service (or any other service) that was worth $100 in 1913 is, *ceteris paribus*, worth $226 today.

Any one conversant with the way in which public utility rates are determined today will know that no such thought is ever entertained. On the contrary, elaborate property valuations are prepared, based sometimes on what the property would have cost if built at some other and more remote period, sometimes based upon no price that ever was. Prices are often used in valuation of public utilities, as many of you know, based on a trend curve, whose author takes the position that while the prices of a certain date were, in fact, what every one knows them to have been, they really ought to have been something else. The record of rate cases in the last two years is full of instances where nobody in the entire proceeding discussed anything in terms of the market prices then actually prevailing and being paid by every individual present at the rate bearing for every commodity necessary to keep himself alive.

### HOW RATE PROCEEDINGS SHOULD BE INFLUENCED BY CURRENT COMMODITY PRICES

Am I, then, to be understood as suggesting that all pre-war valuations be abandoned and only valuations as of the present day be employed? I have been asked this question frequently in the last six months, and the answer to it is that I am not much interested what basis is used for valuing a company's property and for setting up the facts relating thereto so long as both the company and its customers get equal treatment. But if a rate-making body is to carry on its work using the word "dollar" in the same rate case to mean two different things, to mean a dollar of high purchasing power every time one party to the controversy is mentioned, and a dollar of low purchasing power every time the opposing party is mentioned, then there is an elemental injustice present in the proceeding. Can it be claimed as fair that a dollar should have the purchasing power of 1913 ascribed to it whenever a public-utility company's property is being evaluated in dollars, while all parties are fully aware that rates can only be paid in the dollar of today with its vastly different purchasing power?

The investor in a public utility is already severely enough injured by the change in purchasing power of the dollar. To a vast extent his investment has been made out of savings effected in times when the dollar was the fruit of far more labor and material than it is today. One thousand dollars invested in the public-utility business ten years ago represents a much greater amount of labor and a much greater amount of thrift than a corresponding thousand dollars invested today. More of a man's life and toil went into it.

The United States Department of Labor commodity price index, as shown in Fig. 1, puts the figures of the situation in their fair relation. The labor and frugality which will produce 226 dollars today produced only 100 dollars six or seven years ago. We may not be able to remedy this situation; it may be inherent in our economic and financial system, and may be part of the burden that the investor must bear. But there certainly is no use in denying it or covering it up, and certainly none in talking as if the burden and hardship did not exist.

In Fig. 4, in which I have expressed the United States Department of Labor's commodity price index in terms of the shrinking purchasing power of the dollar, I have followed the Department's lead in using the index for 1913 as a base point, calling it 100 cents on the dollar. It will be noted that the purchasing power of each of the six or seven or eight dollars which the investor of 1913 looked to as the income to be derived from each hundred of his savings has shrunk today until each dollar can accomplish only what 44 cents accomplished at the time he made his investment. The chart shows how severely he has been hit.

# The Cost of Organizing and Financing a Public-Utility Project

By F. B. H. PAINE [1]

## INTRODUCTION

HAVING been associated with the late Mr. Paine on important work, and being familiar with his views based on his investigations into somewhat unusual phases of public-utility valuation, it has occurred to me that wider publicity should be given his views than has resulted from his expression of them in reports and before utility commissions. The following paper by him was prepared in July, 1916. It was in fact a report on a particular property. In order to adapt it to the requirements in the present instance I have made a few changes—those which make it real for the general rather than a particular case. No changes in substance have been made. Mr. Paine was an unusual man. His wide experience along the lines of his paper particularly qualified him to express the views set forth. In discussion of them he was most forceful and convincing. It affords me great pleasure to render this service to the memory of so talented a man, so modest and unassuming, and so good a friend. His work ceased just at the time when it had come to fruition.

M. E. COOLEY.

BEFORE the construction of the physical property of a utility can be begun, even before the engineering and other plans for that physical property are begun, there are certain preliminary steps necessary to be taken which involve much effort, time, and expense, and which are incurred at much risk of entire loss before the project as a whole can be said to be finally launched. These steps embrace the company's rights from the public, arrangements for the necessary capital funds with which to reimburse itself for preliminary outlays, and funds with which to construct the physical property necessary to fulfill obligations to the public incurred and incident to its rights to conduct a public-utility business.

I will outline these preliminary and necessary steps, state their reasonable costs, and the basis of my estimates. Each indicated percentage, or their sum, should be applied to the total direct and indirect construction cost, including administration and law expenditures, interest, and taxes during construction. I have placed them in three groups:

A—The Development of the Project......  2½ per cent
B—The Cost of Money..................  5    per cent
C—The Promoter's Remuneration........  5    per cent

Inasmuch as the preliminary effort, time and expense, the cost of money and the promoter's just remuneration vary proportionately with the breadth, the scope, and particularly the magnitude of the property, they may most conveniently be expressed as percentages of the cost of producing a new property or the estimated cost of reproducing an old property. These three groups

---

[1] Deceased.

are as absolutely essential to the service of the public as the physical groups themselves. Indeed, the latter are only made possible by the former.

## A—THE DEVELOPMENT OF THE PROJECT

The items listed below are intended to show in barest outline, by headings, the nature of the preliminary problem. It is not intended to list specifically each and every feature detail, nor can I possibly estimate in detail what each item has or would reasonably cost because the items are all so interrelated and interdependent that no sharp line can be drawn between them. We can, however, estimate upon the cost of this group as a whole, with reasonable accuracy.

1  Conception of the project
2  Determination of its scope and territorial breadth
3  Preliminary outline design and preliminary estimates of cost
4  Estimates of operating expenditures, of operating revenue and resulting profits
5  Prospectus or description of the project
6  Development of the financial plan
7  Legal and other advice
8  Negotiations with bankers and underwriters for capital funds
9  Negotiations with state and community governmental bodies necessary to secure the state charter, franchises, and the approval of the franchises, prospectus and financial plan by the Board of Public Utility Commissioners, and other such matters; but not including any payments to the public for the public rights, nor including any valuation of those public rights.

The old accounting systems were never kept in sufficient detail to show the costs of this group, which I call "Development of the Project." In recent years, however, there have been instances where such costs are approximately known, and there have been many estimates and allowances made by commissions and courts. The percentage allowed on the cost of physical property has, as a rule, fallen within the range of from 2 to 4 per cent, according to the size of the property and state requirements.

The first question, therefore, is how reasonably to estimate the cost of these preliminary requirements as if they were to be incurred as of the present day with the construction of a great public-utility system in view. The necessity for the preliminary work and studies is patent to anybody; the law of the state makes them absolutely essential. That they will cost a great sum of money is obvious; and equally so that the amount involved will be approximately proportional to the magnitude of the project. At this stage of consideration the property being appraised must be looked at as a project; it cannot become a reality until after the preliminary steps have been taken.

The work involved in the development of the project is performed by the same type of men as are concerned subsequently in the administration of the construction of the property and later in its operations. From general experience we know that the effort and time involved would at least quite equal that incident to the physical construction of the projected utility; and it may be fairly assumed that the expenditures involved in the development of the project would be substantially equal to the cost of the administration and legal expenses during construction. Therefore I have so estimated it.

It costs about the same to administer the expenditure of a million dollars properly, for whatever purpose the money be spent. If during a year the expenditures of a utility corporation for operations and for new property amount, say, to ten million dollars, and if in the same year the administration and legal advice costs $250,000, or 2½ per cent of the money expended, that percentage may reasonably be taken to represent the cost of administration and legal advice during the construction of the property; and it may also be taken to represent the cost of the development of the project, for the reasons given in the preceding paragraph. I have found from a study of public-utility companies' accounts that at least 2½ per cent of the money spent for operating and for construction was the cost for administration and legal expenses, as I am here using these terms.

This method of estimating the costs of development of the project would only fall into error where there are some peculiar or unusual conditions, as in a particular state. In some states there are certain conditions which make for a lesser cost of securing the public rights, gaining approval of them and of the financial and other plans; and in other states the public requirements are such as to increase these costs. In some states, for some kinds of utilities, no local franchises are necessary; in other states there is still no controlling body like public utility commissions. Where there is no occasion to obtain approval of any local franchises, or of any project with its prospectus and financial plan, the preliminary costs are correspondingly less. But where separate charters, corporations, and other forms are necessary in each political subdivision of the state, and the necessity exists for securing approval of the controlling body, for each subdivision, the preliminary costs are correspondingly more.

### B—COST OF MONEY

In order that money may be secured for a particular enterprise, such as a public utility, there must be a cost incident to its acquisition; not only a cost incident to the acquisition of the first money needed, but for each enlargement of the property. This is always true of every such enterprise and always will be. This cost of money is sometimes, in fact, is usually, merged in and becomes a part of what is generally termed "discount" on such bonds and stocks as are sold.

I am here referring only to that part of discount which is incident to the expense of procuring the money with which to construct. The corporation might sell its securities "over the counter," which would involve an expense, or they might employ agents for the disposition of their securities (employing the word "security" to include stocks, notes, etc., as well as first-mortgage bonds), commonly known as brokers, underwriters, bankers, etc.

When we wish to rent or sell real estate we employ a real-estate agent to do it for us because that is his business and he can do it and make a profit at a less cost, or to better advantage, than we could ourselves. There is a cost in either case attendant upon each such transaction, i.e., the value of our own time and effort taken from our own ordinary occupation versus the payment of money to an agent whose ordinary business it is.

Funds with which to construct a property of any considerable magnitude are invariably and necessarily secured from the world's money markets. No locality, or no one market, would be sufficient for the needs of a large enterprise. Funds from these world money markets are invariably supplied through bankers or underwriters, who for themselves make an investigation of the property from the engineering, legal, and commercial points of

view. They must bear this expense as well as the expense of selling to the ultimate buyer. They must also assume the risk of changes in the financial situation, or of error in judgment as to the salability of the securities and the price that they will command at such times as the corporation needs the funds. Not infrequently the banker or underwriter is compelled to carry issued securities of a high-grade character for long periods, while the utility gets its money as needed. The public's demand for service cannot be denied whether the money market is favorable or unfavorable. The practical necessities of the case make it wise for a utility company to secure its funds through agents or underwriters who assume the burden of the risks and the expense of reaching the widest, greatest world's money markets. The cost of money is by no means "velvet" to the banker or others who underwrite an "issue."

During the past two years I have been under the necessity, and have had the opportunity and facilities, of informing myself as to the cost of acquiring money in the case of sixteen large and well-known utility companies, with nineteen issues of the highest grade bonds, rated in Moody's Analysis of Public Utilities as "AA" or "A," and marketed within the period from 1908 to 1914. The result of my investigations was to find that the cost incident to the disposition of the senior ("AA" or "A") highest rated securities of these large, well-known utilities was 5·per cent; the discounts were much more than that. The cost incident to selling bonds rated by Moody as "BA" and "BAA," found in the same manner, was 7.9 per cent.

The propriety of allowing this item as one of the proper elements entering into the fair value of a utility has been favorably passed upon by commissions and courts in this country, but the clearest and most complete analysis is that of the English Justice Lawrence in the Railway and Canal Commission, Royal Courts of Justice, in the condemnation case of the National Telephone Company, Limited, vs. His Majesty's Postmaster General. This covers the questions so completely that I quote from that part of the decision dealing with the cost of money.

*The first question then is, would it, in fact, cost anything to provide the necessary capital? The Company have given evidence by way of example, that it cost them 4.41 per cent to raise 5 1-2 million pounds. No one has given evidence that it would not cost anything, nor has that proposition been put forward even in an argument. I know of no commodity and of no service that can be procured as of right for nothing.*

*It has been said that it cannot be an element adding to the value of the plant. The thing transferred here is the plant "in situ," and the cost of construction, less depreciation, is the method by which the value has to be ascertained. It follows that every expense which is necessary in order to construct is an element to be considered, and it has to be considered because it is necessary in the process of construction. The thing to be transferred, say a pole, must be procured, transported, and erected; each of these steps is necessary to the existence of the pole "in situ"; each of these steps cost money, and raising this money is itself an expense and is one as necessary to the existence of the pole as any of the other steps.*

*The cost to be considered is the cost to the hypothetical constructor who is a person in good credit, or, in other words, what it must cost any constructor. This does not involve the conclusion that the cost of raising capital should be added to the price again if the property should be transferred a second or a third time; it is an item of value once and once only, namely on the construction of the plant, and it is merely because it is necessary in order to construct that this item comes into the calculation at all. That it must be included is apparent, if it is tested in a case in which the sale takes place immediately upon the completion of the construction.*

I call special attention to the sentence italicized by me: *The cost to be considered is the cost to the hypothetical constructor who is a person in good credit.* With this part of the decision I personally do not agree when applied to establishing a fair value for another company differently situated. The Board of Public Utility Commissioners does not deal in condemnation or confiscatory matters, but does deal in matters establishing a fair value, among others. "Fair" means more than that which just avoids being confiscatory.

It is obvious that the cost of money is greater on that part of the capital raised on the junior securities than on the senior securities; greater for small undertakings less well situated with respect to the money markets than for large properties very well situated; greater for securities of any class of low rating, than for those of the very highest rating. It does not seem fair to me, therefore, that the allowed cost of the money with which to construct should in all cases be only that allowed in the most favorable circumstances. It follows in my judgment that the fair cost of money necessarily raised for a new enterprise, or for a utility whose securities are not of the highest standing, would be more than 5 per cent—as high as 7.9 per cent. That is the percentage which I found it had cost to sell bonds rated by Moody as "BA" and "BAA."

Securities issued by a corporation under a general mortgage are often classed as "Collateral Trust Bonds." They represent fairly the average, or the mean, of the credit attaching to the various senior and junior security issues of such a corporation necessarily issued to raise all the money needed to construct the whole property. The highest-grade first-mortgage bonds rated in the "A" group by Moody probably could not be sold in amount more than 60 per cent of the total money required. The remaining 40 per cent would have to be acquired through the issue of other forms of junior securities such as common and preferred stocks, second-mortgage bonds and collateral-trust bonds or notes. It costs less to sell the former than it does the latter for the reason that the first mortgage for a limited amount relative to the size of the property has a higher credit than the junior securities. A first-mortgage bond such as I have described absorbs in addition to its own proportion of the corporate credit, all the credit underlying the junior securities as well, and leaves all the risk of the business upon the shoulders of the junior securities in varying degrees according to their degree of juniority. Therefore, when the cost of money with which to construct the whole property is taken as that which it would cost to sell only 60 per cent of the securities, it in my judgment amounts to confiscation of a part of the cost of money involved in raising the rest of the necessary capital funds. That is why I think that usually the fair amount to allow would be more than 5 per cent and might easily be 7.9 per cent of the cost of the property devoted to the public use.

However, under the most liberal interpretation of the phraseology of Mr. Justice Lawrence's decision in the condemnation case referred to, it is the cost of money with the best of credit which is to be considered; therefore, the measure I adopt as the "cost of money" with securities rated "BA" or "BAA," is the cost of capital funds secured for utilities of the highest credit, best salability, such as first-mortgage monds rated "AA" or "A." I take this as the measure in order that my estimate may be regarded as ultraconservative, that it may be accepted as the very minimum.

I arrive at 5 per cent as the cost of money in the following manner: Let us assume that the securities are sold at less than par, or less than their face value. A part of the difference from par is due to the cost of obtaining the money. This part, as I have shown, belongs to the cost of construction and is therefore proper to capitalize. The remaining part is an interest adjustment necessarily made to satisfy the money markets at the moment of sale, which is true discount and may not be a proper capital charge.

Whatever the interest adjustment, whether discount or premium, it is easily determinable as it is the amount less or more than par, or face value, at which the securities are taken by the general public or ultimate purchaser. The cost of money is the difference between the price paid by the ultimate purchaser and the net amount received into the "till" of the corporation. If securities are sold to ultimate purchaser at above par it simply means that the par, or face interest, is higher than necessary to satisfy the money market. The same security is sold sometimes above par and sometimes below par, according to the demands of the money market at the moment of sale. Money is a commodity which varies in its market requirements much as wheat and other commodities vary. The cost of money, however, is distinct from discount or premium and is the difference between that paid by the ultimate purchaser and the amount received into the corporation till.

As it is the cost of raising all the money which it would take to construct at the present time, the per cent of the cost of money is to be applied to the entire estimated cost of construction. Generally speaking, the money with which to construct is secured through the sale of various kinds of securities—common stock, preferred stock, second-mortgage bonds and notes, as well as first-mortgage bonds. Let us assume three examples:

*Example 1:*

| | |
|---|---|
| Par value of bond........................................ | $1000 |
| Interest, per cent....................................... | 5 |
| Bond sold to ultimate purchaser for...................... | 975 |
| Net amount received into corporation till................ | 925 |

In such a case the cost of the money would be the difference between $925 and $975, viz., 5 per cent, which is properly capitalized............................... 50
The interest adjustment required by the money market would be a discount of 2 1-2 per cent.............. 25

*Example 2:*

| | |
|---|---|
| Par value of bond........................................ | $1000 |
| Interest, per cent....................................... | 5 |
| Bond sold to ultimate purchaser for...................... | 1030 |
| Net amount received into corporation till................ | 980 |

In this case the cost of the money would be as in the first example, 5 per cent, properly capitalized............. 50
The interest adjustment required by the money market would be a premium of 3 per cent.................. 30

*Example 3:*

| | |
|---|---|
| Par value of bond........................................ | $1000 |
| Interest, per cent....................................... | 5 |
| Bond sold to ultimate purchaser for...................... | 800 |
| Net amount received into the corporation till............ | 750 |

Here again the cost of money properly to be capitalized would be 5 per cent............................... 50
The interest adjustment required by the money market would be a discount of 20 per cent to adjust the assumed market requirement of 5 per cent interest.... 200

In arriving at my conclusions as to the cost of money I first took the per cent cost incurred in the sale of each of the 19 issues rated at "AA" or "A" of the 16 corporations and found that by adding together the per cent costs and dividing the sum by the number of issues the true average cost was 5.1 per cent. I then multiplied the par value of each issue sold by the per cent cost incurred in that particular sale which gave me the cost in dollars for each case; and by adding all these together I found it had cost $3,164,000 to market $64,845,000 par value of bonds, or 4.9 per cent; 5.1 per cent being the average of the percentages; 4.9 per cent being the weighted average. I took the mean, or 5 per cent, as representing the cost of obtaining money. I found in a similar way that the cost of marketing the "B" group of bonds was 7.9 per cent.

### C—PROMOTER'S REMUNERATION

"I know of no commodity and of no service that can be secured, as of right, for nothing."—Mr. Justice Lawrence. What is it which would induce any promoter or any group of persons to make such a preliminary expenditure as I have described under "development of the project" and then provide the money and assume the two great risks: First, that the project would fall by the wayside and never come to fruition at all; second, its failure afterward? Why should the promoter render such a service as has been rendered to the public in providing most existing utilities? The answer is: Remuneration for service rendered. Promoter's remuneration, as dealt with here, is the cost to the public for the promoter's services. It is entirely distinct from the cost of the preliminary development of the project and the cost of securing the money with which to build the property.

There must be some inducement to bring about the activities of the promoter in the particular enterprise, and that inducement is profit or gain; and for it he renders a valuable service. The service of the promoter is essential at all stages of the utilities' development, not only in the beginning, but with the addition of each increment of capital needed for enlargement or extensions. Enterprise without the service of a promoter is inconceivable. The Railroad Securities Commission appointed by President Taft, of which Arthur T. Hadley, President of Yale University, was Chairman, said on page 30 of their report:

We are told that the profit of the promoter represents a wholly unnecessary burden upon the American public, and that so far as this profit can be done away with it will be good for all parties. Neither of these statements is quite true. The promoters, using the term in a broad sense, may be divided into two classes: constructors who build a railroad whose future is uncertain, in the expectation of selling the stock for more than it cost them; and financiers who induce the public to buy the bonds of such roads. Both of these classes, if they do their work honestly, render useful service to the public. The constructor gives our undeveloped districts the benefit of new roads, which they would not get without his intervention; and if he does his business well he builds the road more economically than anybody else could. The financier renders an equally important service in collecting the capital of the investors to build new railroads or improve old ones.

There are many similar statements made by commissions, courts, and other authorities, and I think all commissions have recognized that the services of a promoter should be properly remunerated.

When a public-service commission is determining the fair value of a property, it should make no difference whether the property is old and has been rendering a valuable service, or whether it is new and much desired by the public; the same elements should enter into the value on which earnings are allowed; and whatever is a proper allowance necessary to induce new and valuable enterprise must also be a proper allowance to give the promoters of an enterprise already in existence and of undoubted public use and in which the risk has been incurred and the service rendered.

The service of a promoter having been essential to a particular utility, the property having been built and having rendered a valuable service to the community, it is entirely proper that a reasonable recompense should be made on account of the service

of promotion. The fact that the promoter might be the present owner does not in any way affect the equity. If now, when the state for the first time fixes the value and exercises control over the property, it denies remuneration to the promoter an injustice will be done. The promoter, indeed, might better have gone elsewhere.

In the application of the Rochester, Corning and Elmira Traction Company for consent to the execution of a mortgage, and authorizing the issuance of capital stock, the Public Service Commission of the State of New York, 2d District (opinion by Chairman Stevens, in Vol. 1 of the report of April 1, 1909, p. 166), dealt with the whole subject very fully (as it has been by many other commissions) and allowed 5 per cent as promoter's remuneration. It has also allowed 5 per cent in the case of other projected utilities.

In the decision by Chairman Stevens, in the case referred to, he speaks of the difficulty of determining the exact amount which the promoter should be paid under the varying conditions. He fixed 5 per cent for a proposed railroad, the desirability of which he had passed upon favorably. My judgment is that his allowance was too small for such a new project (the road was not built), yet he is a man intentionally fair, of good judgment, and has had good opportunities for observation. Five per cent has been accepted in other cases and is a frequent minimum provision in promoting large enterprises. Therefore, it follows, and my judgment is, that five per cent of the cost of a property is the minimum proper capital allowance which should be made, in the appraisal of public utilities

# The Construction Period

By H. C. ANDERSON,[1] ANN ARBOR, MICH.

IN the early days of appraisals, the item of interest during the construction period was frequently ignored. In some of the later appraisals it has been taken at very large figures, large because of the assumption that the construction period extended to the completion of the entire property; that is, no part of it was assumed to go into operation before the completion of the whole. There seems to be very little, if any, disagreement among engineers and commissions as to the annual rate of interest, but in many cases there is a wide variation in the length of the construction period.

If it is assumed the property is to be reproduced "piece-meal" we will have one period of construction. On the other hand, if the property is to be reproduced "wholesale" we will have a different period of construction. If a certain property is to be reproduced "piece-meal" the result would naturally be higher unit cost and lower total interest cost than if reproduced by the "wholesale" method. In some cases the two may balance.

The particular part is to fix, as accurately as possible, the duration of the construction period. Usually some part of the property, a power plant for instance, will be the best guide. Obviously, a street railway, complete in every other respect, must remain idle without power.

The question of the critical speed of construction must be given consideration. Theoretically, a large property could be constructed in a few months, assuming enough men and material could be procured; but to do this it would be necessary to exceed the speed of economical construction, which would result in a very high unit cost, and no doubt would more than offset the decrease in the interest cost. Consideration, also, must be given to the question of obtaining labor and material in large quantities at the time desired, and the fact that no city would consent to the extensive tearing up of its streets that might be necessary with a short period of construction.

One method often used is to assume an interest rate of 6 per cent applied to the entire cost over one-half the construction period. The cost is zero at the beginning and is assumed to increase uniformly to the full amount at the end. A criticism of

this method is that money cannot be borrowed day by day, or month by month, as needed. It is difficult to imagine bankers willing to furnish money in this way.

Another method often used is to assume an interest rate of 6 per cent and 2 per cent allowance on the unexpended balance, assuming the unexpended balance to equal one-half the loan, and the loans to be made at equal intervals during the construction period. The following concrete problems will illustrate the two methods.

PROBLEM NO. 1:

Assume a property costing $12,000,000 and a construction period of three years, interest rate 6 per cent. The expenditures proceeding uniformly from zero at the beginning to the full amount at the end.

Solution:

Six per cent for one-half the construction period will give a total interest charge of 9 per cent, which applied to the $12,000,000 cost will give $1,080,000 as the interest cost.

$$\frac{1,080,000}{12,000,000} = 9 \text{ per cent}$$

PROBLEM NO. 2:

Assume a property costing $12,000,000 and a construction period of three years, interest rate 6 per cent, allowance on balance at the rate of 2 per cent, balances equal one-half the loan. $2,000,000 borrowed each six months.

Solution:

| | | |
|---|---|---|
| First loan ........$2,000,000 @ 6% for 3 yr. | $360,000 | |
| Credit on balance... 1,000,000 @ 2% for ½ yr. | 10,000 | |
| Net interest first loan..................... | | $350,000 |
| Second loan ....... 2,000,000 @ 6% for 2½ yr. | 300,000 | |
| Credit on balance... 1,000,000 @ 2% for ½ yr. | 10,000 | |
| Net interest second loan.................. | | 290,000 |
| Third loan ........ 2,000,000 @ 6% for 2 yr. | 240,000 | |
| Credit on balance... 1,000,000 @ 2% for ½ yr. | 10,000 | |
| Net interest third loan................... | | 230,000 |

[1] Professor of Mechanical Engineering, University of Michigan. Mem. Am. Soc. M. E.

| | | |
|---|---|---|
| Fourth loan ....... 2,000,000 @ 6% for 1½ yr. | 180,000 | |
| Credit on balance... 1,000,000 @ 2% for ½ yr. | 10,000 | |
| Net interest fourth loan.................. | | 170,000 |
| Fifth loan ......... 2,000,000 @ 6% for 1 yr. | 120,000 | |
| Credit on balance... 1,000,000 @ 2% for ½ yr. | 10,000 | |
| Net interest fifth loan.................. | | 110,000 |
| Sixth loan ........ 2,000,000 @ 6% for ½ yr. | 60,000 | |
| Credit on balance... 1,000,000 @ 2% for ½ yr. | 10,000 | |
| Net interest sixth loan.................. | | 50,000 |
| Total interest for three years............ | | $1,200,000 |

$$\frac{12,000,000}{1,200,000} = 10 \text{ per cent}$$

If this money is borrowed $4,000,000 each year for three years instead of $2,000,000 each six months the total interest becomes

$$\frac{1,320,000}{12,000,000} = 11 \text{ per cent}$$

It will be noted, in this analysis, the assumption is made that no part of the money is spent until the day construction starts. In actual practice this would be very difficult, if not impossible. The land for various purposes, such as right of way, power stations, sub-stations, shops, car houses, gas plants, commercial offices, etc., would necessarily have to be purchased one or more years before construction started. The following concrete problem will illustrate this method.

PROBLEM NO. 3:

Assume a property costing $12,000,000. Of this amount the cost of land is $1,200,000 and the structural property $10,800,000. It will require three years to obtain the land, starting one year in advance of construction. Construction period of three years. Interest rate of 6 per cent, allowance on balance at the rate of 2 per cent, balances equal one-half the loan, money borrowed each six months.

Solution:

If the purchase of land starts one year in advance, and money is borrowed every six months there will be 8 periods to consider; following are the amounts borrowed each period.

| | Land | Structural Property | Total |
|---|---|---|---|
| First Period .......... | 200,000 | .......... | $ 200,000 |
| Second Period ........ | 200,000 | .......... | 200,000 |
| Third Period .......... | 200,000 | $1,800,000 | 2,000,000 |
| Fourth Period ......... | 200,000 | 1,800,000 | 2,000,000 |
| Fifth Period .......... | 200,000 | 1,800,000 | 2,000,000 |
| Sixth Period .......... | ...... | 1,800,000 | 2,000,000 |
| Seventh Period ....... | $200,000 | 1,800,000 | 1,800,000 |
| Eighth Period ......... | ...... | 1,800,000 | 1,800,000 |
| Total ............ | $1,200,000 | $10,800,000 | $12,000,000 |

| | | |
|---|---|---|
| First loan ........ $200,000 @ 6% for 4 yr. | $48,000 | |
| Credit on balance.. 100,000 @ 2% for ½ yr. | 1,000 | |
| Net interest second loan................ | | $47,000 |
| Second loan ....... 200,000 @ 6% for 3½ yr. | 42,000 | |
| Credit on balance.. 100,000 @ 2% for ½ yr. | 1,000 | |
| Net interest second loan................ | | 41,000 |
| Third loan ........ 2,000,000 @ 6% for 3 yr. | 360,000 | |
| Credit on balance.. 1,000,000 @ 2% for ½ yr. | 10,000 | |
| Net interest third loan.................. | | 350,000 |
| Fourth loan ...... 2,000,000 @ 6% for 2½ yr. | 300,000 | |
| Credit on balance.. 1,000,000 @ 2% for ½ yr. | 10,000 | |
| Net interest fourth loan ................ | | 290,000 |
| Fifth loan ........ 2,000,000 @ 6% for 2 yr. | 240,000 | |
| Credit on balance.. 1,000,000 @ 2% for ½ yr. | 10,000 | |
| Net interest fifth loan.................. | | 230,000 |

| | | |
|---|---|---|
| Sixth loan ........ 2,000,000 @ 6% for 1½ yr. | 180,000 | |
| Credit on balance.. 1,000,000 @ 2% for ½ yr. | 10,000 | |
| Net interest sixth loan................ | | 170,000 |
| Seventh loan ...... 1,800,000 @ 6% for 1 yr. | 108,000 | |
| Credit on balance... 900,000 @ 2% for ½ yr. | 9,000 | |
| Net interest seventh loan................ | | 99,000 |
| Eighth loan ....... 1,800,000 @ 6% for ½ yr. | 54,000 | |
| Credit on balance .. 900,000 @ 2% for ½ yr. | 9,000 | |
| Net interest eighth loan................ | | 45,000 |
| Total interest cost ..................... | | $1,272,000 |

$$\frac{1,272,000}{12,000,000} = 10.60 \text{ per cent}$$

If this money is borrowed at the beginning of each year instead of every six months the amounts for each period will be as follows:

| | Land | Structural Property | Total |
|---|---|---|---|
| First Period .......... | $400,000 | .......... | $ 400,000 |
| Second Period ........ | 400,000 | $3,600,000 | 4,000,000 |
| Third Period .......... | 400,000 | 3,600,000 | 4,000,000 |
| Fourth Period ........ | ...... | 3,600,000 | 3,600,000 |
| Total ............ | $1,200,000 | $10,800,000 | $12,000,000 |

The total interest cost will be $1,392,000.

$$\frac{1,392,000}{12,000,000} = 11.60 \text{ per cent}$$

I do not know of any one formula that can be applied to every type of utility, and which, when solved, will give accurately the interest cost during construction. Each property will have its own individual problems, and for this reason a thorough study should be made of the particular property under consideration to determine:

1　The magnitude of the property and the difficulties that must be overcome during construction

2　The method used in finding the reproduction cost

3　The proper construction period, taking into consideration the method used in finding the reproduction cost

4　Knowing the construction period, what is the proper method to use in finding the total interest cost?

In the end, after all theories have been considered. it is the judgment of competent engineers that should be the controlling factor.

TAXES DURING CONSTRUCTION

Like interest, taxes during construction is an item of cost incurred during the construction period and must be paid, the only question being, what is the proper amount to allow. It may be argued that the value of a utility depends on its earnings, and there can be no earnings during the construction period, therefore no taxes should be paid. But it does not work that way. The assessor finds the land and puts it on the rolls. He finds the building materials, machinery, and partially erected buildings and they also go on the rolls, perhaps not in full amount, but in some amount at least. The problem, however, is somewhat easier than in the case of interest. It is always possible to go to the company's books and find the amount of taxes the company is actually paying, also the rate, but it is not always possible to find the ratio of assessed to actual value.

It also becomes necessary to estimate just what percentage of the property under construction the assessor will find and put on the rolls. It is also necessary to consider whether the date of starting construction and the date of assessment are the same, or whether the assessment is made in mid-year of each year of the construction period. The final result will not be very different with two methods. The following problem is presented only to illustrate the solution under certain assumed conditions.

PROBLEM NO. 4:

Assume a property costing $12,000,000, of this amount the

(Continued on page 597)

# The Design of Aspirators for Sterilizing Water

By A. E. WALDEN,[1] BALTIMORE, MD.

THE value of ozone as a sterilizing agent in the purification of water is well known. The means for uniting water and ozone or air, however, are not so clearly established, and in an effort to perfect the design of an aspirator which would efficiently accomplish this the writer conducted an extensive series of experiments which form the basis of this paper.

From the operating point of view, it was first of all essential that the heads of water pumped against be kept as low as possible. Data were unobtainable, however, concerning the relation of the volume of water to the volume of air that it was possible to combine or mix at low heads, or the possible variation of the two, which is of greater importance.

To obtain this information numerous different forms of aspirators, ejectors, siphons, jets, and nozzles were accordingly tested. Some of these, of the barometric-condenser type used for both steam and water, failed to give satisfactory results. Some that would work successfully with steam would not work with water; moreover, they would not handle large volumes of air. Others that would work well with large volumes of water could not be manipulated to vary the quantity of water and would not handle large volumes of air, or permit of varying the quantity of air to the quantity of water, and vice versa. Also some that were very satisfactory for use with condensers would not handle large quantities or volumes of air with water without a blowback of air, due to the fact that the column of water and air lacked weight or velocity to compress the air and keep the combination in a forward motion, a pressure equivalent to higher pressure than that in the mixing well or contact cell being required.

Experiments likewise showed that under low heads some aspirators would draw in a continuous draft of air but could not be varied; others, when the draft was heavy, would lower the specific gravity of the mixture to such a value that a blowback occurred and action ceased altogether; and that while it was possible to draw in air and water in a proportion of 1 to 3, by volume, the head should not be reduced much below 12 ft. for satisfactory operation. All of the above conclusions were reached by testing many different types and designs of aspirators and the first results that were at all satisfactory were obtained with a Knowles spirojector. This is shown in Fig. 1.

The Knowles spirojector, however, while giving fairly satisfactory results, did not function at low heads any better than

other types; moreover, it is difficult to construct in either tile or enameled iron, inasmuch as the heat necessary to set the enamel will warp and distort the tube. If a tile-grinding process could be employed at a reasonable cost, it might be possible to eliminate some of the objections to this form of aspirator, such as the lack of flexibility in the variation of quantity of water, etc. In all cases, however, it was impossible materially to vary the quantity of air and ozone to water, and a pumping head of 34 ft. plus a pressure of 3 lb. per sq. in. was of course out of the question.

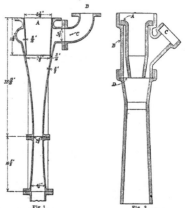

Fig 1    Fig. 3

FIG. 1 SECTIONAL VIEW OF THE KNOWLES SPIROJECTOR when used as an aspirator, air enters at $A$ and water at $B$. When measuring air velocity, opening $A$ is reduced and an anemometer placed in the reduced opening. Pressure gage is located at $C$.

FIG. 3 SECTIONAL VIEW OF ASPIRATOR USED IN TESTS

In this connection it should be pointed out that air velocity tends to reduce the temperature and that if the humidity is

[1] Supt. and Chief Engineer, The Baltimore County Water and Electric Co., Mem. Am. Soc. M. E.

TABLE 1   SUMMARY OF RESULTS OBTAINED IN TESTING VARIOUS TYPES OF ASPIRATORS

| No. | Type of Aspirator[1] | Discharge, cu. ft. water per min. | Air velocity, ft. per min. | Cu. ft. of air per min. | Pressure head, ft.-in. | Negative head, ft.-in. | Bacteria, per cent removal |
|---|---|---|---|---|---|---|---|
| 1 | 3-in. Orifice | 41 | 1100 | 45.0 | 5—9 | 8—0 | 97.58 |
| 2 | 3-in. Orifice | 41 | 1090 | 44.7 | 6—0 | 8—0 | 85 |
| 3 | 2.5-in. Orifice | 41 | 930 | 38.0 | 5—10½ | 8—0 | 95.6 |
| 4 | 3.5-in. Orifice | 41 | 900 | 36.9 | 5—1 | 8—0 | 88.1 |
| 5 | Tile Aspirator | 41 | 1000 | 41.0 | 6—2 | 7—0 | 92.0 |
| 6 | Nozzle 2-A, Throat No. 2, 2½ in. apart | 24.5 | 590 | 24.1 | 5—8½ | 7—0 | 96.2 |
| 7 | Nozzle 2-A, Throat No. 3, 2½ in. apart | 25 | 420 | 17.2 | 6—2 | 7—0 | 96.8 |
| 8 | Nozzle 2-A, Throat No. 4, ½ in. clearance around jet at throat | 24 | 870 | 35.6 | 6—1 | 7—0 | 99.2 |
| 9 | Nozzle No. 2-A, Throat No. 3 | 23 | 630 | 25.8 | 6—3 | 7—6 | 98.6 |
| 10 | Nozzle No. 2-A, Throat No. 5 | 25 | 335 | 17.7 | 6—4 | 7—0 | 95.9 |
| 11 | Nozzle No. 21, Throat No. 25 | 42 | 330 | 13.5 | 5—4 | 7—0 | 97.3 |
| 12 | Nozzle No. 21, Throat No. 22 | 42 | 430 | 17.6 | 5—4 | 7—0 | 99.1 |
| 13 | Nozzle No. 21, Throat No. 23 | 42 | 690 | 28.3 | 5—2 | 7—0 | 99.3 |
| 14 | Nozzle No. 21, Throat No. 24 | 42 | 800 | 32.8 | 5—3 | 7—0 | 97.2 |
| 15 | Nozzle No. 21, Throat No. 6 | 42 | 460 | 18.8 | 5—4 | 7—0 | 98.3 |
| 16 | Nozzle with holes of 5 sets, Throat No. 25 | 22 | 725 | 29.7 | 6—1 | 7—1 | 99 |

[1] Tests Nos. 1-4 were made with a steel-plate orifice inserted at $D$ in Fig. 3.
Test No. 5 was made with a tile nozzle inserted at $D$ in Fig. 3.
Tests Nos. 6-16 were made with nozzles and throats of the shapes shown in Fig. 2.

high and the air impinges on a cold metal surface, such as a cooled tube, the dewpoint is quickly reached and moisture is immediately precipitated on the cool surface. If allowed to

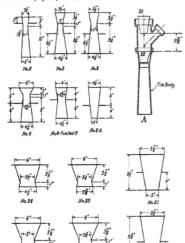

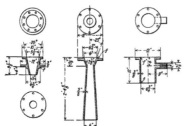

FIG. 2 EXPERIMENTAL ASPIRATOR WITH TILE BODY AND THE VARIOUS NOZZLES AND THROATS USED

FIG. 4 DETAILS OF AN ASPIRATOR HEAD

continue it eventually causes rupture of the dielectric in the ozone generator. Humidity also reduces the quantity of ozone it is possible to produce. The reason therefor is not known at this time, unless it may be that moisture carries the static charge in such a way as not to affect the air itself. The above action, which has never before been mentioned by any experimenters, was accidentally discovered during the series of experiments conducted by the author.

As a result of these preliminary tests a new set of experi-

ments were conducted using various tile nozzles and different throats inserted in a tile body as shown at A in Fig. 2. These nozzles and throats are also shown in Fig. 2. The results of the experiments conducted on these aspirators are given in Table 1, and the varied tests indicated therein show the suc-

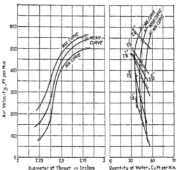

FIG. 5 CURVES SHOWING RELATION OF AIR VELOCITY TO DIAMETER OF THROAT AND TO QUANTITY OF WATER WITH VARYING SIZES OF THROATS (FOR 20-DEG. 3-IN. NOZZLE)

cessive steps which were taken in the selection of the type ultimately chosen.

THE EFFECT OF VARYING HEAD

A series of tests were also made to determine the effect of varying both the positive and negative head, positive head being the elevation above the throat and negative head the distance from the throat to the tailwater. The results of these tests are given in Table 2 and summarized below.

The loss of 4 in. down the pipe, or tail pipe, was estimated to be equal to 0.08436 ft. per foot of pipe.

A test was made for the 2-in. nozzle for diameter of stream at contracted portion, which was found to be 1 13-16 in. at

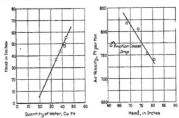

FIG. 6 CURVES SHOWING RELATION OF HEAD TO QUANTITY OF WATER AND AIR VELOCITY, FOR 20-DEG. 3-IN. NOZZLE

one-half the distance of 3 in. from end of nozzle; and discharging 33 cu. ft. under 5.75 ft. positive head and 7 ft. negative head sustained a column of water 81.5 in. high. Roughness or irregularity of the apparatus used, however, will vary these figures materially.

These tests indicate that the length of the jet discharged from the nozzle striking either at the throat or at some point in the diverging tube below the throat gave the best results. As the tests were made the jets were examined by throwing light through the opening C, Fig. 3, so that the jet could be seen, and also through a side opening, which was straight, in a cast-iron special that was employed in place of tile B.

In every case the roughest discharge from the nozzle, if it impinged at or below the throat, drew in the greatest quantity of air. This led to the suggestion that a glass aspirator be constructed, but glass manufacturers would not undertake to make it except at an excessive price.

The appearance of the jet also suggested that possibly a nozzle giving a star-shaped or serrated surface discharge would aspirate a greater quantity of air to a given volume of water.

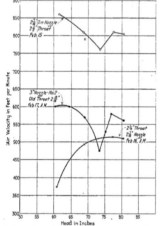

FIG. 7  CURVE SHOWING IRREGULAR CHARACTERISTICS FOUND IN A FEW OF THE LARGE NUMBER OF ASPIRATORS TESTED

These experiments led to the construction of three aspirators, one of which was later adopted and is shown in Fig. 4.

TABLE 2  RESULTS OF TESTS MADE TO DETERMINE EFFECT OF VARYING HEAD

| No. | Type of Aspirator. | Discharge Cu. ft. Water per Min. | Air Velocity Ft. Per Min. | Cu. Ft. Air Per Min. | Pressure Head, Ft. In. | Negative Head, Ft. In. |
|---|---|---|---|---|---|---|
| 1 | Tile Nozzle, Throat No. 22 .. | 42 | 1015 | 41.6 | 7—0 | 7—0 |
| 2 | "        "      Throat No. 25 .. | 42 | 890 | 36.5 | 7—0 | 7—0 |
| 3 | 3-in Nozzle, 2¾-in. Throat... | 47 | 420 | 17.4 | 6—0 | 8—0 |
| 4 | "        "         "      ... | 46 | 380 | 15.6 | 6—0 | 8—0 |
| 5 | "        "         "      ... | 47 | 275 | 11.5 | 6—0 | 7—6 |
| 6 | "        "         "      ... | 47 | 200 | 8.2 | 6—0 | 6—0 |

All of these tests indicated the importance of providing not less than 20 per cent excess air velocity in testing over that in operation in sterilizing, to allow for friction in tail pipe and mixers or devices below the throat, including the change in the specific gravity of the mixture of air and water. This was necessary to keep the mixture in motion.

It was the intent to carry out further experiments to de-

termine what would appear to be the slip of the bubbles of air in the tail pipe and the fact that they do not seem to fall at the same velocity as the column of water, or for some reason do not receive and hold the kinetic energy imparted to them. The bubbles decrease in size with increase of head and increase with a decrease of head, which is as it should be. The bubbles appear to rise at a rate of about 0.75 ft. per sec.

The specific gravity of the mixture of air and water, or ozone, when a certain entrainment is reached, is so low that the energy in a downward direction is lost and a blowback occurs in the air inlet, as before mentioned, which is similar to the air lift, and makes the mixture look like soap bubbles in the down pipe.

### THE TYPE OF ASPIRATORS ADOPTED

The final aspirators were of cast iron, enameled with the best possible white porcelain enamel; but even this is subject to deterioration, due to high ozone concentration, which shows after about five years of operation. There seems to be no question as to the suitability of tile for aspirators when ground to suitable measurements.

The details of an aspirator head are shown in Fig. 4. The curves of Fig. 5 and 6 give its characteristics. Some experiments, however, would seem to indicate that the air intake and the time element of mixing might be changed profitably, considering the cost of handling the water, and that contact in a horizontal pipe beyond the tail pipe for a distance of 20 ft. will turn out a perfectly satisfactory effluent. The curves in Fig. 7 are inserted to show the irregular characteristics found in a few of the great number of aspirators tested before the final type was chosen.

### John Ericsson Memorial Tablets

The members of The American Society of Mechanical Engineers doubtless remember the account published in the January 1920 issue of MECHANICAL ENGINEERING of the memorial meeting held during the Annual Meeting of this Society, on the evening of December 3, 1919, in commemoration of the eightieth anniversary of the arrival in the United States of Captain John Ericsson, his fifty years' association with Cornelius H. DeLamater in engineering work, and finally the thirtieth anniversary of the deaths of these two men in 1889 within one month of each other.

It had been decided previous to the memorial meeting that money left over from subscriptions to it should be set aside as a nucleus for a fund to be devoted to the erection of commemorative tablets to mark the sites of certain buildings which were closely identified with the work of Mr. DeLamater and Captain Ericsson. These buildings comprised: (1), the Phoenix Foundry at Laight and West Streets, New York City, where the first screw-propelled vessel in this country was constructed and where many other original developments were made; (2), Captain Ericsson's residence, 26 Beach Street, where he designed the Monitor and made all his inventions during his later years; (3), the DeLamater Iron Works, foot of West 13th Street, where the engines of the Monitor, Puritan and Dictator were constructed, as well as the first submarine boat, the first torpedo boat, the first torpedo boat destroyer, the first self-propelled torpedoes, the first dynamite gun, the first air compressors, ice machines and many other industrial appliances now in general use throughout the world; (4), the Continental Iron Works at Greenpoint, L. I., where the hulls of the Monitor, the Puritan and other warships were built.

A number of organizations have already subscribed to the tablet fund, either as a body or through individual members. The tablets will be unveiled on March 9, 1922, the sixtieth anniversary of the battle between the Monitor and the Merrimac, on which date also the John Ericsson Memorial Monument will be dedicated in Washington.

The individual members of the Metropolitan Section have been asked to subscribe to this fund, but it is believed that there are many members, as well as industrial firms, organizations and other individuals, located elsewhere who, from sentimental or patriotic motives, would also like to subscribe. Those interested should communicate with the DeLamater-Ericsson Tablet Committee, H. F. J. Porter, Chairman, Room 1100 Engineering Societies Building, 29 West 39th Street, New York, N. Y.

# The Effect of Depth of Copper Plating on Carburization

By F. P. ZIMMERLI,[1] ANN ARBOR, MICH.

ALTHOUGH case hardening has been investigated considerably during the last few years, methods for preventing the cementation of steel in certain areas have received but little attention. Commercial plants which the writer has recently had occasion to visit have called his attention to the importance of the use of non-carburizing materials and the lack of knowledge concerning them. Most of these plants were using copper plating on all parts of the steel not to be cemented and in answer to inquiries regarding the amount used, the general reply was that they put on "enough."

While there are several non-carburizing materials in use, such as fireclay, or waterglass and calcium hydroxide mixed to form a paint, nevertheless copper plating has given the most satisfaction and it is estimated that it is employed by at least 75 per cent of the plants.

The object of the work described in the following paragraphs was to determine the least amount of copper which must be plated on a piece of steel in order that within a specified time no carbon shall have penetrated through the coating and gone into the steel.

The steel chosen for the experimental work was $\frac{3}{4}$-in. bar stock having the following percentage composition: C, 0.112; P, 0.012; S, 0.0365; Si, 0.128; Mn, 0.458. This was cut longitudinally into half-round pieces approximately 0.75 in. by 1.25 in. on the flat side. These pieces were then numbered and the flat sides, approximately 0.94 sq. in. in area, ground smooth and true on an emery wheel. These uniformly smooth surfaces were cleaned and the backs greased preparatory to plating. Pieces Nos. 1 to 10 inclusive were placed in the plating bath at one time and formed the first set, 11 to 20 the next set, etc., the time on each set being different to control the depth of copper plate.

The cyanide bath was made up as follows: the given quantities being for one liter of solution:

| | |
|---|---|
| $Na_2SO_4$ | 20 grams |
| $Na_2CO_3 \cdot 10H_2O$ | 20 grams |
| $NaHSO_3$ | 20 grams |
| $Cu(C_2H_3O_2)_2 \cdot H_2O$ | 20 grams |
| KCN (100 per cent) | 20 grams |

The specific gravity of such a solution is 1.0507 or 7 deg. B. The current density advised for plating is 0.003 ampere per sq. cm. and the copper deposited at 20 deg. cent. 2.3716 grams per ampere-hour. Because of gas evolution and deposition of copper on the backs of the pieces which no grease would seem entirely to keep off it could not be figured beforehand just how much copper would plate on, nor could the current be kept constant at 0.003 ampere per sq. cm. for each set. An effort was made, however, to keep to this value. It was necessary to guess at the percentage of the total surface exposed by the ten specimens in the bath because of gradual copper deposition on the back. The total area of any one piece was 2.867 sq. in. and of this it was desired to plate 0.955 sq. in.; and of a set, ten times that value.

In practice it was believed that 0.005 in. was not an unusual depth of plate. M. T. Lothrop in his paper on Case Carbonizing[2] reports that 0.0005 in. of copper is the least amount sufficient to prevent carbonization due to non-uniform coating giving unplated areas, but he does not state the current density used. This value he obtained by measurement with micrometers. The plating on the specimens used in this work was from 0.00000893 in. to 0.001815 in. thick, which values are both less and greater than that of Mr. Lothrop.

It was impossible either to figure in advance the amount of copper by knowing the current used as before mentioned, or to obtain the true value of copper deposited by weighing before and after plating, the iron displacing the copper in solution even when no current was used and the moisture condensed on the surfaces was appreciable. The tests were therefore continued as follows: The fifty pieces, sets 1, 2, 3, 4, and 5, depending on length of time in copper-plating bath, were divided so that set A consisted of ten samples, two from each of groups 1, 2, 3, 4, and 5. Sets B, C, D, and E were formed likewise. Two were taken from each of groups 1, 2, 3, 4, and 5 because when ten samples were immersed in the plating bath, five on each side of the copper electrode. One piece from each side of the electrode for each time interval in the bath constituted a set. All copper but that on the flat surface was taken off. Several $2\frac{1}{2}$-in. iron pipes 8 in. long with caps were obtained and a small hole drilled in each, thus keeping the pressure atmospheric. Sets A, B, C, D, and E were packed in them with a carburizing mixture which analyzed as follows:

| | |
|---|---|
| Charcoal | 48.70 per cent |
| $BaCO_3$ | 10.56 per cent |
| $Na_2CO_3$ | 4.65 per cent |
| $SiO_2$ | 2.70 per cent |
| $Al_2O_3$ | 33.25 per cent |

The specimens were packed face down in alternate layers with the carburizing material, the container being placed upright, and about an inch of the material was placed over the last specimens placed in a container.

Sets A, B, C, D, and E were carbonized at 1697 ± 3 deg. fahr. in a large gas furnace for 1, 2, 4, 8 and 12 hours, respectively. One-half hour was allowed for the container to reach furnace

TABLE 1 DATA OF TESTS TO SHOW EFFECT OF DEPTH OF COPPER PLATING ON CARBURIZATION OF LOW-CARBON STEEL SPECIMENS

| Piece Number | Time in furnace, at 1700 deg. fahr., hours | Depth of copper deposit, in hundred thousandths of an inch | Depth of case on steel, inches | | Case under copper |
|---|---|---|---|---|---|
| | | | Eutectoid | Hypoeutectoid | |
| 5 | 2 | 1.885 | 0.016 | 0.0186[1] | None |
| 10 | 2 | 1.850 | 0.0106 | 0.0186[1] | None |
| 15 | 2 | 22.79 | 0.0160 | 0.0160[1] | None |
| 16 | 2 | 20.80 | 0.0160 | 0.0106[8] | None |
| 23 | 2 | 89.25 | 0.0106 | 0.0171[1] | None |
| 29 | 2 | 55.25 | 0.0186 | 0.0186 | None |
| 33 | 2 | 50.20 | 0.0106 | 0.0187[1] | None |
| 38 | 2 | 55.60 | 0.0106 | 0.0187[1] | None |
| 45 | 2 | 145.50 | 0.0160 | 0.0214 | None |
| 49 | 2 | 133.50 | 0.0160 | 0.0187 | None |
| 1 | 8 | 1.29 | 0.0453 | 0.0318 | * |
| 8 | 8 | 1.78 | 0.0318 | 0.0318 | None[4] |
| 12 | 8 | 17.8 | 0.0372 | 0.0366 | None |
| 18 | 8 | 13.45 | 0.0453 | 0.0318 | None |
| 24 | 8 | 76.95 | 0.0453 | 0.0318 | None |
| 31 | 8 | 70.00 | 0.0453 | 0.0318 | None |
| 37 | 8 | 53.25 | 0.0453 | 0.0342 | None |
| 41 | 8 | 56.25 | 0.0480 | 0.0342 | None |
| 43 | 8 | 144.50 | 0.0480 | 0.0342 | None |
| 48 | 8 | 158.00 | 0.0480 | 0.0342 | None |
| 2 | 12 | 0.896 | 0.0372 | 0.0532 | * |
| 9 | 12 | 3.02 | 0.0532 | 0.0372 | Cu broken[5] |
| 11 | 12 | 20.30 | 0.0532 | 0.0532 | None |
| 17 | 12 | 11.98 | 0.0318 | 0.0266 | * |
| 25 | 12 | 73.50 | 0.0590 | 0.0372 | * |
| 30 | 12 | 66.00 | 0.0343 | 0.0398 | None |
| 34 | 12 | 88.25 | 0.0372 | 0.0508 | None |
| 39 | 12 | 48.70 | 0.0372 | 0.0560 | * |
| 42 | 12 | 163.10 | 0.0318 | 0.0508 | None |
| 47 | 12 | 124.10 | 0.0560 | 0.0398 | None |

[1] Uneven. [2] Very uneven. [3] Raise of 0.11 to 0.2 per cent C at extreme edge near copper. Copper off in places. [4] Copper off in places. [5] Raise of 0.11 to 0.15 per cent C at extreme edge near copper. Copper mostly off. [6] Copper off in places. [7] Practically none. [8] 0.1 to 0.2 per cent. [9] Raise of 0.11 to 0.2 per cent.

[1] Department of Chemical Engineering, University of Michigan.
[2] Trans. Am. Soc. M. E., vol. 34 (1912), p. 939.

temperature and it took the furnace 10 hr. to cool down to 291 deg. fahr. after being shut off, at which temperature the sets were removed from the furnace.

Upon opening the containers all specimens were examined for oxidation of iron or copper. The exposed iron was first coated with paraffin and nitric acid then used to remove all the copper

The specimens were all sawed in half, polished and etched. They were then examined microscopically for case depth under the copper plating and on the back where there was no copper deposition. The case was always examined for free cementite, which was not found until Set E was reached, where it occurred in some of the specimens; however, it was so very shallow that it

FIG. 1  PHOTOMICROGRAPH OF UNTREATED SPECIMEN ETCHED WITH 5 PER CENT NITRIC ACID (× 100)

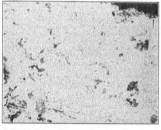

FIG. 2  UNDER COPPER PLATING, SHOWING EDGE OF PIECE. ETCHED WITH 5 PER CENT NITRIC ACID (× 100)

FIG. 3  UNPLATED EDGE, SHOWING CASE. ETCHED WITH 5 PER CENT NITRIC ACID (× 100)

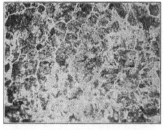

FIG. 4  SAME AS FIG. 3, EXCEPT FARTHER FROM EDGE OF SPECIMEN (× 100)

FIG 5  UNPLATED EDGE, SHOWING CASE. SAME AS FIG. 4 (× 350)

FIG. 6  CASE STILL FARTHER FROM EDGE THAN IN FIG. 4

from the surface. The percentage of copper was determined electrolytically. At the time of this analysis it was realized that the iron might have dissolved some of the copper and some might have been carried off mechanically, but it is felt that in this manner there was less error introduced than in any other.

was figured in the eutectoid. The hypoeutectoid was measured to a point wherein it was estimated that there was not over 0.15 to 2 per cent carbon, since from such a low percentage it decreased rapidly to the 0.11 per cent of the original steel.

(Continued on page 602.)

# Notes on Electric Welding

AT a meeting of the Washington Section of the A.S.M.E. on May 17 of this year, two papers on electric-arc welding were presented, one by Mr. H. S. Rawdon of the Bureau of Standards and the other by Mr. O. H. Eschholz of the Westinghouse Electric & Manufacturing Co. Data are given by these authors on the strength of the joint and the properties of are-fused metal.

## THE ELECTRIC-ARC WELDING OF STEEL: THE PROPERTIES OF THE ARC-FUSED METAL

By HENRY S. RAWDON,[1] WASHINGTON, D. C.

A FUSION weld is fundamentally different from all other types in that the metal of the weld is essentially a casting. The arc-fusion weld has characteristics quite different from other fusion welds. The results of a study of the properties of the arc-fused metal which has been carried out by the Bureau of Standards may be of sufficient interest to warrant presentation before this body. For details the Technologic Paper of the Bureau upon this subject should be consulted.

The application of arc welding in the shipbuilding industry is one of the most important and extensive applications which has ever been proposed for this method. During the year 1918, at the request of and with the coöperation of the Welding Research Sub-Committee of the Emergency Fleet Corporation, an extensive program was outlined by the Bureau of Standards for the study of this type of welding. The principal object of the research was the determination in an empirical way by actual welding tests carried out by skilled operators the relative values and merits of the different types of electrodes or welding pencils which are available commercially for the purpose. Due to changed conditions, however, at the close of the year 1918, the original program was modified and shortened very considerably.

In drawing up the modified program it was decided to make the study of the characteristic properties of the fused-in metal, the primary object of the investigation, the study of the merits of the different types of electrodes being a secondary one. Since

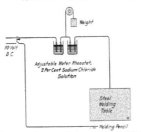

FIG. 1 DIAGRAM OF APPARATUS USED IN THE PREPARATION OF THE SPECIMENS OF ARC-FUSED METAL

the metal of any weld produced by the electric-arc fusion method is essentially a casting, as there is no refinement possible as in some of the other methods, it is apparent that the efficiency of the weld is dependent upon the properties of this arc-fused metal. Hence a knowledge of its properties is of fundamental importance in the study of electric-arc welds.

Though the results available at present are not as conclusive throughout the entire series of examinations made as might be desired, it is believed that they contribute very materially to our knowledge of the properties of the fused-in metal which is ultimately the deciding factor as to the serviceability of the elec-

[1] Physicist, Bureau of Standards.

tric-arc weld. The investigation included a study of the composition changes which accompany the fusion of the metal, the determination of the mechanical properties of the arc-fused metal together with the examination of the structure of the metal after fusion, in the endeavor to explain the observed mechanical and chemical characteristics.

The specimens required for the study of the mechanical properties of the arc-fused iron were prepared for the most part at the Bureau of Standards, direct current being used in all the operations. The apparatus used is shown diagrammatically in Fig. 1. By the use of automatic recorders the voltage and current were measured and records were taken at intervals during the preparation of a specimen.

Since the investigation was concerned primarily with the properties of the arc-fused metal, regular welds were not made. Instead the metal was deposited in the form of blocks (Fig. 2), each one being large enough to permit a tension specimen (0.505

FIG. 2 BLOCKS OF ARC-FUSED METAL WITH TENSION SPECIMEN CUT THEREFROM

in. in diameter, 2 in. gage length) to be machined out of it. Although the opinion is held by some welders that the properties of the metal of an arc weld are affected materially by the adjacent metal by reason of the interpenetration of the two during fusion, it was decided that the change of properties of the added metal induced by fusion alone, was of fundamental importance and should form the basis of any study of arc welding. The method adopted also permitted the use of larger specimens with much less machining than would have been possible had the metal been deposited in the usual form of a weld. The block of arc-fused metal was built up on the end of a section of ½-in. plate of mild steel (ship plate). When a block of sufficient size had been built up, it, together with the portion of the steel plate immediately beneath, was sawed off from the remainder of the steel plate. The tension specimen was turned entirely out of the arc-fused metal. No difficulty whatever was experienced in machining the specimens.

In general, in forming the blocks, the fused metal was deposited as a series of "beads," so arranged that they were parallel to the axis of the tension specimen which was cut later from the block. For purposes of comparison in a few specimens, the metal was deposited in beads at right angles to the length of the block. In all the specimens, after the deposition of each layer, the surface was very carefully and vigorously brushed with a stiff wire brush, to remove the layer of oxide and slag which formed during the fusion. There was found to be but little need to use the chisel for removing this layer.

Tension specimens only were prepared from the arc-fused metal. It is quite generally recognized that the tension test falls very short in completely defining the mechanical properties of any metal; it is believed, however, that the behavior of this material when stressed in tension is so characteristic that its general behavior under other conditions of stress, particularly when subjected to the so-called dynamic tests, i. e., vibration and shock, can be safely predicted from the results obtained. In order to supplement the specimens made at the Bureau a series was also prepared by one of the large manufacturers of equipment for electric welding to be included in the investigation.

Fig. 3 Microstructure of "Pure-Iron" Type of Electrode Used (× 100)

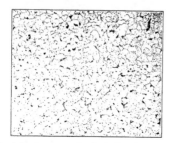

Fig. 4 Microstructure of Low-Carbon Steel Electrode Used (× 100)

Two types of electrodes were used as material to be fused. These differed considerably in composition and were chosen as representative of a "pure" iron and a low-carbon steel (see microstructures in Figs. 3 and 4), the approximate carbon content of the two being respectively 0.02 and 0.15 per cent. Each type was obtained in sizes ranging from ⅛ in. in diameter to ¼ in., so that a considerable range of current densities was possible. The electrodes were used both in the bare condition and after being slightly coated with a paste of an oxidizing and refractory nature, the purpose being to prevent excessive oxidation of the metal of the electrode during fusion and to form also a thin protective coating of slag upon the fused metal.

CHANGES IN COMPOSITION OF ELECTRODES AS A RESULT OF FUSION

In general, the effect of the fusion in the arc is to render the metal of two electrodes quite similar in composition. The data summarized in Table 1 are typical of the changes.

TABLE 1 PERCENTAGE COMPOSITIONS OF ELECTRODES BEFORE AND AFTER FUSION

| Type | A | | A | | B | |
|---|---|---|---|---|---|---|
| Size, in. | ⅛ | | 1/16 | | ¼ | |
| | Before | After | Before | After | Before | After |
| Carbon | 0.056 | 0.031 | 0.022 | 0.010 | 0.15 | 0.024 |
| Silicon | 0.33 | 0.007 | 0.16 | 0.012 | 0.06 | 0.008 |
| Manganese | 0.042 | tr. | 0.038 | tr. | 0.47 | tr. |
| Phosphorus | 0.002 | 0.005 | 0.002 | 0.002 | 0.018 | 0.002 |
| Sulphur | 0.057 | 0.036 | 0.040 | 0.033 | 0.021 | 0.035 |
| Nitrogen[1] | 0.0030 | 0.143 | 0.0040 | 0.126 | 0.0032 | 0.140 |

[1] Average of several determinations on different electrodes of same size and type.

The loss of carbon and of silicon is very marked in each case the nitrogen content of the metal. In general the increase was where these elements exist in considerable amounts. A similar tendency also exists for manganese.

The most noticeable change in composition is the increase in the nitrogen content of the metal. In general the increase was rather uniform for all specimens. In Fig. 5 the average nitrogen contents found for the different conditions of fusion are plotted against the corresponding current densities. Though no definite conclusion seems to be warranted, it may be said that, in general, the percentage of nitrogen taken up by the fused iron increases somewhat as the current density increases. With the

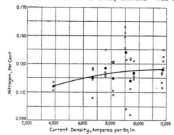

Fig. 5 Relation of the Nitrogen Content of the Arc-Fused Metal to the Current Density

lowest current densities used the amount of nitrogen was found to decrease appreciably. Other investigators in this line have reported an increase in nitrogen content under similar conditions.

MECHANICAL PROPERTIES OF THE ARC-FUSED METAL

When it is considered that the metal which forms a fusion weld is essentially a casting, it is not surprising that there should be a very considerable difference between the mechanical properties of the metal in the form of electrodes or welding pencils and the same after fusion. The average tensile properties of the two types of electrodes are as given in Table 2.

TABLE 2 AVERAGE TENSILE PROPERTIES OF PURE-IRON (A) AND LOW-CARBON STEEL (B) ELECTRODES

| Type of Electrode | Ultimate Strength, lb. per sq. in. | Proportional Limit, lb. per sq. in. | Elongation in 2 in., per cent | Reduction of Area, per cent. |
|---|---|---|---|---|
| A | 61,000 | 40,000 | 14 | 68 |
| B { | 86,000 | 63,000 | 6 | 55 |
| | 67,000 | 37,000 | 15 | 61 |

The size of the electrode tested and the physical state of the metal, that is, the amount of cold working the metal received, cause rather wide variations in the properties. After fusion, however, the properties are very different, as shown in Table 3.

TABLE 3 TENSILE PROPERTIES AND HARDNESS OF ARC-FUSED METAL

| Type of Electrode | Average Ultimate Strength, lb. per sq. in. | Average Yield Point, lb. per sq. in. | Elongation in 2 in., per cent | Reduction of Area per cent | Brinell Hardness | Shore Hardness |
|---|---|---|---|---|---|---|
| A (bare)[1]..... | 48,900 | 32,500 | 7.0 | 10.0 | 103 | 16.6 |
| A (covered)[1].. | 46,600 | 34,600 | 8.5 | 9.0 | 103 | 16.6 |
| B (bare)[1]..... | 48,800 | 34,200 | 8.5 | 10.5 | 103 | 16.0 |
| B (covered)[1].. | 47,400 | 32,200 | 9.9 | 13.8 | 103 | 16.0 |
| C[2]........... | 49,900 | 32,700 | 11.0 | 17.0 | 113 | 17.0 |

[1]"Bare" electrodes were those which were used in the conditions as received "covered" ones were coated as mentioned in the text.
[2]These specimens were prepared by a large welding firm and submitted to the Bureau to be included in the series.

The variations for the specimens of any particular type of electrode are rather wide, usually within the limits of ± 6 per cent. No definite conclusions could be drawn concerning the merits of the various electrodes. All the specimens have properties similar to those of steel castings of a rather inferior grade; the variations noted are of the order which might be expected for such cast steel of inferior quality. There does not appear to be any material advantage so far as the properties of the metal obtained were concerned in coating the electrodes. The specimens

FIG. 6 FRACTURED END OF A TENSION SPECIMEN AFTER TEST; NATURAL SIZE
Nearly all the specimens upon fracturing revealed a defect similar to the one shown above.

FIG. 7 SIDE VIEW (NATURAL SIZE) OF THE TENSION SPECIMEN SHOWN IN FIG. 6

which were prepared by different operators compare very favorably with each other. The most striking feature in the mechanical properties is the low ductility of the metal.

The characteristic appearance of the specimens after testing, illustrating their behavior when stressed in tension until rupture occurs, is shown in Figs. 6 and 7. The features shown are quite typical of the set of specimens. The fracture in all cases revealed interior flaws and defects. Although the results of the tension test appeared to indicate a fair elongation, examination of the specimens showed that the measured elongation does not truly represent a property of the material, but is due rather to the interior defects which indicate lack of perfect union of succeeding additions of metal during the process of fusion. The surface features of the specimens after stressing are very similar to those seen in the type of defective steel familiarly known as "flaky," which was a source of much discussion and delay of production during the progress of the war.

As previously noted, specimens were prepared for the purpose of showing the relation between the direction in which the stress is applied and the manner of deposition of the metal. The metal was deposited so that the beads of fused metal extended across the piece rather than lengthwise, hence the beads were at right angles to the direction in which the tensional stress was applied. The results of the tension tests are given in Table 4 and show that these two specimens (1 and 2) were decidedly inferior to those prepared in the other manner.

## STRUCTURAL FEATURES

The general condition of the metal resulting from the arc fusion is shown in Figs. 8 and 9, which represent longitudinal

FIG. 8 LONGITUDINAL SECTION OF A TENSION SPECIMEN OF ARC-FUSED METAL
Natural size; polished and etched with copper-ammonium chloride solution to accentuate internal defects.

sections of tension specimens. The metal in all of these specimens was found to contain a considerable number of cavities and oxide inclusions, these are best seen after the polished surfaces are etched with a 10 per cent aqueous solution of copper-ammonium chloride. In many of the specimens the successive additions of metal are outlined by a series of very fine inclusions (probably

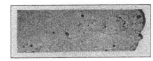

FIG. 9 LONGITUDINAL SECTION OF A TENSION SPECIMEN OF ARC-FUSED METAL
Natural size; polished and etched with copper-ammonium chloride solution to accentuate internal defects.

oxide) which are revealed by the etching. There appears to be no definite relation between the soundness of the metal and the conditions of deposition, i. e., for the range of current density used; nor does either type of electrode used show any decided superiority over the other with respect to porosity of the resulting fusion. The condition of the material prepared by experienced "practical" welders is quite similar to that prepared by the Bureau.

TABLE 4 TENSILE PROPERTIES OF THE SPECIMENS TESTED

| Specimen | Ultimate Strength, lb. per sq. in. | Proportional Limit, lb. per sq. in. | Elongation in 2 in., per cent | Reduction of Area, per cent |
|---|---|---|---|---|
| 1 | 40,450 | 22,500 | 6.5 | 8.5 |
| 2 | 39,500 | 22,500 | 4.0 | 3.0 |

It is to be expected that the microstructure of the material after fusion will be very considerably changed, since the metal is then essentially the same as a casting. It has some features, however, which are not to be found in steel as ordinarily cast. The general type of microstructure was found to vary in the different specimens and to range from a condition which will be designated as "columnar" to that of a uniform fine equi-axed crystalline arrangement. This observation held true for both types of electrodes, whether bare or covered. In the examination of cross-sections of the blocks of arc-fused metal it was noticed that the equi-axed type of structure is prevalent throughout the interior of the piece and the columnar is to be found generally nearer the surface, i. e., in the metal deposited last. It may be inferred from this that the metal of the layers which were deposited during the early part of the preparation of the specimen is refined considerably as to grain size by the successive heatings to which it is subjected as additional layers of molten metal are deposited. The general type of structure of the tension bars cut from the blocks of arc-fused metal will vary considerably according to the amount of refining which has taken place as well as the relative position of the tension specimen within the block. In addition it was noticed that the columnar and coarse equi-axed crystalline condition appears to predominate with fusion at high current densities.

In all of the specimens of arc-fused metal examined micro-

quently deposited over them, i. e., during the formation of this same layer and before any brushing of the surface occurs.

The most characteristic structural feature of the metal after fusion is the presence of numerous needles or plates within the ferrite crystals. The general appearance of this feature is shown in Fig. 11. The number and distribution of these needles were found to vary greatly in different specimens; in general, however, they are most abundant in the columnar and in the coarse equi-axed crystals. Their distribution within the individual crystals indicates that they are arranged along the crystallographic planes.

The usual explanation of these plates is that they are due to the nitrogen which is taken up during the fusion. Other suggestions offered by different investigators at various times are to the effect that they are due to oxide and to carbide. While it is

FIG. 10  MICROSCOPIC EVIDENCE OF UNSOUNDNESS
The globule at the right is surrounded by a film of oxide, and the fracture of specimens when tested in tension originates in such unsound areas.

FIG. 11  CHARACTERISTIC FEATURE OF ARC-FUSED IRON
The "needles" or "plates" are due to the increase which occurs in the nitrogen-content during fusion. (× 500)

FIG. 12  NITROGENIZED IRON, SHOWING CHANGE IN STRUCTURE CAUSED BY HEATING IN AMMONIA
Electrolytic iron was nitrogenized by heating it for several hours (approx. 650 deg. cent.) in ammonia. The change in structure is due to the nitrogen which is taken up. Compare with Fig. 11. (× 500)

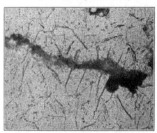

FIG. 13  RELATION OF PATH OF RUPTURE TO THE MICROSTRUCTURE OF ARC-FUSED IRON
The course of the crack or tear in the metal which was produced by stressing in tension does not appear to have been influenced appreciably by the "nitride plates." (× 500)

scopically there were found numerous tiny globules of oxide as shown in Fig. 11. In general they appear to have no definite arrangement but occur indiscriminately throughout the crystals of iron.

A type of unsoundness frequently found is that shown in Fig. 10, which will be referred to as "metallic globule inclusions." In general these globules possess a microstructure similar to that of the surrounding metal but are enveloped by a film, presumably of oxide. It seems probable that they are small metallic particles which were formed as a sort of spray at the tip of the electrode and which were deposited on the solidified crust surrounding the pool of molten metal directly under the arc. These solidified particles apparently are not fused in with the metal which is subse-

not to be denied that the oxide content increases to a very pronounced extent—as is evident from the microscopic examination of any specimen of weld metal whatever, the characteristic appearance of such oxide inclusions, as well as their behavior when the specimen is heated for long periods (e.g., 6 hours at 1000 deg. cent. was used in some of the experiments) makes the suggestion that oxide forms the substance of such plates or needles seem very improbable. Likewise the suggestion of carbide may also be dismissed. The tendency during fusion is for the carbon to be "burnt out" of the metal and only a small amount remains. In practically all of the specimens enough carbide (as pearlite) may be found to account for the carbon content of the metal as revealed by the chemical analysis. The fact that the nitrogen content increases during fusion and also that iron which has been nitrogenized under known conditions has a structure practically identical with that of the arc-fused metal, is very strong evidence in favor of the view that the nitrogen is responsible. See Fig. 12. When once combined with the iron the nitrogen is eliminated only with great difficulty. A heating of six hours in vacuo at 1000 deg. cent. (1830 deg. fahr.) reduced the nitrogen by about only 50 per cent of its initial value.

The pronounced effect of the nitrogen which is gained by the metal during fusion upon the thermal characteristics of the iron in the partial suppression and lowering of the transformation changes is shown in Fig. 14. The suggestion that such a pronounced change in the structure of the metal should have a similar pronounced effect upon the mechanical properties of the metal is a reasonable one. Numerous microscopic examinations were made with the aim of demonstrating to what extent the properties are affected, particularly whether these plates determine to any pronounced extent the course of the fracture in the specimen when broken in tension. It appears, however, as shown in Fig. 13, that the path of rupture is not affected appreciably by these so-called "nitride-plates." The effect of the grosser imperfections of the metal, i. e., oxide inclusions and films, "metallic globule" inclusions, and lack of adhesion between successive layers of the weld metal, is so much greater than any possible effect of the nitride plates in determining the mechanical properties of the specimen as a whole that the conclusion appears to be warranted that this feature of the structure is a matter of relatively minor importance in ordinary arc welds.

In arc-fusion welds in general the mass of weld metal is in intimate contact with the parts which are being welded, so that it is claimed by many that because of the diffusion and intermingling of the metal under repair with that of the electrodes the properties of the latter are considerably improved. The comparison shown in Table 5 somewhat supports this claim. The nearest comparison found available with the Bureau's specimens are some of those of the welds designated as the "Wirt-Jones" series reported by H. M. Hobart of the Research Committee of the American Welding Society. The welds were of the 45-deg. double-V type made in ½-in. ship plate; the specimens for test were of uniform cross-section, 1 in. by ½ in., the projecting metal at the joint having been planed off even with the surface of the plates, and the test bars were so taken that the weld extended transversely across the specimen near the center of its length. The electrodes used were similar to those designated as type B in the Bureau's investigation.

In any consideration of electric-arc welding it should be constantly borne in mind that the weld metal is simply metal which has been melted and has then solidified *in situ*. The weld is essentially a casting, though the conditions for its production are very different from those ordinarily employed in the making of steel castings. The metal loses many of the properties it possesses when in the wrought form, and hence it is not to be expected that a fusion weld made by any process whatever will have all the properties that metal of the same composition would have when in the forged or rolled condition. A knowledge of the characteristic properties of the arc-fused iron is then of fundamental importance in the study of the electric-arc weld.

The peculiar conditions under which the fusion took place also render the weld metal quite different from similar metal melted and cast in the usual manner. It is seemingly impossible to fuse the metal without serious imperfections. The mechanical properties of the arc-fused metal are therefore dependent to an astonishing degree upon the skill, care, and patience of the welding operator. The very low ductility shown by specimens when stressed in tension is the most striking feature observed in the mechanical properties of the material as revealed by the tension test. As explained above, the measured elongation of the tension specimen does not truly indicate a property of the metal. Due to the unsoundness already referred to in the discussion of the structure, the true properties of the metal are not revealed by the tension test to any extent. The test measures, largely for each particular specimen, the adhesion between the successively added

TABLE 5  COMPARISON OF BUREAU OF STANDARDS AND WIRT-JONES TESTS

| Bureau of Standards Tests | | | Wirt-Jones Tests | | |
|---|---|---|---|---|---|
| Electrode | Ultimate Strength, lb. per sq. in. | Elongation in 2 in., per cent | Electrode | Ultimate Strength, lb. per sq. in. | Elongation in 2 in., per cent |
| ¼ in. | 48,700 | 8.0 | ¼ in. | 54,900 | 10.0 |
| ⁵⁄₃₂ in. | 48,800 | 10.0 | ⁵⁄₃₂ in. | 59,800 | 9.0 |

layers, which value varies considerably in different specimens due to the unsoundness caused by imperfect fusion, oxide and other

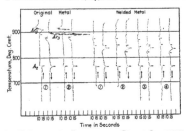

FIG. 14  EFFECT OF ARC FUSION UPON THE THERMAL CHARACTERISTICS OF LOW-CARBON STEEL

The metal originally contained 0.02 per cent carbon; after fusion the composition is very similar except that it has taken up nitrogen to the extent of 0.13 per cent.

inclusions, tiny enclosed cavities and similar undesirable features. The elongation measured for any particular specimen is due largely, if not entirely, to the increase of length resulting from the combined effect of the numerous tiny imperfections which exist throughout the sample.

That the metal is inherently ductile, however, is shown by the behavior upon bending as recorded in the microstructure of the bent specimen. The formation of slip bands within the ferrite grains to a very marked extent occurs under such conditions and is evidence of a high degree of ductility. It appears, however, that the grosser imperfections are sufficient to prevent any accurate measurement of the real mechanical properties of the metal from being made. The conclusion appears to be warranted, therefore, that the changes of composition which the fusion entails, together with the unusual features of microstructure which accompany the composition change, are of minor importance in determining the strength, durability, and other properties of the arc weld.

Since the specimens used in the work described were prepared in a manner quite different from the usual practice of arc welding, no definite recommendations applicable to the latter can be based upon the results. It appears, however, from the results obtained that the two types of electrodes used, i. e., "pure" iron and low-carbon steel, should give very similar results in practical welding. This is due to the changes which occur during the melting, so that the resulting fusions are essentially of the same composition. In fact, the most valuable conclusion which the results of the study warrant is that an arc weld is essentially a casting the properties of which are determined to a very large extent by the skill, ingenuity, and experience of the operator. If this essential characteristic of arc welds is borne in mind, the method will not be used in places where it is evident from the conditions of stresses existing in service that a casting should not be used, and it will also determine in innumerable cases that arc welding is the most efficient method to be considered.

# METALLIC-ELECTRODE ARC-WELDING PROCESS

BY O. H. ESCHHOLZ,[1] EAST PITTSBURGH, PA.

THE metallic-electrode arc-welding process is essentially a method for casting mild steel on a properly fused surface. Upon drawing an arc between a wire electrode and a metal surface, electrical energy is converted into thermal energy which serves to fuse the metal at both terminals. When a short arc length, 1/8 in. or less, is maintained, molten filler electrode metal is conveyed automatically to a fused surface.

The basic phenomena in fusion welding are those affecting metal transition. Upon drawing an arc with bare electrodes in downward welding, a pool of molten metal is formed at the positive or plate terminal and a liquid globule at the end of the negative terminal. This globule enlarges, with the continued application of energy, until the retaining forces, surface tension and cohesion, are exceeded by the globular weight, when the metal drops to the plate. Upon holding a short arc, globule growth is

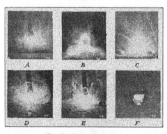

FIG. 1   METAL DEPOSITION

*a, b, c,* illustrate growth, release and deposition of liquid globule: *d* and *e,* show, electrode before and after deposition; *f* shows the appearance of electrode end immediately after interrupting arc.

limited by contact with the fused plate surface, and the forces of surface tension and adhesion at the contact area then assist the force of gravity in removing the molten globule.

It is evident that with a given electrode the rate of metal deposition may be increased, within limits, upon shortening the arc length. Electrodes differ in the speed and size of globule formation. The deposition rate has been observed to vary from two to sixty globules a second and the globule diameter from 1-32 in. to 5-32 in., depending upon the energy density, electrode analysis, preparation and arc length. The appearance of the negative or globular electrode terminal is indicative of its constituents and process of manufacture.

In Fig. 1 are shown high-speed photographs of the arc stream and terminals during metal transition. In this figure, *a, b* and *c* illustrate, respectively, the growth, release and deposition of metal under the exaggerated condition of maintaining a long arc with a large globule-forming electrode of high carbon content; *d* and *e* show respectively the appearance of the arc and a low-carbon steel electrode immediately before and after metal deposition; and *f* gives a view of the globular terminal immediately after interrupting the arc.

The sequence of circuit phenomena[2] accompanying metal deposition is shown graphically in Fig. 2. The initial rush of current at electrode contact serves to accelerate metal fusion. In the succeeding interval, globular growth occurs until contact is made with the plate, short-circuiting the arc. This causes a drop in arc voltage but an increase in current, the added energy ob-

[1] Research Engineer, Westinghouse Electric and Manufacturing Company.
[2] Metal Deposition in Metallic Electrode Arc Welding, by O. H. Eschholz, *Electrical World,* June 26, 1920, and July 17, 1920.

tained in the contact resistance tending to replace the energy developed at the arc terminal and thereby maintain the metal in a fused state.

## PHYSICAL PROPERTIES OF ARC-DEPOSITED METAL

Although this subject would appear to be of primary importance in the design and execution of welded structures, very few data are available concerning it. The refining[3] action of the atmospheric gases on the transferred metal is now, however, recognized.

The following analyses are typical of metal deposited from *bare* wire electrodes:

|               | C     | Mn    | P     | S     | Si    |
|---------------|-------|-------|-------|-------|-------|
| Electrode     | 0.16  | 0.56  | 0.032 | 0.024 | 0.016 |
| Deposited Metal | 0.05 | 0.18 | 0.031 | 0.036 | 0.011 |
| Electrode     | 0.049 | 0.021 | 0.025 | 0.007 | 0.08  |
| Deposited Metal | 0.05 | 0.018 | 0.020 | 0.015 | 0.011 |

The variation in analysis of the conveyed metal is determined by the degree of exposure to oxidation. Upon holding a short arc length and employing special impregnated, covered or coated electrodes, which develop an arc-enveloping screen of inert gases, the analysis of deposited metal will approximate that of the filler electrode.

To determine the physical properties of *arc-deposited metal* test coupons were taken, respectively, from 7-lb. and 14-lb. deposits. In forming these the welder employed a short arc length, an arc current of 175 amperes and a 5-32 in. diameter low-carbon steel bare electrode. The metal was deposited in layers, each succeeding tier being at right angles. The surface of each tier in the 7-lb. deposit was sand-blasted, while in the 14-lb. deposit it was cleaned with a wire brush, which is the practice usually followed.

TABLE 1   RESULTS OF TENSILE TESTS ON MATERIAL IN 7-LB. DEPOSIT OF ARC METAL

| Test No. | U. T. S. | Yield Pt.—Lb. per sq. In. | Elas. Limit[1] | Per cent. Elon. in 2 In. | Per cent Reduction |
|----------|----------|---------------------------|----------------|--------------------------|--------------------|
| 77674-1  | 56,075   | 35,875                    | 29,000         | 16                       | 23.4               |
| 77674-2  | 58,225   | .....                     | .....          | 18                       | 27.8               |

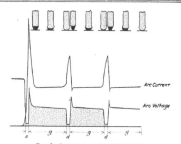

FIG. 2   SEQUENCE OF ARC PHENOMENA

*s,* start of arc; *g,* growth of liquid globule; *d,* deposition of metal.

The fractures displayed the typical structure of a high-grade, partially annealed, low-carbon cast steel. These values exceed those obtainable from metal deposited by the average welder, and may be used as a "bogie" in the development of both welders and electrodes.

The results of tests on the 14-lb. deposit, formed under conditions analogous to those obtaining on welding a locomotive frame, are given in Table 2.

[3] Some Recent Investigations in Arc Welding, by O. H. Eschholz, Proc. A. I. & S. E. E., December, 1917.

TABLE 2 .RESULTS OF TENSILE TESTS ON MATERIAL IN 14-LB. DEPOSIT OF ARC METAL

| Tensile Test No. | U. T. S. | Yield Pt. | Elas. Limit[1] | Per cent Elon. in 2 in. | Per cent Reduction |
|---|---|---|---|---|---|
| | | Lb. per sq. in. | | | |
| 77,651-1 | 58.825 | 41,000 | 40,000 | 8.2 | 19.9 |
| 77,651-2 | 54,650 | 35,000 | 29,000 | 6.5 | 13.4 |

| Compression | | | | | |
|---|---|---|---|---|---|
| | Test No. | | Elas. Limit.[1] Lb. per sq. in. | | |
| | 77651-1A | | 29,450 | | |
| | 77651-2A | | 34,500 | | |

Brinell Hardness, No. 114.
Transverse: Ultimate, 59,400 lb. per sq. in.; elas. limit.[1] 47,600 lb. per sq. in.; supports, 6 in.; deflection, 0.03 in.
Cantilever: Ultimate, 58,200 lb. per sq. in.; elas. limit.[1] 44,750 lb. per sq. in.; supports, 2 in.; deflection, 0.03 in.
Bending: 100 deg. on 1 in. radius, bar ½ in. thick.
Shear: 46,200 and 44,900 lb. per sq. in.
Impact: Izod, unannealed specimens. 2; 1; ft.-lb.

[1] Elastic limits were obtained by Johnson's method.

These data are typical of the characteristics of arc metal when deposited by a conscientious, skilled welder. Similar results will be consistently obtained only when mechanical engineers demand proper instruction and supervision of welding operators. Further improvement in metal characteristics may be secured by the use of special electrodes and heat treatment.

PHYSICAL PROPERTIES OF ARC-WELDED SECTIONS

A weld is formed by the fusion of successive layers of deposited

FIG. 3 TYPICAL SECTION THROUGH A LAYER OF ELECTRODE METAL DEPOSITED ON A LOW-CARBON STEEL PLATE WHEN USING A SHORT ARC, 150 AMPERES AND 5/32-IN. BARE ELECTRODE

Note penetration, small overlap at deposit edges, and extent of structural change caused by thermal disturbance.

FIG. 4 SECTIONS MIDWAY THROUGH WELD SHOWING SHANK METAL, ARC-DEPOSITED METAL AND CHARACTER OF FUSION WITH AND WITHOUT THE USE OF A "FREE DISTANCE" AT VEE APEX

metal. The appearance of the section through a single layer fused to a plate surface is shown in Fig. 3, the characteristic contour, penetration and overlap of which may be noted. On welding sections scarfed to a vee opening it is necessary to separate the shank members by a "free distance" of 1-16 in. to ⅛ in. This separation facilitates the fusion of the initial deposited layer to both members, the most important step in fusion welding. In Fig 4, A and B show respectively sections through plate and deposited metal welded with and without the use of a free distance. In Fig. 4 at A fusion is indicated by the irregular line of recession from the original scarfed surface; in the same figure at B fusion is shown present only at the top of the joint. The clearance between sections, at the apex of the vee, in this weld, was not sufficient to permit arc formation.

The total thermal energy developed at the fused zones is determined by the arc current employed. It is evident, therefore,

[4] A complete discussion of fusion characteristics is contained in a paper by the author on Fusion in Arc Welding. Proc. A. I. E. E., March 1919.

that the arc current must be changed with variation in the thermal capacity, thermal conductivity, and temperature of the weld.[4] In Figs. 5, A and B show sections through welds made with arc currents of 150 and 100 amperes respectively. Fig. 5A shows a fracture at a stress approximately 95 per cent of the U. T. S. of the shank metal, while Fig. 5B shows a break which occurred during machining. Difference in fusion at scarfed surface and in deposit, homogeneity of metal and character of fracture are marked. Fig. 6 summarizes the results of many tests and indicates the variation in weld strength with arc current for skilled and average welders. Fig. 7 gives suggested current values to be used for various plate thicknesses and electrode diameters. It must be recognized, however, that the position of the plate and any unusual condition affecting thermal factors require other current values.

The data in Table 3 have been selected from tests made by the Bureau of Standards on properly executed bare-electrode arc welds prepared for the Welding Committee of the Emergency Fleet Corporation.[5]

TABLE 3 MACHINED, WELDED SHIP PLATES

| Test No. | U. T. S. | Yield Pt. | Elon. in 2 in. | Bending, 1.5 in. diam. Deg. | Vibration.[1] |
|---|---|---|---|---|---|
| | Lb. per sq. in. | | | | |
| Original | 64,700 | 38,400 | 31.5 | 180 | 143,753 |
| 1 | 59,800 | 35,400 | 9.0 | 42 | 845,892 |
| 2 | 62,600 | 39,000 | 11.5 | 44 | 316,825 |
| 5 | 61,100 | 38,650 | 8.5 | 95 | 87,742 |
| 7 | 57,600 | 39,000 | 8.5 | 50 | 41,376 |

[1] Inconclusive. Fiber stress, 25,000 lb. per sq. in.; r.p.m. = 120; tests made on Upton-Lewis machine.

The characteristic average data shown in Tables 4 and 5 were obtained from tests made for Lloyd's Register of Shipping by David Kirkaldy and Sons, London, on ship plates welded with covered electrodes.

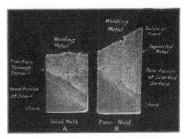

FIG. 5 APPEARANCE OF WELD SECTIONS WHEN USING (A) 150 AMPERES ARC CURRENT AND (B) 100 AMPERES

TABLE 4 TESTS ON SHIP PLATES WELDED WITH COVERED ELECTRODES

| Specimen. | U. T. S. | Yield Pt. | Per cent Elong. in 8 in. | Per cent Reduction per cent | No. of Tests |
|---|---|---|---|---|---|
| | Lb. per sq. in. | | | | |
| Unwelded | 62,000 | | 28 | 45 | 4 |
| Welded plate | 60,400 | | 18 | 31 | 8 |
| Arc deposited metal | 64,500 | 44,000 | 13 | 25 | 2 |

TABLE 5 ALTERNATING-STRESS TEST, 1000 R.P.M., ROTATING WELDED BAR ⅝-IN. DIAMETER

| Specimen. | Stress at Periphery. Lb. per sq. in. | Rotations. | Remarks. |
|---|---|---|---|
| Unwelded | ±23,500 | 5,000,000 | Fractured |
| Welded | ±14,000 | 4,355,000 | Fractured |
| Welded | ±14,000 | 5,000,000 | Unbroken |
| Welded | ±18,800 | 5,000,000 | Unbroken |
| Welded | ±18,800 | 1,705,000 | Fractured |
| Welded | ±20,500 | 1,610,000 | Fractured |

[2] Training of Arc Welders, O. H. Eschholz, Electrical World, September 27, 1919, October 4, 1919.
[5] Welding Mild Steel, H. M. Hobart, Proc. A. I. E. E., March, 1919.

While the data in these tables may be typical now of a skilled welder only, nevertheless they indicate the potentialities of the process and the progress to be obtained in welded structures by the development of proficient operators.

### CONCLUSIONS

The present status of arc welding may be summarized as follows:

a Mainly utilized for the welding of low-carbon steel members subjected to static tension or compression stresses

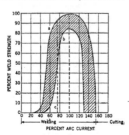

FIG. 6 APPROXIMATE CHANGE IN WELD STRENGTH WITH CHANGE IN ARC CURRENT FOR (A) SKILLED AND (B) AVERAGE WELDERS

Welders tend, as a rule, to employ a current value approximately equal to 75 per cent, that giving best results. The weld strength, as shown by q, will therefore vary from 40 per cent to 90 per cent of maximum, depending on the skill of the operators.

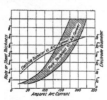

FIG. 7 RELATION OF APPROXIMATE ARC CURRENTS AND ELECTRODE DIAMETERS

b Production and repair welding of parts wherein possible weld failure will not involve life hazard

c Knowledge of welding phenomena and correct welding procedure is not widely disseminated

d Process control and inspection methods, although practicable, are not enforced.

The trend of future activity appears to be as follows:

a Initiation of training schools having a standardized welding course, in large industrial establishments

b Uniform electrode specifications and tests

c Further determination of physical properties of arc-deposited metal with a resulting continued development in electrode manufacture and process control

d Application of inspection methods

e Introduction of automatic arc welding

f Application to non-ferrous metal welding

g Definition of the limitations of the process in the welding of the various alloys, steels and structures subjected to fatigue stresses.

## New York Law Requires Licensing of Engineers

The New York Law for the licensing of professional engineers and land surveyors became effective May 14, 1920. Its administration is in the hands of a board of five licensed engineers appointed by the regents of the University of the State of New York. The personnel of this board has not as yet been named, although it is expected that the appointments will be made on or before October 8. A brief abstract of the law prepared by Engineering Council follows:[1]

APPLICATION FOR CERTIFICATE. Must be made on prescribed form to Regents of the University of the State of New York, Albany, New York. Present practitioners must obtain licenses within two years.

PROFESSIONAL REQUIREMENTS. Law covers all those practicing or offering to practice professional engineering or land surveying.

*Experience Without Degree.* To practice Engineering—Six or more years of active professional engineering work, one of which shall have been in responsible charge, of a character satisfactory to the board. To practice Land Surveying—Four or more years of active engagement in land surveying work of a character satisfactory to the board.

*Educational Allowance.* Each two years of study of engineering in a school of engineering of standing satisfactory to the regents considered as equivalent to one year of active practice.

*Society Membership.* No provision covering membership.

GENERAL REQUIREMENTS. *Citizenship of United States.* To practice Engineering—not necessary. To practice Land Surveying—necessary—or a declaration to become a citizen.

*Age.* At least 21 years.

*Miscellaneous.* Must be of good character and repute. To practice Land Surveying, must speak and write the English language.

EXAMINATIONS. *Conditions Under Which Required.* If evidence presented in application does not appear to board conclusive or warranting issuance of a certificate, applicant may present further evidence, which may include the result of a required examination. Nature and frequency of examinations not prescribed.

FEES. For certificate to practice Engineering or Land Surveying—$25. For certificate to practice both Engineering and Land Surveying—$35. For reissuance of lost or destroyed certificate—$1. If issuance of certificate be denied, fee shall be returned.

PUBLIC WORK. Two years after act takes effect, no county, city, town or village, or other political subdivision shall engage in construction or maintenance of any public work involving professional engineering or land surveying for which plans, specifications and estimates have not been made by, and construction and maintenance supervised by, a licensed engineer or land surveyor, provided contemplated expenditure for completed project does not exceed $2000.

SEALS. Each licensee may obtain a seal of design authorized by board, bearing licensee's name and the legend "licensed professional engineer" or "licensed land surveyor." Plans, specifications, plats, and reports may be stamped with said seal.

EXEMPTIONS. *Employees of Licensed Practitioners.* Employees of licensed engineers or licensed land surveyors, so long as practice does not include responsible charge of design or supervision as principal.

*Contractors and Superintendents.* Not covered by this law.

*Federal, State, and Municipal Employees.* Employees of state or any political subdivision are exempted from the time act becomes effective only until the then existing term of office expires. Officers or employees of United States entirely exempt.

*Corporations or Partnerships* may engage in the practice of engineering or land surveying provided person or persons connected with such corporation or partnership in charge of design or supervision which constitutes such practice is or are licensed. Same exemptions as apply to individuals apply to corporations and partnerships.

*Overlapping Duties.* Engineers not prohibited from making land surveys essential to engineering projects.

RECIPROCITY. *Non-Resident Practitioner* may practice in this state, provided he is legally qualified for such professional service in his own state or country, where the necessary qualifications for which meet the requirements of the board of regents. May be granted a certificate upon payment of the required fee if he holds a like unexpired certificate from another state or territory in the United States in which the requirements for license or registration are of a standard satisfactory to the board.

*New Resident.* Provided he has filed application for license and has paid required fee, may practice for such reasonable time as the board requires in which to consider and grant or deny application.

REVOCATION OF CERTIFICATE. The regents may revoke a certificate for cause only after a hearing at which the accused has the right to be represented by counsel, to introduce evidence and to examine and cross-examine witnesses.

LEGISLATIVE INFORMATION. Law is Article IV-A of Chapter 25, Laws of 1909. Approved May 14, 1920. Introduced in Senate, No. 1104, by Mr. Ferris, March 17, 1920. Article VII-A, Chapter 25, Laws 1909 applies to registration of architects. Laws of 1915, Chapter 454.

[1] Engineering Council does not guarantee the legal accuracy of this abstract.

# SURVEY OF ENGINEERING PROGRESS

## A Review of Attainment in Mechanical Engineering and Related Fields

SUBJECTS OF THIS MONTH'S ABSTRACTS

| | | |
|---|---|---|
| COTTON ROPE FOR POWER TRANSMISSION | FUEL FOR ELECTRIC RAILWAYS | ARMOR-PLATE PENETRATION BY PROJEC-TILES |
| HORSEPOWER TABLE FOR POWER TRANS-MITTED BY COTTON ROPE | SAWMILL REFUSE AS FUEL | STAMPS FOR HOT STAMPING |
| SHEAR STRENGTH OF PLYWOOD WEBS | HARDENING CARBON STEEL AND RATE OF COOLING | BRITISH AND AMERICAN PRACTICE IN STAMPING COMPARED |
| GEARED TURBINES WITH DOUBLE REDUC-TION GEARS | AIR CLASSIFICATION OF PULVERIZED MA-TERIALS | TWIST DRILLS, USE AND ABUSE |
| TURBINES ON S.S. *Corrientes* | RAYMOND SYSTEM OF AIR CLASSIFICATION | MORETTI HYDRAULIC TURBINE LATHE |
| ALCLLA WING | ZEITLIN VARIABLE-STROKE ENGINE | LATHE FOR CRIPPLES, HYDRAULIC |
| DROP-FORGING DEFECTS | AIR-COOLED SLEEVE-VALVE ENGINE | BOWDEN PETROMETER |
| STEEL CASTINGS, LIGHT | BURT SINGLE-SLEEVE VALVE | LANCHESTER CHASSIS AND ENGINE |
| CORE MAKING FOR STEEL CASTINGS | LAWS OF HIGH-SPEED PUNCHING | WATER SOFTENING IN PITTSBURGH |

## Cotton Rope for Power Transmission

COTTON ROPE FOR TRANSMISSION, J. Melville Alison. General discussion of the subject based primarily on English practice, together with data of practical character on the installation and operation if such transmission.

From the article it would appear that cotton has entirely supplanted manila for transmission work in England, the reason for this being the longevity and high factor of efficiency of cotton rope. In fact, the author cites a case where ropes installed in October, 1879, and doing a 24-hour day's work are still in service. This is claimed to be sufficient to overcome the handicap of the higher initial cost of cotton rope. As regards the type of rope, 3- and 4-strand ropes are the most usual. The general advantage seems to lie with the 3-strand rope, because, as shown in Fig. 1,

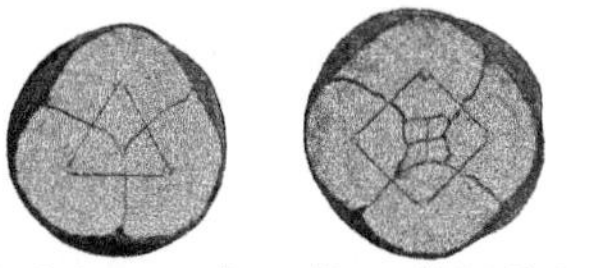

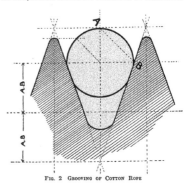

FIG. 1 SECTIONS OF 3- AND 4-STRAND COTTON ROPE

an equilateral triangle is obtained by connecting the centers of the strands, demonstrating the triangulation of the strains, which is a principle adopted by engineers in erecting angular constructions such as bridges. With the 4-strand rope and the same method of connecting the strand centers, a square is obtained, which is a less desirable structure. Furthermore, the 4-strand rope cannot be constructed without a supporting core and its collapse must mean the dislocation of the whole structure. The core, however, represents only about 1/40 of the cross-sectional area of the rope, and it must be reasonably assumed that the stresses exerted by the alternate contraction and extension of the spiral as the rope passes over the sheaves must tend toward the breaking up of this core, which actually takes place.

The relative spiral elongation of the two types of ropes is also such as to favor the 3-strand type.

*Grooving.* In laying out a groove it should be borne in mind that a rope is to all intents and purposes an elastic wedge and reaches its highest driving capacity when it is pressed to the shape of the groove itself. If the sides of the groove are merely tangent to the circle representing a given size of rope, there is a danger of the groove being too shallow, so that when the compression takes place (possibly after a short term of running) the rope is liable to find its way to the bottom of the groove. In this ease slipping takes place, the true indication of which is

heated sheave rims. Grooves with curved sides are undesirable.

The rope should decide the angle of the groove and not vice-versa. The groove may be set out as shown in Fig. 2.

First set a circle of equal diameter to the rope, draw the vertical and horizontal center lines. The chord of the arc $AB$ is then marked off. This becomes the unit measurement and when laid off downward along the vertical line from the center of the circle, indicates the center of the curve of the bottom of the

FIG. 2 GROOVING OF COTTON ROPE

groove. Laid off a second time from this center, the apex of the inverted angle of the groove is determined. This angle always comes out at about 40 deg. whatever the size of the rope may be. Extend the lines of the angle through $BB$, cutting off segment of the circle on the way, until they intersect the horizontal dotted line through $A$, which points fix the radii of the mid-flanges. This groove may be taken as a standard for ropes over 1 in. in diameter and also shows the approximate actual compression which takes place in a rope at work.

*Fixed and Revolving Ropes.* The author is in favor of wedged rope, which with its angular formation makes a better groove contact than the revolving rope. Further, it has less tendency to slippage and a shorter life. Ropes which are underworked will often tend to revolve, a golden rule in rope transmission being that ropes like work to do.

TABLE 1 POWER TRANSMITTED BY VARIOUS ROPES AT VARIOUS SPEEDS

| Velocity, ft. per min. | Diameter of Ropes, In. | | | | | | | | | | | | |
|---|---|---|---|---|---|---|---|---|---|---|---|---|---|
| | ½ | ⅝ | ¾ | ⅞ | 1 | 1⅛ | 1¼ | 1⅜ | 1½ | 1⅝ | 1¾ | 1⅞ | 2 |
| 1000 | 4 | 6 | 9 | 12 | 15 | 19 | 24 | 28 | 34 | 39 | 46 | 53 | 60 |
| 2000 | 4½ | 7 | 10½ | 14 | 18 | 23 | 28 | 31 | 37 | 43 | 51 | 58 | 67 |
| 3000 | 5½ | 8½ | 12½ | 17 | 22 | 28 | 32 | 36 | 40 | 48 | 56 | 65 | 75 |
| 4000 | ... | ... | 14½ | 20 | 26 | 32 | 36 | 40 | 45 | 54 | 64 | 73 | 84 |
| 5000 | ... | ... | ... | 23 | 30 | 38 | 43 | 48 | 55 | 64 | 75 | 87 | 98 |
| 6000 | ... | ... | ... | ... | 35 | 44 | 52 | 58 | 68 | 78 | 92 | 105 | 117 |
| 7000 | ... | ... | ... | ... | ... | .. | 62 | 70 | 84 | 98 | 114 | 131 | 144 |

*Life of the Rope.* All things being equal, durability may be gaged by sectional area. The most economical diameters range from 1½ to 1¾ in., although for rolling-mill work 2-in. ropes are generally used.

Cases are mentioned of cotton ropes having a life of 28 to 40 years. Such cases of longevity lead to the conclusion that well-made cotton ropes are not readily subject to fatigue and have a power of quick recovery, which latter is said to be due to the fact that the ropes pull down on the working side and bulge out to

fashion is recommended. This practice is common in England.

*Sheave Diameters.* In England the minimum standard is taken as approximately 30 times the diameter of the rope—in America forty to fifty times. This, however, is largely a question of speeds to sheave size, as for instance a ½-in. rope will run comfortably over a 4-in. sheave at 1000 r.p.m., a 1-in. rope over a 22-in. sheave at 3000 r.p.m., and it is only when the speed rises to 5000 ft. per min. that we meet the 30-times rule for 1-in. rope.

*Centrifugal Force.* The effect of centrifugal force in rope driving appears to be different from that in belt driving. In this connection Table 1 is of interest, as it would indicate that higher speeds do not detract too much from the ability to transmit power. Cases are mentioned of drives in England and France running at over 7000 ft. per min. and transmitting the highest calculated horsepower. In Cleveland, England, there is a case where the ropes are running at 7800 ft. per min. The driving sheave is 14 ft. 4 in. in diameter, making 168 r.p.m.; the driven sheave is 4 ft. 7 in. diameter, with nineteen 2-in. ropes to transmit 1400 hp. The driven sheave only shows a diameter 27.5 times that of the rope, but is giving comparatively good service.

As regards the effect of moisture, it is stated that cotton ropes are less susceptible to atmospheric changes than manila for the reason that in the first the moisture has a better chance for evaporation. This renders cotton rope practically immune from

TABLE 2 HORSEPOWER TRANSMITTED BY GOOD 3-STRAND COTTON DRIVING ROPE.

| Rope Veloc. ft. per min. | ½ in. | ⅝ in. | ¾ in. | ⅞ in. | 1 in. | 1⅛ in. | 1¼ in. | 1⅜ in. | 1½ in. | 1⅝ in. | 1¾ in. | 1⅞ in. | 2 in. |
|---|---|---|---|---|---|---|---|---|---|---|---|---|---|
| 1000 | 1.0 | 1.6 | 2.3 | 3.2 | 4.2 | 5.3 | 6.6 | 8.0 | 9.5 | 11.2 | 13.0 | 14.9 | 17.0 |
| 1100 | 1.2 | 1.8 | 2.6 | 3.6 | 4.7 | 5.9 | 7.4 | 8.9 | 10.6 | 12.5 | 14.5 | 16.7 | 19.0 |
| 1200 | 1.3 | 2.0 | 2.9 | 4.0 | 5.2 | 6.5 | 8.1 | 9.8 | 11.7 | 13.8 | 16.0 | 18.5 | 20.9 |
| 1300 | 1.4 | 2.2 | 3.2 | 4.4 | 5.7 | 7.1 | 8.8 | 10.7 | 12.8 | 15.1 | 17.6 | 20.2 | 22.8 |
| 1400 | 1.5 | 2.4 | 3.5 | 4.8 | 6.2 | 7.7 | 9.5 | 11.5 | 13.9 | 16.3 | 18.9 | 21.9 | 24.7 |
| 1500 | 1.6 | 2.6 | 3.8 | 5.2 | 6.7 | 8.3 | 10.2 | 12.5 | 15.0 | 17.5 | 20.3 | 23.5 | 26.5 |
| 1600 | 1.7 | 2.8 | 4.1 | 5.5 | 7.1 | 8.9 | 10.9 | 13.4 | 16.0 | 18.7 | 21.7 | 25.0 | 28.3 |
| 1700 | 1.8 | 3.0 | 4.4 | 5.8 | 7.5 | 9.5 | 11.6 | 14.2 | 17.0 | 19.9 | 23.1 | 26.5 | 30.1 |
| 1800 | 1.9 | 3.2 | 4.6 | 6.1 | 7.9 | 10.1 | 12.3 | 15.0 | 18.0 | 21.0 | 24.4 | 28.0 | 31.8 |
| 1900 | 2.0 | 3.4 | 4.8 | 6.4 | 8.3 | 10.7 | 13.0 | 15.8 | 18.9 | 22.1 | 25.7 | 29.5 | 33.5 |
| 2000 | 2.1 | 3.6 | 5.0 | 6.7 | 8.7 | 11.2 | 13.7 | 16.6 | 19.8 | 23.2 | 27.0 | 31.0 | 35.2 |
| 2100 | 2.2 | 3.8 | 5.2 | 7.0 | 9.1 | 11.7 | 14.4 | 17.3 | 20.7 | 24.3 | 28.2 | 32.5 | 36.8 |
| 2200 | 2.3 | 3.9 | 5.4 | 7.3 | 9.5 | 12.2 | 15.1 | 18.0 | 21.6 | 25.0 | 29.4 | 33.9 | 38.4 |
| 2300 | 2.4 | 4.0 | 5.6 | 7.6 | 9.9 | 12.7 | 15.7 | 18.7 | 22.6 | 26.6 | 30.5 | 35.3 | 39.9 |
| 2400 | 2.5 | 4.1 | 5.7 | 7.9 | 10.3 | 13.2 | 16.3 | 19.4 | 23.3 | 27.5 | 31.5 | 36.6 | 41.4 |
| 2500 | 2.6 | 4.2 | 6.0 | 8.2 | 10.7 | 13.7 | 16.9 | 20.1 | 24.1 | 28.5 | 32.9 | 37.9 | 42.9 |
| 2600 | 2.7 | 4.3 | 6.2 | 8.5 | 11.1 | 14.1 | 17.4 | 20.8 | 24.9 | 29.4 | 34.0 | 39.1 | 44.4 |
| 2700 | 2.8 | 4.4 | 6.4 | 8.8 | 11.4 | 14.5 | 17.9 | 21.5 | 25.7 | 30.3 | 35.1 | 40.3 | 45.8 |
| 2800 | 2.9 | 4.5 | 6.6 | 9.1 | 11.7 | 14.9 | 18.4 | 22.2 | 26.5 | 31.1 | 36.1 | 41.4 | 47.1 |
| 2900 | 3.0 | 4.6 | 6.5 | 9.4 | 12.0 | 15.3 | 18.9 | 22.8 | 27.2 | 31.9 | 37.1 | 42.5 | 48.4 |
| 3000 | 3.1 | 4.7 | 7.0 | 9.6 | 12.3 | 15.7 | 19.3 | 23.4 | 27.9 | 32.7 | 38.0 | 43.6 | 49.6 |
| 3100 | ... | 4.8 | 7.2 | 9.8 | 12.6 | 16.0 | 19.7 | 23.9 | 28.5 | 33.4 | 38.9 | 44.6 | 50.7 |
| 3200 | ... | 4.9 | 7.3 | 10.0 | 12.9 | 16.3 | 20.1 | 24.4 | 29.1 | 34.1 | 39.7 | 45.6 | 51.8 |
| 3300 | ... | 5.0 | 7.4 | 10.2 | 13.2 | 16.6 | 20.5 | 24.9 | 29.7 | 34.7 | 40.5 | 46.5 | 52.3 |
| 3400 | ... | 5.1 | 7.5 | 10.4 | 13.4 | 16.9 | 20.9 | 25.3 | 30.3 | 35.3 | 41.2 | 47.4 | 53.8 |
| 3500 | ... | 5.2 | 7.6 | 10.6 | 13.6 | 17.2 | 21.3 | 25.7 | 30.6 | 35.9 | 41.9 | 48.3 | 54.7 |
| 3600 | | | 7.7 | 10.8 | 13.8 | 17.5 | 21.7 | 26.1 | 31.3 | 36.5 | 42.6 | 49.1 | 55.6 |
| 3700 | | | 7.8 | 11.0 | 14.0 | 17.8 | 22.0 | 26.5 | 31.8 | 37.1 | 43.2 | 49.9 | 56.4 |
| 3800 | | | 7.9 | 11.1 | 14.2 | 18.1 | 22.3 | 26.9 | 32.2 | 37.7 | 43.8 | 50.7 | 57.2 |
| 3900 | | | 8.0 | 11.2 | 14.4 | 18.3 | 22.6 | 27.3 | 32.6 | 38.3 | 44.4 | 51.4 | 58.0 |
| 4000 | | | | 11.3 | 14.6 | 18.5 | 22.9 | 27.7 | 33.0 | 38.8 | 45.0 | 52.1 | 58.8 |
| 4100 | | | | 11.4 | 14.8 | 18.7 | 23.2 | 28.1 | 33.4 | 39.3 | 45.5 | 52.7 | 59.6 |
| 4200 | | | | 11.5 | 15.0 | 18.9 | 23.5 | 28.4 | 33.8 | 39.8 | 46.0 | 53.3 | 60.3 |
| 4300 | | | | 11.6 | 15.2 | 19.1 | 23.8 | 28.7 | 34.2 | 40.2 | 46.5 | 53.9 | 61.0 |
| 4400 | | | | 11.7 | 15.4 | 19.3 | 24.1 | 29.0 | 34.6 | 40.6 | 47.0 | 54.4 | 61.7 |
| 4500 | | | | 11.8 | 15.6 | 19.5 | 24.4 | 29.3 | 35.0 | 41.0 | 47.5 | 54.9 | 62.3 |
| 4600 | | | | 11.9 | 15.8 | 19.7 | 24.7 | 29.6 | 35.4 | 41.4 | 48.0 | 55.4 | 62.9 |
| 4700 | | | | 12.0 | 16.0 | 19.9 | 24.9 | 29.9 | 35.8 | 41.8 | 48.5 | 55.9 | 63.5 |
| 4800 | | | | 12.1 | 16.2 | 20.1 | 25.1 | 30.2 | 36.1 | 42.2 | 49.0 | 56.4 | 64.1 |
| 4900 | | | | 12.2 | 16.3 | 20.3 | 25.3 | 30.5 | 36.4 | 42.6 | 49.5 | 56.9 | 64.7 |
| 5000 | | | | 12.3 | 16.4 | 20.5 | 25.5 | 30.8 | 36.7 | 43.0 | 50.0 | 57.4 | 65.3 |
| 5100 | | | | | 16.5 | 20.7 | 25.7 | 31.1 | 37.0 | 43.4 | 50.4 | 57.9 | 65.9 |
| 5200 | | | | | 16.6 | 20.9 | 25.9 | 31.4 | 37.3 | 43.8 | 50.8 | 58.4 | 66.4 |
| 5300 | | | | | 16.7 | 21.1 | 26.1 | 31.7 | 37.6 | 44.2 | 51.2 | 58.9 | 66.9 |
| 5400 | | | | | 16.8 | 21.3 | 26.3 | 32.0 | 37.9 | 44.6 | 51.6 | 59.4 | 67.4 |
| 5500 | | | | | 16.9 | 21.5 | 26.5 | 32.3 | 38.2 | 45.0 | 52.0 | 59.9 | 67.9 |
| 5600 | | | | | | 21.7 | 26.7 | 32.6 | 38.5 | 45.3 | 52.4 | 60.3 | 68.4 |
| 5700 | | | | | | 21.9 | 26.9 | 32.7 | 38.8 | 45.6 | 52.8 | 60.7 | 68.9 |
| 5800 | | | | | | 22.1 | 27.1 | 32.9 | 39.1 | 45.9 | 53.2 | 61.1 | 69.4 |
| 5900 | | | | | | 22.2 | 27.3 | 33.1 | 39.4 | 46.2 | 53.6 | 61.5 | 69.9 |
| 6000 | | | | | | 22.3 | 27.5 | 33.3 | 39.6 | 46.5 | 54.0 | 61.9 | 70.4 |

their normal diameter immediately upon passing to the idle or slack side of the drive

*English and American Systems.* In England the ropes are handled under the multiple system; in America, by the continuous method. The advantage claimed for the English system is immunity from stoppage, since it is seldom that more than one rope gives way at a time.

*Center Distance.* It would appear that driving may be successfully accomplished, even where sheaves are almost touching, by using enough rope. A case is mentioned where sheaves are so close that a finger would scarcely pass between them.

*Over- and Under-Driving.* It is generally accepted that the slack side of the drive should come on top, but this is not universally so. With an erratic or fluctuating load such as obtains in rolling-mill practice, it is reasonable to have the slack on the bottom, as this minimizes the tendency of the rope to wander from its appointed track. Sometimes, however, ropes persist in wandering, in which case the erection of a rope guard in comb

internal mildew which so often is to be found in discarded manila rope.

As regards the arc of contact, formulæ are given in the original article showing methods of proportioning the pulleys for the two cases of slack above and below.

*Splicing.* The subject of splicing is so important to the well-being of the system in general that, rather than have installations spliced by indifferent workmanship, expert splicing mechanics are sent out from England to all parts of the world. Experiments have been carried on with different devices of metal and other couplings with a view to dispensing with splicing, but nothing has yet been found which, in England, would be considered against the long splice. This is usually calculated at about 82 times the diameter of the rope.

*Horsepower.* Attention is called to Table 2, giving horsepower transmitted by good 3-strand cotton driving rope. (*Journal of The Engineering Institute of Canada*, vol. 3, no. 6, June 1920, pp. 292-298, 10 figs.)

# Shear Strength of Plywood Webs

SHEAR STRENGTH OF PLYWOOD WEBS. Tests were conducted by the Forest Products Laboratory, Madison, Wis., to determine the shearing strength of plywood and laminated veneer webs and to determine the influence of the strength of various factors such as the number of plies, the thickness of the various plies, the relative direction of the grain in the plies, the distance between the shearing forces and the direction of the shearing forces relative to the grain of the wood.

The tests were divided into two series. The test specimens were either clamped or glued between two pairs of posts to which the load was applied. The distance between the shear forces was varied from ¼ in. to 6 in., so that the specimens were not sub-

veneer having the grain of all plies parallel is obtained with angles of from 40 to 70 deg. and from 50 to 70 deg., respectively, between the direction of the applied force and the face grain. This shear strength at the optimum angle is usually more than twice that of the shear strength parallel or perpendicular to the face grain. (Figs. 1 and 2.) In Fig. 1 the two curves show the results obtained on 1/12-1/8-1/12-in. birch plywood for each of the two positions in which the specimen may be tested for each angle between face grain and direction of applied load. In other words, the two points for a given angle represent the unit shear strength of the plywood with shear forces in opposite directions.

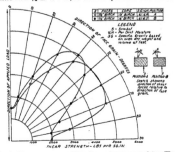

FIG. 1  RELATION BETWEEN UNIT SHEAR STRENGTH OF 3-PLY WOOD AND DIRECTION OF GRAIN

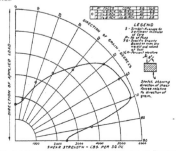

FIG. 2  RELATION BETWEEN UNIT SHEAR STRENGTH OF PARALLEL-PLY LAMINATED VENEER AND DIRECTION OF GRAIN

jected to true shear but to a combination of shear and flexural stresses.

The results of the tests are given in tables and curves. The tests indicate that the shear strength of 3-ply wood of various species, having all plies of one panel of the same thickness and species, is somewhat greater across the face grain than parallel to the face grain.

The unit shear strength of 3-ply birch plywood within the limits of thicknesses tested is slightly less than that for wood having five or more plies.

The unit shear strength of 3-ply wood perpendicular to the face grain and parallel-ply laminated veneer, both perpendicular and parallel to the grain, decreases with an increase in the distance between the shear forces.

The unit shear strength of birch plywood perpendicular to the face grain within the limits of the experiments decreased with an increase in the ratio of the core to total panel thickness.

The maximum shear strength of plywood and of laminated

The data determine that the shear strength of plywood is dependent, in general, upon the specific gravity of the species; and the higher the specific gravity the greater will be the shear strength.

These tests have confirmed the existence of an important difference between plywood and ordinary wood in shear, namely, the nature of the failure. Ordinary wood fails suddenly and completely. In the case of plywood, the failure is more gradual and less complete, as in plywood under shear the plies fail first in shear along the grain but the failures in successive plies are at right angles to each other.

The shear strength of plywood and parallel-ply laminated veneer, as shown in Figs. 1 and 2, is dependent to a very large extent upon the direction of the shear forces relative to the face grain. (Engineering Division, Air Service, McCook Field, *Technical Orders*, No. 14, March 1920, pp. 47-60, 6 figs., *epA*. Abstracted by special permission of the Office of the Director of Air Service, War Department)

# Geared Turbines with Double-Reduction Gear

GEARED TURBINES WITH DOUBLE-REDUCTION GEAR. For the S.S. *Corrientes*, the Parsons Marine Steam Turbine Co. has designed a turbine installation with several novel features.

The total horsepower which the turbines are designed to develop is only 3000 hp., and yet there are three ahead turbines arranged in series, and two astern turbines fitted respectively within the low-pressure and the intermediate-pressure casings. (Fig. 1A.)

The arrangement of gearing is unusual. The high-pressure and intermediate-pressure turbines (Fig. 1B) both engage with one wheel; the intermediate-pressure turbine being arranged vertically above this wheel. The low-pressure turbine engages with

a second wheel on the opposite side of the propeller shaft. Fig. 1C shows the arrangement of the gear box.

In the turbines themselves dummies are absent, the thrust being taken on Michell blocks as shown in Fig. 1C. Carbon glands are used for the high-pressure and intermediate-pressure turbines. The elimination of the dummies reduces to quite low values certain leakage losses which have hitherto served to limit the commercial efficiency of the smaller sizes of the Parsons type of turbine.

The theoretical stage efficiency of a reaction turbine is higher than that of an impulse turbine since each row of blades, whether fixed or moving, constitutes a pressure stage, and the losses in a

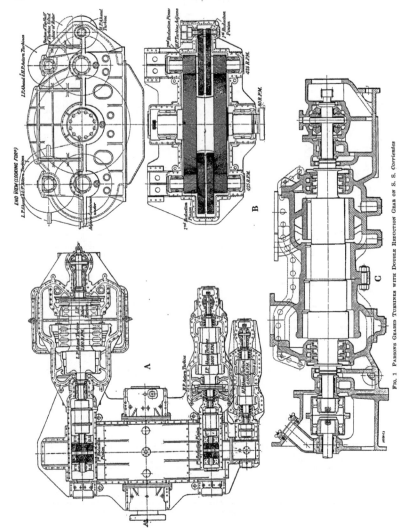

pressure stage thus consist of nozzle friction only, while in each pressure stage of an impulse turbine the bucket losses must be added to the nozzle losses. As a consequence, it is a noteworthy fact that in water-turbine practice the efficiency of the reaction machine is not less than of the impulse type, although the latter is generally free from the leakage losses and the disk friction which are unavoidable with steam turbines of the same type. It is obvious, therefore, that the intrinsic efficiency of the reaction blading is higher than that of impulse blading, but in the past this has been counteracted by the larger leakage losses. It is to reduce these that the dummies have been eliminated and in a further effort in the same direction the "end-tightening" principle has been applied to the high-pressure blading. With this system the turbine is assembled and driven in the shops with no clearance at all at the high-pressure blading, which is thus worn to a uniform bearing and the working clearance is then adjusted to about one-half of the radial clearance which would otherwise be necessary.

The blading is shrouded and the shrouding taken off to a thin edge which abuts on the foundation ring at the root of the following or preceding row of blades. The rotor spindle is adjustable axially and thus the thin edge of the shrouding can be brought into actual contact with the opposing surfaces.

If the turbine is started up in this state it will wear its own clearances, which are therefore as small as could be. It is said that no trouble is experienced from the heat generated by the friction during this period of wear, this being explained by the mechanical principle said to be first formulated by Sir Charles Parsons, namely, that when two unlubricated surfaces are in relative motion and are liable to touch, one of the surfaces should be reduced to a mere fin which is insufficiently rigid to cause distortion of other parts by being heated and is too light to produce much heat through its wearing away.

It is claimed that this type of turbine is particularly adapted for working with superheated steam. The rotors, though of small diameter, are short and stiff, while the use of the end-tightened blading makes large radial clearances possible in the region where the steam temperature is high. (*Engineering*, vol. 110, no. 2850, Aug. 13, 1920, pp. 211, 212, 214 and 2 sheets of plates, 7 figs. besides the plates, *dA*)

## Short Abstracts of the Month

### AERONAUTICS (See also Internal-Combustion Engineering; Measuring Instruments)

ALULA WING. Data on the Alula wing developed by the Blackburn Aeroplane Co. and used in the Pelican 4-ton air truck of the Commercial Aeroplane Wing Syndicate, Ltd.

The Alula wing is claimed to be a new wing type. It is stated that a method of designing a connected wing shape was evolved enabling the designer to vary all features of a wing in unison

FIG. 1   PELICAN 4-TON AIR TRUCK WITH ALULA WING

whenever a particular one had to be changed, in such a manner as not to upset their interrelation. No data as to the mathematical method referred to are, however, presented beyond general statements. Fig. 1 may give some indication as to the character of this wing. (*Aeronautics*, vol. 19, no. 354, July 29, 1920, pp. 96-97, 3 figs., *dg*)

### ARTILLERY (See Machine Shop)

### AUTOMOTIVE ENGINEERING (See Measuring Instruments)

### FORGING

DEFECTS IN DROP FORGINGS, P. Rowley. Rejected drop forgings may owe their rejection to defects in the steel from which they are made. If a billet has a crack or lap which is oxidized inside, the chances are that it will extend during working, especially if the metal is spread by a peg in the die. Cracks which do not open during forging may extend during the oil and water quenching. Piping and ghost lines are also found to be a source of trouble.

Some troubles of the drop forger may be due also to segregation in the metal, especially in alloy steel, of which segregation there are several types. In one case of 3 per cent nickel-chrome steel worked up into connecting rods trouble was encountered during machining. Investigation showed that while the outer layer of a rod cut through the center longitudinally had a carbon content of 0.37 per cent, the inner part approximated to 1 per cent, the segregation of carbon being apparently due to the method adopted by the steel maker to prevent piping. Some of the troubles in drop forging are due to improper design of the parts, traceable to lack of drop-forge knowledge on the part of the designers and draftsmen. These people sometimes fail to differentiate between the flow of a molten metal and the flow of a plastic but at the same time a solid material. In casting, metal may flow from two directions into the same part of the mold to form a solid mass on cooling, but plastic metal cannot run together from two directions to form a good forging owing to the oxidation and decarburization of the exposed surfaces. In this connection, the usual H-section connecting rod as commonly used in internal-combustion engine work is cited as an example of a design but poorly suitable for drop forging. The remedy would be to enlarge the radius in the flute so that the metal might run into the sides of the H-section before flowing into the flash. The adoption of a fish-back design of connecting rod wherever possible would result in saving in die manufacture. (Paper before the Association of Drop Forgers and Stampers at Birmingham, abstracted through *Engineering*, vol 110, no. 2849, Aug. 6, 1920, pp. 185-186, *p*)

### FOUNDRY

#### British Practice in Light Steel Casting

LIGHT STEEL CASTINGS, R. J. Dunderdale. This type of casting was not extensively used in England prior to the war, one of the reasons for it being that the small converters which were then a source of steel for this type of work did not produce an entirely suitable metal. The situation, however, has become entirely different with the introduction of the electric furnace.

The types of furnace chiefly in use in England for light steel castings are the Schneider, Heroult and Electro-Metals in sizes varying from ½ ton to 15 tons a charge. The size suitable for general work in castings ranging from 14 lb. to 140 lb. in weight is the 2-ton furnace, but even then it is desirable to have the furnaces graduated in sizes and used near the type of casting which they would be chiefly employed on.

The sand required for this type of molding differs materially from the usual iron foundry sand. Its physical condition should be hard, sharp and not easily friable. Chemically, it should contain 97 to 98 per cent of silica and as little matter of other kinds as possible. The sand itself should not pass through an 80-mesh screen as its texture will be rendered decidedly finer than this by the clay used for bonding. The best clays for this purpose contain about 30 per cent of alumina and as little organic matter as possible, (a certain amount of iron is said to be an advantage). The moisture of the final mixture which must be good can be as much below 4 per cent as will allow the sand to hold together well.

Baking the mold and painting it with tar from time to time used to be common practice prior to the war. As a matter of fact, however, for a vast majority of light steel castings now made a green mold is entirely satisfactory. A baked mold is still used, however, for very thin work and cored pressure work, providing the mold is made suitably weak and well vented.

A good deal of trouble is experienced from blowholes when the molding is not right, and, in this connection it is stated that the practice of piling is probably not so successful as might be expected, owing to the steam generated in the lower molds getting into the upper molds before the steel does and preventing the exit of gases. This same trouble is apt to occur when two patterns are cast separately in the same box.

Core making in a light-steel-castings foundry has now become a very important part of molding. As regards the sand used in core making, one similar to but rather finer than the facing sand for the molds is desirable.

The causes of the more important failures in light steel castings are discussed in some detail. The three main causes mentioned are short-run castings, slag, and blown and spongy castings.

For repetition work on light steel castings the jolt-ramming type of machine appears to be particularly suitable. In a steel casting it is essential that the face should be hard and the backing moderately weak, in order that the high temperature of the metal and its weight should not break up the face in the first place and the slower-cooling portions shall not be drawn apart by a mold which is uniformly too tightly rammed. The jolt-ramming machine provides this.

As regards fettling, the usual tumbling boxes have not been found sufficient for the hard burning on which takes place in the case of a steel casting, especially in the angles and curves of intricate specimens. Sand blasting has to be resorted to, preferably with steel shot and nozzles.

The development of the light casting industry in England faces the competition of Switzerland, Belgium and Germany, where the same industry appears to be quite firmly established. At present the rate of exchange is in favor of England and it is stated that the future of the industry should be good for some years to come in any event, but there are several weak spots which are apt to be uncovered when competition begins to arise. As a way to help establish this young industry while it is now in a healthy infancy, it is recommended that the light steel castings users combine with the present manufacturers and arrange to parcel out the work to these manufacturers in such a way that one firm is only engaged on two or three standard types of job, this covering the field of repetition work. With regard to the jobbing work, any small quantities of this type should be reserved entirely for two or three foundries suitably equipped with proper range of furnaces. (*Engineering*, vol. 110, no. 2849, Aug. 6, 1920, pp. 167-168, *p*)

## FUELS AND FIRING

### Sawmill Refuse as Fuel for Electric Railways

FUEL SELECTION FOR ELECTRIC RAILWAYS, Darrah Corbet. Data of tests and experience with sawmill refuse as used for fuel by the Portland Railway Light and Power Company of Portland, Ore.

Sawmill refuse ranging from fine sawdust to large blocks of hardwood, at least 1 ft. square, is available for fuel in power stations in many sections of the country, particularly in the Northwest.

It has a fairly high heating value (8500 B.t.u. per lb. of dry non-resinous wood) and can be readily stored for a considerable length of time without very material depreciation in the fuel value and without involving great fire risk. Moreover it can be fed into furnaces not primarily designed for this kind of fuel, although it is probable that better efficiency could be secured with special furnace design.

The original paper gives curves of comparative values of wood and fuel oil (evaporation from and at 212 deg.), Fig. 2. The use of this curve is based on the assumption of oil at $1 a barrel and 12 lb. of evaporation per pound of oil; sawmill-refuse fuel containing 45 per cent moisture and weighing 30 lb. per cu. ft. and efficiency 60 per cent.

Starting with the value of 12 lb. of evaporation per pound of oil, directly above this at the point of intersection with the line indicating the pounds of oil per barrel, we note that on the left

the total evaporation per barrel of oil is 4080 lb. Transferring this below 4080 lb. evaporation across to the right, we note that over the figure $1 per barrel the cost of 1000 lb. of steam is about 24¼ cents.

Starting again with 45 per cent moisture in the sawmill fuel and going to the left to the point of intersection of 60 per cent efficiency, we note that the evaporation per pound of wood is approximately 2.3 lb. water, or with 4000 lb. per unit this is equivalent to about 9200 lb. of water per unit of fuel. Carrying this to the right to the point of intersection with the line indicating cost of steam 24¼ cents per 1000 lb., we note $2.25 as the equivalent cost of a unit of fuel.

On the same basis, if oil cost $2 a barrel the equivalent cost of sawmill refuse would be twice the above, or $4.50 a unit. In this manner a ready reference is secured and the curve can probably be read far more accurately than will be the assumptions made.

The use of powdered coal in several power stations in the Northwest is discussed. At the plant of the Pacific Coal Company at Renton, Wash., a large amount of sludge coal accumulated during the past 15 years. In its present form this fuel is practically useless, but after being pulverized is said to give an average evaporation of 5 lb. of water per pound of coal.

It is also stated that the Milwaukee Electric Railway and Light Company has decided to install a powdered-coal burning equipment for a new 40,000-kw. station.

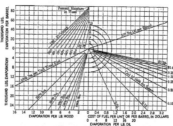

FIG. 2  CURVES OF COMPARATIVE VALUES FOR WOOD AND OIL FUEL
(EVAPORATION FROM AND AT 212 DEG. FAHR.)

On the whole, the author comes to the conclusion that the use of oil fuel will probably be curtailed and more consideration will be given to coal. There is a limited field for sawmill refuse, but better and more efficient means of handling and burning it are desirable. (*Journal of the American Institute of Electrical Engineers*, vol. 39, no. 8, August 1920, abstracted pp. 715-732, 4 figs., 12 tables. Also abstracted in *Electric Railway Journal*, vol. 56, no. 7, Aug. 14, 1920, pp. 314-315, 1 fig., *cp*)

## ENGINEERING MATERIALS

### French Investigation on the Hardening of Carbon Steel

INFLUENCE OF THE VELOCITY OF COOLING ON THE HARDENING OF CARBON STEEL, Portevin and Garvin. The operation of hardening which consists essentially in a more or less rapid cooling from a temperature $T$ (which may be called the temperature of hardening) depends only on the following two variables: (1) the state of the metal at the beginning of the cooling, that is, the temperature $T$; and (2) the law of cooling. It is the law of cooling that the authors discuss in the present paper.

This law of cooling for a given steel depends on the following factors:

A—Temperature at the beginning of cooling, $T$

B—Mass, shape and dimensions of the piece of steel treated

C—Location of the point under consideration in the piece of steel

D—Nature of the bath, that is, the medium by which cooling

is brought about. The action of this medium in its turn depends on its physical properties, namely, its specific heat, its latent heat of vaporization, heat conductivity and viscosity

E—Temperature of the bath, which may vary during the process of cooling depending on the relative masses of the bath itself and of the piece of metal immersed therein

F—Stirring of the bath with respect to the piece of metal

G—State of the surface of the metal piece

H—Temperature at the end of the cooling, or temperature at withdrawal from bath.

For a given point in the piece of metal this law of cooling is completely determined by a curve expressing $\theta = f(t)$, which gives the variation of the temperature $\theta$ as a function of the time $t$. For similar laws of cooling this curve is sufficiently determined by the average velocity of cooling between two given temperatures or by its inverse, namely, the time necessary to traverse a given interval of temperature. The writers adopted as a characteristic value of the velocity of cooling the time necessary for the temperature to change from 700 deg. cent. to 200 deg. cent. which they denote by $\tau$ (700-200), or simply by $\tau$.

The paper investigates the following two questions: First, How does the law of refrigeration, or, in other words, the value of $\tau$, vary in accordance with the variation in the factors enumerated from A to H and second, What relation exists for a given steel and a given initial temperature of the process between $\tau$ and the final state of the steel as determined by the hardness and the micrographic aspect thereof?

The original paper gives the data secured in the form of curves and tables. Only the most important results obtained in this interesting investigation can be reported here.

It has been found that there is a critical region for the velocity of cooling within which the thermal phenomenon denoting the internal transformations in the steel suddenly changes its character and position in the scale of temperatures. The authors call it the critical velocity of hardening in predetermined conditions defining the initial state of the metal and the law of cooling. The following notation is introduced. Ar' is the anomalous state very clearly defined on the curves at about 650 deg. cent., and Ar'' a change in the shape in the curves occurring around 300 deg. cent. The transformation denoted by Ar' and occurring around 650 deg. cent. corresponds to the formation of troostite, and the transformation at the lower temperature Ar'' corresponds to the formation of martensite.

There is therefore a critical velocity denoting the passage from troostitic hardening to martensitic hardening. A parallel phenomenon is observable in measuring hardness. There, there are two regions—one of great hardness around 600 Brinell units and the other of medium hardness around 400 Brinell units. These authors call martensitic hard hardening and troostitic soft hardening.

Extensive data are presented to show the distinction between these two states.

As regards the influence of the chemical composition on the critical velocity of hardening, the authors claim that this critical velocity is at its minimum when the steel is in the neighborhood of the eutectic. Manganese plays an important role from the point of view of critical velocity, and, for example, a sample of steel with 1.10 per cent of carbon and 0.40 per cent of manganese has a martensitic type of hardening, while a similar sample with 1.07 per cent carbon and 0.08 manganese gave a troostitic hardening.

As regards the rate of cooling approaching the critical rate of hardening, the authors claim that the apparent contradictory behavior observed by various experimenters in reference to the influence of the temperature of cooling may be explained by the difference in the size of the sample on which observations were made. If the sample is small or if the examination is limited to the surface of the sample it will be found that the temperature of hardening has no material influence on the result, because under these conditions the process is always carried on at a rate of cooling considerably below the critical. On the contrary, if the sample is large or the interior is taken into consideration and conditions approaching the critical velocity have been reached, a very different conclusion as to the influence of temperature must be arrived at. If the size of the sample and conditions of cooling are such that at the center of the sample soft troostitic hardening takes place, there is, naturally, between the center and the surface a region martensitic-troostitic in which the critical rates have been exceeded. The location of this region shifts with the initial temperature of hardening, and the thickness of the zone of hard martensitic hardening varies. This is why the temperature of hardening becomes of importance as it affects the penetration of the hard hardness and the hardening of large pieces. The other parts of this very interesting paper cannot be abstracted because of lack of space.

Among the conclusions attention is called to the following statement: If, with the same steel and the same initial temperature of cooling the rate of cooling is increased, the lowering of the point of transformation does not occur in a regular manner and there is a discontinuity or at least a variation of a non-gradual character in the location of the point of transformation on cooling. After the point of transformation has gone down gradually, the anomaly on cooling passes from the position Ar' to the position Ar''; whence the notion of critical velocity of hardening.

As a matter of fact, this expression "critical velocity of hardening" is not precise and should be replaced by the designation of the critical zone of the velocity of hardening, taking into consideration the phenomenon of the presence of a double point of transformation which leads to the appearance of troostite and martensite.

For carbon steel 350 deg. cent. appears to be the upper limit of temperature at the withdrawal from the bath in case of repeated hardenings or for medium hardness. Such a temperature will not cause a marked decrease of hardness other than that brought about by the slight anneal occurring during the second period of the operation. (Bulletin de la Société d'Encouragement pour l'Industrie Nationale, vol. 132, no. 3, May-June 1920, pp.

## HANDLING OF MATERIALS

### Comparison of Separation of Pulverized Materials by Screens and Air Currents and Description of Raymond System

AIR CLASSIFICATION OF PULVERIZED MATERIAL, S. B. Kansowitz. The means employed to classify pulverized or ground material have so far been screens or air currents. It is the latter which is described here.

The use of screens or screening cloth is claimed to be limited to a comparatively coarse product. When very fine separation is attempted the screens tend to clog, decreasing the capacity and returning to the mill tailings which contain a high percentage of fine particles. With hard abrasive material the screen openings are rapidly enlarged and then give a product coarser than desired.

Air separation is claimed to avoid all these difficulties. A real air separator should be an integral part of the mill and should be able to remove the ground material from the grinding surfaces as fast as produced. This can be accomplished only by having the air enter underneath the grinding surfaces and blow the ground particles up and away from the rolls.

The separator should be dustless in operation and capable of maintaining uniform product under all reasonable changes of air velocity and air density. Furthermore, it should be so designed that fineness of the finished product could be changed on short notice and without shutting down the equipment.

Figs. 3 and 4 show cross-sections of the Raymond system of air classification. The material from the grinders is drawn into a fan and then blown into a cyclone dust collector. After traveling in the comparatively small-sectioned discharge pipe the material enters the large collector and is compelled to travel in a circular path. The centrifugal force thus produced causes the material to hug the walls of the collector and eventually drop through the bottom as finished product. The air freed from the material passes off through the top of the collector into the return pipe which delivers the air back to the port holes under the grinding chamber. These ports are surrounded with an air-tight casing. In the separator the whirling currents travel in a larger and larger

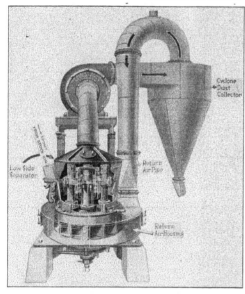

FIG. 3  RAYMOND SYSTEM OF AIR SEPARATION

to be reground, while the finer particles of which the centrifugal force is not great enough to overcome the fan suction, pass into the front intake.

The important feature of this type of separator is that the separation does not depend on the air velocity or air density. Because of this high air velocity can be employed.

Among the advantages of air separation is claimed the fact that a finer product can be obtained than by the use of the finest screens. The full import of this statement may be realized when we bear in mind that there are screens in the market of 350 meshes to the linear inch, or 122,500 holes per square inch. Air separation may also be usefully employed with materials which either are, or tend to become sticky, such as resin, pitch, shellac and various gums. (*Compressed Air Magazine*, vol. 25, no. 7, July 1920, pp. 9722-9723, 3 figs., *dc*)

## INTERNAL-COMBUSTION ENGINEERING

### Rotary Air-Cooled Engine with Variable Piston Stroke

ZEITLIN VARIABLE-STROKE AERO ENGINE. Description of a 220-hp. engine shown at the recent International Aero Exhibition, London.

The engine is of the rotary air-cooled type with nine cylinders and its particular interest lies in the fact that the piston stroke is varied through the cycle. This is accomplished (Fig. 6) by fitting an eccentric to the big end of each connecting rod and rotating the eccentrics around the crankpin at half the engine speed by means of gear-

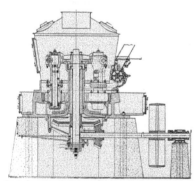

FIG. 4  SECTION OF MILL AND AIR-SEPARATOR CONNECTIONS IN THE RAYMOND SYSTEM OF AIR SEPARATION

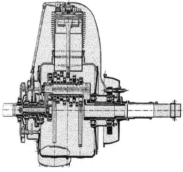

FIG. 5  ZEITLIN VARIABLE-STROKE AERO ENGINE, SHOWING DRIVE OF ECCENTRICS ON CRANKPIN

circular path due to the increased radius of the separator. Because of this the coarser particles are constantly being thrown against the sides of the separator and brought back to the mill

ing. The cycle of operation is as follows: On the working stroke each piston of the engine travels inward a distance of 181 mm. returning on the exhaust stroke for a distance of 203.5 mm. so as to clear the cylinder of the gases of combustion, more thoroughly. The exhaust valve remains open for the greater part of the suction stroke, but after it has closed the piston continues to

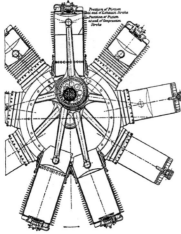

FIG. 6   ZEITLIN VARIABLE-STROKE AERO ENGINE, SHOWING TWO POSITIONS OF PISTON THROUGH THE CYCLE

move for a total distance of 226 mm., uncovering inlet ports in the cylinder liner near the end of the stroke. These inlet ports are connected with the crankcase and the partial vacuum in the

fixed wheel on the engine center and the gear attached to the eccentrics. This, however, would involve certain difficulties and instead of the idler wheel a toothed ring encircling both gears is employed. This toothed ring (Fig. 5) is supported on large ball bearings. The exhaust valves, of which there are nine and which are of the annular type, are operated by three cams driven by epicyclic gearing.

In connection with the exhaust-valve mechanism, the engine is provided with a device to compensate for changes in altitude. The exhaust valve is normally kept open for a certain fraction of the suction stroke arranged, say, to allow the admission of a quantity of air calculated to give the correct compression at a height of 15,000 ft. The quantity admitted at lower altitudes with the same arrangement would, however, be excessive, but reduction of pressure is accomplished by opening the exhaust valve for a certain part of the compression stroke so that some of the air is expelled before it has had time to mix with the gas admitted through the ports.

The engine is said to develop 220 b.hp. and the weight per b.hp. is well below 2 lb. (*Engineering*, vol. 110, no. 2848, July 30, 1920, pp. 138-139, 3 figs., *d*)

## Motor-Car Air-Cooled Engine with Burt Single-Sleeve Valves

AIR-COOLED SLEEVE-VALVE ENGINE. Description of a radial 3-cylinder motor-car air-cooled engine fitted with the Burt single-sleeve valve. The engine as shown in Fig. 7 has a bore and stroke of 80 by 90 mm. (3.14 in. by 3.54 in.), and its most interesting feature is the sleeve valve, a construction which presents unusual difficulties in an air-cooled engine because of the conditions for heat disposal which are present. In this case the sleeve is given a movement which is a compound of up and down and rotational movements, so that a point marked on the sleeve would in a complete cycle trace an ellipse.

The sleeve is operated by means of a small crankpin carrying a short horizontal connecting rod, the small end of which is attached to lugs at the bottom of the sleeve by means of a pin having a vertical axis, that is, an axis at right angles to the axis of the crankpin.

The place of the usual camshaft is taken by a small crankshaft driven at half engine speed with separate crankpins for each cylinder. There is a separate crank for each cylinder driven by

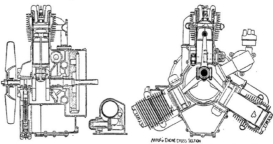

FIG. 7   RADIAL 3-CYLINDER MOTOR-CAR AIR-COOLED ENGINE WITH BURT SINGLE-SLEEVE VALVE

cylinder is sufficient to draw in gas from the crankcase when the ports are uncovered. On the compression stroke the piston travels a distance of 203.5 mm.

The method of driving the eccentrics on the crankpin is of interest. To reduce the friction as much as possible, it is desirable to rotate the eccentrics in the same direction as the engine. One way to do this is to fit a small idler pinion between the

means of skew gearing, the crown wheel of which is mounted at the side of the main crankshaft remote from the fan.

One of the features of the valve system, Fig. 8, is that inlet and exhaust ports are placed alternately round the cylinder, and the ports in the sleeve therefore act alternately for admission and exhaust, which helps to cool the sleeve. The cold incoming mixture circulates round an annular space at the top of the cylinder, cooling the latter and also the cylinder head. (*The Autocar*, vol. 45, no. 1296, Aug. 21, 1920, pp. 317-318, 5 figs., *d*)

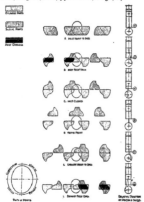

## MACHINE SHOP

### Seriatim Theory of High-Speed Punching Applied also to Perforation of Armor Plate by Projectiles

THE LAWS OF HIGH-SPEED PUNCHING, Capt. T. J. Tresidder. Paper issued by the War Office, Ordnance Committee Press, presenting a novel theory of punching which the author calls the Seriatim Theory of High-Speed Punching.

The theory is of importance also to artillerists as it enables them to determine the remaining velocity of a projectile perforating a given plate at a given striking velocity.

The method of procedure according to this theory is to separate the energy expended in perforating into its three components, namely, that expended on punching, that expended on overcoming friction, and that expended on giving velocity to the expelled plug, the laws governing the punching of a plate by a projectile being claimed to be exactly the same as those known to obtain in the case of plates punched in a punching machine.

The basic argument brought forward is to the effect that mean endlong pressure on the shell due to resistance to punching decreases as the velocity of the shell increases from the bare perforating velocity upward.

There is an important distinction between the blow the projectile can give and that which it does give. Up to bare perforating velocity it gives its all. Beyond the perforation point the blow given is only the maximum the plate can return, and the rest of its possible blow the shell carries on with it.

When a moving body—say, a shell—collides with a stationary one—say, a plate—the action and reaction between them, which are equal to each other, obviously cannot exceed the maximum of which the more feeble is capable. If the blow is arrested, the shot is the more feeble and the mean pressure will depend on its energy. If it is not arrested the plate is the more feeble and the mean pressure will depend on its resistance. If the latter decreases—according to the theory propounded—as speed of attack increases and makes the time available less sufficient for calling into action all the resources of the plate, pressure must decrease with it. Therefore, a shell which encounters a plate with only just enough velocity to perforate it will, according to

this theory, be more stressed by, and will undergo more endlong presure from, the plate's resistance than if its velocity were twice as great.

Consider the force required to punch, as in a punching machine, four plates, each ¼ in. thick, arranged with three thinner hard plates, which have been already perforated, between them. The thin hard plates—that is, the second, fourth and sixth of the pile —will then act as bolders-up for the first, third and fifth, the seventh being held up by the die of the machine as usual. Clearly the pressure that must now be applied to the punch to perforate the four ¼-in. plates—1 in. altogether—is theoretically only the same as would be needed for one ¼-in. plate. That is to say, if a punch can deal with a plate by punching one-fourth of its thickness at a time, it need only have behind it one-fourth the pressure otherwise necessary. Similarly, if a plate is made up of a thousand or ten thousand laminæ which can be separately punched seriatim, the theoretical force or pressure between punch and plate need only be one-thousandth or one ten-thousandth of that required for the solid plate. The distance separating the laminæ need only be sufficient to allow each to fail to shear before the next is seriously engaged, and as this distance will certainly be less as the speed of punching is greater—for the higher the speed the more thoroughly is each lamina held up by its own inertia—it is not inconceivable that as the velocity increases toward the infinite, there will be, in the case of a solid plate, a nearer and nearer approach to a sort of seriatim punching of hypothetical laminæ, whose number approaches infinity and whose thickness and distance apart approach zero.

This is followed by an interesting discussion of the classical case of a soft candle fired as a projectile through a wooden board.

The author's theory was checked by means of tests where measurement was made of the loss of energy suffered by certain projetiles in perforating certain plates at velocities far above those just necessary to carry a shell just through. The bare perforating velocities were known. Knowing the energy lost by the projectile when it is just perforated and calling it *A*, and also knowing the energy lost by the same projectile in perforating the same plate at a higher velocity and calling it *B*, it was only necessary to compare *A* and *B* to test the truth of the main assertion of the Seriatim Theory, namely, that mean pressure due to resistance to punching is less as perforating velocity is greater, which proved to be true in every case.

The approximate net energy absorbed in overcoming the plate's resistance to punching at the high velocity—based on the observed remaining velocity—was compared with that at the low velocity. Their relation to each other was found every time to be, within reasonable limits of observational accuracy, inversely the same as the relation of the corresponding striking velocities.

This amounts to experimental proof of the statement that the net energy absorbed in the punching of a given plate by a given projectile, and also the mean pressure *F* on the projectile due to the plate's resistance to punching, varies inversely as the striking velocity.

It will be seen that, according to the Seriatim Theory, as *V* proceeds from perforation value to infinity, "pressure exerted," "work done," and "time taken" in overcoming the plate's resistance to punching, all diminish toward zero; but "power exerted", which is work divided by time, remains the same at all velocities, if time is fixed, since, however near to zero "work done" and "time taken" may be brought, their relation to each other remains unchanged, both varying inversely as the first power of the velocity.

Bearing in mind, then, that in the perforation of one body by another, energy is expended in three distinct ways, viz., punching, friction overcoming and plug propelling; and that of these we are at the moment dealing with only one, namely, punching, the following new dynamical law seems to result from the arguments of this section:

When any given body, in virtue of kinetic energy, passes completely through another given body without undergoing any change whatever of its own form or dimensions, the power it exerts in effecting the necessary rupture of the material of the perforated body is constant for all velocities.

The report covers many more features of apparently great importance which cannot be abstracted, partly because of lack of space and partly because of insufficient explicit data in the abstract from which this one is taken. (*The Engineer*, vol. 130, no. 3371, Aug. 6, 1920, pp. 126-127, *dtA*)

## Comparison of British and American Practice in Hot Stamping

STEAM AND FRICTION STAMPS FOR HOT STAMPING, W. H. Snow. The author compares American and British practice in the matter of production of stampings and comes to the conclusion that American methods are far superior to the British. He explains this by the fact that the American stamper has had the advantage of a demand for quantities altogether beyond the orders usually met with in England and hence has been able to specialize to a high degree.

The author points out several constructional differences between American and British types.

American stamps, both board drop and steam, are self-contained, i.e., are built entirely on their anvil blocks. The British drop stamp follows ordinary steam-hammer construction in having the anvil block and the supports carrying the lifting gear independent of one another. The requisite alignment of the anvil block and tup is secured by resting guides in the block, the upper ends of the guides being supported laterally by the framework carrying the lifting gear, and being free to move vertically. In this way fewer parts of the stamp are affected by the shock of the blow, and a rigid foundation can be used, making for greater efficiency of blow. It is common to place the block directly on the concrete bed, though it is better to interpose a layer of timber of moderate thickness.

American stamps are generally bedded on timber to a depth of several feet—a rather costly arrangement—an elastic foundation being necessary to soften the shock of the blows on the stamp framework, etc., which carried on and bolted direct to the block would otherwise suffer unduly. In the case of steam stamps, the piston rod is a vulnerable part, and excessive rigidity in the foundation would result in a very short life. Experience has shown that nothing but the best materials are admissible in this type of stamp, if frequent and costly stoppages are to be avoided. In the best American practice, nickel-chrome steel, specially treated, is used for piston rods and heavy bolts, the tup is of specially selected steel, and the frames steel castings.

In British stamps the guides for the tup are generally straight bars of steel, or iron or steel castings of simple form carrying single or multiple "vees" engaged with the tup. The bars rest in holes in the anvil block, and are suitably held in the upper framework, as already mentioned. The lower ends of the bars are generally wedge-shaped, the holes in the block being formed to correspond so that the bars can accommodate themselves to some extent in the event of the block settling unequally and getting tilted. This is a bad arrangement, because the bars jump under the blows and then getting loose momentarily can twist, with the result that the tup is thrown out of alignment with the block.

In American stamps the standards, fitted with adjustable slides for guiding the tup, are made as supports of ample proportions for carrying the cylinder or the headgear, and are spread out at the base and accurately fitted to the block, which is a much better arrangement.

The universal practice as regards the upper die is to dovetail and key it into the block. In American stamps the same plan is adopted for the lower die. A strong steel die holder of ample area is dovetailed into the block, and the die is held by dovetailing in the holder. There are no ready means of altering the position of the dies in relation to one another, and reliance is placed on accurate sinking of the impressions to secure the correct registering.

The general practice among British stampers is to have poppets fitted to the anvil block and to hold the die between screws—usually four in number—working in the poppets and pointing radially. (In France and Belgium six poppets with screws on rectangular axes are favored.)

There is also a material difference in the stamping processes used in the two countries. In American practice preparing for stamping as well as the stamping of the article is done as far as possible under the same stamp. The original British method is to employ a dummying or preparing stamp alongside the main stamp and to rough out the article under it between suitable tools arranged with narrow and wide bases by the ordinary forging process. A variant on this plan is to use fast-running steam or power hammers for preparing quite independently of the stamping, the pieces being forged in numbers and then taken away for stamping. This is a considerable improvement on the older method, as the saving in time by using high-speed hammers more than compensates for the cost of reheating and a much larger output per stamp is obtained.

An interesting feature of the British methods is disclosed by the author when he calls attention to the fact that the use of forging machines for upsetting is unknown in England.

The conclusion to which the author comes is that the British practice needs overhauling. He calls attention, however, to the fact that if American stamps were installed, but no change of methods of working adopted, the full advantage of their construction would not be realized. On the other hand, it would not be feasible to work on American lines with British stamps as their construction is much too weak for this. (*The Engineer*, vol. 130, no. 3373, Aug. 20, 1920, pp. 173-174, *cA*)

## Practical Suggestions on the Use of Twist Drills

THE USE AND ABUSE OF TWIST DRILLS, H. Wills. Twist drills will stand more strain in proportion to their size than almost any other tool. The form of drill point is important because it controls the rate of production, accuracy of the hole, frequency of necessary grinding, and the life of the drill. Speed and feed are also of great importance.

When grinding the drill points the following rules should be observed:

Both cutting lips must be inclined at the same angle with the axis of the drill and must be of equal length. The point angle of 59 deg. has been universally adopted as best suited for average conditions.

The drill point must have the proper clearance or contour of surface back of the cutting edges and this clearance must be identical on both sides. Approximately a 12-deg. clearance angle combined with the center angle of 130 deg., which will give a constantly increasing clearance toward the center, has proven best for average conditions.

If the drill points are not properly ground a number of undesirable conditions will result.

An illustration in the original article shows how a gage can be used to determine approximately the center angle. Although the included angle of the gage is only 118 deg., a center angle of 130 deg. is recommended.

Twist drills are often made with a gradual increase in the thickness of the web about the shank, with the result that as the drill becomes short and the web thinner, greater force is required to drive it. To overcome this, it is good practice to thin the web by grinding away the excess thickness, reducing it to its original dimensions.

In grinding high-speed drills care should be taken not to overheat them, and when heated they should never be plunged into cold water.

As regards speeds and fits, tables are given in the original article for use on various commercial materials. If the drill is properly ground and the corners of the cutting lips begin to show rapid wear, it is an indication that the speed is too great. If the cutting edges roughen or break out in minute particles, it indicates that the feed is too great.

As regards very small drills, attention is called to the fact that they are very frequently run at an excessively low speed with the result that breakage is excessive. These are delicate tools. They must be run true and the cutting edges kept sharp. A fine-grade emery stone is best suited for this purpose. (*Iron Age*, vol. 106, no. 8, Aug. 19, 1920, pp. 461-463, 12 figs., *pA*)

## MEASURING INSTRUMENTS

### Electrical Apparatus for Registering Fuel Consumption on Motor Vehicles

THE BOWDEN PETROMETER. Description of a device acting electrically which registers the fuel consumption on a motor vehicle or aeroplane from moment to moment under all conditions.

As shown in Fig. 9, the petrometer consists essentially of a vessel divided more or less centrally by a diaphragm in which three passages are formed. One of these passages, on the left,

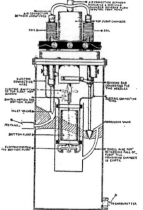

FIG. 9   THE BOWDEN PETROMETER FOR MEASURING FUEL CONSUMP-
TION OF A MOTOR-VEHICLE ENGINE

is connected to the main supply pipe from the tank and can be closed by a needle valve; the second passage, connecting the upper and lower chambers, is opened and closed by another and larger valve. The third passage through the diaphragm, connected by a pipe to a central barrel, is set midway through the diaphragm and is equipped with two disks with a float chamber between them.

Beneath the lower disk are two curved contact pieces, wired through this disk and the upper one—and rigidly, for constructional support—up to the terminals of the coils, which are mounted on a platform above, and outside the fluid-containing vessel, along with an upper float chamber, the armatures—not shown—that carry the needle valves, and the positive terminals from the source of electricity. Each float is mounted in a central spindle that carries a tungsten-tipped contact bar which makes or breaks the electric circuit for the upper or lower "loop," so to say, thus established, as it makes or leaves contact with the stationary contact pieces, making—in the case of the lower float chamber—only while it is empty, but breaking the moment the float lifts, when the lower, or measuring, chamber is full. The upper contact bar, on the contrary, makes contact from the moment the petrol is free to enter the receiving chamber, and lift the upper float, and does not break contact until the upper float chamber is empty. The two needle valves are interconnected by a centrally pivoted rocking bar, which carried a free arm extending downward at a right angle to it, to engage a loop from the spindle of the lower float. The function of this arm and loop, so far as can be seen, would simply steady the fall of the lower float.

All that is apparent from the diagram is that during the "make" of the lower float contact bar, the left-hand coil is energized to lift the inlet needle valve beneath it, which remains lifted until the upper float chamber is sufficiently full from the reception of the petrol for the upper float to lift its contact bar into "make" contact; which action is simultaneous with the "break" of the lower contact bar from the lift of the float in the lower chamber by the entry of a certain amount of petrol through the free connecting or dwell pipe meanwhile. Then at the break of the latter, simultaneously with the make of the former, the left-hand coil is deënergized to drop its needle valve, while the right-hand coil is energized to lift the discharge valve into the measuring chamber, which, as the diagram shows, is in free connection with the carburetor.

The discharge valve is mechanically connected with the dial so that the moment it closes it registers the amount it has admitted into the lower or measuring chamber. The dial has two sets of hands—one, non-adjustable, registering the gross fuel consumption; and the other with a trip hand adjustable for the run con-

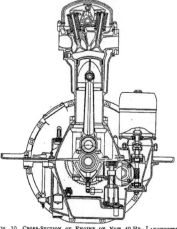

FIG. 10   CROSS-SECTION OF ENGINE ON NEW 40-HP. LANCHESTER
CHASSIS

sumption. In the lower float chamber there is also installed a damping disk, which, acting like a paddle in the constant amount of fluid renders the apparatus proof against jerking or oscillations, which is an important feature. (*Auto-Motor Journal*, vol. 25, no. 31/1021, pp. 799-800, 4 figs., d)

## METALLURGY (See Engineering Materials)

## MOTOR-CAR ENGINEERING

### Lanchester Motor-Car Engine

40-HORSEPOWER LANCHESTER CHASSIS. Description of the new Lanchester chassis of which the description of the engine is of particular interest.

The engine is 4-in. bore by 5-in. stroke and has a brake rating of 89 hp. at 2000 r. p. m. The cylinders are of cast iron and formed in two blocks of three cylinders. The inlet and exhaust ports in the integral cylinder head are so arranged that in each block two inlet and two exhaust valves have common ports, the remaining two valves having separate ports. The two cylinders nearest the carburetor in each block have a common port and the areas and length of the ports and pipes are so arranged that each cylinder obtains an equal amount of mixture.

Two spark plugs per cylinder are fitted in a horizontal position immediately beneath the inlet ports. The spark points are in pockets communicating by small drilled holes with the combustion space which is completely machined, and the spark-plug bosses are completely surrounded by water pockets.

As regards the question of valve seats, the inlet valves only are fitted in a detachable cage, while the exhaust valves are seated directly in the cylinder head. The exhaust valve can, however, be readily removed by dropping it on to the piston head and withdrawing it through the inlet port. Each valve cage is pressed against a taper formed in the cylinder head and held in position by a circular nut slotted for a special wrench, a snug preventing the cage from rotating while the nut is turned.

tank the proper chemicals which either precipitate the scale-forming impurities or change them into impurities that are of non-encrusting character, or both; and also neutralize the free acid content of the water. In the settling tanks, of which there are two (used alternately) the treated water is first stirred and then allowed to stand for two to six hours, depending upon the load being carried.

The reagents used are lime to take care of the bicarbonates and free acid; soda ash for permanent hardness and soluble lime salts; and ferrous sulphate in sugar form to act as a coagulant and facilitate settlement.

Because of the fact that the impurities in the water vary quite irregularly frequent testing becomes necessary. Fig. 11 shows the variation produced in the water by the treatment described above.

The water flows by gravity from the settling tanks, in this case located on the roof of the building, to the filters, the flow being controlled by a float-operated valve so that the filter bed is always

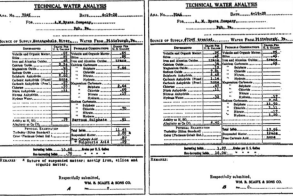

FIG. 11 WATER ANALYSIS BEFORE AND AFTER TREATMENT AT BYERS PLANT

Both inlet and exhaust valves are made from tungsten steel. The cross-section of the engine indicating the general arrangement is shown in Fig. 10. (*The Automobile Engineer*, vol. 11, no. 141, Aug. 1920, pp. 311-318, 20 figs., *d*)

## POWER-PLANT ENGINEERING

WATER-SOFTENING IN PITTSBURGH DISTRICT. The condition existing in the Pittsburgh district with respect to boiler feed-water is that the liquid available for this use in its natural condition is not water but a changing solution of several scale-forming substances with a liberal percentage of free acid and undissolved organic matter. Under these conditions boiler troubles would be inevitable unless some means were taken to control the chemical constitution of feedwater. The present article describes the experience of, and methods developed for such control of feed-water by the plant of the A. M. Byers Company and at the mine and coke plant of W. J. Rainey at Allison.

At the Byers plant the water is taken directly from the Mononghela River and used as condensing water in jet condensers, so that it is delivered to the hot well at a temperature of about 70 deg. From the hot-well part of the water is used in hydraulic presses, while the remainder goes to the softening system where it is treated for boiler use.

This water is treated in an intermittent system. The process in the main consists in adding to the water as it enters the settling

covered but the condensers never overflow. Another float-operated valve prevents flow from the filter to the clear-water tank when it has become filled.

When the plant began using treated water the boilers were heavily encrusted with hard scale. Because of this, only a partial treatment was used to prevent the scale from coming down too rapidly and causing burned tubes. This treatment was gradually increased until the heating surfaces were free from scale. Previous to the introduction of water treatment continual trouble was experienced necessitating replacement of drums and tubes. During the six months that the treating plant has been in operation the tube replacements have been reduced to almost nothing and shutdowns on account of bad water have been entirely eliminated. (*Power Plant Engineering*, vol. 24, no. 16, Aug. 15, 1920, pp. 777-784, 16 figs., *d*)

STAMPING (See Machine Shop)

## CLASSIFICATION OF ARTICLES

Articles appearing in the Survey are classified as *c* comparative; *d* descriptive; *e* experimental; *g* general; *h* historical; *m* mathematical; *p* practical; *s* statistical; *t* theoretical. Articles of especial merit are rated *A* by the reviewer. Opinions expressed are those of the reviewer, not of the Society. The Editor will be pleased to receive inquiries for further information in connection with articles reported in the Survey.

# ENGINEERING RESEARCH

A Department Conducted by the Research Committee of the A. S. M. E.

## A $30,000 Contribution for Coöperative Research

A VERY striking example of how coöperative research may work out in practice is shown by the recent contribution of $30,000 made by the General Electric Company to extend the investigations of the Committee on Fatigue Phenomena of Metals of the Division of Engineering of the National Research Council into the field of nickel steel. This work was originally planned to cover carbon steels only, and was heretofore supported by Engineering Foundation, which appropriated $30,000 for two years' work, and by the University of Illinois, which contributes the equivalent of about $12,000 in the services of Professor Moore and in space, heat, light and equipment.

Although fatigue failure of metal parts subjected to rapidly alternating stresses has been recognized for many years, the recent era of high speeds has yielded cases of great number and importance—in connection with steam-turbine shafts and rotors, airplane-engine crankshafts, the hulls of steel ships, axles and shafts in railway cars, motor cars and trucks, and other machine parts. Such metal parts occasionally fail under ordinary service conditions without showing any general distortion or other symptoms, even when the material is known to be highly ductile. These failures are found only in parts subjected to alternations of stress repeated in some cases millions of times, and therefore are attributed to fatigue of the metal.

Recognizing the value of the results of these investigations to the industries of the country, Engineering Foundation Board about a year ago made a grant of $15,000 a year for a period of two years for the investigation of fatigue phenomena in carbon steels, under a Committee of the Division of Engineering of the National Research Council. Articles of agreement were drawn up between the National Research Council, Engineering Foundation and the University of Illinois whereby the experimental work was to be done at the University of Illinois under the direct supervision of Prof. H. F. Moore, Chairman of the Committee. In addition to the services of Professor Moore, the University furnishes the necessary space and facilities for conducting the work to the best advantage.

Prof. J. B. Kommers, of the University of Wisconsin, a specialist in this subject, has secured a two-year leave of absence and is devoting his full time to the work together with a staff of assistants. The apparatus is all installed and the work progressing according to schedule.

The General Electric Company has recently signed an agreement with Engineering Foundation, National Research Council and the University of Illinois, whereby it agrees to contribute the additional sum of $30,000 to extend the work to include 3 per cent and 3½ per cent nickel steel. This extension is to be considered part of the original program and no restriction is placed by the General Electric Company on the publication of the results.

Although the results of this work on nickel steels will be of immediate commercial value to the General Electric Company, they will also be of value to other manufacturers. The attitude of the company is therefore an unusually broad-minded one. On the other hand, great economies will result from the coöperation agreed to. This is easily appreciated when one considers that the facilities, instead of being erected especially for, and much of the apparatus for conducting these tests are already available and that the consulting services of a group of the foremost experts in the country are furnished without cost, the chief additional cost being some additional apparatus and junior assistants.

Moreover, the coöperative method with its concentration of talent makes much more likely the discovery of the fundamental laws of fatigue, which will be vastly more valuable than the mere empirical information as to the fatigue limits of two varieties of steel.

Thus the broad-gage, far-sighted policy of the General Electric Company demonstrates the commercial feasibility of coöperative research in the fundamentals of engineering, and opens up a large field of usefulness to the National Research Council and Engineering Foundation. It is also an illustration of the large cumulative returns likely to accrue from a wisely placed investment such as that made by the Engineering Foundation Board, which gave the necessary initial impetus to this movement.

## Fuel for Motor Transport

The Fuel Research Board of the Department of Scientific and Industrial Research has prepared an interim memorandum on fuel for motor transport. The memorandum deals chiefly with the possibility of substituting alcohol for petrol and consists of three sections: (a) Historical, (b) The Present Position of Power Alcohol and (c) Alternate Motor Fuels. Copies (at 3d) may be obtained through any bookseller in England or from H. M. Stationery Office, Imperial House, Kingsway, London, W. C. 2.

## Research Laboratories in Industrial Establishments of the United States

The Bulletin of the National Research Council for March, 1920, is devoted to a description of the research laboratories in industrial establishments of the United States compiled by Alfred D. Flinn, assisted by A. J. Porskievies and Ruth Cobb. The list contains the names of over 300 industrial establishments with research laboratories and gives the work done by the laboratories, the research staff, the equipment and the principal product of the company. This Bulletin will be of value to those requiring research work. The Bulletin is published by the National Research Council of the National Academy of Sciences, Washington, D. C., with publication office at 1201 16th St.

## Research Résumé of the Month

### A—RESEARCH RESULTS

The purpose of this section of Engineering Research is to give the origin of research information which has been completed, to give a résumé of research results with formulae or curves where such may be readily given, and to report results of non-extensive researches which in the opinion of the investigators do not warrant a paper.

*Air A1-20* Operation of Blowers in Parallel. Tests were conducted using two motor-driven fans of the single-width, multivane type with double inlet and forward-tipped vanes. Characteristics were obtained for parallel operation at different speeds discharging air against the same head. Certain negative characteristics were determined. The relative advantages of flat and steep characteristic types of forced-draft blowers are discussed. Work was done at the Naval Experiment Station, Annapolis, Md. Address Bureau of Steam Engineering, U. S. Navy, Washington, D. C.

*Apparatus and Instruments A11-20* Gas Analysis. Indicators for carbon dioxide and oxygen in air and flue gas by L. H. Milligan, D. O. Criss and W. S. Wilson are discussed in Technical Paper 238 of the Bureau of Mines. This paper shows the form of apparatus developed by the Bureau of Mines for determining carbon dioxide in large and small quantities. The apparatus is not quite similar in form to other apparatus used for this purpose. Address F. G. Cottrell, Director, Bureau of Mines, Washington, D. C.

*Cement and Other Building Materials A10-20* Aged Cements in Concrete Mixtures. Cement which had been in storage for a year and a half has been tested after lumps had been removed by sieving. Concrete made with the old cement after sieving was compared with that made with new cement and it was found that a 1:1¼:2½ mix of old cement gave concrete equal to that of a 1:2:4 mix with new cement. Similar tests were conducted at Vicksburg, Miss., by the U. S. Engineers' Office. These results show that a 1:2¼:4½ mix of aged cement is equal to a 1:3:6

588

mix of fresh cement. Reference is also made to *Cement and Other Building Materials A7-20*, found in the September number of MECHANICAL ENGINEERING. Bureau of Standards, Washington, D. C. Address S. W. Stratton, Director.

*Chemistry, Inorganic A1-20* Manufacture of Hydrogen. The U. S. Naval Engineering Experiment Station at Annapolis, Md., has reported on an investigation of the possibilities of manufacturing hydrogen gas from carbocoal and steam. Address Bureau of Steam Engineering, U. S. Navy, Washington, D. C.

*Electric Power A4-20* Brake and Stray-Power Methods of Testing Motors. The U. S. Naval Engineering Experiment Station at Annapolis, Md., has reported on a test of comparison of brake and stray-power methods of testing variable-speed motors. Certain losses are not included in stray-power measurement and these tests show that losses amounting to from 1½ to 3 per cent are not determined by the stray-power method. Address Bureau of Steam Engineering, U. S. Navy, Washington, D. C.

*Foods A1-20* Food Investigating Board of Great Britain. Special Report No. 1 on Design of Railway Wagons, price 4d. Report No. 2 on The Literature of Refrigeration, price 5d. Address H. M. Stationery Office, Imperial House, Kingsway, London, WC2.

*Fuel Utilization and Appliances A1-20* Gas Appliances. The Bureau of Standards has made a report on the work of the year in coöperation with the Industrial Fuel Committee of the American Gas Association. The proper design of air shutter, gas orifice and burner throat has been worked out and apparatus has been perfected for measuring the air and gas. The results show that many appliances used at present are low in efficiency and great improvement can readily be made. Some equipment used only 10 to 30 per cent of the heat of the gas and can be made to yield 30 to 50 per cent. Investigations show that burners can be designed to operate natural gas at low pressure for use in the winter season and that the installation of regulators to supply gas at a low and uniform pressure at all times would result in savings. Bureau of Standards, Washington, D. C. Address S. W. Stratton, Director.

*Glass and Ceramics A1-20* Transverse Strength of Fireclay Tile. Fireclay tile subject to temperature of 1350 deg. cent. has been tested to determine modulus of rupture and deformation. Many of these tile cannot support any appreciable load beyond their own weight at this temperature. If loads are to be carried the investigation points to the need of entirely different compositions and combinations of clay and grog. Bureau of Standards. Washington, D. C. Address S. W. Stratton, Director.

*Mining, General A3-20* Output from Coal Miner. Report 2145 by W. W. Adams shows that the output per miner in the U. S. is greater than that in other countries. The amount has increased from 3.37 tons per day in 1901 to 4.40 tons in 1918. New South Wales 4.07 tons per day in 1917, Nova Scotia 3.35 tons per day. Bureau of Mines, Washington, D. C. Address F. G. Cottrell, Director.

*Refrigeration A2-20* Food Investigating Board of Great Britain. Special Report No. 1 on Design of Railway Wagons, price 4d. Report No. 2 on the Literature of Refrigeration, price 5d. Address H. M. Stationery Office, Imperial House, Kingsway, London, WC2.

*Sound A1-20* Submarine Sound Research Work. The U. S. Naval Engineering Experiment Station at Annapolis, Md., has completed for the navy sound research work of the following nature:

(a) Determination of relative efficiency of sound-signaling receiving devices with regard to sensitivity

(b) Determination of relative efficiency of four types of installation with regard to directivity

(c) Depth sounding with listening devices

(d) Comparison of German and United States oscillators

(e) Interfleet communication using listening devices and oscillator

(f) Investigation of screening effect of hull of vessel upon transmission and reception of sound waves

(g) Determination of relative sensitivity and directive qualities of MV electrical blister installation for submarines

(h) Skin installation of MV microphone receiving lines

(i) Investigation of sound apparatus for battleships

(j) Development of range finder

(k) Sound reception in deep water.

Address Bureau of Steam Engineering, U. S. Navy, Washington, D. C.

*Steam Power A1-20* Flow of Steam. The U. S. Naval Engineering Experiment Station has reported a résumé on the flow of steam through pipes and fittings to the Bureau of Steam Engineering of the U. S. Navy. See August, 1920, number of *Journal A. S. N. E.*

*Water, Sewage and Sanitation A3-20* Water Supply and Sewage Disposal for Country Houses. No. 21 of the *Engineering Experiment Station Series* of the University of Missouri is devoted to the water supply and sewage disposal for country homes by E. J. McCaustland, Director. This Bulletin of 36 pages gives construction, drawings and diagrams for the installation of systems of water supply and sewage disposal for country homes. Address E. J. McCaustland, Director. Engineering Experiment Station, Columbia, Mo.

## B—RESEARCH IN PROGRESS

The purpose of this section of Engineering Research is to bring together those who are working on the same problem for coöperation or conference, to prevent unnecessary duplication of work and to inform the profession of the investigators who are engaged upon research problems. The addresses of these investigators are given for the purpose of correspondence.

*Cement and Other Building Materials B3-20* Preserving Coatings for Stone. The Bureau of Standards is investigating a colorless surface treatment for stone for the prevention of disintegration by weathering. Bureau of Standards, Washington, D. C. Address S. W. Stratton, Director.

*Fuels, Gas, Tar and Coke B4-20* Oil Shales. An investigation of the fundamentals of oil-shale retorting is being carried out by the Bureau of Mines. A report No. 2141, by M. J. Gavin, has been made to the Bureau on the plan of this work. In June a report was made by the same author on oil shales and their economic importance. Bureau of Mines, Washington, D. C. Address F. G. Cottrell, Director.

*Glass and Ceramics B2-26* Glass Industry. The Bureau of Standards is endeavoring to develop an improved refractory for the construction of pots for glass making, with particular reference to corrosion. A systematic study is being made of bodies composed of silicious bond clays mixed with aluminous grog and aluminous bond with silicious grog. Crucibles of these mixtures will be fired and then subjected to the action of corrosive barium glass. On breaking the crucibles the depth of penetration will determine the value of the mixture. The transverse strength of the mixture in the dry and fired state, as well as its shrinkage, will be determined. Bureau of Standards, Washington, D. C. Address S. W. Stratton, Director.

*Metallurgy and Metallography B9-20* Bearing Alloys. The tests of babbitt bearing alloys at elevated temperatures have shown that the yield point and ultimate strength decrease rapidly with increasing temperatures. Babbitts containing lead lose their strength more rapidly than those with a tin base. Brinell hardness measurements are being made. Varying quantities of tin are being added to high-grade tin-base babbitt to study the physical properties. A thermostatically controlled air bath is being used for annealing specimens over long periods of time. Bureau of Standards, Washington, D. C. Address S. W. Stratton, Director.

*Metallurgy and Metallography B10-20* Microstructural Change on Tempering Steel. Elaborate series of examinations to determine changes in structure of hardened steels upon tempering has been started at the Bureau of Standards. Magnetic tests and density determinations have indicated pronounced changes. Six different types of steel are being used with carbon content varying from 0.07 per cent to more than 1 per cent. The steel will be hardened by quenching in water from various temperatures, and the length of time at which the material is held at the quenching heat will also be considered. Bureau of Standards, Washington, D. C. Address S. W. Stratton, Director.

*Metallurgy and Metallography B11-20* Materials Investigations. The U. S. Naval Engineering Experiment Station is at work on investigations of failures in shafting, boiler parts and condenser tubes and on new metals and alloys. The work is to determine cause of failure or to obtain data for the revision of specifications. The Bureau is also at work on investigations of impact, resistance and endurance properties of metals. It is also performing experimental work for the Joint Committee on Investigation of the Effect of Phosphorus and Sulphur in Steel. Address Bureau of Steam Engineering, U. S. Navy, Washington, D. C.

*Steam Power B3-20* Boiler Water. The U. S. Naval Engineering Experiment Station is investigating the treatment of boiler water by the use of aluminous plates. Address Bureau of Steam Engineering, U. S. Navy, Washington, D. C.

## C—RESEARCH PROBLEMS

The purpose of this section of Engineering Research is to bring together persons who desire coöperation in research work or to bring together those who have problems and no equipment with those who are equipped to carry on research. It is hoped that those desiring coöperation or aid will state problems for publication in this section.

*Heat C1-20* Heat Transmission to Air at High Pressures. Experimental information regarding the coefficient of heat transfer through steel from gas to compressed air under pressures ranging from 300 to 3000 lb. per sq. in., with a temperature range from —10 deg. fahr. to 60 deg. fahr. is desired. Address W. S. Aldrich, American Bridge Company, Master Mechanics Department, Gary, Ind.

## D—RESEARCH EQUIPMENT

The purpose of this section of Engineering Research is to give in concise form notes regarding the equipment of laboratories for mutual information and for the purpose of informing the profession of the equipment in various laboratories so that persons desiring special investigations may know where such work may be done.

### E—RESEARCH PERSONNEL

The purpose of this section of Engineering Research is to give notes of a personal nature regarding the personnel of various laboratories, methods of procedure for commercial work or notes regarding the conduct of various laboratories.

### F—BIBLIOGRAPHIES

The purpose of this section of Engineering Research is to inform the profession of bibliographies which have been prepared. In general this work is done at the expense of the Society. Extensive bibliographies require the approval of the Research Committee. All bibliographies are loaned for a period of one month

only. Additional copies are available, however, for periods of two weeks to members of the A.S.M.E. or to others recommended by members of the A.S.M.E. These bibliographies are on file in the offices of the Society.

*Fuels, Gas, Tar and Coke F6-20* Coal Storage. Technical Paper 235 on Safe Storage of Coal, by H. H. Stoek, gives instructions for storage and includes bibliography of technical papers on this subject published by the Bureau of Mines, the Engineering Experiment Station of the University of Illinois and articles published in technical journals. Bureau of Mines, Washington, D. C. Address F. G. Cottrell.

*Machine Tools F1-20* Metal-Cutting Tools. A bibliography of five pages on metal-cutting tools from 1918 to June 1920. Search 3061. This supplements *Machine Tools F1-19* from 1903 to 1918. Search 2546. Address A.S.M.E., 29 West 39th St., New York.

*Petroleum, Asphalt and Wood Products F5-20* Recent Articles on Petroleum, and Allied Substances. The Bureau of Mines receives monthly reports on the above subject by E. H. Burroughs, Editorial Assistant. Bureau of Mines, Washington, D. C. Address F. G. Cottrell, Director.

# CORRESPONDENCE

CONTRIBUTIONS to the Correspondence Department of MECHANICAL ENGINEERING are solicited. Contributions particularly welcomed are discussions of papers published in this Journal, brief articles of current interest to mechanical engineers, or suggestions from members of The American Society of Mechanical Engineers as to a better conduct of A. S. M. E. affairs.

## Making the New Federation a Success

To THE EDITOR:

A new machine has been designed, called a "Federation," to enable the engineering profession to act unitedly on matters of public welfare in which the profession as a whole is interested. In the September issue of MECHANICAL ENGINEERING Mr. Morris Llewellyn Cooke finds fault with the design of this machine, which he declares to be "short-sighted almost to the point of insuring failure." He gives as his reason that the Constitution of the new Federation permits the governing boards of the Societies to elect the representatives to the new American Engineering Council instead of having them elected by popular vote of the membership.

But is this distinction, after all, such a very important one? A great many good people labored long and diligently a few years ago to reform the United States Senate. They secured a Constitutional amendment requiring the Senators to be elected by popular vote instead of by the State Legislatures; but does Mr. Cooke think any large improvement has resulted in our Senate? Is it not a fact that representative government sometimes works better than purely democratic methods? What percentage of this Society's members would really take an active interest in the selection of the Society's representatives on the new American Engineering Council?

And whether Mr. Cooke and other critics of the Federation are right or wrong is not important now, *for the time to criticise has passed.* If the profession wants machinery by which it can speak with a united voice on public questions, then the only thing to do is to go in whole-heartedly and make the new Federation a going concern. If the machine proves defective here and there (and it probably will, for few new machines come from the designer perfect), then alter it as experience and wisdom accumulate.

Let us remember the great amount of energy and ability that has been devoted to planning this Federation. If the results of all this work are to be discarded by the profession, it is unthinkable that another set of men would begin over again the toilsome task of harmonizing differences of opinion and evolving a plan in which all existing societies of engineers could coöperate.

The only thing to do, then, for every engineer who believes in coöperative action by the whole profession, is to "boost" the Federation, and find how it may achieve the largest measure of success. Let us remember, too, that success is always a relative

term. Many engineers will have "mighty purposes" in mind, such as Mr. Cooke alludes to, which the new organization will probably be unable to effect. The experience of the present Engineering Council has shown that there are very real limitations on the work *any* such body can do. There is, however, plenty of work within its powers awaiting the new Federation, and it can succeed if the profession will give it the necessary backing.

CHARLES WHITING BAKER.

30 Church Street,
New York, N. Y.

## A Word of Caution Regarding the Federation

To THE EDITOR:

To those of us who went through the harrowing days at the beginning of our participation in the war of 1917, there can be no doubt of the value of an organization harnessing together, for the national security, development and welfare, all those engineering activities which lie at the very foundation of our industrial strength in every line.

I believe I can give no clearer indication of the national need in emergency than to say that within one year after our entry into the war there had been coördinated under the direction of the Council of National Defense, with whose Advisory Commission I had the honor to serve as a member, more than 144,000 distinct organizations of various kinds throughout the United States.

One strong word of caution I should, however, like to voice. We should endeavor at every point to make The Federated American Engineering Societies a truly practical and workable agency for definite accomplishment. We should avoid the pyramiding of new engineering activities, committees, boards, etc., which are so apt to come into being for the furtherance of localized and sometimes selfish interests of individuals or societies and which tend to still further confuse a sufficiently complicated situation. The activities of many of our society members of The Federated American Engineering Societies already overlap, and as these societies endeavor to develop and extend their prerogatives we may naturally expect further complications in authority and in duplication of work.

We have in The Federated American Engineering Societies a coördinating influence which can be made of inestimable practical value, both to its member organizations and to the departments in Washington. Let us see to it, therefore, that this new coördinating body is made to efficiently fulfill those objects for which it has been created.

HOWARD E. COFFIN.

Detroit, Mich.

## Comments on the Proposed Code of Ethics

To THE EDITOR:

The writer is of the opinion that any attempt at a Code of Ethics to be adopted by The American Society of Mechanical Engineers, or by the engineering profession in general, is of doubtful value. He feels, however, that if we are to take such a step, it is worth while to contribute any comment that may aid in putting this in the best shape possible. The form of the report lends itself to a classification under three items, and the writer suggests that the numbered paragraphs be rearranged in accordance with this classification.

The engineer's relation to his own self-respect and to contact with the general public is considered in Pars. 1, 2, 3, 4, 9, and 11, and partly in Par. 12. The engineer's relation to his client or employer is considered in Pars. 5, 6, 7, and 10.

The engineer's relation to his fellow-engineers is considered in Pars. 8, part of 12, 13, and 14. The first three of these have to do with competition for employment.

In his relations to his own self-respect and to the general public, the first four paragraphs are noble and beautiful principles in which we all believe. They are, however, statements of such general character that it would be practically impossible to take any steps for discipline or enforcement regarding them unless the acts involved were so far contrary to all standards that the offender would be expelled from the Society whether there were any formal code of ethics or not.

The subject of reports and expert testimony—Par. 9—is undoubtedly a thing which we will all approve as stated. The writer would wish to state this even more forcibly; that is, that all reports and expert testimony should be given from a judicial standpoint, seeking the truth by including all facts and sound engineering principles. There should be no partisan bias leading to the omission of facts or the distortion of principles in order to reach the desired conclusion. The writer is very doubtful whether it is at all possible to enforce any such ethical principle by action of the Society's authorities on professional conduct.

The writer concurs in the statement in Par. 11 that we should avoid sensational, exaggerated or unwarranted statements about engineering work; but he takes exception to the second sentence of the paragraph to the effect that first descriptions of new inventions, processes, etc., for publication should be furnished only to the technical societies or to the technical press. He would prefer to see it stated that first descriptions, etc., should *preferably* be furnished to the engineering societies or to the technical press. He does not believe that this paragraph should prohibit the first appearance of any proper description in the general press when the engineer concerned finds that advisable.

The writer would prefer to see Par. 12 divided; one paragraph being that, "He should not advertise in an undignified, sensational, or misleading manner." The remainder of the present Par. 12 will be discussed in the comments referring to the engineer's relations with his fellow-engineers.

The statements of the engineer's relations with his clients or employers in Pars. 5, 6, 7, and 10 are all approved. These all register with one ethical principle: namely, that the engineer is not justified in any surreptitious action affecting the relation between himself and his client or employer which would not be approved by them if they were aware of such action.

The engineer's relations with his fellow-engineers when competing with them for business or employment, Par. 8, and parts of Pars. 12 and 13, are more open to question. These set up standards which, if conformed to by the more careful, conscientious members of the profession, would place them at a disadvantage with a considerable percentage of the profession who are not so scrupulous. As stated, they would be very difficult of enforcement. There is at the present time an effort to establish minimum rates for certain classified positions. It would be en-

tirely proper to forbid engineers from seeking such positions at less than the minimum rates in cases where such have been established. We will all agree that no engineer should improperly solicit professional work, but there is likely to be a wide diversity of opinion as to what is and what is not improper.

Everybody will approve Par. 14, regarding engineers assisting each other by exchange of information and experience. This, of course, is ideal. I believe we already approximate to this to a considerable degree. It is not, however, a principle which any authority could be expected to enforce against any individual who was unwilling to conform.

C. M. SPALDING.

Erie, Pa.

To THE EDITOR:

The Code of Ethics published in your September issue of MECHANICAL ENGINEERING is short, general in scope and positive. Furthermore, it has "teeth," for the Committee of Professional Conduct will have very wide powers. The Code of Ethics Committee's report, therefore, meets the recommendation of the Committee on Aims and Organization:

> That it is the sense of this Committee that a short Code of Ethics of broad scope, general in character, and positive rather than negative injunction be prepared and that the same be vigorously enforced.

Some have asked the Committee for more detailed clauses. A book could easily be filled if any attempt were made to be specific in every case. The Committee has therefore felt that the Code should be confined to a few broad, underlying statements, leaving matters of interpretation and elaboration to the "Committee on Professional Practice," whose findings will receive full publicity in our publications.

Others have asked why a Code is necessary. They say that the Committee's recommendations state only principles that engineers have recognized as just and honorable even from school days. All of which is quite true of practically our whole membership, and yet cases have already been presented informally to your Committee for consideration, where engineers of high standing have been in doubt over certain conditions that confronted them in practice. The Code will be particularly helpful to our younger members, who are often more in need of such guidance. In fact, every new member entering the Society will be furnished with a copy of the Code. The Committee on Professional Practice will also act in an advisory capacity to those in doubt on the ethics of certain proceedings.

Still other members have intimated that it is proposed to enforce laws and police duties on our Society. In the Committee's mind these are merely incidents connected with the larger problem, viz., the elevation of engineering to its proper place as one of the most honorable and most respected professions. How can this be attained? We believe that the only sure method is to set up high standards of conduct that will win the respect and confidence of the general public as well as of our own associates and as a Society to see that these standards are rightly upheld. When these conditions prevail, membership in our Society will be an honorable distinction.

Our members will enjoy a well-earned confidence and as men of justice will be sought after for public service which will enable them to render best service as citizens.

But your Committee does not want this code confined only to the mechanical engineers. We hope to establish a joint code and practice in which electrical, civil, mining and chemical engineers can coöperate to the one end, that of adding an honorable distinction and true professional character to engineering.

The Committee proposes to render a final report at the Annual Meeting in December which will contain a few minor changes from the preliminary code. In the meantime, the Committee urges the careful consideration of our first report by every member and will welcome the widest discussion of this most important subject.

A. G. CHRISTIE,
*Chairman Code of Ethics Committee.*

Baltimore, Md.

Published Monthly by The American Society of Mechanical
Engineers at 29 West Thirty-ninth Street, New York

FRED J. MILLER, *President*

WILLIAM H. WILEY, *Treasurer*   CALVIN W. RICE, *Secretary*

PUBLICATION COMMITTEE:
GEORGE A. ORROK, *Chairman*                       J. W. ROE
      H. H. ESSELSTYN          GEORGE J. FORAN
              RALPH E. FLANDERS

PUBLICATION STAFF:
LESTER G. FRENCH, *Editor and Manager*
   FREDERICK LASK, *Advertising Manager*
         WALTER B. SNOW, *Circulation Manager*
                 140 Federal St., Boston

Yearly subscription $4.00, single copies 40 cents. Postage
to Canada, 50 cents additional; to foreign countries $1.00 addi-
tional.

*Contributions of interest to the profession are solicited. Com-
munications should be addressed to the Editor.*

## Extending Ocean Navigation to the Great Lakes

CHARLES WHITING BAKER

TO make the St. Lawrence River navigable so that ocean vessels may reach the Great Lakes, and to develop the water power of the St. Lawrence Rapids, as described by Mr. H. C. Gardner in the last issue of MECHANICAL ENGINEERING, is without doubt by far the most important engineering enterprise now proposed in the world. Indeed, one may well go further and say that this enterprise promises to yield a larger public benefit in proportion to its cost than any engineering work of the first magnitude ever undertaken.

This statement may be criticised on the ground that only very general estimates of the work are as yet available. Enough is known, however, to fully warrant the above claim.

Perhaps a statement of some of the arguments for the St. Lawrence project, in addition to those made by Mr. Gardner, may make still clearer its value and importance.

The economics of transportation have been revolutionized in the past half century, in that *terminal expenses*, and not the cost of hauling between terminals, have become the controlling factor. This is well illustrated, indeed, by some of the figures quoted by Mr. Gardner; but it is a fact which few, even among engineers, thoroughly comprehend.

The great advantage of the St. Lawrence Route is that it would eliminate entirely the enormous terminal expenses in the congested terminals on the Atlantic Seaboard. On traffic moved between the Lake cities and foreign ports, a cargo of dressed beef at Chicago, or steel plates at Cleveland, or wheat at Duluth or Fort William, could be loaded directly on an ocean steamer, whose holds need not be opened until the steamer docked at Liverpool or other foreign port. Contrast this with the present

cost of transfer between railway car and steamer at an Atlantic port, and the great saving by the St. Lawrence Route begins to be evident.

Another matter not mentioned by Mr. Gardner is that the great fleet of Lake vessels, which are now kept idle by the winter ice blockade for nearly half the year, could go down to the sea every fall and take part in the world's carrying trade during the winter months. Of course, a compromise type of vessel, fitted to navigate both the Lakes and the Ocean, would have to be evolved; but that would come about naturally. The United States is destined to become the great coal supply depot of the world. Vessels adapted to the Lake Superior ore trade could also handle to advantage coal to South America and Europe during the winter months.

But it may be asked, if the St. Lawrence project offers the greatest opportunity for profit in the world, why has it not long ago been undertaken? There are many reasons, but the controlling one is this: the profits from the creation of the St. Lawrence Route will go to the public. It is not an enterprise that can possibly be undertaken by private capital to make money. Had this been possible, it would long ago have been completed. Where public welfare is concerned, however, one finds few willing to give the time and energy required to carry an enterprise from the stage of a mere project to practical success.

Further than this, in order to carry out the St. Lawrence project, at least four Governmental bodies must grant consent or take an active part, viz.: the Governments of the United States and of the State of New York, and those of the Dominion of Canada and the Province of Ontario. The various private interests which have reason to oppose the St. Lawrence Route have found it easy, through one or the other of these Governmental organizations, to hinder united action.

The reason why the St. Lawrence project has at last reached a stage where there is promise that it may become a reality, is because an organization has been effected under which fourteen States of the Northwest have united in a league to promote this project, and are doing it officially through their State Governments.

If this organization continues and the work of promotion, public education, and planning the details of the enterprise is carried on with the same energy and ability that would be the case if the project were being furthered as a private enterprise by a powerful financial syndicate, then the success of the project appears certain.

CHARLES WHITING BAKER.

## A Great Work Possible in Industrial Education

As recorded in Section Two of this number, a meeting was recently held of the Committee on Education and Training of The American Society of Mechanical Engineers, at which there was a discussion of more than ordinary importance by a number of eminent authorities who are well-informed on the agencies now existing in the United States for all kinds of industrial training.

This movement for industrial training has taken on a surprising and broad development within recent years and many of the larger corporations already have schools for their men. The Federal Government, the separate states and private firms have made efforts to solve the problem of production by the better training of men to enter the fields of industry and transportation. The Vocational Board in Washington and the National Metal Trades Association have been working systematically towards this end.

At the committee meeting a statement was made of the methods followed by different types of schools, after which the question was raised as to the duty and opportunity of The American Society of Mechanical Engineers in this direction. Its members are well equipped by training and experience to lend a hand. Many of them are already doing this. The constitution of the Society defines its object as the "promotion of the arts and sciences connected with engineering and mechanical construction." It is clear from this that the Society would be entirely within its intended scope and would be building on a firm foundation by taking an active part in the subject of industrial education, which is of such importance to the country.

It is obvious that the Society cannot undertake the instruction of young men for the profession and for the industries, but it *can* assist all the agencies of the United States toward a better understanding of the entire subject and toward greater coöperation. The smaller industries are in real need of information about the training of their men. If the Society should do nothing else than get out a report informing its members of the practice in different types of schools, including those within corporations, it would be contributing a useful work.

It is a year of opportunity in this country and The American Society of Mechanical Engineers ought to assist with all its might toward safeguarding against a lapse into the kind of barbarism masquerading under the name of Bolshevism. We do not belong to a profession confined solely to the technicalities of engineering, for the world has changed and we are part of the vast industrial system in which we must take our part. Hence a committee that will enable the Council of the Society to decide wisely upon the best method of serving the educational interests of the people can do as much for the future of engineering as the Boiler Code Committee is now doing, and that is a great deal.

IRA N. HOLLIS.

## John R. Freeman Addresses Civil Engineers on Engineering Conditions in Japan and China

On the evening of September 15 Past-President John R. Freeman gave an illustrated talk before the members of the American Society of Civil Engineers and invited guests. The subject of Mr. Freeman's talk was Recent Engineering Developments in the Far East as noted during his eight months' sojourn in the Orient. The opening slides were devoted to manufacturing conditions in Japan, and included several pictures of modern factories engaged in the manufacture of electric lamps where the operators, under clean, light, healthy surroundings, worked at the very low wage of from 12 to 20 cents a day. Mr. Freeman felt that his welcome in Japan was sincere and that Americans, generally, were made to feel very much at home there.

Pictures relating to his visit to China covered the hydraulic bore at Hangchow, where a wall of water 15 ft. in height rushes up the bay at the equinoctial tidal season. Mr. Freeman also showed extensive views of the canal system of China, upon which the Chinese delta is dependent almost entirely for its transportation of food and cotton. The canals of China are examples of great engineering skill on the part of the Chinese, some of the masonry walls and locks having been built over 700 years ago and subject to frosts and torrents of destructive power. The bridges over the canals were constructed with great skill and true artistic appreciation.

Mr. Freeman is preparing a report on the proper procedure for removing the possibility of overflow of the Yellow River, which has caused immense loss of life during its flood periods by destroying the transportation system with consequent starvation. He made an extended trip through the Yellow River basin and had surveys of the country and studies of the old records of rainfall and flood channels placed at his disposal in this work. An interesting fact is shown in that the present channel of the Yellow River is located on a ridge, so that the upper normal level of the water is about 15 ft. above that of the surrounding country. The water is held in its course by dikes four or five miles apart, but which are incorrectly placed. These dikes require constant attention, and the Chinese local governments must be ready at all times to make the needed repairs. The water contains about 10 per cent silt by weight, which fills the bed rapidly and makes frequent cleanings necessary. This work is done by coolies, who remove the sediment into baskets at a cost of not more than 10 cents per cubic yard. Great mountains of silt were shown along the channel of the Yellow River. An interesting phenomenon of this river is its tendency to dig a deeper channel at high flood. Mr. Freeman is of the opinion that this tendency may be utilized in safeguarding the country against further floods.

## Development of the Labor Situation in Australia

In general, it may be said that in no country in the world is the welfare of the worker hedged about by more legal safeguards than in Australia. In several states of the Commonwealth of Australia liberal provision is made for workers killed or injured in industrial disputes, old age and invalid pensions are paid, and minimum wages based on statistically-arrived-at cost of living are decreed by courts and boards for the settlement of industrial disputes and strikes.

In the July issue of the *Monthly Labor Review*, published by the Bureau of Labor Statistics in the U. S. Department of Labor, a summary of a report prepared by Trade Commissioner A. W. Ferrin of Melbourne, Australia, is included. This summary explains the development and growth of trade unions in Australia, starting with the increase of population in the cities caused by the influx of immigrants upon the discovery of gold in 1851. The building trade and printing trade unions were those first formed and were followed shortly by the stone masons. Unionism made rapid progress until, in 1919, the membership in trade unions was estimated at 580,000.

Two systems, based upon different principles, are in operation in Australia for the regulation of wages, hours, and general conditions of labor. The " wages board " system, prevailing in Victoria and Tasmania, has as its chief aim the prevention of disputes by the regulation of wages, hours, etc., by a special board appointed for a specific industry on application; while under the " industrial arbitration court " system, prevailing in Western Australia, an industry does not ordinarily come under review until after a dispute has actually arisen. New South Wales, Queensland, and South Australia have both systems.

The " wages board " system calls for the appointment by the Minister of Labor of a special board for a certain industry upon the request of employees or employers in that industry. An order is then issued, constituting a board of not less than four or more than ten selected by the Minister from nominations in the daily press. The constituted board elects a chairman, who votes only in case of a tie. The decisions of the board are submitted to the Minister for approval, and the employers and employees may appeal from the board's decision to a court composed of a judge of the supreme court of the State, appointed for a fixed period, and one representative each of the employers and employees, appointed for the duration of the case in review. The decision of this court is final.

Since under the industrial arbitration court system industries do not come under review until a dispute has arisen, it is quite feasible for the two systems to work together as they do in New South Wales, Queensland, and South Australia. Where wages boards are unable to settle satisfactorily the conditions in a given industry, either side can appeal to the industrial arbitration court and all agreements which have been reached by conferences between employers and employees can be registered with the court, thus acquiring the standing of court awards.

In spite of penalties against breaches of award, striking and locking out, strikes are very numerous. From 1913 through 1918 there have been 2152 disputes affecting 7697 plants, causing a loss of 8,756,389 working days, with a loss of £5,073,346. In 1919 the seamen's strike, causing an estimated loss of wages of £3,000,000, tied up industries in Victoria dependent upon coal from New South Wales. The marine engineers' strike, lasting for three months, ending Feb. 25, 1920, threw 15,000 persons out of work with a wage loss of £1,000,000. A strike of miners at Broken Hill, beginning in May 1919, is still in force. This strike is interesting in the manner in which it has been financed. Strikers understanding various trades, such as cobbling or barbering, contribute their services, the supplies being furnished out of a general fund which is made up of contributions from unions and individuals.

Another feature of Australia's labor disputes is the "go-slow" policy, or "lazy strike," which reduces output to a minimum, with the idea of forcing concessions from capital. It has been adopted with much effect in the Government dockyards, where the riveters reduced their average of rivets to 73 per day per man against a normal of 273, and by telephone operators, who refused to answer more than a given number of calls per day.

In general, it may be said that neither capital nor labor finds the existing legislation for the avoidance and settlement of industrial disputes in Australia satisfactory. Labor is dissatisfied with the wage-board system because it fixes wages for a period, before the completion of which the rapid advance in the cost of living has made the award inadequate; and with arbitration because it believes that more can be obtained by direct action than by resorting to courts. Capital objects to both systems because of the difficulty of inducing labor to abide by awards and the belief that the boards and courts are too radical.

Commercial interests in Australia have expressed the desire that a new system be established whereby the employers and employees can meet on friendly terms and come to a proper understanding of each other's difficulties. In line with this idea, a royal commission has been appointed, sitting at Melbourne, to consider the subject of industrial unrest and to fix, if possible, a basic wage for the whole Commonwealth.

## James Hartness To Be the Next Governor of Vermont

James Hartness, Past-President, The American Society of Mechanical Engineers, who has been a candidate for the Republican nomination for Governor of Vermont, won a sweeping victory at the primaries held on September 14. Mr. Hartness was one of four contestants for the nomination, and won by a vote of nearly two to one over his nearest competitor. Nomination is

JAMES HARTNESS, NOMINEE FOR GOVERNOR

tantamount to election, and he may be safely greeted as "the next Governor of Vermont."

The Burlington (Vt.) *Daily Free Press* considers that the result marked "the triumph of the gospel of hope and progress for Vermont," and in commenting on the result says:

The Hon. James Hartness of Springfield not only won the Republican nomination for Governorship, but he also gained a vote which at the hour of writing seems to have approached proportions of an actual landslide. . . . The people have been educated in this campaign to expect a genuine move toward a greater Vermont. We believe they will not be disappointed. Mr. Hartness will do his part. The only question now is whether we—the people of Vermont—will do our part to match his initiative and energy by helping to carry out his program for new opportunities for Vermont's youths.

The significant thing in Mr. Hartness' nomination is that it was accomplished by an engineer who used an engineer's methods, and

that these methods won out before a jury of an American public. His success should have a potent influence in shaping public opinion in favor of such methods, and should be an inspiration to other engineers who are in a position to render effective service to the state or nation.

To understand the significance of the event, one must trace the history of the industries of Springfield, Vt., Mr. Hartness' home town. Those familiar with the machine-tool industry are acquainted with the remarkable industrial center which has developed in this community, comprising a number of prosperous firms which have sprung from the parent company, the Jones and Lamson Machine Company, which was established there through Mr. Hartness' efforts about 30 years ago. The spirit of the community has been to assist young men of inventive and administrative talent who have originated new tools and devices to establish organizations of their own under the direction of those who by training and experience are competent to conduct them.

As a result, Springfield has grown and prospered in marked contrast in this respect to most other towns of the state of corresponding size. The population of Vermont as a whole does not vary much from year to year. The larger towns and cities increase somewhat in size, but the smaller communities have a tendency to decrease. It is evident, therefore, that the fundamental problem in Vermont relates to those things which concern the organic life and growth of the state. It is evident that it is related only to a very slight degree to programs such as are usually propounded by legislators and politicians.

With characteristic originality, Mr. Hartness made a survey of the conditions of the state, and judged that something must be done to increase its industrial life which, in turn, would provide a better market for the produce of its farms and make the state an attractive abode for its young men. The broad-gaged policy which had been followed at Springfield had proved effective. Why should it not apply equally well elsewhere?

It was on this proposition that Mr. Hartness conducted his campaign, and in his first appeal to the public made the following statement:

We can accomplish much by refining our business methods and in watching expenditures, but that will not make a material change in the conditions that confront us.

I believe that the same spirit and energy that has been successfully displayed by Vermonters in agriculture, dairying, and various other state successes, can be displayed in other desirable industries providing we conduct a drive for industrial plant culture to awaken interest and activity.

It is now proposed to inject this scheme of industrial plant culture into this campaign because it is the only way, it is the best way, it is the big way to solve our problems.

One example of its working is found in the machine-tool industries that have grown from a single plant that brought $30,000 per year into the state in '89 to six plants that brought about $10,000,000 last year. The six plants are all prosperous and filled with young, enthusiastic people who are competent to start other plants. Nearly all of this sum comes into Vermont and is invested here.

As to the efficiency of legislation or administrative action in carrying out such a program, a recent bulletin issued in interest of the movement states: "A popular movement for carrying forward a state's interest in these matters constitutes a real big thing to be done at the present time. We can ignore the trend of this age of mechanism and machinery, or we can wake up to the importance of the plan of industrial plant culture. While we know that legislation cannot produce the skill of the worker or an artist, or the initiative of the industrialist, or the fruit of invention; the people can by popular opinion, expressed by the press and through their representatives in legislature, give encouragement or they can stifle and kill such development."

## Noteworthy Papers on Labor, Management and Production

As repeatedly shown at the meetings of The American Society of Mechanical Engineers, both general and local, there is no single subject in which engineers are so vitally interested at the present time as that of labor and management. Of unusual interest to engineers, therefore, should be the September issue of *The Annals*, the bi-monthly publication of the American

Academy of Political and Social Science.[1] This contains a collection of 30 articles under the general title of Labor, Management and Production, divided under the four following headings: Part I, The Human Factor as the Heart of Industry; Part II, The Drift Toward Science in Industry; Part III, Some Major Problems of Industry; Part IV, The Manager's Part in the Adventure of Industry. .

The series of papers was edited by Morris Llewellyn Cooke, Mem.Am.Soc.M.E., who has written the editor's preface outlining its scope and intent. Collaborating with him as editors were Samuel Gompers, President of the American Federation of Labor, representing organized labor; and Major Fred J. Miller, President, Am.Soc.M.E., representing the managers of labor.

In the selection of contributors, an effort was made to secure papers from two groups of authors of which the names of Mr. Gompers and Major Miller are representative; or, as expressed by the Academy when this issue of *The Annals* was planned, by the "organized workers" on one hand, and "scientists of industry" on the other hand. In selecting the latter, however, the editor states that "no one was asked to contribute to these pages simply because he owned something or employed somebody. Official standing in the labor movement and recognized service in the application of science to the purposes of industry have determined eligibility in every case." The contributors are from the United States, England and France, and it is pleasing to note that one of the French representatives in the list is Charles de Freminville, recently elected Honorary Member, Am.Soc.M.E.

The value to the engineer of such a selection of articles as comprised in this number of *The Annals*, as a source of information upon the problems of the labor movement, is well expressed by Major Miller in his foreword, who says:

"There can be no doubt that much of the misunderstanding between employer and employee may be traced to the fact that each reads, more or less exclusively, the publications that support his views—that indeed must do so, for reasons that are easily understood. Thus each side fails to get the other's viewpoint and it is certain that the industrial situation would be much improved if there could be more of that 'getting together' which accompanies a free interchange of views, to the end that each side may at least comprehend what the other stands for; and why."

Mr. Gompers, whose foreword is entitled The Workers and Production, recognizes in his opening paragraph that "production is the great world problem of today" and other of the labor representatives also take the stand for increased production as in the interest of the wage earner. Mr. Gompers says, however, that "The Workers and Production—the title under which I express these thoughts—means the workers and life. That is to say, the workers and the life of the nation, spiritual and mental as well as physical. To contribute to the thought that is centered upon a better life for those who work, upon a nobler life for the nation and a higher plane for it and the civilization of the world, is a privilege to be sought. The last drag upon the worker's heart and brain, upon his enthusiasm for service, upon his simple efficiency at the bench and forge, is the lingering ideal of production for profit alone. The greatest single achievement for progress possible to this day and this generation is the substitution in industry of the ideal of production for use—for service —and not for profit alone."

Major Miller, writing from the standpoint of industrial management, says:

There are those who, for ten years or more, have recognized and have declared their conviction that the industrial world has been passing through a revolution—for the most part peaceful and constructive, but, nevertheless essentially, a revolution. The Hohenzollern war did not cause this revolution, but only accelerated what was already under way. Among other things it has shown us clearly that the old driver method of industrial management will no longer do. The workmen of the world and as well, the women of every country that participated in the war, have acquired by that participation a new status. Many of the industrial difficulties of the present day are due to the resistance of working men and women everywhere, to being forced back to their former and inferior status. They are insisting that if they are good enough to place their lives at the disposal of the forces of civilization, then they are good enough to have

at least some voice in determining the manner of life they shall lead in the civilization they have striven, they hope successfully, to preserve.

In every country of the world the trend, for years, has been toward democracy; and absolutism, both in governments and in industries, is being generally perceived to be an anachronism in this age of enlightenment. . . . The tendency is to reëxamine our position with respect to our terrestrial environment and to ask why, with the productive powers of man multiplied by myriads of inventions, as they have been within the past century, anyone able and willing to work should, at any time, even temporarily, be without an adequate supply of all that is needed to maintain in good health not only himself, but all who are naturally dependent upon him; and to obtain in addition a fair share of the luxuries which modern civilization is supposed to afford for those who live in civilized countries. . . .

Many of the most profound and disinterested students of this, our greatest problem, believe that the modern tendency of those who labor to plan for action in their common interest is, after all, the best protection society can have against worse things—evils such as have been alluded to above and which history plainly shows have led often to violence, but almost inevitably to degeneration and social decay.

All who are sincerely trying to understand the present course of events will rejoice that, perhaps now more than ever before, men of large affairs, of proven capacity for leadership, heads of important industrial concerns, are giving evidence of their conviction that autocracy in industries is irreconcilable with democracy in governments. Realizing the very great difference between a body of employees all enthusiastically coöperating toward one object and, on the other hand, a body of employees rendering only such service as they think necessary to hold their jobs, and only so long as they wish to hold them, these men are giving this problem their best attention. .

## Favors Printing Code of Ethics in Shape for Framing

The proposed Code of Ethics submitted by the Special Committee to the Spring Meeting at St. Louis (published in the September issue of MECHANICAL ENGINEERING) is worthy of careful consideration by each member of our Society.

The prospect of a definite, concise, straightforward code of ethics adopted by the A.S.M.E. to supersede the present rather cumbersome and lengthy code is gratifying. Any code of ethics that will stand the test is, or should be, merely restatements in detail of the operation of the Golden Rule, which has been the age-old measuring rod that has successfully withstood the assaults of the critic and the cynic and today will withstand the onrush of modern tendency to limit and equalize by restraint the service rendered for compensation or remuneration.

The Code, when adopted, might well be printed in such a shape that it can be framed, so that it may be in constant evidence. By displaying such a code in his office, the engineer subscribes to all of the statements contained therein. Such a use should go far toward inspiring confidence in the individual engineer and in the profession as a whole; for the Code, as proposed, is a statement of high altruistic purpose.—E. S. CARMAN.

## Dr. Samuel Sheldon Dies

Dr. Samuel Sheldon, for the past thirty-one years professor of Physics and Electrical Engineering at the Polytechnic Institute of Brooklyn, died on September 4 at Middlebury, Vt. Dr. Sheldon was born in Middlebury on March 8, 1862. He was graduated from the Middlebury College in 1883 with the degree of A. B. In 1886 he received his A. M. degree. During the next two years he studied at Wurzburg, Germany, and received the degree of Doctor of Philosophy there in 1888. During part of this time he was associated with Kohlrausch, the distinguished physicist, in his celebrated determination of the ohm as the unit of electrical resistance. He was awarded the degree of Doctor of Science from the University of Pennsylvania in 1906, and from Middlebury College in 1911.

Dr. Sheldon was a fellow of the American Electro-Therapeutic Association, fellow and past-president of the American Institute of Electrical Engineers, member of the American Physical Society, member and past-president of the New York Electrical Society, member of the American Electrochemical Society, of the Society for the Promotion of Engineering Education, of the Brooklyn Institute of Arts and Sciences, and president of its department of electricity. He was also a member and past-president of the United Engineering Society and chairman of its Library Committee.

# The Federated American Engineering Societies

First Meeting of Its Governing Body, American Engineering Council, to be Held in November, at Washington, D. C.—The Purposes and Activities of the New Organization— Two More Societies Become Charter Members

THE first meeting of the American Engineering Council, the governing body of The Federated American Engineering Societies, will be held on November 18 and 19, at the New Willard Hotel, Washington, D.C. This will be an event of the greatest importance to the engineering profession.

The first session will be devoted to the election of temporary officers and the appointment of temporary committees, and the second to an address by J. Parke Channing, chairman of the Engineering Council. Mr. Channing's address will review the activities of Engineering Council and will outline particularly the work of its many committees. The second session will afford an opportunity for the discussion of the purposes of the Federation and its field of activities.

The sessions of the second day will be given over to the election of permanent officers and the reports of committees. Herbert C. Hoover, president of the American Institute of Mining and Metallurgical Engineers, will address the convention during this session.

A complete program of the meeting is presented below:

upon matters of common concern to the engineering and allied technical professions.

In the past Engineering Council has been fully alive to the need for concerted effort on the part of engineers in these directions and has accomplished notable results as already fully outlined in these columns. The following are some of the matters now pending before Engineering Council and with all of them the newly created American Engineering Council is destined to be greatly concerned.

(a) Classification and compensation of engineers
(b) Licensing of engineers
(c) Water conservation
(d) National Board for Jurisdictional Awards in the Building Industry
(e) National Public Works Department
(f) Assisting in the preparation of information for the Senate Committee on Reconstruction and Production
(g) Public affairs
(h) Military affairs

---

## Program of First Meeting of American Engineering Council of The Federated American Engineering Societies

*Washington, D. C., November 18-19, 1920: Headquarters, New Willard Hotel.*

**THURSDAY, NOVEMBER 18, 1920**

MORNING SESSION

8:30 a.m. Registration
10:00 a.m. Opening Session of American Engineering Council
1 Call to Order
Richard L. Humphrey, Chairman of the Joint Conference Committee, Consulting Engineer, Philadelphia, Pa.
2 Election of Temporary Chairman
3 Election of Temporary Secretary
4 Appointment of Temporary Committees:
(a) Program
(b) Credentials
(c) Nominations
(d) Constitution and By-Laws
(e) Plan and Scope
(f) Budget
(g) Resolutions.

AFTERNOON SESSION

2:00 p.m. Address—ENGINEERING COUNCIL, by J. Parke Channing, Chairman.
2:30 p.m. Discussion of the field of activity for The Federated American Engineering Societies.

**FRIDAY, NOVEMBER 19, 1920**

MORNING SESSION

9:00 a.m. 1 Report of Committee on Nominations
2 Election of Permanent Officers
3 Report of Committee on Constitution and By-Laws
4 Formal Ratification of Constitution and By-Laws
5 Report of Committee on Plan and Scope

AFTERNOON SESSION

2:00 p.m. 1 Report of Committee on Budget
2 Report of Committee on Resolutions

EVENING SESSION

8:30 p.m. 1 Introductory remarks by the presiding officer, the Chairman of American Engineering Council
2 Address by Herbert C. Hoover, President American Institute of Mining and Metallurgical Engineers (Subject to be announced later)
9:30 p.m. Informal reception and smoker.

**SATURDAY, NOVEMBER 20, 1920**

9:00 a.m. Organization Meeting of Executive Board, American Engineering Council.

---

## The Purposes and Activities of the Federation

Since the future activities of the newly organized Federation is bound to be the central topic under discussion at the forthcoming meeting of the American Engineering Council a statement of the work in prospect and a review of the accomplishment of the present Engineering Council is of timely interest.

The questions are frequently asked, "What will be the field of activity of The Federated American Engineering Societies?" and "What does the organization intend to do?" Briefly, it will so consolidate the engineers of the country that they will present a solid front and join in an united effort for the proper handling of governmental and legislative matters of an engineering character. It will use its power for the service of the community, state and nation in public affairs wherever engineering experience and technical knowledge are involved, and will consider and act

(i) New York state government reorganization
(j) International affiliation of engineers
(k) Curricula of engineering schools
(l) Patents
(m) Payment for estimating
(n) Boston-Washington super-power system
(o) Russian-American engineering coöperation
(p) Types of government contracts

In view of the good work which Engineering Council has accomplished and is now doing the questions arise:

If Engineering Council is a success, why should the proposed Federation supplant it?

If Engineering Council is a failure, why should the Federation be organized along lines so nearly parallel?

Chiefly, in answer to the first question, to establish an organization which shall be representative of engineers everywhere so that all may have a voice and exert their influence in the great affairs of public concern with which a central body would be expected to deal. Engineering Council has six member-societies. At the Organizing Conference for the Federation 71 societies were represented with an aggregate membership of over 80 per cent of that represented by the 110 societies which were invited. As expressed in the bulletin of the Joint Conference Committee, which is now acting for the Federation, the organization of Engineering Council is "from the top downward," whereas the Federation, which came into being in response to a widespread movement throughout the country, has been organized from the "bottom upward" and rests upon a foundation so broad that every engineering society of acknowledged standing may become a supporting element.

The Constitution and By-Laws of the Federated American Engineering Societies for the first time provide a plan for bringing about solidarity in the engineering profession—a plan which assures success because it does not interfere with the existing engineering and allied technical societies.

The answer to the second question is that the Engineering Council has not been a failure, but on the other hand has done a magnificent work. The deficiencies of Engineering Council are not in the quality of the work which it has accomplished, but rather have been due to its organic limitations and to the fact that it is not sufficiently representative of the local, state and regional organizations and affiliations. The Organizing Conference in Washington laid the foundation for a more democratic organization in which the local, state, and regional engineering and allied technical organizations, and affiliations, will be represented, and will have a real voice in the management of its activities.

The Organizing Conference thus recognized the successes and limitations of Engineering Council and has evolved an organization in which all of these successes will be utilized and a broader opportunity afforded for more effective work on behalf of the engineering and allied technical professions.

## Societies Rapidly Organizing for Federation Support

### A. S. M. E. APPOINTS ITS DELEGATES TO AMERICAN ENGINEERING COUNCIL.

The Council of The American Society of Mechanical Engineers recently appointed its representatives on the American Engineering Council, the governing body of The Federated American Engineering Societies. The mechanical engineers will be represented by fourteen delegates who, under the By-Laws of the Federation, are appointed for a period of two years. A list of those named by the A.S.M.E. will be found on page 135 of Section Two.

### AMERICAN INSTITUTE OF ELECTRICAL ENGINEERS BECOMES CHARTER MEMBER.

The American Institute of Electrical Engineers is the second national engineering organization to join the Federation. This action by the Institute was taken at a meeting of the Board of Directors held in New York on August 12 when the following resolution was unanimously adopted:

*Resolved,* That the American Institute of Electrical Engineers accepts the invitation to it to become a Charter Member of The Federated American Engineering Societies, and pledges its hearty coöperation in the work thereof.

### AMERICAN SOCIETY OF CIVIL ENGINEERS TO VOTE ON JOINING FEDERATION.

The American Society of Civil Engineers at a convention held in Portland, Oregon, during August voted to refer the question of joining the Federation to its corporate membership. The resolutions adopted provided:

"That the Board of Direction of the American Society of Civil Engineers be directed to submit at once the question of the American Society of Civil Engineers becoming a charter member of The Federated American Engineering Societies to referendum vote to the Corporate Membership of the American Society of Civil Engineers as recommended by the Joint Conference Committee, said ballot to be accompanied by a copy of the Constitution and By-Laws of said Federation, and

"That the Board of Direction of the American Society of Civil Engineers be further instructed in the event of a favorable vote on said referendum to proceed at once to take such steps as may be necessary for the American Society of Civil Engineers to become affiliated with said Federation."

### CLEVELAND ENGINEERING SOCIETY IS FIFTH MEMBER OF THE FEDERATION.

The Cleveland Engineering Society at its meeting of August 10 voted to become a Charter Member of The Federated American Engineering Societies. The Cleveland society is the fifth organization to join the Federation, the first four in their order of affiliation being, respectively, The American Society of Mechanical Engineers, the Detroit Engineering Society, the Technical Club of Dallas, and the American Institute of Electrical Engineers.

---

## THE CONSTRUCTION PERIOD

(*Continued from page 501.*)

land is $1,200,000 and the structural property $10,800,000. The purchase of land is started one year before construction and continues for three years, the construction period is also three years, the assessment is made in mid-year of each construction period. The assessor will find 70 per cent of the construction cost, the rate is 2 per cent. The amount put into property each year is as follows:

| | Land | Structural Property | Total |
|---|---|---|---|
| First Year ........... | $400,000 | ......... | $ 400,000 |
| Second Year ......... | 400,000 | 3,600,000 | 4,000,000 |
| Third Year .......... | 400,000 | 3,600,000 | 4,000,000 |
| Fourth Year ......... | ........ | 3,600,000 | 3,600,000 |
| Total ........... | $1,200,000 | $10,800,000 | $12,000,000 |

*Solution:*

| | | |
|---|---|---|
| Total cost end of first year................$ | 400,000 | |
| Cost at middle of first year ................. | 200,000 | |
| Taxes paid first year 2% of $140,000 ...... | 140,000 | $ 2,800 |
| Total cost at end of second year............ | 4,400,000 | |
| Cost at middle of second year............. | 2,400,000 | |
| Assessment @ 70% ....................... | 1,680,000 | |
| Taxes paid second year 2% of $1,680,000.... | | 33,600 |
| Total cost at end of third year ............ | 8,400,000 | |
| Cost at middle of third year............. | 6,400,000 | |
| Assessment @ 70% ....................... | 4,480,000 | |
| Taxes paid third year 2% of $4,480,000...... | | 89,600 |
| Total cost at end of fourth year ............ | 12,000,000 | |
| Cost at middle of fourth year................ | 10,200,000 | |
| Assessment @ 70% ....................... | 7,140,000 | |
| Taxes paid fourth year 2% of $7,140,000.... | | 142,800 |
| Total taxes paid ..................... | | $268,800 |

$$\frac{268,800}{12,000,000} = 2.24 \text{ per cent}$$

As in the case of interest, before the final percentage to include for taxes during construction can be determined it will be necessary to know what method was used in finding the reproduction cost, if "piece-meal" the percentage will be one thing, if "wholesale" it will be another. The proper amount to include for taxes finally depends on first, the length of the construction period; second, what per cent of the construction cost will the assessor find; third, the tax rate.

# Engineering Foundation Seeks Large Endowment

### Research Organization Asks for $1,000,000 to Assure it of Adequate Funds for Present and Future Undertakings

THE Engineering Foundation, created in 1914 by the United Engineering Society, is now seeking to increase its endowment to at least a million dollars. The Foundation is based on gifts from Ambrose Swasey, of Cleveland, Ohio, Past-President Am.Soc.M.E., aggregating $300,000 and intended to be the nucleus for a much larger endowment fund. It was established "for the furtherance of research in science and engineering, or for the advancement in any other manner of the profession of engineering and the good of mankind." Its income, however, is entirely inadequate to meet the ever increasing demands, and, in order that the researches now in progress may be continued and those contemplated be undertaken as may be required from time to time, an increase of endowment is thus being sought. The foundation desires gifts of $1000 or more, and it has announced that any person giving $250,000 will be honored as a Founder. Charles F. Rand, of 71 Broadway, New York, Past-President American Institute Mining and Metallurgical Engineers, is chairman of the Foundation. The offices of the Foundation are in the Engineering Societies Building, 29 West 39th Street, New York, Alfred D. Flinn, Secretary.

The endowment funds of the Foundation are owned by United Engineering Society, a New York corporation, for the four national societies. The income is administered by a board composed of thirteen representatives of the American Society of Civil Engineers, American Institute of Mining and Metallurgical Engineers, The American Society of Mechanical Engineers, American Institute of Electrical Engineers, and three members at large. Because of the liberal form of its organization, the Foundation can support researches conducted either by individuals or by organizations; establish and operate research laboratories; aid in applying to engineering the results of the researches in the sciences, and in numerous ways aid in stimulating interest in research among engineers and the industries.

The results to be obtained through research cannot, however, be achieved in a short space of time, but, nevertheless, in its brief existence, Engineering Foundation has accomplished noteworthy results. Since 1916 it has maintained a liaison with scientists through the National Research Council, an organization called into being at the request of the President of the United States and which, because of its relation to the Federal Government, is therefore in a position to be intimately connected with the research problems both of the industries and governmental departments.

At present a most important research is being carried out, in which both the Foundation and the Division of Engineering of National Research Council are interested. It is concerned with the fatigue phenomena in metals. This highly important subject is being investigated at the Engineering Experiment Station of the University of Illinois and is being financed to the extent of $30,000 by the Foundation. Recently the General Electric Company requested that it be permitted to coöperate in the work in order that the program of tests might be extended. The September 1919 issue of MECHANICAL ENGINEERING contained a progress report of the committee appointed to carry on this research, and in it will be found a complete summary of the facts and theories relating to fatigue phenomena so far as at present known.

Another of the Foundation's researches is concerned with the wear of gears. This study begun in 1916 was interrupted by the war and was not resumed until the past summer. The work is being conducted at Leland Stanford Junior University.

In 1917 the Foundation joined with the New York Committee on Submarine Defense in a series of experiments on the concealment of ships by means of spray from special nozzles placed at suitable points on the vessel. This method of protecting ships was found, however, to be impractical.

One of the Foundation's most recent researches has resulted in an improved form of weir for gaging the flow of liquids in open channels. This research was conducted by Clemens Herschel, a noted hydraulic engineer, and a past-president of the American Society of Civil Engineers. The experimental work was conducted in collaboration with the Hydraulic Laboratory of the Massachusetts Institute of Technology and the results obtained were presented as the leading article in MECHANICAL ENGINEERING for February 1920.

Still another phase of research was undertaken in 1919, when the Foundation authorized Dr. E. E. Southard, the late Director of the Massachusetts State Psychiatric Institute, to undertake an investigation of the place in industry of the mentally deficient. The objects of Dr. Southard's research were to develop or discover methods for recognizing and suitably placing persons of slightly abnormal mental characteristics and for adapting psychopathic individuals to usefulness in industry, but his untimely death prevented the conclusion of his work. He had, however, completed three preliminary papers which appeared in the February, April, and June 1920 issues of *Industrial Management* (New York).

Research in mental hygiene of industry deals, however, with but one of the many groups of problems peculiar to industrial personnel relations. The Foundation has, therefore, coöperated with the National Research Council in planning a conference of leading industrial, labor and scientific bodies interested in this field to consider the possibility of comprehensive, unbiased study of the subject. Arrangements for the preliminary conference are well advanced and the meeting will be held this fall in Washington, D. C. Attention will be directed to the non-controversial elements of the subject, capable of scientific study.

In the latter part of 1918, Julius Alsberg, a consulting hydraulic engineer, suggested to the Foundation that there be established a station for the testing of large water wheels and other large hydraulic equipment. The Foundation immediately appointed a committee consisting of Silas H. Woodard, H. Hobart Porter, and Calvert Townley to investigate the matter. In May 1919, the committee reported that such a testing station was not feasible; that it was not necessary to establish a testing flume for small models, as existing flumes met all requirements; but that testing of water wheels now in place would be useful and practicable.

The Engineering Foundation has thus covered several diversified fields and has been limited as to the extent of its investigations solely because of its limited resources. Many important research projects which promised valuable results could not be undertaken because of lack of funds, and it is to provide for these that the Foundation is now engaged in increasing its endowment.

To those who desire more detailed information of its activities, the Foundation will be pleased to send a booklet descriptive of its organization and work.

---

Several calls for assistance have come to the Society from engineers in Austria, which indicate the desperate situation in which many are now placed in that country. The latest appeal comes from Innsbruck from the Technischer Verein Innsbruck, which says:

"To give you an instance of our precarious position, it would require, in consequence of the debasement of our money, the sevenfold amount of the entire funds of our society to procure a typewriter and a copying press, an amount which we would not be able to raise in addition to paying the current expenses of the society. We are in great need of these two machines. In order to circulate items of a technical character among our members, and take the liberty of addressing ourselves to you and of soliciting your assistance in helping us to secure a second-hand typewriter and a copying press. Will you kindly send reply addressed to the Chairman of our Society, Mr. Julius Gruder, Engineer, Innsbruck, Postbox 79."

# Government Activities in Engineering

*Notes Contributed by The National Service Department of Engineering Council [1]*

## Organization and Status of Federal Power Commission

The Federal Power Commission which was created to administer the Water Power Act, passed and signed by the President in the closing days of the last Congress, has been organized along the following lines: engineering, accounting, statistical, regulatory, licensing, legal, and operation. It is apparent that the engineering division will be the most important because it will make general investigation of the electric power industry, power sites, costs, and development.

Unfortunately the actual work of the Commission has been somewhat hampered because the Comptroller of the Treasury has ruled that, under the terms of the act, there is no money available to pay the required personnel. This means that the personnel will have to be assigned to the Commission by the various Departments until such time as appropriations can be made to pay for the required assistance. Thus far Lt-Col. Kelley of the Engineers Corps has been assigned as engineer officer, and Major L. W. Call of the Judge Advocate General's office as chief counsel.

One of the chief duties of the new Commission has been to formulate regulations that would make the applications for waterpower privileges valid. These regulations were drawn up by the Secretary and all interested Government Departments and outside organizations were invited to send their representatives to a Washington conference on August 12 for the purpose of discussion in making further recommendations. This meeting was followed by another conference the next day with representatives of interested financial institutions, so that their recommendations could also be considered before the application was put into final form.

Some regulations covering applications for permits and licenses under the Water Power Act were approved at the September 3d meeting of the Commission, and those covering construction, operation and maintenance of projects are now in process of preparation. It is contemplated that further hearings on these regulations will be held before they are finally adopted.

District offices of the Commission have been opened at St. Paul and St. Louis in the local offices of the Corps of Engineers, and in Denver and San Francisco in the legal offices of the Forest Service.

## Effect of Navy Wage' Award on . Salaries of Technical Men

On September 9 the Secretary of the Navy announced the new schedule of pay which had been made in accordance with the recommendations of the Navy Wage Adjustment Board. This is the first adjustment that has been made since the Macey Board Award in February, 1919.

In general, the various employees of the Navy receive increases of 5 per cent in pay except where the pay is more than eight dollars per day, so that the award does not include the engineers, architects, and draftsmen in the raise. In fact, they are adversely affected because the new schedule requires that all employees of the Department shall observe the standard working day of eight hours. This means that when the order becomes effective technical men will suffer a decrease of about 8 per cent in their present pay because of the loss of one-half day each week. This, combined with the fact that the working day is increased from seven to eight hours, makes an average approximate decrease of 20 per cent in compensation.

[1] Engineering Council is an organization of national technical societies created to consider matters of common concern to engineers as well as those of public welfare in which the profession is interested. The headquarters of Engineering Council are located in the Engineering Societies Building, 29 West 39th Street, New York City. The Council also maintains a Washington office with M. O. Leighton, chairman of the National Service Department, in charge. This office is in the McLachlen Building, 10th and G Streets, Washington, D. C. The officers of Engineering Council are: J. Parke Channing, *Chairman*; Alfred D. Flinn, *Secretary.*

## National Committee for Governmental Economy

There are at least four important agencies carrying on a propaganda for some modification of the administrative branch of the Federal Government. These are, (1), the Budget Committee; (2), National Public Works Department Association; (3), National Education Association; (4), American Public Health Association.

Of these the National Budget Committee has taken the lead in the formation of a National Committee for Governmental Economy which is preparing a general plan for a reorganization of the administrative branch of the Government. This is for the purpose of coördinating the work of any departments which may be formed as a result of the proposed reorganization as, for instance, if a Department of Public Works is created, it is desired that this shall harmonize with other adjustments which must ultimately be made, and the same is true of the Department of Public Health or the Department of Education should they be formed.

The staff of the Committee comprises Stanley Howe, Director of Organization of the Budget Committee, who has the same duties with the new Committee; C. T. Chenery, Executive Secretary for the National Public Works Department Association, is Executive Secretary; and Harold N. Graves, Assistant Chief of the Bureau of Efficiency, is directing the reorganization study. The Committee has offices at 700 Tenth Street, N.W., Washington, D.C., and will issue progress reports from time to time.

## Relative Importance of Mining and Agricultural Appropriations

Appropriations made by the last Congress show only one-thirteenth as much for the promotion of the interests of the mining industry as is shown for promotion of corresponding interests in the agricultural industry. The ratio of economic importance of these two industries is but one to four. Official reports for 1918 show that the total value of mineral products was roughly five and one-half billion dollars, while the value of agricultural products totalled approximately twenty-two and one-half billion dollars. Of course the agricultural industry is more diversified in all of its aspects, but those interested in the various phases of mining work contend that this does not make up for the great difference in ratio of expenditures.

It is now contended that the Bureau of Mines should assist the chemical industries in the same way that it extends its assistance to metal-mining and coal-mining industries, because of the fact that they are so closely allied to the Bureau's present work. It is hoped that the gathering of statistics and the study of problems pertinent to the chemical industries will be taken up by the Bureau of Mines during the next fiscal year.

## Census of Central Electric Light and Power Stations

A partial report dealing with Central Electric Light and Power Stations for 1917 was issued by the Census Bureau on September 7. It is part of a more complete report which covers railways, telephones, telegraphs, and municipal electric fire alarm and police signaling systems, and which is the most complete and valuable census ever taken of American electrical industries.

The report which has just been issued covers the general development of Central Electric Light and Power Stations of both the steam and hydroelectric type. The general record of primary power equipment with special attention to generating equipment, line equipment and substation equipment is also presented, and the output and disposal of current is classified. Very comprehensive financial statistics are also given with special comparative financial and operating summaries of selected groups of electric stations. The report likewise contains some interesting data on employees, salaries, and wages.

# Mining Engineers Visit Lake Superior Copper and Iron Country

## The 122d Meeting of the Institute of Mining and Metallurgical Engineers Held at Points in Michigan and Minnesota, with Inspection Trips to the Great Mining Properties of Lake Superior District

A GENERAL trip of study and inspection of the Lake Superior iron and copper country was a feature of the 122d meeting of the American Institute of Mining and Metallurgical Engineers. Members left Buffalo on August 20, visited the copper mines in and about Houghton, the iron mines of the Marquette and Iron Ranges of Michigan and those of the Mesabi Range in Minnesota, and returned to Buffalo on September 3. Technical sessions were held at Houghton, Minneapolis and Duluth. A banquet was also held at Minneapolis, Herbert Hoover, President of the Institute, delivering the principal address.

### ADDRESS BY HERBERT HOOVER

Mr. Hoover spoke in advocacy of a definite national program in the development of our great engineering problems and emphasized the need for a National Department of Public Works. He said in part:

Our rail and water transport, our water supplies for irrigation, our reclamation, the provision of future fuel resources, the development and distribution of electrical power, all cry out for some broad visioned national guidance. We must create a national engineering sense of provision for the nation as a whole. If we are to develop this national sense of engineering and its relations to our great human problems it must receive the advocacy of such institutions as this.

We, together with our sister engineering societies, represent the engineers of the United States. It is our duty as citizens to give voice to those critical matters of national policy which our daily contact with this, the fundamentally constructive profession, illuminates to us. Just as our medical associations voice the necessity of safeguards to national health; as the bar associations safeguard our judiciary, so the engineers should exert themselves in our national engineering policies. We have none, but we need some, or the next generation will face a lower instead of a higher standard of living than ours.

Continuing, Mr. Hoover discussed the great national engineering problems of coal and petroleum supply, rail and water transportation, mountain water storage, irrigation, lumber supply, etc., and said:

All these problems are much akin, and the time has come when they need some illumination, guidance, and coöperation in their solution from the Federal Government. Nor do I mean a vast extension of Federal bureaucracy in our Government. If, in the first instance, through an agency of the central government, we could have an adequate study and preparation of the plan and method made of these problems of engineering development over the next fifty years, viewed solely in their national aspects, we would have taken the first step toward the adequate provision of an increasing standard of living and a lower cost of living for our descendants.

The second step is to determine that our government will be a government of coöperation, limiting profits surely, but holding to individual initiative as the single hope of human development. In order that we shall have some central point in the Federal government where these problems may be adequately considered, from which they can be ventilated for the verdict of public opinion, where the business brains of the country can be called into conference and coöperation with the government, and therefore with the people, the engineers of the United States have proposed time and again that a Cabinet department should be established in Washington, either new or to replace the Interior Department, to which should be assigned the whole question of public works. . . .

Such a department has become an essential from the point of view of proper consideration and presentation to the American people of those broader national engineering problems upon which the next generation must depend if our country is to march forward.

### TECHNICAL PAPERS

A paper on the Handling and Treatment of Rock-Drill Steel at Copper Range Mines was presented by H. T. Mercer and A. C. Paulson. The shank is formed in a bulldozing machine, a pin 5 in. long being first inserted in the hole to maintain it. After being formed the pin is removed and the shank reheated and quenched in oil. A 1/64-in. clearance is allowed between the shank and drill chuck. All shanks are made without lugs, except in a few cases where the drills are to be used in loose ground. Collars are formed on the jackhammer steel only. Results ob-

tained in heat-treating experiments were also mentioned. Drills were suspended in a chloride bath and the heat regulated by a pyrometer. They were then quenched in tempering oil, a 10 per cent solution of brine, or water, and tests run on each in the mine under uniform conditions. Other tests were made with a further treatment by tempering or drawing. It was concluded that the additional treatment after hardening was altogether unnecessary. No practical difference was found between heating in a chloride bath or in a coke furnace, provided temperature regulation was maintained equally as well in both cases. Water at a temperature of 80 to 100 deg. fahr. was determined to be the best quenching medium for the conditions existing in the mines.

How steam regenerators reduce coal consumption at the mines of the Copper Range Company was told by W. H. Schacht. The hoists furnish the only available supply of exhaust steam. The method of treating a flux of intermittent exhaust steam by the use of steam regenerators consists in passing the steam through a closed vessel, or regenerator, in which it is condensed and reevaporated. Condensation takes place when the steam is in excess of the quantity required to maintain a constant flow of outgoing steam, and reëvaporation takes place when the steam discharged into the regenerator is not sufficient to maintain this constant flow. The steam from the regenerator is delivered to the buildings to be heated with a drop in pressure not exceeding 2 lb. at times of greatest demand and about 1 lb. during average winter conditions. A two-pipe system is used. In the buildings the drainage of mains and return lines is in the direction of the flow, thus eliminating all water hammer. Syphon traps are used on all radiators, direct and indirect, and on all headers and drips. The air and condensation are collected into and returned by a 3-in. line, the end of which is submerged, forming a water seal that is located in the boiler house. At a point about 20 ft. above the overflow of this water seal is located an air separating tank, from which the air and non-condensable vapors are drawn from the system by a Nash No. 0 vacuum pump, requiring for its operation about 1½ hp. This pump is operated by an electric motor. The pressure carried on the regenerator, being also the back pressure on the hoist, varies from 8 to 10 in. vacuum to ½ lb. above the atmosphere; the colder the weather, the lower is the average of the pressure maintained. The return-line pressure is usually constant at about 18 in., making a working pressure available, therefore, of about 8 lb., which is more than sufficient in almost any emergency. There was a slight reduction in the tons hoisted in most plants during the year following the regenerator installations, but this did not affect the coal consumption materially, for some live steam was usually required for heating.

Surface Changes of Carbon Steels Heated in Vacuo, by E. Heaton Hemingway and George R. Ensminger, contained an account of experiments conducted at the Watertown Arsenal. Samples of steel were held at 1000 deg. cent. in vacuo for 10 hours. Three types of markings were discerned: (1) A deeply marked polyhedral structure, which represented the final gamma boundaries; (2) a fine clean-cut structure, which was brought out more clearly by etching, that represented the alpha boundaries; and (3) an indefinite and often partly obliterated structure that represented the boundaries of former gamma crystals that had been absorbed by crystalline growth. The outside layer consisted of ferrite below the temperature at which solid solution exists. Iron oxide reacted with the carbides in steel at temperatures below 1000 deg. cent. to form carbon monoxide and pure iron.

Other papers were presented, among which were the following: Efficiency in the Use of Oil as Fuel, by W. N. Best; Industrial Representation in the Standard Oil Co. (N. J.), by C. J. Hicks; A New Process for Making Fifteen Per Cent Phosphor Copper, by P. E. Demmler; and The Charpy Impact as Applied to Aluminum Alloys, by E. H. Dix, Jr.

# LIBRARY NOTES AND BOOK REVIEWS

A COURSE IN ELECTRICAL ENGINEERING. Vol. I. Direct Currents. By Chester L. Dawes. First edition. McGraw-Hill Book Co., Inc., New York, 1920. Cloth, 6 x 8 in., 498 pp., illus., $4.

The present book is the first of two volumes which are intended to serve as a comprehensive text covering the general field of electrical engineering in a simple manner. It will be useful, the author believes, to students of electrical engineering as an introduction to more advanced texts and also to students who are studying the subject for general training, without the intention of becoming specialists. The treatment is detailed, straightforward and systematic, but avoids much mathematical analysis.

THE DESIGN OF HIGHWAY BRIDGES OF STEEL, TIMBER AND CONCRETE. By Milo S. Ketchum. Second edition, rewritten. McGraw-Hill Book Co., Inc., New York, 1920. Flexible cloth, 6 x 9 in., 548 pp., illus., diagrams, charts, tables, $6.

This new edition has been made necessary by the increase in live loads due to the extensive use of heavy motor trucks and tractors, and the increased use of reinforced concrete in highway bridge construction. The scope of the book has been extended to include the design of concrete and timber bridges as well as steel bridges. The discussion covers all the details of constructing highway bridges. This edition is uniform with the author's Structural Engineers' Handbook.

DREDGING ENGINEERING. By F. Lester Simon. First edition. McGraw-Hill Book Co., Inc., New York, 1920. Cloth, 6 x 9 in., 182 pp., illus., plate, tables, $2.50.

The author of this volume has attempted to supply a work which will describe the principal types of dredges in such a manner as to impart a fundamental working knowledge of their construction and operation. He also takes up concisely the problems that confront the engineer in the conception and accomplishment of dredging projects.

EARTHWORK AND ITS COST. A Handbook of Earth Excavation. By Halbert Powers Gillette. Third edition. McGraw-Hill Book Co., Inc., New York, 1920. Flexible cloth, 5 x 7 in., 1346 pp., illus., tables, $6.

This treatise covers systematically and comprehensively the methods of excavating for engineering work of all kinds. In addition to the results of his own experience, the author has included a summary of the material in American technical literature. Particular attention is given to cost data. The present edition has been revised.

FLUGTECHNIK: GRUNDLAGEN DES KUNSTFLUGES. By Arthur Pröll. R. Oldenbourg, München and Berlin, 1919. Paper, 9 x 12 in., 332 pp., illus. 30.25 marks (33 marks, pound).

This treatise is founded on a manual of the scientific foundations of the subject, prepared for practical self-instruction while the author was director of the aeronautical experiment station of the Austrian war ministry. The notes prepared for this purpose were used later in connection with his lectures on technical aerodynamics in the Hannover High School, from which the present work has been prepared. The book, therefore, presents the subject from the viewpoint of immediate practical usefulness and also emphasizes the connection of aeronautical calculations with the teachings of mechanics. It will serve, the author believes, not only as a text for students, but will also enable the experienced engineer to carry out designs and construction in all details, when combined with practical numerical data. The discussion is confined to monoplanes and biplanes of the usual type.

HOW TO MANAGE MEN. The Principles of Employing Labor. By E. H. Fish. The Engineering Magazine Co., New York, 1920. Cloth, 6 x 9 in., 337 pp., $5.

This volume is a discussion of the relations between workers and their employers, in which an attempt is made to evaluate the things that go to make a satisfactory working organization. The author discusses the establishment of an employment department, the problems of the employment manager, the promotion of industrial relations, and industrial education.

THE IRON AND STEEL INDUSTRY OF THE UNITED KINGDOM UNDER WAR CONDITIONS. A Record of the Work of the Iron and Steel Production Department of the Ministry of Munitions. By F. H. Hatch. Privately printed for Sir John Hunter by Harrison and Sons, London, 1919. Cloth, 6 x 10 in., 167 pp., illus., plates, portraits, charts, tables.

This volume is an interesting account of the methods used by the British Government to obtain the necessary increase in the steel production of Great Britain during the war. The vastness of the field covered, the variety and complexity of the technical problems involved and the far-reaching industrial questions raised, make the subject one of great interest and importance.

MACRAE'S BLUE BOOK. MacRae's Blue Book Co., Chicago and New York, 1920. Cloth, 9 x 11 in., 1853 pp., $10.

An alphabetically arranged directory of American manufacturers, which is especially full with respect to iron and steel products, and the materials used by railways and in building. In addition to the directory of materials it contains a list of addresses, and also of trade names, a standard list of prices and a collection of data used by buyers. Thirty thousand firms are included.

MARINE ENGINEERS' HANDBOOK. Prepared by a staff of specialists, Frank Ward Sterling, Editor-in-Chief. First edition. McGraw-Hill Book Co., Inc., New York, 1920. Flexible cloth, 5 x 7 in., 1486 pp., illus., tables, $7.

This handbook is compiled for the use of operating and designing engineers, and students. To cover the main subjects and the many ramifications of marine engineering, it has been necessary to enlist the services of thirty specialists, each of whom has received the hearty coöperation of others in the same field. The best endeavor has been made to coördinate different practices in order that all data may be reliable and unbiassed. The handbook is a pioneer in its field.

MATHEMATICS FOR ENGINEERS. Part II. By W. N. Rose. E. P. Dutton and Co., New York, 1920. Cloth, 6 x 9 in., 419 pp., illus., tables, $7.

The work of which this is the second volume is intended to occupy a position midway between those written for college students, which are usually academic in character, and those practical books which omit any discussion of the scientific basis underlying them. It is intended to contain all the mathematical work needed by engineers in their practice and by students of all branches of engineering. This volume treats of the differential and integral calculus, spherical trigonometry and mathematical probability.

PERSONNEL ADMINISTRATION. Its Principles and Practice. By Ordway Tead and Henry C. Metcalf. First edition. McGraw-Hill Book Co., Inc., New York, 1920. Cloth, 6 x 9 in., 538 pp., $5.

The purpose of this book is to set forth the principles and the best prevailing practice in the field of the administration of human relations in industry. It is addressed to employers, personnel executives, employment managers and students of personnel administration, but will have value, the authors hope, also for all who are interested in advancing right human relations in industry and in securing a productivity that is due to willing human coöperation, interest and creative power. The field covered includes those efforts usually included in personnel management, employment, health, and safety, training, personnel relations, service features and joint relations. The book also seeks to show the relation of the personnel problems of each corporation to those of its industry as a whole, by considering the activities of employers' associations and organizations of workers.

RAILROAD CURVES AND EARTHWORK. By C. Frank Allen. Sixth edition, revised. N. Y. and London, McGraw-Hill Book Co., Inc., New York, 1920. Flexible cloth, 4 x 7 in., 289 pp., diagrams, charts, tables, $4.

This book has been prepared for the use of the students in the author's classes at the Massachusetts Institute of Technology, but it will also, he believes, be useful to engineers who have to deal with earthwork computation or curves. The new edition is marked by an extension of the treatment of circular arcs; by new methods for the computation of earthwork, additional data on haul, and by general revision.

THE RUDDER MARINE DIRECTORY. A Trade List of Shipbuilding and Marine Industries. Rudder Publishing Co., New York, 1920. Cloth, 6 x 9 in., 438 pp., $5.
The title of this directory is sufficient indication of its contents, and suggests its possible usefulness to those interested in ship construction and operation. The present edition is considerably larger than the first, which appeared in 1919. The lists are conveniently classified and arranged to facilitate their use.

DIE SCHALTUNGSGRUNDLAGEN DER FERNSPRECHANLAGEN MIT WAEHLERBETRIEB. By Fritz Lubberger. R. Oldenbourg, München & Berlin, 1920. Boards, 9 x 13 in., 168 pp., plates, 28 marks.
This book is intended to give engineers acquainted with manually-operated telephone exchanges an introduction to automatic systems, which are frequently so confusing to the beginner. The author has made an extended examination of the systems of wiring used or proposed for this purpose and has selected the essentials of proper procedure from them. Beginning with an easily understood automatic circuit, the book then proceeds to describe generally the principles of the various parts of the circuit, and follows this with explanations of the various systems that have been devised for the practical solution of the problems involved. The book includes bibliographical references, a list of German patents and many plates illustrating the arrangements described.

STORAGE BATTERIES. A Practical Presentation of the Principles of Action, Construction, and Maintenance of Lead and Non-lead Batteries and their Principal Commercial Applications. By Morton Arendt. American Technical Society, Chicago, 1920. Flexible cloth, 6 x 7 in., 136 pp., illus., plate, tables.
Important changes have taken place in the last ten years in the construction of the storage battery, and the principles of the design and manufacture of this " savings bank " of the electrical industry have been studied with the view of making it even more efficient. The importance of proper charging methods, testing, locating and remedying troubles are treated in detail in this volume, which also covers the methods of use by the large generating stations, in the telephone and telegraph industries and in electric vehicles. The work is adapted for home study by beginners as well as for experts.

STRUCTURAL STEELWORK. Relating Principally to the Construction of Steel-framed Buildings. By Ernest G. Beck. Longmans, Green and Co., New York, 1920. Cloth, 6 x 9 in., 462 pp., illus., tables, $7.50.
This book presents information likely to be of use in the design and construction of ordinary steel-framed structures. The principal endeavor has been to be broadly suggestive, rather than particular or exhaustive, to propose commonsense lines of argument based upon straightforward consideration of the facts, instead of attempting to generalize or formulate specific relations from details which cannot be more than typical. The work follows British practice, and is based upon a series of articles contributed to the *Mechanical World*.

TEXTBOOK OF AERO ENGINES. By E. H. Sherbondy and G. Douglas Wardrop. Frederick A. Stokes Co., New York, (Copyright 1920). Cloth, 9x12 in., 363 pp., illus., diagrams, plates, $10.
The present work is a review of the theory and practice of the design of engines for aerial navigation. The first part treats of theory and includes the theory of the cycle, the work available, the effect of altitude, pressure, etc., on the power, and the theory and design of compressors. The second part describes the current types of American, British, French, German and Italian engines. An appendix gives tables of use to designers. The book is fully illustrated with drawings and photographs.

TECHNICAL GAS AND FUEL ANALYSIS. By Alfred H. White. Second edition, revised and enlarged. McGraw-Hill Book Co., Inc., New York, 1920. Cloth, 6x8 in., 319 pp., illus., tables, $3.
The first edition of this work, which appeared in 1913, was intended to present the conclusions of the various committees of

technical societies which had reported on the testing of fuels and their utilization. Since 1913 our knowledge of gas and fuel analysis has been distinctly increased. Standard methods for sampling and analyzing coal and coke and for examining liquid fuels have been adopted by several societies. Methods for gas analysis have been critically studied and new methods developed. These advances have been carefully reviewed for this edition, which is about 20 per cent larger than the former one.

UTILISATION DES VAPEURS D'ECHAPPEMENT DANS LES HOUILLERES en Vue de la Production d'Energie Electrique. By Adrien Barjou, preface by Maurice Soubrier. Gauthier-Villars et Cie., Paris, 1920. Paper, 6x10 in., 90 pp., illus.
The author of this work discusses the utilization of exhaust steam in low-pressure turbines for the production of electricity or compressed air. The particular case studied is that of coal mines which have outgrown their power plants and are faced with the necessity of extending them. The economic and technical results obtained with low-pressure turbines are set forth in this book.

WIRING FOR LIGHT AND POWER. A detailed and fully illustrated commentary on the National Electrical Code. By Terrell Croft. Second edition. McGraw-Hill Book Co., Inc., New York, 1920. Flexible cloth, 5x8 in., 465 pp., illus., tables, $3.
The purpose of this book is to supply explanations, elaborations and illustrations for those sections of the National Electrical Code to which most frequent reference is necessary. For this purpose the most important Code regulations are given verbatim, and their application in practical electrical construction is explained by supplementary text and drawings. This edition has been fully revised to correspond with the 1918 edition of the Code.

# THE EFFECT OF DEPTH OF COPPER PLATING ON CARBURIZATION

(Continued from page 566.)

An ocular micrometer was used to measure the depth of case. The depth $t$ of the plating was calculated from the formula $t$ (inches) $= C/145.845A$, in which $C =$ weight of copper deposited, grams; 145.845 = weight of 1 cu. in. of copper (assuming specific gravity = 8.9); and $A =$ area of surface of deposit, sq. in.

Data from tests of sets B, D and E are given in Table 1. From these it may be seen that the carbon had just started to leak through the copper plating and in all specimens, including those of sets A and C, there was no case below the copper that would ever be noticed in practical work. The amount of copper which analysis does not show and which is due to solution in the iron or any other cause at all is certainly low; and allowing 10 per cent for this would give a safe margin. It is believed that for a 10 to 12 hour heat not over 0.0002 in. need ever be used. In event such an amount of copper is used the only care that should be taken is to have plenty of the carburizing agent present so that it will not be exhausted and the copper oxidized by reaction in which $2CO_2+Cu$ becomes $CuO+2CO$.

In examining the specimens under the microscope the corners of the pieces were given considerable attention. At this point the carbon could enter the steel back of the copper. In some 60 per cent of all specimens, however, such was not the case, and it seemed that the copper exerted some influence to keep carbon from the steel even though it was not directly in contact with it.

The photomicrographs taken—Figs. 1 to 6—show the typical appearance of specimens examined. With the exception of the original, these pictures were taken of piece No. 30; they show very clearly that no carbon penetrated the copper, and that at the same time a good cementation was effected on the back.

These experiments would tend to show that in copper plating with a current of as low density as the one employed:

1 There is no need of using as deep a plating as is now common practice;

2 The depth of copper plating required for ordinary work is not over 0.0002 in., or with a 100 per cent safety factor, 0.0004 in., and very often a lesser thickness would be sufficient.

# MECHANICAL ENGINEERING

| Volume 42 | November 1920 | Number 11 |

## The Constitution and Properties of Boiler Tubes

### Causes Underlying Defective Tubes—Grain Growth under Temperatures below Critical Point a Factor—Higher Carbon Content Suggested

By ALBERT E. WHITE,[1] ANN ARBOR, MICH.

**D**URING the winters of 1913–1914 and 1914–1915 considerable difficulty was experienced by the Park Place Heating Plant of The Detroit Edison Company in maintaining continuity of boiler operation. This difficulty arose because of the frequent shutdowns necessitated when boiler tubes bore evidence of being or becoming defective, requiring in consequence the temporary closing down of the boiler or boilers until the tubes in question could be replaced. Tubes in the front bank of the boilers were particularly prone to develop defects, and since this condition was experienced in its most aggravating form at a time when the boilers were most needed, namely, during the various cold snaps of the winter, investigations were started so that suitable steps might be taken to combat the trouble.

The necessity for obtaining relief was emphasized in June 1915, when No. 7 tube in the third row of No. 7 boiler at the Park Place Heating Plant let go. The appearance of this tube after the accident is given in Fig. 1. This tube failure made necessary the rebuilding of most of the boiler, and especially the replacement of nearly all of the tubes in the boiler. Fortunately no injuries were sustained by any of the men in the plant. Also, the failure occurred at a time of the year when the plant was not operating at full load, so that the customers were not subjected to any inconvenience through lack of steam.

The accident, however, indicated the need for prompt relief and investigations were started along three lines: (a) water softening, (b) rearrangement of baffling, and (c) considerations relating to the composition and constitution of boiler tubes.

### WATER SOFTENING

With reference to water softening, it was noted that practically all of the tubes which were replaced showed a thin but tough scale on the water side. This scale was calcium sulphate and was thin simply because the quantity of this salt in the boiler feedwater was relatively small and in plants operating under normal load conditions would have warranted no consideration. In these boilers, however, which were of the Stirling water-tube type of 750 hp., the long periods of cold often necessitated a load averaging 80 per cent above the nominal rating for periods extending at times into as many as six days. These load conditions, while not apparently unusually exacting, were in reality very severe when cognizance is taken of the fact that 98 per cent of the boiler feedwater was raw water drawn from the city mains.

There was no question but that the thin scale seriously affected heat transmission and was one of the contributing factors leading to tube failure. Much research work was done on this phase of the problem, and it resulted in a decision to treat all of the water in the plant with soda ash. Since this treatment has been in use practically all of the insolubles are caught in the live-steam purifier. There is no scale in any of the tubes, although there are flocculent particles of insoluble sodium carbonate circulating with the water in the boilers. This product, together with the other minor insoluble ones that may be present and the soluble salts, especially sodium sulphate, are kept down to harmless percentages by frequent boiler blow-offs. The Dionic tester is used as a guide, experience indicating that when it reads under 1800 no fear need be had of trouble resulting from foaming and priming.

### CHANGES IN BAFFLING

Steps were also instituted with reference to a study of the baffling. At the time when the trouble was most pronounced, the front baffles were between the first and second rows of tubes. This arrangement had been adopted since it was believed that it would produce higher temperatures with consequent fuel economy and abatement of the smoke nuisance.

The investigations showed that most of the tube replacements were from the front row. For the purpose, therefore, of distrib-

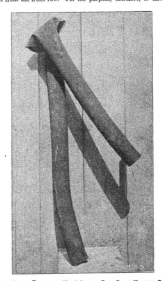

Fig. 1 Burst Tube from No. 7 Boiler, Park Place Heating Plant, Detroit Edison Company

uting the radiant heat among a greater number of tubes, hoping in this way to increase tube life without, however, carrying the same smoke troubles coming from incomplete combustion, the front baffling was changed so as to lie between the second and third rows rather than between the first and second rows. This arrangement has been most beneficial. There has been no decrease in fuel economy nor increase in smoke nuisance and tube replacements have been materially lowered.

[1] Consulting Metallurgical Engineer, Professor of Chemical Engineering, University of Michigan.
Abstract of paper to be presented at the Annual Meeting, New York, December 1920, of The American Society of Mechanical Engineers. All papers are subject to revision.

## CONSTITUTION OF TUBES

Under the third line of investigation listed above—the composition and constitution of boiler tubes—the first step was a study of the constitution of the metal in the tubes. This was undertaken for the purpose of ascertaining the effect of service conditions upon the types of tubes now employed. The metal in a very considerable number of tubes was examined metallographically, the

FIG. 2 AVERAGE BOILER-TUBE STRUCTURE; MAGNIFICATION 100; ETCHED WITH NITRIC ACID

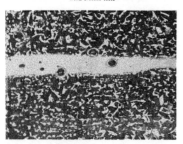

FIG. 3 AVERAGE BOILER-TUBE STRUCTURE; MAGNIFICATION 100; ROSEN-BAIN-ETCHED

etching mediums being the common nitric acid and the so-called Rosenhain. This latter deposits copper over the surface of the metal, giving an even coating when the distribution of the metalloids is uniform; but when such is not the case, disclosing minutest segregations and imperfections most clearly by variations in the depth of the plating.

Two photomicrographs showing the structure of a piece of boiler-tube metal which was selected as typical of the average of the majority of the sections examined, are shown in Figs. 2 and 3. The photomicrograph in Fig. 2 was taken after a sample had been etched in nitric acid, and in Fig. 3 treated with the Rosenhain reagent. It will be noted that the etching is uniformly distributed, indicating that the constituents are evenly dispersed throughout the metal. A clearly defined ghost line is brought out, resulting from phosphorus segregation, although more apparent in one resulting from the Rosenhain etching. There are also strikingly evident cavities or segregations which the writer has detected in many of the samples which he has examined. He considers this condition most serious and will further discuss it below. [In the pamphlet copy of the paper are other photomicrographs of samples representing the average of the best structure and of the poorest structure found among the different specimens.—EDITOR.]

## CAUSES OF TUBE FAILURE

On the completion of this preliminary survey attention was directed to the causes of tube failure. Excluding that due to imperfect heat transmission resulting from scale, the writer believes that the principal causes can be listed under the following heads:

a Failure due to tube brittleness resulting from absorption by the metal of hydrogen and usually attributable to faulty boiler-feedwater treatment

b Failure due to blowholes or other imperfections in the metal

c Failure due to recrystallization of the metal.

*Tube Brittleness Resulting from Hydrogen Absorption.* The first of these causes, namely, that due to tube brittleness resulting from the absorption by the metal of hydrogen, will not be developed in this paper. The facts are outstanding that contained hydrogen in metal makes it extremely brittle. The facts are further outstanding that certain types of water improperly treated, or all water excessively treated with certain types of boiler compounds, will cause the tubes to absorb hydrogen and become brittle. With intelligent treatment, however, there need be no cause for concern over tube failure from this source.

*Failure from Blowholes.* Failure from blowholes and other imperfections, the writer believes, should receive greater consideration in the future than has been accorded it in the past. The matter of course goes back to the steel mill and calls for greater emphasis on quality and less on tonnage.

A photomicrograph of a sample of metal containing blowholes is shown in Fig. 4. The ghost line is evident and even a casual scrutiny reveals radiating lines issuing from the center of the cavity. This condition is manifest in all of the photomicrographs taken of specimens with blowholes and indicates a lack of continuity of the metal. Should service conditions cause the tube to bag and should a cavity of the pipe shown in Fig. 2 be at the nipple of the bag, there is every reason to anticipate a bursting of the tube, with all of the attending dangers and expense.

*Failure from Recrystallization.* This matter merits much consideration. Recrystallization is accompanied by a marked decrease in the elastic limit and fatigue-resisting properties of the metal manifesting this phenomenon. It will occur if steel with a low carbon content which has previously been mechanically deformed at a temperature below the critical is later heated for a sufficient

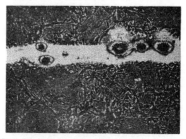

FIG. 4 EXAMPLE OF BLOWHOLE, PREPARED FROM SECTION OF BOILER TUBE; MAGNIFICATION 100; ROSENHAIN-ETCHED

time to any temperature below that at the critical. The common composition for boiler tubes is such that this class of metal is especially susceptible to this phenomenon. Mechanical deformation to some degree is unfortunately assured by present-day methods of handling tubes, for at the mill tubes are straightened and in many cases actually brought to final size when below the critical temperature. Tubes are often bent during fabrication and erection and are universally rolled into the tube sheet when cold, and the methods of cleaning tubes in service often employ forms of apparatus which produce local deformation by repeated hammer blows. Finally, the time and temperature conditions required for recrystallization are present.

### Time-Temperature Criterion for Crystal Growth

Heating of deformed metal to temperatures approaching the critical, causes crystal growth in very short time periods. Corresponding growth occurs more slowly at lower temperatures, the time periods required for the same increasing very rapidly as the temperature to which the material is heated decreases. Values experimentally determined for temperatures from 550 deg. cent. (1022 deg. fahr.) to 675 deg. cent. (1247 deg. fahr.) for one set of conditions, are given by the following equations:

$T$ (minutes) = 8       for 675 deg. cent. (1247 deg. fahr.)
$T$ (minutes) = 8 × 3    for 650 deg. cent. (1202 deg. fahr.)
$T$ (minutes) = 8 × 3$^2$ for 625 deg. cent. (1157 deg. fahr.)
$T$ (minutes) = 8 × 3$^3$ for 600 deg. cent. (1112 deg. fahr.)[1]

or, in general, for temperatures below 675 deg. cent. (1247 deg. fahr.)

$$T = 8 \times 3^n$$

where $T$ = time in minutes
     $t$ = temperature in degrees centigrade
     $n$ = $(675 - t)/25$

That the normal method of handling boiler tubes results in mechanical deformation and that this metal as a result, under the proper conditions of time and temperature, will develop large crystals, is shown in Figs. 5 and 6. Fig. 5. is from a photomicrograph of a specimen of a tube which has been so deformed, and Fig. 6 from a specimen after receiving heat treatment at a temperature below the critical which quickly developed grain growth. A comparison of the grain sizes of the two samples indicates that the tube has been sufficiently deformed to respond to the laws of grain growth when the proper conditions for this development are present.

### Physical Tests

Both tension tests and fatigue tests were made to determine the effect of coarse grains on the physical properties of the tubes. In each of the tension tests three samples per set were employed: the first on metal as received, the second on metal annealed for 10 minutes at 950 deg. cent. or 1742 deg. fahr. and then cooled in the furnace and the third on coarse-grained metal produced by stressing all of the test specimens the same amount and in all cases past the elastic limit and followed by an annealing for three hours at a temperature of 800 deg. cent. or 1472 deg. fahr. In the fatigue tests ten samples per set were used because of the greater difficulties in a test of this kind in securing check results.

The physical properties of the "as received" samples showed satisfactory metal. The average tensile-strength and elastic-limit values of the 'coarse-grain" samples" were 27.2 and 58.9 per cent lower than they were for the "as received" samples and indicate, therefore, a decidedly inferior grade of metal. In the fatigue tests, also, the "coarse-grain" metal was 16.2 per cent poorer than the "as received," and 48.70 per cent poorer than the "annealed" metal.

### Composition of Tubes

In view of all of the conditions above pointed out there arose a question as to whether or not the present composition of boiler tubes, from the consumer's standpoint, was the most acceptable. Would there be a composition as easy to make from the producer's viewpoint, as easy to install, as resistant to the absorption of hydrogen, more strong, as free, if not freer, from blowholes, and above all less subject to recrystallization?

This is a formidable set of conditions, and yet do not tubes with a carbon content between 0.30 to 0.35 per cent more perfectly meet all of the above conditions than tubes with a carbon content ranging between 0.08 to 0.18 per cent?

Tubes with this higher carbon range will not be appreciably more difficult to manufacture or to install and there is nothing to indicate that they will absorb hydrogen more readily. It should be possible to make them as free of blowholes; there is no question but that they are at least 40 per cent stronger as measured by

tensile-strength and elastic-limit tests throughout all working temperatures with no detrimental decrease in elongation or reduction; and finally, and most important of all, tubes with the higher carbon range are not subject to recrystallization.

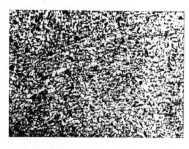

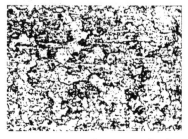

Figs. 5 and 6   Effect of Heat Treatment on the Grain Size of Boiler Tubes; Sections Adjacent to Tube Sheet; Etched in Nitric Acid; Magnification 100; Fig. 5, Untreated; Fig. 6, Heated at 650 Deg. Cent. for 15 Hours

On this last point the literature is suggestive, although there have been no pieces of work as yet published as far as the writer knows which give direct proof.

In view of this condition, therefore, the following test was carried out to ascertain roughly the carbon range in which grain growth on deformed iron[1] when heated at temperatures below the critical range, occurred. Five irons were used with carbon varying from 0.006 to 0.315 per cent. Each of the samples was annealed so as to secure freedom from strains and obtain minimum grain size. This treatment was then followed by impressing a 5-mm. ball on each specimen under a load of 3000 kg. Each sample was then heated for four hours at 675 deg. cent. or 1247 deg. fahr., a temperature considerably below the critical for all of the carbon ranges present. The specimens were then examined and the average grain size in the section undeformed in and in that portion of the deformed area showing maximum grain size compared.

The results given in Table 1 and shown in Fig. 7 indicate that iron in carbon ranges from 0.006 to 0.251 per cent, inclusive, undergoes a perceptible grain growth when treated as just described and that iron with a carbon content of 0.315 per cent is

---

[1] Recrystallization as a Factor in the Failure of Boiler Tubes, White and Wood. Proc. American Society for Testing Materials, vol. xvi, pp. 82-116.

[1] The word "iron" is used in its generic sense and is intended to include steel and what is commonly called ingot iron.

(Continued on page 618)

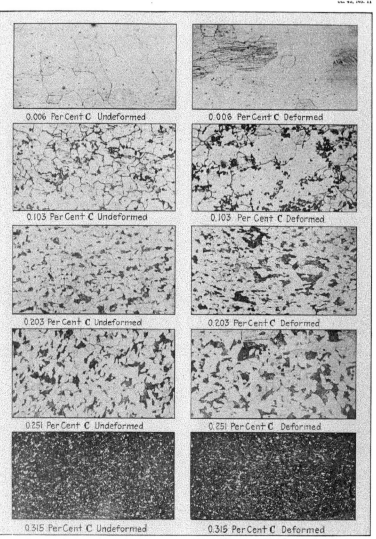

0.006 Per Cent C Undeformed    0.006 Per Cent C Deformed

0.103 Per Cent C Undeformed    0.103 Per Cent C Deformed

0.203 Per Cent C Undeformed    0.203 Per Cent C Deformed

0.251 Per Cent C Undeformed    0.251 Per Cent C Deformed

0.315 Per Cent C Undeformed    0.315 Per Cent C Deformed

FIG. 7  EFFECT OF CARBON CONTENT ON GRAIN GROWTH IN DEFORMED IRON WHEN HEATED BELOW CRITICAL TEMPERATURE;
MAGNIFICATION 100; ETCHED IN NITRIC ACID.  (ACCOMPANYING ARTICLE ON PRECEDING PAGES.)

# Calibration of Nozzles for Measurement of Air Flowing into a Vacuum

By WM. L. DE BAUFRE,[1] LINCOLN, NEB.

*The following paper is based upon an investigation of the flow of an elastic fluid through nozzles having well-rounded entrances, the effects of frictional resistance and of moisture in the atmosphere when the fluid is atmospheric air being taken into account. It is shown that the rate of flow of dry air decreases with increases of moisture, this result being based upon a series of tests on metal nozzles. The paper includes a description of the nozzles, the method of calibrating them, and the results obtained. Formulae employed in the calculations are also derived and the theory involved is outlined.*

**D**URING some tests of air pumps, condensers, etc., conducted at the U. S. Naval Engineering Experiment Station, Annapolis, Md., it was desired to measure quantities of air by means of nozzles, the air flowing from the atmosphere through the nozzle into a vacuum. The problem was therefore to make a number of nozzles having orifices of suitable areas and to calibrate them, rather than to investigate correction factors for nozzles of different shapes. The following paper is based upon a report submitted on this work, and includes not only a description of the nozzles constructed, the method of calibration and the results obtained, but also the formulæ and calculations to be used. The complete paper from which this was abstracted contains a development of the formulæ and the theory involved.

## Description of the Nozzles

Twenty monel-metal nozzle plugs were made of the general shape shown in Fig. 1-A. Monel metal was selected as it is practically non-corrodible and takes a high polish. The plugs were threaded with a 1-in. pipe thread, a standard ring thread gage being used to secure interchangeability. It was intended to use these plugs in a soft brass bushing as shown in Fig. 1-B in order to preserve the threads as long as possible. The threaded joint between plug and bushing was made tight with asphaltum.

Each plug was 1¼ in. long, and the 1-in. pipe thread permitted the use of a maximum diameter of 1 in. for the orifice through the plug. An entrance curve to the orifice was selected as shown in Fig. 1-C, constructed of two arcs approximating an ellipse. This entrance curve takes up three-tenths of the length of the nozzle, leaving in the case of the 1-in. orifice a straight portion seven-tenths of the nozzle length. The cross-section of all nozzles having orifices less than 1-in. nozzle are made geometrically similar to the 1-in. nozzle by counterboring the plug. In all nozzles, therefore, the length of the straight portion is seven-eighths its diameter, and the length of the entrance portion is three-eighths its diameter.

The diameters of the orifices were selected to correspond to standard drill sizes and to give convenient increments in area as given in Table 1. Except for the sizes below ¼ in., they were drilled ¹⁄₆₄ in. smaller in diameter and reamed to size. The entrance surface was rubbed down with emery cloth to templet, giving the nozzle a highly polished appearance.

[EDITOR's NOTE: At this point in the complete paper the author discusses the theory of air flow through nozzles. He first develops equations for frictionless flow when no heat is imparted to or received from the flowing fluid by the nozzle. He next develops equations for flow which is not frictionless and shows that the coefficient of resistance depends upon the diameter and length of the nozzle, and for a given nozzle is therefore constant. He also shows that a constant coefficient of resistance in the relation

$$p_1 v_1{}^n = p v_a{}^n$$

the exponent *n* can be taken as a constant and should evidently be the same for geometrically similar nozzles. The efficiency *N* of the nozzle he defines as the ratio of the kinetic energy actually possessed by

[1] Chairman M. E. Dept., Univ. of Neb., Mem.Am.Soc.M.E.
For presentation at the Annual Meeting, New York, December 7 to 10, 1920, of THE AMERICAN SOCIETY OF MECHANICAL ENGINEERS. The paper is here printed in abstract form and advance copies of the complete paper may be obtained gratis upon application. All papers are subject to revision.

the flowing mass to that theoretically available by adiabatic expansion through the same pressure range. The efficiency is therefore, dependent upon the ratio of expansion $p/p_1$ as well as upon the exponent of the expansion *n*. Equations are developed for calculating rates of flow for all possible values of the expansion ratio $p/p_1$. These equations take the form

$$\frac{G}{F} = \psi \sqrt{p_1 d_1}$$

in which $G$ = rate of flow; $F$ = area of orifice in sq. ft.; $\psi$ = weight-flow factor; $p_1$ = initial pressure; and $d_1$ = initial density. The author also presents additional formulæ and curves for the calculation of the various terms.]

## Description of Apparatus Used

The nozzle to be calibrated was screwed into a brass bushing mounted at $N$ (see Fig. 2) in the side of the sheet-steel tank $A$. This tank was 8 ft. in diameter and 16 ft. high. A special cover or valve was arranged to close the nozzle. This valve, shown in open position in Fig. 3, consisted of a rubber-covered plate mounted

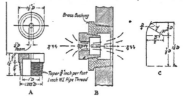

Fig. 1 Details of Nozzle Plugs
A, Cross-section of Nozzle; B, Method of Inserting Nozzle in Tank; C, Construction of Entrance Curve

on trunnions in a flat bar hinged at one end to the tank and having a handle at the other end. With the nozzle plug in place, the two screws at the back of the valve were adjusted so that its rubber-

**TABLE 1 NOZZLE SIZES AND CAPACITIES**

| Nozzle Number | Drill Size | Diam. by Measurement, In. | Capacity, Lb. per Hr. | Capacity, Cu. Ft. per Min. |
|---|---|---|---|---|
| 1 | No. 46 | 0.08537 | 6.79 | 1.51 |
| 2 | No. 33 | 0.11239 | 11.77 | 2.62 |
| 3 | No. 21 | 0.15683 | 22.93 | 5.10 |
| 4 | No. 9 | 0.19653 | 36.01 | 8.01 |
| 5 | No. 1 | 0.22671 | 47.92 | 10.70 |
| 6 | ¼ in. | 0.24944 | 58.00 | 12.90 |
| 7 | 9/32 in. | 0.27721 | 71.63 | 15.90 |
| 8 | 5/16 in. | 0.31527 | 92.65 | 20.60 |
| 9 | 11/32 in. | 0.34385 | 110.20 | 24.50 |
| 10 | 3/8 in. | 0.37490 | 131.00 | 29.10 |
| 11 | 13/32 in. | 0.40600 | 153.70 | 34.20 |
| 12 | 7/16 in. | 0.43822 | 179.00 | 39.80 |
| 13 | 15/32 in. | 0.46755 | 203.80 | 45.30 |
| 14 | 1/2 in. | 0.49895 | 232.10 | 51.60 |
| 15 | 9/16 in. | 0.56823 | 301.00 | 67.00 |
| 16 | 5/8 in. | 0.61981 | 358.10 | 79.70 |
| 17 | 11/16 in. | 0.68733 | 403.90 | 89.90 |
| 18 | 3/4 in. | 0.74950 | 523.70 | 116.50 |
| 19 | 7/8 in. | 0.87595 | 715.20 | 159.10 |
| 20 | 1 in. | 0.99885 | 930.00 | 206.90 |

Note—Capacity based on dry air at 70 deg. fahr. and with an initial pressure of 29.92 in. mercury and a final pressure less than 15 in. mercury.

covered face fitted evenly over the face of the plug. The valve was held tightly closed by a hand nut screwed on a stud in the tank.

The atmospheric pressure was indicated by a mercury barometer hung on a nearby wall, and the vacuum within the tank was measured by a mercury gage $M$ (see Fig. 2). The latter was of the floating-scale type, consisting of a cast-iron float with a brass scale attached thereto. The float rises or falls with the mercury in the cistern so that the zero of the scale always corresponds with the surface of the mercury. The difference between atmospheric pressure and the absolute pressure in the tank was therefore obtained by a single reading—the height of the upper end of the column of mercury in the gage glass, obtained by means of a sliding index. Only one side of the column was used, it being of the duplex type. A thermometer was hung on the column to obtain its temperature.

The temperature within the tank was obtained from the resistance of a copper wire (No. 36 B. & S. Gage) strung from top to bottom of the tank six times and distributed across a diameter as shown at $W$ in Fig. 4. This figure also shows diagrammatically the arrangement of the bridge used to measure the resistance of

FIG. 2 APPARATUS USED FOR TESTS
A, Tank; M, Mercury Gage; N, Nozzle; H, Exhauster

the wire $W$. The leads $L_1$ were of No. 12 copper wire, with compensating leads $L_2$ of the same size and length.

The temperature and humidity of the atmospheric air flowing into the nozzle under calibration were determined from readings of dry- and wet-bulb thermometers placed in the suction duct of a small exhauster $H$ (Fig. 2). The wet bulb was moistened by a cotton wick which dipped into one leg of a U-tube kept filled with water. By raising or lowering this U-tube, the cotton wick could be kept just sufficiently moist around the thermometer bulb, a drop of water occasionally falling from its lower end.

The evacuation of the tank was accomplished by a Westinghouse-Leblanc air ejector connected to the tank at $D$ (Fig. 4). While the tank was being exhausted, valves $C$ and $D$ were open and valve $E$ was closed. When the desired vacuum was attained, valves $C$ and $D$ were closed and valve $E$ was opened to admit water from the vessel $F$ to water-seal valve $D$.

### METHOD OF CONDUCTING TESTS

The nozzle to be calibrated was first mounted in place, the valve adjusted to cover the nozzle squarely and the hand nut screwed down to hold the valve tightly closed. The tank was then exhausted by the ejector to a vacuum of 28 to 29 in. of mercury and the intervening valves closed and water-sealed. The tank was then allowed to stand overnight in this exhausted condition in order to insure that a portion of the water previously put into the tank would have time to evaporate and saturate the interior space. It was found necessary to add water in order to replace the moisture withdrawn with the air by the ejector; and if sufficient time were not allowed for the water to evaporate, the vapor pressure taken from steam tables to correspond to the temperature within the tank would be greater than the actual existing pressure, thus indicating that the space was not saturated with water vapor.

In order to obtain some idea of the weight of water required to be put into the tank, a curve was derived to give the actual volume of air removed at a continuously decreasing pressure according to the relation

$$\frac{v_r}{v_1} = \log_\epsilon \frac{p_1}{p_2} \dots \dots \dots \dots \dots \dots \dots \dots [1]$$

in which $v_r$ = total volume of air removed
$v_1$ = volume of tank
$p_1$ = initial pressure in tank
$p_2$ = final pressure in tank.

The application of this formula may be shown by an example. Thus, if the tank is exhausted from an atmospheric pressure of 30 in. mercury to a vacuum of 29 in. mercury, we shall have

$$\frac{p}{p_1} = \frac{30}{30-29} = 30$$

and from the formula or curve

$$\frac{v_r}{v_1} = 3.41.$$

The volume of the tank as determined by calibration with water was about 865 cu. ft. Hence

$$v_r = 3.41 \times 865 = 2950 \text{ cu. ft.}$$

This will also be the volume of the water vapor removed; and assuming it to be saturated steam at a temperature of 70 deg. fahr., with a corresponding density of 0.001148 lb. per cu. ft. (taken from steam tables), we find 3.4 lb. are withdrawn. Accordingly, about 5 lb. of water was put in the tank before exhausting each time.

Allowing the exhausted tank to stand overnight did not always insure saturation, and before taking readings the nozzle was opened to admit atmospheric air until the absolute pressure within tank doubled; that is, the vacuum dropped from say 29 to 28 in. mercury. The moisture brought in from the atmosphere together with the vapor in the tank insured saturation.

Each calibration run, during which the air nozzle under calibration was open, was preceded and followed by a leakage period. During the initial leakage period, readings of the barometric pres-

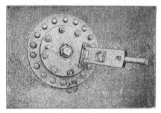

FIG. 3 SPECIAL VALVE FOR CLOSING NOZZLE

sure, the vacuum within the tank, and the temperature within the tank were taken at intervals of 3 min. for a duration of 30 min. Then the nozzle was quickly opened at a given second to start the run and readings of the barometer, mercury column and wet-

TABLE 2 SUMMARY OF RESULTS

| Nozzle Number | Nozzle Diam. In. | Number of Runs | Average Weight-Flow Factor | Average Expansion Exponent | Average Weight-Flow Efficiency Per cent |
|---|---|---|---|---|---|
| 1 | 0.08537 | 9 | 3.646 | 1.349 | 93.86 |
| 2 | 0.11239 | 6 | 3.786 | 1.378 | 97.48 |
| 3 | 0.15683 | 12 | 3.770 | 1.375 | 97.08 |
| 4 | 0.19653 | 4 | 3.824 | 1.387 | 98.46 |
| 5 | 0.22671 | 5 | 3.853 | 1.393 | 99.20 |
| 6 | 0.24944 | 2 | 3.768 | 1.375 | 97.02 |
| 7 | 0.27721 | 4 | 3.834 | 1.389 | 98.72 |
| 8 | 0.31527 | 2 | 3.748 | 1.370 | 96.49 |
| 9 | 0.34385 | 4 | 3.791 | 1.380 | 97.61 |
| 10 | 0.37490 | 2 | 3.776 | 1.376 | 97.22 |
| 11 | 0.40600 | 2 | 3.779 | 1.377 | 97.28 |
| 12 | 0.43822 | 4 | 3.767 | 1.374 | 96.98 |
| 13 | 0.46755 | 2 | 3.774 | 1.376 | 97.16 |
| 14 | 0.49895 | 2 | 3.781 | 1.377 | 97.35 |
| 15 | 0.56823 | 2 | 3.759 | 1.373 | 96.77 |
| 16 | 0.61961 | 2 | 3.831 | 1.388 | 98.64 |
| 17 | 0.68735 | 2 | 3.791 | 1.379 | 97.61 |
| 18 | 0.74950 | 1 | 3.636 | 1.346 | 93.61 |
| 19 | 0.87595 | 1 | 3.790 | 1.379 | 97.58 |
| 20 | 0.99885 | 1 | 3.728 | 1.366 | 95.98 |

and dry-bulb thermometers taken at intervals of 30 sec. The run was ended by quickly closing the nozzle a certain number of minutes after opening it, the length of the run depending upon the size of the nozzle. A short run was necessary with the larger nozzles in order to prevent the pressure within the tank rising above the critical value. The final leakage period began at the time the nozzle was closed and continued for 30 min.

TABLE 3　VALUES OF WEIGHT-FLOW FACTOR FOR VARIOUS UNITS

| Weight Flow $G$ | Orifice Area $F$ | Initial Pressure $p$ | Initial Density $d$ | Numerical Value of Factor $\psi$ |
|---|---|---|---|---|
| lb. per sec. | sq. ft. | lb. per sq. ft. | lb. per cu. ft. | 3.771 |
| lb. per sec. | sq. in. | lb. per sq. in. | lb. per cu. ft. | 0.3143 |
| lb. per sec. | sq. in. | inches mercury | lb. per cu. ft. | 0.2203 |
| lb. per hr. | sq. ft. | lb. per sq. ft. | lb. per cu. ft. | 13,576 |
| lb. per hr. | sq. in. | lb. per sq. in. | lb. per cu. ft. | 1,131.3 |
| lb. per hr. | sq. in. | inches mercury | lb. per cu. ft. | 792.9 |
| kg. per hr. | sq. cm. | mm. mercury | kg. per cu. m. | 11.061 |

TABLE 4　VALUES OF WEIGHT-FLOW FACTOR FOR VARIOUS UNITS

| Weight Flow $G$ | Orifice Area $F$ | Initial Pressure $p$ | Initial Temperature $t$ | Numerical Value of Factor $\psi$ |
|---|---|---|---|---|
| lb. per sec. | sq. ft. | lb. per sq. ft. | deg. fahr. | 3.771 |
| lb. per sec. | sq. in. | lb. per sq. in. | deg. fahr. | 3.771 |
| lb. per sec. | sq. in. | inches mercury | deg. fahr. | 1.852 |
| lb. per hr. | sq. ft. | lb. per sq. ft. | deg. fahr. | 13,576 |
| lb. per hr. | sq. in. | lb. per sq. in. | deg. fahr. | 13,576 |
| lb. per hr. | sq. in. | inches mercury | deg. fahr. | 6,668 |
| kg. per hr. | sq. cm. | mm. mercury | deg. cent. | 18.458 |

### CALCULATION OF RESULTS

The details of the method employed in correcting the data and deducing the results are illustrated and explained in the Appendix of the complete paper for Run 1 of Nozzle 1. To obtain the dry-air density in the tank the temperature and pressure observations for the leakage intervals preceding and following each run were plotted as shown in Fig. 5 and the calculations based on the curves rather than directly upon the observations. A curve was also plotted of weight-flow factor $\psi$ to expansion exponent $n$ so as to enable item 25 being obtained from item 24, and in Table 2 there have been tabulated for each nozzle the average values of the weight-flow factor, the expansion exponent, and the weight-flow efficiency. The number of runs made on each nozzle is also given. It will be noted that the calculations were based upon the pressure and temperature within the tank before the nozzle was opened and after it was closed, when the conditions were static, rather than upon the pressure and temperature, while the air was flowing into the tank.

Assuming all the individual runs to be of equal weight in ap-

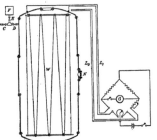

FIG. 4　DIAGRAM SHOWING METHOD EMPLOYED TO MEASURE TEMPERATURE WITHIN TANK

plying the formula for the probable error of the mean value, we obtain the expansion exponent $n = 1.3747 \pm 0.0013$, and assuming the average values in Table 2 for the several nozzles to be of equal weight, we obtain $n = 1.3753 \pm 0.0018$. The probable error for a single run and for a single nozzle are, respectively, $\pm 0.0107$ and $\pm 0.0078$. That is, for carefully made nozzles geometrically similar to those tested, the value of $n$ will probably be within 0.010 of 1.375 if the observations are taken as carefully as during this calibration test. The corresponding value and accuracy of the weight-flow efficiency are 97.1 $\pm$ 1.2 per cent.

Corresponding to the expansion exponent $n = 1.375$, the numerical value of the weight-flow factor $\psi$ in the formula

$$\frac{G}{F} = \psi \sqrt{pd}$$

will depend upon the units in which the several quantities $G$, $F$, $p$ and $d$ are expressed, as indicated in Table 3.

For final pressures greater than the critical pressure, that is, $p/p_1$ greater than 0.533, the factor will be reduced as given by the solid curve Fig. 6, for the units in the sixth line of Table 3.

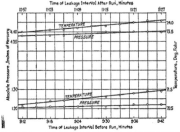

FIG. 5　CURVES SHOWING ABSOLUTE PRESSURE AND TEMPERATURE FOR LEAKAGE INTERVALS BOTH BEFORE AND AFTER RUN NO. 1 ON NOZZLE NO. 1

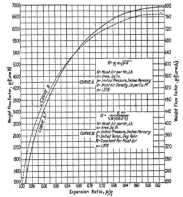

FIG. 6　CURVES OF WEIGHT-FLOW FACTOR WITH EXPANSION RATIO FOR $n = 1.375$

Changing the above formula into the form

$$\frac{G}{F} = \frac{\psi}{\sqrt{R}} \frac{p}{\sqrt{458.5 + t}}$$

also changes in certain cases the numerical value of $\psi$ for the same units of $G$, $F$, $p$ and $t$, as shown in Table 4.

For pressure ratios greater than 0.533, the dotted curve in Fig. 6 gives the reduced values of $\psi$ corresponding to the units in the sixth line of Table 4.

The value of the constant $R$ depends upon the ratio of water vapor to dry air in the fluid flowing through the nozzle and the

(Continued on page 650)

# Experiences with Large Center-Crank Shafts

By LOUIS ILLMER,[1] PHILADELPHIA, PA.

*This paper is descriptive of some disastrous experiences with a number of large gas-engine shafts of the center-crank type. The shafts were mounted upon three-point bearing supports and carried a heavy flywheel between the intermediate and outboard bearings. The stress diagrams show that this mode of support is likely to set up a pernicious interaction of bearing load, culminating in excessive wear in the intermediate main bearing.*

*When the wheel shaft lacks adequate stiffness, the appreciable lifting action at the free end of the web portion of a center-crank shaft may exert a considerable thrust against the cap of the aligned outer main bearing, which in turn reacts upon the intermediate journal, causing it to become overloaded. The resulting rapid wear in the intermediate main bearing reduces the upward cap thrust and thus drop in alignment gradually relieves the intermediate bearing of overload. Further wear causes a portion of the downward load on the intermediate bearing to be transferred to the outer main bearing. It was found that when this readjustment is complete, the downward load upon the two main bearings becomes approximately equalized. The considerable sag required to attain this state of equilibrium as to wear, involved running under stress conditions so severe as to lead to ultimate breakdown of the wheel shaft.*

*In order to keep the wheel-shaft stress within desired limits under load conditions found in these engines, it finally became necessary to enlarge the intermediate main journal to ³/₄ of its original diameter, thus making its dimensions fully as large as required for side-crank construction. It was found to be expedient to make a careful analysis of the underlying stress conditions so as to insure a proper margin of safety in both the web and wheel parts after allowing for the lengthening of the span between the wheel supports due to excess wear in the intermediate main bearing.*

SEVERAL years ago a series of disastrous center-crank shaft failures occurred in a 1200-kw. gas-power plant which the author investigated and in which he found that the wheel shafts were entirely too light for the load they had to carry.

The gas-power plant comprised three 25-in. by 43-in. 500-b.hp. double-acting producer-gas engine units, each driving a 400-kw. 25-cycle three-phase alternator at 100 r.p.m. The engines were of the direct-connected horizontal type, provided with a massive flywheel. The shafts were of the single-throw center-crank type, as indicated in Fig. 1, having a tensile strength of about 65,000 lb. per sq. in.

After being in 24-hour-a-day service for some four or five years, one of the shafts unexpectedly gave way near the flywheel hub, allowing the armature to drop and thus completely wreck its generator and heavy wheel, and in other ways doing serious damage.

In order that the cause of this and similar shaft failures might be intelligently arrived at, a series of sag determinations was conducted on the wheel shaft and a thorough technical investigation made of the underlying stress conditions.

The three-point shaft support for the heavy flywheel was found to set up a pernicious interaction of load which led to excessive wear in the intermediate main bearing. This resulted in the lengthening of the effective shaft span, which in turn produced sag stresses of such magnitude as to cause ultimate rupture of the shaft.

The inadequacy of a center-crank shaft for carrying a heavy wheel is made evident in the accompanying deflection diagrams. An excessive wheel-shaft deflection puts a large upward thrust against the outermost main bearing cap. The reaction of this thrust upon the intermediate main bearing may become so large as to squeeze the babbitt lining out of the bearing shell, in which event the resulting drop in this vital support allows the shaft to run with an excessive sag between this and the outboard bearing. Thus the shaft stresses may be increased far beyond those usually anticipated by the engine builder, and if allowed to continue can readily cause the metal of this shaft to undergo such rapid fatigue as to lead to shaft rupture.

The matters of fixed stress limits and requisite factor of safety underlying good shaft design are rather fully discussed in the Appendix accompanying the complete paper.

## CRANKSHAFT STRESS DIAGRAMS

The shaft outline and load conditions under which the original shafts operated are shown in Fig. 1, in which the line *A-A* indicates the location of the wheel-shaft failure, its relative position being almost identical in three of these shafts. The journal portion of the shaft was carried straight up to the wheel hub but as indicated by the dotted outline in Fig. 1, it was entirely feasible to enlarge the shaft immediately after clearing the camshaft gear. Had this been done originally, it would have obviated the worst of the peak stresses shown in the diagram.

The original shafts were excessively loaded by a massive flywheel weighing about 73,000 lb., and by an armature of about 13,000 lb., to which should be added an assumed magnetic pull of about 10,000 lb., on the basis of ¹/₁₆ in. armature displacement. The location of those weights is shown in Fig. 1, and their relative magnitude is indicated by varying line lengths, extending downward.

In making a preliminary stress analysis, in the conventional manner, it may be assumed that the wheel shaft is pin-supported between the intermediate main bearing I and the outboard bearing III. The resultant wheel-shaft load $P_r = 96,000$ lb., acting downward, causes the shaft to deflect, thus putting the bottom fibers in tension and the top fibers in compression. The rotation of the shaft sets up a reversal of stress with each revolution.

In addition to these alternating stresses, the shaft is subject to a negligible shear stress due principally to the weight of the wheel. In a horizontal engine the effect of the connecting-rod thrust in producing an increase in the wheel-shaft stress may be sufficiently allowed for by taking into account only the twisting stress resulting from the transmission of power to the generator.

The effect of this twisting action may best be combined with the bending moment as indicated in the stress curve in Fig. 1. It will be seen that while the original wheel-shaft strength lacks uniformity, the maximum stress estimated on this preliminary basis reaches a peak value of only about 10,000 lb. per sq. in. near the point of rupture.

Assuming the wheel to be perfectly balanced and the shaft to run true, the basic or net factor of safety for this shaft, as determined by the Appendix, appears to be equal to about 2.2, as against a minimum stipulated factor of three. It is evident that this minor difference is insufficient to account for the wheel-shaft failure, but the diagram does show that this conventional mode of checking center-crank wheel-shaft stresses cannot be relied upon to bring out the cause of rupture.

## EFFECT OF THREE-POINT SHAFT SUPPORT

In order to arrive at the cause of the present shaft failure, it is necessary to take into account the interacting effect of the three-point shaft support. Assuming all the bearings to be in perfect alignment, it will then be found that the excessive wheel-shaft deflection produced by the massive flywheel, puts a 20,000-lb. upward thrust against the cap of the outermost main bearing II, which in turn reacts upon and so overloads the intermediate bearing as to squeeze out its babbitt. As indicated in Fig. 2, the drop in this vital support causes sag stresses to be set up that more than double the peak stress previously determined by the preliminary stress check represented in Fig. 1.

The estimated stress relations shown in Fig. 1 assume the original wheel shaft to be pin-supported between bearings I and III only, a condition that is approximated when the cap is removed from the outer main bearing II.

The cap thrust of 20,000 lb. reacts to effect an increase in the downward load of the intermediate bearing. With the cap II forced into place under conditions of perfect alignment, the resultant downward pressure in the main bearing I is approximately 400 lb. per sq. in., and if to this the downward centrifugal force is added, a maximum pressure of about 500 lb. per sq. in. of net pro-

1 Oil Eng. Expert, Southwark Foundry & Machine Co., Mem. Am. Soc. M.E.
For presentation at the Annual Meeting, New York, December 7 to 10, 1920, of THE AMERICAN SOCIETY OF MECHANICAL ENGINEERS. The paper is here printed in abstract form and advance copies of the complete paper may be obtained gratis upon application. All papers are subject to revision.

jected area is reached. Such a heavy bearing pressure fails to distribute uniformly and the consequence of excessive loading is to

load by lowering its alignment with respect to the end supports of the shaft.

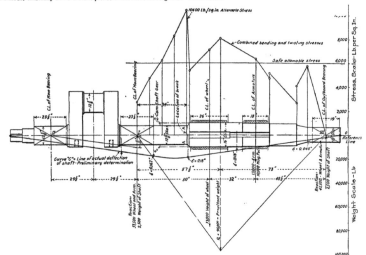

FIG. 1   PRELIMINARY STRESS DIAGRAM FOR PIN-SUPPORTED WHEEL SHAFT

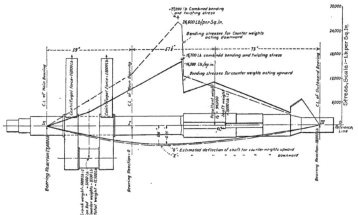

FIG. 2   DIAGRAM FOR SHAFT WORN TO STATE OF EQUILIBRIUM

squeeze the babbitt out of the intermediate shell. The resulting rapid wear serves to relieve the intermediate bearing of its over-

The dropping of the intermediate bearing support has the further effect of reducing the upward thrust against the bearing cap II,

and as the downward forces undergo readjustment by continued wear, the outer main bearing II will gradually assume an increased portion of the downward shaft load. Experience shows that when the load upon each of the main bearings becomes approximately equal, a condition of stable equilibrium will have been established, after which the two main bearings continue to wear down together at a more or less uniform rate.

In order to attain such a state of equilibrium, the shafts must suffer undue deflection as indicated in Fig. 2. Experience further shows that when starting with newly aligned bearings this condition is reached in a remarkably short time, being approximately 30 days of continuous operation under the existing load conditions. The corresponding wear in the intermediate bearing was found to be about $1/8$ in. in this brief period, as against a normal wear of say $1/64$ in. to $1/32$ in. per year for continuous running.

Fig. 2 also shows the marked effect which the dropping of the intermediate bearing support has in enlarging the effective shaft span and thus largely increasing the bending stresses.

These high sag stresses have been checked by comparing the estimated deflection curve $E$ with the actual wheel-shaft sag as taken off one of the original shafts. The deflections were determined by means of a pin gage and these measurements were corrected for the sag of the reference cord. Thus the actual shaft sag, measured at the center of the intermediate main bearing, was

found to be about 0.16 in. lower than the end supports, as against an estimated value of about 0.14 in.

The calculations show further that after reaching the aforesaid state of equilibrium, the estimated maximum pressure acting in the main intermediate bearing is reduced to about 218 lb. per sq. in. of projected area, which can readily be sustained without excessive wear. Under these conditions the shaft suffers a maximum bending stress of about 26,600 lb. per sq. in., and it is found that this peak practically coincides with the point of rupture. This extremely high stress is still further increased when combined with the crank twisting stress.

### DETRIMENTAL EFFECT OF HEAVY COUNTERWEIGHTS

The massive counterweights with which these engines were provided were designed to balance the inertia effects of the heavy reciprocating parts, averaging about 115 lb. per sq. in. of piston area. When the centrifugal force acts upward, a maximum bending stress of 16,300 lb. per sq. in. results, while for downward action the bending stress is increased to about 26,600 lb. per sq. in. Combining these high stresses with the maximum twisting stress without allowing for shock or preignition, an equivalent maximum tension stress of 27,000 lb. per sq. in. is produced in the outer shaft fiber, and when this same fiber turns through 180 deg., it is subjected to a compressive stress of about 16,700 lb. per sq. in.

The corresponding ratio of $f_{min}$ to $f_{max}$ for this partially reversing stress is about 0.62 as defined in the Appendix, and the total stress range that this surface fiber undergoes during each revolution is about 43,700 lb. per sq. in., a limit far in excess of that which the metal can continuously endure. Checking this result on the basis of the formulæ given in the Appendix, the allowable stress range for unit basic factor of safety for 65,000-lb. T.S. steel is about 30,000 lb. per sq. in., which corresponds to a tension stress limit of about $f_{max} = 24,000$ lb. per sq. in. The net or basic factor of safety in the original wheel shaft is thus found to be about 0.9, i.e., less than unity, for which the corresponding excess ten-

sion stress becomes equal to about $f_{exc} = 27,000 - 24,000 = 3,000$ lb. per sq. in.

No record has been kept as to the running time actually required to rupture the original shafts, but judging from the average period of operation of these engines, it is probable that each of the defective shafts made between 100,000,000 and 150,000,000 revolutions prior to failure.

As based upon Stromeyer's fatigue formula as given in the Appendix, the excess stress $f_{exc}$ required to produce rupture in the stated number of revolutions would lie between 2700 and 3000 lb. per sq. in. for an alternating-stress cycle. Accordingly, it would appear that the stress curves shown in Fig. 2 have been closely estimated and that the original shafts must have been subjected to a maximum tension stress approaching the elastic limit of the shaft material.

The character of the ruptured section of these shafts is shown in Fig. 3. The high alternating stresses set up in the outermost fibers undoubtedly produced gradual fatigue, which led to a rupture of the surface metal. This irreparable damage greatly weakened the shaft, and as the rupture crept inward, a point was finally reached where the shaft could no longer bear its load.

The fact that three of the original shafts became defective in practically the same place and that each shaft proved serviceable for a period approximating three years virtually precludes a defect in forging but points rather to improper shaft design as the cause of the mishap. In fact, it now appears that the ultimate failure could have been foretold had the underlying stress relations been analyzed in the manner advocated before deciding upon the final shaft dimensions.

## Transporting Freight in Interchangeable Containers

A long tep toward relieving the tense situation in the transportation lines of the country appears to have been made by the River and Rail Transportation Company of St. Louis, Mo. · This company has developed an interchangeable metal container, termed "Trinity Freight Unit," for facilitating the transfer of goods at terminals. *Railway Age* for September 24 describes this container system of freight transportation. Merchandise is placed in the container at a manufacturing plant or in a warehouse and then, locked and sealed, the container is transported by motor truck to a railroad, or to a waterway, where it is transferred to a flat car or to a boat without rehandling the materials.

The complete unit includes a specially constructed flat car, which differs from those now in use on the railroads of the United States only in providing a means of clamping the containers fast to the platform of the car. Estimates, based on present prices, place the cost of equipping a car with five 10-ton units at $1000. A flat car having these special features may be built for the same cost as that required for an ordinary flat car, or existing flat cars may be altered to carry the containers at an approximate cost of $250.

Containers made in a number of different ways are provided for carrying various kinds of material. A type is designed with side-opening doors for package freight, another with top doors and drop bottom for loose bulk freight, and there are other types for refrigerator service and for carrying liquids.

The units are made in capacities of $2^1/2$ tons and 10 tons each and are proportioned so that five 10-ton units, or twenty $2^1/2$-ton units, or several units of both capacities, and for any of the different classes of freight, can be carried on a flat car of 50 tons capacity. They are rectangular in form and are substantially constructed of steel plate rigidly reinforced with angle irons to withstand the strains due to the weight of the contents and the transferring of the unit from one vehicle to another.

The system has already been tried and proved successful. The United States Railroad Administration adopted it to facilitate water-rail shipments of war supplies. Twenty of the 10-ton package units were built and gave satisfactory service in the New Orleans district in handling package freight. Estimates based on the performance of these units indicate that the handling of freight is not only greatly expedited but can be handled at from 300 to 400 per cent less cost than with any of the present methods.

# Tests on Rear-Axle Worm Drives for Trucks

By KALMAN HEINDLHOFER,[1] PHILADELPHIA, PA.

*In this article is recorded a series of tests on worm drives used in the rear axles of motor trucks to determine the efficiencies under load variations. Efficiency curves showing results with four sets of gears in mesh are included. The effect of the oil temperature in the worm thrust bearing, on the efficiency of the drive is shown, higher efficiencies being derived at higher oil temperatures. The author also presents the record of life of four case-hardened steel worms that failed under load.*

THE various tests on rear-axle worm drives which are to be described were conducted by the author during an investigation of ball-bearing applications. Important data and information were obtained concerning the efficiency of two axles of the same capacity and similar design, the temperature of the worm thrust bearings under full-load conditions, and the life of four worms which broke under load. The testing method employed was that developed by Prof. C. M. Allen and F. W. Roys and described in their paper *Efficiency of Gear Drives.*[2]

Figs. 1 and 2 show the test arrangement, which is represented diagrammatically in Fig. 3. The worm gears to be tested were

and the middle knife edge being supported on the platform scales. A counterweight and rider are mounted on the lever.—EDITOR.]

It will be evident from Fig. 3 that for 100 per cent efficiency the system would be in equilibrium, but since the efficiency is never 100 per cent, the reactive force of the motor will exceed the force of the dynamometer and balance may be restored by shifting the position of the rider weight. The displacement of the rider weight is therefore a measurement of the change of moment, and is a measurement of the power loss.

The motor and transmission, as well as the transmission and worm, were connected by flexible couplings to take care of possible misalignment.

The oil used in the rear axle was Mobil Oil C, and in the transmission, medium mineral oil. The viscosity curves of both oils determined in a Saybolt viscosimeter, are shown in Fig. 4, the vertical scale giving the time, in seconds, required for 60 cc. of oil to pass through a standard orifice.

In order to determine the efficiency of the worm drive the trans-

FIG. 1 FRONT VIEW OF AXLE-TESTING DEVICE     FIG. 2 REAR VIEW OF AXLE-TESTING DEVICE

mounted in a rear-axle housing and driven by an electric induction motor running 1200 r.p.m. under no load and 1160 r.p.m. under full load. It was possible to change the speed by a transmission having gear ratios of 5.93 : 1, 3.24 : 1, 1.74 : 1 and 1 : 1. The power transmitted was absorbed by an Alden dynamometer, which offered a very smooth and readily adjustable torque resistance. To avoid losses in the differential and the use of an extra dynamometer, the differential was locked. At each end the frame of the driving motor was supported by three disk rollers, which allowed the housing to swing on its supports without friction.

[The testing device, as described in the paper above referred to, permits the reactive force of the motor and the force of the dynamometer to act downward through arms of equal length, at the ends of which are knife edges. A lever with three knife edges mounted upon it, has the two outer knife edges adjusted so that the distance between them is equal to the distance (horizontal) between the dynamometer knife edges. The third knife edge divides this distance into segments whose ratio to each other is the same as the gear ratio. The knife edges of the motor and dynamometer are connected to the outer knife edges of the lever, the high-speed motor being coupled to the long arm of the lever

mission efficiency was determined in a separate test, using a similar apparatus to that shown in Fig. 3. The worm efficiency was obtained by subtracting the transmission efficiency from the total.

## EFFICIENCY TESTS

The first two tests were of the rear axles for a 3-ton truck, designated as Nos. 1 and 2, their engines having a rated capacity of 50 hp. at 1000 r.p.m., the worm ratio being $9\frac{1}{2}$ to 1. The efficiency of worm drive and transmission was determined in first, second and third gears, readings being taken at $\frac{1}{3}$, $\frac{2}{3}$ and full load and $\frac{1}{3}$ overload.

The worm is made of low- (0.2 per cent) carbon steel, no nickel, case-hardened and ground, and has four right-hand threads, circular pitch 1.1562, pitch 3.265, lead angle 24 deg. 20 min. The gear is made of phosphor bronze. The worms in the two axles are supported on ball and roller bearings, respectively.

Test results are plotted in Figs. 5 and 6.

In order to ascertain the effect of severe and enduring service on the efficiency, readings were taken on axle No. 1 after running a total of 110 hr. in low gear (196 r.p.m. of worm) and 49 hp. The efficiency reading was taken under the same condition and found to have improved 1 to $1\frac{1}{2}$ per cent.

The efficiencies of the transmission are shown in Fig. 7, while the net efficiencies of the worm are shown in Fig. 8.

It is remarkable that the efficiency of the transmission (Fig. 7) is the lowest in direct drive, that is, when the gears are running

[1] SKF Research Laboratory.
[2] Trans. Am. Soc. M.E., vol. 40, p. 101.

For presentation at the Annual Meeting, New York, December 7 to 10, 1920, of THE AMERICAN SOCIETY OF MECHANICAL ENGINEERS. The paper is here printed in abstract form and advance copies of the complete paper may be obtained gratis upon application. All papers are subject to revision.

idle. This is due to the power absorbed in churning the oil at high speed, in which case the viscous forces opposing motion are high. Oil temperature affects efficiency considerably. In case of the Mobil Oil C, higher temperatures in the rear axle housing improve the efficiency. A heavy oil in winter at light loads and high worm speeds, means low efficiency. The effect of temperature in the beginning disturbed the tests until it was found that by starting with heavy loads and finishing with light loads, the temperature could be maintained constant to a satisfactory degree, and thus consistent readings could be obtained. The small figures alongside the curves indicate the oil temperatures in the worm housing.

In connection with the test on axle No. 2, the author wishes to point out that the efficiency in direct drive at low input would have

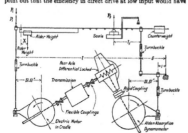

FIG. 3 DIAGRAM OF AXLE-TESTING DEVICE

been somewhat better if a higher oil temperature had existed. These low temperatures were due to precautions being taken to avoid overheating at the end of the test.

### TEMPERATURE OF THE WORM THRUST BEARINGS

On account of the high worm loads carried by the worm thrust bearings, information on the bearing temperature was considered valuable.

To determine maximum temperature conditions, first (low)

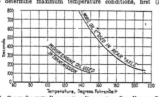

FIG. 4 CURVE SHOWING VARIATION IN VISCOSITY OF OILS USED IN REAR AXLE AND TRANSMISSION

gear and full load was considered the most severe. Under this condition axle No. 1 stood up very well, and the maximum bearing temperature was 185 deg. fahr. with an oil temperature in the housing of 275 deg. fahr. after 5 hr. continuous running under an input of 49 hp. Axle No. 2 under same speed and practically same load gave a bearing temperature of 572 deg. fahr. with an oil temperature in the housing of 400 deg. fahr. after 90 min. continuous running under 52 hp. Due to the excessive heat the worm and the thrust bearing were completely destroyed.

It was then decided to modify the test for axle No. 2 to resemble actual conditions such as are encountered by a loaded truck in snow or on a sandy road, with subsequent hill climbing. Due to the damage done in the preceding test, the worm and thrust bearing were replaced by new ones. Curves plotted in Fig. 9 show the temperature rise with the time. The thrust bearing of axle No. 2 shows a final temperature slightly over 200 deg. fahr. in high gear, full speed and 52 hp. In a comparatively short

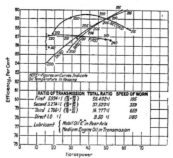

FIG. 5 EFFICIENCY CURVES OF No. 1 REAR-AXLE WORM DRIVE AND TRANSMISSION AT DIFFERENT GEARS

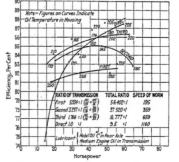

FIG. 6 EFFICIENCY CURVES OF No. 2 REAR-AXLE WORM DRIVE AND TRANSMISSION AT DIFFERENT GEARS

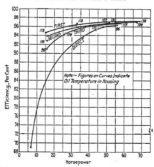

FIG. 7 EFFICIENCY CURVES OF TRUCK TRANSMISSION AT DIFFERENT GEARS

TABLE 1   TABLE SHOWING RESULTS OF FATIGUE TESTS ON WORMS

| | A | B | C | D |
|---|---|---|---|---|
| 1 Worm ratio.................. | 9⅓ : 1 | 9⅓ : 1 | 9⅓ : 1 | 8⅔ : 1 |
| 2 Hp. on motor............... | 49 | 46.8 | 41.2 | 42.2 |
| 3 Hp. transmitted to worm wheel................... | 46.5 | 44.5 | 39 | 40 |
| 4 R.p.m. of worm.............. | 196 | 196 | 196 | 196 |
| 5 Net effective torque, in-lb..... | 14,900 | 14,300 | 12,500 | 12,800 |
| 6 Pitch radius................. | 1.63 | 1.37 | 1.37 | 1.39 |
| 7 Angle of lead............... | 24°20′ | 30°26′ | 30°26′ | 32°12′ |
| 8 Angle of pressure........... | 30° | 27°30′ | 27°30′ | 27°15′ |
| 9 No. of threads.............. | 4 | 4 | 4 | 4 |
| 10 Tangential component of tooth pressure, lb............ | 9,150 | 10,400 | 9,150 | 9,230 |
| 11 Vertical component of tooth pressure, lb............ | 11,700 | 9,200 | 8,100 | 7,550 |
| 12 Item 11 less a percentage for friction loss........... | Less 10% | Less 12% | Less 15% | Less 11% |
| 13 Resultant radial load, lb..... | 10,500 10,500 | 8,100 | 6,900 | 6,700 |
| 14 Worm thrust, lb........... | 14,000 | 13,100 | 11,500 | 11,300 |
| 15 Half span, in.............. | 18,000 | 15,500 | 13,200 | 13,300 |
| 16 Max. bending moment, in-lb. | 6.2 | 7.2 | 7.2 | 6.6 |
| 17 Sectional area of core, sq. in. | 68,300 | 62,600 | 54,300 | 51,000 |
| 18 Life of worm, hr........... | 4.6 | ~2.9 | 2.9 | 2.6 |
| | 118 hr. | 46 min. | 1 hr 30 min. | 8 hr. 40 min. |
| 19 Life of worm, total no. revolutions.................. | 1,390,000 | 9,000 | 17,600 | 102,000 |
| 20 Remarks.................. | Thread broken off; core unbroken | Core broken at center | Core broken at center | Core broken at center |

time the heat in the thrust bearing became excessive. The break in the curve indicates the period of stop needed for changing gears, knife edges on the balance beam, etc. The slightly higher horsepower input in case of axle No. 2 was due to the difficulty in predetermining the correct scale weights.

## FATIGUE TESTS ON WORMS

The running tests conducted under uniform speed and load

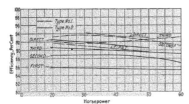

FIG. 8   EFFICIENCY CURVES OF REAR-AXLE WORM DRIVE AT DIFFERENT GEARS

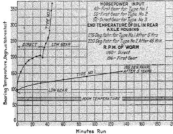

FIG. 9   TEMPERATURE CURVES OF THE WORM THRUST BEARINGS

offered the opportunity to record the life of four worms, three of which were different in design. All these worms broke after a certain number of revolutions under a heavy load. The load, however, did not exceed the maximum which the axle was supposed to carry in low gear.

All worms tested were made of about 0.2 per cent carbon steel. They contained no nickel and were case-hardened, the working surfaces being ground. The test arrangement was the same as shown in Figs. 1 and 2, enabling the input to be accurately determined. During each test the horsepower was kept practically constant.

For the benefit of rear-axle designers a few important endurance data are tabulated in Table 1. Worm designated A is the one whose efficiency is shown in Fig. 5. B and C are identical in design, but different from A. D has a different worm ratio than A, B or C.

Fig. 10 gives an idea of the worm fractures. In case of worms

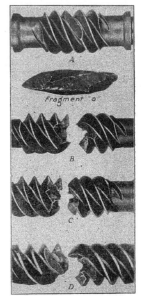

FIG. 10   WORMS THAT FAILED ON FATIGUE TEST

B, C and D their life could probably have been increased considerably by avoiding sharp corners at the root of the thread. This seems to be a small matter, yet it is of great importance. It is well known from theoretical and experimental investigations that sharp corners tend to concentrate or localize stresses. In case of static loads and ductile material this concentration of stresses is much reduced after the elastic limit is transgressed, which is brought about by the permanent stretching of the metal. The stresses are thus redistributed and will be shifted to other parts originally stressed to a smaller extent.

The case, however, is different with stresses often repeated or reversed and failure may occur even within the elastic limit. The relieving of stresses at the peaks will not take place and the corners will form starting points for cracks, which, when once started, will rapidly progress until the section is so weakened that a sudden break occurs. Thus, it is important, that a radius as large as possible be specified by the designer and checked by the inspector.

# Effect of Fittings on Flow of Fluids Through Pipe Lines

By DEAN E. FOSTER,[1] TULSA, OKLA.

IN the layout of piping in power plants, heating systems, re- fineries, and other projects, the engineer is frequently confronted with the problem of making proper allowance for the resistance offered by fittings to the flow of liquids and gases. A number of formulæ and tables have been published giving such information for globe valves and elbows, but none for fittings in general. Konrad Meier, Mem.Am.Soc.M.E., in his excellent treatise en- titled, The Mechanics of Heating and Ventilation, has given the subject of the effect of fittings on resistance to flow a very care- ful study and has presented data thereon in the form of seven charts. In his work of the past several years the writer has found these charts to be very useful in the design of piping for uniform distribution of liquids and gases and in the selection of equip- ment for producing the necessary force for moving fluids, the only objection being the time required when using them. After some study of the equations used in their construction, however, it was found to be a comparatively easy matter to formulate the following relations between the frictional resistance of pipes of various sizes and the various screwed fittings commonly used, and from these relations to establish tables showing the lengths of pipe equivalent in resistance to that offered by fittings.

For the flow of water through pipe lines Mr. Meier uses the formula

$$H_f = \frac{0.0257}{2g} \times v^{1.86} \times \frac{L}{d^{1.25}}$$

where $H_f$ = loss of pressure in ft.
 $v$ = velocity in ft. per sec.
 $d$ = diameter in in.
 $L$ = length in ft.

This equation agrees within 5 per cent with Gardner S. Williams' formula, which is so widely used in this country.

For resistance offered to flow by fittings, Mr. Meier gives the following:

$$H_r = 1.38r \times \frac{v^{1.86}}{2g}$$

where $H_r$ = ft. of pressure loss per fitting
 $r$ = a factor of resistance whose value depends upon the shape of the fitting
 $v$ = velocity of the fluid in ft. per sec.

By combining these two equations we obtain

$$L_e = 53.75r \times d^{1.25}$$

where $L_e$ represents the number of feet of pipe equivalent in re- sistance to a fitting and the other factors as before. Table 1 has been made up from this equation and applies to water, crude or refined petroleum oils of .32 deg. B. gravity or lighter, or other non-viscous liquids.

For the flow of high-pressure steam through pipes Mr. Meier uses the equation

$$P_f = \frac{W}{144} \times 0.0257 \times \frac{v^{1.86}}{2g} \times \frac{L}{d^{1.25}}$$

where $P_f$ = pressure in lb. per sq. in. to overcome frictional loss.

For resistance offered to flow by fittings he gives:

$$P_r = \frac{W}{144} \times 1.12r \times \frac{v^{1.95}}{2g}$$

By combining these two formulæ we obtain

$$L_e = 43.7r d^{1.2}.$$

There is no reason to assume that the flow of gas or air through fittings should occasion losses differing from those caused by the flow of steam. Hence the writer assumes that the equivalents of Table 2 may be used for all three fluids.

[1] Consulting Engineer, Foster and Gilmore, Mem.Am.Soc.M.E.
For Presentation at the Annual Meeting, New York, December 7 to 10, 1920, of THE AMERICAN SOCIETY OF MECHANICAL ENGINEERS. All papers are subject to revision.

TABLE 1 EQUIVALENT LENGTHS OF STANDARD PIPE TO ALLOW FOR VARIOUS SCREW FITTINGS IN CONDUITS CARRYING NON-VISCOUS LIQUIDS[1]

Formula Used: $L_e = 53.75rd^{1.25}$

| Nominal Pipe Size, in. | Actual Inside Diameter, in. | Gate Valve | Long-Sweep Elbow or on Run of Stand-ard Tee | Elbow Medium-Sweep or on Run of Tee Reduced 1/4 | Standard Elbow or on Run of Tee Reduced 1/2 | Angle Valve | Close Return Bend | Tee through Side Outlet | Globe Valve |
|---|---|---|---|---|---|---|---|---|---|
| Factor of Resistance.... | | 0.25 | 0.33 | 0.42 | 0.67 | 0.90 | 1.00 | 1.33 | 2.00 |
| 1/2 | 0.662 | 0.335 | 0.442 | 0.56 | 0.89 | 1.20 | 1.34 | 1.79 | 2.68 |
| 3/4 | 0.824 | 0.475 | 0.627 | 0.79 | 1.27 | 1.71 | 1.90 | 2.52 | 3.80 |
| 1 | 1.049 | 0.640 | 0.844 | 1.07 | 1.72 | 2.30 | 2.56 | 3.40 | 5.12 |
| 1 1/4 | 1.38 | 0.902 | 1.19 | 1.51 | 2.42 | 3.24 | 3.61 | 4.80 | 7.22 |
| 1 1/2 | 1.61 | 1.09 | 1.43 | 1.83 | 2.92 | 3.92 | 4.36 | 5.79 | 8.72 |
| 2 | 2.06 | 1.49 | 1.96 | 2.50 | 3.99 | 5.36 | 5.96 | 7.92 | 11.92 |
| 2 1/2 | 2.46 | 1.86 | 2.46 | 3.13 | 5.00 | 6.72 | 7.47 | 9.93 | 14.94 |
| 3 | 3.06 | 2.46 | 3.25 | 4.11 | 6.66 | 8.87 | 9.86 | 13.11 | 19.72 |
| 3 1/2 | 3.54 | 2.92 | 3.80 | 4.91 | 7.84 | 10.53 | 11.70 | 15.56 | 23.40 |
| 4 | 4.026 | 3.44 | 4.53 | 5.77 | 9.22 | 12.37 | 13.70 | 18.28 | 27.50 |
| 4 1/2 | 4.506 | 3.95 | 5.20 | 6.63 | 10.60 | 14.22 | 15.80 | 21.01 | 31.60 |
| 5 | 5.047 | 4.57 | 6.00 | 7.68 | 12.20 | 16.47 | 18.30 | 24.33 | 36.60 |
| 6 | 6.065 | 5.72 | 7.55 | 9.61 | 15.30 | 20.61 | 22.90 | 30.45 | 45.80 |
| 7 | 7.024 | 6.90 | 9.10 | 11.59 | 18.50 | 24.84 | 27.60 | 36.70 | 55.20 |
| 8 | 7.981 | 8.10 | 10.69 | 13.60 | 21.70 | 29.16 | 32.40 | 43.09 | 64.80 |
| 10 | 10.020 | 10.70 | 14.10 | 17.97 | 28.70 | 38.52 | 42.80 | 56.92 | 85.60 |
| 12 | 12.090 | 12.50 | 17.80 | 22.68 | 36.20 | 48.60 | 54.00 | 71.82 | 108.00 |

TABLE 2 LENGTHS OF STANDARD PIPE TO ALLOW FOR VARIOUS SCREW FITTINGS IN CONDUITS CARRYING STEAM, AIR, OR GAS

Formula Used: $L_e = 43.7rd^{1.2}$

| Nominal Pipe Dis- | Actual Inside Dis- meter, in. | Gate Valve | Long-Sweep Elbow or on Run of Standard Tee | Elbow Medium-Sweep or on Run of Tee Reduced 1/4 | Standard Elbow or on Run of Tee Re- duced in Size 1/2 | Angle Valve | Close Return Bend | Tee through Side Outlet | Globe Valve |
|---|---|---|---|---|---|---|---|---|---|
| Factor of Resistance.... | | 0.25 | 0.33 | 0.42 | 0.67 | 0.90 | 1.00 | 1.33 | 2.00 |
| 1/2 | 0.622 | 0.031 | 0.41 | 0.52 | 0.84 | 1.12 | 1.25 | 1.66 | 2.50 |
| 3/4 | 0.824 | 0.044 | 0.57 | 0.73 | 1.17 | 1.57 | 1.75 | 2.33 | 3.50 |
| 1 | 1.049 | 0.057 | 0.77 | 0.98 | 1.57 | 2.11 | 2.34 | 3.11 | 4.68 |
| 1 1/4 | 1.380 | 0.082 | 1.00 | 1.37 | 2.19 | 2.94 | 3.27 | 4.35 | 6.54 |
| 1 1/2 | 1.610 | 0.098 | 1.29 | 1.64 | 2.63 | 3.52 | 3.92 | 5.21 | 7.84 |
| 2 | 2.067 | 1.32 | 1.74 | 2.23 | 3.53 | 4.77 | 5.30 | 7.05 | 10.60 |
| 2 1/2 | 2.469 | 1.64 | 2.16 | 2.75 | 4.39 | 5.91 | 6.56 | 8.71 | 13.12 |
| 3 | 3.068 | 2.13 | 2.81 | 3.59 | 5.72 | 7.69 | 8.54 | 11.40 | 17.08 |
| 3 1/2 | 3.548 | 2.53 | 3.34 | 4.26 | 6.80 | 9.10 | 10.13 | 13.50 | 20.26 |
| 4 | 4.026 | 2.96 | 3.90 | 4.97 | 7.94 | 10.65 | 11.84 | 15.75 | 23.68 |
| 4 1/2 | 4.506 | 3.27 | 4.43 | 5.66 | 9.05 | 12.14 | 13.50 | 17.95 | 27.00 |
| 5 | 5.047 | 3.88 | 5.11 | 6.42 | 10.40 | 13.95 | 15.51 | 20.60 | 31.02 |
| 6 | 6.068 | 4.81 | 6.35 | 8.09 | 12.90 | 17.35 | 19.27 | 25.60 | 38.54 |
| 7 | 7.023 | 5.75 | 7.59 | 9.66 | 15.40 | 20.70 | 23.02 | 30.60 | 46.08 |
| 8 | 7.981 | 6.70 | 8.85 | 11.20 | 17.90 | 24.10 | 26.80 | 35.60 | 53.60 |
| 10 | 10.02 | 8.35 | 11.54 | 14.70 | 23.40 | 31.50 | 35.00 | 46.60 | 70.00 |
| 12 | 12.09 | 10.90 | 14.40 | 18.35 | 29.30 | 39.30 | 43.70 | 58.10 | 87.40 |

Many charts and diagrams for the quick determination of steam pipe sizes have been published, but they have all been constructed under the assumption that dry saturated steam was the fluid to be carried. The accompanying chart, Fig. 1, however, solves problems involving the carrying of either dry saturated or superheated steam. In order that its use may be clear the following case has been assumed, the pressure loss in lb. per sq. in. per 100 ft. being desired:

| Average steam pressure in line.......... | 140 lb. abs. |
|---|---|
| Superheat............................ | 150 deg. |
| Amount of steam to be delivered........ | 1700 lb. per min. |
| Size of pipe.......................... | 10 in. standard |

In the solution shown in Fig. 1 the chart is entered on the hori- zontal line representing 150 deg. of superheat. This line is fol- lowed to the left until it intersects the curve representing 140 lb. pressure. From this intersection the dashed vertical line is fol- lowed down to its intersection with the horizontal dashed line repre- senting 1700 lb. of steam. From this intersection the inclined dashed line is followed to its intersection with the horizontal line representing 10-in. pipe. The vertical thus determined gives the loss required as 2.08 lb. per 100 ft. Had a 12-in. pipe line been used, the loss would have been only 0.81 lb.

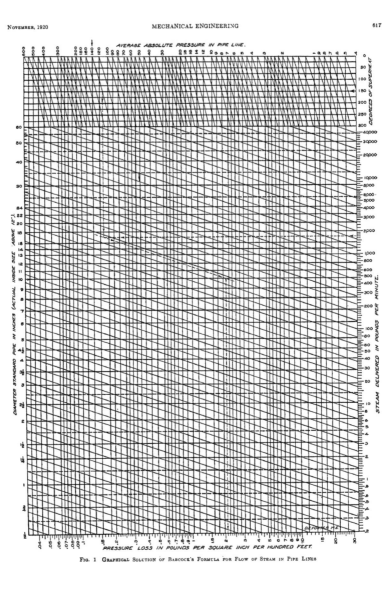

Fig. 1 Graphical Solution of Babcock's Formula for Flow of Steam in Pipe Lines

Fig. 1 can be used for the solution of any case where four of the variables are known and the fifth is to be determined. For the solution of problems involving dry saturated steam, enter the chart at the top at the point representing the pressure assumed and follow a vertical line from this point to its intersection with a horizontal representing the desired quantity of steam. The remainder of the solution will be the same as outlined for superheated steam. If the part representing superheated steam were omitted, this chart would be very similar to one constructed by H. V. Carpenter, Mem.Am.Soc.M.E., and published a number of years ago in *Power*.

To show the use of Table 2 an example has been taken of a 6-in. steam line 1000 ft. long, containing 5 gate valves, 3 angle valves, 20 standard tees, and 10 standard ells. From Table 2 for steam, air, and gas, these fittings are found to be equivalent to 332.10 ft. of 6-in. pipe as follows:

| | |
|---|---|
| 5 Gate valves, each 4.81 | 24.05 |
| 3 Angle valves, each 17.35 | 52.05 |
| 20 Standard tees on run, each 6.35 | 127.00 |
| 10 Standard ells, each 12.90 | 129.00 |
| Total allowance | 332.10 ft. |
| Actual pipe length | 1000.00 |
| Equivalent total length | 1332.10 ft. |

The chart shows that with an initial pressure of 150 lb. and a total loss of 5 lb. pressure or 0.375 lb. per 100 ft. this line will transmit about 225 lb. of steam per min. If the effect of the valves and fittings had been ignored, the calculation would show a capacity of 260 lb. of steam per min. This would have indicated a flow of $11\frac{1}{2}$ per cent more than the real capacity of the line. This error in capacity will be the same, no matter what pressure drop is allowed through the line.

Use of Table 1 is illustrated by the solution of the following problem which assumes that a 4-in. line is to handle 26.2 cu. ft. of water per min. through a horizontal run of 50 ft., at which point the line rises at right angles 35 ft., thence runs horizontally 120 ft. to a tee heading a manifold consisting of five pieces of 4-in. pipe 12 ft. long connected by 4-in. tees, each delivering one-sixth of the total quantity of water. A centrifugal pump is to be selected and properly speeded to handle this quantity of water. The pump is to be located 10 ft. above the water surface.

Since the frictional resistance of pipe varies approximately as the square of the quantity of water passing through it, it will be convenient to combine the last 12 ft. of horizontal run with the first tee. This will give us 6 sections of 4-in. pipe each 12 ft. long and having a connecting tee. From Table 1 for liquids we find a 4-in. tee equivalent to be 4.53 ft. of 4-in. pipe, making each section the equivalent of 16.53 ft. in length. From Cox's tables for flow of water through pipe lines we find a loss of 2.563 ft. per 100 ft. for a 4-in. line handling 26.2 cu. ft. of water per min. The loss for the first section which carries all the water will be 16.53 × 2.563/100 or 0.424 ft. For the second section, which carries ⁵⁄₆ of the total, the loss will be as the square of the quantity carried, or $\frac{5}{6} \times \frac{5}{6} \times 0.424 = 0.295$ ft.

For the entire manifold the losses will be as follows:

| | |
|---|---|
| Section 1 $(6/6)^2 \times 0.424$ | 0.424 |
| Section 2 $(5/6)^2 \times 0.424$ | 0.295 |
| Section 3 $(4/6)^2 \times 0.424$ | 0.198 |
| Section 4 $(3/6)^2 \times 0.424$ | 0.131 |
| Section 5 $(2/6)^2 \times 0.424$ | 0.047 |
| Section 6 $(1/6)^2 \times 0.424$ | 0.012 |
| Total loss for manifold | 1.107 ft. |

For the remainder of the line we have:

| | |
|---|---|
| Suction line | 10 |
| 300 diameters for pump | 100 |
| First horizontal run | 50 |
| Riser | 35 |
| Second horizontal run | 108 |
| Three 4-in. standard elbows, each 9.22 | 27.66 |
| Total equivalent length of pipe to Section 1 of manifold | 330.66 ft. |

and from the foregoing:

| | |
|---|---|
| Loss on 330.66 ft., 2.563/100 per ft. | 8.47 |
| Loss through manifold | 1.11 |
| Total loss | 9.58 ft. |

The total pumping head then will be as follows:

| | |
|---|---|
| Suction head | 10.00 |
| Elevation | 35.00 |
| Friction head | 9.58 |
| Total head | 54.58 ft. |

# CONSTITUTION AND PROPERTIES OF BOILER TUBES

### (Continued from page 605)

not thus visibly subject to grain growth. Not only was there evident a marked growth in the ferrite grains for the irons ranging in carbon content from 0.006 to 0.251 per cent, inclusive, but in the irons in this range where there was a visible quantity of carbon existing as pearlite, very apparent agglomeration or balling up of this constituent was in evidence.

TABLE 1 EFFECT OF CARBON CONTENT ON GRAIN GROWTH IN DEFORMED IRON WHEN HEATED BELOW THE CRITICAL TEMPERATURE

| Carbon Content, Per cent | Number of Ferrite Grains per Square Inch | |
|---|---|---|
| | Undeformed | Deformed |
| 0.006 | 3.9 | 1.9 |
| 0.103 | 25.4 | 13.4 |
| 0.203 | 63.4 | 23.0 |
| 0.251 | 26.3 | 16.0 |
| 0.315 | (a) | (a) |

(a) Grains too small to count. Photomicrographs from both the undeformed and deformed areas show no appreciable difference in grain size.

### SERVICE COMPARISON BETWEEN MEDIUM-HIGH- AND LOW-CARBON BOILER TUBES

Not only do all theoretical considerations point to the procurement of an increased life for boiler tubes through a raising of the carbon content, but the results of some actual tests on which data are now available seem to indicate and confirm this claim.

This test was started in 1916–1917 by placing in the front row of four of the 750-hp. Stirling boilers at the Park Place Heating Plant of the Detroit Edison Co. tubes of a medium-high carbon content averaging around 0.30 per cent carbon, and in four other boilers at the same plant operating under the same loads at about the same time tubes with a carbon content between 0.08 and 0.18 per cent.

The results given in the "Service Test" (Table 2) speak for them-

TABLE 2 SERVICE TEST ON MEDIUM-HIGH- AND LOW-CARBON BOILER TUBES AT PARK PLACE HEATING PLANT, DETROIT, MICH.

| Boiler Number | Installation | | Replacement | |
|---|---|---|---|---|
| | Number | Date | Number | Date |
| HIGH-CARBON TUBES IN FRONT ROW | | | | |
| 3 | 29 | June 19, 1916 | 27 | July 27, 1918 |
| | | | 15 | Summer, 1920 |
| 6 | 29 | May 9, 1916 | 13 | Sept. 11, 1917 |
| | | | 5 | July 15, 1919 |
| | | | 1, | Summer, 1920 |
| 7 | 29 | June 19, 1917 | | |
| 8 | 29 | June 9, 1917 | | |
| Totals, | 116 | | 61 | |
| LOW-CARBON TUBES IN FRONT ROW | | | | |
| 1 | 29 | Sept. 23, 1916 | 27 | July 12, 1918 |
| | | | 5 | Summer, 1920 |
| 2 | 29 | Sept. 18, 1916 | 10 | Sept. 28, 1917 |
| | | | 27 | July 11, 1920 |
| | | | 7 | Summer, 1920 |
| 4 | 29 | June 14, 1916 | 18 | Jan. 6, 1917 |
| | | | 5 | July 27, 1918 |
| | | | 13 | June 17, 1919 |
| | | | 5 | Summer, 1920 |
| 5 | 10 | June 5, 1916 | 10 | August 3, 1917 |
| | 19 | Aug. 3, 1917 | | |
| Totals, | 116 | | 127 | |

selves, for of the medium-high-carbon tubes only about one-half have been replaced, and of the low-carbon tubes, more than a 100 per cent replacement has been necessary.

This paper has been prepared not for the purpose of suggesting radical changes in boiler-tube composition nor for the purpose of criticizing present boiler-tube manufacturing practices, for all things considered, it is on a very high plane with respect to quality. It has been prepared, however, to present certain facts, especially those relating to grain growth, to which tubes of the commonly accepted composition are so subject; and in view of these acts to question whether tubes with a carbon content varying between 0.30 and 0.35 per cent would not insure longer tube life and safer boiler operation than tubes with a carbon range between 0.08 to 0.18 per cent.

# Principles of the Gyro-Compass

By GEORGE B. CROUSE,[1] BLOOMFIELD, N. J.

*The following brief outline covers the main factors in the development of the gyro-compass up to the present time. It explains briefly the principles of the gyroscope and the simple gyro-compass; discusses the disturbing forces acting on the gyro-compass when on shipboard; and finally takes up the three major problems of design, namely, the compensation of the disturbing forces, the methods of suspension, and methods of damping.*

A SUBSTITUTE for magnetism as a directing force has long been sought and in 1852 it appeared that the work of Leon Foucault bid fair to solve this problem by means of the gyroscope. The eminent French scientist found that by the use of the gyroscope, or spinning wheel, a mechanical device could be constructed which, when properly mounted, would indubitably seek the meridian and which depended in no way on the earth's magnetism. The spinning wheel exhibits two properties which distinguish it from inert matter, termed, respectively, "fixity of plane" and "precession."

*Fixity of plane* expresses the fact that a spinning mass acquires by virtue of its rotation a sort of enhanced inertia which strongly resists any attempt to change the plane of rotation. It is a common error to believe that the gyroscope maintains its plane of rotation in space irrespective of any force applied to it. This is de-

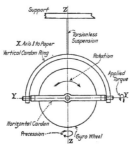

FIG. 1 SIMPLE GYROSCOPIC SYSTEM

cidedly not true. Even a force of very small magnitude will gradually change the plane of rotation, although it does so in a very novel manner.

*Precession* is illustrated in Fig. 1, which shows a simple gyroscopic system. If torque be applied about, say, the $Y$-$Y$ axis, with the intent of forcing the plane of rotation of the wheel to rotate about this $Y$-$Y$ axis, no such result will ensue, but instead the plane will rotate slowly about the $Z$-$Z$ axis, which is perpendicular to the axis of the applied torque. This action is precession.

The angular velocity of the precessional motion, $\Omega$, will depend upon $I$, the moment of inertia of the wheel, and upon $\omega$, the angular velocity of rotation of the wheel, and upon $T$, the torque applied about the $Y$-$Y$ axis as above, the quantities being connected with the precessional velocity $\Omega$ by the equation

$$T = I\omega\Omega$$

The direction of procession will depend on the direction of rotation of the wheel as well as upon the direction of the applied torque, the relation being correctly given in Fig. 1. A reversal either of the direction of the torque or the rotation of the wheel will change

[1] Consulting Engineer.
Abstract of paper to be presented at the Annual Meeting, New York, December 7 to 10, 1920, of THE AMERICAN SOCIETY OF MECHANICAL ENGINEERS. All papers are subject to revision.

the direction of precession, whereas a reversal of both of these will not change the direction of the precession.

## SIMPLE GYRO-COMPASS

With these facts in mind the explanation of the Foucault compass becomes very simple. The elements of the device are shown

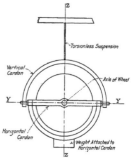

FIG. 2 SIMPLE FOUCAULT COMPASS

in Fig. 2. A gyroscope is so mounted in cardan rings that it may freely turn in any direction and the entire system is perfectly balanced about all three of the mutually-perpendicular axes, save only the axis $Y$-$Y$, about which it is made slightly pendulous.

Let us suppose such a device set up on the equator of the earth,

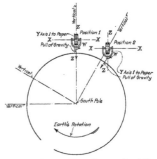

FIG. 3 DIAGRAM TO ILLUSTRATE TILTING EFFECT OF EARTH'S ROTATION ON GYRO-COMPASS

with the axle pointing east-west, as shown in position 1, Fig. 3. It is obvious that the direction of the perpendicular at any point on the equator is continually changing, being rotated in a clockwise direction, when viewed from the south, once in 24 hr. The gyroscope, on the other hand, due to its property of fixity of plane, will resist this rotation, with the result that as the earth rotates the east-pointing end of the wheel will rise with respect to the earth's surface as shown in position 2. As soon, however, as this occurs the weight on the bottom of the wheel case will introduce a couple about the $Y$-$Y$ axis, due to the pull of gravity, and the phenomenon of precession occurs, causing the wheel to rotate about the vertical $Z$-$Z$ axis.

A consideration of the foregoing action will show that the east end of the axle will continue to rise and that the weight will act until the axle has passed through the meridian plane, after which the earth will act to depress this end of the axle, finally bringing it into the horizontal. As soon as the axle is horizontal the gravity couple vanishes and the precession ceases. However, the earth continues to act on the instrument, the axle is again tilted and the wheel again swings back toward the meridian plane. Thus the compass continues to perform a series of oscillations back and

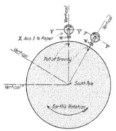

forth across the meridian until damped out by friction or other means. It will at once be apparent from Fig. 4 that when the compass is at rest with the axle in the meridian and horizontal there is no tilt introduced by the earth and the system is in equilibrium. It may also be readily shown that the action is the same

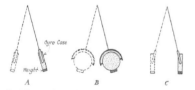

FIG. 5  DIAGRAM SHOWING EFFECT OF INTERCARDINAL ACCELERATIONS

as that described when the compass is located at any other latitude than the equator, with the exception of a small error introduced by the vertical component of the earth's rotation.

It is thus seen that the compass derives its directive force from the rotation of the earth and the force of gravity acting on the pendulous mass of the wheel. In reality an instrument such as that described functions perfectly—granting the necessary degree of mechanical refinement—when placed on a fixed platform where forces other than those mentioned cannot be impressed on the system. However, when the Foucault pendulum is placed on a moving platform, such as a ship at sea, other forces are impressed on it, due to the motion of translation of the ship over the sea and the motions due to wave action. The forces resulting from these causes are in no way to be distinguished from the useful or directive force and are in fact many times greater in magnitude than the useful force, so that under these conditions the instrument becomes inoperative.

It was this fact which delayed the introduction of the gyro-compass in navigation for nearly sixty years after its discovery; and it was the solving of this problem which acted to produce the varied forms of compass which it is the purpose of this paper to describe.

### DISTURBING FORCES ACTING ON THE COMPASS

We may broadly divide the disturbing forces acting on the compass at sea into two classes: First, those due to the motion of translation of the ship; and second, those due to motions of the ship about various axes fixed in relation to the ship: to which may also be added the error introduced by the vertical component of the earth's rotation, and in some cases small errors due to elements of the compass design. Of these the elimination of forces due to the rolling and pitching of the ship has largely dictated the major factors of the compass design.

These rolling forces, acting on the simple compass or pendulous gyroscope, exhibit their effects chiefly on intercardinal headings, being theoretically absent on north-south or east-west headings, and increasingly troublesome up to 45 deg. from these directions.

A convenient method of investigating these rolling forces is to place the complete compass on a pendulum which may swing in a plane making a known angle with the plane of the meridian. In the first case, suppose this plane to coincide with the plane of the meridian. Then, if the compass wheel is not running and the supporting pendulum is swung, the pendulous mass of the wheel case will cause the case to line up with the pendulum, due to the combination of the forces of gravity and acceleration, as shown at $A$ in Fig. 5. When the wheel is spinning, the case will not line up with the pendulum, since it is stabilized about the east-west axis, and the force of acceleration on either end of the swing will therefore be balanced by a gyroscopic reaction. ($C$, Fig. 5). These forces are equal and opposite on each end of the swing and no disturbance of the compass will result. Should the plane of swing of the pen

FIG. 6  SENSITIVE ELEMENT OF THE ANSCHUTZ COMPASS

dulum be east and west, and the case will swing freely about the axle of the wheel, and again no disturbing force will result. ($B$, Fig. 5).

Assume now that the plane of swing makes an angle with the meridian of, say, 45 deg. There will be a component of swing in a north-south direction which will cause a force to be impressed on the pendulous mass, and, as before, it will be equal and opposite on each end of the swing. In addition to this, there will be a movement of the case about the axle of the wheel, due to the east-west component of the swing. This latter motion will cause the point of application of the force of acceleration to be shifted from one side to the other of the vertical axis; and thus, while the impressed force will be equal on each end of the swing, and in an opposite direction, it acts to form a pulsating couple about the vertical whose direction is always the same.

Another way of regarding this phenomenon is to consider that whereas the wheel when spinning is stabilized about an east-west axis, it is perfectly free to rotate about the north-south axis, thus causing the apparent moments of inertia about these two axes to be widely different. The result of this is that when forces tend to rotate the system about any horizontal axis other than the major axes mentioned above, a torque is set up about the vertical which ultimately displaces the compass from the meridian.

### COMPENSATION OF THE DISTURBING FORCES

From this description it is obvious that the simplest way of remedying the difficulty is to stabilize the compass in both directions, or, in other words, about the north-south axis as well as about the east-west axis, and one of the earliest methods was based on this principle. Fig. 6 is from a photographic view of the sensitive element of the Anschutz compass. This compass uses three wheels, all attached to the orienting element, one of which operates in the same manner as the Foucault pendulum, the other two being mounted with their axles at an angle to each other and to the main wheel. By this method the apparent moment of inertia of the system about the north-south axis is greatly increased and at the same time a portion of the useful directive force of both of the auxiliary wheels is utilized to orient the compass on the meridian.

In a later form of the Anschutz compass for use on submarines and other small craft subjected to violent rolling, a fourth wheel is added, which serves only as a stabilizer, the axle being vertical.

It is to be noted, however, that in the Anschutz compass complete stabilization of the sensitive element is not aimed at, but rather the reduction of the disturbing torque to within a limit where its effect on the compass will not be detrimental. On the other hand, a number of attempts have been made to solve the problem by the independent stabilization of the platform on which the sensitive element is mounted. None of the attempts has been

FIG. 7 SPERRY GYRO-COMPASS, SHOWING FLOATING BALLISTIC

so far successful, due, probably, to the difficulty of maintaining the necessary accuracy in the stabilizing apparatus. Numberless trials have demonstrated the fact that it is practically impossible to stabilize a platform on a rolling ship to anywhere near the accuracy required for this purpose.

Another early solution of the problem was the ingenious method developed by the Sperry Gyroscope Company. This method consisted in stabilizing the point of attachment between the gyroscope and the pendulous mass as shown in Fig. 7. The stabilizing device, known as a "floating ballistic" is essentially a pendulum stabilized by a small gyroscope. We find that there again the stabilizing principle is used only to reduce the magnitude of the disturbing forces, and neither the Anschutz principle of three wheels, nor the Sperry floating ballistic, completely compensates.

In 1916 and 1917 a form of compass was developed by Harry L. Tanner, of the Sperry Gyroscope Company, in which complete compensation of the disturbing forces was accomplished. This

compass, a photograph of which is shown in Fig. 8, and a line drawing in Fig. 9, utilizes two wheels of equal size and rotating at equal speeds but in opposite directions. One of these wheels is mounted as in the Foucault pendulum and may be regarded as the compass wheel. The second wheel, located to the east of the meridian and known as the east gyro, precesses in a direction opposite to that of the compass wheel under the influence of disturbing forces, and this fact is utilized to impress a force on the west wheel equal and opposite to the disturbing torque; this torque being transferred between the wheels by means of a yielding connec-

FIG. 8 SPERRY TWIN GYRO-COMPASS

tion between them. At the same time the mounting is such that a large portion of the available directive force of the east wheel is transferred to the west and thus utilized. The compass is completely compensated and was adopted by the U. S. Navy for

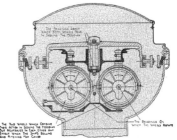

FIG. 9 PARTIAL SECTION OF SPERRY TWIN GYRO-COMPASS

use on destroyers, a class of vessels whose rolling characteristics are notorious.

In the last year a new method of compensation has been developed in the United States by the Sperry Company and in England by Perry and Brown. This consists of the use of a nonpendulous wheel, the gravity couple being introduced in a novel manner. The device, the elements of which are shown in Fig. 10, consists of a gyroscope which is perfectly balanced about all of the three axes. On both the north and south sides of the gyro casing and attached to the casing are bottles containing mercury which are connected by means of a small tube. The action of the device as a compass is then very similar to the simple compass in that the earth acts to tilt the wheel as usual, whereupon the mercury flows from the higher to the lower bottle, thus upsetting the equilibrium about the east-west axis. The torque thus generated causes precession about the vertical axis as before. It should be noted, however, that this torque is oppositely directed to that generated by the usual pendulous mass and therefore it is necessary to rotate the wheel in a direction opposite to that of the ordinary compass. In other words, the wheel must rotate in a counterclockwise direction when viewed from the south.

The explanation of the compensating features of this type of compass lies in the fact that the motion of the mercury from one bottle to the other, under the influence of the forces set up by the

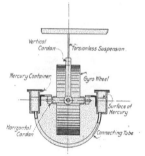

FIG. 10 MERCURY BALLISTIC COMPASS

rolling of the ship, is out of phase with the motion of the compass in its gimbal rings. While the principle involved is new and no great number of compasses built on this principle have been in service, very exhaustive trials of the instrument have been made in

## METHODS OF SUSPENSION

Perhaps of equal importance to the problem of compensation, from the standpoint of the designer of gyro-compasses, is the question of suspension of the sensitive element. While the directive force of the gyro-compass is many times larger than that of the magnetic compass, in most forms of the instrument the masses to be oriented are more than proportionately greater; therefore some means of carrying these masses on a suspension bearing must be devised which is as nearly as possible without friction.

The mercury float forms a very simple method of suspension which immediately suggests' itself from the somewhat similar mounting of the magnetic needle in the type known as the liquid compass. The Anschutz compass utilizes the mercury suspension with considerable success and has done so for a number of years. Its great advantage is its simplicity, but on the other hand a number of practical difficulties present themselves. The mercury collects dust from the air and also oxidizes rapidly unless it is surrounded by some non-oxidizing atmosphere such as hydrogen. When the metal becomes dirty or oxidized the sensitivity of the mounting is greatly decreased.

In the original compass constructed by Foucault a filament suspension was employed similar to that used today for galvanometers, oscillographs and similar instruments. Such a suspension was very sensitive, but, of course, was not suitable for shipboard use since it was not capable of making a complete revolution in azimuth. The Sperry Company applied this torsionless-filament suspension to the practical instrument in a manner which has proved so successful that it has been used on all of the various types of compasses which they have designed. It consists essentially of a suspension head carrying a filament consisting of a bundle of very fine piano wires, to the lower end of which is attached the sensitive element. When the sensitive element of the compass turns in azimuth, or the binnacle turns relatively to the sensitive element, this filament is continuously unwound by means of an electrical follow-up system, the elements of which are shown in Fig. 13. It consists of a so-called phantom element, driven in

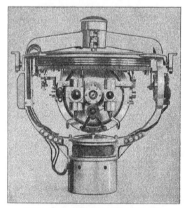

FIG. 11 PERRY-BROWN COMPASS

comparison with other standard compasses. All of these tests show that the compensation is practically perfect, in addition to which the simplicity of the instrument strongly recommends it.

In Fig. 11 is a view of the Perry-Brown compass and Fig. 12 shows the mercury attachment or "mercury ballistic" of the Sperry compass, the mercury bottles being indicated by the arrow.

FIG. 12 MERCURY BALLISTIC ATTACHED TO SPERRY COMPASS

azimuth by an electric motor. As shown in the figure the motor is controlled by trolley contacts operated by the motion of the sensitive element in azimuth relative to the phantom. This suspension is very sensitive and has the additional advantage that a continuous, small oscillation of the phantom relative to the sensitive element is maintained, which acts to reduce the friction of the vertical guide bearings necessary in any type of suspension.

The Perry-Brown compass uses a form of suspension which has also shown great sensitivity. It is similar in principle to the oil-float thrust bearings used on vertical turbines and similar machinery. A small pump is employed to force oil between the surfaces of a step bearing, thus maintaining a perfect film for the separation of the surfaces. The pulsations of the pump give the sensitive element a small up-and-down oscillation and thus acts as a means of reducing the static friction of the vertical guides as in the Sperry instrument.

The jeweled point has frequently been suggested and tried as a gyro-compass suspension, but has never met with any measure of success, due principally to the large weights which must be supported on extremely small areas.

### METHODS OF DAMPING

The third major design element, and one which has been the subject of considerable study and which has resulted in many different solutions, is the means for damping the oscillations of the compass system in azimuth. With the very sensitive suspensions which must be employed, the compass will continue to swing back and forth across the meridian for many hours without coming to rest and some special means must be employed to damp these oscillations in a reasonable length of time. The damping employed is generally very large, usually from 60 to 90 per cent. Since in all compasses so far constructed the period has had to be fixed at 85 min. of time, even with a very high damping factor, several hours are required for the compass to settle on the meridian.

An early solution of the damping problem was that employed in

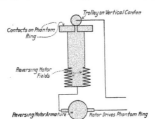

Fig. 13   ELEMENTS OF SPERRY ELECTRICAL FOLLOW-UP SYSTEM

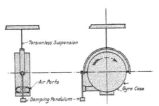

Fig. 14   ANSCHÜTZ AIR-DAMPING DEVICE

the first Anschütz compass, which consisted of two air jets controlled by a pendulum mounted on the case of the wheel, as shown diagrammatically in Fig. 14. The air pressure was supplied by the rapid rotation of the wheel, the nozzles being so arranged that when the wheel was tilted by the earth's rotation one or the other of the nozzles was opened, causing a reaction about the vertical axis which caused precession to bring the axle back into the horizontal plane. It is obvious that this action will slow down the motion of the wheel in azimuth and that thus

the necessary damping will be secured. The controlling pendulum was not stabilized, but this did not seem to interfere with the action of the device, since its average position, of course, was vertical. Various practical difficulties arose with the device in practice, however, and it was early abandoned.

In the Sperry compasses the damping is secured in connection with the power-driven phantom element. In the compass known as the "Mark I" the pendulous mass, instead of being made an

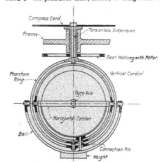

Fig. 15   SPERRY MECHANICAL DAMPING DEVICE

integral part of the wheel case, is supported on the phantom element on horizontal bearings. The attachment of this mass to the wheel is made slightly to the east of the vertical, and thus when the wheel tilts a large reaction is applied about the horizontal

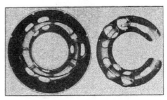

Fig. 16   ANSCHÜTZ DAMPING TANKS

axis to secure the directive action and a small reaction about the vertical axis to effect damping.

In the present Anschütz compasses damping is obtained by the use of a flow of oil from the north to the south side of the horizontal axis through very small orifices, the action being very similar to that of the mercury ballistic described above, except that the flow of oil is made so slow that it is out of phase with the oscillation of the compass in azimuth. The oil thus is continually acting to oppose the torque of the main gravity couple, and the oscillations are thus damped. In the Anschütz compass the oil is allowed to flow through a number of reservoirs in series, the openings between the reservoirs being very small. A photograph of the damping tanks is shown in Fig. 16.

Having solved the three major design problems outlined above, the question of building a successful working instrument resolves itself largely into a question of correct mechanical design and excellency of workmanship. The delicacy of the instrument is such that the utmost care must be exercised in both the design and construction to insure the correct and permanent alignment of the parts and the reduction of friction about all axes, to insure freedom from vibration and similar factors.

# The Heat-Insulating Value of Cork and Lith Board

Results of a Series        ͗nducted at the Engineering Experiment Station of Kansas State
Agricul        ͗wing Cork Board to be Superior to Lith Board
Heat-Insulating Material

By A. A.        ͗WOOD,[2] A. J. MACK,[3] AND L. S. HOBBS[4]

*The following paper treats of a series of tests ͗       ͗       ͗er-
ing Experiment Station of the Kansas State Agricultur͗.       ͗ ͗ the
purpose of determining the conductivity and heat-transmittin͗ ͗v͗rties
of cork board and lith board. The apparatus employed is described in
detail, and the procedure followed in the experimental work is carefully
outlined. Formulae for calculating results are also presented. The results obtained by the authors show that as a heat-insulating material cork
board is slightly better than lith board; that is, for the samples tested the
lith board had a conductivity approximately 5 per cent greater than that
of the cork board.*

IN the following paper are presented the results of a series
of tests which were conducted at the Engineering Experiment
Station of the Kansas State Agricultural College upon the heat-
insulating properties of two refrigerating materials—nonpareil
cork and lith board. The general arrangement of apparatus used
for the tests is shown in Fig. 1. The test box itself is made of nonpareil cork board and is a cube whose internal dimensions are
2 ft. square. One side of this box is removable for the insertion of
specimens of other materials and within it are three resistance

coils which serve as an electric heater during the test. These
coils are arranged so that one, two, or three may be joined in
series, thus making it possible to vary the intensity of the heat
supply. Current is supplied by a 220-volt d.c. line and in order
to maintain a uniform voltage at the heating element, a 32-volt
storage battery is floated across the supply main. A variable resistance is used to regulate the current input.

The electrical energy consumed by the heating coils is measured
by a single millivoltmeter, for through the use of a multiplier, a shunt
and a two-way mercury switch the same instrument is made to
measure both the impressed voltage and amperage supplied to the
coils.

At regular intervals during the test temperature measurements
were taken at the internal and external surfaces of the top, bottom,
front, and sides of the test box, care being taken in inserting the

¹ Dean of Engineering, Purdue University. Mem.Am.Soc.M.E.
² Professor of Mechanical Engineering, Kansas State Agricultural College.
Mem.Am.Soc.M.E.
³ Assistant Professor of Mechanical Engineering, Kansas State Agricultural
College.
⁴ Research Engineer, Dayton, Ohio.
For presentation at the Annual Meeting, New York, December 7 to 10, 1920, of
THE AMERICAN SOCIETY OF MECHANICAL ENGINEERS. The paper is here printed
in abstract form and advance copies of the complete paper may be obtained gratis
upon application. All papers are subject to revision.

thermometer to eliminate all possible influence from the air, both
external or internal. Temperature measurements were also taken
on the front side of the test box, at points 2 in. and 4 in., respectively,
from both internal and external surfaces (see Fig. 3).

These temperature measurements were accomplished by the
use of thermocouples, made from No. 12 gage copper and constantan
wire welded together to form the junction. Since the range in
consequent electromotive force small, the most convenient method
for the measurement of the impressed electromotive force produced
by the thermocouple was by balancing it with a standard electromotive force.

The wiring diagram of the temperature-measuring equipment is
shown in Fig. 2. The various thermocouples are joined by means
of a 14-point switch to the portable potentiometer indicator and
in order to produce a more sensitive potentiometer this instrument
was of a special design, being provided with connections for a
sensitive external galvanometer, which made possible temperature
measurements of great accuracy. The scale division of the instrument gives a direct reading of 0.01 millivolt, so that it is possible
to estimate to 0.001 millivolt.

The thermocouples with their cold ends maintained at 32 deg.
fahr. by an ice bath were calibrated throughout their range of
use against a standard mercury-in-glass thermometer. This
calibration was accurate to 0.5 deg. No calibration of the thermocouples was made below the 32 deg. fahr. point, as the deviation
of the calibration curve above and below this point is probably
negligible.

The refrigerator or thermal testing room was 9 ft. in 12 .͗

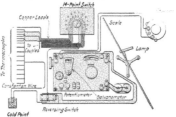

and 10 ft. high. It contained 600 ft. of 2-in. direct-expansion
ammonia piping and its walls, floor, and ceiling were insulated
with cork board. The refrigerating equipment includes a 5-ton
Frick ammonia-compression system, and a 1-ton York machine
used as an auxiliary. The ammonia compressor is steam-driven.
A pressure gage is inserted in the evaporating coils so that the
pressure in the coils can be regulated as desired by opening or
closing the suction valve to the compressor or the expansion valve
to the coils. Since the boiling point of the ammonia in the evaporating coils is dependent upon the pressure, this regulation gave an
indirect means of controlling the temperature in the refrigerator.

## METHOD OF TESTING

The method of testing was as follows: The test box with its
equipment was first placed in the refrigerator. The refrigerating
machine was then started and the pressure in the evaporating
coils so regulated that the desired temperature in the refrigerator

would be produced. No other regulation of the refrigerating equipment was made, except that of maintaining the pressure constant in the evaporating coils. While it was impossible to predict the exact temperature that would result with this regulation, the final temperature when conditions became adjusted remained practically constant at a value of approximately 10 deg. fahr. Lower temperatures were possible but required a longer period of operation before conditions became constant.

As a preliminary, temperature readings of two thermocouples, one inside and the other outside the test box, were taken, and when these became constant the test was then started. The length of test was one hour, and readings of the electrical energy supplied and the temperatures as indicated by the various thermocouples were recorded every 10 min..

Since no standard method of procedure has been suggested for the testing of heat-insulating materials along the lines followed in the present work, preliminary tests were first made in order to establish such a method. These tests were to determine the effect of the temperature of the source of heat supply upon

sulating properties of both cork and lith board were then made. These tests consisted of two series:

*Series 1.* These tests were made to show the heat-insulating properties of cork. The external temperature of the test box was maintained constant at approximately 10 deg. fahr. while internal conditions were varied by means of the electric heater described above. Temperature measurements were taken at the internal and external surfaces of the top, bottom, front and back sides of the test box and also at points 2 and 4 in., respectively, from the internal and external surfaces of the front side (see Fig. 3).

*Series 2.* These tests were identical with those of Series 1, but the front side of the cork-board test box was replaced by a 3-in. lith board.

### METHOD OF CALCULATING RESULTS

In calculating the results from the tests of Series 1, the conductivity of the cork was assumed constant for the various temperatures involved in each test. The heat equivalent of the elec-

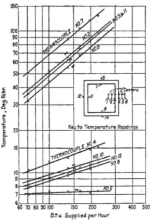

FIG. 3   TEMPERATURE READINGS OBTAINED IN TEST SERIES NO. 1 ON CORK BOARD

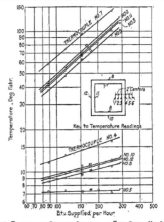

FIG. 4   TEMPERATURE READINGS OBTAINED IN TEST SERIES NO. 2 ON LITH BOARD

the various results, and the probable time that would be required for a variation in the temperature external to the test box to be transmitted to the interior.

From these preliminary results it was concluded that the heating element should consist of the three coils in series as this produced more uniform conditions in the interior of the box, and the insulating effect of the inner air film would approach a condition similar to that found in actual installations. With three coils in series and a current input sufficient to produce a temperature of 70 deg. fahr. in the interior of the box when the external temperature was 10 deg. fahr., the temperature of the heating element was 170 deg. fahr.

It was also shown by these preliminary tests that at least $2\frac{1}{2}$ hr. would be necessary, after conditions external to the test box became uniform, before the internal temperature conditions could be expected to become constant. Furthermore, if after external and internal conditions became constant, the internal temperatures could reasonably be expected to remain uniform for a period of 2 hr.

Following these preliminary tests the investigation of the in-

trical energy was equated to the heat conducted through the various sides as shown by the following equation, the temperature difference of the two measured sides being assumed as the average of the front and back temperature differences.

$$H = 4C\{(t_7 - t_8) + (t_9 - t_{10}) + (t_{11} - t_{12}) + (t_3 - t_4) + 2\left[\frac{(t_{11} - t_{12}) + (t_3 - t_4)}{2}\right]\} \dots\dots\dots [1]$$

where $H$ = the heat equivalent of the electrical input per hr.
$C$ = conductivity of the cork in B.t.u. per sq. ft. per deg. per hr.
$4$ = a constant = sq. ft. of internal surface per side of test box
$t_3$ = temperature, deg. fahr., internal surface, front side
$t_4$ = temperature, deg. fahr., external surface, front side
$t_7$ = temperature, deg. fahr., internal surface, top side
$t_8$ = temperature, deg. fahr., external surface, top side
$t_9$ = temperature, deg. fahr., internal surface, bottom side
$t_{10}$ = temperature, deg. fahr., external surface, bottom side
$t_{11}$ = temperature, deg. fahr., internal surface, back side
$t_{12}$ = temperature, deg. fahr., external surface, back side

The heat transmission of the cork was calculated by dividing the heat conducted through the front surface of the test box by the temperature differences measured 2 in. from the internal and external surfaces of the front side. The heat conducted through the front surfaces was estimated from the determined value of the conductivity and the inner and outer surface temperatures of the front side.

The calculation of the conductivity of the lith board from the tests in Series 2 gave rise to two methods:

*Method 1.* The first method was similar to that used in the calculations for the conductivity of cork. The heat transmitted through the cork of the test box was calculated from the temperature differences of the cork surfaces and the conductivity of the cork as determined in the tests of Series 1. The heat transmitted through the cork was then subtracted from the heat equivalent of the electrical input and the difference represents the heat transmitted through the lith board from which the conductivity may be determined. The equation of this relation is as follows:

$$4C_1(t_3 - t_4) = H - 4C[(t_7 - t_8) + (t_9 - t_{10}) + (t_{11} - t_{12}) + 2k(t_{11} - t_{12})] \dots \dots \dots [2]$$

where $C_1$ = conductivity of lith board in B.t.u. per sq. ft., per deg. per hr.

$C$ = conductivity of cork board in B.t.u. per sq. ft. per deg., per hr. as determined from tests of Series 1

$k$ = the proportion of the temperature difference of the sides to the temperature difference of the back surface

and all other factors have the same significance as in Eq. [1].

The tests of Series 1 showed that the temperature differences of the front and back sides of the test box were dissimilar. This was possibly due to slight differences in temperature at different points in the refrigerator. Therefore, in order to estimate the temperatures of the sides of the box when testing lith board, the relation between the temperatures at the back sides of the test box was determined from the tests in Series 1. From the conductivity of the lith board thus found, the heat transmitted through that surface was calculated. With this heat known, the heat transmission was calculated as in the tests of Series 1.

*Method 2.* The second method for calculating the conductivity of lith board was based on the theory that the conductivity of two different materials is proportional to their temperature differences when subjected to the same conditions. In order to apply this theory to the tests of Series 2, the probable temperature of the front of the test box if cork had been inserted was obtained. Since there was a difference between the external temperatures of the front and back sides of the test box in the tests of Series 1, a constant, $k_2$, was obtained from their proportion. Thus, the external temperature of the back side in Series 2 could be used in estimating the probable external temperature of the front side. This theory is expressed in equation form as follows:

$$C_1 = C \frac{(t_{11} - k_2 t_{12})}{(t_3 - t_4)} \dots \dots \dots \dots \dots [3]$$

where $k_2$ = a constant = the relation between the external temperature of the front and back sides obtained from the tests of Series 1

and all other factors have the same significance as in Eqs. [1] and [2].

### Results of Tests

The results of these tests are given in Tables 1 and 2. Table 1 gives the results of the tests of Series 1, which is based upon the cork-board test box, and Table 2 gives the results of the tests of Series 2, in which the removable cover of the test box was replaced by 3-in. lith board.

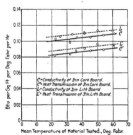

Fig. 5 Curves Showing Conductivity and Heat Transmission of Cork and Lith Board

The conductivity of the lith board as calculated by both methods agreed very closely. The values secured by the Method 2, however, were more consistent and were used in other calculations in preference to those determined by the Method 1. One other feature is noticeable in connection with the Method 2; namely, consistent results of the conductivity could be secured when conditions were not uniform enough to warrant the use of the Method

TABLE 1  RESULTS OF TESTS ON 3-IN. CORK BOARD, SERIES NO. 1

| Test No. | | 1C | 2C | 3C | 4C | 5C | 6C |
|---|---|---|---|---|---|---|---|
| Temperature at thermo couples, deg. fahr. | 1 | 36.95 | 53.69 | 64.94 | 79.79 | 93.20 | 104.53 |
| | 2 | 36.68 | 53.97 | 64.04 | 78.53 | 91.84 | 103.93 |
| | 3 | 33.86 | 50.90 | 60.08 | 75.02 | 87.80 | 100.22 |
| | 4 | 8.33 | 12.65 | 11.20 | 14.18 | 17.96 | 20.12 |
| | 5 | 4.82 | 8.06 | 5.90 | 7.52 | 10.60 | 12.27 |
| | 6 | 4.86 | 7.70 | 5.34 | 7.07 | 10.40 | 11.93 |
| | 7 | 48.36 | 70.25 | 85.53 | 104.18 | 121.37 | 133.16 |
| | 8 | 3.66 | 9.40 | 8.14 | 10.27 | 14.00 | 16.16 |
| | 9 | 29.86 | 45.14 | 53.78 | 67.64 | 78.80 | 90.85 |
| | 10 | 6.56 | 10.40 | 9.23 | 11.30 | 14.90 | 17.39 |
| | 11 | 33.86 | 50.90 | 61.16 | 75.38 | 88.16 | 100.26 |
| | 12 | 6.09 | 10.04 | 8.24 | 10.94 | 14.18 | 16.68 |
| B.t.u. per hr. supplied heater | | 70.62 | 100.00 | 136.26 | 166.16 | 197.69 | 226.56 |
| Average conductivity. B.t.u. per sq. ft. per deg. fahr. per hr. | | 0.102 | 0.098 | 0.105 | 0.104 | 0.108 | 0.109 |
| Average conductivity, B.t.u. per sq. ft. per in. thickness per 24 hr. | | 7.34 | 7.06 | 7.06 | 7.49 | '.78 | 7.85 |
| Average temp. of cork, deg. fahr. | | 21.22 | 32.01 | 36.50 | 45.37 | 53.77 | 61.10 |
| Transmission,[1] B.t.u. per sq. ft. per deg. per hr. | | 0.082 | 0.084 | 0.088 | 0.089 | 0.093 | 0.096 |
| Transmission,[1] B.t.u. per sq. ft. per deg. per 24 hr. | | 1.97 | 1.97 | 2.11 | 2.13 | !.23 | 2.30 |

[1] Based on temperatures 2 in. from surface.
Density of sample, 11.8 lb. per cu. ft.

TABLE 2  RESULTS OF TESTS ON 3-IN. LITH BOARD, SERIES NO. 2

| Test No. | | 4L | 3L | 2L | 2XL | 1L | 5L |
|---|---|---|---|---|---|---|---|
| Temperature at thermocouples, deg. fahr. | 1 | 42.89 | 56.84 | 72.81 | 72.73 | 88.93 | 110.92 |
| | 2 | 42.80 | 56.65 | 72.37 | 72.50 | 88.16 | 111.38 |
| | 3 | 40.19 | 53.06 | 67.91 | 67.91 | 84.79 | 103.38 |
| | 4 | 11.03 | 14.00 | 14.90 | 15.35 | 16.34 | 16.07 |
| | 5 | 6.80 | 8.24 | 7.88 | 8.60 | 8.24 | 6.71 |
| | 6 | 6.53 | 7.88 | 7.70 | 8.33 | 7.79 | 6.62 |
| | 7 | 57.56 | 75.20 | 96.12 | 95.72 | 116.60 | 138.74 |
| | 8 | 8.00 | 9.92 | 10.39 | 10.80 | 11.39 | 10.49 |
| | 9 | 37.22 | 48.56 | 63.20 | 63.33 | 77.18 | 92.21 |
| | 10 | 8.60 | 10.69 | 11.66 | 11.75 | 11.75 | 11.75 |
| | 11 | 41.09 | 54.50 | 69.87 | 69.80 | 84.92 | 105.62 |
| | 12 | 8.60 | 10.54 | 10.99 | 11.64 | 12.20 | 11.73 |
| B.t.u. per hr. supplied heater | | 87.85 | 116.66 | 153.70 | 153.98 | 199.16 | 249.85 |
| Conductivity. B.t.u. per sq. ft. per hr. Method 1 | | | 0.112 | 0.118 | 0.126 | 0.109 | |
| Conductivity. B.t.u. per sq. ft. per hr. Method 2 | | 0.104 | 0.107 | 0.110 | 0.109 | 0.107 | 0.113 |
| Conductivity. B.t.u. per deg. per in. thickness per 24 hr. | | 7.49 | 7.70 | 7.89 | 7.85 | 7.70 | 8.06 |
| Transmission,[1] B.t.u. per sq. ft. per deg. per hr. | | | 0.090 | 0.090 | 0.092 | 0.094 | |
| Transmission,[1] B.t.u. per sq. ft. per deg. per 24 hr. | | | 2.17 | 2.16 | 2.21 | 2.26 | |

[1] Based on temperatures 2 in. from surface.
Density of sample, 10.8 lb. per cu. ft.

1. This fact was made use of in tests Nos. 4L and 3L, in which the temperature difference was small and the time that would have been required to secure uniform conditions extremely long. Since the temperatures were not absolutely constant during these tests, the heat transmissions of tests Nos. 4L and 3L were omitted as they could not be productive of representative results.

The location of the thermometer elements, as indicated in the results will be understood by reference to the key to temperature

*(Continued on page 654)*

# SURVEY OF ENGINEERING PROGRESS

## A Review of Attainment in Mechanical Engineering and Related Fields

### SUBJECTS OF THIS MONTH'S ABSTRACTS

COMBUSTION PROCESS IN THE OIL ENGINE
SOLID INJECTION AND COMPRESSED-AIR INJECTION COMPARED
STEINBECKER PROCESS OF DIESEL-ENGINE INJECTION
HEAT TRANSFER IN FLUES
20,000-R.P.M. SINGLE-STAGE TURBO-BLOWER
AIR FORCES ON CIRCULAR CYLINDERS
CAST IRON FOR LOCOMOTIVE CYLINDER PARTS
CAST IRON, AIR FURNACE AND CUPOLA, COMPARED

CORROSION OF IRON, INFLUENCE OF COPPER, MANGANESE AND CHROMIUM IMPURITIES
PORTLAND CEMENT, LONG-TIME TESTS
HYDRAULIC LIME, LONG-TIME TESTS
VOLCANIC ASHES, LONG-TIME TESTS
PLUTO STOKER AND LOW-GRADE FUELS
FUEL ECONOMY REPORT OF THE BRITISH ASSOCIATION
THERM AS NEW UNIT FOR SALE OF GAS
FLOW OF OIL IN PIPES
FLOW OF OIL AND VISCOSITY
WAVE-POWER TOOLS, DORMAN

CONSTANTINESCO SONIC WAVES APPLIED TO POWER TOOLS
FLEXIBLE PIPE, DORMAN
PUMPS FOR CORROSIVE LIQUIDS
CERATHERM PUMPS
ROTO-PISTON PUMPS
SPRING-SCRAGGING MACHINE
HEAT TRANSFER THROUGH INSULATION AND GEOMETRIC FORM
DYNAMICAL METHOD FOR RAISING GASES TO A HIGH TEMPERATURE
AIR COMPRESSORS, CAUSES OF OVERHEATING

## The Combustion Process in the Oil Engine

THE COMBUSTION PROCESS IN THE OIL ENGINE. Discussion of the various stages constituting the combustion process in an oil engine and criticism of some of the new constructions.

The first of the stages discussed is the fuel injection. For this there are three methods available: namely, injection by compressed air, solid injection without the use of air, and injection by partial combustion in a separate chamber or retort.

As regards air injection, the influence of atomizers, needles and nozzles on the shape of the diagram has been carefully investigated and sufficient data are now available to make it possible to secure a good diagram and a smokeless combustion, no matter whether a medium-pressure, high-speed Diesel engine is desired such as was used on the submarines, or one with good regulation such as a multi-cylinder type engine used for dynamo drive. Likewise, Neumann and others have investigated the influence of the size of the droplets on the performance of the Diesel engine and the size itself has been measured and determined.

The subject of solid injection has been investigated by several concerns in Germany and engines of this kind are being built. Such injection has been used in the first place in hot-bulb engines and is satisfactory for that type. It has been extensively used, however, for Diesel engines by the Vickers concern.

The author does not appear to be in favor of the solid-injection method and gives the following comparison between solid injection and compressed-air injection as calculated for a 1700-hp. submarine-type four-stroke-cycle Diesel engine. In the case of compressed-air injection, an air-fuel volume of 75 cc. (4.57 cu. in.) has to be delivered, per cycle and injection needle, whereas in solid injection there is handled only a volume of 3 cc. (0.18 cu. in.), or one twenty-fifth that in the previous case. But, in the first place, the velocity of injection is in the neighborhood of 300 m. (984 ft.) per sec., and if with solid injection the same constant-pressure diagram is to be obtained, the same amount of fuel has to be injected at the same velocity; and the cross-section of the nozzle must also be only one twenty-fifth that of a corresponding compressed-air injection nozzle, which means that instead of the six nozzles of 2.2 mm. (0.085 in.) diameter used in air injection, there have to be six nozzles likewise used with 0.45 mm. (0.018 in.) in diameter, and in order to introduce the fuel at the same velocity as in the previous case the pressure on the fluid must be somewhere in the neighborhood of 300 atmos. The author claims that the introduction of very small volumes of fuel used makes it an extremely difficult problem to operate at such extremely high pressures.

The author further claims that the difficulties of materializing such high injection pressures are apparent in the Vickers engine in the complicated appliances used for solid injection. Thus, each of the twelve cylinders has its own fuel pump which, with its regulating devices and manometer, is as wide as the cylinder itself. If one should compare a Vickers 12-cylinder engine with a German U-boat-type four-stroke-cycle engine likewise of 1200 hp. but with

six cylinders, it is easy to see how tremendously the difference in the method of injection affects the engine design; and the author fails to see in what way the elimination of an air compressor does anything to simplify the design, particularly as the use of a three- or four-stage compressor would make its operation quite simple.

The author is not aware of any published indicator diagrams of any precision covering the operation of solid-injection engines except the diagrams which appeared in *The Engineer* for Nov. 14, 1919, and apply to a submarine-type Vickers engine. The

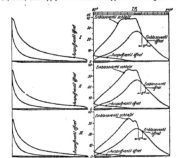

FIG. 1 INDICATOR DIAGRAMS OF A VICKERS SUBMARINE-BOAT TYPE ENGINE WITH SOLID INJECTION

(*Auspufventil oeffnet.* exhaust Valve opens; *Einblaseventil schliest,* inlet Valve closes; *Einblaseventil oeffnet,* inlet Valve opens; *Auspufventil oeffnet,* exhaust Valve opens; *at,* atmospheres; *TP,* dead center.)

author compares these diagrams with similar diagrams obtained from a 1700-hp. U-boat engine built at the Germania Works in Germany. His criticism of the Vickers diagram (Fig. 1) is of interest.

TABLE 1 DATA REFERRING TO DIAGRAMS IN FIG. 1

| Diagram No. | R.p.m. | Pressure, Kg. per Sq. Cm. Pi | Output, I.hp. | Fuel-Injection Pressure, Kg. per Sq. Cm. | Duration of Opening, Deg. | Fuel-Injection Valve — Beginning of Opening Deg. Ahead Dead Center | Beginning of Pressure Rise after Valve Opening — Deg. | Beginning of Pressure Rise after Valve Opening — Seconds |
|---|---|---|---|---|---|---|---|---|
| I..... | 358 | 6.7 | 107 | 302 | 43 | 18 | 8 | 0.0037 |
| II...... | 312 | 5.4 | 76 | 175 | 33 | 16 | 8 | 0.0043 |
| III...... | 182 | 4.5 | 36 | 84 | 21 | 10 | 6 | 0.0055 |

The compression is set at 25 atmos. The explosive character of

627

the combustion is indicated by the substantial rise in pressure during the injection of the fuel. The pressure under which the fuel is introduced is in the first full-load diagram at a level somewhat exceeding 300 atmos. or just about the values computed above. Properly timed introduction or governing of fuel injection, such as one is accustomed to in compressed-air-injection engines, is out of the question here. Furthermore, the low compression cannot be used for computing the dimensions of the parts, such as base-plate, crankshaft, cylinder or cylinder cover, and therefore it is impossible to secure a low-weight engine, this being due to the fact that the compression pressure in low-compression, solid-injection engines rises to the same levels as in compressed-air-injection engines having a compression of from 30 to 33 atmos. Moreover, as is shown by the behavior of the expansion line with partial load, there is a strong after-combustion. It is difficult to understand how under these conditions the Vickers engine can secure the fuel combustion of 170 grams per hp-hr. (0.374 lb.) claimed for it by the makers—all the more so as apparently smokeless combustion is not secured. In proof of this latter statement, the

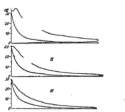

Fig. 2 Indicator Diagrams of a 1700-hp. U-Boat 4-Stroke Cycle Engine of the German Germaniawerft

author quotes the following passage from an article by an English submarine-boat officer (*Motorship and Motorboat*, 1919, p. 175).[1] "They (meaning the Vickers engine) have, however, an important disadvantage, namely, they develop clouds of smoke which stretch for miles behind the submarine. The engines also *vibrate*, no matter how carefully they may be adjusted, and if the personnel in the engine room is not very careful, they may easily produce a smoke-screen effect."

For purposes of comparison, indicator diagrams are reproduced in Fig. 2 which were taken under approximately the same conditions in respect to speed in revolutions and average compression from a 1700-hp. German U-boat engine with compressed-air injection. Here, apparently, explosive combustion and the sudden pressure rise which accompany it are completely eliminated, and the after-burning effect is comparatively slight.

The author comes to the conclusion that solid injection is suitable only for engines with a comparatively small output per cylinder

[1] Passage retranslated into English from the German, as the original publication is not available.—EDITOR.

Table 2 Data Referring to Diagrams in Fig. 2

| Diagram No. | r.p.m. | Pressure, per Sq. Cm. Kg. | Output I.hp. | Fuel-Injection Pressure, Kg. per Sq. Cm. |
|---|---|---|---|---|
| I | 377.6 | 8 | 431 | 85 |
| II | 295.8 | 5 3 | 179 | 55 |
| III | 239 | 3.5 | 109 | 45 |

and operating at fairly constant speed and load. For high-efficiency engines of large output, the solid-injection process must be considerably improved before it can enter into competition with compressed-air injection, and even then it is claimed by the author to have no advantage as compared with the Steinbecker process, which also operates without a compressor.

The main features of this process have already been described. Essentially, it consists of the following: A fuel pump located in the cylinder cover injects the fuel into a special passage about 3 deg. before dead center. A part of the fuel together with air

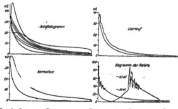

Fig. 3 Indicator Diagrams of an Experimental Steinbecker Diesel Engine
(*Anlassdiagram,* starting curve; *Normallast,* normal load; *Leerlauf,* idling; *Diagram der Retorts,* indicator diagram for the "retort;" *ai,* atmospheres.)

enters a special chamber or retort filled with highly preheated air having a pressure of 30 atmos., which is several atmospheres below the compression pressure and is ignited there so that the pressure reaches 65 atmos. This pressure rise in the retort produces a powerful injection of the fuel into the main cylinder, and it is claimed that the action of this case is better than with solid injection, and the compression as good as with compressed-air injection, while the elimination of the compressor, the fuel-injection valve and the injection compressed-air tank with its valves and accessories are of course of material benefit.

In Fig. 3 are given four diagrams taken on an experimental engine of the Steinbecker type. It is claimed that the combustion was smokeless throughout.

The rest of the article discusses the subjects of vaporization, ignition, and combustion. These may be abstracted in an early issue. (*Zeitschrift des Vereines deutscher Ingenieure,* vol. 64, no. 33, Aug. 14, 1920, pp. 637-643, 13 figs., *tc*)

## Heat Transfer in Flues

HEAT TRANSFER IN FLUES, Lawford H. Fry, Mem.Am.Soc.M.E. Discussion of the laws governing heat transfer in flues, with particular application to locomotive-boiler flues.

The present article is an extension of a former article by the author in *Engineering* for Feb. 19, 1909, and a paper read before The American Society of Mechanical Engineers in 1917 (Trans.Am.Soc.M.E., vol. 39, p. 709), and is based on recent work of the author and others.

The rate of heat transfer is studied in its relation to the various factors on which it depends. These factors are:

1. *Temperature.* The rate of heat transfer depends on the ratio of the temperature of the gas to the temperature of the flue wall.

If the flue wall has a mean temperature $t$, and if the mean gas temperature at one section of the flue is $T_1$ and at another section $x$ ft. distant along the flue is $T_2$, all temperatures being measured from absolute zero in any scale, the relation between the two gas temperatures is given by the equation:

$$\text{lolog } T_1/t - \text{lolog } T_2/t = Mx \dots \dots [1]$$

where "lolog" means "the logarithm of the logarithm," and where $M$ is a constant for a given set of conditions, being determined by the dimensions of the flue and by the rate of gas flow.

2. *Flue Dimensions.* The two dimensions on which the above constant $M$ (and hence the rate of heat transfer) depends, are the perimeter and the mean hydraulic depth. The mean hydraulic depth is the area divided by the perimeter, and in a circular flue

is one-fourth of the diameter. When circular flues only are under consideration it is simpler and equally accurate to speak of the effect of the flue diameter instead of the effect of the mean hydraulic depth. This will be done in the present article, the symbol $d$ being used to represent the flue diameter in inches. If the equations thus established are to be applied to annular or other noncircular flues, it is only necessary to replace $d$ by $4h$, where $h$ is the mean hydraulic depth.

3. *Rate of Flow of Gas.* The effect of the rate of gas flow on the rate of heat transfer is dependent on the weight of gas flowing per hour per inch of flue perimeter. Expressed in symbols this statement is to the effect that the coefficient $M$ in Equation [1] varies with the ratio $W/p$, where $W$ is the weight of gas flowing in pounds per hour and $p$ is the flue perimeter in inches.

The relation between the flue dimensions, the rate of gas flow and the coefficient $M$ is that given by the following equations:

$$\log M = B - m \log W/p \quad\quad\quad [2]$$

where

$$\log (B + 1.3) = \bar{1}.71 - 0.54 \log d \quad\quad [3a]$$

and

$$\log m = \bar{1}.36 + 0.37 \log d \quad\quad\quad [3b]$$

The derivation of these equations from the experimental data is discussed below, but attention is first directed to the conditions of heat transfer disclosed by the formulæ.

Equations [2], [3a] and [3b] show that for a given flue with gas flowing at a constant rate, the coefficient $M$ in Equation [1] is a constant. This means that if distances along the flue $x$ are measured from a point at which the gas temperature is $T$, the gas temperature $T_1$ at any other point is given by equation:

$$\text{lolog } T/t = \text{lolog } T_1/t - Mx \quad\quad\quad [1a]$$

This is to say, the logarithm of the logarithm of the ratio of gas temperature to flue temperature (lolog $T/t$) falls by a straight-line law along the flue. The slope of the line giving this fall of the log of the temperature ratio is measured by the coefficient $M$.

This equation enables determining the heat transfer from gas to flue in any section of the flue by finding the temperature loss of the gas.

Furthermore, the form of Equation [1a] indicates that a large value of $M$ corresponds to a rapid drop of temperature along the flue, that is, a rapid rate of heat transfer.

Table 1 shows how the value of $M$ is affected by the three factors

TABLE 1 VALUE OF COEFFICIENT M FOR VARIOUS FLUE DIAMETERS AND RATES OF GAS FLOW

| In. | 100 Lb./Hr. | 200 Lb./Hr. | 400 Lb./Hr. | 800 Lb./Hr. | 1600 Lb./Hr. |
|---|---|---|---|---|---|
| | | | Values of W— | | |
| 0.5 | 0.133 | 0.117 | 0.104 | 0.092 | 0.081 |
| 1.0 | 0.074 | 0.063 | 0.054 | 0.046 | 0.039 |
| 2.0 | 0.050 | 0.040 | 0.033 | 0.027 | 0.022 |
| 4.0 | 0.040 | 0.030 | 0.023 | 0.018 | 0.014 |
| 8.0 | 0.037 | 0.026 | 0.019 | 0.013 | 0.009 |

on which it depends, namely, flue diameter, flue perimeter and rate of gas flow. Another table (Table 2) in the original article is given to show the length of flue necessary to reduce the gas temperature from an initial value of 2000 deg. to 720 deg. fahr. with a flue-wall temperature of 380 deg. and from this table it appears that with a $^1/_2$-in. flue the above temperature drop will take place in 3.76 ft. of pipe with a flow of 100 lb. per hour, while

a length of 6.15 ft. will be required for the same temperature drop with a flow of 1600 lb. per hour. This is particularly significant as in the first instance the heat transferred from gas to flue is equal to 41,800 B.t.u. per hour, while in the second case the rate of heat transfer will be 16 times as great, or 668,800 B.t.u. per hour.

From the data given in Table 1 it would appear that the rate of heat transfer falls off rapidly as the flue diameter is increased. Other tables are given showing the temperature drop and the amount of heat transferred in flues 5 ft. long when the flue diameter and rate of gas flow are varied. These tables show clearly the effect of increasing flue diameter on increasing the final temperature, that is, on reducing the amount of heat transfer, while the other tables illustrate the effect on the heat transfer of varying the flue length, flue diameter, rate of gas flow and the gas temperature as well as the temperature of the flue wall.

In order to facilitate the use of the formulæ Tables 2 and 3

TABLE 2 VALUES OF COEFFICIENTS m AND B FOR FLUES OF VARIOUS DIAMETERS

| d In. | B | m | d In. | B | m |
|---|---|---|---|---|---|
| 0.38 | $\bar{1}.561$ | 0.160 | 1.50 | $\bar{1}.112$ | 0.266 |
| 0.44 | $\bar{1}.500$ | 0.169 | 2.0 | $\bar{1}.053$ | 0.296 |
| 0.50 | $\bar{1}.444$ | 0.178 | 2.25 | $\bar{1}.031$ | 0.309 |
| 0.58 | $\bar{1}.388$ | 0.188 | 2.5 | $\bar{1}.013$ | 0.321 |
| 0.66 | $\bar{1}.341$ | 0.197 | 3.0 | $\bar{2}.983$ | 0.342 |
| 0.75 | $\bar{1}.300$ | 0.208 | 3.5 | $\bar{2}.961$ | 0.364 |
| 0.87 | $\bar{1}.253$ | 0.218 | 4.0 | $\bar{2}.943$ | 0.382 |
| 1.00 | $\bar{1}.213$ | 0.229 | 5.0 | $\bar{2}.915$ | 0.413 |
| 1.15 | $\bar{1}.175$ | 0.242 | 6.0 | $\bar{2}.895$ | 0.443 |
| 1.30 | $\bar{1}.143$ | 0.252 | 8.0 | $\bar{2}.867$ | 0.494 |

are given, of which the former gives for various flue diameters the intermediate coefficients $m$ and $B$ which are used in establishing the coefficient $M$, while Table 3 gives values of the coefficient $M$ for a number of flue diameters and rates of gas flow. The relation of $m$ and $B$ to the flue diameter is given by the Equations [3a] and [3b]; the values of coefficient $M$ given in Table 3 are calculated from Equation [1], where $W$ is the weight of gas in lb. flowing per hour and $p$ is the flue perimeter in inches.

The data presented above were obtained by the use of methods described in the author's paper in Trans.Am.Soc.M.E., vol. 39, 1917, p. 709, in which he surveyed some of the experimental work done on heat transfer between the gas and the flue and derived experimental formulæ. As he stated, he had tried to harmonize some of the various groups of experiments without success until Hedrick and Fessenden put forward their double logarithmic (lolog) formula (Trans.Am.Soc.M.E., vol. 38, 1916, p. 407). The use of an expression of this type led to the formulæ which have been given above and which give a much wider range than any previous experiences for heat transfer.

The formulæ given by the writer are based on the experiments of Jordan, Nusselt, Josse, The Babcock & Wilcox Company, and Fessenden. In addition to these, the author analyzed five series of boiler tests, two of which were made by the Pennsylvania Railroad with locomotive boilers and the other three with stationary boilers—one by Nicolson on an experimental plug-type boiler, the second by the Bureau of Mines on a Heine boiler, and the third on a return-type boiler. (*Engineering*, vol. 110, no. 2852, Aug. 27, 1920, pp. 265–268, 5 figs., teA)

TABLE 3 VALUES OF COEFFICIENT M

| Pounds of Gas per Inch of Perimeter per Hr. W/p | d = Inside Diameter of Flue in Inches— | | | | | | | | | | | | | | | | |
|---|---|---|---|---|---|---|---|---|---|---|---|---|---|---|---|---|---|
| | 0.38 | 0.44 | 0.50 | 0.58 | 0.66 | 0.75 | 0.87 | 1.00 | 1.15 | 1.30 | 1.50 | 1.75 | 2.00 | 2.25 | 2.50 | 3.00 | 3.50 | 4.00 |
| 2.00 | 0.326 | 0.281 | 0.246 | 0.215 | 0.1915 | 0.1725 | 0.1540 | 0.1390 | 0.1265 | 0.1170 | 0.1080 | 0.0986 | 0.0920 | 0.0866 | 0.0824 | 0.0758 | 0.0710 | 0.0673 |
| 2.83 | 0.310 | 0.265 | 0.231 | 0.201 | 0.1790 | 0.1603 | 0.1420 | 0.1290 | 0.1165 | 0.1075 | 0.0985 | 0.0895 | 0.0831 | 0.0780 | 0.0738 | 0.0674 | 0.0625 | 0.0589 |
| 4.00 | 0.292 | 0.250 | 0.217 | 0.188 | 0.1670 | 0.1493 | 0.1320 | 0.1190 | 0.1070 | 0.0983 | 0.0897 | 0.0811 | 0.0730 | 0.0700 | 0.0659 | 0.0598 | 0.0551 | 0.0517 |
| 3.86 | 0.276 | 0.236 | 0.204 | 0.176 | 0.1563 | 0.1390 | 0.1225 | 0.110 | 0.1009 | 0.0901 | 0.0818 | 0.0736 | 0.0677 | 0.0630 | 0.0590 | 0.0532 | 0.0485 | 0.0452 |
| 8.00 | 0.261 | 0.222 | 0.192 | 0.166 | 0.1460 | 0.1290 | 0.1135 | 0.1015 | 0.0903 | 0.0826 | 0.0746 | 0.0667 | 0.0611 | 0.0565 | 0.0527 | 0.0472 | 0.0427 | 0.0396 |
| 11.3 | 0.246 | 0.210 | 0.180 | 0.155 | 0.1360 | 0.1205 | 0.1053 | 0.0935 | 0.0831 | 0.0756 | 0.0680 | 0.0592 | 0.0552 | 0.0507 | 0.0472 | 0.0420 | 0.0377 | 0.0347 |
| 16.0 | 0.233 | 0.198 | 0.169 | 0.145 | 0.1270 | 0.1120 | 0.0974 | 0.0864 | 0.0763 | 0.0693 | 0.0621 | 0.0548 | 0.0498 | 0.0455 | 0.0422 | 0.0372 | 0.0332 | 0.0304 |
| 22.6 | 0.221 | 0.186 | 0.159 | 0.136 | 0.1187 | 0.1043 | 0.0903 | 0.0800 | 0.0703 | 0.0635 | 0.0566 | 0.0498 | 0.0450 | 0.0409 | 0.0378 | 0.0331 | 0.0292 | 0.0266 |
| 32.0 | 0.209 | 0.176 | 0.150 | 0.127 | 0.1110 | 0.0970 | 0.0837 | 0.0738 | 0.0645 | 0.0582 | 0.0516 | 0.0451 | 0.0405 | 0.0367 | 0.0338 | 0.0294 | 0.0258 | 0.0233 |
| 45.2 | 0.198 | 0.166 | 0.141 | 0.119 | 0.1035 | 0.0903 | 0.0776 | 0.0682 | 0.0594 | 0.0533 | 0.0471 | 0.0409 | 0.0366 | 0.0329 | 0.0303 | 0.0261 | 0.0227 | 0.0200 |
| 64.0 | 0.187 | 0.157 | 0.133 | 0.112 | 0.0965 | 0.0840 | 0.0720 | 0.0631 | 0.0546 | 0.0489 | 0.0430 | 0.0371 | 0.0329 | 0.0296 | 0.0270 | 0.0232 | 0.0200 | 0.0174 |
| 90.5 | 0.177 | 0.148 | 0.125 | 0.1047 | 0.0903 | 0.0781 | 0.0667 | 0.0583 | 0.0502 | 0.0448 | 0.0392 | 0.0336 | 0.0305 | 0.0265 | 0.0242 | 0.0206 | 0.0176 | 0.0159 |
| 128.0 | 0.167 | 0.139 | 0.117 | 0.0981 | 0.0843 | 0.0727 | 0.0618 | 0.0538 | 0.0461 | 0.0410 | 0.0357 | 0.0305 | 0.0268 | 0.0239 | 0.0217 | 0.0183 | 0.0155 | 0.0137 |
| 181.0 | 0.158 | 0.131 | 0.110 | 0.0918 | 0.0788 | 0.0677 | 0.0573 | 0.0498 | 0.0425 | 0.0376 | 0.0326 | 0.0276 | 0.0242 | 0.0215 | 0.0194 | 0.0162 | 0.0137 | 0.0127 |
| 256.0 | 0.150 | 0.124 | 0.104 | 0.0861 | 0.0736 | 0.0630 | 0.0531 | 0.0459 | 0.0390 | 0.0344 | 0.0297 | 0.0250 | 0.0219 | 0.0193 | 0.0173 | 0.0144 | 0.0120 | 0.0106 |
| 362.0 | 0.142 | 0.117 | 0.0974 | 0.0807 | 0.0688 | 0.0586 | 0.0492 | 0.0425 | 0.0359 | 0.0315 | 0.0271 | 0.0227 | 0.0198 | 0.0173 | 0.0155 | 0.0128 | 0.0106 | 0.0092 |
| 512.0 | 0.134 | 0.110 | 0.0916 | 0.0756 | 0.0642 | 0.0546 | 0.0456 | 0.0392 | 0.0329 | 0.0289 | 0.0247 | 0.0206 | 0.0178 | 0.0155 | 0.0139 | 0.0114 | 0.0094 | 0.0081 |
| 724.0 | 0.127 | 0.104 | 0.0862 | 0.0710 | 0.0601 | 0.0508 | 0.0423 | 0.0361 | 0.0303 | 0.0265 | 0.0225 | 0.0187 | 0.0161 | 0.0140 | 0.0124 | 0.0101 | 0.0082 | 0.0071 |
| 1024.0 | 0.120 | 0.098 | 0.0811 | 0.0665 | 0.0561 | 0.0473 | 0.0392 | 0.0334 | 0.0278 | 0.0242 | 0.0205 | 0.0169 | 0.0145 | 0.0125 | 0.0111 | 0.0090 | 0.0072 | 0.0062 |

## Short Abstracts of the Month

### AIR MACHINERY (See Thermodynamics)

### DeLaval Direct-Connected High-Speed Turbo-Blower

22,000-R.P.M. SINGLE-STAGE TURBO-BLOWER. In 1916 the Westinghouse Machine Company built a small blower designed to run at 43,000 r.p.m. which actually ran at speeds as high as 60,000 r.p.m. In one of the tests the rotor of this blower was damaged, and there is no information available whether it was ever actually placed on the market.

It has proved, however, that blowers running at these terrific speeds are fully practicable.

In this connection the unit recently built by the Rateau, Battu, Smoot Co. of New York City and described in *Power* is of par-

FIGS. 1 AND 2 22,000-R.P.M. SINGLE-STAGE TURBO-BLOWER

ticular interest. The unit was built for a South American mining concern and before shipment was operated at 26,000 r.p.m. for test purposes, but is intended for normal operation at 22,000 r.p.m. It is a single-stage turbo-blower which, when operating at its normal speed, will compress 3000 cu. ft. of free air per min. to

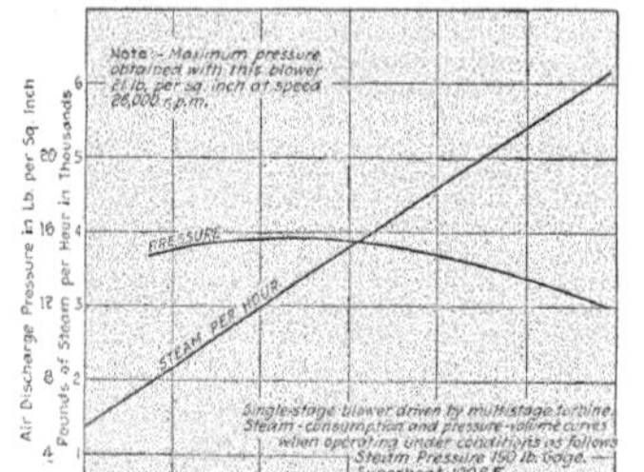

FIG. 3 PERFORMANCE CURVES OF TURBO-BLOWER OF FIG. 1

15 lb. per sq. in. Fig. 1 shows the complete unit assembly, with the turbine at the left.

In the turbine the rotor has three wheels which are machined as an integral part of the shaft which is made of a nickel-chrome-magnesium steel, heat-treated forging, the radial blades being machined as an integral part of the shaft. Fig. 2 shows the rotor removed from the bearing and resting on the lower half of the casing. The 6-in. rule standing vertically on the right-hand end

of the casing gives an idea of the size of the rotor, which weighs only 138 lb., of which 100 lb. is in the turbine rotor, and only 38 lb. in the blower rotor.

The rotor had to be made of a single piece in order to give it the maximum rigidity. It is so designed that at all points the stresses are far below the elastic limit of the metal. In machinery running at the high speeds employed in this case (at 26,000 r.p.m. the tip speed of the blower rotor is 1400 ft. per sec., or greater than the velocity of sound) centrifugal forces become of extreme importance, as one ounce weight on the end of the blade exerts a radial pull of approximately 3.75 tons.

Fig. 3 gives some data as to the normal performance of the turbine and blower as shown by the elaborate tests made before shipment. An interesting feature of the operation is that at full speed and overspeeds the machine operated without the slightest variation. The results of these tests also show that two machines of this type working in series can be made to compress air to 100 lb. pressure in small units, and three machines in series would give a compression as high or higher than 100 lb. for large-size machines. (*Power*, vol. 52, no. 9, Aug. 31, 1920, pp. 327–328, 4 figs., d)

### BUREAU OF STANDARDS

AIR FORCES ON CIRCULAR CYLINDERS, AXES NORMAL TO THE WIND, WITH SPECIAL REFERENCE TO DYNAMICAL SIMILARITY, Hugh L. Dryden. One of the most difficult problems of the airplane designer is to obtain a method of computing forces on full-size machines or full-size machine parts from measurements made on models in a wind tunnel. A certain equation deduced from theoretical considerations has long been known and used, namely, the equation proposed a long time ago, first by Helmholtz, later stated by Reynolds, and developed more fully by Lord Rayleigh and Buckingham. This equation is a logical consequence of certain assumptions and states that, if these assumptions are true, the force of a current of air moving against a solid body may be expressed as $C\rho A V^2$, $\rho$ being the density of the air, $A$ the area of the body projected on a plane normal to the wind, $V$ the velocity of the wind, and $C$ a dimensionless constant depending on a single parameter $VL/\nu$, where $\nu$ is the kinematic viscosity of the air and $L$ is a linear dimension of the body. This equation then implies that if $C$ be plotted against $VL/\nu$, the points will fall on the same curve independent of the individual value of $V$ or $L$ provided the bodies are geometrically similar and presented in the same manner to the wind.

This law is a logical consequence of the assumptions. Although some tests have been made of the validity of the assumptions in the case of wind-tunnel experiments, no extensive investigation of a body of simple geometrical form over a large range of values of $V$ and $L$ has been published. The present paper gives the results of tests on cylinders for different values of $L$. An attempt was made to make the assumptions involved in the derivation of the equation true as closely as is possible in wind-tunnel experiments. Cylinders of 1, $1\frac{1}{4}$, $1\frac{1}{2}$, $1\frac{3}{4}$, 2, $2\frac{1}{2}$, 3, 4, $4\frac{1}{2}$, 5, $5\frac{1}{2}$ and 6 in.

(0.0254 to 0.1524 m.) were used at velocities from 15 to (in the case of the smaller cylinders) 80 m.p.h. (25 to 130 km. per hour, approximately). The range of values of $VL/\nu$ was from 10,000 to 185,000. The cylinders were made of wood with the exception of those less than 2 in. in diameter, which were in brass; but an additional 1-in. wood cylinder and 4-in. brass cylinder were also used. The "guard ring" principle was used to obtain results applicable to infinite cylinders.

The results show that the equation does not represent the facts of wind-tunnel experiments in the case of cylinders. The coefficient $C$ for a 1-in. cylinder is half again as large as that for a 3-in. cylinder at the same value of $VL/\nu$. Above a 3-in. diameter the equation is satisfied closely, the coefficient being practically constant and equal to 0.426. The curve for the 1-in. cylinder checks N.P.L. values closely. The maximum departure of any one observation from the mean is about $2\frac{1}{2}$ per cent. Measurements of the pressure distribution showed that in the case of the small cylinders the ratio of the average decrease in pressure on the back to the maximum increase in pressure on the front is greater than in the case of the large cylinders. (Abstract of *Scientific Paper of the Bureau of Standards*, no. 394, *t*)

CAST IRON FOR LOCOMOTIVE-CYLINDER PARTS, C. H. Strand. Frequent removal of cylinder parts of locomotives results in greatly increased cost of maintenance to the railroads, and consequently the quality of the cast iron entering into their construction is a matter of paramount importance, particularly from the standpoint of wear. These parts include piston-valve bushings, piston-valve packing rings, piston-valve bull rings, cylinder bushings, piston packing rings and piston-head or bull rings. It was found that ordinary high-silicon cast iron gave unsatisfactory wear, particularly in modern superheater locomotives, and the tendency has been toward a harder and stronger iron.

At the request of the U. S. Railroad Administration, the Bureau of Standards has investigated the mechanical, chemical, and microscopical properties of a number of packing rings furnished with service mileage records, as well as arbitration-test bars, chill-test specimens, and miscellaneous examples from different manufacturers. All of this material was cast iron such as used for the various cylinder parts. The Bureau of Standards at the same time made a review of the previous work and specifications on the subject, to ascertain as far as possible the practices of the different foundries, and to suggest such revision of existing specifications as would be warranted by the results of the present and other investigations.

It was found that air-furnace iron is made more uniform in character and in general of somewhat better mechanical properties than cupola iron. The latter, however, often equals or even excels the air-furnace product in mechanical properties. Because of the many variable factors, it was difficult to establish correlation between laboratory and service tests. It was recommended, as a result of the present and other investigations, that the transverse-strength requirements of the Standard American Society for Testing Materials $1\frac{1}{4}$-in. Arbitration Bar be increased from 3200 to 3500 lb. for castings $\frac{1}{2}$ in. or less in thickness, and from 3500 to 3800 lb. for castings over $\frac{1}{2}$ in. in thickness. (Abstract of *Technologic Paper of the Bureau of Standards*, no. 172, *e*)

## ENGINEERING MATERIALS

THE INFLUENCE OF COPPER, MANGANESE AND CHROMIUM AND SOME OF THEIR COMBINATIONS ON THE CORROSION OF IRON AND STEEL, E. A. Richardson and L. T. Richardson. Data of experimental work made upon steel both free from and containing copper, and upon commercial pure iron both free from and containing copper.

These tests brought out a number of deductions, some of which were known before. Steel rusts much faster than iron, but the presence of iron decreases the corrosion of both—more so, however, in the case of steel. It was further found that it was the manganese present in the steel, but not in the iron, which was intensifying the action of copper in reducing corrosion.

Of considerable interest are the remarks of the authors on pure iron. Hitherto the effort was to make manganese in pure iron intended for use as a rust-resisting material as low as possible. However, commercially pure iron contains about 0.04 per cent of copper, which causes it to be red-short. The addition of a small amount of manganese would make this iron slightly more rust-resisting and at the same time remove the tendency to be red-short, while the addition of still more copper and manganese would materially increase its rust-resisting properties.

The cause of the influence of copper on the corrosion of iron has never been satisfactorily explained. The present authors propose what they call the "film" or "intergrain" theory of corrosion resistance. This theory is based on the assumption that iron only becomes rust-resistant when certain impurities are present in certain amounts. Copper creates an intergrain rust-resisting film, which, on the other hand, also assists in creating red-shortness. A brief list of references covering previous research on corrosion, and especially influence of impurities, is appended in the original article. (Paper presented at the general meeting of the *American Electrochemical Society*, Cleveland, Ohio, Sept. 30–Oct. 2, 1920, pp. 123–135, 6 figs., *tp*. Abstracted through advance copy)

## 20-"Year" Tests of Cement and Mortar Materials

LONG-TIME TESTS OF PORTLAND CEMENT, HYDRAULIC LIME AND VOLCANIC ASHES, I. Hiroi. Data of two series of tests covering a period of more than 20 years. Some of the cements used for the first series of tests were manufactured by the old process of burning in so-called "bottled" kilns, while those in the second series were produced in modern rotary kilns. Another series of tests is also under way which it is proposed to extend over a period of 100 years.

The following are some of the conclusions arrived at by the author: In sea water neat-cement briquets attained maximum strength in the course of less than a year, after which they rapidly declined, in some cases completely losing their tensile strength in four or five years. The author ascribes this peculiarity to excessive crystallization, which makes the structure highly brittle. While losing their tensile strength almost entirely, these briquets retain their form and also show considerable amounts of compressive strength, which latter, in fact, possibly even increases while the tensile strength decreases.

Cement-sand mortar mixtures in the proportion of one part of the former to two parts of the latter in air and sea water show progressive increase of strength with age, the air curves running much higher than sea-water ones. The mean results apparently follow more or less closely hyperbolic curves (for which equations are given in the original paper), which would place the eventually attainable tensile strength of such mortar in air and sea water at 85 and 50 kg. per sq. cm. (1210 and 710 lb. per sq. in.), respectively.

Tests were also made to determine the influence of the kind of sand used for mortars, and it was found that while standard and coarse sands produced practically equal strength, fine sand was found to be decidedly inferior in air, fresh water and sea water, both in tension and in compression.

As regards the strength of mortars in sea water, it was found that the lower the proportion of sand in the mortar the stronger the latter will be.

The results, however, have also suggested the possibility that in mortars to be used in sea water an excess of cement is to be avoided as much as a deficiency. A proportion richer than 1 : 1 is too costly and one lower than 1 : 3 may have voids; the proportion of 1 : 2, on the other hand, has little more cement than is sufficient to fill up all the interstices in the sand.

Tests on the effect of curing in fresh water on the strength of mortars kept in sea water have not indicated any definite and material improvement due to curing.

An interesting series of tests was made on the use of volcanic ashes in cements, the ashes used being of Japanese origin exclusively. In air tests it was found that the greater the amount of ashes used the lower the strength of the briquets, and none of the ash-cement mortars had a strength equal to that of the straight cement ones. On the other hand, in sea-water tests the superiority was decidedly on the side of the ash-cement mortars, which are, however, weaker than straight cement mortars in compression.

The action of volcanic ashes when used in a cement mortar appears to be twofold, viz., mechanical and chemical. Mechanically, the ashes increase the density of the mixes, making the latter more or less impermeable to sea water; chemically, the combination of silica with free lime in cement, which makes the latter unassailable by the sulphates contained in sea water, seems to be the most important action. The activity of silica contained in ashes naturally depends on the state in which it is present; and while there is no doubt that the soluble portion is the most active agent, the total amount of silica should also be taken into consideration. Thus, the Otaru ashes, which according to the analysis contain the least amount of soluble silica of the three but the largest amount of the insoluble one (on an average of 61 per cent in the Otaru ashes, 47 per cent in the Yoichi, 34 per cent in the Goto), produced higher strength than either of the other two ashes. That a portion of insoluble silica enters into combination to form soluble compounds in course of time is shown by the several analyses made of (Otaru) ash-cement mortar block kept in sea water, the results of which are given in the following table:

| Time of Analysis | Silica (Total 100) | |
| --- | --- | --- |
| | Soluble | Insoluble |
| Before induration | 43.73 | 56.27 |
| 2 months | 46.28 | 53.72 |
| 7 months | 47.29 | 52.71 |
| 14 months | 50.08 | 49.92 |
| 38 months | 53.95 | 46.05 |

The value of volcanic ashes, as an ingredient in a cement mortar is possible of direct determination by testing the combining power of the ashes with lime. (*Journal of the College of Engineering, Tokyo Imperial University*, vol. 10, no. 7, 1920, pp. 155–172 and 10 plates, *e*)

## FUELS AND FIRING

### German Pluto Stoker and Its Improvements

DEVELOPMENT OF THE PLUTO STOKER FOR THE UTILIZATION OF LOW-GRADE FUELS, Otto Nerger. While attempts were made in Germany and Austria before the war to utilize the poorer grades of fuel, it was under the pressure of war conditions that a really earnest effort was made in this direction. Among the poorer grades of fuel available for industrial use on a large scale by the Central Powers are, first, the waste products of the coal mines, such as coal dust, breeze, products of coal washing, coke breeze

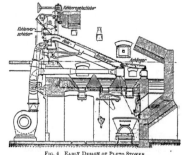

Fig. 4   EARLY DESIGN OF PLUTO STOKER

(*Kohlenregelschieber*, slide plate to govern rate of supply of coal; *Kohlenvorschieber*, slide plate to govern supply of fresh coal to fuel bed; *Hohlrostkörper*, hollow grate bars; *Anhänger*, rocking slag grate.)

and coke ashes. Next to this come such low-grade coals as lignite and peat, the latter fuel being largely in a class by itself, however.

All of these low-grade fuels require special appointments for their utilization, depending to a certain extent on the moisture and ash contents of the fuel, but above all on the content of vola-

tile matter in the fuel, which latter ultimately determines whether the fuel can be utilized at all.

Among other things, low-grade fuels require special grates, and in Germany forced-draft traveling grates were chiefly used, while in Austria and Hungary preference was given to the so-called Pluto stoker.

A Pluto stoker of old design is shown in Fig. 4. Its chief characteristic was the use of a massive hollow grate body and rigid side walls. Early users often found that the hollow grate bars did not last very long and that the fire on them could not be well maintained. In later years, however, improvements made in the design largely eliminated this source of trouble, and it was found that some of the early troubles were due chiefly to the fact that the amount of draft available in the furnaces was entirely insufficient, which led to improper combustion and rapid deterioration of the grate bars.

It is claimed that in those early days there existed an erroneous impression that the availability of forced draft made the further and more important action of the smoke stack unnecessary, which is, however, incorrect, as the action of the forced draft can hardly be sufficient to overcome the resistance of the fuel bed and to produce a slight vacuum in the fire chamber. Even then the question of carrying off the gases of combustion without material turbulence remains still to be taken care of.

In an effort to remedy as far as possible the troubles in the hollow grate bars indicated above, an attempt was made to produce what the original article calls a "removable grate," which makes it possible to take out particularly badly attacked parts of the grate surface and to replace them by fresh elements at a comparatively low expense. Fig. 5 shows such a grate with removable bars. From this it appears that the general shape of the grate has not been changed, but that each section consists of a frame and grate elements inserted therein. This construction has also been of advantage because the bending of the massive hollow grate bars used in the older structure has disappeared in the new, and it is claimed that the new type of grate, of which many thousands were installed during the last five years, has proved to be more economical from the point of view of replacements than chain grates. This is due to the fact that in the Pluto stoker it is only the inserted elements in the firing zone that have to be replaced from time to time, while the frame and other parts are subject to practically no wear.

FIG. 5  PLUTO STOKER GRATE WITH REMOVABLE BARS

In order to reduce as much as possible the dropping through the grate of unconsumed combustible material, the rigid side walls of the older grate were replaced by a springy, adjustable structure such as is shown in Fig. 6. From this it would appear that the structure maintains its adjustment at all times because the springs are located not at the level of the combustion chamber, but in the comparatively cool ashpit. Furthermore, the arrangement of the parts is such that there is always an opportunity to tighten up the springs even while the furnace is being operated.

In many types of furnace it has been found that the residues of combustion clinker together on the short and rigid ash grate and can be removed only by knocking them off with bars through sidewise-located openings. This trouble has been encountered quite often, particularly with fuels inclined toward the formation of clinkering slags. To obviate this the usual rigid plate grate has been replaced by a suspended traveling grate, the construction of which may be seen in Fig. 4. This suspended grate is connected to the hollow bar grate by means of a hinge and, therefore, goes through the same motions as this latter. Because of this the slags are never at rest and have no opportunity to clinker.

Furthermore, the use of the movable suspension grate has the advantage that by the simple displacement of a rear grate support the angle of inclination of the grate may be varied, which, in its turn, varies the rate of sliding of the fuel. In the case of coals of lower ignitibility, which require a longer time for complete combustion, a very slow rate of travel may be secured by an appropriate variation in position of the suspended grate.

A further improvement as compared with the type which was at first used in Germany, consists in the use of a so-called water-column motor with worm-gear drive. This motor is attached to the boiler feed line and makes it possible to vary the rate of travel of the grate within wide limits by throttling more or less the water-admission valve. Furthermore, this type of water-column motor has no fast-moving parts and no uneconomical ratio of speed transformation. It simply transmits the up-and-down movement of the piston directly to the oscillating stoker shaft by means of a connecting rod. The water-column motor replaces not only the electrical driving motor but also transmis-

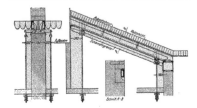

FIG. 6   THE SIDE WALLS OF A PLUTO STOKER WITH SPRING
AUTOMATIC ADJUSTMENT

(*Russboden*, floor; *Asbestostchnur*, asbestos winding; *Schnitt A-B*, section
through *A-B*; *Schamottengrense*, edge of refractory clay.)

sion and governing devices. Tests made with this type of motor have shown that the duration of a double stroke of the stoker shaft can be varied at pleasure from 8 to 80 sec. by proper adjustments on the motor. The details of this motor are not shown in the original article. (*Stahl und Eisen*, vol. 40, no. 29, July 22, 1920, pp. 969–975, 8 figs. and one large table, *de*)

## FUELS AND FIRING

FUEL-ECONOMY REPORT OF THE BRITISH ASSOCIATION. Abstract of the third report of the Committee (Prof. W. A. Bone, Chairman) appointed for the investigation of fuel economy, utilization of coal, and smoke prevention, presented to Section B at the British Association Meeting on Aug. 26. The report gives a considerable amount of data on coal mining and export statistics which cannot be abstracted here, also on future gas standards and alcohol from coke-oven gas.

As regards the gas standards, the Committee looked at the question primarily from the point of view of the national interests as a whole, and particularly from that of domestic and industrial gas consumers. It agreed with the Fuel Research Board that the future basis of charge to the consumer should be the actual number of thermal units supplied to him in the gas which passed through his meter, but desired that the charge should be based upon the "ascertained net calorific value" of the gas supplied rather than its "declared calorific value" as proposed by the Fuel Research Board.

As regards the gas pressure, the Committee recommended not less than 2 in. of water at the exit of the consumer's meter and not any measure of pressure in the mains or service pipes, taking the stand that what mattered to the consumer was the adequacy of the pressure in his pipes rather than in the gas mains outside his premises. The Committee attached great importance to the pressure being maintained as constant as possible, as well as to the requirement that greater attention than ever be paid to the removal of cyanogen and sulphur impurities from the gas.

In this connection some data are mentioned from the Gas Regulation Bill, now under consideration in the British House of Commons. Among other things, this bill introduces a new unit for the sale of gas, namely, the "therm," equal to an amount of gas containing 100,000 B.t.u.

As regards the production of alcohol from coke-oven gas, it is stated that the trials made at the Skinningrove Iron Works have demonstrated the possibility of obtaining on a large scale 1.6 gal. of absolute alcohol per ton of the Durham coal used in the tests. For data on the Skinningrove process see MECHANICAL ENGINEERING, June 1920, p. 352. (*The Engineer*, vol. 130, no. 3374, Aug. 27, 1920, pp. 196 and 198, *g*)

## HYDRAULICS

FLOW OF OIL IN PIPES, Arthur C. Preston. Discussion of the flow of viscous liquids in pipes, having for its aim the development of a method for determining frictional loss as a function of the viscosity and density of the liquid. The flow of mineral oils is primarily considered, but the method is general and may be applied to other fluids. The author claims that the rate of flow of liquid moving through a pipe under a gravity head is influenced by three characteristics of the liquid itself: namely, the density which supplies the pressure-producing flow, the viscosity which retards flow, and inertia, which, under certain conditions, manifests itself in the dissipation of energy by the collision of eddying particles. These three constitute what may be called the internal group of factors affecting flow, while such factors as diameter, length, and roughness of the pipe and the amount of gravity head are external factors and secondary in the order of consideration of the problem as a whole.

Hydraulic tables are claimed to be incapable of furnishing a satisfactory basis for estimating the flow of one liquid when the flow of another has been determined by the same pipe. On the other hand, when a general solution for the problem of liquid flow is obtained, it takes the place of hydraulic formulæ and gives the same results simply by substituting in the general formulæ the physical dimensions of the pipe and the values for the particular density and viscosity of the liquid flowing, such as water.

The first part of the article, which is all that has been published so far, is devoted to a discussion of the questions dealing with the viscosity of the liquid. One of the conclusions to which the author comes is that liquids of equal Saybolt viscosity have equal pipe-friction losses, at least through the range of velocity in which flow is steady and also for all higher velocities. (*Chemical and Metallurgical Engineering*, vol. 23, no. 13, Sept. 29, 1920, pp. 607–612, 3 figs., *tA*)

## MACHINE TOOLS

### Sonic Wave-Power Tools

DORMAN WAVE-POWER TOOLS. In MECHANICAL ENGINEERING, June 1920, p. 359, an abstract was given dealing with the Constantinesco sonic waves and the method of power transmission by rapid impulses in liquids.

Information about this novel method has been scarce and not easily obtainable. The method itself was invented by George Constantinesco, a Roumanian engineer, and was in the early stages of development in England when the world war broke out. As soon as this happened the British government took over the experimental work and made the patents secret under the Defense of the Realm Act. This secrecy was maintained to such an extent that even Walter Haddon, a prominent British manufacturer and associate of the inventor, was forbidden access to his own works, W. H. Dorman & Co., Ltd., and refused information regarding results obtained.

The necessity of such secrecy was due to the fact that the wave-transmission method was early applied to the C.C. gear for the automatic firing of guns on aeroplanes.

Since the close of hostilities, W. H. Dorman & Co., Ltd., restored to the use of their patents, have been at work in the development of power transmission by wave motion and the tools necessary for its application. The following information is obtained from a publication issued by the company, under the title Dorman Wave Power Tools, this source of information being used in this case because of the fact that no other source is available. For the principles of the method, reference is made to the abstract in the June issue of MECHANICAL ENGINEERING referred to in the first paragraph.

The wave-transmission installation consists of three units corresponding to dynamo, transmission line and motor in electrical installations. The first of these units is a wave generator, which consists of one or more metal cylinders, each fitted with a piston connected by a crankshaft to some high-speed prime mover such as a steam or internal-combustion engine, or, considered for this particular purpose, an electric motor; second, a wave-transmission pipe line which may be either rigid or flexible and in which various kinds of fluids may be used, though the highest transmission effi-

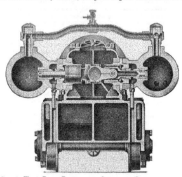

Fig. 7  WAVE POWER-TRANSMISSION GENERATOR (SECTION THROUGH "CAPACITIES")

ciency is obtained with water; third, a wave motor, which consists of one or more metal cylinders, each fitted with a piston designed to receive the power at the intake end, the other end of the piston being suitably connected to the tool or mechanism desired to be operated. The simplest application is found in such appliances as rock drills and riveting hammers, where the piston is used as a floating hammer and strikes directly on to the shank end of the drill or rivet snap.

Essentially, a wave generator is a pump, but since its purpose is not so much to convey the liquid as to convey it in a certain manner, that is, by impulses regularly following each other at predetermined intervals, it has some features not encountered in ordinary pumps. One of these is the so-called "Capacity," shown in Fig. 7. By "Capacity" in wave-transmission terminology is meant two spherical hollow steel castings designed to suit the pressures adopted. These are bolted to the crankcase and located by the crosshead guide. The disposition of these vessels on either side of the crankcase balances the forces of the crankshaft and insures freedom from vibration and quiet running at high speeds. The vessels are connected by a pipe at the top, the function of which is to equalize the pressure in each and enable the whole energy of the generator to be taken from either vessel. At the highest point of the balance pipe is a small needle valve for releasing any air which may get into the system. This is only required for a few seconds when starting up.

Screwed into the left-hand spherical vessel is an inlet charging valve actuated by pressure difference. When the minimum pressure in the capacity is greater than the pressure of the pump, it is closed. But immediately upon the pressure in the capacity being lower than the pump pressure (due to loss of water) it opens.

For the pipe line either rigid or flexible piping may be used. The

construction of the Dorman flexible pipe is shown in Fig. 8. The individual sections are made from solid-drawn steel tube or other metal, with spherical joints at the ends. These joints consist of a piece of steel formed with a spherical recess at each end, into which fits a length of solid-drawn tube upon which is mounted a ball piece which accurately fills the spherical recess. The ball piece is flattened out on one end to receive the special packing ring made from materials suitable for the purpose for which the pipe is used.

A spherically seated nut screwed into the double socket shoulders against the spherical surface of the ball piece and holds the pipe together. It is claimed that this type of piping can be made for pressures up to 10 tons per sq. in. and also that it has been in constant use for the last three years under alternating pressure varying from about 1600 lb. per sq. in. down to atmospheric pressure,

Fig. 8  DORMAN FLEXIBLE STEEL PIPE

the pressure variations taking place between these two extremities forty times per second without causing the slightest trouble.

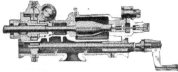

Fig. 9  WAVE POWER-TRANSMISSION HAMMER-TYPE ROCK DRILL

The third element in the wave power-transmission system is the motor, but this is not shown in detail in the publication from which the present abstract is taken. The illustration in Fig. 9, however, will give an idea of the construction of wave-power tools, such as hammer-type rock drills. In this connection, it may be stated that for this type of hammer a rapidity of 2400 to 3000 blows per minute is claimed, which, it is said, gives a higher rate of penetration at a lower power consumption than is possible with the pneumatic drill. (Dorman Wave Power Tools, published by W. H. Dorman & Co., Ltd., Stafford, England, 68 pp., illustrated, d)

## MINING MACHINERY  (See Machine Tools)

## PUMPS

PUMPS FOR CORROSIVE LIQUIDS. The question of pumps for corrosive liquids is of particular, though not exclusive, interest to the chemical industry. Various materials have been tried which, while satisfactory for some purposes, failed under other conditions. Thus, ferrosilicon resists some acids but not hydrochloric acid, and is unsuitable for processes where contamination with iron must be avoided. Lead and regulus metal cannot be used for solutions containing metallic salts. Ebonite will not withstand hot liquids and is attacked by some chemicals.

It would appear that the substance which is most generally suitable for resisting the action of corrosive liquids is some kind of silicious ceramic material, as this can be obtained in forms entirely insoluble in almost any liquid. On the other hand, however, the usual type of ceramic material is difficult to shape and has somewhat unsatisfactory mechanical properties.

During the war, however, a new ware was developed under the necessity of finding some substance suitable for apparatus for condensation of large quantities of acid gas. Like silica ware, it can be plunged when red hot into cold, water without cracking and has also good heat conductivity. This material was known as "ceratherm." Since the war a modified form of this material was adopted by Guthrie & Co. of Accrington, England, in the manufacture of their acid-proof pumps. This modified ceratherm material can be manufactured to accurate dimensions with greater ease than the original material.

The design of ceratherm pumps will be understood from Figs. 10 and 11. The acid-proof material forming the body of the pump is very thick and strong and is cemented into an iron casing. It is so arranged that it is only subjected by the bolts to crushing stress. The gland through which the plunger passes is usually on the section side, so that it is relieved from pressure, and the stuffing box is

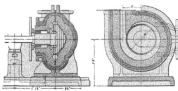

Figs. 10 and 11   Ceratherm Pump for Corrosive Liquids

packed with a small quantity of wool, usually soaked in paraffin wax.

The pumps are built mainly in small sizes to lift, say, from 20 to 100 gal. per min. against a head up to 120 lb., although pumps have been built for a head of 300 lb. and larger sizes have been made. (*Engineering*, vol. 110, no. 2851, Aug. 20, 1920, pp. 253–254, 6 figs.. d)

Roto-Piston Vacuum and Pressure Pumps. In these pumps the motions are produced by two cylinders, one, the Roto-piston, being enclosed in the other and touching it at only one point as they are mounted on different axes (Fig. 12). The inner is revolved at the same angular speed as the outer by cranks connecting the two cylinder heads. The throw of the cranks is such as to allow

Fig. 12   Roto-Piston Pump

the Roto-piston pump to maintain contact with the outer case in an apparently eccentric motion which is, nevertheless, completely balanced. The crescent-shaped space between the cylinders is sealed at its ends by the contact of the cylinders and at any point

by a vane which slides in the Roto piston and maintains contact with the outer cylinder.

As the crescent chamber remains fixed in revolution the cylinders roll past it and the vane moves through with them, displacing the space as positively as in a reciprocating pump although the motion is rotary and continuous. It has a sliding area of a little over 1 in. on the outer case. The main shaft remains stationary during revolution. There are no valves and the pump is air cooled. The pumps are made in both pressure and vacuum types. (*Iron Age*, vol. 106, no. 7, Aug. 12, 1920, p. 393, 1 fig., d. The illustration used in Fig. 12 was obtained through the courtesy of the Crescent Sales and Engineering Co., Detroit, Mich.)

## RAILROAD ENGINEERING (See Thermodynamics)

## REFRIGERATION (See Thermodynamics)

## SPECIAL MACHINERY (See Machine Tools)

### British Machine for Testing Springs by Scragging

Spring-Scragging Machine. Description of a new type of spring-scragging machine recently designed by Samuel Denison and Son, Hunslet Foundry, Leeds, England.

The ordinary scragging test applied to springs consists, in the case of laminated springs, of flattening the camber, or, in the case of springs having no camber, of giving the spring a slight deflection. Helical or volute springs are tested by closing them down hard.

The system adopted in the new machine, Fig. 13, embodies a motor-driven variable-speed rotary oil pump on the Hele-Shaw principle. The scragging gear is virtually a vertical press with a horizontal table or anvil and a vertical inverted cylinder directly overhead.

The energy put into the spring on the down stroke serves to return the tup to its normal position without the use of the pump. Owing to the piston rod on the under surface of the piston, the effective cylinder volumes on the two sides differ. In order to get rid of the surplus oil on the upper side of the piston on the up stroke, therefore, it is bypassed back to the replenishing tank by means of a shuttle valve in the pipe service connecting the top of the cylinder and pump. This valve operates the bypass as soon as the pump is put in the neutral position. While the tup is moving up the underside of the piston is supplied with oil by suction, some of the oil from the top side of the piston being bypassed to the underside for this purpose.

It is possible with this arrangement to obtain about 40 complete strokes of the tup per minute. For this a large pump unit is necessary. In order to assist in rapidly controlling this a small relay pump is introduced into the system as shown in the illustrations. The latter pump is controlled by the operator by means of a lever, to be seen in Fig. 13, and hunting gear to the main pump control, the main pump cutting out the small pump as soon as its work is accomplished. The effort needed to set the small pump in operation is very small, as also is the hand movement necessary to secure the up and down strokes of the tup. (*Engineering*, vol. 110, no. 2851, Aug. 20, 1920, pp. 243–244, and an illustration on p. 242, d)

## THERMODYNAMICS

### Shape of Insulation and Efficiency of Heat Transfer

The Effect of Geometric Form upon the Heat Transfer through Insulation, C. E. Rose. The author investigates the methods for the selection of insulation for various uses: in particular, the relation between the shape of the insulating element to its efficiency as an insulator. This has a particular value for the refrigeration industry. In such applications as the insulation of steam piping, it is of course also desirable to make the insulating element as efficient as possible, but not so important as in refrigeration, because heat is much cheaper to produce than cold and therefore its loss is less important. The writer analyzes three cases, of which the first is quite common, while the other two, as he claims, have not been properly investigated.

The first case is that of cork board used for insulation with the

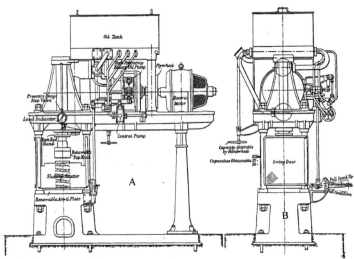

FIG. 13 NEW SPRING-SCRAGGING MACHINE OPERATING AS A PRESS

heat transfer occurring normally to its largest area. The second case is that of circular pipe insulated with a cylinder of insulation (Fig. 14). For this case the writer derives an expression

$$Q = C\,\frac{(1/R_1)(T_2 - T_1)}{\log_e R_2 - \log_e R_1} = \frac{C(T_2 - T_1)}{R_1 \log_e (R_2/R_1)} \ldots\ldots [3]$$

where $Q$ is lb.-deg. fahr. units transferred per sq. ft. of pipe surface per hour; $C$ is a laboratory constant of heat transfer which has to be determined for each material; $R_1$ is the radius of pipe in inches; $R_2$ outside radius of insulation; $T_1$ temperature in deg. fahr., low side of insulation; $T_2$ temperature in deg. fahr., high side of insulation.

The expression

$$Q_1 = \frac{2\pi C(T_2 - T_1)}{\log_e (R_2/R_1)} \ldots\ldots\ldots\ldots [4]$$

gives the rate of absorption for cold piping or radiation for hot piping.

If the rate of heat transfer is known and the thickness of insulation $R_1 - R_2$ must be calculated, the formula

$$R_2 = R_1 e\,\frac{C(T_2 - T_1)}{R_1 Q} \ldots\ldots\ldots\ldots [5]$$

may be used, where $e$ is the Naperian base.

Case 3 is that of a circular pipe enclosed in a square of insulation (Fig. 15), such as a pipe laid in a box or trench filled with granulated cork, the pipe being installed coaxially with the trench.

For this the author derives the following formulæ for the heat transfer:

$$Q_2 = C\,\frac{4}{\pi}\,\frac{R_2}{R_a}\left[\frac{(T_2 - T_1)}{R_1 \log_e (R_a/R_1)}\right] \ldots\ldots\ldots [7]$$

lb-deg. fahr. units per sq. ft. of pipe surface per hour, and

$$Q_2 = 8C\,\frac{R_2}{R_a}\left[\frac{(T_2 - T_1)}{\log_e (R_a/R_1)}\right] \ldots\ldots\ldots\ldots [8]$$

lb-deg. fahr. units per linear foot of pipe per hour, where $R_2$ is the mean integrated radius vector.

The writer takes numerical examples, one for each case, and ascertains what results the changes of form have upon the rate of heat transfer in unit time at constant thickness through 2 in. of formed cork having a heat-transfer constant $C = 0.49$ lb.-deg. fahr. units per sq. ft. per deg. difference per in. of thickness per hour and a temperature difference of 100 deg. fahr. $= T_2 - T_1$.

The actual calculations are given in the article. In case 2, a 2-in. molded cork insulation is applied to a 2-in. pipe, in which case Equation [3] gives $Q = 37.8$ lb.-deg. fahr. units per sq. ft. per hr.

FIG. 14 CIRCULAR PIPE INSULATED WITH A CYLINDER OF INSULATION (DIAGRAMMATIC)

FIG. 15 CIRCULAR PIPE ENCLOSED IN A SQUARE OF INSULATION SUCH AS BOX OR TRENCH (DIAGRAMMATIC)

A similar computation for the case of cork-board insulation and heat transfer normal to its largest area gives for $Q$ a value of 24.5 lb.-deg. fahr. units per sq. ft. per hr., whereas in case 3 with $R_1 = 1$ in. and a thickness of covering at $A_4$ of 2 in., gives 27.6 lb-deg. fahr. units per average sq. ft. per hr.

It would appear, therefore, that the second form transmits 54 per cent and the third form 11 per cent more heat (at the particular thickness of 2 in. with 2-in. pipes) than the first form. (Ice and Refrigeration, vol. 59, no. 3, September 1920, pp 82–84, 4 figs., tp)

## Wire-Drawing Compression Method of Heating Gases —Causes of Overheating in Air Compressors with Leaky Delivery Valves

A DYNAMICAL METHOD FOR RAISING GASES TO A HIGH TEMPERATURE, Prof. W. H. Watkinson. If air be allowed to flow from the atmosphere through a spring-loaded wire-drawing valve adjusted so that the air in the cylinder during the charging stroke is constantly at a pressure of one-quarter of an atmosphere and if this air is subsequently compressed to atmospheric pressure, then, assuming adiabatic and frictionless conditions and neglecting the effect of clearance in the pump cylinder, the absolute temperature of this air at the end of compression will be approximately $1\frac{1}{2}$ times the absolute temperature of the outside atmosphere; that is, $T_e$, Fig. 16, will be approximately equal to $1\frac{1}{2}$ times 288, that is, 432 deg., when the atmospheric temperature is 15 deg. cent. The reason for this is that the constant-pressure line $a\,b$, Fig. 16, is approximately an isothermal line for the varying mass of air in the pump cylinder, and the temperature of the air during the charging stroke is approximately the same as that of the outside atmosphere.

From the equations giving the connection between temperature, pressure and volume of a given mass of gases undergoing expansion and compression, namely,

$$\frac{T_e}{T_b} = \left(\frac{p_e}{p_b}\right)^{\frac{n-1}{n}} = \left(\frac{V_b}{V_e}\right)^{n-1}$$

it is obvious that the absolute-temperature ratio depends on the absolute-pressure ratio and not on the magnitudes of these pressures. If in the above illustration the air had flowed into the cylinder at a pressure of 25 atmos. from a receiver in which the air was at a pressure of 100 atmos. and at a temperature of 15 deg. cent., the temperature of the air after compression in the pump to 100 atmos. would have been 432 deg. abs., i.e., the same as in the above case. Therefore, if a pump arranged as shown in Fig. 17 is used and air from the atmosphere is admitted to the pump cylinder $b$ by means of the wire-drawing valve $d$, and this air after compression is discharged through the valve $e$ into a receiver $f$, its temperature within this receiver will be 432 deg. abs.

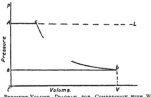

FIG. 16 PRESSURE-VOLUME DIAGRAM FOR COMPRESSION WITH WIRE-DRAWING VALVE

If this air is then admitted to the other end of the cylinder and is wire-drawn during admission, by the valve $g$, to one-quarter of an atmosphere, and then compressed to atmospheric pressure and discharged through the valve $h$, the temperature of the air in the pipe $i$ will be $288 \times 1.5^2 = 648$ deg. abs. and its pressure there will be atmospheric.

With five such double-acting pumps in series, the temperature of the air leaving the last pump at atmospheric pressure would be 16,600 deg. abs., providing the process could be carried out under the conditions assumed for the first stage. Actually, the temperature of the air leaving the second pump would be approximately 4600 deg. abs., and with our present materials and present types of compressors this is probably the upper limit at which the pump could be made to operate. Furthermore, to obtain the enormous increase of temperature stated above by single-stage compression would require the pressure ratio to be about 1,040,000, so that if the initial pressure was at one atmosphere the final would have to be 1,040,000 atmos. The multi-stage wire-drawing method described above might be regarded as a thermodynamic ratchet. One possible application of this principle may be indi-

cated. If the pump shown in Fig. 17 be connected to the compression space of the cylinder $a$ of an internal-combustion engine, the temperature of the air in the pipe $i$ may be raised sufficiently high to effect ignition of fuel or of fuel and air in this pipe. If, for example, the air admitted to the pump cylinder by the valve $d$ is wire-drawn to a pressure of one-half an atmosphere and then compressed to a pressure of two atmospheres and discharged through

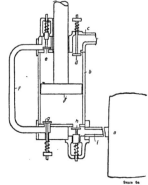

FIG. 17 PUMP ARRANGED TO RAISE GAS TEMPERATURE DYNAMICALLY (WATKINSON METHOD)

the valve $e$ into $f$, where the pressure is one atmosphere, and if the process is repeated on the other side of the piston and the air discharged into the pipe $i$ at a pressure of one atmosphere at the beginning of the compression stroke of the piston of the engine, then the temperature of the air in the pipe $i$ will be the same as if this air had been compressed from one atmosphere to 16 atmos.; and if the compression pressure ratio in the engine cylinder is 10, the air in the pipe or pocket $i$ will at the end of the compression stroke of the engine be at the same temperature as if it had been compressed from one atmosphere to 160 atmos., although its pressure will only be equal to 10 atmos., or that in the engine cylinder with which it is in free communication.

The above discussion would also explain why serious overheating may occur in air compressors having a leaky delivery valve. If, for example, in a single-stage air compressor compressing air to 4 atmos. the delivery valve was held partly open by any means during the charging stroke of the pump, and if it be assumed, as an extreme case, that the backward flow into the cylinder was at such a rate as just to maintain the pressure at one atmosphere during the charging stroke, then after 10 revolutions the air would, under adiabatic conditions, be raised to a temperature of 16,600 deg. cent. abs., and to a very high temperature under actual conditions. The temperature attained owing to such leakage might, in certain cases, be sufficient to cause melting of the valve. In the case of turbo-compressors, backward leakage due to a bent or fractured blade might raise the temperature sufficiently to cause melting of the blades. (Paper before the British Association, Section G, August 1920, abstracted through *The Engineer*, vol. 130, no. 3374, August 27, 1920, p. 108, 2 figs., *tpA*)

## CLASSIFICATION OF ARTICLES

Articles appearing in the Survey are classified as *c* comparative; *d* descriptive; *e* experimental; *g* general; *h* historical; *m* mathematical; *p* practical; *s* statistical; *t* theoretical. Articles of especial merit are rated *A* by the reviewer. Opinions expressed are those of the reviewer, not of the Society. The Editor will be pleased to receive inquiries for further information in connection with articles reported in the Survey.

# ENGINEERING RESEARCH

A Department Conducted by the Research Committee of the A.S.M.E.

## National Physical Laboratory of Great Britain

THE Report of the National Physical Laboratory for the year 1919 has been issued. The report includes the report of the executive committee, a statement of the work proposed for the next year, a list of the papers published by the Laboratory or communications by members of the staff to scientific societies or journals, the report of the director, report of special work done in gage testing and other researches during the war. The heads of the various departments give accounts of the recent work in the laboratory. The subjects covered are as follows:

Physics Department:
- I—Heat:
  - *a* High-Temperature and General Work
  - *b* Thermometer Testing
  - *c* Oil-Apparatus Testing
- II—Optics
- III—Radium and X-Ray Work
- IV—Tide Prediction
- V—Library

Electricity Department
Metrology Department
Engineering Department
Aerodynamics Department
Metallurgy Department
The William Froude National Tank

Copies of the Report at 5s may be obtained from the Imperial House, Kingsway, W.C. 2, London.

## Petroleum Research

The Petroleum Section of the American Institute of Mining and Metallurgical Engineers held meetings in St. Louis on September 21 and 22. An invitation was extended to the Petroleum Division of the American Society for Testing Materials and the American Association of Petroleum Geologists. The subjects of the papers dealing with research matters were as follows: Urgency for Deeper Drilling on the Gulf Coast, A. F. Lucas; Oil-Field Brines, Chester W. Washburne; Efficiency of the Use of Oil as Fuel, W. N. Best; Determination of Pore Space of Oil and Gas Sands, A. F. Melcher; Outline for Analysis of Oil-Field Water Problems, A. W. Ambrose; The Nature of Coal, J. E. Hackford; Extended Life of Wells Due to Rise in the Price of Oil, W. W. Cutler, Jr.; Oil Shales and Petroleum Prospects in Brazil, H. E. Williams.

## Research Associations in Great Britain

In addition to those listed in the March and August issues of MECHANICAL ENGINEERING (pp. 181 and 470), the following Research Associations have received licenses from the Board of Trade under Section 20 of the Companies' Act of 1908:

The British Refractories Research Association, 14 Great George St., S. W. 1; Secretary, R. C. Rann.
The Scottish Shale Oil Scientific and Industrial Research Association, 136 Buchanan St., Glasgow; Secretary, W. Fraser, C.B.E.

The following Research Associations have been approved, but not licensed, in addition to those mentioned previously:

The British Electrical and Allied Industrial Association
The British Silk Association
The British Motorcycle and Cycle Car Research Association
The British Cutlery Association.

The following Associations are preparing Memoranda and Articles of Association in addition to those previously listed:

Jute Spinners and Manufacturers
Gray and Malleable Cast-Iron Founders.

## Research Résumé of the Month

### A—RESEARCH RESULTS

The purpose of this section of Engineering Research is to give the origin of research information which has been completed, to give a résumé of research results with formulæ or curves where such may be readily given, and to report results of non-extensive researches which in the opinion of the investigators do not warrant a paper.

*Air A2-20* Blasting Granite with Compressed Air. Report on Investigation Serial No. 2154 by Oliver Bowles deals with the use of compressed air in the final blasting of granite. In granite quarries where no joints or "sheeting planes" occur, two holes 3 in. in diameter are drilled close together to a depth of about eight feet. A small charge of about a spoonful of black blasting powder is placed in each hole, tamped with clay and fired. This starts a small fracture leading out horizontally from the two holes and by successively larger charges the fracture is enlarged in area until compressed air at 100 lb. can be put in the holes through pipes which have connections made airtight by means of sulphur. This air supplied over an enormous surface exerts an upward pressure and continues the enlargement of the fracture. Address Bureau of Mines, Washington, D. C., F. G. Cottrell, Director.

*Apparatus and Instruments A12-20* Method for Measuring Interior Diameter of Ring Gages. A method employing two steel balls, the sum of the diameter of which is slightly larger than the nominal inside diameter of a ring to be measured, has been devised by the Bureau of Standards. The ring is laid on a surface plate and the larger ball placed in the ring and the smaller ball then rests against the inside of the ring, and on the larger ball. The difference in vertical position between the upper surfaces of the two balls is determined by a micrometer. This distance is easily converted into the vertical distance between the centers of the two balls. From this the horizontal distance between the two centers is determined from the right-angled triangle formed by the vertical distance, horizontal distance and distance between centers of the balls. From the above the diameter of the ring is determined. Bureau of Standards, Washington, D. C. Address S. W. Stratton, Director.

*Apparatus and Instruments A13-20* Quantitative Measurement of Consistency. The Bureau of Standards has developed a method for the quantitative measurement of consistency. This consists in forcing the material under investigation through a capillary tube by means of air pressure. Runs are made at various constant air pressures. Values which are deduced from the readings determine consistency. This does not determine the plasticity, and the Bureau is at work on a plastometer, Scientific Paper No. 276 and Technologic Paper 169, as well as papers in the Proceedings of the A. S. T. M. for 1919–1920, show the application of this apparatus. Bureau of Standards, Washington, D. C. Address S. W. Stratton, Director.

*Apparatus and Instruments A14-20* Etched Balls for Brinell Tests. When specimens are tested in polished state the impressions from etched balls are very distinct, while those from unetched balls are almost invisible from certain angles. The steel ball is etched for about one minute before the test in a 2 per cent alcoholic nitric acid solution. If the specimens are not polished, preliminary etching of balls does not seem to be of much advantage. Bureau of Standards, Washington, D. C. Address S. W. Stratton, Director.

*Cement and Other Building Materials A11-20* Penetration of Water. The Bureau of Standards has devised a new apparatus to determine the rate of penetration of water through building materials. No special form of test piece is necessary if one face of the specimen is fairly smooth. A complete test of materials has not been made but from the preliminary investigation the table below has been prepared:

| Test Specimen | Absorption in 24 Hours, Per cent | Thickness, In. | Water Pressure, Lb. | Time Required for Penetration through Wall |
|---|---|---|---|---|
| Limestone 50 | 5.80 | 1 5/8 | 60 | 10 1/2 min. |
| Limestone 7E | 3.10 | 1 5/16 | 60 | 11 min. |
| Limestone 8907 | 4.40 | 1 5/8 | 60 | 19 min. |
| Limestone 9c | 4.60 | 1 5/8 | 60 | 2 3/4 min. |
| Limestone No. 5G | 3.48 | 1 1/4 | | 1 1/2 min. |
| Sandstone | 5.56 | 2 1/2 | 60 | 10 sec. |
| 1 : 6 Portland-cement mortar | 7.8 | 2 | 60 | 3 1/2 hr. |
| 1 : 1 1/2 : 2 concrete | 5.8 | 2 | 60 | (a) |

(*a*) Did not fail in 24 hr.; when broken through water had penetrated but 1/2 in.

*Fuels, Gas, Tar and Coke A11-20* Gasoline from Natural Gas. Factors in Determining the Gasoline Content in Natural Gas by the Absorption Method is the title of a Report Serial No. 2157 to the Bureau of Mines by D. B. Dow. This report deals with certain factors which should be considered in making absorption tests to determine the gasoline content of natural gas using "mineral seal oil" as the absorbent. The method is described in Technical Paper 232, Bureau of Mines 1919, by Dykema and Neal. The factors to be considered are the saturation in the absorbent oil which should be as high as possible in the first compartment and low in the third and fourth compartment, the loss in distillation, the rate of distillation, the temperature of the condenser and the stopping of the distillation by means of the mineral seal ring rather than by temperature readings. The rise in temperature to 350 deg. fahr. should take between 12 and 30 min. The condenser should be held at 60 deg. fahr. Bureau of Mines, Washington, D. C. Address M. G. Cottrell, Director.

*Fuels, Gas, Tar and Coke A12-20* The Coal Fire. The Department Scientific and Industrial Research through its Fuel Research Board has made a special Report No. 3 on the Coal Fire. This includes the work of Dr. Margaret Fishenden on Domestic Heating, the investigation of the efficiency of open fires and on air pollution from domestic heating. Copies at 4/3 may be obtained from the Imperial House, Kingsway, London, W. C. 2.

*Fuels, Gas, Tar and Coke A13-20* Colorado Oil Shale. Martin J. Gavin, Bureau of Mines and Leslie H. Sharp, State of Colorado, have made report No. 2152 to the Bureau of Mines on some physical and chemical data of Colorado oil shale. The experiments were made on samples which represented an average of the massive type of oil shale encountered in Colorado. The results show a yield of 42.7 gal. of oil of 0.905 specific gravity from 2000 lb. of shale. The specific gravity 0.905 or 24.7 deg. B was determined at 15.56 deg. cent. or 60 deg. fahr. The weight of the shale per cu. ft. is given below:

| Size | Run of mine | —1 in. | —¹/₂ in. | —¹/₄ in. |
|---|---|---|---|---|
| Weight, lb. per cu. ft... | 53.775 | 54.775 | 56.015 | 58.200 |

The apparent specific gravity varied from 1.92 to 2.06. The specific heat from 20 to 90 deg. cent. varied from 0.223 for spent shale to 0.265 for raw shale. The heat of combustion of raw shale was 2460 calories per gram or 4428 B.t.u. per lb. In making this determination it was found very difficult to get complete combustion, even when using oxygen at 500 lb. pressure. The ash residue in most cases contains considerable combustible matter. The thermal conductivity for ranges of 25 to 75 deg. cent. was 0.0032 in absolute units, for which the conductivity of copper is about 0.91. The proximate analysis shows moisture 0.60 per cent, loss on ignition 40.00 per cent, and ash or residue 59.40 per cent. The analysis of the ash gives about 45 per cent silica, 26 per cent iron and alumina, 18 per cent lime, 5.3 per cent magnesium. The heat of combustion of crude shale oil is 10,215 calories per gram or 18,387 B.t.u. per lb. Bureau of Mines, Washington, D. C. Address F. G. Cottrell, Director.

*Glass and Ceramics A2-20* Telescope Objectives. A paper has been prepared by the Ordnance Department which deals with the possible sets of simultaneous values of spherical and chromatic aberration which may be obtained from a two-piece cemented telescope objective of barium crown glass and flint glass. Sets of contours are presented which show in a complete form the manner in which the different aberrations change as the radii are altered. See *Journal of the American Optical Society*. Also S. Tour, Metallurgical Section, Ordnance Dept., Washington, D. C.

*Industrial Management A1-20* Code for Head and Eye Protection. The National Safety Code for the protection of head and eyes of industrial workers has just been sent to the press. For two years the rules have been published in mimeographed copies for criticism and correction. The Code specifies appropriate protectors against mechanical hazards such as flying particles, dust, chemical fumes and excessive light. Bureau of Standards, Washington, D. C. Address S. W. Stratton, Director.

*Iron and Steel A2-20* Boiler Plate at Elevated Temperatures. The tensile properties of boiler plate at temperatures up to and including 563 deg. fahr. are not affected by the rate of loading from very slow loading up to 1.6 in. per min. Bureau of Standards, Washington, D. C. Address S. W. Stratton, Director.

*Mining, General A4-20* Stench Warning in Metal Mines. A. C. Fieldner and S. H. Katz have made a report on this subject in Serial 2153 to the Bureau of Mines. In order to send warnings, messengers, electric lights, telephones, the interruption of the flow of compressed air or the introduction of water into the compressed-air lines have been used. Experiments have just been made on the introduction of a material which would cause a strong odor or stench that could not be mistaken. The liquids suggested have been ethyl mercaptan ($C_2H_5SH$) and amyl acetate. Ethyl mercaptan boils at 98 deg. fahr. and freezes at 228 deg. below zero. It has a disagreeable and characteristic skunk-like odor which will not be mistaken for odors commonly found in mines. It can be obtained from certain chemical manufacturers at $2.25 per lb. Amyl acetate may be obtained from any chemical-supply house or from dealers in paints and lacquers. This has a banana-like odor which is not objectionable but distinctive. Tests have been made in a large number of mines proving the effectiveness of ethyl mercaptan. The amyl acetate tests were not so effective in a short time. 1¹/₄ pints of ethyl mercaptan or 3¹/₂ pints of amyl acetate are required for each 100,000 cu. ft. of free air entering the mine per minute. The injector is constructed of a stout glass cylinder or airtight metal cylinder capable of withstanding the pressure in the line. It is connected above and below the liquid to the

air line with short ¹/₂-in. pipes. Bureau of Mines, Washington, D. C. Address F. G. Cottrell, Director.

*Paints, Varnishes and Resins A1-20* Talc in Fire-Resistant Paint. A fire-resistant paint has been prepared by the Paint Manufacturers' Association started through Henry A. Gardner at the Institute of Industrial Research, Washington, D. C., and Dr. Herman Von Schrenk, of St. Louis. The formula for the paint is as follows: 10.6 lb. basic sulphate white lead, 11 lb. zinc oxide, 33 lb. asbestine (magnesium silicate or talc), 0.5 lb. borax, 0.9 lb. dry lampblack, 24 lb. linseed oil, 2 lb. liquid drier and 10 lb. mineral spirits. It is noted that 33 per cent of the paint is talc. Although called asbestine, this is not a form of asbestos. The material is being tested at present by the Underwriters Laboratory. See Report of the Bureau of Mines, Serial No. 2150. Address Bureau of Mines, Washington, D. C., F. G. Cottrell, Director.

## B—RESEARCH IN PROGRESS

The purpose of this section of Engineering Research is to bring together those who are working on the same problem for coöperation or conference, to prevent unnecessary duplication of work and to inform the profession of the investigators who are engaged upon research problems. The addresses of these investigators are given for the purpose of correspondence.

*Chemistry, Inorganic, B2-20* Ferric Oxide with Small Percentage of Fused Salts. An investigation is being made of the physical and chemical properties of ferric oxide fused with small quantities of various salts. Waltham Watch Co., F. P. Flagg, Chief Chemist, Waltham, Mass.

*Glass and Ceramics B3-20* Enamel for Watch Dials. The Waltham Watch Co. is investigating a satisfactory method of determining the properties of the enamel used on watch dials. Difficulties have been encountered in finding satisfactory methods for determining constituents. Physical properties are being studied. These include the Coefficient of expansion, viscosity, melting point, opacity and color. Address Waltham Watch Company, F. P. Flagg, Chief Chemist, Waltham, Mass.

*Lubricants B1-20* Lubrication at High Temperatures. Tests of viscosities at high temperatures of asphalt and paraffin base oils at 300 deg. fahr. and lower. Address Lockhart Laboratories, L. B. Lockhart, Atlanta, Ga.

*Metallurgy and Metallography B12-20* Furnace Linings and Brass Alloys. The effect of different brass-foundry alloys on different refractory materials used for furnace linings in crucible furnaces, open-flame oil furnaces and electrical furnaces. The Lumen Bearing Company, Address C. H. Bierbaum, Buffalo, N. Y.

*Metallurgy and Metallography B13-20* Bearing Alloys. A study of the methods of eliminating impurities incident to improper foundry methods for bearing alloys. Address Lumen Bearing Company, C. H. Bierbaum, Buffalo, N. Y.

*Metallurgy and Metallography B14-20* Bearing Alloys. The possibility of using new alloys for bearing purposes and limitations of different alloys now in use. Lumen Bearing Company, Address C. H. Bierbaum, Buffalo, N. Y.

*Textile Manufacture and Clothing B3-20* Work of the United States Testing Company. The work in progress at the United States Testing Company is devoted to various phases of textile work.

1 Standardization of Fading Tests. Design and construction of special holder for several samples exposed to artificial light. Comparison of results at Rutgers College, University of Nevada, and in New York City.
2 Mechanical tests as a means of classifying raw silks..
3 The effects on the color in finished knit goods of different ways of doubling and twisting in hosiery tram.
4 Some apparent anomalies in the colorimetric estimation of dyestuffs.
5 Lousiness.
6 A comparison of artificial silks among each other and natural silks.
7 Standardization of chemical methods of determining the per cent content of silk, wool and cotton in fabrics.
8 Weighting and durability of silk fabrics.
9 Relation of laundering processes to fastness of colors.
10 Identification of dyestuffs and intermediaries.
11 Soaps and oils for textile work.
12 Photomicrography of fabrics.
13 Determination of the soap and oil content of thrown silk by extraction method.
14 Chemical and physical characteristics of the sericin and fibroin in so-called hard and soft silks.

Address R. E. Douty, General Manager, U. S. Testing Company, 340 Hudson St., New York.

## C—RESEARCH PROBLEMS

The purpose of this section of Engineering Research is to bring together persons who desire coöperation in research work or to bring together those who have problems and no equipment with those who are equipped to carry on research. It is hoped that those desiring coöperation or aid will state problems for publication in this section.

*Machine Design C2-20* Ball Bearings at High Speeds. The performance of ball bearings at from 7,000 to 25,000 r.p.m. is being investigated theoretically by Panfilo Trombetta. Investigation of a theoretical and experimental nature is desired. Address Panfilo Trombetta, General Electric Company, Schenectady, N. Y.

*Metallurgy and Metallography C2-20* Cartridge Brass. Relation between hardness, tensile strength and amount of cold work on cartridge brass containing 68 to 71 per cent copper, 32 to 29 per cent zinc, and not less than 0.07 per cent lead and not less than 0.05 per cent iron. In the annealed condition the relation between hardness and grain size is known. The relation between the amount of cold work, hardness and tensile strength for metals of different grain size is desired. Address S. Tour, Metallurgical Section, Office of the Chief of Ordnance, Washington, D. C.

*Metallurgy and Metallography C3-20* Overstrain in Steel. The effects of overstrain on steel and the influence of time and temperature on these effects. The statement that elastic limit of steel may be increased by cold work or by the application of excessive stresses is often made. It is desired to know the physical characteristics of steel subjected to cold work or to excessive stresses above the elastic limit and the effect which time and temperatures less than 600 deg. cent. have upon these characteristics. Address S. Tour, Metallurgical Section, Office of the Chief of Ordnance, Washington, D. C.

#### D—RESEARCH EQUIPMENT

The purpose of this section of Engineering Research is to give in concise form notes regarding the equipment of laboratories for mutual information and for the purpose of informing the profession of the equipment in various laboratories so that persons desiring special investigations may know where such work may be done.

*Simonds Steel Mills D1-20* Simonds Manufacturing Company has just completed an addition to its research laboratory at its steel mills in Lockport, N. Y. The building is 40 by 50 ft. New equipment consisting of a special-type electric melting furnace of 300 to 500 lb. capacity and electric heating furnaces with other equipment for extending facilities for research and development work has been added. New formulæ and methods for making special steels are being worked out under the direction of an expert metallurgical staff.

*Doehler Die-Casting Company D1-20* Laboratories equipped for making chemical and physical tests of materials, particularly those used in non-ferrous alloys and especially alloys used in die casting. Doehler Die-Casting Company, Address Charles Pack, Chief Chemist, Court, Ninth and Huntington Sts., Brooklyn, N. Y.

#### E—RESEARCH PERSONNEL

The purpose of this section of Engineering Research is to give notes of a personal nature regarding the personnel of various laboratories, methods of procedure for commercial work or notes regarding the conduct of various laboratories.

#### F—BIBLIOGRAPHIES

The purpose of this section of Engineering Research is to inform the profession of bibliographies which have been prepared. In general this work is done at the expense of the Society. Extensive bibliographies require the approval of the Research Committee. All bibliographies are loaned for a period of one month only. Additional copies are available, however, for periods of two weeks to members of the A.S.M.E. or to others recommended by members of the A.S.M.E. These bibliographies are on file in the offices of the Society.

# CORRESPONDENCE

CONTRIBUTIONS to the Correspondence Department of MECHANICAL ENGINEERING are solicited. Contributions particularly welcomed are discussions of papers published in this Journal, brief articles of current interest to mechanical engineers, or suggestions from members of The American Society of Mechanical Engineers as to a better conduct of A.S.M.E. affairs.

## Can Engineering Students be Given Broad Conception of Production and Management ?

TO THE EDITOR:

Mr. Benedict describes in his paper in the September number what he modestly offers as a partial contribution toward the improvement of engineering schooling. His work exemplifies current systems of shop operation and management. This covers a considerable part of the field of industrial engineering, but not the whole field. Engineers, and particularly mechanical engineers, are too apt to think of machine-shop operation as typical, if not inclusive, of production generally. A result is that in those industries which are quite dissimilar from the shop, engineers class themselves as mechanical men rather than as production men.

If applications of engineering are to be taught in the schools, other industries as well as the shop should have attention. A very few may be mentioned: steel manufacture; the refining of petroleum products; sugar processing; the manufacture of pulp and paper. Shop methods cannot be applied, unmodified, in the production departments of these industries.

Few schools if any can have model paper mills or sugar refineries. They can teach, do teach and should continue to teach, the scientific principles underlying these and all industries. They must now also attempt to develop and teach operation and management principles. There must consequently be some study of operation. This cannot in general be carried on in the experimental way which Mr. Benedict develops for the shop. It must be done by industrial research or survey conducted by means of seminar or thesis assignments, including visits to manufacturing plants. These visits should be of a month or two in duration, not of an hour or two. They will make graduates more immediately useful in the special industries examined; but this is not the main point, or the chief function of the school, which should be concerned with the producing power of its graduate over his whole economic life. The real advantage of the study of typical industries is in its broadening and stimulating effect, and in its confirmation or illustration of fundamental principles.

To carry out such a program requires a rather unusual amount of coöperation from the manufacturers, and the manufacturers must not expect too much immediate or direct benefit. Education is about the longest-range operation we have today.

The new professional course in industrial engineering, which has just been announced by Columbia University, has been prepared with some of these points in mind. Studies are grouped under General Engineering, Business, Machinery, and Industrial. In the last group provision is made for four months, work in a factory, half the time as a student-worker and the balance in directed study and analysis. The major courses in this group are those dealing with manufacturing processes and it is interesting to note that the non-metallic industries are assigned fully half the total time. Processes and mechanical operations are analyzed and the characteristics of machines examined. In the third year of this graduate course each student selects and studies, under guidance, some particular industry. This need not be (although it often will be) a metal-working industry. The study will include a survey of commercial, financial and technical factors, with contemplation of ways and means for improvement. As a test of accomplishment, a final course aims to develop methods of analysis by which the machinery and equipment may be selected for manufacturing an assigned commodity according to an assumed schedule. A part of the job is the preparation of a financial budget for the operation.

The obvious reflection in connection with a program as ambitious as this is that it requires teaching of unusual type and scope. But with the right sort of handling the value of work of this kind should be very great.

132 Nassau St. WILLIAM D. ENNIS.
New York, N. Y.

# The New Ford Plant at River Rouge, Mich.

### Notes of Interest Regarding the Features of This Unique Plant, Gathered Through the Visit of the Cleveland Engineers

MENTION was made in the last number of the recent excursion of the Cleveland engineers to Detroit, and of the visit to the new Ford plant at River Rouge. In *Cleveland Engineering* for September 7, the journal of the Cleveland Engineering Society, is an account of the trip, from which the following notes are taken:

#### ABOUT THE FORD PLANT

Heretofore, iron ore—the rock itself as it is taken from the mine—has never been listed as an automotive part. Even the most completely equipped, self-contained automotive factory, a plant fabricating steel and manufacturing a majority of the units going into its product, has never gone back farther than iron pigs and raw steel.

At River Rouge, one of Detroit's western suburbs, this dream of Henry Ford's is approaching realization. The Clevelanders found River Rouge the scene of an industrial development program of such proportions that its scope is almost incomprehensible.

Points of interest were the waterway, storage yards, by-product plant, blast furnaces, a large body plant which formerly was the Eagle boat plant, power house and foundry, the latter three in course of construction.

The waterway consists of a turning basin large enough for the longest lake vessel, and a slip half a mile long, 250 ft. wide and 25 ft. deep. The Government recently has begun contract work on the widening and deepening of the River Rouge for a distance of two and a half miles from the plant to the Detroit River, which will put the plant on the 22-ft. waterway of the lakes. The visitors had an opportunity to see the hydraulic dredges at work.

The storage yards for coal, stone and ore, just completed, were inspected. The boat-unloading equipment was not operating, pending completion of the Government channel. Meanwhile the company gets its ore by rail.

The by-product plant was in full operation, turning out coke, illuminating gas, light oil, tar and ammonium sulphate.

The plant in general occupies about 23 acres. When Mr. Ford's entire project is realized it will take in a site of nearly a thousand acres. The 120 coke ovens produce 1700 tons of coke, 18,000,000 cu. ft. of gas, 10,000,000 of which is being sold to the city of Detroit, 4000 gal. of benzol, 16,000 gal. of tar and 27 tons of ammonium sulphate, sold as a fertilizer.

Each of the four blast furnaces will have a capacity of 500 tons of iron daily. Every casting and steel part used in the Ford automobile, truck and Fordson tractor will be manufactured at River Rouge when the plant enters into full production.

The power house will consume 1000 tons of coal daily. When complete it will contain turbo-generators of 55,000 kw. capacity and five turbo-blowers of 40,000 ft. capacity a minute. Five 2000-ft.-a-minute air compressors will supply air to the plant for manufacturing purposes. Eight 2640-hp. boilers will be fired by blast-furnace gas and pulverized coal.

The plant is unique in three respects. It will be the first complete consolidated blast furnace, foundry and steel mill in the world. It will be the first large power plant in this country to utilize pulverized coal. It will be the first to make castings from metal coming directly and uncooled from the blast furnaces. This is a feat which engineers have said could not be done.

In engineering circles, the announcement that the Ford company planned to pour hot metal directly from the furnaces to molds, eliminating the re-melting of pig iron in the foundry cupolas, has been received with unusual interest.

It was this feature that the Cleveland visitors were particularly anxious to see. However, pending completion of the foundry, the metal is being made into pig and the engineers had to be satisfied with an explanation of the process. In working this process out the cupolas have not been eliminated entirely, but the metal from the furnaces and the metal from the cupolas will be mixed in definite proportions.

At the time of the excursion pamphlets were distributed by the George T. Ladd Co., Pittsburgh, Pa., on the boilers which they are building for this plant, probably the largest boilers ever constructed, with brief reference, also, to the superheaters and method of firing. The pamphlet states:

#### BOILERS OF THE POWER PLANT

These boilers each contain 26,470 sq. ft. of heating surface, which is exclusive of superheater heating surface, or surface of future economizer. Inasmuch as this heating surface is substantially all in the tubes, it required for each boiler nearly six miles of $3^1/_4$-in. tubing. Steam will be generated at 240 lb. per sq. in. and superheated 200 deg. fahr.

The main steam and water drums are 5 ft. inside diameter by 25 ft. $10^3/_4$ in. long, with $1^3/_{16}$-in. thick shell plates. The steam is led from the two main top drums to a 36-in. steam drum equipped with two 10-in. nozzles, which connect with the two saturated-steam headers of the superheater.

The furnace, while irregularly shaped, is approximately 23 ft. by 24 ft. inside by 55 ft. high above the ashpits. The combustion space allowed by this furnace, exclusive of ashpits, is about 5 cu. ft. per normal rated

horsepower. The total height of the boiler from the ashpit floor to the top of the superheater piping is 82 ft. $9^7/_8$ in.

The settings of the boilers presented many difficulties owing to the extreme size of the furnaces. The lower main walls are $33^1/_4$ in. thick at the haunches with a 9-in. vertical invert at the center. All the high-temperature zones in the furnace are lined with $13^1/_2$ in. of firebrick. A sectional view and side elevation of one of the boilers and its setting are shown in the accompanying illustration.

Economizers are not to be installed at the start, but provision is made in the structure for their future addition. The superheater arrangement, however, permits a full 3-pass gas travel for each side of the boiler, which,

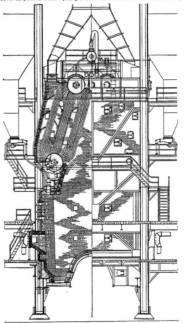

SECTIONAL VIEW OF BOILER UNIT, FORD PLANT

coupled with the fact that the passes can be baffled as closely as is consistent with the ratings desired, seems to assure reasonably low stack temperatures. Mounted over each boiler and supported by the building framing will be a steel stack 11 ft. inside diameter, with the top reaching 327 ft. above the ash-floor level.

#### SUPERHEATERS

Each superheater consists of three headers, two 10 in. in diameter and one 12 in., all approximately 23 ft. long, together with the connecting elements.

The headers are located below the boiler steam drum, with the larger in the center and the two smaller on each side. Steam is taken from two points on the boiler collecting drum and led to the ends of the 10-in. saturated-

*(Continued on page 654)*

# MECHANICAL
# ENGINEERING

The Monthly Journal Published by The American Society of
Mechanical Engineers

FRED J. MILLER, *President*

WILLIAM H. WILEY, *Treasurer*        CALVIN W. RICE, *Secretary*

PUBLICATION COMMITTEE:

GEORGE A. ORROK, *Chairman*                          J. W. ROE

H. H. ESSELSTYN          GEORGE J. FORAN

RALPH E. FLANDERS

PUBLICATION STAFF:

LESTER G. FRENCH, *Editor and Manager*

FREDERICK LASK, *Advertising Manager*

WALTER B. SNOW, *Circulation Manager*

Yearly subscription, $4.00, single copies 40 cents. Postage
to Canada, 50 cents additional; to foreign countries $1.00 addi-
tional.

*Contributions of interest to the profession are solicited. Com-
munications should be addressed to the Editor.*

## Mechanical Engineering Changes Printers

For the past ten years, MECHANICAL ENGINEERING has been printed at the plant of the Williams Printing Company, New York, one of the largest, best-equipped and best-organized printing establishments in the country. The service which this company has rendered has been of the highest order and the careful attention which its employees have given to the execution of the work upon MECHANICAL ENGINEERING has been all that could be desired, and deserves the highest praise. The rapidly rising cost of production in the printing industry, however, has made it seem advisable to the Publication and Papers Committee of The American Society of Mechanical Engineers to have the printing done in a smaller city where the overhead charges and cost of living, and consequently cost of production, are lower than in New York. Accordingly, this journal is now to be printed by the Eschenbach Printing Company, Easton, Pa., which, under the postal laws, becomes the publication office. The editorial and business offices will remain at the headquarters of the Society, 29 West 39th Street, New York, and all communications should be sent to the New York address.

It probably is not realized by the average person that the increases in costs in the publishing business have been greater since the beginning of the war than in almost any other line of industry. We are accustomed to think of the prices of commodities as having advanced about 100 per cent during this period; but if the cost of publishing had advanced *only* 100 per cent, those who issue periodicals would now be supremely happy and would have no cause for concern. The following items give the percentages of increase which have occurred since 1914 in the production of journals like MECHANICAL ENGINEERING and indicate how great has become the expense of issuing a periodical:

PER CENT INCREASE IN PRINTING COSTS SINCE 1914

| | |
|---|---|
| Machine Composition | 105 |
| Hand Composition | 218 |
| Presswork | 170 |
| Binding | 118 |
| Paper | 300 |
| Engravings (inch rate) | 110 |
| Engravings (minimum sizes) | 275 |

The greatest advances have occurred during the past two years, and to a large extent since the printers' strike in New York City

last fall. Nevertheless, the Society has not only maintained MECHANICAL ENGINEERING during this trying period on the same basis and with the same complete presentation of its matter as in previous years, but in addition has made notable improvements, one of which is the inclusion of The Engineering Index.

As for the new printers, the Eschenbach Printing Company has for many years handled the publications of the American Chemical Society and is well equipped for scientific work. Its facilities are to be enlarged to accommodate MECHANICAL ENGINEERING, and while it has not been possible to secure the complete installation of machinery and materials in time for the present number, new presses and composing-room equipment will shortly be in place. For several years past this company has printed Condensed Catalogues, the superior appearance of which is evidence of the workmanlike manner in which its printing has been handled.

In one respect, at least, it is hoped to render better service in the new location than heretofore, namely, in the distribution of MECHANICAL ENGINEERING through the mails. There have been frequent and annoying delays because of the congestion caused by the tremendous amount of printed matter which has to be handled by the New York Post Office, a condition which does not exist to a like extent in the smaller cities.

## The Superpower Survey

WILLIAM S. MURRAY

IN 1905 Mr. W. S. Murray, chairman of the Superpower Survey, was called to the New York, New Haven and Hartford Railroad Company to supervise the electrification of its lines adjacent to New York. This brought forcibly to his attention the economic advantages resulting from the consolidation of power units for the electric distribution of power for industrial and railroad purposes. It is therefore greatly to his satisfaction that Congress has authorized an investigation of the possibilities of the superpower system for the eastern section of the country, which is creating so much interest among engineers and others connected with our industrial plants and transportation systems. The proposed plan has for its central idea the construction of an electrical regional power system applicable to the industries and railroads existing within the territory between Boston and Washington and inland from the coast 150 miles. By the installation of high-powered, high-economy tidewater steam plants, hydroelectric stations, and steam stations located at the mouth of mines, all interconnected with a transmission system, and using also the large central stations now built and located at the larger cities, such as in Boston, Providence, New York, Philadelphia, and Baltimore, there can be integrated one great regional power system from which power can be delivered to the industries and railroads.

The area above-mentioned is the most congested industrial and railroad district in the United States. Here are located hundreds of steam-electric power plants, mostly of small capacity, using on a conservative estimate an average of 40 lb. of steam per kw-hr., as against the probable rate of 15 lb. when supplied from the superpower system. In Mr. Murray's opinion the machine capacity of this district is 17,000,000 hp., of which 10,000,000 is industrial and 7,000,000 railroad. The load factor is not over 15 per cent instead of 50 per cent or more which would result if power were concentrated on a single bus system and distributed therefrom to the industries and railroads.

By such concentration of power there would also result an enormous saving in maintenance through the substitution of electric for steam drive in the factories and on the railroads, and also by the reduction of train-miles on the railroads by virtue of the ability of the electric locomotive to consolidate a greater tonnage in single trains. Automatically, also, labor and maintenance costs at the mines would be greatly reduced and cargo space on the rails now required for useless coal haulage released for other commodities so urgently needed at the present time. Preliminary estimates indicate that the total annual saving from these various results would aggregate $300,000,000.

Facts such as these were brought out at meetings of the American Institute of Electrical Engineers and of other societies, and finally led the Engineering Council, the central body of the national engineering societies of America, to pass a resolution at a meeting held a year ago this month, advocating an appropriation by Congress for a survey of the superpower project. In April, 1920, a committee, of which Mr. Murray was chairman, was appointed by the Council to present the matter before the Appropriations Committee of the House of Representatives. The hearing was held March 18, and testimony was given by Prof. L. P. Breckenridge, Mem.Am.Soc.M.E.; Prof. D. C. Jackson, Mem.Am.Soc.M.E. and Past-President of the A.I.E.E., and Mr. Murray. The case was reported favorably to the House and ably presented there, by Chairman Good of the Appropriations Committee; and $125,000 was appropriated for the power survey.

The work is now being prosecuted by an organization known as the Superpower Survey, under the jurisdiction of the U. S. Geological Survey, with headquarters at 709 Sixth Avenue, New York City. Mr. Murray has divided his engineering staff into three departments of investigation: (1) Power and Transmission; (2) Railways; and (3) Industries, with a division chief at the head of each and an engineer-secretary assigned to the duty of collaborating with these division chiefs in collecting and collating data.

The services of the chief hydraulic engineer of the Geological Survey and of the chief mechanical engineer of the Bureau of Mines are being contributed in part to the engineering staff of the Superpower Survey. Further assistance is being rendered by an Advisory Board of men representative of power producers and power consumers, and of the people as a whole representing a national power policy. Mr. Murray has selected and the Secretary of the Interior has appointed to membership on this Board, gentlemen who would be representative of New England railroads, New York railroads, engineering, technical publicity, the National Electric Light Association, the American Electric Railway Association, and the National Industrial Conference Board. Prof. L. P. Breckenridge, of Yale University, is chairman of this Advisory Board.

The superpower project, in Mr. Murray's opinion, which is in accord with others, marks the beginning of a national power policy and he believes its advantages as developed for the eastern district will be applicable to other districts in other parts of the country, and will constitute one of the most important movements ever developed for the conservation of our fuel and for the economical production of needed products. It will serve to speed up, increase and maintain production in this country at a high point of efficiency, and give to the public a service such as they have a right to demand. [Prepared from notes supplied to MECHANICAL ENGINEERING by Mr. Murray.—EDITOR.]

## Engineers in Public Service

It is gratifying to know that others besides engineers are coming to recognize that the type of mind and the training of the engineer tend to qualify him for responsible executive positions in the service of the public. We all know that, as city managers, members of public-service commissions, etc., they have been for some time rendering valuable and appreciated service.

It is safe to say that Herbert Hoover is perhaps the most widely known and most generally admired public man, the world over, now living.

A member of The American Society of Mechanical Engineers, Mr. Norman J. Gould, is a member of Congress, and James Hartness, Past-President, Am.Soc.M.E., has recently been nominated by the Republicans for Governor of Vermont—a nomination which is generally conceded to be equivalent to election.

Governor Cox has said that when elected he will try to induce Mr. Hoover to enter his Cabinet; and that Governor Cox believes in engineers for executive work was further shown by an incident related to me in a conversation I had with Mr. H. S. Riddle, Mem.

Am.Soc.M.E., of Columbus, Ohio, on June 2, which, it will be noted, was some time before the nominating conventions. Mr. Riddle is President of the Ohio Board of Administration and as such he has general charge of all the institutions of the state except the penitentiary. In the course of a conversation about his work he told me, incidentally, how he came to undertake it.

Shortly after Governor Cox took office for his first term, now nearly six years ago, he asked Mr. Riddle to come to his office. On arrival, Mr. Cox said he had been looking for a man to administer the state institutions—had been informed that he, Mr. Riddle, was the man for that work and he asked him to accept the appointment. Mr. Riddle replied to the effect that he was not looking for a job and besides, said he: "Governor, I don't know as you and I would get along so very well, for you are a Democrat, while I am a Republican." Thereupon Mr. Riddle was assured that what was wanted was a man who could and would manage the state's institutions as they should be managed. Investigation had shown that Mr. Riddle was such a man and what his politics might be made no difference. Mr. Riddle thereupon recognized it as his duty to accept and he still holds the position, which I am sure he fills in a very able manner.

The incident shows that engineers, without regard to their politics, are being more and more sought by public officials for service in the conduct of public work, even when that work is not strictly of an engineering character; and that, I think, is cause for gratification.                                FRED J. MILLER.

## Saving the Natural Gas

The growing apprehension as to the possible (and probable) early exhaustion of the natural-gas supply has led to several important developments for its conservation and economy in its use.

The main influence in this direction has been the effective work of the National Committee on Natural Gas Conservation, appointed last January by former Secretary Lane, which has worked in co-operation with the U. S. Geological Survey of the Department of the Interior. On June 11 this committee made a final report at a conference of Governors and members of public-utility commissions of the natural-gas-producing states, held at Washington, D. C., in which were incorporated rules and regulations for gas production, transmission and utilization.

The spirit of these recommendations has been admirably expressed in an administrative order issued last August by the Public Utilities Commission of Ohio. This order designates two classes of consumers of natural gas: domestic consumers and industrial consumers, and so classifies these as to indicate in what order service is to be discontinued during an emergency in order to conserve a supply of gas for domestic consumers, and in what order service is to be restored when the emergency is past.

Domestic consumers include the users of gas for heating, lighting and cooking in private homes and apartments, but for lighting and cooking only in hotels, restaurants, etc. Industrial consumers are divided into three classes, comprising (1) plants producing food on a large scale, (2) power plants with gas engines, and (3) those not included in the first two classes. When there is not sufficient gas for all consumers, the surplus is to be furnished to them in the order above named. If, after disconnecting all industrial consumers, there is not a sufficient supply for the domestic consumers, there is to be a limited curtailment of use for heating purposes which, as the supply becomes much more restricted, will make it incumbent for householders to install auxiliary heating equipment.

A further recommendation, emphasized also by the National Committee, is that all distributing companies make a study of the best and most economical methods of using natural gas, and that a campaign of education be carried on among consumers. This thought was further advanced in a proclamation issued by the Governor of Ohio.

In the same spirit was a conference held in September at White Sulphur Springs, West Virginia, attended by the public-service commissions of the gas-producing states and representatives of the leading gas companies and gas associations. Definite recommendations were made to the natural-gas companies for the coming winter, calling for more efficient domestic gas appliances and classification of domestic and industrial consumers, and a survey of the territory served by the different companies.

Still another phase of the situation is the law recently passed by the State of West Virginia, known as the Steptoe Law, which provides that the natural gas produced in West Virginia may not be used outside of the state until after all the requirements and demands of home consumers had been supplied. This law has been contested by the states of Ohio and Pennsylvania to enjoin West Virginia from enforcing their law, and the case is now pending in the Supreme Court of the United States. At a recent hearing, Mr. S. S. Wyer, Mem.Am.Soc.M.E., natural gas engineer for the U. S. Bureau of Mines, opposed the West Virginia law.

At the conference at White Sulphur Springs, Prof. I. C. Church, state geologist of West Virginia, testified that from 75 to 90 per cent of gas deposits in Ohio are gone, and three-quarters of the known gas territory in West Virginia is already exhausted. The necessity for conserving the remaining available supply for the benefit of domestic consumers makes it apparent that the use of natural gas for industrial purposes must ultimately be discontinued; even at present there are occasions during periods of extreme cold when the gas supply is insufficient for domestic consumers alone, which indicates that gas for house heating will have to be restricted in order to conserve the supply for cooking and hot-water heating.

## The Water Power Act

With the passage of the Water Power Bill during the last session of Congress, the way has at last been cleared for the development on a large scale of a hitherto neglected source of power. It is true that in the past there has been some utilization of our water-power resources, but for the most part it has been confined to limited sections of the country. While this condition can be traced in part to the low costs of fuel and transportation which formerly prevailed, the controlling factor, nevertheless, has been the limitations imposed by the Government restrictions growing out of the fear of the usurpation of national resources by private interests.

Heretofore Government permits for the development of water-power sites have been revocable at will, and thus the business world has hesitated to embark upon such ventures. The new water-power bill, however, removes this uncertainty, as licenses will now be granted for a term of fifty years and at the end of such time may be renewed; or the Government, if it sees fit, may take over the development, giving, of course, proper compensation to the owners. Such a provision cannot fail to encourage investors.

By the passage of the bill water power now comes to the fore as a national resource of the greatest importance, and to the end that its readers may be kept in touch with the latest developments in this field, MECHANICAL ENGINEERING presents elsewhere in this issue a summary of the regulations governing the administration of the newly enacted law.

## Ground Broken for Vehicular Tunnel Between New York and New Jersey

On October 12, 1920, at Canal and Washington Streets, New York City, the first ground was broken for the vehicular tunnel to be built between New York City and Jersey City. The ceremony was held under the auspices of the New York State Bridge and Tunnel Commission, George B. Dyer, Chairman, and The New Jersey Interstate Bridge and Tunnel Commission, W. H. Noyes, Chairman.

The first work to be done on the tunnel is the construction of two ventilating shafts at the Manhattan end of the tunnel, located respectively at Canal and Spring Streets, near Washington Street. These shafts, which are 60 ft. and 54 ft. deep, respectively, will form permanent ventilating openings connecting with the tubes. The specifications show few material differentiations from the former practice in New York subway work. The chief change is in waterproofing, cotton fabric and asphalt being specified instead of burlap and coal-tar pitch because it is thought that the

hot asphalt in compressed air will give less trouble in the matter of fumes than the coal-tar pitch. C. M. Holland, chief engineer, states that the type of ventilating shaft specified makes it feasible for the subsequent tunnel work to be done either by the shield process or by the trench method.

## High-Speed Machinery

Under the Survey of Engineering Progress in the present issue is described a single-stage air compressor, operating at a normal speed of 22,000 r.p.m., which has been shipped by an American concern to South America. In many respects this is a significant installation. The very fact that a device of this character has been adopted for use in South America, where repair facilities are comparatively scarce, and not always adequate, shows the confidence both of its makers and purchasers in the reliability of the apparatus, notwithstanding the fact that it runs at such great speed.

Great speeds of rotation such as now met with in different types of apparatus have been achieved only by satisfying certain previously prescribed conditions as to materials, workmanship and means for balancing, etc. When it is stated that in the blower referred to, one ounce at the tip of the rotor exerts a stress due to centrifugal forces equal to something like $3\frac{1}{4}$ tons, it will be realized that not only materials capable of withstanding these immense stresses must be used, but also materials not subject to fatigue within extremely severe limits; and still further, methods of producing parts that will insure a very great range of safety.

Special steels, such as nickel-chrome and nickel-chrome-vanadium, together with great skill and care in the manufacture of ferrous alloys, have solved the problem as to the raw material, while, in the instance mentioned, modern methods of forging enabled the shaft and blades to be produced out of a single piece, which helped materially in the direction of necessary excellence of workmanship.

But before these tremendously high-speed machines could be made commercially practicable, a good deal of research in the field of mechanics of fast-revolving parts had to be done. It is only within the most recent time that engineers have succeeded in obtaining a clear comprehension of the laws governing the critical speeds of shafts, whirling phenomena, and the balancing of rotating parts generally, and have evolved methods for observing and measuring dynamic unbalance and for correcting errors therein. This was a somewhat tedious research, requiring an unusually skilful use of both mechanical and mathematical tools of investigation, and the success of modern high-speed machinery shows how well this work has been accomplished.

It should be clearly understood that the use of machinery running at speeds of 20,000 r.p.m. and upward is not merely an engineering "stunt" but is an answer to a very urgent requirement. In some cases, as, for example, with blowers, it permits a more compact design of machine and achieves in a more efficient manner what could also have been done with slower-running machinery. On the other hand, there are cases where a problem can be solved only by the use of such high-speed machinery, as, for example, in the case of the Alexanderson high-frequency generator for wireless telegraphy. There is another class of electric generators, namely, the commutatorless direct-current generator, where the ability to run at these high speeds appears to be a valuable characteristic. The significance of such machines as the DeLaval blower and the Alexanderson high-frequency generator for wireless telegraphy lies just as much, however, in the fact that it shows to what a state of perfection modern engineering has arrived in the production of materials and parts capable of withstanding truly tremendous stresses, and in the comprehension of the hitherto obscure phenomena of dynamic balancing and critical speeds of fast-rotating parts.

———

A natural-gas primer has been issued by the State of Ohio for use in its public schools. This pamphlet, which was originally issued by the U. S. Fuel Administration, deals in the simplest manner possible with the production and distribution of natural gas, the need for its conservation, and its economical use in the home.

# Federal Power Commission Adopts Regulations

## Commission Charged With Administering the New Federal Water Power Act Announces Adoption of Ten Rules Covering Procedure to be Followed in the Securing of Licenses and Permit

THE Federal Power Commission, which is charged with administering the Water Power Act, has adopted a set of comprehensive regulations covering the procedure in connection with the securing of licenses and permits. These regulations, ten in number, are of vital importance to the work of the Commission, as already many applications have been received, and if the present rate is maintained they will total before January 1 at least 4,000,000 horsepower.

The regulations just issued cover such matters as (1) Definition of Terms, (2) Applications—General Requirements, (3) Applications for Preliminary Permits, (4) Applications for Licenses—Major Projects, (5) Applications for Licenses—Minor Projects, (6) Applications for Licenses—Major Projects Already Constructed, (7) Declarations of Intention, (8) Priorities, (9) Permits, and (10) Licenses. Regulations (9) and (10) are of particular importance to those interested, and for that reason are given below in full.

The passage of the Federal Water Power Act concludes twelve years of effort for satisfactory water-power legislation. In accordance with its provisions the Secretary of War, the Secretary of the Interior, and the Secretary of Agriculture became a Commission to administer the Act and at their first meeting the Secretary of War was appointed chairman and O. C. Merrill, who at that time was chief engineer of the Forest Service, was made executive secretary.

The work of the Commission has been divided into engineering, accounting, statistical, regulatory, licensing, legal, and operation. The engineering division is regarded as the most important as it will make general investigations of the electric power industry, power sites, costs and developments. It will also report results of its examinations to Congress preparatory to construction work of the United States; examine and revise plans for development of streams upon which applications for licenses are made; consider construction plans proposed by licensee; and when already existing plants are brought under the Act will fix the necessary stipulations as to maintenance, development, and operation.

(9) PERMITS:

A—Except as hereinafter provided, preliminary permits may be issued on the application of citizens, associations, corporations, States or municipalities desirous of obtaining licenses for the construction, maintenance or operation of dams, water conduits, reservoirs, power houses, transmission lines, or other project works necessary or convenient for the development and improvement of navigation, and for the development, transmission and utilization of power across, along, from or in any of the navigable waters of the United States, or upon any part of the public lands and reservations of the United States (including the Territories), or for the purpose of utilizing the surplus water or water power from any Government dam.

B—Permits will be issued only for the purpose of enabling applicants to maintain their priorities while securing the data required for an application for license and will be for such periods, not exceeding a total of three years, as in the judgment of the Commission will be necessary for studying the proper location and design of the project; for making examinations, surveys, maps, plans, specifications and estimates; for conducting stream measurements; for sinking test pits or making borings to determine foundations for dams or other structures; for securing a market for the power to be developed; for making financial arrangements; or for any other purpose necessary or desirable in the preparation of application for license.

C—Permits will not be issued for projects already constructed; for transmission lines alone; for projects of a power capacity of less than 100 horse power; for projects which, in the judgment of the Commission, do not come within the scope of its authority under the Act, or should be undertaken by the United States itself, or do not propose adequate schemes of development, or would unreasonably interfere with projects under permit, license or other authority theretofore granted; or for projects for which date sufficient for filing application for license are already available. Permit affecting any reservation will be issued only after a finding by the Commission that the proposed use will not interfere with or be inconsistent with the purpose for which such reservation was created or acquired. Permits will not be issued until after the expiration of the publication period prescribed by the Act.

D—In acting upon applications for preliminary permits, and in determining preferences therefore, the Commission may in its discretion upon the request of any applicant or upon its own motion, hold hearing, order testimony to be taken by deposition, summon witnesses, or require the production of documentary evidence.

E—No charges will be made for permits, but permittees will be required as a condition of maintenance of priority, to perform such work and to make such studies and investigations, and such reports thereon, as in the judgment of the Commission may be necessary or desirable to enable both the applicant and the Commission to determine the feasibility, and the character and extent of development, which is proposed or which should be undertaken, which requirements will be expressed in the permit.

F—Upon a satisfactory showing of reasons, therefore, the Commission may authorize permittees to perform such construction work as may be necessary to maintain water rights under State law, or as may be desirable in preparation for the construction of project works; but the granting of such authority shall not be deemed to have created any equities or to have established any rights beyond what would have been created or established had such authority not been given.

H—Each preliminary permit shall set forth the conditions under which priority shall be maintained and a license issued, and shall also set forth the essential terms and conditions of such license.

(10) LICENSES:

A—Except as hereinafter provided, licenses may be issued either in accordance with the provisions of preliminary permits or upon direct application therefore by citizens, associations, corporations, States or municipalities, for the purpose of constructing, operating, and maintaining dams, water conduits, reservoirs, power houses, transmission, and utilization of power across, along, from or in any of the navigable waters of the United States, or upon any part of the public lands and reservations of the United States (including the Territories), or for the purpose of utilizing the surplus water or water power from any Government dam.

B—Licenses will be issued for such periods, not exceeding fifty (50) years, as in the judgment of the Commission, will in each individual case, allow for the satisfactory development and operation of the project and protect the public interest, and shall remain in full force and effect for such periods, unless surrendered or terminated as provided in the Act.

C—Licenses will not be issued for projects which, in the judgment of the Commission, do not come within the scope of its authority under the Act, or should be undertaken by the United States itself, or do not propose adequate schemes of development, or lack satisfactory showing of financial ability, or would unreasonably interfere with projects under permit, license or other authority theretofore granted, or would be opposed to the public interest. No license affecting the navigable capacity of any navigable waters of the United States will be issued until the plans of the dam or other structures affecting navigation have been approved by the Chief of Engineers and the Secretary of War. Licenses within any reservation will be issued only after a finding by the Commission that the license will not interfere or be inconsistent with the purposes for which such reservation was created or acquired. Licenses will not be issued until after the expiration of the publication period prescribed by the Act.

D—Licenses may be altered only upon mutual agreement between the licensee and the Commission. Any such alteration shall be made a part of the license and a substitute for the provision altered, but no such alteration shall operate to alter or amend or in any way whatsoever be a waiver of any other part, condition, or provision of the license.

E—Licenses may be surrendered only upon mutual agreement between the licensee and the Commission, and upon the fulfilment by the licensee of all obligations under the license, with respect to payment or otherwise, existing at the time of such agreement, and, if the project works authorized under the license are constructed in whole or in part, upon such conditions with respect to the disposition of such works as may be determined by the Commission.

F—Licenses may be terminated by written order of the Commission after such reasonable notice, not exceeding ninety (90) days, as the Commission may grant, if there is failure to commence actual construction of the project works within the time prescribed in the license, or as extended by the Commission. Under similar conditions and upon like notice the authority granted under a license may be terminated with respect to any project works or separable part thereof covered by the license, if there is failure to begin construction of such project works or part thereof within the time prescribed in the license or as extended by the Commission; but no part of the project works shall be deemed separable for the purposes of this regulation unless so specified in the license.

G—Licenses may be revoked only through proceedings in equity instituted in a district court of the United States for a district in which some part of the project is situated, and in the manner provided in the Act: (a). In case construction of the project works covered by the license, or of any specified part thereof, has been begun but not completed within the time prescribed in the license, or as extended by the Commission; or (b) in case the terms of the license are violated by the licensee.

# Plans for A.S.M.E. Annual Meet

### Meeting in New York December 7–10 to Consider Vital Transportation Problem—Newly

### cations to Woodworking and a Number of

THE tentative plans for the Forty-first Annual Meeting of The American Society of Mechanical Engineers to be held in New York early in December are prophetic of the best meeting in the history of the Society. The main points which assure success lie in the importance of the topic selected for the Keynote Session and in the fine spirit of coöperation shown by the newly formed Professional Sections, which will present many valuable and interesting papers.

## Keynote Session

Transportation will be the subject of the Keynote Session. Industry in this country is being severely handicapped because of the lack of adequate transportation facilities, and from present indications it will take several years for these facilities to catch up with the increased traffic and the normal development which will take place within the next few years.

Several phases of the transportation question will be considered on a broad, economic basis and with a view to developing practical constructive measures for improving the conditions. Addresses will be made on the railroad situation; water transportation, including rivers, lakes and canals; motor-truck transportation; terminals and freight handling; the Greater New York transportation problem; the development of railroad feeders; and a general address in the nature of a résumé or study of the whole transportation question in a large way.

## Sessions of the Professional Sections

Six of the newly formed Professional Sections have presented plans for vitally interesting sessions at this meeting. This mobilization of engineering skill for the discussion of the problems in the various subdivisions of mechanical engineering has greatly increased the value of the meeting.

### FUEL SECTION

The fuel situation of the United States demands careful attention by every individual, and help in conservation can be secured only by application of engineering principles. These points are clearly brought out in the four papers to be presented by the Fuel Section at the Annual Meeting, the following abstracts of which indicate the broad view-points of the authors.

THE FUEL SUPPLY OF THE WORLD, by L. P. Breckenridge.[1] This paper presents the latest available data on supply, production and use of various fuels of the world, particularly of the United States. It dwells primarily on supplies of kinds of coal: anthracite, bituminous, lignite and peat. The problems of diminishing supply of natural gas and oil are discussed. The writer also touches on the evident impossibility of ever developing any considerable amount of water power in proportion to the total power requirements of that portion of the United States lying east of the Mississippi River. The paper closes with a plea for thrift in the use of coal, setting forth how in a broad way the waste of fuel may be prevented.

FUEL CONSERVATION, by David Moffat Myers.[3] The results secured by the constructive conservation of fuel during the war, a program based on hurriedly formed and necessarily imperfect plans, are an indication of what might be accomplished by thorough and complete plans for constructive conservation in peace time. Some of the main items of waste to be considered are found in (1) boiler plants; (2) coke ovens; (3) domestic heating equipment; and (4) plants and installations misusing steam after its generation. Under the last item the author considers the waste of exhaust steam. He emphasizes the need for coöperation between the private plants needing both heat and power and the central stations, which could supply both instead of wasting some 80 per cent of the heat of the steam into the condenser. In the case of the private plants producing a surplus amount of exhaust steam, more power might easily be generated and sold to the central-station system. Fuel conservation, the author states, can never be satisfied by development of the central-power-plant idea alone; the privately owned plant is essential to economy.

[1] Professor of Mechanical Engineering, Sheffield Scientific School, Yale University, New Haven, Conn.
[3] Consulting Engineer, Griggs and Myers, New York, N. Y.

Touching on the "superpower" system advocated by W. S. Murray and included in a footnote to the paper. Mr. Myers believes that the plan as applied to railroads alone would have many important advantages but that in industrial and building heating the waste of steam and fuel in large central stations would be augmented rather than reduced.

Mr. Myers believes that the need of a definite policy of fuel conservation is urgent. Natural gas is practically gone, it is estimated that the supply of natural fuel will be exhausted in 20 to 30 years, and coal, first anthracite and then bituminous, our last natural fuel resource, will be the next to disappear. The author advocates educating the public to the point where it will demand the coöperation of the Government in the conservation of fuel. In conclusion he presents for consideration and constructive criticism a policy of fuel conservation to be adopted with the coöperation of the Government.

DISTILLATION OF FUELS AS APPLIED TO COAL AND LIGNITE, by O. P. Hood.[1] The necessity for conservation of both capital and raw material are two important factors restricting the distillation of coal. On the other hand, coal distillation is encouraged by the hope that if elements of higher values are recovered from coal by distillation, the remaining fixed carbon can be sold at a price lower than the raw fuel and also that by extensive distillation and other processes, manufactured gas may reinforce our rapidly disappearing natural gas.

Among the processes of distillation which are receiving considerable attention is that resulting from low temperature. The most advanced experiment in this country using low-temperature carbonization and distillation is the carbocoal process. In the production of metallurgical coke, distillation of coal in by-product coke ovens is already the major process.

The most favorable field for the expansion of distillation processes lies in domestic heating service. In lignite areas of the United States where there are large supplies of low-grade fuel and eastern coals are obtained with difficulty, experimental work has shown the possibility of applying distillation methods to this fuel, obtaining gas, tar, ammonium sulphate and a carbonized residue of excellent heating value.

FORM VALUE OF ENERGY IN RELATION TO ITS PRODUCTION, TRANSPORTATION AND APPLICATION, by Chester G. Gilbert[2] and Joseph E. Pogue.[3] Resource energy, as the source of power, heat, and chemical work, is the basis of industrial activity and social advancement. The quantity of energy required in the United States has grown to such stupendous size that its provision in the customary form is becoming increasingly difficult, as reflected in uncertainties of supply and rising cost. Coal, oil and water power—the major sources of energy—need each to be considered in regard to production, transportation and utilization; hence the problem assumes at once the complexity of a mathematical function with nine variables. The field of energy is an economic and technologic checkerboard, no area of which can be measured or appraised without due regard to all its associated components.

The energy situation is therefore reviewed in this paper from two points of view, that of "form value," defined as an intangible quality expressing the broad applicability of the energy form in contrast to its theoretical thermal value as commonly expressed in B.t.u. units; and "resource value," defined as an intangible quality expressing the availability of energy in terms of location and chemical character of its source, and involving the potentiality of chemical control for purposes of multiple production. Form value, in respect to both application and transportation, and resource value are discussed as basic factors in the development of a balanced energy supply. Industrial analogies contained in the paper furnish the reader with practical applications of the theory set forth.

### RAILROAD SECTION

In preparing its program the Railroad Section desires to bring to bear the combined skill of its members in the advancement of ideas and information that will be of benefit to all railroad men of the country in the solution of their immediate problems. The program includes the papers which are abstracted below and gives ample opportunity for the discussion of situations through which railroad problems must pass before solution.

MODERNIZING LOCOMOTIVE TERMINALS, by G. W. Rink.[4] This paper is a discussion of the problem of providing adequate facilities for the proper maintenance of locomotives at engine terminals. The location, size and general layout is dependent on various elements, the two principal factors being character of work to be performed and location of the general locomotive repair shop. The necessity for

[1] Chief Mechanical Engineer, U. S. Bureau of Mines, Washington, D. C.
[2] Consulting Engineer, Arthur D. Little, Inc., Cambridge, Mass.
[3] Industrial Economist and Engineer, Sinclair Consolidated Oil Corporation, New York, N. Y.
[4] Assistant Superintendent Motive Power, Central R. R. of N. J., Jersey City, N. J.

646

# ing Assure Phenomenal Success

## Organized Professional Sections Have Planned Interesting Sessions—Engineering Appli-
## Strong Miscellaneous Papers to be Presented

providing modern facilities is discussed with a view of awakening an interest in this subject, which has an important bearing on the ability of the railroads to handle the increased traffic demands of the country.

The introduction of modern labor-saving facilities not only makes for economy in repairs, but shortens the time necessary to prepare locomotives for service. The various structures which comprise the terminals are treated separately with more or less detail, having in mind that the entire problem must be handled in such a way that it will be of service in modernizing existing locomotive terminals as well as to provide information of value in designing new terminals.

INCREASING THE CAPACITY OF OLD LOCOMOTIVES, by C. B. Smith.[1] The high cost of railroading makes this subject of increased importance at the present time. The usual policy with reference to the purchase

[1] Mechanical Engineer, Boston, Mass.

of new locomotives and the conversion of old ones is not, in the opinion of the writer, a well-formulated policy. The difficulty lies in the fact that shop facilities are inadequate, a large amount of both time and money being consumed unnecessarily. The problems of adapting the old-type locomotives to suburban and local service are discussed and the items which are to be considered in any program for increasing locomotive capacity are listed. Mr. Smith cites examples of satisfactory reconstruction which justify the improvement program advocated by him, and states that the application of all the desirable auxiliaries to old engines is prohibitive without a radical provision for carrying out such a program.

ADJUSTMENT OF VEHICLES ON CURVES, by R. Eksergian.[1] The author gives a suggested outline of important factors that should be con-

[1] Engineer, Baldwin Locomotive Works, Philadelphia, Pa.

---

## TENTATIVE ANNUAL MEETING PROGRAM
### New York, December 7–10, 1920
*(Other subjects or changes to be announced later)*

**Tuesday Afternoon, December 7 (Simultaneous Sessions)**

| FUEL SECTION | WOODWORKING SECTION | MACHINE-SHOP SECTION |
|---|---|---|
| FUEL SUPPLY OF WORLD, L. P. Breckenridge | ENGINEERING IN FURNITURE MANUFACTURE, R. A. Parks | SIDE CUTTING OF THREAD MILLING HOBS, Earle Buckingham |
| LOW-TEMPERATURE DISTILLATION OF COAL, O. P. Hood | USE OF WOOD FOR FREIGHT CARS, H. S. Sackett | |
| FUEL CONSERVATION VERSUS MONEY CONSERVATION, D. M. Myers | WOODWORKING EDUCATORS, F. F. Moon | |
| FORM VALUE OF ENERGY IN RELATION TO ITS PRODUCTION, TRANSPORTATION AND APPLICATION, Chester G. Gilbert and Joseph E. Pogue | WOODEN HOLLOW WARE, John L. Graham | |
| | WOODEN FACTORY FLOORING, L. T. Erickson | |
| | WOOD PRESERVATION, E. S. Park and J. M. Webber | |
| | ELECTRICALLY OPERATED SAWMILLS, A. E. Hall | |

**Tuesday Evening, December 7**

Report of Tellers of Election and Introduction of the President-Elect
Presidential Address
Conferring of Six Honorary Memberships
Presidential Reception and Dance

**Wednesday Morning, December 8**

BUSINESS MEETING. Amendments to the Constitution, Reports of Standing Committees, Committees on Code of Ethics, Industrial Relations, Education, Feedwater-Heater Standardization, etc.

**Wednesday Afternoon, December 8 (Simultaneous Sessions)**

| MANAGEMENT SECTION | RAILROAD SECTION | RESEARCH |
|---|---|---|
| THE LIFE AND WORK OF THE LATE HENRY L. GANTT | STATIC ADJUSTMENT OF TRUCKS ON CURVES, R. Eksergian | CALIBRATION OF NOZZLES FOR THE MEASUREMENT OF AIR FLOWING INTO A VACUUM, Wm. L. DeBaufre |
| AN APPRECIATION FROM FRANCE, M. Ch. de Freminville | INCREASING CAPACITY OF OLD LOCOMOTIVES, C. B. Smith | THE HEAT-INSULATING PROPERTIES OF CORK AND LITH BOARD, A. A. Potter, J. P. Calderwood, A. S. Mack and L. S. Hobbs |
| AN APPRECIATION FROM GREAT BRITAIN, James J. Butterworth | MODERNIZING LOCOMOTIVE TERMINALS, George W. Rink | THE FLOW OF FLUIDS THROUGH PIPE LINES AND THE EFFECT OF PIPE-LINE FITTINGS, D. E. Foster |
| MR. GANTT'S CONTRIBUTION TO INDUSTRY, Fred. J. Miller, Pres. A.S.M.E. | | STEAM FORMULAE, R. C. H. Heck |
| MR. GANTT'S CONTRIBUTION TO SHIPBUILDING, SHIP OPERATION, ORDNANCE AND AIRCRAFT, Marshall Evans | | |
| THE CULMINATION OF MR. GANTT'S WORK, E. A. Lucey | | |
| MR. GANTT'S INDUSTRIAL PHILOSOPHY, W. N. Polakov | | |

**Wednesday Evening, December 8**

BRASHEAR MEMORIAL. Oration will be delivered on the life and work of the late Dr. John A. Brashear, Past-President, A.S.M.E.

**Thursday Morning, December 9**

KEYNOTE SESSION ON TRANSPORTATION. The following phases of the Transportation Problem will be discussed by authorities in this field: Railroads; Waterways; Feeders; Motor Trucks; Terminal Problem of New York City

**Thursday Afternoon, December 9**

Continuation of Keynote Session on Transportation
Ladies' Tea and Dance

**Friday Morning, December 10 (Simultaneous Sessions)**

| DESIGN | TEXTILE SECTION | POWER SECTION |
|---|---|---|
| DISASTROUS EXPERIENCES WITH LARGE CENTER-CRANE SHAFTS, Louis Illmer | HUMIDITY CONTROL IN TEXTILE PLANTS, Author to be announced | SESSION DEVOTED TO CONSIDERATION OF FUTURE DEVELOPMENTS OF POWER |
| TESTS ON REAR-AXLE WORM DRIVES FOR TRUCKS, K. Heindlhofer | POWER APPLICATION TO FINISHING PLANTS, Leo Loeb | MISCELLANEOUS |
| FOUNDATIONS FOR MACHINERY, N. W. Akimoff | TEXTILE FABRICATION FOR MECHANICAL PURPOSES, J. W. Cox | THE CONSTITUTION AND PROPERTIES OF BOILER TUBES, A. E. White |

### Other Papers to be Scheduled Later

sidered in the layout of a locomotive from the aspect of the adaption of the running gear to a plane curve. The paper considers in detail the following points: (a) The geometrical limitations imposed for a given curve with a proper wheelbase and the relation between lateral play and length of wheelbase, etc.; (b) the nature of the lateral reactions induced between the various elements of the running gear for various wheel arrangements and curves; (c) the elementary requirements in the design layout and arrangement of the wheelbase, together with the proper type of guiding trucks, etc.; and (d) the effect of the lateral reactions on the running gear as to strength of axles, etc.

### MANAGEMENT SECTION

The program of the session of the Management Section has been drawn up with the purpose of summarizing the life work of Henry L. Gantt and crystallizing the principles thereof for the guidance of the engineer.

Of the speakers, President Miller has long been an apostle of management as expounded by Mr. Gantt and for several months previous to Mr. Gantt's death was associated with him. Mr. Evans served in the Emergency Fleet Corporation during the war and there had an opportunity to give Mr. Gantt's methods a thorough tryout. Mr. Polakov was intimate with Mr. Gantt and is able to successfully interpret the fundamentals of his philosophy. Mr. Lucey was long a member of Mr. Gantt's organization and directly in touch with his latest plans. Messrs. Butterworth and de Freminville were close students and great admirers of his life and work. The session will have an appeal not only to the disciples of Mr. Gantt but to every engineer in industry, for the engineer has been the first to realize that the industrial problem is a problem of "man," and the understanding is now general that Mr. Gantt was the pioneer in applying the humanizing influence to management problems.

### MACHINE SHOP SECTION

Papers on important points in machine-shop practice will be discussed, and although the valuable paper by Earle Buckingham on the subject Side-Cutting of Thread-Milling Hobs is the only one at present selected by the Committee, we are assured that other valuable contributions will be received which will make up a particularly strong session.

### TEXTILE SECTION

Although this Section has not been formerly organized, its members have shown great interest, and the session to be held at the Annual Meeting will bring out four papers of practical value to the textile manufacturers.

### POWER SECTION

At the time of going to press the plans for this session had not been entirely completed. However, the subject for discussion will be Future Power Developments. Its treatment will be from four points of view, as follows:

(a) The Policy of Future Power Development, covering broad general aspects of future development and placing emphasis on fuel conservation and elimination of uneconomical generating stations.

(b) Effect of Load Factors on Cost, dealing with the economy in operating and capital charges due to diversity of load factors.

(c) Effect of Size of Plant on Cost, treating of high economies possible in large plants.

(d) Financial and Legal Aspects of Future Power Development.

### Miscellaneous

As has been announced previously in MECHANICAL ENGINEERING, a session will be devoted to the discussion of the Principles of Engineering in Woodworking. This session will give detailed information as to the proper application of engineering to varied subjects of the woodworking industry, and will emphasize the great need for further application of engineering principles to this industry.

In addition to the miscellaneous Annual Meeting papers, copious abstracts of which are included in this issue of MECHANICAL ENGI-

NEERING, there will be presented a paper by N. W. Akimoff containing a detailed discussion of Foundations for Machinery. Mr. Roger M. Freeman will also present a paper giving the general layout of the $20,000,000 Naval Ordnance Plant at Charleston, W. Va.

### BRASHEAR MEMORIAL

The greatness of Dr. John A. Brashear is worthy of fitting tribute by The American Society of Mechanical Engineers. With this in mind, Wednesday evening of the meeting week has been set aside for the honoring of a great man. An oration setting forth in form for permanent record the lovable qualities and successful attributes of "Uncle John" will be presented.

### THURSDAY EVENING

Attention is especially called to the action of the Meetings Committee in not scheduling a meeting event on Thursday evening. This program permits the holding of College reunions on Thursday night if desired and gives out-of-town members attending the meeting an opportunity to avail themselves of some of the various amusements which the metropolis offers.

## History of Naval Consulting Board Just Issued

The Navy Department has just issued a 288-page book containing a history of the origin and accomplishment of the Naval Consulting Board. Created in 1915 by the Secretary of the Navy and headed by Thomas A. Edison, this organization played a conspicuous part in the development of war inventions and new devices.

The formation of the Board, it will be recalled, was effected with the aid of the national engineering and allied technical organizations, who selected from their membership men especially qualified to form the personnel of such an important body. The American Society of Mechanical Engineers, for example, were represented by W. R. L. Emmet, a consulting engineer of the General Electric Company and Spencer Miller, of the Lidgerwood Manufacturing Company, whose work has been largely concerned with the design and production of marine apparatus.

The history of the Board was written by Lloyd N. Scott, late Captain, U. S. A., who acted as liaison officer to the Board and the War Committee of Technical Societies from the Invention Section of the General Staff, U. S. A. Special emphasis is laid on the fuel-oil investigation, the new naval station on the Pacific Coast, ship protection, the inventive accomplishments of the members of the board, and the meritorious inventions received from the public. Many of the special problems on which the Board worked in close coöperation with Army and Navy officials are also reviewed.

The preface to the volume comes from the pen of Secretary Daniels, who writes of the work of the Board in the following words:

It would have been impossible for the Navy to have carried on its efficient part in winning the World War without the intelligent and patriotic contribution of civilian thinkers and workers as well as civilians who enlisted in every department of naval effort. Foremost among these civilian patriots stand the members of the Naval Consulting Board. Its members gave themselves fully to the service of their country, bringing scientific and engineering knowledge, with large experience touching the vital problems that confronted the Navy. The membership of this Board was chosen by and from the foremost engineering societies in America, embracing men whose achievements were of world renown. The Navy, indeed the whole country, owes their scientific patriotism a debt not only for what they wrought but quite as large a debt for the stimulus and inspiration they imparted to the naval personnel and to civilians enlisted in national service.

The Astoria Silk Mills of Astoria. L. I., have recently purchased the historic West Point Foundry, in Cold Spring, and will operate it with 700 employees. The foundry closed down a few years ago, after the A. B. & J. M. Cornell Iron Works had operated it for several years. The first locomotive used in New York State, the old wood burner De Witt Clinton, now on exhibition in Grand Central Terminal, was constructed there. The plant was rushed with war orders during the civil war making hundreds of Parrott guns.

# The Federated American Engineering Societies Was Created To Consolidate The Engineering Organizations of America

## Is Your Engineering Society a Member of This Federation?

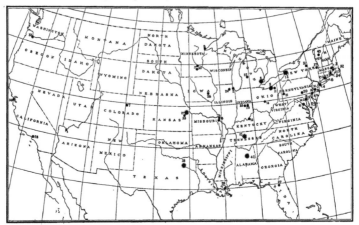

THE FIELD OF THE FEDERATION
Circles Show Local Societies Represented at Washington Conference, June 3 and 4, 1920
Asterisks Show Local Societies Now Members of the Federation

Last June delegates from 30 national and 41 local and regional organizations met and created THE FEDERATED AMERICAN ENGINEERING SOCIETIES. It is a going organization and on November 18 and 19 the first meeting of its governing body, the American Engineering Council, will be held at Washington, D. C. Will your Society be represented? Consult the list below and see. The societies listed in bold-face type are those which are now members of the Federation.

### ORGANIZATIONS REPRESENTED AT WASHINGTON CONFERENCE, JUNE 3 AND 4, 1920

LOCAL, STATE, AND REGIONAL ORGANIZATIONS

| | | |
|---|---|---|
| (1) | Associated Engineering Societies of St. Louis, St. Louis, Mo..... | 1,200 |
| (2) | Associated Engineers of Spokane, Spokane, Wash............... | 200 |
| (3) | Boston Society of Civil Engineers, Boston, Mass.............. | 786 |
| (4) | Brooklyn Engineers' Club, Brooklyn, N. Y................... | 415 |
| (5) | Cleveland Engineering Society, Cleveland, Ohio............... | 1,300 |
| (6) | Connecticut Society of Civil Engineers, New Haven, Conn...... | 385 |
| (7) | Colorado Society of Engineers, Denver, Colo................. | 415 |
| (8) | Detroit Engineering Society, Detroit, Mich................. | 575 |
| (9) | Duluth Engineers' Club, Duluth, Minn...................... | 154 |
| (10) | Engineering Association of Nashville, Nashville, Tenn........ | 125 |
| (11) | Engineering Society of Akron, Akron, Ohio.................. | 436 |
| (12) | Engineering Society of Buffalo, Buffalo, N. Y............... | 600 |
| (13) | Engineers' Club of Baltimore, Baltimore, Md................ | 230 |
| (14) | Engineers' Club of Columbus, Columbus, Ohio................ | 250 |
| (15) | Engineers' and Architects' Club of Louisville, Louisville, Ky... | 225 |
| (16) | Engineers' Club of Philadelphia, Philadelphia, Pa............ | 2,222 |
| (17) | Engineers' Club of Minneapolis, Minneapolis, Minn........... | 179 |
| (18) | Engineers' Club of St. Louis, St. Louis, Mo................ | 625 |
| (19) | Engineers' Club of Trenton, Trenton, N. J.................. | 280 |
| (20) | Engineers' Society of Pennsylvania, Harrisburg, Pa........... | 702 |
| (21) | Engineers' Society of Western Pennsylvania, Pittsburgh, Pa..... | 1,200 |
| (22) | Florida Engineering Society, Gainesville, Fla............... | 162 |
| (23) | Grand Rapids Engineering Society, Grand Rapids, Mich........ | 85 |
| (24) | Illinois Society of Engineers, Wheaton, Ill................. | 262 |
| (25) | Indiana Engineering Society, Indianapolis, Ind.............. | 358 |
| (26) | Iowa Engineering Society, Iowa City, Iowa.................. | 205 |
| (27) | Kansas Engineering Society, Topeka, Kansas................. | 205 |
| (28) | Los Angeles Joint Technical Society, Los Angeles, Cal........ | 1,600 |
| (29) | Louisiana Engineering Society, New Orleans, La.............. | 285 |
| (30) | Minnesota Surveyors and Engineers' Society, Minneapolis, Minn. | 200 |
| (31) | Mohawk Valley Engineers' Club, Utica, N. Y................. | 160 |
| (32) | New England Water Works Association, Boston, Mass.......... | 890 |
| (33) | Oregon Technical Council, Portland, Ore................... | 530 |
| (34) | Providence Engineering Society, Providence, R. I............ | 830 |
| (35) | Scientech Club, Indianapolis, Ind......................... | 120 |
| (36) | San Francisco Joint Council of Engineering Societies, San Francisco, Cal........ | 1,600 |
| (37) | Society of Engineers of Eastern New York, Troy, N. Y........ | 490 |
| (38) | Technical Club of Dallas, Dallas, Texas.................... | 125 |
| (39) | Topeka Engineers' Club, Topeka, Kansas.................... | 63 |
| (40) | Vermont Society of Engineers, Montpelier, Vt............... | 150 |
| (41) | Washington Society of Engineers, Washington, D. C.......... | 482 |
| (42) | Alabama Technical Association, Birmingham, Ala.............. | 110 |

NATIONAL ORGANIZATIONS

| | | |
|---|---|---|
| (1) | American Association of Engineers, Chicago, Ill.............. | 20,000 |
| (2) | American Association of Petroleum Geologists, Norma, Okla..... | 250 |
| (3) | American Ceramic Society, Alfred, N. Y.................... | 1,262 |
| (4) | American Concrete Institute, Boston, Mass.................. | 450 |
| (5) | American Electric Railway Engineering Association, New York, N. Y............ | 800 |
| (6) | American Electrochemical Society, Bethlehem, Pa............. | 2,500 |
| (7) | American Institute of Chemical Engineers, Brooklyn, N. Y..... | 318 |
| (8) | American Institute of Electrical Engineers, New York, N. Y.... | 11,345 |
| (9) | American Institute of Mining and Metallurgical Engineers, New York, N. Y................ | 8,689 |
| (10) | American Railway Engineering Association, Chicago, Ill........ | 1,430 |
| (11) | American Society of Agricultural Engineers, Columbus, Ohio.... | 624 |
| (12) | American Society of Civil Engineers, New York, N. Y......... | 9,652 |
| (13) | American Society of Mechanical Engineers, New York, N. Y.... | 12,705 |
| (14) | American Society of Refrigerating Engineers, New York, N. Y... | 400 |
| (15) | American Society of Safety Engineers, New York, N. Y........ | 456 |
| (16) | American Society of Heating and Ventilating Engineers, New York, N. Y................ | 1,203 |
| (17) | American Society for Testing Materials, Philadelphia, Pa...... | 2,765 |
| (18) | American Water Works Association, New York, N. Y.......... | 2,616 |
| (19) | Association of Railway Electrical Engineers, Chicago, Ill...... | 450 |
| (20) | Illuminating Engineering Society, New York, N. Y........... | 1,250 |
| (21) | Institute of Radio Engineers, New York, N. Y.............. | 1,600 |
| (22) | Mining and Metallurgical Society of America, New York, N. Y. | 505 |
| (23) | National Fire Protection Association, Boston, Mass........... | 3,205 |
| (24) | National Safety Council (Engineering Division), Chicago, Ill.... | 200 |
| (25) | Society of Automotive Engineers, New York, N. Y........... | 4,869 |
| (26) | Society of Industrial Engineers, Chicago, Ill................ | 400 |
| (27) | Society of Naval Architects and Marine Engineers, New York, N. Y. | 1,688 |
| (28) | Society of American Military Engineers, Washington, D. C..... | 1,600 |
| (29) | Society for the Promotion of Engineering Education, Pittsburgh, Pa. | 1,325 |
| (30) | Taylor Society, New York, N. Y.......................... | 370 |

# Engineering Division of National Research Council Issues Report

### Research in the Future to be Conducted by a Special Committee in Each Field—Reorganization of Division Opens Way for Greater Coöperation with Industries and Engineering Organizations—Grouping of Existing and Proposed Committees

THE Division of Engineering of the National Research Council recently rendered to Engineering Foundation, which coöperates with it in its work, a report of its accomplishments to date and the lines along which it expects to develop in the immediate future. As in the case of many other fields of endeavor, engineering research has been greatly handicapped by lack of funds, and Comfort A. Adams, chairman of the Division, emphasizes this fact in his report. He writes:

The opportunities for service of great importance to our industries and to our country are almost appalling in number; but although the past accomplishments of our committees constitute a very considerable contribution toward such service, the progress in any particular case must be discouragingly slow, without substantial financial support. For example, the very substantial accomplishment of the Welding Committee was due in considerable part to the financial support of the Emergency Fleet Corporation, amounting to over $40,000 for one year. If all of our committees were supported on this same scale, by contributions from the interested industries, the annual contribution of any one company would be less than they pay in advertising during one month, and utterly insignificant as compared to the value of the results, as was true in the case of the Welding Committee.

Despite this lack of adequate funds, the Division has carried on research in many fields, and that its future undertakings may be productive, the work now in progress is being reorganized on the group basis; that is, in each general field there will be a General Advisory Committee, with a small, active Executive Committee, under which will be organized the several Research Committees. These General Advisory Committees will also be the agencies through which the Engineering Division will endeavor to stimulate the industrial world to conduct the more commercial forms of research. Furthermore, in order to enlist the active coöperation of the engineering societies, it is hoped to have each society accept as its own research committee the Advisory Committee in its particular field.

The grouping of the existing committees and of those in formation is shown in the accompanying chart. The first group (S) is largely made up of committees organized by Dr. H. M. Howe, formerly chairman of the Division, either during or shortly after the war. Several other committees of this group organized specially for war work, completed their work and were discharged. The committees on Non-Ferrous Metals (NF), Fatigue Phenomena of Metals (F), New Hardness-Testing Machine (HT), and Pulverizing (P), were likewise organized by Dr. Howe, and some of these might have been included in the Steel group, except that they apply also to non-ferrous metals.

All of these committees naturally come within the field of mining and metallurgy, and although they are not at present recognized as officially connected with the American Institute of Mining and Metallurgical Engineers, Professor Adams states in his report that it is expected they will be coördinated in the near future under a general group committee of that Institute.

It is similarly expected that the Electrical Group (E) and the Highway Group (H) will probably be accepted by the American Institute of Electrical Engineers and the American Society of Civil Engineers, respectively. None of the three societies mentioned above has research committees of its own. On the other hand, however, the group dealing with Mechanical Engineering (M) was organized by The American Society of Mechanical Engineers and is being administered and financed by that society.

The Alloy Research Association (AR) was first fathered by the Division of Research Extension and was intended to be an organization of the interested industries and, after it was well started, largely independent of the National Research Council. At present, however, its activities are confined to an information service and its management is attached to the National Research Council, the Engineering Division coöperating in this work.

The Society of Automotive Engineers is very much alive to the importance of research, but although some work has been done under its technical committees, the society is just now feeling its way in respect to the organization and conduct of research.

The Society of Heating and Ventilating Engineers is also actively engaged in research, most of its work being done at the Bureau of Mines' Pittsburgh Laboratory, under the direction of Prof. John R. Allen. The society has collected more than $20,000 per year from the industries for the conduct of this coöperative research, and the Bureau of Mines contributes space, facilities and assistance of about equal value. Several researches are actively under way at the present time.

The Society of Refrigerating Engineers is likewise thoroughly alive to the need for research, and although it has no research committee, it has appointed Professor Rautenstrauch, of Columbia

University, as its representative on the Division, and he is now organizing for research in this field.

Finally, the American Bureau of Welding of the American Welding Society is also about to be reorganized as the General Advisory Committee on Welding to the Division of Engineering, although at present the research committees of the Bureau are officially committees of the Division.

## CALIBRATION OF NOZZLES

*(Continued from page 609)*

greater the amount of moisture, the less the total weight of fluid that flows through the nozzle in a given time. Thus, when air flows through a nozzle from the atmosphere, the usual variations which occur in the amount of moisture contained therein produce appreciable changes in the weight of dry air admitted in a given time.

Based on an expansion exponent of $n = 1.375$, the rates of flow of dry air containing no moisture under the normal conditions of 70 deg. fahr. and 29.92 in. mercury pressure, 0.07492 lb. per cu. ft. density, have been calculated for the nozzles, calibrated, and are tabulated in Table 1.

# Engineering and Industrial Standardization

## The Standardization of Safety Codes

DURING the past ten or fifteen years many states have passed laws providing for the compensation by employers of workmen injured while engaged in the duties of their various occupations. These laws are administered by the state boards, usually called industrial accident boards or compensation commissions, and an organization of these boards has been formed called the International Association of Industrial Accident Boards and Commissions.

Some of these boards have not only administered the state laws respecting compensation but have also issued safety rules for use by the industries of their respective states. Such work is of great importance. Only a few states, however, have funds and technical staffs sufficient to independently prepare safety rules or codes and hence some states copy the rules of others with such changes as seem necessary or advisable, and promulgate them with such instructions and supervision as they are able to give. These rules are revised from time to time to make them more complete, and thus as each state has prepared and revised them considerable divergence has occurred, which has often proved embarrassing to manufacturers furnishing machinery and supplies which are expected to comply with the requirements of states having such rules.

As a step toward the elimination of this duplication the Bureau of Standards coöperated during the war with the War and Navy Departments in the preparation of several industrial safety codes for the use of the Government establishments. Furthermore, in order to get the judgment of all those most active in such work, the Bureau called a conference of approximately 100 organizations interested in safety codes. This conference was held in Washington, January 15, 1919. The special question put before it was how the work of preparing safety codes could be carried out so that they might become national and be generally used throughout the country. The suggestion was made that codes be submitted for the approval of the recently organized American Engineering Standards Committee, and at a second conference held on December 8, 1919, a resolution was passed that safety codes for general use should be prepared according to the procedure of that body.

A second resolution requested the American Engineering Standards Committee to ask the International Association of Industrial Accident Boards and Commissions, the National Safety Council, and the Bureau of Standards to form a Joint Safety Code Committee which should prepare a report giving a list of safety codes which should be prepared and recommending sponsors for the same. On January 3, 1920, the American Engineering Standards Committee took favorable action upon this request and formally asked these three organizations to form a committee.

This Joint Safety Code Committee was organized as an advisory committee to the American Engineering Standards Committee and has rendered valuable service. In its reports so far it has recommended the sponsorships for thirty-seven safety codes. Recent activities of a number of these sub-committees are given below.

*Safety Code on Power Presses.* The National Safety Council which has been assigned by the American Engineering Standards Committee the sponsorship for the Safety Code on Power Presses, has requested The American Society of Mechanical Engineers to appoint two representatives to serve on the Sectional Committee. The Society appointees are E. E. Barney, development engineer, Remington Typewriter Company, New York City, and C. N. Underwood, executive secretary of the Clothiers' Exchange of Rochester.

*Safety Code for Grinding Machinery.* The American Engineering Standards Committee has assigned to the Grinding Wheel Manufacturers' Association of the United States and Canada, and the International Association of Industrial Accident Boards and Commissions, the sponsorship for a Safety Code for Grinding Machinery. At the request of the Grinding Wheel Manufacturers' Association, Walter B. Gardiner, mechanical superintendent, Lincoln Twist Drill Co., Taunton. Mass., has been appointed to serve on this Committee.

*Safety Code on Paper and Pulp Mills.* The National Safety Council as sponsor for this code has requested The American Society of Mechanical Engineers to appoint two representatives to serve on this Committee.

*Safety Code on Logging and Sawmill Machinery.* The Bureau of Standards was appointed sponsor by the American Engineering Standards Committee to formulate a Safety Code on Logging and Sawmill Machinery. The Bureau of Standards has requested The American Society of Mechanical Engineers to designate a representative to serve on the Sectional Committee, and Joseph H. Dickinson, engineer, Lidgerwood Manufacturing Company, New York City, has been appointed.

*Mechanical Transmission of Power.* The International Association of Industrial Accident Boards and Commissions, the National Workmen's Compensation Service Bureau, and The American Society of Mechanical Engineers have been appointed joint sponsors for the Safety Code for the Mechanical Transmission of Power. The A.S.M.E. has designated the following members to serve on this Committee: A. C. Jewett, Winchester Repeating Arms Company, New Haven, Conn.; F. L. Morse, Morse Chain Company, Ithaca, N. Y.; G. M. Naylor, Fairbanks Company, New York City; R. W. Sellew, The Fafnir Bearing Company, New Britain, Conn.; E. D. Wilson, Graton & Knight Manufacturing Company, Worcester, Mass.; D. C. Wright, Dodge Manufacturing Company, Mishawaka, Ind.

## International Aircraft Standardization

One of the most important organizations now concerned with standardization is the International Aircraft Standards Commission. This Commission is composed of representatives of Canada, France, Great Britain, and Italy. The admission of Belgium and Japan awaits formal confirmation, and needless to say the participation of America is greatly desired. The next meeting of the Commission will be held in Paris during November, and for the purpose of discussing America's participation therein a meeting was recently held in Washington, D. C., which was attended by the representatives of Government departments and national engineering organizations.

The meeting was called to order by A. A. Stevenson, chairman of the American Engineering Standards Committee, who stated that the organization which he represented had been requested to arrange for unofficial American representation at the Paris meeting of the Commission, if it were found impossible to secure full and official representation. Mr. Stevenson also stated that the Commission is planning to consider at the November meeting such general matters as the advisability of continuing on its present lines, its relation to whatever organizations should be set up by the Air Convention, and the consideration of reviewing or modifying the references to the Advisory Technical Committee.

After a full discussion of various methods which were proposed for bringing the matter to the attention of the proper Government officials, and what method should be followed in case Government credentials could not be secured, resolutions were unanimously passed authorizing the chairman to appoint a committee of not to exceed five to call on the chairman of the Council of National Defense and to present through him to the President a request for the authorization of official representation at the Paris Conference. The committee of five was also authorized to recommend in the event of favorable consideration of the request for official representation, that the American Engineering Standards Committee nominate members of the delegation who should represent this country.

The conference also went on record as favoring unofficial representatives, in the event that official representation cannot be secured, provided such a delegation would be permitted to sit in at the meetings of the International Commission, participate in the discussion, and be entitled to the report of the proceedings. It is also the sense of the conference that the American Engineering Standards commission, acting on behalf of the Conference, should appoint unofficial representatives.

# NEWS OF THE ENGINEERING SOCIETIES

The Ninth Annual Safety Congress—Meetings of the Iron and Steel Electrical Engineers,
American Society for Steel Treating, and American Chemical Society
—The Sixth Chemical Exposition

## Ninth Annual Safety Congress

That safety must be incorporated into the civic and school life and organized as a community activity is the 1920 slogan of the National Safety Council, under whose auspices the Ninth Annual Safety Congress was held in Milwaukee, September 27 to October 1. Over three thousand men and women delegates attended the congress. Two hundred exhibits in Mechanics' Hall showed every safety device on the market. The nineteen divisions of the council held sectional meetings, the sessions including those of engineering, public utilities, steam railroads, electric railroads, mining, rubber, metals, and the automotive industries. Motion-picture films on safety subjects were shown in the Milwaukee Auditorium during the days of the congress. Actual demonstrations by boy scouts of accident scenes which the safety workers had pictured verbally to the delegates, closed the program. From the point of view both of number of speakers and importance of the subjects discussed, this gathering was the most important in the history of the Council. Making safety a part of the instruction in public schools is the work upon which the Council plans to concentrate its energies in the immediate future. It is expected that the factory operative will thus have learned the principles of safety at an early age, and will be influenced by the instruction throughout his life.

## Association of Iron and Steel Electrical Engineers

Recent developments in electrically driven reversing mills were recorded at the fourteenth annual meeting of the Association of Iron and Steel Electrical Engineers, held at the Hotel Pennsylvania, New York, September 24. A paper by K. A. Pauley on the Electric Reversing Mill Considered from the Standpoint of Tonnage gave an analysis of records on comparative runs in electric and steam mills showing that the electric mill can produce a tonnage equal to or greater than the steam mill. Coupling these records with the advantages of the electric mill from the standpoint of lower power costs, lower maintenance cost, greater flexibility of control, etc., there is little doubt that the steam reversing mill is now as out of date as the non-reversing steam mill has been for many years.

The educational committee suggested educating steel-mill electricians as to means for minimizing delays due to failure of electrical equipment in steel plants. Sample sheets of a practical electrical course being given to electricians in two large steel plants were included in the report of the committee.

The electric-furnace committee pointed out possible improvements in operating details of electric-furnace work. The report was prepared from answers received to a questionnaire sent out to steel manufacturers. The data secured indicate that the greatest need at present is for better electrode economy. Lack of tensile strength seems to be one of the difficulties of electrode trouble caused by the lines of cleavage of the calcined anthracite coal being in a vertical as well as horizontal direction, it being impossible at present to build an electrode where all the cleavage lines are horizontal. One of the chief sources of difficulty in furnace operation is delays due to handling material to and from the furnace. These can be reduced by bringing the scrap on the furnace floor from the rear and adjacent to the furnace in small cars on tracks and using traveling cranes entirely for handling the steel from the furnace.

In a paper on Some Economic Considerations in Design of Power Plants for Steel Mills, T. E. Keating studied in detail the utilization of blast-furnace gas and the efficiencies and costs that may be expected to prevail in such an installation. Power transmission was discussed by A. L. Freret in Underground Transmission,

and by D. M. Petty in Transmission and Distribution of Power in Industrial Plants. A feature of the meeting was an exhibition on the roof of the hotel of representative types of electrical equipment for steel mills.

## American Society for Steel Treating

The second annual convention of the American Society for Steel Treating was held in the Commercial Museum, Philadelphia, September 14–18, 1920, at which meeting the final steps of amalgamation between the Steel Treaters' Research Society and the American Steel Treaters' Society were consummated. Prof. A. E. White, of the University of Michigan, Ann Arbor, is president and W. H. Eisenman, of Cleveland, is secretary. The headquarters of the new society are at Cleveland.

The large floor space of the Commercial Museum was given over to an exhibition of various manufacturers of steel-treating devices and materials. A number of interesting sessions on the various phases of steel treating were held in the Assembly Hall. Among the noteworthy papers was that of H. P. MacDonald, vice-president, Snead & Co., Jersey City, N. J., who described a method of heat-treating aeroplane tubes by passing electric current through a tube to raise the temperature to the required degree, releasing the clamps holding the tube and allowing it to drop into the oil bath at the rate of one tube every 40 sec.

Dr. Charles W. Burrows, magnetic-research engineer, elaborated a method for the magnetic testing of rails, cables, etc., by which blowholes, flaws, seams, cracks, etc., would be revealed. Doctor Burrows showed that the method was commercial, inasmuch as only one minute was required per rail.

T. D. Lynch, of the Westinghouse Co., presented a tentative specification for the manufacture of helical springs required to transmit the power from a motor-driven quill, through a flexible coupling, to the driving wheels of a heavy electric locomotive. The analysis suggested is as follows: C, 0.50 to 0.60; Mn, 0.60 to 0.80; Si, 1.90 to 2.20; P, 0.04 max.; S, 0.04 max.

The properties of a highly resistant alloy steel were explained by Charles M. Johnson, director of research department, Crucible Steel Co. It is a complex alloy, with 0.3 to 0.4 per cent carbon, which reverts to a homogeneous solid solution upon annealing at 1300 deg. cent., and which, upon tempering being continued for several days, allows an excess constituent to divorce itself and become spheroidized. It is quite non-magnetic and resists staining and rusting remarkably. Although the metal can be forged, rolled or sheared in thicknesses 0.1 in. to 1 in., and is machinable, it offers a maximum resistance to prolonged heating at temperatures up to 2000 deg. fahr. and is cut by the oxy-acetylene torch with the utmost difficulty.

Metallurgical features of the manufacture of stainless steel articles were given by W. H. Marble, manager of American Stainless Steel Co. The analysis used is: C, 0.20 to 0.40; Cr, 13; Si, 0.30; Mn, 0.50. Chromium is the essential element furnishing the peculiar resistance to corrosion, tungsten and nickel being sometimes added for an increasing luster in polish. Ingots are box-annealed at 1380 deg. fahr., air-cooled, reheated to 2100 deg. fahr., and hammered carefully, never allowing the temperature to drop below 1650 deg. fahr. Cooling after forging gives hard material, the higher the uninterrupted cooling the harder becoming the material. Rough forgings are now reannealed at 1380 deg. fahr., using all the precautions necessary for high-grade steel, furnace-cooled to 1100 deg. fahr. and then air-cooled. This results in a Brinell hardness of 200. After machining and finishing the piece may be hardened from 1750 deg. fahr. in air, oil or water, depending upon the intricacy of the shape, and tempered at a suitable heat, varying from 280 deg. fahr. for knife blades to 1100 deg. fahr. for exhaust valves.

Other papers covered the Use of Quenching Fluids, the Development of Alloy Steels, and Various Fuels for Heat Treating, and symposiums were held which dealt with Case Carburizing and High-Speed Steel Treatment.

## American Chemical Society

At the meeting of the American Chemical Society in Chicago, September 6 to 10, a fuel symposium was held at which a number of papers were presented of interest to mechanical engineers. The following notes on this session are abstracted from the report in *Power Plant Engineering* for October 1.

Dr. Harry A. Curtis described the process of manufacture of "carbocoal," the new smokeless fuel manufactured from bituminous coal and having certain of the characteristics of anthracite. It is prepared by crushing the soft coal and carbonizing it at the relatively low temperature of 900 deg. fahr. During the carbonization, which is carried out in a horizontal retort lined with carborundum, the coal is continuously stirred and moved slowly through the retorts by means of paddles mounted on steel shafts running lengthwise through the retort. A soft semi-coke is obtained from this low-temperature carbonization. The yield in tar is twice as much as is obtained from the ordinary coking process.

The semi-coke is ground, briquetted with pitch, and the briquets then carbonized for about six hours at 1800 deg. fahr. High-temperature carbonization renders the briquets hard, dense and smokeless. By-products are also obtained from the second carbonization. The first commercial plant manufacturing carbocoal was put into operation at Clinchfield, Va., in June 1920, and is equipped with 24 retorts giving a capacity of 500 tons of coal a day, from which 350 tons of the fuel are made. The process is the result of five years' experimental work at Irvington, N. J., and the large plant at Clinchfield was established as a war project.

Colloidal fuels were discussed by Dr. S. E. Sheppard. He recorded as a signal technical triumph the successful operation from April to July 1918, of the U. S. S. *Gem* on a colloidal fuel containing 30 per cent pulverized coal. Among the advantages he quoted for colloidal fuels were that they contain more heat units per gallon than fuel oils and very little moisture and ash, and that they are immune from spontaneous combustion and may be fireproofed by a "water seal" because they are heavier than water.

The development and increasing use of the by-product coke oven in the United States were dealt with by F. W. Sperr, Jr., and E. H. Bird. Before the war, coke was made largely in the simple beehive oven where tars run to waste and gases are discharged into the air. Now the by-product ovens lead the beehive ovens. The coking of a ton of high-grade coal in a beehive oven requires the consumption of 11,000 cu. ft. of gas, 9 gal. of tar, 4 gal. of light oil and 100 lb. of coke, which in all generate 9,388,000 B.t.u., or the equivalent of 671 lb. of coal. The by-product oven in making a ton of coke uses 4300 cu. ft. of gas, which is equal to 2,408,000 B.t.u., or equivalent to 172 lb. of coal.

Dr. J. B. Garner advocated the enrichment of illuminating water gas with natural gas as a means of economy. In this way, he said, natural gas could be so used as to insure to the public for many years to come a supply of gas at a cost at which it would otherwise be impossible to sell it.

Low-Temperature Carbonization and Its Application to High-Oxygen Coals, by S. W. Parr; Carbonization of Canadian Lignite, by Edgar Stansfield; Gasoline Losses Due to Incomplete Combustion in Motor Vehicles, by A. C. Fieldner, G. W. Jones and A. A. Straub; and Fuel Conservation, Present and Future, by Horace C. Porter, were subjects also considered at the fuel session.

Numerous other professional papers were presented at the other sessions. Attention may be called to the following: Electrometric Method for Detecting Segregation of Dissolved Impurities in Steel, by Edward G. Mahin and R. E. Brewer; Flow of Viscous Liquids Through Pipes, by Robt. E. Wilson and M. Seltzer; and Comparative Study of Vibration Absorbers, by H. C. Howard.

At the meeting of the council of the society the invitation to join The Federated American Engineering Societies and to send delegates to the coming meeting of the American Engineering Council was declined. It was held that the society should offer

the Federation aid and encouragement, but that the fields of chemists and engineers were distinct and separate and that more and better work could be accomplished without interlocking the organizations in these branches.

## Sixth Chemical Exposition

A multitude of products, comprising the most delicate instruments of precision, new chemicals and dyes, and a vast amount of machinery, displayed by nearly 500 exhibitors at the Sixth National Exposition of the Chemical Industries held in the Grand Central Palace, New York, September 20 to 25, afforded tangible evidence of the continued progress of the chemical and dye industries of the United States since the termination of the war. The exposition was formally opened with an address by Dr. Charles H. Herty, who said that during the past fiscal year our exports of chemicals totaled $1,250,000,000, of which $24,000,000 represented dye materials. Dr. Herty also outlined the growth of industrial research. Conservative estimates place the amount to be expended this year on industrial research laboratories, personnel, housing and equipment, at $25,000,000. He cited the example of a plant where industrial research had been instrumental in developing increased production, which suggested to him the possibility of overcoming the present inefficiency of labor through research on improved methods of operation in all lines of industry.

Dr. H. E. Howe, chairman of the Division of Research Extension, National Research Council, emphasized the fact that research plays a vital part in industrial conservation and cited the example of a certain mill in the application of modern bleaching methods. A reaction that was thought to take thirty hours for its completion was found to go forward under properly controlled temperature and pressure in three-quarters of a minute, so that but one-fifteenth of the capital was required to provide apparatus and stock formerly tied up in the process.

Dr. Howe also discussed the conservation of materials and the elimination of industrial waste through scientific control of processes. He related how steel mills were unable to make the type of billet which certain customers wanted until they had made their control more rigid. He took up, also, the question of protecting stores of both raw and finished products and again looked to scientific research for needed improvements. While much has been done there is still much more to be accomplished, and to this end the National Research Council has been organized with the coöperation of the national scientific and technical societies, many independent specialists, educational institutions and scientists employed by the Government.

Symposiums were held on fuel economy, industrial management, materials handling, chemical engineering and ceramics. Interest was enlivened by providing in connection with the papers extensive programs of motion pictures illustrating a wide range of chemical operations. R. C. Beadle, chairman of the Fuel Session, spoke on the duties and responsibilities of the combustion engineer in the present fuel crisis, and W. O. Rankin spoke on the use of pulverized fuel. Other papers at this session were: Saving Fuel by Controlling Chimney Losses, F. F. Uehling; Fluid Heat Transmission, Alexander B. McKechnie; Producer Gas and the Modern Mechanical Producer, W. B. Chapman; Refractory Cement: Life Insurance for a Furnace, F. W. Reisman; Preventing Conduction and Radiation Heat Waste, S. L. Barnes; Increasing Conduction and Reducing Fuel Consumption, W. R. Van Nortwick; The Reason for the Fuel Saving in the Dressler Kiln, Conrad Dressler.

The advance of the United States coal-tar chemical output in 1919 was recorded by Grinnell Jones, chief chemist, United States Tariff Commission. The productive capacity of the by-product coke ovens in the United States increased 17.2 per cent during 1919. With the possible exception of anthracene, adequate supplies of the fundamental raw materials of coal-tar origin will be available from American sources for the growth of the industry. The difficulty with anthracene is that its removal leaves the pitch so hard that it does not find a ready market under American conditions. Active work on the solution of this problem is now under way and the progress already made is encouraging. A comparison of the intermediates produced in 1918 and 1919 shows a considerable increase in the total number and substantial increase in amount in many cases. In 1919 there were about 225 different intermediates produced, against about 140 in 1918.

The session on Materials Handling, of which Roy V. Wright, Chairman Committee on Meetings and Program, Am.Soc.M.E., acted as chairman, showed the influence on the cost of living of the use of costly methods of handling materials in the factory and at the terminal points. In speaking of the transportation problem Mr. Wright, in illustrating the economy resulting from handling goods by machinery, pointed to the fact that the shipper and the consumer have the cars 2.6 times as long as the railroad actually requires for hauling them. If mechanical means were provided to supplement the human factor and increase the average output, he observed, it would be possible to do this work much more quickly, and thus relieve the congestion at the terminals and keep the equipment moving for a proportionately greater time each day.

J. H. Leonard contrasted the perfection of the American railway machine with the "disgraceful" situation in the terminals. The roadbeds are the very best, the tracks are the fullest fruits of metallurgical science, bridges are substantial, rolling stock and motive power are the last words in design and construction, freight trains move expeditiously between points; but there perfection comes to an end. At the terminals trouble and expense begin, congestion quickly follows, with the result that terminal handling costs far exceed the cost of movement. Mr. Leonard saw no relief in sight with the recent announcement by the railroads that they propose spending nearly all of the enormous sums to come from the rate advances in the purchase of freight cars and locomotives. New cars, be added, would mean nothing more than floating storehouses, for the terminals could not handle the freight that would come to them. The solution of the problem, according to Mr. Leonard, must be worked out with the coöperation of almost every branch of engineering, for it is a case of specific adaptations to specific requirements which a survey and study alone can properly determine.

Two other papers were presented at the materials handling section, namely, Bringing the Food to the Table, by Rumsey W. Scott; and Cost Cutting with Conveyors, by W. T. Spivey. Of mechanical engineering interest were also the following papers which were presented at the other sessions: Recoverance: A Physical Property in Material and Its Importance, by Robert G. Guthrie; A Classification of Sheet-Steel Enamels, by R. R. Danielson; and Refractories, by Homer F. Staley.

## HEAT-INSULATING VALUE OF CORK AND LITH BOARD

(Continued from page 626)

reading in Fig. 3. This key represents a section through the test box and shows the points at which temperatures were measured, together with the reference number to the thermometer.

The principal temperature readings obtained in the tests of Series 1 and 2 are shown graphically in Figs. 3 and 4, the results being plotted on logarithmic coördinates. Fig. 5 represents the final results of the conductivity and heat transmission of the two materials tested.

### CONCLUSIONS

The results of these tests are of special interest because of the low value of the external temperature, the average for all tests being approximately 10 deg. fahr. The mean temperature of the materials tested was approximately 40 deg. fahr.

These tests indicate that the heat-insulating property of cork board is slightly better than that of lith board. For the samples tested the lith board has a conductivity approximately 5 per cent greater than that of cork board.

The conductivity and heat transmission of the materials tested increases with the temperature. The change is small, however, between the temperature limits employed in this investigation.

Any discrepancies in the results are more likely to occur at the lower heat inputs with their consequent small temperature differences.

Small differences in temperature materially reduce the heat flow. Under such conditions the rate of change in the adjust-

ment of temperatures is likewise small and it is easily possible to assume conditions constant before they have properly adjusted themselves. This may result in serious error.

The conductivities of the materials as determined in this investigation agree fairly closely with those obtained by other investigators. The United States Bureau of Standards, for example, gives the following values for the conductivity of cork board:

Conductivity, B.t.u. per in. per deg. per sq. ft. per 24 hr. 7.4
Density of material, lb. per cu. ft......................... 11.3
Mean temperature of material, deg. fahr. (25 deg. cent.).. 77.0

## NEW FORD PLANT AT RIVER ROUGE

(Continued from page 641)

steam headers, both connections being on the same side of the boiler. From the saturated-steam headers the steam passes through the elements and is returned to the 12-in. superheated-steam header and discharged at the side opposite that at which it enters the saturated headers, thereby avoiding any possibility of short-circuiting and at the same time attaining correct steam distribution through all the elements.

Probably the most important and unique feature of the superheater is the location of the units. These are placed in the first pass of the boiler, protected by several rows of boiler tubes. The space for the location of the superheater units was selected in order to protect the elements from too high a temperature and at the same time provide for maintaining high superheat without overheating.

A feature in the unit or element design is the use of the forged return bend. These return bends are made on the ends of the units from the metal of the pipe itself by a special mechanical forging process without the use of electric or acetylene welding. The distribution of the metal in the return bends is such that there is an increased section in the bend at the point where the hottest gases impinge. The use of the forged return bend in the unit construction results in a continuous pipe, the bend itself being actually stronger than the body of the pipe. It also affords a sharp return without adding excessive pipe friction and facilitates the convenient and advantageous location of the superheater units.

### PULVERIZED-FUEL EQUIPMENT

The boilers will be fired with a combination of pulverized coal and blast-furnace gas. A pulverizing plant of suitable capacity, using air separation mills, is being installed, and coal will be conveyed to the boilers from this plant by a series of screw conveyors.

Each boiler will be equipped with 12 ' Lopulco" feeders and 4 "Lopulco" triplex burners, firing the coal vertically from the top. They will also be equipped with 8 gas burners for the purpose of injecting blast-furnace gas horizontally through the side, the gas flame and the pulverized-fuel flame so uniting at the proper point in the combustion chamber as to very greatly improve and increase the efficiency of the blast-furnace gas.

The possible capacity ratings when burning powdered coal will be extremely high, as the firing equipment is of such a capacity as to burn completely sufficient coal to produce 400 per cent rating continuously. It is probable that a high percentage of efficiency will be maintained throughout the whole cycle of operation, as the efficiency curve when pulverized fuel is properly handled is very flat from 50 per cent to 400 per cent of normal rating. The day operation of the boilers will be around 250 per cent of rated capacity.

---

A national convention of Italian engineers was held at Naples on May 20–25 under the auspices of the National Association of Italian Engineers. Among the topics discussed was that of Civic Activities of the Engineer.

There were present at the convention numerous government officials, delegations of Italian engineering associations and representative engineers from all sections of Italy.

The tenor of the addresses indicated a determination of the Italian engineers to assume "a leading position" in "technical and political" questions in Italy. It was emphasized by one of the principal speakers that to the engineers, who are by profession dedicated to the cultivation of the exact sciences, belongs an exalted position in all matters related to the general progress of humanity.

An incident of the convention was the reading by the chairman of a letter from the Federation of Italian Engineers, in which this organization announced that it had agreed voluntarily to its own dissolution, leaving the way clear for the National Association of Italian Engineers to engage in the furthering of civic activities of engineers in national questions.

Resolutions were adopted in which the desirability and importance of engineers taking an active interest in civic questions of national character were upheld, and reforms were suggested in regard to the policy followed by the Department of Public Works of Italy.

# LIBRARY NOTES AND BOOK REVIEWS

## The New Fall Books

By HARRISON W. CRAVER, DIRECTOR ENGINEERING
SOCIETIES' LIBRARY

PUBLISHING is to a considerable extent a seasonal business, the busy period beginning with the first days of fall and extending through the months of winter. The present time is a favorable one, therefore, for surveying the probable literary output, as most of the important works that will appear have been arranged for.

▶ In order to ascertain what may be expected, a number of the more important publishers were asked to indicate those of their proposed publications which they expected would appeal to the engineer, and a study of their replies was made.

One is immediately struck, upon looking over the list, with the general belief among publishers and authors that the questions upon which the public wishes enlightenment at present are economic rather than technological. Problems of management, of labor, of accounting and of selling account for a considerable part of the season's output of books. This is what would naturally be expected; the business field has been entirely neglected until recent times and the present demand for business books is extensive and increasing. Conditions prevailing since the war have brought matters of management to the front in a decisive manner.

Among general books for the executive we notice, first, Howard T. Wright's Organization as Applied to Industrial Problems (J. B. Lippincott Company). This, according to the publisher, is a concise general summary of the principles of organization, based on a varied business experience of twenty-five years. Other books are devoted to more specific problems, as Carle M. Bigelow's Management in the Woodworking Industry (*Industrial Management*); Walter N. Polakov's Mastering Power Production (*Industrial Management*), which discusses power-plant operation; and C. U. Carpenter's Increasing Production: Decreasing Costs (*Industrial Management*), which treats of ways and means for utilizing labor, machinery and floor space most efficiently. Two books intended primarily for college use may also be of interest, Horace Secrist's Readings and Problems in Statistical Methods (Macmillan) and O. B. Goldman's Financial Engineering (John Wiley and Sons). The first of these is intended to show students and executives, in a practical manner, how statistics may be used. The second is described as a textbook for consulting, managing and designing engineers and for students.

Among new books on accounting appears one entitled Accounts in Theory and Practice, by Prof. E. A. Saliers, of Yale University (McGraw-Hill), which attempts to emphasize principles more definitely than is customarily done. The Ronald Press Company offers a five-volume work entitled Business Accounting, prepared by a number of experts under the editorial supervision of Harold D. Greeley, the volume on cost accounting being written by D. C. Eggleston. The same publisher has issued also a book entitled Cost Accounting, Its Principles and Practice, by Gould L. Harris, of New York University, and J. P. Jordan, Vice-President of the C. E. Knoeppel Company. This book attempts, in addition to explaining the technique of cost accounting, to show the uses of cost-account data by executives to secure current control of the industry. G. Charter Harrison's Cost Accounting to Aid Production (*Industrial Management*) is also intended to aid the operating department as well as the treasurer, and so discusses cost accounts from an engineering viewpoint.

Technical advertising sees its first textbook in Advertising the Technical Product, by C. A. Sloan, Advertising Manager for the Hyatt Roller Bearing Company and J. D. Mooney, of the General Motors Company (McGraw-Hill), a book which discusses the special problems involved in preparing and placing the advertising of mechanical devices and similar manufactures. The McGraw-Hill Book Company will publish also Essentials of Advertising, by F. L. Blanchard, Editor of *The Fourth Estate*.

Two books on banking appear, intended for the business man rather than the banker. Chester A. Phillips' Bank Credit (Macmillan) explains commercial credit and discusses in detail the analysis of mercantile credit, the interpretation of credit statements and the evaluation of the different items. It should assist the merchant by its explanation of the principles that the bank applies in analyzing his statement. W. H. Kniffen's The Business Man and His Bank (McGraw-Hill) is intended to make the business man familiar with the work of a bank and to show him what he may expect the bank to do for him.

Commerce is represented by two books. The Elements of Marketing, by P. T. Cherington (Macmillan), is based on the author's experience in teaching the subject at Harvard University and is an attempt to formulate the principles underlying the transfer of merchandise from producer to consumer under modern conditions. The Century Company begins a new series on foreign trade with Robert E. Annin's Ocean Shipping: Elements of Practical Steamship Operation, a book dealing with the fundamentals of ship management and operation. Another book in the series that should prove useful is Packing for Export, by H. R. Moody, who was in charge of the packing and shipping of Government supplies overseas during the war.

Labor and labor problems have led to several books treating various phases of the question. Personnel Administration, by Ordway Tead and H. C. Metcalf (McGraw-Hill), is a systematic discussion from the management viewpoint. Daniel Bloomfield has written Labor Maintenance (Ronald), describing how harmony may be maintained between the management and the workers and giving practical illustrations of accomplishment. Nathan W. Shefferman's Employment Methods (Ronald) discusses the functions of the employment department and describes modern methods of selecting and assigning men to obtain a stable labor force. Roy W. Kelly's Training Industrial Workers (Ronald) presents suggestions to be followed by industrial firms in training employees. The Macmillan Company announces John A. Ryan's Social Reconstruction, and The Church and Labor, by John A. Ryan and Joseph Husslein. The former book, compiled from lectures at Fordham University, outlines a comprehensive program, including a minimum wage, collective bargaining, coöperative societies, health insurance and labor participation in management. The second is an authoritative statement of the attitude and teaching of the Roman Catholic Church in relation to labor and society. The same firm also announces Sigmund Mendelsohn's Labor's Crisis, a discussion of some phases of the problem from the employer's point of view, which recommends improvement of housing and modified profit sharing.

Turning from general administrative matters to those more technical, the mechanical engineer finds two additions to the steadily growing library of handbooks. The Marine Engineers' Handbook (McGraw-Hill), edited by Lieut.-Com. F. W. Sterling, U. S. N., is the first American work of the kind and will undoubtedly prove welcome. J. D. Hoffman's Handbook for Heating and Ventilating Engineers (McGraw-Hill) is not new, but appears in a thoroughly revised and modernized form. Welding has been a much-discussed operation in recent days, and three new books are listed: Practical Electric Welding (Spon), by H. B. Swift; Gas and Electric Welding (American Technical Society), by George W. Cravens: and Spot and Arc Welding (Lippincott), by H. A. Horner. Two books deal with lubrication: One, a rewritten and enlarged edition of John R. Battle's Lubricating Engineers' Handbook, is now entitled Lubrication and Industrial Oils, and forms the first volume of a Handbook of Industrial and Oil Engineering to be published by the J. B. Lippincott Company. The other book is by T. C. Thomsen and is entitled Practice of Lubrication (McGraw). Prof. Franklin DeR. Furman, of Stevens Institute of Technology, announces a work entitled Cams: Elementary and Advanced (Wiley). Elements of Engineering Thermodynamics (Wiley) is the joint work of Director James A. Moyer, of the Massachusetts Department of University Extension, Prof. James P. Calderwood, of the Kansas State Agricultural College, and Audrey A. Potter, of Purdue University. For the automotive engineer, E. P. Dutton and Company offer the Theory and Practice of Aeroplane Design, by S. T. G. Andrews and S. F. Benson. John Wiley and Sons have a new edition of Pumping by Compressed Air, by E. C. Ivens.

For some unknown cause, the electrical engineer has little representation in the autumn lists. John Wiley and Sons announces Engineering Electricity, by Prof. Ralph G. Hudson, of Massachusetts Institute of Technology; and Principles of Radio Communication, by Prof. J. H. Morecroft, of Columbia University, Prof. A. Pinto, of Cooper Union, and W. A. Curry, of the New York Edison Company. The McGraw-Hill Book Company announces the third edition of Dr. C. P. Steinmetz's Transient Electric Phenomena. The Vacuum Tube, by Dr. A. J. Van der Bijl, and a Course in Electrical Engineering, by Prof. C. L. Dawes, of Harvard. Spon and Chamberlain present The Strowger Automatic Telephone Exchange, by Mordin, a work of English origin. The Century Company announces two books by M. Luckiesh, of the General Electric Company, treating the subject for non-technical readers. Lighting the Home offers advice on all domestic lighting problems, while Artificial Light: Its Influence on Civilization, is a history of the development of artificial light and its influence on human progress.

The mining engineer is offered various additions to his library. J. R. Finlay has prepared a new edition of his Cost of Mining (McGraw-Hill). Robert McGarraugh writes on Mine Records and Accounts (McGraw-Hill). Steam Shovel Mining (McGraw), by R. Marsh, Jr., is a practical review of methods. Flotation (Wiley), by T. C. Rickard, will be of interest to many. A book that should prove very useful to those who expect to travel or work in Russia is C. W. Purington's Vocabulary of Russian-English and English-Russian Mining Terms (Lippincott). This includes not only technical terms, but also those needed by the traveler and camper, legal and financial terms, and tables of weights. Political and Commercial Geology, and the World's Mineral Resources (McGraw-Hill) is the work of J. E. Spurr, Editor of the *Mining Journal*. The oil producer will find that subject treated in the Geology of Petroleum (McGraw-Hill), by Prof. W. H. Emmons, and in Field Methods in Petroleum Geology (McGraw-Hill) by Professor Cox, Dake and Muilenberg, of the Missouri School of Mines. Prof. W. A. Grabau has prepared the first volume of a Handbook of Salt Geology (McGraw-Hill), dealing with an important question which has hitherto escaped adequate treatment.

The construction engineer finds a new edition of Charles Evan Fowler's Ordinary Foundations (Wiley); Reinforced Concrete Construction (Spon), by Cantell; a Handbook of Building Construction (McGraw-Hill), by G. A. Hool and A. N. Johnson; and Concrete Work, a book to aid self-development of workers in concrete and for students in engineering (Wiley), by William K. Hatt, of Purdue University, and W. C. Voss, of Wentworth Institute. Engineers interested in agricultural matters may find assistance in George Thomas' Development of Institutions Under Irrigation (Macmillan), a history of irrigation in Utah. Studies in French Forestry (Wiley), by Theodore S. Woolsey, Jr., the Executive Member of the Interallied War Timber Committee, 1917-1919, will interest foresters, and C. W. Murphy's Drainage Engineering, (McGraw-Hill), will be welcome. Prof. T. R. Agg, of Iowa State College, is the author of American Rural Highways (McGraw-Hill). A revised edition of the Manual for Testing Materials (McGraw-Hill), by W. K. Hatt and H. H. Scofield, is announced.

The analytical chemist will look for Dr. A. D. Little's Technical Methods of Analysis (McGraw-Hill), the Technical Examination of Crude Petroleum (McGraw-Hill), by W. A. Hamor, and the new edition of Rapid Methods for the Chemical Analysis of Special Steels, Steel-Making Alloys, Their Ores and Graphites (Wiley), by Charles M. Johnson.

The new edition of A Dictionary of Chemical Solubilities—Inorganic (Macmillan), by Arthur A. Comey and Dorothy A. Hahn, will be received with appreciation, as it has been out of print for several years and needed revision and extension. The most noticeable general works are another volume—Volume 9, Part 1—of J. Newton Friend's valuable Textbook of Inorganic Chemistry

(Lippincott), probably the best general English text for reference use, and a thoroughly revised edition of Holleman's Textbook of Organic Chemistry (Wiley), edited by A. Jamieson Walker.

Special books will appear on the chemistry of various industries. C. F. Cross and E. J. Bevan have prepared a new edition of their well-known Textbook of Paper Making (Spon), Edwin Sutermeister, chemist to the D. Warren Company Paper Mills, has written the Chemistry of Pulp and Paper Making (Wiley), and Spon and Chamberlain offer The Recovery and Re-manufacture of Waste Paper, by Strachan. J. Merritt Matthews writes on the Application of Dyestuffs to Textiles, Paper, Leather and Other Materials (Wiley). Alvah H. Sabin has rewritten and enlarged Red-Lead, and How to Use it in Paint (Wiley). Dr. W. B. Bancroft's Applied Colloid Chemistry (McGraw-Hill) is on a subject whose importance is only beginning to be realized. S. L. Hoyt has written Part 2 of his Metallography (McGraw-Hill), treating of the metals and common alloys, and Bonewski's Studies in Alloys is announced by the J. B. Lippincott Company. The Century Company is publishing The New Stone Age, by H. E. Howe, of the National Research Council, a non-technical account of the uses of cement and concrete, intended for the "general" reader.

The same reader is the object of The New World of Science: Its Development During the War (Century), a collection of essays on the scientific advances made in surgery, communication and other arts, edited by Dr. Robert M. Yerkes, of the National Research Council. Another historical work is Hoogard's Modern History of Warships (Spon).

The Century Company announces two biographies of engineers. Rose W. Lane and Charles K. Field have prepared The Making of Herbert Hoover: a Biography, in which special attention is given to his early life; and W. Vaughan has written The Life and Work of Sir William Van Horne, the story of the builder of the Canadian Pacific Railway.

**ADVERTISING THE TECHNICAL PRODUCT.** By Clifford Alexander Sloan and James David Mooney. McGraw-Hill Book Co., Inc., New York. 1920. Cloth, 6 × 9 in., 365 pp., illus., $5.

This book, the work of two men experienced in advertising technical products, is a discussion of the more important factors of the problem. The subjects discussed are the economic elements of such advertising, the instruments available for it, the preparation of technical advertisements and advertising organizations. An interesting and varied collection of actual advertisements, with critical comments, and a brief bibliography are included in the book.

**HYDRAULIC TABLES.** The Elements of Gagings and the Friction of Water Flowing in Pipes, Aqueducts, Sewers, etc.; Flow of Water over Sharp-edged and Irregular Weirs and the Quantity Discharged. By Gardner S. Williams and Allen Hazen. Third edition, revised. John Wiley and Sons. Inc., New York. 1920. Cloth, 6 × 9 in., 115 pp., Illus., plates, tables, $2.

The third edition of these well-known tables has been carefully corrected and revised in minor points, and a new chapter has been added, in which the additional data that have accumulated during the fifteen years since these tables first appeared are examined to ascertain whether changes or adjustments in the formula are needed.

**WINGS OF WAR.** An Account of the Important Contribution of the United States to Aircraft Invention, Engineering, Development and Production During the World War. By Theodore Macfarlane Knappen, with an introduction by Rear-Admiral D. W. Taylor. G. P. Putnam's Sons. Inc., New York. 1920. Cloth, 6 × 8 in., 289 pp., plates, $2.50.

The story of the United States army-aircraft production program is essentially one of confident hopes, bitter disappointments, failures and successes such as inevitably attend the creation from nothing of an immense industrial organization. Until now, the history of this undertaking is chiefly found in the voluminous reports of investigating committees, which overemphasize the failures and undervalue the successes. The present book is an attempt to supply a less one-sided account of the army air effort, one which will give a readable report of the problem, the methods used for its solution and the net results obtained.

## Armor-Plate and Gun-Forging Plant of U. S. Navy

### A Description of the Government-Built and Operated Naval Ordnance Plant at South Charleston, W. Va.—"H"-Type Forge and Furnace Building Among Unique Features of Plant

By ROGER M. FREEMAN,[1] NEW YORK, N. Y.

*The magnitude of such a project as the establishment of the Navy-owned, built and operated Naval Ordnance Plant at South Charleston, W. Va., has made it impossible for the author to give details on all points of the development of the plant. He describes briefly the establishment of the projectile plant and preliminary steps leading up to the actual construction of the armor-plate and gun-forging plant, a general description of which is the main object of the paper. The general layout of the plant and transportation facilities are covered and then the author passes to a more detailed discussion of the unusual features of design and construction. Among these are the arrangement of the stock yard alongside the open-hearth building, the 30-ton electric furnaces for duplexing, the electric traveling cranes to carry 400,000-lb. ingots, the original "H" type of forge shop and treatment plant combined, the 14,000-ton steam-intensified hydraulic press, the twenty 15-ft.-wide Carbottom furnaces, the machine shop with three 106-ft.-wide aisles, and the gun-treatment building, the high portion of which contains a 75-ton crane of 104-ft. span on rails 165 ft. above the floor level which serves a 10- by 105-ft. high electric treatment furnace placed in a concrete pit 55 ft. deep. The 23 electric traveling cranes, twenty of which are 75-ton or larger and three of 250 tons capacity, are discussed. The 25,000-kw. electric substations, the 6500-hp. boiler plant and the 25,000,000-gal. reservoir are also discussed briefly.*

*The paper concludes by summarizing the various steps in the development, outlining the organization of the project and personnel, giving in round numbers construction costs, and stating briefly the present status of work on the plant which is now substantially completed.*

SEVERAL years ago the papers were again full of the historic discussion as to whether the Navy Department should attempt to manufacture armor plate for itself or continue to go to the three non-competing firms who were able to make armor in this country, and pay the price asked. With the strong backing of Secretary Daniels and Senator Tillman, and despite much opposition by the manufacturers, the public at large, and certain Navy circles, Congress directed that the plant be built. Shortly thereafter the war came on and the project was generally forgotten in the rush of ordnance manufacture for the nations of Europe and for this country, when we finally entered.

On August 30, 1917, ground was broken at South Charleston, W. Va. for the construction of a projectile plant as the first unit of the U. S. Naval Ordnance Plant. The main advantages of South Charleston lay in its being in the center of the rich coal districts of West Virginia, in the proximity of iron mines and blast furnaces, and in the availability of natural gas and oil. The reservation consists of somewhat over 200 acres of land located between the Chesapeake & Ohio Railroad and the Great Kanawha River about five miles below Charleston, at South Charleston, W. Va. It is divided into a north unit of approximately 40 acres and a south unit of approximately 160 acres, by Eighth Avenue, which is the main road and which carries the trolley line between Charleston and St. Albans.

The projectile plant, which occupies the north unit, was completed and put into operation in the spring of 1918, since when it has operated satisfactorily, producing principally 6-in. gun forgings, counter-recoil cylinders, 16-in. armor-piercing projectiles, air

[1] Supervising Engineer in charge during design and construction of Naval Ordnance Plant, Mem.Am.Soc.M.E.

For presentation at the Annual Meeting, New York, December 7 to 10, 1920, of THE AMERICAN SOCIETY OF MECHANICAL ENGINEERS. The paper is here printed in abstract form and advance copies of the complete paper may be obtained gratis upon application. All papers are subject to revision.

flasks for torpedoes, steel ingots up to 18 tons, and at the present time the entire demand of the Navy Department for armor bolts. The projectile plant cost about $2,000,000 and consists principally of a forge and foundry, approximately 130 ft. by 560 ft., containing three 6-ton Heroult electric furnaces, two 60-in. cupolas, a small brass foundry, a 3000-ton press and a 500-ton press, eleven forge and five regenerative Carbottom annealing furnaces; a machine shop, 140 ft. by 400 ft., completely equipped for machining minor-caliber gun forgings, large projectiles and general machine-shop work, and a heat-treatment shop, 92 ft. by 153 ft.

In June 1918 it was decided to vigorously push the construction of the main plant for the production of armor plate and major-caliber gun forgings. The writer was summoned from Erie, Pa., where he had supervised the construction of the Erie Forge and Steel Company and other steel plants for the Navy, and placed in charge of the design and construction. A designing and drafting force was built up around his engineering assistants from Erie and his previous work, and a half-dozen draftsmen who had worked on the projectile plant. During the summer of 1918 the layout and the general design for the main buildings were completed, and by November 1 were approved by the bureaus concerned.

It was decided that due to existing conditions in the material and labor markets, the impossibility of writing specifications and getting lump-sum bids without great delay, and on account of the general disrepute into which cost-plus contracts for Government plant construction had fallen, the work should be done directly by the Navy. A construction division was organized and construction work was under way by October 1, 1918.

The armor-plate and gun-forging plant has been designed to produce armor plate of the heaviest type, completely finished, ready to attach to battleships; and major-caliber gun forgings, up to 20-in. 50-caliber in size, rough-machined, which will be sent to the Washington Navy Yard for finish-machining. It will employ about 3000 when completed and in full operation.

The United States Naval Ordnance Plant is not a "war plant" in the modern usage of the term, but a permanent Navy institution which is abundantly justified for the following reasons:

First, additional capacity for the manufacture of armor plate and gun forgings has been provided.

Second, cost of armor manufactured by the Naval Ordnance Plant should be less to the Navy than the price now paid to commercial manufacturers and will afford a definite means of determining what the fair price for armor and gun steel to be paid steel manufacturers by the Navy should be.

Third, the Naval Ordnance Plant will be a large-scale experimental and research laboratory which can continually strive to improve the product and simplify manufacturing methods. In this connection it is interesting to note that the first group of 16-in. armor-piercing projectiles made at the Naval Ordnance Plant passed the test at Indian Head and that the plant is successfully making air flasks for torpedoes.

Fourth, a very great additional advantage in the Government-owned and operated plant is the fact that improvements in product or methods can, when desired, remain the property of the United States.

### GENERAL LAYOUT

The general layout of the south unit was determined first by the requirements of manufacture in modern straight-line methods,

FIG. 1  THE UNITED STATES NAVAL ARMOR-PLATE AND GUN-FORGING PLANT AT SOUTH CHARLES-
Buildings:[1] 1, Machine Shop; 2, Gun-Treating Building; 3, Forge and Furnace Build

and secondly by the topography of the site. Fig. 1 shows the general layout of the armor and gun-forging plant. The four main buildings, namely, the open-hearth, the forge and furnace, the machine shop, and the gun-treatment building are arranged in parallel on a shuttle track which runs at right angles to the major axes of the buildings and connects them all together. This track will be the backbone of the manufacturing processes. All main buildings have been placed so that future expansion to double the capacity of the plant may be readily made.

Scrap and pig iron will be received from the Chesapeake & Ohio Railroad, at the southeast corner of the reservation, sorted from other incoming material in the classification yard, and stored in the stock yard alongside the open-hearth building. Steel will be melted in the open-hearth furnaces, refined in the electric furnaces, and cast into ingots in the pouring pit in the open-hearth building. From there the ingots will be transferred to the forge and furnace building, by way of the shuttle track for heating, preparatory to forging into armor plate or guns under the presses located in the center of the forge and furnace building. The armor-plate forgings will then be carbonized, annealed, tempered, hardened, and ultimately sent over to the machine shop on the shuttle track where the armor plate will be machined and finished, ready for attaching to battleships. Gun forgings will be green-annealed in the south aisle of the forge shop, sent to the north aisle of the machine shop for rough-machining, and thence to the gun-treatment building for heat treatment prior to shipment to the Navy Yard in Washington for finish-machining and assembling.

The service building, which comprises the main electric substation, the air compressors, the water-supply distribution pumps, and the boiler plant, which provides steam for the steam-intensified hydraulic presses and for heating the machine shop, are located as nearly as possible to the center of gravity of their respective loads.

Twin buildings, one of which will be used for general stores and the other for the blacksmith shop, pattern and templet making and storage, have been located in the southeastern portion of the south unit accessible to the railroad system and the forge shop. The maintenance shops, in which the electricians, pipefitters,

riggers and carpenters, are to be located, and the locomotive repair shop and roundhouse, will be combined in one building at the approximate center of the reservation.

The western portion of the south unit as divided by the main railroad track running north and south, contains a large storage yard and the "skull cracker."

The reservoir and settling basin for the industrial water-supply system has been made by throwing two earth-filled dams across the western end of the gully which cuts diagonally across the reservation.

An administration building of sufficient size to house the executives for the entire ordnance plant was built at the same time as the projectile plant and located in the south unit on Eighth Avenue. The balance of the frontage on Eighth Avenue is occupied with temporary buildings for construction offices, stores, the garage, barracks to house about 1000 men on construction work, and Bungalow Park, in which 32 bungalows were erected for the occupancy of the supervisors of the construction division.

## TYPE OF STRUCTURE

In general, the buildings consist of a structural-steel framework, unusually heavy on account of the size of the cranes on concrete foundations, some of which contain upward of 200 cu. yd. of concrete enclosed by walls of specially designed hollow red-tile building block, 5 in. by 12 in. by 8 in. Fifty per cent or more of the surface area of the walls is steel sash. The roof decks are of gypsum composition, cast in place and covered by waterproof roofing. In round numbers about 50,000 cu. yd. of concrete has been placed in the various foundations, 25,000 tons of structural steel erected, and in excavation and grading upward of 500,000 cu. yd. of earth moved. The total area under roof is approximately 700,000 sq. ft. for the main buildings alone.

## TRANSPORTATION FACILITIES

It was evident very early in the design period that one of the most important features of the entire plant would be the arrangements for efficiently handling and moving unusually heavy and awkward loads. For example, one armor-plate ingot which will be made about six times a year will weigh upward of 400,000 lb., and tube forgings for the major-caliber guns will weigh up to possibly 100,000 lb. and run in length up to nearly 100 ft. An armor plate in the making must be handled something like fifty separate times and as the average weight of the finished plates is in the neighborhood of 50 tons, very special attention was required in the design of the electric traveling cranes and in the railroad-track system.

The railroad system, which totals something over seven miles in length, is outlined on Fig. 2. A classification yard of ten tracks including incoming and departure tracks has been provided. Pro-

---
[1] In the background toward the north may be seen the projectile plant with the barracks for the construction forces in the foreground. In the immediate foreground is the main track crossing the upper dam of the reservoir. Immediately in front of the machine shop is the oxy-hydrogen building and gasometers in process of construction. To the right of the machine shop is the concreting plant for the gun-treatment pit, only a portion of the steelwork having been erected. The steelwork for the high portion, when completed, will reach a point three times the height of the adjacent parapet wall of the machine shop, i. e., somewhat over 200 ft., and somewhat higher than the twin chimneys for the boiler plant shown in the middle of the photograph. The fill for the shuttle track is shown leading from the forge shop to the south side of the machine shop. Note the open monitors on the low and high portions of the forge and furnace building. The photograph was taken from a position near the middle of the railroad "Y" which is at the approximate center of the reservation.

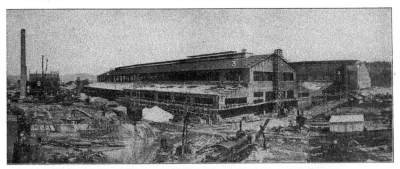

TON, W. VA.; VIEW LOOKING IN A GENERAL EASTWARDLY DIRECTION, AUGUST, 2. 1920

ing: 4, Open-Hearth Building; 6, Boiler House; 7, Foundations for Service Building.

## OPEN-HEARTH BUILDING

visions have been made for tracks entering into either end of practically every crane aisle. A complete loop-within-loop system has been developed so that a wreck or stoppage on any main track will not hold up the operation of the plant or of any individual major building. A three-way reinforced-concrete approach trestle will lead to the charging-floor level of the open-hearth building.

The major specifications of the twenty-three overhead electric traveling cranes are given in Fig. 3. Each main aisle of the four main buildings is provided with erecting girders for the cranes located in the roof trusses.

While by far the larger portion of the traffic at the plant will be by rail, a system of concrete roads 20 ft. in width has been developed in order to provide auto-truck service to all main buildings, and for employees. The main entrances to both the north and south units will be from Eighth Avenue, where the plant railway crosses. A central receiving building with a special siding has been built near this point as shown on Fig. 2, for the unloading of less-than-carload shipments and to permit distribution by auto truck.

### OPEN-HEARTH BUILDING

The unusual features of the open-hearth at the Ordnance Plant consist of the arrangement of the stock yard alongside the building, the 100-ft.-wide pouring aisle with exceptionally heavy cranes, the arrangement for "duplexing" by placing the electric furnaces in the pouring aisle and the dump trestle and storage bins in the crane-served charging aisle.

Figs. 4, 5 and 6, inclusive, show the general arrangement and give the important dimensions of the building which, including the stock yard, is 331 ft. wide by 516 ft. long.

The charging floor of the open-hearth building has been designed to withstand a load of 800 lb. per sq. ft. It extends 30 ft. out under the crane runway of the stock yard to form a charging platform. Two parallel tracks are arranged upon this charging platform with cross-overs conveniently located, and a turn-out leads to the charge track directly in front of the open-hearth furnace doors.

A track connected at both ends to the main system runs through the stock yard at the ground level adjacent to the charging platform. Materials for the charge, such as scrap, pig iron, etc., received on this track can be unloaded directly by the stock-yard cranes by magnet and placed in charging boxes on charging cars on the charging platform. These cars will be made up into trains, each containing a complete charge for an open-hearth furnace, or the incoming cars of material may be unloaded directly into the stock piles. The charge train will be shifted by an electric storage-battery locomotive over the scale, located near the entrance to the charging aisle, directly to the track in front of the open-hearth furnace doors and into position for the charging machine to start

charging the furnaces. The empty train of charging boxes will be shunted back to the charging platform over the same track it entered, or out of the building at the east on to the trestle and back.

The charging floor carries three tracks, two of standard gage which are connected into the three-way approach trestle at the east of the building, and the third of 24 ft. 6 in. gage for the low-type charging machine. The floor is paved with brick.. The charging track is located directly in front of the furnaces and the so-called "hot-metal track" is at the far side of the aisle. This track originates in the approach trestle and continues at the east end of the building over a series of concrete bins arranged so that cars of material can be bottom-dumped. The track then leads to the main charging floor where it will be used for handling materials with which to repair the furnaces and thus avoid interference with operation.

Space has been provided for three 65-ton open-hearth furnaces, two of which are now being completed. An electric traveling crane of 25 tons capacity with an auxiliary hoist of 10 tons capacity serves the entire length of the charging aisle.

Underneath the charging floor and back of the open-hearth furnaces is the pit for the checker chambers enclosed by a reinforced-concrete retaining wall. The lean-to contains the offices, the quick-test laboratory, chimneys, storage areas, lockers, toilets, etc.

The pouring aisle is served by three electric traveling cranes, specifications for which are given in Fig. 3. The distance from the center to center of crane runway rails is 100 ft. Jib cranes of 6 tons capacity with a radius of 25 ft. are mounted on main building columns adjacent to the open-hearth furnaces for handling the spouts.

The unusual width of the pouring aisle was to give space in the open-hearth building for the storage and chipping of ingots, and on account of the large pouring pit in which it is planned to ash-anneal the ingots. The location of the electric furnaces for duplexing in this aisle was an important factor in determining this width.

The pouring pit is divided for convenience in operation into two levels and is of reinforced concrete lined with 13 in. of red brick. The lower level is 16 ft. below the floor, 30 ft. wide and 93 ft. long and the upper level is 8 ft. deep, 30 ft. wide and 46 ft. long.

For open-hearth steel the furnaces will be tapped into the ladles and carried directly to the pouring pit and bottom-poured directly into the ingot molds. In obtaining electric steel by duplexing, the open-hearth furnaces will be used simply for melting the charge, which will be tapped into ladles and then carried down the shop to charge the electric furnaces.

Two 30-ton Heroult electric furnaces have been erected as shown in Fig. 5 on independent concrete foundations, and are surrounded

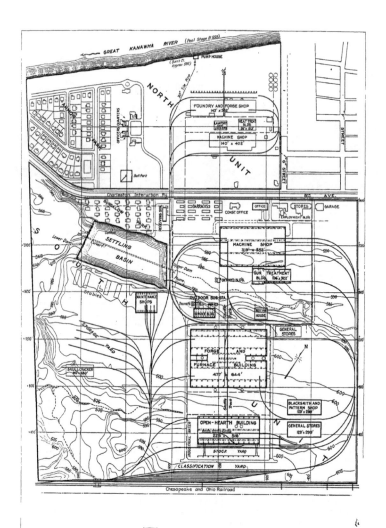

Machine Shop. Gun Treat. Bldg.  Forge and Furnace Bldg.  Open-Hearth Building.

Cross Section Through Buildings

by a charging platform which is an extension of the charging floor and on which are placed the transformer houses. Flux material will be brought over to this floor from the bins in the charging aisle. The furnaces will be charged from the rear. The capacity of one open hearth is sufficient to charge the two large electric furnaces. After the refining process, a ladle will be brought close into the front of the electric furnace by lifting a large trap door, the furnace tilted forward and the contents discharged into the ladle, which will then be carried back to the pouring pit and the molten steel will be bottom-poured into ingots in the usual manner.

In casting the largest ingot, the open-hearth furnaces will be overloaded to 80 tons each and will be brought out at the same time as one or both of the 30-ton electric furnaces overcharged if required.

Tracks enter the pouring aisle at either end for a length of one or two car lengths and the south end of the shuttle track is at the middle of the building adjacent to the pouring pit, conveniently located for the transportation of ingots to the forge shop.

Attention is invited to the runway escape platform for the crane operators at the level of the bottom of the crane cages, also to the provision in the structural design for ventilation. The entire wall area of the building is left open to a distance 8 ft. above the ground, although arrangements are made for vertically lifting doors in case of severe weather. The sides of the large monitor are left entirely open and without louvers, high curbs being provided on the main roof and the overhang of the monitor roof being made excessive for protection against the weather.

### FORGE AND FURNACE BUILDING

The design of the forge and furnace building of the U. S. Naval Ordnance Plant is believed to be a radical departure from anything that has been built for the purpose.

The forge shops of the existing armor plants consist of buildings with one main crane aisle and a lean-to on one or both sides. The furnaces are located in the lean-tos, the presses in the middle or at one end of the main aisle.

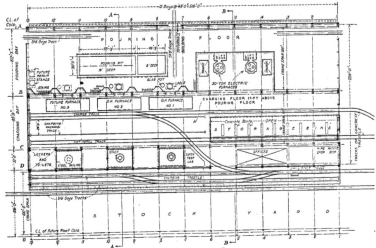

| CRANE NUMBER | OPEN-HEARTH CRANES | | | | FORGE AND FURNACE BUILDING | | | | | | | | MACHINE SHOP | | HEAT TREATM'T | |
|---|---|---|---|---|---|---|---|---|---|---|---|---|---|---|---|---|
| | 1 | 2 | 3 | 4 | 5,6 | 7,8 | 9,10 | 11 | 12 | 13 | 14 | 15 | 16,17 | 18-21 | 22 | 23,24 |
| LOCATION | O.H. SHUTTLE SLIDE | O.H. SHUTTLE SLIDE | O.H. SHUTTLE SLIDE | O.H. SHUTTLE SLIDE | STOCK YARD | MAIN AISLE | MAIN AISLE | PRESS ROOM | | | | PRESS ROOM | | | | |
| CAPACITY (NET TONS) | | | | | | | | | | | | | | | | |
| MAIN HOIST | 250 | 125 | 75 | 25 | 15 | 200 | 100 | 250 | | | | 75 | 150 | 75 | 75 | |
| AUXILIARY HOIST | 40 | 25 | 15 | 10 | NONE | 25 | 25 | 25 | | | | 15 | 25 | 15 | NONE | |
| SPEED (FT PER MIN) | | | | | | | | | | | | | | | | |
| MAIN HOIST | 10 | 10 | 14 | 25 | 50 | 8 | 10 | 7 | | | | 14 | 10 | 14 | | |
| AUXILIARY HOIST | 20 | 23 | 30 | 30 | 0 | 23 | 23 | 30 | | | | 30 | 23 | 30 | 0 | |
| BRIDGE | 200 | 200 | 200 | 250 | 300 | 200 | 200 | 150 | | | | 200 | 200 | 200 | 100 | |
| TROLLEY | 75 | 75 | 100 | 150 | 150 | 75 | 100 | 50 | | | | 70 | 100 | 100 | 50 | |
| CRANE RAILS | 175 LB | 175 LB | 175 LB | 80 LB | 85 LB | 175 LB | | | | | | 175 LB | | | | |
| CLEARANCES | | | | | | | | | | | | | | | | |
| A | | | | | | | | 70'-9" | | | | 70'-0" | | | | |
| B | | | | | | | | 2'-10" | | | | 2'-9" | | | | |
| C | | | | | | | | 46'-8" | | | | 78'-0" | | | | |
| D | | | | | | | | 21'-0" | | | | 3'-6" | | | | |
| G | | | | | | | | 25'-0" | | | | 78'-0" | | | | |
| H | | | | | | | | 2'-6" | | | | 0'-3" | | | | |
| J | | | | | | | | 3'-0" | | | | 7'-0" | | | | |
| K | | | | | | | | 5'-0" | | | | 6'-0" | | | | |
| M | | | | | | | | 26'-6" | | | | 0'-0" | | | | |
| N | | | | | | | | 7'-0" | | | | 3'-8" | | | | |

FIG. 3   SPECIFICATIONS FOR ELECTRIC TRAVELING CRANES

FIG. 5  POURING AISLE OF OPEN-HEARTH BUILDING LOOKING EAST FROM CRANE
GIRDER, JULY 14, 1920

Casting pit is in immediate foreground. Electric furnaces just beyond. Charging floor,
when completed, will extend completely around furnaces. Locations for open-hearth furnaces 1 and 2 are shown to the right just behind middle row of columns. Box cars on charge
track of charging aisle indicate scale. Crane span of charging aisle is 100 ft. wide.

Forging cranes are provided on either side of the press on crane
rails which at one end of the span are supported on the top of
the press and at the other end on separate steel structures.
Forging heats are brought from the furnaces down the main
aisle by the overhead cranes and placed in a position where they
can be picked up by the forging cranes, although in certain cases
the ingot furnaces are in direct reach of the forging cranes. In
order that the presses and the forging cranes may be cleared, the
overhead crane in the main aisle must be placed at a relatively

the building are 644 ft. by 477 ft.  The building consists of two main aisles, each of which has a lean-to
on either side and which are connected by the press
room which forms the cross-bar of the "H."

The 14,000-ton steam-intensified hydraulic forging
press and its driving equipment, including the 2500-
lb. hydraulic-pressure pumps, the 32-in. accumulator
and the steam receivers and intensifiers, is located
in the eastern half of the press room as divided by the
shuttle tracks.  (Fig. 11.)  The 6500-ton press which
will be erected in the future will be located in the
western half.

The direction of the forging is from the south aisle
toward the north aisle.  A 250-ton hydraulic forging
crane with an auxiliary hook of 25 tons capacity is
located on either side of the press on double runway
rails 46 ft. above the floor.  The inside runway rails
are supported directly on top of the press.  These
forging cranes, of which all the motions excepting the
hoisting motions are electrically driven, have a sufficient range of bridge travel to pick up loads from
either of the shuttle tracks.  Water is supplied to
the hoist motions from the pressure system at 2500
lb. per sq. in.

The entire area of the press room proper, which is
approximately 140 ft. wide by 160 ft. long, has been
made entirely free from columns except for the row
necessary to support the inside crane runway rails
for the forging cranes.  The central area over the
presses is served by a 75-ton crane located on rails
approximately 80 ft. above the floor level.  This crane has been
placed to serve in the erection of the presses and of the forging
cranes as well as in their repair.  The span of the runway girders
is 140 ft.  The track scale of 600,000 lb. capacity is located on
the main shuttle track in the northern half of the press room.

Briefly and generally the process for armor-plate manufacture
in this building will be as follows:  The ingot will be received in
the south aisle from the open-hearth building on a flat car by way
of the shuttle track.  It will be lifted by an overhead crane and

FIG. 6  OPEN-HEARTH BUILDING FROM SOUTHEAST, AUGUST 2, 1920

Alongside the Open-Hearth Building is the stockyard with 100 ft. crane span. Classification yard is to extreme left.  Note track doors at
charging-floor level.  Three-way concrete trestle will replace temporary wooden trestle.  In immediate foreground is general storehouse and
blacksmith and pattern shop (No. 3).  Twin buildings each 130 by 300 received through Army salvage.

great height above the ground.  This makes an expensive
building.

The treatment shops of the existing armor plants are located in
separate structures from the forge shops, and as plates must be
returned to the press for bending, straightening, etc., considerable
railroad traffic between the two buildings results, unless two very
large presses are provided.

At the Naval Ordnance Plant the forge shop and the treatment
shop for both armor plate and gun forgings have been consolidated
in the "H"-type forge and furnace building.  The general arrangement is shown by Figs. 7, 8, 9 and 10.  The overall dimensions of

placed in one of the ingot furnaces by means of a porter bar.  To
heat the ingot will require up to 20 hr.  The hot ingot will then be
withdrawn from the furnace and carried down the aisle to either
one of the shuttle tracks and deposited on a car which will then be
pushed through the few feet necessary by an electric locomotive
into the press room.  It will be picked up by a forging crane and
carried under the press.

The first forging operation will last about an hour.  It will
bring the plate down within an inch or so of its finished thickness
and will include cropping the ingot.  The plate, which is then
35 per cent lighter, is returned to the south aisle, deposited on the

car of one of the Carbottom type reforging furnaces, which will then be rolled into the furnace and the plate will be reheated preparatory to reforging. After reforging and rectifying, the plate will come out into the north aisle for the carbonizing, annealing, tempering and hardening operations, and under normal conditions will not return again to the south aisle although returning occasionally to the press room, as required, for bending and straightening. It then goes via shuttle track to the machine shop.

The "H" type of building consolidating the forging plant with the treatment building presents the following main advantages:

(a) A low first cost. An estimated saving of half a million dollars was secured by keeping the height of the runway rails in the main aisles down to 40 ft.

(b) The maximum economy of space secured by consolidating the forge and treatment operations in one building.

(c) The maximum economy in operation should result, due to the compactness of the building itself, the relatively short haul from any furnace to the press room, and the time saved over present methods in operation.

(d) The presses are very accessible to the furnaces and any press can receive work directly from any furnace in either aisle.

(e) Operation of cranes in the furnace aisles will not interfere with or be interfered with by forging operations in the press room.

(f) Operating conditions are ideal in so far as light and ventilation are concerned.

FIG. 8 FORGE AND FURNACE BUILDING, SOUTH AISLE, LOOKING NORTH. EAST FROM CRANE GIRDER, JULY 12, 1920

Deeper crane girders in left foreground and heavy columns are at 70-ft. spans where cross-tracks will lead to press room on left.

(g) The presses and press-drive equipment are concentrated in a central area, thereby avoiding any expensive distribution systems for steam, high-pressure water or electricity.

Furnaces. Twenty-five furnaces are being completed in the

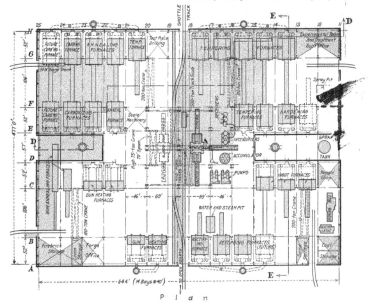

FIG. 7 FLOOR PLAN OF FORGE AND FURNACE BUILDING

forge and furnace building. Of these the three ingot furnaces, the two reforging furnaces and the rectifying furnace in the south aisle; the three carbonizing, three annealing, eight tempering and one hardening furnace in the north aisle are identical in section and differ only in length. They are of the regenerative Carbottom type, the fuel is natural gas, and the air preheated. There are three different lengths of furnace, 50, 42 and 36 ft., respectively. The height inside from the floor line to the top of the underside of the arch is 14 ft. 6 in. and the width inside from face to face of the brick work 15 ft. in each case. The general arrangement is shown in Fig. 10.

An interesting feature of the furnace equipment is the car-pulling mechanism. Having in mind the "electric mules" which

FIG. 9  PRESS ROOM, FORGE AND FURNACE BUILDING, LOOKING EAST
FROM CRANE GIRDER, JULY 16, 1920

Foundations for 14,000-ton press in middle foreground. In immediate foreground is excavation for double column which will support inside crane girders running directly across the top of the press. Note high crane for erection and repair of press and forging crane. Note double rails on outside girders for forging cranes. Intensifiers will be located at position of locomotive crane and an 32-in. accumulator where sand pile is shown.

draw the ships through the locks at Panama, the idea was conceived of using an independent portable electric-driven rack-rail locomotive which would operate on standard-gage tracks placed between the Carbottom tracks and which would draw the Carbottom by means of a standard M.C.B. coupling. A spring reel of cable was arranged for plugging in to the nearest electric power socket and rings were provided for conveniently lifting and carrying the mule to any furnace desired by the overhead crane. The device has subsequently been altered by the operating division to exclude the standard-gage track and wheels, and as built the car-pulling mechanism will be carried on an extension of the girders of the Carbottom furnaces.

The green-annealing furnaces for gun forgings, one of which will be 100 ft. long, are also of the Carbottom type. The balance of the furnaces are of the solid-bottom type. Six radial-brick chimneys with independent firebrick linings are being erected to serve the furnaces by groups.

*Forging Press.* The forging press is of 14,000 gross tons capacity and is of the steam-intensified hydraulic type. The maximum working stroke is 90 in. and the

four press columns are 19 ft. by 8 ft. 6 in. on centers. The columns are each 30 in. in diameter. There are three cylinders each 44 in. in diameter and the working pressure is 7000 lb. per sq. in. There are two double-acting manipulating jacks in pits on either side of the press, each of 150 tons capacity and having a maximum stroke of approximately 18 ft. Valves are arranged so that one forging cylinder alone may be used, giving greater speed for forging gun work under these conditions. Water is supplied at 2500 lb. pressure by three 250-gal.-per-min. pumps. A 32-in. accumulator operates on this system which also supplies the manipulating cylinders, pull-back plungers and the 250-ton hydraulic forging cranes. The triple intensifier operating on steam at 200 lb. pressure from the boiler plant, increases the pressure in the press cylinders to 7000 lb. per sq. in. The general arrangement is shown in Figs. 9 and 11.

As a matter of general interest, the total weight of the press alone is nearly 5,000,000 lb. The total load at the bottom of the foundation, assuming both cranes loaded directly over the press and including the dead weight of the press and of the concrete foundations is approximately 13,000,000 lb. To support this load it was necessary to place the press on four concrete piers, each 9 ft. in diameter, which bear on rock about 60 ft. below the level of the press-room floor.

Under conditions of eccentric loading there is an extremely remote possibility that a very great side thrust may occur at the top of the press. At the insistence of the manufacturer, a structural-steel brace was designed to take this load, as is shown in the section of the press room (Fig. 11).

All motions of the presses and jack are controlled from a pulpit located on the floor of the press room on line with the press and about 15 ft. away. The hoist motions of the 250-ton hydraulic cranes are also controlled at this point. The bridge and trolley motions of the crane are controlled from cages mounted at the far end of each forging crane whence the operator can most readily line up the piece in the press.

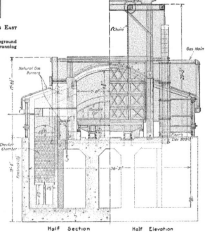

Half Section     Half Elevation

FIG. 10  SECTION AND END VIEW OF CARBOTTOM FURNACES FOR ARMOR PLATE ]

The spray for quenching armor plate is placed at the east end of the north aisle. This consists of a reinforced-concrete pit 16 by 32 ft. by 8 ft. deep. The bottom is covered by a grillage of small pipes perforated with $1/4$-in. holes. A carriage on tracks carrying a like grillage straddles the pit. The plate is taken from the Carbottom hardening furnace nearby by a crane and placed

The building is 320 ft. wide by 552 ft. long. It comprises three main aisles, each of which is 100 ft. wide center to center of crane rails. Crane rails are 40 ft. above the floor level.

The south aisle and the middle aisle are to be used for the finishing of the armor plate. Each is served with a 150-ton and a 75-ton crane with 25-ton and 10-ton auxiliary hoists, respectively. The

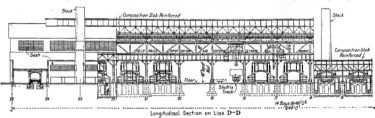

Longitudinal Section on Line D-D

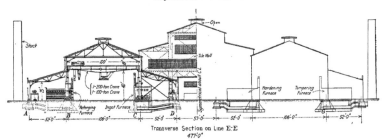

Transverse Section on Line E-E
477'-0"

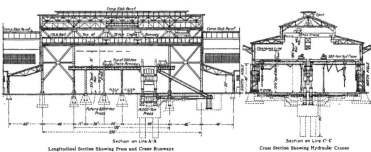

Section on Line A-A
Longitudinal Section Showing Press and Crane Runways

Section on Line C-C
Cross Section Showing Hydraulic Cranes

FIG. 11　SECTIONS OF FORGE AND FURNACE BUILDING

on suitable supports in the pit. The carriage is then moved over it and the plate drenched on either or both sides with water. Approximately 25,000 gal. of water per min. are required during the first twenty minutes of the operations. The water from the spray is wasted into the adjacent gully.

### MACHINE SHOP

The general design of the machine shop is shown by Fig. 12.

principal machine tools in the south aisle are four planers, two universal borers and drillers of the car type, two universal boring drilling and milling machines, a universal radial drill and two armor-plate grinders. At the west end of the south aisle will be located the burning equipment; gas will be piped from a separate oxy-hydrogen plant which has been built 100 ft. south as a matter of safety. At the west end of the south aisle is the erection floor and the surface plates for the fitting up of armor plate prior to

FIG. 12  MACHINE SHOP, MIDDLE AISLE, LOOKING EAST, JULY 20. 1920

Note lighting.  Immediate foreground is location of shuttle track.  Crane span 100 ft.; girders support one 150-ton and one 75-ton crane.  Columns are 45 ft. on centers:

shipment.  The erection floor will be 168 ft. long by 93 ft. wide and will consist of a thick reinforced-concrete slab with rails embedded in the surface.  The surface plates cover an area 85 ft. by 48 ft. and are of steel.  Special arrangements are made in the foundations so that an absolutely level surface may be accurately maintained.

The middle aisle contains an open-side planer, a vertical and horizontal planer, two armor-plate rotary planers and saws, a cutting-off machine and a double armor-plate breast planer.  Offices and a tool room will be located in the farthest bay west.

The north aisle which is served by two 75-ton cranes will be devoted to the machining of gun forgings.  Equipment of ample size to rough-bore and turn 20-in. 50-caliber gun forgings is provided.

The shuttle track ends in the middle aisle of the machine shop.  Gun forgings from the forge shop will arrive by way of the main-track system in order not to make necessary swinging the great length of the forgings through 90 deg., although this can be very conveniently done if required, due to the 100-ft. width of the aisle.  Railroad tracks enter both ends of the north and middle aisles and the east end of the south aisle.  A track scale of 100 tons capacity is placed on the latter track.

Special attention is invited to the method of securing an abundance of daylight.

Somewhat over 60 per cent of the side and end wall area is in steel sash.  Top lighting is provided by longitudinal sawtooth monitors in the side aisles and a double sawtooth monitor in the middle aisle.  The construction is shown in Fig. 12.

### GUN-TREATMENT BUILDING

This building is located alongside the machine shop, the intervening space 40 ft. wide being roofed over to provide for lockers, toilets, offices and a two-story electric substation for the gun-treatment furnaces.  Figs. 13 and 14 show the general arrangement of the building which consists of a low portion and a high portion.  A railroad connection from the main-track system runs through the entire length of the building close to the south wall.

The low portion of the building provides space for a future straightening press, storage area for forgings, while awaiting treatment or result of tests, cutting-off saws and slotting and boring machines for the taking of test specimens.

The high portion provides for the vertical treatment of gun forgings.  A reinforced-concrete pit approximately 80 ft. square and 55 ft. deep has been built and will contain the 10-ft. by 105-ft. vertical electric heat-treatment furnace and a quenching tank 10 ft. in diameter and 105 ft. high.  A 50-ft. furnace is also to be erected on a line with the high furnace and the quenching tank.  Space is provided in the pit and on the floor for a duplication of this equipment.  Stuctural-steel platforms will be erected around the furnaces in galleries 10 ft. apart, for the convenient inspection of the forging in the furnace during operation through peep holes in the side of the furnace.  The steel structure will also support the rolling doors which cover the top of the furnaces and the bridges from which the forgings are hung while in the furnaces.  An interesting detail is the installation of an elevator to carry the operator from the bottom of the pit to any one of the ten galleries in the 105 ft. height of the treatment-furnace structure.

The reinforced-concrete pit is of unusually heavy construction in order to withstand the pressure due to its depth, and due to the additional 20-ft. head of water to which the bottom is subjected.  The design is unique in that no interior cross-bracing whatever is used.  Each of the four sides of the lower half of the pit is designed as a slab supported between the bottom of the pit and a flat horizontal beam at approximately one-half the total depth.  The upper half of the walls are designed as retaining walls, for which the flat beam forms the base.  Special precautions were taken in the construction of the pit to secure watertightness and the entire concrete work was divided into units each consisting of one full day's pouring of concrete, metal water stops being provided at each construction joint and also where the needle beams of the shoring pass through the wall.

The high portion is served by a 75-ton crane with a 104-ft. span which has a 50-ft.-per-min. hoisting speed at full load, and a 100-ft.-per-min. lowering speed.  All motions of this crane are controlled from a pulpit located on line with the quenching tank and at a height just above the top of the furnaces and tanks, on the east wall of the high portion of the building.  The furnaces are placed on line with the tank so that only a hoist and a trolley motion will be required in the quenching operation.

The low portion is served with two cranes which are duplicates of the machine-shop 75-ton cranes.  The runway of the low por-

FIG. 13  GUN-TREATMENT BUILDING. GUN PIT IN FOREGROUND, JULY 12. 1920

Note concreting plant to pour 5000 yd. of concrete involved in pit.  Sheet piles are 40 ft. long.

tion has been extended into the high portion so that pieces may be picked up and carried directly under the area served by the high crane and thus avoid an additional transfer by rail.

### PLANT SERVICES

The service features of the plant include the industrial water supply, the sewerage, electrical power and lighting, steam, air, natural gas, etc. The electric, water and compressed-air systems are centered in the service building, which consists of two component parts, an outdoor substation and the service building proper. Electric power will enter the building at 6600 volts, 3 phase, 60 cycles and be distributed from the main buses to all

buildings, and in addition, the d.c. feeders to the gun-treatment building. The second center is adjacent to the press room of the forge and furnace building and controls the hydraulic pumping equipment and the lighting for the building. The third center is under the charging platform near to the electric furnaces in the open-hearth building and controls the electric furnaces and the lighting in that building.

Adequate provisions have been made for secondary distribution systems, distributing centers for direct and alternating current, lighting, etc.

*Water Supply.* The industrial water supply is taken from the Great Kanawha River which flows west on the north boundary of

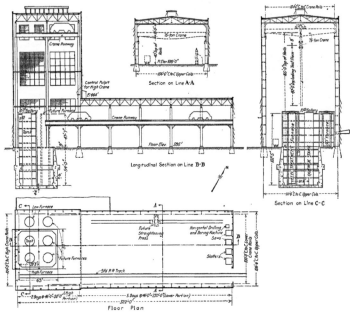

FIG. 14 GUN-TREATMENT BUILDING, PLAN AND SECTIONS

parts of the plant. The building houses four 1500-kw. rotary converters, synchronous condensers, three air compressors, six water pumps, various step-down transformers, necessary auxiliaries and the switchboards for controlling this equipment.

The outdoor electric substation is for the control of the two 44,000-volt and one 66,000-volt three-phase 60-cycle incoming circuits and consists of 3 banks of transformers and the necessary switches, buses, grounds, and supporting framework.

In general, the electrical distribution system may be outlined as follows: For the distribution of alternating currents, three centers in addition to the main substation and the projectile-plant substation are provided. One of these is located adjacent to the gun-treatment building, in the west end of the lean-to between the heat-treatment building and the machine shop, and controls gun-treatment furnaces, lights, and 220-volt a.c. power in both

the reservation. The river pump station is constructed of reinforced concrete directly in the river bank. Two 2000-gal.-per-min. pumps are being installed and space is provided for a third pump. The pump floor is underneath the river level and a positive head thereby assured. Occasional floods bring the river up to 30 ft. or in extreme conditions, 35 ft. The pumps raise the water approximately 50 ft. to a surge tank, whence it flows in a 30-in. tile pipe by gravity to the lower end of the reservoir settling basin. This basin has a capacity of 25,000,000 gal. As the maximum demand of the plant for industrial water is estimated to be approximately 5,000,000 gal. per 24 hr., the reservoir provides at least a five-day settling period for the raw river water.

The water is carried from the reservoir to the service building in a reinforced-concrete flume where the distributing pumps are located in a pump pit which insures a positive head for the pumps.

Distribution is at 60 lb. pressure. As an additional precaution in the remote case that the spray pumps should fail to operate immediately, when the plate was placed in the spray pit for quenching, a standpipe 25 ft. in diameter and 100 ft. high is being provided in the east court of the forge shop.

A separate city water system has been provided to supply water for drinking and lavatory purposes. This water will also be used for 2500-lb. pressure water at the hydraulic press and forging cranes.

*Sewerage.* Sewerage, waste water and in certain cases, roof water, are gathered from branch systems into the trunk line which, along with the other distribution systems of the plant, including gas, water and electrical conduit, follows the general line of the main west road and the main railroad track across the upper dam and ultimately reaches an outfall several hundred feet below the river pump station. The trunk line is 24 in. in diameter.

*Compressed Air.* Compressed air is piped at 100 lb. pressure

Fig. 15   Typical Column Footing for Row F, Forge and Furnace
Building, August 29, 1919
172 cu. yd. concrete. Note man in background.

from the service building to the main buildings in a loop system, large quantities being used in the open-hearth, forge shop and machine shop, and also in the blacksmith shop, where air will be used to operate the hammers. A complete loop system is provided.

*Natural Gas.* Natural gas is obtained from the fields some miles away and will be delivered at the plant at 50 lb. pressure. A loop system is provided with reducer valves located at convenient points throughout the plant. The largest quantities are, of course, used in the open-hearth furnaces and in the furnaces in the forge and furnace building. In the latter building a double loop is provided and meter houses are arranged so that the quantities of gas used in the forge shop and in the treatment shop may be measured separately. The natural-gas supply may be exhausted, it is believed, within the course of the next ten years and with this in view necessary provisions have been made for double checker chambers for the open-hearth furnaces in case producer gas must be resorted to. The use of pulverized coal as a fuel was also in mind when the foundations for the furnaces were designed. Space which was left for double checker chambers can be equally well used for ash baffles.

*Boiler Plant.* The question of providing the best method of drive for the 14,000-ton hydraulic press and additional capacity to operate a stand-by power plant of generous proportions and take care of the heating of the machine-shop and gun-treatment buildings was gone into very thoroughly and a most interesting method of adapting the reversing rolling-mill drive system to a pressure pump to work momentarily against 7000 lb. per sq. in.

Eight 823-hp. Stirling boilers were obtained from the power plant of the Old Hickory Powder Plant at Nashville, Tenn., and also a separate building of sufficient size to house them. They are arranged four on a side facing each other. The automatic

stokers are fed from an overhead concrete bunker of approximately 1000 tons capacity. This in turn is filled by a bucket conveyer from a hopper and crusher located under the railroad siding at the end of the building. The operating floor of the boiler plant is 17 ft. above the ground level so that the grates will be dumped into hopper cars on narrow-gage track and the ash used for fill. Two 12-ft.-diameter by 200-ft.-high radial-brick chimneys have been erected. A 16-in. steam main will run from the boiler plant directly to the press room of the forge shop and an additional line will be run to the service building.

### PROGRESS OF CONSTRUCTION

Actual work for the construction of the plant began in the latter part of September 1918, when ground was broken for temporary storage sheds. Barracks and mess halls were built to accommodate 1000 men and by January 1919 grading for the classification yard and excavation and foundation work for the open-hearth and machine-shop buildings was in full swing. The first steel work was erected in the open-hearth building on May 30, 1919. Excavation and concreting of the foundations for the forge and furnace buildings were undertaken in July and completed during the month of August, about 8000 cu. yd. of concrete being placed over the extended area in less than six weeks. Fig. 15 shows a typical main row column footing for this building.

During the past year the steel work has been completed on all the buildings except the gun-treatment building. All buildings have been enclosed with walls, sash and roof. Foundations have been completed for the 25 furnaces in the forge shop and for the 14,000-ton press. The 30-ton electric furnaces have been erected. The majority of the cranes have been put into operation. The boiler plant is nearing completion, also the service station. The gun pit and the river pump station are practically done.

The construction work remaining to be completed consists chiefly of the distribution systems for the plant services, the concreting of the roads and the installation of lockers, toilets and offices. Bottoms should be burned in the open-hearth furnaces and the first armor ingots cast by the time this paper is presented. The majority of forge furnaces are now completed and the 14,000-ton press should be erected and in operation by the first of the year.

Progress has been considerably delayed by the difficulty of securing and maintaining a sufficient force of qualified engineers and draftsman, and sufficient labor; the uncertainty of deliveries, due to transportation conditions; and the general unsettled conditions in the material market, although the latter situation was relieved by the obtaining of surplus material and used equipment from the Potomac Park Construction Work in Washington, from Navy construction work at Erie, Pa., and from the Army powder plant at Nitro, W. Va. A large quantity of equipment and material for the entire plant was also obtained from Army salvage.

### ORGANIZATION OF PROJECT AND PERSONNEL

The United States Naval Ordnance Plant is operated under the direction of the Bureau of Ordnance of the Navy Department. It was designed and built under the joint direction of that Bureau and the Bureau of Yards and Docks. The entire design and construction of the armor and gun-forging plant was done within the Navy and without recourse to a general contractor.

Admiral Ralph Earle was Chief of Bureau of Ordnance until recently relieved by Admiral McVey. The Armor Plant Division of the Bureau has been under the direction of Commander Logan Cresap.

The Chief of Bureau of Yards and Docks has been Admiral C. M. Parks. The Assistant Chief has been Captain R. E. Bakenhus who has also been Project Manager of the Ordnance Plant Division of the Bureau of Yards and Docks as well as Manager of Shipyard Plants for the Emergency Fleet Corporation.

The original Inspector of Ordnance in charge at the Naval Ordnance Plant was Commander J. B. Rhodes. He was relieved about a year ago by Captain George R. Marvel.

The specifications for and purchasing of the 14,000-ton press and its driving equipment and the detailed design of all furnaces and similar equipment were handled by the steel superintendent, Mr. W. J. Priestly, who is head of the Hot Metal Department.

*(Continued on page 726)*

# Steam Formulas

### A Critical Comparison of the Various Existing Formulations and the Development of a New Set of General Equations

By ROBERT C. H. HECK,[1] NEW BRUNSWICK, N. J.

THIS paper presents the results of an investigation which was at first intended to be no more than a critical comparison of existing formulations, but which soon grew into the task of developing a new set of general equations.

The discussion does not go beyond the four primary quantities temperature, pressure, volume and heat content. The two general or characteristic equations, for superheated steam of course, relate specific volume $v$ and heat $h$, respectively, to temperature $T$ (absolute) and pressure $p$ (also absolute).

In brief summary, the theory of the subject is as follows:

For a perfect or ideal steam gas the volume equation would be

$$pv = BT \dotfill [1]$$

$B$ standing for a constant which is determined by the molecular weight of the substance. According to this formula, product $pv$ would not vary if $T$ were constant, being independent of $p$ as that is varied up or down the isothermal.

With an actual gas, especially near saturation, the product $pv$ shrinks in isothermal compression, after a manner most generally expressed by the equation

$$pv = BT - X \dotfill [2]$$

The contractive term $X$ is a function of both pressure and temperature, increasing with $p$ but diminishing as $t$ is higher and as the gas gets farther away from saturation.

From a study of experimental knowledge and as a result of trial against the data, the general term $X$ is given a specific and detailed form which leads to the $p$-$v$-$T$ equation

$$pv = BT - \frac{E}{T^m} p - \frac{F}{T^n} p^s \dotfill [3]$$

In this formulation the three exponents are fixed as $m = 3.5$, $n = 7.2$ and $s = 2$, while $E$ and $F$ are constant coefficients fixed by trial.

The general formula for heat content is similar to [2], being

$$h = h_0 - X' \dotfill [4]$$

for any temperature $h$ has a maximum value at zero pressure and diminishes as pressure rises along the isotherm. An important theoretical relation whose derivation goes back to Clausius ties $X'$ very definitely to $X$, giving the $h$-$p$-$T$ equation the form

$$h = h_0 - \frac{E'}{T^m} p - \frac{F'}{T^n} p^s \dotfill [5]$$

The coefficients $E'$ and $F'$ bear to $E$ and $F$ fixed ratios which are rather simple functions of the exponents $m$, $n$ and $s$. Zero-pressure heat content $h_0$ is an empirical function of $T$.

Three complete sets of equations have been proposed, namely, Callendar's, used by Mollier for steam tables in 1906 and by Callendar himself in 1915; Goodenough's, which underlie the table published in 1915; and the new equations by the writer. Those submitted to the Society by the writer in 1913 used simple forms for terms $X$ and $X'$ and are not now given serious consideration.

A table in the paper gives all the constants for each of these formulations; and they enter into the comparisons with data, along with other formulations which are listed, notably the Marks and Davis tables.

The theory just described, while rational in its general form, is empirical in the definite valuation of its constants. Unfortunately we have neither method nor knowledge for its synthetic upbuilding from simple elements, so that the only practicable procedure is by trial solutions, testing overall effect against data. A very

[1] Professor of Mechanical Engineering, Rutgers College, Mem.Am.Soc.M.E.
For presentation at the Annual Meeting, December 7 to 10, 1920, of THE AMERICAN SOCIETY OF MECHANICAL ENGINEERS. In this abstract of his paper the author has confined himself to a brief summary of the results of an extended investigation undertaken by him, in which existing steam formulas are compared and a new set of expressions are developed. The complete paper with its abundance of reference data and charts may be obtained gratis upon application by those interested. All papers are subject to revision.

important determinant, especially for the necessary extrapolation far above the pressure limit of experiment, is furnished by Clapeyron's law. This relates latent heat to the increase of volume during vaporization, by a ratio in which the rate of change of pressure with temperature for saturated steam is the variable factor. The saturation pressure-temperature relation is the most accurately and completely known of the properties of steam, the writer accepting Goodenough's formula for this relation as satisfactory and final. The determinant is applied by evaluating assumed general equations for coincident saturation values of $p$ and $t$ and seeing how closely the resulting ratios agree with those from the independent $p$-$t$ relation. The writer's final equations are the outcome of between sixty and a hundred trial solutions.

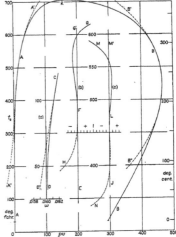

FIG. 1   PRESSURE-VOLUME DIAGRAM FOR SATURATED STEAM

In presenting results and comparisons the paper follows the usual order of the steam tables, considering first saturated steam, then superheated. For the properties of hot water and of dry saturated steam the writer offers new empirical formulas: on the side of saturation these come down from the critical "point" and at about 500 lb. pressure merge with the saturation evaluations of the general equations. Fig. 1 shows a plot of curves for volume, showing product $pv$ rather than the widely variant volume itself. The minor detail diagrams (a), (b) and (c) are concerned with a close study of the merging of formulas, and may be disregarded here. Note that the base (vertical) is saturation temperature $t_s$. In Fig. 2 the base is changed to pressure and comparisons are made with Marks and Davis (MD) and Goodenough (G) values. Remembering that all experimental data for steam are below 250 lb. pressure or 400 deg. fahr. saturation temperature, it is apparent that extension of the steam tables even to 1000 lb. pressure is very much of an extrapolation, and that theoretical relations which

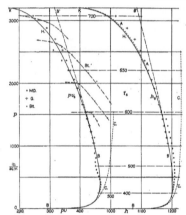

Fig. 2 Diagram of $pv_i$ and $h_i$ on Pressure Base, with High-Range
Comparisons

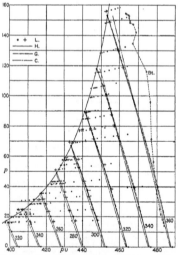

Fig. 3 Comparison of Callendar-Mollier, Goodenough and Heck
Isotherms with the Munich Volume Experiments

will guide in making this extension are highly important and useful.
The paper goes on to make close and magnified comparisons of
the saturation lines with the low-range data, below 250 lb., to show
the exact relative status of all the formulations, from Regnault's

down. Turning now, however, to superheated steam, we have the
three volume equations compared in Fig. 3 with the Linde data;
note how the straight-line isotherms of Callendar (C) fail to conform with the experimental points for the same temperatures,
marked +.

Comparisons with specific-heat data are made in Fig. 4, the experiments all having been made at Munich under the direction of
Knoblauch and the last lot of them constituting the one body of
new data since 1913. There is notably close agreement between
the (G) and (H) curves.

A very practical and direct comparison of results on the side of
heat content is made in Fig. 4, by plotting isotherms for 400,
500, 600 and 800 deg. fahr., up to 600 lb. pressure. Dotted curves,
from the writer's Steam Engine and Turbine are contemporaneous
with the Marks and Davis determinations; for high temperatures
these are not even reasonably correct, being based on data that
have been superseded.

The paper proceeds to discuss the form of the steam equations,
describing the action and effect of the Clapeyron relation as a

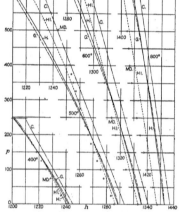

Fig. 4 Comparison of Heat-Content Isotherms

determinant and bringing in use of throttling relations and data
as a test of final form. Noting how closely the writer's curves
agree with those from Goodenough, the question is raised whether
this shall be taken, practically, as an endorsement of the latter.
Certain points in Goodenough's solution are rather sharply criticised, one on the ground of doubtful correctness in theory.

The whole discussion shows how close we are getting to final
results in this department of scientific knowledge. The differences
between the best solutions lie far within the range of experimental
irregularity and uncertainty. The chief need is some accurate
data for high-pressure steam, to confirm, or perhaps slightly to
modify, the highly probable results now reached in this region. To
isolate energy quantities for precise measurement under such pressure conditions would be almost impossible. On the side of volume
the physical quantities are much simpler and easier to handle;
and under the scheme of related formulas there is diminished need
for data in both fields. A comparatively small amount of first-
class physical experimentation would be sufficient to resolve any
present uncertainty. Pending such information, and with the confident expectation that it can call for but little readjustment, the
writer offers his new formulas as the best that can now be done
toward establishing a standard steam table.

# Foundations for Machinery

By N. W. AKIMOFF,[1] PHILADELPHIA, PA.

*The purpose of building foundations is clearly defined and differs radically from that of providing substructures for machinery. Considerable contradiction can be found in blindly applying the principles of the former to the latter, and the results are sometimes so unsatisfactory as to necessitate the use of cushions and other yielding means, thus entirely undermining the very theory on which the foundation was supposedly built. Vibrations arise from either lack of balance or from other causes. The proposed theory contemplates the latter, the problem of balancing being considered as capable of complete solution by suitable treatment. After briefly considering the nature of possible displacements of the foundation as acted upon by various causes leading to vibrations, the author introduces, by way of illustration, a double pendulum, a few experiments with which form the basis of his theory. Means for localizing the expected vibration and of controlling the resulting periods are then illustrated in a working sketch of the proposed arrangement.*

THE weight of the Great Pyramid is approximately 5,274,000 tons, its base is 764 ft. square, and its height about 486 ft. It is built on leveled rock. The piers of Brooklyn Bridge are founded 44 ft. below the bed of the river upon a layer of sand, 2 ft. thick which rests upon bedrock. The massive St. Isaac's Cathedral, Petrograd, is built on a swamp, and the piling has been so carefully proportioned that the exceedingly heavy doors of the cathedral swing easily, whereas the slightest lack of uniformity in settling would doubtless lock them. These well-known structures are here mentioned by way of illustrating the obvious purpose of the foundations upon which they rest, i.e., (1) to distribute the load in as nearly uniform a manner as possible; and (2) to secure uniformity in settling.

Just how all this applies to foundations for all sorts of machinery, and in particular to rotative machinery, is not easy to say. Indeed, the weight, say, of a large pumping engine or of a turbo-generator outfit is generally much lower per square foot of floor space occupied than the limits prescribed by municipal laws or building ordinances, and furthermore, uniformity of settling of a relatively small volume of this nature can be secured without going to the extremes usually observed in designing footings for buildings. What, then, would be the general governing idea in proportioning a foundation, say, for an engine of a given type? Should it be heavy or light? Should it be deep, resting on rock or sand if possible? Should it be independent of the footings of the building, or would it be desirable to tie it to the latter?

By examining the existing records we can find a great variety of rather contradictory answers to each question, but the predominant idea in the mind of the designers appears to be somewhat as follows: Since the engine is likely to vibrate, let us tie it as firmly as we can to the earth itself. The mass of the earth being practically infinite, the amplitude of the resulting vibration will probably be zero. The designer may be utterly unconscious of this reasoning, but he applies it nevertheless and obtains results which sometimes are satisfactory and sometimes exceedingly poor.

## CAUSES OF VIBRATION

A rational basis on which to work is thus seen to be lacking and it is accordingly the object of this paper to point out some definite lines along which a rational theory of substructures for engines and moving machinery can in general be built up. To begin with, it is important to realise that vibrations are caused by two distinct orders of agencies: (1) Those due to unbalance, or, more correctly, lack of running balance; and (2) those due to causes other than unbalance.

As regards unbalance, it may be said that this can be so easily corrected in the construction of machinery that all specifications should invariably call for perfect running balance at all speeds, that is, complete absence of tremor or of "periods" under all condi-

tions. However, there are many causes quite independent of balance, each of which is likely to result in vibrations, as for instance, "whipping" of a slender body (crankshaft, armature, turbine rotor, etc.); water in a steam turbine; peculiarities of the reciprocating mechanism—for instance, a 4-cylinder or an 8-cylinder V-type engine where certain forces do not cancel out and where running balance alone is not conducive to perfect results; or torsional vibrations, which under certain conditions produce an effect very similar to that of unbalance. Our problem, then, is to analyze the effect of these various causes, with the view of designing a substructure for a given machine that will be *least* responsive to these causes, for this is what the "relative" freedom from vibrations really means.

But whatever may be the cause of vibrations, it is safe to say that in general they are always due to *forces*, acting in a plane or planes, perpendicular to a certain axis; also to centrifugal *couples*, located in a plane, rotating about a certain line, usually the axis mentioned just above. We know from elementary mechanics that any motion of a body can be resolved into six distinctly separate motions: three *along* the three mutually perpendicular axes drawn

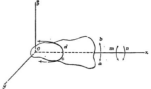

FIG. 1 DIAGRAM OF A BODY WITH ONE POINT FIXED

through any point, within or without the body, and three *about* these same axes. A free body, capable of a displacement in any of these six modes, is said to possess six degrees of freedom.

### VIBRATION AS AFFECTED BY DEGREES OF FREEDOM

If the body is rigidly locked so that no displacement of any kind is possible, we say that all six degrees of freedom have been suppressed. By fixing two points in the body we have the effect of rotation about an axis, and only one degree of freedom, that is, the angular displacement about the axis, characterized by these two points. By fixing one point we suppress all bodily motion along any three axes through this point, but we still have three degrees of freedom. Fig. 1 shows a body whose point $O$ is fixed. Such a body can have only three kinds of displacement: (1) about the axis $y$, as shown by the arrows $a$ or $b$; (2) about the axis $z$, as shown by arrows $c$ or $d$; and (3) about the axis $x$, as shown by arrows $m$ or $n$. It makes absolutely no difference what the forces are which act on the body, no other motion is conceivable.

On the other hand, placing a body upon a thick sheet of yielding material, or for that matter on four springs, means freedom in all six directions; and of course the same applies to cushions or pads. For this reason it appears to be of extremely questionable value to interpose layers of such material between a massive subfoundation and the foundation proper of a machine, even if isolated examples are on record where such an arrangement actually *happened* to give satisfactory results.

### WHAT IS STABILITY?

With the foregoing in mind, let us digress for a moment. Stability with regard to our subject is a somewhat relative term. Why was the Great Pyramid built upon level rock? To insure stability. Why build a massive foundation for an engine? To secure stability. Why provide a layer of yielding material or springs or rubber pads? To secure stability. In other words, it is

[1] Mfr. Balancing Machinery, Mem.Am.Soc.M.E.
Abstract of a paper to be presented at the Annual Meeting, New York, December 7 to 10, 1920, of THE AMERICAN SOCIETY OF MECHANICAL ENGINEERS, 29 West 39th Street, New York. All papers are subject to revision.
NOTE: Patent protection controlled by the General Machinery Foundations Co., Philadelphia, Pa.

quite necessary to define in a more rational way the purpose of a foundation for an engine of a given type. We shall attempt to do this, first stating, however, the well-known effects of vibrations on various types of apparatus.

In large power plants where the main units are of the modern turbo-generating type the steam mains have been known to burst, and subsequent investigation has often revealed no defect in either material or in general arrangement of piping. The accident can thus only be explained as due to "rough" running, that is, vibrations.

The operation of a printing plant or of a leather-working factory is often extremely unpleasant for adjoining dwellings, and sometimes even for buildings located at a rather considerable distance. Inspection often reveals that machinery in such plants is firmly secured to extremely massive foundations and the owners are at loss as to how to remedy the trouble.

An internal-combustion engine often exhibits a certain amount of vibration which can be felt all over the understructure. In fact, owing to violent vibrations some of the tie rods, lamp brackets, etc., on automobiles have been known to snap in two, and in aircraft some of the instruments to drop off the board, yet these were parts of the understructure, to which they, as well as the engine

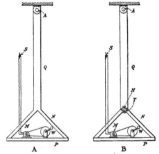

Fig. 2   A—System with One Degree of Freedom; B—System with Two Degrees of Freedom, One of Which May be Suppressed

itself, were firmly secured. The point we wish to emphasize is that in an understructure too much rigidity is as harmful as too much freedom to yield.

### Stability Defined

We are now ready to formulate the new criterion of stability for foundations. Stability is here characterized by remoteness of the operative speed from any one of the several synchronous speeds at which the frequency of the operative speed would be nearly, or exactly, equal to the frequency of the free oscillation of the system, if displaced from natural state of rest and let go.

How many distinct synchronous speeds a system is capable of having depends upon the number of degrees of freedom. An absolutely free system, for instance, placed upon an elastic subfoundation, may have six independent synchronous speeds, or "critical" speeds, as they are sometimes called. A massive foundation resting upon rock is likewise often apt to be, in the larger sense, free in all six degrees, since it occasionally does vibrate and propagate the vibration to other buildings, etc. A system with one fixed point may have only three such synchronous speeds, while a system mounted to rotate about an axis can have only one such speed or "period," as it is often termed. If we could control these synchronous speeds, so as to make sure that none comes anywhere near the actual speed of operation of the machine, we would then have a fairly complete solution of our problem.

### Vibration in a Body with One Degree of Freedom

In order to understand clearly the foregoing as well as the broad methods here proposed, let us consider the following experiment and the general consequences manifestly derived therefrom. Imagine a pendulum (Fig. 2-A) consisting of a platform $P$ rigidly connected to the member $Q$ by means of the side members $N$. The system is free to swing about the axis $A$ in the plane of the figure. A small motor $M$ fastened on the platform operates a countershaft carrying an off-center weight $W$. The motor is fed through a suitable flexible connection and it is always possible to adjust the speed of the countershaft so that the number of revolutions per minute will be equal to the number of double oscillations per minute of the pendular system. If the latter is slightly displaced from its vertical position of equilibrium and let go, the effect of this adjustment of speed will be the so-called "synchronism," and the extent of swing (amplitude) of the pendular system, in general very slight for arbitrary values of the rotative-speed of the weight, will now become violent, in fact, out of all proportion to the magnitude of the weight itself.

This phenomenon of synchronism of cause and effect has been well studied. The amplitude at the exact condition of synchronism should theoretically be infinite, but of course in practice resistances of various kinds are always present, so that instead of *infinite* we have *large* amplitudes. The most curious fact is that in the vicinity of synchronous speed, both above and below, the amplitude drops down to a value almost insignificant, so that if the weight is small the system appears to be practically at rest. Furthermore, any further increase of speed will not produce any effect, contrary to the current opinion of those not very well versed in the matter. For the sake of illustration let this synchronous speed be 100 r.p.m.

### Vibration in a Body with Two Degrees of Freedom

As a modification of the experiment let us now provide another system (Fig. 2-B) identical with the first except that the platform $P$ is not solid with the member $Q$ but is pinned thereto at $H$, the pin used being both frictionless and at the same time so arranged that it can be tightened up by means of the nut $T$, so as to lock the joint, thereby securing the exact effect of the rigid pendular system of Fig. 2-A. Providing the pin in the language of dynamics is the introduction of an additional degree of freedom, thus securing a system of two degrees of freedom; while the tightening of the nut amounts to suppressing one of the degrees of freedom, thus converting a two-degree into a one-degree system. In experimenting we shall first deal with the system of one degree of freedom, tightening the nut $T$ and thus converting the pendulum into a system exactly similar to that discussed above, the synchronous speed being, say, 100 r.p.m. The pendulum will oscillate violently. We now loosen up the nut $T$, introducing an additional degree of freedom, with the apparently surprising result that the amplitude decreases practically to zero. If we reduce the speed considerably, say to 50 r.p.m., violent oscillations of the whole system will reappear, as will likewise be the case in speeding up the countershaft, say to 150 r.p.m. These figures are purely illustrative; whether they will actually correspond to exact facts will of course depend upon the characteristics of the system.

In other words, by introducing an additional degree of freedom we have accomplished this double result: (1) What was synchronous speed for a system with a single degree of freedom is no longer synchronous speed for the same system provided with an additional degree of freedom; (2) the new system has two frequencies of oscillation, at which it is sensitive to disturbing influences, one being below and the other above the value corresponding to that of the same system with the additional degree of freedom suppressed.

It should especially be observed that the oscillations were thus reduced practically to zero, not by steadying the system by something without it, but by some sort of an adjustment wholly within the vibrating system itself. Furthermore, what we did was to increase in a measure the flexibility of the system by breaking it in two; and although at first glance this might have increased the effects of the disturbing agency the actual effect was practically

(Continued on page 699)

# Organization and Construction of Dye Houses

### The Machinery Organization, Location, Construction, Ventilation and Piping of Dye Houses Considered from the Engineering Viewpoint

By A. W. BENOIT,[1] BOSTON, MASS.

THERE are two distinct classes of dye houses, one being the converting plant which does dyeing and finishing for the trade; and the other, the plant dye house, which dyes and finishes for one mill only. The first generally covers a wider range and is prepared to do a great variety of work, while the latter is naturally of a narrower scope and is usually limited to dyeing the product of one plant. The converting plant is sometimes equipped to do bleaching and mercerizing, which add complications to their problems. For the purpose of this paper the dye-house problem will be discussed from the standpoint of the plant dye house.

The operation of dyeing enters largely into the four great branches of the textile industry, cotton, worsted, woolen and silk, and in many cases these lap over each other in the individual plants because of the great variety of combinations of stock used in making cloth. Cotton may be dyed in the raw stock, as yarn, or in the piece. Worsteds are dyed in the top, as yarn, or in the piece. The bulk of the woolen fabrics are dyed in the stock but yarn and piece dyeing is also common. Silk is generally dyed as yarn.

## MACHINERY ORGANIZATION

Following is a list of the kinds of machinery used in cotton, worsted and woolen dye houses. It will be noted that many of the machines appear in more than one of the lists. It must be borne in mind that most dye houses handle more than one kind of stock and in more than one form and will contain machines from more than one group.

*Cotton-Dyeing Machinery:*
For Yarn:
  Boiling-out machines
  Doubling machines
  Chain dyeing machines
  Splitting machines
  Beam dyeing machines.
For Piece Goods:
  Jigs
  Continuous dyeing machines
  Padders.
For Raw Stock:
  Pressure kettles
  Rotary raw-stock kettles
  Hand tubs.
Miscellaneous Machinery:
  Dye cans
  Mangles
  Washers.

*Worsted Dyeing Machinery:*
For Yarn:
  Rotary skein-dyeing kettles
  Spool-dyeing kettles.
For Piece Goods:
  Piece dye kettles.
For Slub Dyeing:
  Top-dyeing machines.

Other Machinery:
  Crabs
  Steamers
  Flat and string washers
  Extractors
  Carbonizing machines
  Dry cans
  Tentering dryers.

*Woolen-Dyeing Machinery:*
For Yarn:
  Rotary skein-dyeing kettles
  Spool-dyeing kettles
  Skein dyeing in machines in which liquor is circulated.
For Piece Goods:
  Piece dye kettles.
For Raw Stock:
  Rotary raw-stock kettles
  Stock dyeing in machines in which dye liquor is circulated
  Hand tubs.

*Silk-Dyeing Machinery:*
For Yarn:
  Rotary kettle for skeins
  Hand tubs.

## LOCATION OF DYE HOUSE

The location of the dye house in the plant is influenced by (1) sequence of operations, (2) water supply, (3) steam supply and (4) drainage. It is impossible to lay down any fixed rule as to the location of the dye house in a plant in its relation to other operations, because most dye houses handle the stock at more than one stage of the process. Each dye house becomes a problem in itself and its location must be decided upon by the particular conditions surrounding it and the purpose for which it is to be used. If a dye house handles its product all in one form such as raw stock,

[1] Assistant Engineer, Charles T. Main.
Abstract of a paper to be presented at the Annual Meeting, New York, December 7 to 10, 1920, of THE AMERICAN SOCIETY OF MECHANICAL ENGINEERS, 29 West 39th Street, New York. All papers are subject to revision.

yarn or cloth, its location is easily fixed, and if it handles stock in more than one form there is generally a preponderance of one kind which determines its location. In general, the position of a dye house for various kinds of plants would be as given below, from which it will be seen that the choice of location depends upon the form of the bulk of the product.

| | | |
|---|---|---|
| Cotton | Dyed yarn | Near warping, dressing and slashing department |
| Cotton | Piece goods | Near gray room and finishing department |
| Worsted | Yarn | Near warp-preparing department |
| Worsted | Piece goods | Near gray room and finishing department |
| Woolen | Raw stock | Near scouring and picking department |
| Woolen | Yarn (unusual) | Near dressing room |

A dye house requires an abundant supply of clean, soft water. If this is available from a canal or river the dye house can often be advantageously located to secure its supply by gravity and thus

FIG. 1 DYE HOUSE WITH VENTILATION

avoid pumping. If the water is taken from a pond, it usually has to be pumped and therefore the location of the supply has no influence on the location of the dye house. When the water is so hard as to require a softening plant, pumping is required and the same thing holds true with regard to location.

*Size of Building.* In determining the general dimensions of the building a machinery organization plan must be made of the equipment to be installed. The width of the bay is not essential and is usually about 10 ft. The length of the span from column to column to wall should be about 25 ft., providing space for one row of kettles and an ample working alley (Fig. 1).

The kettles should be installed in pairs, right and left hand, with the drives whether belt- or motor-driven, outside, and an alley of 4 or 5 ft. between ends of each pair of kettles. If there are but few kettles they can be economically arranged in two rows in the center of the room, the fronts facing the outer walls with a 12- to 16-ft. working alley. The backs should be set 6 ft. apart. If there are a large number of kettles they are best arranged in four rows. The two outer rows facing the walls with a 12- to 16-ft. working alley, the two inner rows facing the center of the room with a 16- to 20-ft. working alley.

The height of the room should be from 16 to 18 ft. at the eaves. If the building is three spans or more in width there should be a monitor in the middle to light the inner rows of kettles as the window light is largely cut off by the machines and ventilators.

*Type of Building.* A one-story building is the best type for ventilation and light and is the most satisfactory unless crowded for room. Slow-burning mill construction consisting of brick walls, wood roof and concrete foundation and trenches, is desirable. The use of exposed steel is to be avoided in every way and for this reason wood sash are preferred to steel, because of the corrosion due to moisture and acid fumes. An ordinary flat roof with a monitor, using hard-pine beams and 4-in. spruce plank, works out satisfactorily. The extra thickness of roof plank is necessary in cold climates for insulating purposes. Spruce or Douglas fir is preferable to hard pine as the latter contains resin which is softened by the heat and is likely to drop on goods waiting to be dyed.

All timber and plank used in the roof and monitor should be kyanized after the framing has been done and before putting into place. This necessitates that much of it be done on the job. It is the best safeguard to employ against decay and well worth the expense.

*Dyer's Office.* An office should be provided for the dyer, located on the north side of the room if possible, having a large window, preferably of plate glass at which he can match samples of work taken from kettles, for shade and color.

*Drug Room.* Space should be provided for the storage of dye stuffs and chemicals used in the dye house. This room should be

FIG. 2 DYE HOUSE WITHOUT VENTILATION

near the dyer's office and partitioned off from the kettle room. It should be well lighted and have a steam and water supply to convert dry colors into liquid form before they are placed in the kettles.

*Laboratory.* The modern and well-equipped dye house of today is provided with a laboratory where the dye stuffs and chemicals purchased, may be tested and small samples of stock dyed. This should be located next to the dyer's office and connected to it.

|Trenches. Practically all types of dye kettles and washers are set so that much of the machine is below the floor level. These should be concrete trenches about 11 ft. wide by 6 ft. deep in which supports are provided for the machines, leaving a clear space underneath for the flow of the waste, and ample room all around to get at all parts requiring attention. In large dye houses and also in those where, because of local ordinances, it is necessary to separate the hot waste dye liquor furnished the rinse water, an auxiliary drain should be provided in this trench. This is done by installing a separate drain pipe on one side or partitioning off part of the trench. A main intercepting drain crosses the dye house, into which all the trenches are connected, and in some cases two such drains are provided to keep the wastes separate.

*Floors.* In a dye house of the above description the trenches are planked over after machines are placed and all spaces between the trenches are backfilled and the floor is laid directly on the ground. The floor consists of a concrete base with either a granolithic or paving-brick top, pitched to drain into the trenches. A granolithic top is only good in small dye houses where the trucking is light, because such floors crack more or less and break down at

the edges under traffic, causing depressions which hold water and make trucking difficult. A granolithic floor that will stand up under heavy trucking can be laid by placing cast-iron grids on the concrete base and putting the granolithic finish over the grids and floating flush with the top. This floor is expensive but will stand a great deal of abuse. The best floor is obtained by using a vitrified paving brick bedded in cement mortar and well grouted. These bricks should be not less than 2 in. thick and thoroughly vitrified. A common hard-burned brick will not answer the purpose.

The chief cause of obscurity in a dye house when kettles are boiling is not due to steam in suspension but to the dense fog caused by the precipitation of the moisture in the saturated air when it comes in contact with cold ceilings, floors, walls, pipes and cold air seeping into the room. Some of the hot vapor rising from the boiling kettles and tubs will condense on coming in contact with the overhead structure and the water will drip to the floor. That which is not expelled by fans or ventilators will cool off to the room temperature, creating a fog which settles toward the floor, gradually filling the whole room.

The simplest form of ventilation is to depend on the natural ventilation from the windows and monitor. In very small dye houses this gives fair results in summer, but in the cold weather it is inadequate. It can be helped by a large heating system designed to keep the temperature of the air above the point where fog will form, or by putting the dryers in the dye house, but at the best it is unsatisfactory.

It is very often attempted to clear the air by using exhaust fans in the monitor, and while this scheme will remove the hot, oppressive vapors, it does not clear the fog which lies in the lower half of the room. The air necessary to replace that driven out by the fans, comes in at the doors, windows or other inlets, and as soon as the cold air comes in contact with the warmer saturated air a fog is produced which cannot be removed by fans.

Another scheme is to put large wooden hoods over the machines, singly or in pairs, with a vent from each hood going up through the roof and well above it. The idea is to catch the hot vapors before they are dispersed and by means of the natural draft due to the difference in temperature, to get them out of the room. This works out fairly well and will give good results under normal conditions in medium-sized dye houses. There is still some fogging due to the cold air in winter being drawn into the room to replace that carried out with the steam but the humidity is much lower than with open kettles so that this is not so serious.

Some large dye houses have been ventilated without the use of hoods over the machines by means of warm air blown in along the ceiling and delivered in large quantities about 7 ft. from the floor toward the machines. The steam, fog and surplus air escape through the monitor windows or ventilators in the roof. By this method the ceiling is kept dry and the fog is eliminated to a distance of about 6 ft. from the floor. The shafting, piping and equipment in the upper part of the room is obscured by the fog which is moving upward. (Fig. 2).

The system of ventilation which has given the best results is to place over the kettles hoods having vents up through the roof and to distribute warm, dry air over the room and in a blanket on the under side of the ceiling in sufficient quantities to produce a slight pressure. This prevents any cold air from coming into the room and causes the steam to discharge directly up through the vents keeping the ceiling free from appreciable condensation and the room free from visible vapor. The temperature of the air to perform this work should not be over 80 deg. fahr.

The air is furnished by a steel-plate fan taking its air through a sectional heater in which any desired number of sections can be heated depending on the temperature, and distributed about the room by a system of air ducts suspended close to the roof, having outlets delivering a blanket of warm air along the ceiling, eliminating all condensation from that source and having branch ducts dropping down to within 7 ft. of the floor with outlets discharging toward the hoods and about the room. During the summer months the windows should be kept closed and the apparatus kept in operation but without steam on the heater. With the room clear of all condensation, vapor or steam, individual or group motor drive for the dye-house machinery may be used with safety.

*(Continued on page 706)*

# Rational Design of Hoisting Drums

By EVERETT O. WATERS,[1] NEW HAVEN, CONN.

*The main problem of hoisting-drum design resolves itself into two parts: determination of flange shape and thickness, and determination of the thickness of the drum body. Other factors, such as brake, clutch, and bearings, have already been investigated and are, therefore not considered in this paper.*

*Flange thickness is a function of (a) rope tension and (b) depth of winding. Previous theoretical investigations have recognized these factors but have failed to take account of the friction between adjacent coils of rope and between rope and drum, which tends to hold the coils in position without the aid of the flanges, and have also disregarded the flattening of the inside coils under pressure from the outer layers.*

*A theoretical formula is deduced for the total pressure against a drum flange caused by the winding-on of rope to a given depth and under a given initial tension. Two other formulae are then deduced, which relate this total pressure to the flange thickness and the maximum allowable tensile and shearing stresses in the material. By means of these formulae a flange of the usual straight-sided or the mushroom type may be designed to withstand safely the pressure of the rope wound upon the drum.*

*The drum body is subjected to combined stresses, since it is under compression from the coils wound upon it, under tension due to the lateral pressure against the flanges, and under combined torsion and bending caused by the load which is to be hoisted. Formulae are derived which may be used to obtain the correct thickness both at the center of the drum, where the normal stresses are greatest, and at the ends of the drum, where the shearing stresses are most prominent.*

*Tests were made with small-size rope wound upon a special drum, which was so arranged that the side thrust against one flange could be directly measured while the rope was being wound on or unwound. These tests show conclusively that formulae for flange pressure which do not take account of rope friction and flattening give excessively large results.*

**S**OMEWHAT over a year ago the writer's attention was called to certain details in the design of rope-winding drums as commonly used in hoisting and conveying machinery. Upon investigation it appeared that little if any study of this subject had been made beyond the private researches of various firms engaged in manufacturing hoisting apparatus. Published matter dealing with this class of machinery makes scanty reference to the design of hoisting drums, and recommends proportions based on careful guesswork or previously accepted practice rather than upon theory or experiment.

Such being the case, it seemed well worth while to make a special study of the hoisting drum, with a view to deducing some logical rules for proportioning the important parts. There are in reality only two such parts that require this analysis: the drum body or core, and the flanges. Other details, such as length of bearing, size of arms connecting the drum core with the hub at either end, and proportions of brake and clutch surfaces, present no novel problems; a discussion of these parts is accordingly omitted from this paper.

## Design of Drum Flanges

In many installations of hoisting machinery the length of rope to be handled is not great and it is quite possible to wind it in a single layer on a drum of moderate diameter and length. This is, of course, the most desirable arrangement, since it insures uniform hoisting speed, obviates the danger of overwinding at either end of the drum if the rope is improperly guided, and prevents the wear on the individual wires which results when two or more coils are wound on the drum in contact with each other. However, it is often necessary to wind great lengths of rope on comparatively small drums, and in such cases the end flanges must be relied upon to support the rope at the ends of the layers of which there may be as many as 14, 16 or more. In addition these flanges are usually

---
[1] Assistant Professor of Machine Design, Yale University. Mem.Am.Soc.M.E.
For presentation at the Annual Meeting. New York, December 7 to 10, 1920, of THE AMERICAN SOCIETY OF MECHANICAL ENGINEERS. The paper is here printed in abstract form and advance copies of the complete paper may be obtained gratis upon application. All papers are subject to revision.

subjected to a stress from brake, friction clutch, or both. It is therefore necessary to know (a) the maximum tangential load on the flanges caused by the brake, (b) the maximum axial or tangential load due to the clutch, and (c) the load due to the axial thrust of the several layers of rope.

## Theoretical Determination of Axial Thrust

*First Method.* The first two of these loads can be found quite readily, as explained in treatises on brakes and clutches. The third is much more difficult to determine, and it has been suspected for some time that the theoretically obtained values for this thrust or pressure do not agree with the facts. The series of tests described in a later paragraph were made for the express purpose of throwing light on this point.

In deriving a simple formula for the axial thrust, it is assumed that the coils of rope are wound uniformly, the coils of each successive layer resting in the spaces between coils of the preceding layer, as shown in Fig. 1. Friction between contiguous coils and between rope and drum surface is neglected. If the drum is grooved, as shown in the figure, all coils in the central part of the drum (those numbered 100 and higher in the figure) will be self supporting, while a certain number of coils, occupying a space o

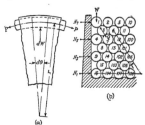

Fig. 1 Diagram of Coils Wound on Grooved Drum

wedge-shaped cross-section at each end of the drum, must rely upon the flanges for their support. Thus in Fig. 1 the coils numbered 1 to 16, inclusive, are the only ones which will cause axial thrust against the left flange. This thrust, which is indicated by the forces $N_1$, $N_3$, $N_5$, etc., in Fig. 1, is computed as follows:

When there are $2m - 1$ layers of rope,

$$N_{2m-1} = bW \cot \gamma = 2 \pi bP \cot \gamma \dots\dots\dots [1a]$$

$$N_1 = (m-1)W \cot \gamma = 2(m-1)\pi P \cot \gamma \dots [1b]$$

When there are $2m$ layers of rope,

$$N_{2m-1} = (b+1)W \cot \pi = 2(b+1)\pi P \cot \gamma \dots [1c]$$

$$N_1 = (m - \tfrac{1}{2})W \cot \gamma = (2m-1)\pi P \cot \gamma \dots [1d]$$

The total axial thrust for an odd or even number $m$ of layers is approximately

$$N_1 + N_3 + N_5 + \dots + N_m = N = \frac{m^2 - 1.5}{2} \pi P \cot \gamma \dots [2]$$

*Second Method.* The foregoing expressions are easy to use, but in the writer's opinion give results which are excessive, this being corroborated by the experiments described toward the end of the paper. Three important factors have been disregarded: friction; cross-over of the rope from a given space between coils on the preceding layer to the second space beyond, due to the fact that alternate layers are wound right and left hand; and compression of the internal layers. All three of these tend to lessen the axial thrust. An approximate expression for this thrust, as modified by these conditions, may be found by considering the end coils as a solid wedge *abcd*, Fig. 2, upon which forces are acting as indicated. The effective slope of this wedge is not the actual slope $\gamma$ of the center lines of the coils, but is $90° - \gamma$, because that is the actual inclination of the contact surfaces between contiguous coils. The coefficient of friction $\gamma$ is probably higher for rope on rope than for rope on drum. For example, some experiments performed by the writer gave 0.201 for the friction of rope on rope and 0.181 for rope on drum, these results having been obtained by wrapping a short length of $^1/_4$-in. rope around a drum, hanging a known weight from one end of the rope, and then suspending enough weight from the other end to just cause slipping. The rope was bright, but not lubricated, except with the grease applied in manufacture; the drum was finish-turned and unlubricated. In view of the closeness of these results, a single value for $\mu$ has been assumed in the following analysis.

The radial force $R_1$ is indeterminate; it is therefore assumed equal to $1/p$ times the vertical component of all forces acting along $dc$ and $cb$, where $p$ is the number of coils between $a$ and $b$. For example, if there are six layers as in Fig. 2,

$$R_1 = \frac{1}{3^1/_2} [R_1 + Q (\sin \gamma + \mu \cos \gamma)]$$

$\Sigma(W)$ is the sum of the radial components of the tensions on all

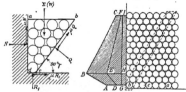

Fig. 2 (Left) Diagram Showing Forces Acting in Rope Wound on Drum

Fig. 3 (Right) Diagram Showing Distribution of Axial Thrust over Surface of Flange

the coils in the wedge; in other words, $\Sigma(W) = 2\pi\Sigma(P - P')$, where $P$ is the initial tension in each coil and $P'$ the tension loss in the same coil due to the pressure from outer layers.

Taking the sum of the horizontal and vertical components of all the forces shown in Fig. 2 and eliminating $R_1$ and $Q$, the following equation for the end thrust against each flange is obtained:

$$N = \frac{(p - \mu^2 - 1) \cos \gamma - \mu p \sin \gamma}{(1 - \mu^2) p \sin \gamma + \mu(2p - \mu^2 - 1) \cos \gamma} 2\pi\Sigma(P - P') . . [3]$$

This equation is to be considered as a refinement of Formula [2], in which no account was taken of $\mu$ or $P'$. It is interesting to note what widely divergent results the two formulæ give, especially when the number of layers is large.

The tension loss $P'$, which to the best of the writer's knowledge has hitherto been disregarded or avoided in all published works dealing with the design of hoisting drums, deserves some discussion at this point. As each successive layer is wound on the drum, the preceding layers are flattened out to a certain extent; this reduces the actual length of each flattened coil and thus relieves it of a part of the initial tension $P$ which was applied to it at the moment when the coil was wound upon the drum. Suppose, for example, that a $^1/_4$-in. rope is being wound upon a 10-in. drum under an initial tension of 1000 lb. First, the bottom layer is completed, each coil having the initial tension of 1000 lb. Then the second layer is wound on, each coil of which has a tension of 1000 lb. But at the same time, the tension in each coil of the first layer drops to 700 lb. When three layers have been completed the outside layer is under the initial tension, while the tension in the second layer has fallen to 500 lb. and that in the first layer has dropped from 700 to 500 lb. With four layers the effect is still more pronounced, the respective tensions being 1000, 400, 250 and 500 lb. In general, the maximum compression occurs near

the middle layer; the inner layers are prevented from much shortening by the unyielding surface of the drum, while the outer layers, being near the surface, are acted upon by comparatively small radial forces.

It is possible to determine the reduction in tension in each layer, and hence the value of $\Sigma(P - P')$, by a simple analysis of the mechanical principles involved. Unfortunately, however, as the number of layers of rope increases the number of unknown quantities increases correspondingly, and the equations which accumulate can only be solved after much laborious computation. But by making the assumption that the radii of all the layers are approximately equal, the formulæ are greatly simplified. Table 1, based on this assumption, gives values of $P'$ for each layer, in terms of the initial tension $P$, for windings of a number of layers.

The quantities $E'$ and $m$, involved in the expression for $h$ in Table 1, deserve a word of explanation, since they represent properties of the rope which determine the extent to which it will yield under lateral pressure. $E'$ may be termed the "lateral modulus of elasticity," and defined as the ratio of lateral pressure per unit length of rope to the decrease in rope diameter measured along the line of action of the pressure. On account of the non-homogeneous character of wire rope, $E'$ bears no direct relation to the true modulus of elasticity $E$ and is best determined by test. The writer obtained the value $E' = 9260$ lb. per sq. in. for new $^1/_4$-in. 6 by 19 hoisting rope with hemp center. Obviously rope with steel center would give higher values of $E'$, while cotton-centered or 8-strand rope would probably give lower values, on account of the greater softness of the core. The coefficient $m$ is a factor of lateral deformation, or ratio of the increase in rope diameter along one axis to the decrease along a perpendicular axis, the lateral pressure causing this deformation being assumed to act along the latter axis. Its value is probably less than 0.3, the value for homogeneous steel, since the pressure which causes deformation is applied at isolated points instead of over the entire surface.

An example was worked out using both formulæ. With the same values in both cases the aggregate tension on all the coils in the wedge, taking the flattening of the coils into account, was found to be 34,460 lb. as compared to an aggregate tension of 63,000 lb. neglecting this flattening effect.

Formula [3] tells nothing about the distribution of the axial thrust over the surface of the flanges. According to the first method for determining it, where friction and coil flattening are disregarded, the distribution over a small sector is as indicated by the line $ABC$, Fig. 3. The effect of friction is probably to reduce the abscissæ of the shaded area a constant amount, to such a line as $DEF$. But the aggregate effect of coil flattening is much more pronounced near the drum body, so that when both friction and coil flattening are considered, the pressure is in all likelihood distributed with a fair degree of uniformity over the sector of flange. This is indicated by the line $GHI$. The experiments described later point toward the same conclusion; and in the absence of more definite information, a uniform distribution of load over a given sector will be assumed in determining the strength of the drum flange.[1]

### EFFECT OF UNGROOVED DRUMS ON AXIAL ROPE PRESSURE

Thus far it has been assumed that the surface of the drum body is grooved. This is undoubtedly the best practice, since it insures the proper spacing of coils, tends to make the winding of all the layers uniform, and reduces wear. In addition it makes the coils of rope in the middle of the drum self-supporting, without the aid of the drum flanges.

With the ungrooved drum, the tendency of the bottom coils to side-slip is counteracted partly by friction, and partly by the fact that a number of coils in the middle part of the drum are "balanced," that is, the pressure from the outer layers which tends to slide one of these coils to the left, is counteracted by an equal and opposite pressure tending to slide the coil to the right. In Fig. 3 the coil $a$ is thus balanced; but all the bottom coils to the left of

---

[1] By "uniform distribution over a sector" is meant a distribution in which (load per unit area) multiplied by (distance from area to drum axis) is constant. This should not be confused with the common meaning of uniform distribution, as used in connection with beams, floors, etc.

receive an excess of pressure from the right, which tends to slide them to the left. This means (remembering that all coils in a layer are in contact with each other) that an additional load in the nature of a shearing force is placed upon the drum flange.

Coil $b$, Fig. 3, presses against the flange with a force $(m-1) \times \pi P \cot \gamma$, neglecting friction, compression of coils, and the pressure of the coil immediately to the right. This force is caused by the coils included in the oblique strip indicated by dotted lines. Denoting by $\Sigma(P')$ the total tension loss in these coils due to compression, and taking account of the friction between coil $b$ and the drum body and flange, the net pressure of this coil against the flange is

$$N_1 = \pi [(m-1)P - \Sigma(P')]\frac{\cot \gamma - \mu}{1 - \mu^2}$$

The coils between $b$ and $a$ will cause a pressure against the flange, of similar form but smaller magnitude, the decrease being due to the partial balance of these coils. Therefore it may be assumed that the total shearing force on a drum flange caused by non-grooving of the body is

$$\Sigma(N_1) = \frac{\pi n}{2}[(m-1)P - \Sigma(P')]\frac{\cot \gamma - \mu}{1 - \mu^2}\ \dots \dots [4]$$

$n$ being the number of unbalanced coils in the first layer.

The problem of deducing a relation between load and stress in the flange is no easy one to solve, owing to the diverse forms of flanges which are embodied in standard designs, and the use of

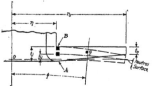

FIG. 4   DIAGRAM FOR STUDY OF STRESSES IN FLANGE OF HOISTING DRUM

stiffening ribs, wide rims, etc. After considerable study the writer selected the mushroom type illustrated in Fig. 4 as being the simplest and at the same time the most widely used. The load-stress-flange-thickness formulæ were then derived by first obtaining a relation between the radial and tangential strains at any point in the flange and the coördinates of that point. This gives a relation between the true radial, tangential, and shearing stresses at the point and the coördinates of the point. By a single integration the total shearing stress on a cylindrical section of the flange containing the point was found; this was set equal to the external load acting on that part of the flange between the cylindrical section and the rim; and after successive integrations the deflection of the point was obtained. By proper substitution the true radial, tangential and shearing stresses at the point could then be found in terms of known quantities, i.e., the flange thickness, inside and outside radii, and external load. By giving this general point certain specific locations, a number of formulæ were deduced.

In determining the constants of integration the following boundary conditions were assumed: The shear at the surface of the flange is zero; the radial stress at the rim of the flange is zero; the slope of the deflected flange is zero at the shoulder (rigid connection between flange and drum body); and the deflection of the flange at the rim is a maximum. In order to simplify the integrations, the following relation was assumed between the thickness of the flange and its radius:

$$t_r = \frac{1}{\sqrt[3]{\alpha + \beta r}}$$

where $t_r$ = thickness of flange at radius $r$, and $\alpha$ and $\beta$ are constants. This gives a shape of flange closely approximating that in Fig. 4. A special case arises when $\beta = 0$; the flange is then of constant thickness.

Formulæ were derived for the radial stress (tension and com-

pression) at shoulder of flange, $p_r$; tangential stress (tension and compression) at rim of flange, $p_t$; and for shearing stress at the shoulder of the flange, $s$. These equations give the stresses caused by the lateral rope pressure. By a similar analysis it is found that the action of the clutch causes similar stresses at the clutch flange, namely, $p'_r$, $p'_t$ and $s'$.

These latter stresses act counter to the stresses $p_r$, $p_t$ and $s$. In other words, when the first layer of rope is being wound on the drum, the clutch flange is stressed in tension in the region $A$, Fig. 4; but as the number of layers increases, this stress becomes less and less until a point is reached at which $p_r$ and $p'_r$ exactly neutralize each other, after which $p_r$ becomes the greater and the flange is under tension in the region $B$. As a rule the net stress due to loads $N$ and $C$ combined will not exceed $p_r$, $p_t$ and $s$, unless the depth of winding is kept down to two or three layers.

To determine the effect of the turning force $S$, consider (1) a small cube in the region $B$, Fig. 4, and (2) a small cube halfway between $A$ and $B$, on the neutral surface. The first cube is subjected to an apparent radial stress $p_1 = (p_r - p'_r)q^2/(q^2 - 1)$, an apparent tangential stress $p_2 = p_1/q$ and a shear $S/(2\pi r_d t_i)$. Combining these in the usual manner to find the principal stresses, and combining the principal stresses to find the true stresses, we have:

$$p' = \frac{1}{2}\left[(p_r - p'_r)^2 + \sqrt{(p_r - p'_r)^2 + \frac{S^2(q + 1)^2}{\pi^2 r_i^2 t_i^2 q^2}}\right]\dots [5]$$

$$S' = \frac{1}{2}\sqrt{(p_r - p'_r)^2 \frac{q^2}{(q + 1)^2} + \frac{S^2}{\pi^2 r_i^2 t_i^2}}\dots \dots [6]$$

$q$ being the factor of the lateral contraction. The second cube is subjected to no normal stresses, but to two shearing stresses at right angles to each other, $S/(2\pi r_d t_i)$ and $0.23qNt_i^2/(r_d t_s^2)$. Hence the total stress is the algebraic sum of these two.

The writer must admit that these formulæ have a formidable appearance and will be likely to discourage the designer who is looking for simple and easily applied rules. The chief difficulty lies with equations for $p_r$, $p_t$, $p'_r$ and $p'_t$; when these have been solved it is a comparatively simple matter to apply Formulæ [5] and [6] as checks to determine whether the addition of a force $S$ will still keep the stresses within safe limits. On the other hand, the equations for $p_r$ and $p_t$ may themselves be much simplified by omitting terms of negligible value and by assuming that the flange is of constant thickness. When the winding space is shallow as compared with the diameter of the drum, the flanges approximate short cantilever beams of depth $t_i$, width $2\pi r_i$ and carrying a uniformly distributed load $N$ and a concentrated load $-C$. The tangential stress vanishes, and the maximum radial stress becomes

$$p_r = 0.477(1 - k)\frac{N - 2C}{t_i}\dots \dots \dots [7]$$

and the maximum shear (due to loads $N$ and $C$) becomes

$$S = \frac{0.23q(N - C)}{r d_i}\dots \dots \dots \dots [8]$$

The additional stress caused by load $S$ may be found as in Formulæ [5] and [6].

### DETERMINATION OF STRESSES IN THE DRUM BODY

The obvious load in this case is one of compression, due to the crushing effect of the rope wound upon the drum under tension. There are, however, additional applied forces which should be considered in a careful analysis of the problem. Besides acting as a cylinder under external pressure, the drum body is also a beam supported at the ends and carrying a moving load, a shaft transmitting a torsional moment and a tensional member serving to tie the flanges together. These forces combine to produce definite stresses which vary in different parts of the drum, so that it is necessary to select certain points at which the stresses are to be determined and to assume certain positions for the moving load $P$. The most suitable points are shown in Fig. 5 by two unit cubes, set "square" with the drum, it being assumed that the rope pull is acting vertically upward. One of these cubes is in the center of the drum body, on the side which is under tension, considering the drum as a beam; the other is alongside one of the points of support of the drum body and in the neutral plane.

Consider the first cube. On each face there is a normal force and in the $x$-$y$ plane there is a shear due to the torque transmitted. $X$ is a tension caused by the rope pressure on the flanges, minus the clutch presure, to which is added the tension caused by the bending effect of the load which is assumed to hang from the middle of the drum. Dividing by the section area for the direct tension, and using the common flexure formula for the bending, we have, approximately,

$$X = \frac{N - C}{2\pi r_i d_d} + \frac{Pl}{4\pi r_i^2 d_d}$$

where $l$ is the length of the drum between bearing centers. $Y$ is

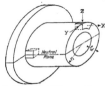

FIG. 5 DIAGRAM FOR STUDY OF STRESSES IN HOISTING-DRUM BODY

a pressure due to the rope tension; on the basis of a uniform distribution of this load over the entire drum body,

$$Y = -\frac{mP - \Sigma(P')}{v l_d},$$

$v$ being the pitch or spacing of coils.

$$Z = \frac{Y l_d}{r_i} = -\frac{mP - \Sigma(P')}{w r_i}$$

The shear due to the twisting moment is a maximum when the rated load $P$ is acting on the outermost layer that can be wound

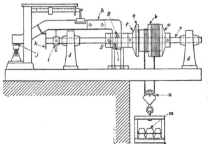

FIG. 6 DIAGRAM OF APPARATUS FOR DETERMINING AXIAL THRUST ON FLANGE OF HOISTING DRUM

on the drum. Denoting the radius to the center of this layer by $r_m$, and substituting in the usual torsion formula, the shear is equal to, approximately,

$$W = \frac{P r_m}{2\pi r_i^2 l_d}$$

The $X$ and $Y$ stresses can be combined with the shear $W$, giving the principal normal stresses and the maximum shearing stress

$$\frac{1}{2}\sqrt{(X - Y)^2 + 4W'^2}\ldots\ldots\ldots\ldots[9]$$

From these are derived the true normal stresses

$$X' = \frac{q - 1}{q} \times \frac{X + Y}{2} + \frac{q + 1}{q} \times \frac{\sqrt{(X - Y)^2 + 4W'^2}}{2} - \frac{Z}{q}\ [10]$$

$$Y' = \frac{q - 1}{q} \times \frac{X + Y}{2} - \frac{q + 1}{q} \times \frac{\sqrt{(X - Y)^2 + 4W'^2}}{2} - \frac{Z}{q}\ [11]$$

$$Z' = Z - \frac{X}{q} - \frac{Y}{q}\ldots\ldots\ldots\ldots[12]$$

To solve for $l_d$ it is only necessary to make the proper substitutions in the equations for $X$, $Y$ and $Z$, then to substitute these in Formulæ [9], [10], [11] and [12], solve, and take the result which is numerically largest as the correct value for the thickness. When the allowable tensile and compressive stresses are different, as in the case of cast iron, it may be assumed that $X'$ represents a tensile, and $Y'$ a compressive, stress. The sign of $Z'$ depends largely upon the relative values of $X$ and $Y$, and in general $Z'$ will be small if based upon a value of $l_d$ obtained by solving Formulæ [10] or [11].

Consider now the second cube, located in the neutral plane of the drum body. The stresses $Y$ and $Z$ are the same as before; but $X$ reduces to

$$X = \frac{N - C}{2\pi r_i d_d}$$

since the bending moment at this point is zero; and $W$ becomes

$$W = \frac{P r_m}{2\pi r_i^2 l_d} + \frac{P}{2\pi r_i d_d}$$

where the first term represents the shear due to torsion, as before, and the second term the maximum shear due to bending. By making the same substitutions as in the previous case, the correct value of $l_d$ may be found for the drum body at either end.

### EXPERIMENTAL INVESTIGATION OF AXIAL THRUST

The apparatus used for this work is shown in Fig. 6, $a$ being the drum, which is fastened rigidly to the shaft $c$ supported in bearings $d$, $d$. An adjustable flange $b$ divides the drum into two parts. The left end of the drum is without a flange, but close to it is placed the separate flange $e$, which has an easy sliding fit on the shaft but is made to turn with it by means of the pin $f$. Any axial load against

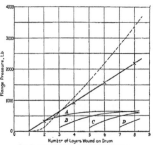

FIG. 7 CURVES SHOWING FLANGE PRESSURE
(Rope pull, 256 lb.; rope, $\frac{1}{4}$ in., 6 by 19; pitch of coils, $\frac{1}{16}$ in.)

this loose flange is transmitted through the knife edge $g$ to the lever $h$, fulcrumed at $i$ and bearing upon platform scales at $j$; thus by balancing the scales a quantitative indication of this side thrust can be obtained for any number of turns of rope upon the drum. The number of turns is shown by the dial $k$, and can be varied by turning the shaft by means of the capstan $l$. The platform and weights $m$ produce whatever rope tension may be desired; and the necessity of a deep shaft in which these weights may rise and fall is obviated by passing the free end of the rope through the sheave $n$ and back on the right-hand end of the drum, so that as it is wound on the left or testing end it is unwound from the right or storage end, and vice versa.

The following procedure was used in making the tests: The

desired size of rope was attached to the drum, a sufficient length being used to fill the left end of the drum to the outside of the flange. One complete layer was wound on the testing end of the drum, the required spacing having been obtained by fiber liners placed between adjacent coils, at close intervals around the circumference of the drum body. It would, of course, have been possible to space the coils by grooving the drum, but this would have limited the apparatus to one size of rope. The desired weight was then placed on the hanging platform, and all was in readiness for a test run. In making a run the shaft was first turned a small amount, varying from 10 to 360 deg., during which time a portion of rope was wound upon the testing end of the drum; the bight of rope hanging below the drum was then given two or three smart hammer blows to cause the new coils to settle well in place, and the scale beam was balanced and reading recorded. This was repeated until the testing end was practically filled with rope. The small turning intervals, 10 deg., 30 deg. and the like, were

place. This means that the arc of cross-over, in which the coils are self-sustaining, is longer for the space than for the closely packed arrangement of coils; hence the flange pressure is less. Furthermore the closely packed coils are in contact with each other at six points, instead of four points, per coil; consequently there is a less pronounced rope flattening in the interior layers, with a correspondingly higher flange pressure as a result.

This latter effect could be taken account of in Formula [3], by selecting the values of $P'/P$ in Table 1 which correspond to higher values of the lateral deformation factor $m$. As the writer did not have the time to obtain experimental values of $m$, he considered it best to omit the theoretical curves from Fig. 8.

At least two conclusions may be drawn from the tests: First, the actual flange pressure increases, roughly, in direct ratio with the number of layers, whereas the theoretical pressure (neglecting friction and rope flattening) is represented by a curve with rising slope; so that at eight layers the actual is only a tenth or less

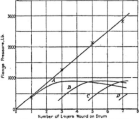

FIG. 8 CURVES SHOWING FLANGE PRESSURE
(Rope pull, 256 lb.; rope, ¼ in., 6 by 19; pitch of coils, ¼ in.)

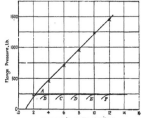

FIG. 9 CURVES SHOWING FLANGE PRESSURE
(Rope pull, 56 lb.; rope, ¼ in., 6 by 7; pitch of coils, ³/₁₆ in.)

used only for the coils near the loose flange, as practically all of the increase in axial thrust occurred when these coils were wound on. For the remaining coils, readings taken for each revolution of the drum were considered sufficient.

A marked periodic rise and fall in the scale readings occurred during each revolution of the drum. This was undoubtedly due to a slight camber in the shaft, with possibly some additional effect caused by untruth in the bearing surfaces $x$ and $y$, Fig. 6, since the phase of this periodic change varied somewhat as the axial thrust increased. On account of its periodic nature, the error could be corrected by simply taking averages of the readings obtained during each revolution of the shaft.

Figs. 7 and 8 show the results of two of these tests in graphic form. In each figure, curve $A$ indicates the flange pressure opposite the first layer, curve $B$ the flange pressure against the third layer, etc. The heavy curve shows the total flange pressure, and the dotted curve gives the total pressure as computed by Formula [3]. A surprising difference in flange pressures is evidenced by a comparison of Fig. 7 with Fig. 8, where the rope pulls are the same, but the coils are wound close instead of being separated. The writer attributes this to a difference in the winding characteristics of the two cases: When the coils are separated to simulate the effect of a grooved drum, they are wound on in each layer in approximately helical form; but when they are closely packed, each coil is a true circle except for an arc of about 30 deg. in which space the advance to the position of the next coil takes

part of the theoretical pressure. The plotted points in Figs. 7 and 8 might be taken to indicate a drooping curve for the actual pressure, but Fig. 9, where four additional layers are shown, points rather to a straight-line relation. Second, the results obtained with Formula [3], and shown in Fig. 7 by a dotted curve, show a fair degree of correspondence with the test curves, and err for the most part on the side of safety. In plotting the dotted curve, $E'$ was taken at 9260 lb. per sq. in., $m$ at 0.3, and $\mu$ at 0.2, all of which are high values. Lower values would be more truthful and would make a still closer agreement of the dotted curve with the test curves.

The question naturally arises: If the tests indicate a straight-line relation between the flange pressure and the number of layers

of rope, what value is there in a complicated expression such as Formula [3], which brings in several indefinite quantities like $\mu$, $p$, $E'$ and $m$? The answer is that the tests show merely a few results from a very wide range of possible conditions; whereas the formula, by taking into account a large number of variables, should be applicable to all conditions, whether the number of layers, rope pull, coefficient of friction, etc., be large or small. The field is still open for further experimentation, using larger rope sizes and greater rope pulls, and the quantities $E'$ and $m$ should be accurately determined for different sizes and lays of rope. The writer hopes that this work can be undertaken in the near future, and that the results will stand as an additional check upon the theoretically derived relation between rope pull and flange pressure embodied in Formula [3].

TABLE 1    VALUES OF $P'$ FOR VARIOUS LAYERS IN TERMS OF THE INITIAL TENSION $P$

| | Number of Layers | | | | | | | | | | | | | | | | | | |
|---|---|---|---|---|---|---|---|---|---|---|---|---|---|---|---|---|---|---|---|
| | | | 2 | | | | 3 | | | | 4 | | | | 5 | | | | 6 |
| $h$ (see note) | 0.6 | 0.7 | 0.8 | 0.9 | 1.0 | 0.6 | 0.7 | 0.8 | 0.9 | 1.0 | 0.6 | 0.7 | 0.8 | 0.9 | 1.0 | 0.6 | 0.7 | 0.8 | 0.9 | 1.0 |
| $P_1'/P$ | 0.00 | 0.00 | 0.00 | 0.00 | 0.00 | 0.30 | 0.35 | 0.40 | 0.45 | 0.50 | 0.13 | 0.13 | 0.13 | 0.13 | 0.13 | 0.05 | 0.04 | 0.04 | 0.03 | 0.03 | 0.02 | 0.02 | 0.01 | 0.01 | 0.01 | 0.00 | 0.00 | 0.00 | 0.00 | 0.00 |
| $P_2'/P$ | | | | | | | | | | | 0.56 | 0.62 | 0.67 | 0.71 | 0.75 | 0.21 | 0.20 | 0.19 | 0.18 | 0.17 | 0.08 | 0.07 | 0.06 | 0.05 | 0.04 | 0.03 | 0.02 | 0.02 | 0.01 | 0.01 |
| $P_3'/P$ | | | | | | | | | | | | | | | | 0.62 | 0.67 | 0.71 | 0.75 | 0.78 | 0.23 | 0.22 | 0.20 | 0.19 | 0.17 | 0.08 | 0.07 | 0.05 | 0.03 | 0.04 |
| $P_4'/P$ | | | | | | | | | | | | | | | | | | | | | 0.64 | 0.68 | 0.72 | 0.75 | 0.78 | 0.23 | 0.22 | 0.20 | 0.19 | 0.17 |
| $P_5'/P$ | | | | | | | | | | | | | | | | | | | | | | | | | | 0.64 | 0.68 | 0.72 | 0.75 | 0.78 |

NOTE: The value of $h$ is given by the following formula., where $a$ = area of wires in rope; $E$ = modulus of elasticity of rope; $E'$ = transverse modulus of elasticity of rope, $m$ = coefficient of lateral deformation of rope, $r$ = average radius of coils under consideration:

$$h = \frac{aE}{E'r^2}\left(\frac{\gamma}{43^\circ} - \frac{90^\circ - \gamma}{43^\circ}m\right)$$

# Steam vs. Electric Locomotives

### Joint Meeting in New York for the Discussion of the Relative Advantages of Modern Steam and Electric Locomotives Has a Record Attendance—Papers and Discussion by Leading Authorities in the Railroad Field

A CONTROVERSIAL subject, if timely, always draws a crowd. On the evening of October 22, in the Engineering Societies Building, New York, the controversial subject was provided and the crowd was there. The meeting was under the joint auspices of the Professional Section on Railroads of the A.S.M.E., the Metropolitan Section of the A.S.M.E. and the New York Section of the A.I.E.E.; and the subject was Steam vs. Electric Locomotives. The attendance was a record-breaking one—every seat in the large auditorium was taken, 200 were standing and 400 or 500 were turned away. The curtain raiser was by Frank J. Sprague, the father of electric railroading, who briefly reviewed certain features connected with the development of electric railroads and the electric locomotive, after which were given the three papers of the evening, followed by the discussion which lasted until nearly midnight. The brief (in the legal sense) for the steam locomotive was presented by John E. Muhlfeld, vice-president, Railway and Industrial Engineers, Inc., New York, and the case for the electric locomotive was presented by A. H. Armstrong and F. H. Shepard, representing respectively the General Electric and Westinghouse companies. The discussion was summed up by George Gibbs, consulting engineer, Pennsylvania Railroad. Space permits the publication of abstracts only of the three papers and a running account of the discussion.

#### REMARKS BY FRANK J. SPRAGUE

Mr. Sprague, in tracing the history of railway electrification in America, read a news item from a copy of the New York Sun published in August 1887, recording the trial of an electric car on Fourth Avenue, New York, which it said "created an amount of surprise and consternation from 32d Street to 117th Street that was something like that caused by the first steamboat on the Hudson." Soon after came the famous Richmond, Va., road, fairly called the forerunner of the modern trolley. Three years later there were in operation, here or abroad, something like 350 electric trolley roads and a quarter of the street-tramway mileage of the United States had been converted to the new power.

The speaker said that so successful had been the early application of electricity to this particular problem that the wildest prophecies were made by imaginative enthusiasts of· the coming debacle of the steam railway, and he found it necessary, as the incoming president of the American Institute of Electrical Engineers, in June 1892, to sound a warning in his inaugural address on the subject of The Coming Development of Electric Railways.

At that time the problem of trunk transportation was still awaiting solution and had to be preceded by another and more pressing development, that of urban rapid transit. In New York the "L" was supreme and the growth of the city was being throttled. The handicap of steam equipment had been only partially removed, and only when it became possible absolutely to disregard it and, instead, to adapt equipment and methods of operation to the possibilities of electric operation, was the success of rapid transit assured. Then were the city's shackles cast off and the modern growth of the great metropolis made possible. It was

even then, however, a long time before the multiple-unit system, the fundamental of electric rapid transit the world over, received recognition.

Preliminary to the larger developments in the broad field of trunk-line operation was the construction in the early 90s by Drs. Duncan and Hutchinson, and Mr. Sprague, with the help of the Baldwin and Westinghouse companies, of a 1000-hp. electric locomotive, which, however, never saw service. Handicapped as was the transmission of power on a large scale and over long distances, until the development of the a. c. transformer and the rotary converter, and by the persistent traditions of steam practice, progress was necessarily impeded, and it was not until about 1902 that because of a grave accident the New York Central Railroad undertook its epoch-making development of a great terminal and the beginning of trunk-line operation.

Looking to the future, Mr. Sprague said: "I wish to record again my unchanging belief in the coming supremacy of electric power for transportation; and, propagandist as I have been these many years, my faith is based not upon any overwhelming claims of superior fuel economy when both are dependent on coal supply, nor upon any material saving in operating expense sufficient to pay the charges incident to the increased capital cost when considered on existing traffic density and methods of operation, but rather upon the broad ground of the overwhelmingly vital demand for *increased capacity*, a demand which ultimately can be met only by the electric system because of characteristics individual to it.

"Of course, the general electrification of trunk lines is not a matter of immediate possibility of accomplishment. It will only come as it is coming, progressively with the accelerating power of example. It will be governed very largely by financial conditions, however justified. Under sufficiently enlightened public policy, railroads might be able to assume the large added burdens incident to electrification, if given sufficient encouragement. But the test of wisdom of railway officials, financiers, and national and state governments is the ability to anticipate and plan against the inevitable before the slowing down and paralysis of traffic, as, for example, as is illustrated by the rapid-transit situation in this city, where the congestion affects the daily life and wellbeing of the whole community.

"I am not unmindful of the many and great difficulties in the way of accomplishment, nor would I have electrical engineers lose themselves in visionary contemplation of unobtainable commercial realities; but I may perhaps repeat the substance of the closing words of that address of 28 years ago, to which I have already referred:

"'Such is the work before you, a work well meriting your best effort. Yet it is well to temper your effort with prudence. Limit your attempts to the solution of those problems which will promise practical benefit. Do not chase rainbows; they are beautiful and poetic, but they have small place in the world's economies.'

"This is said in no spirit of discouragement, for I yield to no man in my confidence in the future of electricity."

## Steam vs. Electric Locomotives—the Steam Side

#### By JOHN E. MUHLFELD, NEW YORK, N. Y.

IN consideration of the existing traffic rates and regulations as established by the Interstate Commerce Commission, and the wages and working conditions as recommended by the Railroad Labor Board, it is assumed that the railroads as a whole will average net operating earnings equal to 6 per cent on their valuation as fixed from time to time by the Interstate Commerce Commission. Some railroads may earn more and some will earn less,

but in every case the minimum fixed charge and the maximum operating and maintenance economy will be required if the stock owners are to receive even a reasonable return on their investments.

In the protection and control of railroad net earnings one of the most involved factors is the kind of motive power to be used, and, unfortunately, in appraising the relative values of steam and electric railway power, comparisons have been made between the opera-

tions of new, up-to-date electrical and of obsolete steam installations; or on the basis of other conditions which are not fairly comparable.

For example, a report was made several years ago on the advisability of electrifying about 275 miles, or a division, of one of
the more prominent western lines, and an erroneous comparison
was made, first, between existing antiquated and uneconomical
steam and an up-to-date electric operation; and second, by omitting
a statement of the investment required to bring the steam operation up to date. When all involved factors were properly adjusted the net capital expenditure of $4,000,000 required for elec-
trification (compared with $1,000,000 needed for modernizing the
steam equipment), and the estimated annual operating saving
from electrification of approximately $750,000, were wiped out
and replaced by a saving of $250,000 from a continuation of the
improved steam operation.

When we discuss the further electrification of the whole or any
part of the 260,000 miles of steam-operated railroad systems in
the United States, which is now making use of about 65,000 steam
and 375 electric locomotives for its passenger, freight and terminal
service, the most essential requisite is a correct and complete
statement of facts, comparing the most up-to-date steam operation
with similar electric operation, after which immediately come the
important factors of the necessary financing and legislation.

*Financing.* Few if any existing steam roads can justify the additional capital investment required per mile of road for electrification, except for short distances under very special conditions
such as prevailed on the Norfolk and Western, where the ventilation and 1.5 per cent grade line features of a five-eighths-mile
single-track tunnel restricted the train movements to a 6-m.p.h.
basis on a congested traffic section of the main line, and even then
only providing the fixed charges and operating expenses are not
too excessive.

The immediate requirements of new money for the more urgent
steam equipment and facilities needed to provide adequate, safe
and expeditious rather than luxurious service in the regeneration
of the railroads, is the obvious reason for the continued utilization
of the overall more economical steam operation; and only after
the possibilities in this direction have been realized can any serious
financial consideration be given to the proposed radical change to
superelectrification.

*Adaptability to Existing Trackage and Facilities.* First and
foremost in the advantages of a continuation of the existing improved steam locomotive for all purposes for which it is permissible,
is its flexibility and adaptability to existing railroad trackage and
terminal and operating facilities, and the relatively low first cost
at which it can be purchased per unit of power developed for the
movement of traffic. Being a self-contained mobile power plant,
it is possible quickly to transfer needed or surplus power from one
part of the line to another and to concentrate it when and where
necessary; whereas, with the electric locomotive this is impossible
unless electrification extends over the entire property or the sources
of power supply have almost prohibitive peak-load capacity.
Furthermore, the various systems of electrification make the interchanging of electric locomotives impracticable without much non-
productive first cost, complication, and maintenance and operating
expense.

*Effectiveness in Increasing Track Capacity.* Without doubt
electrification increases the capacity of a terminal, as evidenced by
the intensified traffic movements at Grand Central Station in New
York, and at Broad Street Station in Philadelphia; but an analysis
of the situation on the New York Central shows that this is not due
to decreasing locomotive movements through the use of multiple
units as is usually made. For example, through-line passenger
trains are handled in and out of Harmon with single steam the same
as with single electric locomotives; and as regards commutation
service, this is largely a motor-car proposition, as indicated by the
fact that the New York Central has 241 motor cars (and no trailers)
as compared with 73 electric locomotives.

Special line conditions, as on the Norfolk and Western, may make
electrification advisable for short distances; but neither the results on that road nor at the New York terminals justify the assumption frequently made that electrification would have obviated
the difficulties experienced in steam-locomotive haulage during the

unprecedentedly cold weather and severe traffic conditions of the
winters of 1917-1918. If otherwise, why did the New Haven not
operate at 100 per cent of its capacity over its electrified zone at
that time? If short of locomotives or motor cars the New York
Central had plenty of surplus that was not in use and which could
not be utilized outside of its electric zone where it was badly needed.
The probable answer is lack of interchangeability, which is still
one of the most discouraging operating factors involved in any
electrification scheme, as brought out in the last report of the A.R.A.
Committee on Design, Maintenance and Operation of Electric
Rolling Stock, wherein the wide variation of current-generating,
transmitting, distributing and contact systems, voltages, types of
locomotives and plenty of general ideas relating to the features sets forth
the present undeveloped state of the art.

Furthermore, in the handling of heavy-tonnage trains by the
unlimited combining of electric-locomotive units, the factors of
peak load, transmission lines, and power-plant capacity must all
be considered, with the probability that permissible modern steam-
locomotive train units can be more economically handled over
dense traffic lines than the electric multiple-unit super-trains.
Although under the multiple-unit system of locomotive and train
operation it is theoretically possible to provide unlimited sustained
hauling capacity at the head of the train, the tonnage to be handled
without rear-end or intermediate helpers is limited by the ability
of the draft rigging on the cars to withstand the pull and shock.
This limitation can be readily met and exceeded in steam-locomotive design and operation, as may be noted from the following
comparison of the St. Paul electric freight and the Virginian steam
freight articulated types of locomotives:

TABLE 1  COMPARISON OF MODERN TYPES OF ELECTRIC AND
STEAM LOCOMOTIVES

|  | St. Paul Electric Articulated | Virginian Steam Articulated |
|---|---|---|
| Tractive power, in simple gear, maximum.......... | 132,500 lb. | 176,600 lb. |
| Tractive power, in compound gear, maximum...... | | 147,200 lb. |
| Tractive power, at 15 miles per hour............. | 71,000 lb. | 108,000 lb. |
| Wheel arrangement (excluding tender)........... | 4-8-8-4 | 2-10-10-2 |
| Length over all (including tender)............... | 112 ft. | 107 ft. |
| Total wheelbase (including tender).............. | 102 ft. 8 in. | 97 ft. 0 in. |
| Driving wheelbase............................. | 75 ft. 0 in. | 64 ft. 3 in. |
| Rigid wheelbase............................... | 10 ft. 6 in. | 10 ft. 10 in. |
| Total weight on driving wheels................. | 448,000 lb. | 617,000 lb. |
| Total weight on truck wheels.................. | 116,000 lb. | 67,000 lb. |
| Total weight of tender (with 1/2 fuel and water | | |
| capacity)................................... | | 148,000 lb. |
| Total weight of locomotive.................... | 564,000 lb. | 832,000 lb. |
| Truck wheels—total No....................... | 8 | 4 |
| Driving wheels—total No..................... | 16 | 20 |
| Driving wheels—diameter..................... | 52 in. | 56 in. |
| Driving wheels—adhesive weight to total....... | 79.4 per cent | 90 per cent |
| Driving wheels—adhesive weight per axle, avg.... | 56,000 lb. | 61,700 lb. |
| Driving wheels—unsprung weight per axle, avg.... | 16,250 lb. | 12,000 lb. |
| Factor of adhesion—maximum tractive effort...... | 3.38 | 3.49 |
| Factor of adhesion—tractive effort at 15 m.p.h.... | 6.31 | 5.70 |
| Source of power.............................. | Electricity from outside hydro-electric plant | Superheated steam from self-contained boiler plant |

*Train Speeds.* The averge freight car is in main-line movement
only about 10 per cent of its life, or 2 hr. 24 min. out of each
24 hr. The balance of its time can be distributed as follows:
55 per cent in the hands of the railroads on account of interchanges,
yard and loading and unloading track movements, surplus cars,
repairing tracks, and road delays; and 35 per cent in the hands of
the shipper and consignee, due to loading and unloading recon-
signment and Sundays and holidays. Therefore, increasing train
speeds beyond established economic limits at the sacrifice of tonnage, and with an increase in fuel, track and equipment upkeep
and danger of operation, is not the solution of the freight-traffic
problem. For example, an increase of 50 per cent over and above
the established economic freight-train speeds would be only 72
min. of the daily life of each freight car; whereas, capital expenditures applied to the reduction of those delays which now involve,
over 21½ hr. per day, or about 90 per cent of the life of the car,
would give much more effective and economical results.

As the electric locomotive is a constant-speed proposition,
whether going up or down grade, and is unable to utilize its rated
capacity and effectiveness through the same range of speed and
tractive-power variations as the more flexible steam locomotive,
the latter can, therefore, be more efficiently operated over the continually changing up and down grades, levels, curves and tangents
traversed by the average freight train in this country.

With respect to passenger-train service, where speed is more of a factor, the steam locomotive performs equally satisfactorily. For example, on the main line of the Baltimore and Ohio, for a distance of 17 miles between Piedmont, West Virginia, and Alta-mont, Maryland, the average gradient is .2.2 per cent. A single Pacific-type locomotive with a tractive power of 43,400 lb. will haul up this grade in 50 min., without helper, or at an average speed of 20 m.p.h., a passenger train consisting of 9 cars weighing 620 tons without locomotive and 830 tons with locomotive. The same train will make the trip down grade in 35 min., or at an aver-age speed of 28¹/₂ m.p.h. The total weight of engine and tender of these Pacifics is 210 tons, which may be compared with a total weight of 265 tons for the St. Paul electric locomotives used to handle similar passenger trains up 17 miles of 2.2 per cent grade from the Columbia River west at an average speed of 25 m.p.h.

*Fuel Consumption.* [Under this heading the author refers to figures which have been quoted in support of electrification on the basis of a coal rate of 2¹/₂ lb. per kw-hr. at the central station in combination with a 40-kw-hr. rate at the point of delivery of the power to the railroad system, resulting in a train movement of 1000 average gross ton-miles for 100 lb. of coal of about 12,000 B.t.u. In commenting on this, he says that in arriving at these data apparently numerous factors were overlooked or disregarded which should be included in any estimate of the power requirements and coal rates for an electrified system. Continuing, he says:]

However, accepting the assertion that the proposed electrifica-tion will produce 1000 gross ton-miles for an average of 40 kw-hr., or 100 lb. of 12,000-B.t.u. coal, as stated and generally approved by electrical engineers, what can the modern steam locomotive do to justify its existence?

Dynamometer-car tests may be cited, made by a Joint Committee of representatives of the New York Central and Pennsylvania railroads and the American Locomotive Company during August 1910, of the first Mallet-type of locomotive put into use on the Pennsylvania Division of the New York Central and operated over the 65 miles of average 0.5 per cent grade line between Avis and Wellsboro Junction. This locomotive was built ten years ago and by no means represents the best practice of the present day when superheat has been increased and a more efficient all-around machine is produced. At that time the average of six runs gave a thermal efficiency of 6.01 for the locomotive, and the following test made on August 27, 1920, is representative:

| | |
|---|---|
| Miles run, about | 65 |
| Cars in train, No | 65 |
| Cars in train, tonnage | 3,734 |
| Running time | 4 hr. 35 min. |
| Time on road | 6 hr. 51¹/₂ min. |
| Average speed, m.p.h | 12.9 |
| Thermal efficiency of locomotive | 6.25 per cent |
| Dry coal per drawbar horsepower-hour | 2.90 lb. |
| B.t.u. in dry coal as fired | 14,053 |
| Cut-off | |
| Drawbar horsepower | 1,270.4 |
| Drawbar pull | 34,071 lb. |
| Steam pressure in branch pipe | 203.3 lb. |
| Superheat in branch pipe | 143.7 lb. |
| Machine efficiency of locomotive | 89.21 per cent |
| Boiler efficiency | 69.07 per cent |

Furthermore, the 2-6-6-2 type Mallet steam locomotives with tractive power of about 88,500 lb. in compound, and 106,200 lb. in simple gear, operated over a distance of about 155 miles of average 0.3 per cent grade line with a ruling grade of 0.6 per cent 13 miles long, between Birmingham, Ala., and Columbus, Ga. During this period steam was used about 50 per cent of the time, the locomotive drifting the balance of the time, and a figure of 1000 gross ton-miles was realized from 100 lb. of coal of approximately 13,000 B.t.u. which compares quite favorably with the foregoing hypo-thetical figures given for electric operation.

The results of some dynamometer-car tests made during 1918 may also be of interest. At that time the steam locomotives tested were of the ordinary superheated Mikado freight type of the following general description:

| | |
|---|---|
| Weight on driving wheels | 110 tons |
| Weight on truck wheels | 32 tons |
| Cylinders, simple | 25 in. by 32 in. |
| Driving wheels, diameter | 56 in. |
| Steam pressure | 200 lb. |
| Tractive power | 50,000 lb. |

One locomotive was fitted for hand firing and burning coal on grates, while another was equipped with the "Lopulco" system for burning powdered coal in suspension, and the tests were made in tonnage freight service handling from 2400 to 2600 tons east-bound and from 1850 to 2250 tons west-bound on the Santa Fe main line between Ft. Madison, Ia., and Marceline, Mo. (the pro-file consisting of 0.8 per cent ruling grades) a distance of 112.7 miles. The coal averaged from 1 to 8 per cent moisture, 33 to 38 per cent volatile, 51 to 41 per cent of fixed carbon, 15 to 12 per cent ash, 4 to 3¹/₂ per cent sulphur, and from 12,055 to 11,050 B.t.u. as fired. The comparative average results are given in Table 2.

TABLE 2 TESTS ON MIKADO FREIGHT-TYPE SUPERHEATER LOCO-MOTIVES

| | Powdered-Coal Locomotive | Hand-Fired Locomotive |
|---|---|---|
| Total trips run (112.7 miles each) | 14 | 10 |
| Total miles run | 1578 | 1127 |
| Average running time, hours | 3.06 | 3.25 |
| Average dead time, hours | 1.25 | 1.01 |
| Average total time, hours | 6.31 | 6.26 |
| Average speed. m.p.h | 22.3 | 21.6 |
| Average trailing tonnage per train | 2278 | 2283 |
| Average gross 1000 ton-miles | 256.5 | 255.4 |
| Average coal per gross 1000 ton-miles | 82.4 | 114.8 |
| Average superheat, deg. fahr | 223 | 173 |
| Average coal per boiler and superheater, coal per hp-hr | 3.74 | 4.99 |
| Average B.t.u. per pound of coal as fired, lb | 12,025 | 11,160 |

As the coal supplied to the grates of the hand-fired locomotive was considerably lower in heat value than that specified in the elec-trification project, and as the tests were run during March and April, it can be assumed from the foregoing that the average yearly performance will approximate 100 lb. of 12,000-B.t.u. coal per 1000 gross ton-miles, or equivalent to what we are promised for the ex-penditure of billions of dollars of new capital and the loss of billions of dollars of investment in existing plant and equipment to inaugurate electrification.

*Efficiency of Locomotive Operation.* It is unquestionably true that when operating under ideal fixed-load conditions, the central power station, either hydroelectric or steam, can produce a horse-power with less initial energy input than is possible on a steam loco-motive. It is also true that the stand-by losses on existing steam locomotives are, in ordinary practice, a serious proportion of the total fuel consumption; but it is likewise a fact that the majority of these can be substantially reduced, if not entirely overcome, by modernizing the present equipment and improving maintenance and operation.

The number of factors entering into an analysis of the net ther-mal efficiency of the electric locomotive, in terms of drawbar pull, are so many as to make it impossible with the lack of dynamometer-car and laboratory test data to arrive at a figure which is not based on a number of assumptions; but as a matter of interest, assuming that *all of the factors are affected equally* in the electric locomotive, the net thermal efficiency at the drawbar, when taking into con-sideration the boiler, engine, generator, step-up transformer, a.c. transmission, step-down transformer, a.c.-d.c. converter, d.c. transmission, motors, and machine efficiencies may, as representa-tive of average existing practice, be illustrated as in Table 3.

TABLE 3 ANALYSIS OF NET THERMAL EFFICIENCY OF ELECTRIC LOCOMOTIVE

| Equipment: | | Load Rating, Per Cent | | |
|---|---|---|---|---|
| | | 100 | 75 | 50 |
| Boiler | Factor | | | |
| | Efficiency | 76.7 | 76 | 72 |
| Engine | Factor | 18.25 | 18.29 | 19.17 |
| | Efficiency | 14 | 13.9 | 13.8 |
| Generator | Factor | 90 | 89.5 | 86 |
| | Efficiency | 12.6 | 12.44 | 11.88 |
| Transformer, Step-Up | Factor | 98 | 96 | 90 |
| | Efficiency | 12.34 | 11.93 | 10.67 |
| Transmission, A.C | Factor | 90 | 95 | 97 |
| | Efficiency | 11.10 | 11.32 | 10.34 |
| Transformer, Step-Down | Factor | 98 | 96 | 90 |
| | Efficiency | 10.87 | 10.85 | 9.30 |
| Converter, A.C. to D.C | Factor | 80 | 75 | 63 |
| | Efficiency | 8.69 | 8.13 | 3.85 |
| Distribution, D.C | Factor | 90 | 95 | 97 |
| | Efficiency | 7.82 | 7.71 | 5.66 |
| Motors, D.C | Factor | 91.5 | 90.8 | 89.5 |
| | Efficiency | 7.15 | 7.00 | 5.05 |
| Machine Efficiency | Factor | 81 | 85 | 90 |
| | Efficiency | 5.79 | 5.95 | 4.54 |

Likewise the net thermal efficiency of existing representative steam locomotives, in terms of drawbar pull, may be illustrated as in Table 4.

TABLE 4 ANALYSIS OF NET THERMAL EFFICIENCY OF STEAM LOCOMOTIVE

| Equipment: | Steam Used: | | Load Rating, Per Cent | | |
|---|---|---|---|---|---|
| | | | 100 | 75 | 50 |
| Boiler......... | Superheated | Factor | | | |
| | | Efficiency | 43.7 | 54.9 | 65.9 |
| | Saturated | Factor | | | |
| | | Efficiency | 45.0 | 57.4 | 70.0 |
| Cylinders....... | Superheated | Factor | 11.9 | 11.0 | 10.5 |
| | | Efficiency | 5.08 | 6.04 | 6.92 |
| | Saturated | Factor | 7.8 | 8.4 | 7.8 |
| | | Efficiency | 3.51 | 4.82 | 5.46 |
| Machine........ | Superheated | Factor | 75 | 80 | 85 |
| | | Efficiency | 3.85 | 4.83 | 5.86 |
| | Saturated | Factor | 77 | 80 | 82 |
| | | Efficiency | 2.70 | 3.86 | 4.47 |

Comparing the electric and steam locomotive figures as illustrated, the relative percentage of power delivered at the track rails to 100 per cent B.t.u. in the coal would be as given in Table 5.

TABLE 5 COMPARISON OF THERMAL EFFICIENCY OF ELECTRIC AND STEAM LOCOMOTIVES

| | Net Thermal Efficiency at Load Ratings of | | |
|---|---|---|---|
| | 100 | 75 | 50 |
| Kind of Locomotive: | Per cent | Per cent | Per cent |
| Electric.......................... | 5.79 | 5.95 | 5.54 |
| Steam, Superheated............. | 3.85 | 4.83 | 5.88 |
| Steam, Saturated................ | 2.70 | 3.86 | 4.47 |

As 100 per cent load rating conditions would, in practice, occur only momentarily and as the majority of the drawbar load represents from 30 to 60 per cent of the locomotive maximum drawbar capacity, comparison should be properly made only of the net thermal efficiencies at 50 per cent load ratings.

[As a check on the foregoing figures upon steam operation, the author presents a table (not here published) in which is given the results of tests[a] on representative types of steam passenger and freight locomotives. In respect to these he says:]

It will be noted that at speeds of from 15 to 75 miles per hour the thermal efficiency of existing superheated-steam locomotives actually ranges from 5.3 to 8.1 per cent as compared with the calculated figures of from 4.83 to 5.88 per cent for 75 and 50 per cent load ratings, respectively. Adding to this an increase of from 15 to 50 per cent in net thermal efficiency that may be produced from the various developments now under way and the steam locomotive of the future will be quite a respectable assembly of engineering efficiency.

*Cost of Enginemen.* Under existing conditions each electric locomotive must carry a motorman, and a second man comparable to the fireman of a steam locomotive, although not functioning as such. The wage of the latter is an added expense without economic return and must be charged to the cost of firing the central power-station boilers or otherwise distributed.

*Cost of Maintenance.* In determining the maintenance cost of the electric locomotive, a true comparison can only be made by including all elements corresponding to those found in the self-contained steam locomotive, which goes back to the upkeep of all facilities having to do with the utilization of the fuel or water power, including the central power-station buildings; boilers; engines; conversion, transmission, distributing and contact-line systems; substations; track rail bonding and insulation; electric disturbance cutouts or neutralizers; extra expense in upkeep of the electric zone trackage; and like auxiliaries, and finally the electric locomotive itself.

With particular reference to the maintenance-cost figures that have been given as applying to the New York Central, Michigan Central, Pennsylvania and St. Paul railroads for the years 1913 to 1918, inclusive, and which range from $3.78 to $10.87 per 100 locomotive-miles run, these no doubt apply to the electric locomotive units only, and if so would appear exceptionally high even for relatively new-built steam locomotives. Until a true reflection of the upkeep cost per electric locomotive, or per 1000 gross ton-miles hauled, can be given by including all the factors and elements of age and mechanism that are embodied in the steam locomotive, comparisons will be worthless.

*Peak-Load Conditions in Relation to Traffic Requirements.* In order to meet the ideal conditions for electrification, the traffic should be uniformly scattered over the 24-hr. period, whereas in the majority of cases train movement is based on traveling and shipping conditions and cannot be advanced or delayed in order to

eliminate peak-load conditions. That this cannot be done in order to maintain a straight-line power demand can be illustrated by citing the condition that exists in any large industrial center where freight accumulates and is switched during the day period and the out and inbound train movements concentrate in fleets, principally in the evening and morning, respectively, and cause peak-load requirements at those times.

*Ease of Starting Trains.* Due to the uniform torque as developed by the electric locomotive, its adherents have laid great stress on its ability to start a heavier train than a steam locomotive of relatively the same tractive power and factor of adhesion. In steam-railroad service the locomotive is seldom required to start "the train," but what it does is to start each car in the train, successively, which nullifies this theoretical advantage of the electric locomotive. In fact, with steam locomotives of the Mallet and other types having compound cylinders equipped with properly designed simpling devices, the starting power is increased about 20 per cent as compared with electric locomotives of equivalent road rating.

*Rate of Acceleration.* In order that the desired running speeds can be reached in the minimum time after the starting of trains, the ability of a locomotive to rapidly accelerate its load is of considerable importance, and in this respect the electric power has had the advantage. The steam-locomotive engineer, however, has not lost sight of this fact and improvements already made in boiler and cylinder horsepower ratios, as well as developments now undergoing for utilizing existing non-productive adhesive weight and increasing the coefficient of friction between the propelling wheels and the track rails will enable the steam locomotive to duplicate the performance of its electric competitor in this regard.

*Train Braking.* Since the development of regenerative braking with the electric locomotive, great emphasis has been laid on the increased security of operation which this permits over heavy-grade lines. Considerable attention has also been directed to the saving brought about through the elimination of the ordinary air braking on such down grades.

The Baltimore and Ohio has successfully and safely handled with steam locomotives its heavy tonnage and dense traffic on the Cumberland and Connellsville Divisions for many years, and this tonnage descends a grade averaging approximately 2.2 per cent for 17 miles, at an average speed of from 15 to 20 miles per hour for freight and from 25 to 30 miles per hour for passenger trains, without slow-downs or stops. This performance is comparable with that on the worst grade conditions encountered in the St. Paul electrified zone.

While the regenerative system of braking can no doubt be developed to the point where it can be safely used, the recent serious accident on the St. Paul wherein a heavy-tonnage freight train made up with an electric locomotive at the head end and a steam Mallet helper locomotive at the rear end, broke away from the latter and derailed the entire train of about 65 cars on a 20-mile grade of 2.2 per cent, due to the failure of the regenerative brake control, makes problematical just what economy will result. When the power so generated cannot be directly used by another pulling locomotive on the line, it must be otherwise absorbed, and it remains for the electrical engineers to prove just how much of it is lost in conversion or by absorption and the resulting net gain as compared with the investment, fixed charge and upkeep and operating cost for the equipment involved.

*Effect of Weather Conditions.* Even though the full steaming capacity, horsepower, and drawbar pull of a modern steam locomotive can be developed during cold-weather conditions, there are the factors of radiation and freezing to be reckoned with, which gives the electric locomotive the advantage in winter, particularly as its effectiveness is greater on account of the lesser tendency for the motors to overheat. This winter advantage, however, is largely overbalanced during the summer when the main motors heat, especially under overloads, and require cooling at terminals; or by overheating lead to insulation breakdowns, burn-outs, or other troubles.

*Road Delays and Tie-Up.* While the electric locomotive has the advantage of not being required to take on fuel and water except for the operation of steam-heating equipment for passenger trains, with the increased capacity of the modern steam-locomotive tenders, and the lower water and fuel rates per drawbar horsepower developed, the delays due to taking on these supplies have been greatly reduced and need not be serious.

Barring collisions, wrecks and like accidents not due to the system of motive power in use, steam operation is not susceptible to complete tie-ups as is the case with electrification, where short-circuits or failures occur due to rains, floods, storms and like causes, and as the result of motor, wiring and insulation heating, deterioration and breakdowns, as the individual mobility of each piece of motive power without regard to any outside source of power enables quick relief.

*Terminal Delays.* The examination of reports of a dense heavy-freight-traffic railroad in the Eastern District shows the time of its steam locomotive for a recent two months' period distributed as follows:

| | | |
|---|---|---|
| 1 | In road service............................. | 50 per cent of total time |
| 2 | At terminals waiting trains and otherwise in hands of transportation department.... } | 26.4 per cent of total time |
| 3 | At terminals in hands of mechanical dept... | 23.6 per cent of total time |

There is no doubt but that the electric has an advantage over the steam locomotive as regards time required for periodical boiler work, fire cleaning and rebuilding, fueling and watering except where fuel oil is used, but where terminal delays occur due to waiting for trains, the time required for such work does not become an expensive determining factor in the daily average miles to be obtained per locomotive.

# Steam vs. Electric Locomotives—the Electric Side

By F. H. SHEPARD,[1] PITTSBURGH, PA.

THE limit to physical expansion of railroad lines and of terminals has been just about reached in many cases, both on account of the prohibitive cost and the inefficiency of terminals of unworkable size. A large measure of relief can still be secured by line and terminal revisions and improvements, but when the inevitable increase in the demand for traffic movement of the future is considered, these improvements savor more or less of expedients to secure relief which can only be temporary and very limited in degree.

The great need of the country (and this is coming to be more and more generally understood) is the free and expeditious movement of our traffic. The yearly average of 22 miles per day for a freight car for the whole country, with monthly averages of as low as five miles on some of our most congested railroads for a single month, emphasizes the fact that this is a problem that somehow, someway, we must solve. The solution lies to a large extent in railroad electrification.

With the present standards of train make-up, classification and terminal handling, electrification will double the capacity of any railroad. With the better equipment we can expect in the future, together with the evolution of improved methods of operation contingent on electric power, this capacity should be doubled again, thus securing four times the present capacity. This should certainly be accepted as a vision of the future, and why not our aim? Unless some broad and consistent program is embraced, the situation, which is serious indeed today, may well be calamitous tomorrow.

The electric locomotive has generally, thus far, been a mere substitute for the steam locomotive, although in some cases, due to the greater power of the electric locomotives, there has been modification of the handling of traffic.

Two conspicuous cases of this have been the Norfolk & Western and the Chicago, Milwaukee & St. Paul. In the case of the Norfolk & Western, two electrics handle the same train as was formerly handled by three Mallet engines, but at twice the speed. In this operation, owing to the great increase in hours of road service as well, one electric locomotive is the practical equivalent of four of the Mallet engines replaced.

On the Chicago, Milwaukee & St. Paul the notable change has been elimination of intermediate terminals on the electrified section between Harlowton, Montana, and Avery, Idaho, 440 miles long. There is at present a single intermediate engine terminal, but the latest passenger locomotives are detached from trains at this terminal for inspection and work only, about once in each eight or ten trips. On regular schedule these engines have a run of 440 miles each day, being taken off for inspection at Deer Lodge after mileage varying from three to five thousand miles. On occasion, when due to a schedule derangement engines have been maintained in continuous road service for periods of 30 hours for a full day of 24 hours, mileage records of up to 766 miles in this mountain service have been established.

[1] Director of Heavy Traction, Westinghouse Electric and Mfg. Co.

The advantage of electric power is its great flexibility and mobility. The difference between locomotives, steam and electric, is fundamental. The steam locomotive carries its own power plant, while the electric locomotive, on the other hand, is simply a transformer of power. The design of the steam locomotive is circumscribed utterly by the necessity of tying up the rest of the machine to a steam boiler. On the other hand, the electric locomotive assembly can differ amazingly as to type, length, axle loading and driving connections. A group of small motors does not differ materially in efficiency from a single large motor. The speed and power of an electric locomotive, therefore, is limited only by conditions of track and construction and condition of car equipment. It thus becomes entirely practicable to build an electric locomotive to take any train which will hold together, over any profile whatever, and at any desired speed. It should easily be practical to increase very greatly the speed of our freight trains so that they would all run at a common speed not very different from that at which the superior trains are operated.

Again, with the retirement of the lighter and weaker car equipments, a material increase in the weight of trains will be realized. Without the limitation in train speed commonly accepted as a handicap to operation of tonnage trains, who can say what the limit to train load will be with electric power? In fact, the character of railroad operation which may be secured with electric power has not yet been visualized. Every other industry that has been electrified has experienced a revolution in methods and service due to electrification. This should be equally true in the case of the movement of our railroad traffic.

Our present methods have been built up entirely under the necessities and limitations of the steam locomotive. This is evidenced by the existence of intermediate terminals at the ends of all the so-called engine districts, where all traffic halts. Again, the steam locomotive requires attention en route, needs supplies of water and coal, and because of its slow movement when hauling our present tonnage trains, it is frequently sidetracked for superior trains, and thus there are more and still more halts to traffic.

Car inspection now takes place at the terminus of each engine district. If under condition of electric operation the engine district can be increased to 200, 400 or even 500 miles, is there any good reason why car inspection should not be eliminated at the present intermediate terminals? In fact, is not the general standard of maintenance of equipment of doubtful value on the present basis of inspection at each 100-mile interval? Cars in subway service, which is certainly full of potential hazard, are economically and reliably maintained, although inspected at intervals of one to three thousand miles. The elimination of these intermediate terminals, with the resultant necessity of keeping the train moving on the main line, would secure an enormous increase in miles per car with a corresponding saving in equipment.

Furthermore, with the dispatch obtained in handling trains, movement could be so marshalled and scheduled that the necessity of storing goods at terminals to protect exports and local consumption would be largely eliminated, and terminals would then become, in fact as in fiction, open instead of closed gateways.

[The author here introduced a general statement upon the comparative performance of steam and electric locomotives which is omitted in this abstract in view of the fact that Mr. Armstrong's paper is devoted largely to this phase of the subject and covers the ground quite thoroughly.—Editor.]

There are a considerable number of different designs of electric locomotives all in successful operation, and each possessing certain advantageous features. Further experience will undoubtedly result in the survival of common types for the different classes of service. The great latitude with which electric locomotives can be designed, while fundamentally most desirable, is in itself at the present time somewhat of a handicap. This is now the subject of intensive analysis and this study will undoubtedly develop a better knowledge of the running characteristics of the steam locomotive.

To state the case briefly, we are all interested in the transportation problem. Electrification is bound to be the most potent factor for its relief. We should therefore invite and embrace the closest coöperation with the engineering and mechanical skill which has been so productive in the steam-locomotive field.

# Steam vs. Electric Locomotives—the Electric Side

By A. H. ARMSTRONG,[1] SCHENECTADY, N. Y.

A COMPARISON of the modern steam and electric locomotive leads immediately to a discussion of the relative fitness of the two types of motive power to meet service conditions. At present railway practice has closely followed steam-engine development, but are we not justified in looking at the transportation problem from the broader standpoint of a more powerful and adaptable type of motive power?

It is not merely a question of replacing a Mikado or Mallet by an electric locomotive of equal capacity. The economies thus effected are in many instances not sufficient in themselves to justify a material increase in capital account. The paramount need of our railways today is improved service, and this can be brought about by introducing the more powerful, flexible and efficient electric locomotive.

## POSSIBILITIES OF DESIGN

Owing to handicap of precedent and prejudice, electricity must take up the railway problem where steam leaves off. In other words, the proof is up to the electrical engineer proposing any marked departure from commonly-accepted standards as established by long years of steam-engine railroading.

The following table gives the commonly accepted constants in locomotive design. While a maximum standing load of 60,000 lb. per axle has been generally accepted for steam engines, it is

COMMONLY ACCEPTED CONSTANTS

| | |
|---|---|
| Limiting gross weight per axle | 60,000 lb. |
| Limiting dead weight per axle | 18,000 lb. |
| Limiting coefficient adhesion running | 18 per cent |
| Limiting coefficient adhesion starting | 25 per cent |
| Ruling gradient | 2 per cent |
| Maximum curvature | 10 deg. |
| Maximum rigidwheel base | 18 ft. |
| Maximum speed on level—passenger | 65–70 m.p.h. |
| Maximum speed on level—freight | 25–30 m.p.h |
| Maximum drawbar pull | 150,000 lb. |

well known that an impact of at least 30 per cent in excess of this figure is delivered to rail and bridges due to unbalanced forces at speed. Impact tests taken on electric locomotives of proper design disclose the feasibility of adopting a materially higher limiting weight per axle than 60,000 lb. without exceeding the destructive effect on track and road-bed now experienced with steam engines. However, owing to the flexibility of electric locomotive design, there is no immediate need of exceeding present steam practice in this respect, although this and other reserves may be called upon in the future.

Accepting the Mikado and Mallet as the highest developments of steam road and helper engines for freight service, the following

COMPARISON OF STEAM AND ELECTRIC LOCOMOTIVES

| Type | Mikado 2-8-2 | Mallet 2-8-8-2 | Electric 6-8-8-6 |
|---|---|---|---|
| Wt. per driving axle | 60,000 lb. | 60,000 lb. | 60,000 lb. |
| No. driving axles | 4 | 8 | 12 |
| Total wt. on drivers | 240,000 lb. | 480,000 lb. | 720,000 lb. |
| Total wt. loco. and tender | 480,000 lb. | 800,000 lb. | 780,000 lb. |
| Tract. eff. at 18 per cent coeff. | 43,200 lb. | 86,400 lb. | 129,600 lb. |
| Gross tons 2 per cent grade | 940 | 1,880 | 2,820 |
| Trailing tons 2 per cent grade | 693 | 1,495 | 2,430 |
| Speed on 2 per cent grade | 14 m.p.h. | 9 m.p.h. | 16 m.p.h. |
| Hp. at driver rims | 1,620 | 2,080 | 5,570 |
| Lbp. at 80 per cent eff. | 2,030 | 2,600 | ... |
| Trailing tons-miles per hr. on 2 per cent gradient | 9,700 | 13,500 | 38,800 |

general comparison is drawn with an electric locomotive that it is entirely practicable to build without in any respect going beyond

the experience embodied in locomotives that are now operating successfully.

The above analysis brings out the fact that to equal the hourly ton-mile performance of one electric locomotive it would require three and four engine crews, respectively, for the Mallet and Mikado types. Furthermore, the electric performance as tabulated above can be obtained with each individual locomotive practically regardless of climatic conditions, efficiency of the crew or time that has elapsed since shopping; and with a demonstrated reliability that has set a new standard in railroading.

## MOUNTAIN SERVICE

The electric locomotive has demonstrated its very great advantages in relieving congestion on single-track mountain-grade divisions. The number of meeting points on a single-track line increases as the square of the number of trains operating at one time and is proportional to the average speed, so it will be appreciated what an advance in mountain railroading is opened up by the adoption of the electric locomotive.

The hazard of mountain operation is greatest on down grades, although the perfection of automatic air brakes has done much to modify its dangers. It is left to electricity, however, to add the completing touch to the safe control of descending trains by supplying regenerative electric braking. Not only are air brakes entirely relieved and held in reserve by this device, but the potential energy in the descending train is actually converted into electricity which is transmitted through the trolley to the aid of the nearest train demanding power. Aside from the power returned from this source (14 per cent of the total on the Chicago, Milwaukee and St. Paul Railway), the chief advantage of electric braking lies in its assurance of greater safety and higher speeds permitted on down grades. The heat now wasted in raising brake shoes and wheel rims often to a red heat is returned to the trolley system and becomes an asset instead of a likely cause of derailment.

## COST OF MAINTENANCE

In order to draw a fair comparison, there should be added to the maintenance charges for locomotives the cost of back shop repairs and all expenses of roundhouse, turntable, ashpit, coal and water stations—in fact the many items contributing to rendering necessary steam-engine service, as most of these charges are eliminated by the adoption of the electric locomotive. Spare parts can be substituted so quickly that, excepting in cases of wreck, there is no need of the back shop for electric locomotives, unless turning tires and painting may be considered heavy repairs. Electric locomotives are now being operated 3000 miles between inspections on at least two electrified railways and the following figures are available:

ELECTRIC-LOCOMOTIVE MAINTENANCE, YEAR 1919

| | N. Y. C. | C. M. & St. P. | B. A. & P. |
|---|---|---|---|
| Number locomotives owned | 73 | 45 | 28 |
| Locomotive weight—tons | 118 | 290 | 84 |
| Annual mileage | 1,946,879 | 2,321,148 | 566,977 |
| Repairs per mile | 6.39 cents | 14.65 cents | 6.48 cents |

On the basis of prewar prices, maintenance costs were approximately 60 per cent of above figures given for the year 1919. In contrast, it can be stated that the present cost of maintaining a type 2-8-8-2 Mallet is fully 60 cents per engine-mile, without including many miscellaneous charges not shared by the electric locomotive. Possibly more direct comparison may be better drawn by expressing maintenance in terms of driver weight.

STEAM AND ELECTRIC REPAIRS

|  | Steam Mallet | C. M. & St. P. Electric |
|---|---|---|
| Cost repairs per mile...................... | 60 cents | 14.65 cents |
| Weight on drivers........................ | 240 tons | 225 tons |
| Cost repairs per 100 tons locomotive weight on drivers........................ | 25 cents | 6.52 cents |

Including all engine service charges, the facts available give foundation for the claim that electric locomotives of the largest type can be maintained for from 25 to 30 per cent of the upkeep cost of steam engines operating in similar service.

## FUEL SAVING

Fuel economy figured prominently among the several reasons leading up to the replacement of the steam engine on the Chicago, Milwaukee and St. Paul Railway as brought out by a careful analysis of the performance of the steam engines then in service.

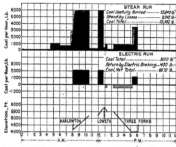

FIG. 1   COAL RECORD FOR STEAM AND ELECTRIC RUNS FROM HARLOWTON TO THREE FORKS, 1000 GROSS TONS MOVED

Although the steam engines tested may now perhaps be considered obsolete and not within the scope of this discussion of the modern engine, nevertheless it is not without value to compare the results of steam and electric locomotives operating over such long distances under identical conditions. The following data as given in Fig. 1 are therefore submitted as applying to a particular equipment only. No claim is made that they are representative of the best modern steam-engine performance, although many thousands of steam engines still in operation will show no greater economies than those given in the following table. The general data applying to the steam and electric locomotives tested are as follows:

C. M. & ST. P. LOCOMOTIVE TEST DATA

| Type | Steam 2–6–2 | Electric 4–4–4–4 |
|---|---|---|
| Weight of engine................... | 206,000 lb. | 568,000 lb. |
| Weight of tender................... | 134,000 lb. | ... |
| Weight total engine and tender....... | 360,000 lb. | 568,000 lb. |
| Weight on drivers.................. | 152,000 lb. | 450,000 lb. |
| Ratio driver weight to total......... | 42.2 per cent | 79.3 per cent |
| Rigid wheelbase................... | 13 ft. | 10 ft. 6 in. |
| Diameter drivers.................. | 63 in. | 52 in. |
| Cylinders........................ | ·21 in. X 28 in. | ... |
| Boiler pressure................... | 200 lb. | ... |
| Heating surface................... | 2346 sq. ft. | ... |
| Grate area....................... | 45 sq. ft. | ... |
| Water capacity................... | 8000 gal. | ... |
| Coal capacity.................... | 14 tons | ... |

Other engines were also tested over other sections of track, but the particular runs recorded in Fig. 1 are chosen for illustration as bringing out most strikingly the inherent disadvantages of operating a steam engine over a single-track mountain-grade division and handicapped by the usual delays attending freight-train service under such conditions. The run of 111.1 miles from Harlowton, elevation 4162 ft., to Three Forks, elevation 4066 ft., over the Belt Mountain Divide at Loweth, elevation 5789 ft., was made by steam with 871 tons trailing in 26 cars and by electric locomotive hauling 64 cars weighing 2762 tons. In order to picture

a direct comparison of the results of the steam and electric runs, test data are reduced to a common basis of 1000 gross tons moved, this unit of measurement including the locomotive and tender weight. The running speed of the electric train was but slightly higher than the steam and the additional correction in the power demand rate of the former is made proportional to the lower speed. Both runs are therefore shown as made in identical time on the basis of 1000 total gross tons moved in each instance. The fuel furnishing power to the steam train was coal having the following analysis:

COAL ANALYSIS

| Fixed Carbon | Volatile Carbon | Ash | Moisture | B.T.U.'s |
|---|---|---|---|---|
| 47.99 | 38.98 | 8.35 | 4.68 | 11,793 |

Electric power was furnished by water and hence no direct coal equivalent is provided by the test result. To afford a common basis of comparison, however, a single assumption seems permissible and a rate of 2$\frac{1}{2}$ lb. of coal per kilowatt-hour is taken as representative of fair electric power-station practice. Power input to the electric locomotive was obtained by carefully calibrated recording wattmeters as well as curve-drawing volt and ampere meters. These values of locomotive input were raised to the value of three-phase power purchased in the ratio of 68 per cent (given by R. Beeuwkes in his A.I.E.E. Paper of July 21, 1920), and the kilowatt-hours so obtained reduced to coal equivalent in the ratio of 2$\frac{1}{2}$ lb. coal per kw-hr.

The run of a more modern steam engine would have effected a material reduction in the 23,640 lb. of coal burned in doing useful work, but the amount of coal wasted in stand-by losses (9042 lb.) might have been duplicated or even possibly increased with larger grate area. As standby losses constitute so large a proportion of the coal total burned (27$\frac{1}{2}$ per cent in this instance) it is apparent that enormous economies over the simple engine tested must be realized in the modern superheater and other improvements since introduced to offset in part the high inherent efficiency of the electric locomotive.

[In his complete paper the author gives a further analysis of the foregoing test in which he deduces comparative figures for coal per hp-hr. at driver rims and coal per rim hp-hr., showing still more favorably for the electric locomotive. He estimates that whatever transmission and conversion losses are interposed between power house and electric locomotive are more than compensated for by the improvement in load factor resulting from averaging the fluctuating demands of many individual locomotives. —EDITOR.]

It would be a simple matter to carry through a series of runs over the electrified zone of the C. M. & St. P. with a modern Mikado equipped with all the up-to-the-minute fuel-saving devices and thus provide the necessary data to draw direct comparisons with the electric locomotive. Such tests with modern steam equipment would undoubtedly discredit the above comparison based upon the economies of six years ago and might lead to something approximating the following blend of fact and theory:

THEORETICAL COMPARISON OF MODERN STEAM AND ELECTRIC LOCOMOTIVES Harlowton–Three Forks

| Type | Mikado 2–8–2 | Electric 4–4–4–4 |
|---|---|---|
| Weight on drivers................... | 240,000 lb. | 450,000 lb. |
| Weight engine and tender............ | 480,000 lb. | 568,000 lb. |
| Tractive efficiency, 18 per cent coefficient... | 43,200 lb. | 81,000 lb. |
| Trailing ton 1 per cent grade......... | 1,420 | 2,836 |
| Hp-hr. at driver rims............... | 4,360 | 8,200 |
| Coal per 1 hp-hr.................... | 3 | ... |
| Coal per driver hp-hr............... | 3.75 | ... |
| Stand-by loss—test result........... | 9,042 lb. | ... |
| Total coal per driver hp-hr.......... | 2.15 lb. | ... |
| Coal at power house hp-hr........... | 5.90 lb. | ... |
| Coal at power house kw-hr........... | ... | 2.5 lb. |
| Coal at locomotive hp-hr............ | ... | 1.86 lb. |
| Coal at locomotive driver hp-hr...... | ... | 3.09 lb. |
| Coal credit due regeneration......... | ... | 0.55 lb. |
| Net coal at driver hp-hr............. | 5.90 lb. | 2.54 lb. |
| Total net coal..................... | 24,800 lb. | 20,900 lb. |
| 1000 trailing ton-miles............. | 157,500 | 314,000 |
| Coal per 1000 ton-miles............ | 138 lb. | 66$\frac{1}{2}$ lb. |
| Ratio coal burned.................. | 2.37 | ... |

The above table is based upon actual electric-locomotive performance, Harlowton–Three Forks, coal taken at 2$\frac{1}{2}$ lb. per kw-hr. at assumed steam power station. Steam-engine values are based upon the known working efficiency of a Mikado equipped with superheaters, but penalized with the same stand-by losses actually determined with simple engine tested on the Harlowton—

Three Forks run. A test run from Harlowton to Three Forks with a modern Mikado engine hauling 1420 tons may possibly show a lower average fuel rate than 3 lb. per i.hp-hr. at drivers, less stand-by waste than 9042 lb. coal, but the average annual performance of many such engines would be most excellent if it reached the net figure arrived at of 5.9 lb. coal per actual hp-hr. work performed at drivers. The electric run, however, is being duplicated daily, as to relation between kilowatt-hours and ton-miles, and it is just this reliability of electric operation that may at times give rise to misunderstanding in the comparison of steam and electric data.

Each individual electric locomotive will reproduce almost exactly the record of all others in similar service, little influenced by either extreme cold or skill of the engineer, while the fireman so-called and still retained, has nothing to do with the matter at all. There is no creeping paralysis gradually impairing the efficiency of an electric locomotive until temporary relief is obtained through frequent boiler washings and roundhouse tinkering, inevitably ending up in the major operations annually performed on the steam engine in the back-shop hospital to keep it going. It is for such reasons that the electrical engineer is slow to accept general statements of average service operation based on the results of tests usually made on steam engines in excellent condition and skillfully handled.

It is with some knowledge of all these facts that the broad statement is made that the general adoption of the electric locomotive would probably result in saving fully two-thirds the fuel now burned on present steam engines and possibly one-half the amount of fuel necessary to steam engines of the most modern construction.

## COMPARATIVE COST

The superior operating advantages of the electric locomotive are admitted by many who believe the first cost to be prohibitive, largely due to the trolley construction, copper feeders, substations, transmission lines, etc., necessary to complete the electrification picture. It is true that such auxiliaries add an amount that may equal the electric-locomotive expense and the task of proving the electric case is not made easier by the fact that steam-engine facilities are already installed and may have little or no salvage value to offset new capital charge for electrification.

Comparing the cost of equivalent steam and electric motive power, it is apparent that on the basis of the same unit prices for labor and material, the first cost is approximately the same. While electric locomotives cost possibly 50 per cent more than steam for equal driver weight, the smaller number required to haul equal tonnage may quite offset this handicap, especially with quantity production of electric locomotives of standard design.

The steam engine also demands a formidable array of facilities peculiar to itself as shown by expenditures made on fourteen railways included in the northwestern group from 1907 to 1919, in which the expense for engines was $68,000,000 and for water stations, shops, enginehouses, shop machinery, turntables, ash-pits, etc., was $42,000,000.

# DISCUSSION

H. B. Oatley[1] suggested that the respective merits of steam and electric locomotives should be considered from the point of view of the return on the capital investment. Modern types of both steam and electric locomotives, he said, must be compared with each other, and the losses in the operation of the one compared with similar losses in the operation of the other. Mr. Oatley expressed his confident belief that the improvements in the steam locomotive, many of which are past the experimental stage and are beginning to be applied as part of regulation equipment, would result in making the steam locomotive so much better, from a fuel and efficiency standpoint, that combined with the other points of advantage which it possesses it would be enabled to retain its superiority.

F. J. Cole[2] called attention to the slow progress which has been made in the United States toward superseding the steam locomo-

tive. Records showed that in April 1919 there were 675 electric locomotives operating in the country. It seems, therefore, that in the 27 years since the building of the first electric locomotive only one per cent of the steam locomotives have been superseded by electric locomotives. Mr. Cole insisted that the relative merits of steam and electric locomotives must be determined largely on the basis of cost. He said that the reports of electrifications do not give costs of installation. The questions that should be answered are: Is it cheaper to operate by steam or by electricity, and can freight per ton-mile be moved at less cost, when all expenditures are considered, by electricity than by steam?

C. H. Quinn[1] presented comparative figures of fuel consumption on the electrified division of the Norfolk & Western Railroad and on modern Mallet engines under similar operating conditions. The figures showed a fuel saving of 29.3 per cent in favor of electric operation.

A. L. Ralston[2] gave data on the thermal efficiency of the 3000-volt direct-current system. On the electrified section of the New York, New Haven and Hartford Railroad the efficiency at full load is 8 per cent; at 75 per cent load, 7.8 per cent; and at 50 per cent, 7.4 per cent. In passenger service, 9.3 lb. of coal are consumed per car-mile on the electrified section, and 19.3 lb. per car-mile on the steam-operated section. The net saving through electrification is thus $393,000 a year, assuming coal to cost $5 a ton. In freight service, electric locomotives consume 84 lb. of coal per 1000 gross ton-miles, while steam locomotives average 199 lb. per 1000 gross ton-miles, the saving being $268,000 a year. The corresponding figures in switching service are 38.3 for electric and 106.8 for steam locomotives, and the saving in this service $94,000. The electrification therefore effects an annual gross saving in the cost of fuel amounting to $755,000.

F. H. Hardin[3] said that the maintenance costs of New York Central Mallet locomotives, including shop and enginehouse repairs, were from 24 to 37 cents per mile, as against the estimate of 60 cents offered by Mr. Armstrong. Also that Mr. Armstrong's estimate of 158 lb. of coal consumption per 1000 gross ton-miles for Mikado locomotives is disproved by the records of locomotives of this type operated by the New York Central Railroad, which have shown a coal consumption of 125 to 130 lb. per 1000 gross ton-miles.

W. L. Bean[4] summed up the comparative advantages and disadvantages of both steam and electric locomotives. Electrical operation requires less coal per unit of traffic handled than steam operation. The mileage per unit of electric equipment is ordinarily greater per unit of time. The first cost of electric locomotives per unit of capacity is greater than in the case of steam. Large electric locomotives cannot be used for heavy local service because of the heating caused by frequent starting. In such service the maximum train which can be handled by an electric locomotive can only approximate what can be handled by a steam engine of about 30,000 lb. tractive effort. As regards design, the discusser contrasted the steady development of the steam locomotive along lines of possible common usage and practice with the vast field of opinion among electrical engineers as to types of installation. He hoped that the lines of development of electrical facilities would tend to converge rather than diverge too widely.

In conclusion, Mr. Bean said that study and comparison of the details, item for item, of any large activity, is necessary in order to get the benefit of real analysis, but satisfactory conclusions as to the merits of the entire project cannot be reached by setting up in a partisan way outstanding advantages on the one hand any more than by listing all the disadvantages on the other. Certain more or less intangibles are important and must be weighed impartially. Among such are the increase in real-estate value through electrification, increased capacity of road, comparative safety and reliability of operation, permanence of type of design, obsolescence and depreciation factors, etc. Tangibles from a money standpoint can and should be segregated and set up in full

---

[1] Chief Electrical Engineer, Norfolk & Western R.R., Roanoke, Va. Mem.Am. Soc.M.E.
[2] Mechanical Superintendent, N. Y., N. H. & H. R.R.
[3] Chief Engineer of Motive Power, N. Y. Central R.R.
[4] Ass't. Mechanical Superintendent, N. Y., N. H. & H. R.R., New Haven, Conn. Mem.Am.Soc.M.E.

---

[1] Chief Engineer, Locomotive Superheater Co., New York. Mem.Am.Soc.M.E.
[2] Chief Consulting Engineer, American Locomotive Co., New York. Mem.Am. Soc.M.E.

scope on both sides of the case and conclusions based on the net result at the bottom line of the balance sheet. If fixed charges on plant, including equipment, plus maintenance charges, plus other outgo, outweigh the savings in fuel plus other operating savings, the net result is a deficit and all manner of proclaiming isolated pecuniary advantages would not induce a careful investor to support the enterprise.

R. Beeuwkes[1] dissented from the statements made by Mr. Muhlfeld in regard to the electrification of the Chicago, Milwaukee and St. Paul Railway. The Virginian steam articulated locomotive, he stated, could not justifiably be compared with the St. Paul electric articulated, as each was designed for an entirely different service and operated under entirely different conditions. The St. Paul locomotive was designed to provide capacity sufficient to handle a 2500-ton trailing load over any portion of the profile without helper on grades of one per cent or less, at a speed of approximately 16 miles per hour. It was not built with the idea of securing the maximum capacity possible, as it appeared to Mr. Bean that Mr. Muhlfeld had implied. In addition, Mr. Bean insisted that fuel economy is not the most important claim for electrification. He referred to decreased cost of engine repairs, enginehouse expenses, train and enginemen wages and increased ton-mile capacity of the electric locomotive as of more importance. Other advantageous features he mentioned were ability to carry overloads and ease of starting trains.

W. F. Kiesel,[2] Jr. submitted experimental data to show that electric operation does not save coal. He said that with the exception of capital investment, coal per drawbar horsepower and stand-by losses, there is too little difference between the two types for operation in open country to warrant further consideration. He also said that, for either steam or electric operation, the length of divisions and locations of terminals are governed by other than locomotive limitations.

E. B. Katte[3] produced a table which showed that fuel-consumption data are apt to be misleading in so far as comparisons are concerned, due to widely different conditions on the various roads in question. He cautioned against placing too much reliance on results obtained from test runs, notwithstanding the fact that they fairly simulate actual conditions. He advised that instead, service records extending over several years must be used, and the data carefully collected, correlated and averaged to give results that may be expected in everyday service.

George Gibbs[4] contrasted the two kinds of power plants for conducting railway transportation from the standpoints of both design and performance. In steam service, he said, the plant is a part of the moving train; in electric it has both stationary and moving elements, viz., a central power-generating plant, various connecting links to bring the power to the train, and means of utilizing it there. As regards simplicity, therefore, the self-contained steam locomotive has an inherent advantage over the combination of elements required for electric propulsion, and the latter must show some peculiar advantages in an *operating*, rather than a *structural*, sense if it is to supersede steam traction. Furthermore, the steam locomotive has been developed to a perfection of detail and a high degree of steam accuracy during the one hundred years of its use; it does wonderful work and is in possession of the field representing a heavy money investment, and therefore can only be displaced, even by something better, by slow degrees. He contended that electrical as well as mechanical engineers must, in the light of evidence furnished by existing installations, concede than an electric system will function in a successful, reliable and efficient manner for any kind of railway service. It is capable of unlimited hauling capacity, is flexible as to speed and has important features conducing to safety in handling trains. But in his opinion the fundamental question affecting the adoption of electric locomotive is whether or not its substitution for steam is warranted by its advantages in the production of more transportation, and if so, whether this is practicable financially. It will be found, he said, that electric traction will pay, directly or indirectly, in special

cases to an extent depending upon the density of traffic and the difficulty of maintaining proper steam operation. As it is unquestionable that an electric installation involves a higher first cost than for steam, its adoption means that more or less existing investment must be scrapped, and therefore the increase in fixed charges must be offset either by the direct operating savings produced or these plus the indirect savings and benefits. The final conclusion arrived at by Mr. Gibbs was that there is always a large saving in fuel with electric traction, generally some saving in maintenance cost of power equipment and often important savings in train-crew costs, enginehouse expenses and minor supplies.

A. W. Gibbs[1] submitted data for a 2-10-0 type steam locomotive of which over 100 are in regular service, which he stated, represent better practice than the locomotives referred to by Mr. Armstrong in the second table of his paper. He said that these locomotives were expressly designed to do all of their work within the economical range of steam distribution, the required power being obtained by increases in steam pressure and size of cylinders. While he had given the power at nearly the speed mentioned by Mr. Armstrong, the performance was excellent as double the speeds given, but the sacrifice in drawbar pull—from nearly 60,000 lb. at 14.7 m.p.h. to about 43,000 lb. at 25.3 m.p.h.—would not be justified. In brief, the improvements in the steam locomotive, if properly availed of, have much narrowed the field of economical electrification.

DATA ON 2-10-0-TYPE STEAM LOCOMOTIVE

| | |
|---|---:|
| Weight in Working Order | 371,000 lb. |
| Weight on Drivers | 342,050 lb. |
| Weight on Engine and Tender | 523,000 lb. |
| Tractive Effort at 14.7 miles per hour, 45 per cent cut-off | 58,900 lb. |
| Gross Tons (2 per cent grade) | 1,280 |
| Trailing Tons | 1,019 |
| Coal per D.H.P. at this speed and cut-off | 2.8 |
| Tractive Effort at 22 miles per hour, 40 per cent cut-off | 42,500 lb. |
| Gross Tons | 923 |
| Trailing Tons | 662 |
| Coal per D.H.P. | 3.2 |
| Tractive Effort at 25.4 miles per hour, 45 per cent cut-off | 43,600 lb. |
| Gross Tons | 948 |
| Trailing Tons | 687 |
| Coal per D.H.P. | 3.8 |

The same arrangement of limited maximum cut-off used in the 2-10-0 locomotive has been embodied in a simple Mallet now running which has the same speed elasticity as the 2-10-0 type.

*Stand-by Losses.* While these losses are actual and large it is very difficult to fix their value. The electric locomotive has the advantage provided there are sufficient trains in motion to smooth out the total demand on the power plant.

*Weather.* The independence of the electric locomotive in severe weather is another undoubted advantage, not so much because of the performance of the motors as from the avoidance of losses and delays due to ashpit work and to frozen pipes and other parts, incidental to the presence of water on the steam locomotive.

*Liability to Interruption.* Electric operation is dependent on uninterrupted connection with the source of power. In the event of breakage of the line, especially of the overhead construction, the trains in the section involved are dead and cannot get themselves out of the way of the repair trains.

*Speeds.* The question of speed has evidently been treated from the freight standpoint, for there has never been any question as to the speed capacity of well-designed electric passenger locomotives, being far beyond that permitted by the rules. As he saw it, the high-speed of trains was less important than uniformity of speeds of different trains. If tonnage trains had the same speed as preference trains, and could thus avoid the great delay due to side tracking of trains of inferior rights, far more would be accomplished than the mere saving in time over the division due to the increased speed.

In the maintenance cost per mile of different electric locomotives, given by Mr. Armstrong, it should be said that any figures of postwar dates are so clouded by the abnormal labor and material costs as to be very doubtful. These locomotives have not run long enough to reach a general level of costs and it would be noted that the average annual mileage is low except in the case of the Milwaukee locomotive. Besides this, there is no evidence as to the maintenance costs of the rest of the outfit, including power plant, transformers, transmission lines, converters or transformers, trolley or third-rail and track circuits, all of which are essential

[1] Electrical Engineer, C. M. & St. P. Ry.
[2] Mechanical Engineer, Pennsylvania R.R., Altoona, Pa.   Mem.Am.Soc.M.E.
[3] Chief Engineer Electric Traction, N. Y. Central R.R., New York.   Mem.Am. Soc.M.E.
[4] Consulting Engineer, Gibbs & Hill, New York.   Mem.Am.Soc.M.E.

[1] Chief Engineer, Pennsylvania Railroad.

(Continued on page 726)

# SURVEY OF ENGINEERING PROGRESS

## A Review of Attainment in Mechanical Engineering and Related Fields

SUBJECTS OF THIS MONTH'S ABSTRACTS

WATER TURBINES, NEW ANALYTICAL THEORY
WATER TURBINES, METHOD OF DESIGN
RATEAU-BATTU-SMOOT HIGH-SPEED BLOWER
COPPER STEEL
OIL-FIRED BOILER ECONOMIES
CENTRIFUGAL PUMPS, DROP IN OUTPUT AFTER STARTING
BLAST-FURNACE GAS CLEANER
WAR-VESSEL DESIGN
SUBMARINES IN NAVAL WARFARE
AIRCRAFT IN NAVAL WARFARE

GROUND VIBRATIONS, TRANSMISSION TO A DISTANCE
GROUND VIBRATIONS, DAMPING
STEAM CONTROL DEVICE, LOW-PRESSURE
TURBINES, STEAM CONTROL DEVICE
WATER-INDICATING DEVICES, TESTS
LOCOMOTIVES, TESTS OF WATER-INDICATING DEVICES
CAREY OIL TRANSMISSION SYSTEM
REGAN AUTOMATIC TRAIN-CONTROL SYSTEM
BRITISH COMMITTEE FOR SCIENTIFIC AND INDUSTRIAL RESEARCH

ZIRCONIUM STEELS
FUEL RESEARCH IN ENGLAND
NATIONAL INSTITUTE OF RESEARCH FOR CANADA
STEAM NOZZLES
STEAM DIFFUSORS
MAGNETIC TESTING APPARATUS
DEFECTOSCOPE
TRUCK TRANSPORTATION IN THE STEEL INDUSTRY

## New Analytical Theory of Water Turbines

NEW ANALYTICAL THEORY OF WATER TURBINES, Prof. A. Petot. Analytically, the problem therefore resolves itself to a search for the maximum value of a function of two independent variables such that their variation occurs within fairly well-restricted intervals. By selecting these variables in a convenient way all the elements of a turbine can be expressed as functions of an auxiliary unknown value given by an equation of the second degree with numerical coefficients. The author proceeds in the following manner.

As a rule in the analytical theory of water turbines it is customary to neglect losses of energy due to viscosity, friction and shocks. It is here shown how these may be accounted for with sufficient approximation and without running into excessive complications. Let $H$ be the head of the water fall; $v$, $u$, $w$ the absolute velocity, the velocity of entrainment and the relative velocity at the entrance to the wheel; $\alpha$ the angle of $v$ with $u$ and $\beta$ the angle of $w$ with $v$, continuation of $u$; $v_1$, $u_1$, $w_1$, $\alpha_1$ and $\beta_1$ are the same elements for the exit from the wheel; $r$ and $r_1$ are the average radii and $m$ is the ratio $r_1/r$; $kv^2/2g$ and $k'w^2/2g$ the losses which have to be taken care of; $\epsilon$ is the degree of reaction of the turbine and $\rho$ is its internal efficiency.

If one neglects certain magnitudes which are given in the problem and others such as the angles $\alpha_1$ and $\beta_1$ which must be as small as possible, the efficiency $\rho$ is merely a function of two variables and it is obvious that these variables should be selected in such a manner that the calculation may be carried through without neglecting any important terms. As a result of his investigations the author has been led to select as such variables $v$ and $u$.

The first question with which one has to deal in a calculation carried on in the manner indicated is that of expressing $w_1$, $v_1$ and $\rho$ as functions of $v$ and $u$. From the two basic equations

$$\rho g H = uv \cos \alpha - u_1 v_1 \cos \alpha_1 \ldots \ldots \ldots [1]$$

$$\rho = 1 - \frac{v_1^2}{2gH} - k \frac{v^2}{2gH} - k' \frac{w_1^2}{2gH} \ldots \ldots \ldots [2]$$

it is found that

$$gH - \frac{v_1^2}{2} - k \frac{v^2}{2} - k' \frac{w_1^2}{2} = uv \cos \alpha - u_1 v_1 \cos \alpha_1 \ldots [2a]$$

Equation [2a], after obvious transformations, becomes

$$(1 + k') w_1^2 = 2gH + m^2 u^2 - 2uv \cos \alpha - kv^2 \ldots \ldots [3]$$

From a consideration of the triangle at the exit, the author obtains

$$v_1^2 = m^2 u^2 + w_1^2 - 2muw_1 \cos \beta_1 \ldots \ldots \ldots \ldots [4]$$

and finally from Equations [4] and [2].

$$\rho g H = uv \cos \alpha + muw_1 \cos \beta_1 - m^2 u^2 \ldots \ldots \ldots [5]$$

The desired equation is thus obtained without neglecting any terms. However, in order not to complicate the computations too much it is advisable to preserve the time being $w_1$ as an auxiliary variable, bearing in mind, however, that in due time it can be gotten rid of by means of Equation [3].

### DETERMINATION OF THE OPTIMA VALUES OF $v$ AND $u$

The two partial derivatives $\dfrac{\partial \rho}{\partial u}$ and $\dfrac{\partial \rho}{\partial v}$ of $\rho$ must be reduced to zero value, which involves the use of Equation [5] and also of $w_1$. These latter may be deduced from Equation [3], which gives

$$(1 + k')w_1 \frac{\partial w_1}{\partial u} = m^2 u - v \cos \alpha \qquad \ldots \ldots (a)$$

$$(1 + k')w_1 \frac{\partial w_1}{\partial v} = - u \cos \alpha - kv$$

Now, equating in succession to zero the expressions $\dfrac{\partial \beta}{\partial u}$ and $\dfrac{\partial \rho}{\partial v}$ there are obtained from these two equations

$$(1 + k')w_1 (v \cos \alpha - 2m^2 u) + m \cos \beta_1 (2gH + 2m^2 u^2 - 3uv \cos \alpha - kv^2) = 0 \quad \Bigg\} \ldots (b)$$

$$(1 + k') w_1 \cos \alpha - m \cos \beta_1 (u \cos \alpha + kv) = 0$$

which together with Equation [3] determine the optima values of $u$ and $v$, and also the corresponding value of $w_1$. If in the first of these two equations $w_1$ be replaced by its value

$$w_1 = \frac{m \cos \beta_1}{(1 + k') \cos \alpha} (u \cos \alpha + kv) \ldots \ldots \ldots \ldots [6]$$

from the second equation there results an equation of state

$$uw \left( \cos \alpha + \frac{m^2 k}{\cos \alpha} \right) = gH \ldots \ldots \ldots \ldots \ldots [7]$$

which differs but little from the equation given by Poncelet.

Proceeding now in the same manner, from Equations [6] and [7] we obtain

$$m^2 u^2 \cos^2 \alpha \ (k' + \sin^2 \beta_1 - kv^2 [m^2 k \cos^2 \beta_1 + (1 + k') \cos^2 \alpha] + \frac{2m^2 kgH \cos^2 \alpha (k' + \sin^2 \beta_1)}{\cos^2 \alpha + m^2 k} = 0 \ldots \ldots \ldots \ldots [8]$$

which thus provides a system of two quadratic equations for determining the two principal unknown values $v$ and $u$.

On the other hand, in order to verify a priori Equation [7], the author places

$$v^2 = \frac{gH \cos \alpha}{\cos^2 \alpha + m^2 k} \times x \ldots \ldots \ldots \ldots [9]$$

and

$$u^2 = \frac{gH \cos \alpha}{\cos^2 \alpha + m^2 k} \times \frac{1}{x} \ldots \ldots \ldots \ldots [10]$$

in which $x$ designates the auxiliary unknown quantity equal to $\dfrac{v}{u}$. If these values of $v^2$ and $u^2$ are inserted in Equation [8], then

$$x^2 \left( k \cos^2 \beta_1 + \frac{1 + k'}{m^2} \cos^2 \alpha \right) - 2x \cos \alpha (k' + \sin^2 \beta_1) - \cos^2 \alpha \frac{k' + \sin^2 \beta_1}{k} = 0 \ldots \ldots \ldots \ldots [11]$$

which is of quadratic character and has numerical coefficients.

689

It has also real roots with opposite signs and it is evident that the positive root $x'$ is the only one that suits the case. From this we can pass to $v$ and $u$, and all the rest can be deduced, as will be shown by comparatively simple numerical computations which can be carried out with such approximation as may be desired.

### Calculation of the Other Elements of the Turbine

The degree of reaction $\epsilon$ is given by the usual formula

$$1 - \epsilon = \frac{(1 + k)v^2}{2gH} \quad\dots\dots\dots\dots\dots [12]$$

or else by

$$1 - \epsilon = \frac{(1 + k)x' \cos \alpha}{2 (\cos^2 \alpha + m^2 k)} \quad\dots\dots\dots [12 \text{ bis}]$$

which can be directly deduced therefrom. In order that the value of $\epsilon$ be positive it is necessary that

$$\frac{(1 + k)x' \cos \alpha}{2 (\cos^2 \alpha + m^2 k)} < 1 \quad\dots\dots\dots\dots [c]$$

or else

$$x' < \frac{2(\cos^2 \alpha + m^2 k)}{(1 + k) \cos \alpha} \quad\dots\dots\dots\dots [13]$$

For this it is necessary that when $x$ is the first member in Equation [11] to be replaced, the result obtained should be positive. Hence, there must be between the elements $\alpha$, $\beta_1$, $m$, $k$ and $k'$ a characteristic inequality of operation with reaction. The author does not attempt to give this inequality explicitly as it is quite complex, but he shows that with the usual value of the elements above referred to it is present.

The angle $\beta$ is deduced as usual from the formula

$$\frac{v}{u} = \frac{\sin \beta}{\sin (\alpha + \beta)} \quad\dots\dots\dots\dots\dots [14]$$

which, bearing in mind the values of $v$ and $u$, gives

$$\tan \beta = \frac{x' \sin \alpha}{1 - x' \cos \alpha} \quad\dots\dots\dots\dots [15]$$

In order to determine the relative velocity $w$ at the entrance to the wheel, we have the relation

$$w^2 = u^2 + v^2 - 2 uv \cos \alpha \quad\dots\dots\dots\dots [16]$$

which may be applied under the above form or replaced by

$$w^2 = \frac{gH \cos \alpha}{\cos^2 \alpha + m^2 k} \left( x' + \frac{1}{x'} - 2 \cos \alpha \right) \dots [16 \text{ bis}]$$

where only the auxiliary unknown quantity $x'$ is used.

As regards the exit velocity $w_1$, it is given by Equation [6], which requires, however, the preliminary computation of $v$ and $u$ by means of Equations [9] and [10]; Equation [3] might also be used in the same manner. From this, we pass to the ratio $w_1/w$, which must be taken into consideration in order to satisfy the conditions of continuity and to obtain the desired degree of reaction.

With $w_1$ known, one might also compute $v_1$ by means of Equation [4] and then the angle $\alpha_1$ by considering the triangle of velocities at the exit. These two elements, however, are not used in the course of this investigation and, therefore, are merely mentioned for the sake of completeness. The former may be determined, if it should be desirable to know how the total loss of efficiency is distributed between the three negative members of Equation [2] and angle $\alpha_1$, to compare its actual value with the values which have been attributed to it in accordance with the investigations of Euler and Poncelet and which are usually accepted today.

It is now necessary to determine only the internal hydraulic efficiency $\rho$ which is obtained by means of Equation [5], where, however, $v'$ is replaced by its value given by Equation [6] and also $w$ by their values obtained from Equation [7]. This gives

$$\rho = 1 - \frac{m^2(k' + \sin^2 \beta_1)}{1 + k'} \left( \frac{u^2}{gH} + \frac{k}{\cos^2 \alpha + m^2 k} \right) \dots [17]$$

where for $u^2$ may be substituted its value from Equation [10], which gives

$$\rho = 1 - \frac{m^2(k' + \sin^2 \beta_1) \left( \frac{1}{x'} \cos \alpha + k \right)}{(1 + k')(\cos^2 \alpha + m^2 k)} \dots [17 \text{ bis}]$$

When the characteristic inequality has been verified, it is necessary to adopt operation with reaction in order to secure maximum efficiency, but the degree of reaction cannot be selected arbitrarily as it is fixed by Equation [12] and varies with the type of turbine. If the inequality is reduced to equality, we have a limit turbine with a zero reaction and a maximum efficiency. Should the inequality be reversed, the maximum efficiency proper to a turbine of that type could be obtained only by making the reaction negative, a case which the author does not consider at all.

Moreover, practice has shown that reaction turbines have as a rule the best efficiency, and it is likely that the characteristic inequality will be found correct in the majority of cases. This should be checked every time, however, by using Equation [13].

Furthermore, if we should admit that the coefficients $k$ and $k'$ both preserve the same values for all the turbines of a given type, which is likely to be true, we shall see that the idea of proportionality of turbines is presented here in as simple a manner as in the usual approximate theory. This makes it possible for us to dispense with the consideration of questions relative to the output $q$ of the waterfall and the corresponding dimensions of turbines, as they can be considered in the same manner as is usually done.

### Approximate Formulæ

Since the values of $k$ and $k'$ do not exceed 0.1, it is evident that in solving Equation [1] in the first approximation one may neglect entirely the term containing $x$ and reduce the coefficient of $x^2$ to its second term. This gives for $x'$ the comparatively simple approximate value

$$\xi = m \sqrt{\frac{k' + \sin^2 \beta_1}{k(1 + k')}} \quad\dots\dots\dots\dots [18]$$

from which can be derived expressions for all the elements of the turbine. For example, with the same degree of approximation one may derive the following expressions:

$$v^2 = \frac{mgH}{\cos \alpha} \sqrt{\frac{k' + \sin^2 \beta_1}{k(1 + k')}} \quad\dots\dots\dots\dots [19]$$

$$u^2 = \frac{gH}{m \cos \alpha} \sqrt{\frac{k(1 + k')}{k' + \sin^2 \beta_1}} \quad\dots\dots\dots [20]$$

$$1 - \epsilon = \frac{m(1 + k)}{2 \cos \alpha} \sqrt{\frac{k' + \sin^2 \beta_1}{k(1 + k')}} \quad\dots\dots\dots [21]$$

$$\rho = 1 - \frac{m}{\cos \alpha} \sqrt{\frac{k(k' + \sin^2 \beta_1)}{1 + k'}} \quad\dots\dots [22]$$

If necessary, the values of $\beta$, $w$, $w_1$, $v_1$ and $\alpha_1$ may be obtained in the same manner.

If we say that $\epsilon$ as obtained from Equation [21] has a positive value, the characteristic inequality will be expressed as

$$\frac{k' + \sin^2 \beta_1}{1 + k'} < \frac{4k \cos^2 \alpha}{m^2(1 + k)^2} \quad\dots\dots\dots\dots [23]$$

which is obviously only an approximation. It appears, however, in view of the factor $4/m^2$ that it must always be fairly well satisfied in the case of centripetal turbines and even in parallel turbines. There is doubt only in reference to centrifugal turbines. This explains why, as has been said above, there is always an advantage in having operation with reaction. These various approximate formulæ cannot take the place of the preceding precise formulæ in the calculation of the elements of turbine design, but they make it possible to view along broad lines the general operation and to give the first idea of the numerical values of certain of the elements which enter therein.

Thus, Equations [21] and [22] show first (because of the factor $m$, the variations of which have a more or less permanent influence than the variations of $k$ and $k'$) that the degree of reaction must be lower for parallel turbines than for centripetal turbines and that these latter have a better efficiency. From the equation

$$\rho = 1 - \frac{2k(1 - \epsilon)}{1 + k} \quad\dots\dots\dots\dots\dots [24]$$

which is a direct outcome of the preceding equations, it follows that the two properties of centripetal turbines, namely, the higher degree of reaction and better efficiency, are correlative with each other.

On the other hand, if we eliminate $k'$ from between Equations [20] and [22], we obtain

$$k = \frac{(1-\rho)u^2 \cos^2 \alpha}{gH} \quad \dots \dots \dots \dots [25]$$

which gives an approximate value of the coefficients $k$ which will be utilized later on in this paper. The approximate value of $k'$ might also be obtained in the same manner, but it does not appear to be of any use.

### EXPERIMENTAL INVESTIGATION OF WATER TURBINES

*First Series of Experiments.* A good many experiments have been made on water turbines, but it would be difficult to deduce with any degree of precision for each particular type of turbine the average value of coefficients $k$ and $k'$ because the machinery on which the experiments were carried out was designed in accordance with old formulæ. Nevertheless, by properly interpreting the results of these experiments we can obtain for $k$ and $k'$ what might be called the first series of values, from which a start may be made to deduce other values more or less closely approaching the true values. To demonstrate how this is done, let us assume that we have a good reaction turbine $M_0$ and that its various elements have been determined as precisely as possible. Equations [10], [11] and [17] are then a system of three equations with three unknowns $x$, $k$ and $k'$, which may be solved in the following manner:

Let us assume that the unknown $k$ has been given an approximate value from Equation [25] and that we designate this value by the letter $a$. Then we can immediately derive from Equations [10] and [17] corresponding values $b$ and $a'$ for the two other unknown quantities $x$ and $k$; since, however, the value $a$ of $k'$ is only approximate, it is evident that Equation [11] cannot be verified by these numbers $a$, $a'$ and $b$, and hence its first term will have a certain value $n$, different from zero. If, however, we repeat the same calculation by giving to $k$ another value $a_1$ in the neighborhood of the value $a$, then numerical values $a'$, $b$ and $n$ will be replaced by other values $a_1$ $b_1$ and $n_1$ lying in the neighborhood of the former values. It is obvious that the method of interpolation can be applied here, and it is easy to see how one can obtain, with as close approximation as may be desired, values $x_0$, $k_0$ and $k_0'$ for the unknown quantities $x$, $k$ and $k'$ of the turbine $M_0$.

These two numbers $k_0$ and $k_0'$ are the first couple of values of the coefficients $k$ and $k'$ and their approximation is the closer the better the turbine $M_0$ is designed. The degree of precision attained in this manner in the determination of the values of the coefficients may be established by comparing $M_0$ to a turbine $M_0'$ designed in accordance with the formulæ given earlier in this article, where $k$ and $k'$ are taken to be respectively equal to $k_0$ and $k_0'$. In this connection one is led to consider simultaneously with the constructional elements $\beta$ and $\lambda = u_1/w$ of the turbine $M_0$, also the corresponding elements $\beta'$ and $\lambda'$ of the turbine $M_0'$ and to compare them with each other.

If it should be found that $\beta'$ and $\lambda'$ are equal to $\beta$ and $\lambda$, respectively, that will mean that the turbine $M_0'$ is nothing else but the turbine $M_0$ and that hence the desired values of $k$ and $k'$ are $k_0$ and $k_0'$ exactly. The question will therefore be immediately solved for all machines of the same type as $M_0$ without there being any need of further experimentation. However, there is but little likelihood that this might be the case and all that one has the right to expect is that the differences $\beta' - \beta$ and $\lambda' - \lambda$ will be very small. In that case the conditions of maximum of efficiency would not be entirely satisfied in the turbine $M_0$ and one would be led to look forward to a turbine $M_1$ which, while being very close to $M_0$ would have a somewhat better efficiency. This is a problem clearly different from that which has been considered above and therefore new experiments would be necessary to solve it.

*Second Series of Experiments.* In the first place, in accordance with what has been stated above, it becomes necessary to give to $k$ and $k'$ numerical values somewhat lower than $k_0$ and $k_0'$. This is, however, only the first indication of what should be done and in order to progress it is necessary to proceed by a series of trial values. Let us imagine, for example, that by taking $k$ and $k'$ to be equal respectively to $0.98k_0$ and $0.98k_0'$ we obtain for $\rho$ by means of Equations [11] and [17 bis] a value which appears to be acceptable. In that case it may be used provisionally;

otherwise, the values of $k$ and $k'$ would have to be either decreased or increased slightly by methods of interpolation, the values of $\beta' - \beta$ and $\lambda' - \lambda$ being the guides as to whether the efficiency values should be accepted.

The new values of $k$ and $k'$, viz., $k_1$ and $k_1'$, having been selected in the above-described manner, let us now imagine that we are computing all the elements of the turbine $M_1$ by means of the general equations given above, that we have built the machine and that we have set it in operation under the same conditions as the turbine $M_0$. Then as regards the maximum efficiency and the operating velocity the distinction should be made between two sets of values—first, $\rho$ and $u$ computed a priori from the Equations [17 bis] and [10] and the values $\rho'$ and $u'$ obtain experimentally by direct measurement. In fact, by comparing $\rho$ with $\rho'$ and $u$ with $u'$ we can see with what degree of approximation the values $k_1$ and $k_1'$ approach the desired values of $k$ and $k'$. The reasoning is here similar to that by means of which we have passed from the first experimental turbine $M_0$ to the second, $M_1$.

Let us assume at first that $\rho'$ and $u'$ are respectively equal to $\rho$ and $u$. The conditions of maximum efficiency having been then satisfied in the turbine $M_1$ the desired values of $k$ and $k'$ are exactly $k_1$ and $k_1'$ and the experiment is satisfactorily computed. In practice it is sufficient for the same purpose that the differences

FIG. 1  DIAGRAM OF THE TURBINE CONSIDERED BY PROFESSOR PETOT

$\rho' - \rho$ and $u' - u$ should be very small, and it is likely that this might be the case if the turbine $M_0$ from which we have started has been well designed.

On the other hand, if these differences be quite large the experiments have to be continued. The author does not attempt, however, more than to indicate the direction of the experiments. As the turbine $M_0$ has been built in accordance with the general formula with values of $k$ and $k'$ lower than $k_0$ and $k_0'$, it is likely that its effective efficiency $\rho'$ will be somewhat higher than that of the turbine $M_0$. This is admitted by the author, with the reservation that should experiment show that this is not the case, the admission might be modified in the further experiments which would be easy because the same method would have to be followed.

Admitting, however, that the turbine $M_1$ represents from the point of view of efficiency a step forward as compared with the turbine $M_0$, the same manipulations are carried out with $M_1$ as were carried out with $M_0$ so as to secure a design for a third turbine $M_2$, which presumably would represent a step ahead as compared with $M_1$, and so on until a turbine is designed in which the differences between $\rho' - \rho$ and $u' - u$ are negligible, which may be considered as the end of the experiments. The important part of this process is that while more and more correct values are obtained for the coefficients $k$ and $k'$, the efficiency of the turbine of a given type is likewise improved.

### CONCLUSIONS

In the study of water turbines, by taking for an auxiliary unknown quantity the ratio $x = \bar{v}/u$ of the absolute velocity and velocity of flow at the entrance to the blades, one may carry to a final conclusion the computations of turbine design by taking into consideration all the charge losses, and without neglecting any of the terms. This unknown quantity $x$ is given by a quadratic equation, which is not surprising, as the various methods, whether analytical or graphical, proposed for the investigation of turbine design (in particular, those of Bodmer and Rateau) come back in the final end to an approximate solution of an equation of this order.

Once, however, we have obtained the value of $x$ for a given set of conditions, we can deduce from it by means of comparatively simple equations all the elements of construction and operation of a turbine of a given type. Also, by neglecting in the quadratic equation of $x$, certain terms which are very small as compared with other terms we are enabled to derive a series of approximate formulæ, which, while not capable of replacing the exact formulæ for the computations, involved in turbine design, still enable us to obtain a preliminary idea of their general operation.

All these considerations obviously would not amount to much if we were unable to determine with sufficient approximation the values which we must attribute for each type of turbine to the coefficients of loss $k$ and $k'$. Ways are, however, described and given showing how this may be done in a practical manner.

The coefficients, $k$ and $k'$, by the way, are not true coefficients, but functions of a large number of variables which it appears to be impossible to express in an analytical form. They may be treated as constants, however, because the functions themselves have a range of variation which is quite restricted. The same applies also to the majority of coefficients which are made use of in hydraulic problems. Because of this, it becomes necessary to determine the average values of $k$ and $k'$ for each type of turbine and for each new device which may be introduced therein, such as new construction of blades, where progress is still possible.

In the present article only simple turbines have been considered, but the same methods may be applied with proper variations to the more complex types. (*La Technique Moderne*, vol. 12, no. 8, Aug. 1920, pp. 332–336, *tA*.)

## Short Abstracts of the Month

### Errata

The direct-connected high-speed turbo-blower described in the November issue of Mechanical Engineering (p. 630) and referred to editorially on p. 644, was erroneously ascribed to the De Laval Company. This turbo-blower was built by the Rateau-Battu-Smoot Company of New York City.

In the original article in *Power*, from which the abstract in Mechanical Engineering was prepared, it was stated that the blower end of the rotor was made of a nickel-chrome-magnesium steel forging. We are now informed by the manufacturers that this forging was made of nickel-chrome *manganese* steel and not as was stated.

### ENGINEERING MATERIALS

Review of the Development of Copper Steel, E. M. Buck. The claim is put forward that tests made by a Committee of the American Society for Testing Materials with the coöperation of the United States Bureau of Standards, together with other data available, proved that by alloying with normal open-hearth or bessemer steel from 0.15 to 0.25 per cent of copper, the rate of corrosion of steel is very much reduced where the products are exposed to alternate attacks of air and moisture. The author also claims that in his investigations he noticed indications which point to a better adherence of paint coatings on copper steel, resulting in a more perfect and longer-continued protection by the paint film.

The manufacture of copper steel has heretofore been largely confined to sheet metal, but it might be extended to other uses. In particular, steel freight cars are mentioned, especially those of the open kind, which suffer greatly from corrosion. (Paper before the American Iron and Steel Institute, New York meeting, Oct. 1920, abstracted through *Iron Age*, vol. 106, no. 18, Oct. 28, 1920, pp. 1100–1110, *g*)

### FUEL AND FIRING

Oil-Fuel Installation, Joseph Pope, Mem.Am.Soc.M.E., and Frank G. Philo. Description of the boiler installation of the Savannah (Ga.) Electric Company which has been recently converted from coal to fuel oil and for which unusual efficiencies are claimed.

Since the adoption of fuel oil the boilers have been operated normally at about 250 per cent of their rate of capacity without a single instance of tube trouble throughout seven months of operation, notwithstanding the fact that the boiler feedwater is not entirely free from solid impurities. In the tests carried out at the plant efficiencies (gross) approximating 80 per cent at 300 per cent of boiler ratings have been obtained. Load conditions at the power station were such as to require the developent of high boiler outputs, which made the question of design of the burner of great importance. This steam-atomizing type of burner was not suitable as it gives excellent results at outputs about the rated capacity but not at the higher overloads.

As regards the mechanical-atomizing type of burner the one which had been in use for marine service had not shown as high an efficiency as even the steam-atomizing burner and did not permit to develop very high boiler ratings. The burner finally adopted was a new type developed by the Babcock & Wilcox Company. The original article does not describe its construction and only says that a specially designed orifice disk or tip is employed in which the oil is led to a central circular perforation through tangential grooves and thus attains a violent rotary motion as it is projected through the orifice. The resulting spray takes the form of a widely diverging cone of extremely fine mist. The burner proper is set in the furnace wall at the center of an air register so designed as to impart a whirling motion to the entering air. The rate of combustion is controlled by the temperature and pressure of the oil. The size of orifice in the burner tip also determines the rating, but the burner tips can be obtained in various sizes and are interchangeable, the larger size having a normal capacity of about 14,000 lb. oil per hour at 200 lb. pressure. (*Electrical World*, vol. 76, no. 15, Oct. 9, 1920, pp. 730–732, 3 figs., *de*)

### HYDRAULICS

Remarkable Drop in Output of Centrifugal Pumps after Starting, J. C. Dijxhoorn. Observations made by the author indicate that many centrifugal pumps show a gradual falling off in efficiency after running for some time. Two characteristic cases are cited in connection with the large pumps used for pumping the Haarlem mere. In the first the output fell to 30 per cent below the normal, and in the second case a marked increase in the energy consumption was noticeable shortly after the pump was started. On stopping and starting the pump, the normal output and energy consumption were again obtained; but the conditions gradually fell back to those previously discovered. After discussing the influence of a possible accumulation of gas (from a peaty soil) in the suction pipe, the author rejects this explanation as being insufficient. He ascribes the cause to the fact that turbine pumps of the Francis type are designed on the assumption that the inflowing water is moving in an axial and not a rotary direction. This assumption is undoubtedly correct during starting, but it can be readily demonstrated that very little would be required to impart a rotary movement to the incoming water column, i.e., the mass of water would move in a spiral through the pipe. This rotary effect would explain the phenomenon and also the fact that in turbine pumps with a high circumferential speed, where little or no rotary motion would be imparted to the incoming column of water, this falling-off in output disappears. The author is of opinion that the transmission of this rotary motion from the rotor of the pump back along the incoming water column is considerably accelerated by the presence of large quantities of suspended matter in the water. The obvious solution of the difficulty for low-speed turbine pumps would be the introduction of a rib or ribs in the suction pipe, to prevent a rotary motion from being imparted to the incoming water column. (*De Ingenieur*. Compare *Mechanical World*, vol. 68, no. 1760, Sept. 24, 1920, p. 216, *c*)

### METALLURGY (See also Engineering Materials)

Cleaning of Blast-Furnace Gas. Description of the Lodge electrostatic process for rough cleaning of blast-furnace gas as used at the Skinningrove Works in England.

The cleaning is applied to gas used in stoves, boilers and other combustion processes. Wet cleaning has the advantage in that, first, the sensible heat in the gas is lost; and further, power has to be expended in pumping the large volumes of water required for the dust removal. The Lodge plant has been installed to take care of this gas, 0.3 gram of dust per cubic meter in the cleaned gas having been fixed as an arbitrary standard. Essentially the method of operation of the Lodge system does not appear to differ very much from the Cottrell system which has been used in this country for a number of years. The plant appears to be working in a satisfactory manner and not only saves the trouble due to uncleaned stoves and boilers, but also has saved gas, which may have something to do with the former. Furthermore, about 50 tons of dust per week is recovered with 27 per cent content of potash. (Paper before The Iron and Steel Institute, Cardiff meeting, Sept. 1920, abstracted through *Iron Age*, vol. 106, no. 15, Oct. 7, 1920, pp. 928-930, *d*)

## MILITARY ENGINEERING

### Battleships, Submarine and Aircraft in Naval Warfare

THE DESIGN OF WAR VESSELS AS AFFECTED BY THE WORLD WAR. Rear-Admiral David Watson Taylor. It is not fully possible as yet to answer the question, "What has been the effect of the World War on warship design?" and the opinion of the service has not yet been crystallized and become definite in this respect.

From such data as are available some facts may, however, be considered as established—one of them being the ability of large, modern, heavy-armored ships not only to survive but to continue in action after the most severe punishment.

Taking the battle of Jutland as a test of the defensive qualities of modern ships, one finds that in the entire action only four such ships, all of them of the battle-cruiser type, were lost as a result of the action. The four outstanding facts of interest to the designer, as they emerge in the author's analysis from the smoke and flame of the battle, are—first, the value of armor protection; second, the necessity for the maximum number of major-caliber guns; third, the tactical value of speed; and fourth, the futility of subjecting all the ships to the attack of modern weapons.

Protection of ships is being recognized as one of the most important problems and both the British and American naval services have solved this problem so far as the torpedo has been developed to date. Moreover, although the solutions differ radically in detail, they do not differ much in underlying ideas.

Of the general phases of the war the submarine campaign is given considerable attention. Nations must in the future be prepared to find the submarine playing an important part in attacking and throttling enemy commerce, even on their own coasts. The proposal brought forward at the Peace Conference to abolish the submarine entirely by international agreement was not adopted, wisely as the author thinks, for so long as the possibility of war remains, progress of science and engineering and their application to the art of war cannot successfully be throttled unless there is complete unanimity of sentiment throughout the civilized world. War experience developed certain facts regarding the submarine. It is essentially an instrument of stealth. Once detected it is at the mercy of a surface vessel, and detection devices, while not perfect yet, have been and will be steadily improved as time goes on. If we had today an accurate device which would locate a submerged submarine with reasonable approximation several miles off and with accuracy when 100 to 200 ft. directly under the surface vessel, the submarine would be obsolete as a weapon of war.

The position of the destroyer as an element of the fighting navy has been particularly enhanced as a result of the war experience, both to attack the enemy and protect its own capital ships. As regards the design of destroyers, the principal demand has been for greater cruising radius and increased shelter and comfort for the personnel.

The rôle played by trawlers, drifters and other small vessels of a similar type has been changed somewhat from that intended at the beginning of the war. Primarily they were to be used as mine sweepers and tenders, but the development of the depth charge made it possible to employ them against submarines, which, in fact, became the principal use to which these small boats were put during the last year and a half of the war. In fact, as time went on, special boats of small sizes were developed for this kind of service, such as the British patrol boats and the United States eagle boats.

Aircraft carriers are particularly mentioned as a new and large type of naval vessel. The actual offensive use of aircraft against naval vessels was little developed during the war.

The author believes that developments in the air will be both along lines of offense against capital ships and of defense of them by auxiliary and offensive aircraft. He states, however, that the big ship which must be protected from projectiles of a ton weight falling at an angle of 30 deg. and fired from ships almost out of sight below the horizon, is not yet in serious danger from bombs carried by present-day aircraft, with chances of hitting small indeed. It seems probable that aircraft will sooner become dangerous to destroyers and light vessels, generally, than to the large ships of the line and even then in order to perform efficiently their functions with the field aircraft must have means for being carried with the fleet, not only on long cruises but actually in battle, which accounts for the development of the aircraft-carrier type of ship.

To sum up, the experience of war, so far as it can be grasped to date, has resulted in demands of every existing type of war-vessel which can only be met by increased size and cost. It has resulted in the introduction of only one new type of major importance, namely, the aircraft carrier, but it has introduced a number of small types which will probably survive but need not be constructed in large quantities in time of peace.

While the present tendency is towards increased size and cost, this very fact, under the present financial economical and political conditions in the world, may actually result in the long run in the disappearance from future building programs of these very types, and the substitution for them of smaller and cheaper units made possible by new developments in size and engineering. (*Journal of the Franklin Institute*, vol. 190, no. 2, Aug. 1920' pp. 157-185, illustrated, *gA*)

## POWER-PLANT ENGINEERING

### Soil Vibration Due to Running Machinery and Methods of Damping It

TRANSMISSION TO A DISTANCE OF GROUND VIBRATIONS DUE TO MACHINERY WITH RECIPROCATING MASSES. Gerb. Discussion of the transmission of ground vibrations due to reciprocating machinery, in which it is found that such transmission to a distance takes place only through plastic subsoil. The vibrations propagate in the form of surface waves. Recommendations are made as to means of avoiding such phenomena.

At the end of 1914 in the Royal Munitions Factory in Danzig a horizontal tandem steam engine of 650 to 850 hp. was installed and when it began to run it set into vibration a number of houses in the neighborhood, some of which were located as far away as 300 m. (say, 1000 ft.). The engine foundation itself was soon split by a horizontal crack in such a manner that the upper and lower parts moved with respect to each other when the engine was running. The free mass forces of the engine in the horizontal and vertical directions amounted to about 10,000 kg. (say, 22,000 lb.) and observation in the houses affected by vibration indicated that the vibrations corresponded to an engine speed of between 160 and 180 r.p.m.

Careful measurements have further indicated that the direction of vibration of the masses did not at all coincide, as might have been expected, with the direction of the reciprocating masses in the engine and that rather the masses vibrated regularly in the direction of the axis of the minimum moment of inertia of the house foundation, hence, at times, normally to the direction of motion of the reciprocating masses of the engine.

The vibrations were of maximum value in the upper floors of the houses but did not grow in proportion to the distance from the ground. An interesting feature lies in the fact that the vibration phenomena were limited to the houses themselves, and the apparatus used for their discovery had not shown any on the ground proper, even in the immediate proximity of the engine. Likewise, it has not proved possible to establish any law for the variation of ·

the accelerations with the distance from the source of vibration.

Only horizontal accelerations were established in the houses, and no measurements of vertical vibrations have been made.

Because of the fact that no vertical accelerations in the houses had been discovered it was assumed that the horizontal accelerations had been transmitted by the plastic stratum of the ground and it has been proposed to isolate the vibrating machine from the surrounding structures by means of a ditch. This means has, however, proved ineffective here, as it did in some other cases, because it is not the solid surface layer of the ground but the plastic subsoil that transmits the vibrations, and this being the case, the only way to prevent the transmission of vibrations is to obviate their transmission from the source to a plastic subsoil body. This cannot be done by insulating layers at the foundation, as materials used for such insulating layers, such as felt, cork, etc., are perfectly capable of taking up vibrations of the higher frequencies propagated as noises, but are not elastic enough as compared with the elastic subsoil and cannot take up elastically horizontal acting forces.

The only truly reliable method of procedure is to attack the evil at the source and to eliminate the generation of vibration by a proper balancing of the reciprocating masses. This can be done in the engine itself only if it has not less than six cylinders. It is,

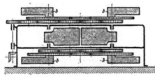

Fig. 1  Device for Balancing Reciprocating Forces in Horizontal Engines of Less Than Six Cylinders

however, possible to do this by special appliances even in a single-cylinder engine. This is based on the fact that horizontally acting free masses in crankshaft drives produce vibrations of the same frequency as the rotative speed of the engine, and, in addition to this, that higher vibrations have a double frequency. As shown in Fig. 1, the basic frequency can be ironed out by two balancing masses a, rotating at the same speed as the engine but in an opposite direction, which eliminates the component of centrifugal force acting at right angles to the direction of motion of the reciprocating masses. For the same reason and also in order to prevent the appearance of moments, the weight which balances out the vibrations of double frequency is divided into four parts b rotating at a speed double that of the engine. In this way the diagram of the centrifugal forces developed coincides at all rotative speeds with the diagram of the free horizontal forces in the engine. Since, however, the direction of the former is opposite to the direction of the latter and occurs in the same space in which the free forces of the engine operate, theoretically no forces or moments can be transmitted to the foundation. The device shown is driven from the camshaft and consumes a very small amount of power as it has only to overcome the comparatively slight friction in the ball bearings and gears. It is, however, necessary that the connection with the engine body should be strong enough to withstand the stresses produced. (*Zeitschrift des Vereines deutscher Ingenieure*, vol. 64, no. 38, Sept. 18, 1920, pp. 750–760, 2 figs., *dp*)

### Device for Cutting Off Last Expansion Stage of Turbine from the Exhaust Main

Low-Pressure Steam Control Device, J. W. Smith. In a mixed-pressure steam-turbine and electric-generator installation placed in the power station of an English factory there was no provision made for automatically cutting off the last expansion stage of the turbine from the exhaust main when the supply of exhaust steam failed.

When the air supply of this steam fell below the quantity which could be passed by the low-pressure starting valve a partial vacuum

was induced in the exhaust main, which, in this case, is about 400 yd. long, with the inevitable result that air was drawn in through some leaky joint or open drain cock. The starting valve could be regulated by hand to pass only the proper quantity of exhaust steam, but such a solution of the problem was troublesome and would mean that one man's whole time would be occupied. Instead it was solved by putting in an automatically operative controller.

First, there is a low-pressure receiver which diminishes the fluctuations in pressure of the exhaust steam from steam-driven machinery, such as steam hammers and non-condensing steam pumps. After leaving the receiver (Fig. 2) the exhaust steam enters the low-pressure chest through pipe C, then passing downward through the starting valve and upward through the emergency stop valve and the governor throttle valve into the turbine. The turbine steam chest is fitted here with an emergency trip-stop valve A which closes upward and the spindle passes vertically through the bottom governor. This valve is of a piston type and is a duplicate of the governor throttle valve B fitted immediately above it. As arranged by the makers it was connected to its opening lever by a pair of short links C. These links were replaced by some longer ones with a distance piece riveted between their ends to form a flat surface D, and the bottom holes of the new links were elongated to allow the valve to be closed without raising the original operating lever. Under this against the wall of the condenser pit, was fitted a lever about 4 ft. 6 in. long. At about 3 in. from its fulcrum a vertical rod was jointed which passed through a suitable guide so that its end was just clear of the flat bottom of the new valve links when the new lever was at the bottom of its travel. The end of the lever had to be raised about 12 in. to close the valve, and suitable guides and stops were provided to prevent all undue stresses. Owing to the 18-to-1 leverage provided, the effect of the friction of the valve and spindle was negligible at the end of the lever. The steam receiver E, the shell of an old Lancashire boiler, has a siphon drain F connected to one end plate about 3 in. above the bottom of the shell. In consequence of the position of this drain there was always a quantity

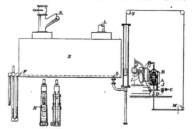

Fig. 2  Diagram of Controller Used on a Mixed-Pressure Steam Turbine at the Gloucester Wagon Company's Central Power Station

of water lying in the bottom of the shell, the surface of which measured roughly 3 ft. by 20 ft. At the other end of the steam receiver the old blow-off cock fitting G on the bottom of the shell had been blanked off. A $2^1/_2$-in. cock was fitted to this blow-off cock branch one week end, and, working back from that the following week, an old 7-ft. length of 8-in. cast iron pipe was rigged up in a vertical position and supported from the floor by a length of $2^1/_2$-in. pipe with a blank flange at the bottom. This pipe formed an excellent float chamber with sufficient head of water to seal the maximum pressure permissible in the receiver. Owing to the wide ratio between the surface of the water exposed to the steam on one leg of the submerged bend, and the surface of the water exposed to atmospheric pressure on the other leg of the submerged bend, about 60 sq. ft. to 50 sq. in., the rise of water in the float chamber was almost 2.3 ft. per lb. pressure in the receiver. A galvanized-iron float was provided, 6 in. diam. by 12 in. deep, with a stretcher bar across its open top to enable it to be suspended correctly. This was filled with water and connected to the new lever by a flexible steel wire rope passing over two guide pulleys. Sufficient weight was then added to the lever to cause the float to be normally half submerged when the lever was clear of both stops.

It was found in tests that the displacement of only 2 in. of water above or below the center of the float is sufficient to operate the valve and that only a $1/2$-lb. variation of steam pressure is required to move the valve from "open" to "shut," or vice versa. Moreover, it is also stated that the device has been in constant use for eight years without any repairs. (Paper read before the Gloucestershire Engineering Society, abstracted through *Power House*, vol. 13, no. 19, Oct. 5, 1920, pp. 437-439, 2 figs., *d*)

### Water-Indicating Devices for Locomotives

GOVERNMENT TESTS OF WATER-INDICATING DEVICES. For the purpose of determining, if possible, the general outline of the flow of water existing at the back head, when high evaporation was taking place, tests were made on one of the U. S. Railroad Administration standardized 2-10-2-type locomotives.

In these tests it was found that the way the top connection to the water column is made affects very materially the general outline assumed by the water on the back head. It also appeared that under certain conditions dry steam was being obtained both at the back knuckle and further ahead, which is believed, however, to be partly due to the exceedingly good water used in this test.

Further tests were made to determine the approximate outline and proportions of the water conditions existing at the back boiler head while the locomotive is being operated with heavy throttle, or when steam is being rapidly generated and simultaneously escaping from the boiler. These tests covered the distance of 808 miles in bad-water districts on approximately level track with a

FIG. 3   EXPLORATION TUBES USED AS WATER-INDICATING DEVICES IN TESTS OF WATER CIRCULATION IN LOCOMOTIVES

locomotive of the heavy 2-8-2-type equipped with superheater and duplex stoker.

The test appliances are shown in Fig. 3, the apparatus consisting of four gage cocks applied directly in the back head near the knuckle; one water column to which three gage cocks and one water glass were attached; one water glass with a 9-in. reading, standard application, with both top and bottom cocks entering boiler back head direct; one water glass applied for experimental purposes with a bottom cock entering the boiler head on back knuckle and one entering 13 in. ahead of the back knuckle, together with four exploration tubes or sliding gage cocks. Fig. 3 shows a side elevation of these exploration tubes.

Here it was found that while generally the same conditions prevail as in other tests, the outline of water reached a higher elevation and greater proportions at the back head than in the previous tests, which is due probably to the poor water used here.

Tests were also made on a switching locomotive in which, in addition to the usual apparatus, a glass tube was inserted in the top of the wrapper sheet which permitted the use of an electric

light inside the boiler, clearly illuminating the steam space over the crown sheets. Five bull's-eye sight glasses were distributed so that the action of water in the back head could be seen while under steam pressure, as shown in Fig. 4.

Both main rods were disconnected, crossheads blocked at end of stroke and valve stems disconnected and so placed that steam was discharged through the exhaust nozzle and stack, creating a forced draft on the fire, representing as nearly operating conditions as possible.

When the throttle was closed and no steam escaping from the boiler the surface of the water was approximately level, with a distinct circulation noted from back to front and from the sides toward the center of the crown sheet. When the safety valves lifted, the water rose with fountain effect, around the edges of the firebox, from 1 in. to 2 in., and the circulation was materially increased.

When the throttle was opened and steam was being generated and escaping from the boiler in greater volume, the level of water

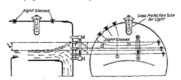

FIG. 4   ARRANGEMENT OF LIGHTS AND OBSERVATION GLASSES IN BOILER TO DETERMINE WATER CONDITIONS

throughout the boiler was seen to rise 1 in. to $1^{1}/_{2}$ in., which rise was registered by the water glass, and a marked flow of water, with fountain effect, was observed rising around the firebox at the back head and wrapper sheets, reaching a height above that over the remaining portion of the crown sheet of approximately 2 in. to 4 in., in proportion to the amount of steam being generated and simultaneously escaping from the boiler.

The important feature to be noted is that this height of water, as seen at the back head, was approximately 4 in. at its maximum, and was registered by the gage cocks, while at the same time it could be seen that the water glass was registering the level further ahead over the crown sheet.

Among the interesting features observed were the size of the steam bubbles which were approximately $1/_4$ in. to $3/_8$ in. in diameter, and the rapidity with which they were seen to rise to the surface and explode. The size and number of these steam bubbles, which were seen rapidly rising next to the back head, explain one of the physical reasons for the increased height of water around the crown sheet and the rapid circulation attained.

These observations establish beyond question that when steam is being generated and escaping there is an upward movement of water at the back head of the locomotive boiler which carries it above that further ahead over the crown sheet, and that the gage cocks, when applied directly in the boiler, register this rise of water and do not indicate the level further ahead, while the water glass registers the level of water further ahead and not the fountain of water at the back head.

Among the general observations made attention may be called to the following: It is recognized that the volume of water in the boiler increases in proportion to the amount of steam being generated and in the same ratio that the steam bubbles below the surface are formed and expanded, the volume of which depends to a very considerable extent upon the purity of the water in the boiler its ability to readily release the steam being generated, consequently increasing the height of water in the same proportion, which height is registered by the water glass.

Gage cocks secured directly in the boiler have been shown to be incapable of correctly indicating the general water level and an arrangement is described and illustrated in the report which is claimed to be more correct and safe than the ordinary methods. (Report of the Bureau of Locomotive Inspection of the Interstate Commerce Commission, abstracted through *Railway Me-*

*chanical Engineer*, vol. 94, no. 10, Oct. 1920, pp. 630–633. This is the second part of the report, the first part having been given in the *Railway Mechanical Engineer*, for Sept. 1920. *dep*)

## POWER TRANSMISSION

### Liquid Variable-Speed Transmission Gear

CAREY OIL TRANSMISSION SYSTEM. Description of the pump used for an oil transmission system. The rotor (Fig. 5) consists of a solid block of steel $R$ in which five recesses are formed to act as cylinders. Each recess is fitted with a glass-hard steel bush in which works an ordinary steel ball acting as a piston and having a total clearance of about one one-thousandth of an inch. These ball pistons were in conjunction with a track $T$ of glass-hard steel, the track being carried by a cast-steel ring $S$ which works between

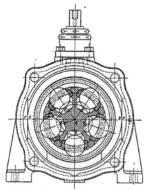

Fig. 5  CROSS-SECTION OF ROTOR OF THE CAREY PUMP

guides $P$ formed in the cast-iron main casing or pump body. This ring may be moved to different positions in a vertical direction between the guides so that varying degrees of eccentricity may be obtained between the center of the track and the axis of the rotor, either above or below the center line. When the centers of both parts coincide, no relative radial motion takes place between the rotor and the ball pistons. As the degree of eccentricity is increased the quantity of oil delivered by the pump increases likewise until the limit of travel is reached. When the track engine is raised above the center, flow takes place in one direction and this direction is reversed when the track is moved to a position below the center line.

In each of the five cylinder recesses there is a port $F$ with a hardened base and a floating valve (not shown in the drawing) having a rocking motion and also a rotating motion so that it can adjust itself to the position of the rotor. This valve carries the inlet and outlet ports.

By combining two machines of this type it is possible to make a variable-speed transmission gear. One machine is driven by a prime mover at constant speed and acting as a variable-stroke pump delivers oil under pressure to the other machine which acts as a hydraulic motor. This can be done by coupling up two Carey machines back to back, but with their respective back covers replaced by a single junction casing which carries the two floating valves and provides the necessary passages for the flow and return of the working fluid. The ratio of speeds between the driving pump and the driven pump is controlled by the lengths of the strokes of the two machines.

It has been found that for speeds up to 500 r.p.m. these trans-

mission gears may be filled with oil, but for higher speeds it is desirable that the rotor should run in air. (*The Engineer*, vol. 130, no. 3377, Sept. 17, 1920, p. 284, 2 figs., *d*)

## PRESSES

AN ADJUSTABLE-POSITION PRESS. Description of a press designed and built by the Toledo Machine and Tool Company, Toledo, Ohio, the most interesting feature of which is, that it is operative in either an upright or an inclined position, a turn of the handwheel permitting a change in position. The press is driven by a direct-connected electric motor so arranged that it stays vertical no matter whether the press is in the vertical or inclined position. (*The Iron Trade Review*, vol. 67, no. 17, Oct. 21, 1920, p. 1140, 2 figs., *d*)

## RAILWAY ENGINEERING (See also Power-Plant Engineering)

REGAN AUTOMATIC TRAIN CONTROL SYSTEM. This system, as installed on the Great Eastern Railway where the Westinghouse brake is employed, consists of a shoe suspended from the center of the rear buffer pin of the engine, an electropneumatic valve behind the steps on the right side, a relay box and release switch on the same side, a battery box and a speed-control device on the right wheel of the controlling nozzle. The equipment on the track consists of a ramp 100 ft. long with inclined ends fixed in the "four-foot" at each signal and electrically connected thereto.

The speed-control device consists of a centrifugal governor mounted upon a baseplate rigidly bolted to the end of the trailing nozzle and accurately centered therewith.

The shoe mechanism consists of a shoe stem, a cylinder and a circuit controller. The upper part of the steam connects to the brake pipe, and as the stem is hollow, it follows that should the shoe be broken off the air would escape and the brake would be automatically applied.

A demonstration of this system was given on the Great Eastern Railway on September 30, 1920, when the speed control was set for 24 miles per hour and the train was stopped from that speed in 225 ft.

In this connection, it may be stated that in an editorial in same issue of *The Engineer* (p. 355–356) automatic train control is advocated for Great Britain and instances are cited of serious accidents which could have been prevented but such a control been available. In fact, it is stated that the Ministry of Transport has appointed a Departmental Committee to consider the whole question.

In discussing the matters to be considered by this new Committee the editorial points out as one having a very serious aspect, the question whether or not the uniform system for the whole of Great Britain should be determined upon. This is particularly important as British locomotives do a considerable amount of running over "foreign" tracks. There is also the question as to what object automatic train control has to achieve; this is, whether or not the driver should have an indication of the state of the running signal which he is unable to see, for example, during fog or falling snow.

An interesting discussion is presented as to the conditions which the track equipment should satisfy. The subject of speed control is also not as simple as it looks, as, for example, it is not easy to insure that a mixed-traffic engine just taken off a fast freight train is properly adjusted for working an express passenger train, and vice versa. (*The Engineer*, vol. 130, no. 3380, Oct. 8, 1920, p. 348, *dg*)

## RESEARCH

### Progress of Scientific and Industrial Research in England During the Past Year

REPORT OF THE COMMITTEE OF THE PRIVY COUNCIL FOR SCIENTIFIC AND INDUSTRIAL RESEARCH FOR THE YEAR 1919–1920. From the report presented it would appear that a very active program for research is being steadily pushed ahead in England and that quite substantial funds are available therefor. In particular, it is pointed out that the British industries with an awakening interest in the value of research and of scientific control of manufactures, are making such demands on the available supply of research workers and offering such salaries that the universities face the danger of losing their teaching staffs in the same manner as they did during the war.

Important grants have been made to students and independent workers as well as to provide professors with research assistants of scientific standing. No conditions are attached to the grants made to workers whose sole aim is the extension of knowledge, either as to the line of their work or as to the use of the results. The only condition is that if they propose to make commercial use of their discoveries they must consult the Committee for Scientific and Industrial Research, because at this point they are leaving the field of pure investigation; otherwise, however, their tenure of the grant is perfectly free.

There are now 18 research associations of various character in England licensed by the Committee and covering many fields of industry, and five other associations have been approved but have not received their licenses.

The Department of Research is committed to a total expenditure of £450,000 on account of the established research associations and to a further expenditure of at least £120,000 on account of those approved but not yet licensed. The total committments may, however, reach £800,000.

It is claimed that the industries fully realize already the importance of research, though they have not yet quite come to a full appreciation of its difficulty or its full worth. The best proof that the value of research is realized is shown that large funds for the endowment of research are being raised privately; for example, the cotton industry is hoping to raise £250,000 for that purpose.

The report mentions more or less briefly the method of organization of research and the subjects which occupy the attention of the various boards and committees.

In this connection it is of interest to note that the question of fuel is considered to be a basic problem of the greatest national importance. As regards the Committee of Research, it is in the hands of a fuel research board. There are numerous other boards covering the vast field of investigation.

As a sample of the work of these committees may be mentioned the following quotation describing the Zirconium Inquiry Committee:

It was frequently suggested during the war that zirconium-containing steels of remarkable hardness had been produced and were being employed by enemy countries in the production of light armor for aeroplanes and tanks. In July 1918, at the request of the Ministry of Munitions, a Zirconium Inquiry Committee was set up with the object of investigating the preparation of ferro-zirconium and, from the latter, of zirconium-containing steels. The experimental difficulties in the way of production of a true alloy of iron and zirconium were numerous and severe, and after much work they were only partly overcome. It was found possible to prepare, in 50-lb. batches, a ferro-zirconium containing, however, a considerable portion of carbon, while much of the zirconium in the material was present as carbide. It was consequently very doubtful whether the zirconium in this material could be transferred into a melt of steel. In the meantime numerous inquiries have failed to discover any source of supply of ferrozirconium containing only small proportions of carbon and silicon, or any evidence that zirconium added to steel conveys to it any beneficial quality. Further, the examination of each armor of enemy origin has not revealed the presence of zirconium.

After the conclusion of the war the importance of the inquiry diminished and in consequence the Committee were of opinion that their investigations should be concluded. We have recently received their report and have recommended that it be communicated to the fighting services.

In an appendix some information is given on the developments of the organization of research in the Overseas Dominions and in the allied and associated countries.

In particular, there is under consideration the proposal for the establishment of a National Institute for Canada, on which the report of the Committee was unanimously adopted by the Canadian House of Commons in May last. The cost of building and equipping the Institute is estimated at $600,000 and approximately $50,000 is required for the salaries of the staff during the first year. The proposed Institute will exercise functions akin to those of the Bureau of Standards in Washington. It is also anticipated that the Institute will play an important part in fostering the movement for coöperative research among the industries of Canada by placing laboratories and (if necessary) staffs at the disposal of

coöperative research guilds when they are not in a position to maintain laboratories of their own.

There are already several organizations in Canada established with a view to industrial research, such as, for example, the Fuel Research Board appointed to deal with the standardization of coals through the Dominion, and the Peat Commission which has already successfully developed a bog at Alfred, Ontario. (Abstract of Fifth Annual Report of the Committee of the Privy Council for Scientific and Industrial Research, presented to Parliament by Command by His Majesty and published in London in 1920, 120 pp., *g*)

# SPECIAL MACHINERY (See Power Transmission)

# STEAM ENGINEERING

## German Tests on Steam Nozzles and Diffusors

STEAM NOZZLES. A German engineer, Friedrich Muller, of Berlin, has recently published certain data of an investigation dealing with tests on a steam diffuser.

Investigations dealing with the inner processes in the conversion of pressure into velocity in nozzles have not entirely cleared up the theory of steam flow; for example, no one has succeeded thus far in giving a complete expression for processes which take place under the assumption of a unidimensional flow, and Professor Stodola in an article in *Zeitschrift des Vereines deutscher Ingenieure* (1919) states that these processes will not be made clear until laws are established on a polydimensional conception of flow.

In the experiments described by Muller, carried out in the laboratory of machine design of the Technical High School, Charlottenburg, live steam of various temperatures and pressures was expanded in a DeLaval nozzle. The nozzle was attached in the usual manner to a movable arm of a reaction device. When at dead center this arm was held rigid and the steam flowed crosswise toward the exit from the nozzle of the diffuser and was then condensed on the pipes of a surface condenser. A valve was provided to make it possible to create the desired back pressure in back of the diffusor. The state of the steam previous to the entrance into the nozzle was determined by measuring the steam pressure by a manometer and likewise by measuring the respective temperatures.

In the tests the following two concepts were followed up: first, the variable state of the steam with the given nozzle; and second, different nozzles with a corresponding initial state of the steam.

The data of the tests are presented in the original publication in eight tables. Among other things, this investigation led to the conclusion that we have not yet the means necessary to produce the designed expansion, but we can by proper devices make the nozzles capable of satisfying given requirements of pressure conditions. On the other hand, these tests have been sufficient to throw a great deal of light on what takes place in the diffusors. Hitherto information available was only sufficient to determine the shape of the diffusor when the given weight of steam expanded in a nozzle had to be compressed to a given back pressure. The author proposes to proceed in the opposite direction, that is, to pay no attention at first to the shape of the diffusor and instead to determine that pressure both in front and back of the diffusor which gives the best efficiency.

By efficiency of the diffusor is, therefore, understood here the ratio of two heat heads, one being the expansion of the steam in the nozzle and the other determinable by the pressures in front and in back of the diffusor. This latter heat head is to be measured on the adiabatic curve of the nozzle expansion.

The most reliable data have been obtained in the tests presented in Fig. 6. The curves *d* and *e* give the most desirable variations of pressure. According to the general theory of nozzles, the following two equations are available: first, equation of continuity:

$$W_{pr} = \frac{G \times v_{14}}{F_m \, Diff} \quad\quad\quad\quad [1]$$

second, equation of sound velocity:

$$W_{th} = g \times k \times p_{14} \times v_{14} \quad\quad\quad\quad [2]$$

in which the following notation is used:

$W_{pr}$ = velocity in meters per second computed on the assumption of continuity

$W_{th}$ = velocity of sound in meters per second

$G$ = weight of steam flowing per second in kg.

$g$ = 9.81 meters per sec. per sec.

$k$ = ratio of specific heat at constant pressure to specific heat at constant volume

$$\psi = \sqrt{g \times k \times \left(\frac{2}{k+1}\right)^{(k+1)/(k-1)}}$$

$p$ = pressure in kg. per sq. m. available at any time

$v$ = available specific volume in cu. m. per kg.

$f_{m\ Düse}$ = minimum cross-section of the nozzle in sq. m. (Düse means nozzle)

$F_{m\ Dif}$ = minimum cross-section of the diffusor in sq. m.

If we insert the experimental values as given in Fig. 6 into Equations [1] and [2], we find that the variation of pressure is free from

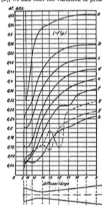

FIG. 6  CURVES OBTAINED IN TESTS ON A STEAM DIFFUSOR AT THE MACHINE CONSTRUCTION LABORATORY OF THE TECHNICAL HIGH SCHOOL IN CHARLOTTENBURG

(at. abs., atmospheres absolute; Diffusorlänge, length of diffusor.)

shocks and the diffusor develops its best efficiency when it is assumed that

$$\frac{W_{th}}{W_{pr}} = 1 \quad\quad\quad\quad\quad [3]$$

Since the maximum, as has been found elsewhere, occurs with a back pressure of about 0.435 atmos. abs., it follows that

$$\frac{Gv_{14}}{F_{m\ Dif}} = \sqrt{g \times k \times p_{14} \times v_{14}} \quad\quad [3a]$$

which when simplified reduces to

$$\left(\frac{G}{F_{m\ Dif}}\right)^2 = (g \times k) - \frac{p_{14}}{v_{14}} \quad\quad [4]$$

Equation [4] becomes similar to the general nozzle equation

$$G = F_{m\ Düse} \times \psi \sqrt{\frac{p_1}{v_1}} \quad\quad\quad [4a]$$

the latter is expressed in the form

$$\left(\frac{G}{F_{m\ Düse}}\right)^2 = (g \times k)\frac{p_m}{v_m} \quad\quad [5]$$

By combining Equations [1] and [2] the conclusion may be arrived at that a definite ratio of pressures may characterize a diffusor just as it does nozzles. It is further found that most likely when the flow in a diffusor is free from shocks, there is a very definite value determining the ratio between pressure in the smallest

cross-section $p_{14}$ to end pressure $p_5$. The tests have shown that this ratio for the given turbine was equal to $p_{14}/p_5 = 0.6$. The author expressed it, however, in another form, namely,

$$\frac{p_{14}}{p_5} = \left(\frac{2}{k+1}\right)^{k/(k-1)} \quad\quad\quad [6]$$

where $k$, in the case of diffusors, has the usual values for saturated steam. Substituting this in Equation [3a] or its derivative, he gets

$$G = F_{m\ Dif} \times \psi \times \sqrt{\frac{p_5}{v}} \quad\quad\quad [6a]$$

Furthermore, as the weight of the steam flowing per unit of time is the same for the nozzle and the diffusor and assuming that $\psi_{Düse}$ for the nozzle is equal to $\psi_{Dif}$ for the diffusor, it appears that the following relation holds good for the ratio between the minimum cross-section to the minimum cross-section to the diffusor.

$$\frac{F_{m\ Dif}}{F_{m\ Düse}} = \frac{\sqrt{\dfrac{p_1}{v_1}}}{\sqrt{\dfrac{p_5}{v_5}}} \quad\quad\quad [7]$$

Moreover, this equation was proved to be approximately correct experimentally and the following values have been found to hold good:

$F_{m\ Düse}$ = 5.11 sq. cm.

$p_1$ = 3 atmos. abs.

$v_1$ = 0.911 cu. m. per kg.

$D_{m\ Dif}$ = 30.2 sq. cm.

$p_5$ = 0.41 atmos. abs.

$v_5$ = 5 cu. m. per kg.

$F_{m\ Dif}/F_{m\ Düse}$ = 1.82/0.287.

Fig. 6 among other things, explains the pressure variations, since with an excessively large ratio of pressures the steam finds in the convergent and divergent parts of the nozzle somewhere cross-sections where oscillations of pressure occur. If, however, the pressure ratio is below the theoretically correct one, then the minimum cross-section must be smaller than that actually available and because of this the steam does not lie at the walls but forms a diffusor of its own. (Mitteilungen des Maschinenbau-Laboratoriums in Berlin, abstracted through Der Praktische Maschinen-Konstrukteur, vol. 53, no. 31, Aug. 5, 1920, pp. 280–282, 1 fig., et)

## TESTING MATERIALS

NEW MAGNETIC TESTING APPARATUS. Description of a new device for the magnetic testing of steel which its inventor, Dr. Charles W. Burrows (Mem.Am.Soc.M.E.), calls the defectoscope.

The big advantage of magnetic testing is, first, that it can be applied to the article in its finished form; and second, that it in no way affects the article or mars it.

When a test is made by means of magnetic defectoscope the bar must first be magnetized. This is effected by means of a relatively short solenoid energized by a direct current of such value that the magnetism is carried well beyond the knee of the induction curve. Next, a detector is used to discover magnetic variations of the bar. This detector consists of two test coils having the same number of turns and surrounding the specimen. Detection is made by shifting the detector from one position to another along the length of the test material, and as it occupies different positions it is threaded by an induction which depends upon the nature of the specimen. If the specimen is not quite uniform the magnetic induction threading one of the coils and the detector is different from the magnetic induction threading the other coil. The result is that the electromotive force generated in one of the test coils is different from that generated in the second test coil and this differential electromotive force indicates lack of homogeneity in the specimen and can be read from the indicator, in this case a heavily damped short-period D'Arsonval galvanometer provided with a recorder which is essentially a photographic film caused to move uniformly across a small slit through whose opening a spot of light is reflected by the galvanometer. The instrument is, of course, provided with the control box containing the necessary electrical switches, rheostats, and instruments.

In the original article it is shown how the defectoscope may be applied to the examination of rails, rods, wire, cable, etc. Among other things, the instrument may be readily applied to the examination of strip material such as is used in the manufacture of various saws and coil springs.

It would appear that for the examination of material which possesses circular symmetry, test methods have to be applied somewhat different from those used for rectilinear material. In fact, small circular objects have a special method of their own for magnetic examination, namely, examination in a rotating magnetic field and measurement of the magnetic torque exerted on the specimen by the rotating magnetic field, so that it is really not the permeability but rather the magnetic hysteresis of the material that is measured. (*Iron Age*, vol. 106, no. 18 Oct. 28, 1920, pp. 1125–1128, 7 figs., *d*)

## TRANSPORTATION

### Motor Transportation of Steel Products

TRUCK TRANSPORTATION IN THE STEEL INDUSTRY. Myers L. Feiser. More steel is said to have been moved by motor truck during the past several months than ever before in this country, due to the demoralized traffic conditions which reached an acute stage last April as a result of the outlaw railroad strike. Steel recently carried in motor trucks from mill to consumer or to cars for shipment to consumer is estimated at more than 100,000 tons monthly, although no reliable records are available to cover the entire industry. Some makers, have, however, tabulated their truck tonnages, and one in the Pittsburgh district reports 25,000 tons in a single month and an average close to this figure for some time.

In normal times distance has been one of the factors limiting the use of trucks in steel hauling, but in the past few months unusually long hauls have been resorted to. One user of sheets in New England, for example, sent a truck all the way to Pittsburgh for sheet bars. Another buyer of sheets sent his truck from Philadelphia to Niles, Ohio, for a quantity of black sheets. These distances are exceptional, however, and most of the trucking of steel recently has been on an average of perhaps 50 miles.

The main sizes of trucks used for steel carrying have been 5-ton and 3-ton. The costs varied widely, but the demand for steel has been such as to make the cost in some cases a minor factor.

The organization of trucking with the various companies has been different, few, if any, steel makers maintaining a trucking service of their own for their customers, though some, as, for example, the Republic Iron and Steel Company, Youngstown, Ohio, have trucks for their own requirements.

Motor-truck transportation apparently appealed particularly to consumers of high-grade steels, which steels are not necessarily required in large quantities, especially that regular motor routes have been established between cities with scheduled stops at plants and towns on the way. It has been found that the charges over these routes vary, but when material is urgently required the extra cost represents a premium which the consumer is willing to pay in order to obtain delivery.

A canvass of steel manufacturers has shown that the increase in the movement of material by truck is held up by the condition of roads, and that otherwise with proper equipment, truck transportation of steel and iron products especially for short distances has apparently a promising future. It would appear, however, that under normal conditions the question of cost will have a decided influence. (*The Iron Trade Review*, vol. 67, no. 17, Oct. 21, 1920, pp. 1127–1129, 5 figs., *dg*)

## CLASSIFICATION OF ARTICLES

Articles appearing in the Survey are classified as *c* comparative; *d* descriptive; *e* experimental; *g* general; *h* historical; *m* mathematical; *p* practical; *s* statistical; *t* theoretical. Articles of especial merit are those of the reviewer, not of the Society. The Editor will be pleased to receive inquiries for further information in connection with articles reported in the Survey.

## FOUNDATIONS FOR MACHINERY

(*Continued from page 672*)

to bring the system to rest. It is still more important, however, to note the fact that we introduced the additional degree of freedom precisely in the sense of action of the disturbing agency, that is, in the sense of the plane of the figure and not at right angles thereto; or, say, in an up-and-down sense, as for instance by providing a coil spring instead of joint *II*.

### THE BASIS OF FOUNDATION DESIGN

This, then, will be taken as basis for our further discussion: In contemplating the design of a foundation we shall always separate those directions (or axes of instantaneous rotation) about which the system cannot (or at least is not likely to) oscillate from those directions (or instantaneous axes) about which the system is more or less certain to vibrate. We next shall select a "steady" point from purely practical considerations, and finally devise such means of controlling the "free periods" of the system as will secure the desired degree of remoteness from synchronism under the actual operative

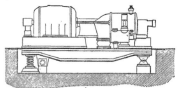

FIG. 3   APPLICATION OF AUTHOR'S DESIGN TO A TURBO-GENERATOR

speed. Such means of course will be springs, exceedingly heavy, and not in the least calculated to allow of any free wabbling of the system. They will also be adjustable so that the desired periods may be readily varied within wide limits; and in general structurally arranged to introduce as few changes as possible.

The author feels that to submit too many particulars as regards the detailed designs of such an arrangement would certainly defeat the purpose of this paper. Considering, therefore, only one type of apparatus, a turbo-generator (Fig. 3), we must start out with the selection of the steady point. We will naturally place it as near the steam main as possible (not to the exclusion, of course, of a suitable expansion joint), as under all conditions, should there be a choice of position, preference should be given to that point as far as possible from the center of gravity of the system, so that any static unbalance (whipping, etc.), would be made to act not as a force *upon*, but as a moment *about*, that steady point. Such point should actually be made as steady as possible and no trouble should be spared in providing suitable piling or digging down to the solid ground and constructing suitable footings.

The next problem is to design a substructure adapted to receive the bedplate of the apparatus and made stiff enough so as to eliminate any "periods" of its own. This bedplate may be made of structural steel or of reinforced concrete, in which latter case the ends thereof may be made of cast iron. The substructure is supported upon the steady point either by a ball-and-socket arrangement, or is simply bolted at that point to the floor plate underneath by a bolt which need not necessarily be very light but which must be arranged in a manner to secure the minimum area of actual contact. Remembering that in apparatus of this sort the tendency to oscillate about the axis *z* (Fig. 1) is always rather negligible, we have practically only two degrees of freedom and only two periods to adjust so as to have them well out of the limits of the operative speed. Hence the two sets of springs, one to take care of the period corresponding to oscillation in the vertical plane, the other to control motion in the horizontal plane. It should not be imagined, however, that these springs will necessarily be very light; they will always have considerable stiffness, but their function is that of being the only members that *can* yield and the whole situation is controlled by the proper choice of these yielding elements.

# ENGINEERING RESEARCH

## A Department Conducted by the Research Committee of the A.S.M.E.

### Department of Scientific and Industrial Research

THE Fifth Annual Report of the Committee of the Privy Council for Scientific and Industrial Research has now been issued (Cmd 905) and may be obtained (price 1—by post 1/2¹/₄) through any bookseller or direct from H. M. Stationery Office at the following addresses: Imperial House, Kingsway, London, W.C. 2; 28 Abingdon Street, London, S.W. 1; 37 Peter Street, Manchester; 1 St. Andrew's Crescent, Cardiff; 23 Forth Street, Edinburgh; or from E. Ponsonby Ltd., 116 Grafton St., Dublin.

Besides a detailed account of the work of the Scientific and Industrial Research Department during the year 1919–1920, this report contains a review of the activities of the Department during the five years that have passed since the establishment of the Committee of the Privy Council for Scientific and Industrial Research statistics of the grants made and sums expended over this period and an outline of the development of coöperative association for industrial research.

### Research Résumé of the Month

#### A—RESEARCH RESULTS

The purpose of this section of Engineering Research is to give the origin of research information which has been completed, to give a résumé of research results with formulæ or curves where such may be readily given, and to report results of non-extensive researches which in the opinion of the investigators do not warrant a paper.

*Apparatus and Instruments A15-20.* Purity of Electrolytic Oxygen. An instrument built by the Leeds & Northrup Company for determining the purity of electrolytic oxygen by thermal conductivity has been calibrated at the Bureau of Standards. A number of manufacturers of chemical products intend to apply this method of manufacture for several of their processes. A 16-point gas analysis prepared for the helium plant at Fort Worth is well under way. Bureau of Standards, Washington, D. C. Address S. W. Stratton, Director.

*Apparatus and Instruments A16-20.* Reflectometer. The Bureau of Standards has recently developed a reflectometer for measuring the light from walls, ceilings and other objects. Serious errors have been made in previous determinations of reflection factors of most materials. Results have been confirmed by other experimenters. The instrument will be described before the convention of the Illuminating Engineering Society during November and in a scientific paper of the Bureau of Standards. An original form of the instrument was described in Scientific Paper 391 and also in the *Journal of the Optical Society* for January, 1920. Bureau of Standards, Washington, D. C. Address S. W. Stratton, Director.

*Chemistry, General and Physical A1-20.* Colloid Chemistry. The Committee of the British Association for the advancement of Science has issued its third report on Colloid Chemistry and its general and industrial applications. This has been published for the Department of Scientific and Industrial Research of Great Britain. The report includes the following sections:

1 Colloid Chemistry of Soap, Part I—Solutions. By Prof. J. W. McBain.
2 Ultramicroscopy. By G. King, M.Sc., F.I.C.
3 Solubility of Gases in Colloidal Solutions. By G. King, M.Sc., F.I.C.
4 Electrical Charge on Colloids. By J. A. Wilson.
5 Imbibition of Gels, Part I. By J. A. Wilson.

The subjects dealt with under the second head are:
6 Imbibition of Gels, Part II—Industrial Applications. By J. A. Wilson.
7 Colloid Problems in Bread Making. By R. Whymper.
8 Colloid Chemistry in Photography. By Dr. R. E. Slade.
9 Collodion in Photography. By H. W. Greenwood.
10 Cellulose Esters. By Foster Sproxton, B.Sc., F.I.C.
11 Colloid Chemistry of Petroleum. By Dr. A. E. Dunstan.
12 Asphalt. By Clifford Richardson, M.Am.Soc., C.E.F.C.S.
13 Varnishes, Paints and Pigments. By Dr. R. S. Morrell.
14 Clays and Clay Products. By A. B. Searle.

The price of the report is 2s.6d. and it may be obtained from Imperial House, Kingsway, London, W.C. 2.

The first report of the Committee issued in 1917 may be obtained from the British Association, Burlington House, London W.1., price 2s., and the second report from Imperial House, Kingsway, London W.C. 2.

*Explosives and Explosions A1-20.* Liquid-Oxygen Explosives in Germany. Mr. George S. Rice has made report No. 2163 to the Bureau of Mines on this subject. The report shows that 136 plants with a total capacity of 3797 liters of liquid oxygen per hour have been installed in the coal, iron, potash and salt mines, as well as for military purposes in various parts of Germany and in the French coal mines. Address F. G. Cottrell, Director, Bureau of Mines, Washington, D. C.

*Foundry Equipment, Materials and Methods A1-20.* Metals Alloyed with Gray-Iron Castings. Experiments have been made to determine if any benefit results by adding such metals as zirconium, aluminum and uranium to gray cast iron. Zirconium in a ferroalloy containing 35 per cent of zirconium was added to the iron in the ladle in the proportions of 2 lb. of alloy per ton of iron. The analysis showed 0.03 per cent zirconium in the castings. Cylinders and test bars were cast from this metal. Transverse tests showed an average of 3000 lb. and tensile tests, above 25,000 lb. per sq. in. These results are approximately the same as those obtained with regular run of castings. The cylinders machined more easily than those without zirconium and were much softer. The castings were close-grained and clear. The addition of ferrozirconium increased the cost about $2 per ton. Uranium was added in the form of an aluminum alloy containing 96 per cent aluminum. Three samples were tried containing 0.21 per cent, 0.50 per cent and 1 per cent uranium. Test bars showed the same transverse strength as regular run of castings with a tensile strength slightly lower than those of the regular iron. The castings showed no better machining qualities but were close-grained and clean. The alloys may have a scavenging effect, but the value of the uranium alloys is questionable. Address H. V. Wille, Baldwin Locomotive Works, Philadelphia, Pa.

*Fuels, Gas, Tar and Coke A14-20.* Water and Petroleum. The Pittsburgh Petroleum Laboratory of the Bureau of Mines has recently developed an improved method for determining water in petroleum emulsions. It is a modification of the familiar procedure of distillation with immiscible solvent. It is described in Report No. 2159 by E. W. Dean and W. A. Jacobs. It is also described by E. W. Dean and V. D. Starke in the *Journal of Industrial and Engineering Chemistry*, vol. 20, May 1920, p. 486. The equipment is made up of standard parts with the exception of the distilling-tube receiver, which can be obtained through the Bureau of Mines, from certain chemical supply houses and an electric heater which may be made from an electric hot plate. Names and addresses of firms may be obtained from the Bureau on request. The method may be used with fuel oils, shale oils, tars, greases and mixtures of powdered coal, oil and water. Address F. G. Cottrell, Director, Bureau of Mines, Washington, D. C.

*Glass and Ceramics A3-20.* Colloidal Kaolin. The Kraus process for the treatment of kaolin is based on the colloidal theory in which colloidal gels already existent but confined or imprisoned by waters of crystallization are released and remaining constituent crystalline and amorphous bodies are partially gelatinized. The colloidal gels are predominant after treatment and there is a change in plasticity, absorption, covering capacity, bonding strength and burning density. There is much improvement in the manufacture of many products. Greater retention is shown when used as a paper filler; china ware is more translucent; paint more adhesive and of greater covering capacity; electric porcelain has higher dielectric strength; oilcloth is more flexible. Colloidal kaolin is made adaptable for new uses as in soap, glass pots, refractories and it may replace silica in steel manufacture. Address Kraus Research Laboratories, Inc., 130 Pearl St., New York.

*Glass and Ceramics A5-20.* Ultramicroscopic Examination of Clays. A study of clays by ultramicroscopic examination indicates that a study along the lines of colloid-chemical principles and analysis will throw much light upon many obscure phenomena met by those who use clay for ceramic and other purposes. See article by Jerome Alexander in the *Journal of the American Ceramic Society*, August 1920.

*Internal-Combustion Motors A8-20.* Preheating Air Fuels. A paper was presented at the June Meeting of the Society of Automotive Engineers by a member of the staff of the Bureau of Standards giving a report on the value of various systems in preheating air fuel used in automobile engines. This work is being continued to ascertain the effect of different degrees of heating. Bureau of Standards, Washington, D. C. Address S. W. Stratton, Director.

*Machine Shop A1-20.* Calibration of End Standards. The Bureau of Standards is using interference methods for the calibration of end standards as precision gages and is obtaining rapid and reliable work and increasing the production of high-precision end standards. An article on this method of calibration is being prepared for publication. Bureau of Standards, Washington, D. C. Address S. W. Stratton, Director.

*Machine Design A2-20.* Shrink Fits. Experiments to determine allow. ance for locomotive reverse shaft arms with bearings 4 in. in diameter and 4 in. long. Ordinary shop practice allows $1/64$ in. for all sizes and styles. Steel specimens prepared in form of round disks 4 in. indi. ameter and 4 in. thick. Steel collars $3/8$ in. thick were machined with holes bored smaller than 4 in. by the following amounts: 0.004 in., 0.008 in., 0.012 in., 0.016 in., 0.020 in. These numbers vary by 0.001 in. per inch of diameter. The collars were shrunk on at a temperature of 600 deg. fahr. and in accordance with common practice, and after cooling, the sleeve was supported by a sleeve with a $4^1/2$-in. hole, while a plug of smaller diameter than 4 in. was placed on the 4-in. disk. After this load was applied from a testing machine to determine the starting pressure and the pressure at each quarter inch. The results of the test are shown by the curve, Fig. 1. The maximum

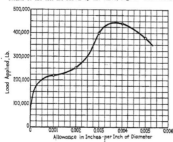

FIG. 1 SHRINK FITS: CURVE OF STARTING PRESSURE

strength seems to occur at about 0.00375 in. per inch of diameter. This result is about the same as that found for axles and wristpins of this size and the same allowance will be made for shrink fits of steel as had formerly been used for forced fits of steel, the latter being the result of careful tests. Address H. V. Wille, Baldwin Locomotive Works, Philadelphia, Pa.

*Mining, General A5-20.* Preserving Mine Timbers. The Forest Products Laboratory installed timbers treated with coal-tar creosote in an Alabama mine ten years ago. During this time all untreated timbers have been removed on account of decay, while 80 per cent of the creosoted timbers are still sound and none have decayed to an extent requiring removal. Other experiments have proved the same thing. Three preservatives are suitable for mine work: coal-tar creosote; zinc chloride, and sodium fluoride. The first of these is the most effective. The fire hazard as shown from long experience is not greatly increased by the use of creosote. Zinc chloride and sodium fluoride are odorless and tend to reduce inflammability. They are cheaper but do not give such permanent protection. Creosote is applied by brush, dipping or by pressure. Zinc chloride and sodium fluoride may be injected by steeping or by pressure. The saving will pay in every case. The wood preservative should be soluble in water if made from inorganic salts. In this way the poisonous substance is carried into the wood. In the case of creosote the poisonous material is carried in the oil. Address Director, Forest Products Laboratory, Madison, Wis.

*Reflectometer A1-20.* Reflectometer. See *Apparatus and Instruments A16-20.*

*Refrigeration A4-20.* Methyl Chloride. A pamphlet describing the properties and uses of methyl chloride has been issued by the Roessler & Hasslacher Chemical Company. This pamphlet gives a historical sketch of its discovery, manufacture and use, its physical properties, its chemical and physiological properties, its purity and its application as a refrigerant. Address The Roessler & Hasslacher Chemical Company, 709 Sixth Ave., New York.

*Textile Manufacture and Clothing A1-20.* Bleaching. Articles in the *Textile World Journal* by Allen Abrams and in *Cotton* by J. Merritt Matthews discuss the use of electrolytic bleaching as compared with the use of lime. In both papers the electrolytic bleach is shown to be more convenient to make, more uniform, less liable to change under storage, less expensive, more rapid, better controlled and the cloth bleached with the electrolytic liquors is more easily freed from residual chlorine and other materials than that bleached with bleaching powders. The tensile strength of the goods does not vary in the two methods of treatment.

*Wood Products A9-20.* Wood Preservation. The Forest Products Laboratory has investigated the possibility of protecting wood by charring. The area of charred wood does not decay or encourage the growth of fungi. This layer, however, is broken in places by checks and infection can enter through these cracks. If charring is deep enough to resist decay the post is weakened. The laboratory has shown that mine timbers could be properly protected by coal-tar creosote, zinc

chloride or sodium fluoride when properly impregnated. Address Director, Forest Products Laboratory, Madison, Wis.

## B—RESEARCH PROBLEMS

The purpose of this section of Engineering Research is to bring together those who are working on the same problem for coöperation or conference, to prevent unnecessary duplication of work and to inform the profession of the investigators who are engaged upon research problems. The addresses of these investigators are given for the purpose of correspondence.

*Automotive Vehicles and Equipment B4-20.* Fan Belts. An investigation to determine the performance of fan belts by using a motor and a typical automobile fan. Revolution counter attached to fan pulley and motor will give slippage. Bureau of Standards, Washington, D. C. Address S. W. Stratton, Director.

*Automotive Vehicles and Equipment B5-20.* International Harvester Co. The Laboratory No. 5 of the Experimental Department of the International Harvester Company is investigating brake lining, clutch facing and fan belting as well as belts, pulleys, pulley coverings and chains. It is making absorption tests on all apparatus and parts of tractors, trucks and stationary engines. Address O. B. Zimmermann, Experimental Department, International Harvester Co., 606 South Michigan Ave., Chicago, Ill.

*Electric Power B3-20.* Farm Lighting. Laboratory No. 5 of the Experimental Department of the International Harvester Co. is investigating storage batteries for the purpose of determining the best type for farm lighting purposes and also the form of electric generator for the same purpose for variation of speed of 600 per cent. Address O. B. Zimmermann, Experimental Department, International Harvester Co., 606 South Michigan Avenue, Chicago, Ill.

*Glass and Ceramics B4-20.* Coal-Gas Retorts. Investigation of the thermal qualities, strength and the other properties of gas retorts. Address Kalmus, Comstock & Westcott, Inc., 110 Brookline Ave., Boston, Mass.

*Glass and Ceramics B5-20.* Glazes. An attempt is being made to differentiate the various glazes applied to clay products by a machine similar to that of Brinell. The steel sphere is replaced by a conical steel point. Penetration is being measured by a micrometer microscope. The pressures are quite high. A study on high-fire porcelain glazes has shown that the range of compositions suitable for high fire increases rapidly with the temperature. Low-silica glazes are found to overfire at temperatures corresponding to cone No. 12. The limits for glazes of temperature up to cone No. 16 have been established quite definitely. Bureau of Standards, Washington, D. C. Address S. W. Stratton, Director.

*Glass and Ceramics B6-20.* Refractories for By-Product Coke Ovens. A study of the properties of a new silica refractory suggested research for by-product coke-oven refractories, considering the effect of carbonic acid gas, the abrasive action of charge; cooling effect; strength of product in cross-breaking and load; testing under heat and the thermal qualities of the refractory employed. This latter points to the advantage in using special refractory material which develops greater efficiency by reducing the time of coking of charge, quantity of gas used in coking, and permitting radical decrease in the temperature of the combustion of the gas. Address Kalmus, Comstock & Westcott, 110 Brookline Ave., Boston, Mass.

*Internal-Combustion Motors B1-20.* International Harvester Co. The following investigations are being undertaken by the Laboratory No. 5 of the International Harvester Company:

1 The development of the intake manifolds for tractor engines using kerosene.

2 The increased temperature of the ingoing mixtures produced by the new hot-spot manifolds increases the average temperature of the gases in the cylinders, which affects heat losses to the cylinder and affects valves, spark plugs. etc.

3 Carburetors and mixers for properly metering fuel to vaporizing chambers.

4 Fuel investigations.

5 Air cleaners, their efficiency in cleaning and possible causes for loss of volumetric efficiency in engines.

6 Proper vaporisers, spraying nozzles and metering pumps for high-compression and semi-Diesel oil engines.

7 The Hvid engine's characteristics and durability.

8 General stationary-engine development.

9 Continuous-combustion-engine developments such as pumps, refractories, proper valve timing, and compression problems in connection with the power cylinder.

10 Overheating problems of engines using kerosene.

11 Proper valve areas consistent with the increased temperature of the mixtures for proper vaporization.

12 Automatic and auxiliary intake valve-spring determination for proper spring design.

13 Governor investigations.

14 Radiator problems concerning trucks and tractors.

15 The proper use of anti-friction bearings for crankshafts and their durability.

16 Durability tests on all engine parts to determine proper design.

17 The action of greases on gears in regard to wear, lubrication and power absorbed.

18 Proper methods and procedure for determining proper lubricating oils for internal-combustion engines.

19 Placing of spark plugs and proper materials of construction for use in heavy-duty kerosene-burning engines.

*Metallurgy and Metallography B15–20.* Decarburization of Steel in Vacuum. A study is being made of steels in vacuo when heated above the $A_1$ point for different periods. The metal loses carbon on the surface, the amount being proportional to the time of heating. The carbon seems to leave as a gas, probably carbon monoxide. Bureau of Standards, Washington, D. C. Address S. W. Stratton, Director.

*Steam Power B4–20.* Unified Power Plants. Laboratory No. 5 of the Experimental Department of the International Harvester Co. is at work on the development of the unified system of burners, boilers, condensers, pumps and controls for steam power plants. Address O. B. Zimmermann, Experimental Department, International Harvester Co., 606 South Michigan Ave., Chicago, Ill.

*Steam Power B5–20.* Condenser Tubes. An investigation is being made by the General Electric Company, of Schenectady, and the Diamond Power Specialty Company, of Detroit, on the life of condenser tubes. The investigation is undertaken to determine the possibility of utilizing calorized copper tubes for condensers. Calorizing has proven its value in preventing oxidation of soot-blower units, pyrometer protection tubes and other equipment. It is not known that calorizing has been utilized successfully as a means for preventing corrosion. Information is desired from power plants and marine engineers respecting the service obtained from condenser tubes, and this coöperation is requested. The communication should state the metal of which tubes are made, size and lengths of tubes and life of tubes. Address Robert June, Diamond Power Specialty Company, Detroit, Mich.

## ˙C—RESEARCH PROBLEMS

The purpose of this section of Engineering Research is to bring together persons who desire coöperation in research work or to bring together those who have problems and no equipment with those who are equipped to carry on research. It is hoped that those desiring coöperation or aid will state problems for publication in this section.

*Machine Design C3–20.* Ball Bearings at High Speeds. The reference to this subject under *Machine Design C2–20* arose from the discussion that as the speed of the ball was increased very materially the centrifugal force became very great, and in such cases large balls are improper because the mass causing centrifugal trouble will vary as the cube of the diameter of the ball, while its bearing power will only increase as the diameter. This will mean centrifugal troubles at high speeds. Inquiry is made on this subject. Has any one studied this particular problem or made experiments which would answer questions arising in this study? Address Panfilo Trombetta, General Electric Company, Schenectady, N. Y.

*Steam Power C1–20.* Condenser Tubes. (See *Steam Power B5–20.*) In this investigation on corrosion of condenser tubes and its prevention, coöperation is asked from engineers of power plants and marine equipment regarding (a) metal used for condenser tubes, (b) size and lengths of tube and (c) life of tubes. Any data will be greatly appreciated for this research. Address Robert June, Diamond Power Specialty Company, Detroit, Mich.

## D—RESEARCH EQUIPMENT

The purpose of this section of Engineering Research is to give in concise form notes regarding the equipment of laboratories for mutual information and for the purpose of informing the profession of the equipment in various laboratories so that persons desiring special investigations may know where such work may be done.

*Henry Souther Engineering Company D1–20.* Metallurgical Laboratory. The equipment of the Henry Souther Engineering Company devoted to metallurgical and sanitary inspection consists of a chemical laboratory equipped to handle routine metallurgical analysis; organic laboratory equipped to handle materials and compounds such as oils, greases, paints, boiler waters and plating compounds. A physical testing laboratory for tensile, transverse and compression tests, hardness machines, impact testers and fatigue testers. The metallurgical laboratory is equipped for practical heat treatment and metallographic studies of steel, iron, brass, bronze and other alloys. Address Henry Souther Engineering Company, 11 Laurel St., Hartford, Conn.

*Hydraulics D1–20.* Hydraulic Laboratory at Purdue University. See *Purdue University D1–20.*

*Kalmus, Comstock & Westcott, Inc. D1–20.* Laboratory for Industrial Research in Ceramics. Laboratory is housed in reinforced-concrete building with water, 110- and 220-volt d.c. current and gas. All power electrical. Especially equipped for refractory and other materials.

*Equipment:*
Muffle furnace, gas- and oil-fired for temperatures up to 1400 deg. cent., atmospheric conditions under control.
Circular pit furnace, gas-fired for temperatures of 1450 to 1500 deg. Load capacity, 6000 lb.

Cross-breaking and compression testing machines.
Leeds & Northrup two-point recording thermometer for base-metal and noble-metal samples.
Special thermal-conductivity-determining apparatus designed and developed at laboratory. Accurate determinations on refractory and insulating materials at 100 to 1000 deg. cent. Apparatus specially designed to guard against side loss.
Small electric muffle furnace for 1000 deg. cent.
36-in. surface grinder, jaw crushers, pulverizers, mixers, driers, air compressors, fans.
Kilns available for standard and special operating conditions. Address Kalmus, Comstock and Westcott, Inc., 110 Brookline Ave., Boston, Mass.

*Minneapolis Steel & Machinery Company D1–20.* Metallurgical Laboratory. This laboratory is equipped for physical, metallurgical and chemical work. Address C. S. Moody, Metallurgical Engineer, Minneapolis Steel & Machinery Company, Minneapolis, Minn.

*Purdue University D1–20.* Hydraulic Laboratory. The laboratory established in 1911 has a maximum flow of 8400 gal. per min. at 34 ft. and 1800 gal. per min. at 231 ft. This is produced by pumping. The equipment is as follows:

(a) One Platt direct-driven single-stage centrifugal pump with a capacity of 8400 g.p.m. against a maximum head of 34 ft.
(b) One American two-stage direct-driven centrifugal pump with a capacity of 1800 g.p.m. against a maximum head of 231 ft.
(c) One Allis-Chalmers direct-driven single-stage centrifugal pump with a capacity of 400 g.p.m. against a maximum si maximum head of 116 ft. The motor is of the d.c. variable-speed type.
(d) One Gould triplex pump with a capacity of 60 g.p.m.
(e) One 2-in. rotary pump.
(f) One Trump 12-in. reaction turbine.
(g) One Fairbanks weighing scales of 20 tons capacity. The balance arm of the scale is equipped with an electric timing device.
(h) One concrete channel, 3 ft. wide, 3 ft. deep, and 80 ft. long, supplied with track and carriage for current-meter investigations.
(i) One large concrete channel, 6 ft. deep and 70 ft. long composed of two sections, one 8 ft. and the other 5 ft. in width. Each section is equipped with a rectangular weir. The flow from the channel passes into the large weighing tank.

The laboratory is 115 ft. by 50 ft. with a main floor and gallery. A concrete reservoir of 82,000 gal. capacity is situated under the main floor. In addition to the apparatus listed above there are a number of small tanks and scales, small impulse wheels, hydraulic ram, orifices, weirs, tubes and water meters. A movable equipment for pitot tubes and current meters up to a maximum range of 5 ft. per sec. is available. The laboratory is in charge of F. W. Greve. Address Dean A. A. Potter, Purdue University, Lafayette, Ind.

*The Westport Mill D1–20.* Chemical Laboratory. Laboratory for chemical and hydrometallurgical work. The laboratory devoted to determining the proper size of Dorr equipments for the Dorr Company, Engineers, New York, Denver and London. Laboratory is equipped for researches dealing with the separation of solutions from finely divided solid substances often of colloidal nature. Purification and preparation of clays and the continuous production of finely ground pigments. Address J. A. Baker, Westport Mill, Westport, Conn.

## E—RESEARCH PERSONNEL

The purpose of this section of Engineering Research is to give notes of a personal nature regarding the personnel of various laboratories, methods of procedure for commercial work or notes regarding the conduct of various laboratories.

## F—BIBLIOGRAPHIES

The purpose of this section of Engineering Research is to inform the profession and especially the members of the A.S.M.E. of bibliographies which have been prepared. These bibliographies have been prepared at the request of members, and where the bibliography is not extensive, this is done at the expense of the Society. For bibliographies of a general nature the Society is prepared to make extensive bibliographies at the expense of the Society on approval of the Research Committee. After these bibliographies are prepared they are loaned to the person requesting them for a period of one month. Additional copies are prepared which are available for periods of two weeks to members of the A.S.M.E. or to others recommended by members of the A.S.M.E. These bibliographies are on file in the offices of the Society and are to be loaned on request. The bibliographies prepared by the staff of the Library of the United Engineering Society which is probably the largest Engineering Library in this country.

*Fuels, Gas, Tar and Coke F7–20.* Kerosene and Carburation. A bibliography of $2\frac{1}{2}$ pages. Search 3117. Address A.S.M.E., 29 West 39th St., New York.

*Pumps F1–20.* Jet Apparatus, Injectors and Ejectors. A bibliography of 3 pages. Search 3126. Address A.S.M.E., 29 West 39th St., New York.

# WORK OF THE A.S.M.E. BOILER CODE COMMITTEE

*THE Boiler Code Committee meets monthly for the purpose of considering communications relative to the Boiler Code. Any one desiring information as to the application of the Code is requested to communicate with the Secretary of the Committee, Mr. C. W. Obert, 29 West 39th St., New York, N. Y.*

THE procedure of the Committee in handling the cases is as follows: All inquiries must be in written form before they are accepted for consideration. Copies are sent by the Secretary of the Committee to all of the members of the Committee. The interpretation, in the form of a reply, is then prepared by the Committee and passed upon at a regular meeting of the Committee. This interpretation is later submitted to the Council of the Society for approval, after which it is issued to the inquirer and simultaneously published in MECHANICAL ENGINEERING, in order that any one interested may readily secure the latest information concerning the interpretation.

Below are given the Interpretations of the Committee in Cases Nos. 259, 285, 305, 307, 310, 314–317, inclusive, as formulated at the meeting September 24, 1920, and approved by the Council. In accordance with the Committee's practice, the names of inquirers have been omitted.

## CASE No. 259 (Reopened)

*Inquiry:* Is it allowable, under the requirements of the A.S.M.E. Boiler Code, to fit a high-pressure steam boiler with a blow-off connection larger than $2^1/_r$-in. in case it is to be used initially for low-pressure steam or hot-water heating, or will it be necessary under the requirements of Par. 308 to apply two or more $2^1/_r$-in. blow-off connections for the return connections to the boiler?

*Reply:* · It is the opinion of the Committee that a boiler could be initially fitted with a blow-off connection larger than $2^1/_r$-in. size when intended for use as a steam or hot-water heating boiler, and when converted into a steam boiler if to operate at over 15 lb. pressure, but not exceeding 100 lb. pressure, a reducing fitting could be used at the opening to reduce to the pipe size required for the blow-off connections by Par. 308.

## CASE No. 285 (Reopened)

*Inquiry:* Is is permissible, under the rules of the Boiler Code, to use standard extra-heavy cast-iron flanges and fittings on pipe connections between boilers and attached-type superheaters, and on the ends of superheater inlet headers for pressures up to 250 lb. per sq. in.? It is pointed out that neither the inlet pipe connections nor the superheater inlet flanges would be subjected to other than saturated steam temperatures.

*Reply:* It is the opinion of the Committee that the flanges and fittings referred to may be made of cast iron, provided the temperature of the steam does not exceed 450 deg. fahr. as specified in Par. 12.

## CASE No. 305

*Inquiry:* Is it permissible, under the requirements of the Boiler Code, to connect sections of wrought-steel headers, one of which is shrunk or forced over the end of the other, making a close fit and secured by means of bolts of ample cross-sectional area, the joint being calked or autogenously welded to insure tightness, the design of the joint to be such that it will be of ample strength, neglecting the holding power due to shrinking or forcing one section over the other and the holding power of the autogenous welding?

*Reply:* It is the opinion of the Committee that while the construction described is not a desirable one, it is not prohibited by the Code.

## CASE No. 307 (Reopened)

*Inquiry:* Is it permissible to so locate the supporting lugs on h.r.t. boilers where more than four lugs are required and under Par. 323 of the Code, that those in each pair come close together, or must the horizontal distance between the center lines of rivets attaching the adjacent lugs to the shell be at least equal to the vertical spacing of rivets that is required for lug attachments in Par. 323, as shown in Fig. 8?

*Reply:* There is no requirement in the Code specifying the distance apart of the lugs forming pairs as required by the last sentence of Par. 323. It is the intent of the Boiler Code that the distance between the lugs should be such as to give at least $1/_2$ in. between the edges of the lugs and not more than 2 in. The load should be equalized between the two lugs.

## CASE No. 310

*Inquiry:* a Can a type of boiler other than the h.r.t. type have lap joints where the courses are over 12 ft. long?

b With butt and double strap construction on longitudinal joints for boilers other than of the h.r.t. type, is it required that the tension test specimens be cut from the shell plate as provided in Par. 190?

*Reply:* a The restriction in length of lap joints to 12 ft. in Par. 190 of the Boiler Code applies specifically to boilers of the h.r.t. type, and there is nothing in the rule which prohibits the construction of shells and drums of lengths exceeding 12 ft. for other types of boilers.

b Inasmuch as the first sentence of Par. 100 applies specifically to h.r.t. boilers, as pointed out in reply (a), it is the opinion of the Committee that this prohibition does not cover other types of boilers.

## CASE No. 314

*Inquiry:* Is it the opinion of the Boiler Code Committee that a lap joint reinforced with a cover plate should be considered as an ordinary lap joint and the requirements for factors of safety given in Pars. 379 and 380 applied accordingly?

*Reply:* It is the opinion of the Committee that a lap joint even though reinforced by a cover plate, should be treated exactly the same as the simple form of lap joint, so that the factors of safety proposed in Pars. 379 and 380 for lap joints would be applicable.

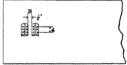

FIG. 8  SPACING OF SUPPORTING LUGS IN PAIRS ON H.R.T. BOILERS

## CASE No. 315

*Inquiry:* Is it to be understood that the requirements of Par. 323 of the Boiler Code, relative to the location of lugs and the distribution of rivets attaching them to the shell, apply to the small sizes of boilers referred to in Par. 324?

*Reply:* The requirements of Par. 323 of the Code are clearly limited to h.r.t. boilers over 78 in. in diameter. It will be found that Par. 325 gives the requirements for attaching the lugs.

## CASE No. 316

*Inquiry:* Is it permissible, under the requirements of the Boiler Code, for a boiler manufacturer to stamp a boiler as A.S.M.E. Code Standard, when it is fitted with a safety-valve nozzle or connection that is adequate only for safety valves operating at high lifts?

*Reply:* The way designated in the Code in which the size of the safety valve or valves that shall be used on any boiler is determined, is by their relieving capacity, and where the safety-valve opening corresponds to that required from such a valve or valves, the boiler is constructed in accordance with the A.S.M.E. Code in this respect, and may be so stamped. The Committee suggests, however, that where boilers are sold for a given operating pressure without knowledge of the type of safety valves that will be used, the safety-valve openings be proportioned for the intermediate lifts and corresponding relieving capacities given in Table 15 of the Code.

### CASE NO. 317

*Inquiry:* If it is permissible, as indicated in Case No. 298, to fit a steam outlet nozzle with a wrought-steel flange screwed to the outer end of the end neck to which the flange is threaded and peened over into a beveled part cut away from the flange, why should not this construction be acceptable for the fastening of the flange at the other end of the nozzle for attachment to the boiler shell?

*Reply:* It is the opinion of the Committee that this construction does not conform to the requirement of the last sentence in Par. 268.

### Data Sheets of the Interpretations of the Boiler Code

As a result of the widespread demand from the steam-boiler field the Boiler Code Committee has inaugurated the practice of reprinting from MECHANICAL ENGINEERING the Interpretations issued at its regular monthly meetings and approved by the Council, and these are now available for general distribution in data-sheet form, trimmed to convenient size and punched for insertion in suitable binders. These data sheets begin with Case No. 200, which is the first Case formulated by the Boiler Code Committee in interpretations of the 1918 edition of the Code; the Cases from Nos. 1–199 have not been reprinted in this form, as with their incorporation in the new edition of the Boiler Code (edition of 1918) they are superseded and thus rendered of no further service.

These data sheets are available upon application to the office of the Secretary of the Boiler Code Committee, at the prevailing rates charged by the Publication Committee for reprinted matter. They may be obtained singly or in complete sets from Case No. 200 up, as may be desired. Should it be desired to preserve them in a convenient form, this may be accomplished by ordering in addition, a boiler code binder, this being fitted with clamping bolts for binding in not only the data sheets but also the Boiler Code, if desired.

In response to numerous calls that have been received for copies of these Interpretation data sheets as issued, preparations have been made for supplying them on regular subscriptions as often as issued where ordered in advance. Further information concerning the terms of such subscriptions will be given upon application to the Secretary's office.

### Promulgation of the Boiler Code

The American Uniform Boiler-Law Society, organized primarily to secure the legal adoption of the A.S.M.E. Boiler Code, has at the same time used its office to disseminate information and news about the Code.

For example, the society has made a survey of the educational institutions and reports that the following are using the Code as a textbook:

Massachusetts Institute of Technology
Sheffield Scientific School
Yale University
Armour Institute of Technology
Case School of Applied Science
University of Vermont
University of Cincinnati
Johns Hopkins University
University of Michigan
University of Minnesota
Rensselaer Polytechnic Institute

Rutgers College and University of Texas, beginning this year, will also use the Code as a textbook.

The following institutions are using the Code as a reference book:

Vanderbilt University (Tennessee)
New Mexico College of Agriculture and Mechanic Arts
Tulane University of Louisiana
University of Colorado
University of Washington
University of Maine
University of Wyoming
Virginia Polytechnic Institute
Washington University (St Louis)

This survey shows that the Code is having a greater sphere of usefulness than was ever anticipated.

---

# CORRESPONDENCE

C ONTRIBUTIONS to the Correspondence Department of MECHANICAL ENGINEERING are solicited. Contributions particularly welcomed are discussions of papers published in this Journal, brief articles of current interest to mechanical engineers, or suggestions from members of The American Society of Mechanical Engineers as to a better conduct of A.S.M.E. affairs.

### Code of Ethics Restated

To THE EDITOR:

The following is a restatement of the Proposed Code of Ethics which was published in the September issue of MECHANICAL ENGINEERING.[1]

In order to remain in good standing in The American Society of Mechanical Engineers a member in any of the grades of the Society shall observe the following rules of conduct:

1 He shall not fail to recognize that the first duty of an engineer is to be a useful citizen, exemplifying in his character and his practice the highest ideals of citizenship and of loyalty to his country.

2 He shall not permit selfish considerations to influence his professional conduct, but shall ever recognize service to society and the public as fundamental in promoting the interests of the engineering profession.

3 He shall not associate himself with any questionable or illegitimate enterprise and shall not countenance such association by a fellow-engineer, but shall endeavor to prevent it, first by friendly counsel and, that failing, by reporting the facts and circumstances to the proper authorities.

[1] Section Two, p. 123.

4 He shall not accept compensation, pecuniary or otherwise, from more than one interested party and shall not receive, directly or indirectly, any royalty, gratuity, or commission on any patented article or process used in the work upon which he is retained, without the knowledge and consent of all interested parties.

5 He shall not undertake any engineering work in which his judgment or the character of his services might be influenced by his other connections or interests, and he shall not allow his clients' or employers' interests to suffer in any way through his professional relations with others.

6 He shall not knowingly compete with a fellow-engineer in the matter of professional charges, or attempt to supplant him after definite steps have been taken toward his employment; and he shall not take over the work of another consulting engineer without first conferring with him and becoming satisfied that ample reasons exist for a change.

7 He shall not advertise in an undignified, sensational or misleading manner or solicit work by dishonest or unseemly methods; and he shall not countenance sensational, exaggerated or unwarranted statements about his own work in the public press; but may publish to the technical world new inventions, processes, and engineering projects and accomplishments through the engineering societies and the technical press in preference to the public press.

8 He shall not misrepresent recognized facts and well-established theories in making reports, or in testifying as an expert; and he shall not confuse newly discovered facts and newly evolved theories with those that have been long accepted; and he shall not jeopardize the good name of the profession by public expression of opinions as an engineer on subjects in which he may not properly qualify as an authority.

9 He shall not make public any information obtained from or through work for a client or employer, nor shall he at any time make such use of it as will embarrass the client or employer in whose service it was obtained, but may use such information as forming part of his professional experience to guide in his own professional practice.

10 He shall not refuse to help his fellow-engineers, by exchange of general information and experience, by personal interest in their welfare or by instruction through engineering societies, schools of applied science, and the technical press, or through the public press under conditions that will insure accuracy and guarantee that such public instruction shall not mislead.

Ann Arbor, Mich. M. E. COOLEY.

## Economy of Passenger Automobiles at Various Speeds

TO THE EDITOR:

The limited supply of fuel for automobiles makes it imperative that the cars be run with the minimum amount of gasoline. It is a duty of every owner of a car to save the fuel, not only because it benefits his own pocketbook, but also because it makes more fuel available for the whole nation.

There are certain speeds at which it is most advantageous and economical to run a car, and those who drive automobiles should be requested to coöperate in saving gasoline by driving their cars as closely as possible at these speeds.

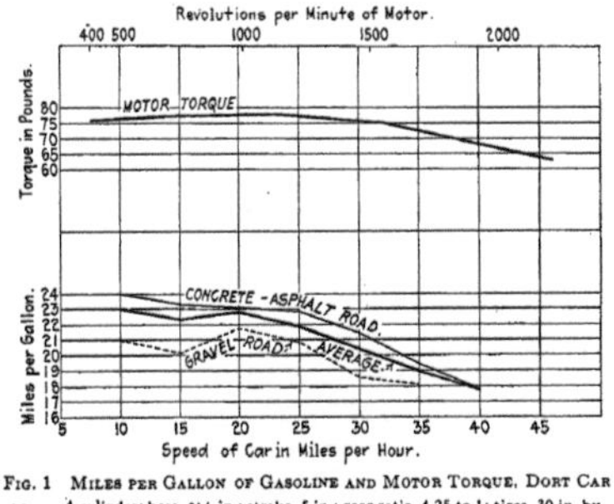


FIG. 1 MILES PER GALLON OF GASOLINE AND MOTOR TORQUE, DORT CAR MOTOR, 4-cylinder; bore, $3^1/_2$ in.; stroke, 5 in.; gear ratio, 4.25 to 1; tires, 30 in. by $3^1/_2$ in.; car weight, 2400 lb.

In order to establish the most economical speed for the average automobile, the author has made some very complete and thorough experiments at speeds of 10, 15, 20, 25, 30, 35 and 40 m.p.h.

These tests were made on concrete roads and gravel roads and in the directions north, south, east and west, in order to offset the effect of the wind, which has a marked influence in increasing or decreasing the consumption of gasoline according to whether the automobile travels with or against it. The car used was a Dort car, with a four-cylinder motor, having a $3^1/_2$-in. bore and 5-in. stroke. The weight of the car is 2400 lb. in running order. The size of the tires is 30 by $3^1/_2$ in. The gear ratio on the rear axle is $4^1/_4$ to 1. The load carried corresponded to a three-passenger load, and the top was up with the windshield closed.

The results of these tests are given in Table 1, and show the average gasoline consumption for each speed in the four directions of the compass. In Fig. 1 the tests are condensed into a single average and a curve is drawn which shows at a glance the most economical speed of the car.

TABLE 1 ECONOMY OF LIGHT CAR AT VARIOUS SPEEDS

4 cyl. motor, $3^1/_2$-in. bore, 5-in. stroke; gear ratio, 4.25 to 1; wheels, 30 in. by $3^1/_2$ in.; car weight, 2400 lb.; 3 passengers.

| Miles per hr. | Road | Direction | Temp. deg. fahr. | Weather | Date | Miles per gal. |
|---|---|---|---|---|---|---|
| 10 | Concrete | South | 80 | Clear | Sept. 8/20 | 23.6 |
| 10 | Concrete | North | 80 | Clear | Sept. 8/20 | 25.6 |
| 10 | Concrete | South | 80 | Clear | Sept. 8/20 | 21.2 |
| 10 | Concrete | North | 80 | Clear | Sept. 8/20 | 25.6 |
| 10 | Gravel | West | 75 | Cloudy | Sept. 9/20 | 21.4 |
| 10 | Gravel | East | 75 | Cloudy | Sept. 9/20 | 20.6 |
| | | | | | Average, | 23.0 |
| 15 | Concrete | South | 75 | Clear | Sept. 3/20 | 23.2 |
| 15 | 2/3 concrete, 1/3 sand | South | 75 | Clear | Sept. 3/20 | 22.4 |
| 15 | 2/3 concrete, 1/3 sand | North | 75 | Clear | Sept. 3/20 | 24.0 |
| 15 | Concrete | North | 75 | Clear | Sept. 3/20 | 25.2 |
| 15 | 2/3 concrete, 1/3 gravel | East | 75 | Clear | Sept. 3/20 | 25.4 |
| 15 | 2/3 concrete, 1/3 gravel | West | 75 | Clear | Sept. 3/20 | 24.4 |
| 15 | Dry gravel | West and south | 75 | Clear | Sept. 3/20 | 21.6 |
| 15 | Loose gravel | East | 75 | Clear | Sept. 3/20 | 18.0 |
| 15 | Loose gravel | West | 75 | Clear | Sept. 3/20 | 19.4 |
| 15 | Concrete | South | 80 | Clear | Sept. 8/20 | 21.2 |
| 15 | Concrete | North | 80 | Clear | Sept. 8/20 | 23.6 |
| 15 | Wet gravel | West | 75 | Cloudy | Sept. 9/20 | 21.2 |
| 15 | Wet gravel | East | 75 | Cloudy | Sept. 9/20 | 20.4 |
| | | | | | Average, | 22.3 |
| 20 | Concrete | South | 75 | Clear | Sept. 3/20 | 21.8 |
| 20 | Concrete | North | 75 | Clear | Sept. 3/20 | 22.4 |
| 20 | Concrete | South | 80 | Clear | Sept. 8/20 | 21.8 |
| 20 | Concrete | North | 80 | Clear | Sept. 8/20 | 24.8 |
| 20 | Wet gravel | West | 75 | Cloudy | Sept. 9/20 | 22.8 |
| 20 | Wet gravel | East | 75 | Cloudy | Sept. 9/20 | 20.8 |
| 20 | Wet concrete | North | 75 | Cloudy | Sept. 9/20 | 23.6 |
| 20 | Wet concrete | South | 75 | Cloudy | Sept. 9/20 | 22.6 |
| 20 | Concrete | South | 75 | Clear | Sept. 13/20 | 23.0 |
| 20 | Concrete | North | 75 | Clear | Sept. 13/20 | 25.0 |
| | | | | | Average, | 22.86 |
| 25 | Concrete | South | 75 | Clear | Sept. 4/20 | 24.4 |
| 25 | Concrete | North | 75 | Clear | Sept. 4/20 | 22.4 |
| 25 | Concrete | East | 80 | Clear | July 1/20 | 22.0 |
| 25 | 1/2 gravel, 1/2 concrete | North | 80 | Clear | July 1/20 | 23.2 |
| 25 | Concrete | South | 80 | Clear | Sept. 8/20 | 21.0 |
| 25 | Concrete | North | 80 | Clear | Sept. 8/20 | 24.0 |
| 25 | Dry gravel | West | 70 | Clear | Sept. 13/20 | 20.4 |
| 25 | Dry gravel | East | 70 | Clear | Sept. 13/20 | 21.2 |
| 25 | 1/2 gravel, 1/2 concrete | East and north | 70 | Clear | Sept. 13/20 | 21.2 |
| 25 | 1/2 gravel, 1/2 concrete | South and west | 70 | Clear | Sept. 13/20 | 18.8 |
| | | | | | Average, | 21.86 |
| 30 | Concrete | North | 80 | Clear | July 1/20 | 24.0 |
| 30 | Concrete | South | 80 | Clear | July 1/20 | 20.4 |
| 30 | Concrete | South | 75 | Clear | Sept. 4/20 | 20.4 |
| 30 | Concrete | North | 75 | Clear | Sept. 4/20 | 21.0 |
| 30 | Concrete | South | 75 | Cloudy | Sept. 9/20 | 20.0 |
| 30 | Concrete | North | 75 | Cloudy | Sept. 9/20 | 24.4 |
| 30 | Dry gravel | West | 70 | Clear | Sept. 13/20 | 19.0 |
| 30 | Dry gravel | East | 70 | Clear | Sept. 13/20 | 20.8 |
| 30 | Concrete | North | 70 | Clear | Sept. 13/20 | 20.0 |
| 30 | Dry gravel | West | 70 | Clear | Sept. 13/20 | 17.6 |
| 30 | Dry gravel | East | 70 | Clear | Sept. 13/20 | 16.8 |
| | | | | | Average, | 20.4 |
| 35 | Concrete | South | 75 | Clear | Sept. 4/20 | 18.0 |
| 35 | Concrete | North | 75 | Clear | Sept. 4/20 | 19.6 |
| 35 | Concrete | South | 75 | Clear | Sept. 9/20 | 18.4 |
| 35 | Concrete | North | 75 | Clear | Sept. 9/20 | 22.0 |
| 35 | Dry gravel | West | 70 | Clear | Sept. 13/20 | 17.4 |
| 35 | Dry gravel | East | 70 | Clear | Sept. 13/20 | 19.0 |
| | | | | | Average, | 19.06 |
| 40 | Concrete | South | 75 | Cloudy | Sept. 11/20 | 16.0 |
| 40 | Concrete | North | 75 | Cloudy | Sept. 11/20 | 19.6 |
| | | | | | Average, | 17.8 |

A curve of the motor torque is plotted above the economy curve in Fig. 1, and an examination of these curves will show that the economy curve follows the torque curve very closely, with a tendency for the former to fall faster above 25 m.p.h., due to the increased wind resistance. It would seem logical, therefore, in order to get the greatest possible economy out of the cars, to lower the gear ratio in the rear axle, thereby reducing the engine speed, or else to have a geared-up fourth speed in the transmission.

It is to be noted that after a speed of 25 m.p.h. is reached the economy curve falls down very fast; therefore it would be to the interest of car owners to drive their vehicles at a speed not exceeding 30 m.p.h. in order to get the best results. A speed of approximately 30 m.p.h. is about the maximum speed at which a car can be driven with safety on our present roads. Of course, it is almost impossible to establish arbitrarily a limit of 30 m.p.h. for the maximum speed of an automobile, as the engineer must take into account the poor roads which are encountered in various parts of the country, where more power is needed to pull the car through, but a car with a maximum possible speed of 40 m.p.h. on good roads would have enough power to go anywhere.

The tests show very plainly that a concrete-asphalt road gives a much better mileage than the ordinary gravel road or the sandy road, and consequently, as the building of good hard roads progresses throughout the country, the economy in gas consumption will become greater.

If these different facts are given wide publicity, not only by the oil industry but by the automotive industry and the public press, it will show to the governments of the states, and to the Federal Government, that the sooner we build up a smooth, solid type of highways all through the country, the quicker will we get better mileage in the operation of motor cars and motor trucks, and the conservation of our natural resources in oil will be greatly enhanced.

Certain emphasis has been laid on the fact that the automotive manufacturers ought to build lighter cars, which would give the necessary comfort with less waste of gas. A number of manufacturers are already producing such cars; but the majority of cars on the market in the light-weight class are great wasters of gasoline as they average only between 15 and 16 miles to the gallon, due to their very inefficient motors.

From the author's experience, a car with a weight not to exceed 2000 lb. in running order, without passengers, should give easily 25 to 30 miles to the gallon; a car weighing between 2000 and 2500 lb. under the same conditions should give, easily, between 20 and 25 miles per gallon; and the heavier cars up to 3000 lb. weight should give 16 to 20 miles to the gallon, without any question. However, the efficiency of the motor alone is not all that is necessary to obtain these figures. The car should be kept in good order, with the brake drums free from dragging brakes, tires should be inflated to their proper pressure, and the lubrication of the complete car well taken care of in order to get the maximum efficiency.

These different points will be brought to the front more and more as the number of automobiles increases and it becomes imperative to use less gasoline per automobile in order to have enough for everybody. E. Planche.

Flint, Mich.

## Effect of Radial Play on Life of Ball Bearings

To the Editor:

Tests have recently been made by the Fafnir Bearing Company to determine the effect of radial play on the life of ball bearings which it may be of interest to report. By radial play is meant the amount of shake or looseness between the inner and outer rings in a direction at right angles to the bore. This play is measured after the bearing is completely assembled ready for shipment, and by the use of a special gage employed in the inspection of Fafnir bearings.

Most users of ball bearings demand that the radial play be very small, some specifying that it shall not exceed 0.001 in. They consider it sufficient that a bearing should run freely before being mounted and that no perceptible play exist.

At first one would believe that the less play the better, as long as the bearing is free to turn. One must consider, however, that the examination or inspection of ball bearings is made before mounting and that the mountings call for a drive fit of the inner ring over the shaft, and sometimes the outer ring is also forced into its housing. Several tests were made to determine what effect this forcing of the bearings into place had. It was found that inner rings expanded from 0.0003 to 0.0004 in., and outer rings contracted from 0.0002 to 0.0003 in., these measurements being made when press fits were used.

These facts make it apparent that the radial play must be considered after the bearing has been mounted. It is of course impossible in most installations to measure radial play after the bearing is mounted, but it is quite possible to measure the expansion and contraction of inner and outer rings. When an average value has been established for these two items, the problem is simplified. For instance, consider a bearing having a play of 0.0005 in. before mounting, and a mounting that causes an expansion of 0.0002 in. of the inner ring and a contraction of 0.0001 in. of the outer ring. The radial play after mounting will be the sum of the expansion plus the contraction (0.0001 + 0.0002 = 0.0003) subtracted from the play before mounting (0.0005 − 0.0003 = 0.0002). Hence the radial play after mounting is very different from that in the unmounted bearing.

It is quite evident that under certain conditions the bearing may be loose enough to turn easily before mounting and may be entirely too tight afterward. In order to determine whether or not a tight

bearing wears out faster than a loose one a series of tests was made, the results of which are tabulated below. It must be remembered that the radial play of these test bearings was measured before mounting in the test block. The test mounting reduced the play in each case by about 0.0002 in.

| Bearing No. | Radial Play, Inches Before Mounting | After Mounting | Hr. Run | Remarks |
|---|---|---|---|---|
| 1 | 0.0000 | Cramped | 347 | |
| 2 | 0.0000 | Cramped | 477 | Failed at this time |
| 3 | 0.0002 | 0.0000 | 323 | |
| 4 | 0.0002 | 0.0000 | 477 | |
| 5 | 0.0004 | 0.0002 | 571 | |
| 6 | 0.0006 | 0.0006 | 803 | |
| 7 | 0.0012 | 0.0010 | 571 | No signs of failure |
| 8 | 0.0012 | 0.0010 | 585 | |

These results indicate that the play should be at least 0.0005 in. after mounting. The necessary play before mounting must be determined for each condition of installation. This must be done by the user under average conditions as he finds them.

In conclusion, it may be well to point out that there is a slight variation of play in all bearings. This is due to the fact that a slight tolerance or variation must be permitted to make bearing manufacture commercially possible. This variation in radial play is usually about 0.0005 in. but for large bearings may exceed this. Thus a specification should read that the radial play must in no case be less than a certain amount. This amount is determined from the known expansion and contraction of the rings as pointed out above. Roland W. Sellew.

New Britain, Conn.

## ORGANIZATION AND CONSTRUCTION OF DYE HOUSES

(Continued from page 674)

### Dye-House Piping

Each kettle has a hot-water, a cold-water and a steam connection which means a good deal of piping to care for and it should be laid out so as to be accessible and permit of changes. All cold-water piping must be below ground and is usually of cast iron run along the inside of the trench on the same supports as the machines. The high- and low-pressure steam and hot-water piping are run overhead and covered.

Steam Supply. The objection to the use of low-pressure steam in dye houses for boiling kettles has about vanished, and where an ample supply of clean low-pressure steam is available, as from either bleeder or non-condensing turbines, a considerable economy can be effected. For worsted piece dyes this steam must be free from oil, but in stock dyeing for woolens the exhaust from reciprocating engines can be used without injurious results if proper precautions are taken. If low-pressure instead of high-pressure steam is used at the kettles, the pipe sizes must be increased. Low-pressure systems have been condensed because of a failure to take this into consideration. A careful study should be made of the steam requirements of the dye house under consideration, especially with reference to peak loads and the pipe sizes proportioned to meet these conditions. The piping should be ample to provide for the future growth of the dye house and the possible rearrangement and changing of machinery. It is difficult because of the varying load in the dye house to maintain an economical heat balance, but much can be done in the way of economy by careful arrangement and the use of surplus steam to heat the water supply.

Warm-Water Supply. There should be provided, either close to the dye house or as part of it, an ample supply of warm water for washers, crabs, rinsing and such dye baths as can start with warm water. The simplest way to accomplish this is by means of wood storage tanks located high enough so that the water will flow by gravity to the machines. The water can be heated by exhaust steam or any other means. If there is a surface condenser in the power plant the circulating water can be used, and with a small sacrifice in vacuum can be delivered at about 120 deg. and pumped from the hot well to the tanks. The effect of this hot-water supply is to speed up the dye-house operations and reduce the peak load on the boilers.

# Appeal to Members of The American Society of Mechanical Engineers on Behalf of Nolan Patent Office Bill H. R. 11,984

MEMBERS of The American Society of Mechanical Engineers are requested by the Committee on Patents of Engineering Council to communicate at once with their Representatives and Senators in Congress, and to Hon. John I. Nolan and Senator George W. Norris, who are the chairmen of the Patent Committees of the House and Senate, respectively, urging action upon the Nolan Patent Bill H. R. 11,984, in accordance with resolutions passed on October 21 by Engineering Council.

This bill has passed both the House of Representatives and the Senate and is shortly to be considered by a Conference Committee of the House and the Senate. As passed by the Senate, amendments were made reducing the patent-office force and salaries of the examining and clerical force in a way to make both seriously inadequate; and further, certain riders were made part of the bill, the effect of which would undoubtedly be to delay its passage.

The specific request of the Committee on Patents is that members ask to have the original values of the figures for the examining and clerical force of the Patent Office and the salaries of the employees restored, and that the riders be eliminated. The text of the Committee's recommendations and of the resolutions of Engineering Council (in abstract) are appended so that any members who desire to take action may have all the facts at their disposal.

The members of the Society will doubtless remember that they, in cooperation with the engineering societies represented in Engineering Council, together with other scientific and industrial organizations, have been giving support to an effort to relieve the present desperate condition of the United States Patent Office by passing Nolan Patent Office Bill H. R. 11,984.

The purpose of this bill is to increase the examining and clerical forces of the Patent Office and to raise their salaries so as to give that office a sufficient force and at salaries that will attract and hold competent men to enable it to make its examinations with that reasonable promptness which is necessary to make it worth while applying for them and with such thoroughness as to reduce the percentage of errors to as low a limit as sufficient time for the work and proper qualifications can possibly effect.

Engineering Council appointed its Patents Committee for the purpose of aiding the Nolan bill and urged the membership of the constituent societies to communicate with the Patent Committees of the House of Representatives and the Senate and with the Representatives and Senators from the districts and states of the respective members on behalf of the said bill. The influence thus exerted, and that of other organizations, was so powerful that, at a hearing before the Rules Committee of the House of Representatives, which was largely attended by officers of members of Engineering Council and of the said societies and organizations, the Nolan bill was ordered made special and the House of Representatives promptly passed it without amendment by a very large majority.

A similar hearing on the bill was held by the Patent Committee of the Senate, but, in order to remove objection to unanimous consent to a special hearing by the Senate, before adjournment of the session, the Patent Committee of the Senate consented to amendments so seriously reducing the force and salaries of the bill as passed by the House of Representatives as to reduce the examining and clerical force below the numbers now actually employed in the Patent Office. The increases of the salaries provided in the bill were also cut down to where they are seriously inadequate to attract or hold a sufficient number of qualified men to enable the Patent Office to do its work. The steady exodus of examiners from the Patent Office, which has been going on for some time, has not been stayed at all by the passage of the bill by the Senate.

The bill was referred by the Senate to a Conference Committee, of which the Senate members are: Senator Geo. W. Norris, of Nebraska, Senator Geo. B. Brandegee, of Connecticut, and Senator William F. Kirby, of Arkansas.

The members of the Conference Committee for the House of Representatives have not been appointed, but Hon. John I. Nolan, of California, is certain to be one.

Engineering Council, regarding the matter as of grave importance, unanimously passed the following resolutions on October 21, 1920:

RESOLUTIONS OF ENGINEERING COUNCIL CONCERNING NOLAN PATENT OFFICE BILL

(*Limited space has made it necessary to give these in abstract*)

WHEREAS, the salaries of examiners, except for a war bonus, have only been increased ten per cent in seventy years and are so low that resignations of examiners are constantly occurring in a steady stream, averaging twenty-five per cent per annum, and resulting in such frequent changes that much inefficiency unavoidably results therefrom, even where examiners are qualified for the work, and many men are necessarily employed as examiners who cannot pass the examination required to qualify for their positions; and the salaries of the clerical forces are considerably below the average of salaries for corresponding work in the governmental departments generally; and

WHEREAS, as a partial remedy for such situation Nolan Bill H. R. 11,984 was introduced into Congress providing for an increase in the examining corps of the Patent Office of but five and eight-tenths per cent and an increase in the clerical force of but three and nine-tenths per cent, and providing increases in the salary for the position of primary examiners from $2700 to $3900 and of assistant examiners in proportion, and providing increases in the salaries of the clerical force only to bring them up approximately to the average corresponding salaries of other Government Departments and Bureaus, and as the cost of the increased salaries and force of the said Nolan Bill was more than met by an increase in the fees for patents provided therein; and

WHEREAS, the United States Senate so amended the said Nolan bill that instead of increasing it decreases the examining corps by fifteen and seven-tenths per cent and reduces the clerical force by about one per cent below the present insufficient numbers of said examining corps and clerical force actually employed in the Patent Office, as well as reduces the salaries, both of the examining and the clerical forces, so that the total present payroll is reduced five and nine-tenths per cent, notwithstanding that the increase in the fees for patents which were made to provide funds for the increased force and salaries were retained in the bill;

NOW, THEREFORE, BE IT RESOLVED: That Engineering Council, representing 45,000 engineers, regards it of large importance that the numbers of the examining and clerical forces for the Patent Office and the salaries therefore in Nolan Patent Bill H. R. 11,984 be restored to those in the bill as it passed the House of Representatives; that the bill be freed from any riders, such as Section 9 thereof, which may delay or jeopardize the passage thereof, and that the bill be made a law at the earliest possible moment.

As the Patent Office is steadily losing more and more of its competent men and is rapidly getting farther and farther behind in its work, and as to get much farther behind would mean for it practically to cease to function, and as the bill, as amended by the Senate, is wholly inadequate to accomplish its purpose, and would be worse than useless, every effort should be made to induce Congress to restore the figures of the bill to their values as passed by the House of Representatives.

The Conference Committee will probably take the bill up for consideration the middle or latter part of December. Each member of the Society is therefore most earnestly requested to write or telegraph to the member of Congress from his District, to the Senators from his state and to the members of the Conference Committee, urging that the figures of the Nolan Patent Office Bill H. R. 11,984 be restored to the values which passed the House of Representatives.

It would also be well to urge that the bill be freed from any riders not related thereto, so that its early enactment will not be hindered by opposition to such riders.

The names of the senators and representatives can be obtained from the World Almanac, or similar publications, and from postmasters.

In connection with this bill, engineers, as a class, for the first time, have exerted a powerful influence in a public matter, and they should see that their work is brought to a successful conclusion.

Yours very truly,
J. PARKE CHANNING
*Chairman, Engineering Council*

EDWIN J. PRINDLE, *Chairman*
J. PARKE CHANNING, *Secretary*
EDWIN J. PRINDLE (*A.S.M.E. Representative*)
D. S. JACOBUS (*A.S.M.E. Representative*)

## DeLamater Collection Presented to A.S.M.E.

Messrs. Sydney and Victor DeLamater Bevin, members of The American Society of Mechanical Engineers, and descendants of Cornelius H. DeLamater, have recently presented to the Society the models, medals and plans formerly belonging to him. The collection consists of:

*a* One model solar engine, 6 in. high, made of brass and steel, in working condition

*b* One model hot-air engine of similar construction, in working condition

*c* One model steam engine

*d* One model cannon, mortar type

*e* Several medals awarded to C. H. DeLamater

*f* Set of miscellaneous plans belonging to the DeLamater Iron Works.

These articles, which are of historic value to the science of mechanical engineering, have been placed in the headquarters of The American Society of Mechanical Engineers at 29 West 39th Street, New York, N. Y.

# MECHANICAL ENGINEERING

The Monthly Journal Published by The American Society of Mechanical Engineers

FRED J. MILLER, *President*

WILLIAM H. WILEY, *Treasurer*       CALVIN W. RICE, *Secretary*

PUBLICATION COMMITTEE:

GEORGE A. ORROK, *Chairman*                              J. W. ROE

H. H. ESSELSTYN          GEORGE J. FORAN

RALPH E. FLANDERS

PUBLICATION STAFF:

LESTER G. FRENCH, *Editor and Manager*

FREDERICK LASK, *Advertising Manager*

WALTER B. SNOW, *Circulating Manager*

Yearly subscription, $4.00, single copies 40 cents. Postage to Canada, 50 cents additional; to foreign countries $1.00 additional.

*Contributions of interest to the profession are solicited. Communications should be addressed to the Editor.*

## Why Not By-Product Producer Gas?

ROBERT H. FERNALD

THE rapid exhaustion of our natural-gas supply through criminal negligence and reckless extravagance resulting in drastic restrictions in order to conservé what little we have left brings us abruptly to a realization of the situation and leads us seriously to query regarding possible substitutions for this remarkable natural fuel.

The situation is acute. It is therefore imperative that we begin exhaustive investigations to determine the most practical solution of the industrial problems that have been for many years past so vitally dependent upon our natural-gas supply. The results reached through such investigations may lead into channels quite different from those under consideration at the present time, as the necessity of the situation may develop unthought-of possibilities.

With the limited amount of study that has been given to this problem to date, the most conspicuous source of relief seems to be through the development of by-product producer gas. This, of course, is practically an unknown field in the United States. Spasmodic attempts to develop interest have come to the attention of engineers from time to time but no serious study of the situation in its relation to the larger commercial developments of the country has as yet been undertaken.

In Europe we find by-product producer gas somewhat extensively used and the large central station near the mines, recovering the by-products from the fuel and distributing the gas under pressure for several miles through the industrial districts, has become a dependable source of supply.

Such plants produce in the neighborhood of 135,000 cu. ft. of 150-B.t.u. gas in addition to some 85 lb. of sulphate of ammonia and 100 lb. of water-free tar as by-products per ton of coal gasified. The sulphate of ammonia alone brings a handsome return to the plant, as it is worth, in normal times, about 3 cents a pound or $60 a net ton. The price for the last few years has, of course, been considerably higher.

The gas from one of the large central by-product plants in England is used for some thirty or forty separate purposes, including melting, welding, annealing, vulcanizing, hardening, enameling, and brazing, outside of a large demand for power. The gas is sold at from $1\frac{1}{2}$ to $2\frac{3}{4}$ pence per 1000 cu. ft., depending upon the volume of consumption. The net cost of the gas per 1000 cu. ft. for the year 1912 based on 40,000 tons of coal gasified, was 1.287 pence. This selling price of $1\frac{1}{2}$ to $2\frac{3}{4}$ pence is equivalent to about 3 to 5.5 cents per 1000 cu. ft. This, of course, is for gas averaging 150 B.t.u. per cu. ft., which is equivalent to from 12 to 22 cents for 600-B.t.u. gas or from 20 to 37 cents for 1000-B.t.u. gas based on the relative B.t.u. values alone.

Through the abundance of our fuel supply and our indifference to the needs of future generations, we have been ruthlessly extravagant and wasteful of our fuel resources during the past decades. An appreciation of the necessity of recovering by-products has but recently received thoughtful consideration from those agencies that are most seriously interested in the problems of fuel conservation.

At the present moment, low-temperature distillation processes are attracting more than casual attention. Investigation may show a combination of low-temperature distillation and the manufacture of by-product producer gas to be a distinctly economical and commercial solution of our natural-gas problems, as these processes, on the one hand, lead to the maximum by-product recovery and on the other, to a supply of gas commercially adaptable to heating and general industrial processes.

Our natural-gas situation is a serious one. The solution requires more than the casual financial interest of the private dividend seeker. It requires systematic study and research involving the expenditure of large sums of money. It is a problem of national importance and must be attacked on a broad basis. It calls for the best thought and most serious consideration of the ablest experts available, financially supported from sources that will guarantee the results of the investigation to be free from industrial or commercial bias.

It has been estimated that two decades will see the total exhaustion of the natural-gas supply. We cannot afford to wait longer.

Should not the engineering profession appeal to the Government in no uncertain terms regarding this necessity and see that the problem is vigorously attacked before the crisis is upon us?

ROBERT H. FERNALD.

## Forty Years Ago

The fortieth Anniversary meetings of The American Society of Mechanical Engineers held last month in various cities make it interesting to recall certain events in the mechanical engineering field which indicate what was the state of the art in 1880 when the Society was founded. A remarkable portrayal of the situation at that time is contained in the presidential address of the late Robert H. Thurston, given before the Society in 1881.

Unquestionably, the absorbing topic of the time was the promise of electricity as a useful agent in industry. The Brush dynamo-electric machine had been invented in 1876 and the Edison carbon-filament incandescent lamp in 1880. Three years later, so rapid had been the progress of the electric light, that there were 400 electric-light stations in operation in this country. The possibilities of this development are told so fascinatingly by Doctor Thurston that his allusion to the subject in his presidential address is here quoted in full:

"That feature of recent progress in engineering which is today attracting most attention and awakening most interest in the minds of the public as well as of the profession, is the introduction of machine-made electricity, and of the electric light, but what seems to me the most important phase of this impending revolution, as I think, not yet generally comprehended. By the ingenuity and skill, the courage and persistence, the energy and enterprise of our brother engineers, Brush and Edison and their condjutors, it seems certain that the dream of the great author of "The Coming Race" will in part be speedily realized, and that for the occasional mild light of the moon, or the yellow sickly flare of the gas flame, will soon be substituted the less uncertain and always available, and always beautiful and mellow, radiance of the electric flame. This is but a beginning, however. A few months ago one of the most earnest and best workers of all who have been with me, at once, friends and pupils, made a very painstaking investigation of the efficiency of a powerful dynamo-electric machine kindly loaned him from Menlo Park. The mean of several series of tests gave, as a result, an efficiency of between 90 and 95 per cent. That is to say, of all the power transmitted to the machine from the steam engine driving it, over 90 per cent appeared on the wire in the form of electrical energy. It follows at once that mechanical power may be transmitted through two such machines, again appearing as mechanical power, with a loss of less than 20 per cent. And it follows from this last fact that the distribution of power by electricity is not unlikely to prove a more important application of this wonderful force than is the electric light."

Inasmuch as we have just emerged from the Great War, it may be noted that it was only in 1880 that the manufacture of modern steel guns was projected in this country, with imported Whitworth steel; and that it was not until several years later that the Whitworth fluid-compression process for steel forgings, both for artillery and for the heavy parts of steam and electric machinery, was introduced in this country. In general, this period was marked by the substitution of high-carbon steel forgings for wrought-iron forgings, and of mild-steel plates for boilers and ships in place of wrought-iron plates, a fact alluded to by Doctor Thurston.

In 1880 the Corliss engine was supreme—it was a period of transition between the big, slow-moving, low-powered Corliss steam engine of the Centennial and the high-speed engine of the nineties. The Otto "silent" gas engine had been in the country four years and held the field until the expiration of the Otto patents in 1890. Referring to steam engineering and the gas engine, Doctor Thurston said in part:

"This figure—16 lb. of steam per hr. and per hp.—(from a Leavitt hoisting engine at Hecla) may be put on record as the very best economy attained by our best engineers at the end of the decade 1870–1880. It is just double the weight which would be required in a perfect engine working steam of the same pressure at maximum efficiency.

"The compressed-air engine, the petroleum engine, and the gas-engine are all just now coming forward. I have no figures that I can rely upon except for the gas engine, which sometimes consumes as little as 28 cu. ft. of gas (1 cubic meter nearly) per hr. per hp.

"Today a steam pressure of six atmospheres (75 lb. per sq. in.) is usual, and 7 atmospheres (90 lb.) is often adopted. Increased pressure has been accompanied by increased speed of piston—from 300 to 500 ft. per min.—and both causes have combined to reduce greatly the size and weight of engines. The consumption of fuel per hour and per horsepower has decreased from 2 lb. to 1.8 lb."

The error which had been fallen into by many writers on the steam engine to the effect that the condensation in the steam-engine cylinder was due solely to the conversion of heat into work, was also pointed out by Doctor Thurston, who said: "It is becoming generally recognized and writers are in a fair way to learn that it is not the fact that the greater part of the liquid water which collects in unjacketed cylinders is produced by liquefaction of steam during its expansion, but that this latter amount is insignificant and that this comes of cylinder condensation, sometimes with a considerable leakage, and often amounts to half or more of all the fluid supplied to the boiler." In other words, the action of steam in the engine cylinder was but beginning to be understood by engineers in general.

Locomotive practice is summed up as follows:

"Locomotives are frequently built weighing 50 tons; 70 tons has been reached, and every builder of engines is ready to guarantee the performance of an engine to draw 2000 tons 20 miles an hour on a level track. In coal consumption we have made some saving of late years. Three pounds of coal per hour and per horsepower is a usual amount, and a consumption of 2.6 lb. of coal, and of 22$\frac{1}{2}$ lb. steam has been reported from recent tests."

Finally, in two trite but effective sentences the labor situation is thus recorded:

"When the last generation was in its prime our factories were in operation 12 or 13 hours: 'Man's work was from sun to sun, and woman's work was never done.' Today man works ten hours, and woman is coming to a stage in which she will work where, when, and how she pleases."

## Finding Jobs for Things

One of the most interesting branches of modern engineering is that which deals with the discovery of new uses for materials and equipment which have lost their applications and remain for a while, so to speak, jobless. The war, in particular, may be regarded as an example of an immensely vast industry which has ceased, and whose raw materials, formerly developed for the destruction of life and property, must now be put to some useful purpose.

For a great many materials "jobs" have already been found, and what follows must therefore be considered merely as examples and possible suggestions of future applications.

During the war, as a protection against the employment of poison gases, highly efficient gas masks were developed, and now that the war is over it is an obvious step to apply them to mine-rescue work and the protection of men engaged in handling poisonous materials. A less obvious matter is that of utilizing ill-smelling gases, a vast number of which were developed for military purposes. It is now suggested that such gases should be used in mines to notify the men, in case of accident, of the necessity for escape; for tests have shown, especially in mines supplied with artificial ventilation, that ill-smelling gases carry their message more rapidly and certainly than sounding devices.

Another war development which is finding application in peace time is the guide cable which was employed by the Germans around Heligoland to permit submarines to return through the mine field to their bases. The cable was laid in such a manner as to indicate safe passage and oscillations sent through it could be caught by an appropriately attuned receiver. At the present time experiments are being carried out by the U. S. Navy with a view to applying the same scheme for guidance of vessels entering harbors in a fog.

But the problem that is of particular importance, however, is in connection with the numberless materials which are becoming available every year as by-products of other manufactures. Selenium is one of these. Before the war it was almost exclusively used for electrical purposes, where its value lay in the fact that its electrical conductivity varies with the intensity of the light to which it is exposed. During the war, when this country experienced a great shortage of manganese, selenium was extensively used to take the place of this material in glass manufacture. The amount which is produced in the copper industry, however, is so large that it has become an important problem to find some peace-time commercial application for this interesting material. It has been recently announced that such an application has been found, due to the discovery of the fact that selenium oxychloride is a powerful solvent for a number of materials such as paraffin, etc.

There is still a third problem associated with this discovery of new uses for materials and it deals with the employment of rare materials. Vanadium, for instance, was for many years known only as a laboratory curiosity. It was known, in general, that it improved the quality of steel, but not enough of it was available to make the work of investigation worth while. Shortly, however, great deposits of it were found among the hitherto inaccessible peaks of the mountain ranges of Peru and in a few years vanadium became a recognized agency in the production of high-grade steel, for which at just about the same time, a big demand was created by the automobile industry.

Another similar material, molybdenum, is still in what might be called a probationary stage. It is well established that under certain conditions molybdenum does improve steel, but just what these conditions are is still a somewhat mooted question; furthermore it is not as yet definitely established whether or not the same results can be obtained at the same cost with such other materials as tungsten or chromium.

There are, however, many materials produced or available in quite substantial quantities for which no suitable use has as yet been discovered. As such may be mentioned, for example, calcium chloride, which is a by-product of soda manufacture.

Another by-product which is still waiting for some one to discover a use for it is straw. Of this many millions of tons are produced a year, and yet, notwithstanding the active work done by the various departments of the Government and in private laboratories, no real use has been found, which may be due partly to the fact that straw is a seasonal material and as such is particularly difficult to handle commercially.

The question of utilization of either waste or idle materials is a large one for our national industries, as it represents one of the most logical ways to cut down the unit costs of production. The packing industry realized this many years ago and boasted for years that everything in the pig except the squeal was put to commercial use. In fact, it is believed that one of the reasons for the tremendous growth of packing establishments and the very generous profits which they have made for their stockholders has been this very fact of utilizing practically 100 per cent of what goes into the stockyards.

The mechanical industries are still very far from this ideal, although doubtless the present high costs of labor, materials, and doing business generally, are proving to be a powerful incentive toward eliminating all sources of waste and putting all by-products possible to useful employment.

## Fuel and Power Conservation on the Pacific Coast

*Special Correspondence*

At the present time engineering affairs on the Pacific Coast are rather quiet, and I think it might be said that the great majority of what might be called engineering or industrial capital is feeling its way very carefully before deciding on any investment of funds. This is partly due to the political situation, but more I think to the great uncertainty as to future moves of labor organizations.

Recent conditions on the Pacific Coast have concentrated the attention of engineers there on the question of power supply and the more economical use of power. Inadequate rainfall for three successive seasons and a very light precipitation of snow in the mountains had, during September of this year, forced the power administrator of California to issue an order that all power users must reduce their consumption twenty per cent and that all electric display signs and unnecessary lighting must be discontinued.

During this same three-year period the cost of fuel oil, the universal industrial fuel of California, had risen by leaps and bounds, and early in the spring of this year managers of industry were thrown into a near panic by the publicity given to the fact that our available supplies of fuel oil were almost exhausted. These facts and circumstances have combined to direct the attention of engineers very emphatically to the problem of power conservation, and much activity is evident in that direction.

One very interesting instance of this activity lies in the careful investigation which is being made by railroad engineers into the economy of electrifying all of their lines and also into the problem of applying Diesel engines to train propulsion, and the suggestion has been made that perhaps these two propulsive methods might be combined.

The great difficulty in applying the Diesel or any other internal-combustion engine to railway propulsion lies in securing sufficient initial torque for starting a train. This, of course, could be readily overcome by using an electric locomotive which would be coupled with a Diesel-electric generating plant on a separate tender or car. This equipment could be used on portions of the rights of way distant from hydroelectric development. The same locomotives could also pick up power from transmission lines where they were readily available.

Probably the greatest benefit which will be derived from this power-conservation movement will come through a careful engineering survey of the sources of supply and a concerted coöperative effort to apply power from those sources in the most efficient and useful manner. Undoubtedly, as has been pointed out in many articles in the technical press, the majority of power users are wasting a large percentage of the power in its application.

This is particularly true in the plants which have sprung up during the emergency shipbuilding program, many of which are now endeavoring to adjust themselves to the changed conditions and to compete in general and industrial engineering practice.

When we consider the economies that can be gained from installation of the best apparatus for the utilization of the energy in fuels as compared with the very low percentage of that energy utilized by the makeshifts often in use, we are almost forced to the conclusion that the use of much of this apparatus should be made an economic crime and that all applications of power should be under proper supervision. As Franklin K. Lane has so well said, every man who uses fuel oil under a steam boiler should be made to show good reason for the apparent waste of a valuable commodity.

Notwithstanding all of this agitation and activity directed toward power conservation, we note the sorry spectacle of the Shipping Board taking a backward step by removing from several of its vessels water-tube boilers and geared turbines and substituting therefor Scotch boilers and reciprocating engines. This leads us to remark that now, as in every period of the history of modern engineering, the greatest enemy to advancement has been prejudice and obstinacy on the part of operating engineers themselves, and this seems to be true to the greatest extent in our own United States where one would least expect it.

During the first three weeks of October there came considerable general rainfall all over the state of California and an abundant snowfall in the mountains, and while the engineers are still laboring on the problem, the popular demand for its solution has vanished overnight, the power commissioner has rescinded the order for a twenty per cent reduction, and display signs, street lighting, etc., are running full blast; but the industrial managers facing the problem of costs find it necessary to trim at every point in order to compete, and in all probability a great deal of permanent benefit will come to the industries of the Pacific slope through this power-shortage scare.

I shall hope in future correspondence to report substantial progress along some of these lines and perhaps be able to make a few suggestions for the benefit of engineering societies and local sections thereof.　　　　　　　　　　　　　　　　　A. J. DICKIE.

## James Hartness Elected Governor of Vermont by Big Majority

James Hartness, past-president Am.Soc.M.E., ran ahead of the presidential ticket in his election as Governor of Vermont. Harding was accorded a majority of 44,300 and Hartness 47,400. The total vote for Hartness was approximately 65,950 and for Martin, the Democratic candidate for governor, 18,550. As previously explained in these columns, Mr. Hartness conducted his campaign for the gubernatorial nomination on a platform calling for the development of the industries of Vermont, similar to that which has taken place at his home town of Springfield. This has become very much of an industrial center through the local encouragement offered to home talent and home firms.

## President Burton Inaugurated at University of Michigan

The inauguration of President Marion Leroy Burton at the University of Michigan took place on the morning of October 14. In addition to an address by President Emeritus Hutchins and President Burton's inaugural address, there were others on the functions of the governing board and the faculty in the administration of a university.

President Burton's inauguration was the occasion for an educational conference which was opened on the afternoon of October 14 by a session devoted to a symposium on educational readjustments. The conference continued on October 15 with symposiums on administrative problems and on constructive measures. There were four speakers at each of these symposiums and many problems of the present-day university were discussed. The conference closed on October 16 with a meeting of the regents of state universities at which the salary problem and student fees and tuition charges were discussed.

# JOHN R. ALLEN

JOHN R. ALLEN, director of the Bureau of Research of the American Society of Heating and Ventilating Engineers in coöperation with the U. S. Bureau of Mines, and vice-president of The American Society of Mechanical Engineers, died on October 26 of pneumonia. Professor Allen contracted a cold when in Detroit to give an address early in October. He recovered from this and went on to Philadelphia where he spoke for an hour and a half the evening of October 14. The next day, Friday, he attended the A.S.M.E. Council meeting in New York and on Saturday participated in a conference on the ventilation of the New York New Jersey Vehicular Tunnel. He arrived home early Sunday morning feeling very ill and went immediately to bed. Toward evening he was again about for a little while, and for the last time. Pneumonia had set in. He died on Tuesday afternoon, October 26. Funeral services were held in the First Baptist Church of Ann Arbor, on Friday, October 29, at which the Society was repre-

JOHN R. ALLEN

sented by Dean M. E. Cooley and Prof. H. C. Anderson, appointed honorary vice-presidents by President Miller to attend the funeral.

Professor Allen was born in Milwaukee on July 23, 1869, and received his preparatory education at Milwaukee and at the Ann Arbor, Mich., high school. In 1892 he was graduated from the University of Michigan with the degree of B.S. (M.E.), receiving his master's degree from the same institution in 1896. For two years he was connected with the L. K. Comstock Construction Co. From 1896 to 1911 he was associated with the University of Michigan, rising from an instructorship to a full professorship. In 1912 he was called to Robert College in Constantinople, Turkey, as dean of the engineering department. Two years later he returned to the University of Michigan as head of the mechanical engineering department, being so connected until 1917, when he became dean of the college of engineering and architecture of the University of Minnesota. In August 1919 Professor Allen accepted the directorship of the Bureau of Research of the American Society of Heating and Ventilating Engineers, with headquarters at the U. S. Bureau of Mines, Pittsburgh.

Professor Allen was extremely active in the work of engineering societies, having been a past-president of the American Society of Heating and Ventilating Engineers and of the Michigan Engineering Society. He was an honorary member of the National District Heating Association, a member of the British Institute of Heating and Ventilating Engineers and of the Society for the Promotion of Engineering Education. He was also a member of the honorary societies, Tau Beta Pi and Sigma Psi. He became a member of our Society in 1894, and at the time of his death held the office of vice-president of the Society.

Professor Allen has been a contributor of important technical papers to these societies, papers which have proved to be reference works of great value. He was the author of textbooks, not only on the subject of steam engines, which he taught, but also on heating and ventilation. His book on the latter subject is recognized throughout the country as a standard. His long connection with the University of Michigan gave him a wide acquaintance and he was called for consulting work to all parts of the world. Notable among his engineering activities during the period of his connection with the University of Michigan were his expeditions to the unexplored districts of western Mexico to report upon the existence of rubber-producing trees; to the Rocky Mountains to report on water-power projects, and his engineering expeditions into northern Michigan and southern Canada.

A great lover of outdoor life, Professor Allen was the owner of two beautiful farms in the outskirts of Ann Arbor, one of which was owned jointly with his adopted son. He brought the boy back with him from his Mexican expedition and adopted him that he might have the opportunity of coming to this country and of being educated here. This boy, who is a native Mexican Indian, has of late years operated both farms, and it was Professor Allen's great pleasure to spend much of his leisure time in farming.

He leaves a wife and daughter residing in Pittsburgh, his daughter attending the Margaret Morrison school of Carnegie Institute.

## WORDS OF APPRECIATION

From Dean M. E. Cooley, a very close friend of Professor Allen's, comes the following appreciation:

Prof. John R. Allen's death was untimely. It came in the midst of his most important life work, apart from his teaching days. He was intensely interested in his heating and ventilating researches, and was trying to establish fundamental laws governing the flow of heat under a variety of practical conditions. A number of colleges were coöperating with him. A year or two more would have seen certain benchmarks firmly established. His death was indeed untimely.

Professor Allen was a man of charming personality. He had a host of friends. In the honor Society of Michigamua here at Michigan his Indian name was "Man of Many Friends." He sat in well at committee meetings and conferences. Always careful of the feelings of others and respectful, difficult situations were overcome and his own views, even when radically different, were well received. He took advice well and was himself a good adviser. He was a splendid teacher. His students respected and admired him. He was a good influence on the young men who sat under him. He was one about whom the older alumni inquired years after their college days. That fact alone marked him with especial honor in the teaching world.

Professor Allen was a well-poised man. Unpleasant things did not upset him. He never lost his temper. His rule was to forget unpleasant things—at least to ignore them. If he could not get away from them in one locality he would move to another. Nothing seemed to ruffle him. At least he did not show it. And yet he was sensitive and easily hurt, but always ready to forgive and start over again.

Professor Allen was resourceful and adaptable. He got on well. In foreign countries he had the same respect and devotion from those with whom he came into contact as at home. He was quick to learn the peculiarities of men and to act accordingly. Rarely did anyone get the better of him for long. His construction gangs frequently embraced Italians, Greeks, Turks, Bulgarians, Armenians. But whatever their nationality, men would work for him cheerfully and well.

Professor Allen loved companionship. He enjoyed his friends and could not do too much for them. He was a good banqueter and could always be relied on for some post-prandial remarks. He enjoyed speaking in public and was more than generous in responding to invitations to address gatherings. His experiences in Mexico when investigating rubber-producing trees, and in Turkey, were favorite non-technical subjects and he made them very attractive with his lantern slides.

The following letter has been received from Prof. O. P. Hood, chief mechanical engineer of the U. S. Bureau of Mines, Washington:

Wherever John Allen made contact with people or work, there will be a keen sense of loss at his going.

Dean Allen came to Pittsburgh in a scientific adventure, the coöperative agreement between the American Society of Heating and Ventilating Engineers and the Bureau of Mines being now and untried as a method of furthering science and technology. His selection to head the work proved so happy that the Bureau had no fears whatever for the character of the output. He dropped into the organization with the greatest accommodation, fitting himself to the situation so graciously as to win instant appreciation from all the members of the Bureau of Mines with whom he was brought in contact. Kindly and helpful in every situation, he was a stimulus to good work. The installation of equipment was just being completed and we all looked forward to an early output of valuable information.

The sense of personal loss to members of the Bureau will be slow to fade.

# Fortieth Anniversary Meeting of A.S.M.E.

Special Meeting in New York with Addresses on the Opportunity and Responsibility of the Engineer
by J. Herbert Case, Federal Reserve Bank of New York, Samuel Gompers, and William B. Dickson

THE special meeting of The American Society of Mechanical Engineers called to celebrate the Fortieth Anniversary of the first meeting of the Society was held in the Engineering Societies Building, New York, on the evening of November 5. Coincident with this general meeting were meetings in different parts of the country of 32 of the 38 sections, all of which contributed to a most appropriate recognition of the growth of the Society, its far-reaching influence and its long period of service.

As a means for "tying-in" the various meetings the plan was carried out at the suggestion of one of the members, Mr. William H. Bristol, of Waterbury, Conn., of having phonographic records made of speeches by officers of the Society to be distributed to the different Sections for use on the evening of the celebration. Mr. Bristol has been working on a device for synchronizing phonograph records and moving-picture films, and in fact has so far perfected the method that a demonstration with films and records was made at Waterbury before the Section meeting there, which was the first public exhibition of the synchronizing device. The records distributed were of addresses by President F. J. Miller and Past-President I. N. Hollis.

Another tying-in feature was effected through the coöperation of the American Radio and Research Corporation, at Medford, Mass. and its vice-president, Mr. H. J. Power. Greetings from Harding and Coolidge, Mayor Peters of Boston, the Boston Section, Dean Anthony of Tufts College, and the company, were radio-phoned to Local Sections throughout New England and as far as Ontario, Canada. President-Elect Harding sent "Greetings and good wishes to The American Society of Mechanical Engineers on the occasion of the celebration of the fortieth anniversary of the organization. The Administration which comes into power next March fourth very much wishes the advice and coöperation of the membership."

The meeting in New York was on the subject of The Opportunities and Responsibilities of the Engineer, with a program arranged by a special committee appointed by the President, consisting of Henry R. Towne, past-president, chairman; Harry A. Hopf, vice-chairman; W. Herman Greul, secretary; William H. Wiley, treasurer of the Society; Frank T. Chapman, representative of the New York Section; and L. B. McMillan, representative of the Committee on Meetings and Program.

The program consisted of preliminary remarks by Henry Towne, the oldest past-president of the Society; A. P. Davis, president A.S.C.E.; William L. Saunders, past-president, A.I.M.E.; and Charles F. Scott, past-president, A.I.E.E. Following these were the three principal addresses: by J. Herbert Case, acting governor, Federal Reserve Bank of New York; Samuel Gompers, president, American Federation of Labor; and William B. Dickson, steel manufacturer. Extended abstracts of these addresses conclude this report.

President Miller presided and said that forty years ago the first regular meeting of the Society was held at the theater of the Union League Club, then at 26th Street and Madison Square. At that time there were enrolled on the books of the Society 189 members and 85 of them were registered as having been in attendance at that first meeting. Continuing, he said:

When a man reaches his fortieth year, he is generally conceded to have reached the zenith of his power of production; but there is no age limit on associations of men—societies of men may be afflicted with some things analogous to old-age infirmities, but when they come they are usually the result either of a lessening opportunity for rendering service, by change of conditions, or by reason of bad management in the affairs of the societies. Our opportunities for service seem to be expanding, and in so far as we can see, they appear to have no limit. With our Local Sections now numbering 38, scattered over the industrial centers of the country, with our student branches now numbering 53, located in the educational institutions at which engineering is taught; with our recently formed Professional Sections now numbering eleven, our participation in the United Engineering Foundation, and in the American Engineering Standards Committee, and in the conduct of our Journal, to mention only the major lines of activity in which the Society is engaged, it can be seen that we are a very active Society.

## Remarks by Representatives of the Four Founder Societies

### REMARKS OF HENRY R. TOWNE

Mr. Towne's remarks consisted chiefly of reminiscences of the history of the Society. He said, in part:

I became a member of this Society in 1882, about two years after its organization. The first meeting which I attended was held in October 1883, under the presidency of Mr. Leavitt, in what was then the house of the American Society of Civil Engineers in East 23d Street, which the older members will remember, and we had at that meeting an attendance of about 150. The total membership of the Society at that time was about 600.

Today, to make the contrast at once, the Society has a membership of 13,500, and is still going strong, and the attendance at its meetings approximates thousands where formerly it was only in the hundreds. In the same connection I may mention the fact that the income of the Society in that year was some $7560, whereas its budget for the coming year, I understand, approximates a half million of dollars.

That meeting was signalized also by the evening session of the second day which was devoted to memorial services for one of the most distinguished American engineers and one who is held in affectionate remembrance by all who knew him—Alexander Lyman Holley. The presiding officer on that occasion was Prof. R. H. Thurston, the orator was Dr. Rossiter W. Raymond, and the mover of the final resolution was Mr. James C. Baylis, names familiar to all the older members.

In that connection I may recall that my predecessors, seven of them, were Prof. R. H. Thurston, the first president, followed by E. D. Leavitt, Prof. John E. Sweet, J. F. Holloway, Coleman Sellers, George H. Babcock, and Horace See, all of whom have joined the great majority.

At the meeting of the Society held in Chicago in 1886, owing to the absence of the President, Mr. Coleman Sellers, because of his unfortunate incapacity due to a very serious operation, I, being the senior available vice-president, was called to serve in his place. At that meeting I had the honor of presenting a paper under the title The Engineer as an Economist, and I believe those who are interested in and familiar with the record of this branch of our professional work credit that paper with being the earliest publication of any kind, in the English language, certainly, to propose that the mechanical engineer in his organized society should take up the study of economics as related to his work, the justification for that being, in the case of our Society, in particular, the fact that probably a majority of mechanical engineers are identified more or less closely with managerial positions, as superintendents or directors of industrial operations. I had noted at that time that there was a gathering volume of data and information and experience in that important field, but no channel for its crystallization, for its diffusion, and for its utilization, no channel either of printing or of meetings for discussion.

I thought, also, that it marked the dawn of a new science, and that The American Society of Mechanical Engineers, more fittingly than any other organization, could sponsor the growth and development of that science. The suggestion was accepted, the Society has acted upon it from that time to this, and the result has been a great and growing volume of material relating to that subject in our discussions, and in our transactions and public papers.

Now, that fact is directly germane to the program for this evening on the subject of The Opportunities and the Responsibilities of the Engineer. In my opinion, the responsibilities of the engineer have become very definite and his opportunities are almost limitless. Mechanical engineers in particular, who are brought so directly in contact with productive operations, have a great opportunity and as equally great responsibility to see that they do their full share in promoting the advance of the new era and the new conditions which it implies and to which it has given origin.

These new conditions are divided under two heads and can be summed up perhaps in two words: efficiency and coöperation. Efficiency of methods of machines and of processes; and coöperation between those who plan and direct the machinery of industry and those who operate it. The old order is passing. Especially is that true in our method of compensation, the compensation for labor. The old methods have served their purpose fairly well, but they are not adapted to present conditions—they are going to be superseded by newer and better methods. We will be called upon to meet a great diversity of conditions in the different industries, and these in turn must be worked out by the collaboration of the industrial engineer, and the labor leader. By coöperation between these two elements the problems which will confront us will be solved, and we are going to have greater efficiency, greater productiveness, and there is no greater thing in the world today than that, and with it will come greater possibilities of friendship and good will.

### REMARKS OF A. P. DAVIS

Mr. A. P. Davis extended the congratulations of the American Society of Civil Engineers to The American Society of Mechanical

Engineers on the completion of its fortieth successful year of opera-tion. He congratulated the Society not only upon its age and upon its growth, but on its activity and its progressiveness. In this connection he said:

I had the honor of being called on by your President recently to dictate a brief greeting to be read at the meetings of the Local Sections of this Society on this occasion, the 40th anniversary of the organization of the parent society, and in that greeting I was glad to state that the usefulness and progressive character of The American Society of Mechanical Engineers was a bright example to be followed and emulated by the older, but smaller and less progressive—though I hope not less useful—organization, the American Society of Civil Engineers.

The American Society of Civil Engineers is now nearing the culmina-tion of an effort to shake off the lethargy of the past, and to become a worthy member of what is known as the Four Founder Societies. The ballot will be opened and counted on next Monday, and on that day we hope that the American Society of Civil Engineers will seal the bond of fellowship with the other three Founder Societies which have entered upon a larger and more useful field than they have occupied heretofore, in joining and leading and forming and molding a federation of the engineers of the United States into one great Federation, which can work together and be really and truly representative of all the engineers of the country, and speak for them as a body.

Mr. Davis pointed out the opportunities for expansion of engi-neering activities, referring to the developments resulting from the World War and concluded by saying:

Among those leaders of thought I place foremost the engineers of the world, the educated, experienced men, whose capacity has been demonstrated in the last few years; and whether or not that influence is to be for good or for evil depends on the activity of the people I see before me, and those corresponding to them in the other Founder Societies.

### ADDRESS OF W. L. SAUNDERS

Mr. Saunders, representing the American Institute of Mining and Metallurgical Engineers, discussed four points regarding engineers. These points are given briefly in the following abstract of his remarks:

I occupy a rather difficult position tonight, in that I am expected to represent Mr. Hoover, but I take pleasure in the thought that the man I represent is particularly representative of the subject which called us to-gether tonight—the engineer, not specifically in his technical work, but in his usefulness to society in general.

There are certain thoughts which I should like you to carry away with you tonight, and the first one which impresses itself upon me is this—that the engineer is moving upward and broadening his influence. When I sit on this platform and look at my old friend Samuel Gompers and see him coöperating with us in a meeting of this kind, I rejoice in the knowledge that the profession to which I belong is now one which recognizes the human side of industry, which recognizes, and as ex-Mayor Seth Low once said, that a man should have the right to have something to say about the conditions under which he spends the major portion of his life. The second thought which occurs to me, which is closely allied with the first, is emphasized in the fact that this meeting is not merely an experience meeting, not merely a meeting of men and women who have come together to say "We are now forty years old, let us talk about the great things we have done in the past—let us be reminiscent," but it is a meeting to call serious at-tention to the fact that this forty years of service of engineers brings upon our shoulders greater responsibilities and duties for the future. Let us not forget that, because we have done things in the past in engineering lines, because we have build up this country as a great industrial nation, we have now to face the responsibility of seeing to it that the engineer not only maintains what he has built up, but that he carries it to greater usefulness.

The third thought which impressed me tonight is this: Mechanical engineering is the basic science of all engineering. And the fourth and last thought that I should like you to carry away with you tonight is that these organizations of engineers, the four Founder Societies, and particularly the mechanicals, represent the spirit of coöperation—the spirit of collectivism as distinguished from the spirit of individualism. By getting together, as these societies do, and matching ideas and doing teamwork, they get re-sults, not only for the societies and for the concerns they may represent, but for each individual. It is teamwork which tells.

### REMARKS BY CHARLES F. SCOTT

Mr. Scott, who was chairman of the committee which had charge of the construction of the Engineering Societies Building, spoke of the services rendered by Mr. Calvin W. Rice, Secretary of The American Society of Mechanical Engineers, whose activities eighteen years ago as chairman of a prior building committee, made this building possible.

Reviewing the achievements of the past forty years, Mr. Scott said:

During this forty years the population of the United States has doubled; the value of the manufactured products of the country have been increased sixfold; the amount of power which is used in manufacturing, twelvefold;

our railway traffic, as measured in ton-miles of freight carried, tenfold. During this same period of forty years other increases have been even greater, for the use of electricity has increased from meager beginnings to the gigantic position, which it now holds. The telephone—the high-speed tool of modern business—has increased from its comparatively small beginnings until now there is a telephone for every ten persons in the United States, and the increase in the number of engineers is typified by the growth of The American Society of Mechanical Engineers, from less than 200, at the beginning, to more than 13,000 at the present time. I understand that the increase in membership during this past year has equalled the total member-ship at the end of the first twelve years.

While the population has been doubling, the fundamental activities of engineering production and transportation and communication have in-creased fivefold, tenfold, one-hundred fold. The annual increment is as great as the total a decade or a score of years ago.

Life today is different from that of our grandfathers, because of the things which engineering has supplied. Relationships are different, and so families and communities are no longer free and independent and self-supporting. The factory and the railroad have made us interdependent for food and clothing, fuel and shelter, and health and pleasure.

Nations are no longer isolated and apart. Steam and electricity have cast aside the barriers of space and time. The whole world today is no longer the original thirteen states when measured in time of travel or facility for the transport of food and materials. This new interdependence of communities or nations is based upon the methods of industrial pro-duction, and of transportation and communication which the engineer has developed, and which he alone understands and can efficiently operate. From these have arisen new customs; the standard of living has been elevated; new ideals and aspirations have been awakened and new possi-bilities for benefiting or harming others have been placed in the hands of small groups of men.

As a result of all this, what are the responsibilities of the engineer? The engineer has responsibilities because he has produced power and has applied it to transportation and industry and because this has given the world new activities and capabilities for doing things on a gigantic scale that have changed the whole character of civilization. He has ushered in a new epoch and it is his responsibility to see that this great mechanism which underlies modern society is kept moving and that the tremendous forces which he has produced are not diverted to its destruction.

What are the opportunities of the engineer? It is his opportunity to extend and multiply the means of production and transportation which have already transformed our mode of life. It is his opportunity to coördinate on a larger and larger scale the engineering activities which have grown up independently. It is his opportunity to coördinate in national scope interests which are separate and overlapping and duplicating. It is his opportunity to apply the principles of physical efficiency to human action. It is the opportunity of the engineer to be a leader in pointing out the direction of future development and to direct the trend of human activities along the paths of the highest ideals. The modern engineer is the director of the human as well as the material and inanimate. He determines the abilities of men as well as the qualities of materials.

Nearly forty years ago David A. Wells wrote that the economic changes due to the introduction of the steam engine and machinery brought about the world-wide depression and panic of 1873. But those changes were small compared with those of recent years. Readjustment to the new order of things which engineering and power are producing in modern life —in industry and social and material affairs—will be far-reaching. It will involve revolution, possibly revolution. The problems are many and complex. But to the engineer a problem is an opportunity.

In the course of the evening President Miller read congratulatory messages from President-Elect Harding, Governor Coolidge, Hon. Andrew J. Peters, Mayor of Boston, Herbert Hoover, E. A. Car-man, President-elect of the A.S.M.E., the American Engineers in Cuba and the Institution of Civil Engineers of London. Greet-ings were also sent by a number of the Sections of the Society and by absent members. Mr. W. H. Patchell, a member of the Coun-cil of the Institution of Mechanical Engineers of London, was present and read a cablegram from that Institution.

## Addresses by J. Herbert Case, Samuel Gompers, and William B. Dickson

### ADDRESS BY J. HERBERT CASE

The Constitution of your Federated American Engineering Societies defines your profession "as the science of controlling the forces and of utilizing the materials of nature for the benefit of man." I should like you to think of banking as a profession with essentially similar aims. I con-ceive that our professions and our functions are not vitally different.

More and more the banker is becoming, in a sense, a financial engineer, and adopting essential engineering methods. I have hopes that in the future we shall be able to deal with banking and financial problems as concretely and with the same sureness and knowledge as the mechanical or electrical engineer does now with his problems.

Ever since the banks took on the function of loaning money, this has been the intent. The mechanical engineer wishes first of all to know the strength of his materials. So, in a way, does the banker. But I think we are presently to go much further than this. The banker of the future will have a variety and definiteness of knowledge that the bankers of the past have not had. Moreover, he will endeavor to do exactly what the engineer endeavors to do in calculating the weight and size of his constructions. He will have to do with stresses and strains and the means of meeting them so that there shall be no breakage and the least possible friction.

It has been the endeavor of more than a generation of thoughtful bankers and economists to devise a banking system which would meet periodic strains so that acute commercial or financial crises might be avoided, exactly as the engineer, for example, plans a bridge to meet every possible strain that can be put upon it, an unusually heavy load, a freshet, an unexpected hurricane, the ordinary dangers of fire and water. I think this endeavor has been successful.

We have witnessed in the past few years a remarkable expansion in credits and during recent months a very remarkable fall in prices. I am not sure that it is not the most remarkable fall in prices, taken as a whole, of which we have any accurate record.

Without wishing to exaggerate the fact, it is easy to see that these prices declines in commodities, whose annual worth runs into the billions of dollars, must in the aggregate have represented an enormous sum. What I wish to point out to you is that all this has taken place without any violent rupture or breakdown, such as has often characterized the past. And I want you to consider the reason why.

The reason why, as I see it, is that, taking a leaf from the basic principles of engineering, there has been introduced into our banking and financial system a greatly augmented factor of safety. We have approximately 10,000 of these banks, representing 70 per cent of America's banking assets, welded together into a coherent and smoothly working system wherein, practically speaking, the resources of these 10,000 banks are more or less pooled, within reasonable limits, and in such a way as to enable them to meet almost any probable strain. Although these 10,000 banks are a trifle less than one-third of the total number, their resources are more than two-thirds of the total banking resources of the nation. Their aggregate deposits are in excess of 20 billions, a fabulous sum even in these days. It is to my mind unthinkable that with 10,000 banks possessing such resources as these, welded together, and pooling their reserves, there can be such a breakdown of our banking system as has come in the past.

My vision goes even further than this. I see our whole banking and financial and currency systems so coördinated and so delicately adjusted to the demands of business that even severe commercial crises shall likewise be largely avoided. I foresee these systems so highly adapted, so well planned, that they will run on almost automatically. I can see our Federal Reserve System and the currency system that goes with it so developed that when the peace of business gets a little too fast, when American enthusiasm gets out of bounds, when the spell of mak ng money rapidly gets the better of men's judgment, this system will almost automatically apply the brakes. And it will apply them, as I imagine, in time, so that perhaps the very high interest rates which we have at the present will no longer be needed. Perhaps some day a general rise in prices will automatically put a check upon credit expansion. It is conceivable that if we should have had a smooth-acting governor automatically controlling our credit policies, the very rapid rise in prices last year might have been in part, at least, avoided and consequently the quite drastic declines that we are witnessing now.

If, in our financial system, we can get a thorough-going introduction of engineering methods, adapted, of course, to financial and banking problems, I foresee a day when these periods of distress will be nearly if not completely eliminated. And I can see that this will do much more. If we can do away entirely with uninvited unemployment; if we can keep the vast industrial machine running like a wonderful Curtis turbine, day in and day out, year in and year out; if we can get rid of these recurrent times of rising prices which affect so severely the cost of living and breed every kind of friction and irritation and discontent; and if, at the same time, we can get rid of periods of business depression; it seems to me that we ought to do away very largely with the prevalent social disturbances, unrest, and insatiate impulse to try impossible schemes of social betterment.

It ought to be pretty clearly evident that a man on an island can eat and have no more than he reaps and makes. What is true of the man on the island is equally true of a nation of a hundred millions, or of the whole wide world. Throughout the last half century the supply of goods, the total of products in this country, has increased at a rate very close to $4^1/_2$ per cent per annum. Our business crises and depressions have been very largely price cycles—if you like, very largely psychological. The vast business of production has gone on pretty much the same. This is the real foundation, I think, for the hope that the financial and banking engineer of the future may so adjust the credit machine that even these price waves and fluctuations, with their attendant dislocations and disasters, may be largely ironed out.

In brief, gentlemen, I feel that the future is full of the richest sort of promise, financial and social as well as industrial. On the engineering side you have done a great and far-reaching work. You have given American engineers and American engineering the highest place in the world. You have set a wonderful, and, I may say, a difficult example to follow. What I wish you to know is that much the same ideas, much the same aims and methods, are at work now in the development of the Federal Reserve System of the United States.

## Address by Samuel Gompers

Because of my high conception of the engineering profession I was glad to accept your invitation to address The American Society of Mechanical Engineers to voice my understanding of the possibilities of your work as engineers of industry, and to suggest what seems to me the responsibility of the engineer to the whole modern industrial system.

One of the difficulties that arises nowadays about our discussion of responsibility is that we fail to realize that professional men, whether doctors, lawyers or engineers, should all be in a very real sense, agents of society and not merely masters in their own particular professions.

We are beginning to realize that just as no nation is isolated from the family of nations, so is it actually true that individually we cannot be isolated from our professions. Every man in his actions, influences in a greater or less degree, other groups or individuals, either for good or evil.

During the past few months engineers have expressed their sense of responsibility in a splendidly stimulating way. In order to accomplish better their purpose of service to others and to contribute to public welfare the best that was in them, all the engineering and technical societies of the United States banded themselves together in a great over-arching organization designated as The Federated American Engineering Societies—a comprehensive organization dedicated to the service of the community, state and nation. This close union makes possible coördination of effort and more efficient progress in achievement of great ideals.

During the past year representatives of engineering organizations as well as individual engineers have come to me, seeking help in getting a better understanding of the human element essential to production and in establishing the proper basis for coöperation with the constructive force which they had come to realize exists in the organizations of human beings engaged in industrial production. For the engineer knows that organization is necessary in order to utilize power—human or material.

It is a tremendously encouraging fact that the engineers throughout the country are coming to appreciate their high calling. It is unnecessary for me to review the mechanical achievements of the engineer, for they speak for themselves, but I do want to point out that concept which has grown up in industry, which many accept, is fundamentally in error. We talk about the production of our factories as being the material, the finished product, in other words, which is sent out in freight ears, and of the individual workmen as by-products. Men, not things, are the true goal of civilization. That civilization fails that does not produce great men and great women, able to create and to use with discrement the material things that serve the spirit. Who can estimate the worth of human beings? I submit that the true ethical point of view of production is that the man himself is the main product and the materials the by-product, and it is in this clearer point of view, it seems to me, the way lies open for joining the forces which the labor movement represents and the forces represented by the activities of your own societies.

The problem involved is not a simple one, for the tendency during the last seventy-five or one hundred years of our western civilization has been to have the machine replace the man. The old feeling of craftsmanship, which existed before the industrial revolution came about, has been greatly modified because of the perfection reached in machine design. This process, however, has been carried entirely too far, for in many places the man has become a human connecting link in a machine and mastered by it instead of controlling the machine himself, as he did with the tools that he used in the old days. The result is that today men's work tends to become mere toil, so it seems to me that the task that lies before us is to develop a definite kind of working environment which will be attractive and which will inspire rather than repulse the workmen. The work itself must become of central concern. This cannot be brought about unless the man finds the opportunity for self-expression in the day's work and a chance to exercise his creative impulses.

During the past 50 years the labor movement has endeavored to protect the workman against the inroads of the machine upon his own life. Our fundamental effort toward this may be epitomized in this declaration. The labor of a human being is not a commodity nor article of commerce. We knew that human labor is inseparable from thinking, living beings, but it took the organized power of our labor organizations to secure recognition of this principle in law and in practice. Workmen's compensation laws and other legislation of a similar nature are a recognition of this fundamental.

I see, however, before the labor movement a great future, as I also see a great future before the engineering profession. If the engineer should join hands with the workman—both devoting their energies to one cause, namely, the development of a kind of industry and a kind of work in which the man will not only learn the processes of production—each day will have increasing opportunities to develop those human functions which are essentially intelligent.

A way has been opened for such coöperation in the declarations of the conventions of the American Federation of Labor, expressing appreciation of the value of the technicians of industry and the desirability of the labor movement's availing itself of scientific aid in all possible ways.

In America our education has been both popular and free. We have had compulsory education for all because we wanted to be sure that all would be prepared for the duties of citizenship. Education, however, is nothing more than the acquiring of greater knowledge of natural law and an opportunity to use this knowledge in the performance of useful work. Is it not logical, therefore, to look forward to the day when our industries will be conducted along educational lines?

It is the deadly monotony of repetitive work that is at the root of most of our troubles, and I, therefore, in the name of the workers, urge you engineers to direct your energies to the solution of this problem. Beware that the machines you create do not become a Frankenstein and enslave the human race.

If you study the laws of humanity with the same degree of intensity that you study the laws of material science, you will render a tremendous service, and as President of the American Federation of Labor it is my firm conviction that the labor movement not only welcomes but invites your coöperation.

### Address by William B. Dickson

After a brief introductory statement, Mr. Dickson analyzed the present showing of signs of instability in the social order. Mechanical developments, which during the past two or three generations have advanced the human race, may nevertheless bring about grave dangers, which must be recognized and counteracted if American ideals are to be preserved.

Mr. Dickson spoke of the danger to be apprehended from a high degree of specialization in modern industry and drew a vivid picture of the artisan, happy in the opportunity for the expression of his creative instinct.

In Mr. Dickson's opinion, the menace of people clothed with political power but stunted in body and soul by their industrial environment presents a problem which will need the most careful consideration by the industrial leader.

Mr. Dickson then proceeded as follows:

The principal theme, however, to which I wish to direct your attention is this: What is the supreme issue confronting mankind today? In my opinion, simply the same issue which runs back through all history, and which we have fondly dreamed was settled once and forever by the American people, namely, aristocracy vs. democracy.

We Americans are so accustomed to think of democracy as the normal system of human government, the very flower of civilization, that the man in our midst who would seriously question this apparently self-evident truth would be looked upon as abnormal, to say the least. We fondly hoped that when Cornwallis surrendered at Yorktown political democracy was achieved. As a matter of fact, democracy is not an achievement; it is an opportunity for further struggle upward.

We must now set our minds to the task of applying democratic principles to industrial relations. I believe there is a grave menace to our American ideals in the highly centralized, autocrati control which is becoming a marked tendency in our great industries. The tendency of our modern industrial system is toward autocratic control of the workers through ownership of what our socialistic friends term "the tools of production," which include not only the natural resources, but also the furnaces, mills, factories and transportation systems.

Instead of indulging in glittering generalities, let me cite an instance of what has happened under the existing system of corporate control. Some years ago, a gentleman at the head of one of our great corporations decided that prices must be maintained in the face of a diminishing demand. In order to accomplish his purpose, he restricted production by shutting down a number of large plants located in different communities, each of which had been built up largely as an adjunct of the plant. Some of these plants were kept closed for about a year, and the result was disaster to the communities. The merchants were driven out of business, real-estate values were depreciated, and the workers were thrown on their own resources and had to break up their homes and seek employment elsewhere. None of these persons had any voice in the momentous decision which was made in a New York office, and which resulted in social paralysis in all of these communities.

History is filled with instances where centralized power has led to conditions inimical to human progress, as that term is usually understood in America. It is the effect of the unconscious insolence of conscious power. If we should read in the paper some morning that a Turkish pasha had exercised his authority in such a way as to deprive a city of its means of subsistence, we would raise our eyes in holy horror and bless our good fortune in living in a more enlightened land. Any manifestation of autocracy is repugnant to the American people, whether it proceeds from a president of a corporation, a president of a labor union, or a president of the United States. What is the answer? Only one, namely, industrial democracy.

In a great national crisis, Lincoln said, "A house divided against itself cannot stand;" "this nation cannot continue to exist half slave and half free." I believe that we are approaching a time when we will need an industrial Lincoln, who will give utterance to the creed of the twentieth century; "A house divided against itself cannot stand; this nation cannot continue to exist politically democratic but economically autocratic." What do I mean by Industrial Democracy? It is exceedingly important that there be no confusion as to this definition. Mr. Carnegie was once asked, "Which is the most important factor in your business, capital, management, or labor?" He replied, "Which is the most important leg on a three-legged stool?" This answer epitomizes my theme, and also what I believe will be the creed of the twentieth century.

In an efficient partnership, such as Mr. Carnegie's answer implied, while each partner may have equal rights, the duties and responsibilities are usually separated so that each exercises his principal functions within his own limited sphere. But where grave questions are to be considered, which vitally affect the organization as a whole, there is general consultation. So, in the new ideal of industrialism, each factor, i.e., capital, management and labor, will continue to have its own separate natural function, as heretofore, but no arbitrary, autocratic decisions affecting the general welfare will be made, either by the directors, the officials, or by the workmen.

Some of you may ask "Did Mr. Carnegie follow this ideal in practice?" My answer is, "No." He did give a larger measure of recognition to management than most of his fellow-manufacturers, but in his attitude toward labor, he was merely a signpost, pointing the right way but never taking it. The Carnegie labor policy was highly autocratic, as is that of its successor, the U. S. Steel Corporation; a benevolent autocracy, if you please, in many splendid ways, although it still maintains that relic of barbarism, the twleve-hour day. But however large you write the word "benevolent," you must always write after it the word "autocracy." The autocratic policy of this great industrial corporation is diametrically opposed to American ideals, and if it and similar organizations in other industries continue to grow and to maintain this autocratic attitude, there can be only one result—industrial feudalism; feudalism with a high degree of comfort and safety for the worker. I grant you, but none the less, feudalism.

One of the most melancholy tasks of the student of history is to observe the insidious ways in which free institutions have been destroyed under the guise of apparently innocent innovations. Rome, after a glorious history as a republic, extending over nearly five centuries, became an empire under the Caesars, and as far as outward forms were concerned, the transition was so gradual that the citizens did not realize that any real change had occurred. This was so apparent that Julius Caesar himself did not dare to accept the title of king. So today I am not imputing blame to any man or class. No man has deliberately said, even to himself, "I will deprive my fellow-citizens of a large measure of liberty in order to enrich myself," nevertheless the things which lead to this condition have been done, and are being done today as we look on.

As a famous American has said: "The essential characteristic of empire in the objectionable sense, is absolutism. Whether or not absolute power be administered benevolently, makes no difference. The evil is in the power itself, not in the nature or manner of its administration.

When a man, or a number of men, for their own ends, create a great industrial unit, they assume an obligation toward the human elements in that unit, and through them to society in general, which cannot be cancelled or suspended arbitrarily. I subscribe to the doctrine that human labor is not a commodity in the ordinary sense of that term. In a completely natural society, every man by reason of close and continuous contact with land and other natural resources would be an independent, self-sustaining unit. When a man has left this natural condition, whether voluntarily or otherwise, and has become the servant of another man, or other men, he has given up a natural right and his employer has assumed an equivalent obligation. The fact that neither the employer nor the employee has been conscious of this exchange, and that both may have acted from purely selfish motives, does not alter the elemental fact, which, in the great national aggregate, constitutes the great unanswered problem of modern times—the elemental fact that is at the base of all social unrest..

I believe that the greatest task to which American employers must address themselves is the devising of practical ways in which labor can be given the full recognition to which, as an equal partner, it is entitled. I make this statement with absolute confidence in the fairmindedness of tho American workingman when he is fully informed and is entirely free to act. If I did not have this confidence, I would despair of the future of our free institutions. I believe, therefore, that one of the first steps necessary to inspire the workman with confidence is the sincerity of the employers recognition of the proper status of labor, in the adoption of a fair system of collective bargaining.

I also believe that in the near future the workmen must become partners through some system of profit-sharing. No scheme could be adopted which would be applicable to all business, as each particular company would have to adopt the general idea to its own particular conditions.

But, some of you may say, "We don't want to have anything to do with your so-called industrial democracy; we are satisfied with the present system and prefer to continue as we are." My answer to this is that the human relations are not static, but dynamic, and unless I am entirely mistaken as to the direction and force of the tide which is now running so strongly in human affairs, your choice will not lie between the present system of industrial control and industrial democracy.

American industry has come to the parting of the ways; on the right, is the road that leads direct to industrial democracy. This road has some heavy grades, and a higher degree of skill will be required to drive on it, but it will bring us out into peace Valley. On the left, is a road also deviating from the old road by which we have come, but it is cunningly camouflaged so as to seem to be the natural continuation of the main highway. It leads directly to industrial feudalism; to that social condition predicted by Hilaire Belloc in his book, The Servile State, in which the workers voluntarily sacrifice freedom in return for comfortable maintenance and safety. From this second road, there is also a by-path which is now being trodden by Russia, and toward which not only our British brethren, but a considerable number of American workmen are being tempted to stray. In other words, the choice lies between democracy on one hand, and serfdom or chaos on the other.

To sum up, what can be done to counteract the tendencies which I have described? These things seem to me to be entirely practicable:

1 Place our industries on a more democratic basis giving recognition to management and labor as, equal partners with capital.
2 Teach democracy in our schools and colleges as thoroughly as we teach arithmetic, so that it will permeate every phase of human life, politically and industrially.

It is a constant source of wonder to me to find so many persons in all walks of life who have no real conception of the vital principles of democracy. Life, in its truest and most virile sense consists largely in making choices, and, like the traveler before the Sphinx, we must answer correctly or be destroyed. I am not looking forward to the new era of industrial democracy as a period of peace and serenity, but rather as a time in which the way has been cleared for a further toilsome climb up the spiral of evolution. I am hopeful that our generation will guess the Sphinx riddle, and that "Out of this nettle, danger, will pluck the flower, safety."

# Engineering and Industrial Standardization

## A Review of Standardization Work Now in Progress, as Reported by the American Engineering Standards Committee

THE American Engineering Standards Committee held its regular quarterly meeting in New York on October 9. Among the more important reports submitted were those from committees on finance, membership, and safety codes. Nomenclature for standards was also discussed, as well as suggestions for standardization in the zinc and other non-ferrous metal industries. Brief notes from the minutes of the meeting follow:

### REPORT OF THE FINANCE COMMITTEE

The report of the Finance Committee, which had been accepted by the Executive Committee and recommended to the Main Committee for approval of the policies therein outlined, had been circulated to the members prior to the meeting. It was accompanied by detailed estimates prepared by the Secretary under instructions of the Executive Committee. The recommendations, in brief, were: (1) That efforts should be made to secure contributions to increase the total annual income of the Committee to $25,000, and (2) that associations of manufacturers rather than individuals or corporations be asked for contributions. These recommendations were very fully discussed, and upon motion by Mr. Robinson it was unanimously resolved that, while it is preferable that contributions be received from associations of manufacturers rather than from corporations, it is entirely proper to accept contributions from corporations provided the appeal for such support be broad enough to include a considerable number of widely different interests, and provided no one company be asked for a relatively large contribution.

### REPORT OF THE MEMBERSHIP COMMITTEE

The chairman of the Membership Committee submitted a report recommending favorable action on the applications for membership from the Department of the Interior, the Gas Group, and the American Electric Railway Association. Upon motion by Mr. Latey, it was resolved that the Department of the Interior be admitted to membership with three representatives; that the Gas Group (comprising, at present, the American Gas Association, the Compressed Gas Manufacturers Association, the International Acetylene Association) be admitted to membership with three representatives; and that the American Electric Railway Association be admitted to membership with one representative at the present time, but with the understanding that this may be increased to two or three, if this should later seem advisable. The action in each case was by unanimous vote.

### REPORT OF JOINT COMMITTEE ON SAFETY CODES

A report of the chairman of the Joint Committee on Safety Codes dated October 5 was submitted and the four recommendations contained therein were approved by the following unanimous actions to the effect

(1) That the Bureau of Standards and Society of Automotive Engineers be requested to assume joint sponsorship for an Aviation Safety Code;

(2) That the American Society of Heating and Ventilating Engineers be requested to assume sponsorship for a code on Ventilation;

(3) That the Electrical Safety Conference be designated as sponsor for safety code on Electrical Power Control; and

(4) That in view of the time which has elapsed since the preparation of the 1917 Foundry Safety Code by the American Foundrymen's Association and the National Founders' Association, and of the progress in safety practices in the meantime, that it be suggested to the sponsor bodies that it is desirable to organize a sectional committee to approve or revise this code before it is approved by the American Engineering Standards Committee.

### COLOR SCHEMES FOR PIPE LINES

S. J. Williams reported that in accordance with the suggestion of the Executive Committee, The American Society of Mechanical Engineers and the National Safety Council had addressed inquiries to a considerable number of organizations in regard to the desirability of attempting the standardization of color schemes. The answers had all been favorable. Considerable information had already been collected. It was thereupon voted that The American Society of Mechanical Engineers and the National Safety Council be requested to assume joint sponsorship for the standardization of Color Schemes for Pipe Lines.

### NOMENCLATURE OF STANDARDS

The following action was taken by unanimous vote:

WHEREAS, The terms, "Tentative American Standard" and "Recommended American Practice" more accurately express the meaning intended to be conveyed by the terms "Tentative Standard" and "Recommended Practice," and

WHEREAS, The latter terms conflict with long-established practices of important organizations, be it

Resolved, That the American Engineering Standards Committee will approve standards as either "Recommended American Practice," "Tentative American Standard," or "American Standard," and will authorize the use of these terms in the publication of standards approved by it.

### ZINC AND OTHER NON-FERROUS METALS

A communication was received from the Association Belge de Standardisation under date of August 31, suggesting international standardization of the following matters pertaining to zinc: grades of spelter, gage thicknesses and tolerances for sheet zinc, methods of weighing and determining moisture content, and methods of analysis of ore and spelter. The Association also suggested that similar work on other non-ferrous metals might follow.

Mr. Stone stated that he had discussed the matter informally with the American Zinc Institute and that the Institute felt that the work proposed on zinc was very desirable. After further discussion, it was unanimously voted that the American Society for Testing Materials and the American Zinc Institute be requested to assume joint sponsorship for the standardization of zinc.

The larger question raised by the Belgian communication of non-ferrous metals in general, was also considered and it was voted that the officers be authorized to call a conference of the bodies which might be interested, to consider whether work on other non-ferrous metals should be undertaken and if so, what the scope of the work should be.

## The Standardization of Elevators

Upon the request of the American Institute of Architects and the Elevator Manufacturers' Association of the United States a conference of the various bodies interested in the question of the standardization of elevators was recently called by the American Engineering Standards Committee for the purpose of deciding whether or not such standardization work should be undertaken, and if so, what its scope should be. The conference was held in New York on September 21, and was attended by representatives of the following organizations:

American Institute of Architects
American Institute of Consulting Engineers
American Institute of Electrical Engineers
American Society of Mechanical Engineers
American Society of Safety Engineers
Bureau of Building (Manhattan)
Bureau of Standards
Electric Power Club
Elevator Manufacturers' Association of New York
Elevator Manufacturers' Association of the United States
National Association of Building Owners and Managers
National Fire Protection Association
Supervising Architect's Office, U. S. Treasury Department

After a full discussion of the subject which touched upon such matters as platform sizes, speeds, accelerations, capacities, methods of test, pit and overhead clearances, well dimensions and batchways, and safety provisions and appliances, it was unanimously

Resolved, That this Conference fully recognizes the need and the desirability of standardizing such features of both passenger and material-handling elevators as capacities, platform sizes, and methods of test.

The relation of the standardization of the fundamental features of elevator to the Elevator Safety Code now being formulated by The American Society of Mechanical Engineers, was also fully considered, and a committee appointed to confer with the Mechanical Engineers and arrange for coöperation in the completion of this code.

The conference also appointed a committee, consisting of one member from each body represented at the meeting, to develop plans for the standardization of elevators. This committee is to report its recommendations to the American Engineering Standards Committee.

## Progress of the Safety Code Program

Considerable progress has been made toward the completion of the program calling for the creation of safety codes. This work has been undertaken by a large number of organizations under the auspices and rules of procedure of the American Engineering Standards Committee, and following the regular procedure, each code is being formulated by a sectional committee, broadly representative of the interests concerned, and composed primarily of representatives designated by the various bodies interested in the particular code. The sectional committee is organized by one or more bodies designated for the purpose by the American Engineering Standards Committee and known as sponsors.

The Head and Eye Protection Code has been completed, and the sponsor, the Bureau of Standards, has submitted the code to the Main Committee for approval. Sponsorships for the additional safety codes have been arranged as follows:

*Construction Work.* National Safety Council
*Electrical Fire Code.* National Fire Protection Association
*Electrical Safety Code.* Bureau of Standards
*Floor Openings, Railways and Toe Boards.* National Association of Mutual Casualty Companies
*Lighting Code.* Illuminating Engineering Society
*Lightning Protection.* American Institute of Electrical Engineers and the Bureau of Standards
*Machine Tools.* National Machine Tool Builders' Association and the National Workmen's Compensation Service Bureau
*Mechanical Transmission of Power.* National Workmen's Compensation Service Bureau, the International Association of Industrial Accident Boards and Commissions, and The American Society of Mechanical Engineers
*Sanitation Code, Industrial.* United States Public Health Service
*Stairways, Fire Escapes and Other Exits.* National Fire Protection Association
*Textiles.* National Safety Council and the National Association of Mutual Casualty Companies

SPONSORSHIPS PREVIOUSLY ANNOUNCED

*Abrasive Wheels.* The Grinding Wheel Manufacturers of the United States and Canada, and the International Association of Industrial Accident Boards and Commissions
*Foundries.* American Foundrymen's Association and the National Founders' Association
*Gas Safety Code.* Bureau of Standards and the American Gas Association
*Head and Eye Protection.* Bureau of Standards
*Paper and Pulp Mills.* National Safety Council
*Power Presses.* National Safety Council
*Pressure Vessels, Non-Fired.* American Society of Mechanical Engineers
*Refrigeration, Mechanical.* American Society of Refrigerating Engineers
*Woodworking Machinery.* International Association of Industrial Accident Boards and Commissions and the National Workmen's Compensation Service Bureau

## Law for Registration of Engineers

On October 21 the Committee on Licensing Engineers reported to Engineering Council the uniform law for registering architects, engineers and land surveyors. The justification for this law is given in Section 1: "In order to safeguard life, health and property, any person practicing architecture, engineering or land surveying shall be required to submit evidence that he or she is qualified so to practice." The law, as drawn, places no limitations on those who may or may not practice the profession of architecture, engineering or surveying provided they refrain from using the title.

The law specifies in detail the personnel of the Board who shall pass upon the registration of the professions coming under the jurisdiction of the law, providing for apportionment among the professions of architecture, engineering and surveying. All the members of the Board shall be citizens, residents of the state, and members in good standing of recognized professional societies. The Board shall each receive a certificate of appointment from the Governor of the State and a certificate of registration and is then authorized to affix its official seal to certificates of registration granted.

The qualifications for registration are based on a study of the complete statement of the applicant's education and a detailed summary of his technical work. These statements should be made under oath and should be supported by the recommendations of not less than two members of his profession. All applicants to be eligible must be over twenty-five years of age, citizens of the United States or Canada, must speak and write English, must be of good character, and must have been actually engaged for six or more years in architectural, engineering or land-surveying work of a character satisfactory to the Board. A year of teaching or of study satisfactorily completed in a school of satisfactory standing, shall be considered equivalent to one year of active engagement.

The functions of the Board are largely administrative and judicial. The burden of presenting evidence of qualifications is placed upon the applicant.

The law states that unless disqualifying evidence be before the Board, the following facts established in the application shall be regarded as prima facie "evidence, satisfactory to the Board," that the applicant is fully qualified to practice architecture, engineering or land surveying:

a Ten or more years of active engagement in architectural, engineering or land surveying work;

b Graduation after a course of not less than four years, in architecture or engineering, from a school or college approved by the Board as of satisfactory standing, and an additional four years of active engagement in architecture, engineering or land-surveying work;

c Full membership in American Institute of Architects, American Institute of Chemical Engineers, American Society of Civil Engineers, American Institute of Electrical Engineers, American Society of Mechanical Engineers, American Institute of Mining and Metallurgical Engineers, Society of Naval Architects and Marine Engineers, or such other national or state architectural or engineering societies as may be approved by the Board, the requirements for full membership of which are not lower than the requirements for full membership in the professional societies or institutes named above.

The Board has the power to revoke certificates of registration if the holder be found guilty of negligence, incompetency or misconduct. The law permits practicing in the state by architects, engineers or land surveyors on a consulting basis, provided the non-resident is qualified for professional service in his or her own state or country.

Corporations are permitted to engage in practice under this law, provided the person or persons connected with such corporations or partnerships, in responsible charge of such practice, is or are registered as herein required as architects, engineers or land surveyors.

The law has been prepared by Engineering Council for an act of legislation in each state. Complete copies of the law may be procured of Alfred D. Flinn, Secretary of Engineering Council, 29 West 39th St., New York City.

## Warner and Swasey Observatory Dedicated at Case School of Applied Science

The Case School of Applied Science, Cleveland, Ohio, on October 12, 1920, dedicated a new astronomical observatory, the gift of Worcester R. Warner and Ambrose Swasey, partners in the noted firm of instrument makers. The building is placed on the crest of a ridge overlooking a beautiful residential section of Cleveland, about two miles from the school campus. The plan of the building is L-shaped, with the telescope tower at the angle. One single-story wing is devoted to a library and conference room, at the inward end of which is constructed a constant-temperature room for two Riefler clocks. The other wing contains two astronomical transits, and a zenith telescope, all Warner and Swasey instruments. A basement provides living quarters for the caretaker, a storeroom, a battery room, and a photographic room. In the tower is a 10-in. telescope, the lens of which was ground by John Brashear, of Pittsburgh.

At the exercises Mr. Swasey related many anecdotes of the firm's astronomical work, and Mr. Warner, who for thirty years has been a trustee of Case School of Applied Science, presented the keys of the observatory to President Charles S. Howe, who made an appropriate response. Prof. D. T. Wilson, professor of astronomy at Case, outlined his work and his plans for the future usefulness of the observatory. The principal address was delivered by Director W. W. Campbell, of Lick Observatory, Cal., on the topic, The Daily Influences of Astronomy.

Some applications of astronomy to daily life, Dr. Campbell said, are the following: supplying the world with correct time by observations through transits, determining latitudes and longitudes for maps and navigators, the fixing of boundaries and the predicting of tides.

# Power Test Codes

## A.S.M.E. Committee on Power Test Codes Submits Preliminary Draft of the First of the Nineteen Test Codes Which It Is Formulating

AS has been mentioned a number of times in MECHANICAL ENGINEERING, The American Society of Mechanical Engineers organized in December 1918 a Committee of 125 specialists to revise the Power Test Codes of 1915. This large Committee is subdivided into a Main Committee of twenty-eight engineers which supervises the work of the nineteen Individual Committees. Each Individual Committee, which is made up of from six to ten men, is revising one particular test code.

The Code which follows is entitled "General Instructions," and as its name indicates, has been formulated for use with each of the test codes.

The Committee will welcome suggestions for corrections or additions. These should be sent to Mr. Fred R. Low, Chairman, before January 1, 1921.

### GENERAL INSTRUCTIONS

#### OBJECT

1 Ascertain the object of the test. This object should be kept in view during preparations for the test and in the conduct of the test itself. The methods to be used and the accuracy sought should be in accord with the object in view.

2 Among the many objects of performance tests, the following may be noted:

Determination of capacity, efficiency or regulation, and the comparison of these with standard or guaranteed results
Comparison of different conditions or methods of operation
Analysis and interpretation of plant performance
Comparison of different kinds of fuel
Determination of the effects of changes of design or proportion upon capacity or efficiency

3 If questions of fulfilment of contract are involved there should be a clear understanding in writing between all parties as to the operating conditions which should obtain during the trial, the methods of testing to be followed, corrections to be made in case the conditions actually existing during the test differ from those specified, and as to all other matters about which dispute may arise, unless these are already expressed in the contract itself.

#### PREPARATIONS

4 Ascertain and record manufacturer's serial number, plant designation or other means of identifying each unit involved in the test in order that later there be no uncertainty as to which unit or units were used in obtaining certain data.

5 Examine and record the general features, arrangement and condition of the apparatus and plant, and, if needed, make sketches to show the arrangement, and any unusual features. When making such examinations operating conditions bearing on the object of the test should be noted. So far as possible examination should be internal as well as external.

6 Determine the principal dimensions of the apparatus to be tested, particular attention being given those bearing on the objects in view. When possible, important dimensions should be determined by actual measurement. When conditions of operation change these dimensions, due allowance should be made therefor.

7 If the object of the test is to determine the highest efficiency or capacity obtainable, any physical defects of operation tending to make the result unfavorable should first be remedied. All fouled parts should be cleaned, the whole being put in first-class condition and evidence that it is in such condition should be secured and recorded. If, on the other hand, the object is to ascertain the performance under existing conditions, no such preparation is required or permissible. If fulfillment of contract is involved changes should not be made in apparatus or operation without the consent of all parties interested.

8 In all tests in which the quantity of gas, vapor or liquid flowing in pipes or ducts is determined or in which such quantity forms an item in calculation of results, precautions, must be taken to guard against leakage. Such precautions are necessary in the case of blow-off connections, drips and drains, cross-connections,

isolating valves, packing and joints of all kinds, etc.. All outlets and inlets other than those actually used in the test should be blanked off or so arranged that any leakage in or out can be observed. A leakage test to determine tightness and to measure unpreventable leaks should be made in every instance.

#### INSTRUMENTS AND TESTING APPARATUS

9 The accompanying main code section headed "Instruments and Testing Apparatus" contains a description of the various instruments and appliances which may be employed for Code tests, together with directions for their application, use and calibration, and a statement of their adaptability and limitations. The choice of instruments and apparatus for any test must be determined by local conditions, adaptability and availability, but in all cases the types called for in the various codes should be used. In making this choice, the degree of accuracy desired should be one of the guiding factors.

10 All instruments and testing apparatus used should have designating numbers or other means of identification firmly attached and such identifications should be entered on the log sheets.

11 All instruments and testing apparatus should be calibrated, and any which may give trouble if not approximately correct and all which may be broken in use should be calibrated in advance. Additional calibrated instruments should be provided to replace any which may be broken during the test. Any instrument which may not maintain its calibration throughout a test should be calibrated before and after the test, and, if necessary, at intervals during its progress.

12 The location and arrangement of instruments and apparatus must be left to the judgment and ingenuity of the person in charge, subject to the directions of the code, the details being largely dependent on the type of apparatus under test, the locality and the facilities available. Instruments and testing apparatus should always be so arranged and installed as to give maximum accessibility and best possible illumination in order to facilitate use.

13 Notebooks or log sheets should be so arranged that all readings of one group of instruments or all readings made byo ne observer appear in one book or on one group of sheets. The pages or sheets upon which entries are to be made should be ruled in advance, should be given proper identification marks, should contain space for the date and the name of the observer and should contain proper headings and subheadings. All such sheets should have at least one column in which time is recorded. When rates are to be determined by differences between observations, a time column should accompany the observation column. Space for remarks should be provided on every sheet.

14 A desk or table from which the tests may be directed and at which the records may be assembled should be provided in tests requiring many observers. This should be located with due regard to the positions they occupy. A signal system or a system of intercommunication is convenient and may be provided.

#### MISCELLANEOUS INSTRUCTIONS

15 The person in charge of a test should have the aid of a sufficient number of assistants, to permit him to give special attention to any part of the work whenever and wherever it may be required. He should make sure that the instruments and testing apparatus continually give reliable indications, and that the readings are correctly recorded. He should also keep in view, at all points, the operation of the plant or part of the plant under test, and see that the operating conditions determined on are maintained and that nothing occurs to vitiate the data.

#### STARTING CONDITIONS

16 When circumstances permit, it is advisable to make one or more preliminary tests for the purpose of determining the adequacy of the instruments and apparatus and of training the personnel. When conditions do not permit such preparatory runs, operations may be started and the time at which conditions become satisfactory can be chosen later as the starting time of the test.

17 Before the test begins or before continuing the test after an important change of conditions during operation, the apparatus should be run under normal conditions for a sufficient length of time to bring about equilibrium with respect to thermal or other factors, except in cases in which the effects of the variation of such conditions are to be determined or included. These exceptions include such effects as result from starting and stopping, carrying units ready for start, and banking fires.

18 In preparation for a test to demonstrate maximum efficiency, it is desirable to make preliminary tests for the purpose of determining the most advantageous conditions.

### OPERATING CONDITIONS

19 In all tests in which the object is to determine the performance under conditions of maximum efficiency, or where it is desired to ascertain the effect of predetermined conditions of operation, all such conditions which have an appreciable effect upon the efficiency should be maintained as nearly uniform during the trial as the limitations of practical work will permit. On the other hand, if the object of the test is to determine the performance under working conditions, no attempt at uniformity should be made unless this uniformity corresponds to the regular practice, and when this is the object the usual working conditions should prevail throughout the trial.

### RECORDS

20 Records of a test are divided into two classes:
(a) Records of preliminary examination such as those outlined in Paragraphs 4 to 8 inclusive, and
(b) Records of observations made, and events occurring during the test.

21 Records of both classes should be made with an original and a sufficient number of carbon copies so that one copy may be given to each party in interest. All records should be signed by the respective observers and countersigned by the person in charge of the test.

22 A log of the observations made during the test should be entered in the notebooks or on the sheets referred to in Paragraph 13. This should be done in such manner that the test may be divided into suitable periods to show the degree of uniformity attained.

23 The readings of instruments from which averages are to be obtained should be recorded at intervals fixed by the extent of their variations. Ordinarily time intervals of from fifteen to thirty minutes are sufficient, but there are cases in which intervals of less than a minute are necessary on account of rapid and irregular variations.

24 Ordinarily the intervals between readings should be regular, and this is particularly true when readings are to be averaged. There are, however, cases in which the purpose of the test is such that regularity is not required and possibly not even desired. In every case the person in charge must determine the requirements in this respect.

25 It is not always necessary to read all instruments at the same intervals, nor is it always necessary to read simultaneously those which are read at the same intervals. In cases in which the readings of given instruments are averaged, respectively, for a long period and these averages are used in calculating results, regularity rather than the attainment of simultaneous readings should be the guiding principle. In cases in which each set of observations is used for calculating a result and the results are then averaged, simultaneous readings are all-important. When observations are made to determine rates by sums or differences, the exact time of making the observation is necessary. Whenever possible, check readings should be made after observation has been recorded.

26 The records made during the test should show the extent of fluctuation of the various instruments in order that data may be available for determining the effect of such fluctuations on the accuracy of the calculated results.

27 A record should never consist of a series of tallies as is sometimes made when loads of given weight are discharged at odd intervals or under similar repeated conditions. In such cases each weight and the time when taken should be recorded.

28 Observations of basic character in a given test should be made and recorded in duplicate by two observers when the importance of the test warrants such precautions. In order that the readings made by these observers may be in accord before it is too late, the readings should be compared and an agreement reached soon after the observations are made.

29 A running graphical log of the principal quantities recorded during a test should be maintained. Such graphical records serve to call attention to undesirable or unintentional lack of uniformity in the conduct of operations and to detect errors of observation or record before it is too late to correct them. Such graphical logs should be arranged so that time is plotted horizontally and quantities vertically.

30 Every event connected with the progress of a test, however unimportant it may appear at the time, should be recorded on the proper log sheets with the time of occurrence and the name of the observer who noted the event and entered the record.

### WORKING UP DATA

31 It is desirable to work up the data approximately during the progress of the test. This tends to call attention to omissions and to irregularities of various parts. Calculations of this kind are, however, to be regarded as auxiliary testing methods and the results obtained should not be used in the preparation of the final report unless the test is intended to be of approximate nature and the result to be so given.

32 When all records have been completed and assembled, they should be reviewed for the purpose of determining how they are to be used in obtaining the results of the test. All parts of the record which show evident irregularities must be either corrected or eliminated in order that the final results may not be knowingly in error, and if satisfactory adjustments cannot be made, the test must be repeated. All adjustments of this type should be noted and explained in the final report.

33 After reviewing the record, the methods of calculation to be adopted must be decided upon. There are two distinctly different methods available and it is necessary to recognize the conditions under which each may be used. These methods are:
(a) The method using primary averages in which all readings of each instrument are averaged for a given period and these averages are then used in calculating final results, and
(b) The method using final averages in which the final result is calculated from each set of coincident observations and the final results are then averaged as a grand average.
The choice between these two methods of calculating results should be determined by the type of formula involved and the degree of accuracy desired. When the formula involves the sum or difference of first powers of the observed quantities, either method will give the same numerical results; when the formula involves the product or the quotient of first powers, the two methods will agree quite closely; but when the formula involves fractional powers or powers greater than the first, the method of primary averages will give approximate results only. The error resulting from the use of this method is due to the fact that averages of fractional or multiple powers are not numerically the same as the fractional or multiple powers of averages.

34 When working up the results graphical logs of the principal quantities and of the principal results should be prepared for inclusion in the report. The Report of the Joint Committee on Standards for Graphic Presentation may be used as a general guide.

35 In working out results and in making reports the principles of the theory of errors may be applied to determine what observations should be rejected, the manner of treating the data obtained and assigning the relative importance of various quantities determined.

### REPORT

36 The report of a test of considerable magnitude should be divided into three distinct parts, as follows:
(a) Brief statement of object, results, conclusions and recommendations.
(b) Complete presentation of the leading facts regarding the entire work.
(c) Appendices giving details which are not included in (b) such as methods of calculation, methods of calibration, descriptions of special testing apparatus, results of preliminary or special tests, etc.

(Continued on page 725)

# First Meeting of the American Engineering Council

### Thirty Societies Represented—Herbert Hoover Elected President—Engineers of the Nation Now Organized for United Action in Public Service

At the first meeting of the American Engineering Council of The Federated American Engineering Societies at Washington, D. C., November 18 to 20, there was an auspicious beginning of activities by the election of Herbert Hoover as president. Five sessions were held to complete the organization of the Federation which is to unite the engineers of America for service to the city, state, and nation. Delegates were present from 21 member societies having an enrollment of 50,000 engineers, and representatives from the nine organizations which are considering membership. The conference is one of the most important and significant events in the history of engineering and portends a broadening vision and wider field of influence for the engineer.

HERBERT HOOVER
President of the Federation

The Council voted to establish permanent headquarters at Washington, D. C., and at a meeting of the Executive Board a committee was appointed to nominate a candidate for executive secretary. Until such executive is appointed, Mr. L. P. Alford, formerly secretary of the Joint Conference Committee, is to act as temporary secretary, with headquarters in the rooms of The American Society of Mechanical Engineers at 29 West 39th Street, New York.

Mr. Hoover gave an address on Industrial Relations, an abstract of which appears below. This resulted in a movement for a committee on Industrial Waste to consider problems of human relations in industry and coöperation between labor and capital.

### THE OPENING SESSION

The call to order was by Richard L. Humphrey, of Philadelphia, Pa., chairman of the Joint Conference Committee. He briefly recounted the history of the movement which resulted in the Federation and of the Organizing Conference last June in which delegates from 72 organizations, representing 100,000 engineers, took part.[1] Since that time he stated, the Joint Conference Committee,

[1] See MECHANICAL ENGINEERING, July 1920.

which has conducted the affairs of the Federation preliminary to the first meeting of its governing body, has kept in touch with the 116 organizations invited to become members of the Federation and has received replies from all but about 3 per cent of the aggregate membership represented. Of this membership, 35 per cent have accepted the invitation; and, including the participating organizations who have given the matter of membership favorable consideration but as yet have not taken final action, 50 per cent were represented in this first meeting. In other words there were delegates in attendance from 21 societies and representatives from 9 participating societies. The largest delegation was from The American Society of Mechanical Engineers.

Mr. Humphrey also spoke briefly of the opportunities and activities that lay before the Federation and expressed his satisfaction that the keynote of the organization centered around the one word—Service. In this connection he said in part:

> You are embarking on a broad field of activity under the critical but hopeful eyes of the entire engineering and allied technical professions. It, therefore, behooves you to maintain the high ideals and add to the traditions of these professions.
>
> It is highly essential to organize and catalogue the engineering resources of this country, to the end that the engineering and allied technical professions shall not again be lacking in preparedness as they were at the beginning of the late war. The best method should be devised for framing a national engineering policy by which The Federated American Engineering Societies can be instrumental in shaping a program for our great national engineering problems. Furthermore the vital need for constructive suggestions based upon careful study of the great national problems, such as transportation, conservation of labor, water, fuel, and other natural resources, should be met through the work of this organization.

Following the remarks by Mr. Humphrey, E. S. Carman, of Cleveland, delegate of The American Society of Mechanical Engineers, was elected temporary chairman, and W. E. Rolfe, delegate of the Associated Engineering Societies of St. Louis, temporary secretary. Mr. Carman, in taking the chair, declared, in accord with Mr. Humphrey, that the keynote of the new organization was embodied in the word "service" and expressed the hope that its aim would always be toward the highest pinnacle.

Resolutions were next passed expressing sincere regret that Mr. Humphrey, who had labored so faithfully for the Federation, and his colleagues of the American Society of Civil Engineers, were not permitted to participate officially.[1] The delegates very graciously accorded the privilege of the floor to these gentlemen by a rising vote.

The meeting then considered the question as to what organizations should be considered charter members and it was voted that all organizations to which invitations had been sent to join the Federation should be considered charter members if acceptances were received by July 1, 1921.

The following temporary committees were then appointed and were instructed to report at the next morning's session: Program, Credentials, Constitution and By-Laws, Nominations, Plan and Scope, Budget, and Resolutions.

L. P. Alford next took the floor and announced the representation on the Executive Board in accordance with the constitution to be by districts for local societies (one delegate from each district) and on the basis of membership for the national societies. The country for the present is to be divided into six districts, as follows: (1) New England and New York State; (2) Minnesota, Wisconsin, and Michigan; (3) Illinois, Indiana, and Ohio; (4) Pennsylvania, New Jersey, Maryland, and Delaware; (5) States south of the Ohio River and east of the Mississippi, and Louisiana; (6) States west of the Mississippi, with the exception of Minnesota, Texas, and Louisiana.

This division allows six representatives from local organizations and 14 from the national societies which are now members of the Federation, the latter number being divided as indicated by

[1] By a recent letter ballot the American Society of Civil Engineers voted against joining the Federation.

the number in parentheses which follows the names of the national societies listed below.

The remainder of the morning session was devoted to a discussion of the location of the Federations' headquarters. Consideration was given to the advantages of locations geographically central and at or near the industrial center of the country; and to the relative advantages of New York, where is located the headquarters of the national engineering societies, and of Washington, D. C. New York from the start was practically counted out. It was considered by many speakers that while there were evident business reasons for selecting New York City, the national character of the organization would be better emphasized by locating elsewhere. Many favored Washington. Besides the governmental departments with which the Federation would need to be in close communication in respect to engineering matters of national importance, Washington is already the headquarters of many organizations of prominence and with whom conference would be needed.

Objection to Washington was that its selection might lead to the impression that the Federation was organized for political purposes and that its officers were engaged in lobbying and other undesirable practices. Nevertheless the trend of opinion was that Washington was the most favorable point for contact with those engaged in promoting the best interests of the nation and in rendering service of various kinds to the citizens of the country. In none of the discussions was a geographically central location for headquarters considered as more than of secondary importance, since it would merely affect the convenience of delegates traveling to headquarters and in so far as the operation of the executive staff is concerned might result in less effective work. In any case it was considered essential that there should be a Washington office.

At the conclusion of this discussion, Philip N. Moore of St. Louis offered a resolution, with the proviso that it be laid on the table for discussion at the afternoon session, that it be the sense of the meeting that the headquarters of the Federation be located in Washington, D. C. This motion was duly seconded, whereupon the morning session was adjourned.

### THE AFTERNOON SESSION

The afternoon session, beginning at two o'clock, was opened by an address on Engineering Council by J. Parke Channing of New York. Mr. Channing has served as chairman of the Council for the past three years and his remarks as to the work accomplished during that time and the tasks before the new American Engineering Council were largely in the nature of recommendations and warnings to the new body. Speaking of the accomplishments of the old Engineering Council, Mr. Channing said in part:

During the war we furnished the Government with the names of 4000 engineers for war service. We aided the Naval Consulting Board and the Army General Staff in examining 135,000 suggestions and inventions for war devices. We assisted the Fuel Administrator and the Bureau of Mines. We supplied Congress with information about Waterpower.

On January 1, 1919, we opened an office in Washington giving varied service to engineers, and have also aided the Government in engineering matters.

In April, 1919, we called a conference at Chicago of seventy-four technical organizations having 105,000 members. This conference is permanently organized and is advocating the establishment of a National Department of Public Works.

Through an appeal to the President, we have brought together a conference of fourteen Government offices engaged in map making, with prospects of getting under one head the completion of the topographical map of the United States.

We now have a Classification and Compensation Committee of Engineers, with sections on Railways, Federal Government, Municipal and State Governments, and we are working in harmony with the Congressional Joint Committee on Reclassification of Salaries.

We have just drafted a typical law for the registration of engineers and we have joined with the National Research Board in a report on an improvement of the patent system as a result of which legislative action is being taken.

---

## Member-Societies and Representatives at the First Meeting of American Engineering Council

*Washington, D. C., November 18–19, 1920*

*Alabama Technical Association, Birmingham, Ala.*
  PAUL WRIGHT, Birmingham, Ala.

*American Institute of Chemical Engineers, Brooklyn, N. Y. (1)*
  ALLERTON S. CUSHMAN, Washington, D. C.
  * HARRISON E. HOWE (alternate), Washington, D. C.

*American Institute of Electrical Engineers, New York, N. Y. (4)*
  * CALVERT TOWNLEY (chairman), New York, N. Y.
  COMFORT A. ADAMS, Cambridge, Mass.
  A. W. BERRESFORD, Milwaukee, Wis.
  * H. W. BUCK, New York, N. Y.
  F. L. HUTCHINSON, New York, N. Y.
  W. A. LYMAN, St. Louis, Mo.
  * WILLIAM McCLELLAN, Philadelphia, Pa.
  L. F. MOREHOUSE, New York, N. Y.
  LEWIS T. ROBINSON, Schenectady, N. Y.
  CHARLES S. RUFFNER, New York, N. Y.
  * CHARLES F. SCOTT, New Haven, Conn.
  * LEWIS B. STILLWELL, New York, N. Y.

*American Institute of Mining and Metallurgical Engineers, New York, N. Y. (8)*
  * HERBERT HOOVER, Palo Alto, Calif.
  * J. PARKE CHANNING, New York, N. Y.
  ARTHUR S. DWIGHT, New York, N. Y.
  EDWIN LUDLOW, New York, N. Y.
  ALLEN H. ROGERS, Boston, Mass.
  PHILIP N. MOORE, St. Louis, Mo.
  JOHN V. W. REYNDERS, New York, N. Y.
  JOSEPH W. RICHARDS, Bethlehem, Pa.

*American Society of Agricultural Engineers, Ames, Iowa (1)*
  SAMUEL H. McCRORY, Washington, D. C.

*American Society of Mechanical Engineers, New York, N. Y. (4)*
  * L. P. ALFORD (chairman), New York, N. Y.
  CHARLES T. MAIN, Boston, Mass.
  * ARTHUR M. GREENE, JR., Troy, N. Y.
  * E. S. CARMAN, Cleveland, Ohio.
  ARTHUR L. RICE, Chicago, Ill.
  * DEXTER S. KIMBALL, Ithaca, N. Y.
  PAUL WRIGHT, Birmingham, Ala.
  W. A. HANLEY, Indianapolis, Ind.
  WILLIAM B. GREGORY, New Orleans, La.
  V. M. PALMER, Rochester, N. Y.
  H. P. PORTER, Tulsa, Okla.
  ROBERT H. FERNALD, Philadelphia, Pa.

  L. C. NOROMEYER, St. Louis, Mo.
  * FRED J. MILLER (alternate), Centre Bridge, Pa.
  ROBERT SIBLEY (alternate), San Francisco, Calif.
  CHARLES WHITING BAKER (alternate), New York, N. Y.

*Associated Engineering Societies of St. Louis, Mo.*
  * WILLIAM E. ROLFE, St. Louis, Mo.

*Detroit Engineering Society, Detroit, Mich.*
  D. J. STARRETT, Detroit, Mich.

*Engineering Association of Nashville, Tenn.*
  C. B. HOWARD, Nashville, Tenn.

*Engineering Society of Buffalo, N. Y.*
  W. B. POWELL, Buffalo, N. Y.

*Grand Rapids Engineering Society, Grand Rapids, Mich.*
  BURRITT A. PARKS, Grand Rapids, Mich.

*Kansas Engineering Society, Topeka, Kan.*
  * LLOYD B. SMITH, Topeka, Kan.

*Louisiana Engineering Society, New Orleans, La.*
  WILLIAM B. GREGORY, New Orleans, La.

*Mohawk Valley Engineers' Club, Utica, N. Y.*
  BYRON E. WHITE, Utica, N. Y.

*Technical Club of Dallas, Tex.*
  * O. H. KOCH, Dallas, Tex.

*The Cleveland Engineering Society, Cleveland, Ohio*
  * JOHN F. OBERLIN, Cleveland, Ohio.

*The Engineers' Club of Baltimore, Md.*
  W. W. VARNEY, Baltimore, Md.

*The Society of Industrial Engineers, Chicago, Ill. (1)*
  * L. W. WALLACE, Baltimore, Md.

*Washington Society of Engineers, Washington, D. C.*
  E. C. BARNARD, Washington, D. C.

*York Engineering Society, York, Pa.*
  WILLIAM J. FISHER, York, Pa.
  H. A. DELANO (alternate), York, Pa.

* Present Members of Executive Board

This shows the general lines along which we have been working and I would suggest that the new American Engineering Council follow in a general way such lines. You will find that you will have to be most rigorous in determining what activities you may undertake and will have to turn down many suggestions which you will find are beyond the province of your Council.

Organized as you are with a broader support than Engineering Council, you will be able to make recommendations on state and local questions through local societies which are members of your body. But you must be careful and not permit yourself to be used for movements which however good in themselves are not especially under the perview of engineers.

Following Mr. Channings' address the chair called upon Alfred D. Flinn, secretary of Engineering Council, Philip N. Moore, and Dr. D. S. Jacobus to express their views as to the field of activity of the Federation. Mr. Moore and Dr. Jacobus have both been active in the work of Engineering Council and Mr. Flinn has served as secretary of the body for the past four years. Their remarks were accordingly received with great interest by the delegates assembled. It was their chief and unanimous opinion that the new American Engineering Council should carry on the work begun by the old Council extending it so as to cover a wider field and thus be of greater service to its member societies and in turn to the individual members. New lines of endeavor would soon develop they also asserted but as to participation in all of these they counseled careful thought before decisions were reached.

Following this discussion the motion of Mr. Moore relative to the location of the headquarters of the Federation, in accordance with the decision of the morning session, was next taken from the table. It was further discussed along the lines followed in the morning. An amendment was finally made to refer it to the Executive Board with power to act. This however was defeated and upon the original motion being put 29 delegates voted in favor and six against. This decision, which thus locates the headquarters of the Federation at the National Capitol, concluded the sessions of the first day.

### THE FRIDAY MORNING SESSION

The Friday session opened at nine o'clock with E. S. Carman, temporary chairman presiding. The chair first announced that the Taylor Society had voted to join the Federation and that their delegate, Morris L. Cooke, was present to take part in the proceedings. The Taylor Society is the twenty-first society to join the Federation.

L. W. Wallace, president of the Society of Industrial Engineers, next addressed the Council on the subject of Labor Conservation. He stressed the importance of safety and welfare work and the necessity of medical advisors in the attainment of such conservation. He also discussed the need of industrial education not alone, as he stated, for the worker but for all, even including the chief executives. Our labor problems, he declared, can only be solved by the earnest coöperation of both employer and employee.

Following Mr. Wallace the Nominating Committee offered its report and at the request of its chairman, D. S. Kimball put before the Council the choice of the committee for president—Herbert Hoover. In his introduction Dean Kimball said that it was important to the profession and the country at large that the standing of the men in the Council should be gaged by their standing among their brothers and that there was one who was thus preëminent. Later he added, 'You have elected a statesman, not an engineer."

The list of officers follows:

*President*

HERBERT HOOVER, Am. Inst. M. & M. Eng.

*Vice-Presidents*

CALVERT TOWNLEY, Am. Inst. Elec. Eng.  (2 year term)
W. E. ROLFE, Assoc. Eng. Soc. of St. Louis (2 year term)
D. S. KIMBALL, Am. Soc. Mech. Eng.   (1 year term)
J. PARKE CHANNING, Am. Inst. M. & M. Eng.  (1 year term)

*Treasurer*

L. W. WALLACE, Soc. of Ind. Eng.

After formal vote had been taken, Mr. Hoover, was escorted to the platform and took the chair. He briefly expressed his appreciation of the action of the Council and then withdrew for a short time to attend a meeting of the Red Cross which was holding a simultaneous session. In his absence Calvert Townley presided and brief acceptance speeches were next made by all the newly elected officers.

The morning session closed with the presentation of the report of the Committee on Constitution and By-Laws, recommending several minor changes in both the Constitution and By-Laws, but of such a character as not to change the import of their main provisions. The changes were discussed and voted upon item by item.

### THE CLOSING SESSION

The closing session of the Council on Friday afternoon saw Mr. Hoover in the chair. The announcement was first made of the members constituting the Executive Board (see page 721) and this was followed by the report of the Committee on Plan and Scope. This report in the main followed the recommendation made by J. Parke Channing in his address at the opening session. It pointed out in particular the importance of service to state and nation by investigation and advice strictly along non-partisan lines, and the necessity of maintaining contact with the local and regional engineering organizations.

Calvert Townley, as chairman of the Budget Committee, next submitted a tentative budget for the Council's consideration. He explained that if the Federation should not add to its member-societies during the year the estimated income would be $59,000 but if the societies who have already made known their desire to join the Federation should do so then an income of $80,000 might be expected. With these minimum and maximum figures in mind the estimated expenses for the maintenance of the organizations' headquarters, for committee expenses, traveling expenses, etc., would be $56,500 as a minimum and $93,500 as a maximum. The report was accepted and referred to the Executive Board for further consideration.

This concluded the official business of the day but prior to adjournment the Council was addressed by Geo. G. Anderson of Los Angeles, Cal. who told of the results of coöperation among the engineers on the western coast. Much has been accomplished he said, through joint effort and he urged that other states organize local engineering councils.

## Address by Herbert Hoover

On Friday evening, November 19, a large audience gathered to listen to an address by Herbert Hoover, president of the Federation, on some Phases of Relationship of Engineering Societies to Public Service. Referring to the great problems which have come as a result of the industrial development of the past fifty years, and the inadequacy of socialism to solve them he said in part:

We have built up our civilization, political, social and economic on the foundation of individualism. We have found in the course of development of large industry upon this system that individual initiative can be destroyed by allowing concentration of industry and service and thus an economic domination of groups over the whole. We have therefore built up public agencies intended to preserve an equality of opportunity through control of possible economic domination. Our regulation of public utilities and of many other types of industry, aiming chiefly to prevent combination in restraint of free enterprise. is a monument to our attempts to limit this economic domination—to give a square deal. While our system of individualism under controlled capitalism may not be perfect, the alternative offers nothing that warrants its abandonment. Our thought, therefore, needs to be directed to the improvement of this structure and not to its destruction.

A profound development in our economic system, apart from control of capital and service during the last score of years, has been the great growth and consolidation of voluntary local and national associations. We have the growth of great employers' associations, great farmers' associations, great merchants' associations, great labor associations—all economic groups striving today by political agitation, propaganda, and other measures to advance group interest. At times they come in sharp conflict with each other and often enough charge each other with crimes against public interest. And to me, one question of the successful development of our economic system rests upon whether we can turn the aspects of these great national associations towards coördination with each other in the solution of national economic problems, or whether they grow into groups for more violent conflict.

The Federated American Engineering Societies stands somewhat apart among these great economic groups. In that it has no special economic interest for its members. Its only interest is the creation of a great national association is public service, to give voice to the thought of the engineers in these questions. And if the engineers with their training in quantitative thought, with their intimate experience in industrial life, can be of service in bringing about coöperation between these great economic groups of special interests, they will have performed an extraordinary service. The engineers should be able to offer expert service in constructive solution of problems to the individuals in any of these groups. And here is a wider vision of this expert service in giving the group service of engineers to group problems.

One of the greatest conflicts rumbling in the distance is that between the employer on one side and organized labor on the other. We hear a great deal from extremists on one side about the domination of organized labor. Probably the tendency to domination exists among extremists on both sides. One of the most perplexing difficulties in all discussion and action in these problems is to eliminate this same extremist. There are certain areas of conflict of interest, but there is between these groups a far greater area of common interest, and if we can find measures by which, through coöperation, the field of common interest can be organized, then the area of conflict can be in the largest degree eliminated.

In this connection the employer sometimes overlooks a fundamental fact in connection with organized labor in the United States. This is that the vast majority of its membership and of its direction are individualists in their attitude of mind and in their social outlook; that the expansion of socialist doctrines finds its most fertile area in the ignorance of many workers, and yet the labor organizations, as they stand today, are the greatest bulwark against socialism. On the other hand, some labor leaders overlook the fact that if we are to maintain our high standards of living, our productivity, it can only be in a society in which we maintain the utmost possible initiative on the part of the employer; and further, that in the long run we can only expand the standard of living by the steady increase of production and the creation of more goods for division over the same numbers.

The American Federation of Labor has publicly stated that it desires the support of the engineering skill of the United States in the development of methods for increasing production and I believe it is the duty of our body to undertake a constructive consideration of these great problems and to give assistance, not only to the Federation of Labor, but also to the other great economic organizations interested in this problem, such as the Employers' Association and the Chambers of Commerce.

It is primary to mention the three-phase waste in production: First, from intermittent employment, second, from unemployment that arises in shifting industrial currents and third, from strikes and lockouts. Beyond this elimination of waste, there is another field of progress in the adoption of measures for positive increase in production.

In the elimination of the great waste and misery of intermittent employment and unemployment, we need at once coördination in economic groups. For example, our engineers have pointed time and again to the bituminous coal industry, where the bad economic functioning of that industry results in an average of but 180 days' employment per annum, where a great measure of solution could be had if a basis of coöperation could be found between the coal operators, the coal miners, the railways, and the great consumers. The combined results would be a higher standard of living to the employees, a reduced risk to the operator, a fundamental expansion of economic life by cheaper fuel. With our necessary legislation against combination and the lack of any organizing force to bring about this coöperation, the industry is helpless unless we can develop some method of governmental interest, not in governmental ownership, but in stimulation to coöperation in better organization.

In help against the misery in the great field of seasonal and other unemployment, we indeed need an expansion and better organization of our local and federal labor exchanges. We have a vast amount of industry, seasonal in character, which must shift its labor complement to other industries. The individual worker is helpless to find the contacts necessary to make this shift unless the machinery for this purpose is provided for him.

In the questions of industrial conflict resulting in lockout and strike, one mitigating measure has been agreed upon in principle by all sections of the community. That is collective bargaining, by which, wherever possible, the parties should settle their difficulties before they start a fight. It is founded not only on the sense of prevention but on the human right to consolidate the worker in a proper balanced position to uphold his rights against the consolidation of capital. This measure, advocated for long years by organized labor, was agreed to by the employers group in the First Industrial Conference. It has been supported in the platform of both political parties. The point where the universal application of collective bargaining has broken down is in the method of its execution. The conflict arises almost wholly over the question of representation and questions of enforcement. The employer in some industries denies the right of men other than his own employees to conduct the negotiations. Labor organizations demand that, as such negotiations require skill, experience and bargaining freedom, they are of more than local application and that thus they only can protect the body of workers by presenting the case on their behalf by skilled negotiators.

The Second Industrial Conference, of which I was a member, proposed a solution to this point by the provision that where there was a conflict over the determination representation should be left to a third party. It also proposed that each party should have the right to summon skill and experience to its assistance. It further proposed that where one of the parties at dispute refuses to enter upon collective bargaining, the entire question should be referred to an independent tribunal for investigation as to the right and wrong of the whole dispute—but only for investigation and report. The conference was convinced that the illumination of the public mind as to the rights and wrongs of these contentions would in itself make for material progress in their solution and that in public education and the condemnation by public opinion of wrong doing lay the root to real progress. The conference did not believe that industrial contention could be cured by compulsory arbitration or any other form of governmental repression.

There are questions in connection with this entire problem of employer and employee relationship, both in its aspects of increased production and in its aspects of wasteful unemployment, that deserve most careful study by our engineers. There lies at the heart of all these questions the great human conception that this is a community working for the benefit of its human members, not for the benefit of its machines or to aggrandize individuals; that if we would build up character and abilities and standard of living in our people, we must have regard to their leisure for citizenship, for recreation, for family life. These considerations, together with protection against strain, must be the fundamentals of determination of hours of labor. These factors being first protected, the maximum production of the country should become the dominating purpose. There is a broad question bearing upon stimulation of self-interest and thus increase in protection that revolves around the method of wage payment. I need not review the advantages, difficulties and weaknesses of bonus, piece-work, profit or saving sharing plans that are in use as a remedy for the deadening results of the same wage payment to good and bad skill alike. The suggestion I wish to put for your consideration is the possible use of another device in encouragement of individual interest and effort by creating two or three levels of wage in agreement for each trade, the position of each man in such scale to be based upon comparative skill and character. This plan should be developed upon the principle of extra graded compensation for added skill and performance, above an agreed basic wage. In order to give confidence, the classification under such scales must be passed upon by representatives of the workers in each shop or department. This plan is now being successfully experimented with.

We must take account of the tendencies of our present repetitive industries to eliminate the creative instinct in its workers, to narrow their field of craftsmanship, to discard entirely the contribution to industry that could be had from their minds as well as from their hands. Indeed, if we are to secure the development of our industries, we could not accomplish increased production without their stimulation. Here again we cannot make an advance unless we can secure coöperation between employer and employee. In large industry this mutuality of interest that existed in small units cannot be restored without definitive organization.

[Mr. Hoover here discussed shop committees as a system of such organization, saying that if organized labor should settle its problems of wage and conditions of labor in general agreement and apply its energies through shop committee organization to development of production as well as to the correction of incidental grievance, another step in coöperation would be accomplished.—EDITOR.]

There is an immediate problem in increased production that is too often overlooked by the theorist. While it is easy to state that increased production will decrease cost and by providing a greater demand for goods secure increased consumption and ultimate greater employment, yet the early stages of this process do result in unemployment and great misery. We usually accomplish these results over long periods of time, but if we would secure coöperation to accomplish them rapidly we must take account of the unemployment and we must say to the community that if it is to benefit by the cheapening costs and thus the increased standard of living, or alternatively if the employer is to take benefits, the entire burden should not be thrust upon the individual who now alone suffers from industrial changes.

In summary, the main point that I wish to make is this, that there is a great area of common interest between the employer and the employee through the reduction of the great waste of voluntary and involuntary unemployment, and in the increase of production. If we are to secure increased production and an increased standard of living, we must keep awake interest in creation, in craftsmanship, and the contribution of the worker's intelligence to management. Battle and destruction are a poor solution to this problem. The growing strength of national organizations on both sides should not be and must not be contemplated as an alignment for battle. Battle quickly loses its rules of sportmanship and adopts the rules of barbarism. These organizations—if our society is to go forward instead of backward—should be considered as the fortunate development of influential groups through which skill and mutual consideration can be assembled for coöperation to the solution of these questions. If we could secure such coöperation, throughout all our economic groups, we should have provided a new economic system, based neither on the capitalism of Adam Smith nor upon the socialism of Karl Marx. We should have provided a third alternative that preserves individual initiative, that stimulates it through protection from denomination. We should have given a priceless gift to the twentieth century.

## Meeting of the Executive Board

The Executive Board, charged with conducting the business of the Federation under direction of the American Engineering Council, held its first meeting on Saturday morning. In the absence of President Hoover, Vice-President Rolfe presided. Charles F. Scott served as secretary.

The first business was the appointment of a committee to recommend a candidate or candidates for Executive Secretary of the American Engineering Council. Following a discussion Dean D. S. Kimball said he had given the matter much thought and without having consulted others would suggest five names representative of the various interests concerned in the conduct of the affairs of the Council. It was understood that President Hoover would act ex-officio. These names were:

W. W. WALLACE, Baltimore, *Chairman*, representing the officers
of the Council.
PHILIP N. MOORE, St. Louis, representing A.I.M.E.
CHARLES F. SCOTT, New Haven, representing A.I.E.E.
L. P. ALFORD, New York, representing A.S.M.E.
JOHN F. OBERLIN, Cleveland, representing the local societies

It was voted that this committee be appointed, with the addition
of the name of Calvert Townley, representing the present Engineering Council, thus constituting with the president, a committee
of seven to report at the next meeting.

The question of whether assessments upon member societies
should be on the basis of voting membership or of total membership in these societies, was referred to the Committee on Membership and Representation.

The Board declared in favor of continuing the efforts of Engineering Council for a National Department of Public Works. This
led to a general discussion of ways and means which could properly
be adopted by the Federation for the promotion of public and
legislative matters. Maj. Fred J. Miller spoke in opposition to
spending money for promotion and to lobbying in any form. He
advocated the formulation of policies by the Council and the use
of publicity methods to bring them to the attention of legislators
and the public. L. B. Stillwell referred to the last French Engineering Congress, held at Paris, as exemplifying a way in which this
could be done. Committees were appointed to present reports
to the Congress on such public questions as highways, hydraulics,
etc., and when formally accepted the reports at once became available for incorporation in legislation and carried with them the
prestige of the Congress.

The Board indorsed the plan, previously suggested by President
Hoover, for a committee to investigate industrial waste and authorized the president to organize for this purpose. The committee
will be the agency through which the Council will take up the question of industrial relations and the losses incident to the lack of
coöperation between capital and labor.

It was voted that the establishment of Washington headquarters,
as directed by vote of the Council, should be deferred until the
appointment of the permanent Executive Secretary, probably in
January. To provide for carrying on the current work, and particularly the preparation and distribution of the minutes of the
present series of meetings, L. P. Alford was elected temporary
secretary with headquarters in the rooms of The American Society of Mechanical Engineers, New York, to serve until the
next meeting of the Board.

The last action of the Board was to request The American Society of Mechanical Engineers to give Mr. Alford such assistance
on the part of its staff as might be required to conduct temporarily
the work of the Federation.

# NEWS OF THE ENGINEERING SOCIETIES

## Conventions of the American Foundrymen's Association, the American Iron and Steel Institute, and the American Electrochemical Society

### American Foundrymen's Association

THE annual convention and exhibition of the American Foundrymen's Association took place at Columbus, Ohio, on October 4
to 8. At the technical sessions 67 papers were scheduled, most
of which were presented. There were 240 exhibitors and the registered attendance exceeded 3800. An outstanding feature of the
convention was the announcement of five donations, four of $5000
each and one of $1000, the income from which will be awarded in
recognition of meritorious work in foundry research and invention.

Dr. Richard Moldenke, Watchung, N. J., presented a paper
on the elements useful for purifying cast iron, with special reference to zirconium.

A paper on the electric furnace and the problem of sulphur in
cast iron was read by George K. Elliot, Lunkenheimer Co., Cincinnati. The tendency of sulphur to segregate, he said, constitutes
the greatest menace of high sulphur, but he claimed that it is
impossible to fix positively any pronounced advantage in iron with
0.015 or 0.030 per cent sulphur over those with twice that amount.
Instead of striving for low sulphur limits the speaker was of the
opinion that the use of more scrap or of pig iron running comparatively high in sulphur will become popular. A. N. Kelly, Modern
Tool Co., Cincinnati, described the methods employed in the production of machine-tool castings. This paper was illustrated with
lantern slides showing the methods employed in the work.

A paper describing the making of high-grade iron castings for
milling-machine tables at the plant of the Brown & Sharpe Mfg.
Co., Providence, R. I., was read by Leroy Sherman. Papers
were also read by E. J. Fowler, San Francisco, on Standardizing
Gray-Iron Castings for Analytical Determination; Electrical
Apparatus in a Modern Iron Foundry, by F. W. Egan, Pittsburgh;
The Foundry of the U.S.S. *Prometheus*, prepared by Lieut. R. F.
Nourse; and Approved Methods of Testing Molding Sands, by
S. W. Stratton, of the U. S. Bureau of Standards, Washington.

The need for developing standards for molding materials was
emphasized and a number of tests for determining the usefulness
of sand and clay in the shop were discussed by R. L. Lindstrom,
Canadian Steel Foundries, Montreal.

The status of the electric furnace in the steel industry as of
September 1, 1920, was surveyed by Edwin F. Cone, associate
editor, *The Iron Age*, New York. There were in the United States
and Canada on January 1, 1920, 363 furnaces of all types operating
or contracted for in the steel industry; on September 1, this figure
had been expanded to 374 or a net increase of 11 furnaces. This
increase was wholly in the United States.

Annealing Steel with Pulverized Coal was the title of a paper
by C. H. Gale, Pressed Steel Car Company, McKees Rocks, Pa.
The author stated that where the investigations were conducted
a marked difference in the physical characteristics of steel was
noticed after pulverized coal was substituted for gas and oil for
the annealing furnaces. The tensile strength became greater and
the carbon and sulphur content was increased.

Other steel foundry papers were: Cleaning Steel Castings, by
A. W. Gregg; Electric Heat Treating of Steel Castings, by E. F.
Collins; Making Acid Electric Steel, by James W. Galvin and Charles
N. Ring; Making Steel Castings to Specifications, by E. R. Young,
Jr.; and Accurate Treatment of Steel Castings, by T. F. Baily,
Alliance, Ohio.

Interesting papers on malleable iron were also presented.
Enrique Touceda, Albany, described some experiments made
with air furnaces, and suggested some changes in open-hearth furnaces for malleable iron. He stated that future advancement in
the malleable-iron industry must come through an improvement
in plant organization and metallurgical apparatus. Fractures and
Microstructures of American Malleable Cast Iron was the subject
of a paper by W. R. Bean, H. W. Highriter and E. S. Davenport,
Eastern Malleable Iron Co., Naugatuck, Conn. Two papers by
H. A. Schwartz, National Malleable Castings Co., Cleveland, described the new research department of that plant, and also the
triplex process for making electric-furnace malleable. E. J.
Lowry, Oliver Chilled Plow Co., South Bend, Ind., presented a
paper on Practical Malleable Annealing.

These notes have been prepared from the extensive account of
the convention published in *The Iron Age* for October 14.

### American Iron and Steel Institute

The American Iron and Steel Institute held its eighteenth
annual convention in New York on October 22. In his presidential address, Judge Elbert H. Gary, recounted his personal observations in a recent tour through France and Belgium, and pre-

dicted speedy recovery by both these nations from the effects of the ravages of war upon industries. In general he took an optimistic view of industry, both abroad and at home.

The technical papers were of the usual high standard. E. A. Wheaton, superintendent, open-hearth department, Bethlehem Steel Co., presented the results of experiments conducted to determine the effect of using a high-manganese iron in the initial basic open-hearth charge. A furnace was operated exclusively on high-manganese iron for a considerable period. The records obtained show that high-manganese iron improves the quality of open-hearth steel without reduction of tonnage or any other injurious effects, and that it is of great assistance in meeting the demands for the better grades of alloy steels. Mr. Wheaton concluded from his data that it is possible, with high-manganese iron, to run lean slags, maintain tonnage, and with sulphur in blast-furnace iron as high as 0.1 per cent to deliver that same iron to the open-hearth furnaces through a mixer with 50 per cent of the sulphur eliminated.

D. M. Buck, metallurgical engineer, American Sheet & Tin Plate Co., Pittsburgh, reviewed the development of copper steel and its non-corrosive properties. Practice has shown that the best results are obtained by alloying with normal open-hearth or bessemer steel, from 0.15 to 0.30 per cent of pure copper. Mr. Buck pointed out that although references are found as early as 1627 to the effect of copper on the physical properties of iron and steel, nothing is found pertaining to the control of corrosion previous to the present century. The relative corrosive properties of untreated and copper-treated iron and steel have been investigated by several authorities from 1900 on. Tests conducted by the American Society for Testing Materials have shown that the life of sheet metal is more than doubled by the copper treatment.

The successful development of the large-hearth blast furnace in the United States was surveyed by Walther Malhesius, superintendent of blast furnaces, Illinois Steel Co., South Chicago, Ill. Hearth dimensions have been increased over the accepted standard without interfering with the height or angle of bosh. The average hearth diameter of all 22-ft. bosh furnaces at the South Chicago plant of the Illinois Steel Co. was increased from $16^1/_2$ ft. in 1911 to $18^1/_2$ ft. in 1919. The best average monthly production of these furnaces equaled 512 tons per furnace per day in December 1911, as against 556 tons in May 1920. Mr. Mathesius attributed the increased output to the adoption of the larger hearths, as no enlargement of the furnaces or important changes of lines above the bosh were made. The fuel consumption per ton of iron for the two periods cited above were 2053 lb. in December 1911, and 2037 lb. in May 1920. Further proof of the successful operation of the larger hearths has been furnished by the operation for some time of a furnace with a hearth diameter of 20 ft. 9 in. This furnace was described at length by Mr. Mathesius, and he presented tables of its performance as well as of the average results of other South Chicago furnaces. The furnace with a 20-ft. 9-in. hearth has consistently worked with a more regular and uniformly lower blast pressure than the other furnaces under comparative conditions. Contrary to what has been claimed as to a large hearth requiring an increase in the number of tuyeres, the furnaces at the South Chicago works have worked well with distances between tuyeres of 4 ft. $7^1/_4$ in., 4 ft. $11^3/_4$ in.; 5 ft. $9^3/_4$ in., 5 ft. $10^3/_8$ in., and 6 ft. $6^1/_4$ in., the last figure corresponding to the furnace with the 20-ft. 9-in. hearth. The enlargement has not imposed additional or increased stresses and duties, while in some respects the demands upon the strength and wearing qualities have actually lessened. Mr. Mathesius pointed out that the successful development of the large-hearth blast furnace, unhampered by any sacrifice of former advantages, is an exclusively American accomplishment, which is evoking the keenest interest abroad.

F. L. Toy, superintendent of open hearths, Homestead Works, Carnegie Steel Co., surveyed the practice in the manufacture of basic open-hearth steel. He referred specially to the improved devices and methods which have been developed for preserving and lengthening the life of the furnace, and discussed in detail the merits of the various fuels for melting the charge, and the efficiency of the furnaces. Other papers were: Recent Developments in the Iron and Steel Industry of India, by C. P. Perrin; Foreign Trade, by A. H. Holliday; and Heat Treatment of Automobile Steels, by R. R. Abbott.

## American Electrochemical Society

The thirty-eighth general meeting of the American Electrochemical Society was held in Cleveland on September 30 and October 1 and 2. The technical papers dealt principally with various phases of the application of electric furnaces in the steel industry. Experiments on heat losses through electrodes were reported by M. R. Wolfe and V. de Wysocki, of Lehigh University. A sixton Heroult furnace was provided with special water-cooled rings. From records obtained by measuring the temperature of the water at intervals, the heat loss through the electrodes was estimated at 18.7 per cent of the total power supplied to the furnace. Data on the influence of copper, manganese, and chromium on the corrosion of their iron alloys, were presented by E. A. and L. T. Richardson, General Electric Co., Cleveland. Tests showed that there is a mutual action between manganese, and copper in their effect upon the atmospheric corrosion of iron. Copper alone reduces the corrosion of pure iron and to a still greater extent the corrosion of steel. This is due to the effect of manganese, which enhances the effect of copper. If manganese is replaced by chromium the effect is still more pronounced. The red-short range in iron due to the presence of copper is removed by either manganese or chromium. It is believed that there is some relation between this red-short range and resistance to atmospheric corrosion. Based upon this the film intergrain hypothesis is suggested to explain the corrosion resistance of these alloys.

A new portable standard cell was described by C. J. Rodman and Thomas Spooner. It is of the cadmium type of special design with the usual ingredients, but it has a container of hard glass with tungsten leads. Instead of the common H-type a compact concentric arrangement has been devised. Measurements of the electrical resistivity of specialized refractories were presented by M. L. Hartman, A. P. Sullivan, and D. E. Allen, research laboratory, Carborundum Co. A progress report of the Söderberg continuous electrode installation at Anniston, Ala. was submitted by Dr. Joseph W. Richards. The Söderberg electrode was introduced into the United States by Elektrokemisk Industri of Christiania, Norway. Ferromanganese is manufactured in an 1800-kw. electric furnace at the Anniston plant. Attempts to use pure nickel wire as a furnace resistor in place of nichrome wire were related by F. A. J. FitzGerald and Grant C. Mayer, FitzGerald Laboratories.

## POWER TEST CODES

(Continued from page 719)

The complete statement referred to in (b) above should be divided as follows:

1 Object, giving authority, preliminary agreements and other relevant information.

2 Description of apparatus tested.

3 Methods of testing, giving conditions of operation, location, arrangement and method of using instruments and testing apparatus, and personnel.

4 Discussion of the data and results both as to accuracy and as bearing on the object of the test.

5 Leading conclusions.

6 Tables of data and calculated results or graphical expositions of the same.

7 Charts not included under item 6.

Reports of tests of lesser magnitude should follow the same general plan.

37 In preparing a report the work should be so done that the report will be self-contained in all respects. The best test of such a requirement is to assume that the report may be brought into court in connection with litigation involving the object for which the test was made. If the report can stand alone under such conditions it may be considered complete.

(*Signed*) WILLIAM. H. KAVANAUGH, *Chairman,*
               CHARLES J. BACON,
               GEO. H. BARRUS,
               ARTHUR M. GREENE, JR.,
               CLARENCE F. HIRSHFELD.

## ARMOR-PLATE AND GUN-FORGING PLANT OF U. S. NAVY

(Continued from page 668)

The machine-tool layout and the specifying and purchasing of the equipment was done by the machine-shop superintendent, Mr. W. E. Hayes, who is the head of the Cold Metal Department.

The general layout, industrial and structural design, design of plant services crane specifications, foundation design for all equipment, etc., were made under the author's direction in the Bureau of Yards and Docks at Washington, and details completed at South Charleston when the engineering force, with the exception of the Electrical Division, was transferred there in the Spring of 1919 to be in closer touch with the work. A Construction and Maintenance Department was organized at the Naval Ordnance Plant to build the new plant and carry on the maintenance of the Projectile Plant. It employed at maximum strength some 40 to 50 designers and draftsmen and somewhat over 1200 men in the field. The principal assistants of the writer, as head of the Construction and Maintenance Department were Mr. J. Hepinstall, supervising engineer of design; Mr. T. H. Callahan, supervising engineer of construction; Mr. W. H. Sears, materials and office, and Mr. H. M. Cogan, supervising engineer of electrical design.

### Costs

Final figures of cost are not yet available. In round numbers it may be stated that the entire plant when completed will represent an investment of about $20,000,000. Construction items of the plant, including cost of all structures completely enclosed and with services, including overhead cranes but exclusive of operating equipment and equipment foundations, will total somewhat over $10,000,000.

The cost records, kept both by the intricate Navy system of accounting which imposes a more or less arbitrary overhead based on the amount of productive labor; and on a man-hour and material basis kept separately by the author, indicate that the work has been done economically. The percentage profit of a general contractor has been saved to the Department. A saving has also been made by the use of Navy cost-plus and Army salvage material and equipment.

The method adopted of handling the entire construction of the armor and gun-forging plant within the Department proved very flexible and made it possible to keep the design work and the construction work in close harmony. It also permitted essential eleventh-hour changes to be made with less inconvenience and delay than would have been possible dealing through a general contractor.

If armor ingots are cast in December, as is confidently expected, it will have taken two years and six months, in exceedingly trying times as to labor and material, to bring the plant from the "open-lot and blank-paper" stage to operation.

## DISCUSSION—STEAM VS. ELECTRIC LOCOMOTIVES

(Continued from page 688)

to the operation of electric locomotives, and are just as much a part of the electric locomotive as the boiler is of the steam locomotive.

*Extent of Electrification.* Where electrification is contemplated a very serious question is: What shall be its extent? Naturally the desire would be to wipe out as many as possible of the extensive accessories to steam operation. If, however, it becomes necessary to operate steam trains over the electrified section, it will obviously be necessary to retain water stations and possibly fuel stations, provided the electrified section is sufficiently long. This operation of steam locomotives under their own power over electrified sections would be necessary in case of redistribution and possibly in case of diversions where the electrified section formed part of the diverted line. Therefore, the claim for economy in doing away

with these features of steam operation would probably not be realized.

In commenting on the paper presented by Mr. Muhlfeld, Mr. Gibbs said that he believed his enthusiasm had carried him too far in minimizing the advantages of electrification. The operation of the electrified roads has undoubtedly been good whether it be terminal or road operation. He thought Mr. Muhlfeld also ignored the fact that the modern improvements which have so added to the performance of the steam locomotives are potential only. For instance, it is very possible and common by indifference to so carry water in the boiler that the superheater becomes merely a steam drier and its value disappears. In many cases, because of neglect of damper mechanism or from dirty flues, little benefit is derived from improved appliances. Modernizing of steam locomotives calls for the intelligent use of the devices, which will come when the old spirit of loyalty returns.

In summarizing his remarks Mr. Gibbs said in part: "It is to be noted that practically all of the electrifications on steam railroads so far have been based on local conditions. In the electrifications in and around cities a moving cause has been the elimination of smoke and other objectionable features incidental to steam operation, and the possibility of increasing the capacities of the passenger terminals. On the Milwaukee road it was the utilization of available water power. On the Norfolk & Western it was to secure increase in capacity on a congested mountain division with tunnel complications. It is fair to assume that other electrifications will be similarly governed by local conditions.

"If, after careful consideration of the road, based on actual train sheets for the heaviest actual or probable congested operation; the capacity and number of active and available locomotives required; crediting the operation with incidental savings which may be effected, and eliminating expenses peculiar to steam operation; it appears that there would be economy in electrification, either from actual savings or better operation, or both, it still remains for the management to decide whether the money required can be spent to better advantage for electrification than for some other features of the general operation."

## Book Notes

CONDENSED CATALOGUES OF MECHANICAL EQUIPMENT. Comprising Condensed, Uniformly Presented and Illustrated Catalogue Information Covering the Products of Manufacturers of Various Classes of Mechanical Equipment, with General Classified Directory and Consulting Engineers' Directory. Tenth annual volume. American Society of Mechanical Engineers, New York, 1920. Cloth, 6 × 9 in., 1004 pp., illus., $4.

The tenth edition of this catalog and directory of mechanical equipment and consulting engineers follows the established lines, although the data on A.S.M.E. standards have been omitted from this issue. Five hundred and nine firms are represented by condensed catalogs of their products, this section being eighty pages longer than in 1919. The directory of mechanical equipment, in which manufacturers are listed under products, is ten per cent larger, and the directory of consulting engineers covers sixty per cent more pages than before. Various improvements have also been made to facilitate convenient reference.

THE COAL TRADE. The Year Book of the Coal and Coke Industry. By Sydney A. Hale. 47th annual edition. The Estate of F. E. Saward. New York, 1920. Cloth, 6 × 8 in., 352 pp., tables, $4.

The present volume is, the compiler states, the largest in size and the widest in scope yet issued. It includes a variety of statistical tables of interest to those in the coal industry, brought to the latest possible dates, and information showing the development of the industry during the past year. The book contains statistics of production, exports, imports, costs, prices, ocean rates, wages and similar matter, as well as reviews of important occurrences.

COMMON SENSE AND LABOUR. By Samuel Crowther. Doubleday, Page & Co., Garden City, N. Y., 1920. Cloth, 5 × 8 in., 284 pp., $2.

The author discusses the causes of present-day dissatisfaction between workmen and employers, the various remedies that have been suggested, and the results obtained in actual cases. Written in readable style, and of interest to employers.

Lightning Source UK Ltd.
Milton Keynes UK
UKHW021309221118
332685UK00010B/1845/P